MANUAL OF CLINICAL MICROBIOLOGY

FOURTH EDITION

MANUAL OF CLINICAL MICROBIOLOGY

FOURTH EDITION

EDITOR IN CHIEF

EDWIN H. LENNETTE

Viral and Rickettsial Disease Laboratory, California Department of Health Services, Berkeley, California

EDITORS

ALBERT BALOWS

Center for Infectious Diseases, Centers for Disease Control, Atlanta, Georgia

WILLIAM J. HAUSLER, JR.

Hygienic Laboratory, University of Iowa, Iowa City, Iowa

H. JEAN SHADOMY

Medical College of Virginia, Richmond, Virginia

AMERICAN SOCIETY FOR MICROBIOLOGY

Washington, D.C. 1985

1913 I St., N.W.
Washington, DC 20006

Library of Congress Cataloging in Publication Data

Main entry under title:

Manual of clinical microbiology.
Includes bibliographies and indexes.
1. Diagnostic microbiology—Handbooks, manuals, etc. 2. Medical microbiology—Handbooks, manuals, etc. I. Lennette, Edwin H., 1908– . II. American Society for Microbiology. [DNLM: 1. Microbiology. QW 4 M294]
QR67.M36 1985 616'.01 84-28304

ISBN 0-914826-65-4 (cloth binding)
ISBN 0-914826-69-7 (flexible binding)

Printed in the United States of America

Contents

Section I. GENERAL
Section Editor: Henry D. Isenberg

Section II. NOSOCOMIAL INFECTION PREVENTION AND CONTROL
Section Editor: James M. Hughes

Section III. AEROBIC BACTERIA
Section Editors: Marie B. Coyle, Josephine A. Morello, P. Byrd Smith

Section IV. ANAEROBIC BACTERIA

Section Editor: Stephen D. Allen

Section V. SPIROCHETES

Section Editor: Russell C. Johnson

Section VI. FUNGI

Section Editor: Billy H. Cooper

Section VII. PARASITES

Section Editor: James W. Smith

Section VIII. VIRUSES, RICKETTSIAE, AND CHLAMYDIAE
Section Editor: C. George Ray

Section IX. SEXUALLY TRANSMITTED DISEASES
Section Editor: Stephen A. Morse

Section X. IMMUNODIAGNOSTIC TESTS
Section Editor: Richard C. Tilton

Section XI. LABORATORY TESTS IN CHEMOTHERAPY

Section Editor: Clyde Thornsberry

Section XII. MOLECULAR METHODS

Section Editor: John C. Feeley

Section XIII. MEDIA, REAGENTS, AND STAINS

Section Editor: Peter Nash

Editorial Board

Contributors

Libero Ajello
Division of Mycotic Diseases, Center for Infectious Diseases, Centers for Disease Control, Atlanta, Georgia 30333

William L. Albritton
Department of Microbiology, University of Saskatchewan, Saskatoon, Saskatchewan, Canada S7N 0W0

Aaron D. Alexander
Department of Microbiology, Chicago College of Osteopathic Medicine, Chicago, Illinois 60615

Stephen D. Allen
Division of Clinical Microbiology, Department of Pathology, Indiana University Medical Center, Indianapolis, Indiana 46223

John P. Anhalt
Section of Clinical Microbiology, Mayo Clinic, Rochester, Minnesota 55905

Lawrence R. Ash
School of Public Health, University of California-Los Angeles, Los Angeles, California 90024

George M. Baer
Centers for Disease Control, Lawrenceville, Georgia 30246

Albert Balows
Center for Infectious Diseases, Centers for Disease Control, Atlanta, Georgia 30333

W. Emmett Barkley
Division of Safety, National Institutes of Health, Bethesda, Maryland 20205

Arthur L. Barry
Clinical Microbiology Institute, Tualatin, Oregon 97062

Marilyn S. Bartlett
Department of Pathology, Indiana University School of Medicine, Indianapolis, Indiana 46223

Raymond C. Bartlett
Department of Pathology, Division of Microbiology, Hartford Hospital, Hartford, Connecticut 06115

Neil R. Blacklow
Division of Infectious Diseases, University of Massachusetts Medical School, Worcester, Massachusetts 01605

Marjorie Bohnhoff
Departments of Pathology and Medicine, University of Chicago Medical Center, Chicago, Illinois 60637

Robert Bortolussi
Izaak Walton Killam Hospital, Halifax, Nova Scotia, Canada B3J 3G9

Edward J. Bottone
Department of Microbiology, The Mount Sinai Hospital, and Mount Sinai School of Medicine, The Mount Sinai Medical Center, New York, New York 10029

Lynda L. Bradford
Microbial Disease Laboratory Section, California Department of Health Services, Berkeley, California 94704

Don J. Brenner
Division of Bacterial Diseases, Center for Infectious Diseases, Centers for Disease Control, Atlanta, Georgia 30333

Frank L. Bryan
Center for Professional Development and Training, Centers for Disease Control, Atlanta, Georgia 30333

Willy Burgdorfer
Rocky Mountain Laboratories, Hamilton, Montana 59840

Roberta B. Carey
Department of Pediatrics, University of Chicago, Chicago, Illinois 60637

Rodney Y. Cartwright
Public Health Laboratory Service, Public Health Laboratory, St. Luke's Hospital, Guildford GU1 3NT, England

William B. Cherry
Center for Infectious Diseases, Centers for Disease Control, Atlanta, Georgia 30333 (Retired)

Julia C. Clark
Diagnostic Virology Laboratory, University of Colorado Health Sciences Center, Denver, Colorado 80220

Marion K. Cooney
Department of Pathobiology, School of Public Health and Community Medicine, University of Washington, Seattle, Washington 98195

Billy H. Cooper
Department of Pathology, Baylor University Medical Center, Dallas, Texas 75246

Marie B. Coyle
Departments of Laboratory Medicine and Microbiology & Immunology, University of Washington, Harborview Medical Center, Seattle, Washington 98104

George Cukor
Infectious Disease Group, DuPont, North Billerica, Massachusetts 01862

Richard F. D'Amato
Division of Microbiology, Department of Pathology, The Catholic Medical Center of Brooklyn and Queens, Jamaica, New York 11432

Jules L. Dienstag
Gastrointestinal Unit, Massachusetts General Hospital, and Department of Medicine, Harvard Medical School, Boston, Massachusetts 02114

Steven D. Douglas
The Children's Hospital of Philadelphia, Philadelphia, Pennsylvania 19104

Walter R. Dowdle
Center for Infectious Diseases, Centers for Disease Control, Atlanta, Georgia 30333

R. J. Doyle
Department of Microbiology and Immunology, University of Louisville Health Sciences Center, Louisville, Kentucky 40292

W. Lawrence Drew
Mount Zion Hospital and Medical Center, San Francisco, California 94120

Martha A. C. Edelstein
Clinical Anaerobic Bacteriology Research Laboratory, Infectious Disease Section, V.A. Wadsworth Medical Center, Los Angeles, California 90073

Paul H. Edelstein
Infectious Disease Section, V.A. Wadsworth Medical Center, and Department of Medicine, University of California-Los Angeles School of Medicine, Los Angeles, California 90073

T. Grace Emori
Hospital Infections Program, Center for Infectious Diseases, Centers for Disease Control, Atlanta, Georgia 30333

Ana Espinel-Ingroff
Department of Medicine, Medical College of Virginia, Virginia Commonwealth University, Richmond, Virginia 23298

M. Essex
Department of Cancer Biology, Harvard School of Public Health, Boston, Massachusetts 02115

J. W. Ezzell
U.S. Army Medical Institute of Infectious Diseases, Ft. Detrick, Frederick, Maryland 21701

Richard R. Facklam
Respiratory and Special Pathogens Laboratory Branch, Centers for Disease Control, Atlanta, Georgia 30333

J. J. Farmer III
Enteric Bacteriology Section, Center for Infectious Diseases, Centers for Disease Control, Atlanta, Georgia 30333

Silas G. Farmer
Department of Pathology, Medical College of Wisconsin, Milwaukee, Wisconsin 53226

Martin S. Favero
Hospital Infections Program, Center for Infectious Diseases, Centers for Disease Control, Atlanta, Georgia 30333

Mary Jane Ferraro
Department of Bacteriology, Massachusetts General Hospital, Boston, Massachusetts 02114

Sydney M. Finegold
Infectious Disease Section, V.A. Wadsworth Medical Center, and Department of Medicine and Department of Microbiology and Immunology, University of California-Los Angeles School of Medicine, Los Angeles, California 90073

Thomas J. Fitzgerald
Department of Medical Microbiology/Immunology, University of Minnesota-Duluth School of Medicine, Duluth, Minnesota 55812

John E. Forney
Laboratory Program Office, Centers for Disease Control, Atlanta, Georgia 30333

Harvey M. Friedman
Department of Medicine, University of Pennsylvania School of Medicine, Philadelphia, Pennsylvania 19104
Joan C. Fung
Department of Laboratory Medicine, University of Connecticut School of Medicine, Farmington, Connecticut 06032
Julia S. Garner
Hospital Infections Program, Center for Infectious Diseases, Centers for Disease Control, Atlanta, Georgia 30333
Thomas L. Gavan
Department of Microbiology, The Cleveland Clinic, Cleveland, Ohio 44106
Gerald L. Gilardi
Department of Laboratories, North General Hospital, New York, New York 10035
Lynn C. Goldstein
Genetic Systems Corporation, Seattle, Washington 98121
Robert C. Good
Mycobacteriology Branch, Centers for Disease Control, Atlanta, Georgia 30333
Norman L. Goodman
Department of Pathology, College of Medicine, Albert B. Chandler Medical Center, University of Kentucky, Lexington, Kentucky 40536-0084
Morris A. Gordon
Wadsworth Center for Laboratories and Research, New York State Department of Health, Albany, New York 12201
Paul A. Granato
Department of Pathology, Veterans Administration Hospital, Syracuse, New York 13210
Donald L. Greer
Department of Pathology/Dermatology, Louisiana State University Medical Center, New Orleans, Louisiana 70112
Neal B. Groman
Department of Microbiology and Immunology, University of Washington, Seattle, Washington 98195
Carlyn Halde
Department of Microbiology and Immunology, School of Medicine, University of California, San Francisco, California 94143
W. J. Hausler, Jr.
Hygienic Laboratory, University of Iowa, Iowa City, Iowa 52242
George R. Healy
Division of Parasitic Diseases, Center for Infectious Diseases, Centers for Disease Control, Atlanta, Georgia 30333
Donald A. Hendrickson
Department of Biology, Ball State University, Muncie, Indiana 47306
Kenneth L. Herrmann
Division of Viral Diseases, Center for Infectious Disease, Centers for Disease Control, Atlanta, Georgia 30333
F. W. Hickman-Brenner
Enteric Bacteriology Section, Center for Infectious Diseases, Centers for Disease Control, Atlanta, Georgia 30333
L. A. Holcomb
University Hygienic Laboratory, University of Iowa, Iowa City, Iowa 52242
F. Blaine Hollinger
Departments of Virology, Epidemiology, and Medicine, Baylor College of Medicine, Houston, Texas 77030
Dannie G. Hollis
Special Bacterial Pathogens Laboratory, Centers for Disease Control, Atlanta, Georgia 30333
James M. Hughes
Hospital Infections Program, Center for Infectious Diseases, Centers for Disease Control, Atlanta, Georgia 30333
Henry D. Isenberg
Division of Microbiology, Long Island Jewish-Hillside Medical Center, New Hyde Park, New York 11042, and Health Sciences Center, State University of New York at Stony Brook, Stony Brook, New York 11794
Peter B. Jahrling
U.S. Army Medical Research Institute of Infectious Diseases, Ft. Detrick, Frederick, Maryland 21701

William M. Janda
Department of Medical Laboratory Sciences, Clinical Microbiology Laboratories, University of Illinois-Chicago, Chicago, Illinois 60612

William R. Jarvis
Hospital Infections Program, Center for Infectious Diseases, Centers for Disease Control, Atlanta, Georgia 30333

Ronald N. Jones
Department of Pathology, Kaiser Permanente Regional Laboratory, Clackamas, Oregon 97015

J. H. Jorgensen
Department of Genetics, North Carolina State University, Raleigh, North Carolina 27695-7614

Leo Kaufman
Division of Mycotic Diseases, Centers for Disease Control, Atlanta, Georgia 30333

K. F. Keller
Department of Microbiology and Immunology, University of Louisville Health Sciences Center, Louisville, Kentucky 40292

Michael T. Kelly
Division of Medical Microbiology, University of British Columbia and Vancouver General Hospital, Vancouver, British Columbia, Canada V5Z 1M9

Alan P. Kendal
Division of Viral Diseases, Center for Infectious Diseases, Centers for Disease Control, Atlanta, Georgia 30333

George E. Kenny
Department of Pathobiology, University of Washington School of Public Health and Community Medicine, Seattle, Washington 98195

Mogens Kilian
Department of Oral Biology, The Royal Dental College, DK-8000 Aarhus C, Denmark

W. E. Kloos
Department of Genetics, North Carolina State University, Raleigh, North Carolina 27695-7614

Stephen J. Kraus
Center for Prevention Services, Centers for Disease Control, Atlanta, Georgia 30333

Donald J. Krogstad
Microbiology Laboratory, Barnes Hospital, and Departments of Medicine and Pathology, Washington University School of Medicine, St. Louis, Missouri 63110

Irene Kwasnik
Department of Laboratory Medicine, University of Connecticut Health Center, Farmington, Connecticut 06032

Geoffrey A. Land
Methodist Hospitals of Dallas, Dallas, Texas 75265

Sandra A. Larsen
Treponema Research Branch, Centers for Disease Control, Atlanta, Georgia 30333

Howard W. Larsh
Missouri State Chest Hospital, Mount Vernon, Missouri 65712

David A. Lennette
Virolab, Inc., Emeryville, California 94608

Evelyne T. Lennette
Virolab, Inc., Emeryville, California 94608

James D. MacLowry
Clinical Pathology Department, Clinical Center, National Institutes of Health, Bethesda, Maryland 20205

Linda M. Marler
Division of Clinical Microbiology, Department of Pathology, Indiana University Medical Center, Indianapolis, Indiana 46223

Michael R. McGinnis
Department of Microbiology and Immunology, University of North Carolina at Chapel Hill, Chapel Hill, North Carolina 27514

John E. McGowan, Jr.
Departments of Pathology and Medicine, Emory University School of Medicine, and Clinical Microbiology Laboratory, Grady Memorial Hospital, Atlanta, Georgia 30333

Kenneth McIntosh
Division of Infectious Diseases, Children's Hospital Medical Center, Boston, Massachusetts 02115

Roger M. McKinney
Center for Infectious Diseases, Centers for Disease Control, Atlanta, Georgia 30333

James C. McLaughlin
Division of Microbiology, Hartford Hospital, Hartford, Connecticut 06115
Joseph L. Melnick
Department of Virology and Epidemiology, Baylor College of Medicine, Houston, Texas 77030
Dorothy M. Melvin
Division of Laboratory Training and Consultation, Laboratory Program Office, Centers for Disease Control, Atlanta, Georgia 30333
Marilyn A. Menegus
University of Rochester Medical Center, Rochester, New York 14642
J. Michael Miller
Laboratory Program Office, Centers for Disease Control, Atlanta, Georgia 30333
Linda L. Minnich
University of Arizona Health Sciences Center, Tucson, Arizona 85724
Josephine A. Morello
Departments of Pathology and Medicine, Clinical Microbiology Laboratories, University of Illinois-Chicago, Chicago, Illinois 60637
George K. Morris
Enteric Bacteriology Section, Center for Infectious Diseases, Centers for Disease Control, Atlanta, Georgia 30333
Stephen A. Morse
Sexually Transmitted Diseases Laboratory Program, Center for Infectious Diseases, Centers for Disease Control, Atlanta, Georgia 30333
C. Wayne Moss
Division of Bacterial Diseases, Center for Infectious Diseases, Centers for Disease Control, Atlanta, Georgia 30333
N. P. Moyer
University Hygienic Laboratory, University of Iowa, Iowa City, Iowa 52242
James H. Nakano
Division of Viral Diseases, Center for Infectious Diseases, Centers for Disease Control, Atlanta, Georgia 30333
Peter Nash
Bio-Metric Systems, Eden Prairie, Minnesota 55344
Gary R. Noble
Office of the Director, Centers for Disease Control, Atlanta, Georgia 30333
Erling Norrby
Department of Virology, Karolinska Institutet, SBL, S-10521, Stockholm, Sweden
Thomas C. Orihel
School of Public Health and Tropical Medicine, Tulane University, New Orleans, Louisiana 70112
Richard A. Ormsbee
Department of Microbiology, University of Montana, Missoula, Montana 59801
Arvind A. Padhye
Division of Mycotic Diseases, Center for Infectious Diseases, Centers for Disease Control, Atlanta, Georgia 30333
Charlotte D. Parker
Department of Microbiology, School of Medicine, University of Missouri-Columbia, Columbia, Missouri 65212
Charlotte M. Patton
Enteric Bacteriology Section, Center for Infectious Diseases, Centers for Disease Control, Atlanta, Georgia 30333
Beverley J. Payne
Microbiology Unit, Missouri Division of Health Laboratory, Jefferson City, Missouri 65101
Elizabeth Phillips
Medical Technology Program, Mankato State University, Mankato, Minnesota 56001
Peter Piot
Department of Microbiology, Institute of Tropical Medicine, B-2000 Antwerp, Belgium
Frank A. Plummer
Department of Medical Microbiology, University of Manitoba, Winnipeg, Manitoba, Canada
Harry D. Pratt
Vector-Borne Disease Control Training Service, Training Branch, Centers for Disease Control, Atlanta, Georgia 30333
William E. Rawls
Department of Pathology, McMaster University, Hamilton, Ontario, Canada L8S 4J9

C. George Ray
University of Arizona Health Sciences Center, Tucson, Arizona 85724
Errol Reiss
Division of Mycotic Diseases, Centers for Disease Control, Atlanta, Georgia 30333
John H. Richardson
Laboratory Consultation and Training Division, Laboratory Program Office, Centers for Disease Control, Atlanta, Georgia 30333
Michael G. Rinaldi
Department of Microbiology, Montana State University, Bozeman, Montana 59717
Glenn D. Roberts
Section of Clinical Microbiology, Mayo Clinic and Mayo Foundation, Rochester, Minnesota 55905
Alvin L. Rogers
Department of Botany and Plant Pathology, Michigan State University, East Lansing, Michigan 48823
Morrison Rogosa
National Institute of Dental Research, Bethesda, Maryland 20205
Jon E. Rosenblatt
The Mayo Clinic and Mayo Foundation, Rochester, Minnesota 55901
Sally Jo Rubin
Department of Pathology and Laboratory Medicine, St. Francis Hospital and Medical Center, Hartford, Connecticut 06105
Kenneth J. Ryan
Clinical Microbiology Laboratories, Department of Pathology, University of Arizona College of Medicine, Tucson, Arizona 85724
Raymond W. Ryan
Department of Laboratory Medicine, University of Connecticut Health Center, Farmington, Connecticut 06032
L. D. Sabath
Department of Medicine, University of Minnesota, Minneapolis, Minnesota 55455
Samuel K. Sarafian
Department of Microbiology and Immunology, Emory University School of Medicine, Atlanta, Georgia 30333
Julius Schachter
Department of Laboratory Medicine, University of California-San Francisco, and San Francisco General Hospital, San Francisco, California 94110
Jack H. Schieble
Viral and Rickettsial Disease Laboratory, California Department of Health Services, Berkeley, California 94704
Walter F. Schlech III
Victoria General Hospital, Halifax, Nova Scotia, Canada B3J 3G9
Nathalie J. Schmidt
Viral and Rickettsial Disease Laboratory, State of California Department of Health Services, Berkeley, California 94704
F. D. Schoenknecht
Clinical Microbiology, University of Washington, Seattle, Washington 98195
H. Jean Shadomy
Department of Microbiology, Medical College of Virginia, Virginia Commonwealth University, Richmond, Virginia 23298
Smith Shadomy
Department of Medicine, Medical College of Virginia, Virginia Commonwealth University, Richmond, Virginia 23298
John C. Sherris
Department of Microbiology and Immunology, University of Washington, Seattle, Washington 98195
Robert E. Shope
Yale Arbovirus Research Unit, Yale University School of Medicine, New Haven, Connecticut 06510
Jean A. Siders
Division of Clinical Microbiology, Department of Pathology, Indiana University Medical Center, Indianapolis, Indiana 46223
Margarita Silva-Hutner
Mycology Laboratory, Columbia-Presbyterian Medical Center, New York, New York 10032
Robert M. Smibert
Department of Anaerobic Microbiology, Virginia Polytechnic Institute and State University, Blacksburg, Virginia 24061

James W. Smith
Department of Pathology, Division of Clinical Microbiology, Indiana University Medical Center, Indianapolis, Indiana 46223

Jean S. Smith
Centers for Disease Control, Lawrenceville, Georgia 30246

Herbert M. Sommers
Clinical Microbiology, Northwestern Memorial Hospital, Chicago, Illinois 60611

Alex C. Sonnenwirth
Department of Microbiology and Immunology and Department of Pathology, Washington University School of Medicine, and Division of Microbiology, The Jewish Hospital of St. Louis, St. Louis, Missouri 63178

Frances O. Sottnek
Center for Infectious Diseases, Centers for Disease Control, Atlanta, Georgia 30333

Stuart E. Starr
Department of Pediatrics, University of Pennsylvania School of Medicine, Philadelphia, Pennsylvania 19104

Vera L. Sutter
Oral Microbiology Research Laboratory, V.A. Wadsworth Medical Center, Los Angeles, California 90073

Frank E. Swatek
Department of Microbiology, California State University, Long Beach, California 90840

Milton R. Tam
Genetic Systems Corporation, Seattle, Washington 98121

Clyde Thornsberry
Hospital Infections Program, Center for Infectious Diseases, Centers for Disease Control, Atlanta, Georgia 30333

Richard C. Tilton
Department of Laboratory Medicine, University of Connecticut School of Medicine, Farmington, Connecticut 06032

Lucy S. Tompkins
Division of Infectious Diseases, Stanford University School of Medicine, and Clinical Microbiology Laboratories, Stanford University Medical Center, Stanford, California 94305

Govinda S. Visvesvara
Parasitology Division, Centers for Disease Control, Atlanta, Georgia 30333

Alexander von Graevenitz
Department of Medical Microbiology, University of Zurich, CH-8028 Zurich, Switzerland

Kenneth W. Walls
Division of Parasitic Diseases, Center for Infectious Diseases, Centers for Disease Control, Atlanta, Georgia 30333

John A. Washington II
Section of Clinical Microbiology, Department of Laboratory Medicine, Mayo Clinic, Rochester, Minnesota 55905

Benedict L. Wasilauskas
Department of Pathology, Bowman-Gray Medical School, Winston-Salem, North Carolina 27103

Robert E. Weaver
Centers for Disease Control, Atlanta, Georgia 30333

Richard P. Wenzel
Department of Internal Medicine, University of Virginia Medical Center, Charlottesville, Virginia 22908

Dale E. Yelton
Genetic Systems Corporation, Seattle, Washington 98121

Robert H. Yolken
Department of Pediatrics, Division of Infectious Diseases, Johns Hopkins University School of Medicine, Baltimore, Maryland 21205

Preface

The first edition of the *Manual of Clinical Microbiology* was published in 1970. In 15 years the Manual has appeared in four editions, each extensively revised to provide readers and workers with an authoritative and useful volume for the bench and the classroom which would be as up to date as current publication methods would permit. The format developed over the years to achieve this purpose has earned the Manual wide acclaim and increasing distribution, both nationally and internationally.

The Editors of this fourth edition were mindful of past success as they undertook the planning and structuring of the book and the selection of the Editorial Board. The authors selected are experts in their respective fields who were willing to give of their time to write chapters representing the best information currently available for clinical, medical, and public health microbiologists and virologists. Each chapter is presented with the assurance that the methods, reagents, and guidance provided are both time tested and completely up to date.

The Editors, the Editorial Board, and the authors have had one guiding principle: to produce a reliable, current, and convenient manual for clinical and public health microbiologists. To achieve these objectives, the Editors retained the general format that has been used in previous editions. Changes have been made, however. Each chapter in the previous edition of the Manual was either thoroughly revised or completely rewritten. In addition, much new material has been added. Some of the new chapters cover automation and rapid diagnostic methods, computer utilization, surveillance and control of nosocomial infections, anaerobic spirochetes, parvoviruses, papovaviruses, and T-cell leukemia viruses. New sections on molecular methods and sexually transmitted diseases have been added. The previous section on immunoserological tests has been completely restructured into a state-of-the-art unit on immunodiagnostic tests, as have the chapters dealing with antimicrobial susceptibility testing. The section on media, reagents, and stains also has been recast and should prove to be of continuing use to the reader.

The Editors recognize that this fourth edition of the *Manual of Clinical Microbiology* would not have been possible without the capable input of the excellent Editorial Board and authors, who all contributed generously of their time, knowledge, and expertise to make this book a reality. The Editors thank the staff of the ASM Publications Department, including Walter Peter, Susan Birch, Ellie Tupper, Sara Joslyn, and Marie Smith, who held up remarkably well under the stress and strain of missed deadlines and were always helpful and facilitative. We are also most grateful to the many individuals who participated in the peer review process to assist us in reaching our goal.

Finally, we express our indebtedness to our secretaries and those of the many authors, who produced several versions of typescript to the satisfaction of the authors, Editorial Board members, reviewers, and Editors. Our burden was lighter and the task more pleasant in this collaborative effort with so many dedicated and cooperative people.

Edwin H. Lennette
Albert Balows
William J. Hausler, Jr.
H. Jean Shadomy

Chapter 1

Section I. General

Taxonomy, Classification, and Nomenclature of Bacteria

DON J. BRENNER

Medical textbooks are the last place to look for definitive taxonomic data. Medical sources are historically the most conservative in keeping abreast of changes in taxonomy and nomenclature. Such nomenclatural anachronisms as "*Vibrio comma*," "noncholera vibrio," and "*Salmonella typhosa*" are examples of the ultraconservative approach of some medical sources to changes in nomenclature.

The most comprehensive treatment of bacterial classification, particularly for nomenclature, type strains, description of taxa, and references to pertinent literature, is found in *Bergey's Manual of Systematic Bacteriology*, vol. 1, and in *Bergey's Manual of Determinative Bacteriology*, 8th ed. (2, 8). These are invaluable reference sources and should be at the desk of every microbiologist. The 8th edition of *Bergey's Manual* was published in 1974. *Bergey's Manual of Systematic Bacteriology* will be published in four subvolumes, of which the first appeared in January 1984 and the others will appear at approximately 1-year intervals. It is therefore necessary to keep the 8th edition until all four subvolumes of the *Manual* are available. New editions of the *Bergey's Manual for Determinative Bacteriology* (formerly the shorter *Bergey's Manual*) will be published covering subvolumes 1 and 2, and 3 and 4, of the new *Manual* and are designed for bench use.

The 8th edition of *Bergey's Manual* is out of date for those taxa in which new species have been described or in which nomenclatural changes have been made. Furthermore, space limitations make it impossible to fully describe many species. The interested party should, therefore, begin a taxonomic search with *Bergey's Manual* and then augment the information obtained there by searching the *International Journal of Systematic Bacteriology*—in which all new species must be described or reference must be made to the journal in which they are described—and by contacting authorities in the specific field. Other journals that publish papers on new species are the *Journal of Clinical Microbiology, Current Microbiology, Annales de Microbiologie* (Institut Pasteur), and *Systematic and Applied Microbiology*.

In *Bergey's Manual* (2, 8) bacteria are placed in the kingdom *Prokaryotae*. They are subdivided into four divisions: *Gracilicutes* for gram-negative-type cell walls, *Firmacutes* for gram-positive cell walls, *Tenericutes* for organisms lacking a cell wall (mycoplasmas), and *Mendosicutes* for bacteria that have faulty cell walls and presumably lack peptidoglycan. Each division is further subdivided into classes. Within each class are orders, and within the orders are families or (if family names are not available) morphological groups that are further subdivided to genera and species.

From a functional standpoint, the bacteria are divided into a number of "sections" ("parts" in the 8th edition) on the basis of Gram reaction, oxygen requirement, spore formation, and metabolic pattern (gram-positive anaerobes; gram-negative, facultatively anaerobic rods; endospore-forming rods; gliding, nonfruiting bacteria; gram-negative heterotrophs; etc.). Each section (part) is further divided to the species level or, where pertinent, to subspecific categories such as biogroups and serotypes (serovars).

The following sections are included in subvolume 1 of *Bergey's Manual*: spirochetes; gram-negative, aerobic, microaerophilic, motile helical or curved bacteria; gram-negative, nonmotile or rarely motile curved bacteria; gram-negative aerobic rods and cocci; gram-negative facultatively anaerobic rods; gram-negative anaerobic rods; gram-negative anaerobic cocci; dissimilatory sulfate- or sulfur-reducing bacteria; rickettsias and chlamydias; mycoplasmas; and unclassified endosymbionts. Subvolume 2 will contain gram-positive cocci; endospore-forming rods; gram-positive, regular, nonsporing rods; gram-positive, irregular, nonsporing rods; mycobacteria; and nocardioform bacteria. Subvolume 3 will contain gliding, nonfruiting bacteria; anoxygenic photosynthetic bacteria; budding and/or appendaged bacteria; archaeobacteria; sheathed bacteria; gliding, fruiting bacteria; chemolithotrophic bacteria; and cyanobacteria. Subvolume 4 will contain the streptomycetes and their allies. Until subvolumes 2 through 4 are published, the reader must depend on the 8th edition of *Bergey's Manual* and alternative sources for information on the groups that they will address.

A detailed description of the classification of bacteria is, of course, beyond the scope of this chapter, but some comments are in order.

Cowan has referred to "the trinity that is taxonomy": classification, identification, and nomenclature (3). Before discussing reasons for nomenclatural and taxonomic changes and the concept of a bacterial species, it is necessary to establish working definitions for these terms.

"Classification" is simply an orderly arrangement of bacteria into groups. There is nothing inherently scientific about classification. Mandel has said that "like cigars, a good species and a good classification is one which satisfies" (10). Cowan correctly observed that classification is purpose oriented; thus, a successful classification is not necessarily good, and a good classification is not necessarily successful (4). Very often specialty groups classify the same organisms in a different manner or to a different level.

"Identification" is the practical use of a classification to isolate and distinguish desirable organisms from undesirable ones, to verify the authenticity or utility of a culture, or to isolate and identify the causative organism of a disease, of a desirable reaction, etc. "Nomenclature" is the means through which the characteristics of a species are defined and communicated among microbiologists. It is essential that a name has the same meaning to all microbiologists, yet there are many names that are defined differently in different parts of the world or by different microbiological specialty groups. *Klebsiella pneumoniae* is defined differently in England than in most other parts of the world, and *Vibrio cholerae* has often been equated with a single serotype by epidemiologists and clinical microbiologists.

Taxonomy is the science of classification. As a science, it is dynamic and subject to change on the basis of available data. New data often necessitate changes in taxonomy. These changes frequently result in changes in the existing classification, in nomenclature, in criteria for identification, and in the recognition of new species.

For eucaryotes the species definition usually stresses the ability of similar organisms to reproduce sexually, with the formation of a zygote, and to produce fertile offspring. Sexual reproduction, in the eucaryotic sense, does not occur in bacteria.

The term species as applied to bacteria has been defined as a distinct kind of organism, having certain distinguishing features, and as a group of organisms which generally bear a close resemblance to one another in the more essential features of their organization. The problem with these definitions is that they are subjective. What is "a close resemblance"? What are "more essential features"? How many "distinguishing features" are sufficient to create a species? Historically, these questions have been answered arbitrarily. Species were often defined solely on the basis of criteria such as host range, pathogenicity, ability or inability to produce gas in the fermentation of a given sugar, and rapid or delayed fermentation of sugars. Since there was no way to devise a single species definition that could be applied to all groups, criteria used to define species were heavily slanted towards the prejudices of the investigators who described the species. For example, Fritz Kauffmann defined a species as "a group of related sero-, biophagotypes" and believed that serology was the ultimate criterion in taxonomy. To him, each serotype of *Salmonella* was a separate species. We now know that all serotypes of *Salmonella* are genetically the same species. These practices probably led Cowan to state that "taxonomy . . . is the most subjective branch of any biological science, and in many ways is more of an art than a science."

Edwards and Ewing, in their monumental studies on *Enterobacteriaceae* (see reference 5), pioneered in establishing the following principles for characterizing, classifying, and identifying organisms:

1. Classification and identification of an organism should be based upon its overall morphological and biochemical pattern. A single characteristic (pathogenicity, host range, biochemical reaction, etc.), regardless of its importance, is not a sufficient basis for classifying or identifying an organism.
2. To accurately determine the biochemical characteristics of a given species, one must test a large diverse strain sample. The reactions of these strains to any test should be expressed in percentages.
3. Atypical strains, when adequately studied, are often perfectly typical members of a given biogroup within an existing species, and sometimes they are typical members of a new species.

In the 1960s, numerical taxonomy (also called computer or phenetic taxonomy) became widely used. In this method a large number of biochemical, morphological, and cultural characteristics (usually between 50 and 200) are used to determine the degrees of similarity between organisms. More recently, susceptibilities to antibiotics and inorganic compounds have been added to the characteristics used in numerical taxonomy.

Classification and identification of bacteria were significantly improved by the numerical approach to taxonomy. Organisms could be classified on the basis of a large number of characteristics. Many new and some previously ignored biochemical tests were assayed as possible aids in classification. Species-specific and genus-specific tests were identified by the numerical approach. Clinical and applied laboratories could then use these tests to help separate specific species and groups of organisms.

In the numerical approach to taxonomy, investigators often calculate the coefficient of similarity or percentage of similarity between strains (for this discussion "strain" refers to a single isolate from a clinical or other specimen). A dendrogram or a similarity matrix is constructed that joins individual strains into groups and joins one group with other groups on the basis of their percentage of similarity. A hypothetical example of a dendrogram is shown in Fig. 1. In this figure, group 1 represents three strains that are about 95% similar and join with a fourth strain at the level of 90% similarity. Group 2 is composed of three strains that are 95% similar, and group 3 contains two strains that are 95% interrelated and a third strain to which they are 90% similar. Similarity between groups 1 and 2 occurs at the 70% level, and group 3 is about 50% similar to groups 1 and 2.

In some studies all the characteristics included in the similarity matrix are given equal weight, and in some studies certain characters are weighted (for

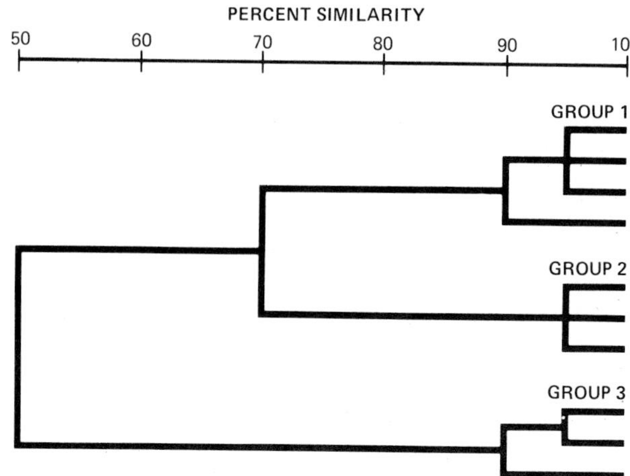

FIG. 1. Hypothetical example of a dendrogram.

example, the presence of spores in *Clostridium* might be weighted more heavily than the organism's ability to utilize a given carbon source). A given level of similarity can be and has been equated with relatedness at the species, genus, and, sometimes, the subspecies level. For instance, strains of a given species may cluster at a 90% similarity level, and species within a given genus may cluster at a 70% level. If these values were applied to Fig. 1, the strains in groups 1, 2, and 3 would each represent a separate species. The species represented by groups 1 and 2 would be placed in the same genus, and the species represented by group 3 would be in a separate genus.

Several problems arise when this approach is used as the sole basis for defining a species:

1. How many and which tests should be chosen?
2. Should the tests be weighted? If so, how?
3. What level of similarity should be chosen to reflect relatedness at the level of species and genus?
4. Is the same level of similarity applicable to all groups?

The molecular weight of DNA in most bacteria is between 1×10^9 and 8×10^9, enough to specify some 1,500 to 6,000 average-sized genes. Therefore, even a battery of 300 tests would, at most, assay only between 5 and 20% of the genetic potential of bacteria. It is almost certainly true that tests which are comparatively simple to carry out and assay (such as tests for carbohydrate utilization and for enzymes whose presence can be assayed colorimetrically) predominate over tests for structural genes, for genes involved in macromolecular synthesis, reproduction, or regulatory functions, and for other genes whose presence is difficult to assay. Additional potential sources of error must be considered when identifying species solely on the basis of phenotype:

1. Different enzymes (specified by different genes) may catalyze the same reaction.
2. Negative reactions can occur when the metabolic gene is present and functional. They can occur through any of several mechanisms, including the inability of the substrate to enter the cell, and a regulatory or suppressor mutation.
3. A negative reaction can occur when the gene is present but not functional because of a mutation in a portion of the gene that is necessary for enzyme activity.
4. The correlation between a reaction and the number of genes (or enzymes) necessary to carry out that reaction is not necessarily one to one. If one assays for the end product, a positive reaction indicates six similar enzymes, whereas a negative reaction can mean the absence or nonfunction of anywhere from one to six enzymes.
5. Fastidious strains will not cluster with nonfastidious strains from the same species. This is often seen with *Escherichia coli* and *K. pneumoniae.* Several other strain characteristics can drastically affect phenotypic characterization. These include slow growth rate, temperature of incubation (*Yersinia, Erwinia*), salt requirement (marine *Vibrio* species), and pH (*Legionella*).
6. Plasmids carrying metabolic genes can enable strains to carry out reactions that are rarely, if ever, seen in the absence of the plasmid (see below).

The same set of "definitive" reactions cannot be used to classify all groups of organisms, and there is no magic number of specific reactions that allows one to define a species. Organisms are identified on the basis of phenotype, but from the taxonomic standpoint, identification to the species level based solely on phenotype is subject to error.

The ideal means of identifying bacterial species would be a "black box" which would separate genes and instantly compare the nucleic acid sequences in a given strain with a standard pattern for every known species—something akin to mass spectrophotometric analysis. Although restriction endonuclease analyses can be done to determine common sequences in isolated genes, we are not at all close to having an appropriate black box, especially one suited for clinical laboratory use. There is, however, a method with which one can compare the total DNA of one organism with that of any other organism. This method, pioneered by Marmur and Doty, Speigelman, Bolton, and McCarthy, and Britten and Kohne, is called nucleic acid hybridization or DNA hybridization. It measures the amount of DNA sequences held in common between any two organisms. One can also approximate the percentage of divergence or unpaired nucleotide bases within related DNA sequences. DNA relatedness studies have been done in yeasts, viruses, bacteriophages, and many groups of bacteria. A partial list of these bacteria includes members of the family *Enterobacteriaceae, Brucella, Bacillus, Pseudomonas, Lactobacillus, Haemophilus, Mycobacterium, Vibrio, Neisseria, Bacteroides* and other anaerobic groups, and *Legionella.*

Five parameters are now used to determine DNA relatedness (1): genome size, guanine-plus-cytosine (G+C) content, DNA relatedness under conditions optimal for DNA reassociation, thermal stability of related DNA sequences, and DNA relatedness under supraoptimal conditions for DNA reassociation. Genetic parameters are presently impractical for clinical laboratories. We therefore must correlate practical biochemical tests with the genetic data. For example, yellow-pigmented strains of *Enterobacter cloacae* were shown to be a separate species genetically, but were not designated as such (*Enterobacter sakazakii*) until there were three practical tests that correlated with the genetic data. The same procedure was followed before designating a number of new species in *Klebsiella, Yersinia,* and *Serratia.*

(i) G+C content (7). The G+C content in bacterial DNA ranges from about 25 to 75%. The G+C percentage is specific for a given species, but is not exclusive for that species. For example, *E. coli,* salmonellae, and *Morganella morganii* have 50 to 52% G+C, and *Bacillus subtilis* and *Pasteurella* spp. have a G+C percentage of about 42%. Therefore, two strains with a similar G+C percentage may or may not belong to the same species. If the G+C content is very different, however, then the strains cannot be members of the same species. G+C content is useful in placing a strain in the proper ballpark for further testing. A good example is a recently isolated organism that is biochemically between *Vibrio* and *Aeromonas.* It was placed in the *Vibrio* genus because its G+C content was 50%, which is within the range for *Vibrio* species and is significantly less than the 57 to 60% G+C found in *Aeromonas* species.

(ii) Genome size (7). True bacterial DNAs have molecular weights between 1×10^9 and 8×10^9 (genome size). Genome size determinations can, in certain circumstances, distinguish between groups that have very different genome sizes. They were used to distinguish *Legionella pneumophila* (the Legionnaires disease bacterium) from *Rochalimaea (Rickettsia) quintana*. *L. pneumophila* has a genome size of about 3×10^9 daltons; that of *R. quintana* is about 1×10^9 daltons.

(iii) DNA relatedness under conditions optimal for DNA reassociation (7). DNA relatedness is determined by allowing single-stranded DNA from one strain to reassociate with single-stranded DNA from a second strain to form a double-stranded DNA molecule. DNA reassociation is a specific, temperature-dependent reaction. The optimal temperature for DNA reassociation is some 25 to 30°C below the temperature at which native double-stranded DNA is denatured into single strands. Studies with a large number of groups indicate that a bacterial species is composed of strains that are usually 70 to 100% related. Relatedness between species is 0% to about 60%. It is important to emphasize that the term "related" does not necessarily mean "identical" or "homologous." Similar nucleic acid sequences can reassociate. This fact and its importance are illustrated below.

(iv) Thermal stability of related DNA sequences (7). It has been shown that each 1% of unpaired nucleotide bases in a double-stranded DNA sequence causes a 1% decrease in the thermal stability of that DNA duplex. By comparing the thermal stability of a control double-stranded molecule in which both strands of DNA are from the same organism with a heteroduplex (DNA strands from two different organisms), we can assess differences in thermal stability. Decreased thermal stability can be thought of as divergence in nucleotide sequences. In practice, strains that are 70% or more related show from 0% to about 5% divergence in related sequences, whereas sequences held in common between different species show 8 to 20% divergence. It is quite important to know the amount of divergence and the relatedness at supraoptimal conditions (see below) when strains are 60 to 70% related.

(v) DNA relatedness under supraoptimal conditions for DNA reassociation (7). When the incubation temperature used for DNA reassociation is raised from 25 to 30°C below the renaturation temperature to only 10 to 14°C below the denaturation temperature, only very closely related (and therefore highly thermally stable) DNA sequences can reassociate. Strains from the same species are 55 to 100% related to supraoptimal incubation temperatures. Strains from different species are 50% or less related. High-temperature reactions are especially important in distinguishing between strains that are 60 to 70% related under optimal reassociation conditions.

By applying these five parameters, we can generate a species definition based on DNA. *E. coli* can be defined as a series of strains with a G+C content of 49 to 52%, a genome size of 2.3×10^9 to 3.0×10^9 daltons, relatedness of 70% or more at an optimal reassociation temperature with 0 to 4% divergence in related sequences, and relatedness of 60% or more at a supraoptimal reassociation temperature. Experience with more than 100 species resulted in an arbitrary molecular definition of species as a group of strains with similar G+C content and genome size that are 70% or more related with 0 to 5% divergence among related sequences and in which relatedness remains at 55% or more at supraoptimal incubation temperature. There is general agreement that 70% or more indicates relatedness at the species level. The 70% species relatedness rule has been occasionally ignored when the existing nomenclature is both very deeply ingrained and useful. One such example is *E. coli* and the four species of *Shigella*. These organisms are all 70% or more related and should therefore be grouped into a single species, instead of the present five species in two genera. This change has not been made because the primary consideration was to avoid the confusion that such a change would create among members of the medical community. Another example is *Yersinia pestis* and *Yersinia pseudotuberculosis*, which are the same species genetically. It has been proposed that they be treated as two subspecies of a single species, but clinically it is extremely important to continue to report *Y. pestis* and to distinguish it from *Y. pseudotuberculosis*.

DNA relatedness provides one species definition that can be equally applied to all organisms. It is not subject to phenotypic variation, to mutations, or to the presence or absence of metabolic or other plasmids. This is because it measures overall relatedness, and atypical biochemical reactions, mutations, and plasmids affect only a very small percentage of the total DNA. This point is illustrated by looking at DNA relatedness data obtained from a large number of biochemically atypical *E. coli* or *E. coli*-like strains (Table 1). Many of these biochemically atypical strains were shown to be genetically typical *E. coli*. The biochemical borders of *E. coli*, therefore, include

TABLE 1. Definition of the biochemical parameters of *E. coli* on the basis of DNA relatedness

Biotypes 80% or more related to typical *E. coli*		
Urea$^+$	KCN$^+$	Mannitol$^-$
Citrate$^+$	Inositol$^+$	Yellow pigment
H_2S^+	Phenylalanine deaminase$^+$	Triple decarboxylase$^-$
Indole$^-$	Adonitol$^+$	Methyl red$^-$
Urea$^+$, mannitol$^-$	H_2S^+, citrate$^+$	H_2S^+, citrate$^+$
Anaerogenic, nonmotile, lactose$^-$		Methyl red$^-$, mannitol$^-$
Biotypes 60% or less related to typical *E. coli*		
KCN$^+$, yellow pigment, cellobiose$^+$ = new species		
Urea$^+$, KCN$^+$, citrate$^+$, cellobiose$^+$ = *Citrobacter amalonaticus*		

all of these biogroups. Two groups of strains were shown not to belong to *E. coli*. These are KCN- and cellobiose-positive, yellow-pigmented strains, which proved to be a new species, and urea-, KCN-, citrate-, and cellobiose-positive strains that were in fact *Citrobacter* spp.

In practice, the approach to bacterial taxonomy should be polyphasic. The first step is phenotypic grouping of strains by biochemical reactions and any other characteristics of interest. In the second step, phenotype groups are tested for DNA relatedness to determine whether the observed phenotypic homogeneity (or heterogeneity) is reflected by genetic homogeneity or heterogeneity. The third and most important step for identification is reexamination of the biochemical characteristics of the DNA relatedness groups. This allows a determination of the biochemical borders of each group and of those reactions that are of diagnostic value for the group. For identification of a given organism, we weight specific tests on the basis of their correlation with DNA results. Occasionally, the commonly used reactions will not totally distinguish between two distinct DNA relatedness groups. In these cases, one must search for other biochemical tests that are of diagnostic value. Heavily weighted tests are of great value for specific organisms (coagulase for *Staphylococcus aureus*, DNase for *Serratia* spp., etc.). We often forget that before these weighted tests are of diagnostic value the organism has been well characterized by growth on selective media, by colonial and cellular morphology, and by other biochemical tests.

Some newer DNA methods are already in use for the detection of pathogenic bacteria. These methods, which include specific gene probes, plasmid profile analyses, and colony DNA hybridization, are covered in detail in Chapter 108 but deserve mention here as tools for identification and for classification at and below the species level (7).

It is possible to detect cholera toxin and both heat-stable and heat-labile enterotoxins in *E. coli* by means of specific gene probes. The genes (or genetic regions including all or part of the genes coding for these toxins) are hybridized with purified DNA from suspected toxigenic strains, with colonies lysed directly on nitrocellulose filter paper, or directly with unprocessed stool material lysed on nitrocellulose filter paper. When stools are used, results are obtained within 48 h. Gene probes are being developed for the identification of toxigenic *Yersinia enterocolitica* and for identification of salmonellae, shigellae, *Neisseria gonorrhoeae*, and legionellae.

Detection of a specific plasmid associated with virulence can be used to detect virulent strains of *Y. enterocolitica* and invasive strains of *E. coli*.

Molecular analysis of plasmids has recently been extensively used to identify epidemic strains of bacteria. Plasmid DNAs are separated electrophoretically (by molecular weight). Plasmids of similar size can be specifically fragmented, by treatment with one or a series of restriction endonucleases which cleave DNA at specific sites, to determine similarities precisely. For plasmid profile and restriction endonuclease analysis to be effective in monitoring epidemic strains, the strains must contain plasmids and these plasmids must be different from those of nonepidemic strains. Fortunately, these conditions are met in the majority of *Enterobacteriaceae* and other bacteria involved in nosocomial disease outbreaks, and in many of the bacteria responsible for food-borne disease outbreaks.

The purpose of classification and identification is to be able to distinguish one organism from another and to group similar organisms on the basis of criteria of interest to all microbiologists or to any single specialty group. The purpose of nomenclature is to provide a convenient system of communication to define an organism without the necessity of listing its characteristics. The most important level of communication is at the species level. A species name should mean the same thing to everyone, regardless of his or her specialty. We cannot have effective communication if strains of the same species are given different names on the basis of source of isolation, serotype, presence or absence of a converting bacteriophage, or the ability to perform a specific function, such as cause a disease, produce an antibiotic, etc. Species have been created on the basis of each of these criteria and many others. These criteria may be extremely important for a specialty group, but, by themselves, they are not a sufficient basis for establishing a species. Species should be established based on the polyphasic criteria mentioned above.

Special-interest groups of microbiologists need to communicate, but their needs can and should be met by designations below the species level as "groups" or "types" on the basis of common serological or biochemical reactions, phage or bacteriocin sensitivity, pathogenicity, or other characteristics. For example, bioserogroup or bioserotype is a group of strains of the same species with common biochemical and serological characteristics that set them apart from other members of the species. Many of these are already commonly used and accepted: serotype, phage type, colicin type, biotype, bioserotype, and pathotype. The ending "var" (phagovar, etc.) can be substituted for "type." We prefer "type." Host specificity has been expressed in many groups by the term "variety." For example, *Erwinia herbicola* var. *ananas* causes rot in pineapple and *Erwinia chrysanthemi* var. *zeae* is pathogenic for corn. By using these designations, there can be communication among all microbiologists at the species level and among all special-interest groups at an intrasubspecific (below species) level.

There is no genetic definition of a genus. If there were, an ideal genus would be composed of a series of species that are similar phenotypically and genetically (50 to 65% related). Some genera that contain phenotypically similar species approach this genetic criterion: *Citrobacter*, *Yersinia*, and *Serratia*, to name a few. More often, the phenotypic similarity is present, but the genetic relatedness is not. *Bacillus*, *Clostridium*, *Vibrio*, *Campylobacter*, *Pseudomonas*, and *Legionella* are good, or at least accepted, phenotypic genera in which relatedness between species is not 50 to 65% but 0 to 65%. When both phenotypic similarity and genetic similarity do not occur, phenotypic similarity should generally be given priority in establishing genera. The reason for phenotypic priority at the genus level is a practical one. When identifying organisms at the bench, it is desirable to have the most phenotypically similar species in the same genus. Primary consideration for a genus is that it contains biochemically similar species that are convenient or important to consider as a group and that must be separated from one another at the bench.

In some cases, generic names have been changed because the original name was not validly published for one of a variety of reasons. These changes are legislated (see below). More often, name changes are made on the basis of data not available when the original name was proposed. More than one name for a genus (or species) can also result from two investigators independently publishing different names for the same group. Even though one of the names will have priority, both will often persist for a long time.

A typical example of a current nomenclatural problem based on the interpretation of data is in the family *Legionellaceae.* Since the first species, *Legionella pneumophila*, was described by workers at the Centers for Disease Control, 21 additional species have been described or are in the process of being described, all of which were placed in the genus *Legionella.* Another group of investigators confirmed the first five species and concluded that two additional genera were necessary on the basis of DNA relatedness. This second group gave precedence to a genetic genus concept, whereas the original investigators argued for a single genus on the basis of phenotypic, pathogenic, and treatment similarities among all of the species. The first group believed that these similarities, and the present inability to separate most species except by serology, made it impractical to create additional genera. The difference in opinion will ultimately be settled by usage in the scientific and medical community. Thus far the single genus *Legionella* has by far the greater usage, but the other generic names are occasionally seen in the literature.

Why should species be named, how are they named, and where are they named? According to the *International Code of Nomenclature of Bacteria* (9) (*Bacteriological Code*), "the primary purpose of giving a name to a taxon is to supply a means of referring to it." In other words, names are to foster communication and to ensure that the description of a given set of characteristics (by using a name to define them) has the same meaning to all scientists, just as it is more meaningful to say Willie Mays or Sean Connery than to describe the greatest center fielder of all time or one of the men who played the role of James Bond.

Species are named in accordance with principles and rules of nomenclature as set forth in the *Bacteriological Code* (9). The first principle of bacterial nomenclature is concerned with creating stability, avoiding or rejecting names that cause error or confusion, and avoiding the useless creation of names. Scientific names are usually taken from Latin or Greek and, regardless of their origin, are treated as Latin. The correct name of a species or higher taxonomic designation is determined by three criteria: valid publication, legitimacy of the name with regard to the rules of nomenclature, and priority of publication (it is the first validly published name for the taxon). "A name has no status under the rules and no claim to recognition unless it is validly published" (9).

Until 1 January 1980, priorities for names dated from 1 May 1753. This caused much confusion since it was difficult to search the literature to ensure that a species had not been previously proposed. The early descriptions were often sketchy and were based on fewer and often different tests than are now used. Furthermore, strains representing the species proposed in the 19th century often were not available, were of uncertain authenticity, or, when tested, did not exactly correspond to the published properties of the species.

Priorities for bacterial names now start as of 1 January 1980. On that date, the *Approved Lists of Bacterial Names* were published in the *International Journal of Systematic Bacteriology* (IJSB) (11). Names not on those lists lost all standing nomenclatural status (*"Arizona hinshawii"* is an example). Now we can quickly determine whether a species has been previously published by consulting the *Approved Lists* and the lists of valid names published periodically in IJSB. The *Approved Lists* and the requirement to search only to 1980 for prior species proposals has removed and will continue to remove much of the past confusion and will preclude many future problems.

To be validly published, a new species proposal must contain the species name, a description of the species, and the designation of a type strain for the species, and the name must be published in IJSB. The proposed name is automatically validly published if the proposal is published in IJSB. If the proposed name is published in another journal, it is not validly published until it appears in IJSB. It is the author's responsibility to send reprints of such publications to the editor and to request publication of the new name(s) in IJSB. In this case, the date of valid publication is the date of publication in IJSB.

Most people seem to believe that, once proposed, a name must go through some formal process leading to its official acceptance. The opposite is true: a validly published name is assumed to be correct unless and until it is officially challenged. This is done by publishing a request for an opinion (to the Judicial Commission of the International Union of Microbiological Societies) in IJSB. The Judicial Commission will then seek advice from appropriate experts and render an opinion. This is done only in cases in which the validity of a name is questioned with respect to compliance with the rules of the *Bacteriological Code.* A question of classification that is based on scientific data (for example, whether a species, on the basis of biochemical or genetic characteristics or both, should be placed in a new genus or an existing genus; see the above discussion of *Legionella*) is not settled by the Judicial Commission but by the preference and usage of the scientific community. This is why generic synonyms exist. Some more recent examples are *Citrobacter (Levinea) amalonaticus, Citrobacter diversus (Levinea malonatica), Morganella (Proteus) morganii, Legionella (Tatlockia) micdadei*, and *Legionella (Fluoribacter) dumoffii.* The reader is invited to consult the *Bacteriological Code*, the *Approved Lists*, and *Bergey's Manual* for further details on bacterial nomenclature.

Taxonomy, albeit far from perfect, is not designed to create confusion in the lives of clinical microbiologists or any other group. The task of the taxonomist is to describe all species, using the best available scientific tools. Newly recognized species have sometimes been masquerading as members of well-known species. Such was the case with *Klebsiella oxytoca* (indole-positive *Klebsiella pneumoniae*), *Enterobacter sakazakii* (yellow-pigmented, sorbitol-negative, delayed DNase-positive *Enterobacter cloacae*), *Vibrio mimicus* (sucrose-negative *Vibrio cholerae*), and *Yersinia intermedia* (rhamnose- and melibiose-positive *Yersinia enterocolitica*). Many others, including the Lyme disease spirochete, legionellae, *Vibrio hollisae, Yersinia ruckeri, Enterobacter gergoviae*, and *Escherichia vulneris*, were not previously recognized.

Clinical microbiologists should keep up with taxonomic changes to know which newly described species present potential clinical problems and which are rarely, if ever, human pathogens. Examples of newly described human pathogens, with an argument that clinical microbiologists should stay abreast of taxonomic changes, were recently given by Farmer (6).

A strong argument has been made for identifying species in the clinical laboratory, and this is not the place to repeat this argument. The probability of correctly identifying an organism, especially an atypical strain, is highly dependent upon the number of tests done. There is no one magic number of tests, because both the number of tests and the specific tests used vary from one group of organisms to another. How far one should go in identifying organisms and which organisms one should screen for are decisions that must be made in each laboratory on the basis of the type of population it serves (Indian, pediatric, cancer or burn hospital, outpatient clinic) and its function (primary laboratory, reference laboratory, etc.).

A final word of caution: you will never see a given organism if you do not use appropriate isolation and enrichment media and do the proper tests for identification. Newly described biochemical tests should be done exactly as described in the literature. Modifications, shortcuts, or any change in media or reagents must not be attempted unless and until these changes have been published or shown to yield comparable results by the individual laboratory. Several years ago, strains of lactose-positive *Salmonella* caused an epidemic in Central and South America. These strains were isolated in some United States laboratories, but other laboratories reported that they had never seen them. Many laboratories pick only lactose-negative colonies from isolation plates; thus, they will never have a problem with "lactose-positive" *Salmonella*. Similarly, many laboratories never isolated *Yersinia enterocolitica, Campylobacter fetus*, many *Vibrio* species, or legionellae until they set up specific procedures for isolating these organisms.

LITERATURE CITED

1. **Brenner, D. J.** 1978. Characterization and clinical identification of *Enterobacteriaceae* by DNA hybridization. Prog. Clin. Pathol. **7**:71–117.
2. **Buchanan, R. E., and N. E. Gibbons (ed.).** 1974. Bergey's manual of determinative bacteriology, 8th ed. The Williams and Wilkins Co., Baltimore.
3. **Cowan, S. T.** 1965. Principles and practice of bacterial taxonomy—a forward look. J. Gen. Microbiol. **37**:43–153.
4. **Cowan, S. T.** 1971. Sense and nonsense in bacterial taxonomy. J. Gen. Microbiol. **66**:1–8.
5. **Edwards, P. R., and W. H. Ewing.** 1971. Identification of *Enterobacteriaceae*, 3rd ed. Burgess Publishing Co., Minneapolis.
6. **Farmer, J. J., III.** 1981. Should clinical microbiology keep up with taxonomy—yes. Clin. Microbiol. Newsl. **1**:5–6.
7. **Johnson, J. L.** 1981. Genetic characterization, p. 450–474. *In* P. Gerhardt, R. G. E. Murray, R. N. Costilow, E. W. Nester, W. A. Wood, N. R. Krieg, and G. B. Phillips (ed.), Manual of methods for general bacteriology. American Society for Microbiology, Washington, D.C.
8. **Krieg, N. R., and J. G. Holt (ed.).** 1984. Bergey's manual of systematic bacteriology, vol. 1. The Williams and Wilkins Co., Baltimore.
9. **Lapage, S. P., P. H. A. Sneath, E. F. Lessel, V. B. D. Skerman, H. P. R. Seeliger, and W. A. Clark (ed.).** 1975. International code of nomenclature of bacteria, 1975 revision. American Society for Microbiology, Washington, D.C.
10. **Mandel, M.** 1969. New approaches to bacterial taxonomy: perspective and prospects. Annu. Rev. Microbiol. **23**:239–275.
11. **Skerman, V. B. D., V. McGowan, and P. H. A. Sneath (ed.).** 1980. Approved lists of bacterial names. Int. J. Syst. Bacteriol. **30**:225–420.

Chapter 2

Microscopy

STEVEN D. DOUGLAS

Microscopy is the science of the interpretive uses and applications of microscopes. The history of the development of microscopy is closely linked to the beginning of microbiology. Descriptions of protozoa and bacteria were recorded in 1683 by Anton van Leeuwenhoek, who used small single lenses ground into convex surfaces; one of these microscopes, which remains at the University of Utrecht, resolves 1.4 μm at 270 magnifications.

Barer has lucidly pointed out that the two main aims of microscopy are formation of a magnified image with as few optical defects as possible and achievement of contrast (1). Contrast is based on the differential absorption of light between the specimen under study and its background.

Both direct visualization and photographic films and plates are sensitive to variations in light intensity, and all methods of microscopy depend on the ability of specimens to absorb, refract, or reflect light, thus producing contrast between details within a specimen and its background.

The resolving power of a lens is defined as the ability to discriminate detail in an object by means of the eye, microscope, camera, or photograph. The limit of resolution is defined as the minimal distance between two points that allows for their discrimination as two separate points. Thus, as derived by Abbé from the law of diffraction: limit of resolution (d) = 0.61 λ/NA. The numerical aperture (NA) = $n \times \sin\alpha$, where n is the refractive index of the medium (which with most optical media does not exceed 1.40), $\sin\alpha$ is the sine of the semiangle of the aperture (which cannot exceed 1.0), and λ is the wavelength of the light (Fig. 1) (1, 2, 7, 15–17).

For the human eye, resolution is achieved when the visual angle is large enough so that light rays from two different points encounter different receptor elements in the retina. The minimal visual angle permitting resolution for the eye is about 1 min of arc. Thus, for the shortest distance of clear vision of 25 cm, the linear distance between objects or resolving power of the unaided eye is about 0.07 mm (Fig. 2a).

The NA is the major determinant of the resolving power of an objective lens. The theoretical limiting resolving power of light microscopes (omitting consideration of refractive aberrations) is calculated with the Abbé formula (e.g., with white light λ = 550 nm, the resolving power is about 250 nm; with monochromatic violet λ = 400 nm, the resolving power is 170 nm). The development of the electron microscope made it possible to greatly increase resolving power by the substitution of highly accelerated electrons for conventional light waves.

The wavelength of electrons is determined by the DeBroglie formula [$\lambda = (12.2\sqrt{V})(0.1 \text{ nm})$], where V is the acceleration voltage (5). Thus, at 50,000 V, λ = 0.005 nm, and the theoretical limit of resolution is 0.2 nm. There are a number of factors, however, which lead to the loss of resolution in the study of biological specimens, particularly in electron microscopy (see below).

The intent of this chapter is to describe various types of light and electron optical techniques which are readily available, indicating their advantages and limitations, and to mention briefly the application of these techniques to clinical microbiology.

LIGHT MICROSCOPY

Simple microscope

A simple microscope is composed of a single magnifying lens which has a large field and produces an upright image (Fig. 3). This type of microscope may be fitted with a micrometer, grid, or other reticle and an appropriate stand and illuminators. Its use is limited by a low NA and it has a resolution of about 10 μm. Simple lenses are useful for dissection, measurement, and examination of agglutination reactions. A widely used modified simple microscope is the bacterial colony counter.

Compound microscope

Compound microscopes are made up of at least two lens systems (Fig. 4a). The standard compound microscope is composed of an objective lens, which has multiple compartments with short focal lengths and which forms a tiny inverted real image in the focal plane of another lens, and the ocular or eyepiece (or projector lens when a camera is attached to the microscope), which magnifies the image so that it can be resolved by the observer.

The objective lens is the most important single determinant of the quality of the image produced by a particular microscope. Unfortunately, no lens is capable of producing a perfect image, and all lenses must be corrected for one or more aberrations or lens errors. Several types of objective lenses are available which have corrections for various types of aberrations; e.g., achromatic lenses are corrected for primary chromatic aberrations, apochromatic objectives are corrected for both chromatic and spherical aberration, and "plan" objectives are multicomponent lenses with these corrections as well as correction for curvature of the field. ("Plan" objectives are the most expensive.) In addition to the degree of correction of a lens, image quality is determined by the NA of the lens. As seen by the application of the Abbé formula, lenses of higher NA (degree of correction being equal) have better resolving power. For high-resolution light microscopy, immersion objectives (NA, 1.2 to 1.4) are useful. The "oil" used in conjunction with immersion objectives is a nondrying, refractively stable, viscous liquid with a refractive index of 1.51. The refractive index of cover glasses and glass slides is comparable. Under most conditions (lenses, illumination, and biological material), the compound light microscope has useful magnifications of up to ×1,000 and resolution

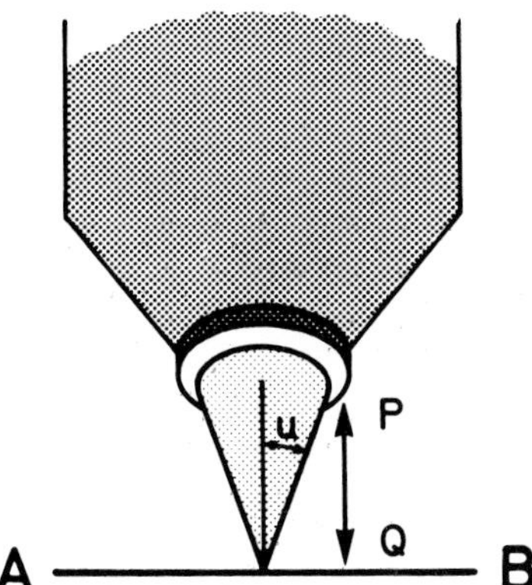

FIG. 1. Schematic view of the aperture cone of an objective lens. Angle (μ) is equivalent to α in the formula NA = $n \times \sin\alpha$. AB, Object plane; PQ = free working distance. From James (7).

of about 200 to 300 nm (Fig. 2) (1, 2, 7, 15–17). Examples of compound microscopes commonly used in the clinical laboratory include the stereoscopic microscope and the bright-field biological microscope.

The stereoscopic microscope combines two compound microscopes which produce separate images for each eye at a slightly different angle. A single or two separate objectives may be used. The single-objective stereoscopic microscope allows simultaneous and complete focus of both images. Either reflected or transmitted light may be used for illumination; however, since there is no substage condenser, the type and intensity of illumination are limiting factors (15). Stereoscopic microscopes are useful in examining colonial morphology of bacteria, fungi, and tissue cultures.

The biological compound microscope utilizes a tungsten filament lamp and bright-field condenser which should have the capability to be centered and focused. Alignment of the condenser by use of critical or Köhler illumination is crucial to obtaining the highest possible resolution. Since both viable and fixed materials have very little contrast when viewed against a bright background, various stains are employed to produce contrast. Vital stains, such as Janus green or methylene blue, are used when observing living cells, and many bacteriological and histochemical stains are available for fixed material. The biological microscope is useful in the examination and identification of microorganisms in smears, tissue fluids, and tissue sections. Another method for producing contrast in unstained material is by the introduction of Rheinberg disks into the light path to produce optical staining. Light of low angular aperture enters the central disk of one color, and light of high angular aperture passing through the specimen enters the lateral portion of the disk, which is of one or more different colors, resulting in a colored image of an unstained specimen (15).

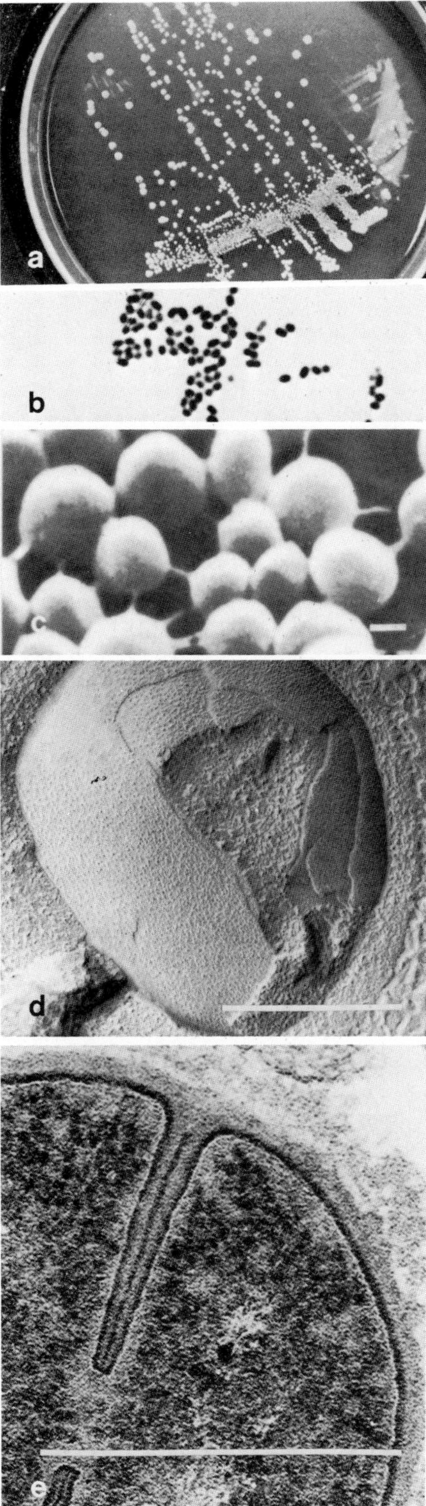

FIG. 2. *Staphylococcus aureus* visualized by different microscopic techniques. (a) Colonies visualized by the unaided eye. (b) Gram-stained organisms viewed with a compound microscope by oil immersion. ×1,000. (c) Scanning electron micrograph. ×10,000. (d) Transmission electron micrograph of freeze-fracture replica. ×50,000. (e) Transmission electron micrograph of thin section. ×100,000. Bar, 0.5 μm. Parts a, b, d, and e are from my laboratory; c is from Klainer and Geiss (9).

Phase-contrast microscopy

When a single-beam light is passed through a specimen, which partially absorbs the light, there is a change in amplitude of the light waves, their phase

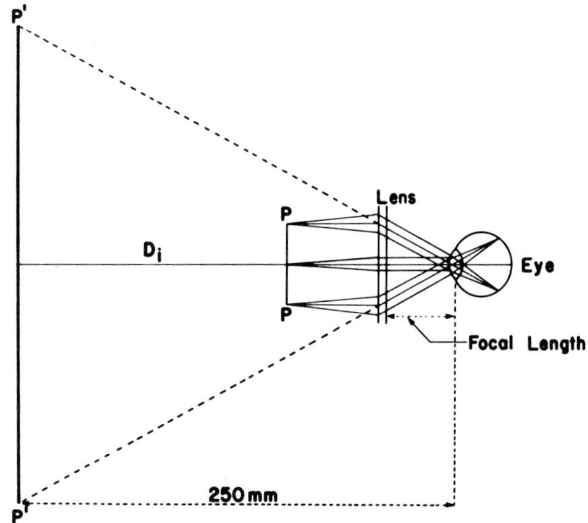

FIG. 3. Simple microscope. From Rochow and Rochow (15).

relationship, or their plane of vibration. Zernike in 1934 developed optical methods to separate the incident and diffracted waves of light and thus to amplify the phase difference between them. In the phase-contrast microscope, the condenser has an annular diaphragm, which produces a hollow core of light, and the objective has a glass disk (phase plate or annular groove) with a thin film of transparent material deposited on it which accentuates phase changes produced in the specimen (Fig. 5) (15–17). This phase change is observed as a difference in light intensity. Phase plates may either retard (positive phase plate) or advance (negative phase plate) the diffracted light relative to the direct light. The phase-contrast microscope with positive phase plates reveals small changes in specimen refractive index as areas of greater darkness. Positive phase-contrast objectives are useful in the examination of granular inclusions and flagella.

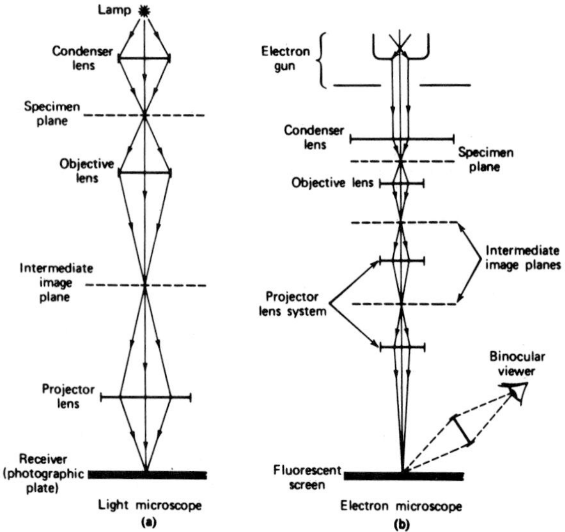

FIG. 4. Comparison of light and electron microscopes. From Slayter (17).

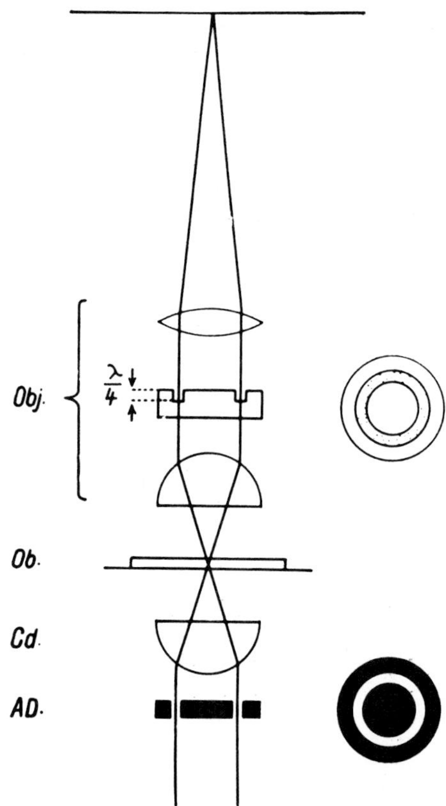

FIG. 5. Path of rays in a phase-contrast microscope. The annular condenser diaphragm and the phase plate are shown at right. From Ruthmann (16).

Phase-contrast microscopes may be modified to produce variable alteration of phase and amplitude (variable phase-amplitude microscopy) or may use oblique polarized illumination (modulation contrast microscopy) to increase contrast and resolution (15).

Dark-field microscopy

The compound microscope may be fitted with a dark-field condenser which has an NA greater than the objective. Light passing through the specimen is diffracted and enters the objective lens, whereas undiffracted light does not, resulting in a bright image against a dark background. Since light objects against a dark background are more easily perceived by the eye than the reverse, dark-field illumination is useful in visualizing bacterial flagella and spirochetes which are poorly defined by bright-field and phase-contrast microscopy. However, no increase in actual resolution is obtained.

Polarization microscopy

Many materials are anisotropic, causing nonrandom oscillations of light within the plane of the structure or molecule. Objects in which the refractive index varies with direction are known as doubly refracting or birefringent.

The types of anisotropy found in living cells include optical anisotropy of crystals, molecular birefringence of long or flat molecules, and form birefringence produced by orientation of molecules or fibers. Anisotropy is detected by illumination with polarized light. Polarizing microscopes have two rotatable polars, one at the light source (polarizer) and one within the objective analyzer (15). Polarization microscopy is a valuable technique for the study of crystals and fibrous proteins. This type of microscopy has been used to study spindle fibers of dividing cells and the orientation of purines and pyrimidines within chromatin fibers (15). Although the polarizing microscope has not been extensively used in the examination of microorganisms, the application of this technique in the clinical laboratory is potentially useful.

Interference microscopy

Both the phase-contrast and interference microscopes are based on interference effects caused by phase changes in the specimen. In interference microscopy, phase changes are converted into amplitude changes by the use of dual light beam systems. Light rays passing through the object interfere with light which has traversed the object in its immediate vicinity. In a system developed by Nomarski, gradients of optical path differences are visible in the object, resulting in differential interference contrast (15).

The advantage of interference microscopy as compared with phase-contrast microscopy is the ability to discern phase changes at the actual point at which they occur, eliminating the halos observed in phase-contrast microscopy. In addition, a shadow-casting effect is produced in the image. Interference microscopy can be used to determine dry mass and thickness of cells. These advantages, however, must be balanced against the high cost of the instrument and the expertise which is required for proper illumination and alignment in order to obtain high resolution.

Fluorescence microscopy

Compounds which have the ability to absorb light of one wavelength and emit light of a longer wavelength are termed fluorochromes. Examples of naturally occurring fluorochromes are certain vitamins, steroids, porphyrins, and nucleic acids. Other compounds which fluoresce include rhodamine, auramine, and fluorescein and acridine dyes, which are used in many fluorescent labeling techniques.

The fluorescence microscope is based on the principle of removal of incident illumination by selective absorption, whereas light which has been absorbed by the specimen and reemitted at an altered wavelength is transmitted. The light source must produce a light beam of appropriate wavelength. An excitation filter removes wavelengths which are not effective in exciting the fluorochrome used. The incident beam is thus of a relatively narrow range of wavelengths. The light fluoresced by the specimen is transmitted through a barrier filter which removes the incident wavelength from the beam (12). Thus, only light which has interacted with the specimen with a change in wavelength contributes to the intensity in the image plane (Fig. 6) (6).

Light sources used in fluorescence microscopy include halogen-quartz lamps, high-pressure mercury arcs, and xenon arcs. In 1967, Ploem developed an epi-illuminator system employing a vertical illuminator and a dichroic mirror which permits passage of light

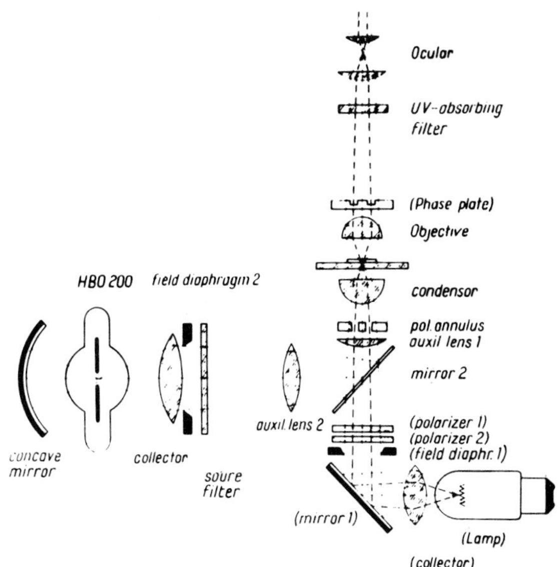

FIG. 6. Path of rays in the phase-contrast fluorescence microscope. Parts in parentheses belong to the phase-contrast equipment. From Ruthmann (16).

of selected wavelengths through the mirror in one direction only. The excitation beam is focused directly on the specimen through the objective lens, and the emitted light is transmitted to the eye through the dichroic mirror.

Fluorescence techniques

Fluorescent dyes have been used to stain many types of biological materials. Truant's acid-fast stain containing auramine and rhodamine is used to detect acid-fast organisms in tissue sections and sputum specimens. Acridine dyes are used to stain nuclei. The specificity of these techniques is determined by the chemical interaction of the stain, the cellular components, and the staining conditions.

Coons developed a technique in which β-anthracene isocyanate, a blue-fluorescing compound, was conjugated to pneumococcal antiserum, and used this reagent to detect bacterial antigen in tissue (12). The specificity of this immunofluorescence technique resides in the conjugated protein, and the fluorochrome serves to localize the reaction. Fluorescein isothiocyanate and lissamine rhodamine are the most commonly used fluorochromes for immunofluorescence, since they have greater stability in neutral, aqueous solutions. At alkaline pH they can be covalently conjugated to proteins through ε-amino and terminal amino groups. Fluorescein has an absorption maximum between 490 and 496 nm and emits a characteristic green color at 517 nm. Rhodamine has an absorption maximum at 550 nm with maximal emission at 580 nm.

The use of fluorescence and immunofluorescence techniques has greatly improved the ability to detect bacterial antigens in tissue smears, sections, and fluids. Fluorescence microscopy frequently requires additional filters to reduce the background fluorescence of naturally fluorescent material in biological specimens. Additional sensitivity is achieved by using the indirect immunofluorescence technique, which employs a nonconjugated primary antibody and a conjugated secondary antibody that amplifies the amount of fluorescence observed. The fluorescent treponemal antibody (FTA-ABS) test used to detect antibodies to *Treponema pallidum* in patient sera is an example of this technique.

Fluorescence may be further enhanced by utilizing the biotin-avidin system. Avidin, a basic glycoprotein derived from egg albumen, has a high affinity for the vitamin biotin. Biotin is easily coupled to antibody, and the subsequent addition of fluorochrome-labeled avidin results in bright fluorescence.

The use of the Ploem epi-illuminator has facilitated the use of two microscopic techniques simultaneously. For example, fluorescence and phase-contrast microscopy are useful in examining living cells which have been stained with fluorescent antibodies or cells which have been stained with two different fluorochromes.

Techniques have recently been developed to regulate the flow of cells into single droplets which are illuminated by monochromatic laser beams (argon ion lasers) and monitored by fluorescence detectors (11). This technique, called flow cytometry, has employed monoclonal antibodies for identification of cell surface antigens, particularly to delineate lymphocyte subpopulations. Flow cytometry will simultaneously analyze fluorescence (amount of monoclonal antibody bound), light scatter (cell size), and cell viability.

ELECTRON MICROSCOPY

The electron microscope and special preparatory techniques developed in conjunction with it have made it possible to resolve cellular detail at the macromolecular level. Two types of electron microscopy and their respective preparatory methods will be considered here: transmission electron microscopy (TEM) and scanning electron microscopy (SEM). Although a detailed consideration of electron optics is beyond the scope of this chapter (16, 17), broad general principles and the utility of electron microscopy in microbiology will be discussed. The high resolution of the electron microscope is made possible by the use of a stream of electrons instead of light rays to produce a magnified image of an object. Recently, high-voltage (15) electron microscopes have been developed which have even greater resolution than standard transmission electron microscopes. Since electron beams can be deflected by magnetic and electric fields, it is possible to design image-forming systems. The basic design of the electron microscope involves a tungsten filament or cathode which, when placed in a vacuum tube and heated, emits electrons which can be accelerated and controlled. The electrons are focused on the plane of the specimen by electromagnetic lenses and metallic apertures. Utilizing electromagnets as condenser, objective, and projector lenses, magnified images are produced which are visualized on a fluorescent screen or recorded on photographic plates (Fig. 2d, 2e, and 4b).

TEM

In typical TEM, a beam of electrons generated by the heating of a tungsten filament is accelerated at a voltage of 25 to 125 kV and passed through a specimen of a thickness between 20 and 80 nm. Except in the case of high-voltage TEM, preparation techniques for TEM are directed toward the production of specimens of sufficient thinness and contrast to permit the passage of electrons through the specimen and onto a photographic film or plate where the specimen image is recorded. High-voltage TEM with accelerating voltages in millions of kilovolts permits the study of thicker specimens, up to approximately 2 μm. The most frequently used preparative techniques for conventional TEM are negative staining, microtomy techniques, and freeze-etching. The role of electron microscopy in microbiology has been summarized by Costerton (3).

Negative staining is based on the observation that specimens completely surrounded by nonpenetrating electron stains will appear bright on a dark background (negative contrast). These negative stains are applied to supporting films (e.g., Formvar and carbon) with the specimen. The most frequently used negative stain is phosphotungstic acid. Negative staining has had wide application in the study of macromolecules, viruses, bacteriophages, and bacterial organelles in a cell-free system (5).

Specimen preparation for thin sectioning (ultramicrotomy) requires fixation, dehydration, and embedding in plastic (5). The most commonly used fixatives for electron microscopy are glutaraldehyde, osmium tetroxide, and uranyl acetate (8, 14). The specimens are then dehydrated, usually in ethanol or acetone. After dehydration, specimens are impregnated with a polymerizable material (e.g., epoxy resins) and then polymerized. The polymerized blocks are sectioned by use of an ultramicrotome which can produce sections as thin as 30 nm. The image contrast of sections may be further enhanced by staining with heavy metal salts (e.g., lead) (Fig. 2e) (13).

It is well known that preparative techniques, particularly those used for thin-sectioned material, produce artifacts within the specimen. Problems encountered with fixation, dehydration, and embedding include loss of both proteins and lipid components from both membranes and cytoplasm, cell shrinkage, and conformational changes in cell organelles (4, 10).

A recently developed technique which makes it possible to study unfixed cells is cryomicrotomy. Specimens are rapidly frozen, usually in propane, and sectioned in a microtome which maintains specimen temperatures between −70 and −170°C. The specimens are then either freeze-dried or directly examined in an electron microscope on a cold stage.

The freeze-fracture and etching technique provides information on the subunit morphology of membranes. Although it is technically difficult and requires specialized instrumentation, it provides valuable cytological data. The technique requires rapid freezing, fracturing, shadowing, and coating. A platinum-carbon replica of the biological material is produced which is examined in the electron microscope. Bilayer membranes fracture symmetrically, and it is thus possible to study the inner and outer membrane faces (Fig. 2d).

The freeze-fracture technique, with its unique capability to demonstrate intramembrane topography, has been used to advantage in the study of antibiotics and their actions on bacterial and fungal plasma membranes. For example, pronounced alterations in conformation of lipid components of *Epidermophyton floccosum* have been observed after application of a variety of polyene antibiotics (8).

Microscopic applications are further extended by immunocytochemistry and autoradiography. Antigens are localized at the electron microscopic level by using ferritin-, peroxidase-, or heavy-atom (e.g., colloidal gold)-labeled antibodies. Light and electron microscopic autoradiography has been useful in studies of macromolecular synthesis and binding sites of small molecules.

SEM

The technique of SEM utilizes a beam of electrons, generated by a tungsten filament, which is rapidly directed over the surface of a specimen. The energy imparted to the metal-coated specimen by this beam of primary electrons results in the emission of secondary electrons from the specimen. This process, utilizing recording and display means similar to those used in television, results in an image. An integrated image thus displayed may be viewed directly or recorded photographically. In contrast to TEM, SEM has the capacity of three-dimensional information transfer, since electrons emitted from a particular area of a three-dimensional specimen vary in direction and energy from those which are emitted from an adjacent area of the same specimen. A limitation of SEM, however, is that its resolution is limited to about 10 nm; surface macromolecules resolvable by TEM (in freeze-fractured or negative-stained specimens) are not visible by SEM.

SEM primarily employs two preparatory techniques: freeze-drying and critical-point drying. The first, freeze-drying, is based on the sublimation of water from a quick-frozen specimen by the application of heat to the specimen within a vacuum chamber. Although this technique is excellent for SEM preparation and produces fewer artifactual changes in the cell than critical-point drying, its use is limited by the necessity of maintaining a good high-vacuum source.

Critical-point drying, on the other hand, may be accomplished with simple equipment. The process is dependent on the same phenomena as freeze-drying, but does not require the use of large vacuum systems. The processes of fixation and dehydration are involved (producing changes similar to those noted for TEM), but the final solvent (not water) is sublimated and exchanged, under pressure, for CO_2 or Freon gas, which is freely miscible with air.

After application of either freeze-drying or critical-point drying, the specimen is coated with a layer of metal, usually gold or platinum, from which secondary electrons are readily emitted.

SEM is particularly suited to the observation of the three-dimensional growth characteristics of microbial colonies as a result of the wide range of magnification it employs and the depth of field possible with the scanning system of beam deflection. The additional uses of SEM are now in the natural sciences, such as X-ray and Auger-electron emission from a surface probed by the electron beam. In the future, these applications may be useful in the analysis of cell walls, cell membranes, and cytoplasmic constituents of a variety of organisms.

CONCLUSION

Advances in the design, resolution, and interpretive use of various types of microscopes have greatly advanced understanding of the form of the bacterial cell. In the future, application of specialized microscopic instrumentation, such as near-infrared microscopy and high-voltage electron microscopy, should provide new insights into microbial structure. The correlation of morphological structure with biochemical composition is the basis for our understanding of the biology of microorganisms.

I am grateful to Karen L. Richards and Carolyn S. Cody for their assistance in the preparation of this chapter and their studies in microscopy.

LITERATURE CITED

1. **Barer, R.** 1964. Microscopy, p. 91–158. *In* G. H. Bourne (ed.), Cytology and cell physiology. Academic Press, Inc., New York.
2. **Chayen, J., and E. F. Denby.** 1968. Biophysical technique as applied to cell biology. Methuen and Co., Ltd., London.
3. **Costerton, J. W.** 1979. The role of electron microscopy in the elucidation of bacterial structure and function. Annu. Rev. Microbiol. **33:**459–479.
4. **Fuller, R., and D. W. Lovelock (ed.).** 1976. Microbial ultrastructure. The use of the electron microscope. Soc. Appl. Bacteriol. Tech. Ser. No. 10.
5. **Glauert, A. M. (ed.).** 1977. Practical methods in electron microscopy. North-Holland Publishing Co., Amsterdam.
6. **Hijmans, W., and M. Schaeffer (ed.).** 1975. Fifth International Conference on Immunofluorescence and Related Staining Techniques. Ann. N.Y. Acad. Sci. **265:**1–627.
7. **James, J.** 1976. Light microscopic techniques in biology and medicine. Martinus Nijhoff Medical Division, Amsterdam.
8. **Kitajima, Y., T. Sekiya, and Y. Nozawa.** 1976. Freeze-fracture ultrastructural alterations induced by filipin, pimaricin, nystatin and amphotericin B in the plasma membranes of epidermophyton, Saccharomyces and red blood cells. A proposal of models for polyene-ergosterol complex-induced membrane lesions. Biochim. Biophys. Acta **445:**452–465.
9. **Klainer, A. S., and I. Geiss.** 1973. Agents of bacterial disease. Harper & Row, Publishers, New York.
10. **Lickfeld, K. G.** 1976. Transmission electron microscopy of bacteria, p. 127–176. *In* J. R. Norris (ed.), Methods in microbiology, vol. 9. Academic Press, Inc., London.
11. **Melamed, M. R., P. F. Mullaney, and M. L. Mendelsohn (ed.).** 1979. Flow cytometry and cell sorting. Wiley, New York.
12. **Nairn, R. C.** 1975. Fluorescent protein tracing, 4th ed. Longman Inc., New York.
13. **Perkins, W. D., and J. K. Koehler.** 1978. Antibody-labeling techniques, p. 39–63. *In* J. K. Koehler (ed.), Advanced techniques in biological electron microscopy. II. Specific ultrastructural probes. Springer-Verlag, Berlin.
14. **Quesnel, L. B.** 1971. Microscopy and microtomy, p. 2–103. *In* J. R. Norris and D. W. Ribbons (ed.), Methods in microbiology, vol. 5a. Academic Press, Inc., London.
15. **Rochow, T. G., and E. G. Rochow.** 1978. An introduction to microscopy by means of light, electrons, X-rays or ultrasound. Plenum Press, New York.
16. **Ruthmann, A.** 1970. Methods in cell research. Cornell University Press, Ithaca, N.Y.
17. **Slayter, E. M.** 1970. Optical methods in biology. Wiley-Interscience, New York.

Chapter 3

Quality Control in Clinical Microbiology

RAYMOND C. BARTLETT

Passage of the Clinical Laboratory Improvement Act in 1967 resulted in publication of federal regulations for mandatory quality control procedures in laboratories involved in interstate commerce (*Federal Register*, vol. 33, 15 October 1968). These regulations gradually were made applicable to all clinical laboratories by mutual agreement between the Centers for Disease Control and the agencies responsible for the Medicare and Medicaid programs. When these regulations were written there had been insufficient experience with quality control in microbiology to establish which procedures and materials demonstrated the most erratic performance. A sense of urgency existed at the time as a result of anecdotal reports and some statistical data suggesting that poor-quality work was widespread in clinical laboratories. Because of this, the regulation stipulated the monitoring of each new batch of culture media and the testing of all reagents on each day of use. This posed a substantial burden for many laboratories and resulted in incomplete compliance.

During the 17 years since the inception of these regulations, increasing numbers of laboratory workers have questioned their need because actual observation has demonstrated that many materials rarely, and some never, have been deficient. There was a great need for systematic collection of such observations to permit an objective reassessment of the need for quality control of equipment, materials, and procedures. We reported from our laboratory the results of over 100 different surveillance procedures which in toto had been conducted over 4,000 times (5). The average number of times that each procedure had been conducted was 41, and only two had been conducted fewer than 25 times. Both the cost of performance and the yield in deficiencies detected were used to establish the cost effectiveness of these surveillance procedures.

Monitoring of personnel functions and equipment was most effective. Monitoring of media overall was less productive; that of susceptibility testing and biochemical reagents was least useful. The media under surveillance had been prepared in our own laboratory. All media to which materials were added after sterilization had demonstrated either contamination, variance in pH, or improper growth characteristics after inoculation with stock cultures at one time or another. A third of the media to which no materials were added after sterilization demonstrated no deficiencies; the remaining two thirds showed either improper performance of stock cultures, variance in pH, or contamination. No deficiencies were detected by the monitoring of antisera and various biochemical tests including coagulase, methyl red, ferric chloride, catalase, Voges-Proskauer, indole, nitrate, and oxidase. Deficient media or disks in use with the diffusion antimicrobial susceptibility (Kirby-Bauer) test were extremely uncommon, although improper performance of this test was frequently detected. The total program had cost slightly over $30,000 per year, about 5% of our bacteriology laboratory budget. Discontinuation of monitoring of materials which had never demonstrated deficiencies, along with reduction of the frequency of monitoring other products from each batch to only each new purchased lot of dehydrated medium or reagents, enabled us to reduce this cost by over $13,000 per year, reducing the percentage of our budget for quality control to about 3%.

Few additional data have been published to assess the experience in other laboratories. A conference conducted under the auspices of the College of American Pathologists in 1979 revealed evidence that a number of commonly performed biochemical tests have never displayed deficiencies during extensive observation in several different laboratories (16). This report also suggested that weekly quality control of the Kirby-Bauer test was appropriate, and this was later confirmed in a separate publication (9). Few data are available regarding the quality of commercially prepared, ready-to-use culture media. Nagel and Kunz reported that deficiencies were not detected through use of stock cultures or monitoring of pH and sterility (13). They indicated that deficiencies were readily observable at the time the plates were put into use.

The regulations established in 1967 may never be amended, but the operational guidelines that are used by inspecting agencies actually establish what types of quality control activities will be expected for accreditation. These guidelines have already undergone some modification, but there are geographical inconsistencies and inconsistencies between inspecting agencies. There is general recognition among laboratory workers, professional societies, and the regulators that there is an urgent need for a new approach to standardization of necessary quality control practices. Currently an effort is being undertaken by the National Committee for Clinical Laboratory Standards to establish such criteria for quality control of commercially prepared, ready-to-use culture media.

Two other factors have influenced reaction to federal regulations. First, the quality of commercially available biological products has improved considerably, although certain types of antisera continue to be of unpredictable quality. Second, controls appropriate for laboratories processing limited numbers and types of microbiological specimens appear to be quite different from those that are appropriate for large laboratories processing hundreds of microbiological specimens of many types each day. Laboratories performing a limited amount of microbiology must expend a greater proportion of their resources on quality control. Recent data from 265 laboratories have shown that those laboratories with one to four workers in microbiology devote, on average, 10% of their time to quality control functions (R. C. Bartlett, Clin. Microbiol. Newsl. **4**:14, 1982). This drops to 7% for laboratories with 5 to 14 workers and only 4% for laboratories with more than 14 workers in microbiology.

Many specimens may be shipped by small laboratories to a referral laboratory; this presents viability problems not encountered by larger laboratories, which are able to process promptly most, if not all, types of specimens. If a technologist examines Gram-stained direct smears of critical specimens such as spinal fluid only once a month, control smears should be stained and examined at the same time. In larger laboratories where technologists examine several specimens of this type each day and regularly observe specimens containing bacteria, such a control measure would be superfluous. This consideration has caused some smaller laboratories to abandon clinical microbiology and to refer such work to larger laboratories which can more economically assure the quality of their work.

These observations suggest that the operational guidelines of inspecting agencies should take into account the relative cost effectiveness of quality control measures in laboratories of various sizes. The following guidelines are provided in the hope that they will assist in resolving discrepancies between the expectations of regulators and the practical and economic capabilities of laboratory workers. A more extensive discussion of quality assurance in clinical microbiology can be found in other publications (3, 4, 6, 12, 16).

MEDICAL USEFULNESS OF AVAILABLE PROCEDURES

Microbiologists should weigh the factors that justify performing microbiological procedures in their own laboratory as opposed to submitting specimens to a referral laboratory. Careful scrutiny of the cost of performance, including quality control, of some procedures that are infrequently requested and are of low medical urgency may not justify their continued use. A careful reassessment should be made of procedures that are offered "after hours." Pressure from physicians for provision of such services may actually result in substandard performance because economics and practicality preclude quality assurance of procedures infrequently performed by workers who lack immediate supervision. Demand for after-hours services often results in unsupervised performance of these tests by house staff, who are increasingly disinclined to accept it as a part of their education and who lack basic training in clinical microbiology. Equally unacceptable is performance of these tests by unsupervised, often part-time, night emergency technical personnel who often perform different kinds of procedures than do daytime staff and may perform some procedures as infrequently as once a month.

CONTROL OF SPECIMEN QUALITY

Test ordering criteria

The continued growth in the average number of laboratory tests per patient day is drawing increasing attention as a factor contributing to rising health care costs. Studies conducted in Hartford, Conn., area hospitals revealed that from 40 to 50% of urine culture requests did not meet fundamental ordering criteria established by a group of infectious disease physicians and microbiologists. Such investigations are difficult to conduct because they require review of charts by trained workers and by a physician to ensure proper classification. These studies are almost always conducted retrospectively and thus have no direct influence on physician ordering practices. The proposal being introduced by the federal government, of reimbursement based on diagnostic-related groupings, will focus greater attention on physician ordering practices. Already many institutions maintain computerized tabulations of laboratory expenditures, by physician, for individual diagnostic-related groupings. We can expect that physician overusers will be questioned about their ordering practices, and progressively more stringent means may be used to control overutilization. This measure has a direct bearing on quality control because constraints on budgets for personnel, space, and equipment will cause deterioration in the quality of work if the volume of specimens is allowed to continue to grow and to include a variable number of submissions that are not justifiable clinically. It has been our practice to call physicians after spot review of medical records fails to support test requests, but it is difficult to conduct this type of control activity on enough requests to have any real impact on ordering practices.

As an example of such control, test ordering at Hartford Hospital will be conducted by on-line entry into a laboratory computer reporting system which will be interfaced with pharmacy records and computerized nursing notes. This will permit a comparison of test requests with antibiotic usage, patient signs and symptoms, and the current diagnosis. As a result, many test requests could be found to require supplemental action such as review with a more senior physician or consultation with laboratory staff. These control measures are in a primitive state of development and implementation at this time but can be expected to evolve rapidly as increasing pressure to control laboratory utilization is exerted and as such measures can be shown to reduce costs.

Collection and transport

There is increasing recognition that the responsibility of the microbiologist for quality assurance goes beyond the accuracy and thoroughness with which the microbial content of specimens is analyzed and reported. Responsibility must include the provision of criteria for proper collection and transport of specimens, methods to ensure adherence to these standards, and their enforcement through close liaison with the nursing and medical staff. More detailed discussions of collection, transport, and processing of specimens can be found in Chapter 8 and *Cumitech 9* (8). Emphasis here is placed on assurance that appropriate standards are established and enforced by monitoring of compliance as specimens are received in the laboratory. The microbiologist should make sure that instructions for the collection of specimens from the lower respiratory, urinary, and genital tracts include methods for excluding incidental indigenous and colonizing bacteria from the specimen. Microbiologists should participate in continuing education programs that repeatedly review these methods with medical and nursing staff. Specimens collected from normally sterile body sites present less of a problem, but standards also should be established for decontamination of the skin with both organic iodide compounds and alcohol before collection of specimens. Appropriate sterile containers must be used that will not risk contamination of the specimen or represent a hazard to laboratory staff. All specimens from patients on isolation must be bagged, with the requisition slip attached to the exterior of the bag. The anticoagulant sodium polyanethol sulfonate should be required for submission of body fluids or joint fluids that may clot. An adequate volume should be collected to assure a final sodium polyanethol sulfonate concentration of no greater than 0.05%. When smaller volumes are submitted, a note should be appended to the report indicating that inhibition of certain bacteria by excess anticoagulant may occur.

Swabs are unsuitable for examination of material for mycobacteria and fungi. In fact, cotton swabs do not yield as much useful information about any type of microorganism as can be obtained from exudate or fluid submitted in a syringe or other appropriate closed sterile container. If swabs must be used, the less toxic Dacron swabs should be employed. They should be submitted in a nonnutrient transport medium such as Cary-Blair or Stuart transport medium or in one of the many commercially available prepackaged devices. Many of these are suitable for isolation of *Neisseria gonorrhoeae* from specimens that are processed within 3 h of collection. Swabs that are dried out when they are received are useless except for those from throats, which can be placed in silica gel strictly for isolation of group A beta-hemolytic streptococci. Calcium alginate swabs have some advantages in being entirely soluble in buffers, which permits better recovery of microorganisms but prevents preparation of direct smears. Tissue can be submitted in a sterile container in which it can be kept moist, but not floating, in water, saline, or anticoagulant. Gassed-out tubes, capped syringes, or commercially available devices such as Port-A-Cul (BBL Microbiology Systems, Cockeysville, Md.) or the B-D anaerobic specimen container (Becton Dickinson & Co., Rutherford, N.J.) should be used whenever culture for anaerobic bacteria is requested.

Swab-collected specimens for virus isolation should be placed in a transport medium such as that described by Leibovitz (10). Direct inoculation of cell culture medium is preferable to shipping swabs in transport medium to a referral laboratory for virus isolation. Fecal specimens submitted for identification of parasites need not be placed in a preservative if examination is performed within 4 h. Otherwise, specimens should be placed in Formalin and polyvinyl alcohol fixatives (see Chapter 55).

Specimens for isolation of bacteria should not be submitted in nutrient medium, with the exception of rectal swabs for culture of intestinal pathogens. Use of nutrient medium in other instances is discouraged because it excludes detection of cell types and bacteria in Gram-stained direct smears.

The diagnosis of infectious disease is commonly compromised by improper handling of tissue removed at surgery or autopsy (see Chapter 55). Culture of tissue is of greater value than blood culture in assisting in the autopsy diagnosis of infection. Histological examination cannot establish a specific diagnosis of infection in many instances, and the use of swabs for microbiological examination enormously reduces the chances that direct examination or isolation in culture will identify the agent. If possible, a laboratory representative should be present in the operating room or autopsy suite to assure that an appropriate portion of tissue is submitted for microbiological study. Occasionally, pathologists and clinicians request that an attempt be made to isolate mycobacteria and fungi from specimens that have been fixed in Formalin. Although successful isolation has been reported, this provides no justification for this request. Discussion between the pathologist or clinician and the microbiologist regarding the futility of attempting such isolation may contribute to ensuring submission of fresh tissue on subsequent occasions.

The time of collection should be stated on all specimens. Clean-catch urine specimens should be processed within 1 h, and all other microbiological specimens should be processed within 3 h. Urine specimens may be refrigerated for up to 24 h if necessary, but other specimens cannot be held, with or without transport medium, at any temperature without risking alterations in microbial population and the loss of fastidious organisms.

Control of unnecessary replication in submission of specimens collected from the same site by the same method of collection on the same day will reduce unnecessary work. Other than blood cultures (in which case three to four specimens per day are appropriate), duplicate specimens should be held at 4°C. Reports should be based on examination of the first specimen received or the specimen of the best apparent quality if more than one specimen are received at once.

Evaluation of specimen quality before culture

Despite efforts to standardize and control collection practices, some specimens are received that are heavily contaminated with indigenous and colonizing bacteria not related to infection (Table 1). Gram-stained direct smears help to detect oropharyngeal contamination of lower respiratory tract specimens (2). This technique will also permit recognition of superficial contamination of specimens of exudate from wounds, the middle ear, and the endometrial cavity. It is wasteful and potentially misleading to the physician to perform and report the results of anaerobic cultures on any specimen when the direct smear demonstrates

TABLE 1. Sites of infection from which specimens commonly become contaminated with indigenous and colonizing bacteria during collection

Site of infection	Source of contaminating and colonizing bacteria
Middle ear	External auditory canal
Lower respiratory tract	Oropharynx
Nasal sinus	Nasopharynx
Bladder	Urethra and perineum
Endometrium	Vagina
Superficial wounds and subcutaneous infections	Skin and mucous membranes
Fistulae	Gastrointestinal tract

squamous epithelial cells, which indicate skin or mucous membrane contamination. It is likewise inadvisable, when squamous cells are seen, to culture for organisms other than *N. gonorrhoeae* from endometrial or endocervical specimens when endometritis is suspected.

Certain types of specimens so consistently yield mixed cultures of indigenous and colonizing flora and organisms of unknown relevance to infection that attempts to produce useful microbiological information from them are more wasteful of laboratory effort than is justified. The medical and nursing staff should be advised in writing and periodically reminded orally that when they submit such specimens they will be requested to consult with laboratory staff regarding the usefulness of processing. This should apply to superficial oral and periodontal lesions, decubiti, varicose ulcers, most burns and superficial gangrenous lesions, and perirectal abscesses, all of which can be effectively treated on an empirical basis. If pathogenic bacteria must be identified in such lesions, they should be debrided and 20 to 500 mg of tissue should be excised and submitted for decontamination and quantitative culture. Alternatively, certain lesions can be aspirated with a needle and syringe to avoid surface contamination. Microbiological examination of bowel content, Foley catheter tips, lochia, vomitus, and discharge from colostomies is not useful for diagnosing infection or guiding therapy. The usefulness of Gram stain and culture of material from the external auditory canal and from gastric aspirate in newborns is of questionable value in diagnosing neonatal sepsis.

Disposition of specimens unsuitable for culture

When specimens are submitted contrary to established standards for collection and transport, when processing appears to be of questionable clinical usefulness, or when specimens are of poor quality based on direct smear examination, they should be held in the laboratory at 4°C. A written report, along with a phone call, should go to the patient care unit, indicating the nature of the deficiency and requesting that another specimen be taken or that the physician consult with the microbiologist if complete processing of the first specimen is for some reason clinically preferable to collection of another. It is desirable to contact the patient care unit instead of the physician for several reasons. Often laboratory personnel will not know which physician ordered specimen collection and submission or who actually collected and transported the specimen. Most instances involve urine and sputum specimens, which patient care unit personnel are responsible for collecting and transporting. When physicians have collected or transported the specimen improperly, the patient care unit personnel can convey the message to the appropriate physician to consult with the microbiologist regarding further processing. Physicians will not take the time to do this unless there is compelling clinical justification. If physicians are contacted directly, many will be inclined to request processing even when the clinical merit of a new specimen would outweigh the inconvenience of collecting it. It is important that the microbiologist convey an attitude of concern for patient care and that his or her action be viewed not as restricting the availability of laboratory services but as attempting to produce the most useful information while minimizing production of potentially misleading information.

Misunderstandings between clinicians and microbiologists will be minimized if standards for collection and transport and for specimen quality are agreed upon in advance and are approved by a medical staff committee. These standards should be placed in both nursing and laboratory procedure manuals. These should be reviewed periodically at conferences with medical and nursing staff to minimize the confrontations that may occur when misunderstanding is allowed to develop.

CONTROL OF EXTENT OF PROCESSING

The microbiologist is also responsible for limiting laboratory effort expended on isolation, identification, and antimicrobial susceptibility testing of microorganisms of questionable clinical significance while concentrating resources and quality control on the processing of those specimens that have a higher probability of yielding clinically useful information. This may be accomplished by controlling the extent of processing of specimens which are likely to be contaminated with indigenous flora and colonizing bacteria. There is little evidence that nasopharyngeal cultures are useful other than for the diagnosis of pertussis. Specimens from the throat should be routinely examined only for *Streptococcus pyogenes*. Search for other organisms in throat specimens may be justified, i.e., in cases of diphtheria, gonorrhea, etc., but this should be based on a specific request from the physician. There is no good evidence that *Staphylococcus aureus, Streptococcus pneumoniae, Haemophilus influenzae, Pseudomonas aeruginosa*, and various members of the *Enterobacteriaceae* family predominating or in pure culture from the nose or throat suggest the presence or etiology of pharyngitis, middle ear, or lower respiratory tract infection in adults or children unless the patients are markedly leukopenic (leukocytes less than 10^9/liter) or lack cellular immunity.

Effort will be conserved and more useful information will be reported if physicians are required to specify whether gonorrhea, vaginitis, endometritis, or surgical wound infection is suspected when specimens are submitted from the cervical-vaginal area.

Despite efforts to control proper collection, transport, and submission of specimens, and the use of examination of Gram-stained direct smears to assess superficial contamination, specimens may yield mixtures of indigenous and colonizing bacteria and multiple species which could be pathogenic for the site from which the specimen was collected. In the absence of knowledge of the clinical circumstances surrounding collection of such specimens, laboratory staff have no way of assessing the diagnostic and therapeutic significance of complete identification and antimicrobial susceptibility testing of potentially pathogenic isolates. Indeed, in highly immunocompromised patients isolates which might ordinarily be considered indigenous flora or harmless colonizers may play a role in infection. We correlate the organisms seen in direct smears of such specimens with those isolated in culture and write a note suggesting that those which are seen only in culture may be less relevant to infection.

It has been recommended that repeat collection be requested when clean-catch urine specimens containing three or more isolates are submitted (1). We request a repeat collection or a consultation from the physician when clean-catch or Foley catheter specimens contain two or more organism strains, to justify complete identification and susceptibility testing. We have found that only about 10% of these specimens represent treatable mixed infections. Most of them represent contamination or colonization. When cultures of other material subject to superficial contamination and colonization, such as lower respiratory tract secretions, wound exudates, and cervicovaginal exudate, contain more than three potential pathogens in addition to indigenous flora, we render a report based on gross colony morphology with a request for repeat collection or a consultation regarding complete identification and susceptibility testing. We hold incompletely processed specimens at 4°C for 5 days. In our experience physicians infrequently request processing. Instead, fresh specimens are submitted, and these often are of superior quality and yield fewer isolates. One of the reasons that physicians do not request complete identification and susceptibility testing when more than three isolates are found is that empirical antimicrobial therapy has usually been initiated and is not likely to be altered by precise identification and susceptibility testing of this many isolates.

In concluding this section, it is reemphasized that effective quality assurance cannot be accomplished unless appropriate clinical criteria exist to support requests and unless specimens of optimal quality are received. Diversion of resources to the processing of poor-quality specimens or specimens for which there is no clinical justification for examination may keep personnel from applying sufficient time to quality control of the processing of specimens that are more capable of producing useful clinical information.

ADMINISTRATION OF QUALITY CONTROL ACTIVITY

A list should be prepared of all equipment and reagents used and of the operations performed by personnel that will be subject to surveillance, with an indication of the frequency of surveillance and criteria for acceptable performance. Responsibility for surveillance activity may be given to one worker, but distribution of this responsibility among all personnel helps to establish a "quality control" consciousness throughout the laboratory, minimizes the amount of time each individual must expend on this activity, and does not require extensive retraining when an individual specialized in quality control leaves employment. If expiration dates and verification of performance characteristics are applicable to any reagent, personnel should be conditioned not to use it if these have not been verified on the label. Similarly, equipment should not be used if established maintenance and surveillance of appropriate operating criteria (temperature, calibration, etc.) have not been conducted according to schedule and documented (Table 2). Records should indicate (i) whether the item was monitored according to schedule and (ii) whether a deficiency was found. These records should be reviewed at least monthly by supervisory personnel. If monitoring is not conducted according to schedule, a reason should be given with corrective action. The corrective action should include either a change in the frequency of monitoring or a redistribution of the workload to permit subsequent surveillance to be conducted on schedule. If deficiencies are observed, their nature should be documented along with corrective action. A visible card file (Fig. 1) is a convenient way of keeping track of scheduled surveillance activities and establishing necessary documentation of corrective action.

TABLE 2. Surveillance of equipment

Item	Conditions	Frequency
Ventilation	Direction of flow; changes per hour; temperature, 23 to 29°C; humidity, 30 to 50%	Semiannual
Temperature-controlled devices	Upper and lower limits; power failure	Daily or continuous
Humidified chambers	Humidity, 30 to 50%	Daily or continuous
Capneic incubators	Concn, 5 to 10%	Daily or continuous
Autoclaves (decontaminating)	Sterilize spores	Weekly
Safety cabinets	Air flow and configuration	Semiannual
Motors	Regular maintenance	Semiannual
Centrifuges	Calibration and maintenance	Semiannual
Microscopes	Inspection and cleaning	Semiannual

Surveillance activity can be divided into several major areas: (i) environment and equipment; (ii) reagents, stains, and biological materials; (iii) performance of personnel; and (iv) use of referral laboratories. Readers are referred to other sources for more detailed recommendations (3, 4, 6, 12, 16).

LABORATORY ENVIRONMENT AND EQUIPMENT

The temperature and humidity of the air in microbiology laboratories should be maintained at between 23 and 29°C and 30 to 50%, respectively. Values outside this range are deleterious to culture media and produce error in serological testing. Microbiology laboratories should be negatively pressured with respect to adjacent corridors, and 100% fresh air should be provided with a minimum of 10 air changes per hour. Temperatures displayed by temperature-controlled equipment should be recorded daily. If recorders are used, the records should be kept on file. Alternatively, a central alarm system can be installed to alert personnel when temperatures exceed allowable limits or when power fails. Trials should be conducted and documented periodically to establish that the system will alert personnel who will act promptly at any hour of the day if power is lost or conditions exceed the acceptable range for the device. Humidity should be maintained at 40 to 50% in

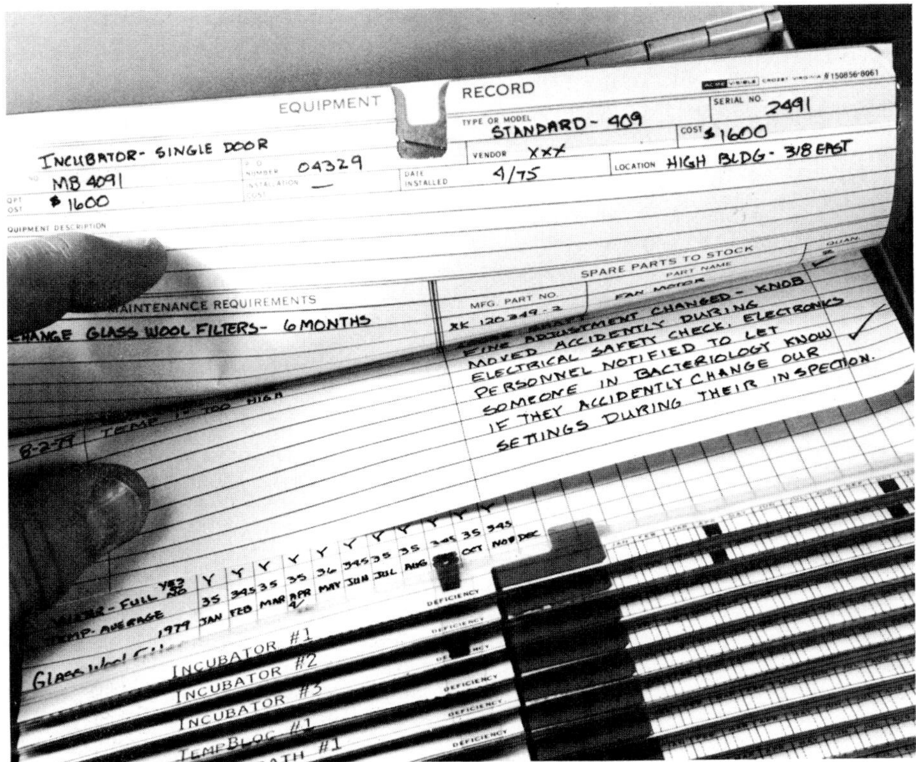

FIG. 1. Visible card file used as a convenient way to organize surveillance activities. Colored plastic indexes can be advanced as surveillance is completed. Tabs may be lowered to signal deficiencies. Cards readily display nature of surveillance measures and expected performance and allow recording of deficiencies and corrective action.

incubators, and carbon dioxide concentrations should be between 5 and 10% in capneic incubators. An oxygen-sensitive indicator such as methylene blue or resazurin or a control culture of *Clostridium novyi* B should be placed in anaerobic jars or chambers to assure that anaerobic conditions have been achieved.

Adequate autoclaving of contaminated media and liquids should be assured by weekly testing of spore solutions which are suspended in the center of the container with the largest volume. Adequate sterilization of culture media should depend on published guidelines related to the volume of medium and on experience with production of media free from contaminants (15). Heating sufficient to sterilize a spore solution is not required in preparation of culture media and may result in damage to its constituents. Biological safety cabinets should be checked twice a year for integrity of high-efficiency air filters and for proper velocity and configuration of air flow. It will be more economical and reliable to establish a contract with the manufacturer, or other firm, which possesses the equipment and special skills required rather than for laboratory staff to attempt this. It should be done every 6 months or whenever any construction or repairs are conducted near the equipment or on any duct work connected to it. Monitoring of appropriate standards of glassware cleanliness may be essential for certain serological and cell culture procedures. Generally, it is not necessary to recheck calibration of volumetric glassware used in microbiology with the exception of automatic diluters and inoculators. A wide range in acceptable tolerances exists; these should be established for each device and checked periodically.

The condition of inoculating wires and loops and calibration of quantitative loops should be checked periodically. Loops should be round and 4 to 5 mm in diameter to assure transfer of an adequate volume of fluid. Both loops and straight wires should be long enough to avoid contamination of the hub of the holder by the deepest tube in use. Nichrome loops should not be used in anaerobic work. Plastic disposable loops help to eliminate most of these problems. Motorized devices must be lubricated, and microscopes should be regularly inspected and cleaned. Centrifuges should be checked with a tachometer, and a conversion chart should be provided to indicate relative centrifugal force (11). All analytical procedures should refer to relative centrifugal force because different centrifugal forces will result when centrifuges with different radii are operated at the same number of revolutions per minute. All specialized instruments require preventive maintenance, calibration, and use of certain control measures. Performance criteria and frequency of surveillance should be documented as for other quality control measures.

A discussion of the use of disinfectants is presented in Chapter 13. Regular application of 0.5% sodium hypochlorite to countertops, or use of disposable plastic covers, is essential to minimize the risk of acquiring hepatitis. Conformance to standards for handling of specimens which may contain unusually hazardous organisms should be monitored. See Chapter 14 for a review of these safety considerations.

REAGENTS, STAINS, AND BIOLOGICAL MATERIALS

There is increasing evidence that the performance and stability of biological materials available from commercial sources have improved in recent years. This has resulted from a voluntary standardization effort within the industry, controls imposed by government, and voluntary participation by manufacturers in evaluation programs such as the one conducted by the Centers for Disease Control for serological reagents. Purchase requisitions for antigens and antisera should always specify that only lots that have been approved by the Centers for Disease Control should be shipped.

There is a wide variation in the stability of reagents, stains, and biological materials used in microbiology. Suggested priorities are given in Table 3, but the reader is cautioned that these recommendations may not conform to the requirements of inspecting agencies. An inventory should be established listing all materials in use, including solid stock and solutions. The method and condition of storage should be indicated, along with expiration dates for opened and unopened containers. Guidelines appearing elsewhere are quite conservative, there being no reliable evidence of actual shelf life in many cases (3). Regulations now require manufacturers to place expiration dates on many reagents, although in many instances the reliable life of the product may be substantially greater and in some instances nearly indefinite. Periodic monitoring should be applied to the proper storage and labeling of all of these materials. Such an inventory will assist in systematic purchasing of these materials and will help assure rotation of supplies. In general, reactivity of any chemical or biological reagents or stains should be established when a new lot is received, and this should be done before old lots are exhausted. Additional surveillance may be applied periodically or each time the material is used (Table 3). In some cases greater emphasis should be placed on proper use of materials and correct interpretation of results by personnel than on the integrity of the materials themselves. Examples include Gram stain reagents or disks and plates used for performance of the diffusion antimicrobial susceptibility (Kirby-Bauer) test (see below).

Culture media

Monitoring of culture media may be placed in two categories: media prepared by user, and commercially prepared media.

(i) Media prepared by the user. Manufacturers of dehydrated media should provide a detailed listing of the components of each lot. The purchaser should be alerted to any change in source or chemical composition of constituents. The pH and results of testing stock cultures with media prepared from the dehydrated material must be reported for the specific lot shipped. Alternatively, if generally accepted standards for verifying media performance become established, the manufacturer may simply indicate that the product meets such standards. Stock cultures used should represent the most fastidious strains that are to be isolated, and for selective media strains should be used that are most likely to be resistant to the inhibition required. Predictability of different reactions should be affirmed by demonstrating positive results with the most weakly reactive strains that might be tested and negative results with strains most likely to yield false-positive results. Lists of stock cultures, methods of preservation, and appropriate cultures to test against specific media are listed elsewhere (3, 4, 6, 12). Media prepared from a new lot or a new bottle should be tested with control cultures for

TABLE 3. Surveillance of materials

Item	Condition	Frequency
Stains	Expected staining characteristics	Each new lot or at least semiannually
Inorganic chemical reagents	Expected reactions	Each new lot
Organic chemical reagents	Expected reactions	Each new lot or at least semiannually
Antigens and antisera[a]	Reactivity with homologous antigen or antisera	Each new lot or at least semiannually
Other biological reagents	Reactivity	Each new lot or at least semiannually
All items	Proper labeling, storage conditions, expiration dates	Monthly
User-prepared media		
Additives after sterilization and all other primary isolation media	Sterility; pH; inhibition and growth of selected control cultures	Each batch prepared
Other media	pH; inhibition and growth of selected control cultures	Each new lot of dehydrated medium
Commercially prepared media		
Additives after sterilization (blood, hemoglobin, enrichment, etc.) and other primary isolation media	Sterility; pH; inhibition and growth of selected control cultures	New supplier: check several sequential lots Change in composition stated by supplier: check affected lot
Other media	pH; inhibition and growth of selected control cultures	Incomplete data from supplier: check each lot

[a] May be tested daily with reference materials to standardize interpretation (see text).

proper pH and for sterility. Media to which materials are added after sterilization, and other primary isolation media, have been found to show deficient performance and contamination often enough to warrant continued monitoring of each batch. Five percent of such batches should be incubated for sterility, and these should not be reused for culture of clinical specimens. Sterility testing may be discontinued after observation of consistently sterile batches. If an incident of contamination of any medium is observed, subsequent batches should be monitored for sterility until the problem is resolved. Other media not containing materials added after sterilization or not being used for primary isolation may be monitored only when new lots of dehydrated material are received, unless problems are encountered in its use, in which case appropriate monitoring should be conducted until the problem is resolved. The amount of monitoring may be reduced by preparing large batches, which are bagged to prevent loss of moisture and are stored at 4°C. Media containing no enrichment or animal supplement are stable for at least 1 year in tubes with screw caps. Similar media in plates may be held for 6 months. Blood, chocolate, and modified Thayer-Martin media can be held for 3 months.

(ii) Commercially prepared media. When a commercial source of prepared medium is used for the first time, the manufacturer should be requested to provide a listing of the source and nature of all components used for preparation of the specific lot shipped, the dates of preparation, shipment, and expiration, the final pH, and the types and sources of control organisms used and results obtained, as outlined above for manufacturers of dehydrated media. Again, if generally accepted standards for verifying media performance become established, the manufacturer may simply indicate that the product meets such standards. In addition, documentation of temperature conditions maintained during transit must be provided. All of this information should be provided with each shipment of each type of medium; otherwise, the purchaser should test the medium as described above. The first time primary isolation media, or media to which components have been added after sterilization, are received, the user should confirm and document sterility, pH, and performance of stock cultures for several sequential lots, as outlined above. This testing could be performed by a referral or centralized laboratory if the user could be assured of receiving the medium within 24 h with documentation of the temperature maintained during that period. If satisfactory results are obtained, it would be unnecessary to retest all subsequent lots or batches of the same medium unless the manufacturer indicated a change in the source or composition of materials. Media not containing materials added after sterilization and not used for primary isolation need not be retested by users if the manufacturer complies with the standards recommended above. These criteria should be applied also to prepackaged microbial identification systems available from commercial sources.

Antimicrobial susceptibility testing materials

Both the uniformity and the stability of dehydrated and prepared media and antimicrobial disks used in the diffusion antimicrobial susceptibility (Kirby-Bauer) test have improved substantially in recent years. Experience in many laboratories has demonstrated that only rare deficiencies occur in materials which could lead to the reporting of erroneous results. The deficiencies are most readily detected by testing new lots of disks and Mueller-Hinton medium before they are put into use. A tentative standard proposed by the National Committee for Clinical Laboratory Standards states that the user must test appropriate control strains every time a new lot of Mueller-Hinton agar or a new lot of antimicrobial disks is introduced (14). Furthermore, the user should monitor the overall performance of the test system by testing all control strains each day the test is performed. However, the user may reduce the frequency of such test monitoring if satisfactory performance in daily control tests can be documented. For this purpose, satisfactory performance is defined as follows: (i) documentation that all control strains were tested each day that samples were processed, for 30 consecutive test days; and (ii) for each drug-microorganism combination, no more than 3 of the 30 zone diameters are outside the control limits.

When these conditions are fulfilled, the user must test each control strain at least once a week and whenever any reagent component is altered. Whenever unacceptable results are observed with the weekly monitoring system, the user must return to daily control tests long enough to define the source of the aberrant results and to document resolution of the problem. The criteria for daily control acceptance also apply to weekly tests, i.e., 1 of every 20 tests may be outside the control limits, and simply repeating the test the following day may be sufficient. Improper interpretation of zone diameters by personnel may be currently the most significant source of error in performance of this test. This is a different problem, and it will be reviewed below (Performance of Personnel). Appropriate controls for proprietary instrumental methods of analysis such as Autobac, AutoMicrobic System, MS-2, etc., must be performed, but these must follow the recommendations of the manufacturer. Dilution methods of susceptibility testing may be controlled by daily or weekly testing of strains for which the minimum inhibitory concentration of each drug is known. Tests employing one or several antimicrobial concentrations short of an entire $\log_2$ series ("break-point" methods) must be controlled by assaying the final concentrations used before testing clinical specimens and then intermittently during use until a stable shelf life under the specific conditions of storage is established.

PERFORMANCE OF PERSONNEL

Procedure manuals should be prepared chiefly to provide guidance to inexperienced technologists in the performance of common procedures. Unusual requests and rare isolates should be brought to the attention of the supervisor, who may use reference materials that need not be incorporated in a basic procedure manual. The most effective means of assessing performance of personnel is the use of internal blind unknowns. Many laboratories have successfully introduced these on a periodic basis for many years and have found that the technique elucidates errors in the use of equipment and materials, reveals misinterpretation of procedure manual guidelines, and detects delays (3). This is a difficult procedure for small laboratories, where covert preparation and distribution of such specimens may be impossible. In these cases, cooperative efforts with other laboratories in a community or region should be considered. Detailed instructions for the preparation and distribution of such specimens, along with examples of results obtained, are discussed elsewhere (3).

Surveillance of performance of materials is frequently confused with surveillance of performance of tests by personnel to detect misuse of materials or incorrect interpretation of results. Use of controls each time tests are performed standardizes the interpretation of testing of clinical specimens by technologists. Positive and negative controls for fluorescent-antibody reagents should be used each day the tests are performed. This is not essential to affirm reactivity of the reagent, which probably could be effectively accomplished when new lots are received, or at most intermittently during the period of use. Similarly, testing *Escherichia coli* (ATCC 25922), *Staphylococcus aureus* (ATCC 25923), and *P. aeruginosa* (ATCC 27855) control strains for the Kirby-Bauer test standardizes correct interpretation of zone diameters more effectively than it detects deficient disks or media (see above). Examples of common microbiological procedures that should be performed with positive and negative control materials primarily to standardize interpretation by personnel are listed in Table 4.

Rubella immunization and skin testing for tuberculosis should be documented for all personnel. Familiarity with fire-fighting equipment, fire blankets, and eye washers should be evaluated through random scheduled drills. Use of chipped or broken glassware, mouth pipetting, inappropriate handling of toxic infectious materials, and incidents involving skin puncture or accidental exposure to infectious materials should be well documented with an indication of corrective action and newly established preventive measures.

WORKLOADS AND SPACE

Insufficient or excessive workloads and inadequate space may be detrimental to the quality of work. Infrequent testing and the poor quality of performance that may result can be resolved by discontinuing procedures and referring them to another laboratory or by having personnel spend enough time in another laboratory performing a larger volume of the procedure in question to gain enough experience to assure proficient performance. Excessive workloads can be controlled in part by limiting personnel effort to procedures that are most likely to produce clinically useful information, as discussed extensively above. Laboratory directors should monitor productivity by measuring relative value units or specimens processed per worker. The most widely applied system of relative values is the Workload Recording Method of the College of American Pathologists (CAP) (7). An average of 45 CAP units per h has been observed in a large number of laboratories. The numbers of bacteriological specimens processed per worker per year observed in a recent survey were 4,100 for large laboratories with more than 14 workers in bacteriology, 4,500 for medium-sized laboratories with 5 to 13 workers in bacteriology, and 5,900 for small laboratories with only 1 to 4 workers in this area (Bartlett, Clin. Microbiol. Newsl. **4**:14, 1982). Corresponding specimen workloads per worker in immunology were 10,500, 13,500, and 11,600. In virology the values were 500 to 1,000 specimens per worker regardless of size. This survey also demonstrated that the average CAP units per specimen were higher in the larger laboratories. For large, medium, and small bacteriology laboratories, respectively, the average CAP units per specimen were: throat, 16, 14, and 11; lower respiratory tract, 25, 21, and 16; wound, 30, 26, and 17; stool, 31, 27, and 19; blood, 30, 26, and 18; anaerobic culture, 31, 22, and 17; spinal fluid, 21, 20, and 16; and genital specimens for *N. gonorrhoeae*, 15, 15, and 11. No ideals or standards have been established for the extent of processing that should be applied to specimens; hence, no acceptable limits have been placed on these values. Laboratory directors must carefully consider, and be prepared to defend, whether higher-than-average productivity represents too low a standard of practice that assumes too much risk in the quality of performance, or that, conversely, below-average productivity represents unjustifiable expense that may result in budget cuts and reduced reimbursement by third-party payers.

Net square feet should be computed by measuring the internal dimensions of all work areas and offices. The average number of workers occupying this space simultaneously should be determined. This will provide a useful means of projecting future space needs. No data are available to establish appropriate limits

TABLE 4. Surveillance of personnel performance

Item	Condition	Frequency
Procedures manual	Complete and corrected	Semiannually
Workload and space	Specimens or units per worker, net square feet per worker	Annually
Continuing education	Hours for each worker	Annually
Referral laboratories	Agreement	Each specimen submitted
Blind unknowns	Adherence to procedures	Monthly
Procedures		
Immunoserological tests	Appropriate reactions of positive and negative controls	Daily
Immunoprecipitin tests		
Fluorescent-antibody test		
Antimicrobial susceptibility test	Achieve target zone diameters or minimal inhibitory concentrations with reference cultures	Weekly

on the amount of space that should be provided to assure quality in clinical microbiology. Some observations and recommendations may be found elsewhere (3, 4; Bartlett, Clin. Microbiol. Newsl. **4**:14, 1982). Among the 265 laboratories recently surveyed, large, medium, and small laboratories revealed the following numbers of net square feet (internal dimensions) per worker; bacteriology, 115, 125, and 127; virology, 153, 50, and 175; and immunology, 135, 150, and 147.

REFERRAL LABORATORIES

As discussed above, referral laboratories can be of use in numerous ways. They will perform tests that may be beyond the capacity or budget of the laboratory that receives the specimen, and they can also provide workers with supplemental experience in certain procedures that may be performed too seldom to assure proficient performance. Furthermore, specimens can be sent to referral laboratories for confirmatory examination to evaluate agreement in results obtained with immunological tests and isolation and identification of bacteria, mycobacteria, and fungi. Comparison of identification and the types of media and identification reactions used, as well as discrepancies in observations obtained with media and biochemical tests, is a very efficient way of detecting deficiencies in processing and misinterpretation of results (3). Referral laboratories often can provide reference cultures of infrequently isolated pathogens and can provide standard reagents and antisera for comparison with those in use in the individual laboratory.

LITERATURE CITED

1. **Barry, A. C., P. B. Smith, and M. Turck.** 1975. Cumitech 2, Laboratory diagnosis of urinary tract infections. Coordinating ed., T. L. Gavan. American Society for Microbiology, Washington, D.C.
2. **Bartlett, J. G., N. S. Brewer, and K. J. Ryan.** 1978. Cumitech 7, Laboratory diagnosis of lower respiratory tract infections. Coordinating ed., J. A. Washington II. American Society for Microbiology, Washington, D.C.
3. **Bartlett, R. C.** 1974. Medical microbiology: quality, cost and clinical relevance. John Wiley & Sons, Inc., New York.
4. **Bartlett, R. C., V. D. Allen, D. J. Blazevic, C. T. Dolan, V. R. Dowell, T. L. Gavan, S. L. Inhorn, G. L. Lombard, J. M. Matsen, D. M. Melvin, H. M. Sommers, M. T. Suggs, and B. S. West.** 1978. Clinical microbiology, p. 871–1005. *In* S. L. Inhorn (ed.), Quality assurance practices for health laboratories. American Public Health Association, Washington, D.C.
5. **Bartlett, R. C., C. A. Rutz, and N. Konopacki.** 1982. Cost effectiveness of quality control in bacteriology. Am. J. Clin. Pathol. **77**:2.
6. **Blazevic, D. J., C. T. Hall, and M. E. Wilson.** 1976. Cumitech 3, Practical quality control procedures for the clinical microbiology laboratory. Coordinating ed., A. Balows. American Society for Microbiology, Washington, D.C.
7. **College of American Pathologists.** 1984. A workload recording method for clinical laboratories. College of American Pathologists, Chicago.
8. **Isenberg, H. D., F. D. Schoenknecht, and A. von Graevenitz.** 1979. Cumitech 9, Collection and processing of bacteriological specimens. Coordinating ed., S. J. Rubin. American Society for Microbiology, Washington, D.C.
9. **Jones, R. N., D. C. Edson, and J. V. Marymont.** 1982. Evaluation of antimicrobial susceptibility test proficiency by the College of American Pathologists Survey Program. Am. J. Clin. Pathol. **78**:168–172.
10. **Leibovitz, A.** 1969. A transport medium for diagnostic virology. Proc. Soc. Exp. Biol. Med. **131**:127–130.
11. **Miale, J. B.** 1977. Laboratory medicine: hematology, 5th ed. C. V. Mosby, St. Louis, Mo.
12. **Miller, J. M.** 1983. Quality control in the microbiology laboratory. Centers for Disease Control, Atlanta, Ga.
13. **Nagel, J. G., and L. J. Kunz.** 1973. Needless testing of quality-assured, commercially prepared media. Appl. Microbiol. **26**:31–37.
14. **National Committee for Clinical Laboratory Standards.** 1983. Tentative Standard M2-T3. Performance standards for antimicrobial disk susceptibility tests. National Committee for Clinical Laboratory Standards, Villanova, Pa.
15. **Perkins, J. J.** 1969. Principles and methods of sterilization in health sciences, 2nd ed., p. 475–479. Charles C Thomas Publisher, Springfield, Ill.
16. **Sommers, H. B. (ed.).** 1979. Clinical relevance in microbiology, CAP Conference/Aspen 1979. College of American Pathologists, Skokie, Ill.

Chapter 4

Indigenous and Pathogenic Microorganisms of Humans

HENRY D. ISENBERG AND RICHARD F. D'AMATO

CHANGING CONCEPTS OF INFECTIOUS DISEASE

Traditionally, the clinical microbiologist has been charged with the responsibility of identifying microorganisms in a clinical specimen as accurately and quickly as possible. There is also a tacit, almost universal agreement between clinicians and microbiologists that the microorganisms of interest are the so-called pathogens. This attitude suggests that a very small number of microorganisms incite infectious disease processes regardless of their quantity, their portal of entry, or the presence of other microorganisms. Most significantly, this view neglects the determinative roles of the host and the environment in the clinically overt manifestations of infectious disease, placing the onus for the disease squarely on the microorganism.

Dubos (5) and Burnet (2) have expanded the appraisal by Theobald Smith (18) of the numerous host and parasite interactions which culminate in clinically overt infectious disease. The studies of these investigators make it evident that the general health of the host, his previous contact with particular microorganisms, his past medical history, and a variety of toxic, traumatic, or iatrogenic insults, not of microbial origin, are significant determinants of infectious disease. Implied is the understanding that indigenous microflora, given the opportunity by an unrelated lowering of host resistance, may become involved in infectious disease. The numerous factors and conditions of microbial and host origins which must be considered in infection have been reviewed recently (8).

The dilemma of the clinical microbiologist is deciding which of the microorganisms isolated from a clinical specimen are involved in the disease. There are very few microorganisms to which the term "pathogenic" can be applied invariably, if pathogenic is defined as causing infectious disease at all times (14). Yet the clinical microbiologist is expected to decide the causal relationship between a microorganism and a disease even though most of the organisms from clinical material are at best only sometimes pathogenic. All that can be presented here is a very limited discussion and outline of some of the factors which affect the categorizing of microorganisms as harmless, potentially hazardous, or actively involved in a lesion.

Clinical microbiologists will have to contend with an additional, albeit artificial, complication. The eighth edition of *Bergey's Manual* ignores the diagnostic designations of bacteria, which are helpful to the clinician as well as the laboratorian in distinguishing between closely related microorganisms. We believe that clinical microbiologists cannot abandon a system of classification that is useful and generally accepted. As but one example, the order which the painstaking investigations of Ewing and his colleagues at the Centers for Disease Control have brought to the confusion hitherto known as the family *Enterobacteriaceae* cannot be ignored for a system which will not allow ease of recognition and confirmation. The clinical microbiologist has responsibilities to the clinician and, most importantly, to the patient. This obligation should never be subjected to the vagaries of taxonomic disagreements.

Microorganisms, i.e., bacteria, yeasts, fungi, protozoa, and viruses, are ubiquitous in the human body (15). Discussion of virus carriage in healthy individuals has been omitted from this presentation because of the paucity of pertinent information. From birth, people live in a microbial biosphere composed of innumerable microorganisms representing types, variants, strains, species, genera, etc. The composition of this microbial environment is by no means static. Numerous additions and deletions, both qualitative and quantitative, take place constantly. Many populated and sterile areas are found in and on the human body, as are sparsely populated areas or regions which harbor transient microbiota. These temporary habitats of microorganisms include the larynx, trachea, bronchi, accessory nasal sinuses, esophagus, stomach and upper portions of the small intestines, upper urinary tract including the posterior urethra, and the corresponding distal areas of the male and female genital organs. The persistent finding of numerous microorganisms in these temporarily inhabited areas or in blood provides, according to Rosebury (14), as reliable a marker as can be found for the imaginary line that divides health from disease. Conclusions concerning the significance of a microorganism isolated from usually sterile areas must be based on properly obtained specimens that were properly handled and transported, that were examined promptly, and that yielded a large number of microorganisms unusual in this locale or recovered occasionally in small numbers.

One other aspect of microbiology cannot be ignored. Not all of the microorganisms that may be present in a given specimen can be cultivated; others are isolated when the physician communicates the relevant clinical information to the microbiologist; still other microorganisms have not yet been cultivated or are very difficult to cultivate. In most instances, the clinical microbiologist insists on the examination of stained preparations from clinical material, especially when the specimen is obtained directly from a pathological lesion or from suspect areas usually populated by a mixed microflora. Soon the innovative techniques of molecular biology and immunochemistry will be applied directly to specimens, augmenting the diagnostic capabilities of clinical microbiologists considerably.

A description of the indigenous and pathogenic microorganisms in various body areas is hampered further by several other considerations. (i) The specimen submitted for microbiological analysis must be obtained from the suspect lesion and not from an adjacent, noninvolved area. (ii) The site of sampling must be properly prepared. (iii) There is little, if any, information concerning the so-called normal host or the variations in microflora from one geographical locale to another. (iv) Socioeconomic background, diet, climate, and other factors may modify the definition and the suitability of a normal host and result in inconsequential changes in the host microbiota.

No hard and fast rule can be presented that divides the microorganisms of people into clear-cut categories of harmless, commensal organisms and pathogenic species. The most commonly encountered microbial species are listed in the various tables of this chapter, along with the anatomical locale in which they are usually encountered and the infectious diseases in which they become involved (9). The list of organisms lacks several that could have been included without stretching the definition of sometimes-pathogenic microorganisms. Also omitted are those microorganisms that are usually restricted to domestic or wild animals, as well as ordinary inhabitants of soil and plants. As an example, the omnipresent bacilli cannot be designated as invariably harmless. Certain *Bacillus* spp. may cause severe eye disease, especially iridocyclitis and panophthalmitis. In debilitated patients, these same organisms have been involved as the causative agents of meningitis and bacteremia. Occasionally, *Bacillus* spp. have complicated the healing of surgical wounds. Other bacilli, which elaborate toxins, may give rise to food poisoning (20).

BODY AREAS

The indigenous or autochthonous and the pathogenic microorganisms of the various body areas have been compiled (3), but the list is constantly growing. All of the organisms listed have been isolated from lesions, although many times the causal relationship between the microbe recovered and the disease was tenuous at best. Despite the dearth of basic information and the very cautious presentation of these listings, they have been accepted as authoritative. Organisms not mentioned as either indigenous or pathogenic are treated with an extreme degree of suspicion or, worse, ignored entirely. This practice still prevails despite considerable improvement in cultural technology and a changed approach toward the basic tenets of infectious disease. Therefore, it must be stated categorically that the following compilations are not exhaustive; the omission or inclusion of a microbial species in any category does not imply that it cannot be isolated from a particular or any other body area or that it cannot incite disease, complicate underlying disease, or colonize anatomical abnormalities of congenital, traumatic, or iatrogenic origin. In the communities and especially in the hospitals of developed countries, antimicrobial drugs exert selective pressures that permit the entry of drug-resistant microorganisms into the intimate human biosphere from an inexhaustible pool in nature.

The areas to be described are divided into two major categories: those that usually bear microorganisms and those that are usually sterile or only occasionally harbor a few microbes but may become heavily colonized if diseased. The microorganisms will be classified as indigenous and pathogenic in various degrees.

RESPIRATORY TRACT

Areas that usually harbor microorganisms

Also see Table 1 (14).

Mouth. The mouth will be considered as composed of the buccal cavity, teeth, tongue, gingivae, palates, and saliva. The following organisms are commonly found in this region. Various pigmented micrococci, *Staphylococcus epidermidis, Staphylococcus aureus,* and *Staphylococcus*-like anaerobic varieties are especially numerous in the saliva and on tooth surfaces, but they are not usually encountered in gingival crevices of healthy individuals. *S. aureus* and the anaerobic micrococci are rare in the predentulous mouth. The viridans streptococci, including both the mitis and salivarius groups, are ubiquitously distributed on all surfaces of the mouth, with certain species favoring specific sites (7). Enterococci are usually present. *Streptococcus pyogenes* is present in a certain small percentage (5 to 10%) of normal mouths, usually in those individuals who yield this organism in throat cultures in small numbers and without symptoms. This finding of group A streptococci in healthy individuals is restricted to adults and their saliva or tooth surfaces. Peptostreptococci are commonly found. *Streptococcus pneumoniae* may be present in the predentulous mouth and has been recovered from saliva and tooth surfaces of as many as 25% of healthy adults. Pigmented *Neisseria* spp., *Branhamella catarrhalis, Veillonella* spp., and aerobic corynebacteria are very common in saliva and in gingival crevices. *Actinomyces bifidus* and *Actinomyces israelii* can be isolated from the gingiva. The lactobacilli are found in saliva, whereas the leptotrichia are more frequently encountered on tooth surfaces. The family *Enterobacteriaceae* is well represented, with *Escherichia coli* and the *Klebsiella-Enterobacter* group the most common, especially in saliva and on teeth and after the initiation of broad-spectrum antimicrobial therapy, when the enteric representatives may be joined by oxidase-producing or nonfermenting gram-negative rods. The hemophiline rods, especially *Haemophilus influenzae* and *Haemophilus parainfluenzae*, are very often recovered from or demonstrated in normal mouths, as are representatives of *Bacteroides* and *Fusobacterium* spp., *Vibrio sputorum*, and a variety of spirochetes, especially *Treponema denticola* and *Treponema refringens*. Several mycoplasmata have been isolated from the saliva of healthy individuals. *Candida albicans* and occasionally other *Candida* spp. apparently exist in the oral cavity without disease production. The protozoa *Entamoeba gingivalis* and *Trichomonas tenax* are found in the gingival crevices of some healthy adults.

This large array of microorganisms found in healthy mouths makes it almost impossible not to find a goodly number represented whenever pathology is present. Undoubtedly, some appear in lesions accidentally. Findings from gum lesions, root canals, caries, etc., reflect this state. Debilitated patients or those on prolonged chemotherapy often have lesions of the tongue from which fungi are isolated; other patients with nutritional deficiencies or possibly hy-

TABLE 1. Microorganisms of the respiratory tract

Organism[a]	Anatomic locale	Infectious disease process
Acinetobacter spp.	Nasopharynx	Meningitis, bacteremia, pneumonia
Actinomyces spp.	Mouth, tonsils	Actinomycosis, salivary calculi
Arachnia propionica	Mouth	Actinomycosis
Bacterionema matruchotii	Tooth surface, gingiva	Not determined unequivocally
Bacteroides spp.	Mouth, tonsils	Lung abscess, lung gangrene
Bacteroides pneumosintes	Throat	Chronic disease of meninges and other organs (rare)
Bifidobacterium spp.	Mouth	Actinomycosis
Campylobacter sputorum	Nasopharynx, tooth surface, gingiva, saliva	Not determined unequivocally
Candida albicans and other yeasts	Mouth, throat	Thrush, pneumonitis
Corynebacterium spp.	Mouth, nose	Subacute bacterial endocarditis, lung abscess
Entamoeba gingivalis	Mouth	Not determined unequivocally
Enterobacteriaceae	Mouth, throat	Pneumonia, lung abscess
Enterococcus	Mouth, tonsils, nose	Bacteremia, meningitis, pneumonia, endocarditis
Eubacterium spp.	Mouth	Not determined unequivocally
Fusobacterium spp.	Mouth, tonsils	? Vincent's angina, lung abscess, complication of human bite
Haemophilus spp.	Mouth, nasopharynx, throat	Laryngotracheobronchitis, meningitis, conjunctivitis, pneumonitis, bacteremia
Lactobacillus spp.	Mouth, saliva	Bacterial endocarditis (very rare), lung abscess (one report)
Leptotrichia buccalis	Mouth, tooth surface	Not determined unequivocally
Micrococcus spp.	Mouth, tonsils	Not determined unequivocally
Moraxella spp.	Nasopharynx, nose	Conjunctivitis
Mycoplasma spp.	Mouth	Primary atypical pneumonia by one species
Neisseria spp.	Mouth, nasopharynx, nose	Meningitis (by one species; very rare by others)
Peptococcus spp.	Tonsils, throat, mouth	Rarely in lung abscess
Peptostreptococcus spp.	Mouth, tonsils	Lung gangrene, lung abscess
Propionibacterium acnes	Nose	Pimples, acne, endocarditis
Rothia dentocariosa	Mouth	Abscess
Selenomonas sputigena	Nasopharynx	Not determined unequivocally
Staphylococcus aureus	Mouth, nasopharynx, tonsils, nose	Pneumonia, otitis, parotitis, abscess
Staphylococcus epidermidis	Mouth, nasopharynx, tonsils, nose	Subacute bacterial endocarditis
Streptococcus pneumoniae	Mouth, nose, tonsils, throat	Pneumonia, conjunctivitis, meningitis, otitis media
Torulopsis glabrata	Mouth	Urinary tract infections
Treponema denticola	Mouth	Not determined unequivocally
Treponema refringens	Mouth	? Vincent's angina
Trichomonas tenax	Mouth	Not determined unequivocally
Veillonella spp.	Mouth, tonsils	Bacterial endocarditis
Vibrio sputorum	Mouth	Not determined unequivocally
Viridans streptococci	Mouth, throat	Subacute bacterial endocarditis

[a] Several species of the genera listed may be found; not all of these species are involved in disease processes.

gienic neglect may have membranous lesions involving the entire oral cavity. Numerous bacteria can be demonstrated in this disease picture, often referred to as Vincent's angina. Only very few of the component bacteria have been cultivated, and the causal relationship between the microbes and the disease is not at all clear.

Throat, including the nasopharynx, oropharynx, and tonsils. Micrococci are among the organisms usually found in the throat, nasopharynx, oropharynx, and tonsils. *S. epidermidis* may be found in the nasopharynx and on the tonsillar area of children older than infants. *S. aureus* is frequently present in the nasopharynx, the oropharynx, and the tonsils. The tonsils may harbor anaerobic micrococci. Viridans streptococci are almost always present in the throat. Various hemolytic streptococci, among them *S. pyogenes*, can be found in the nasopharynx and especially in the tonsillar region of normal, healthy individuals, but in small numbers. Enterococci may be present on tonsils. Anaerobic streptococci are occasionally isolated from tonsils. *S. pneumoniae* and the neisseriae that tolerate 20 to 25°C are present in the healthy throat, accompanied in the nasopharynx, at times, by *Neisseria meningitidis*, usually without any evidence of a disease process or contact with individuals ill with meningococcemia or meningococcal meningitis. *Veillonella* spp., corynebacteria, and *A. israelii* are present, especially on the tonsils. *E. coli*, the *Klebsiella-Enterobacter* group, and various species of *Proteus* can be found with varying frequency in different areas of the throat. A large number of healthy throats harbor *H. influenzae* or *H. parainfluenzae*, and the tonsils especially contain various *Bacteroides* spp., *Fusobacterium* spp., the salivary vibrios, and spirochetes. In addition, culture or special preparations of healthy throat specimens indicate the presence of *Bacteroides pneumosintes*, tiny gram-negative rods capable of passing most bacterial filters, mycoplasmata, *C. albicans* and other *Candida* spp., as well as the same protozoa described for the mouth.

Infectious disease of the throat can be caused by an appreciable number of microorganisms. Some lesions may be initiated by different viruses, only to be followed quickly by bacteria and fungi. Immunization against at least two pathogenic bacteria, *Corynebacterium diphtheriae* and *Bordetella pertussis*, has lessened the incidence of the associated diseases in those geographical areas where such protection is available. Rapid means of transportation and ready interchange with representatives from endemic areas are sufficient reasons to keep clinical microbiology personnel prepared to evaluate the suspicions of alert physicians. In many geographical areas, *S. pyogenes* remains the major pathogenic bacterium involved in throat disease. Its rapid detection is imperative for the initiation of appropriate therapy to prevent sequelae, especially rheumatic fever and acute glomerulonephritis. The recent introduction of latex and coagglutination reagents allow the rapid differentiation of *S. pyogenes* from serogroup C and G streptococci. *S. pneumoniae* may not contribute to pathological processes in the throat. However, large numbers of pneumococci in the nasopharynx, oropharynx, or tonsils may suggest pathology in the lower respiratory tract. Their presence in the locale may also denote sinusitis and, if present in large numbers, otitis media. Although the hemophiline bacteria are present in the throat, *H. influenzae* serogroup B has been involved in epiglottitis and laryngotracheobronchitis, both diseases of young children, rarely of young adults, with grave prognoses. The recent observations of β-lactam antibiotic resistance among the pneumococci and haemophili may adumbrate increases in their involvement in human disease and make their presence in the nasopharynx prognostically and epidemiologically significant. *S. aureus* may be involved in a variety of disease processes in the throat. Its presence is probably the most difficult to interpret. Repeated isolation of this bacterium in large numbers from a carefully cultured lesion diminishes such doubts. As a rule, the number of coagulase-positive staphylococci in this region is small. The opportunistic nature of the organism and the proximity of the throat to its natural habitat allow rapid seeding of any pathological process with these bacteria. They may, thus, secondarily complicate any disease process, and their presence in mixed cultures cannot be interpreted as reflecting contamination only. Should there be infectious disease due to staphylococci in any other part of the body, it is very likely that an inordinately large number of representatives will be recovered from the throat. *Mycoplasma pneumoniae, Klebsiella pneumoniae*, and *Pseudomonas aeruginosa* isolated in large numbers from throat cultures can reflect pathology of the lower respiratory tract; this is true especially of the gram-negative rods which tend to complicate the noninfectious diseases of the lung in cystic fibrosis and otherwise debilitated patients. *P. aeruginosa* may cause sinusitis. This organism has gained notoriety as the most prevalent agent contaminating inhalation equipment and being seeded in large numbers by these pressure machines into the lower respiratory tract (12). Thrush, already described for the mouth, may also involve the throat. Candidal lesions may be found in the throat accompanying a variety of drug regimens, as well as in patients with debilitating and neoplastic diseases and in neonatal infants. The throat and especially the tonsils may be the site for primary lesions of syphilis. Recent reports (6) indicate that *Neisseria gonorrhoeae* may also be detected and may give rise to a local disease in this area.

Nose. The indigenous microflora of the nasal passages and the nares is more limited than that of the adjacent anatomic regions. The nares are the usual habitat of the staphylococci. *S. epidermidis* and *S. aureus* are recovered with great frequency from this site. Although both organisms may not be present at all times in any one individual, their recovery from the nares is to be expected. On rare occasions, viridans streptococci can be isolated from the nasal passages. In children, and especially in infants, enterococci and other bacteria reflecting the fecal microbiota are not uncommon. On occasion, the noses of healthy persons have also yielded *S. pyogenes* and *S. pneumoniae*. The nonpathogenic neisseriae are transients in this location as well. Healthy contacts of patients with meningococcal disease may harbor *N. meningitidis*. Additional residents of the nose are various corynebacteria and occasionally *Moraxella lacunata*.

Actual disease of the nasal passages must be differentiated carefully from disease of adjacent areas, including the skin, nasopharynx, oropharynx, tonsils, and sinuses. It is not uncommon to recover from the nasal passages great numbers of microorganisms that reflect the diseased state of neighboring or more distant areas. Such isolations are of significance and represent, to a large degree, the rationale for this frequently performed microbiological analysis. However, the nasal passages themselves do become involved. Regardless of the primary causative agents, viral or chemical, staphylococci, especially *S. aureus*, will complicate, as a secondary invader, any disease process in this area. Independent staphylococcal disease in the form of boils, furunculosis, etc., may be present. In premature and newborn infants, lesions due to hemolytic *E. coli, P. aeruginosa*, and *C. albicans* may be encountered, but these usually represent more generalized disease, as is augured by the isolation of *Acinetobacter calcoaceticus* subsp. *lwoffi, Moraxella* spp., *A. calcoaceticus* subsp. *antitratus*, and various flavobacteria. Ozaena, a disease characterized by atrophy of the nasal mucosa (20), is caused by gram-negative rods. An infective granuloma, rhinoscleroma, most often involves the nose, although the pharynx and the remaining upper respiratory tract may be involved as well. Certain distinct klebsiellae, of the species *Klebsiella rhinoscleromatis*, are regarded by most investigators as the etiological agent. The disease is uncommon in the United States. Mild rhinitis due to *N. meningitidis* has been reported.

Usually sterile areas

The larynx, trachea, bronchi, bronchioles, alveoli, and the accessory nasal sinuses are usually sterile. Contamination by occasional microorganisms is usual, but the various defense mechanisms of these organs remove such offenders quickly and efficiently. The usual specimen submitted to the clinical microbiological laboratory for establishing infectious disease of the respiratory tract is the sputum specimen. Bronchial washings, bronchoscopy specimens, thoracentesis fluid, transtracheal aspirates, and aspirates from tracheostomies or lung lesions are gaining in frequency of submission for microbiological analysis. Sputum and some aspirates are contaminated by the microbiota of the throat, nose, and mouth. Rapid, efficient processing, including examination of smears to evaluate leukocytes and epithelial cells, is imperative to prevent these contaminating microorganisms from obscuring etiological agents. Transtracheal aspirates are preferred for accurate microbiological diagnosis of lower respiratory tract infections. The most commonly encountered bacteria in the sputum are *S. aureus*, *H. influenzae*, representatives of the *Enterobacteriaceae*, *P. aeruginosa*, and *C. albicans*. These organisms may be involved in the disease, they may be contaminants, or they may reflect superinfection after therapy. It must also be remembered (13) that in any infective lesion of the lungs, the demonstration of the causative organisms in sputum can hardly be expected unless escape of exudate into the bronchus has occurred.

It is perhaps most suitable to list the various diagnoses of chest disease and those etiological agents most frequently recovered. Acute infectious bronchitis, most frequently seen during winter in children and the aged and associated with viruses of the adenovirus-influenza groups, may be caused or complicated by *S. pneumoniae*, *S. pyogenes* and related serogroups, *S. aureus*, and *H. influenzae* (*B. pertussis* when whooping cough is present). Chronic bronchitis, of unknown etiology and associated frequently with pulmonary emphysema, is assuming increasing importance. Sputum from such patients displays a variety of microorganisms, not necessarily the same species on repeat examinations. The presence of bacteria associated with pathology of the respiratory tract, such as pneumococci or group A streptococci, *H. influenzae*, and staphylococci, may be significant. *P. aeruginosa*, flavobacteria, and other environmental microorganisms that contaminate inhalation equipment may also be important. The laboratory work-up for chronic bronchitis should invariably include cultures for mycobacteria, actinomycetes, and those fungi involved in pneumonic disease. Acute mediastinitis is usually the result of perforation of the esophagus in conjunction with instrumentation, obstruction, external wounds, downward propagation of deep cellulitis of the neck, forceful vomiting, and occasionally the extension of infectious disease of the lungs, pleural cavity, or pericardium. The bacteria usually found are those common to the mouth and pharynx, which are drained into the area. The bacteria which contribute most frequently to the disease picture are the peptostreptococci, *Bacteroides* spp., fusobacteria, and occasionally clostridia. Pneumonia may be caused by *S. pneumoniae*, *S. pyogenes*, *H. influenzae*, *S. aureus*, *K. pneumoniae*, *Francisella tularensis*, *C. albicans*, *E. coli* (usually the hemolytic serogroups), *P. aeruginosa*, *Proteus* spp., and *Legionella pneumophila* or similar bacteria (11). In newborn or very young infants chlamydiae should be ruled out as etiological agents. Pulmonary abscesses may result from staphylococcal or Friedlander pneumonias or the aspiration of particulate matter. The organisms most commonly cultured from such lesions are *Bacteroides* spp., *Fusobacterium* spp., peptostreptococci, enterococci, staphylococci, clostridia, klebsiellae, escherichiae, and pseudomonads. The examination of such abscesses should include a search for acid-fast organisms, aerobic and anaerobic actinomycetes, and the fungi associated with pulmonary lesions. Empyema is usually a secondary disease, i.e., an extension of the primary disease. Some special attention must be accorded certain clinical variants of empyema, especially empyema of infants usually caused by *S. aureus*, empyema due to *Entamoeba histolytica*, and empyema due to *Actinomyces* spp. or *Nocardia* spp.

GASTROINTESTINAL TRACT

Areas that usually harbor microorganisms

See Table 2.

The part of the gastrointestinal tract that invariably harbors microorganisms is the large intestine, although fecal organisms can be recovered in aspirates of the lower ileum of normal individuals (14). Once past the ileocecal valve, the contents of the large intestine reflect the microbiota of feces. It is surprising that information concerning the fecal microflora is so meager (4, 14). It seems almost superfluous to list the organisms which can be encountered; many of them are ignored in the search for enteric pathogens, i.e., salmonellae and shigellae, which certainly do not represent the only microbes endowed with the potential of disease production. Rosebury (14) lists the following microorganisms as indigenous: *S. epidermidis*, *S. aureus*, viridans streptococci, enterococci, occasionally *S. pyogenes* and related serogroups, peptostreptococci, lactobacilli (especially in infants), corynebacteria, some mycobacteria, clostridia, actinomycetes, the various members of the family *Enterobacteriaceae*, *P. aeruginosa*, *Alcaligenes faecalis*, flavobacteria, *Bacteroides* spp., *Fusobacterium* spp., *Eubacterium* spp., *Propionibacterium* spp., *Bifidobacterium* spp., mycoplasmata, *C. albicans*, various other yeasts, and a large variety of protozoa such as *Entamoeba coli*, *Endolimax nana*, *Dientameba fragilis*, *Iodameba butschlii*, *Trichomonas hominis*, *Giardia lamblia*, and *Chilomastix mesnili*. Not all microbial groups are present in each and every individual at all times. Undoubtedly, many other genera can be found with frequency. The dominance of the anaerobic bacteria in the normal adult colon must be remembered despite the present, justified disregard for these bacteria during routine microbiological analysis of fecal specimens. Geographical distribution, dietary habits, and sanitary habits will also be active in the selection of a resident microflora. It is unfortunate that the paucity of information does not permit even a short statement concerning the role of the intestinal microbiota in the economy of the host and as a first line of defense against colonization by microorganisms less adapted to the human environment. A similar state of ignorance shrouds the host-parasite relationship between humans and this large microbial aggregate, and no really authoritative treatment has been accorded the fecal microflora as potential etiological agents of infectious disease in other areas of the body. Recently, a role of the normal intestinal microbiota in

TABLE 2. Microorganisms of the gastrointestinal tract

Organism[a]	Anatomic locale	Infectious disease process
Achromobacter spp.	Large intestine, lower ileum	Postoperative and posttraumatic wound infectious complications
Acidaminococcus fermentans	Large intestine	Not determined unequivocally
Acinetobacter calcoaceticus	Large intestine, ileum	Postoperative complications, various complications in immunocompromised individuals
Aeromonas spp.	Large intestine, lower ileum	Diarrhea (rare), septicemia (rare), osteomyelitis, postoperative complications
Alcaligenes faecalis	Large intestine, lower ileum	Gastroenteritis (rare), bacteremia
Bacillus spp.	Large intestine	Food poisoning, wound infection
Bacteroides spp.	Large intestine, lower ileum	Peritonitis, abscess, cholecystitis, enteritis
Bifidobacterium spp.	Large intestine	Diverticulitis, peritonitis
Butyrivibrio fibrosolvens	Large intestine	Not determined unequivocally
Campylobacter spp.	Large intestine	Diarrhea
Candida albicans and other yeasts	Large intestine, lower ileum	Postoperative complications
Clostridium spp.	Large intestine, lower ileum	Food poisoning, choledochitis, cholecystitis, pseudomembranous enterocolitis
Corynebacterium spp.	Large intestine, lower ileum	Not known
Enterobacteriaceae	Large intestine, lower ileum	Abscess, peritonitis, bacteremia, diarrhea, enteric fevers, typhoid fever, postoperative and posttraumatic complications, meningitis, endocarditis, intoxication
Enterococcus	Large intestine, lower ileum	Peritonitis, cholecystitis, postoperative complications
Eubacterium spp.	Large intestine	Diverticulitis, peritonitis
Flavobacterium spp.	Large intestine, lower ileum	Meningitis, bacteremia
Fusobacterium spp.	Large intestine, lower ileum	Abscess, bacteremia
Lactobacillus spp.	Large intestine, lower ileum	Not known
Mycobacteria spp.	Large intestine, lower ileum	Not by indigenous bacteria
Mycoplasma spp.	Large intestine, lower ileum	Not determined unequivocally
Peptococcus spp.	Large intestine	Various abscess, peritonitis, myonecrosis
Peptostreptococcus spp.	Large intestine, lower ileum	Cholecystitis, abscess
Propionibacterium spp.	Large intestine	Endocarditis
Pseudomonas aeruginosa	Large intestine, lower ileum	Gastroenteritis, meningitis, bacteremia, postoperative complications
Ruminococcus bromii	Large intestine	Not determined unequivocally
Sarcina spp.	Large intestine	Not determined unequivocally
Staphylococcus aureus	Large intestine, lower ileum	Pancreatic abscess, enteritis, complication of pseudomembranous enterocolitis, food poisoning
Veillonella spp.	Large intestine	Not determined unequivocally
Viridans streptococci	Large intestine, lower ileum	Not known
Vibrio spp.	Large intestine, lower ileum	Not by indigenous bacteria

[a] Several species of the genera listed may be found; not all of these species are involved in disease processes.

the production of quantities of carcinogens has been advocated. This approach may possibly bring new dimensions to clinical microbiology and our understanding of disease (4).

The most readily recognized diseases of the gastrointestinal tract, usually initiated in the colon or ileum but often involving other parts of the tract or other organs, are salmonellosis and shigellosis. All members of the genus *Salmonella* are capable of evoking the various clinical symptoms of salmonellosis and its complications. The various shigellae can cause diarrhea or the syndrome known as bacillary dysentery. *Vibrio cholerae* and the attendant disease are encountered increasingly in many parts of the world. Enterotoxigenic and invasive *Escherichia coli* are known to cause pathology in the human intestine. Efforts to develop ready means to recognize these strains should be successful in the near future. *S. aureus, Proteus mirabilis,* and occasionally *C. albicans* may complicate pseudomembranous enterocolitis caused by *Clostridium difficile*. Staphylococcal food poisoning must also be considered as affecting the gastrointestinal tract, as does food poisoning due to clostridial toxins, in addition to botulism and pseudomembranous enterocolitis. The toxins, rather than the microorganisms, cause the symptomatology, and the bacteria need not be demonstrable in feces or vomitus. Acute diarrhea may at times be associated with *P. aeruginosa, Proteus* spp., or *C. albicans. Entamoeba histolytica* is the etiological agent of amoebic dysentery and its complications. Large numbers of *G. lamblia* may reflect a subacute infectious colitis due to this organism in susceptible individuals. Balantidiasis, quite rare in the United States, can occasionally be suspected when biopsies of the rectal mucosa reveal the ciliate *Balantidium coli*. Localized pathology of the large intestine may involve *Aeromonas* spp., *Plesiomonas* spp., or *Yersinia enterocolitica*, the latter causing not only occasional gastroenteritis but also the symptoms of appendicitis. Hemolytic serotypes as well as enterotoxigenic and invasive varieties of *E. coli* have been isolated from intestinal disease processes. In addition, *Vibrio parahaemolyticus* and related bacteria may be involved in food-poisoning episodes after the ingestion of seafood. Several species of *Bacillus* and *Campylobacter* are responsible for many, but mild, diarrheas (16). Anal carriage of *S. pyogenes* in the nosocomial spread of the disease and of the gonococcus in clinical or cryptic disease must not be overlooked in specific investigations of these infections. The special considerations that bear on the diagnosis of patients suspected of AIDS (acquired immune deficiency syndrome) cannot be adequately covered in this chapter.

Usually sterile areas

The esophagus and the stomach are contaminated with bacteria whenever food is ingested. The microbial population does not survive well in these two sections of the gastrointestinal tract. Similarly, the small intestine (except the distal ileum), the liver, and the gall bladder are usually free from microbial contamination or harbor transient microbial populations. This also applies to the peritoneum. Microorganisms in these areas usually are secondary to underlying diseases such as carcinoma or reach sites because the large intestine has been punctured or ruptured. Peritonitis, regardless of its initial cause, can be caused by any of the fecal organisms listed. They occur as part of the accompanying bacteremia, but usually mixed aerobic and anaerobic, gram-positive and gram-negative microorganisms participate. Peritonitis may give rise to intraabdominal abscesses, especially pelvic, paracolic, intermesenteric, subphrenic, and retroperitoneal. All may show the mixed microbial population or the single causative agents *S. aureus*, *Bacteroides* spp., enterococci, *P. aeruginosa*, and *E. coli*. In an appreciable number of cholecystitis cases, bacteria have been isolated. These were possibly involved as secondary opportunists. *E. coli*, enterococci, peptostreptococci, and clostridia are the most frequently encountered, but many other common and uncommon representatives of the fecal microflora in health and disease may be found. Bacterial cholangitis usually is secondary to intra- or extrahepatic obstruction in the area of the bile duct. Cholangitis caused by salmonellae, staphylococci, and streptococci may be accompanied by septicemia due to these bacteria (13). Again, the organisms usually recovered reflect the microbiota of the intestinal tract. Amoebic abscesses of the liver constitute a complication of the primary amoebic dysentery. Pyogenic hepatic abscesses in the antibiotic era yield *E. coli* most frequently. Pancreatic abscess formation is a secondary complication of pancreatitis. The microorganisms most frequently found are *S. aureus* and *E. coli*. Clostridial diseases are uncommon but not rare (13). They include acute gaseous cholecystitis due to *Clostridium perfringens*, but other clostridia, *E. coli*, and aerobic and anaerobic streptococci have been associated with this entity. Invasion of the bile ducts from the intestinal tract results in clostridial choledochitis, from which *C. perfringens* has been isolated exclusively. Enteritis necroticans is caused by heat-resistant *C. perfringens* type F. Clostridial cellulitis of the abdominal wall is a complication of the perforation of intestinal neoplasm with local peritonitis or of colon or biliary tract surgery.

GENITOURINARY TRACT

Areas that usually harbor microorganisms

See Table 3.

External genitalia. Rosebury (14) and Skinner and Carr (15) have listed the following organisms as present on the surface of genitalia: *S. epidermidis*, viridans streptococci, enterococci, peptostreptococci, corynebacteria, mycobacteria, various members of the *Enterobacteriaceae*, *Bacteroides* spp., *Fusobacterium* spp., mycoplasmata, and *C. albicans* and other yeasts.

The external genitalia are subject to the same infectious diseases as other skin areas. The special lesions of the external genitalia are venereal in nature and include the lesions of syphilis, chancroid or soft chancre, granuloma inguinale, which is primarily seen in the tropics and subtropics and is caused by an agent known as *Calymmatobacterium granulomatis*, and lymphogranuloma venereum resulting from infection with certain serogroups of *Chlamydia trachomatis*.

Anterior urethra. An appreciable number and variety of microorganisms can usually be recovered from the anterior urethra in normal healthy individuals of both sexes: coagulase-negative and occasionally coagulase-positive staphylococci, enterococci, various nonpathogenic neisseriae, aerobic corynebacteria, rarely certain mycobacteria, the various enteric gram-negative rods, chlamydiae, *A. calcoaceticus* subsp. *lwoffi*, *Gardnerella vaginalis*, mycoplasmata, and *Candida* spp. including *C. albicans*. *Trichomonas vaginalis* may on occasion gain access to this part of the urethra without overt disease.

It is difficult to delineate exactly where the anterior portion of the urethra ends, especially when disease is present. In the male, it is indicated to consider disease involving the entire urethra. Urethritis may be caused specifically by the gonococcus or nonspecifically by a variety of bacteria, including the staphylococci, chlamydiae, mycoplasmata, the fecal gram-negative rods, and *Listeria* spp. In the female, contamination by vaginolabial microbiota of the anterior portion of the urethra is also unavoidable. It is appropriate to underline that *N. gonorrhoeae* can be recovered from the anterior urethra of the female and male with cryptic disease. Bacteriuria of young women due to *Staphylococcus saprophyticus* is reported with increasing frequency.

Vagina. The usual microflora of the vagina from the menarche to the menopause is dominated by lactobacilli, designated as Doderlein bacillus and probably glycogen-fermenting *Lactobacillus acidophilus*. Prepubescent females from shortly after birth and postmenopausal women harbor skin microbiota in this region. Despite the control over the vaginal environment exerted by the lactobacilli, many other microorganisms can be cultivated from vaginal samples of healthy women. Thus, *S. aureus* and *S. epidermidis*, viridans streptococci, enterococci, peptostreptococci, group B streptococci, the low-temperature-tolerant neisseriae, corynebacteria, some actinomycetes, members of the family *Enterobacteriaceae*, the acinetobacters, chlamydiae, *G. vaginalis*, occasionally clostridia and other anaerobic rods, mycoplasmata, *C. albicans*, other yeasts, and *T. vaginalis* may be isolated.

The major infectious diseases of the female pudendum are the venereal diseases syphilis and gonorrhea as well as candidiasis, trichomoniasis, and nonspecific vaginitis caused by a variety of organisms that may act as opportunistic secondary invaders reflecting disturbances in the microbiological balance as a complication of disease elsewhere and its treatment. *G.*

TABLE 3. Microorganisms of the genitourinary tract

Organism[a]	Anatomic locale	Infectious disease process
Acinetobacter spp.	Anterior urethra, vagina	Urethritis, disease of newborn, complication of instrumentation and surgery, complication of burns
Bacteroides spp.	External genitalia	Complication of surgical procedures, especially in females
Bifidobacterium spp.	Vagina	Not known
Candida albicans and other yeasts	External genitalia, anterior urethra, vagina	Candidiasis
Chlamydia spp.	Urethra, vagina	Urethritis, cervicitis, neonatal disease, lymphogranuloma venereum
Clostridium spp.	Vagina	Complication of surgical procedures, abortion
Corynebacterium spp.	External genitalia, anterior urethra, vagina	Not determined unequivocally
Enterobacteriaceae	External genitalia, anterior urethra, vagina	Pyelonephritis, cystitis, bacteriuria
Enterococcus	External genitalia, anterior urethra, vagina	Pyelonephritis, cystitis, bacteriuria
Fusobacterium spp.	External genitalia, vagina	Not determined unequivocally
Gardnerella vaginalis	Anterior urethra, vagina	Vaginitis
Lactobacillus spp.	Vagina	None
Moraxella spp.	Vagina	Postpartum, postoperative, and neonatal complications (rare)
Mycobacterium spp.	External genitalia, anterior urethra, vagina	Not by indigenous organisms
Mycoplasma spp.	External genitalia, anterior urethra, vagina	? Nonspecific urethritis
Neisseria spp.	External genitalia, anterior urethra, vagina	Not by indigenous organisms excluding cryptic carriage
Peptococcus spp.	External genitalia, vagina	Postpartum or postoperative complications
Peptostreptococcus spp.	External genitalia, vagina	Puerperal fever
Sarcina spp.	External genitalia, vagina	Postpartum or postoperative complications (rare)
Staphylococcus aureus	External genitalia (rare), anterior urethra, vagina	Urethritis, furunculosis
Coagulase-negative staphylococci	External genitalia (rare), anterior urethra, vagina	Not clearly established; *S. saprophyticus* causes urinary tract infections in young women
Streptococcus agalactiae	Vagina	Various diseases of the neonate, endocarditis, abscess, meningitis, myocarditis, osteomyelitis, septicemia (19)
Trichomonas vaginalis	Anterior urethra, vagina	Vaginitis
Viridans streptococci	External genitalia (rare), anterior urethra, vagina	None

[a] Several species of the genera listed may be found; not all of these species are involved in disease processes.

vaginalis and chlamydial infections rank high among the so-called nonspecific vaginitides, but fecal and skin organisms, including staphylococci, enterococci, *Listeria* spp., etc., may contribute to clinically overt disease (14). *Campylobacter fetus* and related unclassified organisms have been isolated from the vaginas of women who have aborted repeatedly. The vaginal microflora, especially the anaerobic bacteria, may contribute to the complications of septic abortions. Postpartum sepsis and salpingitis, frequently caused by facultatively and obligately anaerobic enteric gram-negative rods and to a lesser degree by clostridia and staphylococci, probably reflect the microbiota of the vagina. Similarly, colonization and infectious disease of the newborn reflect the microbial population of the vagina.

Usually sterile areas

As a rule, the remaining structures of the genitourinary tract are without permanent microbiota. Infectious disease of the kidneys is still not completely understood, with two major theories advanced to explain the seeding of this organ by infectious organisms. There are many predisposing factors for pyelonephritis, especially host factors and underlying diseases, which appear mandatory for the establishment of infectious disease. The major bacterial offenders are the gram-negative rods, especially *E. coli*, *Proteus* spp., *P. aeruginosa*, and the *Klebsiella-Enterobacter* group; enterococci are not uncommon and can be demonstrated in adequate numbers repeatedly in untreated acute pyelonephritis. *Mycobacterium tuberculosis* may infect the kidney and be demonstrated in urine. The prostate in the male may become a secondary focus of gonorrheal disease. However, other microorganisms, including staphylococci, enterococci, listeriae, *E. coli*, mycoplasmata, pseudomonads, achromobacters, and rarely trichomonads, may lodge in this organ, especially in middle-aged and older individuals.

SKIN, EAR, AND EYE

Areas that usually harbor microorganisms

It seems superfluous to remark that the outermost border of the body is constantly populated with microorganisms that reflect the contacts, habits, profession, etc., of the individual. Still, the organisms which are universally present on the skin number far fewer than one would expect (Table 4). They are the staphylococci, among which *S. epidermidis* outnumbers *S. aureus*, occasionally *S. pyogenes*, aerobic corynebacteria, *Propionibacterium acnes*, mycobacteria, and a variety of usually harmless yeasts. Certain areas, especially those adjoining the various body openings, reflect the microbiota of these adjacent sites.

The most common and most obvious skin diseases are caused by the staphylococci. *S. aureus* is involved most frequently in boils and furuncles, and *S. epidermidis* is found in pimples and acne, often accompanied by *P. acnes*. Pustulosis, especially of the newborn, also has *S. aureus* as the etiological agent. The most common agents of impetigo are *S. pyogenes* and *S. aureus*. Coagulase-positive staphylococci are also involved in superficial folliculitis and sycosis barbae. Although streptococci and other oral bacteria may also be found, *S. aureus* is the major etiological agent of acute suppurative parotitis. Actinomycosis, as the name implies, and clinically similar lesions are usually caused by members of the order *Actinomycetales* and represent a chronic suppurative or granulomatous disease, culminating in abscesses. Deep cellulitis of the neck is most frequently caused by streptococci, both aerobic and anaerobic, as well as staphylococci. *S. aureus* is responsible for hydradenitis suppurativa, a disease more common in women and usually involving the axillae, and it is the most frequent disease-producing agent in acute mastitis and breast abscess. Paronychia most often involve staphylococci and streptococci and less commonly involve *C. albicans*. These former organisms are also frequently recovered from tenosynovitis. Clostridia are usually responsible for anaerobic cellulitis, mixtures of bacteria are responsible for crepitant cellulitis, and *S. pyogenes* is involved in necrotizing fascitis or hemolytic streptococcal gangrene. Erysipelas is a subcutaneous form of streptococcal cellulitis. Erysipeloid, an acute, cutaneous, bluish-red inflammation of the hands and wrists, which is usually seen in males in certain vocations, is caused by *Erysipelothrix insidiosa*. The eruptions of secondary syphilis may be most obvious on the skin. The various skin disorders caused by the dermatophyte fungi are well known, and the skin infections due to *C. albicans* have already been mentioned.

The external auditory canal usually reflects the microbiota of skin. Perhaps *S. pneumoniae* and the gram-negative rods including *P. aeruginosa* have been recovered with greater frequency from this site than from other skin areas. The middle and inner ear are usually sterile. Diseases of the outer ear reflect the disorders of the skin. Other parts of the auditory organ may be involved during generalized disease or may be affected only locally by *S. aureus*, *S. pneumoniae*, *S. pyogenes*, *P. aeruginosa*, *H. influenzae*, and occasionally other microorganisms.

The healthy conjuctiva of the eye may harbor a number of skin organisms. Among those most frequently recovered from the healthy eye are the staphylococci; viridans streptococci, *S. pyogenes*, the pneumococcus, neisseriae, and corynebacteria are recovered rarely. The fecal gram-negative rods have been isolated even less frequently from this site, but *H. influenzae* has an incidence second only to the coagulase-negative staphylococci. Conjunctival cultures have also yielded *C. albicans* and other yeasts. Eye disease has been caused by *S. aureus* and *P.*

TABLE 4. Microorganisms of the skin, ear, and eye

Organism[a]	Anatomic locale	Infectious disease process
Acinetobacter calcoaceticus	Skin	Complications of compromised patients
Bacillus spp.	Skin	Iridocyclitis, panophthalmitis; meningitis and bacteremia in compromised patients
Candida albicans and other yeasts	Skin	Paronychia (rare)
Chlamydia trachomatis	Uncertain	Trachoma, inclusion conjunctivitis
Corynebacterium spp.	Skin, ear, eye	Bacterial endocarditis, complication of cardiac surgery
Epidermophyton floccosum	Skin	Skin infections, athlete's foot
Haemophilus aegyptius	Eye	Eye disease
Haemophilus influenzae	Eye	Eye disease
Micrococcus spp.	Skin	Not determined unequivocally
Moraxella spp.	Eye	Eye disease
Mycobacterium spp.	Skin	Mycobacteriosis (skin)
Neisseria spp.	Eye, skin	Not known for indigenously occurring representatives
Peptococcus spp.	Skin	Complications of compromised patients
Peptostreptococcus spp.	Skin	Not known
Pityosporum ovale	Skin	Dandruff (?)
Propionibacterium acnes	Skin	Pimples, acne, bacterial endocarditis
Sarcina spp.	Skin	Complications of compromised patients
Staphylococcus aureus	Skin, ear, eye (rarely)	Boils, furuncles, impetigo, pustulosis, mastitis
Staphylococcus epidermidis	Skin, ear, eye	Pimples, acne, endocarditis, complication of cardiac surgery, thrombophlebitis
Trichophyton spp.	Skin	Skin diseases
Viridans streptococci	Skin, eye	None

[a] Several species of the genera listed may be found; not all of these species are involved in disease processes.

aeruginosa (very often iatrogenically) and occasionally by *S. pneumoniae*. More often, infectious diseases peculiar to the eye have yielded *M. lacunata, H. influenzae, Haemophilus aegyptius*, and, on rare occasions, *Bacillus* spp.

BLOOD, CEREBROSPINAL FLUID, EXUDATES, AND TRANSUDATES

The blood and spinal fluid of healthy individuals are usually sterile. Occasionally, bacteria may be isolated from blood cultures of normal, asymptomatic individuals. Often these bacteria represent low-level contamination of the circulation from inapparent skin lesions or other sources. Such findings complicate the interpretation of laboratory results but are usually of no consequence. However, during an appreciable number of infectious diseases or during infectious complications of a primary disease, microorganisms may be recovered from blood. They may be present transiently or persistently. Positive blood cultures reflect a variety of conditions and diseases and encompass a large array of microorganisms. Positive blood cultures may result from traumatic and surgical wounds, as well as from burns, injury to bones and joints, brain abscesses, furunculosis, cellulitis, meningitis, pneumonia, lung abscesses, empyema, mucoviscidosis, mycotic aneurysm, cardiac anomalies, peritonitis, intestinal or biliary obstruction, cholangitis, carcinoma, urinary obstruction, nephropathies, postpartum endometritis, and septic abortion. Immunosuppressive, cytotoxic, or radiation therapy and conditions such as arteriosclerosis, chronic debility, diabetic acidosis, hematological diseases, hepatic insufficiency, and malignancies of all sorts may lead to positive blood culture findings. No hard and fast rule concerning the types of microorganisms of significance can be stated. Certainly staphylococci have been recovered repeatedly, not only *S. aureus* but also *S. epidermidis*. The latter can be involved in bacterial endocarditis and has complicated cardiac catheterization and prostheses procedures. The gram-negative rods, especially *E. coli*, the obligately anaerobic bacteria of the colon, *Klebsiella-Enterobacter* organisms, *P. aeruginosa, Proteus* spp., and *Providencia* spp., have been recovered after trauma or surgery of contaminated body areas or as nosocomially acquired microorganisms complicating protracted hospitalization of patients with a variety of diseases. The recovery of salmonellae from the blood of individuals with systemic salmonellosis is not uncommon. Other so-called fevers of unknown origin may yield brucellae, pasteurellae, pneumococci, *N. meningitidis, Listeria monocytogenes*, etc. Bacterial endocarditis patients may harbor various viridans streptococci, coagulase-negative staphylococci, enterococci, corynebacteria, *Bacteroides* spp., *C. albicans*, etc. The rare recovery of environmental or so-called commensal microbes from the blood of patients with bacterial endocarditis need not be surprising. Fungi have been found in blood cultures from patients with lung abscesses, mycotic aneurysms, and hematological disorders. Clostridia have been recovered in the blood of accident victims and from patients with cholangitis and postpartum complications. In the newborn with bacteremia or septicemia during the first week of life, the organisms most often responsible are *Streptococcus agalactiae* (19) and *S. aureus*.

Positive cultures obtained from cerebrospinal fluids also reflect a variety of conditions. Besides meningitis, trauma, infectious complications of surgery, cranial and spinal epidural abscesses, subdural abscess, septic thrombophlebitis of the venous sinuses, and brain abscesses contribute to positive findings. There are systemic infectious diseases which may afflict the meninges, including severe pneumococcal pneumonia, salmonellosis, *H. influenzae* pneumonia and bacteremia, tuberculosis, listeriosis, *A. calcoaceticus* subsp. *lwoffi* septicemia of the newborn, generalized candidiasis, and advanced sepsis with every type of gram-negative rod, especially *E. coli* and *P. aeruginosa*. Meningitis is caused primarily by *H. influenzae* group b in children and by the meningococcus and pneumococcus in a majority of adults; the aged are most frequently involved with the latter organism. *S. aureus, E. coli, P. aeruginosa, Salmonella* spp., and *L. monocytogenes* are recovered not too uncommonly. *Flavobacterium meningosepticum* has been isolated occasionally. Traumatic or surgical injury may be complicated by any one of the many microorganisms, but the staphylococci, enterococci, and gram-negative rods predominate. In intracranial abscesses, anaerobic microorganisms are most common. These include peptostreptococci, *Bacteroides* spp., *Actinomyces* spp., *P. acnes*, and, less frequently, *Veillonella* spp. Cranial epidural abscesses have yielded staphylococci, streptococci, pneumococci, and occasionally some gram-negative rods. Spinal epidural abscesses have involved *S. aureus* and *P. aeruginosa*. Other gram-positive cocci have been isolated, as have *Salmonella* spp., which have caused epidural abscesses while infecting vertebrae (13). Subdural abscess is a comparatively rare complication of *H. influenzae* meningitis in children. *Cryptococcus neoformans, M. tuberculosis*, and leptospiral disease of the meninges must also be considered.

Transudates and exudates accompany a large variety of clinical conditions. They may be sterile or contain a varied microflora. Any and all microorganisms that involve the afflicted organs may be recovered. The organisms found reflect the anatomical site of the disease nidus or adjacent areas. The organisms outlined in the foregoing sections may also be encountered. These include the anaerobic bacteria, fungi, *M. tuberculosis*, and *L. pneumophila*.

WOUNDS AND BURNS

The microbiota of wounds reflects their anatomical site, the mode of infliction, i.e., traumatic or surgical, the environment in which the wounds were inflicted, and the degree of microbial contamination of adjacent areas that were perforated in the process. These considerations supplement rather than substitute for the general considerations which maintain an adequate host-parasite equilibrium. Traumatic wounds are usually complicated by aerobic indigenous microorganisms, especially *S. aureus*, group A streptococci, enterococci, *P. aeruginosa, E. coli, Proteus* spp., flavobacteria, and *Acinetobacter* spp. Among the anaerobic bacteria associated with traumatic wounds, the histotoxic and neurotoxic clostridia are most prominent, leading, under proper conditions, to gas gangrene or tetanus. The most common clostridia of gas gangrene are *C. perfringens* type A, *Clostridium septicum*, and *Clostridium novyii*, but these organisms may be harmless contaminants of wounds. The diagnosis of gas gangrene is strictly a clinical one; the isolation of clostridia may alert a clinician but does not constitute the diagnosis of clostridial cellulitis or clostridial myonecrosis. *Clostridium tetani* should not present a problem in a community of properly immunized individuals. Criminally inflicted stab wounds, often intentionally contaminated with feces, require not only expert surgical but also exquisite microbiological attention.

Infections complicating surgery may be of two kinds. One, the so-called wound infection, usually denotes a complication of a clean surgical procedure, a procedure performed under the best available aseptic conditions on tissues usually sterile and not found grossly contaminated during surgery. These wounds may yield *S. aureus*, enterococci, or gram-negative rods, and rarely microorganisms such as *S. pyogenes*, corynebacteria, pneumococci, *Bacillus subtilis*, etc. The second kind of infection occurs when the surgeon performs all or part of the surgical procedures in a contaminated area, resulting in infectious disease complications in a certain percentage, and often reflecting the state of health of the patient. The microorganisms mirror the microbiota of the particular anatomical site, but frequently minority members of the microflora attain a majority in the usually sterile tissue subjected to surgery and may gravely complicate the recovery of the patient. Among the bacteria involved in this complication are *E. coli*, *P. aeruginosa*, *Proteus* spp., *Providencia* spp., the *Klebsiella-Enterobacter-Serratia* group, flavobacteria, *Acinetobacter* spp., and *Bacteroides* spp.

Microbial contamination of severe burns can be a life-threatening complication. The organism most commonly encountered and most difficult to control is *P. aeruginosa*, often in conjunction with flavobacteria and other gram-negative rods and staphylococci, which abound in the institutional environment or on the uninvolved skin of the patient. Early microbiological analysis of burns usually shows a large number of a great variety of microorganisms, many of which are eventually supplemented by the bacteria cited above.

SURGICAL SPECIMENS

The microbiology of surgical specimens depends on the anatomical site and the underlying disease, the care exercised at the time of excision and during subsequent handling, and the available information that will help to narrow the number of etiological agents. The usual pyogenic organisms may be recovered, often as single components. Systemic fungi, nocardiae, actinomycetes, and various mycobacteria, including, of course, *M. tuberculosis*, as well as mixtures of anaerobic bacteria, have been found. In addition to bacteria such as *S. aureus*, primarily infected bone has yielded peptostreptococci, *Salmonella* spp. (especially in children with sickle-cell disease), gonococci, *Veillonella* spp., representatives of the fecal gram-negative rods, and rarely actinomycetes and blastomycetes. Foreign bodies, including prostheses, old sutures, and indwelling catheters, have been sources of many species of organisms, among them *P. aeruginosa*, salmonellae, escherichiae, and staphylococci.

AUTOPSY SPECIMENS

Autopsy specimens can yield invaluable information when the examination is performed promptly and aseptically and when cultures are plated immediately and accompanied by smears. Routinely, the kidney, spleen, liver, and lung are negative unless distinct lesions are cultured or an appreciable portion of the organ is diseased. If more than 6 h has elapsed between death and the examination, heart blood cultures and the tissues will reflect postmortem or perimortem contamination by organisms from the large bowel. An experienced and interested prosector will culture the areas of tissues possibly involved in an infectious disease process. A large variety of suspected and totally unexpected microorganisms can be recovered and demonstrated subsequently in tissue sections. The microorganisms range from the usual bacteria to lesions populated by brucellae, nocardiae, penicillia, aspergilli, pneumococci, group A streptococci, zygomycetes, clostridia, salmonellae, and shigellae, to name but a few.

MICROBIAL NUMBERS AND CLINICAL DISEASE

The preceding considerations underline the lack of understanding, in quantitative terms, of the many factors which contribute to overt symptoms of infectious disease in any particular person. The notion that disease production in the host resides solely in the pathogenic and virulent properties of the microorganisms can no longer be accepted. The application of molecular biology has initiated a better understanding of the mechanisms of microbial attachment to host cells, the advantages of certain plasmids and episomes in the expression of filaments, fimbriae, and fibrillae, the utility of some lipoteichoic acids, and the protection of colonizing microorganisms by exopolysaccharides (1). However, the numerous specific and nonspecific defense mechanisms of a particular host, the general health of this individual, and the stresses and insults to which he has been subjected recently have considerable influence, if not the determinant role, in the initiation and progression of an infectious disease process. The quality and quantity of the microbiota in the intimate biosphere of the individual at any given moment play significant roles in the process. These roles are as multifarious as the genera and species of microorganisms that constitute the microbial ecosystem of the person. The understanding of microbial ecological relationships among the constituents of this large pool of microbes is lacking, especially with regard to the health of the host. We know very little about the selectivity exercised by the tissues and organs of the host in the establishment of a local, autochthonous microbiota. There are but a few indications that age, physical environment, and nutrition may affect the minimal infectious dose of some microorganisms with more pronounced pathogenic proclivities. As with other areas of human biology, it is the abnormal condition that permits an intimation of understanding. If infectious disease is viewed as the disturbance of the equilibrium between two very complex systems, the host and his or her microbiota, at a specific moment, then the information derived from the study of the compromised patient and nosocomially significant organisms may be useful in the comprehension of the many factors operative in the maintenance of the host-parasite equilibrium. Microorganisms of nosocomial significance are widely distributed in nature but are involved in disease complications in medical facilities only. They may be minority members of an individual's microbiota, or they may be residents of the hospital, where they are concentrated as a result of antibiotic-associated selective pressures (10). Patients, compromised in their capacity to resist microorganisms for various medical and surgical reasons, most often find themselves under an umbrella of antimicrobial therapy. Since their humoral and cellular defense mechanisms may be suppressed pharmacologically, antibiotic-resistant endogenous or institutional organisms establish themselves quite readily as a majority in usually contaminated anatomic sites or even in tissues considered sterile. The original number of such microbes may be very small, certainly much less than required with healthy individuals, if the environmental surveillance and epidemiological studies can be applied as indicators of the distribution of such nosocomial microorganisms. The impaired patient mechanisms and probably the disturbed microbial ecology resulting from antimicrobial therapy allow these hospital organisms to reach critical levels with attendant overt clinical disease. It is not completely clear whether these microorganisms must attain a certain critical number or whether the access of just a few suffices as long as host impairment and lack of competition of normal microbiota persist. These observations emphasize that pathogenicity and virulence cannot be applied to a microbial genus or species without reference to the host and the host environment or to special circumstance. Although it is obvious that much must be learned before a clear understanding of infectious disease is established (17), it is equally clear that the concern and responsibility of the clinical microbiologist must transcend the narrow borders of microbial diagnosis or recognition and encompass aspects of the history and condition, the treatment, and the present and past environment and community of the patient. This more encompassing view is required to allow the proper distinction of significant microorganisms isolated from pathological specimens. Without such information, the unclear lines of separation between the indigenous and pathogenic microbiota of humans will be confused still more, depriving the clinician of information and the capability to intercede effectively against microorganisms for the benefit of his patients (8).

LITERATURE CITED

1. **Beachey, E. H.** 1980. Bacterial adherence. Chapman and Hall, London.
2. **Burnet, F. M.** 1962. Natural history of infectious disease, 3rd ed. Cambridge University Press, Cambridge.
3. **D'Amato, R. F., M. F. Sierra, and M. R. McGinnis.** 1977. Infectious diseases and etiologic or associated bacterial/fungal agents, p. 11–38. *In* D. Seligson (ed.), CRC Handbook series in clinical laboratory science. CRC Press, Inc., Cleveland.
4. **Drasar, B. S., and M. J. Hill.** 1974. Human intestinal flora. Academic Press, London.
5. **Dubos, R. J.** 1958. The evolution and the ecology of microbial diseases, p. 14–27. *In* R. J. Dubos (ed.), Bacterial and mycotic infections of man, 3rd ed. J. B. Lippincott Co., Philadelphia.
6. **Fiumara, N. J., H. M. Wise, Jr., and M. Many.** 1967. Gonorrheal pharyngitis. N. Engl. J. Med. **276:**1248–1250.
7. **Gibbons, R. J.** 1975. Attachment of oral streptococci to mucosal surfaces, p. 127–131. *In* D. Schlessinger (ed.), Microbiology—1975. American Society for Microbiology, Washington, D.C.
8. **Isenberg, H. D., and A. Balows.** 1981. Bacterial pathogenicity in man and animals, p. 83–122. *In* M. P. Starr, H. Stolp, H. G. Truper, A. Balows, and H. G. Schlegel (ed.), The prokaryotes. Springer-Verlag, New York.
9. **Isenberg, H. D., and J. I. Berkman.** 1966. Recent practices in diagnostic bacteriology, p. 237–317. *In* M. Stefanini (ed.), Progress in clinical pathology, vol. 1. Grune and Stratton, New York.
10. **Isenberg, H. D., and J. I. Berkman.** 1971. The role of drug-resistant and drug-selected bacteria in nosocomial disease. Ann. N.Y. Acad. Sci. **182:**52–58.
11. **Jones, G. L., and G. A. Hébert.** 1979. "Legionnaires": the disease, the bacterium and methodology. Center for Disease Control, Atlanta.
12. **NcNamara, M. J., M. C. Hill, A. Balows, and E. B. Tucker.** 1967. A study of hospital infections. Ann. Intern. Med. **66:**480–488.
13. **Pulaski, E. J.** 1954. Common bacterial infections. The W. B. Saunders Co., Philadelphia.
14. **Rosebury, T.** 1961. Microorganisms indigenous to man. McGraw-Hill Book Co., Inc., New York.
15. **Skinner, F. A., and J. G. Carr.** 1974. The normal microflora of man. Academic Press, London.
16. **Smibert, R. M.** 1978. The genus *Campylobacter*. Annu. Rev. Microbiol. **32:**673–709.
17. **Smith, H.** 1972. The little-known determinants of microbial pathogenicity, p. 1–24. *In* H. Smith and J. H. Pearce (ed.), Microbial pathogenicity in man and animals. Cambridge University Press, Cambridge.
18. **Smith, T.** 1934. Parasitism and disease. Princeton University Press, Princeton, N.J.
19. **Wilkinson, H. W.** 1978. Group B streptococcal infections. Annu. Rev. Microbiol. **32:**41–57.
20. **Wilson, G. S., and A. A. Miles.** 1975. Topley and Wilson's principles of bacteriology and immunology, 6th ed. The Williams & Wilkins Co., Baltimore.

Chapter 5

Procedures to Use During Outbreaks of Food-Borne Disease

FRANK L. BRYAN

Clinical microbiological laboratories are sometimes called upon to examine foods during outbreaks of disease that might be food borne. The laboratories are often the first agency either to determine that patients may be ill from food-borne disease or to be sent clinical specimens or food samples during investigation of a disease outbreak. During an epidemic investigation, the laboratory can help medical practitioners by verifying presumptive diagnoses. It can aid epidemiologists and field investigators by identifying the vehicle, by determining the degree of contamination, by determining the source of the etiological agent and tracing its spread, and by confirming the factors that contributed to the outbreak. Laboratory personnel must often advise investigators on taking suitable specimens from patients, controls, and food handlers and on collecting samples of foods that are apt to yield pathogens.

Depending on the probable disease under investigation, specimens and samples can include (i) serum, stools, vomitus, and urine from patients and controls; (ii) blood, spleen and liver tissue, and intestinal contents from fatal cases; (iii) stool or rectal swabs, blood, nasal or throat swabs, and exudate or pus from lesions of persons who handled the implicated or suspect food; (iv) food from implicated lots of processed foods; (v) leftover food from an incriminated or suspect meal; (vi) swabs of equipment used to process epidemiologically implicated foods; (vii) portions, scraps, or swabs of raw foods that may have introduced pathogens into the kitchen or plant environments; and (viii) when applicable, rectal or cloacal swabs of animals, swabs of animal droppings, samples of feed, or swabs of environmental contacts of animals. This chapter is limited to procedures for collecting and processing food and environmental samples and for interpreting results of laboratory examinations of these. Information for collecting specimens is presented in Chapter 8.

SELECTION AND COLLECTION OF SAMPLES

Suspect foods are those that are implicated by an attack-rate table or other epidemiological data or that have a history of being mishandled or mistreated. Potentially hazardous foods are those that readily support rapid and progressive growth because of their properties (pH, nutrients, water activity [a_w]) or that have a history of being vehicles in outbreaks of food-borne disease.

Samples of suspect foods should be collected at patients' domiciles and at establishments where the patients ate the suspect food or meal or from which they purchased the suspect food. Ordinarily, samples should be taken at patients' domiciles of any leftover food or beverage that was eaten within the last 72 h or of ingredients used in suspect foods. Samples of suspect foods or potentially hazardous foods from the suspect meal and samples of food from an allegedly contaminated lot should be collected at food service or processing establishments.

Representative samples of foods chosen randomly from unopened, original packages should be collected during surveys or quality control evaluations (7). Sampling during outbreak investigations, however, is done quite differently. On the basis of professional judgment, sample units of food or from equipment are collected at the point of operation (such as after possible contamination, survival, or growth) that is most likely to yield food-borne pathogens. In general, sample units should be taken from the geometric center of a food, the area where multiplication would most likely have occurred and consequently the most likely to yield positive results and where the highest counts should be found. An exception would be fermented sausage, which should be sampled a short distance below the surface, or side layers of refrigerated foods (where cooling is more rapid than in the center of the mass) for an indication of what the counts were soon after serving and before refrigeration.

Storage facilities should be checked for foods that could have been served previously. If no foods are left from the suspect meal or lot, collect sample units of food prepared in a similar manner to the suspect food, raw foods (particularly those of animal origin), ingredients used in the suspect food, or foods from the same suspect source.

Collect portions of each component of a food (such as meringue and filling for custard pies) that was prepared by different persons or prepared or stored in different ways before final assembly.

If the mode of spread or source of etiological agents is sought, and if definitive typing of isolates is contemplated, swabs of food-contact surfaces of equipment and environmental samples can be taken. When collecting swab sample units, swab or rub sterile sponges randomly over large areas of foods of animal origin and over contact surfaces of utensils and equipment used to process, store, or prepare suspect foods.

Collect samples aseptically with sterile implements or sterile gloves and put them into sterile containers as described in Table 1. Equipment for collecting, holding, and preserving samples is listed in Table 2. Ordinarily, sampling implements should be wrapped and sterilized in the laboratory before going to the site where the food was allegedly mishandled or mistreated. Field disinfection, when necessary, of sampling equipment can be accomplished as described in Table

TABLE 1. Methods of collecting, preserving, packing, and shipping samples

Sample	Methods of collecting and preserving	Methods of packing and shipping
Solid food or mixture of two or more food items	Cut or separate portions of food with sterile knife or other implement if necessary. Aseptically collect at least 200 g of sample from geometric center or other locations as deemed necessary, using a sterile implement, transfer to a sterile plastic bag or wide-mouth glass jar, and refrigerate.	Label. Put refrigerant around sample container. Do not freeze or use dry ice. Take sample to laboratory or ship it by most rapid means.
Liquid food or beverage	Stir or shake. Take sample in one of the following ways:	
	1. Pour, or ladle with sterile implement, at least 200 ml into sterile container. Refrigerate sample.	As above.
	2. Put long sterile tube into liquid and then cover top opening with finger. Transfer liquid to sterile jar or bag. Refrigerate sample.	As above.
	3. Immerse Moore swab in vat of liquid food or insert it into pipeline and allow liquid to flow through. Keep it in place for several hours if possible. Transfer swab to a jar containing enrichment broth.	Take sample to laboratory as soon as practicable; refrigeration may not be needed.
	4. If liquid is not viscous, pass 1 to 2 liters through membrane filter. Transfer filter pad aseptically into a jar of enrichment broth.	As above.
Frozen food	Use one of the following procedures: 1. Ship or take small volumes of frozen food to the laboratory without thawing or opening. 2. Drill with large-diameter, sterile auger from top of container diagonally through center to bottom at opposite side. Repeat from other side until at least 200 g is collected. 3. Chip frozen material with hammer and sterile chisel and collect chips with sterile implement; transfer at least 200 g of chips into sterile container.	Keep frozen. Use dry ice if necessary. Take or ship in insulated container.
Raw meat or poultry	Sample in one of the following ways: 1. With sterile implement or sterile plastic glove put chicken carcass, poultry part, or large cut of meat into a large, sterile plastic bag. Add 100 to 300 ml of enrichment broth and shake. For turkeys use 500 ml of enrichment broth. Remove sample and close bag. 2. Wipe sterile sponge over large area of carcass or cut of meat. Put swab into jar of enrichment broth. 3. Moisten swab with buffered, distilled water or 0.1% peptone water. Swab large portion of carcass or cut of meat. Put swab into enrichment broth. 4. With sterile, plastic glove, wipe carcass with sterile gauze squares; put gauze into bottle of enrichment broth. 5. Aseptically cut portions of meat or skin from different areas of carcass or cut of meat, or remove portion of carcass. Put at least 200 g into sterile, plastic bag or glass jar. Refrigerate. 6. Aseptically cut 200 g of neck skin and put into sterile container.	Same as with solid or liquid food, or if in enrichment broth take to laboratory as soon as practicable.
Dehydrated foods	Insert sterile, hollow tube from top of one side of container, diagonally through center to bottom of opposite side. Hold top and transfer to sterile container. Repeat from	Keep in tightly sealed, moisture-resistant container. Take or ship to laboratory.

Continued

TABLE 1—*Continued*

Sample	Methods of collecting and preserving	Methods of packing and shipping
	opposite side until at least 200 g is collected. An alternative method is to scoop material with a sterile spoon, spatula, tongue depressor, or similar implement. Transfer material to sterile container.	
Scrap material, air filters, sweepings, dust, litter, etc.	Cut with sterile knife if necessary. Collect at least 200 g of material with sterile tongue depressor, spatula, spoon, or tongs and place in sterile plastic bags or wide-mouth jars.	Same as above, depending on material.
Environmental or equipment-surface swab	Moisten swab with sterile 0.1% peptone water or buffered distilled water and swab contact surfaces of equipment or environmental surfaces. Put swab in enrichment broth.	Package, label, and ship as fecal swab.
Air	Impinge on plate or in liquid with air sampling device.	Tape collection container closed, label, and take to laboratory. Refrigerate liquid samples.
Water	Take historical samples, including water in bottles in refrigerators, ice cubes, and water in tanks. Take line-water samples after turning on tap for 10 s. Take water-source samples after running water for 5 min. Hold sterile bottle under tap and fill to 1 in. (ca. 2.5 cm) below lip. Collect 1 to 5 liters. Alternatively, membrane filters can be used. Moore swabs can be used to sample water in streams or pipelines; keep them in place up to 48 h, then transfer to jars of enrichment broth.	Tape collection container closed; label. Pack with absorbent material. Box and take or ship to laboratory. Refrigeration is usually not required.

TABLE 2. Equipment useful for investigation[a]

Sterile sample containers	Plastic bags (disposable or Whirl-Pak type), wide-mouth jars (0.3 to 1.5-liter capacity) with screw caps, water sample bottles (bottles for chlorinated water should contain enough sodium thiosulfate to provide a concentration of 100 mg of this compound per ml of sample), foil or heavy wrapping paper (wrapped), metal cans.
Sterile and wrapped sample collection implements	Spoons, scoops, tongue depressor blades, butcher knife, forceps, tongs, spatula, drill bits, metal tubes (1.25 to 2.5 cm in diameter, 30 to 60 cm long), pipettes, scissors, swabs, Moore swabs (compact pads of gauze made from strips 120 by 15 cm tied in the center with a long, stout twine or wire; for sewer, drain, stream, or pipeline samples).
Specimen-collecting equipment	Cartons (with lids) for stool specimens, bottles containing a preservative and transport medium, stool specimen protective canisters and cartons, sterile swabs, rectal swab outfits, sterile gauze pads (10 by 10 cm), tubes of transport medium.
Temperature-recording devices	Thermocouples (needle point, button type, soldered end) with either recording potentiometer, data logger, or digital indicator; bayonet-type thermometer (either 8 or 5 in. [ca. 20.3 or 12.7 cm, respectively] in length) with dial (0 to 220°F), or bulb-type thermometer. Optionally, thermosisters and indicator thermometers with platinum probes can be used. Reflecting thermometers can be used to measure temperatures of surfaces.
Supporting equipment	Fine-point felt-tip marking pen, roll of adhesive or masking tape, labels, waterproof cardboard tags with eyelets and wire ties, flashlight, electric drill, matches, 0.1% peptone water or buffered distilled water (5 ml in screw-capped tubes), test tube rack, insulated chest, pH meter, a_w hygrometer, investigational forms.
Sterilizing agents	95% ethyl alcohol, propane torch.
Refrigerants	Canned ice, refrigerant in plastic bags, liquid in cans, rubber or plastic bags or jars which can be filled with water and frozen, heavy-duty plastic bags for ice.
Clothing	White laboratory coat, paper hats, disposable plastic gloves, disposable plastic boots. (These are optional.)

[a] At least 15 sterile plastic bags or wide-mouth jars, 15 sterile spoons, 6 specimen collection containers or devices, temperature-measuring devices, one each of the supporting equipment, and sterilizing equipment should be preassembled in a kit which is kept by the agency responsible for investigating food-borne illness. Periodic resterilization or replacement of sterile supplies, media, and transport media is required to maintain kit in a ready-to-use condition.

TABLE 3. Field disinfection of equipment

1. Expose to steam at 100°C for 1 h in enclosed chamber, if available. (Spores may survive.)
2. Flame thoroughly with propane torch or Bunsen burner.
3. Wash, rinse, dry, and then immerse in 95% alcohol, remove, and flame; repeat two or three times. (Spores may survive if procedure is inadequately performed.)
4. Wash and then immerse in boiling water (spores may survive if not removed by washing).
5. Wash, rinse, immerse for at least 30 s in solution containing solute equivalent to not less than 100 ppm hypochlorite, rinse in sterile water; if necessary, wipe dry with sterile cloth (spores may survive if not removed by washing).

3. Bryan (1), Bryan et al. (3, 4), Gabis et al. (6), and the International Commission on Microbiological Specifications for Foods (7) give additional information on sampling.

HANDLING SAMPLES AFTER COLLECTION

As soon as a sample unit has been collected, it should be marked by an identifying code. Its description and any features that might be useful in interpreting results (such as the temperature of the food and the environment from which it was taken) should be recorded in a notebook or on a sample collection form, such as that illustrated in Fig. 1 (3).

Frequently, sample units are held for a while before they are taken or sent to the laboratory. Frozen foods should be kept frozen either in freezers in the establishment where they are collected or by contact with dry ice. Perishable foods should be kept at 0 to 4.4°C, either in refrigerators at the establishment where they are collected or in an insulated box. Sample units of hot foods or beverages in sealed containers should be cooled rapidly under cold running water or in ice baths and then held at 0 to 4.4°C.

As soon as sampling has been completed, samples should be packed appropriately to avoid spillage or breakage and to maintain desired temperatures. Frozen foods should be packed with dry ice. Sample units of perishable or chilled food should be kept cold with ice in plastic bags. If dry ice is used (which is not recommended), insulate the sample units to prevent their freezing. When it is not possible to deliver the sample personally, it should be sent to the laboratory by the most rapid and economical means feasible. The laboratory should be advised of the forthcoming samples, the method of shipment, shipment numbers, and the expected time of arrival. (See Table 1 for procedures for preserving and transporting specific types of samples.)

RECEIVING AND PREPARING SAMPLE UNITS FOR ANALYSIS

The physical appearance of each sample unit and its container should be observed and recorded upon arrival. Also, if a temperature-control sample is submitted, the temperature should be recorded at this time. Otherwise, the temperature of perishable foods should be recorded immediately after a portion has been removed for analysis. When appropriate, pH and water activity (a_w) determinations should be made at this time also.

Samples should be analyzed as soon as practicable after they arrive at the laboratory. It is sometimes necessary, however, to store them for a short time before work can begin. Those that arrive frozen should be either kept frozen or thawed in a refrigerator. Perishable foods should be refrigerated at 0 to 4.4°C. Low-moisture and canned foods can be stored at room temperature. Sample units for bacteriological analysis should be analyzed within 30 h, if possible.

Analytical samples should be removed aseptically from their containers. A Gram or other appropriate stain should be made of a drop of liquid food or beverage or from a well-mixed (1:10) homogenate of each analytical sample of solid food. Methods for preparing homogenates and examination for specific pathogens are described in Chapter 8 and by Speck (10) and the International Commission on Microbiological Specifications for Foods (8).

DETERMINATION OF APPROPRIATE TESTS

The types of tests and the order in which they are to be done should be determined by (i) the clinical signs and symptoms and the incubation periods (if known) of the affected persons; (ii) results of Gram or other appropriate stain of liquid foods or the homogenate of solid foods; and (iii) the type of food (9). Pathogens to look for depend on the symptoms, signs, and incubation periods of the patients and other appropriate clinical or epidemiological data. Six categories (according to signs and symptoms) of food-borne diseases are listed in Table 4, as follows: (i) upper gastrointestinal tract signs and symptoms (nausea, vomiting) occur first or predominate; (ii) lower gastrointestinal tract signs and symptoms (abdominal cramps, diarrhea) occur first or predominate; (iii) sore throat and respiratory tract signs and symptoms occur; (iv) neurological signs and symptoms (visual disturbances, vertigo, tingling, paralysis) occur; (v) allergic signs and symptoms (facial flushings, itching) occur; and (vi) signs and symptoms associated with general infection (fever, chills, malaise, prostration, aches, swollen lymph nodes) occur.

Usually, the syndrome can readily be classified in one of the above categories. If the patient's incubation period is known, further classification can be done. This classification may take the form of <1 h, 1 to 6 h, 7 to 12 h, 13 to 72 h, and >72 h. Initially, portions of the sample units should be tested for the etiological agents most likely associated with the syndrome. If the results are negative, remaining portions should be tested for other possible agents.

The Gram reaction and cellular morphology of predominant organisms provide information indicative of the kind of microorganisms to seek by culture techniques. (See Table 5 for a guide for the interpretation of staining results.) An analytical sample, however, will usually have to contain a rather high number of microorganisms before one will be seen in a microscopic field.

Bacteriological studies and epidemiological investigations have established the usual sources, reservoirs,

FOOD/ENVIRONMENTAL SAMPLE COLLECTION REPORT

Complaint Number	Sample Number[1]

Form E

Place Collected	Address	Phone
Person-In-Charge	Sample	Date/Hour Collected

Reason for Collecting Sample:
☐ Food from alleged outbreak, ☐ Food ingredient, ☐ Similar food prepared in similar manner to that involved in outbreak,
☐ Special survey, ☐ Routine, ☐ Environmental, ☐ Other (specify) ____________

Method of Collecting and Shipping Sample:
Method of Sterilizing: Container[2] | Collection Utensil[2]

Location Food Stored When Sampled	Temperature: Food	Storage Unit	Time Between Serving and Sampling

Shipped: ☐ Refrigerated, ☐ Frozen, ☐ Ambient	Identification Marks	Cost of Sample
Product Identification: Name	Brand	Lot Number

Manufacturer's Name	Address	Container Size or Weight

Symptoms of Victims:
☐ Nausea, ☐ Vomiting, ☐ Abdominal Cramps, ☐ Fever, ☐ Diarrhea ☐ Other (specify) ____________

Time of Eating Suspect Food/Meal: Date Hour	Time of Onset: Date Hour	Incubation Period	Duration of Illness
Investigator	Title	Agency	Date

Test Requested	Presence/Absence	Count/Concentration	Definitive Type
☐ Staphylococci			
☐ Staphyloenterotoxin			
☐ *C. perfringens*			
☐ *B. cereus*			
☐ *Salmonella*			
☐ *Shigella*			
☐ *E. coli*			
☐ *V. parahaemolyticus*			
☐ *C. botulinum*			
☐ Botulinus toxin			
☐ Chemical			
☐ Aerobic Colony Count			
☐ Coliform			
☐ Enterococci			
☐ Other (specify)			

Condition of Food	pH	a_w	Temperature: When received

Comments and interpretations

Laboratory Analyst	Agency	Date/Hour: Received	Started	Completed	Agent Identified

[1]Attach a list of number, sample, and tests desired for other samples collected at the same establishment during the same investigation.
[2]Specify only if unusual (such as field) method of sterilizing or sanitizing collection container or utensil or collecting sample is used.

FIG. 1. Food/environmental sample collection report form.

TABLE 4. Guide for laboratory tests indicated by certain signs and symptoms and incubation periods

Category of illness	Incubation period	Predominant signs and symptoms	Specimens to analyze from patients	Specimens to analyze from food workers	Test for:
Upper gastrointestinal tract (nausea, vomiting); signs and symptoms occurring first or predominating	Less than 1 h	Nausea, vomiting, unusual taste, burning of mouth	Vomitus, urine, blood, stool		Antimony, arsenic, cadmium, copper, lead, zinc
		Nausea, vomiting, retching, diarrhea, abdominal pain	Vomitus		Gastroenteritis-type mushrooms
	1 to 2 h	Nausea, vomiting, cyanosis, headache, dizziness, dyspnea, trembling, weakness, loss of consciousness	Blood		Nitrites
	1 to 6 h; mean, 2 to 4 h	Nausea, vomiting, retching, diarrhea, abdominal pain, prostration	Vomitus, stool	Nasal swab, swab of lesion on skin	*Staphylococcus aureus* and its enterotoxins, *Bacillus cereus*
	6 to 24 h	Nausea, vomiting, diarrhea, thirst, dilation of pupils, collapse, coma	Urine, blood (SGOT, SGPT[a] enzyme tests), vomitus		*Amanita, Phallodin, Gyromitrin* toxin group of mushrooms
Lower gastrointestinal tract (abdominal cramps, diarrhea); signs and symptoms occurring first or predominating	8 to 22 h; mean, 10 to 12 h	Abdominal cramps, diarrhea	Stool	Stool, rectal swab	*Clostridium perfringens, Bacillus cereus, Streptococcus faecalis* (?), *Streptococcus faecium* (?)
	12 to 72 h; mean, 18 to 36 h	Abdominal cramps, diarrhea, vomiting, fever, chills, malaise	Stool	Stool, rectal swab	*Salmonella, Arizona, Shigella* species, pathogenic *Escherichia coli,* other *Enterobacteriaceae, Vibrio parahaemolyticus, Yersinia enterocolitica, Campylobacter jejuni, Aeromonas* spp. (?), *Pseudomonas aeruginosa* (?)
	1.5 to 3 days	Diarrhea, fever, vomiting, abdominal pain; possibly respiratory symptoms	Stool	Stool	Enteric viruses, Norwalk agent, rotaviruses
	1 to 6 weeks	Mucoid diarrhea (fatty stools), abdominal pain, weight loss	Stool	Stool	*Giardia lamblia*
	1 to several weeks; mean, 3 to 4 weeks	Abdominal pain, diarrhea, constipation, headache, drowsiness, ulcers, variable; often asymptomatic	Stool	Stool	*Entamoeba histolytica*

Continued

TABLE 4. *Continued*

Category of illness	Incubation period	Predominant signs and symptoms	Specimens to analyze from patients	Specimens to analyze from food workers	Test for:
	3 to 6 months	Nervousness, insomnia, hunger pains, anorexia, weight loss, abdominal pain, sometimes gastroenteritis	Stool	Stool	*Taenia saginata, Taenia solium*
Sore throat and respiratory	12 to 72 h	Sore throat, fever, nausea, vomiting, rhinorrhea, sometimes a rash	Throat swab, blood	Throat swab, swab of lesions on skin	*Streptococcus pyogenes*
	2 to 5 days	Inflamed throat and nose, spreading grayish exudate, fever, chills, sore throat, malaise, difficulty swallowing, edema of cervical lymph node	Throat swab, blood	Throat swab, swab of lesions	*Corynebacterium diphtheriae*
Neurological (visual disturbances, vertigo, tingling, paralysis)	Less than 1 h	Tingling and numbness, giddiness, staggering, drowsiness, tightness of throat, incoherent speech, respiratory paralysis			Shellfish toxin
		Gastroenteritis, nervousness, blurred vision, chest pain, cyanosis, twitching, convulsions	Blood, urine, fat biopsy		Organic phosphate
		Excessive salivation, perspiration, and tearing; gastroenteritis, irregular pulse, pupils constricted, asthmatic breathing	Urine		Muscaria-type mushrooms
		Lightheadedness, drowsiness, followed by state of excitement, confusion, delirium, visual disturbances	Urine		Ibotonic acid and muscimol groups of mushrooms
		Tingling and numbness, dizziness, pallor, gastroenteritis, hemorrhage, desquamation of skin, eyes fixed, loss of reflexes, twitching, paralysis			Tetraodon toxin
	1 to 6 h	Tingling and numbness, gastroenteritis, dizziness, dry			Ciguatera toxin

Continued

TABLE 4. *Continued*

Category of illness	Incubation period	Predominant signs and symptoms	Specimens to analyze from patients	Specimens to analyze from food workers	Test for:
		mouth, muscular aches, dilated pupils, blurred vision, paralysis			
		Nausea, vomiting, tingling, dizziness, weakness, anorexia, weight loss, confusion	Blood, urine, stool, gastric washings		Chlorinated hydrocarbons
	12 to 72 h	Vertigo, double or blurred vision; loss of reflex to light; difficulty swallowing, speaking, breathing; dry mouth; weakness; respiratory paralysis	Blood, stool		*Clostridium botulinum* and its neurotoxins
	More than 72 h	Numbness, weakness of legs, spastic paralysis, impairment of vision, blindness, coma	Urine, blood, stool, hair		Organic mercury
		Gastroenteritis, leg pain, ungainly high-stepping gait, foot and wrist drop	Biopsy of gastrocnemius muscle		Triothocresyl phosphate
Allergic type (facial flushing, itching)	Less than 1 h	Headache, dizziness, nausea, vomiting, peppery taste, burning of throat, facial swelling and flushing, stomach pain, itching of skin	Vomitus		Histamine
		Numbness around mouth, tingling sensation, flushing, dizziness, headache, nausea			Monosodium glutamate
		Flushing, sensation of warmth, itching, abdominal pain, puffing of face and knees	Blood		Nicotinic acid
Generalized infection (fever, chills, malaise, prostration, aches, swollen lymph nodes)	4 to 28 days; mean, 9 days	Gastroenteritis, fever, edema about eyes, perspiration, muscular pain, chills, prostration, labored breathing	Muscle biopsy		*Trichinella spiralis*
	7 to 28 days; mean, 14 days	Malaise, headache, fever, cough, nausea, vomiting, con-	Stool, blood, urine	Stool, rectal swab	*Salmonella typhi*

Continued

TABLE 4. *Continued*

Category of illness	Incubation period	Predominant signs and symptoms	Specimens to analyze from patients	Specimens to analyze from food workers	Test for:
		stipation, abdominal pain, chills, rose spots, bloody stools			
	10 to 13 days	Fever, headache, myalgia, rash	Lymph node biopsy, blood		*Toxoplasma gondii*
	10 to 50 days; mean, 25 to 30 days	Fever, malaise, lassitude, anorexia, nausea, abdominal pain, jaundice	Urine, blood (SGOT, SGPT enzyme tests)		Serological evidence of hepatitis A virus (etiological agent not yet isolated)
	Varying periods (depends on specific illness)	Fever, chills, headache or joint ache, prostration, malaise, swollen lymph nodes, and other specific symptoms of disease in question	Blood, stool, urine, sputum, lymph node, gastric washings (one or more, depending on organism)		*Bacillus anthracis*, *Brucella melitensis*, *Brucella abortus*, *Brucella suis*, *Coxiella burnetii*, *Francisella tularensis*, *Listeria monocytogenes*, *Mycobacterium tuberculosis*, other *Mycobacterium* spp., *Pasteurella multocida*, *Streptobacillus moniliformis*, other pathogens as deemed necessary

[a] SGOT, Serum glutamic oxalacetic transaminase; SGPT, serum glutamic pyruvic transaminase.

and vehicles of many food-borne pathogens. When a food that has a history of having been a vehicle of a specific illness has been ingested and when the signs and symptoms and incubation period of the ill are compatible with that illness (Table 4), or when examination indicates characteristics of specific agents (Table 5), examine the suspect food either for agents that cause the suggested illness or for toxins produced by the agents. Table 6 lists tests to run when examining foods that have been suspected to be or have been epidemiologically implicated as vehicles of food-borne illness.

INTERPRETATION OF RESULTS

Food products have diverse chemical components and properties. They are processed, stored, and prepared in a variety of ways that affect the quantity and type of microorganisms that may have been introduced into and have survived and grew in them. Microorganisms in foods are in a dynamic state, and the numbers and the predominant types vary with time, with the nature of the substrate, and with environmental conditions. These same factors influence the methods used to isolate pathogens in the foods, as well as to interpret results of the laboratory tests.

Pathogens are often easier to isolate from clinical specimens than from foods. For example, pathogenic members of the *Enterobacteriaceae* in the human gut are in an environment that provides greater enrichment than that provided by foods. These same pathogens sometimes occur in small numbers in foods because of injury caused by processing, die-off during storage, or overgrowth by spoilage organisms. Therefore, routine clinical laboratory methods may be inadequate to detect pathogens in foods. (See references 8 and 10 for appropriate procedures for foods.)

An agent thought to be responsible for an illness can be confirmed if a pathogen that causes a syndrome similar to that observed is isolated from appropriate specimens from the patients. An agent also can be confirmed if toxins are identified in specimens from patients or if there is evidence of rise in antibody titer in serum specimens from patients.

To confirm actual involvement of a food, the same type of pathogen (serotype, phage type, or other definitive type) or the same toxin as was found in specimens from patients must be found in the epidemiologically implicated food. Even when clinical specimens are not available, a food is confirmed as a vehicle if toxic substances (such as zinc, staphylococ-

TABLE 5. Microscopic observations for determining morphology of predominant organism[a] in suspect foods

Gram reaction	Cell shape	Cell groupings	Flagella arrangement[b]	Spore[b]	Test for[c]:
Positive	Rods	Singly, in pairs, or short chains	Peritrichous	Avoid; central, subterminal, terminal spores swelling the cell	*Clostridium botulinum*
Positive	Rods	Singly, in pairs, or short chains	Atrichous	Avoid; subterminal spores, nonswelling cell	*Clostridum perfringens*
Positive	Rods	Singly and in pairs, but frequently form tangled chain	Peritrichous	Ellipsoidal; central or paracentral spores	*Bacillus cereus*
Positive	Cocci	Clusters	Atrichous	None	*Staphylococcus aureus*
Positive	Cocci	Long chains	Atrichous	None	*Streptococcus pyogenes*
Positive	Cocci	Short chains	Atrichous	None	*Streptococcus faecalis, Streptococcus faecium*
Negative	Rods	Singly and in pairs	Peritrichous (most types)	None	*Salmonella* spp., *Arizona* spp., *Escherichia coli, Yersinia enterocolitica*
Negative	Rods	Singly and in pairs	Atrichous	None	*Salmonella pullorum, Salmonella gallinarum, Shigella* spp., *Escherichia coli* (few types atrichous)
Negative	Slightly curved rods	Singly and in pairs	Monotrichous (polar)	None	*Vibrio parahaemolyticus*
Negative	Short curved rods, pleomorphic	Singly and in spiral chains	Monotrichous (polar)	None	*Vibrio cholerae* and related vibros
Negative	Comma and S shaped, pleomorphic	Singly or in short or long spiral chains	Usually monotrichous, but some bipolar or atrichous	None	*Campylobacter jejuni*

[a] As disclosed by Gram or other stain of food sample or food sample homogenate (1:10 dilution); large numbers (frequently 10^5 or more) of cells must be present before readily detected by microscopic examination.
[b] Features are not always seen in Gram stain.
[c] See Bryan (1, 2) for other organisms to test for.

cal enterotoxin, or botulinal toxin) are detected in it. A vehicle can be epidemiologically suspect if food-specific attack rates are high in a group of persons who have eaten a specific food and low in a group of persons who have not eaten the food. Further association can be made if a significant number of specific pathogens (such as 10^5 or more organisms of *Staphylococcus aureus, Clostridium perfringens*, or *Bacillus cereus* per g of food) that cause a syndrome similar to that of the patients are isolated from the food or if enteric pathogens (such as *Salmonella* spp. or *Shigella* spp.) are recovered from the food. Criteria for classifying outbreaks of specific food-borne diseases as "confirmed," according to the Centers for Disease Control (5), are given in Table 7.

The source or mode of spread of the causative agent can often be ascertained if the agent is isolated from raw foods, food ingredients, equipment, food workers, or live animals or their environment. Definitive typing of isolates is always required to make such associations. An account of the preparation of a food that is suspected as being a vehicle must contain reference to appropriate opportunities for pathogens or toxins to contaminate the food and, where applicable, opportunities for survival and growth of pathogens. Otherwise, the account is incomplete or in error. When necessary, field studies with appropriate laboratory backup should be done to provide evidence of contamination, survival of microorganisms or toxins, or growth of pathogenic bacteria.

TABLE 6. Tests to examine foods alleged, suspected, or epidemiologically implicated as vehicles of food-borne illness

Food	Test for:
Canned foods (primarily home-canned)	*Clostridium botulinum* and its neurotoxins, pH
Cereals and foods containing cornstarch	*Bacillus cereus*, mycotoxins
Cheese	*Staphylococcus aureus* and its enterotoxins, *Brucella* spp., pathogenic *Escherichia coli*
Confectionery products	*Salmonella* spp., a_w
Cream-filled baked goods, custards	*Staphylococcus aureus* and its enterotoxins, *Salmonella* spp., *Bacillus cereus*, pH, a_w
Crustacea	*Vibrio parahaemolyticus*, crustacean toxins
Dry milk	*Salmonella* spp., *Staphylococcus aureus* and its enterotoxins
Egg and egg products	*Salmonella* spp., beta-hemolytic streptococci
Fermented meats	*Staphylococcus aureus* and its enterotoxins
Fish	*Vibrio parahaemolyticus*, histamine, *Proteus* spp., fish toxins
Ham	*Staphylococcus aureus* and its enterotoxins
Mayonnaise	pH
Meat, meat products, and foods containing meat	*Salmonella* spp., *Clostridium perfringens*, *Staphylococcus aureus* and its enterotoxins, *Campylobacter jejuni*, *Yersinia enterocolitica*
Mexican-style foods	*Clostridium perfringens*, *Bacillus cereus*, *Shigella* spp., *Salmonella* spp., *Staphylococcus aureus* and its enterotoxins
Molluscan shellfish	*Vibrio parahaemolyticus*, shellfish toxin (saxotoxin), *Vibrio cholerae* O1 and non-O1, epidemiological and serological implication of hepatitis A or Norwalk agent
Oriental-style foods	*Vibrio parahaemolyticus*, *Bacillus cereus*, monosodium glutamate
Pinto beans	*Clostridium perfringens*, *Bacillus cereus*
Potato	*Bacillus cereus*, *Clostridium botulinum* and its neurotoxin
Poultry, poultry products, and mixed foods containing poultry	*Salmonella* spp., *Clostridium perfringens*, *Staphylococcus aureus* and its enterotoxins, *Campylobacter jejuni*, *Yersinia enterocolitica*
Raw fruits and vegetables	Parasites, *Shigella* spp., pathogenic *Escherichia coli*, *Listeria monocytogenes*
Raw milk	*Salmonella* spp., *Staphylococcus aureus* and its enterotoxin, *Campylobacter jejuni*, *Yersinia enterocolitica*, beta-hemolytic streptococci
Rice	*Bacillus cereus*
Salads of mixed vegetables, meat, poultry, or fish	*Staphylococcus aureus* and its enterotoxins, *Salmonella* spp., beta-hemolytic streptococci, *Shigella* spp., pathogenic *Escherichia coli*, pH, epidemiological and serological evidence of hepatitis A or Norwalk agent
Smoked meat, poultry, fish products	*Salmonella* spp., *Staphylococcus aureus* and its enterotoxins, *Clostridium botulinum* and its neurotoxins, a_w
Soft drinks, fruit juices, and concentrates (previously in metallic containers or vending machines)	Chemicals such as copper, zinc, cadmium, lead, antimony, tin; pH
Soups, stews, or gumbos	*Bacillus cereus*, *Clostridium perfringens*

TABLE 7. Criteria for confirming an outbreak of food-borne disease when cases have the typical syndrome characteristics of the disease

Disease	Laboratory or epidemiological criteria[a]					
	Isolation of pathogen	Serotype association	Titer increase	No. recovered	Toxin detection	Other criteria
Bacillus cereus gastroenteritis		Same serotype of *B. cereus* from stool specimens from most ill persons but not from controls Same serotype of *B. cereus* from ill persons and epidemiologically implicated food		$\geq 10^5$ *B. cereus* per g from epidemiologically incriminated foods		
Brucellosis	*Brucella* spp. in blood of ill persons		Fourfold or greater increase in agglutination titer between blood			

Continued

TABLE 7—*Continued*

Disease	Laboratory or epidemiological criteria[a]					
	Isolation of pathogen	Serotype association	Titer increase	No. recovered	Toxin detection	Other criteria
			specimens taken during acute illness and 3 to 6 weeks after onset of illness			
Botulism	*Clostridium botulinum* from stool, or ill from epidemiologically implicated food				Detection of botulinal toxin in sera, feces, or food	Frequently a history of eating home-canned or home-fermented fish, roe, or sea mammal meat
Clostridium perfringens enteritis		Same serotype of *C. perfringens* from specimens from most ill persons but not from controls		≥10^5 *C. perfringens* per g in epidemiologically implicated food	Demonstration of toxin in feces	
		Same serotype of *C. perfringens* from ill persons and epidemiologically implicated food		≥10^4 *C. perfringens* from ill persons is presumptive evidence		
Escherichia coli diarrhea		Same serotype of *E. coli* from most ill persons but not from controls			Demonstration of culture either to be enterotoxigenic (by gut loop, infant mouse, cell culture, or other biological technique) or to be invasive by production of conjunctivitis in guinea pig eye or other technique (e.g., ELISA[b])	
		Same serotype of *E. coli* from ill persons and from epidemiologically implicated food				
Salmonellosis	*Salmonella* from stool, rectal swab (urine or blood if septicemic symptoms occur) of ill persons, or epidemiologically implicated food	Same serotype of *Salmonella* from ill persons and from epidemiologically implicated food				
Shigellosis	*Shigella* spp. from stool or rectal	Same serotype from ill person and				

Continued

TABLE 7—*Continued*

Disease	Laboratory or epidemiological criteria[a]					
	Isolation of pathogen	Serotype association	Titer increase	No. recovered	Toxin detection	Other criteria
	swab of ill persons or epidemiologically implicated food	from epidemiologically implicated food (and from stool of food worker)				
Staphylococcal enterotoxicosis		Same phage type of specimen from ill persons and from epidemiologically implicated food (and skin, nose, or lesion of food worker)		≥10^5 *S. aureus* per g from epidemiologically implicated food	Detection of enterotoxin in epidemiologically implicated food	
Streptococcal sore throat or scarlet fever		Same M and T types of group A or G streptococci from ill persons and epidemiologically implicated food				
Cholera	Isolation of *Vibrio cholerae* from vomitus or stool of ill persons or from epidemiologically implicated food		Rise of serum titer during acute or early convalescent phases of illness and fall of titer during late convalescent phase in unimmunized persons		Demonstration of culture or filtrate to be enterotoxigenic by gut loop, infant mouse, cell culture, or other biological technique	
Vibrio parahaemolyticus gastroenteritis		Isolation of Kanagawa-positive *V. parahaemolyticus* of same serotype from stool of most ill persons		≥10^5 *V. parahaemolyticus* from epidemiologically implicated food		
Yersiniosis	Isolation of *Yersinia enterocoliticus* or *Y. pseudotuberculosis* from stool of most ill persons or from epidemiologically implicated food		Fourfold or greater rise of agglutination titer between blood specimens taken during acute illness and 2 to 4 weeks after onset of illness			
Campylobacteriosis	Isolation of *Campylobacter jejuni* from stools of most ill		Fourfold or greater rise of agglutination titer between blood			

Continued

TABLE 7—*Continued*

Disease	Laboratory or epidemiological criteria[a]					
	Isolation of pathogen	Serotype association	Titer increase	No. recovered	Toxin detection	Other criteria
	persons or from epidemiologically implicated food		specimens taken during acute illness and 2 to 4 weeks after onset of illness			
Other bacterial diseases	Variable, depending on clinical and laboratory appraisal of individual circumstances.					
Hepatitis A		Serological evidence of virus				Liver function tests; frequently a history of eating raw shellfish
Norwalk and related viral disease		Serological evidence of virus				Syndrome, incubation period, duration of illness
Trichinosis	Demonstration of larva in food	Serological evidence of infection				History of eating pork, bear, or arctic mammals
	Demonstration of cyst from muscle biopsy specimen					
Paralytic shellfish poisoning				Detection of large numbers of toxigenic species of dinoflagellates in water from which epidemiologically implicated mollusks were harvested	Detection of saxitoxin in epidemiologically implicated mollusks	History of eating shellfish; red tides
Ciguatera					Demonstration of ciguatoxin in epidemiologically implicated fish	History of ciguatera-associated fish
Puffer fish poisoning					Demonstration of tetrodotoxin in puffer fish	History of eating puffer fish
Scombroid poisoning					Detection of histamine levels of ≥100 mg/100 g of fish muscle	History of eating *Scombroidea* fish (tuna, mackerel)
Gastroenteritis-group mushroom poisoning					Demonstration of toxic chemicals in epidemiologically implicated mushrooms	History of eating gathered mushrooms

Continued

TABLE 7—*Continued*

Disease	Laboratory or epidemiological criteria[a]					
	Isolation of pathogen	Serotype association	Titer increase	No. recovered	Toxin detection	Other criteria
Mushroom-alcohol intolerance					Demonstration of toxic chemicals in epidemiologically implicated mushrooms or urine	History of eating gathered mushrooms
Muscarine-group mushroom poisoning					Demonstration of muscarine in epidemiologically implicated mushrooms or urine	History of eating gathered mushrooms
Ibotenic acid and muscimol groups of mushroom poisoning					Demonstration of ibotenic acid or muscimol in epidemiologically implicated mushrooms	History of eating gathered mushrooms
Amatoxin, phallotoxin, or gyromitrin groups of mushroom poisoning					Demonstration of amanitatoxin, phallotoxin, gyromitrin in epidemiologically implicated mushrooms or urine	History of eating gathered mushrooms
Plant poisoning					Demonstration of toxic chemical in epidemiologically implicated plant	History of eating toxic species of plant
Heavy metal poisoning					Demonstration of high concentration of metallic ion in epidemiologically implicated food or beverage	History of storing high-acid food or beverage in metal container or pipeline
Other chemical poisoning					Demonstration of high concentration of chemical substances in epidemiologically implicated food or beverage	History of use or suspect chemical in food or environment

[a] One or more of these criteria are demonstrated.
[b] ELISA, Enzyme-linked immunosorbent assay.

LITERATURE CITED

1. **Bryan, F. L.** 1979. Infections and intoxications due to other bacteria, p. 211–297. *In* H. Riemann and F. L. Bryan (ed.), Food-borne infections and intoxications, 2nd ed. Academic Press, Inc., New York.
2. **Bryan, F. L.** 1982. Diseases transmitted by foods. A classification and summary, 2nd ed. Centers for Disease Control, Atlanta.
3. **Bryan, F. L., H. W. Anderson, R. K. Anderson, K. J. Baker, H. Matsuura, T. W. McKinley, R. C. Swanson, and E. C. D. Todd.** 1976. Procedures to investigate food-borne illness, 3rd ed. International Association of Milk, Food, and Environmental Sanitarians, Ames, Iowa.
4. **Bryan, F. L., H. W. Anderson, K. J. Baker, G. F. Craun, W. Duel, K. H. Lewis, T. W. McKinley, R. A. Robinson, R. C. Swanson, and E. C. D. Todd.** 1979. Procedures to investigate waterborne illness. International Association of Milk, Food, and Environmental Sanitarians, Ames, Iowa.
5. **Centers for Disease Control.** 1983. Foodborne disease outbreaks, annual summary 1981. Centers for Disease Control, Atlanta.
6. **Gabis, D. A., A. R. Brazis, and T. F. Midura.** 1984. Sample collection, shipment and preparation for analysis, p. 3–9. *In* M. L. Speck (ed.), Compendium of methods for the microbiological examination of foods, 2nd ed. American Public Health Association, Washington, D.C.
7. **International Commission on Microbiological Specifications for Foods.** 1974. Microorganisms in foods. 2. Sampling for microbiological analysis: principles and specific applications. University of Toronto Press, Toronto.
8. **International Commission on Microbiological Specifications for Foods.** 1978. Microorganisms in foods. 1. Their significance and methods of enumeration, 2nd ed. University of Toronto Press, Toronto.
9. **Sanders, A. C., F. L. Bryan, and J. C. Olson, Jr.** 1984. Foodborne illness—suggested approaches for the analysis of foods and specimens obtained in outbreaks. *In* M. L. Speck (ed.), Compendium of methods for the microbiological examination of foods, 2nd ed. American Public Health Association, Washington, D.C.
10. **Speck, M. L. (ed.).** 1984. Compendium of methods for the microbiological examination of foods, 2nd ed. American Public Health Association, Washington, D.C.

Chapter 6

Rapid Manual and Mechanized/Automated Methods for the Detection and Identification of Bacteria and Yeasts

RICHARD F. D'AMATO, JAMES C. McLAUGHLIN, AND MARY JANE FERRARO

Clinical microbiology, the traditional stepchild of pathology, has been very slow to benefit from the technological advances that have had such a positive influence on other areas of laboratory medicine. It would not have been unusual to walk into a clinical microbiology laboratory in the late 1960s to find that little had changed regarding microbial detection and identification in the nearly 80 years since Koch's original utilization of agar. In the transition from classical to contemporary methodologies used for microbial identification, the reference point has shifted from multistep methods to unitary procedures with emphasis on standardization, speed, reproducibility, miniaturization, mechanization, and automation. Advances in molecular biology and immunochemistry are providing microbiologists with tools to detect rapidly the presence of specific microorganisms in clinical specimens. In reality, the time required for reaction endpoints for rapid tests ranges from immediately for agglutination techniques to 1 week for auxanographic methods for yeast identification. The commercialization of many rapid and mechanized/automated microbial identification and detection methods has increased the ability of the clinical microbiology laboratory to provide rapid results with accuracy. Additionally, rapid methods could reduce laboratory costs and decrease the time a patient is hospitalized. In light of the fiscal constraints on hospitals in the current era of prospective payments, any cost savings become important.

The literature contains numerous evaluations of detection and identification systems, but comparisons of accuracy for many of the newer systems are available only in abstract form and therefore are excluded from this chapter. There is a trend in the laboratory evaluations of commercial identification systems. The earliest efforts compared the percentage agreement of individual biochemical reactions versus a "standard" method. Later investigators, aware that percentage agreements vary with media formulations, compared only final identifications. The inconvenience and expense of preparing a battery of specially formulated media as utilized by reference centers such as the Centers for Disease Control has led many investigators now to compare the accuracy of the identification of one system approach with that of another. The many factors to consider in performing a comparison of one or more identification systems have been reviewed by D'Amato et al. (50) and Edberg and Konowe (61). A review of the systems approach to microbial identification is presented by D'Amato et al. (50).

The purpose of this chapter is to review commercially available rapid, manual, and mechanized/automated systems for bacterial and fungal detection and identification. Because of the many commercial systems currently available, only those that are established approaches in clinical laboratories are reviewed. The systems are divided into three major categories: (i) manual biochemical systems for microbial identification; (ii) mechanized/automated systems; and (iii) immunological systems. Within each category, identification systems and detection methods (some of which are also used for rapid identification) are discussed in alphabetical order and according to the type of microorganism they are designed to identify. We do not address the issues of cost (which varies greatly with test volume), convenience of use, or the need for all laboratories to avail themselves of the comprehensive identifications offered by some systems. Evaluations of each system can be found in the references listed after the manufacturer's name and address. Methods for microorganisms other than bacteria and fungi and for antimicrobial susceptibility testing are discussed in other chapters in this volume.

MANUAL BIOCHEMICAL SYSTEMS FOR MICROBIAL IDENTIFICATION

Enterobacteriaceae

API 20E (Analytab Products, Plainview, N.Y.) (2, 3, 19, 20, 23, 25, 26, 29, 38, 55, 60, 70, 74, 78, 86, 91–93, 99, 102, 111, 117, 126, 131, 143, 144, 151, 163, 164, 166, 175, 180)

The API 20E is a widely used identification system, complemented by an extensive and diverse computer data base. Bacterial identification can be achieved with the same series of test substrates in 5 h or after overnight incubation. The system consists of a plastic strip with 20 microtubes containing dehydrated substrates as listed in Table 1. The test substrates are reconstituted when inoculated with a suspension of the test organism in sterile physiologic saline. After inoculation, the cupule sections of the arginine, lysine, ornithine, H_2S, and urea microtubes are overlaid with mineral oil, and the strip is placed in a disposable incubation tray containing water and incubated at 35 to 37°C for 5 h for rapid identification or for 18 to 24 h. Reagents are added to detect tryptophan deamination, indole, and acetoin. The 21 test results, which include oxidase detection, are converted to a seven-digit profile using the octal coding system. Identification is made through the Analytical Profile Index, by use of a differential chart, or by computer through the offices of the manufacturer. An API Same-Day Analytical Profile Index for profile numbers obtained after 5 h of incubation is also available.

Enteric-Tek (Flow Laboratories, Inc., McLean, Va.) (31, 63, 73)

The Enteric-Tek system comprises a round, multicompartmented plastic tray containing 11 peripheral

TABLE 1. Tests included in manual systems for identification of gram-negative bacteria[a]

Test	API 20E	Rapid E	Entero-tube	Micro-ID	Micro-Media	Minitek	Enteric-Tek
Acetoin production (Voges-Proskauer)	+	+	+	+	+	−	−
Acid from:							
Adonitol	−	+	+	+	+	−	+
Amygdalin	+	−	−	−	−	−	−
Arabinose	+	+	+	+	+	+	+
Cellobiose	−	+	−	−	−	−	−
Dulcitol	−	−	+	−	−	−	−
Glucose	+	+	+	−	+	+	+
Inositol	+	−	−	+	+	+	−
Lactose	−	−	+	−	+	−	+
Mannitol	+	−	−	−	−	−	−
Melibiose	+	+	−	−	−	−	−
Raffinose	−	+	−	−	+	−	−
Rhamnose	+	+	−	−	+	+	+
Sorbitol	+	−	+	+	+	−	+
Sucrose	+	+	−	−	+	−	−
Trehalose	−	+	−	−	−	−	−
Xylose	−	+	−	−	−	−	−
Arginine dihydrolase	+	−	−	−	+	−	−
β-Galactosidase (ONPG)	+	+	−	+	+	+	−
Citrate utilization	+	+	+	−	+	+	+
Cytochrome oxidase production	+	+	−	−	−	−	−
Deamination of phenylalanine or tryptophan	+	+	+	+	+	+	+
Esculin hydrolysis	−	+	−	+	−	−	−
Gas from glucose	−	−	+	−	−	−	−
Gelatinase production	+	−	−	−	−	−	−
H_2S production	+	−	+	+	+	+	+
Indole production	+	+	+	+	+	+	+
Lysine decarboxylase	+	+	+	+	+	+	+
Malonate utilization	−	+	−	+	+	+	+
Nitrate reduction	+	−	−	+	−	+	−
Ornithine decarboxylase	+	+	+	+	+	+	+
Urease production	+	+	+	+	+	+	+

[a] Adapted from reference 50.

wells and 1 center well, each of which contains a specific agar-based medium. The system permits the determination of 14 biochemical parameters (Table 1). Each well is inoculated with a suspension of the test organism in sterile water, and the plate is then incubated at 35 to 37°C for 18 to 24 h. After incubation, Kovacs reagent is added to the center well. The reactions in each well are recorded. A five-digit profile number is generated, and identification is accomplished with the Enteric Computer Code Book or with a differential chart. Computer-assisted identification is also available from the manufacturer.

Enterotube II (Roche Diagnostics, Nutley, N.J.) (16, 47, 52, 59, 62, 72, 77, 86, 92, 111, 116, 125, 131, 137, 163, 164, 173, 180, 182)

The Enterotube II is a 12-compartment, plastic tube containing agar-based media which permits the determination of 15 biochemical parameters (Table 1). It is the easiest and fastest multitest system to inoculate. Each end of the tube contains a screw cap which is removed before inoculation. A single colony is picked up with the tip of an inoculating needle which extends through the body of the tube. The media are inoculated by withdrawing the needle through the compartments, completely reinserting the needle, and again withdrawing it until the tip is in the H_2S/indole compartment. The end of the needle is severed at a prescored position, and both caps are replaced. The tube is incubated for 18 to 24 h at 35 to 37°C, and reagents are then added to the appropriate chambers for indole and acetoin. Test results are converted into a five-digit profile number according to the octal coding system. Identification is made using a differential chart, an identification manual, the Computer Coding and Identification System, or the manufacturer's computer.

Micro-ID System (General Diagnostics, Morris Plains, N.J.) (33, 60, 74, 101, 102)

The Micro-ID System represents a departure from the traditional methodologies of bacterial identification in that it relies on the presence of preformed enzymes in the bacterial inoculum to permit a rapid 4-h identification. It consists of a molded styrene tray containing 15 reaction chambers, each containing an individual test substrate (Table 1). The first 5 chambers contain individual substrate and detection disks, and the remaining 10 contain combination substrate/detection disks. These chambers are inoculated with a bacterial suspension in sterile physiologic saline equivalent to a McFarland no. 1 turbidity standard. The system is then incubated for 4 h at 35 to 37°C. After incubation, potassium hydroxide is added to the Voges-Proskauer chamber. Then the unit is rotated to moisten the upper detection disks in the first five wells. Test results are converted into a five-digit profile number using the octal coding system, and the profile number may be interpreted by using the Micro-ID Identification Manual, by computer through the offices of the manufacturer, or with a differential chart. If individual laboratory requirements dictate, the system may be refrigerated and read the following day.

Micro-Media Quad Enteric Panel (Micro-Media Systems, Inc., San Jose, Calif.) (20, 26, 102, 104)

The Quad Enteric Panel consists of four individual sets of 20 tests (Table 1) furnished in a frozen microtiter tray. After the tray is thawed, a suspension of the test organisms is poured into a four-compartment seed trough. An 80-pronged transfer lid is lowered from the seed trough into the test panel and simultaneously inoculates each well. The tray is then covered and incubated for 18 to 24 h at 35 to 37°C. Appropriate reagents are added, and test results are converted to a seven-digit profile number. Identification is accomplished with an index, the MMS Biotype Code Book, or a differential chart or by computer through the office of the manufacturer.

Minitek (BBL Microbiology Systems, Cockeysville, Md.) (67, 78, 81, 84, 106, 118, 153, 163, 164)

The Minitek system utilizes 20-well plastic plates to which paper disks impregnated with various biochemical substrates are added. The system permits the evaluation of 36 possible tests, and unlike most other systems, the number and choice of tests are left to the user. However, an index has been introduced which requires the use of a fixed set of 14 tests (Table 1). The disks are inoculated with an appropriate bacterial suspension, and mineral oil is added to the urea and H_2S/indole wells. After incubation for 18 to 24 h at 35 to 37°C, reagents are added to assay for production of indole and acetoin, phenylalanine deaminase activity, and nitrate reduction. As with the other systems, reactions are read and, using the octal coding system, converted to a five-digit profile number which is then matched in an index, the Enteric Code Book. Profile numbers not found in the index can be identified by computer through the offices of the manufacturer. The Enterobacteriaceae III System is presently available and provides identification after 4 h of incubation.

Rapid E (Analytab Products, Plainview, N.Y.) (8)

The Rapid E system is physically similar to the API 20E. It consists of a plastic strip of 20 microtubes containing dehydrated substrates (Table 1). Substrates are reconstituted upon the addition of a suspension of the test organism in medium provided with the system, and reactions are read after 4 h of incubation at 35 to 37°C. Profile numbers are constructed and interpreted with the Identification Code Book.

Nonfermenters and Oxidase-Positive Fermenters

API 20E (Analytab Products, Plainview, N.Y.) (35, 57, 88, 123, 132, 133, 136, 155, 174)

Developed for identification of *Enterobacteriaceae*, the API 20E system as described above is quite suitable for the identification of oxidase-positive fermenters. However, the carbohydrate reactions are of limited value for the identification of nonfermenting gram-negative rods, regardless of their oxidase activity. Therefore, six supplementary tests (nitrate and nitrite reduction, motility, acid production from glucose, and growth on MacConkey agar) must be performed. After incubation for 48 h at 35 to 37°C, appropriate reagents are added, and a nine-digit profile number is generated. Identification is accomplished with the same index used for the identification of *Enterobacteriaceae* or with a differential chart. Computer-assisted identification is also available.

Minitek (BBL Microbiology Systems, Cockeysville, Md.) (13, 35, 41, 133, 178)

The Minitek system described for the identification of *Enterobacteriaceae* can also be used for the identification of nonfermenters and oxidase-positive fermenters. Disks containing the following substrates are recommended for the identification of these organisms: maltose, sucrose, mannitol, xylose, glucose (aerobic and anaerobic), citrate, *o*-nitrophenyl-β-D-galactopyranoside (ONPG), nitrate, phenylalanine, urea, lysine, ornithine, arginine, and tryptophan. The plates containing the disks are inspected after 24 and 48 h of incubation in a humidity chamber at 35 to 37°C. After appropriate reagents are added, tests are read and converted into a profile number which is consulted in the Numerical Identification Code Book. Identification can also be made with a differential chart.

N/F System (Flow Laboratories, Inc., McLean, Va.) (18, 35, 108, 174)

The N/F System was designed specifically for the identification of nonfermenters and oxidase-positive fermenters. It consists of two tubes (42P tube and GNF tube), used for initial screening, and the Uni-N/F-Tek plate, which is physically similar to the plate used for the identification of *Enterobacteriaceae*. After incubation for 18 to 24 h, the following reactions are provided by the screening tubes: glucose fermentation, nitrate reduction, fluorescein production, growth at 42°C, and pyocyanin production. These reactions are used for the identification of *Pseudomonas aeruginosa* and *Pseudomonas fluorescens*/*Pseudomonas putida*. When an isolate cannot be identified by the screen, the Uni-N/F-Tek plate is inoculated. The plate contains agar-based media for the determination of indole and H_2S production in the center well and glucose, xylose, mannitol, lactose, and maltose oxidation, acetamide assimilation, esculin hydrolysis, and urease, DNase, and β-galactosidase production in the peripheral wells. After inoculation, the plate is incubated at 35 to 37°C for 24 to 48 h. Results may be interpreted with a differential chart or an index, the N/F Computer Code Book.

Oxi-Ferm (Roche Diagnostics, Nutley, N.J.) (35, 57, 88, 90, 95, 108, 129, 132–134, 136, 146, 154)

The Oxi-Ferm system is similar in design to the Enterotube II. It consists of eight compartments that contain agar-based media for testing for glucose fermentation, oxidation of xylose and glucose, production of urease, H_2S, indole, and nitrogen gas, arginine dihyrolase, and citrate utilization. The substrates are inoculated in the same way as for the Enterotube II. After 48 h of incubation at 35 to 37°C, results are recorded and a four-digit profile number is generated. Identification is made with a differential chart or with the "Computer Coding and Identification System for Oxi-Ferm Tube." This is not a probabilistic or other computer-derived identification method, but essentially a register which lists profile numbers.

Rapid NFT (Analytab Products, Plainview, N.Y.)

The DMS Rapid NFT consists of a plastic strip of 20 microcupules containing dehydrated substrates. Included are tests for the assimilation of glucose, arabinose, mannose, mannitol, *N*-acetylglucosamine, maltose, gluconate, caprate, adipate, malate, citrate, and phenylacetate; glucose fermentation; arginine dihydrolase; and hydrolysis of urea, esculin, gelatin, and ONPG. The strip is placed in an incubation tray, and the substrates are inoculated with a suspension of the test isolate in medium supplied with the system. The unit is then incubated for 24 h. However, unlike other systems which are incubated at 35°C, the Rapid NFT is incubated at 29 to 31°C. Reactions are read after appropriate reagent additions, and results are converted to a seven-digit profile number. Identification is made with the Identification Code Book or a differential table.

Anaerobes

API 20A (Analytab Products, Plainview, N.Y.) (82, 83, 121, 130, 171)

The API 20A system is similar in design to the API 20E system. The following tests are included: fermentation of glucose, mannitol, lactose, sucrose, maltose, salicin, xylose, arabinose, glycerol, cellobiose, mannose, melezitose, raffinose, sorbitol, rhamnose, and trehalose; production of indole, urease, gelatinase, and catalase; and esculin hydrolysis. After the microtubes in the strip are inoculated with the test organism suspended in Lombard-Dowell broth (58), the strip is incubated anaerobically for 24 h at 35 to 37°C. After incubation, appropriate reagents are added, the tests are read, and a seven-digit profile number is generated. Identification is made with the API 20A Analytical Profile Index, with a differential chart, or by computer through the offices of the manufacturer.

A new API system is the AN-IDENT, which consists of a plastic strip with 20 microcupules containing 20 dehydrated substrates. These conventional and nonconventional (chromogenic) substrates provide results after 4 h of *aerobic* incubation at 35 to 37°C. The nonconventional substrates are based upon the release of a chromogen when substrate hydrolysis takes place. Some of the chromogens are colorless, and their detection is accomplished by the addition of a detection reagent such as cinnamaldehyde. Others require no reagent addition. Included are tests for indole production, *N*-acetyl-glucosaminidase, α-glucosidase, α-arabinosidase, β-glucosidase, α-fucosidase, phosphatase, α- and β-galactosidase, indoxyl-acetate, arginine dihydrolase, catalase, pyroglutamic acid arylamidase, and leucine, proline, tyrosine, arginine, alanine, histidine, phenylalanine, and glycine aminopeptidases. After incubation and the addition of appropriate reagents, results are converted into a seven-digit profile number and are interpreted with the AN-IDENT Profile Index, with a differential chart, or by the manufacturer's computer.

Anaerobe-Tek (A/T) (Flow Laboratories, Inc., McLean, Va.) (34, 114)

Anaerobe-Tek is similar to the other identification systems manufactured by Flow Laboratories. It consists of a multicompartmented plate containing agar media. The tests provided include starch, esculin and gelatin hydrolysis, milk proteolysis, bile tolerance, lecithinase, lipase, catalase, DNase, H_2S and indole production, and fermentation of glucose, trehalose, lactose, and mannitol. After inoculation, the system is incubated anaerobically for 24 to 48 h at 35 to 37°C. Test results are converted into a six-digit profile number, and identification is accomplished with a computerized numerical code book or a differential chart.

Minitek (BBL Microbiology Systems, Cockeysville, Md.) (11, 82, 83, 169, 170)

The Minitek system with an appropriate battery of disks incubated under anaerobic conditions is also used for the identification of anaerobic bacteria. The recommended test disks are esculin, nitrate, arabinose, glucose, glycerol, lactose, maltose, mannitol, rhamnose, salicin, sucrose, trehalose, and xylose. The disks are inoculated with a suspension of test organism made in Lombard-Dowell broth (58). Plates are incubated for 48 h, and appropriate reagents are added. Results are recorded and interpreted by means of an index or a differential chart.

IDS RapID ANA (Innovative Diagnostic Systems, Inc., Decatur, Ga.)

The IDS RapID ANA consists of a plastic panel which contains 10 test cavities with the following conventional and chromogenic tests: hydrolysis of *p*-nitrophenylphosphate, ONPG, *p*-nitrophenyl-α-D-glucoside, *p*-nitrophenyl-β-D-glucoside, *p*-nitrophenyl-α-D-galactoside, *p*-nitrophenyl-α-D-fucoside, *p*-nitrophenyl-*N*-acetylglucosaminide, leucylglycl-β-napthylamide, glycyl-β-naphthylamide, prolyl-β-naphthylamide, phenylalanyl-β-naphthylamide, arginyl-β-naphthylamide, seryl-β-naphthylamide, and pyrrolidonyl-β-naphthylamide, triphenyltetrazolium reduction, arginine dihydrolase, trehalose fermentation, and indole production. The panel is structured so that the test organism suspension simultaneously inoculates the substrates. After inoculation, the panel is incubated *aerobically* for 4 h at 35 to 37°C. Reactions in some test cavities are read before and after appropriate reagent addition. Test results are scored, and identification is made with the aid of a differential chart or an index, the RapID ANA Code Compendium.

Streptococci

API 20S (Analytab Products, Plainview, N.Y.) (10, 98, 105, 128)

The API 20S, designed for the identification of streptococci and aerococci, consists of a strip with 20 microcupules containing dehydrated conventional and chromogenic substrates. Included are tests for esculin hydrolysis in the presence of bile, fermentation of mannitol, sorbitol, glycerol, sorbose, raffinose, lactose, sucrose, and trehalose, β-glucosidase, *N*-acetylglucosaminidase, and β-galactosidase, alkaline phosphatase, pyroglutamic acid arylamidase, leucine, serine, and arginine aminopeptidases, and hippurate and indoxyl-acetate hydrolysis. After the substrates are inoculated with a suspension of the test organism in sterile physiologic saline, the strip is incubated for 4 h at 35 to 37°C. Appropriate reagents are added, and test results are converted to a seven-digit profile number. Identification is made with the Streptococcus Analytical Profile Index or with a differential table.

Rapid STREP (Analytab Products, Plainview, N.Y.) (149, 150)

The Rapid STREP system also consists of a strip containing 20 cupules and makes use of conventional and chromogenic tests. The system provides tests for the fermentation of ribose, L-arabinose, mannitol, sorbitol, lactose, trehalose, insulin, raffinose, starch, and glycogen, acetoin production, hippurate and esculin hydrolysis, pyrrolidonylarylamidase, α-galactosidase, β-glucuronidase, β-galactosidase, alkaline phosphatase, leucine arylamidase, and arginine dehydrolase. After inoculation of the dehydrated substrates with a suspension of the test organism in medium supplied with the system, the strip is incubated aerobically for 4 h at 35 to 37°C. Appropriate reagents are added, and test results are converted to a seven-digit profile number. If the profile number cannot be found in the Identification Codebook, the strip is reincubated for an additional 14 to 20 h. Identification can also be accomplished with a differential table.

Staphylococci

Staph-Ident (Analytab Products, Plainview, N.Y.) (4–6, 42, 56, 71, 107)

The Staph-Ident system was designed to identify *Staphylococcus aureus* and coagulase-negative staphylococci. Dehydrated conventional and chromogenic test substrates are contained in microcupules on a plastic strip similar to other API manual identification systems. Included are tests for phosphatase, urease, β-glucosidase, β-glucuronidase, β-galactosidase, arginine dihydrolase, and mannose, mannitol, trehalose, and salicin fermentation. The substrates are inoculated with a suspension of bacteria in sterile physiologic saline, and the strip is incubated at 35 to 37°C for 5 h. A color-developing reagent is added to the β-galactosidase test, and reactions are recorded. Identification is made with a differential chart or the Staph-Ident Profile Register.

Staph-Trac (Analytab Products, Plainview, N.Y.) (71)

The Staph-Trac system consists of a plastic strip with cupules and permits the performance of 20 tests for the identification of *S. aureus* and coagulase-negative staphylococci. Included are tests for the fermentation of glucose, fructose, mannose, maltose, lactose, trehalose, mannitol, melibiose, xylitol, raffinose, xylose, sucrose, α-methyl-glucoside, and *N*-acetylglucosamine, nitrate reduction, alkaline phosphatase, production of acetyl-methyl-carbinol, arginine dihydrolase, and urease. The dehydrated substrates are inoculated with the test organism suspended in medium supplied with the system. The strip is incubated in an incubation tray for 24 to 48 h at 35 to 37°C. After reagent addition, tests are read and results are converted into a seven-digit profile number. Identification is accomplished with a differential table or the Identification Code Book, which contains a limited number of profiles.

Neisseria

Minitek (BBL Microbiology Systems, Cockeysville, Md.) (12, 80, 124, 141)

The Minitek system for the identification of *Neisseria*, like the other Minitek systems, uses paper disks impregnated with substrates. The following tests are used: glucose, maltose, sucrose, and ONPG. Disks are dispensed onto a plate and are inoculated with a suspension of the test isolate in medium supplied with the system. The plate is incubated in a humidity chamber for 4 h at 35 to 37°C. Reactions are read and interpreted with a differential table.

Rapid N/H System (Innovative Diagnostic Systems, Inc., Decatur, Ga.) (145)

The Rapid N/H system was developed for the 4-h identification of *Neisseria* and *Haemophilus* spp. The plastic test panel, which is similar in design and operation to the RapID ANA panel, has 10 test cavities with 12 dehydrated conventional and chromogenic substrates for the performance of the following tests: phosphatase, nitrate reduction, hydrolysis of ONPG, prolyl-*p*-nitroanilid, and gamma-glutamyl-*p*-nitroanilid, resazurin reduction, glucose and sucrose utilization, indole, urease, orinithine decarboxylase, and β-lactamase. A suspension of the test isolate equivalent to a McFarland no. 3 turbidity standard is made with medium supplied with the system. The substrates are inoculated and the panel is incubated at 35 to 37°C for 4 h. Reactions in some test cavities are read before and after appropriate reagent addition. Test results are scored, and identification is made with a differential chart or the IDS Code Compendium.

RIM-N (Austin Biological Laboratories, Inc., Austin, Tex.)

The RIM-N system is based upon the rapid carbohydrate degradation tests of Kellogg and Turner (100) and Brown (30). To perform the test, 3 drops each of glucose, lactose, sucrose, and maltose are added to separate tubes. Inocula of the test organism are collected on individual harvester loops. The loops are placed in the tubes, and the tubes are vigorously mixed. The shaft of each loop is snapped off, and the remaining portion serves as a stopper. The tubes are incubated for 30 min at 35 to 37°C. Reactions are recorded, and identification is made with a differential chart.

Yeasts

API 20C (Analytab Products, Plainview, N.Y.) (28, 32, 68, 109, 120, 142, 156, 179, 183)

The API 20C system for yeast identification consists of a strip with 20 microcupules, each of which contains carbon source substrates for assimilation reactions. Included are glucose, glycerol, 2-keto-D-gluconate, arabinose, xylose, adonitol, xylitol, galactose, inositol, sorbitol, methyl-D-glucoside, *N*-acetyl-D-glucosamine, cellobiose, lactose, maltose, sucrose, trehalose, melezitose, and raffinose. The strip is placed in an incubation tray, and a light suspension of the test organism in medium supplied with the system is used to inoculate the substrates. The strip is incubated, preferably, at 30°C for 72 h, and results are recorded daily. Results are converted into a seven-digit profile number which is consulted in the API 20C Analytical

Profile Index, or identification may be made with a differential chart or through the manufacturer's computer. Definitive identification with all yeast identification systems must be made in conjunction with microscopic and macroscopic morphology observations of the test isolate.

Minitek (BBL Microbiology Systems, Cockeysville, Md.)

The Minitek Yeast Identification System is an auxanographic method utilizing 12 carbohydrate-impregnated paper disks. A yeast suspension is made in yeast carbon assimilation agar and poured into a 150-mm petri dish. Disks containing glucose, maltose, sucrose, lactose, galactose, melibiose, cellobiose, inositol, xylose, raffinose, trehalose, and dulcitol are then placed on the surface of the solidified agar. Positive assimilation is determined by observing growth around substrate disks after incubation at 25 to 30°C for 24 to 48 h. A Yeast Code Book or a differential chart is available to assist in identifying an isolate after an assimilation profile has been determined.

Uni-Yeast-Tek System (Flow Laboratories, Inc., McLean, Va.) (27, 28, 44, 45, 68)

The Uni-Yeast-Tek System is similar to the other identification systems produced by Flow Laboratories. The central well on the plate contains cornmeal-Tween 80 agar for the observation of microscopic morphology. The 11 peripheral wells permit the determination of urease production, nitrate reduction, and the assimilation of sucrose, lactose, maltose, raffinose, cellobiose, soluble starch, and trehalose. Three tubes provided with the system are the C/N Screen for the presumptive identification of *Cryptococcus neoformans* by the detection of phenol oxidase activity, the GBE tube for germ tube production, and the SAM tube for sucrose assimilation. If germ tubes are present, then sucrose assimilation is assessed. If they are absent, then the plate and C/N screen are inoculated. The test substrates are incubated at 25 to 30°C for 6 days and examined daily. Results are interpreted by means of a "Logic Wheel" which is based on a dichotomous key approach, or by means of a conventional dichotomous key.

MECHANIZED/AUTOMATED SYSTEMS

Autobac IDX System (General Diagnostics, Morris Plains, N.J.) (22, 48, 103, 158, 159)

The Autobac IDX System provides the user with the capability to identify *Enterobacteriaceae* and glucose-nonfermenting gram-negative bacilli, to screen urine specimens for the presence of significant bacteriuria, and to determine antimicrobial susceptibilities. This semiautomated system consists of a light-scattering photometer, an incubator-shaker, a data terminal, a disk dispenser, and a multichamber optical cuvette. The approach of the system is unique, basing identification of gram-negative bacteria on differential growth inhibition by antibacterial compounds. The panel of 18 compounds includes antimicrobial agents (carbenicillin, cephalothin, colistin, cycloserine, kanamycin, novobiocin), dyes (acriflavine, brilliant green, malachite green, methylene blue), and other chemicals (cobalt chloride, 3,5-dibromosalicylic acid, dodecylamine hydrochloride, floxuridine, omadine disulfide, sodium azide, thallous acetate). The effects of these inhibitors, along with six manually obtained parameters (growth, lactose fermentation and bile precipitation on MacConkey agar, swarming, spot indole and oxidase tests), are used to identify isolates. To perform an identification, a standardized inoculum is prepared and dispensed into an Autobac cuvette containing the 18 different antibacterial disks. The cuvette is incubated in the incubator-shaker for 3 h, after which it is placed in the photometer which computes a light-scatter index value for each chamber. If growth is insufficient at 3 h the cuvette is returned to the incubator-shaker, and additional readings are taken at any interval up to 6 h. The light-scatter index values and six primary plate observations/tests are utilized in a two-stage quadratic discriminant analysis to arrive at an identification. The two most probable identifications, along with their respective relative probabilities, are printed automatically.

For the urine screen, a sample of urine is transferred to a chamber in the cuvette. The cuvette is incubated, and light-scatter readings are taken at various times between 1 and 6 h (usually at 3 to 4 h). Bacterial multiplication is detected by a density increase which results in a voltage decrease. The initial reading is determined after 15 min. Density increases resulting in a voltage decrease equal to or greater than 0.2 V indicate a bacterial density of $\geq 10^5$ CFU/ml.

AutoMicrobic System (Vitek Systems, Inc., Hazelwood, Mo.) (1, 7, 10, 21, 52, 65, 75, 85, 93, 94, 97, 102, 135, 148, 165, 167, 168)

The AutoMicrobic System is an automated system designed to identify gram-negative and gram-positive bacteria and yeasts. It is also capable of detecting, enumerating, and identifying microorganisms directly from urine specimens and performing antimicrobial susceptibility tests. The system comprises a filler/sealer module, a reader incubator, a computer module, a data terminal, and a printer.

The Gram Negative Identification test kit consists of a disposable 30-well card which contains the test substrates listed in Table 2. The Gram Positive Identification card contains 28 test media and two control wells. The substrates or differential substances used in the test wells are bacitracin, optochin, NaCl (6%), bile (10 and 40%), esculin, arginine, urea, tetrazolium red, novobiocin, and 18 carbohydrates. The Yeast Card contains 26 biochemical media and four negative control wells. The tests include utilization or assimilation of carbohydrates, urea hydrolysis, nitrate reduction, and inhibition by cycloheximide.

To use the system for microbial identification, a standardized suspension is prepared in a test tube from a pure culture of the bacterium or yeast; the test tube is then connected to the card by a capillary straw. The cards are filled automatically in the filler/sealer module and are then incubated at 35 to 37°C in the reader/incubator module for bacterial identification, or off-line at 30°C for yeast identification. The bacterial identification card wells are monitored hourly for changes in optical density, with final automated interpretation and identification available within 4 to 6 h for *Enterobacteriaceae*, 6 to 18 h for glucose-nonfermenting gram-negative bacilli, and 4 to 15 h for gram-positive bacteria. At the end of a 22- to 24-h off-line incubation, the yeast card is placed in the

TABLE 2. Tests included in mechanized/automated systems for identification of gram-negative bacteria[a]

Test	AMS	Sceptor	AutoSCAN-4	Avantage Microbiology Center	UniScept Plus	Autobac IDX
Acetoin production (Voges-Proskauer)	−	−	+	−	See API 20E (Table 1)	See text
Acid from						
Adonitol	+	+	+	+		
Arabinose	+	+	+	+		
Cellobiose	−	−	+	−		
Glucose	+	+	+[b]	+		
Glucose in presence of *p*-coumaric acid	+	−	−	−		
Glucose in presence of irgasan	+	−	−	−		
Inositol	+	+	+	+		
Lactose	+	−	−	+		
Maltose	+	+	+[b]	−		
Mannitol	+	+	−	+		
Melibiose	+	+	−	−		
Raffinose	+	−	+	−		
Rhamnose	+	+	+	+		
Sorbitol	+	+	+	+		
Sucrose	+	+	+	+		
Xylose	+	+	+[b]	+		
Acetamide utilization	+	+	+	+		
Arginine dihydrolase	+	+	+	−		
β-Galactosidase (ONPG)	+	+	+	−		
β-Glucosidase	+	−	−	−		
Citrate utilization	+	+	+	+		
Deamination of phenylalanine or tryptophan	+	+	+	−		
Esculin hydrolysis	−	+	+	+		
Growth in cetrimide	+	+	−	−		
H_2S production	+	+	+	−		
Indole production	−	+	+	−		
Lysine decarboxylase	+	+	+	+		
Malonate utilization	+	+	+	+		
Nitrate reduction	−	−	+	−		
Ornithine decarboxylase	+	+	+	+		
Polymyxin B susceptibility	−	−	−	+		
Starch hydrolysis	−	−	+	−		
Tartrate utilization	−	−	+	−		
Urease production	+	+	+	+		

[a] Adapted from reference 50.
[b] Carbohydrates present in enteric or oxidation-fermentation base media or both.

reader/incubator module for automatic reading, analysis of results, and identification. The computer printout of primary and alternative choices of identification is accompanied by an estimate of the probability of the correct choice as calculated from a probability matrix.

Urines are processed with a 20-well card, the urine Identi-Pak. An appropriately diluted sample of the patient's urine is automatically introduced into the card. Ten of the card wells contain selective media which permit the identification of nine of the most frequently encountered bacterial and yeast urinary tract pathogens. Five of the wells are used for enumeration, and the remaining five are blank. After filling, the cards are placed in the reader/incubator, where each well is evaluated hourly by light-emitting diodes and phototransistor detectors. Positive results can be obtained as early as 4 h. However, a 13-h incubation period is required before a specimen can be considered negative. The urine Identi-Pak is being modified to permit testing three urine specimens per card.

AutoSCAN-4 (American Microscan, Mahwah, N.J.) (181)

The AutoSCAN-4 is a computer-controlled instrument for automated reading and interpretation of the MicroScan gram-negative bacilli or gram-positive cocci identification panels. Each test tray contains frozen, conventional test substrates (Table 2) for bacterial identification, available separately or in combination with MIC panels. After manual inoculation with a 95-prong transfer lid and incubation for 18 to 24 h at 35 to 37°C, the panels are inserted into the AutoSCAN reader, which simultaneously examines spectrophotometrically all of the wells in a specific panel at up to six different wavelengths. Data analysis and printing of the most probable identifications with their respective likelihood percentages occur automatically.

Avantage Microbiology Center (Abbott Laboratories, Diagnostic Division, Irving, Tex.) (46, 55, 119)

The Avantage Microbiology Center is a semiautomated system designed to identify gram-negative bacteria and yeasts, to screen urine specimens, and to perform antimicrobial susceptibility tests. The system consists of one to several analysis modules coupled to a control center and an inoculum dispenser. Separate disposable plastic test cartridges, comprising 20 optically clear chambers, are available for bacterial or yeast identification.

The gram-negative bacterial identification cartridge contains 20 lyophilized conventional biochemical substrates (Table 2). The yeast identification cartridge also contains 20 lyophilized substrates which test for the utilization of sugars, *p*-hydroxy benzoic acid, and protocatechuic acid, urease production, nitrate reduction, and a negative control well. The result of a conventional germ tube test is necessary to complete the yeast identification. In either case an inoculum is prepared from a pure culture and manually dispensed into each of the 20 reaction chambers of the test cartridge. After insertion of the cartridge into the analysis module for an initial light transmission reading, the bacterial or yeast cartridges are incubated off-line at 35°C for 5 to 24 h, respectively. The cartridge is then reinserted into the analysis module for the final reading, which is compared with the base-line reading stored in the control center. Test interpretation and identification based on a probability matrix are performed and printed automatically.

For screening urines, a sample of the specimen is introduced into individual vials containing special nutrient broth. The vials are placed in a special holder that is introduced into an incubator/shaker. Light-emitting diodes scan the vials at 5-min intervals and record turbidity changes. Microbial concentrations of 10^5 CFU/ml can be detected in 4 h.

BACTEC (Johnston Laboratories, Cockeysville, Md.) (12, 49, 122, 140, 172)

The BACTEC system utilizes the generation of $^{14}CO_2$ from substrates containing ^{14}C to detect microorganisms in normally sterile body fluids such as blood. Samples are introduced into aerobic or anaerobic vials or both, containing ^{14}C-labeled substrates. The vials are incubated and periodically monitored for microbial metabolism by assaying for $^{14}CO_2$ in the head space of the vial. The amount of $^{14}CO_2$ present is converted to a growth index which must exceed an established base line for a positive reading. The culture vial(s) must then be subcultured for microbial isolation. Depending on laboratory needs, semiautomated and fully automated instruments are available.

The system is also capable of identifying *Neisseria* spp. and analyzing specimens for the presence of mycobacteria in approximately 50% of the time required by conventional methods. *Mycobacterium tuberculosis* can be differentiated from other mycobacteria in approximately 3 days after initial detection, and mycobacterial susceptibility test results can be made available in approximately 6 days after detection.

Bac-T-Screen (Marion Scientific, Kansas City, Mo.) (51, 139)

The Bac-T-Screen uses a disposable filter card through which a sample of a urine specimen is forced under pressure. Microorganisms present in the urine are retained on the filter and are stained with a patented dye. The intensity of the stain on the filter is then compared with a color chart to ascertain whether significant bacteriuria is present. Results are available in approximately 2 min.

Lumac (3M Co., St. Paul, Minn.)
Monolight (Analytical Luminescence Laboratory, Inc., San Diego, Calif.)

Both 3M Co. and Analytical Luminescence Laboratory, Inc., manufacture instrumented urine screens based on the luciferin-luciferase approach for relating the quantity of microbial ATP to the number of CFU per milliliter in the urine specimen. Before analysis, nonbacterial ATP present in mammalian cells in the urine specimen is released and hydrolyzed enzymatically during a short incubation period. Microbial ATP is then chemically released and reacts with the luciferin-luciferase reagents, resulting in light emission. The amount of light emitted is measured with a photon counter, and the reading is related to the level of microbial ATP present in the urine specimen. This measurement is in relative light units and is used to determine the number of CFU per milliliter in the urine. Results are available in less than 1 h.

Sceptor System (BBL Microbiology Systems, Cockeysville, Md.) (181)

The Sceptor System is a semiautomated system designed to identify gram-positive, gram-negative, and anaerobic organisms. The system comprises an automated prep station, a reader/recorder, and a data management center. The microdilution test panels contain desiccated biochemical substrates (Table 2) which are rehydrated during autoinoculation of a standardized suspension of the test organism. After 18 to 20 h of off-line incubation at 35 to 37°C, the test panel is placed in the reader/recorder, which features a controlled light source for color interpretation of each reaction. The reactions are then compared with positive and negative color controls. The test results are analyzed in the data management center, which computes the organism identification.

UniScept Plus (Analytab Products, Inc., Plainview, N.Y.)

The UniScept Plus System is a semiautomated system that can be applied to the identification of *Enterobacteriaceae* and other gram-negative bacteria and for qualitative and quantitative antimicrobial susceptibility testing. The system consists of an autoinoculator, an autoreader, and a data system computer. The test substrates (Table 2), while biochemically identical to those in the API 20E, are arranged in a 2-by-10 format in cupules with an optically clear backing.

A standardized suspension of organisms aspirated in a syringe allows for autoinoculation of the test plate. After 18 to 24 h of incubation at 35 to 37°C, the plates are placed in the photometric autoreader, where ratio readings at up to five different wave lengths are taken in multiple positions along the cupule. The computer provides an automated interpretation, with a percent probability estimate given for the top three choices.

IMMUNOLOGICAL METHODS

Immunological procedures for the detection or identification of microorganisms have the desirable characteristics of ease of use, relatively rapid reaction endpoints, and, ideally, the required sensitivity and specificity. The coagglutination procedure employs suspensions of staphylococci with specific antibody attached by its Fc segment to the protein A component of the staphylococcal cell wall. The Fab fragment is left free to react with homologous antigen. In latex agglutination methods, latex particles are substituted for staphylococci.

Bactigen (Wampole Laboratories, Cranburg, N.J.) (43, 115, 177)

Bactigen uses the latex agglutination approach for the detection of *Streptococcus pneumoniae, Haemophilus influenzae* type b, and *Neisseria meningitidis* groups A, B, C, and Y in cerebrospinal fluid, serum, and urine. A drop of reagent and the clinical specimen are placed on a glass slide, and reaction endpoints are reached after 10 min of agitation on a rotator.

Culturette Group A Strep ID Kit (Marion Scientific, Kansas City, Mo.)

The Culturette Kit is used for the direct detection of group A streptococci in throat swabs. After specimen collection, the swab is placed in a tube and three extraction reagents are added. A portion of the extracted specimen is placed on a glass slide and mixed with a latex suspension sensitized with anti-group A *Streptococcus* antibodies. The slide is placed on a rotator, and results are available 10 min after specimen processing.

Directigen (Hynson, Westcott, and Dunning, Baltimore, Md.)

The Directigen meningitis kits are based on latex agglutination reactions using monoclonal antibodies. The Directigen procedure is similar to that of Bactigen. The Meningitis Test Kit permits the detection of *H. influenzae* type b, *S. pneumoniae*, and *N. meningitidis* groups A and C. Kits are also available for *N. meningitidis* group B and group B streptococci. The Directigen Group A Strep Test Kit is used for the direct detection of group A streptococci from throat swabs. It is similar to Culturette, and reaction endpoints are reached 60 min after test initiation.

Fluortec-F and -M (General Diagnostics, Morris Plains, N.J.) (53, 89, 127)

Fluortec-F and Fluortec-M are direct immunofluorescence test kits for the rapid detection and identification of members of the *Bacteroides fragilis* and *Bacteroides melaninogenicus* groups in clinical specimens. The test kit includes polyvalent, fluorescein isothiocyanate-labeled antibody pools against members of the *B. fragilis* and *B. melaninogenicus* groups. Fixed smears of the specimen are prestained with rhodamine. Excess stain is removed, and the smears are reacted with one of the labeled antibody pools. After an incubation period and washing, the smears are dried and observed with a fluorescent microscope. The reagents can also be used to identify members of the *B. fragilis* and *B. melaninogenicus* groups from culture.

Gono Gen (Micro-Media Systems, Inc., San Jose, Calif.) (110)

The Gono Gen is a coagglutination slide test kit utilizing seven pooled monoclonal antibodies against the principal outer membrane protein (protein I) antigen prepared from several strains of *Neisseria gonorrhoeae*. A suspension of the test organism is boiled for 10 min, and a drop is placed on a glass slide and mixed with a drop of the reagent. A positive test is evidenced by agglutination within 2 min.

Gonozyme (Abbott Laboratories, North Chicago, Ill.) (54, 152)

Gonozyme is an enzyme immunoassay for detection of *N. gonorrhoeae* antigens in urethral and endocervical swab specimens. To perform the test, the clinical specimen is incubated with a specially treated bead which adsorbs gonococci and gonococcal antigens. The bead is washed to remove unbound material and then reacted with rabbit antigonococcal antibody. Unbound antibody is removed by washing, and anti-rabbit goat antibody conjugated with horseradish peroxidase is added to the bead. The bead is washed again, and a peroxidase substrate is added. After a final incubation step, a yellow-orange color develops in proportion to the amount of gonococcal antigen adsorbed to the bead. The reaction can be quantified with a spectrophotometer.

Phadebact (Pharmacia Diagnostics, Piscataway, N.J.) (9, 14, 15, 17, 36, 37, 39, 43, 66, 69, 76, 79, 87, 112, 113, 115, 147, 157, 160–162, 176, 177)

Phadebact utilizes the coagglutination principle. Reagents are available for the detection of *Streptococcus agalactiae, Streptococcus pneumoniae, H. influenzae*, and *N. meningitidis* in cerebrospinal fluid and for the identification of *N. gonorrhoeae, Streptococcus* groups A, B, C, D, and G, *H. influenzae*, and *S. pneumoniae* from isolation plates. Reactions are carried out with glass slides, and results are available in approximately 1 min. Cerebrospinal fluids must be heated to 80°C for 5 min before testing.

Streptex/Wellcogen (Wellcome Diagnostics, Research Triangle Park, N.C.) (24, 40, 64, 138, 162)

The Streptex/Wellcogen procedure uses antibody-coated latex particles to detect homologous antigen. In the Streptex kit, reagents are available for identifying group A, B, C, D, F, and G streptococci after a 1-h enzyme extraction of antigen. The Wellcogen kit can be used for the detection of group B streptococci in cerebrospinal fluid, serum, and urine.

LITERATURE CITED

1. **Aldridge, C., P. W. Jones, S. Gibson, J. Lanham, M. Meyer, R. Vannest, and R. Charles.** 1977. Automated microbiological detection/identification system. J. Clin. Microbiol. **6**:406–413.
2. **Aldridge, K. E., B. B. Gardner, S. J. Clark, and J. M. Matsen.** 1978. Comparison of Micro-ID, API 20E, and conventional media systems in identification of *Enterobacteriaceae*. J. Clin. Microbiol. **7**:507–513.
3. **Aldridge, K. E., and R. L. Hodges.** 1981. Correlation studies of Entero-Set 20, API 20E, and conventional media systems for *Enterobacteriaceae* identification. J. Clin. Microbiol. **13**:120–125.
4. **Aldridge, K. E., C. Kogos, C. V. Sanders, and R. L. Marier.** 1984. Comparison of rapid identification assays for *Staphylococcus aureus*. J. Clin. Microbiol. **19**:703–704.
5. **Aldridge, K. E., C. W. Stratton, L. S. Patterson, M. E. Evans, and R. L. Hodges.** 1983. Comparison of the Staph-Ident system with a conventional method for species identification of urine and blood isolates of coagulase-negative staphylococci. J. Clin. Microbiol. **17**:516–520.
6. **Almeida, R. J., and J. H. Jorgensen.** 1983. Identification of coagulase-negative staphylococci with the API Staph-Ident system. J. Clin. Microbiol. **18**:254–257.
7. **Almeida, R. J., J. H. Jorgensen, and J. E. Johnson.** 1983. Evaluation of the AutoMicrobic system gram-positive identification card for species identification of coagulase-negative staphylococci. J. Clin. Microbiol. **18**:438–439.

8. **Altwegg, M.** 1983. Performance of two four-hour identification systems with atypical strains of *Enterobacteriaceae*. Eur. J. Clin. Microbiol. **2**:529–533.
9. **Anand, C. M., and E. M. Kadis.** 1980. Evaluation of the Phadebact Gonococcus Test for confirmation of *Neisseria gonorrhoeae*. J. Clin. Microbiol. **12**:15–17.
10. **Appelbaum, P. C., M. R. Jacobs, J. I. Heald, W. M. Palko, A. Duffett, R. Crist, and P. A. Naugle.** 1984. Comparative evaluation of the API 20S system and the AutoMicrobic system gram-positive identification card for species identification of streptococci. J. Clin. Microbiol. **19**:164–168.
11. **Appelbaum, P. C., C. S. Kaufmann, J. C. Keifer, and H. J. Venbrux.** 1983. Comparison of three methods for anaerobe identification. J. Clin. Microbiol. **18**:614–621.
12. **Appelbaum, P. C., and R. B. Lawrence.** 1979. Comparison of three methods for identification of pathogenic *Neisseria* species. J. Clin. Microbiol. **9**:598–600.
13. **Appelbaum, P. C., J. Stavitz, M. S. Bentz, and L. C. von Kuster.** 1980. Four methods for identification of gram-negative nonfermenting rods: organisms more commonly encountered in clinical specimens. J. Clin. Microbiol. **12**:271–278.
14. **Arvilommi, H.** 1976. Grouping of beta-haemolytic streptococci by using coagglutination, precipitation, or bactitracin sensitivity. Acta Pathol. Microbiol. Scand. Sect. B **84**:79–84.
15. **Arvilommi, H., O. Uurasmaa, and A. Nurkkala.** 1978. Rapid identification of group A, B, C, and G beta-haemolytic streptococci by a modification of the co-agglutination technique. Comparison of results obtained by co-agglutination, fluorescent antibody test, counterimmunoelectrophoresis, and precipitin technique. Acta Pathol. Microbiol. Scand. Sect. B **86**:107–111.
16. **Barnes, W., C. Broers, J. Farrow, and L. Potter.** 1980. Reducing the cost of urinary isolate identification with Enterotube II. J. Am. Med. Technol. **42**:213–214.
17. **Barnham, M., and A. A. Glynn.** 1978. Identification of clinical isolates of *Neisseria gonorrhoeae* by a coagglutination test. J. Clin. Pathol. **31**:189–193.
18. **Barnishan, J., and L. W. Ayers.** 1979. Rapid identification of nonfermentative gram-negative rods by the Corning N/F system. J. Clin. Microbiol. **9**:239–243.
19. **Barry, A. L., and R. E. Badal.** 1979. Rapid identification of *Enterobacteriaceae* with the Micro-ID system versus API 20E and conventional media. J. Clin. Microbiol. **10**:293–298.
20. **Barry, A. L., R. E. Badal, and L. J. Effinger.** 1979. Identification of *Enterobacteriaceae* in frozen microdilution trays prepared by Micro-Media Systems. J. Clin. Microbiol. **10**:492–496.
21. **Barry, A. L., T. L. Gavan, R. E. Badal, and M. J. Telenson.** 1982. Sensitivity, specificity, and reproducibility of the AutoMicrobic system (with the *Enterobacteriaceae*-Plus biochemical card) for identifying clinical isolates of gram-negative bacilli. J. Clin. Microbiol. **15**:582–588.
22. **Barry, A. L., T. L. Gavan, P. B. Smith, J. M. Matsen, J. A. Morello, and B. H. Sielaff.** 1982. Accuracy and precision of the Autobac system for rapid identification of gram-negative bacilli: a collaborative evaluation. J. Clin. Microbiol. **15**:1111–1119.
23. **Bartlett, R. C., T. S. Kohan, and C. Rutz.** 1979. Comparative costs of microbial identification employing conventional and prepackaged commercial systems. Am. J. Clin. Pathol. **71**:194–200.
24. **Bixler-Forell, E., W. J. Martin, and M. D. Moody.** 1984. Clinical evaluation of the improved Streptex method for grouping streptococci. Diagn. Microbiol. Infect. Dis. **2**:113–118.
25. **Blazevic, D. J., D. L. Mackay, and N. M. Warwood.** 1979. Comparison of Micro-ID and API 20E systems for identification of *Enterobacteriaceae*. J. Clin. Microbiol. **9**:605–608.
26. **Borchardt, K. A., and J. Gibson.** 1977. Comparison of enteric identification systems. Health Lab. Sci. **14**:5–10.
27. **Bowman, P. I., and D. G. Ahearn.** 1975. Evaluation of the Uni-Yeast-Tek kit for the identification of medically important yeasts. J. Clin. Microbiol. **2**:354–358.
28. **Bowman, P. I., and D. G. Ahearn.** 1976. Evaluation of commercial systems for the identification of clinical yeast isolates. J. Clin. Microbiol. **4**:49–53.
29. **Brooks, K. A., M. Jens, and T. M. Sodeman.** 1974. A clinical evaluation on the API microtube system for identification of *Enterobacteriaceae*. Am. J. Med. Technol. **40**:55–61.
30. **Brown, W. J.** 1974. Modification of the Rapid Fermentation test for *Neisseria gonorrhoeae*. Appl. Microbiol. **27**:1027–1030.
31. **Bruckner, D. A., V. Clark, and W. J. Martin.** 1982. Comparison of Enteric-Tek with API 20E and conventional methods for identification of *Enterobacteriaceae*. J. Clin. Microbiol. **15**:16–18.
32. **Buesching, W. J., K. Kurek, and G. D. Roberts.** 1979. Evaluation of the modified API 20C system for identification of clinically important yeasts. J. Clin. Microbiol. **9**:565–569.
33. **Buesching, W. J., D. L. Rhoden, A. O. Esaias, P. B. Smith, and J. A. Washington II.** 1979. Evaluation of the modified Micro-ID system for identification of *Enterobacteriaceae*. J. Clin. Microbiol. **10**:454–458.
34. **Buesching, W. J., J. R. Svirbely, and L. W. Ayers.** 1983. Evaluation of the Anaerobe-Tek system for identification of anaerobic bacteria. J. Clin. Microbiol. **17**:824–829.
35. **Burdash, N. M., E. R. Bannister, J. P. Manos, and M. E. West.** 1980. A comparison of four commercial systems for the identification of nonfermentative gram-negative bacilli. Am. J. Clin. Pathol. **73**:564–569.
36. **Burdash, N. M., and M. E. West.** 1982. Identification of *Streptococcus pneumoniae* by the Phadebact coagglutination test. J. Clin. Microbiol. **15**:391–394.
37. **Burdash, N. M., M. E. West, R. T. Newell, and G. Teti.** 1981. Group identification of streptococci. Evaluation of three rapid agglutination methods. Am. J. Clin. Pathol. **76**:819–822.
38. **Butler, D. A., C. M. Lobregat, and T. L. Gavan.** 1975. Reproducibility of the Analytab (API 20E) system. J. Clin. Microbiol. **2**:322–326.
39. **Carlson, B. L., M. S. Haley, J. R. Kelly, and W. M. McCormack.** 1982. Evaluation of the Phadebact test for identification of *Neisseria gonorrhoeae*. J. Clin. Microbiol. **15**:231–234.
40. **Castle, D., S. Kessock-Philip, and C. S. F. Easmon.** 1982. Evaluation of an improved Streptex kit for the grouping of beta-hemolytic streptococci by agglutination. J. Clin. Pathol. **35**:719–722.
41. **Chester, B., and T. J. Cleary.** 1980. Evaluation of the Minitek system for identification of nonfermentative and nonenteric fermentative gram-negative bacteria. J. Clin. Microbiol. **12**:509–516.
42. **Christensen, G. D., J. T. Parisi, A. L. Bisno, W. A. Simpson, and E. H. Beachey.** 1983. Characterization of clinically significant strains of coagulase-negative staphylococci. J. Clin. Microbiol. **18**:258–269.
43. **Collins, J. K., and M. T. Kelly.** 1983. Comparison of Phadebact coagglutination, Bactogen latex agglutination, and counterimmunoelectrophoresis for detection of *Haemophilus influenzae* type b antigens in cerebrospinal fluid. J. Clin. Microbiol. **17**:1005–1008.
44. **Cooper, B. H.** 1980. Clinical laboratory evaluation of a screening medium (CN screen) for *Cryptococcus neoformans*. J. Clin. Microbiol. **11**:672–674.
45. **Cooper, B. H., J. B. Johnson, and E. S. Thaxton.** 1978. Clinical evaluation of the Uni-Yeast-Tek system for rapid presumptive identification of medically impor-

tant yeasts. J. Clin. Microbiol. **7**:349–355.

46. **Cooper, B. H., S. Prowant, B. Alexander, and D. H. Brunson.** 1984. Collaborative evaluation of the Abbott yeast identification system. J. Clin. Microbiol. **19**:853–856.
47. **Coppel, S. P., and I. G. Coppel.** 1974. Comparison of the R-B system and the Enterotube for the identification of *Enterobacteriaceae*. Am. J. Clin. Pathol. **61**:218–222.
48. **Costigan, W. J., and G. E. Hollick.** 1984. Use of the Autobac IDX system for rapid identification of *Enterobacteriaceae* and nonfermentative gram-negative bacilli. J. Clin. Microbiol. **19**:301–302.
49. **Damato, J. J., M. T. Collins, M. V. Rothlauf, and J. K. McClatchy.** 1983. Detection of mycobacteria by radiometric and standard plate procedures. J. Clin. Microbiol. **17**:1066–1073.
50. **D'Amato, R. F., B. Holmes, and E. J. Bottone.** 1981. The systems approach to diagnostic microbiology. Crit. Rev. Microbiol. **9**:1–44.
51. **Davis, J. R., C. E. Stager, and G. F. Araj.** 1984. Clinical laboratory evaluation of a bacteriuria detection device for urine screening. Am. J. Clin. Pathol. **81**:48–53.
52. **Davis, J. R., C. E. Stager, R. D. Wende, and S. M. H. Qadri.** 1981. Clinical laboratory evaluation of the AutoMicrobic system *Enterobacteriaceae* biochemical card. J. Clin. Microbiol. **14**:370–375.
53. **DeGirolami, P., and C. P. Mepani.** 1981. Evaluation of a direct fluorescent antibody staining method for rapid identification of members of the *Bacteroides fragilis* group. Am. J. Clin. Pathol. **76**:78–82.
54. **Demetriou, E., R. Sackett, D. F. Welch, and D. W. Kaplan.** 1984. Evaluation of an enzyme immunoassay for detection of *Neisseria gonorrhoeae* in an adolescent population. J. Am. Med. Assoc. **252**:247–250.
55. **DiPersio, J. R., J. W. Dyke, and R. D. Vannest.** 1983. Evaluation of the updated MS-2 bacterial identification system in comparison with the API 20E system. J. Clin. Microbiol. **18**:128–135.
56. **Doern, G. V., J. E. Earls, P. A. Jeznach, and D. S. Parker.** 1983. Species identification and biotyping of staphylococci by the API Staph-Ident system. J. Clin. Microbiol. **17**:260–263.
57. **Dowda, H.** 1977. Evaluation of two rapid methods for identification of commonly encountered nonfermenting or oxidase-positive, gram-negative rods. J. Clin. Microbiol. **6**:605–609.
58. **Dowell, V. R., Jr., G. L. Lombard, F. S. Thompson, and A. Y. Armfield.** 1977. Media for isolation, characterization, and identification of obligately anaerobic bacteria. U.S. Department of Health, Education and Welfare, Washington, D.C.
59. **Durzi, S. S.** 1979. Evaluation of Enterotube and Minitek rapid systems for the identification of *Enterobacteriaceae*. Can. Inst. Food Sci. Technol. J. **12**:61–65.
60. **Edberg, S. C., B. Atkinson, C. Chambers, M. H. Moore, L. Palumbo, C. F. Zorzon, and J. M. Singer.** 1979. Clinical evaluation of the Micro-ID, API 20E, and conventional media systems for identification of *Enterobacteriaceae*. J. Clin. Microbiol. **10**:161–167.
61. **Edberg, S. C., and L. S. Konowe.** 1982. A systematic means to conduct a microbiology evaluation, p. 268–299. *In* V. Lorian (ed.), Significance of medical microbiology in the care of patients, 2nd ed. The Williams and Wilkins Co., Baltimore.
62. **Elston, H. R., J. A. Baudo, J. P. Stanek, and M. Schaab.** 1971. Multibiochemical test system for distinguishing enteric and other gram-negative bacilli. Appl. Microbiol. **22**:408–414.
63. **Esaias, A. O., D. L. Rhoden, and P. B. Smith.** 1982. Evaluation of the Enteric-Tek system for identifying *Enterobacteriaceae*. J. Clin. Microbiol. **15**:419–424.
64. **Facklam, R. R., R. C. Cooksey, and E. C. Wortham.** 1979. Evaluation of commercial latex agglutination reagents for grouping streptococci. J. Clin. Microbiol. **10**:641–646.
65. **Ferraro, M. J. B., M. A. Edelblut, and L. J. Kunz.** 1981. Accurate automated identification of selected *Enterobacteriaceae* at four hours. J. Clin. Microbiol. **13**:151–157.
66. **Finch, R. G., and I. Phillips.** 1977. Serological grouping of streptococci by a slide coagglutination method. J. Clin. Pathol. **30**:168–170.
67. **Finklea, P. J., M. S. Cole, and T. M. Sodeman.** 1976. Clinical evaluation of the Minitek differential system for identification of *Enterobacteriaceae*. J. Clin. Microbiol. **4**:400–404.
68. **Fuchs, P. F., and C. T. Dolan.** 1982. Performance of yeast identification systems. An analysis of the College of American Pathologists special mycology survey data. Am. J. Clin. Pathol. **78**:664–667.
69. **Futrovsky, S. L., C. A. Gaydos, and J. Keiser.** 1981. Comparison of the Phadebact Gonococcus Test with the rapid fermentation method. J. Clin. Microbiol. **14**:89–93.
70. **Gardner, J. M., B. A. Snyder, and D. Gröschel.** 1972. Experiences with the Analytab system for the identification of *Enterobacteriaceae*. Pathol. Microbiol. (Basel) **38**:103–106.
71. **Giger, O., C. C. Charilaou, and K. R. Cundy.** 1984. Comparison of the API Staph-Ident and DMS Staph-Trac systems with conventional methods used for the identification of coagulase-negative staphylococci. J. Clin. Microbiol. **19**:68–72.
72. **Goldin, M.** 1972. A comparison of multiple-test systems for the presumptive identification of *Enterobacteriaceae*. Am. J. Med. Technol. **38**:288–291.
73. **Goldstein, J., J. J. Guarneri, P. Della-Latta, and J. Scherer.** 1982. Use of the AutoMicrobic and Enteric-Tek systems for identification of *Enterobacteriaceae*. J. Clin. Microbiol. **15**:654–659.
74. **Gooch, W. M., III, and G. A. Hill.** 1982. Comparison of Micro-ID and API 20E in rapid identification of *Enterobacteriaceae*. J. Clin. Microbiol. **15**:885–890.
75. **Grasmick, A. E., N. Naito, and D. A. Bruckner.** 1983. Clinical comparison of the AutoMicrobic system gram-positive identification card, API Staph-Ident, and conventional methods in the identification of coagulase-negative *Staphylococcus* spp. J. Clin. Microbiol. **18**:1323–1328.
76. **Grasso, R. J., L. A. West, N. J. Holbrook, D. G. Halkias, L. J. Paradise, and H. Friedman.** 1981. Increased sensitivity of a new coagglutination test for rapid identification of *Haemophilus influenzae* type b. J. Clin. Microbiol. **13**:1122–1124.
77. **Grunberg, E., E. Titsworth, G. Beskid, R. Cleeland, Jr., and W. F. Delorenzo.** 1969. Efficiency of a multitest system (Enterotube) for rapid identification of *Enterobacteriaceae*. Appl. Microbiol. **18**:207–213.
78. **Guthertz, L. S., and R. L. Okoluk.** 1978. Comparison of miniaturized multitest systems with conventional methodology for identification of *Enterobacteriaceae* from foods. Appl. Environ. Microbiol. **35**:109–112.
79. **Hahn, G., and I. Nyberg.** 1976. Identification of streptococcal groups A, B, C, and G by slide coagglutination of antibody-sensitized protein A-containing staphylococci. J. Clin. Microbiol. **4**:99–101.
80. **Hampton, K. D., R. A. Stallings, and B. L. Wasilauskas.** 1979. Comparison of a slide coagglutination technique with the Minitek system for confirmation of *Neisseria gonorrhoeae*. J. Clin. Microbiol. **10**:290–292.
81. **Hansen, S. L., D. R. Hardesty, and B. M. Myers.** 1974. Evaluation of the BBL Minitek system for the identification of *Enterobacteriaceae*. Appl. Microbiol. **28**:798–801.
82. **Hansen, S. L., and B. J. Stewart.** 1976. Comparison of API and Minitek with Center for Disease Control methods for the biochemical characterization of anaerobes. J. Clin. Microbiol. **4**:227–231.

83. **Hanson, C. W., R. Cassorla, and W. J. Martin.** 1979. API and Minitek systems in identification of clinical isolates of anaerobic gram-negative bacilli and *Clostridium* species. J. Clin. Microbiol. **10:**14–18.
84. **Hanson, C. W., E. Marso, and W. J. Martin.** 1978. Comparison of the Minitek test system with a conventional screening procedure for identification of *Enterobacteriaceae*. Health Lab. Sci. **15:**3–8.
85. **Hasyn, J. J., and H. R. Buckley.** 1982. Evaluation of the AutoMicrobic system for identification of yeasts. J. Clin. Microbiol. **16:**901–904.
86. **Hayek, L. J., and G. W. Willis.** 1976. A comparison of two commercial methods for the identification of the *Enterobacteriaceae*—API 20E and the Enterotube—with conventional methods. J. Clin. Pathol. **29:**158–161.
87. **Helstad, A. G., and M. K. Bruns.** 1980. Rapid laboratory identification of *Neisseria gonorrhoeae* by coagglutination. J. Clin. Microbiol. **11:**753–754.
88. **Hofherr, L., H. Votava, and D. J. Blazevic.** 1978. Comparison of three methods for identifying non-fermenting gram-negative rods. Can. J. Microbiol. **24:**1140–1144.
89. **Holland, J. W., L. R. Stauffer, and W. A. Altemeier.** 1979. Fluorescent antibody test kit for rapid detection and identification of members of the *Bacteroides fragilis* and *Bacteroides melaninogenicus* groups in clinical specimens. J. Clin. Microbiol. **10:**121–127.
90. **Holmes, B., J. Dowling, and S. P. Lapage.** 1979. Identification of gram-negative nonfermenters and oxidase-positive fermenters by the Oxi/Ferm tube. J. Clin. Pathol. **32:**78–85.
91. **Holmes, B., W. R. Willcox, and S. P. Lapage.** 1978. Identification of *Enterobacteriaceae* by the API 20E system. J. Clin. Pathol. **31:**22–30.
92. **Holmes, B., W. R. Willcox, S. P. Lapage, and H. Mainick.** 1977. Test reproducibility of the API(20E), Enterotube, and Pathotec systems. J. Clin. Pathol. **30:**381–387.
93. **Isenberg, H. D., T. L. Gavan, P. B. Smith, A. Sonnenwirth, W. Taylor, W. J. Martin, D. Rhoden, and A. Balows.** 1980. Collaborative investigation of the AutoMicrobic system *Enterobacteriaceae* biochemical card. J. Clin. Microbiol. **11:**694–702.
94. **Isenberg, H. D., T. L. Gavan, A. Sonnenwirth, W. I. Taylor, and J. A. Washington II.** 1979. Clinical laboratory evaluation of automated microbial detection/identification system in analysis of clinical urine specimens. J. Clin. Microbiol. **10:**226–230.
95. **Isenberg, H. D., and J. Sampson-Scherer.** 1977. Clinical laboratory evaluation of a system approach to the recognition of nonfermentative or oxidase-producing gram-negative, rod-shaped bacteria. J. Clin. Microbiol. **5:**336–340.
96. **Janda, W. M., J. A. Morello, and M. Bohnhoff.** 1984. Use of the API NeIdent system for identification of pathogenic *Neisseria* spp. and *Branhamella catarrhalis*. J. Clin. Microbiol. **19:**338–341.
97. **Johnson, J. E., and A. W. Brinkley.** 1982. Comparison of the AutoMicrobic system and a conventional tube system for identification of nonfermentative and oxidase-positive gram-negative bacilli. J. Clin. Microbiol. **15:**25–27.
98. **Jorgensen, J. H., S. A. Crawford, and G. A. Alexander.** 1983. Rapid identification of group D streptococci with the API 20S system. J. Clin. Microbiol. **17:**1096–1098.
99. **Jorgensen, J. H., J. E. Johnson, G. A. Alexander, R. Paxson, and G. L. Alderson.** 1983. Comparison of automated and rapid manual methods for the same-day identification of *Enterobacteriaceae*. Am. J. Clin. Pathol. **79:**683–687.
100. **Kellogg, D. S., Jr., and E. M. Turner.** 1973. Rapid fermentation confirmation of *Neisseria gonorrhoeae*. Appl. Microbiol. **25:**500–552.
101. **Kelly, M. T., D. C. Hale, and J. M. Matsen.** 1981. Rapid identification by the Micro-ID system of *Enterobacteriaceae* detected by urine screening. J. Clin. Microbiol. **14:**295–297.
102. **Kelly, M. T., and J. M. Latimer.** 1980. Comparison of the AutoMicrobic system with API, Enterotube, Micro-ID, Micro-Media systems, and conventional methods for identification of *Enterobacteriaceae*. J. Clin. Microbiol. **12:**659–662.
103. **Kelly, M. T., J. M. Matsen, J. A. Morello, P. B. Smith, and R. C. Tilton.** 1984. Collaborative clinical evaluation of the Autobac IDX system for identification of gram-negative bacilli. J. Clin. Microbiol. **19:**529–533.
104. **Kelly, S. A., and J. A. Washington II.** 1979. Evaluation of Micro-Media Quad Panels for identification of the *Enterobacteriaceae*. J. Clin. Microbiol. **10:**515–518.
105. **Keville, M. W., and G. V. Doern.** 1982. Comparison of the API 20S *Streptococcus* identification system with an immunorheophoresis procedure and two commercial latex agglutination tests for identifying beta-hemolytic streptococci. J. Clin. Microbiol. **16:**92–95.
106. **Kiehn, T. E., K. Brennan, and P. D. Ellner.** 1974. Evaluation of the Minitek system for identification of *Enterobacteriaceae*. Appl. Microbiol. **28:**668–671.
107. **Kloos, W. E., and J. F. Wolfshohl.** 1982. Identification of *Staphylococcus* species with the API Staph-Ident system. J. Clin. Microbiol. **16:**509–516.
108. **Koestenblatt, E. K., D. H. Larone, and K. J. Pavletich.** 1982. Comparison of the Oxi/Ferm and N/F Systems for identification of infrequently encountered nonfermentative oxidase-positive fermentative bacilli. J. Clin. Microbiol. **15:**384–390.
109. **Land, G. A., B. A. Harrison, K. L. Hulme, B. H. Cooper, and J. C. Byrd.** 1979. Evaluation of the new API 20C strip for yeast identification against a conventional method. J. Clin. Microbiol. **10:**357–364.
110. **Lawton, W. D., and G. J. Battaglioli.** 1983. Gono Gen coagglutination test for confirmation of *Neisseria gonorrhoeae*. J. Clin. Microbiol. **18:**1264–1265.
111. **Leighton, P. M., and J. A. Little.** 1983. Clinical comparison of the Enterotube II and API 20E systems for bacterial identification. Am. J. Clin. Pathol. **79:**367–370.
112. **Lewis, J. S., and J. E. Martin, Jr.** 1980. Evaluation of the Phadebact Gonococcus Test, a coagglutination procedure for confirmation of *Neisseria gonorrhoeae*. J. Clin. Microbiol. **11:**153–156.
113. **Lim, D. V., R. D. Smith, and S. Day.** 1979. Evaluation of an improved rapid coagglutination method for the serological grouping of beta-hemolytic streptococci. Can. J. Microbiol. **25:**40–43.
114. **Lombard, G. L., D. N. Whaley, and V. R. Dowell, Jr.** 1982. Comparison of media in the Anaerobe-Tek and Presumpto Plate Systems and evaluation of the Anaerobe-Tek System for identification of commonly encountered anaerobes. J. Clin. Microbiol. **16:**1066–1072.
115. **Marcon, M. J., A. C. Hamoudi, and H. J. Cannon.** 1984. Comparative laboratory evaluation of three antigen detection methods for diagnosis of *Haemophilus influenzae* type b disease. J. Clin. Microbiol. **19:**333–337.
116. **Martin, W. J., P. K. W. Yu, and J. A. Washington II.** 1971. Evaluation of the Enterotube system for identification of members of the family *Enterobacteriaceae*. Appl. Microbiol. **22:**96–99.
117. **Marymont, J. H., III, J. H. Marymont, Jr., and T. L. Gavan.** 1978. Performance of *Enterobacteriaceae* identification systems. Am. J. Clin. Pathol. **70:**539–547.
118. **McCarthy, L. R., J. B. Mayo, G. Bell, and D. Armstrong.** 1978. Comparison of a commercial identification kit and conventional biochemical tests used for the identification of enteric gram-negative rods. Am. J. Clin. Pathol. **69:**161–164.
119. **McCracken, A. W., W. J. Martin, L. R. McCarthy, D. A. Schwab, B. H. Cooper, N. G. P. Helgeson, S. Prowant, and J. Robson.** 1980. Evaluation of the MS-2 system for rapid identification of *Enterobacteriaceae*. J. Clin. Mi-

crobiol. **12:**684–689.
120. **Miller, R. E., Jr., and L. P. Lu.** 1976. Evaluation of a multitest microtechnique for yeast identification. Am. J. Med. Technol. **42:**238–242.
121. **Moore, H. B., V. L. Sutter, and S. M. Finegold.** 1975. Comparison of three procedures for biochemical testing of anaerobic bacteria. J. Clin. Microbiol. **1:**15–24.
122. **Morgan, M. A., C. D. Horstmeier, D. R. DeYoung, and G. D. Roberts.** 1983. Comparison of a radiometric method (BACTEC) and conventional culture media for recovery of mycobacteria from smear-negative specimens. J. Clin. Microbiol. **18:**384–388.
123. **Morris, M. J., V. M. Young, and M. R. Moody.** 1978. Evaluation of a multitest system for identification of saccharolytic pseudomonads. Am. J. Clin. Pathol. **69:**41–47.
124. **Morse, S. A., and L. Bartenstein.** 1976. Adaptation of the Minitek system for the rapid identification of *Neisseria gonorrhoeae*. J. Clin. Microbiol. **3:**8–13.
125. **Morton, H. E., and M. A. J. Monaco.** 1971. Comparison of Enterotubes and routine media for the identification of enteric bacteria. Am. J. Clin. Pathol. **56:**64–66.
126. **Moussa, R. S.** 1975. Evaluation of the API, the PathoTec, and the improved Enterotube systems for the identification of *Enterobacteriaceae*, p. 407–420. *In* C. G. Hedèn and T. Illèni (ed.), New approaches to the identification of micro-organisms. John Wiley & Sons, New York.
127. **Mouton, C., P. Hammond, J. Slots, and R. J. Genco.** 1980. Evaluation of Fluoretec-M for detection of oral strains of *Bacteroides asaccharolyticus* and *Bacteroides melaninogenicus*. J. Clin. Microbiol. **11:**682–686.
128. **Nachamkin, I., J. R. Lynch, and H. P. Dalton.** 1982. Evaluation of a rapid system for species identification of alpha-hemolytic streptococci. J. Clin. Microbiol. **16:**521–524.
129. **Nadler, H., H. George, and J. Barr.** 1978. Accuracy and reproducibility of the Oxi/Ferm system in identifying a select group of unusual gram-negative bacilli. J. Clin. Microbiol. **9:**180–185.
130. **Nord, C.-E., A. Dahlbäck, and T. Wadström.** 1975. Evaluation of a test kit for identification of anaerobic bacteria. Med. Microbiol. Immunol. **161:**239–242.
131. **Nord, C.-E., A. A. Lindberg, and A. Dahlbäck.** 1974. Evaluation of five test kits—API, AuxoTab, Enterotube, PathoTec and R/B—for identification of *Enterobacteriaceae*. Med. Microbiol. Immunol. **159:**211–220.
132. **Oberhofer, T. R.** 1979. Comparison of the API 20E and Oxi/Ferm systems in identification of nonfermentative and oxidase-positive fermentative bacteria. J. Clin. Microbiol. **9:**220–226.
133. **Oberhofer, T. R.** 1983. Use of the API 20E, Oxi/Ferm, and Minitek system to identify nonfermentative and oxidase-positive fermentative bacteria: seven years of experience. Diagn. Microbiol. Infect. Dis. **1:**241–256.
134. **Oberhofer, T. R., J. W. Rowen, G. F. Cunningham, and J. W. Higbee.** 1977. Evaluation of the Oxi/Ferm tube system with selected gram-negative bacteria. J. Clin. Microbiol. **6:**559–566.
135. **Oblack, D. L., J. C. Rhodes, and W. J. Martin.** 1981. Clinical evaluation of the AutoMicrobic System Yeast Biochemical Card for rapid identification of medically important yeasts. J. Clin. Microbiol. **13:**351–355.
136. **Otto, L. A., and U. Blachman.** 1979. Nonfermentative bacilli: evaluation of three systems for identification. J. Clin. Microbiol. **10:**147–154.
137. **Painter, B. G., and H. D. Isenberg.** 1973. Clinical laboratory experience with the improved Enterotube. Appl. Microbiol. **25:**896–899.
138. **Petts, D. N.** 1984. Early detection of streptococci in swabs by latex agglutination before culture. J. Clin. Microbiol. **19:**432–433.
139. **Pezzlo, M. T., M. A. Wetkowski, E. M. Peterson, and L. M. De La Maza.** 1983. Evaluation of a two-minute test for urine screening. J. Clin. Microbiol. **18:**697–701.
140. **Pizzuto, D. J., and J. A. Washington II.** 1980. Evaluation of rapid carbohydrate degradation tests for identification of pathogenic *Neisseria*. J. Clin. Microbiol. **11:**394–397.
141. **Reddick, A.** 1975. A simple carbohydrate fermentation test for identification of the pathogenic *Neisseria*. J. Clin. Microbiol. **2:**72–73.
142. **Roberts, G. D., H. S. Wange, and G. E. Hollick.** 1976. Evaluation of the API 20C microtube system for the identification of clinically important yeasts. J. Clin. Microbiol. **3:**302–305.
143. **Robertson, E. A., G. C. Macks, and J. D. MacLowry.** 1976. Analysis of cost and accuracy of alternative strategies for *Enterobacteriaceae* identification. J. Clin. Microbiol. **3:**421–424.
144. **Robertson, E. A., and J. D. MacLowry.** 1974. Mathematical analysis of the API Enteric 20 Profile Register using a computer diagnostic model. Appl. Microbiol. **28:**691–695.
145. **Robinson, M. J., and T. R. Oberhofer.** 1983. Identification of pathogenic *Neisseria* species with the RapID NH system. J. Clin. Microbiol. **17:**400–404.
146. **Rosenthal, S. L., L. F. Freundlich, and W. Washington.** 1978. Laboratory evaluation of a multitest system for identification of gram-negative organisms. Am. J. Clin. Pathol. **70:**914–917.
147. **Rosner, R.** 1977. Laboratory evaluation of a rapid four-hour serological grouping of groups A, B, C, and G beta-streptococci by the Phadebact Streptococcus Test. J. Clin. Microbiol. **6:**23–26.
148. **Ruoff, K. L., M. J. Ferraro, M. E. Jerz, and J. Kissling.** 1982. Automated identification of gram-positive bacteria. J. Clin. Microbiol. **16:**1091–1095.
149. **Ruoff, K. L., and L. J. Kunz.** 1982. Identification of viridans streptococci isolated from clinical specimens. J. Clin. Microbiol. **15:**920–925.
150. **Ruoff, K. L., and L. J. Kunz.** 1983. Use of the Rapid STREP system for identification of viridans streptococcal species. J. Clin. Microbiol. **18:**1138–1140.
151. **Rutherford, I., V. Moody, T. L. Gavan, L. W. Ayers, and D. L. Taylor.** 1977. Comparative study of three methods of identification of *Enterobacteriaceae*. J. Clin. Microbiol. **5:**458–464.
152. **Schachter, J., W. M. McCormack, R. F. Smith, R. M. Parks, R. Bailey, and A. C. Ohlin.** 1984. Enzyme immunoassay for diagnosis of gonorrhea. J. Clin. Microbiol. **19:**57–59.
153. **Shayegani, M., M. E. Hubbard, T. Hiscott, and D. McGlynn.** 1975. Evaluation of the R/B and Minitek systems for identification of *Enterobacteriaceae*. J. Clin. Microbiol. **1:**504–508.
154. **Shayegani, M., A. M. Lee, and D. M. McGlynn.** 1978. Evaluation of the Oxi/Ferm tube system for identification of nonfermentative gram-negative bacilli. J. Clin. Microbiol. **7:**533–538.
155. **Shayegani, M., P. S. Maupin, and D. M. McGlynn.** 1978. Evaluation of the API 20E system for identification of nonfermentative gram-negative bacteria. J. Clin. Microbiol. **7:**539–545.
156. **Shinoda, T., L. Kaufman, and A. A. Padhye.** 1981. Comparative evaluation of the Iatron serological *Candida* Check kit and the API 20C kit for identification of medically important *Candida* species. J. Clin. Microbiol. **13:**513–518.
157. **Shively, R. G., J. T. Shigei, E. M. Peterson, and L. M. De La Maza.** 1981. Typing of *Haemophilus influenzae* by coagglutination and conventional slide agglutination. J. Clin. Microbiol. **14:**706–708.
158. **Sielaff, B. H., E. A. Johnson, and J. M. Matsen.** 1976. Computer-assisted bacterial identification utilizing antimicrobial susceptibility profiles generated by Autobac I. J. Clin. Microbiol. **3:**105–109.
159. **Sielaff, B. H., J. M. Matsen, and J. E. McKie.** 1982.

Novel approach to bacterial identification that uses the Autobac system. J. Clin. Microbiol. **15**:1103–1110.

160. **Slifkin, M., C. Engwall, and G. R. Pouchet.** 1978. Direct-plate serological grouping of beta-hemolytic streptococci from primary isolation plates with the Phadebact Streptococcus Test. J. Clin. Microbiol. **7**:356–360.

161. **Slifkin, M., and G. Interval.** 1980. Serogrouping single colonies of beta-hemolytic streptococci from primary throat culture plates with nitrous acid extraction and Phadebact streptococcal reagents. J. Clin. Microbiol. **12**:541–545.

162. **Slifkin, M., and G. R. Pouchet-Melvin.** 1980. Evaluation of three commercially available test products for serogrouping beta-hemolytic streptococci. J. Clin. Microbiol. **11**:249–255.

163. **Smith, K. E.** 1975. An investigation into three rapid multiple-test systems for identification of *Enterobacteriaceae* and their possible application in a clinical pathology laboratory. Lab. Med. **6**:25–28.

164. **Smith, P. B.** 1975. Performance of six bacterial identification systems. U.S. Department of Health, Education and Welfare, Center for Disease Control, Atlanta, Ga.

165. **Smith, P. B., T. L. Gavan, H. D. Isenberg, A. Sonnenwirth, W. I. Taylor, J. A. Washington II, and A. Balows.** 1978. Multilaboratory evaluation of an automated microbial detection/identification system. J. Clin. Microbiol. **8**:657–666.

166. **Smith, P. B., K. M. Tomfohrde, D. L. Rhoden, and A. Balows.** 1972. API system: a multitube micromethod for identification of *Enterobacteriaceae*. Appl. Microbiol. **24**:449–452.

167. **Smith, S. M., K. R. Cundy, G. L. Gilardi, and W. Wong.** 1982. Evaluation of the AutoMicrobic system for identification of glucose-nonfermenting gram-negative rods. J. Clin. Microbiol. **15**:302–307.

168. **Sonnenwirth, A. C.** 1977. Preprototype of an automated microbial detection and identification system: a developmental investigation. J. Clin. Microbiol. **6**:400–405.

169. **Stargel, M. D., G. L. Lombard, and V. R. Dowell, Jr.** 1978. Alternative procedures for identification of anaerobic bacteria. Am. J. Med. Technol. **44**:709–722.

170. **Stargel, M. D., F. S. Thompson, S. E. Phillips, G. L. Lombard, and V. R. Dowell, Jr.** 1976. Modification of the Minitek miniaturized differentiation system for characterization of anaerobic bacteria. J. Clin. Microbiol. **3**:291–301.

171. **Starr, S. E., F. S. Thompson, V. R. Dowell, Jr., and A. Balows.** 1973. Micro-method system for identification of anaerobic bacteria. Appl. Microbiol. **25**:713–717.

172. **Strauss, R. R., J. Holderbach, and H. Friedman.** 1978. Comparison of a radiometric procedure with conventional methods for identification of *Neisseria*. J. Clin. Microbiol. **7**:419–422.

173. **Tomfohrde, K. M., D. L. Rhoden, P. B. Smith, and A. Balows.** 1973. Evaluation of the redesigned Enterotube—a system for the identification of *Enterobacteriaceae*. Appl. Microbiol. **25**:301–304.

174. **Warwood, N. M., D. J. Blazevic, and L. Hofherr.** 1979. Comparison of the API 20E and Corning N/F systems for identification of nonfermentative gram-negative rods. J. Clin. Microbiol. **10**:175–179.

175. **Washington, J. A., II, P. K. W. Yu, and W. J. Martin.** 1971. Evaluation of accuracy of multitest micromethod system for identification of *Enterobacteriaceae*. Appl. Microbiol. **22**:267–269.

176. **Wasilauskas, B. L., and K. D. Hampton.** 1982. Determination of bacterial meningitis: a retrospective study of 80 cerebrospinal fluid specimens evaluated by four in vitro methods. J. Clin. Microbiol. **16**:531–535.

177. **Welch, D. F., and D. Hensel.** 1982. Evaluation of Bactogen and Phadebact for detection of *Haemophilus influenzae* type b antigen in cerebrospinal fluid. J. Clin. Microbiol. **16**:905–908.

178. **Wellstood-Nuesse, S.** 1979. Comparison of the Minitek system with conventional methods for identification of nonfermentative and oxidase-positive fermentative gram-negative bacilli. J. Clin. Microbiol. **9**:511–516.

179. **Weymann, L. H., C. E. Stager, S. G. M. Qadri, A. Villarreal, and S. M. H. Qadri.** 1979. Evaluation of a modified dye pour-plate auxanographic method for the rapid identification of clinically significant yeasts: comparison with two commercial systems. Med. Microbiol. Immunol. **167**:11–20.

180. **Willis, G., and I. J. Y. Cook.** 1975. A comparative study of API, ENCISE and conventional methods. Med. Technol. **5**:4–9.

181. **Woolfrey, B. F., R. T. Lally, and C. O. Quall.** 1983. Evaluation of the AutoSCAN-3 and Sceptor systems for *Enterobacteriaceae* identification. J. Clin. Microbiol. **17**:807–813.

182. **Wüst, J., and F. H. Kayser.** 1974. Evaluation of the redesigned Enterotube and its interpretation systems. Pathol. Microbiol. **40**:316–325.

183. **Zwadyk, P., Jr., R. A. Tariton, and A. Proctor.** 1977. Evaluation of the API 20C for identification of yeasts. Am. J. Clin. Pathol. **67**:269–271.

Chapter 7

Computers in Clinical Microbiology

JAMES D. MacLOWRY AND KENNETH J. RYAN

HISTORICAL PERSPECTIVE

Enthusiasm for the use of data processing equipment in clinical laboratories first gained momentum in the early 1960s with the introduction of electronic accounting machine equipment that was used primarily for billing purposes. Soon thereafter, computers were introduced into the laboratory, particularly in the area of clinical chemistry and to a lesser degree in hematology, for the purpose of data acquisition either from on-line instruments or by various off-line data entry modes. There were a number of reasons for this initial developmental activity in the field of chemistry, not the least of which was the fact that instruments were available that had the potential for being interfaced with a computer, allowing data to be acquired on-line with very little human intervention. Also, virtually all of the data generated in chemistry were numeric and, therefore, fit the existing thought constructs regarding the manipulation of data within a computer. A number of companies talked in very grandiose terms about the revolution that would occur with the use of computerized data processing systems in the laboratory and spoke glibly of the addition of clinical microbiology data to the existing and developing chemistry and hematology systems. Few of the companies which started out in the field of laboratory computers in the mid- to late 1960s are now in existence, partly as a consequence of the overselling of the ability of the then-existing computers. Specifically, the software available for those computers and the restrictive system concepts created considerable user frustration and disappointment. It is not inappropriate to draw an analogy between the great enthusiasm of the mid- and late 1960s regarding the utility of computers in the clinical laboratory and the present explosion of computer hardware and software engendered by the use of microcomputers, and to relate the disillusionment which is gradually becoming evident in the early 1980s to the overselling of computer applications.

The thrust of this chapter on the clinical applications of computers in diagnostic microbiology laboratories will be to attempt to delineate those situations in which the computer has obvious advantages and to try to distinguish them from situations in which the computer plays a very minimal role and from possible future situations which may or may not ever come to fruition. The overselling of both actual and imagined capabilities of computer systems in the clinical laboratory has been a depressing saga interspersed with some significant contributions. It has, unfortunately, become almost axiomatic that any job that can be done by a computer should be done by a computer. Furthermore, it is often suggested that if a particular task has already been computerized, it can be performed admirably in any laboratory environment. It is our purpose to attempt to alert the inexperienced and unwary to this technological quagmire and also to present those accomplishments which are appropriate to consider in a given laboratory situation.

It is critical to make a distinction between what can be done with computers in the clinical laboratory, what has already been done, and what exists in a form which is appropriate to a given laboratory situation. Furthermore, it is also necessary when one is considering the possibility of utilizing a computer for some or many functions in the laboratory to know what has actually been accomplished, whether that which has been accomplished is appropriate for the given laboratory situation, and whether the hardware and, more importantly, the software which accomplished the task is available to a potential user. Some very effective and innovative systems have been designed by creative programmers working in concert with interested laboratory personnel, but sometimes the systems which they have developed are not available for anyone outside their particular environment (10).

There has been a considerable tendency to view each laboratory as a unique entity and, therefore, not able to use the experience of other laboratories. This tempts workers in each environment to try to design their own data processing system, which creates the situation of the proverbial wheel being reinvented endlessly. In fact, close scrutiny from an organizational standpoint suggests that there is little difference in the ways most laboratories operate and that existing operational modes could be altered to accommodate an already existing computer system. The temptation to build a specialized data processing system is particularly great if a computer group is available for consultation within the institution. These computer groups have been notorious for underestimating the problems of developing a system for the laboratory. Most will admit that they have had no experience in this regard, but they are willing to make grandiose projections regarding the simplicity of a particular project. Many of these attempts have ended disastrously because there was not the wisdom to realize that the task was much more difficult than originally envisioned. This particular tendency is probably even greater presently, as we are bombarded with suggestions that any high school student is able to write programs that can be utilized for almost any purpose. Therefore, the conclusion drawn is that it is really quite simple for an interested person in a laboratory setting to develop the appropriate programs for operating a laboratory system. It is very popular at present to discuss networking of computers so that they will freely communicate with each other. This concept is able to be developed because of the availability and relatively low cost of microcomputers. We would gently suggest, as will be discussed in other parts of this chapter, that this particular enthusiasm conceals a number of pitfalls and that the simplicity of this approach is illusory.

The remainder of this chapter will develop the following four main themes regarding the use of computers in clinical laboratories and, specifically, clinical microbiology laboratories: (i) the expectations, requirements, and evaluation of laboratory systems in general; (ii) the utility of the computer, particularly in the area of diagnostic capability in microbiology; (iii) the use of the computer for epidemiology functions; and (iv) the computer as a part of the rapidly growing area of instrumentation in clinical microbiology. The areas are not meant to be mutually exclusive, and we recognize that there can be considerable overlap among some of these aspects of computer functioning in a laboratory, but we believe that certain points can be made more effectively with this somewhat arbitrary division of computer function and use within the laboratory.

LABORATORY SYSTEMS

Clinical microbiology computer systems are designed to replace the handwritten mode of information transmission. The reasons for making such a change are to improve the efficiency of the laboratory operation and to increase the speed and clarity of presentation of results to the physician. An additional benefit should be the availability of analytical reports which are either unavailable or time-consuming in a manual system. Laboratory directors who wish to use a computer must decide between designing their own system and purchasing one from a vendor. Although a full discussion of the factors involved in these decisions is beyond the scope of this chapter, we will address a number of considerations which are fundamental to any microbiology information system. As an underlying principle, the computer system should adapt to and facilitate the daily activities of the two people who use it most: the bench microbiologist or technologist and the physician managing the patient. From a work flow standpoint, the major functions involved are test ordering, result entry, and reporting.

Test ordering

Test orders may be entered into the computer in the laboratory or on the hospital unit. Ordering directly from hospital units requires a more sophisticated laboratory system interfaced to a hospital information system (10). The test-ordering routines themselves are similar to those used in computerized chemistry laboratories but, in addition, require the ability to more completely characterize the timing and describe the specimens submitted for culture. Ordering routines should also allow the placement of specific physician requests (e.g., rule out nocardia, rule out diphtheria) in the record. The inclusion of clinical information or antibiotic therapy should be weighed against the time required for entry and the probable accuracy of the data. Once tests are ordered, the system should allow the usual work flow to proceed, with the additional support of aids such as labels containing the patient's name, date, test order, and computer accession number.

Result entry

The speed, ease, and accessibility of the computer for result entry will ultimately determine its success in laboratory use. The complexities of microbiological results are considerable compared with the numeric results typical for other sections of the clinical laboratory. A single bacteriology accession number may require multiple result entries, including Gram smear, taxonomic terms, English text statements, and antimicrobial susceptibility results. Each of these may be in multiples and require quantitative modifiers or interpretations (e.g., 4+, many, 4.0 μg/ml, resistant). Adding to the complexity, not all of the information is generated at the same time, and some of it will be sequentially modified as a matter of routine; for example, a blood culture result may be initially reported as gram-negative rods, later updated to non-lactose-fermenting gram-negative rods, and finally identified as *Serratia marcescens*.

To meet the needs of microbiological entry, the computer system must be rapid and flexible in its entry modes. Modifications must be possible without backtracking and with the same ease as for the original entries. Entry modes which have been used to accomplish these goals have been mark sense documents (10) and computer keyboards using a combination of codes and free text (8). More recently the latter has been enhanced to allow the keys on a standard keyboard to be programmed to allow the entry of an organism name or narrative statement with a single key stroke (11). Entry may also be simplified or speeded up by the use of light pens or the movement of cursors on a preformatted cathode ray tube screen. Result entry must also include a verification step in which it is determined that data have been received and placed in the proper patient file by the computer. This is most rapidly done by interactive programs which are able to display the data entered to the person who entered them for immediate verification. This entry mode usually does not provide a hard copy of the verification transaction. Hard-copy verification lists are time-consuming to use but do provide a specific audit trail.

Another decision on result entry is whether the actual entry will be done by the technologist or by someone else. It is preferable that the technologist performing the work enter the results, particularly if this can be done as part of the work process itself with direct cathode ray tube verification. This method requires a terminal at each work station. The number of terminals necessary can be reduced by the use of automatic answering programs at work stations with a high proportion of negative results, such as those for blood cultures. This procedure requires entry of positive results only, with the negative results being automatically updated by the computer (8). Less efficient entry systems may require separation of the work and information flow, leading to entry by a data clerk. This approach requires additional steps for transfer of records for both entry and verification. This is not only slower but more error prone.

Reporting

Printed reports and the other means by which physicians gain access to patient results are the end product of a clinical laboratory computer system. Print programs should allow flexible formatting of such reports so they are at least as good as the preprinted forms used in manual systems. Columnar formats designed for the presentation of numeric data are generally not suitable. It is also undesirable to use any but the most familiar abbreviations or codes because the physician should not have to solve a puzzle to extract the clinical meaning from the report. The computer system should be able to generate cumulative reports which summarize days to weeks of an inpatient stay in a manner specified by the user. For example, it is useful to have all blood cultures printed together in chronological order. It is also reasonable to expect the system to allow authorized users both in and outside the laboratory immediate access to verified patient files via remote terminals. These terminals may be located in data centers, outpatient clinics, inpatient units, intensive care units, and emergency rooms and occasionally in physicians' offices. It should be noted that most systems are limited in the number of remote terminals (whether cathode ray tubes or printers) that can be added.

Management tools

In addition to reporting routine laboratory results, a clinical microbiology system can also be expected to provide a number of reports designed to improve the overall efficiency of the laboratory. These include specimen logs, reports of overdue tests, quality control statistics, antimicrobial susceptibility probabilities, work load statistics, hospital epidemiology, and many other reports, including those designed for specialized reviewing needs of laboratory supervisors, laboratory directors, and infectious disease specialists. To be useful, such reports need to be available on demand by laboratory personnel. It is critical to specify exactly the type of report desired since a system may be able to generate an "epidemiology report" which may not be at all useful for one's particular needs. It may be very expensive or not possible to program the system to provide the exact report desired.

A few specific comments on different reports that can be produced for management purposes may be useful. The specimen logs generated usually are provided daily and are organized either alphabetically or by specimen number. Both of these formats are useful. A listing of incomplete or overdue test results is very helpful in keeping track of a complex system. It is very easy to incompletely finalize all or part of a computer result file, and unless the file is automatically reviewed periodically, these oversights can unnecessarily clutter the files.

Work load statistics can be very helpful, but it is essential that the exact type of counting be determined at the outset. Some systems may only count the test type, e.g., fungus culture, and not the source of the specimen. Conversely, only the source may be counted and not the test type. It is also possible that the types of organisms identified may be counted, but that the patients from whom those isolates were derived may not be identified. Occasionally it may be useful to know the number of different patients from whom samples have been taken, with duplicate cultures excluded from analysis. Such statistics may not be possible on many systems.

The antimicrobial susceptibility profiles which have considerable epidemiological interest to most hospitals should be available for retrieval at reasonable intervals. Often it is desired to review them at 1- or 6-month intervals. Some systems may not be able to store the data long enough to provide a 6-month or 1-year survey. It is also of interest to be able to identify the patient and hospital location in the case of unusual profiles, and this ability should exist in the system as it is often very difficult to program later.

Many of these report functions can be very useful for management, epidemiology, and therapy, but it is necessary that it be demonstrated for a prospective system that the necessary reports are available. One should carefully define what is useful or necessary and then insist that it be produced, not just promised, in a functioning system.

DIAGNOSTIC FUNCTIONS

Traditionally, microbiologists have utilized at least two fundamentally different approaches for the identification of bacteria. The taxonomists utilize a broader number of tests in their studies of microorganisms, and often, but not always, these represent a specific set of tests. These tests were used to develop a data base which contained the test results for each distinct taxonomic grouping, and for each test a percentage of positivity was determined. Some tests were considered to be more significant than others in the identification of certain organisms and therefore weighed more heavily in the minds of the investigator attempting to identify or categorize organisms.

The clinical microbiologists, usually because of the pressures of time and financial constraints, tended to utilize some type of dichotomous branching scheme whereby a single or perhaps a few initial tests were used; they would then, depending on the reactions derived, utilize additional tests if necessary. Once again, some of these tests weighed much more heavily than others in the identification of certain organisms. The dichotomous branching scheme, although it can be extraordinarily efficient at times, has some very great pitfalls, particularly when a specific test does not perform well or the strain happens to be unusual and gives an atypical result for a critical test at a branch point. It was for these reasons that a number of investigators in the early and mid-1960s, quite independently of each other, began to look at the possibility of utilizing some of the already existing mathematical models, which were being applied to a wide variety of data bases, as a possible way of more efficiently utilizing the already existing biochemical data related to organism identification (7, 9, 16). Some of the most extensive clinical microbiology data related to members of the family *Enterobacteriaceae* (3), and it was this particular group of bacteria that initially occupied the attention of a number of authors (5, 9). It was found that computer manipulation of a large data base could be done with a considerable amount of facility and with far more accuracy than most microbiologists were able to achieve. Some of these early studies not only dealt with the traditional biochemical tests for a variety of organisms but also utilized the increasing amount of antimicrobial susceptibility data for a wide variety of organisms as another type of biochemical reactivity of organisms (2, 6). These initial studies were very encouraging in their level of diagnostic accuracy.

Fortuitously at approximately the same time in the early 1970s, commercially available test kits with a specific concatenation of tests became available. The number of tests (20) in one kit was considerable, and it became obvious that if the routine clinical laboratory wished to use this type of approach for identification of microorganisms, a more effective interpretive scheme for all of these tests was necessary. The joining of computer mathematical modeling with the data base that had been derived proved very quickly to be an extraordinarily powerful tool in utilizing and interpreting the biochemical tests (13). This scheme did not weigh any test more heavily than any other and relied only on the extensiveness of the data base for its accuracy. This particular kit very rapidly became the identification standard in the field for members of the family *Enterobacteriaceae*, and it very considerably enhanced the ability of laboratories to accurately identify a wide variety of isolates, particularly within the family *Enterobacteriaceae*. This scheme and a number of others which are presently commercially available have, by utilizing computer manipulation of the data, increased the level of sophistication of the routine laboratory to a level which was not imaginable even 10 years ago.

Almost all of these systems utilize some sort of code book which contains a listing of numbers that are in some way derived from the concatenation of biochemical tests used in the system. For each number, an identification or series of identifications is listed, usually with an evaluation of how accurate that identification might be. Most of the systems now have reached the level of sophistication whereby they are also able to comment on which tests in the coded number have given unusual results for a particular identification, thus alerting the laboratory to the possibility of a misreading of the test result. Some systems now have the capability to read these particular test results in an automated or semiautomated fashion. As they are read, the data are automatically fed into the computer, and by using a probabilistic mathematical model, identification of an organism is made. Most users who see only the code book derivation of the identifications may not be aware of the computer modeling that underlies the particular identification, but it is there nevertheless. Most of these manufacturers also make available the opportunity for users to call them with numbers that are not in the code books. They run these numbers through their data base and determine a probability of identification which can be communicated back to the laboratory in a short period of time. It should be mentioned that these programs by and large are mathematical manipulations of a data matrix, usually using some form of a Bayesian probability model, rather than a searching of the data base for a complete match of test results (4, 12).

The use of the computer in the general area of taxonomy is obviously of at least indirect interest to the clinical microbiology laboratory, but this will not be discussed in detail here since the usual clinical laboratory is not involved with the development of a data base for new or unusual organisms or concerned with the specific relationship of one poorly described organism to another. There is an excellent summary of this area for the interested reader (17).

Another area of at least theoretical interest concerns the use of antimicrobial susceptibility information in the identification of bacterial isolates. This approach has a number of different ramifications. First, it has been shown that one can identify bacteria with a fair degree of accuracy by using only the antimicrobial susceptibility pattern of an isolate (6). The accuracy is less than that obtained by using the more traditional biochemical tests, but most microbiologists are aware that with some isolates certain characteristic susceptibility patterns can be useful in identification. As a consequence, some of the commercially available systems will in fact use some antimicrobial agents as part of their identification schemes. This latter aspect of using antimicrobial susceptibility testing as another biochemical test has proven to be considerably useful in a number of identification schemes.

Another aspect of the potential utility of antimicrobial susceptibility testing which has been talked about with great enthusiasm but which has rarely been implemented within computer systems has to do with using the susceptibility information as a check upon the biochemical identification. What one would ideally prefer to have would be a system whereby once an identification is made from the biochemical information, there would be an automatic checking routine using the antimicrobial susceptibility data, and any discrepancies between the expected susceptibilities for the identification derived from the biochemical tests would be flagged in some way. At present, this kind of an error-checking routine is rarely used, usually because of software constraints. Computer programmers quite readily admit that it can be done, but when they are pressed to make this type of a modification in already existing programs, they display a great deal of reluctance. This particular feature, even if it is not used specifically in an error-checking mode, could be very useful if unusual susceptibility patterns were flagged by utilizing the hospital's expected antimicrobial spectrum. It is possible that this type of flagging of unusual results might promote more rapid epidemiological observations. However, this does not mean that these types of observations cannot be made in a manual fashion by utilizing very carefully drawn guidelines. It is somewhat unrealistic in laboratories where many different individuals rotate through the susceptibility test area to try to keep in mind some predetermined guidelines or to check each isolate against them, and this usually does not occur.

One additional interesting mathematical manipulation of these large data bases when a specific combination of tests has been used allows one to reduce the set of biochemical tests to a minimum number without appreciably reducing the accuracy of the identification. Use of an algorithm, such as a stepwise linear discriminate function algorithm, allows one at each point to determine which test is most efficacious for the identification of a variety of different species, and then by building a new test set one can derive an optimal set which represents a subset of the original data base (14). This has at present primarily a theoretical interest, but as manufacturers examine more biochemical reactions for wider varieties of organisms, it is entirely possible that it will be necessary to use some type of algorithm to reduce the tests to a manageable number. These studies cannot be done manually and require the tremendous power of a large computer, which is able to do virtually countless calculations. Usually this kind of an analysis cannot be done on a microcomputer because of insufficient storage capabilities.

EPIDEMIOLOGICAL FUNCTIONS

The use of microbiological data for epidemiological purposes is an often-cited benefit of laboratory computerization. Laboratories with microbiology information systems should be able to use the data already entered, while those without a computer system could still set up epidemiology programs on a microcomputer. In either case, efficiency requires decisions on how much of the patient report needs to be saved. Setting up a file separate from that used for patient reports is useful for a number of reasons. The data file can be organized to facilitate rapid on-line searches, and computer storage space can be more effectively used so that long-term storage (up to 1 year or more) is possible. For infection control purposes, an epidemiology file needs to be kept on-line for at least 6 months, which is longer than most laboratory systems can keep clinical patient files available for immediate retrieval.

The most useful epidemiological function is the ability to scan the epidemiology file and select cases with user-specified combinations of epidemiological parameters. The information stored should include the patient name, hospital number, location at the time of specimen collection, specimen dates and times, culture results, accession numbers, and susceptibility patterns. The scanning program should allow selection of any combination of organism(s), specimen(s), and hospital location(s) over a specified time period. Searches of this kind are extremely useful for epidemiological problem solving, particularly if they can be produced the same day the request is made. It is also important to be able to identify specific patients rather than just numbers of patients with particular isolates. A number of programs are able to count the number of patients in a particular hospital unit who have a specific isolate but are unable to identify those patients. These programs are not very helpful to the epidemiologist.

Scans by susceptibility pattern are more complicated and may not be needed if the program prints susceptibility results in a format that is readily examined visually. Antimicrobial susceptibility statistics are often considered an epidemiological function, although they actually serve a more clinical purpose as a guide to physicians in antimicrobic selection. As with epidemiology programs, susceptibility statistics should be available without re-entry of the primary data in a laboratory computer system, or they could be entered into an independent microcomputer. It is tempting to combine the epidemiological and susceptibility statistic functions since many of the features of time, specimen, location, and patient identification are common to both. Although this may be possible, it is not wise to restrict the epidemiological file to those organisms with susceptibility tests since this may exclude epidemiologically significant isolates. Susceptibility statistic programs need not have sophisticated means of analyzing pattern changes to be useful. A summary should be available at 3- to 6-month intervals without additional work on the part of the laboratory staff.

Microcomputers can be very useful if they are used as locally distributed computers which maintain a segment of the central computer data base. There are many problems involved with using the integrated central computer data base, whether it is a centralized laboratory computer or a total hospital information system for retrievals other than for a specific patient. These systems are often severely restricted in their abilities to retrieve across patients or for more than short time intervals. Regular, perhaps daily, interaction with a local microcomputer which develops a specific data base, e.g., for epidemiological purposes, can be a powerful and rapid retrieval capability. It currently is difficult to program local microcomputers to communicate with each other, but software applications are being developed to facilitate this type of interaction.

INTEGRATION OF COMPUTERS INTO INSTRUMENTATION

The greatest potential for rapid and accurate handling of clinical microbiological data is by direct transfer of data from the microcomputer in an automated instrument to the laboratory computer. Placing an instrument "on line" means that this is accomplished at the time the information is generated by the instrument, with no manual entry. The interface process can be viewed as occurring at two levels. The first is the sending of data by the instrument and their recognition by the laboratory computer. This includes matching to the specific patient accession and organism numbers. The instrument vendor can be expected to provide an electronic interface board which sends the information in a generally recognizable form, such as ASCII code. Translating this code back into microbiological terminology is the job of the laboratory computer system programmer or vendor. For this to be possible, the instrument vendor must be willing to provide the programmer with accurate and complete documentation of all signals to be transmitted across the interface. For example, the instrument may list bacterial identifications by a numeric code which must then be translated into whatever code system is used by the laboratory computer. Matching the results to the patient's culture can be accomplished through the use of the laboratory accession number in the instrument's accessioning system. The general process is essentially the same as that used for automated chemistry instruments.

There are a number of aspects of this particular problem which need to be addressed separately. Does the instrument contain its own microcomputer which is involved with the direct control of the instrument and the manipulation of data which is generated? If the instrument has its own computer, a number of questions have to be asked regarding its utility to the laboratory. Is the computer used only to control the instrument and, therefore, to produce a quantitation, an identification, or an antimicrobial susceptibility result? If the computer produces data in a form which is readable by humans, can these data be sorted and manipulated by the user in predetermined or readily programmable ways? Is the computer within the instrument able to perform functions other than those related to the control of the instrument? These questions are all quite separate, and each of them, unfortunately, may require a considerable amount of inquiry to the manufacturer regarding the possibility of accomplishing any or all of the tasks. For example, a number of instruments contain computers used for control of the instrument, and the manufacturer states that it has supplied an interface port which can be utilized to transfer data to a laboratory system. One then must ask very direct questions regarding whether the interface itself actually exists. If the hardware and software for the interface exist, do they exist for the specific type of laboratory computer in question? It is unlikely that the interface already exists for one's particular laboratory application. The question then becomes: how can this interface, both the hardware and the software, be developed and at what cost and in what time frame? At that juncture, when the questions are posed with that particular emphasis, the manufacturer may have very little interest in developing the interface. The manufacturer's interest is inversely related to the price quoted for the interface. Sometimes the answer is given that you can do this yourself with a small amount of programming. This is a flag which should mean that one should exercise utmost caution, because a small amount of programming often extends into an extremely cumbersome problem, part of which is aggravated by the manufacturer being unwilling, because of proprietary concerns, to give information which is required for the construction of the interface.

Once the interface data are recognized and matched to the patient accession number, transferring them into the active patient report file could be considered a second step. The verification process required at this point presents unique problems with microbiology instruments. A multichannel chemistry instrument usually completes a test run minutes after the specimen is introduced. A verification step need not add significantly to the total time required from the start of the test run until the result is available to the physician. Automated microbiology instruments may produce results 4 to 12 h after the specimen is introduced. This may require verification outside regular working hours. A decision simply to wait until the next day sacrifices the speed capability possessed by both the instrument and the computer. One approach to this problem has been the use of editing criteria automatically applied by the laboratory computer to the results at the time of transmission by the instrument. This reduces the probability of erroneous results being placed in the patient's file (15).

Once an on-line interface has been established, care must be taken to prevent software changes from causing it to fail. One source of such failures is a decision by the instrument vendor to update its software without informing the user of such changes. It is unwise, for example, for users of an on-line instrument to install a new software disk sent by the instrument manufacturer without checking the effect that the changes have on the interface. As with most bits of sophisticated technology, the installation and maintenance of an on-line interface involves a significant degree of cooperation between microbiologists and the laboratory or vendor computer scientists. Such an effort is worthwhile if it leads to a genuine link between automated microbiology and automated (computerized) information handling.

If it is not possible to interface the instrument with a laboratory system or if a laboratory system does not exist, there may still be compelling reasons for using such an instrument. It is entirely possible that the instrument itself may have a sufficient amount of storage capability that the type of manipulation that one wishes to do with the data can be done as an extra function of the computer, but this is something which must be discussed in great detail and, in fact, must be demonstrated before one can be assured that it is possible. Another aspect of this problem has to do with whether a particular function exists within the computer software or whether it is a matter of writing additional software to complement or add functions which may be required. It is entirely possible that there may be a sufficient amount of storage to do other functions and that the computer is sufficiently user friendly to allow one to perform these functions, but this may not be the case. Also, it must be realized that alteration of programs in some of these instruments is specifically forbidden by the manufacturer and that if one has the temerity to proceed, this may void any real or implied contracts regarding the servicing of these instruments.

Some of the instrumentation, particularly if it is not interfaced with the laboratory system, requires a great deal of entry of demographic data about the patient. One has to critically evaluate prospectively to make sure that an inordinate amount of work is not necessary to identify specimens. Occasionally the comment is made by the manufacturer that one can use a particular instrument if one has a printer to generate patient reports. This is obviously a potential advantage, but the amount of data that one needs to input may be excessive and may nullify any particular advantage of the system. One always has to be somewhat skeptical of the usefulness of generating a report which may only be useful for a certain subset of all of the isolates found in the patient population. Specifically, one may have problems with the medical record department concerning prohibitions on the types of reports allowed in the patient record.

A potential advantage of some of the instruments, particularly those that are involved with automated reading of antimicrobial susceptibility tests, has to do with the selection of antimicrobial agents to report to the physician. Although this selection is not done presently in most hospitals, it is consistent with the cost restraints of the new regulations regarding payment for various diagnosis-related groups. Many of the newer antimicrobial agents are extraordinarily expensive and may or may not have advantages over older, less expensive agents. The information derived from consensus evaluation of use of antimicrobial agents is something which could be incorporated into each antimicrobial susceptibility report (1). Specifically, one should not be reporting large numbers of antibiotics for each isolate but rather should be reporting antimicrobial agents which are less toxic and less expensive. This selective reporting could be accomplished with a computer by comparing the source of the isolate with the identification of the isolate and then reporting only those antimicrobial agents which would be considered to be the treatment of choice or the alternate treatment for these particular isolates. This concept of very selective antimicrobic reporting is important in a constructive, unified approach to try to reduce the continually increasing cost of medical care. However, the mere fact that an automated instrument has a computer which is reporting susceptibility data does not necessarily mean that these kinds of selective programs can be incorporated into the instrumentation. These kinds of questions must be asked of each manufacturer, and the manufacturers need to be encouraged to develop software which allows each user to modify the selection of antimicrobial agents, depending either on local usage or on agreement between the infectious disease consultants, the pharmacy, and possibly even the hospital administration.

One final consideration needs to be mentioned regarding the utilization of computers that are part of a specific instrument. The temptation may be great to attempt to utilize such a computer as a stand-alone microbiology laboratory computer. It is possible that under some circumstances this could be accomplished. It should be noted that these microcomputers are very limited in their ability to interact with other microcomputers. Therefore, claims by a manufacturer that a particular system associated with an instrument can be easily used for the entire laboratory need to be greeted with considerable skepticism. It is entirely possible, for example, that a number of instruments, all of which have computers, may exist within the microbiology laboratory environment, and that one would be tempted to interface these computers to a larger computer on one specific instrument. It is unlikely that it would be possible to perform this task without a very considerable amount of programming capability. Also, different manufacturers would be very reluctant to encourage the interfacing of their instrument with other instruments. Although the general comment will be made that this can be done, one must realize that to perform this task one needs specific information regarding internal aspects of the software of a specific manufacturer, which may come under the category of privileged information and hence not be available for the system planned.

As mentioned above, there is presently a great deal of enthusiasm for the networking of microcomputers, and computer scientists tend to make light of the existing problems by arguing that the hardware is quite inexpensive and that the software is easy to deal with. This overselling and probably misrepresentation of the capabilities of systems is very reminiscent of that which occurred in the late 1960s and early 1970s, except that now every laboratory worker feels that he or she has some interest and capability in dealing with home computers and, therefore, that he or she should be able to accomplish these tasks. In the late 1960s and early 1970s, one had to deal with large systems; therefore one was more constrained in the ability to modify the systems and had a built-in form of reluctance to become involved. This particular barrier has now been broken and will certainly lead some laboratory directors and personnel into areas of considerable confusion. Another factor that needs to be considered has to do with what represents adequate backup for a particular system; if a particular instrument serves as a core for a laboratory system and there is no appropriate backup for that instrument, this can cause a great deal of difficulty for the laboratory. It should be noted that systems developed along the lines outlined above may be very difficult to interface with total laboratory systems, and therefore one would be at a considerable disadvantage compared with buying a more expensive but coherent total laboratory system which would include chemistry, hematology, and perhaps the blood bank.

LITERATURE CITED

1. **Anonymous.** 1984. The choice of antimicrobial drugs. Med. Lett. **26:**19–26.
2. **Darland, G.** 1975. Discriminant analysis of antibiotic susceptibility as a means of bacterial identification. J. Clin. Microbiol. **2:**391–396.
3. **Edwards, P. R., and W. H. Ewing.** 1962. Identification of Enterobacteriaceae, 2nd ed. Burgess Publishing Co., Minneapolis.
4. **Fisher, R. A.** 1950. Contributions to mathematical statistics. John Wiley & Sons, Inc., New York.
5. **Friedman, R. B., D. Bruce, J. D. MacLowry, and V. Brenner.** 1973. Computer-assisted identification of bacteria. Am. J. Clin. Pathol. **60:**396–403.
6. **Friedman, R., and J. MacLowry.** 1973. Computer identification of bacteria on the basis of their antibiotic susceptibility patterns. Appl. Microbiol. **26:**314–317.
7. **Gyllenberg, H. G.** 1965. A model for computer identification of microorganisms. J. Gen. Microbiol. **39:**401–405.
8. **Kunz, L. J., J. W. Poitras, J. Kissling, B. A. Mercier, M. Cameron, C. Lazarus, R. C. Moellering, Jr., and G. O. Barnett.** 1976. The role of the computer in microbiology, p. 181–193. *In* J. E. Prier, J. Bartola, and H. Friedman (ed.), Modern methods in medical microbiology: systems and trends. University Park Press, Baltimore.
9. **Lapage, S. P., S. Boscomb, W. R. Wilcox, and M. A. Curtis.** 1973. Identification of bacteria by computer: general aspects and perspectives. J. Gen. Microbiol. **77:**273–290.
10. **Lawrie, D. J., R. J. Elin, V. J. Gill, T. L. Lewis, J. D. MacLowry, and F. G. Witebsky.** 1979. Microbiology subsystem of a total, dedicated laboratory computer system. J. Clin. Microbiol. **10:**861–875.
11. **Peebles, J. E., and K. J. Ryan.** 1980. A microbiology information system, p. 534–538. *In* J. T. O'Neil (ed.), Proceedings: the Fourth Annual Symposium on Computer Applications in Medical Care. IEEE Computer Society, Long Beach.
12. **Pratt, J. W., H. Raiffa, and R. Schlaifer.** 1965. Introduction to statistical decision theory, p. 10.1–10.15. McGraw-Hill Book Co., New York.
13. **Robertson, E. A., and J. D. MacLowry.** 1974. Mathematical analysis of the API Enteric 20 Profile Register using a computer diagnostic model. Appl. Microbiol. **28:**691–695.
14. **Robertson, E. A., and J. D. MacLowry.** 1975. Construction of an interpretive pattern directory for the API 10 S kit and analysis of its diagnostic accuracy. J. Clin. Microbiol. **1:**515–520.
15. **Ryan, K. J., and J. E. Peebles.** 1982. On-line computer entry of routine and AutoMicrobic System bacteriology results, p. 23–27. *In* R. C. Tilton (ed.), Rapid methods and automation in microbiology. American Society for Microbiology, Washington, D.C.
16. **Sneath, P. H. A.** 1957. The application of computers to taxonomy. J. Gen. Microbiol. **17:**201–226.
17. **Sneath, P. H. A., and R. R. Sokal.** 1973. Numerical taxonomy: the principles and practice of numerical taxonomy. W. H. Freeman and Co., San Francisco.

Chapter 8

Collection, Handling, and Processing of Specimens†

HENRY D. ISENBERG, JOHN A. WASHINGTON II, ALBERT BALOWS, AND ALEX C. SONNENWIRTH

A major concern of clinical microbiologists has been the rapid identification of significant microorganisms isolated from clinical specimens. The scientific literature contains many publications directed at the identification of genus, species, or both in the least possible time; yet the subject of this chapter has been almost totally ignored, despite the fact that it constitutes the foundation for all subsequent work. It is not possible to present in the space available an all-encompassing guide to the procurement, proper management, and primary inoculation procedures of clinical specimens (see Table 1). Our interest is to present a working outline that is admittedly not all inclusive, but is an approach that knowledgeable clinical microbiologists should pursue in the handling of clinical specimens in their laboratories. There are many variations on the approach presented here, and many situations will arise that are not covered by this chapter.

It is necessary to establish certain ground rules for clinical microbiology. The purpose of the clinical microbiology laboratory is to rapidly and accurately provide the clinician with information concerning the presence or absence of a microbial agent that may be involved in an infectious disease process. When indicated, this information should be supplemented with an antibiotic profile. It is more important for a physician to know that a gram-negative rod susceptible to various antimicrobial agents is present than to wait several days or weeks for a properly identified microorganism (27, 61). Clinical microbiologists must admit that, when identifying a microorganism, they base the identification on a number of salient characteristics. Clinical microbiologists are not taxonomists in the strict sense; they do not weigh each microbial character on an equal scale and, after collecting an appreciable number of characteristics, determine the identity of the organism according to Adansonian precepts (1). It is not surprising to find that clinical microbiologists use a variety of identification schemes that do not necessarily reflect the classification schemes of taxonomists or even follow *Bergey's Manual of Determinative Bacteriology.* To bridge any communication gap that may occur with the medical staff, the staff of the clinical microbiology laboratory should be aware of the synonyms of the organisms that are identified by the accepted genus and species designation. The approaches outlined in this chapter require the modifications deemed applicable by each clinical microbiologist for the needs of a particular laboratory. They also must be modified to accommodate the existing requirements of prospective payment policies. A thorough knowledge of microorganisms expected in specimens from nonsterile body sites or common contaminants of sterile material, acquaintance with the recent history of the patient and the clinical impression of the physician, and the occurrence and antibiotic profiles of the more usual isolates involved in nosocomial complications must bear on decisions that affect the workup of specimens and the identification of the microorganisms they contain. Judicious decisions on the amount of analytical effort to be expended are now mandatory to provide high-quality microbiology for the patient in need of such care. The discriminate pruning of test procedures, elimination of unimportant procedures, rejection of unnecessary or improperly submitted specimens, and concerted educational effort in the ordering of tests will help to lessen the cost of health care.

† The authors dedicate this chapter to the memory of Alex C. Sonnenwirth, who passed away after the completion of this revision.

COLLECTION OF SPECIMENS

Abscesses

See section on Wounds.

Anatomical and surgical pathology

The microbiological examination of tissues obtained at the time of surgery or autopsy is of considerable importance because the specimen may represent the entire pathological process. Surgical specimens are obtained at considerable expense and some risk to the patient. Further surgery to obtain more material may be contraindicated or refused; additional autopsy specimens often are not available because the remains have already been embalmed and buried. The microbiologist must be prepared to do whatever is necessary to establish the diagnosis, frequently with assistance from the pathologist. Many of the necessary procedures have been described in detail (53, 65, 66, 67) and form the basis of the discussion which follows.

Selecting the proper specimen and collecting an adequate sample for examination are essential. When the lesion is large or when there are several lesions, multiple specimens from different sites must be collected. Samples from an abscess should include pus and a portion of the wall of the abscess. The use of a cotton swab to collect a small amount of pus from an abscess cavity is tantamount to malpractice! It must be remembered that abscesses of the brain, lung, pleural space, peritoneal cavity, liver, pelvis, and wounds commonly contain anaerobes or a mixed flora and that an adequate sample of infected material should be collected and transported to the laboratory in an anaerobic container. When granules are observed in the pus, they should suggest infection with actinomycetes.

Gross surgical specimens submitted for histopathological examination are ideal for microbiological study since portions of the specimen may be carefully selected for analysis before placing the tissue in fixative. A close rapport between microbiologist and surgical pathologist is invaluable. Histological examination reveals whether or not the lesion is malignant and, if it is inflammatory, whether it is granulomatous or suppurative. If the lesion is malignant, it may not be necessary to perform cultures. If it is inflammatory, the type of reaction may indicate what types of culture are necessary; however, it is well to bear in mind that the histopathology of infections due to mycobacteria, fungi, and brucellae is quite variable, but the diseases are not mutually exclusive. Histopathological diagnosis of an infection depends on the visualization of a sufficient number of organisms in characteristic form. Tissue stains for acid-fast bacilli provide positive results in only 30 to 40% of those specimens with positive cultures, whereas special stains for fungi often show organisms not recovered in cultures, or they may show structures resembling the organisms but lacking definitive characteristics (66). Cultures, therefore, are absolutely necessary to establish or confirm the diagnosis of an infectious process.

A gross surgical specimen may be bisected aseptically by the surgeon. One half is submitted to surgical pathology and the other half is placed in a sterile, wide-mouthed, screw-capped jar and sent to the microbiology laboratory. Lung biopsy tissue impressions are made by imprinting the freshly cut surface several times on each of two or three glass slides for staining with Giemsa, Gomori methenamine-silver, or toluidine blue for *Pneumocystis carinii*. The tissue is finely minced with sterile scissors and ground aseptically with a pestle in a sterile mortar, with 60-mesh aluminum oxide (alundum) as an abrasive. Alternatively, tissue may be homogenized in a sealed plastic bag by using a Stomacher Lab Blender (Spiral System Instruments, Inc., Bethesda, Md.). The lung homogenate may be centrifuged and the sediment used to prepare smears for *P. carinii*. This procedure yields improved detection of *P. carinii* cysts compared with impression smears (62). Slides for Gram and special stains may be prepared from the homogenate to determine the presence of microorganisms and guide the culture approach. Antigen detection methods may also be applied if appropriate and if reagents are available. A 20% suspension of the ground material is made in broth and placed in a sterile dropper bottle to be used for inoculating culture media and for storage. Preparing and examining special histological stains may take several days, and further microbiological studies also may be indicated; therefore, it is advisable to store any residual tissue emulsion at 5°C for 1 or 2 weeks before discarding it. It is recommended that as much material as possible be examined microbiologically by the inoculation of multiple plates or tubes containing the appropriate media. Inoculation of appropriate cell cultures for virus isolation should be done if indicated. If it is anticipated that virus isolation may be attempted later, some of the body secretions or tissue (whole or ground) should be stored at −70°C. If there is any intention to consider electron microscopy study of the tissue for viruses, a portion of the involved tissue should be preserved in glutaraldehyde (see Chapter 61).

It is generally thought that cultures of embalmed tissue are useless; however, Weed and Bagenstoss were able to isolate tubercle bacilli, *Histoplasma capsulatum*, *Nocardia asteroides*, and various species of bacteria from tissues that had been embalmed for 24 to 48 h (67). Embalming will diminish considerably the probability of recovering an organism; however, in certain cases, cultures of tissues embalmed for as long as 48 h may be worthwhile, provided nothing else is available and the area of the embalmed tissue selected for cultures is centrally located.

The value of postmortem microbiology remains controversial. In some institutions the heart blood is routinely cultured; in others, cultures are limited to cases and specific organs in which an infectious disease is suspected. In recent years, attempts have been made to determine whether clinical or autopsy evidence of infection is an appropriate mechanism for selecting cases to culture, whether sampling should be limited to tissues or organs suspected of being infected, and whether the microbiologist should attempt to correlate postmortem culture results with those obtained antemortem (18, 35, 72). In general, the results of these studies have demonstrated that humans possess an indigenous tissue flora, that there does not appear to be postmortem transmigration of organisms, that the frequency of positive cultures is poorly related to clinical or autopsy evidence of infectious disease, that there is a substantial lack of correlation between ante- and postmortem culture results, that cultures performed on a single postmortem tissue only are rarely of value, and that in some cases postmortem cultures of multiple tissues may be of value in identifying the etiological agent of the infection, especially if these represent cases with well-recognized clinical entities caused by a single organism or cases with an overwhelming infection. The value of postmortem cultures is, therefore, limited except in selected cases.

When postmortem cultures are indicated, the prosector should obtain at least a 6-cm^3 portion of tissue, with one serosal or capsular surface intact. The tissue should then be placed in a sterile, sealed, plastic container and sent to the laboratory for immediate processing or storage at 5°C. Subsequent procedures are based on those described by Dolan (17). In the laboratory, the capsular or serosal surface is thoroughly seared with a soldering iron and incised with a sterile instrument. A 1-cm^3 portion of tissue is removed aseptically from the core of the tissue block; it is used to prepare impression smears and is then ground, as described above, to provide a 10 to 20% emulsion to be used as inoculum for cultures. When a small specimen is received, it may be immersed in boiling water for 3 to 5 s to decontaminate its surface. Any residual portion of surgical or anatomical specimens should be retained in a refrigerator by the microbiologist until it is reasonably well established that there is no longer any need to retain the tissue.

The routine described by Isenberg and Berkman for postmortem examination (27) has also been found useful by a number of pathologists and entails qualitatively surveying lungs, liver, spleen, and kidney, as well as obtaining heart blood. Smears and cultures are made of the material aseptically obtained at the time of autopsy. Comparison between the findings from the direct smears and especially the inoculated blood plates usually indicates correlation. One of the major points which must be stressed is the need to obtain microbiological cultures from areas grossly appearing infected or involved in an infectious process. The past history of the patient, the course of therapy, and the agonal events must also be considered so that media to culture organisms that do not grow readily on routinely used media can be obtained. Such examinations may serve also to delineate the most significant nosocomial microbes in the environment of a specific hospital (28).

Blood cultures

The prompt and accurate isolation and identification of the etiological agents of septicemia remain among the most important functions performed by the clinical microbiology laboratory. The indications for obtaining blood cultures are a sudden relative increase in the pulse rate and temperature of the patient, a change in sensorium, and the onset of chills, prostration, and hypotension. Other indications include a prolonged, mild, and intermittent fever in association with a heart murmur. Bacteremia is continuous in endocarditis, endarteritis, uncontrolled infections, typhoid fever, and brucellosis, and it is usually intermittent in other infections. Timing in the collection of cultures in endocarditis, for example, may not be critical; however, in other cases, timing of the collection is important because the bacteremia is usually intermittent and may precede the onset of fever or chills by as much as 1 h.

In patients with suspected bacterial endocarditis, three blood cultures are sufficient to isolate the etiological agent in nearly all instances. These should be collected separately and, the condition of the patient permitting, at no less than hourly intervals within a 24-h period. In intermittent bacteremias, three separate blood cultures within 24 to 48 h are usually sufficient to isolate the etiological agent. In such instances, the time interval between cultures is frequently determined by clinical circumstances and the urgency to initiate antimicrobial therapy. In patients who have received antimicrobial agents before blood collection, a total of four to six separate blood cultures may be necessary to isolate the etiological agent, or consideration should be given to using special techniques to adsorb or inactivate the antibiotic(s) present in the blood.

It is essential that blood for culture be collected aseptically, first by cleansing the skin with 80 to 95% alcohol and then by applying 2% iodine in concentric fashion to the venipuncture site. Because of the lower incidence of skin hypersensitivity to iodophors, they may be used in lieu of iodine. Instant antisepsis never occurs, and the iodine or iodophor should remain intact on the skin for at least 1 min. The intended venipuncture site should not then be touched unless the fingers used for palpation are similarly disinfected. After the venipuncture, any residual iodine should be removed with an alcohol sponge or pad.

It is recommended that 10 to 20 ml of blood be collected for each culture. Culture of a lesser volume may result in lower recovery rates because of the relatively low numbers of organisms present in most bacteremias. In infants and children, collection of 1 to 5 ml appears to be satisfactory. Optimally, the blood is inoculated directly into culture media at the bedside of the patient, with either a syringe and needle or a transfer set. Alternatively, the blood may be transported to the laboratory in a sterile, evacuated tube containing sodium polyanetholesulfonate (SPS) and then inoculated into culture media (72). Several new approaches have become available recently and are summarized by Reller (48). These may prove to be useful substitutes for the traditional methods and may decrease the time-to-positive interval as well as the technical attention requirements.

It is essential that blood be inoculated into culture media on a 10% (vol/vol) basis to counteract the normal bactericidal activities of chemical and cellular mediators of immunity. Any residual bactericidal effects remaining after this dilution of blood in the culture media have been shown to be abolished by the presence of 0.05 and 0.025% SPS, which is a polyanionic anticoagulant that is also anticomplementary and antiphagocytic.

Any general-purpose, commercially available nutrient broth medium may be used for the culture of blood. Soybean casein digests, such as tryptic soy broth (Difco Laboratories, Detroit, Mich.), Trypticase soy broth (BBL Microbiology Systems, Cockeysville, Md.), Columbia broth, and brain heart infusion broth, have been found to be satisfactory. Strict reliance on fluid thioglycolate, Thiol (Difco), or supplemented, prereduced anaerobically sterilized media is not recommended because of lower isolation rates of aerobic or facultatively anaerobic bacteria in these media. Commercially available liquid media are generally bottled under vacuum with CO_2 and contain 0.025% SPS. As such, they are satisfactory for cultivation of anaerobes from blood (see Chapter 37 for more details about anaerobic blood cultures). Bottles containing 50 or 100 ml of medium should be employed. The value of liquid media made hyperosmotic, generally by the addition of sucrose, remains uncertain. In general, such media have not been found to increase the detection rates or to decrease the time interval for detection of bacteremia.

Because of the mutually exclusive atmospheric requirements of several groups of microorganisms encountered in septicemia, it is desirable to use two bottles, one which is not vented and one which is vented. Venting is performed by aseptically inserting a sterile, cotton-plugged needle through the rubber stopper of the bottle and then withdrawing the needle after the vacuum within the bottle has been released. If the venting unit is left in place, the bottle should be incubated in an atmosphere containing 10% CO_2.

Cultures are incubated at 35°C and are examined macroscopically or radiometrically later on the day of their collection and daily thereafter for at least 7 days for evidence of growth. Clinically significant bacteria are recovered within this period in at least 95% of instances; however, longer periods of incubation may be necessary with specimens from seriously ill patients who are receiving and appear not to be responding to antimicrobial agents or from patients with endocarditis or endarteritis presumably due to fastidious microorganisms. Recovery of such bacteria beyond the first 7 days of incubation may provide the opportunity for administration of more specific and effective antimicrobial agents.

Gram- or acridine-orange-stained smears and aerobic and anaerobic subcultures of obvious or suspected positive cultures should be prepared immediately, and the results of the microscopic examination of the smear, if positive, should be reported by phone and in writing to the physician of record as quickly as possible. Subcultures for direct antibiotic susceptibility testing may be performed; however, these results should be considered tentative and require confirmation by retesting the isolate with a standardized method.

Routine (blind) subcultures of grossly negative cultures should be performed between 6 and 14 h after blood collection and after 48 h of incubation. Subcultures are inoculated onto quadrants of chocolate blood agar plates which are incubated in 3 to 6% CO_2 for 48 h. Subcultures are necessary to ensure the recovery of *Pseudomonas aeruginosa*, *Haemophilus* spp., meningococci, gonococci, and yeasts, particularly when blood culture bottles are not vented. Routine anaerobic subcultures are unnecessary. In some instances a routine Gram- or acridine-orange-stained (48) smear made within 24 h after blood collection has been helpful.

The recovery of diphtheroids (aerobic and anaerobic), *Bacillus* spp., and *Staphylococcus epidermidis* usually signifies contamination, unless they are present in multiple cultures. Nonhemolytic streptococci, excluding group D streptococci, and alpha-hemolytic streptococci in single cultures are of uncertain significance. *Bacteroidaceae, Enterobacteriaceae, P. aeruginosa, Haemophilus* spp., *Staphylococcus aureus*, pneumococci, and yeasts are nearly always clinically significant.

Polymicrobial bacteremia has been reported in as many as 18.4% of microbiologically proven septic episodes (69) and has been associated with higher mortality than has unimicrobial infection (68).

The routine addition of penicillinase to blood culture media does not appear to be justified, except in selected patients who are receiving high doses of a penicillin or cephalosporin at the time of blood collection. In such instances, it is essential to prepare concurrent cultures of the penicillinase solution, contamination of which has been associated with outbreaks of pseudosepticemia. In general, antimicrobial agents are inhibited or neutralized by dilution of the blood in broth, addition of SPS to the medium, and lysis-concentration techniques. Although the advantages of antibiotic-adsorbent resins are controversial and may be system dependent, it is generally agreed that such devices should always be used in conjunction with conventional blood culture systems.

The isolation of brucellae and many other organisms is facilitated by inoculation of the blood into Castaneda double medium or into soybean casein digest broth, which must be subcultured at least twice weekly for 4 weeks. In either case, prolonged capneic incubation is required. Castaneda double medium without SPS is useful in recovering pathogenic neisseriae from blood.

Leptospiremia is usually present only during the first week of illness. One to three drops of freshly drawn blood are inoculated into each of several tubes containing 5 ml of a suitable semisolid medium, such as Fletcher or Ellinghausen medium, and the cultures are incubated in the dark at 30°C for 28 days. A portion of each culture is examined weekly by darkfield or fluorescence microscopy.

Although it has been recommended that detection of fungemia should be accomplished by inoculation of blood on a 10% (vol/vol) basis into a soybean casein digest broth with subcultures to fungal isolation media at 48-h intervals or by inoculation of the blood into a Castaneda double medium, it now appears that fungal isolation rates are substantially improved by lysis-concentration techniques.

For more details regarding blood cultures, readers should consult *Cumitech 1* (9), *Cumitech 1A* (49), and monographs (2, 63) devoted to blood cultures.

Bone

Specimens from orthopedic procedures or postoperative complications must be transported to the laboratory quickly and in media which prevent drying. Microbiologists should insist that each culture be accompanied by a smear procured from the same site as the culture. It is important to insist that information on the patient be provided to permit analysis for various microorganisms not included in the routine screening of surgical specimens. Thus, specimens from animal bites which penetrated to the periosteum require inoculation of media which support *Haemophilus haemoglobinophilus* and *Pasteurella multocida*. Special care and consultation with the surgeon of record must be exercised with traumatic and nosocomially complicated osteomyelitides. Microbiologists must do their utmost to help the surgeon differentiate between skin organisms and those from the deeper recesses of tissues and bone. Multiple specimens, obtained with great care and accompanied by smears, can help provide this information. Blood cultures are important adjunct procedures in monitoring such patients.

Ear

The bacteriological findings from persons with acute otitis media have been sufficiently consistent in many reported studies that there would seem to be little justification for performing tympanocentesis, except in therapeutic failures or in neonates, in whom the bacterial etiology differs from that of older children. Since the correlation between nasopharyngeal and middle ear cultures is poor, cultures of the nasopharynx may provide misleading results. Most studies show that *Streptococcus pneumoniae, Haemophilus influenzae*, and *Streptococcus pyogenes* are the most common etiological agents of acute otitis media; however, *Branhamella catarrhalis, P. aeruginosa, Escherichia coli, Klebsiella pneumoniae*, and *S. aureus* may also be isolated from middle ear effusions of infants (12, 42, 55), and in therapeutic failures *S. aureus* and *P. aeruginosa* are predominant (10, 12).

Eye

In patients with conjunctivitis and, especially, with keratitis, there are special problems in specimen collection and processing. Swabs are generally inadequate for establishing the presence of microorganisms because of the small sample size. An additional problem is the antimicrobial activity of topical anesthetics (51). Microbes responsible for keratoconjunctivitis include bacteria, actinomycetes, fungi, and viruses. It is, therefore, recommended that swabs for cultures be taken before topical anesthetics are applied and that corneal scrapings be taken after they are applied.

Ideally, a laboratory technologist should assist the ophthalmologist in collecting the specimen and preparing the cultures.

Examination of Gram- and Giemsa-stained smears of the corneal scrapings may provide preliminary clues to the nature of the disease. If necessary, stains for acid-fast bacteria or a potassium hydroxide wet mount can be made. Because of the limited amount of material available, it is recommended that the scrapings be spot-inoculated directly onto each one of the following media: chocolate blood agar, brain heart infusion agar with 5% sheep blood, inhibitory mold agar, and, if indicated, Lowenstein-Jensen agar. Swab cultures may be inoculated onto one half of each agar plate and the scrapings onto the other half. See Chapter 85 for the handling of specimens to be cultured for *Chlamydia* spp. A more detailed description of pertinent procedures may be found in *Cumitech 13* (30).

Feces

Proper collection and preservation of feces is a frequently neglected but important requirement for the isolation of microorganisms contributing to intestinal disease. Unless the specimen can be taken immediately to the laboratory and properly handled upon delivery, a number of important microorganisms will not survive the changes in pH which occur with a drop in temperature. This is especially true of most shigellae and an appreciable number of salmonellae. When delays are unavoidable, it should be a standard rule that all stool specimens be submitted in a stool preservative, such as 0.033 M phosphate buffer mixed with equal volumes of glycerol. An indicator may be added which will assure an approximate pH of 7. The indicator added to the specimen provides visual proof that the pH drop has been inordinate. Stool specimens are introduced into a screw-capped glass container with stool preservative immediately after they have been passed. A 0.5- to 2-g quantity is sufficient. If sterile swabs are used in obtaining the specimen, they should be passed beyond the anal sphincter, carefully rotated, and withdrawn. The swabs may be added to a screw-capped tube containing preservative and transported to the laboratory for culture. Difficulties are occasionally encountered when patients have diarrhea and are unable to use a bedpan. Under those circumstances, paper towelling may be deposited into the toilet bowl, and the diarrheal stool may be obtained above the water level. Toilet paper, which usually has been impregnated with bismuth salts in the manufacturing process, should not be used, since a number of organisms found in feces are inhibited or killed by these salts.

If the patient is hospitalized, personnel obtaining the specimen should be instructed explicitly to choose portions of the stool that display either mucus, blood, or both. These areas usually harbor a large number of the organisms that are involved in the disease process.

It cannot be stated too often that the vast majority of bacteria found in feces are anaerobic gram-negative rods. They are usually ignored in the attempt to diagnose infectious intestinal disorders. This is not to imply that such organisms may not be involved in disease processes of the intestinal tract. However, to date, such a causal relationship has not been established, with the exception of food-borne *Clostridium perfringens* intoxication and pseudomembranous colitis after antimicrobial therapy resulting in the implantation of *Clostridium difficile*. It is important to remember that the biochemical activities of the anaerobic microbiota contribute to the detriment of *Enterobacteriaceae*. Therefore, the facultatively anaerobic enteric bacteria require the type of buffering provided by a stool preservative.

There are times when surveys for carriers as well as institutional outbreaks indicate a need to use rectal swabs to obtain specimens. Under these circumstances, the stool preservative should not be used; the swab should immediately be placed, instead, into a medium such as Gram-Negative broth (BBL Microbiology Systems), especially if the responsible organism has already been identified and contacts or carriers are being sought. Enrichments such as tetrathionate broth or selenite F may be used, particularly if salmonellae are involved.

As with most other specimens, a single stool specimen with little or no accompanying information is inadequate. The laboratory can be much more helpful to the clinician if a brief history is provided. If necessary, the physician of record should be called for this information so that the clinical microbiologist can be particularly attentive to culturing the stool on media most likely to yield the responsible bacteria, fungi, or parasites. Without such information, procedures cannot be modified to accommodate as complete a microbiological analysis as is required. Of equal concern are the increased cost and the time lost in searching for impossible organisms.

A single negative stool culture cannot be regarded as sufficient for laboratory confirmation of noninvolvement of infectious bacteria. Although the number three is not inviolate, repeated cultures are indicated if the clinical picture suggests the involvement of bacteria and the first two cultures are unrewarding despite the symptoms. Similarly, after the diagnosis has been made, microbiological surveillance of the convalescent individual and of contacts who may have become carriers should be conducted at regular intervals until at least three negative specimens have been obtained consecutively.

When the history or early examination suggests the possibility that anaerobes, *Bacillus* sp., or staphylococci; *Vibrio cholerae, Vibrio parahaemolyticus*, and related bacteria; *Campylobacter* sp.; toxigenic or invasive *E. coli*; and certain viral agents may be involved, special media should be used for direct inoculation of the stool. When gastrointestinal mycobacteriosis is suspected, the clinical microbiology laboratory should follow the suggestions in Chapter 22. Specimens collected for the demonstration of parasites and ova are discussed in the section on parasitology (Chapters 55 through 60). Gram-stained smears for the demonstration of bacteria have not been used frequently in recent times. In the past, examination of such smears was popular for determining the ratio of gram-negative to gram-positive bacteria. It is not possible to determine the type of microorganisms present simply by observing the morphology and staining characteristics of bacteria in a stool specimen. The usefulness of such a method is restricted to cases such as pseudomembranous colitis, staphylococcal enterocolitis, or monilial disease, where the overwhelming number of gram-positive cocci or yeastlike cells would easily be demonstrated in a properly made Gram stain of a portion of the stool specimen or rectal swab; extraordinary numbers of pus cells can be detected in this fashion also. For further details, see *Cumitech 12* (50).

Body fluids

Cerebrospinal fluid. The examination of cerebrospinal fluid (CSF) from patients suspected of having meningitis represents one of the major emergency procedures faced by personnel in the clinical microbiology laboratory. The reasons for this urgency are that bacterial meningitis is a rapidly fatal disease if untreated or inadequately treated and that appropriate antimicrobial therapy often requires prompt identification of the etiological agent. Lumbar puncture and examination of the CSF should be undertaken whenever the physician suspects meningitis or wants to rule it out. It must be remembered that the typical signs of meningeal irritation in the adult, such as fever, headache, vomiting, nuchal rigidity, hyperreflexia, etc., are usually absent in infants and neonates, in whom the clinical manifestations of meningitis are often vague and nonspecific. An unexplained febrile illness in an irritable infant who is doing poorly should lead one to suspect meningitis. Meningitides due to mycobacteria, fungi, leptospires, or protozoa are generally insidious in onset. The diagnosis of viral meningoencephalitis is frequently established by exclusion and by serological means. Although some viruses may be isolated from CSF, viral isolates are more likely to be obtained from other clinical specimens.

Lumbar puncture must be performed under conditions of strict asepsis, since contamination of the specimen can occur readily and confuse the identification of the etiological agent. The skin should be disinfected with povidone-iodine. Specimens should be collected in sterile containers which can be sealed with a screw cap to preclude leakage and loss or contamination of the contents. Cotton-plugged or rubber-stoppered tubes should not be used, and snap-top containers should be checked to ensure that a tight seal does occur and that some of the contents are not aerosolized on opening. The sterility and absence of microorganisms in CSF specimen containers should be periodically confirmed by culture and also presumptively by Gram stain (40). Methods for the detection of microbial antigens in CSF are indicated now that several types of reagents are available commercially. These approaches are recommended especially if no organisms can be detected microscopically or to confirm the identity of the common etiological agents seen on smears or both (32). Prompt transport of the specimen to the laboratory is mandatory, since fastidious organisms such as *H. influenzae* and *Neisseria meningitidis* may not survive storage or variations in temperature. For these reasons, some advocate that smears be prepared and cultures be inoculated at the bedside of the patient when the CSF is obtained. Though such a practice may be ideal, it is seldom practical. The clinical microbiologist, therefore, should (i) examine the lumbar puncture tray routinely used in the hospital to ensure that the CSF containers are of satisfactory quality (not all commercially available trays have airtight specimen containers), (ii) attempt to establish a standardized skin preparation, and (iii) develop the systems whereby the specimen can be transported promptly to the laboratory. An adequate sample (as much as possible) of CSF should be available for microbiological examination, particularly when a diagnosis of tuberculous or fungal meningitis is considered, since the numbers of microorganisms present are often small. If only one specimen container is filled, it should be submitted to the microbiology laboratory first so that it can be opened aseptically and samples for chemical and cytological studies can be removed at the time cultures are inoculated. The microbiology laboratory should process the fluid immediately by preparing Gram-stained smears and inoculating the appropriate media.

Since usually only small numbers of microorganisms are present in infected CSF, some procedure for the concentration of any organisms present should generally be performed. The simplest of these procedures is centrifugation of the specimen at $1{,}000 \times g$ for 15 min; then the supernatant fluid is removed for chemical or serological studies, and the sediment is used for both smear and culture purposes. This centrifugation is adequate for those fluids in which there is an increase in inflammatory cells; however, its limitation is that in very early cases of bacterial meningitis and in fungal meningitis the cell count may be normal despite a positive culture. Furthermore, a force of $10{,}000 \times g$ for 10 min is required to sediment *H. influenzae*, which means that with most standard clinical laboratory centrifuges, which develop a maximum of $1{,}000 \times g$, 60 min may be necessary (56). For these reasons, prolonged (30-min) examination of the Gram-stained smear under oil immersion may be necessary. The same sampling problem occurs when attempts are made to culture the specimen from a nonexistent sediment. In instances in which there is no cellular response or a poor one, it is recommended that the specimen be concentrated instead by filtration through a 0.45-μm membrane filter which is directly placed on an appropriate culture medium. Sterile, disposable, 13- and 25-mm filter units (Swinnex; Millipore Corp., Bedford, Mass.) are satisfactory and convenient for this purpose.

A variety of immunological procedures have been developed for the rapid diagnosis of bacterial meningitis; however, standardized reagents and some special equipment are requird for some of these procedures. Fluorescence microscopy, capsular swelling, counterimmunoelectrophoresis, coagglutination, and latex particle agglutination are some of the principal techniques used. The usefulness of these tests is in those instances in which the patient has received antibiotic therapy and in life-threatening situations when a fast and reasonably presumptive diagnosis is required. These tests should be well controlled to be reliable (46). In most laboratories, however, the Gram-stained smear remains the most reliable and convenient rapid diagnostic tool.

The inflammatory and noninflammatory responses of the host may be helpful in the differential diagnosis of meningitis. Whereas the leukocytic response in the CSF in acute bacterial meningitis is usually polymorphonuclear, that in tuberculous, fungal, leptospiral, or protozoan meningitis is usually lymphocytic and less intense. Although polymorphonuclear leukocytes may be predominant in the CSF early in the course of aseptic meningitis, there is usually a clear shift to mononuclear cells within 8 h (21). The CSF glucose is usually depressed in cases of acute bacterial or tuberculous meningitis and normal in the other meningitides; CSF protein is usually elevated. Cytological and chemical changes in the CSF may occur in patients with brain abscess; however, smears and cultures of the CSF are generally negative in these cases unless the abscess ruptures into the subarachnoid space or the ventricles.

One final point in partially treated cases of bacterial meningitis deserves emphasis, and that is that cellular and chemical findings and the recovery rates of bacteria in cultures may be altered (15, 16). Furthermore, a tendency has been noted for gram-positive organisms to appear to be gram negative in such cases (15), so that findings in a Gram-stained smear must be interpreted cautiously.

When fungal meningitis is suspected, a drop of the CSF sediment should be mixed with a drop of India ink (Pelikan, Gunther, and Wagner, Hanover, Federal Republic of Germany) or nigrosin (Harleco, Div. of Harman-Leddon Co., Philadelphia, Pa.) solution on a clean glass slide, covered with a cover slip, and examined with a decreased intensity of light. Nigrosin is preferable to India ink because it is free from discernible particulate matter. The presence of encapsulated, budding, yeastlike cells in the wet preparation is virtually diagnostic of cryptococcal meningitis; however, special care must be exercised in differentiating nonencapsulated yeasts from erythrocytes or leukocytes, air bubbles, or even talc. In some cases of cryptococcal meningitis, there is little or no cellular response or depression of the sugar value in the CSF (46).

Cryptococcal meningitis is frequently associated with disseminated infection, and it is not uncommon to recover *Cryptococcus neoformans* from sources other than CSF (urine, blood, sputum, bone marrow, etc.). In patients from whom *C. neoformans* has been recovered initially from such other sources, serious consideration must be given to lumbar puncture and examination of the CSF, since symptoms referable to the central nervous system may be absent. In patients with suspected cryptococcal meningitis, performance of the latex-cryptococcal antigen test with CSF is often helpful because the India ink or nigrosin preparation is negative in roughly 50% of instances.

The development of meningoencephalitis of obscure etiology in patients with a recent history of swimming in stagnant fresh water before the onset of symptoms should lead to the prompt examination of the CSF for motile amoebae (19).

Fluids other than CSF. As with the lumbar puncture, the percutaneous aspiration of pleural, pericardial, peritoneal, and synovial fluids must be performed aseptically to avoid contamination of the specimen and to prevent the accidental introduction of microorganisms into these anatomical spaces. The specimen should immediately be injected into a sterile tube or bottle. Since infection of these spaces may be due to anaerobes, it is recommended that fluid or pus be collected with a sterile syringe and needle, that any air bubbles present in the syringe be expelled, and that the material then be injected into an anaerobic transport tube or vial containing CO_2. Again, as much material as it is feasible and practical to collect should be submitted to the laboratory. A small amount of sterile heparin may be added to the fluid to prevent coagulation, since clots may trap microorganisms. Coagulated material should be emulsified and cultured along with a portion of its surrounding fluid. The special procedures and media advocated for the isolation of *Legionella pneumophila* should be applied to pleural fluids (31).

Gram-stained smears of the centrifuged sediment of clear or slightly cloudy fluids should be examined carefully; however, frankly purulent material should be smeared directly and examined after it has been stained for the presence of bacteria.

Respiratory tract

Cultures of the respiratory tract (3, 6, 64) must be interpreted cautiously because of the microflora normally present in the nose, oral cavity, and pharynx and because of the frequency of nosocomial acquisition of potentially pathogenic microorganisms by seriously ill patients. Since potential pathogens such as *S. aureus*, *H. influenzae*, *Streptococcus pneumoniae*, *P. aeruginosa*, *Enterobacteriaceae*, and yeasts may be present in the oropharynx, their isolation from cultures of the respiratory secretions does not represent a priori evidence of their etiological role in respiratory infections.

Nasopharyngeal cultures should be performed to detect carrier states of *N. meningitidis*, *Corynebacterium diphtheriae*, and *S. pyogenes*. In addition, such cultures aid in the diagnosis of whooping cough and croup. Nasopharyngeal specimens should be obtained with a Dacron, cotton, or calcium alginate swab on a flexible wire which is gently passed through the nose into the nasopharynx, rotated, removed, and placed into a suitable transport medium for isolation. Nasopharyngeal aspirates are superior to swabs for isolating *Bordetella pertussis* from children suspected of having whooping cough.

Although nasopharyngeal cultures are sometimes recommended in comatose patients unable to expectorate a sputum sample, the frequency of aspiration and subsequent anaerobic pleuropulmonary infection in these patients and the misleading results provided by oropharyngeal contamination make percutaneous transtracheal aspiration mandatory (7).

H. influenzae type B may cause a form of croup known as acute epiglottitis. The distinctive appearance of the epiglottis and the rapidly progressive and fulminating course, which may lead to death within 24 h, demand prompt initiation of therapy. Nasopharyngeal and blood specimens should be obtained and cultured for *H. influenzae* once an airway has been assured.

It should be kept in mind that orolaryngeal involvement is not uncommon in both acute and chronic disseminated forms of histoplasmosis and blastomycosis (11), in tuberculosis, and in leishmaniasis. Cultures for these agents and biopsy with histological demonstration of the organisms are necessary to establish the correct diagnosis. It is imperative that the clinician communicate his or her clinical suspicions to the microbiologist immediately. Otherwise, valuable time and specimens may be lost.

Throat cultures are obtained most frequently for the diagnosis of streptococcal pharyngitis and less commonly for the diagnosis of pertussis, diphtheria, and pharyngitis due to gonococci or viruses. There is no need to identify other microorganisms in routinely submitted throat cultures, since there is little or no evidence to document their role in producing pharyngitis. Exudative pharyngitis, enlarged cervical nodes, headache, nausea, vomiting, and abdominal pain are commonly associated with streptococcal pharyngitis, whereas cough, rhinorrhea, and hoarseness are commonly associated with viral pharyngitis (24). There are, however, many viral infections which are confused clinically with streptococcal pharyngitis. Acute tonsillopharyngitis with vesicles or shallow ulcers on the anterior fauces, palate, and buccal mucosa is usually due to herpes simplex virus or to coxsackievirus A. A history of an incomplete immunization series or no immunization during childhood should alert the clinician to the possibility of infection with *B. pertussis* or *C. diphtheriae*, particularly when incomplete immunization is coupled with the characteristic signs and symptoms of these diseases. Gonococcal pharyngeal infection should be suspected in patients with gonococcal infections at other sites, especially among those practicing fellatio (70).

Requisitions for culture or specimens from the upper respiratory tract should specify the suspected etiological agent so that laboratory personnel can take appropriate steps to ensure its isolation. The culture should be obtained under direct visualization with a Dacron, cotton, or calcium alginate swab by vigorously swabbing both tonsillar areas, the posterior pharynx, and any areas of inflammation, ulceration, exudation, or capsule formation. The tongue should be depressed with a tongue blade or spoon to minimize contamination of the swab with oral secretions which may dilute, overgrow, or inhibit the growth of pharyngeal flora.

Bacterial culturing of sputum is fraught with error, and clear-cut results are seldom obtained. Specimens are frequently collected haphazardly by paramedical personnel who are not aware of the necessity for a fresh, clean specimen resulting from a deep cough and who fail to transport the specimen to the laboratory promptly. Expectorated sputum is frequently contaminated with oropharyngeal flora, and it is difficult to determine which of the many different potential pathogens isolated is responsible for pulmonary infection. Rarely, a potential pathogen is isolated in pure culture and may be presumed to represent the etiological agent. To quote Barrett-Connor (4), who found no pneumococci in 45% of the sputum and nasopharyngeal cultures examined from patients with pneumococcal bacteremia, "The routine sputum culture for the diagnosis of acute bacterial pneumonia may be a sacred cow. Not only can the results lead to serious mismanagement of the patient, but also sputum cultures represent one of the largest workloads and expenses for the hospital bacteriology laboratory." Correlation between results of cultures of transtracheal aspirates and results of qualitative and quantitative cultures of the sputum has been poor (25), except in patients producing sputum containing fewer than 25 squamous epithelial cells per ×100 field microscopically (23). There is, therefore, no completely satisfactory method for isolating bacterial pathogens from expectorated sputum at the present time. Under no circumstances is anaerobic bacterial culturing of sputum performed. A Gram-stained preparation should precede all sputum culture efforts. Fewer than 25 epithelial cells per low-power field or at least equal numbers of pus cells and epithelial cells, preferably with a predominance of neutrophiles, are an absolute requirement of sputum specimens for cultural analyses.

Lower respiratory secretions collected by nasotracheal aspiration with a catheter may also become contaminated with upper respiratory flora. Another method, which bypasses oropharyngeal contamination completely, is percutaneous transtracheal aspiration, a relatively harmless procedure that can be performed rapidly with local anesthesia (7). This procedure is recommended for seriously ill or comatose patients who have pneumonia and cannot raise sputum or who are suspected of having anaerobic pleuropulmonary infection. Serious complications are rare when an experienced and qualified physician performs this procedure; nonetheless, patients must be followed carefully. Since the saline itself may be bactericidal, any specimen so collected should be cultured promptly; if this is not feasible, a balanced salt solution, such as lactated Ringer solution, should be used instead (47). Specimens should be transported promptly to the laboratory or injected into an anaerobic transport tube or vial. It is recommended that a Gram-stained smear of the material collected by transtracheal aspiration be examined and that aerobic bacterial cultures be performed with all possible speed. In some instances, culturing for anaerobic bacteria, mycobacteria, and fungi may also be indicated. Direct staining of such aspirates with specific fluorescent antibodies may hasten the diagnosis of diseases such as legionellosis; a search for *P. carinii* may also be indicated, provided the appropriate clinical impression has been communicated to the laboratory staff.

Culture of early-morning freshly expectorated sputum or of expectorated sputum induced by a heated aqueous aerosol of 10% glycerol and 15% sodium chloride, followed in about 1 h by a gastric washing (13), is useful for recovery of mycobacteria and fungi. Gastric washings obtained after induced coughing may have more mycobacteria (13). A series of such collections may be desirable. Ideally, the specimens should be cultured on the day they are collected. Pooled specimens are sometimes collected for culture; however, to prevent overgrowth of contaminants, it is necessary to refrigerate each specimen until the pool is processed. Twenty-four-hour sputum collections are unnecessary and, in fact, undesirable.

In some instances, the etiology of pulmonary infections may be diagnosed by percutaneous transthoracic needle biopsy of the lung, thoracentesis, needle aspiration of an abscess or empyema cavity, or open lung biopsy. When open biopsy is performed, multiple specimens should be obtained from different sites if the lesion is large or if multiple lesions are present. In addition, a portion of an abscess wall should be removed, as well as a sample of the pus within (66).

Histological examination serves two purposes: to determine whether the lesion is inflammatory and, if so, to provide clues as to the nature of the infectious agent. These clues may, in turn, lead to further histological studies with special stains and to additional special cultures. Open lung biopsy may be necessary to establish the diagnosis of infection due to *P. carinii* or *L. pneumophila* in immunocompromised hosts, and it may be helpful in either establishing or confirming infections due to mycobacteria, fungi, and viruses. In experienced hands, the diagnosis of pneumocystosis may be made rapidly by appropriately staining material obtained by bronchial or transbronchial biopsy (14).

Specimens obtained by bronchoscopy have generally not been suitable for bacterial culture because during the procedure the lower respiratory tract is inevitably contaminated with flora indigenous to the upper respiratory tract. It has been suggested that by inserting a distally occluded catheter with telescoping cannulas it is possible to minimize bacterial contamination of upper respiratory origin (73). At this time, it appears that the results of quantitative cultures of the protected catheter brush are more clinically useful than those obtained from qualitative cultures; however, it would appear prudent to culture bronchial material for bacteria on a selective basis only and to restrict anaerobic cultures to special situations.

Skin

Caution must be exercised with specimens alleged to originate from skin (64). In the absence of lesions that can be cultured readily, these specimens are of little help, at best. Frequently they are misleading, especially if the request does not specify examination for dermatophytes. In that instance, proper scrapings should be obtained (see Chapter 48). Routine cultures of cellulitis or rashes are usually of little help. Specimens from various and sundry lesions must be obtained after the area has been properly disinfected. Cultures and smears as well as media for virus isolation should be inoculated, especially if the skin lesions are closed and aspiration with a sterile syringe and needle can be performed by a qualified clinician. When dealing with productive lesions, surface material should be discarded, and exudates from the interior should be used for microbiological analysis. Submission of smears of such material should be mandatory. Microbiological monitoring of burns requires identification of the usually polymicrobic microbiota and should be accompanied by frequent blood cultures, especially if the injury is extensive.

Urine

Urinary tract infections encompass diseases that involve areas ranging from the kidney to the urethra, with the urethra and bladder most commonly affected. Infections are classified as uncomplicated (medical) when there are no anatomical or neurological abnormalities present and as complicated (surgical) when residual inflammatory changes, obstructive uropathy, calculi, or neurological lesions are present (37). Infections usually arise via the urethra by the ascending route and, less commonly, hematogenously. In most hospitals, urinary tract infections represent the most common form of nosocomial infection (60).

Females of childbearing age seen for the first time with symptoms of uncomplicated urinary tract infection are nearly always infected with *E. coli*. These strains are usually susceptible to antimicrobial agents, including sulfonamides, which are active against gram-negative bacilli. Because a high degree of correlation exists between urinary sterilization and in vitro susceptibilities (59), some argue that it is unnecessary to obtain cultures from these patients. However, the prevailing viewpoint is that it is essential to obtain urine cultures during therapy. According to Stamey et al., "Better medicine is practiced, at less expense to the patient, if a culture is repeated 48 to 72 hours after starting therapy than if a useless antimicrobial agent is continued for ten days in the face of ineffective therapy" (58). Urine cultures obtained during therapy should be sterile, since the symptomatic response of the patient is a poor indicator of successful therapy. Pyuria is not a totally reliable index of the presence of disease because it may be absent during the process or may occur in disorders other than bacterial infection, such as extreme dehydration, trauma secondary to instrumentation or calculi, chemical inflammation, renal tuberculosis, acute glomerulonephritis, and nonbacterial gastroenteritis and respiratory infections.

The reliability of a culture of a single, clean-voided specimen is about 80% in the female and virtually 100% in the adult male if he is circumcised or has carefully retracted the foreskin and cleansed the glans. In females, the reliability of a clean-voided specimen increases to 90% and nearly 100% if two and three specimens, respectively, are obtained and contain the same organism (37). Obviously, in symptomatic patients one specimen only is usually collected before therapy is initiated; however, in asymptomatic patients two or three specimens should be collected to document the presence of bacteriuria.

Guidelines for the diagnosis of urinary tract infections and minimal diagnostic criteria have been presented in detail by Kunin (37). In most females, the clean-voided midstream method of specimen collection is satisfactory, provided that the patient is capable of cleansing herself and collecting her own specimen or that the nursing personnel, in the case of the bedridden patient, clearly understand printed or stated instructions (37). All too frequently, instructions are given casually. The result is a grossly contaminated specimen. In addition, the task of obtaining a specimen from the bedridden patient is frequently assigned to the most inexperienced hospital aide or orderly. Hospitals should be encouraged to establish urine collection teams or select individuals who are trained in the proper methods of urine collection and transport.

Urine collection for culture by urethral catheterization is seldom indicated, except in those cases in which catheterization must otherwise be done for diagnostic or therapeutic reasons. There is a small risk of infection after urethral catheterization which varies according to the type of patient catheterized. Because of normal urethral colonization with bacteria, it may be difficult to determine whether organisms isolated from a catheterized urine sample are of urinary or urethral origin.

Contamination of the urine by urethral or introital microflora may be obviated by suprapubic aspiration. This is a rapid, safe, and simple technique when performed by a qualified physician. Suprapubic aspiration is indicated in patients with clinical evidence of urinary tract infection but in whom bacterial counts in clean-voided specimens are low and, therefore, indeterminate, in neonates, in young infants, in patients in whom catheterization may be contraindicated, and in those with suspected anaerobic bacteriuria. The patient should have a full bladder at the time the procedure is performed. After the skin has been properly disinfected, a 19- or 20-gauge needle attached to a syringe is passed through the skin in the midline at a point approximately one-third the distance from the symphysis pubis to the umbilicus. Urine is aspirated into the syringe.

In patients with chronic indwelling urethral catheters attached to closed drainage, urine is collected for culture by disinfecting with a suitable agent the wall of the catheter at its juncture with the drainage tube and puncturing it with a 21-gauge needle attached to a syringe into which the urine is aspirated. It is possible to puncture such a catheter repeatedly without leakage occurring from the puncture sites. The connection between the catheter and the drainage tube should not be broken for specimen collection, nor should material for culture be taken from the drainage bag.

During the course of cystoscopy, ureteral catheterization, or retrograde pyelography, urine may be collected for culture. In the case of an obstructed ureter, bladder urine may be sterile, whereas the urine proximal to the obstruction may be infected; hence, the urologist will often collect several specimens and request that each be handled separately.

Transport and storage of urine specimens are important adjuncts to the reliability of culture results. A sterile, screw-capped container or tube should be used. Urine is an excellent culture medium, and a small or insignificant number of bacteria can multiply rapidly to a significant number unless certain precautions are taken. If the urine cannot be cultured within 1 h after its collection, it can be refrigerated with satisfactory results for at least 24 h and, according to one report (39), for as long as 5 to 10 days. The wisdom of prolonged refrigeration has recently been questioned, and refrigeration should be used only if absolutely necessary.

Significant bacteriuria occurs when there are 100,000 colonies or more per ml in a clean-voided, midstream specimen obtained from asymptomatic patients. However, for acute dysuria and frequency in young, sexually active women, a colony count of 100 CFU/ml may be a useful criterion (44). In such cases, it is imperative to notify the laboratory so that procedures appropriate to the detection of as few as 100 CFU/ml are used. It must be emphasized that this criterion should not be extrapolated to populations that have either a substantially greater likelihood of contamination of voided urine specimens or a lower prevalence of infection (44). Lower counts in a catheterized specimen require reculturing for confirmation of the results. Any count in a suprapubic aspirate is significant, although recovery of skin contaminants, such as diphtheroids and staphylococci, may not be significant and may require repetition of the procedure to confirm or rule out their presence in the urine. It must be recognized, however, that bacteriuria may be low on a diurnal basis when the patient is well hydrated, if the urine pH is below 5.0 or its specific gravity is less than 1.003, in the presence of an antimicrobial agent, with complete ureteral obstruction, and in chronic pyelonephritis. The role of cell wall-defective variants in urinary tract infections has not been established clearly. Anaerobic bacteriuria, although not uncommon in clean-voided, midstream specimens, has been only rarely confirmed by suprapubic aspiration (54), which represents the only reliable means by which to establish this diagnosis.

Chemical and other newer screening tests notwithstanding, significant bacteriuria may be established with roughly 90% accuracy by microscopic examination of a Gram-stained smear of well-mixed, unspun urine or of a wet preparation of unstained urinary sediment. The presence of at least two bacteria per ×1,000 microscopic field of the Gram-stained smear signifies bacteriuria of 100,000 or more colonies per ml (45).

Candiduria is abnormal; its quantitation is of little importance and may even be misleading if standards currently applicable to bacteriuria are applied (52). What is important is its relationship to the clinical picture, and it may serve not only as a warning of generalized candidiasis but also as an indication of the presence of an underlying disease. *Candida* spp. are frequently grown on conventional bacteriological media; however, their recovery is improved, particularly when they are present in small numbers mixed with bacteria, on fungal agar media containing antibiotics, e.g., inhibitory mold agar. A microscopic examination of a wet preparation of urinary sediment may serve as easily and inexpensively to demonstrate the presence of yeasts and pseudohyphal elements.

Pyuria without bacteriuria indicates the possibility of renal tuberculosis. There may be gross or microscopic hematuria, and secondary bladder involvement may produce symptoms of cystitis. Three consecutive clean-voided, early-morning specimens should be collected for cultures. Twenty-four-hour urine collections are undesirable because of the frequent overgrowth of bacteria other than mycobacteria.

Finally, a urinary calculus should be split so that cultures can be made of its interior, as well as of its surface, since it is not uncommon to isolate bacteria from within the stone which are not present on its surface (41).

Venereal disease

Until recently, concern with gonorrhea has dominated the field of venereal disease (33, 64). At this time, the only diagnostic test is the microbiological demonstration of viable gonococci in the various exudates and lesions which are present in persons with this disease. In the male, the microscopic observation of diplococci resembling gonococci in a Gram-stained smear of a urethral discharge remains unproblematic. There is excellent correlation between the presumptive diagnosis of uncomplicated gonorrhea in the male based on the microscopic examination of a Gram-stained smear of urethral exudate and the confirmation of the diagnosis by the culture and identification of the organisms by standard techniques. On occasion, individuals attending a clinic or consulting their physicians may exhibit a purulent discharge with little or no evidence of microorganisms. In these instances, the discharge or the urethral canal should be sampled for culture inoculation. Direct inoculation of Thayer-Martin (TM) medium or modified TM medium (38) is advisable in situations where an inoculated plate can be immediately transported to the laboratory. If the history and symptoms suggest exposure but only a small urethral discharge can be obtained, urethral scrapings should be obtained by the clinician with a small swab or a smooth, small wire loop; the secretions should be cultured appropriately, and a smear should be examined. If it is known or suspected that the patient is homosexual, oropharyngeal and rectal cultures are indicated. TM agar or modified TM agar plates should be inoculated directly. Recent reports of cryptic carriage of gonococci in the distal urethra of clinically asymptomatic men exposed to infected women make the culturing of this site with bacteriological loops inserted into the urethra for a short distance a significant test.

Obtaining an adequate specimen in female patients may present greater problems (20). The presence of a cervical discharge is helpful in the diagnosis. However, the Gram-stained smear obtained from such a discharge cannot be relied upon to establish the absence or presence of the gonococcus, and cultures should be made without exception. A cervical culture and, when possible, an anal culture have been recommended. If pelvic examination suggests that the vaginal glands or the urethra is involved, swabs from these areas should also be cultured for the gonococcus on modified TM medium. Should transport to the laboratory be delayed, modified TM medium may be employed.

Skin lesions, especially in the female, have been observed with increasing frequency. Very often they appear at the same time as pudendal involvement. Cultures of scrapings from the base of such lesions have been successful, but they are positive in a comparatively low percentage of the cases. Careful uprooting of the skin lesion with a sterile needle should be employed for obtaining specimens from these sites. Staining of smears of material from such sites with *Neisseria gonorrhoeae* fluorescent-antibody conjugate is often helpful in establishing a rapid, presumptive diagnosis. The report should state how this presumptive diagnosis was determined.

When a diagnosis of gonococcal arthritis is suspected, properly obtained specimens should be cultured on a blood agar plate and a noninhibitory plate of chocolate agar. Gas chromatography of synovial fluid appears to be a rapid and useful technique in the laboratory diagnosis of gonococcal and other arthritides. Caution must be exercised before needling such joints, with particular emphasis on adequate skin preparation. It is advisable that smears be made from the aspirated material and that it be cultured also for organisms other than the gonococcus. The use of hypertonic media for the detection of cell wall-deficient variants is advisable. Finally, when gonococcemia is suspected, the blood culture bottles should be subcultured minimally at 1-, 2-, and 4-day intervals on noninhibitory chocolate agar.

Nonselective enriched chocolate agar, TM medium, or modified TM medium is advocated for cultures of the exudate in ophthalmia neonatorum. Endocervical cultures of the mother should also be obtained immediately.

The immediate inoculation and proper incubation of the culture medium should be stressed by the clinical microbiology laboratory. It is comparatively easy to indoctrinate clinic and hospital personnel in the proper fashion of carrying out this procedure. Illustrative materials, available from the Centers for Disease Control, are most helpful in presenting the proper procedures. The microscopic observation of bacteria on smears prepared carelessly is often obscured by the thickness of the material allowed to dry on a slide. Ward and clinic personnel should be reminded continuously that thinner smears are more helpful, instructed how to prepare slides in a proper fashion, and shown the difference between thick and thin smears for the purposes of recognizing gonococci.

It is also necessary to instruct and inform physicians, nurses, and attendants that the gonococcus can remain viable for some time in certain body fluids and that precautions to protect themselves and laboratory personnel handling the material are required. Containers contaminated on the outside, swabs not properly transported in enclosures, and smears dripping with discharge should not be handled in the laboratory without all individuals using every precaution to prevent personal contamination.

Dark-field demonstration of *Treponema pallidum* from primary and secondary lesions of syphilis or the demonstration of *T. pallidum* with specific direct immunofluorescent conjugate are the direct microbiological examinations that may be requested of a clinical laboratory. These procedures are useless if performed on material obtained and transported to the laboratory. Patients with such lesions must be examined within the vicinity of a dark-field microscope. Proper preparation of these lesions, the preparation of the slide, and the recognition of the treponemes are not within the scope of this chapter. However, it should be evident that, because of the infrequency with which this examination is performed in the routine clinical microbiology laboratory, only very experienced personnel should be assigned the task of identifying *T. pallidum* in the dark-field preparations made from lesions. It may even be advisable to send patients to a laboratory where the test is performed routinely. The misdiagnosis of syphilis can be a traumatic and damaging experience to the patient and must be avoided at all costs.

Chancroid is rarely reported. Recently, Hammond et al. have shown that the isolation of *Haemophilus ducreyi* from lesions can be readily accomplished by applying a swab sample to supplemented chocolate agar containing 3 μg of vancomycin per ml. Proper attention to incubation is necessary (26). Instructions and transport vessels for the isolation of significant viral agents such as hepatitis virus and herpesvirus hominis II, chlamydiae, or ureaplasmas are given in Chapters 81, 64, 85, and 36, respectively.

Wounds, boils, abscesses, etc.

See reference 64.

Material from a previously undrained abscess, if properly collected and transported to the laboratory, should contain the etiological agent(s) of disease in most instances. An opened wound, ulcer, or sinus tract, however, frequently becomes contaminated with skin, mucosal, or airborne microorganisms. In general, the use of a swab to collect material from these sites is of limited value, since the amount of material supplied for examination is not only very small but also likely to represent an inadequate sample. Moreover, nothing is more worthless than a dry swab from a dry lesion. In chronic, localized lesions the number of organisms present may be small. A sterile needle and syringe should be used to collect a generous quantity of liquid material. Since anaerobic bacteria are commonly recovered from certain wounds, such material should be sent to the laboratory immediately or, preferably, injected into an anaerobic tube or vial for transport. Since sinus tracts often originate in bone or lymph nodes, the orifice of the tract should be cleaned thoroughly with a suitable antiseptic; then curettings of the lining of the tract should be taken as close to the base of the tract as possible. Ulcerative lesions of the skin and mucosa should be biopsied or, if they are small, excised for histopathological and microbiological studies. Irrigation, intravenous, or intraarterial catheters should be carefully removed after their entry site in the skin is disinfected; the tips should be cut off aseptically and submitted in a sterile container to the laboratory for culture.

Repeated cultures of open draining wounds or of large areas of devitalized tissue will frequently yield a profusion of microorganisms. Their significance is uncertain, their numbers and variety tax the laboratory, and their presence is unlikely to be affected by rational systemic antimicrobial therapy. Repetitive antimicrobial susceptibility testing of isolates from such cultures not only wastes the laboratory's time and the patient's money, but may also provide misleading results, inasmuch as attempts at therapy may simply favor superinfection by resistant microorganisms.

Although acute infections are generally recognizable clinically and their etiology is usually established by conventional bacteriological procedures, the recognition and establishment of the etiology of chronic localized infections are not so readily accomplished. Whereas staphylococci, streptococci, *Enterobacteriaceae*, and *Pseudomonadaceae* predominate in the former type of infection, actinomycetes, brucellae, mycobacteria, and fungi must be considered in the differential diagnosis of the latter. As we have already emphasized, biopsy and histopathological examination are valuable adjuncts in the diagnosis of the latter type of infection. The importance of microscopic study of the tissue is emphasized further by the occasional resemblance between neoplasms and infectious processes.

Burn wounds represent a somewhat unique problem because they are initially contaminated with staphylococci, streptococci, and clostridia and subsequently with such opportunists as *Pseudomonas* spp., *Klebsiella* spp., *Providencia* spp., *Candida albicans*, zygomycetes, and viruses. The burn eschar is a nonviable, devascularized structure which provides a milieu favorable for microbial growth. Quantitative cultures of biopsies of black, degenerated, or unhealthy areas of the burn or of drainage from burn wounds may be helpful in establishing the etiology of local or invasive infections. Gram-stained smears of any drainage may be helpful in providing an early clue to the nature of the infectious process. Such patients must be monitored with blood cultures at frequent intervals.

Quantitative smears and cultures of tissue from traumatically acquired wounds, taken after surgical debridement, are useful in predicting the risk of wound sepsis and, therefore, in helping the surgeon decide whether or not to close the wound primarily.

TRANSPORT OF SPECIMENS TO THE LABORATORY

Inefficiency in transporting specimens to the laboratory (29) after they have been obtained from the patient is a major problem and becomes a convenient excuse of laboratory workers for not finding organisms whose presence is suspected by the clinician. When a specimen is lost, the loss becomes a convenient excuse for the clinician who has left the specimen with allegedly responsible personnel who, in turn, have instructed others to transport it to the laboratory. Although this activity is not directly under the jurisdiction of a clinical microbiology laboratory, every effort must be made to control its operation. The best possible way to obtain acceptable results is to insist that transportation time be as short as possible. Some laboratories refuse to accept specimens if they have been in transport too long. Certain specimens should be transported in a medium or vehicle which preserves the organisms in the specimen and helps to maintain the ratio of one organism to the other in the specimen. This is especially important for those specimens in which normal microbiota may be admixed with bacteria or other microorganisms foreign to the location. It becomes even more necessary to use a transport medium if significant microorganisms are present in very low numbers. House medical staff and nursing supervisors should remind appropriate personnel as often as required that prompt and proper transport of specimens from the clinics or patient floors to the laboratory is important. Many laboratories have established a policy of refusing to process specimens that have been handled improperly.

Microbiological specimens may be transported to the laboratory by various means. A variety of media may be used directly for this purpose. A nonproliferating, buffer type of transport medium such as that of Stuart, Toshach, and Patfula may be employed; however, the charcoal transport medium is preferred by some (29). Ordinary nutrient broth or anaerobic broth may be used when swabs or aspirates are involved. In other instances, microbiologists insist that, for example, throat cultures for streptococci be inoculated just as the swab is obtained onto blood agar plates and that the inoculations be made near the periphery over an area ca. 2 cm in diameter. The swab used for obtaining the specimen is then broken into a fluid transport medium and submitted in conjunction with the blood agar plate. Modified TM medium (38) is recommended for gonococcal specimens when they are to be transported to a laboratory. For the isolation of *B. pertussis*, direct inoculation of Bordet-Gengou agar is advocated. Many laboratories do not accept stool specimens for culture unless they are transported in various buffered preservatives. Preparation of these transport media is described in various manuals (see Chapter 111). One of the best and most up-to-date references, which should be consulted when problems arise concerning transport of specimens, is Public Health Publication number 876, revised September 1973, entitled *Collection, Handling and Shipment of Microbiological Specimens*. The major points advocated in this manual may be summarized as follows. (i) Aspirated specimens, such as fluids, etc., should be collected with a sterile needle and syringe instead of with swabs which permit rapid drying and subsequent loss of viability of the bacteria. Such specimens should be transported promptly to the laboratory or injected into an anaerobic transport tube or vial. (ii) Microorganisms survive better if 0.5% agar is added to media such as thioglycolate, chopped meat, etc. (iii) Smears for staining should be prepared at once, especially when the material is from pathological lesions which have not been aspirated with a syringe and needle and from which a large quantity of material cannot be obtained. It is important for adequate and proper smears to be prepared at the bedside by clinicians who fully understand the significance of this procedure and are aware that thick smears are of very little help. Thin smears can be prepared from material that seems thick and tenacious by firmly pressing two slides together over the areas containing the applied specimen. This method is recommended for urethral discharges, wound exudates, aspirated material from abscesses, etc. By simply sliding the slides apart, thin areas of smear are made which are useful as a guide for the clinical microbiologist in the selection of media for primary culturing.

Special treatment is required for specimens when the etiological agent is thought to be anaerobic. The material is aspirated with a needle and syringe. It may be practical to remove the needle, cap the syringe with its original seal, and bring the specimen directly to the laboratory. No more than 10 min ought to be allowed for the transport of such specimens; otherwise, such specimens must be injected immediately into an anaerobic transport tube or vial. The laboratory must be alerted that a specimen for anaerobic cultures is on the way. If prereduced media are available, they must be inoculated as soon as the specimen is obtained. It is the responsibility of the microbiology laboratory to ensure that personnel using this approach fully understand the method of inoculation. Swabs for anaerobic culture are usually less satisfactory than aspirates, even when transported in an anaerobic tube.

Transport of urine specimens to the laboratory presents some problems. The urine must be obtained properly and transported and processed as soon as possible. No more than 1 h should elapse between the time the specimen is obtained and the time of incubation. If this time schedule cannot be followed, the urine specimen must be refrigerated immediately. If transport to the laboratory takes more than a few minutes, the urine specimen should be placed in an insulated bag containing scotch ice. Upon delivery to the laboratory, the urine specimen should be cultured immediately or refrigerated again. In any event, it is not advisable to refrigerate urine for longer than 18 to 25 h before culture under routine circumstances (see below, Inoculation of Media and Colony Isolation Methodology) (5).

All laboratory personnel must be aware that submitted material which is not in a supportive environment must be handled immediately, especially spinal fluid, urine, pus, or any material not transported in a preservative. All smears should be fixed and stained immediately. In addition, laboratory personnel doing the primary culturing of specimens should acquaint themselves with the clinical diagnosis and any other pertinent laboratory findings, especially with select specimens such as blood, spinal fluid, wounds, burns, etc. Thus, hematological and chemical data, immunochemical findings on levels of immunoglobulins, complement, etc., may frequently be helpful in interpreting the findings of the laboratory and planning for the isolation of certain specific microorganisms. Laboratory personnel must also remember to handle all clinical specimens and cultures with appropriate attention to laboratory safety for themselves and coworkers.

RECORDING OF LABORATORY SPECIMENS

The person collecting the specimen is responsible for ensuring that the name, registration, and location of the patient are correctly (and legibly) written or imprinted on the culture request form and that this information corresponds with that written or imprinted on a label affixed to the specimen container (29). The type or source of the sample and the physician's choice of tests to be performed must be specified on the request form. Errors in specimen identification may have disastrous consequences and must be avoided. Correct identification of the specimen, however, continues in the laboratory, where different types of cultures, subcultures, and other appropriate tests and procedures may proliferate and, ultimately, all results are compiled into one or more reports. Multiple specimens from the same patient may be collected for diagnostic purposes or for following the progress of therapy; for these specimens, sample source, timing, and type of test performed must be clearly recorded.

In most laboratories, a system of accession numbers expedites record keeping. There are undoubtedly many variations of the system at all levels of sophistication, depending on whether human or machine systems are employed. Since off- or on-line computer systems are still relatively uncommon in clinical microbiology laboratories, most accessioning systems are human and, therefore, subject to an incidence of random errors from 1 per 100 operations to 1 per 1,000 operations (36).

One approach to accessioning is the use of sequentially numbered self-adherent labels produced with sufficient replicates so that one can be affixed to each copy of the culture request form, the specimen container, work card(s), each of the primary plates or tubes of media inoculated, and the average number of plates or tubes of media used in subcultures. Further categorization according to the type of specimen and the year it was received can be achieved by color-coding the labels and using a numbered prefix to indicate the year of accession, followed by one or two letters to designate the type of specimen received. A list of accessions by laboratory and specimen type is helpful in documenting whether or not the specimen was received and can be used for rapid location and tracking of work in progress. Such a listing may also be used to tabulate the monthly and annual workloads received by the laboratory. These figures enable the laboratory director to anticipate requirements for technical assistance and space (8).

In addition to the specimen records discussed above, the dates and times the specimen was collected and received should be recorded on the request forms. If the specimen was received in an unsatisfactory condition, this fact should be noted in the accession list, on the request forms, and in an incidence record. The clinician of record must be notified by phone and in writing. Compliance with these standards not only is required by laboratory inspection and accreditation agencies but also provides data that are useful in the delivery of proper patient care.

Most laboratories use some type of work card on which a record of the work in progress is noted and updated as each step is completed. If the work card format is systematized, an individual who is unfamiliar with a particular culture can answer inquiries about it before the final report is ready. Inspection of the work card to determine the appropriateness of the tests performed and the accuracy of their interpretation is an important responsibility of the laboratory director when the final report is prepared.

Final reports must be retained in the laboratory at least until the results are recorded in the history of the patient and in keeping with legal and accreditation requirements. Retention beyond this point is not practical unless it fulfills a specific purpose, be it epidemiological, investigational, or medico-legal, since record storage is expensive; retention of records is unnecessary if the data are also a part of other permanent records. Microbiological results provide the data base for the epidemiology program of a hospital and should be organized in a manner which facilitates analysis of infection rates by medical or surgical services and hospital location. This organization will enhance detection of hospital endemics or epidemics. Antimicrobial susceptibility data should be tabulated by organism and by antimicrobial agent on a yearly basis to detect trends in susceptibility and provide clinicians with the information needed for selecting initial therapy for serious infections. Several different computer systems address these problems. Unfortunately, experience to date is still too limited to warrant discussion or recommendation.

Many variations of specimen recording are extant. The essential features of the approach described must be part of all record-keeping efforts. It should be stressed that the record must reflect, in addition, the various steps which comprised the analytical process, to ensure accuracy and detection of aberrant reactions while enabling other laboratorians to follow the decision-making process at a later date. All quality controls exercised in the interest of the particular specimens must be recorded. Although it may not be an absolute requirement that each specimen record reflect this concern, the various controls exercised must be logged to provide the reassurance of proper materials, reagents, media, etc.

PREPARATION OF SPECIMENS FOR PRIMARY CULTURING

A variety of considerations must be entertained by the technical personnel when any specimen is delivered to a laboratory: the source of the specimen, the transport medium or status of the specimen, the accompanying smear or the nature of the specimen from which a smear is to be made, and the established routine for the laboratory in general. All of these matters must be brought into play for the technical personnel to determine the steps to be taken in preparing specimens for primary culture (29).

The section on Collection of Specimens (above) lists a variety of anatomical sources from which laboratory specimens may be obtained. It seems reasonable that in any laboratory all of the specimens should be cultured on a certain series of microbiological media. Which media will be used in any given laboratory is left to the experience of the director or supervisor of the clinical microbiology laboratory. The consideration of possible anaerobic microorganisms in a particular specimen must be accompanied by the use of media capable of supporting growth of anaerobic bacteria. Another important factor is the antimicrobial therapy history of the patient from whom the specimen was obtained. The physician submitting the specimen should provide the microbiologist with a working diagnosis, so that the latter may then consider the addition of special media which might help isolate those microorganisms consistent with the clinical impression. If the specimen contains large or inhibitory amounts of the antimicrobial agent administered, it may be necessary to dilute the specimen.

If the organisms suspected of being present are not frequently encountered in the particular laboratory, a representative of such microorganisms should be cultured in duplicate media to ensure that proper support and conditions for their proliferation are provided and to permit comparison of cultural characteristics when the unknown organism has grown. The use of such control cultures has been advocated by some for years (27). It is a helpful measure which frequently aids in a speedier recognition of organisms encountered only rarely.

As mentioned earlier, certain specimens require dilution. This is especially true of urine specimens submitted for quantitative microbiological analysis. There are many ways of ascertaining the number of bacteria per milliliter of urine. Whichever the method chosen, it must be of proven efficacy and accuracy when performed in the laboratory. The most common approach is the use of a calibrated loop that delivers a specified volume of urine to an agar plate, which is then streaked. Another approach is diluting the specimen through several tubes until a 1:100 dilution has been obtained. A small portion, usually 0.1 ml, is then delivered to the plates used for assessing the number of bacteria, and this small volume is then spread across the plates with a glass spreader, with a bacteriological loop or needle, or with the pipette itself. Finally, dilution by the shake-agar technique is used occasionally. Various dilutions of the urine specimen are added to melted, measured volumes of agar. The agar and urine are mixed, and the mixture is delivered into a petri dish where it is allowed to harden (5). A variety of automated, semiautomated, and screening devices are available for the detection of bacteria in urine; their advantages have recently been reviewed (43).

When a piece of tissue is submitted for microbiological analysis, it should be cut into small pieces or, if available, a tissue grinder should be used for maceration (see above, Collection of Specimens). If the material submitted is very viscous, it may be treated with sterile *N*-acetylcysteine or an equal volume of 1.5% amyl acetate, both of which have been shown not to inhibit microorganisms. This treatment is very helpful with certain body fluids and sputum.

Before primary culturing is begun, all media and equipment must be well organized on a clean bench top. A sequence of the steps to be performed should be established so that the inoculation procedure can be executed smoothly. Proper entries in workbooks, on worksheets, or on patient records should be made to indicate that each particular step has been executed. Adequate safeguards should be within easy reach; a disinfectant solution must always be on hand. The proper placement of a Bunsen burner, a microincinerator, loop carriers, or sterile swabs, if they are to be used, is a prerequisite for efficient working. Whenever possible, safety cabinets should be used. These are mandatory if aerosolization may occur during the processing of the specimen.

INOCULATION OF MEDIA AND COLONY ISOLATION METHODOLOGY

Many methods of inoculating various media are available to the laboratory (29). The bacteriological inoculating loop is used most often, particularly when a homogeneous specimen is available. When broth cultures or fluid specimens are transferred to media, sterile cotton or synthetic fiber swabs, Pasteur pipettes, calibrated pipettes, syringes, and needles may be used for distributing them to various preparations of several solid and liquid media. In general, it is unwise to utilize these various transfer devices for spreading the culture across the entire agar plate. Most laboratories, therefore, plant a portion of specimens on an area near the periphery of the plate(s). After the specimen has been delivered to all of the various substrates, it is then further distributed by streaking it across the agar in a fashion that permits an ongoing dilution of the specimen. This is best accomplished with a bacteriological loop. The specimen is spread back and forth over the surface of perhaps one-quarter to one-third of the plate, with the bacteriological loop held loosely between the thumb and index finger. The weight of the bacteriological loop is allowed to exert its own pressure. There is no need to exert additional pressure with the hand. After the first quadrant has been inoculated, the plate is turned 90°, and the first few cross-streaks touch the original inoculum. After that, the laboratory worker continues to inoculate the surface of the agar, avoiding the first inoculated quadrant. When another quarter of the plate is covered, it is once more turned 90° at the end, thereby permitting a small area to be inoculated after the fourth turn has been made. This is to ensure the presence of isolated colonies. An important caveat must be stated. Familiarity with the inoculation procedure often leads to a rapid performance of this task. Personnel must contain their enthusiasm sufficiently to avoid hitting the sides of the petri dish, since such collisions create aerosols, potentially dangerous to personnel and laboratory reagents.

The microbiologist must keep in mind the purpose of the medium being inoculated. For example, highly selective media require a heavier inoculum, and one drop of the specimen may be inadequate to yield growth of the microorganism in question. Therefore, for those media, it may be appropriate to utilize a swab to inoculate the first quadrant of the plate. Resorting to a bacteriological loop thereafter is helpful in obtaining isolated colonies. Familiarity with the composition and purpose of each medium plus experience is the ultimate way to master this aspect of the procedure.

If the purpose of the procedure is to obtain an evenly distributed inoculum of a diluted specimen, a bent glass rod may be used. This rod is designed to make approximately a 45° angle. The side used for inoculation usually covers the radius of a standard petri dish. This sterile glass applicator is placed over the inoculum, and the plate carrying the agar is rotated until an even layer of liquid is above the agar. Care must be exercised that the inoculating volume is proper (usually 0.5 to 1.0 ml).

The special requirements for inoculating specimens for anaerobic cultivation are discussed in Chapter 37 and *Cumitech 5* (22). Animal inoculations are performed with decreasing frequency in the average clinical microbiology laboratory. It would be useful to consult standard texts describing this procedure should the need arise. As a word of caution, individuals not experienced in performing animal inoculations should send the specimens to central or reference laboratories which are equipped to provide this service. It is also important that animal facilities and care meet prescribed regulations.

Occasionally, there is a need to use the pour plate technique. Pour plates are used infrequently in the quantitation of urinary microorganisms, but the method is often used to quantitate the microorganisms in environmental cultures. For this type of inoculation, sterile pipettes are required. If dilutions are to be made, tubes containing measured amounts of sterile diluent are needed. A Bunsen burner is usually required. The other types of inoculations can all be accomplished with microincinerators, a device safer and easier to use when pathological material is involved.

If the specimen for primary inoculation is of very tenacious or friable material and is to be cultured for fungi, such as skin scrapings, nail scrapings, hairs, etc., it may be necessary to use a small sterile scalpel to cut the specimen into smaller pieces on the surface of an agar medium. This will permit the pathological specimen to be submerged into the agar and further inoculated onto the plate by using the scalpel to make additional cuts in the medium.

The inoculation methods described are to be used only for the inoculation of primary isolation media.

Other methods are appropriate for transferring laboratory-grown microorganisms; these are not covered here.

MEDIA FOR PRIMARY ISOLATION

At the risk of redundancy, it must be stated once more that the purpose of using a number of different solid and broth media is twofold. It permits the isolation of many of the various microbial agents in a specimen. At the same time, a number of media select specific members of the microbial population by suppressing other microorganisms present in the specimen.

Table 1 represents the recommended media and Gram stain requirements for primary bacteriological cultures as used at the Mayo Clinic, Rochester, Minn. It is reproduced here as a guide to the selection of media as practiced in one of the outstanding institutions in the United States. It is obvious that for each medium chosen, others may be substituted. It is equally obvious that the choice of media may be expanded considerably. This recommendation encompasses the requirements for the isolation of microorganisms usually looked for in these specimens and is in no way all encompassing. Additional media must be chosen when information concerning specific microorganisms is available from the clinician. This guide is similar to others that have been published by various medium manufacturers such as BBL Microbiology Systems and Difco Laboratories or advocated in texts (27, 28, 64, 71), etc.

Detailed discussion of the many media for primary isolation cannot be included here. It must be remembered that the choice of which medium to use for a specific purpose represents perhaps best the mystique of clinical microbiology, a mystique compounded by habit, preferences of favorite teachers, personal experience, and occasionally solid data. The choice is also determined to a degree by the availability of commercially prepackaged agar plates and broth.

INCUBATION OF INOCULATED MEDIA

After inoculation on proper substrates, cultures should be incubated at an optimal temperature as quickly as possible. Delay in bringing certain organisms to the desired temperature may hamper their ability to grow in the desired period of time. Turbidity or other manifestations of growth in broth will be impaired. Overgrowth by undesirable microorganisms will be encouraged, especially in liquid media. The question of proper temperature has preoccupied clinical microbiologists for some time. Although it is theoretically desirable to come close to body temperature for many of the organisms involved in disease production, a number of microorganisms cannot tolerate the optimal temperature for humans. Therefore, most incubators in clinical laboratories are set at 35°C, which permits the growth of virtually all organisms capable of existing at 37°C. A single incubator set at 35°C is obviously inadequate for many of the temperature requirements of a large clinical laboratory. Provision for changes of incubation temperature should be available and can be provided even in the smallest laboratory by the judicious use of temperature blocks, water baths, or small incubators. If only one major incubator is available, it is recommended that the temperature be set at 35°C. A drawer or a closed cabinet can be used in laboratories that do not have multiple incubators for room temperature incubation. A water bath may be adjusted to a temperature beyond that of the main incubator, such as 42, 56, or 60°C.

It is desirable that most cultures be inspected after 15 to 18 h of incubation, the so-called overnight incubation. Not only are most cultures readable after this time, but also it is important to the clinician and the patient that a preliminary report be issued within the first 24 h. Even a negative report on certain cultures, especially those of body fluids, has significance to the clinician. However, many of the bacteria in clinical specimens require additional incubation for visible growth. Good clinical microbiologists routinely inspect all cultures that were negative during the first 18- to 24-h period on a daily basis for a period of 1 to 2 weeks. In fact, recovery of bacteria after 4 to 5 days is not a frequent occurrence. However, this observation is offset by the isolation of bacteria after a 1- to 2-week incubation period, especially from blood cultures. Although the percentage of such delayed isolations is low, it nevertheless constitutes isolation of organisms important to the diagnosis and treatment of the patient. Only broth media containing portions of the clinical specimen are so treated. Incubation of agar media in excess of 3 days is unnecessary unless specific information for slow-growing microorganisms or a suspicion of a prolonged lag phase exists. Should the clinical information suggest that bacterial L-forms, the so-called cell wall-defective variants, are suspected in the specimen submitted, hypertonic conditions should be established in the media used for primary and subculture purposes. This may be accomplished even after ordinary media have been inoculated by the addition to fluid media of suitable hypertonic solutions.

Adequate moisture is a further requirement of all incubation chambers. Even the most primitive incubator should be supplied with such a source. Many of the new incubators have water reservoirs attached or are equipped to regulate the evaporation of water into the incubator. Pans of water may be placed on the lowest shelf of an incubator not adequately equipped for this purpose. When automatic regulation of moisture is present, a relative humidity of 70 to 80% should be provided. Provision and control of atmospheric conditions are other important aspects for proper incubation for primary isolation. Many bacteria prefer a carbon dioxide atmosphere for growth. A capnophilic environment may be provided in special incubators equipped with a CO_2 gas tank which regulates adequate carbon dioxide content in the atmosphere. Burning a candle in a tightly closed jar, the candle extinction method, is still considered quite adequate in providing the increased CO_2 required by all such bacteria. Another method of providing an increased CO_2 environment is to tip a 1% HCl solution into a beaker containing $NaHCO_3$ previously placed in a wide-mouthed jar. The special atmospheric requirements of mycobacteria and anaerobic microorganisms are described in Chapters 22 and 37.

All incubators must be equipped with a thermometer. Whereas many of the newer models have mechan-

TABLE 1. Recommended media and Gram stain requirements for various specimens[a]

Specimen	Media										
	Supplemented thioglycolate[b]	Blood agar	EMB or McC	PEA	CBA	TM[c]	HE, XLD, GNB	Actinomyces	Anaerobic[d]	Mueller-Hinton broth	Gram stain
Autopsy tissue	X	X	X	X							X
Bowel drainage by mechanical means (Gomco, sump, etc.)	X	X	X	X			X (not gastric)				X
Ear, nose, parotid, antrum, mouth, bronchial, sinus (head)	X	X	X	X	X			Mouth, nose tissue only	Tissue only		
Eye	X	X	X	X	X				Blood agar only		X
Fluids: dialysis, chest, abdominal, pericardial, peritoneal	X	X	X	X	X				X		X
Fluids: spinal, scalp flap, subgaleal, ventricular	X	If not filtered			X				Blood agar only	X	X
Prostate	X	X	X	X	X	If swab					X
Stools and rectal swabs		X	X	X			X				X
Surgical tissue from lung, lymph nodes, spleen	X	X	X	X	Lung only			X	X		X
Suprapubic aspirate	1 ml of urine	X	X								X
Throat, sputum, tonsil, nasopharynx, tracheal		X	X	X	X						
Throat swab for *N. meningitidis*					X	X					
Transtracheal aspirate, synovial and joint fluids	X	X	X	X	X				X		X
Urethra, vagina, cervix, prostate for *N. gonorrhoeae* ("GC") only[c]					X	X					X
Urine		X	X								
Vagina, lochia, cervix, urethra, pelvic area, perineum	X	X	X	X	X	If MTM not received[c]			Pelvic and perineum		Pelvic and perineum
Vaginal swabs for *Gardnerella vaginalis*	X	X			X						X
Wounds, abscesses, ulcers, exudates, tissues, etc.	X	X	X	X	Brain		Perianal, neck and rectal tissue abscesses only		X		X

[a] Adapted from Washington (64). Abbreviations: EMB, eosin methylene blue agar; PEA, phenylethyl alcohol-blood agar; CBA, supplemented chocolate blood agar; TM, Thayer-Martin agar; HE, Hektoen enteric agar; XLD, xylose-lysine-deoxycholate agar; GNB, gram-negative broth (Hajna); McC, MacConkey agar; MTM, modified Thayer-Martin medium.

[b] Thioglycolate-135C (BBL Microbiology Systems) supplemented with either 0.5 to 1.0 ml of rabbit serum (Thioglycolate-135C with 1.0 ml of serum, used for all anaerobic cultures) or 1.0 ml of ascitic fluid.

[c] Cultures for *N. gonorrhoeae* may be inoculated directly onto modified Thayer-Martin medium before being submitted to the laboratory.

[d] Supplemented thioglycolate, blood agar, kanamycin-vancomycin-blood agar, and PEA.

ical recording devices, those incubators which do not should at least be monitored by a high- and low-indicator thermometer, a device which will record the lowest and highest temperature achieved between inspections. Daily checking and recording of the temperature are necessary. Similarly, the sanitary care of walk-in incubators must be of the highest order. Accidental spills, collections of old, overlooked specimens, dust, and other soil must be avoided by meticulous attention to a cleaning schedule. Even the smaller table-top or closet-type incubators should be inspected periodically to ensure proper sanitary conditions. All personnel with access to the incubator must be aware that accidental spillage of specimens, culture media, etc., represents a serious hazard that must be quickly disinfected and cleaned up.

EXAMINATION OF CULTURES FOR PRESUMPTIVE RECOGNITION AND IDENTIFICATION OF ISOLATES

One of the most decisive steps in clinical microbiology is the preliminary inspection of growth on primary isolation media after the first period of incubation. Once growth has been recognized or its identity suggested by the various manifestations in and on isolation media, an accepted set of standardized procedures should be followed to initiate presumptive and definitive identification. The recognition that a potentially significant microorganism is present in a clinical specimen must result from consideration of the source of the specimen; namely, is it from a body area in which microorganisms are rarely, if ever, found in the normal state, or does the organism come from an anatomic site which has a normal microbial flora? It is important to consider not only the appearance of certain organisms on generally supportive media, but also the numbers and kind of bacteria which manifest themselves on various selective media used for the cultivation of the original specimen.

The experienced microbiologist or the technical staff require only a glance at a series of selective and general media to gain an appreciation of the organisms present. However, no one should misunderstand or underestimate the importance of this first approach to acquaint oneself with the organisms present as being sufficient for further action. It is imperative that this first scanning be followed by close inspection of each primary isolation medium employed. More often than not, such agar plates should be inspected with the aid of a hand lens, an illuminated magnifying lens, if available, or, best of all, a stereoscopic microscope. Much of the morphology of bacteria, fungi, and yeasts on primary isolation media is a self-taught pattern recognition system which combines the aspects of colonial morphology on one medium with the appearance of the same organism on various other selective media. Even the absence of detectable colonial growth on one medium is important to the identification. It is customary for laboratorians in the clinical microbiology laboratory to utilize magnification to search out and inspect beta-hemolytic streptococcal and possibly pneumococcal colonies. However, an inspection of all plates with some magnification helps to acquaint workers in the clinical microbiology laboratory with subtle differences in colonial morphology on a variety of media. Colony inspection with magnification should be encouraged in all laboratories. Even the laboratory which does not receive large numbers of clinical specimens can afford to train neophytes in clinical microbiology by providing them with known cultures which can be studied on the various media used in that laboratory. Learning to describe the morphological characteristics of each significant bacterium on each medium will enhance the ability to recognize these organisms in mixtures that are common from contaminated body areas and also make possible a prompt preliminary decision when the bacteria appear in cultures of blood and other body fluids.

The characterization of colonial morphology follows well-established guidelines. There are many authoritative texts which will acquaint clinical microbiologists with the easiest manner of describing the appearance of a colony. Many methods are useful in recording these particular differences. Some prefer to sketch the appearance of the colony in their notebooks. Others utilize terms that describe the appearance. It is also helpful to establish comparable size differences of the colonies. Although "eye-balling" and guessing the size of a colony lack all quantitative aspects, it is helpful to note these impressions. Recording the findings of colonial morphology aids in the final recognition of organisms if such data are recorded immediately and permanently, rather than attempting to reconstruct the reasons for further studies long after the original impressions of the primary colonies have become blurred in the mind of the worker.

The properties of surface colonies on solid media can be described very simply as summarized in many of the standard texts of microbiology. It is important to note whether the shape of a colony is circular, irregular, radiate, or rhizoid. The size of the colony should be expressed in millimeters. Those who use calipers or millimeter rules for the measurement of zones of inhibition in antimicrobial susceptibility tests have learned to make such measurements quickly and accurately. The elevation of the colony must be described either by sketching or by words such as raised, low convex, convex or dome shaped, umbonate, umbilicate, and with or without beveled margins. Wilson and Miles (71) add to this description the structure of the colony, which they describe as amorphous, fine, medium, or coarsely granular, filamentous, or curved. The surface may be described as smooth, contoured, beaten copper, rough, fine, medium, or coarsely granular, ringed, papillate, dull, or glistening. The word matt has been used to describe the appearance of colonies somewhere between the dull and glistening states. The edge of the colony may be described as entire, undulate, lobate, crenated, erose, fimbriate, curved, or effuse. The pigmentation of a colony should be detected by reflected or transmitted light. Pigment may be described as fluorescent, iridescent, or opalescent, as well as by color description. It should also be noted whether these pigments are carotenoid or water insoluble, i.e., confined to the colony itself or diffused into the surrounding medium. Another variable is the opacity of the colony. Opacity may be classified as transparent, translucent, or opaque. The consistency of the colony becomes apparent to the experienced microbiologist; therefore, descriptions such as butyrous, viscid, friable, and membranous may help in characterizing the colonial morphology as well as serving as a guide to the handling of a colony should it be necessary to subculture it. The manner in which a colony behaves when it is picked for making a Gram stain may serve as a distinctive characteristic with a particular medium. This is referred to as emulsifiability and is either easy or difficult; it may be noted that the colony forms a homogeneous or granular suspension or that it remains membranous when rubbed in a drop of water with an inoculating needle. These notations are helpful, especially when training individuals to describe the appearance of colonies, and continue to aid in the recognition of organisms by their colonial morphology. Certain organisms will produce characteristic odors on various media. The ability to distinguish some bacteria by the odor produced can be of great help. However, odor detection is to be done with care and caution and is not totally reliable.

Growth in fluid media helps in the presumptive identification of microorganisms. Fluid media may be evaluated by first noting the degree of growth present. It is reported as none, scant, moderate, or abundant. Another variable is the degree of turbidity. This may be recorded as absent or, if present, as slight, moderate, or dense. Another useful descriptive observation is whether the turbidity is uniform, granular, or flocculent. Variation in these properties will occur in different fluid media. It is not possible to compare the appearance of an organism in thioglycolate broth with its appearance in brain heart infusion broth. The presence or absence of precipitates in broth media is also important. The degree of precipitate—slight, moderate, or abundant—should be noted. The precipitates may also be described as powdery, granular, flocculent, membranous, or viscid. In some instances,

TABLE 2. Growth on primary plate culture[a]

Blood agar (aerobic) after overnight incubation			
Colonies 1 mm or more in diameter		Colonies less than 1 mm in diameter	
Gram-positive cocci	*Staphylococcus* spp. *Micrococcus* spp. (*Streptococcus* spp.)	Gram-positive cocci in pairs or chains (short)	
		1. Growth on MacConkey medium	*Enterococcus* spp. (*Streptococcus* spp. group D) *Streptococcus* group B, C, G
Gram-negative cocci	*Neisseria* spp.		
Gram-positive rods	*Corynebacterium* spp. *Bacillus* spp.	2. No growth on MacConkey medium[b]	Group A (beta-hemolytic *Streptococcus*) *S. pneumoniae* (pneumococcus) Viridans streptococci
Gram-negative rods			
1. Growth on MacConkey agar	*Enterobacteriaceae* including *Yersinia* spp., *Aeromonas* spp., *Vibrio* spp., *Flavobacterium* spp., *Acinetobacter* spp. *Pseudomonas* spp., (*Brucella* spp., *Pasteurella* spp., *Actinobacillus* spp.)		Gram-positive cocci in clumps (*Staphylococcus*) *Micrococcus* spp. Gram-positive rods *Corynebacterium* spp. *Lactobacillus* spp. *Nocardia* spp. *Erysipelothrix* spp. *Listeria* spp.
2. No growth on MacConkey agar[b]	(*P. multocida*, few other *Pasteurella* spp., *Haemophilus* spp.)		Gram-negative rods *Haemophilus*[c] spp. *Brucella*[c] spp. *Pasteurella*[c] spp. *Bordetella*[c] spp.
Blood agar (anaerobic + 10% CO_2) after overnight incubation (no equivalent growth on aerobic culture)			
Colonies about 1 mm or spreading		Gram-positive rods	*Lactobacillus* spp. *Propionibacterium* spp. *Actinomyces* spp. *Eubacterium* spp.
Gram-positive rods	*Clostridium* spp.		
Colonies minute[d]			
Gram-positive cocci	*Peptostreptococcus* spp. *Peptococcus* spp.		
Gram-negative cocci	*Veillonella* spp.	Gram-negative rods	*Bacteroides*[e] spp. *Fusobacterium* spp.

[a] Revised from Sonnenwirth (57).
[b] Sensitive to bile salts.
[c] These organisms often require special media and conditions for primary isolation; occasionally they will grow on the media shown above. *Brucella abortus* does not grow on primary isolation without CO_2; *Francisella tularensis* usually does not grow on ordinary media; *P. multocida* does not grow on MacConkey agar.
[d] Many of these small colonies fail to appear for several days. Some may prove to be macroaerophilic, not true anaerobes.
[e] Occasionally 1 mm or larger.

TABLE 3. Growth characteristics of frequently isolated bacteria on some commonly used agars[a]

Organism	Growth on indicated agar			Growth on indicated agar			
	EMB	MacConkey	Hektoen enteric	SS	Bismuth sulfite	XLD	SEA
Citrobacter spp.	Translucent colonies, greenish metallic sheen (LF)	Uncolored, transparent; red (LF)	Usually inhibited; when present, colonies are small and bluish-green	Similar to *Salmonella* spp.	Black; green-brown	Opaque, yellow	Inhibited
Enterobacter spp., *Serratia* spp.	Metallic sheen, similar to *E. coli* but somewhat larger	Red-pink	Green centers with yellow to brown periphery	White or cream colored, opaque, mucoid	Raised mucoid colonies, silvery sheen	Opaque, yellow	Inhibited
E. coli (rapid lactose fermenters)	Dark center; greenish metallic sheen	Red or pink, may be surrounded by a zone of precipitated bile	Moderately inhibited; orange to salmon-pink	Red to pink; colorless with a pink center	Mostly inhibited; black-brown, greenish surface; no metallic sheen	Opaque, yellow	Inhibited
Klebsiella spp.	Larger than *E. coli*, mucoid, brownish, tend to coalesce, often convex	Pink, mucoid	Yellow centers, periphery orange	Red to pink; colorless with a pink center	Mostly inhibited	Opaque, yellow	Inhibited
Proteus spp.	Translucent, colorless	Uncolored, transparent	Most strains are inhibited; dark center, greenish (H_2S producers), similar to *Salmonella* spp.	Black center, clear periphery	Green; black (H_2S producers), mostly inhibited	Opaque, yellow (*P. mirabilis*, *P. vulgaris*); red (*P. rettgeri*, *Morganella morganii*)	Small gray colonies (few)
Pseudomonas spp.	Translucent, colorless; amber	Uncolored, transparent	Most strains are inhibited; colonies are small, flat, and green to brown	Mostly inhibited, transparent, colorless colonies	Inhibited	Sometimes red colonies	Inhibited
Salmonella spp.	Translucent, amber colonies; colorless	Uncolored, transparent	Blue to blue-green; most colonies have black centers (H_2S producers)	Opaque; transparent; uncolored; black center, clear periphery	*S. typhi* black with sheen or dotted black or greenish-gray; other species black or green	Black-centered red (H_2S producers); red (no H_2S)	Inhibited
Shigella spp.	Translucent, amber colonies; colorless	Uncolored, transparent	Blue to blue-green, periphery of colonies lighter than center portion	Opaque; transparent	Mostly inhibited; *S. flexneri* and *S. sonnei* are brown, raised, and craterlike	Red	Inhibited

Organism	SEA	110	Mannitol-salt	MS	Chocolate agar	Thayer-Martin
Enterococci	Translucent to whitish colonies surrounded by dark-brown to black zones	Mostly inhibited	Mostly inhibited	Blue-black, shiny center, clear periphery	White to gray	Mostly inhibited
Listeria spp.	Pinpoint colonies with reddish to black-brown zones	Inhibited	Inhibited	Inhibited	Gray	Inhibited
Neisseria spp.	Inhibited	Inhibited	Inhibited	Inhibited	Opaque, grayish white	Mostly inhibited
N. gonorrhoeae	Inhibited	Inhibited	Inhibited	Inhibited	Opaque, grayish white	Gray
N. meningitidis	Inhibited	Inhibited	Inhibited	Inhibited	Opaque, grayish white	Gray
Staphylococcus spp.	Small, white-gray colonies	White; orange to yellow	Colonies with yellow zones (mannitol fermenters); colonies with red or purple zones (mannitol not fermented)	Mostly inhibited	White to gray	White to gray, mostly inhibited
Streptococcus spp.						
Beta-hemolytic	Tiny colonies	Mostly inhibited	Mostly inhibited	Small blue colonies	White to gray	Mostly inhibited
Alpha-hemolytic	Tiny colonies	Mostly inhibited	Mostly inhibited	Various sizes	White to gray	Mostly inhibited
Nonhemolytic	Tiny colonies	Mostly inhibited	Mostly inhibited	Various sizes	White to gray	Mostly inhibited
S. salivarius	Tiny colonies	Mostly inhibited	Mostly inhibited	Blue gumdrop colonies	White to gray	Mostly inhibited
S. mitis	Tiny colonies	Mostly inhibited	Mostly inhibited	Small blue colonies	White to gray	Mostly inhibited

[a] Abbreviations: EMB, eosin methylene blue agar; SS, salmonella-shigella agar; XLD, xylose-lysine-deoxycholate agar; SEA, selective enterococcus agar; 110, *Staphylococcus* 110 agar; MS, mitis-salivarius agar; LF, lactose fermenter.

slight shaking will resuspend the precipitates formed, whereas no effect will be noted with others. The formation of a pellicle on the surface of the broth is still another characteristic that should be observed. If a pellicle is present, it may have a ring of growth around the inner surfaces of the tube. The pellicle may be thick, thin, smooth, granular, or rough. This manifestation may also be destroyed by shaking or may remain intact.

Growth on selective media may constitute a clue to the identification of certain microorganisms. Tables 2, 3, and 4 present summaries of some of the bacteria which may be recognized presumptively at this stage. It must be emphasized that it is not the purpose of these tables to serve as other than a guide to help the microbiologist decide whether to identify a particular organism further. This decision is influenced by the source of the specimen, the microorganism recovered, and the history of the patient. As an illustration, consider the recovery of *E. coli* from a stool specimen of a 30-year-old male. If all the colonial morphology patterns indicate that this organism is *E. coli* and that

TABLE 4. Simplified guide to the presumptive recognition of common groups of bacteria[a]

AEROBIC CULTURES

I. Gram-positive cocci
  A. Catalase positive
    1. Arranged in clusters, large colonies—*Staphylococcus* spp. (Chapter 15) (perform coagulase test); glucose fermented anaerobically
    2. In pairs, fours, or small clusters; use glucose oxidatively or not at all—*Micrococcus* spp. (Chapter 15)
  B. Catalase negative
    1. Short and long chains and even pairs; fermentative (anaerobic) utilization of sugars—*Streptococcus* spp. (Chapter 16)
      Note: Beta-hemolytic streptococci may belong to pyogenes or enterococcus group; alpha-hemolytic streptococci may belong to viridans or enterococcus group; nonhemolytic streptococci (gamma) may be viridans or enterococcus group
      a. Pyogenes group: usually, but not always, beta-hemolysis; do not grow at 45°C; do not survive 30 min at 60°C; serological groups A, B, C, F, G, H, K, L, M, and O; Lancefield grouping or fluorescent-antibody technique; most strains, especially group A, susceptible to bacitracin
      b. Viridans group: alpha-hemolysis or none; not soluble in bile; not inhibited by Optochin; usually grow at 45°C (*Aerococcus* group of Cowan and Steel included here)
      c. Enterococcus group: some beta-hemolytic, others alpha- or gamma-hemolytic; grow in 0.1% methylene blue milk, in 6.5% NaCl, and at pH 9.6; grow at 45°C and survive 30 min at 60°C; usually grow on MacConkey agar; bile-esculin positive
      d. Usually lancet shaped, pairs, single, or short chains; bile soluble; inhibited by Optochin; alpha-hemolysis; no growth at 45°C; virulent for mouse—pneumococci (*S. pneumoniae*)
    2. Primarily clumps and tetrads; no acid from glucose anaerobically—*Aerococcus* spp.

II. Gram-negative cocci, mostly in pairs
  A. *Neisseria* spp. (all oxidase positive) (Chapter 17)
    1. No growth at 22°C or on nutrient agar; growth on modified Thayer-Martin agar (MTM)
      a. Requires enriched medium; acid from glucose and maltose; agglutination by antimeningococcal serum—*N. meningitidis*
      b. Requires enriched medium; acid from glucose only—*N. gonorrhoeae*
    2. Growth on MTM; ferments glucose, maltose, and lactose—*N. lactamica*
    3. Growth at 22°C and on nutrient agar; light inoculum yields no growth on MTM
      a. Growth on ordinary media; no acid from glucose, maltose, sucrose, or lactose—*Branhamella catarrhalis*
      b. Yellow pigment; no acid from glucose, maltose, sucrose, or lactose—*N. flavescens*
      c. Good growth on ordinary media; acid from glucose and maltose—pharyngeal group
  B. Rule out *Acinetobacter* (former *Mima-Herellea* group, *Achromobacter* spp., *Bacterium anitratum*) (see III.A.5)

III. Gram-negative rods
  A. Good growth on ordinary media, including MacConkey agar[b]
    1. Fermentative,[c] oxidase negative,[d] nitrate reduced to nitrite
      a. Lactose usually fermented,[e] phenylalanine deaminase not produced
        (1) Voges-Proskauer (V-P), citrate, urease, and usually H_2S negative; indole and methyl red positive—*E. coli* (Chapter 24)
        (2) V-P and citrate positive; growth in KCN; usually indole negative, urease positive or delayed; motile—*Enterobacter* spp.; nonmotile—*Klebsiella* spp. (Chapter 24)
        (3) H_2S and citrate positive; lysine decarboxylase not produced; V-P, urease, and indole negative; growth in KCN—*Citrobacter* spp. (Chapter 24)
      b. Lactose usually not fermented; phenylalanine deaminase not produced
        (1) H_2S, citrate, and lysine decarboxylase positive; indole, V-P, urease, and KCN negative—*Salmonella* spp. (Chapter 24) (exceptions: no gas, little H_2S—*S. typhi* and others) (use specific antisera)
        (2) H_2S, citrate, V-P, and urease negative; nonmotile; usually no gas—*Shigella* spp. (Chapter 24) (or nonmotile *Escherichia* spp.); some *Shigella* spp. ferment lactose slowly
        (3) V-P and citrate positive; urease delayed positive; motile; pigment often formed, especially at room temperature—*Serratia* spp. (Chapter 24)
        (4) H_2S, indole, lysine, and methyl red positive; urease, V-P, citrate, and KCN negative; most sugars (except glucose and maltose) not fermented—*Edwardsiella* spp. (Chapter 24)

Continued

TABLE 4—*Continued*

AEROBIC CULTURES

III. Gram-negative rods *continued*
- c. Phenylalanine deaminase produced
  - (1) Urease positive—*Proteus* spp. (Chapter 24)
    - (a) H_2S positive; indole positive—*P. vulgaris*
    - (b) H_2S positive; indole negative—*P. mirabilis*
    - (c) H_2S negative; citrate negative—*Morganella morganii*
    - (d) H_2S negative; citrate positive—*Providencia rettgeri*
  - (2) Urease negative—*Providencia* spp.
- 2. Fermentative; oxidase negative; growth on MacConkey agar, usually on salmonella-shigella (SS) agar (exceptions); motile at 20 to 25°C but not at 35°C; urease positive, phenylalanine negative—*Yersinia pseudotuberculosis, Y. enterocolitica* (Chapter 24)
- 3. Fermentative; oxidase positive
  - a. Catalase positive; nitrate reduced; usually motile; arginine dihydrolase produced—*Aeromonas* spp. (Chapter 25)
  - b. Cells spiral or comma shaped; motile; fermentative; lysine and ornithine decarboxylase produced—*Vibrio* spp. (Chapter 26) (identify *V. cholerae* with specific antisera, biochemical tests)
  - c. No growth on MacConkey agar—see *Pasteurella* spp. (*P. multocida, P. ureae,* etc.); *Cardiobacterium* spp. (Chapter 28)
  - d. Growth on MacConkey agar—*Pasteurella haemolytica* (Chapter 28)
- 4. Oxidative utilization of sugars (oxidation-fermentation [O-F] medium, no fermentation)[f]
  - a. Oxidase usually positive; no gas from sugars; motile; grow on MacConkey, SS, and cetrimide agars; many have soluble pigments (green or yellow)—*Pseudomonas* spp. (Chapters 29 and 30)
  - b. Oxidase positive (or variable); growth variable on MacConkey agar; oxidative or no utilization of sugars; yellow pigment; no reduction of nitrate—*Flavobacterium* spp. (Chapter 29)
  - c. Oxidase negative; growth on MacConkey agar; malonate negative; nonmotile; no reduction of nitrate; no decarboxylases; 10% lactose, citrate positive; majority malonate positive—*Acinetobacter* spp. (other former names and groups—*Achromobacter anitratus, B. anitratus, Mima-Herellea* group) (Chapter 29)
- 5. Carbohydrates not attacked (no oxidation or fermentation, O-F medium)
  - a. Oxidase positive; growth on MacConkey and cetrimide agars; nitrate reduced to nitrite; urease negative; no decarboxylases; motile—*Alcaligenes* spp. (Chapter 29)
  - b. Oxidase positive; nonmotile; usually no growth on MacConkey agar (exceptions); penicillin susceptible; citrate and urease negative—*Moraxella* spp. (Chapter 29)
  - c. Oxidase negative; growth on MacConkey agar
    - (1) *Acinetobacter calcoaceticus* subsp. *lwoffi* (Chapter 29)
    - (2) *Pseudomonas maltophilia* (Chapter 30)
    - (3) *Bordetella parapertussis* (Chapter 34)
  - d. Oxidase positive; growth on MacConkey agar; microaerophilic; slow growth—*Campylobacter fetus* (Chapter 27)
- B. Some grow on ordinary media, others need enriched media; fermentative; no gas produced; oxidase usually positive, nitrate reduced, catalase positive—*Pasteurella* spp. (Chapter 28)
- C. Slow growth; pleomorphic cells; nonmotile; fermentative; no gas; often polar staining; oxidase and catalase variable; slow coagulation of milk; no decarboxylases—*Actinobacillus* spp. (Chapter 28); growth on MacConkey agar; oxidase positive—*A. lignieresii, A. equuli*; no growth on MacConkey agar; oxidase negative—*A. actinomycetemcomitans*; oxidative; no growth on MacConkey agar; oxidase negative—*P. mallei* (Chapter 28)
- D. Some requirement of special media (no growth on MacConkey agar) and conditions; no capsule; nonmotile—*Brucella* spp. (members of this group have to be differentiated by CO_2 requirement, H_2S production, dye inhibition test, and agglutination tests) (Chapter 32)
- E. No growth or poor growth without special factors in media; capsule variable, nonmotile
  - 1. Require factors X and/or V—*Haemophilus* spp. (Chapter 33); characteristic satellitism along *Staphylococcus* streak or other colonies; no growth on plain agar; encapsulated strains identified by type-specific antisera—*H. influenzae*; grows on plain agar along a *Staphylococcus* streak or other colonies providing factor V—*H. parainfluenzae; H. ducreyi* cultivation rare; *H. aphrophilus*
  - 2. Do not require factors X and/or V; growth improved by addition of serum or ascitic fluid; sugars not attacked; oxidase positive—*Moraxella* spp. (Chapter 29); oxidase negative—*Gardnerella vaginalis* (Chapter 18)
- F. Primary isolation best on complex media with blood; oxidase positive; shows characteristic colonies on Bordet-Gengou agar; agglutination by specific antiserum—*B. pertussis* (rough variant of *B. pertussis* grows on ordinary media); *B. parapertussis* and *B. bronchiseptica* grow on MacConkey agar; closely related antigenically (Chapter 34)
- G. No growth on ordinary media; requires special media (cystine-glucose-blood agar)—*Francisella tularensis* (Chapter 28)
- H. Grow best on enriched media, crescent-shaped or spiral cells (long screws or portions of a turn)—*Spirillum* spp. (Chapter 35); no growth on artificial media—*S. minus* and *S. volutans*
- I. Primary isolation on complex media with cysteine and iron, 5% CO_2; difficult to stain with Gram stain in tissues but reacts with Dieterle technique or direct fluorescent antibodies—*Legionella* spp. (Chapter 31)

IV. Gram-positive rods
- A. Catalase positive; no spores formed; no growth on MacConkey agar
  - 1. Nonmotile; arranged in Chinese figures; stain unevenly with bands and granules—*Corynebacterium* spp. (Chapter 18); toxin production; fermentative—*C. diphtheriae* and *C. ulcerans*; other corynebacteria and diphtheroids differentiated by biochemical and toxigenicity tests; some fermentative, others do not attack sugars at all
  - 2. Motile (at 20°C but usually not at 37°C); short, diphtheroid-like rods; often narrow zone of beta-hemolysis—*Listeria monocytogenes* (use biochemical and pathogenicity tests; Chapter 19)
  - 3. Motile; do not attack sugars; indole production and nitrate reduction negative; long rods; some form filaments and coccoid bodies in broth—*Kurthia* spp. (see *Bergey's Manual*, 8th ed.)

Continued

TABLE 4—*Continued*

AEROBIC CULTURES

IV. Gram-positive rods *continued*
  B. Catalase negative; no spores formed
    1. Nonmotile; frequently form long filaments; usually no growth in litmus milk; fermentative (glucose, lactose); H_2S in butt of triple sugar iron agar; nonbranching—*Erysipelothrix* spp. (Chapter 20)
    2. Form chains; growth is better on tomato juice agar; H_2S negative in butt of triple sugar iron agar—*Lactobacillus* spp. (Chapter 41)
  C. Acid fast; no branching; no spores
    1. Nonmotile; no branching; no hyphae—*Mycobacterium* spp. (Chapter 22)
      a. Special media required; slow growth—*M. tuberculosis* (distinguish by biochemical and cultural characteristics); nontuberculous mycobacteria (formerly atypical) (use cultural characteristics, pigment formation, physiological and biochemical tests)
      b. Rapid growth—"saprophytes"
  D. Nonmotile; some branching; some acid fast; hyphae, but no true conidia produced; oxidative—*Nocardia* spp. (Chapter 23)
  E. Catalase positive; spores formed, many motile—*Bacillus* spp. (Chapter 21); characteristic colonies, nonmotile, usually nonhemolytic—*B. anthracis* (use biochemical differentiation)

ANAEROBIC CULTURES

I. Gram-positive cocci
  A. Occurring mainly in clusters but also in pairs—*Peptococcus* spp. (gas chromatography, biochemical reactions; Chapter 39)
  B. Mainly in pairs and chains—*Peptostreptococcus* spp. (identification by gas chromatography and biochemical reactions; Chapter 39)

II. Gram-negative cocci occurring in irregular masses, small cocci—*Veillonella* spp., *Acidaminococcus* spp., *Megasphaera* spp. (Chapter 39)

III. Gram-negative rods
  A. Usually nonmotile of varying sizes and shapes; nonsporeforming; often foul smelling; cells greater than 0.6 μm; do not produce butyric acid—*Bacteroides* spp. (Chapter 40)
  B. Some with pointed ends, effuse colonies; many very pleomorphic, produce butyric acid—*Fusobacterium* spp. (Chapter 40)
  C. Motile; polar flagella; curved—*Butyrivibrio* spp., *Succinimonas* spp. (Chapter 40)
  D. Spiral organisms—*Borrelia* spp., *Treponema* spp. (Chapters 43 and 44)

IV. Gram-positive rods; no spores
  A. Vegetative mycelium produced, which fragments; true branching; non-acid fast—*Actinomyces* spp. (identified by cultural and biochemical characteristics; catalase negative; slow-growing, dry, crumbly colonies on solid media, resembling tubercle bacillus colonies; in fluid medium granules adherent to walls of tube; unstained preparation of crushed granule shows typical clubs; must be differentiated from anaerobic diphtheroids and lactobacilli; Chapter 41)
  B. No catalase or indole produced, no spores—*Bifidobacterium* spp. (Chapter 41)
  C. Catalase and indole positive; nitrate reduced; gas from glucose; produce propionic acid—*Propionibacterium* spp. (Chapter 41)
  D. Nonmotile; nonbranching; do not ferment lactose; produce butyric acid—*Eubacterium* spp. (Chapter 41)

V. Gram-positive rods; spores formed; some species may appear gram negative; motile and nonmotile; forming endospores that distort the cells; some species microaerophilic; catalase negative—*Clostridium* spp. (many members of this genus are saprophytes, but some are highly pathogenic for humans and animals; differentiation by gas-liquid chromatography, biochemical reactions, pathogenicity tests, and specific toxins; Chapter 38)

[a] Revised from Sonnenwirth (57). For the identification of yeasts, fungi, parasites, and viruses, consult the appropriate sections of this Manual.

[b] Some do not grow on MacConkey agar; exceptions listed.

[c] O-F medium; fermentation observable in triple sugar iron medium.

[d] Cytochrome (indophenol) oxidase test.

[e] Lactose is valuable in the case of prompt fermenters (on differential plates overnight or in 24 to 48 h in fermentation media); some strains of the groups listed show delayed or no fermentation of lactose. See *Enterobacteriaceae* (Chapter 24).

[f] Usually no change in triple sugar iron medium or on carbohydrate-containing differential plate media.

it is not being confused with other members of the *Enterobacteriaceae* family, especially that it is not a *Shigella* species, or that such an identification error is of no consequence in this specimen, then no further work needs to be done. This is true especially if the history of the patient does not indicate any gastrointestinal symptoms. The same organism isolated from a urine specimen from the same individual would require additional tests to identify it.

The clinical microbiologist must decide which of the organisms isolated on primary culture must be identified. There are no easy guidelines which would cover all eventualities. If, after careful inspection, the organisms represent the normal microbiota of the part of the body from which the specimen was obtained, further identification of these organisms by additional testing is superfluous. If the specimen was obtained from an anatomical site which is normally sterile, the laboratory must proceed to establish the identification of genus and species, but not at the expense of presumptively reporting the initial impressions.

This need for presumptive identification is in response to the major responsibility of the clinical microbiology laboratory. As mentioned at the onset, it is the function of the clinical microbiology laboratory to report as accurately and quickly as possible the presence of a significant microorganism in a clinical specimen and to provide the clinician with a guide to antimicrobial therapy. This aim can be met by the judicious application of the principles outlined in this chapter and by the early and quick use of diagnostic tests available for the recognition of certain microorganisms. An example is the isolation and recognition of a *Staphylococcus* sp. from a specimen. A coagulase test would quickly confirm that this organism is *S. aureus* without requiring additional testing. This does not rule out the possible future need to perform phage typing for epidemiological purposes. Similarly, the recognition of either *Salmonella* spp. or *Shigella* spp. in a stool specimen may require the use of grouping antisera to confirm the identity of the organism. The species or serotype designation may be obtained, and biochemical tests to confirm the identification of the organism may be carried out. However, the certainty and authority with which one can report the presence of such an organism in a specimen are enhanced largely by the ability to perform rapid tests leading to strong presumptive identification.

The special circumstances of the patient frequently require the testing of a microorganism further than would appear necessary at first. Also, the persistent isolation of an organism from a usually sterile body area ought to lead to identification tests, even if under ordinary circumstances such an organism would be described as nonpathogenic. Early presumptive reporting is imperative. It is the practice in many laboratories to render a verbal report to the clinician. This is followed by a written report reiterating the preliminary nature of the verbal report. Laboratory personnel (and clinicians) must also be aware that occasionally the preliminary report may prove to be incorrect. One should never hesitate to retract the initial report and indicate the correctness of the final report. This is done with explanations if required. Clinical microbiologists must not hesitate to alert the physician and the infection committee of the institution when they have presumptive evidence of organisms that are potentially dangerous to the institution and the community at large.

The classical purpose of primary isolation is the separation of microorganisms into groups which can be identified according to the principles of sound microbiology. Primary isolation constitutes the starting point for the detective work which leads to the diagnosis, if not the identification, of microorganisms in a manner expounded by the experts to guide the reader to the level of genus and species recognition. The dynamic state of the intimate human biosphere makes it imperative that the clinical microbiologist remain aware of the changes which bring new populations into this environment and that the role of these newcomers in disease production or complication be recognized with dispatch. This responsibility can be met only by the judicious use of the chapters which follow.

LITERATURE CITED

1. **Balows, A., and H. D. Isenberg.** 1978. Biotyping in the clinical microbiology laboratory. Charles C Thomas, Publisher, Springfield, Ill.
2. **Balows, A., and A. C. Sonnenwirth.** 1983. Bacteremia: laboratory and clinical aspects. Charles C Thomas, Publisher, Springfield, Ill.
3. **Bannatyne, R. M., C. Clausen, and L. R. McCarthy.** 1979. Cumitech 10, Laboratory diagnosis of upper respiratory tract infections. Coordinating ed., I. B. R. Duncan. American Society for Microbiology, Washington, D.C.
4. **Barrett-Connor, E.** 1971. The nonvalue of sputum culture in the diagnosis of pneumococcal pneumonia. Am. Rev. Respir. Dis. **103:**845–848.
5. **Barry, A. L., P. B. Smith, and M. Turck.** 1975. Cumitech 2, Laboratory diagnosis of urinary tract infections. Coordinating ed., T. L. Gavan. American Society for Microbiology, Washington, D.C.
6. **Bartlett, J. G., N. S. Brewer, and K. J. Ryan.** 1978. Cumitech 7, Laboratory diagnosis of lower respiratory tract infections. Coordinating ed., J. A. Washington II. American Society for Microbiology, Washington, D.C.
7. **Bartlett, J. G., J. E. Rosenblatt, and S. M. Finegold.** 1973. Percutaneous transtracheal aspiration in the diagnosis of anaerobic pulmonary infection. Ann. Intern. Med. **79:**535–540.
8. **Bartlett, R. C., G. O. Carrington, and C. Mielert.** 1968. Quality control in clinical microbiology (revised). Commission on Continuing Education, Council on Microbiology, American Society of Clinical Pathology, Chicago.
9. **Bartlett, R. C., P. D. Ellner, and J. A. Washington II.** 1974. Cumitech 1, Blood cultures. Coordinating ed., J. C. Sherris. American Society for Microbiology, Washington, D.C.
10. **Bass, J. W., T. M. Cashman, A. L. Frostad, R. M. Yamaoka, R. A. Schooler, and E. P. Dierdorff.** 1973. Antimicrobials in the treatment of acute otitis media. Am. J. Dis. Child. **125:** 397–402.
11. **Bennett, D. E.** 1967. Histoplasmosis of the oral cavity and larynx: a clinicopathologic study. Arch. Intern. Med. **120:**417–427.
12. **Bland, R. D.** 1972. Otitis media in the first six weeks of life: diagnosis, bacteriology, and management. Pediatrics **49:**187–197.
13. **Carr, D. T., A. G. Karlson, and G. G. Stillwell.** 1967. A comparison of cultures of induced sputum and gastric washings in the diagnosis of tuberculosis. Mayo Clin. Proc. **42:**23–25.
14. **Chalvardjian, A. M., and L. A. Grawe.** 1963. A new procedure for the identification of *Pneumocystis carinii* cysts in tissue sections and smears. J. Clin. Pathol. **16:**383–384.
15. **Converse, G. M., J. M. Gwaltney, D. A. Strassburg, and J. Q. Hendley.** 1973. Alteration of CSF findings by partial treatment of bacterial meningitis. Clin. Res. **21:**120.
16. **Dalton, H. P., and M. J. Allison.** 1968. Modification of laboratory results by partial treatment of bacterial meningitis. Am. J. Clin. Pathol. **49:**410–413.
17. **Dolan, C. T.** 1979. Postmortem microbiology, p. 138–145. *In* J. Ludwig (ed.), Current methods of autopsy practice, 2nd ed. The W. B. Saunders Co., Philadelphia.
18. **Dolan, C. T., A. L. Brown, and R. E. Ritts, Jr.** 1971. Microbiological examination of postmortem tissues. Arch. Pathol. **92:**206–211.
19. **Duma, R. J., H. W. Ferrell, C. Nelson, and M. Jones.** 1969. Primary amebic meningoencephalitis. N. Engl. J. Med. **24:**1315–1323.
20. **Eschenbach, D., H. M. Pollock, and J. Schachter.** 1983.

Cumitech 17, Laboratory diagnosis of female genital tract infections. Coordinating ed., S. J. Rubin. American Society for Microbiology, Washington, D.C.

21. **Feigin, R. D., and P. G. Shackelford.** 1973. Value of repeat lumbar puncture in the differential diagnosis of meningitis. N. Engl. J. Med. **289:**571–574.
22. **Finegold, S. M., W. E. Shepherd, and E. H. Spaulding.** 1977. Cumitech 5, Practical anaerobic bacteriology. Coordinating ed., W. E. Shepherd. American Society for Microbiology, Washington, D.C.
23. **Geckler, R. W., D. H. Gremillion, C. K. McAllister, and C. Ellenbogen.** 1977. Microscopic and bacteriological comparison of paired sputa and transtracheal aspirates. J. Clin. Microbiol. **6:**396–399.
24. **Hable, K. A., J. A. Washington II, and E. C. Hermann, Jr.** 1971. Bacterial and viral throat flora: comparison of findings in children with acute upper respiratory tract disease and in healthy controls during winter. Clin. Pediatr. **10:**199–203.
25. **Hahn, H. H., and H. N. Beaty.** 1970. Transtracheal aspiration in the evaluation of patients with pneumonia. Ann. Intern. Med. **72:**183–187.
26. **Hammond, G. W., C. J. Lian, J. C. Wilt, and A. R. Ronald.** 1978. Comparison of specimen collection and laboratory techniques for isolation of *Haemophilus ducreyi*. J. Clin. Microbiol. **7:**39–43.
27. **Isenberg, H. D., and J. I. Berkman.** 1966. Recent practices in diagnostic bacteriology. Prog. Clin. Pathol. **1:**237–317.
28. **Isenberg, H. D., B. G. Painter, J. I. Berkman, L. Philipson, and V. Tucci.** 1974. The post-mortem microbiological analysis as an indicator of nosocomially-significant microorganisms in the hospital environment. Health Lab. Sci. **11:**85–89.
29. **Isenberg, H. D., F. D. Schoenknecht, and A. von Graevenitz.** 1979. Cumitech 9, Collection and processing of bacteriological specimens. Coordinating ed., S. J. Rubin. American Society for Microbiology, Washington, D.C.
30. **Jones, D. B., T. J. Liesegang, and N. M. Robinson.** 1981. Cumitech 13, Laboratory diagnosis of ocular infections. Coordinating ed., J. A. Washington II. American Society for Microbiology, Washington, D.C.
31. **Jones, G. L., and G. L. Hébert.** 1979. "Legionnaires' ": the disease, the bacterium and methodology. Centers for Disease Control, Atlanta, Ga.
32. **Kaplan, S. L.** 1983. Antigen detection in cerebrospinal fluid—pros and cons. Am. J. Med. **75**(Suppl. 1B):109–118.
33. **Kellogg, D. S., Jr., K. K. Holmes, and G. A. Hill.** 1976. Cumitech 4, Laboratory diagnosis of gonorrhea. Coordinating ed., S. Marcus and J. C. Sherris. American Society for Microbiology, Washington, D.C.
34. **Kiani, D., E. L. Quinn, K. H. Burch, T. Madhavon, L. D. Saravolatz, and T. R. Neglett.** 1979. The increasing importance of polymicrobial bacteremia. J. Am. Med. Assoc. **242:**1044–1047.
35. **Koneman, E. W., T. M. Minckler, D. B. Shires, and D. S. de Jongh.** 1971. Postmortem bacteriology. II. Selection of cases for culture. Am. J. Clin. Pathol. **55:**17–23.
36. **Krieg, A. J., T. J. Johnson, C. McDonald, and E. Cottore.** 1971. Clinical laboratory computerization. University Park Press, Baltimore.
37. **Kunin, C. M.** 1979. Detection, prevention and management of urinary tract infections, 3rd ed. Lea & Febiger, Philadelphia.
38. **Martin, J. E., J. H. Armstrong, and P. B. Smith.** 1974. New system for cultivation of *Neisseria gonorrhoeae*. Appl. Microbiol. **27:**802–805.
39. **Mou, T. W., and H. A. Feldman.** 1961. The enumeration and presentation of bacteria in urine. Am. J. Clin. Pathol. **35:**572–575.
40. **Musher, D. M., and R. F. Schell.** 1973. False-positive Gram stains of cerebrospinal fluid (letters). Ann. Intern. Med. **79:**603–604.
41. **Nemoy, N. J., and T. A. Stamey.** 1971. Surgical, bacteriological, and biochemical management of "infection" stones. J. Am. Med. Assoc. **215:**1470–1476.
42. **Ostfeld, E., M. Harell, D. Michaeli, and E. Rubinstein.** 1978. Acute gram-negative bacillary otitis media. Am. J. Dis. Child. **132:**721–722.
43. **Pezzlo, M. T.** 1983. Automated methods for detection of bacteriuria. Am. J. Med. **75**(Suppl. 1B):71–78.
44. **Platt, R.** 1983. Quantitative definition of bacteriuria. Am. J. Med. **75**(Suppl. 1B):44–52.
45. **Pollock, H. M.** 1983. Laboratory techniques for detection of urinary tract infection and assessment of value. Am. J. Med. **75**(Suppl. 1B):79–84.
46. **Ray, C. G., B. L. Wasilauskas, and R. Zabransky.** 1982. Cumitech 14, Laboratory diagnosis of central nervous system infections. Coordinating ed., L. R. McCarthy. American Society for Microbiology, Washington, D.C.
47. **Rein, M. F., and G. L. Mandell.** 1973. Bacterial killing by bacteriostatic saline solutions—potential for diagnostic error. N. Engl. J. Med. **289:**794–795.
48. **Reller, L. B.** 1983. Recent and innovative methods for detection of bacteremia and fungemia. Am. J. Med. **75**(Suppl. 1B):26–30.
49. **Reller, L. B., P. R. Murray, and J. D. MacLowry.** 1982. Cumitech 1A, Blood cultures II. Coordinating ed., J. A. Washington II. American Society for Microbiology, Washington, D.C.
50. **Sack, R. B., R. C. Tilton, and A. S. Weissfeld.** 1980. Cumitech 12, Laboratory diagnosis of bacterial diarrhea. Coordinating ed., S. J. Rubin. American Society for Microbiology, Washington, D.C.
51. **Schmidt, R. M., and H. S. Rosenkranz.** 1970. Antimicrobial activity of local anesthetics: lidocaine and procaine. J. Infect. Dis. **121:**597–607.
52. **Schonebeck, J.** 1972. Studies on Candida infection of the urinary tract and on the antimycotic drug 5-fluorocytosine. Scand. J. Urol. Nephrol. Suppl. **11:**1–48.
53. **Segal, E. L., G. E. Starr, and L. A. Weed.** 1959. Study of surgically excised pulmonary granulomas. J. Am. Med. Assoc. **170:**515–522.
54. **Segura, J. W., P. P. Kelalis, W. J. Martin, and L. H. Smith.** 1972. Anaerobic bacteria in the urinary tract. Mayo Clin. Proc. **47:**30–33.
55. **Shurin, P. A., V. M. Howie, S. I. Pelton, J. H. Ploussard, and J. O. Klein.** 1978. Bacterial etiology of otitis media during the first six weeks of life. J. Pediatr. **92:**893–896.
56. **Smith, A. L.** 1973. Diagnosis of bacterial meningitis. Pediatrics **42:**589–592.
57. **Sonnenwirth, A. C.** 1979. Collection and culture of specimens and guides for bacterial identification, p. 1122–1169. *In* S. Frankel, S. Reitman, and A. C. Sonnenwirth (ed.), Gradwohl's clinical laboratory methods and diagnosis, 8th ed. C. V. Mosby Co., St. Louis.
58. **Stamey, T. A., D. E. Govan, and J. M. Palmer.** 1965. The localization and treatment of urinary tract infections: the role of bactericidal urine levels as opposed to serum levels. Medicine **44:**1–35.
59. **Stamey, T. A., and A. Pfan.** 1970. Urinary infections: a selective review and some observations. Calif. Med. **113:**16–35.
60. **Stamm, W. E.** 1983. Measurement of pyuria and its relation to bacteriuria. Am. J. Med. **75**(Suppl. 1B):53–58.
61. **Steel, K. J.** 1962. The practice of bacterial identification. Symp. Soc. Gen. Microbiol. **12:**405–432.
62. **Thomson, R. B., Jr., T. F. Smith, and W. R. Wilson.** 1982. Comparison of two methods used to prepare smears of mouse lung tissue for detection of *Pneumocystis carinii*. J. Clin. Microbiol. **16:**303–306.
63. **Washington, J. A., II (ed.).** 1978. The detection of septicemia. CRC Press, West Palm Beach, Fla.
64. **Washington, J. A., II.** 1981. Laboratory procedures in clinical microbiology. Springer-Verlag, New York.
65. **Weed, L. A.** 1954. Microbiologic methods in surgical pathology. Mayo Clin. Proc. **29:**393–399.

66. **Weed, L. A.** 1958. Technics for the isolation of fungi from tissues obtained at operation and necropsy. Am. J. Clin. Pathol. **29:**496–502.
67. **Weed, L. A., and A. H. Bagenstoss.** 1951. The isolation of pathogens from tissues of embalmed human bodies. Am. J. Clin. Pathol. **23:**1114–1120.
68. **Weinstein, M. P., J. R. Murphy, L. B. Reller, and K. A. Lichtenstein.** 1983. The clinical significance of positive blood cultures: a comprehensive analysis of 500 episodes of bacteremia and fungemia in adults. II. Clinical observations, with special reference to factors influencing prognosis. Rev. Infect. Dis. **5:**54–70.
69. **Weinstein, M. P., L. B. Reller, J. R. Murphy, and K. A. Lichtenstein.** 1983. The clinical significance of positive blood cultures: a comprehensive analysis of 500 episodes of bacteremia and fungemia in adults. I. Laboratory and epidemiologic observations. Rev. Infect. Dis. **5:**35–53.
70. **Wiesner, P. J., E. Tronca, T. Bonin, A. H. B. Pedersen, and K. K. Holmes.** 1973. Clinical spectrum of pharyngeal gonococcal infection. N. Engl. J. Med. **268:**181–185.
71. **Wilson, G. S., and A. A. Miles (ed.).** 1964. Topley & Wilson's principles of bacteriology and immunity, 5th ed. The Williams & Wilkins Co., Baltimore.
72. **Wilson, W. R., C. T. Dolan, J. A. Washington II, A. L. Brown, and R. E. Ritts, Jr.** 1972. Clinical significance of postmortem cultures. Arch. Pathol. **94:**244–249.
73. **Wimberley, N., L. J. Faling, and J. G. Bartlett.** 1979. A fiberoptic bronchoscopy technique to obtain uncontaminated lower airway secretions for bacterial culture. Am. Rev. Respir. Dis. **119:**337–343.
74. **Winn, W. R., M. L. White, W. T. Carter, A. B. Miller, and S. M. Finegold.** 1966. Rapid diagnosis of bacteremia and quantitative differential membrane filtration culture. J. Am. Med. Assoc. **197:**539–548.

Chapter 9

Section II. Nosocomial Infection Prevention and Control

Epidemiology of Nosocomial Infections

JAMES M. HUGHES AND WILLIAM R. JARVIS

Nosocomial infections result in substantial morbidity, prolongation of hospital stay, increases in direct patient care costs, and mortality. In 1958, the Advisory Committee on Infections within Hospitals to the American Hospital Association recommended the establishment of hospital-wide infection control committees (5). Our understanding of the epidemiology of nosocomial infections has increased dramatically during the last 20 years. The 1960s were characterized by an appreciation of the fact that nosocomial infections were a significant problem in hospitals and by the initiation of systematic approaches to the surveillance and control of nosocomial infections. In 1976, the Joint Commission on the Accreditation of Hospitals made the existence of a hospital-wide infection control program and a multidisciplinary infection control committee requirements for certification (19). This requirement has resulted in the organization of hospital infection control programs in nearly all hospitals in the United States. The complex interactions between these programs and local, state, and national organizations and agencies are discussed in Chapter 10. The microbiology laboratory plays a critical role in both the surveillance and control components of an infection control program. These important roles are detailed in Chapter 11. Surveillance conducted by these infection control programs has demonstrated that degrees of risk for the development of nosocomial infections differ among patients. Among the patients at highest risk are those in critical-care units. The special problems which these patients pose are addressed in Chapter 12. Some serious nosocomial infections are associated with medical devices (see Chapter 12); many others are associated with poor antiseptic practices. The role of sterilization and disinfection of patient care devices and equipment and the role of antiseptic practices in the prevention of nosocomial infections are summarized in Chapter 13. Finally, the microbiology laboratory can be a potential source of infections for hospital personnel. Measures for the prevention of laboratory-acquired infections are outlined in Chapter 14.

The purposes of this introductory chapter are to provide background information on the magnitude of the problem of nosocomial infections in acute-care hospitals and extended-care facilities in the United States, to present descriptive epidemiologic data on nosocomial infections, to review the modes of transmission of nosocomial pathogens, to discuss the epidemiology of pediatric and viral nosocomial infections, and to summarize data on the prolongation of stay, increases in direct patient care costs, and mortality associated with nosocomial infections.

MAGNITUDE OF THE PROBLEM

Definition

A nosocomial infection is an infection that is not present or incubating when a patient is admitted to a hospital. Bacterial nosocomial infections generally have onset more than 48 to 72 h after hospital admission. In determining whether a given infection is nosocomial or community acquired, the incubation period of the specific infection (e.g., varicella or hepatitis B) must be considered. Some nosocomial infections (e.g., surgical wound infections) may have onset after a patient is discharged from the hospital. Such infections are less likely to be recognized by infection control personnel than infections with onset during hospitalization.

Nosocomial infections may be either endogenous or exogenous (21). Endogenous infections are caused by those organisms that are present as part of the normal flora of the patient; exogenous infections are those caused by organisms acquired by exposure to hospital personnel, medical devices, or the hospital environment.

Incidence

It has been estimated that approximately 5.5% of hospitalized patients develop a nosocomial infection (4). Combining this incidence with the number of patients admitted annually to hospitals yields an estimate of approximately 2×10^6 nosocomial infections occurring each year (1, 4). Many variables influence the nosocomial infection rate in individual institutions. Examples of such variables include the average level of patient risk (i.e., the case mix) and the sensitivity and specificity of surveillance techniques. Among randomly selected patients on general medical and surgical services in a representative sample of U.S. hospitals in 1975 and 1976, the nosocomial infection rate was 5.2 per 100 discharges, or 5.2% (14). This study identified a number of factors, including age, sex, service, duration of hospitalization, presence of previous infections, underlying illness, operation, duration of surgery, urinary catheters, continuous ventilatory support, and immunosuppressive therapy, as determinants of patient risk. Intrinsic or host factors, such as age, sex, underlying disease, and immunosuppressive therapy, increase the risk of infection at all sites. The importance of specific extrinsic risk factors varies with the kind of infection (Table 1).

During a 3-year period from 1972 to 1975, the nosocomial infection rate at the University of Virginia Hospital was 6 per 100 admissions, or 6% (35). The

TABLE 1. Important extrinsic risk factors for major nosocomial infections

Infection	Risk factor
Urinary tract	Indwelling catheter Duration of catheterization Instrumentation
Pneumonia	Endotracheal tube Mechanical ventilation Thoracoabdominal surgery Nasogastric tube
Surgical wound	Preoperative stay Preoperative shaving Duration of surgery Degree of wound contamination Presence of foreign body
Primary bacteremia	Intravascular cannulas Duration of cannulation

sensitivity of this surveillance system was 82% compared with prospective daily chart review and 94% compared with retrospective chart review. Monthly infection rates during this period varied from 4 to 9%. Among hospitals participating in a statewide nosocomial infection surveillance system in Virginia from 1974 to 1977, the nosocomial infection rate was 3.3% (36). The sensitivity of surveillance in these hospitals was 69% compared with prospective daily chart review, and the specificity was 99%. Annual nosocomial infection rates varied from 2% in community hospitals to 9% in university hospitals. Although these studies report data collected nearly 10 years ago, they provide data on infection rates in different types of hospitals in which the sensitivity of the surveillance system was known.

The National Nosocomial Infections Study (NNIS) is a nationwide surveillance system organized by the Center for Disease Control (CDC) in 1970. NNIS represents the only source of current national data on the incidence of and trends in the rates of nosocomial infections. NNIS members are a nonrandom sample of U.S. hospitals, but the members are reasonably representative of teaching hospitals affiliated with medical schools and large nonteaching hospitals not affiliated with medical schools (18). The sensitivity of surveillance for nosocomial infections in these hospitals has not been systematically evaluated, but it has been estimated to be approximately 65% (4). During 1980 through 1983, the reported nosocomial infection rate was 3.3% (18). The vast majority of reported infections are sporadic or endemic. A much smaller proportion are related to outbreaks.

Institutionally acquired infections also affect patients in extended-care facilities. Data on the incidence of such infections are extremely limited, and no national data on the magnitude of the problem exist. One study of infections in 532 residents of seven skilled nursing homes in Utah revealed a prevalence rate of 16.2% (9).

Descriptive epidemiology

Nosocomial infection rates vary by service and by level of patient risk. During 1980 through 1983, the nosocomial infection rate was highest on the surgery, medicine, and gynecology services (Table 2). Rates of nosocomial infection also vary by surgical subspeciality. For example, at the University of Virginia Hospital between 1972 and 1975, the nosocomial infection rate ranged from a low of 0.2% for admissions to ophthalmology to a high of 11% for admissions to general surgery (32).

The urinary tract is the most common site of nosocomial infection. Urinary tract infections accounted for approximately 41% of all nosocomial infections reported to CDC during 1980 through 1983. During this period, surgical wound infections accounted for 19% of infections, lower respiratory tract infections accounted for 16%, and primary bacteremia accounted for 6%. Infections at all other sites accounted for the remaining 18% of infections.

The most common pathogen causing nosocomial infections during this period was *Escherichia coli*, followed by *Staphylococcus aureus*, enterococci, and *Pseudomonas aeruginosa*. However, the distribution of pathogens varied by service and by site of infection. *E. coli* was the leading pathogen on adult services, and *S. aureus* was the leading pathogen on pediatric and newborn services. *E. coli* was the most frequent cause of urinary tract infections, and *S. aureus* was the most frequent cause of surgical wound and cutaneous infections and primary bacteremia.

Infection rates also vary by the type of hospital. NNIS hospitals are stratified into nonteaching, small teaching (≤500 beds), or large teaching (>500 beds) hospitals. Infection rates were lowest for nonteaching hospitals (2.4 per 100 discharges), intermediate for small teaching hospitals (3.1 per 100 discharges), and highest for large teaching hospitals (4.0 per 100 discharges). These differential rates among the three hospital categories are probably accounted for by differences in the case mix among these categories, with the highest average patient risk among patients in the large teaching hospitals.

A review of 223 epidemics of nosocomial infection investigated by epidemiologists from CDC between 1956 and 1979 suggests that the epidemiology of nosocomial infection outbreaks differs from that of endemic infection. The gastrointestinal tract, skin, and bloodstream were the most common sites of infection, and *S. aureus*, *Salmonella* sp., and hepatitis B virus were the most common pathogens (27). A contaminated common vehicle was the mode of transmission in 23% of these outbreaks. During the 1970s, outbreaks investigated by CDC and others were often caused by multiply resistant organisms (6, 27). Infections with multiply resistant organisms are particularly common in critical-care areas (see Chapter 12).

Eleven percent of the outbreaks described by Stamm et al. were pseudoepidemics (27). In 11 (55%) of 20 pseudoepidemics investigated between 1956 and 1975, errors in the collection, handling, or processing of specimens were responsible for the outbreak (see Chapter 11). Of these 11 outbreaks, 9 were caused by gram-negative bacilli (30).

Common sites of institutionally acquired infections in residents of extended-care facilities differ somewhat from those of endemic infections in acute-care hospitals. In the Utah prevalence study, the most frequent infections were infected decubitus ulcers and conjunctivitis, followed by symptomatic urinary tract

TABLE 2. Nosocomial infection rate, by service, in NNIS hospitals, 1980 through 1983

Service	Rate[a]
Surgery	4.5
Medicine	3.6
Gynecology	2.9
Obstetric	1.8
Newborn[b]	1.2
Pediatric	1.1

[a] Per 100 discharges.
[b] Includes normal newborn, intermediate care, and NICUs.

infection, infections of the upper and lower respiratory tracts, and diarrhea (9). The prevalence of asymptomatic bacteriuria in patients with indwelling catheters was 85%. Antimicrobial resistance was common among these isolates. The importance of extended-care facilities as a reservoir of resistant organisms for patients in acute-care hospitals needs to be assessed.

Modes of transmission

Four modes of transmission of nosocomial pathogens exist (2). By far the most common mode of transmission of nosocomial pathogens in the United States is contact transmission, which may result from direct contact between patients or between patients and patient care personnel. Indirect contact transmission occurs when inanimate objects in the environment (e.g., endoscopes) become contaminated and are not adequately disinfected or sterilized between patients. Droplet transmission, another form of contact spread, occurs by means of large droplets, which can spread over a few feet.

The second most common mode of transmission is common vehicle transmission. Examples of contaminated common vehicles implicated in the transmission of nosocomial infections include food, blood and blood products, diagnostic reagents, and medications.

The third most common mode of transmission is airborne transmission. In such instances, infectious agents (e.g., agents for smallpox, aspergillosis, and legionellosis) have been transmitted over great distances. Vectorborne transmission of nosocomial pathogens is rarely, if ever, documented in the United States, but it can be important in hospitals in developing countries.

NOSOCOMIAL INFECTIONS IN PEDIATRIC PATIENTS

Nosocomial infections are an important cause of morbidity and mortality in hospitalized infants and children. The estimated incidence of nosocomial infections in infants and children varies widely and depends upon the age of the patient and the presence of other factors that determine patient risk. Nosocomial infections develop in <1% of infants admitted to well-baby nurseries (WBN) (20), 1 to 5% of infants and children admitted to general pediatric wards (8, 34), and up to 29% of infants admitted to neonatal intensive care units (NICU) (17, 32).

Data on the incidence of nosocomial infections in pediatric patients are limited. Overall infection rates at hospitals for children range from 2.8 to 6.5 infections per 100 discharges (3, 8, 26). Although the overall rate of infection in pediatric patients appears to be lower than the rate reported for adults, several factors might result in an underestimate of the present nosocomial infection rate in pediatric patients. In these studies, no attempt was made to identify viral nosocomial pathogens. Furthermore, these studies were conducted over 10 years ago, and the number of immunocompromised patients has increased substantially over this period as a result of the intense immunosuppressive therapy used in cancer chemotherapy and organ transplantation. The emergence of pediatric intensive care units has also led to an increase in the number of patients at high risk for nosocomial infection.

Between 1980 and 1983, the nosocomial infection rate on pediatric and newborn services in NNIS hospitals was 1.2 per 100 discharges, or 1.2%. Differences in nosocomial infection rates are again apparent among the three hospital categories, with the highest rates in the large teaching hospitals (Table 3). These data suggest that patients with severe underlying disease who are hospitalized at large tertiary-care centers are at the greatest risk of developing a nosocomial infection.

NNIS newborn service data cannot be stratified by the type of unit (i.e., WBN or NICU). However, Maguire et al. have shown that the risk of a nosocomial infection in an infant admitted to an NICU is approximately 28 times greater than that in an infant admitted to a WBN (20). Although the risk of a nosocomial infection is low in WBNs, conditions such as overcrowding or understaffing have been shown to be associated with a greater risk of infection (13). These data suggest that hospitals with overcrowded or understaffed WBNs and those with NICUs have higher rates of infection.

The epidemiology of nosocomial infections in infants and children is unique; in particular, the rates of infection by site and pathogen differ from those reported in adults. Unlike adult services, where the urinary tract is the most common site of infection, cutaneous infections were the most frequent infections on the combined NNIS pediatric and newborn services, followed by lower respiratory tract infections, primary bacteremia, urinary tract infections, and surgical wound infections (Table 4). Differences are also apparent when the pediatric and newborn services are analyzed separately. On the newborn service, cutaneous infections predominate, lower respiratory tract infections and primary bacteremia are

TABLE 3. Nosocomial infection rate for newborn and pediatric services, by hospital category, in NNIS hospitals, 1980 through 1983

Hospital category[a]	Nosocomial infection rate[b] by service		
	Newborn	Pediatric	Both
LMSA	16.3	15.3	15.9
SMSA	11.1	11.7	11.3
NMSA	7.3	2.7	5.7
All	12.2	11.2	11.8

[a] LMSA, Large medical school-affiliated hospitals; SMSA, small medical school-affiliated hospitals; NMSA, hospitals not affiliated with a medical school.
[b] Per 1,000 discharges.

TABLE 4. Nosocomial infection rate for pediatric and newborn services, by infection, in NNIS hospitals, 1980 through 1983

Infection	Nosocomial infection rate[a] by service		
	Newborn	Pediatric	Both
Cutaneous	46.1	14.8	34.3
Lower respiratory tract	13.8	18.5	15.6
Primary bacteremia	14.4	14.4	14.4
Urinary tract	5.2	17.4	9.8
Surgical wound	4.1	11.2	6.8

[a] Per 10,000 discharges.

less frequent, and urinary tract and surgical wound infections are least frequent. In contrast, on the pediatric services, urinary tract and lower respiratory tract infections are the most frequent, followed by cutaneous infections, primary bacteremia, and surgical wound infections. The proportion of newborn and pediatric patients included in the analysis, therefore, may influence the distribution of infections by site and may explain why the predominant site of infection varies in different hospitals (3, 8, 26, 34).

The distribution of pathogens causing nosocomial infections in pediatric patients is also different from that in adults. In contrast to adult services, where *E. coli* is the organism most frequently causing nosocomial infections, on the individual and combined NNIS pediatric and newborn services, *S. aureus* is the most common etiologic agent (Table 5). These findings are similar to those in other reports (3, 8, 26), except that the NNIS data reflect the recent emergence of *Staphylococcus epidermidis* as an important pathogen in the pediatric population (23). NNIS data reveal further differences between pediatric and newborn services. *S. aureus* accounts for nearly 40% of the pathogens isolated from patients with nosocomial infections on the newborn services, but it accounts for only 17% of the pathogens isolated from such patients on the pediatric service. Group B streptococcal infections were reported nearly seven times more frequently from the newborn services than from the pediatric services.

NOSOCOMIAL VIRAL INFECTIONS

The recognition of viral nosocomial infections depends on active surveillance for these infections and the availability of a viral diagnostic laboratory for the processing of specimens. Recent developments in rapid diagnostic viral techniques have resulted in better estimates of the incidence and relative importance of nosocomial viral infections. Estimates of the incidence of these infections vary widely and depend on the population studied and the surveillance and diagnostic techniques employed. NNIS data from 1980 to 1983 indicate that approximately 1 of 10,000 hospitalized patients developed a nosocomial viral infection and that 0.3% of nosocomial infections had a viral etiology. However, since many NNIS hospitals do not have diagnostic virology laboratories, these figures certainly underestimate the magnitude of the problem. Valenti et al. estimate that viruses cause over 5% of nosocomial infections and that over 75,000 nosocomial viral infections occur in the United States each year (28).

The epidemiology of viral nosocomial infections differs from that of bacterial nosocomial infections. Over 90% of nosocomial viral infections involve the respiratory or gastrointestinal tract, whereas fewer than 15% of bacterial nosocomial infections occur at these sites (28, 33). In pediatric patients, the incidence of nosocomial upper and lower respiratory tract infections caused by viruses far exceeds that caused by bacteria; in one study, nearly one of every six pediatric patients under 4 years of age admitted to the hospital during the winter and spring developed a nosocomial viral respiratory tract infection (33). The age- and service-specific attack rates for viral nosocomial infections also differ from those seen for bacterial infections. Whereas nosocomial bacterial infections occur most frequently in older patients on the medicine and surgery services, nosocomial viral infections occur most frequently in younger patients on the newborn and pediatric services. In NNIS hospitals between 1980 and 1983, 0.3% of all infections on the medicine and surgery services were caused by viruses, whereas 1.3% of all infections on the pediatric and newborn services were caused by viruses. Valenti et al. also found that viral nosocomial infections occurred more frequently on pediatric services than on adult services; 72.4 per 10,000 patients admitted to the pediatric service developed a nosocomial viral infection, while 13.8 and 6.6 per 10,000 adults admitted to the medicine and surgery services, respectively, developed a viral infection (28). Observations by Welliver and McLaughlin that viral agents were responsible for 14.3% of all nosocomial infections at their hospital for children provide further evidence that viruses are an important cause of nosocomial infections in infants and children (31).

A number of viruses have been associated with nosocomial infections. Respiratory viruses include influenza, parainfluenza, and respiratory syncytial viruses, adenoviruses, and rhinoviruses. Hepatitis viruses include hepatitis A, B, and non-A, non-B. Herpesviruses include cytomegalovirus, varicella-zoster virus, and the herpes simplex viruses. Rotaviruses

TABLE 5. Nosocomial infection rate and percent distribution of the 10 most frequently identified pathogens on pediatric and newborn services in NNIS hospitals, 1980 through 1983

Pathogen	Service					
	Newborn		Pediatric		Both	
	Rate[a]	%[b]	Rate	%	Rate	%
S. aureus	44.3	38	17.8	17	34.3	31
E. coli	10.3	9	14.6	14	12.0	11
S. epidermidis	12.2	10	9.1	9	11.0	10
Klebsiella spp.	6.8	6	8.5	8	7.4	7
Enterococci	6.8	6	5.9	6	6.4	6
Candida spp.	4.6	4	7.6	7	5.8	5
Group B *Streptococcus* sp.	8.0	7	1.2	1	5.4	5
P. aeruginosa	2.8	2	8.1	8	4.8	4
Enterobacter spp.	3.2	3	2.5	2	2.9	3
Proteus spp.	1.4	1	2.6	3	1.8	3

[a] Per 10,000 discharges.
[b] Percentage of all pathogens isolated from patients with nosocomial infections.

are also associated with nosocomial infections. The source of viral nosocomial infections can be either endogenous or exogenous. Endogenous infection, or reactivation of latent virus, is most commonly identified in adults, in whom nosocomial infections caused by cytomegalovirus, herpes simplex virus, and varicella-zoster virus account for most nosocomial viral infections (28). However, exogenous infections, in particular those caused by influenza and hepatitis viruses, also are identified frequently in adults. On pediatric services, most nosocomial viral infections are exogenous. The etiologic agents associated with hospital-acquired, exogenous viral infections are similar to those circulating in the community; each year, respiratory syncytial virus and rotavirus outbreaks occur on pediatric services, and influenza outbreaks occur on adult services (28). In most cases, nosocomial spread of these viruses occurs through contact transmission involving patients and patient-care personnel. The interruption of viral transmission requires an increased awareness of the importance of viruses as nosocomial pathogens, rapid identification of the etiologic agent, initiation of isolation precautions for infected patients, vaccination or prophylactic therapy when appropriate (e.g., for influenza), and strict handwashing before and between all patient contacts.

RAMIFICATIONS OF NOSOCOMIAL INFECTIONS

Prolongation of stay

In a study done in 1975 and 1976 in three hospitals, physicians used standardized techniques to assess the extent to which nosocomial infections prolonged hospital stay (15). Nosocomial infections prolonged hospitalization by an average of 3.1 to 4.5 days in the three hospitals. Prolongation of stay varied by site and was greatest for nosocomial pneumonia, bacteremia, and surgical wound infections. Comparisons of infected patients with matched or unmatched controls have yielded even higher estimates of prolongation of stay (16, 22). All studies of prolongation of stay have a number of methodologic problems, and their results must be interpreted cautiously (22). However, these figures appear to be reasonable minimum estimates.

Direct patient care costs

In the study conducted in three hospitals in 1975 and 1976, the increase in direct patient care charges, excluding fees for physicians, resulting from nosocomial infections ranged from an average of $590 to $641 per infection (15). Studies that compared costs for infected patients with costs for matched or unmatched controls have yielded even higher estimates (16, 22). As with prolongation of stay, the average increase in direct patient care charges varied according to the site of infection, with lower respiratory tract infections, bacteremia, and surgical wound infections associated with the largest amount of extra charges. Surgical wound and lower respiratory tract infections account for a disproportionately large share of the costs of nosocomial infections (24).

Mortality

Among infections reported by NNIS hospitals during 1980 through 1983, 1% were reported to cause the death of the patient and 3% to contribute to death. Applied to the estimated 2×10^6 annual nosocomial infections nationwide, these figures suggest that nosocomial infections may cause approximately 20,000 deaths and contribute to approximately 60,000 deaths each year. These data should, however, be interpreted cautiously, since the relationship of infection to death in individual cases may be quite difficult to assess. In addition, many patients dying as a direct result of a nosocomial infection have severe underlying diseases and are considered to be in the terminal phase of illness at the time of admission (10). In fact, in patients who were considered terminal on admission, the frequency of nosocomial infection was similar in those who died and those who survived. However, in those patients who were not considered terminal on admission, nosocomial infections were significantly more common in those who died than in those who survived (11). Nosocomial pneumonia was the most frequent infection causing death; however, primary bacteremia and meningitis were most likely to cause death (10). Although nosocomial urinary tract infections have infrequently been reported to cause death, a recent study suggests that a nosocomial urinary tract infection occurring in a catheterized patient results in a threefold increase in risk of death (25). Further studies are required to document the contribution of nosocomial infections to mortality.

INFECTION SURVEILLANCE AND CONTROL PROGRAMS

The results of the recently completed Study on the Efficacy of Nosocomial Infection Control (SENIC Project) conducted by CDC indicate that well-organized hospital infection surveillance and control programs can be expected to prevent approximately one-third of all nosocomial infections (R. W. Haley, D. H. Culver, J. W. White, W. M. Morgan, T. G. Emori, V. P. Munn, and T. M. Hooton, Am. J. Epidemiol., in press). The preventability of infections varies somewhat among the major sites, ranging from a low of 22% for pneumonia to a high of 35% for surgical wound infections and bacteremia. The critical components of effective programs have been identified and include both surveillance and control strategies (Haley et al., in press). An analysis of the costs of infection control programs and the potential savings in direct patient care costs associated with prevented infections suggests that infection surveillance and control programs that prevent more than 5 to 6% of infections result in a net savings to the hospital (7, 12). This figure should be considered by hospital administrators when they evaluate potential cost containment strategies as hospitals enter the era of prospective reimbursement based on diagnosis-related groups (29).

LITERATURE CITED

1. **Bennett, J. V.** 1978. Human infections: economic implications and prevention. Ann. Intern. Med. **89:**761–763.
2. **Brachman, P. S.** 1979. Epidemiology of nosocomial infections, p. 9. *In* J. V. Bennett and P. S. Brachman (ed.), Hospital infections, 1st ed. Little, Brown & Co., Boston.
3. **Cooper, R. G., and C. Sumner.** 1970. Hospital infection data from a children's hospital. Med. J. Aust. **2:**1110–1113.
4. **Dixon, R. E.** 1978. Effect of infections on hospital care. Ann. Intern. Med. **89:**749–753.

5. **Eickhoff, T. C.** 1978. Standards for hospital infection control. Ann. Intern. Med. **89:**829–831.
6. **Eickhoff, T. C.** 1982. Nosocomial infections. N. Engl. J. Med. **306:**1545–1546.
7. **Eickhoff, T. C.** 1984. Hospital epidemiology: an emerging discipline, p. 241. *In* J. S. Remington and M. N. Swartz (ed.), Current clinical topics in infectious diseases, vol. 4. McGraw-Hill Book Co., New York.
8. **Gardner, P., and D. G. Carles.** 1972. Infections acquired in a pediatric hospital. J. Pediatr. **81:**1205–1210.
9. **Garibaldi, R. A., S. Brodine, and S. Matsumiya.** 1981. Infections among patients in nursing homes: policies, prevalence, and problems. N. Engl. J. Med. **305:**731–735.
10. **Gross, P. A., H. C. Neu, P. Aswapokee, C. Van Antwerpen, and N. Aswapokee.** 1980. Deaths from nosocomial infections: experience in a university hospital and a community hospital. Am. J. Med. **68:**219–223.
11. **Gross, P. A., and C. Van Antwerpen.** 1983. Nosocomial infections and hospital deaths: a case-control study. Am. J. Med. **75:**658–662.
12. **Haley, R. W.** 1978. Preliminary cost-benefit analysis of hospital infection control programs (The SENIC Project), p. 93–95. *In* F. Daschner (ed.), Proceedings of an International Workshop at Baiersbronn. Gustav Fischer Verlag, Stuttgart.
13. **Haley, R. W., and D. A. Bregman.** 1982. The role of understaffing and overcrowding in recurrent outbreaks of staphylococcal infection in a neonatal special-care unit. J. Infect. Dis. **145:**875–885.
14. **Haley, R. W., T. M. Hooton, D. H. Culver, R. C. Stanley, T. G. Emori, C. D. Hardison, D. Quade, R. H. Shachtman, D. R. Schaberg, B. V. Shah, and G. D. Schatz.** 1981. Nosocomial infections in U.S. hospitals, 1975–1976: estimated frequency by selected characteristics of patients. Am. J. Med. **70:**947–959.
15. **Haley, R. W., D. R. Schaberg, K. B. Crossley, S. D. Von Allmen, and J. E. McGowan.** 1981. Extra charges and prolongation of stay attributable to nosocomial infections: a prospective interhospital comparison. Am. J. Med. **70:**51–58.
16. **Haley, R. W., D. R. Schaberg, S. D. Von Allmen, and J. E. McGowan.** 1980. Estimating the extra charges and prolongation of hospitalization due to nosocomial infections: a comparison of methods. J. Infect. Dis. **141:**248–257.
17. **Hemming, V. G., J. C. Overall, and M. R. Britt.** 1976. Nosocomial infections in a newborn intensive care unit: results of forty-one months of surveillance. N. Engl. J. Med. **294:**1310–1316.
18. **Hughes, J. M., D. H. Culver, J. W. White, W. R. Jarvis, W. M. Morgan, V. P. Munn, J. L. Mosser, and T. G. Emori.** 1983. Nosocomial infections surveillance, 1980–1982. Morbid. Mortal. Weekly Rep. CDC Surveillance Summaries **32**(Spec. Suppl. 4):1SS–16SS.
19. **Joint Commission on Accreditation of Hospitals.** 1976. Accreditation manual for hospitals, p. 49. Joint Commission on Accreditation of Hospitals, Chicago.
20. **Maguire, G. C., J. Nordin, M. G. Myers, F. P. Koontz, W. Heirholzer, and E. Nassif.** 1981. Infections acquired by young infants. Am. J. Dis. Child. **135:**693–698.
21. **Maki, D. G.** 1978. Control of colonization and transmission of pathogenic bacteria in the hospital. Ann. Intern. Med. **89:**777–780.
22. **McGowan, J. E., Jr.** 1982. Cost and benefit: a critical issue for hospital infection control. Am. J. Infect. Control **10:**100–108.
23. **Munson, D. P., T. R. Thompson, D. E. Johnson, F. S. Rhame, N. VanDrunen, and P. Ferrieri.** 1982. Coagulase-negative staphylococcal septicemia: experience in a newborn intensive care unit. J. Pediatr. **101:**602–605.
24. **Pinner, R. W., R. W. Haley, B. A. Blumenstein, D. R. Schaberg, S. D. Von Allmen, and J. E. McGowan, Jr.** 1982. High cost nosocomial infections. Infect. Control **3:**143–149.
25. **Platt, R., B. F. Polk, B. Murdock, and B. Rosner.** 1982. Mortality associated with nosocomial urinary-tract infection. N. Engl. J. Med. **307:**637–642.
26. **Roy, T. E., S. McDonald, M. L. Patrick, and J. A. Keddy.** 1962. A survey of hospital infections in a pediatric hospital. I. Description of hospital, organization of survey, population studied, and some general findings. Can. Med. Assoc. J. **87:**531–538.
27. **Stamm, W. E., R. A. Weinstein, and R. E. Dixon.** 1981. Comparison of endemic and epidemic nosocomial infections. Am. J. Med. **70:**393–397.
28. **Valenti, W. M., M. A. Menegus, C. B. Hall, P. H. Pincus, and R. G. Douglas, Jr.** 1980. Nosocomial viral infections. I. Epidemiology and significance. Infect. Control **1:**33–37.
29. **Vladeck, B. C.** 1984. Medicare hospital payment by diagnosis-related groups. Ann. Intern. Med. **100:**576–591.
30. **Weinstein, R. A., and W. E. Stamm.** 1977. Pseudoepidemics in hospital. Lancet **ii:**862–864.
31. **Welliver, R. C., and S. McLaughlin.** 1984. Unique epidemiology of nosocomial infection in a children's hospital. Am. J. Dis. Child. **138:**131–135.
32. **Wenzel, R. P.** 1981. Surveillance and reporting of hospital-acquired infections, p. 35. *In* R. P. Wenzel (ed.), CRC handbook of hospital-acquired infections, 1st ed. CRC Press, Inc., Boca Raton, Fla.
33. **Wenzel, R. P., E. C. Deal, and J. P. Hendley.** 1977. Hospital-acquired viral respiratory illness on a pediatric ward. Pediatrics **60:**367–371.
34. **Wenzel, R. P., C. A. Osterman, and K. J. Hunting.** 1976. Hospital-acquired infections. II. Infection rates by site, service and common procedures in a university hospital. Am. J. Epidemiol. **104:**645–651.
35. **Wenzel, R. P., C. A. Osterman, K. J. Hunting, and J. M. Gwaltney, Jr.** 1976. Hospital-acquired infections. I. Surveillance in a university hospital. Am. J. Epidemiol. **103:**251–260.
36. **Wenzel, R. P., C. A. Osterman, T. R. Townsend, J. M. Veazey, Jr., K. H. Servis, L. S. Miller, R. B. Craven, G. B. Miller, Jr., and R. S. Jackson.** 1979. Development of a statewide program for surveillance and reporting of hospital-acquired infections. J. Infect. Dis. **140:**741–746.

Chapter 10

Nosocomial Infection Surveillance and Control Programs

JULIA S. GARNER AND T. GRACE EMORI

Nosocomial infection surveillance and control programs in acute-care hospitals in the United States have evolved largely over the last 15 years. Before 1970, activities performed in the name of infection control in many hospitals consisted primarily of routine microbiologic culturing of air and environmental surfaces, such as floors, walls, and tabletops in various areas of the hospital. Only a few hospitals had infection control committees, even though such committees had been recommended by the American Hospital Association as early as 1958, and almost none were conducting surveillance of nosocomial infections (13).

In 1970, hospital infection control in the United States altered its course, shifting its emphasis away from the environment to the surveillance of infections in patients. The impetus for the change was the first International Conference on Nosocomial Infections (9) held at the Centers for Disease Control in Atlanta, Ga. The conference brought together persons working in the field of nosocomial infections to share their experience and knowledge of the epidemiology of nosocomial infections and to identify effective systems for the reduction of these infections. In a summary of the conference, R. E. O. Williams stated, "Quite clearly, the first message from this conference is the need for surveillance of nosocomial infections" (16). Two other statements by Williams indicated that infection control personnel had begun to abandon their preoccupation with the hospital environment: "Environmental sampling is very rarely relevant," and "It is a mistake to rely on microbiological tests just to see whether a shelf is clean or not" (16). The importance of the microbiology laboratory as a source of information about nosocomial infections, however, was stressed during the conference (12) (see Chapter 11).

In the brief span from 1970 to 1976, the vast majority of U.S. hospitals established infection surveillance and control programs (13). By 1976, about 25% of hospitals had reduced or discontinued routine culturing of the environment; 87% of hospitals were conducting some form of infection surveillance, with half having very active surveillance programs. Almost without exception, hospitals had an infection control committee, and 42% had a nurse working in infection control at least half-time (13).

Most of the early infection control efforts and changes in program direction occurred voluntarily in hospitals, since no regulations or standards for infection control programs were in force until 1976, when infection control became a requirement for hospitals seeking accreditation from the Joint Commission on Accreditation of Hospitals (14). Moreover, the early effect of accreditation bodies on infection control has primarily been to promote and extend what hospitals had already started voluntarily. Thus, the most important determinants of the direction, structure, and activities of the infection control program in each hospital are the characteristics of the hospital. Hospitals set the priorities for infection control based on an assessment of their own infection problems, patient risks, and availability of resources.

INFECTION CONTROL PROGRAMS IN HOSPITALS

The infection control committee in most hospitals is the central decision- and policy-making body for infection control in the hospital. Ideally, the regular members of the committee are interested and influential representatives from major hospital departments, including clinical microbiology. Some committees also have ad hoc members representing ancillary departments that occasionally deal with infection control problems. In addition, most committees have a working staff who provide surveillance data and technical information for the committee. The staff usually consists of a physician who serves as chairperson of the committee and an infection control coordinator, usually a nurse, who works full- or part-time, depending on the size of the hospital. In addition to having a physician chairperson of the infection control committee, an increasing number of hospitals (usually those affiliated with medical schools) have a physician committee member with special training in infectious diseases and epidemiology who serves as the hospital epidemiologist. Infection control committees usually meet every 1 to 2 months to review hospital-specific information on the occurrence of nosocomial infections, evaluate technical information, and formulate policy. In an ideal administrative structure, hospital by-laws permit infection control committee decisions to receive serious attention rapidly through deliberation or approval by the medical board or board of trustees.

The number, type, and administrative placement of infection control personnel are usually determined by the size of the hospital and whether it is affiliated with a medical school. In large, medical school-affiliated hospitals, infection control is often a small but separate hospital department under the direction of a physician-hospital epidemiologist, who also serves as a member of the faculty. The number of infection control personnel working in the department varies according to program activities. In general, one full-time-equivalent infection control coordinator is recommended for surveillance and control activities for every 250 hospital beds. In medium-sized community hospitals, an infection control coordinator may be the only person in the hospital with specific full-time and continuous responsibility for infection control. In such situations, there are various options for the placement of the coordinator in the structure of the hospital. Above all, administrative placement should provide maximum flexibility to facilitate the crossing of departmental lines, since the infection control coor-

TABLE 1. Names and addresses of agencies and organizations involved in the prevention and control of nosocomial infections

American Academy of Pediatrics (AAP)
P.O. Box 1034
Evanston, IL 60204

American College of Surgeons (ACS)
55 East Erie Street
Chicago, IL 60611

American Hospital Association (AHA)
840 North Lake Shore Drive
Chicago, IL 60611

American Society for Microbiology (ASM)
1913 I Street, N.W.
Washington, DC 20006

Association of Operating Room Nurses (AORN)
10170 East Mississippi Avenue
Denver, CO 80231

Association for Practitioners in Infection Control (APIC)
505 Hawley Street
Mundelein, IL 60060

Centers for Disease Control (CDC)
Public Health Service
U.S. Department of Health and Human Services
1600 Clifton Road, N.E.
Atlanta, GA 30333

Certification Board for Infection Control (CBIC)
P.O. Box 5428
Willowick, OH 44094

Environmental Protection Agency (EPA)
401 M Street, S.W.
Washington, DC 20460

Food and Drug Administration (FDA)
U.S. Department of Health and Human Services
5600 Fishers Lane
Rockville, MD 20857

Health Care Financing Administration (HCFA)
U.S. Department of Health and Human Services
6325 Security Boulevard
Baltimore, MD 21207

Joint Commission on Accreditation of Hospitals (JCAH)
875 North Michigan Avenue
Chicago, IL 60611

dinator acts as the liaison between the infection control committee and the hospital departments when infection control policies and procedures are proposed, reviewed, implemented, and evaluated.

A wide range of activities is performed by infection control coordinators in U.S. hospitals. On the average, in 1976 they were spending 46% of their time performing surveillance, 23% researching and developing infection control policies, 13% training hospital personnel about infection control, 10% consulting with doctors, nurses, and other personnel or hospital departments, and 8% investigating suspected outbreaks of infection (10). Most infection control coordinators are nurses, but some have training in other fields, such as medical technology and microbiology. Regardless of professional background, infection control coordinators perform their activities similarly, except that nurses tend to emphasize infection control training for hospital personnel, and coordinators with laboratory backgrounds spend more of their time conducting surveillance and investigating outbreaks (11).

Through their various activities, the infection control coordinators come into contact with all levels of clinical, administrative, and support service personnel working throughout the hospital and are expected to be able to interact effectively with all of them. These contacts with hospital personnel range from informal discussions of specific patient care practices with nurses on the ward or conducting formal teaching sessions with nurses, doctors, or other members of the hospital staff to holding conferences with administrators, department heads, or chiefs of medical services.

Several committees or programs within the hospital also may collect data and discuss problems of interest to the infection control program. For example, the Pharmacy and Therapeutics Committee may collect data on the use of systemic antimicrobial agents in the hospital. In addition, programs such as utilization review, risk management, and quality assurance may collect and analyze data on high-risk patients. Since high-risk patients often have nosocomial infections, these data may be useful for infection control program activities. Thus, infection control personnel must have a sound technical knowledge of infection control and an awareness of other committees or programs within the hospital as well as organizations outside the hospital that can influence infection control.

ORGANIZATIONS THAT INFLUENCE INFECTION CONTROL IN HOSPITALS

Various accreditation bodies, governmental agencies, and national organizations are involved in infection control (Table 1). The influence of these organizations (especially the Joint Commission on Accreditation of Hospitals and the Health Care Financing Administration) on infection control in hospitals can be significant. Some national organizations assist hospital infection control programs by providing information and education resources.

Accreditation and Governmental Agencies

Joint Commission on Accreditation of Hospitals

The Joint Commission on Accreditation of Hospitals (JCAH), an organization sponsored by the American College of Physicians, American College of Surgeons, American Dental Association, American Hospital Association, and American Medical Association, establishes various standards for the operation of hospitals and offers a voluntary accreditation program for hospitals that comply with the standards. The standards and their interpretation, which are revised every few years, can be found in the JCAH publication *Accreditation Manual for Hospitals* (15).

An important requirement for obtaining JCAH accreditation is an on-site hospital visit conducted by a JCAH survey team. The survey team usually consists of a hospital administrator, a physician who is often a surgeon, and a nurse who has had experience in nursing administration. At the completion of the visit, the surveyors and hospital officials hold a summation conference during which survey findings are discussed and clarified; hospital officials are given full opportunity to comment on any adverse or other findings noted by the surveyors. Infection control personnel should expect to participate in the survey and the summation conference. Hospitals may be accredited for 1 or 2 years. One-year accreditation means that the survey findings necessitate an on-site visit within 1 year to assess progress in the correction of deficiencies. Those hospitals receiving the maximum 2-year accreditation must also demonstrate progress in the correction of any deficiencies. Hospitals with JCAH accreditation are eligible to participate in Medicare and Medicaid programs without an additional survey from the state health department.

Health Care Financing Administration

The Health Care Financing Administration (HCFA), an agency of the U.S. Department of Health and Human Services, administers reimbursements to hospitals under Medicare and Medicaid. HCFA also sets the conditions for participation in the government reimbursement program for hospitals and long-term care facilities that are not accredited by the JCAH or American Osteopathic Association. Proposed HCFA standards for hospital infection control are found in 42 CFR, Section 482.42, which was published in the *Federal Register* (11a). Copies of HCFA standards can be purchased from the U.S. Government Printing Office, Washington, DC 20402.

Food and Drug Administration

The Food and Drug Administration (FDA), also an agency of the U.S. Department of Health and Human Services, has several activities that influence hospital infection control programs. The Center for Drugs and Biologics develops FDA policy for the safety, effectiveness, and labeling of all drugs and biologics for human use. The FDA also develops and implements standards for the safety and effectiveness of all over-the-counter drugs and products that contain antimicrobial ingredients, such as handwashes for health care personnel, skin antiseptics, and preoperative skin preparations. Persons responsible for the selection of such products for use in hospitals should be guided by information from the FDA on the safety and effectiveness of the ingredients. The Center for Devices and Radiologic Health develops FDA policy and priorities for the safety, effectiveness, and labeling of medical devices for human use. Problems or potential problems with drugs, biologics, and medical devices should be reported promptly to the nearest regional office of the FDA.

Environmental Protection Agency

Two activities of the Environmental Protection Agency influence hospital infection control programs. The Office of Solid Waste develops hazardous waste standards, including guidelines for the disposal of infective or potentially infective solid waste from laboratories and other areas of the hospital. The Environmental Protection Agency also registers liquid chemical formulations that are used for disinfection or sterilization in hospitals. Persons responsible for the selection of sterilants and disinfectants should be guided by information obtained from the Environmental Protection Agency on the safety and effectiveness of the ingredients.

Centers for Disease Control

The Centers for Disease Control, an agency of the U.S. Department of Health and Human Services, has an organizational unit, the Hospital Infections Program (HIP) in the Center for Infectious Diseases, that is involved exclusively with nosocomial infections. HIP personnel conduct studies to evaluate methods for the prevention and control of nosocomial infections (e.g., the Study on the Efficacy of Nosocomial Infection Control and evaluation of antimicrobial susceptibility testing methodology [see Section XII]). HIP provides epidemic assistance and consultation upon request from state health departments to hospitals regarding the identification and control of nosocomial infections, including those related to dialysis systems. HIP also conducts surveillance of nosocomial infections through the National Nosocomial Infections Study (see Chapter 9). Guidelines for the prevention and control of the most common nosocomial infections (urinary tract and surgical wound infections, pneumonia, and bacteremia), as well as guidelines for isolation precautions, personnel health, and sterilization and disinfection of patient-care equipment, are issued periodically by HIP, and copies are sent to all acute-care hospitals in the United States. Additional copies of the guidelines can be purchased from the National Technical Information Service, U.S. Department of Commerce, 5285 Port Royal Road, Springfield, VA 22161. HIP does not conduct training courses on the surveillance and control of nosocomial infections; however, such courses are offered by a separate organizational unit at the Centers for Disease Control, the Center for Professional Development and Training. HIP personnel participate as speakers in these and other national and regional training programs and educational conferences.

Local and state health departments

Local and state health departments in the United States vary greatly in their range of services, expertise, and interest in nosocomial infection control. Few of them, except those in large cities, have personnel primarily involved in nosocomial infection control activities. Local health departments are often responsible for the inspection of hospital kitchens and can usually provide information about rodent and pest control, local ordinances for solid waste disposal from hospitals, and the sanitation of hydrotherapy pools in hospitals.

Most state health departments, in addition to having regulatory authority, have programs or personnel involved in nosocomial infection control activities. Virtually all state health departments have an organizational unit directed by a trained epidemiologist who can provide epidemiologic advice and assistance in the investigation of outbreaks of nosocomial infections. Moreover, health department personnel are familiar with other state or federal agencies whose knowledge and resources might be helpful during an investigation. In addition, state health departments with an organizational unit devoted to nosocomial infection control may sponsor or cosponsor training sessions on various aspects of nosocomial infections.

National Associations and Organizations

American Academy of Pediatrics

The American Academy of Pediatrics (AAP) has published recommendations on the care of pediatric patients with infectious diseases in hospitals. The infuence of the AAP can be very strong in hospitals with a large pediatric service. These recommendations are published every few years by the AAP Committee on Infectious Diseases in a handbook entitled *Report of the Committee on Infectious Diseases* (4), commonly referred to as the Red Book. The Red Book contains information on the clinical manifestations and epidemiology of pediatric infectious diseases ranging from adenovirus infection to varicella, as well as recommendations for the treatment and control of pediatric infectious diseases in hospitals and clinics. In addition, the Red Book is an excellent resource for information about incubation periods and various childhood viral diseases.

The AAP Committee on the Fetus and Newborn, together with the Committee on Obstetrics: Maternal and Fetal Medicine, American College of Obstetricians and Gynecologists, publishes *Guidelines for Perinatal Care* (5). A chapter on the control of infections in obstetric and nursery areas contains recommendations for preventive measures such as handwashing, personnel dress codes, and skin care of newborns, as well as recommendations for environmental cleaning and disinfection and the sterilization of equipment used for perinatal care. Recommendations for isolation precautions for mothers and neonates with infections and for the management of nursery outbreaks of disease are included.

American College of Surgeons

The recommendations of the American College of Surgeons influence surgeons and operating room practices in U.S. hospitals. The Committee on Control of Surgical Infections publishes the *Manual on Control of Infection in Surgical Patients* (6). The manual discusses all aspects of the prevention of surgical wound infections and contains various recommendations concerning those infections, including classification of wounds, immediate preoperative preparation of the patient, preparation of the operating team and supporting personnel, and preparation and maintenance of the operating room environment.

American Hospital Association

The American Hospital Association (AHA) has a long history of involvement in several areas of nosocomial infection control activities. It sponsors the Advisory Committee on Infections Within Hospitals, which meets several times a year to discuss and advise AHA about current or emerging infection control problems facing hospitals. The Committee has recently issued recommendations for the control of rubella in hospitals (1), for the use of hepatitis B vaccine in hospital employees (2), and for a hospital-wide approach to acquired immunodeficiency syndrome (3). Members of the Committee, along with other experts, also appear in AHA-sponsored teleconferences that allow question-and-answer sessions between Committee members and personnel in hospitals. The Committee writes a handbook, *Infection Control in the Hospital* (7), that contains information on various aspects of nosocomial infections.

AHA also sponsors several societies that influence hospital departments indirectly involved with infection control. For example, the American Society of Hospital Central Service Personnel, an affiliate of AHA, conducts educational programs for hospital central service personnel and publishes manuals for the operation of central service departments in hospitals.

American Society for Microbiology

The American Society for Microbiology is a large, multidisciplinary organization whose members have various interests in laboratory and clinical infectious diseases, including nosocomial infections. Division L (Nosocomial Infections) is responsible for the part of the annual meeting of the Society which includes lectures, symposiums, and slide and poster presentations on epidemiologic and microbiologic aspects of nosocomial infections. In addition to providing a forum for the presentation of scientific data, Division L sessions serve as a link between nosocomial infection control and clinical microbiology.

Association of Operating Room Nurses

The Association of Operating Room Nurses (AORN) is a professional organization of more than 29,000 registered, professional nurses engaged in operating room nursing at the supervisory, teaching, and staff levels in hospitals. The AORN annual meeting usually includes one or more sessions on infection control in the operating room. The Recommended Practices Subcommittee of AORN formulates the recommendations of the Association for operating room nursing and publishes them periodically in the *AORN Journal*. Many of these practices involve infection control in the operating room.

Association for Practitioners in Infection Control

The Association for Practitioners in Infection Control (APIC) was founded in 1972 to enhance communication among and develop educational programs for infection control personnel in hospitals. About 75% of the approximately 7,000 members of APIC are infection control nurses. APIC holds an annual educational conference that includes sessions for beginning and advanced infection control personnel. The APIC Curriculum Committee publishes a large two-part manual containing a curriculum (8) designed to serve as a study guide and to assist in the design of educational programs for infection control personnel. The APIC Standards Committee reviews or participates in the development of infection control standards by other national associations or governmental agencies. As an outgrowth of the APIC Certification Committee, the Certification Board for Infection Control was established in late 1981 and offered its first certification examination in 1983. Additional information about the examination can be obtained from the Certification Board.

THE FUTURE

In addition to the above organizations and associations, new organizations are emerging that will influence hospital infection control programs in the 1980s. For example, the Society for Hospital Epidemiologists of America is beginning to offer educational programs for hospital epidemiologists. The Institute for Health Policy Analysis of the Georgetown University Medical Center is also involved in the complex issues related to the reuse of disposable medical devices. Within the next few years we can expect other important changes in hospital infection control programs. Some of the changes will undoubtedly occur as a result of the new prospective payment plans based on diagnostic-related groups and increased emphasis on cost containment. Other changes will be prompted by the publication of results of new studies on the epidemiology and control of nosocomial infections. Therefore, hospitals must continue to set priorities for their own infection control programs based on an assessment of their own infection problems, patient risks, and availability of resources.

LITERATURE CITED

1. **Advisory Committee on Infections Within Hospitals, American Hospital Association.** 1981. Recommendations for the control of rubella within hospitals. Infect. Control **2:**410–411, 424.
2. **Advisory Committee on Infections Within Hospitals, American Hospital Association.** 1983. Hepatitis B vaccine recommendations for hospital employees. Infect. Control **4:**41–44.
3. **Advisory Committee on Infections Within Hospitals, American Hospital Association.** 1984. A hospitalwide approach to AIDS. Infect. Control **5:**242–248.
4. **American Academy of Pediatrics.** 1982. Report of the Committee on Infectious Diseases, 19th ed. American Academy of Pediatrics, Evanston, Ill.
5. **American Academy of Pediatrics and American College of Obstetricians and Gynecologists.** 1983. Guidelines for perinatal care. American Academy of Pediatrics, Evanston, Ill., and American College of Obstetricians and Gynecologists, Washington, D.C.
6. **American College of Surgeons.** 1984. Manual on control of infection in surgical patients, 2nd ed. J. B. Lippincott Co., Philadelphia.
7. **American Hospital Association.** 1979. Infection control in the hospital, 4th ed. American Hospital Association, Chicago.
8. **Association for Practitioners in Infection Control.** 1983. The APIC curriculum for infection control. Kendall/Hunt Publishing Co., Dubuque, Iowa.
9. **Brachman, P. S., and T. C. Eickhoff (ed.).** 1971. Proceedings of the International Conference on Nosocomial Infections, Atlanta, Georgia, Center for Disease Control, August 3–6, 1970. American Hospital Association, Chicago.
10. **Emori, T. G., R. W. Haley, and J. S. Garner.** 1981. Techniques and uses of nosocomial infection surveillance in U.S. hospitals, 1976–1977. Am. J. Med. **70:**933–940.
11. **Emori, T. G., R. W. Haley, and J. S. Garner.** 1982. Comparison of surveillance and control activities of infection control nurses and infection control laboratorians in United States hospitals, 1976–1977. Am. J. Infect. Control **10:**3–16.

11a. **Federal Register.** 1983. Standards and certification for participation in Medicare and Medicaid. Fed. Reg. **48:**311–312.

12. **Garner, J. S., J. V. Bennett, W. E. Scheckler, D. G. Maki, and P. S. Brachman.** 1971. Surveillance of nosocomial infections, p. 277. *In* P. S. Brachman and T. C. Eickhoff (ed.), Proceedings of the International Conference on Nosocomial Infections, Atlanta, Georgia, Center for Disease Control, August 3–6, 1970. American Hospital Association, Chicago.
13. **Haley, R. W., and R. H. Schachtman.** 1980. The emergence of infection surveillance and control programs in U.S. hospitals: an assessment, 1976. Am. J. Epidemiol. **111**(Suppl. 5):574–591.
14. **Joint Commission on Accreditation of Hospitals.** 1976. Accreditation manual for hospitals. Joint Commission on Accreditation of Hospitals, Chicago.
15. **Joint Commission on Accreditation of Hospitals.** 1984. Accreditation manual for hospitals. Joint Commission on Accreditation of Hospitals, Chicago.
16. **Williams, R. E. O.** 1971. Summary of conference, p. 318–321. *In* P. S. Brachman and T. C. Eickhoff (ed.), Proceedings of the International Conference on Nosocomial Infections, Atlanta, Georgia, Center for Disease Control, August 3–6, 1970. American Hospital Association, Chicago.

Chapter 11

Role of the Microbiology Laboratory in Prevention and Control of Nosocomial Infections

JOHN E. McGOWAN, JR.

The hospital laboratory has important responsibilities in the surveillance, control, and prevention of nosocomial infections. These duties encompass six aspects: (i) working with other hospital personnel in infection control activities, especially those of the infection control committee; (ii) accurately identifying organisms responsible for nosocomial infections; (iii) timely reporting of laboratory data relevant to infection control; (iv) supporting investigations of specific hospital infection problems as they arise; (v) providing additional studies, when necessary, to establish similarity or difference of organisms; and (vi) conducting certain routine microbiologic studies of hospital personnel or the hospital environment. Each of these will be examined in turn.

PARTICIPATION IN HOSPITAL-WIDE INFECTION CONTROL ACTIVITIES

The laboratory can make major contributions to infection control when the persons responsible for infection control efforts and those in charge of the clinical microbiology laboratory cooperate closely to attack this problem. Often, the microbiologist is placed in charge of both efforts, making communication perfect (62).

Infection Control Committee activities

A representative from the microbiology laboratory staff should be an active member of the hospital's Infection Control Committee. This contributes significantly to close cooperation between clinical, infection control, and microbiology personnel.

In the typical hospital, most members of the infection control committee will not have a background in microbiology. Thus, the laboratory representative will provide needed microbiologic expertise to the group for determining the significance of culture data and for laboratory aspects of investigations undertaken by the group.

Culture results from the microbiology laboratory are crucial data for successful infection control. Laboratory activities affecting the identification of the etiology of nosocomial infection (for example, adequacy of the basic techniques for primary isolation, species identification, and antimicrobial susceptibility testing) should be reviewed by the microbiologist for the Infection Control Committee.

Clinical microbiology personnel often have to make cost-benefit judgments about use of laboratory equipment, instruments, and procedures (12, 59). The insights and methodology used for such laboratory activities should be useful to the Infection Control Committee's review of current activities and attempts to improve productivity. Such review efforts can be expected to become more frequent, in view of the major changes currently occurring in hospital reimbursement in the United States (60).

Teaching microbiologic aspects of nosocomial infection to infection control personnel

The persons responsible for infection control usually are not trained in clinical laboratory procedures (33, 95). Since the key to success in infection control efforts is communication, it is necessary that all involved speak the same language. For this purpose, some training of epidemiology personnel in clinical microbiology is important. The goal of such teaching is not necessarily to make the infection control staff accomplished laboratory workers, but rather to familiarize them with the procedures and practices of the laboratory, with the microorganisms involved in nosocomial infection, with the validity of test procedures used in identifying these pathogens, and with the strengths and limitations of the resulting data.

Similarly, it is important for the microbiologist to learn some of the concepts of the epidemiologist, since few laboratory directors or technologists have adequate grounding in epidemiology (33). Especially important is exposure to techniques used for measuring frequency of infection and to the concept of colonization versus infection.

Such joint efforts permit ready communication between the two groups of colleagues. Teaching of this type can be done formally or as part of the day-to-day contacts between the infection control staff and laboratory personnel.

ACCURATE IDENTIFICATION OF ORGANISMS INVOLVED IN NOSOCOMIAL INFECTION

Any information that permits the successful tracing of organism movements within the hospital can be of use to the hospital infection control team. Thus, the ability of the laboratory to isolate and identify responsible microorganisms is as crucial to infection control as it is to the individual patient (33). This is true whether the positive cultures represent episodes of infection or indicate colonization of the patient (1).

Monitoring specimen quality

Laboratory information presented to the hospital epidemiologist must reflect organisms actually associated with the patient's site of culture rather than contaminants. Specimens that are not collected or transported properly, even when handled as well as possible once they reach the laboratory, are likely to reflect contamination rather than true pathogens. For example, failure to isolate organisms from deep wounds or abscesses of patients who are not on antibiotics, or inability to recover pathogens seen on Gram stain in cases of presumed anaerobic infections, suggests inadequate anaerobic transport media, delay or inappropriate refrigeration of specimens in transit, or use of inadequate isolation techniques for isolating anaerobes (7, 95).

The laboratory must monitor specimen handling continually and work closely with the wards and clinics to make sure that the possibility of contaminated specimens or missed organisms is minimized. The frequency with which probable contaminants are isolated from clinical specimens can measure the quality of specimen collection in a specific hospital area. For example, periodic review of the relative incidence of false-positive smears for acid-fast bacilli may highlight problems in sputum collection and processing (9).

Assessment of the specimens at the time they are received in the laboratory is one of the best ways to evaluate their suitability; specimens identified as inadequate are not processed further and do not confuse either clinician or epidemiologist. For example, microscopic review of Gram stain of sputum specimens remains an excellent way to determine whether or not these specimens are contaminated (9, 100). Culture or smear results for other types of specimens may also suggest contamination at the time of collection (8). Application of such criteria will ensure that the information generated from the specimens that are processed completely will provide accurate data to the epidemiologist.

Examination of Gram stain for morphology can identify organisms that might be epidemiologically important, but not reflected by culture. Thus, presence of a mixed flora in Gram stain of a sputum specimen, when coupled with an aerobic culture yielding only *Haemophilus influenzae*, may indicate possible mixed aerobic-anaerobic infection rather than pneumonia due to *Haemophilus* sp. As infection control implications of these two etiologies may differ, evaluations of this type by the laboratory can be important. Likewise, nonculture methods for identifying the presence of rotavirus (demonstration by electron microscopy, immunologic assays, etc.) have helped us learn more about this organism as a cause of nosocomial infection in older age groups (61).

Efforts such as those listed above are undertaken routinely by laboratory personnel as a part of their service to patients to reduce errors in diagnosis and unnecessary antimicrobial therapy. Such an approach also will improve the specificity of infection surveillance data, which otherwise might be confounded by isolates of questionable etiologic significance.

Providing adequate techniques for recovery of nosocomial pathogens

Often it is difficult to determine the causative agents in nosocomial infection. The majority of cases today for which the etiology is known involve gram-negative aerobic bacilli (61). Most frequent among these gram-negative rods are *Klebsiella* spp., *Enterobacter* spp., *Pseudomonas* spp., *Serratia* spp., *Proteus* spp., and *Escherichia coli* (in approximately that order). In recent years, *Acinetobacter, Flavobacterium*, and *Pseudomonas* species other than *Pseudomonas aeruginosa* have become increasingly prominent (61). In the past few years, anaerobic bacterial organisms (usually found in mixed aerobic-anaerobic infections) and new bacterial agents like the now numerous variants of *Legionella* have been implicated in both epidemic and endemic nosocomial disease (30). In addition, viral agents (e.g., rotavirus), fungi (especially *Candida* spp. and other non-*Candida* yeasts), and parasites like *Pneumocystis* have been identified as important causes of nosocomial infection (61).

This expansion of the list of possible microbial pathogens for hospitalized patients has made it more difficult for both microbiologists and clinicians to deal effectively with hospital infection. Effective handling of such problems requires the laboratory to keep up with the steadily unfolding panorama of organisms that are important causes of nosocomial infection, and to implement and maintain culture and other techniques that will bring these to light.

Pinpointing nosocomial pathogens

Infection control personnel constantly are searching for evidence that a common organism has spread from patient to patient (33). The ability to detect such an event is enhanced by laboratory procedures leading to identification of the organism to the level of species. For example, identifying an isolate as *Pseudomonas cepacia* can help the epidemiologist because this organism has a characteristic hospital reservoir; it frequently is associated with illness or pseudoepidemics caused by contaminated water or other solutions (58). By contrast, identifying the organism only as "*Pseudomonas* species" lumps the isolate with a group of organisms that do not have as characteristic a hospital niche.

Since incomplete identification of organisms may obscure real problems and make retrospective epidemiologic investigation impossible, the epidemiologist may ask the laboratory to identify most isolates to the species level. Whether this can be done routinely or only by request depends on the resources of the individual laboratory (53). At a minimum, the laboratory should be capable of identifying gram-positive cocci and gram-negative aerobic bacilli to the species level when special or recurring problems in a given institution make such information useful for dealing with nosocomial infection problems (8, 95). Commercial, multiple test media for biochemical testing can provide this degree of characterization (53, 83). Laboratory personnel also should be familiar with procedures for obtaining additional assistance from state and national reference laboratories in identifying unusual isolates beyond the expertise of an individual laboratory.

When organisms are identified, it is important that standard criteria and nomenclature be consistently applied. The changes in nomenclature that result from better knowledge of organism relationships can cause confusion. For example, the renaming as *Acinetobacter calcoaceticus* subsp. *anitratus* of the organism formerly called *Herrelea vaginicola* gave false alarm to institutions not used to seeing or dealing with what appeared to be a new intruder (95).

Sometimes it is the pattern of susceptibility to antimicrobial agents that discriminates epidemiologically significant organisms from other apparently similar hospital organisms. For example, many United States hospitals currently encounter nosocomial infections due to *Staphylococcus aureus* strains resistant to methicillin (89). Such organisms can be the subject of infection control activities only if the laboratory maintains effective and efficient means for their identification.

Reporting of "biotyping" information (pattern of response to biochemical testing) can be of value in differentiating organisms that are frequently encountered, but the need for clinical laboratories to provide this identification on a routine basis has not been established (1, 33).

Adding procedures for infection control purposes

The laboratory constantly must consider whether laboratory techniques in addition to those used for organism recovery and identification can aid infection control goals. For example, the semiquantitative method for culture of intravenous catheters (55) can help distinguish between contamination at the time of catheter removal and true infection of the intravascular device. This special technique generates information that can be quite helpful to the infection control officer's assessment of whether nosocomial infection was present. In addition, of course, the data aid the clinician who must decide whether antimicrobial therapy is needed.

Identification and quality control

Many spurious outbreaks of hospital-acquired infection have been traced to inaccurate or inconsistent microbiologic procedures. For example, incorrect reading of coagulase tests might result in misidentification of coagulase-negative organisms as being coagulase positive and cause a "pseudooutbreak" of *S. aureus* infection. Adequate quality control is essential to avoidance of such problems (8). Periodic review of selected laboratory materials, media, and other equipment should be performed. On occasion, erroneous microbiologic results related to the inadvertent use of contaminated or faulty materials may occur. For example, nonviable contaminants have been found in specimen tubes in commercial lumbar puncture trays; these resulted in the assumption that an outbreak of nosocomial meningitis was occurring when the contaminants were seen in Gram stains of cerebrospinal fluid (94). Such "pseudoepidemics" (J. E. McGowan, Jr., G. F. Mallison, and R. A. Weinstein, *in* J. V. Bennett and P. S. Brachman, ed., *Hospital Infections*, in press) must be considered when laboratory culture or stain results do not correlate with clinical or epidemiologic findings.

TIMELY REPORTING OF LABORATORY DATA

Laboratory records are an important tool for surveillance of infections (36). More than 80% of infections defined by other criteria as nosocomial may be identified by review of positive cultures from the microbiology laboratory (31). Thus, data gathered by infection control personnel during laboratory visits form an important base to which further surveillance data from clinical rounds must be added.

Both sources must be used to obtain an accurate estimate of the true rate of occurrence of nosocomial infection in a given hospital. The laboratory can indicate only which organisms were present in culture. The epidemiologist then must supplement this with clinical information to determine whether organisms found in culture indicate infection or colonization. If colonization is present, the identification of organisms in the culture may be of little help to the clinician. To the epidemiologist, however, both the organisms involved in epidsodes of colonization and those of infection are of interest. Either may be evidence of spread of organisms from one site to another; such spread may indicate the opportunity for control measures to halt transmission.

To deal with individual problems of nosocomial infection in the hospital as they arise, control measures must be taken as quickly as possible and must be based on accurate assessment of the problem and its causes (34). Without rapid identification and reporting of the organisms involved, control measures cannot be efficiently and rationally designed and implemented.

Even in the absence of an outbreak, microbiologic and immunologic reports may be the starting point for further epidemiologic investigations. Often, these investigations also require information about attributes of the patient, the personnel involved in care, or the diagnostic and therapeutic procedures that were provided to the patient. Obtaining these nonlaboratory data usually is easier when the patient is still present in the hospital, or at least is still fresh in the minds of hospital personnel. Prompt reporting of pertinent laboratory results facilitates information retrieval of this type.

Reporting procedures

Prompt reporting to both clinicians and infection control personnel is essential when isolates of nosocomial significance are presumptively identified (43). Examples of such isolates include *Neisseria meningitidis*, salmonellae or shigellae from stool specimens, positive smears and cultures for *Mycobacterium tuberculosis* from any patient or employee, the isolation of *S. aureus* from lesions of a newborn, and organisms resistant to an unusually large number of antimicrobial agents. Any of these situations require quicker action than a "final report," as important epidemiologic investigations often can be triggered by preliminary data from the laboratory. At the same time, "early warning" must not be requested for so many situations that this becomes an unreasonable burden for the laboratory personnel.

Results can be brought to the attention of infection control colleagues by telephone or page, if urgent; if not, a mention during the daily visits of the infection control staff to the laboratory usually suffices.

Retention of laboratory records

In addition to instituting control measures, infection control workers often need to analyze laboratory data from various time periods to try to detect patterns of infection (62). To assist this effort, it is helpful if the laboratory can provide an "archival" summary of organisms on a periodic basis. The specific laboratory data that can aid epidemiologic analyses will vary from hospital to hospital; the information to be included, and the frequency with which such summaries are made, should be determined by the individuals providing and working with the data in each hospital. As a general guide, the source of each specimen, date of collection, patient identification, hospital number, hospital service, ward, and the organisms identified in the final report should be recorded (93). Records also should be kept of results of antimicrobial susceptibility tests and of any special biochemical or typing reactions.

All cultures should be recorded so that results are readily available by date, by type of specimen, and by pathogens isolated for both inpatients and outpatients. Computer storage and retrieval of all results is optimal, but noncomputerized rapid retrieval and sorting systems also may be useful (78). For example, records could be maintained simply and inexpensively in bound log books kept chronologically for each major type of specimen (i.e., blood, wound, skin, cerebrospinal fluid, urine, stool, sputum, etc.). Sole reliance on a filing system of loose lab slips is not desirable because specific data are difficult to retrieve and are easily lost.

The permanent records of the microbiology laboratory should include dates and other details of any major changes in culturing techniques or laboratory procedures. Dates of changes in the criteria for identification and taxonomic designations applied to isolates should be recorded as well.

The length of time such records can be maintained depends on hospital size, work volume of the laboratory, and available storage facilities as well as on infection control needs. Thus, storage time should be determined by laboratory personnel after consultation with the hospital infection control staff. Six months is a reasonable minimum (McGowan et al., in press).

Summary reports for clinical use

Profiles of susceptibility of frequently tested nosocomial pathogens to antimicrobial agents can be of considerable assistance in guiding therapy for sepsis of unclear cause and for other infections. Testing of other organisms (e.g., slow-growing bacteria or organisms requiring special test procedures) may be performed at intervals to develop a profile of their susceptibility. As long as susceptibility patterns can be presumed to remain stable, such testing may be a useful substitute for testing each isolate at the time of recovery.

In liaison with the Infection Control Committee, the laboratory should provide at regular intervals a summary of susceptibility patterns and make it available to the clinicians on the staff. Tabulations that may be of particular use include frequency of susceptibility to individual drugs by site of infection (which may provide guidance to the clinician for empiric therapy of infection before the causative organism has been identified) and tabulation of frequency of susceptibility to individual antimicrobial agents by pathogen (this may be used to direct therapy after an organism has been identified but before susceptibility tests have been completed). The Infection Control Committee, a medical staff committee, a quality assurance committee, or more than one of these may wish to receive such data to guide their review of antibiotic utilization (50).

A listing of the relative costs of the currently employed antimicrobial compounds may be developed with cooperation of the pharmacy; inclusion of this information with susceptibility summaries may aid efforts to reduce costs of antimicrobial usage (62).

PROVIDING SUPPORT FOR INVESTIGATIONS OF SPECIFIC HOSPITAL INFECTION PROBLEMS

Outbreaks of nosocomial infection must be dealt with as rapidly as possible (34). This means that the laboratory may face exceptional demands for service at the beginning of, and throughout, an epidemic period (95). Advance preparation for such situations makes response easier in time of need. The laboratory should prepare contingency plans for the types of outbreaks that have occurred most frequently in the past in its hospital so that it is ready to deal with these exceptional requests in smooth fashion.

Investigation of an outbreak of nosocomial infection may require isolation and identification of isolates not only in specimens from patients, but also in those taken from personnel who might be colonized with the outbreak strain and from environmental objects implicated by epidemiologic investigation that might be similarly contaminated (34, 80). Such activity may require the laboratory to process and evaluate large numbers of cultures. Special techniques may be necessary to accomplish such projects. For example, reliable detection of *Salmonella* carriage, rather than infection with this organism, requires enhancement of growth by use of selective media (33). The laboratory and the infection control team can process this work efficiently by making careful assessment of the sites to be cultured and determining which culture media and techniques will be employed.

Costs for laboratory procedures that are not related directly to care of the individual patient (e.g., bacteriologic sampling of personnel and the environment) should be borne by a budget separate from that of the laboratory. To facilitate all of the microbiologic activities necessitated by an outbreak, the laboratory (or the hospital epidemiologist or the Infection Control Committee, depending on the organizational structure in a given hospital) should have a contingency fund to enable personnel, materials, and space to be temporarily assigned to epidemic aid support. An investigation of an outbreak should not be financed by charging the individual patients for cultures taken during the study.

Hospital-supported continuing education is essential for ensuring that investigation support can be provided in the microbiology laboratory. Fortunately, a number of organizations, including the American Society for Microbiology, the Association for Practitioners in Infection Control, and the American Society of Clinical Pathology, provide frequent programs on nosocomial infection topics.

Culture techniques for outbreak investigation (also see Chapter 13)

A wide variety of items and substances can be responsible for cross-infection. Thus, surveys of patients, personnel, or the hospital environment may be useful during investigation of specific problems within a hospital. These should be instituted in response to, and should specifically address, epidemiologic findings (59). Without this focus, considerable expense may be incurred in production of information that is worthless or misleading. Thus, approach requests for such cultures with caution, and be clear how the culture results will affect patient care or epidemiologic control measures before undertaking such tasks (59).

The infection control workers and laboratorian must be familiar with the general aspects of culture procedures discussed below. Detailed description of suitable culture techniques for every possible vehicle of cross-infection is beyond the scope of this chapter. Methods for sampling to recover specific pathogens are described in other chapters of this manual.

Culture of blood products after transfusion reaction. Infection after transfusion of blood or blood products remains a problem (48). If a transfusion reaction occurs, the administered unit should be discontinued immediately and disconnected from the administration catheter. Sterile caps should be placed over the exposed ends of needles or tubing to prevent subsequent contamination. A 5-ml sample should be removed aseptically from the administration tubing in the laboratory and processed according to the method recommended by the American Association of Blood Banks described below (2). It is desirable as well to obtain blood culture specimens at this time by venipuncture from the patient.

Culture of parenteral fluids and intravascular therapy equipment. Bacteremia associated with parenteral therapy may require culture of the needle, the catheter itself, portions of the administration set, the fluid being administered, portions of the cap or closure provided with the fluid, or some combination of the above. Blood culture specimens should be collected simultaneously from the patient. Lot numbers of all commercial materials involved should be recorded on the patient's chart as well as on all pertinent laboratory records, as some episodes have resulted from "intrinsic" (at time of manufacture) contamination of commercial products (54).

Catheters and needles may be cultured by a swab-rinse technique (as described below), but the semiquantitative culture method of Maki et al. (55) is preferable because information of greater clinical use is obtained. When needles and catheters are cultured, avoid sampling the hubs and other portions of the administration set that may have been exposed to superficial contamination.

If portions of the administration set are suspected, these must be properly capped to exclude contamination during transport and laboratory processing. The bottle and administration set should remain connected and be placed in a plastic bag to minimize contamination during delivery to the laboratory. Both broth and membrane filter methods have been used for culture of fluid. The membrane filter method (52) appears simpler and more rapid.

Culture of surfaces. Cultures of external surfaces or internal cavities (e.g., tubes and containers) may be conducted by a swab-rinse technique (9, 95). Brain heart infusion broth supplemented with 0.5% beef extract is used. If the cultured objects are likely to contain residual disinfectants, the broth should contain 0.07% lecithin and 0.5% Polysorbate 80 as neutralizers. A cotton applicator swab that has been immersed in this broth and then wrung out is used to swab the surface to be sampled. The swab is returned to a tube with a cap after sampling; the portion of the stick handled by the operator should be broken off.

Culture of tubes and containers. While containers and the lumens of tubular structures may be sampled by the swab-rinse method described above, a rinse technique often is more convenient (95; McGowan et al., in press). A suitable quantity of brain heart infusion broth is poured into the lumen of tubular structures (40 to 50 ml for respiratory therapy tubing), and the object is tilted to rinse the sides. Up to 50 agitations are desirable. A sample of the broth then is placed in a screw-cap tube. Bottles and containers such as nebulizer reservoirs may be sampled by adding 10 to 15 ml of rinse broth, inserting a sterile stopper if required, shaking vigorously for about 30 s, and then removing a portion for culture.

The broth is decanted or pipetted into a sterile container and agitated to ensure a homogeneous suspension. Plate counts are made by preparing a series of 10-fold solutions in tubes of tryptic digest-casein-soy broth, using 1 ml of the test sample for the first dilution. From each tube of the dilution series, 0.5 ml is pipetted onto the surface of a plate containing tryptic digest-casein-soy agar supplemented with 5% sheep or rabbit blood. The plate is tilted until the inoculum is thoroughly dispersed and is then allowed to dry at 35°C (95°F). The plates then are inverted and incubated. If colonies appear to be coalescing or spreading after 24 h of incubation, colony counts are made at that time. Otherwise, counts are made after 48 h of incubation.

For specific identification of isolates, the original broth sample should be subcultured at 4 and 24 h after initial sampling. For each subculture, 0.5 ml is pipetted onto the surfaces of tryptic digest-casein-soy blood agar, MacConkey agar, and cetrimide agar plates (or the identification media chosen by a given laboratory). The plates are tilted to distribute the inoculum, allowed to dry, and then incubated at 35°C. After the original sample has been incubated at 35°C for 24 h, these same media are inoculated again. At this time, if growth seems apparent on the basis of turbidity, a loopful should be subcultured and streaked to produce isolated colonies. At least two colonies of each morphologic type present should be picked for identification.

Culture of disinfectants and antiseptics. (Also see Chapter 13). Contamination of disinfectants and antiseptics has been implicated in nosocomial infection on a number of occasions (69). The organisms usually involved are nonfermenting gram-negative aerobic bacilli. Plating of serial dilutions of the product to dilute its antimicrobial activity has been of value (69).

Culture of respiratory therapy equipment. In situations of high endemic levels or epidemic levels of occurence of nosocomial respiratory infections, sampling of respiratory therapy apparatus may be helpful (81). Protocols for culturing tubes or containers (21, 90) are similar to methods listed above. A direct-dilution method of sampling has been reported by one group of investigators to be satisfactory for detection of organisms in the range of 10 to 10^6 CFU/ml (56). Ventilator tubing often becomes contaminated within the first 2 h of use, and the number of bacteria in the circuit may not be appreciably different in samples at 48 h compared with those taken at 24 h of use (21).

Cultures of air. Airborne spread of nosocomial bacterial or viral infection is known to occur, but is probably uncommon (35). Air sampling should be required infrequently; routine sampling is not recommended (25). When sampling is required, it may be performed either with settling plates or with more sophisticated equipment (35).

Particles suspended in hospital air vary greatly in size and in the number of microorganisms they contain. The surface of a standard 100-mm petri dish represents an area of about 1/15 square foot. Assuming that the air in the study area contains particles of average size, an open petri dish in still air will sample microbial particles from about a cubic foot of air during 15 min of exposure (95). Brain heart infusion agar or Trypticase soy agar can be used for such sampling (McGowan et al., in press).

Although settling plates are an inexpensive way to evaluate airborne microbial contamination, quantitative results may correlate poorly with those obtained by mechanical, volumetric air samplers because of variation in particle size and unpredictable influences of air turbulence and atmospheric conditions. For example, under low humidity conditions, droplet nuclei of about 3 μm can remain suspended indefinitely and can be collected only with high-velocity, volumetric air samplers. In the hospital, the range of particle size averages from 10 to 15 μm (35). Furthermore, the colonies present after settling on an agar plate will not precisely measure the total number of airborne microbial "particles" because such particles may contain more than one viable cell. This is also why air sampling techniques in which volumetric samples are taken by bubbling air through collection fluid, which breaks up airborne particulate matter, better reflect the total number of organisms than will air samplers that impinge contaminated particles on agar.

A slit sampler is suitable for most microbial air-sampling applications (95). A centrifugal sampler can also serve well if detection of very small particles is not essential (13). Brain heart infusion or Trypticase soy agar should be used in the sampling plates. A staged sampler (35, 95) should be needed only when there is some reason to determine the size distribution of the particles. This should be an extremely infrequent event in most United States hospitals. Efficient vacuum sources must be used for many samplers, and the rate of flow of air must be properly calibrated to ensure accurate results.

Standards for levels of microbiologic contamination of air have been suggested (96), but no uniform agreement has been achieved because of the lack of correlation between these levels and occurrence of clinical infection.

Cultures of floors and other surfaces. (See McGowan et al., in press.) A number of methods for sampling of floors and other surfaces have been described in detail (9, 22). One method in common use (95) is an adaptation of the swab technique previously described. In this application, a square hole (2 by 2 cm) is cut from the center of a sheet of heavy paper, which subsequently is sterilized and wrapped. The culture is collected by placing the paper on the surface to be sampled. The swab is slowly rubbed in close, parallel streaks across the exposed area. The swab is then moved in a direction perpendicular to the first streaks, and the procedure is repeated.

Rodac plates (9) are designed to permit direct contact of agar with flat surfaces. Such plates should be filled with 16.5 ml of Trypticase soy agar containing 0.07% lecithin and 0.5% Polysorbate 80 as neutralizers of disinfectant. The number of plates to use will depend on the size of the surface being sampled and the level of statistical confidence that is required. Both the plates and the surface to be sampled should be dry at the time the sample is collected. Plates are pressed firmly against the surface, avoiding a rotary or sliding motion. Colonies are counted after incubation at 35 to 37°C (95 to 99°F) for 48 h. The use of various types of automatic or semiautomatic colony counters will save time in counting large numbers of plates.

Standards for acceptable levels of contamination of floors and bedside tables as sampled by the Rodac plate technique described above have been suggested (19). There is no evidence, however, that any particular level of contamination is directly correlated with an increased risk of infection, so routine sampling of this type is not indicated.

Cultures of water and ice. (See McGowan et al., in press.) Water that meets U.S. Public Health Service standards for drinking water contains 10^6 or more microorganisms per ml (9); some of these organisms are potential pathogens. Ice likewise can contain organisms that can pose a threat of infection, especially in patients with compromised host defenses. However, a correlation of these culture findings with occurrence of patient illness has been found only infrequently.

Samples of water (or melted ice) can be cultured by passing large quantities through a 0.45-μm (or 0.22-μm) membrane filter (Millipore) and then culturing the filter in broth or directly on agar. Four or more colonies per 100 ml by this membrane filter test is considered abnormal (9).

Maximum sensitivity is achieved when the previously described 24-h brain heart infusion broth dilution method is used. By this technique, substantial numbers of bacteria may be isolated, and a most probable number of 2.2 or more colonies by this test is considered abnormal (9), but there is no evidence that this finding by itself indicates a hazard.

Use of selective media

To reduce the work load in the laboratory and to expedite the processing of specimens, selective media should be used whenever possible for culturing specimens during outbreak investigations (McGowan et al., in press). Susceptibility testing of known or suspected epidemic strains may be used to identify an appropriate selective medium. Pretesting of the medium is essential because of possible synergy or antagonism between the added antimicrobial agents or between these drugs and the medium; such interactions could cause inhibition of growth of the epidemic strain or failure to inhibit growth of nonimplicated organisms. Once the implicated organism is found to grow on a selective medium, this formulation then can be used to exclude numerous bacteria unrelated to the outbreak. This may accelerate the detection of contaminated equipment or infected patients.

Examples of appropriate selective media are numerous. Differential primary plating media, such as a modified MacConkey agar, may be helpful in selectively isolating *P. aeruginosa* from contaminated material or mixed cultures (24). Similarly, tetrathionate broth is an excellent medium for selective preenrichment of *Salmonella* cultures. Mueller-Hinton agar containing sorbitol, a pH indicator, and antibiotics (vancomycin, colistin, and nystatin) provides selective differentiation of *Serratia* species. Many epidemic strains of *S. aureus* are resistant to mercuric chloride, and the incorporation of small amounts of this compound in Trypticase soy agar can be helpful in inhibiting nonepidemic strains of *S. aureus*, *Staphylococcus epidermidis*, and most gram-negative organisms except *Pseudomonas* spp. Recently, methicillin-resistant strains of *S. aureus* have become a problem in some United States hospitals; the use of agar containing small concentrations of methicillin has been helpful in investigating hospital problems due to these strains (89).

ADDITIONAL STUDIES TO ESTABLISH SIMILARITY OR DIFFERENCE OF ORGANISMS

On occasion, the epidemiologist will suspect that a group of nosocomial infections with organisms of the same species have a common origin. To investigate whether strains in this "cluster" (77) are common or different, the usual practice is to examine results of routine biochemical tests and patterns of susceptibility to routinely tested antimicrobial agents. However, for organisms commonly encountered in the hospital (e.g., *S. aureus* or *Klebsiella* spp.), the general pattern of these results may be similar on the basis of chance alone. Conversely, for other organisms (e.g., *P. aeruginosa*) the difference in these characteristics from strain to strain is so small that the tests provide little information about similarity or difference of tested strains (1). In situations where no differences can be shown for the above tests, examination ("typing") of further organism characteristics ("markers") can be of great assistance (1, 33, 34; McGowan et al., in press).

Typing techniques

Typing systems of value in investigating nosocomial infection problems vary with the organism involved (Table 1). Although hospital laboratory personnel may not have the facilities to perform specialized typing procedures, they should know which organisms can be typed, and which cannot, and should be aware of centers where specific procedures can be performed. When epidemiologically important isolates must be forwarded to other laboratories, materials should be packaged for air transport in conformance with federal regulations (74).

Susceptibility to additional antimicrobial agents or to heavy metals. Further testing of susceptibility, either to antimicrobial agents not routinely included or to other antibacterial substances (e.g., silver), may differentiate strains in some cases. Benefits must be weighed against the extra time and cost required for testing and recording of the additional studies. Testing susceptibility to heavy metals is sometimes called "resistotyping" when used in this context (27).

The pattern of susceptibility or resistance is reported most simply by specifying the group ("susceptible," "resistant," etc.) in which the test value falls. Some microorganisms can be further differentiated by indicating the relative degree of susceptibility or resistance that is present (29). This can be done by noting the absolute value of zone size in agar diffusion testing or by providing MIC or MBC (45). The laboratory and infection control team jointly should identify situations in which this more quantitative information would be useful.

Biotyping. The use of certain characteristic biochemical reactions to identify subgroups of bacteria is widely practiced (83). Differentiation schemes based on this method have been devised for a variety of bacterial organisms, both anaerobes and aerobes. Lack of precision in the commercial systems can be a hindrance (83), but these chemical profiles are still often used in conjunction with pattern analysis of antimicrobial susceptibility testing to attempt discrimination of isolates (Table 1).

Phage typing. Susceptibility to bacteriophages (phage typing) is a characteristic used for a number of organisms of nosocomial importance. The technique is especially handy for grouping strains of *S. aureus* (82). This procedure usually is available only in reference laboratories. Plasmid transfer of phage characteristics apparently can confound results. Interpretation of results of phage typing can be difficult; guidelines are available (14).

Serotyping. Measuring immunologic reactions is a frequently employed technique for typing of many gram-negative aerobic bacilli, especially *Klebsiella pneumoniae* and *P. aeruginosa* (Table 1). Methods for typing capsular antigens of *S. aureus* recently have been enhanced (5). Serotyping can be of great help for other organisms shown in the table as well, both in outbreak situations and in research investigations (1). However, routine serotyping of isolates in the absence of an outbreak has not been cost effective.

Bacteriocin typing. Bacteriocins are products manufactured by some bacteria that can inhibit the growth of other organisms. Production of such bacteriocins by an epidemic strain, or susceptibility of the organism in question to those produced by other bacteria, can be used as a typing tool for a number of organisms (Table 1). The method requires careful use of controls, and widespread agreement on standards for reagents and interpretation is unusual. Bacteriophage typing and pyocin typing often permit further subdivision of *P. aeruginosa* strains of the same serotype; for maximum usefulness, all three procedures probably should be performed (1).

Other typing systems. A number of other systems for typing are employed on occasion for organisms found to be important in cross-infection (Table 1). The use of plasmid analysis or nucleic acid homology in nosocomial infection studies is discussed elsewhere. Production of enzymes or marker proteins has been employed on occasion. One method of typing for swarming strains of *Proteus mirabilis*, known as Dienes typing, can readily be performed by hospital laboratories (40). A line of demarcation develops at the junction between two swarming strains when unrelated strains simultaneously are inoculated on the same agar plate, but the line does not appear when the inoculated strains are similar (46).

Use of the other systems shown in Table 1 has been reported infrequently, and the relative usefulness of these methods for investigation of nosocomial infection is not yet determined.

TABLE 1. Some typing systems used for epidemiologic investigation and selected microorganisms for which the system has been employed

Typing system	Organisms investigated	Reference no.[a]	Typing system	Organisms investigated	Reference no.[a]
Pattern of susceptibility to antimicrobial agents or heavy metals	*Clostridium difficile*	101	Serotyping	*Campylobacter jejuni*	76
	Diphtheroids	66		*Escherichia coli*	23
	Enterobacter spp.	68		*Haemophilus* spp.	92
	Escherichia coli	23		*Klebsiella* spp.	73
	Haemophilus influenzae	92		*Legionella* spp.	98
	Klebsiella spp.	34		*Listeria* spp.	84
	Mycobacterium fortuitum	16		*Proteus* spp.	4
	Proteus spp.	4		*Providencia rettgeri*	70
	Pseudomonas aeruginosa	77		*Pseudomonas aeruginosa*	77
	Salmonella spp.	63		*Pseudomonas cepacia*	39
	Serratia marcescens	71		*Salmonella* spp.	26
	Shigella spp.	27		*Serratia* spp.	71
	Staphylococcus aureus	67		*Shigella* spp.	26
	Staphylococcus coagulase-negative	15		*Staphylococcus aureus*	5, 17
	Yersinia spp.	6		*Staphylococcus* coagulase-negative	17
				Streptococci	38
				Yersinia spp.	6
Biotyping	Diphtheroids	66	Bacteriocin production or susceptibility	*Clostridium difficile*	79
	Enterobacter spp.	68		*Enterobacter* spp.	10
	Escherichia coli	23		*Escherichia coli*	41
	Haemophilus influenzae	92		*Klebsiella* spp.	11
	Klebsiella spp.	73		*Proteus* spp.	4
	Mycobacterium fortuitum	16		*Pseudomonas aeruginosa*	32
	Proteus spp.	4		*Serratia* spp.	71
	Serratia marcescens	71		*Shigella* spp.	64
	Staphylococcus aureus	33		Group A *Streptococcus*	87
	Staphylococcus coagulase-negative	15	Plasmid profile or nucleic acid homology	Multiple organisms (see Chapter 108)	
	Yersinia spp.	6			
Phage susceptibility	*Clostridium difficile*	79	Analysis of enzyme production	*Pseudomonas aeruginosa*	44
	Escherichia coli	57		*Staphylococcus aureus*	1
	Klebsiella spp.	73	Analysis of marker proteins	*Clostridium difficile*	86
	Listeria spp.	88		*Haemophilus influenzae*	65
	Proteus spp.	42		*Staphylococcus aureus*	18
	Pseudomonas aeruginosa	51	Dienes reaction	*Proteus mirabilis*	40, 42
	Salmonella spp.	47	Serum opacity	Streptococci	97
	Serratia marcescens	71	Colony morphology	Diphtheroids	66
	Shigella spp.	46	RNA electrophoresis	Rotavirus	75
	Staphylococcus aureus	82	Killer system analysis	*Candida albicans*	72
	Staphylococcus coagulase-negative	15	Cytotoxicity assay	*Clostridium difficile*	101
	Yersinia spp.	6			

[a] Number of reference using the indicated typing scheme for the organism listed.

Saving nosocomial pathogens

For the supplemental tests described above to be performed, the laboratory must retain strains that may relate to nosocomial infection for a given period while it is determined whether further testing by the methods listed above is needed or not. In cooperation with infection control personnel, the laboratory should subculture and save epidemiologically important isolates, whether such isolates are from outbreaks or from single cases of unusual or potentially epidemic diseases. How long a storage period is required for this purpose will vary from hospital to hospital and should be agreed upon between epidemiologist and clinical laboratory supervisor (62). A system for reviewing and periodically discarding these isolates also must be established.

The technique used to ensure the viability of the organisms (freezing, lyophilization, etc.) should be determined by the laboratory after considering the equipment and personnel that are available. For example, one method for storing *S. aureus* strains involves placing a small amount of growth on a blank paper disk, which is put in a 2-ml glass screw-cap vial containing a few granules of silica gel. If the vial is kept tightly closed, isolates may be held for up to 6 months. The disk can be immersed in broth to retrieve the isolate (McGowan et al., in press).

MICROBIOLOGIC STUDIES OF HOSPITAL PERSONNEL OR THE HOSPITAL ENVIRONMENT

Environmental or employee culture surveys are not recommended as a routine practice. In the past, culture of various animate and inanimate sites on a routine basis was advocated, but data now suggest that high cost outweighs any clinical or epidemiologic benefits for all except a few situations (49). This aspect of a control program recently has been reviewed in detail (59, 80).

A few general microbiologic studies of this type should be done routinely. These procedures should be considered one of the costs of the hospital's infection control program, and charges for the cultures should not be billed to individual patients (33, 59, 95). Included are monitoring of sterilization, sampling of infant formula and other hospital-prepared products, culture of blood components and dialysis fluid, and periodic sampling of disinfected equipment, as discussed below.

Monitoring of sterilization (also see Chapter 13)

All steam and ethylene oxide gas sterilizers should be checked at least once each week with a suitable live-spore preparation (80). In addition, each load in either type of sterilizer that contains implantable objects (prosthetic cardiac valves, hip prostheses, etc.) should be monitored with a spore test (the implantable objects should not be used until the spore test is reported as negative, usually at 48 h of incubation). Dry-heat sterilizers should be monitored at least once each month (80).

Sterilization of laboratory supplies and equipment represents a special case. Microbiologic culture media do not require the duration of exposure that is required for sterilization of material known to contain large populations of bacteria. Ampoules containing spore solutions are not an appropriate way to check sterilization of laboratory media, as heating culture media to temperatures sufficient to sterilize a test strip or spore ampoule will damage the medium through overheating.

Microorganisms chosen for spore tests are more resistant to sterilization than are most naturally occurring pathogens (85). The test spores are provided in relatively high concentrations to ensure a margin of safety. For steam sterilization, spores of the thermophile *Bacillus stearothermophilus* are used. Ethylene oxide and dry-heat sterilizers are tested with spores of *Bacillus subtilis* (subsp. *niger*, strain *globigii*). Spores of both species are incorporated in some of the available commercial test kits.

The spores usually are provided either in impregnated filter-paper strips or in solution in glass ampoules. Other types of spore preparations are commercially available and require different handling. In each case, the manufacturer's directions should be followed closely.

Test strips or spore solutions should always be placed in the center of the specimen to be tested, never on an open shelf in the autoclave. The center of a pack located near the bottom front exhaust valve will be exposed to the least adequate duration and temperature of sterilization, and thus provides the best location for a test measurement. Adequacy of sterilization of fluids is tested by placing an ampoule containing a spore solution in the largest vessel.

Spore strips. Most spore strip preparations are packaged in envelopes that contain one or two test strips and a control strip. The test strips are packaged in separate envelopes, which are removed and sterilized at the time other material is processed. Handling of the strip after sterilization should be minimized to reduce the likelihood of cross-contamination. Such handling can be reduced by removing the strips from their envelopes before use and placing them in sterile glass tubes. The tubes are then sterilized with the screw cap removed or with other closures permeable to steam in place.

When they reach the laboratory, the test strips and control strip are cultured by placing the strips in a tube of tryptic digest-casein-soy broth. The transfer from tube to culture medium should be made in a laminar-flow cabinet (when available), using sterile forceps and scissors. Common sources of contamination include the forceps and scissors, which may be insufficiently sterilized by flaming or wiping with alcohol. Since alcohol solutions are not sporocidal, they may contain viable spores that might not be killed by flaming, and brief flaming may be insufficient to heat instruments to a temperature that will destroy viable spores. Care should be taken not to cross-contaminate the sterilized spore strips with the control strips.

The broth is incubated at 37°C for spores used to control gas sterilization (*B. subtilis*) and 56°C for those used to monitor steam sterilization (*B. stearothermophilus*). It is not necessary to culture a positive control strip for each test; if strips are obtained from a single lot, only 10% of the positive control strips need to be tested (McGowan et al., in press).

Spore solutions. Spore solutions are available commercially in sealed glass ampoules for testing the adequacy of sterilization of fluids. These ampoules should be incubated at 56°C (133°F) in a water bath. If there is no change in the indicator by 7 days, the test is reported as negative.

Interpreting results. When positive spore cultures are obtained, Gram stain and subculture should be performed. If organisms other than gram-positive bacilli are observed, the test should be repeated and reported as "possible laboratory contamination, test being repeated."

For each group of tests, uninoculated culture media should be incubated at room temperature and at 35 or 56°C (whichever is to be used for incubation of the inoculated media) to ensure that medium contamination has not led to false-positive reports (37). Condensation on the cover of a 56°C water bath may cause contamination of the caps and closures of tubes. A heating block may be used to avoid this, or the bath may be left uncovered; the latter will make it necessary to provide a reservoir to maintain the water level, as the evaporation rate is high at this temperature.

Whenever positive results are obtained, the sterilizers should be checked immediately for proper use and function (80). Careful examination must be made of thermometer and pressure gauge readings as well as review of recent time and temperature records. If any deficiency is observed, or if the repeated sterility test still results in growth, engineering personnel and experts in autoclave maintenance and function should be consulted promptly. Objects other than those used for implants do not need to be recalled at this point unless defects are discovered in the sterilizer or its use; if spore tests remain positive after proper use of the sterilizer is documented, the machine should be removed from service until the defects are corrected (80).

Sampling of infant formula and specific other products prepared in the hospital

It previously was recommended that infant formula, if prepared in the hospital kitchen, should be monitored on a weekly basis (80). However, this restriction is not being stressed in newer Centers for Disease Control guidelines and has become a matter for individual hospital choice.

Certain other products prepared in a hospital have been demonstrated to have a potential for transmission of nosocomial infection. These should be monitored until it is clear that any problem has been rectified. At one hospital, these consisted of hyperalimentation fluid prepared in the hospital and breast milk pools collected for group use (59). Whether any products fit into this category, the criteria for acceptability, and the frequency of monitoring for any problem materials identified should be determined by the Infection Control Committee in individual hospitals.

Culture of blood components

The American Association of Blood Banks does not recommend random culturing of units of blood to ensure sterility. However, periodic culturing of products prepared in an "open" system is advocated (3). An open system is one in which the transfer container "is not integrally attached to the blood container" (91). Ten-milliliter samples of the component should be cultured both aerobically and anaerobically for up to 10 days. Incubation at 30 to 32°C is advised (2).

Culture of dialysis fluid (also see Chapter 13)

The water used for preparation of dialysis fluid and the dialysis fluid itself at the end of treatment should be tested by colony count at least once per month, according to one set of recommendations (80). The water should contain less than 200 CFU/ml. In addition, dialysis fluid at the end of treatment should contain <2,000 CFU/ml. Defined methods for such testing generally follow Centers for Disease Control guidelines (28).

Periodic sampling of disinfected equipment (also see Chapter 13)

If chemical disinfection or pasteurization rather than sterilization is used on equipment such as cystoscopes and other endoscopes or anaesthesia equipment, then some authorities recommend that periodic microbiologic sampling be done to ensure the absence of vegetative cells after processing. By contrast, current Centers for Disease Control guidelines recommend against such testing for respiratory therapy equipment (81).

The frequency of sampling of disinfected devices of the above types, if done at all, will depend on any evidence that nosocomial infection is associated with their use. There is little agreement on which items of this type should be tested, or how often. Standards for acceptability are lacking. If such a monitoring program is begun, it may be possible to cut back on the frequency of culturing after a period of time in which cultures are negative, as long as no changes are made in equipment and techniques being used (59).

Sampling of unused sterile disposable equipment is not necessary. Samples for culture of reusable equipment, if this testing is done at all, should be taken after the product has been disinfected and made ready for use on patients.

Environmental sampling that is not recommended (also see Chapter 13)

Certain sampling activities are specifically singled out as unneeded, as follows. (i) Routine culturing of patients or hospital personnel is not recommended (80, 99) for reasons discussed above. Such surveys may be useful during the investigation of specific problems within a hospital and should be instituted in response to, and should specifically address, epidemiologic findings. (ii) Routine sampling of commercial patient-care items that are labeled "sterile" (e.g., intravascular catheters and fluids) is not recommended (80). On occasion, these have been contaminated with organisms that can cause disease in patients, but the low frequency of contamination makes it difficult and expensive to perform sterility testing adequately. (iii) In-use testing of antiseptics and disinfectants should not be a routine procedure for hospital microbiology laboratories (80). If contamination of commercial products sold as sterile is suspected, infection control personnel should be notified and the nearest office of the U.S. Food and Drug Administration should be contacted immediately (95). State regulations may require immediate notification to state health authorities as well. (iv) The American Association of Blood Banks does not recommend random culture of blood units to ensure sterility (3). (v) In the absence of an epidemic or high endemic rate of nosocomial pulmonary infections, routine culturing to monitor the disinfection process for respiratory therapy equipment should not be done (81). Likewise, routine sampling of respiratory therapy equipment while it is being used by a patient is not recommended (81). (vi) Routine culture of dialysate is not useful in detecting peritonitis in patients receiving peritoneal dialysis (20). (vii) Routine air sampling of wards or operating rooms is not worthwhile (25).

LITERATURE CITED

1. **Aber, R. C., and D. C. Mackel.** 1981. Epidemiologic typing of nosocomial microorganisms. Am. J. Med. **70:**899–905.
2. **American Association of Blood Banks.** 1981. Technical

Manual of the American Association of Blood Banks, 8th ed., p. 56–57. American Association of Blood Banks, Washington, D.C.
3. **American Association of Banks.** 1981. Technical manual of the American Association of Blood Banks, 8th ed., p. 378–379. American Association of Blood Banks, Washington, D.C.
4. **Anderson, R. L., and F. B. Engley, Jr.** 1978. Typing methods for *Proteus rettgeri*: comparison of biotype, antibiograms, serotype, and bacteriocin production. J. Clin. Microbiol. **8:**715–724.
5. **Arbeit, R. D., W. W. Karakawa, W. F. Vann, and J. B. Robbins.** 1984. Predominance of two newly described capsular polysaccharide types among clinical isolates of *Staphylococcus aureus*. Diagn. Microbiol. Infect. Dis. **2:**85–91.
6. **Baker, P. M., and J. J. Farmer III.** 1982. New bacteriophage typing system for *Yersinia enterocolitica, Yersinia kristensenii, Yersinia frederiksenii*, and *Yersinia intermedia*: correlation with serotyping, biotyping, and antibiotic susceptibility. J. Clin. Microbiol. **15:**491–502.
7. **Bartlett, J. G.** 1979. Anaerobic bacterial pneumonitis. Am. Rev. Respir. Dis. **119:**19–23.
8. **Bartlett, R. C.** 1974. Medical microbiology: quality, cost and clinical relevance, p. 9–42. John Wiley & Sons, New York.
9. **Bartlett, R. C., D. H. M. Gröschel, D. C. Mackel, G. F. Mallison, and E. H. Spaulding.** 1974. Control of hospital-associated infections, p. 841–857. *In* E. H. Lennette, E. H. Spaulding, and J. P. Truant (ed.), Manual of clinical microbiology, 2nd ed. American Society for Microbiology, Washington, D.C.
10. **Bauernfeind, A., and C. Petermuller.** 1984. Typing of *Enterobacter* spp. by bacteriocin susceptibility and its use in epidemiological analysis. J. Clin. Microbiol. **20:**70–73.
11. **Bauernfeind, A., C. Petermuller, and R. Schneider.** 1981. Bacteriocins as tools in analysis of nosocomial *Klebsiella pneumoniae* infections. J. Clin. Microbiol. **14:**15–19.
12. **Broughton, P. M. G., and F. P. Woodford.** 1983. Benefits of costing in the clinical laboratory. J. Clin. Pathol. **36:**1028–1035.
13. **Casewell, M. W., P. G. Fermie, C. Thomas, and N. A. Simmons.** 1984. Bacterial air counts obtained with a centrifugal (RCS) sampler and a slit sampler—the influence of aerosols. J. Hosp. Infect. **5:**76–82.
14. **Centers for Disease Control.** 1981. Phage typing *Staphylococcus aureus*: understanding and interpreting the pattern, p. 40–42. *In* National nosocomial infections study report, annual summary 1978. Centers for Disease Control, Atlanta, Ga.
15. **Christensen, G. D., J. T. Parisi, A. L. Bisno, W. A. Simpson, and E. H. Beachey.** 1983. Characterization of clinically significant strains of coagulase-negative staphylococci. J. Clin. Microbiol. **18:**258–269.
16. **Clegg, H. W., M. T. Foster, W. E. Sanders, Jr., and W. B. Baine.** 1983. Infections due to organisms of the *Mycobacterium fortuitum* complex after augmentation mammaplasty: clinical and epidemiologic features. J. Infect. Dis. **147:**427–433.
17. **Cohen, J. O.** 1972. Serotyping of staphylococci, p. 419–430. *In* J. O. Cohen (ed.), The staphylococci. Wiley Interscience, New York.
18. **Cohen, M. L., L. M. Graves, P. S. Hayes, R. J. Gibson, J. K. Rasheed, and J. C. Feeley.** 1983. Toxic shock syndrome: modification and comparison of methods for detecting marker proteins in *Staphylococcus aureus*. J. Clin. Microbiol. **18:**372–375.
19. **Committee on Microbial Contamination of Surfaces, Laboratory Section, American Public Health Association.** 1970. A comparative microbiological evaluation of floor-cleaning procedures in hospital patient rooms. Health Lab. Sci. **7:**3–7.
20. **Cooper, G. L., J. A. White, J. A. D'Elia, P. C. DeGirolami, C. Arkin, A. Kaldany, and R. Platt.** 1984. Lack of utility of routine screening tests for early detection of peritonitis in patients requiring intermittent peritoneal dialysis. Infect. Control **5:**321–325.
21. **Craven, D. E., T. A. Goularte, and B. A. Make.** 1984. Contamination of condensate in ventilator circuits: a risk factor for nosocomial pneumonia. Am. Rev. Respir. Dis. **129:**625–628.
22. **Craythorn, J. M., A. G. Barbour, J. M. Matsen, M. R. Britt, and R. A. Garibaldi.** 1980. Membrane filter contact technique for bacteriological sampling of moist surfaces. J. Clin. Microbiol. **12:**250–255.
23. **Crichton, P. B., and D. C. Old.** 1980. Differentiation of strains of *Escherichia coli*: multiple typing approach. J. Clin. Microbiol. **11:**635–640.
24. **Daly, J. A., R. Boshard, and J. M. Matsen.** 1984. Differential primary plating medium for enhancement of pigment production by *Pseudomonas aeruginosa*. J. Clin. Microbiol. **19:**742–743.
25. **Editorial.** 1984. Air sampling in operating theatres. J. Hosp. Infect. **5:**1–2.
26. **Edwards, P. R., and W. H. Ewing.** 1972. Identification of *Enterobacteriaceae*. Burgess Publishing, Minneapolis.
27. **Elek, S. D., J. R. Davies, and R. Miles.** 1973. Resistotyping of *Shigella sonnei*. J. Med. Microbiol. **6:**329–345.
28. **Favero, M. S., and N. J. Peterson.** 1977. Microbiologic guidelines for hemodialysis systems. Dial. Transplant. **6:**34–36.
29. **Flournoy, D. J.** 1982. Quantitative antibiogram as a potential tool for epidemiological typing. Infect. Control **3:**384–387.
30. **Fraser, D. W.** 1981. Bacteria newly recognized as nosocomial pathogens. Am. J. Med. **70:**432–438.
31. **Freeman, J., and J. E. McGowan, Jr.** 1981. Methodologic issues in hospital epidemiology. I. Rates, case-finding, and interpretation. Rev. Infect. Dis. **3:**658–667.
32. **Fyfe, J. A. M., G. Harris, and J. R. W. Govan.** 1984. Revised pyocin typing method for *Pseudomonas aeruginosa*. J. Clin. Microbiol. **20:**47–50.
33. **Goldmann, D. A.** 1980. Laboratory procedures for infection control, p. 939–951. *In* E. H. Lennette, A. Balows, W. J. Hausler, Jr., and J. P. Truant (ed.), Manual of clinical microbiology, 3rd ed. American Society for Microbiology, Washington, D.C.
34. **Goldmann, D. A., and A. B. Macone.** 1980. A microbiologic approach to the investigation of bacterial nosocomial infection outbreaks. Infect. Control **1:**391–400.
35. **Gröschel, D. H.** 1980. Air sampling in hospitals. Ann. N.Y. Acad. Sci. **353:**230–239.
36. **Gross, P. A., A. Beaugard, and C. Van Antwerpen.** 1980. Surveillance for nosocomial infections: can the sources of data be reduced? Infect. Control **1:**233–236.
37. **Gurevich, I., J. E. Holmes, and B. A. Cunha.** 1982. Presumed autoclave failure due to false-positive spore strip tests. Infect. Control **3:**388–392.
38. **Hahn, G., and I. Nyberg.** 1976. Identification of streptococcal groups A, B, C, and G by slide co-agglutination of antibody-sensitized protein A-containing staphylococci. J. Clin. Microbiol. **4:**99–101.
39. **Heidt, A., H. Monteil, and C. Richard.** 1983. O and H serotyping of *Pseudomonas cepacia*. J. Clin. Microbiol. **18:**738–740.
40. **Herruzo-Cabrera, R., J. Garcia-Caballero, J. Garcia-Reneses, F. Garcia-Caballero, and J. Rey-Calero.** 1984. The use of the Dienes test in the epidemiology of proteus urinary tract infections in patients with spinal cord disease. J. Hosp. Infect. **5:**92–95.
41. **Hettiaratchy, I. G. T., E. M. Cooke, and R. A. Shooter.** 1973. Colicine production as an epidemiologic marker of *Escherichia coli*. J. Med. Microbiol. **6:**1–11.
42. **Hickman, F. W., and J. J. Farmer III.** 1976. Differentiation of *Proteus mirabilis* by bacteriophage typing and the Dienes reaction. J. Clin. Microbiol. **3:**350–358.

43. **Isenberg, H. D.** 1982. Microbiology and the ailing patient, p. 1–11. *In* V. Lorian (ed.), Significance of medical microbiology in the care of patients, 2nd ed. The Williams & Wilkins Co., Baltimore.
44. **Janda, J. M., and E. J. Bottone.** 1981. *Pseudomonas aeruginosa* enzyme profiling: predictor of potential invasiveness and use as an epidemiological tool. J. Clin. Microbiol. **14:**55–60.
45. **Jones, R. N.** 1982. The antimicrobial susceptibility test: rapid and overnight, agar and broth, automated and conventional, interpretation and trend analysis, p. 341–369. *In* V. Lorian (ed.), Significance of medical microbiology in the care of patients, 2nd ed. The Williams & Wilkins Co., Baltimore.
46. **Kallings, L. O., A. A. Lindberg, and L. Sjoberg.** 1968. Phage typing of *Shigella sonnei*. Arch. Immunol. Ther. Exp. **16:**280–287.
47. **Kasatiya, S., T. Caprioli, and S. Champoux.** 1978. Bacteriophage typing scheme for *Salmonella infantis*. J. Clin. Microbiol. **10:**637–640.
48. **Khabbaz, R. F., P. M. Arnow, A. K. Highsmith, L. A. Herwaldt, T. Chou, W. R. Jarvis, N. W. Lerche, and J. R. Allen.** 1984. *Pseudomonas fluorescens* bacteremia from blood transfusion. Am. J. Med. **76:**62–68.
49. **Kramer, B. S., P. A. Pizzo, K. J. Robichaud, F. Witebsky, and R. Wesley.** 1982. Role of serial microbiologic surveillance and clinical evaluation in the management of cancer patients with fever and granulocytopenia. Am. J. Med. **72:**561–568.
50. **Kunin, C. M.** 1981. Evaluation of antibiotic usage: a comprehensive look at alternative approaches. Rev. Infect. Dis. **3:**745–753.
51. **Lindberg, R. B., and R. L. Latta.** 1974. Phage typing of *Pseudomonas aeruginosa*: clinical and epidemiologic considerations. J. Infect. Dis. **130**(Suppl.):S33–S42.
52. **Longfield, J. N., P. Charache, E. L. Diamond, and T. R. Townsend.** 1982. Comparison of broth and filtration methods for culturing of intravenous fluids. Infect. Control **3:**397–400.
53. **Lorian, V., and D. B. Louria.** 1984. Speciation polemic. J. Infect. Dis. **149:**661–662.
54. **Maki, D. G.** 1980. Through the glass darkly—nosocomial pseudoepidemics and pseudobacteremias. Arch. Intern. Med. **140:**26–28.
55. **Maki, D. G., C. E. Weise, and H. W. Safafin.** 1977. A semi-quantitative culture method for identifying intravenous-catheter-related infection. N. Engl. J. Med. **296:**1305–1309.
56. **Malecka-Griggs, B., and D. J. Reinhardt.** 1983. Direct dilution sampling, quantitation, and microbial assessment of open-system ventilation circuits in intensive care. J. Clin. Microbiol. **17:**870–877.
57. **Marsik, F. J., and J. T. Parisi.** 1971. Bacteriophage types and O antigen groups of *Escherichia coli* from urine. Appl. Microbiol. **22:**26–31.
58. **Martone, W. J., C. A. Osterman, K. A. Fisher, and R. P. Wenzel.** 1981. *Pseudomonas cepacia*: implications and control of epidemic nosocomial colonization. Rev. Infect. Dis. **3:**708–715.
59. **McGowan, J. E., Jr.** 1981. Environmental factors in nosocomial infection—a selective focus. Rev. Infect. Dis. **3:**760–769.
60. **McGowan, J. E., Jr.** 1982. Cost and benefit—a critical issue for hospital infection control. Fifth Annual National Foundation for Infectious Diseases Lecture. Am. J. Infect. Control **10:**100–108.
61. **McGowan, J. E., Jr.** 1983. Antimicrobial resistance in hospital organisms and its relation to antibiotic use. Rev. Infect. Dis. **5:**1033–1048.
62. **McGowan, J. E., Jr.** 1984. Topics in clinical microbiology: the role of the laboratory in control of nosocomial infection. Infect. Control **5:**144–148.
63. **McHugh, G. L., R. C. Moellering, C. Ç. Hopkins, and M. N. Swartz.** 1975. *Salmonella typhimurium* resistant to silver nitrate, chloramphenicol, and ampicillin. Lancet **i:**235–239.
64. **Morris, G. K., and J. G. Wells.** 1974. Colicin typing of *Shigella sonnei*. Appl. Microbiol. **27:**312–316.
65. **Murphy, T. F., K. C. Dudas, J. M. Mylotte, and M. A. Apicella.** 1983. A subtyping system for nontypable *Haemophilus influenzae* based on outer-membrane proteins. J. Infect. Dis. **147:**838–846.
66. **Murray, B. E., A. W. Karchmer, and R. C. Moellering, Jr.** 1980. Diphtheroid prosthetic valve endocarditis: a study of clinical features and infecting organisms. Am. J. Med. **69:**838–848.
67. **Nakashima, A. K., J. R. Allen, W. J. Martone, B. D. Plikaytis, B. Storer, L. M. Cook, and S. P. Wright.** 1984. Epidemic bullous impetigo in a nursery due to a nasal carrier of *Staphylococcus aureus*: role of epidemiology and control measures. Infect. Control **5:**326–331.
68. **Old, D. C.** 1982. Biotyping of *Enterobacter cloacae*. J. Clin. Pathol. **35:**875–878.
69. **Parrott, P. L., P. M. Terry, E. N. Whitworth, L. W. Frawley, R. S. Coble, I. K. Wachsmuth, and J. E. McGowan, Jr.** 1982. *Pseudomonas aeruginosa* peritonitis associated with contaminated poloxamer-iodine solution. Lancet **ii:**683–685.
70. **Penner, J. L., and J. N. Hennessy.** 1979. Application of O-serotyping in a study of *Providencia rettgeri (Proteus rettgeri)* isolated from human and nonhuman sources. J. Clin. Microbiol. **10:**834–840.
71. **Pitt, T. L.** 1982. State of the art: typing of *Serratia marcescens*. J. Hosp. Infect. **3:**9–14.
72. **Polonelli, L., C. Archibusacci, M. Sestito, and G. Morace.** 1983. Killer system: a simple method for differentiating *Candida albicans* strains. J. Clin. Microbiol. **17:**774–780.
73. **Rennie, R. P., C. E. Nord, L. Sjoberg, and I. B. R. Duncan.** 1978. Comparison of bacteriophage typing, serotyping, and biotyping as aids in epidemiological surveillance of *Klebsiella* infections. J. Clin. Microbiol. **8:**638–642.
74. **Richardson, J. H., and W. E. Barkley (ed.).** 1983. Biosafety in microbiological and biomedical laboratories. Publication no. 1983-646-010/8285. U.S. Government Printing Office, Washington, D.C.
75. **Rodriguez, W. J., H. W. Kim, C. D. Brandt, M. K. Gardner, and R. H. Parrott.** 1983. Use of electrophoresis of RNA from human rotavirus to establish the identity of strains involved in outbreaks in a tertiary care nursery. J. Infect. Dis. **148:**34–40.
76. **Rogol, M., I. Sechter, I. Braunstein, and C. B. Gerichter.** 1983. Extended scheme for serotyping *Campylobacter jejeuni*. J. Clin. Microbiol. **18:**283–286.
77. **Schaberg, D. R., R. W. Haley, A. K. Highsmith, R. L. Anderson, and J. E. McGowan, Jr.** 1980. Nosocomial bacteriuria: a prospective study of case clustering and antimicrobial resistance. Ann. Intern. Med. **93:**420–424.
78. **Schneierson, S. S., and D. Amsterdam.** 1967. A manual punch card system for recording, filing, and analyzing antibiotic sensitivity test results. Am. J. Clin. Pathol. **47:**818–820.
79. **Sell, T. L., D. R. Schaberg, and F. R. Fekety.** 1983. Bacteriophage and bacteriocin-typing scheme for *Clostridium difficile*. J. Clin. Microbiol. **17:**1148–1152.
80. **Simmons, B. P.** 1981. Centers for Disease Control guideline for hospital environmental control—microbiologic surveillance of the environment and of personnel in the hospital. Infect. Control **2:**145–146.
81. **Simmons, B. P., and E. S. Wong.** 1982. Guideline for prevention of nosocomial pneumonia. Infect. Control **3:**327–333.
82. **Smith, P. B.** 1972. Bacteriophage typing of *Staphylococcus aureus*, p. 431–441. *In* J. O. Cohen (ed.), The staphylococci. Wiley Interscience, New York.
83. **Smith, P. B.** 1983. Biotyping—its value as an epidemio-

logic tool. Clin. Microbiol. Newsl. **5**:165–166.

84. **Stamm, A. M., W. E. Dismukes, B. P. Simmons, C. G. Cobbs, A. Elliott, P. Budrich, and J. Harmon.** 1982. Listeriosis in renal transplant recipients: report of an outbreak and review of 102 cases. Rev. Infect. Dis. **4**:665–682.
85. **Starkey, D. H.** 1980. The use of indicators for quality control of sterilizing processes in hospital practice: a review. Am. J. Infect. Control **8**:79–84.
86. **Tabaqchali, S., S. O'Farrell, D. Holland, and R. Silman.** 1984. Typing scheme for *Clostridium difficile*: its application in clinical and epidemiological studies. Lancet **i**:935–938.
87. **Tagg, J. R.** 1984. Production of bacteriocin-like inhibitors by group A streptococci of nephritogenic M types. J. Clin. Microbiol. **19**:884–887.
88. **Taylor, A. G., J. McLauchlin, H. T. Green, M. B. Macaulay, and A. Audurier.** 1981. Hospital cross-infection with *Listeria monocytogenes* confirmed by phage-typing. Lancet **ii**:1106.
89. **Thompson, R. L., I. Cabezudo, and R. P. Wenzel.** 1982. Epidemiology of nosocomial infections caused by methicillin-resistant *Staphylococcus aureus*. Ann. Intern. Med. **97**:309–317.
90. **Townsend, T. R., S. Wee, and B. Koblin.** 1982. An efficacy evaluation of a synergized glutaraldehyde-phenate solution in disinfecting respiratory therapy equipment contaminated during use. Infect. Control **3**:240–244.
91. **U.S. Food and Drug Administration.** 1983. Code of Federal Regulations—Food and Drugs, Title 21, Parts 600 to 799, revised as of April 1, 1983, p. 116. U.S. Government Printing Office, Washington, D.C.
92. **Wallace, R. J., Jr., D. M. Musher, E. J. Septimus, J. E. McGowan, Jr., F. J. Quinones, K. Wiss, P. H. Vance, and P. A. Trier.** 1981. *Haemophilus influenzae* infections in adults: characterization of strains by serotypes, biotypes, and beta-lactamase production. J. Infect. Dis. **144**:101–106.
93. **Washington, J. A., II.** 1975. Utilization of microbiologic data. Human Pathol. **6**:267–270.
94. **Weinstein, R. A., F. W. Bauer, R. D. Hoffman, P. G. Tyler, R. L. Anderson, and W. E. Stamm.** 1975. Factitious meningitis. J. Am. Med. Assoc. **233**:878–879.
95. **Weinstein, R. A., and G. F. Mallison.** 1978. The role of the microbiology laboratory in surveillance and control of nosocomial infections. Am. J. Clin. Pathol. **69**:130–136.
96. **Whyte, W., O. M. Lidwell, E. J. L. Lowbury, and R. Blowers.** 1983. Suggested bacteriologic standards for air in ultraclean operating rooms. J. Hosp. Infect. **4**:133–139.
97. **Wiesenthal, A. M.** 1984. A maternal-neonatal outbreak of infections due to an unusual group A beta-hemolytic streptococcus. Infect. Control **5**:271–274.
98. **Wilkinson, H. W., A. L. Reingold, B. J. Brake, D. L. McGiboney, G. W. Gorman, and C. V. Broome.** 1983. Reactivity of serum from patients with suspected legionellosis against 29 antigens of *Legionellaceae* and *Legionella*-like organisms by indirect immunofluorescence assay. J. Infect. Dis. **147**:23–31.
99. **Williams, W. W.** 1983. CDC guidelines for the prevention and control of nosocomial infections. Guideline for infection control in hospital personnel. Infect. Control **4**:326–349.
100. **Wong, L. K., A. L. Barry, and S. M. Horgan.** 1982. Comparison of six different criteria for judging the acceptability of sputum specimens. J. Clin. Microbiol. **16**:627–631.
101. **Wust, J., N. M. Sullivan, U. Hardegger, and T. D. Wilkins.** 1982. Investigation of an outbreak of antibiotic-associated colitis by various typing methods. J. Clin. Microbiol. **16**:1096–1101.

Chapter 12

Infection Control Priorities in Critical Care Medicine: Device-Associated Intravascular Infections

RICHARD P. WENZEL

Nosocomial infections occur in 5 to 10% of patients admitted to hospitals in the United States (9, 10, 38) and result in increased morbidity, mortality, and costs (11, 14, 16–18, 31, 33, 34, 38, 45). As a result, surveillance has been advocated by the Centers for Disease Control to develop priorities for research and infection control activities (1, 15). Efforts to streamline surveillance methods have been advanced (41, 43, 44), and a concept of identifying preventable infections has evolved (22). In general, preventable infections are those related to a device or a specific procedure (36, 40), in contrast to infections that often occur in immunosuppressed hosts.

It has recently been suggested that a high priority for infection control be the identification of procedure-related infections in patients in intensive care units (ICUs), since a significant proportion of such infections may be preventable (42). In general, patients admitted to the ICU have a higher risk of nosocomial infection than other hospitalized patients. In a 2-year study, Donowitz et al. showed that ward patients had a 6% rate of infection and that ICU patients had an 18% rate (12). Except for infection at the operative site, there was a significantly greater rate of infection for patients in the ICU (Table 1), a fact which attests to their serious underlying illnesses and need for invasive monitoring devices. The major organisms causing hospital-acquired bloodstream infections were *Staphylococcus epidermidis* (7/100 admissions) and *Pseudomonas* species (5.3/100). Importantly, there was a tendency for a higher proportion of *Staphylococcus aureus* bloodstream isolates to be methicillin resistant and for a higher proportion of gram-negative rod bloodstream isolates to be aminoglycoside resistant in ICU patients as compared with ward patients.

In addition to endemic infections, epidemic infections are also significantly increased in critical care areas (46). Most epidemics occur in critical care units and are related to contaminated hospital equipment or products. By definition, all are preventable. In a recent report from the University of Virginia, the reservoir in 5 of the 11 identified outbreaks involved devices, and 8 of the 11 cases involved infections of the bloodstream (Table 2) (46).

The purpose of this chapter is to emphasize the importance of preventable vascular-device-related infections in patients hospitalized in critical care units. Special emphasis will be placed on the arterial and pulmonary artery catheters and the increasingly used Hickman-Broviac catheters.

ARTERIAL CATHETERS

The vast majority of adult patients in critical care units have arterial catheters in place for monitoring blood pressure and for providing access to arterial blood for blood gas analyses. Although a number of studies have reported the infectious complications of the transducer assembly, comparatively little information is available on the risk of infection deriving from the arterial catheter itself.

The most useful information came from a study by Band and Maki in 1979 in which they stressed the importance of certain risk factors for line-associated infection: (i) a surgical cutdown approach instead of a percutaneous puncture; (ii) placement of the catheter for more than 4 days; and (iii) the presence of any inflammation (pus or erythema) around the catheter entrance site (3). Both contamination of the catheter and development of arterial-catheter-related bloodstream infection were more frequent in the presence of these risk factors (Table 3). As a result, it is generally recommended that arterial catheters be placed by percutaneous puncture, that the site be changed if there is any inflammation, and that the catheters be routinely changed at 72 to 96 h.

In a subsequent study, Maki and Band examined the relative risk of catheter contamination in patients who had antibacterial ointments applied to the catheter site at 48-h intervals (21). They compared an iodophor ointment to one containing polymyxin, neomycin, and bacitracin and utilized a control group with no topical agent applied. Although various types of catheters were studied, the authors concluded that if an ointment is to be used, the iodophor ointment may be preferable for arterial catheter care. Their conclusion was based on the fact that staphylococcal infections occurred less frequently with either antibacterial ointment compared with controls and that three of the four candida infections occurred in the group treated with the triple-antibiotic ointment (21).

PULMONARY ARTERY CATHETERS

In addition to arterial catheters, Swan-Ganz pulmonary artery catheters are also being utilized with increasing frequency to monitor ICU patients to aid in the diagnosis and treatment of congestive heart failure. Little information is available on the risk of catheter-related bloodstream infections, but some information was summarized by Michel et al. in 1981 (24). In three studies, the rate of catheter-associated sepsis was 0 to 2/100 patients; a fourth study, performed at a time when there was an outbreak of infections in the critical care unit, indicated a 10% rate (Table 4). Most clinicians think that the true endemic infection rate is less than 2%.

Importantly, as one examines the data, one can appreciate that the patient population in the four studies varied and that the culture techniques employed to detect catheter-related sepsis also varied among the studies (Table 4). Thus, we are left with the impression, based on limited data, that the rate of infection related to pulmonary artery catheters is low.

TABLE 1. Nosocomial infection rates for patients on hospital wards compared with those in ICUs[a]

Site of infection	Infection rate[b]	
	Ward	ICU
Bloodstream	0.7	5.2
Lung	0.7	4.6
Urinary tract	2.4	4.6
Operative wound	1.1	1.4
Other	1.0	2.3

[a] Adapted from data of Donowitz et al. (12).
[b] Number of infections per 100 admissions.

In general, the subclavian approach is preferred for placing pulmonary artery catheters. Nevertheless, in a small study by Singh et al., there was no difference in the rate of contamination of catheters placed by the internal jugular versus the subclavian route (35). However, it is the impression of most critical care nurses and physicians that contamination of the catheters with airway secretions is greater when the pulmonary artery catheter is inserted by the internal jugular route.

The optimal time for changing pulmonary artery catheters has not been determined. However, by analogy with the study on arterial catheters, many ICUs have a protocol to change the pulmonary artery catheter site at 4-day intervals. An additional unanswered question is whether the catheters should be changed over a wire and maintained at the same site or whether an entirely new catheter should be placed at a new site. Studies are in progress to attempt to answer that question, which focuses on the potential risk of infection versus the potential trauma of repeated percutaneous punctures to place a new line. In the meantime, it should be emphasized that manufacturers' directions should be followed in using pulmonary artery catheters. Specifically, the disposable catheters should not be resterilized and reused since such procedures could lead to malfunction and complications, including major clotting within the lumen (26).

HICKMAN-BROVIAC CATHETERS

The Hickman-Broviac catheters have become increasingly popular as a method of long-term infusion of fluids and medications as well as a method for providing access to blood specimens for analysis. Originally used for hyperalimentation, particularly in patients with serious small bowel disease, catheters are currently used for bone marrow transplant patients and patients with malignancies for continued

TABLE 2. Eleven outbreaks of nosocomial infections, illustrating the frequent locations in ICUs[a]

Organism	Anatomic site of infections	Reservoir	No. of deaths/ no. of patients[b]	Unit[c]
Serratia marcescens	Bloodstream	Pressure monitor transducer head	10/19	SICU
Klebsiella pneumoniae and *Klebsiella oxytoca*	Bloodstream	Banked breast milk (contaminated electric pump)	1/5	NICU
Staphylococcus aureus	Bloodstream	Swan-Ganz catheter	0/4	CCU
Staphylococcus epidermidis	Bloodstream	Broviac catheter	0/5	NICU
Staphylococcus epidermidis, Candida albicans, and several gram-negative rods	Bloodstream	Swan-Ganz and arterial catheters	3/6	MICU
Pseudomonas cepacia	Multiple, including bloodstream	Contaminated cocaine	5/56[d]	SICU
Pseudomonas putida	Bloodstream	Probably blood	3/3	SICU
Staphylococcus aureus	Skin	Nursing personnel with dermatitis[e]	0/6	NICU
Methicillin-resistant *Staphylococcus aureus*	Multiple, including bloodstream	Patients (ward or roommates)	5/54	BU/SICU
Viral? (gastrointestinal)	Gastrointestinal tract	Patients (ward or roommates)	0/8	NICU
Salmonella enteritidis	Gastrointestinal tract	Patients (ward or roommates)	0/3	Pediatrics ward

[a] Infections were identified by routine surveillance between 1 January 1978 and 1 December 1982 at the University of Virginia Hospital. Data are from reference 46.
[b] These are crude case/fatality ratios, which do not imply a cause-effect relationship.
[c] Infections occurred solely or predominantly in the indicated unit. Abbreviations: SICU, surgical ICU; NICU, newborn ICU; CCU, coronary care unit; MICU, medical ICU; BU, burn unit.
[d] The outbreak was a combination of epidemic and pseudoepidemic.
[e] The dermatitis in seven nurses was related to excessive use of a newly introduced antiseptic handwashing agent containing chlorhexidine.

TABLE 3. Infections related to arterial catheter placement in ICU patients[a]

Factor examined	No. of patients	Catheters contaminated with ≥15 organisms (%)	Catheter-related bloodstream infections (%)
Percutaneous puncture	112	16	2
Surgical cutdown	18	33	17
<4 days[b]	17	10	0
≥4 days[b]	59	27	9
No inflammation	116	16	2
Inflammation at catheter site	14	35	21

[a] Adapted from Band and Maki (3).
[b] Duration of arterial catheter use.

access to the venous circulation. Most clinicians describe their use as a humane addition to the management of such patients, since repeated venipunctures are avoided. More recently, the catheters have become popular in neonatal ICUs, in some adult ICUs, and occasionally in a ward setting in patients who require long-term antibiotic therapy. The latter have included patients with bacterial osteomyelitis and those with systemic fungal infections. The catheters are tunnelled under the skin of the chest wall, with an entrance site halfway between the nipple and sternum. After entering the vascular system via the subclavian vein, the tip of the catheter lies in the superior vena cava or right atrium. A Dacron cuff wrapped around a small segment of the catheter beneath the chest wall excites a fibrotic reaction which prevents extraluminal migration of bacteria.

Despite the frequent use of these catheters, there are no data to show that they are less likely to be associated with infections than standard intravenous catheters. A number of uncontrolled studies show that patients frequently become infected but that the infection rate, defined as the number of infections per 1,000 days of use, is low even though these catheters are often left in place for prolonged periods. From 10 to 52% of patients in reported series became infected, but the number of infections per 1,000 catheter days ranged from 0.49 to 4.8 (2, 4–8, 19, 20, 23, 27, 30, 32, 37, 39). In addition, the risk of infection was greater in those with altered immune conditions than in those with only gastrointestinal disease (Table 5).

When one examines the organisms reported to have caused bloodstream infections related to Hickman-Broviac catheters (Table 6), one notes the presence of gram-positive organisms, especially *S. epidermidis* and *S. aureus*. Fungal species as a group are also important pathogens in this setting, and of the gram-negative organisms, *Escherichia coli* is the most common agent.

Current information suggests that organisms traverse the catheter from the skin of the patient or the skin of those touching the patient by the intraluminal route. Thus, meticulous attention to aseptic technique during insertion and manipulation of the catheter appears to be an extremely important infection control measure.

One issue facing clinicians is how one can interpret results of blood cultures drawn through the central catheter. In general, most clinicians have preferred to use only the information from peripheral venous cultures. However, a recent study suggested the utility of quantitative blood cultures in the evaluation of septicemia in children with Broviac catheters (29). The authors studied 30 febrile episodes in 14 patients. In all cases of septicemia due to catheter contamination, quantitative blood cultures drawn through the central catheter had a concentration of pathogens ≥10 times that in blood obtained from a peripheral vein.

A major clinical question arises when a patient with a Hickman-Broviac catheter in place develops a bloodstream infection: should the catheter be removed? Current data do not permit a definite answer, and many clinicians prefer to leave the catheter in place while antibiotic therapy is initiated. If, after several days of effective antibiotics, the infection is not controlled, then the catheter is removed. On the other hand, there is some clinical information which suggests that if the offending organism is a corynebacterium or a fungal species, the chances of successful antibiotic therapy alone are very low unless the catheter is removed.

CONTROL OF DEVICE-ASSOCIATED INFECTIONS

Problems in critical care areas of the hospital are or should be a high priority for infection control research and service. ICUs represent 5 to 10% of beds in the hospital, yet patients housed there account for at least one-third of all life-threatening infections: nosocomial bloodstream infections and pneumonias (42). In large part, such data reflect the serious underlying problems of the patients and the use of multiple devices which are associated with infections. Little is known about optimal management of devices despite the fact that improved care may prevent serious endemic and epidemic problems. The problem is compounded by the fact that there exist few prospective studies to define specific risk factors for infections associated with devices.

In addition to the use of devices which predispose patients to or increase the risk of infections, it is likely that high usage of broad-spectrum antibiotics selects for the types of organisms involved. The recent introduction of second- and third-generation cephalosporins has coincidentally been associated with the rec-

TABLE 4. Bloodstream infections related to Swan-Ganz catheters[a]

Year (reference)	Type of ICU	No. of cases	Catheter culture method	Catheter-related infection rate (%)
1978 (28)	Coronary care	152	Not stated	1
1979 (13)	Respiratory-medical-surgical	92	Broth	2
1980 (25)	Critical care-trauma	153	Semiquantitative	10
1981 (24)	Medical-surgical	60	Broth	0

[a] Adapted from Michel et al. (24).

TABLE 5. Bloodstream infection associated with Broviac-Hickman catheters

Hospital	Reference	No. of infections per 1,000 catheter days	No. of patients infected/ no. with catheters (%)	Underlying disease in study population[a]
University of Washington	7	0.76	2/18 (11)	GI
	30	0.49	8/38 (21)	GI
	19	?	24/75 (32)	MALIG
UCLA	8	0.87	1/6 (17)	GI (PED)
	4	4.0	9/88 (10)	PED
	23	2.6	2/18 (10)	MALIG (PED)
	27	0.87	33/198 (17)	GI, MALIG (all ages)
Auckland, New Zealand	5	?	11/25 (44)	MALIG
University of Kansas	37	1.3	8/84 (10)	GI, MALIG (all ages)
Balt, Calif.	39	4.4	22/47 (47)	MALIG
Genoa, Italy	2	4.7	10/40 (25)	? (all ages)
University of Tasmania	6	4.8	4/15 (27)	MALIG
Copenhagen, Denmark	20	0.81	6/23 (26)	GI
University of Pittsburgh	34	2.7	14/27 (52)	MALIG (PED)

[a] Abbreviations: GI, gastrointestinal disease; MALIG, malignancy; PED, pediatric patients.

ognition of infections caused by *S. epidermidis*, enterococci, more resistant gram-negative rods, and yeasts. To many in infection control, the birthplace of antibiotic resistance is the ICU, and such areas should be under close surveillance.

The following are some suggested approaches, which need to be tested in prospective studies, to preventing ICU-related infections. (i) The design of ICUs should include plans to separate patients by wall partitions to minimize the movement of personnel from patient to patient without proper handwashing. Sinks placed at the entrances to the unit or in cubicles (not corners) in ICUs may improve compliance with protocols for handwashing. (ii) One-to-one nurse-to-patient ratios 24 h a day may significantly reduce cross-infection, particularly if clinical specimens are obtained properly by the one nurse caring for the patient rather than by several technicians. In addition, management of medical devices may be optimal if limited to the primary care nurse. (iii) Isolation precautions for ICU patients who require two or more systemic antibiotics may minimize transmission of antibiotic-resistant organisms to susceptible roommates in the ICU. (iv) Recognition and prevention of device-related infections by clinicians may improve with use of individual patient "flow sheets" listing all devices used, dates of insertion, and documentation of the need for continuation. (v) Surveillance and reporting of hospital-acquired infections should include a separate category for device-related infections. The introduction of all new devices and products, including handwashing agents, should be reported to the infection control team. (vi) Strict protocols should be used for the management of all devices in the ICU.

TABLE 6. Bloodstream isolates associated with Hickman-Broviac catheter sepsis

Organism	No. of isolates reported in reference (year):						
	8 (1977)	31 (1980)	5 (1980)	2 (1981)	39 (1981)	27 (1981)	32 (1982)
Staphylococcus epidermidis	0	4	2	0	8	6	3
Staphylococcus aureus	0	1	0	3	3	15	1
Alpha-hemolytic streptococcus	1	0	0	0	2	2	0
Enterococcus	0	2	3	0	2	2	0
Corynebacterium spp.	0	0	0	0	1	1	0
Clostridium spp.	0	0	0	0	1	0	0
Other streptococci	0	0	2	0	0	0	0
Escherichia coli	0	0	1	2	3	2	3
Klebsiella pneumoniae	0	0	1	0	2	1	2
Pseudomonas aeruginosa	0	1	0	0	2	0	1
Enterobacter spp.	0	0	0	0	1	0	3
Bacteroides fragilis	0	0	0	0	1	0	0
Pseudomonas stutzeri	0	0	0	0	0	0	1
Serratia spp.	0	0	0	0	0	1	0
Candida albicans	0	1	0	5	2	4	0
Candida tropicalis	0	0	0	0	1	0	1
Aspergillus spp.	0	0	0	0	1	0	0

LITERATURE CITED

1. **Aber, R. C., and J. B. Bennett.** 1979. Surveillance of nosocomial infections, p. 53–61. *In* J. V. Bennett and P. Brachman (ed.), Hospital infections. Little, Brown & Co., Boston.
2. **Adami, G. F., A. Bicagalupo, U. Bonalumi, M. T. Van Lindt, and F. Griffanti-Bartali.** 1981. Use of Hickman right atrial catheter for vascular access in marrow transplant recipients. Arch. Surg. **116:**1099.
3. **Band, J. D., and D. G. Maki.** 1979. Infections caused by arterial catheters used for hemodynamic monitoring. Am. J. Med. **67:**735–741.
4. **Begala, J. E., K. Maher, and J. D. Cherry.** 1982. Risk of infection associated with the use of Broviac and Hickman catheters. Am. J. Infect. Control **10:**17–23.
5. **Blacklock, H. A., R. S. Hill, A. G. Clark, M. V. Pillai, J. R. D. Matthews, and J. F. Wade.** 1980. Use of modified right atrial catheter for venous access in leukaemic patients. Lancet **i:**993–994.
6. **Braithwaite, P., V. J. Rust, and R. M. Lowenthal.** 1981. The value of the Hickman-type right atrial catheter for venous access in cancer patients. Aust. N. Z. J. Surg. **51:**359–363.
7. **Broviac, J. W., J. J. Cole, and B. H. Scribner.** 1973. A silicone rubber atrial catheter for prolonged parenteral alimentation. Surg. Gynecol. Obstet. **136:**602–606.
8. **Byrne, W. J., T. C. Halpin, M. J. Asch, E. W. Fonkalsrud, and M. E. Amest.** 1977. Home total parenteral nutrition: an alternative approach to the management of children with severe chronic small bowel disease. J. Pediatr. Surg. **12:**359–366.
9. **Center for Disease Control.** 1979. National nosocomial infections study report, p. 1–3. Center for Disease Control, Atlanta, Ga.
10. **Dixon, R. E.** 1978. Effect of infections on hospital care. Ann. Intern. Med. **89:**749–753.
11. **Donowitz, L. G., and R. P. Wenzel.** 1980. Endometritis following cesarean section: a controlled study of the increased duration of hospital stay and direct cost of hospitalization. Am. J. Obstet. Gynecol. **137:**467–469.
12. **Donowitz, L. G., R. P. Wenzel, and J. W. Hoyt.** 1982. High risk of hospital-acquired infections in the ICU patient. Crit. Care Med. **10:**355–357.
13. **Elliott, C. G., G. A. Zimmerman, and T. P. Clemmer.** 1979. Complications of pulmonary artery catheterization in the care of critically ill patients. Chest **76:**647–652.
14. **Freeman, J. B., A. Rosner, and J. E. McGowan, Jr.** 1979. Adverse effects of nosocomial infection. J. Infect. Dis. **140:**732–740.
15. **Garner, J. S., J. C. Bennett, W. E. Scheckler, D. G. Maki, and P. S. Brachman.** 1971. Surveillance of nosocomial infections, p. 277–281. *In* Proceedings of the International Conference on Nosocomial Infections. American Hospital Association, Chicago.
16. **Givens, C. D., and R. P. Wenzel.** 1980. Catheter-associated urinary tract infections in surgical patients: a controlled study on the excess morbidity and costs. J. Urol. **124:**646–648.
17. **Green, J. W., and R. P. Wenzel.** 1977. Postoperative wound infection: a controlled study of the increased duration of hospital stay and direct cost of hospitalization. Ann. Surg. **185:**264–268.
18. **Haley, R. W., D. R. Schaberg, S. D. Von Allmen, and J. E. McGowan.** 1980. Estimating the extra charges and prolongation of hospitalization due to nosocomial infections: a comparison of methods. J. Infect. Dis. **141:**248–257.
19. **Hickman, R. O., C. D. Buckner, R. A. Clift, J. E. Sanders, P. Stewart, and E. D. Thomas.** 1979. A modified right atrial catheter for access to the venous system in marrow transplant recipients. Surg. Gynecol. Obstet. **148:**871–875.
20. **Ladefoged, K., F. Efsen, J. K. Christoffersen, and S. Jarnum.** 1981. Long-term parenteral nutrition. II. Catheter related complications. Scand. J. Gastroenterol. **16:**913–919.
21. **Maki, D. G., and J. D. Band.** 1981. A comparative study of polyantibiotic and iodophor ointments in prevention of catheter-related infection. Am. J. Med. **70:**739–744.
22. **McGowan, J. E., Jr., P. L. Parrott, and V. P. Duty.** 1977. Nosocomial bacteremia. Potential for prevention of procedure-related cases. J. Am. Med. Assoc. **237:**2727–2729.
23. **Merritt, R. J., C. E. Ennis, R. J. Andrassy, D. M. Hays, F. R. Sinatra, D. W. Thomas, and S. E. Siegel.** 1981. Use of Hickman right atrial catheter in pediatric oncology patients. J. Parenter. Enter. Nutr. **5:**83–85.
24. **Michel, L., M. Marsh, J. C. McMichan, P. A. Southorn, and N. S. Brewer.** 1981. Infection of pulmonary artery catheters in critically ill patients. J. Am. Med. Assoc. **245:**1032–1036.
25. **Myers, M. L., G. Ravindar, T. Austin, and W. Sibbald.** 1980. Swan Ganz (SG) catheterization and sepsis: a prospective study. Crit. Care Med. (April) 272.
26. **Park, G. R., and D. H. T. Scott.** 1982. Complication of the reuse of flow-directed pulmonary artery catheters. Br. Med. J. **284:**258–259.
27. **Pollack, P. F., M. Kadden, and W. J. Byrne.** 1981. 100 patient years' experience with the Broviac elastic catheter for central venous nutrition. J. Parenter. Enter. Nutr. **5:**32–30.
28. **Prachar, H., M. Dittel, C. Jobst, E. Kiss, E. Machacek, and H. Nobis.** 1978. Bacterial contamination of pulmonary artery catheters. Intensive Care Med. **4:**79–82.
29. **Raucher, H. S., A. C. Hyatt, A. Barzilai, M. D. Harris, M. A. Weiner, N. S. Leiko, and D. S. Hodes.** 1984. Quantitative blood cultures in the evaluation of septicemia in children with Broviac catheters. J. Pediatr. **104:**29–31.
30. **Riella, M. C., and B. H. Scribner.** 1976. Five years' experience with a right atrial catheter for prolonged parenteral nutrition at home. Surg. Gynecol. Obstet. **143:**205–208.
31. **Rose, R., K. J. Hunting, T. R. Townsend, and R. P. Wenzel.** 1977. The morbidity, mortality and economics of hospital-acquired bloodstream infections: a controlled study. South. Med. J. **70:**1267–1269.
32. **Scheckler, W. E.** 1978. Septicemia and nosocomial infections in a community hospital. Ann. Intern. Med. **89:**754–756.
33. **Scheckler, W. E.** 1980. Hospital costs of nosocomial infections. A prospective three month study in a community hospital. Infect. Control **1:**150–152.
34. **Shapiro, E. D., E. R. Wald, and K. A. Nelson.** 1982. Broviac catheter-related bacteremia in oncology patients. Am. J. Dis. Child. **136:**679–681.
35. **Singh, S., N. Nelson, I. Acosta, F. E. Check, and V. K. Puri.** 1982. Catheter colonization and bacteremia with pulmonary and arterial catheters. Crit. Care Med. **10:**736–739.
36. **Stamm, W. E.** 1978. Infections related to medical devices. Ann. Intern. Med. **89:**764–769.
37. **Thomas, J. H., R. I. MacArthur, G. E. Pierce, and A. S. Hermreck.** 1980. Hickman-Broviac catheters, indications and results. Am. J. Surg. **140:**791–796.
38. **Townsend, T. R., and R. P. Wenzel.** 1981. Nosocomial bloodstream infections in a newborn intensive care unit: a case matched control study of morbidity, mortality, and risk. Am. J. Epidemiol. **114:**73–80.
39. **Wade, J. C., K. A. Newman, S. S. Schimpff, D. A. Van Echo, R. A. Gelbu, W. A. Reed, and P. H. Wiernik.** 1981. Two methods for improved venous access in acute leukemia patients. J. Am. Med. Assoc. **246:**140–144.
40. **Weinstein, R. A., and L. S. Young.** 1979. Other procedure related infections, p. 489–505. *In* J. V. Bennett and P. Brachman (ed.), Hospital infections. Little, Brown & Co, Boston.

41. **Wenzel, R. P., K. J. Hunting, and C. A. Osterman.** 1977. Post-operative wound infection rates. Surg. Gynecol. Obstet. **144**:749–752.
42. **Wenzel, R. P., C. A. Osterman, L. G. Donowitz, J. W. Hoyt, M. A. Sande, W. J. Martone, J. E. Peacock, Jr., J. I. Levine, and G. B. Miller, Jr.** 1981. Identification of procedure-related nosocomial infections in high-risk patients. Rev. Infect. Dis. **3**:701–707.
43. **Wenzel, R. P., C. A. Osterman, and K. J. Hunting.** 1976. Hospital-acquired infections. II. Infection rates by site, service and common procedures in a university hospital. Am. J. Epidemiol. **104**:645–651.
44. **Wenzel, R. P., C. A. Osterman, K. J. Hunting, and J. M. Gwaltney, Jr.** 1976. Hospital-acquired infections. I. Surveillance in a university hospital. Am. J. Epidemiol. **103**:251–260.
45. **Wenzel, R. P., C. A. Osterman, T. R. Townsend, J. M. Veazey, Jr., K. H. Servis, L. S. Miller, R. B. Miller, G. B. Miller, Jr., and R. S. Jackson.** 1979. Development of a statewide program for surveillance and reporting of hospital-acquired infections. J. Infect. Dis. **140**:741–746.
46. **Wenzel, R. P., R. L. Thompson, S. M. Landry, B. S. Russell, P. J. Miller, S. Ponce de Leon, and G. B. Miller, Jr.** 1983. Hospital-acquired infection in intensive care unit patients: an overview with emphasis on epidemics. Infect. Control **4**:371–375.

Chapter 13

Sterilization, Disinfection, and Antisepsis in the Hospital

MARTIN S. FAVERO

The effective use of antiseptics, disinfectants, and sterilization procedures is important in the prevention of nosocomial infections. Physical agents, such as moist or dry heat, play the major role in sterilization, and chemical germicides are used primarily for disinfection and antisepsis. In recent years, there has been an explosion in the number of germicidal products available to hospitals in the United States. In 1973, the American Society for Microbiology Ad Hoc Committee on Microbiological Standards of Disinfection in Hospitals surveyed 16 U.S. hospitals with a combined bed capacity of more than 9,000. The survey showed that the average number of different formulations used per hospital was 14.5, with a range of 8 to 22. A total of 224 products were used in the 16 hospitals, and 125 of the products were proprietary products.

The choice of agents and procedures to be used for hospital environmental sanitation and antisepsis depends on a variety of factors, and no single agent or procedure is adequate for all purposes. Factors to be considered in the selection of procedures include the degree of microbial killing required, the nature of the item to be treated, and the cost and ease of using the available agents. This chapter discusses each of these factors and practical methods for evaluating the effectiveness of the various agents and procedures.

REGULATION OF CHEMICAL GERMICIDES IN THE UNITED STATES

Chemical germicides that are formulated as disinfectants are regulated and registered by the Environmental Protection Agency (EPA). The authority for this responsibility comes under the Federal Insecticide, Fungicide, and Rodenticide Act (FIFRA). The EPA requires manufacturers of chemical germicides formulated as general disinfectants, hospital disinfectants, and disinfectants applied in other environments, such as the food industry, to test these formulations by using specific protocols for microbicidal efficiency, stability, and toxicity to humans. The decision to register a disinfectant is based on data provided to the EPA by the manufacturer. Although in past years the EPA has reserved the right to independently test and verify formulations of chemical germicides for their specified efficacy, in practice only those formulations to be registered as sterilants or sporicides were actually tested. In 1982, the EPA discontinued this testing, so that current registration of all chemical germicides is based solely on efficacy data provided to the EPA by the manufacturer. Formulations of chemical germicides currently are registered by the EPA based on data obtained from the manufacturer without independent verification except as these formulations are tested by investigators and reported in the scientific literature. This lack of independent verification has caused concern among health professionals who believed that EPA verification of sporicidal or sterilant claims was a significant consideration in the selection of a sterilant or high-level disinfectant. Unfortunately, EPA verification is no longer available.

Users should keep in mind that the field of chemical germicides is highly competitive and that exaggerated claims are often made about the germicidal efficiency of specific formulations. New formulations of chemical germicides that become commercially available should be scrutinized by health professionals. It may also be necessary to consult with the Disinfectants Branch, Registration Division, Office of Pesticides, EPA, when questions regarding specific claims or use arise. Studies published in the scientific literature and presented at scientific meetings also provide important information on the efficacy of these formulations.

Chemical germicides that are formulated as antiseptics, preservatives, or drugs to be used on or in the human body or as preparations to be used to inhibit or kill organisms on the skin are regulated by the Food and Drug Administration (FDA). However, the FDA method of regulating these formulations is significantly different from that of the EPA. The FDA has an advisory review panel on nonprescription, antimicrobial drug products. Manufacturers of such formulations voluntarily submit data to the panel, which in turn categorizes the products for their intended use, i.e., antimicrobial soaps, health care personnel hand washes, patient preoperation skin preparations, skin antiseptics, skin wound cleansers, skin wound protectants, and surgical hand scrubs. Generic chemical germicides for each use are further divided into category I, safe and efficacious; category II, not safe or efficacious; and category III, insufficient data to categorize. Consequently, chemical germicides formulated as antiseptics and regulated by the FDA are categorized basically by use and efficacy and are not regulated or registered in the same fashion as EPA regulates and registers a chemical germicide. For a more extensive discussion of these subjects, the reader is referred to papers by Block (5) and Zanowiak and Jacobs (26).

ANTIMICROBIAL EFFECTIVENESS

Although the definitions of sterilization, disinfection, and antisepsis (14) have been generally accepted, it is common to see all three terms misused. The precise definitions of the three terms and the basic knowledge of how to achieve and monitor each state are extremely important if long-known principles are to be effectively applied.

Sterilization

Sterilization is defined as the use of a physical or chemical procedure to destroy all microbial life, including highly resistant bacterial endospores. In the hospital, sterilization particularly pertains to those organisms that may exist on inanimate objects. Moist heat under pressure (steam autoclaving) and ethylene oxide gas are the major sterilizing agents used in hospitals. However, when used appropriately, some chemical germicides normally considered disinfectants can also be used for sterilization.

Disinfection

Disinfection is generally less lethal than sterilization. A disinfection procedure inactivates virtually all recognized pathogenic microorganisms but not necessarily all microbial forms (e.g., bacterial endospores) on inanimate objects. As can be seen by this definition, disinfection does not ensure an "overkill," and disinfection lacks the margin of safety achieved by sterilization. The effectiveness of a disinfection procedure depends on a number of factors, each of which may significantly affect the end result. Among these factors are the nature and number of contaminating microorganisms (especially the presence of bacterial endospores), the concentration of the chemical, the length of exposure to the chemical, the amount of organic matter (soil, feces, blood, etc.) present, the type and condition of the material to be disinfected, and the temperature. Thus, disinfection is a procedure which reduces the level of microbial contamination, but there is a broad range of activity which extends from sterility at one extreme to a minimal reduction in the number of contaminating microorganisms at the other.

Absolute sterility is difficult to prove, and as a result, it is common to define sterility in terms of the probability that a contaminating microorganism will survive treatment. For example, sterilizing processes are usually challenged and verified with 10^6 to 10^9 dried bacterial endospores, and sterilization is defined as the state in which the probability of any one spore surviving is 10^{-6} or lower. This rationale has been used to establish cycles for steam sterilizers and ethylene oxide gas sterilizers, and it produces a degree of overkill as well as a quantitative assurance of true sterilization. It is difficult to evaluate liquid-chemical disinfection processes by using these criteria, however, and disinfection processes cannot be assumed to have the same reliability as sterilization.

Antisepsis

An antiseptic is a substance that is used on or in living tissue to inhibit or destroy microorganisms. Quite often, the distinction between an antiseptic and a disinfectant is not made; a disinfectant is a chemical germicide that is used solely for destroying microorganisms on inanimate objects, whereas an antiseptic is used on or in living tissue. Some chemical agents, iodophors for example, are used as active agents in chemical germicides that are formulated both as disinfectants and as antiseptics. However, the precise formulations are usually significantly different depending on intended use, and germicidal efficacy differs substantially. Thus, disinfectants should never be used as antiseptics and vice versa.

DISINFECTANT ACTIVITY

The classification of chemical germicides in various concentrations according to their levels of microbicidal potency is important but arbitrary. I have decided to retain the system originally proposed by Spaulding (23), rather than the system used by the EPA (5). For the purposes of this chapter, three levels of disinfection are defined: high, intermediate, and low (Table 1). In the EPA classification (5), chemical germicides that are registered as sporicides would be equivalent to high-level disinfectants. Germicides registered as sanitizers would probably fall into the category of low-level disinfectants, but there are numerous formulations that would be classified as either low- or intermediate-level disinfectants, depending on the specific label claims. For example, some chemical germicide formulations are claimed to be efficacious against *Mycobacterium tuberculosis*. By the Spaulding system, such a formulation would probably be classified as an intermediate-level disinfectant. However, chemical germicide formulations with specific label claims for *Salmonella cholerae-suis*, *Staphylococcus aureus*, and *Pseudomonas aeruginosa* could fall into intermediate- or low-level disinfectant categories. Efficacy against these challenge organisms is required for EPA classification as a hospital disinfectant.

Some high-level disinfectants can kill large numbers of resistant bacterial endospores under severe test conditions but may require as long as 24 h to do so. However, most disinfectants cannot achieve this

TABLE 1. Levels of germicidal action

Level	Efficacy against					
	Bacteria			Fungi[a]	Viruses	
	Vegetative bacteria	Tubercle bacillus	Spores		Lipid and medium-sized	Nonlipid and small
High	+[b]	+	+[c]	+	+	+
Intermediate	+	+	±[d]	+	+	±[e]
Low	+	−	−	±	+	−

[a] Includes asexual spores but not necessarily chlamydospores or sexual spores.

[b] +, Killing effect can be expected when the normal-use concentrations of chemical disinfectants or pasteurization are properly employed; −, little or no killing effect.

[c] Only with extended exposure times are high-level disinfectants capable of actual sterilization.

[d] Some intermediate-level disinfectants, e.g., iodophors, formaldehyde, tincture of iodine, and chlorine compounds, can be expected to exhibit some sporicidal action.

[e] Some intermediate-level disinfectants, e.g., alcohols and phenolic compounds, may have limited virucidal activity.

level of antimicrobial activity. In practical terms, high-level disinfection procedures, if done properly, can be considered almost equivalent to sterilization without the added insurance of overkill; since a number of critical patient care items are damaged by high temperatures and cannot be heat sterilized, they must be disinfected with chemical germicides. High-level disinfectants are used fairly often to process medical and surgical materials. In the absence of bacterial spores, these germicides are rapidly effective; however, the absence of spores cannot usually be assured. Although the number of spores will generally be small (22), sporicidal capacity is nevertheless an essential property of high-level disinfection, and the sporicidal activity depends both on the agent and how it is used.

A good example of a germicide whose effectiveness depends on how it is used is 2% aqueous glutaraldehyde, which is in fact capable of sterilizing, but only after extended contact time in the absence of extraneous organic material. Some materials are not physically able to withstand immersion in fluid for 6 to 10 h. Even if prolonged contact were possible, the treated materials would have to be rinsed thoroughly with sterile water, dried in a special cabinet with sterile air, and stored in a sterile container to ensure that the material remained sterile. Thus, glutaraldehyde-based chemical germicides are capable of sterilizing under a strict set of circumstances (i.e., precleaned items, 6 to 10 h of contact time, and room temperature). However, users often soak items in a high-level disinfectant for 10 to 30 min, rinse them in nonsterile water, and refer to the items as "sterile." This situation indicates a misunderstanding of the terms sterile and disinfected, as well as overconfidence in a product and overestimation of the safety of the processed item.

Intermediate-level disinfectants do not necessarily kill bacterial endospores, but they do inactivate the tubercle bacillus, which is significantly more resistant to aqueous germicides than are ordinary vegetative bacteria. These disinfectants are also effective against fungi (asexual spores, but not necessarily dried chlamydospores or sexual spores) as well as lipid, nonlipid, medium-sized, and small viruses (Table 1). The tubercle bacillus, lipid and nonlipid viruses, and other microbial types listed in Table 1 are used as indicator organisms because of the various degrees of their resistance to chemical germicides and not necessarily because of their importance in causing nosocomial infections. For example, cells of *M. tuberculosis* are among the most resistant vegetative microorganisms known, and after bacterial endospores, these cells constitute a most severe challenge to a chemical germicide. This high-level chemical germicide may be used as a high- or intermediate-level disinfectant targeted to control many types of nosocomial pathogens but not specifically to control respiratory tuberculosis, which in the United States is no longer a major infection control problem in hospitals.

Although intermediate-level disinfectants are considered effective against viruses, there are some exceptions. Klein and Deforest (19) have shown that the resistance of viruses to chemical disinfectants can vary significantly. They reported that small nonlipid viruses were significantly more resistant than medium-sized viruses with lipid in their protein coats. Some of the most widely used chemical germicides failed to destroy picornaviruses, among which are the enterovirus group and the rhinoviruses of the common cold. Consequently, it is not necessarily true that intermediate-level disinfectants that have good tuberculocidal activity also destroy all viruses. The human hepatitis viruses (A, B, and non-A, non-B) have been difficult to study because some (B and non-A, non-B) have not yet been cultured in the laboratory. However, there is no evidence that any of these viruses is unusually resistant to physical or chemical agents (8, 20), and the hepatitis B virus has been shown to be inactivated by several intermediate- to high-level disinfectants including two glutaraldehyde-based formulations, free chlorine (500 ppm), an iodophor disinfectant, and 70% isopropanol (6).

Low-level disinfectants are those that cannot be relied on to destroy bacterial spores, tubercle bacilli, or small nonlipid viruses within a practical time period. These disinfectants may be useful in actual practice because they can rapidly kill vegetative forms of bacteria and fungi as well as medium-sized lipid-containing viruses. Examples of low-level germicides formulated as disinfectants or antiseptics are: aqueous quaternary ammonium compounds, hexachlorophene, chlorhexidine, and parachlorometaxylenol (PCMX). Relatively high concentrations of mercurial compounds are required to achieve significant bactericidal activity. Mercurial compounds are fairly low-level disinfectants that are not considered efficient and are of very little use in modern disinfection strategies in hospitals.

FACTORS INFLUENCING GERMICIDAL PROCEDURES

Microorganisms vary widely in their responses to physical and chemical stresses, but it is generally agreed that few, if any, organisms approach the resistance of bacterial endospores (14). Because of this resistance, bacterial spores are used as biological indicators for sterilization cycles. In a broad descending order of relative resistance, considerably below that of bacterial spores, are the tubercle bacilli, nontuberculous mycobacteria, fungal spores, small or nonlipid viruses, vegetative fungi, medium-sized or lipid viruses, and vegetative bacterial cells.

The differences in chemical resistance exhibited by various vegetative bacteria are relatively minor except for the differences in tubercle bacilli and some nontuberculous mycobacteria (10), which, presumably because of their hydrophobic cell surfaces, are comparatively resistant to a variety of chemical germicides. Among the ordinary vegetative bacteria, staphylococci and enterococci are somewhat more resistant than most other gram-positive bacteria. Antibiotic-resistant hospital strains of *Staphylococcus* are not discernibly more resistant to germicides than are susceptible strains. A number of gram-negative bacteria such as *Pseudomonas*, *Klebsiella*, *Enterobacter*, and *Serratia* species may also show comparative resistance to some disinfectants and antiseptics. This resistance is noteworthy since these species are among the emerging pathogens responsible for outbreaks of hospital infection.

Gram-negative water bacteria that can grow well and achieve levels of 10^3 to 10^7 organisms per ml in distilled, deionized, or reverse osmosis water have been shown to be significantly more resistant to a variety of disinfectants when the bacteria are isolated and grown in pure culture in water without subculturing on laboratory media than when cells are subcultured in the normal fashion (16, 18). The same phenomenon has been shown for nontuberculous mycobacteria (10). These differences in resistance become important when low-level disinfectants are used, particularly at marginal or dilute concentrations. In addition, naturally occurring spores are significantly more resistant to dry heat than are those that are subcultured (7). Some gram-negative water bacteria can attach to and colonize surfaces and may form a film or glycocalyx (11). Microbial cells can thus be shielded and survive for significant periods in the presence of chemical germicides that ordinarily inactivate these organisms rapidly (2, 13, 15). These phenomena emphasize the importance of actual use tests when chemical germicides are evaluated for application in specific environments, as well as the importance of precleaning items before disinfection.

Under a given set of circumstances, the higher the level of microbial contamination, the longer the exposure to the inactivating agent must be. Consequently, the lack of good physical cleaning of an item before subjecting it to disinfection or sterilization may easily cause the process to fall far short of its intended goal. Feces, blood, mucus, or soil may also contribute to the failure of a given disinfection or sterilization process. Organic soil may occlude microorganisms and prevent penetration by physical and chemical agents, or the soil may directly inactivate certain germicidal chemicals. Cleaning is particularly important in disinfection that does not include the overkill factor of sterilization. Indeed, in one report, a flexible fiberoptic endoscope was implicated in an outbreak of *Serratia* septicemia in a hospital (25). This instrument had been sterilized with ethylene oxide gas, but it had not been properly cleaned before the procedure. Thus, a rigorous sterilization cycle capable of killing exposed bacterial spores may not kill even relatively delicate vegetative bacterial cells that are protected by organic material.

Generally speaking, with all other variables constant, the higher the concentration of chemical agent or the longer a process is continued, the greater is its effectiveness. For temperature-based procedures, an increase in temperature during exposure will usually significantly increase the efficacy of the chemical germicide.

DIRECTIONS FOR THE USE OF COMMERCIALLY AVAILABLE CHEMICAL GERMICIDES

Another important factor that should be kept in mind when a procedure for sterilization, disinfection, or antisepsis is being formulated is the necessity to read and follow the directions of the manufacturer. This caution applies to sterilization equipment, such as steam sterilizers and ethylene oxide gas sterilizers, as well as to commercially available chemical germicides. For disinfectants and antiseptics, general guidelines for use and contraindications, as well as a listing of the active ingredients, are found on the label and in the literature supplied with a particular product. When these directions are not taken into consideration, significant errors can be made in use dilutions as well as in the application of certain chemical germicides. A good example is iodophors.

An iodophor is a combination of iodine and a solubilizing agent or carrier in which the resulting complex or combination acts as a reservoir of iodine and liberates small amounts of free iodine when the iodophor is diluted with water. Examples of carriers are quaternary ammonium compounds, detergents, and polyvinylpyrrolidone. Formulations containing iodophors usually list certain percentages of available iodine and do not specify the amount of free iodine, which is the chemical form responsible for killing microorganisms. Available iodine content often is used as an indicator of germicidal potency, but this approach is incorrect. Many aspects related to the physical and organic chemistry of iodine complexes are not fully understood. For example, a povidone-iodine germicide formulated as an antiseptic may contain 10% povidone-iodine and 1% available iodine. The term "available iodine" refers to the amount of iodine that is titratable with sodium thiosulfate. When a solution contains 1% available iodine, almost all of it is in a complexed form, and very little exists as free iodine; but during the chemical assay, sodium thiosulfate reacts with both the complexed and the free iodine. The amount of free iodine in these types of solutions is approximately 1 ppm and is significantly controlled by the amount of potassium iodide present as well as by the amount of water. Concentrated solutions of iodophor contain less free iodine in undiluted solutions than do solutions diluted up to a specific point. It is virtually impossible to chemically assay free iodine in the presence of complexed iodine without resorting to an extraction technique with solvent. Thus, one can readily appreciate that the label direction of the manufacturer for an iodophor disinfectant that calls for a 1:213 aqueous dilution of a concentrated formulation is designed to give the maximum degree of microbicidal efficiency for the amount of free iodine present.

The amount of available iodine noted on the label of a chemical germicide formulated as an antiseptic may be very similar to that listed on a germicide formulated as a disinfectant. This similarity does not alter the rationale for classifying disinfectants as intermediate level, but it does present a problem in defining the appropriate use concentration. Since it is complicated to assay for free iodine in the presence of iodophor solutions and since it is the current practice of manufacturers to include the amount of available iodine on product labels as an implication of potency, I have elected to retain the amount of available iodine to denote strength (Table 2). With iodophors, the directions of the manufacturer are much more critical with respect to actual use dilutions with water than are the directions with most other disinfectants, and care should be taken to follow the label directions closely. Furthermore, iodophors formulated as antiseptics contain significantly less free iodine than those formulated as disinfectants, so that iodophor antiseptics should not be used as disinfectants and vice versa (2, 13, 14).

TABLE 2. Methods of sterilization or disinfection[a]

Method	Concn or level[b]	Activity level
STERILIZATION		
Heat		
Moist heat (steam under pressure)	250°F (121°C) or above; prevacuum cycle, 271°F (132°C)	
Dry heat	171°C for 1 h; 160°C for 2 h; 121°C for 16 h or longer	
Gas		
Ethylene oxide	450–500 mg/liter at 55–60°C[c]	
Liquid		
Glutaraldehyde, aqueous	Variable[d]	
Hydrogen peroxide, stabilized	6–25%	
Formaldehyde, aqueous[e]	6–8%	
DISINFECTION		
Heat		
Moist heat[f]	75–100°C	High
Liquid		
Glutaraldehyde, aqueous	2%	High
Hydrogen peroxide, stabilized	3–6%	High
Formaldehyde, aqueous	1–8%	High to low
Iodophors[g]	30–50 mg of free iodine per liter; 70–150 mg of available iodine per liter	Intermediate
Chlorine compounds[h]	500–5,000 mg of free available chlorine per liter	Intermediate
Alcohol (ethyl; isopropyl)[i]	70%	Intermediate
Iodine and alcohol	0.5% + 70%	Intermediate
Phenolic compounds, aqueous	0.5–3%	Intermediate to low
Quaternary ammonium compounds, aqueous	0.1–0.2%	Low
ANTISEPSIS[j]		
Alcohol (ethanol; isopropanol)	70%	Intermediate
Iodophor	1–2 mg of free iodine per liter; 1–2% available iodine	Intermediate to low
Chlorhexidine	0.75–4.0%	Low
Hexachlorophene	1–3%	Low
Parachlorometaxylenol	0.5–4%	Low
Mercurial compound	0.1–0.2%	Low

[a] This list of chemical germicides contains generic formulations. Other commercially available formulations can also be considered for use. Users should ensure that the formulations are registered with the EPA or categorized by the FDA. Information in the scientific literature or presented at symposia or scientific meetings can also be considered in determining the suitability of certain formulations. Adequate precleaning of surfaces is the first prerequisite for any disinfecting or sterilizing procedure. The longer the exposure to a physical or chemical agent, the more likely it is that all pertinent microorganisms will be eliminated. An exposure of 10 min may not be adequate to disinfect many objects, especially those that are difficult to clean because of narrow channels or other areas that can harbor organic material as well as microorganisms; thus, longer exposure times, i.e., 20 to 30 min, may be necessary. This is especially true when high-level disinfection is to be achieved.

[b] For sterilization, see the recommendation of the manufacturer for exposure times and conditions. For disinfection, exposure times should be 20 to 30 min or longer.

[c] In a humidified sterilizer designed for ethylene oxide at 55 to 60°C.

[d] There are several glutaraldehyde-based proprietary formulations on the U.S. market, i.e., low-, neutral-, or high-pH formulations recommended for use at normal or raised temperatures with or without ultrasonic energy, and also a formulation containing 2% glutaraldehyde and 7% phenol. Instructions of the manufacturer regarding use as a sterilant or disinfectant or regarding anticipated dilution during use should be closely followed.

[e] Because of the ongoing controversy of the role of formaldehyde as a potential occupational carcinogen, the use of formaldehyde is recommended only in limited circumstances under carefully controlled conditions, i.e., for the disinfection of certain hemodialysis equipment.

[f] Includes hot water pasteurization.

[g] Only those iodophors registered with the EPA as hard-surface disinfectants should be used, and the instructions of the manufacturer regarding proper use dilution and product stability should be closely followed. Antiseptic iodophors are not suitable for use as disinfectants.

[h] There currently is a formulation registered with the EPA as a sterilant and disinfectant, depending on contact time, whose active ingredient is chlorine dioxide. The instructions of the manufacturer regarding use as a sterilant or disinfectant or regarding anticipated dilution during use should be closely followed.

[i] With volatile products such as alcohols, careful attention should be given to proper contact time during a disinfection protocol.

[j] This list includes those antiseptics which are commonly used in hospitals; the list is not complete in that not all formulations categorized by the FDA are mentioned. For more detail, the reader is referred to the paper by Zanowiak and Jacobs (26).

INTENDED USES OF GERMICIDAL AGENTS

Patient care equipment can be categorized into critical, semicritical, and noncritical items (14) (Table 3). Critical items include instruments or objects, e.g., scalpels, cardiac catheters, implants, and hemodialyzers, that are introduced directly into the body, either into the bloodstream or into normally sterile areas. In this instance, sterility is required, and all contaminating microorganisms must be destroyed.

Semicritical items come into contact with mucous membranes, but they do not ordinarily enter normally sterile tissues. Local host defense mechanisms can be expected to protect against challenges from small numbers of exogenous microorganisms, but for safety, semicritical items should not be contaminated with vegetative bacteria. Although sterilization of these items is desirable and quite often the cheapest and fastest procedure available, it is not absolutely essential. For semicritical items that do not tolerate heat or cannot withstand long periods of immersion in chemical germicides or exposure to ethylene oxide gas, it is reasonable to use a high-level disinfection process, a procedure designed to destroy ordinary vegetative bacteria, most fungal spores, tubercle bacilli, and small nonlipid viruses.

Noncritical items offer little risk of transmitting infectious agents. Such items include face masks, carafes, electrocardiogram electrodes, walls, floors, furniture, and other environmental surfaces that do not ordinarily come into contact with mucous membranes. Many persons rely on hot water or detergent cleansing, but some workers use low-level disinfectants either alone or in addition to the washing.

EVALUATION OF ACTUAL GERMICIDAL EFFECTIVENESS

Microbiologic assays

A hospital staff can make a rational choice from the various sterilization, disinfection, and antisepsis processes that are available by considering the intended uses for the product (is sterility required, or may

TABLE 3. Methods of ensuring adequate processing and safe use of medical devices

Patient care object and classification	Example	Method	Comment
Critical			
Sterilized in the hospital	Surgical instruments and devices; trays and sets	1. Thoroughly clean objects and wrap or package them for sterilization. 2. Follow instructions of the manufacturer for use of each sterilizer or use recommended protocol. 3. Monitor time-temperature charts. 4. Use commercial spore preparations to monitor sterilizers. 5. Inspect package for integrity and for exposure of sterility indicator before use. 6. Use before maximum safe time has expired.	Sterilization processes are designed to have a wide margin of safety. If spores are not killed, the sterilizer should be checked for proper use and function; if spore tests remain positive, discontinue use of the sterilizer until it is properly serviced. Maximum safe storage time of items processed in the hospital varies according to type of package or wrapping material(s) used; follow instructions of the manufacturer for use and storage times.
Purchased as sterile	Intravenous fluids; irrigation fluids; normal saline; trays and sets	1. Store in safe, clean area. 2. Inspect package for integrity before use. 3. Use before expiration date if one is given.	Notify the U.S. FDA if intrinsic contamination is suspected.
Semicritical			
Should be free of vegetative bacteria; may be subjected to high-level disinfection rather than sterilization	Respiratory therapy equipment and instruments that will touch mucous membranes	1. Sterilize if possible; if not, follow a protocol for high-level disinfection. 2. Bag and store in safe, clean area. 3. Conduct quality control monitoring after any important changes in the disinfection process.	Bacterial spores may survive after high-level disinfection but are usually not pathogenic. Microbiologic sampling can verify that a high-level disinfection process has resulted in destruction of vegetative bacteria.
Noncritical			
Usually contaminated with some bacteria	Bedpans; crutches; rails; linens; food utensils; electrocardiogram leads	1. Follow a protocol for cleaning or low-level disinfection process or both.	
Water-produced or -treated items	Water used for hemodialysis fluids	1. Assay water and dialysis fluids monthly. 2. Water should not have more than 200 bacteria per ml and dialysis fluids not more than 2,000 bacteria per ml.	Gram-negative water bacteria can grow rapidly in water and dialysis fluids and can place dialysis patients at risk of pyrogenic reactions or septicemia. These water sources and pathways should be disinfected routinely.

high-, intermediate-, or low-level disinfection be adequate?) and by understanding the factors that influence germicidal effectiveness, as discussed above. It is not practical for hospital laboratories to test the antimicrobial effectiveness of commercially available chemical germicides unless such testing is part of a well-designed research project. As mentioned in the first part of this chapter, in past years the hospital could rely on testing performed by the EPA for chemical germicides classified as sporicides. Formulations registered with the EPA were thought to meet standard test criteria for effectiveness. Currently, hospitals should rely on scientific literature, scientific meetings, and scientific data provided by manufacturers in addition to EPA registration and FDA categorization. Testing of antiseptics (9, 21, 24) and disinfectants (4, 12, 14) is a complex and expensive process, and few clinical microbiology laboratories will devote their resources to such testing.

The actual effectiveness of a germicide is influenced only in part by the nature of the agent. Of equal, or perhaps greater, importance is the way the germicide is used in the hospital. Many disinfectants, especially low- and intermediate-level disinfectants, have little margin of safety; misuse by hospital personnel may lead to germicidal failure. Thus, workers in a hospital infection control program may decide to use microbiologic cultures in a limited program to monitor the effectiveness of disinfection and sterilization (see Chapter 11). Routine or widespread environmental culturing is generally discouraged, because it offers few data of use to an infection control program. Moreover, an environmental monitoring program must be well designed with a specific objective in mind. It makes little sense, for example, to evaluate items or areas that are unlikely to play a role in disease transmission. Thus, floors, furniture, or other noncritical items should not be tested even to evaluate the effectiveness of hospital housekeeping personnel. Environmental assays, to the extent that they are used, should be limited to high-risk (critical or semicritical) items. Even then, such assays should not take the place of scrupulous attention to the actual performance of the sterilization or disinfection procedures. With these cautions in mind, the following guidelines are offered for the microbiologic monitoring of selected high-risk procedures.

Steam autoclaves

Steam sterilizers (autoclaves) monitor several physical parameters during the sterilization process. For example, autoclaves equipped with time and temperature recorders generally ensure an adequate sterilization cycle. However, assurance that the process has actually sterilized objects and liquids can be inferred only after a number of factors have been controlled. These factors include proper loading of the sterilizer, elimination of residual air, proper bleeding of the autoclave so that superheating does not produce false readings, and proper operation of the physical gauges. Commercially prepared spores (*Bacillus stearothermophilus*) should be used at least once a week to confirm that a routine cycle achieves actual sterilization. Where the biological indicator is placed is important and depends on the type of load to be monitored. As a general rule, spore carriers should be placed in the center of the largest pack in the largest load or in an area of the load that is least likely to reach sterilizing temperature. The biological indicator should not be placed on an open shelf or on the peripheral exterior surface of a pack. The indicator should be processed according to the recommendations of the manufacturer (see Chapter 11).

Gas sterilizers

Sterilization by gaseous ethylene oxide provides a much narrower margin of safety than sterilization by steam, and malfunctions can occur that are not detectable by the physical gauges associated with the commercial units. When ethylene oxide gas is used to treat implantable devices that require sterility, each load should be monitored; otherwise the sterilizer should be monitored weekly. Spores of *Bacillus subtilis* var. *niger* (*globigii*), also commercially available, should be used. Here again, where the spore carrier is placed is important. It should be in the middle of the largest pack in an area of the load that is least likely to receive adequate exposure to ethylene oxide. The specific assay procedure should be performed according to the directions of the manufacturer.

Dry-heat sterilizers

Dry-heat sterilizers should be checked at least once weekly by the same basic strategies described above and with spores of *B. subtilis* var. *niger* (*globigii*).

Respiratory therapy breathing circuits and anesthesia equipment

Proper cleaning and disinfection procedures are the most important part of an environmental control program to reduce infections transmitted by respiratory therapy and anesthesia equipment breathing circuits. Many items may be sterilized with steam or ethylene oxide, but if they are subjected to chemical disinfection, they may be spot checked every few months or when disinfection or usage procedures change. Routine or scheduled microbiologic assays are not required. There is no adequate microbiologic guideline for this equipment that is supported by epidemiologic studies. The most widely used criterion of acceptability is the absence of vegetative bacteria on components of the breathing circuits after disinfection.

Hemodialysis systems

Gram-negative water bacteria can multiply relatively quickly in fluids associated with hemodialysis systems such as distilled, softened, deionized, and reverse osmosis water, as well as in the dialysis fluid itself. Although these fluids do not need to be sterile, excessive levels of contamination by gram-negative bacteria pose a risk of pyogenic reactions and septicemia. An epidemiologic-based, quantitative microbiologic guideline for levels of contamination has been proposed (3, 17). It is suggested that dialysis fluid and the water used to prepare dialysis fluids be checked microbiologically at least once a month. Microbiologic guidelines for these procedures include sampling the water used to prepare dialysis fluid at the point where it is mixed with concentrated dialysis fluid. The level of bacterial contamination should not exceed 200 organisms per ml. Dialysis fluid should be sampled at the end of a dialysis treatment, and the level of bacterial contamination should not exceed 2,000 organisms per ml. In both instances, routine standard plate counts can be done by pour or surface plating or by membrane filter procedures with appropriate culture media such as tryptic soy or standard methods agar.

Arterial pressure transducers

Arterial pressure transducers have been incriminated in disease transmission, and the best means of disease control are adequate cleaning and sterilization as well as proper placement. Disposable domes should not be reused. Scheduled microbiologic sampling is not required. If a microbiologic assay is performed, the criterion of acceptability is sterility.

Miscellaneous procedures and equipment

Numerous items of patient care equipment pose various degrees of risk of infection. These items make direct contact with skin, mucous membranes, or body orifices, but not with deep tissue. Items in this category are flexible fiberoptics, antiseptic solutions, soaps, hydrotherapy equipment, and nonsterile solutions prepared in the hospitals. However, with most of these materials, the most important element in environment control is not microbiologic sampling, but rather adherence to tested protocols for their cleaning, preparation, disinfection or sterilization, length of use, and maintenance. Even spot checking these items and procedures in most instances is not recommended because of the absence of meaningful microbiologic guidelines supported by epidemiologic criteria.

REUSE OF DISPOSABLE MEDICAL DEVICES

In the 1970s, many medical devices that were reusable and thus could be cleaned and resterilized were replaced by disposable medical devices. This change was thought to be cost-effective, because the personnel and operational costs of cleaning, sterilizing, and repackaging reusable devices could be eliminated. However, in recent efforts to save money, some hospitals are reusing disposable medical devices. The number and types of disposable medical devices are varied and range from critical medical devices to noncritical ones. The question arises: can medical devices labeled "single-use, sterile, and disposable" be safely reprocessed and reused in a hospital? Currently, no reported studies have shown that specific disposable medical devices can be reprocessed, sterilized, or disinfected and safely function as originally intended. A notable exception is the disposable hollow-fiber hemodialyzer. Approximately 43% of dialysis centers in the United States process disposable hemodialyzers and reuse them on the same patient (1). When standardized protocols and adequate sterilization procedures are used, these procedures do not appear to be associated with problems of disease transmission. However, dialysis centers that do reuse dialyzers use standardized cleaning and sterilization protocols and quality control measures that determine the efficacy of the dialyzer. With other medical devices, there are basically no data available that show that the device after reprocessing and sterilization still functions safely and does not retain toxic levels of the chemical germicides used in the reprocessing procedure.

Hospitals should be aware that if they reuse disposable medical devices, the liability associated with those devices shifts from the manufacturer to the user. Furthermore, there are some medical devices that can be sterilized but will be discernibly damaged, and the function of the device can be affected, which in turn affects the safety of the patients. Unless there are overriding cost considerations, disposable critical and semicritical medical devices meant for one-time use probably should not be reused. If they are reused, it should be determined that they can be cleaned, sterilized, or disinfected without altering their function and that problems of residual toxicity and overall safety will not be involved with their reuse.

UNNECESSARY MICROBIOLOGIC ASSAYS

There are a number of items and procedures in the hospital environment for which microbiologic sampling either on a scheduled or periodic basis is not cost-effective or rational. These items include sterile intravenous solutions, injectables, disposable syringes, disposable blood lines, artificial kidneys, and all other items that are received in a sterile state (see Chapter 11). Equipment and solutions sterilized in the hospitals do not need to be sampled microbiologically. Rather, quality assurance testing associated with sterilization procedures as described above should be used to verify that the sterilization process per se is performing to specifications and that the associated procedures are being performed correctly.

Although it is recognized that inanimate surfaces and air associated with critical areas such as surgical suites and intensive care areas may contain reservoirs of microorganisms, the chance for disease transmission in environments that are adequately cleaned and maintained is remote. Environmental control procedures associated with housekeeping and engineering services should use and adhere to tested cleaning, disinfection, and maintenance protocols. Consequently, microbiologic sampling should not be necessary for intramural air or inanimate environmental surfaces. However, in an outbreak of hospital-acquired disease, if a certain part of the environment, such as the air ventilation system, appears to be associated with disease transmission, appropriate sampling should be initiated.

ENVIRONMENTAL MICROBIOLOGIC SAMPLING DURING OUTBREAKS OF INFECTIONS

The strategy that should be used during an outbreak of infection with respect to environmental microbiologic sampling depends on several things. First, the epidemiologist must determine whether certain procedures, equipment, instruments, or other parts of the environment may be playing a direct or indirect role in the outbreak. An outbreak of nosocomial infections does not automatically mean that environmental microbiologic sampling at any level is required. Second, if environmental microbiologic sampling is believed to be necessary, the microbiologist and epidemiologist should coordinate the sampling strategy and determine the implicated procedures, items, or parts of the environment requiring microbiologic assay.

A microbiologic guideline applied in this context differs from scheduled or periodic sampling. During the investigation of an outbreak of nosocomial infections, environmental sampling is usually directed toward a specific pathogen. Consequently, if the outbreak is due to *P. aeruginosa*, this organism and perhaps specific serotypes are sought in the various environmental items that are sampled. In this respect, the guideline tends to be more qualitative than quantitative, although in some instances one must rely on established guidelines. For example, if water or ice in a hospital is incriminated in an outbreak of nosocomial salmonellosis, procedures should be used to determine fecal coliform bacteria and the total number of microorganisms, in addition to searching for *Salmonella* species. The guideline here, then, is flexible and basically is determined by the nature of the outbreak and the results of the epidemiologic investigation.

LITERATURE CITED

1. **Alter, M. J., M. S. Favero, N. J. Petersen, I. L. Doto, R. T. Leger, and J. E. Maynard.** 1983. National surveillance of dialysis-associated hepatitis and other diseases, 1976 and 1980. Dial. Transplant. **12:**860–865.
2. **Anderson, R. L., R. L. Berkelman, D. C. Mackel, B. J. Davis, B. W. Holland, and W. J. Martone.** 1984. Investigations into the survival of *Pseudomonas aeruginosa* in poloxamer-iodine. Appl. Environ. Microbiol. **47:**757–762.
3. **Association for the Advancement of Medical Instrumentation.** 1981. American national standard for hemodialysis systems. Association for the Advancement of Medical Instrumentation, Arlington, Va.
4. **Association of Official Analytical Chemists.** 1970. Official methods of analysis, 10th ed. Association of Analytical Chemists, Washington, D.C.
5. **Block, S. S.** 1983. Federal regulation of disinfectants in the United States, p. 831–844. *In* S. S. Block (ed.), Disinfection, sterilization and preservation, 3rd ed. Lea & Febiger, Philadelphia.
6. **Bond, W. W., M. S. Favero, N. J. Petersen, and J. W. Ebert.** 1983. Inactivation of hepatitis B virus by intermediate-to-high-level disinfectant chemicals. J. Clin. Microbiol. **18:**535–538.
7. **Bond, W. W., M. S. Favero, N. J. Petersen, and J. H. Marshall.** 1970. Dry-heat inactivation kinetics of naturally occurring spore populations. Appl. Microbiol. **20:**573–578.
8. **Bond, W. W., N. J. Petersen, and M. S. Favero.** 1977. Viral hepatitis B: aspects of environmental control. Health Lab. Sci. **14:**235–252.
9. **Bruch, M. K.** 1983. Methods of testing antiseptics and antimicrobials used topically in humans, p. 946–963. *In* S. S. Block (ed.), Disinfection, sterilization and preservation, 3rd ed. Lea & Febiger, Philadelphia.
10. **Carson, L. A., N. J. Petersen, M. S. Favero, and S. M. Aguero.** 1978. Growth characteristics of atypical mycobacteria in water and their comparative resistance to disinfectants. Appl. Environ. Microbiol. **36:**839–846.
11. **Costerton, J. W., G. G. Geesey, and K.-J. Cheng.** 1978. How bacteria stick. Sci. Am. **238:**86–95.
12. **Engley, F. B., and B. P. Dey.** 1970. A universal neutralizing medium for antimicrobial chemicals, p. 100–106. *In* Proceedings of the 56th Mid-Year Meeting. Chem. Spec. Manuf. Assoc., New York.
13. **Favero, M. S.** 1982. Iodine—champagne in a tin cup. Infect. Control **3:**30–32.
14. **Favero, M. S.** 1983. Chemical disinfection of medical and surgical materials, p. 469–492. *In* S. S. Block (ed.), Disinfection, sterilization and preservation, 3rd ed. Lea & Febiger, Philadelphia.
15. **Favero, M. S., W. W. Bond, N. J. Petersen, and E. H. Cook.** 1983. Scanning electron microscopic observations of bacteria resistant to iodophor solutions, p. 158–177. *In* G. A. Digenis and J. Ansell (ed.), Proceedings: international symposium on povidone. University of Kentucky, Lexington.
16. **Favero, M. S., L. A. Carson, W. W. Bond, and N. J. Petersen.** 1971. *Pseudomonas aeruginosa*: growth in distilled water from hospitals. Science **173:**836–838.
17. **Favero, M. S., and N. J. Petersen.** 1977. Microbiologic guidelines for hemodialysis systems. Dial. Transplant. **6:**34–36.
18. **Favero, M. S., N. J. Petersen, L. A. Carson, W. W. Bond, and S. H. Hindman.** 1975. Gram-negative water bacteria in hemodialysis systems. Health Lab. Sci. **12:**321–334.
19. **Klein, M., and A. Deforest.** 1963. Antiviral action of germicides. Soap. Chem. Spec. **39:**70–72, 95–97.
20. **Miner, N. A.** 1978. Viral hepatitis: prevention and control. Postgrad. Med. **60:**19–22.
21. **Price, P. R.** 1938. The bacteriology of normal skin: a new quantitative test applied to a study of the bacterial flora and the disinfectant action of mechanical cleansing. J. Infect. Dis. **63:**301–318.
22. **Spaulding, E. H.** 1939. Chemical sterilization of surgical instruments. Surg. Gynecol. Obstet. **69:**738–744.
23. **Spaulding, E. H.** 1972. Chemical disinfection and antisepsis in the hospital. J. Hosp. Res. **9:**5–31.
24. **Ulrich, J.** 1964. Technics for skin sampling for microbial contaminants. Health Lab. Sci. **1:**133–136.
25. **Webb, S. F., and A. Vall-Spinosa.** 1975. Outbreak of *Serratia marcescens* associated with the flexible fiberbronchoscope. Chest **68:**703–705.
26. **Zanowiak, P., and M. R. Jacobs.** 1982. Topical anti-infective products, p. 525–542. *In* Handbook of nonprescription drugs, 7th ed. American Pharmaceutical Association, Washington, D.C.

Chapter 14

Biological Safety in the Clinical Laboratory

JOHN H. RICHARDSON AND W. EMMETT BARKLEY

The safe operation of a clinical laboratory is a facility management responsibility. This responsibility can be delegated, reassigned, abandoned, or ignored, but when an accident occurs, it ultimately returns, unforgivingly, to the laboratory director. The laboratory director must develop and institute an operational safety program that will effectively minimize the potential for overt laboratory-associated disease or physical harm to all who have direct or indirect exposure to biological, chemical, or radioactive hazards and to the physical laboratory facility. This chapter, however, will deal primarily with biological hazards related to infectious agents in clinical microbiology laboratories.

A comprehensive laboratory safety program must address the considerations of storage, use, and disposal of hazardous chemicals and radioactive materials, facility operation and maintenance, staff training, and medical surveillance. A number of useful publications are available to assist laboratory managers in developing operational safety policies and practices appropriate for their specific laboratory (3, 13, 14).

It should be emphasized that risk of exposure to infectious agents is not restricted to clinical microbiology laboratory personnel. Hepatitis B infection, a recognized hazard to laboratory personnel working with blood and blood products (11), has a recorded incidence in clinical chemistry personnel seven times that expected in the general population (10).

EVALUATION OF RISKS

It is difficult to quantitate the risk of working with an infectious agent in the laboratory. Individual risk predictably increases with the frequency and level of contact with the agent. Community risk relates to the level of immunity to the agent and the effectiveness of secondary barrier systems (e.g., HEPA [high-efficiency particulate air] filters) within the laboratory.

Of primary concern in the clinical laboratory working with specimens and agents of domestic origin is the risk of exposure and infection to laboratory and support personnel working in or entering the laboratory. Risks are ultimately influenced by an often unpredictable interrelationship of the agent, the host, and the activity. Factors applicable to the agent include virulence, infectious dose, routes of infection, toxigenicity, and host spectrum. The variables influencing risk to the host may include age, sex, race, pregnancy, presence and level of antibody or cellular immunity, and the availability and use of vaccines, specific therapeutic compounds, or medications. Finally, the type of laboratory activity (e.g., diagnostic, production, research) may significantly affect the risk to personnel because of the type, quantity, and concentration of agents handled, the manipulations utilized in working with the agent, and the effectiveness of primary and secondary containment equipment and laboratory practices. The facility descriptions, safety equipment, and operational practices described and referenced in this chapter all address the objective of minimizing the risk of exposure, infection, and disease in laboratory personnel.

LABORATORY DESIGN CONSIDERATIONS

Laboratory design considerations must be based on the function of the laboratory (e.g., diagnostic, research, production), the intrinsic hazard of the agents usually handled, and the procedures utilized in the laboratory operation. From a practical standpoint, clinical laboratories can realistically divide diagnostic activities with indigenous infectious agents into two general categories based on the degree of risk associated with aerosol transmission of the disease. Hepatitis B virus and *Mycobacterium tuberculosis* are two microorganisms which characterize this concept. A summary of the recommended practices and techniques, safety equipment, and facilities for work involving infectious agents is provided in Table 1 (14).

Hepatitis B virus is representative of the broad spectrum of indigenous disease agents present in the community and in diagnostic specimens received and handled in clinical microbiology laboratories. Such agents are a "moderate risk" in terms of infection hazard to laboratory personnel. Laboratory-acquired infection is associated primarily with ingestion, exposure of mucous membranes of the eyes, nose, or mouth to infectious fluids, and parenteral inoculation. Manipulations of small volumes and low concentrations of infectious materials containing hepatitis B virus ordinarily do not pose a risk of infection via the respiratory route. The standard and special microbiological practices, safety equipment, and facilities recommended for Biosafety Level 2 (see Table 1) are appropriate for hepatitis B virus and agents which pose similar occupational risks (14). Human T-cell leukemia virus type III (HTLV-III), the prime etiological candidate for AIDS (acquired immune deficiency syndrome), may pose occupational infection risks similar to those of hepatitis B virus. Those precautions recommended for hepatitis B virus are also applicable to the AIDS agent. Good microbiological technique and judicious use of primary containment equipment, such as biological safety cabinets, constitute the primary safeguards when working with agents in this risk category. Recommended laboratory facilities do not require special engineering or architectural provisions for ventilation or access control. The design and layout of the laboratory should facilitate housekeeping and provide for decontamination of infectious laboratory wastes.

M. tuberculosis is representative of indigenous infectious risks associated with transmission via infectious aerosols generated by common laboratory manipulations of clinical materials and cultures. The human 50% infective dose (respiratory) for *M. tuberculosis* is on the order of 10^1 organisms (8). Ingestion, exposure of mucous membranes to infectious fluids, and parenteral inoculation also pose occupational infection

TABLE 1. Summary of recommended biosafety levels for infectious agents[a]

Biosafety level	Practices and techniques	Safety equipment	Facilities
1	Standard microbiological practices.	None: primary containment provided by adherence to standard laboratory practices during open bench operations.	Basic
2	Level 1 practices plus: laboratory coats; decontamination of all infectious wastes; limited access; protective gloves and biohazard warning signs as indicated.	Partial containment equipment (i.e., Class I or II biological safety cabinets) used to conduct mechanical and manipulative procedures which have high aerosol potential that may increase the risk of exposure to personnel.	Basic
3	Level 2 practices plus: special laboratory clothing; controlled access.	Partial containment equipment used for all manipulations of infectious material.	Containment
4	Level 3 practices plus: entrance through change room where street clothing is removed and laboratory clothing is put on; shower on exit; all wastes are decontaminated on exit from facility.	Maximum containment equipment (i.e., Class III biological safety cabinet or partial containment equipment in combination with full-body, air-supplied, positive-pressure personnel suit) used for all procedures and activities.	Maximum containment

[a] From reference 14.

risks to laboratory personnel. The inhalation risk can be appropriately managed by applying safeguards described under Biosafety Level 3 (14).

Because of the high risk of disease transmission via infectious aerosols, all manipulations of known or potentially infectious clinical materials and cultures should be conducted in open-fronted negative-pressure or vertical laminar flow (Class I or II) biological safety cabinets (14) or other primary containment devices. Special requirements for the laboratory facility include an access control space through which work areas are entered and a ventilation system which maintains the laboratory under negative air pressure relative to adjacent "clean" areas and exhausts air from the laboratory without recirculation to other areas of the building. Although BCG vaccine immunization of laboratory personnel working with tuberculosis is not a common or recommended practice in the United States, vaccination against other agents (e.g., Q fever, tularemia, certain arboviruses) that pose demonstrated aerosol infection hazards may further reduce individual risks.

RESTRICTED ACCESS

Access to laboratories handling hazardous materials (e.g., infectious, toxic, radioactive, flammable) should be restricted, except for personnel working in the laboratory or providing regular support services. All others should enter only with specific authorization of the laboratory supervisor. All persons entering the laboratory should be aware of the hazards present and must meet applicable entry requirements, which may include specific immunizations and the wearing of dosimeters, safety glasses, and protective clothing or devices.

As the level of hazard increases, based on the materials handled or the laboratory function, so does the need for stricter access control. Physical control methods may include lock and key, card-operated locks, and entry through clothing change areas.

A uniform system of hazard warning signs should be utilized to indicate the hazards present, special entry requirements, and the name of the person responsible for the laboratory facility. The international biohazard symbol (1) should be incorporated in signs warning of infectious hazards.

IMMUNIZATION

Persons working with infectious agents or entering areas where such materials are handled are at increased risk of infection (6, 7). Vaccination may provide moderate to substantial levels of protection to such at-risk individuals. The biological products, either licensed or available as investigational or new drugs, which may be required or recommended (14) for prophylactic immunization of laboratory personnel include:

Anthrax vaccine
Botulism toxoid (ABCDE)
Cholera vaccine
Hepatitis B virus vaccine
Diphtheria-tetanus toxoid
Eastern equine encephalitis vaccine
Japanese encephalitis vaccine
Measles vaccine
Plaque vaccine
Poliomyelitis vaccine
Q fever vaccine
Rabies vaccine
Rift Valley fever vaccine
Rubella vaccine
Smallpox vaccine
Tularemia vaccine
Typhoid vaccine
Venezuelan equine encephalitis vaccine
Western equine encephalitis vaccine
Yellow fever vaccine

Immunity is relative and variable in terms of individual response to antigens, duration and level of measurable antibody, and level of protection. For this reason, immunization of personnel should be considered as an adjunct to good microbiological practices and appropriate laboratory facilities—never as a substitute for either.

Clinical laboratory personnel working with measles, rubella, poliomyelitis, plaque, botulism, and rabies should be examined serologically for evidence of antibodies or immunized in accordance with applicable recommendations of the Advisory Committee on Immunization Practices (2).

Immunization with less efficacious vaccines (e.g., cholera) or with vaccines which may produce increasing levels of sensitization with repeated use (e.g., typhoid) may be recommended for at-risk personnel but should not be required as a condition of employment. Immunization of laboratory personnel should be specifically targeted at those agents that are known or are likely to be handled in the laboratory. A complete record of all immunizations, both required and recommended, should be maintained in the employee's personal medical records.

LABORATORY INFECTIONS

In Pike's summary (6) of almost 4,000 cases of laboratory-associated infections, fewer than 20% resulted from known overt exposures or accidents. The overwhelming majority of cases had a common history of having "worked with the agent" in the laboratory. Infectious aerosols, generated as a result of common and perhaps, on occasion, poorly executed laboratory procedures and manipulations, may have contributed to a substantial number of these cases either by creating an unrecognized inhalation exposure or through indirect contact with surfaces contaminated by settling of infectious particles. Of those cases resulting from known accidents, oral aspiration through a pipette, autoinoculation with a needle and syringe, aerosol exposures resulting from needle and syringe sprays or separations of needles from syringes, centrifuge accidents, and animal bites were the leading causes.

Most of the known accidents were preventable by a combination of good microbiological technique, appropriate safety equipment, and awareness on the part of the laboratory worker. For example, mouth-pipetting accidents are avoided by requiring automatic or bulb pipetting devices for the manipulation of all infectious or toxic liquids. Mouth pipetting of all fluids, in fact, should be prohibited in all laboratories handling toxic or infectious materials. The use of needles and syringes should be avoided except for the parenteral injection of fluids or the aspiration of fluids from humans or animals. Needles and syringes should never be used as a substitute for a pipette. When required for parenteral injections or aspiration of potentially infectious fluids, only syringes with locking hubs should be used. Needle and syringe size and length must be appropriate for the intended use, and personnel must be aware of the potential hazard and properly trained in the use of this equipment. After use, needles and syringes should be placed immediately in a puncture-resistant container and autoclaved before reuse or disposal. Needles should not be bent, sheared, replaced in the sheath or guard, or removed from the syringe before decontamination. Personnel must be trained and equipped to handle experimental animals. Anesthetics or tranquilizers should be routinely used in the restraint of larger experimental species to minimize injuries to both personnel and animals.

Centrifuge accidents, which may produce both infectious aerosols and flying fragments (e.g., glass, metal), are also avoidable. Centrifuge safety involves a combination of knowledgeable and proper use of the equipment (e.g., selection of compatible centrifuge tubes and cups, proper balancing and acceleration and deceleration procedures) and selection and use of appropriate physical containment devices. Centrifuge safety cups with sealed covers are the most widely applicable containment devices for centrifuges. An alternative to the use of safety cups is the placement of centrifuges in exhaust chambers or biological safety cabinets (4) which operate under negative pressure and exhaust through HEPA filters.

Primary containment equipment (centrifuge safety cups, safety blenders, biological safety cabinets) and facility design features, such as directional air flow from clean areas to areas of contamination and exhaust of laboratory air without recirculation, can significantly reduce the hazard of infectious aerosols.

ANIMAL CARE

As a general principle, the containment requirements and good microbiological practices used for handling infectious agents in the laboratory are also applicable to experimental animals infected with the same agent. The "Guide for the Care and Use of Laboratory Animals" [Department of Health, Education and Welfare Publication no. (NIH) 78-23, revised 1978] is one of a number of useful references available on laboratory animal management and use.

Facilities for laboratory animals used for experiments with both infectious and noninfectious materials should be physically separated from other locations such as animal production and quarantine areas, clinical laboratories, and especially facilities providing patient care. Special design features of animal facilities include: floors which are impervious to water; wall and ceiling surfaces which are resistant to water; directional air flow from clean areas to areas of contamination; and direct discharge, without any recirculation, of exhaust air from the animal facility. Exhaust air may be recirculated within the animal facility if it has first been passed through HEPA filters. The design and layout of animal facilities should facilitate cleaning. A "clean hall/dirty hall" design is especially useful. Floor drains in animal rooms should be provided only if specifically needed. If floor drains are provided, the drain trap should be kept filled with a suitable chemical disinfectant. Housekeeping and operational practices should provide clean and uncluttered hallways, animal rooms, and service areas. Cages used to house infected animals should be autoclaved before litter is removed and before being washed.

In most cases, animal care personnel must be considered to be at the same level of risk as laboratory personnel working with the same agent. With some agents, infected laboratory animals are unlikely to constitute a direct exposure risk to animal care personnel (e.g., malaria). For other agents, activities with infected laboratory animals (e.g., carnivores infected with rabies) may represent a higher level of hazard to animal care personnel than that experienced by laboratory staff during routine diagnostic procedures on the open bench.

Animal care personnel should be immunized against infectious agents on the basis of the same risk assessment criteria applied to laboratory personnel. Biohazard warning signs should be posted in the animal facility indicating the specific hazards and special entry requirements. Access to animal areas should be controlled at a level commensurate with that utilized in the laboratory handling the same infectious agent. Protective clothing and devices for laboratory and animal care staff should be based on a risk assessment of the agent and of the specific activity.

EMERGENCY PLANS

Laboratories working with infectious agents should develop emergency plans in anticipation of the type of accident or emergency that is most likely to develop within the facility. Basic to almost every clinical laboratory is the possibility of spills of infectious materials and of fires. It is the responsibility of the laboratory supervisor to assess the probability of such an accident or emergency, to instruct personnel in applicable procedures, and to conduct drills in the use of emergency procedures and equipment.

The components of an emergency plan include: establishing criteria for immediate evacuation of the laboratory in the event of accidents or emergencies which might contaminate the area with infectious agents or toxic chemicals; providing a system to report accidents or emergencies to the supervisor, safety personnel, or both; developing an accident risk assessment capability by identifying resource personnel; arranging for medical evaluation, surveillance, and treatment of exposed personnel; and developing or identifying procedures and resources to perform necessary decontamination or cleanup of the facility.

With regard to fire prevention, special attention must be given to the use, storage, and disposal of flammable solvents, use of flammable compressed gases, and electrical safety. The Occupational Safety and Health Administration General Industry Standard (Code of Federal Regulations, Title 29, Part 1910) provides guidelines applicable to fire safety. The appropriate number and type of fire extinguishers should be placed in laboratories, and personnel should be trained in their use.

DECONTAMINATION PROCEDURES

Basic to the operation of clinical laboratories is the capability of efficiently and effectively handling and decontaminating potentially infectious materials and wastes received or generated by the laboratory. Such materials and wastes include diagnostic specimens, inoculated media and cell cultures, all viable cultures of microbial agents, experimental animals (including embryonated hen eggs), and all glassware, instruments, equipments, surfaces, and spaces used to handle such materials.

In the clinical laboratory there is no practical substitute for a steam autoclave for decontaminating many types of infectious wastes. The steam autoclave, however, is not a fail-safe device. Its effectiveness depends on a number of variables, including personnel knowledgeable in proper use and loading procedures, regular biological and physical monitoring, and appropriate maintenance. Each facility must develop operating procedures based on the type of materials to be autoclaved and on the proven efficacy of the time-temperature-pressure settings against the target or simulant microorganisms. Commercially available preparations of microorganisms with defined heat sensitivity (e.g., spores of *Bacillus stearothermophilus*) are recommended for regular monitoring of autoclaves (see Chapter 13).

Other decontamination methods include the use of physical (e.g., dry heat, boiling, irradiation, UV rays) and chemical (e.g., phenolics, hypochlorites, formaldehyde) procedures. All methods have inherent advantages and limitations. It is imperative that the user be acquainted with both the good and bad aspects of a method before applying it to the day-to-day situation. Decontamination procedures and their application to various laboratory operations have been extensively reviewed and discussed elsewhere (5; see Chapter 13).

If, as a result of laboratory operations, accidents, or spills of infectious materials, there is need for decontamination of the entire laboratory, gas fumigation may be effectively used. Formaldehyde gas, generated by heating or atomizing Formalin (2 ml of Formalin per ft^3 [0.028 m^3] of space), is among the most readily available and effective of the gaseous fumigants. Formaldehyde has a broad range of activity and is quite economical to use. It is most effective under controlled temperatures (70°F, 21°C) and relative humidity (70%). Formaldehyde gas has the obvious disadvantage of a pungent, suffocating, irritating odor and can polymerize on surfaces when misused. Formaldehyde is explosive in concentrations of 7.0 to 73%; however, the recommended use rate is approximately 10% of the minimal explosive concentration. Care in the application of the fumigant is also required because formaldehyde may be a potential carcinogen.

RECEIVING AND HANDLING DIAGNOSTIC SPECIMENS

It is logical to assume that all clinical specimens are potentially infectious. It is prudent to handle such materials in a manner which will reasonably preclude overt exposure of personnel to pathogens most likely to be present. The ubiquitous hepatitis B virus is the most consistent infection hazard in the widest variety of human clinical specimens of domestic origin (9).

Accordingly, as a minimal precaution, clinical laboratories should utilize a number of commonsense practices in the receipt and initial handling of all diagnostic specimens. For example, personnel should be trained and an area should be designated to receive, open, record, and distribute incoming specimens within the laboratory. The receiving area should be well lighted and provided with work surfaces that are impervious to water and readily cleanable. Personnel must be alert to the probability of improperly packaged, damaged, or leaking specimens, advised of the possibility of exposure to infectious agents in such specimens, and instructed in decontamination procedures to follow in the event of leaking or spilled specimens in the receiving area. Personnel should wear laboratory coats or uniforms to protect skin and street clothing from exposure to infectious agents.

Rubber gloves provide an additional level of protection. Personal hygiene practices should include handwashing after the handling of specimen containers and should forbid eating, drinking, smoking, and storage of food in the area. Housekeeping practices should assure a neat uncluttered work area and cleaning and disinfection of work surfaces both when activities are completed and immediately after spills of potentially infectious materials. Plastic-backed absorbent paper toweling facilitates cleanup of the work surface after regular use or in the event of spills. Uncontaminated and contaminated wastes should be removed from the receiving area daily and discarded or recycled after appropriate decontamination.

When followed, these practices may substantially reduce the risk of exposure to the broad spectrum of infectious agents routinely received in clinical laboratories.

LITERATURE CITED

1. **Baldwin, C. L., and R. S. Runkle.** 1967. Biohazards symbol: development of a biological hazards warning signal. Science **158:**264–265.
2. **Department of Health, Education and Welfare, Centers for Disease Control.** 1981. Selected recommendations of the Public Health Service Advisory Committee on Immunization Practices. Centers for Disease Control, Atlanta, Ga.
3. **Fucaldo, A. A., B. J. Erlick, and B. Hindman (ed).** 1980. Laboratory safety theory and practice. Academic Press, Inc., New York.
4. **Hall, C. B.** 1975. A biological safety centrifuge. Health Lab. Sci. **12:**104–106.
5. **Lawrence, C. A., and S. S. Block (ed.).** 1971. Disinfection, sterilization and preservation. Lea & Febiger, Philadelphia.
6. **Pike, R. M.** 1976. Laboratory-associated infections: summary and analysis of 3,921 cases. Health Lab. Sci. **13:**105–114.
7. **Pike, R. M.** 1979. Laboratory-associated infections: incidence, fatalities, causes and prevention. Annu. Rev. Microbiol. **33:**41–66.
8. **Riley, R. L.** 1957. Aerial dissemination of pulmonary tuberculosis. Am. Rev. Tuberc. **76:**931–941.
9. **Schmidt, N. J., and E. H. Lennette.** 1972. Safety precautions for performing tests for hepatitis-associated "Australia" antigen and antibodies. Am. J. Clin. Pathol. **57:**526–530.
10. **Skinhøj, P.** 1974. Occupational risks in Danish clinical chemistry laboratories. II. Infections. Scand. J. Clin. Lab. Invest. **33:**27–29.
11. **Syndman, D. R., J. A. Bryan, and R. A. Dixon.** 1975. Prevention of nosocomial viral hepatitis type B (hepatitis B). Ann. Intern. Med. **83:**838–845.
12. **U.S. Public Health Service.** 1974. Laboratory safety at the Center for Disease Control. Department of Health, Education and Welfare Publication no. CDC 76-8118. Center for Disease Control, Atlanta, Ga.
13. **U.S. Public Health Service.** 1978. NIH laboratory safety monograph. Office of Research Safety, National Cancer Institute, Bethesda, Md.
14. **U.S. Public Health Service, Centers for Disease Control and National Institutes of Health.** 1984. Biosafety in microbiological and biomedical laboratories. Department of Health and Human Services, Publication no. (CDC) 84-8345.

Chapter 15

Section III. Aerobic Bacteria

Staphylococci

W. E. KLOOS AND J. H. JORGENSEN

CHARACTERIZATION

Staphylococci are gram-positive cocci (0.5 to 1.5 μm in diameter) that occur singly and in pairs, tetrads, short chains (three or four cells), and irregular "grape-like" clusters. They are nonmotile, nonsporeforming, and usually catalase positive and unencapsulated. Most species are facultative anaerobes. With the exception of the anaerobic species *S. saccharolyticus* and some unusual strains of *S. aureus*, growth is more rapid and abundant under aerobic conditions. Some uncommon strains may require the presence of CO_2 for good growth. Most strains grow well on noninhibitory media. Cell wall-deficient forms may require a hypertonic environment.

The cell wall contains peptidoglycan and teichoic acid. The diamino acid present in the peptidoglycan is L-lysine. Staphylococci are susceptible to lysis by the endopeptidase lysostaphin, which cleaves glycine-glycine linkages present in the interpeptide bridges of the peptidoglycan. They are generally resistant to lysis by the muramidase lysozyme. Micrococci, on the other hand, are generally resistant to lysostaphin (58). Staphylococci and micrococci can be further differentiated on the basis of (i) anaerobic acid production from glucose (positive for most *Staphylococcus* species and negative for most *Micrococcus* species) (66), (ii) production of acid aerobically from glycerol in the presence of 0.4 μg of erythromycin per ml (positive for most staphylococci and negative for most micrococci) (58), (iii) growth on selective media (e.g., the SK medium of Schleifer and Kramer [59], supporting the growth of staphylococci, or the FTO medium of Curry and Borovian [11], supporting the growth of micrococci), (iv) growth rate and morphology of colonies on noninhibitory media (34), and (v) modified oxidase and benzidine tests designed to identify the presence of cytochrome *c* (negative in most *Staphylococcus* species and positive in micrococci) (20). Unsaturated menaquinones and cytochromes *a* and *b* (and cytochrome *c* in *S. caseolyticus*, *S. lentus*, and *S. sciuri*) form the electron transport system in staphylococci. Due to the presence of cytochromes, staphylococci are benzidine positive by the standard test (14), whereas streptococci are benzidine negative.

Staphylococci are currently included together with micrococci and planococci in the family *Micrococcaceae*, as a matter of historical and practical convenience (14). All three genera represented are catalase-positive, gram-positive cocci with a cell wall peptidoglycan containing L-lysine as the diamino acid. The guanine-cytosine content of DNA (10), nucleic acid hybridization (60), and 16S rRNA cataloging (65), however, indicate that the three genera are not closely related phylogenetically and should not be combined in one family. *Staphylococcus* species are most closely related to the broad *Bacillus-Lactobacillus–Streptococcus* cluster.

The genus *Staphylococcus* is currently composed of 20 species. Those found on humans include *S. aureus*, *S. epidermidis*, *S. hominis*, *S. haemolyticus*, *S. warneri*, *S. capitis*, *S. saccharolyticus*, *S. auricularis*, *S. simulans*, *S. saprophyticus*, *S. cohnii*, and *S. xylosus* (32, 37, 38, 57). Other primates carry *S. aureus*, *S. saprophyticus*, *S. xylosus*, and different subspecies of *S. auricularis*, *S. warneri*, *S. haemolyticus*, and *S. cohnii* than are usually found on humans (33). *S. intermedius* is found frequently on carnivora and has also been isolated from certain other mammals and birds (23, 52). *S. hyicus* subsp. *hyicus* is found frequently on pigs, whereas *S. hyicus* subsp. *chromogenes* is found predominantly on cattle (15, 17). *S. sciuri* is found on rodents and certain other mammals (39), *S. lentus* on sheep and goats (33), *S. caprae* occasionally on goats (18), and *S. gallinarum* on poultry (18). *S. carnosus* is being used as a starter culture in the processing of fermented meats, such as sausage and salami (55). *S. caseolyticus* has been isolated from milk and dairy products (56).

Most *Staphylococcus* species are common inhabitants of the skin and mucous membranes. Some exhibit habitat or niche preferences. For example, *S. capitis* is found as large populations on the human head, especially the scalp and forehead where sebaceous glands are numerous and well developed. *S. auricularis* has a strong preference for the external auditory meatus. *S. hominis* and *S. haemolyticus* generally produce larger populations in areas of the skin where apocrine glands are numerous, such as the axillae and pubic areas. *S. aureus* prefers the anterior nares as a habitat.

Certain *Staphylococcus* species are found frequently as etiological agents of a variety of human and animal infections. In this chapter, we will be concerned primarily with the identification of *S. aureus*, *S. epidermidis*, and *S. saprophyticus*, species most commonly associated with human infections, and *S. intermedius* and *S. hyicus*, species of special veterinary interest.

CLINICAL SIGNIFICANCE

Skin infections caused by *S. aureus* are the most common human staphylococcal infections. These include cellulitis, pustules, boils, carbuncles, impetigo, and postoperative wound infections of various sites. A common community-acquired disorder is food poisoning caused by thermostable enterotoxins elaborated in foods during growth of *S. aureus*. Serious processes such as bacteremia, endocarditis, meningitis, pneumonia, and osteomyelitis continue to be seen as both community- and hospital-acquired infections.

In the late 1950s and early 1960s *S. aureus* caused considerable morbidity and mortality as a nosocomial pathogen of hospitalized patients. The advent and use of penicillinase-resistant, semisynthetic penicillins in the intervening years have provided successful antimicrobial therapy of serious *S. aureus* infections. However, methicillin-resistant *S. aureus* strains (MRSA) have recently emerged as a major clinical and epidemiological problem in U.S. hospitals (9, 24). While MRSA have most often been isolated from extremely ill patients in large, tertiary care hospitals, they have also been seen in patients in smaller community hospitals and rehabilitation facilities (2, 54). The majority of these strains are resistant to several of the most commonly used antimicrobial agents, including macrolides, aminoglycosides, and the β-lactam antibiotics in current use, including the latest generation of cephalosporins (45). Serious infections due to MRSA have most often been successfully treated with an older, potentially toxic antibiotic, vancomycin (64).

A community-acquired disease of potentially serious consequences, toxic shock syndrome, also has been attributed to infection or colonization with *S. aureus* (61). Most of the patients with toxic shock syndrome have been young, menstruating females who use certain types of highly absorbent tampons during menses (46). However, the toxin(s) associated with the syndrome may also be elaborated by *S. aureus* present at sites other than the genital in nonmenstruating women and in men (13). Methods for recognizing these strains or their toxins are not currently available for routine clinical laboratory use.

During the past 10 years, there has been increased concern for the other staphylococci that constitute a major component of our normal microflora, the coagulase-negative staphylococci. These species are endogenous to humans and certain animals and have previously been considered to be saprophytic or of low pathogenicity. The human coagulase-negative species include *S. epidermidis, S. hominis, S. haemolyticus, S. warneri, S. capitis, S. saccharolyticus, S. auricularis, S. saprophyticus, S. cohnii, S. xylosus* (often obtained by contact with animals), and *S. simulans*. Several of these species were previously included in the designation of a single species, *S. epidermidis* or *S. albus*. Others were noted as micrococci, and *S. saccharolyticus* was previously designated *Peptococcus saccharolyticus*. Such isolates have often been regarded as skin or mucous membrane contaminants when encountered in clinical specimens. However, more recent information has documented examples of infections by coagulase-negative species; these infections include bacteremia and infective endocarditis; infection or colonization of ventriculoatrial or other cerebrospinal fluid shunts, intravenous catheters, and joint prostheses; peritonitis associated with dialysis; osteomyelitis or pyoarthritis; and, prominently, genitourinary tract infections (19, 26, 31, 49, 53, 62, 68). Many of these infections arise as complications of invasive procedures common in modern medical care. *S. epidermidis* sensu stricto is prominent among species causing such infections, although careful species identification of clinical isolates has demonstrated the importance of several other species as probable or potential pathogens (Table 1). *S. saprophyticus* has now been clearly identified as one of the most common causes of urinary tract infections in young, sexually active females; recent studies indicate that it may be the second most common cause of such infections in this group (6, 28, 44). The possibility that *S. saprophyticus* is an agent of nongonococcal urethritis in males or causes other sexually transmitted diseases has also been proposed by European investigators (27).

Recently, methicillin-resistant *S. epidermidis* strains also have emerged as a clinical problem, especially in patients with prosthetic heart valves (31) or those who have undergone other forms of cardiac surgery (7). Resistance to methicillin is heterogeneic and may extend to the cephalosporin antibiotics. Difficulties in performing in vitro tests that adequately recognize cephalosporin resistance of these strains continue to be a problem (67). Serious infections due to methicillin-resistant *S. epidermidis* have been successfully treated with combination therapy which includes vancomycin plus rifampin or an aminoglycoside (31).

The coagulase-positive species *S. intermedius* and the coagulase-variable species *S. hyicus* are of particular importance to veterinary medicine. These species and *S. aureus* are serious opportunistic pathogens of animals. *S. intermedius* has been implicated in a variety of dog infections such as otitis externa, pyoderma, abscesses, reproductive tract infections, mastitis, and wound infections (16, 51, 52). *S. hyicus* subsp. *hyicus* (a coagulase-variable subspecies) has been implicated in infectious exudative epidermitis and septic polyarthritis of pigs and is occasionally isolated from the milk of cows suffering from mastitis

TABLE 1. Clinical importance of *Staphylococcus* species

Species group	Common pathogen	Questionable or uncommon pathogen	Undetermined or rare pathogen
S. aureus	*S. aureus*		
S. epidermidis	*S. epidermidis*	*S. haemolyticus, S. hominis, S. warneri, S. saccharolyticus*	*S. capitis, S. caprae*[a]
S. auricularis			*S. auricularis*
S. saprophyticus	*S. saprophyticus*	*S. cohnii*	*S. xylosus*
S. simulans		*S. simulans*	*S. carnosus*
S. intermedius	*S. intermedius*		
S. hyicus	*S. hyicus* subsp. *hyicus*	*S. hyicus* subsp. *chromogenes*	
S. sciuri			*S. sciuri, S. lentus*
S. caseolyticus			*S. caseolyticus*
S. gallinarum			*S. gallinarum*

[a] Species group designation is questionable.

(15, 50). *S. hyicus* subsp. *chromogenes* (a coagulase-negative subspecies) is commonly isolated from the milk of cows suffering from mastitis, although its role as an etiological agent is questionable.

COLLECTION, TRANSPORT, AND STORAGE OF SPECIMENS

The general principles of collection, transport, and storage of specimens, as described in Chapter 8, are applicable to staphylococci. No special methods or precautions are usually required because staphylococci are easily obtained from clinical material from most infection sites and are relatively resistant to drying and to moderate temperature changes. Special hypertonic media may be required for the transport and growth of cell wall-deficient forms encountered under certain circumstances (30). Some strains may require anaerobic conditions or CO_2 supplementation for satisfactory growth, but these strains survive transport and limited storage in air.

DIRECT EXAMINATION

Because of the rapidity of performing a Gram stain, the direct examination of normally sterile fluids such as cerebrospinal fluid, joint aspirates, or pulmonary secretions collected by transtracheal aspiration can be of great value. Direct examination of nonsterile fluids may also be very useful if the microscopist carefully evaluates the specimen by noting the presence of inflammatory cells versus epithelial cells. Even if large numbers of gram-positive cocci are present, only a presumptive report of "gram-positive cocci resembling staphylococci" should be made. This report must be confirmed by culture and appropriate identification techniques. It must also be emphasized that microscopy by itself cannot adequately differentiate various species of staphylococci from one another or from micrococci, planococci, some streptococci, aerococci, or various anaerobic cocci.

CULTURE AND ISOLATION

The basic procedures for culture and isolation described in Chapter 8 should be followed. Regardless of the source of the specimen, blood agar (preferably sheep blood agar) and a fluid medium such as thioglycolate broth should be inoculated. On blood agar, abundant growth of most staphylococcal species occurs within 18 to 24 h. By this time, colonies will be 1 to 3 mm in diameter, and they are usually circular, smooth, and raised, with a butyrous consistency. Most species cannot be distinguished from one another on the basis of colony morphology within a 24-h incubation period. Colonies conforming to the description above should be Gram stained, subcultured, and tested for genus and species properties. Most staphylococci of major medical interest produce growth in the upper as well as the lower anaerobic portions of thioglycolate broth or semisolid agar (36).

Specimens from heavily contaminated sources such as feces should be streaked also onto a selective medium such as SK agar (59), mannitol-salt agar, Columbia CNA agar, lipase-salt-mannitol agar, or phenylethyl alcohol agar (see Chapter 111). These media inhibit the growth of gram-negative organisms but allow staphylococci and certain other gram-positive cocci to grow. On selective media, incubation should be extended to at least 48 h for good colony development.

IDENTIFICATION

Staphylococcus species can be identified on the basis of colony morphology, coagulase production, oxygen requirements, hemolysis, novobiocin resistance, acetylmethylcarbinol (acetoin) production, aerobic acid production from certain carbohydrates, and certain enzyme activities (Table 2). The most clinically significant species can be identified on the basis of a few key characteristics (Table 3).

Correct identification of the coagulase-positive species as etiological agents or indicators of potential health risk is of prime importance. It should be noted here, however, that although *S. intermedius* and *S. hyicus* may be etiological agents of animal infections, they have not yet been implicated in human infections. Of the coagulase-negative species, *S. epidermidis* and *S. saprophyticus* have been associated with human infections most often. These species can be distinguished from each other and from related species with a high degree of accuracy. The coagulase-negative species *S. hominis*, *S. haemolyticus*, and *S. simulans* have each been isolated from more than 5% but less than 20% of the total human infections attributed to coagulase-negative staphylococci and therefore are of some concern to the clinical microbiologist. *S. warneri*, *S. saccharolyticus*, and *S. cohnii* have been only occasionally isolated from infections.

Colonial appearance

On nonselective blood agar, nutrient agar, tryptic soy agar, brain heart infusion agar, or P agar (41, 57), abundant growth of most species occurs in 18 to 24 h at 34 to 37°C. The exceptional species *S. lentus* and *S. auricularis* grow much more slowly than other staphylococci and usually require 24 to 36 h for detectable colony development. Growth of micrococci also occurs more slowly on these media. Well-isolated colonies of most staphylococci will generally be 1 to 3 mm in diameter within 24 h and 3 to 10 mm in diameter by 5 days of incubation, depending on the species.

Colony morphology can be a useful supplementary character in the identification of species and strains when time is available. Colonies of *S. aureus*, *S. saprophyticus*, *S. hominis*, *S. haemolyticus*, and *S. hyicus* subsp. *chromogenes* are often pigmented on primary isolation, whereas colonies of *S. epidermidis*, *S. simulans*, *S. intermedius*, and *S. hyicus* subsp. *hyicus* are very rarely pigmented (17, 23, 37, 57). Most pigments are carotenoids, and the colors range from yellow to orange. Colonies of *S. aureus* and, to a somewhat lesser extent, those of *S. intermedius* and *S. hyicus* become translucent to nearly transparent with continued incubation. Colonies of the other species mentioned above are generally opaque, although under crowded conditions *S. epidermidis* colonies may have dark, translucent centers. On P agar, colonies of *S. epidermidis* are about 3 to 5 mm in diameter after incubation at 34 to 35°C for 3 days followed by storage at room temperature for an additional 2 days. Colonies of this species are slightly raised, becoming sticky with age, and usually gray or gray-white. Colonies of

TABLE 2. Differentiation of *Staphylococcus* species

Character	*S. aureus*	*S. epidermidis*	*S. capitis*	*S. caprae*	*S. warneri*	*S. haemolyticus*	*S. hominis*	*S. saccharolyticus*	*S. auricularis*	*S. saprophyticus*
Colony size (large)[c]	+[d]	−	−	+	d	+	−	−	−	+
Colony pigment[e]	+	−	−	−	d	d	d	−	−	d
Anaerobic growth[f]	+	+	(+)	(+)	+	(+)	−	+	(±)	(+)
Aerobic growth	+	+	+	+	+	+	+	−	+	+
Coagulase	+	−	−	−	−	−	−	−	−	−
Hemolysis[g]	+	−	−	(d)	(d)	(+)	−	−	−	−
Nitrate reduction	+	+	d	+	−	d	d	+	(d)	−
Acetoin	+	+	d	+	+	d	d	ND	d	+
Cytochrome *c*[h]	−	−	−	−	−	−	−	−	−	−
Phosphatase	+	+	−	(+)	−	−	−	d	−	−
Urease[i]	+	+	−	+	+	−	+	ND	−	+
Arginine utilization[i]	+	+	d	+	d	+	d	+	d	−
β-Glucosidase[i]	+	(d)	−	−	+	d	−	ND	−	d
β-Glucuronidase[i]	−	−	−	−	d	d	−	ND	−	−
β-Galactosidase[i]	−	−	−	−	−	−	−	ND	(d)	d
Novobiocin resistance[j]	−	−	−	−	−	−	−	−	−	+
Acid (aerobically) from:										
Maltose	+	+	−	d	(+)	+	+	−	(+)	+
D-Trehalose	+	−	−	+	+	+	d	−	(+)	+
D-Mannitol	+	−	+	−	d	d	−	−	−	d
D-Xylose	−	−	−	−	−	−	−	−	−	−
Xylitol	−	−	−	−	−	−	−	−	−	d
D-Cellobiose	−	−	−	−	−	−	−	−	−	−
Sucrose	+	+	(+)	−	+	+	(+)	−	d	+
D-Turanose	+	d	−	−	d	d	d	ND	(d)	+
D-Mannose	+	(+)	+	+	−	−	−	(+)	−	−
D-Ribose	+	d	−	−	d	d	−	ND	−	−
Raffinose	−	−	−	−	−	−	−	−	−	−
α-Lactose	+	d	−	+	d	d	d	−	−	d
β-D-Fructose	+	+	+	−	+	d	+	(+)	+	+

[a] *S. cohnii* is composed of several subspecies (33). Characteristics of the human subspecies are noted in the left column, and those of the primate (including human and lower primate) subspecies are noted in the right column. The subspecies are currently unnamed.

[b] *S. hyicus* is composed of several subspecies (17, 51). Characteristics of *S. hyicus* subsp. *hyicus* are noted in the left column, and those of *S. hyicus* subsp. *chromogenes* are noted in the right column. Two additional characters (not listed), hydrolysis of polysorbate (Tween 80) and hyaluronidase activity (17), have been shown to be useful in distinguishing these two subspecies. *S. hyicus* subsp. *hyicus* is positive for both Tween 80 and hyaluronidase, whereas *S. hyicus* subsp. *chromogenes* is negative for these characters.

[c] Positive is defined as a colony diameter of ≥6 mm after incubation on P agar (41) at 34 to 35°C for 3 days and at room temperature (ca. 25°C) for an additional 2 days (37).

[d] +, 90% or more strains positive; −, 90% or more strains negative; d, 11 to 80% of strains positive; ND, not determined. Parentheses indicate a delayed reaction.

[e] Positive is defined as the visual detection of carotenoid pigments (e.g., yellow, yellow-orange, orange) during colony development at normal incubation or room temperatures.

[f] Anaerobic growth in a semisolid thioglycolate medium. Symbols: +, moderate or heavy growth down tube within 18 to 24 h; ±, heavier growth in the upper portion and weaker growth in the lower portion of tube; −, no visible growth within 48 h, but by 72 h very weak diffuse growth or a few scattered colonies may be observed in the anaerobic portion of tube. Parentheses indicate delayed growth appearing within 24 to 72 h, sometimes noted as large discrete colonies in the anaerobic portion of tube.

[g] Hemolysis on bovine blood agar. Symbols: +, wide zone of hemolysis within 24 to 36 h; (+), delayed moderate to wide zone of hemolysis within 48 to 72 h; −, narrow zone (≤1 mm) or no detection of hemolysis within 72 h.

[h] Cytochrome *c* is determined by the modified oxidase test of Faller and Schleifer (20).

[i] Urease and arginine utilization are characters which were shown to be useful in distinguishing *Staphylococcus* species by the dms STAPH Trac (API SYSTEM S.A., Montalieu-Vercieu, France, and DMS Laboratories, White House, N.J.) or API STAPHI-DENT (Analytab Products, Plainview, N.Y.) staphylococcal systems; β-glucosidase, β-glucuronidase, and β-galactosidase activities were shown to be useful characters in distinguishing species by the API STAPH-IDENT staphylococcal system.

[j] Positive is defined as a MIC of ≥1.6 μg/ml or a growth inhibition zone diameter of ≤16 mm with a 5-μg novobiocin disk.

[k] Acid production from maltose is typically very weak. After incubation for 48 to 72 h a slight yellowish to yellow-green color is produced under the culture streak, especially with canine *S. intermedius* strains. Strains isolated from pigeons or horses are somewhat different from carnivora strains and are more variable with respect to acid production.

S. hominis are small, like those of *S. epidermidis*, but remain smooth and butyrous. The more common pigmented *S. hominis* colonies have a yellow or orange-yellow center with alternating light and dark (e.g., whitish and grayish) concentric rings proceeding from near the center to the edge. Unpigmented *S. hominis* colonies exhibit the light and dark concentric ring pattern. Colonies of *S. epidermidis* grow significantly larger on P agar supplemented with 5% blood than on unsupplemented P agar, whereas the reverse

TABLE 2—*Continued*

S. cohnii[a]		*S. xylosus*	*S. simulans*	*S. carnosus*	*S. intermedius*	*S. hyicus*[b]		*S. sciuri*	*S. lentus*	*S. caseolyticus*	*S. gallinarum*
Human	Primate					Subsp. *hyicus*	Subsp. *chromogenes*				
d	+	+	+	+	+	+	+	+	−	−	+
−	d	d	−	−	−	−	+	d	d	d	d
d	(+)	d	+	+	(+)	+	+	(+)	(±)	(±)	(+)
+	+	+	+	+	+	+	+	+	+	+	+
−	−	−	−	−	+	d	−	−	−	−	−
(d)	(d)	−	(d)	−	d	−	−	−	−	−	(d)
−	−	d	+	+	+	+	+	+	+	+	+
d	d	d	−	+	−	−	−	−	−	−	−
−	−	−	−	−	−	−	−	+	+	+	−
−	+	d	(d)	+	+	+	+	+	+	+	(+)
−	+	+	+	−	+	d	d	−	−	ND	+
−	−	−	+	+	d	+	+	−	−	ND	−
−	−	+	−	−	d	d	d	+	+	ND	+
−	+	d	d	−	−	d	−	−	−	ND	−
−	+	+	+	+	+	−	−	−	−	ND	−
+	+	+	−	−	−	−	−	+	+	−	+
(d)	(+)	+	−	−	(±)[k]	−	d	(d)	d	+	+
+	+	+	d	d	+	+	+	+	+	d	+
d	d	d	+	+	(d)	−	d	+	+	−	+
−	−	+	−	−	−	−	−	−	(d)	−	+
(d)	(d)	(d)	−	−	−	−	−	−	−	−	d
−	−	−	−	−	−	−	−	+	+	ND	+
−	−	+	+	−	+	+	+	+	+	d	+
−	−	d	−	−	d	−	d	−	−	−	+
(d)	+	+	d	+	+	+	+	(d)	(+)	−	+
−	−	d	d	ND	+	+	+	+	+	+	+
−	−	−	−	−	−	−	−	−	+	ND	+
−	+	d	+	d	d	+	+	(d)	d	+	d
+	+	+	+	+	+	+	+	+	(+)	+	+

is usually the case for colonies of *S. hominis*. Colonies of *S. haemolyticus* are larger than those of *S. epidermidis* or *S. hominis* and are usually 5 to 8 mm in diameter. Like *S. hominis* colonies, they are smooth and butyrous. Colonies of *S. saprophyticus* are large (5 to 8 mm in diameter), entire, usually glistening, opaque, butyrous, and more convex than colonies of the above species in the *S. epidermidis* group. Colonies of *S. intermedius* and *S. hyicus* are usually more convex than those of *S. aureus*.

On selective mannitol-salt agar, the acid-producing species *S. aureus*, *S. simulans*, and *S. saprophyticus* may be distinguished from the non-acid-producing species *S. epidermidis* and most strains of *S. hominis* and *S. hyicus* subsp. *hyicus*. Strains capable of producing acid from mannitol usually do so within 24 to 48 h. Their colonies and the surrounding medium are yellow as a result of acidification in the presence of phenol red indicator. On lipase-salt-mannitol agar, most strains of *S. aureus* produce a yellow (acid) zone due to acidification of mannitol and, in addition, produce an opaque zone around the colony due to the presence of lipovitellin-lipase activity.

Some staphylococci which are cell wall deficient or require supplementation with specific metabolites (e.g., hemin, menadione, CO_2, etc.) may produce dwarf colonies on routine laboratory media (1, 30, 63). Rare encapsulated strains produce colonies that are smaller and more convex in profile than those of unencapsulated strains and that have a glistening-wet appearance. Upon storage of these, growth may become slimy and run down the surface of agar under normal gravitational force.

Coagulase production

The ability to clot plasma continues to be the most widely used and generally accepted criterion for the identification of frankly pathogenic staphylococci, i.e., *S. aureus* in humans and animals and *S. intermedius* and *S. hyicus* subsp. *hyicus* in animals (10, 17, 23). Two different coagulase tests can be performed: a tube test for free coagulase and a slide test for bound coagulase, or clumping factor. While the tube test is more definitive, the slide test may be used as a rapid screening technique. A variety of plasmas may be used for either test; however, dehydrated rabbit plasma containing citrate or EDTA is commercially available and most satisfactory. Human plasma should not be used unless it has been carefully tested for clotting capability and for lack of inhibitors.

The slide coagulase test is performed by making a heavy suspension of growth in distilled water, adding 1 drop of plasma, stirring the mixture, and observing for clumping within 10 s. The slide test is very rapid and more economical of plasma than the tube test. However, 10 to 15% of *S. aureus* strains may yield a negative test, thus requiring that isolates yielding negative slide tests be reexamined by the tube test. Slide test reactions must be read quickly because false-positive results may appear with reaction times longer than 10 s. In addition, colonies for testing must

TABLE 3. Minimum tests for identification of the most clinically significant species

Character	*S. aureus*[a,b]	*S. epidermidis*[a]	*S. saprophyticus*[a]	*S. intermedius*[b]	*S. hyicus* subsp. *hyicus*[b]
Coagulase	+[c]	−	−	+	d
Novobiocin resistance[d]	−	−	+	−	−
D-Mannitol	+	−	d	(d)	−
Colony pigment[e]	+			−	−
Acetoin	+			−	−
β-Galactosidase[f]	−			+	−
Maltose[g]	+			(±)	−
Phosphatase[h]	+	+	−		
D-Trehalose[i]	+	−	+		
D-Xylose[j]	−	−	−		
Sucrose[j]	+	+	+		

[a] Pathogenic for humans. *S. epidermidis* may produce a mild or subclinical mastitis in the bovine and caprine udder and therefore is of some limited veterinary interest.

[b] Pathogenic for animals.

[c] +, 90% or more strains positive; −, 90% or more strains negative; d, 11 to 80% of strains positive. Parentheses indicate a delayed reaction. Minimum tests results are shown for species of major veterinary or human clinical interest. Omitted symbols for extra test results are shown in Table 2.

[d] Positive is defined as a MIC of ≥1.6 μg/ml or a growth inhibition zone diameter of ≤16 mm with a 5-μg novobiocin disk.

[e] Positive is defined as the visual detection of carotenoid pigments (e.g., yellow, yellow-orange, orange) during colony development at normal incubation or room temperatures. Some strains of *S. aureus* may be unpigmented. Some strains of *S. intermedius* produce a pale violet pigment in the center of colonies.

[f] Activity as determined in the API STAPH-IDENT staphylococcal system.

[g] Some bovine strains of *S. aureus* may produce weak acid from maltose (J. L. Watts, personal communication). For *S. intermedius*, acid production from maltose is typically very weak. After incubation for 48 to 72 h, a slight yellowish to yellow-green color is produced under the culture streak, especially with canine strains. Strains isolated from pigeons or horses are somewhat different from carnivora strains and are more variable with respect to acid production.

[h] Some strains of *S. epidermidis* are phosphatase negative. These can be distinguished from *S. hominis* on the basis of their strong anaerobic growth in a thioglycolate medium within 18 to 24 h.

[i] Some strains of *S. saprophyticus* may not produce acid from D-trehalose, and rare strains of *S. epidermidis* may produce acid from D-trehalose.

[j] *S. saprophyticus* can be distinguished from *S. xylosus* and *S. cohnii* on the basis of no acid from D-xylose and acid from sucrose, respectively. For additional verification it should be noted that *S. saprophyticus* produces acid from D-turanose but does not produce acid from D-mannose.

not be picked from media containing high concentrations of salt (e.g., mannitol-salt agar), because autoagglutination and false-positive results may occur. Most strains of *S. intermedius* and *S. hyicus* yield negative slide tests.

The tube coagulase test is best performed by mixing 0.1 ml of an overnight culture in brain heart infusion broth with 0.5 ml of reconstituted plasma, incubating the mixture at 37°C in a water bath or heat block for 4 h, and observing the tube for clot formation by slowly tilting the tube 90°. Alternatively, a large, well-isolated colony on a noninhibitory agar can be transferred into 0.5 ml of reconstituted plasma and incubated as described above. Any degree of clotting constitutes a positive test. However, a flocculent or fibrous precipitate is not a true clot and should be recorded as negative. Incubation of the test overnight has also been recommended for *S. aureus* (29). For veterinary clinical laboratories it is important to note that some strains of *S. intermedius* and most coagulase-producing strains of *S. hyicus* subsp. *hyicus* require more than 4 h for a positive coagulase test. Clot formation by these species may require up to 12 to 24 h of incubation. If incubation exceeds 4 h, the following points must be considered: (i) staphylokinase produced by some strains may lyse the clot after prolonged incubation, yielding false-negative results; (ii) if the plasma used is not sterile (and some are not), either false-positive or false-negative results may occur; and (iii) an inoculum from an agar-grown colony may not be pure, and a contaminant may produce false results after prolonged incubation. In this regard, plasma containing EDTA is superior to citrated plasma because citrate-utilizing organisms (e.g., some streptococci) may produce clot formation by consuming the citrate. For those uncommon *S. aureus* strains requiring a longer clotting period, other characters should also be tested to confirm identity. Additional characters are also required to identify coagulase-negative mutants and some encapsulated strains of *S. aureus*.

Other identification methods for *S. aureus*

Newer, alternative methods for identification of *S. aureus* include a rapid commercial hemagglutination slide test for clumping factor (BBL Microbiology Systems, Cockeysville, Md.); commercial latex agglutination tests that detect both clumping factor and protein A, properties of the cell wall of *S. aureus* (Scott Laboratories, Inc., Fiskeville, R.I.; I-M, Inc., Operating Unit of American Micro Scan, Lexington, Ky.; and Carr-Scarborough Microbiologicals, Inc., Stone Mountain, Ga.); and commercial biochemical test systems for species identification of staphylococci, including *S. aureus* (Analytab Products, Plainview, N.Y.; DMS Laboratories, Inc., White House, N.J.; American Micro Scan, Mahwah, N.J.; BBL Microbiology Systems, Cockeysville, Md.; and Vitek Systems, Inc., Hazelwood, Mo.). Certain of these methods may be chosen by individual laboratories for either primary or alternative identification of presumed *S. aureus* isolates. The biochemical test systems are described in more detail in a later section of this chapter.

Hemolysis

Differences in the hemolytic activity of certain *Staphylococcus* species are best demonstrated with bovine blood agar (5% blood in P agar); alternatively, human or sheep blood agar can be used. If human blood is used, it must be carefully controlled and known not to contain antibiotics inhibitory to staphylococci. The moderate to strong hemolysis of blood by most strains of *S. haemolyticus* within 48 to 72 h at 34 to 37°C is very useful in distinguishing this species from other coagulase-negative species (36, 37, 57). Most *S. aureus* strains demonstrate rapid and strong hemolysis, i.e., within 24 to 36 h. Hemolysis is variable in *S. intermedius*, although many strains produce moderate to strong hemolysis. Most strains of *S. hyicus* and other species produce no, or only weak, hemolysis. Hemolytic staphylococci generally produce stronger hemolysis with human blood and weaker hemolysis with sheep blood than with bovine blood.

Novobiocin resistance

The novobiocin-resistant species, including *S. saprophyticus, S. cohnii, S. xylosus, S. sciuri, S. lentus*, and *S. gallinarum* are generally resistant to novobiocin concentrations of up to 1.6 μg/ml of agar (MIC ≥ 1.6 μg/ml) (18, 36, 39, 57). A simple disk diffusion procedure for estimating novobiocin susceptibility and distinguishing *S. saprophyticus* from other clinically important species can be performed by use of a 5-μg novobiocin susceptibility disk and either P agar (36), Mueller-Hinton agar (3), or tryptic soy sheep blood agar (22). With an inoculum suspension equivalent in turbidity to the 0.5 McFarland opacity standard, a 5-μg novobiocin disk, and incubation at 35 to 37°C for 16 to 20 h, novobiocin resistance is indicated by a zone diameter of ≤16 mm with any of the above-described media. Rapid disk-elution procedures using either manual or automated instrument interpretation have also been reported to reliably predict novobiocin resistance after only 4 to 5 h of incubation (4, 25, 43). Novobiocin resistance is uncommon in the other clinically important species.

Phosphatase activity

Phosphatase activity may be determined rapidly by using a modification of the technique of Pennock and Huddy (48), in which a 0.005 M solution of phenolphthalein monophosphate (sodium salt in 0.01 M citric acid–sodium citrate buffer, pH 5.8) is used (36). Before use, all glassware should be rinsed thoroughly with deionized water to remove contaminating phosphates. Tubes containing 0.5 ml of the above buffer should be inoculated with a loopful of an overnight culture to a density of approximately 10^9 CFU per ml. After incubation at 37°C for 4 h, the reaction is stopped by adding 0.5 ml of 0.5 N sodium hydroxide and 0.5 ml of 0.5 M sodium bicarbonate. Color is developed by adding 0.5 ml of 4-aminoantipyrine solution (0.6 g/100 ml) and 0.5 ml of potassium ferricyanide solution (2.4 g/100 ml). Phosphatase activity is indicated by the development of a red color. Strains of *S. aureus* and most strains of *S. epidermidis, S. intermedius*, and *S. hyicus* produce a deep red color, whereas strains of *S. simulans* generally produce a pink color indicating weak activity. Approximately 10 to 20% of *S. epidermidis* strains isolated from clinical specimens are phosphatase negative (yellow). These can be distinguished from *S. hominis* on the basis of their strong anaerobic growth in thioglycolate within 18 to 24 h. *S. hominis, S. haemolyticus*, and *S. saprophyticus* are usually phosphatase negative.

A newer, alternative method for determining phosphatase activity has been incorporated into several of the commercial biochemical test systems for species identification of staphylococci. The method is based on the hydrolysis of *p*-nitrophenylphosphate into P_i and *p*-nitrophenol by alkaline phosphatase. Phosphatase activity is indicated by the release of yellow *p*-nitrophenol from the colorless substrate.

Acetoin production

Acetoin production from glucose or pyruvate is a useful alternative character to distinguish *S. aureus* (positive) from the other coagulase-positive species *S. intermedius* (negative) and coagulase-positive strains of *S. hyicus* (negative) (17, 23). The rapid paper disk method of Davis and Hoyling (12) is recommended for this test. Cultures are inoculated as a patch or smear onto tryptone-yeast extract-glucose agar (12). After 48 h of incubation at 30°C, a 1-cm disk of Whatman 3M paper freshly soaked in 10% sodium pyruvate solution is placed on each growth patch, and the plates are reincubated for 3 h. Acetoin production is detected by spotting 1 drop each of 40% potassium hydroxide, 1% creatinine, and 1% α-naphthol (alcoholic) solution onto each disk and observing a pink-red color change within 1 h at room temperature. The accuracy of the disk test is comparable to that of conventional tube acetoin tests requiring longer incubation periods.

Acid production from carbohydrates

Acid production from carbohydrates may be easily detected by using an agar plate method (36). Carbohydrate agars are prepared by adding an appropriate amount of filter-sterilized carbohydrate stock solution to an autoclave-sterilized purple agar base medium (containing bromocresol purple indicator) to give a final carbohydrate concentration of 1%. Culture streaks are prepared by lightly inoculating a 0.5- to 1-cm long streak on the surface of an agar plate with a loopful of an overnight culture. Partitioned quadrant plates having two widely separated culture streaks per quadrant are generally more satisfactory than standard nonpartitioned plates having eight equidistant culture streaks when acid production is strong (e.g., with maltose, β-D-fructose, D-mannitol, or sucrose). Cultures are incubated at 34 to 37°C in air and examined at 24 and 72 h. Reactions are interpreted as follows: +, moderate to strong acid, yellow indicator color extends out from the culture streak into the surrounding medium within 72 h; ±, weak acid, distinct yellow indicator color under culture streak, but not extending into the surrounding medium within 72 h; −, no acid, very faint or no yellow indicator color under the culture streak within 72 h. Strong acid production can be detected as early as 12 to 24 h, and moderate acid production can be detected within 48 to 72 h.

Some variation in acid production from certain carbohydrates can be expected within certain species. However, when reactions of several key carbohydrates are taken together, accuracy in identification is greatly increased.

S. saprophyticus can be distinguished from other novobiocin-resistant species by its production of acid from sucrose and turanose and absence of acid production from xylose, arabinose, mannose, cellobiose, and raffinose. *S. epidermidis* can be distinguished from other novobiocin-susceptible species by its production of acid from maltose, sucrose, and fructose and absence of acid production from mannitol and trehalose. Some rare strains of *S. epidermidis* may produce acid from trehalose. These can be distinguished from other species on the basis of phosphatase activity, anaerobic growth in thioglycolate, and colony morphology.

Anaerobic growth in thioglycolate medium

Tubes (150 by 13 mm) containing 8 ml of cooled (50 to 52°C) thioglycolate semisolid medium (Brewer's fluid thioglycolate medium plus an additional 0.3% agar) (36) are inoculated and mixed gently with a 0.1-ml saline suspension of an overnight culture, giving a final density of 10^6 to 10^7 CFU per ml. Cultures are incubated without shaking for 5 days at 34 to 37°C and observed daily for anaerobic growth characteristics. Strains demonstrating dense, uniform anaerobic growth, such as is typical for *S. aureus*, *S. epidermidis*, and *S. simulans*, show this property within 16 to 24 h and do not require additional incubation (37). However, cultures producing a gradient of dense to light growth down the tube or those showing very weak or no anaerobic growth within 24 to 48 h should be incubated for at least 72 h. *S. hominis* usually shows no or very weak anaerobic growth and may produce a few tiny colonies in the anaerobic portion of the medium. *S. haemolyticus* often produces relatively large, discrete colonies in the anaerobic portion of the medium after 24 to 72 h of incubation. The anaerobic growth of *S. intermedius* is slower and weaker than that observed for *S. aureus*.

Identification of species by using commercial biochemical test systems

Some of the conventional methods described in this chapter may not be suitable for routine use in certain clinical laboratories, because a number of specialized media are required. Several manufacturers of commercial kit identification systems or automated instruments have recently released products that can identify a number of the *Staphylococcus* species with an accuracy of 70 to >90% with relative speed and simplicity (5, 21, 42). For some of these systems additional testing is recommended to increase accuracy, e.g., testing conventional coagulase production or novobiocin resistance (5, 42). Identification systems that are currently available include API STAPH-IDENT (Analytab), dms STAPH Trac (DMS), Pos Combo panel (American Micro Scan), Minitek Gram-Positive System (BBL), and the Vitek Gram-Positive Identification Card (Vitek).

The API STAPH-IDENT strip consists of a series of 10 microcupules containing dehydrated substrates, nutrient media, or both. The microcupules provide a reaction vessel and support for different carbohydrate utilization and enzymatic tests. After inoculation, strips are incubated for 5 h at 35 to 37°C in air and then interpreted. Positive color reactions are converted to a four-digit profile for species identification according to instructions of the manufacturer. It is recommended that the additional tests of novobiocin susceptibility (using a 5-μg disk) and coagulase activity be used to more accurately identify certain strains of *S. saprophyticus*, *S. cohnii*, *S. hominis*, and *S. aureus*. Anaerobic growth in thioglycolate should be used as an additional test to distinguish strains of *S. epidermidis* that are phosphatase negative on the strip from *S. hominis*. Analytab has developed recently a similar type of rapid identification system for both staphylococcal and streptococcal species; this system is designated API 20 GP and consists of 20 tests.

The dms STAPH Trac kit is a rapid identification system for staphylococcal and certain micrococcal species. The STAPH Trac strip consists of a series of 20 microcupules containing substrates for the performance of 19 different biochemical tests and a negative control. After inoculation, the strips are incubated for 24 h at 37°C in air and then interpreted. Positive reactions are converted to a seven-figure profile according to instructions of the manufacturer. It is recommended that the additional test of lysostaphin susceptibility be used to differentiate staphylococci from micrococci.

The American Micro Scan Pos Combo panel is designed for the rapid identification of different staphylococcal and streptococcal species as well as certain species of several other gram-positive genera. This system consists of a series of 27 miniaturized tests. Inoculated panels are incubated for 18 h at 35°C in air and then interpreted. Positive reactions are translated into seven- or eight-digit "biotype numbers." A portion of the Pos Combo panel may also be used for antibiotic susceptibility testing.

The Minitek Gram-Positive System is designed for the rapid identification of staphylococcal and streptococcal species and certain micrococcal species. The Minitek system consists of plates with 20 wells, each containing a paper disk impregnated with a different substrate or biochemical. A total of 21 tests are included in the system; one of the wells performs two different tests. Inoculated plates are incubated for 18 to 24 h at 35 to 37°C in air and then interpreted. Positive reactions are translated into a seven-digit profile. Several supplemental tests (e.g., coagulase activity, hemolysis, and novobiocin susceptibility) are recommended for more accurate identification of *S. saprophyticus*, *S. haemolyticus*, *S. hominis*, and *S. aureus*.

The Vitek AutoMicrobic System utilizes a Gram-Positive Identification Card to identify a variety of gram-positive bacteria, including, among others, certain staphylococcal and streptococcal species. Based on tests currently used in this system, it is possible to identify *S. aureus*, *S. epidermidis*, *S. saprophyticus*, and *S. hominis*. The Gram-Positive Identification Card contains wells for 27 different biochemical tests. With some modification it may be possible to identify other staphylococcal species with this system. The incubation period for the Gram-Positive Identification Card varies according to the growth rate of the inoculated organism. Identification of most staphylococci will be made after incubation for 4 h, while slower-growing bacteria may require up to 13 h of incubation. At the end of the incubation period, the data terminal prints a status report including a species identification.

Until the clinical significance of routine species identification of all coagulase-negative isolates can be clearly established, clinical laboratories may choose to restrict complete species identification to isolates from normally sterile sites such as blood or cerebrospinal fluid, and to routinely distinguish (i) *S. sapro-*

phyticus from other coagulase-negative staphylococci isolated from urine and (ii) *S. epidermidis* from other coagulase-negative staphylococci isolated from colonized shunts, catheters, or prosthetic devices.

Recognition of MRSA

For proper selection of antimicrobial agents for therapy and for hospital infection control, it is extremely important that MRSA be quickly and accurately recognized. During the past several years numerous reports have documented the technical intricacies of detecting MRSA by conventional or instrument susceptibility test methods. MRSA are characterized as demonstrating heteroresistance toward beta-lactam antibiotics, in that two subpopulations (one susceptible and the other resistant) coexist within a culture. Successful detection of MRSA depends largely on fostering the growth of the more resistant subpopulation, which is favored by neutral pH (7.0 to 7.4), cooler temperatures (30 to 35°C), prolonged incubation (up to 48 h), and addition of sodium chloride to susceptibility test media. At this time, reliable recognition of MRSA appears most likely through the use of one of two dilution methods. Microdilution MIC tests using Mueller-Hinton broth supplemented with divalent cations and 2% sodium chloride and incubated at 35°C for at least 24 h appear to accurately detect MRSA. With this procedure, MRSA demonstrate a methicillin MIC of ≥16 μg/ml and an oxacillin or nafcillin MIC of ≥8 μg/ml (67). An alternative screening procedure that also appears reliable is the use of Mueller-Hinton agar supplemented with 4% NaCl and containing either methicillin (10 μg/ml), oxacillin (6 μg/ml), or nafcillin (6 μg/ml). An inoculum of 10^4 CFU is placed on the surface of the agar, and the plates are incubated at 35°C for 24 to 48 h. MRSA grow on these plates, whereas susceptible strains fail to grow under these conditions (67).

EPIDEMIOLOGICAL TYPING SYSTEMS

Because of their ubiquity, it is sometimes desirable to separate strains of staphylococci into types or groups (40). Typing is usually indicated to detect the source of an outbreak of staphylococcal food poisoning or nosocomial infections. A variety of techniques have been used, including serological typing, biochemical reactions (biotyping), antibiotic susceptibility patterns, bacteriophage typing, and plasmid profiles. Each of these has its practical limitations.

Serological typing has been shown to be of value, but because of the difficulties of preparing specific antisera, it is rarely used today except in a few European research laboratories. Biotyping is used primarily with coagulase-negative *Staphylococcus* species because *S. aureus* strains are relatively homogenous biochemically. Antibiograms can be used epidemiologically since they are commonly determined in the laboratory, and highly standardized procedures have been established. The emergence of a *Staphylococcus* strain with a unique susceptibility pattern can be a valuable signal to the clinical microbiologist and can provide a marker by which similar strains can be detected. It must be recognized, however, that susceptibility patterns can be strongly influenced by the scope of antibiotic use within a given locale, so that results can vary considerably from one community to another. Thus, antibiograms are of most use within a confined area rather than on a larger geographical scale. Susceptibility profiles may even change within an institution over time or at different locations within the institution, thus making strain identification more difficult.

The most established system for epidemiological typing is bateriophage typing. Since 1952 this technique has found widespread use, and an international system has been established to control both the typing phages and the procedures. It is not, however, a static system, as it must respond to the emergence of new strains of staphylococci that are nontypable. Thus, the bacteriophages used in the standard typing set vary over an extended period of years. This factor must be considered when comparing phage patterns of previous years with those of today. *S. aureus* is the *Staphylococcus* species most commonly typed, as it is the most commonly encountered pathogen in the genus. Many larger hospital laboratories and a number of state health department laboratories offer this service or can forward cultures to the Centers for Disease Control, Atlanta, Ga., where a reference center is maintained. A set of bacteriophages for typing strains of *S. epidermidis* exists, but only a few laboratories have this capability. The normal variations existing in phage typing patterns of *S. epidermidis* are not well defined, and typing this species is usually less productive than typing *S. aureus*.

Determination of plasmid profiles of isolates by agarose gel electrophoresis is proving to be a promising method of identifying strains or clonal populations of staphylococci (8, 35, 47); however, this approach is currently beyond the scope of many clinical laboratories. On occasion, clonal variation in plasmid profile may be observed within a strain. This is most often noted by the addition or deletion of a specific plasmid or modifications resulting in a change of plasmid size. Many of the *Staphylococcus* species demonstrate complex plasmid profiles, thus allowing for considerable variation in profile between strains (35). Some strains may carry as many as 5 to 10 different plasmids.

LITERATURE CITED

1. **Acar, J. F., F. W. Goldstein, and P. Lagrange.** 1978. Human infections caused by thiamine- or menadione-requiring *Staphylococcus aureus*. J. Clin. Microbiol. **8:**142–147.
2. **Aeilts, G. D., F. L. Sapico, H. N. Canawati, G. M. Malik, and J. Z. Montgomerie.** 1982. Methicillin-resistant *Staphylococcus aureus* colonization and infection in a rehabilitation facility. J. Clin. Microbiol. **16:**218–223.
3. **Almeida, R. J., and J. H. Jorgensen.** 1982. Use of Mueller-Hinton agar to determine novobiocin susceptibility of coagulase-negative staphylococci. J. Clin. Microbiol. **16:**1155–1156.
4. **Almeida, R. J., and J. H. Jorgensen.** 1983. Rapid determination of novobiocin resistance of coagulase-negative staphylococci with the MS-2 system. J. Clin. Microbiol. **17:**558–560.
5. **Almeida, R. J., J. H. Jorgensen, and J. E. Johnson.** 1983. Evaluation of the AutoMicrobic system Gram-Positive Identification Card for species identification of coagulase-negative staphylococci. J. Clin. Microbiol. **18:**438–439.
6. **Anderson, J. D., A. M. Clarke, M. E. Anderson, J. L. Isaac-Renton, and M. G. McLoughlin.** 1981. Urinary

tract infections due to *Staphylococcus saprophyticus* biotype 3. Can. Med. Assoc. J. **124**:415–418.
7. **Archer, G. L., and M. J. Tenenbaum.** 1980. Antibiotic-resistant *Staphylococcus epidermidis* in patients undergoing cardiac surgery. Antimicrob. Agents Chemother. **17**:269–272.
8. **Archer, G. L., N. Vishniavsky, and H. G. Stiver.** 1982. Plasmid pattern analysis of *Staphylococcus epidermidis* isolates from patients with prosthetic valve endocarditis. Infect. Immun. **35**:627–632.
9. **Boyce, J. M., and W. A. Causey.** 1982. Increasing occurrence of methicillin-resistant *Staphylococcus aureus* in the United States. Infect. Control **3**:377–383.
10. **Buchanan, R. E., and N. E. Gibbons (ed.).** 1974. Bergey's manual of determinative bacteriology, 8th ed., p. 478–490. The Williams & Wilkins Co., Baltimore.
11. **Curry, J. C., and G. E. Borovian.** 1976. Selective medium for distinguishing micrococci from staphylococci in the clinical laboratory. J. Clin. Microbiol. **4**:455–457.
12. **Davis, G. H. G., and B. Hoyling.** 1973. Use of a rapid acetoin test in the identification of staphylococci and micrococci. Int. J. Syst. Bacteriol. **23**:281–282.
13. **Davis, J. P., M. T. Osterholm, C. M. Helms, J. M. Vergeront, L. A. Wintermeyer, J. C. Forfang, L. A. Judy, J. Rondeau, and W. L. Schell.** 1982. Tri-state toxic-shock syndrome study. II. Clinical and laboratory findings. J. Infect. Dis. **145**:441–448.
14. **Deibel, R. H., and J. B. Evans.** 1960. Modified benzidine test for the detection of cytochrome-containing respiratory systems in microorganisms. J. Bacteriol. **79**:356–360.
15. **Devriese, L. A.** 1977. Isolation and identification of *Staphylococcus hyicus*. Am. J. Vet. Res. **38**:787–792.
16. **Devriese, L. A., and V. Hájek.** 1980. A review. Identification of pathogenic staphylococci isolated from animals and foods derived from animals. J. Appl. Bacteriol. **49**:1–11.
17. **Devriese, L. A., V. Hájek, P. Oeding, S. A. Meyer, and K. H. Schleifer.** 1978. *Staphylococcus hyicus* (Sompolinsky 1953) comb. nov. and *Staphylococcus hyicus* subsp. *chromogenes* subsp. nov. Int. J. Syst. Bacteriol. **28**:482–490.
18. **Devriese, L. A., B. Poutrel, R. Kilpper-Bälz, and K. H. Schleifer.** 1983. *Staphylococcus gallinarum* and *Staphylococcus caprae*, two new species from animals. Int. J. Syst. Bacteriol. **33**:480–486.
19. **Eng, R. H. K., C. Wang, A. Person, T. E. Kiehn, and D. Armstrong.** 1982. Species identification of coagulase-negative staphylococcal isolates from blood cultures. J. Clin. Microbiol. **15**:439–442.
20. **Faller, A., and K. H. Schleifer.** 1981. Modified oxidase and benzidine tests for separation of staphylococci from micrococci. J. Clin. Microbiol. **13**:1031–1035.
21. **Giger, O., C. C. Charilaou, and K. R. Cundy.** 1984. Comparison of API Staph-Ident and DMS Staph-Trac systems with conventional methods used for the identification of coagulase-negative staphylococci. J. Clin. Microbiol. **19**:68–72.
22. **Goldstein, J., R. Schulman, E. Kelly, G. McKinley, and J. Fung.** 1983. Effect of different media on determination of novobiocin resistance for differentiation of coagulase-negative staphylococci. J. Clin. Microbiol. **18**:592–595.
23. **Hájek, V.** 1976. *Staphylococcus intermedius*, a new species isolated from animals. Int. J. Syst. Bacteriol. **26**:401–408.
24. **Haley, R. W., A. W. Hightower, R. F. Khabbaz, C. Thornsberry, W. J. Martone, J. R. Allen, and J. M. Hughes.** 1982. The emergence of methicillin-resistant *Staphylococcus aureus* infections in United States hospitals. Ann. Intern. Med. **97**:297–308.
25. **Harrington, B. J., and J. M. Gaydos.** 1984. Five-hour novobiocin test for differentiation of coagulase-negative staphylococci. J. Clin. Microbiol. **19**:279–280.
26. **Holt, R. J.** 1971. The colonization of ventriculo-atrial shunts by coagulase-negative staphylococci, p. 81–87. *In* M. Finland, W. Marget, and K. Bartmann (ed.), Bacterial infections: changes in their causative agents, trends, and possible basis. Springer-Verlag, Berlin.
27. **Hovelius, B., I. Thelin, and P. A. Mardh.** 1979. *Staphylococcus saprophyticus* in the aetiology of nongonococcal urethritis. Br. J. Vener. Dis. **55**:369–374.
28. **Jordan, P. A., A. Iravani, G. A. Richard, and H. Baer.** 1980. Urinary tract infections caused by *Staphylococcus saprophyticus*. J. Infect. Dis. **142**:510–515.
29. **Jungkind, D. L., N. J. Torhan, K. E. Corman, and J. M. Bondi.** 1984. Comparison of two commercially available test methods with conventional coagulase tests for identification of *Staphylococcus aureus*. J. Clin. Microbiol. **19**:191–193.
30. **Kagan, B. M.** 1972. L-forms, p. 65–74. *In* J. O. Cohen (ed.), The staphylococci. Wiley-Interscience, New York.
31. **Karchmer, A. W., G. L. Archer, and W. E. Dismukes.** 1983. *Staphylococcus epidermidis* causing prosthetic valve endocarditis: microbiologic and clinical observations as guides to therapy. Ann. Intern. Med. **98**:447–455.
32. **Kilpper-Bälz, R., and K. H. Schleifer.** 1981. Transfer of *Peptococcus saccharolyticus* Foubert and Douglas to the genus *Staphylococcus*: *Staphylococcus saccharolyticus* (Foubert and Douglas) comb. nov. Zentralbl. Bakteriol. Parasitenkd. Infektionskr. Hyg. Abt. 1 Orig. Reihe C **2**:324–331.
33. **Kloos, W. E.** 1980. Natural populations of the genus *Staphylococcus*. Annu. Rev. Microbiol. **34**:559–592.
34. **Kloos, W. E.** 1981. The identification of *Staphylococcus* and *Micrococcus* species isolated from human skin, p. 3–12. *In* H. I. Maibach and R. Aly (ed.), Skin microbiology: relevance to clinical infection. Springer-Verlag, New York.
35. **Kloos, W. E., B. S. Orban, and D. D. Walker.** 1981. Plasmid composition of *Staphylococcus* species. Can. J. Microbiol. **27**:271–278.
36. **Kloos, W. E., and K. H. Schleifer.** 1975. Simplified scheme for routine identification of human *Staphylococcus* species. J. Clin. Microbiol. **1**:82–88.
37. **Kloos, W. E., and K. H. Schleifer.** 1975. Isolation and characterization of staphylococci from human skin. II. Descriptions of four new species: *Staphylococcus warneri*, *Staphylococcus capitis*, *Staphylococcus hominis*, and *Staphylococcus simulans*. Int. J. Syst. Bacteriol. **25**:62–79.
38. **Kloos, W. E., and K. H. Schleifer.** 1983. *Staphylococcus auricularis* sp. nov.: an inhabitant of the human external ear. Int. J. Syst. Bacteriol. **33**:9–14.
39. **Kloos, W. E., K. H. Schleifer, and R. F. Smith.** 1976. Characterization of *Staphylococcus sciuri* sp. nov. and its subspecies. Int. J. Syst. Bacteriol. **26**:22–37.
40. **Kloos, W. E., and P. B. Smith.** 1980. Staphylococci, p. 83–87. *In* E. H. Lennette, A. Balows, W. J. Hausler, Jr., and J. P. Truant (ed.), Manual of clinical microbiology, 3rd ed. American Society for Microbiology, Washington, D.C.
41. **Kloos, W. E., T. G. Tornabene, and K. H. Schleifer.** 1974. Isolation and characterization of micrococci from human skin, including two new species: *Micrococcus lylae* and *Micrococcus kristinae*. Int. J. Syst. Bacteriol. **24**:79–101.
42. **Kloos, W. E., and J. F. Wolfshohl.** 1982. Identification of *Staphylococcus* species with the API STAPH-IDENT system. J. Clin. Microbiol. **16**:509–516.
43. **Leighton, P. M., and M. White.** 1983. Rapid determination of novobiocin susceptibility for the identification of *Staphylococcus saprophyticus*. Diagn. Microbiol. Infect. Dis. **1**:261–264.
44. **Marrie, T. J., C. Kwan, M. A. Noble, A. West, and L. Duffield.** 1982. *Staphylococcus saprophyticus* as a cause of urinary tract infections. J. Clin. Microbiol. **16**:427–431.
45. **Neu, H. C.** 1982. The new beta-lactamase-stable cephalo-

sporins. Ann. Intern. Med. **97:**408–419.

46. **Osterholm, M. T., J. P. Davis, R. W. Gibson, J. S. Mandel, L. A. Wintermeyer, C. M. Helms, J. C. Forfang, J. Randeau, and J. M. Vergeront.** 1982. Tri-state toxic-shock syndrome study. I. Epidemiologic findings. J. Infect. Dis. **145:**431–440.
47. **Parisi, J. T., and D. W. Hecth.** 1980. Plasmid profiles in epidemiologic studies of infections of *Staphylococcus epidermidis*. J. Infect. Dis. **141:**637–643.
48. **Pennock, C. A., and R. B. Huddy.** 1967. Phosphatase reaction of coagulase-negative staphylococci and micrococci. J. Pathol. Bacteriol. **93:**685–688.
49. **Peters, G., R. Locci, and G. Pulverer.** 1982. Adherence and growth of coagulase-negative staphylococci on surfaces of intravenous catheters. J. Infect. Dis. **146:**479–482.
50. **Phillips, W. E., Jr., R. E. King, and W. E. Kloos.** 1980. Isolation of *Staphylococcus hyicus* subsp. *hyicus* from a pig with septic polyarthritis. Am. J. Vet. Res. **41:**274–276.
51. **Phillips, W. E., Jr., and W. E. Kloos.** 1981. Identification of coagulase-positive *Staphylococcus intermedius* and *Staphylococcus hyicus* subsp. *hyicus* isolates from veterinary clinical specimens. J. Clin. Microbiol. **14:**671–673.
52. **Raus, J., and D. N. Love.** 1983. Characterization of coagulase-positive *Staphylococcus intermedius* and *Staphylococcus aureus* isolated from veterinary clinical specimens. J. Clin. Microbiol. **18:**789–792.
53. **Rubin, J., W. A. Rodgers, H. M. Taylor, E. D. Everett, F. F. Prowant, L. V. Fruto, and K. D. Nolph.** 1980. Peritonitis during continuous ambulatory dialysis. Ann. Intern. Med. **92:**7–13.
54. **Saravolatz, L. D., D. J. Pohlod, and L. M. Arking.** 1982. Community-acquired methicillin-resistant *Staphylococcus aureus* infections: a new source for nosocomial outbreaks. Ann. Intern. Med. **97:**325–329.
55. **Schleifer, K. H., and U. Fischer.** 1982. Description of a new species of the genus *Staphylococcus: Staphylococcus carnosus*. Int. J. Syst. Bacteriol. **32:**153–156.
56. **Schleifer, K. H., R. Kilpper-Bälz, U. Fischer, A. Faller, and J. Endl.** 1982. Identification of *"Micrococcus candidus"* ATCC 14852 as a strain of *Staphylococcus epidermidis* and of *"Micrococcus caseolyticus"* ATCC 13548 and *Micrococcus varians* ATCC 29750 as members of a new species, *Staphylococcus caseolyticus*. Int. J. Syst. Bacteriol. **32:**15–20.
57. **Schleifer, K. H., and W. E. Kloos.** 1975. Isolation and characterization of staphylococci from human skin. I. Amended descriptions of *Staphylococcus epidermidis* and *Staphylococcus saprophyticus* and descriptions of three new species: *Staphylococcus cohnii, Staphylococcus haemolyticus*, and *Staphylococcus xylosus*. Int. J. Syst. Bacteriol. **25:**50–61.
58. **Schleifer, K. H., and W. E. Kloos.** 1975. A simple test system for the separation of staphylococci from micrococci. J. Clin. Microbiol. **1:**337–338.
59. **Schleifer, K. H., and E. Krämer.** 1980. Selective medium for isolating staphylococci. Zentralbl. Bakteriol. Parasitenkd. Infektionskr. Hyg. Abt. 1 Orig. Reihe C **1:**270–280.
60. **Schleifer, K. H., S. A. Meyer, and M. Rupprecht.** 1979. Relatedness among coagulase-negative staphylococci: deoxyribonucleic acid reassociation and comparative immunological studies. Arch. Microbiol. **122:**93–101.
61. **Schlievert, P. M., K. N. Shands, B. B. Dan, G. P. Schmid, and R. D. Nishimura.** 1981. Identification and characterization of an exotoxin from *Staphylococcus aureus* associated with toxic shock syndrome. J. Infect. Dis. **143:**509–516.
62. **Sewell, C. M., J. E. Clarridge, E. J. Young, and R. K. Guthrie.** 1982. Clinical significance of coagulase-negative staphylococci. J. Clin. Microbiol. **16:**236–239.
63. **Slifkin, M., L. P. Merkow, S. A. Krevzberger, C. Engwall, and M. Pardo.** 1971. Characterization of CO_2 dependent microcolony variants of *Staphylococcus aureus*. Am. J. Clin. Pathol. **56:**584–592.
64. **Sorrell, T. C., D. R. Packham, S. Shanker, M. Foldes, and R. Munro.** 1982. Vancomycin therapy for methicillin-resistant *Staphylococcus aureus*. Ann. Intern. Med. **97:**344–350.
65. **Stackebrandt, E., and C. R. Woese.** 1979. A phylogenetic dissection of the family *Micrococcaceae*. Curr. Microbiol. **2:**317–322.
66. **Subcommittee on Taxonomy of Staphylococci and Micrococci.** 1965. Recommendations. Int. Bull. Bacteriol. Nomencl. Taxon. **15:**107–110.
67. **Thornsberry, C., and L. K. McDougal.** 1983. Successful use of broth microdilution in susceptibility tests for methicillin-resistant (heteroresistant) staphylococci. J. Clin. Microbiol. **18:**1084–1091.
68. **Wilson, P. D., E. A. Salvati, P. Aglietti, and L. J. Kutner.** 1973. The problem of infection in endoprosthetic surgery of the hip joint. Clin. Orthop. Relat. Res. **96:**213–221.

Chapter 16

Streptococci and Aerococci

RICHARD R. FACKLAM AND ROBERTA B. CAREY

CHARACTERIZATION

Description

The streptococci and aerococci belong to the family *Streptococcaceae*, which is made up of gram-positive, cytochrome-negative, coccoidal bacteria that usually grow in chains of various lengths but sometimes form tetrads. Under certain growth conditions, these organisms are elongated and therefore appear rodlike when Gram stained. Although morphological characteristics are useful, they cannot always be used to differentiate between members of the families *Streptococcaceae* and *Micrococcaceae*. The most definitive difference between the two families is that catalase is present in cultures of *Micrococceae* and absent in those of *Streptococcaceae*. Unfortunately, simply adding hydrogen peroxide to an agar slant culture is not specific for catalase, and therefore the occasional peroxidase-producing *Streptococcus* or *Aerococcus* species and the catalase-negative *Micrococcus* and *Staphylococcus* species are not differentiated with the relatively simple catalase test. They can, however, be differentiated with the benzidine test. A positive benzidine reaction indicates that the strain contains cytochromes and thus is a member of the *Micrococcaceae*. A negative benzidine reaction indicates that a gram-positive coccus is a member of the *Streptococcaceae*.

In *Bergey's Manual of Determinative Bacteriology* (29), the *Streptococcaceae* are divided into five genera: *Aerococcus, Gemella, Leuconostoc, Pediococcus*, and *Streptococcus*. The taxonomic status of the genera *Gemella* and *Pediococcus* is, however, uncertain. *Gemella* species may actually be viridans streptococci. Sufficient information is not available to differentiate between *Gemella hemolysans* and *Streptococcus morbillorum* (40). *Pediococcus* species are probably either aerococci or streptococci (111). *Leuconostoc* strains have not been found in humans. Therefore, the *Streptococcaceae* isolated from humans need only be identified as streptococci or aerococci (40).

Streptococci are facultative with respect to oxygen. Many strains grow better anaerobically than aerobically and thus are sometimes called "microaerophilic" or "aerotolerant anaerobic." Because the medically important streptococci and aerococci are homofermentative, clinical laboratories with gas-liquid chromatography equipment can use an analysis of spent media to determine whether the end product of glucose fermentation is primarily lactic acid (if so, homofermentation is taking place, and the genus is *Streptococcus* or *Aerococcus*; if not, heterofermentation is taking place, and the genus is *Peptococcus*) (51). Unfortunately, agreement has not been reached on the proper way to classify exceptional strains such as homofermentative, anaerobic gram-positive cocci. Streptococci are oxidase negative. This property, together with the Gram stain characteristics and cellular morphology, differentiates streptococci from *Neisseria* spp. *Neisseria* spp. are also catalase positive. Some streptococcal species contain specific polysaccharide antigens. These group-specific antigens are useful in the differentiation and classification of streptococci. Antigenic analysis of the aerococci has not been reported. Not all streptococci found in pathological conditions have group-specific antigens, and other methods of identifying these strains are necessary. The different types of hemolytic action are an aid in the identification of species. There is a wide range of physiological characteristics among streptococci. Microbiologists can take advantage of useful physiological characteristics to help identify the pathogenic streptococci.

A combination of antigenic, hemolytic, and physiological characteristics must be determined to identify the pathogenic streptococci and aerococci.

Clinical significance

Humans are the natural reservoir for the beta-hemolytic group A streptococcus, *S. pyogenes*. Transmission from person to person is frequently associated with close contact with an asymptomatic carrier who is colonized in the nasopharynx, skin, vagina, or rectum (103). Contaminated food may also serve as a vehicle for infecting a large number of people. The group A streptococci may be serotyped according to their cell wall M and T antigens, which are helpful in tracing common-source outbreaks.

S. pyogenes is a known cause of pharyngitis, tonsillitis, sinusitis, lymphadenitis, pyoderma, impetigo, cellulitis, erysipelas, bacteremia, arthritis, osteomyelitis, endocarditis, and meningitis. Pneumonia is a rare occurrence that is associated with a preceding viral illness. In the child <4 years of age, upper respiratory infections may be subacute, with rhinorrhea as the only manifestation. School age children are acutely ill with fever, sore throat, exudative tonsillitis, and cervical adenitis. The carrier state develops in 25% of the patients despite appropriate penicillin therapy. Semi-quantitative cultures do not reliably differentiate true streptococcal infection from the carrier state (107). Antibiotic therapy given early in the course of the disease may abort a rise in anti-streptolysin O or anti-DNase B titers, so that failure to demonstrate a serological response does not exclude a streptococcal etiology (105). Erythrogenic strains produce scarlet fever, in which the streptococcal infection is accompanied by a rash. Antiserum injected intradermally in the area of the rash produces a blanching response (the Schultz-Charlton reaction) that is diagnostic for scarlet fever (17).

Approximately 5 weeks after a pharyngeal infection with *S. pyogenes*, rheumatic fever may occur. The clinical manifestations include carditis, polyarthritis, chorea, erythema marginatum, and subcutaneous nodules. Unlike rheumatic fever, acute glomerulonephritis may follow either a pharyngeal or a skin infection with group A streptococci. A limited number of M and T serotypes are recognized as nephritogenic strains and are responsible for the majority of cases of acute glomerulonephritis. The latent period for nephritis is 10 days after throat infections and 18 to 21 days after skin infections. Hot, humid weather favors pyoderma, which is usually located on exposed skin surfaces and is preceded by minor skin trauma. In patients with skin infections, elevated anti-DNase B titers occur more frequently than a rise in antistreptolysin O titer (30). The pathogenesis of rheumatic fever and glomerulonephritis is unknown. Cell-mediated and humoral autoimmune mechanisms, as well as a direct toxic effect of the bacteria on the host tissue (55), have been suggested. While a decline in the incidence and the severity of group A streptococcal disease and its sequelae has been reported in the United States and western Europe (81, 107), the disease and its effects remain a major problem in the developing nations of the world, where overcrowding and poor hygiene are more prevalent.

S. agalactiae, first recognized as a pathogen in cattle, is currently one of the most significant human pathogens in the neonatal period. The organism is harbored in the genital and gastrointestinal tracts of 25% of healthy human adults. Because of the high pharyngeal carriage rate, its significance as a cause of pharyngitis is uncertain (43, 77). The most common mode of acquisition by the neonate is exposure to the maternal genital flora in utero through ruptured membranes or by contamination during passage through the birth canal. Nosocomial transmission has been documented, especially in hospitals with high maternal colonization rates (75). The incidence in the United States is 2 to 5 of every 1,000 live births (44). Early-onset disease (at <7 days of age) occurs predominantly in low-birth-weight babies who are born to mothers with obstetric complications. The initial clinical presentation includes respiratory distress due to pneumonia and also sepsis. The mortality rate is 55%. When infection occurs later (at >7 days of age), it usually presents as meningitis with a mortality rate of 14 to 23% (1, 77).

The group B streptococci elaborate a group-specific carbohydrate that can be detected in the body fluids of infected persons. For the rapid diagnosis of group B streptococcal infection (Chapter 90), cerebrospinal fluid, serum, and urine can be examined for the presence of bacterial antigen. The group B streptococci are classified into four serotypes (Ia, Ib, II, and III) which are based on capsular polysaccharide antigens. A protein antigen, formerly called Ic and now called c, is used to further characterize some strains, especially Ia/c and II/c. An equal distribution of all serotypes causes early-onset disease; however, 89% of the late-onset infections are caused by serotype III strains (112). Low levels of maternal antibody to the type III polysaccharide antigen correlate with increased susceptibility of the neonate to disease (6). Immunoprophylaxis and chemoprophylaxis are being evaluated as possible methods of preventing neonatal disease (44).

Unusual manifestations of group B streptococcal disease in the infant and young child include otitis media, ethmoiditis, conjunctivitis, abscess, pericarditis, omphalitis, and osteomyelitis (1, 54). The organism is an opportunistic pathogen in adults who are compromised by chemotherapy, diabetes mellitus, liver disease, or pregnancy. The spectrum of group B infections in the adult includes postpartum infection, bacteremia, pneumonia, empyema, pyelonephritis, septic arthritis, gangrene, osteomyelitis, endocarditis, and meningitis (63, 77).

The four species of beta-hemolytic group C streptococci are primarily important as veterinary pathogens. *S. equisimilis* is the species that most commonly affects humans. It has been isolated from the throat, nose, vagina, and rectum of asymptomatic carriers and cited as a pathogen causing puerperal sepsis, pharyngitis, cellulitis, pneumonia, osteomyelitis, bacteremia, endocarditis, brain abscess, and meningitis (93, 102). *S. zooepidemicus*, which causes septicemia in cattle, horses, sheep, foxes, and guinea pigs, was isolated from two outbreaks of human pharyngitis that were followed by cases of glomerulonephritis. In both instances the organism was most likely acquired by the consumption of unpasteurized milk (9, 32). Nephritis has also been documented as a sequela of group C skin infection (80), but rheumatic fever has not been reported after either respiratory or skin infection with these organisms. *S. equi*, which causes an acute, contagious respiratory disease known as strangles in horses, is rarely found in humans. A minute-colony form of beta-hemolytic group C streptococci (*S. milleri* or *S. anginosus*, group C) is also found in human infections.

The group G streptococci have been described as human and animal pathogens. These organisms are found as normal flora of the vagina, gastrointestinal tract, and skin. Pharyngeal carriage has been reported in 23% of healthy persons and in 12% of those with pharyngitis; however, the nonsuppurative sequelae associated with the group A streptococci have not been documented for group G streptococci. Serious group G infections include puerperal and neonatal sepsis, otitis media, pneumonia, empyema, peritonitis, cellulitis, and meningitis. Bacteremia and endocarditis are more likely to occur in patients with malignancies or preexisting valvular disease (4, 12).

The group D streptococci are normal inhabitants of the human gastrointestinal tract, and they may spread from this site to cause bacteremia, cholecystitis, and wound infections. Approximately 10% of urinary tract infections and 20% of cases of endocarditis are caused by these organisms. The medically important group D streptococci are divided into the enterococcal species (*S. faecalis*, *S. faecium*, and *S. durans*) and the nonenterococcal species (*S. bovis*). In the past, emphasis has been placed on differentiating these two groups, for it was assumed that the nonenterococcal strains were unequivocally susceptible to penicillin, whereas the enterococci required both a penicillin and an aminoglycoside for the treatment of serious infections. This distinction cannot reliably predict the susceptibility pattern of the organism, because occasional *S. bovis* strains are relatively resistant to the lethal action of penicillin and other antibiotics (83, 89).

When septicemia with the group D streptococci occurs in the neonatal period, its presentation may be identical to early-onset group B streptococcal disease. Anatomical defects of the central nervous system, neurological intervention, endocarditis, and urinary tract infections with the group D streptococci may predispose the patient to meningitis, which has a mortality rate of 33% (10). Sepsis with *S. bovis* is highly associated with colonic carcinoma, and the recovery of this organism may serve as the first indication of underlying gastrointestinal lesions (57). *S. equinus*, which colonizes the alimentary tract of horses, and *S. avium*, commonly isolated from chicken feces, rarely cause infections in humans.

The group F beta-hemolytic streptococci and other beta-hemolytic streptococci that form minute colonies are called *S. anginosus* in this text but are termed *S. milleri* by the British taxonomists. In some of the medical literature, these streptococci are considered a cause of serious suppurative infection. Clinical syndromes include abscess formation in the major organs and dentition, purulent involvement of the body cavities, bacteremia, endocarditis, and osteomyelitis (70). Significant underlying disease or local tissue injury has been reported in most patients who were infected with these organisms (92).

Human infections rarely occur with the beta-hemolytic streptococci belonging to groups E, L, M, U, and V, although their importance as veterinary pathogens has been well noted. Streptococci possessing group H, K, and O antigens are not beta-hemolytic and are included in the viridans group. Streptococci possessing group R, S, and T antigens are collectively known as *S. suis*. Although these organisms are primarily associated with disease in swine, the group R streptococci have been known to cause meningitis and sepsis in humans (21, 113).

S. pneumoniae is the most frequent cause of otitis media and bacteremia in infants and children. It is the major cause of community-acquired bacterial pneumonia, with approximately a half-million cases a year reported in the United States. The pneumococcus is a common cause of meningitis, with an incidence of 1.5 per 100,000 persons per year. The organism is frequently isolated from the upper respiratory tract of healthy individuals, especially preschool children (56). The 83 serotypes are based on the polysaccharide capsular antigens. Pneumococcal vaccines are available that contain 12, 14, or 23 antigens of the most common serotypes. These types represent 68 to 97% of the strains causing disease (18, 82). Infants have the highest rate of pneumococcal infection; however, individuals at risk also include those who are postsplenectomy or who have asplenia, Hodgkin's disease, sickle cell disease, chronic bronchitis, or obstructive pulmonary disease. *S. pneumoniae* has been isolated from patients with conjunctivitis, sinusitis, mastoiditis, pericarditis, occult bacteremia, arthritis, and endocarditis.

Facklam (36) has designated 10 species for the heterogeneous group of alpha- and nonhemolytic streptococci, known as the viridans streptococci. Colman and co-workers (23, 24) describe eight species of viridans streptococci. Since species designations are different in each classification system, these streptococci may bear more than one name in the literature. The viridans streptococci are normal flora of the human oral cavity, and some species, such as *S. mutans*, play an important role in the production of dental caries and plaque. In the 1970s, the streptococci were responsible for 58% of all subacute bacterial endocarditis, which is a decrease from previous decades. Of all streptococcal endocarditis, 50% is attributed to the viridans group; the beta-hemolytic streptococci and *S. pneumoniae* are rarely isolated from endocarditis patients at the present time. The nutritionally variant viridans streptococci make up 5 to 6% of the streptococcal endocarditis isolates (83). Also known as thiol-requiring, vitamin B_6-dependent, or satelliting streptococci, these nutritional variants are part of the normal throat flora and have been isolated from pleural fluid, pancreatic abscesses, and the blood cultures of patients with postpartum sepsis or abortion. Because the viridans streptococci are infrequently identified to the species level, it is difficult to associate a particular species with a unique pattern of disease.

Aerococci, which resemble the group D streptococci in their biochemical reactions, have been isolated from patients with urinary tract infections or endocarditis (22, 76). More often the aerococcus has been cited as an environmental contaminant.

COLLECTION, TRANSPORT, AND STORAGE OF SPECIMENS

Methods of collection

Throat. The technique of swabbing the throat is as important in the isolation of streptococci as is cultivation of the specimens obtained. The two most common pitfalls that result in inadequate specimens are (i) swabbing the tongue or uvula tissues rather than the pharynx; and (ii) inadequately exposing the pharynx. The pharynx must be adequately exposed and illuminated. The tonsils and pharynx should be rubbed with a cotton- or Dacron-tipped applicator (swab), and the tongue and uvula tissues should be avoided. Any exudate should be touched with the swab.

Nose. Nasal cultures should be taken with a sterile, cotton-tipped, flexible wire. The swab may be moistened with sterile water or saline before it is introduced into the patient's nose. The tip of the nose is raised with one hand, and the swab is introduced gently along the floor of the nasal cavity, under the middle turbinate, until the pharyngeal wall is reached. Force should not be used; if any obstruction is encountered, the nasopharyngeal culture cannot be taken on that side.

Skin. Culture specimens are best obtained from skin lesions by removing the crusts of the pustule or vesicle cap. The sterile swab should be firmly rubbed into the lesion. This rubbing may cause the patient some discomfort, but the procedure is necessary to ensure the maximum recovery of streptococci.

Wound cultures should be treated in the same manner. If the lesions or wounds are dry, a moistened swab should be used.

Blood, cerebrospinal fluid, sputum, urine, and other body fluids. The methods used for the collection and processing of blood, cerebrospinal fluid, sputum, urine, and other body fluids are described in Chapter 8.

Transport

The method of transporting the swab to the bacteriology laboratory depends upon (i) the source of the specimen, (ii) the length of time the specimen is expected to be in transit, and (iii) the philosophies of the physician submitting the specimen and the laboratory director, that is, which bacteria they consider upper respiratory tract pathogens. Before laboratory procedures are begun, the submitting physician and the laboratory director should decide on the extent of bacteriological examination needed for each specimen from each source. Special procedures and media may be used in some situations, but not in others. If the specimen is taken from the throat, only the beta-hemolytic streptococci and *Corynebacterium diphtheriae* are usually considered pathogens. On occasion, *Haemophilus influenzae* is considered pathogenic, especially in young children, and some physicians request special bacteriological workups of specimens taken from patients at risk.

If no more than 2 h is expected to elapse between the time the swab specimen is collected and the time it is examined in the laboratory, no special precautions need be taken. The streptococci survive well in a dry environment. The swab is returned to the paper envelope or a sterile test tube for transit to the laboratory. If the swab is not to be processed until the next day or if other pathogens, such as those from wound infections, must be considered, a holding medium (for example, Stuart or Amies medium) should be used. If the swab is to be in transit for more than 1 day, then the silica gel or the dry filter paper transport system should be used. These systems can be used for both throat and skin swabs. The materials needed for these systems are available commercially (Carter, Rice, Storrs & Bement, Inc., East Hartford, Conn.). A modified silica gel transport system can be made by placing enough silica gel crystals in a screw cap tube (15 by 125 mm) to cover the cotton tip of the swab. The tube and crystals are autoclaved and dried in a hot-air oven.

Storage

Most streptococci survive for several months on tightly capped blood agar slants stored at 4°C. The exceptions are the pneumococcal and some viridans strains that do not survive more than a week. None of the streptococcal strains survives well in broth cultures; some strains die after only 3 or 4 days. All streptococci survive 1 to 2 years frozen in blood at −70°C and 20 or more years lyophilized. Sand desiccation can also be used to store most strains (pneumococci do not survive) for 20 or more years.

DIRECT EXAMINATION

Gram stains

Direct examination of throat, nose, and skin specimens is of little value in the identification of pathogenic streptococci. Nonpathogenic streptococci are normal inhabitants of these areas and do not differ from the pathogenic streptococci in their staining characteristics or cellular morphology. However, Gram-stained smears of blood, cerebrospinal fluid, and other body fluids are of some help in the identification of the pathogen as a streptococcus.

Gram-stained smears of sputum are of value in the identification of pneumococci when increased numbers of polymorphonuclear leukocytes are also noted. The sputum should be gently homogenized with 1 to 2 ml of sterile saline. This homogenization can be done by refluxing the sputum-saline mixture in a small syringe without a needle attached. Sputum so treated should be placed on a glass slide, air dried, heat fixed, and stained by the Gram technique. Gram-positive cocci found singly, in pairs, or in short chains are indicative of pneumococci.

Quellung reactions

Pneumococci can also be identified by directly examining body fluids with the Quellung test. This procedure is described below in the section on serological identification.

Antigen detection

Throat cultures. Studies have shown that the group A streptococcal antigen can be detected in throat swabs containing sufficient numbers of group A streptococci (46, 97). A 5-min micro-nitrous acid extraction procedure is used in combination with a slide coagglutination (COA) group A antibody reagent (Phadebact) to yield identification within 20 to 30 min of obtaining the swab. The procedure is outlined in the Appendix to this chapter. The results of studies comparing this procedure with conventional procedures have shown a high specificity (95%) but somewhat lower than desired sensitivity (75%) (46, 97). The rapid extraction procedure does not detect group A antigen on swabs containing low numbers of streptococci. Although no published data are available for similar studies with swabs of skin or nasal origin, it can be assumed that the same results will be observed.

Several commercial kits for the direct detection of group A antigen from throat swabs are, or soon will be, available. These commercial kits contain materials for the extraction of the group A streptococcus antigen from the swab by enzymatic or chemical means. Generally, the enzymatic extractions require longer times (30 to 60 min) than chemical extractions (2 to 10 min). The antigen detection system included in the kits is either an enzyme-linked immunosorbent assay or a group of agglutination (latex or COA) reagents. The agglutination tests are generally less complex and require less time to perform than the enzyme-linked immunosorbent assay. Data are available that compare the Directigen Group A Strep Test (Hynson, Westcott & Dunning) and the Culturette Brand Ten-Minute Group A Strep ID Technique (Marion Scientific) with conventional blood agar plate methods for detecting group A streptococci from throat swabs (66, 67, 98). Generally, the direct antigen tests compared very favorably with the conventional identification tests for the detection of group A streptococci. Specificities of 98 to 99%, sensitivities of 91 to 95%, and overall correlations of 95 to 98% compared with the conventional procedures have been reported for the direct antigen tests (66, 67, 98). Each of these procedures has advantages and disadvantages. Before one of them is adopted, be sure to evaluate the complexity and turn-around time to ensure that these procedures will fit the needs of the laboratory. Obtain and study the package insert instructions for each of the products under consideration.

Urogenital swabs. Studies have shown that group B streptococcal antigen can be extracted from cervical, uterine, and placental swabs in the same manner as group A streptococcal antigen is extracted from throat swabs (96).

Blood cultures. Growth in blood cultures can be identified if Gram stain characteristics are determined along with the streptococcal antigenic characteristics. Group A, B, C, D, F, and G streptococci can be identified by the procedure of Wellstood (109) or Wetkowski et al. (110). Caution must be exercised in the interpretation of positive slide agglutination reactions of such cultures because cross-reactions may occur; i.e., pneumococci react with group C reagent, and some viridans streptococci may agglutinate with group A, C, F, and G reagents. Generally these procedures yield identification within 2 to 4 h of observing growth.

Cerebospinal, peritoneal, and pleural fluids. Several commercially available kits contain reagents for the detection of the common bacteria that cause meningitis, including *S. pneumoniae, S. agalactiae, Neisseria meningitidis,* and *H. influenzae.* These reagents detect the carbohydrate capsular antigens of these bacteria. Several kinds of test methods are available, including COA (31), latex agglutination (LA) (13), and counterimmunoelectrophoresis (13, 61). The technical details of the procedures should be obtained from the package insert of the manufacturer.

CULTURE AND ISOLATION

Recommendations for primary throat cultures

Streptococci are fastidious in their nutritional requirements. Enriched infusion agar and broth, such as tryptic soy, heart infusion, Todd-Hewitt, or proteose peptone, should be used. These media are free of reducing sugars, substances that influence the expression of beta-hemolysis by streptococci. The pH of the medium should be 7.3 to 7.4. For some strains of *S. faecalis* (group D), different animal bloods affect hemolysis. These strains are beta-hemolytic on horse, human, and rabbit blood agar but alpha-hemolytic on sheep blood agar. Microscopic examination of the area surrounding growth is necessary to determine the type of hemolysis, regardless of the blood source. If human blood is used, each lot must be tested with control group A strains to ensure that the inhibitors present in some samples (such as type-specific antibodies, antistreptolysin O antibodies, antibiotics, and citrates) inhibit neither the growth nor the hemolysins of the organisms. Because colonies of *Haemophilus haemolyticus* are indistinguishable from those of beta-hemolytic streptococci, sheep blood is recommended for throat cultures. This blood lacks sufficient amounts of pyridine nucleotides (V factor) to support the growth of *H. haemolyticus*.

Different concentrations of blood affect the size of the area of erythrocyte (RBC) destruction (zone size) and may affect the decision as to the type of hemolysis. If streak plates are used, lower concentrations of blood may make it difficult to distinguish alpha- from beta-hemolysis. Higher concentrations of blood may cause beta-hemolytic strains to appear nonhemolytic, unless the agar is cut or stabbed with inoculum. The best blood agar plates for primary isolation contain 5% defibrinated blood in agar approximately 4 mm deep.

Pour-streak plate

1. Place a throat swab in 1 ml of broth, and incubate it at 37°C for 2 h. Specimens received in the laboratory within 2 to 4 h after they are taken may be cultured after 0 to 2 h of incubation in broth. Specimens that have been in transit for 4 to 8 h should be incubated in broth for a minimum of 2 h, whereas those in transit over 8 h should be incubated in broth for 4 to 5 h.
2. Remove the swab from the 1 ml of broth, drain it against the inside of the tube, and place the swab in a sterile tube.
3. Melt 15 to 20 ml of blood agar base in a tube and hold it in a water bath at 50°C.
4. Add 0.8 to 1.0 ml of sterile defibrinated blood to the melted and cooled agar.
5. Pick up a loopful of broth containing swab washings, drain the loop against the side of the tube, and then transfer the inoculum to the blood agar tube.
6. Mix the contents thoroughly, flame the tube lip, and pour the inoculated medium into a sterile petri dish.
7. When the agar is hard, rotate the specimen swab over a small section of the surface. With an inoculating loop, spread the inoculum over half of the plate, and crosshatch the plate for isolation. Stab into the agar after each crosshatch series.

Overnight broth swabs

1. Insert the specimen swab into a tube of broth, incubate the culture overnight, and then mix the contents of the tube to get an even suspension of organisms.
2. Transfer 1 loopful of broth culture to 15 ml of sterile saline and mix well. (If growth is light, it may be necessary to use 2 to 3 loopfuls.)
3. Follow the procedure for steps 3 through 7 of the pour-streak plate procedure described above, except in step 5 use a loopful of the saline dilution, not a drained loop.

Streak plate. Swabs may be cultured on blood agar plates immediately after collection or after any period of enrichment or selective enrichment. If the swab has been enriched, press the swab against the wall of the tube to remove excess moisture before transferring the inoculum to the agar plate. Firmly roll the swab over one-sixth of the plate. Use a sterile wire loop to crosshatch the remainder of the plate. Stab the agar several times with the wire loop. Make two or three stabs in an area of the plate that has not been streaked. The wire loop does not have to be resterilized at any stage of the streaking or stabbing of the plate.

Recommendations for primary cultures other than throat

Sputum. Sputum should be homogenized as described above in the direct examination section. A loopful of the sputum should be streaked onto the surface of the blood agar infusion base medium as described above for streak plates.

Other body fluids. Other body fluids, such as blood, cerebrospinal fluid, etc., may be processed by either the pour plate or the streak plate method. Use a loopful of the body fluid and proceed as described above.

Incubation atmospheres and temperatures

Pour plates. Incubate pour plates under any atmosphere. They are anaerobic by the nature of their preparation.

Streak plates. Streak plates for pneumococci should be incubated in a candle extinction jar or a CO_2 incubator with 5 to 10% CO_2.

Preferably, streak plates for streptococci should be incubated under anaerobic conditions with 5 to 10% CO_2 and 85 to 90% N_2. These plates can be incubated under normal aerobic conditions with very little loss in the recovery rates of group A streptococci. This procedure, however, has not met with equal success in all laboratories (33, 72, 79). Some non-group A streptococci may fail to grow in normal atmospheres. If stabs are made with a wire loop in the streak plates, these plates can be incubated in candle extinction jars or CO_2 incubators. If this last method is used, the stabbed area of the plate must be used to determine streptococcal hemolysis as described below.

Incubation temperature. All plates and broths should be incubated at between 35 and 37°C.

Other culture methods

When quantitative information is not sought, enrichment, selection, or selective-enrichment techniques are being used with increasing frequency for the primary isolation of streptococci. There are contradictory opinions as to whether patients should be treated regardless of the numbers of infecting organisms. If colony counts on primary isolation plates are needed, enrichment is unnecessary. If, on the other hand, the objective is to detect even low numbers of streptococci, enrichment (incubation of the inoculum in broth), selection (incubation in an environment more conducive to the growth of streptococci than to unwanted organisms), or selective enrichment may be advantageous.

Enrichment. Any of the aforementioned infusion broths can be used for enrichment. Group A and B streptococci can be identified by immunofluorescence within 4 h, even in the presence of large numbers of contaminating organisms. Otherwise, swabs can be incubated in Todd-Hewitt broth overnight. Blood agar plates are then inoculated with the broth culture, and if contaminants overgrow the culture on nonselective medium, selective blood agar (i.e., Columbia colistin [10 mg/ml]-nalidixic acid [15 mg/ml], phenylethyl alcohol [0.25%], or colistin [10 mg/liter]-oxolinic [5 mg/liter]) can be used for the isolation of the streptococci (27, 34, 78).

Selective technique. Because high concentrations of some inhibitors (e.g., gentamicin) may also inhibit streptococcal growth, a nonselective plate should be used simultaneously with the selective media. The source of the specimen may dictate which selective agent to use. For example, inhibitors of gram-negative rods (neomycin [30 μg/ml] or nalidixic acid [15 μg/ml] and polymyxin [10 μg/ml]) are not usually necessary for throat swabs, but such inhibitors are very useful for rectal, vaginal, or wound specimens. Conversely, crystal violet (1 μg/ml), an inhibitor of staphylococci, which are often found in the throat, would be useful in streptococcal throat swab cultures. Sulfamethoxazole (23.75 μg/ml)-trimethoprim (1.25 μg/ml) (SXT) has also been used in tryptic soy agar to inhibit staphylococci, viridans streptococci, and gram-negative rods (50). SXT has been used in primary blood agar plates with various degrees of success (33, 60). Investigators attempting to recover beta-hemolytic streptococci have usually reported higher isolation rates with gentamicin (5.5 μg/ml in Columbia agar or 5.0 μg/ml in tryptic soy agar) blood agar than with nonselective media (11, 72). The investigators reported that the improved recovery rate was due to better growth of non-group A rather than of group A streptococci. Selective agar (tryptic soy agar) containing gentamicin (5.0 μg/ml) has been reported to substantially improve the recovery rates of pneumococci from the oropharynx (25).

Selective-enrichment technique. As the name suggests, selective-enrichment broths provide the advantages of enrichment and selection by providing optimal conditions for streptococcal growth while inhibiting the growth of competitors. The most frequently used selective agents have been sodium azide (1:16,000) to inhibit gram-negative rods and crystal violet (1:500,000) to inhibit staphylococci. Several modifications of this broth have been described (74, 88). Todd-Hewitt broth has been modified for the selection of group B streptococci (nalidixic acid [15 μg/ml] and gentamicin [8 μg/ml] or nalidixic acid [15 μg/ml], polymyxin [1 μg/ml], and crystal violet [0.1 μg/ml]). A common mistake in the preparation of selective-enrichment broths is the use of the same concentration of inhibitors for broth as for selective agar. The fact that some inhibitors diffuse more widely in broth may dictate that their concentrations be reduced (47).

Regardless of whether an enrichment, selection, or selective-enrichment technique is used, there is no substitute for a well-prepared streak plate with stabs into the agar or a pour plate for determining hemolysis from the primary plates.

IDENTIFICATION

Recognition of the colonies

Typically, after 18 to 24 h of incubation on blood agar, the colonies of group A streptococci are about 0.5 mm in diameter, transparent or translucent, and domed; they have a smooth or semimatt surface and an entire edge. They are surrounded by a well-defined zone of complete hemolysis, usually 2 to 4 times the diameter of the colony; however, considerable variations occur. The appearance of the colonies depends markedly on the medium used and, to some extent, on the atmosphere of incubation. All colonial characteristics are not manifested on a single medium or in a single atmosphere. Subsurface colonies also vary. Some colonies are lancet shaped, whereas others are oval or round. The appearance of surface or subsurface beta-hemolytic group C or G streptococcal colonies does not differ sufficiently from that of group A colonies to be of any value in identification. Group B streptococcal colonies may be somewhat larger than group A colonies, but both are smooth, with entire edges. The group B colonies are surrounded by a much smaller zone of complete hemolysis, and some strains do not lyse RBCs at all. Group D streptococcal colonies are somewhat larger than group A colonies on the surface of blood agar (0.5 to 1.0 mm). The group D colonies are less opaque, and some strains are glossy white, resembling staphylococcal colonies on blood agar. Group D streptococci may exhibit beta-, alpha-, or no hemolytic action of RBCs. The beta zones produced by the hemolytic action of group D streptococci are usually larger than the beta zones produced by other streptococci. Group F streptococci generally form "minute colonies." Zones of hemolysis similar in size to those produced by group A streptococci surround these minute colonies. This characteristic has little diagnostic value, however, since some strains of groups C and G and even some strains of group A also form minute colonies.

Some strains of group B streptococci form dull, brick-red, pigmented colonies after anaerobic incubation. Again, this characteristic is of little value for identification, because not all strains of group B streptococci form this pigment. The most important cultural characteristic for the recognition of streptococci is their hemolytic action on the RBCs in the medium.

The viridans streptococcal colonies vary in size from pinpoint (0.1 mm) to equal to or larger than the colonies of group A streptococci (0.5 mm). The colonies are usually considerably smaller than those of the pneumococci. They may appear mucoidal and translucent or glossy and nontranslucent. The colony size and appearance are affected by medium composition and the atmosphere of incubation. The colonies may be surrounded by a small zone of alpha-hemolysis (partial destruction of RBCs) or no zone of hemolysis. Under anaerobic incubation, viridans streptococci are usually nonhemolytic. Pneumococcal colonies are round with entire edges, mucoid, and about 1 mm in diameter. When the culture has been incubated in candle extinction jars or CO_2 incubators, the colonies are surrounded by a fairly large zone of alpha-hemolysis. Microscopic examination (magnification of ×40 to ×50) of the colonies is a useful aid in differentiating the pneumococci from the alpha-hemolytic streptococci. Young pneumococcal colonies are raised (like alpha-hemolytic streptococci), but as the culture ages, the colonies become flattened and the central part of each colony may become depressed (unlike alpha-hemolytic streptococci). The colonies of aerococci are similar to those of the viridans streptococci, although some strains may form colonies nearly equal in size to colonies of group D streptococci. The colonies are usually surrounded by a zone of alpha-hemolysis, although some strains are nonhemolytic.

Hemolysis

Hemolysis is the most useful characteristic for the identification of streptococci. The hemolytic action of streptococci on RBCs was described and defined by Brown in 1919. These definitions are as follows.

(i) **Alpha-hemolysis.** An indistinct zone of partial destruction of RBCs appears around the colony, often accompanied by a greenish to brownish discoloration of the medium.

(ii) **Beta-hemolysis.** A clear, colorless zone appears around the streptococcus colonies, in which the RBCs have undergone complete discoloration.

(iii) **No hemolysis.** No apparent hemolytic activity or discoloration is produced by the colony.

(iv) **Alpha-prime- or wide-zone alpha-hemolysis.** A small halo or envelope of intact or partially lysed RBCs lies adjacent to the bacterial colony, with a zone of complete hemolysis extending farther out into the medium. When examined macroscopically, alpha-prime-hemolysis can be confused with beta-hemolysis.

Brown's observations were based on a microscopic examination of subsurface colonies in blood agar pour plates. Through the years, these definitions have been used to characterize colonies growing on the surface of streaked blood plates. This extended application has not been accomplished smoothly because of the character of the hemolysins responsible for beta-hemolysis and because of the misinterpretation of alpha-prime- as beta-hemolysis.

Among the streptococci, at least three distinct hemolysins are responsible for beta-hemolytic activity. The hemolysins of group A streptococci are differentiated on the basis of antigenicity and susceptibility to inactivation by oxidation. Streptolysin O is antigenic and oxygen labile; streptolysin S is nonantigenic and oxygen stable. Oxygen-sensitive streptolysin O can be reactivated in the presence of reducing agents.

Furthermore, streptolysin S is not produced in serum-free broth, and its production is inhibited in media rich in fermentable carbohydrate. When these restrictive properties are considered, it becomes obvious that aerobic incubation of streaked blood plates inhibits the hemolytic activity of streptolysin O and limits the characterization of beta-hemolytic streptococci to streptolysin S activity, which may vary from strain to strain. When streaked blood plates are incubated in the presence of atmospheric oxygen, the investigator imposes limitations upon the hemolytic expression of the organism, which could cause the beta-hemolytic characteristic to be overlooked.

Peroxide-producing beta-hemolytic streptococci can appear alpha-hemolytic on the surface on blood agar plates incubated aerobically or in atmospheres of increased CO_2. In addition, peroxide-producing alpha-hemolytic streptococci may inhibit the expression of beta-hemolysis produced by group A streptococci on the surface of blood agar plates (53, 62). This phenomenon is especially important to consider when determining hemolysis on primary isolation plates from throat swabs, because alpha-hemolytic streptococci are part of the normal throat flora and may obscure the potentially pathogenic beta-hemolytic streptococci.

If streak plates are used as the method of choice, at the very least, stab the streaked plate with the inoculating loop to get subsurface growth and to permit the detection of both O and S streptolysins. At a magnification of approximately ×60, beta-, alpha-, and alpha-prime-hemolysis can be differentiated around the subsurface growth in blood agar pour and stabbed plates.

Identification of major categories

The major differential characteristics of the streptococci and aerococci are listed in Table 1. Once isolated in pure culture, a catalase-negative, gram-positive coccus is tested for hemolysis in pour plates (preferable) or streak-stab plates. If the strain is beta-hemolytic, it is tested for serological reaction with group A, B, C, D, F, and G antisera. If the strain is not beta-

TABLE 1. Characteristics of value for differentiating the major categories of streptococci and aerococci

Category	Most common cellular arrangement	Hemolysis	Streptococcal group antigen	Hydrolysis of BE[a]	Growth in NaCl (6.5%)	Bile solubility
Beta-hemolytic "groupable" streptococci	Chains	Beta	A, B, C, F, and G	−	−*	−
Group D streptococci						
Enterococci	Short chains, diplo	Alpha, beta, none	D	+	+	−
Nonenterococci	Short chains, diplo	Alpha, none	D	+	−	−
Viridans streptococci	Chains	Alpha, none	None[b]	−*	−	−
Pneumococci	Diplo, short chains	Alpha	None	−	−	+
Aerococci	Single cells, tetrads	Alpha, none	None	v	+	−

[a] BE, Bile-esculin; v, variable; +, positive; −, negative; *, occasional exception.
[b] No group B or D antigens present; occasionally group A, C, F or G antigens present.

hemolytic, it is tested only with group B and D antisera. Viridans streptococci may contain cross-reacting antigens that react with group A, C, F, and G antisera, and thus non-beta-hemolytic streptococcal strains should not be tested for these antigens (36).

Streptococci that are neither beta-hemolytic nor members of group B or D are identified with physiological tests. Group D streptococci can also be identified presumptively with the bile-esculin (BE) test. The 6.5% NaCl tolerance test differentiates enterococci from group D streptococci that are not enterococci (37, 38). Viridans streptococci and pneumococci are not beta-hemolytic, do not react with group B or D antiserum, and do not grow in 6.5% NaCl broth. Pneumococci are soluble in bile salts, whereas viridans streptococci are not. The aerococci are not beta-hemolytic, do not react with streptococcal grouping antisera, are insoluble in bile, and grow in 6.5% NaCl broth. Some aerococci blacken BE and must be differentiated from enterococci on the basis of cellular arrangement and other distinguishing physiological characteristics.

Strains in each category in Table 1 are tested further for species identification (see below).

Serological grouping procedures

Nearly all the beta-hemolytic streptococci isolated from human infections possess specific carbohydrate antigens. These carbohydrate antigens are called streptococcal group antigens, and they can be demonstrated by a variety of techniques. Group-specific precipitating, agglutinating, and fluorescent-antibody (FA) sera, which can be used with extracts, cell suspensions, and spent broth media, are commercially available. Attempts to extend these procedures to the non-beta-hemolytic streptococci have been unsuccessful except for groups B, D, and N (36). Group N streptococci are not found in human infections. We recommend the use of group A, B, C, D, F, and G antisera for the identification of beta-hemolytic streptococci and the use of only group B and D antisera for identifying non-beta-hemolytic streptococci.

Procedures for the extraction of streptococcal antigens are outlined in the Appendix. These procedures include the Lancefield hot-HCl, hot formamide, autoclave, nitrous acid, *Streptomyces albus* enzyme, pronase B enzyme, and *S. albus*-lysozyme extraction procedures.

Each extraction method has certain advantages and disadvantages. For example, the Lancefield hot-acid technique is the standard for grouping and must be used for typing group A and B streptococci. It is the only technique available for extracting the protein type-specific antigens as well as the carbohydrate (group A, B, C, F, and G) and teichoic acid (group D and N) antigens. However, it is somewhat more complex and time consuming than are other methods. The hot-formamide technique is also relatively complex and time consuming. Like the Lancefield method, it can be used to extract all group antigens; however, this technique cannot be used for typing group A and B streptococci because it destroys their protein type-specific antigens.

The autoclave technique is relatively simple and can be used for grouping. When used with the Centers for Disease Control (CDC) grouping antisera for the identification of group A, B, C, D, F, and G streptococci, the hot-acid, formamide, and autoclave techniques were equally effective.

The standard and micro-nitrous acid techniques are simple and effective for the identification of group A, B, C, F, and G streptococci. Although the technique is easy to perform, the *S. albus* enzyme extraction can be used to group only A, B, C, F, and G streptococci. The group D antigen is not extracted satisfactorily by the *S. albus* enzyme.

The pronase B enzyme extraction technique, like the *S. albus* enzyme technique, is easy to perform but does not extract group D or F antigen as satisfactorily as does the hot-acid, formamide, autoclave, or *S. albus*-lysozyme enzyme technique. In CDC tests, satisfactory extractions were obtained 100% of the time only for group A streptococci.

In contrast, the *S. albus*-lysozyme enzyme technique extracts the group antigen of group A, B, C, D, F, and G streptococci. The reagents for this technique are more expensive than are those for any of the other techniques.

Capillary precipitin test

The effectiveness of all extraction techniques depends largely on the quality of the antisera used in the precipitin test. With potent, specific antisera, all techniques work well within the limits of the extraction procedures described above. Control streptococcal strains should be used to test each new lot of commercial antiserum. Some are of notoriously poor quality. Cost, complexity, time requirements, and efficiency of extraction are all factors that influence the choice of extraction technique. The Lancefield precipitin technique originally involved layering the extract under the antiserum, a method that was satisfactory for hyperimmune antisera. Because CDC and commercial antisera are usually not as potent as those specified for the Lancefield procedure, however, we layer the extract over the antiserum. The procedure for performing the CDC capillary precipitin test is described in the Appendix.

Slide agglutination tests

Two slide agglutination tests in which carrier particles for the group-specific antisera are used have been described. The reagents for these tests are now available from several commercial sources. In the COA test, specially prepared protein A-rich staphylococcal cells conjugated to group-specific streptococcal antisera are used. Latex particles conjugated to group-specific streptococcal antisera are used in the LA test. The carrier particles in these tests (staphylococcal cells and latex particles) are so large that agglutination can be seen without the aid of magnification.

The most distinctive characteristic of these reagents is that they shorten the identification time over that of conventional extraction and capillary precipitation tests and yet are equal in accuracy. These reagents can be used to detect streptococcal group antigens directly on cells in culture, in extracts, or from spent broth medium. To ensure accurate results, the package insert instructions that accompany each product should be strictly followed. These instructions are generally simple and often offer alternative procedures to ensure accurate identification. The COA and LA reagents have been used in combination with the micro-nitrous acid extraction procedure to identify group A streptococci directly from throat swabs (46, 66, 67, 97, 98). The COA and LA reagents have been used to identify beta-hemolytic colonies from primary throat culture blood agar plates. In some cases single colonies (99) have been used, whereas other investigators have used either four or five beta-hemolytic colonies per grouping reagent (100) or mixed flora (20). When these direct testing procedures fail to identify the beta-hemolytic colonies, 4-h and overnight broth cultures and culture supernatants have been used as antigens. Most of these reagents are of high quality, and the choice of which one to use is a matter of personal preference (19, 49, 100). Because potential errors arise when group D streptococci are identified with the COA and LA tests, we recommend a battery of two physiological tests to aid in the identification of these streptococci. In addition, since enterococcal and nonenterococcal group D streptococci differ in their susceptibility to antibiotics, microbiologists must have a means of differentiating them. We recommend testing all suspected group D streptococci for BE reactions and tolerance to 6.5% NaCl broth (NaCl). These tests are described later in this chapter. To limit the number of BE and NaCl tests performed, we suggest that oropharyngeal isolates not be tested. Group D streptococci are seldom found in throat specimens and are of little clinical significance at that site. When sheep blood agar is used to determine streptococcal hemolysis, most group D streptococci are non-beta-hemolytic. Thus, the use of sheep blood for the processing of throat swabs eliminates most group D strains from consideration for serological testing. Non-beta-hemolytic streptococci are considered normal oral flora. If the culture is from tissue or from a body fluid, the BE and NaCl tests must be performed to minimize the potential errors in the COA and LA procedures.

Identification of groups A and B by IF staining

Until the recent advances in slide agglutination grouping of extracted throat and vaginal swabs, IF staining had advantages over conventional identification techniques (69). These advantages included rapidity (several hours rather than several days), sensitivity (detection of small numbers of streptococci in mixed cultures), and detection of nonhemolytic group B streptococci (often missed with conventional techniques). The IF reagent, or "conjugate," is composed of appropriate dilutions of anti-group carbohydrate immunoglobulin attached to fluorescein isothiocyanate. Nonspecific staining of *Staphylococcus aureus* or members of other streptococcal groups, which cannot be distinguished from group A streptococci morphologically, can be avoided by the addition to the conjugate of either unlabeled streptococcal group C antiserum or unlabeled nonimmune serum. The "blocking" of nonspecific staining is explained theoretically by the demonstrated nonimmune binding of unlabeled immunoglobulin (by the Fc portion) to staphylococcal protein A or certain streptococcal receptors (58), which can prevent the nonspecific binding of the IF reagent to these receptors (i.e., by the Fc of the conjugated antiserum).

The group B streptococcal IF procedure, described by Romero and Wilkinson (85), is less widely used because specific reagent is not commercially available. If all serotypes are to be stained, the group B conjugate must contain, in addition to unlabeled nonimmune serum (for the same reasons as above), both group B and type-specific fluorescein isothiocyanate-labeled antibodies. Specific group A and B conjugates should not contain antibodies reactive with the several R-protein antigens found among many streptococcal groups. For staining, streptococcal cells can be grown on blood agar, suspended in buffer, and either placed directly on glass slides or grown in Todd-Hewitt broth and then placed on glass slides. The latter enrichment is often preferred for pharyngeal or vaginal swabs because of the short incubation time (2 to 6 h) required for detection. Group B cells are usually washed only briefly to avoid the removal of surface antigens needed for correct identification. Specimens can be fixed by heat, ethanol, or Formalin. Although several modifications have been described, only the most commonly used procedure for group A and B IF staining is presented in the Appendix.

Serological identification of the pneumococci

Tentatively identified pneumococcal colonies can be serologically identified by the Quellung reaction. Cells from a single colony suspended in a drop of physiological saline or various body fluids (including cerebrospinal fluid, peritoneal fluid, transtracheal aspirates, and sputa) can be examined directly with the Quellung test, or a broth culture can be examined. Suitable infusion broths include tryptose-phosphate, Todd-Hewitt, and thioglycolate.

Quellung test

1. Place a small drop of culture, cell suspension, or body fluid on a glass slide.
2. Add a loopful (1 mm) of antiserum and mix well.
3. Add a small loopful of saturated aqueous methylene blue dye and mix.
4. Place a cover slip over the mixture, and after 10 min, examine the mixture microscopically with the oil immersion lens.
5. To avoid antigen excess, which may cause negative reactions, prepare the slides so that each microscopic field contains 50 to 100 cells.
6. To obtain the oblique illumination needed to examine the slide, adjust the iris diaphragm so that only about one-third of the light passes through the condenser at low power (×10). Minor modifications are given in detail by Austrian (5) and Lund (65). A positive Quellung reaction is the result of the binding of the pneumococcal capsular polysaccharide with type-specific antiserum. The corresponding change in the refractive index causes the capsule to appear swollen. Actually, it becomes more visible. The pneumococcal cell stains dark blue and is surrounded by a sharply demarcated halo which represents the outer edge of the capsule. The light transmitted through the capsule appears brighter than either the pneumococcal cell or the background of the slide. Single cells, pairs, chains, and even clumps of organisms may have positive Quelling reactions. Pneumococci are identified with the Quelling test and a battery of antisera. The Statens Serum Institute, Copenhagen, Denmark, produces polyvalent (omniserum) antiserum with 83 type-specific antibodies. The CDC, Atlanta, Ga., and the Statens Serum Institute produce polyvalent pooled antisera (nine pools, A through I). Each pool contains 7 to 11 type- or group-specific antisera, the latter containing from one to four types. Other commercial type, group, and pooled antisera are available, and the antibody composition of the pools is different in the different sources. Therefore, instructions supplied with each antiserum must be followed. Unless typing is required for epidemiological purposes, omniserum and pooled sera are used to identify pneumococci. The presence of cross-reactive antibodies in omniserum (5) means that colonial and Gram-stain morphology must also be relied upon.

Since serological grouping is the definitive procedure used to identify the beta-hemolytic streptococcal species, the results of the serological procedures can be reported as indicated in Table 2. That is, if a group A serological reaction is found, an identification of *S. pyogenes* can be reported; exceptions to this policy are discussed later. If a group B serological reaction is found, *S. agalactiae* can be reported. However, group C and D serological reactions are not species specific. To identify the three group C species, the investigator

TABLE 2. Differentiation of streptococci found in human infections by serological characteristics

Hemolysis and group	*Streptococcus* sp.
Beta-hemolytic (use group A, B, C, D, F, and G antisera)	
Group A	*S. pyogenes*
Group B	*S. agalactiae*
Group C[a]	*S. equisimilis*, *S. zooepidemicus*, or *S. equi*
Group D[b]	
Group F	*S. anginosus*
Group G	Lancefield group G
None	Beta-hemolytic; unable to demonstrate group antigen
Not beta-hemolytic (use group B and D antisera only)	
Group B	*S. agalactiae*
Group D[b]	
Not group B or D	*S. pneumoniae*[c], viridans sp.[b,d]

[a] *S. equisimilis* forms acid in trehalose but not sorbitol broth, *S. zooepidemicus* forms acid in sorbitol but not trehalose broth, and *S. equi* does not form acid in either trehalose or sorbitol broth. See Table 4 for the identification of the minute-colony form.

[b] See Table 3.

[c] Quellung positive with omniserum or pooled pneumococcal antisera. Bile solubility and optochin susceptibility can be substituted for the Quellung test.

[d] Quellung negative with omniserum or pooled pneumococcal antisera. Bile solubility and optochin susceptibility can be substituted for the Quellung test.

must test for the formation of acid in trehalose and sorbitol broth (see footnote *a* in Table 2). The group D and viridans streptococcal species are differentiated according to 18 physiological tests (Table 3). The group F streptococci are *S. anginosus*, but the group G streptococci do not have a specific epithet. They should be reported as Lancefield group G streptococci. If a positive Quellung reaction is obtained with omniserum or pneumococcal pool antisera, the report of *S. pneumoniae* may be made.

Physiological differentiation of group D and viridans streptococci and aerococci

Once an isolate is identified as a group D or viridans streptococcus or presumptive aerococcus, the physiological characteristics (Table 3) can be used to differentiate the various species. The formulas and methods for determining these characteristics are included in Chapter 111 of this Manual. Because some strains of each of the species listed in Table 3 react atypically in one or more of the tests, the "best fit" is determined, or as described by Deibel (28), a "spectrum analysis" is done. The atypical reactions are those listed as occasional exceptions (+* or −*) in Table 3.

Approximately 75% of the strains will have no atypical reactions; i.e., the test results will conform to those listed as positive or negative in Table 3.

Approximately 20% of the strains will react atypically in one test; the remaining 5% will react atypically in two or more tests and will therefore be difficult to identify at the species level (35, 36). It is absolutely

TABLE 3. Differentiation of group D and viridans streptococci and aerococci found in human infections[a]

Species	Hemolysis			Physiological test																
	Alpha	Beta	None	Bile-esculin	Growth in 6.5% NaCl	Growth at 10°C	Pyruvate	Arginine	Esculin	Starch	Hippurate	Sucrose	Lactose	Mannitol	Sorbitol	Arabinose	Sorbose	Inulin	Raffinose	Glucan
S. faecalis	+	+	+	+	+	+	+	+*	+	−	v	+*	+*	+	+*	−*	−	−	−*	N
S. faecium	+	−	+	+	+	+	−	+	+	−	v	+	+	+	−	+*	−	−*	+*	N
S. avium	+	−	+	+	+	−	+	−	+	−	−	+	+	+	+	+	+	+*	+	N
S. durans	+	+	+	+	+	+	−	+	+	−	v	−*	+	−	−	−	−	−	−*	N
Aerococci	+	−	−*	v	+	−	−	−	v	−	+*	+	+*	v	−	−	−	−*	v	N
S. bovis	−*	−	+	+	−	−	−	−	+	+	−	+	+	+	−	−	−	+*	+*	L
S. mutans	−*	+	+	−*	−	−	−	−*	+*	−	−	+	+	+	+	−	−	+*	+*	D
S. uberis	+	−	+	−	−*	−	−	−	+	−	+*	+	+	+	+	−*	−	+*	+*	N
S. intermedius	−*	−	+	−*	−	−	−	+*	+	−*	−	+	+	−	−	−	−	−	v	N*
S. bovis (var.)	−*	−	+	+	−	−	−	−	+	−*	−	+	+	−	−	−	−	−	v	N*
S. constellatus	+	−	−*	−*	−	−	−	+*	+*	−	−	+	−	−	−	−	−	−	−*	N
S. equinus	+	−	−	+	−	−	−	−	+	−	−	+	−	−	−	−*	−	−	−*	N
S. sanguis I	+	−	−*	−*	−	−	−*	+	+*	+*	−	+	+	−	−*	−	−	+	−*	D*
S. salivarius	−*	−	+	−*	−	−	−	−	+*	−*	−	+	+*	−	−	−	−	+	+*	L*
S. mitis	+	−	−*	−	−	−	−	−	−	−	−	+	+	−	−	−	−	−	−	N
S. sanguis II	+	−	−*	−	−	−	−	−*	−	−*	−	+	+	−	−	−	−	−	+	D*
S. morbillorum	−*	−	+	−	−	−	−	−	−	−	−	+	−	−	−	−	−	−	−	N
S. acidominimus	+	−	−*	−	−	−	−	−	−	−	+*	+	−	−	−	−		−	−	N

[a] Symbols: +, positive reaction; −, negative reaction; v, variable reactions; *, occasional exceptions occur. For glucans: D, dextran; L, levans; N, no glucans.

necessary to demonstrate that the *S. bovis* variant and *S. equinus* have the group D antigen (preferably by extraction and serogrouping). These two strains cannot be distinguished physiologically from *S. intermedius* and *S. constellatus*, respectively. Although serogrouping is preferable for the other group D species, some strains of *S. faecium, S. durans, S. avium*, or *S. bovis* can be identified on the basis of physiological characteristics if no or only one exceptional reaction is noted.

It is also necessary to demonstrate that aerococci do not possess the group D antigen, because their physiological characteristics resemble those of *S. durans*. No single characteristic is typical of the aerococci, which makes identifying them difficult. The two most nearly typical characteristics are the arrangement of cells in tetrads and their growth in 6.5% NaCl broth, but some streptococci react comparably. The tests listed in Table 3 can be used to identify the aerococci. They are usually alpha-hemolytic, grow in 6.5% NaCl broth but not in broth incubated at 10 or 45°C, and hydrolyze hippurate but not arginine. Atypical strains do not hydrolyze hippurate. Two species of aerococci may be found in humans (40), i.e., one that hydrolyzes hippurate and one that does not. On blood agar, the aerococci resemble the viridans streptococci, whereas in BE and salt tolerance tests, they resemble the group D streptococci. Most aerococci are susceptible to bacitracin, whereas most group D and viridans streptococci are not. This characteristic is a useful adjunct to those listed in Table 3.

Two major schemes for the identification and differentiation of the streptococci have evolved in the past few years (8, 23, 24, 36, 76), one from Great Britain and the other from CDC. There is general agreement between the schemes, but there are also areas of disagreement. Both systems recognize the typical group A, B, C, and G streptococci, including the four species within group C. However, agreement is not apparent with group F, the minute-colony groups A, C, and G, or the beta-hemolytic streptococci without group antigens.

The minute-colony forms of group A, C, and G streptococci appear identical to the beta-hemolytic group F streptococci and the beta-hemolytic streptococci without group antigens. The colony size is small (<0.5 mm in diameter) after 18 h of incubation in an increased CO_2 atmosphere. Many of the minute-colony group A, C, and G strains do not grow in normal atmosphere. As with the group F and the nongroupable beta-hemolytic strains, increased CO_2 concentrations are required in the incubation atmosphere for their growth. At CDC we have recognized all of the latter beta-hemolytic strains as separate entities, but not necessarily separate species. The beta-hemolytic group F, minute-colony groups A, C, and G, and beta-hemolytic strains without group antigens are all classified as *S. milleri* by the British taxonomists (8, 23, 76). The British include two additional species in the *S. milleri* classification. *S. constellatus* and *S. intermedius*, formerly *S. anginosus-constellatus* and *S.* MG-*intermedius*, respectively, are non-beta-hemolytic strains included in the CDC viridans classification. There is general agreement in recognition of the viridans species, other than *S. constellatus* and *S. intermedius*. Both systems have descriptions of *S. mutans*, *S. salivarius*, *S. sanguis*, *S. uberis*, and *S. acidominimus* (23, 36). The British taxonomists have used the terms *S. mitior* and dextran-positive *S. mitior*, whereas we have used *S. mitis* and *S. sanguis* biotype II, respectively, to describe two viridans species. Agreement is apparent in the overall description of these species, but the final tests for differentiation are different. The British taxonomists use dextran formation, whereas at CDC, we use acid formation in raffinose broth to differentiate *S. mitis* from *S. sanguis* biotype II (*S. mitior* and dextran-positive *S. mitior*, respectively). British taxonomists have not included *S. morbillorum* (*G. hemolysans*) in their reports. Both systems recognize the pneumococci and aerococci as distinct species and related genera, respectively.

Table 4 summarizes the taxons recognized in the *S. milleri* classification at CDC. We have also listed the major serological and physiological characteristics and a proposed identification that is consistent with the current list of accepted bacterial species (95). The three taxons *S. anginosus*, *S. constellatus*, and *S. intermedius* are included on that list; therefore, legitimately these taxons can be used. The taxon *S. anginosus* has been used to designate the beta-hemolytic group F streptococci and the beta-hemolytic strains without detectable group antigens (29). Subspecies of the beta-hemolytic streptococci that have the minute-colony form and group A, C, and G antigens may be designated as *S. anginosus* group A, C, or G. This designation would convey useful information implying that these strains are not typical group A, C, or G streptococci but are beta-hemolytic.

The name *S. intermedius* should be used for strains previously called *S.* MG-*intermedius*. *S. intermedius* is suggested rather than *S. milleri* only because it is on the list of accepted species (95). The lactose-negative, non-beta-hemolytic strains that were previously called *S. anginosus-constellatus* should be called *S. constellatus*. Mannitol-positive variants of these strains have been recognized by Ruoff and Kunz (87). Although several physiological characteristics of the beta-hemolytic strains are listed in Table 4, these characteristics should not be used to identify the species. Rather, the colony morphology and hemolytic activity on blood agar plates and the serological reaction are all that need be used to characterize them. The physiological tests listed in Table 4 are used to differentiate the two viridans species. The miniaturized, automated or semiautomated systems for the identification of the group D and viridans streptococcal species will be an asset to most laboratories. These systems will utilize a pure subculture of a streptococcus strain grown overnight for inocula. An additional 4 to 24 h is required for final identification in some cases. Thus, these systems have little value for the identification of the beta-hemolytic group A, B, C, F, or G streptococci or the pneumococci, which can be identified more rapidly by serological methods. However, these devices improve the turn-around time and work load of laboratories that identify group D and viridans streptococci by conventional techniques. Several miniaturized and rapid identification systems for the streptococci are now commercially available. The Minitek system (BBL Microbiology Systems, Cockeysville, Md.) incorporates various substrates for enzyme reactions into paper disks. The disks are placed into wells in plastic plates, and the inoculum is then introduced into each well. The manufacturer suggests that the plate be incubated anaerobically if viridans streptococci are suspected. The reported results have indicated a high degree of correlation between the Minitek and conventional tests when substrates are essentially the same (52, 91). However, some tests (esculin and arginine) had to be incubated 5 to 7 days. Thus the system is miniaturized but not necessarily rapid. Although the package insert indicates that the Minitek system can be used to identify the beta-hemolytic and group D *Streptococcus* species, evaluation of the product has been limited to the viridans streptococci (52, 91). The API 20S system (Analytab Products, Ayerst Laboratories, Plainview, N.J.) and the API Rapid Strep system (DMS Laboratories, Inc., Flemington, N.J.) products are similar to each other. Although the tests contained in the two systems are not exactly the same, both contain 20 tests and are read in similar fashion. The API 20S

TABLE 4. Differentiation of the *S. milleri* group of streptococci at CDC[a]

Current identification	Serological and physiological characteristics						Proposed identification
	Hemolysis	Group antigen	Lactose	Arginine and esculin	Mannitol and raffinose	Hippurate, starch, sorbitol, inulin, and glucans	
S. anginosus	Beta	F	−*	+	−	−	*S. anginosus*, group F
Minute-colony group A	Beta	A	−*	+	−	−	*S. anginosus*, group A
Minute-colony group C	Beta	C	−*	+	−	−	*S. anginosus*, group C
Minute-colony group G	Beta	G	−*	+	−	−	*S. anginosus*, group G
Beta-hemolytic streptococcus not group A, B, C, D, F, or G	Beta	None	−*	+	−	−	*S. anginosus*, no group
S. MG-*intermedius*	None, alpha	NA	+	+	−*	−	*S. intermedius*
S. anginosus-constellatus	Alpha, none	NA	−	+	−	−	*S. constellatus*

[a] Symbols: +, positive reaction; −, negative reaction; *, occasional exceptions occur; NA, not applicable.

system has a 4-h incubation period, and the Rapid Strep system has 4- and 24-h incubation periods. In the latter system, if identification is not achieved at the 4-h reading, the user is instructed to incubate the test overnight. For both systems, the tests in dehydrated form are incorporated into microcupules attached to stiff cardboard holders (strips). The inoculum rehydrates the substances in the tests. Both systems identify the group D species with a high degree of accuracy (95 to 98%) (2, 3, 39, 73, 106; unpublished data). The rates of identification of the viridans streptococci to the species level vary between 39 and 85% (2, 3, 39, 73, 106; unpublished data). Some viridans species are identified more accurately than others, and in addition, some of the identifications by the API 20S can be made only after supplemental tests are performed.

The AutoMicrobic system (AMS; Vitek Systems, Inc., Hazelwood, Mo.) is also available for identifying the streptococci. The Gram-Positive Identification (GPI) card, one component of the system, is designed to identify several other common gram-positive bacteria, as well as the streptococci. The card contains dehydrated substances for 27 tests and two controls. The inoculum, which is automatically inserted into the test card by the AMS inoculator module, rehydrates the substances for the tests. The GPI cards are then placed in an AMS reader-incubator module, and a computer-assisted identification is possible after 4 h, but some strains require a 13-h incubation period. The identification of the group D *Streptococcus* species by the GPI system is about 95% accurate (3, 86; unpublished data). The identification rates for the differentiation of viridans streptococcal species have been lower; however, some species were quite accurately identified (e.g., *S. mitis*, 100% [3] and 79% [86]), while others were very poorly identified (e.g., *S. sanguis* biotype I, 29% [3] and 43% [86]). Results obtained at CDC (R. Facklam, G. S. Bosley, D. Rhoden, A. R. Franklin, N. Weaver, and R. Schulman, submitted for publication) indicate that approximately 80% of the viridans strains are identified by the GPI system.

The overall identification rates by the API 20S, Rapid Strep, and AMS-GPI systems for the viridans streptococci are affected by different species and the number of each species tested. If we analyze our data from the evaluations of the three systems for only the seven most commonly occurring viridans species (*S. constellatus*, *S. intermedius*, *S. mitis*, *S. mutans*, *S. salivarius*, *S. sanguis* I, and *S. sanguis* II), we find that the API 20S system identified 84% of 192 strains tested, and the AMS-GPI system identified exactly the same percentage of the same strains. These figures are comparable to those obtained with the Rapid Strep system (85% of the same seven species). Thus, under our test conditions, there is no real difference in overall identification rates by any of the three systems. Supplemental testing is required by the API 20S system, but it is not required by either the Rapid Strep or GPI system.

Examine the package insert instructions and the publications mentioned above before choosing a system that meets your needs.

No uniform pattern of physiological characteristics has been observed among the nutritionally variant strains of streptococci. These strains appear to resemble strains of *S. mitis*, *S. morbillorum*, and *S. salivarius* (26). All the nutritionally variant strains tested at CDC (more than 100 strains) have been successfully grown in broth containing 0.001% pyridoxal (vitamin B_6). All strains can be grown as satellite colonies adjacent to staphylococcus growth on 5% sheep blood-tryptic soy agar. Either of these procedures can be used to identify the nutritionally variant streptococci.

Presumptive identification of streptococci

The beta-hemolytic streptococci, pneumococci, and alpha- or nonhemolytic group B and D streptococci are most accurately and completely identified by serolgoical procedures. The group D and other non-beta-hemolytic strains (viridans) are differentiated to the species level by physiological tests. However, all the strains of streptococci can be presumptively identified or categorized by a minimum of simple tests. Small laboratories may not have sufficient time or funds for serological groupings and complete physiological testing and may find the alternative methods listed in Table 5 especially useful. The source of the infecting strain, its colonial morphology, and its hemolytic activity on primary blood agar media are the characteristics used in determining which tests to perform. For example, results of hemolysis and bacitracin susceptibility or hydrolysis of L-pyrrolidonyl-β-naphthylamide (PYR) are usually sufficient to presumptively differentiate the beta-hemolytic group A streptococci from the beta-hemolytic non-group A streptococci isolated from throat and skin lesions. Similarly, determining optochin susceptibility or bile solubility on suspected pneumococcal colonies is sufficient for the presumptive identification of pneumococci regardless of source. Beta-hemolytic group B streptococci are presumptively identified by hemolysis, hippurate hydrolysis, the CAMP test, or a combination of reactions to bacitracin and SXT disks. The group D nonenterococcal and viridans streptococci are identified by their hemolytic reactions on blood agar, reactions in BE, and salt tolerance or by PYR tests. These identifications are usually necessary for isolates from systemic infections; however, if a non-beta-hemolytic streptococcus is obtained from the upper respiratory tract, skin, or feces, it is generally considered part of the normal flora and does not need to be identified.

Hemolysis. Methods for determining streptococcal hemolysis were discussed above. Hemolysis is the most important characteristic to determine for the presumptive identification of streptococci.

Bacitracin test. The bacitracin susceptibility test can be used for the presumptive differentiation of beta-hemolytic group A and non-group A streptococci.

Several important factors affect bacitracin test results. Commercially available disks can be used to differentiate beta-hemolytic group A and other beta-hemolytic streptococci, but only if the disks are differential (0.04 U) rather than susceptibility disks (which are too concentrated). A heavy inoculum is advisable, because if the inoculum is too light, non-group A streptococci will appear to be susceptible to bacitracin. A pure culture must be used. Bacitracin differential disks placed on nonselective primary plates inoculated with throat swabs presumptively identify only

TABLE 5. Presumptive identification of streptococci[a]

Category	Hemolysis	Susceptibility to: Bacitracin	Susceptibility to: SXT[b]	Hydrolysis of: Hippurate	Hydrolysis of: PYR[c]	CAMP	Bile-esculin	Growth in 6.5% NaCl	Optochin and bile[d]
Group A	Beta	+	−	−	+	−	−	−	−
Group B	Beta*	−*	−	+	−	+	−	+*	−
Beta-hemolytic streptococci not group A, B, or D	Beta	−*	+	−	−	−	−	−	−
Group D, enterococcus	Alpha, beta, none	−	−	−*	+	−	+	+	−
Group D, not enterococcus	Alpha, none	−	+*	−	−	−	+	−	−
Viridans group	Alpha, none	−*	+	−*	−	−	−*	−	−
Pneumococcus	Alpha	±	?	−	−	−	−	−	+

[a] Symbols: +, positive reaction or susceptible; −, negative reaction or resistant; *, exceptions occasionally occur.
[b] SXT, Sulfamethoxazole and trimethoprim.
[c] PYR, L-Pyrrolidonyl-β-naphthylamide.
[d] Optochin susceptibility and bile solubility.

50 to 65% of the group A streptococci (71, 101). However, one investigator has shown that by placing a bacitracin disk on a highly selective primary plate medium, 97% of the group A streptococci can be presumptively identified (59). To be certain that the inhibitors in the medium and the bacitracin in the disk do not act synergistically to inhibit group A streptococci, a second nonselective medium should also be used.

Only beta-hemolytic streptococci should be tested, because many alpha-hemolytic streptococci, including pneumococci, are inhibited by the bacitracin differential disk. Each new lot of disks should be tested with known strains of group A and non-group A streptococci, so that lot-to-lot variation and stability can be determined.

Any zone of inhibition must be interpreted as positive, even though several authors and a technical bulletin recommend measuring zone size. No experimental data are available to show that zone diameters must be measured to differentiate group A from non-group A streptococci. Furthermore, false-positive results are potentially less harmful than are false-negatives. Approximately 5% of the group A streptococci have zones less than 10 mm in diameter. At CDC, we (37, 38) identified 99.5% of the group A streptococci without considering zone size. To perform the test, a pure beta-hemolytic culture (three or four colonies or a loopful of an overnight broth culture) is streaked onto a blood agar plate (15 by 100 mm). The bacitracin disk is placed in the center of the streaked area, and the plate is incubated overnight at 35 to 37°C. A zone of inhibition of streptococcal growth around the bacitracin disk indicates that the strain can be reported as "beta-hemolytic, presumptive group A by bacitracin," or if no inhibition zone is present, the strain can be reported as "beta-hemolytic *Streptococcus* species, presumptively not group A by bacitracin."

PYR. PYR is a substrate that is hydrolyzed by 100% of *S. pyogenes* strains and the enterococci. All other streptococci are unable to hydrolyze PYR. An agar medium and a broth medium containing PYR have been described (14, 41), but the manufacturer (Carr-Scarborough Microbiologicals, Inc., Stone Mountain, Ga.) of these media has not released the exact formulas. Our evaluations of these media have shown them to be highly specific and sensitive (14, 41). For the PYR agar test, an organism is inoculated onto the agar, and the medium is incubated overnight at 35°C in a normal atmosphere. The PYR reagent (*N,N*-dimethylaminocinnamaldehyde) is added to the surface growth. A red color develops on the growth of *S. pyogenes* and enterococci, whereas a yellow color or no color change occurs with the other streptococci. For the PYR broth test, three or four colonies of streptococci are inoculated into the broth. The broth is incubated at 35°C for 4 to 5 h, after which 1 drop of the PYR reagent is added. After 1 min, a red color develops if *S. pyogenes* or enterococci are present. A yellow color or no color change in 1 min indicates a negative reaction.

Although we have not tested the PYR broth test extensively with beta-hemolytic streptococci, we feel that PYR-positive beta-hemolytic streptococci can be presumptively identified as group A streptococci.

The enterococci can be identified with the 4-h PYR broth test if the strains have been serogrouped with group D reagent or if the BE reaction is positive. That is, positive group D or BE reactions and positive PYR tests identify the streptococci as enterococci (14). *S. bovis* is presumptively identified by positive BE and negative PYR tests or definitively identified by positive group D serological and negative PYR tests. Viridans streptococci are identified by negative reactions in BE, group D serological, and PYR tests.

SXT test. Disks containing 1.25 mg of trimethoprim and 23.75 mg of sulfamethoxazole can be used to presumptively separate group A and B streptococci from the other beta-hemolytic groups (38). Groups A and B must then be differentiated from each other by additional testing. SXT disks are available commercially. Sheep RBCs and tryptic soy agar base must be used to make the plates for the SXT test. The same criteria for inoculum size and the interpretation of zones of inhibition that are used for bacitracin are used for SXT. Most group A and B streptococci are resistant to SXT when cultures are incubated in a normal atmosphere.

Beta-hemolytic, SXT-resistant, bacitracin-susceptible streptococci are presumptively identified as group

A, and streptococci that are bacitracin resistant are presumptively identified as group B. Beta-hemolytic, SXT-susceptible, bacitracin-resistant strains are labeled non-group A or B beta-hemolytic streptococci. Our data indicate that SXT- and bacitracin-susceptible strains should also be labeled non-group A or B beta-hemolytic streptococci (38).

CAMP test. A more specific test for the presumptive identification of group B streptococci is the CAMP test. Two procedures are used to determine the CAMP reaction of streptococci, i.e., a conventional test in which a beta-lysin-producing *Staphylococcus aureus* strain is used and a disk test in which a disk containing the staphylococcal beta-lysin is used (38). This latter test is performed on a sheep or bovine blood agar plate by making a single streak of the *Streptococcus* species perpendicular to but not touching (3 to 4 mm apart) a streak of beta-lysin-producing *S. aureus* strain or the disk containing beta-lysin. Washed sheep RBCs that have been suspended in sterile physiological saline and tryptic soy agar are recommended for the medium. Control strains of group A, B, C, and G streptococci should be used, especially when plates are obtained commercially. The inoculated plates should be incubated in a candle jar or a normal atmosphere but not anaerobically. Some group A streptococci are CAMP positive when incubated in the candle jar, and even more strains are positive when incubated anaerobically. Group B streptococci produce a substance (CAMP factor) that enlarges the zone of lysis produced by the *Staphylococcus* beta-lysin to form a typical arrowhead or flame-shaped clearing at the juncture of the two organisms or a crescent-shaped clearing at the juncture of the beta-lysin-containing disk and the *Streptococcus* streak. Bacitracin- or PYR-negative, CAMP-positive, beta-hemolytic streptococci can be reported as "presumptive group B streptococci by the CAMP test."

Bacitracin-positive, CAMP-positive, beta-hemolytic streptococci are group A or B. The two groups can be differentiated by hemolysis. Group B streptococci have smaller hemolysis zones on the surface and in the stab than do group A streptococci. Bacitracin-positive, CAMP-positive group A and B streptococci are found frequently enough to make this distinction necessary.

Bacitracin- or PYR-positive, CAMP-negative, beta-hemolytic streptococci are presumptive group A streptococci. Bacitracin- or PYR-negative, CAMP-negative, beta-hemolytic streptococci are beta-hemolytic streptococci, not group A or B.

Nonhemolytic group B streptococci are usually CAMP positive; therefore, nonhemolytic streptococci that are BE negative (test to be discussed below) and CAMP positive can be presumptively identified as nonhemolytic group B streptococci.

Sodium hippurate test. A third presumptive test that can be used to identify group B streptococci is the hippurate hydrolysis test (37). The formulas for the medium and reagents are described in Chapters 111 and 112. Inoculate the broth with two or three colonies of beta-hemolytic streptococci. Incubate the broth overnight at 35°C, and then centrifuge the broth tube. To 0.8 ml of supernatant fluid, add 0.2 ml of ferric chloride reagent and mix well. If a heavy precipitate forms and remains longer than 10 min, the test is positive, and the *Streptococcus* species is presumptively group B. If the test is negative or weakly positive, the growth tube should be incubated for another 24 h, and the test should be repeated.

Nearly all (99%) group B streptococci, whether they are beta-hemolytic or nonhemolytic, hydrolyze hippurate and therefore react positively. Group D streptococci may also react positively but are distinguished by the BE test. Beta-hemolytic streptococci other than group B (groups A, C, G, and F) do not hydrolyze hippurate. The alpha-hemolytic streptococci, *S. uberis* and *S. acidominimus*, which also hydrolyze hippurate, are rarely found in human infections and thus may be misidentified as non-beta-hemolytic group B streptococci. Additional tests, such as those listed in Table 3, must be used to confirm the identity of these strains. The BE-negative, hippurate-positive, beta-hemolytic streptococci can be reported as "presumptive group B streptococci by hippurate hydrolysis."

BE test. Many tests have been described to identify the group D streptococci presumptively, but the most accurate test is probably the BE test (37, 38, 48). The formula for the medium is given in Chapter 111.

The BE medium can be used in agar slants or plates. For agar slants, inoculate the BE medium with two or three colonies and incubate the slant at 35°C. If more than half of the slant is blackened within 24 to 48 h, the test is positive, and the *Streptococcus* species is presumptively identified as group D. If less than half of the slant is blackened or no blackening occurs within 24 to 48 h, the test is negative. For agar plates, inoculate the BE medium with two or three colonies and incubate the cultures overnight in a normal atmosphere or an atmosphere containing increased CO_2 at 35 to 37°C. The test is positive when any blackening of the medium occurs. If no blackening occurs, the test is negative.

All group D streptococci are BE positive; therefore this test cannot be used to differentiate enterococci, which are penicillin resistant, from nonenterococcal strains, which are usually penicillin susceptible. The strains must be differentiated so that patients with enterococcal infections can be treated with a combination of antibiotics.

Salt tolerance (6.5% NaCl) test. The enterococcal species (*S. faecalis*, *S. faecium*, *S. durans*, and *S. avium*) are easily differentiated from the nonenterococcal species (*S. bovis* and *S. equinus*) by the 6.5% NaCl tolerance test (37). The formulas for the media are given in Chapter 111. The salt tolerance of streptococci can be determined by either the broth test or the agar plate test. For the broth test, inoculate two or three streptococcal colonies into 6.5% NaCl broth and incubate the culture at 35°C. Examine the culture for growth (indicated by turbidity and sometimes a change in the indicator color) after 24 and 48 h. Growth within 48 h indicates that the strain is salt tolerant, i.e., positive, and it can be identified as an enterococcus if it is also BE positive or if it is serologically group D. For the agar plate test, inoculate two or three streptococcal colonies onto the 6.5% NaCl agar and incubate the culture overnight in a normal or increased-CO_2 atmosphere. The test is positive when growth is visible on the agar surface. If no growth is visible after 24 h, the test is negative.

The enterococci usually grow heavily in 6.5% NaCl

and cause an indicator change within 24 h, although some strains grow only after 48 h, and some do not cause an indicator change even after 72 h. About 80% of group B streptococci also grow in the broth medium; fewer grow on the agar plate medium. Beta-hemolytic groups A, C, F, and G usually do not grow in either medium, although salt-tolerant group A streptococci do grow occasionally. The alpha-hemolytic, nongroupable streptococci (viridans) do not grow in either the broth or the agar medium containing 6.5% NaCl, nor do group D species *S. bovis* and *S. equinus*. Cultures that are positive in salt tolerance tests and that are not presumptive group B or D streptococci should be tested for purity by streaking the growth from the salt tolerance test broth onto a blood agar plate and comparing the morphology with that of the original strain. If the morphology differs, a Gram stain and a catalase test should be performed.

An alpha- or nonhemolytic strain that grows in 6.5% NaCl broth and reacts negatively in the BE test may be an *Aerococcus* species. It is confirmed as such by Gram stain morphology (tetrad cellular arrangement), negative arginine hydrolysis, and 10 and 45°C tests (Table 3).

Bile solubility test. The bile solubility test can be used for the presumptive differentiation of pneumococci from the other viridans streptococci (64). Inoculate a broth (Todd-Hewitt, tryptic soy, etc.) with two or three suspect colonies, and incubate it overnight at 35°C in an atmosphere of increased CO_2. Place 0.5 ml of the broth culture in each of two test tubes (13 by 100 mm). Add 1 or 2 drops of phenol red indicator, and neutralize the broths with 1 N NaOH. Add 0.5 ml of 2% sodium deoxycholate (bile) to one tube and 0.5 ml of saline to the other tube. Incubate the tubes at 35°C and examine them periodically for up to 2 h. A clearing of turbidity in the bile tube but not in the saline control tube indicates a positive test; that is, the pneumococcal cells have been lysed ("solubilized").

The bile solubility test can also be performed by making a saline suspension of cells from growth on an agar plate. A turbidity equal to that of an 0.5 to 1.0 McFarland density standard should be used. The test is performed in the same manner as that described above except that the indicator and pH adjustment are generally not necessary.

Occasionally, the turbid broth or saline suspension clears only partially. For example, 83% of 263 pneumococcal strains were completely lysed by 2% sodium deoxycholate, and the remainder were partially lysed, whereas when the concentration of deoxycholate was increased to 10%, 86% of the strains were completely lysed (unpublished data). Therefore, either 2 or 10% deoxycholate can be used, but if lysis is not complete, an additional test (optochin) is needed to allow a more accurate, presumptive identification of pneumococci. Partially soluble, alpha-hemolytic, optochin-susceptible streptococci are presumptively identified as pneumococci; optochin-resistant strains are viridans streptococci.

Optochin test. As noted above, optochin susceptibility is used for the presumptive differentiation of alpha-hemolytic viridans streptococci and pneumococci. Two or three suspect colonies are streaked onto a quarter of a blood agar plate, and the optochin disk is placed in the upper third of the streaked area. The plate is incubated overnight at 35°C in a candle jar or CO_2 incubator. Cultures do not grow as well in a normal atmosphere, and larger zones of inhibition occur. If a 6-mm disk is used, a zone of inhibition of at least 14 mm in diameter is considered positive for pneumococci. A diameter of between 6 and 14 mm is questionable for pneumococci, and the strain is presumptively identified as a pneumococcus only if it is bile soluble. For 10-mm optochin disks, a zone of inhibition of at least 16 mm in diameter is positive, and strains with inhibition zones of between 10 and 16 mm should be tested for bile solubility.

Several combinations of tests (Table 5) can be used to presumptively identify the streptococci. The most simple, convenient, and accurate combination is of PYR, CAMP, and BE tests on agar plates, along with the hemolytic reactions. The battery of tests should be interpreted as a whole; i.e., results obtained with individual tests should be evaluated as they relate to those obtained in all tests run. We suggest that the streptococci can be presumptively identified by the following combinations of tests in addition to that mentioned above: (i) bacitracin, SXT, CAMP, BE, and NaCl tests; (ii) bacitracin, SXT, hippurate, BE, and NaCl tests; and (iii) PYR, hippurate, and BE tests. Table 5 lists the positive and negative reactions of each of the six categories of streptococci to each of the presumptive tests.

Susceptibility to antimicrobial agents

The group A streptococci are universally susceptible to penicillin G, a fact that precludes antimicrobial susceptibility testing for these organisms unless the patient is allergic to penicillin. In these cases, erythromycin or tetracycline is substituted; however, 3 to 5% of the strains tested are resistant to these antibiotics. Excellent in vitro activity has been reported for the newer beta-lactam antibiotics (15), except moxalactam, which is less effective in vitro for all streptococcal species. Routine susceptibility tests for group A streptococci isolated from the throat or skin are not necessary or recommended.

Penicillin G remains the drug of choice for the treatment of infections caused by the other beta-hemolytic streptococci. Generally, the group B, C, F, and G streptococci are susceptible to ampicillin, cephalothin, and chloramphenicol and resistant to tetracycline. Vancomycin or erythromycin is an alternative antibiotic for the penicillin-allergic patient, although occasional strains demonstrate resistance (4, 7). When a beta-hemolytic streptococcus is confirmed as the cause of an infectious process, single-drug therapy with penicillin or ampicillin is usually given. However, evidence for accelerated killing of the group B streptococci with the combination of ampicillin and gentamicin (90) and reports of penicillin-tolerant strains may indicate the need to review single-drug regimens in patients with group B streptococcal infections (84, 94).

The group D enterococci are penicillin resistant and are treated with penicillin or ampicillin in combination with an aminoglycoside. When high-level resistance to streptomycin is documented (MIC of >2,000 μg/ml), the combination of penicillin and streptomycin is unlikely to be synergistic. Penicillin and gentamicin, or vancomycin and gentamicin, remain effective against these strains. Combinations of penicillin with tobramycin, netilmicin, or sisomycin are effective in vitro against *S. faecalis*, but not *S. faecium*. This absence of synergy is not predicated by high-level resistance to these agents, as it is with streptomycin (68). The group D enterococci are uniformly resistant to the cephalosporins, which may contribute to the emergence of these enterococci as opportunistic pathogens in patients receiving cephalosporins. Unlike the enterococci, the aerococci are susceptible to penicillin and other antimicrobial agents.

The majority of the strains of *S. bovis* and the viridans streptococci are susceptible to penicillin, which is the drug of choice for treating infections with these organisms. However, 17% of the viridans streptococci, including the vitamin B_6-dependent strains, are moderately resistant to penicillin, with MICs of >0.1 μg/ml (45, 83). In general, *S. mitis* (*S. mitior*) and *S. sanguis* require higher concentrations of antibiotics for inhibition and killing than do other viridans streptococci (16, 104). Similarly, *S. bovis* has demonstrated tolerance to penicillin, vancomycin, and cephalothin (89). Combination therapy may be required to eradicate these streptococci in serious infections. Antibiotic susceptibility tests should be performed on all alpha- and nonhemolytic streptococci isolated from systemic infection.

The pneumococci are normally inhibited by ≤0.02 μg of penicillin G per ml. Penicillin-insensitive strains (MIC of 0.1 to 1.0 μg/ml) and penicillin-resistant strains (MIC of ≥2.0 μg/ml) have been reported (108). All pneumococcal isolates from serious infections should be screened for their penicillin susceptibility as described in Chapter 107. Relative resistance to penicillin may not preclude successful treatment with large doses of penicillin G, but other penicillins that achieve only low concentrations in blood may not be suitable for treating infections caused by relatively resistant pneumococci. Alternatively, chloramphenicol may be used. Of greater concern is the emergence of multiply resistant strains of *S. pneumoniae* that are resistant to all beta-lactam antibiotics, erythromycin, clindamycin, tetracycline, chloramphenicol, streptomycin, rifampin, and SXT.

Evaluation of cultures

The laboratorian must convey information that best helps the physicians manage their patients. In the case of throat cultures, the report should include: (i) the identity of the organism isolated, either presumptive or confirmed, and (ii) the relative number of organisms present. A convenient method for indicating the estimated number of organisms present on the culture plate is as follows: 1+, 10 colonies or less; 2+, 10 to 50 colonies; 3+, more than 50 colonies; and 4+, predominant or pure culture.

Laboratory results of skin cultures should be reported in the same manner as results of throat cultures. Physicians are interested primarily in beta-hemolytic group A streptococci from throat and skin swabs. Reports of results with blood, cerebrospinal fluid, and other body fluids should include the species or group identity of the organism if possible. The very minimum that should be reported is the hemolytic reaction and whether or not the streptococcus is an enterococcus. Reports on urine samples must include the number of organisms per milliliter. Feingold et al. (42) evaluated both clinical and laboratory findings in an attempt to assess the importance of determining the various serological groups of streptococci isolated from extrarespiratory sources. They found that the clinical picture and the laboratory findings were valuable in assessing the importance of streptococci isolated from blood and urine specimens. Readers are urged to study Feingold's criteria and to communicate with the attending physician about laboratory findings. The laboratorian's responsibility does not end with a simple report of findings; he or she is obliged to know the clinical significance of the bacteria identified. In this way, the microbiologists can at least tell the physician what other scientists have reported and relate their findings to these reports.

APPENDIX

METHODS AND MATERIALS FOR IDENTIFICATION OF STREPTOCOCCI

Antigen Preparation for Precipitin Grouping of Beta-Hemolytic Streptococci

Hot HCl extraction (Lancefield extract)

1. Grow pure cultures in 30 ml of Todd-Hewitt or other suitable broth at 35 to 37°C.
2. Pack the cells by centrifugation.
3. Discard the supernatant fluid, and add 1 drop of 0.04% metacresol purple and about 0.3 ml of 0.2 N HCl (in 0.85% NaCl) to the sedimented cells. Mix the fluids well and transfer the suspension to a small tube. If the suspension is not definitely pink (pH 2.0 to 2.4), add another drop of 0.2 N HCl.
4. Place the tube in a boiling-water bath for 10 min. Shake the tube several times.
5. Remove the tube from the water bath and centrifuge the tube.
6. Decant the supernatant fluid into a small clean tube.
7. Neutralize the supernatant by adding 0.2 N NaOH (in distilled water) drop by drop until the extract is slightly purple (pH 7.4 to 7.8). A deep purple indicates that the pH is too high, a condition that may cause nonspecific cross-reactions. Although it is better not to add more salts or to increase the volume, a back titration with 0.2 N HCl may be necessary.
8. Centrifuge the extract and decant the supernatant fluid into a small screw-capped vial. Add 1 drop of a 1:500 dilution of Merthiolate (1% in 1.4% sodium borate), and store the solution at 4 or −20°C.

Hot formamide extraction

1. Grow the strains in 5 ml of Todd-Hewitt or other suitable broth overnight at 35 to 37°C.
2. Pack the cells by centrifugation.
3. Discard the supernatant fluid, add 0.1 ml of formamide to the cells, and mix.
4. Place the tube in a 150°C oil bath for 15 min.
5. Cool the tube in running tap water, and add to it 0.25 ml of acid-alcohol (95 parts anhydrous alcohol and 5 parts 2 N HCl). Shake the tube to mix the solution.

6. Centrifuge the tube and decant the supernatant into a small tube.
7. Add 0.25 ml of acetone. Shake the tube to mix the solution.
8. Centrifuge the tube and discard the supernatant fluid.
9. Add 1 ml of saline and 1 drop of phenol red indicator to the precipitate. Shake the mixture and neutralize it with a trace of sodium carbonate powder. Remove any insoluble precipitate by centrifugation.

Autoclave extraction

1. Grow the cells in 30 ml of Todd-Hewitt or other suitable broth at 35 to 37°C.
2. Pack the cells by centrifugation.
3. Discard the supernatant fluid, add 0.5 ml of 0.85% NaCl solution to the cells, and shake the mixture to suspend the cells.
4. Autoclave the tube for 15 min at 121°C.
5. Centrifuge the tube.
6. Decant the supernatant fluid into a clean, sterile container.

Nitrous acid extraction

1. Grow the cells on the surface of a blood agar plate or in 5 ml of Todd-Hewitt or other suitable broth overnight at 35 to 37°C.
2. If 5 ml of broth is used, recover the cells by centrifugation, and add 1 drop of saline to the packed cells. If an agar plate is used, recover the cells by adding 1 or 2 drops of saline to the plate, scraping the growth free, and transferring the suspension to a small tube.
3. Add 2 drops of 4 M $NaNO_2$ solution (276 g of $NaNO_2$ per liter of distilled water) to the cell suspension.
4. Add 1 drop of glacial acetic acid and mix well.
5. Allow the solution to react for 15 min at room temperature.
6. Add 1 drop of metacresol purple indicator, and adjust the pH to 7.4 with 1 N NaOH.
7. Centrifuge the tube to clarify the solution.
8. Decant the supernatant.

Micro-nitrous acid extraction

1. Streptococcal colonies for extraction can be taken from the primary blood agar plate culture.
2. Transfer one or more colonies to a Durham tube (3 by 30 mm) containing 20 μl of a 2 M sodium nitrate solution (138 g of $NaNO_2$ per liter of distilled water).
3. Add 3 μl of glacial acetic acid or 10 μl of 33% glacial acetic acid and mix well.
4. Allow the mixture to react for 15 min at room temperature.
5. Add 5 μl of phenol red indicator and adjust the mixture to pH 7.2 to 7.4 with 10 N NaOH. It is convenient to make the crude adjustment with 10 N NaOH and the fine pH adjustments with 0.2 N NaOH.
6. Add distilled water to a final volume of 50 μl.
7. Centrifuge the solution to clarify it.

Micro-nitrous acid extraction of swabs

1. Place the swab in a 1.0-ml wheaten vial containing 0.3 ml of distilled H_2O.
2. Vigorously roll and press the swab against the inner wall of the vial.
3. Centrifuge the vial for 10 min at 3,000 × *g* at room temperature.
4. Decant the supernatant, and drain the fluid through filter paper.
5. Add 20 μl of 4 N sodium acetate.
6. Add 2.5 μl of glacial acetic acid (diluted 1:1 with distilled H_2O).
7. Mix the sedimented material into the solution and incubate for 5 min at room temperature.
8. With a microspatula add 18 to 24 mg of sodium bicarbonate to the vial and mix it in.
9. Place 10 μl of neutralized extract on a glass slide.
10. Add 10 μl of Phadebact group A reagent and examine the solution under indirect light for agglutination.

***S. albus* enzyme extraction**

1. Grow streptococcal strains on a blood agar plate overnight at 37°C.
2. Pipette 0.25 ml of *S. albus* enzyme solution (available commercially) into a small test tube (12 by 75 mm or smaller).
3. Scrape a large loopful of growth from the blood agar plate, and suspend the growth in the enzyme solution.
4. Place the tube in a 45°C water bath until the solution is clear (about 90 min).
5. Cool the tube to room temperature and centrifuge it for 10 min.
6. Decant the supernatant fluid into a clean container. *S. albus* enzyme should be resuspended in the volume specified by the manufacturer and stored at −20 to −70°C.

Pronase B enzyme extraction

1. Grow the strains on a blood agar plate overnight at 35 to 37°C.
2. Prepare borate buffer by adding 525 ml of borate solution (12.404 g of boric acid dissolved in 100 ml of 1 N NaOH and diluted to 1 liter with distilled water) to 475 ml of 0.1 N HCl and 10 ml of 1 M $CaCl_2$.
3. Prepare the buffered enzyme solution (20 mg of pronase B per ml of borate buffer). Dispense the solution in 0.5-ml portions in tubes (13 by 75 mm); cork the tubes and store them at −20 or −70°C.
4. Remove all the growth from the plate with a swab. Place the swab in a tube containing the buffered enzyme solution, and squeeze the swab as dry as possible by rotating and squeezing it against the side of the tube. The suspension should be cloudy.
5. Incubate the suspension at 35 to 45°C for 2 h.
6. Centrifuge the tube for 15 to 30 min.
7. Decant the supernatant into a clean, sterile container.

***S. albus* enzyme-lysozyme enzyme extraction**

1. Grow the strains on a blood agar plate overnight at 35 to 37°C.
2. Prepare the enzyme mixture by adding the *S. albus* enzyme to 5 ml of a solution of lysozyme (5 mg/ml of distilled water). Centrifuge the solution to clari-

fy it, and store it in 0.5-ml quantities in cork-stoppered tubes (10 by 75 mm) at −20°C.

3. Transfer the growth from the blood agar plate to the enzyme solution (0.5 ml) with a sterile swab. Mix the swab in the solution, and then rotate it against the side of the tube to remove as much of the moisture as possible. Discard the swab.

4. Incubate the enzyme-cell mixture in a water bath at 45 to 50°C for 90 min.

5. Centrifuge the mixture to clarify it, and decant the supernatant into a clean container.

CDC Capillary Precipitin Test for Detecting Streptococcal Group Antigens

1. Dip the capillary tube (vaccine capillary tube with outside diameter of 1.2 to 1.5 mm, Kimble borosilicate glass, both ends open, and lightly fire polished) into antiserum (in a screw-capped vial) until a column about 1 cm long has been drawn in by capillary action. (To maintain the sterility of the sera, sterilize the capillary tubes and keep them sterile at the lower end until after the serum is taken up.)

2. Holding the tube carefully so that air does not enter it, wipe it with facial tissue.

3. Dip the tube into the streptococcal extract until an amount equal to that in the serum column is drawn up. If an air bubble separates the serum and the extract, discard the tube and repeat the procedure.

4. Wipe the tube carefully. Fingerprints, serum, or extracts on the outside of the tube may simulate or obscure a positive reaction.

5. Plunge the lower end of the tube into plasticine until a small plug fills the opening. Do not let the reactants mix. The plasticine plug (at the same end of the tube as the reactants) will hold the reactants in place while the tube is inverted. Alternatively, hold a finger over the end of the tube until step 6 is completed.

6. Invert the tube and insert it gently into the plasticine-filled groove of a capillary holding rack.

7. Examine the tube in bright light against a dark background. If a white precipitate appears within 5 min, the reaction is strongly positive; weaker reactions develop more slowly. Precipitates that appear after 30 min should be disregarded.

FA Testing for Group A and B Streptococci

Procedure for group A streptococci

Prepare 0.01 M phosphate-buffered saline (PBS), pH 7.6, by diluting stock solution (12.36 g of Na_2HPO_4, 1.80 g of $NaH_2PO_4 \cdot H_2O$, and 85.00 g of NaCl per liter of distilled water) 1:10 in distilled water. Prepare buffered-glycerol mounting fluid by mixing 9 parts glycerol with 1 part carbonate buffer, pH 9.0. The latter is prepared by adding 100 ml of a solution containing 4.2 g of $NaHCO_3$ to a solution of 5.3 g of Na_2CO_3 in 100 ml of water until the pH is adjusted to 9.0. Inoculate the test organisms into 1 ml of Todd-Hewett broth, and incubate the culture at 35 to 37°C until the broth is turbid (2 to 6 h is usually required for positive throat swabs). Centrifuge the turbid broth, suspend the cell sediment in approximately 1 ml of PBS, centrifuge the suspension, resuspend the cells in several drops of PBS, and prepare a smear of the cell suspension on an IF slide. Allow the smear to air dry, and add several drops of IF reagent to it. Incubate the slide for 0.5 h at room temperature in a moist chamber (easily prepared by fitting half of a 15-cm petri dish with moist filter paper). Tap off the excess conjugate, and then rinse the slide first in PBS (in a container for 10 min or briefly with running PBS) and then briefly in distilled water. Allow the slide to air dry or gently blot it dry, and add 1 drop of mounting fluid and a cover slip. Examine the smears with a ×95 to ×100 oil immersion lens on a fluorescence microscope equipped for transmitted light with an HBO-200 mercury arc lamp, a 3-mm Schotts BG-12 primary filter, and a 1-mm Schotts OG-1 barrier filter (or comparable fluorescence assembly). Group A streptococcal cells should stain with a 3 to 4+ intensity and appear as fluorescent, yellow-green "doughnuts." Other streptococcal groups and *S. aureus* cells should be invisible or stain no more than 2+.

Procedure for group B streptococci

Prepare PBS, pH 7.6, and mounting fluid, pH 9.0, as described above in the procedure for group A streptococci. Inoculate the test organisms into 1 ml of Todd-Hewitt broth, and incubate the broth at 35°C until the broth is turbid (often in 2 to 6 h). Prepare a smear of the cell suspension on an IF slide, and allow it to air dry at room temperature. Add several drops of 95% ethanol as fixative and allow it to evaporate, or add several drops of undiluted Formalin, and after 5 min rinse the slide with PBS and then with distilled water (see the group A procedure explained above). Air dry or blot dry, apply IF reagent to the smear, and incubate the slide in a moist chamber for 0.5 h at room temperature. Tap off the excess conjugate, rinse and blot the slide as described above, add a drop of mounting fluid and a cover slip to the smear, and look for IF-positive streptococcal cells that fluoresce in UV light. (See the description above under group A procedure.) All group B serotypes should stain equally well. The IF (intensity of stained cells) for group A and B streptococci is estimated as follows:

- 4+ = maximal fluorescence; brilliant yellow-green, clear-cut cell outline; sharply defined nonstaining center of cell
- 3+ = less brilliant yellow-green fluorescence; clear-cut cell outline; sharply defined nonstaining center of cell
- 2+ = less brilliant, but definite fluorescence; less clear-cut cell outline; nonstaining center area fuzzy
- 1+ = definite fluorescence, but very subdued; peripheral and center staining at same intensity

A satisfactory working dilution of the test reagent is a dilution that stains all of the group A strains at 3+ or 4+ fluorescence intensity and the other strains at not more than 1+ or 2+. A satisfactory control conjugate should not stain group A streptococci. Occasionally, *S. aureus*, group C and G streptococci, and rare group A streptococci stain at low levels. In reporting the FA results, the laboratorian must report only his or her findings. If the hemolysis is not known before the FA results, the report should be "group A streptococci by FA" or "not group A streptococci by FA." If the hemolysis is known, then a report of beta-hemolytic streptococci group A by FA, beta-hemolytic streptococci not group A by FA, or no beta-hemolytic streptococci isolated can be made.

LITERATURE CITED

1. **Anthony, B. F., and D. M. Okada.** 1977. The emergence of group B streptococci infections of the newborn infant. Annu. Rev. Med. **28**:355–369.
2. **Appelbaum, P. C., P. S. Chaurushiya, M. R. Jacobs, and A. Duffett.** 1984. Evaluation of the Rapid Strep system for species identification of streptococci. J. Clin. Microbiol. **19**:588–591.
3. **Applebaum, P. C., M. R. Jacobs, J. I. Heald, W. M. Palko, A. Duffett, R. Crist, and P. A. Naugle.** 1984. Comparative evaluation of the API 20S system and the AutoMicrobic system Gram-Positive Identification Card for species identification of streptococci. J. Clin. Microbiol. **19**:164–168.
4. **Auckenthaler, R., P. E. Hermans, and J. A. Washington II.** 1983. Group G streptococcal bacteremia: clinical study and review of the literature. Rev. Infect. Dis. **5**:196–204.
5. **Austrian, R.** 1976. The Quellung reaction, a neglected microbiologic technique. M. Sinai J. Med. **43**:699–709.
6. **Baker, C. J., and D. L. Kasper.** 1976. Correlation of maternal antibody deficiency with susceptibility to neonatal group B streptococcal infections. N. Engl. J. Med. **294**:752–756.
7. **Baker, C. J., B. J. Webb, and F. F. Barrett.** 1976. Antimicrobial susceptibility of group B streptococci isolated from a variety of clinical sources. Antimicrob. Agents Chemother. **10**:128–131.
8. **Ball, L. C., and M. T. Parker.** 1979. The cultural and biochemical characters of *Streptococcus milleri* strains isolated from human sources. J. Hyg. **82**:63–78.
9. **Barnham, M., T. J. Thornton, and K. Lange.** 1983. Nephritis caused by *Streptococcus zooepidemicus* (Lancefield group C). Lancet **i**:945–948.
10. **Bayer, A. S., J. S. Seidel, and T. T. Yoshekawa.** 1976. Group D enterococcal meningitis. Arch. Intern. Med. **136**:883–889.
11. **Black, W. A., and F. Van Buskirk.** 1973. Gentamicin blood agar used as a general-purpose selective medium. Appl. Microbiol. **25**:905–907.
12. **Blair, D. C., and D. B. Martin.** 1978. Beta hemolytic streptococcal endocarditis: predominance of non-group A organisms. Am. J. Med. Sci. **276**:269–277.
13. **Bortolussi, R., A. J. Wort, and S. Casey.** 1982. The latex agglutination test versus counterimmunoelectrophoresis for rapid diagnosis of bacterial meningitis. Can. Med. Assoc. J. **127**:489–493.
14. **Bosley, G. S., R. R. Facklam, and D. Grossman.** 1983. Rapid identification of enterococci. J. Clin. Microbiol. **18**:1275–1277.
15. **Bourbeau, P., and J. M. Campos.** 1982. Current antibiotic susceptibility of group A beta hemolytic streptococci. J. Infect. Dis. **145**:916.
16. **Bourgault, A. M., W. R. Wilson, and J. A. Washington II.** 1979. Antimicrobial susceptibilities of species of viridans streptococci. J. Infect. Dis. **140**:316–321.
17. **Breese, B. B.** 1978. Streptococcal pharyngitis and scarlet fever. Am. J. Dis. Child. **132**:612–616.
18. **Broome, C. V., and R. R. Facklam.** 1981. Epidemiology of clinically significant isolates of *Streptococcus pneumoniae*. Rev. Infect. Dis. **3**:277–280.
19. **Burdash, N. M., M. E. West, R. T. Newell, and G. Teti.** 1981. Group identification of streptococci. Evaluation of three rapid agglutination methods. Am. J. Clin. Pathol. **76**:819–822.
20. **Castle, D., S. Dessock-Philip, and C. S. F. Easmon.** 1982. Evaluation of an improved Streptex kit for the grouping of beta-haemolytic streptococci by agglutination. J. Clin. Pathol. **35**:719–722.
21. **Chau, P. Y., and R. Kay.** 1983. *Streptococcus suis* meningitis, an important underdiagnosed disease in Hong Kong. Med. J. Aust. **1**:414–417.
22. **Colman, G.** 1967. Aerococcus-like organisms isolated from human infections. J. Clin. Pathol. **20**:294–297.
23. **Colman, G., and L. C. Ball.** 1984. Identification of streptococci in a medical laboratory. J. Appl. Bacteriol. **57**:1–14.
24. **Colman, G., and R. E. O. Williams.** 1972. Taxonomy of some human viridans streptococci, p. 281–299. *In* L. W. Wannamaker and J. M. Matsen (ed.), Streptococci and streptococcal diseases, recognition, understanding, and management. Academic Press, Inc., New York.
25. **Converse, G. M., III, and H. C. Dillon, Jr.** 1977. Epidemiological studies of *Streptococcus pneumoniae* in infants: methods of isolating pneumococci. J. Clin. Microbiol. **5**:293–296.
26. **Cooksey, R. C., F. S. Thompson, and R. R. Facklam.** 1979. Physiological characterization of nutritionally variant streptococci. J. Clin. Microbiol. **10**:326–330.
27. **Dayton, S. L., D. D. Chipps, D. Blasi, and R. F. Smith.** 1974. Evaluation of three media for selective isolation of gram-positive bacteria from burn wounds. Appl. Microbiol. **27**:420–422.
28. **Deibel, R. H.** 1964. The group D streptococci. Bacteriol. Rev. **28**:330–366
29. **Deibel, R. H., and H. W. Seeley, Jr.** 1974. Family II. *Steptococcaceae* fam. nov., p. 490–517. *In* R. E. Buchanan and N. E. Gibbons (ed.), Bergey's manual of determinative bacteriology, 8th ed. The Williams & Wilkins Co., Baltimore.
30. **Dillon, H. C.** 1979. Post-streptococcal glomerulonephritis following pyoderma. Rev. Infect. Dis. **1**:935–943.
31. **Drow, D. L., D. F. Welch, D. Hensel, K. Eisenach, E. Long, and M. Slifkin.** 1983. Evaluation of the Phadebact CSF test for detection of the four most common causes of bacterial meningitis. J. Clin. Microbiol. **18**:1358–1361.
32. **Duca, E., G. Teodorovicil, C. Radu, A. Vita, P. Talasman-Niculescu, E. Bernescu, C. Feldi, and V. Rosca.** 1969. A new nephritogenic streptococcus. J. Hyg. **67**:691–698.
33. **Dykstra, M. A., J. C. McLaughlin, and R. C. Bartlett.** 1979. Comparison of media and techniques for detection of group A streptococci in throat swab specimens. J. Clin. Microbiol. **9**:236–238.
34. **Ellner, P. D., C. J. Stoessel, E. Drakeford, and F. Vasi.** 1966. A new culture medium for medical microbiology. Am. J. Clin. Pathol. **45**:502–504.
35. **Facklam, R. R.** 1972. Recognition of group D streptococcal species of human origin by biochemical and physiological tests. Appl. Microbiol. **23**:1131–1139.
36. **Facklam, R. R.** 1977. Physiological differentiation of viridans streptococci. J. Clin. Microbiol. **5**:184–201.
37. **Facklam, R. R., J. F. Padula, L. G. Thacker, E. C. Wortham, and B. J. Sconyers.** 1974. Presumptive identification of group A, B, and D streptococci. Appl. Microbiol. **27**:107–113.
38. **Facklam, R. R., J. F. Padula, E. C. Wortham, R. C. Cooksey, and H. A. Rountree.** 1979. Presumptive identification of group A, B, and D streptococci on agar plate media. J. Clin. Microbiol. **9**:665–672.
39. **Facklam, R. R., D. L. Rhoden, and P. B. Smith.** 1984. Evaluation of the Rapid Strep system for the identification of clinical isolates of *Streptococcus* species. J. Clin. Microbiol. **20**:894–898.
40. **Facklam, R. R., and P. B. Smith.** 1976. The gram positive cocci. Hum. Pathol. **7**:187–194.
41. **Facklam, R. R., L. G. Thacker, B. Fox, and L. Eriquez.** 1982. Presumptive identification of streptococci with a new test system. J. Clin. Microbiol. **15**:987–990.
42. **Feingold, D. S., N. L. Stagg, and L. J. Kunz.** 1966. Extrarespiratory streptococcal infections. Importance of the various streptococcal groups. N. Engl. J. Med. **275**:356–361.

43. **Ferrieri, P., and L. L. Blair.** 1977. Pharyngeal carriage of group B streptococci: detection by three methods. J. Clin. Microbiol. **6:**136–139.
44. **Fischer, G., R. E. Horton, and R. Edelman.** 1983. Summary of the National Institutes of Health Workshop on Group B Streptococcal Infection. J. Infect. Dis. **148:**163–166.
45. **Gephart, J. F., and J. A. Washington.** 1982. Antimicrobial susceptibilities of nutritionally variant streptococci. J. Infect. Dis. **146:**536–539.
46. **Gerber, M. A.** 1983. Micronitrous acid extraction-coagglutination test for rapid diagnosis of streptococcal pharyngitis. J. Clin. Microbiol. **17:**170–171.
47. **Gray, B. M., M. A. Pass, and H. C. Dillon, Jr.** 1979. Laboratory and field evaluation of selective media for isolation of group B streptococci. J. Clin. Microbiol. **9:**466–470.
48. **Gross, K. C., M. P. Houghton, and L. B. Senterfit.** 1975. Presumptive speciation of *Streptococcus bovis* and other group D streptococci from human sources by using arginine and pyruvate tests. J. Clin. Microbiol. **1:**54–60.
49. **Guinet, R., J. Andre, G. Barbe, M. Boude, G. Carret, A. M. Freydiere, Y. Gille, J. P. Marcel, and M. A. Mazoyer.** 1982. Serological grouping of streptococci: a collective evaluation in six laboratories of three rapid methods. Med. Microbiol. Immunol. **171:**23–32.
50. **Gunn, B. A., D. K. Ohashi, C. A. Gaydos, and E. S. Holt.** 1977. Selective and enhanced recovery of group A and B streptococci from throat cultures with sheep blood agar containing sulfamethoxazole and trimethoprim. J. Clin. Microbiol. **5:**650–655.
51. **Holdeman, L. V., and W. E. C. Moore.** 1974. New genus, *Coprococcus*, twelve new species, and emended descriptions of four previously described species of bacteria from human feces. Int. J. Syst. Bacteriol. **24:**260–277.
52. **Holloway, Y., M. Schaareman, and J. Dankert.** 1979. Identification of viridans streptococci on the Minitek miniaturised differentiation system. J. Clin. Pathol. **32:**1168–1173.
53. **Holmberg, K., and H. O. Hallander.** 1973. Production of bactericidal concentrations of hydrogen peroxide by *Streptococcus sanguis*. Arch. Oral Biol. **18:**423–434.
54. **Howard, J. B., and G. H. McCracken, Jr.** 1974. The spectrum of group B streptococcal infections in infancy. Am. J. Dis. Child. **128:**815–818.
55. **Kaplan, M. H.** 1979. Rheumatic fever, rheumatic heart disease and the streptococcal connection: the role of streptococcal antigens cross-reactive with heart tissue. Rev. Infect. Dis. **1:**988–996.
56. **Klein, J. O.** 1981. The epidemiology of pneumococcal disease in infants and children. Rev. Infect. Dis. **3:**246–253.
57. **Klein, R. S., R. A. Recco, M. T. Catalano, S. C. Edberg, J. I. Casey, and N. H. Steigbigel.** 1977. Association of *Streptococcus bovis* with carcinoma of the colon. N. Engl. J. Med. **296:**800–802.
58. **Kronvall, G.** 1973. A surface component in group A, C, and G streptococci with non-immune reactivity for immunoglobulin G. J. Immunol. **111:**1401–1406.
59. **Kurzynski, T., C. Meise, R. Daggs, and A. Helstad.** 1979. Improved reliability of the primary plate bacitracin test on throat cultures with sulfamethoxazole-trimethoprim blood agar plates. J. Clin. Microbiol. **9:**144–146.
60. **Kurzynski, T. A., and C. M. Van Holten.** 1981. Evaluation of techniques for isolation of group A streptococci from throat cultures. J. Clin. Microbiol. **13:**891–894.
61. **Lauwers, S., and N. Clumiuk.** 1981. Rapid diagnosis of bacterial meningitis. J. Infect. **3:**27–32.
62. **LeBien, T. W., and M. C. Bromel.** 1975. Antibacterial properties of a peroxidogenic strains of *Streptococcus mitior* (mitis). Can. J. Microbiol. **21:**101–103.
63. **Lerner, P. I., K. V. Gopalakrishna, E. Wolinsky, M. C. McHenry, J. S. Tan, and M. Rosenthal.** 1977. Group B streptococcus (*S. agalactiae*) bacteremia in adults: analysis of 32 cases and review of the literature. Medicine **56:**457–473.
64. **Lund, E.** 1959. Diagnosis of pneumococci by the optochin and bile tests. Acta Pathol. Microbiol. Scand. **47:**308–315.
65. **Lund, E.** 1960. Laboratory diagnosis of pneumococcus infections. Bull. W.H.O. **23:**5–13.
66. **McCusker, J. J., E. L. McCoy, C. L. Young, R. Alamares, and L. S. Hirsch.** 1984. Comparison of Directigen Group A Strep Test with a traditional culture technique for detection of group A beta-hemolytic streptococci. J. Clin. Microbiol. **20:**824–825.
67. **Miller, J. M., H. L. Phillips, R. K. Graves, R. R. Facklam.** 1984. Evaluation of the Directigen Group A Strep test kit. J. Clin. Microbiol. **20:**846–848.
68. **Moellering, R. C., Jr., O. M. Korzeniowski, M. A. Sande, and C. B. Wennersten.** 1979. Species-specific resistance to antimicrobial synergism in *S. faecium* and *S. faecalis*. J. Infect. Dis. **140:**203–208.
69. **Moody, M. D., A. C. Siegel, B. Pittman, and C. C. Winter.** 1963. Fluorescent-antibody identification of group A streptococci from throat swabs. Am. J. Public Health **53:**1083–1092.
70. **Murray, H. W., K. C. Gross, H. Masur, and R. B. Roberts.** 1978. Serious infections caused by *Streptococcus milleri*. Am. J. Med. **64:**759–764.
71. **Murray, P. R., A. D. Wold, M. Marsha, and J. A. Washington.** 1976. Bacitracin differentiation for presumptive identification of group A streptococci: comparison of primary and purified plate testing. J. Pediatr. **89:**576–579.
72. **Murray, P. R., A. D. Wold, C. A. Schreck, and J. A. Washington II.** 1976. Effects of selective media and atmosphere of incubation on the isolation of group A streptococci. J. Clin. Microbiol. **4:**54–56.
73. **Nachamkin, I., J. R. Lynch, and H. P. Dalton.** 1982. Evaluation of a rapid system for species identification of alpha-hemolytic streptococci. J. Clin. Microbiol. **16:**521–524.
74. **Nakamizo, Y., and M. Sato.** 1972. New selective media for the isolation of *Streptococcus hemolyticus*. Am. J. Clin. Pathol. **57:**228–235.
75. **Parades, A., P. Wong, E. O. Manson, L. H. Tabor, and F. F. Barrett.** 1977. Nosocomial transmission of group B streptococci in a newborn nursery. Pediatrics **59:**679–682.
76. **Parker, M. T., and L. Y. Ball.** 1976. Streptococci and aerococci associated with systemic infections in man. J. Med. Microbiol. **9:**275–302.
77. **Patterson, M. J., and A. E. B. Hafeez.** 1976. Group B streptococci in human disease. Bacteriol. Rev. **40:**774–792.
78. **Petts, D. N.** 1984. Colistin-oxolinic acid-blood agar: a new selective medium for streptococci. J. Clin. Microbiol. **19:**4–7.
79. **Pien, F. D., C. L. Ow, N. S. Isaacson, N. T. Goto, and R. C. Rudoy.** 1979. Evaluation of anaerobic incubation for recovery of group A streptococci from throat cultures. J. Clin. Microbiol. **10:**392–393.
80. **Poon-King, T., I Mohammed, R. Cox, E. V. Potter, N. M. Simon, A. C. Siegal, and D. P. Earle.** 1967. Recurrent epidemic nephritis in South Trinidad. N. Engl. J. Med. **277:**728–733.
81. **Quinn, R. W.** 1982. Epidemiology of group A streptococcal infections—their changing frequency and severity. Yale J. Biol. Med. **55:**265–270.
82. **Robbins, J. B., R. Austrian, C. J. Lee, S. C. Rastogi, G. Schiffman, J. Henrichsen, P. H. Makela, C. V. Broome, R. R. Facklam, R. H. Tiesjema, and J. C. Park.** 1983. Considerations for formulating the second generation pneumococcal capsular polysaccharidal vaccine with emphasis on the cross reactive types within groups. J. Infect. Dis. **148:**1136–1159.

83. **Roberts, R. B., A. G. Krieger, N. L. Schiller, and K. C. Gross.** 1979. Viridans streptococcal endocarditis: the role of various species, including pyridoxal-dependent streptococci. Rev. Infect. Dis. **1:**955–965.
84. **Rolston, K. V. I., J. L. LeFrock, and R. F. Schell.** 1982. Activity of nine antimicrobial agents against Lancefield group C and group G streptococci. Antimicrob. Agents Chemother. **22:**930–932.
85. **Romero, R., and H. W. Wilkinson.** 1974. Identification of group B streptococci by immunofluorescence staining. Appl. Microbiol. **28:**199–204.
86. **Ruoff, K. L., M. J. Ferraro, M. E. Jerz, and J. Kissling.** 1982. Automated identification of gram-positive bacteria. J. Clin. Microbiol. **16:**1091–1095.
87. **Ruoff, K. L., and L. J. Kunz.** 1983. Use of the Rapid STREP system for identification of viridans streptococcal species. J. Clin. Microbiol. **18:**1138–1140.
88. **Sato, M.** 1972. A new selective enrichment broth for detecting beta-hemolytic streptococci in throat cultures: quinoline derivate and three percent salt as an additional agent to Pike's inhibitors. Jpn. J. Microbiol. **16:**538–540.
89. **Savitch, C. B., A. L. Barry, and P. D. Hoeprich.** 1978. Infective endocarditis caused by *Streptococcus bovis* resistant to the lethal effect of penicillin G. Arch. Intern. Med. **138:**931–934.
90. **Schauf, V., A. Deveikis, L. Riff, and A. Serota.** 1976. Antibiotic killing kinetics of group B streptococci. J. Pediatr. **89:**194–198.
91. **Setterstrom, J. A., A. Gross, and R. S. Stanko.** 1979. Comparison of Minitek and conventional methods for the biochemical characterization of oral streptococci. J. Clin. Microbiol. **10:**409–414.
92. **Shlaes, D. M., P. I. Lerner, E. Wolinsky, and K. V. Gopalakrishna.** 1981. Infections due to Lancefield group F and related streptococci (*S. milleri, S. anginosus*). Medicine **60:**197–207.
93. **Siefkin, A. D., D. L. Peterson, and B. Hansen.** 1983. *Streptococcus equisimilis* pneumonia in a compromised host. J. Clin. Microbiol. **17:**386–388.
94. **Siegal, J. D., K. M. Shannon, and B. M. DePasse.** 1981. Recurrent infection associated with penicillin tolerant group B streptococci: a report of two cases. J. Pediatr. **99:**920–924.
95. **Skerman, V. B. D., V. McGowan, and P. H. A. Sneath (ed.).** 1980. Approved lists of bacterial names. Int. J. Syst. Bacteriol. **30:**225–420.
96. **Slifkin, M., D. Freedel, and G. M. Gil.** 1982. Direct serogrouping of group B streptococci from urogenital and gastric swabs with nitrous acid extraction and the Phadebact streptococcus test. J. Clin. Pathol. **78:**850–853.
97. **Slifkin, M., and G. M. Gil.** 1982. Serogrouping of beta-hemolytic streptococci from throat swabs with nitrous acid extraction and the Phadebact Streptococcus Test. J. Clin. Microbiol. **15:**187–189.
98. **Slifkin, M., and G. M. Gil.** 1984. Evaluation of the Culturette Brand Ten-Minute Group A Strep ID technique. J. Clin. Microbiol. **20:**12–14.
99. **Slifkin, M., and G. Interval.** 1980. Serogrouping single colonies of beta-hemolytic streptococci from primary throat culture plates with nitrous acid extraction and Phadebact streptococcal reagents. J. Clin. Microbiol. **12:**541–545.
100. **Slifkin, M., and G. R. Pouchet-Melvin.** 1980. Evaluation of three commercially available test products for serogrouping beta-hemolytic streptococci. J. Clin. Microbiol. **11:**249–255.
101. **Sprunt, K., D. Vail, and R. S. Asnes.** 1974. Identification of streptococcus pyogenes in a pediatric outpatient department: a practical system designed for rapid results and resident teaching. Pediatrics **54:**718–723.
102. **Stamm, A. M., and C. G. Cobbs.** 1980. Group C streptococcal pneumonia: report of a fatal case and review of the literature. Rev. Infect. Dis. **2:**889–898.
103. **Stamm, W. E., J. C. Feeley, and R. R. Facklam.** 1978. Wound infection due to group A streptococcus traced to a vaginal carrier. J. Infect. Dis. **138:**287–292.
104. **Thornsberry, C., C. N. Baker, and R. R. Facklam.** 1974. Antibiotic susceptibility of *Streptococcus bovis* and other group D streptococci causing endocarditis. Antimicrob. Agents Chemother. **5:**228–233.
105. **Vosti, K. L.** 1977. Streptococcal diseases, p. 235–246. *In* P. D. Hoeprich (ed.), Infectious diseases, 2nd ed. Harper & Row, Publishers, Hagerstown, Md.
106. **Waitkins, S. A., D. R. Anderson, and P. K. Todd.** 1981. An evaluation of the API-Strep identification system. Med. Lab. Sci. **38:**35–39.
107. **Wannamaker, L. W.** 1979. Changes and changing concepts in the biology of group A streptococci and in epidemiology of streptococcal infections. Rev. Infect. Dis. **1:**967–973.
108. **Ward, J.** 1981. Antibiotic-resistant *Streptococcus pneumoniae*: clinical and epidemiologic aspects. Rev. Infect. Dis. **3:**254–266.
109. **Wellstood, S.** 1982. Evaluation of Phadebact and Streptex kits for rapid grouping of streptococci directly from blood cultures. J. Clin. Microbiol. **15:**226–230.
110. **Wetkowski, M. A., E. M. Peterson, and L. M. De La Maza.** 1982. Direct testing of blood cultures for detection of streptococcal antigens. J. Clin. Microbiol. **16:**86–91.
111. **Whittenbury, R.** 1965. A study of some pedicocci and their relationship to *Aerococcus viridans* and the enterococci. J. Gen. Microbiol. **40:**97–106.
112. **Wilkinson, H. W.** 1978. Analysis of group B streptococcal types associated with disease in human infants and adults. J. Clin. Microbiol. **7:**176–179.
113. **Zanen, H. C., and H. W. B. Engel.** 1975. Porcine streptococci causing septicemia in man. Lancet **i:**1286–1288.

Chapter 17

Neisseria and *Branhamella*

JOSEPHINE A. MORELLO, WILLIAM M. JANDA, AND MARJORIE BOHNHOFF

CHARACTERIZATION

Neisseria and *Branhamella* species are classified in the family *Neisseriaceae*, along with *Kingella*, *Moraxella*, and *Acinetobacter* species. Genetic studies have indicated that *Branhamella* species, formerly classified as *Neisseria* species, are more closely related to members of the genus *Moraxella*. In current taxonomic literature, therefore, branhamellae are considered to belong in a subgenus of *Moraxella*, and the most common species found in humans is referred to as *Moraxella (Branhamella) catarrhalis* (2). Because clinical microbiology terminology sometimes lags behind taxonomy, the branhamellae will be considered in this chapter under the genus name *Branhamella*.

Species of *Neisseria* and *Branhamella* are alike with regard to their morphology, limited metabolic activity, and usual habitat, but differ in their pathogenic potential and genetic relatedness (21). These organisms are gram-negative cocci that may be somewhat refractory to decolorization. The cells usually occur in pairs with flattened adjacent sides, a characteristic that is responsible for their kidney or coffee bean appearance in microscopic preparations. Cell division is in two planes at right angles, sometimes resulting in tetrad formation. Individual cells may vary in size from 0.6 to 1.5 μm, depending on the species, source of the isolate (clinical specimen or culture), and age of the culture. One species, *Neisseria elongata*, is a short rod, 0.5 μm wide, which often occurs as a diplobacillus (50). Older cells of some species tend to autolyze in culture, with resulting ballooning of the organisms. The cells produce no endospores and are nonmotile. Some species are encapsulated, and some produce a yellow-green carotenoid pigment.

Physiologically, most strains of *Neisseria* and *Branhamella* have complex growth requirements, have an optimum growth temperature of 35 to 37°C, and degrade few or no carbohydrates. All species are aerobes that produce the enzymes cytochrome oxidase and, except for *N. elongata*, catalase.

The natural habitat of these organisms is the mucous membranes of warm-blooded animals. Only two species, *Neisseria gonorrhoeae* and *Neisseria meningitidis*, are considered to be primary pathogens, and both infect humans only. Although *N. gonorrhoeae* is isolated from many patients with asymptomatic infection, it is always considered to be a disease producer. On the other hand, *N. meningitidis* may be isolated from the nasopharynx and throat of a variable proportion of healthy humans as well as from patients with meningococcal disease. Most other *Neisseria* species and *B. catarrhalis* have been isolated from serious infections, but they are considered to be opportunistic pathogens.

Since the isolation and identification of *N. gonorrhoeae* and *N. meningitidis* are of primary interest in the clinical microbiology laboratory, the major portion of this chapter deals with the clinical and diagnostic features of these two organisms. The chapter concludes with a brief discussion of other *Neisseria* species and *B. catarrhalis*.

NEISSERIA GONORRHOEAE

Clinical significance

Gonorrhea is the most commonly reported bacterial infection in the United States, occurring in an estimated 3 million individuals each year. Beginning in 1964, the incidence of gonorrhea increased ca. 10% each year until 1977, when the rate of infection began to level off. Two important factors responsible for the gonorrhea "epidemic" include the increasing resistance of *N. gonorrhoeae* to available antibiotics and a large reservoir of asymptomatic females and males who unknowingly transmit the disease to their sexual partner(s).

N. gonorrhoeae is highly susceptible to adverse environmental influences such as temperature extremes and drying, and it does not survive long outside its natural host. Gonorrhea is transmitted between individuals by direct, close, usually sexual contact. In males the most common manifestation of infection is acute urethritis, characterized by the abrupt onset of dysuria and a purulent urethral discharge. The incubation period is ca. 3 to 4 days. Complications such as prostatitis, epididymitis, and urethral stricture occur less frequently now than in the preantibiotic era, presumably because of successful therapy of initial uncomplicated infections. The overall incidence of asymptomatic gonorrhea in men has been estimated to be from 3 to 10%, although it may be 40% or higher in contacts of symptomatic females (20).

The primary site of urogenital gonorrhea in females is the endocervix. Infection may be accompanied by a purulent vaginal discharge, menstrual irregularity, and dysuria. A majority of women with uncomplicated gonorrhea, however, remain asymptomatic or do not develop severe enough symptoms to seek medical help. In ca. 15% of females with gonococcal cervicitis, the infection ascends, resulting in salpingitis, pelvic peritonitis, or both, complications referred to as pelvic inflammatory disease, or PID (20). Although other organisms may cause PID, gonococci are often isolated from the cervices of affected women, indicating that gonococci probably play a major role at least in initial episodes of this disease. Repeated episodes of PID commonly involve anaerobes from the vaginal flora. Gonococcal PID usually is manifest during the first few days of the menstrual cycle and occurs more frequently in women using an intrauterine device for contraception. The inflammatory reaction of PID may cause tubal scarring and blockage, with resulting infertility. Some women with gonococcal PID develop perihepatitis (Fitz-Hugh–Curtis syndrome), presumably by spread of gonococci through the peritoneal cavity. Gonococci from cervical secretions may also contaminate the female perineum, resulting in anorectal and periurethral gland infections.

Although anorectal and oropharyngeal gonorrhea can occur in women as a result of rectal intercourse and fellatio, infections at these sites are more common in homosexual males. The majority of cases of both anorectal and oropharyngeal gonorrhea are asymptomatic, but patients may present with proctitis or acute pharyngitis and tonsillitis (25).

In a small percentage of individuals with gonorrhea, the organisms spread hematogenously to produce disseminated gonococcal infection (DGI). DGI is characterized by the presence of a sparse rash on the extremities and arthritis in one or more joints. At present, gonococci are the major cause of septic arthritis in young adults. Although *N. gonorrhoeae* may be isolated from the blood, skin lesions, or joint fluid of patients with DGI, often the organism is isolated only from genital, anorectal, or pharyngeal cultures of affected individuals with signs and symptoms of DGI. Disseminated infection occurs more commonly in females than males, perhaps because there is a greater tendency for the organisms to disseminate during menstruation and pregnancy. In rare instances, DGI may result in gonococcal endocarditis or meningitis.

In some areas of the United States, the unusual AHU strain of *N. gonorrhoeae*, which requires arginine, hypoxanthine, and uracil for growth on defined media, has been isolated from a disproportionate number of patients with DGI (38) and from males with asymptomatic urethral gonorrhea (9). The ability of these AHU strains to withstand the bactericidal action of human serum may be responsible, in part, for their increased tendency to disseminate (13).

Infants born to infected mothers may develop gonococcal ophthalmia (ophthalmia neonatorum) if the prophylactic eyedrops or ointment usually administered at birth are not used or, as happens rarely, are not effective. Gonococcal ophthalmia must be treated immediately, or blindness can result. Nonsexually transmitted infections may occur in very young children, usually after accidental contamination with discharges from an infected parent or relative. In older children, however, gonococcal infections are acquired primarily by precocious sexual activity or sexual molestation. The primary syndrome in prepubertal girls is vulvovaginitis; otherwise, the clinical picture is similar to that seen in adults.

Specimen collection

For diagnosing gonorrhea, appropriate sites for specimen collection depend to some extent on the age, sex, and sexual practices of the individual and the clinical features of the infection. When specimens are taken, lubricate any instruments used (vaginal speculums, anoscopes) with warm water only, as other lubricants may be toxic for gonococci. Specimens may be collected on swabs or, in some instances, with spur-free, platinum-iridium bacteriological loops. Some cotton swabs contain unsaturated fatty acids that are inhibitory for gonococci (28). Use these only if the specimen can be plated immediately onto culture medium or sent to the laboratory in transport medium containing activated charcoal to adsorb the inhibitory substances. Preferably, use swabs composed of nontoxic synthetic material such as calcium alginate or Dacron. Collect material on separate swabs for Gram stain and for culture. Appropriate culture sites and collection techniques are described below (and also in Table 1).

TABLE 1. Body sites to culture for *N. gonorrhoeae*

Patient	Primary site	Secondary site
Female	Endocervix	Rectum, (urethra), pharynx[a]
Male heterosexual	Urethra	
Male homosexual	Urethra, rectum, pharynx	
DGI, female	Blood, endocervix, rectum	Pharynx,[a] skin lesions,[b] joint fluid[b]
DGI, male	Blood, urethra	Pharynx,[a] skin lesions,[b] joint fluid,[b] rectum[c]

[a] If history of oral-genital contact.
[b] If present.
[c] If history of anal-genital contact.

Endocervix. For suspected gonorrhea or when screening for infection in women, always collect endocervical specimens because they most often yield gonococci in both symptomatic and asymptomatic infections. After inserting the speculum, wipe away cervical mucus with a cotton ball; then carefully swab the endocervical canal, moving the swab from side to side. Do not use any disinfectant before collecting the specimen, and avoid contaminating the swab with vaginal flora.

Urethra. Collect urethral specimens from symptomatic and asymptomatic males at least 1 h after they have urinated. Purulent discharge can be collected directly on a swab, or if there is no discharge, obtain a specimen by gently scraping the mucosa of the anterior urethra with a sterile loop. Preferably, insert a calcium alginate urethral or nasopharyngeal swab ca. 2 cm into the urethra and rotate the swab gently as it is withdrawn. Collect female urethral specimens in a similar manner from symptomatic females with urethral exudate.

Culture of first-voided urine from both males (33) and females may yield the organism as well. Centrifugation of the specimen may increase gonococcal recovery.

Anorectal specimens. Collect anorectal specimens from all patients who may have DGI, homosexual males, females with urogenital symptoms, and asymptomatic females likely to be infected (e.g., contacts of known or suspected infected males). In some instances, gonococci may be isolated from an anal canal culture when the endocervical culture is negative, especially after antimicrobial therapy (i.e., test of cure). Obtain the specimen by inserting a swab ca. 4 to 5 cm into the anal canal. Move the swab from side to side to sample the crypts. If fecal contamination occurs, discard the swab and use another to obtain the specimen. An anoscope may be used to visualize and collect mucopurulent material directly, but it is not mandatory.

Oropharynx. When oropharyngeal infection is suspected, swab the posterior pharynx and the region of the tonsillar crypts.

Conjunctiva. In infants (and occasionally other age groups) with conjunctivitis, obtain swabs of conjunctival exudate for Gram stain and culture.

Bartholin's gland. Collect any draining pus on swabs, or aspirate material from closed abscesses with a syringe and needle. In the latter instance especially, infection with anaerobes as well as gonococci should be considered, and the specimens should be handled appropriately.

Blood. Obtain blood for culture from any patient with suspected DGI, and inoculate it directly into a suitable culture medium. Blood may be sent to the laboratory in a sterile, evacuated blood collection tube containing sodium polyanetholsulfonate (SPS). SPS is toxic for some strains of gonococci (see blood culture section); therefore, if this method is used, the blood specimen should be inoculated into broth as soon as possible, preferably within 1 h after collection.

Joint fluid. In patients with DGI, aspirate material from possibly infected joints. Transport the material directly in a capped syringe or in a sterile specimen container to the laboratory for Gram stain and culture.

Skin lesions. A small punch biopsy is preferable to material aspirated from the lesion. Place the specimen on a chocolate agar plate or other sterile, moist container to prevent drying during transport. The positive culture rate is low.

Specimen transport

Gonococci are highly susceptible to adverse environmental conditions, and often only small numbers of organisms are present in clinical material. When specimens are sent to the laboratory, they must be transported in a manner that preserves organism viability. Appropriate transport conditions depend to some extent on the type of facility in which the specimen is obtained, for example, public health clinic, private physician's office, hospital room or clinic, or emergency room. In each of these situations, the time elapsed between specimen collection and receipt at the laboratory differs and may be unpredictable.

Direct plating. The greatest number of gonococcal cultures are positive when specimens are inoculated in the patient care area directly onto a nutritive growth medium such as modified Thayer-Martin (MTM) (34), Martin-Lewis (ML) (35), or New York City (NYC) (16), incubated immediately at 35 to 37°C in an atmosphere of 3 to 10% CO_2, and examined within 48 h. This is the method of choice whenever an incubator and a candle extinction jar are available, and the jars can be transported readily to the laboratory after 24 to 48 h of incubation. These conditions seldom exist, however, and alternative transport methods are most commonly used.

Nonnutritive transport media. Buffered holding media such as that of Stuart and Amies (Chapter 111) were originally devised to maintain gonococci during transit, although at present they are widely used to preserve a great variety of clinically significant bacteria. Gonococci survive well in these media for at least 6 to 12 h, provided that they are not exposed to temperature extremes. By 24 h, however, the numbers of gonococci decrease to an extent that may prevent their recovery, especially if small numbers were present initially in the specimen. Do not use this type of transport medium if the specimen cannot be delivered to the laboratory and plated onto growth medium within 12 h. The advantages of these holding media are that they are simple to use, require no special equipment, are relatively inexpensive, and have a long shelf life at room temperature.

Nutritive (growth) transport systems. When prolonged delays in transit are expected, systems that incorporate culture medium and a method for increasing CO_2 concentration are required. A commonly used approach is inoculation of a biological environment chamber such as the JEMBEC plate (36). This plate is a rectangular, polystyrene dish with a removable cover and a molded inner well that holds a CO_2-generating tablet (composed of sodium bicarbonate and citric acid). The plates are prepared with one of the gonococcal selective media, such as MTM, ML, or NYC. For inoculation, roll the specimen swab over one-third to one-half of the agar surface; then cross-streak the remainder of the plate with a bacteriological loop or a fresh swab. Insert a CO_2-generating tablet in the well, replace the cover, and place the entire unit into a small, plastic, zip-locked bag. Moisture evaporated from the medium is sufficient to activate the tablet and provide a CO_2 atmosphere during incubation. For maximum recovery and survival of gonococci, incubate the inoculated plates for ca. 18 to 24 h before delivering or mailing them to the laboratory. If the plates are preincubated, they may be inspected for the presence of gonococcal colonies immediately upon arrival, and negative cultures may be incubated for an additional 24 h. If further incubation in the laboratory is necessary, remove the plates from the plastic bag and incubate them under increased CO_2 tension. Colonies growing on media in the JEMBEC plates are easy to observe and subculture because the plates are handled exactly like petri plates.

Media in JEMBEC plates may be stored refrigerated for 4 to 6 weeks if the plates are sealed in plastic to prevent drying. The media should be brought to room temperature before they are inoculated.

In other nutritive transport systems, for example, the Bio-Bag (Marion Scientific, Kansas City, Mo.) and the Gono-Pak System (Nasco, Modesto, Calif.), conventional culture plates are inoculated and then placed into individual, plastic, zip-locked bags, each of which contains a CO_2-generating ampoule or tablet. These bags can be handled in the same manner as JEMBEC plates.

Transport containers for specimens not collected on swabs are discussed in the section above on specimen collection. All of these containers must be delivered to the laboratory as quickly as possible for proper plating and incubation.

Direct examination

Gram stain. The Gram stain is the method of choice for direct examination of genital specimens. When smears of urethral exudate from males are properly prepared and stained and correctly interpreted, correlation with culture results is greater than 98% (20). In many sexually transmitted disease clinics, such specimens are not cultured if the Gram-stained smear is positive. In females, however, smears of endocervical secretions detect only 50 to 70% of culture-positive specimens (32). Thus, both Gram-stained smears and cultures of endocervical secretions should be done. Examining Gram-stained smears of intraurethral specimens from asymptomatic males is not a sensitive diagnostic test; therefore, cultures are required as well. Smears prepared from blind rectal swabs are difficult to interpret, so smear examination is not recommended. Direct visualization of the rectal mucosa and sampling of purulent material through an anoscope frequently yield easily interpretable Gram-stained preparations (Fig. 1), but culture should also be performed. In our experience, smear examination has worked well for rapid diagnosis of symptomatic gonococcal proctitis in homosexual men. Gram-stained smears for detecting *N. gonorrhoeae* in oropharyngeal specimens are not recommended because other *Neisseria* and *Branhamella* species of the upper respiratory tract flora are morphologically indistinguishable from gonococci.

Smear preparation. For optimal results, smears should be prepared directly from the specimen immediately after collection. In smears prepared from swabs sent in an aqueous or semisolid transport medium, organisms may appear atypical. In addition, dilution of the specimen by the medium may affect the sensitivity of smear examination results, particularly if the discharge is not copious. If both a culture and a smear are required, two swab specimens should be collected. The swab should be rolled gently over the surface of a glass microscope slide in one direction only. This method minimizes distortion and breakage of polymorphonuclear leukocytes (PMN) and preserves the characteristic appearance of the microorganisms. Three or four complete, nonoverlapping rolls of the swab on the slide are sufficient. Excessively thick smears do not decolorize properly during the Gram stain procedure and may decrease its diagnostic utility.

Smear interpretation and reporting. Gram-stained smears of the urethral discharge from men with gonococcal urethritis typically contain gram-negative diplococci located intracellulary in PMN. Usually, two or more pairs of organisms are observed within some PMN, whereas many other leukocytes contain no bacteria at all (Fig. 2). Few, if any, other bacteria are seen on the smear. Smears from men with early symptomatic disease frequently show rare intracellular gram-negative diplococci and many extracellular organisms trapped in stringy mucous material. In Gram-stained smears of rectal mucopurulent material collected under direct visualization through an anoscope, gram-negative intracellular diplococci are usually seen.

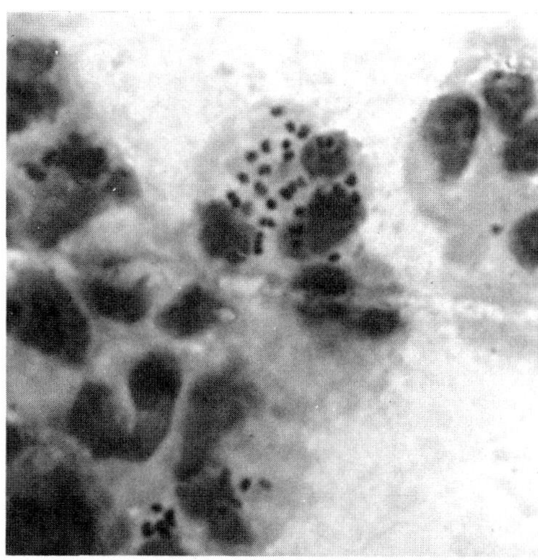

FIG. 2. Gram stain of male urethral exudate. Some PMN contain many diplococci; others contain none. ×1,500.

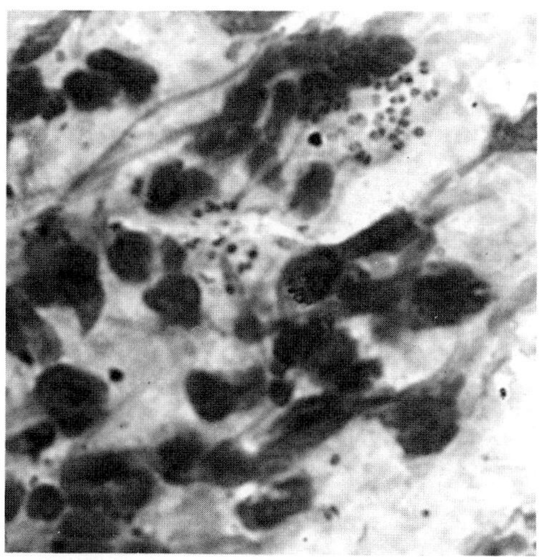

FIG. 1. Gram stain of mucopurulent rectal exudate showing many diplococci inside PMN. ×1,500.

Gram-stained smears from properly collected cervical discharge may show intracellular gram-negative diplococci, but accurate smear interpretation may be complicated by the presence of normal female genital flora that resemble *Neisseria* species. The tendency to overinterpret endocervical and rectal smears, which usually contain gram-negative coccobacilli and bipolarly staining enteric bacilli, must be avoided. The observations should be considered presumptive until culture results are known.

Report the numbers, Gram-stain properties, and morphology of all organisms present and the numbers and types of cells (PMN, squamous epithelial) seen in smears. Organisms resembling *Neisseria* species that are located within PMN should be quantified and reported as "gram-negative intracellular diplococci." The quantity may be described in words (few, moderate, many) or scored from 1+ to 4+ as described in *Cumitech 4* (28).

Gram stains prepared from conjunctival pus in ophthalmia and from petechial lesions and joint fluid of patients with DGI may aid diagnosis. In the latter two specimen types especially, organisms may not grow in culture even though they are seen in the smear.

Fluorescent-antibody test. Direct fluorescent-antibody techniques are not recommended for examining smears prepared from genital or extragenital specimens and should not be used as a replacement for either the Gram stain or culture. Commercially available reagents lack sufficient sensitivity and specificity to be of value. Not all strains of *N. gonorrhoeae* react with available conjugates.

TABLE 2. Antimicrobial agents present in selective media for pathogenic *Neisseria* species[a]

Antimicrobial agent	Function	Concn (μg/ml) in medium		
		MTM	ML	NYC
Vancomycin	Inhibits gram-positive bacteria	3	4	2
Colistin	Inhibits gram-negative bacteria, including saprophytic *Neisseria*	7.5	7.5	5.5
Nystatin	Inhibits airborne molds and some yeasts	12.5		
Anisomycin	Inhibits molds and some yeasts		20	
Amphotericin B	Inhibits molds and yeasts			1.2
Trimethoprim lactate	Inhibits swarming *Proteus*	5	5	3

[a] From J. A. Morello, API Species **5**:1–7, 1981; used by permission.

Culture and isolation

Specimens transported on selective agar media. Many specimens arrive at the laboratory previously plated onto selective agar media such as MTM, ML, and NYC. Basically, MTM and ML media are chocolate agar with four antibiotics added (Table 2). If they are correctly prepared and used while fresh, these media afford excellent growth of most strains of pathogenic *Neisseria* species, while inhibiting the growth of saprophytic *Neisseria* species and other commensals that are likely to contaminate oropharyngeal and anogenital specimens. In NYC medium (formulated at the New York City Department of Health), hemolyzed horse erythrocytes, horse plasma, and yeast dialysate are used in place of the hemoglobin and defined supplements present in chocolate, MTM, and ML media. Selectivity is provided by adding the antibiotics listed in Table 2. NYC medium has also been prepared without the hemolyzed horse cells and used successfully for gonococcal isolation (19). NYC medium is a clear medium that supports luxuriant growth of pathogenic *Neisseria* species, large-colony mycoplasma, and T-strain mycoplasma *(Ureaplasma urealyticum)*.

If the cultures have been incubated at 35 to 37°C for 18 h or more before they arrive at the laboratory, examine them immediately to detect typical colonies of *N. gonorrhoeae*. If they have not been preincubated or if no growth is observed, incubate them in a CO_2 atmosphere and observe them daily for at least 2, but preferably 3, days.

From 3 to 10% or more of gonococci (usually AHU strain organisms) are susceptible to the concentrations of vancomycin used in the selective media and, therefore, will not be recovered (37). Some gonococcal strains are also susceptible to trimethoprim, but the concentrations used in these media do not appear to affect their recovery as much as vancomycin does. Nevertheless, because the normal flora of the pharynx, rectum, and female genital tract grows more rapidly than gonococci and in some instances inhibits gonococcal growth (43), antibiotic-containing media that suppress these commensals usually permit better isolation of gonococci than do nonselective media.

Other specimens. As soon as possible after their arrival in the laboratory, inoculate onto growth medium all specimens received in nonnutritive transport medium or sterile containers. Conjunctival swabs, skin biopsies, and joint fluids can be plated onto chocolate agar rather than a selective medium because they are not likely to be contaminated with saprophytes. Whenever possible, inoculate male and female genital exudate specimens onto chocolate agar as well as onto a selective medium to help detect vancomycin-susceptible organisms.

Inoculate chocolate agar by rolling the swab across one quadrant of the plate, making certain that all areas of the swab are sampled. Streak the remainder of the plate with an inoculating loop to obtain isolated colonies after incubation. Heavily inoculate selective media by rolling the swab across one-third to one-half of the plate surface; then streak to obtain isolated colonies.

Most of the commonly used blood culture broths (e.g., tryptic soy, Columbia, brain heart infusion) support the growth of gonococci, provided that the broth is vented and incubated under increased CO_2 tension. SPS, a common blood culture broth additive, inhibits the growth of some gonococci and meningococci (15, 46). Gelatin appears to neutralize this inhibitory effect (15, 46) and can be added to the broth (in a 1% final concentration) when either of these organisms is suspected. Joint fluids may be inoculated into liquid as well as onto solid media. The BACTEC system (Johnston Laboratories, Inc., Towson, Md.) is satisfactory for the growth and detection of *N. gonorrhoeae*.

Incubation conditions. Although not all gonococci have an absolute requirement for CO_2, their growth is stimulated by it. Therefore, always incubate specimens in a CO_2 incubator or candle extinction jar. Use white wax candles because colored candles may produce toxic substances while burning. The level of CO_2 in the jar is ca. 3% when the candle flame dies out. Alternatively, use commercially available CO_2-generating envelopes (similar to envelopes that produce an anaerobic environment) (BBL Microbiology Systems, Cockeysville, Md.) in a tightly closed chamber such as a candle jar or an anaerobic jar with the catalyst removed.

High humidity is also essential for good gonococcal growth. In a candle jar, humidity is provided by evaporated moisture from the culture medium. In CO_2 incubators that do not have built-in humidifiers, pans of water on the lower shelf can provide moisture during incubation. The surface of commercially prepared agar plates often appears dry. Their performance can be improved by adding a few drops of sterile culture broth to the plate surface and allowing it to soak in before the plate is streaked.

Incubate agar media at 35 to 37°C for at least 72 h before sending a negative culture report. For blood broths that do not show microbial growth after 24 to

48 h of incubation, make blind subcultures on chocolate agar. Hold blood cultures for at least 7 days before discarding them.

Identification

Gonococcal colonies are best observed with the aid of a magnifying glass or, preferably, a stereoscopic microscope. After 24 h of incubation, the colonies are ca. 0.5 to 1 mm in diameter and appear gray to white, opaque, raised, and glistening. Isolated colonies may increase in size to 3 mm with further incubation.

In culture, gonococci produce at least five different colony types, termed types 1 through 5 (T1 through T5) (4, 29). T1 and T2 colonies are small and raised and predominate in cultures of clinical material. When colonies growing on chocolate agar are viewed microscopically with incident light (striking their surface from above), T1 and T2 colonies reflect the light and appear to have bright highlights. Cultures can be maintained in this state by selectively subculturing T1 and T2 colonies. With nonselective subculture, however, these colony forms dissociate to types 3 to 5, which are larger, flatter, and less opaque and do not reflect the light (Fig. 3). Even gross observation of gonococcal cultures, without a stereoscopic microscope, will reveal colonies of different size and morphology, especially if the culture has been transferred two or three times. This colony variation is sometimes mistaken for a mixed culture but can serve as a clue to the presence of *N. gonorrhoeae*. Most organisms that may be confused with gonococci produce colonies of uniform size.

So-called atypical or AHU strains of gonococci are nutritionally fastidious and require arginine, hypoxanthine, and uracil for growth on defined media. These strains grow slowly, produce smaller than normal colonies on agar plates, and are often difficult to identify biochemically (38). Microbiologists must be aware of their existence because, in certain areas of the United States, they are isolated from a majority of patients with DGI and asymptomatic infections. AHU strains also produce several colony types in culture, but all forms are smaller than those produced by normal strains (Fig. 4).

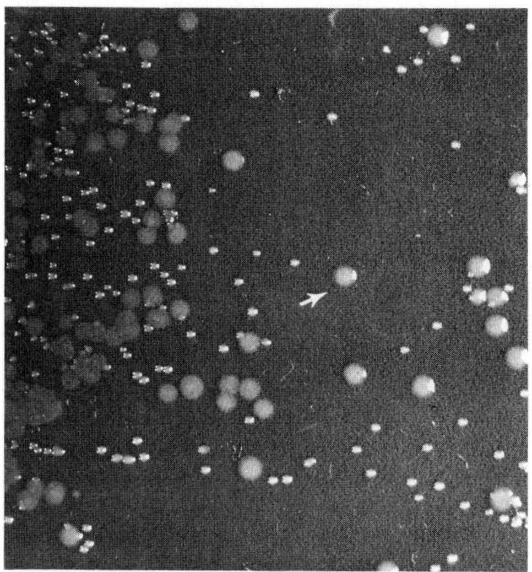

FIG. 3. *N. gonorrhoeae* on chocolate agar. Small colonies with bright highlights are types T1-T2; larger colonies (arrow) are types T3-T4. ×2.5.

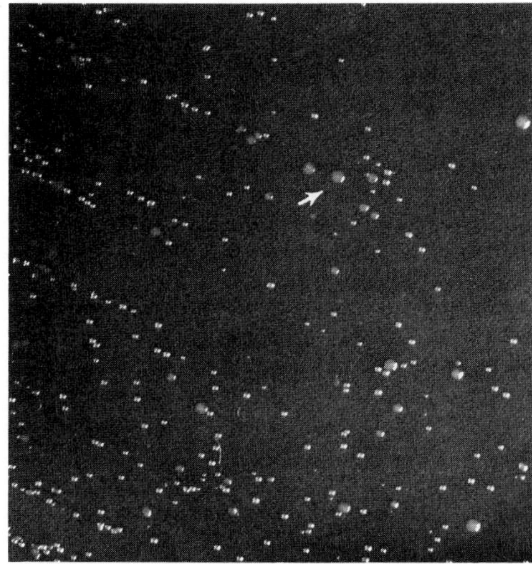

FIG. 4. AHU strain of *N. gonorrhoeae* on chocolate agar. Compare the sizes of types T1-T2 and T3-T4 (arrow) with the sizes shown in Fig. 3. ×2.5.

Presumptive identification. Examine the culture plates daily for at least 72 h. Colonies of the slow-growing AHU strain may not be visible before 48 h. Flooding the plates with oxidase reagent after 48 to 72 h may help detect very small numbers of colonies, but these must be subcultured immediately because the reagent kills the organisms. Once plates have grown out, do not continue incubating them because gonococci autolyze as they age, and the culture becomes nonviable. If identification tests cannot be performed when the plates are removed from the incubator, place them in a candle extinction jar, and store the jar at room temperature. Although further work-up should be performed as soon as possible, the organisms will survive for 2 or 3 days under these conditions. If they are left this long, however, prepare a fresh subculture before proceeding with biochemical tests.

Before attempting further identification of gonococci, perform an oxidase test and Gram stain on all suspicious colonies. Place a few drops of oxidase reagent (1% solution of di- or tetramethyl-*p*-phenylenediamine dihydrochloride) on a piece of filter paper, pick up a portion of a colony with a platinum inoculating loop or straight wire, and rub the growth into the impregnated area of the filter paper. With the tetramethyl reagent, the dark purple of a positive test appears rapidly, usually within 10 s if the test culture is fresh. The dimethyl reagent is more toxic, and the color change, proceeding through pink and maroon to black, is slower. Use of Nichrome wire instead of platinum inoculating needles may cause false-positive reactions. Commercially available paper strips impregnated with oxidase reagent (Pathotec-CO; General Diagnostics, Warner-Lambert Co., Morris Plains, N.J.) are satisfactory for this test.

Even when the oxidase test is positive, a Gram stain is essential because certain oxidase-positive, gram-negative bacilli, such as *Kingella denitrificans*, grow on antibiotic-containing selective media and produce colonies resembling those of gonococci (40). Oxidase-positive, gram-negative diplococci isolated from urogenital specimens may be identified presumptively as *N. gonorrhoeae*.

Confirmatory identification. Confirmatory tests are recommended for all isolates from urogenital sites and are required for isolates from other body sites. Methods available include assays for carbohydrate degradation that depend on growth, preformed enzymes, or radiometric detection; chromogenic enzyme substrate tests; and serological methods such as fluorescent-antibody and coagglutination tests.

Carbohydrate degradation tests in CTA base medium. The standard method for identifying *Neisseria* species is to determine their acid production from carbohydrates in a cystine-tryptic digest agar (CTA) base medium (51). Conventional CTA-carbohydrate media are semisolid agar deeps containing 1% filter-sterilized carbohydrate. The test battery includes glucose, maltose, sucrose, fructose, lactose, and a carbohydrate-free control tube. The test for enzymatic breakdown of *o*-nitrophenyl-β-D-galactopyranoside (ONPG) may be substituted for the lactose tube.

Many laboratory workers have difficulty obtaining satisfactory results with CTA media, but when the test is performed properly, an identification usually can be made within 24 h. The major pitfalls are failure to (i) use a heavy inoculum, (ii) work with pure cultures, and (iii) use reagent-grade carbohydrates. To avoid the first two problems, do not attempt to inoculate carbohydrate tubes with growth from the selective primary isolation plates, which may contain only a few colonies of gonococci as well as inhibited contaminants. Instead, first subculture several typical colonies onto one or two chocolate agar plates (not selective medium), and incubate them in CO_2 for ca. 18 h. With longer incubation, the organisms begin to autolyze, the inoculum becomes gummy (related to released nucleoprotein from autolyzed gonococci), and biochemical reactions are delayed. AHU strains of gonococci, however, may require incubation for 24 h or longer to achieve sufficient growth on chocolate agar.

Examine the subculture plates carefully to ascertain that the cultures are pure (use magnification), and then perform an oxidase test and Gram stain to confirm the presence of gram-negative diplococci. The carbohydrate medium may be inoculated by either of the following methods. (i) Prepare a dense suspension of the organisms in 0.5 ml of saline. With a capillary pipette, dispense 1 drop of suspension onto the surface of each agar deep; then stab the inoculum into the upper third of the medium. (ii) For each tube, scrape a full 3-mm loopful of growth from the surface of the chocolate agar plate and deposit this inoculum a few millimeters below the surface of the medium. In our experience, the latter method provides more rapid and reliable results. A change in the color of the phenol red indicator from orange-pink to yellow, signifying carbohydrate degradation, often occurs within 1 or 2 h.

In our laboratory, we use a modified CTA-carbohydrate medium which consists of CTA agar (1.5%) slants with 2% carbohydrates. A full loopful of growth from a chocolate agar plate is deposited on the surface of the agar slant, which provides the aerobic neisseriae with a larger surface area than do the agar deeps. Carbohydrate breakdown is detected first in the area of the slant directly under the inoculum, usually within 1 h and almost invariably by 4 h.

Incubate carbohydrate tubes at 35 to 37°C in a standard, not CO_2, incubator. Inspect the tubes at periodic intervals for 24 h. If the inoculum is sufficient, positive reactions should be seen within this time, and further incubation is unnecessary. A few strains may require more prolonged incubation. Characteristically, gonococci produce acid from glucose, but not from maltose, sucrose, fructose, or lactose (Table 3).

Problems that occur when CTA-carbohydrate media are used for gonococcal identification include: (i) no carbohydrate degradation observed—this may be related to lack of sufficient inoculum or use of an old culture; (ii) degradation of both glucose and maltose—this is usually related to contamination of the maltose reagent with glucose or to the presence of a meningococcus rather than a gonococcus; (iii) all carbohydrates appear degraded—this is related either to incubation of the tubes in a CO_2 atmosphere (faint color change) or to a contaminated inoculum (usually a bright yellow throughout the tube). Any tube that appears bright yellow should be considered possibly contaminated, and a smear should be prepared for Gram stain.

When either of the first two problems listed above is encountered, it is best to confirm the identification with one of the alternative tests to be described. Until confirmatory identification is completed, suspected gonococci should be subcultured daily onto chocolate agar so that a viable culture is always available for repeat testing.

Rapid fermentation test. A nongrowth carbohydrate degradation test is described in detail in *Cumitech 4* (28). Basically, small volumes of phosphate-buffered carbohydrate solutions with phenol red indicator are inoculated heavily with the test organism and incubated at 35°C. Results are available within 1 to 4 h. This test is very economical, the reagents are easy to prepare and inoculate, and the results are clear-cut.

The RIM-Neisseria kit (American Micro Scan, Campbell, Calif.) is a commercially available adaptation of the rapid fermentation test. Small quantities of buffered carbohydrate-phenol red solutions are added to special buffered tubes included in the kit and are inoculated with the harvesting loops provided. Results are easily interpreted and often may be observed within 30 min of the 1-h incubation period.

Both of these systems may be inoculated directly from the primary isolation plate if sufficient colonies are present. Because bacterial growth does not occur, small numbers of contaminants that may be present do not affect the results, but incubation cannot be continued overnight.

Minitek method. In the Minitek method (BBL), paper disks impregnated with high concentrations of the appropriate carbohydrates are placed in individual wells of a plastic plate. A small volume of a heavy suspension of the organism prepared in broth is pipetted into each disk-containing well. Plates are incubat-

TABLE 3. Characteristics of *Neisseria* species and *B. catarrhalis*[a]

Species	Colony morphology	Growth on: MTM, ML, or NYC medium	Growth on: Chocolate or blood agar at 22°C	Growth on: Nutrient agar at 35°C	Acid production from: Glucose	Acid production from: Maltose	Acid production from: Lactose	Acid production from: Sucrose	Acid production from: Fructose	Reduction of: NO_3	Reduction of: NO_2	Polysaccharide synthesis	DNase
N. gonorrhoeae	Gray to white, smooth, five colony types on subculture from primary	+	0	0	+	0	0	0	0	0	0	0	0
N. meningitidis	Nonpigmented or gray to white, some yellowish, smooth, transparent, encapsulated strains mucoid	+	0	0	+	+	0	0	0	0	V	0	0
N. lactamica	Nonpigmented or yellowish, smooth, transparent	+	V	+	+	+	+	0	0	0	+	0	0
N. sicca	Nonpigmented, wrinkled, coarse and dry, adherent	0	V	+	+	+	0	+	+	0	+	+	0
N. subflava	Greenish, yellow, smooth, often adherent	0	V	V	+	+	0	V	V	0	+	V	0
N. mucosa	Sometimes yellowish, mucoid appearance due to capsule production	0	+	+	+	+	0	+	+	+	+	+	0
N. flavescens	Yellow, opaque, smooth	0	+	+	0	0	0	0	0	0	+	+	0
N. cinerea	Grayish-white, slightly granular	V	V	+	0	0	0	0	0	0	+	0	0
N. elongata[b]	Grayish-white, slight yellowish tinge, flat, glistening, dry, claylike consistency	0	+	+	0	0	0	0	0	0	+	0	0
B. catarrhalis	Nonpigmented or gray, opaque, smooth	0	+	+	0	0	0	0	0	+	+	0	+

[a] Symbols: +, strains typically positive, but genetic mutants that lack the requisite enzyme activity are occasionally encountered; 0, most strains negative; V, variable characteristic.
[b] Weakly positive or negative catalase test, in contrast to other *Neisseria* species.

ed without CO_2 and observed hourly. Most positive reactions occur within 4 h (39); the remainder are read after overnight incubation. Use of this system requires Minitek equipment (plates, pipette gun and tips, humidor, and disk dispenser) and is most convenient for those laboratories that use the Minitek system for the identification of other microorganisms.

BACTEC method. The BACTEC *Neisseria* Differentiation Kit (Johnston Laboratories) is an additional capability for users of the BACTEC radiometric instrument for blood cultures. The three vials in the kit contain small volumes of ^{14}C-labeled glucose, maltose, and fructose in deionized water. The vials are heavily inoculated with a broth suspension of the organism to be identified and then are incubated for 3 h at 35°C. Carbohydrate utilization is detected by the BACTEC instrument as $^{14}CO_2$ evolved from the vials. An ONPG test is performed concurrently to detect β-galactosidase (characteristic of *Neisseria lactamica*).

The BACTEC method is highly sensitive and may identify those strains of gonococci that do not produce detectable acid in growth-dependent identification systems (47). In addition, however, some strains of saprophytic *Neisseria* that do not degrade glucose in conventional tests may yield a low-positive glucose reading in the BACTEC system. Regardless of the identification system used, always be suspicious of "gonococci" isolated from unusual sources or unlikely clinical situations, and perform more than one confirmatory test with them.

Chromogenic enzyme substrate systems. Recently, two kits have become available that use both conventional biochemical reactions and novel bacterial enzyme substrates to identify *Neisseria* and *Branhamella* species. The API NeIdent panel (Analytab Products, Inc., Plainview, N.Y.) consists of 11 miniaturized conventional and chromogenic enzyme substrate tests. A heavy suspension (McFarland no. 3) of organisms is prepared from either selective or nonselective isolation medium in a special buffer and inoculated onto the strip. After 4 h of incubation at 35°C, cinnamaldehyde reagent is added to four of the wells to develop the tests, and all reactions are read and recorded. A four-digit code is generated that is used with the NeIdent Profile Register to identify *Neisseria* species, *B. catarrhalis*, and the *Neisseria*-like genera *Moraxella*, *Kingella*, and CDC groups M-5 and M-6. Early evaluations of the NeIdent system indicate a need for expansion and improvements in the data base (26).

The RapID NH system (Innovative Diagnostics Systems, Inc., Decatur, Ga.) is a plastic cuvette containing dehydrated substrates for 12 tests. The cuvette is inoculated with a McFarland no. 3 suspension of the organism to be identified, and it is incubated at 35°C for 4 h. Conventional tests and two aminopeptidase substrate tests are included in the system. After reagent addition to selected tests, positive and negative color reactions are recorded, and a code number is generated to aid identification. In one evaluation, the RapID NH system provided accurate identification of pathogenic *Neisseria* species (42). The system can also be used to identify *Haemophilus* species and other fastidious gram-negative bacilli.

Serological methods

Fluorescent-antibody test. The Centers for Disease Control now considers the fluorescent-antibody technique to be a presumptive rather than a confirmatory test for *N. gonorrhoeae* identification. Reagents are available from only one commercial vendor (Difco Laboratories, Detroit, Mich.), and some gonococcal strains react weakly or not at all with the currently available conjugate. In addition, cross-reactions with other *Neisseria* species occur. Newer, faster, and more sensitive immunological methods are now available for gonococcal identification; therefore, the fluorescent-antibody test is no longer recommended.

Coagglutination. Protein A-containing strains of *Staphylococcus aureus* bind the Fc region of antibody molecules at the bacterial cell surfaces. When such strains are coated with gonococcal antibody and mixed with gonococcal cells, visible clumps form (a positive coagglutination test). This is the basis for two commercially available kits for the identification of *N. gonorrhoeae*: the Phadebact Gonococcus Test (Pharmacia Diagnostics, Piscataway, N.J.) and the Gono Gen Test (Micro-Media Systems, Potomac, Md.).

For the Phadebact test, a dense, aqueous suspension of the organisms to be tested is boiled for 5 to 10 min and then cooled and mixed with the rabbit antibody-coated reagent and a control reagent coated with gamma globulin from nonimmune rabbits. Strong clumping in the test, but not the control, reagent within 2 min is a positive test and provides definitive identification of *N. gonorrhoeae* (5). Although the results are usually clear-cut, for some strains of gonococci nonspecific clumping and stringing may be observed in both the test and control reagents, making interpretation difficult. Reactivity depends on the strain, the age of the culture, and the medium from which the inoculum is taken. Cross-reactions with other *Neisseria* species, especially *N. lactamica*, occur; therefore, an ONPG test should be performed in conjunction with the Phadebact test on suspicious organisms derived from the oropharynx. Because *N. lactamica* is being isolated more frequently from nonpharyngeal sites, it is prudent to routinely perform an ONPG test when noncarbohydrate methods are used for the identification of gonococci.

The Gono Gen system differs from the Phadebact test in that the staphylococci are coated with murine monoclonal antibodies directed against the principal outer membrane proteins of several gonococcal strains (31). The test is performed essentially as described above. Both positive and negative controls are included in the kit. Reactions are clear-cut and usually appear within 30 s of mixing the organism suspension with the coagglutination reagent.

Comments. The choice of method for the identification of gonococci depends to some extent on the number of isolates to be identified and the required turn-around time. When large numbers of isolates need to be processed, the use of CTA-carbohydrate media or buffered, rapid-fermentation carbohydrate substrates is economical. The reagents can be prepared inexpensively in the laboratory, provided that personnel are available to do so. When these test systems are heavily inoculated, a result usually can be obtained within 4 h. Although kits and other commercially available systems that base identification on biochemical or serological assays provide an equally or even more rapid result, the cost per test is usually higher. When only small numbers of gonococcal isolates are to be identified, however, factors such as the convenience and long shelf life of the kits may outweigh the increased cost.

***N. gonorrhoeae* detection by enzyme immunoassay.** The Gonozyme test (Abbott Laboratories, North Chicago, Ill.) detects gonococci directly in urethral and endocervical specimens. Swabs collected from these sites are immersed in a diluent containing a specially treated plastic bead that adsorbs gonococci and gonococcal antigens. By a double-antibody sandwich technique, antigens bound to the bead are labeled with antibody conjugated to the enzyme, horseradish peroxidase. When the enzyme-labeled bead is incubated with the enzyme substrate, a yellow color develops. The intensity of the color is measured on a Quantum spectrophotometer (Abbott Laboratories) that reads and interprets the results. If no gonococcal antigen is present, no color develops. Results are available in ca. 2 h.

For male urethral specimens, Gonozyme has been reported to have a sensitivity and specificity equivalent to the Gram stain, but for endocervical specimens, the sensitivity and specificity of Gonozyme is not as high as culture (45). Although this method is useful for detecting nonviable organisms and vancomycin-susceptible gonococci that might be missed in culture, its use precludes other tests that rely on the presence of viable gonococci, such as β-lactamase testing for penicillinase-producing *N. gonorrhoeae* (PPNG). In addition, expensive proprietary reagents and equipment are required.

Antimicrobial susceptibility testing

Antimicrobial susceptibility tests are not routinely performed for *N. gonorrhoeae* isolates. Although gonococci have become increasingly resistant to penicillin and other antimicrobial agents, the cure rate of uncomplicated urogenital infections is more than 95% when the dosages of penicillin currently recommended by the U.S. Public Health Service are administered. In 1976, strains of PPNG imported from the Philippines were reported in the United States. PPNG strains possess a plasmid that codes for the production of β-lactamase, an enzyme that hydrolyzes penicillin. These resistant strains have been found in most states in this country, and a dramatic increase in incidence has been reported from a few large cities, including New York, Los Angeles, and San Francisco.

For surveillance purposes and optimal treatment, all *N. gonorrhoeae* isolates should be tested for β-lactamase production. The rapid acidometric, iodometric, or chromogenic cephalosporin method recommended in *Cumitech 6* for detecting ampicillin-resistant *Haemophilus influenzae* strains may be used (48). Currently, the chromogenic cephalosporin, nitrocefin, is available for use as a disk test (Cefinase; BBL). Several strains can be tested on a single disk. Disk tests that rely on the acidometric method are also commercially available (Marion Laboratories, Kansas City, Mo.). Whenever possible, β-lactamase tests should be performed directly from the primary isolation plates because the plasmid encoding for resistance is unstable and may be lost during subculture.

In some regions of the United States, the prevalence of PPNG has increased to such an extent that susceptibility tests with alternative therapeutic agents may be necessary. At present, a single dose of spectinomycin is recommended for the treatment of PPNG infections. Susceptibility tests for both penicillin and spectinomycin may be performed with the following disk tests. In Trypticase soy, brain heart infusion, or Mueller-Hinton broth, prepare a suspension of the test organism to equal the turbidity of the 0.5 McFarland Standard. As in the agar disk diffusion susceptibility test (National Committee for Clinical Laboratory Standards), swab the organism over the surface of an agar plate containing GC agar base supplemented with 1% IsoVitaleX (BBL). Place a 10-U penicillin disk and a 100-μg spectinomycin disk widely separated on the surface of the agar. Incubate at 35°C for 24 h under 5 to 7% CO_2 or in a candle jar. Measure the diameters of the zones of inhibition and interpret as follows: (i) for penicillin, less than 20 mm indicates that the organism is resistant and probably produces β-lactamase, and 20 mm or more indicates that the organism is susceptible and probably does not produce β-lactamase; (ii) for spectinomycin (tentative), less than 14 mm indicates that the organism is resistant, 15 to 17 mm indicates intermediate resistance, and 18 mm or more indicates that it is susceptible.

Strains that appear resistant to penicillin by the disk test should be tested with a specific β-lactamase method to confirm the result. Most penicillin-susceptible organisms yield zone diameters of 30 mm or greater around the disk. Penicillin-resistant organisms have zones of less than 20 mm, and small colonies of β-lactamase-producing organisms may be found within the zones of inhibition. Most spectinomycin resistance is of a high level, with resistant organisms growing up to the edge of the disk (6-mm zone).

If susceptibility tests for other recommended antimicrobial agents (i.e., tetracycline, cefoxitin, cefotaxime, or trimethoprim-sulfamethoxazole) are required, an agar dilution method should be used. Proteose peptone no. 3 agar (Difco) supplemented with 1% hemoglobin and 1% IsoVitaleX is recommended for testing all antibiotics except trimethoprim-sulfamethoxazole (14). For the latter drug, supplemented Diagnostic Sensitivity Test agar (Oxoid, London, England) is preferable (3).

In a few areas of the United States, β-lactamase-negative, penicillin-resistant *N. gonorrhoeae* strains have been isolated. These gonococci appear to have chromosomally mediated rather than plasmid-mediated resistance. In agar dilution susceptibility tests, they were resistant to penicillin, ampicillin, and tetracycline, but susceptible to spectinomycin, cefoxitin, cefuroxime, and cefotaxime. Zone diameter interpretive standards for disk agar diffusion susceptibility tests of these strains are under investigation.

Except for β-lactamase testing, routine antimicrobial susceptibility tests for gonococci are not required and should be reserved for special clinical situations. These situations include isolates from patients who fail to respond to therapy, isolates from patients who acquired gonorrhea in areas where resistant strains are prevalent, and isolates from contacts of these patients. Information about current resistance patterns in the United States is available in the Centers for Disease Control *Morbidity and Mortality Weekly Reports*.

Gonococcal typing

Although several methods for differentiating strains of *N. gonorrhoeae* have been investigated, auxotyping (7) and serological tests (44) have received the most attention. Auxotyping has been used more extensively but is too complex for routine use. Auxotyping separates gonococcal strains according to their requirement for one or more metabolites. The organisms are inoculated onto an array of synthetic media from which various substances (primarily amino acids and nitrogenous bases) have been individually omitted. If a strain cannot grow on medium from which arginine, for example, has been omitted, it requires arginine for growth. Auxotyping has permitted unusual strains of gonococci to be characterized, such as the small-colony AHU strains that often cause DGI. Most gonococci belong to one of only three or four auxotypes, but they can be further separated by their antimicrobial susceptibilities, for example.

Methods for serotyping gonococci have included coagglutination, microimmunofluorescence, and enzyme-linked immunosorbent assays. Although the results in each system do not correspond completely, it appears that gonococci belong to two major serological groups, which are based on studies of their principal outer membrane proteins. At least nine antigenic subsets occur within these two groups. As has been found with the auxotyping method, specific gonococcal characteristics are associated with the serological groups. For example, one serogroup is highly correlated with DGI and resistance to the bactericidal action of normal human serum. Coagglutination reagents for gonococcal serotyping are available commercially (Micro-Media).

NEISSERIA MENINGITIDIS

Clinical significance

Meningococci are isolated most commonly from the throat or nasopharynx of asymptomatic persons who carry the organism for variable time periods, usually only a few weeks. The organism survives poorly in the environment and has no other host; therefore, human carriers are the major reservoir. Person-to-person transmission occurs by direct contact with respiratory secretions or by airborne droplets contaminated with the organism. The usual meningococcal carriage rate is ca. 5%, but it may rise to higher levels in confined populations, such as among military recruits. Homosexual men have a 40% or higher incidence of meningococcal oropharyngeal carriage (25).

In a small percentage of colonized individuals, the organisms disseminate from the nasopharynx through the bloodstream to produce meningococcemia, meningitis, or both. The disease may be mild, or it may progress extremely rapidly, resulting in the death of a previously healthy person within a few hours. The incidence of disease is highest in infants between 6 months and 1 year of age and in adolescents.

Meningococcemia usually is characterized by profound vascular effects, the most visible being a petechial or purpuric skin rash that occurs in about 75% of patients with the disease. In fulminant infection (Waterhouse-Friderichsen syndrome), there is widespread intravascular coagulation with resulting shock and usually death. The pathogenetic mechanism of these effects is not entirely clear, but presumably it is related to or initiated by endotoxin in the meningococcal cell wall.

Although the symptoms of meningococcal meningitis are similar to those of other purulent meningitides, when the rash is present, a presumptive diagnosis can be made since this combination of findings is unusual in other infections. Meningococci have also been implicated as agents of arthritis, which is a relatively common complication of meningococcemia, and, rarely, of purulent conjunctivitis, sinusitis, endocarditis, and primary pneumonia.

Screening programs for gonococci have incidentally revealed that meningococci are found in the cervix and vagina, where they may cause serious pelvic disease (17), and they also have been isolated from the anal canal and male urethra (25). Although the clinical implications are not always clear, these findings emphasize the need for species identification of neisserial isolates from these sites.

Meningococci are subdivided into serological groups depending on the presence of either capsular or cell wall antigens. Currently recognized groups include A, B, C, D, X, Y, Z, 29E, and W135, but other new groups have been reported (1). Classically, group A is responsible for epidemic meningococcal disease, and groups B and C are responsible for infections during interepidemic periods, although the latter groups have also been implicated in epidemics. During outbreaks of disease, carriers tend to be colonized with the prevailing serogroup. Group D meningococci are seen only rarely, and the other serotypes are isolated sporadically from both carriers and patients with disease.

Specimen collection and transport

Meningococci are isolated most readily when specimens are obtained before antibiotic therapy is initiated and are transported rapidly to the laboratory with protection from drying and temperature extremes, especially cold. Depending on the clinical presentation, appropriate specimens include blood, cerebrospinal fluid (CSF), petechial aspirates or biopsy, joint fluid, conjunctival swabs, sputum or transtracheal aspirates, and nasopharyngeal swabs.

CSF. The first sample of CSF collected is most likely to be uncontaminated and is best for culture. Subsequent samples can be used for hematological and chemical determinations or for detecting meningococcal antigen by counterimmunoelectrophoresis (CIE), coagglutination, or latex agglutination. Send between 2 and 5 ml of CSF to the microbiology laboratory; less than 2 ml may not be adequate for testing.

Blood. Draw at least two blood specimens, preferably at separate intervals when the patient is experiencing chills or fever. Inoculate blood directly into culture broth. Since there is evidence that the anticoagulant SPS is toxic for some strains of meningococci (15), evacuated blood collection tubes containing this substance probably should not be used when meningococcemia is suspected.

Skin lesions. For petechial cultures, a small amount of sterile saline may be injected into the lesion and then carefully aspirated with a fine hypodermic needle. If possible, appropriate media should be inoculated at the patient's bedside. Alternatively, because the volume of fluid collected is usually small, recap the end of the needle and deliver the syringe to the laboratory rather than transfer the fluid to another container. If petechial biopsy specimens are obtained, they must be kept moist by transporting them immediately to the laboratory on a chocolate agar plate or in a sterile, closed container with a few drops of saline inside.

Nasopharyngeal swabs. Nasopharyngeal specimens are most easily obtained by passing a bent swab behind the uvula and wiping across the posterior pharynx. Send these specimens and other material on swabs (e.g., conjunctival pus) to the laboratory in nonnutritive, semisolid media such as Amies or Stuart medium. Alternatively, if the specimen must be sent through the mail, plant it directly onto medium in a JEMBEC plate or other CO_2-containing system. Whenever possible, incubate these cultures for 24 h before sending them to the laboratory. Incubated specimens must be shipped in accordance with U.S. postal regulations. Although meningococci can be isolated from the throat, a higher yield is obtained from the nasopharynx, which is the preferred site for specimens.

Direct examination

Gram stain: smear preparation. Because only a few meningococci may be present during the early stages of infection, concentrate CSF specimens by centrifugation or filtration as described in Chapter 8. Decant all except 1 ml of the supernatant, and save it for an antigen detection test or for culture. Suspend the sediment in the remaining fluid and prepare one or two smears for Gram stain. For noncloudy CSF specimens, it may be necessary to layer several drops to see any organisms after staining. During the staining procedure, rinse the slide on the back surface with gentle rocking to direct the stream of water over the stained area on the front surface. This maneuver prevents the smear from washing away. Little information is gained from Gram-stained smears of sputum or nasopharyngeal swabs because meningococci cannot be distinguished from saprophytic neisseriae and other morphologically similar bacteria of the upper respiratory tract.

Smear interpretation and reporting. In clinical specimens, meningococci are gram-negative, bean-shaped diplococci, ca. 0.8 by 0.6 μm in diameter. In some instances, they resist decolorization and may appear very deep pink, almost gram positive. The organisms are usually found both extracellularly and intracellularly in PMN (Fig. 5). The presence of PMN is correlated with a good host response and favorable prognosis; therefore, in addition to reporting the numbers, staining characteristics, and morphology of microorganisms present, report the type(s) of inflammatory cells seen and whether the organisms are located intra- or extracellularly. Smears may require examination for 10 min or longer before organisms are seen.

In patients with meningococcal pneumonia, Gram-stained smears of tracheal aspirate reveal PMN and a predominance of gram-negative diplococci. In contrast, in patients with aspiration pneumonia, a myriad of other organisms is seen along with gram-negative diplococci. The latter most likely are saprophytic *Neisseria* or *Veillonella* species rather than meningococci. This important distinction must be made because initial patient management is often based on the Gram-stained smear report. Interpretation of smears in meningococcal infections at unusual sites may be further complicated by the occurrence of gram-negative coccobacilli (*Acinetobacter* or *Moraxella* species) in fresh clinical specimens.

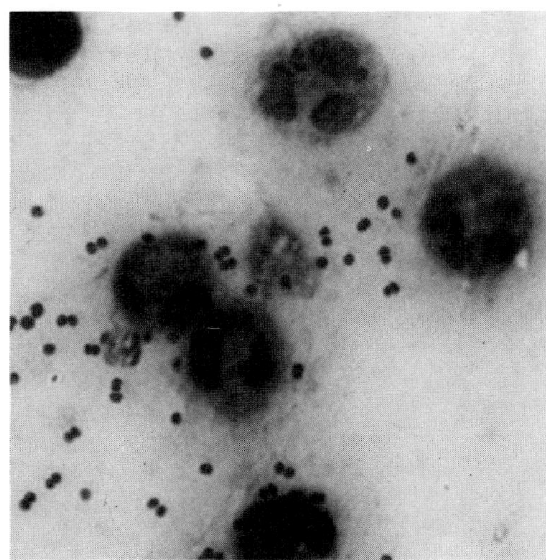

FIG. 5. Gram stain of CSF from a child with meningococcal meningitis. Many intra- and extracellular diplococci are present. ×1,500.

CIE. CIE depends on the migration of soluble antigen and antibody through an agar block under the influence of an electric field. If the antibody is specific for the antigen, a line of precipitate forms where they meet in optimal proportions. Group A and C meningococci produce a soluble carbohydrate capsule. As a result, the CSF (and sometimes other fluids) of patients infected with these groups may contain a significant amount of antigen, and CIE is often positive. CIE results for other meningococcal groups are much less satisfactory, and meningococcal disease should not be ruled out if the test is negative. In addition, cross-reactions with other bacteria occur; e.g., *Escherichia coli* K1 contains antigens that cross-react with group B meningococci.

Latex agglutination and coagglutination tests. Both latex agglutination and coagglutination tests for detecting meningococcal antigens in body fluids are commercially available (latex agglutination test available from Hynson, Westcott and Dunning, Baltimore, Md., and Burroughs-Wellcome Corp., Research Triangle Park, N.C.; coagglutination test available from Pharmacia). All kits contain polyvalent reagent for groups A, B, and C meningococci, and some include reagents for groups Y and W135 as well. These two methods are considerably more sensitive for detecting antigen than is CIE (12, 24). In the clinical laboratory, antigen detection tests should always be used in conjunction with Gram stain and culture. Positive results provide a rapid, presumptive diagnosis and allow early administration of appropriate therapy. Negative results, however, do not exclude a bacterial etiology of the disease. Further evaluations are needed to determine the sensitivity and specificity of commercial latex agglutination and coagglutination reagents for the detection of antigens in CSF, serum, and urine as well as their use for the diagnosis of meningococcal infections other than meningitis.

Culture and isolation

Meningococci are not as nutritionally fastidious as gonococci, but specimens from patients with suspected meningococcal disease must be cultured on enriched medium. Plate all specimens except those likely to contain mixed flora, e.g., nasopharyngeal swabs, onto blood and chocolate agar plates. Streak to obtain isolated colonies. Although meningococci (and gonococci) do not grow well in liquid medium, specimens from sterile sites should also be inoculated into aerobic and anaerobic broths such as tryptic soy broth with blood and supplemented thioglycolate broth to aid recovery of other organisms. Incubate all aerobic media at 35 to 37°C in a 3 to 10% CO_2 environment. Be certain that CO_2 incubators have a 50% or higher humidity.

Plate nasopharyngeal swabs onto blood agar and MTM, ML, or NYC medium. Rub the swab over one-quarter of the area of the blood agar plate and one-half of the selective agar medium; then streak the remainder of the plate to obtain isolated colonies. The antibiotics present in the selective media inhibit most, but not all, nasopharyngeal flora except *N. meningitidis*, *N. gonorrhoeae*, and *N. lactamica*. Coexistence of any of these three organisms in a single specimen may be difficult to determine unless the colonies are well separated by streaking. Incubate all media at 35 to 37°C under increased CO_2 and humidity.

Standard blood culture broths such as tryptic soy, Columbia, and brain heart infusion support the growth of meningococci. Commercially available blood culture bottles must be vented to establish appropriate atmospheric conditions. The concentration of SPS present in most commercial broths (0.02 to 0.05%) inhibits some strains of meningococci. As described for gonococci, the addition of gelatin to a final concentration of 1% neutralizes this adverse effect. Meningococci grow and are detected in the BACTEC blood culture system.

Identification

Colony morphology. On agar media, well-isolated meningococcal colonies are much larger than gonococcal colonies, usually attaining a diameter of 1 mm or greater after 18 h of incubation (Fig. 6). The colonies are round and convex, with a smooth, moist, glistening surface and an entire edge. Encapsulated groups A and C may appear mucoid. Although meningococcal colonies are usually not described as being pigmented, on sheep blood agar they often appear gray, whereas some groups (notably 29E and W135) are creamy white in the area of heavy inoculum. In addition, the medium beneath and adjacent to the colonies exhibits a greenish cast. Young cultures have a butyrous consistency, and the colonies emulsify easily in saline. With continued incubation, however, the organisms autolyze, nucleoprotein is released from the cells, and the colonies become sticky and rubbery. At this point, most of the cells are nonviable. Close examination of the culture plate may reveal small daughter colonies growing from the edges of the larger colonies. Often, the culture can be recovered only by subculturing from these small colonies, which revert to normal size on fresh agar medium. Colony variation such as that found with *N. gonorrhoeae* strains has not been described for *N. meningitidis*.

Presumptive identification. Meningococcal isolates from CSF and skin lesions are usually present in pure culture. *N. gonorrhoeae* and *N. lactamica* may coexist with meningococci in the nasopharynx. They will grow on the selective ML, MTM, and NYC media commonly used for nasopharyngeal specimens, as may certain gram-negative bacilli that commonly inhabit the nasopharynx (strains of *Kingella, Moraxella*, and *Eikenella*). Therefore, carefully examine primary isolation plates, preferably with increased magnification, to determine whether all colonies appear similar. Screen all suspicious colonies with an oxidase test and a Gram stain, as described for the presumptive identification of gonococci. If the primary culture is not pure or is more than 36 h old, prepare a subculture of oxidase-positive, gram-negative diplococci on fresh blood or chocolate agar plates. Incubate for 18 to 24 h at 35°C under increased CO_2 tension; then perform a confirmatory test.

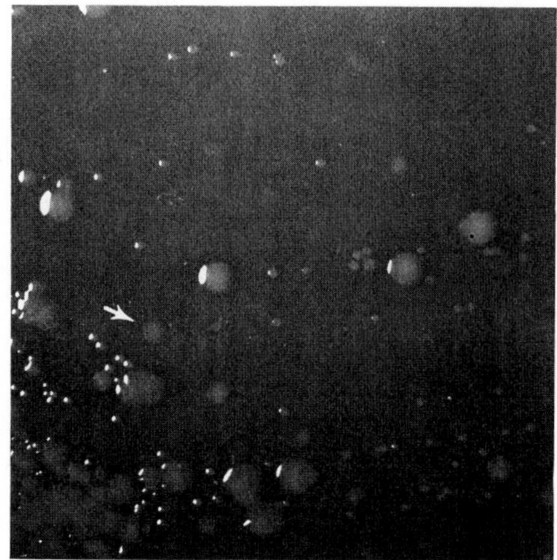

FIG. 6. *N. meningitidis* and *N. gonorrhoeae* on chocolate agar. Meningococcal colonies are larger than gonococcal T3-T4 colonies (arrow). ×2.5.

Confirmatory identification. Characteristically, meningococci degrade glucose and maltose, but not sucrose, fructose, or lactose (Table 3). For the most part, the methods described for the identification of gonococci are used to identify meningococci. Since meningococci grow more rapidly and luxuriantly, however, fewer problems arise.

Carbohydrate degradation tests. The use of CTA-base carbohydrate media, rapid fermentation tests, and the Minitek and BACTEC systems for *Neisseria* identification are discussed above. When meningococci are heavily inoculated in the carbohydrate media, reactions generally are quite pronounced after 1 to 4 h. A bright-yellow color signifies acid production, and negative reactions are indicated by an alkaline (pink) or neutral (orange) reaction. During growth, meningococci degrade peptones in the medium to yield alkaline products which, with prolonged incubation, may obscure the acid reactions in the glucose and maltose tubes. Therefore, examine reactions after 1, 4, and, if necessary, 24 h to detect acid production. If the inoculum was light, further incubation may be necessary. When using the BACTEC system, read the reactions only after the full 3-h incubation period. The maltose vial contains only one-fourth as much ^{14}C as the other vials, and significant levels of $^{14}CO_2$ may not evolve in shorter time periods. Occasionally a false-positive fructose result is obtained when meningococci are tested on the BACTEC system, the growth index value registering slightly above the threshold reading.

On initial isolation some strains of meningococci, primarily those that are resistant to sulfonamides, do not produce acid from maltose (30). Repeated subculture on noninhibitory medium often restores maltose-degrading capability; thus, suspected meningococcal strains that are maltose negative should be retested after several transfers on blood or chocolate agar. The occurrence of glucose-negative meningococci has also been reported. Whenever a suspicious organism is isolated that does not behave as expected biochemically, use an alternative identification system.

Susceptibility testing

In the clinical laboratory, routine antimicrobial susceptibility testing of meningococci is not required since the organisms remain highly susceptible to penicillin. A meningococcus harboring the same β-lactamase plasmid as PPNG has been reported from Canada (10). It was isolated with a PPNG strain from a clinical specimen. Although as yet an uncommon event, this demonstration that antibiotic resistance plasmids can be transferred from gonococci to meningococci in vivo should alert microbiologists to the possible occurrence of β-lactamase-producing meningococci. Isolates from patients who are not responding well to antimicrobial therapy warrant an agar dilution susceptibility or β-lactamase test to detect possible resistance.

Physicians may request sulfadiazine susceptibility tests to determine whether this drug will be effective for prophylactic treatment of close contacts. The agar dilution method (Chapter 100), performed with a limited series of concentrations, is satisfactory for this purpose. Plate 10^5 to 10^6 organisms on Mueller-Hinton agar plates containing 1, 5, or 10 μg of sulfadiazine per ml. Isolates growing at this last concentration are considered resistant. Control plates, including known susceptible and resistant meningococci, as well as an antimicrobial agent-free plate, must be run at the same time. For rifampin, growth on Mueller-Hinton agar plates containing 0.25 μg of rifampin is considered evidence of rifampin resistance.

Meningococcal grouping

A number of serological methods have been used to group meningococci; however, agglutination is the most reliable and routinely used technique (49). In 0.5 ml of phosphate-buffered saline, pH 7.2, prepare a milky suspension of cells grown for 8 to 12 h on blood agar. Allow any clumps of organisms to settle for 1 min; then, with a Pasteur pipette, mix 1 drop of the suspension with 1 drop of antiserum on a slide or in the well of a plastic tray. Rotate the mixture for 2 to 4 min, observing the reaction with indirect lighting. High-titer, specific antiserum will usually produce a strong reaction within 2 min. Almost all meningococci isolated from systemic infections will be groupable by this technique. Nasopharyngeal isolates from carriers, however, may not be groupable; i.e., they will agglutinate in all (autoagglutinable) or in none of the antisera. Rapid passage of the organism on blood agar or the use of pH 6.8 buffer for the agglutination test may resolve the problem. Meningococcal grouping sera are available commercially from Difco Laboratories and Burroughs-Wellcome Corp.

The auxotyping technique is not useful for differentiating strains of meningococci because they are biosynthetically more competent than gonococci (see the section on gonococcal typing). Thus, meningococci usually grow on the complete set of auxotyping media, even though metabolites that gonococci require are lacking. All gonococci require cysteine (or cystine) for growth, whereas ca. 90% of meningococci do not. Because of this, a nutritionally complete medium lacking only cysteine can be used to differentiate gonococci from most strains of meningococci.

NEISSERIA LACTAMICA

Clinical significance

N. lactamica has been implicated in rare instances of meningitis, bacteremia, and respiratory infections and has also been isolated from the urogenital tract (27). Its primary habitat is the nasopharynx of infants and young children, 3 months to 6 years of age. This contrasts with meningococcal carriage, which is more prevalent in teenagers and young adults. Nasopharyngeal carriage of *N. lactamica* may stimulate the formation of bactericidal antibodies that protect against meningococcal disease, since there is serological cross-reaction between the two organisms (18). *N. meningitidis* and *N. lactamica* must be differentiated in culture, because the clinical and epidemiological implications of each are distinct.

Identification

N. lactamica produces the enzyme β-galactosidase, which degrades lactose and ONPG. Except for this characteristic, the organism is remarkably similar to *N. meningitidis* in its colony morphology, ability to grow on MTM, ML, and NYC media, and acid production from glucose and maltose (Table 3). On occasion, the colonies have a yellowish tint and may be smaller and less moist than meningococcal colonies. Most strains of *N. lactamica*, unlike gonococci and meningococci, grow at 35 to 37°C on simple nutrient agar and reduce nitrite with the production of nitrogen gas; the latter is characteristic of some meningococcal strains.

To identify *N. lactamica*, use the same carbohydrate degradation methods as for gonococci and meningococci. With some strains, ONPG is hydrolyzed so rapidly that, if this test is inoculated first, a positive result occurs before the remaining inoculations are completed. The BACTEC system may not detect positive glucose readings for all *N. lactamica* strains, especially if the inoculum is too light. However, the organisms should produce adequate $^{14}CO_2$ from maltose and will be ONPG positive.

Many strains of *N. lactamica* are nongroupable because they autoagglutinate. Since they share cross-reacting antigens, a number of strains will agglutinate in specific meningococcal antisera, especially group B.

NEISSERIA SPECIES AND *BRANHAMELLA CATARRHALIS*

Clinical significance

The primary habitat of *B. catarrhalis* and of *Neisseria* species that colonize humans is the nasopharynx. The latter species include *Neisseria sicca*, *Neisseria subflava*, *Neisseria flavescens*, *Neisseria cinerea*, *Neisseria mucosa*, and *N. elongata* (50). Although they are considered to be of low virulence, these organisms have been implicated rarely as the etiological agent of one or more of the following infections: meningitis, bacteremia, endocarditis, empyema, pericarditis, and pneumonia (22). When they are isolated from blood, CSF, or other normally sterile body fluids and secretions, these organisms must be identified and distinguished from *N. meningitidis* or even *N. gonorrhoeae*.

Identification

Table 3 lists the features that differentiate the pathogenic *Neisseria* species and *B. catarrhalis*. Observations of colony morphology, presence or absence of pigment, and ability to grow on MTM, ML, or NYC medium are important characteristics for initial separation. Pigment is often more visible when several colonies are picked from the plate with an inoculating

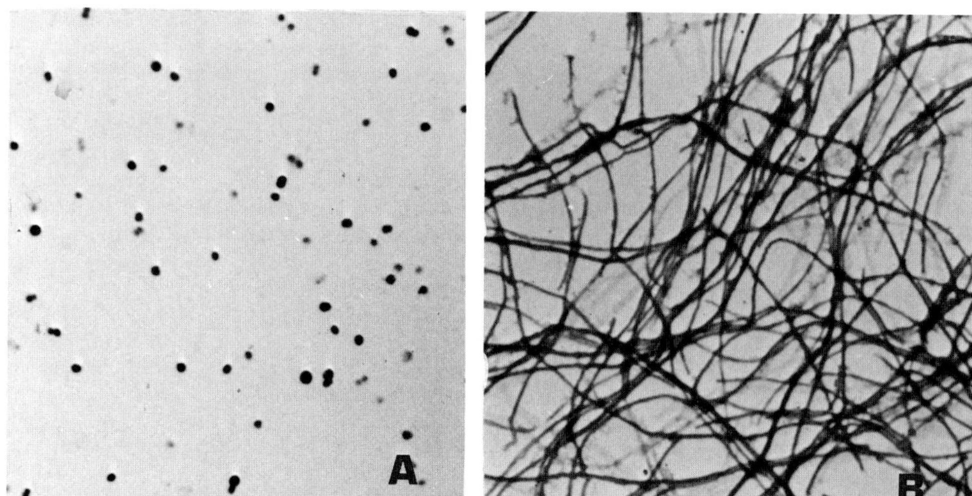

FIG. 7. Cocci (A) and bacilli (B) exposed to subinhibitory concentrations of penicillin. Some cocci are swollen, but still coccoid; bacilli form long strings. ×1,000.

loop and held against a white background. Most of the saprophytes grow at room temperature (ca. 22°C) on enriched media and also at 35°C on simple nutrient agar (beef extract, peptone, agar); enriched agar base media such as tryptic soy, brain heart infusion, and Columbia are not satisfactory for this latter test. Occasionally, stock culture isolates (such as those used in proficiency tests) of gonococci, and especially meningococci, grow sparsely at the low temperature. Clinically significant strains of *B. catarrhalis* usually produce β-lactamase and may have certain atypical characteristics, for example, the ability to grow on MTM, but not on nutrient agar at 22°C (11).

Carbohydrate reactions readily separate gonococci, meningococci, and *N. lactamica* from one another and from the saprophytic, gram-negative cocci, but they do not always distinguish between the saprophytes. Tests for reduction of nitrate and nitrite may be useful in this respect; however, for optimal organism growth, the media may require supplementation with 10% serum. Use a 0.05 to 0.1% concentration of KNO_2 and KNO_3 in the medium (Chapter 111). The ability to synthesize polysaccharide from sucrose may help to differentiate some of the saprophytic *Neisseria* species from pathogenic species (23). Prepare brain heart infusion agar with 5% sucrose. Inoculate plates with the organism, and incubate for 48 h at 35 to 37°C. Add a 1:4 dilution of fresh Lugol iodine to the plate. A positive test is indicated by the formation of a blue color around colonies that synthesize the polysaccharide. Tests for DNA hydrolysis (as performed with gram-negative bacilli) have been used to distinguish *B. catarrhalis* from *Neisseria* species (41).

Coccobacillary gram-negative rods in the family *Neisseriaceae* closely resemble *Neisseria* species in Gram-stained smears and culture, but there is a simple test available to determine their true bacillary morphology. Streak the organism in question on a blood agar plate, and place a 10-U penicillin disk on the inoculated area. Incubate for 18 to 24 h; then examine growth from the edge of the zone of inhibition by preparing a crystal violet-stained smear. As illustrated in Fig. 7, under the influence of subinhibitory concentrations of penicillin, bacilli form long, stringy cells, whereas cocci retain their coccal morphology (6). In this test, *N. elongata* forms long cells and must be differentiated from gram-negative bacilli in the family *Neisseriaceae* by additional tests (2).

PRESERVATION OF STOCK CULTURES

In culture, meningococci and gonococci remain viable only when they are transferred every 48 to 72 h, and they will not survive more prolonged incubation or refrigeration. Cultures of *Neisseria* species are preserved most satisfactorily by lyophilization or by freezing at −40°C or lower. To prepare stock cultures by freezing, scrape off the entire growth from a 24-h blood or chocolate agar plate (AHU strains may require two plates) and suspend it in broth containing 10% glycerol or 10 to 50% serum. We have preserved cultures of gonococci and meningococci for at least 12 years at −70°C in 1-dram screw-capped vials containing a solution of 50% tryptic soy broth and 50% decomplemented horse serum. The vials may be entered a number of times and refrozen, but when growth on subculture appears diminished, prepare a fresh vial.

A method has been described for preserving gonococci and meningococci for several months by overlaying chocolate agar slants on which the organisms have grown with sterile mineral oil to 2 cm beyond the upper tip of the slant. The tubes are closed tightly, wrapped with plastic tape, and stored at 30°C (8).

We are grateful to Leon LeBeau, University of Illinois at Chicago, for help with the photographs.

LITERATURE CITED

1. **Ashton, F. E., A. Ryan, B. Diena, and H. J. Jennings.** 1983. A new serogroup (L) of *Neisseria meningitidis*. J. Clin. Microbiol. **17**:722–727.
2. **Bøvre, K.** 1984. *Moraxella*, p. 296–303. *In* N. R. Krieg (ed.), Bergey's manual of systematic bacteriology, vol. 1. Williams & Wilkins Co., Baltimore.
3. **Brown, S. T., S. E. Thompson, J. W. Biddle, S. J. Kraus, A. A. Zaidi, and G. S. Kleris.** 1982. Treatment of uncom-

plicated gonococcal infection with trimethoprim-sulfamethoxazole. Sex. Transm. Dis. **9**:9–14.

4. **Brown, W. J., and S. J. Kraus.** 1974. Gonococcal colony types. J. Am. Med. Assoc. **228**:862–863.
5. **Carlson, B. L., M. S. Haley, J. R. Kelly, and W. M. McCormack.** 1982. Evaluation of the Phadebact test for identification of *Neisseria gonorrhoeae*. J. Clin. Microbiol. **15**:231–234.
6. **Catlin, B. W.** 1975. Cellular elongation under the influence of antibacterial agents: way to differentiate coccobacilli from cocci. J. Clin. Microbiol. **1**:102–105.
7. **Catlin, B. W.** 1978. Characteristics and auxotyping of *Neisseria gonorrhoeae*, p. 345–380. *In* T. Bergen and J. R. Norris (ed.), Methods in microbiology, vol. 10. Academic Press, Inc., New York.
8. **Cody, R. M.** 1978. Preservation and storage of pathogenic *Neisseria*. Health Lab. Sci. **15**:206–209.
9. **Crawford, G., J. S. Knapp, J. Hale, and K. K. Holmes.** 1977. Asymptomatic gonorrhea in men: caused by gonococci with unique nutritional requirements. Science **196**:1352–1353.
10. **Dillon, J. R., M. Pauze, and K.-H. Yeung.** 1983. Spread of penicillinase-producing and transfer plasmids from the gonococcus to *Neisseria meningitidis*. Lancet **i**:779–781.
11. **Doern, G. V., and S. A. Morse.** 1980. *Branhamella (Neisseria) catarrhalis*: criteria for laboratory identification. J. Clin. Microbiol. **11**:193–195.
12. **Drow, D. L., D. F. Welch, D. Hensel, K. Eisenach, E. Long, and M. Slifkin.** 1983. Evaluation of the Phadebact CSF test for detection of the four most common causes of bacterial meningitis. J. Clin. Microbiol. **18**:1358–1361.
13. **Eisenstein, B. I., T. J. Lee, and P. F. Sparling.** 1977. Penicillin sensitivity and serum resistance are independent attributes of strains of *Neisseria gonorrhoeae* causing disseminated gonococcal infection. Infect. Immun. **15**:834–841.
14. **Elliott, W. C., G. Reynolds, C. Thornsberry, D. S. Kellogg, H. W. Jaffe, S. T. Brown, J. Armstrong, and M. R. Rein.** 1977. Treatment of gonorrhea with trimethoprim-sulfamethoxazole. J. Infect. Dis. **135**:939–943.
15. **Eng, J., and E. Holten.** 1977. Gelatin neutralization of the inhibitory effect of sodium polyanethol sulfonate on *Neisseria meningitidis* in blood culture media. J. Clin. Microbiol. **6**:1–3.
16. **Faur, Y. C., M. H. Weisburd, M. E. Wilson, and P. S. May.** 1973. A new medium for the isolation of pathogenic *Neisseria* (NYC medium). I. Formulation and comparisons with standard media. Health Lab. Sci. **10**:44–54.
17. **Givan, K. F., B. W. Thomas, and A. G. Johnston.** 1977. Isolation of *Neisseria meningitidis* from the urethra, cervix and anal canal: further observations. Br. J. Vener. Dis. **53**:109–112.
18. **Gold, R., I. Goldschneider, M. L. Lepow, T. F. Draper, and M. Randolph.** 1978. Carriage of *Neisseria meningitidis* and *Neisseria lactamica* in infants and children. J. Infect. Dis. **137**:112–121.
19. **Granato, P. A., C. Schneible-Smith, and L. B. Weiner.** 1981. Primary isolation of *Neisseria gonorrhoeae* on hemoglobin-free New York City medium. J. Clin. Microbiol. **14**:206–209.
20. **Handsfield, H. H.** 1977. Clinical aspects of gonococcal infection, p. 57–79. *In* R. B. Roberts (ed.), The gonococcus. John Wiley & Sons, Inc., New York.
21. **Henriksen, S. D.** 1976. *Moraxella, Neisseria, Branhamella*, and *Acinetobacter*. Annu. Rev. Microbiol. **30**:63–83.
22. **Herbert, D. A., and J. Ruskin.** 1981. Are the "nonpathogenic" neisseriae pathogenic? Am. J. Clin. Pathol. **75**:739–742.
23. **Hoke, C., and N. A. Vedros.** 1982. Characterization of atypical aerobic gram-negative cocci isolated from humans. J. Clin. Microbiol. **15**:906–914.
24. **Ingram, D. L., A. W. Pearson, and A. R. Occhiuti.** 1983. Detection of bacterial antigens in body fluids with the Wellcogen *Haemophilus influenzae* b, *Streptococcus pneumoniae*, and *Neisseria meningitidis* (ACYW135) latex agglutination tests. J. Clin. Microbiol. **18**:1119–1121.
25. **Janda, W. M., M. Bohnhoff, J. A. Morello, and S. A. Lerner.** 1980. Prevalence and site-pathogen studies of *Neisseria meningitidis* and *N. gonorrhoeae* in homosexual men. J. Am. Med. Assoc. **244**:2060–2064.
26. **Janda, W. M., J. A. Morello, and M. Bohnhoff.** 1984. Use of the API NeIdent system for identification of pathogenic *Neisseria* spp. and *Branhamella catarrhalis*. J. Clin. Microbiol. **19**:338–341.
27. **Jephcott, A. E., and R. S. Morton.** 1972. Isolation of *Neisseria lactamica* from a genital site. Lancet **ii**:739–740.
28. **Kellogg, D. S., Jr., K. K. Holmes, and G. A. Hill.** 1976. Cumitech 4, Laboratory diagnosis of gonorrhea. Coordinating ed., S. Marcus and J. C. Sherris. American Society for Microbiology, Washington, D.C.
29. **Kellogg, D. S., Jr., W. L. Peacock, Jr., W. E. Deacon, L. Brown, and C. I. Pirkle.** 1963. *Neisseria gonorrhoeae*. I. Virulence genetically linked to clonal variation. J. Bacteriol. **85**:1274–1279.
30. **Kingsbury, D. T.** 1967. Relationship between sulfadiazine resistance and the failure to ferment maltose in *Neisseria meningitidis*. J. Bacteriol. **94**:557–561.
31. **Lawton, W. D., and G. J. Battaglioli.** 1983. Gono Gen coagglutination test for confirmation of *Neisseria gonorrhoeae*. J. Clin. Microbiol. **18**:1264–1265.
32. **Lossick, J. G., M. P. Smeltzer, and J. W. Curran.** 1982. The value of the cervical Gram stain in the diagnosis and treatment of gonorrhea in women in a venereal disease clinic. Sex. Transm. Dis. **9**:124–127.
33. **Luciano, A. A., and L. Grubin.** 1980. Gonorrhea screening: comparison of three techniques. J. Am. Med. Assoc. **243**:680–681.
34. **Martin, J. E., J. H. Armstrong, and P. B. Smith.** 1974. New system for cultivation of *Neisseria gonorrhoeae*. Appl. Microbiol. **27**:802–805.
35. **Martin, J. E., and J. S. Lewis.** 1977. Anisomycin: improved antimycotic activity in modified Thayer-Martin medium. Public Health Lab. **35**:53–62.
36. **Martin, J. E., Jr., and R. L. Jackson.** 1975. A biological environment chamber for the culture of *Neisseria gonorrhoeae*. J. Am. Vener. Dis. Assoc. **2**:28–30.
37. **Mirrett, S., L. B. Reller, and J. S. Knapp.** 1981. *Neisseria gonorrhoeae* strains inhibited by vancomycin in selective media and correlation with auxotype. J. Clin. Microbiol. **14**:94–99.
38. **Morello, J. A., S. A. Lerner, and M. Bohnhoff.** 1976. Characteristics of atypical *Neisseria gonorrhoeae* from disseminated and localized infections. Infect. Immun. **13**:1510–1516.
39. **Morse, S. A., and L. Bartenstein.** 1976. Adaptation of the Minitek system for the rapid identification of *Neisseria gonorrhoeae*. J. Clin. Microbiol. **3**:8–13.
40. **Odugbemi, T., and R. J. Arko.** 1983. Differentiation of *Kingella denitrificans* from *Neisseria gonorrhoeae* by growth on a semisolid medium and sensitivity to amylase. J. Clin. Microbiol. **17**:389–391.
41. **Riou, J.-Y.** 1977. Diagnostic bacteriologique des espèces des genres Neisseria et Branhamella. Ann. Biol. Clin. **35**:73–87.
42. **Robinson, M. J., and T. R. Oberhofer.** 1983. Identification of pathogenic *Neisseria* species with the RapID NH system. J. Clin. Microbiol. **17**:400–404.
43. **Saigh, J. H., C. C. Sanders, and W. E. Sanders, Jr.** 1978. Inhibition of *Neisseria gonorrhoeae* by aerobic and facultatively anaerobic components of the endocervical flora: evidence for a protective effect against infection. Infect. Immun. **19**:704–710.
44. **Sandstrom, E. G., K. C. S. Chen, and T. M. Buchanan.** 1982. Serology of *Neisseria gonorrhoeae*: coagglutination serogroups WI and WII/III correspond to different outer membrane protein I molecules. Infect. Immun. **38**:462–470.

45. **Schachter, J., W. M. McCormack, R. F. Smith, R. M. Parks, R. Bailey, and A. C. Ohlin.** 1984. Enzyme immunoassay for diagnosis of gonorrhea. J. Clin. Microbiol. **19:**57–59.
46. **Staneck, J. L., and S. Vincent.** 1981. Inhibition of *Neisseria gonorrhoeae* by sodium polyanetholsulfonate. J. Clin. Microbiol. **13:**463–467.
47. **Strauss, R. R., J. Holderbach, and H. Friedman.** 1978. Comparison of a radiometric procedure with conventional methods for identification of *Neisseria*. J. Clin. Microbiol. **7:**419–422.
48. **Thornsberry, C., T. L. Gavan, and E. H. Gerlach.** 1977. Cumitech 6, New developments in antimicrobial agent susceptibility testing. Coordinating ed., J. C. Sherris. American Society for Microbiology, Washington, D.C.
49. **Vedros, N. A.** 1978. Serology of the meningococcus, p. 293–314. *In* T. Bergen and J. R. Norris (ed.), Methods in microbiology, vol. 10. Academic Press, Inc., New York.
50. **Vedros, N. A.** 1984. *Neisseria*, p. 290–296. *In* N. R. Krieg (ed.), Bergey's manual of systematic bacteriology, vol. 1. Williams & Wilkins Co., Baltimore.
51. **Vera, H. D.** 1948. A simple medium for identification and maintenance of the gonococcus and other bacteria. J. Bacteriol. **55:**531–536.

Chapter 18

Corynebacterium spp. and Other Coryneform Organisms

MARIE B. COYLE, DANNIE G. HOLLIS, AND NEAL B. GROMAN

CHARACTERIZATION

The genus *Corynebacterium* was originally established to accommodate the diphtheria bacillus and later included a few other species, from animals, which had very similar morphological features. Since morphological similarity was believed to indicate relatedness, coryneform organisms from sources other than animals and humans were subsequently assigned to this genus, which now includes a heterogeneous collection of morphologically similar plant pathogens and saprophytic species from a wide range of habitats. The coryneforms have been defined as pleomorphic gram-positive rods occurring in angular arrangements and developing a variable proportion of coccoid cells in stationary cultures. Rudimentary branching may occur, but definite mycelia are not formed. These organisms do not produce endospores, are not acid fast, and may be motile.

When the chemical composition of the cell wall was introduced as a taxonomic tool, it was found that, with few exceptions, the species that were isolated from humans and animals were similar to each other and that they differed from most of the plant pathogens and saprophytic species. In view of all the chemotaxonomic information currently available, most taxonomists agree that the genus *Corynebacterium* should be restricted to those organisms with cell walls that contain *meso*-diaminopimelic acid, arabinose, galactose, and relatively short-chain mycolic acids with 22 to 38 carbon atoms (3, 23, 35). The G+C content of their DNA is between 51 and 59 mol%. The majority of species are facultative anaerobes, and these contain dihydrogenated menaquinones with either eight or nine isoprene units. Although many of the *Corynebacterium* species that are plant pathogens and saprophytes do not fit this definition, they have not yet been assigned to other genera. Conversely, there are species that are currently included in other genera that fit the chemotaxonomic definition of the genus *Corynebacterium* (35, 36).

The currently proposed genera for the coryneform group are *Corynebacterium, Brevibacterium, Arthrobacter, Cellulomonas, Curtobacterium, Microbacterium, Caseobacter, Rhodococcus,* and *Oerskovia*. Some authors include the genus *Propionibacterium*, since many strains of *P. granulosum, P. avidum,* and occasionally *P. acnes* are facultative anaerobes and resemble the coryneform group (23, 36). There is very limited information in the literature on the conventional biochemical reactions for most coryneform genera other than *Corynebacterium* and *Propionibacterium*. The reliable identification of isolates outside of these two genera usually requires the analysis of cell wall composition and information on DNA composition (35).

The close relationship between the corynebacteria of humans and animals and the mycobacteria and nocardiae is evidenced by striking similarities in their cell walls, particularly the presence of arabinose, galactose, *meso*-diaminopimelic acid, and mycolic acids (3).

The term diphtheroid has been used in medical bacteriology for gram-positive rods that resemble and may be confused with *C. diphtheriae* and are presumably a species of the genus *Corynebacterium*. The diphtheroids are pleomorphic gram-positive rods which stain irregularly and may contain metachromatic (polyphosphate) granules. The cells may be arranged in palisades and V-shapes resembling cuneiform configurations. They are usually nonmotile, catalase positive, nonsporeforming, and non-acid fast, with little tendency to branch. This conventional morphological definition of the corynebacteria of humans and animals falls within the general definition of the coryneforms. When aerobic coryneform bacteria from human skin were characterized on the basis of cell wall amino acids and sugars, only 60% were found to be *Corynebacterium* species, another 20% were closely related to *Brevibacterium* species, 5% resembled aerobic *Propionibacterium* species, and 15% appeared to be environmental isolates from a variety of other unnamed genera (45, 48).

The incidence of opportunistic infections due to coryneform bacteria has continued to increase with the increasing survival of severely compromised patients. The corynebacteria which have been reported as opportunistic pathogens include *C. xerosis, C. pseudodiphtheriticum* (*C. hofmannii*), *C. equi* (*Rhodococcus equi*), and *Bacterionema matruchotii* (*C. matruchotii*) as well as the JK, D-2, A-4, and G-2 groups as defined by the Special Pathogens Section of the Centers for Disease Control (CDC), Atlanta, Ga. (30). In contrast to the increased reports of opportunistic coryneform infections, the number of reported diphtheria cases has steadily declined from 435 in 1970 to 5 in 1981 (9). However, it is essential that clinical laboratories retain the capability of recognizing this pathogen, since the mortality from diphtheria in this country continues to range between 7 and 11%. Other *Corynebacterium* species which can cause infections in noncompromised individuals are *C. ulcerans, C. haemolyticum* (*Arcanobacterium haemolyticum*), *C. pseudotuberculosis* (*C. ovis*), and "*C. minutissimum*." Reference 37 provides an extensive review of infections caused by nondiphtheria corynebacteria and includes a bibliography that can be used as a supplement to this chapter.

In response to the increasing frequency with which laboratorians encounter coryneform bacteria as significant pathogens, this chapter has been expanded to provide some information on almost all of the groups of coryneform bacteria that have been identified by the Special Pathogens Section of the CDC (30). Many groups have not yet been analyzed by chemotaxonomic methods for generic assignments. Other groups have already been assigned to a genus other than *Corynebacterium*. Of the 29 different species or groups presented in Table 1, only 10 are included in the genus *Corynebacterium* in the Approved Lists of Bacterial Names from the International Committee of Systematic Bacteriology (55). *C. haemolyticum* has been proposed as the only species in the newly defined genus *Arcanobacterium* (11), and *C. pyogenes* has been proposed as a new member of the genus *Actinomyces* (49). *C. equi* is currently found under two genera in the Approved Lists, i.e., the genera *Corynebacterium* and *Rhodococcus*. The genus *Oerskovia* includes two of the CDC coryneform groups, A-1 and A-2 (57). Some of the *Corynebacterium* species listed in Table 1 have not been included in the Approved Lists of Bacterial Names. *C. ulcerans* is not yet a recognized species, but

TABLE 1. Identification of medically significant corynebacteria and other coryneform organisms[a]

Species and group[b]	Catalase	Beta-hemolysis	Nitrate reduction	Urease	Gelatin hydrolysis	Motility	Esculin hydrolysis	Carbohydrate utilization				
								Glucose	Maltose	Sucrose	Mannitol	Xylose
Corynebacterium diphtheriae[c,d]	+	v[c]	+[c]	−	−	−	−	+	+	−[c]	−	−
C. ulcerans[d,f,g]	+	+[e]	−	+	−/+[h]	−	−	+	+	−	−	−
C. pseudotuberculosis[d,f,i]	+	+[e]	v[j]	+	−[k]	−	−	+	+	−	−	−
C. xerosis[l]	+	−	+	−	−	−	−	+	v	+	−	−
C. striatum	+	−	+	−	−	−	−	+	−	+	−	−
C. kutscheri	+	v	+	+	−	−	+	+	+	+	−	−
C. renale	+	−	−	+	−	−	−	+	−	−	−	−
C. bovis[m]	+	−	−	−	−	−	−	+	−	−	−	−
C. pseudodiphtheriticum	+	−	+	+	−	−	−	−	−	−	−	−
C. equi[l]	+	−	v	v	−	−	−	−[n]	−	−	−	−
C. aquaticum[o]	+	−	−	−	−	+	+	+	+	+	+	+
"*C. minutissimum*"	+	−	−	−	−	−	−	+	+	v	−	−
"*C. genitalium*"[p]	+	−	−	−	−	−	−	+	w	−	−	−
C. pyogenes (Actinomyces pyogenes)	−	+	−	−	+	−	−	+	+	v	v	+
C. haemolyticum (Arcanobacterium haemolyticum)	−	+[e]	−	−	−	−	−	+	+	v	−	−
Bacterionema matruchotii	+	−	v	−	−	−	−	+	v	v	v	−
Oerskovia turbata	+	−	v	v	+	+	+	+	+	+	−	+
O. xanthineolytica	+	−	+	v	+	+	+	+	+	+	−	+
Group A-3	+	−	+	−	−	+	+	+	+	+	−	+
Group A-4	+	−	v	−	v	v	v	+	+	+	+	+
Group A-5	+	−	v	−	v	v	v	+	+	+	+	−
Group F-1	+	−	v	+	−	−	−	+	v	+	−	−
Group F-2	+	−	v	+	−	−	−	+	+	−	−	−
Group G-1[q]	+	−	+	−	−	−	−	+ or (+)	v	+ or (+)	−	−
Group G-2[q]	+	−	−	−	−	−	−	+ or (+)	v	+ or (+)	−	−
Group JK[r]	+	−	−	−	−	−	−	+	v	−	−	−
Group I[l]	+	−	+	−	−	−	−	+	v	−	−	−
Group E[s]	−	−	−	−	−	−	v	+	+	+	−	+
Group D-2	+	−	−	+	−	−	−	−	−	−	−	−

[a] Symbols: +, 90% or more positive within 4 days; −, 90% or more negative; + or (+), 90% or more positive, with some strains positive after 4 or more days; v, more than 10% and less than 90% positive; w, weakly positive.
[b] CDC coryneform groups.
[c] Biotype *mitis* is weakly beta-hemolytic; it includes sucrose-positive strains that are rare in the United States and nitrate-negative strains that are the variety belfanti. Biotype *gravis* attacks glycogen and starch and includes a few weakly hemolytic strains and rare isolates that are sucrose positive.
[d] Produces halos on Tinsdale medium and does not hydrolyze pyrazinamide.
[e] Narrow zones of slight hemolysis.
[f] PLD positive.
[g] Also ferments glycogen and usually trehalose and starch, which should be read for 7 days.
[h] Negative or weak at 37°C, positive at 25°C.
[i] Does not attack glycogen or trehalose and usually not starch.
[j] Usually negative in strains from sheep and goats and positive in strains from horses and cattle.
[k] Negative at 35 and 25°C but positive at 30°C after 14 days.
[l] Halos not produced on Tinsdale medium.
[m] *o*-Nitrophenyl-β-D-galactopyranoside positive.
[n] Weak reaction after 7 days.
[o] Glucose oxidizer. Reactions of reference strain ATCC 14655.
[p] Reactions of biotype 2 (ATCC 33031) and biotype 4 (ATCC 33033).
[q] Fastidious. Indistinguishable from *B. matruchotii* in these tests.
[r] *o*-Nitrophenyl-β-D-galactopyranoside negative.
[s] Fastidious. Major end products are succinic and acetic acids.

it is treated as such in this chapter on the basis of both tradition and DNA homology studies (25). The epithets *C. genitalium* (19) and *C. minutissimum* (56) have each been applied to biochemically diverse collections of organisms that probably include more than one taxon of the coryneform bacteria, as discussed below. As the technology for biochemically distinguishing other diphtheroids advances, the list of identifiable opportunistic *Corynebacterium* species and other coryneforms will undoubtedly expand.

CORYNEBACTERIUM DIPHTHERIAE

Clinical significance

Diphtheria is an acute infectious disease caused by toxinogenic strains of *C. diphtheriae*. The toxin synthesized in the local lesion is absorbed and carried by the blood to all parts of the body, but the toxic effects involve primarily the heart and peripheral nerves. The usual portal of entry for *C. diphtheriae* is the upper respiratory tract, where the organism multiplies in the superficial layers of the mucous membrane. There, it elaborates its exotoxin, which causes necrosis of the neighboring tissues. The inflammatory response results in formation of the diphtheritic pseudomembrane composed of bacteria, necrotic epithelium, phagocytes, and fibrin. The diphtheritic membrane usually appears first on the tonsils or posterior pharynx and may then extend either upward into the soft and hard palates and the nasopharynx or downward into the larynx and trachea. Laryngeal diphtheria is particularly hazardous because of the possibility of airway occlusion.

Cutaneous diphtheria is common in tropical areas and also accounts for a sizable proportion of recent *C. diphtheriae* isolates from the northwestern United States (46) and the western provinces of Canada (34). It may be that routine culturing of skin lesions during diphtheria outbreaks will reveal a higher incidence of *C. diphtheriae* than was previously suspected (5). In North America the skin lesions yielding *C. diphtheriae* are usually secondarily infected abrasions or insect bites which also contain beta-hemolytic streptococci or *Staphylococcus aureus* or both. The major epidemiological significance of *C. diphtheriae* in skin lesions may be its roles as a reservoir for these pathogens and as an enhancer of their distribution during epidemics. The biological significance may be the opportunity for genetic exchange with coryneforms of the skin (53).

Diphtheritic lesions also occur in the anterior nares, inner nose, mouth, eye, middle ear, and in rare cases, the vagina. Endocarditis due to both toxinogenic and nontoxinogenic *C. diphtheriae* has also been reported (14).

Collection of specimens

In both patients and contacts, cultures should be taken from the nasopharynx as well as the throat, since 20% of positive cultures can be missed when only one site is cultured (38). Nasopharyngeal cultures should be obtained with a flexible alginate swab that reaches deep into the posterior nares. In suspected cases of diphtheria, throat cultures are taken with a cotton swab which is firmly applied to any area with a membrane or inflammation. For asymptomatic patients, the tonsillar fossae, posterior pharynx, and retrouvular areas should be sampled as well as the nasopharynx. Before cultures of wounds are taken, the lesions should be cleansed with sterile normal saline, and the crusted material should be removed. A cotton-tipped applicator is then firmly applied to the base of the wound.

Transport

When transport time is less than 24 h, the swab should be streaked onto the surface of a Loeffler (or Pai) slant and left on the slant for shipment to the laboratory. When longer transit times are unavoidable, swabs probably should be shipped in the aluminum foil packets of silica gel which are in general use for mailing specimens for streptococcal cultures. Upon receipt in the laboratory, swabs from silica gel transporters are moistened with a few drops of broth, rubbed gently on the surface of a Loeffler or Pai slant, and then placed in a tube of broth for overnight incubation at 35°C. It is essential that desiccated swabs be incubated overnight in a broth supplemented with plasma or blood before they are plated on the routine media recommended below. Overnight incubation in broth may increase the yield of *C. diphtheriae* from desiccated swabs by 70% (16).

Direct examination

The diagnosis of diphtheria based on microscopic examination of a direct smear should not be attempted, since both false-positive and false-negative reports are likely to occur.

Culture and isolation

Specimens are inoculated onto a Loeffler or Pai slant, a cystine-tellurite agar plate, and a 5% sheep blood agar plate (Fig. 1). After overnight incubation at 35°C, growth from the Loeffler or Pai slant may be stained with methylene blue for rapid presumptive diagnosis of diphtheria. When the microscopic morphology is typical of *C. diphtheriae* (see below), a preliminary report of organisms resembling *C. diphtheriae* observed on methylene blue smears may be sent to the physician. Overnight growth from the Loeffler or Pai slant is also subcultured to a blood agar and a cystine-tellurite plate. In an early evaluation of cystine-tellurite medium, Frobisher noted that specimens containing small numbers of *C. diphtheriae* are more likely to yield positive cultures if they are first incubated overnight on Loeffler medium and then inoculated onto cystine-tellurite agar (18). Colonies of *C. diphtheriae* are characteristically black or gunmetal gray on cystine-tellurite after 24 to 48 h at 35°C. Plates should be held for 48 h before they are considered negative. Some strains of other organisms, particularly diphtheroids, staphylococci, and occasionally streptococci, can also reduce tellurite to tellurium and produce black colonies on this medium. A representative of each colony type which has coryneform morphology on Gram stain is subcultured to a blood agar plate for purity before biochemical testing to confirm or rule out *C. diphtheriae*.

The blood agar plate is also used to screen for beta-hemolytic streptococci and *S. aureus*, which may be present in the lesions being cultured for *C. diphtheriae*. Since occasional strains of *C. diphtheriae* are com-

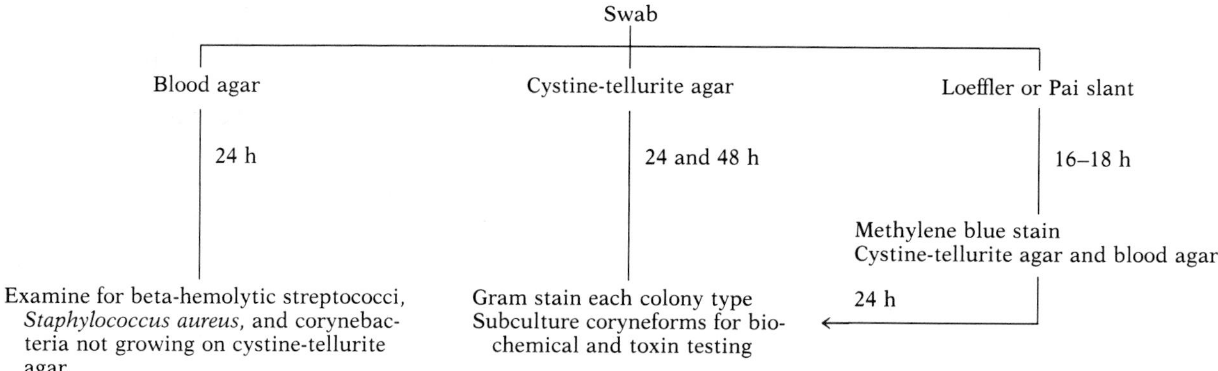

FIG. 1. Method of processing specimens to rule out *C. diphtheriae*.

pletely inhibited by cystine-tellurite agar, blood agar plates should be examined for colonies of coryneforms.

Modified Tinsdale medium is now used less often than cystine-tellurite medium. It was originally recommended as the most reliable medium that could be used by inexperienced personnel for the rapid isolation of *C. diphtheriae* (43). Dark-brown halos around gray-black colonies are usually produced only by *C. diphtheriae, C. ulcerans*, and *C. pseudotuberculosis*. Sometimes, however, very rare strains of diphtheroids, staphylococci, or streptococci may also produce halos around dark colonies. If the plates are stabbed when inoculated, halos usually are produced around the stabs in less than 24 h, even by strains which require 48 h to produce halos around individual colonies. Incubation in an atmosphere of 5 to 10% carbon dioxide is reported to retard the development of the characteristic halos. The major disadvantage of modified Tinsdale medium is its very short shelf life of only 4 days. Even fresh plates may not support the growth of some *C. diphtheriae* strains from primary cultures. The instability of this medium renders it unsuitable for use as the only differential medium for *C. diphtheriae*. Any laboratory that relies on modified Tinsdale medium for diphtheria screening should test each new lot of medium with at least two strains of each biotype. The quantity of growth should be comparable to that on blood agar, and all strains should produce dark-brown halos within 48 h. Each new batch of medium from an acceptable lot need be performance tested with only a single strain, preferably the strain which is most likely to be inhibited by an unsatisfactory medium.

Although modified Tinsdale medium is not recommended as a primary plating medium, it can be useful as a differential medium on which only *C. diphtheriae, C. ulcerans*, and *C. pseudotuberculosis* are expected to produce characteristic halos (58).

Identification

Colonial morphology. The three cultural types of *C. diphtheriae* (commonly referred to as biotypes) are designated *gravis, intermedius*, and *mitis*; these epithets reflect the severity of disease produced by each type during major European epidemics in the early 1900s (42). Although correlation between these types and the severity of disease is no longer clear, typing remains a useful tool in the epidemiology of diphtheria outbreaks. The definitive identification of cultural types should be the domain of reference laboratories. Awareness of the three distinct types, however, will enable technologists to recognize unfamiliar strains which do not resemble the laboratory's quality control strain.

Most strains of the *intermedius* type are lipophilic, which accounts for their limited growth on routine medium that does not contain blood, serum, or Tween 80 (61). The *intermedius* strains produce flat, creamy, transparent colonies on blood agar, whereas the *gravis* and *mitis* strains produce larger, convex colonies. Weak beta-hemolysis is observed with most *mitis* strains and a few *gravis* strains, but not with *intermedius* strains. The *intermedius* strains produce smaller colonies than the other types on cystine-tellurite agar. All types produce halos on Tinsdale medium.

Microscopic morphology. The three cultural types of *C. diphtheriae* may have distinctive microscopic morphologies when they are grown on Loeffler slants and stained with Loeffler methylene blue. Microscopic morphology can vary on different batches of Loeffler slants. Type *mitis* meets the textbook description of *C. diphtheriae*, consisting of fairly long, pleomorphic but rigid rods in cuneiform arrangements; metachromatic granules are intensely stained. Type *intermedius* is highly pleomorphic, ranging from very long to very short rods. The long rods may be irregularly barred. Cells are frequently swollen at the ends containing large terminal granules. The *gravis* type is least likely to be recognized as *C. diphtheriae* on microscopic examination. The cells are usually short, with many being coccoid or pyriform in shape. They tend to stain uniformly and resemble *C. pseudodiphtheriticum (C. hofmannii)*.

The cellular morphology of *C. diphtheriae* when grown on Pai slants generally is that of pleomorphic rods whose ends may be swollen, producing a club shape. Groups of cells in V- and L-shaped arrangements that create the "Chinese character" effect are observed. When stained with Loeffler alkaline methylene blue, reddish-purple polyphosphate granules, which impart a beaded or barred appearance, may be observed in the cells. There is no consistent difference in the microscopic morphology of the biotypes of *C. diphtheriae* from Pai slants. Typically, *C. diphtheriae* does not grow heavily on this medium.

Diphtheroid strains can closely resemble any of the

cultural types of *C. diphtheriae*, and, conversely, atypical strains of *C. diphtheriae* may be mistaken for diphtheroids. When diphtheria was a common infectious disease, many technologists were highly skilled in distinguishing diphtheroids from *C. diphtheriae* solely on the basis of colonial and microscopic morphology. Today, however, since most technologists have encountered only a few stock strains of *C. diphtheriae* in their entire careers, the use of morphological criteria should be restricted to reference laboratories in areas where diphtheria is endemic. Cultures should not be considered negative until coryneforms have been biochemically tested to rule out *C. diphtheriae*. Obviously, the identification of coryneforms from respiratory or skin specimens should be undertaken only when there is clinical or epidemiological reason to suspect the presence of *C. diphtheriae*.

Biochemical characterization

C. diphtheriae is catalase and usually nitrate positive. Urea is not hydrolyzed. Acid, without gas, is produced from glucose and maltose. Trehalose and sucrose are negative, with the exception of sucrose-fermenting *mitis* and *gravis* strains, which are unusual in the United States. Another rare variety of type *mitis* is belfanti, which is nitrate negative. Only isolates of type *gravis* ferment starch and glycogen. The enteric fermentation base containing Andrade indicator is satisfactory for biochemical testing of corynebacteria. One or two drops of rabbit serum must be added to each tube of fermentation medium since most *intermedius* strains, as well as occasional isolates of the other biotypes, will not grow enough to change the pH in unsupplemented broth. Biochemical tests can be read in 24 h, but negative media should be held for 72 h. When identifying cultural types of *C. diphtheriae* isolates, the Special Pathogens Section of the CDC uses a heart infusion medium with bromcresol purple indicator and a pH adjustment of 7.8.

A rapid pyrazine carboxylamidase test can distinguish *C. diphtheriae* and the closely related *C. ulcerans* and *C. pseudotuberculosis* from all other *Corynebacterium* species and other gram-positive rods that might be isolated from the pharynx. In a large study (59), it was found that the only *Corynebacterium* species that did not hydrolyze pyrazinamide were *C. diphtheriae*, *C. ulcerans*, *C. pseudotuberculosis*, and *C. flavescens* (a yellow organism isolated from cheese). Other *Corynebacterium* species and members of the genera *Actinomyces*, *Nocardia*, *Brevibacterium*, and *Bacterionema* were positive in this test.

Toxinogenicity testing

All isolates of *C. diphtheriae* are tested for toxin production. In most instances isolates are sent to a reference laboratory, which may use either an in vitro or an in vivo method for toxin testing. The in vitro immunodiffusion test is more rapid than the guinea pig lethality test, but the technical skills required for accurate results in the immunodiffusion test limit its use to highly experienced personnel.

In vivo test. The most reliable method for the detection of toxin production is the subcutaneous test; this test requires two guinea pigs for each isolate tested. The inoculum is prepared from the entire growth on a 24-h Loeffler or Pai slant suspended in 12 ml of sterile broth or a heavy broth culture (24 to 48 h). Saline cannot be substituted for broth since *C. diphtheriae* rapidly loses viability in saline suspensions (31). The suspensions should be at least as dense as a McFarland no. 3 nephelometer standard, as some strains produce only small amounts of toxin. The abdomen of each guinea pig is shaved or clipped, and the area is disinfected with alcohol. The control animal is intraperitoneally injected with 1,000 U of diphtheria antitoxin. One to two hours later, both the control and test guinea pigs are subcutaneously injected in the prepared area with 5 ml of the culture suspension.

A toxinogenic strain will usually cause death in the test animal within 1 to 4 days, and the control animal will be unaffected. If the culture is nontoxinogenic, neither guinea pig will show evidence of local necrosis or intoxication within 10 days. Test guinea pigs that die within 10 days should be autopsied. The most striking finding is swelling and congestion of the adrenal glands, with scattered hemorrhages in the medulla or cortex or both. Both the abdominal and pleural cavities may contain excess fluid which is clear, cloudy, or bloody. The subcutaneous tissues at the site of inoculation show hemorrhagic, gelatinous edema, and the regional lymph nodes are usually enlarged.

In vitro test. The modified Elek immunodiffusion test is used by many reference laboratories for the rapid detection of toxinogenic strains (7). The presence of potassium tellurite in the agar allows mixed primary cultures to be tested directly from Loeffler or Pai slants. Melted Elek agar base is cooled to 50°C, and 10 ml of the agar is poured into a petri dish containing 2 ml of sterile rabbit serum and 1.0 ml of a sterile 0.3% potassium tellurite solution. The plate contents are thoroughly mixed by gentle rotation and allowed to harden partially. A strip of sterile filter paper (Whatman no. 3, 1.5 by 7 cm) is dipped into a tube containing diphtheria antitoxin diluted to contain about 100 U/ml. The excess fluid is drained off, and the strip is gently pressed onto the surface of the center of the plate. The surface is allowed to dry with the lid partially open for about 1 h in an incubator.

A plate can be inoculated with two to three isolates plus two quality control strains, i.e., a weakly positive *intermedius* strain and a negative *mitis* strain. A loop containing a heavy inoculum from the 24-h growth on a Loeffler slant is used to streak the plate perpendicular to and on both sides of the strip without touching the strip. Plates are incubated at 35°C. After 18 to 24 h, they are read with a hand lens against a dark background with transmitted light. A positive reaction is seen as a precipitin line forming a 45° angle with the inoculum streak. Many toxinogenic strains of *C. diphtheriae* and *C. ulcerans* give a positive reaction within 24 h; plates should also be read at 48 and 72 h. False-positive reactions may be seen after 24 h as weak bands near the filter strip, but these bands can be recognized by comparison with the positive control. Negative in vitro tests should be spot checked by the guinea pig test. Numerous reports of technical problems due to variable reagents (7) indicate that the Elek test should be done only in reference laboratories.

Antibiotic susceptibility

Antibiotics have little or no effect on the clinical course of diphtheria, but treatment of both patients and carriers is a major means of limiting the size of an outbreak. *C. diphtheriae* is susceptible to penicillin, erythromycin, and most of the usual antibiotics used against gram-positive organisms. A notable exception to this susceptibility is methicillin resistance, which is common in *C. diphtheriae* and most other coryneform bacteria. The possibility of erythromycin resistance in nontoxinogenic strains probably warrants susceptibility testing of all *C. diphtheriae* isolates (13). Mueller-Hinton agar supplemented with 5% sheep blood should be used for standard disk diffusion tests or agar dilution tests of MICs, since most *intermedius* strains and some *mitis* strains will not grow on plain Mueller-Hinton agar (41). Trypticase soy broth (BBL Microbiology Systems, Cockeysville, Md.) or brain heart infusion broth supplemented with 10% fetal calf serum or Tween 80 may be used for broth dilution tests.

It should be noted that the clinical diagnosis and treatment of diphtheria should not await laboratory confirmation.

OTHER *CORYNEBACTERIUM* SPECIES AND CORYNEFORM BACTERIA OF MEDICAL IMPORTANCE

Several species of corynebacteria other than *C. diphtheriae* have been reported to cause tissue and bloodstream infections in humans, including bacteremia, endocarditis, osteomyelitis, respiratory tract infections, and wound infections. In many instances, nondiphtherial corynebacterial infections have occurred in immunosuppressed hosts and often are related to the use of indwelling catheters or prosthetic heart valves. Some of the species listed in Table 1 have been isolated only from humans sources, but *C. pyogenes*, *C. pseudotuberculosis (C. ovis)*, *C. equi*, and *C. renale* are well-recognized pathogens of animals. Blood agar supports the growth of all these coryneform bacteria. Certain organisms, e.g., *C. bovis*, *C. pyogenes*, group JK, and group E, as well as some strains of *C. diphtheriae*, require the addition of 1 to 2 drops of sterile rabbit serum to carbohydrate fermentation broth, since growth is poor in the absence of lipid supplementation. Tween-80 supplement has been reported to interfere with interpretations of sugar reactions (33). A rapid fermentation test which provides 4-h carbohydrate reactions of fastidious bacteria is particularly useful for the identification of lipophilic corynebacteria (29).

C. ulcerans

C. ulcerans is not a recognized species, but it appears from recent DNA homology studies (25) that this species is distinct from *C. diphtheriae* and *C. pseudotuberculosis*, though members of all three species produce diphtheria toxin.

C. ulcerans has been implicated as a cause of exudative pharyngitis and diphtherialike disease; it can also be isolated from the nasopharynx of carriers (12). A few cases of other tissue infections have been reported, including an ulcer on the hand of a dairy worker. This organism has been isolated from raw milk and can cause mastitis in cows (12). Microscopically, the organism is pleomorphic, with predominantly coccoid forms and some rod forms. Few metachromatic granules are seen. On Tinsdale medium the brownish-black colonies with distinct halos are indistinguishable from *C. diphtheriae* colonies. *C. ulcerans* is urease positive and nitrate negative and ferments glucose, maltose, trehalose, and starch in peptone broth. The fermentation of trehalose is generally slow and should be read for 7 days. Gelatin hydrolysis is slow or absent at 37°C but usually occurs within 2 to 3 days at 25°C. The ability to hydrolyze urea and the lack of nitrate reduction distinguish this species from *C. diphtheriae* type *gravis*. Some strains can produce two toxins: one is neutralized by diphtheria antitoxin and can usually be detected by the Elek test, whereas the activity of the second toxin, probably phospholipase D (PLD), resembles that of *C. pseudotuberculosis* and is unaffected by diphtheria antitoxin. The organism produces a local abscess at the site of intradermal injection in a guinea pig which is protected with diphtheria antitoxin. The in vivo determination of diphtheria toxin production in the presence of PLD requires special growth conditions and considerable experience in reading the intradermal reaction in a rabbit (40).

C. pseudotuberculosis (C. ovis)

C. pseudotuberculosis causes chronic disease in domesticated animals, particularly caseous lymphadenitis in sheep and abscesses or ulcerative lymphangitis in horses. There have been occasional reports of *C. pseudotuberculosis* infections in otherwise normal humans; most were cases of granulomatous lymphadenitis, but one case was an eosinophilic pneumonia (37). Gram stains of direct smears may show very few diphtheroid bacilli. The microscopic morphology of in vitro growth may resemble that of streptococci, coccobacilli, or pleomorphic bacilli with metachromatic granules. Pinpoint colonies are produced after 24 h of incubation on sheep or horse blood agar. After 48 h, the colonies are small (1 to 2 mm), dry, and yellowish, and usually produce slight beta-hemolysis. Broth cultures develop granular sediments and pellicles with no turbidity. A brown halo develops around black colonies on Tinsdale agar. There is an unusual amount of variation in the reported biochemical reactions of this species. Nitrate is reduced by the vast majority of isolates from horses and cattle but is seldom found with isolates from sheep and goats (6). The conflicting reports on gelatin hydrolysis may be due to technical differences. Laboratories using nutrient gelatin stabs incubated at 35 or 20 to 22°C for up to 30 days have found no hydrolysis, but incubation at 30°C has resulted in digestion of the gelatin after 14 days (21). The differentiation of *C. pseudotuberculosis* from *C. ulcerans* is primarily based on the inability of the former to ferment glycogen, trehalose, and, usually, starch.

A simple test for PLD activity which readily distinguishes *C. pseudotuberculosis* and *C. ulcerans* from all other *Corynebacterium* species (4) has recently been described. Cooperative production of hemolysis by the PLD produced by these two species and the phospholipase C produced by *C. equi* can be readily demonstrated on sheep blood agar plates. The test is performed by using swabs to spot heavy inocula of overnight growth of *C. equi* onto the center of the agar plate and then streaking *C. pseudotuberculosis* or *C. ulcerans* as two or four radiating spokes from the *C. equi*. After incubation for 48 h at 37°C, there is marked hemolysis around *C. equi* growth when it is incubated in the presence of either *C. pseudotuberculosis* or *C. ulcerans*. Hemolysis is not produced in comparable tests of other *Corynebacterium* species (excluding *C. haemolyticum* or *C. pyogenes*).

C. pyogenes (*Actinomyces pyogenes*)

It has recently been proposed that *C. pyogenes* be transferred to the genus *Actinomyces* as *A. pyogenes* on the basis of glucose metabolism, numerical taxonomy, cell wall peptidoglycans, absence of mycolic acids, and the G+C content of the DNA (49). Like most of the *Actinomyces*, *C. pyogenes* is an aerotolerant anaerobe and is biochemically very similar to *A. bovis* except that it is hemolytic and actively proteolytic.

C. pyogenes is frequently isolated from purulent lesions of cattle, sheep, and pigs. Human infections have been reported, but none of the isolate descriptions distinguished *C. pyogenes* from *C. haemolyticum* on the basis of proteolytic activity. On sheep blood agar plates, *C. pyogenes* produces pinpoint, whitish, beta-hemolytic colonies; the zone of hemolysis is at least twice the diameter of the colony. Gram stains contain both coccal and diphtheroid forms. Gelatin hydrolysis is rapid, occurring in less than 48 h. Serum must be added to carbohydrate broths for growth. The catalase test is usually negative, but a report of catalase-positive strains has been confirmed (52).

C. haemolyticum (*Arcanobacterium haemolyticum*)

It has recently been proposed that *C. haemolyticum* be reclassified in a new genus, *Arcanobacterium*, which comprises a single species, *A. haemolyticum* (11). This organism appears distinct from all coryneforms and actinomycetes on the basis of numerical taxonomy and data from the analysis of fatty acids, peptidoglycans, menaquinones, and DNA.

C. haemolyticum has been reported to be a cause of pharyngitis and skin ulcers. Occasionally, pharyngeal infections mimic diphtheria, including membranous exudates and peritonsillar abscesses (24, 37). There are rare reports of the recovery of *C. haemolyticum* from blood, bone, lungs, sinuses, and abscesses, including a brain abscess. The microscopic morphology of cells from Loeffler medium is similar to that of *C. diphtheriae*; however, growth on tellurite medium is poor. The biochemical tests for identification are presented in Table 1. The discrepancies between Table 1 and other sources are due to differences in methodology. The description of *C. haemolyticum* by Hermann, which is based on conventional methods, is most useful for the clinical laboratory (28). Colonies on sheep blood agar are small (0.1 mm at 24 h and 0.5 mm at 48 h), with a narrow zone of hemolysis which may not be present within the first 24 h of incubation. Colonies are larger on rabbit and human blood, and zones of hemolysis are three to five times wider than the colony diameters. The catalase test is negative when done by conventional methods. This organism does not liquify gelatin unless it is tested under the special conditions used by MacLean et al., who first described the species (39). Fermentation of glucose and maltose usually occurs within 24 to 48 h, but some strains require 5 days. The addition of serum stimulates growth in broth; however, serum addition is not recommended for fermentation tests since false-positive reactions may occur. The key tests which distinguish *C. haemolyticum* from *C. pyogenes* are the failure of *C. haemolyticum* to hydrolyze gelatin and to peptonize litmus milk and its inability to ferment xylose.

C. equi (*Rhodococcus equi*)

C. equi has been tentatively assigned to the newly proposed genus *Rhodococcus*, which includes nine species, most of which were formerly in the genus *Nocardia* (22). Rhodococci are aerobic actinomycetes that often form a primary mycelium that soon fragments into irregular elements. When inoculated into fresh nutrient medium, the rod-shaped and coccoid cells give rise to a primary mycelium.

C. equi is pathogenic for horses, swine, and cattle, It can be found in soil near livestock. It has been reported as causing pulmonary infections in patients with neoplastic diseases or renal transplants, some of whom had contact with livestock. On Gram stain, the rods vary in length, some assuming coccal configuration. The organism grows well on ordinary media, where it produces large, mucoid, pale salmon-pink colonies. It is differentiated from other pathogenic corynebacteria by its inability to ferment carbohydrates. Some investigators have reported it to be acid fast (37).

C. xerosis

C. xerosis is part of the normal flora of the nasopharynx and skin. It has caused endocarditis in patients with prosthetic cardiac valves and intravenous shunts and in one patient with a previously normal valve (15). Bacteremia, pneumonia, and surgical wound infections may occur in compromised hosts (37). Microscopically, *C. xerosis* may resemble *C. diphtheriae*, but there is a preponderance of barred forms over granular forms. It is grown readily on ordinary media; lipid does not enhance growth. Colonies may be yellow to tan. The distinctive biochemical features include the abilities to reduce nitrate, ferment several carbohydrates, and hydrolyze pyrazinamide. Urease activity is absent, and no halo is produced on Tinsdale medium.

C. pseudodiphtheriticum (*C. hofmannii*)

The diphtheroid *C. pseudodiphtheriticum* occurs as part of the normal flora of the pharynx. There have been reports of endocarditis in both natural and prosthetic valves as well as a fatal urinary tract infection in a renal transplant patient (37, 44). The colonial morphology can mimic that of *C. diphtheriae*. It is inert in carbohydrate fermentation tests; urease and nitrate reactions are positive.

C. striatum

C. striatum is a member of the normal flora of the nasal mucous membranes. It has been reported as the causative agent in a pleuropulmonary infection in a leukemic patient (8). Colonies grow slowly, with diameters of approximately 1 mm after 48 h. Some strains produce a soluble yellow pigment. Microscopically the organisms are gram-variable rods which are short and thick with a striped or barred appearance due to the arrangement of metachromatic granules. The biochemical reactions presented in Table 1 reflect the 53 human isolates that have been identified at the CDC. Bovine strains are included in other collections and may account for different biochemical patterns described in other publications (52).

Corynebacterium group JK

Corynebacterium group JK, a recently characterized group of diphtheroid isolates, has been cultured from blood, tissue, cerebrospinal fluid, peritoneal fluid, genitourinary sites, and wound infections. Most infections have occurred in patients with granulocytopenia or previous cardiac surgery and are often associated with the use of indwelling intravenous catheters (37). These organisms appear to be part of the normal skin flora, particularly in inguinal, toe web, axillary, and rectal areas. Colonization rates are low in healthy individuals, but 25 to 35% of hospitalized patients have been found to carry this organism (20, 60). The Gram stain morphology of the group JK bacillus shows gram-positive coccobacillary or coccal forms resembling streptococci. Colonies are slow growing; they are small, gray to white, glistening, and usually nonhemolytic on sheep blood agar. Biochemical reactions which differentiate the group JK bacillus from other corynebacteria are the inability to produce urease, reduce nitrate, or readily ferment most carbohydrates. Fermentation of sugars in serum-supplemented peptone broth occurs after prolonged incubation in glucose and galactose for most isolates. Fermentation of starch, maltose, fructose, mannose, dextrin, and glycogen occurs variably, often after incubation of up to 3 weeks (51). *o*-Nitrophenyl-β-D-galactopyranoside is not hydrolyzed, and no halo is produced on Tinsdale medium. The antibiogram in some settings can be a distinguishing feature, since most isolates from leukopenic patients with extended antibiotic exposure have been multiply resistant. Isolates from other patients and healthy volunteers were frequently susceptible to cephalosporins and aminoglycosides. All isolates have been susceptible to vancomycin.

C. bovis

C. bovis is a common commensal in the bovine udder, but it may be a cause of bovine mastitis. There have been seven reported cases of human infections due to this organism (37), but no biochemical descriptions were provided in the original publications. Most of the isolates were from blood or cerebrospinal fluid and were associated with prosthetic devices or trauma. *C. bovis* is a lipophilic organism that requires the addition of serum to the fermentation medium. The urea reaction described in *Bergey's Manual of Determinative Bacteriology*, 8th edition (52), is positive; however, the urea reactions of the type strain and two other reference strains examined at the CDC were negative. The carbohydrate reactions presented in Table 1 are those observed with conventional tests and differ from the reactions described in Bergey's 8th edition, which were obtained from washed cells inoculated into rapid sugar tests in unbuffered water with bromthymol blue (33). The *o*-nitrophenyl-β-D-galactopyranoside test is positive.

C. kutscheri

C. kutscheri frequently causes latent infections in colonies of laboratory rodents. When resistance is lowered, the overt disease is characterized by pseudotuberculous visceral lesions. Rodent infections in the wild have also been reported. There has been a report of chorioamnionitis in a premature infant (17). The morphology is not distinctive, consisting of small, gray-white to yellowish colonies with serrate edges and irregularly staining rods which are slender, clubbed, and in cuneiform arrangements containing metachromatic granules. Esculin, nitrate, and urea reactions are positive.

C. renale group

C. renale, a frequent isolate from the vagina of healthy cows, can cause cystitis and pyelonephritis in cattle. It produces dry, white to pale-yellow colonies. The bacilli are relatively large, 0.7 by 3.0 μm or more, and contain metachromatic granules. Three groups, originally described as immunological types I, II, and III of *C. renale*, are now designated as *C. renale*, *C. pilosum*, and *C. cystitidis*, respectively, on the basis of DNA hybridization and numerical analyses of phenotypic characters (62). Both *C. pilosum* and *C. cystitidis* can be distinguished from *C. renale* by their hydrolysis of starch and absence of casein hydrolysis. Each can be distinguished from *C. renale*, as presented in Table 1, since *C. cystitidis* produces acid from xylose in serum broth and *C. pilosum* reduces nitrate. Both reports of human infections due to the *C. renale* group have been from France; one infection was a rectal abscess, and one infection was a breast abscess (47).

Coryneform groups A-1 through A-5

Coryneform group A comprises motile, yellow-pigmented organisms that have been divided into five subgroups primarily on the basis of carbohydrate reactions (57). Members of groups A-1 and A-2 can be assigned to the genus *Oerskovia*, characterized by the production of filaments that break up into motile rods that are usually monotrichous when they are small coccobacilli and peritrichous (with one to five flagella) when they are longer rods. All *Oerskovia* isolates hydrolyze both gelatin and casein, whereas a majority of isolates in groups A-3, A-4, and A-5 are negative in these tests. *Oerskovia turbata* has been a contaminant of homograft heart valves and caused one case of endocarditis after the replacement of an aortic valve (50). Coryneform group A-4 was isolated from the vitreous humor of a patient after injury with a metallic foreign body (27).

Coryneform group D-2

Coryneform group D-2 has been isolated from transtracheal aspirates from an elderly patient with pneumonia (32). During a survey for carriage of antibiotic-resistant diphtheroids, a cancer research center found 6 of the 52 isolates to be urease positive and biochemically less reactive than the group JK isolates (63). The six isolates were subsequently identified as belonging to group D-2, which can be distinguished from *C. pseudodiphtheriticum* by the inability to reduce nitrate.

Coryneform group E

Coryneform group E is a heterogeneous group which includes some organisms that have been recently recognized as aerotolerant *Bifidobacterium adolescentis* strains. One reported infection due to coryneform group E (*B. adolescentis*) was in a patient who had pyelonephritis and septicemia with no obvious predisposing condition (26). This group is negative for catalase, urease, and usually nitrate reduction. In serum-supplemented broths, it attacks all the carbohydrates listed in Table 1 except mannitol. End products are useful in distinguishing members of the *Bifidobacterium* genus since they produce large amounts of lactic and acetic acids and small amounts of formic and succinic acids, whereas the remaining group E isolates produce large amounts of succinic acid and lesser amounts of acetic acid.

Coryneform group F

Coryneform group F, a group of fermentative, catalase-positive, urease-positive organisms, has been divided into two subgroups on the basis of the carbohydrate reactions. Some strains of group F-1 are fastidious. Group F-2 can be distinguished from *C. ulcerans* and *C. pseudotuberculosis* by the lack of beta-hemolysis, the inability to produce a halo on Tinsdale agar, and a negative PLD test. These bacteria have been cultured from a variety of human sources. The most frequent source of isolation of group F-1 has been genitourinary sites.

Coryneform group G

Coryneform group G is a fastidious group of diphtheroidlike organisms which is very similar to *Bacterionema matruchotii (C. matruchotii)*; however, none of the strains received in the Special Bacterial Pathogens Laboratory at the CDC have stained with the *B. matruchotii* fluorescent-antibody conjugate. The group was divided into two subgroups on the basis of the nitrate reaction. The most frequent source of isolation, particularly of group G-1, has been the eye. Some other sources of group G were blood, cerebrospinal fluid, and throat and genitourinary sites. A case of endocarditis due to group G-2 has been published (1).

Coryneform group I

Coryneform group I organisms are quite similar to *C. diphtheriae*. These bacteria can be distinguished from *C. diphtheriae* by the hydrolysis of pyrazinamide, the inability to produce a halo on Tinsdale agar, and sometimes the maltose reaction. These bacteria have been cultured from wound, throat, blood, ear, eye, and other sites.

C. aquaticum

C. aquaticum has been isolated from blood cultures of a young male with endocarditis and from a diabetic elderly female (37). Many isolates have been received at the CDC, but their clinical significance has not been documented. The organism is common in distilled water and probably in natural fresh water. It is small, coccobacillary, and weakly gram positive. Motility is due to a few peritrichous flagella, some of which may be curly. Growth is aerobic. After overnight incubation on 5% sheep blood agar plates, the slightly yellow colonies are convex and nonhemolytic. The organism hydrolyzes esculin, and acid is produced aerobically, but not anaerobically, from glucose, maltose, sucrose, mannitol, and xylose. It does not attack starch or lactose.

"C. genitalium"

"C. genitalium" includes a diverse group of lipophilic corynebacteria that are recovered from urogenital sites and cultured either on Trypticase soy agar containing 0.1% Tween 80 or on blood agar plates that have been partially converted to chocolate agar by heating them at 65°C for 30 to 40 min. They have been distinguished from other diphtheroids by the fact that subcultures will grow on blood agar but not on commercially prepared chocolate agar. The suggestion that these organisms are a common cause of nonspecific urethritis (19) has not yet been confirmed. Some of the biotypes of *C. genitalium* resemble lipid-dependent strains of previously recognized corynebacterial groups such as *C. bovis*, group JK, and group D-2.

"C. minutissimum"

The term *"C. minutissimum"* is commonly used in the dermatology literature to refer to fluorescent diphtheroids associated with erythrasma, a mild, chronic, localized infection of the stratum corneum that is characterized by reddish-brown, scaly patches. Lesions usually involve intertriginous sites, particularly the groin, toe webs, and axillae, which characteristically fluoresce coral red when examined by Wood's light. The diagnosis does not require culture, but it has been found that if skin scrapings are cultured in a medium containing porphyrin precursors, young colonies will show coral-red fluorescence. A taxonomic scheme that relied heavily on a lipolysis test found that these fluorescent diphtheroids comprised at least eight different biochemical groups (56). They are present in larger concentrations in erythrasma lesions than in normal skin. Fluorescence due to the production of porphyrins is not a unique characteristic and is found in many bacteria, including a number of the recognized *Corynebacterium* species. *"C. minutissimum"* as described in Table 1 is a well-defined entity, which presumably represents one of the eight different groups isolated from erythrasma. It can be distinguished from *C. xerosis*, another fluorescing diphtheroid, by its inability to reduce nitrate. There is no report of *"C. minutissimum"* infections other than erythrasma, but isolates from many body sites, including blood and cerebrospinal fluid, have been submitted to the CDC for identification.

Other coryneform genera

Saprophytic coryneforms from soil, dairy products, and other sources, as well as a few coryneform plant pathogens, have been assigned to genera distinct from the genus *Corynebacterium*. The primary criteria for

TABLE 2. Criteria for genera containing coryneforms[a]

Diamino acid and type of peptidoglycan	Genus	G+C base ratio (mol%)	Arabinose	Mycolic acids	Isolation sites
meso-Diaminopimelic acid					
A[b]	*Corynebacterium*	51–59	+[c]	+	Animals, soil
?[d]	*Caseobacter*	60–67	+	+	Cheese
?	*Rhodococcus*	59–69	+	+	Soil
A	*Brevibacterium*	60–64	−	−	Cheese, skin
Ornithine					
A	*Cellulomonas*	71–75	−	−	Soil
B	*Curtobacterium*	67–71	−	−	Plants
Lysine					
A	*Arthrobacter*	59–66	−	−	Soil
B	*Microbacterium*	69–70	−	−	Dairy sources

[a] Compiled from information assembled by Keddie and colleagues (35, 36).

[b] Peptidoglycan A: direct cross-linkage between the distal amino group of the diamino acid in position 3 and the carboxyl group of D-alanine in position 4 of adjacent peptide subunits. Peptidoglycan B: cross-linkage between the δ carboxyl groups of D-glutamic acid in position 2 and the C terminus of D-alanine in position 4 of adjacent peptide subunits. A diamino acid is present in the interpeptide bridge.

[c] Galactose is present in the cell wall of all genera except *Cellulomonas*.

[d] Data not available.

these genera have emerged from an analysis of cell wall constituents and the determination of DNA base ratios and, in some cases, DNA homologies. The currently recognized coryneform-containing genera are given in Table 2, which shows that the diamino acid contained in the peptidoglycan is the primary criterion of separation. Other useful traits include the types of peptidoglycan (23), the presence or absence of arabino-galactan and mycolic acids, and the types of fatty acids, phospholipids, glycolipids, and isoprenoid quinones. The unique presence of teichoic acids in *Brevibacterium linens*, the heat resistance of *Microbacterium* spp., and the cellulolytic activity of *Cellulomonas* spp. are also useful. There is still much to be done in the classification of the coryneforms, but considerable progress has been made in sifting out inappropriately named strains and establishing delimiting criteria for these genera. Because so many species described in the literature are inappropriately named, the reader is urged to consult the detailed analysis of saprophytic aerobic coryneforms given by Keddie and Jones (36).

Currently available information does not enable clinical microbiologists to assign most coryneform isolates to a genus outside of *Corynebacterium* without the benefit of cell wall analysis or, in some cases, G+C content of the DNA. Once the coryneform bacteria listed in Table 1 have been assigned to their respective genera on the basis of the chemotaxonomic features presented in Table 2, it will be possible to develop a data base for each species with conventional or other biochemical tests that are readily available to the clinical microbiologist. Meanwhile, the majority of clinical isolates can be distinguished by the scheme presented in Table 1.

LITERATURE CITED

1. **Austin, G. E., and E. O. Hill.** 1983. Endocarditis due to *Corynebacterium* group G-2. J. Infect. Dis. **147:**1106.
2. **Barksdale, L.** 1970. *Corynebacterium diphtheriae* and its relatives. Bacteriol. Rev. **34:**378–422.
3. **Barksdale, L.** 1981. The genus *Corynebacterium*, p. 1827–1837. *In* M. P. Starr, H. Stolp, H. G. Truper, A. Balows, and H. G. Schlegel (ed.), The prokaryotes. A handbook on habitats, isolation and identification of bacteria. Springer-Verlag, New York.
4. **Barksdale, L., R. Linder, I. T. Sulea, and M. Pollice.** 1981. Phospholipase D activity of *Corynebacterium pseudotuberculosis* (*Corynebacterium ovis*) and *Corynebacterium ulcerans*, a distinctive marker within the genus *Corynebacterium*. J. Clin. Microbiol. **3:**335–343.
5. **Belsey, M. A., M. Sinclair, M. R. Roder, and D. R. LeBlanc.** 1969. *Corynebacterium diphtheriae* skin infections in Alabama and Louisiana. N. Engl. J. Med. **280:**125–141.
6. **Biberstein, E. L., H. D. Knight, and S. Jang.** 1971. Two biotypes of *Corynebacterium pseudotuberculosis*. Vet. Rec. **89:**691–692.
7. **Bickham, S. T., and W. L. Jones.** 1972. Problems in the use of the *in vitro* toxigenicity test for *Corynebacterium diphtheriae*. Am. J. Clin. Pathol. **57:**244–246.
8. **Bowstead, T. T., and S. M. Santiago, Jr.** 1980. Pleuropulmonary infection due to *Corynebacterium striatum*. Br. J. Dis. Chest **74:**198–200.
9. **Centers for Disease Control.** 1982. Fatal diphtheria—Wisconsin 1982. Morbid. Mortal. Weekly Rep. **31:**553–555.
10. **Collins, M. D.** 1982. Reclassification of *Bacterionema matruchotii* in the genus *Corynebacterium* as *Corynebacterium matruchotii* comb. nov. Zentrabl. Bakteriol. Parasitenkd. Infektionskr. Hyg. Abt. 1 Orig. C **3:**364–367.
11. **Collins, M. D., D. Jones, and G. M. Schofield.** 1982. Reclassification of *Corynebacterium haemolyticum* (MacLean, Liebow and Rosenberg) in the genus *Arcanobacterium* gen. nov. as *Arcanobacterium haemolyticum* nom. rev., comb. nov. J. Gen. Microbiol. **128:**1279–1281.
12. **Cook, G. T., and W. H. H. Jebb.** 1952. Starch-fermenting, gelatin-liquefying corynebacteria and their differentiation from *C. diphtheriae gravis*. J. Clin. Pathol. **5:**161–164.
13. **Coyle, M. B., B. H. Minshew, J. A. Bland, and P. C. Hsu.** 1979. Erythromycin and clindamycin resistance in *Corynebacterium diphtheriae* from skin lesions. Antimicrob. Agents Chemother. **16:**525–527.
14. **Davidson, S., Y. Rotem, B. Bogkowski, and E. Rubinstein.** 1976. *Corynebacterium diphtheriae* endocarditis. Am. J. Med. Sci. **271:**351–353.
15. **Eliakim, R., P. Silkoff, G. Lugassy, J. Michel.** 1983. *Corynebacterium xerosis* endocarditis. Arch. Intern. Med. **143:**1995.
16. **Facklam, R. R., D. N. Lawrence, and F. O. Sottnek.** 1978.

Modified culture technique for *Corynebacterium diphtheriae* isolation from desiccated swabs. J. Clin. Microbiol. **7**:137–138.

17. **Fitter, W. F., D. J. Se Sa, and R. Richardson.** 1979. Chorioamnionitis and funisitis due to *Corynebacterium kutscheri*. Archiv. Dis. Child. **55**:710–712.
18. **Frobisher, M.** 1937. Cystine-tellurite agar for *C. diphtheriae*. J. Infect. Dis. **10**:99–105.
19. **Furness, G., A. T. Evangelista, and Z. Kaminski.** 1977. *Corynebacterium genitalium* (nonspecific urethritis corynebacteria). Biologic reactions differentiating commensals of the urogenital tract from the pathogens responsible for urethritis. Invest. Urol. **15**:23–27.
20. **Gill, V. J., C. Manning, M. Lamson, P. Woltering, and P. A. Pizzo.** 1980. Antibiotic-resistant group JK bacteria in hospitals. J. Clin. Microbiol. **13**:472–477.
21. **Goldberger, A. C., B. A. Lipsky, and J. J. Plorde.** 1981. Suppurative granulomatous lymphadenitis caused by *Corynebacterium ovis (pseudotuberculosis)*. Am. J. Clin. Pathol. **76**:486–490.
22. **Goodfellow, M., and G. Alderson.** 1977. The actinomycete-genus *Rhodococcus*: a home for the "rhodochrous" complex. J. Gen. Microbiol. **100**:99–122.
23. **Goodfellow, M., and D. E. Minnikin.** 1981. Introduction to the coryneform bacteria, p. 1811–1826. *In* M. P. Starr, H. Stolp, H. G. Truper, A. Balows, and H. G. Schlegel (ed.), The prokaryotes. A handbook on habitats, isolation and identification of bacteria. Springer-Verlag, New York.
24. **Green, S. L., and K. S. LaPeter.** 1981. Pseudodiphtheritic membranous pharyngitis caused by *Corynebacterium hemolyticum*. J. Am. Med. Assoc. **245**:2330–2331.
25. **Groman, N., J. Schiller, and J. Russell.** 1984. *Corynebacterium ulcerans* and *Corynebacterium pseudotuberculosis* responses to DNA probes derived from corynephage β and *Corynebacterium diphtheriae*. Infect. Immun. **45**:511–517.
26. **Guillard, F., P. C. Applebaum, and F. B. Sparrow.** 1980. Pyelonephritis and septicemia due to gram-positive rods similar to *Corynebacterium* group E (aerotolerant *Bifidobacterium adolescentis*). Ann. Intern. Med. **92**:635–636.
27. **Hanscom, T., and W. A. Maxwell.** 1979. *Corynebacterium* endophthalmitis. Laboratory studies and report of a case treated by vitrectomy. Arch. Ophthalmol. **97**:500–502.
28. **Hermann, G. J.** 1961. The laboratory recognition of *Corynebacterium hemolyticum*. Am. J. Med. Technol. **27**:61–66.
29. **Hollis, D. G., F. O. Sottnek, W. J. Brown, and R. E. Weaver.** 1980. Use of the rapid fermentation test in determining carbohydrate reactions of fastidious bacteria in clinical laboratories. J. Clin. Microbiol. **12**:620–623.
30. **Hollis, D. G., and R. E. Weaver.** 1981. Gram-positive organisms: a guide to identification. Special Bacteriology Laboratory, Centers for Disease Control, Atlanta.
31. **Holt, H. D., and H. D. Wright.** 1940. The importance of the suspending fluid used in virulence tests upon *C. diphtheriae*. J. Pathol. Bacteriol. **51**:287–296.
32. **Jakobes, N. F., and C. A. Perlino.** 1979. "Diphtheroid" pneumonia. South. Med. J. **72**:475–476.
33. **Jayne-Williams, J. J., and T. M. Skerman.** 1966. Comparative studies on coryneform bacteria from milk and dairy sources. J. Appl. Bacteriol. **1**:72–92.
34. **Jellard, C. H.** 1972. Diphtheria infections in North West Canada, 1969, 1970 and 1971. J. Hyg. **70**:503–510.
35. **Keddie, R. M., and I. J. Bousfield.** 1980. Cell wall composition in the classification and identification of coryneform bacteria. Soc. Appl. Bacteriol. Symp. Ser. **8**:167–188.
36. **Keddie, R. M., and D. Jones.** 1981. Saprophytic, aerobic coryneform bacteria, p. 1838–1878. *In* M. P. Starr, H. Stolp, H. G. Truper, A. Balows, and H. G. Schlegel (ed.), The prokaryotes. A handbook on habitats, isolation and identification of bacteria. Springer-Verlag, New York.
37. **Lipsky, B. A., A. C. Goldberger, L. S. Tompkins, and J. J. Plorde.** 1982. Infections caused by nondiphtheria corynebacteria. Rev. Infect. Dis. **4**:1220–1235.
38. **Lyman, E. D., and J. A. Youngstrom.** 1956. Diphtheria cases and contacts: is it necessary to take cultures from both nose and throat? Nebr. State Med. J. **41**:361–362.
39. **MacLean, P. D., A. A. Liebow, and A. A. Rosenberg.** 1946. A hemolytic corynebacterium resembling *Corynebacterium ovis* and *Corynebacterium pyogenes* in man. J. Infect. Dis. **79**:69–90.
40. **Maximescu, P., A. L. Pop, A. Oprisan, and E. Potorac.** 1968. Relations biologues entre *C. ulcerans*, *C. ovis* et *C. diphtheriae*. Etude experimentale. Arch. Roum. Pathol. Exp. Microbiol. **27**:733–750.
41. **McLaughlin, J. V., S. T. Bickham, G. L. Wiggins, S. A. Larsen, A. Balows, and W. L. Jones.** 1971. Antibiotic susceptibility patterns of recent isolates of *Corynebacterium diphtheriae*. Appl. Microbiol. **21**:844–851.
42. **McLeod, J. W.** 1943. The types *mitis*, *intermedius* and *gravis* of *Corynebacterium diphtheriae*. Bacteriol. Rev. **7**:1–41.
43. **Moore, M. S., and E. I. Parsons.** 1958. A study of modified Tinsdale's medium for the primary isolation of *Corynebacterium diphtheriae*. J. Infect. Dis. **102**:88–93.
44. **Nathan, A. W., D. R. Turner, C. Aubrey, J. C. Cameron, D. G. Williams, C. S. Ogg, and M. Bewick.** 1982. *Corynebacterium hofmannii* after renal transplantation. Clin. Nephrol. **17**:315–318.
45. **Noble, W. C.** 1984. Skin microbiology: coming of age. J. Med. Microbiol. **17**:1–12.
46. **Pedersen, A. H. B., J. Spearman, E. Tronca, M. Bader, and J. Harnisch.** 1977. Diphtheria on Skid Road, Seattle, Wash., 1972–1978. Public Health Rep. **92**:336–342.
47. **Peloux, Y., R. Chatelain, and R. Erny.** 1981. Abces du sein provoqué par une corynebacterie du group "renale." Pathol. Biol. **29**:299–300.
48. **Pitcher, D. G.** 1977. Rapid identification of cell wall components as a guide to the classification of aerobic coryneform bacteria from human skin. J. Med. Microbiol. **10**:439–445.
49. **Reddy, C. A., C. P. Cornell, and A. M. Fraga.** 1982. Transfer of *Corynebacterium pyogenes* (Glage) Eberson to the genus *Actinomyces* as *Actinomyces pyogenes* (Glage) comb. nov. Int. J. Syst. Bacteriol. **32**:419–429.
50. **Reller, L. B., G. L. Maddoux, M. R. Eckman, and G. Pappas.** 1975. Bacterial endocarditis caused by *Oerskovia turbata*. Ann. Intern. Med. **83**:664–666.
51. **Riley, P. S., D. G. Hollis, G. B. Utter, R. E. Weaver, and C. N. Baker.** 1979. Characterization and identification of 95 diphtheroid (group JK) cultures isolated from clinical specimens. J. Clin. Microbiol. **9**:418–424.
52. **Rogosa, M., C. S. Cummins, R. A. Lelliott, and R. M. Keddie.** 1974. Coryneform group of bacteria, p. 599–632. *In* R. E. Buchanan and N. E. Gibbons (ed.), Bergey's manual of determinative bacteriology, 8th ed. The Williams & Wilkins Co., Baltimore.
53. **Schiller, J., M. Strom, N. Groman, and M. Coyle.** 1983. Relationship between pNG2, and Em^r plasmid in *Corynebacterium diphtheriae*, and plasmids in aerobic skin coryneforms. Antimicrob. Agents Chemother. **24**:892–901.
54. **Schleifer, K. H., and E. Stackbrandt.** 1983. Molecular systematics of prokaryotes. Annu. Rev. Microbiol. **37**:143–187.
55. **Skerman, V. B. D., V. McGowan, and P. H. A. Sneath.** 1980. Approved lists of bacterial names. Int. J. Syst. Bacteriol. **30**:225–420.
56. **Somerville, D. A.** 1972. A quantitative study of erythrasma lesions. Br. J. Dermatol. **87**:130–137.
57. **Sottnek, F. O., J. M. Brown, R. E. Weaver, and G. F. Carroll.** 1977. Recognition of *Oerskovia* species in the clinical laboratory: characterization of 35 isolates. Int. J. Syst. Bacteriol. **27**:263–270.
58. **Sottnek, F. O., and J. N. Miller.** 1979. Isolation and

identification of *Corynebacterium diphtheriae*. Center for Disease Control, Atlanta.

59. **Sulea, I. T., M. C. Pollice, and L. Barksdale.** 1980. Pyrazine carboxylamidase activity in *Corynebacterium*. Int. J. Syst. Bacteriol. **30:**466–472.
60. **Tompkins, L. S., F. Juffali, and W. E. Stamm.** 1982. Use of selective broth enrichment to determine the prevalence of multiply resistant JK corynebacteria on skin. J. Clin. Microbiol. **15:**350–351.
61. **Ward, K. W.** 1948. Effect of Tween 80 on certain strains of *C. diphtheriae*. Proc. Soc. Exp. Biol. Med. **67:**527–528.
62. **Yanagawa, R., and E. Honda.** 1978. *Corynebacterium pilosum* and *Corynebacterium cystitidis*, two new species from cows. Int. J. Syst. Bacteriol. **28:**209–216.
63. **Young, V. M., W. F. Meyers, M. R. Moody, and S. C. Schimpff.** 1981. The emergence of coryneform bacteria as a cause of nosocomial infections in compromised hosts. Am. J. Med. **70:**646–650.

Chapter 19

Listeria

ROBERT BORTOLUSSI, WALTER F. SCHLECH III, AND WILLIAM L. ALBRITTON

CHARACTERIZATION

Listeria monocytogenes is a short, motile, gram-positive, nonsporeforming rod that exhibits characteristic gram-positive cell wall structure by electron microscopy. The organism is a facultative anaerobe, produces catalase, and can grow over a wide temperature range on ordinary media. Isolates from patients with clinical infections produce a narrow zone of beta-hemolysis around or beneath the colonies on blood agar. Characteristic tumbling motility in a hanging-drop preparation is more pronounced after growth at 20 to 25°C than at 37°C, and a typical umbrella type of growth is observed in semisolid motility medium.

The genus *Listeria* was previously classified in the family *Corynebacteriaceae*, but the 8th edition of *Bergey's Manual of Determinative Bacteriology* includes the genus *Listeria* with the genera *Erysipelothrix* and *Caryophanon* as "genera of uncertain affiliation" (16). Stuart and Welshimer (19) have suggested that a new family, *Listeriaceae*, be created to accommodate the monospecific genus *Listeria* and a new monospecific genus, *Murraya*, including *M. grayi* subsp. *grayi* (formerly *L. grayi*) and *M. grayi* subsp. *murrayi* (formerly *L. murrayi*). Rocourt et al. (11) have suggested that the genus *Listeria* be divided into five groups based on similarities of biochemical properties and DNA relatedness. Genomic group 1 contains the type strain *L. monocytogenes* and related strains with serovars including 1/2a, 1/2b, 1/2c, 4a, 4b, and others. *L. innocua*, *L. welshimeri*, and *L. seeligeri* have been accepted as names for genomic groups 3, 4, and 5, respectively (7, 12), while "*L. bulgarica*" (serovar 5) has not yet been validated as the species designation for genomic group 2. Virulence of the strains in the genomic groups differs widely. Only strains of genomic group 1 and 2 are pathogenic in mammals.

L. monocytogenes is antigenically related to a number of organisms, especially *Staphylococcus aureus* and *Streptococcus faecalis*, and definitive serological identification requires serotyping of both somatic (O) and flagellar (H) antigens (14, 15).

CLINICAL SIGNIFICANCE

Pathogenic and nonpathogenic *Listeria* spp. are ubiquitous in nature and, with selective media, can frequently be cultured from soil, vegetation, and water (6, 21). *L. monocytogenes* has been isolated in association with disease in a wide variety of fish, birds, and mammals (6). It is an important veterinary pathogen associated with abortion and encephalitis in sheep and cattle, and veterinarians have acquired primary cutaneous listeriosis from handling infected abortuses. A wide variety of other clinical illnesses due to *L. monocytogenes* have been described, but most human infections are accounted for by the following: neonatal sepsis or meningitis, sepsis or meningitis in immunocompromised patients, and sepsis or nonspecific flulike illness in healthy women during pregnancy, which can lead to infection of the fetus.

The incidence of listeriosis in humans is unknown, but *L. monocytogenes* remains an uncommon pathogen. In the United States, *L. monocytogenes* is the fifth most common cause of bacterial meningitis after *Haemophilus influenzae*, *Neisseria meningitidis*, *Streptococcus pneumoniae*, and the group B *Streptococcus* spp. Between 1978 and 1981, 265 cases of meningitis due to *L. monocytogenes* were reported from 27 states participating in the National Bacterial Meningitis Surveillance Study. In a few regions, *L. monocytogenes* was reported to be the second most common cause of neonatal meningitis after group B streptococci and the second most common single cause of bacterial meningitis in persons over 60 after *S. pneumoniae*.

Listeriosis occurs sporadically year-round, with an increase during the summer months. Community-acquired and nosocomial epidemics of listeriosis have been described in Europe, the United States, and Canada. The source for these outbreaks has been obscure, but recent evidence suggests that food-borne transmission may be important and may be the source of sporadic cases as well (13). Fecal carriage of *L. monocytogenes* in healthy individuals is common, particularly among those with occupations in animal husbandry and slaughter (8).

Neonatal listeriosis shares clinical similarities with group B streptococcal illness. Two distinct clinical syndromes are seen. Early-onset septicemic infection (during the first week of life) is associated with prematurity, low birth weight, and obstetrical complications known to predispose to infection. This form may occur with transplacental infection from maternal sepsis or may be due to colonization of the genital tract with *Listeria* spp. Neonatal disease after the first week of life (late-onset disease) usually presents as clinical meningitis. After 1 month of age, infections are primarily associated with advanced age and immune deficiency states. However, bacterial meningitis due to *L. monocytogenes* can occur occasionally in healthy individuals.

COLLECTION, TRANSPORT, AND STORAGE OF SPECIMENS

Appropriate specimens for culture vary with the clinical syndrome, but *L. monocytogenes* is readily isolated from the usual sources, such as blood, cerebrospinal fluid, amniotic fluid, and genital tract secretions. Primary isolation from biopsy material and placental or fetal tissue has been successful. Subcultures may be transported on nutrient agar slants at ambient temperatures or at 4°C and remain stable for days to weeks. Primary isolation rates from contaminated specimens such as stool can be improved by a period of cold enrichment for up to 3 months. Stool specimens are preferable to rectal swabs when carriage studies for epidemiological purposes are undertaken. Specimens are mixed in tryptose broth and held at 4°C, with weekly subcultures for 4 weeks and

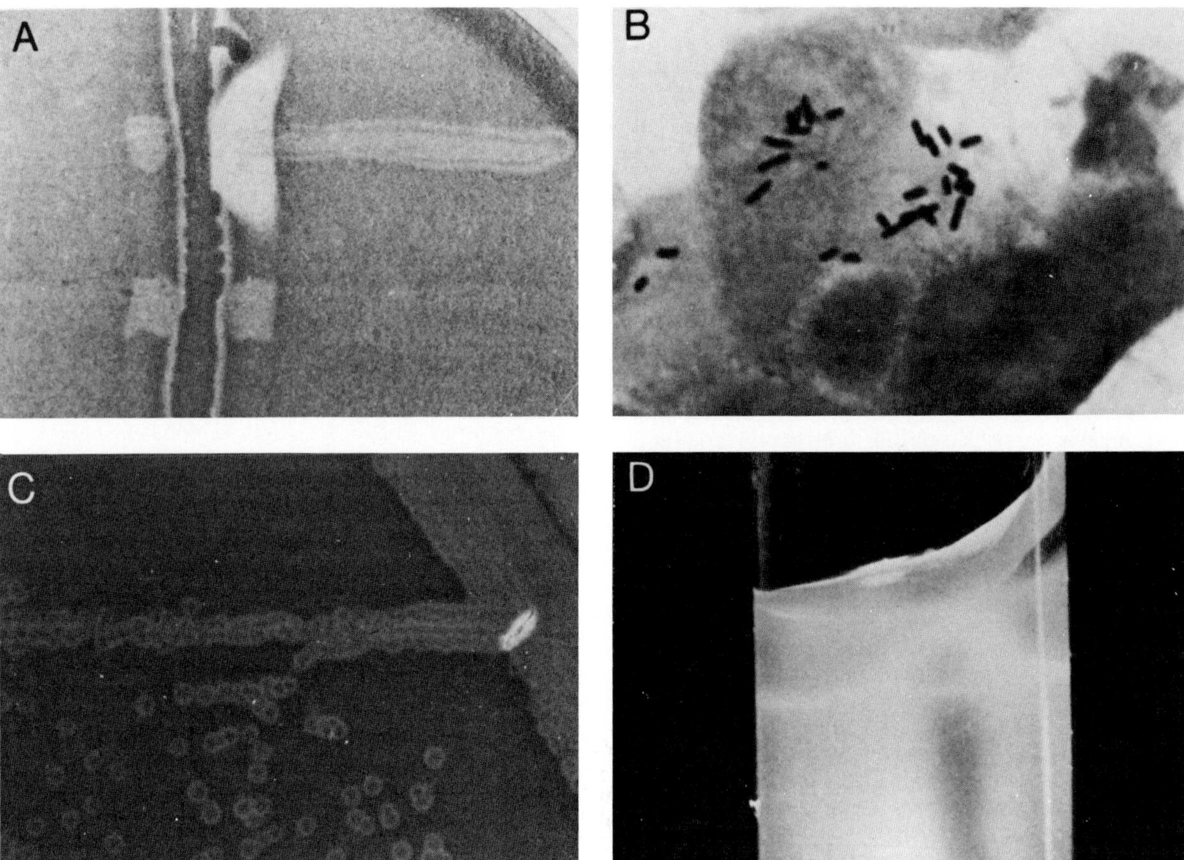

FIG. 1. (A) CAMP test (enhancement of beta-hemolysis of *S. aureus*), showing wing-shaped hemolysis of group B streptococci (top right) and rounded hemolysis of *Listeria* spp. (bottom and left). A beta-hemolytic strain of *S. aureus* runs vertically. (B) Gram stain of peritoneal exudate. (C) Beta-hemolysis of *L. monocytogenes* on blood agar. (D) Umbrella motility on soft agar.

then monthly subcultures for 2 months. Pure cultures may be maintained indefinitely at −70°C in defibrinated rabbit blood.

DIRECT EXAMINATION

Direct examination of the Gram-stained sediment of normally sterile fluids, such as cerebrospinal or amniotic fluid, is useful in establishing a tentative diagnosis. In cerebrospinal fluid smears, *Listeria* organisms are found extra- and intracellularly and may be confused with corynebacteria, pneumococci, streptococci, and, if excessively decolorized, *H. influenzae*. A marked predominance of gram-positive bacilli in specimens from sources contaminated with mixed flora may also be helpful. Specific immunochemical staining with fluorescent antibody has been described (3) and may be useful in the preliminary identification of organisms, especially if the organisms are in contaminated specimens. Very specialized studies have described the specific detection of *L. monocytogenes* in tissue by nucleic acid hybridization (18), but neither hybridization nor fluorescent-antibody methods are likely to become widely used in routine clinical laboratories.

CULTURE AND ISOLATION

Most clinical isolations of *L. monocytogenes* are made on conventional media. Heart infusion agar plates containing 5% sheep, horse, or rabbit blood, brain heart infusion broth, and various commercial blood culture media have all been satisfactory. On blood agar, beta-hemolysis is usually marked within 18 to 24 h in an agar stab, whereas hemolysis surrounding individual colonies may be weak. On blood-free media, such as tryptose agar, *Listeria* colonies give the appearance of a blue-green iridescence when observed with oblique illumination. Investigators have recommended a variety of selective media for the isolation of *L. monocytogenes* (10), especially for specimens collected from environmental sources and feces. As previously mentioned, cold enrichment procedures may improve the isolation rates from tissue sources and contaminated specimens. In general, such media and conditions are best used in epidemiological surveys and have limited application in the routine diagnostic laboratory.

IDENTIFICATION

Cultural characteristics

L. monocytogenes strains from clinical sources appear as round, translucent colonies showing narrow-zone beta-hemolysis (serotype 5 strains show wide-zone hemolysis) on blood agar plates (Fig. 1) or blue-green iridescence on clear agar when examined with oblique light.

Microscopic examination

Gram stains of clinical specimens may reveal short, intra- and extracellular gram-positive bacilli (Fig. 1); however, morphology is variable. Coccobacilli with occasional short chains which could be confused with streptococci may also be seen. Gram stains of broth cultures generally show longer gram-positive bacilli with the more typical palisade formation characteristic of diphtheroids. Occasionally, uneven staining leads to confusion with gram-negative, pleomorphic organisms such as *Haemophilus* spp.

Motility is most pronounced at 25°C. Hanging-drop preparations show a typical, end-over-end, tumbling motility. Broth cultures used for H-antigen preparation are satisfactory only if grown at 25°C. Although motility is diminished at 35°C, an umbrella type of formation is seen in the subsurface, reduced-oxygen environment of motility medium when strains are grown at either 25 or 30°C (Fig. 1).

Biochemical reaction

Listeria strains have been evaluated by many of the biochemical tests available in the research and clinical laboratory (2, 5, 9, 11, 22, 23). Table 1 gives a limited number of characteristics which will allow a presumptive species and group identification. Strains of *L. monocytogenes* isolated from clinical cases are a biochemically homogeneous group. *L. bulgarica* strains have been reported from several countries and are isolated predominantly from veterinary sources, especially aborted sheep. Genomic groups 3, 4, and 5 and *M. grayi* subsp. *grayi* and subsp. *murrayi* strains have been found in animal feces and vegetation. *L. innocua*, *L. welshimeri*, and *L. seeligeri* appear to be the most common species in nature.

In recent years, the use of selective media for isolation from specimens with mixed flora has resulted in the isolation of many nonhemolytic, nonvirulent strains. Many of these strains belong to the newly validated species *L. innocua* (7). These strains have not been encountered in human cases, and the serotypes are often different from those found in diseased animals. The CAMP phenomenon (hemolysis-accentuating factor) has been used as a simple method of identifying group B hemolytic streptococci. Strains of *S. aureus* which produce double zones of hemolysis and at least one strain of *Rhodococcus equi* (NCTC 1621) are capable of producing a similar phenomenon with *Listeria* species (Fig. 1) (11, 15, 17). Additional studies may be helpful in characterizing some strains (5, 9, 22).

Serological methods

Definitive serological identification requires serotyping of both O and H antigens (4, 14, 15). Isolates presumptively identified as *L. monocytogenes* should be submitted through the state health department to the attention of the National Listeria Typing Center, Centers for Disease Control, Atlanta, GA 30333. The production of specific typing sera is tedious and not warranted for diagnostic laboratories. Rapid slide and tube agglutination tests are available commercially, but in general, more definitive serotyping is necessary for epidemiological studies. Serotypes Ia, Ib, and IVb account for over 92% of all isolates, although some regional differences occur.

DIFFERENTIATION FROM RELATED SPECIES

L. monocytogenes is most frequently misidentified as a streptococcus and may be discarded as a contaminating diphtheroid. Demonstration of motility distinguishes *Listeria* strains from *Erysipelothrix* species and many *Corynebacterium* species. *L. monocytogenes* strains ferment glucose, produce catalase, are methyl red and Voges-Proskauer positive, and do not reduce nitrates or produce urease. No H_2S is produced in the butt of triple sugar iron agar. *L. monocytogenes* can be separated from *Corynebacterium aquaticum*, a motile species, since this species produces acid from glucose aerobically only and is Voges-Proskauer negative. As previously discussed, infections with group B streptococci share many of the clinical features of listeriosis, and the infecting organisms resemble each other in some biochemical characteristics (both hydrolyze hippurate) (9). *Listeria* isolates may occasionally be confused with enterococci, since both are positive in bile-esculin and salt tolerance tests (22). Gram-stain morphology, motility, and catalase activity are the most practical laboratory tests for the differentiation of *Listeria* spp. from group B streptococci and enterococci.

ANIMAL INOCULATION

Virulence can be confirmed experimentally by ophthalmic pathogenicity in rabbits (Anton test), intraperitoneal inoculation of mice, and inoculation of the chorioallantoic membrane in embryonated eggs, but

TABLE 1. Biochemical characteristics of *Listeria* strains

Genomic group	Species	Umbrella motility	CAMP test[a]		Acid production in enteric base from:		
			Staphylococcus aureus	*Rhodococcus equi*	Xylose	Rhamnose	Mannitol
1	*L. monocytogenes*	+	+	−	−	+	−
2	*L. bulgarica*[b]	+/−	−	+	+[c]	−[c]	−[c]
3	*L. innocua*	+	−	−	−	+/−	−
4	*L. welshimeri*	+	−	−	+	+/−	−
5	*L. seeligeri*	+	+	−	+	−	−
	M. grayi subsp. *grayi*	+	−	−	−	−	+
	M. grayi subsp. *murrayi*	+	−	−	−	−	+

[a] As described by Roucourt and Grimont (12).
[b] Taxonomic status not firmly established (14, 18).
[c] Serum must be added to base to enhance growth.

such tests are not required for the routine isolation and identification of clinical isolates.

OTHER METHODS

Phage typing of *L. monocytogenes* is currently under investigation for international use. It has been used in a presumptive typing system, and some studies indicate that it is potentially useful in epidemiological and taxonomic investigations (1, 20).

LITERATURE CITED

1. **Audurier, A., R. Chatelain, F. Chalons, and M. Piechaud.** 1979. Lysotypie de 823 souches de *Listeria monocytogenes* isolées en France de 1958 à 1978. Ann. Microbiol. **130:**179–189.
2. **Bojsen-Moller, J.** 1972. Human listeriosis: diagnostic epidemiological and clinical studies. Acta Pathol. Microbiol. Scand. Suppl. **229:**1–157.
3. **Cherry, W. B., and M. D. Moody.** 1965. Fluorescent-antibody techniques in diagnostic microbiology. Bacteriol. Rev. **29:**222–250.
4. **Donker-Voet, J.** 1959. A serological study of some strains of *Listeria monocytogenes*, isolated in Michigan. Am. J. Vet. Res. **20:**176–179.
5. **Emody, L., and B. Ralovich.** 1972. Ability of *Listeria monocytogenes* strains to attack carbohydrates and related compounds. Acta Microbiol. Acad. Sci. Hung. **19:**287–291.
6. **Gray, M. L., and A. H. Killinger.** 1966. *Listeria monocytogenes* and listeric infections. Bacteriol. Rev. **30:**309–382.
7. **International Journal of Systematic Bacteriology.** 1983. Validation of the publication of new names and new combinations previously effectively published outside the IJSB. List no. 10. Int. J. Syst. Bacteriol. **33:**438–440.
8. **Kampelmacher, E. H., and L. M. van Noorle Jansen.** 1969. Isolation of *Listeria monocytogenes* from feces of clinically healthy humans and animals. Zentralbl. Bakteriol. Parisitenkd. Infektionskr. Hyg. Abt. 1 Orig. **211:**353–359.
9. **Kontnick, C., A. von Graevenitz, and V. Piscitelli.** 1977. Differential diagnosis between *Streptococcus agalactiae* and *Listeria monocytogenes* in the clinical laboratory. Ann. Clin. Lab. Sci. **7:**269–276.
10. **Kramer, P. A., and D. Jones.** 1969. Media selection for *Listeria monocytogenes* J. Appl. Bacteriol. **32:**381–394.
11. **Rocourt, J., F. Grimont, P. A. D. Grimont, and H. P. R. Seeliger.** 1982. DNA relatedness among serovars of *Listeria monocytogenes sensu lato*. Curr. Microbiol. **7:**383–388.
12. **Rocourt, J., and P. A. D. Grimont.** 1983. *Listeria welshimeri* sp. nov. and *Listeria seeligeri*. Int. J. Syst. Bacteriol. **33:**866–869.
13. **Schlech, W. F., P. M. Lavigne, R. A. Bortolussi, A. C. Allen, E. V. Haldane, A. J. Wort, A. W. Hightower, S. E. Johnson, S. H. King, E. S. Nicholls, and C. V. Broome.** 1983. Epidemic listeriosis—evidence for transmission by food. N. Engl. J. Med. **308:**203–206.
14. **Seeliger, H. P. R.** 1975. Serovariants of *Listeria monocytogenes* and other *Listeria* species. Acta Microbiol. Acad. Sci. Hung. **22:**179–181.
15. **Seeliger, H. P. R., and K. Hohne.** 1979. Serotyping of *Listeria monocytogenes* and related species. Methods Microbiol. **13:**31–49.
16. **Seeliger, H. P. R., and H. J. Welshimer.** 1974. Genus *Listeria*, p. 593–596. *In* R. E. Buchanan and N. E. Gibbons (ed.), Bergey's manual of determinative bacteriology, 8th ed. The Williams & Wilkins Co., Baltimore.
17. **Skalka, B., J. Smola, and K. Elischerová.** 1982. Routine test for in vitro differentiation of pathogenic and apathogenic *Listeria monocytogenes* strains. J. Clin. Microbiol. **15:**503–507.
18. **Steinman, C. R.** 1975. Specific detection and semiquantitation of microorganisms in tissue by nucleic acid hybridization. I. Characterization of the method and application of model systems. J. Lab. Clin. Med. **86:**164–174.
19. **Stuart, S. E., and H. J. Welshimer.** 1974. Taxonomic reexamination of *Listeria* Pirie and transfer of *Listeria grayi* and *Listeria murrayi* to a new genus, *Murraya*. Int. J. Syst. Bacteriol. **24:**177–185.
20. **Watson, B. B., and W. C. Eveland.** 1965. The application of phage fluorescent antiphage staining system in specific identification of *Listeria monocytogenes*. I. Species specificity and immunofluorescent sensitivity of *Listeria monocytogenes* phage observed in smear preparations. J. Infect. Dis. **115:**363–369.
21. **Weis, J., and H. P. R. Seeliger.** 1975. Incidence of *Listeria monocytogenes* in nature. Appl. Microbiol. **30:**29–32.
22. **Wetzler, T. F., N. R. Freeman, M. L. French, L. A. Renkowski, W. C. Eveland, and O. J. Carver.** 1968. Biological characterization of *Listeria monocytogenes*. Health Lab. Sci. **5:**46–62.
23. **Wilkinson, B. J., and D. Jones.** 1977. A numerical taxonomic survey of *Listeria* and related bacteria. J. Gen. Microbiol. **98:**399–421.

Chapter 20

Erysipelothrix

ROBERT E. WEAVER

CHARACTERIZATION

Erysipelothrix rhusiopathiae is a facultatively anaerobic, gram-positive, rod-shaped bacterium which decolorizes readily and may appear to be gram negative. Organisms from smooth colonies are 0.2 to 0.4 μm by 0.5 to 2.5 μm. In smears from rough colonies, long, nonbranching filaments which may be granular are observed. Production of H_2S in the butt portion of triple sugar iron agar (TSI) slants is a characteristic which readily distinguishes most strains of this organism from other morphologically similar gram-positive bacteria.

The organism is primarily a pathogen of swine and other animals, producing the disease referred to as swine erysipelas. These infections may be chronic or acute and include septicemia, endocarditis, arthritis, and urticarial or cutaneous forms of infection.

The most common form of infection in humans, erysipeloid, is an inflammatory lesion of the skin which usually involves the hands and fingers. The lesions are characterized by an elevated erythematous edge which spreads peripherally as discoloration of the central area fades. Septicemia is an infrequent complication of erysipeloid and may develop in the absence of a preceding cutaneous lesion. Endocarditis has occurred in persons in whom there was no indication of previous valvular damage.

Acquisition of *E. rhusiopathiae* by humans is usually the result of contact with tissue of infected animals or contaminated animal products. The persons at greatest risk of infection include abattoir workers, butchers, veterinarians, fishermen, sellers of fish, and housewives. Outbreaks of erysipeloid have been described among workers who produce buttons from bone.

Contaminated soil is a potential source of *E. rhusiopathiae*. The organism has been isolated from the soil of pig pens as long as 5 years after swine erysipelas was observed on the farm (8).

COLLECTION

Successful isolation of *E. rhusiopathiae* from erysipeloid lesions can be made from biopsy specimens (3). Cleanse and disinfect the skin with cetrimide. After the skin has dried, inject 1% procaine to produce an intradermal wheal in the advancing edge of the lesion. Excise a full-thickness portion of skin from the edge of the lesion by the "pinch" method. The organism may also be obtained from the erysipeloid lesion by injecting saline into the edge of the lesion and then reaspirating without withdrawing the needle (6). In patients with septicemia or endocarditis, the organism can be isolated by routinely used blood culture techniques (4).

TRANSPORT

Biopsy or aspiration specimens should be placed in an infusion broth containing 1% glucose immediately after collection. They should be sent promptly to the laboratory. The inoculated broth may be kept at room or refrigerator temperature until it reaches the laboratory.

DIRECT EXAMINATION

There is no evidence that direct examination of specimens is of any value in the diagnosis of human infections.

CULTURE AND ISOLATION

Primary culture

For isolation of *E. rhusiopathiae* from skin lesions, biopsy specimens or tissue aspirates should be placed in an infusion broth containing 1% glucose, incubated aerobically or in CO_2 at 35 to 37°C, and subcultured to a blood agar plate at 24-h intervals. Routinely used blood culture media are satisfactory for primary isolation from patients with septicemia or endocarditis.

If the skin is properly cleansed and treated with a disinfectant before a biopsy specimen is obtained, the use of selective media is not necessary. However, selective media have been used to isolate *E. rhusiopathiae* from soil, manure, and animal tissues which were heavily contaminated with other bacteria (7, 8). The reader is referred to the first edition of this Manual for descriptions of media and procedures for isolation of this organism from contaminated materials (1).

IDENTIFICATION

Cultural characteristics

At 18 to 24 h on Trypticase soy agar (BBL Microbiology Systems, Cockeysville, Md.) containing 5% sheep blood, smooth colonies are usually 0.1 to 0.5 mm in diameter. The colonies are convex, circular, and transparent. Rough colonies are larger and have a matt surface and a fimbriated edge. Frequently there is a greenish discoloration of the blood underneath the colonies. The discoloration is more prominent in areas of the medium stabbed with the inoculation loop. The appearance of the organism on heart infusion agar containing 5% rabbit blood is similar except that the greenish discoloration under the colonies may be less distinct. Growth on heart infusion or Trypticase soy agar slants is light at 18 to 24 h.

Microscope examination

Smears should be prepared from cultures and Gram stained. Their appearance is as described above.

Biochemical characterization (Table 1)

E. rhusiopathiae is catalase and oxidase negative. In TSI slants it produces an acid slant and butt, and most strains form sufficient H_2S to blacken the butt. The formation of H_2S by a gram-positive bacterium is

TABLE 1. Characteristics of *E. rhusiopathiae*

Test or compound	Reaction	Test or compound	Reaction
Catalase	−	Glucose	A[a]
Oxidase	−	Xylose	−
Motility	−	Mannitol	−
H_2S in TSI	+	Lactose	A
Nitrate reduction	−	Sucrose	−
		Maltose	−[b]

[a] A, Acid produced within 48 h.
[b] A few strains produce acid in 6 to 7 days.

almost always indicative of the presence of *E. rhusiopathiae* because very few of the gram-positive organisms encountered in the clinical laboratory form H_2S. Some *Streptococcus* and *Bacillus* species do so, but they can be differentiated from *E. rhusiopathiae* by cellular morphology and formation of spores, respectively.

Acid is regularly produced from glucose and lactose within 48 h in a fermentation base with Andrade indicator and supplemented with 2 drops of rabbit serum per 4 ml of medium. A few human strains referred to the Division of Bacterial Diseases, Special Bacterial Pathogens Laboratory, Center for Infectious Diseases, Centers for Disease Control, have produced a weakly acidic reaction in maltose by 6 to 7 days. Acid is not formed from xylose, mannitol, or sucrose. In the rapid fermentation test, carbohydrate reactions of *E. rhusiopathiae* have been obtained within 4 h (2).

Although not required for identification, an additional reaction that is highly characteristic of this organism is the "test tube brush" pattern of growth in gelatin stab cultures. Because the gelatin must be incubated at a temperature sufficiently low (25 to 28°C) to maintain it in the solidified state, most clinical laboratories will not find it convenient to perform this test.

Serological methods

There are no serological procedures that are practicable for routine use in a clinical laboratory for the identification of the organism or for the demonstration of antibodies in patients' sera.

Differentiation from related species

The catalase-negative gram-positive species *Actinomyces (Corynebacterium) pyogenes* and *Arcanobacterium (Corynebacterium) haemolyticum* can be differentiated from *E. rhusiopathiae* because the former produce a beta-like hemolysis on blood agar and do not produce H_2S in the butt of TSI slants. *Listeria monocytogenes* can be differentiated by catalase formation, motility, hemolysis, and H_2S production.

Animal inoculation

Most strains of *E. rhusiopathiae* are virulent for mice, and their identification can be confirmed by use of the following mouse protection test. Into two or more mice, inject 0.1 ml of an 18- to 24-h broth culture subcutaneously on one side and 0.3 ml of commercial equine hyperimmune *E. rhusiopathiae* antiserum on the opposite side. An additional group of two or more mice is injected in a similar fashion with 0.1 ml of the broth culture. If the culture is *E. rhusiopathiae*, the mice that did not receive antiserum will die in 5 to 6 days. Those receiving antiserum will be protected.

ANTIBIOTIC SUSCEPTIBILITY

E. rhusiopathiae is susceptible to penicillin, cephalosporin, erythromycin, and clindamycin. It is resistant to novobiocin and the aminoglycoside antibiotics (5).

EVALUATION

Isolation of *E. rhusiopathiae* from blood or skin lesions is a finding that should be quickly reported. It is well established that this organism is the cause of erysipeloid and can produce destructive endocarditis.

LITERATURE CITED

1. **Blair, J. E., E. H. Lennette, and J. P. Truant (ed.).** 1970. Manual of clinical microbiology. American Society for Microbiology, Bethesda, Md.
2. **Hollis, D. G., F. O. Sottnek, W. J. Brown, and R. E. Weaver.** 1980. Use of the rapid fermentation test in determining carbohydrate reactions of fastidious bacteria in clinical laboratories. J. Clin. Microbiol. **12:**620–623.
3. **Price, J. E. L., and W. E. J. Bennett.** 1951. The erysipeloid of Rosenbach. Br. Med. J. **2:**1060–1062.
4. **Muirhead, N., and T. M. S. Reid.** 1980. *Erysipelothrix rhusiopathiae* endocarditis. J. Infect. **2:**83–85.
5. **Simerkoff, M. S., and J. J. Rahal, Jr.** 1973. Acute and subacute endocarditis due to *Erysipelothrix rhusiopathiae*. Am. J. Med. Sci. **266:**53–57.
6. **Wilson, G. S., and A. A. Miles.** 1975. Topley and Wilson's principles of bacteriology, virology and immunity, 6th ed., vol. 2, p. 1716. The Williams & Wilkins Co., Baltimore.
7. **Wood, R. L.** 1965. A selective liquid medium utilizing antibiotics for isolation of *Erysipelothrix insidiosa*. Am. J. Vet. Res. **26:**1303–1308.
8. **Wood, R. L., and R. A. Packer.** 1972. Isolation of *Erysipelothrix rhusiopathiae* from soil and manure of swine raising premises. Am. J. Vet. Res. **33:**1611–1620.

Chapter 21

Bacillus

R. J. DOYLE, K. F. KELLER, AND J. W. EZZELL

CHARACTERISTICS

Members of the genus *Bacillus* are characterized as bacteria which are gram-positive rods, endospore forming, aerobic or facultative, catalase positive, and generally motile (8). In practice, it has been difficult to precisely place *Bacillus* species in well-defined taxonomic positions. This difficulty seems paradoxical, considering the importance this genus has had in the understanding of infectious diseases, contemporary molecular genetics, cellular differentiation, and agricultural and industrial microbiology.

Difficulty in the definition of *Bacillus* species is reflected in the treatment of this genus in the 7th and 8th editions of *Bergey's Manual of Determinative Bacteriology* (8). In the 7th edition (1939), the total number of species listed was 25, whereas in the 8th edition, 22 species were listed, but 26 other taxa were considered closely related. Many of the traits which are customarily used to construct a species are variable in certain *Bacillus* species. Some of these traits include pH and temperature limits of growth, resistance to lysozyme, bacteriophage susceptibility, profile of surface antigens, hemolysis of blood agar, AT/GC ratio in the DNA, and carbohydrate utilization.

CLINICAL SIGNIFICANCE

Because most members of the genus *Bacillus* are saprophytic, it has been traditional to minimize their significance in clinical microbiology. Numerous reports during recent years, however, suggest that it is no longer tenable to consider *Bacillus anthracis*, the etiologic agent of anthrax, the only human pathogen of the genus. *Bacillus* species other than the anthrax bacillus have been implicated in serious infections associated with immunosuppression, traumatic wounds and burns, operative procedures, hemodialysis, parenteral drug abuse, and food poisoning. Several noteworthy reviews have been published which describe cases of bacteremia, meningitis, meningoencephalitis, pneumonia, endocarditis, urinary tract infections, peritonitis, pericarditis, pleuritis, and ocular infections due to nonanthrax *Bacillus* species (11, 19–21, 25). Many animal infections are also attributable to members of this genus, and *B. thuringiensis*, *B. alvei*, and *B. larvae* are important insect pathogens in agriculture. Small rodents are highly susceptible to experimental infections with *B. anthracis*. White mice, rabbits, and guinea pigs develop fatal infections as a result of subcutaneous inoculations of very small numbers of the virulent organisms (23).

In humans, anthrax most commonly occurs as a cutaneous lesion or boil which usually progresses to a black eschar despite antibiotic treatment. Since humans are relatively resistant to the cutaneous infections of anthrax, the lesions could heal without medical intervention. On the other hand, the lesions may lead to septicemia and meningitis unless given proper treatment, which usually consists of antimicrobial therapy with penicillin or tetracycline and typically defers surgical incision and drainage. Pulmonary anthrax, although less frequent in occurrence, always requires treatment and is often fatal. A third form of the disease, gastrointestinal anthrax, usually results from the ingestion of infected meat or spore-contaminated food, and a high mortality (25 to 50%) is observed.

As a matter of historical interest, Koch's postulates were developed as a result of his studies on the pathogen *B. anthracis*. In addition, Pasteur studied the use of heat-attenuated vaccine strains of this organism to protect animals against anthrax.

The remainder of this chapter provides an outline for the identification of the most commonly isolated members of the genus *Bacillus* as well as detailed methods for the identification of *B. anthracis*, *B. cereus*, *B. mycoides* (*B. cereus* var. *mycoides*), and *B. thuringiensis*. Table 1 reviews the major characteristics of these closely related bacteria.

SAFETY PRECAUTIONS

Strict safety considerations must be given to specimens and cultures which may contain *B. anthracis*. It is imperative to work in a biological safety hood, to disinfect all bench space (1% [wt/vol] sodium hypochlorite is acceptable), and to wear gloves while processing clinical specimens and performing subsequent identification tests. Caution must be taken to prevent aerosolization or self-inoculation with specimens or cultures. Discarded or contaminated items should be autoclaved or soaked in 1% sodium hypochlorite.

COLLECTION, TRANSPORT, AND STORAGE OF SPECIMENS

The direct sampling of cutaneous lesions of patients suspected of having an infection of *B. anthracis* is recommended. Dry sterile swabs should be used to obtain vesicular exudate material. One swab is required for the transport and subsequent recovery of organisms, and another swab is required for the preparation of smears for Gram stains and fluorescent-antibody staining. The edges of lesions devoid of serous exudate should be gently massaged with sterile swabs moistened with saline, broth media, or sterile water. The collection and transport of other clinical specimens such as blood, spinal fluids, etc., should be done according to the procedures described in Chapter 8. In addition, the preparation of smears from blood and lesions should also follow the general procedures outlined in Chapter 8. *Bacillus* species in specimens may be frozen to −70°C; however, repeated freezing and thawing should be avoided.

BACILLUS ANTHRACIS

Table 2 and Fig. 1 outline the principal procedures for the identification of *B. anthracis*. On microscopic examination of Gram-stained smears, anthrax bacilli

TABLE 1. Summary of properties of *B. anthracis* and related *Bacillus* species

Property	Response of				Reference
	B. anthracis	*B. cereus*	*B. mycoides*	*B. thuringiensis*	
Resistance to lysozyme	+[a]	Variable	+	+	1, 9, 19
Resistance to mutanolysin	Weak	Weak	Weak	Weak	27
Capsule formation	Variable	−	−	−	16, 20
Agglutinated by *Glycine max* lectin	+	−	+	−	6
Agglutinated by *Helix pomatia* lectin	−	−	+	−	6
Hydrolysis of *p*-nitrophenyl-β-D-*N*-acetylglucosamine	−	+	−	+	24
Sensitivity to penicillin	Sensitive	Resistant	Resistant	Resistant	1, 7–10
Motility	−	+	−	+	1, 3–5, 8, 9
Beta-hemolysis on sheep blood agar	Variable	Variable	Variable	Variable	3, 4, 20
Growth at 45°C	Slight	Rapid	None	Slight	1, 8–10, 19
Enhanced hydrolysis of *p*-nitrophenyl-α-D-glucoside in presence of 1% Triton X-100	+	−	−	−	24
% GC	32–40	32–40	32–40	32–40	1
Lysis of bacteriophage	+	−	Variable	−	3, 7
Pathogenesis in animals (skin reactions)	+	+	−	+	20
Colonial morphology	MH[b]	MH	MH	MH	20

[a] +, Greater than 90% of the reactions are positive; −, less than 10% of the reactions are positive; Variable, 10 to 90% of the reactions are positive.

[b] MH, Medusa head or serrated outgrowth from the periphery of the colony. See Parry et al. (20) for a detailed description of the colonial morphologies of *Bacillus* species.

appear as large rods (3 to 5 μm long and 1 to 1.25 μm wide) with flattened ends. Vegetative cells usually occur as short chains in clinical specimens, whereas spores are rarely found.

In culture, however, long chains are typically formed with an appearance often described as resembling a jointed bamboo rod. Endospores are central to subterminal and nonswollen, and they frequently occur in a free state in culture. Sporulation is most extensive in well-aerated cultures grown at 32 to 35°C.

TABLE 2. Characteristics of members of the genus *Bacillus*[a]

	Morphology				Anaerobic growth	Gas from glucose	Arabinose and xylose	Starch hydrolysis	V-P	$NO_3 \rightarrow NO_2$	Growth in 7% NaCl	Growth at 50°C	Lecithinase	Additional differential tests
	Spore swells sporangium	Vacuoles	Motility	Rods 1 μm or wider										
B. anthracis	−	+	−	+	+	−	−	+	+	+	+	−	+	See Table 1
B. cereus	−	+	V	+	+	−	−	+	+	+	+	−	+	See Table 1
B. mycoides	−	+	−	+	+	−	−	+	+	+	+	−	+	See Table 1
B. thuringiensis	−	+	V	+	+	−	−	+	+	+	+	−	+	See Table 1
B. megaterium	−	−	V	+	−	−	V	+	−	V	+	−	−	Ferments mannitol
B. subtilis	−	−	+	−	−	−	+	+	+	+	+	+	−	Ferments mannitol; citrate positive
B. pumilus	−	−	+	−	−	−	+	−	+	−	+	+	−	Ferments galactose
B. licheniformis	−	−	+	−	+	+	+	+	+	+	+	+	−	Ferments rhamnose
B. firmus	−	−	V	−	−	−	V	+	−	+	+	−	−	Citrate negative; ferments mannitol
B. circulans	+	−	V	−	V	−	+	+	−	V	V	+	−	Citrate negative but ONPG positive[b]
B. polymyxa	+	−	+	−	+	+	+	+	+	+	−	−	−	Gluconate positive
B. laterosporus	+	−	+	−	+	−	−	−	−	+	−	+	+	Canoe-shaped parasporal bodies on spores
B. brevis	+	−	+	−	−	−	−	−	−	V	−	+	−	Ferments glucose
B. sphaericus	+	−	+	V	−	−	−	−	−	−	V	−	−	Does not ferment glucose
B. alvei	+	−	+	V	+	−	−	+	+	−	−	−	−	Ferments adonitol
B. stearothermophilus	V	−	+	V	−	−	V	+	−	V	−	+	−	Growth at 65°C
B. coagulans	V	−	+	V	+	−	V	+	V	V	−	+	−	Ferments lactose
B. macerans	+	−	+	−	+	+	+	+	−	+	−	+	−	Casein not hydrolyzed

[a] Key: +, 90% of isolates positive; −, 10% of isolates positive; V, variable reactions, 10 to 90% of isolates positive; V-P, Voges-Proskauer reaction.

[b] ONPG, *o*-Nitrophenyl-β-D-galactopyranoside.

1. Culture the specimen on sheep blood agar. The selective medium of Knisely is recommended for highly contaminated specimens, although some strains of *B. anthracis* are inhibited by this medium.
2. Subculture isolates on bicarbonate- and serum-containing medium in the presence of 5% CO_2. Determine the presence of a capsule macroscopically, with fluorescein-labeled anti-poly(D-glutamic acid), or by M'Fadyean stain. Both stains can be directly used in tissue sections or blood smears.
3. Use the toxin-antitoxin reaction. A positive reaction confirms *B. anthracis*. A negative reaction does not rule out *B. anthracis*.
4. Determine the susceptibility of the isolate to penicillin. *B. anthracis* is generally susceptible to penicillin G and typically forms the "string of pearls" morphology, whereas other *Bacillus* species are characteristically resistant to the drug.
5. Determine the rates of hydrolysis of *p*-nitrophenyl-α-D-glucopyranoside and *p*-nitrophenyl-α-D-maltoside in the presence and absence of 1% (vol/vol) Triton X-100. The detergent enhances chromophore formation in *B. anthracis* but retards hydrolysis of the substrates by *B. cereus*, *B. mycoides*, and *B. thuringiensis*.
6. Note whether there is agglutination of cells by *Glycine max* lectin but not by *Helix pomatia* lectin.
7. Determine γ phage susceptibility.

FIG. 1. Recommended procedures for the identification of *Bacillus anthracis*, listed in terms of importance as established by our laboratories.

On sheep blood agar after 18 to 24 h at 35°C, the colonies are 2 to 5 mm in diameter and nonhemolytic, with weak hemolysis by some strains in areas of heavy growth. Aging (>36 h) colonies on laboratory media such as blood agar, tryptic soy agar, or nutrient agar have a rough texture (ground-glass appearance) and a serrated edge, they often form curled peripheral projections to produce the classical medusa head appearance (20), and they typically lack a capsule. Specimens should be simultaneously inoculated (if at all possible) onto medium (nutrient agar or brain heart infusion agar) containing 0.5% sodium bicarbonate and incubated in 5% CO_2 or a candle jar. Virulent strains become encapsulated, resulting in mucoid colonies. Capsule formation, which is plasmid mediated (C. B. Thorne, personal communication), may be enhanced on these media by the incorporation of 0.7% (wt/vol) bovine serum albumin. The formation of mucoid colonies only in the presence of elevated CO_2 is characteristic of virulent *B. anthracis* strains (16). No other bacillus species does this. The capsule is composed of poly(D-glutamic acid) and, in addition to the tripartite toxin (see below), serves as a virulence determinant. The capsule can be detected microscopically by using capsule stains. The M'Fadyean stain (17) as described by Parry et al. (20) stains the bacterial cell blue, whereas the surrounding capsule is pink. Fluorescent antibody may also be used to detect encapsulated organisms in tissues or from culture. Fluorescein conjugates of heterologous sera and monoclonal antibody to capsule are available through the Centers for Disease Control (CDC), Atlanta, Ga., but there are cross-reactions with *B. megaterium*. It should be noted that the amount of capsule surrounding the cells may be diminished in aged cultures, after freezing and thawing, or upon extensive washing.

B. anthracis produces a tripartite toxin, mediated by a second plasmid (18), which is composed of an edema factor (extracytoplasmic adenylate cyclase; 14), protective antigen, and lethal factor. Protective antigen is required for the activity of the edema and lethal factors and is the predominant protein secreted by *B. anthracis* under a variety of cultural conditions. *B. anthracis* colonies can thus be identified by placing antibody in a well cut approximately 5 mm from a suspect colony. When this procedure is performed with appropriate media (i.e., R-medium plus 1.5% agar, [22]), precipitin bands can be observed after overnight incubation at 4°C. Both heterologous antisera and monoclonal antibody mixtures directed against the toxin components have been produced for this double diffusion assay (CDC).

Clinical specimens and environmental samples should be mixed with sterile nutrient broth. Macerating, mincing, or vortexing the specimens to effect a suspension of cells is recommended. These procedures must be carried out with safety precautions which prevent the aerosolization of potentially infectious agents. Dilutions are made and plated onto the selective medium of Knisely (12) and onto sheep blood agar. It is not recommended that a presumptive diagnosis of *B. anthracis* be based solely on colonial morphology or a lack of hemolysis. In a study by Brown et al. (3) in which strains were collected from around the world, a total of 122 strains were examined, and 36% were found to produce some degree of hemolysis. This report, however, did not reveal the ages of the cultures. Plates should be examined for hemolysis after 18 h.

Although hemolysis on blood agar is not a reliable index for identification, colonies which are strongly beta-hemolytic have never been reported for *B. anthracis*. Beta-hemolysis among *B. cereus* and *B. thuringiensis* strains is variable, and caution must be exercised not to confuse nonhemolytic strains with *B. anthracis*.

The medusa head colony is exhibited not only by *B. anthracis*, but also often by *B. cereus*, *B. mycoides*, and *B. thuringiensis* (20). Viscidity of colonies is reported to be an unreliable parameter for *B. anthracis*; however, all freshly isolated strains from clinical specimens submitted to the CDC have had this characteristic.

A motility test is one of the most useful and reliable procedures for the preliminary screening of *Bacillus* isolates for *B. anthracis*. For determination of motility, young colonies are first inoculated into Penassay (or other rich) broth and shaken aerobically at 35°C. When exponential growth has been achieved in the broth, samples are removed for determination of motility by the hanging-drop method. It is important that motility determinations be carried out as expeditiously as possible upon removal from the culture, as motility depends on an energized membrane. The motility of *B. cereus* and *B. thuringiensis* strains is in most cases easily detected, whereas *B. anthracis* is nonmotile. Any *Bacillus* isolate exhibiting motility may be safely assumed to be a species other than *B. anthracis*.

For glycosidase assays, colonies from either the selective medium or blood agar may be suspended in

50 mM phosphate buffer (pH 7.2) to an optical density of 0.5 to 0.7 (1-cm path length, 450 nm). Incubation (35°C) of cell suspensions with 0.5 to 1.0 mM *p*-nitrophenyl-α-D-glucopyranoside in the presence or absence of 1% (vol/vol) Triton X-100 can be used as a further diagnostic test for *B. anthracis* (24). α-Glucosidase activity is increased in *B. anthracis* by incubation in the detergent, whereas glycosidase activity in *B. cereus*, *B. mycoides*, and *B. thuringiensis* is diminished. Visible detection of the chromophore is generally possible in 30 to 90 min. Commercial paper strips impregnated with the nitrophenyl saccharides may be used in qualitative glycosidase assays. Quantitative assessments of α-glucosidase activity can be determined by measuring absorbance changes at 420 nm (24).

The phosphate-buffered suspensions can also be used in lectin agglutination assays. To 50 μl of cell suspension (0.2 to 0.5 optical density units, 1-cm path length, 450 nm) is added 50 μl of *Glycine max* (soybean) lectin (200 μg/ml). Controls (with no lectin) are run in parallel. The agglutinations can be performed on regular microscope slides, but Boerner plates with concave wells are preferred. Slides should be rocked gently and continuously at room temperature. A positive agglutination within 1 to 2 min is indicative of *B. anthracis* or *B. mycoides* and may be readily visualized without magnification. A second lectin agglutination test with *Helix pomatia* (snail) lectin is positive only for *B. mycoides*. The control slide is used to check for autoagglutination, which may bias the results if highly positive (6).

The growth of *B. anthracis* on a rich agar (tryptic blood agar base, Mueller-Hinton, or other is acceptable) impregnated with a disk containing 10 U of penicillin G results in a clearly defined zone of inhibition. A Kirby-Bauer type of plate may be used, but standards for zone diameters have not been established. *B. anthracis* strains are generally highly susceptible to penicillin, whereas *B. cereus*, *B. mycoides*, and *B. thuringiensis* are resistant. Cells growing at the leading edge of the zone of inhibition may be removed and examined directly as a wet mount under oil immersion. During growth inhibition by penicillin, the cell cylinders of *B. anthracis* tend to bulge, resulting in the classical "string of pearls" reaction.

Finally, plates streaked for luxuriant growth may be seeded with serially diluted volumes of bacteriophage γ (CDC). Upon incubation (35 to 37°C, overnight), plaques will appear if the suspect culture is *B. anthracis*. It is usually unnecessary to observe the plaques with a microscope, as the clear zones may be detected visibly. *B. cereus* and *B. thuringiensis* do not normally lyse in the presence of γ phage, but a large number of *B. mycoides* strains have been reported to be susceptible (4, 5, 20).

BACILLUS CEREUS

B. cereus food poisoning is a well-established disease entity that has two distinct clinical manifestations: an emetic type and a diarrheal type. The diarrheal type is most commonly associated with meats, sauces, etc., whereas the emetic type has been reported almost exclusively in association with rice dishes. Most investigators consider *B. cereus* food poisoning a true intoxication rather than a food-borne infection. Several enterotoxins have been partially characterized, and in addition, the clinical picture is compatible with that of classical food intoxication. For example, there is usually a rapid onset of symptoms and an absence of fever, and the illness is of a relatively short duration. The incubation period for the emetic type varies from 1 to 6 h, whereas the diarrheal form, which is more common, has an incubation period of 6 to 24 h.

The isolation of *B. cereus* from a patient's stool is not sufficient for a diagnosis because the organism may be present in normal stool samples. The laboratory diagnosis of *B. cereus* food poisoning is usually made by quantitative cultures to determine the number of organisms per gram of food sample. The finding of at least 10^5 organisms per g of the suspect food is considered adequate for a diagnosis.

We have found the following procedures suitable for the isolation and presumptive identification of *B. cereus* strains. The methods are essentially those of Kramer et al. (13), who provide additional details on the preparation of media and reagents.

1. Samples (foodstuffs, environmental samples, vomitus, or fecal specimens) are macerated in sterile 0.5% (wt/vol) peptone broth. Serial dilutions are made of the macerate, and 0.1-ml volumes are plated onto blood agar base containing polymyxin (final concentration of 25 U/ml [2, 11]).

2. After an overnight incubation at 37°C, colonies are enumerated and examined for suspected *B. cereus*. Most *B. cereus* colonies isolated from foods appear flattened and slightly grey, with edges that are typically irregular. Mucoid colonies have not been reported. Parry et al. (20) provide a series of excellent photographs of colonies of various *Bacillus* species, including *B. cereus*.

3. Suspect colonies are inoculated on mannitol-egg yolk agar (13) by stabbing (4 to 6 mm) various sites on the plate. After an overnight incubation at 35 to 37°C, the plates are examined for halos around the sites of inoculation and for evidence of fermentation by a change in the violet background to yellow. *B. cereus* is lecithinase positive and mannitol fermentation negative.

4. The foregoing steps are sufficient to provide a presumptive identification of *B. cereus*. Further steps are needed for confirmation. Suspect colonies should be subcultured in the appropriate media and checked for motility (most *B. cereus* organisms are motile; see the previous section), hemolysis (*B. cereus* strains may be alpha- or beta-hemolytic), and susceptibility to penicillin (*B. cereus* is resistant). *B. cereus* is readily differentiated from *B. mycoides* because the latter organism is agglutinated by *G. max* or *H. pomatia* lectins. Confirmation of *B. cereus* requires serotyping with antibody directed against flagellar antigens (13). Experience has shown that most *Bacillus* isolates which grow on blood agar base-polymyxin and on mannitol-egg yolk agars are indeed *B. cereus*, and unless there are compelling reasons, further confirmation is not required.

OPPORTUNISTIC *BACILLUS* SPECIES

In recent years, opportunistic infections have become a major infectious disease problem. The marked increase in infections caused by microorganisms previously considered only harmless commensals will most likely continue to rise as a result of the increasing use of immunosuppressant drugs and organ transplants and the fact that many patients with neoplastic diseases live longer today. The *Bacillus* species are probably the least reported opportunistic bacteria, as they have been by tradition summarily dismissed by many laboratories as skin or air contaminants. This problem emphasizes the need for close liaison between the clinician and the laboratory.

In addition to causing food poisoning, *B. cereus* is also the most frequently reported of the *Bacillus* species as an etiologic agent for a wide variety of clinical infections. Other *Bacillus* species that have been implicated in human infections are *B. circulans*, *B. macerans*, *B. coagulans*, *B. brevis*, *B. pumilus*, *B. sphaericus*, *B. thuringiensis*, and even *B. subtilis* (11, 15, 21, 25, 26). Although there is no single criterion for assessing the clinical relevance of a *Bacillus* isolate, there are certain conditions or circumstances which may warrant additional study. These include: (i) specimen obtained from a compromised host or main-line drug addict; (ii) colonies located only on the original streak lines; (iii) isolate cultured from a normally sterile area, such as blood, spinal fluid, etc.; (iv) finding of gram-positive rods on primary smears of clinical material; (v) repeated isolations.

The most frequent isolations of bacilli are usually obtained in mixed cultures from local drainage sites, and the clinical significance of these isolates is usually difficult to assess. However, it is reasonable to assume that the isolates represent superficial colonization or contamination until proven otherwise.

LITERATURE CITED

1. **Berkeley, R. C. W., and M. Goodfellow (ed.).** 1981. The aerobic endospore-forming bacteria: classification and identification. Academic Press, Inc., New York.
2. **Bouwer-Hertzberger, S. A., and D. A. A. Mossil.** 1982. Quantitative isolation and identification of *Bacillus cereus*, p. 255–259. In J. E. L. Corry, D. Roberts, and F. A. Skinner (ed.), Isolation and identification methods for food poisoning organisms. Academic Press, Ltd., London.
3. **Brown, E. R., M. D. Moody, E. L. Treece, and C. W. Smith.** 1958. Differential diagnosis of *Bacillus cereus*, *Bacillus anthracis*, and *Bacillus cereus* var. *mycoides*. J. Bacteriol. **75**:499–509.
4. **Burdon, K. D.** 1956. Useful criteria for the identification of *Bacillus anthracis* and related species. J. Bacteriol. **71**:25–42.
5. **Burdon, K. L., and R. D. Wende.** 1960. On the differentiation of anthrax bacilli from *Bacillus cereus*. J. Infect. Dis. **107**:224–234.
6. **Cole, H. B., J. W. Ezzell, Jr., K. F. Keller, and R. J. Doyle.** 1984. Differentiation of *Bacillus anthracis* and other *Bacillus* species by lectins. J. Clin. Microbiol. **19**:48–53.
7. **Feeley, J. C., and C. M. Patton.** 1980. *Bacillus*, p. 145–149. *In* E. H. Lennette, A. Balows, W. J. Hausler, Jr., and J. P. Truant (ed.). Manual for clinical microbiology, 3rd ed. American Society for Microbiology, Washington, D.C.
8. **Gibson, T., and R. E. Gordon.** 1974. Genus I. *Bacillus*, 529–550. *In* R. E. Buchanan and N. E. Gibbons (ed.), Bergey's manual of determinative bacteriology, 8th ed. The Williams & Wilkins Co., Baltimore.
9. **Gordon, R. E.** 1981. One hundred and seven years of the genus *Bacillus*, p. 1–15. *In* R. C. W. Berkeley and M. Goodfellow (ed.), The aerobic endospore-forming bacteria: classification and identification. Academic Press, Inc., New York.
10. **Gordon, R. E., W. C. Haynes, and C. H.-N. Pang.** 1973. The genus *Bacillus*. U.S. Department of Agriculture agricultural handbook no. 427. U.S. Department of Agriculture, Washington, D.C.
11. **Ihde, D. C., and D. Armstrong.** 1973. Clinical spectrum of infections due to *Bacillus* species. Am. J. Med. **55**:839.
12. **Knisely, R. F.** 1966. Selective medium for *Bacillus anthracis*. J. Bacteriol. **92**:784–786.
13. **Kramer, J. M., P. C. B. Turnbull, G. Munshi, and R. J. Gilbert.** 1982. Identification and characterization of *Bacillus cereus* and other *Bacillus* species associated with food poisoning, p. 261–286. *In* J. E. L. Corry, D. Roberts, and F. A. Skinner (ed.), Isolation and identification methods for food poisoning organisms. Academic Press, Ltd., London.
14. **Leppla, S. H.** 1982. Anthrax toxin edema factor: a bacterial adenylate cyclase that increases cyclic AMP concentrations in eukaryotic cells. Proc. Natl. Acad. Sci. U.S.A. **79**:3162–3166.
15. **Melles, Z., I. Nikodemusz, and A. Abel.** 1969. Die pathogene Wirkung aerobes sporenbildendes Bakterien. Zentralbl. Bakteriol. Parasitenk. Infektionskr. Hyg. **212**:174–177.
16. **Meynell, E., and G. G. Meynell.** 1964. The roles of serum and carbon dioxide in capsule formation by *Bacillus anthracis*. J. Gen. Microbiol. **34**:153–164.
17. **M'Fadyean, J.** 1903. A peculiar staining reaction of the blood of animals dead of anthrax. J. Comp. Pathol. **16**:35–41.
18. **Mikesell, P., B. E. Ivins, J. D. Ristroph, and T. M. Dreier.** 1983. Evidence for plasmid-mediated toxin production in *Bacillus anthracis*. Infect. Immun. **39**:371–376.
19. **Norris, J. R., R. C. W. Berkeley, N. A. Logan, and A. G. O'Dannell.** 1981. The genera *Bacillus* and *Sporolactobacillus*, p. 1711–1742. *In* M. Starr, N. Stolp, A. Balows, H. Schlegel, and H. Trueper (ed.), The prokaryotes. A handbook on habitats, isolation, and identification of bacteria, vol. 2. Springer-Verlag, New York.
20. **Parry, J. A., P. C. B. Turnbull, and J. R. Gibson.** 1983. A colour atlas of *Bacillus* species. Wolfe Medical Publications, Ltd., London.
21. **Pearson, H. E.** 1970. Human infections caused by organisms of the *Bacillus* species. Am. J. Clin. Pathol. **53**:506–515.
22. **Ristroph, J. D., and B. E. Ivins.** 1983. Elaboration of *Bacillus anthracis* antigens in a new, defined culture medium. Infect. Immun. **39**:483–486.
23. **Ross, J. M.** 1955. On the histopathology of experimental anthrax in the guinea pig. Br. J. Exp. Pathol. **36**:336–339.
24. **Sadler, D. F., J. W. Ezzell, Jr., K. F. Keller, and R. J. Doyle.** 1984. Glycosidase activities of *Bacillus anthracis*. J. Clin. Microbiol. **19**:594–598.
25. **Tuazon, C. U., H. W. Murray, C. Levy, M. N. Solny, J. A. Curtin, and J. N. Sheagren.** 1979. Serious infections from *Bacillus* sp. J. Am. Med. Assoc. **241**:1137–1140.
26. **Turnbull, P. C., B. K. Jorgensen, J. M. Kramer, R. J. Gilbert, and J. M. Parry.** 1979. Severe clinical conditions associated with *Bacillus cereus* and the apparent involvement of exotoxins. J. Clin. Pathol. **32**:289–293.
27. **Zipperle, G. F., J. W. Ezzell, and R. J. Doyle.** 1984. Glucosamine substitution and muramidase susceptibility in *Bacillus anthracis*. Can. J. Microbiol. **30**:553–559.

Chapter 22

Mycobacterium

HERBERT M. SOMMERS AND ROBERT C. GOOD

The contribution of the clinical laboratory to the diagnosis of mycobacterial disease may be conveniently considered in three phases: detection and isolation of mycobacteria, identification of the mycobacteria isolated, and determination of the drug susceptibility of these organisms. A laboratory report suggesting mycobacterial disease may have serious consequences for the patient and affect his life for many months or even years. At the present time, tuberculosis is treated in a general hospital or on an outpatient basis, rather than in a tuberculosis sanatorium. With the decline in the incidence of tuberculosis and the decentralization of treatment, the volume of tuberculosis bacteriology is distributed among many laboratories in a community, with few laboratories receiving large numbers of such specimens. Therefore, the clinical microbiologist should become familiar with the procedures necessary for good mycobacterial services and decide whether the number of specimens received is large enough to warrant maintaining the expertise and materials necessary for optimal performance. In some cases there may be an inadequate volume to permit maintaining full proficiency, and it may be more practical to delegate one or more phases of this work to another laboratory (either private or public) where adequate services are available. Suggestions for limiting the extent or level of services offered by different laboratories have been published by the College of American Pathologists and the American Thoracic Society (Table 1).

SOME GENERAL CHARACTERISTICS OF DISEASE-PRODUCING MYCOBACTERIA

Mycobacteria are acid-fast, alcohol-fast, aerobic, nonsporeforming, nonmotile bacilli. Their lipid content is high. Growth is slow. The most rapidly growing species require 2 to 3 days on simple media at temperatures of 20 to 40°C, and most disease-associated mycobacteria require 2 to 6 weeks on complex media at very specific temperatures. *Mycobacterium leprae* fails to grow in vitro. *M. ulcerans, M. haemophilum,* and *M. marinum* require a temperature of 30 to 32°C, similar to that of the skin; *M. xenopi* and *M. avium* are favored by temperatures above 37°C. Colonies may be rough, with the bacilli compacted in dense coils (e.g., *M. tuberculosis*); they may be smooth and transparent, with the bacilli in no discernible pattern (e.g., the *M. avium* complex); or they may be intermediate in roughness (e.g., *M. kansasii*). Colonies of some species (*M. xenopi* and some rapid growers) form fragile, branching, filamentous extensions in a radial pattern on the surface of culture medium; other extensions may penetrate the medium or even project into the air. Pigmentation may be photochromogenic (light is required for the formation of pigment), scotochromogenic (pigment is formed in either the dark or the light), absent, or sporadic, according to the species.

Nineteen *Mycobacterium* species are associated with disease in humans, producing slowly developing, destructive granulomas that may undergo necrosis with ulceration or cavitation. Disease may be confined to cooler, more superficial parts of the body, or it may invade internal organs. Tuberculosis of the lungs may disseminate to other parts of the body via the blood, the lymphatics, or the intestinal tract. Although useful in the past, animals are not generally recommended for the isolation of mycobacteria from clinical specimens, as many organisms causing disease in humans may not cause disease in animals. With the more gentle digestion techniques and multiple in vitro tests currently available, cultural methods alone are adequate. Serologic diagnosis of mycobacterial disease in patients has not become a reliable clinical laboratory procedure, although recent developments with enzyme-linked immunosorbent-antibody tests have suggested a role for the detection of specific circulating antibody to mycobacteria (5). Several *Mycobacterium* species rarely cause disease in humans but may occur in clinical specimens as saprophytes. Mycobacteria can occur in soil, water, food, and several species of animals, where they can be a potential danger to humans.

Nontuberculous mycobacteria

The designation "atypical" has been commonly used for mycobacteria other than *M. tuberculosis* or *M. bovis*. Since these bacteria in reality are not atypical, but are characteristic of their particular species, it is better to avoid this designation. The terms "mycobacteria other than tubercle bacilli" and "nontuberculous mycobacteria" are preferable.

Groups (Runyon)

In 1959, Runyon proposed a grouping of mycobacteria exclusive of *M. tuberculosis* or *M. bovis* that occur in clinical specimens (67). The groups are not species; rather, each group comprises several species. Group I consists of the photochromogenic species of slow growers. Members of group II are scotochromogenic slow growers. Group III contains the nonphotochromogenic and often, at least initially, nonchromogenic slow growers. Group IV consists of rapid growers, defined as maturing in less than 1 week at 25°C (room temperature) or at 37°C. Thus, although group I is generally and correctly said to consist of the species *M. kansasii, M. simiae, M. marinum,* and *M. asiaticum*, very rarely a strain of *M. kansasii* is encountered which qualifies for group II or III. It should be noted that mycobacteria in the *M. avium* complex are the major disease-associated strains of group III, but many of these organisms may be lightly or moderately pigmented and may mislead the laboratory worker into considering them group II scotochromogenic organisms. *M. szulgai* is photochromogenic at 25°C and scotochromogenic at 37°C, thereby illustrating the need for the identification of the species of each isolate and the limitations of using the Runyon groups for

TABLE 1. Suggestions for limiting the extent or level of service offered by clinical laboratories

Extent of services (College of American Pathologists)	Levels of laboratory services for mycobacterial diseases (American Thoracic Society)[a]
Extent 1 No mycobacteriological procedures performed	
Extent 2 Acid-fast stain of exudates, effusions, and body fluids, etc., with inoculation and referral of cultures to reference laboratories for further identification	Level I a. Collect adequate clinical specimens b. Transport specimens to a higher-level laboratory for isolation and identification c. May prepare and examine smears for presumptive diagnosis or as a means of monitoring the progress of diagnosed patients on chemotherapy or both
Extent 3 Isolation of mycobacteria; identification of *M. tuberculosis* and preliminary identification of the nontuberculous forms as photochromogens, scotochromogens, nonphotochromogens, and rapid growers; drug susceptibility testing may or may not be performed	Level II Perform functions of level I laboratories, including preparation of smears, and a. Process specimens as necessary for culture on standard egg-based media or egg- and agar-based media b. Identify *M. tuberculosis* c. May perform drug susceptibility studies of *M. tuberculosis* d. Retain mycobacterial cultures for additional or repeat tests; 6 months is recommended
Extent 4 Definitive identification of mycobacteria isolated to the extent required to establish a correct clinical diagnosis and to aid in the selection of safe and effective therapy; drug susceptibility testing may or may not be performed	Level III Perform all functions of level I and II laboratories, and a. Identify all *Mycobacterium* species from clinical specimens; cultures with known reaction patterns for growth and biochemical activities should be tested routinely with clinical isolates for quality control b. Should perform drug susceptibility studies of mycobacteria c. May conduct research and provide training

[a] For the complete text of levels of services with applicable references, see Am. Rev. Respir. Dis. **128:**213, 1983.

classification. If species names are available, reference to groups is undesirable.

Species

The species of mycobacteria that may be encountered in humans and should be distinguished in clinical laboratories are listed in Table 2. Besides the most important species, *M. tuberculosis*, the other tubercle bacillus, *M. bovis*, must still be recognized as a threat to humans. Although in the United States *M. bovis* now rarely causes human disease, it may occasionally be isolated from immigrants. *M. bovis* still occurs sporadically in cattle and other animals. It is not uncommon for some laboratories to isolate bacillus Calmette-Guérin (BCG) from patients receiving immunotherapy for cancer. BCG was derived from *M. bovis*.

Recent investigations have indicated that some previously named species may be complexes of three or more related species. Since the taxonomic status and distinctive clinical importance of some of these newly named species are not yet established, it may be sufficient to identify occasional isolates only as far as their complex. *M. szulgai* is most often associated with pulmonary infection but may also be found in extrapulmonary sites (16). *M. simiae* was originally found in monkeys, but it has also been isolated from humans in Cuba, the United States, and other countries around the world (66). *M. asiaticum* has now been shown to be associated with disease in humans in Australia (7). *M. malmoense* has recently been described in Sweden, England, Australia, and the United States (22, 72). *M. haemophilum*, a hemoglobin- or hemin-requiring species, has been reported from Israel, Australia, and the United States, where it has caused infection of the skin (17, 78). *M. avium* and *M. intracellulare*, which cause disease in humans and animals, are groups in the *M. avium* complex. The two species are closely related and can only be distinguished by type-specific antisera in an agglutination test. Species listed as not commonly causing disease (Table 2) may be found to cause disease in especially susceptible patients. These species include *M. gordonae, M. gastri, M. terrae, M. triviale*, and others.

Characteristics of species in the *M. fortuitum* complex (Table 3) can be used for further classification. Subdivision of the species is helpful for epidemiologic studies and planning a therapeutic approach, since drug susceptibility varies among the subclasses. The spectrum of disease manifestations resulting from infection with rapidly growing mycobacteria has recently been summarized (90). Cutaneous infections accounted for 75 (60%) of 125 cases observed over a 4-year period, with 40 of these infections following surgical procedures (principally, median sternotomy and augmentation mammaplasty), whereas 34 infections followed accidental trauma. Twenty-four cases (19%) of pulmonary disease were found. Disseminated disease, cervical lymphadenitis, keratitis, or endocarditis were seen in the other 27 (21%) patients. Response to therapeutic measures was good; only 9% of the patients died. *M. fortuitum* and *M. chelonae* were each responsible for about half of the infections, but

TABLE 2. Distinctive properties of cultivable mycobacteria encountered in clinical specimens[a]

Runyon group	Complex[b]	Species	Clinical significance[c]	Growth rate[d] at				Colony type			Niacin	Susceptibility to T2H[g] (5 μg/ml)	Nitrate reduction
				45°C	37°C	31°C	24°C	Usual colony morphology[e]	See fig. no.	Pigmentation[f]			
		M. ulcerans	1	–	–	S	–	R		N	–	–	–
	TB	*M. tuberculosis*	1	–	S	S	–	R	2, 3A, B, C, E, G	N	+	–	+
		M. bovis	1	–	S		–	Rt		N	–	+	–
I		*M. marinum*	2		∓	M	M	S/SR	3D	P	∓	–	–
		M. kansasii	2		S	S	S	SR/S	3F, H	P	–	–	+
		M. simiae	3–2	–	S			S		P	+	–	–
		M. asiaticum		–	S		S			P	–	–	–
II	*M. scrofulaceum*	*M. scrofulaceum*	3–2		S	S	S	S		S	–	–	–
		M. szulgai	1		S	S	S	S or R		S/P	–	–	+
		M. gordonae	4		S		S	S		S	–	–	–
		M. flavescens	4		M		M	S		S[j]	–	–	+
		M. xenopi	3	S	S			Sf		S	–	–	–
III	*M. avium*	*M. avium*	2	–/+	S		±	St/R		N	–	–	–
		M. intracellulare	2	–/+	S		±	St/R		N	–	–	–
		M. gastri	4		S		S	S/SR/R		N	–	–	–
		M. malmoense	1		S	S	S	S		N	–	–	–
		M. haemophilum	1	–	–	S[k]	S	R		N	–	–	–
		M. nonchromogenicum	4		S		S	SR		N	–	–	+
	M. terrae	*M. terrae*	4		S		S	SR		N	–	–	+
		M. triviale	4		M		S	R		N	–	–	+
IV		*M. fortuitum*	4–3	–	R		R	Sf/Rf		N	–	–	+
	M. fortuitum	*M. chelonae*	4–3	–	R		R	S/R		N	V	–	–
		M. phlei	4	R	R		R	R		S			+
		M. smegmatis	4	R	R		R	R/S		N			+
		M. vaccae	4		R		R	S		S			+

[a] Plus and minus signs indicate the presence or absence, respectively, of the feature; blank spaces indicate either that the information is not currently available or that the property is unimportant. V, Variable; ±, usually present; ∓, usually absent.
[b] For most clinical laboratories, designation to the complex is usually sufficient.
[c] Potential clinical significance: 1, present only as pathogens; 2, present usually as pathogens; 3, present commonly as nonpathogens; 4, present usually as nonpathogens.
[d] S, Slow; M, moderate; R, rapid.
[e] R, Rough, S, smooth; SR, intermediate in roughness; t, thin or transparent; f, filamentous extensions.
[f] P, Photochromogenic; S, scotochromogenic; N, nonphotochromogenic. Note: *M. szulgai* is scotochromogenic at 37°C and photochromogenic at 25°C.
[g] T2H, Thiophene-2-carboxylic acid hydrazide.
[h] Tween hydrolysis may be + at 10 days.
[i] Arylsulfatase, 14 days, is +.
[j] Young cultures may be nonchromogenic or possess only pale pigment which may intensify with age.
[k] Requires hemin as a growth factor.
[l] *M. chelonae* subsp. *chelonae* is –, *M. chelonae* subsp. *abscessus* is + (Table 3).

80% of the isolates of *M. chelonae* were subspecies *abscessus*, and 83% of the isolates of *M. fortuitum* were biovar *fortuitum*.

SAFETY PRECAUTIONS IN THE MYCOBACTERIAL LABORATORY

Sources of equipment and supplies required in mycobacterial laboratories are well detailed in brochures and manuals available from the Centers for Disease Control, Atlanta, Ga. (86).

Close adherence to a few rules and precautions makes the hazard of working in a well-equipped mycobacterial laboratory small. Minimization of the dispersal of mycobacteria into the air and avoidance of the inhalation of airborne bacilli are of utmost importance. Any direct contact with tubercle bacilli is to be avoided. Maintenance of optimal health is of primary importance. The pathogenic potential of mycobacteria demands unwavering respect.

Personnel assigned to work with mycobacteria should have a regularly scheduled physical examination at least once per year. An annual skin test with Tween-stabilized, purified protein derivative should be routine in persons negative to tuberculin. An initial

TABLE 2—*Continued*

Species	Semiquantitative catalase (>45 mm)	68°C catalase	Tween hydrolysis, 5 days	Tellurite reduction	Tolerance to 5% NaCl	Iron uptake	Arylsulfatase, 3 days	MacConkey agar	Urease	Pyrazinamidase, 4 days	Agglutination tests available
M. ulcerans	−	+	−		−		−		−	−	
M. tuberculosis	−	−	−[h]	∓	−	−	−	−	+	+	
M. bovis	−	−	−	∓	−	−	−	−	+	−	
M. marinum	−	−	+	∓	−	−	∓[i]	−	+	+	+
M. kansasii	+	+	+	∓	−	−	−	−	+	−	+
M. simiae	+	+	−	+	−	−	−		+	+	+
M. asiaticum	+	+	+	−	−	−	−		−	−	
M. scrofulaceum	+	+	−	∓	−	−	V	−	+	±	+
M. szulgai	+	+	∓[h]	±	−	−	V	−	+	+	+
M. gordonae	+	+	+	−	−	−	V	−	−	∓	+
M. flavescens	+	+	+	∓	+	−	−	−	+	+	
M. xenopi	−	+	−	∓	−	−	+	−	−	V	+
M. avium	−	±	−	+	−	−	−	∓	−	+	+
M. intracellulare	−	±	−	+	−	−	−	∓	−	+	+
M. gastri	−	−	+	∓	−	−	−	−	+	−	
M. malmoense	−	±	+	+	−	−	−		V	+	
M. haemophilum	−	−	−	−	−	−	−	−	−	+	
M. nonchromogenicum	+	+	+	−	−	−	−	V	−	V	
M. terrae	+	+	+	−	−	−	−	V	−	V	
M. triviale	+	+	+	−	+	−	∓	−	−	V	
M. fortuitum	+	+	V	+	+	+	+	+	+	+	+
M. chelonae	+	V	V	+	V[l]	−	+	+	+	+	+
M. phlei	+	+	+	+	+	+	−	−			
M. smegmatis	+	+	+	+	+	+	−	−			
M. vaccae	+	+	+	+	V	+	−	−			

chest roentgenograph should be obtained on all those positive to the skin test. More frequent examinations should be obtained when indicated.

Special precautions

Avoid the production of infectious aerosols or dust; avoid the inhalation of contaminated air. The following preventive devices or measures should be intelligently applied. A minimum of biosafety level 2 is recommended for all procedures used in the mycobacteriology laboratory, and work should be performed in a class I or IIA biologic safety cabinet. Biosafety level 3, which includes a restricted access laboratory and special protective clothing, is recommended for the handling of cultures of *M. tuberculosis* and *M. bovis* (J. H. Richardson and W. E. Barkley, *Biosafety in Microbiological and Biomedical Laboratories*, in press).

Safety transfer hoods

Laminar-airflow biologic safety cabinets equipped either with high-efficiency filter exhaust (HEPA) or with recirculation through a HEPA filter are recom-

TABLE 3. Distinctive characteristics of the *M. fortuitum* complex

Species	Biovariant or subspecies	Tolerance to 5% NaCl (28°C)	Color after iron uptake		Utilization of			Inhibition by	
			Rust	Tan	Sodium citrate	Mannitol	Inositol	Pipemidic acid	Polymyxin B
M. fortuitum	Biovar *fortuitum*	+	+		−	−	−	+	+
	Biovar *perigrinum*	+	+		−	+	−	+	+
	Unnamed biovariant	+	+		−	+	+		
M. chelonae	Subsp. *chelonae*	−	−	−	+	−	−	−	−
	Subsp. *abscessus*	+	−	−	−	−	−	−	−
	Unnamed subspecies	−	−	+[a]	+	+	−		

[a] See the text for explanation.

mended for all processing, with the possible exception of centrifugation. Mycobacterial laboratory procedures should not be permitted in a laboratory lacking an adequate safety hood [see the brochure *Class II (Laminar Flow) Biohazard Cabinetry*, 1976, available from the National Sanitation Foundation, Ann Arbor, MI 48105].

The transfer of specimens from one container to another, mixing of specimens for concentration, preparation of smears, and inoculation and transfer of cultures should be conducted in a biologic safety hood. A satisfactory biologic safety hood should maintain a flow of air of at least 75 linear ft (22.9 m)/min over the materials and directed away from the worker. An approved procedure for safety operation, maintenance, and monitoring should be established. Consultant engineers are available to monitor the airflow and filter efficiency of safety cabinets on a periodic basis. Before carrying out any transfer procedures in a safety cabinet, the technologist should cover the base of the working surface with cheesecloth, cloth towels, or absorbent paper towels that have been thoroughly moistened with phenolic disinfectant. This cover will control inadvertent surface contamination from droplets or spillage from containers. A phenol- or alcohol-soaked gauze sponge should be used to wipe the edge of centrifuge tubes and other containers after potentially contaminated solutions have been transferred. The room in which the safety hood is located should be kept scrupulously clean; the work surface of the hood should be swabbed with phenolic germicide before and after work, and it should be irradiated with UV light when not in use. A filter change, as needed, must be done with due precaution by an individual who is specially trained. It should be recognized that working in a hood with the exhaust turned off and vented to the outside may cause an airflow into the face of the operator, constituting a greater hazard than working in an open laboratory. Hoods vented to the outside must be continually inspected for malfunction of the backflow prevention device.

Proper ventilation

The ventilation system should create a directional airflow that draws air into the laboratory through the entry area. Exhaust air from the laboratory room need not be filtered before it is discharged to the outside, but it should be dispersed away from occupied areas and air intakes. Centrifuges and shaking machines, if not designed with safety features, should be in safety hoods or in specially ventilated enclosures. Air movement in rooms should be gentle enough not to cause dust dispersal.

Aerosol-proof containers

For material subjected to shaking or centrifugation, use containers sealed with rubber-, plastic-, or Teflon-lined screw caps. Closed centrifuge tubes containing clinical specimens should be placed in sealed centrifuge cups.

Effective masks

Effective masks are available, but too infrequently used. Properly fitting masks of a kind which has been demonstrated to minimize the passage of airborne bacteria are highly recommended. They continue to be effective for many hours and are not appreciably affected by moisture from the breath. The masks should not be touched with contaminated hands. Clean caps and gowns, as well as disposable rubber gloves, should be used when working with specimens that might contain viable mycobacteria. When one is finished, caps and gowns should be placed in a separate laundry bag to be processed with contaminated or potentially contaminated linen.

UV irradiation

UV irradiation is a useful adjunct to surface decontamination procedures and is helpful for the control of airborne contaminants in restricted areas. Keep UV lamps clean, and test the output of germicidal wavelengths each month with a meter (available from Ultra-Violet Products, Inc., San Gabriel, CA 91778).

Sand-alcohol bottle

A 250- to 500-ml Erlenmeyer screw top flask filled with washed sea sand and 95% ethyl alcohol may be used to wipe off inoculating wires, loops, and spades before they are flamed (Fig. 1); an alternative is the glass-enclosed incinerator attached to some Bunsen burners. Electric induction incinerators are also available for the sterilization of needles and transfer spades after use with mycobacteria, and these incinerators can be conveniently placed in a biohazard hood. The use of disposable cotton swabs, applicator sticks, Pasteur pipettes, or paper straws for the transfer of inoculum eliminates the necessity of flaming an inoculating wire. Before the sterilization of applicator sticks, their ends may be sliced to make effective spades for colony transfer.

The use of inoculating loops may be associated with several serious hazards. First, breaking the air-fluid interface of a specimen or broth may create a very fine aerosol capable of dispersing mycobacteria into the adjacent environment, and second, the premature rupture of a loopful of inoculum, broth, or other *Mycobacterium*-containing fluid can similarly result in a very fine aerosol of potentially dangerous organisms.

Splash-proof discard containers

Do not pour contaminated fluids down the drain. Rather, discard them into autoclavable, covered containers, preferably of stainless-steel construction (Fig. 2).

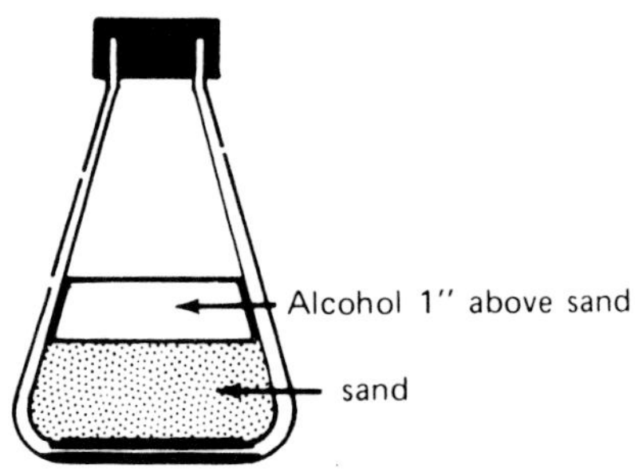

FIG. 1. Alcohol-sand flask. Reproduced from Vestal (86).

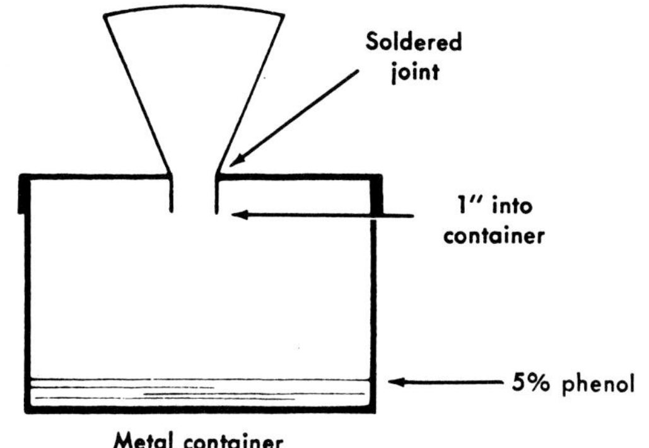

FIG. 2. Splash-proof container. Reproduced from Vestal (86).

Manually operated pipettes

Put nothing into the mouth in the mycobacteriology laboratory. A number of safety pipettors are available which obviate the need for mouth pipetting.

Germicides

Avoid skin, mouth, or other contact with contaminated surfaces. Any persons leaving a mycobacteriology laboratory should wash their hands thoroughly. While the hands are still wet, flood them with 70% isopropyl alcohol and allow them to air dry. Provide handy dispensers of effective germicides at every work area.

Not all so-called germicides are tuberculocidal. Avoid any that have not been demonstrated to be effective against tubercle bacilli. Most quaternary ammonium disinfectants do not kill tubercle bacilli. Suitable disinfectants are as follows:

1. Isopropyl or ethyl alcohol, 70%. Contact should be for 5 min. Residual action, none. (Only these or some dilute phenolics should be used on the skin.)
2. Phenol-soap mixtures employing *o*-phenylphenol or other phenol derivatives with contact periods of 10 to 30 min are adequate. Residual action, 2 to 3 days. Various proprietary products are available, e.g., Amphyl (Lehn and Fink, Toledo, Ohio), Osyl (National Laboratories, Toledo, Ohio), and Staphene and Vesphene (Vestal Laboratories, St. Louis, Mo.). Phenols are irritating to the skin.
3. Sodium hypochlorite, 1:200 to 1:1,000. Contact should be for 10 to 30 min. Residual action, none. Clorox or other household products may be used.
4. Formaldehyde, 3 to 8%; alkaline glutaraldehyde, 2%. Contact time should be for at least 30 min, preferably longer.
5. Phenol, 5%. Contact time of 10 to 30 min is adequate.

COLLECTION OF SPECIMENS

The greatest single problem in the recovery of mycobacteria from clinical specimens is the presence of large numbers of contaminating microorganisms. This problem is partially solved by obtaining a fresh specimen and by the refrigeration of any specimen that cannot be processed promptly. Any negative or doubtful result on a poor specimen should be prominently labeled "inadequate specimen." The provision of instruction sheets for the clinical staff, including nurses and assistants, and another sheet for the patients will assist in the collection of proper specimens.

Successful culture requires getting the best possible specimen.

Sputum

Sterile 50-ml, screw-capped centrifuge tubes or sputum cups, preferably of the one-use disposable type, are suitable containers for sputum specimens. If acetyl-cysteine decontamination is used, it is desirable that the containers be graduated and that the 10-ml volume be clearly marked, since this is the maximal amount of sputum to be used in this type of container. If it is necessary to examine a larger volume, the collection of two or more 10-ml specimens in separate tubes facilitates decontamination. A good sputum specimen is 5 to 10 ml of recently discharged material from the brochial tree, with minimal amounts of oral or nasal secretions. Three such small specimens, collected on successive days, are desirable. If there is any delay in the rapid delivery of the specimens to the laboratory, they should be refrigerated. An early-morning sputum specimen is preferred, since there is a tendency for bronchial secretions to pool in the lungs when the patient is recumbent during the night.

Aerosol-induced sputum and gastric lavage

For patients who have neither a cough nor spontaneous expectoration, suitable specimens may be obtained by the induction of a cough by the inhalation of warm, aerosolized, sterile sodium chloride solution (10%). More recently, ultrasonic nebulizers have been found to be very effective, since more of the solution is aerosolized in less time. The equipment used for the detection of cancer cells may also be used, but propylene glycol should not be in the aerosol, as it is inhibitory to mycobacteria. Patients usually prefer the induction of sputum to the aspiration of gastric contents. Induced sputum is usually superior to gastric lavage for the recovery of tubercle bacilli. Gastric lavage done 0.5 h after the aerosol induction of sputum may be profitable, especially in patients with minimal disease, and may be the only way to collect specimens from infants, some children, and adults who are otherwise unable to cooperate with sputum induction procedures. The combination of induced sputum and gastric lavage will yield more positive results than either procedure alone (11). The optimal time for sputum induction and gastric lavage is early in the morning before meals. Since the objective of gastric lavage is to obtain swallowed sputum, the specimen should be obtained at least 8 h after the patient has eaten or taken oral drugs.

If the processing of gastric lavage is to be delayed for more than a few hours, the collection bottle should contain sodium carbonate powder (about 100 mg) or another alkaline buffer salt. If the specimen is delivered to the laboratory and processed promptly, the buffer is not necessary or desirable, as it interferes with decontamination treatment.

Urine

The preferred urine specimen is the early-morning, cleanly voided, midstream portion. Multiple specimens of this kind are superior to a 24-h pooled specimen. Keep the specimen refrigerated before processing, and as with all other specimens, process it as rapidly as possible. Do not use bottles containing preservatives, as they can kill mycobacteria.

Pleural, spinal, joint, and other fluids, exudates, and tissues

Specimens of body fluids, exudates, and tissues should be submitted in sterile containers. If appropriate, add citrate or heparin anticoagulant. Large amounts of specimen are preferable. Small amounts of exudate may require moistening with sterile water. Examine the specimen before it dries, or refrigerate it. Because of the high lipid content in the mycobacterial cell wall, the specific gravity of mycobacteria is close to unity. For that reason, pleural and other body fluids should be diluted with saline or buffer before centrifugation to improve the chances for sedimentation of the relatively buoyant mycobacterial cells. The recovery or demonstration of mycobacteria by pleural biopsy is frequently a more sensitive way to establish the diagnosis of pleural tuberculosis than is trying to recover the organism from a pleural exudate with a potentially higher specific gravity than that of the organism.

In most cases, cerebrospinal fluid is received in very small amounts, even less than 1 ml, and must be used sparingly for each of the several tests. Since any acid-fast organism found can be assumed to be the etiologic agent and because of the urgency, extra consideration must be given to microscopic examination. Smears prepared in several layers in an area of not more than 1 cm^2 allow the examination of much of the preparation within 30 min.

Most samples of cerebrospinal fluid readily pass through a membrane filter with a 0.45-μm pore size, and this method may be used to collect all the acid-fast bacilli in the specimen. The filter may then be cut into pieces and placed on the surface of several different types of culture media or into a tube of 7H9 or other suitable broth (92).

Occasionally it is necessary to culture a draining sinus or some other site best sampled by a swab. The hydrophobic characteristics of the lipid-containing cell walls of mycobacteria may make it difficult to transfer the organisms from swab to culture media, so that the best growth may occur by resting the tip of the swab on the slanted surface of egg- or agar-based medium and allowing time for growth of the organisms in the interstices of the swab.

MICROSCOPY: DETECTION OF ACID-FAST BACILLI

All clinical specimens submitted for the determination of possible mycobacterial infection should be examined for acid-fast bacilli. The unique acid-fast characteristic of mycobacteria makes microscopy of primary importance. Whether the method used is the classic stain represented by the Ziehl-Neelsen procedure or the fluorochrome technique, the identical property is determined, i.e., bacterial retention of dye after exposure to acid-alcohol. Fluorescence with the latter technique is equivalent to acid-fastness. Microscopy helps in the detection of new cases of mycobacterial infection, serves as an adjunct to culture by determining the acid-fast characteristics of bacterial growth, provides the indication for direct drug susceptibility tests for given specimens and the appropriate dilution of inocula for the tests, gives an indication of the progress of the disease in individual patients from whom a series of specimens are examined, and may be used as a criterion for discharge from the hospital after the initiation of therapy.

Report forms should specify that the finding of acid-fast bacilli indicates only the presence of mycobacteria and that species identification by culture is required.

The proper performance of microscopy for mycobacteria and the interpretation of results require close attention to the following considerations. (i) Occasionally, the careful selection of portions of a specimen to prepare a smear may provide much better results than the use of a concentrate. (ii) Many sputum or other types of specimens containing tubercle bacilli as shown by culture will be negative by smear examination because of the low sensitivity of the method, related in part to the number of organisms present. (iii) It is sometimes difficult to distinguish acid-fast artifacts from bacilli. Therefore, as a rule, the observation of one or only a few acid-fast bodies should be reported as suspicious, and a recommendation should be made that further studies are indicated. Recommendations for reporting the numbers of bacilli seen by microscopy are given in Table 4. The irregular release of mycobacteria from endobronchial foci may result in a variable pattern of positive and negative smears from the same patient. Occasionally, microscopy may be positive when subsequent culture is negative. This difference may result from the inactivation of tubercle bacilli by drugs or by overly toxic decontamination procedures.

TABLE 4. Method for reporting numbers of acid-fast bacilli observed in stained smears[a]

No. of AFB observed	CDC method report
0	Negative for AFB (−)
1–2/300 F	No. seen[b] (±)
1–9/100 F	Avg no./100 F (1+)
9–9/10 F	Avg no./10 F (2+)
1–9/F	Avg no./F (3+)
>9/F	>9/F (4+)

[a] Examination at 800 to 1,000× is assumed. Magnifications less than 800× should be clearly stated. If the microscopist uses a consistent procedure for smear examination, relative comparisons of multiple specimens should be easy for the clinician, regardless of the magnification used. To equate the numbers of bacilli observed at less than 800× with those seen under oil immersion, adjust the counts as follows: for magnifications of about 650×, divide the count by 2; for magnifications near 450×, divide the count by 4; for magnifications near 250×, divide the count by 10; e.g., if eight acid-fast bacilli (AFB) per 10 fields (F) are seen at 450×, the count at 1,000× would be about 2 AFB per 10 F (8 ÷ 4). Taken from American Thoracic Society, "Diagnostic Standards and Classification of Tuberculosis and Other Mycobacterial Diseases," Am. Rev. Respir. Dis. **123**:343–358, 1981.

[b] Counts of less than 3 AFB per 300 F at 800 to 1,000× are not considered positive; another specimen (or a repeat smear of the same specimen) should be processed if one is available.

Laboratory workers should be aware of sources of error which can lead to false-negative or false-positive reports. Inadequate destaining is a common cause of poor smears. Tubercle bacilli and other mycobacteria are strongly acid-fast and cannot easily be destained. Large portions of insufficiently destained material on a slide will make the recognition of tubercle bacilli unlikely. If smears are too thick, they cannot be properly destained and are prone to flake off. If only one or very few acid-fast bacilli are found, another smear should be prepared and examined.

Poor contrast also makes interpretation difficult. The stain and counterstain should be sharply contrasting; the counterstain should be relatively weak so as not to hide acid-fast bacilli. Stains from different sources vary in dye content or brightness. Brilliant-green counterstain may be better for slightly color-blind individuals, although fluorochrome staining is usually preferred.

Acid-fast organisms not originally in the specimen may come from contaminated stain solutions, tap water, distilled-water delivery tubes, or the transfer of material from specimen to specimen or slide to slide in the careless preparation and staining of smears. Transfer from a positive smear to others may occur by the vehicle of immersion oil (use a cover glass or Diaphane). The flaking of smears from microscope slides is an important source of difficulty. Vigilance is required to minimize the possibility of cross-transfer of flakes. Observe the following precautions: make smears on new, clean slides, and fix them thoroughly with heat (2 h at 65°C); whenever possible, handle each slide separately, especially during washing, which must be gentle; do not use common staining jars.

Microscopy without Culture

For physicians or small hospital laboratories lacking facilities for culture, microscopy remains a most valuable procedure. The specimen to be examined should initially be treated with an equal volume of 5% sodium hypochlorite (Clorox; see reference 41). This agent will cause disintegration of the bacilli if it is allowed to act for too long, so smears should be prepared promptly. This procedure can lack sensitivity if the patient has early or minimal disease, and it will be positive only when relatively large numbers of organisms are being shed, unless the specimen can be concentrated (see below).

Recently the concentration of a sodium hypochlorite-treated sputum specimen by filtration through a polycarbonate filter of 0.45-μm pore size was found to increase the sensitivity of the acid-fast stain procedure compared with a similarly treated specimen concentrated by centrifugation. The filtration method can be used as a rapid alternative for the screening of acid-fast bacilli when culture for mycobacteria is impractical (75).

Fluorescence Microscopy with UV Light Source

The great advantages of fluorescence microscopy are in the ease, speed, and completeness of observation. A 25× objective may be used to scan the smear, permitting inspection of a large area in a short time. Other advantages are better contrast, minimal eyestrain, and the relative unimportance of the color acuity of microscopists. The technique of Truant et al. (84) with auramine-rhodamine staining has been demonstrated to be highly satisfactory in many laboratories. Smears positive for acid-fast bacilli are interpreted as described by Truant (Fig. 3A).

Reagents

(i) Auramine-rhodamine stain: auramine, 1.5 g; rhodamine, 0.75 g; glycerol, 75 ml; phenol, 10 ml; distilled water, 50 ml. Clarify the solution by filtration through glass wool. Store the solution at room temperature. (ii) Acid-alcohol decolorizer: 0.5 ml of HCl in 100 ml of 70% ethyl alcohol. (iii) Counterstain: potassium permanganate, 0.5 g; distilled water, 100 ml. Store the solution at room temperature in a dark bottle.

Staining procedure

Fix the smears by heating them on a slide warmer; 65 to 75°C for 2 h or overnight is acceptable, but it may be necessary to cover the slide warmer with a fitted top to achieve a temperature of 65 to 75°C for 2 h. Fixing the smear by heating it with a Bunsen burner on a staining rack is also acceptable. Flood the smear with the fluorochrome stain solution, and allow the smear to stand for 15 to 20 min at room temperature (20 to 25°C); then rinse it in tap water. Decolorize the smear in 0.5% HCl in ethyl alcohol (acid-alcohol), and again rinse it thoroughly in tap water. Flood the smears with counterstain, and allow them to stand for 2 to 4 min. This treatment quenches the background fluorescence of tissue debris. Excessive treatment with the counterstain (i.e., more than 5 min) should be avoided, as it may lower the intensity of fluorescence of stained bacilli. Rinse the smear with water and air dry it. A microscopic examination will demonstrate acid-fast bacilli as yellow or golden yellow (Fig. 3A).

Equipment

The Zeiss BG-12 or Corning 5113 primary (excitation light) filters and the Corning OG-1 barrier filter are satisfactory. Although most fluorescence work is done with oil on the condenser, a dry, dark-field condenser will enable the microscopist to obtain a larger field of vision completely filled with light so that smears can be scanned at a total magnification as low as 60×. For a magnification of 400× with a dry dark-field condenser, it is helpful to place a drop of immersion oil or glycerol over the smear and then cover it with a cover slip.

Fluorescence Microscopy with Blue Light Source

Fluorochrome-stained mycobacteria are seen very well with a microscope equipped with a quartz-halogen illuminator and the proper combination of primary (exciter) and secondary (barrier) filters. This system provides the same benefits as UV apparatus, with the added advantages that the blue-light apparatus does not require a special darkroom (although subdued lighting is recommended), does not require oil on the condenser or slide, is simpler and more economical, and does not present radiation hazards.

The equipment needed is a standard binocular mi-

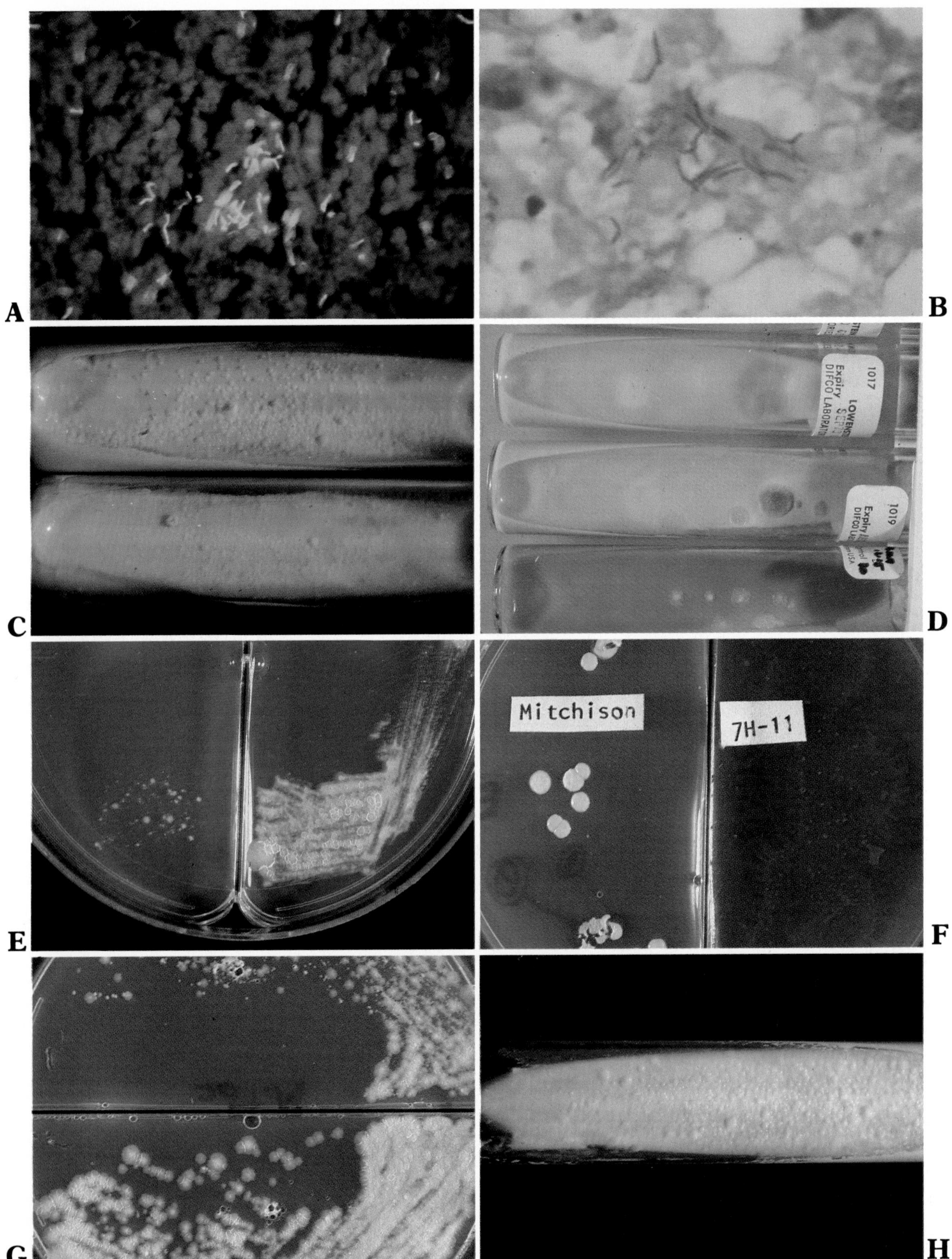

FIG. 3. (A) Truant auramine-rhodamine stain with barrier filter wavelength of 500 nm. (B) Ziehl-Neelsen stain of tuberculous lymphadenitis. (C) *M. tuberculosis* H37Ra on American Thoracic Society medium (top) and Lowenstein-Jensen medium (bottom). Note the better growth on American Thoracic Society medium. (D) *M. marinum* on Petragnani medium (top), American Thoracic Society medium (middle), and Lowenstein-Jensen medium (bottom). The amount of growth varies inversely with the concentration of malachite green in the medium. (E) Isoniazid-resistant *M. tuberculosis* on 7H10 medium (left) and 7H11 medium (right). Not all isoniazid-resistant strains show this amount of growth stimulation on 7H11 medium. (F) Selective 7H11 medium (Mitchison medium) (left) and 7H11 medium (right). Contaminants have been suppressed with selective 7H11 but have overgrown *M. kansasii* on 7H11 without antibiotics. (G) *M. tuberculosis* on selective 7H11 (top) and 7H11 (bottom). The inclusion of antibiotics in selective 7H11 may cause a partial suppression of growth of *M. tuberculosis* and other mycobacteria when compared with the same medium without antibiotics. (H) *M. kansasii* exposed to light at the top of the slant to demonstrate photochromogenicity.

croscope with an illuminator containing a collector lens and a 12-V, 100-W quartz-halogen lamp or high-intensity tungsten bulb, front-surface reflecting mirrors, bright-field condenser, low-power objectives (10 or 25×) for scanning and a high-dry (63×) planachromat (flat field; if the high-dry objective is corrected for a cover slip, then a cover slip must be placed, not mounted, over the smear), a 100× oil-immersion objective for a more critical examination of fluorescent bodies, and 10× compensating eyepieces. A turret or intermediate tube with a holder for secondary filters located in the tube body between the objectives and the eyepieces facilitates filter changing. The optics and light path must be precisely aligned to avoid a loss of light intensity. The light source is adjusted for Kohler illumination.

Combinations of primary and secondary filters are selected to provide good contrast between a dark background and the fluorescing, yellow bacillus (Fig. 3A). The background must be sufficiently light so that nonfluorescing debris can be seen while the focus is maintained and the slide is scanned. A BG-12 primary filter transmitting only wavelengths less than about 50 nm (peak, 404 nm) in combination with secondary filters which transmit only wavelengths above 500 or 530 nm, such as Zeiss no. 50 or 53, respectively, provides a satisfactory demonstration of fluorescing mycobacteria.

Reagents

(i) Auramine O stain is prepared essentially as described by Richards and Miller (64). This stain differs from the Truant procedure listed above (84) in that auramine is used as a single staining agent without the incorporation of rhodamine. Completely dissolve 0.1 g of auramine O in 10 ml of 95% ethyl alcohol. Mix 3 ml of liquefied phenol in 87 ml of water. Combine the alcoholic auramine O solution with the phenol-water. Store the solution at room temperature in amber bottles. (ii) Acid-alcohol: 0.5 ml of HCl in 100 ml of 70% ethyl alcohol. (iii) Potassium permanganate: 0.5% aqueous solution.

Staining procedure

Heat fix smears of sputum or body fluids on a slide warmer at 65 to 75°C for 2 h. Use a cover for the slide warmer if necessary to achieve the proper temperature. Deparaffinize tissue sections and bring them down to water. Stain the smears at room temperature for 15 min (stain tissues for 30 min); rinse them with tap water. Decolorize the smears for 2 min (decolorize tissues for 20 min) with acid-alcohol, and rinse the slides with tap water. Counterstain the smears with potassium permanganate for 3 min. Rinse them with water and air dry them.

Cover the smear with a cover slip of the correct thickness for the high-dry objectives to be used; noncorrected objectives are available for most microscopes, obviating the need for cover slips. The use of mounting medium with the cover slip is not absolutely necessary, but if used, it must be nonfluorescing. Stained tissue sections should be mounted in the usual manner. Include controls of known positive tissue and smears prepared from positive-sputum specimens.

Nonfluorescence Microscopy

Ziehl-Neelsen stain

Reagents. (i) Carbolfuchsin: a saturated solution of basic fuchsin (3 g of basic fuchsin in 100 ml of 95% ethyl alcohol), 10 ml; 5% aqueous solution of phenol, 90 ml. (ii) Acid-alcohol: 3 ml of HCl with 95% ethyl alcohol to bring the volume to 100 ml. (iii) Counterstain: methylene blue, water soluble, 0.3% aqueous.

Staining procedure. Fix the smears by gentle heating over a Bunsen flame (or on a covered electric slide warmer, 65°C for 2 h). Place a piece of filter paper, slightly larger than the size of the smear, on the slide. Flood the slide with carbolfuchsin solution, and heat the slide to steaming with a flame; allow the slide to stand for 5 min without further heating. (If an electric staining rack is used, allow the slides to stain for 15 min.) Remove the filter paper strips, and wash the slides in tap water. Decolorize the smears in several successive portions of acid-alcohol until no more color appears in the washings (about 2 min; longer times may be required for thicker smears). Wash the slide with tap water; to reduce the contamination of organisms from one slide to another, use a staining rack. Do not use a common staining jar or dish. Counterstain the smear with methylene blue for about 30 s. Wash the slide with water and air dry it. For a positive control, use a smear prepared from a specimen from a patient with a strongly positive sputum.

Kinyoun acid-fast stain

Reagents. (i) Kinyoun carbolfuchsin: basic fuchsin, 4 g; phenol, 8 g; alcohol (95%), 20 ml; distilled water, 100 ml. Dissolve the fuchsin in alcohol. Add phenol and water. (ii) Acid-alcohol: 3 ml of concentrated HCl in 97 ml of 95% ethyl alcohol. (iii) Counterstain: malachite green (stock solution, 4 g of malachite green and 50 ml of 95% alcohol; working solution, 20 ml of stock solution and 180 ml of distilled water).

Staining procedure. Prepare the smear; fix it with gentle heat. Stain the smear with Kinyoun carbolfuchsin for 3 min (do not heat it), and then wash the slide gently in running water. Decolorize the smear with acid-alcohol until no more color appears in the washing (about 2 min); wash the slide gently in running water. Counterstain the smear with malachite green for 30 s; wash the slide gently in running water, and dry it in air. For a positive control, use a smear from a strongly positive, patient-derived specimen.

Stain for tissue sections. For acid-fast bacilli in tissues, the Fite-Faraco stain is recommended (1), to be modified only by the use of hematoxylin as a counterstain rather than methylene blue. The tissue should be cut at a thickness of 5 μm.

Observation of carbolfuchsin-stained preparations

Conventional acid-fast stained smears are observed with the oil-immersion lens (900 to 1,000×), and mycobacteria appear as red-stained bacilli against a blue or green background, depending upon the counterstain (Fig. 3B). The acid-fast organisms may be coccobacillary to long, slightly curved, bacillary forms (0.5 to 5.0 by 0.2 to 0.6 μm). Occasionally, the microscopist will see bacilli whose shape may be characteristic of a given species, such as *M. kansasii* (long, often broad-banded cells). It is well to make note of these observations. Identification to the species level on the basis of bacillary morphology cannot be done with any degree of accuracy.

The microscopist responsible for reading preparations stained with either fluorochrome or carbolfuchsin stains must make quantitative notations of the numbers of organisms observed per field or per slide (Table 4). If the number of bacilli seen is only one or two per slide, make and examine another preparation. The preparation of inocula from digested, concentrated sputum specimens for direct drug susceptibility studies is based upon the numbers of bacilli observed by microscopy (see below under Drug Susceptibility Testing).

SPECIMEN PROCESSING AND CULTURE

Homogenization and Decontamination

The best yield of mycobacteria may be expected to result from the use of the mildest procedure which gives sufficient control of contaminants. Agents which liquefy secretions may or may not also control contaminants. Sodium hydroxide, the most commonly used decontaminant, will serve both functions but, like acids and some other decontaminants, is only somewhat less harmful to tubercle bacilli than it is to contaminating organisms. The stronger the alkali, the higher its temperature during the time it acts on the specimen, and the longer it is allowed to act, the greater will be the killing action on both contaminants and mycobacteria (Fig. 4).

Acetylcysteine, dithiothreitol, and several other enzymes effectively liquefy tenacious sputum. Although these agents do not inhibit contaminants, some evidence indicates that their use permits a milder alkali treatment and thereby improves the recovery of mycobacteria. Recently, the use of cetylpyridinium chloride in specimens mailed from remote collection stations to a central processing station has shown a good recovery of *M. tuberculosis* without a significant overgrowth by contaminating oral-pharyngeal bacteria

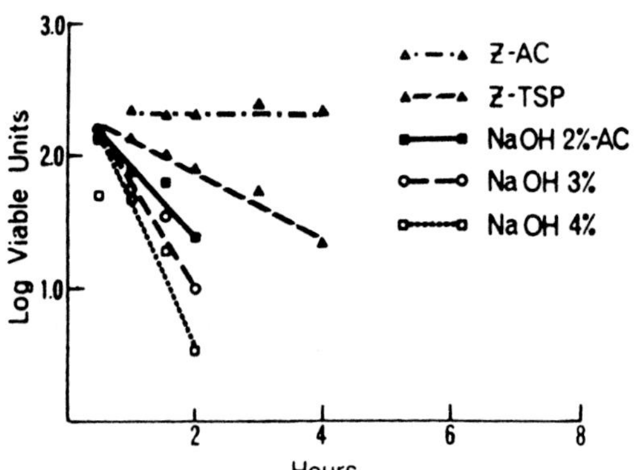

FIG. 4. Death rates of tubercle bacilli in digested sputum. Z-AC, Zephiran-NALC; Z-TSP, Zephiran-trisodium phosphate; NaOH 2%-Ac, 2% NaOH plus NALC; 3% and 4% NaOH do not contain acetylcysteine. Reprinted with permission from Krasnow and Wayne, Am. J. Clin. Pathol. **45**:352, 1966.

(76). Trisodium phosphate liquefies sputum rapidly, but it requires a long exposure for decontamination of the specimen when used alone. Benzalkonium chloride (Zephiran) with trisodium phosphate as described in a method given below shortens the required period of exposure and selectively destroys many contaminants with little bactericidal action on tubercle bacilli.

Specimens differ greatly in their need for decontamination. The agent selected for this purpose ideally should reflect these differences and be modified in strength according to the type and amount of contaminants. The freshness of a specimen and the use of refrigeration before it is processed to reduce overgrowth by commensals are important factors in the need for decontamination.

Some specimens may not need decontamination. Specimens that are not usually decontaminated are aseptically obtained urine; surgical specimens; spinal, synovial, or other internal body fluids; and biopsies of liver, lung, and kidney tissue. Tissues are processed by cutting them into pieces with sterile scissors and reducing the pieces to pulp in a glass-Teflon tissue grinder. The addition of 10 parts of sterile water results in a soupy fluid that can be transferred with a pipette. Although clean surgical tissues do not require decontamination, autopsy tissues usually do. For specimens which are difficult or impossible to duplicate, it is good practice to hold a portion in a freezer pending results on the first portion cultured. Whenever doubt concerning contamination of a specimen exists, a portion of the specimen may be inoculated without prior treatment into liquid medium (Dubos, Proskauer-Beck, or 7H9) while the remainder is kept refrigerated. Inspect the liquid culture daily, and if contamination develops, as shown by Gram and acid-fast stains, subject both this culture and a portion of the specimen to an appropriate decontamination procedure. If no contamination appears after 7 to 10 days and no mycobacterial growth is apparent (by smear examination of the sediment), inoculate the remaining portion of the specimen without treatment. Excessive contamination, usually defined as a rate exceeding 5% of the specimens cultured, is sometimes encountered in specimens from individual patients, from certain localities, or at certain times of the year. The following treatments are suggested. (It must be recognized that harsh treatment to control contamination is deleterious to mycobacteria.)

1. Cautiously increase the strength, duration, or temperature of the alkali treatment; maximal limits are arbitrary, but 4% NaOH at 37°C for more than 60 min will probably kill most tubercle bacilli.
2. Use a selective as well as a primary culture medium such as selective 7H11 (54), Mycobactosel (BBL Microbiology Systems, Cockeysville, Md.) (60), or the Gruft modification of Lowenstein-Jensen medium (25) containing antimicrobial agents which are inhibitory primarily against non-acid-fast bacteria and fungi (Tables 5 and 6).
3. The oxalic acid method of Corper and Uyei (14) is reported to be superior to alkali for the elimination of *Pseudomonas aeruginosa* and some other contaminants. Digest the sputum with an equal volume of 5% oxalic acid. Agitate the solution on a mixer, and then

TABLE 5. Primary mycobacterial isolation media[a]

Medium	Components	Malachite green[b] (g/100 ml)
American Thoracic Society	Fresh egg yolks, potato flour, glycerol	0.02
Lowenstein-Jensen	Fresh whole eggs, defined salts, glycerol, potato flour	0.025
Petragnani	Fresh whole eggs, egg yolks, whole milk, potato, potato flour, glycerol	0.052
Middlebrook 7H10	Defined salts, vitamins, cofactors, oleic acid, albumin, catalase, glycerol, dextrose	0.0025
Middlebrook 7H11	Defined salts, vitamins, cofactors, oleic acid, albumin, catalase, glycerol, 0.1% casein hydrolysate	0.0025

[a] Reproduced from H. M. Sommers, "Mycobacterial Diseases," *in* J. B. Henry (ed.), *Clinical Diagnosis and Management by Laboratory Methods*, 16th ed., W. B. Saunders, Philadelphia, 1979.
[b] Inhibitory agent.

allow it to stand at room temperature for 30 min, with occasional shaking. Add sterile physiologic saline. Centrifuge the solution in a sealed cup, decant the supernatant fluid, and bring the sediment to pH 7 with 4% NaOH containing a pH indicator.

4. Liquefaction, including viscosity reduction, in the presence or even in the absence of a digestant, is facilitated by the vigorous mixing of a solution in a sealed container with a Vortex-type mixer. If such mixers are properly used, aerosol production is minimized. The tube should be held on the vibrating base in such a way that churning, splashing, and foaming of the mixture are avoided. Homogenization should occur not by vibratory agitation but by centrifugal swirling, and this swirling should not be vigorous enough to permit material to rise to the cap. Wait at least 15 min after agitation before opening the tube to allow any fine aerosol droplets formed during the mixing to settle. All such procedures should be carried out in a biologic safety hood, and a mask, gown, and disposable gloves should be worn.

Concentration of Bacilli by Centrifugation

The concentration of mycobacteria by centrifugation is affected by the adequacy of prior homogenization to reduce the viscosity of the specimen, the relative specific gravity of the bacilli versus the suspending fluid, the relative centrifugal force (RCF) employed, and the duration of centrifugation. All of these factors can be controlled to some extent. Since the density of tubercle bacilli is only slightly greater than that of liquefied sputum, centrifugation should be at a high RCF and should be as long as possible, within the limits of excessive exposure to the digestant; 2,000 × *g* for 30 min is minimal. (On a no. 2 IEC centrifuge [International Equipment Co., Needham Heights, Mass.] with a head radius of 25 cm, 2,500 rpm gives an RCF of 2,000 × *g*.) The application of an RCF of up to 3,800 × *g* has been shown to be associated with an improved correlation between the finding of acid-fast organisms by smear and positive cultures from the same specimen (Table 7). (Note: Revolutions per minute does not have the same meaning as RCF when gravitational force is measured.) When specimens are centrifuged at an RCF above 3,000 × *g*, care should be taken to ensure that the centrifuge tubes will be able to withstand the gravitational stress and not collapse. All specimen-containing glass or plastic centrifuge tubes must be placed in sealed cups for centrifugation to reduce the chance of aerosolization should a specimen tube break or collapse. Since centrifugation at high speeds may be associated with the generation of heat, there should be careful monitoring to ensure that no increase in temperature occurs in the clinical specimens that could injure or kill the mycobacteria. For this reason, refrigerated centrifuges are recommended when high RCFs (≥3,000 × *g*) are used. Patients with minimal roent-

TABLE 6. Selective mycobacterial isolation media[a]

Medium	Components	Inhibitory agents
Gruft modification of Lowenstein-Jensen	Fresh whole eggs, defined salts, glycerol, potato flour, RNA (5 mg/100 ml)	Malachite green, 0.025 g/100 ml Penicillin, 50 U/ml Nalidixic acid, 35 μg/ml
Mycobactosel[b]		
Lowenstein-Jensen	Fresh whole eggs, defined salts, glycerol, potato flour	Malachite green, 0.025 g/100 ml Cycloheximide, 400 μg/ml Lincomycin, 2 μg/ml Nalidixic acid, 35 μg/ml
Middlebrook 7H10	Defined salts, vitamins, cofactors, oleic acid, albumin, catalase, glycerol, glucose	Malachite green, 0.0025 g/100 ml Cycloheximide, 360 μg/ml Lincomycin, 2 μg/ml Nalidixic acid, 20 μg/ml
Selective 7H11 (Mitchison medium)	Defined salts, vitamins, cofactors, oleic acid, albumin, catalase, glycerol, glucose, casein hydrolysate	Carbenicillin, 50 μg/ml Amphotericin B, 10 μg/ml Polymyxin B, 200 U/ml Trimethoprim lactate, 20 μg/ml

[a] Reproduced from H. M. Sommers, "Mycobacterial Diseases," *in* J. B. Henry (ed.), *Clinical Diagnosis and Management by Laboratory Methods*, 16th ed., W. B. Saunders, Philadelphia, 1979.
[b] BBL.

TABLE 7. Effect of increasing centrifugal force on positive smears and cultures for mycobacteria[a]

Relative centrifugal force	Smears positive for acid fast[b] (%)	Cultures positive for mycobacteria[b] (%)	Correlation of positive smears/cultures[c] (%)
1,260	1.8	7.1	25
3,000	4.5	11.2	40
3,800	9.6	11.6	82

[a] Adapted from T. W. Rickman and N. P. Moyer, J. Clin. Microbiol. **11**:618–620, 1980.

[b] Percent of total specimens examined.

[c] Percent of culture-positive specimens that were smear positive.

genographic evidence of pulmonary disease are more likely to have sputum specimens culture positive and smear negative for mycobacteria than are patients with well-established tuberculosis showing cavity formation (38). Conversely, patients with chronic cavitary disease are more likely to have smear-positive, culture-negative specimens after the institution of therapy, especially when rifampin is used. Because it is known that some bacilli may be buoyant in centrifugation, the most thorough culture programs have included inoculation onto culture media of a portion of the supernatant fluid as well as the sediment; however, the advantage of this procedure as a routine practice has not been established. All centrifugation should be carried out in sealed safety cups (e.g., sealed dome shields, no. 1124, International Equipment).

For some fluids, particularly cerebrospinal fluids, the use of bacterium-withholding filters may be an effective concentration procedure (92).

Culture Media for Isolation of Mycobacteria

Culture media selected should include both a primary, nonselective, inspissated egg medium (such as Lowenstein-Jensen, American Thoracic Society, or Petragnani medium) and a nonselective agar medium (Middlebrook 7H10 or, preferably, Middlebrook 7H11) (Table 5). At least one selective medium of either egg or agar base (Table 6) should also be used. Middlebrook 7H11 medium is preferred over Middlebrook 7H10, as the addition of 0.1% casein hydrolysate improves the recovery of isoniazid-resistant *M. tuberculosis* (Fig. 3E). Lowenstein-Jensen medium is usually preferred over Petragnani or the American Thoracic Society inspissated egg-based medium. Petragnani medium can cause a partial inhibition of mycobacteria due in part to an increased content of malachite green, but this medium is good for highly contaminated specimens (Fig. 3C and D). Similarly, American Thoracic Society medium can be easily overgrown by contaminants because it contains a smaller amount of malachite green; it is best used for lightly contaminated or clean specimens such as cerebrospinal and pleural fluids, as it will result in larger, less-inhibited colony growth. Care should be exercised in preparation and storage of Middlebrook 7H10 and 7H11 media because excessive heat or light exposure may result in the release of formaldehyde, which can kill or inhibit mycobacteria (56). Both egg- and agar-based media may be somewhat improved by enrichment. The use of an additional medium supplemented with 0.2% pyruvate is recommended when *M. bovis* is suspected (18). The addition of 0.25% L-asparagine or 0.1% potassium aspartate to 7H10 medium (37) affords a maximal production of niacin. Cultures plated on Middlebrook 7H10 or 7H11, each a clear agar medium, and examined inverted with a low-power (10×) objective on a microscope provide the advantages of (i) the earliest demonstration of growth, (ii) evidence of species (for example, a smooth-colony mycobacterium can be definitely and immediately recognized as not *M. tuberculosis*), and (iii) the early and definite recognition of contaminants (by colony selection and transfer, pure cultures often may be obtained readily). Although the clear agar media permit earlier detection of mycobacteria, on prolonged incubation, Lowenstein-Jensen medium often yields a greater number of positive results (41). Therefore, the use of both types of medium gives optimal results in terms of speed and total yield. Similarly, the use of a selective medium containing antimicrobial agents can be helpful in the isolation of mycobacteria from contaminated specimens without the further use of strong decontaminating agents (Fig. 3F) (54). The inclusion of antibiotics in selective media may decrease the growth of mycobacteria (Fig. 3G).

CO_2 enrichment to 5 to 10% is essential for the growth of mycobacteria on 7H10 and 7H11 agar and is also stimulatory to growth on egg media (Fig. 5) (4). Maximal growth stimulation of mycobacteria by CO_2 occurs during days 4 to 10 of incubation.

Acceptable Digestant Methods

Acetylcysteine-alkali procedure for sputum

Reagents. (i) Acetylcysteine-alkali digestant. Combine 50 ml of 2.94% trisodium citrate · $3H_2O$ (= 0.1 M) with 50 ml of 4% NaOH. To this solution add 0.5 g of powdered *N*-acetyl-L-cysteine (NALC) just before use. Discard the solution after 24 or 48 h. If used for 2 days, the solution should be stoppered and refrigerated. (ii) Phosphate buffer, 0.067 M, pH 6.8. (iii) Sterile 0.2%

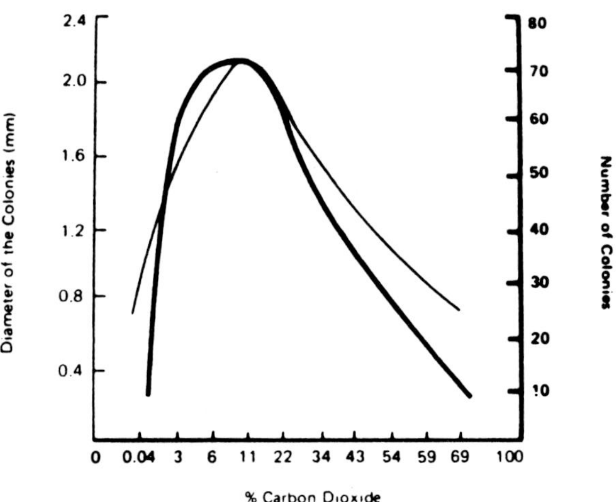

FIG. 5. Effect of CO_2 on the growth (colony size and number of colonies) of *M. tuberculosis* on primary isolation from sputum. Reproduced from H. L. David, U.S. Public Health Service Publ. 76-8316, Center for Disease Control, Atlanta, Ga., 1976.

solution of bovine albumin fraction V adjusted to pH 6.8.

Procedure. Note and, if requested, record the volume and nature of the specimen, i.e., purulent, mucopurulent, mucoid, serous, bloody, etc. If the amount received is more than 10 ml, select with a pipette about 10 ml of purulent-appearing material, and place the material in a sterile, 50-ml, aerosol-free, screw-cap centrifuge tube. Add an equal volume of acetylcysteine-alkali digestant (the final concentration of NaOH is 1%). The digestant should be dispensed with a fresh, sterile pipette for each specimen. Take every precaution to avoid contamination of the digestant. Stopper the tube tightly, and mix the solution for not more than 30 s on a Vortex-type test-tube mixer until the solution is liquefied. If liquefaction is not complete in this time, agitate the solution at intervals during the following decontamination period. Avoid the kind of movement which causes aeration of the specimen, since the acetylcysteine is readily inactivated by oxidation, and mucoid material may repolymerize. A small pinch of crystalline NALC may be added to especially viscous specimens for liquefaction.

Allow the mixture to stand for 15 min at room temperature (20 to 25°C), with occasional gentle shaking if needed, and then fill the tube within 1 or 2 cm of the top with sterile 0.067 M phosphate buffer (pH 6.8). This buffer provides a gentle neutralization of the digestion mixture and lowers the specific gravity of the specimen. Beware of cross-contamination; i.e., do not touch the specimen containers with the buffer dispenser. Centrifuge the solution for at least 15 min at 2,200 to 2,500 × *g* or at a higher RCF if suitable equipment and supplies are available. Wipe the top of the tube with cotton or a sponge moistened with 70% alcohol. Decant the supernatant fluid into a splash-proof discard can containing 5% phenol or other germicide (Fig. 2). Again, wipe the lip of the tube with alcohol.

Prepare the smears on new microscope slides for Ziehl-Neelsen or fluorochrome staining. Use either a sterile applicator stick or a flamed, 3-mm-diameter bacteriologic loop, and smear a portion of the sediment over an area 1 by 2 cm. If the quantity of sediment is very small, delay making smears until after the next step, i.e., the addition of albumin. Add to the sediment 1 ml of sterile 0.2% bovine albumin fraction V adjusted to pH 6.8.

Make a 1:10 dilution of 0.5 ml of the albumin suspension in sterile water. Occasionally, growth occurs only from this diluted inoculum. Inoculate the appropriate media. Inoculate each of the suspensions (diluted and not diluted) onto the surface of an egg-based medium (such as Lowenstein-Jensen), a 7H10 or 7H11 agar plate, and a selective culture medium (Table 5). If biplates are used, both diluted and undiluted suspensions may be in the same plate. Inoculation may be done with disposable capillary pipettes delivering 3 drops to each medium. Plates should be tilted or the drops should be streaked to ensure isolated colonies and the detection of more than one colony type, if present. Use the remainder of the inoculum for the inoculation of drug-containing media as indicated below, or store it in a refrigerator for further treatment if contamination becomes evident.

Place all the inoculated plates into CO_2-permeable polyethylene (not Mylar) bags, to prevent drying, and then into incubators with a CO_2-enriched atmosphere. Candle jars are not acceptable as a source of CO_2.

Sodium hydroxide procedure

For the sodium hydroxide procedure, follow the steps described for the acetylcysteine-alkali method, substituting sodium hydroxide, preferably not stronger than 2%, for the acetylcysteine-alkali digestant. In all NaOH procedures, timing must be rigidly controlled. If it is absolutely necessary to reduce excessive contamination, increase the NaOH concentration (e.g., to 3 or 4%) rather than the time of exposure to the alkali (Fig. 4).

Zephiran-trisodium phosphate digestion procedure

Reagents. (i) Zephiran-trisodium phosphate digestant. Dissolve 1 kg of trisodium phosphate ($Na_3PO_4 \cdot 12H_2O$) in 4 liters of hot distilled water. To this solution add 7.5 ml of Zephiran concentrate (17% benzalkonium chloride; Winthrop Laboratories, New York, N.Y.) and mix. (ii) Neutralizing buffer (Difco Laboratories, Detroit, Mich.).

Procedure. Add an equal volume of Zephiran-trisodium phosphate digestant to the specimen in a screw-capped jar or bottle. Agitate the mixture vigorously for 30 min on a mechanical shaker. Permit the material to stand, without shaking, for an additional 30 min. In a safety hood, transfer all or a portion of the specimen to a screw-capped, 50-ml centrifuge tube. Centrifuge the specimen, and collect the sediment as specified in the description of the acetylcysteine-alkali method. Thoroughly suspend the sediment in 20 ml of neutralizing buffer, and centrifuge the specimen again for 20 min. The neutralizing buffer will serve to inactivate traces of Zephiran in the sediment (40). The recovery of mycobacteria on egg-based media is usually slightly better than on agar media, as lecithin in the eggs will neutralize the residual benzalkonium chloride.

Discard the supernatant fluids; there will be sufficient residual buffer to permit resuspension of the sediment. With a disposable Pasteur pipette, inoculate 3 drops of the sediment onto each medium to be used, and streak the drops to distribute the inoculum for isolated colonies.

Incubation and Inspection of Cultures

Tube or bottle cultures

Tube or bottle cultures should be examined weekly for a minimum of 6 weeks. They may be held for an additional 4 weeks; if they are still negative at that time, they may be discarded. Reports are made when growth of mycobacteria is noted, identification has been made, or 6 weeks of incubation without growth have elapsed. If subsequent growth occurs, a corrected report is issued.

As soon as growth is macroscopically evident on any culture, prepare smears to examine for acid-fast bacilli, and examine the culture for pigmentation (see below) and other properties.

Plate cultures

Plates may be placed in individual, plastic, CO_2-permeable bags (see above under Acetylcysteine-alkali procedure for sputum), or as many as 10 to 12 plates may be placed in a large bag, with rubber bands or tape used to hold down the twisted and folded-over top. Place the bagged plates so that no moisture condenses on the lids. Ordinarily, this condensation will be avoided by the inversion of the plates when the heat source is below the incubator shelf; in some incubators or in some locations in a given incubator, conditions may require a different placement. Examine the plates at 5 to 7 days and each subsequent week with a microscope with 100× magnification to screen for growth on culture medium transparent to transmitted light. A dissecting microscope may also be used.

PROPERTIES USEFUL FOR IDENTIFICATION OF MYCOBACTERIAL PATHOGENS

Most disease-associated mycobacteria may be identified by their rate of growth, colonial pigmentation and morphology, and a few biochemical properties. One should never rely on the results of a single test for identification. Based on the initial observations of the type of growth, appropriate tests to establish the species may be selected from Table 2. It should be recognized that individual strains may deviate from the characteristics of the species. Where no entry is present in Table 2, the property does not contribute to identification, or insufficient information is available for the inclusion of the property. A more complete characterization of different species is given in the following sections.

Although the properties characterizing non-disease-associated mycobacteria are also included in Table 2, complete identification of these species in the clinical laboratory is usually unnecessary. However, it is necessary to differentiate organisms which may appear similar to disease-associated mycobacteria.

Microscopy is the first step in species identification. In addition to an examination of the smear made from the specimen or its concentrate, observation must be made of stained preparations of colonies obtained on culture. This is to confirm that observed growth is mycobacterial and to determine the possible contamination of the culture. Although certain morphologic characteristics of mycobacteria have been associated with different species, bacillary morphology should not be the basis of species identification.

Growth in Relation to Temperature

Usually, it is unnecessary to obtain more complete data regarding growth-temperature relations than will have been seen in the routine procedures for primary culture isolation. Digests from specimens obtained from external areas of the body (rather than from sputum) should be inoculated for incubation at 24°C or, preferably, at 32°C, as well as at 35 to 37°C in a CO_2 atmosphere. Evidence of rapidly growing mycobacteria will commonly be seen at the first weekly inspection of cultures, although some strains exhibit their characteristic growth rate only on subculture. When more definite and complete information on growth rates and temperature preference is needed for species identification, prepare a barely turbid suspension of the bacilli. From a 10^{-2} dilution of this suspension, inoculate equal amounts onto subcultures for incubation at 24, 32, 37, and 42°C. Examine the subcultures for the initial appearance of grossly visible colonies. It is important to streak for isolated colonies, as heavy inocula can give a false impression of rapid growth.

Rapid growth (in <1 week) at 24 or 35 to 37°C

If growth occurs also at 42°C, the strain will usually be other than *M. fortuitum* or *M. chelonae*; inoculate nitrate, arylsulfatase, MacConkey agar without crystal violet, Lowenstein-Jensen medium containing 5% NaCl, and iron uptake tests.

Slow growth (2 or more weeks)

(i) Growth at 35 to 37°C, none at 24 or 42°C: *M. tuberculosis* and *M. bovis*. (ii) Growth at 35 to 37 and 42°C, none at 24°C: *M. xenopi* and some *M. avium* complex strains. (iii) Growth at 35 to 37°C, slower at 24°C, negative at 42°C: *M, kansasii*. (iv) Growth at 32 and 24°C in 2 weeks, none or slower at 35 to 37°C: *M. marinum*. (v) Growth at 32°C in 2 to 4 weeks, double that time at 25 or 35°C, no growth at 37°C: *M. haemophilum*. (vi) Growth at 32°C in >3 weeks, but not at 24 or 35 to 37°C: *M. ulcerans*.

The temperature-growth rate relationships are important in the identification of *M. marinum*, *M. ulcerans*, *M. haemophilum*, and *M. xenopi*. To demonstrate photoreactivity in *M. szulgai*, it is necessary to incubate test cultures at 25 and 35 to 37°C. The growth of *M. szulgai* will be considerably slower at 25°C.

Colonies

The characteristics of colonies described are those seen on plates inverted on the stage of a stereomicroscope with a 10× objective and transmitted light (68). For most species, growth is better on Middlebrook 7H11 agar than on cornmeal-glycerol agar (glycerol, 3%), but some species differences may be more readily seen on the latter. On cornmeal-glycerol agar, tubercle bacilli fail to grow, and *M. kansasii* grows poorly unless seeded heavily. Filamentous extensions from bacillary colonies, e.g., *M. xenopi* and *M. fortuitum*, are usually better developed and more persistent on cornmeal-glycerol agar.

The essence of roughness is the cohesion of bacteria, usually in cords or strands; these cords or strands are commonly serpentine. Observation with the 10× objective of a microscope reveals curving strands in thin colonies. If the colonies become thicker, light is diffracted so much by the stranding that the colonies appear dark or opaque. Smooth colonies of comparable thickness have a homogeneous texture, permitting light transmission; hence, they are translucent.

Colony characteristics to be noted for species identification

Note that most strains will show more than one colony type; e.g., predominantly smooth strains may show a few rough colonies.

Roughness
- Rough, smooth, intermediate
- Kinds
  - Strands or cording; serpentine or other pattern
  - Nonstranded roughness

Shape
- Thin, umbonate, domed; eugonic, dysgonic

Rhizodes
- Filamentous or more or less massive growth of the colony into medium either from the colony center or dispersed

Filamentous extensions
- Some of these are rhizodes; some are on the surface of the medium; others may be short aerial hyphae; some may be branched

Crystals or other extrabacillary formations

Pigmentation and photoreactivity

Colonies of mycobacteria, if definitely colored, owe this characteristic principally to carotenoid pigments which range from yellow to red. Scotochromogenic strains form pigment in the dark and also in the light; commonly, more pigment is produced if growth occurs in a lighted incubator. Photochromogenic strains are stimulated to produce pigment by light exposure and, ordinarily, do not show yellow pigmentation unless exposed to light under proper conditions, i.e., during their early growth and with good aeration of the culture surface (96). *M. kansasii* becomes definitely yellow 6 to 12 h after 1 h of exposure to bright light. *M. szulgai* is unique in being scotochromogenic if grown at 37°C, but photochromogenic if grown at 25°C. During the routine examination of cultures, if the first colonies to appear are nonpigmented, leave the cultures close to a bright, incandescent or fluorescent light in the laboratory for 1 h or more. During this period, smears may be made and examined. Return the cultures, with loosened caps, to the incubator, and examine the colonies the next morning for yellow pigmentation. Often color changes are pale and may even be questionable. For a more systematic determination of photoreactivity, the following procedure is useful. A broth culture of the test organism, diluted sufficiently so that isolated colonies can be obtained, is inoculated into three tubes of Lowenstein-Jensen medium. Two of the tubes are wrapped with aluminum foil, and the third is left exposed to any ambient light in the incubator. Cultures thought to be photoreactive should be incubated at 30 and 35 to 37°C; those considered scotochromogenic should be inoculated in duplicate and incubated at both 24 and 35 to 37°C.

Several days after growth is first noted in the light-exposed control cultures, the foil-wrapped tubes are examined for growth. If there is evidence of colony formation, one of the two foil-wrapped tubes is exposed to strong light. A 100-W tungsten bulb or fluorescent equivalent is adequate. The cap of the culture tube should be loosened during the 3 to 5 h of illumination. Return all cultures to their respective incubators, and examine the cultures after 24, 48, and 72 h (cultures grown at 24°C will respond less rapidly than those grown at 37°C).

Carotene crystals are regularly formed only by photochromogenic *M. kansasii*, *M. marinum*, *M. simiae*, and *M. asiaticum*. Crystals may be seen as early as 2 to 3 weeks if growth has been in continuous light. The use of 100× magnification is required, or somewhat lower magnification will serve if a stereomicroscope is used.

Properties related to pigmentation are as follows:

- None; weak, not definitely yellow—*M. tuberculosis*
- Scotochromogenic—all scotochromogenic species (Table 2) and some rapidly growing species
- Photochromogenic—*M. kansasii*, *M. marinum*, *M. simiae*, and *M. asiaticum* (Fig. 3H)
- Crystalline β-carotene—*M. kansasii* and the photochromogenic species (Table 2)
- Small colonies nonpigmented, older and larger colonies yellow, not light induced—some strains of the *M. avium* complex
- Some strains pigmented, some not; variable—*M. smegmatis*
- Colonies green on media containing malachite green—*M. fortuitum*

Arylsulfatase

The 3-day arylsulfatase test (93) is used mainly to differentiate clinically significant rapid growers. With few exceptions, only *M. fortuitum* and *M. chelonae* split phenolphthalein from tripotassium phenolphthalein sulfate within 3 days. The 14-day test may be useful in the identification of slowly growing mycobacteria such as *M. marinum*, *M. szulgai*, *M. xenopi*, *M. triviale*, and *M. flavescens*. Two procedures for the determination of arylsulfatase production are available: (i) Wayne Arylsulfatase Agar and (ii) the 3- and 14-day broth tests of Kubica and Rigdon (43). For most purposes, the 3-day test given below is adequate and easier to perform.

Inoculum

Prepare a barely turbid suspension.

Reagents

(i) Substrate. Incorporate 1 ml of glycerol and 65 mg of tripotassium phenolphthalein disulfate (Nutritional Biochemicals Corp., Cleveland, Ohio) into 100 ml of melted Dubos oleic agar base (Wayne Arylsulfatase Agar, BBL). Dispense the mixture in 2-ml amounts into screw-capped vials (18 by 60 mm). Autoclave the vials. Permit the mixture to harden in an upright position. (ii) Na_2CO_3, 1 M (10.6 g in water to make 100 ml).

Procedure

Inoculate the medium with 1 drop of the bacillary suspension. Incubate the medium at 37°C for 3 days. Add 1 ml of Na_2CO_3 solution, and observe the medium for pinkness, indicating free phenolphthalein and a positive test.

Controls

Use cultures of *M. fortuitum* and *M. avium* complex as positive and negative controls, respectively.

Catalase Drop Method

Essentially all mycobacteria are catalase positive. The only exceptions are some isoniazid-resistant mutants of *M. tuberculosis* and *M. gastri* and some nonpathogenic, isoniazid-resistant strains of *M. kansasii*. The catalase drop test is very useful for the quick and easy determination of significant isoniazid resistance of *M. tuberculosis*, which ordinarily reflects prior contact with this drug.

Cultures

Any mycobacterial culture may be tested, but the test is usually limited to those cultures suspected or known to be *M. tuberculosis*. Only colonies on medium without drugs are tested.

Reagents

Use a 1:1 mixture of 10% Tween 80 and 30% hydrogen peroxide (Superoxol; Merck & Co., Inc., Rahway, N.J.). Prepare a fresh mixture for use each day.

Procedure

Add a drop of the reagent to growth on a slant or plate. Observe for the formation of bubbles (O_2) around the colonies. Note that some colonies may be positive and others negative.

Catalase after Heating to 68°C

The test for catalase after heating the cells to 68°C (42) is valuable in conjunction with the niacin test for the recognition of tubercle bacilli. A positive test definitely indicates a species other than *M. bovis, M. tuberculosis, M. gastri*, or *M. haemophilum*, which are always negative. Strains of other species (Table 2) may be negative for this test.

Cultures

Well-developed, isolated colonies, preferably from egg-based culture medium, e.g., Lowenstein-Jensen, are tested.

Reagents

(i) A 1:1 mixture of 10% Tween 80 and 30% hydrogen peroxide (freshly prepared). (ii) Phosphate buffer, 0.067 M, pH 7: 61.1 ml of 0.067 M disodium phosphate, Na_2HPO_4 (9.47 g/liter), 38.9 of 0.067 M potassium acid phosphate, KH_2PO_4 (9.07 g/liter).

Procedure

Suspend several colonies in 0.5 ml of phosphate buffer (0.067 M), pH 7, in a screw-capped tube (16 by 125 mm). Place the tube in a water bath at 68°C for 20 min. Cool the suspension to room temperature, and add 0.5 ml of the Tween-peroxide mixture. Observe the mixture for bubbling, and record the result as positive or negative. Hold the tubes for 20 min before discarding them as negative.

Controls

Use cultures of *M. tuberculosis* and *M. kansasii* as negative and positive controls, respectively.

Semiquantitative Catalase Test

The semiquantitative test for catalase has proved valuable in the separation of some species of mycobacteria. Two subgroups of *M. kansasii* have been recognized: one produces <45 mm of bubbles, whereas those strains more commonly associated with disease produce >45 mm of bubbles. *M. tuberculosis, M. bovis, M. marinum*, the *M. avium* complex, *M. gastri, M. malmoense, M. xenopi*, and *M. haemophilum* produce a column of bubbles <45 mm high, but other species usually produce a higher column.

Inoculum

Use a 7-day broth culture or a cell suspension of comparable turbidity.

Reagent and medium

(i) Freshly prepared 1:1 mixture of 10% Tween 80 and 30% hydrogen peroxide. (ii) Lowenstein-Jensen deep tubes (available commercially). Dispense 5 ml of Lowenstein-Jensen medium into screw-capped tubes (20 by 150 mm). Inspissate the medium with tubes in an upright position in a water bath at 85°C for 60 min. Do not substitute agar medium. Remove the tubes and incubate them at 35 to 37°C overnight to check for sterility.

Procedure

Inoculate a Lowenstein-Jensen medium deep tube with 0.2 ml of the bacterial suspension. Incubate the medium for 2 weeks at 35°C with the cap loosened. Add 1 ml of the Tween-peroxide mixture to the Lowenstein-Jensen culture, and measure the column of bubbles in millimeters after 5 min in an upright position at room temperature. Note the two categories: those that produce more and those that produce less than 45 mm of bubbles.

Controls

Use *M. tuberculosis* (which produces a column <45 mm high) and *M. kansasii* (which produces a column >45 mm high) as positive controls, and use uninoculated medium as a negative control.

Inhibition Tests for Identification

Casal and Rodriguez (12) and Wallace et al. (89) have proposed disk susceptibility tests with pipemidic acid and polymyxin, respectively, for the differentiation of *M. fortuitum* and *M. chelonae*. The value of the pipemidic acid test has been confirmed by Lévy-Frébault et al. (50).

Procedure

For pipemidic acid susceptibility tests, suspend the cells from a 7-day culture on Middlebrook 7H10 medium in saline or water to a turbidity equivalent to a no. 1 McFarland standard. Dilute the suspension 1:1,000, and evenly inoculate the plates of Mueller-Hinton agar. Place a disk containing 20 μg of pipemidic acid (available from Robert Good) on the plate, and incubate the culture. Examine the plates after 2 and 5 days, and measure the zone of inhibition. A zone of ≥10 mm is interpreted as inhibition. Strains of *M. chelonae* are resistant to pipemidic acid, whereas strains of *M. fortuitum* are susceptible.

Procedure

For polymyxin susceptibility tests, dilute growth in Middlebrook 7H9 liquid medium to one-half the turbidity of the no. 1 McFarland standard. Swab the suspension onto a Mueller-Hinton agar plate that has been poured to a depth of 4 mm and previously swabbed on the surface with 10% Middlebrook OADC (oleic acid, albumin, dextrose, catalase; added to Middlebrook 7H10 and 7H11 agar after autoclaving). Place a disk containing 300 U of polymyxin on the surface of the agar, and incubate the culture at 35°C for 72 h. Measure the zones of frank inhibition around the disks, but ignore a fine haze of growth that may appear. Strains of *M. fortuitum* are inhibited with zones of ≥9 mm, whereas strains of *M. chelonae* are completely resistant to the drug.

Iron Uptake

In the iron uptake test (97), only rapid growers, such as *M. fortuitum* and *M. phlei*, are positive. *M. chelonae* does not take up iron except for the unnamed subspecies (see below and Table 3) detected by the alternate procedure.

Cultures

Inoculate Lowenstein-Jensen slants with 1 drop of barely turbid aqueous suspension of the strain to be tested.

Reagent

Use aqueous ferric ammonium citrate, 20%. Dispense the reagent into small containers. Autoclave the containers.

Procedure

Incubate the Lowenstein-Jensen slants at 37°C until definite growth appears. Add about 1 drop of the sterile citrate solution for each 1 ml of Lowenstein-Jensen medium. Incubate the culture at 35 to 37°C for a maximum of 21 days. Record the appearance of a rusty-brown color in the colonies and a tan discoloration of the medium as positive.

Controls

Use *M. fortuitum* as a positive control and *M. chelonae* as a negative control.

Alternate procedure

See reference 74 also.

Prepare Lowenstein-Jensen medium, and add 2.5% (wt/vol) ferric ammonium citrate. Inspissate the medium in a slanted position. Inoculate the medium with 1 drop of a barely turbid suspension of the culture to be tested, and incubate the culture for 10 to 24 days at 28°C in a slanted position with the caps loosened. Colonies appear rusty brown if iron is taken up (+). When a colony is tan and a rusty-brown color appears around the edge of the slant, record the reaction as ±. Record a negative reaction if growth appears the same as in medium without added ferric ammonium citrate. Only the unnamed subspecies of *M. chelonae* has been found to give the ± reaction.

MacConkey Agar Growth Test

Growth on MacConkey agar without crystal violet is used to distinguish the species of the *M. fortuitum* complex from other species (28). Only species of this complex grow within 5 days.

Inoculum

Use Tween-albumin broth or 7H9 broth cultures 7 to 10 days old.

Medium

Prepare medium from dehydrated MacConkey agar without crystal violet (Difco) base; pour about 15 ml into each plate. This medium is not that commonly used for gram-negative bacteria. Since crystal violet inhibits mycobacteria, check to ensure that no crystal violet is present.

Procedure

Inoculate a plate of MacConkey agar without crystal violet with a 3-mm loopful of the broth culture, streaking to obtain isolated colonies (a spiral inoculation from the center outward is best). Examine for growth after 5 and 11 days. Only strains of the *M. fortuitum* complex (Table 2) grow to the end of the streak. Other mycobacteria may show some growth where the inoculum is very heavy.

Controls

Use *M. fortuitum* as a positive control and *M. phlei* as a negative control.

Niacin Test

The ability of *M. tuberculosis* to produce abundant niacin has been widely demonstrated. Niacin is secreted from bacilli into the medium, and if the medium is not liquid, an aqueous extract of the medium around the colonies is tested. If the test is negative but other properties indicate *M. tuberculosis* (such as nitrate strongly reduced and catalase destroyed after heating at 68°C), the niacin test should be repeated. Cultures less than 3 weeks old or containing only small numbers of colonies may not yet have produced enough niacin to be detected, so they should be tested for up to 6 weeks. Niacin-negative *M. tuberculosis* strains are exceedingly rare. Positive niacin tests with strains of other *Mycobacterium* species are also rare except for *M. simiae* and some BCG strains of *M. bovis*. Most of these strains may be immediately recognized as other than tubercle bacilli by rapid growth, pigmentation, or the clinical history. *M. simiae* cultures may produce niacin slowly and become positive after prolonged incubation. In a clinical laboratory, there is little advantage in doing niacin tests on strains which are scotochromogenic or rapidly growing. Photochromogenic isolates should be studied to exclude or establish *M. simiae*. The possibility of individual strains or nonpigmented slow growers other than *M. tuberculosis* being niacin positive must not be ignored.

Some laboratory workers avoid doing niacin tests because of the toxicity of one of the reagents, cyanogen bromide (tear gas). Its easily recognized odor is a warning. The less obvious danger of pipetting tubercle bacilli is probably greater. Both dangers are well controlled by the use of a biologic safety cabinet vented to the outside. A paper strip method for niacin determination has the advantages of simplicity and of reduced exposure to tear gas (no handling of cyanogen bromide). The paper strip method is fully dependable (34, 101).

Cultures

Luxuriant growth will occur on egg medium or on Middlebrook 7H10 or 7H11 medium enriched with 0.25% L-asparagine or 0.1% potassium aspartate (37).

Include a known culture of *M. tuberculosis* and an uninoculated medium control.

Reagents

(i) Cyanogen bromide (CNBr), 10%. Working in a chemical fume hood and wearing gloves, place about 2 tablespoons of pure CNBr in a tared Erlenmeyer flask. Weigh the flask. Add enough water to make a 10% solution. When the CNBr is fully dissolved, transfer it to a brown bottle, and cap the bottle tightly. Include on the label the date prepared and a caution to avoid inhalation of fumes and contact with skin or mucous membranes. Keep both the solution and the stock bottle in a refrigerator. (ii) Aniline, 4%, in alcohol. Add 4 ml of aniline to 96 ml of 95% ethyl alcohol. Keep the mixture in a refrigerator in a brown bottle labeled with the contents and the date of preparation. If pure, this solution is clear and colorless. If it is brown or becomes so, discard it. Pure aniline may be obtained by redistillation.

Tube or spot plate test

See reference 70 also.

Working in a hood, layer 0.5 to 1.5 ml of sterile water over the medium around the colonies. If growth is confluent, scrape it to one side so that the water comes in direct contact with the underlying medium. Let the water stand at 36°C for 10 to 15 min (rarely, an extended extraction time of 60 min may be necessary), and then remove a portion of the water (0.5 ml or 2 drops) to a small test tube or to a porcelain spot plate. Add an equal volume of the aniline solution, and make note of any color; no yellow should be evident. Add cyanogen bromide in the same volume as that of aniline. An immediately developing yellow color indicates niacin is present. Add an excess of NaOH to eliminate toxicity before discarding the tests. The porcelain plate may be reconditioned by placing it in boiling water for a few minutes. Always discard cyanogen bromide in NaOH to prevent the unintentional formation of HCN.

Paper strip method for niacin

See references 34 and 101 also.

Niacin test strips are commercially available (Niacin Test Strips, TB, Difco; Remel, Lenexa, Kans.). Follow the directions provided with the strips. Some investigators have found that 2 h for the extraction of niacin (rather than the recommended 30 min) yields more strongly positive reactions for *M. tuberculosis* without causing false-positive reactions for other species.

Controls

Use *M. tuberculosis* as a positive control and the *M. avium* complex and uninoculated medium as negative controls.

Nitrate Reduction

The nitrate reduction test (87) is valuable for the identification of *M. tuberculosis, M. kansasii, M. szulgai*, certain non-disease-associated strains of nonphotochromogens, and *M. fortuitum*, which are nitrate reductase positive. *M. bovis, M. marinum, M. simiae*, the *M. avium* complex, *M. xenopi, M. gastri, M. malmoense*, and *M. chelonae* are negative or only very weakly positive.

Cultures

Cultures on solid medium should be 3 to 4 weeks old except for rapid growers, which may be 2 to 4 weeks old.

Reagents

(i) A 1:2 dilution of concentrated HCl. (ii) 0.2% aqueous solution of sulfanilamide. (iii) 0.1% aqueous *N*-(1-naphthyl)ethylenediamine dihydrochloride. (iv) 0.01 M solution of $NaNO_3$ in 0.022 M phosphate buffer, pH 7 (Nitrate Broth, Difco, may be substituted). (v) Powdered zinc.

Store the solutions at room temperature. They remain stable for several weeks. Record the date of preparation. If either reagent ii or iii changes color, discard it and prepare a fresh solution.

Procedure

Place a few drops of sterile distilled water in a screw-capped tube (16 by 125 mm). Grind in one loopful or one spadeful of mycobacterial growth. Add 2 ml of $NaNO_3$ solution. Shake the mixture and incubate it in a water bath at 37°C for 2 h. Add 1 drop of reagent i, 2 drops of reagent ii, and 2 drops of reagent iii. Examine the solution immediately for the development of a pink-to-red color contrasting with the reagent control. Color intensity should be related to separately prepared color standards, since only a definite red is considered positive. Add a pinch of powdered zinc to all the negative tubes to reduce nitrate to nitrite. The formation of a red color here indicates that the negative reading was valid. Recently, a nitrate reductase test that uses dry, crystalline reagents has been described (91). The initial response by laboratory workers using the procedure has been very favorable. The test is simple, easily performed, and thought to be more sensitive than the standard tube test (see above). Further evaluation of the test is under way.

Paper strip method for nitrate reduction

See reference 61 also.

Commercially available paper strips (Nitrate Test Strips, Difco) also are satisfactory for the nitrate reduction test. Follow the directions provided with the strips. If a strip is negative and other procedures suggest that the test should be positive, the result should be confirmed by the tube test.

Controls

Use *M. tuberculosis* as a positive control and uninoculated medium as a negative control.

Pyrazinamidase

The deamidation of pyrazinamide (PZA) to pyrazinoic acid in 4 days is useful for the differentiation of *M. marinum* (positive) from *M. kansasii* (negative) and of weakly niacin-positive *M. bovis* (negative) from *M. tuberculosis* (positive) and members of the *M. avium* complex (positive) (95).

Reagents

(i) Dubos broth base containing 0.1 g of PZA, 2.0 g of pyruvic acid, and 15.0 g of agar per liter. Dispense the base in 15-ml amounts into screw-capped tubes. Autoclave the tubes for 15 min at 121°C. Solidify the agar with the tubes in an upright position. (ii) Aqueous ferrous ammonium sulfate, 1%, freshly prepared.

Procedure

Heavily inoculate the agar with growth from a 2- to 3-week-old culture. The inoculum should be visible. Incubate the culture at 37°C for 4 days. Add 1 ml of freshly prepared ferrous ammonium sulfate to the tubes, and place them in a refrigerator for 4 h. Examine the tubes for a pink band in the agar.

Controls

Use *M. avium* and uninoculated medium as positive and negative controls, respectively.

Sodium Chloride Tolerance

Of the slowly growing mycobacteria, only *M. triviale* grows in the presence of 5% NaCl, and of the medically significant, rapidly growing mycobacteria, only *M. chelonae* subsp. *chelonae* and the unnamed subspecies fail to grow in the presence of 5% NaCl (33) (Table 3).

Inoculum

Use a barely turbid suspension.

Substrate

American Thoracic Society or Lowenstein-Jensen medium containing 5% NaCl (Difco and Remel) is used. The same medium without salt should be used for controls.

Procedure

Inoculate the medium with 0.1 ml of the bacterial suspension. Incubate the culture at 37°C. Read the culture for growth or no growth at 4 weeks.

Controls

Inoculate the media both with and without 5% NaCl with *M. fortuitum* (positive) and *M. tuberculosis* (negative).

Substrate Utilization

See reference 85 also.

It has been proposed that these tests be used for the differentiation of biovariants and subspecies of strains in the *M. fortuitum* complex (74).

Inoculum

Use 7-day-old cultures incubated at 35°C in 5 ml of Middlebrook 7H9 liquid medium in screw-capped tubes. Dilute the cultures 1:10 in physiologic saline, and inoculate slants of complete medium with 0.1 ml of the diluted suspension.

Media

Basal medium contains $(NH_4)_2SO_4$, 2.4 g; KH_2PO_4, 0.5 g; $MgSO_4 \cdot 7H_2O$, 0.5 g; and distilled water, 950 ml. This salt solution is adjusted to pH 7.0 with 10% KOH for tests with sodium citrate and to pH 7.2 for tests with mannitol, inositol, or control medium. Add 20 g of purified agar per 1,000 ml of salt solution. After sterilization at 121°C for 20 min, cool the solution to 56°C and add 50 ml of filter-sterilized substrate. Substrate solutions contain 5.0 g of mannitol or inositol or 5.6 g of sodium citrate per 50 ml of distilled water. Add sterile water without substrate for the control medium. Dispense the medium in 8-ml amounts into screw-capped tubes (20 by 150 mm), and allow the medium to solidify in a slanted position.

Procedure

Incubate the inoculated tubes at 25 to 28°C for 2 weeks. Growth on medium with substrate is interpreted as positive, whereas no growth, as on the control medium, is interpreted as negative.

Controls

Use *M. fortuitum* biovar *perigrinum* as a positive control for mannitol and *M. chelonae* subsp. *chelonae* as a positive control for citrate. Negative controls are medium without substrate.

Tellurite Reduction

The reduction of colorless potassium tellurite to black metallic tellurium within 3 to 4 days is the distinctive property of *M. avium* complex strains (36). Of other mycobacteria tested, only rapid growers are also positive within this period.

Cultures

Use 7-day-old cultures in 5 ml of Middlebrook 7H9 liquid medium in screw-capped tubes. The broth medium should be fairly turbid as an indication of active growth.

Reagent

Use a 0.2% aqueous solution (0.1 g in 50 ml of distilled water) of potassium tellurite. After the salt is dissolved in water, dispense the solution in 2- to 5-ml amounts, and sterilize the tubes in an autoclave at 121°C for 10 min. To avoid contamination, use only one tube of tellurite solution for a series of tests performed on 1 day, and discard the remainder of the solution.

Procedure

Add 2 drops of the tellurite solution to each culture, and return the cultures to the incubator. Examine the cultures daily for 4 or more days. A positive test is reflected by a jet-black precipitate.

Controls

Use the *M. avium* complex as a positive control and *M. kansasii* as a negative control.

Thiophene-2-Carboxylic Acid Hydrazide

The thiophene-2-carboxylic acid hydrazide (T2H) test is used for distinguishing *M. bovis* from *M. tuberculosis* and other species. Only *M. bovis* is susceptible to low concentrations of this compound (5 μg/ml) (8). Some strains of *M. tuberculosis* are inhibited by 10 μg/ml, and some strains of *M. bovis* may show minimal growth at 1 μg/ml.

Substrate

Use Dubos oleic acid agar with albumin complex or Middlebrook 7H11 medium. Incorporate 5 μg of T2H (Aldrich Chemical Co., Inc., Milwaukee, Wis.) per ml into the agar. Dispense the substrate in 5-ml amounts onto slants or onto Felsen disposable plastic quadrant plates, alternating with the same medium without T2H.

Procedure

Prepare a barely turbid suspension of the organisms to be tested in sterile water. Dilute this suspension 1:1,000 with sterile water. Inoculate 3 drops of the 1:1,000 suspension onto the T2H-containing and control media, and incubate the media at 37°C. Record the time when definite growth is observed on the control medium. Maintain the T2H-containing medium for an additional 3 weeks unless definite growth appears earlier. Record the organism as resistant if growth on the T2H medium is greater than 1% of the growth on the control. If the control plate has confluent growth, assume that this equals 10^4 organisms.

Controls

Use *M. tuberculosis* as a positive control and *M. bovis* as a negative control.

Tween 80 Hydrolysis

The enzymatic hydrolysis of Tween 80 (a polyethylene derivative of sorbitan monooleate) releases complexed neutral red, resulting in a change in color of the test substrate (98). The change in color is not due to a pH shift from the formation of oleic acid, but to the destruction (hydrolysis) of Tween 80. The test is helpful in the identification of scotochromogenic and nonphotochromogenic mycobacteria. Species of these two groups that hydrolyze Tween 80 readily are seldom of clinical significance (with rare exceptions). *M. scrofulaceum* strains are negative; *M. gordonae* and *M. malmoense* strains are positive. The *M. avium* complex, *M. xenopi*, and *M. haemophilum* are negative; other nonphotochromogens are positive.

Inoculum

Use actively growing colonies obtained from solid medium.

Reagents

(i) Phosphate buffer, 0.067 M, pH 7, 100 ml. (ii) Tween 80, 0.5 ml. (iii) Neutral red stock solution, 2 ml: 0.1% aqueous solutions of actual dye content (e.g., if actual dye content is 85%, dissolve 0.1 g in 85 rather than 100 ml of water). Mix the three reagents, and note that the color is amber as the dye is complexed. Dispense the mixture in 4-ml amounts in screw-capped tubes (16 by 125 mm), and autoclave the tubes at 121°C for 15 min. Incubate the mixture at 35 to 37°C overnight to ensure sterility. Store the mixture in a refrigerator, and protect it from light. The substrate is not stable longer than 2 weeks at 4 to 10°C. A concentrate for the preparation of the test substrate is available and can be obtained from Difco (35).

Procedure

Emulsify a 3-mm loopful of growth in a tube of substrate. Incubate the culture at 35 to 37°C. Observe the tubes for a color change from amber (straw) to red after 1, 5, and 10 days.

Interpretation

Record the number of days required for the first appearance of pink. If this time is less than 5 days, the test is positive; if the time is 5 to 10 days, the test is doubtful (±); if no change in color occurs by day 10, the test is negative.

Controls

Use *M. kansasii* as a positive control and *M. scrofulaceum* as a negative control.

Urease

The determination of the ability of an isolate to hydrolyze urea often helps in the characterization of mycobacterial strains aberrant in some other property. For example, *M. scrofulaceum* is urease positive, whereas members of the *M. avium* complex are urease negative. The urease test will help in the recognition of the occasionally encountered pigmented strain of the *M. avium* complex. Similarly, *M. bovis* is urease positive, and this test helps distinguish a drug-resistant *M. bovis* strain from the *M. avium* complex. The method described here is a modification of the procedure of Toda et al. (83).

Inoculum

Use actively growing colonies obtained from solid medium.

Reagents

Mix 1 part of urea agar-base concentrate (Difco or an equivalent product) with 9 parts of sterile water. Do not add agar. Dispense the mixture in 4-ml amounts into screw-capped tubes (16 by 125 mm), and store the tubes at 4°C.

Procedure

Emulsify the equivalent of a 3-mm loop of growth in a tube of substrate. Incubate the culture at 35 to 37°C, and observe the culture for a color change from amber to pink or red. Discard the culture after 3 days.

Interpretation

A change in the color of the medium to pink or red within 3 days is recorded as a positive reaction.

Disk Test for Urea

Murphy and Hawkins (58) have described a procedure to detect urease production with a urea-impregnated paper disk (available from Difco). This procedure has been reported to yield results more quickly than the test described above, with a more consistent separation into positive and negative responses by species.

Controls

Use *M. kansasii* as a positive control and *M. gordonae* as a negative control.

Steadham (79) has described a carefully buffered urea broth which is simple to prepare and reliable in test results obtained after 7 days of incubation.

IMMUNOLOGY: PHAGE SUSCEPTIBILITY

Fluorescent-antibody, double diffusion in gel, erythrocyte or particle agglutination, and complement fixation tests have been used with some success in the differentiation of *Mycobacterium* species and in the characterization of antibodies in the sera of patients. Serologic procedures for routine use in clinical laboratories have not been standardized, and none are recommended at this time.

Simultaneously performed skin tests with standardized, purified protein derivatives from different *Mycobacterium* species have proved useful for the identification of bacilli inoculated into guinea pigs (21, 52). In contrast to experimental animals, in which single infecting agents can be documented, the possibility of multiple sensitizing infections in humans often results in unpredictable cross-reactions which are difficult to interpret (59). At present, the availability of skin test antigens prepared from different species or groups of organisms (complexes) has been curtailed until such antigens can be standardized in the same manner as the skin test antigen for *M. tuberculosis*, PPD-S.

Enzyme-Linked Immunosorbent Assay for Serum Antibodies

Application of the enzyme-linked immunosorbent assay procedure to the sera of patients with active tuberculosis has shown elevated, specific titers in the immunoglobulin G fraction when PPD-S is used as an antigen (5, 32). The demonstration of an elevated immunoglobulin G antibody against *M. tuberculosis* could be very useful for the identification of patients with active tuberculosis whose sputum smears and cultures are negative. Initial studies have shown that patients with tuberculosis have a significantly greater mean level of specific antibody than patients with nontuberculous mycobacterial infections, sarcoidosis, histoplasmosis, blastomycosis, and cryptococcosis; patients who have received BCG vaccination; those who have a history of treated tuberculosis; and a series of control patients who were both PPD skin test positive and skin test negative. The availability of a reliable enzyme-linked immunosorbent assay for specific immunoglobulin G against *M. tuberculosis* could have a highly significant clinical role in the initiation of specific therapy.

Thin-Layer Chromatography

Recently the use of thin-layer chromatography has been very helpful in selected cases in distinguishing differences in lipid content of mycobacterial cell walls (9). Thin-layer chromatography has also been useful in the demonstration of the unique lipid characteristics of previously unrecognized species, e.g., *M. szulgai* (53).

Mycobacterial Phage Susceptibility

M. tuberculosis can be divided into eight major phage types with four mycobacteriophages. Each major type can be subdivided into 64 subtypes with six additional mycobacteriophages (30). Overall population differences in phage types can be detected, but the greatest advantage of the determination of phage types resides in the ability to support epidemiologic studies (31). Although only one such study has been published (63), phage typing has been used to define the association of strains from epidemiologically linked cases and to detect instances of probable laboratory contamination by carry-over. Phage susceptibility has also been used successfully for the differentiation of strains of BCG from *M. tuberculosis* and virulent strains of *M. bovis* (29). Phage typing procedures are specialized and require the maintenance of proficiency; therefore, studies with phages should be conducted only in level III laboratories. Further use of this technique will provide greater depth to future studies of the epidemiology of tuberculosis.

Agglutination tests with rabbit antisera are useful for the differentiation of most strains of *M. kansasii*, *M. marinum*, *M. simiae*, the *M. avium* complex, *M. scrofulaceum*, *M. szulgai*, *M. gordonae*, *M. fortuitum*, and *M. chelonae* (100). Agglutination reactions with the last two species may not be fully reliable. Seroagglutination and phage typing are used primarily for epidemiologic purposes rather than taxonomy.

Gas-Liquid Chromatography

Gas-liquid chromatography has been used in the past and recently in the detection and identification of characteristic long-chain fatty acids from mycobacteria. The early studies of Reiner used pyrolysis gas-liquid chromatography (62). Using trifluoroacetylation of whole-cell methanolysates, Larsson and Mårdh consistently showed species-specific patterns of fatty-acid methyl esters and methyl glycosides among 10 strains of *M. avium*, *M. kansasii*, and *M. tuberculosis* (46). Tisdall et al. developed a procedure for the saponification of organisms in methanolic NaOH and were able to identify correctly most *Mycobacterium* species by using chromatograms and colony characteristics. When the procedure was applied to routine species identification, good correlation was found with stock cultures and clinical isolates (82). Subsequently, it has been suggested that acid methanolysis is more efficient than basic saponification and that mycolic acid separation can be improved by using a higher temperature in the column. Pyrolysis separates long-chain mycolic acids into characteristic C_{22}, C_{24}, and C_{26} cleavage products (26). Under appropriate conditions, if the instrument, culture, and trained technologist are ready, the time for gas-liquid chromatography analysis is approximately 2 h. At present, all such programs should be considered developmental, as both the instrumentation and the procedures used to detect the various products are not yet standardized.

Another approach to the diagnosis of tuberculosis is to detect tuberculostearic acid in clinical specimens by mass spectrometry (47). Such systems are actively under study but are not yet ready for use in the clinical laboratory.

SPECIES IDENTIFICATION

M. leprae, M. ulcerans, M. marinum, and *M. haemophilum*

M. leprae, M. ulcerans, M. marinum, and *M. haemophilum* may be present in specimens from a superficial body area. Since these bacteria rarely grow at 37°C or above, cultures inoculated with such specimens should be incubated at 32 ± 1°C, as well as (other cultures) at 37°C. *M. leprae* does not grow in vitro. The diagnosis of *M. leprae* is presumptive and dependent on the correlation of typical histopathologic lesions with their content of acid-fast bacilli. These lesions are abundant in patients with lepromatous leprosy and very sparse in patients with tuberculoid leprosy (6). *M. leprae* may be present in discharge from the mouth and nose.

One should not expect to isolate *M. ulcerans, M. marinum,* or *M. haemophilum* from specimens of sputum or gastric lavage. *M. marinum* grows readily at 30 to 32°C on any medium commonly employed for mycobacteria. *M. ulcerans* grows much more slowly and may require 6 to 12 weeks for the initial isolation. *M. ulcerans* may fail to grow on some media. *M. marinum* resembles *M. kansasii* in photochromogenicity and in cell and colony morphology, but the species is easily distinguished by its source (superficial lesion), its more rapid growth at 31 than at 37°C or no growth at 37°C, and usually by a negative nitrate reduction test. *M. haemophilum* is a recently described species isolated from skin lesions. It requires hemin for growth and may be isolated on blood or chocolate agar, Middlebrook 7H10 agar containing hemolyzed but not whole sheep erythrocytes, or Lowenstein-Jensen medium containing 1% ferric ammonium citrate. Incubation should be at both 20 and 32°C for a minimum of 2 to 7 weeks (71). No growth occurs at 37°C.

M. tuberculosis

M. tuberculosis is recognized by slowly developing, rough, eugonic colonies of a characteristic buff tint and by the production of niacin. On the most favorable culture media and with other optimal conditions, colonies are recognizable in less than 3 weeks, but some strains, especially those resistant to isoniazid and those growing on suboptimal media after decontamination with harsh digestion agents, may require 4 to 6 weeks or longer. Characteristic microcolonies may be readily recognized in less than 1 week by the microscopic observation of growth on Middlebrook 7H11 agar plates (68). The variations in hue and texture of colonies of tubercle bacilli on a given medium are readily learned but difficult to describe. In the preparation of smears from cultures, poor dispersibility is recognized. Serpentine cord formation (best seen in growth from a liquid medium lacking a wetting agent) is typical, but it must be noted that rough strains of many mycobacterial species will exhibit a similar stranding of bacilli. If the niacin test is positive and the other properties mentioned are in conformity, *M. tuberculosis* may be reported. If the niacin test is negative, a number of other tests should be made, including a repetition of the niacin test after further incubation. Other confirmatory properties are that *M. tuberculosis* reduces nitrate, loses catalase activity after being heated at 68°C, and usually is susceptible to antituberculosis drugs, unless the organism is from long-treated patients. If a strain of *M. tuberculosis* is isoniazid resistant, it may be found to be catalase negative and have reduced or no virulence for guinea pigs.

M. bovis

M. bovis grows poorly or not at all on some media favorable for *M. tuberculosis*. The Jensen modification of Lowenstein egg medium usually supports better growth of *M. bovis* than does 7H10 or 7H11 agar. The media most favorable for *M. bovis* contain 0.4% pyruvate, as does $NaCO_2COCH_3$ without glycerol (18). In contrast to *M. tuberculosis*, poor or no growth on medium containing more than 1% glycerol is one of the distinguishing properties of *M. bovis*, especially on primary isolation. Colonies are buff, low, and small and may appear either smooth or rough on egg medium. On Middlebrook 7H11 agar, colonies are very thin and often show little or no stranding, but if pyruvate has been added to the medium, colonies show serpentine cords like those of eugonic *M. tuberculosis*. The following properties of *M. bovis* are useful for distinguishing it from *M. tuberculosis*. Nitrate is not reduced, and the niacin test is usually negative. *M. bovis* is resistant to PZA but susceptible to 5 μg of T2H per ml (unless it is isoniazid resistant).

M. kansasii

M. kansasii occurs in several forms: photochromogenic, high catalase; photochromogenic, low catalase; scotochromogenic; and nonchromogenic. All of these forms except the first are very rare in clinical specimens. Finding the very distinctive dependence upon light exposure for carotene formation, usually coupled with carotene crystal production, in a strain having a growth rate similar to (or slightly more rapid than) that of *M. tuberculosis* at 37°C is presumptive evidence for reporting *M. kansasii*. Additionally, growth occurs at 25°C (tubercle bacilli show no growth), but very slowly (*M. marinum* grows faster). Young cells are characteristically long, and some are broad and cross-banded (structural evidence of the utilization of the fatty material of the medium). Colonies are characteristically intermediate between fully rough and fully smooth. The centers are elevated. In the thinner margins, curving strands of bacilli may usually be seen with a low-power microscope. Some strains are almost totally smooth, and others are completely rough. Nitrate is reduced; Tween 80 and urea are hydrolyzed. The catalase activity of disease-related strains is vigorous: >45 mm in the semiquantitative test. Rarely, strains not related to disease are encountered. These strains have usually been found to be low in catalase activity (94).

M. simiae

M. simiae was first isolated from nonhuman primates and since then has been found associated with human disease in Europe, Cuba, and the United States (66). The organism produces niacin, is photochromogenic (a characteristic that may be unstable), hydrolyzes Tween 80 slowly (>10 days), and produces thermostable catalase. It has usually been isolated from patients with pulmonary disease, but it has been isolated from environmental specimens also.

M. asiaticum

M. asiaticum was first isolated in Hungary from the lymph nodes and viscera of healthy monkeys. Although originally the distinction was not clear, it is now known that *M. asiaticum* is a separate species from *M. simiae*, which was isolated from monkeys at the same time. *M. asiaticum* has been isolated from five patients in Australia, although the organism was considered responsible for disease in only two (7). Good and Snider have reported two strains isolated in the United States in 1979 and another four in 1980, but it isn't known whether or not these strains were associated with disease (22). *M. asiaticum* is photochromogenic but, in contrast to *M. simiae*, is negative for niacin production.

M. scrofulaceum, *M. szulgai*, and *M. gordonae*

Subdivisions of the slowly growing scotochromogens are not well defined. Some strains of the *M. avium* complex may be scotochromogenic, especially after prolonged incubation. The Tween 80 hydrolysis test provides an initial separation of these organisms: *M. gordonae* is usually positive in 5 days, and *M. szulgai* is slowly positive (usually >7 days), whereas *M. scrofulaceum* is not positive even after 10 days. *M. szulgai* is the only one of this group to give a positive nitrate reduction test, although at times this characteristic may be weak. The urease test will also separate *M. scrofulaceum* both from *M. gordonae* and from occasional pigmented strains of the *M. avium* complex. *M. gordonae* is rarely associated with disease. *M. scrofulaceum* may be associated with cervical adenitis in children, but less commonly than are the *M. avium* complex and *M. tuberculosis* (45). *M. szulgai* has been found to cause pulmonary disease, cervical adenitis, olecranon bursitis, and tenosynovitis of the hand. Most isolates have been associated with disease (16). Both *M. gordonae* and *M. scrofulaceum* commonly occur as nonpathogens.

M. xenopi

M. xenopi can be recognized by its small, erect colonies, slow growth, characteristic yellow color, more rapid growth at 42 than at 37°C, and failure to grow at 25°C. This species has a greater susceptibility to antituberculosis drugs than is seen with other nontuberculous mycobacteria, e.g., it is susceptible to 1 μg of isoniazid per ml. Although this species has previously been placed with the nonphotochromogenic mycobacteria, the brightly pigmented yellow colonies usually found on primary isolation suggest that the organism would be better considered with the scotochromogenic mycobacteria. Occasional colonies of *M. xenopi* are nonpigmented. Persistent branching and filamentous extensions (as seen with a low-power microscope) around the circular colonies on cornmeal-glycerol agar or other plated media are very distinctive. Rough colonies usually exhibit aerial hyphae, as can be seen with a stereomicroscope. Microscopic colonial morphology shows a distinctive "bird's nest" appearance with sticklike projections in young colonies. Disease due to *M. xenopi* is not common. When present, the disease appears to be more frequent in middle-aged men with other predisposing pulmonary diseases.

M. avium and *M. intracellulare* (the *M. avium* complex)

Species in the *M. avium* complex are not usually distinguishable from one another and need not be. They are characterized at the time of primary isolation by slowly growing, thin, transparent, homogeneous, smooth colonies. A small proportion of *M. intracellulare* colonies may be partially or completely rough; for *M. avium*, the proportion of rough colonies may often be greater. After subculture, a transformation to more eugonic type colonies may occur, with the centers becoming prominently domed. Eventually all colonies may be hemispherical. Usually nonpigmented, the colonies, particularly the dome-shaped colonies, may become yellow with age; rarely are colonies pigmented from the onset of detectable growth. Correlated with these colony variations are changes in other properties, notably, pathogenicity (19). Cells are commonly short, although under certain conditions, long, thin bacilli are seen. All strains are Tween hydrolysis negative, and many strains reduce tellurite in 3 days. Although the *M. avium* complex may cause a serious tuberculosislike disease that is difficult to treat, the complex may also occur in clinical specimens as a non-disease-associated organism. The *M. avium* complex is now the most frequent mycobacterium other than *M. tuberculosis* isolated in the United States (22). Although the isolation of the organism does not establish infection, the patient from whom the organism is isolated should be carefully evaluated to distinguish between colonization and infection. During the past several years, disseminated infection with *M. avium* complex has become almost common in patients with acquired immune deficiency syndrome (AIDS) (23). In these patients, the normal cellular response to such a challenge does not always occur, and little inflammatory response may be found at the time of death. The recognition that fever without pulmonary chest X-ray findings may be associated with disseminated mycobacteriosis has resulted in increasing attempts, many successful, to isolate mycobacteria from the blood. The use of the lysis-centrifugation system or the radiometric detection of ^{14}C-metabolites has been shown to be helpful (51).

Two groups of organisms which appear intermediate between the *M. avium* complex and *M. scrofulaceum* have been recognized (27). Both are pigmented, but one group produces a column of foam ≥45 mm in the semiquantitative catalase test and is urease negative, whereas the second group produces <45 mm of foam and is urease positive. These organisms have been assigned the acronym MAIS, and those with the latter reactions are much less common that the first group.

Other species of slowly growing nonphotochromogens not known as pathogens but encountered in the clinical laboratory differ from members of the *M. avium* complex in hydrolyzing Tween in 5 days or less. Some of these organisms have distinctive colonies. Strains of the *M. avium* complex and *M. triviale* may have colonies so rough as to be confused with tubercle bacilli, but these strains are niacin negative. *M. triviale* is Tween positive, whereas the *M. avium* complex is Tween negative. Strains of *M. triviale* are distinguished from other nonphotochromogenic bacteria by their ability to grow on media containing 5% sodium chloride. For other properties of the nonpathogenic species, including *M. gastri* and *M. terrae*, see Table 2. In general, strains of *M. gastri* and *M. terrae* complex are not associated with disease, although occasional cases have been reported (44).

M. malmoense

In 1977, Schröder and Juhlin described a nonphotochromogenic mycobacterium associated with pulmonary disease in Sweden (72). This organism grows slowly, being seen after 2 to 3 weeks of incubation at 37°C and after up to 6 weeks of incubation at 22°C. Colonies are grayish white and appear smooth, glistening, opaque, domed, and circular, 0.5 to 1.5 mm in diameter. This species does not produce niacin, is nitrate negative, hydrolyzes Tween 80, and usually produces heat-labile catalase (Table 2). Although resistant to isoniazid, streptomycin, *p*-aminosalicylic acid, and rifampin, it is susceptible to ethambutol (1 μg/ml) and cycloserine (16 μg/ml). Since 1977, additional isolates have been found in Wales and Australia, where all the isolates were associated with human pulmonary disease. Twelve strains of *M. malmoense* were isolated in the United States in 1980 (22).

M. fortuitum and *M. chelonae*

M. fortuitum and *M. chelonae* are recognized by growth in <7 days, rough or smooth colonies, and the absence of a definite yellow pigment (but sometimes mature colonies of *M. fortuitum* may become green by taking up the malachite green in egg-based medium). Both species are characteristically arylsulfatase positive and will grow on MacConkey agar without crystal violet. Because they share a number of similar metabolic characteristics and are not infrequently found in the same types of infection, they have been referred to as the *M. fortuitum* complex in the past. However, recent studies have shown different patterns of drug susceptibility and associations with sites of infection. Susceptibility studies with antimicrobial agents other than antituberculosis drugs have suggested that strains of *M. fortuitum* are more likely to be susceptible to these agents than is *M. chelonae*. For these and other reasons, it has been suggested that species and subspecies names be used to distinguish between different taxonomic groups of these two organisms which might otherwise be obscured by the grouping of both species into the *M. fortuitum* complex (90). *M. fortuitum* characteristically has branching, filamentous extensions from 1- to 2-day-old colonies on cornmeal-glycerol or Middlebrook 7H11 agar, and short aerial hyphae from rough colonies may be seen with a stereomicroscope. Both species show full resistance to all antituberculosis drugs, except that many isolates of *M. fortuitum* are susceptible to capreomycin. *M. chelonae* lacks the filamentous extensions from colonies at 1 to 2 days that are typical of *M. fortuitum* (68). Of further value, *M. fortuitum* is positive for nitrate reductase, whereas *M. chelonae* is negative, and most *M. fortuitum* strains are positive for iron uptake, whereas *M. chelonae* strains are commonly negative or have a modified reaction (Table 3). Taxonomically, two subspecies of *M. chelonae* have been recognized, i.e., *M. chelonae* subsp. *chelonae* and *M. chelonae* subsp. *abscessus*. The latter subspecies can grow on Lowenstein-Jensen medium containing 5% NaCl at 28°C, whereas *M. chelonae* subsp. *chelonae* cannot. Recently, the recognition of several outbreaks of infection caused by these organisms coupled with their known resistance to standard antituberculosis drugs resulted in the finding that therapy with other antimicrobial agents may be more effective (90). Such antimicrobial agents as cefoxitin, sulfonamides, members of the tetracycline group, and several aminoglycosides may be helpful.

DRUG SUSCEPTIBILITY TESTING

Most newly diagnosed patients with tuberculosis will be treated with two or more of the primary antituberculosis drugs, which are isoniazid, streptomycin, ethambutol, rifampin, and PZA. Many isolates need not be tested for susceptibility if the patient has not had prior chemotherapy. In any case, it is usually not necessary to await the results of these tests to institute treatment. By contrast, if the patient has previously received antituberculosis drugs, it is important to determine the susceptibility pattern, as it is in patients with recurrent tuberculosis that drug resistance is most commonly seen. Clinically, if the drug regimen selected has not converted the sputum of the patient to an acid-fast, smear-negative status within 4 to 5 months, there is a distinct possibility that a drug-resistant organism is emerging, and susceptibility tests should be repeated.

Conventional diffusion techniques for susceptibility tests, which rely on the size of a zone of inhibition surrounding a drug-containing disk, are not suitable for the slowly growing mycobacteria because the drug diffuses throughout the medium before the organism has a chance to grow.

The generally accepted methods for the determination of the drug susceptibility of mycobacteria are based on the growth of the organism on a solid medium. Several important principles must be recognized to understand the difference between the susceptibility testing of mycobacteria and that of the more rapidly growing bacteria.

1. The composition of the medium should have a minimal effect on the inactivation of the drug. The use of Middlebrook 7H10 or 7H11 medium in place of Lowenstein-Jensen or other egg media eliminates the binding of drugs to large-molecular-weight proteins and also eliminates the problems associated with drug inactivation that can occur at the high temperature associated with the inspissation of egg-based media.

2. Drugs vary in stability. Drugs in media may be modified on exposure to extremes of pH or temperature, prolonged storage, or drying. Therefore, drug-

containing media should be stored in a refrigerator, shielded from light, and kept in plastic bags to protect them from evaporation. No drug-containing medium should be used longer than 4 weeks after preparation. In view of the simplicity of preparing test media by the use of paper disks (see below), it should not be necessary to stockpile large amounts. Include control organisms with every set of susceptibility tests to demonstrate the activity of drug-containing media. The drug-susceptible strain H37Rv of *M. tuberculosis* is satisfactory, but controls should also include strains resistant to one or more drugs as well as a selected strain of *M. kansasii* that is known to be resistant to the low concentration of isoniazid but susceptible to the high concentration.

3. Standardization of the inoculum is important. Since drug susceptibility or resistance depends on the presence or absence of growth on control and drug-containing media, the inoculum for each culture must be carefully controlled. This standardization is done by determining the growth and numbers of colonies on control cultures seeded with different dilutions of inoculum. Homogenization of the inoculum to eliminate large clumps of cells is essential. The inoculum must be heavy enough to result in at least 200 colonies on a control medium to provide for statistically significant data, but not so heavy that confluent growth covers the surface of the control culture; therefore, two inoculum dilutions are used.

The frequency of a single, drug-resistant bacillus in a culture or an infectious lesion of tubercle bacilli has been estimated at 1 in 10^5 bacteria for isoniazid and 1 in 10^6 bacteria for streptomycin. Because an open pulmonary cavity in a patient may contain a bacillary population as high as 10^7 to 10^9 bacteria, it is apparent that excessive inoculation with a population which is primarily drug-susceptible can result in sufficient growth to suggest more resistance than is actually present. For this reason, it is necessary to employ an inoculum which will permit the expression of susceptibility or resistance as less or greater than 1% of the total population of cells examined. The 1% endpoint has been proposed on the basis of clinical studies that have shown that when more than about 1% of a bacillary population has become resistant to a drug, the drug does not continue to be useful; i.e., the resistant cells soon become the predominant form, nearly 100% of the population.

4. If susceptibility tests are performed on cultures that have more than one type of colony, the inoculum must be prepared from a representative selection. Drug-resistant tubercle bacilli usually grow more slowly and may produce smaller colonies than do drug-susceptible bacilli.

5. Since a modification of the proportion of susceptible to resistant bacteria may occur with in vitro subculture, direct susceptibility tests, with a portion of the decontaminated and concentrated specimen of the patient as an inoculum, are preferred whenever smear examination reveals acid-fast bacilli. Direct susceptibility tests also have the advantage of earlier results, but frequently the results may be unsatisfactory because of contamination or low numbers of colony-forming units.

6. Serious consequences may result from an inaccurate report of drug resistance. This inaccuracy may lead to the substitution of secondary drugs which can involve more discomfort and toxicity to the patient as well as greatly increased expense.

TABLE 8. Distribution of drug-containing disks for susceptibility tests

Plate no.	Quadrant no.	Drug	Amt (μg) per disk	Final drug concn (μg/ml)
1	I	Control no. 1		0
	II	Isoniazid	1	0.2
	III	Isoniazid	5	1.0
	IV	Ethambutol	25	5.0
2	I	Control no. 2		0
	II	Streptomycin	10	2.0
	III	Streptomycin	50	10.0
	IV	Rifampin	5	1.0

Preparation of susceptibility test media

In the past, it was necessary to prepare large amounts of media and incorporate multiple concentrations of the drugs to be tested. This procedure is cumbersome and expensive. A more practical procedure for many laboratories is the method of Wayne and Krasnow (99), in which paper disks (BBL) containing standardized amounts of drugs are placed in individual quadrants of plastic petri dishes as indicated below. This simplified procedure allows each laboratory to make just the number of plates needed for immediate use and eliminates labeling errors, as the disks carry identification codes. The reliability of this method has been established, and results have been shown to be comparable to those achieved by standard methods. To prepare the plates for susceptibility tests, dispense the disks aseptically to the individual quadrants of the petri dishes, centering the disks in the quadrant (Table 8). A 40-ml amount of complete Middlebrook 7H10 or 7H11 agar medium will be needed for the eight quadrants to be used. Dispense 5.0 ml of medium to all four quadrants of both plates. Incubate the plates at 4°C overnight to permit the uniform diffusion of the drug from the paper disk. The plates may then be used immediately or stored for a maximum of 4 weeks in closed plastic bags. Some evaporation from the test plates will occur, even on refrigeration, resulting in some change in drug concentration. For this reason, use fresh medium whenever possible.

Direct method of susceptibility testing

The direct method of susceptibility testing may be used if acid-fast bacilli are seen on the smear of the concentrated, decontaminated clinical specimen. Make dilutions of the sediment according to the schedule given in Table 9, and inoculate each quadrant with 0.1 ml of the dilution.

TABLE 9. Dilution of concentrate for inocula

No. of acid-fast bacilli per oil-immersion field	Control quadrant 1	Control quadrant 2	Drug quadrants
Less than 1	Undiluted	10^{-2}	Undiluted
1–10	10^{-1}	10^{-3}	10^{-1}
More than 10	10^{-2}	10^{-4}	10^{-2}

Indirect method

Susceptibility tests are usually performed with cultures rather than specimens. If the culture is fairly young, the inoculum may be prepared by scraping colonies from the surface of the medium, taking care to sample all parts of the culture. Suspend the bacterial mass in 4 ml of Dubos Tween-albumin or Middlebrook 7H9 broth containing three or four small, sterile glass beads. Place the mixture on a Vortex mixer for about 1 min, and take precautions to prevent aerosol production. Let the tube stand for 15 min or longer. Dilute the suspension with broth until it is just barely turbid. Dilute this stock suspension 10^{-2} and 10^{-4} in broth. Inoculate 0.1 ml (2 to 3 drops from a Pasteur pipette) of the 10^{-2} dilution onto control and drug-containing media. Similarly, inoculate the control medium with 0.1 ml of the 10^{-4} dilution (control 2).

If the culture to be tested is old, it may be necessary to grow it in broth to prepare a viable inoculum. After good growth has occurred, proceed as above.

Incubation and interpretation

Incubate the plates at 35 to 37°C in air supplemented with 5 to 10% CO_2, and read the plates weekly for 3 weeks. Record the amount to growth as follows: innumerable to confluent colonies, +++ to ++++; approximately 100 to 200 colonies, ++; 50 to 100 colonies, +; fewer than 50 colonies, give the actual count. Report the results within 3 weeks after inoculation. Longer incubation periods may yield growth of even susceptible organisms because of drug deterioration. The supplementing of air with carbon dioxide may stimulate the growth of tubercle bacilli, and the effects of the drug may be counteracted, e.g., with PZA and low concentrations of streptomycin.

In most cases, the proportion of resistant colonies can be estimated as greater or fewer than 1% of the control population. Thus, if growth on a drug quadrant exceeds the growth on the second dilution control (i.e., control 2), then more than 1% of the population is resistant. In only rare instances should there be an ambiguous result, and in such cases, the tests should be repeated. Report the results at 2 to 3 weeks or earlier if clearly recognizable growth occurs on drug-containing as well as drug-free control media.

Secondary drug susceptibility tests

It is recommended that the determination of resistance to PZA and secondary drugs (capreomycin, *p*-aminosalicylic acid, cycloserine, ethionamide, and kanamycin) be limited to a few reference laboratories. In general, the media and methods for testing susceptibility to the secondary drugs are similar to those employed for the primary drugs.

Antibiotic disks are available for only *p*-aminosalicylic acid, kanamycin, and ethionamide. Cycloserine, ethionamide, and the primary drug ethambutol are relatively unstable in culture media, and good quality control procedures must be followed to prevent misleading susceptibility test results. Appropriate drug concentrations for the susceptibility testing of mycobacteria are shown in Table 10.

In the preparation of media to test for susceptibility to PZA, the agar-based medium must be buffered at pH 5.5 since the drug is active only at an acid pH (acid-buffered 7H10 medium is available from GIBCO Diagnostics, Madison, Wis.). Albumin-glucose-catalase supplement is used with the acid medium rather than oleic acid-albumin-glucose-catalase, since free oleic acid causes a significant inhibition of growth of *M. tuberculosis* at the low pH. Some lots of albumin-glucose-catalase supplement still failed to support growth of *M. tuberculosis* at this low pH. Butler and Kilburn (10) have developed a rapid turbidimetric test to determine the growth-supporting potential of albumin-glucose-catalase enrichment for *M. tuberculosis* at pH 5.5. When an atmosphere of air supplemented with 5 to 10% CO_2 is used, growth of tubercle bacilli on 7H10 medium is stimulated, and the growth-inhibiting effects of PZA are counteracted. Due to the potential problems, it is suggested that only level III laboratories should perform the test.

TABLE 10. Drug concentrations for susceptibility testing

Drug	Drug concn (μg/ml) on Middlebrook medium	
	7H10	7H11
Isoniazid	0.2, 1.0	0.2, 1.0
p-Aminosalicylic acid	2.0	8.0 or 10.0
Streptomycin	2.0, 10.0	2.0, 10.0
Rifampin	1.0	1.0
Ethambutol	5.0	7.5 or 10.0
Ethionamide	5.0	10.0
Kanamycin	5.0	6.0
Capreomycin	10.0	10.0
Cycloserine	30.0	30.0
PZA	25.0[a]	

[a] At pH 5.5.

Susceptibility tests with nontuberculous mycobacteria

Most strains of *M. kansasii* are susceptible to somewhat higher concentrations of isoniazid, streptomycin, and ethambutol than are usually tested, and the strains are quite susceptible to rifampin. Strains of many of the other slowly growing mycobacteria are resistant to antituberculosis drugs in vitro, even though patients may respond to multiple-drug antituberculosis chemotherapy. Thus, in vitro susceptibility results do not correspond completely with treatment results.

For further information on drug susceptibility testing for mycobacteria, see the position paper published by the Scientific Assembly on Microbiology, Tuberculosis and Pulmonary Infections (3).

For the rapidly growing mycobacteria, methods have been described for susceptibility testing that are comparable to the disk diffusion methods for non-acid-fast bacteria. Diffusion susceptibility testing of *M. fortuitum* and *M. chelonae* with amikacin, doxycycline, minocycline, kanamycin, sulfonamides, and trimethoprim may be of clinical use (88). However, MICs may be determined with relative ease by the method of Swenson et al. (80).

PRIMARY DRUG RESISTANCE

Tuberculosis patients who have never received antituberculosis drugs and from whose secretions drug-resistant tubercle bacilli can be isolated represent tuberculosis primary drug resistance (PDR). For many years the PDR rate was thought not to exceed 2 to 5% of all cases. However, in 1975 the U.S. Public Health Service initiated an ongoing surveillance program designed to sample different populations and geographic areas that had not been represented in previous surveys. The first (39) and third (13) summaries of this program have shown considerable variation in the incidence of PDR among patients in different geographic areas of the country as well as in various ethnic groups, ranging from PDR rates as high as 15% in Los Angeles and 19% in Harlingen, Tex., to 3.4% in Detroit. The PDR rate for all ethnic and racial groups in 1977 was 8.6% (39), but by 1982 the rate had dropped to 6.9% (13). High rates were found in Hispanics and Asians in their characteristic geographic locations in Texas and California (Fig. 6). The decision to offer drug susceptibility testing for mycobacteria may therefore vary from one laboratory to another, depending on geographic location and the population served as well as on the volume of strains and the availability of special facilities. If recent mycobacterial isolates are not tested for drug susceptibility, cultures should be retained in a refrigerator for a minimum of 6 to 12 months in the event that a question of drug resistance subsequently appears.

RECENT DEVELOPMENTS IN THE DRUG THERAPY OF TUBERCULOSIS

Recently, a better understanding of the mode of action of isoniazid and rifampin has led to dramatic improvements in drug therapy for tuberculosis. For many years it has been known that tubercle bacilli occur in the largest numbers in cavitary lesions, in which the milieu is neutral or slightly alkaline. The organisms are also present in small numbers in small, closed lesions and within macrophages, in which the milieu is more acid. The bacilli multiply actively in cavitary lesions, whereas they replicate slowly and even intermittently in the other sites (24). Fortunately, both rifampin and isoniazid are fully bactericidal in both types of sites and penetrate well into closed lesions and macrophages. These facts are considered the reason these two drugs are able to eradicate tuberculous infection in most patients within 9 months, a time that had not been achieved with any previous combination. Immediate success in the elimination of organisms from the sputum is achieved by killing the large extracellular, intracavitary population. The smaller population within closed lesions and macrophages must also be killed, because they can persist and may give rise to a relapse months or years later.

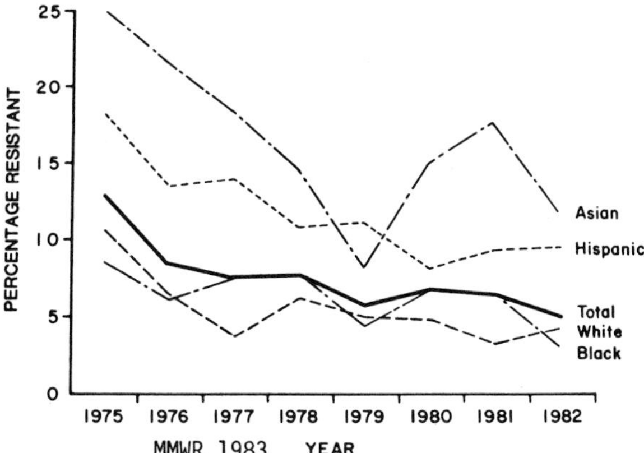

FIG. 6. Percentage of tuberculosis patients with primary drug resistance by race and ethnic group in the United States, 1975–1982. Reproduced from Morbid. Mortal. Weekly Rep. **32**:521, 1983.

The possibility of the reduction of the duration of chemotherapy of tuberculosis is traced to two key animal experiments carried out 12 and 25 years ago and involving rifampin and PZA. Both these drugs and isoniazid were shown to have the ability to kill tubercle bacilli in a tuberculous lesion. Additional experiments have revealed that streptomycin is bactericidal for extracellular bacilli in the neutral or slightly alkaline medium of cavitary lesions. PZA is inactive in this medium but is bactericidal for bacilli in the acid milieu within macrophages, where streptomycin is inactive. For these reasons, the combination of streptomycin and PZA is considered equivalent to either rifampin or isoniazid. Since isoniazid and rifampin kill *M. tuberculosis* in either environment, the duration of therapy can be shortened to 9 instead of the previously standard 18 to 24 months of therapy without a significant change in the relapse rate. Dutt and Stead (20) have found that after the first month of therapy, when a third drug is added (usually either streptomycin or ethambutol), isoniazid and rifampin can be given only two times a week without significantly changing the relapse rates. This intermittent therapy offers tuberculosis control programs several important advantages, including the ability to supervise administration of the drugs, thereby improving patient compliance, and a reduction in the cost of medication to less than one-third the amount necessary for daily therapy.

Recommendations for the therapy of mycobacterial diseases have been summarized in a current position paper by the American Thoracic Society (2).

RADIOMETRIC PROCEDURES FOR THE MYCOBACTERIOLOGY LABORATORY

In 1977, Middlebrook et al. reported the formulation of a medium (7H12) containing [1-^{14}C]palmitic acid which could be used for the detection of the growth of *M. tuberculosis* (55). The method relied upon the measurement of labeled carbon dioxide released by the metabolism of palmitic acid in an automatable ion chamber system (BACTEC, Johnston Laboratories, Towson, Md.). These initial studies indicated that an inoculum of 200 viable units of *M. tuberculosis* could be detected in 12 to 14 days, and only 20 viable units could be detected in 14 to 17 days. These promising results led to further studies for the application of the technique to more routine laboratory procedures for detection, identification, and performance of drug susceptibility tests with primary antituberculosis drugs (Fig. 7).

A multicenter, collaborative study reported that the results of drug susceptibility tests of *M. tuberculosis* with the radiometric and standard methods were similar, but there was better agreement with drug-susceptible strains than with drug-resistant strains (77). Susceptibility to *p*-aminosalicylic acid could not

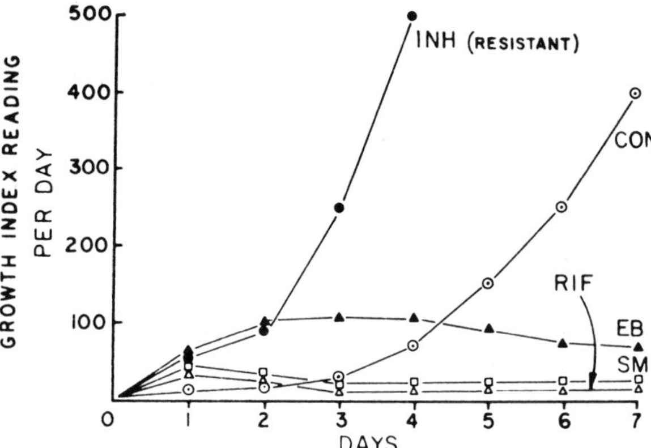

FIG. 7. Radiometric drug susceptibility test pattern of an INH-resistant strain of *M. tuberculosis*. CON, Control. Reproduced from reference 73.

be determined by the radiometric procedure, and overall, results with the new procedure were better from a specialty laboratory than from a routine clinical laboratory. Siddiqi et al. (73) and Laszlo et al. (48) have reported that the overall agreement between radiometric and standard methods for *M. tuberculosis* was >95% and that results were reportable on 98% of the tests in 5 days and on 95% of the tests in 7 days, respectively, in the two studies. The problems encountered with the determinations of susceptibility to ethambutol were believed to be due to tests with an inappropriate concentration of the drug. In a study designed to measure both intra- and interlaboratory reproducibility, susceptibility test results obtained by radiometric procedures were as good as or better than those obtained by conventional procedures (R. C. Good, G. M. Cauthen, C. L. Woodley, G. D. Kelly, and J. O. Kilburn, Abstr. Annu. Meet. Am. Soc. Microbiol. 1984, U5, p. 87).

M. tuberculosis can be detected rapidly in decontaminated clinical specimens by inoculation onto a selective Middlebrook 7H12 medium containing polymyxin B, amphotericin B, carbenicillin, and trimethoprim (PACT). Damato et al. (15) found that >70% of smear-positive specimens were culture positive in the radiometric procedure in 14 days, with or without the addition of PACT to the medium, compared with 21 days by the standard procedure. Similarly, detection times for *M. tuberculosis* from smear-negative specimens with radiometric and conventional culture systems were 13.7 and 26.3 days, respectively (57). A larger number of positive specimens were detected when Lowenstein-Jensen, Middlebrook 7H10, and selective Middlebrook 7H11 media were all used than with a single medium or by the radiometric procedure alone. Radiometric and conventional culture procedures were approximately equivalent for the recovery of *M. tuberculosis* from 5,375 clinical specimens, but the recovery of *M. avium* complex was better by the radiometric procedure (81).

In a collaborative study involving five laboratories, recovery and drug susceptibility tests of *M. tuberculosis* were completed in 18 days by the radiometric procedure as opposed to 38.5 days for the conventional method (65). Throughout these developmental tests, the *Mycobacterium* species isolated was unknown until routine identification procedures were completed. Therefore, the initial time saved in detection was of little benefit, since 4 to 8 weeks was still needed for species identification. Another problem noted by several laboratories is an increase in the isolation of nonpathogenic mycobacteria (*M. gordonae*) by the radiometric procedure. These isolates must be subcultured before individual colonies can be obtained for identification by standard methods.

The rapid identification of *M. tuberculosis* isolated from clinical specimens in contrast to nontuberculous mycobacteria can be accomplished by measuring their susceptibility to *p*-nitro-α-acetylamino-β-hydroxypropiophenone (NAP) (49; S. H. Siddiqi, C. C. Hwangbo, V. A. Silcox, R. C. Good, D. E. Snider, Jr., and G. Middlebrook, Am. Rev. Respir. Dis., in press). Both the rate of $^{14}CO_2$ evolution and susceptibility to NAP are used as identification criteria. The rate of CO_2 evolution is of help as it is directly proportional to the number of cells in culture.

The currently suggested radiometric procedure for the detection, identification, and drug susceptibility testing of *M. tuberculosis* from clinical specimens is as follows (proper safety precautions must be used in the inoculation and reading steps; BACTEC 460, which is constructed for aerosol containment, must be used in these procedures).

1. Digest and decontaminate the specimen by the NALC-NaOH procedure. Suspend the specimen in 1 to 2 ml of buffer so that clumps of bacteria are dispersed.
2. Prepare two vials with 2 ml each of 12A medium (prepared with 2 μCi of ^{14}C-labeled palmitic acid per vial) containing PACT (Johnston Laboratories). Inoculate 0.1 ml of specimen into vial 1, 0.5 ml into vial 2, and 0.1 ml onto a slant of Lowenstein-Jensen medium. Use a part of the suspension to prepare smears for staining. If an organism with a lower optimal growth temperature is suspected, inoculate another vial for incubation at that temperature.
3. Incubate the inoculated vials at 37°C without shaking, and read them on the BACTEC (Johnston Laboratories) daily. Inspect the vials visually to rule out contamination.
4. When the growth index (GI; the value which indicates radioactivity in the gas phase of the culture; 100 GI units ≃ 0.025 μCi of $^{14}CO_2$) reaches 50 to 100 U, inoculate 1 ml of the culture into vial 2 containing 12A medium plus *p*-nitro-α-acetylamino-β-hydroxypropiophenone and into vial 3 containing 12A medium only. Vials 2 and 3 are read daily for 2 to 5 days. If growth is inhibited, the isolate is either *M. tuberculosis* or *M. bovis*, whereas no inhibition of growth indicates that the isolate is another *Mycobacterium* species.
5. When the GI reaches 300, examine the culture microscopically. Serpentine cords are an indication of tubercle bacilli, whereas small clumps or individual cells indicate other species.
6. When the GI reaches 500 or more, the susceptibility of *M. tuberculosis* strains to drugs can be tested. Procedures for tests against antimycobacterial drugs with other *Mycobacterium* species have not been described and are not recommended at this time. Using drugs supplied for the procedure (Johnston Laboratories), prepare the inoculate the drug-containing media with 0.1 ml of the positive culture and with

representative control strains. Inoculate one vial of 12A medium prepared without drugs with 0.1 ml of a 1:100 dilution of the positive culture (which will be equivalent to 1% growth in the proportion method for drug susceptibility testing). Incubate the vial and read it daily until the GI of the control vial reaches 30. If the increase in GI (ΔGI) in drug-containing medium is more than the ΔGI in the control, the culture is considered resistant to that drug. Drugs that can be tested are isoniazid, streptomycin, rifampin, and ethambutol. The manufacturer should be consulted regarding the recommended test concentrations for each drug, since these are currently under review; however, final drug concentrations in the test vial are 90% of the stated concentrations because of the dilution factor (2.0 ml of medium plus 0.1 ml of drug solution plus 0.1 ml of inoculum yields 2.2 ml).

Although the radiometric procedure appears simple and straightforward, problems arise that can be handled only by experienced personnel. Therefore, the procedure should only be used in American Thoracic Society level II or level III or College of American Pathologists extent 3 or 4 laboratories.

QUALITY CONTROL

An ongoing, active quality control program is essential in the mycobacteriology laboratory. Control organisms are available through proficiency testing programs, the American Type Culture Collection, and the Trudeau Mycobacterial Culture Collection based at the National Jewish Hospital and Research Center in Denver, Colo. The Trudeau Mycobacterial Culture Collection has prepared a mycobacterial Standard Taxonomic Characters Kit which is available at no charge by writing on letterhead stationery to the Curator, Mycobacterial Culture Collection, National Jewish Hospital and Research Center, 3800 East Colfax Avenue, Denver, CO 80206.

Mycobacterial cultures can be maintained on egg-based medium slants at 4°C for up to 12 months. Alternatively, the growth from several slants can be suspended in skim milk, placed in small vials, rapidly frozen in a slurry of dry ice and alcohol, and stored at −20 or, preferably, −70°C. More frequently needed mycobacterial species can be maintained in Middlebrook 7H9 broth at 37°C or room temperature and can be subcultured monthly or bimonthly.

Positive control smears should be included with each set of patient specimen slides to ensure that the stain, staining procedure, and microscope are functioning properly. Smears containing mycobacteria can be made from broth cultures. The best method is to prepare a large set of smears from concentrated sputum specimens collected from patients with active tuberculosis. The staining and morphological features of these organisms are more characteristic of organisms causing disease than are the features of organisms grown in culture. Care should be exercised in the handling of such slides, as heat fixation at 65°C for 2 h does not always inactivate mycobacteria.

Upon preparation or receipt in the laboratory, each lot of culture medium should be inoculated with stock strains of *M. tuberculosis*, *M. kansasii*, the *M. avium* complex, and *M. fortuitum*. In most instances, media can be stored for 10 to 14 days, after which quality control results are available. In the case of Middlebrook 7H10 and 7H11 agars, the media can be inoculated with clinical specimen concentrates within a week of the time at which quality control organisms are inoculated, as long as clinical specimen concentrates are maintained in the refrigerator for 2 to 3 weeks afterward. Should the quality control strain not perform well on the new medium, the refrigerated concentrates can be reinoculated to another batch, or a different type, of medium. Storage of Middlebrook 7H10 and 7H11 media for more than 5 weeks may be associated with poor growth characteristics, as formaldehyde may form after prolonged storage and direct exposure to strong light (56).

Appropriate positive and negative control organisms should be included when testing each identification characteristic. Similarly, a selection of resistant and susceptible organisms should be inoculated onto each batch of susceptibility test medium, including a strain of *M. tuberculosis* susceptible to all antimycobacterial agents, strains resistant to some of the agents, and a selected strain of *M. kansasii* that is known to be resistant to the low concentration of isoniazid but susceptible to the high concentration.

In addition to internal quality control measures, including the use of authenticated strains of *Mycobacterium* species, laboratories should participate in mycobacterial proficiency-testing surveys. Surveys are extremely helpful in providing feedback on the ability of the laboratory to isolate and identify mycobacteria and to test for susceptibility to antimycobacterial drugs. The surveys also provide the opportunity to compare the capabilities of the laboratory with those of its peers. Proficiency surveys are usually designed for laboratories to participate to the extent or level of service normally offered by that laboratory. This flexibility provides an opportunity for small laboratories to improve their ability to work with different species and further develop their ability to offer more advanced services without placing them at risk for licensure during a training period. Such surveys also provide a good opportunity to obtain strains of mycobacteria showing specific characteristics, e.g., unusual species or those with highly selective drug resistance patterns.

ANIMAL TESTING

Animal inoculation for the isolation of *M. tuberculosis* is no longer as important as it was in the past. Although the use of animals is still a very important part of research studies, animals are seldom used in clinical laboratories today. For details on the use of animals in the mycobacteriology laboratory, refer to the third edition of this Manual (69).

LITERATURE CITED

1. **Armed Forces Institute of Pathology.** 1957. Manual of histologic and special staining technics. Armed Forces Institute of Pathology, Washington, D.C.
2. **Bailey, W. C., R. K. Albert, P. T. Davidson, L. S. Farer, J. Glassroth, E. Kendig, R. G. Loudon, and L. S. Inselman.** 1983. Treatment of tuberculosis and other mycobacterial diseases. Am. Rev. Respir. Dis. **127:**790–796.
3. **Bailey, W. C., J. B. Bass, J. E. Hawkins, G. P. Kubica, and R. S. Wallace.** 1984. Drug susceptibility testing for mycobacteria. Am. Thorac. Soc. News **10:**9–10.
4. **Beam, E. R., and G. P. Kubica.** 1968. Stimulatory ef-

fects of carbon dioxide on the primary isolation of tubercle bacilli on agar containing medium. Am. J. Clin. Pathol. **50:**395–397.

5. **Benjamin, R. G., and T. M. Daniel.** 1982. Serodiagnosis of tuberculosis using the enzyme-linked immunoabsorbent assay (ELISA) of antibody to *Mycobacterium tuberculosis* antigen 5^{1-3}. Am. Rev. Respir. Dis. **126:**1013–1016.
6. **Binford, C. H.** 1977. Leprosy, p. 415. *In* W. A. D. Anderson and J. M. Kissave (ed.), Pathology, 7th ed. C. V. Mosby Co., St. Louis.
7. **Blacklock, Z. M., D. J. Dawson, D. W. Kane, and D. McEvoy.** 1983. *Mycobacterium asiaticum* as a potential pulmonary pathogen for humans. Am. Rev. Respir. Dis. **127:**241–244.
8. **Bönicke, R.** 1958. Die Differenzierung humaner und boviner Tuberkelterien mit Hilfe von Thiophen-2-carbonsaure-hydrazid. Naturwissenschaften **46:**392–393.
9. **Brennan, P. J., M. Heifets, and B. P. Ullom.** 1982. Thin-layer chromatography of lipid antigens as a means of identifying nontuberculous mycobacteria. J. Clin. Microbiol. **15:**447–455.
10. **Butler, W. R., and J. O. Kilburn.** 1982. Improved method for testing susceptibility of *Mycobacterium tuberculosis* to pyrazinamide. J. Clin. Microbiol. **16:**1106–1109.
11. **Carr, D. T., A. G. Karlson, and G. G. Stilwell.** 1967. A comparison of cultures of induced sputum and gastric washings in the diagnosis of tuberculosis. Mayo Clin. Proc. **42:**23–25.
12. **Casal, M. J., and F. C. Rodriguez.** 1981. Simple, new test for rapid differentiation of the *Mycobacterium fortuitum* complex. J. Clin. Microbiol. **13:**989–990.
13. **Centers for Disease Control.** 1983. Primary resistance to antituberculous drugs. Morbid. Mortal. Weekly Rep. **32:**521–523.
14. **Corper, H. J., and N. Uyei.** 1930. Oxalic acid as a reagent for isolating tubercle bacilli and a study of the growth of acid-fast nonpathogens on different mediums with their reactions to chemical reagents. J. Lab. Clin. Med. **15:**348–369.
15. **Damato, J. J., M. T. Collins, M. V. Rothlauf, and J. K. McClatchy.** 1983. Detection of mycobacteria by radiometric and standard plate procedures. J. Clin. Microbiol. **17:**1066–1073.
16. **Davidson, P. T.** 1976. *Mycobacterium szulgai,* a new pathogen causing infection in the lung. Chest **69:**799–801.
17. **Davis, B. R., J. Brumbach, W. J. Sanders, and E. Wolinsky.** 1982. Skin lesions caused by *Mycobacterium haemophilum.* Ann. Intern. Med. **97:**723–724.
18. **Dixon, J. M. S., and E. H. Cuthbert.** 1967. Isolation of tubercle bacilli from uncentrifuged sputum on pyruvic acid medium. Am. Rev. Respir. Dis. **96:**119–122.
19. **Dunbar, F. P., I. Pejovic, R. Cacciatore, L. Peric-Golia, and E. H. Runyon.** 1969. *Mycobacterium intracellulare*: maintenance of pathogenicity in relationship to lyophilization and colony form. Scand. J. Respir. Dis. **49:**153–162.
20. **Dutt, A. K., and W. W. Stead.** 1982. Present chemotherapy for tuberculosis. J. Infect. Dis. **146:**698–704.
21. **Edwards, L. B., L. Hopwood, and C. E. Palmer.** 1965. Identification of mycobacterial infections. Bull. W.H.O. **33:**405–412.
22. **Good, R. C., and D. E. Snider.** 1982. Isolation of nontuberculous mycobacteria in the United States, 1980. J. Infect. Dis. **146:**829–833.
23. **Greene, J. B., G. S. Sidhu, S. Leasin, S. F. Levine, H. Masur, M. S. Simberkoff, P. Nicholas, R. C. Good, S. B. Zolla-Pazner, A. A. Pollock, M. C. Tapper, and R. S. Holzman.** 1982. *Mycobacterium avium-intracellulare*: a cause of disseminated life threatening infection in homosexuals and drug abusers. Ann. Intern. Med. **97:**539–546.
24. **Grosset, J.** 1980. Bacteriologic basis of short-course chemotherapy for tuberculosis. Clin. Chest Med. **1:**231–241.
25. **Gruft, H.** 1971. Isolation of acid-fast bacilli from contaminated specimens. Health Lab. Sci. **8:**79–82.
26. **Guerrant, G. O., M. A. Lambert, and C. W. Moss.** 1981. Gas-chromatographic analysis of mycolic acid cleavage products in mycobacteria. J. Clin. Microbiol. **13:**899–907.
27. **Hawkins, J. E.** 1977. Scotochromogenic mycobacteria which appear intermediate between *Mycobacterium avium-intracellulare* and *Mycobacterium scrofulaceum.* Am. Rev. Respir. Dis. **116:**963–964.
28. **Jones, W. D., and G. P. Kubica.** 1964. The use of MacConkey's agar for differential typing of *Mycobacterium fortuitum.* Am. J. Med. Technol. **30:**187–195.
29. **Jones, W. D., Jr.** 1979. Further studies of mycobacteriophage 33D (Warsaw) for differentiation of BCG from *M. bovis* and *M. tuberculosis.* Tubercle **60:**55–58.
30. **Jones, W. D., Jr., R. C. Good, N. J. Thompson, and G. D. Kelly.** 1982. Bacteriophage types of *Mycobacterium tuberculosis* in the United States. Am. Rev. Respir. Dis. **125:**640–643.
31. **Jones, W. D., Jr., and C. L. Woodley.** 1983. Phage type patterns of *Mycobacterium tuberculosis* from Southeast Asian immigrants. Am. Rev. Respir. Dis. **127:**348–349.
32. **Kalish, S. B., R. C. Radin, J. P. Phair, D. Levitz, C. R. Zeiss, and E. Metzger.** 1983. Use of an enzyme-linked immunosorbent assay technique in the differential diagnosis of active pulmonary tuberculosis in humans. J. Infect. Dis. **147:**523–530.
33. **Kestle, D. G., V. D. Abbott, and G. P. Kubica.** 1967. Differential identification of mycobacteria. II. Subgroups of Groups II and III (Runyon) with different clinical significance. Am. Rev. Respir. Dis. **95:**1041–1052.
34. **Kilburn, J. O., and G. P. Kubica.** 1968. Reagent-impregnated paper strips for detection of niacin. Am. J. Clin. Pathol. **50:**53–532.
35. **Kilburn, J. O., K. F. O'Donnell, V. A. Silcox, and H. L. David.** 1973. Preparation of a stable mycobacterial Tween hydrolysis test substrate. Appl. Microbiol. **26:**826.
36. **Kilburn, J. O., V. A. Silcox, and G. P. Kubica.** 1969. Differential identification of mycobacteria. V. The tellurite reduction test. Am. Rev. Respir. Dis. **99:**94–100.
37. **Kilburn, J. O., K. D. Stottmeier, and G. P. Kubica.** 1968. Aspartic acid as a precursor for niacin synthesis by tubercle bacilli grown on 7H10 agar medium. Am. J. Clin. Pathol. **50:**582–586.
38. **Kim, T. C., R. S. Blackman, K. M. Heatwole, T. Kim, and D. F. Rochester.** 1984. Acid-fast bacilli in sputum smears of patients with pulmonary tuberculosis. Am. Rev. Respir. Dis. **129:**264–268.
39. **Kopanoff, D. E., J. O. Kilburn, J. L. Glassroth, D. E. Snider, L. S. Farer, and R. C. Good.** 1978. A continuing survey of tuberculosis primary drug resistance in the United States: March 1975 to November 1977. Am. Rev. Respir. Dis. **118:**835–842.
40. **Krasnow, I., and G. C. Kidd.** 1965. The effect of a buffer wash of sputum sediments digested with Zephiran trisodium phosphate on the recovery of acid-fast bacilli. Am. J. Clin. Pathol. **44:**238–240.
41. **Krasnow, I., and L. G. Wayne.** 1969. Comparison of methods for tuberculosis bacteriology. Appl. Microbiol. **18:**915–917.
42. **Kubica, G. P., and G. L. Pool.** 1960. Studies on the catalase activity of acid-fast bacilli. I. An attempt to subgroup these organisms on the basis of their catalase activities at different temperatures and pH. Am. Rev. Respir. Dis. **81:**387–391.
43. **Kubica, G. P., and A. L. Rigdon.** 1961. The arylsulfatase activity of acid-fast bacilli. III. Preliminary investigation of rapidly growing acid-fast bacilli. Am. Rev. Respir. Dis. **83:**737–740.

44. **Kuze, F., A. Mitsuoka, W. Chiba, Y. Shimizu, M. Ito, T. Teramatsu, N. Maekawa, and Y. Suzuki.** 1983. Chronic pulmonary infection caused by *Mycobacterium terrae* complex: a resected case. Am. Rev. Respir. Dis. **128:**561–565.
45. **Kwan, K. L., K. D. Stottmeier, I. H. Sherman, and W. R. McCabe.** 1984. Mycobacterial cervical lymphadenopathy: relation of etiologic agents to age. J. Am. Med. Assoc. **251:**1286–1288.
46. **Larsson, L., and P.-A. Mårdh.** 1976. Gas chromatographic characterization of mycobacteria: analysis of fatty acids and trifluoroacetylated whole-cell methanolysates. J. Clin. Microbiol. **3:**81–85.
47. **Larsson, L., and P. A. Mårdh.** 1980. Detection of tuberculostearic acid in biological specimens by means of glass capillary gas chromatography: electron and chemical ionization mass spectrometry, utilizing selected ion monitoring. J. Chromatogr. **182:**402–408.
48. **Laszlo, A., P. Gill, V. Handzel, M. M. Hodgkin, and D. M. Helbecque.** 1983. Conventional and radiometric drug susceptibility testing of *Mycobacterium tuberculosis* complex. J. Clin. Microbiol. **18:**1335–1339.
49. **Laszlo, A., and S. H. Siddiqi.** 1984. Evaluation of a rapid radiometric differentiation test for the *Mycobacterium tuberculosis* complex by selective inhibition with *p*-nitro-α-acetylamino-β-hydroxypropiophenone. J. Clin. Microbiol. **19:**694–698.
50. **Lévy-Frébault, V., M. Daffé, K. S. Goh, M.-A. Lanéelle, C. Asselineau, and H. L. David.** 1983. Identification of *Mycobacterium fortuitum* and *Mycobacterium chelonei*. J. Clin. Microbiol. **17:**744–752.
51. **Macher, A. M., J. A. Kovacs, V. Gill, G. Roberts, J. Ames, C. H. Park, S. Straus, H. C. Lane, J. E. Parrillo, A. S. Fauci, and H. Masur.** 1983. Bacteremia due to *Mycobacterium avium-intracellulare* in the acquired immunodeficiency syndrome. Ann. Intern. Med. **99:**762–765.
52. **Magnusson, M.** 1961. Specificity of mycobacterial sensitins. I. Studies on guinea pigs with purified "tuberculin" prepared from mammalian and avian tubercle bacilli, *Mycobacterium balnei* and other acid-fast bacilli. Am. Rev. Respir. Dis. **83:**57–68.
53. **Marks, J., and T. Szulga.** 1965. Thin-layer chromatography of mycobacterial lipids as an aid to classification: technical procedures. *Mycobacterium fortuitum*. Tubercle **46:**400–411.
54. **McClatchy, J. K., R. F. Waggoner, W. Kanes, M. S. Cernick, and T. L. Bolton.** 1976. Isolation of mycobacteria from clinical specimens by use of selective 7H11 medium. Am. J. Clin. Pathol. **65:**412–416.
55. **Middlebrook, G., Z. Reggiardo, and W. D. Tigert.** 1977. Automatable radiometric detection of growth of *Mycobacterium tuberculosis* in selective media. Am. Rev. Respir. Dis. **115:**1066–1069.
56. **Miliner, R. A., K. D. Stottmeier, and G. P. Kubica.** 1969. Formaldehyde: a photothermal activated toxic substance produced in Middlebrook 7H10 medium. Am. Rev. Respir. Dis. **99:**603–607.
57. **Morgan, M. A., C. D. Horstmeier, D. R. DeYoung, and G. D. Roberts.** 1983. Comparison of a radiometric method (BACTEC) and conventional culture media for recovery of mycobacteria from smear-negative specimens. J. Clin. Microbiol. **18:**384–388.
58. **Murphy, D. B., and J. E. Hawkins.** 1975. Use of urease test disks in the identification of mycobacteria. J. Clin. Microbiol. **1:**465–468.
59. **Palmer, C. E., and L. B. Edwards.** 1968. Identifying the tuberculous infected: the dual test technique. J. Am. Med. Assoc. **205:**167–169.
60. **Petran, E. L., and H. D. Vera.** 1971. Media for selective isolation of mycobacteria. Health Lab. Sci. **8:**225–230.
61. **Quigley, H. S., and H. R. Elston.** 1970. Nitrite strips for detection of nitrate reduction by mycobacteria. Am. J. Clin. Pathol. **53:**663–665.
62. **Reiner, E.** 1965. Identification of bacterial strains by pyrolysis-gas-liquid chromatography. Nature (London) **206:**1272–1274.
63. **Reves, R., D. Blakey, D. E. Snider, Jr., and L. S. Farer.** 1981. Transmission of multiple drug resistant tuberculosis: report of a school and community outbreak. Am. J. Epidemiol. **113:**423–435.
64. **Richards, O. W., and D. K. Miller.** 1941. An efficient method for the identification of tuberculosis bacteria with a simple fluorescence microscope. Am. J. Clin. Pathol. Tech. Suppl. **5:**1–8.
65. **Roberts, G. D., N. L. Goodman, L. Heifets, H. W. Larsh, T. H. Lindner, J. K. McClatchy, M. R. McGinnis, S. H. Siddiqi, and P. Wright.** 1983. Evaluation of the BACTEC radiometric method for recovery of mycobacteria and drug susceptibility testing of *Mycobacterium tuberculosis* from acid-fast smear-positive specimens. J. Clin. Microbiol. **18:**689–696.
66. **Rose, H. D., G. J. Dorff, M. Lauwasser, and N. K. Sheth.** 1982. Pulmonary and disseminated *Mycobacterium simiae* infection in humans. Am. Rev. Respir. Dis. **126:**1110–1113.
67. **Runyon, E. H.** 1959. Anonymous mycobacteria in pulmonary disease. Med. Clin. N. Am. **43:**273–290.
68. **Runyon, E. H.** 1972. Identification of acid-fast pathogens utilizing colony characteristics, 3rd ed. Veterans Administration, Salt Lake City.
69. **Runyon, E. H., A. G. Karlson, G. P. Kubica, and L. G. Wayne.** 1980. Mycobacterium. *In* E. H. Lennette, A. Balows, W. L. Hausler, Jr., and J. P. Truant (ed.), Manual of clinical microbiology, 3rd ed. American Society for Microbiology, Washington, D.C.
70. **Runyon, E. H., M. J. Selin, and H. W. Harris.** 1959. Distinguishing mycobacteria by the niacin test. Am. Rev. Tuberc. Pulm. Dis. **79:**663–665.
71. **Ryan, C. G., and B. W. Dwyer.** 1983. New characteristics of *Mycobacterium haemophilum*. J. Clin. Microbiol. **18:**976–977.
72. **Schröder, K. H., and I. Juhlin.** 1977. *Mycobacterium malmoense* sp. nov. Int. J. Syst. Bacteriol. **27:**241–246.
73. **Siddiqi, S. H., J. P. Libonati, and G. Middlebrook.** 1981. Evaluation of a rapid radiometric method for drug susceptibility testing of *Mycobacterium tuberculosis*. J. Clin. Microbiol. **13:**908–912.
74. **Silcox, V. A., R. C. Good, and M. M. Floyd.** 1981. Identification of clinically significant *Mycobacterium fortuitum* complex isolates. J. Clin. Microbiol. **14:**686–691.
75. **Smithwick, R. W., and C. B. Stratigos.** 1981. Acid-fast microscopy on polycarbonate membrane filter sputum sediments. J. Clin. Microbiol. **13:**1109–1113.
76. **Smithwick, R. W., C. B. Stratigos, and H. L. David.** 1975. Use of cetylpyridinium chloride and sodium chloride for the decontamination of sputum specimens that are transported to the laboratory for the isolation of *Mycobacterium tuberculosis*. J. Clin. Microbiol. **1:**411–413.
77. **Snider, D. E., Jr., R. C. Good, J. O. Kilburn, L. F. Laskowski, R. H. Lusk, J. J. Marr, Z. Reggiardo, and G. Middlebrook.** 1981. Rapid drug susceptibility testing of *Mycobacterium tuberculosis*. Am. Rev. Respir. Dis. **123:**402–406.
78. **Sompolinsky, D., A. Lagziel, D. Naveh, and T. Yankilevitz.** 1978. *Mycobacterium haemophilum* sp. nov., a new pathogen of humans. Int. J. Syst. Bacteriol. **28:**67–75.
79. **Steadham, J. E.** 1979. Reliable urease test for identification of mycobacteria. J. Clin. Microbiol. **10:**134–137.
80. **Swenson, J. M., C. Thornsberry, and V. A. Silcox.** 1982. Rapidly growing mycobacteria: testing of susceptibility of 34 antimicrobial agents by broth microdilution. Antimicrob. Agents Chemother. **22:**186–192.
81. **Takahashi, H., and V. Foster.** 1983. Detection and recovery of mycobacteria by a radiometric procedure. J. Clin. Microbiol. **17:**380–381.
82. **Tisdall, P. A., D. R. DeYoung, G. D. Roberts, and J. P.**

Anholt. 1982. Identification of clinical isolates of mycobacteria with gas-liquid chromatography: a 10-month follow-up study. J. Clin. Microbiol. **16:**400–402.

83. **Toda, T., Y. Hagihara, and K. Takeya.** 1960. A simple urease test for the classification of mycobacteria. Am. Rev. Respir. Dis. **83:**757–761.
84. **Truant, J. P., W. A. Brett, and W. Thomas, Jr.** 1962. Fluorescence microscopy of tubercle bacilli stained with auramine and rhodamine. Henry Ford Hosp. Med. Bull. **10:**287–296.
85. **Tsukamura, M.** 1966. Adasonian classification of mycobacteria. J. Gen. Microbiol. **45:**253–273.
86. **Vestal, A. L.** 1975. Procedures for the isolation and identification of mycobacteria. U.S. Public Health Service Publ. 75-8230, Center for Disease Control, Atlanta.
87. **Virtanen, S.** 1960. A study of nitrate reduction by mycobacteria. Acta Tuberc. Scand. Suppl. **48:**1–119.
88. **Wallace, R. J., Jr., J. R. Dalovisio, and G. A. Pankey.** 1979. Disk diffusion testing of susceptibility of *Mycobacterium fortuitum* and *Mycobacterium chelonei* to antibacterial agents. Antimicrob. Agents Chemother. **16:**611–614.
89. **Wallace, R. J., Jr., J. M. Swenson, V. A. Silcox, and R. C. Good.** 1982. Disk diffusion testing with polymyxin and amikacin for differentiation of *Mycobacterium fortuitum* and *Mycobacterium chelonei*. J. Clin. Microbiol. **16:**1003–1006.
90. **Wallace, R. J., Jr., J. M. Swenson, V. A. Silcox, R. C. Good, J. A. Tschen, and M. S. Stone.** 1983. Spectrum of disease due to rapidly growing mycobacteria. Rev. Infect. Dis. **5:**657–679.
91. **Warren, N. G., B. A. Body, and H. P. Dalton.** 1983. An improved reagent for mycobacterial nitrate reductase tests. J. Clin. Microbiol. **18:**546–549.
92. **Wayne, L. G.** 1957. The use of millipore filters in clinical laboratories. Am. J. Clin. Pathol. **28:**565–567.
93. **Wayne, L. G.** 1961. Recognition of *Mycobacterium fortuitum* by means of the three-day phenophthalein sulfatase test. Am. J. Clin. Pathol. **36:**185–187.
94. **Wayne, L. G.** 1962. Two varieties of *Mycobacterium kansasii* with different clinical significance. Am. Rev. Respir. Dis. **86:**651–656.
95. **Wayne, L. G.** 1974. Simple pyrazinamidase and urease tests for routine identification of mycobacteria. Am. Rev. Respir. Dis. **109:**147–151.
96. **Wayne, L. G., and J. R. Doubek.** 1964. The role of air in the photochromogenic behaviour of *Mycobacterium kansasii*. Am. J. Clin. Pathol. **42:**431–435.
97. **Wayne, L. G., and J. R. Doubek.** 1968. Diagnostic key to mycobacteria encountered in clinical laboratories. Appl. Microbiol. **16:**925–931.
98. **Wayne, L. G., J. R. Doubek, and R. L. Russell.** 1964. Tests employing Tween 80 as substrate. Am. Rev. Respir. Dis. **90:**588–597.
99. **Wayne, L. G., and I. Krasnow.** 1966. Preparation of tuberculosis susceptibility testing mediums by means of impregnated discs. Am. J. Clin. Pathol. **45:**769–771.
100. **Wolinsky, E.** 1959. Nontuberculous mycobacteria and associated diseases. Am. Rev. Respir. Dis. **119:**107–159.
101. **Young, W. D., Jr., A. Maslansky, M. S. Lefar, and D. P. Kronish.** 1970. Development of a paper strip test for detection of niacin produced by mycobacteria. Appl. Microbiol. **20:**939–945.

Chapter 23

Aerobic Pathogenic *Actinomycetaceae*

MORRIS A. GORDON

The aerobic pathogenic actinomycetes are branched filamentous bacteria related to the mycobacteria and to saprophytic *Streptomyces* species. Some species tend to fragment into bacillary and coccoid forms, and some form aerial chains of either conidia or arthrospores. They are gram positive, but some tend to stain irregularly; some species are weakly and partially acid-fast. Their cell wall constitution is of type I, III, or IV, according to genus and species (8). *Oerskovia turbata*, an actinomycete characterized by a soft yellow colony and fragmentation of branched substrate hyphae into flagellated motile elements, has a cell wall of type VI and is in many other respects an exception within this group. It will therefore not be further characterized here. A description of this and other *Oerskovia* species is given by Lechevalier (31). *O. turbata* has been reported as an agent of bacterial endocarditis (43), and an *Oerskovia* sp. was cultured from a kidney abscess (13).

NOCARDIA, NOCARDIOPSIS, ACTINOMADURA, AND *STREPTOMYCES* SPECIES

Species to be considered are *Nocardia asteroides, Nocardia brasiliensis, Nocardia caviae, Actinomadura madurae, Actinomadura pelletieri, Nocardiopsis dassonvillei, Streptomyces paraguayensis*, and *Streptomyces somaliensis*.

Characterization

Species of the genera *Nocardia, Nocardiopsis, Actinomadura*, and *Streptomyces* form colonies varying from glabrous and waxy, raised or heaped, and variously pigmented to densely mycelial, white, tough, and moldlike. Growth rates are comparable to those of the rapidly growing mycobacteria; colonies of *N. asteroides* and *O. turbata* are usually apparent within 3 days, but others may take 1 week or more to develop. The optimal temperature for growth is 30 to 37°C, with a range of 10 to 50°C, which varies with the species. All species are nonencapsulated. All are catalase positive. The species are differentiated biochemically by their action upon casein, tyrosine, xanthine, and starch agar plates, urea agar slants, and tubes of gelatin and bromocresol purple milk, by their sensitivity to lysozyme, and by their ability to produce acid from certain sugars (Table 1).

Cell wall analysis. The four genera in this group can be separated into cell wall types I, III, and IV by the presence in or absence from their cell walls of glycine, arabinose, galactose, and either of two isomeric forms of 2,6-diaminopimelic acid. Type III is further divided by the presence or absence of madurose (Table 1). (Mycobacteria have cell walls of type IV.)

Clinical significance

The spectrum of disease encompasses severe, suppurative or cavitary, pulmonary infection, often simulating tuberculosis (most commonly *N. asteroides*, but also *N. brasiliensis* and *N. caviae*); cutaneous and subcutaneous abscesses (*N. asteroides* and *N. brasiliensis*); bloodstream invasion with secondary, often fatal, involvement of meninges and brain, kidney, and other organs (primarily *N. asteroides*); and actinomycetic mycetoma, i.e., swollen, indurated lesions which eventually involve the bone and which discharge granules in pus from sinuses (*N. asteroides* only rarely) (34). *N. asteroides* is also an occasional agent of keratitis (30). The etiologic agents are part of the normal soil microflora, and infection normally is acquired either by the inhalation of contaminated dust or by direct, traumatic contact with contaminated soil or vegetation. Person-to-person transmission may also occur (see below). Pulmonary colonization by *N. asteroides* may occur in the presence of underlying disease or defective airway clearance. Among animals, *N. asteroides* infection is most common in cattle and dogs, but it has been reported in various other mammals as well as in birds and fish (39).

As a result of a survey in 1976 (5), it was estimated that 500 to 1,000 human cases of nocardiosis are recognized each year in the United States. Of these, 85% are serious pulmonary or systemic infections, and of these serious infections, between 8.6 and 18.8% are caused by *Nocardia* species other than *N. asteroides*. The survey confirmed that nocardial infections are usually opportunistic in the human host but found that at least 15% of the cases in that series occurred in patients with no known predisposing condition. As has been observed in other surveys, male patients outnumbered female patients by 3:1.

Pulmonary or generalized nocardiosis has become increasingly prevalent, especially in patients with severe underlying diseases, preponderantly neoplastic, who are receiving intensive medical therapy. Particularly implicated are steroids and antineoplastic drugs. In addition to lymphomas and leukemias, the predisposing conditions include pulmonary alveolar proteinosis, chronic pulmonary disease, chronic intestinal disease, cirrhosis of the liver, and severe wounds (38). Severe infections have also occurred in renal and cardiac transplant patients being treated with prednisone and azathioprine. Although nocardiosis has been associated with mycoses in several cases, almost all of the associated mycoses were themselves opportunistic infections (particularly candidiasis and aspergillosis); most of the patients suffered from neoplasms and other disorders mentioned above and had been treated with steroids or cytotoxic agents. There is no substantial evidence of opportunism in bovine mastitis.

Beaman and Sugar (6) reviewed naturally acquired nocardiosis in animals and provided a summary of extensive investigations into the pathogenesis of experimental nocardial infections.

The species in this group are not so geographically limited as some of their names imply. *N. brasiliensis* is most common in Mexico but is also found in the United States and elsewhere; *S. somaliensis* has been

TABLE 1. Physiologic characteristics of pathogenic *Nocardia, Nocardiopsis, Actinomadura,* and *Streptomyces* species[a]

Species	Cell wall type[b]	Decomposition[c] of							Acid from		Resistance to lysozyme
		Casein	Tyrosine	Xanthine	Starch	Gelatin	BCP milk[d]	Urea	Lactose	Xylose	
N. asteroides	IV	−	−	−	−[e]	−[f]	−[g]	+	−	−	+
N. brasiliensis	IV	+	+	−	−[e]	+	+	+	−	−	+
N. caviae	IV	−	−	+	−[e]	−	−[g]	+	−	−	+
Nocardiopsis dassonvillei	IIIC	+	+	+	+	+	+ (rapid)	± (35%+)	−	+	−
A. madurae	IIIB	+	+ 14%−[h]	−	+	+	±	−	± 55%+	+	−
A. pelletieri	IIIB	+	+	−	− 13%+[h]	+	±	−	−	−	−
S. somaliensis	I	+	+	−	±	+	+	−	−	−	−
S. paraguayensis	I	+	+	+	−	+	+	+		−	
Streptomyces spp. (nonpathogenic)	I	+	+	−	+[i]	+[i]	+[i]	±	+	+	−[j]
		+	+	+				50%+			
		−	+	+							

[a] For further information, see references 25, 27, 33, 36, and 37.
[b] Cell walls of all actinomycetes contain glucosamine, muramic acid, glutamic acid, and alanine. In addition, major amounts of the following components are found in the respective groups: (I) L-diaminopimelic acid (DAP) and glycine; (III) *meso*-DAP; and (IV) *meso*-DAP, arabinose, and galactose (8). Whole-cell hydrolysates of subgroup IIIB contain 3-*O-methyl*-D-galactose (madurose), absent in IIIC. *Nocardia* species can be differentiated from *Actinomadura* and *Mycobacterium* species by chromatographic lipid analysis (29) and from the latter by the arylsulfatase test (49).
[c] Within 2 weeks at 27°C.
[d] Bromocresol purple milk.
[e] About 50% of strains positive by a different method (27).
[f] Some strains reportedly liquefy certain gelatin media.
[g] Usually turns alkaline.
[h] See reference 25.
[i] Most species give positive results.
[j] Most *Streptomyces* species are sensitive to lysozyme, as are *Mycobacterium phlei* and *Mycobacterium smegmatis; Mycobacterium fortuitum* and *Mycobacterium marinum* are resistant (26).

found in Mexico as well as in Africa; *A. madurae* is cosmopolitan; and mycetomas, although preponderantly tropical, occur in all parts of the world. Mycetoma is seen more frequently in males than in females and mostly in young adults and in rural areas, apparently reflecting the risk of traumatic exposure to the microorganisms on thorns and splinters and in soil. There are several additional so-called, but inadequately defined, *Nocardia* species which, along with most *Streptomyces* species, are not known to incite human or animal disease and are not ordinarily found in clinical specimens. These species are not treated here. Aside from the acid-fastness of the three *Nocardia* species shown in Table 2, there is no general method for distinguishing disease-related from nonpathogenic species of the two genera.

Nocardiosis is generally considered to be noncontagious among humans, although herd infections among dairy cattle indicate that it is passed from animal to animal by mechanical vectors (42). Development of a typing scheme, with serum from hyperimmune rabbits in an immunodiffusion test (40), made epidemiologic studies possible and showed that most isolates from each of several outbreaks were of a single serotype (41). Subsequently, an epidemic of seven cases of nocardiosis within 9 months among renal transplant patients in a London hospital afforded an opportunity to apply this epidemiologic tool to human patients, with results that suggest transmission from a common source (48). Serotyping, together with physiologic characterization of the *N. asteroides* isolates, indicated that isolates from patients and from the unit environment were identical but they were different from some of those obtained outside the unit. Epidemiologic studies suggested that the probable source was a patient with pulmonary nocardiosis who had been admitted to the intensive care area in which the six subsequent cases developed.

Collection of specimens

Types of specimens. For pulmonary cases, fresh, single (not cumulative) specimens of sputum or bronchial washings should be submitted in sterile screwcap containers. Volumes must be adequate for both direct microscopic examination and culture. Biopsy and autopsy specimens, exudate, pus, and scrapings from ulcers should be submitted in tightly stoppered, sterile vials, tubes, or bottles; desiccation of these specimens should be avoided. Swabs are not recommended, but if there is no alternative, they should be premoistened with sterile water and submitted in a sealed tube. They should not be coated with additives, such as charcoal, that may obscure direct examination. Spinal fluid or blood specimens for culture should be submitted in sterile sealed tubes.

Transport and storage. Specimens should be conveyed to the examining laboratory as quickly as possible. Specimens in storage or delayed in transit should be kept at approximately 4°C. Preserving fluids are not recommended. If the laboratory is distant, isolation media are best inoculated locally.

TABLE 2. Morphology and staining of pathogenic *Nocardia, Nocardiopsis, Actinomadura*, and *Streptomyces* species

Species	Appearance in host tissue[a]		Sabouraud dextrose agar culture (27–37°C)	Acid-fast
	Pus	Stained sections		
N. asteroides	Granules small (25–150 μm), soft, white to yellowish, lobulated, sometimes clubbed; occurring only rarely, generally in mycetomas; usually occurs as clumped or scattered, acid-fast, branched filaments, undergoing fragmentation (Fig. 13)	Granules rare, irregularly oval, staining lightly with hematoxylin, often with a distinct eosinophilic periphery, fringed and sometimes clubbed; Gram stain gives a similar pattern (Fig. 5); more commonly, either colonies of a loose mycelium or scattered, fragmenting filaments sharply delineated by Grocott method (Fig. 4) but poorly or not at all by H & E[b]	Colonies orange, glabrous, heaped and folded (Fig. 6) to white or pink, raised and chalky with aerial hyphae (Fig. 7); crumbly or leathery and adherent; hyphae fragment into bacillary and coccoid elements (Fig. 14); chains of arthroconidia-like elements (Fig. 16) occasionally form on aerial hyphae; grows at 46°C	+
N. brasiliensis	As above, but granules are common in mycetomas; branched filaments in cutaneous abscesses	As above, but granules are common in mycetomas	As above; no growth at 46°C	+
N. caviae	As in *N. brasiliensis*	As in *N. brasiliensis*	As above; growth at 46°C variable	+
Nocardiopsis dassonvillei	Insufficient information; granules not reported	Insufficient information	Densely filamentous colonies (Fig. 8) (*Streptomyces*-like) with abundant aerial mycelium and long aerial chains of conidia (Fig. 15); vegetative mycelium may fragment to some extent into rods; growth (80% of strains) at 40°C; no growth at 45°C	−
A. madurae	Large (1–5 mm), soft, white to yellowish or reddish granules, irregularly oval, serpiginous, or lobulated	H & E: Center of granule is hollow or tenuous, surrounded by denser network staining dark purple; wide, dense, pink border with long fringes and usually clubs (Fig. 1)	Optimal temperature, 30–37°C; colonies are waxy, heaped and folded (Fig. 9), membranous and tough; white to tan, pale orange, pink or red; nonfragmenting, but sparse aerial hyphae may form short chains of conidia	−
A. pelletieri	Granules soft, small (300–500 μm), deep red, smooth-edged or finely denticulate, irregularly spherical, sometimes with large lobes	H & E: Granules round, sharply delimited, characteristically fracturing into large segments; heavily and homogeneously purple, with a lighter purple peripheral band (Fig. 2)	Optimal temperature, 30–37°C; colony resembles a crushed cranberry, with areas of bright and dark red; slow-growing, heaped irregularly, waxy granular (Fig. 10); may have sparse aerial hyphae; pigment soluble in mineral oil	−
S. somaliensis	Hard, yellow to brown, round to oval granules, 1–2 mm wide	H & E: Granules both large and tiny, round, dense and homogeneous, but staining light purple and often partially pink in patches; tend to rupture into parallel strips; smooth, sharply defined, nonbanded border (Fig. 3)	Optimal temperature, 30°C; slow growing, leathery, eventually heaped and folded; cream colored to brown or black, glabrous or with whitish aerial hyphae (Fig. 12); nonfragmenting; aerial hyphae may form chains of conidia characteristic of the genus	−
S. paraguayensis	Black granules		There is doubt as to the existence of this species as a distinct, pathogenic agent. Cultures designated *N. paraguayensis* resemble saprophytic *Streptomyces* spp. (white or cream to gray, tough colonies with raised center [Fig. 11], often bearing short aerial hyphae)	−
Streptomyces spp.	Nonpathogenic		Colonies tough; glabrous, velvety or (most often) chalky; of various colors, commonly white or grayish; often with earthy odor; nonfragmenting, but specialized branches of aerial mycelium characteristically form medium to long, curved or spiral chains of conidia	− (Spores often +)

[a] See references 33 and 36.
[b] H & E, Hematoxylin and eosin stain.

Direct examination

Wet mount. Pus and other exudates from patients with suspected mycetoma should be diluted in sterile water, if necessary, and examined macroscopically for the presence of granules (microcolonies) to be crushed and stained. These and other body fluids, including sputum, bronchial washings, and the centrifuged sediment from spinal fluid or urine, should be examined microscopically in an unstained wet mount between slide and cover slip, under greatly reduced light or phase contrast, for branching filaments 1 μm or less in diameter. Differentiation of granules according to species is described in Table 2. Branched hyphal filaments of this size may belong to any of these four genera or to *Actinomyces* species.

Gram stain. All forms of these species, whether filamentous, bacillary, or coccoid, are gram positive, but *Nocardia* species tend to stain irregularly, and their filaments are generally beaded (see Fig. 14). Filamentous forms may not be differentiable from those of *Actinomyces* species, and bacillary elements may resemble those of *Actinomyces, Mycobacterium,* or *Corynebacterium* species. Intact granules reveal little detail and should be crushed before staining to expose the component filaments, which would distinguish them from microcolonies of *Staphylococcus* species or other agents of botryomycosis.

Acid-fast stain. Acid-fastness is variable in *N. asteroides, N. brasiliensis,* and *N. caviae,* both in clinical specimens and in culture, depending upon strain, culture conditions, and staining method; but when present and properly controlled, acid-fastness is diagnostic for this group. Individual filaments and fragments in a smear may vary from partially or completely acid-fast (see Fig. 13) to entirely non-acid-fast. The usual Ziehl-Neelsen and Kinyoun staining procedures may be employed, but the period of decolorization with acid alcohol must not exceed 5 to 10 s. Alternatively, 2% H_2SO_4 may be applied for 1 min. Acid-fastness is sometimes enhanced by prolonged growth on proteinaceous media, e.g., on bromocresol purple milk for 1 month.

Tissue sections. The various species may appear in sections of biopsy or autopsy specimens, generally within abscesses, either as granules (in a mycetoma), as radially dendritic colonies, or as scattered fine, branching filaments. Granules show up well in sections stained with hematoxylin and eosin (Fig. 1 through 3), but this stain often fails completely to demonstrate other morphologic forms of *Nocardia* species, even when they are shown by other stains to be present. The Grocott-Gomori methenamine-silver method also stains granules very well and is probably the best for disclosing dendritic colonies (Fig. 4) and individual filaments and fragments. The Gram-Weigert technique is effective for both granules (Fig. 5) and filaments if done properly, as are acid-fast stains for the appropriate species. Periodic acid-Schiff stains are not recommended. (See Table 2 for the differentiation of species by morphologic and staining characteristics of their granules or filaments.)

Culture and isolation

Specimens are planted on duplicate or triplicate Sabouraud dextrose agar slants (without antibiotics) for incubation at 25 to 30°C and streaked on beef heart infusion-blood agar plates for incubation at 37°C. Since clinical manifestations of some forms of nocardiosis and actinomycosis may be indistinguishable from one another, and since branching, gram-positive filaments in tissues and exudates may be those of either aerobic or anaerobic actinomycetes, it is often appropriate to inoculate media for anaerobic culture as well (see Chapters 37 and 41). Beef infusion broth or thioglycolate broth, to be incubated at 37°C, may be inoculated as an enrichment medium. Since *N. asteroides* grows well at 45°C, initial incubation at temperatures above 37°C may help to separate this species from other bacteria. Cultures are examined at intervals from 48 h to 2 weeks or more. *N. asteroides* often survives mycobacterial concentration procedures, including sodium hydroxide methods, and treatment of sputum with dithiothreitol or *N*-acetyl-L-cysteine has been recommended. Species of all four genera grow especially well on Lowenstein-Jensen medium and 7H10 (oleic acid-albumin) agar, and these or yeast-dextrose agar should be used for primary isolation, at 27 to 30°C, of *Actinomadura* spp. Even at these optimal temperatures, *A. madurae* or *A. pelletieri* may develop visible growth only after 4 weeks. It is advisable to wash and crush granules for inoculation of media.

Identification

Cultural characteristics. Some cultural characteristics are shown in Fig. 6 through 12. Glabrous, orange colonies of *Nocardia* species may resemble those of rapidly growing mycobacteria. In questionable cases, the presence of aerial hyphae in the actinomycetes, if not obvious to the naked eye (Fig. 7), may be observed by placing plate cultures (e.g., on tyrosine agar) of mature growth under a compound microscope or by preparing slide cultures (27) with Bennett or Hickey and Tresner medium. Aerial hyphae may be distinguished as arising from the substrate hyphae and appearing wider than the latter, with a heavier, black outline (they are more refractile; see Fig. 18). Their formation may be inhibited by exposure to light. Some strains give off a musty odor ("newly plowed earth") or produce the so-called tyrosinase reaction (clear, brownish discoloration of the medium) or both, but these two properties are typical also of many saprophytic *Streptomyces* species, generally with chalky colonies. The formation of pigment and spores is encouraged by growth on Czapek-Dox solution agar. Growth on liquid media, including bromocresol purple milk, occurs as a thick surface pellicle.

The term *Nocardia farcinica* has historically been associated with isolates from lesions of bovine farcy, formerly prevalent in Guadeloupe, French West Indies, as well as in northern Africa. Cultures so labeled have proven to be of two types, one not readily distinguishable from *N. asteroides* and the other a mycobacterium. Since the original isolate from Guadeloupe, upon which this binomial was based in 1889, has been lost and no one knows which organism was isolated at that time, *N. farcinica* must be regarded as a nomen dubium. The disease has been eradicated from Guadeloupe, and the strains currently being isolated from cases of bovine farcy in Africa are mycobacteria, described by Chamoiseau in 1973 (12) under the generally accepted name of *Mycobacterium*

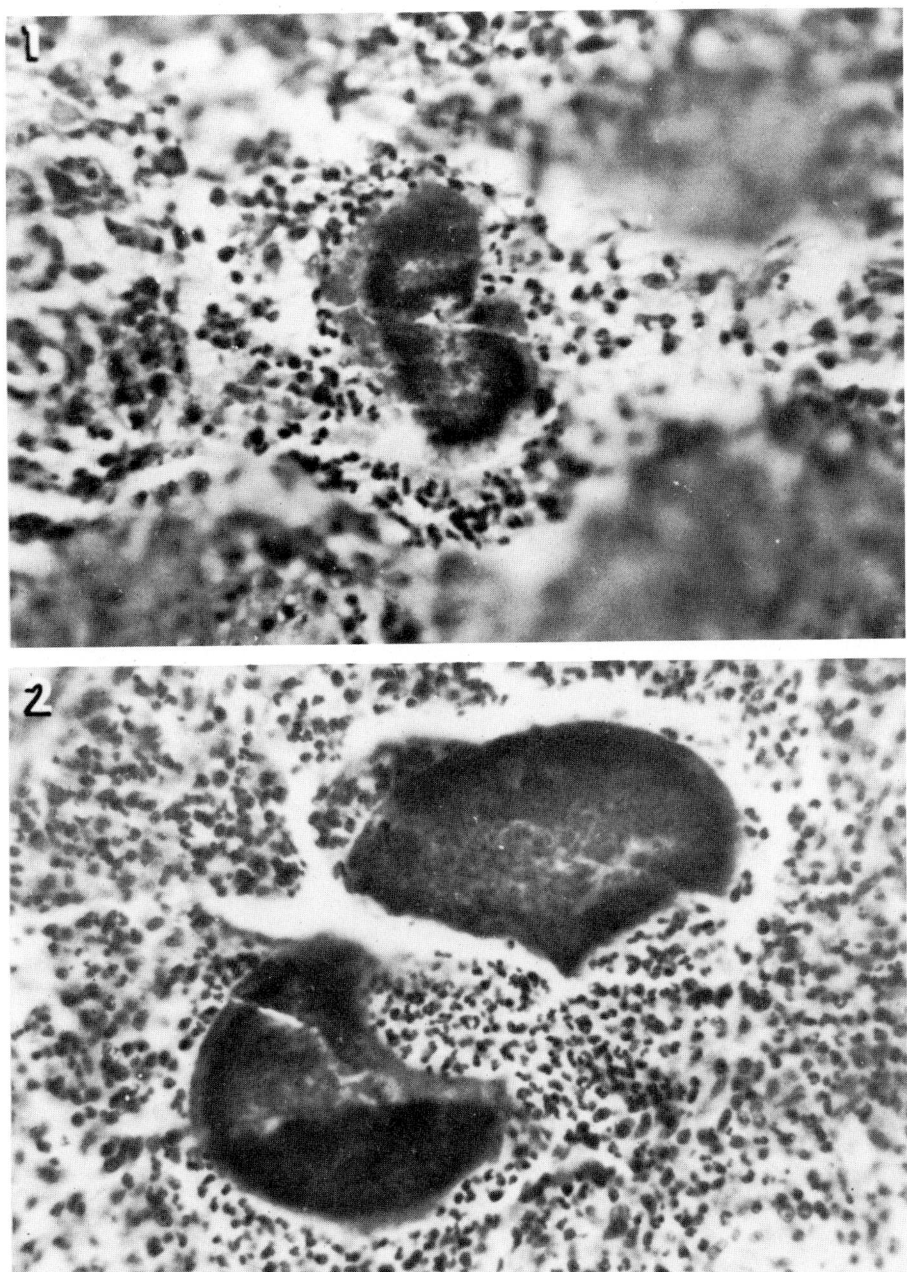

FIG. 1. *A. madurae* in a section from a human mycetoma. Note the tenuous center of the granule, the denser periphery stained with hematoxylin, and the wide, fringed, clubbed eosinophilic border. Hematoxylin and eosin stain. ×378.

FIG. 2. *A. pelletieri* in a section from a human mycetoma. Round granule, tending to break into large lobes; heavily hematoxylin-stained with somewhat lighter border. Hematoxylin and eosin stain. ×378.

farcinogenes. This species resembles *Nocardia* species in being completely mycelial, but is nonfragmenting and fully acid-fast.

Microscopic examination of culture growth. Microscopic examination of cultures (Fig. 13 through 18), whether by wet mount, Gram stain, or acid-fast stain, reveals very fine, branching filaments (less than 1 μm in diameter) with a tendency, in the acid-fast species, to fragment into bacillary and coccoid forms (Fig. 14). *Actinomadura* species and pathogenic *Streptomyces* species, as well as saprophytic species of the latter, tend not to fragment; but the aerial mycelium of most saprophytic *Streptomyces* species and of *Nocardiopsis dassonvillei* is characterized by medium to long chains of conidia (Fig. 15), often in spirals or whorls, which are sometimes acid-fast. Short chains of aerial arthrospores may be seen in some pathogenic strains (Fig. 16), and conidia may be seen in others (Table 2). To detect spontaneous fragmentation, as distinct from breaking up of the mycelium during the preparation of slides (Fig. 17), smears should be made with as little traumatization as possible.

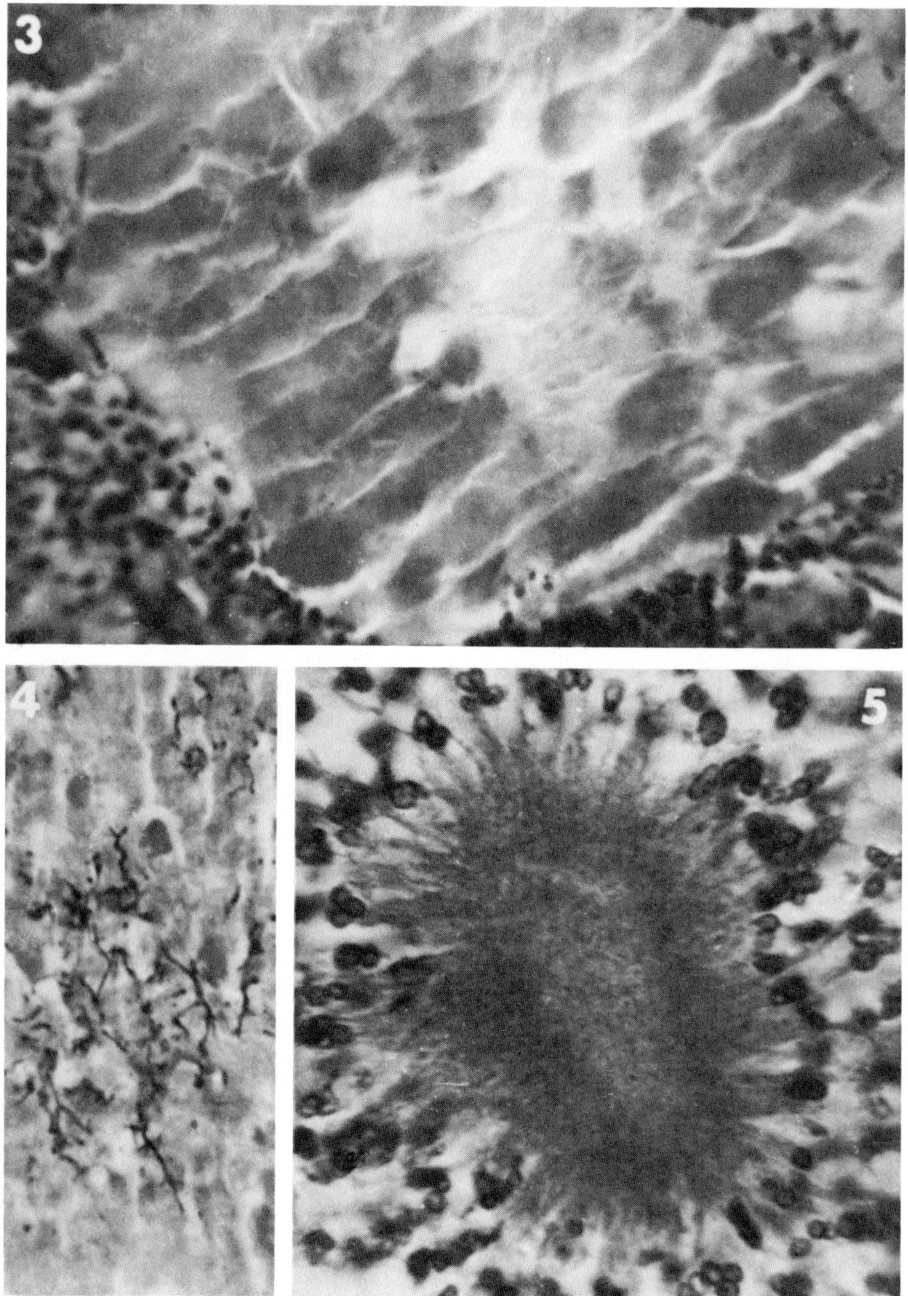

FIG. 3. *S. somaliensis* in a section from a human mycetoma. Round, dense granule, lightly stained with hematoxylin; fracturing into strips. Hematoxylin and eosin stain. ×378.

FIG. 4. *N. asteroides* colonies in a section of human brain. Grocott stain. ×850.

FIG. 5. *N. asteroides* granule in a section of abscessed omentum from an experimentally infected guinea pig. Gram-positive interior; fringed, clubbed, gram-negative border. Gram-Weigert stain. ×850.

Biochemical reactions. Media for the testing of biochemical reactions are incubated at 27 to 30°C and are observed at 2-day intervals for 2 weeks or until good growth occurs. The reactions of the various species are given in Table 1.

Pathogenicity. There is some variation in virulence for guinea pigs among strains of *N. asteroides*, but the species is in general regularly pathogenic when tested in the following manner. All of the material on the surface of four Sabouraud dextrose agar slants, each of which is at least half-covered with heavy growth, is finely emulsified in an equal volume of 5% hog gastric mucin by grinding with a mortar and pestle. Each of two guinea pigs, weighing 250 to 300 g, is injected intraperitoneally with 1 ml of the suspension. The animals generally die within 7 to 10 days and display

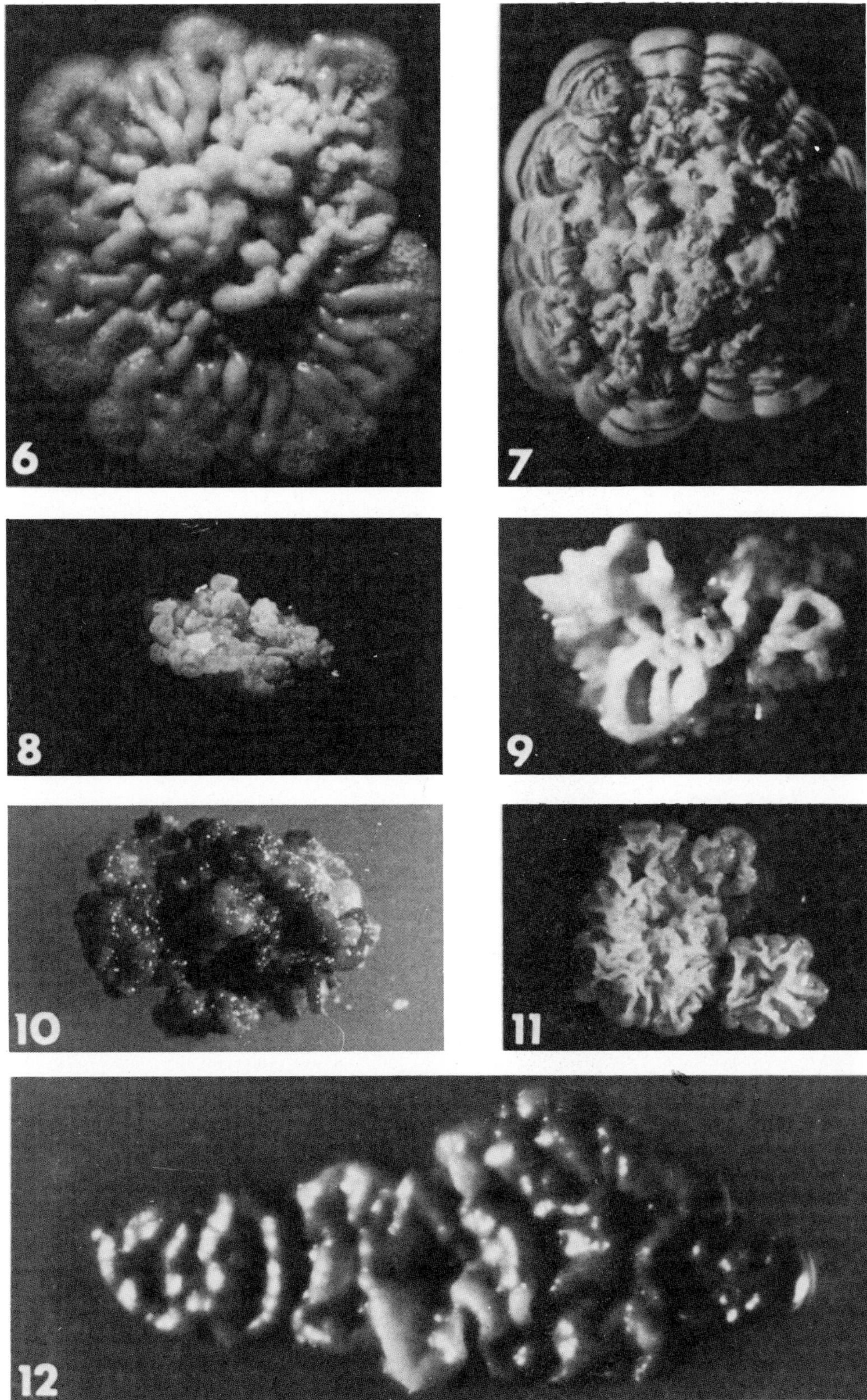

FIG. 6–11. "Giant" colonies of various species on Sabouraud dextrose agar plates, after 9 days at 27°C. ×6. Fig. 6: *N. asteroides*, glabrous orange form. Fig. 7: *N. asteroides*, gypsoides (white, chalky) variety (*N. brasiliensis* and *N. caviae* are similar to *N. asteroides*). Fig. 8: *Nocardiopsis dassonvillei*. Fig. 9: *A. madurae*. Fig. 10: *A pelletieri*. Fig. 11: *S. paraguayensis*.

FIG. 12. *S. somaliensis*, streaked heavily on blood agar slant, after 3 days at 27°C. ×6.

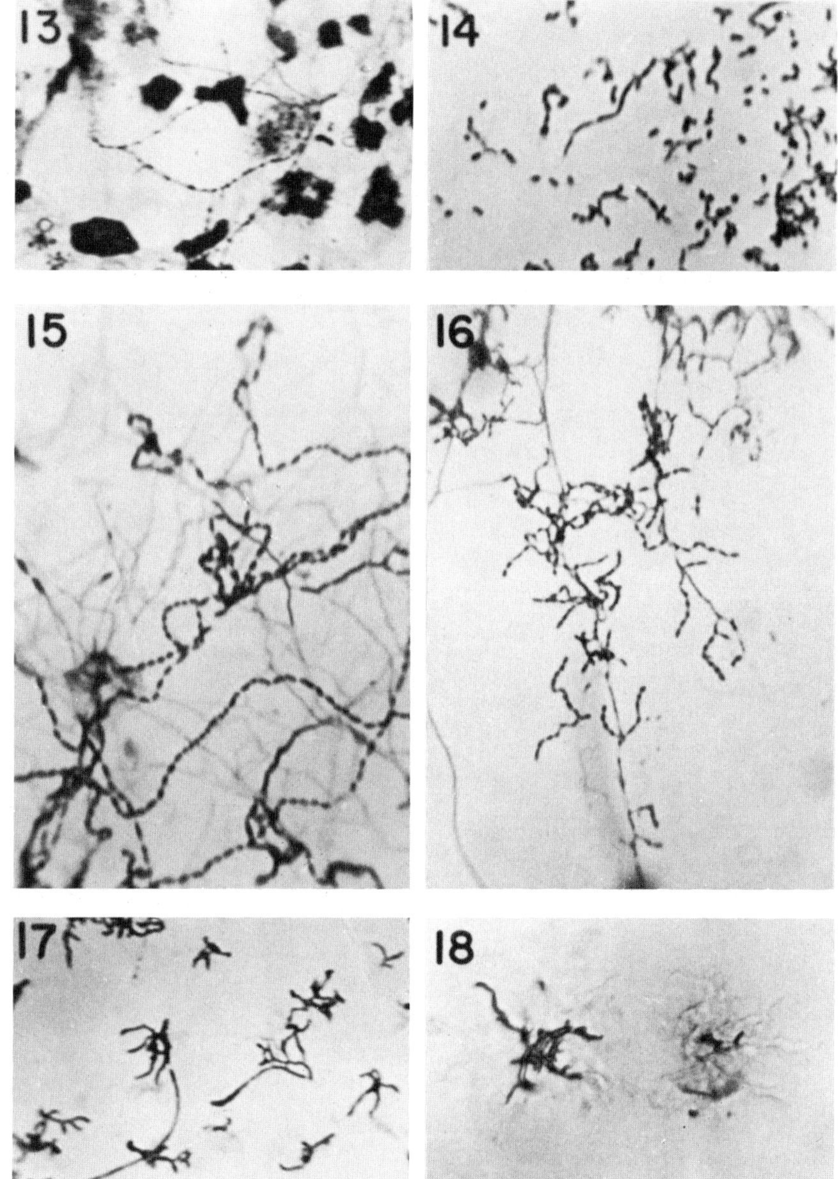

FIG. 13. *N. asteroides* in a smear of abscessed omentum from an experimentally infected guinea pig. Modified Kinyoun acid-fast stain. ×972.

FIG. 14. *N. asteroides* in a smear from a glabrous orange colony on Sabouraud dextrose agar, after 3 weeks at 27°C; extensive fragmentation. Gram stain. ×972.

FIG. 15. *Nocardiopsis dassonvillei*, showing long chains of ovate conidia. Slide culture on Hickey and Tresner medium, incubated for 10 days at 25°C. Giemsa stain. ×972.

FIG. 16. *N. asteroides*, showing chains of aerial arthroconidia. Slide culture on Hickey and Tresner medium. Giemsa stain. ×972.

FIG. 17. *A. pelletieri* smear from a colony on Sabouraud dextrose agar, 27°C; prepared in the manner of a blood film, causing moderate traumatic fragmentation. Gram stain. ×972.

FIG. 18. Pour-slide culture of *N. asteroides* on Hickey and Tresner medium, after 3 days at 27°C; photographed in situ with a 43× objective to illustrate the contrast between aerial (darker) and substrate hyphae. ×432.

more or less massive, often coalescing, caseous abscesses involving the organs of the abdominal cavity, particularly the omentum. If the animals survive this period and are not obviously ill, the abdomen should be palpated for adhesions or hard masses, the presence of which dictates sacrifice of the animal. If neither animal becomes ill, one is sacrificed at 2 weeks and the other at 4 weeks, and the omentum, abdominal viscera, and lungs are examined for lesions.

Smears should be prepared from pus, stained by both Gram and modified acid-fast methods, and examined for fragmenting, gram-positive, partially acid-fast, branching filaments (Fig. 13), which under the circumstances are diagnostic for one of the acid-fast *Nocardia* species. Pus or diseased tissue should also be cultured on Sabouraud dextrose agar at both room temperature and 37°C. A negative result, i.e., with no virulence for the guinea pig, does not rule out *N. asteroides*, since some strains are relatively avirulent under these conditions. Most of those strains, however, will display pathogenicity upon repetition of the test.

Most strains of *N. caviae* and *N. brasiliensis* are also pathogenic for the guinea pig. *Actinomadura* and *Streptomyces* species have been reported as nonpathogenic for this animal. In mice inoculated intraperitoneally, *N. asteroides* tends to induce acute suppurative abscesses, whereas *N. brasiliensis* usually causes formation of granulomata characterized by foamy macrophages.

Immunoserologic diagnosis. *Nocardia* spp. share some of their antigens with mycobacteria, *Actinomyces* spp., and *Streptomyces* spp., as determined by both immunodiffusion and immunofluorescence tests. However, antigenic differences distinguish these genera as well as the three *Nocardia* species. Practical advantage has been taken of these differences primarily in cases of mycetoma, in which immunodiffusion with specific antigens permits identification of the individual actinomycete species and their differentiation from the agents of eumycetoma, a capability which is extremely important in guiding therapy. Complement fixation and counterimmunoelectrophoresis are also used in the diagnosis of mycetomas (35).

Although reliable immunodiffusion tests have long been established for systemic nocardiosis in dogs and nocardial mastitis in cattle, only recently have promising steps been taken in the development of diagnostic serology for human nocardiosis. Two antigens employed in an immunodiffusion test showed sensitivities, based upon 71 cases, of 50 and 70%, but tended to cross-react with sera from patients with actinomycosis and tuberculosis (10). A complement fixation test has been reported as having an 81% sensitivity in a series of 16 cases, with cross-reactions in tuberculosis and leprosy (46). For details of the serologic tests see the *Manual of Clinical Immunology*, second edition (22).

Susceptibility to antimicrobial agents. Sulfonamides, particularly sulfadiazine and sulfisoxazole, have long been considered the drugs of choice for the treatment of nocardial infections, but they have given way in recent years to the combination of trimethoprim with sulfamethoxazole (TMP-SMX). In vitro testing of isolates of *Nocardia* species for antibiotic susceptibility has never been a popular procedure, not only because of technical difficulties, which are compounded in testing for synergy, but also because in vitro sensitivity of the sulfonamides has shown little relationship to clinical success. In one study of TMP-SMX synergism (9), results were found to depend critically upon the nocardial strain, the duration of incubation, and, particularly, the ratio of the two drugs. It was felt that the commercial fixed-dose combination contained too little trimethoprim for the best synergism. However, another study (7) found synergism over a wide range of concentrations of both antifungal agents. In 19 patients at one medical center who received the fixed combination of TMP-SMX (47), the overall cure or improvement rate was 89%; and this experience, together with other reports in the literature, was taken to support this combination as the therapeutic drug of choice for nocardial infections. Those authors recommended antibiotic susceptibility testing on all *Nocardia* isolates, although all isolates they tested were sensitive in vitro. In another study (2), by agar disk and paper strip diffusion methods, synergistic action was shown for 12 of 18 isolates of *N. asteroides* from various sources. Five patients with nocardial pneumonia were treated with TMP-SMX, and all were cured; all of the four isolates tested from these patients showed synergistic patterns of inhibition, so that these in vitro tests appeared to correlate with clinical usefulness.

Alternative treatments must be considered in cases of sulfonamide or TMP-SMX resistance or for patients who react adversely to sulfonamides. Extensive in vitro studies (4, 15, 17, 28) have shown *N. asteroides* to be sensitive particularly to amikacin and minocycline, and these drugs have been effective in both experimental nocardiosis in mice (50) and a limited number of human cases (3, 14, 51, 52). A combination of ampicillin and erythromycin has also been effective (3). Of a number of β-lactam antibiotics tested in vitro, only cefotaxime, cefmenoxime, ceftriaxone, cefuroxime, and *N*-formimidoyl thienamycin (imipenem) showed good activity against *N. asteroides* (17, 28). Imipenem showed synergism in vitro with TMP-SMX and with cefotaxime, as did amikacin with TMP-SMX (18). It has been recommended (11) that nocardial infections routinely be treated with sulfonamide (now TMP-SMX) and that in vitro sensitivity tests, which in any case are not readily available, be done only if the clinical response is poor or if the patient proves to be sensitive to sulfonamides.

Treatment with TMP-SMX has been effective in actinomycetic mycetoma as well as in other forms of nocardiosis. The addition of streptomycin to TMP-SMX or to diaminodiphenylsulfone (Dapsone), long the standard of chemotherapy for this type of mycetoma, is reported to improve the effectiveness of these drugs (16).

Essential criteria. Actinomycetes can be differentiated from fungi by the very fine diameter of their branching filaments, by their cell wall composition, and by the absence of a nuclear membrane. They differ from mycobacteria in their more extensive filamentation, with true branching and usually aerial hyphae, in the tenacity of their colonies, and in being only weakly acid-fast or non-acid-fast. Mycobacteria have cell walls of type IV, but they differ from *Nocardia* species in their mycolic acids and, for some species, in sensitivity to lysozyme.

Valuable presumptive tests. Presumptive identification of certain *Nocardia* species can be made as follows (27). An actinomycete that forms aerial hyphae but fails to hydrolyze casein, L-tyrosine, or xanthine can presumptively be identified as *N. asteroides*. An acid-fast culture with aerial hyphae which dissolves casein and tyrosine but not xanthine can safely be presumed to be *N. brasiliensis*. Aerial-hyphae-forming strains that decompose xanthine but not casein or tyrosine can be presumed with reasonable certainty to be *N. caviae*. A lack of demonstrable acid-fastness does not rule out any of these three species.

DERMATOPHILUS CONGOLENSIS

Characterization

D. congolensis is unique among bacteria in that it forms branching filaments which divide in both transverse and multiple longitudinal planes to form packets of cuboid or coccoid cells which become motile spores (19). Another species, *Geodermatophilus obscurus*, which together with *D. congolensis* makes up the family *Dermatophilaceae* (20), also forms clusters of cuboid cells by dividing in both transverse and longitudinal planes; but filamentation in *G. obscurus* is rudimentary or absent, the thallus often consisting simply of a muriform, tuber-shaped structure. *G. obscurus* differs from *D. congolensis* in several additional fundamental respects, including nutritional requirements, DNA base composition (45), and lack of pathogenicity (23), and will not be discussed further here. *Frankia* sp. CpI1, a root nodule bacterium characterized by filaments which at a certain stage of its life cycle form sporangium-like vesicles with some resemblance to structures of *D. congolensis*, has also proven to be nonpathogenic in the *Dermatophilus* pathogenicity test (21). With respect to actinomycete cell wall composition, *D. congolensis* is type IIIB, *G. obscurus* is type IIIC, and *Frankia* sp. is related to types II and IV.

Germination of the *D. congolensis* spore gives rise to a mycelium of narrow, tapering, encapsulated filaments with lateral branching at right angles. As longitudinal septa are laid down centrifugally, the filaments broaden from an initial diameter of 0.5 to 1.5 μm up to 5 μm, always tapering distally (Fig. 19). The formation of swarms of motile, coccoid spores completes the cycle (Fig. 20). The spores are isodiametric, with tufts of 5 to more than 50 flagella. All stages of the complex life cycle are gram positive and non-acid-fast.

Clinical significance

D. congolensis is the etiologic agent of dermatophilosis (streptotrichosis) (32), an exudative, pustular dermatitis of worldwide distribution which affects mainly cattle, sheep (as mycotic dermatitis, lumpy wool, or strawberry foot rot), and horses, but also goats, other domesticated and many feral mammals, lizards, and occasionally humans (24). Human infection, in the form of pustules, furuncles, or desquamative eczema of the hands or forearms, is acquired through contact with diseased animals. Structures indistinguishable from *D. congolensis* are seen in scrapings and biopsies from patients with pitted keratolysis, which attacks the soles of the feet, but this microorganism has not been cultured from such material. In animals, damage to hides and wool results from crusting and scab formation, and extensive infection often leads to fatalities in lambs and cattle (32), apparently through concurrent infection or cachexia. There are several reports of subcutaneous and lymph node infections in animals (16). The microorganism has been isolated only from clinical specimens, although it may remain viable for prolonged periods in detached dried scabs.

Collection of specimens

D. congolensis can be isolated in pure culture from clinical materials by streaking scabs or exudate, preferably from unopened pustules, directly on blood agar plates and incubating them at 37°C. Exudate should be submitted to the laboratory in sterile tubes or on swabs moistened with sterile water. These samples should be refrigerated if culturing is to be delayed. Scrapings, crusts, and scabs may be transmitted dry, in sterile tubes or jars, and stored at ambient temperature. Biopsy specimens for culture should be transmitted in sterile, screw cap jars without preservative.

Direct examination

D. congolensis may be seen in any stage of its life cycle. The most diagnostic (and pathognomonic) stage is marked by the presence of branched filaments, 2 to 5 μm in diameter, dividing both transversely and longitudinally into packets of coccoid forms (Fig. 21). Smears and unstained wet mounts can be made from exudate, from the underside of crusts or the roof of pustules, and from scabs ground and suspended in saline. Bacterial stains, particularly methylene blue, are effective in the demonstration of the structures in smears, but the Gram stain tends to be too dark and obscures details. The Giemsa method is probably best. Giemsa and Grocott methenamine-silver stains are best for paraffin sections, which often reveal the microorganism abundantly in hair follicles, accompanied by large numbers of eosinophilic leukocytes (Fig. 21).

Culture and isolation

Clinical materials, preferably the underside of scabs, should be streaked, with the use of optimal isolation techniques, on beef infusion-blood agar plates for aerobic incubation at 37°C. For highly contaminated specimens, animal passage, although uneconomical, is very effective. Crusts and scabs are ground and applied to the shaved and scarified ear or flank of a rabbit; lesions appear in 2 to 7 days, and from these the organism is readily isolated in pure culture (23). Another method (1), designed to minimize contamination of primary cultures, entails grinding of dry scabs or of wool tufts containing dry scabs to a fine powder. The powder is extracted with 2 volumes of distilled water (shake thoroughly, wet the wool, and press out the fluid). The suspension is allowed to settle for 15 min, and the supernatant is

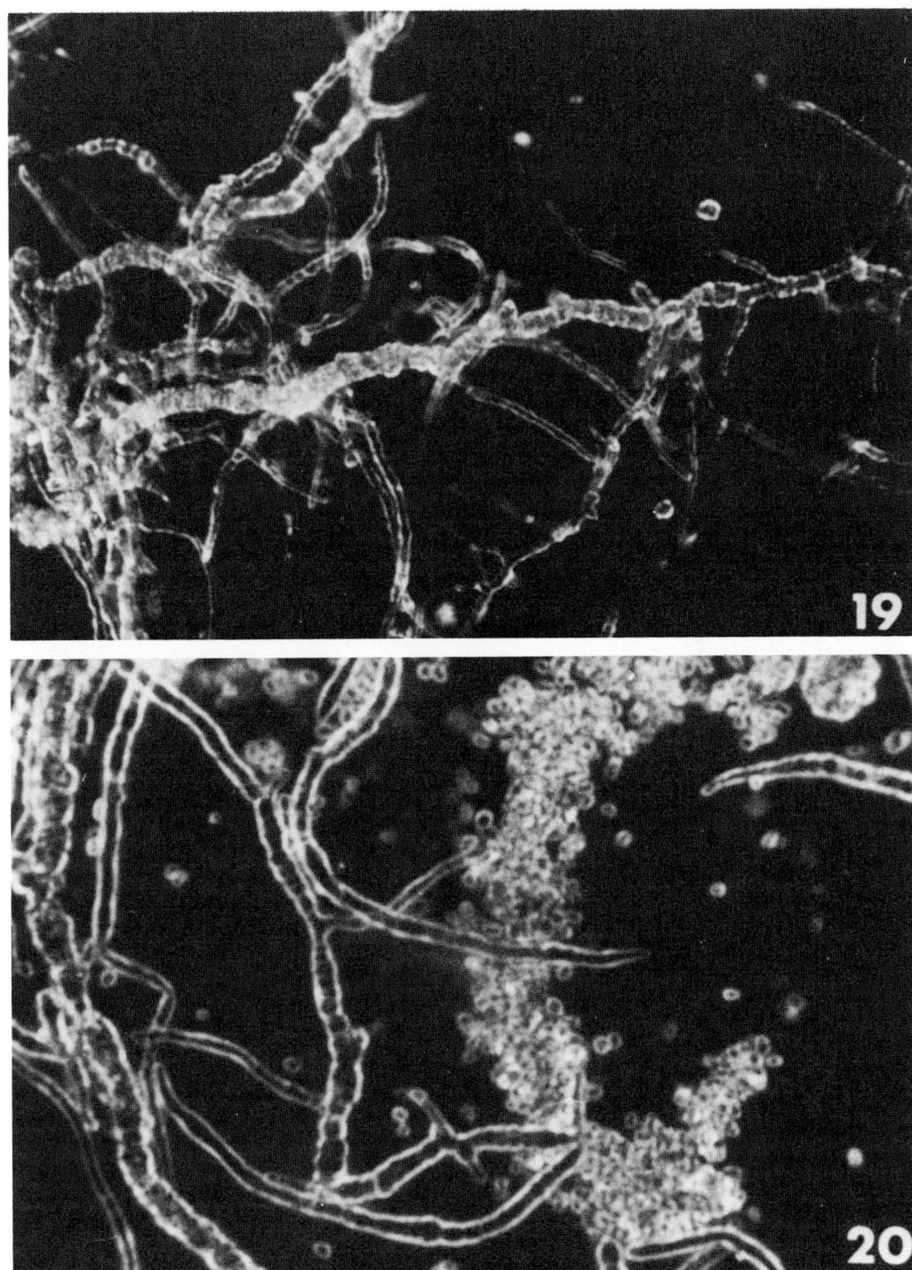

FIG. 19. *D. congolensis*, showing branched filaments which taper from coarse multiseptate bases to fine, nonseptate apical and lateral hyphae. Wet mount from a broth culture; dark field. ×720. Reproduced with permission from the *New York State Journal of Medicine*.

FIG. 20. *D. congolensis*, showing the final transformation of mature hyphae into an agglomeration of motile spores. Dark field. ×1,750. Reproduced with permission from the *New York State Journal of Medicine*.

removed with a Pasteur pipette and streaked on beef infusion-blood agar plates containing 1,000 IU of polymyxin B sulfate per ml. Incubation is at 37°C for 48 h.

Identification

Cultural characteristics. When *D. congolensis* is incubated on beef heart infusion-horse blood agar plates for 24 h at 37°C, aerobically or with 5 to 10% CO_2, the colonies are tiny (0.5 to 1.0 mm); round, square, or irregular; grayish white; raised, rough, glabrous, hard, and adherent; and typically pitting the medium. In 2 to 5 days (Fig. 22), they characteristically develop orange pigment (more rapidly where colonies are crowded), but occasional variants remain white or gray. Beta-hemolysis, particularly prominent on horse blood, develops earliest in crowded areas. Isolated colonies grow to 4 to 6 mm in diameter in 1 week, and many become mucoid at the apex. At 27°C, growth is similar but slower; the colonies are pinpoint

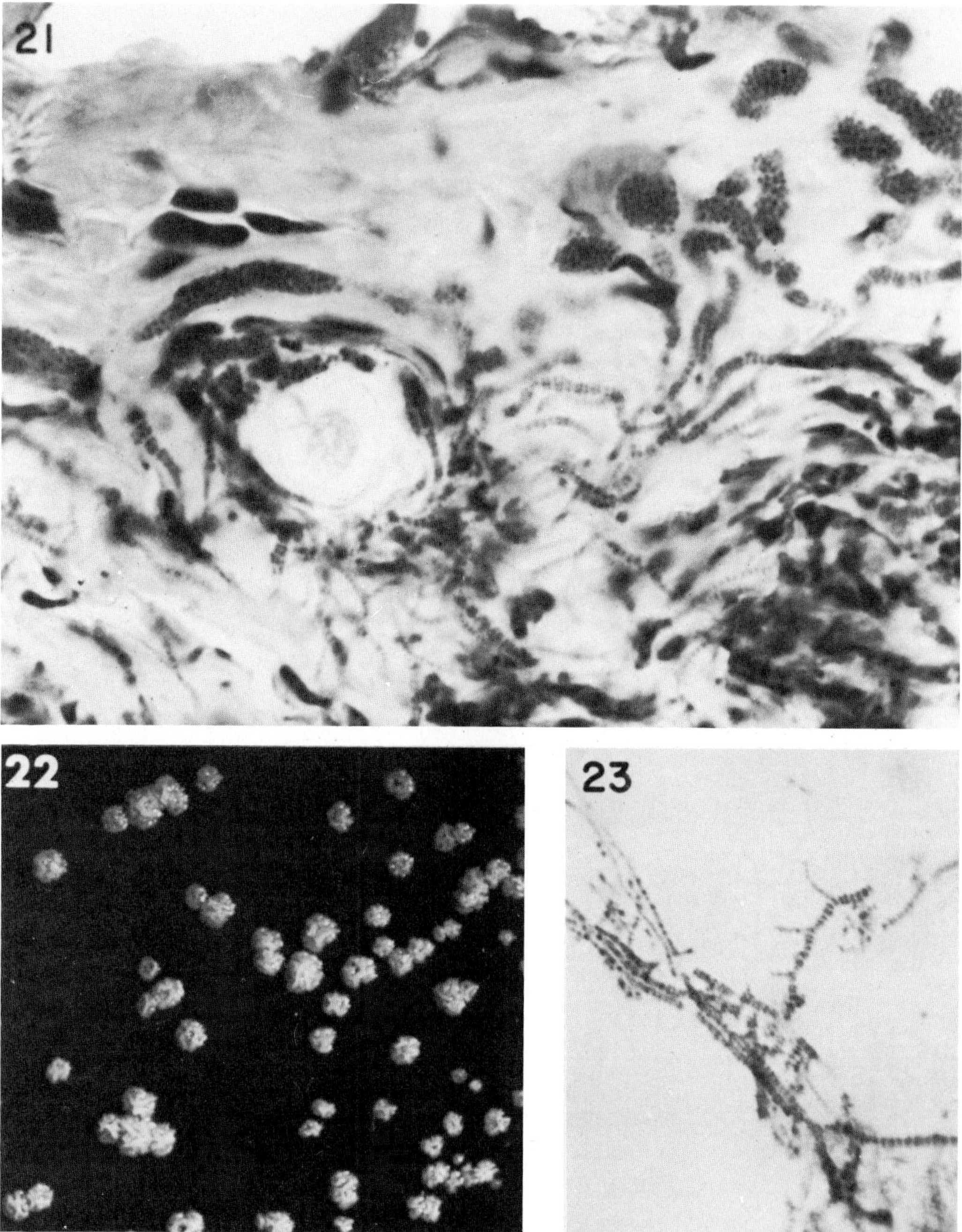

FIG. 21. *D. congolensis*, showing various developmental stages in a section of mouse skin, concentrated around a hair follicle. Hematoxylin and eosin stain. ×1,800. Reproduced with permission from the *New York State Journal of Medicine*.

FIG. 22. Colonies of *D. congolensis* on blood agar streak plate, after 3 days at 37°C. ×3.6.

FIG. 23. *D. congolensis*, showing the early segmentation stage from brain heart infusion agar, after 3 days at 37°C. Methylene blue stain. ×972. Reproduced with permission from the American Society for Microbiology.

in size at 48 h but reach 1 to 2 mm, possibly up to 3 mm, at 1 week. Anaerobically the colonies reach a diameter of 0.2 to 1 mm in 48 h but are less voluminous than aerobic colonies, white to translucent, and umbonate. Hemolysis develops later, if at all.

Heavy streak inoculation of pure cultures onto brain heart infusion slants results, after 3 days, in confluent growth which varies in color, according to the strain, from gray to orange and which may be granular or membranous. Growth is generally markedly adherent but often with a butyrous or caseous overlay. In beef infusion-peptone broth culture at 37°C, there is a thick sediment and a clear supernatant fluid, occasionally with a surface ring. There is no growth on Sabouraud dextrose agar or Czapek-Dox solution agar.

Microscopic examination of culture growth. The characteristic microscopic appearance of *D. congolensis* is of thickened, branched filaments dividing both transversely and longitudinally (Fig. 23). However, there is extreme variation in morphology. Depending upon the strain and the age of the culture, one may see a completely micrococcoid picture, with cocci arranged irregularly or in cubical packets, clusters of germinating spores, or branched filaments, segmenting or not. Motility is usually evident in recent isolates. If only cocci are seen and *D. congolensis* is suspected, cultures should be examined at an earlier age for hyphae.

Biochemical reactions. *D. congolensis* is catalase positive, as determined by a spot test. A half-loopful of growth from brain heart infusion agar (without blood, which itself gives a positive catalase test) is emulsified in a drop of 30% H_2O_2 and observed for effervescence. Other tests are incubated at 37°C and read at 48 h and at 5, 7, and 14 days. The urease test is regularly positive within 24 h. Casein and starch are hydrolyzed, but not tyrosine and xanthine. Bromocresol purple milk is peptonized. Gelatin and Loeffler coagulated serum medium are liquefied by almost all strains. Nitrate reduction, methyl red, and Voges-Proskauer test results are negative; indole is not produced. Acid but no gas is produced in proteose-peptone broth from glucose and fructose (in 48 h); it is produced transitorily from galactose (acid in 48 h, negative in 2 weeks), and, by some strains, it is produced belatedly (1 to 2 weeks) from maltose. No acid is produced from sucrose, lactose, xylose, dulcitol, mannitol, sorbitol, or salicin.

Pathogenicity. A demonstration of virulence for animals is not necessary for the identification of *D. congolensis*, but infection is readily established in rabbits and may serve to differentiate this species from filamentous forms of *G. obscurus* (23). The experimental inoculation of abraded skin results in acute ulcerative pustular dermatitis involving principally the hair follicles (Fig. 21).

Immunoserologic diagnosis. There is a close serologic relationship among strains of *D. congolensis* from diverse geographic locations and animal species, and there is little or no cross-reactivity with other microorganisms, including actinomycetes (19, 44). Circulating antibodies have been detected by immunodiffusion in chronically infected cattle (44) and in experimental animals. Nevertheless, diagnostic serology has not been developed for routine use. Immunofluorescence has been applied to the identification of *D. congolensis* in animal tissues, being particularly useful where only coccal forms are seen.

Susceptibility to antimicrobial agents. The growth of *D. congolensis* in vitro is inhibited by many antibacterial drugs, including penicillin, streptomycin, chloramphenicol, tetracyclines, erythromycin, kanamycin, nitrofurantoin, and sulfonamides, but not by griseofulvin. A combination of penicillin G with streptomycin is more effective than either antibiotic alone for the treatment of dermatophilosis in sheep.

Essential criteria. The microscopic morphology of *D. congolensis* is unique among known microorganisms, and it can hardly fail to be recognized when seen in all of its manifestations, although *G. obscurus* resembles it in some respects. The typical biochemical properties serve as confirmatory characteristics.

LITERATURE CITED

1. **Abu-Samra, M. T., and G. S. Walton.** 1977. Modified techniques for the isolation of *Dermatophilus* spp. from infected material. Sabouraudia **15:**23–27.
2. **Adams, H. G., B. A. Beeler, L. S. Wann, C. K. Chin, and G. F. Brooks.** 1984. Synergistic action of trimethoprim and sulfamethoxazole for *Nocardia asteroides*: efficacious therapy in five patients. Am. J. Med. Sci. **287:**8–12.
3. **Bach, M. C., A. P. Monaco, and M. Finland.** 1973. Pulmonary nocardiosis: therapy with minocycline and with erythromycin plus ampicillin. J. Am. Med. Assoc. **224:**1378–1381.
4. **Bach, M. C., L. D. Sabath, and M. Finland.** 1973. Susceptibility of *Nocardia asteroides* to 45 antimicrobial agents in vitro. Antimicrob. Agents Chemother. **3:**1–8.
5. **Beaman, B. L., J. Burnside, B. Edwards, and W. Causey.** 1976. Nocardial infections in the United States, 1972–1974. J. Infect. Dis. **134:**286–289.
6. **Beaman, B. L., and A. M. Sugar.** 1983. *Nocardia* in naturally acquired and experimental infections in animals. J. Hyg. **91:**393–419.
7. **Beaumont, R. J.** 1971. Trimethoprim as a possible therapy for nocardiosis and melioidosis. Med. J. Aust. **2:**1123–1127.
8. **Becker, B., M. P. Lechevalier, and H. A. Lechevalier.** 1965. Chemical composition of cell-wall preparations from strains of various form-genera of aerobic actinomycetes. Appl. Microbiol. **13:**236–243.
9. **Bennett, J. E., and A. E. Jennings.** 1978. Factors influencing susceptibility of *Nocardia* species to trimethoprim-sulfamethoxazole. Antimicrob. Agents Chemother. **13:**624–627.
10. **Blumer, S. O., and L. Kaufman.** 1979. Microimmunodiffusion test for nocardiosis. J. Clin. Microbiol. **10:**308–312.
11. **Carroll, G. F., J. M. Brown, and L. D. Haley.** 1977. A method for determining *in vitro* drug susceptibilities of some nocardiae and actinomadurae. Results with 17 antimicrobial agents. Am. J. Clin. Pathol. **68:**279–283.
12. **Chamoiseau, G.** 1973. *Mycobacterium farcinogenes* agent causal du farcin du boeuf en Afrique. Ann. Microbiol. (Paris) **124:**215–222.
13. **Cruickshank, J. G., A. H. Gawler, and C. Shaldon.** 1979. *Oerskovia* species: rare opportunistic pathogens. J. Med. Microbiol. **12:**513–515.
14. **Curry, W. A.** 1980. Human nocardiosis. A clinical review with selected case reports. Arch. Intern. Med. **140:**818–826.
15. **Dalovisio, J. R., and G. A. Pankey.** 1978. In vitro susceptibility of *Nocardia asteroides* to amikacin. Antimicrob. Agents Chemother. **13:**128–129.
16. **Gibson, J. A., R. J. Thomas, and R. L. Domjahn.** 1983. Subcutaneous and lymph node granulomas due to *Dermatophilus congolensis* in a steer. Vet. Pathol. **20:**120–122.
17. **Gombert, M. E.** 1982. Susceptibility of *Nocardia asteroides* to various antibiotics, including newer beta-lactams, trimethoprim-sulfamethoxazole, amikacin, and *N*-formimidoyl thienamycin. Antimicrob. Agents Chemother. **21:**1011–1012.
18. **Gombert, M. E., and T. M. Aulicino.** 1983. Synergism of imipenem and amikacin in combination with other antibiotics against *Nocardia asteroides*. Antimicrob. Agents Chemother. **24:**810–811.
19. **Gordon, M. A.** 1964. The genus *Dermatophilus*. J. Bacteriol. **88:**509–522.
20. **Gordon, M. A.** 1974. Family V. *Dermatophilaceae* Austwick 1958, 42, *emend. mut. char.* Gordon 1964, 521,

p. 723–726. *In* R. E. Buchanan and N. E. Gibbons (ed.), Bergey's manual of determinative bacteriology, 8th ed. The Williams & Wilkins Co., Baltimore.
21. **Gordon, M. A., M. P. Lechevalier, and E. W. Lapa.** 1983. Nonpathogenicity of *Frankia* sp. CpI1 in the *Dermatophilus* pathogenicity test. Actinomycetes **18:**50–53.
22. **Gordon, M. A., and E. S. Mahgoub.** 1980. Immune response to aerobic pathogenic *Actinomycetaceae*, p. 526–529. *In* N. R. Rose and H. Friedman (ed.). Manual of clinical immunology, 2nd ed. American Society for Microbiology, Washington, D.C.
23. **Gordon, M. A., and U. Perrin.** 1971. Pathogenicity of *Dermatophilus* and *Geodermatophilus*. Infect. Immun. **4:**29–33.
24. **Gordon, M. A., I. F. Salkin, and W. B. Stone.** 1977. *Dermatophilus* dermatitis enzootic in deer in New York State and vicinity. J. Wildl. Dis. **13:**184–190.
25. **Gordon, R. E.** 1966. Some criteria for the recognition of *Nocardia madurae* (Vincent) Blanchard. J. Gen. Microbiol. **45:**355–364.
26. **Gordon, R. E., and D. A. Barnett.** 1977. Resistance to rifampin and lysozyme of strains of some species of *Mycobacterium* and *Nocardia* as a taxonomic tool. Int. J. Syst. Bacteriol. **27:**176–178.
27. **Gordon, R. E., and J. M. Mihm.** 1962. Identification of *Nocardia caviae* (Erikson) *Nov. Comb.* Ann. N.Y. Acad. Sci. **98:**628–636.
28. **Gutmann, L., F. W. Goldstein, M. D. Kitzis, B. Hautefort, C. Darmon, and J. F. Acar.** 1983. Susceptibility of *Nocardia asteroides* to 46 antibiotics, including 22 β-lactams. Antimicrob. Agents Chemother. **23:**248–251.
29. **Hecht, S. T., and W. A. Causey.** 1976. Rapid method for the detection and identification of mycolic acids in aerobic actinomycetes and related bacteria. J. Clin. Microbiol. **4:**284–287.
30. **Hirst, L. W., G. K. Harrison, W. G. Merz, and W. J. Stark.** 1979. *Nocardia asteroides* keratitis. Br. J. Ophthalmol. **63:**449–454.
31. **Lechevalier, M. P.** 1972. Description of a new species, *Oerskovia xanthineolytica*, and emendation of *Oerskovia* Prauser et al. Int. J. Syst. Bacteriol. **22:**260–264.
32. **Lloyd, D. H., and K. C. Sellers (ed.).** 1976. *Dermatophilus* infection in animals and man. Academic Press, Ltd., London.
33. **Mackinnon, J. E., and R. C. Artagaveytia-Allende.** 1956. The main species of pathogenic aerobic actinomycetes causing mycetomas. Trans. R. Soc. Trop. Med. Hyg. **50:**31–40.
34. **Mahgoub, E. S.** 1976. Medical management of mycetoma. Bull. W.H.O. **54:**303–310.
35. **Mahgoub, E. S., and I. G. Murray.** 1973. Mycetoma. William Heinemann Medical Books Ltd., London.
36. **Mariat, F.** 1962. Critères de détermination des principales espèces d'*Actinomycètes* aérobies pathogènes. Ann. Soc. Belge. Med. Trop. **4:**651–672.
37. **Meyer, J.** 1976. *Nocardiopsis*, a new genus of the order *Actinomycetales*. Int. J. Syst. Bacteriol. **26:**487–493.
38. **Murray, J. F., S. M. Finegold, S. Froman, and D. W. Will.** 1961. The changing spectrum of nocardiosis. Am. Rev. Respir. Dis. **83:**315–330.
39. **Parnell, M. J., G. B. Hubbard, K. C. Fletcher, and R. F. Schmidt.** 1983. *Nocardia asteroides* infection in a purple throated sunbird (*Nectarinia sperapa*). Vet. Pathol. **20:**497–500.
40. **Pier, A. C., and R. E. Fichtner.** 1971. Serologic typing of *Nocardia asteroides* by immunodiffusion. Am. Rev. Respir. Dis. **103:**698–707.
41. **Pier, A. C., and R. E. Fichtner.** 1981. Distribution of serotypes of *Nocardia asteroides* from animal, human, and environmental sources. J. Clin. Microbiol. **13:**548–553.
42. **Pier, A. C., E. H. Willers, and M. J. Mejia.** 1961. *Nocardia asteroides* as a mammary pathogen of cattle. II. The sources of nocardial infection and experimental reproduction of the disease. Am. J. Vet. Res. **22:**698–703.
43. **Reller, L. B., G. L. Maddoux, M. R. Eckman, and G. Pappas.** 1975. Bacterial endocarditis caused by *Oerskovia turbata*. Ann. Intern. Med. **83:**664–666.
44. **Richard, J. L., J. R. Thurston, and A. C. Pier.** 1976. Comparison of antigens of *Dermatophilus congolensis* isolates and their use in experimental and natural infections, p. 216–228. *In* D. H. Lloyd and K. C. Sellers (ed.), *Dermatophilus* infection in animals and man. Academic Press, Ltd., London.
45. **Samsonoff, W. A., M. A. Detlefsen, A. F. Fonseca, and M. R. Edwards.** 1977. Deoxyribonucleic acid base composition of *Dermatophilus congolensis* and *Geodermatophilus obscurus*. Int. J. Syst. Bacteriol. **27:**22–25.
46. **Shainhouse, J. Z., A. C. Pier, and D. A. Stevens.** 1978. Complement fixation antibody test for human nocardiosis. J. Clin. Microbiol. **8:**516–519.
47. **Smego, R. A., Jr., M. B. Moeller, and H. A. Gallis.** 1983. Trimethoprim-sulfamethoxazole therapy for *Nocardia* infections. Arch. Intern. Med. **143:**711–718.
48. **Stevens, D. A., A. C. Pier, B. L. Beaman, P. A. Morozumi, I. S. Lovett, and E. T. Houang.** 1981. Laboratory evaluation of an outbreak of nocardiosis in immunocompromised hosts. Am. J. Med. **71:**928–934.
49. **Tsukamura, M.** 1970. Relationship betwen *Mycobacterium* and *Nocardia*. Jpn. J. Microbiol. **14:**187–195.
50. **Wallace, R. J., Jr., E. J. Septimus, D. M. Musher, M. B. Berger, and R. R. Martin.** 1979. Treatment of experimental nocardiosis in mice: comparison of amikacin and sulfonamide. J. Infect. Dis. **140:**244–248.
51. **Wren, M. V., A. M. Savage, and R. H. Alford.** 1979. Apparent cure of intracranial *Nocardia asteroides* infection by minocycline. Arch. Intern. Med. **139:**249–250.
52. **Yogev, R., T. Greenslade, C. F. Firlit, and P. Lewy.** 1980. Successful treatment of *Nocardia asteroides* infection with amikacin. J. Pediatr. **96:**771–773.

Chapter 24

Enterobacteriaceae

MICHAEL T. KELLY, DON J. BRENNER, AND J. J. FARMER III

Organisms in the family *Enterobacteriaceae* are facultative gram-negative rods that ferment glucose. They are oxidase negative, reduce nitrates to nitrites (except some strains of the genera *Enterobacter, Erwinia, Klebsiella, Xenorhabdus,* and *Yersinia*), and do not require nor is their growth enhanced by NaCl. They may be motile by peritrichous flagella (except *Tatumella*) or nonmotile. These characteristics serve to differentiate members of the family *Enterobacteriaceae* from other glucose-fermenting, facultative gram-negative rods which may fall in the family *Vibrionaceae* or *Pasteurellaceae*. Negative oxidase and/or lack of NaCl requirement separate the *Enterobacteriaceae* from the *Vibrionaceae*, and negative oxidase and/or lack of requirement for organic nitrogen sources distinguish the *Enterobacteriaceae* from the *Pasteurellaceae*.

Enterobacteriaceae are widely distributed on plants, in soil, and in the intestine of humans and animals. They are associated with many types of human infections, including abscesses, pneumonia, meningitis, septicemia, and intestinal, urinary, and wound infections. *Enterobacteriaceae* are a major component of the normal human intestinal flora, but they are relatively uncommon in the normal flora of other body sites. Hospitalized patients often become colonized with *Enterobacteriaceae*, and these organisms are a major cause of nosocomial infections. *Enterobacteriaceae* may account for 80% of clinically significant isolates of gram-negative bacilli in clinical microbiology laboratories, and they may account for 50% of all clinically significant isolates. These bacteria are particularly common in certain types of specimens. For example, they account for nearly 50% of septicemia cases, 60 to 70% of bacterial enteritis cases, and more than 70% of urinary tract infections.

Since the last edition of this Manual, a number of changes have occurred in the nomenclature and classification of members of the family *Enterobacteriaceae*. The family now contains over 20 genera and more than 100 species (5). Recent changes in, and additions to, the family have been based on a multiphasic approach to taxonomy centered in genetic relatedness studies by DNA-DNA hybridization and tempered by phenotypic characteristics, including biochemical features and antimicrobial susceptibility. Many of the newly recognized genera and species are of no or unknown clinical significance and will very rarely be encountered in clinical microbiology laboratories. Approximately 49 species or groups are currently recognized as definite or probable human pathogens. The current nomenclature of these organisms and their corresponding designations in the previous edition of this Manual are presented in Table 1. The reader is referred to other sources on the *Enterobacteriaceae* (5, 10) for discussion of other species or groups that are not known to be human pathogens, and it should be recognized that any of these organisms may infect humans in selected circumstances (compromised hosts, trauma, etc).

INFECTIONS DUE TO *ENTEROBACTERIACEAE*

Intestinal infections

Members of the *Enterobacteriaceae* account for a major portion of intestinal infections in humans and animals in the United States and throughout the world. Although many of the *Enterobacteriaceae* have been implicated as a cause of diarrhea, only members of the genera *Escherichia, Salmonella, Shigella,* and *Yersinia* are clearly established as enteric pathogens.

Escherichia coli. *E. coli* is associated with at least four types of human enteric disease, including enteropathogenic, enterotoxigenic, enteroinvasive, and hemorrhagic. Enteropathogenic strains cause diarrhea, mostly in infants, by unknown mechanisms. These organisms are recognized by serological testing, and they belong to a limited group of specific serotypes. Although many of these strains produce no known factors of pathogenesis, they are capable of causing diarrhea in normal volunteers (21). The occurrence of diarrhea due to enteropathogenic *E. coli* seems to have decreased markedly over the past 20 years (12).

Enterotoxigenic strains of *E. coli* cause secretory diarrhea by the elaboration of heat-labile or heat-stable enterotoxins or both (32). Surface attachment factors are also important and appear to be required for the organisms to attach to the intestinal mucosa, where they then elaborate toxin(s). Enterotoxigenic *E. coli* strains typically cause profuse, watery diarrhea, and they are often implicated in cases of traveler's diarrhea (13). The importance of enterotoxigenic *E. coli* in diarrheal disease in the United States is largely unknown due to the lack of practical methods for enterotoxin detection (see Enterotoxin Testing section below).

Enteroinvasive strains of *E. coli* cause a dysentery-like illness similar to *Shigella* infection (35). These *E. coli* strains invade the mucosal cells, causing destruction of the cells followed by sloughing of areas of the mucosal lining. Typically, patients with diarrhea due to enteroinvasive *E. coli* have stools containing blood, mucus, and polymorphonuclear leukocytes. Enteroinvasive strains of *E. coli* are thought to occur less commonly than enterotoxigenic strains, but the true importance of enteroinvasive *E. coli* in the United States is unknown because a suitable assay for use in clinical laboratories to detect these strains has not been available. The recent association of a high-molecular-weight plasmid with invasiveness may provide a useful marker for the detection of these organisms.

Hemorrhagic colitis is a recently recognized enteric infection due to *E. coli* strains of a specific serotype, O157:H7. These strains cause a severe diarrhea characterized by grossly bloody stools (30). Most of the reported cases have been acquired from the consumption of partially cooked hamburger. *E. coli* strains associated with hemorrhagic colitis produce a cytotoxin for Vero and HeLa cells that appears to be identical to the shiga toxin of *Shigella dysenteriae*. It is believed that this toxin damages vascular endothelial cells, resulting in hemorrhage into the bowel lumen (20).

Salmonella. Salmonellae cause a wide range of human enteric disease, from self-limited gastroenteritis with mild symptoms of short duration, to severe gastroenteritis with or without bacteremia, to typhoid fever, which is a severe, debilitating, and potentially life-threatening illness. *S. choleraesuis*, *S. paratyphi* A, and *S. typhi* are particularly important because of their frequent association with severe disease and bacteremia. Other *Salmonella* serotypes may also cause bacteremia, but they are more likely to cause gastroenteritis without bacteremia. Salmonellae typically invade through the bowel mucosa and multiply in the submucosa. Depending on the virulence of particular strains and the host response, they may invade and multiply in the bloodstream or the lymphatic tissue or both. Salmonellae are the most common cause of bacterial diarrhea in the United States.

Shigella. *Shigella* species cause classical bacillary dysentery characterized by severe, cramping abdominal pain and diarrhea with blood and mucus. These organisms invade mucosal cells, causing death and sloughing of the cells into the bowel lumen. However, they seldom invade beyond the mucosa. *Shigella* infections remain one of the most commonly recognized causes of bacterial diarrhea in the United States.

Yersinia. Over the past 10 years, *Y. enterocolitica* has been clearly demonstrated to be a significant cause of gastroenteritis (37). It is more often recognized in northern than southern parts of the northern hemisphere and is a major cause of enteric illness in Europe, Canada, and the northern United States. Although the organism may produce an enterotoxin that closely resembles that of *E. coli* ST (25), *Y. enterocolitica* is considered primarily an invasive pathogen, and involvement of intestinal lymph nodes is common. Such infections are difficult to distinguish from acute appendicitis. It appears that, like *E. coli*, only strains of *Y. enterocolitica* that possess essential virulence factors are capable of causing human intestinal disease.

Extraintestinal infections

Except for *Shigella* species, which rarely cause infections outside the gastrointestinal tract, most of the *Enterobacteriaceae* are capable of causing a variety of extraintestinal infections. However, a small number of species, including *Enterobacter aerogenes*, *E. cloacae*, *Escherichia coli*, *Klebsiella pneumoniae*, *Proteus mirabilis*, and *Serratia marcescens*, account for the vast majority of infections. Urinary tract infections, primarily cystitis, are the most common, followed by respiratory, wound, bloodstream, and central nervous system infections. Urinary tract, respiratory, and wound infections due to *Enterobacteriaceae* are each encountered more commonly than intestinal infections in most clinical microbiology laboratories. Many of these infections, especially sepsis and meningitis, are life threatening, and many infections of any type due to *Enterobacteriaceae* are hospital acquired. Because of the severe nature of many of these infections, prompt isolation, identification, and susceptibility testing of *Enterobacteriaceae* are essential.

SPECIMEN COLLECTION AND TRANSPORT

Enterobacteriaceae may be recovered from infectious processes in virtually any body site, and good practices for specimen collection in general should be followed for the collection of blood, body fluid, respiratory, wound, and other specimens for the culture of *Enterobacteriaceae*. Urine specimens should be collected by methods that reduce the possibility of contamination of the specimen. Such methods include the clean-catch, mid-stream urine technique, catheterization under aseptic conditions, and in certain instances, suprapubic aspiration. Most *Enterobacteriaceae* multiply rapidly in media such as urine, and it is important to transport specimens to the laboratory as quickly as possible to avoid overgrowth of such organisms. This is especially true for urine specimens because of the need for quantitation, and specimens should be refrigerated unless they can be cultured within 2 h of collection. Alternatively, urine preservative systems may be used in situations where specimen transportation is often delayed. The importance of rapid processing of specimens after collection cannot be overemphasized, and inadequate specimen delivery systems may reduce the quality of otherwise well-collected specimens.

Stool specimens also require special attention in both collection and transportation. Specimens should be collected early in the course of illness when the causative agent is likely to be present in largest numbers. With the exception of typhoid fever, enteric pathogens are more difficult to detect as the course of illness progresses, and cultures collected in the recovery phase may often be negative. Freshly passed stool is much preferred over rectal swabs for the culture of *Enterobacteriaceae*, and stool provides the highest recovery of enteric pathogens. Rectal swab specimens may be used in selected circumstances, such as for infants or for mass screening in investigation of epidemics, but these specimens must be collected properly by sampling the mucosa within the rectal vault to be of any value. Ideally, enteric specimens should be cultured within 2 h of collection, and if delays in processing are unavoidable, a transport medium may be used. Buffered glycerol-saline solution (pH 7.4) is often used for this purpose, but rapid processing rather than a transport medium should be used whenever possible.

CULTURE OF *ENTEROBACTERIACEAE* FROM CLINICAL SPECIMENS

Most *Enterobacteriaceae* grow readily on media commonly used in clinical microbiology laboratories, and specimens other than stool can be cultured on blood or chocolate agar if the specimens are from a normally sterile body site. Specimens such as urine and respiratory and wound samples that are likely to contain mixtures of organisms may be cultured on selective media to enhance the recovery of *Enterobacteriaceae*. MacConkey agar or eosin methylene blue agar may be used for this purpose, and these media offer the added advantage of differential features which allow a preliminary grouping of enteric and other gram-negative bacteria. In addition, broth enrichment cultures may be of value, especially when small numbers of *Enterobacteriaceae* may be present in the specimen.

Stool and rectal swab specimens require special culture procedures involving selective-differential media and enrichment broths. Selective media are those that allow the growth of *Enterobacteriaceae* while inhibiting the growth of normal stool flora organisms, and differential media are those that allow a preliminary characterization of organisms based on colony morphology and selected biochemical reactions. Media may be highly selective, allowing the growth of only a limited group of organisms (e.g., bismuth sulfite agar for *Salmonella*), or only moderately selective, allowing the growth of most gram-negative organisms and inhibiting predominantly gram-positive organisms (e.g., MacConkey agar). Selective-differential media are essential for the efficient recovery of enteric pathogens from stool specimens because these media reduce the overgrowth of pathogens by normal flora organisms and aid in the recognition of pathogens in primary cultures. Broth enrichments are also useful for the culture of stool specimens because they provide for the growth of enteric pathogens while retarding the growth of normal flora organisms. Such media are particularly useful when pathogens may be present in low numbers. In the acute phases of gastrointestinal illness, broth enrichments are probably not essential because enteric pathogens are usually present in high numbers, but specimens are often not submitted to the laboratory until the later stages of illness, and broth enrichment may be the only means of detecting the infecting organism in such cases. Broth enrichment may also be useful in epidemiological studies to enhance the detection of low numbers of organisms from asymptomatic or convalescing patients. Gram-negative broth and selenite broth are enrichment media that have been found useful in many laboratories through the years. Cold enrichment in phosphate-buffered saline has also been found to enhance the recovery of *Y. enterocolitica* from stool specimens, but this technique is apparently useful only in cultures of convalescent or asymptomatic subjects.

A battery of media for routine use in the culture of *Enterobacteriaceae* from stool specimens should include a nonselective medium such as blood agar, a moderately selective differential medium such as MacConkey agar, a more selective differential medium such as xylose-lysine-deoxycholate agar or Hektoen enteric agar, and a broth enrichment. The use of such a battery will provide for the effective detection of the *Enterobacteriaceae* most commonly associated with enteric infection, including *Salmonella* and *Shigella* species and *E. coli*. In areas where *Y. enterocolitica* is encountered, a selective-differential medium such as *Yersinia* selective agar may also be added. A complete stool culture procedure will also include media for the isolation of *Campylobacter* species and possibly media for the isolation of *Vibrio* species in areas where cholera and other *Vibrio* infections occur.

TABLE 1. Clinically significant members of the family *Enterobacteriaceae* and changes in their taxonomy and nomenclature since the last edition of the Manual

Current classification	Previous designation
Cedecea davisae[a]	None
C. neteri[a]	None
Citrobacter freundii	Same
C. diversus	Same
C. amalonaticus	Same
Edwardsiella tarda	Same
Enterobacter cloacae	Same
E. aerogenes	Same
E. agglomerans	Same
E. gergoviae	Same
E. sakazakii	Same
Escherichia coli	Same
Ewingella americana	None
Hafnia alvei	Same
Klebsiella pneumoniae	Same
K. oxytoca	Same
K. ozaenae	Same
K. rhinoscleromatis	Same
K. planticola[a]	None
Kluyvera ascorbata	None
K. cryocrescens	None
Morganella morganii	Same
Proteus mirabilis	Same
P. penneri	None
P. vulgaris	Same
Providencia stuartii	Same
P. rettgeri	Same
P. alcalifaciens	Same
Salmonella typhi	Same
S. choleraesuis	Same
S. paratyphi A	*S. enteritidis* bioserotype paratyphi A
"Arizona group"	*Arizona hinshawii*
Salmonella serotype	*S. enteritidis* bioserotype
Serratia marcescens	Same
S. liquefaciens group[a]	*S. liquefaciens*
S. odorifera[a]	None
S. rubidaea[a]	Same
Shigella dysenteriae	Same
S. flexneri	Same
S. boydii	Same
S. sonnei	Same
Tatumella ptyseos[a]	None
Yersinia enterocolitica	Same
Y. frederiksenii[a]	Same
Y. intermedia[a]	Same
Y. kristensenii[a]	None
Y. pestis	Same
Y. pseudotuberculosis	Same

[a] Questionable clinical significance.

BIOCHEMICAL TESTING

Members of the family *Enterobacteriaceae* are identified primarily by testing for biochemical reactions. Several methods are available for the performance of biochemical tests, including a variety of commercial systems and the classical tubed media. In addition, various combinations of classical tests are available, and simple screening tests for identification have been proposed. Only the classical biochemical tests and rapid screening tests will be considered here because they are the basis for the taxonomy and identification of the *Enterobacteriaceae* (see Chapter 6 for a discussion of commercial systems).

Classical biochemical tests

Of the many biochemical tests that are available, a battery of 28 tests is sufficient to differentiate all the currently recognized, clinically significant *Enterobac-*

TABLE 2. Biochemical characteristics of clinically significant *Enterobacteriacea*

Characteristic	% Positive reactions																						
	Cedecea davisae	*C. neteri*	*Citrobacter amalonaticus*	*C. diversus*	*C. freundii*	*Edwardsiella tarda*	*Enterobacter aerogenes*	*E. agglomerans*	*E. cloacae*	*E. gergoviae*	*E. sakazakii*	*Escherichia coli*	*E. fergusonii*	*Ewingella americana*	*Hafnia alvei*	*Klebsiella oxytoca*	*K. ozaenae*	*K. planticola*	*K. pneumoniae*	*K. rhinoscleromatis*	*Kluyvera ascorbata*	*K. cryocrescens*	*Morganella morganii*	*Proteus mirabilis*
Primary test battery																								
D-Adonitol	0	0	0	98	0	0	100	7	25	0	0	5	96	0	0	100	97	100	90	100	0	0	0	0
Arginine dihydrolase	50	100	85	65	65	0	0	0	97	0	100	17	4	0	6	0	6	0	0	0	0	0	0	0
Citrate utilization	95	100	85	100	95	0	95	50	100	98	100	0	19	97	10	95	30	100	98	0	96	76	0	65
DNase	0	0	0	0	0	0	0	0	0	0	0	0	0	0	0	0	0	0	0	0	0	0	0	0
D-Glucose, gas production	70	100	97	98	95	100	100	20	100	98	98	90	92	0	98	97	50	100	97	0	93	95	87	96
Hydrogen sulfide (TSI)[a]	0	0	0	0	80	100	0	0	0	0	0	0	0	0	0	0	0	0	0	0	0	0	2	95
Indole production	0	0	100	100	5	100	0	20	0	0	11	95	96	0	0	100	0	20	0	0	89	86	98	2
Lysine decarboxylase	0	0	0	0	0	100	98	0	0	90	0	90	92	0	100	98	40	100	98	0	97	35	0	0
Motility	95	100	98	95	95	98	97	85	95	90	91	80	92	60	90	0	0	0	0	0	98	86	95	95
Ornithine decarboxylase	95	0	95	100	20	100	98	0	96	100	91	65	100	0	98	2	3	0	0	0	100	100	98	100
Phenylalanine deaminase	0	0	0	0	0	0	0	20	0	0	50	0	0	0	0	2	0	0	0	0	0	0	95	98
Sucrose	100	100	15	20	30	0	100	75	97	98	100	50	0	0	10	100	20	100	100	75	98	81	0	15
Urease	0	0	80	75	70	0	2	20	65	93	0	0	0	0	4	90	10	98	95	0	0	0	98	90
Voges-Proskauer	50	50	0	0	0	0	98	70	100	100	100	0	0	97	85	90	0	98	95	0	0	0	0	25
Secondary test battery																								
L-Arabinose	0	0	100	100	100	9	100	95	100	98	100	100	96	0	95	98	98	100	100	100	100	100	0	0
myo-Inositol	0	0	0	0	3	0	100	15	15	0	75	0	0	0	0	98	55	100	95	95	0	0	0	0
KCN, growth in	86	65	95	0	96	0	98	35	98	0	100	3	0	7	95	97	88	100	98	90	92	86	100	98
Lactose	19	35	50	35	50	0	95	40	93	55	100	90	0	70	5	100	30	100	98	0	98	95	0	2
Malonate	91	100	0	90	15	0	85	65	75	96	18	0	30	0	70	98	3	100	93	95	96	86	2	2
D-Mannitol	100	100	100	100	100	0	100	100	100	100	100	98	100	100	100	100	100	100	100	100	100	95	0	0
Melibiose	0	0	5	0	50	0	100	50	90	97	100	75	0	0	0	100	97	100	100	100	100	100	0	0
ONPG (β-galactosidase)	90	100	100	96	95	0	100	90	100	97	100	95	83	96	90	100	80	100	100	0	100	100	5	0
Raffinose	10	0	5	0	30	0	96	30	97	97	100	50	0	0	2	100	90	100	100	75	98	100	0	0
L-Rhamnose	0	0	100	100	100	0	100	85	92	98	100	80	92	23	97	100	55	100	100	96	100	100	0	2
Salicin	100	100	40	20	5	0	100	65	75	98	100	40	54	90	13	100	97	100	100	98	100	100	0	0
D-Sorbitol	0	100	100	98	98	0	100	30	95	0	0	94	0	0	0	100	75	100	100	100	40	45	0	0
Trehalose	100	100	100	100	100	0	100	97	100	100	100	98	96	100	100	100	98	100	100	100	100	100	13	98
D-Xylose	100	100	100	100	100	0	100	93	100	98	100	95	96	13	98	100	95	100	100	100	100	91	0	98

[a] TSI, Triple sugar iron agar.

teriaceae (Table 2). A primary battery of 14 tests, including adonitol fermentation, arginine dihyrolase, citrate utilization, DNase, gas from glucose, hydrogen sulfide, indole production, lysine decarboxylase, motility, ornithine decarboxylase, phenylalanine deaminase, sucrose fermentation, urease, and Voges-Proskauer, is sufficient for the identification of most isolates to the genus level and of many isolates to the species level. A secondary battery of 14 additional tests may be used as a supplementary battery to identify more difficult isolates. The secondary test battery consists mainly of carbohydrate fermentation tests plus tests for the hydrolysis of *ortho*-nitrophenyl-galactoside (ONPG) and for growth in KCN (because of the potential hazard in the preparation of the KCN medium, it is recommended that clinical laboratories purchase the medium from commercial suppliers). Several media are available for combined testing of the primary test battery reactions. Triple sugar iron agar provides results for gas and H_2S production and

TABLE 2. *Continued*

	% Positive reactions																							
	P. penneri	*P. vulgaris*	*Providencia alcalifaciens*	*P. rettgeri*	*P. stuartii*	*Salmonella choleraesuis*	*S. paratyphi* A	*S. typhi*	*Salmonella*, other serotypes	*Serratia liquefaciens* group	*S. marcescens*	*S. odorifera*	*S. rubidaea*	*Shigella boydii*	*S. dysenteriae*	*S. flexneri*	*S. sonnei*	*Tatumella ptyseos*	*Yersinia enterocolitica*	*Y. frederiksenii*	*Y. kristensenii*	*Y. pestis*	*Y. pseudotuberculosis*	*Y. intermedia*
Primary test battery																								
D-Adonitol	0	0	100	100	5	0	0	0	0	5	40	55	98	0	0	0	0	0	0	0	0	0	0	0
Arginine dihydrolase	0	0	0	0	0	55	15	3	70	0	0	0	0	18	3	2	2	0	0	0	0	0	0	1
Citrate utilization	0	15	98	95	93	25	0	0	95	90	98	98	95	0	0	0	0	2	0	15	0	0	0	0
DNase	25	60	0	0	2	0	0	0	2	85	98	100	98	0	0	0	0	0	9	0	0	0	0	0
D-Glucose, gas production	45	85	85	10	0	95	100	0	96	75	55	8	30	0	0	3	0	0	5	40	23	0	0	18
Hydrogen sulfide (TSI)[a]	10	95	0	0	0	50	10	97	95	0	0	0	0	0	0	0	0	0	0	0	0	0	0	0
Indole production	0	98	100	100	98	0	0	0	0	2	2	55	2	25	45	50	0	0	50	100	30	0	0	99
Lysine decarboxylase	0	0	0	0	0	95	0	98	98	90	100	96	55	0	0	0	0	0	0	0	0	0	0	0
Motility	85	95	96	94	85	95	95	97	95	95	97	100	85	0	0	0	0	0	2	0	8	0	0	6
Ornithine decarboxylase	0	0	0	0	0	100	95	0	97	100	100	40	0	2	0	0	98	0	95	92	92	0	0	100
Phenylalanine deaminase	100	100	98	98	95	0	0	0	0	0	0	0	0	0	0	0	0	90	0	0	0	0	0	0
Sucrose	100	97	15	15	50	0	0	0	2	98	100	40	98	0	0	2	2	98	95	100	0	0	0	100
Urease	100	95	0	98	30	0	0	0	0	3	15	2	2	0	0	0	0	0	75	60	77	5	95	77
Voges-Proskauer	0	0	0	0	0	0	0	0	0	93	98	81	100	0	0	0	0	5	2	0	0	0	0	12
Secondary test battery																								
L-Arabinose	0	0	0	0	0	0	100	2	100	98	0	100	100	94	45	60	95	0	98	100	77	100	50	100
myo-Inositol	0	0	0	90	95	0	0	0	35	60	75	100	20	0	0	0	0	0	30	7	15	0	0	18
KCN, growth in	100	100	100	97	100	0	0	0	0	90	95	35	25	0	0	0	0	0	2	0	0	0	0	6
Lactose	0	2	0	5	2	0	0	2	2	10	2	87	100	2	0	2	2	0	5	40	8	0	0	35
Malonate	0	0	0	0	0	0	0	0	0	2	3	0	94	0	0	0	0	0	0	0	0	0	0	6
D-Mannitol	0	0	2	100	10	98	100	100	100	100	100	98	100	97	0	95	100	0	98	100	100	97	100	100
Melibiose	0	0	0	5	0	45	95	100	95	75	0	98	98	15	0	55	25	25	0	0	0	50	70	80
ONPG (β-galactosidase)	0	10	0	5	10	0	0	0	2	93	95	100	100	10	40	2	90	0	95	100	70	50	70	93
Raffinose	0	5	0	5	7	0	0	0	2	85	2	45	98	2	0	40	3	11	5	30	0	0	15	45
L-Rhamnose	0	5	0	70	0	100	100	0	95	15	0	94	2	0	30	5	75	0	0	100	0	2	70	100
Salicin	0	75	0	50	2	0	0	0	0	97	95	65	98	0	0	0	0	55	20	92	15	70	25	100
D-Sorbitol	0	0	0	0	0	90	95	98	95	95	100	100	2	45	30	30	2	0	98	100	100	50	0	100
Trehalose	55	30	0	0	98	0	100	100	100	100	100	100	100	85	85	65	100	93	98	100	100	100	100	100
D-Xylose	100	95	0	10	7	98	0	82	97	100	7	100	98	11	4	2	2	9	70	100	85	90	100	100

also for the fermentation of glucose. Motility-indole-ornithine medium provides results in a single medium for motility, indole production, and ornithine decarboxylase. Phenylalanine-urease broth provides testing for phenylalanine deaminase and urease. Thus, the 14 primary test battery reactions can be obtained from 10 tubed biochemical media. However, each of the secondary test battery reactions must be performed individually. Clinically significant isolates that cannot be identified by the primary and secondary test batteries should be sent to a reference laboratory for further testing.

Screening tests

Test batteries that are rapid or abbreviated or both may be used to identify or aid in the identification of *Enterobacteriaceae*. A spot test for oxidase is an essen-

TABLE 3. Biochemical characteristics of clinically significant genera of the family *Enterobacteriaceae*

Characteristic	Biochemical reaction of genus[a]:																
	Cedecea	*Citrobacter*	*Edwardsiella*	*Enterobacter*[b]	*Escherichia*	*Ewingella*	*Hafnia*	*Klebsiella*	*Kluyvera*	*Morganella*	*Proteus*	*Providencia*	*Salmonella*	*Serratia*	*Shigella*	*Tatumella*	*Yersinia*
Primary test battery																	
Adonitol	−	V	−	V	V	−	−	+	−	−	−	V	−	V	−	−	−
Arginine	V	V	−	V	V	−	−	−	−	−	−	−	V	−	V	−	−
Citrate	+	+	−	+	V	+	V	V	V	−	V	+	V	+	−	−	−
DNase	−	−	−	−	−	−	−	−	−	−	V	−	−	+	−	−	−
Gas	V	+	+	+	+	−	+	V	+	V	V	V	V	V	−	−	V
H_2S	−	V	+	−	−	−	−	−	−	−	V	−	V	−	−	−	−
Indole	−	V	+	−	+	−	−	V	V	+	V	+	−	V	V	−	V
Lysine	−	−	+	V	+	−	+	V	V	−	−	−	V	V	−	−	−
Motility	+	+	+	+	V	V	+	−	V	+	+	+	+	+	−	−	−
Ornithine	V	V	+	+	V	−	+	−	+	+	V	−	V	V	V	−	V
Phenylalanine	−	−	−	−	−	−	−	−	−	+	+	+	−	−	−	+	−
Sucrose	+	V	−	+	V	−	V	V	V	−	V	V	−	V	−	+	V
Urease	−	V	−	V	−	−	−	V	−	+	+	V	−	−	−	−	V
VP	V	−	−	+	−	+	V	V	−	−	V	−	−	+	−	−	−
Secondary test battery																	
Arabinose	−	+	−	+	+	−	+	+	+	−	−	−	V	V	V	−	V
Inositol	−	−	−	V	−	−	−	V	−	−	−	V	V	V	−	−	V
KCN	V	V	−	V	−	−	+	+	V	+	+	+	−	V	−	−	−
Lactose	V	V	−	V	V	V	−	V	+	−	−	−	−	V	−	−	V
Malonate	+	V	−	V	V	−	V	V	V	−	−	−	−	V	−	−	−
Mannitol	+	+	−	+	+	+	+	+	+	−	−	V	+	+	V	−	+
Melibiose	−	V	−	+	V	−	−	+	+	−	−	−	V	V	V	V	V
ONPG	+	+	−	+	+	+	+	V	+	−	−	−	−	+	V	−	V
Raffinose	−	V	−	+	V	−	−	+	+	−	−	−	−	V	V	V	V
Rhamnose	−	+	−	+	V	V	+	V	+	−	−	V	V	V	V	−	V
Salicin	+	V	−	V	V	+	V	+	+	−	V	V	−	V	−	V	V
Sorbitol	V	+	−	V	V	−	−	+	V	−	−	−	+	V	V	−	V
Trehalose	+	+	−	+	+	+	+	+	+	V	V	V	V	+	V	+	+
Xylose	+	+	−	+	+	V	+	+	+	−	+	−	V	V	−	−	V

[a] Symbols: −, ≤9% of strains positive; V, 10 to 89% of strains positive; +, ≥90% of strains positive. Only clinically significant species are included in the percentage values for each genus.
[b] Not including *E. agglomerans*.

tial preliminary screening test that should be performed early in the identification of all isolates. Since all *Enterobacteriaceae* are oxidase negative, this simple test can save much unnecessary work by the elimination of oxidase-positive organisms. However, caution should be exercised in the use of this test because false-negative reactions may occur with isolates tested from selective-differential media.

E. coli and swarming *Proteus* species make up a substantial percentage of the *Enterobacteriaceae* encountered in clinical laboratories, and these organisms can be accurately identified by experienced clinical microbiologists by colony morphology and a spot indole test. Flat, lactose-positive colonies surrounded by a zone of precipitated bile on MacConkey agar can be identified as *E. coli* if they give a positive spot indole test. For typical growth of swarming *Proteus*, spot indole-negative isolates can be identified as *P. mirabilis*. Many of the other commonly encountered organisms can be readily identified by using the 10 tubed media described above for the primary test battery.

IDENTIFICATION

Cedecea species

Members of the genus *Cedecea* resemble *Serratia* species in many of the commonly tested biochemical characteristics, but *Cedecea* species are DNase negative (Table 3). Members of this genus are also lipase positive (corn oil) and gelatin hydrolysis negative (15). Members of this genus are rarely recovered from clinical specimens, but when encountered, isolates of *C. davisae* and *C. neteri* may be of clinical significance. More than 90% of these strains give positive reactions for citrate utilization, motility, ONPG, and acid production from sucrose, malonate, mannitol, salicin, trehalose, and xylose. More than 90% of the strains are also negative for H_2S, indole, lysine decarboxylase, phenylalanine deaminase, and urease. The two potentially significant species are distinguished by sorbitol fermentation and ornithine decarboxylase. *C. davisae* is sorbitol negative and ornithine positive, whereas *C. neteri* is sorbitol positive and ornithine negative.

Citrobacter species

Members of the genus *Citrobacter* typically give positive tests for citrate, gas from glucose fermentation, motility, ONPG, and fermentation of arabinose, mannitol, rhamnose, sorbitol, trehalose, and xylose (Table 3). *Citrobacter* species are also negative for DNase, lysine decarboxylase, phenylalanine deaminase, Voges-Proskauer, and inositol fermentation. The three currently recognized *Citrobacter* species, *C. amalonaticus, C. diversus*, and *C. freundii*, are all well-recognized human pathogens that cause a variety of infections. *C. freundii* strains often produce H_2S and may be confused with *Salmonella* species unless adequate biochemical testing is performed. *Citrobacter* isolates may also give false-positive agglutination tests in *Salmonella* grouping sera.

Adonitol, H_2S, indole, KCN, and malonate are useful tests for the identification of *Citrobacter* isolates (Table 4). *C. amalonaticus* isolates are adonitol, H_2S, and malonate negative but indole and KCN positive. *C. diversus* isolates are adonitol, indole, and malonate positive but H_2S and KCN negative. *C. freundii* isolates are adonitol, indole, and usually malonate negative.

Edwardsiella species

E. tarda is the single clinically significant species of the genus *Edwardsiella*, and this organism is a well-established cause of a variety of extraintestinal human infections. The role of *E. tarda* as an intestinal pathogen is yet to be firmly established. Recently, new species and groups of the genus *Edwardsiella* have been described (17), but there is no evidence that they cause human infections. Clinically significant *Edwardsiella* strains are positive for gas from glucose fermentation, H_2S, indole, lysine decarboxylase, motility, and ornithine decarboxylase (Table 3). They are negative for other biochemical tests commonly performed.

Enterobacter species

The genus *Enterobacter* is a large and heterogeneous group of organisms. Most *Enterobacter* species are characterized by positive tests for citrate, gas from glucose, motility, ornithine decarboxylase, Voges-Proskauer, ONPG, and fermentation of sucrose, arabinose, mannitol, melibiose, raffinose, rhamnose, trehalose, and xylose (Table 3). Most members of the genus are negative for DNase, H_2S, indole, and phenylalanine deaminase. Although there are eight species of *Enterobacter* currently recognized taxonomically, only five species are clinically significant in human infections. *E. aerogenes* and *E. cloacae* are the members of the genus most often encountered in clinical laboratories, and these are species clearly associated with pulmonary, bloodstream, central nervous system, soft tissue, genitourinary, and other extraintestinal infections of humans. *E. agglomerans* is also a well-established human pathogen that has particularly been associated with nosocomial infections (27). *E. gergoviae* and *E. sakazakii* are more recently recognized species. *E. gergoviae* has been associated with urinary tract, pulmonary, and bloodstream infections (6, 29), and *E. sakazakii* has been associated with several types of infections, most notably with neonatal sepsis and meningitis (9).

Tests useful in the differentiation of *Enterobacter* species include adonitol and sorbitol fermentation, arginine dihydrolase, KCN, and lysine and ornithine decarboxylase (Table 5). *E. aerogenes* is distinguished by positive reactions for all tests except arginine, and *E. agglomerans* is negative or variable for the reactions listed. This latter species is poorly defined taxonomically and consists of several genetically distinct groups as determined by DNA hybridization (4). This diversity makes exact biochemical identification difficult. For practical purposes, *Enterobacter* strains that are arginine, lysine, and ornithine negative can be reported as *E. agglomerans*. Most *E. agglomerans* isolates are also yellow pigmented, which may be useful in their identification. *E. cloacae* isolates are arginine, ornithine, KCN, and sorbitol positive, but lysine negative. Most isolates are also adonitol negative. *E. gergoviae* is lysine and ornithine positive, but negative for adonitol, arginine, KCN, and sorbitol. *E. sakazakii* is arginine, ornithine, and KCN positive, but adonitol, lysine, and sorbitol negative. This species is also yellow pigmented and gives a delayed-positive DNase test.

TABLE 4. Biochemical differentiation of *Citrobacter* species

Test	Reaction of[a]:		
	C. amalonaticus	*C. diversus*	*C. freundii*
Adonitol fermentation	−	+	−
H_2S production (triple sugar iron)	−	−	+
Indole production	+	+	−
KCN, growth in	+	−	+
Malonate utilization	−	+	V

[a] Symbols: −, ≤9% of strains positive; V, 10 to 89% of strains positive; +, ≥90% of strains positive.

Escherichia species

For many years, *E. coli* was the only species in the genus *Escherichia*, but four additional species are now recognized (10). Three of the four species, i.e., *E. fergusonii, E. hermannii*, and *E. vulneris*, have been isolated from potentially significant human sources, but at this time they are not yet proven to be human pathogens. It should also be recognized that the new *Escherichia* species are uncommon, and *E. coli* accounts for more than 99% of *Escherichia* isolates. *E. coli* is typically positive for gas production from glucose, indole, lysine, arabinose, mannitol, ONPG, trehalose, and xylose (Table 3). Isolates are also generally negative for DNase, H_2S, phenylalanine deaminase, urease, Voges-Proskauer, inositol, and KCN. Other fermentation tests, not included in the recommended routine batteries, that may be useful in differentiating *E. coli* from the other species include cellobiose, arabitol, and mucate. *E. coli* is mannitol and mucate positive, but cellobiose and arabitol negative. These additional tests may be needed when dealing with isolates of *E. coli* from hemorrhagic colitis cases (see Serological Testing below) because these isolates are typically sorbitol negative.

TABLE 5. Biochemical differentiation of *Enterobacter* species

Test	Reaction of[a]:				
	E. aerogenes	*E. agglomerans*	*E. cloacae*	*E. gergoviae*	*E. sakazakii*
Adonitol fermentation	+	−	V	−	−
Arginine dihydrolase	−	−	+	−	+
KCN, growth in	+	V	+	−	+
Lysine decarboxylase	+	−	−	+	−
Ornithine decarboxylase	+	−	+	+	+
Sorbitol fermentation	+	V	+	−	−

[a] Symbols: −, ≤9% of strains positive; V, 10 to 89% of strains positive; +, ≥90% of strains positive.

Ewingella americana

The recently named genus *Ewingella* has a single species, *E. americana* (10, 14). These organisms are typically citrate, Voges-Proskauer, and ONPG positive, and they ferment mannitol, salicin, and trehalose (Table 3). *Ewingella* isolates are typically decarboxylase and phenylalanine deaminase negative, and they give negative or variable reactions for other commonly employed biochemical tests. Approximately one-fourth of *E. americana* isolates studied to date have been from blood cultures.

Hafnia alvei

The genus *Hafnia* contains a single species, *H. alvei*. Organisms classified as *H. alvei* were previously known as "*Enterobacter hafniae.*" *H. alvei* is characterized by positive reactions for lysine decarboxylase, motility, ornithine decarboxylase, growth in KCN, ONPG, and fermentation of arabinose, mannitol, rhamnose, trehalose, and xylose (Table 3). *H. alvei* isolates are negative for adonitol, arginine, indole, and urease. Although infrequently encountered compared with other *Enterobacteriaceae*, isolates of *H. alvei* may cause a variety of extraintestinal infections.

Klebsiella species

Members of the genus *Klebsiella* give negative or variable reactions for most of the primary battery of biochemical tests. *Klebsiella* species are characteristically adonitol positive, but arginine, DNase, H_2S, motility, ornithine, and phenylalanine negative (Table 3). They are also active in carbohydrate fermentations, giving positive tests for arabinose, mannitol, melibiose, raffinose, salicin, sorbitol, trehalose, and xylose. There are seven currently recognized *Klebsiella* species (10), but only five species are known to be clinically significant. Two of these five species, *K. ozaenae* and *K. rhinoscleromatis*, are variants of *K. pneumoniae*, but they are considered separate species because of their association with specific human diseases. These five clinically significant species (*K. oxytoca, K. ozaenae, K. planticola, K. pneumoniae*, and *K. rhinoscleromatis*) can be distinguished by tests for indole, malonate, ONPG, and Voges-Proskauer (Table 6). *K. oxytoca* is distinguished by a positive indole reaction; *K. ozaenae* is negative for malonate and Voges-Proskauer; *K. pneumoniae* is indole negative, but malonate, ONPG, and Voges-Proskauer positive; and *K. rhinoscleromatis* is indole and ONPG negative. *K. planticola* (1), the most recently recognized species, is similar to *K. pneumoniae* or *K. oxytoca* but can be distinguished from these species by its ability to ferment glucose at 5 but not at 45°C. In addition, an ornithine-positive biogroup of *K. planticola* has been recognized, and isolates in this group are often clinically significant. *Klebsiella* species account for a significant proportion of urinary tract, bloodstream, and other extraintestinal infections. *K. pneumoniae* and *K. oxytoca* are the most commonly encountered members of the genus, and although the other species are rarely isolated, they may cause significant infections.

Kluyvera species

Kluyvera species are positive for ornithine decarboxylase, ONPG, and fermentation of arabinose, lactose, mannitol, melibiose, raffinose, rhamnose, salicin, trehalose, and xylose (Table 3). There are two currently recognized species of *Kluyvera*, i.e., *K. ascorbata* and *K. cryocrescens* (11). These species are very similar biochemically, and it is recommended that clinical laboratories report the genus identification without attempting to identify the species. However, the species

TABLE 6. Biochemical differentiation of *Klebsiella* species

Test	Reaction of[a]:				
	K. oxytoca	*K. ozaenae*	*K. planticola*	*K. pneumoniae*	*K. rhinoscleromatis*
Glucose fermentation at 5°C	−	−	+	−	−
Indole production	+	−	V	−	−
Malonate utilization	+	−	+	+	+
ONPG	+	V	+	+	−
Voges-Proskauer	+	−	+	+	+

[a] Symbols: −, ≤9% of strains positive; V, 10 to 89% of strains positive; +, ≥90% of strains positive.

can be separated by the ascorbate test (10), with *K. ascorbata* giving a positive test. In addition, *K. ascorbata* is resistant to cephalothin and carbenicillin, whereas *K. cryocrescens* is susceptible. *Kluyvera* isolates have occasionally been associated with significant human infections, and the two species seem to have equal pathogenic potential. However, *K. ascorbata* appears to be numerically the more important of the two.

Morganella morganii

The genus *Morganella* contains a single species, *M. morganii*, which was previously classified in the genus *Proteus* (10). *M. morganii* is a well-recognized human pathogen that has been associated primarily with urinary tract infections but also causes a variety of other human infections. The organism is characteristically positive for indole, motility, ornithine decarboxylase, phenylalanine deaminase, urease, and growth in KCN (Table 3). It is generally inactive in carbohydrate fermentations and rarely ferments sugars other than glucose. Tests for H_2S, indole, ornithine, and xylose fermentation are useful for separating *M. morganii* from members of the genus *Proteus*.

Proteus species

Members of the genus *Proteus* are positive for motility, phenylalanine deaminase, urease, growth in KCN, and xylose fermentation (Table 3). *Proteus* species are also characteristically negative for adonitol, arginine, lysine, and most carbohydrate fermentations other than glucose and xylose. There are four currently recognized *Proteus* species, but only three, *P. mirabilis*, *P. penneri*, and *P. vulgaris*, are known to cause human disease (10). *Proteus* species are best known as urinary tract pathogens, but they are also well-documented agents of various systemic and localized extraintestinal infections. The three clinically significant *Proteus* species can be differentiated by tests for indole and ornithine decarboxylase. *P. mirabilis* is indole negative and ornithine positive; *P. penneri* is indole negative and ornithine negative; and *P. vulgaris* is indole positive and ornithine negative. *P. penneri* is also resistant to chloramphenicol, whereas *P. mirabilis* and most *P. vulgaris* strains are susceptible.

Providencia species

Clinically significant members of the genus *Providencia* are characterized by positive tests for citrate utilization, indole production, motility, phenylalanine deaminase, and growth in KCN (Table 3). *Providencia* species are also negative for arginine dihydrolase, DNase, H_2S, lysine and ornithine decarboxylases, Voges-Proskauer, and ONPG. They frequently fail to ferment carbohydrates other than glucose. Four species are currently recognized in the *Providencia* genus, but only three, i.e., *P. alcalifaciens*, *P. rettgeri*, and *P. stuartii*, are known to be human pathogens (10, 26). These species are most commonly associated with urinary tract infections, but they may also cause a variety of other infections. The *Providencia* species can be distinguished biochemically by testing for adonitol fermentation, trehalose fermentation, and urease (Table 7). *P. alcalifaciens* is adonitol positive, but trehalose and urease negative; *P. rettgeri* is adonitol and urease positive, but trehalose negative; and *P. stuartii* is adonitol negative, trehalose positive, and either positive or negative for urease.

TABLE 7. Biochemical differentiation of *Providencia* species

Test	Reaction of[a]:		
	P. alcalifaciens	*P. rettgeri*	*P. stuartii*
Adonitol fermentation	+	+	−
Trehalose fermentation	−	−	+
Urea hydrolysis	−	+	V

[a] Symbols: −, ≤9% of strains positive; V, 10 to 89% of strains positive; +, ≥90% of strains positive.

Salmonella species

Members of the genus *Salmonella* are causative agents of a variety of infections ranging from simple gastroenteritis to severe illnesses such as enteric fever with bacteremia. Typhoid fever is an example of the life-threatening illnesses that may be associated with members of the genus *Salmonella*. These organisms may also cause local infections of virtually any organ as a result of seeding during bacteremia.

Most *Salmonella* isolates are characterized by positive reactions for motility and fermentation of mannitol and sorbitol (Table 3). They are also characteristically negative for DNase, indole, phenylalanine deaminase, urease, Voges-Proskauer, growth in KCN, ONPG, and fermentation of adonitol, sucrose, lactose, raffinose, and salicin, and they fail to utilize malonate. Most clinical *Salmonella* isolates are also H_2S positive. However, H_2S production is considered a variable characteristic for the genus overall because only 50% of *S. choleraesuis* and 10% of *S. paratyphi* A are H_2S positive (Table 2).

The classification of *Salmonella* species has been controversial for several years. Until recently, the Centers for Disease Control reported three biochemically distinct species of *Salmonella*, i.e., *S. typhi*, *S. choleraesuis*, and *S. enteritidis*, along with a separate genus for "*Arizona*," and this approach was followed in the previous edition of this Manual. However, all salmonellae appear to be very closely related, and the Centers for Disease Control have now revised this approach to be more consistent with the traditional reporting of serotypes rather than species. Reporting of specific serotypes requires the availability of an extensive battery of antisera and the labor resources to use them. Most clinical laboratories are unable to identify *Salmonella* isolates to specific serotypes, and serogroup reporting is more common. Isolates are then referred to reference laboratories for specific serotyping. This approach is sufficient for most *Salmonella* isolates, but *S. typhi*, *S. choleraesuis*, and *S. paratyphi* A should be specifically identified in clinical laboratories because of their special clinical significance. Hence, in this edition of the Manual, we recommend that *Salmonella* isolates be identified biochemically as *S. choleraesuis*, *S. typhi*, or *S. paratyphi* A, with serogroup confirmation. Other *Salmonella* isolates should be confirmed by serogrouping and reported as a certain *Salmonella* serogroup, with specific serotypes to be reported subsequently, after reference laboratory testing. The final report for these isolates should include the genus and serotype name, e.g., *Salmonella* serotype *typhimurium* or, for simplicity, *S. typhimurium*.

S. choleraesuis, *S. paratyphi* A, and *S. typhi* can be separated by tests for arabinose fermentation, citrate utilization, gas from glucose, lysine decarboxylase, ornithine decarboxylase, rhamnose fermentation, and trehalose fermentation (Table 8). *S. choleraesuis* is arabinose and trehalose negative, occasionally positive for citrate, and positive for the other tests. *S. paratyphi* A is characteristically citrate and lysine negative, but positive for the other reactions. *S. typhi* is positive only for lysine and trehalose, and other *Salmonella* serotypes are positive for all the tests. Isolates of *S. typhi* are also recognizable by their typical reactions in triple sugar iron agar including acid butt, alkaline slant, a band of H_2S formation at the slant-butt interface, and lack of gas production. Specific identification of *S. typhi*, *S. paratyphi* A, and *S. choleraesuis* should be accomplished as soon as possible after isolation in the clinical laboratory because of the propensity of these organisms to cause life-threatening infections of considerable epidemiological significance. It is also important to identify other *Salmonella* isolates at the genus and serogroup level, but specific serotype identification can await results from a reference laboratory. The identity of all *Salmonella* isolates, whether at the genus or species level, should be confirmed both biochemically and serologically (see below).

The "Arizona group" was once considered a separate genus, but these isolates are now known to be very closely related to strains of *Salmonella*. They are now considered members of this genus (7). These organisms will be reported by reference laboratories as *Salmonella* serotypes, e.g., *Salmonella* serotype 47:r:z (formerly *Arizona hinshawii* 23:24:31).

Serratia species

Members of the genus *Serratia* are characterized by negative reactions for arginine dihydrolase, H_2S production, phenylalanine deaminase, and urease (Table 3). *Serratia* isolates are unusual among the *Enterobacteriaceae* in their production of three extracellular hydrolytic enzymes, i.e., DNase, gelatinase, and lipase. In addition, members of the genus characteristically give positive reactions for citrate, motility, Voges-Proskauer, ONPG, and fermentation of mannitol and trehalose. Members of the genus originally were often red pigmented, but most strains isolated in clinical laboratories today are nonpigmented.

Three species and one group of the genus *Serratia* are currently recognized as potential human pathogens (10). *S. marcescens* has a colorful history of use in experiments during a time when it was thought to be a harmless commensal (39). However, it is now recognized as a frequent cause of extraintestinal human infections ranging from simple cystitis to life-threatening bloodstream and central nervous system infections. It is particularly noted as a cause of nosocomial infections, and many hospitals harbor multiply antibiotic-resistant strains of *S. marcescens*. *S. liquefaciens* was, until recently, considered a single distinct species. However, it is currently recognized as a collection of closely related species that cannot be readily separated in clinical laboratories. Therefore, these organisms should be reported as "*S. liquefaciens* group." Members of this group are only occasionally encountered in clinical laboratories in the United States, and they are rarely associated with human infections. *S. odorifera* is a recently recognized species that is rarely encountered in clinical laboratories but has been associated with central nervous system and bloodstream infections. *S. rubidaea* is a well-known species that is occasionally associated with significant human infections but is most often an insignificant isolate from sputum, wound, or stool cultures. At least three other *Serratia* species are currently recognized, but there is no evidence that they are of any clinical significance.

Differentiation of the clinically significant members of the *Serratia* genus can be accomplished by testing for ornithine decarboxylase and fermentation of adonitol, arabinose, and sorbitol (Table 9). Members of the *S. liquefaciens* group are arabinose, ornithine, and sorbitol positive, but adonitol negative. *S. marcescens* is ornithine and sorbitol positive, but arabinose negative. *S. odorifera* is arabinose and sorbitol positive and variably positive for adonitol and ornithine. Differentiation of *S. odorifera* from the *S. liquefaciens* group may be difficult biochemically, but *S. odorifera* has a characteristic pungent odor that has been variously described as musty, potatolike, or vegetablelike (16). *S. rubidaea* is identified by positive reactions for adonitol and arabinose, but negative reactions for ornithine and sorbitol.

TABLE 8. Biochemical differentiation of selected members of the *Salmonella* group

Test	Reaction of[a]:			
	S. choleraesuis	*S. paratyphi* A	*S. typhi*	Other[b]
Arabinose fermentation	−	+	−	+
Citrate utilization	V	−	−	+
Glucose gas production	+	+	−	+
Lysine decarboxylase	+	−	+	+
Ornithine decarboxylase	+	+	−	+
Rhamnose fermentation	+	+	−	+
Trehalose fermentation	−	+	+	+

[a] Symbols: −, ≤9% of strains positive; V, 10 to 89% of strains positive; +, ≥90% of strains positive.

[b] Typical strains in serogroups A through E.

Shigella species

Members of the genus *Shigella* have long been recognized as causative agents for bacillary dysentery. Although these organisms are very important as a cause of gastrointestinal infections, they rarely cause other types of infections. *Shigella* species are genetically very closely related to *E. coli* and would be considered taxonomically as members of the *Escherichia* genus (3). However, because of their distinctive ability to cause dysentery, *Shigella* species should continue to be reported as a distinct genus in clinical laboratories. Members of the genus *Shigella* are among the biochemically least-reactive members of the *Enterobacteriaceae*. They are characteristically negative for adonitol, citrate, DNase, gas from glu-

TABLE 9. Biochemical differentiation of *Serratia* species

Test	Reaction of[a]:			
	S. liquefaciens group	*S. marcescens*	*S. odorifera*	*S. rubidaea*
Adonitol fermentation	−	V	V	+
Arabinose fermentation	+	−	+	+
Ornithine decarboxylase	+	+	V	−
Sorbitol fermentation	+	+	+	−

[a] Symbols: −, ≤9% of strains positive; V, 10 to 89% of strains positive; +, ≥90% of strains positive.

cose, H_2S, lysine, motility, phenylalanine, sucrose, urease, Voges-Proskauer, inositol, KCN, lactose, malonate, salicin, and xylose (Table 3). Members of the genus are no more than variably positive for any of the commonly employed biochemical tests other than glucose fermentation and nitrate reduction.

There are four currently recognized *Shigella* species, i.e., *S. dysenteriae*, *S. flexneri*, *S. boydii*, and *S. sonnei*. All are capable of causing dysentery, and *S. dysenteriae* has been associated with a particularly severe form of illness thought to be related to toxin production. The identification of *Shigella* species is important for both clinical and epidemiological implications. Three biochemical tests are helpful in the separation of the *Shigella* species (Table 10). *S. dysenteriae* is mannitol and ornithine negative, and 40% of the isolates are ONPG positive. *S. flexneri* is mannitol positive, but ONPG and ornithine negative, and *S. boydii* has the same reactions except for occasional (10%) ONPG-positive strains. *S. sonnei* gives positive reactions for mannitol, ONPG (90%), and ornithine and is the only *Shigella* species that is ornithine positive. Serological testing is also needed for the identification of *Shigella* isolates. Although *S. dysenteriae* and *S. sonnei* are biochemically distinct, *S. flexneri* and *S. boydii* are often biochemically identical. Serological grouping is essential for the separation of *S. flexneri* and *S. boydii* and is also required for the confirmation of *S. dysenteriae* and *S. sonnei*.

Tatumella ptyseos

Genus *Tatumella* is a new genus with a single species, *T. ptyseos*, which was previously designated CDC group EF-9 (18). Strains of *T. ptyseos* have been placed in the *Enterobacteriaceae*, based on DNA hybridization studies, despite some unusual characteristics, including susceptibility to penicillin and polar or lateral rather than peritrichous flagella (10, 18). The organism has been recovered from a variety of clinical specimens, including blood cultures. Therefore, it is considered a rare but potentially significant human pathogen. Isolates of *T. ptyseos* are relatively inactive biochemically at 35 to 37°C but are more active at 25°C. They are typically negative for all of the commonly employed biochemical tests except phenylalanine deaminase and fermentation of sucrose and trehalose (Table 3). Isolates may occasionally give positive reactions for fermentation of melibiose, raffinose, and salicin. A positive test for phenylalanine deaminase with negative indole and urease reactions is highly suggestive of *Tatumella*.

Yersinia species

Members of the genus *Yersinia* are characterized by negative or variably positive reactions for most of the commonly used biochemical tests (Table 3). All members of the genus are negative for arginine dihydrolase, citrate utilization, DNase, H_2S, motility, phenylalanine deaminase, Voges-Proskauer, and malonate utilization. They are also positive for fermentation of mannitol and trehalose. Some strains give additional positive reactions (especially for motility and Voges-Proskauer) if incubated at 25 instead of 35 to 37°C. Six clinically significant species of the genus *Yersinia* are currently recognized, and there are at least two other species and groups that will not be discussed here because they are of no known clinical significance (8).

It is well established that some strains of *Y. enterocolitica* can cause gastrointestinal infections, and they may cause a severe form of gastroenteritis that mimics appendicitis. These infections may occur as sporadic cases or as food- or waterborne outbreaks. The organism may also less commonly cause extraintestinal infections. The characteristics that separate *Y. enterocolitica* from other *Yersinia* species include positive tests for ornithine decarboxylase and sucrose fermentation and negative tests for melibiose, raffinose, and rhamnose fermentation (Table 11). More than 70% of the isolates are also positive for urease.

Y. frederiksenii is a recently named organism that was previously a biogroup of *Y. enterocolitica* (36). Its role in human infections is questionable, but it has rarely been isolated from blood, wound, and sputum cultures and must be considered a potential human pathogen (10). It has also been isolated from stool specimens, but its significance as an intestinal pathogen is yet to be established. The organism is commonly found in the environment. *Y. frederiksenii* is characterized biochemically by positive tests for ornithine decarboxylase and fermentation of rhamnose and sucrose.

Y. kristensenii and *Y. intermedia* are also new species that were formerly biogroups of *Y. enterocolitica* (2). These organisms have been recovered from significant human sources, but their role in human infections has not been firmly established. *Y. kristensenii* is characterized by positive tests for ornithine decarboxylase and negative reactions for raffinose, rhamnose, and sucrose fermentation. *Y. intermedia* is ornithine, rhamnose, and sucrose positive. It is also positive for raffinose and melibiose fermentation at 25°C, and these reactions are helpful in distinguishing *Y. intermedia* from *Y. frederiksenii*.

TABLE 10. Biochemical and serological differentiation of *Shigella* species

Test	Reaction of[a]:			
	S. dysenteriae	*S. flexneri*	*S. boydii*	*S. sonnei*
Mannitol fermentation	−	+	+	+
ONPG	V	−	V	+
Ornithine decarboxylase	−	−	−	+
Serogroup	A	B	C	D

[a] Symbols: −, ≤9% of strains positive; V, 10 to 89% of strains positive; +, ≥90% of strains positive.

TABLE 11. Biochemical differentiation of *Yersinia* species

Test	Reaction of[a]:					
	Y. enterocolitica	*Y. frederikensii*	*Y. kristensenii*	*Y. pestis*	*Y. pseudotuberculosis*	*Y. intermedia*
Melibiose fermentation	−	−	−	V	V	V
Ornithine decarboxylase	+	+	+	−	−	+
Raffinose fermentation	−	V	−	−	V	V
Rhamnose fermentation	−	+	−	−	V	+
Sucrose fermentation	+	+	−	−	−	+
Urease	V	V	V	−	+	V

[a] Symbols: −, ≤9% of strains positive; V, 10 to 89% of strains positive; +, ≥90% of strains positive. Results are for tests done at 37°C; test results may differ if done at 25°C (see text).

Y. pestis is the well-known cause of the severe systemic illness, plague, which may be acquired directly or indirectly from infected animals (bubonic) or via aerosols from infected patients (pneumonic). Most infections in the United States are from accidental contact with an infected animal. *Y. pestis* is a dangerous organism, and extreme caution should be used in handling it in the laboratory. A controversy has existed over the nomenclature of this organism because of the recent proposal that *Y. pestis* should be considered a subspecies of *Y. pseudotuberculosis*. However, because of the extreme danger of serious infection from *Y. pestis* compared with the danger from *Y. pseudotuberculosis*, clinical and public health laboratories should continue to report separate species (28). *Y. pestis* is a fastidious organism, and it is characterized by mostly negative biochemical tests (Tables 2 and 11). However, more than 90% of the strains ferment arabinose, mannitol, trehalose, and xylose. A negative urease test is important for distinguishing *Y. pestis* from *Y. pseudotuberculosis*.

SEROLOGICAL TESTING

Serological confirmation is an essential part of the identification of *Salmonella* and *Shigella* isolates, and it is the only means of recognizing certain pathogenic *E. coli* strains. Serological identification of these organisms is performed by slide agglutination techniques with polyvalent or specific somatic (O)-antigen grouping sera followed, in some cases, by testing with individual sera for specific serotype identification. Serogroup testing should be performed by all clinical microbiology laboratories, at least for *Salmonella* and *Shigella* isolates. Serotype identification based on flagellar (H) antigen testing can be referred to public health laboratories in most instances. Because of the possibility of serological cross-reactions, it is essential that isolates be subjected to rigorous biochemical testing before serological analysis is performed. Biochemical and serological testing are complementary, and both are required for identification of *Salmonella* and *Shigella* species and certain *E. coli* isolates.

Slide agglutination tests are done by making a dense suspension of the organism to be tested in a drop of saline (treated with phenol or mercuric iodine). Growth from a triple sugar iron slant is a convenient source for the preparation of the suspension. A drop of properly diluted antiserum is added to the organism suspension and mixed with a wooden stick. The slide is then rocked back and forth and observed for clumping of the suspension. A positive test is indicated by rapid, complete agglutination of the bacteria. Control suspensions without added antisera should also be tested to detect autoagglutination.

Salmonella species

Isolates confirmed biochemically as members of the genus *Salmonella* should be tested with polyvalent antiserum. These reagents contain agglutinins for *Salmonella* O antigens and Vi antigen, representing approximately 95% of the *Salmonella* isolates likely to be encountered in clinical laboratories. If agglutination occurs with the polyvalent reagent, testing with individual grouping reagents should be done. Isolates giving a positive test may be reported as a certain *Salmonella* serogroup and referred to a reference laboratory for serotyping. If an isolate is biochemically a *Salmonella* species but no agglutination is obtained with the polyvalent reagent, a suspension of the organism should be heated in a boiling-water bath for 15 min, cooled, and then retested. Heating removes capsular antigens that may block the reaction with the somatic (O) antigen. In addition, up to 5% of strains will not agglutinate because they belong to a serogroup not represented in the reagents used. Such strains should be sent to a reference laboratory for further testing. Finally, the serological reactions of *S. typhi* are of special note. Unheated *S. typhi* suspensions will typically agglutinate in Vi antisera but not in D-grouping sera, whereas after heating, the reactions are reversed. Agglutination will occur in most polyvalent antisera with both heated and unheated suspensions because the polyvalent reagents contain agglutinins to both O group D and Vi antigens.

Shigella species

Serological confirmation of *Shigella* isolates is based only on O-antigen testing and is therefore less complicated than *Salmonella* testing. *Shigella* isolates fall into four serogroups, A, B, C, and D, which correspond to the four recognized *Shigella* species, *S. dysenteriae*, *S. flexneri*, *S. boydii*, and *S. sonnei*, respectively. The O-grouping sera for testing *Shigella* isolates are polyvalent (A, 10 of the 12 serotypes; B, 6 serotypes and two variants; and C, 15 of the 18 serotypes) except for the group-D reagent, which has a single serotype. Isolates biochemically identified as *Shigella* species should be tested by slide agglutination, as described above, for reactivity with the four grouping sera. Agglutination in one of the four sera plus the biochemical reactions of the isolate can be used for species identification, and results can be reported as *S. sonnei*, *S. flexneri*, etc. If the isolate is biochemically a *Shigella* species but fails to agglutinate in the grouping sera, it should be retested after boiling as described above. If heated suspensions fail to agglutinate and the isolate is well characterized biochemically, it should be sent to a reference laboratory for further testing. Particular care must be taken in the biochemical identification of isolates because of serological cross-reactions with *E. coli* (including the former Alkalescens-Dispar group) and other *Enterobacteriaceae*.

E. coli

Serotyping for the identification of *E. coli* strains capable of causing enteric disease has been applied mostly to the group of organisms known as "enteropathogenic." A limited number of the approximately 162 O serogroups of *E. coli* have been associated with outbreaks of diarrhea in nurseries. However, these organisms have been infrequently recognized in recent years. Serotyping of *E. coli* strains is a complicated and difficult process and very seldom yields definitive etiological information. Accordingly, routine testing in hospital laboratories for the detection of enteropathogenic serotypes is no longer recommended. When nursery or other outbreaks of diarrheal disease occur and no specific pathogen can be identified, *E. coli* isolates may be sent to a reference laboratory for serotyping to possibly identify a cause of the outbreak. Specific serotypes have also been associated with enterotoxigenic and enteroinvasive strains of *E. coli*, but reagents have not yet become commercially available for routine testing in clinical laboratories (22).

Recently, a new enteric infection has been ascribed to *E. coli*. Hemorrhagic colitis is a distinctive form of severe, bloody diarrhea that often requires hospitalization (30). *E. coli* O157:H7 strains have been implicated as the cause of this disease, and they can be definitively identified only by serotyping. Such serotyping is best performed at reference centers. Several representative *E. coli* colonies should be picked for testing, and a negative test for sorbitol fermentation may be useful in the selection of isolates for serotyping (38).

ENTEROTOXIN TESTING

Laboratory diagnosis of diarrhea due to *E. coli* poses a particular problem because of the ability of this organism to cause diarrhea by several different mechanisms. This problem is further complicated by the inability of traditional cultural and serological methods to detect many pathogenic strains. The detection of most pathogenic *E. coli* strains requires methods for toxin testing. Until recently, the available methods were not applicable to routine use in clinical laboratories because they required the use of animals or cell cultures. However, newly described methods can be used in clinical laboratories, especially if the reagents become commercially available. These methods are based on immunologic and DNA hybridization technologies.

Promising immunologic techniques include enzyme-linked immunosorbent assays and particle agglutination methods. Although a gel immunodiffusion technique is commercially available, it requires several days to perform and is technically demanding (19). Enzyme-linked immunosorbent assay methods involve the binding of toxin in culture filtrates to ganglioside-coated wells of plastic microtiter plates. The presence of the toxin is detected by the addition of specific antibody, followed by the addition of enzyme-conjugated antibody. A positive test is indicated by a color reaction. Such tests have been found to have up to 100% sensitivity and specificity when compared with traditional cell culture assays (34). The second promising immunologic technique is coagglutination with protein A-bearing *Staphylococcus aureus* conjugated with antibody against *E. coli* enterotoxins. This technique has been found to be effective for the detection of enterotoxin in colony lysates of *E. coli* (31). Both techniques also have promise for the direct detection of enterotoxins in stool specimens without culture (23).

DNA hybridization has been successfully applied to the detection of *E. coli* enterotoxin from isolated colonies and directly in stool samples (24). The technique employs radioactively labeled DNA fragments that encode for the heat-stable and the heat-labile enterotoxins of *E. coli*. The application of these genetic probes to isolated colonies involves growing the organism to be tested on nitrocellulose filters, lysing the colonies, and hybridizing the colony DNA with the probe DNA. Hybridization is detected by autoradiography, and positive colonies are indicated by a black spot. For the direct detection of enterotoxins, stool specimens are spotted onto filter paper, incubated overnight, and tested as described above. These techniques have been very effective for the detection of both the heat-stable and the heat-labile toxins (33).

The availability of these methods makes enterotoxin testing feasible for many laboratories, and when these or similar methods become commercially available in the near future, it will be possible for all but the smallest laboratories to detect *E. coli* enterotoxins. Due to a lack until now of enterotoxin detection methods applicable to clinical laboratories, the importance of *E. coli* as a cause of diarrhea in the United States has been unclear. The availability of these newer methods should clarify the role of *E. coli* and lead to better diagnosis of bacterial diarrhea.

ANTIBIOTIC SUSCEPTIBILITY

The antibiotic susceptibility of the *Enterobacteriaceae* is variable according to the genus and species and, to a certain extent, within species. Accordingly, susceptibility testing should be done on individual isolates when antibiotic therapy is indicated. However, most genera and some species have susceptibility patterns that are predictable in general, and useful information may be gained from such patterns to guide antimicrobial therapy until susceptibility testing results are available. However, susceptibility patterns may vary among institutions, and profiles should be generated for each laboratory if possible. In addition, susceptibility patterns for individual species have been unstable in recent years. For example, until recently, *E. coli* was considered susceptible to most of the commonly tested drugs for gram-negative bacilli, but increasing numbers of isolates in recent years have been resistant to ampicillin, cephalothin, and other drugs. Therefore, it is important for laboratories to update their susceptibility profiles regularly.

The following general patterns of susceptibility hold true for most hospitals, although individual variation occurs as mentioned above. *Citrobacter diversus* is characteristically susceptible to all drugs commonly tested for gram-negative bacilli except ampicillin and carbenicillin. *C. freundii* is more resistant overall, but two-thirds of its isolates are susceptible to carbenicillin and more than one-third are susceptible to ampicillin. The most commonly encountered *Enterobacter* species, *E. aerogenes* and *E. cloacae*, are resistant to ampicillin and cephalothin. As mentioned above, most *E. coli* isolates are susceptible to the commonly tested drugs, but an increasing proportion are resistant, especially to ampicillin and cephalothin. In our experience, more than 40% of *E. coli* isolates are resistant to ampicillin, and more than 20% are resistant to cephalothin. Such *E. coli* strains resemble *Enterobacter* species in their susceptibility profiles. *K. pneumoniae* isolates are characteristically resistant to ampicillin and carbenicillin, but most isolates are susceptible to other drugs commonly tested. *K. oxytoca* isolates are more often susceptible to carbenicillin. *M. morganii* isolates are predictably resistant to ampicillin and cephalothin, and they are characteristically resistant to colistin. In addition, more than 40% of the isolates are resistant to tetracycline. Isolates of *P. mirabilis* are among the most susceptible of the *Enterobacteriaceae*, but they are classically resistant to colistin and tetracycline. Isolates of *P. vulgaris* have a susceptibility pattern very similar to that of *M. morganii*. *Providencia* species tend to be relatively resistant to antimicrobial agents, and *Salmonella* and *Shigella* isolates in the United States are susceptible to most drugs, although multiply resistant strains are common in other countries. *S. marcescens*, especially hospital strains, is often drug resistant, and it is classically resistant to ampicillin, cephalothin, and colistin.

Some of the less commonly encountered species also have characteristic susceptibility patterns. *Cedecea* strains are usually colistin and cephalothin resistant. *K. cryocrescens* is usually susceptible to carbenicillin and cephalothin, but *K. ascorbata* is generally resistant. *P. penneri* strains were initially recognized in part because of resistance to chloramphenicol, and *T. ptyseos* isolates are unique among the *Enterobacteriaceae* in their marked susceptibility to penicillin.

LITERATURE CITED

1. **Bagley, T., R. J. Seidler, and D. J. Brenner.** 1981. *Klebsiella planticola* sp. nov.: a new species of *Enterobacteriaceae* found primarily in non-clinical environments. Curr. Microbiol. **6:**105–109.
2. **Bercovier, H., J. Ursing, D. J. Brenner, A. G. Steigerwalt, G. R. Fanning, G. P. Carter, and H. H. Mollaret.** 1980. *Yersinia kristensenii*: a new species of *Enterobacteriaceae* composed of sucrose-negative strains (formerly called *Yersinia enterocolitica* or *Yersinia enterocolitica*-like). Curr. Microbiol. **4:**219–224.
3. **Brenner, D. J.** 1981. Introduction to the *Enterobacteriaceae*, p. 1105–1127. *In* M. P. Starr, H. Stolp, H. G. Truper, A. Balows, and H. G. Schlegel (ed.), The prokaryotes. Springer-Verlag, New York.
4. **Brenner, D. J.** 1981. The genus *Enterobacter*, p. 1173–1180. *In* M. P. Starr, H. Stolp, H. G. Truper, A. Balows, and H. G. Schlegel (ed.), The prokaryotes. Springer-Verlag, New York.
5. **Brenner, D. J.** 1984. Family I. *Enterobacteriaceae* Rahn 1937, p. 408–516. *In* N. R. Krieg and J. G. Holt (ed.), Bergey's manual of systematic bacteriology, vol. 1. The Williams & Wilkins Co., Baltimore.
6. **Brenner, D. J., C. Richard, A. G. Steigerwalt, M. A. Asbury, and M. Mandel.** 1980. *Enterobacter gergoviae* sp. nov.: a new species of *Enterobacteriaceae* found in clinical specimens and the environment. Int. J. Syst. Bacteriol. **30:**1–6.
7. **Crosa, J. H., D. J. Brenner, W. H. Ewing, and S. Falkow.** 1973. Molecular relationships among the Salmonelleae. J. Bacteriol. **115:**307–315.
8. **Ewing, W. H., A. J. Ross, D. J. Brenner, and G. R. Fanning.** 1978. *Yersinia ruckeri* sp. nov., the redmouth (RM) bacterium. Int. J. Syst. Bacteriol. **28:**37–44.
9. **Farmer, J. J., III, M. A. Asbury, F. W. Hickman, D. J. Brenner, and The *Enterobacteriaceae* Study Group.** 1980. *Enterobacter sakazakii*: a new species of "*Enterobacteriaceae*" isolated from clinical specimens. Int. J. Syst. Bacteriol. **30:**569–584.
10. **Farmer, J. J., III, B. R. Davis, F. W. Hickman-Brenner, A. McWhorter, G. P. Huntley-Carter, M. A. Asbury, C. Riddle, H. G. Wathen-Grady, C. Elias, G. R. Fanning, A. G. Steigerwalt, C. M. O'Hara, G. K. Morris, P. B. Smith, and D. J. Brenner.** 1985. Biochemical identification of new species and biogroups of *Enterobacteriaceae* isolated from clinical specimens. J. Clin. Microbiol. **21:**46–76.
11. **Farmer, J. J., III, G. R. Fanning, G. P. Huntley-Carter, B. Holmes, F. W. Hickman, C. Richard, and D. J. Brenner.** 1981. *Kluyvera*, a new (redefined) genus in the family *Enterobacteriaceae*: identification of *Kluyvera ascorbata* sp. nov. and *Kluyvera cryocrescens* sp. nov. in clinical specimens. J. Clin. Microbiol. **13:**919–933.
12. **Gangarosa, E. J., and M. H. Merson.** 1977. Epidemiologic assessment of the relevance of the so-called enteropathogenic serogroups of *Escherichia coli* in diarrhea. N. Engl. J. Med. **296:**1210–1213.
13. **Gorbach, S. L., B. H. Kean, D. G. Evans, D. J. Evans, Jr., and D. Bessudo.** 1975. Travelers' diarrhea and toxigenic *Escherichia coli*. N. Engl. J. Med. **292:**933–936.
14. **Grimont, P. A. D., J. J. Farmer III, F. Grimont, M. A. Asbury, D. J. Brenner, and C. DeVal.** 1983. *Ewingella americana* gen. nov., sp. nov., a new *Enterobacteriaceae* isolated from clinical specimens. Ann. Inst. Pasteur **134:**39–52.
15. **Grimont, P. A. D., F. Grimont, J. J. Farmer III, and M. A. Asbury.** 1981. *Cedecea davisae* gen. nov., sp. nov. and *Cedecea lapagei* sp. nov., new *Enterobacteriaceae* from clinical specimens. Int. J. Syst. Bacteriol. **31:**317–326.
16. **Grimont, P. A. D., F. Grimont, C. Richard, B. R. Davis, A. G. Steigerwalt, and D. J. Brenner.** 1978. Deoxyribonucleic acid relatedness between *Serratia plymuthica* and other *Serratia* species, with a description of *Serratia odorifera* sp. nov. (type strain: ICPB 3995). Int. J. Syst. Bacteriol. **28:**453–463.
17. **Grimont, P. A. D., F. Grimont, C. Richard, and R. Sakazaki.** 1980. *Edwardsiella hoshinae*, a new species of *Enterobacteriaceae*. Curr. Microbiol. **4:**347–352.
18. **Hollis, D. G., F. W. Hickman, G. R. Fanning, J. J. Farmer III, R. E. Weaver, and D. J. Brenner.** 1981. *Tatumella ptyseos* gen. nov., sp. nov., a member of the family *Enterobacteriaceae* found in clinical specimens. J. Clin. Microbiol. **14:**79–88.
19. **Honda, T., S. Taga, Y. Takeda, and T. Miwatani.** 1981. Modified Elek test for detection of heat-labile enterotoxin of enterotoxigenic *Escherichia coli*. J. Clin. Microbiol. **13:**1–5.
20. **Karmali, M. A., M. Petric, C. Lim, P. C. Fleming, and B. T. Steele.** 1983. *Escherichia coli* cytotoxin, hemolytic-uremic syndrome, and hemorrhagic colitis. Lancet **ii:**1299–1300.

21. **Levine, M. M., E. J. Bergquist, D. R. Nalin, D. H. Waterman, R. B. Hornick, C. R. Young, S. Satman, and B. Rowe.** 1978. *Escherichia coli* strains that cause diarrhea but do not produce heat-labile or heat-stable enterotoxins and are non-invasive. Lancet **i:**1119–1122.
22. **Merson, M. H., B. Rowe, R. E. Black, I. Huq, R. J. Gross, and A. Eusof.** 1980. Use of antisera for identification of enterotoxigenic *Escherichia coli*. Lancet **ii:**222–224.
23. **Morgan, D. R., H. L. DuPont, L. V. Wood, and C. D. Ericsson.** 1983. Comparison of methods to detect *Escherichia coli* heat-labile enterotoxin in stool and cell-free culture supernatants. J. Clin. Microbiol. **18:**798–802.
24. **Moseley, S. L., P. Echeverria, J. Seriwatana, C. Tirapat, W. Chaicumpa, T. Sakuldaipeara, and S. Falkow.** 1982. Identification of enterotoxigenic *Escherichia coli* by colony hybridization using three enterotoxin gene probes. J. Infect. Dis. **145:**863–869.
25. **Nunes, M. P., and I. D. Ricciardi.** 1981. Detection of *Yersinia enterocolitica* heat-stable enterotoxin by suckling mouse bioassay. J. Clin. Microbiol. **13:**783–786.
26. **Penner, J. L.** 1981. The tribe *Proteeae*, p. 1204–1224. *In* M. P. Starr, H. Stolp, H. G. Truper, A. Balows, and H. G. Schlegel (ed.), The prokaryotes. Springer-Verlag, New York.
27. **Pien, F. D., W. J. Martin, P. E. Hermans, and J. A. Washington.** 1972. Clinical and bacteriologic observations on the proposed species, *Enterobacter agglomerans* (the Herbicola-Lathyri bacteria). Mayo Clin. Proc. **47:**739–745.
28. **Quan, T. J., J. D. Poland, and A. M. Barnes.** 1984. Where has *pestis* gone? Dangerously egregious taxonomic recommendations. ASM News **50:**193–194.
29. **Richard, C. B., B. Joly, J. Sirot, G. H. Stoleru, and M. Popoff.** 1976. Étude de souches de *Enterobacter* appartenent à un groupe particulier proche de *E. aerogenes*. Ann. Inst. Pasteur **127:**545–548.
30. **Riley, L. W., R. S. Remis, S. D. Helgerson, H. B. McGee, J. G. Wells, B. R. Davis, R. J. Hebert, E. S. Olcott, L. M. Johnson, N. T. Hargrett, P. A. Blake, and M. L. Cohen.** 1983. Hemorrhagic colitis associated with a rare *Escherichia coli* serotype. N. Engl. J. Med. **308:**681–685.
31. **Rönnberg, B., and T. Wadström.** 1983. Rapid detection by a coagglutination test of heat-labile enterotoxin in cell lysates from blood agar-grown *Escherichia coli*. J. Clin. Microbiol. **17:**1021–1025.
32. **Sack, R. B.** 1975. Human diarrheal disease caused by enterotoxigenic *Escherichia coli*. Annu. Rev. Microbiol. **29:**333–353.
33. **Seriwatana, J., P. Echeverria, J. Escamilla, R. Glass, I. Huq, R. Rockhill, and B. J. Stoll.** 1983. Identification of enterotoxigenic *Escherichia coli* in patients with diarrhea in Asia with three enterotoxin gene probes. Infect. Immun. **42:**152–155.
34. **Svennerholm, A.-M., and G. Wiklund.** 1983. Rapid GM1-enzyme-linked immunosorbent assay with visual reading for identification of *Escherichia coli* heat-labile enterotoxin. J. Clin. Microbiol. **17:**596–600.
35. **Tullock, E. F., Jr., K. J. Ryan, S. B. Formal, and F. A. Franklin.** 1973. Invasive enteropathogenic *Escherichia coli* dysentery. Ann. Intern. Med. **79:**13–17.
36. **Ursing, J., D. J. Brenner, H. Bercovier, G. R. Fanning, A. G. Steigerwalt, J. Brault, and H. H. Mollaret.** 1980. *Yersinia frederiksenii*: a new species of *Enterobacteriaceae* composed of rhamnose positive strains (formerly called atypical *Yersinia enterocolitica* or *Yersinia enterocolitica*-like). Curr. Microbiol. **4:**213–218.
37. **Weissfeld, A. S.** 1981. *Yersinia enterocolitica*. Clin. Microbiol. Newsl. **3:**91–93.
38. **Wells, J. G., B. R. Davis, I. K. Wachsmuth, L. W. Riley, R. S. Remis, R. Sokolow, and G. K. Morris.** 1983. Laboratory investigation of hemorrhagic colitis outbreaks associated with a rare *Escherichia coli* serotype. J. Clin. Microbiol. **18:**512–520.
39. **Yu, V. L.** 1979. *Serratia marcescens*. N. Engl. J. Med. **300:**887–892.

Chapter 25

Aeromonas and *Plesiomonas*

ALEXANDER VON GRAEVENITZ

The genera *Aeromonas* and *Plesiomonas* belong to the family *Vibrionaceae* by virtue of their morphology (gram-negative rod shape with predominantly polar flagellation), facultative anaerobiosis (respiratory and fermentative metabolism), and guanine-plus-cytosine ratio (24, 31). Their normal habitats are water sources. Occasionally, cold-blooded animals and humans are infected; natural infection in warm-blooded animals is very rare (28).

DESCRIPTION OF THE GENERA

Aeromonas

Members of *Aeromonas* are facultatively anaerobic, asporogenous, gram-negative rods, 1.0 to 4.4 μm long and 0.4 to 1.0 μm wide, possessing polar, usually monotrichous flagella with a wavelength of 1.7 μm (exceptions are the nonmotile species *A. salmonicida* and the proposed species *A. media*). In 2- to 4-h-old cultures, short lateral flagella with a wavelength of less than 1.7 μm may be seen. Lophotrichous flagellation is exceptional. Aeromonads are heterotrophic, produce oxidase and catalase, and ferment glucose and other carbohydrates with the production of acid or acid and gas. Nitrates are reduced to nitrites. Many exoenzymes are produced, e.g., amylase, DNase, lipase, phosphatases, proteinases, etc. The temperature range for growth is between 0 and 41°C; for human strains it usually lies between 10 and 41°C; *A. salmonicida* grows only below 37°C. Some strains are biochemically more active at 22°C than at 37°C (9, 20, 24). The pH range for growth is 5.5 to 9.0. The guanine-plus-cytosine content of the DNA is 57 to 63 mol% (24). Aeromonads are not susceptible to the compound O/129 (2, 4-diamino-6,7-diisopropylpteridine); i.e., on Mueller-Hinton agar, there is no zone around disks containing 10 or 150 μg of O/129 phosphate (Oxoid Ltd., Basingstoke, U.K.) and a zone of ≤15 mm appears around disks containing 570 μg of O/129 phosphate (Institut Pasteur Production, Paris, France) (25).

Bergey's Manual of Systematic Bacteriology (24) recognizes four species of *Aeromonas*: *A. hydrophila*, *A. caviae*, *A. sobria*, and *A. salmonicida* (with the subspecies *salmonicida*, *achromogenes*, and *masoucida*). Very recently, a new aquatic species, *A. media*, nonmotile like *A. salmonicida* and not found in human specimens, has been proposed (1). The latter two species will be disregarded here. The biochemical reactions listed in Table 1 are a composite from all three motile species. Table 2 lists the differential reactions of the three species. There have been contradictory data as to whether further tests like fermentation of cellobiose and arabinose, beta-hemolysis, and production of lysine decarboxylase are of differential value as well (6, 15). The need for identification to species level in the clinical laboratory should be decided on grounds of epidemiology and, perhaps, enteropathogenicity (see below).

Plesiomonas

There is only one species in the genus *Plesiomonas*: *P. shigelloides* (formerly C27 or *Aeromonas shigelloides*). Like aeromonads, *P. shigelloides* is a facultatively anaerobic, asporogenous, gram-negative rod measuring 2.0 to 3.0 μm by 0.1 to 1.0 μm, with polar, lophotrichous flagella (generally two to five) with a wavelength of 3.5 to 4.0 μm; 2- to 4-h-old cultures may show lateral flagella with a shorter wavelength. Monotrichous cells as well as nonmotile strains occur. *P. shigelloides* is heterotrophic, produces oxidase and catalase, and ferments glucose and only a few other carbohydrates (Table 1) without gas formation (31). Nitrates are reduced to nitrites. No exoenzymes are formed. The minimum growth temperature is 8°C; optimum and maximum temperatures are 37°C and between 40 and 44°C, respectively. The pH range for growth is 5.0 to 7.7. Most strains are susceptible to O/129 (25). The guanine-plus-cytosine content of the DNA is 51 mol%. A few strains share a common antigen with *Shigella sonnei* phase 1 (25); cross-agglutination with *Shigella* spp. A and C antisera has also been reported (23).

NATURAL HABITAT AND CLINICAL SIGNIFICANCE

Aeromonas

Aeromonas spp. are widely distributed in stagnant and flowing fresh waters, in salt waters which interface with fresh water, and in sewage, with densities ranging from less than one to several thousand cells per milliliter (12, 27). They have been found at pH values of 5.2 to 9.8 and at temperatures between 4 and 45°C and are not considered halophilic since their NaCl tolerance ranges only from 0 to 4% (25). They have also been isolated from soil and foodstuffs (9).

Natural disease due to *Aeromonas* spp. has been found in amphibians, e.g., as "red leg disease" in frogs (not specific for aeromonads) (28). Reptiles or fish may also be infected or may carry the organisms. Experimentally, these animals as well as guinea pigs and mice can be infected intravenously or intraperitoneally (28).

Human infections with *Aeromonas* spp. occur predominantly during the period from May to November (14), probably due to the aquatic origin of the bacteria, and have been the subject of several recent reviews (7, 14). Four categories of infections have been described:

1. Cellulitis or wound infection related to exposure to water or soil;
2. Acute diarrheal disease of short duration, sometimes bloody or choleriform, occurring worldwide and affecting any age (although it seems particularly frequent in patients with traveler's diarrhea and in those less than 5 years of age [14, 23]);

TABLE 1. Biochemical characteristics of motile *Aeromonas* spp. and *Plesiomonas shigelloides*[a]

Test	*Aeromonas* spp.	*P. shigelloides*
H_2S (triple sugar iron or Kligler iron agar)	− (0)	− (0)
Urease	− (4w)	− (0)
Indole	d (87)	+ (100)
Methyl red		
37°C	+ (95)	+ (100)
26°C	d (57)	+ (100)
Voges-Proskauer		
37°C	d (33)	− (0)
26°C	d (66)	− (0)
Citrate (Simmons)	d (52/26)	− (0)
Growth in KCN medium	d (58)	− (2)
Motility	+ (98)	d (85)
Gelatin, 22°C	+ (78/21)	− (0)
Lysine decarboxylase	d[b]	+ (96/4)
Arginine dihydrolase	d (75/10)	+ (93/2)
Ornithine decarboxylase	− (0)	+ (100)[c]
Phenylalanine deaminase	d (25w)	d (41w)
Glucose		
Acid	+ (100)	+ (100)
Gas	d (46)	− (0)
Lactose	d (9/28)	+ (65/26)
Sucrose	d (83/4)	− (0/6)
Arabinose	d (52/2)	− (0)
Mannitol	+ (99)	− (0)
Dulcitol	− (0)	− (0)
Salicin	d (41/22)	d (32)
Adonitol	− (0)	− (0)
Inositol	d[b]	+ (100)
Sorbitol	d (13/1)	− (0)
Raffinose	− (2w)	− (0)
Rhamnose	− (4)	− (0)
Maltose	+ (98/1)	+ (100)[c]
Xylose	− (0)	− (0)
Trehalose	+ (100)	+ (96/2)
Cellobiose	d (37/2)	− (0)
Glycerol	+ (80/9)	d (15/68)
Esculin	d (60)	− (0)
Melezitose	− (0)	− (0)
Melibiose	− (4/2)	d (48/9)
Mannose	+ (93)	d (14/3)
Malonate	− (0)	− (0)
Mucate	− (0)	− (0)
Citrate (Christensen)	d (52/26)	− (0)
Sodium acetate	d (74/8)	d (2/33)
Lipase (corn oil)	+ (98)	− (0)
Nitrate		
To nitrite	+ (99)	+ (100)
To gas	− (0)	− (0)
Oxidase	+ (100)	+ (98)
Catalase	+ (100)	+ (100)
Growth on:		
Cetrimide	− (2w)	− (0)
MacConkey agar	+ (100)	+ (91)
Salmonella-shigella agar	d (85)	d (87)
String test	− (10)	− (0)
DNase	+ (99)[d]	− (0/4)[c]
Caseinase	+ (98)[d]	− (0)[c]
o-Nitrophenyl-β-D-galactopyranoside	+ (100)[d]	d (75)[c]
Growth in 6.5% NaCl broth	− (0)[d]	− (0)[c]
Amylase	+ (100)[d]	− (0)[c]

[a] Adapted from Ewing and Hugh (8). Strains tested at 37°C (unless indicated otherwise). Symbols: +, positive; −, negative; d, different reactions; w, weak reaction. Numbers in parentheses show the percentage of strains positive in 1 to 2 days/percentage of strains positive in ≥3 days.

[b] Different reactions; no recent data (see text).

[c] Data from Richard et al. (25) on 57 strains tested at 37°C.

[d] Data from McCarthy (20) on 91 strains tested at 22°C.

3. Septicemia, mostly in association with hepatic, biliary, or pancreatic disease or with malignancy, particularly acute leukemia;
4. Other infections, e.g., various soft tissue infections, and rare cases of urinary tract infections, meningitis, peritonitis, otitis, and endocarditis.

Each species has been isolated in each disease category (6, 14). Nosocomial strains are in the minority. Pure and mixed cultures have been observed. Their eventual source seems to have been water, soil, foodstuffs, or the intestinal flora. There are animal as well as human carriers. With selective techniques, carrier rates of 2 to 3% have been observed in England (21, 33), the United States (37), and Australia (4); a recent study from Thailand yielded carrier rates of 8 to 16% in children and 27% in adults (23). Several toxic substances have been isolated from cultures of *Aeromonas* spp., i.e., hemolysins, leucocidin, proteases, and at least one enterotoxin. This cytotonic, heat-labile enterotoxin can be separated in vitro from the cytotoxic alpha- and beta-hemolysins; it shows a positive ileal loop test (19). Whether a heat-stable enterotoxin is also produced has not been clearly established, since positive suckling mouse tests (4) may have been due to the action of beta-hemolysin (19). Cytotoxicity to Y1 or HeLa cells has been found associated at least with positive lysine decarboxylase and Voges-Proskauer tests (6, 16, 23) but was seen in strains from diarrheic individuals and carriers alike (23). Diarrhea was, however, associated with strains whose agglutination of human group O cells was not inhibited by fucose, glactose, or mannose (3). Recently, enteroinvasive *Aeromonas* strains were found (23), but efforts to induce diarrhea in animals have failed (19, 23).

Plesiomonas

The aquatic distribution of *P. shigelloides* is limited by its minimum growth temperature of 8°C (30) and the fact that it does not survive well in seawater (25). It has been isolated from the gut of freshwater fish, from surface water, and from many animals (snakes, monkeys, dogs, cats, goats, cattle, toads; 2, 30, 35). The bulk of human *P. shigelloides* strains have been isolated from stools of diarrheic patients living in subtropical and tropical areas (Mali, Zaire, Kenya, Madagascar, Tahiti, Cuba, Thailand, India), Japan, Australia, and only rarely Europe and the United States (2, 23, 25, 35). Most of these infections were probably waterborne. Carriers are very rare (2), except in endemic areas (2 to 24% in Thailand [23]) and in association with occasionally observed epidemics

TABLE 2. Important characteristics differentiating *Aeromonas caviae*, *A. hydrophila*, and *A. sobria*[a]

Test	*A. hydrophila*	*A. caviae*	*A. sobria*
Esculin hydrolysis	+	+	−
Growth in KCN	+	+	−
Acid from salicin	+	+	−
Gas from glucose	+	−	+
Voges-Proskauer reaction	+	−	d
Elastase production	+	−	−

[a] Data from Popoff (24). Symbols: +, typically positive; −, typically negative; d, different reactions.

(34). Tests for enteropathogenicity have mostly been negative (18, 23). The few extraintestinal infections have recently been summarized (cellulitis, septicemia, neonatal meningitis) (22).

COLLECTION, TRANSPORT, AND STORAGE OF SPECIMENS

Survival of *Aeromonas* and *Plesiomonas* spp. in various transport media has not been specifically investigated. There is no reason to assume that methods of collection, transport, and storage for specimens suspected to contain these organisms should differ from those outlined in Chapter 8.

CULTURE AND ISOLATION

Primary culture of *Aeromonas* and *Plesiomonas* spp. is accomplished on the usual laboratory media (blood agar, enteric differential agars, and tryptic digest broths). However, not every strain will grow on enteric differential or selective agars (see Table 1).

Selective techniques to detect *Aeromonas* spp. in stools

A recent study has compared nine solid and two liquid media for their suitability to select *Aeromonas* spp. and *P. shigelloides* from human stools (36). Media with optimal sensitivity and specificity for *Aeromonas* and *Plesiomonas* spp. were alkaline peptone-water and inositol-brillant green-bile salts agar (29) (Plesiomonas Differential-Agar; E. Merck AG, Darmstadt, Federal Republic of Germany). For *Aeromonas* spp., dextrin-fuchsin sulfite agar (27), xylose-sodium desoxycholate-citrate agar (33), and Pril-xylose-ampicillin agar (26; Pril is a quaternary ammonium detergent compounded of primary alkyl sulfate, alkyl-benzyl sulfonate, and salts; Böhme Fettchemie GmbH, Düsseldorf, Federal Republic of Germany) could also be recommended. Blood agar with 43 g of α-*p*-nitrophenol glycerine per liter (4), MacConkey agar with trehalose instead of lactose, and an enrichment broth with bile salts and novobiocin have also been used, the latter two for water samples (16). Other media, including TCBS agar, are of insufficient specificity and sensitivity for diagnostic purposes, as are colony characteristics like beta-hemolysis, oxidase positivity, or lactose negativity (36). For *P. shigelloides*, growth on various media selective for *Salmonella* or *Shigella* spp. (25) has been reported, with no data on sensitivity.

IDENTIFICATION

Recommended Procedures

Aeromonas

On blood agar, many strains of *Aeromonas* spp. show a large zone of beta-hemolysis, but nonhemolytic strains do occur (18, 20, 24). On enteric differential agars, many strains show no signs of lactose fermentation, but a minority of strains yield lactose-fermenting colonies (13).

In the routine diagnostic laboratory, the most important characteristics that should lead to a presumptive diagnosis of *Aeromonas* spp. are growth on MacConkey agar, a positive oxidase reaction, and fermentation of carbohydrates. The oxidase test in aeromonads is positive if performed with any reagent on colonies from blood agar, but it may be negative if performed on colonies from enteric agars which show signs of fermentation (13). On triple sugar iron agar or Kligler iron agar, most strains yield acid butts and slants; alkaline slants, gas formation, or both may occur, but never H_2S formation. If oxidation-fermentation media with carbohydrates are used, acidification occurs in the "open" and "closed" tubes. These tests separate *Aeromonas* spp. from the *Enterobacteriaceae* and from nonfermentative and fastidious fermentative gram-negative rods, and should, in conjunction with traditional or commercial kit systems, lead to a correct diagnosis.

Separation from other *Vibrionaceae* may be more difficult in light of the multitude of new *Vibrio* species. Resistance to O/129 is not characteristic of *Vibrio* spp. *Aeromonas* spp. from humans, moreover, do not grow in 6% NaCl broth and do not possess an ornithine decarboxylase, features that separate them from halophilic and nonhalophilic *Vibrio* spp., respectively.

Differentiation of *Aeromonas* spp. from *P. shigelloides* chiefly involves tests for ornithine decarboxylase, DNase, and O/129 resistance (see Table 2). Recently, inositol-positive *Aeromonas* strains have been reported (30). Flagellar straining is generally not necessary for confirmation. Table 1 should be consulted for identification to species.

Plesiomonas

P. shigelloides colonies are not beta-hemolytic on blood agar and generally do not ferment lactose on enteric agars. Dissociation into colonies with large (2 mm) and small (0.5 mm) diameters or into lactose-positive and lactose-negative ones is not rare (25, 35). Key tests for a presumptive diagnosis are those listed for *Aeromonas*. Triple sugar iron and Kligler iron agar yield an alkaline slant and an acid butt with neither gas nor H_2S formation. For differentiation from *Aeromonas* spp., see above. *Vibrio* spp. are mostly DNase positive, mannitol positive, and inositol negative (see Chapter 26).

Serological Investigations

Aeromonas

Agglutination and indirect hemagglutination studies (9, 17) have led to serotyping of only one-quarter of the known *Aeromonas* strains. Cross-reactivity with *P. shigelloides* and between *A. hydrophila* and *A. sobria* has been observed (9, 17). In patients with deep infections due to *A. hydrophila*, antibodies (antihemolysin, agglutinins, precipitins) have been detected (5).

Plesiomonas

A recent schema has put together 30 O and 11 H antigens (32). It proved useful for the investigation of two diarrhea epidemics but could not type all environmental strains (34).

Antimicrobial Susceptibility Studies

Aeromonas

Aeromonas strains are mostly resistant to penicillin, ampicillin, carbenicillin, and ticarcillin. Conversely, most are susceptible to azlocillin, mezlocillin, piperacillin, and the second- and third-generation cephalosporins, but development of resistance to these drugs has been observed. More than half are resistant to cephalothin. Most strains are susceptible to the aminoglycosides, chloramphenicol, tetracycline, nalidixic acid, nitrofurantoin, and trimethoprim-sulfamethoxazole (10, 11).

Plesiomonas

Plesiomonas strains can be resistant to ampicillin and carbenicillin, but they are mostly susceptible to the other drugs mentioned (22, 25).

LITERATURE CITED

1. **Allen, D. A., B. Austin, and R. R. Colwell.** 1983. *Aeromonas media*, a new species isolated from river water. Int. J. Syst. Bacteriol. **33:**599–604.
2. **Arai, T., N. Ikejima, T. Itoh, S. Sakai, T. Shimada, and R. Sakazaki.** 1980. A survey of *Plesiomonas shigelloides* from aquatic environments, domestic animals, pets and humans. J. Hyg. **84:**203–211.
3. **Burke, V., M. Cooper, J. Robinson, M. Gracey, M. Lesmana, P. Echeverria, and J. M. Janda.** 1984. Hemagglutination patterns of *Aeromonas* spp. in relation to biotype and source. J. Clin. Microbiol. **19:**39–43.
4. **Burke, V., M. Gracey, J. Robinson, D. Peck, J. Beaman, and C. Bundell.** 1983. The microbiology of childhood gastroenteritis: *Aeromonas* species and other infective agents. J. Infect. Dis. **148:**68–74.
5. **Caselitz, F.-H., V. Freitag, and G. Jannasch.** 1975. Spezifischer Antikoerpernachweis bei Aeromonasinfektionen. Zentralbl. Bakteriol. Parasitenkd. Infektionskr. Hyg. Abt. 1 Orig. Reihe A **233:**347–354.
6. **Daily, O. P., S. W. Joseph, J. C. Coolbaugh, R. I. Walker, B. R. Merrell, D. M. Rollins, R. J. Seidler, R. R. Colwell, and C. R. Lissner.** 1981. Association of *Aeromonas sobria* with human infection. J. Clin. Microbiol. **13:**769–777.
7. **Davis, W. A., J. G. Kane, and V. F. Garagusi.** 1978. Human *Aeromonas* infections: a review of the literature and a case report of endocarditis. Medicine (Baltimore) **57:**267–277.
8. **Ewing, W. H., and R. Hugh.** 1974. *Aeromonas*, p. 230–237. *In* E. H. Lennette, E. H. Spaulding, and J. P. Truant (ed.), Manual of clinical microbiology, 2nd ed. American Society for Microbiology, Washington, D.C.
9. **Ewing, W. H., R. Hugh, and J. G. Johnson.** 1961. Studies on the *Aeromonas* group. Center for Disease Control, Atlanta, Ga.
10. **Fainstein, V., S. Weaver, and G. P. Bodey.** 1982. In vitro susceptibilities of *Aeromonas hydrophila* against new antibiotics. Antimicrob. Agents Chemother. **22:**513–514.
11. **Fass, R. J., and J. Barnishan.** 1981. In vitro susceptibilities of *Aeromonas hydrophila* to 32 antimicrobial agents. Antimicrob. Agents Chemother. **19:**357–358.
12. **Hazen, T. C., C. B. Fliermans, R. P. Hirsch, and G. W. Esch.** 1978. Prevalence and distribution of *Aeromonas hydrophila* in the United States. Appl. Environ. Microbiol. **36:**731–738.
13. **Hunt, L. K., T. L. Overman, and R. B. Otero.** 1981. Role of pH in oxidase variability of *Aeromonas hydrophila*. J. Clin. Microbiol. **13:**1054–1059.
14. **Janda, J. M., E. J. Bottone, and M. Reitano.** 1983. *Aeromonas* species in clinical microbiology: significance, epidemiology, and speciation. Diagn. Microbiol. Infect. Dis. **1:**221–228.
15. **Janda, J. M., M. Reitano, and E. J. Bottone.** 1984. Biotyping of *Aeromonas* isolates as a correlate to delineating a species-associated disease spectrum. J. Clin. Microbiol. **19:**44–47.
16. **Kaper, J. B., H. Lockman, R. R. Colwell, and S. W. Joseph.** 1981. *Aeromonas hydrophila*: ecology and toxigenicity of isolates from an estuary. J. Appl. Bacteriol. **50:**359–377.
17. **Leblanc, D., K. R. Mittal, G. Olivier, and R. Lallier.** 1981. Serogrouping of motile *Aeromonas* species isolated from healthy and moribund fish. Appl. Environ. Microbiol. **42:**56–60.
18. **Ljungh, A., M. Popoff, and T. Wadström.** 1977. *Aeromonas hydrophila* in acute diarrheal disease: detection of enterotoxin and biotyping of strains. J. Clin. Microbiol. **6:**96–100.
19. **Ljungh, A., and T. Wadström.** 1982–83. Toxins of *Vibrio parahaemolyticus* and *Aeromonas hydrophila*. J. Toxicol. Toxin Rev. **1:**257–307.
20. **McCarthy, D. H.** 1975. The bacteriology and taxonomy of *Aeromonas liquefaciens*. Technical Report Series no. 2. Fish Diseases Laboratory, Weymouth, Dorset, U.K.
21. **Millership, S. E., S. R. Curnow, and B. Chattopadhyay.** 1983. Faecal carriage rate of *Aeromonas hydrophila*. J. Clin. Pathol. **36:**920–923.
22. **Pathak, A., J. R. Custer, and J. Levy.** 1983. Neonatal septicemia and meningitis due to *Plesiomonas shigelloides*. Pediatrics **71:**389–391.
23. **Pitarangsi, C., P. Echeverria, R. Whitmire, C. Tirapat, S. Formal, G. J. Dammin, and M. Tingtalapong.** 1982. Enteropathogenicity of *Aeromonas hydrophila* and *Plesiomonas shigelloides*: prevalence among individuals with and without diarrhea in Thailand. Infect. Immun. **35:**666–673.
24. **Popoff, M.** 1984. Genus III. *Aeromonas* Kluyver and van Niel 1936, 398, p. 545–548. *In* N. R. Krieg and J. G. Holt (ed.), Bergey's manual of systematic bacteriology, vol. 1. The Williams & Wilkins Co., Baltimore.
25. **Richard, C., M. Lhuillier, and B. Laurent.** 1978. *Plesiomonas shigelloides*: une vibrionacée entéropathogène exotique. Bull. Inst. Pasteur **76:**187–200.
26. **Rogol, M., I. Sechter, L. Grinberg, and C. B. Gerichter.** 1979. Pril-Xylose-Ampicillin agar, a new selective medium for the isolation of *Aeromonas hydrophila*. J. Med. Microbiol. **12:**229–231.
27. **Schubert, R. H. W.** 1967. Das Vorkommen der Aeromonaden in oberirdischen Gewässern. Arch. Hyg. **150:**688–708.
28. **Schubert, R. H. W.** 1967. Die Pathogenität der Aeromonaden für Mensch und Tier. Arch. Hyg. **150:**709–716.
29. **Schubert, R. H. W.** 1977. Ueber den Nachweis von *Plesiomonas shigelloides* Habs und Schubert, 1962, und ein Elektivmedium, den Inositol-Brillantgrün-Gallesalz-Agar. E. Rodenwaldt-Arch. **4:**97–103.
30. **Schubert, R. H. W.** 1981. Zur Oekologie von *Plesiomonas shigelloides*. Zentralbl. Bakteriol. Parasitenkd. Infektionskr. Hyg. Abt. 1 Orig. Reihe B **172:**528–533.
31. **Schubert, R. H. W.** 1984. Genus IV. *Plesiomonas* Habs and Schubert 1962, 324, p. 548–550. *In* N. R. Krieg and J. G. Holt (ed.), Bergey's manual of systematic bacteriology, vol. 1. The Williams & Wilkins Co., Baltimore.
32. **Shimada, T., and R. Sakazaki.** 1978. On the serology of *Plesiomonas shigelloides*. Jpn. J. Med. Sci. Biol. **31:**135–142.
33. **Shread, P., T. J. Donovan, and J. V. Lee.** 1981. A survey of the incidence of *Aeromonas* in human faeces. Soc. Gen. Microbiol. Q. **8:**184.
34. **Tsukamoto, T., Y. Kinoshita, T. Shimada, and R. Sakazaki.** 1978. Two epidemics of diarrhoeal disease possibly caused by *Plesiomonas shigelloides*. J. Hyg. **80:**275–280.
35. **Vandepitte, J., A. Makulu, and F. Gatti.** 1974. *Plesiomonas shigelloides*: survey and possible association with diarrhoea in Zaïre. Ann. Soc. Belge Med. Trop. **54:**503–513.
36. **von Graevenitz, A., and C. Bucher.** 1983. Evaluation of differential and selective media for isolation of *Aeromonas* and *Plesiomonas* spp. from human feces. J. Clin. Microbiol. **17:**16–21.
37. **von Graevenitz, A., and L. Zinterhofer.** 1970. The detection of *Aeromonas hydrophila* in stool specimens. Health Lab. Sci. **7:**124–127.

Chapter 26

Vibrio

J. J. FARMER III, F. W. HICKMAN-BRENNER, AND MICHAEL T. KELLY

Most clinical laboratories isolate many strains of *Enterobacteriaceae* each year, and workers feel secure with the family (23). Unfortunately, this is not the case with the genus *Vibrio*, which was once relegated to the category "strangers in the world of clinical microbiology." *Vibrio* species are probably quite rare in many inland laboratories in the United States, where contact with the ocean or seafood is rare. Laboratories which serve coastal areas are more likely to encounter *Vibrio* strains (15). This situation in the United States is in contrast to some undeveloped areas of the world where cholera and its causative organism, *Vibrio cholerae*, are very common. *Vibrio* cultures are also encountered in countries where there are cultural differences in food preference. For example, in Japan, where fish and other seafoods are eaten raw, *Vibrio parahaemolyticus* is one of the most common causes of gastroenteritis. In this chapter we will discuss methods which will apply to this diverse spectrum of *Vibrio* distribution. The material which applies to *Vibrio* species as a group is discussed first and is followed by a discussion of each of the 10 *Vibrio* species which occur in clinical specimens.

Because of space limitations, many primary references cannot be given. However, several excellent reviews are available on the family *Vibrionaceae* (9), the marine vibrios (6, 7, 30), the genus *Vibrio* (8, 11, 14, 19, 52, 55, 58), *V. cholerae* (2, 5, 26, 34, 48, 50, 62), *V. parahaemolyticus* (3, 28, 37, 41, 45, 63), the vibrios not associated with human disease (1, 6, 7, 19, 30), and *Vibrio* toxins (59). For the other species, there is much less literature, and we will give some primary references.

NOMENCLATURE AND CLASSIFICATION

In the last few years the genus *Vibrio* has changed from a poorly characterized heterogeneous group of organisms to a well-understood natural group. This has been due to the removal of "nonfermentative vibrios," "anaerobic vibrios," and "microaerophilic vibrios" to other genera such as *Campylobacter (Vibrio fetus), Wolinella (Vibrio succinogenes), Pseudomonas*, and *Alteromonas* (6). The genus *Vibrio* is now much more homogeneous. The species currently recognized as belonging to genus *Vibrio* are listed in Table 1. The genus *Vibrio* is classified in the family *Vibrionaceae* (Table 2) along with three other genera, i.e., *Photobacterium, Aeromonas*, and *Plesiomonas*. Table 3 gives the guanine-plus-cytosine ratios of the genera of *Vibrionaceae*, which can be used as a first step in determining the relatedness of genera. The genus *Vibrio* is closely related to the genus *Photobacterium*, which would be expected because of their phenotypic similarities, but it is more distantly related to the genus *Aeromonas* and to the family *Enterobacteriaceae* (27). More distant relatives include the purple photosynthetic bacterium genus *Chromatium* and the nonfermentative genera *Pseudomonas* and *Acinetobacter* (27). Table 2 gives some of the properties of organisms belonging to the families *Vibrionaceae, Enterobacteriaceae*, and *Pseudomonadaceae*. Table 3 gives some properties which differentiate the four genera in the family *Vibrionaceae*. Most of the *Vibrio* species are not closely related to each other in a phylogenetic sense (6, 9, 17); thus *Vibrio* is a heterogeneous genus, similar to the genus *Pseudomonas*. The genus *Vibrio* is maintained as a matter of convenience, since most *Vibrio* strains share common phenotypic properties (Table 3).

Vibrio classification for the clinical laboratory

The identification of *Vibrio-Photobacterium* species can be greatly simplified from the 31 species given in Table 1 if we consider only the species which cause human disease or occur in human clinical specimens. This simplified classification for the genus *Vibrio* recognizes the 10 named species in the top part of Table 1 and considers all the other named species of *Vibrio* and *Photobacterium* as "marine vibrio species" (synonym, marine vibrios).

We define a "marine vibrio" as a *Vibrio* or *Photobacterium* strain which is oxidase positive, ferments D-glucose, does not grow in nutrient broth without added NaCl, but does grow in nutrient broth with added NaCl or "sea salts." Its properties do not fit any of the 10 "clinical species" of the genus *Vibrio*. Many of these marine vibrios require 1 to 2% NaCl for growth and grow better at a lower temperature than at 36°C. Some are bioluminescent.

Many of the organisms isolated from ocean or estuary water or from the plants and animals which live in this water will belong to the marine vibrio group. These organisms can be identified to species (6–8), but this task is much more difficult because the identification methods are very different from those used in clinical laboratories (52).

Reporting *Vibrio* species

Based on the above discussion, we recommend that clinical laboratories report the 10 *Vibrio* species given in the top of Table 1 by genus and species name and report the other species as "marine vibrios." Reference laboratories may also wish to adopt this simplified classification until all the named *Vibrio* and *Photobacterium* species have been characterized by the same methods and an identification matrix becomes available. We are collecting these data for a data matrix similar to the one for all the species of *Enterobacteriaceae* (23).

CLINICAL SIGNIFICANCE

Vibrio species cause a number of human infections (14, 19) which can be classified as intestinal or extraintestinal. Each of these infections is described briefly below but is discussed in more detail in the section for each species.

TABLE 1. Classification of the genus *Vibrio* and some related organisms

Vibrio species which occur in human clinical specimens		
V. alginolyticus	*V. hollisae*	
V. cholerae	*V. metschnikovii*	
V. damsela	*V. mimicus*	
V. fluvialis	*V. parahaemolyticus*	
V. furnissii	*V. vulnificus*	
Vibrio and *Photobacterium* species which do not occur in human clinical specimens		
V. aestuarianus	*V. natriegens*	*P. angustum*
V. anguillarum	*V. nereis*	*P. leiognathi*
V. campbellii	*V. nigripulchritudo*	*P. phosphoreum*
V. costicola	*V. ordalii*	
V. diazotrophicus	*V. orientalis*	
V. fischeri	*V. pelagius*	
V. gazogenes	*V. proteolyticus*	
V. harveyi	*V. splendidus*	
V. logei	*V. tubiashii*	

Infections of the intestinal tract: cholera and other diarrheas

Five *Vibrio* species have been shown to cause or be associated with diarrhea. *V. cholerae* is well known as the cause of cholera (5, 26, 50), which is a distinct clinical entity in its most severe form but in its milder form may be hard to distinguish from other secretory diarrheas. *V. parahaemolyticus* is a well-documented cause of acute gastroenteritis (28, 45). More recently, *Vibrio fluvialis* (35, 40), *V. hollisae* (31, 47), and *V. mimicus* (20, 54) have been implicated as causes of diarrhea. *Vibrio furnissii* (18) has been isolated from a few cases of diarrhea, but there is no real evidence that it can actually cause diarrhea.

Extraintestinal infections

Vibrio strains have been isolated frequently from certain extraintestinal infections but are rarely isolated from others. Table 4 is a summary of the sources of *Vibrio* cultures studied at the Centers for Disease Control (CDC). Vibrios are often isolated from blood, wounds of arms and legs, infected eyes and ears, and the gallbladder. They are rarely reported from patients with meningitis, pneumonia, and infection of the reproductive organs or urinary tract, but they can occasionally occur (58). Within the genus *Vibrio* there is a division between the species which cause intestinal infection and those which cause extraintestinal infections; however, this division is not absolute. The pandemic strain of *V. cholerae* (serogroup O1, cholera toxin$^+$) seems to be well adapted to the human intestinal tract; seldom is it found at other sites (Table 4). Similarly, many of the wound infections and septicemia reportedly due to *V. parahaemolyticus* were really due to *Vibrio vulnificus*, which was not described until recently. *V. vulnificus* is well established as an important cause of septicemia (often fatal) and wound infections (13, 14, 32). *Vibrio damsela* (43) is associated with human wound infections (47), but its causal role needs strengthening. *Vibrio alginolyticus* (14) has been isolated from several types of soft-tissue infections. *Vibrio metschnikovii* is usually an environmental organism (39), but it has been isolated in a single case of peritonitis in a patient with an inflamed gallbladder (36). Most, but not all, of these infections are associated with seawater contact.

TABLE 2. Some characteristics[a] of the families *Enterobacteriaceae*, *Vibrionaceae*, and *Pseudomonadaceae*

Characteristic	Property of family (typical genus)		
	Enterobacteriaceae (Escherichia)	*Vibrionaceae (Vibrio)*	*Pseudomonadaceae (Pseudomonas)*
Grow aerobically	+	+	+
Grow anaerobically	+	+	−
Ferment D-glucose	+	+	−
Flagella locations	Peritrichous	Polar[b]	Polar
Oxidase	−	+	+
Often bioluminescent	−	+	−
Na^+ required by many species	−	+	−
Inhibited by vibriostatic compound O129	−	+	−

[a] These data apply to most species in the families, but there are exceptions.

[b] Some *Vibrio* species have a sheathed polar flagellum when grown in a liquid medium but peritrichous flagella when grown on a solid medium.

TABLE 3. Properties of the four genera of the family *Vibrionaceae*[a]

Property	Property of genus			
	Vibrio	*Photobacterium*	*Aeromonas*	*Plesiomonas*
Guanine + cytosine content of DNA (mol%)	38–51	40–44	57–63	51
Associated with human infections	+	−	+	+
Na^+ required for growth (or stimulates growth)	+	+	−	−
Sensitive to vibriostatic compound O129	+	+	−	+
Lipase	+	V	+	−
D-Mannitol fermentation	+	−	+	−
Sheathed polar flagella	+	−	−	−
Accumulate poly-β-hydroxybutyrate, but do not utilize β-hydroxybutyrate	−	+		

[a] Apply to most species in the genus, but there are exceptions. A blank space indicates that data are not available. Adapted from Baumann and Schubert (9). V, Variable reaction for the genus.

LABORATORY: *VIBRIO* SPECIES IN RELATION TO USUAL LABORATORY TECHNIQUES

Many of the techniques used in clinical microbiology and enteric bacteriology laboratories work well

TABLE 4. Sources of 1,230 *Vibrio* isolates[a]

Source	No. of *Vibrio* isolates										
	V. cholerae		*V. mimicus*	*V. metschnikovii*	*V. hollisae*	*V. damsela*	*V. fluvialis*	*V. furnissii*	*V. alginolyticus*	*V. parahaemolyticus*	*V. vulnificus*
	O1	Non O1									
Human											
Feces or intestinal	118	94	39	0	30	0	15	16	4	114	6
Spinal fluid	0	1	0	0	0	0	0	0	0	0	2
Blood	1	42	0	1	2	0	0	0	0	2	63
Wound											
Hand or arm	0	2	0	0	0	1	0	0	1	3	12
Foot or leg	1	9	1	0	0	9	0	0	12	7	10
Other or unknown	1	3	1	0	0	0	0	1	9	1	1
Ear	0	24	6	0	0	0	0	0	18	0	0
Eye	0	0	0	0	0	0	0	0	2	0	0
Gallbladder	1	1	1	0	0	0	0	0	1	0	0
Urine	0	4	0	2	0	0	0	0	0	0	1
Respiratory tract	0	10	0	0	0	0	0	0	6	0	1
Other or unknown	4	12	5	4	1	0	0	0	6	0	8
Nonhuman											
Animals, nonmarine											
Primate	0	1	0	0	0	0	0	0	0	0	0
Pet or farm	0	6	0	0	0	0	0	1	0	0	0
Other	0	1	0	0	0	1	0	1	0	0	0
Animals, marine											
Fish	0	7	0	0	0	7	0	0	2	1	0
Oyster	8	29	9	0	0	1	5	0	0	11	4
Clam	0	5	0	0	0	0	0	0	2	0	0
Shrimp	1	5	1	0	0	0	0	0	0	5	0
Crab	2	0	0	1	0	0	0	0	3	1	0
Bird	0	0	1	2	0	0	0	0	0	0	0
Other	0	1	0	2	0	0	0	0	0	2	0
Water, unspecified	38	80	9	3	0	0	1	0	1	1	3
Ocean, or estuary	0	4	3	0	0	0	0	0	3	1	9
Lake or stream	0	3	2	0	0	0	0	1	0	0	0
Sewage	28	20	0	3	0	1	0	0	0	0	1
Food	0	18	1	3	0	0	0	0	1	4	0
Culture collections	0	0	0	0	0	0	0	0	0	2	0
Other or unknown	43	18	13	2	1	2	9	3	3	7	3

[a] Studied at the Vibrio Reference Laboratory, Enteric Bacteriology Section, CDC.

with the genus *Vibrio*. However, there are some specialized items that are recommended for laboratories which often isolate and identify *Vibrio* species (4, 10, 33, 46, 60, 61) and for clinical laboratories that want to increase their capability. Two factors can complicate the isolation and identification of *Vibrio* species. Sometimes *Vibrio* cultures will not grow well on the highly selective media used to isolate enteric pathogens. The other factor is that the halophilic species of *Vibrio* need added NaCl for optimum growth and activity (see Fig. 5.17 in reference 8), and several laboratory media have suboptimum amounts of Na^+ (less than 0.5% NaCl). Solutions to these problems are discussed below. Four approaches to the isolation of *Vibrio* species are as follows.

1. Use normal procedures, and make no special effort to search for *Vibrio* species.
2. Use normal procedures and plating media, and look for oxidase-positive colonies.
3. Incorporate TCBS agar as an extra plate for stool cultures and also for other likely specimens such as those from wounds, blood, eye, and ear.
4. Use other special procedures to enhance the isolation of *V. cholerae*, *V. parahaemolyticus*, and other *Vibrio* species.

The approach adopted by a particular laboratory will probably depend on the frequency with which *Vibrio* cultures are encountered. For example, a small laboratory in the American Midwest will rarely see a *Vibrio* culture, and approach 1 or 2 will probably be adequate. A large clinical laboratory near the ocean may wish to use TCBS agar routinely as a plating medium (15). There are both advantages and disadvantages to the routine use of TCBS medium. On the positive side, this medium brings immediate attention to a possible *Vibrio* isolate, whose colonies may not be obvious on other plating media. On the negative side, the routine use of TCBS may not be cost-

TABLE 5. Growth of *Vibrio* cultures on TCBS medium

Organism	Colony appearance on TCBS[a] (%):		Growth-plating efficiency
	Green	Yellow	
V. cholerae	<1	>99	Good
V. mimicus	100	0	Good
V. metschnikovii	0	100	May be reduced
V. hollisae	100	0	Very poor
V. damsela	95	5	Reduced at 36°C
V. fluvialis	0	100	Good
V. furnissii	0	100	Good
V. alginolyticus	0	100	Good
V. parahaemolyticus	99	1	Good
V. vulnificus	90[b]	10[b]	Good
Marine vibrios	Variable	Variable	Variable

[a] Percentage of strains which produce the given kind of colony.

[b] The original report of Hollis et al. (32) gave the percentage positive for sucrose fermentation as 3%. Since this time, the number of sucrose-positive strains we have received has increased dramatically (to 15%).

effective. In an 18-month study, one of us (M.T.K.) found that every *Vibrio* isolate found on TCBS medium was also detected on the other plating media used (which were screened for oxidase-positive colonies). Other negative factors include the facts that different lots of commercial TCBS medium vary in their selectivity and that some *Vibrio* species or strains do not grow well on TCBS (Table 5). These positive and negative aspects of the routine use of TCBS medium must be considered.

An active laboratory on the Indian subcontinent, where cholera is epidemic or endemic, will probably want to look specifically for *V. cholerae*. A laboratory interested in the environmental aspects of the genus *Vibrio* may want to use the specialized procedures in approach 4 to increase isolation rate.

COLLECTION, TRANSPORT, AND STORAGE OF SPECIMENS

Extraintestinal

The usual procedure for the collection and processing of an extraintestinal specimen is followed. There are no special procedures for *Vibrio* species.

Feces collection

Stool specimens should be collected early, preferably within the first 24 h of illness, and before the patient has received any antimicrobial agents. Fluid stool may be collected by inserting a petrolatum-lubricated soft-rubber catheter into the rectum. Rectal swabs may also be used. Although highly efficient in the acute phase of illness, rectal swabs are probably less satisfactory for convalescent patients or transiently infected asymptomatic persons. The administration of purgatives has been reported to increase the detection of persons excreting small numbers of vibrios. Vomitus, if available, may also be collected for culture.

Transport

Whenever possible, stool or rectal swab specimens should be inoculated onto isolation plates with minimal delay. The viability of *Vibrio* species is well maintained at the alkaline pH of rice-water stool but is unpredictable in formed stools. Vibrios are very susceptible to desiccation; hence, specimens must not be allowed to dry.

When there will be a delay in plating a culture (especially when it must be transported by courier), rectal swabs or fecal material should be placed in the semisolid transport medium of Cary and Blair, which maintains the viability of *Vibrio* cultures for up to 4 weeks. Buffered glycerol-saline, often used in enteric bacteriology, is an unsatisfactory transport medium even for short periods. Tellurite-taurocholate-peptone broth has been extensively used with success as an enrichment transport medium at the Cholera Research Laboratory in Dacca, Bangladesh, where specimens collected in the field are generally plated within 12 to 24 h. In the absence of available suitable transport media, strips of blotting paper may be soaked in liquid stool and inserted into airtight plastic bags. Specimens collected in this way may remain viable for up to 5 weeks.

Storage

Specimens in transport medium may be shipped to the laboratory without refrigeration.

Direct examination

Direct examination of stool material is not recommended for general purposes. However, stools may be examined by dark-field microscopy for the characteristic size, shape, and darting motility of vibrios, especially after a brief incubation in broth (10).

CULTURE AND ISOLATION

Figure 1 gives an overall plan for the use of special methods to enhance the *Vibrio* isolation rate. A simple procedure will probably result in a higher isolation rate for *Vibrio* species. Hemolytic colonies on sheep blood agar plates are tested for their oxidase reaction. This procedure will detect strongly hemolytic colonies of *Aeromonas*, some *Vibrio* species, and also some weakly hemolytic *Vibrio* and *Plesiomonas* cultures. Oxidase testing can also be done on nonhemolytic colonies (see below).

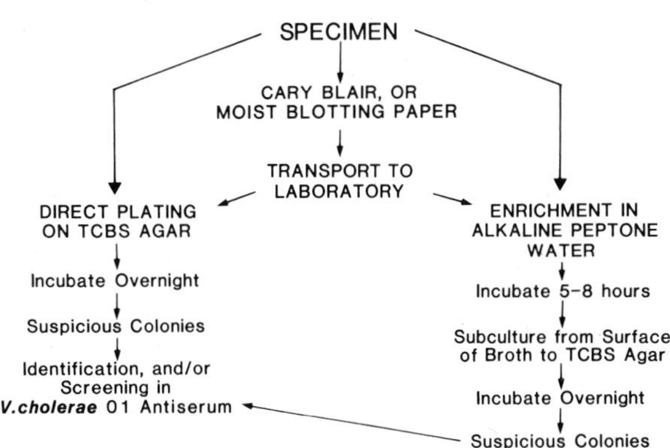

FIG. 1. Overall plan for using special methods for the isolation and identification of *Vibrio* cultures.

Extraintestinal specimens

Extraintestinal specimens are usually processed with no particular attention to *Vibrio*. The *Vibrio* species of medical importance grow well on blood agar and may also grow on MacConkey agar. However, a more thorough search for *Vibrio* isolates can be done with oxidase testing (see below) or by including a plate of TCBS medium.

Stool specimens

Stool specimens will usually be plated (see section on Specimen Processing, Chapter 8) onto a nonselective medium such as blood agar; onto selective media such as MacConkey agar, phenylethyl alcohol-blood agar, xylose-lysine-deoxycholate agar, and Hektoen enteric agar; and into gram-negative broth (Hajna). Specialized plating media for *Campylobacter*, *Yersinia*, and *Salmonella* species are used in some laboratories. *Vibrio* cultures will grow well on blood agar, where they will be beta-hemolytic (*V. cholerae* non O1 and *V. cholerae* O1 of the eltor biotype), alpha-hemolytic (*V. vulnificus* and many others), or nonhemolytic. *Vibrio* cultures usually grow on MacConkey agar (but sometimes with a reduced plating efficiency) and will appear as colorless (lactose-negative) colonies. Oxidase testing can be done on colonies grown on blood agar and usually on lactose-negative colonies on selective media; however, lactose-positive colonies from selective media can give a false-negative oxidase reaction. *Vibrio* cultures often do not grow well on the more selective enteric plating media.

Oxidase testing of colonies on primary plates

Many laboratories have adopted oxidase testing of colonies on primary plates as an alternative to adding selective media or special procedures for *Vibrio* species. All the *Vibrio* species which occur in clinical specimens are strongly oxidase positive in contrast to species in the family *Enterobacteriaceae*, which are oxidase negative. The lone exception is the oxidase-negative species *V. metschnikovii*, which is extremely rare in clinical specimens. This difference in oxidase reaction makes it easy to spot *Vibrio* colonies among colonies of *Enterobacteriaceae*. Two approaches have been used to spot *Vibrio* cultures this way. (The oxidase-positive species *Pseudomonas aeruginosa* would cause many false positives, but it can usually be recognized by its typical odor and colony morphology on plating media.)

Method 1. Five to 10 colonies from the blood agar are individually touched and spot tested for their oxidase reaction by Kovac's method.

Method 2. A drop of Kovac's oxidase reagent is added to an area of the blood agar plate where the colonies are crowded but still separated. The drop should cover 50 to 100 colonies. An alternative is to pick growth from a crowded area and do the Kovac's oxidase test. If oxidase-positive colonies are present, individual isolated colonies (not exposed to the reagent) from an area of less crowding are tested for their oxidase reactions as described above. Since most specimens will have no oxidase-positive colonies (other than *P. aeruginosa*), this method reduces the number of colonies which require individual screening.

Oxidase-positive colonies (other than *P. aeruginosa*) detected by the above method are then identified to species. This method has another advantage because it also detects cultures of *Aeromonas* and *Plesiomonas*, which are now being looked for in diarrheal stools. In many areas of the United States this method has proved successful but not cost-effective, because of the low yield of *Vibrio*, *Plesiomonas*, and *Aeromonas* strains.

SPECIALIZED MEDIA AND METHODS FOR *VIBRIO* ISOLATION

Special efforts to isolate *V. cholerae* and other *Vibrio* species from feces

Figure 1 is a schema adapted from Wachsmuth et al. (60) and Furniss et al. (29) for the isolation of *V. cholerae* from feces. This method should also be useful in the isolation of other *Vibrio* cultures from feces and other specimens. TCBS agar and enrichment in alkaline peptone water select for *Vibrio* species. As indicated in Fig. 1, these special procedures are integrated into the normal laboratory procedures, but they are used in addition to the normal isolation media.

Microscopic examination of growth

In 1883, Koch noted the appearance of small curved rods in the rice-water stool of cholera patients. These typical curved rods were not seen in the feces of patients without cholera. A number of other authors have made the same observation and commented on the possible usefulness of this observation. *V. cholerae* has considerable variability in its cellular morphology (Fig. 2), which includes typical curved rods, straight rods, short noncurved rods, and involution forms (Fig. 2). Often these forms can be seen in the

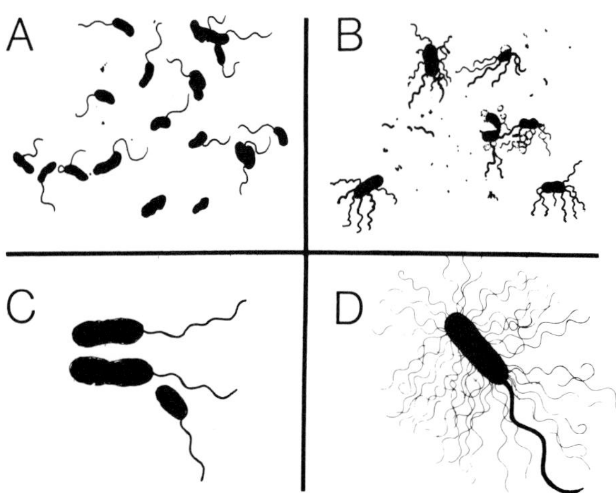

FIG. 2. Cellular morphology of *V. cholerae* and *V. parahaemolyticus*. (A) *V. cholerae* from 18-h culture on nutrient agar (Van Ermengen cilia stain; redrawn from Fig. 16 of reference 50). (B) Flagella stain of *V. parahaemolyticus* (redrawn from Fig. 2 of reference 45). (C) Electron micrograph of *V. parahaemolyticus* grown in liquid medium. Note the single, sheathed polar flagellum (redrawn from Fig. 1 of reference 45). (D) Electron micrograph of *V. parahaemolyticus* grown on a solid medium. Polar and peritrichous flagella are present; note the different size and shape of peritrichous flagella compared with the sheathed polar flagellum (composite drawing based on several original photographs).

same culture. For this reason, the microscopic examination of feces or cultures does not have a prominent role today. However, the finding of typical curved forms can be used as presumptive evidence for the presence of *Vibrio* strains. *Vibrio* cultures grown in liquid media have polar flagella (Fig. 2C), but many strains have peritrichous flagella when grown on solid media (Fig. 2D). Flagella stains have no place in the routine identification of vibrios.

IDENTIFICATION

Because pathogenicity and clinical significance vary among different species, a *Vibrio* strain should be identified to species. There are many different approaches to identification in general (for a review, see reference 24). One important question needs an immediate answer. Is the *Vibrio*-like organism isolated from a case of choleralike diarrhea the pandemic strain of *V. cholerae*?

Ruling out *V. cholerae* serogroup O1

It is very important to give a definitive answer to the above question as quickly as possible. Observation of a large number of sucrose-positive (yellow) colonies on TCBS agar from a patient with severe watery diarrhea gives warning of the possible presence of *V. cholerae* serogroup O1. Suspicious colonies are subcultured early in the day by heavy inoculation onto a nonselective medium such as blood agar or tryptic soy agar which does not contain carbohydrates. After 5 to 8 h, good growth should be present, which can be tested in polyvalent antisera to *V. cholerae* serogroup O1 (21). (Experienced workers often use growth from the TCBS plate to do the agglutination.) Positive agglutination is presumptive evidence of *V. cholerae* serogroup O1, and the result should be reported immediately to the physician. Biochemical testing is then necessary to confirm the identification as *V. cholerae*. Local health authorities should be notified immediately, and the culture, along with information about the case, should be forwarded to the state health laboratory.

Biochemical identification

Most laboratories use media and tests designed to identify cultures of the family *Enterobacteriaceae*. These procedures work well for *V. cholerae* and *V. mimicus* because these species have only a slight requirement for Na^+, which is fulfilled by the amount of NaCl in the medium. However, most of the halophilic *Vibrio* species require much more Na^+ for growth and expression of their metabolic pathways. Some of the media for doing biochemical tests do not contain enough NaCl for these halophilic species (Table 6). The standard enteric test is listed first, and the percentage of positive responses for each species is given. The percentage of positive responses for each species is then given for a modified medium with 1% NaCl added. More *Vibrio* species are indole positive

TABLE 6. The effect of media (especially Na^+ content) and test conditions on biochemical test results

Test	% Positive for *Vibrio* species[a]:									
	V. cholerae	*V. mimicus*	*V. metschnikovii*	*V. hollisae*	*V. damsela*	*V. fluvialis*	*V. furnissii*	*V. alginolyticus*	*V. parahaemolyticus*	*V. vulnificus*
Indole production										
Peptone water	86	94	22	30	0	0	0	14	35	39
Peptone water + 1% NaCl	86	98	26	93	0	0	0	24	80	65
Heart infusion + 1% NaCl	97	95	17	97	0	13	11	42	89	94
Methyl red test										
Standard	25	14	17	NG	NG	0	0	NG	NG	NG
1% NaCl	99	99	96	0	100	96	100	77	78	79
Voges-Proskauer										
Standard	74	2	26	NG	NG	0	0	NG	NG	NG
Standard + 1% NaCl	73	1	24	0	33	0	0	8	0	0
1% NaCl–Barritt method	93	9	96	0	95	0	0	83	0	0
Arginine dihydrolase										
Moeller's	0	0	0	0	81	52	0	0	0	0
Moeller's + 1% NaCl	0	0	59	0	95	93	100	0	0	0
Lysine decarboxylase										
Moeller's	99+	99	13	0	0	0	0	29	63	85
Moeller's + 1% NaCl	99+	100	36	0	52	0	0	99	100	99
Ornithine decarboxylase										
Moeller's	99+	99	0	0	0	0	0	3	71	47
Moeller's + 1% NaCl	99+	99	0	0	0	0	0	53	89	53
Gelatin hydrolysis										
Standard	52	60	65	0	5	41	56	45	66	55
Standard + 1% NaCl	62	63	38	0	6	85	86	76	89	75
Esculin hydrolysis										
Standard	0	0	4	0	0	0	0	0	0	0
Standard + 1% NaCl	2	0	59	0	0	8	0	2	1	39
Nitrate→nitrite										
Standard	99+	99+	0	NG	NG	59	33	NG	0	2
Standard + 1% NaCl	99+	100	0	100	100	100	100	100	100	100

[a] After 48 h of incubation at 36°C. NG, Most cultures do not grow in the medium.

TABLE 7. Biochemical test results and other properties of the 10 *Vibrio* species which occur in clinical specimens[a]

Property[b]	% Positive for *Vibrio* species[a]:									
	V. cholerae	*V. mimicus*	*V. metschnikovii*	*V. hollisae*	*V. damsela*	*V. fluvialis*	*V. furnissii*	*V. alginolyticus*	*V. parahaemolyticus*	*V. vulnificus*
*Indole production (HIB + 1% NaCl)	99	98	20	97	0	13	11	85	98	97
Methyl red (+1% NaCl)	99	99	96	0	100	96	100	73	80	80
*Voges-Proskauer (+1% NaCl, Barritt)	75	9	96	0	95	0	0	95	0	0
Citrate, Simmons'	97	99	75	0	0	93	100	1	3	75
H_2S on TSI	0	0	0	0	0	0	0	0	0	0
Urea hydrolysis	0	1	0	0	0	0	0	0	15	1
Phenylalanine deaminase	0	0	0	0	0	0	0	1	1	35
*Arginine, Moeller's + 1% NaCl	0	0	60	0	95	93	100	0	0	0
*Lysine, Moeller's + 1% NaCl	99	100	35	0	50	0	0	99	100	99
*Ornithine, Moeller's + 1% NaCl	99	99	0	0	0	0	0	50	95	55
Motility at 36°C	99	98	74	0	25	70	89	99	99	99
Gelatin hydrolysis + 1% NaCl (22°C)	90	65	65	0	6	85	86	90	95	75
KCN, growth in	10	2	0	0	5	65	89	15	20	1
Malonate utilization	1	0	0	0	0	0	11	0	0	0
*D-Glucose, acid production	100	100	100	100	100	100	100	100	100	100
*D-Glucose, gas production	0	0	0	0	10	0	100	0	0	0
Acid production from:										
Adonitol	0	0	0	0	0	0	0	1	0	0
*L-Arabinose	0	1	0	97	0	93	100	1	80	0
*D-Arabitol	0	0	0	0	0	65	89	0	0	0
*Cellobiose	8	0	9	0	0	30	11	3	5	99
Dulcitol	0	0	0	0	0	0	0	0	3	0
Erythritol	0	0	0	0	0	0	0	0	0	0
D-Galactose	90	82	45	100	90	96	100	20	92	96
Glycerol	30	13	100	0	0	7	55	80	50	1
myo-Inositol	0	0	40	0	0	0	0	0	0	0
*Lactose	7	21	50	0	0	3	0	0	1	85
*Maltose	99	99	100	0	100	100	100	100	99	100
*D-Mannitol	99	99	96	0	0	97	100	100	100	45
D-Mannose	78	99	100	100	100	100	100	99	100	98
Melibiose	1	0	0	0	0	3	11	1	1	40
α-Methyl-D-glucoside	0	0	25	0	5	0	0	1	0	0
Raffinose	0	0	0	0	0	0	11	0	0	0
L-Rhamnose	0	0	0	0	0	0	45	0	1	0
*Salicin	1	0	9	0	0	0	0	4	1	95
D-Sorbitol	1	0	45	0	0	3	0	1	1	0
*Sucrose	100	0	100	0	5	100	100	99	1	15
Trehalose	99	94	100	0	86	100	100	100	99	100
D-Xylose	0	0	0	0	0	0	0	0	0	0
Mucate, acid production	1	0	0	0	0	0	0	0	0	0
Tartrate, Jordan	75	12	35	65	0	35	22	95	93	84
Esculin hydrolysis (+1% NaCl)	0	0	60	0	0	8	0	3	1	40
Acetate utilization	92	78	25	0	0	70	65	0	1	7
*Nitrate→nitrite (+1% NaCl)	99	100	0	100	100	100	100	100	100	100
*Oxidase	100	100	0	100	95	100	100	100	100	100
DNase (25°C)	93	35	50	0	75	100	100	95	92	50
*Lipase, corn oil	92	17	100	0	0	90	89	85	90	92
*ONPG test	94	90	50	0	0	40	35	0	5	75
Yellow pigment at 25°C	0	0	0	0	0	0	0	0	0	0
Tyrosine clearing	13	30	5	3	0	65	45	70	77	75
Growth in nutrient broth with:										
*0% NaCl	100	100	0	0	0	0	0	0	0	0
*1% NaCl	100	100	100	99	100	99	99	99	100	99

Continued

TABLE 7—*Continued*

Property[b]	% Positive for *Vibrio* species[a]:									
	V. cholerae	*V. mimicus*	*V. metschnikovii*	*V. hollisae*	*V. damsela*	*V. fluvialis*	*V. furnissii*	*V. alginolyticus*	*V. parahaemolyticus*	*V. vulnificus*
6% NaCl	53	49	78	83	95	96	100	100	99	65
*8% NaCl	1	0	44	0	0	71	78	94	80	0
*10% NaCl	0	0	4	0	0	4	0	69	2	0
12% NaCl	0	0	0	0	0	0	0	17	1	0
String test	100	100	100	100	80	100	100	91	64	100
0129, zone	99	95	90	40	90	31	0	19	20	98
Polymyxin B, zone	22	88	100	100	85	100	89	63	54	3

[a] After 48 h of incubation at 36 ± 1°C.

[b] An * indicates that the test is recommended as part of the routine set for *Vibrio* identification. HIB, Heart infusion broth; TSI, triple sugar iron; ONPG, *o*-nitrophenyl-β-D-galactopyranoside.

when the medium is changed from peptone water to heart infusion broth, and more strains are Voges-Proskauer positive when the reagent for detecting acetyl methyl carbinol contains alpha-naphthol (Table 6). Table 6 shows the pronounced effect of media and methods on the results of biochemical tests. Table 7 lists the biochemical reactions for 10 *Vibrio* species which occur in clinical specimens. The vast majority of *Vibrio* cultures isolated from clinical specimens will be easily identified as one of the species listed. The six tests which we find very helpful in dividing the 10 species into 5 groups are: requirement for Na^+, oxidase, nitrate reduction to nitrite, arginine dihydrolase, lysine decarboxylase, and ornithine decarboxylase (Table 8).

Growth in the absence of added Na^+ (growth in nutrient broth with 1% NaCl, but no growth in nutrient broth) is the essential test for the differentiation of *V. cholerae* and *V. mimicus* from the other eight *Vibrio* species and from the marine vibrios. Table 9 gives the key characteristics for *V. cholerae* and *V. mimicus* and the key tests for differentiating these two species, which are very similar biochemically. Table 10 gives the tests which are used to define the classical and eltor biogroups of *V. cholerae*. *V. metschnikovii* is easily differentiated because it is oxidase negative and nitrate negative. All of the other vibrios listed in Table 8 are positive for these two tests. *V. hollisae* is negative for arginine dihydrolase and for lysine and ornithine decarboxylases (triple decarboxylase negative), which differentiates it from the remaining *Vibrio* species (Table 8). The remaining six species are subdivided into two groups, i.e., the group which is arginine dihydrolase positive and includes *V. damsela*, *V. fluvialis*, and *V. furnissii* and the group which is arginine dihydrolase negative but lysine decarboxylase positive and includes *V. alginolyticus*, *V. parahaemolyticus*, and *V. vulnificus* (Table 8). Tests useful for further differentiation of the species in these groups are given in Tables 11 through 13.

Commercial identification systems

Many laboratories use commercial kits to identify clinical isolates. Changes were necessary in some of the original procedures, e.g., suspension of the organism in NaCl solution (0.85 to 1%) rather than distilled water. The halophilic *Vibrio* species grew very poorly without the added NaCl and were often lysed in the distilled water. MacDonald et al. (44) found that artificial seawater was better (gave more positive reactions) as a suspending medium for the halophilic *Vibrio* species. One problem has been that some of the commercial products confuse *Aeromonas* species with *V. fluvialis*. These two organisms can easily be differentiated with one additional test: *V. fluvialis* is a

TABLE 8. Seven key tests to divide the 10 *Vibrio* species which occur in clinical specimens into five groups[a]

Test	Reaction of *Vibrio* species in:									
	Group 1		Group 2	Group 3	Group 4			Group 5		
	V. cholerae	*V. mimicus*	*V. metschnikovii*	*V. hollisae*	*V. damsela*	*V. fluvialis*	*V. furnissii*	*V. alginolyticus*	*V. parahaemolyticus*	*V. vulnificus*
Growth in nutrient broth										
0% NaCl	+	+	−	−	−	−	−	−	−	−
1% NaCl	+	+	+	+	+	+	+	+	+	+
Oxidase			−	+	+	+	+	+	+	+
Nitrate→nitrite			−	+	+	+	+	+	+	+
Arginine dihydrolase				−	+	+	+	−	−	−
Lysine decarboxylase				−				+	+	+
Ornithine decarboxylase				−						

[a] Symbols: +, almost all strains positive, usually 90% or more; −, almost no strains positive, usually 0 to 10% positive. All results are for tests incubated at 36°C for 48 h.

TABLE 9. Key characteristics of *V. cholerae* and *V. mimicus*[a]

Properties	Reactions of	
	V. cholerae	*V. mimicus*
Shared by both species		
Oxidase	+	+
Growth in nutrient broth		
No NaCl added	+	+
1% NaCl added	+	+
Arginine dihydrolase	−	−
Lysine decarboxylase	+	+
Ornithine decarboxylase	+	+
Fermentation of		
Lactose (1 to 2 days)	9	79
Lactose (3 to 7 days)	24	56
To differentiate the two species		
Sucrose fermentation	>99	0
Lipase (corn oil)	95	10
Voges-Proskauer	65[b]	0

[a] For symbols, see table 8, footnote *a*.
[b] The eltor biogroup is usually positive, and the classical biogroup is usually negative.

halophilic *Vibrio* species and does not grow in nutrient broth without added NaCl, but *Aeromonas* strains grow in this medium. In summary, the kits have improved in their ability to identify *Vibrio* species, but misidentifications are still a problem. One of us (M.T.K.), who has tried several different kits, prefers conventional tube tests for *Vibrio* identification.

ANTIBIOTIC SUSCEPTIBILITY

Vibrio strains important in clinical microbiology usually grow well on Mueller-Hinton agar, which is used in the disk susceptibility test. Although Mueller-Hinton agar contains no added NaCl (see the commercial formula), it contains a hydrochloric acid hydrolysate of casein, which apparently has enough NaCl to allow good growth of the halophilic *Vibrio* species. Some of the environmental marine vibrios grow poorly on Mueller-Hinton agar, presumably because of their higher requirement for Na^+. Broth dilution susceptibilities in Mueller-Hinton broth can be done without modification for most *Vibrio* species because the broth contains sufficient Na^+ (32).

Antibiotic resistance is rare in *Vibrio* species compared with species of the family *Enterobacteriaceae*. Table 14 summarizes the results of isolates received at CDC. In most cases, resistance is probably intrinsic to the species rather than acquired through plasmid transfer or antibiotic exposure. The one exception to this generalization is the antibiotic resistance found in some outbreaks of *V. cholerae* which have become resistant through the acquisition of R factors. In the United States, strains of *V. cholerae* and other *Vibrio* species have rarely had this type of resistance. Resistance to polymyxin antibiotics (polymyxin B and colistin) can be useful in spotting a culture of the eltor biogroup of *V. cholerae* or a culture of *V. vulnificus*. Most other *Vibrio* species are more susceptible.

TABLE 10. Differentiation of the classical and eltor biogroups of *V. cholerae* serogroup O1[a]

Property	Biogroup	
	Classical	Eltor
Voges-Proskauer	−	+
Zone around polymyxin B (50 U)	+	−
Hemolysis of erythrocytes	−	+
Agglutination of chicken erythrocytes	−	+
Lysis by bacteriophage		
Classical IV	+	−
Eltor 5	−	+

[a] For symbols, see Table 8, footnote *a*.

TABLE 11. Differentiation of the arginine-positive species *V. damsela*, *V. fluvialis*, and *V. furnissii*[a]

Test	Reaction of *Vibrio* species		
	V. damsela	*V. fluvialis*	*V. furnissii*
Voges-Proskauer	+	−	−
Citrate, Simmons'	−	+	+
L-Arabinose fermentation	−	+	+
D-Mannitol fermentation	−	+	+
Sucrose fermentation	−	+	+
Gas production during fermentation	V	−	+

[a] For symbols, see Table 8, footnote *a*. V, 26 to 74% positive.

SEROLOGICAL TYPING OF ISOLATES

The serological typing of isolates is usually done by reference laboratories rather than by clinical laboratories. The one exception is the use of *V. cholerae* O1 antisera (Table 15) which can be used in clinical laboratories to test a strain of *V. cholerae* to determine whether the strain is O1 positive or negative. This procedure was discussed above. Strains of *V. cholerae* which are not O group 1 can be further differentiated by serotyping. Two systems for typing non-O1 *V. cholerae* have been described (16, 53, 56).

In a system described by Sakazaki and Shimada (53), antisera are made against heated cells, and the O antigen is determined by agglutination. There are

TABLE 12. Differentiation of *V. fluvialis* and *V. furnissii*

Property	% Positive for property	
	V. fluvialis	*V. furnissii*
Glucose, gas production	0	99
Esculin hydrolysis	72	0
Carbon sources, growth on[a]		
Citrulline	97	4
Glucuronate (D−)	94	7
Putrescine	31	100
Aminovalerate (delta)	0	63
Cellobiose	63	4
Glutarate	−[b]	+[b]

[a] Data from Lee et al. (40) except that for glutarate, which is from Baumann et al. (8). These carbon source utilization tests are usually done in reference laboratories rather than clinical laboratories.
[b] Symbols: see Table 8, footnote *a*.

TABLE 13. Differentiation of the arginine-negative, lysine-positive species *V. alginolyticus*, *V. parahaemolyticus*, and *V. vulnificus*[a]

Property	Reaction of		
	V. alginolyticus	*V. parahaemolyticus*	*V. vulnificus*
Fermentation of:			
Cellobiose	–	–	+
Lactose	–	–	[+]
Salicin	–	–	+
Zone sizes around:			
Colistin	Large	Large	Small
Ampicillin	Small	Small	Large
Carbenicillin	Small	Small	Large
Growth in:			
8% NaCl	+	[+]	–
10% NaCl	V	–	–
Voges-Proskauer	+	–	–
Sucrose fermentation	+	–	[–]
L-Arabinose fermentation	–	[+]	–

[a] Symbols: +, almost all strains positive, usually 90% or more; [+], most strains positive, usually 75 to 89%; V, strain-to-strain variation, 26 to 74% positive; [–], few strains positive, usually 11 to 25% positive; –, almost no strains positive, usually 0 to 10% positive.

over 60 O antigens in this schema. In a system described by Smith (56), antisera are made against unheated cells and the "Smith type" is determined by slide agglutination of living cultures. There are over 72 different types in the Smith schema. Serological typing of non-O1 *V. cholerae* should be done only to answer specific epidemiological questions, and such typing is done by only a few reference laboratories. An international working group will study the two systems and make recommendations for standardization.

There is also a serological typing schema for *V. parahaemolyticus*. This schema includes 11 numbered O antigens and 55 numbered K antigens (Table 16). The O antigen is first determined by slide agglutination with an autoclaved antigen. The K antigen is then determined by slide agglutination with an unheated suspension. A complete set of O and K antisera (individuals and pools) is produced commercially by Toshiba Kagaku Kogyo Co., Inc., Maruishi, Bldg. 2, 1-Chrome Kanda-Kajicho, Chiyode-KU, Tokyo, Japan, and is available in the United States from Nichimen Company, Inc., 1185 Avenue of the Americas, New York, NY 10036. Serotyping of *V. parahaemolyticus* is a technique which should be done in reference laboratories only to answer specific questions.

DIAGNOSIS OF *VIBRIO* INFECTIONS BASED ON A PATIENT'S ANTIBODY RESPONSE

The best method for the diagnosis of infections due to *Vibrio* species is the isolation of the causative organism during the acute phase of illness. However, it is not always possible to obtain the necessary cultures, so an alternative is to measure the patient's antibody response to the *Vibrio* species which is suspected. This approach has been used frequently for cholera, and the methods have been discussed in detail by Feeley and DeWitt in the *Manual of Clinical Immunology* (25).

A retrospective diagnosis of cholera can be established with a high degree of certainty by the titration of acute- and convalescent-phase sera in agglutination, vibriocidal, or antitoxin tests. Acute-phase sera should be collected at 0 to 3 days, and convalescent-phase sera should be collected at 10 to 21 days after onset of illness. Paired sera should be submitted to laboratories familiar with these specialized tests. A fourfold or greater rise in vibriocidal or agglutination titer is diagnostic, provided recent (within 2 weeks) cholera immunization can be ruled out. Antitoxin determinations may be of special value in persons recently immunized. The titration of single convalescent sera is usually not worthwhile, except that the absence of significant levels of vibriocidal antibody would nearly exclude the diagnosis of cholera.

Similar serological tests can be used with the other *Vibrio* species and are especially recommended to further document the etiological role of some of the new species. Antibody responses against the patient's particular strain can be particularly useful in distinguishing colonization from actual infection and can be determined in both intestinal and extraintestinal infections.

TABLE 14. Antibiotic susceptibility (Kirby-Bauer disk method[a]) of 1,009 *Vibrio* strains

Antibiotic	Zone size (mm)			% of *Vibrio* strains susceptible (no. of strains studied)									
	Resistant	Intermediate	Susceptible	*V. cholerae* (480)	*V. mimicus* (75)	*V. metschnikovii* (22)	*V. hollisae* (34)	*V. damsela* (21)	*V. fluvialis* (25)	*V. furnissii* (9)	*V. alginolyticus* (69)	*V. parahaemolyticus* (144)	*V. vulnificus* (130)
Penicillin G	6–11	12–21	≥22	2	3	9	97	0	0	0	0	0	2
Ampicillin	8–11	12–13	≥14	87	97	31	100	52	32	11	0	12	99
Carbenicillin	8–17	18–22	≥23	64	83	27	100	14	16	0	0	1	54
Cephalothin	6–14	15–17	≥18	98	100	100	100	76	40	0	32	17	65
Colistin	6–8	9–10	≥11	4	61	91	100	76	100	100	25	11	2
Tetracycline	6–14	15–18	≥19	98	100	73	97	86	88	89	94	98	99
Sulfadiazine	6–12	13–16	≥17	26	17	5	56	71	36	11	16	3	28
Chloramphenicol	6–12	13–17	≥18	99	100	100	100	100	88	100	100	100	100
Streptomycin	6–11	12–14	≥15	60	61	32	100	24	84	100	54	17	42
Kanamycin	6–13	14–17	≥18	92	89	14	100	43	88	100	62	37	53
Gentamicin	6–12	13–14	≥15	98	99	100	100	100	100	100	100	97	100
Nalidixic acid	6–13	14–18	≥19	99	99	100	100	100	100	100	97	99	99

[a] Done on Mueller-Hinton Agar (no added NaCl) at 36 ± 1°C.

TABLE 15. Antigenic relationships in *V. cholerae* O1 subtypes Ogawa, Inaba, and Hikojima

Serogroup O1 subtype	O factors present in culture	Agglutination in absorbed serum[a]	
		Ogawa	Inaba
Ogawa	A, B	+	−
Inaba	A, C	−	+
Hikojima	A, B, C	+	+

[a] Specific factor sera are prepared by absorption. For example, an Ogawa antiserum is prepared by injecting an Ogawa culture and then absorbing the resulting antiserum with an Inaba culture, which removes the antibodies to O antigen factor A, leaving antibodies to O factor B.

V. CHOLERAE SEROGROUPS O1 AND NON O1

In the mid-1930s it was recognized that the vast majority of vibrios isolated from cholera cases agglutinated in a single antiserum (50). Agglutination in this antiserum (later called *V. cholerae* O1 antiserum) became the main criterion for identifying an organism as *V. cholerae*. Organisms which did not agglutinate in this serum were given the vernacular names "nonagglutinating vibrios," "NAGs," or "noncholera vibrios." These names included organisms known today as *V. cholerae* non O1, *V. parahaemolyticus*, other *Vibrio* species, and even *Aeromonas* or *Plesiomonas* species. It was soon recognized that some of the NAGs were identical with *V. cholerae* except for their agglutination in the antiserum to serogroup O1 (52). However, it has been only in the last few years that these vibrios have been classified in the species *V. cholerae*. Today it is accepted that there are two groups of *V. cholerae* strains, i.e., serogroups O1 and non O1. It is usually convenient to discuss these serogroups separately. *V. cholerae* O1 is the serogroup isolated from cholera cases, in which it produces severe, watery diarrhea through the action of cholera toxin. *V. cholerae* non O1 can cause a choleralike illness, but also causes a much wider spectrum of disease (48), including extraintestinal infections (34).

V. cholerae Serogroup O1

V. cholerae serogroup O1 is the group of *V. cholerae* strains which cause pandemic cholera. These strains are defined as strains of *V. cholerae* which agglutinate in O1 serum, include serogroups Ogawa and Inaba (and Hikojima), and include biogroups classical and eltor.

TABLE 16. Association of K antigens with certain O antigens of *V. parahaemolyticus*

O group	K antigens
1	1, 25, 26, 32, 28, 41, 55, 56
2	3, 28
3	4, 5, 6, 7, 29, 30, 31, 33, 37, 43, 45, 48, 54, 57
4	4, 8, 9, 10, 11, 12, 13, 34, 42, 49, 53, 55
5	15, 27, 30, 47
6	18, 46
7	19
8	20, 21, 22, 39
9	23, 44
10	24, 52
11	36, 40, 50, 51

History

Historians do not always agree on the history of cholera, perhaps because early descriptions made it difficult to differentiate cholera from other secretory diarrheas. The cholera bacillus was isolated by Koch in 1884, so bacteriological confirmation was possible only after this date. Cholera, or Asiatic cholera as it was often called in early writings, has probably always existed in India. It may have been found outside of India before 1800, but not in large epidemics. The spread of Asiatic cholera has come in seven large pandemics. The first pandemic began in 1816 to 1817; six other pandemics followed, beginning in 1829, 1852, 1863, 1881, 1889, and 1961 (50). Until recently, the classical biogroup has caused pandemic cholera. However, the seventh pandemic was caused by the eltor biogroup. Recently the classical biogroup has displaced the eltor biogroup in some areas of India.

Clinical significance

In an area in which cholera is endemic, many individuals will ingest *V. cholerae*. Most of these people will have either a mild diarrhea or no symptoms at all (asymptomatic colonization of the intestine with recovery of the organism from formed stool). This pattern has been much more common in the seventh cholera pandemic caused by the eltor biogroup of *V. cholerae*. At the other extreme, some individuals develop a dramatic, acute diarrhea which has been called cholera gravis. This is the terrifying disease which caused such fear in the previous century when the word cholera was mentioned.

In severe cholera there is massive diarrhea, with large volumes of rice-water stool (clear fluid with flecks of mucus) passed painlessly. The amount of fluid passed can be 1 liter per hour. In 4 to 6 days this fluid would amount to over twice the body weight. There is usually vomiting and little desire to maintain nutrition through normal eating. If the patient is untreated, there will be prostration with symptoms of severe dehydration: painful muscle cramps, watery eyes, loss of skin elasticity, and anuria (absence of urine excretion). Death can occur very quickly after onset of symptoms because of the severe dehydration.

The mainstay of treatment is intravenous therapy, which restores water and electrolyte balance and prevents acidosis. Cases with milder illness are often treated with oral electrolyte solutions. Tetracycline therapy reduces the period of excretion but is not a substitute for rehydration.

Milder forms of diarrhea due to *V. cholerae* O1 are more difficult to distinguish from other mild diarrheas. These cases can last from 1 to 5 days with several liquid stools per day. Cramping and vomiting can also occur. The role of *V. cholerae* O1 is shown by the isolation and identification of the causative organism. With the eltor biogroup of *V. cholerae* O1, this milder disease is seven times more common than severe cholera.

Pathogenesis

The pathogenesis of diarrhea due to *V. cholerae* O1 is well understood. The organism is ingested, and some cells survive the acid pH of the stomach and pass into the small intestine. The organism colonizes the small intestine and begins to grow and produce cholera toxin. Cholera toxin (molecular weight, 84,000) is composed of an A1 subunit (molecular weight, 21,000), an A2 subunit (molecular weight, 7,000), and five B subunits (molecular weight, 10,000 each). The B subunit of intact cholera toxin attaches to a specific receptor, ganglioside G1, on the cell membrane of the host's tissue. The A1 subunit of cholera toxin then activates the enzyme adenylate cyclase in the host cells, which then increases the level of cyclic AMP and the hypersecretion of salt and water. The net results are the massive outpouring of liquid stool and resulting dehydration which are typical of cholera.

Sources

V. cholerae serogroup O1 is well adapted to the human intestinal tract. The vast majority of strains come from human feces (Table 4). Only a few human isolates are extraintestinal. *V. cholerae* serogroup O1 can also be isolated from environments and from other animals that come in contact with feces of cholera cases. There have been reports of *V. cholerae* serogroup O1 from environmental sources which are unlikely to have come from cholera cases. These strains do not produce cholera toxin and probably represent different clones of *V. cholerae*. They are probably similar to the *V. cholerae* non-O1 strains in clinical and public health importance.

Isolation

In rice-water stools from cholera cases, *V. cholerae* is usually present essentially as a pure culture and in very high numbers (10^6 to 10^8 organisms per ml of stool). Isolation will present no problem in these cases. In formed feces, the organism will probably be present in much lower numbers, and normal enteric flora will also be present. The approach given in Fig. 1 will yield more positives in this situation.

Biochemical reactions

In countries in which cholera is common, there is no need to do extensive biochemical studies to confirm a culture as *V. cholerae*. However, complete biochemical testing is recommended in industrialized countries where cholera is rare. Tables 7 through 9 give the properties of *V. cholerae*. The test for Na^+ requirement differentiates *V. cholerae* from the halophilic *Vibrio* species. Sucrose fermentation differentiates *V. cholerae* and *V. mimicus*. The decarboxylase pattern of arginine negative, lysine positive, and ornithine positive differentiates *V. cholerae* from the oxidase-positive fermentation species in the genera *Aeromonas* and *Plesiomonas*.

Antibiotic susceptibility

Antibiotic resistance is rare in *V. cholerae*. Most strains are susceptible to tetracycline, the drug of choice. Occasionally a strain of *V. cholerae* becomes resistant and can spread. This event usually happens where cholera is endemic and there is more chance for a strain to acquire an antibiotic resistance plasmid.

Susceptibility to the polymyxin group of antibiotics is a property used to differentiate the classical and eltor biogroups.

Geographical distribution

The World Health Organization regularly reports on the worldwide distribution of cholera. In its survey for 1982, 54,856 cases were reported from 37 countries. The majority of cases were in Africa (37,427 cases), Asia (15,191 cases), and the Trust Territories in the Pacific Ocean (2,217 cases).

Cholera (defined as the disease caused by *V. cholerae* serogroup O1, cholera toxin positive) in the United States is quite interesting (12, 38). In the United States, the last case of cholera from a pandemic occurred before 1900. Then, unexpectedly, an isolated case occurred in Texas in 1973. This case was followed by an outbreak in Louisiana in 1978, two sporadic cases in Texas in 1982, and an outbreak on an oil-drilling platform on a bayou near the Texas coast in 1982. These strains have been thoroughly studied and have the following properties in common: serogroup Inaba, cholera toxin positive, strongly hemolytic, and a unique bacteriophage lysis pattern. When compared with other cultures from the seventh pandemic of cholera, the American cultures are quite different, suggesting that they are a different strain (clone). The origin and history of this strain are now being studied.

V. cholerae Serogroup Non O1

V. cholerae serogroup non O1 can cause a cholera-like disease, but it is also isolated from patients with mild diarrhea and extraintestinal infection. It is also isolated from the aquatic environment.

History

Until recently, the strains of serogroup non O1 were not classified in the species *V. cholerae* but were reported as "nonagglutinating vibrios," "NAGs," or "noncholera vibrios." Today they are classified as *V. cholerae* non O1, and the literature on their ecology (19) and role in human disease (34, 48) is expanding rapidly.

Clinical significance: diarrhea

These organisms have been isolated from patients with severe dehydrating (choleralike) gastroenteritis. Other strains have been isolated from patients with mild diarrhea. Unlike *V. cholerae* O1, the non-O1 strains have also been found in patients with diarrhea when feces have contained blood or mucus or both. Fever is also a symptom in some of the cases. These different responses may correlate with the different types of pathogenesis described below. The duration of illness is usually less than 3 days.

Pathogenesis of intestinal infections

Although the pathogenesis of intestinal infections is not as well known as is that of *V. cholerae* O1, three pathogenic mechanisms have been postulated: (i) production of cholera (choleralike?) toxin, (ii) production of a heat-stable enterotoxin (positive result in the infant mouse assay used to detect *Escherichia coli* heat-stable enterotoxin), and (iii) invasive disease (positive response to whole cells but not to cell filtrates in the ligated rabbit ileal loop, or infant mouse assay). Some strains of *V. cholerae* non O1 have no positive response in any of the above assays. The correlation of the possible virulence factors with the different types of gastroenteritis-invasive diarrhea needs further investigation.

Extraintestinal infections

V. cholerae non O1 has a wider spectrum of disease than does *V. cholerae* O1. Non-O1 strains have been isolated from patients with septicemia who had cirrhosis or other underlying disease. In our series (Table 4), there were 42 blood isolates, compared with only 1 blood isolate for *V. cholerae* O1. The non-O1 strains were also isolated from ears, wounds, respiratory tract, and urine. There were 52 isolates of serogroup non O1 from these sources, compared with only 2 isolates for *V. cholerae* O1 (Table 4).

Laboratory

The discussion of laboratory procedures for *V. cholerae* O1 (see above) also applies to the non-O1 strains. The essential difference is that the O1 strains agglutinate in *V. cholerae* serogroup O1 antiserum, but the non-O1 strains do not. This difference emphasizes the need for reference laboratories to maintain this antiserum, even though it will be needed only occasionally. The biochemical reactions of both *V. cholerae* groups are almost identical.

Geographical distribution

V. cholerae non-O1 strains have been isolated worldwide both from patients and from the environment. In the United States, many strains are found in sporadic cases of diarrhea. Most of these cases have occurred after raw oysters were eaten. In the United States, fewer than 5% of the isolates have been positive for cholera toxin, a situation in contrast to that in Bangladesh, where about one-third of the isolates have been positive (11).

VIBRIO MIMICUS

V. mimicus is a new *Vibrio* species (20) which is associated with diarrhea, usually after the consumption of uncooked seafood, particularly raw oysters (54). *V. mimicus* has been isolated in many countries (Canada, Mexico, Philippines, New Zealand, Guam, and Bangladesh), which suggests a worldwide distribution in countries situated on an ocean. The isolation of *V. mimicus* from water, oysters, and shrimp and its distribution in coastal areas indicate that its ecology may be similar to the ecology of *V. cholerae* non O1, which has been well studied.

History

There have been a number of reports on *V. cholerae* strains with atypical biochemical reactions. Traditionally, these strains have been reported with designations such as "*V. cholerae* lysine decarboxylase negative," "*V. cholerae* mannitol negative," or "*V. cholerae* sucrose negative." Davis et al. (20) studied representative strains from six of these unusual biogroups by both DNA hybridization and phenotypic analysis. Five of the six groups were very highly related to *V. cholerae* by DNA hybridization. These groups had been identified correctly as atypical strains of *V. cholerae*. However, the group of sucrose-negative strains was only 24 to 54% related to *V. cholerae*. Based on this low relatedness and phenotypic differences, Davis et al. proposed a new species, *V. mimicus*. The species name (specific epithet) *mimicus* refers to the fact that the strains "mimic" *V. cholerae*.

Clinical significance

Based on the information submitted with cultures, Davis et al. (20) felt that *V. mimicus* was associated with diarrhea and was probably linked to the eating of shellfish. Three strains of *V. mimicus* were positive for heat-stable enterotoxin in the infant mouse assay, and five strains were positive for heat-labile enterotoxin in the Y-1 adrenal cell assay or the enzyme-linked immunosorbent assay. However, other strains from diarrhea cases were negative in the assays. The enterotoxin data were very similar to those for non-O1 *V. cholerae* strains in that some produced heat-labile or heat-stable enterotoxin, but most strains from patients with diarrhea produce neither toxin.

Shandera et al. (54) did a retrospective case-control study on 17 patients whose cultures of *V. mimicus* had been sent to CDC from 1979 to 1981. This study further strengthened the association of *V. mimicus* with gastroenteritis-diarrhea after the consumption of seafood, particularly raw oysters. The role of *V. mimicus* as an actual cause of diarrhea needs stronger documentation.

Ear infection

Two of the *V. mimicus* isolates studied by Shandera et al. were from patients' ears (54). Case 1 was a 39-year-old person with chronic otitis media since the age of 20. Case 2 was a 13-year-old child who had Kawasaki's syndrome at age 5. Both patients had seawater exposure during the week before onset of illness, one person wading in a North Carolina estuary and the other swimming off the coast of Hawaii. These data for *V. mimicus* are similar to those for other *Vibrio* species from ear infections. *V. mimicus* has been recovered from ears after salt- or brackish-water exposure, but its causal role in ear infections needs further documentation.

Isolation

Most of the *V. mimicus* strains were isolated during a search for *V. cholerae* and other *Vibrio* species. *V. mimicus* is sucrose negative, so colonies on TCBS agar are green rather than yellow (Table 5). It grows well on normal enteric media, so isolation and identification should pose no special problems.

Biochemical reactions

Tables 7 through 10 give the reactions of *V. mimicus* along with those of *V. cholerae*. These two species are very similar in most of the tests normally needed for identification; however, Table 9 gives reactions which are useful for differentiation. *V. mimicus* is negative for lipase production and sucrose fermentation. *V. cholerae* usually has the opposite pattern, although a few strains are lipase negative. *V. mimicus* resembles the classical biogroup of *V. cholerae* because it is Voges-Proskauer negative and susceptible to polymyxin. The eltor biogroup of *V. cholerae* is usually Voges-Proskauer positive and resistant to polymyxin. *V. mimicus* grows in nutrient broth without NaCl, a reaction which differentiates it along with *V. cholerae* from the halophilic *Vibrio* species which require a higher NaCl concentration for growth.

Antibiotic susceptibility

V. mimicus is generally susceptible to antibiotics except for sulfadiazine (87% resistant). A few strains are resistant to penicillins, colistin, nalidixic acid, and kanamycin (20; Table 14). *V. mimicus* is very similar to *V. cholerae* in its susceptibility.

VIBRIO PARAHAEMOLYTICUS

V. parahaemolyticus has been known as a cause of acute gastroenteritis since 1950. Food-borne outbreaks and sporadic cases occur worldwide and are usually associated with the consumption of contaminated seafood. Two books and a recent review article summarize much of the information known about this organism (28, 37, 45).

History

On 20 and 21 October 1950, there was an outbreak of food poisoning in Osaka, Japan. There were 272 patients with acute gastroenteritis, 20 of whom died. These deaths led to an extensive investigation of the outbreak, and eventually an organism was isolated which was shown to be the etiological agent (see reference 45, p. 1–5, for the complete, fascinating story). The properties of the organism were studied, and it was named as a new bacterial species, *Pasteurella parahaemolytica*. The halophilic nature of the new organism was not discovered until 1955 (45), and it was then reclassified as a halophilic *Vibrio* species, *V. parahaemolyticus*.

Clinical significance in gastroenteritis

V. parahaemolyticus causes gastroenteritis with nausea, vomiting, abdominal cramps, low-grade fever, and chills. The diarrhea is usually watery but sometimes bloody. The disease is usually mild and self-limiting but can be fatal (a 7% fatality rate in the first reported outbreak). There is a good correlation between pathogenicity and a positive Kanagawa test which measures the ability of the strain to produce a hemolysin for human erythrocytes when it is grown on a special medium. About 96% of the *V. parahaemolyticus* strains from well-documented cases of human gastroenteritis are Kanagawa positive, but only about 1% of the strains isolated from the environment are positive. Although there is excellent correlation between this hemolysin and human disease, the mechanisms of pathogenesis are still being debated, since other toxins or virulence factors may be involved. Rehydration is usually the only treatment needed, but in some severe cases, the patient will require hospital admission. Antimicrobial therapy may be beneficial, and tetracycline appears to be the drug of choice.

Outbreaks of gastroenteritis due to *V. parahaemolyticus* occur worldwide, but they are not as common in the United States as in some other countries. In Japan, *V. parahaemolyticus* is an extremely important diarrheal agent and causes 50 to 70% of the cases of food-borne enteritis (52). All of the outbreaks are associated directly or indirectly with seafood. Direct infection comes from the ingestion of raw fish or shellfish which are contaminated with *V. parahaemolyticus*. This mechanism of transmission is apparently quite common in Japan because of the national custom of eating raw fish (52). Contamination after cooking is apparently the mechanism for indirect infection. This mechanism is important in outbreaks where the food has been cooked. Kitchen utensils and water probably introduce the organism into the cooked food, where it can then multiply if correct food-handling procedures are not used (52).

Extraintestinal infections

There is confusion about the role of *V. parahaemolyticus* in infection outside the intestinal tract. Many authors refer to rapidly fatal septicemia or to wound infections due to this organism. However, in many of the original reports the organism was actually *V. vulnificus* rather than *V. parahaemolyticus*. This error is due to the fact that *V. vulnificus* was incorrectly identified as *V. parahaemolyticus*. All reports of severe extraintestinal infections due to *V. parahaemolyticus* must be examined critically to be sure the organism in question is not *V. vulnificus*, another *Vibrio* species, or a marine vibrio. Two cases of eye infections (58) provide evidence that *V. parahaemolyticus* can be a pathogen outside the intestinal tract. However, these cases are apparently rare. The vast majority of the isolates in our series are from feces, but 2 were from blood and 11 were from wounds (Table 4). Isolates from the marine environment are quite common (Table 4), a fact which conforms to the known distribution of the organism.

Isolation

Unlike the situation with most of the *Vibrio* species, much has been written about the isolation of *V. parahaemolyticus*. The fact that this species was not detected as an important cause of gastroenteritis until 1950 indicates that isolation with the usual enteric media can pose problems. Oxidase testing of colonies on a nonselective medium such as blood agar should provide a high isolation rate. A plate of TCBS medium can be included in those geographical areas where *V. parahaemolyticus* is most common. The isolation methods designed for *V. cholerae* are also efficient for *V. parahaemolyticus*. *V. parahaemolyticus* grows well on TCBS agar as 2- to 3-mm green colonies. A number of selective media have also been described elsewhere (14), but these media are most useful with environmental specimens or for detecting carriers and are almost never used in the routine processing of clinical specimens.

Identification

V. parahaemolyticus is one of the halophilic *Vibrio* species and belongs in the lysine-positive, arginine-negative group (Table 8). It is negative for the Voges-Proskauer test and for the fermentation of lactose and salicin. These and other tests are useful in differentiation (Table 13).

Urea-positive strains of *V. parahaemolyticus*

Until recently, *V. parahaemolyticus* was considered negative for urea hydrolysis. Sakazaki and Balows reported that none of their 2,354 strains was urea positive (52). However, in the last few years we have received strains which were otherwise typical of *V. parahaemolyticus* but were urea positive. These urea-positive strains were confirmed by DNA hybridization as *V. parahaemolyticus*. The number of urea-positive strains has increased steadily: in 1977, 17% were urea positive; in 1978, 14%; in 1979, 36%; in 1980, 12%; in 1981, 47%; in 1982, 73%; and in 1983, 83%. This sample is biased because laboratories refer atypical strains rather than typical ones. However, 50% or more of *V. parahaemolyticus* strains from California and from outbreaks investigated by CDC have also been urea positive. These samples are less biased and show that the increase in urea-positive strains is real.

Antibiotic susceptibility

V. parahaemolyticus is usually resistant to ampicillin and carbenicillin but susceptible to colistin. These properties are shared with *V. alginolyticus*, but *V. vulnificus* has the opposite susceptibilities (15). Table 14 gives more details on antibiotic susceptibility.

VIBRIO HOLLISAE

V. hollisae is a recently described halophilic species in the genus *Vibrio* which is associated with diarrhea after the consumption of raw seafood.

History

V. hollisae was named as a new species in 1982 (31). Two laboratories had independently studied this organism under the vernacular names "EF13" and "Enteric Group 42." Data from DNA hybridization indicated that all these strains were highly related and distinct from other named *Vibrio* species (31). Hickman et al. named this organism *V. hollisae.*

Clinical significance: sources

V. hollisae is a probable cause of human diarrhea. Its association with diarrhea is very strong, but additional evidence is needed to document its causal role and pathogenesis. Of the original 16 strains, 15 were from feces, and many of the patients had diarrhea (31). Morris et al. (47) described the clinical and epidemiological features of 11 diarrhea cases that were culture positive for *V. hollisae.*

They also reported one isolate from blood. It was from a patient with hepatic cirrhosis, hepatic encephalopathy, bronchopneumonia, and sepsis due to the yeast *Cryptococcus*. The patient was comatose and died 2 days after hospital admission (47).

Isolation

Laboratory strains of *V. hollisae* did not grow (2 days, 36°C) on TCBS (Table 5) or MacConkey agar (31). Since these two media are frequently used for the isolation of *Vibrio* species and species of the family *Enterobacteriaceae* from stools, it is possible that *V. hollisae* is often missed, even though it may be the predominant organism. *V. hollisae* grows well on sheep blood agar (36°C), so it can be detected by the use of the oxidase reagent as discussed above.

Identification

V. hollisae does not grow in nutrient broth without added NaCl; thus it is one of the halophilic *Vibrio* species. Tables 6 through 8 give the biochemical reactions; only a few reactions need be mentioned. *V. hollisae* is indole positive when tested in heart infusion broth with 1% NaCl added, but only 38% positive in peptone water with only 0.5% NaCl. Moeller's lysine, arginine, and ornithine are all negative. Motility in *V. hollisae* is very unusual. None of the strains was motile (semisolid medium, 36°C) after 48 h of incubation, but after 7 days, 88% were motile. This characteristic is unusual for the genus *Vibrio*. *V. hollisae* ferments only D-glucose, L-arabinose, D-galactose, and D-mannose, a characteristic fermentation pattern among *Vibrio* species. Recently we confirmed a urea-positive strain of *V. hollisae*, which is the first one with this unusual property.

Antibiotic susceptibility

Stains of *V. hollisae* have a very characteristic antibiogram. There are very large zones around all antibiotics tested, including penicillin (Table 14).

VIBRIO FLUVIALIS

V. fluvialis is a new *Vibrio* species which seems to cause sporadic cases of diarrhea worldwide (40, 57) and has been implicated in outbreaks of diarrhea in Bangladesh (35).

History

In 1981, Lee et al. (40) gave the name *V. fluvialis* to a group of halophilic vibrios which had been previously known as "group F vibrios" and as "group EF6." These organisms had been isolated from a number of environmental sources throughout the world. They had also been isolated from humans with diarrhea. Lee et al. defined two biogroups (synonym, biovars) in this species which correlated with source and certain biochemical tests. *V. fluvialis* biogroup I did not produce gas during fermentation (it was anaerogenic) and was isolated from human diarrhea as well as from the environment. *V. fluvialis* biogroup II produced gas during fermentation (it was aerogenic) and was isolated from the environment, but not from humans with diarrhea. Brenner et al. (18) later showed that strains of *V. fluvialis* biogroup II were more distantly related to strains of *V. fluvialis* biogroup I and proposed that these strains be classified in a new *Vibrio* species, *V. furnissii.*

Clinical significance: sources

The sources of the early clinical isolates of *V. fluvialis* showed a marked association with diarrhea (40). Further studies have provided additional evidence for its causal role. One difficulty in assessing the clinical significance and epidemiology of *V. fluvialis* has been the presence of other possible pathogens. Most of the reports have come from geographical areas where several possible pathogens are often present in feces.

Gastroenteritis caused by *V. fluvialis* is usually described as choleralike (35). Patients typically have watery diarrhea with vomiting (97%), abdominal pain (75%), moderate to severe dehydration (67%), and often fever (35%). Usually infants, children, and young adults are affected. Frank blood is found in a small percentage of stool samples, but erythrocytes or leukocytes are found in most (75%).

The pathogenesis of *V. fluvialis* diarrhea is not completely known. Assays for heat-labile and heat-stable enterotoxins and invasiveness have generally been negative. About 20% give a positive rabbit ileal loop test for enterotoxin. Recent studies have provided additional evidence for enterotoxin or enterotoxin-like molecules (42, 49).

Isolation

V. fluvialis grows on TCBS agar as 2- to 3-mm yellow colonies (Table 5). Although this species requires Na^+, its requirement is much less than some of the more halophilic *Vibrio* species, which makes isolation and identification easier. Nishibuchi and Seidler recently described a selective broth for isolating this species from water (see reference 49).

Identification

The most striking aspect of *V. fluvialis* identification is the possible confusion with *Aeromonas* strains, since both species are arginine dihydrolase positive. Many cultures we receive as "*V. fluvialis*" turn out to be *Aeromonas hydrophila*. The converse is also true. A simple test to differentiate these two organisms is growth in nutrient broth with 0 and 1% NaCl. *V. fluvialis* is a halophilic vibrio and will not grow without the added NaCl. *A. hydrophila* grows in both media. Table 7 gives the biochemical reaction for the identification of *V. fluvialis*. *V. fluvialis* is generally susceptible to antibiotics (Table 14).

VIBRIO FURNISSII

V. furnissii is a new *Vibrio* species which was formerly known as "*V. fluvialis* biovar II," "*V. fluvialis* aerogenic," or "*V. fluvialis* gas$^+$." It is now recognized as a separate species and is named *V. furnissii* (18).

Clinical significance: sources

The vast majority of the isolates of the *V. fluvialis-V. furnissii* group isolated from patients with diarrhea have not produced gas during fermentation. Thus, they are *V. fluvialis*. One exception was a case of human diarrhea in Indonesia (58). Further investigation (18) into case histories of the isolates studied at CDC indicated that *V. furnissii* had been isolated from two outbreaks of acute gastroenteritis. These outbreaks were among American tourists returning from the Orient in 1969. Other enteric pathogens were also isolated (*V. parahaemolyticus*, *V. cholerae* non O1, *V. fluvialis*, *Salmonella* spp., and *Plesiomonas* spp.), so the causal role of *V. furnissii* was not clear. There have been no documented cases of diarrhea caused by this organism. Thus we have not included it with the other *Vibrio* species as a probable cause of diarrhea. However, microbiologists should be alert to its presence and for evidence of its causal role in diarrhea. *V. furnissii* is apparently rare in human clinical specimens. Table 4 indicates that human stool was the most common source (16 isolates), but 1 isolate was from a wound. Lee et al. (40) mention that *V. furnissii* (listed as *V. fluvialis* biovar II) is widespread in the aquatic environment and is common in estuaries.

Laboratory

V. furnissii resembles *V. fluvialis* very closely (Table 7). The two species are so close phenotypically that only a few tests are available to differentiate them (Table 12). Gas production will be the key differential test in most clinical laboratories (Table 12). *V. furnissii* can be confused with strains of *Aeromonas* if its halophilic nature is not recognized.

VIBRIO VULNIFICUS

V. vulnificus has been recognized as a distinct species of *Vibrio* since 1976 (8). It causes wound infections and life-threatening septicemia (13). In the last few years, there has been intensive study of this organism (see references 8, 11, 19, 52, and 58 for primary references).

History

In 1976, Hollis et al. (32) described a salt-requiring organism that appeared to be different from other *Vibrio* species. Its properties were similar to those of *V. parahaemolyticus* and *V. alginolyticus*, but the new organism fermented lactose. The vernacular name "lactose-positive *Vibrio*" was given to this organism. In the literature it has been referred to as "L$^+$ *Vibrio*" and "Lac$^+$ *Vibrio*." This organism was studied by several laboratories and given the scientific name *Beneckea vulnifica* by Reichelt et al. in 1976 (8). Its classification in the genus *Beneckea* was not widely accepted, so it was subsequently classified in the genus *Vibrio* (22). Today, it is almost universally known as *V. vulnificus*.

Clinical significance: sources

V. vulnificus has been associated with two disease syndromes, primary septicemia and wound infection (13).

Primary septicemia is a very serious infection with a fatality rate of about 50%. Most patients with primary septicemia due to *V. vulnificus* have preexisting liver disease (13), but other patients have been normal individuals (58). In most cases the disease begins several days after the patient has eaten raw oysters. Individuals develop malaise which progresses rapidly to fever, chills, and prostration and often to severe hypotension and death. The progression can be very rapid, from asymptomatic to death within 24 h. At autopsy, there are findings of gram-negative shock, hemorrhage, and necrosis of involved muscles and skin (with organisms culturable from the lesions). Cultures of blood and skin lesions are usually positive for *V. vulnificus*.

V. vulnificus also causes severe wound infections, usually after trauma and exposure to marine animals or the marine environment (13). The mortality rate is not nearly as high as in primary septicemia caused by this organism (about 7% compared with about 50%, respectively). Wound infections progress rapidly with swelling and erythema, often followed by the development of bullae or vesicles leading to tissue necrosis. Symptoms include a very painful wound, fever, and chills. Surgical intervention, including amputation, is often required. Antimicrobial therapy is often required; the most effective drugs appear to be tetracycline and aminoglycosides, based on laboratory susceptibilities (32). About one-third of the patients with wound infection had underlying disease.

Isolation

V. vulnificus grows well on blood agar and TCBS agar (Table 5). Although most strains of *V. vulnificus* are sucrose negative (3% positive in the original report of Hollis et al. [32]), the number of sucrose-positive strains in our collection has increased markedly over the years. Thus, on TCBS agar, *V. vulnificus* can be either green (most) or yellow (some) (Table 5). This fact has not always been considered in searches for this organism. Most isolations of *V. vulnificus* in clinical laboratories will be made on blood agar, since blood and wounds are the usual sources of this species. Commercial blood culture bottles usually support growth of *V. vulnificus*; however, the Na^+ content is probably suboptimal, which may yield some cells with aberrant sizes and shapes. *V. vulnificus* does not survive well on media with an NaCl content of only 0.5%. Some cultures are dead by the time they reach our laboratory.

Identification

V. vulnificus is a halophilic *Vibrio* species. It is lysine positive and arginine negative, which puts it into the same group as *V. parahaemolyticus* and *V. alginolyticus* (Tables 8 and 13). It is differentiated by lactose fermentation, the ONPG test, salicin fermentation, and several other tests (Table 13). The antibiogram (disk diffusion) is also helpful in the differentiation of these three species (Table 14). *V. vulnificus* has either no zone or a small zone around the colistin disk, but large zones around ampicillin and carbenicillin. *V. parahaemolyticus* and *V. alginolyticus* have the opposite pattern (Table 13).

Antibiotic susceptibility

V. vulnificus is susceptible to most antibiotics except the polymyxin group (Table 14). Hollis et al. tested 33 strains by both the disk and agar dilution methods (MICs on Mueller-Hinton medium with no added salt). All strains were susceptible to penicillin, ampicillin, carbenicillin, cephalothin, chloramphenicol, gentamicin, tetracycline, rifampin, nitrofurantoin, and sulfisoxazole. Although *V. vulnificus* is susceptible to most antibiotics used against gram-negative bacteria, animal studies indicate that tetracycline may be most effective in vivo (58). Early administration of aminoglycoside or tetracycline antibiotics before the onset of hypotension appears to be critical to successful therapy.

Other infections

Other infections from which *V. vulnificus* has been isolated include pneumonia in a drowning victim and endometritis which developed in a woman after exposure to seawater (58).

Sources

Table 4 gives the sources of our 124 isolates. The vast majority were from blood or wounds. Unusual isolates included one from urine and five from stool. *V. vulnificus* is also isolated from the marine environment.

Pathogenesis

Animal studies have shown that *V. vulnificus* can cause severe local infections (51) with gross edema leading to tissue necrosis and death. Iron availability has been implicated in the pathogenesis of *V. vulnificus* infections (58). A toxin produced by *V. vulnificus* has been demonstrated, although its role in the pathogenesis of infections is still unclear.

VIBRIO DAMSELA

V. damsela is a new *Vibrio* species which is associated with human wound infections. It is found in the marine environment and causes skin lesions on certain marine fish.

History

V. damsela was described in 1981 by Love et al. (43). The original isolates were from skin lesions on marine fish from off the California coast. Koch's postulates were fulfilled when it was documented that this organism caused the lesions on the damselfish. Strains of *V. damsela* were compared with unidentified human strains of *Vibrio*, and several *V. damsela* strains were identified (43). The strains from marine fish, human clinical specimens, and the environment led to the proposal of this new species in the genus *Vibrio*.

Clinical significance: sources

Morris et al. (47) reviewed the case histories of six patients with wound infections from which isolates had been sent to CDC from 1971 to 1981. All of the infections had been acquired in coastal areas.

The sources of our *V. damsela* isolates are given in Table 4. All of the clinical isolates were from wounds; 9 of the 10 wounds were leg or foot wounds. Marine animals, particularly marine fish, made up the majority of the remaining isolates. A number of fish from Dakar, Senegal, yielded this organism. Other isolates included sewage (one isolate), oysters (one isolate), and a wound on a raccoon (one isolate). The role of *V. damsela* as an actual cause of human wound infection needs further documentation.

Isolation

V. damsela grows on blood agar and other noninhibitory media, but the Na^+ content is less than optimum. Strains grow on TCBS agar as 2- to 3-mm green colonies (Table 5). Love et al. noted that the damselfish isolate grew much better on TCBS at 25 than at 36°C. There have been few details on the isolation of this organism from human clinical specimens.

Identification

The complete biochemical reactions of *V. damsela* are given in Table 7. *V. damsela* is a typical halophilic *Vibrio* species in many respects. Strains do not grow in nutrient broth without added NaCl, which indicates its requirement for Na^+. *V. damsela* is arginine positive and Voges-Proskauer positive and has a characteristic fermentation pattern. It ferments D-glucose, D-mannitol, and D-mannose, and most strains ferment

D-galactose and trehalose. The characteristics which differentiate it from other *Vibrio* species are given in Tables 8 and 11.

Antibiotic susceptibility

V. damsela is generally susceptible to antibiotics, but less so than many other *Vibrio* species. Resistance is often seen for ampicillin, carbenicillin, penicillin, streptomycin, and sulfadiazine (Table 14).

VIBRIO ALGINOLYTICUS

V. alginolyticus is a relatively new species in the genus *Vibrio*. Most clinical isolates have come from infected wounds, ears, or eyes after trauma and seawater exposure. It is very common in the marine environment.

History

V. parahaemolyticus was the first of the halophilic *Vibrio* species to be studied in detail. An organism similar to, but different from, *V. parahaemolyticus* was often isolated from the marine environment, but not from patients with gastroenteritis. This organism was originally called "biotype 2 of *V. parahaemolyticus*." However, a number of studies have shown it to be a distinct species, i.e., *V. alginolyticus* (52).

Clinical significance: sources

There is now a considerable amount of literature reporting the isolation of *V. alginolyticus* from soft-tissue infections. Wound and ear infections are usually mentioned, with eye infections mentioned much less frequently. The data show the definite association of *V. alginolyticus* with infection at these sites; however, the etiological role and pathogenesis of this organism have not been thoroughly shown. However, most authors list *V. alginolyticus* as a pathogenic *Vibrio* species, particularly of wound and ear infections. Antibiotic treatment has been used in most cases, and surgical debridement has been used in some.

Table 4 shows the sources of the 74 isolates of *V. alginolyticus* in our series. Wounds and ears are the most common sources. Interestingly, there were 12 isolates from foot or leg wounds, but only 1 isolate from hand or arm wounds. *V. alginolyticus* is also widely distributed in the marine environment (29, 52). Most clinical isolates probably come from patients exposed to the marine environment, particularly after trauma.

Isolation

V. alginolyticus is a marine *Vibrio* species and requires added NaCl for optimum growth. However, it grows well on many of the enteric media. It is sucrose positive and occurs as 2- to 3-mm yellow colonies on TCBS agar (Table 5).

Identification

Table 6 lists the properties of this species. It is lysine positive and arginine negative, which puts it into the group with *V. parahaemolyticus* and *V. vulnificus* (Table 8). *V. alginolyticus* is usually Voges-Proskauer positive and usually grows in 8 and 10% NaCl, which differentiates it from the other two species. Table 13 gives some other differential characteristics.

Antibiotic susceptibility

V. alginolyticus is usually susceptible to ampicillin and carbenicillin but resistant to colistin (Table 14). This pattern is similar to that of *V. parahaemolyticus; V. vulnificus* has the opposite pattern.

VIBRIO METSCHNIKOVII

V. metschnikovii is an old species in the genus *Vibrio* (39) that has been frequently isolated from fresh, brackish, and marine waters. It has been isolated from clinical specimens only rarely.

History

In 1884, Gamalria reported the isolation of a new organism, *V. metschnikovii*, from a fowl that had died of a choleralike disease (39). Little was written about the organism until Lee et al. began isolating similar organisms on TCBS agar from marine and freshwater environments in the United Kingdom (39). These strains were oxidase negative and did not reduce nitrate to nitrite, but otherwise they were quite typical halophilic *Vibrio* species. Lee et al. proposed that the strains be classified as *V. metschnikovii* and wrote an emended description of this organism (39).

Clinical significance: sources

Lee et al. described 40 strains of *V. metschnikovii*, but none were from human clinical specimens; however, in 1981 Jean-Jacques et al. (36) reported the first clinically significant isolate of this newly redefined species. An 82-year-old woman with peritonitis and an inflamed gallbladder yielded *V. metschnikovii* from a positive blood culture. The isolate was considered clinically significant.

Table 4 gives the sources of the CDC isolates. One unusual finding was the two strains from urine. One was from the midstream urine of an 80-year-old woman in Canada. The other was from a routine urine culture of a 65-year-old male with chronic alcoholism, poorly controlled diabetes, and incontinence and urinary frequency. No information was available on the clinical significance of these two isolates. The first wound culture was from a 64-year-old female who was on chronic renal dialysis and had peripheral vascular disease. She developed an ulcer around a bunion on her foot which was treated by a wedge resection, followed by a foot amputation and finally a below-the-knee amputation. The wound yielded *V. metschnikovii* along with five other possible pathogens. The patient went fishing frequently, which may have been her source of *V. metschnikovii*. Further study is needed to determine the clinical significance of *V. metschnikovii* from sources such as urine and wound.

Nonhuman sources (taken from Lee et al. [39] and our own data) include rivers (11 isolates), sewage (9 isolates), cockles (4 isolates), shrimp (3 isolates), lobster (2 isolates), crab (1 isolate), and fowl (2 isolates). These data suggest that *V. metschnikovii* is widely distributed in the environment and that humans occasionally become colonized or infected from these sources.

Laboratory

V. metschnikovii grows on TCBS agar as 2- to 3-mm yellow colonies (Table 5) and apparently grows well on other laboratory media. It requires Na^+ for growth but in much smaller amounts (5 to 15 mM) than the other Na^+-requiring *Vibrio* species (8). Table 7 gives the complete characteristics of *V. metschnikovii*. This species is very easy to identify because it is oxidase negative, does not reduce nitrate to nitrite, and requires Na^+ for growth (Table 8). These three properties differentiate it from all the other species of the families *Enterobacteriaceae* and *Vibrionaceae*. *V. metschnikovii* is generally susceptible to antibiotics (Table 14).

LITERATURE CITED

1. **Anderson, J. I. W., and D. A. Conroy.** 1970. *Vibrio* disease in marine fishes, p. 266–272. *In* S. F. Snieszko (ed.), A symposium on diseases of fishes and shell fishes. Special publication no. 5, American Fisheries Society, Washington, D.C.
2. **Balows, A., G. H. Hermann, and W. E. DeWitt.** 1971. The isolation and identification of *Vibrio* cholerae—a review. Health Lab. Sci. **8:**167–175.
3. **Barker, W. H., Jr., and E. J. Gangarosa.** 1974. Food poisoning due to *Vibrio parahaemolyticus*. Annu. Rev. Med. **25:**75–81.
4. **Barua, D.** 1970. Laboratory diagnosis of cholera cases and carriers. Public Health Pap. **40:**47–52.
5. **Barua, D., and W. Burrows (ed.).** 1974. Cholera. The W. B. Saunders Co., Philadelphia.
6. **Baumann, P., and L. Baumann.** 1981. The marine gram-negative eubacteria: genera *Photobacterium, Beneckea, Alteromonas, Pseudomonas*, and *Alcaligenes*, p. 1302–1331. *In* M. P. Starr, H. Stolp, H. G. Trüper, A. Balows, and H. G. Schlegel (ed.), The prokaryotes. Springer-Verlag, Berlin.
7. **Baumann, P., and L. Baumann.** 1984. Genus II. *Photobacterium* Beijerinck 1889, 401[AL], p. 539–545. *In* N. R. Krieg and J. G. Holt (ed.), Bergey's manual of systematic bacteriology, vol. 1. Williams & Wilkins, Baltimore.
8. **Baumann, P., A. L. Furniss, and J. V. Lee.** 1984. Genus I. *Vibrio* Pacini 1854, 411[AL], p. 518–538. *In* N. R. Krieg and J. G. Holt (ed.), Bergey's manual of systematic bacteriology, vol. 1. Williams & Wilkins, Baltimore.
9. **Baumann, P., and R. H. W. Schubert.** 1984. Family II. *Vibrionaceae* Vernon 1965, 5245[AL], p. 516–517. *In* N. R. Krieg and J. G. Holt (ed.), Bergey's manual of systematic bacteriology, vol. 1. Williams & Wilkins, Baltimore.
10. **Benenson, A. S., M. R. Islam, and W. B. Greenough III.** 1964. Rapid identification of *Vibrio cholerae* by darkfield microscopy. Bull. W.H.O. **30:**827–831.
11. **Blake, P. A.** 1981. New information on the epidemiology of *Vibrio* infections, p. 107–117. *In* T. Holme, J. Holmgren, M. H. Merson, and R. Möllby (ed.), Acute enteric infections in children. New prospects for treatment and prevention. Elsevier/North Holland Publishing Co., Amsterdam.
12. **Blake, P. A., D. T. Allegra, J. D. Snyder, T. J. Barrett, L. McFarland, C. T. Caraway, J. C. Feeley, J. P. Craig, J. V. Lee, N. D. Puhr, and R. A. Feldman.** 1980. Cholera—a possible endemic focus in the United States. N. Engl. J. Med. **302:**305–309.
13. **Blake, P. A., M. H. Merson, R. E. Weaver, D. G. Hollis, and P. C. Heublein.** 1979. Disease caused by a marine *Vibrio*: clinical characteristics and epidemiology. N. Engl. J. Med. **300:**1–5.
14. **Blake, P. A., R. E. Weaver, and D. G. Hollis.** 1980. Diseases of humans (other than cholera) caused by vibrios. Annu. Rev. Microbiol. **34:**341–367.
15. **Bonner, J. R., A. S. Coker, C. R. Berryman, and H. M. Pollock.** 1983. Spectrum of *Vibrio* infections in a Gulf Coast community. Ann. Intern. Med. **98:**464–469.
16. **Brenner, D. J., B. R. Davis, Y. Kudoh, M. Ohashi, R. Sakazaki, T. Shimada, and H. L. Smith, Jr.** 1982. Serological comparison of two collections of *Vibrio cholerae* non O1. J. Clin. Microbiol. **16:**319–323.
17. **Brenner, D. J., G. R. Fanning, F. W. Hickman-Brenner, A. G. Steigerwalt, B. R. Davis, and J. J. Farmer III.** 1983. DNA relatedness among *Vibrionaceae* with emphasis on the *Vibrio* species associated with human infection. Colloq. INSERM (Inst. Nat. Sante Rech. Med.) **114:**175–184.
18. **Brenner, D. J., F. W. Hickman-Brenner, J. V. Lee, A. G. Steigerwalt, G. R. Fanning, D. G. Hollis, J. J. Farmer III, R. E. Weaver, S. W. Joseph, and R. J. Seidler.** 1983. *Vibrio furnissii* (formerly aerogenic biogroup of *Vibrio fluvialis*), a new species isolated from human feces and the environment. J. Clin. Microbiol. **18:**816–824.
19. **Colwell, R. R. (ed.).** 1984. *Vibrios* in the environment. John Wiley & Sons, New York.
20. **Davis, B. R., G. R. Fanning, J. M. Madden, A. G. Steigerwalt, H. B. Bradford, Jr., H. L. Smith, Jr., and D. J. Brenner.** 1981. Characterization of biochemically atypical *Vibrio cholerae* strains and designation of a new pathogenic species, *Vibrio mimicus*. J. Clin. Microbiol. **14:**631–639.
21. **Donovan, T. J., and A. L. Furniss.** 1982. Quality of antisera used in the diagnosis of cholera. Lancet **ii:**866–868.
22. **Farmer, J. J., III.** 1979. *Vibrio ("Beneckea") vulnificus*, the bacterium associated with sepsis, septicemia, and the sea. Lancet **ii:**903.
23. **Farmer, J. J., III, B. R. Davis, F. W. Hickman-Brenner, A. McWhorter, G. P. Huntley-Carter, M. A. Asbury, C. Riddle, H. G. Wathen-Grady, C. Elias, G. R. Fanning, A. G. Steigerwalt, C. M. O'Hara, G. K. Morris, P. B. Smith, and D. J. Brenner.** 1985. Biochemical identification of new species and biogroups of *Enterobacteriaceae* isolated from clinical specimens. J. Clin. Microbiol. **21:**46–76.
24. **Farmer, J. J., III, J. G. Wells, W. Terranova, M. L. Cohen, and C. M. West.** 1981. *Enterobacteriaceae*, p. 341–392. *In* A. Balows and W. J. Hausler (ed.), Diagnostic procedures for bacterial, mycotic, and parasitic infections. 6th ed. American Public Health Association, Washington, D.C.
25. **Feeley, J. C., and W. E. DeWitt.** 1976. Immune response to *Vibrio cholerae*, p. 289–295. *In* N. R. Rose and H. Friedman (ed.), Manual of clinical immunology. American Society for Microbiology, Washington, D.C.
26. **Finkelstein, R. A.** 1973. Cholera. Crit. Rev. Microbiol. **2:**553–623.
27. **Fox, G. E., E. Stackebrandt, R. B. Hespell, J. Gibson, J. Maniloff, T. A. Dyer, R. S. Wolfe, W. E. Balch, R. S. Tanner, L. J. Magrum, L. B. Zablen, R. Blakemore, R. Gupta, L. Bonen, B. J. Lewis, D. A. Stahl, K. R. Luehrsen, K. N. Chen, and C. R. Woese.** 1980. The phylogeny of procaryotes. Science **209:**457–463.
28. **Fujino, T., G. Sakaguchi, R. Sakazaki, and Y. Takeda (ed.).** 1974. International symposium on *Vibrio parahaemolyticus*. Saikon Publishing Company, Ltd., Tokyo.
29. **Furniss, A. L., J. V. Lee, and T. J. Donovan.** 1978. The *Vibrios*. Public Health Laboratory Service Monograph Series no. 11. Maidstone Public Health Laboratory. Her Majesty's Stationery Office, London.
30. **Hastings, J. W., and K. H. Nealson.** 1981. The symbiotic luminous bacteria, p. 1332–1345. *In* M. P. Starr, H. Stolp, H. G. Trüper, A. Balows, and H. G. Schlegel (ed.), The prokaryotes. Springer-Verlag, Berlin.
31. **Hickman, F. W., J. J. Farmer III, D. G. Hollis, G. R. Fanning, A. G. Steigerwalt, R. E. Weaver, and D. J. Brenner.** 1982. Identification of *Vibrio hollisae* sp. nov. from patients with diarrhea. J. Clin. Microbiol. **15:**395–401.
32. **Hollis, D. G., R. E. Weaver, C. N. Baker, and C. Thorns-**

berry. 1976. Halophilic *Vibrio* species isolated from blood cultures. J. Clin. Microbiol. **3**:425–431.

33. **Hugh, R., and R. Sakazaki.** 1972. Minimal number of characters for the identification of *Vibrio* species, *Vibrio cholerae*, and *Vibrio parahaemolyticus*. J. Conf. Public Health Lab. Dir. **30**:133–137.
34. **Hughes, J. M., D. G. Hollis, E. J. Gangarosa, and R. E. Weaver.** 1978. Non-cholera vibrio infections in the United States. Ann. Intern. Med. **88**:602–606.
35. **Huq, M. I., A. K. M. J. Alam, D. J. Brenner, and G. K. Morris.** 1980. Isolation of *Vibrio*-like group, EF-6, from patients with diarrhea. J. Clin. Microbiol. **11**:621–624.
36. **Jean-Jacques, W., K. R. Rajashekaraiah, J. J. Farmer III, F. W. Hickman, J. G. Morris, and C. A. Kallick.** 1981. *Vibrio metschnikovii* bacteremia in a patient with cholecystitis. J. Clin. Microbiol. **14**:711–712.
37. **Joseph, S. W., J. B. Kaper, and R. R. Colwell.** 1982. *Vibrio parahaemolyticus* and related halophilic vibrios. CRC Crit. Rev. Microbiol. **10**:77–124.
38. **Kelly, M. T., J. W. Peterson, W. E. Sarles, M. Romanko, D. Martin, and B. Hafkin.** 1982. Cholera on the Texas gulf coast. J. Am. Med. Assoc. **247**:1598–1599.
39. **Lee, J. V., T. J. Donovan, and A. L. Furniss.** 1978. Characterization, taxonomy, and emended description of *Vibrio metschnikovii*. Int. J. Syst. Bacteriol. **28**:99–111.
40. **Lee, J. V., P. Shread, A. L. Furniss, and T. N. Bryant.** 1981. Taxonomy and description of *Vibrio fluvialis* sp. nov. (synonym group F vibrios, group EF6). J. Appl. Bacteriol. **50**:73–94.
41. **Ljungh, A., and T. Wadstrom.** 1983. Toxins of *Vibrio parahaemolyticus* and *Aeromonas hydrophila*. J. Toxicol. Toxin Rev. **1**:257–307.
42. **Lockwood, D. E., A. S. Kreger, and S. H. Richardson.** 1982. Detection of toxins produced by *Vibrio fluvialis*. Infect. Immun. **35**:702–708.
43. **Love, M., D. Teebken-Fisher, J. E. Hose, J. J. Farmer III, F. W. Hickman, and G. R. Fanning.** 1981. *Vibrio damsela*, a marine bacterium, causes skin lesions on the damselfish *Chromis punctipinnis*. Science **214**:1139–1140.
44. **MacDonell, M. T., F. L. Singleton, and M. A. Hood.** 1982. Diluent composition for use of API 20E in characterizing marine and estuarine bacteria. Appl. Environ. Microbiol. **44**:423–427.
45. **Miwatani, T., and Y. Takeda.** 1976. *Vibrio parahaemolyticus*: a causative bacterium of food poisoning. Saikon Publishing Co., Tokyo. (Available from Maruzen Company, Ltd., Import and Export Dept., P.O. Box 5050, Tokyo International 100-31 Japan.)
46. **Morris, G. K., M. H. Merson, I. Huq, A. K. M. G. Kibrya, and R. Black.** 1979. Comparison of four plating media for isolating *Vibrio cholerae*. J. Clin. Microbiol. **9**:79–83.
47. **Morris, J. G., H. G. Miller, R. Wilson, C. O. Tacket, D. G. Hollis, F. W. Hickman, R. E. Weaver, and P. A. Blake.** 1982. Illness caused by *Vibrio damsela* and *Vibrio hollisae*. Lancet **i**:1294–1297.
48. **Morris, J. G., R. Wilson, B. R. Davis, I. K. Wachsmuth, C. F. Riddle, H. G. Wathen, R. A. Pollard, and P. A. Blake.** 1981. Non-O group 1 *Vibrio cholerae* gastroenteritis in the United States: clinical, epidemiologic, and laboratory characteristics of sporadic cases. Ann. Intern. Med. **94**:656–658.
49. **Nishibuchi, M., and R. J. Seidler.** 1983. Medium-dependent production of extracellular enterotoxins by non-O-1 *Vibrio cholerae*, *Vibrio mimicus*, and *Vibrio fluvialis*. Appl. Environ. Microbiol. **45**:228–231.
50. **Pollitzer, R.** 1959. Cholera. World Health Organization Monograph Series no. 43. World Health Organization, Geneva.
51. **Poole, M. D., and J. D. Oliver.** 1978. Experimental pathogenicity and mortality in ligated ileal loop studies of the newly reported halophilic lactose-positive *Vibrio* sp. Infect. Immun. **20**:126–129.
52. **Sakazaki, R., and A. Balows.** 1981. The genera *Vibrio*, *Plesiomonas* and *Aeromonas*, p. 1272–1301. *In* M. P. Starr, H. Stolp, H. G. Trüper, A. Balows, and H. G. Schlegel (ed.), The prokaryotes. Springer-Verlag, Berlin.
53. **Sakazaki, R., and T. Shimada.** 1977. Serovars of *Vibrio cholerae* identified during 1970–1975. Jpn. J. Med. Sci. Biol. **30**:279–282.
54. **Shandera, W. X., J. M. Johnston, B. R. Davis, and P. A. Blake.** 1983. Disease from infection with *Vibrio mimicus*, a newly recognized *Vibrio* species. Ann. Intern. Med. **99**:169–171.
55. **Shewan, J. M., and M. Vérnon.** 1974. Genus I. *Vibrio* Pacini 1854, 411, p. 340–345. *In* R. E. Buchanan and N. E. Gibbons (ed.), Bergey's manual of determinative bacteriology, 8th ed. The Williams & Wilkins Co., Baltimore.
56. **Smith, H. L., Jr.** 1979. Serotyping of non-cholera vibrios. J. Clin. Microbiol. **10**:85–90.
57. **Tacket, C. O., F. Hickman, G. V. Pierce, and L. F. Mendoza.** 1982. Diarrhea associated with *Vibrio fluvialis* in the United States. J. Clin. Microbiol. **16**:991–992.
58. **Tison, D. L., and M. T. Kelly.** 1984. *Vibrio* species of medical importance. Diagn. Microbiol. Infect. Dis. **2**:263–276.
59. **Wachsmuth, I. K.** 1984. Laboratory detection of enterotoxin, p. 93–115. *In* P. Ellner (ed.), Infectious diarrheal diseases: current concepts and laboratory procedures. Marcel Dekker, Inc., New York.
60. **Wachsmuth, I. K., G. K. Morris, and J. C. Feeley.** 1980. *Vibrio*, p. 226–234. *In* E. H. Lennette, A. Balows, W. J. Hausler, Jr., and J. P. Truant (ed.), Manual of clinical microbiology, 3rd ed. American Society for Microbiology, Washington, D.C.
61. **World Health Organization.** 1983. Manual for laboratory investigation of acute enteric infections. Diarrhoeal Diseases Control Program, World Health Organization, Geneva.
62. **World Health Organization Scientific Working Group.** 1980. Cholera and other vibrio-associated diarrhoeas. Bull. W.H.O. **58**:353–374.
63. **Zen-Yoji, H., S. Sakai, T. Terayama, Y. Kudo, T. Ito, M. Benoki, and M. Nagasaki.** 1965. Epidemiology, enteropathogenicity, and classification of *Vibrio parahaemolyticus*. J. Infect. Dis. **115**:436–444.

Chapter 27

Campylobacter

GEORGE K. MORRIS AND CHARLOTTE M. PATTON

The genus *Campylobacter* consists of a well-defined group of bacteria. They are slender, spirally curved rods, 0.2 to 0.5 μm wide and 0.5 to 5 μm long. Campylobacters may be comma, S, or gull wing in shape and may occur in short or occasionally long chains. Cells may become spherical or coccoid, especially in old cultures. The cells are nonsporeforming. These bacteria are motile by a single polar flagellum at one or both ends, and occasionally strains are nonmotile. They are microaerophilic and have a respiratory type of metabolism. Carbohydrates are neither fermented nor oxidized. The classification at the species level has been undergoing changes and expansion during the past decade. Taxonomy used herein is the system officially approved by the International Committee on Systematic Bacteriology, except for those species for which the names have not yet been officially approved (see reference 37 and later listings in *International Journal of Systematic Bacteriology*). In such cases the names are enclosed in quotation marks. The names are listed in Table 1 along with previously published synonyms and diseases associated with the species.

CLINICAL SIGNIFICANCE

Campylobacter jejuni, a newly recognized enteric pathogen, appears to be by far the most important species from the standpoint of human diseases. In humans the disease usually occurs as diarrhea (sometimes with blood in the stool), abdominal pain, fever, nausea, and sometimes vomiting (4). The organism can sometimes be isolated from blood specimens in addition to the stool. It is estimated that as many as 2 million cases of *Campylobacter* infections occur in the United States each year. This number is equivalent to the estimated number of cases of salmonellosis and exceeds the number of shigellosis cases. In the United States the disease appears to be especially prevalent among college students. Non-ill carriers are rare in the United States but are not uncommon in some developing countries. The source of human infections is usually food of animal origin, especially raw (unpasteurized) milk. The organisms are excreted in the feces of healthy domestic animals. Studies with chickens, turkeys, and cattle have shown that as many as 50 to 100% of a flock or herd of these animals excrete *C. jejuni*. The organisms can also be isolated with high frequency from surface waters.

C. coli causes diseases in humans similar to those caused by *C. jejuni*, but the frequency is not known because most clinical laboratories do not distinguish between *C. jejuni* and *C. coli*. Studies in Canada and England indicate that *C. coli* accounts for 3 to 5% of human cases of *Campylobacter* infections (19). *C. coli* is very often found in healthy pigs.

C. laridis is a newly described thermophilic species isolated from human, avian, and mammalian species but most frequently from sea gulls (3). Five of six clinical isolates referred to the Campylobacter Reference Laboratory at the Centers for Disease Control (CDC) were from patients with a secretory type of diarrhea or abdominal pain or both, symptoms similar to those of infections with *C. jejuni*. Sea gulls did not appear to be a source of these human infections.

C. fetus subsp. *fetus* sometimes causes disease in humans, but the symptoms are usually extraintestinal, especially septicemias, and in patients with preexisting disease. Most of these patients do not have diarrheal disease. Infection in animals results in abortion in sheep and cattle.

C. fetus subsp. *veneralis* probably does not cause disease in humans, but it causes abortion and infertility in cattle and is transmitted venereally.

The three subspecies of *C. sputorum* have not been associated with human disease, although *C. sputorum* subsp. *sputorum* has been isolated from the oral cavity of humans (14, 24a). *C. sputorum* subsp. *bubulus* has been isolated from the uterine and genital sites of cattle and sheep but has not been implicated in disease (42). *C. sputorum* subsp. *mucosalis* has been associated with intestinal diseases in pigs (22, 23).

C. concisus has been isolated from the gingival crevices of humans with periodontal disease, but its pathogenicity is unknown (40).

C. nitrofigilis is associated with saltwater plants in nature and is not pathogenic for animals or humans (28).

"*C. fecalis*" has not been isolated from humans but has been found in feces of healthy sheep and in cattle with enteric disease (1, 12).

"*C. hyointestinalis*" has been associated with proliferative ileitis in swine (13).

"*C. cinaedi*" and "*C. fennelliae*" have been isolated from rectal swabs, and "*C. cinaedi*" has also been isolated from the blood of homosexual men with intestinal symptoms. The role of these species in human disease is unknown (11; P. A. Totten, C. L. Fennell, F. C. Tenover, J. M. Wezenberg, P. L. Perine, W. E. Stamm, and K. K. Holmes, J. Infect. Dis., in press).

"Pyloric campylobacters" are *Campylobacter*-like organisms isolated from biopsy specimens of humans with gastritis and peptic ulcerations of the stomach and duodenum (26).

"Aerotolerant *Campylobacter* species" have been associated with abortion in cattle and pigs and mastitis in cows. Primary isolation is made in a microaerophilic atmosphere at 30°C on *Leptospira* EMJH medium. Growth occurs aerobically after subculture (25, 31).

PATHOGENESIS

Little is known about the pathogenic mechanisms of *C. jejuni*, *C. coli*, and *C. laridis*. The presence of blood and leukocytes in the stools of infected persons indicates that *C. jejuni* can be invasive; however, a secretory form of diarrhea also occurs, especially in children, suggesting the involvement of an enterotoxin. Several workers have reported enterotoxinlike sub-

TABLE 1. Bacteria in the genus *Campylobacter*

Name[a]	Published synonyms	Diseases and conditions caused in humans and animals
C. jejuni	*C. fetus* subsp. *jejuni*, related vibrios, *Vibrio jejuni*	Diarrhea and sometimes septicemia in humans; abortion in sheep; enteritis in cattle
C. coli	*Vibrio coli*	Diarrhea in humans
C. laridis	Nalidixic acid-resistant thermophilic *Campylobacter* group	Associated with diarrhea in humans
C. fetus subsp. *fetus*	*C. fetus* subsp. *intestinalis, Vibrio fetus* var. *intestinalis*	Abortion in sheep and cattle; septicemia in humans, usually in compromised hosts
C. fetus subsp. *venerealis*	*C. fetus* subsp. *fetus, Vibrio fetus* var. *venerealis*	Abortion and infertility in cattle
C. sputorum subsp. *sputorum*	*C. sputorum, Vibrio sputorum*	None
C. sputorum subsp. *bubulus*	*C. bubulus, Vibrio sputorum* var. *bubulum, V. bubulus*	None
C. sputorum subsp. *mucosalis*		Intestinal tumors in pigs
C. concisus		None
C. nitrofigilis		None (nitrogen-fixing, marine origin)
"*C. fecalis*"	*Vibrio fecalis*	Diarrhea in cattle
"*C. hyointestinalis*"		Proliferative ileitis in swine
"*C. cinaedi*"	CLO-1	Associated with diarrhea in homosexual men
"*C. fennelliae*"	CLO-2	Associated with diarrhea in homosexual men
"Pyloric Campylobacter"		Associated with gastritis and pyloric ulcers in humans
"Aerotolerant Campylobacter"		Associated with abortion in cattle and pigs; mastitis in cows

[a] Proposed names not officially approved by the International Committee on Systematic Bacteriology are in quotes.

stances produced by *C. jejuni*. Ruiz-Palacios et al. (36) reported the production of a choleralike enterotoxin by *C. jejuni*. Klipstein and Engert (20) confirmed the finding by both biological and immunological assays and described the properties of the heat-labile enterotoxin. Although the enterotoxin appeared to be immunologically similar to cholera toxin and *Escherichia coli* heat-labile enterotoxin, Olsvik et al. (32), in studies with genetic probing, did not detect genes for production of cholera toxin and heat-labile enterotoxin. Clinical observations suggest that the pathogenesis of *C. coli* and *C. laridis* is similar to that of *C. jejuni*.

COLLECTION, TRANSPORT, AND STORAGE OF SPECIMENS

Feces or rectal swabs should be collected from patients with diarrhea. Patients who have had diarrhea may still be excreting *Campylobacter* species even if they are having formed stools at the time of collection. Inoculate specimens promptly onto selective plating media, preferably within 2 h. If this time limit cannot be met, rectal swabs or swabs prepared from a fresh stool should be placed in Cary-Blair transport medium (8). *C. jejuni* does not survive well in buffered glycerol-saline, a transport medium commonly used for other enteric pathogens, and the medium is not recommended for *Campylobacter* species. Specimens should be held in transport media at refrigeration temperature. The specimens can usually be held for 1 to 2 weeks at 4°C before the recovery of *Campylobacter* species is severely reduced.

For the transport of pure cultures of *Campylobacter*, either Wang *Campylobacter* transport medium (43) or Cary-Blair medium may be used. Inoculate Wang medium with the pure culture, incubate it overnight in a microaerophilic atmosphere, and tighten the cap. To use Cary-Blair medium for transport, remove 24- to 48-h growth from a blood agar plate (BAP) with a cotton-tipped swab. Immerse and leave the swab in the medium, and seal the tube. Transport or hold the cultures at refrigeration temperature.

For the storage of *Campylobacter* cultures for long periods, suspend 24- to 48-h growth from a BAP in tryptic soy broth containing 20% glycerol or in defribrinated rabbit blood, and store the culture at −70°C or in liquid nitrogen.

Lyophilization is sometimes necessary for shipping cultures. To lyophilize, we recommend making a heavy inoculum of 24- to 48-h blood agar growth in serum-inositol suspending medium (2). The serum-inositol medium is 5% *myo*-inositol in filter-sterilized newborn calf serum (9 parts serum to 1 part 50% *myo*-inositol [wt/vol]). Store lyophilized cultures at 4°C.

DIRECT EXAMINATION

Although we do not routinely examine stool specimens by direct microscopic examination, some workers find that, with sufficient experience, such examination is useful for rapid presumptive identification in cases of acute diarrhea. The most effective technique is to use a phase-contrast or dark-field microscope and observe for the darting, tumbling motility of spiral rods. Brucella broth or a similar medium is recommended as the suspending fluid for the stools. If dark-field microscopy is not available, observing for small curved rods by crystal violet stain may be useful.

CULTURE AND ISOLATION

Three factors are important in the isolation of thermophilic *Campylobacter* species: (i) the use of selective plating media, (ii) the incubation of plates in reduced oxygen with added CO_2, and (iii) the incubation of primary isolation plates at 42°C. Several selective plating media have been developed for isolating *Campylobacter* species, and numerous methods are available to achieve suitable atmospheric conditions.

Media

Skirrow (38), Butzler (7), and Campy BAP (4) selective plating media are widely used. These media are available from commercial sources. Minor changes have been made in both the Skirrow and the Butzler medium formulations since they were first introduced. The recently suggested formulations (15) of the media are given in the chapter on culture media (Chapter 111). Comparative evaluations of these media for the isolation of *C. jejuni* from human stools show a comparable number of positives with each when the plates are incubated for 24, 48, and 72 h (29, 44). If economy dictates the use of only one plating medium and only one reading time, we recommend Campy BAP medium and a 48-h incubation time. The use of two different plating media is preferable to a single plating medium in that the yield is increased by approximately 10% with two plating media (44).

We do not recommend the routine use of selective enrichment broth for the isolation of *Campylobacter* species from fecal specimens; however, numerous selective enrichment broths have been developed to improve the recovery of *C. jejuni* from both clinical and environmental specimens (4, 5, 9, 10). Kaplan obtained a 10% increase in the isolation rate of *C. jejuni* from human fecal specimens with overnight cold enrichment in Blaser Campy Thio (R. L. Kaplan, J. E. Barrett, W. Landau, and L. J. Goodman, Abstr. Annu. Meet. Am. Soc. Microbiol. 1984, C77, p. 249), a medium made with thioglycolate broth containing 0.16% agar and the antimicrobial agents listed for Blaser Campy BAP (4). This increase in the yield of positive specimens is similar to the increase observed with a combination of two different plating media.

The usefulness of an enrichment broth may vary with the type of specimen being processed (27). Enrichment may be beneficial with mishandled fecal specimens, such as those delayed in transit or held at ambient temperature too long. Other specimens possibly benefiting from enrichment are fecal or environmental specimens thought to contain low numbers of *Campylobacter* organisms, such as specimens from carriers, postsymptomatic cases, patients who have been treated with antibiotics, and samples of food and water. Enrichment broth is especially advantageous for the isolation of *C. jejuni* and *C. coli* from foods (10).

Inoculation of media

Rectal swabs are inoculated directly onto plates of selective media by spreading the feces in a circular area approximately 1 in. (2.54 cm) in diameter. Formed stools can be processed by taking up portions of the stool on a cotton-tipped swab and inoculating plates as for rectal swabs or by emulsifying a small portion (approximately 0.5 g) in 2 to 3 ml of phosphate-buffered saline or broth and inoculating 1 to 2 drops from a Pasteur pipette onto each plate. Liquid stools may also be inoculated directly with 1 to 2 drops from a Pasteur pipette, or a swab can be dipped in the stool, with care being taken to sample in areas of blood or mucus, if present. Streak the plate for isolation. If enrichment broth is to be used, it should be used in conjunction with, and not in lieu of, direct plating.

Incubation of the plates

The inoculated plates are incubated at 42 to 43°C in an atmosphere containing 5% O_2, 10% CO_2, and 85% N_2. The thermophilic campylobacters will grow at 36°C, but the higher temperature suppresses the normal competing fecal flora while permitting the thermophilic campylobacters to grow.

The atmosphere suitable for the growth of *Campylobacter* species may be produced in several ways. The method of choice will depend on laboratory size, work load, and relative cost. The evacuation-replacement system is an accurate and reproducible means of promptly attaining the proper atmosphere and is suitable for large or small work volumes. The plates are placed in an anaerobic container without catalyst, and the container is evacuated twice to 38 cm (ca. 15 in.) of Hg and refilled each time with a gas mixture of 10% CO_2 and 90% N_2. Good results have been reported by workers using similar gas mixtures.

A frequently used alternative is the disposable gas-generating envelope which is available commercially. The anaerobe jar can be opened to add new specimens, but a new envelope must be used each time. Plates should be stacked in the jar no more than six high for the maximal isolation of *C. jejuni*.

The polybag method is recommended by some workers. It consists of a plastic bag (8 by 15 in. [ca. 20 by 38 cm]) which is collapsed two or three times by hand or vacuum and refilled each time with a gas mixture from a tank of 5% O_2–10% CO_2–85% nitrogen. The bag is sealed with a rubber band. If no more than six to eight plates are placed in a bag, good results are obtained.

The candle jar provides an atmosphere of approximately 17 to 19% oxygen and 2 to 3% CO_2. This atmosphere is not ideal, and some *Campylobacter* strains will not grow in it. Several investigators have shown that candle jars incubated at 42°C give significantly better results than those incubated at 36°C. FBP supplement (consisting of 0.05% each of ferrous sulfate, sodium metabisulfite, and sodium pyruvate) added to the medium also improves the isolation rate. This supplement enhances the organism's tolerance for oxygen by quenching the toxic forms of oxygen such as superoxide anions and hydrogen peroxide. Such substances are produced when the medium is exposed to light and air. The supplement is available commercially.

Examination of plates

For maximal isolation, plates should be examined at 24, 48, and 72 h. If economy permits only one reading, we recommend one at 48 h. Colonies of thermophilic campylobacters are nonhemolytic and gray or colorless. They may be flat and watery with irregular edges, or round and convex with entire edges. Colonies may be pinpoint in size or spreading over large areas of the plate. Suspect colonies should be screened with three presumptive tests: motility and morphology by observation under dark-field or phase-contrast microscopy, oxidase, and Gram reaction. Use 0.3% carbol fuchsin as a counterstain in the Gram stain (see Identification section below for details). *Campylobacter* species are motile, with a darting, tumbling motility; they are oxidase positive; and they appear as gram-negative spiral rods. Presumptively positive colonies are confirmed by further biochemical identification tests described below. If a dark-field or phase-contrast microscope is not available, colonies may be rapidly screened for typical cell morphology by staining with crystal violet.

Since *C. fetus* is most likely to be involved in extraintestinal infections, the isolation of these organisms from blood and other normally sterile body fluids does not represent the same problem as the isolation of the thermophilic bacteria from feces. The important point is to culture the specimens under microaerophilic conditions. These organisms are usually isolated in pure culture on nonselective BAPs or from liquid blood culture medium. Colonies that are positive in the presumptive tests described above should be confirmed by further biochemical tests. If patients show diarrheal symptoms and fecal specimens need to be cultured for *C. fetus* subsp. *fetus*, specimens should be cultured on selective media without cephalothin or cephazolin and incubated at 36°C. Cephalothin and cephazolin inhibit the growth of some strains of *C. fetus* subsp. *fetus*, and most strains of *C. fetus* subsp. *fetus* do not grow at 42°C. If only selective media containing cephalothin or cephazolin are available, suspensions of fecal specimens can be passed through filters (0.65-μm pore size), and the filtrates can be inoculated onto noninhibitory blood media. Most campylobacters will pass through a 0.65-μm filter, and other bacteria will not.

IDENTIFICATION

Presumptively positive colonies should be definitively identified by the tests described below. The inoculum for the tests is 24- to 48-h growth from a BAP suspended in heart infusion broth to a turbidity matching a McFarland no. 1 standard. Occasionally, stains may require 3 days of incubation for sufficient growth for the preparation of the inoculum. Test media are inoculated with approximately 0.1 ml (2 drops from a Pasteur pipette) and incubated for 3 days except where noted.

Growth conditions

Growth in aerobic and anaerobic atmospheres is determined on heart infusion agar slants. Growth in 1% glycine and 3.5% NaCl is determined in brucella albimi broth medium (GIBCO Laboratories) containing 0.16% agar. Growth at 25, 36, and 42°C is determined on BAPs. Plates should be inoculated with a 6-mm loop. Make a single streak of the broth suspension across separate plates for incubation at each temperature. For economy, three or more strains can be inoculated onto each plate.

Catalase

The 3% hydrogen peroxide reagent is added to *Campylobacter* growth on a heart infusion agar slant. Any formation of bubbles is considered positive.

Gram stain

Use 0.3% carbol fuchsin as a counterstain because *Campylobacter* species stain poorly with safranin. The carbol fuchsin solution is prepared as follows. For solution A, add 0.3 g of basic fuchsin to 10 ml of ethanol. For solution B, add 5 ml of phenol (melted crystal) to 95 ml of distilled water. Mix solutions A and B.

Nitrate reduction

The test for reduction of nitrate is determined in a special nitrate broth medium prepared from 25 g of heart infusion broth (Difco Laboratories), 2 g of potassium nitrate, and 1,000 ml of distilled water. To a 3-day culture in 4 ml of nitrate broth, add 0.25 ml each of nitrate reagents 1 (8 g of sulfanilic acid in 1,000 ml of 5 N acetic acid) and 2 (6 ml of dimethyl-α-napthylamine in 1,000 ml of 5 N acetic acid). Development of a red color is a positive reaction and indicates the production of nitrate reductase by the organism. If no red color develops, add powdered zinc to the tube to determine whether nitrates are still present.

Hydrogen sulfide production

The production of hydrogen sulfide is tested in triple sugar iron agar and on a lead acetate paper strip suspended over a tube of brucella albimi broth (GIBCO) containing 0.02% L-cysteine. Triple sugar iron agar may be inoculated directly with a needle by streaking the slant and stabbing the butt with growth from a 24- to 48-h BAP.

Tolerance to antimicrobial agents

Inoculate the surface of a BAP with heart infusion broth inoculum in a manner to yield confluent growth. Place a 30-μg nalidixic acid disk and a 30-μg cephalothin disk on the surface of the agar. Incubate the culture at 36°C overnight and read it. If no growth is seen on the plate, incubate it and read it at 48 h. Any zone of inhibition is considered a susceptible reaction.

Hippurate test

Hippurate hydrolysis is performed by a modification of previously described methods (16, 18). A loopful of an overnight growth from heart infusion agar with 5% defibrinated rabbit blood is emulsified in 0.4 ml of 1% aqueous sodium hippurate, and the dense suspension is incubated in a 37°C water bath for 2 h. After incubation, use a pipette to slowly add 0.2 ml of ninhydrin solution (3.5% ninhydrin in a 1:1 mixture of acetone and butanol) down the side of the tube to form an overlay. Incubate for 10 min more in the 37°C water bath, and examine the suspension for color. Deep purple (crystal violet) is considered positive. Weak color reactions are considered negative. For convenience, the 0.4-ml tubes of sodium hippurate can be prepared ahead of time and stored at −20°C. The ninhydrin solution is stored in the dark at room temperature.

The tests are interpreted for identification of the campylobacters to species and subspecies level by referring to reactions in Table 2. Very few practical tests for distinguishing *Campylobacter* species at the species and subspecies levels are available. As a result,

TABLE 2. Phenotypic characteristics of *Campylobacter* species[a]

Organism	Growth					Biochemical reaction								Other test responses			
	25°C	36°C	42°C	1% Glycine	3.5% NaCl	Oxidase	Catalase	Glucose[b]	Nitrate reduction	Nitrite reduction	H_2S in TSI	H_2S on lead acetate paper	Hippurate hydrolysis[c]	Nalidixic acid[d]	Cephalothin[d]	Hydrogen[e]	C-19:0 fatty acid
C. jejuni	–	+	+	+	–	+	+	–	+	–	–	+	+	S	R	–	+
C. coli	–	+	+	+	–	+	+	–	+	–	–	+	–	S	R	–	d
C. laridis	–	+	+	+	–	+	+	–	+		–	+	–	R	R	–	–
"*C. fecalis*"	–	+	+	+	+	+	+	–	+		+	+	–	d	S	–	–
C. fetus subsp. *fetus*	+	+	d	+	–	+	+	–	+	–	–	d	–	R	S	–	–
C. fetus subsp. *venerealis*	+	+	–	–	–	+	+	–	+	–	–	d	–	d	S	–	–
"*C. hyointestinalis*"	d	+	d	+	–	+	+	–	+	–	+	+	–	R	S		
C. sputorum subsp. *sputorum*	–	+	+	+	–	+	–	–	+	+	+	+	–	R	S	–	–
C. sputorum subsp. *bubulus*	d	+	–	+	+	+	–	–	+	+	+	+	–	d	S	–	–
C. sputorum subsp. *mucosalis*	–	+	+	–	–	+	–	–	+	+	+	+	–	d	S	+	–
C. concisus	–	+	–	–	–	+	–	–	+	+	+	+		R		+	
"*C. cinaedi*"	–	+	–	+	–	+	+	–	+		–	+	–	S	S		
"*C. fennelliae*"	–	+	–	+	–	+	+	–	–		–	+	–	S	S		
"Pyloric Campylobacter"		+				+	+	–[f]	–			+		R			
"Aerotolerant Campylobacter"	+	+	–	d	–	+	+	–	–		–	–	–	S	R		

[a] Symbols: +, 90% or more of the strains are positive; –, 90% or more of the strains are negative; d, 11 to 89% of the strains are positive; R, resistant; S, susceptible. TSI, Triple sugar iron. Table compiled from references 3, 12, 13, 16, 17, 22, 23, 26, 30, 31, 35, 39, 40, and 42; from Totten et al., in press; and from data from the Campylobacter Reference Laboratory, CDC.

[b] Ferments or oxidizes glucose.

[c] Ninhydrin test.

[d] 30-μg disk.

[e] Hydrogen required for microaerophilic growth.

[f] Oxidation of glucose not indicated.

few phenotypic characters assume critical importance for taxonomic work with this genus. With some strains it will be impossible to definitively identify them by species with the laboratory tests routinely available in the clinical laboratory.

At this time, the genus *Campylobacter* is composed of seven species with names officially recognized by the International Committee on Systematic Bacteriology (Table 1). *C. fetus* is composed of two subspecies, and *C. sputorum* has three subspecies. There are at least six *Campylobacter* or *Campylobacter*-like groups for which the name has not been chosen or is not official.

Campylobacter species are generally placed into two broad groups, the catalase-positive and catalase-negative campylobacters. The *Campylobacter* strains most often associated with disease in humans are catalase positive. The catalase-positive campylobacters can be separated into groups according to the range of temperatures at which they will grow. The thermophilic campylobacters find the optimum temperature for growth at 42°C and do not grow at 25°C. The thermophilic campylobacters associated with disease in humans are *C. jejuni*, *C. coli*, and *C. laridis*. Resistance to nalidixic acid will usually distinguish *C. laridis* from *C. jejuni* and *C. coli*. The key test for distinguishing *C. jejuni* from *C. coli* is the hippurate hydrolysis test. *C. jejuni* can hydrolyze hippurate to glycine and benzoic acid. *C. coli* and other campylobacters do not hydrolyze hippurate. Certain strains of *C. jejuni* can give variable results in the ninhydrin method for the hydrolysis of hippurate; therefore, for these strains it is necessary to use the more sensitive gas-liquid chromatography method (21) for hippurate hydrolysis.

The catalase-positive strains of *Campylobacter* that grow at 25°C and usually not at 42°C are *C. fetus* and "*C. hyointestinalis.*" A few thermotolerant strains of *C. fetus* subsp. *fetus* and "*C. hyointestinalis*" grow at 42°C; however, growth at 25°C is the key differential test to distinguish the thermotolerant strains from the thermophilic *Campylobacter* species. *C. fetus* subsp. *fetus* can cause disease in humans and can be distinguished from *C. fetus* subsp. *venerealis* by the 1% glycine test and from "*C. hyointestinalis*" by the test for H_2S in triple sugar iron (Table 2). "*C. cinaedi*" and "*C. fennelliae*" are catalase-positive *Campylobacter* species associated with human disease. These species grow at 37 and not at 25 or 42°C.

SEROTYPING OF STRAINS

C. jejuni and *C. coli* can be serotyped for epidemiologic evaluation. Two methods representative of those now in common use are the Penner method (34) for soluble heat-stable antigens with an indirect hemagglutination technique and the Lior method (24) for heat-labile antigens with a slide agglutination technique and absorbed antisera. Absorbed antisera are not required for the Penner method, making it the less difficult method to implement. The Lior method is simpler to perform and gives more rapid results than the Penner method. Cultures often react in multiple antisera in the Penner method, whereas multiple reactions are infrequent with the Lior method.

Both systems are equally useful for the serotyping of strains from case clusters and epidemics (33).

SEROLOGIC DIAGNOSIS

Antibody responses in patients with culture-positive *Campylobacter* enteritis have been demonstrated by a variety of techniques. Serologic diagnosis has proven most useful in bacteriologically negative patients involved in outbreaks in which an isolate of *C. jejuni* available from at least one patient involved in the outbreak is used as an antigen. Serologic techniques are not practical for the diagnosis of sporadic cases of *Campylobacter* infection because of the antigenic heterogeneity of the organism.

ANTIMICROBIAL SUSCEPTIBILITY

The method recommended for antimicrobial susceptibility analysis is the MIC method with Mueller-Hinton broth with 5% lysed horse blood as described in Chapter 101. The results of the in vitro antimicrobial susceptibility testing of *Campylobacter* strains analyzed at CDC are summarized in Table 3.

PLASMIDS

Many *Campylobacter* strains contain plasmids, and plasmid profiles have been used as markers for the epidemiologic analysis of outbreak-associated strains (6, 41).

TABLE 3. Antimicrobial susceptibility of *Campylobacter* species[a]

Antimicrobial agent	Response of		
	C. jejuni (n = 45)	*C. coli* (n = 7)	*C. fetus* (n = 50)
Amikacin	S	S	
Ampicillin	MS	S	MS
Carbenicillin	S	S	
Cefotaxime	S	S	
Cefoxitin	R	R	
Cefoperazone	R	R	
Cephalothin	R	R	MS
Chloramphenicol	S	S	S
Clindamycin	S	S	MS
Erythromycin	MS	MS	MS
Gentamicin	S	S	S
Metronidazole	S	S	
Moxalactam	MS	R	
Nafcillin	R	R	
Nalidixic acid	S	S	R
Nitrofurantoin	S	S	
Penicillin G	MS	MS	MS
Polymyxin B	S	S	
Sulfamethoxazole-trimethoprim (cotrimoxazole)	S	S	S
Sulfamethoxazole	S	R	
Tetracyclines	S	S	
Tobramycin	S	S	
Trimethoprim	R	R	
Vancomycin	R	R	R

[a] R, Strains usually resistant; S, strains usually susceptible; and MS, strains are moderately susceptible. The results with *C. coli* should be used with caution because of the small number of strains tested. This table is based on unpublished data of C. Thornsberry, CDC, Atlanta, Ga. The interpretation breakpoints are those published by the National Committee for Clinical Laboratory Standards (M7-T), Villanova, Pa.

EVALUATION

C. jejuni probably causes diarrhea in humans more frequently than other bacterial pathogens such as *Salmonella* and *Shigella* species. Clinical laboratories examining fecal specimens for bacterial enteric pathogens should include tests for *Campylobacter* species in the laboratory regimen. The most common source of human infections is food of animal origin, especially unpasteurized milk.

LITERATURE CITED

1. **Al-Masha, R. R., and D. J. Taylor.** 1980. *Campylobacter* spp. in enteric lesions in cattle. Vet. Rec. **107:**31–34.
2. **Barbaree, J. M., and A. Sanchez.** 1982. Cross-contamination during lyophilization. Cryobiology **19:**443–447.
3. **Benjamin, J., S. Leaper, R. J. Owen, and M. B. Skirrow.** 1983. Description of *Campylobacter laridis*, a new species comprising the nalidixic acid resistant thermophilic *Campylobacter* (NARTC) group. Curr. Microbiol. **8:**231–238.
4. **Blaser, M. J., I. D. Berkowitz, F. M. LaForce, J. Cravens, L. B. Reller, and W. L. Wang.** 1979. *Campylobacter* enteritis: clinical and epidemiological features. Ann. Intern. Med. **91:**179–185.
5. **Bolton, F. J., and L. Robertson.** 1982. A selective medium for isolating *Campylobacter jejuni/coli*. J. Clin. Pathol. **35:**462–467.
6. **Bopp, C. A., K. A. Birkness, I. K. Wachsmuth, and T. J. Barrett.** 1985. In vitro antimicrobial susceptibility, plasmid analysis, and serotyping of epidemic-associated *Campylobacter jejuni*. J. Clin. Microbiol. **21:**4–7.
7. **Butzler, J. P., P. Dekeyser, M. Detrain, and F. Dehaen.** 1973. Related *Vibrio* in stools. J. Pediatr. **82:**493–495.
8. **Cary, S. G., and E. B. Blair.** 1964. New transport medium for shipment of clinical specimens. I. Fecal specimens. J. Bacteriol. **88:**96–98.
9. **Chan, F. T. H., and A. M. R. MacKenzie.** 1984. Advantage of using enrichment-culture techniques to isolate *Campylobacter jejuni* from stools. J. Infect. Dis. **149:**481–482.
10. **Doyle, M. P., and D. J. Roman.** 1982. Recovery of *Campylobacter jejuni* and *Campylobacter coli* from inoculated foods by selective enrichment. Appl. Environ. Microbiol. **43:**1343–1353.
11. **Fennell, C. L., P. A. Totten, T. C. Quinn, D. L. Patton, K. K. Holmes, and W. E. Stamm.** 1984. Characterization of *Campylobacter*-like organisms isolated from homosexual men. J. Infect. Dis. **149:**58–66.
12. **Firehammer, B. D.** 1965. The isolation of vibrios from ovine feces. Cornell Vet. **55:**482–494.
13. **Gebhart, C. J., G. E. Ward, K. Chang, and H. J. Kurtz.** 1983. *Campylobacter hyointestinalis* (new species) isolated from swine with lesions of proliferative ileitis. Am. J. Vet. Res. **44:**361–367.
14. **Gibbons, R. J., S. S. Socransky, S. Sawyer, B. Kapsimalis, and J. B. MacDonald.** 1963. The microbiota of the gingival crevice area of man. II. The predominant cultivable organisms. Arch. Oral Biol. **8:**281–289.
15. **Goossens, H., M. DeBoeck, H. Van Landuyt, and J. P. Butzler.** 1984. Isolation of *Campylobacter jejuni* from human feces, p. 39–50. *In* J. P. Butzler (ed.), *Campylobacter* infection in man and animals. CRC Press, Inc., Boca Raton, Fla.
16. **Harvey, S. M.** 1980. Hippurate hydrolysis by *Campylobacter fetus*. J. Clin. Microbiol. **11:**435–437.
17. **Harvey, S. M., and J. R. Greenwood.** 1983. Relationships among catalase-positive campylobacters determined by deoxyribonucleic acid-deoxyribonucleic acid hybridization. Int. J. Syst. Bacteriol. **33:**275–284.
18. **Hwang, M.-N., and G. M. Ederer.** 1975. Rapid hippurate hydrolysis method for presumptive identification of

group B streptococci. J. Clin. Microbiol. **1**:114–115.

19. **Karmali, M. A., and M. B. Skirrow.** 1984. Taxonomy of the genus *Campylobacter*, p. 1–20. *In* J. P. Butzler (ed.), *Campylobacter* infection in man and animals. CRC Press, Inc., Boca Raton, Fla.
20. **Klipstein, F. A., and R. F. Engert.** 1984. Properties of crude *Campylobacter jejuni* heat-labile enterotoxin. Infect. Immun. **45**:314–319.
21. **Kodaka, H., G. L. Lombard, and V. R. Dowell, Jr.** 1982. Gas-liquid chromatography technique for detection of hippurate hydrolysis and conversion of fumarate to succinate by microorganisms. J. Clin. Microbiol. **16**:962–964.
22. **Lawson, G. H. K., J. L. Leaver, G. W. Pettigrew, and A. C. Rowland.** 1981. Some features of *Campylobacter sputorum* subsp. *mucosalis* subsp. nov., nom. rev. and their taxonomic significance. Int. J. Syst. Bacteriol. **31**:385–391.
23. **Lawson, G. H. K., A. C. Rowland, and P. Wooding.** 1975. The characterization of *Campylobacter sputorum* subspecies *mucosalis* isolated from pigs. Res. Vet. Sci. **18**:121–126.
24. **Lior, H., D. L. Woodward, J. A. Edgar, L. J. Laroche and P. Gill.** 1982. Serotyping of *Campylobacter jejuni* by slide agglutination based on heat-labile antigenic factors. J. Clin. Microbiol. **15**:761–768.

24a. **Loesche, W. J., R. J. Gibbons, and S. S. Socransky.** 1965. Biochemical characteristics of *Vibrio sputorum* and relationship to *Vibrio bubulus* and *Vibrio fetus*. J. Bacteriol. **89**:1109–1116.

25. **Logan, E. F., S. D. Neill, and D. P. Mackie.** 1982. Mastitis in dairy cows associated with an aerotolerant *Campylobacter*. Vet. Rec. **110**:229–230.
26. **Marshall, B. J., and J. R. Warren.** 1984. Unidentified curved bacilli in the stomach of patients with gastritis and peptic ulceration. Lancet **i**:1311–1315.
27. **Martin, W. T., C. M. Patton, G. K. Morris, M. E. Potter, and N. D. Puhr.** 1983. Selective enrichment broth medium for isolation of *Campylobacter jejuni*. J. Clin. Microbiol. **17**:853–855.
28. **McClung, C. R., D. G. Patriquin, and R. E. Davis.** 1983. *Campylobacter nitrofigilis* sp. nov., a nitrogen-fixing bacterium associated with roots of *Spartina alterniflora* Loisel. Int. J. Syst. Bacteriol. **33**:605–612.
29. **Morris, G. K., C. A. Bopp, C. M. Patton, and J. G. Wells.** 1981. Media for isolating *Campylobacter*. Arch. Lebensmittelhyg. **33**:151–153.
30. **Moss, C. W., A. Kai, M. A. Lambert, and C. Patton.** 1984. Isoprenoid quinone content and cellular fatty acid composition of *Campylobacter* species. J. Clin. Microbiol. **19**:772–776.
31. **Neill, S. D., W. A. Ellis, and J. J. O'Brien.** 1978. The biochemical characteristics of *Campylobacter*-like organisms from cattle and pigs. Res. Vet. Sci. **25**:368–372.
32. **Olsvik, O., K. Wachsmuth, G. Morris, and J. C. Feeley.** 1984. Genetic probing of *Campylobacter jejuni* for cholera toxin and *Escherichia coli* heat-labile enterotoxin. Lancet **i**:449.
33. **Patton, C. M., T. J. Barrett, and G. K. Morris.** 1983. Serotyping *Campylobacter jejuni/coli* by two systems: the CDC experience, p. 96–97. *In* A. D. Pearson, M. B. Skirrow, B. Rowe, J. R. Davies, and D. M. Jones (ed.), *Campylobacter* II. Proceedings of the Second International Workshop on Campylobacter Infections, Brussels, 6–9 September 1983. Public Health Laboratory Service, London.
34. **Penner, J. L., and J. N. Hennessy.** 1980. Passive hemagglutination technique for serotyping *Campylobacter fetus* subsp. *jejuni* on the basis of soluble heat-stable antigens. J. Clin. Microbiol. **12**:732–737.
35. **Roop, R. M., R. M. Smibert, J. L. Johnson, and N. R. Kreig.** 1984. Differential characteristics of catalase-positive campylobacters correlated with DNA homology groups. Can. J. Microbiol. **30**:938–951.
36. **Ruiz-Palacios, G. M., J. Torres, N. I. Torres, E. Escamilla, B. R. Ruiz-Palacios, and J. Tamayo.** 1983. Cholera-like enterotoxin produced by *Campylobacter jejuni*. Lancet **ii**:250–253.
37. **Skerman, V. B. D., V. McGowan, and P. H. A. Sneath (ed.).** 1980. Approved lists of bacterial names. Int. J. Syst. Bacteriol. **30**:225–420.
38. **Skirrow, M. B.** 1977. *Campylobacter* enteritis: a "new" disease. Br. Med. J. **2**:9–11.
39. **Smibert, R. M.** 1984. Genus *Campylobacter* Sebald and Véron 1963, 907^{AL}, p. 111–118. *In* N. R. Krieg and J. G. Holt (ed.), Bergey's manual of systematic bacteriology, vol. 1. Williams & Wilkins, Baltimore.
40. **Tanner, A. C. R., S. Badger, C.-H. Lai, M. A. Listgarten, R. A. Visconti, and S. S. Socransky.** 1981. *Wolinella* gen. nov., *Wolinella succinogenes* (*Vibrio succinogenes* Wolin et al.) comb. nov., and description of *Bacteroides gracilis* sp. nov., *Wolinella recta* sp. nov., *Campylobacter concisus* sp. nov., and *Eikenella corrodens* from humans with periodontal disease. Int. J. Syst. Bacteriol. **31**:432–445.
41. **Taylor, D. E.** 1984. Plasmids in *Campylobacter*, p. 87–96. *In* J. P. Butzler (ed.), *Campylobacter* infection in man and animals. CRC Press, Inc., Boca Raton, Fla.
42. **Véron, M., and R. Chatelain.** 1973. Taxonomic study of the genus *Campylobacter* Sebald and Véron and designation of the neotype strain for the type species, *Campylobacter fetus* (Smith and Taylor) Sebald and Véron. Int. J. Syst. Bacteriol. **23**:122–134.
43. **Wang, W.-L. L., N. W. Luechtefeld, L. B. Reller, and M. J. Blaser.** 1980. Enriched brucella medium for storage and transport of cultures of *Campylobacter fetus* subsp. *jejuni*. J. Clin. Microbiol. **12**:479–480.
44. **Wells, J. G., C. A. Bopp, and M. J. Blaser.** 1981. Evaluation of selective media for the isolation of *Campylobacter jejuni*, p. 80–83. *In* D. G. Newell (ed.), Campylobacter. MTD Press Unlimited, Boston.

Chapter 28

Gram-Negative Fermentative Bacteria and *Francisella tularensis*

ROBERT E. WEAVER, DANNIE G. HOLLIS, AND EDWARD J. BOTTONE

The gram-negative fermentative microbial species that are the central theme of this chapter are infrequently encountered as human pathogens. Nevertheless, when causing human infection, they engender dire consequences for the infected person, pose considerable therapeutic dilemmas, and challenge the microbiologist regarding their isolation and subsequent characterization.

Many of the species produce organ-specific infections such as endocarditis; others are closely linked epidemiologically to animal reservoirs from which they are introduced into human tissue through a variety of mechanisms, both direct, e.g., animal or insect bites (*Pasteurella multocida* and *Francisella tularensis*), and indirect, e.g., consumption of contaminated foods and water (*F. tularensis* and *Chromobacterium violaceum*). Several species such as *Actinobacillus actinomycetemcomitans*, *Cardiobacterium hominis*, and *Capnocytophaga* species constitute part of the normal human microbial flora. *Actinobacillus* and *Capnocytophaga* species are involved primarily in periodontal disease, but in specific settings, these organisms as well as *C. hominis* may produce soft-tissue infections or bacteremia. Each species presents a particular diagnostic challenge. This chapter describes the basic microbiologic attributes distinguishing these somewhat unusual entities.

With the exception of *F. tularensis*, special selective procedures are not used to isolate these bacteria. They are isolated on standard media such as blood agar and chocolate agar, which are usually used to culture the types of specimens in which these microorganisms are found. Gram stains and oxidase tests may be done from colonies on nonselective media. If the organism has not been isolated on blood agar or if there is any question about the action of the organism on the blood cells, it should be streaked to a blood agar plate (heart infusion agar containing 5% defibrinated rabbit blood). The plates are streaked in a manner that ensures well-isolated colonies, and the agar is stabbed several times. This procedure provides information pertaining to the action on blood cells not only of surface colonies but also of subsurface growth. Plates are incubated for 18 to 24 h at 35 to 37°C in an initial atmosphere of about 1 to 3% carbon dioxide (a candle jar is satisfactory). Growth on the plates is examined with a ×12 dissection microscope or a hand lens, and data on colonial morphology, pigment, and action on blood cells are recorded. The oxidase test is performed by adding 1 or 2 drops of reagent (0.5% aqueous solution of tetramethyl-*p*-phenylenediamine dihydrochloride) to the growth on a portion of a plate held in a slightly tilted position. If the reaction is negative or cannot be readily interpreted, the test is performed by the method of Kovacs (39). The latter method is very sensitive, and final readings of tests are made at 10 s.

A well-isolated colony is picked and inoculated into a tube each of slanted heart infusion agar, heart infusion broth, and triple sugar iron (TSI) agar (Fig. 1). The TSI medium is inoculated over the entire surface of the slant, and the butt of the medium is stabbed to the bottom of the tube. These three tubes are incubated overnight (18 to 24 h), and a record is made of characteristics such as growth, pellicle formation, pigmentation, and hydrogen sulfide formation. Gram stains made of growth from heart infusion agar and heart infusion broth cultures are prepared and examined.

Preliminary information as to whether a microorganism uses carbohydrate fermentatively or is nonfermentative is obtained from reactions in TSI medium. If acidity is produced in the butt of the TSI medium, the bacterium should be considered fermentative. Nonfermentative bacteria rarely produce this reaction. An alkaline slant and a neutral pH in the butt of the TSI medium are characteristic of a nonfermenter. Acidity of the slant but not the butt portion of TSI medium may be observed with *Pseudomonas pseudomallei* and an occasional strain of *Acinetobacter calcoaceticus*, both of which use carbohydrates oxidatively and produce heavy growth on the surface of the slant. A microorganism that produces acidity only in the slant portion of the medium and grows poorly on or in the medium, however, should be considered fermentative. Fastidious microorganisms which do not produce a change in the pH of either the slant or the butt of the medium should be tested for their ability to use carbohydrates in both fermentation and oxidation-fermentation (OF) media.

When it is not certain whether an organism uses carbohydrates fermentatively, both open and sealed tubes of OF medium containing 1% glucose are inoculated. This test does not always provide the answer, because some organisms either do not grow or grow poorly in this medium.

Of the bacteria discussed in this chapter, *Kingella kingae*, *Kingella denitrificans*, and dysgonic fermenter DF-2 rarely produce acid reactions in TSI medium. About 12% of the strains of *C. hominis* and 45% of the strains of *Capnocytophaga* species also do not produce acidity in TSI medium. *F. tularensis* does not produce acid from carbohydrates unless special media supplemented with cysteine or cystine are used.

The set of media used for the examination of fermentative bacteria (Fig. 1) contains six carbohydrate substrates in liquid peptone basal medium with Andrade indicator. It is sometimes necessary to enrich this basal medium by the addition of 2 drops of rabbit serum to each tube containing 3 ml of medium to ensure or enhance the growth of fastidious microorganisms such as *K. kingae* and DF-2. All media are incubated at 35 to 37°C for 48 h in an ordinary incubator or a candle jar, depending upon which

Blood agar plate (5% defibrinated rabbit blood)

|

Oxidase tests, Gram stains

|

Heart infusion agar slant
Heart infusion broth
Triple sugar iron agar

|

Substrates in liquid peptone basal medium with:
- Glucose
- Xylose
- Mannitol
- Lactose
- Sucrose
- Maltose

|

MacConkey agar slant
Simmons citrate agar
Methyl red/Voges-Proskauer medium
Motility medium
Christensen urea agar
Nitrate medium (peptone base)
Nitrate medium (heart infusion base)
2% Tryptone medium
Nutrient broth
Nutrient broth with 6% sodium chloride
Gelatin
Tryptone-glucose-yeast extract agar slant (3 tubes)
Esculin agar

FIG. 1. Substrates recommended for the identification of miscellaneous gram-negative fermentative bacteria.

environment is considered optimal for the particular microorganism under investigation. In addition to the carbohydrate media mentioned above, each of the sets of media includes media and test procedures which permit determination of the most important characteristics of the bacteria discussed in this chapter and which provide sufficient data for identifying the majority of these bacteria at 48 h.

Tests for nitrate reduction and indole production are made after 48 h of incubation. The Durham inserts in the nitrate medium should be examined carefully for evidence of gas formation. Tests for indole are made by extracting the tryptone water with xylol before adding Ehrlich-Boehme reagent. Extraction is a prerequisite because some of these bacteria (e.g., *C. hominis* and HB-5) produce minute quantities of indole. Heart infusion broth cultures also may be used for indole determinations, especially with dysgonic microorganisms.

The methyl red test is done after 4 to 5 days of incubation. The Coblentz method is recommended for the detection of acetoin (acetyl methyl carbinol). A tube of nutrient broth and a tube of nutrient broth containing 6% sodium chloride are included to determine the sodium chloride requirements of various microorganisms. The halophilic species, such as *Vibrio parahaemolyticus* and *V. alginolyticus*, grow in nutrient broth only when it is supplemented with sodium chloride. None of the organisms discussed in this chapter requires the supplementation with sodium chloride.

Three tryptone-glucose-yeast extract agar slants are incubated (one each at 25, 35, and 42°C) to determine the optimal growth temperature. The catalase test is performed by flooding the surface of the slant of the tryptone-glucose-yeast extract agar culture that exhibits optimal growth with 3% hydrogen peroxide. If a bacterium does not grow on this agar, the temperature studies are repeated with heart infusion agar slants. If the catalase test is negative on the slant with optimal growth, the test is repeated on the slants that were incubated at other temperatures and supported growth of the microorganism. In some instances, additional tests and substrates must be used for accurate identification. These include tests for decarboxylase activities (Moeller method) and for the utilization of additional carbohydrate substrates.

A key to the identification of these miscellaneous fermentative gram-negative bacteria is given in Fig. 2. The key should be used only as a guide; variable reactions may be encountered. *F. tularensis* has not been included in the key. Cellular morphology and the requirement for or the stimulation of growth by cysteine or cystine are key characteristics for the recognition of this bacterium.

ACTINOBACILLUS SPECIES

In *Bergey's Manual of Systematic Bacteriology*, vol. 1 (40), five species of *Actinobacillus* are recognized: *A. lignieresii*, *A. equuli*, *A. suis*, *A. capsulatus*, and *A. actinomycetemcomitans*. *A. capsulatus*, which has been described as an etiological agent of arthritis in rabbits

- Growth on MacConkey agar
  - Oxidase negative
    - Motile: *Chromobacterium violaceum*
    - Nonmotile: HB-5
  - Oxidase positive
    - Motile: *Chromobacterium violaceum*
    - Nonmotile, urea positive
      - Indole positive: *Pasteurella pneumotropica*
      - Indole negative: *Actinobacillus equuli*, *Actinobacillus lignieresii*, *Actinobacillus suis*, *Pasteurella aerogenes*
    - Nonmotile, urea negative
      - Indole positive: HB-5
      - Indole negative: *Kingella kingae*, *Pasteurella haemolytica*, EF-4
- No growth on MacConkey agar
  - Oxidase negative
    - Catalase positive: *Actinobacillus actinomycetemcomitans*
    - Catalase negative, indole positive: HB-5
    - Catalase negative, indole negative: *Capnocytophaga* species
  - Oxidase positive
    - Urea positive, indole positive: *Pasteurella pneumotropica*, *Pasteurella* sp. new species 1
    - Urea positive, indole negative: *Actinobacillus equuli*, *Actinobacillus lignieresii*, *Actinobacillus suis*
    - Urea negative
      - Indole positive, catalase positive: *Pasteurella multocida*, *Pasteurella* sp. new species 1
      - Indole positive, catalase negative: *Cardiobacterium hominis*, *Kingella indologenes*, HB-5
      - Indole negative, catalase positive: *Actinobacillus actinomycetemcomitans*, *Pasteurella haemolytica*, DF-2, EF-4
      - Indole negative, catalase negative: *Kingella denitrificans*, *Kingella kingae*

FIG. 2. Key to identification of miscellaneous gram-negative fermentative bacteria.

TABLE 1. Characteristics of species of *Actinobacillus*

Characteristic	*A. equuli* (*n* = 19)		*A. lignieresii* (*n* = 30)		*A. hominis* (*n* = 1), sign	*A. suis* (*n* = 33)		*A. actinomycetem-comitans* (*n* = 120)	
	Sign[a]	%+[b]	Sign	%+		Sign	%+	Sign	%+
Hemolysis (clear zone)	−	0	−	4	−	v	76	−	0
Motility	−	0	−	0	−	−	0	−	0
Gas from glucose	−	0	−	0	−	−	0	v	28
Acid from:									
Glucose	+	100	+	96 (4)	+	+	94 (6)	+ or (+)	83 (16)
Xylose	+	100	+ or (+)	87 (13)	(+)	+	94 (6)	v	33 (9)
Mannitol	+	100	+ or (+)	91 (9)	+	v	54[c]	v	66 (16)
Lactose	+	95	v	17 (61)	(+)	+ or (+)	79 (18)	−	0[d]
Sucrose	+	100	+	96 (4)	+	+	94 (6)	−	0[d]
Maltose	+	95	+ or (+)	83 (17)	+	+	94 (6)	+ or (+)	81 (15)
Trehalose	+	100	−	0	+	+	96		
Melibiose	+ or (+)	79 (21)	−	0	(+)	+	91		
Raffinose	+	100	v	25 (46)	+	+	96		
Catalase	v	68 (5)	v	89	+	v	85	+	99
Oxidase	+	100	+	100	+	+	100	v	19 (2)
Growth on MacConkey agar	v	84 (5)	v	67 (8)	+	+ or (+)	82 (12)	−	4 (1)
Simmons citrate	−	0	−	0	−	−	0	−	0
Urease	+	100	+	100	+	+	97 (3)	−	0
Nitrate reduction	+	100	+	100	+	+	100	+	100
Gas from nitrate	−	0	−	0	−	−	0	−	0
Indole	−	0	−	0	−	−	0	−	0
Gelatin hydrolysis	v	70	−	0	−	−	3	−	0
TSI[e] slant, acid	+	100	+	100	+	+	100	+	100
TSI butt, acid	+	100	+	100	+	+	100	+	100
Esculin hydrolysis	−	0	−	0	(+)	+	100	−	0
Lysine decarboxylase	−	0	−	0	−	−	0		
Arginine dihydrolase	−	0	−	0	−	−	0		
Ornithine decarboxylase	−	0	−	0	−	−	0		

[a] Sign: +, 90% or more positive in 1 or 2 days; −, no reaction, 90% or more; + or (+), 90% or more positive, some strains positive after 3 or more days; v, more than 10% and less than 90% positive.
[b] Numbers in parentheses indicate percentage of delayed reactions (3 days or more).
[c] Strains fermenting mannitol may not be *A. suis*; see the text.
[d] *Haemophilus aphrophilus* ferments lactose and sucrose but not xylose.
[e] TSI, Triple sugar iron.

but has not been reported for humans, has not been included in this chapter.

A. lignieresii and *A. equuli*

A. lignieresii and *A. equuli* have similar characteristics, and the two species may belong together in only a single species (66).

Both species are gram-negative rods that are oxidase positive, urease positive, and nonmotile and do not form indole. Nitrate is reduced. *A. lignieresii* and *A. equuli* are differentiated on the basis of trehalose and melibiose reactions. *A. equuli* ferments both substrates; *A. lignieresii* does not ferment either substrate. None of the strains of either species hydrolyzes esculin (Table 1).

Both *A. lignieresii* and *A. equuli* have been associated with human infections. Primarily, however, they are animal pathogens. *A. lignieresii* produces granulomatous lesions in the upper alimentary tract of cattle and suppurative lesions in the skin and lungs of sheep. *A. equuli* causes septicemia, arthritis, and nephritis in foals and pigs. *A. lignieresii* has been isolated from the mouths of normal cattle and sheep. *A. equuli* has been isolated from the mouths of normal horses and swine.

The Centers for Disease Control (CDC) has received three strains of *A. equuli* from human sources: donkey bite wound, joint fluid, and leg wound. Three strains of *A. lignieresii* isolated from humans were from wounds for which a cause was not stated. A fourth strain was isolated from a horse bite wound.

On rabbit blood agar, neither species hemolyzes the erythrocytes. Some strains cause a greenish discoloration of the blood. After incubation for 18 to 24 h, colonies are 1 to 2 mm in diameter. They are low convex with an entire edge and translucent to semi-opaque. In Gram-stained smears, small coccoid forms and short rods are usually observed.

A. suis

In the 8th edition of *Bergey's Manual of Determinative Bacteriology* (8), *A. suis* was listed as *species incertae sedis*. In the current edition of *Bergey's* (40), *A. suis* has been included in the genus *Actinobacillus*. *A. suis* has been isolated from septicemic disease and a variety of lesions in piglets.

The reactions of *A. suis* listed in Table 1 are the reaction of strains of *Actinobacillus*-like organisms received at CDC. Unlike *A. lignieresii* and *A. equuli*, these organisms usually produce a clear zone of hemolysis. Another reaction that differentiates these strains from *A. lignieresii* and *A. equuli* is esculin hydrolysis. The remaining characteristics are similar to those of the latter two species. Of these esculin-positive strains, 54% ferment mannitol and, therefore, differ from the description of *A. suis* (40). Although the mannitol-positive strains may belong to another species, we have chosen to group together the esculin-positive strains of *Actinobacillus*-like organisms. In addition to the hemolytic and esculin reactions, the key characteristics include the fermentation of carbohydrates without the formation of gas, a positive oxidase reaction, the hydrolysis of urea, the reduction of nitrate, a negative indole reaction, and no motility. The colonial and cellular morphologies are similar to those of *A. lignieresii* and *A. equuli*.

We have received strains of *A. suis* isolated from horses, pigs, and cattle. The sources of 17 human isolates include: blood, 2; upper respiratory tract and sputum, 5; wound, 5; animal bite wounds (donkey, zebra, and hamster), 3; laceration of the humerus, 1; and finger lesion, 1.

A. hominis

Friis-Møller isolated a *Pasteurella/Actinobacillus*-like organism from the sputa of 17 patients with chronic lung disease in Copenhagen and proposed the name *A. hominis* (16). This species has not been included in *Bergey's Manual of Systematic Bacteriology*, vol. 1 (40).

The major characteristics of the type strain are listed in Table 1. In our laboratory the type strain hydrolyzed esculin in 3 to 7 days. Further studies are needed to determine whether some of the mannitol-positive strains we have included with *A. suis* might be *A. hominis*. Friis-Møller reported that *A. hominis* does not ferment D-mannose, D-arabinose, or cellobiose. All three carbohydrates are fermented by *A. suis*. *A. equuli* and *A. lignieresii* ferment D-mannose (16).

A. actinomycetemcomitans

Clinical significance. *A. actinomycetemcomitans* is primarily isolated from actinomycotic or actinomycotic-like lesions and from the blood of patients with endocarditis. From 1912, when *A. actinomycetemcomitans* was first described, until 1962, most isolates reported were from lesions in which the organism was associated with other types of organisms, primarily *Actinomyces israelii*. In 1962 King and Tatum (37) described 33 isolates, 27 of which had been isolated from blood. Page and King (51) later summarized the case histories of 32 patients from whom *A. actinomycetemcomitans* had been isolated. Twenty-five of the isolates were from blood cultures. The diagnosis of bacterial endocarditis was made in 23 of the 25 patients. Five cultures were from abscesses, and one culture each was obtained from a sinus tract and pleural fluid. Affias et al. (1) reported two cases of endocarditis and summarized 15 earlier reports of *A. actinomycetemcomitans*-associated endocarditis. The total number of cases summarized in this report was 39, and 12 of the patients died as a result of the infections.

The role of *A. actinomycetemcomitans* or other bacteria present along with *A. israelii* in actinomycotic lesions is not clear. Holm (28), however, has presented evidence that *A. actinomycetemcomitans* may be able to maintain actinomycotic-like lesions after the *Actinomyces* sp. has been eliminated by antibiotic therapy.

A. actinomycetemcomitans has been shown to be part of the normal mouth flora (Heinrich and Pulverer, quoted by Pulverer and Ko [52]). Infections caused by the organism are probably of endogenous origin. Recently, *A. actinomycetemcomitans* has been linked to periodontal disease. Slots and colleagues (59) recovered this species from the periodontal pockets, cheek mucosa, tongue, and saliva of 6 of 12 adult periodontitis patients and from 9 of 12 juvenile periodontitis patients.

Direct examination. Direct examination of *A. actinomycetemcomitans* is not of value. Bacteria are frequently observed in Gram-stained smears of material from actinomycotic lesions. However, the identification of *A. actinomycetemcomitans* cannot be made by direct examination of specimens.

Culture and isolation. Recommended procedures for primary culture follow.

(i) Soft-tissue lesions. Specimens from soft-tissue lesions should be inoculated onto blood or chocolate agar. Incubate the culture at 35 to 37°C in a candle jar or a 5% CO_2 atmosphere. Examine the culture at 18 to 24 h. Pick various colony types and incubate them for an additional 18 to 24 h; examine these cultures for any development of additional colony types.

(ii) Blood cultures. The isolation of *A. actinomycetemcomitans* in several blood culture media has been reported. The media have included: Trypticase soy broth (BBL Microbiology Systems, Cockeysville, Md.) (1), brain heart infusion broth with *p*-aminobenzoic acid and agar (63), and Schaedler broth with 0.05% sodium polyanetholsulfonate (63).

Growth in blood culture bottles has been observed as early as day 3 of incubation. In some reports, growth was not detected until 6 or more days after the blood cultures were inoculated. Growth may be noted first as delicate granules in the sedimented blood (Fig. 3). Growth also may appear as granules adherent to

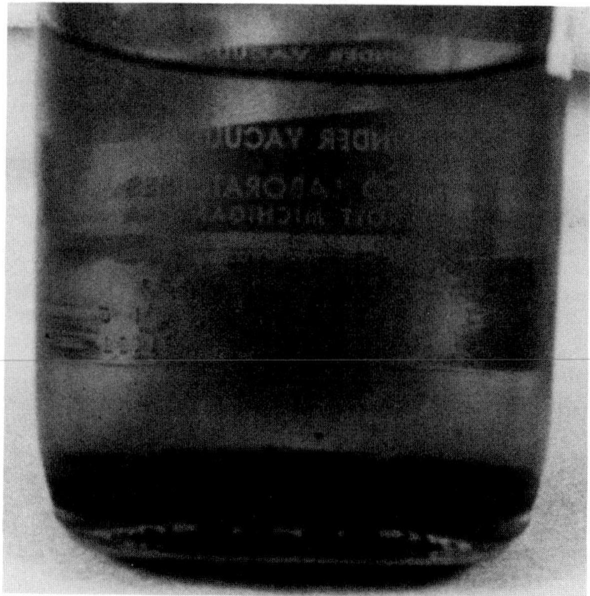

FIG. 3. Blood culture showing discrete floccules of growth on sedimented erythrocyte layer; often noted with *Actinobacillus actinomycetemcomitans* and *Haemophilus aphrophilus*.

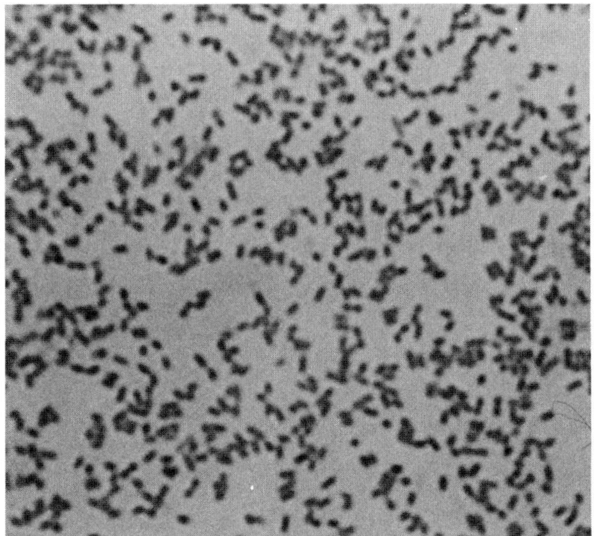

FIG. 4. Gram-stained smear of *Actinobacillus actinomycetemcomitans* showing small coccoid to coccobacillary morphology (×1,000).

the sides of the bottles, especially near the surface of the broth. A ring of growth is formed that is more easily observed when the bottle is tipped (45). The blood-broth is subcultured to a blood or chocolate agar plate, which is then incubated in a candle jar or CO_2 incubator.

Identification. *A. actinomycetemcomitans* is a fastidious, small, gram-negative, coccoid to coccobacillary organism (Fig. 4). Its morphology is similar to that of *Haemophilus aphrophilus*. Its size is comparable to that of *Brucella* species. Some strains become more rod shaped after several transfers in the laboratory. An occasional filament may be observed.

In broth media the organism forms granules that adhere to the walls of the tube (Fig. 5). The broth remains clear. The organism tends to maintain this characteristic, but a few strains produce a light turbidity and a slightly ropy sediment after several transfers.

After 24 h of incubation at 35 to 37°C on blood agar, colonies range from punctate to 0.5 mm in diameter. If incubation is continued, well-isolated colonies will be 2 to 3 mm in diameter by 7 days. More than one colonial form may be seen. The majority of colonies are smooth, slightly domed, and translucent, and they have a slightly wrinkled surface. In the center is a very opaque spot. With further incubation, a four- to six-pointed star structure forms (Fig. 6). The star formation grows into the agar, leaving an impression that can be seen when the colony is scraped away. This phenomenon is seen better on a clear medium than on blood agar. Some strains have mixtures of the two colony types.

A. actinomycetemcomitans frequently will not grow on solid medium unless the medium is incubated in a candle jar or a CO_2 incubator. Those strains that grow aerobically are enhanced by incubation in an increased CO_2 atmosphere.

The biochemical characteristics of *A. actinomycetemcomitans* are listed in Table 1. Characteristics important for differentiating this organism from species with similar characteristics are no growth (or, rarely, poor growth) on MacConkey agar, positive catalase reaction, negative urease and indole reactions, no motility, and fermentation of glucose but not of lactose and sucrose. Of 120 strains studied at CDC, 28% produced gas in the fermentation of glucose. The gas production is weak and may not be observed in the insert tubes placed in glucose broth. Gas production by some strains was detected, as described by King and Tatum (37), by plunging a hot needle into a 48-h thioglycolate broth culture. If a strain is capable of producing gas from glucose, bubbles form along the path of the hot needle. This procedure should be carried out only in a biological safety cabinet.

The fermentation of some carbohydrates, particularly xylose and mannitol, is variable. The majority of strains examined in our laboratory could be placed into five biotypes based on reactions in these two carbohydrates (Table 2). Pulverer and Ko (52) described eight biotypes based on the fermentation of xylose, mannitol, and galactose by 140 strains.

Serological methods. Serological studies have shown that several serotypes can be determined by agglutination and precipitation reactions (37, 53). Serological typing, however, does not appear to provide clinically relevant information. There are no serological tests for demonstrating past or present infection with *A. actinomycetemcomitans*.

CAPNOCYTOPHAGA SPECIES

Capnocytophaga is the genus name proposed by Leadbetter et al. (41) for fastidious gram-negative gliding bacteria isolated from both healthy and diseased sites in the oral cavity of humans. These bacteria are fusiform, have a fermentative metabolism, and usually grow under either anaerobic or aerobic conditions in the presence of CO_2. Little or no growth is

FIG. 5. Glucose growth culture of *Actinobacillus actinomycetemcomitans* showing floccular growth adherence to sides of tube. Similar phenomenon may be observed with *Haemophilus aphrophilus* and *Eikenella corrodens*.

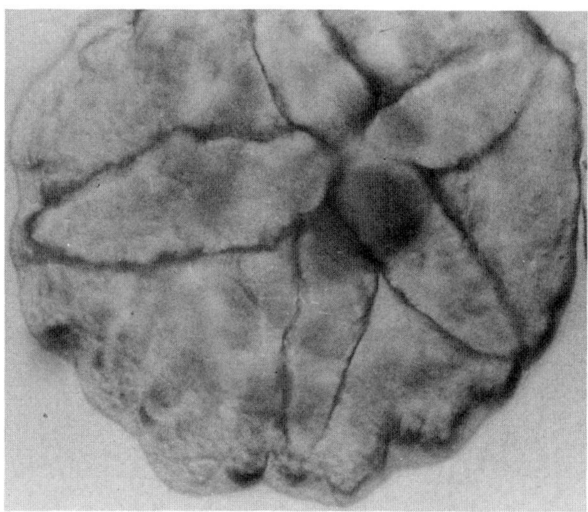

FIG. 6. Star-shaped configuration often noted in colonies of *Actinobacillus actinomycetemcomitans* and *Haemophilus aphrophilus* growing on tryptic soy agar (×100).

obtained under aerobic conditions without an increased CO_2 content. Three species have been described (41): *C. ochracea* (synonymous with *Bacteroides ochraceus*), *C. gingivalis*, and *C. sputigena*. On the basis of DNA homology studies (67), it has been shown that the organisms designated as the DF-1 (dysgonic fermenter 1) group in the early 1960s by Elizabeth O. King, CDC, are *Capnocytophaga* species.

In addition to being isolated in association with disease of the oral cavity, particularly periodontal disease, *Capnocytophaga* species have been isolated in association with systemic disease in compromised hosts (47). Of 172 strains received at CDC, 58 (34%) were isolated from sputum or throat, and 40 (23%) were isolated from blood. Seven strains were isolated from spinal fluid, and seven strains were isolated from bronchial specimens. The remainder were isolated from various sources, including vagina, cervix, amniotic fluid, trachea, pleural fluid, and eye.

The colonial morphology is varied. After incubation for 18 to 24 h at 35 to 37°C on rabbit blood agar, the most characteristic colony is usually quite small or not visible to the unaided eye; by 2 to 4 days it is usually 2 to 3 mm in diameter and slightly convex, with a flat, spreading, fringelike edge (Fig. 7). The colonies may appear to "pit" the agar. Another colonial form is round with a smooth edge. Some strains produce colonies that adhere firmly to agar media. The color of the bacterial growth on blood agar may be gray to white, pink, or yellow. When scraped from the agar surface, the cell mass is usually yellow. Leadbetter et al. (41) observed that the composition of the agar medium influences the degree of spreading of colonies. An agar concentration of about 3% (wt/vol) favors maximum spreading.

In smears from blood agar, *Capnocytophaga* species are medium to long, thin rods with tapered ends (Fig. 8). They closely resemble the cells of the CDC DF-2 group.

The gliding motility of log-phase cultures was observed by dark-field microscopy for all the strains studied by Williams et al. (67). The motility of *Capnocytophaga* species usually cannot be detected in motility medium. We have observed, however, that 40% of the strains grow out slightly from the stab line in motility medium after incubation for 3 to 7 days. Flagellum stains of these strains show an occasional cell with a single polar or lateral flagellumlike structure.

The most useful characteristics for the identification of this genus are: requirement for CO_2, slow growth, cellular morphology, fermentation of carbohydrates, and negative catalase and oxidase reactions. The biochemical reactions listed in Table 3 are those of the DF-1 strains collected at CDC and probably represent the combined reactions of all three species. Of eight CDC strains of the DF-1 group studied for DNA homology by Williams et al. (67), seven strains were shown to be *C. ochracea*, and one strain was *C. gingivalis*. From the study of Socransky et al. (61), it appears that the fermentation of lactose and galactose and the reduction of nitrate are the most useful reactions for the differentiation of *C. ochracea*, *C. sputigena*, and *C. gingivalis* (Table 4).

Several of the strains that were isolated from throats and referred to our laboratory were reported to have been isolated on Thayer-Martin selective medium.

The acid products of glucose fermentation are acetic and succinic acids.

CARDIOBACTERIUM HOMINIS

Members of *C. hominis* were designated group IId by King (36) and were named by Slotnick and Dougherty (57). This species is composed of gram-negative, rod-shaped bacteria which are pleomorphic with bulbous ends and have a tendency to retain some of the crystal violet stain. Frequently, the cells occur in clusters resembling a rosette (Fig. 9).

When cultivated on agar plates that contain 5% rabbit blood, cultures of *C. hominis* do not hemolyze erythrocytes. Colonies on blood agar after 24 h of incubation are minute, but they attain a size of 1.0 mm after 48 h. They are convex, circular with entire edges, smooth, glossy, and butyrous. Some strains produce colonies which pit the agar. The recommended conditions of incubation are 35 to 37°C with increased carbon dioxide tension (a candle extinction jar is satisfactory).

C. hominis is fermentative and usually produces acid throughout TSI medium, but there is no evidence of hydrogen sulfide production in the medium. The production of hydrogen sulfide may, however, be detected by suspending lead acetate strips over inoculated TSI or Kligler iron agar slants (69).

TABLE 2. Characteristics of the biotypes of *Actinobacillus actinomycetemcomitans*

Acid from:	Biotype				
	n = 63	n = 30	n = 15	n = 6	n = 4
Glucose	+	+	+	+	+
Xylose	−	+	+	−	+
Mannitol	+	+	−	−	+
Lactose	−	−	−	−	−
Sucrose	−	−	−	−	−
Maltose	+	+	+	+	−

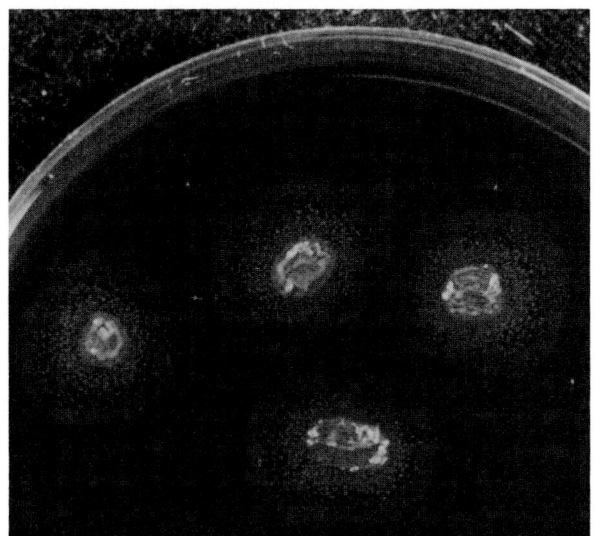

FIG. 7. Blood agar culture of *Capnocytophaga* species showing flat, concentrically spreading growth from a central point inoculation. Fingerlike projections emanating from colony periphery may also be observed (×1.5).

Carbohydrate utilization may be determined in liquid peptone medium to which about 2 drops of rabbit serum per 3 ml of medium has been added to ensure adequate growth. Various degrees of acidity are produced from glucose, mannitol, sucrose, and maltose within 2 to 7 days. Xylose and lactose are not fermented (Table 3).

These microorganisms are oxidase positive, catalase negative, and nonmotile. They do not grow on MacConkey or Simmons citrate medium. Urea is not hydrolyzed, and nitrate is not reduced. Indole is formed, but the amount produced within 48 h is small, and extraction with xylol generally is necessary for its demonstration.

C. hominis is an etiological agent of endocarditis (57, 69) and has been shown to be part of the normal flora of the nose and throat in humans (58). Most strains referred to CDC were isolated from blood. One strain was isolated from spinal fluid. The organism is quite susceptible to a variety of antimicrobial agents, including various penicillins, cephalothin, tetracycline, chloramphenicol, aminoglycosides, and colistin (55).

CHROMOBACTERIUM VIOLACEUM

C. violaceum is a gram-negative, facultatively anaerobic, rod-shaped bacterium. It is motile with polar and lateral flagella. A violet pigment (violacein), insoluble in water and chloroform but soluble in ethanol, is produced by most strains. On blood agar, the colonies may appear black. Some strains produce both pigmented and nonpigmented colonies. Carbohydrates are fermented, usually without the formation of gas. Aerogenic strains were described first by Sivendra (56).

C. violaceum is a normal inhabitant of soil and water and usually is considered nonpathogenic. However, there are several reports of infections in humans (34) and animals (18). The types of human infections include urinary tract infections, localized abscesses, diarrhea, and systemic infections with abscesses in multiple organs. One route of infection has been through wounds contaminated with soil or water. Ingestion may have been the route of infection in at least one case. In most instances the route of infection has not been demonstrated clearly. Most cases have occurred in tropical and subtropical climates. In the United States, human infections have been reported from Florida, Louisiana, and South Carolina. The patient in South Carolina may have acquired the infection by a puncture wound sustained in Florida.

C. violaceum is susceptible to aminoglycosides, chloramphenicol, and tetracycline.

Identification

Most strains can be identified by pigmentation and the fermentative use of carbohydrates, which is indicated by an acid reaction in the butt portion of TSI or Kligler iron agar slants within 18 to 24 h. The other organism that produces violacein, *C. lividum*, uses carbohydrates oxidatively.

The characteristics of *C. violaceum* are listed in Table 3. *C. violaceum* grows on MacConkey agar. The oxidase reaction is variable; therefore, nonpigmented strains must be differentiated from members of the family *Enterobacteriaceae* or the *Vibrio, Aeromonas,* and *Plesiomonas* species. The carbohydrate, decarboxylase, and dihydrolase reactions are usually sufficient for this differentiation. Glucose is fermented, usually without the production of gas; the fermentation of sucrose is variable; only 1 of 36 strains we have examined fermented maltose; xylose, mannitol, and lactose are not fermented. The arginine dihydrolase reaction is positive, and the lysine and ornithine decarboxylase reactions are negative. Urease and Simmons citrate reactions are variable. Nitrate is reduced by most strains; esculin is not hydrolyzed. Some of the nonpigmented strains have produced indole.

In stained smears *C. violaceum* is a short to medium-length rod. The pigmented strains usually give a bipolar staining reaction. After incubation for 18 to 24 h on blood agar, colonies are approximately 0.5 to 1.5

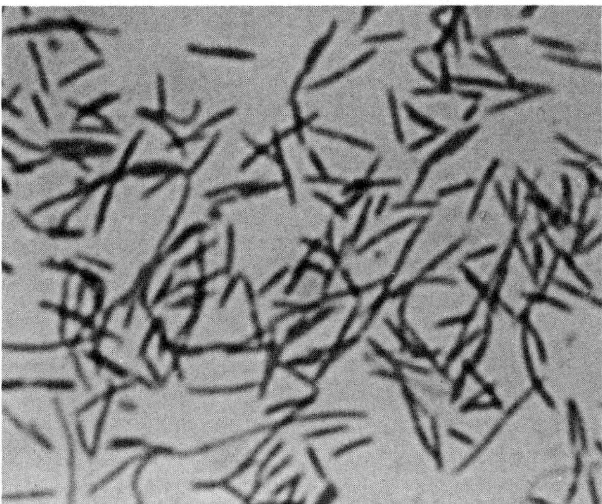

FIG. 8. Smear of blood agar culture of *Capnocytophaga* species showing characteristic fusiform morphology (×1,000).

TABLE 3. Characteristics of *Capnocytophaga* species, *Cardiobacterium hominis*, and *Chromobacterium violaceum*

Characteristic	*Capnocytophaga* species (DF-1) (*n* = 155)		*Cardiobacterium hominis* (*n* = 32)		*Chromobacterium violaceum*[a] (*n* = 36)	
	Sign[b]	%+[c]	Sign	%+	Sign	%+
Hemolysis (clear zone)	−	0	−	0	v	48
Motility	v	5 (40)	−	0	+	100
Gas from glucose	−	0	−	0	−	0[d]
Acid from:						
Glucose	+	90 (10)	+ or (+)	78 (22)	+	100
Xylose	−	0	−	0	−	0
Mannitol	−	0	+ or (+)	50 (50)	−	0
Lactose	v	75 (11)	−	0	−	0
Sucrose	+	90 (9)	+ or (+)	61 (39)	v	20 (6)
Maltose	+ or (+)	86 (14)	+ or (+)	72 (28)	−	0 (3)
Catalase	−	7	−	3	+	97
Oxidase	−	7	+	100	v	67
Growth on MacConkey agar	−	0	−	0	+	100
Simmons citrate	−	0	−	0	v	68 (9)
Urease	−	0	−	0	v	5 (14)
Nitrate reduction	v	63	−	0	+	97
Gas from nitrate	−	0	−	0	−	0
Indole	−	0	+	97	v	21
Gelatin hydrolysis	−	0	−	0	v	86
TSI[e] slant, acid	v	73	+	96	−	8
TSI butt, acid	v	55	v	83	+	94
Esculin hydrolysis	v	81 (2)	−	4	−	0
Lysine decarboxylase	−	0			−	0
Arginine dihydrolase	−	0			+	100
Ornithine decarboxylase	−	0			−	0

[a] 92% produce a violet pigment.
[b] Sign: +, 90% or more positive in 1 or 2 days; −, no reaction, 90% or more; + or (+), 90% or more positive, some strains positive after 3 or more days; v, more than 10% and less than 90% positive.
[c] Numbers in parentheses indicate percentage of delayed reactions (3 days or more).
[d] Aerogenic strains have been described (56).
[e] TSI, Triple sugar iron.

mm in diameter. They are convex and smooth. About half of the cultures we have examined produced a clear zone of hemolysis on rabbit blood agar.

The sources of cultures isolated from humans and referred to CDC have included: blood, wound, lung, liver, sputum, spleen, brain, skin lesion, and ear.

FRANCISELLA TULARENSIS

(A major portion of this section is taken from the chapter by Henry T. Eigelsbach in the 2nd edition of this Manual.)

F. tularensis, originally named *Bacterium tularense* in 1911, was designated *Pasteurella tularensis* in the 7th edition (1957) of *Bergey's Manual of Determinative Bacteriology* (8). Upon recommendation of U.S. and U.S.S.R. investigators, this species was listed in the 8th edition under "Part 7, Gram-Negative Aerobic Rods and Cocci, Genera of Uncertain Affiliation, Genus *Francisella*" to honor the late Edward Francis and the U.S. Public Health Service for pioneering investigations on tularemia and its etiological agent. Two biovars are recognized. *F. tularensis* biovar *tularensis* (type A) has been isolated in nature in North America only. It is more virulent for the domestic rabbit and produces more severe disease in humans than does the second variety. *F. tularensis* biovar *palaearctica* (type B) is found wherever tularemia occurs throughout the northern hemisphere.

F. tularensis is a particularly small, singly occurring, nonmotile, nonencapsulated, nonsporulating, poorly staining, bipolar, aerobic rod that usually requires special media for isolation and growth. In young cultures (propagated for 18 to 24 h on solid medium), morphology is relatively uniform, but older cultures exhibit pleomorphism. The organism is distinguished from other bacteria by its small size (0.3 to 0.5 by 0.2 μm), faint bipolar staining with aniline dyes, inability to grow or poor growth on ordinary media, stimulation of growth by cystine or cysteine, fluorescent-antibody reaction, and agglutination with a specific antiserum.

Biochemical characterization is of little value in identification. Only one serotype has been described. In the agglutination test, no difficulty should be encountered in distinguishing the high-titer homologous reaction from low-titer (usually <1:20) cross-reactions that occur with heterologous antigens of the *Brucella* species.

TABLE 4. Differentiation of the species of *Capnocytophaga*[a]

Characteristic	*C. ochracea* (*n* = 27)		*C. sputigena* (*n* = 6)		*C. gingivalis* (*n* = 5)	
	Sign[b]	%+	Sign	%+	Sign	%+
Acid from:						
Lactose	+	92	v	40	−	8
Galactose	v	83	−	0	−	0
Nitrate reduction	−	8	v	83	−	4

[a] Data from Socransky et al. (61).
[b] Sign: +, 90% or more positive reactions; −, 90% or more negative reactions; v, 11 to 89% positive reactions.

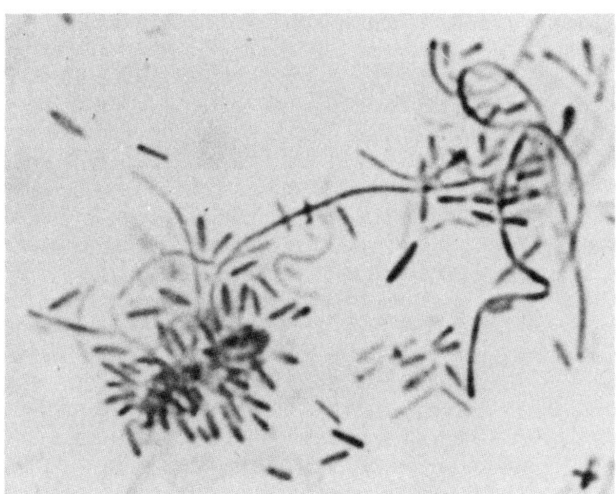

FIG. 9. Smear of agar culture of *Cardiobacterium hominis* showing pleomorphic morphology consisting of filament forms, bacilli with swollen ends resembling "teardrops," and characteristic rosette formation (×1,000).

Clinical significance

Human tularemia in North America is an acute, usually moderately severe, febrile, granulomatous, infectious, zoonotic disease. The clinical picture and severity vary appreciably according to the route of infection and the virulence of the organism. In North America, before the advent of antibiotic therapy, glandular, ulceroglandular, and oculoglandular tularemia in untreated patients resulted in a case mortality rate of approximately 5%. A substantially higher mortality, approximately 30%, resulted from pulmonary (inhalational) tularemia or from the "typhoidal" type of infection (characterized by the absence of an obvious portal of infection and usually resulting from the ingestion of contaminated water or undercooked infected meat). Some lung involvement may occur in all forms of tularemia and is attributed to a transient bacteremia. Severe pulmonary disease initiated by circumstances other than inhalation of the organism is termed secondary pneumonia. Mortality rates in Eurasia are lower for all clinical types (averaging less than 1%) because of the innate lower virulence of strains common to that area. Therapeutically the drug of choice is streptomycin or gentamicin (31).

Tularemia is transmitted to humans by a variety of animals, including wild rabbits, muskrats, beavers, squirrels, woodchucks, sheep, and game birds, or by biting insects (usually ticks, deerflies, or mosquitoes). Infection follows the handling of infected animal carcasses, insect bites, ingestion of improperly cooked meat or contaminated water, or inhalation of airborne organisms, e.g., dust from hay contaminated with infected carcasses and the feces of infected rodents. Bites or scratches by resistant wild or domestic carnivores whose mouth parts have been contaminated by eating infected animals may also result in human infection. Although humans of all ages, sexes, and races are susceptible, human-to-human transmission is extremely rare. The incubation period in humans is usually 3 to 4 days but ranges from 2 to 10 days, depending primarily upon the dose. Most cases are characterized by the formation of a slightly tender, erythematous papule at the site of entry of the organism that progresses to pustule formation and then to a focal ulcer with surrounding erythema. Regional lymph nodes become enlarged and tender, and they often suppurate. These local manifestations accompany or precede the sudden onset of the usual constitutional reactions: fever, chills, severe headache, myalgia, malaise, anorexia, nausea, and sometimes prostration. Secondary pneumonia may occur and is usually accompanied by substernal chest pain, cough, and dyspnea. The clinical diagnosis can be confirmed by the isolation of *F. tularensis* from local lesions, regional lymph nodes, sputum, gastric aspirates, or nasopharyngeal washings. In oculoglandular tularemia, conjunctival scrapings frequently yield the organisms. *F. tularensis* is rarely recovered from the blood except during the first few days after infection and in untreated fulminating disease.

Collection, transport, and storage of specimens

F. tularensis is infrequently isolated from blood; nonetheless, at least 3 ml of blood, without anticoagulant, should be cultured on one of the recommended agar media subsequently described. A primary lesion or enlarged, draining, regional lymph nodes usually contain sufficient *F. tularensis* cells to ensure abundant growth. The pustule or crusted sites of draining lymph nodes and areas immediately surrounding the ulcer are cleansed with 70% alcohol and allowed to dry; the pustule should be incised with a sterile scalpel. A drop or more of fluid is gently expressed and collected in a sterile capillary tube fitted with a rubber bulb; single samples usually suffice. In oculoglandular tularemia, conjunctival scrapings may be obtained by traversing the area with a sterile swab under gentle pressure. A portion of each of these clinical specimens should be inoculated, at bedside, directly onto culture medium; the remainder should be expressed into a small tube containing 0.5 ml of sterile nutrient broth (BBL 11478, Bacto 0003, or equivalent) at approximately neutral pH and frozen for further reference. If no primary lesion is present or if the organism is not isolated from the primary lesion, 2 ml of sterile saline may be injected into enlarged, intact lymph nodes, and the material can be withdrawn and cultured. As subsequently described, a portion of the sample should be frozen and stored. Biopsy specimens can be cultured directly; however, excision of enlarged lymph nodes, even during therapy, may cause a severe constitutional reaction (54). Viable *F. tularensis* may be isolated from sputum, gastric aspirates, and pharyngeal washings during the systemic phase of tularemia regardless of the type of clinical disease; culturing these materials as well as pleural fluid and bronchial secretions is recommended primarily when typhoidal or pulmonary tularemia is suspected. Sputum, gastric aspirates, and pharyngeal washes are obtained in the early morning before the patient drinks water or brushes his or her teeth. To secure pharyngeal washes, the patient is requested to gargle about 15 ml of sterile nutrient broth and expectorate into a sterile container with cover. If possible, several daily samples should be examined. More consistent isolation has been obtained from sputum samples and gastric aspirates than from pharyngeal washes (49).

After collection, specimens must be kept at 10°C or lower to prevent overgrowth by normal flora; procedures for isolation should be performed immediately or as soon as possible. Because direct culture procedures might fail, a portion of the clinical specimens should be held at −30 to −70°C. The rate of freezing is unimportant, but thawing should be accomplished rapidly at 37°C (43). The size of the container should limit the volume of air as much as is practical. Specimens for transmittal to another laboratory for testing should be packaged with dry ice in a container that conforms to postal regulations.

Direct examination

Direct or indirect fluorescent-antibody techniques are considered the best tools for rapid and specific identification of *F. tularensis* in exudates, tissue impressions, and tissue sections; even preparations preserved in Formalin or paraffin may be employed. Ordinary light microscope examination of dye-stained clinical specimens is unproductive.

Culture and isolation

When enriched with defibrinated rabbit blood or outdated, packed, human blood cells, commercially prepared glucose-cysteine agar with thiamine or cystine heart agar is satisfactory for growing *F. tularensis* from most clinical specimens. Chocolate agars which are enriched with supplements containing cystine or cysteine also may be satisfactory for growth of the organism. Cultures isolated from blood have been received for identification with the statement that the organism could be subcultured from the blood culture medium only on chocolate agar. The incorporation of antibiotics (penicillin, polymyxin B, and cycloheximide) is required when clinical specimens containing normal flora are cultured for *F. tularensis*. One strain of *F. tularensis* referred to CDC had been isolated from a mixed bacterial population in a finger lesion by inoculation on Thayer-Martin medium. Others also report that media used for the isolation of gonococci can be used for the isolation of *F. tularensis* (3).

Three strains of *F. tularensis* isolated from pulmonary specimens on charcoal-yeast extract agar supplemented with α-ketoglutarate have been referred to CDC. These isolates did not grow on unsupplemented charcoal-yeast extract agar. In addition, some strains of *F. tularensis* have grown on regular blood agar (65). The clinical microbiologist, therefore, should not assume that a culture is not *F. tularensis* because it is able to grow on blood agar.

Formulas for the recommended media can be found in Chapter 111. Occasionally a marginal lot of medium can be improved to yield more rapid and abundant growth by the addition of 0.5 g of ferrous sulfate per liter of medium. Every precaution should be observed to restrict the formation of surface moisture on plates, a condition that is deleterious to the growth of *F. tularensis*. Each lot of media should be quality control tested by inoculation with avirulent *F. tularensis* ATCC 6223, a strain with more fastidious growth requirements than virulent strains.

Laboratory personnel must use a syringe (needle removed) or a pipette fitted with a safety suction device to transfer potentially highly infectious material to the medium. The volume of inoculum should not exceed 0.2 to 0.3 ml per plate. Small volumes of clinical specimens that have a potentially high concentration of *F. tularensis* as well as normal flora (ulcer, draining lymph nodes, or conjunctival scrapings) should be streaked with a wire loop to ensure that isolated colonies are obtained. Gastric aspirates, pharyngeal washes, pleural fluid, lymph node perfusions, or excised tissue (processed in Ten Broeck grinders with 2 ml of sterile saline) should be transferred to the surface of the medium at the center of the plate and thoroughly spread with a sterile U-shaped glass rod. At least 3 ml of gastric aspirates or pharyngeal washes should be cultured because of the relatively low concentration of *F. tularensis* cells. Plates are inverted and incubated at 35 to 37°C. Increased CO_2 is not required but is not harmful.

Direct culture of adequate amounts of clinical specimens on appropriate growth medium is usually sufficient for the isolation of *F. tularensis* and has the advantage of being more rapid and less hazardous than the inoculation of laboratory animals. However, if adequate facilities (including restricted personnel entry and the use of clear plastic cages fitted with filter tops to be opened only in a vented hood) are available for housing highly infectious animals, appropriate samples of clinical specimens that fail to yield isolates on direct culture can be inoculated intraperitoneally into guinea pigs. The presence of one to five viable cells of *F. tularensis* usually results in death within 5 to 10 days; the spleen pathology (enlargement four to five times normal size and studding with numerous, minute, gray foci of necrosis) is pathognomic of tularemia. Moribund animals should be cultured immediately; *F. tularensis* can be readily isolated from heart blood, spleen, and liver.

Identification

Clinical specimens containing massive numbers of viable *F. tularensis* cells or tissues from infected guinea pigs yield confluent, smooth, gray growth within 18 h when cultured on recommended media. When relatively few organisms are present, pinpoint colonies may appear as early as 24 h after inoculation and develop into 1.0- to 1.5-mm, smooth, gray colonies within 48 h. Incubation for 72 to 96 h results in colonies 3 to 4 mm in diameter. The medium immediately surrounding confluent growth or isolated colonies is characteristically green in appearance. A Gram stain of the growth usually reveals closely aligned but individual, faintly staining, minute, gram-negative, coccoidal forms. Closer inspection of well-separated organisms at the periphery of the stained area will demonstrate that the coccoidal forms are actually the bipolarly staining ends of the rod-shaped organism separated by an even more faintly staining central area. Characteristic morphology is more readily observed in Giemsa-stained preparations.

F. tularensis produces acid but not gas from glucose, maltose, mannose, and fructose. Acid production from glycerol (50) and the possession of a citrulline ureidase system (42) serve to distinguish *F. tularensis* biovar *tularensis* from the less virulent *F. tularensis* biovar *palaearctica*. Biochemical characterization, however, is not necessary or recommended for identification.

Important differential criteria for the identification of *F. tularensis* are: absence of growth on ordinary media and little or no growth on blood-enriched media that lack added cysteine or cystine, relatively slow colony growth on special media, distinctive cellular morphology (small size, in particular) and staining properties, specific fluorescent antibody reaction, and slide agglutination with specific antiserum.

Serological confirmation is routinely and rapidly accomplished by slide agglutination. All strains of *F. tularensis* agglutinate with specific antiserum. Commercially available concentrated antigen can be used undiluted to prepare a positive control; the negative control should show no clumping. Cross-reactions are rare and are characterized by slow and incomplete clumping.

Animal inoculation is not required or recommended for the routine identification of *F. tularensis* isolates.

Immunoserological diagnosis

The agglutination test performed with patient serum is standard and reliable and usually becomes positive early in week 2 of infection. In the absence of a previously known infection, a titer of 1:40 is considered diagnostic, but diagnosis must be confirmed by a rising agglutinin titer. By week 3, titers usually reach 1:320 or higher. Antibiotic therapy, initiated before serological confirmation of the diagnosis, does not prevent the development of a diagnostic titer.

Sera from patients with tularemia or brucellosis may show minor cross-reactions; therefore, a control with *Brucella* antigen should be included. Serum that agglutinates both organisms to the same or a similar titer should be subjected to agglutinin absorption tests (13).

Serologically, even in the absence of other data, rising agglutinin titers in successive tests are diagnostic. If acute-phase serum is not available but relatively high agglutinin titers are observed during convalescence and a negative history of previous tularemia infection or vaccination is ascertained, data are reported as indicative of tularemia.

Laboratory safety

Although numerous laboratory infections have occurred in nonvaccinated laboratory personnel routinely working with high concentrations of *F. tularensis* in cultures and in tissues, trained technicians using reasonable precautions and readily available safety equipment can perform diagnostic procedures with minimal risk. It is mandatory that precautions be taken to prevent the creation of aerosols and the contamination of the skin. All work with potentially infectious clinical materials and cultures of *F. tularensis* should be performed in a vented hood; surgical gloves should be worn to prevent skin contact with the organism. Also, special care must be observed in the housing of animals inoculated with clinical specimens and in the examination of their tissues.

An innocuous but highly effective live vaccine has been developed for the immunization of individuals at high risk, including laboratory animal caretaker personnel. The live vaccine is not directly available commercially but can be obtained by consultation with the CDC, Attention: Immunobiologics Activity, Atlanta, GA 30333.

KINGELLA SPECIES

The proposal to create the genus *Kingella* was made in 1976 by Henriksen and Bøvre (23). The first species to be named in the genus was *K. kingae*. Snell and Lapage (60) proposed two additional species, *K. denitrificans* and *K. indologenes*. All three species were shown to utilize glucose fermentatively in OF medium supplemented with 5% (vol/vol) horse serum (60).

K. kingae

E. O. King first listed the characteristics of *K. kingae* and designated it as *Moraxella*, new species 1. Subsequently it was named *Moraxella kingii* (22) (corrected to *M. kingae*) before being moved to the genus *Kingella*. The cells of *K. kingae* are coccoid to medium-sized rods with square ends and occur in pairs and short

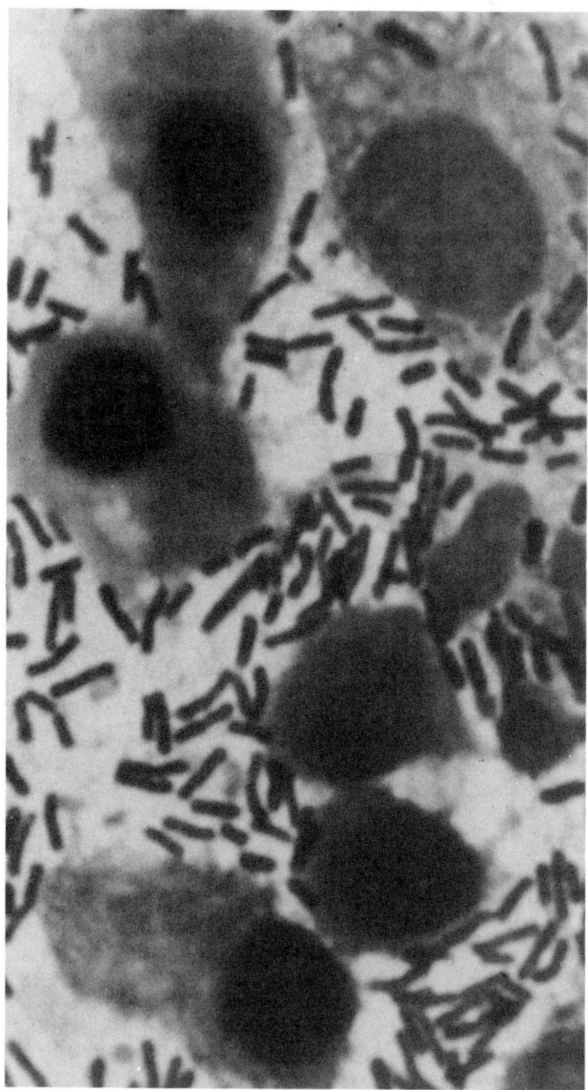

FIG. 10. Smear of purulent exudate showing characteristic morphology of *Kingella kingae*, consisting of plump rods with square ends. Diplococcal and diplobacillary forms predominate in smears of agar-grown cultures (×1,000).

chains (Fig. 10). They are gram negative but have a tendency to resist decolorization. Colonies on blood agar plates have a zone of clear, betalike hemolysis around them. After 18 to 24 h of incubation, the colonies are 0.5 mm or less in diameter. One or both of two colony types may be present. One type pits or corrodes the agar and may have a thin, spreading zone of growth around the edge of the colony. The second type does not have a spreading edge and does not pit the agar.

K. kingae does not grow well on TSI agar and usually does not acidify either the slant or the butt. Acid is produced from glucose and maltose in liquid peptone medium. No acid is produced from xylose, mannitol, lactose, or sucrose. For some strains, the carbohydrate broths have to be supplemented with serum to obtain sufficient growth for the acid to be detected.

These microorganisms are oxidase positive, catalase negative, urease negative, and nonmotile (Table 5). About one-third grow on MacConkey agar. Simmons citrate medium is not alkalinized. Nitrate is reduced by less than 10% of the strains, and indole is not formed. The key reactions for the identification of *K. kingae* are catalase, oxidase, fermentation of glucose and maltose, hemolysis, urease, and indole.

The natural habitat is probably the human oropharynx. The sources of isolation of cultures referred to CDC indicate that clinically significant infections occur. Of 75 isolates, 35 were from blood, 21 were from bone- or joint-associated sites, and 14 were from the throat. There has been one report of *K. kingae* as the cause of endocarditis in a 4-year-old boy with a ventricular septal defect (10). *K. kingae* is very susceptible to penicillin, erythromycin, chloramphenicol, oxytetracycline, ampicillin, streptomycin, carbenicillin, and gentamicin (5).

K. denitrificans

K. denitrificans was first described by Hollis et al. (27) and was designated TM-1. In 1976, Snell and Lapage (60) proposed the name *K. denitrificans*, noting, as had Hollis et al. (27), similarities between it and *K. kingae*.

The designation TM-1 was given to the organism because it was first recognized on Thayer-Martin medium inoculated with throat swabs taken in a survey for carriers of *Neisseria meningitidis* and *Neisseria lactamica*. The microorganism probably is rarely the cause of an infection. Over 80% of the strains submitted to CDC have been isolated from the throat and other sites in the upper respiratory tract. Five isolates were from rectal and genitourinary sources, and two isolates were from blood.

Clinical microbiologists should become aware of this bacterium because it may be mistakenly identified as *Neisseria gonorrhoeae* on the basis of its ability to grow on Thayer-Martin medium, colonial morphology, oxidase reaction, and carbohydrate reactions. In some strains, coccoid forms predominate, increasing the difficulty in differentiating it from *N. gonorrhoeae*.

In a Gram stain, *K. denitrificans* is composed of gram-negative, coccoid to short, straight rod-shaped cells. Pairs and short chains may be observed, and occasionally a long rod-shaped bacterium is seen. After 24 h of incubation on blood agar, the colonies are usually 0.5 mm or less in diameter, circular, low convex, and semitranslucent. Colonies of some strains pit the surface of agar media. The colonial morphology of strains that do not pit the agar surface is similar to the colonial morphology of *N. gonorrhoeae* and *N. lactamica*.

K. denitrificans is oxidase positive, catalase negative, nonproteolytic, and nonmotile (Table 5). The erythrocytes in blood agar medium usually are unaffected, but occasionally a slight green discoloration may be seen. The microorganism does not grow well on TSI medium, and an acid reaction is not produced in the butt of the medium.

Acid is produced slowly (usually 3 to 7 days) in liquid peptone medium containing glucose. The addition of 2 drops of rabbit serum to each tube may be necessary to ensure the adequate growth of some strains. Acid is not produced from xylose, mannitol, lactose, sucrose, or maltose. False-positive reactions have been obtained in maltose when horse serum was used to supplement the medium (26). Maltase in the serum converts maltose to glucose. Of an additional 21 carbohydrates tested (27), galactose was the only other carbohydrate utilized (three of eight strains).

Indole is not produced; decarboxylation of lysine and ornithine and dihydrolation of arginine do not occur. Nitrate is reduced; gas is produced by 88% of the strains.

TABLE 5. Characteristics of *Kingella* species

Characteristic	*K. denitrificans* (n = 60)		*K. kingae* (n = 33)	
	Sign[a]	%+[b]	Sign	%+
Hemolysis (clear zone)	−	0	v	84
Motility	−	0	−	0
Gas from glucose	−	0	−	0
Acid from:				
Glucose	+ or (+)	30 (62)	+ or (+)	42 (48)
Xylose	−	0	−	0
Mannitol	−	0	−	0
Lactose	−	0	−	0
Sucrose	−	0	−	0
Maltose	−	0	+ or (+)	76 (24)
Catalase	−	10	−	0
Oxidase	+	100	+	100
Growth on MacConkey agar	−	0	v	33 (3)
Simmons citrate	−	0	−	0
Urease	−	0	−	0
Nitrate reduction	+	93	−	3 (3)
Gas from nitrate	v	88	−	0
Indole	−	0	−	0
Gelatin hydrolysis	−	0	−	0
TSI[c] slant, acid	−	2	v	28 (4)
TSI butt, acid	−	0	v	12
Esculin hydrolysis	−	0	−	0

[a] Sign: +, 90% or more positive in 1 or 2 days; −, no reaction, 90% or more; + or (+), 90% or more positive, some strains positive after 3 or more days; v, more than 10% and less than 90% positive.

[b] Numbers in parentheses indicate percentage of delayed reactions (3 days or more).

[c] TSI, Triple sugar iron.

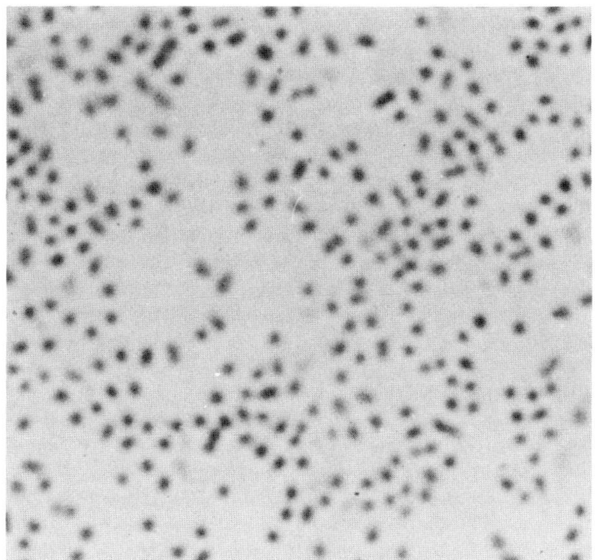

FIG. 11. Smear of *Pasteurella multocida* showing characteristic small coccoid cells with occasional bacillary forms (×1,200).

The key reactions for the identification of *K. denitrificans* are growth on Thayer-Martin medium, acid formation from glucose, oxidase, catalase, nitrate reduction, and indole. Differentiation from the phenotypically similar organism EF-4 is made on the basis of the catalase and arginine dihydrolase reactions. Nitrate reduction is a key reaction for differentiating between strains of *K. denitrificans*, with a coccoid cellular morphology, and *N. gonorrhoeae*. The resistance of *K. denitrificans* to amylase is another characteristic that differentiates it from *N. gonorrhoeae* (48).

K. indologenes

K. indologenes has never been identified in our laboratory. The two strains examined by Snell and Lapage (60) were isolated from eye infections. *K. indologenes* is catalase negative, oxidase positive, and nonmotile, as are the other *Kingella* species. Indole is formed, and acid is produced from glucose, sucrose, and maltose. Xylose and mannitol are not fermented.

PASTEURELLA SPECIES

The genus *Pasteurella* is listed along with the genera *Haemophilus* and *Actinobacillus* in the family *Pasteurellaceae* in *Bergey's Manual of Systematic Bacteriology* (40). Antibiotic susceptibility results show these organisms to be quite susceptible to penicillin. Tetracycline and chloramphenicol also are effective, particularly against strains of *P. multocida*.

P. multocida

Characterization. *P. multocida*, the *Pasteurella* species most frequently associated with human infections, is a small coccobacillary and rod-shaped organism (Fig. 11) which can be differentiated from other *Pasteurella* species on the basis of indole, urease, and ornithine decarboxylase reactions.

P. multocida is a pathogen of several species of animals as well as of humans. Infections in humans generally can be included in one of three categories (62): (i) localized infections consisting of cellulitis, abscesses, or adenitis which may progress to osteomyelitis of the underlying bone when they have been produced by penetrating dog or cat bites; (ii) chronic pulmonary infections in which the organism is the primary pathogen or is associated with other organisms; and (iii) systemic disease or meningitis.

Types of infections from which *P. multocida* has been isolated include appendiceal abscess, peritonitis, septic arthritis, liver abscess, puerperal sepsis, brain abscess, and renal infections.

The majority of infected animal bite or scratch wounds have been inflicted by dogs or cats. In two reports (12, 29) cat-associated wounds were approximately two to three times as frequent as wounds inflicted by dogs. Other animals that have caused bite wounds from which *P. multocida* was isolated include lion, opossum, rat, panther, and rabbit.

In most instances, patients from whom *P. multocida* was isolated had a history of contact with animals or animal tissue. However, in an investigation of 100 isolates, no history of contact with these sources could be obtained from 31% of the patients (30). Johnson and Rumans (33) have published a report of unusual infections caused by *P. multocida* and a review of earlier reports.

P. multocida apparently is a commensal in the mouths, throats, and noses of a variety of wild mammals and birds. It also has been isolated from the respiratory tract of humans in the absence of apparent infection.

Identification. (i) Cultural characteristics. The majority of cultures examined in our laboratory have produced smooth convex colonies of butyrous consistency on blood agar (Fig. 12). Some strains, particularly from the respiratory tract, produce watery to mucoid colonies (Fig. 13). The diameter of the colonies after 18 to 24 h of incubation is approximately 1 to 2

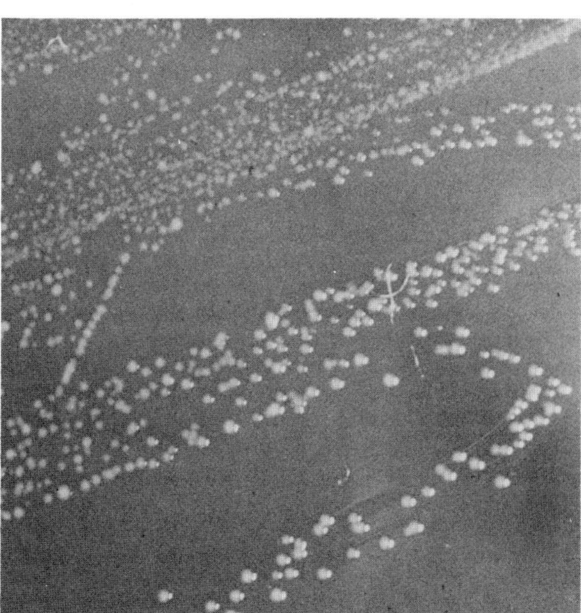

FIG. 12. Blood agar culture (48 h) of *Pasteurella multocida* showing small (1- to 2-mm), smooth, easily emulsifiable colonies (×10).

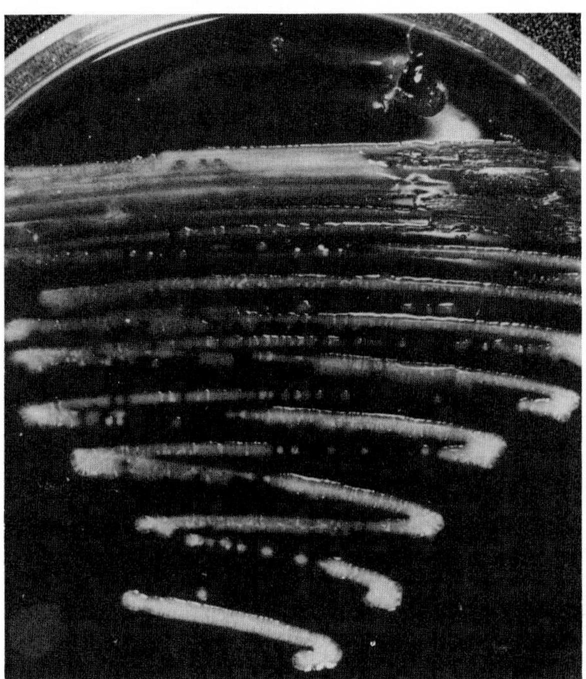

FIG. 13. Mucoid colonies of *Pasteurella multocida* often seen with isolates recovered from the respiratory tract of patients with chronic pulmonary disease. Blood agar culture (×3).

mm. There is no lysis of the blood cells, but there may be a greenish discoloration of the blood agar. A distinctive odor, possibly related to indole, is usually present.

In Gram-stained smears from heart infusion agar or blood agar with a heart infusion base, the organisms are predominately coccoid and very small. From dextrose-starch agar, rods ranging from 0.6 to 5.0 mm have been observed (20). The rod forms may have a bipolar staining reaction. However, this reaction is not as prominent in Gram-stained smears as it is in smears stained with Wright and Giemsa stains. The cells may be single or in pairs and occasionally in short chains. Capsules may be discerned in smears of purulent exudates (Fig. 14).

(ii) Biochemical reactions. Biochemical reactions for *P. multocida* are listed in Table 6. *P. multocida* strains produce acid reactions in the butts of Kligler or TSI slants as a result of fermentative activity. However, fermentation of glucose frequently cannot be demonstrated in OF medium containing glucose because many strains do not grow sufficiently well in the medium.

The oxidase reaction is positive when growth on an 18- to 24-h blood plate is tested with a 5% aqueous solution of tetramethyl-*p*-phenylenediamine dihydrochloride. If commercially available oxidase test strips or the tetramethyl reagent containing stabilizing agents is used, the reaction may be negative. Each laboratory should quality control its oxidase testing

TABLE 6. Biochemical characteristics of species of *Pasteurella*

Characteristic	*P. multocida* (*n* = 306)		*Pasteurella* sp. new species 1 (*n* = 91)		*P. pneumotropica* (*n* = 107)		*P. ureae* (*n* = 97)		*P. haemolytica* (*n* = 67)		*P. aerogenes* (*n* = 16)	
	Sign[a]	%+[b]	Sign	%+	Sign	%+	Sign	%+	Sign	%+	Sign	%+
Hemolysis (clear zone)	−	0	−	0	−	0	−	0	v	72	−	0
Motility	−	0	−	0	−	0	−	0	−	0	−	0
Gas from glucose	−	0	v	16	−	0	−	0	−	0	+	100
Acid from:												
Glucose	+	100	+	100	+	97 (3)	+	100	+	96 (4)	+	100
Xylose	v	67	−	0	+ or (+)	76 (19)	−	0	v	66 (3)	v	81
Mannitol	v	78	−	0	−	2 (1)	+	99 (1)	v	30 (9)	−	6
Lactose	−	8	−	3	v	14 (39)	−	0	v	7 (34)	v	19 (38)
Sucrose	+	100	+	99	+	97 (3)	+	99 (1)	+	97 (3)	+	94
Maltose	−	2	+	100	+	97 (3)	+	91 (5)	+ or (+)	85 (13)	+	100
Catalase	+	98	+	96	+	100	v	63	v	84 (2)	+	100
Oxidase	+	97	+	98	+	99	+	99	+	91	+	100
Growth on MacConkey agar	−	2 (1)	−	1	v	36 (17)	−	0	v	79 (6)	+	100
Simmons citrate	−	0	−	0	−	0	−	0	−	0	−	0
Urease	−	0	v	78 (3)	+	95 (1)	+	100	−	0	+	100
Nitrate reduction	+	99	+	100	+	100	+	99	+	100	+	100
Gas from nitrate	−	0	−	0	−	0	−	0	−	0	−	0
Indole	+	99	+	100	+	90	−	0	−	0	−	0
Gelatin hydrolysis	−	0	v	13	−	0	−	0	v	5 (7)	−	0
TSI[c] slant, acid	+	98 (1)	+	100	+	100	+	100	+	100	+	94
TSI butt, acid	+	99	+	100	+	97	+	99	+	100	+	100
Esculin hydrolysis	−	0	−	0	−	0	−	0	v	23	−	0
Lysine decarboxylase	−	0	−	2	v	33	−	0	−	3	−	0
Arginine dihydrolase	−	0	−	0	−	0	−	0	−	0	−	0
Ornithine decarboxylase	+	94	−	0	+	100	−	0	−	10	v	88

[a] Sign: +, 90% or more positive in 1 or 2 days; −, no reaction, 90% or more; + or (+), 90% or more positive, some strains positive after 3 or more days; v, more than 10% and less than 90% positive.

[b] Numbers in parentheses indicate percentage of delayed reactions (3 days or more).

[c] TSI, Triple sugar iron.

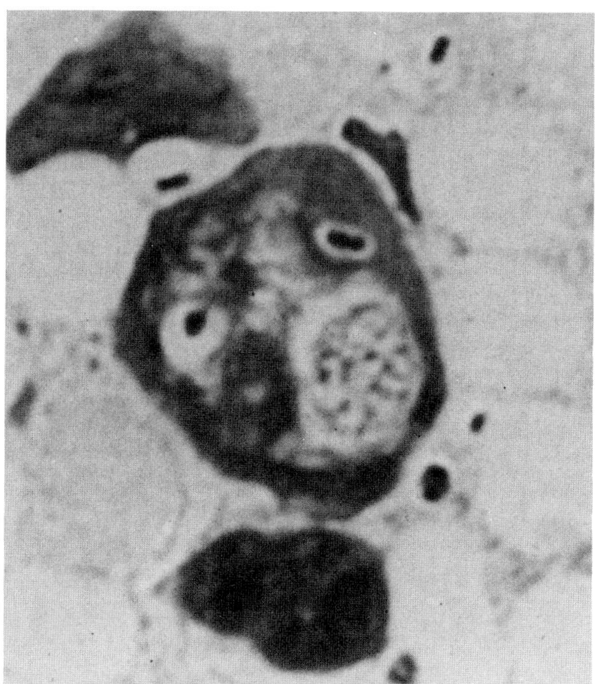

FIG. 14. Direct smear of purulent exudate after cat bite showing small, encapsulated, coccobacillary forms of *Pasteurella multocida*.

method with documented strains of *P. multocida*.

Additional major biochemical characteristics include: nitrate reduction without the formation of gas, indole production, fermentation of glucose and sucrose without gas formation, usually negative fermentation reactions in lactose and maltose, absence of growth on MacConkey agar, a negative urease reaction, and a positive ornithine decarboxylase reaction.

There is strain-to-strain variability in the fermentation of xylose, mannitol, lactose, and maltose. The various patterns of fermentation observed at CDC are presented in Table 7. The reactions were determined in a peptone base with Andrade indicator. Similar results have been obtained in a study of 30 isolates from humans and over 1,000 isolates from nonhuman sources, with phenol red broth (Difco Laboratories) as the base for carbohydrate media (20).

Serological typing has been done on the bases of mouse protection tests, capsular or surface polysaccharides in agglutination and passive hemagglutination procedures, and O antigens in gel diffusion techniques. Serological typing has been performed primarily in veterinary microbiology laboratories.

As indicated above, *P. multocida* can be differentiated from other *Pasteurella* species on the basis of indole, urease, and ornithine decarboxylase reactions. The catalase, nitrate, sucrose, indole, and urea reactions differentiate it from other genera of nonmotile, fermentative, gram-negative bacteria which may be oxidase positive and do not grow on MacConkey agar.

P. pneumotropica group

P. pneumotropica is the name proposed by Jawetz (32) for a group of organisms isolated from pneumonic lesions produced in laboratory mice by rapid, serial intranasal passage of mouse lung tissue. Frederiksen (14), however, has indicated that at least three biotypes of organisms have been identified as *P. pneumotropica*. He studied strains from Heyl (25), Jawetz (32), and Henriksen (21). At CDC we have identified organisms having the characteristics of the cultures described by Jawetz and by Heyl as *P. pneumotropica*. Cultures that appear to be the same as those described by Henriksen are identified as *Pasteurella* sp. new species 1 or "gas," a term used by the late Elizabeth O. King. The two designations *P. pneumotropica* (Jawetz and Heyl) and *Pasteurella* sp. new species 1 will be used in this section. A better definition of the relationships of these organisms probably will not be obtained until DNA homology studies are performed.

***P. pneumotropica* (Jawetz and Heyl).** *P. pneumotropica* (Jawetz and Heyl) appears to be of little importance as a cause of disease in humans. The strains examined by Jawetz and by Heyl were isolated from mice. Of 95 cultures referred to CDC, 62 were isolated from mice, 11 from rats, and 13 from other small animals. Nine cultures were isolated from human sources. These sources included foot wound, urine, throat, pleural fluid, wound caused by a rabbit bite, lymph node, and cellulitis. The specific source was not provided for one of the human isolates.

P. pneumotropica is a small, gram-negative, nonmotile rod. After 18 to 24 h of incubation at 35°C on blood agar, colonies are low convex and may vary in diameter from approximately 0.5 to 1.5 mm. *P. pneumotropica* can be differentiated from *P. multocida* on the basis of urease, maltose, and mannitol reactions (Table 6). Differentiation from *P. ureae*, *P. haemolytica*, and *P. aerogenes* can be made on the basis of indole, gas from glucose, and urease reactions. The most useful reactions differentiating *P. pneumotropica* and *Pasteurella* sp. new species 1 are gas from glucose, ornithine decarboxylase, and xylose.

***Pasteurella* sp. new species 1.** *Pasteurella* sp. new species 1 has also been called *Pasteurella* "gas" and the Henriksen biotype of *P. pneumotropica* (15). The sources from which this organism has been isolated indicate it produces infections in humans and is transmitted by contact with animals, primarily dog and cat bites. The sources provided for 81 cultures referred to CDC were: dog bite, 24; wound, 17; abscess, 3; hand, 3; foot, 3; animal bite, 3; lung, 2; cat bite, 2; throat, 2; blood, 2; cat mouth, 3; dog, 2; miscellaneous, 15; and unknown, 1. On the basis of the description, it is apparent that Winton and Mair (68) isolated this organism from a dog bite wound. Gump and Holden (19) isolated the organism from five blood cultures obtained from a patient who had been bitten

TABLE 7. Characteristics of the biotypes of *Pasteurella multocida*

Acid from:	Biotype				
	n = 168	n = 66	n = 32	n = 24	n = 7
Glucose	+	+	+	+	+
Xylose	+	−	−	+	+
Mannitol	+	−	+	+	+
Lactose	−	−	−	+	−
Sucrose	+	+	+	+	+
Maltose	−	−	−	−	+

on the hand by a cat and subsequently developed endocarditis.

Identification. *Pasteurella* sp. new species 1 grows readily on blood agar and infusion agar. After 18 to 24 h at 35°C, colonies are convex, translucent, and approximately 1 to 2 mm in diameter. Blood cells are not hemolyzed. On Gram stain, gram-negative, short, and coccoidal rods will be observed.

The major biochemical characteristics are listed in Table 6. A unique characteristic for a *Pasteurella*-like organism is the formation of gas from glucose and other carbohydrates. Only 16% of the strains we examined produced a volume of gas large enough to be detected in the fermentation vial. Thirty-seven percent of the strains (15 of 43) were shown to produce gas when cultures in thioglycolate broth containing glucose were tested by the hot-needle technique (37). The major characteristics are fermentation of glucose, positive oxidase reaction, no growth on MacConkey agar, reduction of nitrate, formation of indole, usually hydrolysis of urea, and a negative ornithine decarboxylase reaction. The indole, gas from glucose, ornithine decarboxylase, and urease reactions differentiate *Pasteurella* sp. new species 1 from the other *Pasteurella* species. It can be differentiated from *Actinobacillus* species on the basis of the indole reaction. Serological relationships to other *Pasteurella* and *Actinobacillus* species have not been studied.

P. ureae

In a study of *Pasteurella*-like organisms isolated from the respiratory tract of humans, Henriksen and Jyssum (24) described three strains of an apparently new variety or species and proposed the name *P. haemolytica* var. *ureae*. After studying an additional 18 strains, Jones (35) concluded that they represented a distinct species and proposed that the name be *Pasteurella ureae*, which is the name used in the current edition of *Bergey's Manual of Systematic Bacteriology* (40).

No host other than humans has been identified. Most of the strains listed in published reports have been isolated from the respiratory tract. These strains have been isolated from patients with ozena, rhinosinusitis, and bronchitis. In most instances, other bacteria also have been present, and the role, if any, of *P. ureae* in the pathological conditions has not been clear. In one study (35), *P. ureae* was isolated from about 1% of sputum specimens received in a period of 8 months. Of 97 cultures received in our laboratory, 65 (67%) were isolated from sputum. Other sources included bronchus, throat, trachea, nose, and sinus. Two strains were isolated from blood. Reports of the isolation of *P. ureae* from sites other than the respiratory tract are rare. It has been isolated in pure culture, however, from the blood of a patient with septicemia (17), from the spinal fluid of two patients who developed meningitis secondary to a skull fracture (38, 64), and from the blood and spinal fluid of a patient who developed an infection after intracranial surgery (4).

Identification. *P. ureae* is a pleomorphic gram-negative rod. It tends to be wider than most species of *Pasteurella*. The cells of cultures grown on solid media vary from short to long rods with a tendency to form filaments. Usually some of the cells contain unstained vacuoles. After incubation for 18 to 24 h, well-isolated colonies on blood agar are approximately 1 to 1.5 mm in diameter. They are smooth and low convex. There is no hemolysis of the blood cell.

The biochemical characteristics are listed in Table 6. A key characteristic is the urease reaction. The Christensen slant should be inoculated heavily with growth from a blood plate. The color change begins within a few minutes. *P. ureae* is oxidase positive and does not grow on MacConkey agar medium. The reactions that differentiate it from other urease-positive *Pasteurella* species are the fermentation of mannitol and the inability to form indole or decarboxylate ornithine. Differentiation from the MacConkey-negative strains of the animal-associated *Actinobacillus* species is based on the greater saccharolytic activity of the *Actinobacillus* species.

P. haemolytica

P. haemolytica has several characteristics that are common to all of the *Pasteurella* species that were recognized in the 8th edition of *Bergey's Manual* (40). These characteristics include fermentation of carbohydrates without the formation of gas, positive oxidase reaction, and negative motility (Table 6). *P. haemolytica* is the only *Pasteurella* species, however, which does not produce either indole or urease. The indole and urease reactions also differentiate *P. haemolytica* from the recently proposed species *P. aerogenes* and the organisms we have referred to as *Pasteurella* sp. new species 1. In addition, the urease reaction is the major characteristic for differentiating *P. haemolytica* from the animal-associated species of *Actinobacillus*.

Freshly isolated strains produce a zone of clear hemolysis on blood agar, a characteristic that may be lost or reduced after a few transfers on laboratory media. When received at CDC, 48 of 62 strains were hemolytic. Two biotypes, A and T, are recognized. Some of the characteristics that differentiate the two biotypes are listed in Table 8.

P. haemolytica causes infections primarily in cattle and sheep. It probably is part of the normal flora of these animals. We have received 11 strains as human isolates. One was isolated from a woman who developed an ulceroglandular infection after she shot and dressed a deer (46). The isolation of *P. haemolytica* from the blood of a patient with endocarditis has also been reported (11).

TABLE 8. Differentiation of the biotypes of *Pasteurella haemolytica*

Characteristic	Type A (n = 46)		Type T (n = 21)	
	Sign[a]	%+[b]	Sign	%+
Acid from:				
Xylose	+	96 (4)	−	0
Lactose	v	11 (50)	−	0
Trehalose	−	0	v	80
Mannose	−	3	+	95 (5)
Esculin hydrolysis	−	0	v	83

[a] Sign: +, 90% or more positive reactions; −, 90% or more negative reactions; v, 11 to 89% positive reactions.

[b] Numbers in parentheses indicate percentage of delayed reactions (3 days or more).

P. aerogenes

P. aerogenes was proposed by McAllister and Carter (44) as the name for 25 strains of a gram-negative bacterium that produced gas from carbohydrates and were isolated from the intestines and organs of swine with various diseases. In Gram stains from colonies on blood agar, cells vary in size from 0.5 to 1.0 μm wide and 1.1 to 2.0 μm long. Occasionally, filaments up to 15 mm long are seen. Colonies on blood agar range in size from approximately 0.5 to 1.0 mm in diameter after 24 h at 35 to 37°C. They are nonhemolytic, circular, smooth, entire, convex, and translucent.

The key biochemical tests for identifying *P. aerogenes* (Table 6) are: oxidase (positive), growth on MacConkey agar (positive), motility (negative), urease (positive), nitrate reduction (positive), indole formation (negative), and ornithine decarboxylase (positive).

The pathogenicity of *P. aerogenes* in swine is uncertain. Only 1 of the 25 strains studied by McAllister and Carter was thought to be a primary pathogen. The organism was recovered in large numbers from several organs of aborted fetuses and was considered the probable cause of the abortion.

We have received eight human isolates at CDC. Four of the isolates were from wounds caused by swine bites, and two isolates were from wounds for which the cause was not stated. The other two isolates were from urine and peritoneal fluid.

GROUP DF-2

Characterization

DF-2 is the designation assigned in the Special Bacterial Pathogens Laboratory, CDC, to an unidentified group of gram-negative, rod-shaped bacteria collected in this laboratory and described by Butler et al. (9). All but one of the isolates from humans referred to CDC had been isolated from blood or spinal fluid. The one exception was isolated from a wound. Cellular morphology varies from strain to strain; however, the predominant forms are usually long thin rods, frequently curved and exhibiting a variable degree of pleomorphism. Growth on heart infusion agar is enhanced by blood or serum and usually by incubation in a candle jar.

The DF-2 organism apparently is transmitted by contact with the mouth or oral secretions of dogs. Of the 17 patients with DF-2 infections reviewed by Butler et al. (9), 10 had histories of recent dog bites. Bailie et al. (2) have demonstrated the presence of the organism in the oral flora of 4 of 50 dogs (selective isolation procedures were not used). Patients with underlying diseases appear to be at greater risk of infection than "normal" individuals. Underlying diseases were identified in 14 of the 17 patients. Splenectomy appears to be a significant risk factor. Five of the 17 patients had had splenectomies, a higher incidence of splenectomy than is found in the general population. In the splenectomized patients, the infection tended to be more severe. In two of them, the organisms were numerous enough to be observed in smears of peripheral blood. Three patients died. Clinical features observed included endocarditis in three patients and meningitis in four.

Information received with subsequent cultures submitted to the Special Bacterial Pathogens Laboratory has continued to indicate splenectomy as an important predisposing factor. There have been, however, infections and even deaths of individuals in whom no underlying conditions could be demonstrated.

Direct examination

The presence of long thin rods in blood smears should alert one to the possibility of a DF-2 infection. Specific identification, however, cannot be made in this manner.

Culture and isolation

It is not known whether some blood culture media are superior to others for supporting the growth of this organism. Difficulties, however, in subculturing from blood cultures to agar media have been reported. In one instance, initial subcultures on agar media grew only under anaerobic conditions. After a few transfers under anaerobic conditions, the strain could be grown in a candle extinction jar. Another laboratory isolated the organism in tryptic soy broth but could not obtain growth on an agar medium. At CDC the organism was successfully subcultured on heart infusion agar with 5% rabbit blood in a candle extinction jar. By comparison, growth on tryptic soy agar with 5% sheep or rabbit blood was very poor. At present, therefore, we would recommend heart infusion agar with 5% rabbit or sheep blood for subculturing blood cultures containing organisms with morphology comparable to that of DF-2. Incubation should be in a candle extinction jar or CO_2 incubator at 35 to 37°C. Growth occurs, but to a lesser degree, under anaerobic conditions (GasPak, BBL).

Identification

The microscopic morphology and the appearance of growth on heart infusion agar with 5% rabbit blood should suggest the identity of this organism. At 18 to 24 h, the majority of cells are from 1 to 4 μm in length. Longer rods and filaments are formed by many strains. The longer forms are frequently curved. Most rods are uniform in diameter and have tapered ends. In addition, spindle- and cigar-shaped rods may be observed (Fig. 15). An occasional thickened cell is crescent shaped.

At 18 to 24 h on heart infusion agar with rabbit blood, growth will be moderate to heavy and will have a purplish cast. The colonies are punctate. During an additional 18 to 24 h of incubation, well-isolated colonies enlarge to 2 to 3 mm in diameter.

Colonies are convex, smooth, and circular. Because DF-2 does not grow well in TSI agar or OF glucose medium, reactions in these two media usually do not indicate that the organism produces acid from carbohydrates. To obtain consistent carbohydrate results, we supplement 3 ml of carbohydrate broth with 0.1 ml of rabbit serum. The use of a relatively heavy inoculum is also recommended. The carbohydrate reactions can be demonstrated within 4 h by the rapid fermentation test (7).

The production of acid from glucose in the broth medium occurs even if the inoculated broth is covered with a layer of stiff petrolatum. This result indicates that carbohydrates are probably fermented. Acid is

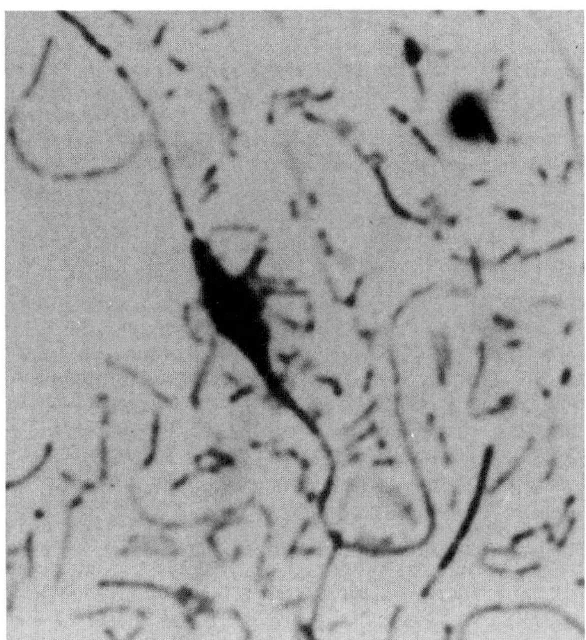

FIG. 15. Smear of DF-2 showing pleomorphic morphology consisting of irregularly stained rods and filament forms with funduliform swellings.

produced from glucose, lactose, and maltose, but not from xylose, sucrose, or mannitol. All strains that have been tested are *o*-nitrophenyl-β-D-galactopyranoside positive.

Other important reactions are (Table 9): catalase positive, oxidase positive, no growth on MacConkey agar or Simmons citrate medium, urease negative, nitrate not reduced, and motility negative. The arginine dihydrolase reaction is usually positive. The best results have been obtained in this test when 1-ml volumes of the Moeller medium contained in tubes (13 by 100 mm) were inoculated very heavily with growth from an 18- to 24-h blood agar plate.

Based on cellular morphology, the aerobic organisms that DF-2 most closely resembles are the *Capnocytophaga* species. Differentiation from these organisms can be made on the basis of the oxidase, catalase, sucrose, nitrate, and arginine reactions (Tables 3 and 9).

DF-2 is susceptible to several antimicrobial drugs, including penicillin and erythromycin. It is resistant to gentamicin, colistin, and kanamycin (9).

GROUPS EF-4a AND EF-4b

Group EF-4a is composed of gram-negative, short, rod-shaped bacteria. Small coccoid forms, long rods, and chains of four to seven cells also may be observed. These bacteria are eugonic, and their metabolism of carbohydrates is fermentative.

Bacteria of group EF-4 may have no effect on the cells in blood agar plates, or they may produce an indeterminate reaction. A slight green discoloration of the blood cells in this medium may occur. After 24 h of incubation, the colonies average 1.0 mm in diameter and are convex, entire, circular, semiopaque, and smooth. A popcornlike odor is usually present. The growth may have a slight yellow to orange pigmentation. The slant of TSI medium is alkalinized after 24 to 48 h of incubation, and most cultures (73%) produce acid in the butt of the medium. There is no evidence of hydrogen sulfide production in TSI medium.

Members of this group are nonmotile and are oxidase and catalase positive (Table 9). Some strains produce a slight, yellow to tan, soluble pigment. Strains vary in their ability to grow on MacConkey agar. When they do grow, they usually grow lightly. Simmons citrate medium usually is not alkalinized. Urea is not hydrolyzed in Christensen medium, and indole is not formed. Arginine is dihydrolyzed by 79% of the cultures; lysine and ornithine are not decarboxylated. Gelatin is liquefied by 79% of the strains. When tests are made after 48 h of incubation, 97% of the strains reduce nitrate. Gas is produced from nitrate by 62% of the strains, and 13% completely reduce both nitrate and nitrite without producing gas. If peptone is used rather than heart infusion as the basal medium for the nitrate test, the complete reduction of both nitrate and nitrite is less likely to occur. A distinctive characteristic of group EF-4a cultures is

TABLE 9. Characteristics of unclassified groups DF-2, EF-4, and HB-5

Characteristic	DF-2 (n = 27)		EF-4 (n = 97)		HB-5 (n = 44)	
	Sign[a]	%+	Sign	%+	Sign	%+
Hemolysis (clear zone)	−	0	−	0	−	0
Motility	−	0	−	0	−	0
Gas from glucose	−	0	−	0	+	100
Acid from:						
Glucose	v	85	+	100	+	100
Xylose	−	0	−	0	−	0
Mannitol	−	0	−	0	−	0
Lactose	+	100	−	0	−	0
Sucrose	−	0	−	0	−	0
Maltose	+	100	−	0	−	0
Catalase	+	100	+	100	−	2
Oxidase	+	96	+	100	v	54
Growth on MacConkey agar	−	0	v	42 (8)[b]	v	34 (23)
Simmons citrate	−	0	−	3 (1)	−	0
Urease	−	0	−	0	−	0
Nitrate reduction	−	0	+	97	+	100
Gas from nitrate	−	0	v	62	−	0
Indole	−	0	−	0	+	100
Gelatin hydrolysis	−	0	v	79	−	0
TSI[c] slant, acid	v	17	−	3	+	100
TSI butt, acid	v	15	v	73	+	95
Esculin hydrolysis	v	77	−	0	−	0
Lysine decarboxylase	−	0	−	0	−	0
Arginine dihydrolase	v	85	v	77 (2)	−	0
Ornithine decarboxylase	−	0	−	0	−	0

[a] Sign: +, 90% or more positive in 1 or 2 days; −, no reaction, 90% or more; v, more than 10% and less than 90% positive.

[b] Numbers in parentheses indicate percentage of delayed reactions (3 days or more).

[c] TSI, Triple sugar iron.

the inability to ferment any carbohydrates other than glucose.

Of 97 strains of EF-4a microorganisms, 77 were isolated from human wounds. At least 32 of the wounds were caused by dog bites, and 7 wounds were caused by cat bites. Information concerning the cause of the other wounds was not received. A few cultures were isolated from the oropharynx of dogs and cats. We have not received any strains isolated from blood or spinal fluid.

Group EF-4b consists of a collection of 35 strains that have growth characteristics and cellular morphology which are similar to those of group EF-4a strains. The major difference is that these strains do not ferment glucose. Acid is produced from glucose, however, in the open tube of OF medium. Gas is not produced in the reduction of nitrate, and arginine is not dihydrolyzed. Members of this group of organisms have also been isolated from human wounds caused by dog and cat bites. Further studies are needed to elucidate the relationship of this group of organisms to group EF-4a.

GROUP HB-5

The arbitrary designation "group HB-5" is applied to a group of fermentative microorganisms which are quite uniform in their biochemical characteristics. The group is composed of gram-negative, coccoid to rod-shaped bacteria of medium length.

Recommended conditions for growth are incubation for 18 to 24 h under increased carbon dioxide tension at 35 to 37°C. Under such conditions, colonies of blood agar plates usually are 0.5 to 1.0 mm in diameter, smooth, entire, and convex in appearance. Sometimes colonies are mottled. There is no hemolysis of the blood cells. Occasionally, however, there is a slight green discoloration of the cells, particularly in stabbed areas.

HB-5 bacteria are fermentative and produce acid throughout in TSI medium in 18 to 24 h, although neither lactose nor sucrose is fermented. Apparently there is not sufficient growth of the organism to cause reversion of the pH of the slant. Some strains produce a small amount of gas in the butt of this medium. Hydrogen sulfide is not produced in the medium.

The oxidase reaction is variable (when positive, it is usually weak), and the catalase reaction is negative (Table 9). About 34% of the strains grow lightly on MacConkey medium within 48 h, and an additional 23% grow in 3 to 7 days. Growth does not occur on Simmons citrate medium. Urea is not hydrolyzed, lysine and ornithine are not decarboxylated, and arginine is not dihydrolyzed. Indole is produced in small amounts; hence, extraction with xylol is necessary before tests are made. Nitrate is reduced to nitrite.

Glucose, fructose, and mannose are fermented with the production of small volumes (10% or less of Durham insert) of gas. Some isolates also ferment galactose and glycerol, but acid production is weak.

The major source of isolation of 59 HB-5 cultures referred to CDC has been the genitourinary tract (urine, vagina, cervix, penis, Bartholin gland, placenta, amniotic fluid, and urethra). Five isolates were from blood. Other sources of isolation included finger lesions, rectal abscess, perianal furuncle, surgical incision infection (appendectomy), and leg abscess.

LITERATURE CITED

1. **Affias, S., A. West, J. Stewart, and E. V. Haldane.** 1978. *Actinobacillus actinomycetemcomitans* endocarditis. CMA Journal **118**:1256–1260.
2. **Bailie, W. E., E. C. Stowe, and A. M. Schmitt.** 1978. Aerobic bacterial flora of oral and nasal fluid of canines with reference to bacteria associated with bites. J. Clin. Microbiol. **7**:223–231.
3. **Berdal, B. P., and E. Søderlund.** 1977. Cultivation and isolation of *Francisella tularensis* on selective chocolate agar, as used routinely for the isolation of gonococci. Acta Pathol. Microbiol. Scand. Sec. B **85**:108–109.
4. **Bia, F., R. Marier, W. F. Collins, Jr., and A. von Graevenitz.** 1978. Meningitis and bacteremia caused by *Pasteurella ureae*. Report of a case following intracranial surgery. Scand. J. Infect. Dis. **10**:251–253.
5. **Bosworth, D. E.** 1983. *Kingella (Moraxella) kingae* infections in children. Am. J. Dis. Child. **137**:650–653.
6. **Breed, R. S., E. G. D. Murray, and N. R. Smith (ed.).** 1957. Bergey's manual of determinative bacteriology, 7th ed. The Williams & Wilkins Co., Baltimore.
7. **Brown, W. J.** 1974. Modification of the rapid fermentation test for *Neisseria gonorrhoeae*. Appl. Microbiol. **27**:1027–1030.
8. **Buchanan, R. E., and N. E. Gibbons (ed.).** 1974. Bergey's manual of determinative bacteriology, 8th ed. The Williams & Wilkins Co., Baltimore.
9. **Butler, T., R. E. Weaver, T. K. V. Ramani, C. T. Uyeda, R. A. Bobo, J. S. Ryu, and R. B. Kohler.** 1977. Unidentified gram-negative rod infection. A new disease of man. Ann. Intern. Med. **86**:1–5.
10. **Christensen, C. E., and G. C. Emmanouilides.** 1967. Bacterial endocarditis due to "*Moraxella* new species 1." N. Engl. J. Med. **277**:803–804.
11. **Doty, G. L., G. N. Loomus, and P. L. Wolf.** 1963. *Pasteurella* endocarditis. N. Engl. J. Med. **268**:830–832.
12. **Francis, D. P., M. A. Holmes, and G. Brandon.** 1975. Infections after domestic animal bites and scratches. J. Am. Med. Assoc. **233**:42–45.
13. **Francis, E., and A. C. Evans.** 1926. Agglutination, cross agglutination and agglutinin absorption in tularemia. Public Health Rep. **41**:1273–1295.
14. **Frederiksen, W.** 1973. *Pasteurella* taxonomy and nomenclature. Contrib. Microbiol. Immunol. **2**:170–176.
15. **Frederiksen, W.** 1981. Gas-producing species within *Pasteurella* and *Actinobacillus*, p. 185–196. *In* M. Kilian, W. Frederiksen, and E. L. Biberstein (ed.) *Haemophilus, Pasteurella* and *Actinobacillus*. Academic Press, Inc., New York.
16. **Friis-Møller, A.** 1981. A new *Actinobacillus* species from the human respiratory tract: *Actinobacillus hominis* nov. sp., p. 151–157. *In* M. Kilian, W. Frederiksen, and E. L. Biberstein (ed.), *Haemophilus, Pasteurella* and *Actinobacillus*. Academic Press, Inc., New York.
17. **Gatti, F., V. Seynhaeve, and R. Weaver.** 1968. Première description d'un cas de septicémie humaine due à *Pasteurella ureae*. Ann. Soc. Belge Med. Trop. **48**:463–468.
18. **Graves, M. G., J. M. Strauss, V. Abbas, and C. E. Davis.** 1969. Natural infection of gibbons with a bacterium producing violet pigment (*Chromobacterium violaceum*). J. Infect. Dis. **120**:609–610.
19. **Gump, D. W., and R. A. Holden.** 1972. Endocarditis caused by a new species of *Pasteurella*. Ann. Intern. Med. **76**:275–278.
20. **Heddleston, K. L., and G. Wessman.** 1975. Characteristics of *Pasteurella multocida* of human origin. J. Clin. Microbiol. **1**:377–383.
21. **Henriksen, S. D.** 1962. Some *Pasteurella* strains from the human respiratory tract. A correction and supplement. Acta Pathol. Microbiol. Scand. **55**:355–356.
22. **Henriksen, S. D., and K. Bøvre.** 1968. *Moraxella kingii* sp. nov., a hemolytic, saccharolytic species of the genus *Moraxella*. J. Gen. Microbiol. **51**:377–385.

23. **Henriksen, S. D., and K. Bøvre.** 1976. Transfer of *Moraxella kingae* Henriksen and Bøvre to the genus *Kingella* gen. nov. in the family *Neisseriaceae*. Int. J. Syst. Bacteriol. **26**:447–450.
24. **Henriksen, S. D., and K. Jyssum.** 1960. A new variety of *Pasteurella haemolytica* from the human respiratory tract. Acta Pathol. Microbiol. Scand. **50**:443.
25. **Heyl, J. G.** 1963. A study of *Pasteurella* strains from animal sources. Antonie van Leeuwenhoek J. Microbiol. Serol. **29**:79–83.
26. **Hollis, D. G., R. E. Weaver, and P. S. Riley.** 1983. Emended description of *Kingella denitrificans* (Snell and Lapage 1976): correction of the maltose reaction. J. Clin. Microbiol. **18**:1174–1176.
27. **Hollis, D. G., G. L. Wiggins, and R. E. Weaver.** 1972. An unclassified gram-negative rod isolated from the pharynx on Thayer-Martin medium (selective agar). Appl. Microbiol. **24**:772–777.
28. **Holm, P.** 1951. Studies on the aetiology of human actinomycosis. II. Do the "other microbes" of actinomycosis possess virulence? Acta Pathol. Microbiol. Scand. **28**:391–406.
29. **Hubbert, W. T., and M. N. Rosen.** 1970. 1. *Pasteurella multocida* infection due to animal bite. Am. J. Public Health **60**:1103–1108.
30. **Hubbert, W. T., and M. N. Rosen.** 1970. II. *Pasteurella multocida* infection in man unrelated to animal bite. Am. J. Public Health **60**:1109–1117.
31. **Jacobs, R. F., and J. P. Narain.** 1983. Tularemia in children. Pediatr. Infect. Dis. **2**:487–491.
32. **Jawetz, E.** 1950. A pneumotropic *Pasteurella* of laboratory animals. I. Bacteriological and serological characteristics of the organism. J. Infect. Dis. **86**:172–183.
33. **Johnson, R. H., and L. W. Rumans.** 1977. Unusual infections caused by *Pasteurella multocida*. J. Am. Med. Assoc. **237**:146–147.
34. **Johnson, W. M., A. F. Disalvo, and R. R. Steuer.** 1971. Fatal *Chromobacterium violaceum* septicemia. Am. J. Clin. Pathol. **56**:400–406.
35. **Jones, D. M.** 1962. A *Pasteurella*-like organism from the human respiratory tract. J. Pathol. Bacteriol. **83**:143–151.
36. **King, E. O.** 1964. The identification of unusual pathogenic Gram-negative bacteria. Center for Disease Control, Atlanta, Ga.
37. **King, E. O., and H. W. Tatum.** 1962. *Actinobacillus actinomycetemcomitans* and *Haemophilis aphrophillus*. J. Infect. Dis. **111**:85–94.
38. **Kolyvas, E., S. Sorger, M. I. Marks, and C. H. Pai.** 1978. *Pasteurella ureae* meningitis. J. Pediatr. **92**:81–82.
39. **Kovacs, N.** 1956. Identification of *Pseudomonas pyocyanea* by the oxidase reaction. Nature (London) **178**:703.
40. **Krieg, N. R., and J. G. Holt (ed).** 1984. Bergey's manual of systematic bacteriology, vol. 1. Williams & Wilkins, Baltimore.
41. **Leadbetter, E. R., S. C. Holt, and S. S. Socransky.** 1979. *Capnocytophaga*: new genus of gram-negative gliding bacteria. 1. General characteristics, taxonomic considerations and significance. Arch. Microbiol. **122**:9–16.
42. **Marchette, N. J., and P. S. Nicholes.** 1961. Virulence and citrulline ureidase activity of *Pasteurella tularensis*. J. Bacteriol. **82**:26–32.
43. **Mazur, P., M. A. Rhian, and B. G. Mahlandt.** 1957. Survival of *Pasteurella tularensis* in sugar solutions after cooling and warming at sub-zero temperatures. J. Bacteriol. **73**:394–397.
44. **McAllister, H. A., and G. R. Carter.** 1974. An aerogenic *Pasteurella*-like organism recovered from swine. Am. J. Vet. Res. **35**:917–922.
45. **Meyers, B. R., E. J. Bottone, S. Z. Hirschman, S. S. Schneierson, and K. Gershengorn.** 1971. Infection due to *Actinobacillus actinomycetemcomitans*. J. Clin. Pathol. **56**:204–211.
46. **Muraschi, T. F., C. K. Smith, and J. K. Miller.** 1962. Primary cutaneous (ulceroglandular) infection due to *Pasteurella hemolytica*. N.Y. State J. Med. **62**:3137–3139.
47. **Newman, M. G., V. L. Sutter, M. J. Pickett, U. Blanchman, J. R. Greenwood, V. Grinenko, and D. Citron.** 1979. Detection, identification, and comparison of *Capnocytophaga, Bacteroides ochraceus*, and DF-1. J. Clin. Microbiol. **10**:557–562.
48. **Odugbemi, T., and R. J. Arko.** 1983. Differentiation of *Kingella dentrificans* from *Neisseria gonorrhoeae* by growth on a semisolid medium and sensitivity to amylase. J. Clin. Microbiol. **17**:389–391.
49. **Overholt, E. L., W. D. Tigertt, P. J. Kadull, and M. K. Ward.** 1961. An analysis of forty-two cases of laboratory-acquired tularemia. Am. J. Med. **30**:785–806.
50. **Owen, C. R.** 1970. *Francisella* infections, p. 468–483. *In* H. L. Bodily, E. L. Updyke, and J. O. Mason (ed.), Diagnostic procedures for bacterial, mycotic and parasitic infection, 5th ed. American Public Health Association, New York.
51. **Page, M. I., and E. O. King.** 1966. Infection due to *Actinobacillus actinomycetemcomitans* and *Haemophilis aphrophillus*. N. Engl. J. Med. **275**:181–188.
52. **Pulverer, G., and H. L. Ko.** 1970. *Actinobacillus actinomycetemcomitans*: fermentative capabilities of 140 strains. Appl. Microbiol. **20**:693–695.
53. **Pulverer, G., and H. L. Ko.** 1972. Serological studies on *Actinobacillus actinomycetem-comitans*. Appl. Microbiol. **23**:207–210.
54. **Saslaw, S., H. T. Eigelsbach, H. E. Wilson, J. A. Prior, and S. Carhart.** 1961. Tularemia vaccine study. I. Intercutaneous challenge. Arch. Intern. Med. **107**:689–701.
55. **Savage, D. D., R. L. Kagan, N. A. Young, and A. E. Horvath.** 1977. *Cardiobacterium hominis* endocarditis: description of two patients and characterization of the organism. J. Clin. Microbiol. **5**:75–80.
56. **Sivendra, R.** 1976. Unusual *Chromobacterium violaceum*: aerogenic strains. J. Clin. Microbiol. **3**:70–71.
57. **Slotnick, I. J., and M. Dougherty.** 1964. Further characterization of an unclassified group of bacteria causing endocarditis in man: *Cardiobacterium hominis* gen. et sp. n. Antonie van Leeuwenhoek J. Microbiol. Serol. **30**:261–272.
58. **Slotnick, I. J., J. A. Mertz, and M. Dougherty.** 1964. Fluorescent antibody detection of human occurrence of an unclassified bacterial group causing endocarditis. J. Infect. Dis. **114**:503–505.
59. **Slots, J., H. S. Reynolds, and R. J. Genco.** 1980. *Actinobacillus actinomycetemcomitans* in human periodontal disease: a cross-sectional microbiological investigation. Infect. Immun. **29**:1013–1020.
60. **Snell, J. J. S., and S. P. Lapage.** 1976. Transfer of some saccharolytic *Moraxella* species of *Kingella* Henriksen and Bøvre 1976, with descriptions of *Kingella indologenes* sp. nov. and *Kingella dentrificans* sp. nov. Int. J. Syst. Bacteriol. **26**:451–458.
61. **Socransky, S. S., S. C. Holt, E. R. Leadbetter, A. C. R. Tanner, E. Savitt, and B. F. Hammond.** 1979. *Capnocytophaga*: new genus of gram-negative gliding bacteria. III. Physiological characterization. Arch. Microbiol. **122**:29–33.
62. **Swartz, M. N., and L. J. Kunz.** 1959. *Pasteurella multocida* infections in man. Report of two cases—meningitis and infected cat bite. N. Engl. J. Med. **261**:889–893.
63. **Vandepitte, J., H. deGeest, and P. Jousten.** 1977. Subacute bacterial endocarditis due to *Actinobacillus actinomycetemcomitans*. J. Clin. Pathol. **30**:842–846.
64. **Wang, W. L. L., and G. Haiby.** 1966. Meningitis caused by *Pasteurella ureae*. Am. J. Clin. Pathol. **45**:562–565.
65. **Weber, M. L., N. J. Sawka, S. A. Halperin, and P. Ferrieri.** 1984. Oculoglandular tularemia. Clin. Microbiol. Newsl. **6**:36–37.
66. **Wetmore, P. W., J. F. Thiel, Y. F. Herman, and J. R. Harr.** 1963. Comparison of selected *Actinobacillus* spe-

cies with a hemolytic variety of *Actinobacillus* from irradiated swine. J. Infect. Dis. **113:**186–194.

67. **Williams, B. L., D. Hollis, and L. V. Holdeman.** 1979. Synonymy of strains of Center for Disease Control group DF-1 with species of *Capnocytophaga*. J. Clin. Microbiol. **10:**550–556.

68. **Winton, F. W., and N. S. Mair.** 1969. *Pasteurella pneumotropica* isolated from a dog-bite wound. Microbios **2:**155–162.

69. **Wormser, G. P., and E. J. Bottone.** 1983. *Cardiobacterium hominis*: review of microbiologic and clinical features. Rev. Infect. Dis. **5:**680–691.

Chapter 29

Glucose-Nonfermenting Gram-Negative Bacteria

SALLY JO RUBIN, PAUL A. GRANATO, AND BENEDICT L. WASILAUSKAS

About 15% of all isolates encountered in clinical microbiology laboratories are glucose-nonfermenting gram-negative bacteria (NFB). About two-thirds are *Pseudomonas* spp. The remaining 5% include named and unnamed groups that are a minority of significant isolates, but are important because of their role in hospital-acquired infections and frequent antimicrobial resistance.

The majority of the bacteria described in this chapter were characterized at the Centers for Disease Control (CDC) by the late Elizabeth O. King and her successors. These bacteria are a heterogeneous group that many laboratories find difficult to identify because they are frequently inert in currently used biochemical tests. In the future, carbon source utilization determinations, which test the ability of a bacterium to grow using a single, specific carbon source, may simplify identification. All of these organisms are gram-negative bacteria that do not ferment glucose, although a few may be weak fermenters. All are asporogenous and Voges-Proskauer negative. Cellular morphology, as well as the ability to oxidize glucose, varies. Each of the following groups will be discussed separately:

Eikenella corrodens
Acinetobacter
Alcaligenes
Achromobacter
CDC groups IVe and IVc-2
CDC group EO-2
CDC group EF-4b
Agrobacterium radiobacter
Flavobacterium, *Flavobacterium* spp. IIe, IIf, IIh, IIi, and IIj
Moraxella
CDC groups M5 and M6

Materials and Methods

A number of protocols have been suggested for the identification of NFB. The most widely used, however, is the one described by Miss King and her successors (40). Formulas for the media and reagents for identification of NFB are found in Chapter 111.

Primary plating media for isolation of NFB depend on the specimen source but must include a blood agar (BA) medium and a selective enteric agar, preferably MacConkey agar. Some workers use rabbit blood, whereas others prefer sheep blood.

Some NFB fail to grow well at 35°C. Primary isolation media incubated for 24 h at 35°C may be held for an additional day at room temperature (18 to 22°C) or at 30°C to permit growth of NFB. All other culture media used to study and characterize NFB should be incubated at 30°C (or room temperature) unless otherwise stated.

The growth of some NFB is enhanced by CO_2; therefore, blood-containing media should be incubated in an atmosphere of 3 to 10% CO_2 (a candle jar is satisfactory).

The identification of NFB is often delayed 24 h because enteric identification systems are inoculated first. Most NFB, however, are oxidase positive. Therefore, time can be saved by screening colonies with either 1 to 2 drops of a 0.5% aqueous solution of tetramethyl-*p*-phenylenediamine dihydrochloride or a commercially available oxidase strip (Pathotec-CO; General Diagnostics, Morris Plains, N.J.).

Identification steps necessarily differ between the clinical and the reference laboratory. In the clinical situation, time and cost are critical and the goal is an accurate identification, whereas the reference laboratory may be required to perform many more tests for taxonomic studies and to identify referred atypical strains.

A well-isolated colony from the primary BA medium is selected and used to inoculate the seven media listed in Table 1. These media allow determination of characteristics that can then be used to select appropriate further tests for specific identification. In the clinical laboratory, usually only five to seven additional tests are necessary for an accurate identification. The reference laboratory may use a large battery of tests for all unusual organisms.

Some gram-positive bacilli easily decolorize with the Gram stain. These gram-negative-appearing rods may be confused with NFB. String formation in 3% KOH (a positive test is string formation of 24-h growth from BA within 60 s) and inhibition of growth (any zone) around a 5-μg vancomycin disk may be used to confirm the Gram stain reaction of NFB (38). Gram-negative bacilli give a positive string test and are resistant to vancomycin, whereas gram-positive bacilli are negative in the string test and are susceptible to vancomycin. Some *Bacillus* spp. may give a false-positive string test but are susceptible to vancomycin. *Achromobacter* (group Vd) and *Agrobacterium radiobacter* give a false-negative string test but almost always stain gram negative and are vancomycin negative. *Moraxella* spp. and *Acinetobacter* spp. may be false-negative in both tests, but their presence should be suspected because they have a characteristic morphology and are nonmotile (38).

Cellular morphology of NFB varies considerably, and sometimes determining whether an organism is a coccobacillus or a coccus is difficult. Examining growth from the edge of the zone of inhibition around the ampicillin or penicillin disk from the susceptibility plate may help make this differentiation. Coccobacilli such as *Moraxella* are spindle shaped or form filaments, whereas gram-negative cocci granulate, enlarge, and may undergo lysis (7).

Fermentative organisms usually produce acid in the butt of a triple sugar iron agar (TSI) or Kligler iron agar slant. Oxidative organisms produce less acid and, often, enough alkaline products (by oxidative peptone metabolism) to neutralize the acid. Therefore, oxidation-fermentation (O-F) basal medium with

TABLE 1. Media for initial grouping of NFB

Medium	Inoculation procedure	Characteristic determined
Heart infusion broth		Motility, Gram reaction; inoculum for other tests (1 drop per tube)
BA	Streak and stab. Examine growth with hand lens.	Colonial morphology, odor, pigment, action in blood, indophenol oxidase
MacConkey agar	Streak.	Growth/no growth
TSI or Kligler iron agar slant	Stab butt and streak entire slant.	Hydrogen sulfide production, fermentation of glucose
O-F–glucose, O-F–maltose, O-F–lactose	Stab four times to about 0.5 in. (ca. 1.25 cm) below the surface.	Oxidation of glucose, lactose, maltose

1% carbohydrate is used to detect oxidative activity. This medium contains less peptone than TSI and 0.3% agar. The agar prevents dilution of any acid produced as well as keeping it concentrated in the medium at the site of production. The CDC laboratory prefers phenol red as an indicator, whereas others use bromthymol blue, which may be more sensitive but may also be somewhat inhibitory. In either case, the initial pH of the medium is critical; it should be 7.3 when phenol red is the indicator and 6.8 for bromthymol blue. Before inoculating a commercial O-F medium, boil and cool it to assure the proper agar colloid system. Before incubating, loosen screw-cap tops.

If an organism is a questionable fermenter, include an inoculated tube of O-F–glucose with a stiff petrolatum seal. Paraffin and liquid mineral oil seals are not as effective.

Motility can be determined by direct microscopic examination of a wet mount of a young culture or may be demonstrated more graphically using a semisolid agar medium. The amount of agar in motility medium must be no greater than 0.3%. Higher concentrations of agar impede spreading and prevent growth of strict aerobic bacteria at the bottom of the tube. Motility medium should be incubated at 18 to 20°C. Results in motility medium must be interpreted carefully. If motility is used as the sole criterion to delineate a species, a flagella stain is required.

After the characteristics listed in Table 1 have been determined, initial grouping of an organism can be made by following the key in Fig. 1. Once the isolate can be assigned to a group of organisms, other characteristics may be determined as necessary to determine the genus and species. Key reactions are shown in brackets in the tables.

Commercial systems for identification of NFB have been developed and evaluated. Such systems include API 20E (Analytab Products, Inc., Plainview, N.Y.), Minitek (BBL Microbiology Systems, Cockeysville, Md.), Oxi/Ferm (Roche, Inc., Nutley, N.J.), and Flow NF (Flow Laboratories, Roslyn, N.Y.). Results from evaluations of the same product vary. Some variation is probably due to whether stock strains, fresh clinical isolates, or both were used in the study. Often only small numbers of less frequently isolated species were tested. All four systems correctly identify at least 90% of the *Pseudomonas aeruginosa* and *Acinetobacter* spp. correctly, with few misidentifications or supplemen-

GLUCOSE NONOXIDIZERS
- MacConkey positive
  - Oxidase negative
    - Nonmotile: *Acinetobacter calcoaceticus* subsp. *lwoffi*
  - Oxidase positive
    - Motile: *Alcaligenes faecalis*, *Alcaligenes odorans*, *Alcaligenes denitrificans*, *Bordetella bronchiseptica*, CDC group IVe, CDC group IVc-2, *Achromobacter xylosoxidans* (Glu v)
    - Nonmotile: *Flavobacterium odoratum*, *Moraxella nonliquefaciens* (Mac r), *Moraxella osloensis* (Mac v), *Moraxella phenylpyruvica*, *Moraxella urethralis*, *Moraxella atlantae*, CDC group M-5 (Mac v), CDC group M-6 (Mac v)
- MacConkey negative
  - Oxidase positive
    - Nonmotile: *Eikenella corrodens*, *Flavobacterium* sp. IIf (Mac r), *Flavobacterium* sp. IIj, *Moraxella lacunata*, *Moraxella nonliquefaciens* (Mac r), *Moraxella bovis*, *Moraxella osloensis* (Mac v), CDC group M-5 (Mac v), CDC group M-6 (Mac v)

GLUCOSE OXIDIZERS
- MacConkey positive
  - Oxidase negative
    - Nonmotile: *Acinetobacter calcoaceticus* subsp. *anitratus*
  - Oxidase positive
    - Motile: *Achromobacter* spp. Vd-1 and Vd-2, *Achromobacter xylosoxidans* (Glu v), *Agrobacterium* spp.
    - Nonmotile: CDC group EO-2 (Mac v), CDC group EF-4b (Mac v), *Flavobacterium meningosepticum*, *Flavobacterium* sp. IIb (Mac v), *Flavobacterium breve*, *Flavobacterium* sp. IIf (Mac r), *Flavobacterium multivorum*, *Flavobacterium spiritovorum* (Mac v)
- MacConkey negative
  - Oxidase positive
    - Nonmotile: CDC group EO-2 (Mac v); CDC group EF-4b (Mac v); *Flavobacterium* spp. IIb (Mac v), IIe, IIh, and IIi; *Flavobacterium spiritovorum* (Mac v)

FIG. 1. Key to the identification of NFB (see Chapter 30 for *Pseudomonas* spp.). Abbreviations: Glu, glucose; Mac, MacConkey agar; r, 10 to 25% positive; v, 26 to 80% positive.

tal tests needed (2, 13, 25, 27, 34, 35, 39). With other NFB, results in all four systems were generally poor. Misidentifications were more common and supplemental tests were necessary more often.

A new system, API 20NE, consists of 8 conventional tests and 12 carbon source assimilation tests. No evaluations of the system have been reported; however, a preliminary study indicates it may be an improvement over current systems (A. von Graevenitz, personal communication).

Future studies of the biochemical capabilities of these organisms, such as carbon source utilization, should lead to improved identification schemes and improved, accurate commercial systems. Until then, diagnostic microbiology laboratories must rely on conventional methods such as those described in this chapter for identifying NFB.

EIKENELLA CORRODENS

Eikenella corrodens, formerly known as HB-1 (32), is a facultative anaerobic microorganism that should not be confused with the obligate anaerobic bacterium *Bacteroides corrodens* (now known as *Bacteroides ureolyticus*). Although each of these microbial species can produce colonies that pit or "corrode" the surface of agar media, careful studies have shown that *E. corrodens* and *B. corrodens* are distinct genetically, biochemically, and serologically and warrant separate taxonomic classification.

E. corrodens is a small (0.5 by 1 to 3 μm), nonsporeforming, nonencapsulated, nonmotile, microaerophilic, gram-negative coccobacillus that usually fails to grow on MacConkey agar, eosin methylene blue agar, or other similar selective media. In air, *E. corrodens* grows slowly on hemin-containing media such as BA or chocolate agar, but growth is enhanced by incubation in a 3 to 10% CO_2 atmosphere. Although not essential for primary isolation, CO_2 does stimulate growth. When grown aerobically, the organism appears to have an obligate requirement for hemin, but upon subculture variants that do not require hemin may be selected. When grown anaerobically, the organism does not require hemin, and growth on nonheme-containing media such as brain heart infusion agar is comparable to that observed on BA incubated in CO_2 or under anaerobiosis.

In humans, *E. corrodens* is considered part of the resident microflora of mucous membrane surfaces. Most frequently the organism is recovered as a commensal from specimens that originate or pass through the upper respiratory tract, but it may also be isolated from gastrointestinal or genitourinary tract specimens.

In recent years, *E. corrodens* infections have become increasingly recognized. Human infections usually result from predisposing factors such as trauma to a mucous membrane surface which compromises normal host defense mechanisms and allows the organism to gain access to surrounding tissue. Once primary infection occurs, hematologic dissemination may establish infectious foci in other parts of the body.

E. corrodens is usually involved in mixed bacterial infections, especially with streptococci and *Enterobacteriaceae*. Infections frequently involve the head and neck or abdominal area. *E. corrodens* may also be the sole infecting pathogen and has been reported in cases of endocarditis, meningitis, subdural empyema, septic arthritis, pneumonia, postsurgical infections, and soft-tissue abscesses. Recent reports have suggested that *E. corrodens* is a common pathogen in human bites and clenched-fist injuries that are frequently complicated by bone resorption and osteomyelitis.

E. corrodens can be isolated from a wide variety of clinical specimens. Most commonly the organism is recovered as a commensal from expectorated sputum and pharyngeal, mouth, and gingival specimens, but it may also be isolated from lower respiratory tract specimens, abscesses, wounds, blood, spinal fluid, and joint aspirates. Specimens should be inoculated onto BA or chocolate agar and incubated at 35°C in a 5 to 10% CO_2 environment. Recovery of *E. corrodens* from polymicrobic respiratory tract specimens may be improved by adding clindamycin (5 μg/ml) to a blood-supplemented medium (37).

Within 18 to 24 h of incubation, tiny, pinpoint colonies may be detected. Occasionally, prolonged incubation may be required before growth becomes apparent. Because *E. corrodens* is usually isolated from specimens with faster-growing organisms, plates should be inspected carefully. On BA or chocolate agar plates, colonies of *E. corrodens* may appear either smooth or rough. A distinguishing feature and a useful identification characteristic is the organism's ability to pit or "corrode" the surface of agar medium. However, the corroding phenomenon is not universally present, being observed only in about 45% of the isolates. When present, pitting of the agar surface is easily detected by observing the plate from different angles with oblique lighting. Occasionally, several colonies may have to be moved to detect the depression caused by their growth.

On BA plates, the colonial morphology of *E. corrodens* is distinctive. The organism produces a dry, flat, radially spreading colony with an irregular periphery and a moist central core. Upon prolonged incubation, the medium surrounding the colonies becomes slightly green. When observed under a stereoscopic microscope, three distinct zones of colonial growth are evident: (i) a clear, moist, glistening center zone; (ii) a highly refractile, speckled, pearllike circle of growth resembling mercury droplets; and (iii) an outer, nonrefractile perimeter of spreading growth. The typical colony morphology is shown in Fig. 2 and 3. On chocolate agar plates, the three zones of growth are less pronounced than on BA plates. Pure cultures of *E. corrodens* have been variously described as producing an odor similar to hypochlorite bleach or a musty odor similar to that produced by certain *Pasteurella* and *Haemophilus*-like organisms.

In liquid medium, microbial growth may vary with individual strains. Adding 0.1 to 0.2% agar or 10 μg of cholesterol per ml to the medium may improve growth. In thioglycolate broth containing yeast extract, a band of growth usually occurs a few centimeters below the surface after 3 days. In glucose broth supplemented with yeast extract, growth may develop as a uniform turbidity or as discrete granules adherent to the side of the tube (Fig. 4).

The biochemical characteristics that distinguish *E. corrodens* are listed in Table 2. The organism may be regarded as biochemically inactive, lacking oxidative

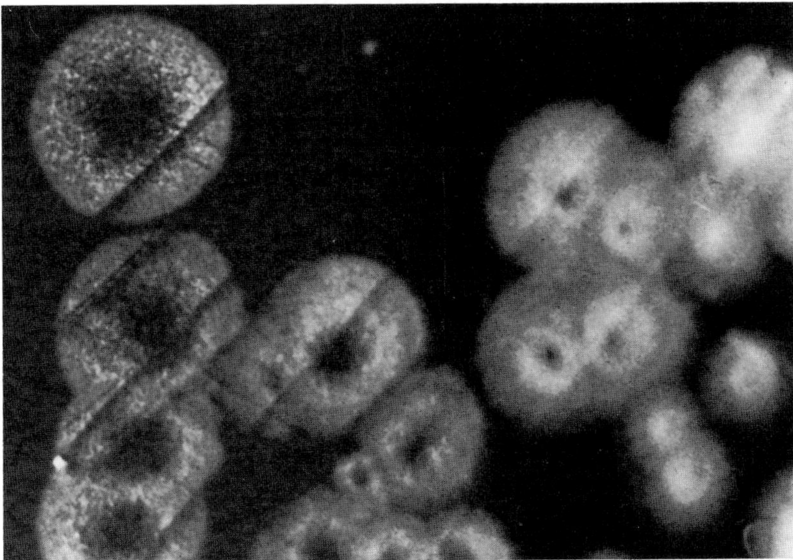

FIG. 2. Characteristic colonies of *E. corrodens* after growth on sheep blood agar after 48 h. ×2.5. From Bottone et al. (4) with permission.

or fermentative capabilities and failing to produce catalase, urease, indole, or H_2S. The organism is oxidase and nitrate positive, which are important distinguishing characteristics. Rare isolates may be weakly catalase positive or nitrate negative.

Four major antigenic components of *E. corrodens* have been recognized. All strains appear to be closely related antigenically, although some strains lack one or two of the major antigenic components. Presently, specific antisera are not available for serologic confirmation of suspected isolates.

In in vitro antibiotic susceptibility testing *E. corro-*

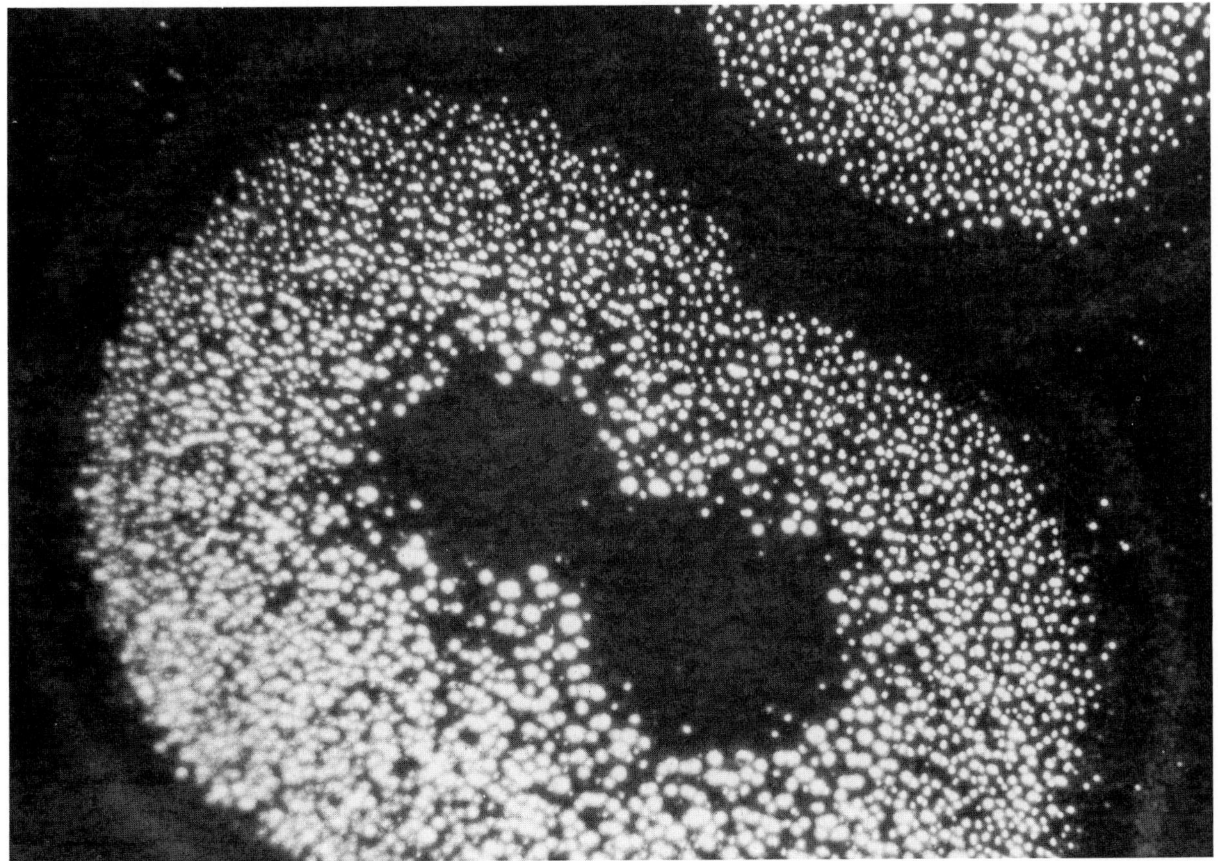

FIG. 3. Higher magnification of *E. corrodens* colony showing three characteristic zones of growth. ×112.5. From Bottone et al. (4) with permission.

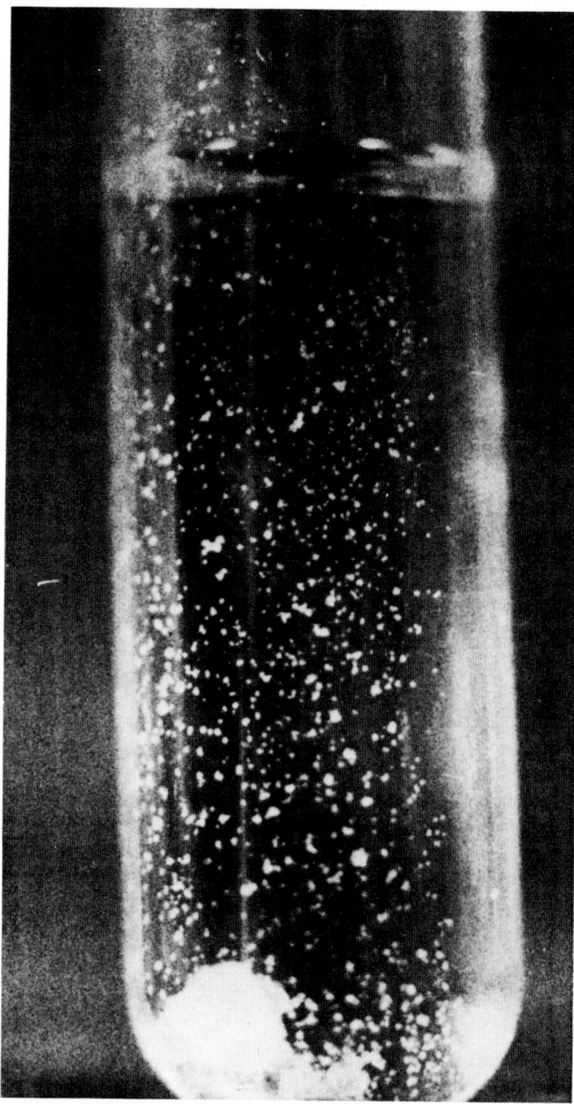

FIG. 4. Granular growth of *E. corrodens* with adherence to side of tube in supplemented glucose broth. Photomicrograph kindly provided by E. J. Bottone, Mt. Sinai Hospital, New York, N.Y.

dens is usually susceptible to penicillin, ampicillin, carbenicillin, chloramphenicol, tetracycline, cefoxitin, moxalactam, and *N*-formimidoyl thienamycin and is usually resistant to clindamycin, which may be a useful aid in its identification. Performing antibiotic susceptibility tests with this organism may be difficult because it grows poorly in liquid media, preventing proper inoculum standardization. This problem can be minimized by adding cholesterol or agar to the medium as previously mentioned. Once uniformly turbid growth is achieved, antibiotic susceptibility testing should be performed by the agar dilution or microtiter broth dilution method (17).

Animal inoculation studies are not helpful in determining identity or pathogenicity because *E. corrodens* does not produce disease in laboratory animals.

ACINETOBACTER

Acinetobacter consists of a single species, *calcoaceticus*, with two subspecies, *anitratus* and *lwoffi*. Many epithets (23) have been used to describe these subspecies, including *Mima* and *Herellea*.

The organisms are widely dispersed in nature and have been isolated from soil, water, and sewage. Both *Acinetobacter* subspecies are part of the normal flora of moist areas of human skin, such as the groin, axilla, and antecubital fossa. Occasionally they have been isolated from saliva and the pharynx. Community-acquired *Acinetobacter* pneumonia occurs uncommonly. Typically, patients are middle-aged men with some underlying disease, and most have been bacteremic (9). *Acinetobacter* has been a cause of both hospital epidemics and pseudoepidemics. The most common *Acinetobacter* nosocomial infection is pneumonia or tracheobronchitis, although infections at all body sites have been reported. Most patients are not bacteremic, unlike those with community-acquired pneumonia. Typical patients are middle-aged or older, in an intensive care unit, and on a respirator (16). These organisms are often transferred from the hands of hospital personnel and result in significant patient morbidity (22). Additionally, a variety of support care equipment, such as respirators and instruments used for invasive procedures, have been contaminated with *Acinetobacter* and continue to be a source for outbreaks of infection (26). Most infections have been caused by *A. calcoaceticus* subsp. *anitratus*.

On Gram-stained smears of primary specimens, *Acinetobacter* often appears as diplococci measuring 1.0 by 0.7 μm. Smears made from broth or agar cultures demonstrate more rodlike morphology, with cells approximately 2.0 by 1.2 μm. On ordinary labo-

TABLE 2. Biochemical characteristics of *Eikenella corrodens* (595 strains)[a]

Test performed	Sign[b]	% Positive
Oxidase	+	100
Catalase	−	9 (w)[c]
Growth on:		
MacConkey agar	−	0.8
SS agar	−	0
Cetrimide agar	−	0
Hydrogen sulfide on:		
TSI	−	0
Lead acetate paper	(+) or −	0 (64)[d]
Oxidation-fermentation	I	100
Urease	−	0
Indole	−	0
Methyl red/Voges-Proskauer	−	0
Citrate (Simmons), alkaline reaction	−	0
Motility	−	0
Gelatin	−	0
Acid from:		
Glucose	−	0
Xylose	−	0
Mannitol	−	0
Lactose	−	0
Sucrose	−	0
Maltose	−	0
Esculin	−	0
Nitrate to nitrite only	+	99.7
Pigment (pale yellow)	+	100

[a] See text for other characteristics.

[b] Sign: +, 90% or more positive in 1 or 2 days; −, no reaction (90% or more); (+) or −, most strains positive after 3 days or longer but some strains negative; I, inactive.

[c] (w), Weakly positive reaction.

[d] Percent positive reactions in 3 or more days.

ratory media these organisms grow well, producing gray-white colonies 2 to 3 mm in size after 18 to 24 h of incubation.

Both subspecies are gram-negative, nonmotile, nonsporeforming, oxidase-negative, aerobic bacteria. They are differentiated by their ability to attack carbohydrates: *A. calcoaceticus* subsp. *anitratus* oxidizes a variety of sugars, whereas *A. calcoaceticus* subsp. *lwoffi* is inert. *Acinetobacter* can be easily distinguished from other genera of NFB on the basis of negative motility and oxidase reactions. A complete list of biochemical reactions is found in Table 3. A serotyping system has been developed at CDC and may be useful for typing outbreak strains.

Both *Acinetobacter* subspecies are resistant to most antimicrobial agents and present a therapeutic challenge when encountered in acutely ill patients. Most strains are susceptible to aminoglycosides, carbenicillin, and trimethoprim-sulfamethoxazole.

ALCALIGENES

Bacteria of the genus *Alcaligenes* are nonsaccharolytic, oxidase positive, motile with peritrichous flagella, and obligately aerobic. *Bordetella bronchiseptica* (*bronchicanis*), CDC group IVc-2, and CDC group IVe are also nonsaccharolytic and peritrichously flagellated but can hydrolyze urea. Although *Achromobacter* and *Alcaligenes* are very similar, the peritrichously flagellated *Achromobacter* species produce acid from xylose and, usually, glucose. The nonsaccharolytic, oxidase-positive, motile *Pseudomonas* species have polar rather than peritrichous flagella. The nomenclature of *Alcaligenes* and *Achromobacter* is controversial.

TABLE 3. Biochemical characteristics of *Acinetobacter* spp.[a] (40)

Test performed	*A. calcoaceticus* subsp. *anitratus* (501)[b]		*A. calcoaceticus* subsp. *lwoffi* (253)[b]	
	Sign	% Positive	Sign	% Positive
Morphology	cc		cc	
Motility: flagella	[nm]		[nm]	
Action on BA	v	45 ly	v	48 ly
Fermentative or oxidative	[O]		[n-o]	
Carbohydrate base	O-F		O-F	
Acid from:				
Glucose	[+]	100	[−]	0
Xylose	+	99	−	0
Mannitol	[−]	2	−	0
Lactose	[+]	97 (2)	−	0
Sucrose	−	0	−	0
Maltose	v	63 (6)	−	0
Esculin hydrolysis	−	0	−	0
Nutrient broth:				
0% NaCl	+	100	[+]	97
6% NaCl	v	22	−	9
Catalase	+	100	+	100
Oxidase	[−]	0	[−]	0
Growth on:				
MacConkey agar	[+]	99 (1)	[+]	90 (7)
SS agar	v	20 (1)	v	7 (5)
Cetrimide agar	−	3	−	0
Simmons citrate	+	93	v	42
Urea, Christensen	v	28 (18)	−	5 (4)
Nitrate reduction	[−]	1	−	3
Gas from nitrate	−	0	−	0
Indole	−	0	−	0
TSI acid:				
Slant	−	5	−	0
Butt	−	0	−	0
H_2S:				
TSI butt	−	0	−	0
Lead acetate paper	v	32	v	58
Gelatin hydrolysis	−	9	−	4
Litmus milk	v	37 (10) A	v	86 k
Pigment	−	7 yel-ta sol	v	13 yel-br sol
Growth at:				
25°C	+	97	+	100
35°C	+	97	+	100
42°C	v	87	v	63

[a] Symbols: [ ], key reactions; ( ), late reaction, 3 to 7 days; +, positive reaction (90% or more strains tested were positive) within 48 h, except for gelatin, reaction within 14 days; −, negative reaction (10% or fewer strains tested were positive), except for action on BA, where it indicates no reaction; + or (+), positives and late positives together total 90% or more. Abbreviations (alphabetically): A, acid; cc, coccobacillus; k, alkaline; ly, diffuse lysis of erythrocytes extending out from heavy growth, no action by individual colonies; nm, nonmotile; n-o, nonoxidizer; O, oxidative; O-F, oxidation-fermentation; sol, soluble; v, variable (11 to 89% positive); yel-br, yellow-brown; yel-ta, yellow-tan.

[b] Number of strains tested.

TABLE 4. Biochemical characteristics of *Alcaligenes* spp.[a] (40)

Test performed	*A. faecalis* (69)[b]		*A. odorans* (49)[b]		*A. denitrificans* (34)[b]	
	Sign	% Positive	Sign	% Positive	Sign	% Positive
Morphology	mrs		mrs		mrs	
Motility: flagella	[m:pe]		[m:pe]		[m:pe]	
Action on BA	v	48 ly	v	48 br, gr	v	35 ly
Fermentative or oxidative	[n-o]		[n-o]		[n-o]	
Carbohydrate base	O-F		O-F		O-F	
Acid from:						
Glucose	[−]	0	[−]	0	[−]	0
Xylose	[−]	0	[−]	0	[−]	0
Mannitol	−	0	−	0	−	0
Lactose	−	0	−	0	−	0
Sucrose	−	0	−	0	−	0
Maltose	−	0	−	0	−	0
Catalase	+	100	+	98	+	100
Oxidase	[+]	100	[+]	100	[+]	100
Growth on:						
MacConkey agar	[+]	100	[+]	100	[+]	100
SS agar	[v]	78 (2)	[+]	100	[v]	65
Cetrimide agar	v	28 (1)	v	59	v	26 (12)
Simmons citrate	+	98 (1)	+	100	v	85 (3)
Urea, Christensen	[−]	0 (1)	[−]	2	v	12 (3)
Nitrate reduction	[v]	45	[−]	0	[+]	100
Gas from nitrate	[−]	0	−	0	[+]	100
Nitrite reduction	[−]	0	[+]	100	[+]	100
Indole	−	0	−	0	−	0
TSI acid:						
Slant	−	0	−	0	−	0
Butt	−	0	−	0	−	0
H_2S:						
TSI butt	[−]	0	−	0	[−]	0
Lead acetate paper	v	52	−	8	v	32
Gelatin hydrolysis	−	0	v	22	−	0
Litmus milk	k	98	k	96	k or (k)	88 (12)
Pigment	v	13 yel, br sol	v	22 yel sol	−	9 br sol
Growth at:						
25°C	+	100	+	100	+	100
35°C	+	100	+	100	+	100
42°C	v	75	v	18	v	56
Esculin hydrolysis	−	0	−	0	−	0
Lysine decarboxylase	−	0	−	0[c]	−	0[c]
Arginine dihydrolase	−	0	−	0[c]	−	0[c]
Ornithine decarboxylase	−	0	−	0[c]	−	0[c]
Nutrient broth:						
0% NaCl	+	100	+	100	+	100
6% NaCl	v	54 (6)	+	98 (2)	v	29

[a] Symbols: See Table 3, footnote *a*. Abbreviations (alphabetically): br, brown (pigment), browning of blood; gr, greening of blood; k, alkaline; ly, diffuse lysis of erythrocytes extending out from heavy growth; mrs, medium straight rods; m, motile; n-o, nonoxidative; O-F, oxidation-fermentation; pe, peritrichous (flagella); sol, soluble; v, variable (11 to 89% positive); yel, yellow.
[b] Number of strains tested.
[c] Less than 10 strains tested.

Bergey's Manual of Systematic Bacteriology (21) lists two clinically significant species, *Alcaligenes faecalis* and *Alcaligenes denitrificans*, and two subspecies of *A. denitrificans* (subsp. *denitrificans* and subsp. *xylosoxidans*). The subspecies *xylosoxidans* is a synonym for *Achromobacter xylosoxidans*. However, the Special Bacteriology Laboratory of the CDC lists three different species of clinically significant *Alcaligenes: A. faecalis, A. odorans*, and *A. denitrificans* (40). *Bergey's Manual* lists *A. faecalis* and *A. odorans* as synonyms (21). Tests to determine the characteristics listed in *Bergey's Manual* are not usually available in clinical laboratories, and therefore the Special Bacteriology Laboratory nomenclature is used in this chapter.

A. faecalis is the most frequently isolated species. Its habitat seems similar to that of *Pseudomonas* spp. Outside the hospital it is found in water and soil; within the hospital it may be found in moist items such as respirators, hemodialysis systems, and intravenous solutions. Some persons may carry this species as part of their normal skin flora. Little ecological information is available for *A. odorans* and *A. denitrificans*.

A. faecalis has been isolated from the blood of patients with and without symptoms of septicemia. *A. faecalis* bacteremia is generally associated with various predisposing factors. Infection is usually hospital acquired and is often associated with sources such as contaminated hemodialysis fluid and contaminated intravenous drugs. This species has also been isolated from sputum and urine.

In humans *A. odorans* has been isolated most often

TABLE 5. Biochemical characteristics of *Achromobacter* spp.[a] (40)

Test performed	*A. xylosoxidans* (135)[b]		Vd (*Achromobacter* sp.) (71)[b]	
	Sign	% Positive	Sign	% Positive
Morphology	mrs		mrs	
Motility: flagella	[m:pe]		[m:pe][c]	
Action on BA	v	42 ly	v	23 LG
Fermentative or oxidative	[O]		[O]	
Carbohydrate base	O-F		O-F	
Acid from:				
Glucose	[v]	78	[+ or (+)]	86 (13)
Xylose	[+]	99	+	96 (4)
Mannitol	−	0	v	46 (34)
Lactose	−	0	[−]	0
Sucrose	−	0	v	28 (25)
Maltose	−	0	v	32 (25)
Catalase	+	98	+	100
Oxidase	[+]	100	[+]	100
Growth on:				
MacConkey agar	[+]	100	[+]	100
SS agar	[+]	98	[+]	96 (1)
Cetrimide agar	+	95 (1)	−	0 (3)
Simmons citrate	+	95	v	54 (13)
Urea, Christensen	[−]	0	[+]	92 (8)
Nitrate reduction	[+]	100	+	100
Gas from nitrate	[v]	60	[+]	99
Indole	−	0	[−]	0
TSI acid:				
Slant	−	0	−	0
Butt	−	0	−	0
H_2S:				
TSI butt	[−]	0	v	49
Lead acetate paper	−	0	+	100
Gelatin hydrolysis	−	0	−	0
Litmus milk	k	95	v	85 k
Pigment	−	5 br sol	v	11 br-yel sol
Growth at:				
25°C	+	98	+	100
35°C	+	100	+	100
42°C	v	84	v	56
Esculin hydrolysis	−	0	v	38 (11)
Lysine decarboxylase	−	0	−	0
Arginine dihydrolase	v	13	v	68
Ornithine decarboxylase	−	0	−	0
Nutrient broth:				
0% NaCl	+	100	+	100
6% NaCl	v	69	v	45
3-Ketolactose production (3)			−	0[d]

[a] Symbols: See Table 3, footnote *a*. Abbreviations (alphabetically): br, brown; LG, lavender-green under heavy growth; m, motile; mrs, medium straight rods; O, oxidative; O-F, oxidation-fermentation; pe, peritrichous; v, variable (11 to 89% positive); yel, yellow; sol, soluble.
[b] Number of strains tested.
[c] Frequently, individual cells have only a single flagellum, either polar, subpolar, or lateral.
[d] Less than 10 strains tested.

from urine, but also from ear discharges, wounds, sputa, and feces. Mixed cultures are frequent. *A. denitrificans* has been isolated from blood, ear discharges, spinal fluid, and urine.

All three species grow well on MacConkey agar and use citrate as a sole carbon source. They do not produce indole or, in most cases, hydrolyze esculin or gelatin. Colonies vary in size after 24 h of incubation. *A. denitrificans* colonies are about 0.5 mm in diameter, low, convex, and glistening, and have an entire edge. Colonies of *A. odorans* range from 1 to 1.5 mm and are umbonate with a flat spreading edge. *A. faecalis* forms two morphologically different colony types. Type I colonies resemble those of *A. denitrificans*, and type II colonies are similar to *A. odorans*.

The three species are basically inert in currently used tests and can be differentiated from each other by only a few characteristics. About half of the *A. faecalis* strains grow in 6% NaCl broth and are able to reduce nitrate. Gas is not produced. *A. odorans* grows in the presence of 6% NaCl, does not reduce nitrate, but does reduce nitrite to gas. *A. odorans* produces a sweetish, fruity odor described as resembling strawberries or pared apples. Less than a third of *A. denitrificans* strains grow in 6% NaCl broth, but all reduce nitrate and nitrite to gas. Characteristics of these species are listed in Table 4.

Carbon substrate utilization profiles may provide additional means of differentiating these species. Rarick et al. (30) tested the ability of 162 *Alcaligenes*

TABLE 6. Differentiation of group IVe, group IVc-2, *Alcaligenes denitrificans*, *Bordetella bronchiseptica*, and *Brucella* spp.[a] (12, 40)

Test performed	Group IVe	Group IVc-2	*Alcaligenes denitrificans*	*Bordetella bronchiseptica*	*Brucella* spp.
Motility	+[b]	+[c]	+[c]	+[c]	−
Oxidase	+	+	+	+	+
Urea, Christensen	+	+	V[d]	+	+
Gas from nitrate	V[e]	−	+	−	V
Acid in O-F medium from:					
Glucose	−	−	−	−	V
Xylose	−	−	−	−	+

[a] For symbols and abbreviations, see Table 3, footnote *a*.
[b] One to two polar flagella and long lateral flagella; weakly motile; a few are nonmotile.
[c] Peritrichous flagella.
[d] Total of 12% positive at <48 h; 3% at ≥3 days.
[e] Total of 60% positive.

isolates to assimilate each of 188 different carbon sources. They found that cultures identified as *A. denitrificans* and *A. odorans* had unique carbon substrate profiles, whereas *A. faecalis* cultures could be divided into two different types. Type I strains were similar both morphologically and biochemically to the *A. denitrificans* cultures, whereas type II strains were similar morphologically to the *A. odorans* cultures except that they did not produce a dark green color on rabbit blood agar.

Alcaligenes is generally susceptible to trimethoprim-sulfamethoxazole. Susceptibility to aminoglycosides, ampicillin, chloramphenicol, carbenicillin, and cephalothin is variable. *A. denitrificans* tends to be more resistant to antimicrobial agents, especially aminoglycosides.

Animal pathogenicity testing has been performed only with *A. faecalis*. Guinea pigs and rabbits inoculated by various routes are not susceptible to infection, but if mucin is added to the inoculum, mice inoculated intraperitoneally die.

ACHROMOBACTER

The generic term *Achromobacter* is used to describe gram-negative, peritrichously flagellated bacteria that are oxidase positive, attack carbohydrates aerobically, and do not produce 3-ketolactose from lactose. Among taxonomists, the designation *Achromobacter* has been controversial and *Bergey's Manual of Systematic Bacteriology* does not accept *Achromobacter* as a genus name (21). Yabuuchi and Yano (43) proposed that *Achromobacter* be revised with *A. xylosoxidans* as the type species. This taxonomy is now recognized by the CDC, but in *Bergey's* it is listed as a synonym for *Alcaligenes denitrificans* biotype *xylosoxydans* (21). The name *Achromobacter*, however, is recognized by most clinical microbiologists and, therefore, is used in this chapter. *Achromobacter* spp. can be readily separated from similar bacteria. Other oxidase-positive, motile glucose oxidizers either have polar flagella (*Pseudomonas*) or produce 3-ketolactose from lactose (*Agrobacterium radiobacter*).

Presently two species, *A. xylosoxidans* and an unnamed species designated Vd, are recognized by the Special Bacteriology Laboratory of the CDC (Table 5). Cellular fatty acid analysis supports the division of *Achromobacter* sp. Vd from *A. xylosoxidans* (10).

The natural habitat of *Achromobacter* spp. has not been studied. They may be part of the indigenous flora of the large intestine, and they have been isolated from various hospital and environmental water sources. *A. xylosoxidans* has been isolated from many clinical sources including blood, urine, wounds, and sputa. Although many isolates of *A. xylosoxidans* are colonizers, some are clearly clinically significant. Reported infections include a case of recurrent purulent meningitis and several fatal bacteremias. Nosocomial outbreaks are usually associated with an aqueous source (nonbacteriostatic saline, hemodialysis fluid) (24, 31). Infected patients in these outbreaks all had compromised host defenses. In a number of instances *A. xylosoxidans* was isolated from purulent ear discharges but was frequently mixed with other gram-negative bacilli (28). If flagella stains are not done and *Pseudomonas* species other than *P. aeruginosa* are not determined, isolates of *Achromobacter* may be easily misidentified as *Pseudomonas* spp. Pien and Higa (28) suggest that some of the *Pseudomonas* spp. reported in acute otitis externa may actually be *Achromobacter*. *Achromobacter* group Vd has also been isolated from various clinical sources, but only one confirmed infection, a pancreatic abscess in a compromised patient, has been reported (1).

Yabuuchi and Yano (43) described in detail the type species *A. xylosoxidans* (formerly designated groups IIIa and IIIb). Colonies on BA are about 1 mm in diameter, smooth, and glistening, with an entire edge. Although oxidation of glucose may be weak, delayed, or negative, 99% of strains oxidize xylose. Almost all strains grow on cetrimide agar. Urease is not produced.

The unnamed *Achromobacter* strains (group Vd) differ from *A. xylosoxidans* in that all hydrolyze urea and none grow on cetrimide agar. Many reactions require at least 2 to 3 days to become positive. Some investigators recognize two Vd biotypes on the basis of their patterns of oxidative metabolism of certain carbohydrates. Biotype 2 produces acid from sucrose, maltose, and mannitol, whereas biotype 1 does not (14). Further characteristics of *Achromobacter* are listed in Table 5.

Antimicrobial susceptibility data for *Achromobacter* are limited. Most *A. xylosoxidans* strains are susceptible to carbenicillin and trimethoprim-sulfamethoxazole and are resistant to aminoglycosides. Susceptibility to chloramphenicol, colistin, and tetracycline is variable. Among the newer antibiotics, moxalactam and ceftazidime inhibit *A. xylosoxidans* in vitro. Re-

TABLE 7. Biochemical characteristics of unnamed groups: EO-2, EF-4b, IVe, IVc-2[a] (40)

Test performed	EO-2[b] (93)[c]		EF-4b (34)[c]		IVe (37)[c]		IVc-2 (1)[c]	
	Sign	% Positive	Sign	% Positive	Sign	% Positive	Sign	% Positive
Morphology	cc		cc, srs		srs		srs	
Motility: flagella	[nm]		[nm]		[m][d]		[m:pe]	
Action on BA	v	23 al or gr	v	41 ly	v	71 ly	v	53 ly
Fermentative or oxidative	[O]		[O]		[n-o]		[n-o]	
Carbohydrate base	O-F		O-F		O-F		O-F	
Acid from:								
Glucose	[+]	100	+ or (+)	70 (26)	[−]	0	[−]	0
Xylose	[+]	99 (1)	−	0	[−]	0	[−]	0
Mannitol	v	10 (10)	−	0	−	0	−	0
Lactose	+	91 (6)	[−]	0	−	0	−	0
Sucrose	[−]	0	−	0	−	0	−	0
Maltose	v	20 (1)	[−]	0	−	0	−	0
Catalase	+	98	[+]	100	+	100	+	100
Oxidase	[+]	100	+	100	[+]	100	[+]	100
Growth on:								
MacConkey agar	v	63 (11)	v	65 (6)	[v]	62 (27)	[+]	94 (6)
SS agar	−	4 (1)	−	0	[−]	5	[−]	3 (6)
Cetrimide agar	−	0			−	0 (3)	−	0 (3)
Simmons citrate	v	22 (3)	v	14 (6)	v	14 (16)	+	100
Urea, Christensen	v	74 (3)	[−]	0	[+]	97	[+]	100
Nitrate reduction	v	85	[+][e]	97	[+]	100	[v]	11
Gas from nitrate	[−]	2	[−]	0	[v]	60	[−]	0
Indole	[−]	0	[−]	0	−	0	−	0
TSI acid:								
Slant	−	0	−	0	−	0	−	0
Butt	−	0	−	6[f]	−	0	−	0
H_2S:								
TSI butt	−	0	−	0	[−]	0	[−]	0
Lead acetate paper	v	75	v	88	v	38	v	51
Gelatin hydrolysis	−	0	−	9	−	0	−	0
Litmus milk	v	32 k	−	6 k	v	65 k	k	97
Pigment	v	17 yel sol			−	3 yel sol	v	27 yel-ta sol
Growth at:								
25°C	v	83	v	88	v	67	+	100
35°C	+	95	+	100	v	88	v	86
42°C	v	35	v	69	v	18	−	0
Esculin hydrolysis	−	0	−	0	−	0	−	0
Lysine decarboxylase	[−][g]	0	[−]	0	NA[h]		NA	
Arginine dihydrolase	[−][g]	0	[−]	0	NA		NA	
Ornithine decarboxylase	−[g]	0	−	0	NA		NA	
Nutrient broth:								
0% NaCl	v	66	+ or (+)	89 (7)	v	19 (3)	+	100
6% NaCl	v	34	−	0	v	14 (5)	v	11

[a] Symbols: See Table 3, footnote *a*. Abbreviations (alphabetically): al, alpha-hemolysis; cc, coccobacillus; gr, greening of blood; k, alkaline; ly, diffuse lysis of erythrocytes extending out from heavy growth; m, motile; nm, nonmotile; n-o, nonoxidizer; O, oxidative; O-F, oxidation-fermentation; pe, peritrichous (flagella); sol, soluble; srs, short straight rods; v, variable (11 to 89% positive); yel, yellow; yel-ta, yellow-tan.

[b] Colonies are frequently mucoid.

[c] Number of strains tested.

[d] Some group IVe strains are weakly motile with one or two polar and long lateral flagella. Occasional strains are nonmotile.

[e] Some strains reduce both nitrate and nitrite without gas formation.

[f] Weak reaction.

[g] Less than 10 strains tested.

[h] NA, Not available.

sults with azlocillin, piperacillin, cefoperazone, and cefamandole vary (29).

Chester and Cooper (8) tested 23 isolates designated *Achromobacter* group Vd. Most isolates were resistant to ampicillin, carbenicillin, penicillin G, cephalothin, and chloramphenicol. About half were gentamicin and tobramycin susceptible, and 70 to 75% were susceptible to amikacin, tetracycline, and trimethoprim-sulfamethoxazole. Most strains are resistant to second-generation cephalosporins.

UNNAMED NFB GROUPS IVe, IVc-2, EO-2, AND EF-4b

Group IVe is an unclassified, nonsaccharolytic, gram-negative bacterium that closely resembles *Alcaligenes*. Cellular morphology may be coccoid, rod shaped, or filamentous. Like *Alcaligenes* it is oxidase positive and motile, and most strains grow on MacConkey agar. Colonies on BA media at 48 h are 1.0 mm in diameter, convex, and circular and have an entire edge. Group IVe differs from *Alcaligenes* by its ability to hydrolyze urea almost immediately in Christensen urea agar if the medium is warmed to incubator temperature before inoculation. Motile cultures of group IVe have both polar and long lateral flagella with long wavelength and low amplitude. Motility may be weak and difficult to demonstrate. Occasional strains are nonmotile.

Group IVe strains also closely resemble *Bordetella bronchiseptica*. They may be differentiated by nitrate reduction and a flagella stain. Group IVe strains usually produce nitrite and nitrogen gas from nitrate, whereas less than half of the *B. bronchiseptica* strains produce nitrite and none produce nitrogen gas. *B. bronchiseptica* has peritrichous flagella rather than polar and long lateral flagella. Weakly motile strains of group IVe may be confused with *Brucella* spp. (nonmotile), especially if xylose oxidation is not tested.

All clinical laboratory group IVe strains have been isolated from human urine except for one from a patient with IVe bacteremia associated with obstruction uropathy (33). This blood isolate was susceptible to kanamycin, gentamicin, cephalothin, nalidixic acid, trimethoprim-sulfamethoxazole, and tetracycline and resistant to ampicillin, carbenicillin, chloramphenicol, colistin, and nitrofurantoin. Urine isolates are susceptible to aminoglycosides, cephalosporins, and colistin at achievable serum levels. At antibiotic concentrations found in urine they are also susceptible to erythromycin, tetracycline, and nitrofurantoin (41).

Another closely related organism is CDC group IVc-2, which is biochemically almost identical to *B. bronchiseptica*. Nitrate reduction and growth on salmonella-shigella (SS) agar were considered differentiating characteristics in that *B. bronchiseptica* grows on SS agar and reduces nitrate to nitrite, whereas group IVc-2 rarely either grows on SS agar or reduces nitrate. Group IVe, group IVc-2, and *B. bronchiseptica* have different carbon substrate utilization profiles (30) as well as distinct fatty acid compositions (12).

Tests for differentiating these organisms from each other and from *Brucella* are found in Table 6. Further characteristics of groups IVe and IVc-2 are listed in Table 7.

Group IVc-2 strains are usually resistant to penicillin, ampicillin, cephalothin, chloramphenicol, and aminoglycosides. Most strains are susceptible to carbenicillin, cefoxitin, cefamandole, tetracycline, polymyxin, and trimethoprim-sulfamethoxazole (15).

Organisms previously designated group EF-4 (eugonic fermenter; Chapter 28) are now divided into two groups, EF-4a and EF-4b. They are biochemically and morphologically similar except that EF-4a ferments glucose whereas EF-4b can only oxidize glucose. EF-4b strains cannot hydrolyze arginine, whereas many (77%) EF-4a strains can. Also the fatty acid compositions of the two groups differ (11).

EF-4b is a nonmotile, oxidase-positive glucose oxidizer. No other sugars are oxidized. Growth on MacConkey agar is variable. Nitrate is reduced, but no gas is produced. Some strains can reduce both nitrate and nitrite without gas formation. Further characteristics of EF-4b are found in Table 7. Both EF-4a and EF-4b have been isolated from wounds associated with cat or dog bites.

Group EO-2 (eugonic oxidizer) is also a nonmotile, oxidase-positive glucose oxidizer. Many (74%) strains hydrolyze urea. Other biochemical reactions of EO-2 are found in Table 7. EO-2 colonies are frequently mucoid.

EO-2 is usually susceptible to tetracycline, chloramphenicol, aminoglycosides, polymyxin, and trimethoprim-sulfamethoxazole and resistant to cephalothin. Results with other antimicrobial agents are variable (15).

TABLE 8. Biochemical characteristics of *Agrobacterium radiobacter*[a] (40)

Test performed	*A. radiobacter* (38)[b]	
	Sign	% Positive
Morphology	spr	
Motility: flagella	[m:pe][c]	
Action on BA	v	18 ly
Fermentative or oxidative	[O]	
Carbohydrate base	O-F	
Acid from:		
Glucose	[+]	100
Xylose	[+]	100
Mannitol	+	100
Lactose	[+]	100
Sucrose	+	100
Maltose	[+]	100
Catalase	+	100
Oxidase	[+]	100
Growth on:		
MacConkey agar	[+]	100
SS agar	v	26 (3)
Cetrimide agar	−	0
Simmons citrate	+	100
Urea, Christensen	[+ or (+)]	89 (11)
Nitrate reduction	v	87
Gas from nitrate	[−]	5
Indole	[−]	0
TSI acid:		
Slant	−	0
Butt	−	0
H_2S:		
TSI butt	v	8 (5)
Lead acetate paper	+	100
Gelatin hydrolysis	−	3
Litmus milk	szf	100
Pigment	−	0
Growth at:		
25°C	+	100
35°C	+	100
42°C	v	34
Esculin hydrolysis	+	100
Lysine decarboxylase	[−]	0
Arginine dihydrolase	[−]	3
Ornithine decarboxylase	−	0
Nutrient broth:		
0% NaCl	+	100
6% NaCl	−	0
3-Ketolactose production (3)	[+]	100

[a] Symbols: See Table 3, footnote *a*. Abbreviations (alphabetically): ly, diffuse lysis of erythrocytes extending out from heavy growth, no action by individual colonies; m, motile; O, oxidative; O-F, oxidation-fermentation; pe, peritrichous; szf, serum formation (clear top in milk); v, variable (11 to 89% positive).

[b] Number of strains tested.

[c] Hooked flagella.

AGROBACTERIUM

The genus *Agrobacterium* has long been known as a plant pathogen causing crown gall and hairy root disease on a variety of vegetation. The organisms inhabit soil and are isolated from human specimens only rarely.

Four species have been proposed: *A. tumefaciens, A. rhizogenes, A. rubri,* and *A. radiobacter*. This last species is biochemically, morphologically, culturally, and serologically identical to *A. tumefaciens* except that *A. radiobacter* is not a plant pathogen. Whether or not *A. radiobacter* should be considered a separate species remains to be determined. Perhaps newer

TABLE 9. Biochemical characteristics of *Flavobacterium* spp.[a] (14, 40)

Test performed	*F. meningo-septicum* (148)[b]		*Flavobacterium* sp. group IIb[c] (155)[b]		*F. breve* (3)[b]		*F. odoratum* (74)[b]		*F. multivorum*[d] (22)[b]		*F. spiritivorum*[e] (11)[b]	
	Sign	% Positive	Sign	% Positive	Sign	% Positive	Sign	% Positive	Sign	% Positive	Sign	% Positive
Morphology	lrs II		mrs II		s-lr fc		lr		srs		srs	
Motility: flagella	[nm][f]		[nm]		[nm]		[nm]		[nm]		[nm]	
Action on BA	v	75 LG	v	26 LG	sl ly	100	v	77 LG	v	gr	–	0
Fermentative or oxidative	[O]		[O]		[O]		[n-o]		[O]		O	
Carbohydrate base	O-F		O-F		O-F		O-F		O-F		O-F	
Acid from:												
Glucose	[+]	95 (4)	[+]	92 (6)	[+]	100	[–]	0	[+]	100	[+]	100
Xylose	–	2 (1)	v	30 (1)	–	0	[–]	0	+	100	+	91 (9)
Mannitol	[+]	91 (8)	[–]	10	[–]	0	–	0	[–]	0	[+]	100
Lactose	v	42 (15)	–	0	–	0	–	0	+	100	+	100
Sucrose	–	0	v	13 (1)	–	0	–	0	+	100	+	100
Maltose	+	93 (7)	+	92 (6)	+	100	–	0	[+]	100	+	100
Starch	–	0	+	100	+	100	NA[g]		NA		NA	
Trehalose	+	100	[+]	100	[–]	0	NA		NA		NA	
ONPG	+	100	v	57	[–]	0	–	0	+	100	NA	
Catalase	+	100	+	99	+	100	+	100	+	100	+	100
Oxidase	[+]	99	[+]	96	[+]	100	[+]	99	[+]	100	[+]	100
Growth on:												
MacConkey agar	[+ or (+)]	89 (3)	v	54 (9)	[+]	100	[+]	91 (5)	[+]	100	[v]	0 (55)
SS agar	–	1	–	0	–	0	v	30 (11)	–	0	–	0
Simmons citrate	v	9 (3)	–	2 (1)	–	0	–	0	–	0	–	0
Urea, Christensen	–	3 (5)	v	14 (28)	–	0	[+]	100	+	95	[+ or (+)]	53 (36)
Nitrate reduction	[–]	0	v	22	–	0	[–]	0	–	0	–	0
Gas from nitrate	–	0	–	0	–	0	–	0	–	0	–	0
Nitrite reduction	NA		v	20	NA		v	83	v	38	NA	
Indole	[+]	100	[+]	98	[+]	100	[–]	0	[–]	0	[–]	0
TSI acid:												
Slant	–	0	–	1	–	0	–	0	v	57 (5)	–	0[j]
Butt	–	0 (3)	–	10	–	0	–	0	v	5 (76)	–	0[j]
H_2S:												
TSI butt	–	3	–	1	–	0	[–]	0	–	0	–	0[j]
Lead acetate paper	+	98	+	99	+ or (+)	33 (67)	v	16	+	86 (5)	v	56[j]
Gelatin hydrolysis[h]	+	91	v	78	+	100	[+]	96	–	0	v	11[j]
Litmus milk	[pep]	98	[pep]	93	v	67 k	pep	93	–	0 (10) Aw	–	0[j]
Pigment:									v	57 sl yel	v	36 pale yel
Insoluble	—[i]		[yel]	99	sl yel	100	v	85 yel				
Soluble	v	0 (37) ta-br	v	11 yel-br	(br)	0 (100)	v	28 ta-br	–	5 ta	NA	
Growth at:												
25°C	+	100	+	100	+	100	+	100	+	100	+	100[j]
35°C	+	100	+	100	+	100	+	100	+	100	+	100[j]
42°C	v	45	v	42	–	0	v	31	–	0	v	11

Continued

TABLE 9—*Continued*

Test performed	*F. meningo-septicum* (148)[b]		*Flavobacterium* sp. group IIb[c] (155)[b]		*F. breve* (3)[b]		*F. odoratum* (74)[b]		*F. multivorum*[d] (22)[b]		*F. spiritivorum*[e] (11)[b]	
	Sign	% Positive	Sign	% Positive	Sign	% Positive	Sign	% Positive	Sign	% Positive	Sign	% Positive
Esculin hydrolysis	[+]	99	v	70	[−]	0	−	0	[+]	100	+	100
Lysine decarboxylase	−	0	−	0	−	0	−	0	[−]	0	−	0[j]
Arginine dihydrolase	−	0	−	0	−	0	−	0 (9)	[−]	0	v	25[j]
Ornithine decarboxylase	−	0	−	0	−	0	−	0	−	0	−	0[j]
Nutrient broth:												
0% NaCl	+	100	+	100	+	100	+	100	+	100	+	100[j]
6% NaCl	−	7	−	0	−	0	v	20 (5)	−	0	−	0[j]
ONPG	+	100	v	43	−	0	−	0	+	100	NA	
DNA hydrolysis	+	100	−	4	+	100	+	95	−	0	NA	

[a] Symbols: See Table 3, footnote *a*. Abbreviations (alphabetically): Aw, weakly acid; br, brown; fc, filaments, curved; gr, greening of blood, usually accompanied by lysis; k, alkaline; II, "II forms," pleomorphism, where rod-shaped cells often appear thin to very thin in the central region, with thicker ends; LG, lavender-green coloration under heavy growth, indication of proteolysis; lr, long rods; lrs, long straight rods; ly diffuse lysis of erythrocytes extending out from heavy growth, no action by individual colonies; mrs, medium straight rods; nm, nonmotile; n-o, nonoxidative; O, oxidative; O-F, oxidation-fermentation; pep, peptonization; s-lr, short to long rods; srs, short straight rods; ta, tan; v, variable (11 to 89% positive); yel, yellow.

[b] Number of strains tested.

[c] Proposed name, *F. indologenes*.

[d] Proposed name, *Sphingobacterium multivorum*.

[e] Proposed name, *Sphingobacterium spiritivorum*.

[f] Polar and lateral flagella have been demonstrated on some strains.

[g] NA, Not available.

[h] Seven- to 14-day incubation.

[i] Some strains exhibit slight yellow pigment.

[j] Less than 10 strains tested.

identification methods such as guanine-cytosine ratio analysis and DNA homology studies will resolve this question.

A. radiobacter has been recovered from various human specimens, but until recently its clinical significance has been somewhat difficult to interpret. In 1980, Plotkin (29) reported a case of prosthetic valve endocarditis caused by *A. radiobacter* in which there was little doubt concerning its pathogenicity. *A. radiobacter* is the only species that has ever been isolated from clinical specimens. Microbiologists should include it in their differential identification schemes because it may be confused with species of *Pseudomonas* and *Achromobacter*.

The organism grows on ordinary laboratory media, is oxidase and nitrate positive, oxidizes glucose, and is motile with peritrichous flagella. The key biochemical test that differentiates *A. radiobacter* from *Achromobacter* and other species of *Agrobacterium* is oxidation of lactose at carbon 3 of the glycosyl moiety to produce 3-ketolactose (3). Because 3-ketolactose is a reducing sugar, it can be detected with Benedict reagent. The oxidation of various other carbohydrates and flagellar arrangement distinguishes *A. radiobacter* from the pseudomonads. A complete list of appropriate biochemical and physiologic parameters used in identification is found in Table 8.

Antibiotic susceptibility data dealing with this organism are sparse. Plotkin (29) reported that his isolate was resistant to most antimicrobial agents except trimethoprim-sulfamethoxazole. His patient was successfully treated with the synergistic combination of trimethoprim-sulfamethoxazole and polymyxin.

FLAVOBACTERIUM

The genus *Flavobacterium* has undergone considerable taxonomic revision since the last edition of this book. Although some bacterial groups have been reclassified into this genus and several new species have been proposed, the exact taxonomic position of some of these organisms is still unsettled. With improved characterization of these organisms, we can expect that some of these *Flavobacterium* species will be transferred into other genera, such as *Cytophaga* or *Sphingobacterium*, as has been recently proposed (5, 6, 42).

The genus *Flavobacterium* includes nonmotile, asporogenous, gram-negative bacilli that are oxidase positive and produce chromogenic colonies. The flavobacteria use glucose and other carbohydrates fermentatively, but the acid production in O-F medium sealed with petrolatum is often weak and may not become apparent until after more than 48 h of incubation. Furthermore, if liquid peptone is used as the basal medium, the reactions may be delayed for as long as 14 to 21 days. For these reasons, it is advisable to consider flavobacteria as oxidizers or nonfermenters and to employ O-F medium (unsealed) for

TABLE 10. Biochemical characteristics of *Flavobacterium* sp. groups IIe, IIf, IIh, IIi, and IIj[a] (14, 40)

Test performed	IIe (18)[b]		IIf (87)[b]		IIh (21)[b]		IIi (23)[b]		IIj (41)[b]	
	Sign	% Positive	Sign	% Positive	Sign	% Positive	Sign	% Positive	Sign	% Positive
Morphology	srs		srs		srs, II		srs		s-lr	
Motility: flagella	[nm]		[nm]		[nm]		[nm]		[nm]	
Action on BA	v	13 LG	v	76 LG	v	32 LG	v	39 LG	v	36 LG
Fermetative or oxidative	[O]		[n-o]		[O]		[O]		[n-o]	
Carbohydrate base	O-F		O-F		O-F		O-F		O-F	
Acid from:										
Glucose	[+]	100	[−]	0	[+ or (+)]	85 (15)	[+]	91 (9)	[−]	0
Xylose	−	0	[−]	0	−	5	+ or (+)	87 (13)	[−]	0
Mannitol	[−]	0	−	0	[−]	0	[−]	0	−	0
Lactose	[−]	0	−	0	[−]	0	+	91 (9)	−	0
Sucrose	−	0	−	0	−	0	[+]	91 (9)	−	0
Maltose	[+]	100	−	0	[+]	95	+	91 (9)	−	0
Catalase	v	88	+	98	+	100	[+]	100	+	100
Oxidase	v	88	[+]	100	[+]	100	[+]	100	[+]	100
Growth on:										
MacConkey agar	[−]	7	[−]	0 (10)	[−]	0	[−]	0	[−]	2
SS agar	−	0	−	0	−	0	−	0	−	0
Simmons citrate	−	0	−	0	−	0	−	0	−	0
Urea, Christensen	−	0	[−][c]	0	−	0	v	14 (18)	[+]	100
Nitrate reduction	[−]	0	−	0	[−]	0	−	0	[−]	0
Indole	[+]	100	[+]	100	[+]	100	[+]	100	[+]	98
TSI acid:										
Slant	−	0	−	0	−	0	−	0	−	0
Butt	−	0	−	0	−	5	[−]	0	−	0
H_2S:										
TSI butt	−	0	−	0	−	0	−	0	−	0
Lead acetate paper	v	78	+	95	+	100	v	70	v	59
Gelatin hydrolysis[d]	−	0	[+]	100	[−]	7	[−]	0	+	98
Litmus milk	v	12 IR	v	39 pep	−	7 IR	−	4 A	v	18 pep
Pigment										
Insoluble	[−]	0	−	0	−	0	v	22 yel	−	0
Soluble	br-yel	82	br-ta	98	br	100	v	48 yel-ta	ta-yel	100
Growth at:										
25°C	+	100	v	58	+	100	+	100	v	30
35°C	+	100	+	100	+	100	+	100	+	95
42°C	−	0	v	70	−	5	v	36	−	10
Esculin hydrolysis	[−]	0	−	0	[+]	100	+	96	−	0
Lysine decarboxylase	NA[e]		−	0[f]	NA		−	0[f]	−	0[f]
Arginine dihydrolase	NA		−	0[f]	NA		−	0[f]	+	100[f]
Ornithine decarboxylase	NA		−	0[f]	NA		−	0[f]	−	0[f]
Nutrient broth:										
0% NaCl	v	88	+	99	v	86	+	100	v	15
6% NaCl	−	0	−	7	−	5	−	9	−	0
ONPG	−	0	−	0	v	50	+	100	−	0
DNA hydrolysis	+	95	−	3	+	100	−	0	−	0

[a] Symbols: See Table 3, footnote *a*. Abbreviations (alphabetically): A, acid; br, brown; II, "II forms," pleomorphism where rod-shaped cells often appear thin to very thin in the central region, with thicker ends; IR, indicator reduced; LG, lavender-green coloration under heavy growth, indication of proteolysis; nm, nonmotile; n-o, nonoxidative; O, oxidative; O-F, oxidative-fermentative; pep, peptonization; s-lr, short to long rods; srs, short straight rods; ta, tan; v, variable (11 to 89% positive); yel, yellow.

[b] Number of strains tested.

[c] In 3 to 7 days, 27 of 86 strains produced pink reactions which might be interpreted as weakly positive.

[d] In 7 to 14 days of incubation.

[e] NA, Not available.

[f] Less than 10 strains tested.

determining carbohydrate utilization.

Currently, the clinically important members of the genus *Flavobacterium* include *F. meningosepticum*, *Flavobacterium* sp. group IIb, *F. breve*, *F. odoratum* (formerly known by the CDC classification M-4F; 20), and *F. multivorum*, previously known as group IIk, biotype 2 (19). In addition, CDC groups IIe, IIf, IIh, IIi, and IIj have been reclassified as *Flavobacterium* spp. IIe, IIf, IIh, IIi, and IIj, respectively. A recent study proposed that *Flavobacterium* sp. group IIb be designated *F. indologenes* and that *F. multivorum* be transferred to a new genus, *Sphingobacterium*, due to its

TABLE 11. Biochemical characteristics of *Moraxella* spp.[a] (40)

Test performed	*M. phenylpyruvica* (50)[b]		*M. urethralis* (22)[b]		*M. atlantae* (23)[b]	
	Sign	% Positive	Sign	% Positive	Sign	% Positive
Morphology	[cc, br]		[cc]		sbr, cc	
Motility: flagella	[nm]		[nm]		[nm]	
Action on BA	v	16 ly	v	18 ly	–	9 al or ly
Fermentative or oxidative	n-o		n-o		[n-o]	
Carbohydrate base	O-F		O-F		O-F	
Acid from:						
Glucose	[–]	0	[–]	0	–[c]	0
Xylose	–	0	[–]	0	–	0
Mannitol	–	0	–	0	–	0
Lactose	–	0	–	0	–	0
Sucrose	–	0	–	0	–	0
Maltose	–	0	–	0	–	0
Catalase	+	90	+	100	+	91
Oxidase	[+]	100	[+]	100	[+]	100
Growth on:						
MacConkey agar	v	80 (6)	[+]	96	[+ or (+)]	87 (13)
SS agar	–	0	–	9	–	0
Sodium acetate	v	43	v	60	v	21 (21)
Simmons citrate	–	0	v	46	[–]	0
Urea, Christensen	[+]	100	[–]	0	–	0
Phenylalanine deaminase	[+]	97	[+]	100	–	0
Nitrate reduction	v	68	[–]	0	[–]	5
Nitrite reduction	–	0	[+]	100	v	20
Indole	–	0	–	0	–	0
TSI acid:						
Slant	–	0	–	0	–	0
Butt	–	0	–	0	–	0
H_2S:						
TSI butt	–	0	–	0	–	0
Lead acetate paper	v	47	–	9	v	71
Gelatin hydrolysis	–	0	–	0	–	0
Litmus milk	v	22 (20) k	v	36 k	–	9 IR
Pigment	–	8 yel-br sol	–	4 amb sol	NA[d]	
Growth at:						
25°C	v	85	v	50	v	68
35°C	+	100	+	100	+	100
42°C	v	29	v	59	v	50
Esculin hydrolysis	–	0	–	0	–	0
Nutrient broth:						
0% NaCl	v	53	+	96	v	36
6% NaCl	v	19	v	59	–	0
Penicillin susceptibility[e]	v	73	+	100	NA	

[a] Symbols: See Table 3, footnote *a*. Abbreviations (alphabetically): al, alpha hemolysis; amb, amber; br, broad rods; cc, coccobacillus; IR, indicator reduced; k, alkaline; ly, diffuse lysis of erythrocytes extending out from heavy growth, no action by individual colonies; nm, nonmotile; n-o, nonoxidative; O-F, oxidation-fermentation; pep, peptonization; sbr, short broad rods; sol, soluble; v, variable (11 to 89% positive); yel, yellow.

[b] Number of strains tested.

[c] Usually does not grow in O-F medium; if it does grow, it usually does not change pH.

[d] NA, Not available.

[e] A BA plate is streaked with growth from an 18- to 36-h culture. A 10-U penicillin disk is placed on the streaked area. A zone of inhibition is positive.

high cellular concentration of sphingophospholipids (42).

The natural habitat of flavobacteria is soil and water. In hospitals the organisms may contaminate water systems and have been recovered from nebulizers, water baths, distilled water lines, sink faucets, drinking fountains, cold humidifiers, incubators, saline irrigation solutions, and hemodialysis systems. Normally, flavobacteria are not part of the resident microflora of humans, yet humans may become colonized and occasionally infected as a result of hospitalization.

F. meningosepticum is the species most commonly involved in nosocomial infections. Infants, particularly those born prematurely, have the greatest incidence of infection, and several epidemics of meningitis with high mortality rates have occurred in hospital nurseries. In meningitis, blood cultures are usually positive. *F. odoratum*, *F. multivorum*, and *Flavobacterium* sp. groups IIb and IIf have been recovered from clinical specimens such as blood, urine, and wound exudates, but they have not been reported in nosocomial outbreaks. Isolates of IIj have been most frequently recovered from infected lesions that have resulted from bites or scratches by dogs or cats. In addition, the microorganism has been recovered from a variety of different specimens, including spinal fluid, blood, and sputum. Some strains have also been isolated from

TABLE 11—*Continued*

M. nonliquefaciens (50)[b]		*M. osloensis* (22)[b]		*M. lacunata* (23)[b]	
Sign	% Positive	Sign	% Positive	Sign	% Positive
cc, sbr		cc, sbr		cc	
[nm]		[nm]		[nm]	
v	10 ly	v	7 al	v	24 al
n-o		n-o		n-o	
O-F[c]		O-F		O-F[c]	
[−]	0	[−]	0	[−]	0
−	0	−	0	−	0
−	0	−	0	−	0
−	0	−	0	−	0
−	0	−	0	−	0
−	0	−	0	−	0
+	95	+	95	+	100
[+]	100	[+]	100	[+]	100
[−]	8 (2)	v	70	−	4
−	0	−	0	−	0
−	0	[+]	100	[−]	7
[−]	0	−	0	−	0
[−]	0	[−]	0	[−]	0
−	0	v	14	v	31
[+]	95	v	24	[+]	100
		[−]	0		
−	0	−	0	−	0
−	0	−	0	−	0
−	0	−	0	−	0
−	0	−	0	−	0
v	83	v	74	−	0
−	0	−	0	v	74
−	2 (1) k	−	9 k	v	58 pep
−	3 yel sol	−	6 yel-br sol	−	4 br sol
+	93	+	96	v	47
v	88	+	98	v	87
v	15	v	51	−	0
−	0	−	0	−	0
v	22	[+]	98	[−]	0
−	0	v	12	−	0
+	99	+	92	+	100

animals, including dogs and cats. *F. breve* and the remaining species of flavobacteria are recovered infrequently from clinical material, and little is known about their involvement in human disease.

Flavobacteria may also be recovered from a wide variety of hospital environmental samples, such as those from nebulizers. Growth on MacConkey agar is variable, and therefore BA or chocolate agar is recommended for primary isolation. The appearance of oxidase-positive, yellow-pigmented colonies on BA after 24 to 48 h of incubation is suggestive of flavobacteria. Biochemical characterization is necessary to distinguish *Flavobacterium* from similarly pigmented *Pseudomonas* spp. and *Pseudomonas*-like bacteria. Also flavobacteria are nonmotile, whereas other similarly pigmented bacteria are usually motile.

Most *Flavobacterium* strains produce colonies on nutrient agar with a bright-yellow carotenoid pigment, fail to grow on SS and cetrimide agars, and are usually resistant to polymyxin. The organisms are proteolytic and produce a lavender-green discoloration of erythrocytes on BA within 18 to 24 h of incubation. Gram-stain examination of flavobacteria from solid media reveals long, thin, gram-negative rods that usually exhibit slightly bulbous ends. With the exception of *F. odoratum* and *F. multivorum*, most species encountered in clinical specimens produce small amounts of indole when tested with Ehrlich reagent after xylene extraction.

Characteristics of flavobacteria are shown in Tables 9 and 10. Strains of *F. meningosepticum* grow well on BA after 24 h of incubation and produce colonies that are 1 to 1.5 mm in diameter with an entire edge and are convex, smooth or slightly mottled, glistening, and butyrous. Typically, colonies produce a slightly yellow pigment, and isolates produce acid from glucose, fructose, maltose, and mannitol and produce alkali from xylose and sucrose. Urea is not usually hydrolyzed, nitrate is not reduced, growth on MacConkey agar is variable, and all strains produce β-galactosidase and extracellular DNase. Strains of *F. odoratum* also produce colonies with a pale-yellow pigment, but hydrolyze urea, may or may not produce DNase, and do not hydrolyze *o*-nitrophenyl-β-D-galactopyranoside (ONPG). Cultures of *F. odoratum* produce a fruity odor similar to that produced by *Alcaligenes odorans*, which allows *F. odoratum* to be readily distinguished from other species of *Flavobacterium*.

F. meningosepticum forms a homogeneous species with respect to phenotypic characters, DNA base composition, and DNA-DNA reassociation studies. The species can be subdivided into six distinct serotypes (A through F). Most human infections due to *F. meningosepticum* are caused by serotype C, followed in frequency by serotypes B, F, A, and E.

Flavobacterium sp. group IIb and *Flavobacterium* sp. group IIf are similar to *F. meningosepticum* in phenotypic characteristics and DNA base composition, but the three species can be readily differentiated. *Flavobacterium* sp. group IIb isolates produce colonies with a bright-yellow intracellular pigment on BA plates and are variable for urea, ONPG, and reduction of nitrate to nitrite. Also, none of these strains produces an extracellular DNase, and growth on MacConkey agar is variable. *Flavobacterium* sp. group IIf strains produce luxuriant growth on BA and peptone media with colonies that are 1 to 1.5 mm in diameter, convex with entire edges, and very moist. Colonies are pale yellow and produce a tan to amber-to-brown soluble pigment into the surrounding medium. Growth does not occur on SS or cetrimide agar, and the majority of strains do not grow on MacConkey agar. If growth from heart infusion agar is examined, the microorganisms appear as coccoid forms and short, rod-shaped bacteria, whereas those from broth culture appear as rods of medium length with some filaments. Good growth occurs in O-F medium with various carbohydrates, but alkali, not acid, is produced. Members of group IIf hydrolyze gelatin but fail to hydrolyze urea and ONPG. They are susceptible to penicillin and polymyxin, which helps to differentiate them from strains of *F. meningosepticum*. *F. breve* produces a pale-yellow pigment but is urea and ONPG negative.

F. multivorum grows on BA and after 24 h of incubation produces pinpoint colonies that may be pale yellow in color or nonpigmented. Within 2 to 3 days of incubation the colonies are usually 1 mm in diameter, circular with an entire edge, convex, and glossy. These organisms do not grow on SS or cetrimide agar and do not produce indole. All strains are oxidase positive. *F. multivorum* strains produce variable reactions on TSI. Most strains alkalinize the slant of this medium, but some produce an acid reaction. In each instance the butt of the TSI medium remains unchanged. Occasional isolates produce acid throughout the TSI medium, which is indicative of a fermentative microorganism. When members of this latter group are inoculated into O-F medium containing glucose and sealed with petrolatum, the results are equivocal. Whether these particular strains are oxidizers or latent fermenters is uncertain. The majority of *F. multivorum* strains, however, produce distinct, clear-cut reactions in O-F medium, and it is advisable to use this medium for determining the oxidative or fermentative ability of the organism as well as its ability to utilize various carbohydrates. Although *F. multivorum* does not produce H_2S in TSI and Kligler iron agar, it produces 1+ to 4+ H_2S reactions on lead acetate paper strips. Some *F. multivorum* isolates are capable of producing acid from ethanol and mannitol, which has resulted in a proposal to create a new species of flavobacteria, *F. spiritivorum* (18).

Flavobacterium sp. group IIj strains grow moderately well on BA after 24 h of incubation. Colonies are no more than 0.5 mm in diameter and are convex, butyrous, and sticky, as well as difficult to remove from the agar surface. These growth and colonial characteristics are useful criteria in helping to differentiate *Flavobacterium* sp. group IIj from group IIf. In addition, the colonies are not usually pigmented, but a tan-to-brown soluble pigment may be produced. Gram stain morphology shows medium-to-long, thin rods. Some pleomorphic cells may be observed which have bulbous ends and thin centers. Most strains fail to grow in O-F medium. When *Flavobacterium* sp. group IIj isolates are inoculated heavily onto Christensen medium, urea is hydrolyzed rapidly, which is a key characteristic in its identification. Also, most strains of group IIj are susceptible to penicillin and resistant to polymyxin.

Flavobacteria isolated from human sources are usually resistant to most antimicrobial agents, and no one particular agent is routinely successful in treatment. In vitro studies indicate that many isolates may be susceptible to erythromycin, novobiocin, nalidixic acid, rifampin, and trimethoprim-sulfamethoxazole.

F. meningosepticum and other species of *Flavobacterium* do not appear to be pathogenic for laboratory animals.

MORAXELLA

The genus *Moraxella* comprises six clinically significant species: *osloensis*, *lacunata*, *nonliquefaciens*, *phenylpyruvica*, *atlantae*, and *urethralis*. The last two species were previously referred to as CDC groups M3 and M4, respectively. Members of this genus are found on the mucous membranes of humans and lower animals. Most are considered opportunists, although all species have been involved in a variety of primary infections. All species are nonsaccharolytic, oxidase positive, and nonmotile. MacConkey agar supports the growth of most species. Because *Moraxella* spp. are small gram-negative bacilli, 1.0 by 2.0 μm, they appear coccobacillary on Gram-stained smears. Occasionally *M. lacunata* may require rabbit serum enrichment for sufficient growth. Most species are susceptible to penicillin. The genus can be distinguished from other NFB on the basis of reactions in O-F–glucose, oxidase, and motility. Biochemical and growth reactions are found in Table 11.

M. osloensis grows well on BA, producing colonies of 2.0 to 2.5 mm in diameter. Most strains grow on MacConkey agar, many at 42°C. All strains can use sodium acetate as their sole carbon source, a feature differentiating them from *M. nonliquefaciens*.

M. lacunata may require rabbit serum added to peptone media for adequate growth. The serum is thought to neutralize toxic factors found in peptone media. Colonies on BA are rather small, measuring 0.1 to 0.3 mm even after several days of incubation. Most strains fail to grow on MacConkey agar or to use sodium acetate as a sole carbon source. All strains reduce nitrates, digest Loeffler slants, and are susceptible to penicillin in vitro.

Strains of *M. nonliquefaciens* are similar to *M. osloensis*. Some isolates appear mucoid on primary

TABLE 12. Biochemical characteristics of CDC groups M-6 and M-5[a] (40)

Test performed	M-6 (40)[b]		M-5 (59)[b]	
	Sign	% Positive	Sign	% Positive
Morphology	mbr		[mrs]	
Motility: flagella	[nm]		[nm]	
Action on BA	−	5 al or gr	v	44 ly
Fermentative or oxidative	n-o		n-o	
Carbohydrate base	O-F		O-F	
Acid from:				
Glucose	[−]	0	[−]	0
Xylose	[−]	0	[−]	0
Mannitol	−	0	−	0
Lactose	−	0	−	0
Sucrose	−	0	−	0
Maltose	−	0	−	0
Catalase	[−]	8	+	100
Oxidase	[+]	100	[+]	100
Growth on:				
MacConkey agar	v	22 (28)	v	42 (20)
SS agar	−	0	−	0
Sodium acetate	v	83[c]	v	25[c]
Simmons citrate	−	0	−	0
Urea, Christensen	[−]	0	−	0
Phenylalanine deaminase	−	0[c]	v	73
Nitrate reduction	[+]	100	[−]	0
Nitrite reduction	[+]	100	[v]	84
Indole	−	0	−	0
TSI acid:				
Slant	−	0	−	0
Butt	−	0	−	0
H_2S:				
TSI butt	−	0	−	0
Lead acetate paper	v	88	v	86
Gelatin hydrolysis	−	0	−	0
Litmus milk	−	0	−	3 k
Pigment	v	18 yel-ta sol	v	48 yel-ta sol
Growth at:				
25°C	v	78	+	95
35°C	+	100	+	100
42°C	v	47	v	63
Esculin hydrolysis	−	0	−	0
Nutrient broth:				
0% NaCl	+	93	+ or (+)	65 (29)
6% NaCl	−	3	−	8
Penicillin susceptibility[d]	v	88	+	100

[a] Symbols: See Table 3, footnote *a*. Abbreviations (alphabetically): al, alpha hemolysis; gr, greening of blood usually accompanied by lysis; ly, diffuse lysis of erythrocytes extending out from heavy growth; mbr, medium-length broad rods; mrs, medium straight rods; nm, nonmotile; n-o, nonoxidizer; sol, soluble; ta, tan; v, variable (11 to 89% positive); yel, yellow.

[b] Number of strains tested.

[c] Less than 10 strains tested.

[d] See Table 11, footnote *d*.

media. Colonies appear to be approximately 2 to 3 mm after 18 to 24 h of incubation. Most strains reduce nitrate and produce H_2S with lead acetate paper. This species appears to be a normal inhabitant of the human respiratory tract.

M. phenylpyruvica grows well on ordinary laboratory media without enrichment. Colonies are somewhat smaller than those of other species, measuring only 0.5 to 1.0 mm in diameter. Virtually all isolates produce phenylalanine deaminase and urease, which are key reactions in differentiating this species from other *Moraxella* spp.

M. atlantae (old CDC group M3) grows poorly on most laboratory media. Small spreading colonies 0.2 to 0.5 mm in diameter are common. All strains grow on MacConkey agar, although some grow rather slowly. Most strains are nitrate and nitrite negative.

M. urethralis (old CDC group M4) was formerly referred to as *Mima polymorpha* subsp. *oxidans*. Most strains grow well on BA and MacConkey agar. All isolates produce phenylalanine deaminase, although the reaction may be weak. Most strains will reduce nitrite, but are nitrate negative. All strains appear to be susceptible to penicillin in vitro. These organisms, as the species name suggests, have been isolated primarily from the human genitourinary tract.

GROUPS M5 AND M6

CDC groups M5 and M6 are quite similar to *Moraxella* spp. and consequently are categorized with this genus. Both organisms are gram negative, oxidase positive, nonsaccharolytic, and nonmotile.

Group M5 isolates frequently have been associated with dog bites. Most strains grow well on BA and

MacConkey agar. All are nitrate negative and susceptible to penicillin in vitro. They most closely resemble *M. osloensis*, but can be differentiated from this species on the basis of nitrite reduction and sodium acetate utilization.

Group M6 has been associated with a wider variety of specimens, including wounds, urine, sputum, and blood (36). Most strains lack catalase activity, which distinguishes them from group M5 and other *Moraxella* species. All isolates reduce nitrate and nitrite. Most strains are also susceptible to penicillin in vitro. Biochemical and physiologic parameters are shown in Table 12.

LITERATURE CITED

1. **Appelbaum, P. C., and D. B. Campbell.** 1980. Pancreatic abscess associated with *Achromobacter* group Vd biovar 1. J. Clin. Microbiol. **12:**282–283.
2. **Appelbaum, P. C., J. Stavitz, M. S. Bentz, and L. C. von Kuster.** 1980. Four methods for identification of gram-negative nonfermenting rods: organisms more commonly encountered in clinical specimens. J. Clin. Microbiol. **12:**271–278.
3. **Bernaerts, M. J., and J. DeLey.** 1963. A biochemical test for crown gall bacteria. Nature (London) **197:**406–407.
4. **Bottone, E. J., J. Kittick, Jr., and S. S. Schneierson.** 1973. Isolation of bacillus HB-1 from human clinical sources. Am. J. Clin. Pathol. **59:**560–566.
5. **Bruun, B.** 1982. Studies on a collection of strains of the genus *Flavobacterium*. Acta Pathol. Microbiol. Immunol. Scand. Sect. B **90:**415–421.
6. **Bruun, B.** 1983. Studies on a collection of strains of the genus *Flavobacterium*. 2. Nutritional studies. Acta Pathol. Microbiol. Immunol. Scand. Sect. B **91:**35–41.
7. **Catlin, B. W.** 1975. Cellular elongation under the influence of antibacterial agents: way to differentiate coccobacilli from cocci. J. Clin. Microbiol. **1:**102–105.
8. **Chester, B., and L. H. Cooper.** 1979. *Achromobacter* species (CDC group Vd): morphological and biochemical characterization. J. Clin. Microbiol. **9:**425–436.
9. **Cordes, L. E., E. W. Brink, P. J. Checko, A. Lentnick, R. W. Lyons, P. S. Hayes, T. C. Wu, D. W. Tharr, and D. W. Fraser.** 1981. A cluster of acinetobacter pneumonia in foundry workers. Ann. Intern. Med. **95:**688–693.
10. **Dees, S. B., and C. W. Moss.** 1978. Identification of *Achromobacter* species by cellular fatty acid and by production of keto acids. J. Clin. Microbiol. **8:**61–66.
11. **Dees, S. B., J. Powell, C. W. Moss, D. G. Hollis, and R. E. Weaver.** 1981. Cellular fatty acid composition of organisms frequently associated with human infections resulting from dog bites: *Pasturella multocida* and groups EF-4, IIj, M-5, and DF-2. J. Clin. Microbiol. **14:**612–616.
12. **Dees, S., S. Thanabalasundrum, C. W. Moss, D. G. Hollis, and R. E. Weaver.** 1980. Cellular fatty acid composition of group IVe, a nonsaccharolytic organism from clinical sources. J. Clin. Microbiol. **11:**664–668.
13. **Dowda, H.** 1977. Evaluation of two rapid methods for identification of commonly encountered nonfermenting or oxidase-positive gram-negative rods. J. Clin. Microbiol. **6:**605–609.
14. **Gilardi, G. L.** 1984. Identification of glucose-nonfermenting gram-negative bacilli. Clin. Microbiol. Newsl. **6:**111–113.
15. **Gilardi, G. L.** 1984. Susceptibility patterns of glucose-nonfermenting gram-negative bacilli. Clin. Microbiol. Newsl. **6:**149–152.
16. **Glew, R. H., R. C. Moellering, Jr., and L. J. Kunz.** 1977. Infections with *Acinetobacter calcoaceticus* (*Herellea vaginicola*): clinical and laboratory studies. Medicine (Baltimore) **56:**79–97.
17. **Goldstein, E. J. C., C. E. Cherubin, and M. Shulman.** 1983. Comparison of microtiter broth dilution and agar dilution methods for susceptibility testing of *Eikenella corrodens*. Antimicrob. Agents Chemother. **23:**42–45.
18. **Holmes, B., R. J. Owen, and D. G. Hollis.** 1982. *Flavobacterium spiritovorum*, a new species isolated from human clinical specimens. Int. J. Syst. Bacteriol. **32:**157–165.
19. **Holmes, B., R. J. Owen, and R. E. Weaver.** 1981. *Flavobacterium multivorum*, a new species isolated from human clinical specimens and previously known as group IIk, biotype 2. Int. J. Syst. Bacteriol. **31:**21–34.
20. **Holmes, B., J. J. S. Snell, and S. P. Lapage.** 1977. Revised description, from clinical isolates, of *Flavobacterium odoratum* Stutzer and Kwaschnina 1929, and designation of the neotype strain. Int. J. Syst. Bacteriol. **27:**330–336.
21. **Kersters, K., and J. DeLey.** 1984. Genus *Alcaligenes* Castellani and Chalmers 1919, 936, p. 361–373. *In* N. R. Krieg and J. G. Holt (ed.), Bergey's manual of systematic bacteriology, vol. 1. Williams and Wilkins Co., Baltimore.
22. **Larsen, E. L.** 1981. Persistent carriage of gram-negative bacteria on hands. Am. J. Infect. Control **9:**112–119.
23. **Lautrop, H.** 1974. Genus IV. *Acinetobacter*. Brisou and Prevot 1954, 727, p. 436–438. *In* R. E. Buchanan and N. E. Gibbons (ed.), Bergey's manual of determinative bacteriology, 8th ed. Williams and Wilkins Co., Baltimore.
24. **McGuckin, M. B., R. J. Thrope, K. M. Koch, A. Alavi, M. Staum, and E. Abrutyn.** 1982. An outbreak of *Achromobacter xylosoxidans* related to diagnostic tracer procedures. Am. J. Epidemiol. **115:**785–793.
25. **Oberhofer, T. R.** 1983. Use of the API 20E, Oxi/Ferm, and Minitek Systems to identify nonfermentative and oxidase-positive fermentative bacteria: seven years of experience. Diag. Microbiol. Infect. Dis. **1:**241–256.
26. **O'Connell, C. J., and R. Hamilton.** 1981. Gram negative rod infections. 2. *Acinetobacter calocaceticus* infection in general hospital. N.Y. State J. Med. **81:**750–753.
27. **Otto, L. A., and U. Blachman.** 1979. Nonfermentative bacilli: evaluation of three systems for identification. J. Clin. Microbiol. **10:**147–154.
28. **Pien, F. D., and H. Y. Higa.** 1978. *Achromobacter xylosoxidans* isolates in Hawaii. J. Clin. Microbiol. **7:**239–241.
29. **Plotkin, G. R.** 1980. *Agrobacterium radiobacter* prosthetic valve endocarditis. Ann. Intern. Med. **93:**839–840.
30. **Rarick, H. R., P. W. Riley, and R. Martin.** 1978. Carbon substrate utilization studies of some cultures of *Alcaligenes denitrificans*, *Alcaligenes faecalis*, and *Alcaligenes odorans* isolated from clinical specimens. J. Clin. Microbiol. **8:**313–319.
31. **Reverdy, M. E., J. Freney, J. Fleurette, M. Coulet, M. Surgot, D. Marmet, and C. Ploton.** 1984. Nosocomial colonization and infection by *Achromobacter xylosoxidans*. J. Clin. Microbiol. **19:**140–143.
32. **Riley, P. S., H. W. Tatum, and R. E. Weaver.** 1973. Identity of HB-1 of King and *Eikenella corrodens* (Eiken) Jackson and Goodman. Int. J. Syst. Bacteriol. **23:**75–76.
33. **Rockhill, R. C., and L. I. Lutwick.** 1978. Group IVe-like gram-negative bacillemia in a patient with obstructive uropathy. J. Clin. Microbiol. **8:**108–109.
34. **Shayegani, M., A. M. Lee, and D. M. McGlynn.** 1978. Evaluation of the Oxi/Ferm tube system for identification of nonfermentative gram-negative bacilli. J. Clin. Microbiol. **7:**533–538.
35. **Shayegani, M., P. S. Maupin, and D. M. McGlynn.** 1978. Evaluation of the API 20E system for identification of nonfermentative gram-negative bacteria. J. Clin. Microbiol. **7:**539–545.
36. **Simor, A. E., and I. E. Salit.** 1983. Endocarditis caused by M6. J. Clin. Microbiol. **17:**931–933.
37. **Slee, A. M., and J. M. Tanzer.** 1978. Selective medium for isolation of *Eikenella corrodens* from periodontal lesions. J. Clin. Microbiol. **8:**459–462.

38. **von Graevenitz, A., and C. Bucher.** 1983. Accuracy of KOH and vancomycin tests in determining the Gram reaction of nonenterobacterial rods. J. Clin. Microbiol. **18:**983–985.
39. **Warwood, N. M., D. J. Blazevic, and L. Hofherr.** 1979. Comparison of the API 20E and Corning N/F systems for identification of nonfermentative gram-negative rods. J. Clin. Microbiol. **10:**175–179.
40. **Weaver, R. E., D. G. Hollis, W. A. Clark, and P. Riley.** 1983. The identification of unusual pathogenic gram negative bacteria. Centers for Disease Control, Atlanta, Ga.
41. **Welch, W. D., R. K. Porschen, and B. Luttrell.** 1983. Minimal inhibitory concentrations of 19 antimicrobial agents for 96 clinical isolates of group IVe bacteria. Antimicrob. Agents Chemother. **24:**432–433.
42. **Yabuuchi, E., T. Kaneko, I. Yano, C. W. Moss, and N. Miyoski.** 1983. *Sphingobacterium* gen. nov., *Sphingobacterium multivorum* comb. nov., *Sphingobacterium mizutae* sp. nov., and *Flavobacterium indologenes* sp. nov.: glucose-nonfermenting gram-negative rods in CDC groups IIk-2 and IIb. Int. J. Syst. Bacteriol. **33:**580–598.
43. **Yabuuchi, E., and I. Yano.** 1981. *Achromobacter* gen. nov. and *Achromobacter xylosoxidans* (ex Yabuuchi and Ohyama 1971) nom. rev. Int. J. Syst. Bacteriol. **31:**477–478.

Chapter 30

Pseudomonas

GERALD L. GILARDI

Of the numerous species of pseudomonads, until recently only *Pseudomonas aeruginosa*, *Pseudomonas pseudomallei*, and *Pseudomonas mallei* were considered important human pathogens. Since the early 1960s, however, other pseudomonads such as *Pseudomonas maltophilia* and *Pseudomonas cepacia* have been associated with infections, primarily as opportunistic pathogens of the compromised host. Prompt diagnosis of *Pseudomonas* infections, identification of the clinical isolate, and determination of specific therapy (because of the unpredictable resistance to antimicrobial agents) are necessary for a favorable prognosis.

A system is presented for the isolation and progressive identification of named and unnamed *Pseudomonas* species listed in Table 1 and their differentiation from other glucose-nonfermenting gram-negative rods encountered in clinical specimens. The current classification is based on the internal division of *Pseudomonas* into various RNA homology groups (50) which represent natural genetic arrangements. Within the RNA groups are subgeneric groups, e.g., the fluorescent group, which contain species that share many common phenotypic properties.

METHODS FOR ISOLATION AND IDENTIFICATION OF *PSEUDOMONAS* SPECIES

Isolation media

Pseudomonads grow easily on routine primary isolation media used in bacteriology and are readily isolated from clinical specimens and from hospital environments. They are usually isolated on peptone agar media with or without infusion containing 5% sheep or rabbit blood, although media enriched with blood are not essential for their isolation. In addition to blood agar, one of the less inhibitory selective-differential media, e.g., Leifson deoxycholate, MacConkey, or eosin-methylene blue agar medium, should be used. Although the less inhibitory selective-differential media may increase the probability of recovering pseudomonads from specimens containing numerous other bacterial species, they do not support the growth of all pseudomonads. Pseudomonads are less frequently isolated on the more inhibitory selective-differential media, e.g., Leifson deoxycholate citrate or salmonella-shigella agar. Media such as cetrimide agar (Difco Laboratories, Detroit, Mich.), pseudomonas agar F (Difco Laboratories), Pseudosel agar (BBL Microbiology Systems, Cockeysville, Md.), and pseudomonas isolation agar (Difco Laboratories) containing cetyltrimethylammonium bromide (cetrimide), 2,4,4-trichloro-2-hydroxydiphenyl ether (irgasan), 9-chloro-9-(4-diethylaminophenyl)-10-phenylacridan (C-390), or similar compounds are used for the selective isolation of *P. aeruginosa* and other *Pseudomonas* species.

Identification methods

Definitive taxonomic and biochemical clarification of glucose-nonfermenting gram-negative rods allows for easy recognition of most species by simple routine diagnostic tests. The battery of media and tests recommended in Table 2 has been modified over a number of years and may need further revision as additional glucose-nonfermenting gram-negative rods are recognized.

Screening procedure

A screening procedure designed to separate similar genera and to distinguish species within the genus *Pseudomonas* should collect the following information: odor, pigmentation, and colonial morphology; Gram reaction, somatic shape, and spore formation; motility and flagellar anatomy; mode of glucose utilization; production of hydrogen sulfide, arginine dihydrolase, and indophenol oxidase; growth at 42°C; and oxidation of glucose, xylose, lactose, and maltose.

Somatic morphology

Pseudomonas species are straight or slightly curved rods. *Acinetobacter* species are usually very short plump rods often approaching coccoid shape, and *Moraxella* species are usually very short plump rods, predominantly in pairs and short chains and occasionally pleomorphic. *Achromobacter* species are noticeably barred when stains are made from broth cultures. The Gram reaction and spore formation of an isolate suspected of being a pseudomonad should be carefully studied. Troublesome isolates presumptively considered to be *Pseudomonas* species may be gram-negative-staining *Bacillus* species or gram-positive coryneforms.

Presumptive identification and screening media

Kligler iron or triple sugar iron agar. Suspect *Pseudomonas* colonies on primary isolation media are transferred to Kligler iron agar (KIA) or triple sugar iron agar (TSIA) slants. Glucose-fermenting gram-negative rods produce an acid butt in these media. Glucose-nonfermenting gram-negative rods usually fail to produce visible growth in the butt, and there is no change in the color of the indicator in the butt after 24 h of incubation. Most glucose-nonfermenting gram-negative rods produce an alkaline (red) slant. Strains of some species may produce no change in the color of the slant. Some *P. pseudomallei* strains and other *Pseudomonas* isolates may produce an acid (yellow) slant. *P. putrefaciens* causes the entire butt of KIA or TSIA to become black, without gas.

Isolates selected for further study should be streaked on an agar medium to verify the purity since a colony removed from primary isolation agar medium may not give rise to a pure culture.

TABLE 1. *Pseudomonas* species encountered in clinical specimens

Name	Synonyms
RNA group I	
Fluorescent group	
Pseudomonas aeruginosa	*Bacterium aeruginosum, Bacillus pyocyaneus, Pseudomonas polycolor*
Pseudomonas fluorescens	*Pseudomonas chlororaphis, Pseudomonas aureofaciens, Pseudomonas lemonnieri*
Pseudomonas putida	*Pseudomonas ovalis*
Stutzeri group	
Pseudomonas stutzeri	*Bacillus denitrificans* II, *Bacterium stutzeri, Pseudomonas stanieri*, group Vb-1, group Vb-3
Pseudomonas mendocina	Group Vb-2
Alcaligenes group	
Pseudomonas alcaligenes	
Pseudomonas pseudoalcaligenes	*Pseudomonas alcaligenes* biovar B
Pseudomonas sp. group 1	*Pseudomonas denitrificans*
RNA group II	
Pseudomallei group	
Pseudomonas mallei	*Bacillus mallei, Actinobacillus mallei*
Pseudomonas pseudomallei	*Bacillus pseudomallei, Malleomyces pseudomallei*
Pseudomonas cepacia	*Pseudomonas multivorans, Pseudomonas kingae*, group EO-1
Pseudomonas gladioli	*Pseudomonas marginata, Pseudomonas alliicola*
Pseudomonas pickettii	*Pseudomonas thomasii, Pseudomonas pseudoalcaligenes* type 2, group Va-1, group Va-2
RNA group III	
Acidovorans group	
Pseudomonas acidovorans[a]	
Pseudomonas testosteroni[a]	
Pseudomonas delafieldii	
RNA group IV	
Diminuta group	Group Ia
Pseudomonas diminuta	*Corynebacterium vesiculare*
Pseudomonas vesicularis	
RNA group V	
Pseudomonas maltophilia	*Xanthomonas maltophilia, Pseudomonas melanogena*, group I
Species of uncertain affiliation	
Pseudomonas mesophilica	*Bacillus extorquens, Vibrio extorquens, Pseudomonas extorquens, Methylobacterium extorquens, Protomonas extorquens*
Pseudomonas paucimobilis	*Flavobacterium devorans, Flavobacterium capsulatum*, group IIk-1
Pseudomonas pertucinogena	
Pseudomonas putrefaciens	*Achromobacter putrefaciens, Pseudomonas rubescens, Alteromonas putrefaciens*, group Ib-1, group Ib-2
Pseudomonas sp. group Ve-1	
Pseudomonas sp. group Ve-2	
Pseudomonas-like group 2	

[a] *Pseudomonas terrigena, Vibrio terrigenus, Vibrio percolans, Lophomonas alcaligenes, Pseudomonas indoloxidans, Pseudomonas desmolytica, Pseudomonas testosteroni*, and *Pseudomonas acidovorans* are synonyms of *Comamonas terrigena*.

Oxidative-fermentative basal medium. An isolate presumptively identified in KIA or TSIA as a glucose-nonfermenting gram-negative rod should be confirmed as such by inoculating oxidative-fermentative (OF) basal medium (Difco Laboratories no. 1688) with 1% glucose. The relatively high carbohydrate concentration of OF glucose medium permits the detection of oxidative acidity in the presence of ammonia and other bases derived from the relatively small quantity of peptone in the medium. OF maltose, lactose, and xylose media should also be inoculated in the screening procedure. The media should be pH 6.8 (green, not blue) at the time of inoculation. The color of the indicator changes to green-yellow or yellow during an acid reaction and to blue during an alkaline reaction. No color change indicates that the organism failed to grow.

Indophenol oxidase. To detect indophenol oxidase (cytochrome oxidase, cytochrome *c* oxidase) activity, growth can be picked with a platinum loop and rubbed into a paper strip impregnated with oxidase reagent; the dark-blue color which appears within 30 s indicates indophenol oxidase activity. The test can also be performed by adding tetramethyl-*p*-phenylenediamine dihydrochloride or α-naphthol followed by dimethyl-*p*-phenylenediamine monohydrochloride reagents to colonies. The tetramethyl reagent tends to be more sensitive than the dimethyl reagent. Oxidase tests should be performed with 18- to 24-h-old growth by either method, using colonies present on noninhibitory medium, e.g., infusion agar. They should not be performed using colonies present on selective-differential agar media or Mueller-Hinton agar.

Arginine dihydrolase; lysine and ornithine decarboxylases. The activities of arginine dihydrolase and lysine and ornithine decarboxylases should be determined in Moeller decarboxylase base medium (Difco Laboratories no. 0890) containing the indicated amino acid. After the media and the base medium control are inoculated, sterile mineral oil is aseptically overlaid. Enzymatic activity is indicated by an alkaline (dark purple) reaction when compared with the inoculated base medium control (light slate color).

Growth at 42°C. Growth at 42°C on an infusion agar slant is used in the differentiation of fluorescent *Pseudomonas* species which fail to produce pyocyanin.

Motility. Motility can be determined in a hanging-drop preparation which is prepared by placing a loopful of a broth culture in the center of a cover glass edged with petrolatum and inverting the cover glass

TABLE 2. Media and tests for isolation and identification of *Pseudomonas* species

For isolation, presumptive identification, and screening
Blood, infusion, or peptone agar
MacConkey, Leifson deoxycholate, or eosin-methylene blue agar
Kligler iron or triple sugar iron agar
Motility
OF glucose, xylose, lactose, and maltose media
Indophenol oxidase
Arginine dihydrolase (Moeller)
Growth at 42°C
Odor, colonial morphology, pigmentation
Somatic morphology (Gram stain)
Flagella stain
Antibiogram
For further identification
Christensen urea agar
Phenylalanine deaminase test agar
Nitrate broth
Lysine and ornithine decarboxylases
DNase test medium
OF fructose and mannitol media
Esculin hydrolysis agar
Nutrient gelatin
Tryptone broth
Acetate assimilation test medium

on a concave glass slide. The preparations are examined with low- and high-dry objectives of the microscope.

Some glucose nonfermenters may fail to grow or spread in media used to detect motility of *Enterobacteriaceae*. Motility medium designed for glucose-nonfermenting gram-negative rods contains a critical quantity of agar and is also a convenient stock culture medium. The Gard plate, motility medium in a petri dish, is used to select a vigorously motile strain from a population of cells which appears nonmotile or sluggishly motile or which appears to contain few motile cells. The medium is stab inoculated in the center of the plate, and motile inoculum is later removed from the periphery of the giant spreading colony.

Flagellar morphology. Flagellar morphology is readily observed with a light microscope after staining, for example, with the tannic acid-fuchsin method described by Leifson (36). Although many *Pseudomonas* species can be identified and differentiated from other motile glucose-nonfermenting rods without determining their flagellar morphology, it should be determined for the identification of apyocyanogenic or melanogenic *P. aeruginosa* strains, motile glucose nonoxidizers (e.g., *P. alcaligenes, P. diminuta, P. testosteroni, Alcaligenes faecalis, Bordetella bronchicanis*), and enzymatically mutilated strains. Some glucose-nonfermenting gram-negative rods cannot be identified without this information.

Most flagellated glucose-nonfermenting gram-negative rods can be divided into three groups based on attachment of the flagellum or flagella to the bacterial soma. Cells of strains of polar monotrichous species usually have one flagellum per pole, and a few cells in the population may have two flagella at one pole. The number of flagella per pole on cells of strains of species with a tuft of polar flagella varies from zero to six or more. Cells with a flagellum or flagella at both poles often appear to be dividing. Cells of strains of peritrichous species have one or more flagella arranged on a cell. The flagellar shape and the site of attachment of the flagellum or flagella to the bacterial cell are usually uniform, constant, and characteristic of the species. Cells without flagella often occur in flagellated populations. Under certain conditions of growth, some strains produce morphovars with lateral flagella. The wavelength of lateral flagella is shorter than that of the terminal flagellum or flagella.

When *Pseudomonas* species are flagellated, the flagellum or flagella are attached at a pole. *Alcaligenes* species, *Bordetella bronchicanis, Agrobacterium* species, and *Achromobacter* species are peritrichous when flagellated, produce indophenol oxidase, and are asporogenous. *Acinetobacter* species are asporogenous nonmotile rods and do not produce indophenol oxidase, whereas *Moraxella* species are asporogenous nonmotile rods that produce indophenol oxidase. *Flavobacterium* species are asporogenous nonmotile rods that produce indophenol oxidase, indole, and an intracellular yellow pigment. *Sphingobacterium* species do not produce indole but otherwise are similar to flavobacteria.

Additional identification methods

Approximately 15% of all gram-negative isolates from clinical specimens are glucose-nonfermenting gram-negative rods. Pyocyanogenic *P. aeruginosa* isolates account for about 70% of these isolates. *Acinetobacter* is the second most frequently encountered glucose-nonfermenting gram-negative rod, followed by *P. maltophilia*. Other frequently encountered glucose nonfermenters include *P. fluorescens, P. putida, P. cepacia, P. stutzeri, Flavobacterium indologenes,* and *Alcaligenes* species.

Although some *Pseudomonas* species and other glucose-nonfermenting gram-negative rods can be identified from information collected during the screening procedure described, it is necessary to detect additional characteristics (media listed in Table 2) to identify other isolates.

Since *Pseudomonas* species do not produce indole, *Flavobacterium meningosepticum* and related species which produce indole can be distinguished. The synthesis of indole by glucose-fermenting gram-negative rods is usually detected by adding Kovacs reagent to a 24- or 48-h culture containing tryptone. Indole production by *Flavobacterium* species can be detected by adding Ehrlich reagent, after xylene extraction, to a 24-h broth culture. Urease activity can be detected quickly if the urea medium is warmed to incubator temperature before inoculation. The ability of pseudomonads to utilize an organic compound (e.g., acetate, DL-norleucine, pelargonate, testosterone, DL-β-hydroxybutyrate, *p*-hydroxybenzoate, etc.) as sole source of carbon and energy in a mineral base medium is employed in identification schemes (51).

Incubation temperature

Although primary isolation media are usually incubated at 35 or 37°C, some *Pseudomonas* species in clinical specimens grow slowly at these temperatures and may be masked by other bacteria. Primary isolation media initially incubated at 35°C for 24 to 48 h should be reincubated at 30°C or room temperature

TABLE 3. Type, neotype, and lectotype strains for identification of *Pseudomonas* species

Species	ATCC[a] no.	NCTC[b] no.	Strain status	Mol% G+C
Pseudomonas acidovorans	15668		Type	67
P. aeruginosa	10145	10332	Neotype	
P. alcaligenes	14909	10367	Neotype	66
P. cepacia	25416	10743	Type	67
P. diminuta	11568	8545	Type	67
P. fluorescens	13525	10038	Neotype	
P. gladioli	10248		Lectotype	68
P. mallei	23344		Neotype	69
P. maltophilia	13637	10257	Type	67
P. mendocina	25411		Type	64
P. mesophilica	29983		Type	66
P. paucimobilis	29837	11030	Type	65
P. pertucinogena	190		Type	60
P. pickettii	27511		Type	64
P. pseudoalcaligenes	17440		Type	63
P. pseudomallei		1691[c]	Neotype	
P. putida	12633		Neotype	
P. putrefaciens	8071		Type	44
P. stutzeri	17588		Neotype	64
P. testosteroni	11996	10698	Type	62
P. vesicularis	11426		Type	66
Comamonas terrigena	8461	1937	Neotype	65

[a] American Type Culture Collection, Rockville, Md.
[b] National Collection of Type Cultures, London, England.
[c] Ragaviah strain (61).

(18 to 22°C) for an additional day or 2 to permit the growth of the temperature-sensitive bacteria. Media used to detect enzymatic activity of glucose-nonfermenting gram-negative rods should be incubated at 30°C unless otherwise stated or indicated. Since flagellar proteins are optimally synthesized at low temperatures, motility medium and cultures to be stained for flagella should be incubated at room temperature.

Each laboratory should establish in-house standards and data for identification of glucose-nonfermenting gram-negative rods. The identification procedure should be standardized to include a comparative study of the corresponding type, neotype, or holotype or a recognized reference strain of the species concerned. A list of such strains is recorded in Table 3. Some of the methods used in the study of pseudomonads have been described in greater detail (22, 60).

A number of factors determine the extent to which a laboratory should proceed with identification of glucose-nonfermenting gram-negative rods, including frequency of isolation, need for epidemiological surveillance, special interests of the infectious disease staff, and resources available to the diagnostic bacteriology laboratory, e.g., physical facilities, budget, knowledgeable personnel, and time. Some diagnostic laboratories may find it expedient to use one or more of the commercial micromethod identification kits designed for glucose nonfermenters (2); others may find it expedient to send suspected *Pseudomonas* species and other glucose-nonfermenting gram-negative rods to a reference laboratory for further identification.

PSEUDOMONAS SPECIES

The minimal characters for identification of most strains of *Pseudomonas* species are recorded in Table 4. The species are obligate aerobes and have a respiratory type of metabolism with oxygen as the terminal electron acceptor. Some species can use nitrate as an alternative electron acceptor and grow under anaerobic conditions in the presence of nitrate. Energy is not obtained by fermentative or photosynthetic metabolism. Some species do not produce acid from glucose. Indole and acetylmethylcarbinol are not produced, and the methyl red test is negative. *Pseudomonas* species produce various fluorescent and nonfluorescent pigments. Most species grow in a mineral base medium, without organic growth factors, containing ammonium ion as sole source of nitrogen and glucose as sole source of carbon and energy. A few species require vitamins or amino acids as growth factors. Most pseudomonads contain 57 to 70 mol% guanine plus cytosine (G+C) in their DNA. The characters of *Pseudomonas* species and the minimal characters for identification of each, based on a study of 2,102 strains, are recorded in Tables 5 through 29.

FLUORESCENT GROUP (*PSEUDOMONAS AERUGINOSA, PSEUDOMONAS FLUORESCENS,* AND *PSEUDOMONAS PUTIDA*)

Most strains of the three species of the fluorescent group produce water-soluble pigments (pyoverdins) which fluoresce when exposed to UV light of short wavelength (ca. 254 nm). Fluorescent pigments may be yellow-green, yellow-brown, or without color. They are not soluble in chloroform. Production of pyoverdins is influenced by nutritional factors, and media which support growth may not promote their synthesis. Most strains produce indophenol oxidase and arginine dihydrolase. Those strains which lack these features may be more closely related to phytopathogenic fluorescent pseudomonads such as *P. syringae*. The majority of strains grow in mineral base medium containing ammonium ion as sole source of nitrogen and glucose as sole source of carbon and energy.

Pseudomonas aeruginosa

Identification

P. aeruginosa has the characters of the genus *Pseudomonas* as recorded in Table 4. The characteristics of 84 apyocyanogenic strains of *P. aeruginosa* are recorded in Table 5. Minimal characters for identification of most apyocyanogenic *P. aeruginosa* strains are given in Table 6. The neotype strain of *P. aeruginosa* has been described (28, 59).

TABLE 4. Minimal characters for identification of strains of *Pseudomonas* species

Character	Sign
Gram-negative, straight or slightly curved rod	+
Asporogenous	+
Polar monotrichous or polar tuft of flagella[a]	+
Motility[a]	+
OF glucose medium open, acid	+ or −
OF glucose medium sealed, acid	−
Indophenol oxidase	+ or −
Catalase	+
Photosynthetic pigment	−

[a] *P. mallei* is nonmotile and without flagella.

TABLE 5. Characters used in identification of fluorescent group *Pseudomonas* species[a]

Test, substrate, or morphology	*P. aeruginosa* (84 strains)		*P. fluorescens* (194 strains)		*P. putida* (273 strains)	
	Sign	% Positive	Sign	% Positive	Sign	% Positive
Acid:						
Glucose, 1% (OFBM)	+	98	+	100	+	100
Fructose	+ or −	89	+	99	+	99
Galactose	+ or −	80	+	97	+	99
Mannose	+ or −	77	+	99	+	99
Rhamnose	− or +	20	− or +	48	− or +	32
Xylose	+ or −	85	+ or −	97	+ or −	98
Lactose	−	0	− or +	12	− or +	14
Sucrose	−	0	− or +	47	−	8
Maltose	− or +	11	− or +	36	− or +	21
Mannitol	+ or −	68	+	93	− or +	18
Lactose, 10% (PAB)	− or +	14	+ or −	62	− or +	42
ONPG	−	4	−	1	−	1
Pyoverdin	+ or −	71	+	95	+ or −	84
Hydrogen sulfide (KIA)	−	0	−	0	−	0
Nitrate reduction	+ or −	75	− or +	19	−	0
Gas from nitrate	+ or −	61	−	5	−	0
Indophenol oxidase	+	100	+	100	+	100
Arginine dihydrolase (DBM)	+	99	+	99	+	99
Lysine decarboxylase	−	0	−	0	−	0
Ornithine decarboxylase	−	0	−	0	−	0
Phenylalanine deaminase	−	7	−	3	−	0
Hydrolysis:						
Urea	+ or −	67	− or +	41	− or +	43
Esculin	−	0	−	0	−	0
Tween 80	+ or −	71	+ or −	63	−	0
Starch	−	7	− or +	19	− or +	18
DNA	− or +	10	−	0	−	0
Lecithin	−	8	+	90	−	0
Gelatin	+ or −	50	+	100	−	0
Acetamide[b]	− or +	37	−	1	−	4
Hemolysis	− or +	39	− or +	14	−	0
Growth on:						
SS	+ or −	88	+	97	+	98
MacConkey	+	99	+	100	+	100
6.5% NaCl	−	7	−	2	− or +	10
Cetrimide (PA)	+	91	+	96	+	92
MBM + acetate	+	96	+	100	+	100
Growth at 42°C	+	100	−	0	−	0
Polymyxin susceptible	+	98	+	97	+	99
Motile	+	97	+	100	+	100
No. of flagella[b]	1		>1		>1	

[a] Except where indicated, cultures were incubated at 30°C. All strains were indole negative. *P. aeruginosa* strains did not produce pyocyanin. Abbreviations for all tables: OFBM, OF basal medium; PAB, purple agar base; DBM, decarboxylase base, Moeller; SS, salmonella-shigella agar; PA, Pseudosel agar; MBM, mineral base medium. Signs: +, 90% or more positive within 2 days; −, no reaction (90% or more); + or −, most cultures positive, some strains negative; − or +, most strains negative, some cultures positive; 1, polar monotrichous flagella; and >1, polar multitrichous flagella.

[b] Not all strains were examined; results are based on the fraction of strains tested.

Most *P. aeruginosa* strains are identified on the basis of the characteristic grapelike odor of aminoacetophenone, colonial morphology, and production of pyocyanin, a blue, water-soluble, nonfluorescent, phenazine pigment that is soluble in chloroform. *P. aeruginosa* is the only *Pseudomonas* species or gram-negative rod known to excrete pyocyanin. Strains which produce pyocyanin and the yellowish pyoverdins often impart a greenish color to culture media.

P. aeruginosa strains may synthesize various combinations of pyocyanin, pyoverdins, pyorubin (red, water-soluble pigment), and pyomelanin (brown to black, water-soluble pigment). Pyocyanin may be masked by other pigments. Strains which produce pyorubin or pyomelanin, as well as highly mucoid strains, may fail to oxidize carbohydrates, as do most strains of *P. aeruginosa*. Highly mucoid strains may also lack motility.

A number of colony types (smooth, coliform-type, rough, mucoid, gelatinous, dwarf) may be observed. Colonial variants arising by dissociation of a single strain give the false impression that different bacterial species are present. The variant forms may demonstrate altered exoenzyme activity, bacteriophage pattern, serovar, antibiograms, and antigenicity.

Apyocyanogenic strains may produce aberrant biochemical character patterns, and their identification may pose a problem. The following universal characters of pyocyanogenic strains are unreliable for identification of apyocyanogenic strains: grapelike odor;

TABLE 6. Minimal characters for identification of apyocyanogenic *P. aeruginosa* strains

Character	Sign	% Positive[a]
Polar monotrichous, fewer than three flagella per pole	+	97
Motility	+	97
OF glucose medium open, acid	+	98
OF lactose, sucrose media, acid	−	0
Indophenol oxidase	+	100
Pyoverdin	+ or −	71
L-Arginine dihydrolase	+	99
Growth at 42°C	+	100

[a] From Table 5.

nitratase, nitritase, gelatinase, caseinase, and lipase activity; acid production from mannitol in OF medium; utilization of acetamide; tolerance to cetrimide; and susceptibility to carbenicillin. Apyocyanogenic strains are more often susceptible to tetracycline, streptomycin, and kanamycin than are pyocyanin-producing strains. Generally, nonpigmented strains are not typable by antisera usually used to determine the serovar of *P. aeruginosa* strains, and they show little or no aeruginocin (pyocin) activity. Apyocyanogenic strains can be recognized as variants of *P. aeruginosa* by several uniform characters (see Table 6), namely, the presence of polar monotrichous flagella, growth at 42°C, and failure to produce acid from lactose and sucrose. Strains designated as unidentified fluorescent *Pseudomonas* species are similar to apyocyanogenic *P. aeruginosa* strains in that they grow at 42°C, but they possess a polar tuft of flagella (1).

Chemical constituents of bacterial cells, such as cellular fatty acids (8, 9, 44), respiratory quinone systems (48, 49), and biosynthetic enzymes (66), are potentially useful tools for identification of *P. aeruginosa* and other *Pseudomonas* species. Serological (41), bacteriophage (6), aeruginocin typing (16), and enzyme profile (31) systems have been used as epidemiological markers.

Sources of isolation and relationship to disease

P. aeruginosa is a physiologically versatile organism, widely distributed in soil, water, sewage, the mammalian gut, and plants. It may be pathogenic for humans, as well as for certain animals, insects, and plants. *P. aeruginosa* is susceptible to drying but will survive for many months in water at ambient temperature. Simple nutritional requirements and the ability to metabolize a variety of organic substances may enable *P. aeruginosa* to survive and multiply in fluids and moist environments found in hospital wards.

Infection with *P. aeruginosa* is generally restricted to hospitalized patients. Predisposing factors for nosocomial infections include metabolic, hematologic, and malignant diseases. Hospital-acquired infections occur in patients who have had prior instrumentation or manipulative procedures such as urethral catheterizations, tracheostomies, lumbar punctures, and intravenous infusions of medications and fluids. Patients become susceptible to *P. aeruginosa* infections after prolonged treatment with immunosuppressive agents, corticosteroids, antimetabolites, antibiotics, and radiation. Hospitalized patients may acquire the bacteria from common environmental sources by contact with human or inanimate vectors. Sources in the hospital environment shown to be transitory vectors include infusion fluids, ophthalmic solutions, dilute phenolic solutions, benzalkonium chloride solution, hexachlorophene soap, hand creams, body lotions, bedside water decanters, ornamental plants, flower vase water, forceps, shaving brushes, oral thermometers, suction apparatus, respiratory ventilators, nebulizing humidifiers, incubators, whirlpools, syringes, ward utensils, dish mops, sponges, floors, baths, sinks, and water taps.

Alternatively, patients may develop an endogenous infection from bacteria present on their skin or in their respiratory or intestinal tract. The proportion of the general population carrying *P. aeruginosa* in the intestinal tract is low, but the administration and selective influence of antibiotic therapy among hospitalized patients have increased the carriage rate among these subjects. The most important human sources for *P. aeruginosa* are infected wounds, urine, and lesions producing an exudate. These moist sources contaminate dressings, bedsheets, patient garments, and equipment, from which the organism may be transferred to other patients by the hands of hospital personnel or visitors. Person-to-person transfer appears to be the most important mode of transmission. *P. aeruginosa* is a cause of severe epidemic diarrhea of infants, ocular infections, burn infections, cystic fibrosis, hot tub- and whirlpool-associated folliculitis, osteomyelitis, and malignant external otitis. Susceptibility tests of *P. aeruginosa* by broth dilution methods may yield low MICs with the aminoglycosides because of the low concentration of magnesium and calcium ions in Mueller-Hinton and many other broth media. Agar diffusion procedures yield more reliable results because of the optimal concentration of divalent cations contributed by most agar media.

P. aeruginosa produces a variety of toxins and enzymes such as surface slime, hemolysin, fibrinolysin, lipase, esterase, lecithinase, elastase, coagulase, DNase, phospholipase, endotoxin, enterotoxin, and exotoxin, some of which may contribute to its patho-

TABLE 7. Minimal characters for identification of *P. fluorescens* and *P. putida* strains[a]

Character	*P. fluorescens*		*P. putida*	
	Sign	% Positive	Sign	% Positive
Polar tuft of three or more flagella per pole	+	100	+	100
Motility	+	100	+	100
OF glucose medium open, acid	+	100	+	100
OF lactose medium, acid	− or +	12	− or +	14
Indophenol oxidase	+	100	+	100
Pyocyanin	−	0	−	0
Pyoverdin	+	95	+ or −	84
L-Arginine dihydrolase	+	99	+	99
Gelatin hydrolysis	+	100	−	0
Growth at 42°C	−	0	−	0

[a] From Table 5.

genesis. The role of virulence factors in the pathogenesis of *P. aeruginosa* has been demonstrated in animal models. A number of literature reviews summarize the relationship of *P. aeruginosa* and other pseudomonads to human diseases (12–14, 42, 45, 55, 64).

Pseudomonas fluorescens and *Pseudomonas putida*

Identification

P. fluorescens and *P. putida* have the characters of the genus *Pseudomonas* as recorded in Table 4. The characteristics of 194 strains of *P. fluorescens* and 273 strains of *P. putida* are listed in Table 5. Minimal characters for identification of most *P. fluorescens* and *P. putida* strains are recorded in Table 7. The neotype strains of *P. fluorescens* and *P. putida* have been described (25, 59, 60).

Strains of *P. fluorescens* and *P. putida* have a polar tuft of flagella. The number of flagella per cell varies from zero to six or more. Most strains contain some cells with three or more flagella per pole. The number of flagella per pole is a valuable criterion for distinguishing *P. aeruginosa* from *P. fluorescens* and *P. putida*. *P. aeruginosa* strains are polar monotrichous and very rarely contain cells with three flagella per pole. *P. fluorescens* and *P. putida* do not produce pyocyanin, and some strains fail to produce fluorescent pigments. They do not grow at 42°C, and many strains fail to grow at 35°C.

These two species historically have been distinguished by proteolytic activity. *P. fluorescens*, but not *P. putida*, produces an extracellular protease, a metalloenzyme, which hydrolyzes gelatin and casein; there is some question, however, concerning the reliability of distinguishing these strains on this basis alone. The variable production of lipase (formation of calcium oleate from Tween 80) and hydrolysis of phosphatidylcholine (egg yolk reaction) by *P. fluorescens* help distinguish it from *P. putida*. *P. putida* does not reduce nitrate to nitrogen gas. *P. fluorescens* and *P. putida* are distinguished from apyocyanogenic strains of *P. aeruginosa* by flagellar morphology, growth at 42°C, and acid production in OF medium from such disaccharides as lactose, maltose, and sucrose. Whereas *P. fluorescens* and *P. putida* are usually susceptible to kanamycin and resistant to carbenicillin, the reverse is true for most *P. aeruginosa* strains.

P. fluorescens strains can be subdivided into seven biovars (A through G), and *P. putida* strains can be subdivided into two biovars on the basis of denitrification, pigment production, levan synthesis from sucrose, and utilization of certain organic compounds (60). Three phenazine-producing biovars of *P. fluorescens* (*P. chlororaphis*, *P. aureofaciens*, *P. lemonnieri*) phenotypically resemble *P. fluorescens* but are recognized as distinct species by some workers.

Sources of isolation and relationship to disease

P. fluorescens and *P. putida* are common inhabitants of soil, water, and plants. They have been recovered from the skin and ears of divers; cosmetics; aviation turbine fuel; hard and soft lens wetting, cleaning, and saline solutions; and contaminated foodstuffs including beef, pork, and milk. They have been frequently isolated from the hospital environment, i.e., water sources, sink drains, floors, benzalkonium chloride solution, fluid from an ultrasonic nebulizer, humidifiers, intravenous catheter, and contaminated blood and blood products stored in a refrigerator.

The species are part of the normal oropharyngeal flora, and most strains are isolated in the clinical laboratory from respiratory tract specimens. A number of these isolates are from cystic fibrosis patients. These organisms are usually environmental contaminants and are rarely opportunistic pathogens for humans. *P. fluorescens* can release endotoxins in contaminated blood and blood products. The phospholipid moiety of the endotoxin produces irreversible shock. These bacteria have been associated with empyema, urinary tract infection, pelviperitonitis, infection in a chronic bronchitic, wound infections, septicemia, and septic arthritis.

STUTZERI GROUP (*PSEUDOMONAS STUTZERI* AND *PSEUDOMONAS MENDOCINA*)

The two species in the stutzeri group are salt tolerant and nonhalophilic but require sodium cation for growth. They grow under strictly anaerobic conditions in nitrate-containing media, as do *P. aeruginosa*, *P. pseudomallei*, and other bacterial species which reduce nitrate to nitrogen gas. The strains are motile with polar monotrichous flagella. Lateral flagella are produced by some cells in some strains. Indophenol oxidase and oxidative acidity from glucose are produced. These two species grow with ammonium ion as sole source of nitrogen and acetate as sole source of carbon and energy. They are susceptible to polymyxins.

Pseudomonas stutzeri

Identification

P. stutzeri has the characters of the genus *Pseudomonas* as recorded in Table 4. The characteristics of 168 strains of *P. stutzeri* are recorded in Table 8. Minimal characters for identification of most *P. stutzeri* strains are recorded in Table 9. A neotype strain has been designated and described (52, 60).

Most freshly isolated strains produce dry, wrinkled, leathery, adherent colonies as well as smooth colonies and various intermediate types. Colonies may become buff to light-brown in color due to high concentrations of cytochrome *c* in the cells. Smooth colonies predominate after repeated subculture. Rare strains produce only smooth, spreading colonies, similar to strains of *Proteus* species, or colonies which pit blood agar medium.

Ability to produce nitrogen gas may be lost on repeated subculture. Strains which no longer denitrify or produce wrinkled colonies will regain these features after serial passage in nitrate-containing media. Arginine dihydrolase is produced only by rare, biovar Vb-3 strains.

Maltose and, usually, starch are utilized. These and other reactions distinguish *P. stutzeri* from *P. mendocina*. *P. stutzeri* may be confused with *P. pseudomallei*, since both produce rough, wrinkled, corrugated colonies. Flagellar morphology; susceptibility to polymixin; starch, arginine, and lactose reactions; and other characters distinguish *P. stutzeri* from *P. pseudomallei*.

TABLE 8. Characters used in identification of stutzeri group *Pseudomonas* species[a]

Test, substrate, or morphology	*P. stutzeri* (168 strains)		*P. mendocina* (6 strains)	
	Sign[b]	% Positive	Sign	% Positive
Acid:				
Glucose, 1% (OFBM)	+	100	+	100
Fructose	+	94	+	100
Galactose	+	91	+	100
Mannose	+ or −	88	+	100
Rhamnose	− or +	24	−	0
Xylose	+	94	+	100
Lactose	−	0	−	0
Sucrose	−	0	−	0
Maltose	+	98	−	0
Mannitol	+ or −	69	−	0
Lactose, 10% (PAB)	−	0	−	0
ONPG	−	0	−	0
Pyoverdin	−	0	−	0
Hydrogen sulfide (KIA)	−	0	−	0
Nitrate reduction	+	100	+	100
Gas from nitrate	+	100	+	100
Indophenol oxidase	+	100	+	100
Arginine dihydrolase (DBM)	−	5	+	100
Lysine decarboxylase	−	0	−	0
Ornithine decarboxylase	−	0	−	0
Phenylalanine deaminase	+ or −	51	+ or −	50
Hydrolysis:				
Urea	− or +	15	+ or −	50
Esculin	−	0	−	0
Tween 80	+	97	+ or −	83
Starch	+	93	−	0
DNA	−	0	−	0
Lecithin	−	8	−	0
Gelatin	−	1	−	0
Acetamide[b]	−	0	−	0
Hemolysis	−	0	−	0
Growth on:				
SS	+ or −	82	+ or −	83
MacConkey	+	100	+	100
6.5% NaCl	+	100	+	100
Cetrimide (PA)	−	2	+	100
MBM + acetate	+	100	+	100
Growth at 42°C	+ or −	89	+	100
Polymyxin susceptible	+	99	+	100
Motile	+	100	+	100
No. of flagella[b]	1		1	

[a] Except where indicated, cultures were incubated at 30°C. All strains were indole negative. For abbreviations and signs, see Table 5, footnote *a*.

[b] Not all strains were examined; results are based on the fraction of the strains tested.

P. stutzeri demonstrates marked nutritional versatility in that only 17 organic compounds serve as substrates for all strains, whereas 72 compounds can be used by some strains. The wide variation in the DNA base composition, 60.7 to 66.3 mol% G+C, of *P. stutzeri* may be related to the high degree of nutritional heterogeneity. Since the phenotypic characters of *P. stutzeri* are heterogeneous, it may include several biologically distinct species.

Sources of isolation and relationship to disease

P. stutzeri is ubiquitous in soil and water and has been found in manure, humus, straw, sewage, stagnant water, distilled water, deionized water, sink drains, tap water aerator, baby formula, aerosolization equipment, hemodialysis machine, inanimate surfaces, aqueous green soap, hard lens wetting solution, eye cosmetics, and various clinical specimens. On a few occasions *P. stutzeri* has been directly associated with an infectious process, e.g., endocarditis, septicemia, septic arthritis, postoperative and posttraumatic infections of the extremities, otitis media, conjunctivitis, and urinary tract infections.

Pseudomonas mendocina

Identification

P. mendocina has the characters of the genus *Pseudomonas* as recorded in Table 4. The characteristics of six strains of *P. mendocina* are recorded in Table 8. Minimal characters for identification of most *P. mendocina* strains are recorded in Table 9. *P. mendocina* ATCC 25411 (CH20) (52) should be recognized as the type (24).

Colonies of *P. mendocina* are flat, smooth, and butyrous and produce a brown-yellow, intracellular,

TABLE 9. Minimal characters for identification of *P. stutzeri* and *P. mendocina* strains[a]

Character	*P. stutzeri*		*P. mendocina*	
	Sign	% Positive	Sign	% Positive
Polar monotrichous, fewer than three flagella per pole	+	100	+	100
Motility	+	100	+	100
OF glucose medium open, acid	+	100	+	100
OF lactose medium, acid	−	0	−	0
OF maltose medium, acid	+	98	−	0
Indophenol oxidase	+	100	+	100
Nitrate to gas	+	100	+	100
L-Arginine dihydrolase	−	5	+	100

[a] From Table 8.

carotenoid pigment. Wrinkled colonies are not produced. Water-insoluble, alcohol-soluble carotenoid and carotenoidlike pigments, such as those found in *P. mendocina*, characteristically remain associated with the cells. Arginine dihydrolase is produced. Starch and maltose are not utilized. The base variation of the DNA in *P. mendocina* shows little variation, with 62.8 to 64.3 mol% G+C content.

P. mendocina may be confused with denitrifying biovars of *P. fluorescens*. Polar monotrichous morphology, growth at 42°C, and other characters distinguish it from nitrogen gas-producing strains of *P. fluorescens.*

Sources of isolation and relationship to disease

P. mendocina has been isolated from soil, water, and urine. It has not been associated with infections in humans.

ALCALIGENES GROUP (*PSEUDOMONAS ALCALIGENES* AND *PSEUDOMONAS PSEUDOALCALIGENES*)

The two species in the alcaligenes group have the characters of the genus *Pseudomonas* as recorded in Table 4. The cells of most strains have no somatic curvature, and some strains contain a few cells which are distinctly curved. They are polar monotrichous when flagellated, and the wavelength of the flagellum is approximately 1.6 μm. The flagellar morphology of the species in the alcaligenes group is similar to that of *P. aeruginosa*.

Most strains grow in mineral base medium containing ammonium ion as sole source of nitrogen and acetate as sole source of carbon and energy. The neotype strain of *P. alcaligenes* does not utilize glucose, and alkali accumulates in unsealed OF glucose medium and OF basal medium. Indophenol oxidase is produced. Pyocyanin and pyoverdins are not produced.

Pseudomonas alcaligenes and *Pseudomonas pseudoalcaligenes*

Identification

The characteristics of 51 strains of *P. alcaligenes* and 62 strains of *P. pseudoalcaligenes* are recorded in Table 10. Minimal characters for identification of most *P. alcaligenes* and *P. pseudoalcaligenes* strains are recorded in Table 11. The neotype strain of *P. alcaligenes* (26) and the type strain of *P. pseudoalcaligenes* (60) have been described. The flagellar morphology of the neotype strain of *P. alcaligenes* is illustrated in reference 26.

Strains assigned to the alcaligenes group can be accommodated in two named species, *P. alcaligenes* (*P. alcaligenes* biovar A) and *P. pseudoalcaligenes* (*P. alcaligenes* biovar B). *P. pseudoalcaligenes* produces a weak acid reaction from fructose in OF basal medium and is nutritionally more versatile than *P. alcaligenes*. Fructose is the only substrate used by all strains of *P. pseudoalcaligenes* and none of the strains of *P. alcaligenes*. An additional five substrates not used by any of the latter are used by most of the more versatile *P. pseudoalcaligenes* strains. Phenotypic characters do not easily distinguish the two species. Hybridization experiments indicate a high degree of homology within each of the two species and a clear separation between them.

Sources of isolation and relationship to disease

P. alcaligenes and *P. pseudoalcaligenes* have been isolated from pond, river, swimming pool, and turtle aquarium waters; cosmetics; cell culture; raw milk; frozen fish; liver of piglet; rabbit blood culture; feces of rabbits, frogs, and humans; and numerous clinical sources. They are occasionally opportunistic agents. *P. alcaligenes* has been associated with endocarditis, neonatal septicemia, empyema, and eye infections. *P. pseudoalcaligenes* has been identified as the etiological agent of pneumonitis, a postoperative knee infection, septicemia, and meningitis.

Pseudomonas sp. group 1, which contains some strains previously designated *Pseudomonas denitrificans* (65), is distinguished from *P. alcaligenes* by the reduction of nitrate to nitrogen gas. The species name *P. denitrificans* (Christensen) Bergey et al. 1923 has been rejected by the Judicial Commission of the International Committee on Systematic Bacteriology, since strains designated *P. denitrificans* appear to belong to species in several genera. An isolate referred to as *P. denitrificans* caused bacteremia and meningitis.

PSEUDOMALLEI GROUP (*PSEUDOMONAS PSEUDOMALLEI, PSEUDOMONAS MALLEI, PSEUDOMONAS CEPACIA,* AND *PSEUDOMONAS PICKETTII*)

The principal feature shared by the species in the pseudomallei group is nutritional versatility in the type and number of organic compounds utilized as sole sources of carbon and energy, which include carbohydrates, mono- and dicarboxylic acids, mono- and polyalcohols, aromatic compounds, amino acids, and amines. Growth occurs in mineral base medium containing ammonium ion as sole source of nitrogen and glucose as sole source of carbon and energy. A few strains fail to produce indophenol oxidase, and others produce a slow and very weak indophenol oxidase reaction. The majority of strains are not susceptible to antibiotics of the polymyxin class.

P. pickettii, the animal pathogens *P. mallei* and *P. pseudomallei*, and the plant pathogens *P. cepacia, P.*

TABLE 10. Characters used in identification of alcaligenes group *Pseudomonas* species[a]

Test, substrate, or morphology	*P. alcaligenes* (51 strains)		*P. pseudoalcaligenes* (62 strains)	
	Sign	% Positive	Sign	% Positive
Acid:				
Glucose, 1% (OFBM)	−	0	− or +	12
Fructose	−	0	+	100
Galactose	−	0	−	2
Mannose	−	0	−	2
Rhamnose	−	0	−	0
Xylose	−	0	− or +	10
Lactose	−	0	−	0
Sucrose	−	0	−	0
Maltose	−	0	− or +	13
Mannitol	−	0	−	2
Lactose, 10% (PAB)	−	0	−	0
ONPG	−	0	−	0
Pyoverdin	−	0	−	0
Hydrogen sulfide (KIA)	−	0	−	0
Nitrate reduction	+ or −	59	+	95
Gas from nitrate	− or +	16	−	5
Indophenol oxidase	+	100	+	100
Arginine dihydrolase	− or +	10	− or +	34
Lysine decarboxylase	−	0	−	0
Ornithine decarboxylase	−	0	−	0
Phenylalanine deaminase	− or +	20	− or +	16
Hydrolysis:				
Urea	− or +	20	−	1
Esculin	−	0	−	0
Tween 80	− or +	49	− or +	10
Starch	− or +	10	−	5
DNA	−	0	−	0
Lecithin	−	0	−	0
Gelatin	−	2	−	2
Acetamide[b]	−	0	−	3
Hemolysis	− or +	14	−	0
Growth on:				
SS	− or +	16	+ or −	73
MacConkey	+	100	+	94
6.5% NaCl	− or +	12	−	2
Cetrimide (PA)	−	8	+ or −	61
MBM + acetate	+	90	+	99
Growth at 42°C	+ or −	59	+ or −	76
Polymyxin susceptible	+	96	+	95
Motile	+	100	+	90
No. of flagella[b]	1		1	

[a] Except where indicated, cultures were incubated at 30°C. All strains were indole negative. For abbreviations and signs, see Table 5, footnote *a*.

[b] Not all strains were examined; results are based on the fraction of strains tested.

gladioli (P. marginata, P. alliicola), and *P. caryophylli* are enzymologically, phenotypically, and genotypically related (4, 53, 66).

Pseudomonas pseudomallei

Identification

P. pseudomallei has the characters of the genus *Pseudomonas* as recorded in Table 4. The characteristics of seven strains of *P. pseudomallei* are recorded in Table 12. Characters useful for the identification of *P. pseudomallei* strains are recorded in Table 13. The neotype strain (61) was isolated in 1921.

Colonies vary from mucoid and smooth to rough and wrinkled in texture and from bright orange to cream color. Colonial dissociation occurs with the same frequency as in *P. aeruginosa*. *P. pseudomallei*, *P. mallei*, *P. stutzeri*, and *P. aeruginosa*, as well as other

TABLE 11. Minimal characters for identification of *P. alcaligenes* and *P. pseudoalcaligenes*[a]

Character	*P. alcaligenes*		*P. pseudoalcaligenes*	
	Sign	% Positive	Sign	% Positive
Polar monotrichous, fewer than three flagella per pole, and normal wavelength	+	100	+	100
Motility	+	100	+	100
OF glucose medium open, acid	−	0	− or +	12
OF fructose medium, acid	−	0	+	100
Indophenol oxidase	+	100	+	100
Hydrogen sulfide, black butt in KIA	−	0	−	0

[a] From Table 10.

TABLE 12. Characters used in identification of pseudomallei group *Pseudomonas* species[a]

Test, substrate, or morphology	*P. pseudomallei* (7 strains)		*P. mallei* (6 strains)[b]		*P. cepacia* (201 strains)		*P. pickettii* biovar 1 (26 strains)		*P. pickettii* biovar 2 (24 strains)	
	Sign[c]	% Positive	Sign	% Positive	Sign	% Positive	Sign	% Positive	Sign	% Positive
Acid:										
Glucose, 1% (OFBM)	+	100	+	100	+	100	+ or (+)	100	+ or (+)	100
Fructose	+	100	+ or (+)	100	+	100	+ or (+)	100	+ or (+)	100
Galactose	+	100	+	100	+	100	+ or (+)	100	+ or (+)	100
Mannose	+	100	+ or (+)	100	+	100	+ or (+)	100	+ or (+)	100
Rhamnose	+ or −	71	−	0	−	0	−	0	−	0
Xylose	+ or −	86	−	0	+	99	+ or (+)	100	+ or (+)	100
Lactose	+	100	(+) or +	100	+	98	+ or (+)	100	−	0
Sucrose	+ or −	86	−	0	+ or −	81	−	0	−	0
Maltose	+	100	(+)	100	+	97	+ or (+)	100	−	0
Mannitol	+	100	(+) or −	83	+	100	−	0	−	0
Lactose, 10% (PAB)	+	100		NT[d]	+	97	+, (+), or −	77	−	0
ONPG	−	0		NT	+ or −	78	−	0	−	0
Pyoverdin	−	0	−	0	−	0	−	0	−	0
Hydrogen sulfide (KIA)	−	0	−	0	−	0	−	0	−	0
Nitrate reduction	+ or −	86		NT	− or +	37	+	92	+	100
Gas from nitrate	+	100	−	0	−	0	+, (+), or −	88	+ or (+)	100
Indophenol oxidase	+ or +(w)	100	+, +(w), or −	67	+ or +(w)	92	+	100	+	100
Arginine dihydrolase (DBM)	+	100	+ or (+)	100	−	0	−	0	−	0
Lysine decarboxylase	−	0	−	0	+	90	−	0	−	0
Ornithine decarboxylase	−	0	−	0	+ or −	65	−	0	−	0
Phenylalanine deaminase	−	0		NT	−	2	−	4	− or +	38
Hydrolysis:										
Urea	− or +	43	− or +	17	− or +	43	+	100	+	100
Esculin	+ or −	57		NT	+ or −	65	−	0	−	0
Tween 80	+ or −	86		NT	+	100	+	100	+	100
Starch	−	0		NT	−	1	+ or −	50	−	8
DNA	−	0		NT	−	0	−	0	−	0
Lecithin	+ or −	86		NT	+ or −	50	−	0	−	0
Gelatin	+	100	−	0	+ or −	74	+ or −	77	− or +	38
Acetamide[e]	−	0		NT	+ or −	60	−	4	−	0
Hemolysis	− or +	43		NT	−	5	−	0	−	0
Growth on:										
SS	−	0	+ or −	67	−	9	−	0	−	4
MacConkey	+	100		NT	+	95	+ or −	77	+	100
6.5% NaCl	−	0		NT	−	0	−	0	−	0
Cetrimide (PA)	−	0		NT	+ or −	68	−	0	−	0
MBM + acetate	+	100	+	100	+	99	+	100	+	100
Growth at 42°C	+	100	−	0	+ or −	58	− or +	27	+ or −	63
Polymyxin susceptible	−	0		NT	−	0	−	0	−	0
Motile	+	100	−	0	+	99	+	96	+	100
No. of flagella[e]	>1				>1		1		1	

[a] Except where indicated, cultures were incubated at 30°C. All strains were indole negative. For abbreviations, see Table 5, footnote *a*.

[b] Data from Hugh (21).

[c] Signs: +, 90% or more positive within 2 days; −, no reaction (90% or more); (+), reactions delayed 3 or more days; (+) or +, most reactions delayed 3 or more days, some occur within 2 days; + or (+), most reactions occur within 2 days, some delayed 3 or more days; +, (+), or −, most reactions occur within 2 days, some delayed 3 or more days, some negative; (+) or −, most reactions delayed 3 or more days, some negative; + or −, most cultures positive, some strains negative; − or +, most strains negative, some cultures positive; (w), weakly reactive; 1, polar monotrichous flagella; >1, polar multitrichous flagella.

[d] NT, Not tested.

[e] Not all strains were examined; results are based on the fraction of strains tested.

pseudomonads, may produce wrinkled, rough, corrugated colonial variants. Initial growth of *P. pseudomallei* is accompanied by an odor of putrefaction supplemented by a distinctive, aromatic, pungent odor. Strains are motile with a tuft of three or more flagella per pole. Abundant growth occurs under anaerobic conditions in nitrate-containing media. Gelatin is hydrolyzed, lactose is utilized, and growth

TABLE 13. Characters for identification of *P. pseudomallei* and *P. mallei* strains

Character	*P. pseudomallei*[a]		*P. mallei*[b]	
	Sign	% Positive	Sign	% Positive
Polar tuft of three or more flagella per pole	+	100	−	0
Motility	+	100	−	0
OF glucose medium open, acid	+	100	+	100
OF maltose medium, acid	+	100	(+)	100
Citrate, Simmons	+	96[b]	−	0
Nitrate to gas	+	100	−	0
L-Lysine decarboxylase	−	0	−	0
L-Arginine dihydrolase	+	100	+ or (+)	100
Growth at 42°C	+	100	−	0

[a] From Table 12.
[b] Data from Hugh (21). For signs, see Table 12, footnote *c*.

occurs at 42°C. Identification should be confirmed by agglutination or fluorescent-antibody reaction utilizing commercially available antiserum for *P. pseudomallei*. Flagellar morphology and arginine, lactose, and other reactions distinguish it from *P. stutzeri*.

Sources of isolation and relationship to disease

P. pseudomallei causes melioidosis, an endemic glanders-like disease of humans and animals, in Southeast Asia and northern Australia. Human melioidosis rarely occurs in the Western Hemisphere. It usually occurs in the United States among those who have returned from countries where it is endemic. The free-living soil organism usually infects the host through a skin opening or by inhalation after the host has had contact with soil or water. A soiled pelvic wound received by a man in a farming accident in Oklahoma became infected with an organism identified by cultural and biochemical characteristics as *P. pseudomallei*. It is rarely transmitted from person to person. Prostatic melioidosis can be the source of human-to-human sexual transmission of *P. pseudomallei*. The organism may remain dormant, and asymptomatic infection can persist in humans for many years before the appearance of clinical disease. The incidence of latent infections in humans in endemic areas appears high. Pulmonary melioidosis is a relatively benign disease with good prognosis. Untreated septicemic melioidosis progresses rapidly and has a high mortality rate. Patients in Australia developed nosocomial infection after urethral catheterization; the source of the organism was probably soil on the hospital grounds. Laboratory-acquired infections have been reported. Because the host spectrum is broad and the clinical manifestations of melioidosis in humans are varied, diagnosis is dependent on isolation and identification of the agent of infection as *P. pseudomallei*.

Pseudomonas mallei

Identification

P. mallei has the characters for the genus *Pseudomonas* as recorded in Table 4. It is the only nonmotile species in the genus *Pseudomonas*. The characteristics of six strains of *P. mallei* are recorded in Table 12. Characters useful for identification of *P. mallei* strains are recorded in Table 13. The neotype strain has been described (58).

Colonies are smooth and range from white to cream color. *P. mallei* grows slowly as compared with *P. aeruginosa* or *P. pseudomallei*, and the cell crop yield is lower. It fails to produce nitrogen gas, hydrolyze gelatin, or grow at 42°C. Although *P. mallei* vigorously reacts with *P. pseudomallei* antiserum, it is distinguished from *P. pseudomallei* by the characters recorded in Table 13.

Sources of isolation and relationship to disease

P. mallei is the cause of glanders (farcy), a natural disease primarily of equines. Occasionally, it is transmitted from equine hosts to humans by direct contact through abraded skin and inhalation. It can be transmitted from person to person. Glanders has been eradicated in the United States and Canada. *P. mallei* can be isolated from blood, sputum, or pus from patients with glanders. It is thought to be the only highly adapted parasite of animals in the genus *Pseudomonas*.

Pseudomonas cepacia

Identification

P. cepacia has the characters of the genus *Pseudomonas* as recorded in Table 4. The characteristics of 201 strains of *P. cepacia* are recorded in Table 12. Minimal characters for identification of most *P. cepacia* strains are recorded in Table 14. The type strain has been designated (53).

Some *P. cepacia* strains produce a conspicuous sulfur-yellow or green, water-soluble, chloroform-soluble, nonfluorescent phenazine pigment. On chemically defined media, the pigments may show a great variety of colors including brown, red, and purple, depending on the carbon sources used for growth. A strain may produce one or more pigments. These pigments occur in the bacterial growth and often diffuse into the agar medium. A diffusible fluorescent pigment may be produced which is violet in UV light and is distinguished from the bluish-green fluorescence of *P. fluorescens*. Many strains of clinical origin are nonpigmented.

Colonies on blood agar are opaque, glistening, occasionally mottled, convex in elevation, and butyrous in consistency. A sweet odor characteristically associat-

TABLE 14. Minimal characters for identification of *P. cepacia* strains[a]

Character	Sign	% Positive
Polar tuft of three or more flagella per pole	+	99
Motility	+	99
OF glucose medium open, acid	+	100
OF lactose medium, acid	+	98
OF maltose medium, acid	+	97
OF mannitol medium, acid	+	100
Nitrate to gas	−	0
L-Lysine decarboxylase	+	90
L-Arginine dihydrolase	−	0

[a] From Table 12.

TABLE 15. Minimal characters for identification of *P. pickettii* strains[a]

Character	Biovar 1		Biovar 2	
	Sign	% Positive	Sign	% Positive
Polar monotrichous, fewer than three flagella per pole	+	96	+	100
Motility	+	96	+	100
OF glucose medium open, acid	+ or (+)	100	+ or (+)	100
OF lactose medium, acid	+ or (+)	100	−	0
OF maltose medium, acid	+ or (+)	100	−	0
OF mannitol medium, acid	−	0	−	0
Indophenol oxidase	+	100	+	100
Nitrate to gas	+, (+) or −	88	+ or (+)	100
L-Arginine dihydrolase	−	0	−	0

[a] From Table 12.

ed with *P. aeruginosa* is produced by some strains. Isolates should be subcultured frequently since they tend to become nonviable after 2 to 3 days on blood agar medium. Optimal growth occurs at 30°C, and some strains grow poorly at 35°C.

P. cepacia is motile with a tuft of three to eight polar flagella. Some cells in a population may have no flagella, and some may have only one flagellum per pole. Cells with lateral flagella occur. A few strains are atrichous. *o*-Nitrophenyl-β-D-galactopyranoside (ONPG), lysine decarboxylase, esculin hydrolysis, proteolytic activity, acidification of carbohydrates, and other reactions distinguish *P. cepacia* from other pseudomonads. A classification into at least eight biovars based on ONPG, nitratase, and esculin hydrolysis permits epidemiological marking of the strains. Bacteriocin patterns and plasmid profiles distinguish strains of clinical and plant origin. A clinical serotyping scheme is based on 12 somatic and flagellar antigens.

It is likely that some clinical isolates identified as *P. cepacia* more closely resemble *P. gladioli* (8). The latter species gives negative reactions for indophenol oxidase, oxidation of sucrose and maltose, and lysine decarboxylase tests, and a positive reaction for urease test.

Sources of isolation and relationship to disease

P. cepacia was first described as a phytopathogen that causes sour skin, an onion bulb rot. It has a wide geographic distribution and has been isolated from raw and pasteurized milk; hard lens and soft lens wetting and saline solutions; nasal sprays; rotting tree trunk; forest soil, muck soil, and soil enrichments; and natural, river, and tap waters. In hospital environments in Malaysia, the United States, Denmark, Holland, France, Canada, and the United Kingdom, *P. cepacia* has been isolated from distilled water, flower vases, disinfectants, detergent solutions, saline solution, aerosol polymyxin, spray atomizer, anesthetics, thermometers, ultrasonic nebulizers, incubators, respirators, humidifiers, water bath, sink drains, pressure transducers, intravenous cannula, hemodialysis coils, hand dip, urinary catheter kits, baby lotion, intravascular inserts, and intravenous fluids.

P. cepacia has been isolated from a variety of clinical materials. It is an opportunistic pathogen, and most infections are of nosocomial origin. *P. cepacia* has been associated wth septicemia, meningitis, endocarditis, urinary tract infections, pneumonia, postoperative wound infections, abscesses, septic arthritis, peritonitis, conjunctivitis, chronic granulomatous disease, and cystic fibrosis. *P. cepacia* demonstrates a multiple antibiotic resistance pattern, especially to the aminoglycosides usually active against *P. aeruginosa*.

Intraperitoneal injection of *P. cepacia* cells in mice resulted in 50% lethal dose values of 3.8×10^8 cells. Endotoxin can be detected by the *Limulus* amoebocyte lysate assay. Animal inoculation is not reliable in the differentiation of *P. cepacia* from *P. pseudomallei*, since some isolates of *P. cepacia* produce pathological changes in animals similar to those produced by *P. pseudomallei*.

Pseudomonas pickettii

Identification

P. pickettii has the characters of the genus *Pseudomonas* as recorded in Table 4. The characteristics of 50 strains of *P. pickettii* are recorded in Table 12. Minimal characters for identification of most *P. pickettii* strains are recorded in Table 15. The type strain has been described (57).

Cells of strains of *P. pickettii* are polar monotrichous. Occasional cells with two polar flagella are observed. The development of growth on blood agar medium characteristically is slow. Growth is not pigmented, and soluble pigments are not produced. Growth on blood agar medium tends to become nonviable after 3 or 4 days.

The development of acid reactions from carbohydrates in OF basal medium and the reduction of nitrate to nitrogen gas characteristically are slow and may require 48 h of incubation to detect. The optimum temperature of incubation for the detection of nitrogen gas is 30°C, not 35°C. Strains which reduce nitrate to nitrogen gas grow anaerobically in the presence of nitrate ion and an oxidizable substrate.

Urease and indophenol oxidase are produced. Arginine dihydrolase and ornithine and lysine decarboxylases are not produced. Strains grow in mineral base medium containing ammonium ion as sole source of nitrogen and glucose as sole source of carbon and energy.

The morphology and the delayed denitrification and delayed carbohydrate oxidation reactions may have contributed to the misidentification of these strains as *P. alcaligenes* or *P. pseudoalcaligenes*.

P. pickettii is a distinct but heterogeneous species and comprises up to seven biovars based on different oxidation patterns and other phenotypic features, i.e., acid production from lactose, maltose, and mannitol and gelatinase production (33, 56). *P. pickettii* biovar 1 (CDC group Va-1) produces acid from lactose and maltose, whereas biovar 2 (CDC group Va-2) does not.

Sources of isolation and relationship to disease

P. pickettii has been isolated from patient sources and from environmental sources including water from an artificial kidney machine, autoclave-cooling water, intravenous fluids, umbilical catheter tip, intravenous catheter, respiratory therapy saline solution, dialysis fluid, nebulizers, cosmetics, paper mill effluent, and teething rings. *P. pickettii* has been associated with septicemia, a case of acute, nonfatal meningitis, and infections of the urinary and respiratory tracts.

ACIDOVORANS GROUP (*PSEUDOMONAS ACIDOVORANS* AND *PSEUDOMONAS TESTOSTERONI*) (*COMAMONAS TERRIGENA*)

The species in the acidovorans group have the characters of the genus *Pseudomonas* as recorded in Table 4, and they also share other attributes. Although most strains are populations of cells with no somatic curvature, an occasional cell of some strains is distinctly curved. Cells are motile with a polar tuft of flagella. The number of flagella at one pole varies from one to six. Strains usually contain some cells with tufts which possess three or more flagella. The tuft of flagella is usually at one pole of the cell; cells with tufts at both poles appear to be dividing. Polar flagella have a mean wavelength of 3.1 μm and an amplitude of 1.08 μm. This distinctive flagellar shape has been described and illustrated in photomicrographs (37, 38). Some strains produce cells with lateral flagella.

Growth is not pigmented, but colonies of some strains are surrounded by a tan or brown zone of discolored agar medium. Most strains grow in mineral base medium containing ammonium ion as sole source of nitrogen and acetate as sole source of carbon and energy.

Alkali accumulates at the surface of OF glucose medium. Acid is not produced from glucose in OF basal medium, and most carbohydrates are not oxidized. Indophenol oxidase is produced. Although indole is not produced, some strains produce anthranilic acid and kynurenine in tryptone broth. These and other tryptophan derivatives cause Kovacs reagent to become orange (R. Hugh and J. D. Welch, Bacteriol. Proc., p. 148, 1969).

Pseudomonads in the acidovorans group, e.g., *P. acidovorans* and *P. testosteroni*, have the characteristics which Guenther (1894) attributed to *Vibrio terrigenus*. During the past 90 years, numerous designations (7, 18–20, 54, 60), including *Pseudomonas terrigena*, have been used for this species; they are listed in Table 1. The specific epithet in the name *Vibrio terrigenus* has priority. The curved soma, particularly the multiple curves of rough and filamentous forms, flagellar morphology, and physiology suggest a relationship of this taxon to certain spirilla. Electron microscopy shows a layer of regularly arranged subunits on the outer surface of the cell wall of *P. acidovorans* (34). The subunits are similar to those of *Spirillum* species, but they do not occur in other pseudomonads. Hugh (18) proposed the new combination *Comamonas terrigena* for this taxon. ATCC 8461 is the neotype strain.

Pseudomonas acidovorans and *Pseudomonas testosteroni*

The characteristics of 95 strains of *P. acidovorans* and 18 strains of *P. testosteroni* are recorded in Table 16. Minimal characters for identification of most *P. acidovorans* and *P. testosteroni* strains are recorded in Table 17. The type strains of *P. acidovorans* (60) and *P. testosteroni* (19, 43) have been described.

TABLE 16. Characters used in identification of acidovorans group *Pseudomonas* species[a]

Test, substrate, or morphology	*P. acidovorans* (95 strains)		*P. testosteroni* (18 strains)	
	Sign	% Positive	Sign	% Positive
Acid:				
Glucose, 1% (OFBM)	−	0	−	0
Fructose	+	100	−	0
Galactose	−	0	−	0
Mannose	−	0	−	0
Rhamnose	−	0	−	0
Xylose	−	0	−	0
Lactose	−	0	−	0
Sucrose	−	0	−	0
Maltose	−	0	−	0
Mannitol	+	95	−	0
Lactose, 10% (PAB)	−	0	−	0
ONPG	−	0	−	0
Pyoverdin	−	0	−	0
Hydrogen sulfide (KIA)	−	0	−	0
Nitrate reduction	+	95	+	94
Gas from nitrate	−	0	−	0
Indophenol oxidase	+	100	+	100
Arginine dihydrolase	−	0	−	0
Lysine decarboxylase	−	0	−	0
Ornithine decarboxylase	−	0	−	0
Phenylalanine deaminase	−	3	−	6
Hydrolysis:				
Urea	−	0	−	0
Esculin	−	0	−	0
Tween 80	− or +	20	− or +	17
Starch	−	0	−	0
DNA	−	1	−	0
Lecithin	−	0	−	0
Gelatin	−	2	−	0
Acetamide[b]	+	100	−	0
Hemolysis	−	0	−	0
Growth on:				
SS	+ or −	66	−	6
MacConkey	+	99	+ or −	89
6.5% NaCl	−	0	−	0
Cetrimide (PA)	− or +	12	−	6
MBM + acetate	+	100	+ or −	67
Growth at 42°C	− or +	10	+ or −	50
Polymyxin susceptible	+ or −	61	+	94
Motile	+	100	+	100
No. of flagella[b]	>1		>1	

[a] Except where indicated, cultures were incubated at 30°C. All strains were indole negative. For abbreviations and signs, see Table 5, footnote *a*.

[b] Not all strains were examined; results are based on the fraction of strains tested.

TABLE 17. Minimal characters for identification of *P. acidovorans* and *P. testosteroni* strains[a]

Character	*P. acidovorans*		*P. testosteroni*	
	Sign	% Positive	Sign	% Positive
Polar tuft of three or more flagella per pole, and long wavelength	+	100	+	100
Motility	+	100	+	100
OF glucose medium open, acid	−	0	−	0
OF fructose medium, acid	+	100	−	0
OF mannitol medium, acid	+	95	−	0
Indophenol oxidase	+	100	+	100
Hydrogen sulfide, black butt in KIA	−	0	−	0

[a] From Table 16.

Pseudomonads in the acidovorans group are divided into two species on the basis of utilization of fructose and additional phenotypic and genotypic differences (60). Non-fructose-utilizing *P. testosteroni* strains are distinguished from fructose-utilizing *P. acidovorans* strains. Acetamide hydrolysis, mediated by an aliphatic amidase, is a practical reaction for identifying *P. acidovorans* strains. This enzyme occurs also in pyocyanogenic, but few apyocyanogenic, *P. aeruginosa* strains and most *P. cepacia* strains.

Sources of isolation and relationship to disease

P. acidovorans and *P. testosteroni* have been isolated from clinical specimens; intravenous tubing; urinary catheter; ultrasonic nebulizer; serum; feces of human, rabbit, turtle, frog, and cobra; sewer, ditch, river, well, and fish aquarium waters; raw milk; frozen shrimp; eye wash; and soil. They have a wide geographic distribution and have been isolated in the United States, Australia, England, Italy, Japan, India, France, the Netherlands, and Java. They are rarely opportunistic pathogens. *P. testosteroni* was associated with septicemia and conjunctivitis. *P. acidovorans* (along with *Enterobacter cloacae*) was the etiologic agent of septicemia in five patients undergoing cardiovascular monitoring.

DIMINUTA GROUP (*PSEUDOMONAS DIMINUTA* AND *PSEUDOMONAS VESICULARIS*)

The two species in the diminuta group have the characters of the genus *Pseudomonas* as recorded in Table 4. The distinctive characteristic of the strains of these species is the very tightly coiled monotrichous flagellum, with a wavelength that varies from 0.62 to 0.98 μm. The wavelength of most polar monotrichous pseudomonads is approximately 2 μm. Specific growth factors are required; however, the species in the diminuta group grow in most peptone media without growth factor supplement. Indophenol oxidase is produced. Acid is produced from primary alcohols by all strains that can utilize alcohols, and some strains hydrolyze gelatin and starch. Members of the diminuta group in general have a restricted range of biochemical activities.

P. diminuta and *P. vesicularis* constitute a distinct group that is only distantly related to other pseudomonads and is more closely related to *Acetobacter* because of the production of acid from primary alcohols and the multiple requirements for growth factors. This group probably deserves a separate generic rank (10, 66).

Pseudomonas diminuta and *Pseudomonas vesicularis*

Identification

The characteristics of 42 strains of *P. diminuta* and 32 strains of *P. vesicularis* are recorded in Table 18. Minimal characters for identification of most *P. diminuta* and *P. vesicularis* strains are recorded in Table 19. The flagellar morphologies of the type strains of *P. diminuta* (39) and *P. vesicularis* (11) have been described and illustrated.

Most carbohydrates in OF basal medium are not oxidized by *P. diminuta*. Pantothenate, biotin, cyanocobalamin, and cystine are required for growth. Carotenoid pigments are not produced. *P. diminuta* strains produce a pellicle in broth culture.

P. vesicularis strains may oxidize glucose, maltose, and other carbohydrates in OF basal medium; acid production, however, may be very weak, equivocal, or absent. The hydrolysis of esculin by *P. vesicularis* is a useful method for distinguishing this species from *P. diminuta*. Pantothenate, biotin, and cyanocobalamin are required for growth. Colonies of some strains produce orange-red carotenoid pigments. A brown discoloration may accumulate in the agar medium surrounding colonies. *P. vesicularis* does not produce a pellicle in broth culture. *P. vesicularis* and most *P. diminuta* strains produce acid from ethanol.

Source of isolation and relationship to disease

P. diminuta has been isolated from stream and ditch waters, contaminated cell culture, respiratory equipment, and various patient sources. *P. vesicularis* has been isolated from the urinary bladder epithelium of a medicinal leech, whooping cough plate, tap water aerator, and clinical sources. *P. diminuta* was the etiologic agent in one case of septicemia.

PSEUDOMONAS MALTOPHILIA (*XANTHOMONAS MALTOPHILIA*)

Identification

P. maltophilia has the characters of the genus *Pseudomonas* as recorded in Table 4. The characteristics of 527 strains of *P. maltophilia* are recorded in Table 20. Minimal characters for identification of most *P. maltophilia* strains are recorded in Table 21. The type strain has been described (27), and the name has been revived (23).

The morphological and biochemical reaction pattern of *P. maltophilia* is remarkably uniform. Colonies develop a characteristic lavender-green color on blood agar media. Brown-colored by-products of metabolism may accumulate in certain agar media. The intensity of the brown color is enhanced by such factors as elevated incubation temperature. A similar reaction is observed among other glucose-nonfermenting gram-negative rods. Some strains slowly produce a very faint yellow, intracellular pigment which does not diffuse out of the colony into the agar

TABLE 18. Characters used in identification of diminuta group *Pseudomonas* species[a]

Test, substrate, or morphology	*P. diminuta* (42 strains)		*P. vesicularis* (32 strains)	
	Sign	% Positive	Sign	% Positive
Acid:				
Glucose, 1% (OFBM)	−	0	− or +(w)	41
Fructose	−	0	−	0
Galactose	−	0	− or +(w)	19
Mannose	−	0	−	0
Rhamnose	−	0	− or +(w)	16
Xylose	−	0	− or +(w)	16
Lactose	−	0	−	0
Sucrose	−	0	−	0
Maltose	−	0	− or +(w)	47
Mannitol	−	0	−	0
Lactose, 10% (PAB)	−	0	−	0
ONPG	−	0	+ or −	50
Pyoverdin	−	0	−	0
Hydrogen sulfide (KIA)	−	0	−	0
Nitrate reduction	−	2	−	3
Gas from nitrate	−	0	−	0
Indophenol oxidase	+	100	+	100
Arginine dihydrolase (DBM)	−	0	−	0
Lysine decarboxylase	−	0	−	0
Ornithine decarboxylase	−	0	−	0
Phenylalanine deaminase	−	7	−	0
Hydrolysis:				
Urea	−	0	−	0
Esculin	−	0	+	100
Tween 80	− or +	12	− or +	16
Starch	−	0	+	94
DNA	− or +	14	−	0
Lecithin	−	0	−	0
Gelatin	+ or −	62	+ or −	56
Acetamide[b]	−	0	−	0
Hemolysis	−	0	−	0
Growth on:				
SS	−	0	−	0
MacConkey	+	98	− or +	19
6.5% NaCl	−	0	−	0
Cetrimide (PA)	−	0	−	0
MBM + acetate	−	0	−	0
Growth at 42°C	− or +	21	−	0
Polymyxin susceptible	+ or −	50	+ or −	88
Motile	+	100	+	97
No. of flagella	1		1	

[a] Except where indicated, cultures were incubated at 30°C. All strains were indole negative. For abbreviations and signs, see Table 5, footnote *a*. (w), Weakly reactive.

[b] Not all strains were examined; results are based on the fraction of strains tested.

medium. This water-soluble, yellow pigment may be due to flavins. Water-soluble melanin is not produced in tyrosine-containing media. Hemolysis does not occur around discrete colonies, but there is a greenish discoloration of erythrocytes around confluent growth. Growth on blood agar is accompanied by a strong odor of ammonia.

Some cells in all strains of *P. maltophilia* have a polar tuft of three or more flagella. Oxidative acidity accumulates in OF maltose and glucose media; however, the acid reaction is more pronounced in maltose than in glucose. OF glucose medium may remain neutral or become weakly acid after 18 to 24 h. An ONPG reaction is produced by most strains. *P. maltophilia* produces high levels of lactose dehydrogenase which, rather than beta-galactosidase, is probably responsible for the metabolism of lactose. Lysine decarboxylase, esculin, DNase, and gelatin reactions are produced. Although a weak indophenol oxidase reaction is produced by a few *P. maltophilia* strains, the species is usually described as negative for this test. *P. maltophilia, P. syringae,* and other phytopathogenic, fluorescent pseudomonads appear to lack cytochrome *c*, which may be necessary for the indophenol oxidase reaction.

Most, but not all, strains of *P. maltophilia* require either methionine or cystine plus glycine for growth: however, peptone-containing media, without growth factor supplement, support growth of all strains. Strains that do not require growth factors (biovar 2) differ from methionine-requiring strains (biovar 1) by the manner in which they assimilate select carbon compounds and other properties (30).

Proposals have been made (10, 62, 66) to transfer *P. maltophilia* to the genus *Xanthomonas* as a separate species, *Xanthomonas maltophilia*. This is based partly on the following common reactions: acid reactions from OF glucose, lactose, and maltose; alkaline reactions from rhamnose and mannitol; negative indophenol oxidase reaction; hydrolysis of esculin and ONPG; and a requirement for amino acids as growth factors.

Sources of isolation and relationship to disease

P. maltophilia is a free-living, ubiquitous organism with a wide geographic distribution. It has been found in raw and pasteurized milk; distilled, well, stagnant, and river waters; sewage; external ear canals and skin of divers; spray atomizer; hard lens cleaning solution; frozen fish; feces of snakes, lizards, frogs, and rabbits; rotten eggs; and soil in petroleum zones. It has been recovered from the internal tissue of banana pseudostem, decaying banana sucker, cotton seed, bean pod, tobacco seedlings, and rhizosphere of cabbage and other cultured plants. In the hospital environment, it has been isolated from contaminated cell culture,

TABLE 19. Minimal characters for identification of *P. diminuta* and *P. vesicularis* strains[a]

Character	*P. diminuta*		*P. vesicularis*	
	Sign	% Positive	Sign	% Positive
Polar monotrichous, fewer than three flagella per pole, and short wavelength	+	100	+	97
Motility	+	100	+	97
OF glucose medium open, acid	−	0	− or +(w)	41
OF lactose medium, acid	−	0	−	0
Esculin hydrolysis	−	0	+	100
Indophenol oxidase	+	100	+	100
Hydrogen sulfide, black butt in KIA	−	0	−	0

[a] From Table 18.

TABLE 20. Characters used in identification of *P. maltophilia*[a]

Test, substrate, or morphology	*P. maltophilia* (527 strains)	
	Sign	% Positive
Acid:		
Glucose, 1% (OFBM)	+	100
Fructose	+	99
Galactose	− or +	25
Mannose	+	98
Rhamnose	−	0
Xylose	+ or −	54
Lactose	+ or −	88
Sucrose	+	92
Maltose	+	100
Mannitol	−	0
Lactose, 10% (PAB)	−	0
ONPG	+	95
Pyoverdin	−	0
Hydrogen sulfide (KIA)	−	0
Nitrate reduction	− or +	40
Gas from nitrate	−	0
Indophenol oxidase	−	2
Arginine dihydrolase	−	0
Lysine decarboxylase	+	99
Ornithine decarboxylase	−	0
Phenylalanine deaminase	−	0
Hydrolysis:		
Urea	−	0
Esculin	+	100
Tween 80	+	100
Starch	−	0
DNA	+	100
Lecithin	−	0
Gelatin	+	100
Acetamide[b]	−	0
Hemolysis	−	0
Growth on:		
SS	−	0
MacConkey	+	100
6.5% NaCl	−	0
Cetrimide (PA)	−	0
MBM + acetate	−	2
Growth at 42°C	− or +	48
Polymyxin susceptible	+	93
Motile	+	100
No. of flagella[b]	>1	

[a] Except where indicated, cultures were incubated at 30°C. All strains were indole negative. For abbreviations and signs, see Table 5, footnote *a*.

[b] Not all strains were examined; results are based on the fraction of strains tested.

streptomycin solution, contaminated deionized water, sink drains, water bath, faucet aerator, incubator reservoirs, respirators, nebulizers, intravenous catheters, evacuated blood collection tubes, blood monitoring transducers, and cardiopulmonary bypass equipment.

P. maltophilia is usually a commensal or contaminant and is part of the transient flora of hospitalized patients. It has been recovered from almost every site on or within the body. Since 1967, this organism has been associated with primary pneumonia and opportunistic infections including endocarditis, septicemia, broncho-, lobar, and aspiration pneumonia, urinary tract infections, cholangitis, conjunctivitis, acute mastoiditis, meningitis, wound infections, and abscesses. Enzymes including esterases, lipases, mucinase, hyaluronidase, RNase, and hemolysins may be potential virulence factors.

PSEUDOMONAS MESOPHILICA

Identification

P. mesophilica has the characters of the genus *Pseudomonas* as recorded in Table 4. The characteristics of six strains of *P. mesophilica* are recorded in Table 22. Minimal characters for identification of most *P. mesophilica* strains are recorded in Table 23. The type strain has been described (3).

Colonies are shiny, smooth, round, raised, entire, and rarely visible before 3 days of incubation at 30°C. An intracellular, nondiffusible, water-insoluble, pink, oxocarotenoid pigment is produced. Pink pigmentation is a characteristic of most methanol-utilizing bacteria. Acid is produced slowly, within 2 to 4 days, from fructose, xylose, and methanol in OF basal medium. Indophenol oxidase, urease, and amylase reactions are produced. Strains are motile, with polar monotrichous flagella. They have 65.8 mol% G+C content in their DNA.

Sources of isolation and relationship to disease

P. mesophilica has been isolated from tobacco leaves; perennial rye grass leaves; methane-oxidizing enrichment cultures from soil, sewage, and rumen of a cow; water; nebulizer; and human clinical materials.

PSEUDOMONAS PAUCIMOBILIS

Identification

P. paucimobilis has the characters of the genus *Pseudomonas* as recorded in Table 4. The characteristics of 110 strains of *P. paucimobilis* are recorded in Table 24. Minimal characters for identification of most *P. paucimobilis* strains are recorded in Table 25. The type strain has been described (17).

Colonies are either butyrous, or mucoid and viscid, and develop an intracellular, nondiffusible, carotenoid (nostoxanthin) pigment on most media. This yellow pigment is distinct from the brominated arylpolyenes (xanthomonadins) present in *Xanthomonas* species. No brown pigments are produced. The name *paucimobilis* indicates that only a few cells of a strain in a population may be motile. Motility is best demonstrated at room temperature. When motile, they are polar monotrichous. Indophenol oxidase is produced,

TABLE 21. Minimal characters for identification of *P. maltophilia* strains[a]

Character	Sign	% Positive
Polar tuft of three or more flagella per pole	+	100
Motility	+	100
OF glucose medium open, acid	+	100
OF maltose medium, acid	+	100
OF maltose medium sealed, acid	−	0
L-Lysine decarboxylase	+	99
Extracellular DNase	+	100

[a] From Table 20.

TABLE 22. Characters used in identification of *P. mesophilica*[a]

Test, substrate, or morphology	*P. mesophilica* (six strains)	
	Sign	% Positive
Acid:		
Glucose, 1% (OFBM)	−	0
Fructose	(+)	100
Galactose	−	0
Mannose	−	0
Rhamnose	− or (+)	17
Xylose	(+)	100
Lactose	−	0
Sucrose	−	0
Maltose	−	0
Mannitol	−	0
Lactose, 10% (PAB)	−	0
ONPG	−	0
Pyoverdin	−	0
Hydrogen sulfide (KIA)	−	0
Nitrate reduction	+ or −	50
Gas from nitrate	−	0
Indophenol oxidase	+	100
Arginine dihydrolase (DBM)	−	0
Lysine decarboxylase	−	0
Ornithine decarboxylase	−	0
Phenylalanine deaminase	−	0
Hydrolysis:		
Urea	+	100
Esculin	−	0
Tween 80	−	0
Starch	+	100
DNA	−	0
Lecithin	−	0
Gelatin	−	0
Acetamide[b]	−	0
Hemolysis	−	0
Growth on:		
SS	−	0
MacConkey	−	0
6.5% NaCl	−	0
Cetrimide (PA)	−	0
MBM + acetate	+	100
Growth at 42°C	−	0
Polymyxin susceptible	− or +	33
Motile	+	100
No. of flagella[b]	1	

[a] Except where indicated, cultures were incubated at 30°C. All strains were indole negative. For abbreviations and signs, see Table 5, footnote *a*. (+), Most reactions delayed 3 or more days.

[b] Not all strains were examined; results are based on the fraction of strains tested.

and ONPG and esculin are hydrolyzed. Acid is produced in OF basal medium from a number of carbohydrates but not from sugar alcohols. *P. paucimobilis* grows in mineral base medium with ammonium ion as sole source of nitrogen and acetate as sole source of carbon and energy. A negative test for urease distinguishes this species from phenotypically similar strains of *Sphingobacterium (Flavobacterium) multivorum*.

Evidence from DNA hybridization results, rRNA cistron comparisons, and cellular fatty acid profiles indicates that *P. paucimobilis* is not an authentic member of the genus *Pseudomonas* and that a new genus designation may be justified (46, 49).

TABLE 23. Minimal characters for identification of *P. mesophilica* strains[a]

Character	Sign	% Positive
Polar monotrichous, fewer than three flagella per pole	+	100
Motility	+	100
OF xylose medium open, acid	(+)	100
OF maltose medium, acid	−	0
Indophenol oxidase	+	100
Urease	+	100
Nitrate to gas	−	0
Intracellular pink pigment	+	100

[a] From Table 22 and text.

Sources of isolation and relationship to disease

P. paucimobilis has been isolated from clinical specimens; respirators; humidifiers; ultrasonic nebulizers; infusate and dialysis fluids; water bath; tap water aerator; incubator, tap, distilled, boiled, bottled mineral, and swimming pool waters; blow bottles; water bottles; contaminated cell culture; and surgical unit air. *P. paucimobilis* has been implicated in an infected leg ulcer after trauma, postoperative chronic cellulitis, septicemia, acute meningitis, and urinary tract infection. DNase, esterases, phosphatases, lipase, and endotoxin may contribute to its virulence.

PSEUDOMONAS PUTREFACIENS

Identification

P. putrefaciens has the characters of the genus *Pseudomonas* as recorded in Table 4. The characteristics of 72 strains of *P. putrefaciens* are recorded in Table 26. Minimal characters for identification of most *P. putrefaciens* strains are recorded in Table 27. ATCC 8071 was designated (35) and acknowledged (47) as the "type strain" of *Alteromonas (Pseudomonas) putrefaciens*.

Cells of *P. putrefaciens* strains possess polar monotrichous flagella. Some strains may contain cells with lateral flagella. Colonies on agar media are slightly viscous or mucoid and usually reddish-brown or pink.

P. putrefaciens is the only glucose-nonfermenting gram-negative rod known to produce hydrogen sulfide in KIA and TSIA. For this reason strains have been mistaken for rare anaerogenic glucose-fermenting isolates of *Salmonella* and *Proteus* species (see Fig. 13 through 15 of reference 29). Occasionally, strains produce only a small quantity of or no hydrogen sulfide in the butts of KIA or TSIA.

The production of acid from glucose is oxygen dependent and may be weak and delayed for several days. Ornithine decarboxylase and DNase tests are positive. Most strains grow at 30°C, many fail to grow at 35°C, and some are psychrophilic. Growth usually develops in mineral base medium containing ammonium ion as sole source of nitrogen and acetate as sole source of carbon and energy. Sodium cations are required for growth of some strains.

Two distinct groups (40) of strains exist within *P. putrefaciens*, and additional groups have been described (47). Biovar 1 (CDC group Ib-1) has a relatively low G+C content of 49 mol% in its DNA and consists

TABLE 24. Characters used in identification of *P. paucimobilis*[a]

Test, substrate, or morphology	*P. paucimobilis* (110 strains)	
	Sign	% Positive
Acid:		
Glucose, 1% (OFBM)	+	100
Fructose	+	100
Galactose	+	100
Mannose	+	100
Rhamnose	+ or −	64
Xylose	+	100
Lactose	+	100
Sucrose	+	100
Maltose	+	100
Mannitol	−	0
Lactose, 10% (PAB)	+ or −	86
ONPG	+	100
Pyoverdin	−	0
Hydrogen sulfide (KIA)	−	0
Nitrate reduction	−	0
Gas from nitrate	−	0
Indophenol oxidase	+	90
Arginine dihydrolase (DBM)	−	0
Lysine decarboxylase	−	0
Ornithine decarboxylase	−	0
Phenylalanine deaminase	− or +	35
Hydrolysis:		
Urea	−	0
Esculin	+	100
Tween 80	− or +	30
Starch	+ or −	50
DNA	−	0
Lecithin	−	0
Gelatin	−	0
Acetamide[b]	−	0
Hemolysis	−	0
Growth on:		
SS	−	0
MacConkey	−	5
6.5% NaCl	−	0
Cetrimide (PA)	−	0
MBM + acetate	+	100
Growth at 42°C	−	0
Polymyxin susceptible	+ or −	82
Motile	+ or −	89
No. of flagella[b]	1	

[a] Except where indicated, cultures were incubated at 30°C. All strains were indole negative. For abbreviations and signs, see Table 5, footnote *a*.

[b] Not all strains were examined; results are based on the fraction of strains tested.

primarily of strains from environmental sources. These strains produce acid from OF sucrose and maltose and fail to grow in 6.5% NaCl or on salmonella-shigella agar. Biovar 2 (CDC group Ib-2), with a high G+C content of 58 mol% in its DNA, includes mostly strains from human clinical specimens. These strains grow in 6.5% NaCl and on salmonella-shigella agar but do not produce acid from disaccharides.

Bacteria of low G+C content have been excluded (50) from the genus *Pseudomonas* and transferred to the new genus *Alteromonas* created (5) precisely to accommodate *Pseudomonas*-like bacteria of low G+C content. The transfer of this group of *P. putrefaciens* strains to the genus *Alteromonas* as *Alteromonas putrefaciens* has not found uniform support (67).

Sources of isolation and relationship to disease

P. putrefaciens is found in soil and water and has a wide geographic distribution. It has been found in raw and sweet milk and cream; eggs; snake feces; sewage; stagnant, sea, well, lake, and river waters; contaminated cell culture; external ear canals of divers; oil emulsion; and natural gas and petroleum brines. It is responsible for putrid deterioration of butter, hydrogen sulfide spoilage in haddock fillets, and green discoloration of fresh meat and vacuum-packed beef, and it has been isolated from fresh, frozen, and stored cod fillets, ground beef, and frozen poultry carcasses.

Isolates of *P. putrefaciens* from clinical specimens of human origin in most instances represent colonization in the absence of clinical disease. *P. putrefaciens* has been etiologically associated with septicemia, an intraabdominal abscess, otitis media, an infected tibia and overlying ulcer after trauma, and purulent skin ulcerations of the legs of diabetics and debilitated patients. Intraperitoneal injection of cells of *P. putrefaciens* in mice resulted in 50% lethal dose values of 1.2×10^9 cells, indicating that death was more likely due to endotoxin and not to invasiveness.

Ve GROUP OF *PSEUDOMONAS* SPECIES (*PSEUDOMONAS* SP. Ve-1 AND *PSEUDOMONAS* SP. Ve-2)

The species in the Ve group have the characters of the genus *Pseudomonas* as recorded in Table 4. Strains in the Ve group may produce smooth colonies or wrinkled, rough, and adherent as well as smooth colonies. The colonies produce an intracellular, nondiffusible, water-insoluble, yellow pigment. Aggregates of cells in sausage-shape chain formation (symplasmata) occur in rough colony forms only. Symplasmata are seen with the low-power objective of a microscope in a hanging-drop preparation of a 24-h broth culture or condensate from an agar slant culture.

Acid is produced from a number of carbohydrates and polyhydric alcohols in OF basal medium. These species grow in mineral base medium containing ammonium ion as sole source of nitrogen and glucose as sole source of carbon and energy. Ve group strains are motile, use polyhydric alcohols, and do not produce indophenol oxidase. These characteristics distinguish Ve from *P. paucimobilis* strains. Two subdivisions of Ve group strains are recognized (65).

TABLE 25. Minimal characters for identification of *P. paucimobilis* strains[a]

Character	Sign	% Positive
Polar monotrichous, fewer than three flagella per pole	+ or −	89
Motility	+ or −	89
OF glucose medium open, acid	+	100
OF maltose medium, acid	+	100
OF mannitol medium, acid	−	0
ONPG reaction	+	100
Urease	−	0
Indophenol oxidase	+	90
Esculin hydrolysis	+	100
L-Lysine decarboxylase	−	0

[a] From Table 24.

TABLE 26. Characters used in identification of *P. putrefaciens*[a]

Test, substrate, or morphology	Biovar 1 (11 strains)		Biovar 2 (52 strains)		Biovar 3 (9 strains)	
	Sign	% Positive	Sign	% Positive	Sign	% Positive
Acid:						
Glucose, 1% (OFBM)	+ or (+)	100	+ or (+)	100	+ or (+)	100
Fructose	−, +, or (+)	46	+, (+), or −	50	+, (+), or −	78
Galactose	−	9	−	0	−	0
Mannose	−	9	−	0	−	0
Rhamnose	−	0	−	0	−	0
Xylose	−	9	−	0	−	0
Lactose	−, +, or (+)	18	−	0	−	0
Sucrose	+ or (+)	100	−	0	−	0
Maltose	+ or (+)	100	−	0	−	0
Mannitol	−	0	−	0	−	0
Lactose, 10% (PAB)	−	9	−	0	−	0
ONPG	−	0	−	0	−	0
Pyoverdin	−	0	−	0	−	0
Hydrogen sulfide (KIA)	+	100	+	100	+	100
Nitrate reduction	+	100	+	100	+	100
Gas from nitrate	−	0	−	0	−	0
Indophenol oxidase	+	100	+	100	+	100
Arginine dihydrolase	−	0	−	0	−	0
Lysine decarboxylase	−	0	−	0	−	0
Ornithine decarboxylase	+	100	+	100	+	100
Phenylalanine deaminase	−	0	−	0	−	0
Hydrolysis:						
Urea	−	0	−	6	−	0
Esculin	− or +	27	−	0	−	0
Tween 80	+	100	+	100	+	100
Starch	−	0	−	0	− or +	11
DNA	+	100	+	100	+	100
Lecithin	−	0	+	100	−	0
Gelatin	+	100	+	100	+ or −	78
Acetamide[b]	−	0	−	0	−	0
Hemolysis	−	0	−	0	−	0
Growth on:						
SS	−	0	+	100	−	0
MacConkey	+	100	+	100	+	100
6.5% NaCl	−	0	+	100	−	0
Cetrimide (PA)	−	0	−	0	−	0
MBM + acetate	+	100	+	100	+ or −	67
Growth at 42°C	−	0	+	90	−	0
Polymyxin susceptible	+	100	+	100	+	100
Motile	+	100	+	100	+	100
No. of flagella[b]	1		1		1	

[a] Except where indicated, cultures were incubated at 30°C. All strains were indole negative. For abbreviations, see Table 5, footnote *a*. Signs: +, 90% or more positive within 2 days; −, no reaction (90% or more); + or (+), most reactions occur within 2 days, some delayed 3 or more days; +, (+), or −, most reactions occur within 2 days, some delayed 3 or more days, some negative; + or −, most cultures positive, some strains negative; − or +, most strains negative, some cultures positive; −, +, or (+), most strains negative, some reactions occur within 2 days, some delayed 3 or more days; 1, polar monotrichous flagella.

[b] Not all strains were examined; results are based on the fraction of strains tested.

TABLE 27. Minimal characters for identification of *P. putrefaciens* strains[a]

Character	Biovar 1[b]		Biovar 2		Biovar 3	
	Sign	% Positive	Sign	% Positive	Sign	% Positive
Polar monotrichous, fewer than three flagella per pole	+	100	+	100	+	100
Motility	+	100	+	100	+	100
OF sucrose medium open, acid	+ or (+)	100	−	0	−	0
OF maltose medium, acid	+ or (+)	100	−	0	−	0
Indophenol oxidase	+	100	+	100	+	100
Extracellular DNase	+	100	+	100	+	100
Hydrogen sulfide, black butt in KIA	+	100	+	100	+	100
Growth on SS agar	−	0	+	100	−	0
Growth on 6.5% NaCl agar	−	0	+	100	−	0

[a] From Table 26.

[b] Proposed designation, *Alteromonas putrefaciens* (35, 47).

TABLE 28. Characters used in identification of group Ve *Pseudomonas* species[a]

Test, substrate, or morphology	*Pseudomonas* sp. Ve-1 (21 strains) Sign	% Positive	*Pseudomonas* sp. Ve-2 (77 strains) Sign	% Positive
Acid:				
Glucose, 1% (OFBM)	+	100	+	100
Fructose	+	100	+	100
Galactose	+	100	+	100
Mannose	+	100	+	100
Rhamnose	+ or −	86	+	90
Xylose	+	100	+	100
Lactose	−	0	−	0
Sucrose	−	0	− or +	25
Maltose	+	100	+	99
Mannitol	+	100	+	100
Lactose, 10% (PAB)	+	100	+	95
ONPG	+	100	−	9
Pyoverdin	−	0	−	0
Hydrogen sulfide (KIA)	−	0	−	0
Nitrate reduction	+	95	−	0
Gas from nitrate	−	0	−	0
Indophenol oxidase	−	0	−	0
Arginine dihydrolase (DBM)	+ or −	76	−	0
Lysine decarboxylase	−	0	−	0
Ornithine decarboxylase	−	0	−	0
Phenylalanine deaminase	−	5	− or +	31
Hydrolysis:				
Urea	− or +	10	− or +	30
Esculin	+	100	−	0
Tween 80	−	0	−	8
Starch	+ or −	57	+ or −	79
DNA	−	0	−	0
Lecithin	−	0	−	0
Gelatin	+ or −	52	−	7
Acetamide[b]	−	0	−	0
Hemolysis	−	0	−	0
Growth on:				
SS	+ or −	62	− or +	45
MacConkey	+	100	+	100
6.5% NaCl	+ or −	76	− or +	21
Cetrimide (PA)	− or +	19	+ or −	79
MBM + acetate	+	100	+	100
Growth at 42°C	+ or −	81	− or +	17
Polymyxin susceptible	+	100	+	100
Motile	+	100	+	100
No. of flagella[b]	>1		1	

[a] Except where indicated, cultures were incubated at 30°C. All strains were indole negative. For abbreviations and signs, see Table 5, footnote *a*.

[b] Not all strains were examined; results are based on the fraction of strains tested.

Pseudomonas sp. Ve-1 and *Pseudomonas* sp. Ve-2

Identification

The characteristics of 21 strains of *Pseudomonas* sp. Ve-1 and 77 strains of *Pseudomonas* sp. Ve-2 are recorded in Table 28. Minimal characters for identification of most *Pseudomonas* spp. Ve-1 and Ve-2 strains are recorded in Table 29.

Pseudomonas sp. Ve-1 strains have a polar tuft of flagella and hydrolyze esculin. *Pseudomonas* sp. Ve-2 strains are polar monotrichous and do not hydrolyze esculin. Strains of the former may produce arginine dihydrolase and nitratase. Cellular fatty acid composition shows group Ve strains to be similar to other species of *Pseudomonas* (9). *Pseudomonas* sp. Ve-1 strains have 56.8 mol% G+C content in their DNA, which is at the lower limit of the range for *Pseudomonas* species. The 66.9 mol% G+C content in the DNA of *Pseudomonas* sp. Ve-2 strains is near the G+C values of *Chromobacterium lividum*, *Chromobacterium violaceum*, and many *Pseudomonas* species. Each warrants species recognition (15, 63).

Sources of isolation and relationship to disease

Pseudomonas spp. Ve-1 and Ve-2 have been isolated from a number of clinical specimens. In the hospital environment, *Pseudomonas* sp. Ve-2 has been recovered from respiratory therapy equipment and sink drains. Bacteremia caused by *Pseudomonas* sp. Ve-1 occurred in a patient after skull fracture and surgery. An immunocompromised patient developed septicemia attributed to *Pseudomonas* sp. Ve-1.

OTHER *PSEUDOMONAS* SPECIES

Pseudomonas pertucinogena (32), identified as rough phase IV *Bordetella pertussis*, has the characters of the genus *Pseudomonas* as recorded in Table 4. The orga-

TABLE 29. Minimal characters for identification of group Ve *Pseudomonas* strains[a]

Character	*Pseudomonas* sp. Ve-1 Sign	% Positive	*Pseudomonas* sp. Ve-2 Sign	% Positive
Polar monotrichous, fewer than three flagella per pole	−	0	+	100
Polar tuft of three or more flagella per pole	+	100	−	0
Motility	+	100	+	100
OF glucose medium open, acid	+	100	+	100
OF mannitol medium, acid	+	100	+	100
Indophenol oxidase	−	0	−	0
Esculin hydrolysis	+	100	−	0
L-Lysine decarboxylase	−	0	−	0

[a] From Table 28.

nism is polar monotrichous and produces indophenol oxidase, phenylalanine deaminase, and pertucin, a bacteriocin that inhibits the growth of *B. pertussis.*

Recently described (8) *Pseudomonas*-like group 2 strains have features similar to species in the pseudomallei group but are distinguished from the latter by cellular fatty acid composition and select conventional biochemical tests.

LITERATURE CITED

1. **Ajello, G. W., and A. W. Hoadley.** 1976. Fluorescent pseudomonads capable of growth at 41°C but distinct from *Pseudomonas aeruginosa*. J. Clin. Microbiol. **4:**443–449.
2. **Appelbaum, P. C., J. Stavitz, M. S. Bentz, and L. C. von Kuster.** 1981. Comparison of four methods for identification of gram-negative non-fermenters: organisms less commonly encountered in clinical specimens. Med. Microbiol. Immunol. **169:**163–168.
3. **Austin, B., and M. Goodfellow.** 1979. *Pseudomonas mesophilica*, a new species of pink bacteria isolated from leaf surfaces. Int. J. Syst. Bacteriol. **29:**373–378.
4. **Ballard, R. W., N. J. Palleroni, M. Doudoroff, R. Y. Stanier, and M. Mandel.** 1970. Taxonomy of the aerobic pseudomonads: *Pseudomonas cepacia, P. marginata, P. alliicola*, and *P. caryophylli*. J. Gen. Microbiol. **60:**199–214.
5. **Baumann, L., P. Baumann, M. Mandel, and R. D. Allen.** 1972. Taxonomy of aerobic marine eubacteria. J. Bacteriol. **110:**402–429.
6. **Bergan, T.** 1978. Phage typing of *Pseudomonas aeruginosa*, p. 169–199. *In* T. Bergan and J. R. Norris (ed.), Methods in microbiology, vol. 10. Academic Press, London.
7. **Davis, G. H. G., and R. W. A. Park.** 1962. A taxonomic study of certain bacteria currently classified as *Vibrio* species. J. Gen. Microbiol. **27:**101–119.
8. **Dees, S. B., D. G. Hollis, R. E. Weaver, and C. W. Moss.** 1983. Cellular fatty acid composition of *Pseudomonas marginata* and closely associated bacteria. J. Clin. Microbiol. **18:**1073–1078.
9. **Dees, S. B., C. W. Moss, R. E. Weaver, and D. G. Hollis.** 1979. Cellular fatty acid composition of *Pseudomonas paucimobilis* and groups IIK-2, Ve-1, and Ve-2. J. Clin. Microbiol. **10:**206–209.
10. **De Vos, P., and J. De Ley.** 1983. Intra- and intergeneric similarities of *Pseudomonas* and *Xanthomonas* ribosomal ribonucleic acid cistrons. Int. J. Syst. Bacteriol. **33:**487–509.
11. **Galarneault, T. P., and E. Leifson.** 1964. *Pseudomonas vesiculare* (Büsing et al.) comb. nov. Int. Bull. Bacteriol. Nomencl. Taxon. **14:**165–168.
12. **Gardner, P., W. B. Griffin, M. N. Swartz, and L. J. Kunz.** 1970. Nonfermentative gram-negative bacilli of nosocomial interest. Am. J. Med. **48:**735–749.
13. **Gilardi, G. L.** 1972. Infrequently encountered *Pseudomonas* species causing infection in humans. Ann. Intern. Med. **77:**211–215.
14. **Gilardi, G. L.** 1976. *Pseudomonas* species in clinical microbiology. Mt. Sinai J. Med. **43:**710–726.
15. **Gilardi, G. L., S. Hirschl, and M. Mandel.** 1975. Characteristics of yellow-pigmented nonfermentative bacilli (groups Ve-1 and Ve-2) encountered in clinical bacteriology. J. Clin. Microbiol. **1:**384–389.
16. **Govan, J. R. W.** 1978. Pyocin typing of *Pseudomonas aeruginosa*, p. 61–91. *In* T. Bergan and J. R. Norris (ed.), Methods in microbiology, vol. 10. Academic Press, London.
17. **Holmes, B., R. J. Owen, A. Evans, H. Malnick, and W. R. Willcox.** 1977. *Pseudomonas paucimobilis*, a new species isolated from human clinical specimens, the hospital environment, and other sources. Int. J. Syst. Bacteriol. **27:**133–146.
18. **Hugh, R.** 1962. *Comamonas terrigena* comb. nov. with proposal of a neotype and request for an opinion. Int. Bull. Bacteriol. Nomencl. Taxon. **12:**33–35.
19. **Hugh, R.** 1965. A comparison of *Pseudomonas testosteroni* and *Comamonas terrigena*. Int. Bull. Bacteriol. Nomencl. Taxon. **15:**125–132.
20. **Hugh, R.** 1970. A practical approach to the identification of certain nonfermentative gram-negative rods encountered in clinical specimens. J. Conf. Public Health Lab. Directors **28:**168–187.
21. **Hugh, R.** 1970. *Pseudomonas* and *Aeromonas*, p. 175–190. *In* J. E. Blair, E. H. Lennette, and J. P. Truant (ed.), Manual of clinical microbiology. American Society for Microbiology, Washington, D.C.
22. **Hugh, R.** 1978. Classical methods for isolation and identification of glucose nonfermenting gram-negative rods, p. 1–13. *In* G. L. Gilardi (ed.), Glucose nonfermenting gram-negative bacteria in clinical microbiology. CRC Press, Boca Raton, Fla.
23. **Hugh, R.** 1981. *Pseudomonas maltophilia* sp. nov., nom. rev. Int. J. Syst. Bacteriol. **31:**195.
24. **Hugh, R., and G. L. Gilardi.** 1980. *Pseudomonas*, p. 288–317. *In* E. H. Lennette, A. Balows, W. Hausler, Jr., and J. P. Truant (ed.), Manual of clinical microbiology, 3rd ed. American Society for Microbiology, Washington, D.C.
25. **Hugh, R., L. Guarraia, and H. Hatt.** 1964. The proposed neotype strains of *Pseudomonas fluorescens* (Trevisan) Migula 1895. Int. Bull. Bacteriol. Nomencl. Taxon. **14:**145–155.
26. **Hugh, R., and P. Ikari.** 1964. The proposed neotype strain of *Pseudomonas alcaligenes* Monias 1928. Int. Bull. Bacteriol. Nomencl. Taxon. **14:**103–107.
27. **Hugh, R., and E. Leifson.** 1963. A description of the type strain of *Pseudomonas maltophilia*. Int. Bull. Bacteriol. Nomencl. Taxon. **13:**133–138.
28. **Hugh, R., and E. Leifson.** 1964. The proposed neotype strains of *Pseudomonas aeruginosa* (Schroeter 1872) Migula 1900. Int. Bull. Bacteriol. Nomencl. Taxon. **14:**69–84.
29. **Hugh, R., and J. A. Webster.** 1978. Photographic survey of glucose-nonfermenting gram-negative rods. American Society for Microbiology, Washington, D.C.
30. **Ikemoto, S., K. Suzuki, T. Kaneko, and K. Komagata.** 1980. Characterization of strains of *Pseudomonas maltophilia* which do not require methionine. Int. J. Syst. Bacteriol. **30:**437–447.
31. **Janda, J. M., and E. J. Bottone.** 1981. *Pseudomonas aeruginosa* enzyme profiling: predictor of potential invasiveness and use as an epidemiological tool. J. Clin. Microbiol. **14:**55–60.
32. **Kawai, Y., and E. Yabuuchi.** 1975. *Pseudomonas pertucinogena* sp. nov., an organism previously misidentified as *Bordetella pertussis*. Int. J. Syst. Bacteriol. **25:**317–323.
33. **King, A., B. Holmes, I. Phillips, and S. P. Lapage.** 1979. A taxonomic study of clinical isolates of *Pseudomonas pickettii, "P. thomasii,"* and "group IVd" bacteria. J. Gen. Microbiol. **114:**137–147.
34. **Lapchine, L.** 1979. Regularly arranged structures on the surface of some *Pseudomonas* sp. FEMS Microbiol. Lett. **5:**223–225.
35. **Lee, J. V., D. M. Gibson, and J. M. Shewan.** 1977. A numerical taxonomic study of some *Pseudomonas*-like marine bacteria. J. Gen. Microbiol. **98:**439–451.
36. **Leifson, E.** 1951. Staining, shape, and arrangement of bacterial flagella. J. Bacteriol. **62:**377–389.
37. **Leifson, E.** 1960. Atlas of bacterial flagellation. Academic Press, Inc., New York.
38. **Leifson, R., and R. Hugh.** 1953. Variation in shape and arrangement of bacterial flagella. J. Bacteriol. **65:**263–271.
39. **Leifson, R., and R. Hugh.** 1954. A new type of polar

monotrichous flagellation. J. Gen. Microbiol. **10:**68–70.
40. **Levin, R. E.** 1972. Correlation of DNA base composition and metabolism of *Pseudomonas putrefaciens* isolates from food, human clinical specimens, and other sources. Antonie van Leeuwenhoek J. Microbiol. Serol. **38:**121–127.
41. **Liu, P. V., H. Matsumoto, H. Kusama, and T. Bergen.** 1983. Survey of heat-stable, major somatic antigens of *Pseudomonas aeruginosa*. Int. J. Syst. Bacteriol. **33:**256–264.
42. **Lowbury, E. J. L.** 1975. Ecological importance of *Pseudomonas aeruginosa*: medical aspects, p. 37–65. *In* P. H. Clarke and M. H. Richmond (ed.), Genetics and biochemistry of *Pseudomonas*. John Wiley and Sons, New York.
43. **Marcus, P. I., and P. Talalay.** 1956. Induction and purification of alpha- and beta-hydroxysteroid dehydrogenase. J. Biol. Chem. **218:**661–674.
44. **Moss, C. W.** 1978. New methodology for identification of nonfermenters: gas-liquid chromatographic chemotaxonomy, p. 171–201. *In* G. L. Gilardi (ed.), Glucose nonfermenting gram-negative bacteria in clinical microbiology. CRC Press, Boca Raton, Fla.
45. **Neu, H. C.** 1978. Clinical role of *Pseudomonas aeruginosa*, p. 83–104. *In* G. L. Gilardi (ed.), Glucose nonfermenting gram-negative bacteria in clinical microbiology. CRC Press, Boca Raton, Fla.
46. **Owen, R. J., and P. J. H. Jackman.** 1982. The similarities between *Pseudomonas paucimobilis* and allied bacteria derived from analysis of deoxyribonucleic acids and electrophoretic protein patterns. J. Gen. Microbiol. **128:**2945–2954.
47. **Owen, R. J., R. M. Legros, and S. P. Lapage.** 1978. Base composition, size and sequence similarities of genome deoxyribonucleic acids from clinical isolates of *Pseudomonas putrefaciens*. J. Gen. Microbiol. **104:**127–138.
48. **Oyaizu, H., and K. Komagata.** 1981. Chemotaxonomic and phenotypic characterization of the strains of species in the *Flavobacterium-Cytophaga* complex. J. Gen. Appl. Microbiol. **27:**57–107.
49. **Oyaizu, H., and K. Komagata.** 1983. Grouping of *Pseudomonas* species on the basis of cellular fatty acid composition and the quinone system with special reference to the existence of 3-hydroxy fatty acids. J. Gen. Appl. Microbiol. **29:**17–40.
50. **Palleroni, N. J.** 1984. Genus I. *Pseudomonas* Migula 1984, p. 141–199. *In* N. R. Krieg and J. G. Holt (ed.), Bergey's manual of systematic bacteriology, vol. 1. The Williams and Wilkins Co., Baltimore.
51. **Palleroni, N. J., and M. Doudoroff.** 1972. Some properties and taxonomic subdivisions of the genus *Pseudomonas*. Annu. Rev. Phytopathol. **10:**73–100.
52. **Palleroni, N. J., M. Doudoroff, and R. Y. Stanier.** 1970. Taxonomy of the aerobic pseudomonads: the properties of the *Pseudomonas stutzeri* group. J. Gen. Microbiol. **60:**215–231.
53. **Palleroni, N. J., and B. Holmes.** 1981. *Pseudomonas cepacia* sp. nov., nom. rev. Int. J. Syst. Bacteriol. **31:**479–481.
54. **Park, R. W. A.** 1962. A study of certain heterotrophic polarly flagellate water bacteria: *Aeromonas, Pseudomonas*, and *Comamonas*. J. Gen. Microbiol. **27:**121–133.
55. **Pedersen, M. M., E. Marso, and M. J. Pickett.** 1970. Nonfermentative bacilli associated with man. III. Pathogenicity and antibiotic susceptibility. Am. J. Clin. Pathol. **54:**178–192.
56. **Pickett, M. J., and J. R. Greenwood.** 1980. A study of the Va-1 group of pseudomonads and its relationship to *Pseudomonas pickettii*. J. Gen. Microbiol. **120:**439–446.
57. **Ralston, E., N. J. Palleroni, and M. Doudoroff.** 1973. *Pseudomonas pickettii*, a new species of clinical origin related to *Pseudomonas solanacearum*. Int. J. Syst. Bacteriol. **23:**15–19.
58. **Redfearn, M. S., N. J. Palleroni, and R. Y. Stanier.** 1966. A comparative study of *Pseudomonas pseudomallei* and *Bacillus mallei*. J. Gen. Microbiol. **43:**293–313.
59. **Shewan, J. M., and W. C. Haynes.** 1967. Report of the Subcommittee on *Pseudomonas* and related organisms (1962–1966). Int. J. Syst. Bacteriol. **17:**255–259.
60. **Stanier, R. Y., N. J. Palleroni, and M. Doudoroff.** 1966. The aerobic pseudomonads: a taxonomic study. J. Gen. Microbiol. **43:**159–271.
61. **Stanton, A. T., and W. Fletcher.** 1932. Melioidosis. Studies from the Institute for Medical Research Federated Malay States, no. 21. John Balo, Sons and Danielsson, Ltd., London.
62. **Swings, J., P. De Vos, M. Van den Mooter, and J. De Ley.** 1983. Transfer of *Pseudomonas maltophilia* Hugh 1981 to the genus *Xanthomonas* as *Xanthomonas maltophilia* (Hugh 1981) comb. nov. Int. J. Syst. Bacteriol. **33:**409–413.
63. **Tatum, H. W., W. H. Ewing, and R. E. Weaver.** 1974. Miscellaneous gram-negative bacteria, p. 270–294. *In* E. H. Lennette, E. H. Spaulding, and J. P. Truant (ed.), Manual of clinical microbiology, 2nd ed. American Society for Microbiology, Washington, D.C.
64. **von Graevenitz, A.** 1978. Clinical role of infrequently encountered nonfermenters, p. 119–153. *In* G. L. Gilardi (ed.), Glucose nonfermenting gram-negative bacteria in clinical microbiology. CRC Press, Boca Raton, Fla.
65. **Weaver, R. E., D. G. Hollis, W. A. Clark, and P. Riley.** 1983. The identification of unusual pathogenic gram negative bacteria. Centers for Disease Control, Atlanta, Ga.
66. **Whitaker, R. J., G. S. Byng, R. L. Gherna, and R. A. Jensen.** 1981. Comparative allostery of 3-deoxy-D-arabino-heptulosonate 7-phosphate synthetase as an indicator of taxonomic relatedness in pseudomonad genera. J. Bacteriol. **145:**752–759.
67. **Wilkinson, S. G., and P. F. Caudwell.** 1980. Lipid composition and chemotaxonomy of *Pseudomonas putrefaciens* (*Alteromonas putrefaciens*). J. Gen. Microbiol. **118:**329–341.

Chapter 31

Legionella

PAUL H. EDELSTEIN

CHARACTERIZATION

The genus *Legionella*, family *Legionellaceae*, contains at least 22 species, most of which have yet to be isolated from clinical specimens (3, 4, 8, 9, 20, 21, 25, 28, 31; D. J. Brenner, personal communication, March 1984). These organisms are primarily aquatic saprophytes which occasionally cause clinical disease. The major reservoirs of *Legionella* appear to be fresh-water sites, air-conditioning units, and various potable-water plumbing fixtures (16). These organisms are nutritionally fastidious, gram-negative, nonsporeforming bacilli. They are variable in size and length, depending upon source and growth conditions. In infected tissue and secretions, they are small uniform bacilli or coccobacilli, approximately 0.5 by 1.0 to 2.0 μm. When grown on culture plates, they range in size and shape from uniform small bacilli, 0.5 by 2.0 μm, to pleomorphic filamentous bacilli up to 100 μm in length. Most, but not all, species are motile by virtue of one or two monopolar flagellae. Sugars are neither oxidized nor fermented; amino acids serve as the primary source of energy. Despite their diversity, all *Legionella* spp. have an absolute nutritional requirement for L-cysteine on primary isolation, and all except one species (*L. oakridgensis*) require the presence of L-cysteine in growth media for optimal growth on subsequent passage (9). Oxidase reactions are variable. The catalase reaction is variable and reflects peroxidase rather than catalase activity in some species (29). Most species grow best in humidified air, but some require 2 to 5% CO_2 for optimal growth; whether this represents a metabolic requirement for CO_2 or is a reflection of medium pH requirement is unknown. The optimal pH of growth media for *L. pneumophila* is 6.85 to 6.95. The cellular fatty acids of all species are distinctive and are unique for gram-negative bacilli, containing primarily branched-chain fatty acids (21, 27).

Taxonomy of the *Legionellaceae* is in a state of constant change. New species and serogroups are being discovered at a dizzying rate, seemingly almost monthly. Many authorities now feel that the complexity of the genus may approach that of *Salmonella*. There are now eleven published species (*L. pneumophila, L. micdadei, L. bozemanii, L. dumoffii, L. feelei, L. gormanii, L. longbeachae, L. jordanis, L. wadsworthii, L. oakridgensis*, and *L. sainthelensi*), but there are at least 11 more recognized species not yet published (D. J. Brenner, personal communication, April 1984). Ten serogroups of *L. pneumophila* and two each of *L. longbeachae* and *L. bozemanii* are recognized; this status also changes frequently, however. Many of the newly recognized species have been isolated only from environmental sources, but it is too soon to know whether this indicates relative pathogenicity.

The only criterion used to distinguish species of *Legionella* has been DNA hybridization studies (21). Phenotypic traits such as pigmentation, autofluorescence, gelatinase production, hippurate hydrolysis, and cellular fatty acid and ubiquinone analyses can be used to differentiate presumptively among species or groups of species. Serotyping plays a major role in species and serogroup identification but is not an absolute arbiter because of shared serologic reactivity by some genetically distinct strains (3, 11, 17, 31).

Differentiating the legionellae can be epidemiologically important in that potential environmental sources of outbreaks can be defined and small clusters of cases can possibly be distinguished from a background of sporadic cases (26). However, there appears to be no major clinical difference between pneumonias caused by different *Legionella* spp. Therefore it is not of major importance that a clinical laboratory be able to accurately distinguish between species, which in some cases is now impossible by serological techniques alone (4, 8, 17; G. W. Gorman, personal communication, April 1984).

Clinical significance

The legionellae are the agents of Legionnaires disease, a multisystem disease manifested primarily as pneumonia (26, 44). The clinical importance of Legionnaires disease is twofold: (i) antimicrobial agents that are usually used to treat pneumonia are inefficacious against legionellosis, whereas prompt administration of effective antimicrobial agents such as erythromycin often leads to cure; (ii) the disease can occur in epidemic form, with a high fatality rate if untreated, and it is being increasingly recognized as a major cause of nosocomial pneumonia. Rare isolations have also been made in cases other than pneumonia, such as wound abscesses. In addition, a systemic, febrile, nonfatal, and nonpneumonic illness known as Pontiac fever is caused by, or at least is very closely associated with, several of the *Legionella* spp. Finally, there is very suggestive clinical and serological evidence that *Legionella* spp. may cause encephalopathy unassociated with pneumonia, as well as native valve endocarditis. There is no good evidence for a human carrier state or for colonization with *Legionella*. However, extensive studies to answer these questions have not been performed (6).

COLLECTION, TRANSPORT, AND STORAGE OF SPECIMENS

Appropriate specimens for culture

Culture of respiratory tract secretions, tissues, and fluids has the highest yield in the culture diagnosis of Legionnaires disease. Occasionally blood cultures are positive, and very rarely other extrapulmonary sites such as wound abscesses, peritoneal fluid, pericardial fluid, prosthetic heart valves, and bowel abscesses yield positive results (1, 5, 12, 24, 45). Cultures of kidney, liver, and spleen also sometimes are positive in fatal cases of Legionnaires disease. Culture diagnosis of Pontiac fever or encephalopathy without pneumonia has not been achieved. Use of invasive techniques to obtain respiratory tract secretions, tissues, or fluids is not usually required, as sputum is ordinarily an adequate specimen when selective media and techniques are used (7, 11, 12, 14, 46). However, the patient's clinical condition sometimes mandates that invasive techniques be used, primarily to exclude other causes of pneumonia in the seriously ill immunocompromised host. Also, it is unknown how sensitive selective culture media and techniques are for the recovery of *Legionella* spp. other than *L. pneumophila*. Thus it is possible that invasive techniques may be required to obtain culture material for some *Legionella* infections. Of these types of specimens, transtracheal aspirates and open-lung biopsies have the highest yield for *Legionella*, and pleural fluid has the lowest yield (12). Specimens obtained during bronchoscopy, such as washings, appear to be almost as useful as sputum, whereas transbronchoscopic biopsies appear to have a low yield.

Specimen collection

All samples should be collected and transported in a sterile, dry screw-top container; if there is a possibility of specimen desiccation, then sterile nonbacteriostatic water should be added in a small amount (13). Use of saline should be avoided, as this is occasionally inhibitory to *Legionella*. Similarly, clinicians should be instructed to use a minimum of saline when obtaining bronchoscopy or transtracheal aspirate samples or, better yet, to use sterile nonbacteriostatic water. Also, it is likely that local anesthetics such as lidocaine are inhibitory to *Legionella*, as they are to other bacteria; only a minimum amount should be used during bronchoscopy for optimal culture yield.

Specimen transportation

Special holding media are not required, as legionellae are generally very hardy, providing that desiccation and exposure to saline and high-molarity (>0.01 M) inorganic buffers are avoided. All specimens should be transported to the laboratory as soon as possible to avoid overgrowth by contaminating flora. If more than a 30-min transport delay is anticipated, then it is best to refrigerate the sample at 5°C until it is delivered to the laboratory. The specimen should be frozen at −20°C or lower if more than a 24- to 48-h transport delay is anticipated. Freezing should be avoided for short-term storage, as freeze-thaw cycles usually result in a drop in viable count. It is also likely that freezing at −70°C is better than at higher temperatures. Specimens should be transported at 5°C unless already frozen, in which case they should be transported in the frozen state.

Collection, storage, and transport of specimens for direct fluorescent-antibody (DFA) examination is no different than for specimens which are to be cultured, as DFA smears are made from the same specimens. Submission of samples just for DFA examination is strongly discouraged, as DFA is a less sensitive test than culture in most instances. If only Formalin-fixed material is available, it is best transported as the wet tissue in Formalin; a small piece (1 by 1 by 1 cm) of consolidated lung is sufficient. Formalin-fixed tissue impression smears and unstained paraffinized tissue can be transported in clean containers at ambient temperatures.

DIRECT EXAMINATION

Legionellae can be visualized in specimens by using a number of specific and nonspecific staining techniques (13, 18, 21, 30, 34, 35, 44). In general the nonspecific methods, such as the Gram, Gimenez, and Dieterle stains, are less sensitive than DFA examination, which is specific. The major exception to this is when organisms in the specimen are from a serogroup or species not included in the DFA test reagents. In this case, the nonspecific stains can be very helpful.

Specimens suitable for staining by both nonspecific and specific stains include all normally sterile respiratory tract and extrapulmonary tissues and fluids, with the exception of blood and urine. Formalin-fixed lung and other organs, as well as paraffin-fixed tissue, are also suitable specimens for specific and nonspecific staining. Use of the nonspecific stains for examination of sputum and other normally nonsterile respiratory tract specimens is not generally helpful; *Legionella* does not differ significantly in its microscopic morphology from some of the normal upper respiratory tract flora.

The Gimenez stain appears to be about as sensitive as silver impregnation methods and is more sensitive than and just as rapid as the Gram stain (Fig. 1). When present in high numbers in lung, *Legionella* is visualized with a modified Gram stain in which 0.1% basic fuchsin is used as the counterstain. The only advantage of the laborious silver impregnation stains, such as the Dieterle, Warthin-Starry, and Tseng-Renner methods, is that organisms can be visualized in paraffin-fixed tissues. Neither the Gram, Gimenez, nor their variations will stain *Legionella* in embedded tissues. Modified acid-fast stains can also be used to detect *L. micdadei* in paraffin-fixed tissue.

Direct immunofluorescent staining of clinical samples has the advantage of specificity (99.9%) and rapidity (7, 12–14, 21, 25, 45, 46). A sample can be screened by DFA testing within hours after receipt, and if positive, it is almost always diagnostic of Legionnaires disease. The major disadvantage of DFA testing is its low sensitivity in relation to culture yield; this is estimated to be between 25 and 80%, depending upon the study. The low sensitivity is due to two major factors: the inherently high concentration of cells necessary to visualize by microscopic means ($\cong 10^4$ cells per ml) and the limited number of cross-reactions between different antigenic types. Thus a negative DFA test does not exclude disease. Because of the limited cross-reactivity between different antigenic types, multiple antisera must be used for screening purposes. Three polyvalent pools (Centers for Disease Control pools A, B, and C) are used which are reactive with *L. pneumophila* serogroups 1 through 6, *L. longbeachae* serogroups 1 and 2, *L. bozemanii, L. micdadei, L. dumoffii*, and *L. gormanii*. Only the specificity and sensitivity of pool A (*L. pneumophila* serogroups 1 through 4) are well known. Pools B and C seem to be as specific, but need further study; their sensitivity is unstudied. The quality of commercially made DFA reagents has been extremely variable, and the potential user should demand de-

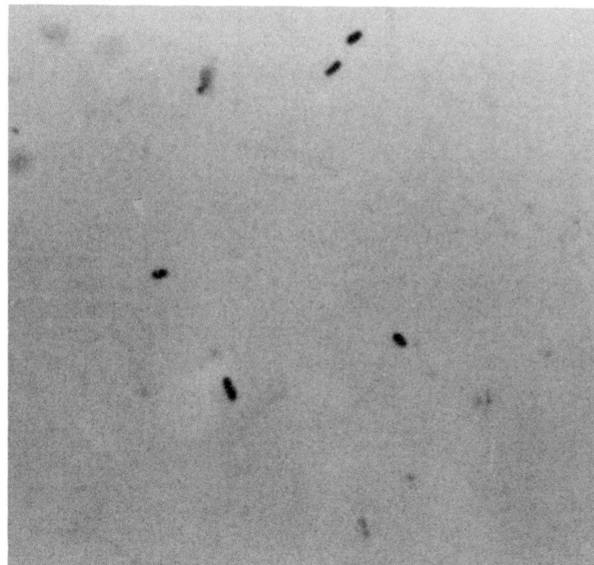

FIG. 1. Gimenez stain of lung homogenate from a guinea pig with experimentally induced Legionnaires disease; bacterial morphology in human lung is identical. Note that the bacilli are much smaller than those taken from a culture plate. In color, *Legionella* cells stain red and the background stains green. Magnification, ×1,200.

tails of quality control testing by the supplier before purchase.

False-positive DFA tests are occasionally noted. In most instances, this is because of laboratory-based problems and not because of cross-reacting bacteria in the specimen. Laboratory-related causes include the use of nonfiltered reagents containing cross-reactive organisms and the transfer of positively staining *Legionella* from a positive to a negative slide by contamination of staining racks or bench top surfaces. Cross-reactions with some *Pseudomonas* spp., *Bacteroides fragilis, Escherichia coli,* and *Streptococcus pneumoniae* are very rarely noted when testing large numbers of organisms and have been an insignificant problem in actual use situations. Regardless, the DFA test requires great skill and experience; it should not be utilized by a laboratory which plans to perform the test infrequently.

Smears for DFA examination and Gimenez stain are made in the same way except that the slide used for DFA should be a glass slide with two 10-mm Teflon circles (Cel-Line Associates, Newfield, N.J.) and that for Gimenez stain should be a single frosted-end microslide. Both types of slides should be cleaned with alcohol before use. If fresh tissue is used, impression smears rather than smears of homogenates provide the best results. Formalin-fixed tissue is scraped with a sterile scalpel, and the scraped material is spread thinly onto slides. Pleural fluid is first centrifuged (3,000 × *g* for 20 min), and the sediment is spread thinly onto the slides. Purulent or bloody areas of sputum or a transtracheal aspirate are teased from other material with sterile wooden sticks, and a thin smear of this material is then made.

The DFA staining procedure requires careful attention to reagents, staining dishes, wash solutions, and drying areas to avoid bacterial contamination and transfer of organisms from slide to slide. Formalin fixation and staining with the DFA conjugate is performed in a moisture chamber, conveniently made with disposable 150-mm petri dishes lined with moistened (with filter-sterilized water) filter paper and using glass rod slide supports. The moisture chambers are replaced weekly to avoid buildup of environmental contaminants. Reagents requiring filter sterilization can most conveniently be handled by using a disposable 0.2-μm sterile filter attached to a 50-ml syringe; the filters should be changed at least weekly. Coplin jars are used for the buffer soaking step, with only a single slide used per jar. Jars are rinsed after each use with tap water several times and a final distilled-water rinse, then dried on a rack. Distilled water used for rinsing the jars should be relatively free of contaminating organisms (<1,000 CFU per ml). The area used for drying slides should be covered with absorbent material which is changed daily, and special efforts should be made to avoid slide-to-slide contact or transfer of draining liquid from one slide to another.

The slides are air dried and then gently heat fixed. Slides for DFA examination are fixed in filter-sterilized 10% neutral Formalin in a moisture chamber for 10 min, washed with filter-sterilized distilled water, and allowed to air dry. The appropriate conjugate is then added to the slides in small volume (≅0.05 ml) and is evenly distributed over the circle material with sterile applicator sticks. The slides are then incubated at room temperature in the moisture chamber for 20 min, after which they are washed in filter-sterilized 0.01 M phosphate-buffered saline (pH 7.6). Then they are immersed in non-filter-sterilized phosphate-buffered saline for 5 min at room temperature in a Coplin jar. The slides are rinsed with nonfiltered distilled water and allowed to air dry before mounting with carbonate-buffered glycerol (pH 9.0).

Examination of DFA slides is performed with a microscope equipped with a UV light source and the proper filters for fluorescein isothiocyanate. Each circle is scanned at ×400 to ×500 over its complete area, much as a differential leukocyte smear is read, from top to bottom or side to side; this should take about 5 to 10 min per circle. Legionellae are seen as small coccobacilli with intense peripheral green fluorescence; they are almost never found in clumps, which may be an indication of cross-reaction or contamination. Atypical bacterial forms may be seen after the initiation of erythromycin therapy, such as coccal and tear-drop shapes; bacterial clumping can also occur late in the course of therapy (14). High-power (×800 to ×1,000) confirmation of the fluorescent forms should be performed. There is no agreed-upon criterion for what the minimum number of organisms observed should be to call a smear positive (13, 14, 21). The Centers for Disease Control criterion is ≥5 per smear; my own is either ≥5 per smear or at least 1 organism in each of four circles. It is usually possible to serotype specimens by using monospecific antisera. False-positive smears due to cross-reacting organisms can sometimes be identified if fluorescein isothiocyanate-conjugated normal rabbit serum stains the organisms. Cross-reacting bacteria can also be detected by failure of at least one monospecific serum to stain organisms.

CULTURE AND ISOLATION

Isolation media

Buffered charcoal yeast extract supplemented with α-ketoglutaric acid (BCYEα) medium is the primary growth medium used (11, 13, 21). It is made selective by the addition of either cefamandole, polymyxin B, and anisomycin (BMPAα medium), or glycine, vancomycin, polymyxin B, and anisomycin (MWY medium) (11, 13, 37). All three media are used for plating clinical samples, as each has its own benefits of sensitivity and selectivity. A differential medium has been described that is useful in isolating more than one species of *Legionella* from the same specimen; this contains bromcresol purple and bromthymol blue (36). There is no good evidence that use of a differential medium improves overall plate sensitivity. Nonetheless, many commercial firms incorporate dyes into their *Legionella* media, as does our laboratory with MWY medium.

Production of these media requires strict attention to the sources of ingredients, the order of adding them, medium pH, and quality control. The medium pH is critical and must be in the range of 6.85 to 6.95 at its use temperature. Unfortunately, adequate quality control testing cannot be performed with a stock strain; it should be done using lung tissue either from a human patient with Legionnaires disease or, preferably, from a guinea pig with experimentally induced Legionnaires disease. Prepared and dehydrated media are available from commercial sources, but because they may be variable in quality, documentation of adequate quality control from the manufacturer is essential.

Specimen decontamination, plating, and medium incubation

Lung and sputum contain substances inhibitory for *Legionella*, so dilution of samples is a requisite for optimal yield; pleural fluid concentration by centrifugation may increase yield, however. A 1:10 dilution of sample in tryptic soy broth is used; other diluents may be satisfactory as long as they do not contain excess sodium. Lung is ground in an all-glass tissue grinder with a small amount of tryptic soy broth and then diluted before plating. Sputum, endotracheal tube aspirates, and other such samples are placed in a sterile petri dish. More bloody or purulent-appearing areas are transferred to another petri dish with sterile wooden sticks, and then about 0.1 ml of this material is diluted in 0.9 ml of tryptic soy broth before plating; the remainder of the sample is used for making smears. Specimens which have already been diluted as a result of the collection process, such as some transtracheal aspirates and bronchial lavage samples, are not diluted before plating. In all cases, it is wise to homogenize samples with sterile glass beads before plating.

Since none of the selective media is highly selective, preplating decontamination is useful for normally nonsterile specimens, such as sputum, bronchial lavage samples, and endotracheal tube aspirates (7, 13). Decontamination is performed by diluting the specimen 1:10 in a KCl-HCl (pH 2.2) solution; this is mixed, incubated at room temperature for 4 min, and then plated directly onto both selective and nonselective media. Since this preplating treatment may reduce specimen yield, it is best to plate specimens with and without prior treatment.

All plates should be inoculated with ≅0.1 ml of specimen and streaked for isolation. The streaked plates are incubated in a humidified air incubator at 35°C and 80 to 90% relative humidity. Incubation can also be performed in a humidified CO_2 incubator if the CO_2 content is less than or equal to 5%.

Two diphasic BCYEα blood culture bottles are each inoculated with 5 ml of blood (13, 21). They are incubated upright under the same conditions as BCYEα plates. The bottles should be tilted every other day so that the broth transiently coats the agar slant.

Isolation of *Legionella*

Recognition of *Legionella* colonies on plates is facilitated by their slow growth rate and by their characteristic colonial morphology. Daily observation of plates with a dissecting microscope is the key to successful isolation. Colonies which appear on or after day 2 of incubation should be suspect; the average time for appearance of microscopic (≅0.5-mm) *Legionella* colonies is 2 days, and that for macroscopic (1- to 2-mm) colonies is 3 days. *Legionella* grows very slowly in blood culture and generally takes about 2 weeks of incubation to become positive; specimens should be held for 1 month before being discarded as negative. *Legionella* may sometimes take up to 2 weeks to grow from respiratory tract samples; all plates should be held routinely for 2 weeks and examined daily for the first week and again at 14 days. Very rarely, microscopic *Legionella* colonies may appear after only overnight incubation, especially from samples of lung from patients who died of Legionnaires disease without receiving antimicrobial therapy. Regardless, macroscopic colonies that appear after overnight incubation are not *Legionella*, especially if they are 2.0 mm or greater in diameter. Microscopic *Legionella* colonies vary in color from colorless to iridescent blue or pink and are round and flat with entire edges; they may be either translucent or speckled. Macroscopic colonies, on their first day of appearance, are usually convex, round, speckled colonies with entire edges and are blue-white to pinkish in color. Within 24 h these colonies may lose their characteristic speckled appearance and color. They then increase in size to a maximum of 3 mm in diameter and may flatten, often more in the center than on the edges, producing a heaped-up edge. The colonies are often pleomorphic in appearance, with large and small colony types commonly occurring. The most characteristic colony type is usually seen on the first day of appearance of macroscopic, convex colonies; after that, the colonies may be mistaken for other organisms, especially if careful daily records are not kept of plate growth and colonial morphology. When touched with a loop, *Legionella* colonies are usually tenacious and form a string when the loop is lifted; the colonies do not lift freely in one piece from the agar surface. Young colonies of some other organisms may cause confusion with *Legionella*, especially *Eikenella corrodens*, some *Pseudomonas* spp., and some *Bacillus* spp., all of which may occasionally grow on BMPAα or MWY medium.

Legionella growing in diphasic blood culture bottles can be detected by visualizing growth on the bottle slant. There is no need for subculture unless growth is noted on the slant. Rarely, BACTEC (Johnston Laboratories, Cockeysville, Md.) blood bottles may contain viable legionellae which can be detected by a slight increase in the growth index and by subsequent subculture to BCYEα medium (10).

Some *Legionella* sp. colonies can be identified quickly by their electric blue-white color when illuminated by a long-wave UV (Woods) lamp (Table 1). The use of long-wave UV light may be more helpful than differential dye media, since many bacteria commonly present in sputum assume the same colors on dye-containing media that *Legionella* colonies do.

IDENTIFICATION

Identifying an organism as a *Legionella* species is relatively easy, requiring only simple techniques and growth tests (Fig. 2). Identification of the species may be more complex and sometimes may be impossible except for the extremely sophisticated laboratory (13, 21). There is no good evidence that identification to the species level is important for other than epidemiologic purposes, and each laboratory must decide for itself whether its academic and epidemiologic interests justify the time and expense sometimes required for the process.

A Gram stain should be performed of suspect *Legionella* colonies. This should always be done using a small amount of a colony emulsified in distilled water to avoid false-positive staining cells. Basic fuchsin (0.1%), rather than safranin, is used as a counterstain. *Legionella* taken from a culture plate stains negatively and is usually pleomorphic, varying in length from 2 to over 100 μm, but usually has a constant width of 0.5 to 1.0 μm (Fig. 3).

Two major types of tests are used to identify and differentiate *Legionella* spp. The most important is a growth test for L-cysteine requirement, which can be used alone to identify *Legionella* in most cases. Serotyping is the primary tool used for presumptive species identification and confirmation of growth tests, but may be either false-negative or false-positive. Biochemical testing can help confirm identity, but is not specific enough to definitively identify a nonserotypable strain. More definitive tests include cellular fatty acid analysis by gas-liquid chromatography and cellular ubiquinone analysis by reverse-phase thin-layer liquid chromatography. The final confirmation of species identity is made by DNA homology testing, but this is not considered a routinely required test.

A flow diagram for the presumptive identification of legionellae is shown in Fig. 2. The media used to test for a nutritional requirement for L-cysteine are BCYEα medium without L-cysteine (BCYEα-Lcys medium) and tryptic soy–5% sheep blood agar (BAP). Both of these medium types should be inoculated with suspect *Legionella* colonies, as the results obtained provide some help in the presumptive identification of *L. oakridgensis* and also because some fastidious non-*Legionella* organisms will not grow on BAP, but will grow on BCYEα-Lcys medium. Non-*Legionella* organisms will grow on BCYEα-Lcys as luxuriantly as on BCYEα, whereas there will be no or extremely scant growth of *Legionella* on BCYEα-Lcys in comparison with that observed on BCYEα medium. The occasional sparse growth of *Legionella* observed on BCYEα-Lcys is attributed to nutrient carry-over from BCYEα, except in the case of *L. oakridgensis*, which appears to require L-cysteine only for primary isolation. *L. jordanis* will become medium-adapted to BCYEα-Lcys, but only after multiple passages, and even then will still grow better on BCYEα medium. Microbiologists

TABLE 1. *Legionella* sp. identification[a]

Species	Characteristics										
	Long-wave UV fluorescence	Tyrosine-yeast extract browning	Oxidase	Catalase	Gelatinase	β-Lactamase	Hippurate hydrolysis	Major cellular fatty acid(s)	Contains a17:1	Requires L-cysteine	Major ubiquinone(s)[b]
L. pneumophila	Yellow-green	+	$+^{w}$ [c,d]	$+^{w}$	+	+	+[e]	i16:0, 16:1	−	+	Q12
L. longbeachae	Yellow-green	+	$+^{w}$	+	+	+[c]	−	i16:0, 16:1	−	+	Q9–12
L. oakridgensis	Yellow-green	+	−	+	+	±	−	i16:0, 16:1	−	−[f]	Q10
L. micdadei	Yellow-green	−	±	+	−	−	−	a15:0	+	+	Q12,13
L. dumoffii	Blue-white	+[c]	−	$+^{w}$	+	+	−	a15:0	−	+	Q9–12
L. gormanii	Blue-white	+	−	$+^{w}$	+	+	−	a15:0	−	+	Q9–12
L. bozemanii	Blue-white	+	−	+	+	+	−	a15:0	−	+	Q9–12
L. jordanis	Yellow-green	+	+	+	+	+	−	a15:0	−	±[f]	Q12,13
L. wadsworthii	Yellow-green	−	−	+	+	+	−	a15:0	−	+	Q10
L. anisa	Blue-white/ yellow-green[g]	+	+	+	+	+	−	a15:0	−	+	Q10–12
L. feelei	None	+	−	+	−	−	+	a15:0, i16:0, 16:1	−	+	Q13
L. sainthelensi	Yellow-green	+	+	+	+	+	−	i16:0, 16:1	−	+	Q9–12

[a] Modified from reference 13 with permission of the publisher. References: 3, 4, 8, 9, 19–22, 27, 28, 31, 45; C. W. Moss, G. W. Gorman, and D. J. Brenner, personal communications (March 1984).

[b] 3+ or greater on a scale of 0–4.

[c] Some strains negative.

[d] $+^{w}$, Weak reaction.

[e] Some serogroup 4 strains are negative.

[f] Stock strains only.

[g] Some strains fluoresce blue-white, and others are yellow-green.

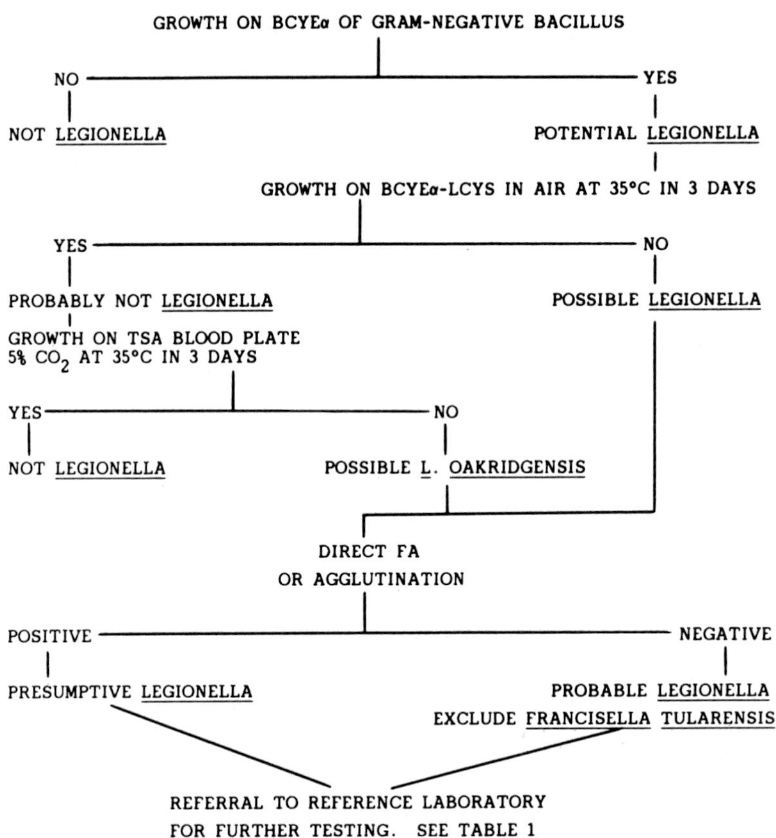

FIG. 2. Flow scheme for identification of *Legionella* spp. TSA, Tryptic soy agar.

working in areas endemic for tularemia should be aware that *Francisella tularensis* can mimic the growth requirements of *Legionella*, and this organism should be considered for nonserotypable strains which meet these growth requirements (38).

Biochemical testing is sometimes useful in characterizing *Legionella*, which is asaccharolytic and urease and nitrate negative (13, 21). Oxidase testing should be performed using 1% *N,N,N′,N′*-tetramethyl-*p*-phenylenediamine dihydrochloride made just before use without ascorbic acid stabilization; results of this test are variable from species to species. Catalase testing, performed with 3% H_2O_2 in 10% Tween 80, gives variable results with *Legionella*. Production of a diffusible melanin-like brown pigment can be demonstrated by inoculating tyrosine-supplemented yeast extract medium; pigment production is seen after heavy growth appears on the plate. A fluorescent pigment is also made by some *Legionella* spp. and is very helpful in distinguishing between species. The fluorescent pigments are best demonstrated by illuminating areas of heavy growth on tyrosine-yeast extract medium with a long-wave UV lamp, although individual colonies growing on any medium will show the same characteristics.

Gelatinase testing, which should be performed with a special BCYEα-based medium, relies on the fact that a gelatinase-positive strain liquefies the tubed medium which is ordinarily solid at 5°C (13). The gelatinase medium is inoculated heavily with test strain, incubated at 35°C, and checked daily at 5°C for 7 days for liquefaction. Positive and negative gelatinase controls should be run simultaneously. Other important biochemical tests include those for β-lactamase production (performed with nitrocefin-containing disks) and hippurate hydrolysis (performed by a rapid method) (19). Endopeptidase analysis is a promising new technique which has been used by one group of investigators to separate the *Legionellaceae* into biochemically related groups (2).

Cellular fatty acid analysis by gas-liquid chromatography is a major taxonomic tool, as the general pattern seen with the *Legionellaceae* is genus specific for mesophilic gram-negative bacilli (27). One pitfall of fatty acid analysis is that thermophilic *Bacillus* spp. can mimic *Legionella* to the extent of a similar gas-liquid chromatographic pattern (33). These bacilli can be excluded on the basis of their growth at 55°C. The method is somewhat cumbersome and perhaps is better left to a reference laboratory. Fatty acid composition falls into two major patterns for *Legionella*, a predominance either of methyl-12-methyltetradecanoic (a15:0) acid or of methyl-14-methylpentadecanoic (i16:0) and methylhexadecanoic (16:1) acids. Thus fatty acid analysis is a powerful tool to confirm identification of *Legionella*, but does not provide

TABLE 2. Significant serologic cross-reactions observed between the *Legionellaceae* with DFA conjugates[a]

Antiserum to:	Reacts with antigen(s) of[b]:
L. pneumophila SG 3	*L. pneumophila* SG 2, 6
L. pneumophila SG 6	*L. pneumophila* SG 3
L. pneumophila SG 8	*L. pneumophila* SG 4, 5
L. longbeachae SG 2	*L. anisa, L. sainthelensi,* and *L. bozemanii* SG 2
L. bozemanii SG 1	*L. jordanis*
L. jordanis	*L. bozemanii* SG 1
L. sainthelensi	*L. oakridgensis* and *L. longbeachae* SG 1
L. oakridgensis	*L. sainthelensi*

[a] SG, Serogroup. References: 3, 4, 8, 9, 31.

[b] Most of these cross-reactions can be eliminated by absorption with the cross-reacting antigen or, in many cases, by using a high working dilution of antibody.

enough information by itself to differentiate among species.

Cellular isoprenoid quinone (ubiquinone) analysis is a promising technique which is less cumbersome than gas-liquid chromatography, as it does not require the use of expensive equipment. The *Legionellaceae* contain unique ubiquinones and can be divided into five different groups on the basis of qualitative and quantitative differences in ubiquinone content. This analysis is performed using reverse-phase thin-layer liquid chromatography (21; C. W. Moss, W. F. Bibb, D. E. Karr, and G. O. Guerrant, Proc. Int. Symp. Bacteriol., Lille, France, 25–26 May 1983).

Serotyping can be performed using either agglutination or immunofluorescent techniques (13, 21, 25, 42). It is not helpful if negative, and it is only presumptive if positive. This is because of the likelihood that not all *Legionella* serotypes have been discovered, making it possible that nonserotypable isolates are new serotypes of *Legionella*. Serological cross-reactions exist within the *Legionellaceae* and may not respect species boundaries (Table 2). Presently there is at least one strain of *L. anisa* which can only be distinguished phenotypically from *L. longbeachae* serogroup 2 on the basis of cellular fatty acid or ubiquinone analysis because of otherwise identical phenotypic traits, including serotype (G. W. Gorman, personal communication, April 1984). Also *L. sainthelensi,* which is phenotypically identical to some strains of *L. longbeachae,* shows serologic cross-reactions with *L. longbeachae* antisera, making separation difficult (8).

Serotyping can be performed by using either a DFA method or an agglutination method. In either case, a suspension of organisms in 10% buffered Formalin is made first. For DFA testing, the best results seem to be obtained with a 1:100 dilution (in 10% buffered Formalin) of a bacterial suspension equal in density to that of a no. 1 MacFarland barium sulfate standard. This eliminates a prozone phenomenon which occasionally produces false-negative results. The diluted suspension is allowed to air dry on a microscope fluoro-slide, heat fixed, and then stained and examined exactly as are clinical specimens. Cross-reactions (Table 2) can be a hindrance to species identification and serogrouping, making other phenotypic typing tests necessary. None of the non-*Legionella* organisms described to cross-react with *Legionella* antibody conjugate shares identical growth requirements with *Legionella,* so false-positive serotyping should not be a problem if suspect organisms are tested for growth requirements before serotyping. Agglutination testing is performed by adding bacterial suspension to the agglutinating reagent and observing for agglutination in 1 min. Agglutination reagents are unavailable commercially, but DFA reagents are available from the Centers for Disease Control as well as a number of commercial sources.

Soluble antigen detection in urine

Kohler and Sathapatayarongs have established that detection of soluble *L. pneumophila* antigen in the urine can be used to diagnose Legionnaires disease (23). This test appears to have a sensitivity of 80% and a specificity of 99%, but it is serogroup specific. Reagents for this testing method are not available commercially.

Serological diagnosis

Diagnosis of Legionnaires disease can be made using an indirect immunofluorescent method for detecting changes in serum antibody titers (21, 39, 40). A rise in titer level from <1:32 to ≥1:128 is considered evidence for seroconversion. This testing method is only well standardized for *L. pneumophila* serogroup 1 infections and has an estimated sensitivity of 60 to 80% and specificity of 95 to 99% (40, 43). Since the overall prevalence of sporadic Legionnaires disease is in the 5 to 10% range, the positive predictive accuracy of this method of testing is fairly low (40 to 90%), making serological diagnosis more helpful in epidemiological studies than for diagnosis of sporadic disease. Other methods of serological diagnosis have been described, such as enzyme-linked immunosorbent as-

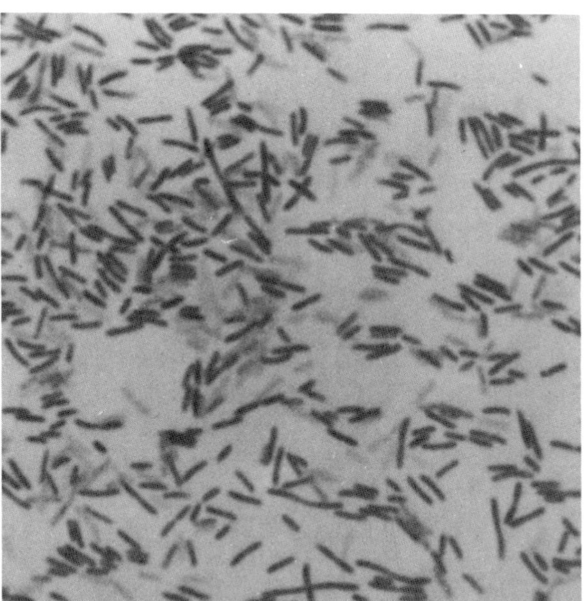

FIG. 3. Gram stain, with 0.1% basic fuchsin counterstain, of *L. pneumophila* grown on a BCYEα plate for 48 h. Note the pleomorphic forms, including filamentous ones. This is in contrast to the microscopic morphology of *Legionella* in tissues (Fig. 1). Magnification, ×1,200.

say and microagglutination; these alternative methods need better standardization before they can be regarded as substitutes for the indirect immunofluorescent method (15, 21, 39,45). Different methods of antigen preparation have led to claims of superiority of one preparation over the other, but more comparative studies are needed (32, 41).

ANTIMICROBIAL SUSCEPTIBILITY TESTING

There is no indication for performing antimicrobial susceptibility testing for *Legionella*. This is because of the lack of correlation of in vitro susceptibility and antimicrobial efficacy, as well as the lack of a well-standardized in vitro method which does not inactivate antimicrobial agents.

LABORATORY SAFETY

Laboratory-acquired Legionnaires disease is exceptionally rare. No case of culture-documented laboratory-associated disease has been reported. Most laboratories work with *Legionella* on the open bench, except when generating aerosols or when inoculating animals. *Coccidioides immitis, Blastomyces dermatitidis,* and *F. tularensis* do grow readily on BCYEα, arguing for extreme caution when handling these plates if a mold or unusual gram-negative bacillus is growing on them.

LITERATURE CITED

1. **Arnow, P. M., E. J. Boyko, and E. L. Friedman.** 1983. Perirectal abscess caused by *Legionella pneumophila* and mixed anaerobic bacteria. Ann. Intern. Med. **98:**184–185.
2. **Berdal, B. P., K. Bøvre, Ø. Olsvik, and T. Omland.** 1983. Patterns of extracellular proline-specific endopeptidases in *Legionella* and *Flavobacterium* spp. demonstrated by use of chromogenic peptides. J. Clin. Microbiol. **17:**970–974.
3. **Bibb, W. F., R. J. Sorg, B. M. Thomason, M. D. Hicklin, A. G. Steigerwalt, D. J. Brenner, and M. R. Wulf.** 1981. Recognition of a second serogroup of *Legionella longbeachae*. J. Clin. Microbiol. **14:**674–677.
4. **Bissett, M. L., J. O. Lee, and D. S. Lindquist.** 1983. New serogroup of *Legionella pneumophila*, serogroup 8. J. Clin. Microbiol. **17:**887–891.
5. **Brabender, W., D. R. Hinthorn, M. Asher, N. J. Lindsey, and C. Liu.** 1983. *Legionella pneumophila* wound infection. J. Am. Med. Assoc. **250:**3091–3092.
6. **Bridge, J. A., and P. H. Edelstein.** 1983. Oropharyngeal colonization with *Legionella pneumophila*. J. Clin. Microbiol. **18:**1108–1112.
7. **Buesching, W. J., R. A. Brust, and L. W. Ayers.** 1983. Enhanced primary isolation of *Legionella pneumophila* from clinical specimens by low-pH treatment. J. Clin. Microbiol. **17:**1153–1155.
8. **Campbell, J., W. F. Bibb, M. A. Lambert, S. Eng, A. G. Steigerwalt, J. Allard, C. W. Moss, and D. J. Brenner.** 1984. *Legionella sainthelensi*: a new species of *Legionella* isolated from water near Mt. St. Helens. Appl. Environ. Microbiol. **47:**369–373.
9. **Cherry, W. B., G. W. Gorman, L. H. Orrison, C. W. Moss, A. G. Steigerwalt, H. W. Wilkinson, S. E. Johnson, R. M. McKinney, and D. J. Brenner.** 1983. *Legionella jordanis*: a new species of *Legionella* isolated from water and sewage. J. Clin. Microbiol. **15:**290–297.
10. **Chester, B., E. G. Poulous, M. J. Demarry, E. Albin, and T. Prilucik.** 1983. Isolation of *Legionella pneumophila* serogroup 1 from blood with nonsupplemented blood culture bottles. J. Clin. Microbiol. **17:**195–197.
11. **Edelstein, P. H.** 1981. Improved semiselective medium for isolation of *Legionella pneumophila* from contaminated clinical and environmental specimens. J. Clin. Microbiol. **14:**298–303.
12. **Edelstein, P. H.** 1983. Culture diagnosis of *Legionella* infections. Zentralbl. Bakteriol. Parasitenkd. Infektionskr. Hyg. Abt. 1 Orig. Reihe A **255:**96–101.
13. **Edelstein, P. H.** 1984. Legionnaires' disease laboratory manual. National Technical Information Service, Springfield, Va.
14. **Edelstein, P. H., R. D. Meyer, and S. M. Finegold.** 1980. Laboratory diagnosis of Legionnaires' disease. Am. Rev. Respir. Dis. **121:**317–327.
15. **Elder, E. M., A. Brown, J. S. Remington, J. Shonnard, and Y. Naot.** 1983. Microenzyme-linked immunosorbent assay for detection of immunoglobulin G and immunoglobulin M antibodies to *Legionella pneumophila*. J. Clin. Microbiol. **17:**112–121.
16. **Fliermans, C. B.** 1983. Autoecology of *Legionella pneumophila*. Zentralbl. Bakteriol. Parasitenkd. Infektionskr. Hyg. Abt. 1 Orig. Reihe A **255:**58–63.
17. **Garrity, G. M., E. M. Elder, B. Davis, R. M. Vickers, and A. Brown.** 1982. Serological and genotypic diversity among serogroup 5-reacting environmental *Legionella* isolates. J. Clin. Microbiol. **15:**646–653.
18. **Greer, P. W., F. W. Chandler, and M. D. Hicklin.** 1980. Rapid demonstration of *Legionella pneumophila* in unembedded tissue. An adaptation of the Giménez stain. Am. J. Clin. Pathol. **73:**788–790.
19. **Hébert, G. A.** 1981. Hippurate hydrolysis by *Legionella pneumophila*. J. Clin. Microbiol. **13:**240–242.
20. **Herwaldt, L. A., G. W. Gorman, T. McGrath, S. Toma, B. Brake, A. W. Hightower, J. Jones, A. L. Reingold, P. A. Boxer, P. W. Tang, C. W. Moss, H. Wilkinson, D. J. Brenner, A. G. Steigerwalt, and C. V. Broome.** 1984. A new *Legionella* species, *Legionella feelei* species nova, causes Pontiac fever in an automobile plant. Ann. Intern. Med. **100:**333–338.
21. **Jones, G. L., and G. A. Hébert (ed.).** 1979. "Legionnaires'": the disease, the bacterium, and methodology. Centers for Disease Control, Atlanta, Ga. National Technical Information Service, Springfield, Va.
22. **Karr, D. E., W. F. Bibb, and C. W. Moss.** 1982. Isoprenoid quinones of the genus *Legionella*. J. Clin. Microbiol. **15:**1044–1048.
23. **Kohler, R. B., and B. Sathapatayavongs.** 1983. Recent advances in the diagnosis of serogroup 1 *L. pneumophila* pneumonia by detection of urinary antigen. Zentralbl. Bakteriol. Parasitenkd. Infektionskr. Hyg. Abt. 1 Orig. Reihe A **255:**102–107.
24. **McCabe, R. E., J. C. Baldwin, C. A. McGregor, C. Miller, and K. L. Vosti.** 1984. Prosthetic valve endocarditis caused by *Legionella pneumophila*. Ann. Intern. Med. **100:**525–527.
25. **McKinney, R. M., L. Thacker, P. P. Harris, K. R. Lewallen, G. A. Hébert, P. H. Edelstein, and B. M. Thomason.** 1979. Four serogroups of Legionnaires' disease bacteria defined by direct immunofluorescence. Ann. Intern. Med. **90:**621–624.
26. **Meyer, R. D.** 1983. Legionella infections: a review of five years of research. Rev. Infect. Dis. **5:**258–278.
27. **Moss, C. W., and S. B. Dees.** 1979. Further studies of the cellular fatty acid composition of Legionnaires disease bacteria. J. Clin. Microbiol. **9:**648–649.
28. **Orrison, L. H., W. B. Cherry, R. L. Tyndall, C. B. Fliermans, S. B. Gough, M. A. Lambert, L. K. McDougal, W. F. Bibb, and D. J. Brenner.** 1983. *Legionella oakridgensis*: unusual new species isolated from cooling tower water. Appl. Environ. Microbiol. **45:**536–545.
29. **Pine, L., P. S. Hoffman, G. B. Malcolm, R. F. Benson, and G. W. Gorman.** 1984. Whole-cell peroxidase test for identification of *Legionella pneumophila*. J. Clin. Microbiol. **19:**286–290.
30. **Pounder, D. J., and S. Stevens.** 1981. Legionnaires' disease at autopsy. Am. J. Forensic Med. Pathol. **2:**139–142.

31. **Tang, P. W., S. Toma, C. W. Moss, A. G. Steigerwalt, T. G. Cooligan, and D. J. Brenner.** 1984. *Legionella bozemanii* serogroup 2: a new etiological agent. J. Clin. Microbiol. **19:**30–33.
32. **Taylor, A. G., and T. G. Harrison.** 1983. Serological tests for *Legionella pneumophila* serogroup 1 infections. Zentralbl. Bakteriol. Parasitenkd. Infektionskr. Hyg. Abt. 1 Orig. Reihe A **255:**20–26.
33. **Thacker, L., R. M. McKinney, C. W. Moss, H. M. Sommers, M. L. Spivack, and T. F. O'Brien.** 1981. Thermophilic sporeforming bacilli that mimic fastidious growth characteristics and colonial morphology of legionellae. J. Clin. Microbiol. **13:**794–797.
34. **Tseng, C. H., and E. D. Renner.** 1983. A new staining method for *Legionella pneumophila*. Am. J. Clin. Pathol. **79:**377–378.
35. **Van Orden, A. E., and P. W. Greer.** 1977. Modification of the Dieterle spirochete stain. J. Histotechnol. **1:**51–53.
36. **Vickers, R. M., A. Brown, and G. M. Garrity.** 1981. Dye-containing buffered charcoal-yeast extract medium for differentiation of members of the family *Legionellaceae*. J. Clin. Microbiol. **13:**380–382.
37. **Wadowsky, R. M., and R. B. Yee.** 1981. A glycine-containing selective medium for isolation of *Legionellaceae* from environmental specimens. Appl. Environ. Microbiol. **42:**768–772.
38. **Westerman, E. L., and J. McDonald.** 1983. Tularemia pneumonia mimicking Legionnaires' disease: isolation of organism on CYE agar and successful treatment with erythromycin. South. Med. J. **76:**1169–1170.
39. **Wilkinson, H. W.** 1982. Serologic diagnosis of Legionellosis. Lab. Med. **13:**151–157.
40. **Wilkinson, H. W.** 1983. Status of serological tests for *Legionella* antigen and antibody at the Centers for Disease Control. Zentralbl. Bakteriol. Parasitenkd. Infektionskr. Hyg. Abt. 1 Orig. Reihe A **255:**3–7.
41. **Wilkinson, H. W., and B. J. Brake.** 1982. Formalin-killed versus heat-killed *Legionella pneumophila* serogroup 1 antigen in the indirect immunofluorescence assay for legionellosis. J. Clin. Microbiol. **16:**979–981.
42. **Wilkinson, H. W., and B. J. Fikes.** 1980. Slide agglutination test for serogrouping *Legionella pneumophila* and atypical *Legionella*-like organisms. J. Clin. Microbiol. **11:**99–101.
43. **Wilkinson, H. W., A. L. Reingold, B. J. Brake, D. L. McGiboney, G. W. Gorman, and C. V. Broome.** 1983. Reactivity of serum from patients with suspected legionellosis against 29 antigens of *Legionellaceae* and *Legionella*-like organisms by indirect immunofluorescence assay. J. Infect. Dis. **147:**23–31.
44. **Winn, W. C., Jr., and R. L. Myerowitz.** 1981. The pathology of the *Legionella* pneumonias. Human Pathol. **12:**401–422.
45. **Winn, W. C., Jr., and A. W. Pasculle.** 1982. Laboratory diagnosis of infections caused by *Legionella* species. Clin. Lab. Med. **2:**343–369.
46. **Zuravleff, J. J., V. L. Yu, J. W. Shonnard, B. K. Davis, and J. D. Rihs.** 1983. Diagnosis of Legionnaires' disease. An update of laboratory methods with new emphasis on isolation by culture. J. Am. Med. Assoc. **250:**1981–1985.

Chapter 32

Brucella

W. J. HAUSLER, JR., N. P. MOYER, AND L. A. HOLCOMB

CHARACTERIZATION

Members of the genus *Brucella* are intracellular parasites which cause epizootic abortions in a variety of animals and septicemic febrile illness or localized infection of bone, tissue, or organ systems in humans. Organisms are isolated from unpasteurized dairy products, infected animals, and clinical specimens such as blood, tissues, and occasionally abscesses of organs. The species infective for humans are *B. suis*, which usually infects swine, *B. abortus*, predominantly a pathogen for cattle, *B. melitensis*, found in goats and sheep, and *B. canis*, a pathogen of dogs. *B. neotomae*, which occurs in the desert wood rat, and *B. ovis*, which is pathogenic for sheep, are not known to cause disease in humans.

Organism

Brucellae are small, nonmotile, nonsporulating, gram-negative coccobacilli or short rods (0.5 to 0.7 μm by 0.5 to 1.5 μm) arranged singly or in pairs and short chains. Capsules, if present, are small.

Growth

Growth occurs aerobically, often enhanced by CO_2, but no growth occurs under strict anaerobic conditions. Thiamine, niacin, and biotin are required for growth, and some strains require the addition of serum to the medium for growth. Calcium panthenate and *meso*-erythritol stimulate growth (9). Optimal growth temperature is 37°C, with a temperature range of 10 to 40°C. Optimal pH range is 6.6 to 7.4. Colonies appear on agar surface after 2 to 3 days of incubation and reach 2 to 3 mm after 4 or 5 days. Growth is slower on selective media. Colonies are nonhemolytic and nonpigmented and have a smooth glistening surface. *B. ovis* and *B. canis* occur normally in the rough form, whereas the other four species have been encountered only in the smooth form. Dissociation readily occurs in the laboratory. A cell-wall-defective variant of bovine origin has been reported (10).

Physiology

Brucellae are catalase-positive, oxidase-positive (except *B. ovis* and *B. neotomae*), urease-variable organisms which reduce nitrate to nitrite. They are citrate and methyl red negative, do not produce acetylmethylcarbinol or gelatinase, and do not release *o*-nitrophenol from *o*-nitrophenol-β-D-galactoside. Metabolism is mainly oxidative with little fermentative action on carbohydrates in conventional media. Production of H_2S, resistance to Thionin and basic fuchsin, and urea hydrolysis help differentiate the species. Complete biotype or strain identification requires oxidative metabolism studies (20).

Serological identification

Smooth strains of *Brucella* are agglutinated with unabsorbed antisera to smooth *Brucella* (commercially available) with known cross-reactions to *Francisella tularensis*, *Vibrio cholerae*, and *Yersinia enterocolitica* serotype 9 (13, 14). Monospecific antisera prepared by differential absorption with *B. melitensis* and *B. abortus* can be used to determine the predominant lipopolysaccharide, M. or A. *B. ovis*, *B. canis*, and rough variants of the other *Brucella* species cross-agglutinate when unabsorbed anti-rough *Brucella* sera are used, but no cross-reactions occur when these organisms are agglutinated with unabsorbed anti-smooth *Brucella* sera. Fluorescent antibody techniques are useful for genus identification only (17).

Laboratory animal pathogenicity

Smooth strains of *Brucella* produce subclinical to lethal infections in guinea pigs, hamsters, and rabbits on intraperitoneal inoculation. Rough variants of smooth *Brucella* strains have decreased pathogenicity for laboratory animals.

CLINICAL SIGNIFICANCE

Brucellosis in humans has a variable incubation time, an insidious or abrupt onset, and no pathogenomonic symptoms or signs. For these reasons the majority of laboratory-confirmed cases of brucellosis are based on serological tests rather than isolation of the organism (17). In 10 to 15% of patients with brucellosis, various complications of an articular, osseous, visceral, or neurological nature occur. Osteomyelitis is the most frequent complication in humans (24).

In the United States, brucellosis is largely job-related. Most of the 2,238 reported cases during the period 1972–1981 occurred in persons with occupational or avocational exposure to animals or laboratory cultures of *Brucella* (5, 24). Only 183 cases were reported in 1983 (7). *B. canis*, a pathogen of dogs, has accounted for more than 30 cases of brucellosis in humans (24). Sporadic episodes of food-associated brucellosis have occurred in recent years. In most instances the organism responsible was *B. melitensis* and the vehicle was Mexican or Mediterranean goat cheese (2, 6, 24). Brucellosis has occurred in hunters butchering elk, moose, and bison in the United States and in persons ingesting raw meat, bone marrow, or liver of reindeer, moose, and caribou in Alaska (19).

COLLECTION AND TRANSPORT OF SPECIMENS

Multiple blood cultures should be obtained when brucellosis is suspected. Acute-phase serum should be obtained as soon as possible after onset of the disease, followed by a convalescent-phase serum collected 14 to 21 days after onset. Additional sera may be required to establish a diagnosis when blocking antibody is present or when immunoglobulin G studies are indicated. Infected tissues and abscesses should be cultured, as should bone marrow and liver biopsies if available. Rarely, cerebrospinal fluids, pleural fluid, peritoneal fluid, urine, and other specimens may be collected for isolation of *Brucella*. Although *Brucella* organisms are relatively resistant to adverse environmental conditions, specimens should be cultured as soon as possible. If a delay is expected or specimens are to be sent to a central laboratory for culturing, they should be refrigerated.

DIRECT EXAMINATION

The direct fluorescent-antibody technique is a rapid method for demonstrating the organism in specimens (17), but the conjugate is not commercially available at this time. Smears are prepared by swab method, from buffy coat or by tissue impression. The smear is dried, heat fixed, and stained (1). The direct fluorescent-antibody test result should be considered a presumptive report; therefore all specimens should be cultured. There is no evidence that the demonstration of organisms in tissue by the modified Koster or Ziehl-Neelsen staining methods is of any value in the diagnosis of brucellosis in humans.

CULTURE AND ISOLATION

Infection with *Brucella* is readily acquired by laboratory workers through skin contact, inhalation of aerosols, eye and mouth contact, and accidental inoculation by needle and syringe. Culture work should be done in a biological safety cabinet, and laboratory workers should wear protective clothing (8, 9).

Blood

When blood and other body fluids are cultured, the Castaneda technique is recommended. This technique utilizes a biphasic (solid and liquid) medium in the same bottle. The solid phase is prepared by adding agar to Trypticase soy agar (BBL Microbiology Systems, Cockeysville, Md.), tryptose agar, or brucella agar to a final concentration of 2.5%. The liquid phase is prepared from the same basal medium without agar and added aseptically after the agar has solidified. The Castaneda bottle should also have an atmosphere of 5 to 10% CO_2. This procedure using the Castaneda bottle is described by Alton et al. (1). If Castaneda bottles are not available, commercial blood culture bottles can be used for culturing *Brucella* spp.; however, subcultures must be made every 4 to 5 days. It is important to use media with added CO_2 and to vent the blood culture bottle. Blood cultures should be incubated at 35 to 37°C and examined for 30 days.

Abscesses or tissues

Fibrous clots, exudates, and tissue are aseptically ground in a mortar with sterile sand. Add ground material to broth, allow the sand to settle, decant, centrifuge, and inoculate media and guinea pigs, if available. This material is inoculated onto 5% sheep blood agar, brucella agar containing 5% serum, or serum dextrose agar (Oxoid USA Ltd., Columbia, Md.). When significant contamination is likely, a selective medium should be inoculated in addition to the basal medium. Overgrowth of cultures can be controlled by adding 1.4 ml of 0.1% aqueous crystal violet (certified) per liter of medium before sterilization, but this concentration of crystal violet may inhibit small numbers of *B. suis* and *B. melitensis*. The medium developed by Kuzdas and Morse (16), which contains bacitracin, polymyxin B, cycloheximide, and circulin, has been proven effective in clinical studies (23). Other selective media used successfully include Farrell's medium (11) containing bacitracin, polymyxin B, nalidixic acid, vancomycin, cycloheximide, nystatin (Oxoid), and modified Thayer-Martin medium (3).

Incubation

Inoculated plates are incubated at 35 to 37°C in an atmosphere of 5 to 10% CO_2 for 10 days. The high humidity which develops in sealed chambers or jars under these conditions promotes rapid growth of molds which may contaminate cultures and render them valueless. A layer or tray of dry $CaCl_2$ placed in the bottom of the chamber or jar will help to control humidity.

Examination of cultures

Examine plates for signs of growth. After 4 to 5 days, *Brucella* colonies are 2 to 7 mm in diameter, spheroidal in shape, moist, slightly opalescent in appearance, translucent, and bluish-white in reflected light. These characteristics may vary somewhat with pH and available moisture. Transfer isolated colonies to several tryptose agar slants or slants of similar media and incubate them under 10% CO_2 tension at 37°C for 48 h. If sufficient growth is not obtained, reincubate for another 24-h period. *Brucella* spp. yield a fine, clear, translucent growth with a slight amber tinge.

IDENTIFICATION

Preliminary identification

Gram stain and examine for gram-negative pleomorphic coccobacilli. Since brucellae take the counterstain poorly, apply counterstain for 1 to 3 min instead of the usual 30 s. Nonhemolytic colonies that are gram-negative coccobacilli, do not ferment lactose or glucose, are obligate aerobes, and are oxidase positive are tested for agglutination in anti-smooth *Brucella* serum (Difco Laboratories). The direct fluorescent-antibody technique (17) may be performed as a presumptive test in place of or in conjunction with agglutination tests (1).

Suspected *Brucella* colonies are emulsified in 2 separate drops of saline. A drop of anti-smooth *Brucella* serum is added to the first drop, and normal serum is added to the second. The suspensions and sera are mixed and examined for agglutination, which should be rapid and complete unless dissociation has occurred. *B. canis* and *B. ovis* will not agglutinate in anti-smooth *Brucella* serum. Control cultures of known *Brucella* spp. should be used in conjunction with the suspected culture.

A semiquantitative urease test can be performed by preparing a dense suspension of the organisms and placing a loopful of the suspension on a slant of Christensen urea agar. *B. suis* and *B. canis* will turn the indicator pink in less than 1 min, whereas *B. abortus* will usually take 5 to 10 min.

The above morphological, biochemical, and serological results are sufficient criteria to place the isolate in the genus *Brucella*. Negative cultures do not

TABLE 1. Differential characteristics of the species and biotypes in the genus *Brucella*[a]

Species	Biotype	CO_2 required	H_2S production	Urease activity	Growth on dyes[b]: Basic fuchsin II	Basic fuchsin III	Thionin I	Thionin II	Thionin III	Thionin blue (1:500,000)	Erythritol (1 mg/ml)	Penicillin (5 U)	Agglutination in sera[c]: A	M	R	Lysis by phage Tbilisi: RTD	$10^4\times$ RTD	Most common host reservoir
B. melitensis	1	−	−	Var	+	+	+	+	+	+	+	+	−	+	−	−	−	Sheep, goats
	2	−	−	Var	+	+	+	+	+	+	+	+	+	−	−	−	−	Sheep, goats
	3	−	−	Var	+	+	+	+	+	+	+	+	+	+	−	−	−	Sheep, goats
B. abortus	1	±	+	1–2 h	+	+	−	−	−	+	+	+	+	−	−	+	+	Cattle
	2	+	+	1–2 h	−	−	−	−	−	−	±	−	+	−	−	+	+	Cattle
	3	±	+	1–2 h	+	+	−	−	+	+	+	+	+	−	−	+	+	Cattle
	4	±	+	1–2 h	+	+	−	−	−	+	+	+	−	+	−	+	+	Cattle
	5	−	−	1–2 h	+	+	−	+	+	+	+	+	−	+	−	+	+	Cattle
	6	−	±	1–2 h	+	+	−	+	+	+	+	+	+	−	−	+	+	Cattle
	7	−	±	1–2 h	+	+	−	+	+	+	+	+	+	+	−	+	+	Cattle
	9	±	+	1–2 h	+	+	−	+	+	+	+	+	−	+	−	+	+	Cattle
Strain 19	1	−	+	1–2 h	+	+	−	−	−	−	−	−	+	−	−	+	+	Vaccine
B. suis	1	−	+	0–30 min	−	−	+	+	+	±	+	−	+	−	−	−	+	Pigs
	2	−	−	0–30 min	−	−	−	+	+	−	+	−	+	−	−	−	+	Pigs, horses
	3	−	−	0–30 min	+	+	+	+	+	+	+	±	+	−	−	−	+	Pigs
	4	−	−	0–30 min	+	+	+	+	+	−	+	−	+	+	−	−	+	Reindeer
B. canis		−	−	0–30 min	−	−	+	+	+	±	+	−	−	−	+	−	−	Dogs
B. neotomae		−	+	0–30 min	−	−	−	−	+	+	+	−	+	−	−	−	+	Wood rat
B. ovis		+	−	Neg	+	+	+	+	+	−	−	−	−	−	+	−	−	Sheep (rams)

[a] Data courtesy of Diagnostic Bacteriology Laboratory, National Veterinary Services Laboratory, Ames, Iowa.

[b] Species differentiation is obtained on Trypticase soy or tryptose agar with the following graded concentrations of dyes: 1:25,000 (I), 1:50,000 (II), 1:100,000 (III). Other concentrations may be preferable with other growth media. Interpretation of results should be controlled with the reference strains of each species. Tests should be conducted in CO_2 for those strains requiring CO_2.

[c] A, Monospecific *B. abortus* antiserum; M, monospecific *B. melitensis* antiserum; R, anti-rough serum.

rule out brucellosis. Preliminary identification is sufficient laboratory evidence for the physician to initiate therapy.

Definitive identification and biotyping

There are now 6 species and 15 biotypes comprising the genus *Brucella*. There are three recognized biotypes of *B. melitensis*, eight biotypes of *B. abortus*, and four biotypes of *B. suis* (15). Biotyping of *Brucella* strains aids in understanding the epidemiology of brucellosis. The decreasing incidence of this infection, however, does not allow most laboratories to maintain proficiency in identifying species and biotypes. Common criteria for the definitive identification of the species and biotypes of the genus *Brucella* are (i) production of H_2S, (ii) requirement of increased CO_2 for growth, (iii) agglutination in monospecific sera, (iv) urease production, and (v) growth in the presence of basic fuchsin and Thionin in solid media (Table 1).

H_2S production. Production of H_2S is detected by suspending a lead acetate paper strip directly over but not touching the inoculated surface of a tryptose agar slant. The strip is examined daily for 4 days and is replaced with a fresh strip each day. Hydrogen sulfide production blackens the strip.

Requirements for additional CO_2. The strain under examination is inoculated on duplicate tryptose agar slants; one slant is incubated in air, and the other is incubated in an atmosphere of 5 to 10% CO_2. Since mutants occur that are no longer dependent on additional CO_2, this test should be carried out immediately after initial isolation of the organism. Good growth in CO_2 and scanty or no growth in air indicates a requirement for CO_2.

Agglutination in monospecific serum. A dense suspension of the organism to be tested is prepared in 0.5% phenolized saline and heated at 60°C for 1 h. A drop of the suspension is added to a drop of each monospecific antiserum and mixed. Agglutination should occur within 1 min with one of the sera. Control cultures of *B. abortus* biotype 1, *B. melitensis* biotype 1, and *B. ovis* or *B. canis* should be used for this test. These control cultures should agglutinate in their homologous serum within 1 min without agglutinating in the other sera.

Tube agglutination test. A culture suspension is prepared by harvesting a fresh slant culture with 0.5% phenolized saline. The harvest is then heated at 60°C for 1 h, diluted, and standardized to 78% light transmission at a wavelength of 650 nm in a spectrophotometer. The monospecific tube test is performed in duplicate, i.e., one set of tubes for each of the monospecific sera. The serum is diluted just beyond its known titer by the double-dilution method, starting at 1:5. An equal amount of the antigen suspension, prepared as above, is then added to each tube plus the saline control. The tubes are incubated for 24 h at 37°C and read. Usually the strain being studied will be agglutinated by one of the sera to its known titer but not at all by the other serum. Some strains are agglutinated by both monospecific sera at varying titers. *B. melitensis* biotype 3 and *B. abortus* biotype 7 produce this type of result (1).

Growth in presence of Thionin and basic fuchsin. The concentration of basic fuchsin and Thionin (Allied Chemical and Dye Co., New York, N.Y.) used in these tests is between 10 and 40 μg of dye per ml of medium (1:25,000 and 1:100,000). The actual concentration of dye is that which will differentiate among control cultures of *B. melitensis*, *B. abortus*, and *B. suis* biotypes. This will depend upon the basal medium used, as well as the bacteriostatic action and purity of the dye, and must be determined by using control cultures.

The medium is prepared by heating a 0.1% dye solution in a boiling-water bath for 20 min and then adding the required amount to the melted agar base (tryptose agar or Trypticase soy agar). The agar is mixed and poured into petri dishes. The surface of the plates should be dry before inoculation. From each pure culture and control strain to be tested, a suspension is made by suspending a loopful of the organism in 1.0 ml of sterile saline. The suspensions should be of equal density. A sterile cotton swab can be used to inoculate the medium. Six cultures, including reference strains, may be tested on each plate. The plates are incubated under 10% CO_2 at 37°C for 3 to 4 days and examined for growth. *B. abortus* 19 cannot be distinguished from other CO_2-independent strains of *B. abortus* biotype 1 by the routine identification procedures. Growth inhibition tests utilizing Thionin blue (1:500,000), erythritol (1 mg/ml), and penicillin (5-U disk), however, will allow differentiation (9, 21).

Results from the routine typing tests mentioned above will identify almost all *Brucella* strains as a particular biotype. Phage typing and oxidative-metabolic tests are required to identify occasional strains (1).

Phage typing. The phage test is particularly useful in distinguishing *B. abortus* biotypes 3 to 9 from *B. melitensis*. The routine test dilution (RTD) of bacteriophage Tbilisi will completely lyse smooth cultures of *B. abortus*, but *B. suis* and *B. melitensis* cultures are not affected by this phage dilution (1). *B. suis* is partially lysed by a phage concentration of $10^4\times$ RTD, whereas most strains of *B. melitensis* are not lysed by this concentration. Test cultures are grown for 24 h on agar slants, then washed off with enough normal saline solution to produce a suspension containing 10^9 cells per ml. Inoculate a well-dried agar plate with a cotton swab that has been soaked in the bacterial suspension being tested. Using a Pasteur pipette delivering 50 drops/ml, add 1 drop of the RTD of phage on the inoculated areas of the test cultures. Repeat the process for $10^4\times$ RTD on the same agar plate. Allow to dry and incubate in 5 to 10% CO_2 at 37°C for 48 h. Smooth and smooth-intermediate *B. abortus* cultures are lysed by the RTD and $10^4\times$ RTD; rough and other nonsmooth phases are not lysed. *B. suis* cultures are usually lysed by $10^4\times$ RTD, but not the RTD. *B. melitensis* is not lysed by either.

SUSCEPTIBILITY TO ANTIMICROBIAL AGENTS

Since intracellular parasites are relatively inaccessible to antibiotics, prolonged treatment is advised. Tetracyclines alone are effective for treatment of brucellosis, but the combination of one of the tetracyclines with streptomycin is currently regarded as the best treatment for this disease since the relapse rate for patients receiving antibiotic combinations is significantly lower (24). The combination of cotrimoxazole with rifampin or tetracycline and streptomycin with rifampin is also effective (24). Most strains of *Brucella* are highly resistant to the penicillins and cephalosporins. Kanamycin and gentamicin can replace streptomycin if resistance develops (24).

Routine susceptibility testing of *Brucella* isolates is not necessary. Relapses that occur in brucellosis and the progression of some acute cases to chronic brucellosis have not been due to the emergence of resistant strains. If resistance to tetracycline or streptomycin is suspected, a standard broth dilution method may be used to determine the susceptibility of the organism (12).

IMMUNOSEROLOGICAL DIAGNOSIS

Since isolation of the organisms from infected patients is difficult, serological tests are relied upon for the routine diagnosis of brucellosis in the majority of cases. The agglutination test is the principal test used clinically in the United States (22, 24), although limitations are noted. The standard antigen is prepared from *B. abortus* 1119-3 and is available to state and federal laboratories from the U.S. Department of Agriculture, National Animal Disease Laboratory, Ames, IA 50010. The tube agglutination test as described by McCullough (18) is recommended. Laboratories receiving an occasional request for this test should refer the specimen to their state public health laboratory.

The microagglutination test correlates well with the tube agglutination test and is used by public health laboratories as a quantitative screening test for large numbers of sera (4). Serum titers equal to or greater than 1:80 should be retested using the tube agglutination test. The 2-mercaptoethanol test for immunoglobulin G is useful in evaluating response to antimicrobial therapy and in diagnosing chronic brucellosis (24). Other tests have been introduced, but none have been widely accepted for clinical diagnosis of brucellosis.

If brucellosis is suspected in the absence of isolation of the organism, rising agglutinin titers are diagnostic. Unfortunately, as a result of the nature of the disease, acute-phase serum is seldom available. In the absence of a previously known infection, a single titer of 1:160 does not exclude the possibility of *Brucella* infection, and titers of 1:160 lose diagnostic significance in groups repeatedly exposed to brucellae, such as abattoir employees and veterinarians. Serological results must be critically assessed along with clinical findings and occupational and other epidemiological factors before a diagnosis is made.

LITERATURE CITED

1. **Alton, G. G., L. M. Jones, and D. E. Pietz.** 1975. W.H.O. laboratory techniques in brucellosis. W.H.O. Monogr. Ser. 55.
2. **Arnow, P. M., M. Smaron, and V. Ormiste.** 1984. Brucellosis in a group of travelers to Spain. J. Am. Med. Assoc. **251:**505–507.
3. **Brown, G. M., C. R. Ranger, and D. J. Kelley.** 1970. Selective medium for the isolation of *Brucella ovis*. Cornell Vet. **61:**265–280.
4. **Brown, S. L., G. C. Klein, F. T. McKinney, and W. L. Jones.** 1981. Safranin O-stained antigen microagglutination test for detection of *Brucella* antibodies. J. Clin. Microbiol. **13:**398–400.
5. **Centers for Disease Control.** 1982. Annual summary 1981: reported morbidity and mortality in the United States. Morbid. Mortal. Weekly Rep. **30:**14.
6. **Centers for Disease Control.** 1983. Brucellosis—Texas. Morbid. Mortal. Weekly Rep. **32:**548–553.
7. **Centers for Disease Control.** 1984. Summary—cases of specified notifiable diseases, United States. Morbid. Mortal. Weekly Rep. **32:**677–692.
8. **Centers for Disease Control and National Institutes of Health.** 1984. Biosafety in microbiological and biomedical laboratories. HHS Publication no. CDC 84–8395. U.S. Department of Health and Human Services, Washington, D.C.
9. **Corbel, M. J., C. D. Bracewell, E. L. Thomas, and K. P. W. Gill.** 1979. Techniques in the identification and classification of *Brucella*. *In* F. A. Skinner and D. W. Lovelock (ed.), Identification methods for microbiologists. Society for Applied Bacteriology Technical Series no. 14, 2nd ed. Academic Press, Inc., New York.
10. **Corbel, M. J., A. C. Scott, and H. M. Ross.** 1980. Properties of a cell-wall defective variant of *Brucella abortus* of bovine origin. J. Hyg. **85:**103–113.
11. **Farrell, I. D.** 1974. The development of a new selective medium for the isolation of *Brucella abortus* from contaminated sources. Res. Vet. Sci. **16:**280–286.
12. **Hall, W. E., and R. E. Manion.** 1970. In vitro susceptibility of *Brucella* to various antibiotics. Appl. Microbiol. **20:**600–604.
13. **Hurvel, B.** 1972. Serological cross-reactions between different *Brucella* species and *Yersinia enterocolitica*. Immunodiffusion and immunoelectrophoresis. Acta Vet. Scand. **13:**472–483.
14. **Hurvel, B., P. Ahvonen, and E. Thal.** 1971. Serological cross-reactions between different *Brucella* species and *Yersinia enterocolitica*. Agglutination and complement fixation. Acta Vet. Scand. **12:**86–94.
15. **International Committee on Systematic Bacteriology. Subcommittee on Taxonomy of *Brucella*.** 1982. Minutes of the Meeting September 4–5, 1978. Int. J. Syst. Bacteriol. **32:**260–261.
16. **Kuzdas, C. D., and E. V. Morse.** 1953. A selective medium for the isolation of brucellae from contaminated materials. J. Bacteriol. **66:**502–503.
17. **McAllister, T. A.** 1976. Laboratory diagnosis of human brucellosis. Scott. Med. J. **21:**129–131.
18. **McCullough, N. B.** 1976. Immune response to *Brucella*, p. 304–311. *In* N. R. Rose and H. Friedman (ed.), Manual of clinical immunology. American Society for Microbiology, Washington, D.C.
19. **Meyer, M. E.** 1974. Advances in research on brucellosis, 1957–1972. Adv. Vet. Sci. Comp. Med. **128:**231–246.
20. **Meyer, M. E., and H. S. Cameron.** 1961. Metabolic characterization of the genus *Brucella*. II. Oxidative metabolic patterns of the described biotypes. J. Bacteriol. **82:**396–400.
21. **Morgan, W. J. B.** 1961. The use of the thionin blue sensitivity test in the examination of *Brucella*. J. Gen. Microbiol. **25:**135–139.
22. **Spink, W. W., N. D. McCullough, L. M. Hutchings, and C. K. Mingle.** 1954. A standardized antigen for agglutination technique for human brucellosis. Report no. 3 of the National Research Council, Committee on Public Health Aspects of Brucellosis. Am. J. Pathol. **24:**496–498.
23. **Weed, L. A.** 1957. Use of a selective medium for isolation of *Brucella* from contaminated surgical specimens. Am. J. Clin. Pathol. **27:**482–485.
24. **Young, E. J.** 1983. Human brucellosis. Rev. Infect. Dis. **5:**821–842.

Chapter 33

Haemophilus

MOGENS KILIAN

CHARACTERIZATION

Members of the genus *Haemophilus* are obligate parasites which constitute part of the normal flora of the respiratory tract of humans and many animal species. The type species *Haemophilus influenzae* is responsible for a variety of diseases in humans, ranging from chronic respiratory infection to meningitis. Other species are implicated in venereal disease and conjunctivitis, and some are occasional causes of endocarditis and abscess formation.

In morphology, *Haemophilus* organisms range from coccobacilli to filamentous rods. They are gram negative, non-acid fast, nonmotile, and non-spore forming. Strains isolated from invasive infections are usually encapsulated.

Members of the genus *Haemophilus* are facultatively anaerobic. In vitro growth requires accessory growth factors: X factor (hemin) and V factor (nicotinamide-adenine dinucleotide). *H. influenzae* requires both of these compounds, whereas some of the other species of the genus require only one of them (Table 1). Growth to sizable colonies on blood agar occurs only around bacteria secreting the V factor, e.g., staphylococci. Preferred media are chocolate agar (blood agar base with 10% heated, defibrinated horse or bovine blood) or Levinthal agar. *H. ducreyi*, which is notably difficult to cultivate, requires special media (see below). Most strains grow better in a humid atmosphere with added 5 to 10% CO_2. The optimal temperature is about 33 to 37°C (see below).

Biochemical characteristics

Carbohydrates are fermented. End products from the fermentation of glucose are succinic, lactic, and acetic acids. Strains of some species produce gas in fermentation media. Only *H. ducreyi* has not been shown to attack carbohydrates. All strains reduce nitrate. The majority of *H. influenzae* strains are oxidase and catalase positive.

Antibiotic susceptibility

Most strains are susceptible to penicillin and its derivatives and to chloramphenicol, sulfonamides, and the tetracyclines. However, due to the spread of conjugative plasmids, an increasing number of clinical *Haemophilus* isolates are now resistant to the penicillins, and occasional strains are resistant to chloramphenicol. The likelihood that a *H. influenzae* isolate will be resistant to penicillin is, in most countries, 10 to 15% and may, in certain areas, exceed 40%. A recent study disclosed a 76% frequency of β-lactamase-producing *H. parainfluenzae* in throat cultures from ambulatory children attending a Canadian hospital (29). *Haemophilus* organisms are resistant to bacitracin, which may be employed as a selective agent in isolation media.

CLINICAL SIGNIFICANCE

H. influenzae is one of the three leading causes of bacterial meningitis. In the United States, the estimated annual number of cases of meningitis due to this organism is 10,000, of which the majority occur in young children. Virtually all of these cases are caused by organisms which possess a serotype b capsule, and about 93% of the strains belong to biotype I (14, 30). The same sero- and biotype is also the major etiological agent in acute epiglottitis (obstructive laryngitis). Occasional cases of meningitis caused by nonencapsulated strains or strains possessing a capsule of one of the other five serotypes usually have a different pathogenesis than the typical *H. influenzae* meningitis. Such cases are often secondary to trauma or occur in patients with impaired host defenses. A considerable number of cases of *H. influenzae* type b meningitis are associated with purulent otitis media. Invasion of the blood stream by encapsulated strains of *H. influenzae* (usually serotype b) may result in suppurative arthritis, osteomyelitis, cellulitis, and pericarditis. Primary *H. influenzae* pneumonia occurs in children as well as in adults, but is relatively uncommon.

Although a primary pathogen, *H. influenzae* serotype b may be isolated from upper respiratory tract cultures of healthy individuals. The frequency of pharyngeal *H. influenzae* serotype b colonization is below 1% during the first 6 months of life but averages 3 to 5% throughout the rest of childhood, although it may be considerably higher in selected populations (16, 23, 31).

Haemophilus species constitute approximately 10% of the constant bacterial flora of the healthy upper respiratory tract. The predominant species is *H. parainfluenzae*, which accounts for three-fourths of the *Haemophilus* flora. Nonencapsulated *H. influenzae*, usually of multiple biotypes, is present in the pharynx of the majority of healthy children but normally constitutes less than 2% of the total bacterial flora. With increasing age of the patient, *H. influenzae* becomes less frequent as a pharyngeal commensal (16). *H. haemolyticus* is rarely encountered in the human respiratory tract.

Nonencapsulated *H. influenzae* of the biotypes found in the healthy respiratory tract (predominantly biotypes II and III) may be isolated from cases of sinusitis, otitis media, chronic or acute exacerbations of lower respiratory tract infections including cystic fibrosis, and chronic conjunctivitis. Cases of *H. influenzae* otitis media are associated with significantly increased proportions (≥ 50% of total bacterial flora) of the same biotype in the nasopharynx (21). Although the implication of nonencapsulated *H. influenzae* in such infections is probably secondary to viral or trachomatous infection or to the obstruction of normal passages, *H. influenzae* is likely to play an important role in the pathogenesis of these infections, partly because the bacterium is capable of destroying the

TABLE 1. Principal differential characteristics of *Haemophilus* species

Species	Factor requirement		Hemolysis	Fermentation of:			Presence of catalase	CO_2 enhances growth
	X[a]	V		Glucose	Sucrose	Lactose		
H. influenzae (*H. aegyptius*)[b]	+	+	−	+	−	−	+	−
H. haemolyticus	+	+	+	+	−	−	+	−
H. ducreyi	+	−	−	−	−	−	−	−
H. parainfluenzae[b]	−	+	−	+	+	−	D[c]	−
H. parahaemolyticus[b,d]	−	+	+	+	+	−	D	D
H. segnis[b]	−	+	−	W[e]	W	−	D	−
H. paraphrophilus[b]	−	+	−	+	+	+	−	+
H. aphrophilus	−	−	−	+	+	+	−	+

[a] As determined by the porphyrin test.
[b] For further characteristics see Table 2.
[c] D, Difference encountered.
[d] Strains requiring extra CO_2 in the incubation atmosphere have been labeled *H. paraphrohaemolyticus* (34).
[e] W, Weak fermentation reaction.

normal immune protection of mucous membranes (17).

An organism closely resembling *H. influenzae* is associated with an acute purulent and contagious form of conjunctivitis which occurs as a seasonal endemic, especially in hot climates. This organism, which bears the name *H. aegyptius* (Koch-Weeks bacillus) (27), is more difficult to cultivate in vitro than is *H. influenzae,* but the two species are notably difficult to differentiate in the laboratory (see below).

H. ducreyi is the cause of the venereal disease soft chancre or chancroid. Morbidity figures from the United States show an annual incidence of approximately 1,000 cases. In recent years, significant epidemics have occurred in several parts of the world. Symptomless cervical carriage of *H. ducreyi* may occur (28).

The species *H. parainfluenzae, H. parahaemolyticus, H. aphrophilus, H. paraphrophilus,* and *H. segnis* all form part of the normal oral microflora. The oral *Haemophilus* species, which very rarely encompass *H. influenzae,* amount to a mean number of 4×10^7 organisms per ml of saliva (16, 30). Some of these species (*H. aphrophilus, H. paraphrophilus,* and *H. segnis*) are predominantly found in dental plaque. *H. haemolyticus* is occasionally encountered in dental plaque collected from gingival crevices. Like several other oral bacteria, these species are occasionally implicated in endocarditis and abscesses of internal organs (1). Cases of meningitis ascribed to *H. parainfluenzae* can probably be explained by the misidentification of *H. influenzae* isolates (see below).

There are no reports of human infections due to the many animal pathogenic species (16). An exception is a single communication on otitis media caused by *H. haemoglobinophilus,* which is part of the normal mucosal flora of dogs.

COLLECTION OF SPECIMENS

Since most infections caused by *H. influenzae* serotype b are associated with bacteremia, blood samples are important sources for isolation of this organism. This is also true in case of suspected bacterial endocarditis. Likewise, cerebrospinal fluid (CSF) should be collected for examination whenever the physician suspects meningitis or wants to rule it out. Specimens of the blood and CSF of the patient are obtained and processed as described in Chapter 8. Prompt transport of these samples to the laboratory is mandatory to ensure the fastest possible diagnosis and survival of microorganisms in the sample. Additional specimens from which *Haemophilus* organisms may be isolated include fluid aspirated from infected joints, pus, nasopharyngeal or throat swabs, and occasionally vaginal swabs. As *Haemophilus* species are normal inhabitants of the upper respiratory tract, it is important that samples taken from foci in this location remain representative of the infecting flora by avoiding contamination with the commensal flora as far as possible. Thus, for sampling of the lower respiratory tract, any method which bypasses the upper respiratory tract, such as bronchial washings, transtracheal aspiration, etc., is preferable.

Cotton swabs should be moistened in broth before the sample is taken and should be transported to the laboratory in a transport medium. Samples should be kept at room temperature. Viability of *Haemophilus* organisms is readily lost as a result of drying out or chilling, and as a rule, they do not survive more than a few days in clinical samples. This is particularly true for the fastidious species which may be isolated from conjunctivae or genital ulcers. Hence, whenever possible, such specimens should be directly inoculated onto isolation media.

Before being sampled, chancroid lesions are cleaned with sterile saline, and the ulcer base is sampled with a broth-moistened cotton swab. If possible, cultivation from the ulcer should be supplemented by aspirating pus from infected bubonic lymph nodes.

As a supplement to cultivation, it is advisable to take scrapings of conjunctivae and genital ulcers for direct microscope examination.

Isolated *Haemophilus* cultures may be stored for many years after lyophilization in skim milk. An alternative method of storage is freezing of 24-h broth cultures or suspensions of freshly grown cells in 10% glycerol at −60 to −70°C. Most strains may also be maintained by weekly transfers on agar media. However, strains of *H. aegyptius* and *H. ducreyi* will rapidly die out.

DIRECT MICROSCOPE EXAMINATIONS

Direct microscope examination of CSF is particularly important to obtain the fastest possible diagnosis. A film of CSF concentrated by centrifugation or filtration (see Chapter 8) is prepared on a microscope slide, allowed to air dry, fixed in absolute methanol for 10 min, and stained by the Gram method. In the preparation of the Gram stain, care should be exercised in decoloration, as the coccobacillary form of *H. influenzae* may resemble pneumococci in morphology. The bacteria may be few in number but are usually detected by careful examination for 10 to 30 min. *H. influenzae* type b will appear as coccobacilli or short rods with rounded ends. The type b capsule, which is present in virtually all strains found in CSF, can be demonstrated by the capsular swelling reaction (see Chapter 16). If a sufficient number of organisms are present, a drop of CSF is mixed on a microscope slide with a drop of an antiserum against the type b capsule. The type b capsule will appear swollen and sharply delineated when compared with a control smear of the organism. Immunofluorescence staining of the capsule with fluorescein-conjugated anti-type b serum provides another excellent means of identifying type b encapsulated strains in clinical samples.

Several immunological techniques for the detection of capsular antigen in CSF and other body fluids have recently been developed. These include countercurrentimmunoelectrophoresis, latex agglutination, staphylococcal coagglutination, and enzyme- or radioimmunoassay.

The reported sensitivities and specificities of these methods have varied probably because of differences in test reagents, antisera, conditions of testing, and other aspects. However, several of the methods are now available as commercial ready-to-use kits. Recent comparisons of commercial latex agglutination (Bactogen; Wampole Laboratories, Div. Carter-Wallace, Inc., Cranbury, N.J.) and coagglutination (Phadebact Haemophilus Test; Pharmacia Fine Chemicals, Inc., Piscataway, N.J.) diagnostic kits with countercurrent immunoelectrophoresis for the detection of type b capsular antigen in CSF found the two agglutination tests to be superior with regard to both sensitivity and specificity (6, 19, 22, 33). Both enzyme- and radioimmunoassays (7, 19) are highly sensitive, but they require specialized equipment not generally available in bacteriological laboratories.

Microscope examination of Gram- or Giemsa-stained smears from conjunctivae or genital ulcers is an important supplement because cultivation is often inadequate for establishing the presence of the implicated microorganisms. Even direct plating of swabs will yield positive cultivation in only 70 to 80% of cases. *H. aegyptius* will appear as slender, gram-negative rods in smears of conjunctival scrapings. *H. ducreyi* will typically appear in smears prepared from chancroid lesions as short chains of gram-negative bacilli which may be located extra- or intracellularly. Although not diagnostic, such findings may provide clues to the nature of the disease.

CULTIVATION

When the isolation of infectious microorganisms from sources which may yield *Haemophilus* organisms is attempted, it is important to keep in mind that most conventional agar media do not support growth of these organisms. Although the growth factors X and V are contained in blood cells, only the X factor is directly available in ordinary blood agar. To release the V factor, the blood cells have to be broken up by heating, as in chocolate agar or in Levinthal medium. In addition to liberating the V factor, this heat treatment is necessary to inactivate V factor-destroying enzymes present in blood.

Cultivation of samples of spinal fluid or pus or subcultivation from blood cultures should be performed on both chocolate agar and blood agar. After inoculation, the blood agar plate is cross-inoculated with a streak of a staphylococcal strain. The staphylococci provide the V factor and allow the detection of *Haemophilus* organisms growing as satellite colonies.

Swabs from the respiratory tract should be inoculated onto blood agar plates which have likewise been cross-inoculated with a staphylococcal strain. If *Haemophilus* organisms are of particular interest, it is advisable to use a selective medium. Even when present in considerable numbers, *Haemophilus* organisms may easily escape detection after incubation because of overgrowth by the remaining flora. Chocolate agar supplemented with bacitracin (300 mg/liter) (11) has been found to give excellent recoveries.

Because of the particularly fastidious nature of *H. aegyptius* and *H. ducreyi*, special media are recommended for their isolation. Chocolate agar supplemented with 1% IsoVitaleX (BBL Microbiology Systems, Cockeysville, Md.) provides a good medium for isolation of both of these organisms, although primary growth from clinical samples is never luxuriant (9, 32). For the cultivation of conjunctival swabs, it is also recommended that the sample be spread on ordinary blood agar cross-inoculated with a staphylococcal strain. The latter medium facilitates the identification of other pathogens that may be found in this location. Since chancroid lesions usually contain a mixed flora, it is advisable to add vancomycin (5 mg/liter) to the chocolate-IsoVitaleX medium as a semiselective agent (9).

Agar plates used for the isolation of *Haemophilus* organisms should not be allowed to dry before use. A wrinkled surface on the medium indicates a degree of dryness likely to inhibit or delay the growth of *Haemophilus* organisms. Cultures should be incubated at 35 to 37°C. An exception is cultures of *H. ducreyi*, which grow significantly better at 33°C. A moist atmosphere supplemented with 5 to 10% CO_2 is preferred by most strains and is mandatory for the isolation of *H. ducreyi*. Given optimal conditions, most *Haemophilus* species grow to at least 1- to 2-mm colonies after incubation for 18 to 24 h. However, *H. aegyptius* requires incubation for 2 to 4 days, and growth of *H. ducreyi* is notoriously slow, often requiring incubation for up to 9 days (9).

With the exception of *H. aphrophilus* and *H. ducreyi*, all *Haemophilus* species which may be isolated from humans require V factor. On blood agar these species will, therefore, grow as satellite colonies around the staphylococcus streak. Some *Haemophilus* strains (*H. haemolyticus*, *H. parahaemolyticus*, and *H. paraphrohaemolyticus*) show beta-hemolytic zones around colonies, which thereby may resemble those of pyogenic streptococci. The satellite growth of these hemolytic *Haemophilus* species is often less pronounced because of the release of V factor from lysed blood cells.

H. influenzae colonies on chocolate agar are grayish, semiopaque, smooth, flat, and convex, usually with an entire edge. In dense areas of the plate, encapsulated strains tend to grow confluently, in contrast to colonies of nonencapsulated strains, which remain separate. On clear agar media, like Levinthal agar, colonies of encapsulated strains show a bright iridescence when light is obliquely transmitted from behind. The bluish-green color of nonencapsulated strains examined in the same way should not be taken as an indication of capsulation. The iridescence of encapsulated strains is most clearly detected in young cultures (10 to 18 h) and will gradually disappear during prolonged incubation. In some strains with capsules of serotypes other than type b, the iridescence is not always clear-cut. Strains from cases of meningitis and epiglottitis virtually always produce indole, which gives the growth on agar media a characteristic pungent smell.

Colonies of *H. parainfluenzae* may be up to 3 mm in diameter after incubation for 24 h and appear either smooth or rough and wrinkled. Most colonies are flat, grayish, and semiopaque. *H. aphrophilus* and *H. paraphrophilus* grow as rough, raised colonies which rarely attain a diameter exceeding 1 to 2 mm. When incubated in air without extra CO_2, these species will grow, if at all, in colonies of various sizes, which give the culture a contaminated look. *H. aegyptius*, *H. segnis*, and *H. ducreyi* grow as small (<1-mm), smooth colonies. The last species may also grow in colonies of different sizes.

IDENTIFICATION OF CULTURES

The satellite phenomenon, which may be detected in primary agar plate cultures, provides a convenient means for a tentative genus identification of all the V factor-requiring species. However, it should be emphasized that other bacteria, such as some streptococci and corynebacteria, may show satellite growth as well.

The prerequisite for the detection of the satellite phenomenon is an agar medium which lacks V factor. Since ordinary blood agar contains various amounts of free V factor, depending on the method of preparation and length of storage, it may sometimes be difficult to achieve convincing satellite growth on this medium. The special problem associated with hemolytic strains has already been mentioned. In case of doubtful reactions, far better results are obtained on a blood agar medium to which the blood (5 to 10%) is added before autoclaving. Since nicotinamide-adenine dinucleotide (V factor) is heat labile, this medium is completely devoid of V factor, but otherwise it satisfies all the growth requirements of *Haemophilus* species, including that for X factor. Instead of a staphylococcus streak, the V factor may also be provided by a nicotinamide-adenine dinucleotide-containing paper disk placed on the surface of the plate after inoculation.

Once the V-factor requirement has been identified, the most important means for further differentiation is determination of the X-factor requirement. This determination has often been made by demonstrating growth around an X factor-containing paper disk placed on an inoculated agar medium or by comparing growth on agar media with and without added blood. The inherent problem of these methods is that probably no complex medium which will otherwise satisfy all growth requirements of *Haemophilus* organisms is totally free of X factor. Therefore, even when particular care is being exercised to avoid carrying over X factor with inoculum, this method will lead to an erroneous result in about 18% of cases (M. Kilian and K. R. Eriksen, unpublished data). If the identity is not being confirmed by biochemical tests, *H. influenzae* strains may be misidentified as *H. parainfluenzae* and vice versa. This undoubtedly explains some of the reported cases of meningitis ascribed to *H. parainfluenzae*.

The porphyrin test (14) provides a more accurate and rapid means of determining the X-factor requirement. This method is based on the observation (5) that hemin-independent *Haemophilus* strains excrete porphobilinogen and porphyrins, all of which are intermediates in the hemin biosynthetic pathway (Fig. 1) when supplied with δ-aminolevulinic acid. X-requiring strains do not excrete these compounds because of a lack of all the enzymes involved in the biosynthesis of heme. Both porphobilinogen and porphyrins may be visualized by simple methods which form the basis for the test.

Porphyrin test

Substrate. The substrate consists of 2 mM δ-aminolevulinic acid–hydrochloride (Sigma Chemical Co., St. Louis, Mo.) and 0.8 mM $MgSO_4$ in 0.1 M phosphate buffer at pH 6.9. The substrate is distributed in 0.5-ml quantities in small glass tubes and may be stored for several months in a refrigerator.

Inoculation. Suspend a heavy loopful of bacteria from an agar plate culture in the substrate. Incubate for 4 h.

Reading. After incubation, expose the substrate to a Wood light (wavelength, approximately 360 nm), preferably in a dark room. A red fluorescence from the bacterial cells or from the fluid indicates porphyrins; i.e., the strain is independent of the X factor. In cases of doubtful reactions, the tubes may be reincubated for up to 24 h.

Alternative method for reading. After incubation, add 0.5 ml of Kovacs reagent (*p*-dimethylaminobenzaldehyde, 5 g; amyl alcohol, 75 ml; and concentrated HCl, 25 ml), shake the mixture vigorously, and allow the phases to separate. A red color in the lower water

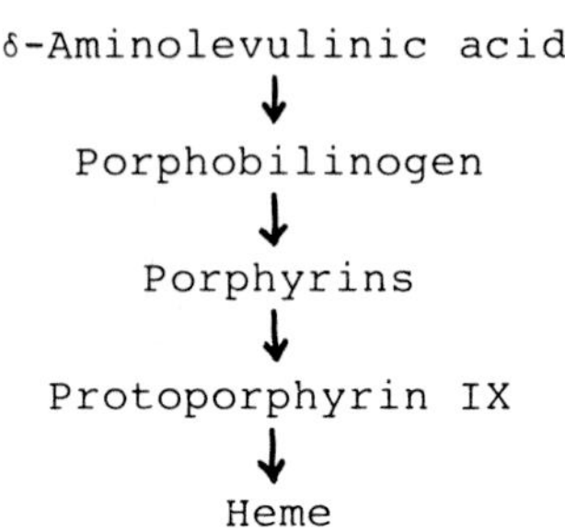

FIG. 1. Principal steps of the heme biosynthetic pathway.

phase is indicative of porphobilinogen, which means that the strain is independent of the X factor. When this method of reading is employed, it is advisable to include an inoculated tube without δ-aminolevulinic acid as a negative control. Kovacs reagent also gives a red color reaction with indole. Although an indole reaction will be present in the upper alcohol phase, indole-positive strains of *H. influenzae* may erroneously be identified as X factor independent in the absence of an appropriate control.

Other biochemical tests

Table 1 gives the key tests which are valuable in further differentiation of the *Haemophilus* species that may be expected in human samples. Criteria for the identification of species primarily associated with animal diseases may be found elsewhere (15). Tests of the fermentation of glucose, sucrose, and lactose are important for species identification. Fermentation reactions are performed in 1% solutions of the respective carbohydrates in phenol red broth base (Difco Laboratories, Detroit, Mich.) supplemented with X and V factors (10 mg of each per liter) after autoclaving (14). Reactions are usually clear-cut after 24 h of incubation, but some species, such as *H. segnis* and *H. aegyptius*, show weak reactions. Fermentation tests may also be performed by the use of commercial substrate-containing disks (3). However, studies have not yet been carried out to ensure comparable results, in particular with respect to the lactose fermentation tests, where differences may be expected as a result of different pH indicators.

H. influenzae and *H. parainfluenzae* may be subdivided into a number of biotypes (8, 14, 25) on the basis of three biochemical reactions: indole production, urease activity, and ornithine decarboxylase activity (Table 2). The biotypes of *H. influenzae* in particular have shown a relationship to the source of isolation (2, 14, 16, 26). There is a certain correlation between biotype and capsular serotype of *H. influenzae* strains (16). For epidemiologic purposes, subtyping on the basis of outer membrane proteins or lipopolysaccharides is a more sensitive tool than biotyping (4, 12, 20).

The three reactions used for the subdivision of these biotypes are performed as rapid tests. In addition to the rapidity, the advantage of this type of test is that growth is not required. Hence, the media do not have to include growth factors. All three test media (0.5-ml quantities) are inoculated with a heavy loopful of bacteria from an agar culture, and the results are read after incubation for 4 h. The test for ornithine decarboxylase may in some cases require additional incubation for 18 to 20 h. Recommended media are given below.

Indole test. The substrate is 0.1% L-tryptophan in 0.05 M phosphate buffer at pH 6.8. After incubation for 4 h with bacteria, add 1 volume of Kovacs reagent, and shake the mixture. Red color in the upper alcohol phase indicates the presence of indole.

Urease test. The substrate is 0.1 g of KH_2PO_4, 0.1 g of K_2HPO_4, 0.5 g of NaCl, and 0.5 ml of 1:500 phenol red in 100 ml of distilled water. Adjust the pH to 7.0 with NaOH, autoclave, and add 10.4 ml of a 20% (wt/vol) filter-sterilized aqueous solution of urea. (For 1:500 phenol red, dissolve 0.2 g of phenol red in NaOH, and add distilled water to 100 ml.) Red color within 4 h indicates urease activity.

Ornithine decarboxylase test. The substrate is the medium used regularly for other bacteria. However, for *Haemophilus* organisms it is inoculated with a heavy loopful of bacteria. The medium is commercially available from Difco. A purple color developing within 4 to 24 h indicates ornithine decarboxylase activity.

Alternative methods

Several commercial systems now available give comparable results when used for biotyping of *H. influenzae* and *H. parainfluenzae*: PathoTec strips (General Diagnostics, Morris Plains, N.J.), API 10S or API 20E strips (Analytab Products, Plainview, N.Y.), and the Minitek System (BBL) (3, 10, 13). However, for species identification, some of the available kits are less satisfactory, because they do not allow a differentiation of the V factor-requiring species.

If media other than the above are used for the identification and biotyping of *Haemophilus* strains, it is advisable to include strains of known identity as a reference.

H. aegyptius has the same key biochemical characteristics as *H. influenzae* biotype III (Table 2). Characteristics which may be of use in the separation of the two taxa include poorer in vitro growth of *H. aegyptius*, its more slender, rodlike shape, its ability to agglutinate erythrocytes, its inability to ferment xylose, and its susceptibility to troleandomycin (5 μg;

TABLE 2. Key to the differentiation of the biotypes of *H. influenzae* and *H. parainfluenzae*, *H. aegyptius*, *H. parahaemolyticus*, and *H. segnis*

Species and biotype	Indole production	Urease activity	Ornithine decarboxylase activity
H. influenzae			
Biotype I	+	+	+
Biotype II	+	+	−
Biotype III	−	+	−
Biotype IV	−	+	+
Biotype V	+	−	+
Biotype VI	−	−	+
Biotype VII[a]	+	−	−
H. aegyptius[b]	−	+	−
H. parainfluenzae			
Biotype I	−	−	+
Biotype II	−	+	+
Biotype III	−	+	−
Biotype IV	+	+	+
H. parahaemolyticus	−	+	D[c]
H. segnis	−	−	−

[a] In the previous edition of this Manual, the reaction pattern here used for biotype VII was used for biotype VI. However, at the same time, Oberhofer and Back (25) used the label biotype VI for strains with a different reaction pattern. Their definition has been adopted here, and biotype VII is characterized as proposed by Gratten (8).

[b] For tests differentiating *H. aegyptius* and *H. influenzae* biotype III, see the text.

[c] D, Difference encountered.

TABLE 3. Differential tests for *H. aphrophilus*, *H. paraphrophilus*, and some related species

Species	V factor required	Indole production	Urease activity	Ornithine decarboxylase activity	Lysine decarboxylase activity	Fermentation of:			Nitrate reduction	Presence of catalase
						Glucose	Sucrose	Lactose		
H. aphrophilus	–	–	–	–	–	+	+	+	+	–
H. paraphrophilus	+	–	–	–	–	+	+	+	+	–
A. actinomycetemcomitans	–	–	–	–	–	+	–	–	+	+
Eikenella corrodens	–	–	–	+	+	–	–	–	+	–
Cardiobacterium hominis	–	+	–	–	–	+	+	–	–	–

Roerig-Pfizer, Inc., New York, N.Y.) (14, 27). However, none of these characteristics will unequivocally differentiate the two species. Although at present it may seem taxonomically unjustified to maintain the separation, clinical data indicate that the two species are different.

The name *H. paraphrohaemolyticus* has been used for strains which require extra CO_2 in the incubation atmosphere but are otherwise identical to *H. parahaemolyticus* (34). However, so far, it is questionable whether or not there is any clinical or taxonomic justification for maintaining these two groups as separate species.

The species *H. aphrophilus* and *H. paraphrophilus* are closely related to *Actinobacillus actinomycetemcomitans*. On primary isolation, *H. aphrophilus* may require the X factor. However, this is usually not the case upon subcultivation (14, 18). Therefore, it may be argued that *H. aphrophilus* does not meet the criteria for inclusion in the genus *Haemophilus* as presently defined. Biochemical reactions, which are valuable in separating *H. aphrophilus* from related organisms, are provided in Table 3.

SEROLOGICAL IDENTIFICATION

The use of serological methods for species identification of *Haemophilus* strains has not been adequately evaluated. Therefore, serological identification is of value only for encapsulated strains of *H. influenzae*. Such strains can be separated into serotypes a through f based on six serologically distinct capsular polysaccharides (26; M. Pittman, unpublished data). A subdivision of strains on the basis of capsular serotypes is highly relevant, since severe pathogenicity is almost exclusively associated with strains possessing the serotype b capsule. Serotypes a, d, e, and f strains are occasionally isolated from infections as well as from the healthy respiratory tract, whereas serotype c strains are notoriously rare.

Capsular serotypes may be determined by agglutination, capsular swelling test, immunofluorescence microscopy, and countercurrentimmunoelectrophoresis. The main value of the last three methods is for the direct identification of strains in clinical samples as described above. Once the organism is isolated, the most practical method for serotyping encapsulated strains is by slide agglutination.

The cell suspension used for the slide agglutination test must be prepared from a young (6- to 18-h) agar culture, as the capsular structure tends to deteriorate in older cultures. A smooth suspension of bacteria is made in normal saline containing Formalin (0.5% vol/vol). The cell suspension must be of sufficient density to permit the antigen-antibody reaction to proceed to completion within 1 min. In a strong positive reaction, all bacteria are agglutinated, and the fluid between the clusters is clear.

Antisera for serotyping encapsulated *H. influenzae* are available from Difco, Burroughs Wellcome Corp. (Research Triangle Park, N.C.), and some state laboratories. Since such antisera also contain antibodies to some somatic antigens, agglutination may occur as a result of reaction with antigens other than capsules. Therefore, it is important that only strong reactions occurring within 1 min be counted as positive anticapsular reactions.

For the screening of nasopharyngeal swab cultures for the presence of encapsulated strains, Levinthal agar containing antiserum is a valuable and sensitive tool (24). However, the method is too slow and expensive for general use for the identification of single strains.

Antigenic similarities to the six capsular polysaccharides of *H. influenzae* have been found in a number of unrelated bacteria. However, these cross-reactions should not create practical problems for laboratory diagnosis. Bacteria possessing antigens cross-reactive with the type b capsule include *Streptococcus pneumoniae* serotypes 6, 15a, 29, and 35a, *Escherichia coli* K100, *Staphylococcus aureus*, *Staphylococcus epidermidis*, *Streptococcus pyogenes*, *Streptococcus faecalis*, *Bacillus alvei*, and *Bacillus pumilus*.

LITERATURE CITED

1. **Albritton, W. L.** 1982. Infections due to *Haemophilus* species other than *H. influenzae*. Annu. Rev. Microbiol. **36:**199–216.
2. **Albritton, W. L., S. Penner, L. Slaney, and J. Brunton.** 1978. Biochemical characteristics of *Haemophilus influenzae* in relationship to source of isolation and antibiotic resistance. J. Clin. Microbiol. **7:**519–523.
3. **Back, A. E., and T. R. Oberhofer.** 1978. Use of the Minitek system for biotyping *Haemophilus* species. J. Clin. Microbiol. **7:**312–313.
4. **Barenkamp, S. J., R. S. Munson, Jr., and D. M. Granoff.** 1981. Subtyping isolates of *Haemophilus influenzae* type b by outer-membrane protein profiles. J. Infect. Dis. **143:**668–676.
5. **Biberstein, E. L., P. D. Mini, and M. G. Gills.** 1963. Action of *Haemophilus* cultures on δ-aminolevulinic acid. J. Bacteriol. **86:**814–819.
6. **Collins, J. K., and M. T. Kelly.** 1983. Comparison of Phadebact coagglutination, Bactogen latex agglutination, and counterimmunoelectrophoresis for detection of *Haemophilus influenzae* type b antigens in cerebrospinal fluid. J. Clin. Microbiol. **17:**1005–1008.
7. **Drow, D. L., D. G. Maki, and D. D. Manning.** 1979. Indirect sandwich enzyme-linked immunosorbent assay for rapid detection of *Haemophilus influenzae* type b infection. J. Clin. Microbiol. **10:**442–450.
8. **Gratten, M.** 1983. *Haemophilus influenzae* biotype VII. J. Clin. Microbiol. **18:**1015–1016.

9. **Hammond, G. W., C. J. Lian, J. C. Wilt, and A. R. Ronald.** 1978. Comparison of specimen collection and laboratory techniques for isolation of *Haemophilus ducreyi*. J. Clin. Microbiol. **7:**39–43.
10. **Holländer, R.** 1981. Die biochemische Characterisierung von *Haemophilus*-Stämmen mit Hilfe der API 20E- und API 50E-Testsysteme. Zentralbl. Bakteriol. Hyg., Abt. 1 Orig. Reihe A **250:**322–329.
11. **Hovig, B., and E. H. Aandahl.** 1969. A selective method for the isolation of *Haemophilus* in material from the respiratory tract. Acta Pathol. Microbiol. Scand. **77:**677–684.
12. **Inzana, T. J.** 1983. Electrophoretic heterogeneity and interstrain variation of the lipopolysaccharide of *Haemophilus influenzae*. J. Infect. Dis. **148:**492–499.
13. **Juni, B. A., J. M. Rysavy, and D. J. Blazevic.** 1982. Rapid biotyping of *Haemophilus influenzae* and *Haemophilus parainfluenzae* with PathoTec strips and spot biochemical tests. J. Clin. Microbiol. **15:**976–978.
14. **Kilian, M.** 1976. A taxonomic study of the genus *Haemophilus*, with the proposal of a new species. J. Gen. Microbiol. **93:**9–62.
15. **Kilian, M., and E. L. Biberstein.** 1984. *Haemophilus* Winslow, Broadhurst, Buchanan, Krumwiede, Rogers and Smith 1917, p. 558–569. *In* N. R. Krieg and J. R. Holt (ed.), Bergey's manual of systematic bacteriology, vol. 1. The Williams & Wilkins Co., Baltimore.
16. **Kilian, M., W. Frederiksen, and E. L. Biberstein (ed.).** 1981. *Haemophilus, Pasteurella* and *Actinobacillus*. Academic Press, London.
17. **Kilian, M., J. Reinholdt, S. B. Mortensen, and C. H. Sørensen.** 1983. Perturbation of mucosal immune defence mechanisms by bacterial IgA proteases. Clin. Respir. Physiol. **19:**99–104.
18. **King, E. O., and H. W. Tatum.** 1962. *Actinobacillus actinomycetemcomitans* and *Haemophilus aphrophilus*. J. Infect. Dis. **111:**85–94.
19. **Leinonen, M., and H. Käyhty.** 1978. Comparison of counter-current immunoelectrophoresis, latex agglutination, and radioimmunoassay in detection of soluble capsular polysaccharide antigens of *Haemophilus influenzae* type b and *Neisseria meningitidis* of groups A or C. J. Clin. Pathol. **31:**1172–1176.
20. **Loeb, M. R., and D. H. Smith.** 1980. Outer membrane protein composition in disease isolates of *Haemophilus influenzae*: pathogenic and epidemiological implications. Infect. Immun. **30:**709–717.
21. **Long, S. S., F. M. Henretig, M. J. Teter, and K. L. McGowan.** 1983. Nasopharyngeal flora and acute otitis media. Infect. Immun. **41:**987–991.
22. **Marcon, M. J., A. C. Hamoudi, and H. J. Cannon.** 1984. Comparative laboratory evaluation of three antigen detection methods for diagnosis of *Haemophilus influenzae* type b disease. J. Clin. Microbiol. **19:**333–337.
23. **Michaels, R. H., C. S. Poziviak, F. E. Stonebraker, and C. W. Norden.** 1976. Factors affecting pharyngeal *Haemophilus influenzae* type b colonization rates in children. J. Clin. Microbiol. **4:**413–417.
24. **Michaels, R. H., F. E. Stonebraker, and J. B. Robbins.** 1975. Use of antiserum agar for detection of *Haemophilus influenzae* type b in the pharynx. Pediatr. Res. **9:**513–516.
25. **Oberhofer, T. R., and A. E. Back.** 1979. Biotypes of *Haemophilus* encountered in clinical laboratories. J. Clin. Microbiol. **10:**168–174.
26. **Pittman, M.** 1931. Variation and type specificity of the bacterial species *Haemophilus influenzae*. J. Exp. Med. **53:**471–492.
27. **Pittman, M., and D. J. Davis.** 1950. Identification of the Koch-Weeks bacillus (Hemophilus aegyptius). J. Bacteriol. **59:**413–426.
28. **Plummer, F. A., L. J. D'Costa, H. Nsanze, J. Dylewski, P. Karasira, and A. R. Ronald.** 1983. Epidemiology of *Haemophilus ducreyi* in Nairobi, Kenya. Lancet **ii:**1293–1295.
29. **Scheifele, D. W., and S. J. Fussell.** 1981. Frequency of ampicillin-resistant *Haemophilus parainfluenzae* in children. J. Infect. Dis. **143:**495–498.
30. **Sims, W.** 1970. Oral haemophili. J. Med. Microbiol. **3:**615–625.
31. **Turk, D. C., and J. R. May.** 1967. *Haemophilus influenzae*. Its clinical importance. English Universities Press, London.
32. **Vastine, D. W., C. R. Dawson, I. Hoshiwara, C. Yoneda, T. Daghfous, and M. Messadi.** 1974. Comparison of media for the isolation of *Haemophilus* species from cases of seasonal conjunctivitis associated with severe endemic trachoma. Appl. Microbiol. **28:**688–690.
33. **Welch, D. F., and D. Hensel.** 1982. Evaluation of Bactogen and Phadebact for detection of *Haemophilus influenzae* type b antigen in cerebrospinal fluid. J. Clin. Microbiol. **16:**905–908.
34. **Zinnemann, K., K. B. Rogers, J. Frazer, and S. K. Devaraj.** 1971. A haemolytic V-dependent CO_2-preferring *Haemophilus* species *Haemophilus paraphrohaemolyticus* nov. spec. J. Med. Microbiol. **4:**139–143.

Chapter 34

Bordetella

CHARLOTTE D. PARKER AND BEVERLEY J. PAYNE

CHARACTERIZATION

Bordetella spp. are small, gram-negative coccobacilli which are obligate aerobes, fail to attack carbohydrates, grow slowly in vitro, parasitize the respiratory tracts of animals, and show striking interspecies DNA homology (12–14, 21). *Bordetella pertussis* was first isolated by Bordet and Gengou in 1906 as the causative agent of whooping cough. It is among the most fastidious bacteria known, and clinical isolates can be cultured only by special techniques (20). *B. pertussis* is inhibited by a number of common medium constituents, such as fatty acids, metal ions, sulfides, and peroxides. Starch, activated charcoal, serum albumin, blood, cyclodextrins, or other protective substances must be added to isolation medium. On special media, such as Bordet-Gengou (BG) agar (potato infusion + 20% sheep blood) or Regan-Lowe (RL) agar (Oxoid CM119 charcoal agar + 10% horse blood + cephalexin), *B. pertussis* forms tiny colonies within 3 to 5 days. No growth occurs on infusion agar or on ordinary blood agar.

Two other *Bordetella* species currently recognized share antigens with *B. pertussis* and may cause human infections, but are less fastidious. *B. parapertussis* also causes whooping cough, and *B. bronchiseptica* causes respiratory infections in rabbits, dogs, swine, horses, and other animals. *B. bronchiseptica* may cause human respiratory or wound infections, but apparently does not cause whooping cough. Both *B. parapertussis* and *B. bronchiseptica* grow on infusion agar and on blood agar within 2 to 3 days, and *B. bronchiseptica* usually grows on MacConkey agar. A fourth species, *B. avium*, recently proposed by Kersters et al. (13), has been found to cause rhinotracheitis in turkeys, but it has not yet been associated with human infections. It is not fastidious and will grow on MacConkey agar.

The frequency of *B. pertussis* infection has been underestimated in recent years because of unfamiliarity with typical whooping cough and the general unavailability of convenient and reliable laboratory diagnostic techniques. Typical whooping cough begins, after an incubation period of 7 to 10 days, with a mild respiratory infection resembling a cold. This catarrhal stage progresses, with a sporadic cough which worsens and becomes paroxysmal over 1 to 2 weeks, until the typical "whoop," a forced inspiration over a partially closed glottis, appears. Coughing is often worse at night and frequently accompanied by vomiting. Lymphocytosis is usually marked during this period. After several weeks, the paroxysms of coughing and vomiting decrease in frequency, and the patient enters a convalescent stage.

Although typical disease can occur at any age, very young infants frequently have choking or apnea and cyanosis rather than whooping, while immunized children or adults may have only catarrhal symptoms or a mild bronchitis. Apparent asymptomatic cases occur occasionally in adults, but true carriers have not been documented. A single case of typical whooping cough may indicate a cluster of infected individuals. *B. pertussis* infection should be suspected in any young child with a severe respiratory infection, and the diagnosis should also be considered in older children and adults who have been exposed to pertussis and have respiratory symptoms or who have prolonged or paroxysmal coughing (7, 16, 19).

B. pertussis infection is localized to the ciliated epithelial cells of the respiratory tract, and the organism may be recovered by posterior nasopharyngeal culture. Such cultures may remain positive for 5 weeks or more after the paroxysmal cough develops, but the percentage of positive cultures decreases steadily with time (17). *B. pertussis* has been shown to be susceptible to erythromycin, tetracycline, and chloramphenicol, and these drugs are effective for therapy (2, 3). Ampicillin is ineffective for therapy (1, 7). While well-documented studies are lacking, the administration of appropriate antibiotics usually suppresses catarrhal symptoms and may lead to rapid improvement of paroxysmal cough. However, the administration of antibiotics does not appear to affect the course of illness in hospitalized children diagnosed during the paroxysmal stage (1). Antibiotics appear to shorten the duration of shedding of the organisms in respiratory secretions from weeks to days (1), which may limit the spread of infection to susceptible contacts. Although controlled studies of chemoprophylaxis have not been reported, administration of erythromycin may also be useful in preventing infection in susceptible contacts of patients with pertussis (11).

COLLECTION AND TRANSPORT OF SPECIMENS

Collection of specimens

Diagnosis of pertussis requires collection of suitable respiratory tract samples for direct fluorescent-antibody (DFA) test and culture. The specimen of choice is duplicate pernasal nasopharyngeal swabs (see Fig. 1), obtained by using tiny calcium alginate swabs on fine, flexible wire (Ultrafine Calgiswab [Inolex Corp., Glenville, Ill.] or equivalent). Both swabs should be taken in a similar manner. With the patient's head immobilized, a swab should be gently inserted into the nostril until it reaches the posterior nares and is then left in place for a few seconds. The tickling sensation of the swab usually induces a cough. (If resistance is encountered during insertion of the swab, the other side should be tried, as some persons have a deviated septum or a large turbinate.) Bronchial or nasopharyngeal secretions obtained by aspiration or washing may provide superior specimens (M. Roe, personal communication) but are collected conveniently only in a hospital setting. Collection of throat swabs in addition to nasopharyngeal swabs has been reported to result in a higher rate of positive cultures. However, calcium alginate swabs and special pertussis media

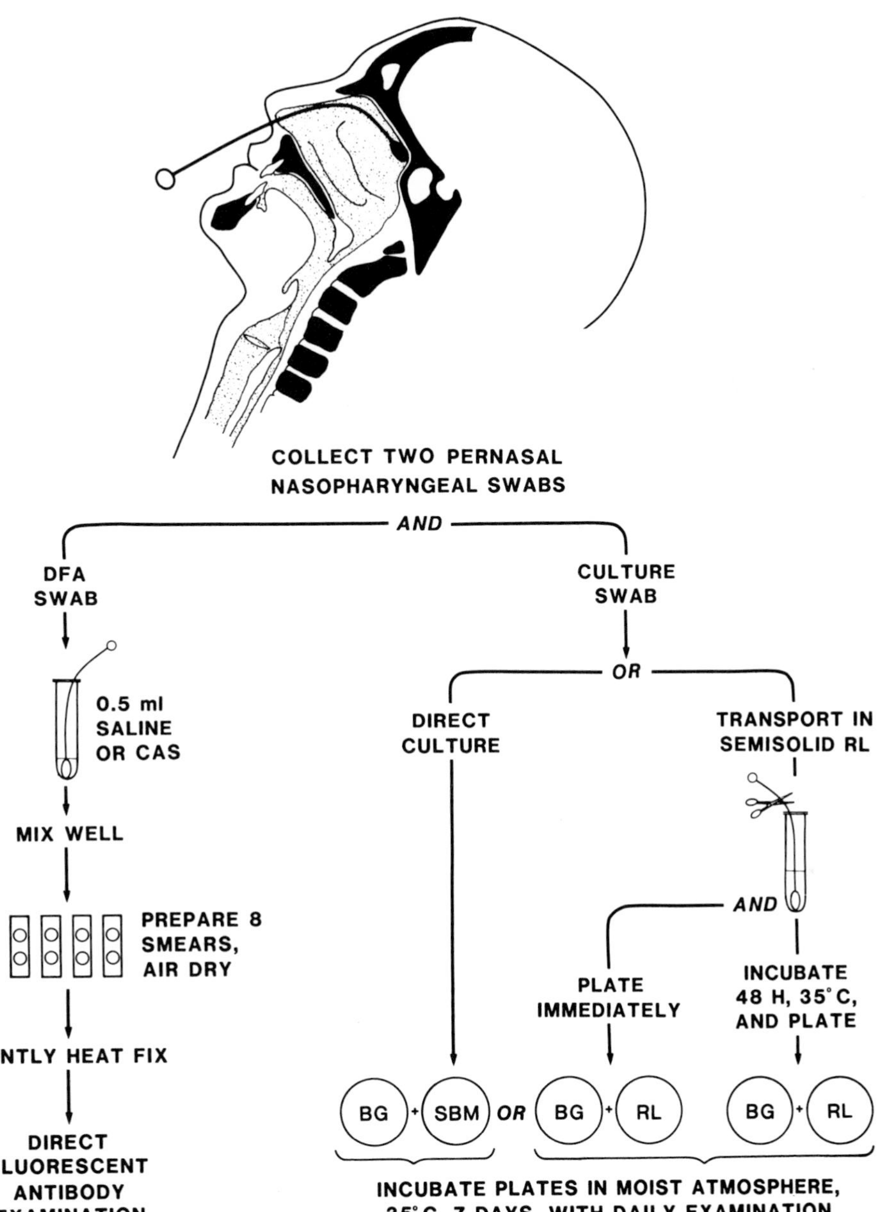

FIG. 1. Recommended protocol for diagnosis of whooping cough. The proper placement of the swab for collection of a nasopharyngeal specimen is shown, although the size of the calcium alginate tip is exaggerated for clarity. Two swabs should be obtained, one for DFA examination and one for culture. CAS, 0.1% casein hydrolysate; SBM, selective *Bordetella* medium (BG containing 2.5 μg of methicillin or 40 μg of cephalexin per ml; or RL).

are also required for throat cultures (15, 22). Obtaining serial cultures and culturing all household contacts who have upper respiratory symptoms also will increase the frequency of laboratory-confirmed cases.

Alternative specimens to nasopharyngeal swabs are nasopharyngeal or bronchial washings, suction aspirates, transtracheal aspirates, etc. Because of the difficulty of obtaining such specimens, every effort should be made to inoculate them directly onto fresh BG plates, as well as to inoculate RL transport medium and prepare smears for DFA. When feasible, a nasopharyngeal swab should also be obtained and processed. For "cough plate" requests, nasopharyngeal swabs should be obtained.

Transport of specimens

Direct plating at bedside on freshly made BG agar plates is the most reliable culture technique and should be used if possible.

In situations where direct plating is impossible, RL transport medium should be utilized. This semisolid transport medium developed by Regan and Lowe (23) has proved satisfactory in use in the United States, provided the exact formulation and usage directions are followed (E. Dodd and M. Wallace, personal communications). RL transport medium consists of one-half strength Oxoid CM119 charcoal agar with 10% defibrinated horse blood and 40 μg of cephalexin per ml (with or without 50 μg of amphotericin B per ml), dispensed so as to half-fill small vials. RL transport medium is inoculated by pushing one of the specimen swabs into the medium and leaving it submerged there while the tube is transported to the laboratory or mailed to a reference laboratory. RL transport medium is stable for 2 months if stored refrigerated.

The State of New York Department of Health utilizes slants made of full-strength RL medium for transport, by streaking the surface and then pushing the calcium alginate swab into the depth of the agar and leaving it submerged for transport (M. Shayegani, personal communication). The slant technique has the advantage that the surface growth can be screened by fluorescent-antibody staining. Shelf life and other details are the same as for semisolid RL.

DIRECT EXAMINATION

DFA is valuable, especially for serious respiratory infections in children, since rapid results may be obtained. DFA is the most frequent laboratory diagnostic test used (5, 6), but should be accompanied by culture. Because of difficulties in interpreting DFA examination of clinical material, success is directly proportional to the experience of the microscopist. For laboratories not routinely performing this procedure, we recommend preparing the slides "in-house" and then referring them (or duplicates) to a reference laboratory for examination.

In outbreaks where DFA and good culture techniques have been used in parallel, DFA has shown about 50% false-negative results (4, 7). A recent study using isolation medium similar to RL, except containing sheep blood, found 19 positive DFA tests among 28 culture-proven patients (8). False-positive DFA results may occur due to unreliable antiserum or failure of the microscopist to recognize characteristic *B. pertussis* morphology (4).

DFA test

One of the two pernasal nasopharyngeal swabs should be placed in a tube containing 0.5 ml of sterile 0.1% casein hydrolysate or sterile saline, sealed tightly, and agitated with a Vortex mixer or shaken vigorously for 20 s. (We use casein hydrolysate because *B. pertussis* survives in it for about 2 h, especially if kept chilled. Thus, if only one swab can be obtained, half the fluid can be used to prepare smears and the other half can be used for direct inoculation of BG plates or to inoculate RL transport medium. *B. pertussis* may not survive exposure to saline.) The fluid should be used to prepare a minimum of eight smears 10 to 15 mm in diameter. We prepare smears on commercially available FA slides which have two etched circles each 10 mm in diameter. Alternatively, circles may be drawn on slides with a permanent marking pen. Several drops of liquid from the tube should be used for each smear, and smears should be air dried. (Swabs rolled directly onto slides yield smears of poorer quality.) The smears, which should be plainly visible after drying, should be gently heat fixed. Smears so prepared may be mailed to a reference laboratory for testing or may be stored desiccated in the cold for at least 6 months.

Our recommended technique uses Difco antiserum (the only commercially available serum obtainable in the United States in March 1984). Difco 2359-56 chicken antiserum to *B. pertussis*, conjugated to fluorescein isothiocyanate, will be referred to as pertussis-conjugate, and Difco 2378-56 chicken antiserum to *B. parapertussis*, conjugated to fluorescein isothiocyanate, will be called parapertussis-conjugate. The following techniques should be modified appropriately for use with other sera. Because fluorescein isothiocyanate-conjugated negative Difco antiserum is unavailable, and because both pertussis-conjugate and parapertussis-conjugate show specificity, we routinely read the pertussis-conjugate stained smear against the parapertussis-conjugate stained smear as the control for nonspecific fluorescence. The correct working dilution for each conjugate should be determined, following directions in the package insert, against thin smears of stock *B. pertussis* and *B. parapertussis* isolates.

Staining protocol

1. Add a drop of the working dilution of the appropriate conjugate to each smear and spread the conjugate to cover. Stain two patient smears with pertussis-conjugate (to permit examination of adequate material) and one patient smear with parapertussis-conjugate.
2. Place slides in a moist chamber for 30 min.
3. Tap off excess conjugate, rinse gently with phosphate-buffered saline (1.24 g of Na_2HPO_4, 0.18 g of $NaH_2PO_4 \cdot H_2O$, and 8.5 g of NaCl, distilled water to 1 liter; pH 7.6), and then allow the slides to stand for 10 min in individual containers of phosphate-buffered saline to prevent cross-contamination. (Copland jars work well.)
4. Remove slides, rinse gently with distilled water, and allow to air dry.
5. Add 1 drop of mounting fluid (1 part carbonate buffer [0.29 g of $Na_2CO_3 \cdot 10H_2O$ and 0.76 g of $NaHCO_3$, distilled water to 1 liter; pH 9.0] plus 9 parts neutral glycerol) and a cover slip to each smear and examine under oil immersion.
6. Hold duplicate smears for possible Gram stain, repeat DFA, or both.

Each time testing is done, three control slides should be stained. Thin smears (very few organisms) of stock *B. pertussis* cells should be stained with pertussis-conjugate and with parapertussis-conjugate. A thin smear of stock *B. parapertussis* should be stained with parapertussis-conjugate. For validity, the *B. pertussis*/pertussis-conjugate and the *B. parapertussis*/parapertussis-conjugate slides must each be strongly positive, and the *B. pertussis*/parapertussis-conjugate slide must be negative.

Stock strains of *B. pertussis* and *B. parapertussis* show the most typical coccobacillary morphology when young cultures on antibiotic-free BG medium are examined. Stock strains should be maintained by lyophilization, or frozen at −70°C in blood. (Subcultures of stock strains on BG plates retain viability and show typical morphology for up to 1 month if stored sealed in the refrigerator.) Control smears may be prepared either from cultures on BG plates or from recent clinical isolates, and smears may be stored with a desiccant in the cold for at least 6 months. Stock cultures readily become adapted to in vitro growth conditions. Thus, the number of transfers should be minimized.

Patient smears should be examined carefully under oil immersion for very small coccobacilli occurring singly, in pairs, or occasionally in small groups. With Difco conjugates, *B. pertussis* cells are a brightly fluorescent yellow-green with a clear-cut periphery and a nonstaining or dimmer-staining cell center (doughnut appearance). Organisms showing incorrect morphology or organisms that fluoresce with both conjugates should be reported DFA negative. (Questionable morphology can, at times, be resolved with a Gram stain of one of the remaining smears. We have observed nonspecific staining of gram-negative diplococci, gram-positive cocci, and diphtheroid-like rods. True cross-reactions, extraneous antibodies found in the conjugate, or nonspecific binding of immunoglobulins by staphylococci and streptococci may account for nonspecific staining.) Patient smears which are positive with pertussis-conjugate and negative with parapertussis-conjugate should be reported as positive for *B. pertussis* by DFA test.

CULTURE AND ISOLATION

Isolation techniques

A number of factors influence the successful isolation of *B. pertussis*. Proper collection of nasopharyngeal specimens by using calcium alginate swabs, and use of RL transport medium or direct plating at bedside on fresh BG plates, are the most critical. For consistent isolation of the organism from clinical cases of whooping cough, every detail of specimen collection and culture technique must be optimal. Careful preparation of media, use of both selective and nonselective isolation media, and incubation at 35°C with high humidity are minimum culture requirements. We use a candle jar to provide a convenient moist chamber, although *B. pertussis* has no CO_2 requirement. Allowing swab specimens or isolation plates to become dry during incubation is very detrimental.

For directly plated specimens, BG plates are recommended. Two freshly prepared BG plates (one plain and one containing either 2.5 μg of methicillin or 40 μg of cephalexin per ml) should be inoculated heavily over one-third of their surface and then streaked for isolation (4, 20, 25). For direct plating we recommend immersing the swab in 0.5 ml of sterile casein hydrolysate at the bedside and returning it immediately to the laboratory for agitation and culture of 0.1 to 0.2 ml of the fluid per BG plate. The additional moisture provided by this method may enhance primary isolation. Plates so inoculated should be incubated right side up for the first day until the fluid is absorbed.

Protocol for processing specimens received in RL transport medium

Both nonselective plates (BG) and selective plates (RL, or BG with antibiotics) should be inoculated.

1. The swab should be removed from the transport medium with forceps and rolled over one-third of the surface of each plate.
2. The swab should be returned to the semisolid RL transport medium and incubated in the transport medium for 48 h at 35°C.
3. The primary plates should be streaked for good isolation and incubated at 35°C for 7 days in a humid atmosphere.
4. After 48 h, a second set of plates should be inoculated from the swab and incubated as before.
5. All plates should be examined daily for 7 days.

Several United States workers have reported success in isolating *B. pertussis* on RL medium (8; F. Chan, E. Rossier, A. M. R. MacKenzie, and A. Camus, Abstr. Annu. Meet. Am. Soc. Microbiol. 1983, C260, p. 355; C. Gullans, K. Bromberg, M. Sierra, and M. Yudlowitz, Abstr. Annu. Meet. Am. Soc. Microbiol. 1983, C262, p. 355; E. Dodd, D. Schoonmaker, and M. Wallace, personal communications). Amphotericin B at a final concentration of 50 μg/ml may be added to plates inoculated from RL transport medium at 48 h to combat yeast or fungal contamination when deemed necessary by observation of initial isolation plates (23). Antibiotic-free media are usually more rapidly positive, but selective media are invaluable for specimens that contain few *B. pertussis* organisms. Sheep blood has been reported as a successful substitute for horse blood in RL plates in a single study (8). Quality control should be performed on new lots of RL medium by showing that they support growth of small inocula of properly preserved clinical isolates of *B. pertussis*. Stock laboratory strains of *B. pertussis* may grow well on media (including blood agar) which will not support the growth of clinical isolates.

The plates should be examined daily by eye and with a dissecting microscope under oblique illumination. Occasionally, microscopically visible colonies of *B. pertussis* appear after 48 h, but colonies visible to the naked eye rarely occur earlier than 3 days on either BG or RL medium. On BG medium, colonies are tiny and glistening, resemble a bisected pearl, and usually exhibit tiny zones of hazy hemolysis. Tiny colonies with a characteristic compact, glistening appearance are typical on RL agar (Fig. 2).

Typical colonies should be Gram stained. If small gram-negative rods or coccobacilli are seen, fluorescent-antibody tests and growth tests should be performed. *B. parapertussis* and *B. bronchiseptica* should be confirmed by biochemical testing, although the

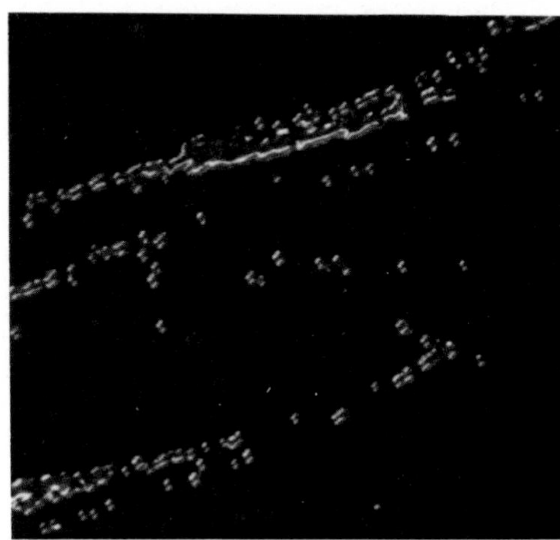

FIG. 2. *B. pertussis* colonies on RL plate. Original magnification, ×4.

fluorescent-antibody stain can also be utilized with *B. parapertussis*.

IDENTIFICATION

Cultural characteristics

In reflected light on BG agar, colonies of *B. pertussis* appear as half pearls, domed, smooth, transparent, and glistening; they have an entire circular edge and are not usually over 1 mm in diameter. *B. parapertussis* colonies on BG are similar, but appear sooner and become larger. The surface of the colonies is duller, and a slight brownish color may be present. *B. parapertussis* grows on blood agar plates within 1 to 2 days, showing an opaque, smooth colony. *B. bronchiseptica* grows most rapidly, with colonies evident on BG agar under a dissecting microscope within 24 h and well developed within 48 h. The colonies resemble *B. pertussis* at first, but become larger and develop a rough surface that is pitted like beaten metal. On blood agar, *B. bronchiseptica* forms larger colonies than on BG agar, and the colonies are grayish-white and somewhat flat and lack the metallic appearance that they have on BG agar. All three species may produce hazy zones of hemolysis on BG agar. *B. parapertussis* is usually hemolytic on blood agar, whereas *B. bronchiseptica* tends to show hazy hemolysis.

On RL agar, *B. pertussis* colonies are round, smooth, shiny, domed, very compact appearing, of small but variable size, and white to gray in color (Fig. 2). Colonies may run together slightly. *B. parapertussis* is similar, but is grayer and less domed and grows more rapidly. *B. bronchiseptica* colonies are much larger, are less variable in size, have a nonhomogeneous appearance and a putrid odor, and grow rapidly.

Microscopic examination

Bordetellae are small, gram-negative bacilli which occur singly, in pairs, or in small clumps and are usually about 1.0 by 0.3 to 0.5 μm in size. Because of the slow uptake of the safranin used in the Gram stain, pale staining is characteristic. We recommend allowing the safranin counterstain to remain on the smear for a full 2 min; otherwise, it is nearly impossible to see the cells at times. In old cultures or cultures on medium containing antibiotics, cells become pleomorphic and may have threadlike filaments and thick bacillary forms. *B. parapertussis* is typically the most rod shaped and may sometimes be seen in a palisade arrangement. *B. bronchiseptica* is motile by peritrichous flagella; the others are not motile.

Biochemical reactions

Members of the genus *Bordetella* are relatively inert biochemically. None ferment sugars; gelatin is not liquefied; indole and acetoin are not produced; H_2S is not produced; most strains are weakly catalase positive; and litmus milk is usually rendered alkaline. The production of an alkaline reaction in most media is characteristic. Urease tests on Christensen urea slants are rapidly positive for *B. bronchiseptica* and more slowly positive for *B. parapertussis*. *B. parapertussis* produces a brown soluble pigment on aerobically incubated tyrosine agar or in defined liquid medium (24) prepared as described by Manclark and Meade (18). *B. bronchiseptica* reduces nitrates to nitrites.

Table 1 presents a summary of the differential features of the three species.

Serological methods

Fluorescent-antibody tests are the preferred method for serological confirmation of the identity of *B. pertussis*. A presumptive positive culture can be reported as early as 2 to 3 days after culture, if a sweep of growth from the primary plate is examined. If antiserum is available, agglutination testing of pure cultures is also satisfactory. Heavy growth of suspected *B. pertussis* from the primary plate or from a subculture is suspended in saline and adjusted to a turbidity equivalent to a reading of 100 Klett units (green filter). Several small drops of antipertussis serum at the appropriate dilution are placed on a slide and mixed with a drop of the saline-suspended cells. Clumping of the cells indicates a positive reaction. Controls should include a known *B. pertussis* culture and a mixture of unknown cells and normal serum.

Serological tests to demonstrate circulating antibodies in patients' sera are not routinely used for

TABLE 1. Characteristics of human clinical isolates of *Bordetella* species (12, 21)

Characteristic	*B. pertussis*	*B. parapertussis*	*B. bronchiseptica*
Involvement in respiratory infections	Moderately common, humans only	Rare, humans only	Frequent in animals, uncommon in humans
Recovery from wounds, abscesses, etc.	−	−	+
Visible growth on BG agar	In 3 to 4 days, occasionally longer	In 1 to 2 days	In 1 to 1.5 days
Growth on blood agar on primary isolation	−	+	+
Growth on heart infusion agar	−	+	+
Brown soluble pigment (heart infusion-tyrosine agar or defined liquid medium)	−	+	−
Motility	−	−	+ (peritrichous flagella)
Urease (Christensen) heavy inoculum	−	+ (24 h)	+ (1–4 h)
Nitrate	−	−	+
Carbohydrate utilization	−	−	−
Litmus milk	Alkaline, 12–14 days	Alkaline, 1–4 days	Alkaline, 1–2 days
Specific heat-labile antigen (for genus and species)	1, 7	7, 14	7, 12
Other heat-labile antigens	2, 3, 4, 5, 6	8, 9, 10	8, 9, 10, 11, 13

diagnosis. Agglutination tests to detect antibody are cumbersome (18). However, detection of *B. pertussis*-specific antibodies in nasal wash fluids or serum offers promise (9, 10), and other serological tests are likely to be developed.

Maria Wallace (Missouri Division of Health Laboratory, Jefferson City) gave us the benefit of her long experience in reading DFA smears and culturing for *B. pertussis*, and we express special thanks to her. The following persons very generously shared their current and unpublished results with us: Everett D. Dodd (Oklahoma State Department of Health, Laboratory Service, Oklahoma City), Mary Ellen Endo (Illinois Department of Public Health, Division of Laboratories, Chicago), Peter H. Gilligan (St. Christopher's Hospital for Children, Philadelphia, Pennsylvania), M. Magus (Ministry of Health, Ontario, Canada), Martha H. Roe (The Children's Hospital, Denver, Colo.), and M. Shayegani and Dianna Schoonmaker (State of New York Department of Health, Albany).

LITERATURE CITED

1. **Baraff, L. J., J. Wilkins, and P. F. Wehrle.** 1978. The role of antibiotics, immunizations, and adenoviruses in pertussis. Pediatrics **61**:224–230.
2. **Bass, J. W., F. W. Crast, J. B. Kotheimer, and I. A. Mitchell.** 1969. Susceptibility of *Bordetella pertussis* to nine antimicrobial agents. Am. J. Dis. Child. **117**:276–280.
3. **Bass, J. W., E. L. Klenk, J. B. Kotheimer, C. C. Linnemann, and M. H. D. Smith.** 1969. Antimicrobial treatment of pertussis. J. Pediatr. **75**:768–781.
4. **Broome, C. V., D. W. Fraser, and J. W. English.** 1979. Pertussis—diagnostic methods and surveillance, p. 19–22. *In* C. Manclark and J. Hill (ed.), International Symposium on Pertussis. DHEW Publication no. (NIH) 79-1830. U.S. Government Printing Office, Washington, D.C.
5. **Centers for Disease Control.** 1982. Pertussis surveillance, 1979–1981. Morbid. Mortal. Weekly Rep. **31**:333–336.
6. **Centers for Disease Control.** 1984. Pertussis outbreak—Oklahoma. Morbid. Mortal. Weekly Rep. **33**:2–10.
7. **Field, L. H., and C. D. Parker.** 1977. Pertussis outbreak in Austin and Travis County, Texas, 1975. J. Clin. Microbiol. **6**:154–160.
8. **Gilligan, P.** 1983. Laboratory diagnosis of *Bordetella pertussis* infection. Clin. Microbiol. Newsl. **5**:115–117.
9. **Goodman, Y. E., A. J. Wort, and F. L. Jackson.** 1981. Enzyme-linked immunosorbent assay for detection of pertussis immunoglobulin A in nasopharyngeal secretions as an indicator of recent infection. J. Clin. Microbiol. **13**:286–292.
10. **Granstrom, M., G. Granstrom, A. Lindfors, and P. Askelof.** 1982. Serologic diagnosis of whooping cough by an enzyme-linked immunosorbent assay using fimbrial hemagglutinin as antigen. J. Infect. Dis. **146**:741–745.
11. **Immunization Practices Advisory Committee.** 1981. Diphtheria, tetanus, and pertussis: guidelines for vaccine prophylaxis and other preventive measures. Morbid. Mortal. Weekly Rep. **30**:392–407.
12. **Johnson, R., and P. H. A. Sneath.** 1973. Taxonomy of *Bordetella* and related organisms of the families *Achromobacteriaceae, Brucellaceae*, and *Neisseriaceae*. Int. J. Syst. Bacteriol. **23**:381–404.
13. **Kersters, K., K.-H. Hinz, A. Hertle, P. Seegers, A. Lievens, O. Siegmann, and J. DeLey.** 1984. *Bordetella avium*, sp. nov., isolated from the respiratory tracts of turkeys and other birds. Int. J. Syst. Bacteriol. **34**:56–70.
14. **Kloos, W. E., N. Mohapatra, W. J. Dobrogrosz, J. W. Ezzell, and C. R. Manclark.** 1981. Deoxyribonucleotide sequence relationships among *Bordetella* species. Int. J. Syst. Bacteriol. **31**:173–176.
15. **Lautrop, H., and B. W. Lacey.** 1960. Laboratory diagnosis of whooping cough or *Bordetella* infections. Bull. W.H.O. **23**:15–35.
16. **Linnemann, C. C.** 1979. Host-parasite interactions in pertussis, p. 3–18. *In* C. Manclark and J. Hill (ed.), International Symposium on Pertussis. DHEW Publication no. (NIH) 79-1830. U.S. Government Printing Office, Washington, D.C.
17. **Linnemann, C. C. and J. W. Bass.** 1981. *Bordetella* infections, p. 249–260. *In* A. Balows and W. J. Hausler, Jr. (ed.), Diagnostic procedures for bacterial, mycotic and parasitic infections, 6th ed. American Public Health Association, Washington, D.C.
18. **Manclark, C. R., and B. D. Meade.** 1980. Serological response to *Bordetella pertussis*, p. 496–499. *In* N. R. Rose and H. Friedman (ed.), Manual of clinical immunology, 2nd ed. American Society for Microbiology, Washington, D.C.
19. **Nelson, J. D.** 1978. The changing epidemiology of pertussis in young infants; the role of adults as reservoirs of infection. Am. J. Dis. Child. **132**:371–373.
20. **Parker, C. D., and C. C. Linnemann, Jr.** 1980. Bordetella, p. 337–343. *In* E. H. Lennette, A. Balows, W. J. Hausler, Jr., and J. P. Truant (ed.), Manual of clinical microbiology, 3rd ed. American Society for Microbiology, Washington, D.C.
21. **Pittman, M.** 1974. Genus *Bordetella*, p. 282–283. *In* R. E. Buchanan and N. E. Gibbons (ed.), Bergey's manual of determinative bacteriology, 8th ed. The Williams and Wilkins Co., Baltimore.
22. **Regan, J.** 1980. The laboratory diagnosis of whooping cough. Clin. Microbiol. Newsl. **2**:1–3.
23. **Regan, J., and F. Lowe.** 1977. Enrichment medium for the isolation of *Bordetella pertussis*. J. Clin. Microbiol. **6**:303–309.
24. **Stainer, D. W., and M. J. Scholte.** 1970. A simple chemically defined medium for the production of phase I *Bordetella pertussis*. J. Gen. Microbiol. **63**:211–220.
25. **Sutcliffe, E. M., and J. D. Abbott.** 1972. Selective medium for the isolation of *Bordetella pertussis* and *parapertussis*. J. Clin. Pathol. **25**:732–733.

Chapter 35

Streptobacillus moniliformis and *Spirillum minus*

MORRISON ROGOSA

STREPTOBACILLUS MONILIFORMIS

Characteristics

Rods, nonsporing, less than 1 μm wide by 1 to 5 μm long, with rounded or pointed ends, are highly pleomorphic, forming curved and looped filaments as long as 100 to 150 μm. In young cultures, filaments are relatively homogeneous, appearing to consist of single cells. As the culture ages, fine granules and alternate light and dark bands often appear in the filaments, and fragmentation into irregular, coccobacillary elements occurs. The filaments may consist of a series of oval to elongated bulbous swellings, each 1 to 3 μm in diameter, giving the appearance of a string of beads, and may contain numerous granules randomly distributed. True branching does not occur. Also, cell, membrane-associated, and extracellular oil-like, ether-soluble droplets containing cholesterol and other lipids are very characteristic. Morphology is influenced considerably by media, cultural conditions, and age of culture. Under favorable conditions, such as in young cultures in favorable media or in smears from pathological blood, joint fluids, etc., cells generally appear to be more rodlike and uniform, with occasional short filaments randomly arranged. There is considerable irregularity in retention of stains: the monilia-like swellings stain more intensely than the filaments. *S. moniliformis* is not acid fast by Ziehl-Neelsen techniques and is gram negative. Giemsa or Wayson stain may be more satisfactory for the demonstration of the organisms than is Gram stain.

Older synonyms of *S. moniliformis* are listed in *Index Bergeyana* (4). The 8th edition of *Bergey's Manual of Determinative Bacteriology* (35) placed the genus *Streptobacillus* in "Part VIII: Gram-negative, Facultatively Anaerobic Rods" in a group containing eight "genera of uncertain affiliation." In the current *Bergey's Manual of Systematic Bacteriology, Vol. 1* (31), published in 1984, the genus *Streptobacillus* has been placed in Section 5, which is comprised of facultatively anaerobic gram-negative rods. The excellent description by Savage (31) of the genus and *S. moniliformis* agrees in all important respects with the description in the eighth edition. Minor differences in the fermentation of some carbohydrates in the case of an organism whose fermentative capacity is generally weak and technically difficult to determine are of no great significance and do not add to the burden of recognizing the organism.

Clinical significance

S. moniliformis infections are usually acquired after the bite of a rat, mouse, or cat. The nasopharynx and infected middle ears of wild rats and mice very frequently harbor the organism, and epizootic incidents characterized by naturally occuring polyarthritis have occurred in laboratory mice. Isolations have also been made from the tendon sheath and sternal bursa of arthritic turkeys and from cervical abscesses in guinea pigs. Human cases resulting from the ingestion of milk to which rats have had access have been reported as Haverhill fever or erythema arthriticum epidemicum. Infrequently, there is no history of a rodent bit or animal contact, and the disease may appear after traumatic injury.

Correct diagnosis of rat-bite fever may often be difficult. In patients who present with fever, headache, and a rash, the symptoms are strikingly similar to those of Rocky Mountain spotted fever. A patient of Portnoy et al. (23) lived in a wooded, tick-infested area from which cases of Rocky Mountain spotted fever had been previously reported. In this case, negative serological results were obtained for leptospirosis, tularemia, murine typhus, Q fever, Rocky Mountain spotted fever, and infectious mononucleosis before *S. moniliformis* was recovered from blood cultures 48 h after inoculation. Although the strain was very susceptible to penicillins, the patient was allergic to penicillin and was successfully treated with chloramphenicol (50 mg/kg per day) for 10 days.

S. moniliformis and *Spirillum minus* have been associated with rat-bite fever acquired in laboratories where rodents are handled. Collins (7) cites 16 references from the years 1925 through 1983. Unlike the relatively high mortality of untreated cases in the general population, there are 20 cases of laboratory-acquired infections with no deaths. Very probably the insight of laboratory personnel results in quick diagnosis and treatment.

Some relatively recent cases of streptobacillary rat-bite fever with unusual clinical aspects have been reported. Raffin and Freemark (24) describe a unique clinical course in a boy who developed a subglottic mass and bilateral parotid swelling. The patient responded initially to tetracycline but relapsed after therapy; cure was ultimately achieved with a 3-week course of penicillin VK (250,000 U orally every 6 h). Faro et al. (8) and Renaut et al. (25) each report a case of generalized *S. moniliformis* septicemia successfully treated with penicillin G.

Hopkinson and Lloyd (12) describe an outbreak of an acute, septicemic, sudden death syndrome in Spinifex hopping mice (*Notomys alexis*). Rats had repeatedly broken into the cage housing 20 mice and had inflicted bite wounds on 2 of them. Autopsies on seven mice revealed swollen livers, and *S. moniliformis* was recovered from the livers and lungs. Although *S. moniliformis* has not been shown to be an inhabitant of the nasopharynx of normal mice, it has been isolated from the nasopharynx of mice with clinical signs of epidemic arthritis, indicating that *S. moniliformis* can also be transmitted by mouse bites. Organisms identical to or resembling *S. moniliformis* have been isolated from 21% of pneumonic lungs of 56 calves examined by Gourlay et al. (10). Although these authors agree that these organisms are members of the genus *Streptobacillus*, they stated that "it would seem more appropriate to designate them *Streptobacillus actin-*

oides rather than *Bacillus actinoides* or *Actinobacillus actinoides* as has been done previously." The name *S. moniliformis* was not accepted by them.

The distribution of human infections due to *S. moniliformis* appears to be worldwide. Because the usual rodent vector suggests a poor or primitive living environment, it is highly likely that the relatively small number of reported cases seriously understates the true incidence of these infections. Very probably, most cases throughout the world are never reported because they are undiagnosed and untreated and because they occur where reporting facilities are inadequate or lacking.

Although some inflammation may result from a bite wound caused by an animal harboring the organism, there is usually normal healing without induration. Unlike the lesion of Sodoku, there is no subsequent reactivation of the initial lesion or uncomplicated lymphadenitis. The incubation period is usually less than 10 days and is followed by abrupt, prostrating illness with chills, fever, vomiting, and severe headache. Alternate remissions and febrile incidents may persist for weeks or months. Usually, there is an accompanying cutaneous eruption (rubellaform, morbilliform, or petechial). Arthritic symptoms are frequent, with excess joint fluid and painful swellings. Reported mortality varies from 0 to 10% in untreated cases. Endocarditis or pneumonia may be serious sequelae. *S. moniliformis* has been recovered from blood, joint fluids, cutaneous eruptions, and a brain abscess. Bacterial-phase infections have been successfully treated with penicillin. Penicillin-resistant infections have been effectively treated with streptomycin. Two cases of L-phase variant infections (in which the organisms were resistant to penicillin) were eliminated by chlortetracycline.

Otherwise normal patients with bacterial-phase infections have been successfully treated with daily penicillin doses varying from 60,000 to 1,700,000 U over periods of 4 to 19 days of therapy. However, because there is a good correlation between drug dosage and response to therapy and also because penicillin-resistant transitional and L-forms are not uncommon (9), it is wise to prescribe daily penicillin dosages of at least 10^6 U combined with tolerable therapeutic doses of streptomycin for a period of at least 7 days. In successful therapy, a dramatic drop to normal body temperature, accompanied by signs of returning well-being, occurs within 1 to 2 days. Exposed patients with rheumatic fever, valvular heart diseases, etc., are especially susceptible to endocarditis. Treatment for *S. moniliformis* endocarditis should be the same as for streptococcal endocarditis, with high daily doses of penicillin or other appropriate antibiotic (depending on the sensitivity of the patient) for 2 weeks or more.

Culture and isolation

Media. Bacterial-phase organisms have been isolated and maintained on a number of common basal media enriched with 15% sterile, defibrinated rabbit blood. However, it is preferable to employ a clear medium which will enhance the isolation and maintenance of the more fragile and nutritionally fastidious L-phase variants (6). The following recommended medium is very useful because it favors the growth of a wide variety of L-phase and bacterial-phase nutritionally fastidious organisms (1, 6). Dissolve 40 g of dehydrated heart infusion agar (Difco Laboratories, Detroit, Mich.) in 850 ml of deionized, distilled water, and adjust the pH to 7.6 with 5 N NaOH. Dispense 85-ml volumes of the solution into screw-cap bottles, and sterilize the preparation by autoclaving for 15 min at 121°C. Just before pouring the plates, add 10 ml of sterile horse serum (previously heated for 30 min at 56 to 60°C) and 5 ml of a sterile 10% (wt/vol) solution of yeast extract (Oxoid Ltd., London, England; Difco; or BBL Microbiology Systems, Cockeysville, Md.) previously adjusted to pH 7.0 and sterilized by filtration through a Seitz-type pad (0.01-μm pore size) or a membrane filter (0.45-μm pore size, Millipore Corp., Bedford, Mass., or equivalent). Pour the plates in sterile disposable plastic petri dishes (60 by 15 mm).

A similar broth medium containing 25 g of dehydrated heart infusion broth (Difco) instead of 40 g of dehydrated heart infusion agar (Difco), but with all other ingredients and medium preparation identical with those of the agar medium, is recommended.

Isolation. A blood specimen is citrated by the addition of 10 ml of blood to 10 ml of sterile 2.5% sodium citrate in a small sterile flask and mixing the fluids as usual. Prepare three separate films, and stain each one with Gram, Wayson, or Giemsa stain. Centrifuge the flask contents for 30 to 45 min to pack the cells. Discard the supernatant fluid. Use the sedimented cells to inoculate one freshly poured, solidified agar plate and two tubes of broth. Mix an estimated 0.1 ml of sedimented cells with 0.1 ml of broth; inoculate 0.1 ml of the mixed, suspended cells on the surface of an agar plate, and distribute the inoculum by gently tilting the plate in a number of directions. Use the remaining cells to inoculate the two cotton-plugged tubes of broth. Place tubes and plates upright in a wide-mouthed, screw-cap glass jar, include a lighted candle, tighten the screw-cap cover, and incubate at 35 to 37°C. Incubate the plates for 2 to 3 days. Examine the tubes daily for the appearance of characteristic fluff balls, particularly on the surface of sedimented cells. Use a good light for this inspection, and do not shake or disturb the tubes. If no growth is apparent, transfer 1 ml to fresh broth tubes daily for at least 3 successive days. If growth is observed, remove a fluff ball by pipette for subsequent transfer. Also, transfer a small drop to each of three separate clean slides; spread, dry, and gently fix the drop; and stain it with Gram, Giemsa, or Wayson stain. Examine the preparations for characteristic microscopic morphology.

Joint fluids should also be citrated by mixing equal volumes of fluid and 2.5% citrate. Otherwise, joint fluids may often clot. Stain smears with Gram, Giemsa, or Wayson stain. Inoculate two tubes of recommended broth with 1 ml or more of fluid per tube (as much as the sample volume will permit). Also, inoculate 0.1 ml onto fresh agar plates, and distribute the inoculum by carefully tilting the plate in a number of directions. Incubate and examine the preparation in the same way as blood cultures.

Pus or exudate from a wound, cutaneous eruption, or abscess should be smeared, stained, and examined microscopically. Also, inoculate one tube of broth and one agar plate (by swabbing, if the sample is limited). Incubate and examine the preparation as already described.

Identification

Growth. *S. moniliformis* is aerobic and facultative. Although growth occurs at reduced partial pressures of oxygen, in candle jars, etc., such closed environments are not strictly anaerobic but supply increased CO_2 and conserve essential moisture at the surface of agar media. Blood, serum, or ascitic fluid is required for growth. However, growth on Loeffler serum agar slants is poor. Growth does not occur or cannot be sustained in infusion, nutrient, and other broths and agars, unless they are supplemented with one of these body fluids. The pH optimum is 7.4 to 7.6, with a range of 7.0 to 8.0. Colonies generally 1 to 2.5 mm in diameter develop within 3 days on serum agar at 35 to 37°C and may be viable for as long as 1 week. There is no growth at 23°C. Bacterial-phase colonies tend to be round with a discrete edge, low convex or slightly raised, and glistening, with a butyrous consistency; smaller, granular colonies and rougher colonies intermediate in size between the first type and L_1 colonies are sometimes seen. Within, beneath, or adjacent to the primary colonies, microscopic L-phase colonies may be found. The L-phase colonies are embedded in the agar and have a fried-egg configuration, with a relatively dark center surrounded by a translucent zone containing swollen bodies and what appear to be oil globules. Thus, the spontaneously appearing L-phase variant colonies of *S. moniliformis* are indistinguishable from other L-phase, pleuropneumonia, or *Mycoplasma* colonies and must be examined by the same microscopic methods (see Chapter 36). Bacterial-phase cells have normal cell walls and are susceptible to penicillin (1- or 10-U disks), whereas L-phase cells lack a cell wall and are resistant to potassium penicillin G (1,000 U or more per ml). The latter have various revertant tendencies in the absence of penicillin. The incorporation of penicillin in a medium stabilizes the culture in the L phase if such cells are present. In fluid culture containing 10% blood, growth generally appears as fluffy balls resembling bread crumbs and may be confined to the bottom portion of the tube or the surface of sedimented cells and stroma.

For the transfer of L-phase colonies, a small agar block is cut out with a sterile spade or spatula from an area of dense growth (6, 17). Invert this block on a fresh agar plate so that the growth area contacts the surface of the agar plate. Gently push the agar block halfway across the surface with a sterile spatula, taking care to leave the block in the same inverted position. Incubate the plates as described above.

In purification and transfer procedures, a wire needle or loop should never be used. Large inocula obtained by agar block cutouts from plates or by pipette transfers (10%, vol/vol) from broth are essential (6, 17, 21). This is often the case with fastidious organisms.

A laboratory specimen may contain several species of L-phase variants because they are naturally occurring or because of previous antibiotic therapy. L-phase variants of coryneform bacteria are not uncommon (36), and some of these may have been mistaken in the past for *S. moniliformis*. Some incorrect statements in earlier literature that *S. moniliformis* is gram positive in young broth cultures and an attempt to classify *S. moniliformis* as *Actinobacillus muris* (34) or *Asterococcus muris* (11) may stem from this confusion. It is therefore important to purify these cultures by replating colonies on the maintenance or isolation agar medium by the agar block technique. The purification procedure should be repeated, if necessary, until there is reasonable assurance of culture purity before biochemical tests are performed.

In isolated pure cultures, a marked decrease of pH in broth cultures occurs in 1 to 2 days, and cultures easily become nonviable. It is therefore necessary to subculture daily. However, young broth cultures and organisms in infected tissues and body fluids frozen at −20 to −70°C remain viable for several years. Cells are killed when held at 55 to 56°C for 30 min.

Biochemical characteristics. This section is based largely on the studies of Aluotto et al. (1) and Cohen et al. (6) and on previous descriptions of the genus *Streptobacillus* (11, 19, 31, 35). The basal medium is the same as the agar medium described for isolation and maintenance except that, for each liter of complete test agar, the basal constituents are dissolved in only 750 ml of water to allow for the addition of test solutions without diluting the strength of basic components. For phenylalanine deamination and esculin hydrolysis tests only, the respective substrates and reagents for testing the reactions are incorporated in the basal medium before autoclaving. The remaining test substrate solutions are added aseptically after being sterilized by filtration through Swinnex-25 plastic filter units fitted with MF-Millipore type GS membrane filters (25-mm diameter, 0.22-μm pore size). Table 1 outlines the protocol of substrates and reagent quantities used in a variety of tests.

The basal agar is dispensed in 19-ml volumes into 2-oz (59-ml) prescription bottles. Just before pouring the plates, add 2.5 ml of heated horse serum and 1.2 ml of 10% (wt/vol) yeast extract to each bottle of molten agar held at 50°C. Since horse serum contains an active maltase, rabbit serum must be used only in maltose test plates. Also, add the individual sterile substrates, reagent, or indicator stock solutions to separately prepared bottles of medium (Table 1). Test controls must include an uninoculated plate containing substrate and an inoculated plate in which distilled water is substituted for substrate solution.

Prepare bacterial-phase inocula by swabbing off a single culture plate and suspending the swab in 2 ml of 0.85% NaCl; use 1 drop to inoculate each test plate. L-phase variants are inoculated onto test plates by the agar block method as already described. All cultures are incubated in candle jars (see previous description).

The pH of plates containing phenol red can be estimated by using indicator standards (La Motte Chemical Products Co., Baltimore, Md.). Incubate media containing carbohydrate for 3 weeks, if necessary. Reactions may be considered positive if the pH throughout the plate shows a reduction of 0.4 more than that shown by controls. Changes in pH which differ from changes in pH of controls by less than 0.4 or localized changes confined to areas of heavy growth are weak reactions. The other tests listed in Table 1

TABLE 1. Concentrations of substrates, indicators, and reagents employed in tests for biochemical reactions[a]

Test	Substrate	Indicator or reagent	Stock solution (%, wt/vol)	Final concn (%, wt/vol)	Amt added per 25 ml of medium
Oxidase activity		*N,N*-Dimethyl-*p*-phenylenediamine hydrochloride	1.0		
Catalase activity		Hydrogen peroxide	30		
Phosphatase activity	Phenolphthalein diphosphate Na salt		1.0	0.01	0.25 ml
		NaOH	20		
Oxidation-fermentation	Glucose		10	1.0	2.5 ml
		Phenol red[b]	0.25	0.0025	0.25 ml
Phenylalanine	DL-Phenylalanine			0.2	0.05 g
		Ferric chloride	10		
Esculin hydrolysis	Esculin			0.1	0.025 g
		Ferric citrate		0.05	0.0125 g
Arginine hydrolysis	L-Arginine hydrochloride		10	1.0	2.5 ml
		Phenol red	0.25	0.0025	0.25 ml
Urea hydrolysis	Urea		20	2.0	2.5 ml
		Phenol red	0.25	0.0025	0.25 ml
Nitrate reduction	Potassium nitrate		1.0	0.1	2.5 ml
		Sulfanilic acid[c]	0.8		
		α-Naphthylamine[c]	0.5		
Tetrazolium reduction	2,3,5-Triphenyltetrazolium chloride		1.0	0.005	0.125 ml
Tellurite reduction	Potassium tellurite		1.0	0.005	0.125 ml
Acetylmethylcarbinol production	Glucose		1.0	0.5	1.25 ml
		Creatine	1.0		
		Potassium hydroxide	40		
Carbohydrate breakdown	Inulin and salicin		5	0.5	2.5 ml
	All other carbohydrates		10	1.0	2.5 ml
		Phenol red	0.25	0.0025	0.25 ml

[a] Reproduced from Cohen et al. (6) with the permission of the American Society for Microbiology.
[b] Prepared by dissolving 0.25 g of phenol red in 70.5 ml of 0.1 N NaOH and then bringing the volume to 100 ml with water.
[c] In 5 N acetic acid.

are performed and evaluated as described by Aluotto et al. (1) and Cohen et al. (6).

Comprehensive studies of the biochemical activities of *S. moniliformis* are few (6, 27, 28, 31, 35). The following reactions are negative: gelatin liquefaction; casein or serum digestion; indole and acetyl methyl carbinol production; phenylalanine deamination; urea hydrolysis; nitrate reduction; oxidase, catalase, and benzidine reactions; and gluconate oxidation. Esculin may be hydrolyzed slightly, and H_2S may be produced in slight amounts. Alkaline phosphatase is produced, methylene blue is reduced anaerobically, and potassium tellurite and 2,3,5-triphenyltetrazolium chloride are reduced aerobically and anaerobically. Phosphatase and tetrazolium reduction tests may be weakly positive.

The following carbohydrates and polyols are not attacked; adonitol, cellobiose, dulcitol, glycerol, inositol, inulin, mannitol, melezitose, melibiose, raffinose, rhamnose, sorbitol, sorbose, and trehalose (6, 27, 28, 35). There may be variable acid production from arabinose, lactose, salicin, sucrose, and xylose. Acid without gas is produced from dextrin, fructose, galactose, glucose, glycogen, maltose, mannose, and starch. Galactose, maltose, and mannose fermentations may be weak.

Glucose is catabolized fermentatively. The most reliable of the tests (i.e., reproducible under the test conditions) are those for oxidase, catalase, phosphatase, glucose oxidation or fermentation, phenylalanine deamination, esculin hydrolysis, carbohydrate utilization, and nitrate, tetrazolium, and tellurite reduction (1, 6). Variable results are sometimes obtained in the arginine, urea, and acetyl methyl carbinol tests because of probably marginal sensitivities of the demonstrating reactions (6). The characteristics of *S. moniliformis* are summarized by Wittler and Cary (35), Rogosa (27–29), and Savage (31).

The moles percent of guanine plus cytosine in the DNA of an L-phase variant was a remarkably low 23.9% as determined by thermal denaturation (33).

Serological tests. Agglutinins in patient serum may be demonstrated in infections with *S. moniliformis*. A titer of 1:80 is regarded as diagnostic, and titers of 1:5,120 have been reported (3, 26). Two or more serum specimens at intervals of 5 days should be tested, especially if the titer of the initial sample is 1:80. The antigen of the bacillary-phase culture may be difficult to prepare because of clumping of cells. However, if glycerol (final concentration, 3%) is added to broth media before inoculation, this tendency to agglutinate spontaneously may be overcome (26). Cultures are killed with Formalin (1.2 ml per 10-ml tube), centrifuged, washed three times with 0.85% NaCl, and suspended in saline to an optical density of approximately 0.5. Perform standard agglutination procedures. Incubate the tubes at 56°C overnight; then refrigerate the tubes for 3 h, and read the results as usual. In addition to the usual controls, a normal human and known agglutinating rabbit antiserum should be included for security of diagnosis. Strain ATCC 14647 may be used in the preparation of agglutinogen in the agglutination test and for antigen in the preparation of rabbit antiserum. Consistently low titers of approximately 1:80 are not diagnostic of a recent infection because titers in this range may persist for at least 2 years (3).

The L-phase variant shares a common antigen with the bacillus but lacks a second antigen present in the bacillary form (18). L-phase antisera protect against challenge with homologous L-phase antigen but do not protect against the bacillus (9).

Wasserman tests in *S. moniliformis* infections have been reported positive. However, these results do not now appear to be correct. In the past, this disease may have been clinically confused with Sodoku, in which about 50% of sera have given positive Wasserman reactions (13).

SPIRILLUM MINUS

Characteristics

Although Roughgarden (30) cites two claims that *S. minus* has been cultivated in laboratory media, these claims have never been confirmed, and it seems safe to say that *S. minus* has not been cultivated in vitro (19, 20). Considering the morphology, pathogenicity, and sources of *S. minus*, Krieg (20) suggests that "serious attention should be given to the possibility that the organism might belong to, or be related to, the genus *Campylobacter*, and the microaerophilic techniques employed for campylobacters might also prove useful for *S. minus*."

The cells are gram negative. They appear short and thick and may vary in dimensions but are generally 0.5 μm wide and 1.7 to 5 μm long; two to six spirals and bipolar tufts of flagella are present with active motility.

Clinical significance

Much of the earlier literature has referred to the disease as Soduku (sic). Sodoku more nearly reproduces Japanese phonetics. The organism has been named *Spirillum minor* (5, 28, 35) or *Spirillum minus* (27). Krieg (20) states that the specific epithet *minor* is grammatically incorrect and the correct form is *minus*. The organism can be transferred to mice, rats, guinea pigs, and monkeys. Usually, infection follows the bite of a rat, mouse, or rodent-ingesting animal (2, 5, 13–15, 32). The incubation period is about 2 weeks. The major clinical features differentiating Sodoku from *S. moniliformis* infections are (i) a recrudescence of the initial wound with inflammation, induration, and an occasional chancre-like ulceration; (ii) associated lymphangitis and regional lymphadenitis; (iii) a different rash appearing as a maculopapular, erythematous, or dark-purple eruption spreading from the initial lesion; (iv) occasional palpable liver; and (v) failure to cultivate the etiological agent (2). The remaining major clinical signs are indistinguishable in *S. minus* and *S. moniliformis* infections (2). Mortality appears to be 6 to 10% (14, 30). The disease has been successfully treated with streptomycin (14, 15, 30) and also with penicillin (14, 30).

Kiefer et al. (16) describe a case occurring in Germany with recurrent fever episodes, lymphadenopathy, liver and spleen enlargement, and severe headache associated with an abnormal brain scan revealing diffuse cerebral ischemias with meningoencephalic involvement. After a 10-day course of large penicillin doses (3×10^7 U per day), recovery was eventually achieved. Mittermayer et al. (22) report a case occurring in Czechoslovakia in a laboratory worker bitten on a finger by a white laboratory rat. *S. minus* was detected in the peritoneal content of guinea pigs infected with blood from the diseased man. Penicillin treatment was highly effective.

Identification

Microscopic tests. In humans, microscopic demonstration of *S. minus* should be attempted in exudates from the initial lesion, in adjacent lymph nodes, in cutaneous eruptions, or in blood. Wet mounts should be examined by dark-field or phase-contrast microscopy. Care should be taken not to confuse numerous fibrils and stroma with *S. minus*. Blood films should also be stained with Giemsa or Wright stain.

Cultural studies are sufficient to make a laboratory diagnosis of *S. moniliformis* infections, and animal inoculations are not necessary, but in suspected Sodoku, there is no recourse other than animal studies when initial microscopy fails to demonstrate *S. minus*.

Before laboratory animals are inoculated, their blood should be examined for naturally occurring organisms resembling *S. minus*. Such organisms are sometimes present, and animals harboring them should not be employed. Four mice should be injected intraperitoneally with 1 ml of patient blood, and one guinea pig should be injected with 2 ml of blood by the same route. In mice, the organism is most abundant in blood within 2 to 3 weeks. Wet mounts and smears of mouse peritoneal fluid should be examined weekly for 4 weeks. Guinea pig blood should be defibrinated and lightly centrifuged, and the fibrin-free supernatant blood should be examined in wet mounts by dark-field or phase-contrast microscopy. Slide films should also be stained and examined by usual light microscopy. *S. minus* is usually most numerous in infected guinea pig blood 1 to 3 weeks after injection. As controls, an equivalent number of animals should be injected with a patient's blood specimen previously heated to 52°C for 1 h and examined as already described.

Often, when organisms are scarce or not demonstrable in the peritoneal fluid or blood of inoculated animals, impression films of the heart muscle may reveal numerous cells (5 to 50 per microscopic field). The heart of a dead or sacrificed animal is sliced in half, and a clean glass microscope slide is passed with pressure across the cut surface. Dry, fix, and flood the slide with a dilute Giemsa stain. This stain consists of 1 drop of Giemsa spirochete stain (Hynson, Westcott, and Dunning, Inc., Baltimore, Md.) added to 1 ml of distilled water. After 20 min, the slide is drained and rinsed with acetone for about 15 s (34). In these cases, heart tissue sections stained by silver impregnation methods also reveal the organism. Impression smears of crushed heart tissue preparations may also be stained with Thedan blue solution T-5 (Allied Chemical Co., New York, N.Y.). The saponin present in this stain destroys erythrocytes and many blood elements, and *S. minus* is stained within 3 min (13).

Serological tests. Live organisms in a fresh preparation have been immobilized by immune serum. However, this test has been questioned (13) and is not routinely used. An active lytic principle is present in immune sera from rabbits and monkeys. An infectious suspension of mouse heart is neutralized within 3 h at 35°C and protects inoculated mice from infection. Human sera have not been evaluated in this mouse protection test (13). Complement fixation tests for syphilis using patients' sera have been reported positive (2, 13, 14).

Sera from rabbits infected with *S. minus* have given positive Weil-Felix reactions with *Proteus* OX strains (13).

LITERATURE CITED

1. **Aluotto, B. B., R. G. Wittler, C. O. Williams, and J. E. Faber.** 1970. Standardized bacteriologic techniques for the characterization of *Mycoplasma* species. Int. J. Syst. Bacteriol. **20:**35–58.
2. **Anderson, W. A. D.** 1961. Pathology, 4th ed., p. 285–286. The C. V. Mosby Co., St. Louis, Mo.
3. **Brown, T. M., and J. C. Nunemaker.** 1942. Rat-bite fever: a review of the American cases with reevaluation of etiology: report of cases. Bull. Johns Hopkins Hosp. **70:**201–328.
4. **Buchanan, R. E., J. G. Holt, and E. F. Lessel, Jr.** 1966. Index Bergeyana. An annotated alphabetic listing of names of the taxa of bacteria, p. 1035, 1068–1069. The Williams & Wilkins Co., Baltimore.
5. **Carter, H. V.** 1888. Note on the occurrence of the minute blood-spirillum in an Indian rat, p. 45–48. Sci. Mem. Med. Offrs. Army India Part 3.
6. **Cohen, R. L., R. G. Wittler, and J. E. Faber.** 1968. Modified biochemical tests for characterization of L-phase variants of bacteria. Appl. Microbiol. **16:**1655–1662.
7. **Collins, C. H.** 1983. Laboratory-acquired infections. Butterworths, London.
8. **Faro, S., C. Walker, and R. L. Pierson.** 1980. Amnionitis with intact amniotic membranes involving *Streptobacillus moniliformis*. Obstet. Gynecol. **55**(Suppl. 3):9S–11S.
9. **Freundt, E. A.** 1956. Experimental investigations into the pathogenicity of the L-phase variant of *Streptobacillus moniliformis*. Acta Pathol. Microbiol. Scand. **38:**246–258.
10. **Gourlay, R. N., B. F. Flanagan, and S. G. Wyld.** 1982. *Streptobacillus actinoides (Bacillus actinoides)*: isolation from pneumonic lungs of calves and pathogenicity studies in gnotobiotic calves. Res. Vet. Sci. **32:**27–34.
11. **Heilman, F. R.** 1941. A study of *Asterococcus muris (Streptobacillus moniliformis)*. I. Morphologic aspects and nomenclature. J. Infect. Dis. **69:**32–44.
12. **Hopkinson, W. I., and J. M. Lloyd.** 1981. *Streptobacillus moniliformis* septicaemia in Spinifex hopping mice (*Notomys alexis*). Aust. Vet. J. **57:**533–534.
13. **Jellison, W. L.** 1963. Soduku, p. 640–641. *In* A. H. Harris and M. B. Coleman (ed.), Diagnostic procedures and reagents. American Public Health Association, Inc., New York.
14. **Jellison, W. L.** 1963. Rat-bite fever (Soduku and Haverhill fever), p. 652–667. *In* T. G. Hull (ed.), Diseases transmitted from animals to man, 5th ed. Charles C Thomas, Publisher, Springfield, Ill.
15. **Jellison, W. L., P. L. Eneboe, R. R. Parker, and L. E. Hughes.** 1949. Rat-bite fever in Montana. Public Health Rep. **64:**1661–1665.
16. **Kiefer, H., W. Froscher, and H. P. Mohr.** 1981. Rattenbisskrankheit mit meningoenzephaler Beteiligung. Med. Klin. **76:**653–655.
17. **Klieneberger, E.** 1935. The natural occurrence of pleuropneumonia-like organisms in apparent symbiosis with *Streptobacillus moniliformis* and other bacteria. J. Pathol. Bacteriol. **40:**93–105.
18. **Klieneberger, E.** 1942. Some new observations bearing on the nature of the pleuropneumonia-like organism known as L_1 associated with *Streptobacillus moniliformis*. J. Hyg. **42:**485–497.
19. **Krieg, N. R.** 1974. Genus I. *Spirillum* Ehrenberg 1832, 38, p. 196–207. *In* R. E. Buchanan and N. E. Gibbons (ed.), Bergey's manual of determinative bacteriology, 8th ed. The Williams & Wilkins Co., Baltimore.
20. **Krieg, N. R.** 1984. Genus *Aquaspirillum* Hylemon, Wells, Krieg and Jannasch, 1973, 361[AL], p. 72–90. *In* N. R. Krieg and J. G. Holt (ed.), Bergey's manual of systematic bacteriology, vol. 1. The Williams & Wilkins Co., Baltimore.
21. **Levaditi, C., S. Nicolau, and P. Poincloux.** 1925. Sur le role etiologique de *Streptobacillus moniliformis* (nov. spec.) dans l'erytheme polymorphe aigu septicemique. C. R. Acad. Sci. **180:**1188–1190.
22. **Mittermayer, T., Z. Koppel, and V. Fazekas.** 1975. Pripad Sodoku v Kosiach (A case of Sodoku in Kosice). Bratisl. Lek. Listy **64:**475–479. (In Czech; summary in English and German.)
23. **Portnoy, B. L., T. K. Satterwhite, and J. D. Dyckman.** 1979. Rat bite fever misdiagnosed as Rocky Mountain spotted fever. South. Med. J. **72:**607–609.
24. **Raffin, B. J., and M. Freemark.** 1979. Streptobacillary rat-bite fever: a pediatric problem. Pediatrics **64:**214–217.
25. **Renaut, J. J., C. Pecquet, C. Verlingue, H. Barriere, M. Deriemnic, and A. L. Courtieu.** 1982. Septicemie a *Streptobacillus moniliformis*. Nouv. Presse Med. **11:**1143.
26. **Robinson, L. B.** 1963. *Streptobacillus moniliformis* infections, p. 642–651. *In* A. H. Harris and M. B. Coleman (ed.), Diagnostic procedures and reagents, 4th ed. American Public Health Association, Inc., New York.
27. **Rogosa, M.** 1970. *Streptobacillus moniliformis* and *Spirillum minus*, the causative agents of two distinct rat-bite fevers, p. 226–231. *In* J. E. Blair, E. H. Lennette, and J. P. Truant (ed.), Manual of clinical microbiology, 1st ed. American Society for Microbiology, Bethesda, Md.
28. **Rogosa, M.** 1974. *Streptobacillus moniliformis* and *Spirillum minor*, p. 326–332. *In* E. H. Lennette, E. H. Spaulding, and J. P. Truant (ed.), Manual of clinical microbiology, 2nd ed. American Society for Microbiology, Washington, D.C.
29. **Rogosa, M.** 1980. *Streptobacillus moniliformis* and *Spirillum minor*, p. 350–356. *In* E. H. Lennette, A. Balows, W. J. Hausler, Jr., and J. P. Truant (ed.), Manual of clinical microbiology, 3rd ed. American Society for Microbiology, Washington, D.C.
30. **Roughgarden, J. W.** 1965. Antimicrobial therapy of rat bite fever, a review. Arch. Intern. Med. **116:**39–54.
31. **Savage, N.** 1984. Genus *Streptobacillus* Levaditi, Nicolau and Poincloux 1925, 1188[AL], p. 598–600. *In* N. R. Krieg and J. G. Holt (ed.), Bergey's manual of systematic bacteriology, vol. 1. The Williams & Wilkins Co., Baltimore.
32. **Schottmüller, H.** 1914. Zur Atiologie und Klinik der Bisskrankheit (Rattin-, Katzen-, Eichornchen-Bisskrankheit). Dermatol. Wochenschr. **58**(Suppl.):77–103.
33. **Williams, C. O., R. G. Wittler, and C. Burris.** 1969. Deoxyribonucleic acid base compositions of selected mycoplasmas and L-phase variants. J. Bacteriol. **99:**341–343.
34. **Wilson, G. S., and A. A. Miles.** 1964. Topley and Wilson's principles of bacteriology and immunity, 5th ed., p. 518–520. The Williams & Wilkins Co., Baltimore.

35. **Wittler, R. G., and S. G. Cary.** 1974. Genus *Streptobacillus* Levaditi, Nicolau and Poincloux 1925, 1188, p. 378–381. *In* R. E. Buchanan and N. E. Gibbons (ed.), Bergey's manual of determinative bacteriology, 8th ed. The Williams & Wilkins Co., Baltimore.

36. **Wittler, R. G., W. F. Malizia, P. E. Kramer, J. D. Tucket, H. N. Pritchard, and H. J. Baker.** 1960. Isolation of a corynebacterium and its transitional forms from a case of subacute bacterial endocarditis treated with antibiotics. J. Gen. Microbiol. **23**:315–333.

Chapter 36

Mycoplasmas

GEORGE E. KENNY

CHARACTERIZATION

Mycoplasmas (mycoplasmata) are small (0.2- to 0.3-μm) pleomorphic organisms bounded only by a cell membrane with no evidence of a cell wall (13); hence, they are insensitive to penicillin. The morphology is variable: cells appear as coccoid bodies and as filamentous and star-shaped forms. Cells do not Gram stain but can be stained (poorly) with Giemsa stain. Cells are best visualized in broth cultures by dark-field or phase-contrast microscopy, but the pleomorphic nature of the organisms makes difficult the distinction of organisms from artifacts in the medium. Consequently, organisms are recognized by their formation on solid medium of typical, small, "fried-egg" colonies (Fig. 1) which are 10 to 300 μm in diameter and visible only with magnification. The demonstration of typical colonies is the sine qua non for the identification of *Mycoplasma* spp.

GROWTH

Because the organisms have a small genome (5×10^8 daltons for *Mycoplasma* species and *Ureaplasma urealyticum* and 10^9 daltons for *Acholeplasma laidlawii*), they are highly fastidious and require enriched medium ordinarily containing a peptone, yeast extract, and animal serum (10 to 20%). Unusual growth requirements include cholesterol (supplied by the serum supplement), preformed nucleic acid precursors (supplied by the yeast extract), and urea for *U. urealyticum*. Growth rates vary, ranging from as little as 1 h to as much as 6 h for *M. pneumoniae*. The yield of organisms from broth culture is small (1 to 20 mg of protein per liter), and cultures ordinarily show at most a faint haze which can best be recognized in comparison with uninoculated medium. Groups of species can be distinguished by their ability to ferment glucose, utilize arginine, or hydrolyze urea (Table 1).

TAXONOMY

The organisms associated with humans include *U. urealyticum*, *A. laidlawii*, and eight species in the genus *Mycoplasma* (Table 1). The lack of a cholesterol requirement distinguishes *A. laidlawii*, and the ability to hydrolyze urea is specific to the genus *Ureaplasma*. Since the heterogeneity observed in the genus *Mycoplasma* (7, 13) far exceeds that of a conventional genus, generalizations concerning the characteristics of a given species should be extrapolated with caution to other species. The human *Mycoplasma* species belong to three different serological clusters of related species (7). The arginine-utilizing, nonglycolytic organisms represent the major cluster of serologically related organisms. *M. pneumoniae* is distinct from the other organisms both biologically and serologically. *M. fermentans* also is unrelated to the other species. *U. urealyticum* is unique not only among mycoplasmas but also among bacteria because it requires urea even in highly complex media (8). Individual species are identified by the inhibition of growth on agar by disks impregnated with rabbit antiserum (1, 2). *Mycoplasma* species have no known relationship to bacteria, but the organisms can be confused with cell-wall-deficient organisms.

NOMENCLATURE

Although the correct trivial name for the organisms in genus *Mycoplasma* is mycoplasmata, the term mycoplasmas, the English common name, is used more commonly. The term mycoplasmas is also used to generally include all membrane-bounded procaryotes, since no more general term has yet been accepted. The use of the term mycoplasma (a singular noun) as a plural noun or an adjective is grammatically incorrect (the terms mycoplasmal and mycoplasmic are correct adjectival forms).

CLINICAL SIGNIFICANCE

M. pneumoniae is a major cause of primary atypical pneumonia and accounts for as much as 10% of total X-ray-proven pneumonia. Four additional species are found in the oral cavity. *M. salivarium* is a common inhabitant of the gingival crevice and is not observed in edentulous persons. It is present in higher concentrations in persons with periodontal disease (4) but has no definite role in this complex disease. *M. orale*, *M. faucium*, and *M. buccale* appear to be normal oral flora. Thus, for respiratory specimens, *M. pneumoniae* is the only organism sought. *M. pneumoniae* is not a normal flora, but it can persist in the respiratory tract for 1 to several months after infection. Subclinical infections are common; the disease is ordinarily mild and is frequently termed "walking pneumonia." In contrast to most other respiratory infections, *M. pneumoniae* infections are not seasonal and may be found as frequently in summer as in winter. In large populations, the disease is endemic year-around, with periodic increases in incidence, but in small populations the outbreaks may appear as local epidemics. The disease spreads slowly in families, and transmission apparently requires close contact.

In the genital tract the situation is more complex. At least 60% of apparently healthy women carry *U. urealyticum*, and some 20% carry *M. hominis* in the vagina (19). The incidence is less in sexually circumspect populations. However, both organisms appear to be effective opportunists when present in internal sites. Both organisms have been recovered from the bloodstream of women with mild postpartum fever (12). *M. hominis* and possibly *U. urealyticum* may have a role in pelvic inflammatory disease. If *U. urealyticum* has a role in infertility, this role is probably small and likely depends upon the infection of internal sites. In men, the situation is more complex; *U. urealyticum* may have a minor role in nonspecific urethritis, but it is not nearly as important as chlamydiae. However, *U. urealyticum* may have a role in prostatitis. In both

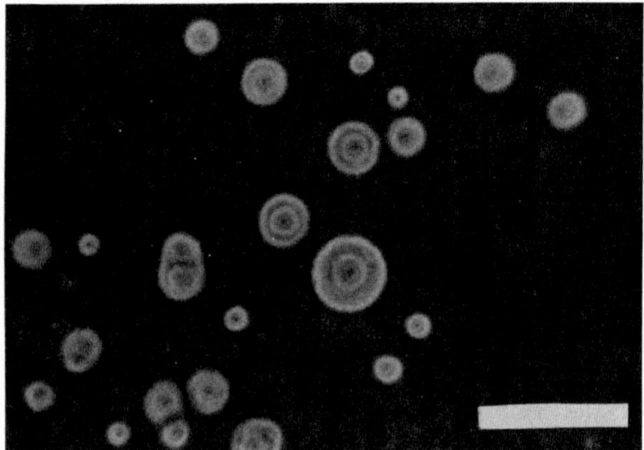

FIG. 1. Colonies of *M. pneumoniae* AP-164 on H agar (10 days of incubation). Bar, 100 μm.

men and women, evidence for a role in upper urinary tract disease is increasing because the organisms can be isolated from the upper urinary tract in the absence of other microbes. It also appears that *U. urealyticum* can infect the lungs of infants (19), but the organism is also found as a passenger in infants. Both *M. hominis* and *U. urealyticum* are occasionally isolated from the throats of adults. *M. fermentans* is rarely isolated. A new organism from the genital tract is *M. genitalium* (20). This organism resembles *M. pneumoniae* in that it grows slowly and is aerobic. It also shows serological cross-reactions with *M. pneumoniae*. The organism is most difficult to isolate, and only a few isolates have been made thus far.

SPECIAL CONSIDERATIONS FOR CULTURE OF MYCOPLASMAS

The unusual features of mycoplasmas and their heterogeneity indicate the necessity for a more detailed consideration of their cultural properties. Since mycoplasmas grow into the agar with only slight surface growth, transfer of the organisms with a loop is ineffective. To transfer a colony to a broth culture, simply excise the agar around a colony and place it in broth. For transfer to agar, the piece of agar is emulsified in broth and transferred to a plate. To observe agar plates, a stereoscopic microscope with oblique lighting is employed at ×20 to ×60 magnification because this equipment provides the best working distance and resolution for observing the colonies on the agar surface through the bottom of the plate without opening the plate. This method of observation is rapid and avoids contamination of the agar surface not only with bacteria but also with mycoplasmas during the repeated observations necessary for the detection of growth. Except for the specialized stains for *U. urealyticum*, colonies of individual species cannot be definitely distinguished. A major concern is the isolation of cell-wall-deficient bacteria which resemble mycoplasmas. These organisms ordinarily form large fried-egg colonies which transfer poorly and usually revert to bacteria on penicillin-free medium. Fortunately, these organisms are rarely isolated.

Because of the heterogeneity of members of the order *Mycoplasmatales*, no one medium will suffice to isolate all organisms. Media with different pHs are required (Table 1). *U. urealyticum* requires media with a pH of 6.0 for optimum growth, with little growth above pH 7.0, whereas most other species will grow on media with a pH of 7.0. The most pH-tolerant organism is *M. hominis*, which grows on media with pHs from 5.5 to 8.0, whereas *M. pneumoniae* grows poorly on media with pHs below 6.5 or above 8.0. Atmosphere is also important. Both *M. pneumoniae* and *M. genitalium* are obligate aerobes, and both appear to be enhanced by the presence of CO_2 in the atmosphere (Table 1). *M. salivarium, M. orale, M. faucium*, and *M. buccale* grow best anaerobically and grow poorly on H agar aerobically. *M. hominis, U. urealyticum*, and possibly *M. fermentans* appear indifferent to atmosphere. The requirement for CO_2 is much less clear because the presence of CO_2 in the atmosphere also affects the pH of the medium. This is a particular problem because the media always contain serum which contains HCO_3^-. The difference in the final pH of agar medium containing 20% serum may be as much as 1 pH unit lower when the pH is measured in 2.5% CO_2 rather than in air. Thus, aerobic organisms such as *M. pneumoniae* may appear to grow better in microaerophilic environments (95% N_2, 5% CO_2) than in air because of the more favorable neutral pH achieved in such atmospheres. Therefore, it is important to measure the final pH of agar media with a surface electrode after incubation in an appropriate atmosphere or to include phenol red in all media as

TABLE 1. Properties of human mycoplasmas and ureaplasmas

Species	Serovar	Metabolic marker			Optimal pH	Growth in atmosphere			Time for appearance of colonies
		Arginine	Urea	Glucose		Aerobic	Anaerobic[a]	Candle jar	
Genital organisms									
Ureaplasma urealyticum	14	−	+	−	5.5–6.5	++	++	++	1–4 days
Mycoplasma hominis	?	+	−	−	5.5–8.0	++	++	++	1–4 days
M. fermentans	1	+	−	+	~7.3[c]	+	++	?	3–20 days
M. genitalium	?	−	−	+	~7.3[c]	?	−	++	Slow
Respiratory organisms									
M. pneumoniae	1	−	−	+	6.5–7.5	++	−	++	3–20 days
M. salivarium	1	+	−	−	6.0–7.0	+	++	?	2–5 days
M. orale[b]	1	+	−	−	~7.0[c]	+	++	?	3–10 days
Acholeplasma laidlawii	1	−	−	+	6.0–8.0	++	++	++	1–5 days

[a] Hydrogen plus carbon dioxide with catalyst (Gaspak).
[b] Properties of *M. faucium* and *M. buccale* are similar.
[c] Limited data available.

recommended in Chapter 111 so that pH may be monitored visually.

In addition to pH and atmosphere, medium components are most important to the isolation of mycoplasmas. Although a significant literature now exists on the comparison of one medium formulation with another for isolation, one of the most important factors is the variability of the peptones, yeast extracts, and sera used for isolation, in addition to the final pH problems mentioned above. Peptones are biological products with limited standardization. The soy peptone used in the media described in Chapter 111 is strikingly variable. Certain lot numbers may yield few colonies when inoculated with specimens of *M. pneumoniae*, even though these lots will provide for good growth of prototypic strains. All other medium formulations contain various peptones and digests, all of which show equal variability. Fresh yeast extract and horse sera appear to show less variability. Horse serum or other sera used for culture should be obtained from vendors who have a testing program for mycoplasmas, since sera occasionally contain these organisms. "Agamma" sera are less effective than whole sera; certain commercial lots are essentially dialyzed and thus contain no bicarbonate. These lots may yield a more favorable medium pH for *M. pneumoniae* but are otherwise deficient for the isolation of *M. pneumoniae*. Overall, the preparation of medium requires careful controls. It is imperative that laboratories have available wild strains of all species to be tested for, preferably either specimen materials or isolates in the earliest possible transfer. Reference cultures grow readily on media which will not provide for isolation of the organism from the patient.

A number of different medium systems are widely employed (3, 14, 16, 18, 20), but I have had good experience in the isolation of both genital and respiratory mycoplasmas with the media described in Chapter 111. No doubt media will be improved and standardized further in the future.

Artifacts are a major problem in the detection of mycoplasmas in clinical specimens. Cells and debris resemble the fried-egg colonies; thus it is imperative that the laboratory demonstrate that the agent is a transferable entity with specific properties. Certain serum lots induce the formation of pseudocolonies, which are whorls of lipid crystals which appear on the surface of agar plates. These crystals transfer, but they will not produce metabolic products. Large amounts of cellular materials in inocula may produce acid from glucose by normal mammalian enzymes. Excessive moisture on the surfaces of agar plates can render colonies invisible because the organisms replicate in the surface film of fluid and form numerous tiny colonies which are too small to be seen. This problem can be avoided by pouring plates 24 h before use and holding them at room temperature overnight to eliminate excessive surface moisture. Plates should be wrapped in plastic to prevent desiccation and stored at −4°C for no more than 2 weeks before use.

Penicillin is ordinarily employed in medium to inhibit bacterial overgrowth of the slowly growing mycoplasmas. Thallium acetate is also used to inhibit bacterial growth but should be avoided in isolation attempts for *U. urealyticum* and *M. genitalium* (thallium acetate is toxic towards humans, and care should be taken in its use). Amphotericin (1:2,000) is useful for the inhibition of yeast cells.

COLLECTION, TRANSPORT, AND STORAGE OF SPECIMENS

Specimens collected on swabs should be placed in transport medium; 2 ml of Trypticase soy broth (BBL Microbiology Systems, Cockeysville, Md.) with 0.5% bovine albumin has been used satisfactorily. Mycoplasmas are remarkably susceptible to drying; thus, specimens or swabs should be placed in transport medium or tested promptly. Penicillin is usually incorporated in transport medium to suppress bacterial overgrowth. Tissues and sputum specimens are submitted directly to the laboratory. If cultures cannot be inoculated immediately, the transport medium should be frozen at −70°C where organisms will survive indefinitely. Storage at −20°C is effective for only several weeks and is not recommended. Specimens may be held for 24 to 48 h at 4°C with little loss in titer. Mycoplasmas are unique in that they withstand freezing and thawing very well, provided that the medium contains protein. Other transport media may be used, provided that they contain protein and have been tested.

DIRECT EXAMINATION

No methods for direct examination of mycoplasmas are yet known. Studies are in progress in various laboratories to attempt to develop direct immunofluorescence tests with monoclonal antibodies or antigen capture assays.

ISOLATION OF *M. PNEUMONIAE*

Specimens from the respiratory tract are those usually examined. A 0.1-ml amount of transport medium (in which the swab has been extracted) is inoculated both into diphasic broth and onto an H-agar plate. Sputum, body fluids, and disrupted tissue materials should be diluted 1:10 and 1:100 (with respect to their original amounts) before being used to inoculate broth and plates; this dilution reduces the amounts of inhibiting substances normally present in tissues. Agar plates are incubated aerobically at 37°C in a sealed container and examined microscopically at 2, 5, 10, 15, and 21 days for the presence of typical colonies. Diphasic cultures should be examined microscopically by viewing the broth through the side of the tube for the presence of spherules (fluid medium colonies), which may appear as early as 5 days. Diphasic cultures should also be observed for a decrease in pH as indicated by phenol red indicator. Diphasic cultures should be transferred to H agar at 21 days, and these cultures should be observed for 21 days. Most positive specimens will show typical small colonies on agar and spherule and acid production in fluid medium by 10 to 12 days, although some specimens may turn positive only after longer intervals (up to 30 days). The fact that an organism has been recovered which grows slowly on agar, ferments glucose, and forms spherules provides a presumptive diagnosis. This presumptive diagnosis is confirmed by a test for hemolysis (3) which can be done on the original isolation plate if sufficient colonies exist. The only other organisms isolated from the respiratory tract under the recommended conditions are *M. hominis*, which grows rapidly and forms large colonies, and rarely *A. laidlawii*, which will also hemolyze guinea pig erythrocytes. However, *A. laidlawii* grows rapidly and will transfer and form large colonies in several days. The hemolysis test is carried out by overlaying the colonies with a thin layer of 8% guinea pig erythrocytes in saline agar. Incubation is carried out overnight at 37°C, and a zone of hemolysis will be observed surrounding individual colonies and groups of colonies. At this point, specimens may be reported as positive for an organism with cultural characteristics resembling *M. pneumoniae*. Absolute identification is carried out by the inhibition of colonial growth with specific antiserum. Portions (0.1 ml) of serial 10-fold dilutions of the broth culture (1:1, 1:10, 1:100) are spread uniformly over H-agar plates and allowed to adsorb; then paper disks (2 to 3 mm in diameter) soaked with antiserum are placed on the plates. Incubation is carried out for 4 to 6 days, and the plates are examined for zones of inhibition of colonial growth. The test is quite sensitive to excessive numbers of organisms.

M. pneumoniae is relatively easy to isolate since patients carry the organism for 1 to 2 months after infection, and isolation is little hampered by antibiotic therapy (6). However, because of the slowness of isolation, testing for the organism has little immediate diagnostic usefulness, but testing is useful epidemiologically and in family infections (6). Although *M. orale* and *M. salivarium* are abundant in throat swabs, H agar is selective in that these organisms will rarely be recovered aerobically. Other media are less selective, and 0.002% methylene blue has been employed in both agar and broth media to inhibit these organisms; however, methylene blue also inhibits *M. pneumoniae* to some extent, so this practice is not recommended.

ISOLATION OF *M. HOMINIS* AND *U. UREALYTICUM*

Since both *M. hominis* and *U. urealyticum* are ordinarily sought in the genital tract, methods for the isolation of these two organisms will be considered together even though their properties are quite different. *U. urealyticum* poses significant problems during isolation because of its steep death phase caused apparently by the exhaustion of urea (8) as well as by an elevated pH from urease activity. In broth cultures, it is essential to buffer the medium properly and supply an optimum but not an excessive amount of urea. In agar plates, the situation is similar. In the media described in Chapter 111, *U. urealyticum* will survive for about 12 h after color change, and colonies on plates will remain viable for about 2 days after their appearance. The optimum buffer appears to be MES [2-(*N*-morpholino)ethanesulfonic acid], and molar ratios of buffer and urea for broth cultures are 2:1 to 4:1 (i.e., 10 or 20 mM MES [pH 6.3] to 5 mM urea). For agar, a ratio of 10:1 buffer to urea (i.e., 30 mM MES [pH 6.3] to 3 mM urea) is sufficient to prevent drastic increases in pH in the vicinity of the colony and subsequent rapid loss of viability. The addition of a reducing agent (i.e., 1 mM sodium sulfite) to U broth reduces the lag phase for some, but not all, isolates. *M. hominis* is the least fastidious of the potential human pathogens and grows well on H agar and in H broth. *M. hominis* strains will show a faint but distinct haze in broth culture, whereas *U. urealyticum* produces no haze at all, but produces a readily recognizable pH change on the hydrolysis of urea.

For the isolation of the organisms, 0.1-ml portions of specimen are inoculated into H broth and onto H agar, U agar, and U diphasic medium. Agar plates are incubated aerobically or in a CO_2-gassed incubator at 37°C. Anaerobic incubation gives no increase in the number of isolates of *M. hominis* or *U. urealyticum* provided that the medium has a suitable final pH. Broth cultures are incubated aerobically. Plates should be observed at 1, 2, 3, and 4 days. The U broth should be observed twice each day for color change. Broth cultures should be transferred into a fresh U broth immediately and to a U agar plate. Colonies of *U. urealyticum* can be identified by the $CaCl_2$ stain (0.1 M $CaCl_2$ and 0.1 M urea; see Chapter 111) or the single-reagent test of Shepard (16, 17). One drop of reagent is added to the agar plate with colonies. Colonies of *U. urealyticum* show a brown halo under transmitted light within 1 to 5 min, whereas *M. hominis* colonies are unaffected. Isolates which stain are considered confirmed if they prove to be transferable entities which utilize urea. A medium which contains $CaCl_2$ as an intrinsic stain in the agar has been previously described (16). This medium can directly distinguish *U. urealyticum* from *M. hominis*, but the difficulties are that colonies are smaller (because of excessive urea) and viability is lost quickly because of high pH in the vicinity of the colony. Although human *U. urealyticum* can be divided into as many as 14 serotypes (15), antisera are not available. *M. hominis* will produce large colonies on both H and U agars (unless lincomycin is included at 20 to 30 μg/ml to select for *U. urealyticum*; erythromycin can be used to select for *M. hominis*). H-broth cultures should be transferred to H agar at 2 and 4 days (broth cultures remain viable for about 48 h after growth begins). Nearly all large-colony isolates are *M. hominis*, but occasionally bacteria are recovered which produce colonies which resemble but are not identical to those of *M. hominis*. These ordinarily grow in bacteriological media without serum. Large-colony forms can be tentatively identified as *M. hominis* if they are transferable entities, utilize arginine (H broth supplemented with 50 mM arginine), and produce typical large colonies. Their identity can be verified by growth inhibition in a manner similar to that for *M. pneumoniae*.

Tissues should be ground in transport medium and diluted 1:10 and 1:100 with respect to the original tissue and inoculated into media as above. Since both organisms have been recovered from the blood (12, 19), requests for isolation from that source should increase. *M. hominis* is quite easy to isolate, but *U. urealyticum* is more difficult. Blood should be collected without coagulant and immediately inoculated into broth. For *M. hominis*, H broth in a volume 10 times the blood volume should be sufficient (i.e., 5 ml of blood into 50 to 100 ml of medium), and the organism has been recovered in conventional bacterial blood cultures. Cultures should be subcultured daily for 5 days. For *U. urealyticum*, the situation is less clear. We have used a rather arbitrary medium which recovers both *M. hominis* and *U. urealyticum* apparently effectively (12).

M. fermentans is rarely isolated. It is not known if the organism is rare or if present media are inadequate. From current information, the recommended media and incubation times will not permit the detection of *M. genitalium*, which has been isolated only on SP-4 medium (20). Present information indicates that the organism appears to require CO_2 and a pH around 7.3 and is inhibited by thallium acetate.

SIGNIFICANCE OF ISOLATIONS

The isolation of *M. pneumoniae* always provides significant clinical information because the organism is not normal flora. For the genital organisms, the situation is different, because the organisms colonize the vagina and the urethra without symptoms. Isolations from internal sources may be more important, provided that other microorganisms are sought also. Isolation from internal sources has little significance unless precautions are taken to avoid vaginal or urethral contamination.

SERODIAGNOSIS

For the diagnosis of *M. pneumoniae* infections, the complement fixation test with either whole organisms or lipid antigen (10) has proven effective for the detection of infections in pneumonia patients. However, the fact that the lipid antigen of the organism is a mixture of relatively simple glycolipids which appear broadly distributed in nature (being found even in plants) (7) suggested that false positives could be generated to this antigen. This has turned out to be true, as persons with proven bacterial meningitis frequently show significant antibody increases to *M. pneumoniae* antigens which are clearly false-positive (11). For this reason, enzyme-linked immunosorbent assays (ELISA) have been sought for *M. pneumoniae*. The tests so far developed are relatively insensitive and appear to have a specificity little different from that of the lipid antigen. These tests do have the advantage of permitting the detection of immunoglobulins G and M. However, it is clear that it is not possible to use solubilized whole organisms or unfractionated extracts because the active components will be represented in such small quantities in the antigenic mixture as to be indetectable by ELISA (9). Clearly, more purified and more generally commercially available antigens will be required for ELISA to be more than a tool of research laboratories. The only test that might be generally usable for the detection of infections with *U. urealyticum* is the immunofluorescence test (5), which may also be useful for the detection of infections by other mycoplasmas. Developments with ELISA look promising for both *U. urealyticum* and *M. hominis*, but commercially available antigens are not yet available.

LITERATURE CITED

1. **Clyde, W. A., Jr.** 1964. Mycoplasma species identification based upon growth inhibition by specific antisera. J. Immunol. **92:**958–965.
2. **Clyde, W. A., Jr.** 1983. Serological identification of mycoplasmas from humans, p. 37–45. *In* J. G. Tully and S. Razin (ed.), Methods in mycoplasmology, vol. 2. Academic Press, Inc., New York.
3. **Clyde, W. A., Jr.** 1983. Recovery of mycoplasmas from the respiratory tract, p. 9–17. *In* J. G. Tully and S. Razin (ed.), Methods in mycoplasmology, vol. 2. Academic Press, Inc., New York.
4. **Engel, L. D., and G. E. Kenny.** 1970. *Mycoplasma salivarium* in human gingival sulci. J. Peridontal Res. **5:**163–171.
5. **Gallo, D., K. W. Dupuis, N. J. Schmidt, and G. E. Kenny.** 1983. Broadly reactive immunofluorescence test for measurement of immunoglobulin M and G antibodies to *Ureaplasma urealyticum* in infant and adult sera. J. Clin. Microbiol. **17:**614–618.
6. **Grayston, J. T., H. M. Foy, and G. E. Kenny.** 1969. The epidemiology of mycoplasma infections of the human respiratory tract, p. 651–682. *In* L. Hayflick (ed.), The Mycoplasmatales and L-phase of bacteria. Appleton-Century-Crofts, New York.
7. **Kenny, G. E.** 1979. Antigenic determinants, p. 351–384. *In* M. F. Barile and S. Razin (ed.), The mycoplasmas, vol. 1. Academic Press, Inc., New York.
8. **Kenny, G. E., and F. D. Cartwright.** 1977. Effect of urea concentration on growth of *Ureaplasma urealyticum* (T-strain mycoplasma). J. Bacteriol. **132:**144–150.
9. **Kenny, G. E., and C. L. Dunsmoor.** 1983. Principles, problems, and strategies in the use of antigenic mixtures for the enzyme-linked immunosorbent assay. J. Clin. Microbiol. **17:**655–665.
10. **Kenny, G. E., and J. T. Grayston.** 1965. Eaton PPLO (*Mycoplasma pneumoniae*) complement-fixing antigen: extraction with organic solvents. J. Immunol. **95:**19–25.
11. **Kleemola, M., and H. Kayhty.** 1983. Increase in titers of antibodies to *Mycoplasma pneumoniae* in patients with purulent meningitis. J. Infect. Dis. **146:**284–288.
12. **Lamey, J. R., D. A. Eschenbach, S. H. Mitchell, J. M. Blumhagen, H. M. Foy, and G. E. Kenny.** 1982. Isolation of mycoplasmas and bacteria from the blood of postpartum women. Am. J. Obstet. Gynecol. **143:**104–112.
13. **Razin, S., and E. A. Freundt.** 1984. The mycoplasmas, p. 740–793. *In* N. R. Krieg and J. G. Holt (ed.), Bergey's manual of systematic bacteriology, vol. 1. The Williams & Wilkins Co., Baltimore.
14. **Robertson, J. A.** 1978. Bromothymol blue broth: improved medium for detection of *Ureaplasma urealyticum* (T-strain mycoplasma). J. Clin. Microbiol. **7:**127–132.
15. **Robertson, J. A., and G. W. Stemke.** 1982. Expanded serotyping scheme for *Ureaplasma urealyticum* strains isolated from humans. J. Clin. Microbiol. **15:**873–878.
16. **Shepard, M. C.** 1983. Culture medium for ureaplasmas, p. 137–146. *In* S. Razin and J. Tully (ed.), Methods in mycoplasmology, vol. 1. Academic Press, Inc., New York.
17. **Shepard, M. C., and G. K. Masover.** 1979. Special features of ureaplasmas, p. 451–494. *In* M. F. Barile and S. Razin (ed.), The mycoplasmas, vol. 1. Academic Press, Inc., New York.
18. **Taylor-Robinson, D.** 1983. Recovery of mycoplasmas from the genitourinary tract, p. 19–26. *In* J. G. Tully and S. Razin (ed.), Methods in mycoplasmology, vol. 2. Academic Press, Inc., New York.
19. **Taylor-Robinson, D., and W. M. McCormack.** 1979. Mycoplasmas in human genitourinary infections, p. 307–366. *In* J. G. Tully and R. F. Whitcomb (ed.), The mycoplasmas, vol. 2. Academic Press, Inc., New York.
20. **Tully, J. G., R. M. Cole, D. Taylor-Robinson, and D. L. Rose.** 1981. A newly discovered mycoplasma in the human genital tract. Lancet **i:**1288–1291.

Chapter 37

Section VI. Anaerobic Bacteria

Isolation and Examination of Anaerobic Bacteria

STEPHEN D. ALLEN, JEAN A. SIDERS, AND LINDA M. MARLER

Anaerobic bacteria cause essentially every type of infection that aerobic or facultatively anaerobic bacteria cause and will be overlooked unless appropriate procedures for primary isolation and anaerobic culture are used. Infections involving anaerobes are often polymicrobial, i.e., the anaerobes are mixed with aerobes, facultative anaerobes, or other anaerobes, or there may be a single species in pure culture. It is necessary for the laboratory to isolate and identify anaerobic bacteria from properly selected and collected specimens because (i) anaerobic infections are common, (ii) these infections are associated with high morbidity and mortality, (iii) clinical clues suggestive of anaerobic infections are not specific, and (iv) the treatment of these infections varies with the species of bacteria involved and often differs from that of infections not involving anaerobes (1a, 17).

Knowledge of the species identified can aid the physician in the determination of the probable clinical significance of isolates and in the selection of antimicrobial agents for therapy. In addition, the anaerobes have shown changing patterns in their susceptibility to the three or four antimicrobial agents that were commonly used to treat anaerobic infections during the past decade. Recently, several additional antimicrobial agents were marketed that are being used to treat anaerobic infections. These compounds vary in cost, pharmacology, and toxicity, and their activities against anaerobe isolates are unpredictable (18, 39, 47). Thus, isolation, identification, and results of susceptibility testing are now required for clinically significant anaerobic bacteria, just as these procedures are required for bacteria that grow aerobically (1a, 18).

The chapters in this section deal with the laboratory diagnosis of infectious diseases involving anaerobic bacteria. Procedures for the isolation and initial characterization of anaerobes from properly collected specimens are presented in this chapter. Descriptions of medically important genera and species of anaerobes and their identification characteristics are covered in the four chapters which follow. The anaerobic spirochetes are discussed in Chapter 45. Antimicrobial susceptibility testing methods are presented in Chapter 103.

CHARACTERIZATION

Relation to oxygen

Medically encountered bacteria are broadly categorized as aerobes or anaerobes, depending on their relationships to oxygen at different pressures and on the kinds of metabolic reactions they use to generate energy for growth and various activities. Although it is convenient to speak of only two categories of bacteria, aerobes and anaerobes, with respect to oxygen requirements and types of oxidation-reduction reactions required to gain energy for growth and metabolism, this categorization is an oversimplification. There is in fact a continuous spectrum of bacteria ranging from those that grow only in the presence of oxygen to those that grow only in its absence. Some of the terms that have been applied to this spectrum and examples of bacteria in the various categories are given in Table 1.

Obligately aerobic bacteria. Obligately aerobic bacteria require molecular oxygen as a terminal electron acceptor and cannot grow without it. They generate energy oxidatively and are unable to generate sufficient energy for growth by fermentation reactions. The formation of unsaturated fatty acids and sterols by certain aerobic microorganisms are examples of other activities that require O_2 (5). Although *Pseudomonas aeruginosa* reduces oxygen and would seem to fit the above description of an obligate aerobe, it is not rare to find this species growing on anaerobically incubated media. As an alternative to oxygen, *P. aeruginosa* uses nitrate as a terminal electron acceptor (by anaerobic respiration); its growth on anaerobic plates does not indicate failure of the anaerobic incubation system to exclude oxygen. Thus, *P. aeruginosa* and other nitrate-reducing bacteria are facultative anaerobes because they do not require O_2 for growth (5).

Microaerophilic bacteria. Microaerophilic bacteria require oxygen as a terminal electron acceptor for growth, but they do not grow on the surface of solid medium in an aerobic incubator (which has 20 to 21% O_2). They grow minimally, if at all, under anaerobic conditions. It is likely that oxygen at 0.2 atm (ca. 20.3 kPa) is toxic to these bacteria, and they are unable to generate sufficient energy for growth anaerobically (5).

Anaerobic bacteria. Anaerobic bacteria are those bacteria that do not require oxygen as a terminal electron acceptor for growth or metabolic activities. Although they may possess certain cytochromes, the anaerobes lack those cytochromes that are required to transfer electrons to molecular oxygen. Thus, their energy comes from fermentation reactions. In addition, anaerobic respiration is used by some anaerobes, such as *Desulfovibrio* spp., which reduce sulfate to H_2S; sulfate serves as a terminal electron acceptor in this case.

Facultatively anaerobic bacteria. Facultatively anaerobic bacteria are able to obtain energy and grow oxidatively by using oxygen as a terminal electron acceptor (via aerobic respiration), or they can grow under anaerobic conditions, obtaining energy by fermentative pathways in which organic compounds serve as terminal electron acceptors. In clinical labo-

TABLE 1. Relative growth of various groups of bacteria on the basis of their relationships to oxygen[a]

Group[b]	Examples	Relative growth on anaerobe blood agar in:			
		Ambient air	Candle jar or 5–10% CO_2-air incubator	Mixture of 5% O_2, 10% CO_2, and 85% N_2	Anaerobic system without O_2
Aerobes					
Obligate	*Micrococcus lutea*	++++	++	±	–
	Nocardia asteroides	++++	++	±	–
Microaerophiles	*Campylobacter jejuni*	–	+ or ++	++++	–
Anaerobes					
Facultative	*Escherichia coli*	++++	++++	++++	+++
	Pseudomonas aeruginosa	++++	+++	++	+ or ±
Aerotolerant	*Clostridium tertium*	+	+ or ++	+++	++++
	Clostridium histolyticum	+	+ or ++	+++	++++
Obligate	*Bacteroides fragilis*	–	–	–	++++
	Peptostreptococcus magnus	–	–	–	++++

[a] Symbols: ++++, best growth; +++ or ++, degrees of moderate growth; +, poor growth; ± scant growth; –, no growth.
[b] See the text for definitions of these terms.

ratories, most of the bacteria that grow in air or in a 5 to 10% CO_2-air incubator are facultative organisms; these bacteria also grow well in anaerobic systems. In the absence of oxygen, certain facultative anaerobes (as indicated above) can use nitrate via anaerobic respiration.

Aerotolerant anaerobes. The term "aerotolerant anaerobes" is applied to those bacteria that do not use molecular oxygen and are inhibited by it to some extent. These bacteria grow best under anaerobic conditions but grow scantily in an aerobic incubator.

Obligately anaerobic bacteria. The obligately anaerobic bacteria do not use molecular oxygen; their growth is inhibited by it, probably because oxygen is directly toxic to them. Obligately anaerobic bacteria do not grow on nutritionally adequate blood agar, chocolate agar, or other solid media in an ambient air incubator, a candle extinction jar, or a 5 to 10% CO_2-air incubator. Most of the obligate anaerobes associated with infections of humans are moderate anaerobes. According to Loesche (31), who coined the term, "moderate obligate anaerobes" are capable of growth at oxygen levels ranging from 2 through 8% (mean, 3% O_2). Loesche used the term "strict obligate anaerobes" to describe bacteria that did not form colonies on agar surfaces exposed to 0.5% or more oxygen. Strict obligate anaerobes are members of the indigenous flora of humans but are rarely isolated from properly collected and transported specimens from ill patients.

The growth of strict anaerobes in culture (or in nature) is favored by the exclusion of oxygen from the system and by a low oxidation-reduction potential (E_h) within the medium. There is now abundant evidence that oxygen, and not a high E_h, is toxic to anaerobic bacteria in an oxidized culture medium (or in a natural or pathogenic habitat). The addition to media of reducing agents such as amorphous ferrous sulfide, thioglycolate, cysteine, and other sulfur-containing amino acids results in both a lowering of the E_h and a removal of oxygen (4, 48). Experiments have shown that the growth of obligate anaerobes occurs over a wide range of redox potentials, provided that oxygen is excluded or kept at very low levels (34, 48). Thus, reducing agents probably enhance the growth of anaerobes in media, primarily by the removal of oxygen (34).

Little is known about why anaerobic bacteria are inhibited or killed by oxygen. The mechanisms of oxygen toxicity appear to be multiple, and many factors have been developed by microorganisms that can protect them from the lethal effects of oxygen and toxic oxygen derivatives. A number of microorganisms that use oxygen as a terminal electron acceptor generate toxic oxygen reduction products, including hydrogen peroxide, the hydroxyl radical, singlet oxygen, and superoxide anions, and release these products into the medium. Many aerobes and facultative anaerobes produce catalases or peroxidases which eliminate hydrogen peroxide or other peroxides. It was once reasoned that anaerobes are killed by H_2O_2 and organic peroxides that might accumulate in media because it was presumed that anaerobes could not produce catalases or peroxidases. Now, several anaerobes are known to produce catalases (e.g., *Bacteroides fragilis*, certain anaerobic cocci, *Propionibacterium acnes*, and others). Another theory has it that various anaerobic bacteria form a superoxide anion during growth; the bacteria lack a means of eliminating the anion and thus are killed by this toxic radical. However, it is now well established that several anaerobe species produce superoxide dismutase, an enzyme which converts superoxide to less toxic O_2 and H_2O_2. According to Gregory et al., the levels of superoxide dismutase produced by different strains of anaerobes correlated directly with degrees of oxygen sensitivity but did not correlate with the sources of isolates or with pathogenicity (23). Undoubtedly, there must be other factors that influence the harmful relationships between oxygen and anaerobic bacteria.

Anaerobes as normal flora

Anaerobic bacteria are widely distributed in natural environments that have a low oxygen tension and redox potential. In humans, anaerobic bacteria reside as normal flora on the skin and mucous membrane surfaces of the nasopharynx, oropharynx, mouth, gastrointestinal tract, orifices of the external genitalia, urethra, and vagina (Table 2). With few exceptions (e.g., *Clostridium botulinum*), all of the pathogenic anaerobic bacteria are part of the normal flora in one or more of the above locations, and they are all

TABLE 2. Principal habitats of selected anaerobes of the normal flora of humans

Organism	Mouth or pharynx	Intestines	Urogenital tract	Skin
Actinomyces israelii	+			
Arachnia propionica	+			
Bacteroides fragilis group		+		
B. melaninogenicus group	+	+	+	
B. bivius	+		+	
B. disiens	+		+	
Bifidobacterium dentium	+	+	+	
Clostridium perfringens		+		
C. ramosum		+		
C. septicum		+		
Eubacterium lentum	+	+		
Fusobacterium nucleatum	+	+		
F. necrophorum	+			
Lactobacillus catenaforme	+	+		
Peptostreptococcus anaerobius	+	+	+	
P. magnus	+	+	+	
Propionibacterium acnes	+	+	+	+
Veillonella parvula	+	+	+	

opportunistic pathogens. *Actinomyces* spp. exist in tonsils without causing harm. *B. fragilis* is without consequence as long as it stays in the lower gastrointestinal tract. *Bacteroides melaninogenicus* and *Fusobacterium nucleatum* are harmless in the mouth and intestines of healthy individuals. *Clostridium perfringens* lives in the gastrointestinal tract of 30 to 50% of us and may be found on the perianal skin of healthy individuals. Most isolates of *P. acnes* from the skin, mouth, nose, and urethra have no pathogenic significance. However, when anaerobes of the indigenous flora find conditions suitable for growth in tissues outside their habitats, they can cause disease. It is important for the clinical microbiologist to be aware of the normal flora of various body sites. This knowledge can aid in anticipating what organisms are most commonly associated with infections involving a given portal or anatomic source and can be useful in determining the potential clinical significance of isolates. It is also important to recognize that the normal flora may be modified by several factors, including antibiotics, chemotherapeutic agents, obstruction, and various diseases. The indigenous microorganisms of humans are discussed in further detail in Chapter 4.

Anaerobic infections

Anaerobic infections can involve any region of the body, provided that conditions in the tissues are suitable (9, 17). Anaerobes are commonly involved in abscesses of any organ (e.g., brain, lung, liver, and tuboovarian abscesses), actinomycosis, antibiotic-associated colitis, appendicitis, bacteremia, cholecystitis, chronic otitis media, crepitant and noncrepitant cellulitis, dental and oral infections, endocarditis, endometritis, meningitis, myonecrosis, neutropenic enterocolitis due to *Clostridium septicum*, osteomyelitis, peritonitis, thoracic empyema, salpingitis, septic arthritis, sinusitis, subdural empyema, and other infections. The majority of anaerobic infections are caused by bacteria from endogenous sources (i.e., the normal flora of the oropharynx, nasopharynx, gastrointestinal tract, genitourinary tract, and skin). Exogenous diseases caused by anaerobes include foodborne, wound, and infant botulism, gastroenteritis and enteritis necroticans due to *C. perfringens*, tetanus, clostridial myonecrosis (gas gangrene), crepitant cellulitis, infected animal bites, and septic abortion. The incidence of obligate anaerobes in infectious diseases has been reviewed elsewhere (17, 21, 22) and is summarized in Table 3. Anaerobes are important in several other types of infections that are not listed here but have been recently reviewed elsewhere (19, 27).

Although most deep abscesses and other necrotizing lesions containing anaerobes are polymicrobial, it is not uncommon to isolate a single species of an anaerobe in pure culture. The major species of anaerobes encountered in human diseases are shown in Table 4.

Predisposing factors involved in the pathogenesis of anaerobic infections include surgical or other trauma, poor blood supply, tissue necrosis, malignancy, various diseases, and growth of aerobes or facultative anaerobes in tissue. All of these factors tend to lower the oxidation-reduction potential, oxygen tension, or both and help provide a favorable environment for anaerobic growth. In addition, the anaerobes produce enzymes, endotoxins, exotoxins, capsules, and other virulence factors, reviewed by others elsewhere (17, 19, 21, 27).

TABLE 3. Relative incidence of anaerobic bacteria in various infections[a]

Type of infection	Incidence (%)
Bacteremia	10–20
Central nervous system	
Brain abscess	89
Subdural empyema	~50
Meningitis	Low[b]
Head and neck	
Chronic sinusitis	50
Chronic otitis media	[b]
Periodontal abscess	100
Other oral infections	[b]
Pleuropulmonary	
Aspiration pneumonia	85–90
Lung abscess	93
Necrotizing pneumonia	85
Empyema	76
Chronic bronchitis, bronchiectasis	[b]
Intraabdominal	
Peritonitis and abscess	90–95
Liver abscess	>50
Female genital tract	
Salpingitis, pelvic peritonitis	56 or higher
Tuboovarian abscess	92
Vulvovaginal abscess	74
Septic abortion and endometritis	73
Soft tissue	
Gas gangrene (myonecrosis)	100
Crepitant cellulitis	High[b]
Necrotizing fasciitis	High[b]
Urinary tract	
Cystitis	1
Urethritis	<1

[a] Adapted from Finegold (17), Gorbach (21), Gorbach and Bartlett (22), and Koneman et al. (29).

[b] Percentage data not available.

TABLE 4. Percentage distribution of major anaerobic bacteria involved in human clinical infections[a]

Group or species	% of isolates
Gram-negative nonsporeforming bacilli	
Bacteroides spp.	39
B. fragilis group (*B. fragilis* and *B. thetaiotaomicron* much more common than *B. vulgatus*, *B. distasonis*, or *B. uniformis*; *B. ovatus* rare)	
B. melaninogenicus group (especially *B. intermedius* and *B. asaccharolyticus*, but includes additional species[b])	
B. oralis group (*B. oralis*, *B. buccalis*, *B. veroralis*[b])	
B. bivius-B. disiens group[b]	
B. oris-B. buccae group[b,c]	
B. ureolyticus	
Fusobacterium spp.	4
F. nucleatum	
F. necrophorum	
F. mortiferum	
F. varium	
Gram-positive cocci	
Peptostreptococcus spp.	22
P. magnus	
P. asaccharolyticus	
P. prevotii	
P. anaerobius	
Staphylococcus sp.	1
S. saccharolyticus[d]	
Streptococcus sp.	1
S. intermedius[e]	
Gram-positive nonsporeforming bacilli	
Actinomyces spp. (especially *A. israelii*; others isolated include *A. naeslundii*, *A. odontolyticus*, and *A. viscosus*	<1
Arachnia sp.	<1
A. propionica	
Bifidobacterium spp.	<1
B. dentium[f]	
B. adolescentis	
Propionibacterium sp.	17
P. acnes	
Eubacterium sp.	4
E. lentum	
Sporeforming bacilli	
Clostridium spp.	11
C. perfringens	
C. ramosum	
C. difficile	
C. septicum	
C. paraputrificum	
C. tertium	
C. sporogenes	
C. histolyticum	
C. novyi	
C. tetani	
C. botulinum	

[a] Based on data from the Indiana University Medical Center from 1972 through 1980.

[b] Also called the pigmented *Bacteroides* group. The different groups of *Bacteroides* spp. are discussed in Chapter 40.

[c] Formerly *B. ruminicola*.

[d] *S. saccharolyticus* (formerly *Peptococcus saccharolyticus*) varies in its aerotolerance; some strains are obligate anaerobes but most are aerotolerant anaerobes. See Chapter 39.

[e] *S. intermedius* (formerly *Peptostreptococcus intermedius*) is not an obligate anaerobe.

[f] Formerly *B. eriksonii*.

Clinical features or clues which suggest infection with anaerobic bacteria include the following: foul odor, lesion in close proximity to a mucosal surface, underlying disease with tissue necrosis or impaired blood supply or both (e.g., malignancy), gangrenous necrosis, abscess, previous antibiotic treatment, septic thrombophlebitis, infection after bites, penetrating wounds to the abdomen or pelvis, aspiration, infection after gastrointestinal surgery, and septic abortion (17, 18). None of these clinical clues is specific. Although foul odor is one of the most characteristic clinical features, it lacks specificity or may be absent. Thus, clinicians can only suspect whether anaerobes, nonanaerobic microorganisms, or both are involved, and specific diagnosis and treatment require both aerobic and anaerobic processing of properly selected, collected, and transported specimens.

COLLECTION OF SPECIMENS

Information on the collection of specimens for the culture of various microorganisms, including anaerobic bacteria, is given in Chapter 8. Again, it is emphasized that proper selection, collection, and transport of clinical specimens are crucial for the laboratory diagnosis of infections caused by anaerobic bacteria.

Selection of specimens for anaerobic culture

As a general rule, we recommend that all materials from areas of the body not likely to be contaminated with normal flora should be cultured for anaerobic bacteria as well as for facultative anaerobes and aerobic microorganisms. A standard aerobic and anaerobic bacterial culture, along with identification and susceptibility testing of isolates as clinically indicated, should be done on the following types of specimens:

- Wounds and abscesses
- Aspirated pus
- Tissue (biopsy, surgical, and autopsy)
- Body fluids (cerebrospinal, pleural, paracentesis, pericardial, and synovial)
- Blood
- Bone marrow
- Transtracheal aspirates
- Direct lung aspirates (also aspirates from other body sites)
- Sulfur granules from patients with suspected actinomycosis
- Urine (suprapubic aspirate, ureterostomy, dialysate, ureter, cystoscopy, and nephrostomy)
- Urogenital (only from sites not having anaerobes as usual flora)
- Small intestine contents (for workup of blind-loop and similar malabsorption syndromes)
- Stool or gastrointestinal contents (for toxin assays and cultures in diseases suspected of involving *Clostridium difficile*, *C. perfringens*, or *C. botulinum*)

Specimens not listed above should receive only aerobic cultures (along with aerobic identification and susceptibility tests if indicated) when received in the laboratory with a physician's request for routine bacteriology (or a similar request). The laboratory should provide the physicians, wards, and nursing stations with guidelines for the proper selection of specimens, along with anaerobic transport containers and directions for their use (9). Laboratory personnel should communicate closely with physicians, nurses, students, and others who are involved in the selection of specimens for anaerobic culture.

Specimens which are ordinarily unacceptable for anaerobic culture include the following:

- Throat or nasopharyngeal swabs
- Gingival swabs
- Expectorated sputum
- Bronchoscopic specimens not collected by a protective, double-lumen catheter (2, 51)
- Gastric contents, small bowel contents (except in blind-loop and similar syndromes), feces (except in the workup of toxicoinfectious diseases due to certain clostridia), rectal swabs, colocutaneous fistulae, and colostomy stomata
- Surface material from decubitus ulcers, swab samples of other surfaces, sinus tracts, and eschars
- Material adjacent to skin or mucous membranes other than the above which have not been properly decontaminated
- Voided urine
- Vaginal or cervical swabs

At times it is necessary to inform physicians and nurses in a spirit of helpfulness and cooperation that specimens contaminated with normal flora will not be cultured for anaerobic bacteria because (i) these specimens predictably yield numerous isolates of doubtful clinical significance, (ii) the laboratory results may be misleading to the clinician, and (iii) such added work is an unnecessary, costly burden on the resources of both the patient (or a health care provider) and the laboratory. The microbiologist or laboratory director should be prepared to suggest appropriate ways to make a laboratory diagnosis and, if need be, to aid with specimen collection.

Collection

Recommended procedures for the collection of specimens for anaerobic culture are outlined in Table 5. If possible, aspirates collected by needle and syringe or tissue samples should be collected for anaerobic culture. The use of swabs is least desirable because swabs are easily contaminated, expose anaerobes to ambient oxygen, allow the specimen to dry out, permit the collection of comparatively small specimen volumes, and are less satisfactory than aspirates for the preparation of smears for direct microscopic examination.

The microbiologic investigation of certain female genital tract infections poses special problems related to anaerobic bacteriology. Puerperal or postpartum endometritis occurs after 1 to 4% of childbirths and, in most instances, involves microorganisms of the normal cervicovaginal microflora (37, 46). Anaerobes, including *B. fragilis*, can be very significant in these infections. *Clostridium* spp., including *C. perfringens*, may be isolated from the endometrial cavity, but their importance can be interpreted only with a full knowledge of the clinical setting. Most patients do not have myonecrosis of the uterus and have a relatively benign hospital course (46). In addition to anaerobes, the most significant organisms include non-group-A streptococci (particularly, group B streptococci), *Escherichia coli*, other *Enterobacteriaceae*, and, rarely, *Listeria monocytogenes* and group A streptococci. The problem in the interpretation of postpartum endometrial cultures is that cervicovaginal flora is present in

TABLE 5. Recommended procedures for collection of specimens for anaerobic culture[a]

Site	Specimens and methods of collection
Central nervous system	Cerebrospinal fluid (especially when turbid) Abscess material Tissue biopsy
Dental area, ear, nose, throat, and sinuses	Carefully aspirated or biopsied material from abscesses after surface decontamination with povidone-iodine Needle aspirates and surgical specimens from sinuses in chronic sinusitis
Pulmonary area	Transtracheal aspiration Percutaneous lung puncture Thoracotomy specimen Thoracentesis (pleural fluid) Bronchoscopic specimen obtained with protective, double-lumen catheter[b]
Abdominal area	Paracentesis fluid Needle-and-syringe aspiration of deep abscesses under ultrasound or at surgery Surgical removal specimen if not contaminated with intestinal flora Bile
Female genital tract	Culdocentesis after surface decontamination of the vagina with povidone-iodine Laparoscopy specimens Surgical specimens Endometrial cavity specimen with double-lumen catheter and microbiologic brush[c] after cervical os is decontaminated
Urinary tract	Suprapubic aspirate of urine
Bone and joint	Aspirate of joint (in suppurative arthritis) Deep aspirate of drainage material after surgery (e.g., in osteomyelitis)
Soft tissue	Open wounds—deep aspirate of margin or biopsy of the depths of wound only after careful surface decontamination with povidone-iodine Sinus tracts—aspiration by syringe and small plastic catheter after careful decontamination of skin orifice Deep abscess, anaerobic cellulitis, infected vascular gangrene, clostridial myonecrosis—needle aspirate after surface decontamination Surgical specimens, including curettings and biopsy material Decubiti and other surface ulcers—thoroughly cleanse area with povidone-iodine by surgical scrub technique, and aspirate pus from deep pockets or obtain biopsy from deep tissue at margin

[a] Adapted from Dowell and Allen (10).

[b] Reviewed elsewhere by Allen and Siders (2).

[c] Technique uses a telescoping double-catheter assembly similar to that available for bronchoscopy. This procedure is inadequate to diagnose postpartum endometritis (see the text).

the uterine cavities of postpartum women whether or not they have endometrial infection (37, 46). If postpartum endometritis is suspected, Gram stains of uterine contents should be prepared, and cultures of blood and uterine contents should be processed aerobically and anaerobically. Bacteremia may be caused by bacteria that invade the blood from the infected uterus. There are no quantitative differences in the bacteriology of endometrial specimens from infected versus noninfected uteri (46). Specific bacteria, such as *B. fragilis*, *C. perfringens*, and the nonanaerobic bacteria mentioned above, should be sought and identified by rapid procedures. This disease is a situation in which a discussion of the clinical circumstances with the clinician can help the microbiologist to interpret Gram-stained smear and culture results and establish the extent of culture workup. The microbiologic investigation of uterine contents in suspected infections after abortions poses similar problems. Again, aerobic and anaerobic cultures of endometrial specimens and blood are indicated, but interpretations of results must be made in the context of all available clinical information.

Amniotic fluid cultures may be obtained from women in labor or with ruptured amniotic membranes who develop fever, maternal or fetal tachycardia, uterine tenderness, foul odor, and leukocytosis (20). Clinicians usually call this condition intraamniotic infection, amnionitis, chorioamnionitis, or intrapartum infection. Intraamniotic infection is associated with maternal sepsis, shock, and death, neonatal sepsis and death, and stillbirth (20). The pathogenesis of intraamniotic infection and the role of anaerobes in this condition are not clearly established. Aerobic and anaerobic cultures should be done on amniotic fluid collected by amniocentesis, at Cesarean section, or by aspiration with a transcervical intrauterine catheter, and cultures should be done on blood. Amniotic fluid collected by amniocentesis or at Cesarean section is less likely to be contaminated than are catheterized specimens which pass through the cervix (20). Gram-stained smears may aid in determining the quality of these specimens. Anaerobes, particularly the anaerobic gram-positive cocci, *Bacteroides bivius*, certain other *Bacteroides* spp., *Fusobacterium necrophorum*, *F. nucleatum*, and *Clostridium* spp. are involved in about half the women with intraamniotic infection (20).

Any laboratory that accepts blood for bacterial culture should be prepared to culture both aerobically and anaerobically. Current recommendations for blood culture procedures in Chapter 8 and in a *Cumitech* (38) include guidelines for anaerobic culture.

C. difficile, *C. perfringens*, *C. septicum*, and *C. botulinum* are major pathogens which may be sought in intestinal and fecal specimens. Guidelines for the collection of specimens in the diagnosis of clostridial diseases are given in Chapter 38.

Transport of specimens

A major challenge for the bacteriology laboratory is to minimize the amount of time between the collection and receipt of specimens. The laboratory should provide appropriate transport containers for use in examining, operating, and emergency rooms, in intensive care units, and on wards. Clinicians should notify the laboratory when transtracheal aspirates or similar hard-to-obtain specimens are to be collected; similarly, the laboratory needs to be aware that surgeons plan to perform an invasive diagnostic procedure before the procedure is done (e.g., open lung biopsy for microbiologic examination). This kind of communication can minimize errors in the transport and processing of crucial specimens. Furthermore, the usual hospital delivery services should not be relied upon for the transportation of transtracheal aspirates, bronchial brush or endometrial brush samples, tissue, pleural fluid, central nervous system aspirates, and other specimens of similar importance. The physician, nurse, or other trusted individual should hand deliver such specimens to the laboratory as soon as possible after collection. Specimens just aspirated with a needle and syringe should have any bubbles of air cleared from the syringe and needle into a piece of sterile gauze. The needle can be capped with a rubber stopper, and then the syringe can become a transport container. A small-volume sample (e.g., a few drops obtained by transtracheal aspiration) may be subject to aeration during the procedure and should ideally be received by the laboratory within 10 min after collection. Larger-volume samples with 1 ml or more of sample should be received by the laboratory within about 1 h. Alternatively, oxygen-free vials or tubes can be used, preferably containing a redox indicator such as resazurin (which is colorless when reduced, but pinkish when oxidized). Anaerobic transport containers should be received by the laboratory within 2 to 3 h of specimen collection. The Vacutainer Anaerobic Specimen Collector (Becton Dickinson and Co., Cockeysville, Md.), the Anaport system (Scott Laboratories, Inc., Fiskeville, R.I.), and the Bio-Bag type A (Marion Laboratories, Inc., Kansas City, Mo.) are suitable containers to transport tissue biopsy specimens (29). As indicated above, swabs are the least desirable method for specimen collection. However, at times it is impossible to obtain an aspirate or tissue sample, and a swab is used. The swab should be placed in tubes that will protect the specimens from drying, exposure to oxygen, or oxidizing conditions. Commercial two-container sets are available (Carr-Scarborough Microbiologicals, Inc., Stone Mountain, Ga.; GIBCO Diagnostics, Lawrence, Mass.; Scott; Port-A-Cul, BBL Microbiology Systems, Cockeysville, Md.; and other sources) containing a swab in an anaerobic atmosphere and a second tube containing a semisolid deep of transport medium (29). When the specimen is collected quickly with the swab from the gassed-out anaerobic container, the swab containing the specimen is then immediately placed in the second tube, which contains the semisolid reduced transport medium (such as modified Cary-Blair or Stuart medium), and this tube should be recieved by the laboratory within 2 to 3 h of collection (29).

DIRECT EXAMINATION OF SPECIMENS

Gross examination of clinical specimens may provide information about the nature and quality of the material collected. Foul odor, purulent appearance, necrosis of tissue, gas in tissue, or sulfur granules should suggest the possibility of anaerobes.

The direct microscopic examination of clinical specimens shows the type of cells present and the number and morphologic features of microorganisms. This information may help guide the clinician in the selection of therapy until culture and susceptibility results are available. In addition, the microscopic findings may aid in the choice of isolation media.

The Gram stain is an important means of quality control. If morphotypes observed in the smears are not recovered in culture, the procedures may have been defective for one or more of the following reasons: (i) the specimen was improperly collected or handled, (ii) the isolation media were defective, (iii) the anaerobic incubation system was defective, or (iv) anaerobic subculture was performed improperly. It is important to report the microscopic findings promptly to the attending physician since many anaerobes require 48 h or more before growth can be seen and additional time for identification and susceptibility testing.

Giemsa and acridine orange stains are often helpful for the observation of bacterial forms that Gram stain poorly. Phase microscopy and dark-field examination of wet mounts may be useful for the demonstration of motility, spirochetes, and endospores. As discussed in Chapter 112, a modified Kinyoun acid-fast stain may aid in distinguishing non-acid-fast actinomycetes from *Nocardia* spp. Fluorescent-antibody stains are used in certain research and reference laboratories for the rapid identification of *C. septicum*, *Clostridium novyi*, *Clostridium haemolyticum*, *F. necrophorum*, *Actinomyces israelii*, *Arachnia propionica*, and certain anaerobic cocci (10). Commercial fluorescent-antibody reagents for the rapid detection and identification of members of the *B. fragilis* and *B. melaninogenicus* groups may be useful in urgent situations (49). However, some of the species in these groups are not reactive with the conjugates (e.g., *Bacteroides gingivalis*), whereas the conjugate for pigmenting bacteroides cross-reacts with *B. bivius* and *Bacteroides disiens*.

Findings from the direct microscopic examination of clinical materials consistent with the presence of anaerobes include the following.

(i) Large, broad, gram-positive rods with blunted ends in a necrotic background with few or rare leukocytes in a smear from a patient with suspected gas gangrene: possibly *C. perfringens* (spores of this organism are rarely present)

(ii) Pale, irregularly staining, pleomorphic gram-negative rods with bipolar staining in a smear from an abscess: possibly *Bacteroides* or *Fusobacterium* species

(iii) Pale, gram-negative, filamentous, slim rods with tapered ends: possibly *F. nucleatum*

(iv) Clusters and chains of gram-positive cocci within neutrophilic exudate from a postoperative intraabdominal wound: although *Staphylococcus* and *Streptococcus* species should be considered, the anaerobic genus *Peptostreptococcus* would be a major possibility

(v) Sulfur granules showing peripheral clubs with branched filamentous rods from a cervicofacial lesion: suggests one of the actinomycetes (e.g., *A. israelii*, *A. propionica*, *Bifidobacterium dentium*, and others)

The direct analysis of short-chain fatty acids by gas-liquid chromatography (GLC) can aid in the rapid presumptive diagnosis of anaerobes in blood cultures (52). However, metabolic product analysis by itself suffers from lack of specificity. For example, a major amount of propionic acid in a sample might suggest *Propionibacterium* species but could also have been produced by *A. propionica* or *Veillonella* spp. The examination of Gram-stained smears of specimens that are analyzed by GLC should be done to aid in interpretations.

Except for the direct detection of *C. botulinum* and *C. difficile* toxins, the use of non-fluorescent-antibody serologic procedures for the direct detection of anaerobes has been limited. A counterimmunoelectrophoresis procedure, previously described for the detection of *C. difficile* toxin (41), lacks sensitivity and cross-reacts with antigens of other species (25). Commercially available immunologic procedures for the rapid detection of certain anaerobe antigens have long been needed. The introduction by manufacturers of reliable packaged kits to detect *B. fragilis* and clostridial toxins in clinical specimens would likely be well received by clinical laboratories.

SELECTION AND USE OF MEDIA FOR PRIMARY ISOLATION OF ANAEROBIC BACTERIA

The primary isolation of obligately anaerobic bacteria from clinical specimens involves the use of nonselective, selective, and enrichment media. The choice of media will depend on the anatomic source of the specimen and on the findings from direct microscopic examination. Various microbiologists have different preferences, but it is generally agreed that media of all three types are needed. Table 6 lists media used at the Indiana University Medical Center for the isolation of obligate anaerobes and other bacteria from wounds, abscesses, and body fluids other than blood. A guide to the selection of media according to specimen source is given in Table 7.

Nonselective media

The use of nutritionally adequate media supplemented with hemin and vitamin K_1 is of key importance. A variety of anaerobe blood agar media have been described. The formulation developed at the Centers for Disease Control (CDC) by Dowell et al. (16; also see Chapter 111) is used in our laboratory. CDC anaerobe blood agar contains Trypticase soy agar base (BBL) supplemented with yeast extract, hemin, vitamin K_1, L-cystine, and 5% sheep blood (or rabbit blood). This medium supports much better growth of *B. melaninogenicus*, *F. necrophorum*, *C. haemolyticum*, certain strains of *A. israelii* and *Bacteroides thetaiotaomicron*, and certain thiol-dependent streptococci (6) than does the supplemented rabbit blood-Trypticase soy agar medium formerly recommended by the CDC anaerobe laboratory (11; S. D. Allen, G. L. Lombard, V. R. Dowell, Jr., A. Y. Armfield, F. S. Thompson, and M. D. Stargel, unpublished data). Both CDC anaerobe blood agar and Schaedler blood agar contain additional L-cystine for the improved growth of *C. novyi* type B, *C. haemolyticum*, *F. necrophorum*, and certain cocci. However, there is more smooth-to-rough variation of colonies on Schaedler blood agar (45) than on CDC anaerobe blood agar, probably due to the increased carbohydrate content in the Schaedler medium. After preparation, plates of CDC

TABLE 6. Basic set of primary isolation media for recovery of anaerobes and other bacteria from wounds and abscesses

Medium	Incubation[a]	Purpose
Sheep blood agar	CO_2	Nonselective plating medium for general use
MacConkey agar	O_2	Recovery of aerobic and facultatively anaerobic gram-negative bacilli
Colistin-nalidixic acid-Columbia base	CO_2	Recovery of all aerobic and facultative gram-positive bacteria; inhibits most gram-negative bacteria
Chocolate blood agar	CO_2	Used primarily for recovery of *Haemophilus* spp. and *Neisseria* spp.; may be required for other fastidious bacteria
Modified Thayer-Martin agar	CO_2	Used when *Neisseria gonorrhoeae, Neisseria meningitidis*, or *Neisseria lactamica* is suspected
CDC anaerobe blood agar	An	Nonselective plating medium for anaerobic bacteria (contains yeast extract, vitamin K_1, hemin, and L-cystine)
Anaerobe phenylethyl alcohol blood agar	An	Recovery of most anaerobic bacteria; inhibits facultatively anaerobic gram-negative bacilli
Anaerobe paromomycin-vancomycin blood agar	An	Selective recovery of anaerobic gram-negative bacilli and *Veillonella* spp.
Thioglycolate medium (BBL-0135C without indicator, enriched with hemin)	An	An enrichment broth used as a supplement or backup to plating media; especially useful when actinomycetes are suspected

[a] O_2, In air or aerobic incubation; CO_2, in 5 to 10% CO_2 or candle jar; An, incubation in an anaerobic system.

anaerobe blood agar are wrapped in cellophane bags to retard dehydration. They can be stored in refrigerator (2 to 4°C) for up to 6 weeks (G. L. Lombard, A. Y. Armfield, M. D. Stargel, and J. B. Fox, Abstr. Annu. Meet. Am. Soc. Microbiol. 1976, C95, p. 41). Brain heart infusion, brucella, and Columbia agar bases have been recommended by others (24, 46). Murray found no significant differences in the quantitative recovery of 10 different anaerobes on commercially prepared brain heart infusion, brucella, Columbia, Schaedler, and tryptic soy agar plates which were stored in cellophane bags in a refrigerator for up to 4 weeks (36). CDC anaerobe blood agar was not included in that study.

Use of liquid media during primary isolation

Liquid media provide a backup to solid media during primary isolation. Enriched thioglycolate and chopped-meat–glucose media support the growth of many fastidious anaerobes. Enriched thioglycolate medium is particularly useful for the isolation of slow-growing anaerobes such as *A. israelii*, which may take several days to develop colonies on solid media. Chopped-meat–glucose medium is especially useful for the isolation of *Clostridium* species by a heat shock or ethanol spore selection procedure (30) and is an excellent holding medium for cultures (16). In addition, both media are used during the identification of isolates. Enriched thioglycolate medium should be prepared from thioglycolate medium without indicator (BBL-0135C, BBL). Both thioglycolate and chopped-meat–glucose media should be supplemented with vitamin K_1 and hemin.

Before inoculation, the liquid media are either held in an anaerobic environment (e.g., 85% N_2, 10% H_2, 5% CO_2) as described elsewhere (10; also see Chapter 111) or heated for 10 min in a boiling-water bath to drive off oxygen, followed by cooling before inoculation. Enriched thioglycolate and chopped-meat–glucose media, prepared as recommended elsewhere (Chapter 111), are commercially available. After inoculation, liquid media in screw cap tubes should be incubated in an anaerobic system with the caps loosened. Alternatively, prereduced and anaerobically sterilized (PRAS) liquid media, prepared by the directions of the Virginia Polytechnic Institute and State University (VPI) *Anaerobe Laboratory Manual*, should be incubated after gassing the butyl rubber-stoppered tubes with CO_2 as described by Holdeman et al. (24).

Selective media

Because abscesses and wound infections frequently contain mixtures of obligate anaerobes, facultative anaerobes, and aerobes, species present in small numbers are likely to be missed if only nonselective media

TABLE 7. Selection of primary isolation media for culture of wound and abscess specimens from various anatomical sites

Specimen source	Media[a] and incubation atmosphere at 35°C		
	Air	5–10% CO_2	Anaerobic
Central nervous system		SAP, Choc	AnBAP, THIO
Eye, ear, oropharyngeal area	MAC	SAP, Choc, CNA	AnBAP, PEA, THIO
Pulmonary area	MAC	SAP, Choc	AnBAP, PEA, (PV), THIO
Intraabdominal area	MAC	SAP, CNA, (MTM)	AnBAP, PEA, PV, THIO
Genitourinary area	MAC	SAP, CNA, MTM	AnBAP, PEA, PV, THIO
Muscle and other soft tissue	MAC	SAP, CNA	AnBAP, PEA, (NEY), CMG
Bone marrow		SAP, Choc	AnBAP, THIO
Body fluids (other than blood and urine)		SAP, Choc	AnBAP, THIO

[a] MAC, MacConkey agar; SAP, sheep blood agar; Choc, chocolate blood agar; CNA, colistin-nalidixic acid blood agar; MTM, modified Thayer-Martin medium; AnBAP, CDC anaerobe blood agar; PEA, phenylethyl alcohol blood agar; PV, paromomycin-vancomycin blood agar; NEY, neomycin egg yolk agar; THIO, enriched thioglycolate medium; CMG, chopped-meat–glucose medium. Media shown in parentheses are optional.

are used. The use of appropriate selective and nonselective media is thus advantageous (9).

A variety of selective media have been described. The use of phenylethyl alcohol blood agar and either paromomycin-vancomycin or kanamycin-vancomycin blood agar is recommended. Phenylethyl alcohol blood agar, prepared by supplementing CDC anaerobe blood agar with 0.25% phenylethyl alcohol (16), supports the growth of most anaerobic bacteria (gram negative and positive) but inhibits the growth of facultatively anaerobic gram-negative bacilli, including *Proteus* spp. Paromomycin-vancomycin and kanamycin-vancomycin media (see Chapter 111) contain 100 μg of either paromomycin or kanamycin and 7.5 μg of vancomycin per ml of CDC anaerobe blood agar. This kind of medium is useful for the selective isolation of obligately anaerobic gram-negative bacteria including *Bacteroides* and *Fusobacterium* species. Most gram-positive bacteria encountered in clinical specimens are inhibited by the vancomycin, whereas most gram-negative facultative anaerobes are inhibited by the aminoglycoside. Gentamicin-vancomycin blood agar is used instead of kanamycin-vancomycin or paromomycin-vancomycin blood agar at the Mayo Clinic (J. E. Rosenblatt, personal communication). The rationale for media containing either paromomycin or gentamicin instead of kanamycin is that some facultatively anaerobic gram-negative bacilli are occasionally resistant to kanamycin but are susceptible to these other aminoglycosides.

Egg yolk-neomycin agar may aid in the isolation of *Clostridium* species from mixed microbial populations. The lipase and lecithinase reactions provide useful differential characteristics. The medium inhibits various facultatively anaerobic gram-negative bacteria but allows the growth of many gram-negative and -positive anaerobes, including clostridia (29).

Inoculation procedures

Use capillary pipettes for the inoculation of primary isolation media with liquid specimens. Inoculate tubes of liquid media near the bottom with 1 or 2 drops of inoculum. Place 1 drop on each plating medium, and then streak the drop with a platinum or stainless steel loop, using a quadrant plating technique to obtain well-isolated colonies. Before discarding the pipette, prepare a smear for Gram-stain examination.

Mince solid-tissue specimens with sterile scissors. Add 1 part enriched thioglycolate medium or 1 part buffered-gelatin diluent (11) per volume of solid tissue, and grind the mixture with a sterile tissue grinder. Inoculate the primary isolation media as described above.

If two swab samples are received as one specimen, inoculate one-quarter of each plate of agar medium with the first swab, and then inoculate the liquid medium. Again, streak the plates to obtain isolated colonies. Prepare a smear for Gram-stain examination with the second swab. If only one swab is received, scrub the material from the swab in a small volume (0.5 to 1 ml) of thioglycolate; then inoculate the media as described above for liquid specimens. With 1 drop of the suspension, prepare a smear for microscopic examination.

ANAEROBIC HOLDING JAR PROCEDURE

Anaerobic bacteria vary in their tolerance to oxygen (31). Therefore, care must be taken to avoid prolonged oxygen exposure of freshly inoculated plating media. It is not always practical in busy clinical laboratories to incubate plates in an anaerobic system immediately after they have been inoculated. The use of an anaerobic holding jar procedure is a practical alternative to immediate incubation. The holding jar procedure described below, a modification of the Martin procedure (35), allows primary plating, inspection of colonies, and subculture of colonies at the open laboratory bench without undue exposure of anaerobes to oxygen (S. D. Allen, G. L. Lombard, A. Y. Armfield, F. S. Thompson, and M. D. Stargel, Abstr. Annu. Meet. Am. Soc. Microbiol. 1977, C142, p. 59). Three holding jars can be used (Fig. 1). One jar is for reduced, uninoculated media, a second jar is for freshly streaked plates which are awaiting incubation, and a third jar is for plates containing colonies to be subcultured. As an alternative to jars, rectangular holding boxes (14 in. [36 cm] wide by 8.5 in. [22 cm] deep by 9 in. [23 cm] high), constructed from 0.5-in. (1.3-cm) Plexiglas with lids hinged down the center (Fig. 2) are used in our laboratory. They are now commercially available (Carr-Scarborough). The plating media described above can be used. The procedure for jars is as follows.

1. The uninoculated plates to be used during the work day should be held in an anaerobic glove box or an anaerobe jar for 6 to 24 h before inoculation.
2. Place the reduced plating media described above in the first holding jar, and flush the jar continuously with a gentle stream of nitrogen or CO_2.
3. Inoculate the surface of each plate on the open laboratory bench. Immediately after each plate is streaked, place it in the second holding jar, which also is flushed with nitrogen or CO_2.
4. Use a third holding jar for any plates containing colonies for subculture which were removed from an anaerobic system after incubation.

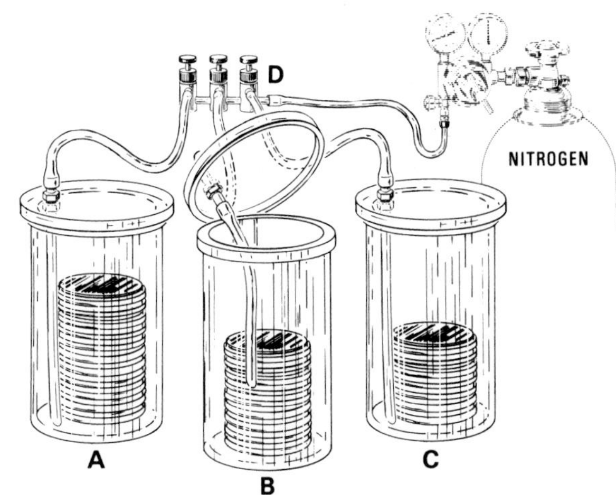

FIG. 1. Anaerobic holding jar system. Jar A contains uninoculated plates; jar B contains freshly inoculated plates. Plates with colonies to be subcultured are being held in jar C. The flow rate of nitrogen to each jar is regulated with the gang valve (D) and the regulator on the tank.

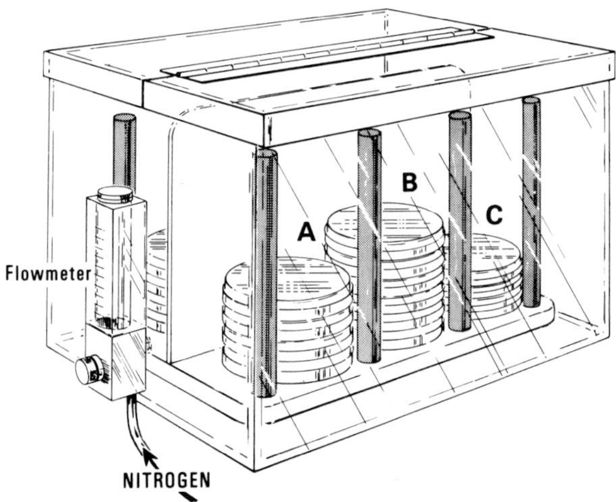

FIG. 2. Plexiglas holding box. This alternative to the three-jar system is commercially available but is simple to construct. The plastic petri dish holder inside the box is available from several commercial sources. The gas flow is regulated with a flow meter and the regulator on the tank (see the text).

5. After the holding jar is filled with freshly inoculated plates, remove the holding jar lid, and incubate the plates with either the GasPak (BBL) or the evacuation-replacement (ER) jar technique, or place the plates in an anaerobic glove box.

Either inexpensive commercial-grade N_2 or commercial-grade CO_2 can be used with the holding jar system. Regulate the flow rate of gas to the holding jars as follows: open the small needle valve on the gas manifold (Fig. 1), and adjust the gas tank regulator to 4 lb/in^2 for 1 to 2 min to purge the jar of air. Then decrease the flow rate to 1 or 2 bubbles per s. The flow rate can be measured by holding the rubber tubing (0.25-in. [0.6-cm] diameter) (which ordinarily goes into the jar) just under the surface of the water in a small beaker. Alternatively, a flow meter (Dwyer Co., Michigan City, Ind.) can be used (the flow rate should be 0.25 ft^3/h or 50 to 100 cm^3/min) (Fig. 2). The gang valve with three or four needle valves which is used to regulate the flow of gas to each jar can be purchased at stores in which aquarium supplies are sold. Commercial-grade N_2 and commercial-grade CO_2 (the least expensive available) are both suitable. A heated copper catalyst (Sargent furnace) has been recommended (35) to remove traces of O_2 from CO_2, but this step is not necessary if N_2 is used.

USE OF ANAEROBIC SYSTEMS

The primary isolation of obligately anaerobic bacteria from clinical specimens requires the incubation of inoculated media in an anaerobic system. Anaerobe jars (with either a hydrogen-carbon dioxide generator or ER technique), glove boxes, and roll-streak tubes with PRAS media yield comparable recoveries of anaerobic bacteria commonly encountered in properly collected and transported clinical specimens (28, 40). Success with these systems requires that the media be nutritionally adequate and either fresh or properly reduced and that the system be properly used and have an active catalyst (in jars and glove boxes). Factors to consider in the selection of a system to use in a particular laboratory include available space, specimen work load, cost of the equipment, cost of the media used with the system, and the technical capabilities of the personnel.

Anaerobe jars

An anaerobe jar is a cylindrical container made of plastic, glass, or metal. A metal or plastic lid (usually with an O-ring gasket) is clamped to a flange at the top of the jar to create an airtight seal. Some jar lids have vents or valves through which air can be evacuated and an anaerobe gas mixture can be added. A vented lid is not required if an H_2-CO_2 generator (GasPak, BBL; Oxoid Anaerobic Jar, Oxoid, U.S.A., Inc., Columbia, Md.; and Scott) is used. After the jar is sealed, the addition of a gas mixture containing H_2 in the presence of a catalyst reduces oxygen inside the jar to form water. Several different jars have been used for the cultivation of anaerobic bacteria (29). In this country, the GasPak jar is still the most widely used anaerobic system.

The GasPak jar uses a "cold" catalyst consisting of palladium-coated alumina pellets which is active at room temperature. This catalyst is inactivated by H_2S, by other volatile metabolic products of bacteria, and by excessive moisture. Optimal activity can be assured if the catalyst is replaced with new or rejuvenated pellets before each use (11). Heating in a dry-heat oven at 160 to 170°C for at least 2 h restores the catalyst to full activity. After being heated, the pellets should be stored in a clean, dry, airtight container.

Anaerobic conditions can be established in jars either by the ER technique or with a disposable H_2-CO_2 generator. Although both methods are simple to use and effective, the ER technique is more economical and establishes anaerobic conditions more rapidly than the H_2-CO_2 generator.

ER technique. A suitable gas mixture for the ER procedure (10% H_2, 5% CO_2, 85% N_2) is available from several suppliers. An inexpensive setup for the ER procedure (50) is shown in Fig. 3. In-house vacuum is sufficient if it permits evacuation of the jar to 20 to 24 in. (51 to 61 cm) of mercury; otherwise, a vacuum pump is needed. To perform the procedure, replace the used catalyst in the lid of the jar with fresh or rejuvenated pellets. Put the materials to be incubated inside the jar. Caps of screw cap tubes must be loosened to permit the exchange of gases. Place a methylene blue indicator in the jar. An indicator strip (BBL) or a tube of methylene blue-$NaHCO_3$-glucose mixture can be used (11). After fastening the lid to the jar, connect the vent on the lid to an ER device as illustrated in Fig. 3. Evacuate the jar to 20 to 24 in. of mercury, and fill the jar with commercial-grade N_2. Repeat this cycle. Evacuate the jar for the third time, but instead of N_2, fill the jar with the anaerobe gas mixture. Clamp the rubber tubing attached to the vented jar, disconnect the jar from the vacuum and gas line, and then place the jar in an incubator.

GasPak generator procedure. The disposable GasPak H_2-CO_2 generator consists of a sealed foil envelope containing two tablets. One contains citric acid and sodium bicarbonate, and the other contains sodium borohydride. When water is introduced into the enve-

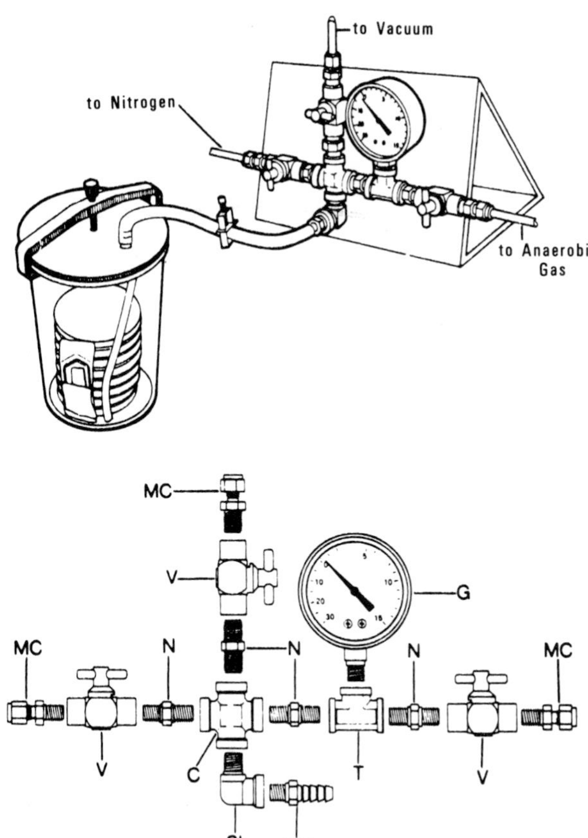

FIG. 3. ER system (from Whaley and Gorman [50], reproduced with permission). The lower drawing shows the parts needed for construction: G, Ashcroft compound ± gauge; N, 0.25-in. (0.6-cm) brass hex nipple: T, 0.25-in. brass tee female; C, 0.25-in. brass cross female; HC, 0.25-in. brass male hose connector; SL, 0.25-in. street L; MC, 0.25-in. pipe to 0.25-in. Swagelok; and V, 0.25-in. brass female circle seal valve.

lope, the former tablet releases CO_2, while the sodium borohydride tablet releases H_2. The GasPak generator procedure is described below.

Replace the used catalyst in the jar with new or rejuvenated catalyst. Place the materials to be incubated into the jar (remember to loosen the caps of any screw cap tubes), and include methylene blue indicator. Snip the corner of the generator envelope, and place the envelope in an upright position in the jar with the materials to be incubated. Add 10 ml of tap water or distilled water with a syringe or pipette according to the directions of the manufacturer. Clamp the lid on the jar. Place the jar in an incubator. Examine the jar to make sure it is working. The lid with catalyst on its undersurface should be warm to the touch, and condensation should appear on the inside surface of the jar within 15 to 30 min after the generator is activated. If condensation does not appear, remove the lid and check the jar for leaks; repeat the setup. The most common causes of GasPak jar failure include failure to use an active catalyst, a faulty or defective gas generator, and a poor sealing gasket in the lid. Occasionally, water fails to reach the two tablets inside the GasPak generator envelope. Use a new envelope when this occurs.

An oxidation-reduction potential indicator should always be included in the jar. Disposable methylene blue indicator strips (BBL) are blue when oxidized and colorless when reduced. Although 5 h or more usually elapse before the indicator becomes colorless after the jar is placed in a 35°C incubator, the indicator should remain colorless as long as the system stays anaerobic. When functioning properly, GasPak jars with generators usually contain $<1\%$ oxygen within 0.5 to 1 h after the lid has been sealed and the jar has been placed in a 35°C incubator (unpublished data).

Recently, Seip and Evans (42) reported that at 20 to 25°C, the O_2 concentration in the GasPak 100 Anaerobic System was 0.2 to 0.6% within 60 min after activation of the generator. At the same time, the CO_2 concentration was 4.6 to 6.2%, and the E_h values of three different media were −30 to −229 mV. These results indicated a rapid lowering of the oxygen concentration and the establishment of reducing conditions in the media, even though the methylene blue indicator did not become decolorized in less than 6 h.

Anaerobic glove box

An anaerobic glove box consists of a gas-tight chamber with glove portals and an entry lock for the transfer of materials in or out of the chamber. The operator of the chamber places his or her hands and arms in gloves to handle the culture materials which have been placed inside. A hydrogen-containing atmosphere is recirculated through a palladium catalyst to remove oxygen from inside the chamber. Recently, glove boxes of several different designs have been described and made commercially available.

The flexible-vinyl plastic glove box, originally developed by R. G. Freter and colleagues at the University of Michigan, has undergone several modifications (26, 29) and is now a practical system for the cultivation of anaerobic bacteria in the clinical laboratory. It is commercially available in two sizes (Coy Laboratory Products, Inc., Ann Arbor, Mich.). The type A chamber has a bag 65 in. (165 cm) wide by 32 in. (81 cm) deep by 40 in. (102 cm) high and two pairs of sleeves and gloves. Equipment is placed in the chamber through a 25-in. (64-cm) aluminum equipment port on one end of the chamber. A rigid metal entry lock is attached to the other end of the vinyl chamber. A vacuum pump, a gas mixture tank (containing 85% N_2, 10% H_2, and 5% CO_2), and a tank of commercial-grade N_2 are connected to the entry lock. An ER procedure similar to that described above for jar systems is used for passing materials in and out. The N_2 is not absolutely essential for either the ER jar procedure or the glove box, but it helps to conserve the more expensive gas mixture. The entry lock is evacuated to 20 in. (51 cm) of mercury and refilled with N_2. The cycle is repeated, and after a third evacuation, the entry lock is filled with the gas mixture. The inner door of the lock can then be opened. Recently, the manufacturer has marketed an optional device which automates the ER procedure.

Media are incubated within an incubator placed inside the chamber, or the entire chamber can be maintained at 35°C (i.e., by using heated catalyst boxes). An incubator permits the interior of the glove box to remain at room temperature. More space is available for the incubation of cultures in a heated glove box than if an incubator is placed inside. Unfortunately, a heated box is too hot to work in comfortably for more than short times, media tend to dry out, and condensation is more difficult to control than in an unheated chamber.

Glove boxes of different design are available from other manufacturers. Cox and Mangels (7) described a small portable chamber (30 in. [76 cm] wide by 18 in. [46 cm] deep by 18 in. [46 cm] high) which does not require the microbiologist to wear gloves. This chamber is commercially available (Anaerobe Systems, Santa Clara, Calif.). A glove box of another design has been described by Dickman et al. (8) and has been marketed (Heinicke Instruments, Inc., Hollywood, Fla.). This is a small (44.5 in. [113 cm] wide by 20.5 in. [52 cm] deep by 40 in. [102 cm] high), flexible-vinyl glove box attached to a metal base. The chamber is heated. Materials are passed in and out by use of a detachable entry module. We at the Indiana University Medical Center and the CDC (V. R. Dowell and G. L. Lombard, personal communication) have found that these glove boxes and the Forma model 1024 glove box (Forma Scientific, Marietta, Ohio) function satisfactorily.

The Forma glove box has a stainless steel cabinet (60 in. [152 cm] wide by 29 in. [74 cm] deep by 30 in. [76 cm] high) and will fit on a standard laboratory bench. The front panel of the cabinet is flexible vinyl and has one pair of sleeves and gloves. An incubator with sliding doors is built into one portion of the rear wall, and shelves for the storage of media and cultures are built into another portion. The outer door of the rectangular entry port opens to the front. Thus, the glove box can be placed adjacent to other equipment or in a corner of the laboratory. The entry port has an automated ER cycle. An anaerobic cabinet module (without an entry lock) can be added to the system to increase (more than double) the incubator and storage space.

Using a model GP Oxygen Analyzer (Lockwood-McLorie, Inc., Horsham, Pa.), we have found that the oxygen concentration in Coy and Forma anaerobic chambers can be maintained at 6 ppm or lower provided that the catalyst is changed weekly and there are no leaks in the systems. A less expensive, more practical way to monitor the atmosphere in an anaerobic glove box is to continuously circulate a portion of the gaseous atmosphere into a bottle of methylene blue indicator by using a small vibrator pump (29). Tears or leaks occasionally occur in any glove box. The location of the tear can be detected with a hydrogen leak detector (Coy) or by releasing a very small amount of Freon-12 within the glove box and scanning the exterior for leaks with a detector (e.g., General Electric Halogen Leak Detector, type H-10) as described elsewhere (26). Leaks can often be repaired with electrical tape, a silicone caulking compound, or a vinyl repair kit (Coy).

Relative humidity within a glove box should be maintained at 70 to 85%. The humidity can be determined with a hygrometer. Silica gel desiccant (Tell-Tale, grade H, type IV silica gel desiccant, Davidson Chemical Corp., Baltimore, Md.) placed in trays inside the chamber will decrease the humidity when it is excessive (e.g., ≥95%). The silica gel can be reused after being heated for 2 to 3 h at 150 to 200°C (it is blue when dehydrated).

PRAS media and the roll-streak tube technique

The roll tube, roll-streak system with PRAS media was developed by W. E. C. Moore and associates of the VPI Anaerobe Laboratory (24). Their system is based on the roll tube method of R. E. Hungate (1950) and has been used extensively in research studies, particularly on the digestive tract flora. Roll tubes and PRAS media are also used in some clinical laboratories. During preparation, the constituents of PRAS media are combined, boiled to remove dissolved oxygen, and then tubed, autoclaved, and stored in butyl rubber-stoppered tubes under oxygen-free gas (24). The roll tubes (25-mm diameter) usually contain PRAS-supplemented brain heart infusion agar medium in a thin transparent layer on the inner surface of the tube. These can be purchased (Scott and Carr-Scarborough) or prepared in the laboratory. After the tubes have been autoclaved, a rolling machine is used to solidify the agar on the surface of the glass. Both PRAS agar and PRAS liquid media require a reducing agent (i.e., L-cysteine hydrochloride) to maintain a low oxidation-reduction potential. The reducing agent is added before the medium is sterilized. The solid and liquid media are inoculated and subcultured in a stream of oxygen-free carbon dioxide with the aid of a VPI anaerobic culture system (Bellco Glass, Inc., Vineland, N.J.) or a Kontes Transoflex anaerobic culture system (Kontes Co., Vineland, N.J.) (Fig. 4). In essence, the roll tube or broth tube becomes its own anaerobic culture chamber and is then placed in an ambient air incubator for incubation.

Three kinds of commercially available carbon dioxide can be used for gassing roll tubes and liquid

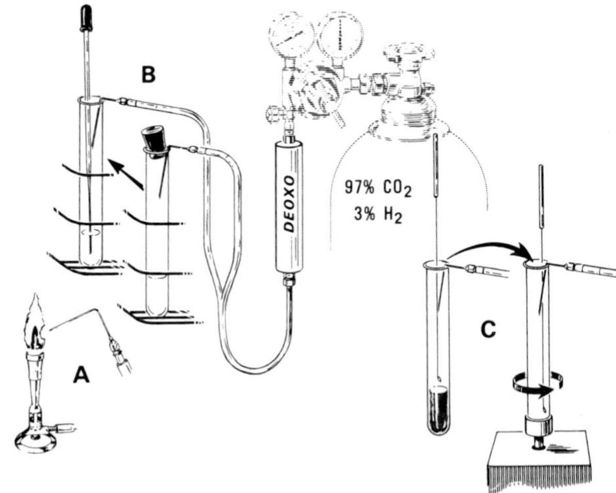

FIG. 4. Anaerobic tube culture technique (PRAS media). The Deoxo catalyst (palladium) is used if the CO_2 contains 3% H_2. (A) The flow from the gas cannula should indent the Bunsen flame. (B) Illustration of connected gassing apparatus. Necks of stoppered tubes are flamed; then the stoppers are removed with a modified hemostat (jaws bent to conform to stopper circumference) which is flame sterilized. A flame-sterilized CO_2 cannula is immediately inserted. Cultures are transferred with glass capillary pipettes or stainless steel or platinum loops. (C) Apparatus for streaking agar roll tube.

media. The first is anaerobe-grade (O_2-free) CO_2, which does not have to be passed through a catalyst. Second, commercial-grade CO_2 can be rendered oxygen-free by passing it through a heated copper catalyst (Sargent furnace). With use, the copper turnings in the Sargent furnace become oxidized and must be periodically reduced by passing hydrogen through the turnings. The hydrogen causes the black, oxidized copper to become reduced to its original bright copper color. The source of hydrogen to reduce the copper can be a mixture containing 97% CO_2 and 3% H_2. Third, this 97% CO_2-3% H_2 mixture can be used for tube culture if it is passed through a Deoxo Hydrogen Purifier (Fisher Scientific Co., Pittsburgh, Pa.) to remove traces of O_2. The Deoxo catalyst contains the same kind of palladium-coated alumina pellets used in the jar and glove box systems. The catalyst does not work unless the gas mixture contains a minimum of 3% H_2.

Liquid PRAS media can be inoculated with an inoculating loop or Pasteur pipettes (Fig. 4). A syringe and needle should be used if screw cap tubes with butyl rubber diaphragms are purchased (Scott and Carr-Scarborough). Agar roll tubes are inoculated according to the roll-streak technique of Holdeman et al. (24).

Like the anaerobic glove box, roll tubes permit the inspection and subculture of colonies at any time without an undue exposure of anaerobes to atmospheric oxygen. This protection is a major advantage over jar techniques.

Other anaerobic systems

The Bio-Bag type A Anaerobic Culture Set (Marion) consists of a gas-impermeable, transparent plastic bag and two ampoules. When the ampoules are crushed, one ampoule generates hydrogen and the other releases resazurin indicator. The Bio-Bag can be used as an anaerobic transport container for tissue (or a syringe containing pus), or it can be used to transport or incubate inoculated anaerobic plates. Tissue should be placed in a sterile, 100-mm petri dish within premoistened sterile gauze or in a loosely capped screw top container. The container of tissue is then put inside the bag, and the top of the bag is sealed with a heat sealer. In addition to its use as a transport device, a Bio-Bag can be used to incubate one or two plates for the cultivation of anaerobes. A slightly larger bag has recently been marketed which can hold three plates. When the gas generator is crushed and the bag is sealed, hydrogen released from the generator combines with oxygen in the presence of a catalyst to produce water vapor. Provided they are not obscured by moisture condensed on the bag, colonies can be observed without removing a plate from its bag.

Anagel (BBL) is a dehydrated, agar-containing powder which, when rehydrated and placed within the lid of a petri dish, is designed to reduce the concentration of oxygen, generate carbon dioxide, and produce an atmosphere suitable for the reduction of the culture medium and the cultivation of anaerobic bacteria. The rehydrated Anagel forms a seal with the edge of the bottom half of the petri dish. According to the manufacturer, oxygen is removed from both the culture medium and the headspace gas (between Anagel and the culture medium surface).

INCUBATION

The incubation temperature ordinarily used for the primary isolation and cultivation of most anaerobic bacteria in the clinical laboratory is 35°C. However, *C. perfringens* grows more rapidly at 42 to 47°C. It can be grown within 4 to 6 h in enriched cooked-meat–glucose and thioglycolate broth cultures incubated at 46°C (10). Chopped-meat agar for the demonstration of the spores of *Clostridium* species which do not sporulate readily is incubated at 30°C (16).

Inoculated media should ordinarily be incubated in an anaerobe jar for at least 48 h before the jar is opened (43). Duplicate jars can be used in urgent situations. Thus, one jar is opened after only 6 to 18 h of incubation, whereas the second jar remains in the incubator for 48 h or longer before it is opened. This early examination permits the rapid isolation and recognition of *C. perfringens* when gas gangrene is suspected. Of course, a jar can be opened inside a glove box at any time for an early examination of growth, resealed in the glove box, taken out, and reincubated.

If roll tubes are used or if plates are in an anaerobic glove box or a Bio-Bag type A (Marion), media can be inspected daily for growth without disturbing the developing colonies (i.e., by exposing them to oxygen). Solid primary isolation media should be reincubated for at least 5 to 7 days. This reincubation allows the recovery of slow-growing anaerobes.

Broth cultures should be incubated in an anaerobic system rather than in ambient air. Hold the thioglycolate and cooked-meat–glucose broth cultures for a minimum of 1 week before discarding them as negative. For suspected actinomycosis, osteomyelitis, endocarditis, and other serious infections, we recommend holding the broth cultures for at least 2 weeks of incubation.

EXAMINATION OF ANAEROBIC CULTURES

Inspection and subculture of colonies on primary isolation media

Obligate anaerobes vary on how long they can tolerate exposure to oxygen (31), as is discussed above. Some anaerobes (e.g., *B. fragilis* and *C. perfringens*) may survive on agar media exposed to room air for several hours, whereas other anaerobes lose viability after only a few minutes in air. Therefore, we recommend use of the holding jar procedure (see above) during inspection and subculture procedures to avoid a prolonged exposure of colonies to oxygen. Alternatively, both the anaerobic glove box and the VPI roll-streak techniques prevent oxygen exposure during these procedures. Although the Bio-Bag permits the inspection of plates for growth without exposure of the colonies to O_2, subculture from a Bio-Bag is analogous to using an anaerobic jar system.

Inspect the colonies on plates with a dissecting microscope (×7 to ×15 magnification), with both transmitted and reflected light. The dissecting microscope will help to ensure the isolation of single colonies and is also useful because the distinctive colonial features of many anaerobes which are not evident without magnification are seen. A hand lens can be used, but this method is less desirable.

Record the characteristics of each colony type, i.e., action on blood, egg yolk reactions, etc. Transfer each colony type to another anaerobe blood agar plate, and streak the plate to purify the isolate. Also, inoculate an aerobic blood agar plate to determine the ability of the isolate to grow in 5 to 10% CO_2-air (see below, under Determination of relationship to oxygen). If the colony is large enough and well separated, inoculate a tube of enriched thioglycolate or cooked-meat–glucose medium for biochemical inoculation and a tube of peptone-yeast-glucose medium for GLC, and prepare a Gram-stained smear. The thioglycolate and cooked-meat–glucose subcultures are incubated anaerobically for 24 to 48 h or until good growth occurs. At this time a Gram stain should be made. If the subculture looks pure, it can be used as a holding medium or to perform additional differential tests.

After incubation, check for the purity of each colony that was subcultured. Examine Gram stains prepared from the broth culture and the colonies on the blood agar plates incubated anaerobically. Record the relationship of each colony type to oxygen, as described below.

Reexamine the primary isolation plates after they have been incubated for 96 h. Observe for any slow-growing anaerobes which may have appeared. Look at the first quadrant where the specimen was inoculated onto the plate. This is the time and place at which pigmented colonies of the *B. melaninogenicus-B. asaccharolyticus* group will often be found. Isolate any new colony types as done previously.

Lastly, examine the thioglycolate or cooked-meat–glucose primary enrichment culture that was originally inoculated with the clinical specimen at the same time the primary isolation plates were inoculated. If no growth is present on the primary anaerobic plating media or if the colonies seen do not account for all the morphologic forms found in the direct Gram-stained smear of the clinical specimen, then subculture the broth medium to an anaerobe blood agar plate for anaerobic incubation and to a blood agar plate for CO_2-air incubation. Also, examine a Gram-stained smear prepared from the broth culture.

Morphologic considerations

The appearance of colonies on anaerobe blood agar, observed with a dissecting microscope, and the Gram reaction and microscopic features of isolates are especially useful in rapid presumptive identification. Morphology is particularly valuable for the rapid differentiation of the *B. fragilis* group, the *B. melaninogenicus* group, *Bacteroides ureolyticus, F. nucleatum, F. necrophorum, Fusobacterium mortiferum*, other fusobacteria, the actinomycetes, bifidobacteria, *C. perfringens, C. difficile, C. septicum, Clostridium sordellii, Clostridium ramosum, Clostridium sporogenes*, and other clostridia (29).

Microscopic features. The Gram reaction, morphology of vegetative cells, presence or absence of spores, motility, and flagella are all key features in the classification and identification of anaerobic bacteria. Gram-stained smears should be examined from both solid and liquid cultures that show good growth. Older cultures may be better for the detection of spores.

Many obligate anaerobes classified as gram positive are gram variable. Certain species (e.g., *C. ramosum*) are usually gram negative even though they are classified with gram-positive organisms.

Record the shape of the cells, size, intracellular features (e.g., vacuoles, granules, etc.), presence of branching, pleomorphism (e.g., filaments, swollen bodies, spherical forms, bifids, etc.), and the arrangement of cells (e.g., singly, pairs, tetrads, chains, clusters, picket fence arrangements).

Bacterial endospores can usually be observed with the Gram stain. Spores will be unstained. An examination of wet mounts by phase microscopy may aid in the search for spores. Free spores and endospores are much more refractile than vegetative cells. A spore stain may also be helpful at times. It is best to examine chopped-meat medium or another medium without carbohydrate when searching for spores of an isolate suspected of being a *Clostridium* species. It may be necessary to inoculate a chopped-meat agar slant (incubated for 5 days at 30°C) and to perform a heat shock or alcohol spore selection technique to demonstrate spores (30). Spores of some clostridia (e.g., *C. perfringens*) can seldom be demonstrated in laboratory culture. If spores are seen, note their shape, size, and position in cells (subterminal or terminal) and whether they cause swelling of the cells.

Flagellum stains are of most value for the characterization of motile, gram-negative anaerobic bacilli, e.g., *Succinimonas* spp., *Anaerovibrio* spp., *Anaerobiospirillum* spp., *Wolinella* spp., *Selenomonas* spp., and members of certain other gram-negative genera described in Chapter 40.

Colonial characteristics. Inspect the colonies on the primary plating media with a dissecting microscope, and describe them. Examine well-isolated colonies to study colonial morphology, and examine areas of heavy growth to detect microorganisms present in small numbers. Observe the agar surface for pitting colonies and spreading growth, and check for hemolysis. Examine the plate cultures under long-wave UV light (365 nm) to detect fluorescent colonies (e.g., *B. melaninogenicus* and *B. asaccharolyticus* often show brick red fluorescence). Describe the characteristics of each colony type. Record the medium, age (incubation time), color or pigment, and diameter, and describe the form, elevation, edge, surface, density, and consistency. Cultural characteristics will vary on different formulations of anaerobe blood agar.

Determination of relationship to oxygen

Determine the relationship to oxygen of each colony type present on anaerobic solid media. Inoculate the following media:

1. One anaerobe blood agar plate to be incubated anaerobically
2. One aerobic blood agar (or chocolate agar) plate to be incubated in a candle extinction jar or in a 5 to 10% CO_2 incubator (the chocolate agar is particularly needed to distinguish nutritionally fastidious *Haemophilus* spp. and other bacteria which will grow on anaerobe

blood agar incubated anaerobically and on chocolate agar under increased CO_2 tension but which fail to grow on blood agar in a 5 to 10% CO_2-air incubator or in air)

3. One aerobic blood agar plate to be incubated in room air

Incubate each plate at 35°C for a minimum of 24 h, or longer (at least 5 to 7 days) for slow-growing isolates. The length of incubation can be estimated from the time required for growth on the primary plates. (In addition, subculture each colony type onto either cooked-meat–glucose or enriched thioglycolate medium and into a tube of peptone-yeast-glucose medium for GLC if this was not done previously.)

Record the relationship to oxygen as obligate anaerobe or nonanaerobe (e.g., aerotolerant anaerobe, microaerophile, or facultative anaerobe) (Table 1). As defined above, obligately anaerobic bacteria do not grow on the surface of nutritionally adequate blood agar or chocolate agar incubated in room air, a candle jar, or a 5 to 10% CO_2-air incubator. For further definitions of terms and descriptions regarding the relationships of isolates to O_2, see the discussion above and Table 1.

USE OF GLC TO DETERMINE VOLATILE AND NONVOLATILE ACID METABOLIC PRODUCTS

The volatile and nonvolatile acid fermentation products of anaerobic bacteria are key characteristics for identification. Once the relationship to oxygen has been determined, the majority of clinically encountered anaerobes can be identified to genus on the basis of Gram reaction, spores, cellular morphology, and metabolic products. Gas chromatographic procedures for the determination of these products with inexpensive instruments with thermal conductivity detectors are simple, rapid, accurate, reproducible, and practical for the clinical laboratory. Although analysis by GLC is not mandatory for preliminary or presumptive identification of the *B. fragilis* group, the *B. melaninogenicus-B. asaccharolyticus* group, *F. nucleatum*, *C. perfringens*, *Peptostreptococcus anaerobius*, and certain other anaerobes commonly recovered from clinical specimens (1), it is required for the definitive identification of many *Bacteroides* and *Fusobacterium* species and most species of the genera *Clostridium*, *Peptococcus*, *Peptostreptococcus*, *Actinomyces*, *Arachnia*, *Bifidobacterium*, *Eubacterium*, *Lactobacillus*, and *Propionibacterium*.

Ether extract procedure for analysis of volatile fatty acids

The ether extract procedure is for the identification of short-chain volatile fatty acids that are soluble in ether. The acids detected include acetic, propionic, isobutyric, butyric, isovaleric, valeric, isocaproic, and caproic acids. Pyruvic, lactic, and succinic acids are not volatile and therefore are not detected by this procedure, but they can be identified after the preparation of volatile methylated derivatives.

1. Inoculate 7- to 8-ml tubes of prereduced peptone-yeast extract-glucose (11, 16) broth with a few drops (0.05 to 0.1 ml) of an actively growing culture or a few isolated colonies from an anaerobic blood agar plate.
2. Incubate the culture under anaerobic conditions for 48 h or until adequate growth is obtained.
3. Transfer 2.0 ml of the culture to a clean screw cap tube (13 by 100 mm).
4. Acidify the culture to pH 2.0 or below by adding 0.2 ml of 50% aqueous H_2SO_4.
5. Add 1 ml of ethyl ether, cap the tube, and invert the tube 20 times to mix the solution.
6. Centrifuge the solution briefly (750 to 1,000 × *g*) to break the ether-culture emulsion.
7. Place the ether-culture mixture in an alcohol-dry ice bath to freeze the aqueous (bottom) portion. Pour or pipette off the ether (top) layer into a clean tube. We prefer to use this freezing technique to avoid the pipetting of water with the ether layer.
8. Add one or two anhydrous $CaCl_2$ pellets to remove residual water.
9. Inject 14 μl of the extract into the SP-1220 or SP-1000 column of a gas chromatograph.
10. Identify the volatile acids by comparing the elution times of the products in the extracts with the elution times of a known acid mixture (volatile-fatty-acid standard) chromatographed on the same day.

Procedure for the analysis of nonvolatile acids (pyruvate, lactate, succinate, and phenylacetate)

1. Transfer 1 ml of the peptone-yeast extract-glucose culture to a clean screw cap tube.
2. Add 0.4 ml of H_2SO_4 and 2 ml of methanol. Place the tube in a 55°C water bath overnight.
3. Add 1 ml of distilled water and 0.5 ml of chloroform; then centrifuge the solution briefly to break the emulsion (chloroform will be in the bottom of the tube).
4. After placing the tip of the needle in the chloroform layer, fill a syringe with the chloroform extract.
5. Wipe the outside of the needle with a clean tissue, and inject 14 μl of the extract into the SP-1000 column. (The SP-1220 column does not permit the separation of pyruvate from lactate.)
6. Identify nonvolatile or methylated acids by comparing the elution times of products with the elution times of the nonvolatile-acid standard solution chromatographed on the same day.
7. After testing approximately 20 methylated samples, recondition the packing material by injecting 14 μl of methanol into the column of the gas chromatograph.

GLC standards and controls

Standard solutions containing 1 meq of each volatile and nonvolatile acid per ml should be examined each time unknowns are tested. The volatile-acid standard should contain at least the following: acetic, propionic, isobutyric, butyric, isovaleric, valeric, isocaproic, and caproic acids. The nonvolatile-acid standard should contain pyruvic, lactic, succinic, and phenylacetic acids.

It is also important to examine a tube of uninoculated medium in the same manner, since different lots of peptone-yeast extract-glucose broth may contain significant quantities of some acids.

Equipment and operating conditions

Gas chromatographs are relatively inexpensive, safe, simple to operate, and reliable. They are commercially available from various manufacturing companies. The operating conditions for two of the chromatographs with thermal conductivity detectors currently available are summarized in Table 8.

Stainless steel or aluminum columns (6 ft. by 0.25 in. [1.8 m by 0.6 cm]) are used. Column packing materials currently recommended for the determination of metabolic products include (i) 15% SP-1220–1% H_3PO_4 on 100/120 Chromosorb W AW (Supelco Inc., Bellefonte, Pa.), which is used only for determining volatile metabolic products since it does not permit resolution of pyruvate from lactate under the above operating conditions, and (ii) 10% SP-1000–1% H_3PO_4 on 100/120 Chromosorb W AW (Supelco Inc.), which can be used for either volatile or nonvolatile metabolic product analysis but resolves the volatile acids more slowly than does the SP-1220 packing.

Several other liquid media which support good growth of anaerobes (e.g., chopped meat-glucose, Schaedler broth, and Lombard-Dowell glucose) can be used instead of peptone-yeast extract-glucose broth for GLC. However, the metabolic products of an organism grown in another medium may differ from those produced in peptone-yeast extract-glucose medium. For example, some anaerobes produce isoacids (e.g., isobutyric and isovaleric acids) in peptone-yeast extract-glucose and other media of high peptone content but fail to produce these acids in media of lower peptone content, such as Lombard-Dowell glucose medium (G. L. Lombard, F. S. Thompson, and A. Y. Armfield, Abstr. Annu. Meet. Am. Soc. Microbiol. 1978, C29, p. 282). Lombard and associates have also found that hemin, vitamin K, yeast extract, L-cystine, and other supplements which affect growth can have a profound influence on fermentation products. Thus, caution should be exercised in the interpretation of the results of GLC from a different medium when the identification tables referred to were prepared from a peptone-yeast extract-glucose data base. The peptone-yeast extract-glucose medium formulation used at CDC (11, 16) is different from the medium recommended in the fourth edition of the VPI manual (24). If chopped-meat–glucose medium is used, the amount of lactic acid present in the uninoculated medium makes it difficult, if not impossible, to determine whether lactic acid was produced by an unknown isolate.

Further details on equipment and procedures for the determination of metabolic products by GLC are given elsewhere (11, 24, 29, 32, 46).

OVERVIEW OF SYSTEMS FOR BIOCHEMICAL CHARACTERIZATION

The biochemical characteristics that are most useful for the identification of anaerobe isolates vary, depending on whether one is dealing with clostridia, anaerobic cocci, gram-negative nonsporeforming anaerobic bacilli, gram-positive nonsporeforming anaerobic bacilli, or anaerobic spirochetes. It is beyond the scope of this section to include all the approaches and procedures that are currently used for the biochemical characterization of anaerobic bacteria. Some of the most commonly used approaches are discussed below and reviewed elsewhere (3a, 29, 44); other approaches are included in the chapters on anaerobe identification that follow.

Conventional systems

Conventional systems for the characterization of isolates with biochemical test media in large tubes are mentioned briefly. For those workers interested in further details on conventional test procedures, refer to the VPI *Anaerobe Laboratory Manual* by Holdeman et al. (24), the CDC laboratory manuals by Dowell and colleagues (11, 16, 32), the CDC tables by Dowell and Lombard (14), and the *Wadsworth Anaerobic Bacteriology Manual* by Sutter et al. (46).

PRAS media, for determining the biochemical characteristics of anaerobes, have long been used for phenotypic work in studies of taxonomic classification and for reference identification (24). For the PRAS media approach, tubes containing medium for carbohydrate fermentation tests are inoculated with a special gassing apparatus (Fig. 4) or with syringe and needle, and the pH is measured after incubation and growth of the organism. Other tests are for growth in the presence of bile; production of indole, H_2S, urease, lipase, and lecithinase; proteolysis and catalase activity; hydrolysis of gelatin, esculin, and starch; and the reduction of nitrate. The media can be prepared in the laboratory according to the directions of Holdeman et al. (24), or they can be purchased from commercial manufacturers. Recently, the results of identifications with PRAS media prepared by two different manufacturers (Carr-Scarborough and the PRAS II system, Scott) were compared in our laboratory. The Carr-Scarborough PRAS media and the PRAS II system showed 95% agreement for identifications with the VPI manual (S. D. Allen, J. A. Siders, and N. A. Baker, Abstr. Annu. Meet. Am. Soc. Microbiol. 1983, C425, p. 382).

The conventional media of Dowell et al. (16), used by the CDC and other reference laboratories, are formulated differently than PY-base PRAS media and are commercially available from Carr-Scarborough and Nolan Laboratories (Atlanta, Ga.). Fermentation tests in the CDC CHO-based media are read with bromthymol blue indicator (yellow at pH 6.0). Recently, we found 99% agreement between identifications based on using the PRAS media procedure of Holdeman et al. (24) and identifications based on the CDC

TABLE 8. Operating conditions for two commercially available gas chromatographs with thermal conductivity detectors

Gas chromatograph[a]	Temp (°C)			Attenuation	Gas flow rate (cm^3/min)	Chart speed (cm/min)	Bridge current (mA)
	Detector	Injection port	Column				
CAPCO	Preset	Preset	145	1	120	1	100
GOW-MAC	152	152	115	1	120	2	200

[a] Manufacturers: CAPCO, Dodeca, Fremont, Calif., and GOW-MAC, Gow-Mac Instrument Co., Bound Brook, N.J.

media using the procedures of Dowell and Hawkins (11). In a separate study, the reproducibility of tests in each system was greater than 99% (S. D. Allen et al., Abstr. Annu. Meet. Am. Soc. Microbiol. 1983, C425, p. 382). Even with the close agreement of identification results obtained with these two conventional systems, our experience supports the concept that VPI tables (24) should be used for PY-base PRAS media and that CDC tables (11, 14) should be used for identifications when thioglycolate or CHO-based biochemicals are used according to the formulations of Dowell et al. (16). A computer program is commercially available for use with the PRAS II system (Scott).

Minitek and API 20A packaged microsystems

Two commercially available, packaged micrometh-od systems have been widely used for anaerobe identification. These are the API 20A (Analytab Products, Inc., Plainview, N.Y.) and the Minitek (BBL) systems. Both systems have been evaluated in a number of laboratories, and reviews of these studies have been published (29, 44).

The API 20A system consists of a plastic strip with 20 microtubules containing dehydrated substrates to determine the following: indole, urease, and catalase production, gelatin and esculin hydrolysis, and the fermentation of glucose plus 15 additional carbohydrates. The microtubules of the strip are inoculated with a turbid suspension of fresh colonies. There is now a data base for reading the API 20A reactions at 24 h, or the strips can be read at 48 h. The indicator to detect carbohydrate fermentation is bromocresol purple. An acid (positive) reaction is yellow (pH 5.2) or yellow-green, and a negative reaction is purple. Various anaerobes (mainly clostridia) reduce the indicator to colorless, straw yellow, muddy green-yellow, or pale (bleached-out) purple. If this change occurs, it is necessary to add back indicator to each carbohydrate tubule before reading the reactions. However, reactions still may not be clear-cut. Further details of the procedure are supplied by the manufacturer along with identification tables and a numerical analytical profile index.

The Minitek anaerobe differentiation system uses paper disks impregnated with various biochemical substrates. The disks are dispensed into wells of a special disposable plastic plate. Biochemical characteristics which can be determined with the system include the fermentation of glucose and several other carbohydrates plus esculin hydrolysis, nitrate reduction, and production of indole. The inoculum is prepared from fresh colonies on blood agar to achieve a dense suspension. Plates are incubated anaerobically for 48 h. Color reactions in this system may also be difficult to interpret because of reduction of the phenol red indicator. A positive reaction for carbohydrate fermentation is indicated by yellow (phenol red is yellow at pH 6.8 and lower). Any orange to red-orange is negative. As for the API 20A system, exposure of the esculin well to long-wave UV light is recommended to test for esculin hydrolysis. The directions of the manufacturer should be consulted for further details. A Minitek numerical identification system is available.

Laboratories using either of these micromethod systems should use only the identification tables or numerical systems provided by the manufacturers. Both microsystems work best with selected saccharolytic gram-negative bacilli. The systems have been criticized for a lack of sufficient tests to characterize and identify asaccharolytic or weakly saccharolytic anaerobes, such as anaerobic cocci, and both systems should be supplemented with several additional tests. Tests that are often required for the identification of anaerobes that are not determined with these kits per se include the determination of the relationship of isolates to oxygen, colony characteristics, reactions on blood agar, Gram reaction, microscopic features, appearance and rapidity of growth in liquid medium, lecithinase and lipase activities on egg yolk agar, growth in the presence of 20% bile, action on milk, inhibition of growth by selected antibiotics and by sodium polyanetholsulfonate, and metabolic products determined by GLC.

Presumpto plate system

The Presumpto quadrant plate system that was developed by Dowell, Lombard, and associates of the CDC (15) is a practical and accurate way to characterize and identify anaerobes. This system has differential tests in Lombard-Dowell agar, with antibiotic disks, three quadrant plates, aerotolerance testing, observation of colonies, cellular morphology, and other criteria for identification (15, 29). The commercially available Presumpto plate packaged system (Nolan Laboratories; Carr-Scarborough; and Remel, Lenexa, Kans.) consists of three quadrant plates containing 12 differential agar media. The following characteristics are determined: indole, indole derivatives, growth on plain Lombard-Dowell agar, esculin hydrolysis, H_2S, catalase, lipase, lecithinase, proteolysis, growth in presence of 20% bile, and an insoluble precipitate in bile agar in Presumpto plate 1; glucose fermentation, stimulation of growth by fermentable carbohydrate, starch hydrolysis, casein hydrolysis, and DNAse activity in Presumpto plate 2; and mannitol, lactose, and rhamnose fermentation and gelatin hydrolysis in Presumpto plate 3. Presumpto plate 1 can be used with a CDC anaerobe blood agar plate on which disks are placed to determine inhibition by penicillin (2-U disk), kanamycin (1,000-μg disk), and rifampin (15-μg disk) for the presumptive identification of gram-negative nonsporeforming anaerobic bacilli (12). Identification tables for use with the Presumpto system are published elsewhere (12, 13, 29).

AT system

The Anaerobe-Tek (AT) system (Flow Laboratories, Inc., McLean, Va.), another packaged system that was recently marketed for anaerobe identification, consists of a round plastic plate divided into 11 peripheral compartments and a central well. This system permits the detection of 15 different characteristics within differential agar media. Although there is some similarity to the Presumpto quadrant plate system, there are several differences, particularly with medium composition and the performance of certain tests (33). The AT system permitted correct identification of only about 50% of the clinically isolated anaerobes tested in two collaborating laboratories (S. D. Allen, J. Snyder, J. A. Siders, and N. A. Baker, Abstr. Annu. Meet. Am. Soc. Microbiol. 1982, C294, p. 320), and its reproducibility was between 92 and 99% in the two laboratories. At this time, the data base of the AT system needs further work and expansion; media and procedural modifications are also needed.

Multiple-well microtube plate procedures for biochemical characterization and susceptibility testing of anaerobes

Recently, we have been developing a practical microtube plate system that permits the determination of up to 50 differential tests for the expanded biochemical characterization of anaerobe isolates (3). A data base and a computer program for data storage with numerical profiles and tables for identification have been developed. Media can be distributed into plates containing 96 wells (Bellco) with an automated Quick Spense dispensing device (Bellco) and then frozen at −80°C until needed. In addition to biochemical testing, broth microdilution antimicrobial susceptibility tests done with this system show a high correlation with the agar dilution reference method (3; see Chapter 100). This system is rapid to inoculate and read, and it offers potential cost savings to laboratories that have the resources to prepare and store their own plates. Prepackaged microtube plates for microtube susceptibility testing are now commercially available from at least two companies; these plates should be of interest to laboratories that lack the equipment and personnel to manufacture their own plates.

Rapid enzyme systems

An exciting new development in clinical anaerobic bacteriology is the introduction of commercially packaged microsystems for the detection of enzymes within a few hours after the systems have been inoculated (3a). One enzyme detection system, API ZYM (Analytab), has been marketed for several years as a research tool but is without a data base for anaerobe identification. Recent work with the API ZYM system has indicated that enzyme profiles of clinically encountered anaerobes and selected stock reference strains are sufficiently distinctive to merit future development of this kit for the rapid identification of anaerobes (33a). In addition to API ZYM, a second rapid system, called the An-Ident, has been recently marketed by Analytab. Like API ZYM, An-Ident requires only 4 h of aerobic incubation. Unlike the situation with API ZYM, a data base and numerical identification profile for An-Ident are being developed by the manufacturer, and the system differs from the API ZYM system with regard to the dehydrated substrates for enzyme detection. It also uses some miniaturized conventional tests and requires different reagents. The An-Ident system is currently being evaluated in a number of laboratories. Recently, the IDS RapID ANA (Innovative Diagnostic Systems, Inc., Atlanta, Ga.) micromethod was introduced, which is also a rapid enzyme kit for the identification of anaerobes within 4 h of aerobic incubation after inoculation with a turbid culture suspension. A comparison in our laboratory of IDS RapID ANA with conventional identifications with PRAS media demonstrated correct identifications of 90% of more than 300 fresh clinical isolates of anaerobic bacteria (L. M. Marler, N. B. O'Bryan, J. A. Siders, and S. D. Allen, Abstr. Annu. Meet. Am. Soc. Microbiol. 1984, C149, p. 261).

EXTENT OF PROCESSING AND EXPEDITED REPORTING OF RESULTS

The results of the direct microscopic examination should be reported rapidly, during the same day the specimen was received. This report should include a phone call in urgent situations, in addition to sending a report of direct Gram stain results to be put on the patient's chart. A preliminary culture report should be sent when an isolate is shown to be an obligate anaerobe. This preliminary report is often 24 h after performance of the anaerobic and aerobic subcultures from the primary isolation plate, but slow-growing anaerobes may require more time (e.g., 2 to 7 days). The report should describe the Gram reaction, cellular morphology, relation to oxygen, and relative number (e.g., rare, few, moderate, many, or 1+, 2+, 3+, or 4+) of each organism on the primary isolation plates. At this time it may be possible to report the presumptive identity to species of *C. perfringens*, *F. nucleatum*, the pigmented or fluorescent *B. melaninogenicus*-*B. asaccharolyticus* group, and certain others. Members of the *B. fragilis* group can often be recognized by an experienced microbiologist on the basis of Gram stain and colony characteristics (as observed through a dissecting microscope) on CDC anaerobe blood agar (29), phenylethyl alcohol, and paromomycin-vancomycin or kanamycin-vancomycin medium on the basis of the relationship to oxygen (i.e., obligate anaerobe; Table 1). Identification tests should then be performed for each isolate as described in Chapters 38 through 41, depending on the morphologic group of the anaerobic isolate.

Wounds and abscesses frequently contain multiple species of bacteria (9), among which anaerobes are mixed with other anaerobes, facultative anaerobes, or aerobes. There may be 6, 8, 10, or more colony types on the primary isolation plates, and it is likely that species present in small numbers may be missed. The cost (in terms of time and supplies) of identification and susceptibility testing of multiple isolates is considerable. There are no established rules on whether to limit the number of isolates identified (1). The relative proportion of squamous epithelial cells and polymorphonuclear leukocytes and the relative number of different morphotypes of microorganisms present in the Gram-stained smear of the specimen aid in assessing specimen quality. When numerous morphologic forms of bacteria are seen (e.g., in a wound, peritoneal fluid, endometrial specimen, etc.) but no inflammatory cells are present, it is our practice to contact the physician, discuss the clinical setting, attempt to interpret the quality of the specimen, and limit the extent of culture workup. In most instances there is no useful information to be gained by culture and identification when the direct smear suggests or confirms that the specimen was of poor quality. In some instances, the physician may wish to know only if certain anaerobe species are present (e.g., *B. fragilis* or *C. perfringens*). On the other hand, in certain clinical situations (e.g., bacteremia, brain abscess, lung abscess, tuboovarian abscess, crepitant cellulitis, myonecrosis, and other serious infections) (29), it would be a serious mistake to limit identification of the anaerobe species. It is desirable to hold cultures for several days under anaerobic conditions when isolates are reported to a preliminary group level. This holding will allow more complete identification if it becomes desirable or is requested by the physician.

It is emphasized that biochemical reactions in different test systems depend on the ratio of inoculum to substrate, the substrate concentration, the composition and buffering capacity of the basal medium, the pH endpoint as determined by a pH meter or pH indicator, and other variables (29, 44). Thus, the results of differential tests in one system will not necessarily agree with those obtained in another system. The tables referred to for the identification of isolates should be those based on the media and methods used for the characterization of an isolate. For example, you should not use VPI tables if you use Minitek differential tests. The procedures and references recommended by the authors of Chapters 38 through 41 should be followed if their identification tables are to be interpreted accurately.

The new rapid enzyme systems, discussed above, are of considerable interest. As we enter the era of prospective payment and diagnosis-related groups, these systems offer promising approaches to microbiologists who are faced with providing reasonably accurate identification results to clinicians within a time frame that is relevant to immediate patient care, while minimizing laboratory costs (1a, 3a). The potential clinical usefulness of anaerobe identifications should be much greater with a system that identifies isolates on the same day that colonies are available in pure culture than with an older system that requires 1, 2, or more days of anaerobic incubation before identifications can be made. It appears that many of the asaccharolytic anaerobes, or weak fermenters, that "do nothing" or are nonreactive in traditional carbohydrate tests are reactive in the new enzyme kits. Conventional systems for the identification of anaerobes will remain the reference standard methods for phenotypic characterization and will be necessary for taxonomic classification and other research. Unfortunately, conventional systems for anaerobe identification are labor intensive and have a relatively high materials cost. The overall time required for laboratory personnel to inoculate, read, and record results with the rapid enzyme kits is about 2 to 3 min, which is less time than for the conventional systems. Also, materials costs of the new systems, thus far, are competitive. Gram reaction, morphology, determination of relationships to oxygen, and a few additional tests are required to supplement the rapid enzyme systems, but the need to use GLC and a battery of other tests is less than that required for older packaged micromethod kits (e.g., MiniTek and API 20A) that depend on the growth and fermentation of carbohydrates. As a note of caution, the data bases of these enzyme systems will need to be expanded in some areas (e.g., additional data are needed for nonsporeforming gram-positive anaerobic bacilli and clostridia) before these systems can be recommended for routine use (1a, 3a).

It is often difficult to know how far to go in the identification of a given anaerobe isolate (1, 9). The time required will depend both on the growth rate of the isolate and on the kinds of tests used (or required) for its differentiation. The results of presumptive identification should be reported as soon as they are available. Definitive identification (i.e., by a conventional approach) should be carried out as dictated by clinical circumstances, the needs of the physician, and research interests. Definitive identification may take longer but is still required to further define the role of obligate anaerobes in health and disease, to educate both clinicians and microbiologists, and to assist the physician in providing optimal patient care. In an individual laboratory, the choice of identification procedures and the extent of anaerobe identification will depend on the technical competence of the personnel, the resources available, the patient population being served, and the needs of the physician staff (1).

Use of reference laboratories

Laboratories are encouraged to utilize the services of reference laboratories for assistance or confirmation of identification, especially in serious illnesses such as anaerobe septicemia. Reference laboratories are also useful for the performance of anaerobe susceptibility tests if the laboratory does not have this capability. Isolates should be submitted in agar deeps (e.g., motility medium, Thiogel [BBL], etc.), plain cooked-meat medium, or a transport medium (e.g., Cary-Blair or the Port-A-Cul; see references 29 and 46). Inoculate the agar deep or cooked-meat medium to the bottom with a capillary pipette, and incubate the medium until good growth has occurred before shipment. Inoculate the Cary-Blair or Port-A-Cul transport medium with a swab from fresh colonial growth, but do not incubate it before shipment. All tubes should be tightly sealed with a screw cap (11, 29).

Antimicrobial susceptibility testing of anaerobic bacteria

Refer to Chapter 103 for a discussion of methodology and the current options on antimicrobial susceptibility testing of anaerobic bacteria for the clinical microbiology laboratory.

LITERATURE CITED

1. **Allen, S. D.** 1979. Identification of anaerobic bacteria: how far to go. Clin. Microbiol. Newsl. **1**:3–5.

1a. **Allen, S. D.** 1984. Current relevance of anaerobic bacteria. Clin. Microbiol. Newsl. **6**:147–149.

2. **Allen, S. D., and J. A. Siders.** 1982. An approach to the diagnosis of anaerobic pleuropulmonary infections. Clin. Lab. Med. **2**:285–303.

3. **Allen, S. D., J. A. Siders, K. S. Johnson, and E. H. Gerlach.** 1982. Simultaneous biochemical characterization and antimicrobial susceptibility testing of anaerobic bacteria in microdilution trays, p. 266–270. *In* R. C. Tilton (ed.), Rapid methods and automation in microbiology. American Society for Microbiology, Washington, D.C.

3a. **Allen, S. D., J. Siders, L. Marler, and N. O'Bryan.** 1984. Rapid identification of anaerobes, p. 233–240. *In* A. Sanna and G. Morace (ed.), New horizons in microbiology. Elsevier Science Publishers B.V., New York.

4. **Brock, T. D., and K. O'Dea.** 1977. Amorphous ferrous sulfide as a reducing agent for culture of anaerobes. Appl. Environ. Microbiol. **33:**254–256.
5. **Brock, T. D., D. W. Smith, and M. T. Madigan.** 1984. Biology of microorganisms, 4th ed. Prentice-Hall, Inc., Englewood Cliffs, N.J.
6. **Cooksey, R. C., F. S. Thompson, and R. R. Facklam.** 1979. Physiological characterization of nutritionally variant streptococci. J. Clin. Microbiol. **10:**326–330.
7. **Cox, M. E., and J. I. Mangels.** 1976. Improved chamber for the isolation of anaerobic microorganisms. J. Clin. Microbiol. **4:**40–45.
8. **Dickman, M. D., A. R. Chappelka, C. Aff, J. Gerhard, and R. P. Orcutt.** 1979. Compact anaerobic glove box for hospitals and research laboratories. J. Clin. Microbiol. **9:**294–296.
9. **Dowell, V. R., Jr.** 1975. Wound and abscess specimens, p. 70–81. *In* A. Balows (ed.), Clinical microbiology. How to start and when to stop. Charles C Thomas, Publisher, Springfield, Ill.
10. **Dowell, V. R., Jr., and S. D. Allen.** 1981. Anaerobic bacterial infections, p. 171–214. *In* A. Balows and W. J. Hausler, Jr. (ed.), Diagnostic procedures for bacterial, mycotic, and parasitic infections, 6th ed. American Public Health Association, Washington, D.C.
11. **Dowell, V. R., Jr., and T. M. Hawkins.** 1977. Laboratory methods in anaerobic bacteriology. CDC laboratory manual. Department of Health, Education, and Welfare publication no. (CDC) 78-8272. Center for Disease Control, Atlanta.
12. **Dowell, V. R., Jr., and G. L. Lombard.** 1977. Presumptive identification of anaerobic nonsporeforming gram-negative bacilli. Center for Disease Control, Atlanta.
13. **Dowell, V. R., Jr., and G. L. Lombard.** 1981. CDC tables: reactions of anaerobic bacteria in differential agar media. U.S. Department of Health and Human Services, Public Health Service, Centers for Disease Control, Atlanta.
14. **Dowell, V. R., Jr., and G. L. Lombard.** 1982. CDC tables: differential characteristics of anaerobic bacteria. U.S. Department of Health and Human Services, Public Health Service, Centers for Disease Control, Atlanta.
15. **Dowell, V. R., Jr., and G. L. Lombard.** 1982. Differential agar media for identification of anaerobic bacteria, p. 258–262. *In* R. C. Tilton (ed.), Rapid methods and automation in microbiology. American Society for Microbiology, Washington, D.C.
16. **Dowell, V. R., Jr., G. L. Lombard, F. S. Thompson, and A. Y. Armfield.** 1977. Media for isolation, characterization, and identification of obligately anaerobic bacteria. CDC laboratory manual. Center for Disease Control, Atlanta.
17. **Finegold, S. M.** 1977. Anaerobic bacteria in human disease. Academic Press, Inc., New York.
18. **Finegold, S. M.** 1983. Antimicrobial therapy of anaerobic infections, p. 39–55. *In* J. E. Yetiv and J. R. Bianchine (ed.), Recent advances in clinical therapeutics, vol. 3. Antivirals and antimicrobials, anticancer agents, cardiovascular therapy. Grune and Stratton, Inc., New York.
19. **Finegold, S. M. (ed.).** 1984. International symposium on anaerobic bacteria and their role in disease. Rev. Infect. Dis. **6**(Suppl. 1)**:**S1–S299.
20. **Gibbs, R. S., J. D. Blanco, P. J. St. Clair, and Y. S. Castaneda.** 1982. Quantitative bacteriology of amniotic fluid from women with clinical intraamniotic infection at term. J. Infect. Dis. **145:**1–8.
21. **Gorbach, S. L.** 1979. General concepts, p. 1854–1863. *In* G. L. Mandell, R. G. Douglas, Jr., and J. E. Bennett (ed.), Principles and practice of infectious diseases. John Wiley & Sons, Inc., New York.
22. **Gorbach, S. L., and J. G. Bartlett.** 1974. Anaerobic infections. N. Engl. J. Med. **290:**1177–1184, 1237–1245, 1289–1294.
23. **Gregory, E. M., W. E. C. Moore, and L. V. Holdeman.** 1978. Superoxide dismutase in anaerobes: survey. Appl. Environ. Microbiol. **35:**988–991.
24. **Holdeman, L. V., E. P. Cato, and W. E. C. Moore (ed.).** 1977. Anaerobe laboratory manual, 4th ed. Virginia Polytechnic Institute and State University, Blacksburg.
25. **Jarvis, W., O. Nunez-Montiel, F. Thompson, V. Dowell, M. Towns, G. Morris, and E. Hill.** 1983. Comparison of bacterial isolation, cytotoxicity assay and counterimmunoelectrophoresis for the detection of *Clostridium difficile* and its toxin. J. Infect. Dis. **147:**778.
26. **Jones, G. L., D. M. Whaley, and S. M. Dever.** 1977. Use of the flexible anaerobic glove box. Center for Disease Control, Atlanta.
27. **Kasper, D. L., and S. M. Finegold (ed.).** 1979. Virulence factors of anaerobic bacteria. Rev. Infect. Dis. **1:**245–400.
28. **Killgore, G. E., S. E. Starr, V. E. Delbene, D. M. Whaley, and V. R. Dowell, Jr.** 1973. Comparison of three anaerobic systems for the isolation of anaerobic bacteria from clinical specimens. Am. J. Clin. Pathol. **59:**552–559.
29. **Koneman, E. W., S. D. Allen, V. R. Dowell, Jr., and H. M. Sommers.** 1983. Color atlas and textbook of diagnostic microbiology, 2nd ed. J. B. Lippincott Co., Philadelphia.
30. **Koransky, J. R., S. D. Allen, and V. R. Dowell, Jr.** 1978. Use of ethanol for selective isolation of sporeforming microorganisms. Appl. Environ. Microbiol. **35:**762–765.
31. **Loesche, W. J.** 1969. Oxygen sensitivity of various anaerobic bacteria. Appl. Microbiol. **18:**723–727.
32. **Lombard, G. L., and V. R. Dowell, Jr.** 1982. Gas-liquid chromatography, CDC laboratory manual. U.S. Department of Health and Human Services, Public Health Service, Centers for Disease Control, Atlanta.
33. **Lombard, G. L., D. N. Whaley, and V. R. Dowell, Jr.** 1982. Comparison of media in the Anaerobe-Tek and Presumpto plate systems and evaluation of the Anaerobe-Tek system for identification of commonly encountered anaerobes. J. Clin. Microbiol. **16:**1066–1072.
33a. **Marler, L. M., S. D. Allen, and J. A. Siders.** 1984. Rapid enzymatic characterization of clinically encountered anaerobic bacteria with the API ZYM system. Eur. J. Clin. Microbiol. **3:**294–300.
34. **Marounek, M., and R. J. Wallace.** 1984. Influence of culture E_h on the growth and metabolism of the rumen bacteria *Selenomonas ruminantium, Bacteroides amylophilus, Bacteroides succinogenes* and *Streptococcus bovis* in batch culture. J. Gen. Microbiol. **130:**223–229.
35. **Martin, W. J.** 1971. Practical method for isolation of anaerobic bacteria in the clinical laboratory. Appl. Microbiol. **22:**1168–1171.
36. **Murray, P. R.** 1978. Growth of clinical isolates of anaerobic bacteria on agar media: effects of media composition, storage conditions, and reduction under anaerobic conditions. J. Clin. Microbiol. **8:**708–714.
37. **Penn, R. L.** 1983. Gynecological and obstetrical infections, p. 555–593. *In* R. E. Reese and R. G. Douglas, Jr. (ed.), A practical approach to infectious diseases. Little, Brown & Co., Boston.
38. **Reller, L. B., P. R. Murray, and J. D. MacLowry.** 1982. Cumitech 1A, Blood cultures II. Coordinating ed., J. A. Washington II. American Society for Microbiology, Washington, D.C.
39. **Rosenblatt, J. E.** 1984. Antimicrobial susceptibility testing of anaerobic bacteria. Rev. Infect. Dis. **6:**S242–S248.
40. **Rosenblatt, J. E., A. Fallon, and S. M. Finegold.** 1973. Comparison of methods for isolation of anaerobic bacteria from clinical specimens. Appl. Microbiol. **25:**77–85.
41. **Ryan, R. W., I. Kwasnik, and R. C. Tilton.** 1980. Rapid detection of *Clostridium difficile* toxin in human feces. J. Clin. Microbiol. **12:**776–779.
42. **Seip, W. F., and G. L. Evans.** 1980. Atmospheric analysis and redox potentials of culture media in the GasPak system. J. Clin. Microbiol. **11:**226–233.
43. **Spaulding, E. H., V. Vargo, T. C. Michaelson, M. Kor-**

zeniowski, and R. M. Swenson. 1974. Anaerobic bacteria: culture and identification, p. 87–103. *In* J. E. Prier and H. Friedman (ed.), Opportunistic infections. University Park Press, Baltimore.

44. **Stargel, M. D., G. L. Lombard, and V. R. Dowell, Jr.** 1978. Alternative approaches to biochemical differentiation of anaerobic bacteria. Am. J. Med. Technol. **44:**709–722.
45. **Starr, S. E., G. E. Killgore, and V. R. Dowell, Jr.** 1971. Comparison of Schaedler agar and Trypticase soy-yeast extract agar for the cultivation of anaerobic bacteria. Appl. Microbiol. **22:**655–658.
46. **Sutter, V. L., D. M. Citron, and S. M. Finegold.** 1980. Wadsworth anaerobic bacteriology manual, 3rd ed. The C. V. Mosby Co., St. Louis.
47. **Tally, F. P., G. J. Cuchural, N. V. Jacobus, S. L. Gorbach, K. E. Aldridge, T. J. Cleary, S. M. Finegold, G. B. Hill, P. B. Iannini, R. V. McCloskey, J. P. O'Keefe, and C. L. Pierson.** 1983. Susceptibility of the *Bacteroides fragilis* group in the United States in 1981. Antimicrob. Agents Chemother. **23:**536–540.
48. **Walden, W. C., and D. J. Hentges.** 1975. Differential effects of oxygen and oxidation-reduction potential on the multiplication of three species of anaerobic intestinal bacteria. Appl. Microbiol. **30:**781–785.
49. **Weissfeld, A. S., and A. C. Sonnenwirth.** 1981. Rapid detection and identification of *Bacteroides fragilis* and *Bacteroides melaninogenicus* by immunofluorescence. J. Clin. Microbiol. **13:**798–800.
50. **Whaley, D. N., and G. W. Gorman.** 1977. An inexpensive device for evacuating the gassing anaerobic systems with in-house vacuum. J. Clin. Microbiol. **5:**668–669.
51. **Wimberley, N., L. J. Faling, and J. G. Bartlett.** 1979. A fiberoptic bronchoscopy technique to obtain uncontaminated lower airway secretions for bacterial culture. Am. Rev. Respir. Dis. **119:**337–343.
52. **Wüst, J.** 1977. Presumptive diagnosis of anaerobic bacteremia by gas-liquid chromatography of blood cultures. J. Clin. Microbiol. **6:**586–590.

Chapter 38

Clostridium

STEPHEN D. ALLEN

CHARACTERIZATION

The anaerobic sporeforming bacilli encountered in clinical materials from humans belong to the genus *Clostridium*. The spores are usually ovoid to spherical and distend the vegetative cells. Certain species, e.g., *C. perfringens*, produce spores only under special culture conditions. The majority of *Clostridium* species are obligate anaerobes, and a few are aerotolerant. Most species are gram positive, at least early in culture, but some, e.g., *C. ramosum* and *C. clostridiiforme*, are almost always gram negative in overnight culture. Several species, such as *C. tetani*, are usually gram negative by the time the spores are formed. Although most vegetative cells are rod shaped, they vary from coccoid to filamentous. All but a few species are characteristically motile by means of peritrichous flagella. In most instances, the nonmotile species isolated from clinical materials are *C. perfringens, C. ramosum*, and *C. innocuum*. Catalase is not produced except in rare circumstances. If produced, it is only weak or in a small amount. Clostridia produce a range of short-chain acid metabolic products; they are usually fermentative, proteolytic, or both, but some are asaccharolytic and nonproteolytic. Their metabolisms and energy requirements are extremely diverse (41, 44).

The clostridia are distributed widely in soil and in freshwater and marine sediments throughout the world. Some are psychrophilic and some are thermophilic, but most are mesophilic. Genus *Desulfotomaculum* is an additional genus of sporeforming anaerobic bacilli that is encountered in nature, but the genus *Clostridium* is the only genus of sporeforming anaerobes found in human clinical specimens (41). Several *Clostridium* species inhabit the lower intestinal tract of humans and other animals as part of the normal flora. Most species are harmless saprophytes.

There have been few changes in the nomenclature and taxonomy of clostridia since the text of the last edition of this Manual was prepared. In 1979, two interesting new species, *C. cocleatum* and *C. spiroforme*, were designated. They are helically coiled, sporeforming anaerobes that have been isolated from the intestinal materials of humans and other animals (28). They have not been associated with diseases of humans. Another new species, *C. villosum*, was isolated from subcutaneous abscesses of cats (35) but has not been found in humans. The organisms formerly called *C. perenne* and *C. paraperfringens* are now classified as *C. barati* (9). From 1980 through 1984, nine new *Clostridium* species from nonhuman sources were recognized, and an additional species was reinstated. A total of 83 *Clostridium* species are now officially recognized. Only a limited number of these species are encountered with any frequency in clinical specimens from humans (Table 1).

Since some aerotolerant clostridia (e.g., *C. tertium, C. histolyticum*, and *C. carnis*) may grow on the surface of a fresh agar medium under aerobic conditions, it is possible to confuse these species with certain facultative *Bacillus* species. However, members of the genus *Clostridium* form spores under anaerobic conditions and almost never produce catalase. Also, aerotolerant clostridia show much better growth (i.e., they form larger colonies) under anaerobic conditions than in air, whereas *Bacillus* species often form larger colonies on aerobically incubated media than on media incubated anaerobically.

Endogenous infections

Although exogenous infections including botulism, *C. perfringens* food-borne illness, gas gangrene associated with traumatic wounds, and tetanus are historically well known and important when they occur, endogenous infections involving clostridia of the indigenous flora are much more common. As for other endogenous infections involving anaerobes, special circumstances are required for the development of infection with the clostridia. Common predisposing factors include trauma, operative procedures, vascular stasis, obstruction, treatment with immunosuppressive agents, chemotherapeutic agents used in the treatment of malignancy, prior treatment with antimicrobial agents (as in pseudomembranous colitis), and underlying illness such as leukemia, carcinoma, or diabetes mellitus. Under the right conditions, clostridia can invade and multiply in essentially any area of the body. Since clostridia are members of the normal intestinal microflora, their presence in a clinical specimen does not necessarily imply association with a pathological condition. Some clostridia, apparently nonpathogenic for humans, can colonize and grow in tissue with an impaired blood supply, although they are unable progressively to invade healthy tissue. Moreover, strains of pathogenic species may react in the same way, for their pathogenic properties are manifested only under special circumstances. Therefore, close liaison between the attending physician and the clinical microbiologist is essential for an assessment of the clinical significance of clostridial isolates and the establishment of the correct diagnosis.

C. perfringens and other histotoxic clostridia

C. perfringens is encountered in a wide variety of clinical settings, ranging from simple contamination of a wound to traumatic or nontraumatic myonecrosis, gangrenous cholecystitis, postabortion infection with devastating septicemia and intravascular hemolysis, necrotizing pneumonia, and empyema (17, 21, 22, 41, 49). *C. perfringens* is commonly considered in a group of organisms known as the histotoxic clostridia. The histotoxic clostridia most commonly involved in gas gangrene (myonecrosis) are *C. perfringens* (80%),

TABLE 1. *Clostridium* species most frequently encountered in clinical specimens at the Indiana University Medical Center Anaerobe Laboratory, 1979 through 1983[a]

Species	No. of isolates	% of total isolates
C. perfringens	272	32.8
C. ramosum	105	12.7
C. innocuum	86	10.4
C. difficile	49	5.9
C. butyricum	46	5.5
C. sporogenes	44	5.3
C. clostridiiforme	40	4.8
C. bifermentans	36	4.3
C. septicum	30	3.6
C. cadaveris	30	3.6
C. tertium	26	3.1
C. paraputrificum	18	2.1
C. glycolicum	17	2.0
Other recognized species[b]	31	3.6

[a] Based on the data of J. A. Siders and S. D. Allen. The total of 830 isolates does not include 135 isolates (13.9% of 965 isolates) which did not belong to a recognized species. All of the isolates were from properly collected specimens. No isolates were from feces, intestinal materials, or other contaminated sources.

[b] Includes one to seven isolates of each of the following: *C. barati, C. sordellii, C. subterminale, C. sphenoides, C. putrificum, C. carnis, C. indolis, C. celatum*, and *C. hastiforme*.

C. novyi (40%), and *C. septicum* (20%), followed by *C. histolyticum* and *C. bifermentans*. Other species such as *C. sordellii, C. fallax, C. sporogenes*, and *C. tertium* have been encountered in myonecrosis, but their pathogenic significance is not certain. Clostridial myonecrosis is not a bacteriological entity. It is a rapidly progressive, invasive, clinicopathological condition with liquefactive necrosis of muscle, gas formation, and associated clinical signs of toxemia. Blood cultures are positive in about 15% of patients (21). The clinical and pathological features are well described elsewhere (17, 21, 41, 49).

C. perfringens food-borne illness

C. perfringens generally ranks behind *Salmonella* spp. and *Staphylococcus aureus* as the third most common cause of food posioning in the United States (2). The organism produces 12 different toxins. It has been divided into five toxin types (A, B, C, D, and E) based on the production of four major lethal toxins (alpha, beta, epsilon, and iota) (41). Almost all outbreaks and cases of *C. perfringens* food-borne gastroenteritis in the United States appear to be due to type A strains (40). *C. perfringens* type C is the only other toxin type encountered in human illness. Type C is associated with a disease called enteritis necroticans (discussed below). In *C. perfringens* type A food-borne disease, the food vehicle is almost always an improperly cooked meat (e.g., beef, turkey, chicken, or pork) or a meat product, such as gravy, that has cooled slowly after cooking or may have been reheated to a moderate temperature (2). Sporulation allows the organism to survive the initial cooking; the spores then germinate, and vegetative cells proliferate during slow cooling or reheating. *C. perfringens* type A food-borne illness should be suspected when there is an outbreak of diarrhea with crampy abdominal pain within about 7 to 15 h after the consumption of a suspected food (40). However, the incubation period has ranged up to 30 h. Most patients are afebrile; nausea and vomiting occur in less than a third of patients (2), and the stools are frequently foamy and foul smelling. Illness results from the ingestion of food with about 10^8 viable vegetative cells which, in the alkaline environment of the intestine, undergo sporulation, with production of an enterotoxin while the spores are being formed (40). The enterotoxin, a 34,000-molecular-weight protein which is a structural component of *C. perfringens* spores, produces fluid and electrolyte secretion in the ileum of mice and other animal models (50). Usually, the illness is mild, and most patients recover within 2 to 3 days after onset. The diagnosis is confirmed by the culture of at least 10^5 organisms per g from epidemiologically implicated food and by the demonstration, by a quantitative spore selection technique, of median spore counts of at least 10^6 *C. perfringens* spores per g of feces collected within 24 h of the onset of illness (2, 40). Laboratory testing for *C. perfringens* food-borne illness is a public health laboratory function which should be done concomitantly with epidemiologic support. Ideally, isolates of the same serotype should be cultured from epidemiologically incriminated food and ill persons, but not from controls. Furthermore, the required serologic reagents are not available to clinical laboratories. Unfortunately, the experience with serotyping of *C. perfringens* at the Centers for Disease Control (CDC) has not been highly successful (40).

C. perfringens in enteritis necroticans

C. perfringens produces a much more severe necrotizing disease of the small bowel known as enteritis necroticans, which can occur sporadically or in an epidemic form. The syndrome has been called Darmbrand in Germany and pig-bel in Papua, New Guinea (17, 21, 41, 49). In this condition, seen mostly in children, there is evidence that *C. perfringens* type C, which is either part of the normal intestinal flora or is ingested with contaminated pork or other meat during a feast, proliferates in the small intestine and produces beta toxin. This toxin production probably leads to focal paralysis, inflammation, hemorrhage, and segmental gangrenous necrosis of the intestine, particularly the jejunum. Among the factors likely to be involved in the pathogenesis of enteritis necroticans are (i) overeating, which might distend the bowel and cause partial obstruction; (ii) poor nutrition, which leads to low levels of production of the pancreatic proteases (particularly trypsin, which ordinarily destroys beta toxin); and (iii) a diet that is rich in trypsin inhibitors (such as the semicooked sweet potatoes that are often eaten at pig feasts) (47). The incidence of this condition in the United States is unknown, probably because the disease is rather obscure and would be easy to overlook clinically.

Another type of gastrointestinal illness that may be caused by *C. perfringens* in some instances is antibiotic-associated diarrhea (8) (discussed below). Although the etiology of neonatal necrotizing enterocolitis is unknown, epidemiologic data support a direct role for microorganisms and their toxins in the pathogenesis of this disease. Of the implicated organisms, clostridia, particularly *C. perfringens*, are among the organisms most likely to be involved (30).

C. difficile gastrointestinal illness

C. difficile is a major cause of antibiotic-associated diarrhea and pseudomembranous colitis (7, 11). Treatment specifically directed against *C. difficile* in these conditions has usually been successful. The habitats of *C. difficile* include soil, water, and the intestinal contents of various lower animals. The organism is part of the transient or permanent fecal flora of many healthy infants and about 3% of healthy adult volunteers. It has been found in the feces of up to 30% of hospitalized patients who were colonized but were without gastrointestinal illness or a history of recent antibiotic treatment. Numerous antimicrobial agents of all the major classes of antibiotics have been associated with diarrhea and colitis (7). In addition to the association between antibiotic treatment and gastrointestinal disease, *C. difficile* is probably involved in some instances in diarrhea or colitis related to the treatment of cancer patients with chemotherapeutic agents, pseudomembranous colitis that does not follow the use of chemotherapeutic agents or antibiotics, a similar disease after obstruction or vascular compromise of the bowel, and the relapse of nonspecific inflammatory bowel disease (31). Also, *C. difficile* has been isolated in a few cases of sudden infant death syndrome. The clinical features of *C. difficile* gastrointestinal illness have been reviewed elsewhere (7). Most (but not all) strains of *C. difficile* isolated from patients with characteristic intestinal disease are toxigenic (10, 34). Virtually all the toxigenic strains produce two toxins in vitro (and probably in vivo) designated toxin A (called an enterotoxin) and toxin B (called a cytotoxin) (34). Although data are lacking regarding the pathogenesis of *C. difficile* disease, diagnosis and successful treatment have been correlated directly with an assay for the cytotoxin in cell culture. Isolation and identification of the organism can also be done as an aid in diagnosis (discussed below).

Although many cases of antibiotic-associated diarrhea are now caused by *C. difficile, S. aureus* was thought to be the primary cause in the 1950s (7) and is still a potential cause. Recently, evidence was reported by Borriello et al. that enterotoxigenic *C. perfringens* is another possible cause of antibiotic-associated diarrhea in some patients (8). Fecal suspensions from 11 patients with antibiotic-associated diarrhea produced a cytopathic effect (CPE) in Vero cell cultures that was neutralized by an antiserum raised against *C. perfringens* enterotoxin. This CPE was not neutralized by gas gangrene polyvalent antitoxin or by *C. sordellii* antitoxin (which neutralizes the CPE of *C. difficile* toxin), nor was it detected in the feces of 16 patients of similar age who did not have diarrhea, 29 patients who had diarrhea due to inflammatory bowel disease, or 12 patients with infective diarrhea (due to nonclostridial organisms). The clinical manifestations of the 11 patients with positive enterotoxin neutralization tests were reported to be more severe and prolonged than are usually seen in patients with *C. perfringens* food-borne illness due to type A organisms (8). Assays for *C. perfringens* enterotoxin are not readily available to clinical laboratories and currently are done in only a few research and referral laboratories.

C. septicum disease

In addition to causing at least 20% of the cases of gas gangrene, *C. septicum* may be a major pathogen in some patients who have intestinal disease involving the ileocecal region. Fatal *C. septicum* bacteremia is associated with malignancy, particularly carcinoma of the bowel arising in the cecum or ascending colon, and leukemia (33). Neutropenic enterocolitis, also known as the ileocecal syndrome or typhlitis, is a distinct clinicopathological entity that is encountered in leukemic patients who have neutropenia and in patients without leukemia who have cyclic neutropenia, neutropenia caused by drug-induced agranulocytosis, or idiopathic agranulocytosis. Recent reports suggest that *C. septicum* plays a primary role in the pathogenesis of ulceration, hemorrhage, and necrosis of the distal ileum or cecum in at least some of these patients (29, 37).

Tetanus

Tetanus is an extremely dramatic illness produced by the action of a potent neurotoxin, tetanospasmin, which is elaborated by *C. tetani*. Tetanospasmin is a heat-labile protoplasmic protein that is released after the autolysis of *C. tetani*. The toxin becomes bound to gangliosides in the central nervous system and blocks inhibitory impulses to the motor neurons, thus producing prolonged muscle spasms of both flexor and extensor muscles. Like the botulinal toxins, tetanospasmin acts at myoneural junctions to inhibit the release of acetylcholine. The binding sites for tetanospasmin and botulinal toxins differ; spasticity is characteristic of tetanus, and paralysis is characteristic of botulism (12). Tetanus is still largely a disease of the unimmunized in the United States and is reported most frequently from areas of the rural South. It would be unusual for the clinical laboratory to be requested to isolate *C. tetani* from a wound, since tetanus usually presents little diagnostic problem for the clinician. For more details on tetanus, see Finegold (17), Smith (41), Willis (49), and the recent review by Dowell (12).

Botulism

Botulism, a neuroparalytic disease produced by the neurotoxins of *C. botulinum*, is currently classified in four categories: (i) classical food-borne botulism, an intoxication caused by the ingestion of preformed botulinal toxin in contaminated food; (ii) wound botulism, the rarest form of botulism, which results from the elaboration of botulinal toxin in vivo after growth of *C. botulinum* in an infected wound; (iii) infant botulism, in which botulinal toxin is elaborated in vivo in the intestinal tract of an infant who has been colonized with *C. botulinum*; and (iv) an undetermined classification of botulism for those cases involving individuals older than 12 months in which no food or wound source is implicated (1). There are seven toxigenic types of *C. botulinum* (A, B, C, D, E, F, and G) based on antigenically distinct toxins produced by different strains of the organism. Types A, B, E, and F are the principal causes of botulism in humans; types C and D have been associated with botulism in birds and mammals (42). Type G organisms have been isolated from soil in Argentina and from autopsy materials from five individuals who died suddenly, but the type G organisms have not been clearly implicated in cases of botulism (12). The potent neurotoxins of *C. botulinum* are protoplasmic proteins which are synthesized during growth of the organism and released during lysis (42). Botulinal toxins cause paralysis by irreversibly binding to peripheral nerve endings, including the neuromuscular junction, thereby blocking the release of acetylcholine at these synaptic sites (1, 42). The characteristic clinical hallmark of botulism is an acute flaccid paralysis, which begins with bilateral cranial nerve impairment involving muscles of the face, head, and pharynx and then descends symmetrically to involve muscles of the thorax and extremities. Death may result from respiratory failure caused by a paralysis of the tongue or muscles of the pharynx that occludes the upper airway or from paralysis of the diaphragm and intercostal muscles. Patients diagnosed with food-borne or wound botulism should immediately receive trivalent (type ABE) antitoxin and promptly receive intensive respiratory care. For further information on botulism in general, see Smith (42), the CDC handbook (1), and Dowell (12).

Infant botulism is the most frequently encountered form of botulism. A total of 395 cases was reported in the United States between 1976 and 1983 (3). Infant botulism has now been reported from 36 states in 11 geographic regions, including Alaska and Hawaii; nearly half of the cases were reported in California. Infant botulism has also been reported from Canada, England, Czechoslovakia, and Australia. The botulinal toxin type was A in 50% and B in 49% of the cases, and there was one case of type F and one case of B/F (3). In general, the geographical distribution of toxin types in infant botulism cases has paralleled the distribution of *C. botulinum* toxin types in soil sampled from across the United States (43). Infants have ranged from 2 to 38 weeks old. The ingestion of spores by the infant during the first weeks of life is probably a prerequisite for the development of the disease. Preformed toxin has never been detected in any food or liquid ingested by the babies. To date, the only clearly defined risk factor is exposure to honey, a potential source of spores. The CDC has recommended that honey not be fed to infants less than 1 year old (3). Within the intestine, *C. botulinum* multiplies and elaborates toxin, but the pathophysiology of toxin production and absorption within the human gut remains to be determined. The first symptom is invariably constipation, although this symptom is usually mild and often overlooked. Patients who are ultimately hospitalized usually develop lethargy and mild weakness with feeding difficulties, pooled oral secretions, and an altered cry. The baby eventually becomes floppy, loses head control, and may go on to develop ophthalmoplegia, ptosis, flaccid facial expression, dysphagia, other signs of cranial nerve deficits, and generalized muscular weakness. Respiratory insufficiency necessitating respiratory therapy also may occur, as in other forms of botulism (4, 5). There is a spectrum of clinical features in infant botulism ranging from mild illness not requiring hospitalization to sudden death, and this syndrome accounts for a small percentage of cases of sudden infant death syndrome (5, 6). The differential diagnosis of infant botulism has included sepsis, myasthenia gravis, failure to thrive, benign congenital hypotonia, and a variety of other conditions.

COLLECTION AND TRANSPORT OF CLINICAL SPECIMENS

As with other anaerobic bacteria, the proper selection, collection, and transport of clinical specimens are extremely important for the laboratory diagnosis of clostridial infections. For recommended collection and transport procedures in general, refer to Chapters 8 and 37. Several tissue specimens should be taken from the active site of infection when gas gangrene is suspected, because the clostridia are often not distributed uniformly in pathological lesions.

Specimens for confirmation of *C. perfringens* food-borne illness

For a laboratory confirmation of *C. perfringens* food-borne illness, most clinical laboratories will need to use the services of a reference laboratory (e.g., local or state public health laboratory). For the shipment of food and fecal specimens to a reference laboratory, the samples should be collected in sterile, leakproof containers and transported at 4°C. Rectal swabs should be placed in an anaerobic transport container (Chapter 37) and shipped at 4°C (40).

Specimens for *C. difficile* culture and toxin assay

A single, freshly passed fecal specimen (about 25 g or 25 to 50 ml, if liquid) is the preferred specimen for *C. difficile* culture and toxin assay. Swab specimens are inadequate for the toxin assay because the volume of sample obtained is too small. Other appropriate specimens include lumen contents and surgical removal or autopsy samples of the large bowel. Specimens should be transported in tightly sealed, leakproof plastic containers. If specimens can be processed by the laboratory within 3 to 4 h after collection, transportation at room temperature will suffice. Alternatively, the toxin appears to be stable for up to 24 h at 2 to 4°C, and specimens can be transported on ice or held in the refrigerator for that length of time. Specimens should be frozen if a delay in processing greater than 24 h is anticipated, and they should be shipped to a reference laboratory on dry ice.

Suspected neutropenic enterocolitis involving *C. septicum*

The specimens of choice for suspected neutropenic enterocolitis involving *C. septicum* are three blood cultures collected from three different venipuncture sites, stool (i.e., 25 g or 25 ml), and lumen contents or tissue from the involved ileocecal area collected at surgery or autopsy and transported in tightly sealed leakproof containers. In addition, a biopsy sample of muscle (or an aspirate of fluid from the involved area, taken by needle and syringe) should be collected if the patient is also suspected of having myonecrosis.

Specimens for *C. botulinum* culture and toxin assay

The clinical diagnosis of food-borne botulism can be confirmed by the demonstration of botulinal toxin in serum, feces, gastric contents, or vomitus or by the recovery of *C. botulinum* from the feces of the patient. The organism has been isolated only rarely from individuals who were not victims of the illness or who had eaten food contaminated with *C. botulinum* that had caused botulism in others during an outbreak (1). The demonstration of botulinal toxin and *C. botulinum* in suspect foods aids in determining the food item responsible for an outbreak, but it provides only indirect evidence to support a clinical diagnosis of botulism. Ideally, 15 to 20 ml of serum (not whole blood), 25 to 50 g of stool, and the suspect food(s) should be collected. Specimens to collect from patients with suspected wound botulism include serum, feces, tissue, exudate, or swab samples from the wound. When infant botulism is suspected, collect serum (2 ml) and as much passed stool as possible. In most instances, the diagnosis of infant botulism has been confirmed by the detection of the toxin or *C. botulinum* or both in feces. Toxin has been detected in serum only on rare occasions in infants with this diagnosis.

Most hospital laboratories are not properly equipped to process specimens from patients suspected of having botulism. The CDC, Atlanta, Ga., provides epidemiologic aid and emergency laboratory services 24 h a day, every day of the week. The attending physician or state epidemiologist should notify the CDC at any time of the day when there is a suspected case of botulism, so that appropriate action can be taken to establish the diagnosis, initiate treatment, and investigate the potential outbreak. Appropriate telephone numbers are (404) 329-3753 on weekdays and (404) 329-3644 at night (after 4:30 p.m.) and on weekends and holidays.

DIRECT EXAMINATION

The direct microscopic examination of clinical materials can provide extremely useful information for the physician in the diagnosis and treatment of clostridial infections. Gas gangrene is an extremely urgent situation, requiring a rapid clinical diagnosis. The direct examination of a Gram-stained smear of the wound may be of special aid to the clinician in establishing the diagnosis. Characteristic findings are the absence of inflammatory cells and other cellular outlines and the presence of clostridia in smears prepared from the central areas of the lesion. It is important to distinguish clostridial myonecrosis, which requires radical surgical excision or amputation of a limb, from anaerobic cellulitis or spreading fasciitis (17, 21). The muscle is not necrotic in these last two conditions and does not need to be removed. In addition to the involvement of clostridia, anaerobic cellulitis may involve anaerobic cocci, facultatively anaerobic cocci, *Bacteroides* spp., *Fusobacterium* spp., and other microorganisms. Also, cell outlines of striated muscle cells and granulocytes remain intact in the latter condition.

The usual Gram stain is satisfactory for the direct examination of a specimen. Special note should be taken of gram-positive rods with or without spores because sporulation in tissue is not common with the two species most frequently encountered, *C. perfringens* and *C. ramosum*. *C. perfringens* usually appears as large, relatively short, fat, gram-positive rods in tissue smears; the cells of *C. ramosum* are more slender, longer, and often curved (31). *C. perfringens* may or may not be encapsulated in smears from wounds; capsules usually are present in smears of endometrial specimens from postabortion *C. perfringens* infections. Gram stains are also usually sufficient for the demonstration of spores. Examination with a phase microscope is helpful if the spores are mature or nearly so, but special spore stains offer no advantage. If spores are present, note their shape (spherical or oval) and position (terminal or subterminal to central) in the cells. The best single medium for the demonstration of spores is cooked-meat medium made up as an agar slant. Incubate the culture anaerobically at 5 to 7°C below the optimum for growth of the clostridia. For most species, 30°C is satisfactory, but 37°C is better for the sporulation of *C. perfringens*.

CULTURE AND ISOLATION

The general procedures described in Chapter 37 for the collection and transport of specimens and for the isolation and examination of cultures apply to the clostridia. Clostridia usually produce good growth on CDC anaerobe blood agar and phenylethyl alcohol blood agar (PEA) after 1 to 2 days of incubation. A few species, for example, *C. perfringens*, form colonies after overnight incubation. When clostridia are suspected in wound or abscess specimens (e.g., from gas gangrene), it is recommended that egg yolk agar (modified McClung-Toabe formula) or neomycin egg yolk (NEY) agar be inoculated in addition to blood agar and PEA. To prepare NEY medium, neomycin is added to achieve a final concentration of 100 μg/ml. Neomycin is heat stable and can be added before the medium is autoclaved. Neomycin can also be added to anaerobe blood agar in the same concentration. The purpose of neomycin is to inhibit some of the facultatively anaerobic gram-negative bacilli; thus, NEY medium is moderately selective.

After incubation, examine the blood agar and PEA cultures with a dissecting microscope, noting particularly the hemolysis pattern, colony structure, and any evidence of swarming or of motile colonies. Examine the egg yolk agar or NEY culture for evidence of lecithinase (phospholipase C) or lipase production. Lecithinase activity is indicated in either medium by the development of an insoluble, opaque, whitish precipitate within the agar. An iridescent sheen or oil-on-water appearance (pearly layer) on the surface growth indicates lipase activity. Proteolysis, the third reaction that can be seen on egg yolk agar or NEY medium, is indicated by a zone of translucent clearing in the medium around the colonies. In addition to the modified McClung-Toabe egg yolk agar formulation (13), these same reactions can be determined on the hemin-supplemented egg yolk agar formulation recommended by Finegold and Edelstein (Chapter 40) or on Lombard-Dowell egg yolk agar (Presumpto plate) (16, 31).

If swarming growth has covered the surface of the agar medium, inoculate another blood agar plate, and incubate it anaerobically only overnight. Subculture from the colonies as soon as the plates are taken from the anaerobe jar. If swarming is again observed, inoculate a plate of anaerobe blood agar made up with 4 to 6% agar. This mixture is known as stiff blood agar (13, 41). When isolated colonies can be picked, subculture them to cooked-meat medium, incubate the culture overnight, and use it for the inoculation of differential media. Also inoculate peptone-yeast extract-glucose broth for gas-liquid chromatography.

Spore selection techniques

Most clinical specimens other than blood in which clostridia may be sought (e.g., feces, material from wounds and abscesses, muscle and other soft tissue, surgical removal specimens, necropsy materials, etc.) yield a mixture of nonsporeforming bacteria. The possibilities include virtually any nonsporeforming bacteria that grow aerobically or anaerobically. A spore selection technique, with heat or alcohol treatment, is a useful selective means of isolating sporeformers while inhibiting nonsporeformers. Spores of clostridia (or of *Bacillus* spp.) resist the heat or alcohol treatment, whereas vegetative cells are killed by it. After these treatments, spores will germinate and produce growth after appropriate solid or liquid media are inoculated and incubated. Heat treatment alone should not be relied upon, because the spores of various *Clostridium* spp. and strains of species vary in their degree of resistance to heat. The treatment of specimens with at least 50% ethanol for 1 h aids in the selective isolation of clostridia from mixed infections and circumvents the problem of different spore tolerances to heat (32). The major problem to avoid with alcohol treatment is the presence of solid specimen particles that are not adequately penetrated by the alcohol. When the specimen is not sufficiently homogenized, the vegetative cells of nonsporeformers may not be inhibited.

Alcohol treatment. For alcohol treatment, to a 1-ml sample of a fecal suspension, a homogenate of a wound or exudate, etc., in a sterile screw-cap tube, add an equal volume of absolute (or 95%) ethyl alcohol (32). Mix the specimen gently at room temperature (22 to 25°C) for 1 h. An Ames Aliquot Mixer (Miles Laboratories, Inc., Elkhart, Ind.) is a convenient way to provide continuous mixing. Subculture the treated material, and inoculate chopped-meat–glucose (or chopped-meat–glucose–starch) medium, anaerobe blood agar, or egg yolk agar. Incubate the culture, and inspect it for growth as described above (and in Chapter 37). (For stool specimens it is often advantageous to alcohol treat separate 1-ml samples from a series of 1:10 dilutions. This treatment helps the alcohol penetrate solid particles.)

Heat treatment. For heat treatment (13), preheat a tube of chopped-meat–glucose–starch medium in an 80°C water bath for 5 min, and then add 1 ml of sample suspension. Heat the culture for 10 min, and then remove the tube and cool it in cold water. Subculture the treated sample suspension into an unheated tube of chopped-meat–glucose–starch medium and onto an anaerobe blood agar and egg yolk agar plate. Incubate the culture anaerobically, and examine it for growth as described above.

Laboratory investigation of *C. perfringens* foodborne illness

Methods for the enumeration of *C. perfringens* in foods and *C. perfringens* spores in feces, with egg yolk-free tryptose-sulfite-cycloserine agar, are described in detail by Hauschild and colleagues (23–25). Although methods for the detection of *C. perfringens* enterotoxin in feces have been described previously and are of considerable interest, these assays are still considered experimental (40).

Primary isolation of *C. difficile* from feces

The isolation of *C. difficile* from fecal samples can be accomplished by an alcohol or heat shock spore selection technique, as described above, and by using selective plating media. If a spore selection technique is used, inoculate the alcohol- or heat-treated sample onto CDC anaerobe blood agar and an egg yolk plate after treatment. After 48 h of incubation on anaerobe blood agar, colonies of *C. difficile* are nonhemolytic, 2 to 4 mm in diameter, slightly raised, flat, and spreading and have a rhizoid edge. They are gray-translucent and, under the dissecting microscope, show an iridescent, crystalline, internal specking (31). They are negative for lecithinase and lipase on egg yolk agar. In addition, two selective media (PEA and cycloserine-cefoxitin-egg yolk-fructose agar [19]) should be inoculated with untreated fecal material. Colonies of *C. difficile* on PEA are similar in appearance to those on anaerobe blood agar. On cycloserine-cefoxitin-egg yolk-fructose agar after 48 h of incubation, colonies of *C. difficile* are about 5 to 8 mm in diameter and show a yellow-green fluorescence under long-wavelength UV light. Under regular lighting, the colonies are yellowish, circular to irregular, and flat, with a rhizoid edge and a ground-glass appearance (19). They have a distinctive odor like that of *p*-cresol. They are gram positive to gram variable, produce subterminal spores, and have vegetative cells about 0.5 by 3 to 6 μm. Although they are adjuncts to the cytotoxin neutralization test, the isolation and identification of the organism alone do not prove that a patient has *C. difficile*-mediated gastrointestinal illness, since *C. difficile* can be isolated from the feces of healthy individuals and of hospitalized patients who have no evidence of gastrointestinal illness. The *C. sordellii* antitoxin, gas gangrene polyvalent antitoxin, and *C. difficile* antitoxins that are currently available (see below for sources) have the potential for cross-reactions with antigens of other clostridia. Furthermore, it is possible that the cytotoxin that is detected by current cell culture assays may not be responsible for the pathogenesis of *C. difficile* intestinal illness. For these reasons, it is recommended that both cultures for the organism and assays for the toxin be performed, at least until practical, sensitive, and specific assays for both the enterotoxin and the cytotoxin become commercially available.

PRESUMPTIVE IDENTIFICATION

Presumptive identification of a few species can be accomplished fairly rapidly. Fluorescent-antibody reagents for *C. novyi*, *C. septicum*, *C. chauvoei* (found only in infections of herbivores), and *C. sordellii* (Wellcome Research Laboratories, Beckenham, England) permit the rapid presumptive identification of these species.

C. perfringens is signaled by colonies on blood agar plates that are surrounded by an inner zone of complete hemolysis and an outer zone of discoloration and incomplete hemolysis (with rabbit, human, or sheep blood) and are composed of short to intermediate gram-positive rods without spores. Subculture such a colony to an egg yolk agar plate, one-half of which has been spread with *C. perfringens* antitoxin, and incubate the culture anaerobically overnight. *C. perfringens* will produce a zone of precipitation around colonies on the control side of the plate and no precipitation, or little, on the side spread with antitoxin (this response is known as the Nagler reaction). Similar reactions will be given by *C. bifermentans, C. sordellii*, and *C. barati* (formerly *C. paraperfringens*), but these species should not cause difficulty. *C. bifermentans* does not form a double zone of hemolysis, sporulates readily, is motile, is more proteolytic than *C. perfringens*, and varies from *C. perfringens* in other cultural characteristics. *C. sordellii* resembles *C. bifermentans*, but *C. sordellii* is urease positive. None of the other clostridia listed in Table 2 are urease positive. *C. barati* is so seldom encountered in clinical material (Table 1) that it is not likely to be an appreciable source of error. *C. perfringens* liquifies gelatin, but *C. barati* does not.

C. difficile is relatively easy to recognize and identify. As indicated above, it has a distinctive odor in culture and produces characteristic colonies on anaerobe blood agar, PEA, and cycloserine-cefoxitin-egg yolk-fructose agar (31). It produces subterminal spores which are readily seen on Gram stains of colonies from 2- to 3-day-old blood agar plates. *C. difficile* is motile, which can be shown by the preparation of a wet mount from a fresh colony or a broth culture. Metabolic product analysis typically shows acetic, propionic, isobutyric, butyric, isovaleric, valeric, and isocaproic acids. A large peak of isocaproic acid is a key characteristic. Other commonly encountered *Clostridium* species with subterminal spores that may produce isocaproic acid include *C. sporogenes, C. bifermentans, C. sordellii*, and *C. subterminale. C. sporogenes* produces Medusa head colonies on anaerobe blood agar that are distinct from those of *C. difficile. C. bifermentans* and *C. sordellii* are both indole positive, but *C. difficile* is indole negative. *C. subterminale* differs in several biochemical characteristics (Table 2). The Presumpto plate system permits the rapid differentiation of clostridia in general and is a practical way to separate *C. difficile* from other clostridia (14, 16, 31). In this system, *C. difficile* hydrolyzes esculin and gelatin but is negative for indole and nitrate (analyses are done by rapid disk tests). *C. difficile* is saccharolytic and is one of the few clostridia that ferments mannitol. A new rapid test which aids in the rapid identification of *C. difficile* is the performance of gas-liquid chromatography on a norleucine-tyrosine broth culture (36). All 120 strains of *C. difficile* tested produced caproic acid and *p*-cresol in the norleucine-tyrosine medium; none of the other clostridia or other bacteria examined produced both products.

It is often difficult to isolate *C. tetani* from a suspected lesion. When this organism is being sought, a freshly poured blood agar plate should be inoculated lightly. Incubate the culture for 1 day, and examine it carefully for swarming, which may be in the form of a very thin layer. Transfer cells from the edge of the swarming area to a tube of broth, and streak a plate of medium containing 5% agar; incubate the culture, and pick an isolated colony. *C. tetani* may sometimes be demonstrated in a specimen more readily than it can be isolated. Mix a small amount of material from the lesion in sterile broth, and inject 0.1 ml of this material into each of four mice, beside the base of the tail. Inject two of the mice with 0.1 ml of tetanus antitoxin. Death or symptoms of tetanus in the unprotected but not the protected mice indicate the presence of *C. tetani* in the specimen. Somewhat different approaches to presumptive identification have been described elsewhere (31, 45).

IDENTIFICATION

To identify clostridia, inoculate the media indicated in Table 2, prepared according to Dowell and Hawkins (13) (see Chapter 111 on media). Incubate the cultures for 2 to 7 days at 35 to 37°C; 2 days are sufficient if growth is prompt and adequate. Examine Gram stains of the cooked-meat culture to determine the presence, position, and shape of the spores. If spores are not found, inoculate a tube of cooked-meat medium, heat the culture at 70°C for 10 min, and incubate it. Growth in this heated tube usually indicates the presence of spores, although none may be apparent microscopically. Alternatively, an alcohol spore selection technique may be helpful for those clostridia with heat-sensitive spores (32).

Determine which of the carbohydrates were fermented by using a pH meter or by adding indicator to the tube as described elsewhere (13). Determine the metabolic products if at all possible with a peptone-yeast extract-glucose broth culture and a gas chromatograph (Chapter 37). The identification of species can be made without gas chromatography, but such identification usually involves more time. The information listed in Table 2 will serve to identify most of the clostridia commonly isolated from clinical specimens. Information on additional tests and descriptions of differential characteristics of additional species can be found elsewhere (9, 13–15, 26, 31, 45).

Toxin tests are necessary for the identification of a few species. *C. sporogenes* cannot be differentiated with certainty from the proteolytic group I strains of *C. botulinum* unless toxin tests are used. Although *C. botulinum* is rarely encountered in clinical material, at least 27 cases of wound botulism in humans have been reported (12). A few strains of group III *C. botulinum* produce lecithinase as well as lipase and are difficult to distinguish from *C. novyi* type A except by toxin tests or by the use of *C. novyi* fluorescent-antibody conjugate. To test for toxin, inoculate two tubes of cooked-meat–glucose medium; incubate one tube at 37°C overnight, and incubate the other tube at 37°C for 3 days. Test the overnight culture first; if no toxin is found, test the 3-day culture. Centrifuge the culture, remove the liquid, and place 1.2-ml amounts in several tubes. Prepare mixtures with 0.3 ml of appropriate antiserum per tube for the various species suspected. Let the mixture stand for 30 min at room temperature or at 37°C, and inject 0.5-ml portions intraperitoneally into each of two mice. Observe the

TABLE 2. Differential characteristics of commonly encountered clostridia[a]

Species	Spores	Egg yolk agar		Growth on aerobic blood agar	Gelatin hydrolysis	Milk digestion	Indole production	Carbohydrate fermentation						Principal metabolic products	Other
		LEC	LIP					Glucose	Maltose	Lactose	Sucrose	Salicin	Mannitol		
C. bifermentans	OS	+	−	−	+	+	+	+	+	−	−	V	−	A, (p), (ib), (b), (iv), (v), (ic)	Urease negative
C. botulinum[b]															
Group I[c]	OS	−	+	−	+	+	−	+	+	−	−	V	−	A, P, IB, B, IV, (v), (ic)	
Group II[c]	OS	−	+	−	+	−	−	+	+	−	−	V	−	A, P, B	
Group III[c]	OS	V	+	−	+	−	−	+	V	−	−	−	−	A, B	
C. butyricum	OS	−	−	−	−	−	−	+	+	+	+	+	−	A, (p), B	
C. cadaveris	OT	−	−	−	V	−	+	+	−	−	−	−	−	A, (p), (ib), B, (iv)	
C. chauvoei[d]	OS	−	−	−	+/−	−	−	+	+	+	+	−	−	A, B	
C. clostridiiforme	OS	−	−	−	−	−	−/+	+	+	V	+	V	−	A	Spores seldom observed; usually gram negative
C. difficile	OS	−	−	−	V	−	−	+	−	−	−	V	+	A, p, IB, B, IV, V, IC	
C. histolyticum	OS	−	−	V	+	+	−	−	−	−	−	−	−	A	
C. innocuum[e]	OT	−	−	−	−	−	−	+	−	−	V	+	+	A, B	
C. limosum	OS	+	−	−	+	+	−	−	−	−	−	−	−	A	
C. novyi A[c]	OS	+	+	−	+	−	−	+	V	−	−	−	−	A, P, B, (v)	
C. novyi B[c]	OS	+	−	−	+	+	−	+	+	−	−	−	−	A, P, B, (v)	
C. paraputrificum	OT	−	−	−	−	−	−	+	+	+	+	+	−	A, (p), B	
C. perfringens[e]	OS	+	−	−	+	+	−	+	+	+	+	V	−	A, (p), B	Spores seldom observed; double zone of hemolysis
C. ramosum[e]	R/OT	−	−	−	−	−	−	+	+	+	+	+	V	A	Spores seldom observed; frequently gram negative
C. septicum	OS	−	−	−	+	−	−	+	+	+	−	+	−	A, B	
C. sordellii	OS	+/−	−	−	+	+	+	+	+	−	−	−	−	A, (p), (ib), (b), (iv), (ic)	Urease positive
C. sphenoides	RS/T	−	−	−	−/+	−	+/−	+	+	V	+/−	+	+	A	Usually gram negative
C. sporogenes	OS	−	+	−	+	+	−	+	+	−	−	V	−	A, (p), (ib), B, IV, (ic)	
C. subterminale	OS	−	−	−	+	+	−	−	−	−	−	−	−	A, (p), IB, B, IV	
C. tertium	OT	−	−	+	−	−	−	+	+	+	+	+	+	A, B	
C. tetani	RT	−	−	−	+	−	V	−	−	−	−	−	−	A, p, B	

[a] Key: +, positive reaction; −, negative reaction; V, variable reaction; /, either/or; O, oval; R, round; S, subterminal; T, terminal; LEC, lecithinase production; LIP, lipase production. Fermentation products: A, acetic; B, butyric; F, formic; IB, isobutyric; IC, isocaproic; IV, isovaleric; P, propionic; V, valeric. ( ), May or may not be present; capital letters indicate major peaks; lowercase letters indicate minor peaks.

[b] Group I contains proteolytic strains (A, B, F, and G), group II contains types C and D, and group III contains nonproteolytic strains (B, E, and F).

[c] Toxin neutralization test required for identification (13).

[d] Pathogenic for herbivores.

[e] *C. innocuum*, *C. perfringens*, and *C. ramosum* are nonmotile.

mice for 3 days, and record the deaths that occur. Only specific sera for laboratory testing should be used for toxin identification; therapeutic sera are often unsatisfactory because they may contain antibodies to toxins of species other than those listed on the label. Diagnostic clostridial antisera are available from Wellcome Reagents Ltd., Wellcome Research Laboratories, Beckenham, England BR3 3BS.

If it is necessary to determine the toxin type of an isolate of *C. perfringens* or *C. botulinum*, it is best to send the isolate to a reference laboratory. *C. perfringens* types other than type A seldom are encountered in clinical specimens from humans. Veterinary clinical microbiology laboratories, however, should be familiar with the technique for the determination of the toxin type of *C. perfringens* isolates.

C. difficile cytotoxin assay

C. difficile has been implicated in the pathogenesis of many cases of antibiotic-associated diarrhea and colitis and in certain other gastrointestinal illnesses discussed previously. It produces two toxins, at least one of which, possibly both, is thought to mediate the disease. It appears that most of the toxigenic *C. difficile* isolates produce a cytotoxin that causes CPEs in certain kinds of cell cultures that can be neutralized with antitoxin in a toxin neutralization assay (10, 34). This assay does not test for the enterotoxin of *C. difficile*, which can be detected with a hamster model. In general, the presence of cytotoxin in a cell-free extract of feces correlates with *C. difficile*-mediated intestinal illness. However, the cytotoxin may be detected in the feces of healthy individuals, particularly in children or infants under 1 year, who are asymptomatically colonized with the organism.

Procedure

See references 10 and 31 also.

1. Centrifuge liquid stool at 2,000 × *g* for 20 min or at 10,000 × *g* for 10 min. Discard the pellet; save the supernatant. If the specimen is semisolid or formed, add an equal volume of buffered gelatin diluent (pH 7.0) (13) or phosphate-buffered saline, and allow the fecal suspension to extract overnight at 4°C. Then centrifuge the specimens as for liquid stool, and save the supernatant.
2. Filter through a membrane filter (0.45-μm pore size).
3. Inoculate a cell culture tube containing a confluent monolayer of human diploid lung fibroblasts (WI-38 cells; Flow Laboratories, McLean, Va.) with 0.1 ml of cell-free fecal supernatant and 0.1 ml of buffered gelatin diluent (or phosphate-buffered saline). Other cell lines can be used instead of WI-38 cell lines (10).
4. Observe the cells for cytotoxicity after 4, 24, and 48 h of incubation at 35°C. Most cells are positive by 18 to 24 h of incubation.
5. Set up the antitoxin neutralization test as follows: add 0.2 ml of gas gangrene polyvalent antitoxin (Lederle Laboratories, American Cyanamid Co., Pearl River, N.Y.), *C. sordellii* antitoxin (Lederle or Food and Drug Administration Bureau of Biologics, Rockville, Md.), or *C. difficile* antitoxin (available from T. D. Wilkins, Virginia Polytechnic Institute and State University, Blacksburg) to 0.2 ml of cell-free fecal supernatant, and mix them.
6. Inoculate a tube of WI-38 cells with 0.2 ml of the supernatant-antitoxin mixture described above, and then incubate and observe the cultures for cytotoxic effects or CPE.
7. In a positive toxin neutralization test, the following should be observed: (i) a rounding of WI-38 cells with radiating cytoplasmic processes or other CPE in the tube containing fecal supernatant without antitoxin and (ii) an absence of CPE in the tube containing the supernatant plus antitoxin.
8. Although a toxin titer can be determined by serial 2- or 10-fold dilutions of supernatant (diluted in buffered gelatin diluent), the results of toxin titers correlate very poorly with the severity of the illness. Toxin titers are usually not clinically useful and need not be done routinely.

Currently described counterimmunoelectrophoresis procedures for the identification and detection of *C. difficile* cytotoxin (38) lack sufficient sensitivity and specificity for routine clinical use (27). Recently, Laughon et al. described an enzyme immunoassay which is designed to detect both the cytotoxin and the enterotoxin of *C. difficile* in fecal samples (34). This experimental procedure appears promising but is not available commercially.

SUSCEPTIBILITY TO ANTIMICROBIAL AGENTS

Penicillin G shows excellent activity against most strains of *C. perfringens* and is the antibiotic of choice for the clostridia in general (39, 46). However, occasional strains of *C. perfringens*, some *C. ramosum* strains, and certain other *Clostridium* species may be resistant (17, 18).

Although clindamycin is highly active against most commonly encountered anaerobic bacteria, a number of clostridia are frequently resistant to it. These clostridia include strains of the following species: *C. ramosum*, *C. difficile*, *C. tertium*, *C. subterminale*, *C. innocuum*, *C. sporogenes*, and some strains of *C. perfringens* (48).

Chloramphenicol, carbenicillin, and metronidazole are active against nearly all the clostridia, with only a few exceptions (17, 18). The clostridia have shown variable resistance to the cephalosporins and tetracyclines, and they are usually resistant to aminoglycosides. Many clostridia other than *C. perfringens* are resistant to cefoxitin, moxalactam, cefoperazone, and other third-generation beta-lactam drugs (31).

In vitro susceptibility studies of *C. difficile* have indicated that most strains are susceptible to penicillin G, ampicillin, and vancomycin but resistant to various cephalosporins. The activity of clindamycin and the tetracyclines against this organism has been variable (7, 20).

EVALUATION

The isolation of *Clostridium* species from a specimen is without meaning unless it is considered in relation to the clinical condition of the patient. Because clostridia are ubiquitous, they are likely to be found in any area that is directly or indirectly contaminated with feces, soil, or dust. Even the toxigenic species are only opportunistic pathogens, and conditions suitable for progressive infection occur only rarely.

This situation is particularly the case with *C. perfringens*. Although this species may cause a host of pathological conditions, many isolates may have no clinical significance. *C. septicum*, on the other hand, is rarely isolated except from serious, often fatal clinical conditions. The important association of *C. septicum* bacteremia with malignancy was mentioned elsewhere in this chapter (33). *C. septicum* bacteremia is associated with high mortality. *C. novyi* is seldom isolated in civilian hospital laboratories. During wartime it has been responsible for gas gangrene as the result of wounds that were contaminated with soil when inflicted. Most *C. novyi* strains encountered in wounds belong to type A; only a few belong to type B. The isolation of either type from a wound would be regarded with concern.

The isolation of a pathogenic strain of *C. sordellii* from a human infection is rare; the great majority of isolates appear to lack toxigenicity. However, this situation is somewhat unclear, for this species is notorious for the rapidity with which organisms can lose their pathogenicity in the laboratory.

The isolation of *C. botulinum*, particularly type A, indicates the possibility of wound botulism. Nevertheless, toxigenic strains of *C. botulinum* have been isolated from wounds in the absence of clinical evidence for botulism. In addition to the culturing of wound samples, serum testing is particularly important (to demonstrate toxin) when wound botulism is suspected (42). A similar situation holds for *C. tetani*. Since most people have been immunized with tetanus toxoid, the isolation of this organism from a wound may be insignificant clinically. Even in unimmunized persons, it is not uncommon to find this organism in wounds without the patient showing symptoms and signs of tetanus.

C. histolyticum is encountered in far less than 1% of the wounds of civilian life, even when it is searched for. Its presence is without clear significance unless a progressive anaerobic infection of muscle is in progress, and then the prognosis is poor.

LITERATURE CITED

1. **Anonymous.** 1979. Botulism in the United States, 1899–1977. Handbook for epidemiologists, clinicians and laboratory workers. Center for Disease Control, Atlanta.
2. **Anonymous.** 1983. Foodborne disease outbreaks annual summary 1981. Center for Disease Control, Atlanta.
3. **Anonymous.** 1984. Infant botulism. Massachusetts. Morbid. Mortal. Weekly Rep. **33**:165–166.
4. **Arnon, S. S.** 1980. Infant botulism. Annu. Rev. Med. **31**:541–560.
5. **Arnon, S. S., and J. Chin.** 1979. The clinical spectrum of infant botulism. Rev. Infect. Dis. **1**:614–624.
6. **Arnon, S. S., K. Damus, and J. Chin.** 1981. Infant botulism: epidemiology and relation to sudden infant death syndrome. Epidemiol. Rev. **3**:45–66.
7. **Bartlett, J. G.** 1979. Antibiotic-associated colitis. Clin. Gastroenterol. **8**:783–801.
8. **Borriello, S. P., H. E. Larson, A. R. Welch, F. Barclay, M. F. Stringer, and B. A. Bartholomew.** 1984. Enterotoxigenic *Clostridium perfringens*: a possible cause of antibiotic-associated diarrhea. Lancet **i**:305–307.
9. **Cato, E. P., L. V. Holdeman, and W. E. C. Moore.** 1982. *Clostridium perenne* and *Clostridium paraperfringens*: later subjective synonyms of *Clostridium barati*. Int. J. Syst. Bacteriol. **32**:77–81.
10. **Chang, T.-W., M. Lauerman, and J. G. Bartlett.** 1979. Cytotoxicity assay in antibiotic-associated colitis. J. Infect. Dis. **140**:765–770.
11. **Dowell, V. R., Jr.** 1979. Antibiotic-associated colitis. Hosp. Pract. **14**:75–80.
12. **Dowell, V. R., Jr.** 1984. Botulism and tetanus: selected epidemiologic and microbiologic aspects. Rev. Infect. Dis. **6**(Suppl. 1):S202–S207.
13. **Dowell, V. R., Jr., and T. M. Hawkins.** 1977. Laboratory methods in anaerobic bacteriology, CDC laboratory manual. Department of Health, Education, and Welfare publication no. (CDC) 78-8272. Center for Disease Control, Atlanta.
14. **Dowell, V. R., Jr., and G. L. Lombard.** 1981. Reactions of anaerobic bacteria in differential agar media, CDC tables. U.S. Department of Health and Human Services, Public Health Service, Centers for Disease Control, Atlanta.
15. **Dowell, V. R., Jr., and G. L. Lombard.** 1982. CDC tables: differential characteristics of anaerobic bacteria. U.S. Department of Health and Human Services, Public Health Service, Centers for Disease Control, Atlanta.
16. **Dowell, V. R., Jr., and G. L. Lombard.** 1982. Differential agar media for identification of anaerobic bacteria, p. 258–262. *In* R. C. Tilton (ed.), Rapid methods and automation in microbiology. American Society for Microbiology, Washington, D.C.
17. **Finegold, S. M.** 1977. Anaerobic bacteria in human disease. Academic Press, Inc., New York.
18. **Finegold, S. M.** 1977. Therapy for infections due to anaerobic bacteria: an overview. J. Infect. Dis. **135**(Suppl.):525–529.
19. **George, W. L., V. L. Sutter, D. Citron, and S. M. Finegold.** 1979. Selective and differential medium for isolation of *Clostridium difficile*. J. Clin. Microbiol. **9**:214–219.
20. **George, W. L., V. L. Sutter, and S. M. Finegold.** 1978. Toxigenicity and antimicrobial susceptibility of *Clostridium difficile*, a cause of antimicrobial agent-associated colitis. Curr. Microbiol. **1**:55–58.
21. **Gorbach, S. L.** 1979. Other *Clostridium* species (including gas gangrene), p. 1876–1885. *In* G. L. Mandell, R. G. Douglas, Jr., and J. E. Bennett (ed.), Principles and practice of infectious diseases. John Wiley & Sons, New York.
22. **Gorbach, S. L., and H. Thadepalli.** 1975. Isolation of *Clostridium* in human infections: evaluation of 114 cases. J. Infect. Dis. **131**:S81–S85.
23. **Hauschild, A. H. W.** 1975. Criteria and procedures for implicating *Clostridium perfringens* in food-borne outbreaks. Can. J. Public Health **66**:388–392.
24. **Hauschild, A. H. W., and R. Hilsheimer.** 1974. Enumeration of food-borne *Clostridium perfringens* in egg yolk-free tryptose-sulfite-cycloserine agar. Appl. Microbiol. **27**:521–526.
25. **Hauschild, A. H. W., R. Hilsheimer, and D. W. Griffith.** 1974. Enumeration of fecal *Clostridium perfringens* spores in egg yolk-free tryptose-sulfite-cycloserine agar. Appl. Microbiol. **27**:527–530.
26. **Holdeman, L. V., E. P. Cato, and W. E. C. Moore (ed.).** 1977. Anaerobe laboratory manual, 4th ed. Virginia Polytechnic Institute and State University, Blacksburg.
27. **Jarvis, W., O. Nunez-Montiel, F. Thompson, V. Dowell, M. Towns, G. Morris, and E. Hill.** 1983. Comparison of bacterial isolation, cytotoxicity assay, and counterimmunoelectrophoresis for the detection of *Clostridium difficile* and its toxin. J. Infect. Dis. **147**:778.
28. **Kaneuchi, C., T. Miyazato, T. Shinjo, and T. Mitsuoka.** 1979. Taxonomic study of helically coiled, sporeforming anaerobes isolated from the intestines of humans and other animals: *Clostridium cocleatum* sp. nov. and *Clostridium spiroforme* sp. nov. Int. J. Syst. Bacteriol. **29**:1–12.
29. **King, A., A. Rampling. D. G. D. Wright, and R. E. Warren.** 1984. Neutropenic enterocolitis due to *Clostridium septicum* infection. J. Clin. Pathol. **37**:335–343.
30. **Kliegman, R. M., and A. A. Fanaroff.** 1984. Necrotizing enterocolitis. N. Engl. J. Med. **310**:1093–1103.
31. **Koneman, E. W., S. D. Allen, V. R. Dowell, Jr., and H. M. Sommers.** 1983. Color atlas and textbook of diagnostic microbiology, 2nd ed. J. B. Lippincott Co., Philadelphia.
32. **Koransky, J. R., S. D. Allen, and V. R. Dowell, Jr.** 1978. Use of ethanol for selective isolation of sporeforming microorganisms. Appl. Environ. Microbiol. **35**:762–765.

33. **Koransky, J. R., M. D. Stargel, and V. R. Dowell, Jr.** 1979. *Clostridium septicum* bacteremia: its clinical significance. Am. J. Med. **66:**63–66.
34. **Laughon, B. E., R. P. Viscidi, S. L. Gdovin, R. H. Yolken, and J. G. Bartlett.** 1984. Enzyme immunoassays for detection of *Clostridium difficile* toxins A and B in fecal specimens. J. Infect. Dis. **149:**781–788.
35. **Love, D. N., R. F. Jones, and M. Bailey.** 1979. *Clostridium villosum* sp. nov. from subcutaneous abscesses in cats. Int. J. Syst. Bacteriol. **29:**241–244.
36. **Nunez-Montiel, O. L., F. S. Thompson, and V. R. Dowell, Jr.** 1983. Norleucine-tyrosine broth for rapid identification of *Clostridium difficile* by gas-liquid chromatography. J. Clin. Microbiol. **17:**382–385.
37. **Rifkin, G. D.** 1980. Neutropenic enterocolitis and *Clostridium septicum* infection in patients with agranulocytosis. Arch. Intern. Med. **140:**834–835.
38. **Ryan, R. W., I. Kwasnik, and R. C. Tilton.** 1980. Rapid detection of *Clostridium difficile* toxin in human feces. J. Clin. Microbiol. **12:**776–779.
39. **Schwartzman, J. D., L. B. Reller, and W.-L. L. Wang.** 1977. Susceptibility of *Clostridium perfringens* isolated from human infection to twenty antibiotics. Antimicrob. Agents Chemother. **11:**695–697.
40. **Shandera, W. X., C. O. Tacket, and P. A. Blake.** 1983. Food poisoning due to *Clostridium perfringens* in the United States. J. Infect. Dis. **147:**163–170.
41. **Smith, L. D.** 1975. The pathogenic anaerobic bacteria, 2nd ed. Charles C Thomas, Publisher, Springfield, Ill.
42. **Smith, L. D.** 1977. Botulism: the organism, its toxins, the disease. Charles C Thomas, Publisher, Springfield, Ill.
43. **Smith, L. D.** 1978. The occurrence of *Clostridium botulinum* and *Clostridium tetani* in the soil of the United States. Health Lab. Sci. **15:**74–80.
44. **Smith, L. D., and G. Hobbs.** 1974. Genus III. *Clostridium* Prazmowski 1880, 23, p. 551–572. *In* R. E. Buchanan and N. E. Gibbons (ed.), Bergey's manual of determinative bacteriology, 8th ed. Williams & Wilkins Co., Baltimore.
45. **Sutter, V. L., D. M. Citron, and S. M. Finegold.** 1980. Wadsworth anaerobic bacteriology manual, 3rd ed. The C. V. Mosby Co., St. Louis.
46. **Sutter, V. L., and S. M. Finegold.** 1976. Susceptibility of anaerobic bacteria to 23 antimicrobial agents. Antimicrob. Agents Chemother. **10:**736–752.
47. **Walker, P. D., T. G. C. Murrell, and L. K. Nagy.** 1980. Scanning electron microscopy of the jejunum in enteritis necroticans. J. Med. Microbiol. **13:**445–450.
48. **Wilkins, T. D., and T. Thiel.** 1973. Resistance of some species of *Clostridium* to clindamycin. Antimicrob. Agents Chemother. **3:**136–137.
49. **Willis, A. T.** 1969. Clostridia of wound infection. Butterworths, London.
50. **Yamamoto, K., I. Ohishi, and G. Sakaguchi.** 1979. Fluid accumulation in mouse ligated intestine inoculated with *Clostridium perfringens* enterotoxin. Appl. Environ. Microbiol. **37:**181–186.

Chapter 39

Anaerobic Cocci

JON E. ROSENBLATT

CHARACTERIZATION

A recent major taxonomic revision based on DNA guanine-plus-cytosine content studies has occurred involving the anaerobic gram-positive cocci (8). All former species of *Peptococcus*, with the exception of *Peptococcus niger*, have been placed in the genus *Peptostreptococcus*, which now includes *Peptostreptococcus anaerobius*, *Peptostreptococcus asaccharolyticus*, *Peptostreptococcus indolicus*, *Peptostreptococcus magnus*, *Peptostreptococcus micros*, *Peptostreptococcus prevotii*, and *Peptostreptococcus productus*. In addition, a new species, *Peptostreptococcus tetradius*, has been proposed to include those organisms formerly designated *"Gaffkya anaerobia"* (8). A subsequent study disagrees somewhat with this taxonomic reorganization (16). These data suggested that the anaerobic gram-positive cocci could be separated into seven distinct groups. *Peptococcus niger* and *Peptostreptococcus parvulus* were said to have similar guanine-plus-cytosine contents, which were distinct from those of the other gram-positive cocci; some of the other cocci were closely related, and others were not. For the moment it is more reasonable to recognize the classification of Ezaki et al. (8).

It has also been proposed that *Peptostreptococcus parvulus* be transferred to the genus *Streptococcus* (4). Anaerobic gram-positive cocci have been included in the genus *Streptococcus* if they ferment carbohydrates with the production of major amounts of lactic acid. This genus includes three other organisms (*Streptococcus intermedius*, *Streptococcus constellatus*, and *Streptococcus morbillorum*) which were formerly classified in either the genus *Peptococcus* or *Peptostreptococcus*. Many strains of these three species are actually capnophiles (formerly referred to as microaerophiles), which grow best, either initially or after one or more subcultures, in a 5 to 10% CO_2–air incubator. These bacteria characteristically produce pairs and chains of cocci when grown in liquid media, but cells tend to be elongated and of unequal size when grown on solid media. Some strains are obligate anaerobes, while others grow on the surface of plates incubated in 5 to 10% CO_2–air. Many strains may not grow in CO_2 or show aerotolerance until several subcultures have been made. At the Mayo Clinic these organisms account for approximately 9% of all "streptococci" (including anaerobes, capnophiles, and facultative organisms) isolated from anaerobic cultures. Most of 48 isolates could be placed in the viridans group of streptococci with the classification scheme of Facklam (9). *Streptococcus* strain MG, *Streptococcus anginosus*, *Streptococcus sanguis*, and *S. morbillorum* were the most common corresponding identifications. As noted by Facklam (9), "microaerophilic streptococci" share certain physiologic characteristics with the viridans group of streptococci. They are important human pathogens and will be isolated most often in mixed culture from intra-abdominal and biliary tract sepsis as well as from infections of the respiratory and female genital tracts. Since only a small minority of organisms in the genus *Streptococcus* are obligately anaerobic, these will not be included in the approach to identification used here.

The organism formerly called *Peptococcus saccharolyticus* has been transferred to the genus *Staphylococcus* (17). Most strains reduce nitrate to nitrite, are weakly catalase positive and indole negative, weakly ferment glucose, mannose, and fructose, and produce a major amount of acetic acid (with only minor amounts of lactate and succinate) in peptone-yeast extract-glucose broth. The majority of strains appear to be obligate anaerobes in aerotolerance tests with incubation for 48 h, but most will grow aerobically within 4 to 7 days of incubation. Because only a small percentage of strains are obligately anaerobic (5, 7), they will not be discussed further in this chapter. Other genera of anaerobic gram-positive cocci are *Ruminococcus*, *Coprococcus*, *Gemmiger*, and *Sarcina* (14, 15); these organisms are so infrequently encountered in human clinical infections that they do not warrant further consideration here.

The anaerobic gram-negative cocci include *Veillonella parvula*, *Veillonella atypica*, *Veillonella dispar*, *Acidaminococcus fermentans*, and *Megasphaera elsdenii* (15). These organisms have rarely been documented as being significant human pathogens and account for a very small minority of the anaerobic cocci isolated from human specimens. Of these, *V. parvula* is the only species encountered with any frequency in properly selected and collected clinical specimens.

The anaerobic cocci are a prominent part of the normal human flora, especially in the mouth, upper respiratory tract, and large intestine (10). The gram-positive cocci are important human pathogens, and at the Mayo Clinic they accounted for approximately 30% of the most frequent anaerobic isolates during 1983 (Table 1). The most common isolate was *P. magnus*, accounting for 13% of the most frequent isolates and 51% of the anaerobic gram-positive cocci; following in order of frequency were *P. asaccharolyticus*, *P. anaerobius*, *P. micros*, and *P. prevotii*. Wren et al. (24) have reported a similar frequency of isolation of gram-positive cocci and also found *P. magnus* to be the most common isolate. The anaerobic cocci have been isolated from many different infections, including abscesses of the breast and other soft tissues, surgical wounds, and infections involving the neck and dento-alveolar tissues, respiratory tract, and female genital tract (3, 10). At the Mayo Clinic during 1983, the anaerobic cocci were involved in 4% of anaerobic bacteremias, and *P. magnus* was present in the majority. *P. magnus* is isolated from a variety of sites, most often as part of a polymicrobial infection (3). However, approximately 15% of the time it is present in pure culture. Table 2 illustrates the infections from which *P. magnus* was recovered (bone and joint, intra-abdominal sepsis, soft tissue, and bacter-

TABLE 1. Anaerobic bacteria most frequently isolated from clinical specimens at the Mayo Clinic during 1983[a]

Organism	No. (%) of isolates[b]
Bacteroides fragilis group	458 (34)
Peptostreptococcus magnus	176 (13)
Bacteroides melaninogenicus	127 (10)
Clostridium perfringens	97 (7)
Peptostreptococcus asaccharolyticus	72 (5)
Veillonella parvula	55 (4)
Bacteroides ruminicola subsp. *brevis*	49 (4)
Bacteroides oralis	48 (4)
Bacteroides bivius	42 (3)
Clostridium ramosum	41 (3)
Eubacterium lentum	41 (3)
Fusobacterium nucleatum	38 (3)
Peptostreptococcus anaerobius	35 (3)
Peptostreptococcus micros	31 (2)
Peptostreptococcus prevotii	30 (2)

[a] Excluding *Propionibacterium acnes*.

[b] Percentages based only on total of most frequent isolates.

emia). The recognition of anaerobic cocci, especially *P. magnus*, as significant causes of osteomyelitis and septic arthritis, especially when foreign bodies are present, is an important recent finding. Sixty percent of 92 anaerobes isolated from 40 patients with osteomyelitis were cocci; 30% of the cocci were *P. magnus* (13). Fifty-three percent of 72 anaerobes from 43 patients with septic arthritis were cocci; 45% were *P. magnus*. Half of these patients had infected total joint arthroplasties (11). *P. indolicus*, previously isolated only from animal sources, has recently been recovered for the first time from a human infection (finger of a sheepherder) (2).

COLLECTION, TRANSPORT, AND STORAGE OF SPECIMENS

See Chapters 8 and 37 for a discussion of collection and transport of specimens for anaerobic culture. Beyond these there are no further unique requirements for the cultivation of anaerobic cocci.

DIRECT EXAMINATION

The anaerobic cocci may be separated microscopically into those that are gram positive and those that are gram negative. Counterstaining with either safranin or basic fuchsin is suitable. There are no unique microscopic characteristics which will suggest that these are anaerobic rather than facultative cocci. The gram-positive cocci are usually distinctive, although occasionally, coccobacillary forms are seen. In particular, the cells of *P. anaerobius* and *P. productus* are often elongated and appear in pairs or chains. Generally the peptococci occur singly or in small groups and are larger than the peptostreptococci, which frequently appear in chains; however, there is considerable overlap, especially in view of the new taxonomy, and this is not a reliable method for separating these genera. *P. magnus* and *P. micros* are differentiated primarily by the larger size of the former. Examination by polyacrylamide gel electrophoresis has shown these species to be distinctive and easily separated (8). The observation of unusually large cocci (which do not resemble yeasts), especially when found in groups or "packets," may suggest the presence of *Megasphaera* spp., *P. tetradius*, or *Sarcina* spp.

CULTURE AND ISOLATION

The anaerobic cocci grow well on the nonselective plating media usually recommended for the anaerobic cultivation of clinical specimens. The base medium may be Centers for Disease Control anaerobe blood agar, brain heart infusion agar, brucella agar, or Schaedler agar supplemented with 5% sheep blood, vitamin K_1, and hemin, although the anaerobic cocci are not known to have a specific requirement for the latter two supplements. Since most gram-positive cocci are inhibited by vancomycin, one cannot expect to recover these organisms from selective media containing this antimicrobial agent (usually found in combination with kanamycin, gentamicin, or paromomycin plus other supplements). However, blood agar plates containing phenylethyl alcohol or neomycin will not inhibit gram-positive cocci, and these media aid in recovery of these organisms from mixed cultures by suppressing overgrowth of facultative gram-negative bacilli, especially spreading *Proteus* spp. The anaerobic cocci grow well in a number of broth media. At the Mayo Clinic a thioglycolate broth (BBL-0135C; BBL Microbiology Systems, Cockeysville, Md.) supplemented with hemin (5 μg/ml), vitamin K_1 (0.1 μg/ml), rabbit serum (5%), $NaHCO_3$ (1 mg/ml), and a $CaCO_3$ chip is used for both inoculation of specimens and growth or maintenance of individual isolates.

A strain of *P. anaerobius* has been shown to grow better on blood agar plates coincubated in anaerobic jars with other cultures than on plates incubated in an anaerobic glove box (20). This coincubation phenomenon probably is common to many other anaerobes and can be easily demonstrated by inclusion of several plates heavily inoculated with *Clostridium perfringens* in the same jar as the test isolate. The mechanism for growth stimulation is not known but appears to involve volatile gases rather than medium Eh changes. In my view, anaerobic jar incubation may be preferable to the glove box, not only for certain fastidious organisms but for primary specimen cultures as well.

IDENTIFICATION

Growth of the anaerobic cocci is usually slower than that of *Bacteroides* or *Clostridium* spp. The generally

TABLE 2. Human infections from which *Peptostreptococcus magnus* was isolated[a]

Infection	% of total infections: Pure culture[b]	% of total infections: Mixed culture[c]
Septic arthritis and osteomyelitis	56	21
Soft tissue	38	38
Intra-abdominal sepsis	0	11
Miscellaneous[d]	0	30
Bacteremia	6	0

[a] From Bourgault et al. (3).

[b] Total of 32 infections; only *P. magnus* was isolated.

[c] Total of 151 infections; *P. magnus* was isolated.

[d] Includes foot ulcers; pleural fluid, female genital tract, lung, and sinus infection; and brain abscesses.

TABLE 3. Identifying characteristics of anaerobic cocci[a]

Organism	Coagu-lase	In-dole	Ni-trate	Escu-lin	Gela-tin	Urease	Fermentation of:					GLC analysis[b]
							Cello-biose	Glu-cose	Lac-tose	Mal-tose	Su-crose	
Peptococcus niger	−	−	−	−	−	−	−	−	−	−	−	B, C, iv, a
Peptostreptococcus												
P. anaerobius[c]	−	−	−	−	−	−	−	+	−	−	−	A, IC, ib, b, iv
P. magnus	−	−	−	−	V	−	−	−	−	−	−	A, (1), (s)
P. micros	−	−	−	−	−	−	−	−	−	−	−	A, (1), (s)
P. indolicus	+	+	+	−	−	−	−	−	−	−	−	A, B, p, (1), (s)
P. asaccharolyticus	−	+	−	−	−	−	−	−	−	−	−	A, B, (1), (p), (s)
P. prevotii	−	−	±	−	−	−	−	−	−	−	−	A, B, (1), (p), (s)
P. tetradius	−	−	−	−	−	+	−	+	−	+	+	B, L, (a), (p)
P productus	−	−	−	+	−	−	+	+	+	+	+	A, (1), (s)
Veillonella parvula	−	−	+	−	−	−	−	−	−	−	−	A, P, (s)

[a] Symbols: +, 90% or more positive; −, 90% or more negative; ±, usually negative, some strains positive; V, variable reaction. See the text for descriptions of tests.

[b] GLC analysis of fatty acid products in peptone-yeast extract-glucose. Capital letters indicate major metabolic products, whereas lowercase letters indicate minor products. Parentheses indicate a variable reaction. A, Acetic acid; P, propionic acid; IB, isobutyric acid; B, butyric acid; IV, isovaleric acid; V, valeric acid; IC, isocaproic acid; C, caproic acid; L, lactic acid; S, succinic acid.

[c] Inhibited by sodium polyanethol sulfonate.

small colonies are not apparent on blood agar plates until after a full 48 h of incubation. Growth in broth media is often slow also and will occur in aggregates or clumps rather than as diffuse turbidity. Further supplementation of thioglycolate broth (see above) with 0.5% yeast extract may improve growth of the cocci (19). This slow growth may necessitate prolonged incubation (3 to 5 days) for biochemical reactions or antibiotic susceptibility testing. Results of such tests should not be read unless good growth is present. The gram-positive cocci produce small convex colonies which are a grayish white and opaque. The edges are entire, and the surface may appear stippled or pock marked. *Veillonella* colonies are convex and translucent with an entire edge; such colonies on blood agar plates have been shown to exhibit red fluorescence under long-wave (365-nm) UV light. A gram-positive coccus which produces black colonies on blood agar plates has been designated *Peptococcus niger*; this organism is extremely rare and has been isolated primarily from human normal flora (23).

The biochemical tests that are useful for identification of the anaerobic cocci are coagulase production, indole production, nitrate reduction, esculin hydrolysis, gelatin liquefaction, urease production, fermentation of carbohydrates (cellobiose, glucose, lactose, maltose, and sucrose), and gas chromatographic analysis of acid metabolic products in peptone-yeast extract-glucose broth culture. Most of these tests may be performed with either prereduced peptone-yeast extract-based broth media (14, 22) or thioglycolate broth-based media (Chapter 111). Commercially prepared rapid kits (Minitek [BBL]; API 20A strip [Analytab Products, Plainview, N.Y.]) offer a practical alternative, especially to conventional fermentation tests, provided that the directions of the manufacturer are followed and results are interpreted by using the tables provided. The species of gram-positive cocci are differentiated primarily on the basis of fermentation reactions as well as end products of metabolism detected by gas-liquid chromatography (GLC) (14). *P. anaerobius* may also be identified on the basis of its unique susceptibility to sodium polyanethol sulfonate; a zone of inhibition of growth (12 to 18 mm or greater) on an anaerobe blood agar plate is produced around a sodium polyanethol sulfonate-saturated paper disk after anaerobic incubation (12). Another rapid test for the identification of *P. anaerobius* involves the ability of this organism to degrade tyrosine on an agar plate; other anaerobic gram-positive cocci are unable to do so (1). Tyrosine degradation results in production of *p*-hydroxyhydrocinnamic acid which can be detected by GLC (18). Should the Gram reaction be equivocal, gram-negative anaerobes may be separated from gram-positive anaerobes by the susceptibility of the former to colistin and their resistance to vancomycin when tested by a special set of antibiotic-containing paper disks (22). The susceptibility pattern is reversed for gram-positive anaerobes.

Table 3 lists the identifying characteristics of the anaerobic cocci. *P. niger* is quite inactive biochemically but produces major butyric and caproic acid peaks on GLC analysis. *P. magnus*, the most commonly isolated anaerobic coccus, ferments no carbohydrates and is indole, nitrate, and esculin negative. Gelatin liquefaction is variable, but the reaction is enhanced by incorporation of Tween 80 (0.02%) in the medium, as is growth of this organism in general. Major amounts of acetic acid and minor (or trace) amounts of lactic and succinic acids are found on GLC analysis of metabolic products in peptone-yeast extract-glucose broth cultures. The large size of *P. magnus* cocci (1 to 2 μm in diameter) helps to differentiate this organism from the smaller *P. micros*, which has similar biochemical reactions. *P. indolicus* is also nonfermentative but is unique in that it produces indole and coagulase and reduces nitrate. It forms major acetic and butyric acid peaks on GLC analysis. *P. asaccharolyticus* and *P. prevotii* are the remaining nonfermentative anaerobic gram-positive cocci. The former is indole positive and the latter is indole negative, while both organisms produce major acetic and butyric acid peaks.

P. anaerobius ferments glucose only and produces major acetic and isocaproic acid GLC peaks and minor propionic, isobutyric, butyric, and isovaleric acid peaks. Production of isocaproic acid is a key differential characteristic of this species. *P. productus* is the most fermentative of the anaerobic cocci; it ferments cellobiose, glucose, lactose, maltose, mannose, sorbitol, sucrose, and xylose. *P. productus* hydrolyzes esculin and produces a major amount of acetic acid. *P. tetradius* is also quite fermentative, producing acid from glucose, maltose, sucrose, and mannose (but not lactose). This is the only clinically encountered anaerobic coccus which produces urease. Major lactic and butyric acid peaks are detected by GLC analysis of peptone-yeast extract-glucose fermentation products. *V. parvula*, the only anaerobic gram-negative coccus of clinical significance, is inactive biochemically except for its reduction of nitrate. It produces major amounts of acetic and propionic acids and a minor or trace amount of succinic acid.

The anaerobic cocci are susceptible to a variety of antimicrobial agents. Although there are some differences in degree of susceptibility between the data of Sutter (21) and studies carried out at the Mayo Clinic (6) (Table 4), most of the gram-positive cocci are inhibited by readily achievable concentrations in serum of the antimicrobial agents tested. The only exceptions are tetracycline (12.5 μg/ml required to inhibit 90% of the strains) and metronidazole for which differing results have been obtained. Whereas Sutter (21) found an MIC for 90% of the strains of 4.0 μg/ml, Edson et al. (6) obtained a value of >25 μg/ml (although 70% of the strains were inhibited by 3.12 μg/ml). The reasons for this discrepancy are not clear. Suffice it to say that metronidazole should not be considered one of the most active agents against the anaerobic cocci. Penicillin G is the drug of choice for treating infections involving anaerobic cocci. Alternate choices would be a cephalosporin, clindamycin, or chloramphenicol. Susceptibility data for the anaerobic gram-negative cocci are similar to those given above, but rarely if ever would specific therapy need to be directed against this component of an infection.

Given the predictable susceptibility patterns of the anaerobic cocci, one can legitimately ask whether specific identification of organisms in this group is necessary in the clinical laboratory and whether susceptibility testing ever should be done. The answer is no for the laboratory with limited facilities, personnel, and interest in anaerobic bacteriology. However, there is clearly much still to be learned about the role of these organisms in clinical infection—witness the recent first isolation of *P. indolicus* from a human infection (2) and the recognition of the importance of *P. magnus* in a variety of infections, including septic arthritis associated with total joint arthroplasties (3). Therefore, laboratories that are capable of performing the tests required for specific identification of the anaerobic cocci should probably do so. Antimicrobial susceptibility testing should be restricted to those organisms isolated from blood, brain abscesses, or bone and joint infections, where specific susceptibility data may be of help in the management of these serious and often chronic infections.

TABLE 4. Antimicrobial susceptibility of anaerobic gram-positive cocci

Antibiotic	MIC_{90}[a] (μg/ml)	
	Sutter[b]	Edson et al.[c]
Penicillin	1.0	≤0.8
Cephalothin	—	≤0.8
Cefoperazone	4.0	—
Cefoxitin	4.0	1.56
Chloramphenicol	4.0	6.25
Clindamycin	8.0	≤0.8
Tetracycline	—	12.5
Metronidazole	4.0	≥25

[a] Concentration at which 90% of the strains were inhibited.

[b] Data of Sutter (21).

[c] Mayo Clinic data (6) obtained by the broth microdilution method.

LITERATURE CITED

1. **Babcock, J. B.** 1979. Tyrosine degradation in presumptive identification of *Peptostreptococcus anaerobius*. J. Clin. Microbiol. **9**:358–361.
2. **Bourgault, A. M., and J. E. Rosenblatt.** 1979. First isolation of *Peptococcus indolicus* from a human clinical specimen. J. Clin. Microbiol. **9**:549–550.
3. **Bourgault, A.-M., J. E. Rosenblatt, and R. H. Fitzgerald.** 1980. *Peptococcus magnus*: a significant human pathogen. Ann. Intern. Med. **93**:244–248.
4. **Cato, E. P.** 1983. Transfer of *Peptostreptococcus parvulus* (Weinberg, Nativelle, and Prévot 1937) Smith 1957 to the genus *Streptococcus: Streptococcus parvulus* (Weinberg, Nativelle, and Prévot 1937) comb. nov., nom. rev., emend. Int. J. Syst. Bacteriol. **33**:82–84.
5. **Crosa, J. H., B. L. Williams, J. J. Jorgensen, and C. A. Evans.** 1979. Comparative study of deoxyribonucleic acid homology and physiological characteristics of strains of *Peptococcus saccharolyticus*. Int. J. Syst. Bacteriol. **29**:328–332.
6. **Edson, R. S., J. E. Rosenblatt, D. T. Lee, and E. A. McVey III.** 1982. Recent experience with antimicrobial susceptibility of anaerobic bacteria. Increasing resistance to penicillin. Mayo Clin. Proc. **57**:737–741.
7. **Evans, C. A., K. L. Mattern, and S. L. Hallam.** 1978. Isolation and identification of *Peptococcus saccharolyticus* from human skin. J. Clin. Microbiol. **7**:261–264.
8. **Ezaki, T., N. Yamamoto, K. Ninomiya, S. Suzuki, and E. Yabuuchi.** 1983. Transfer of *Peptococcus indolicus, Peptococcus asaccharolyticus, Peptococcus prevotii*, and *Peptococcus magnus* to the genus *Peptostreptococcus* and proposal of *Peptostreptococcus tetradius* sp. nov. Int. J. Syst. Bacteriol. **33**:683–698.
9. **Facklam, R. R.** 1977. Physiological differentiation of viridans streptococci. J. Clin. Microbiol. **5**:184–201.
10. **Finegold, S. M.** 1977. Anaerobic bacteria in human disease. Academic Press, Inc., New York.
11. **Fitzgerald, R. H., J. E. Rosenblatt, J. H. Tenny, and A.-M. Bourgault.** 1982. Anaerobic septic arthritis. Clin. Orthop. Relat. Res. **164**:141–148.
12. **Graves, M. H., J. A. Morello, and F. E. Kocka.** 1974. Sodium polyanethol sulfonate sensitivity of anaerobic cocci. Appl. Microbiol. **27**:1131–1133.
13. **Hall, B. B., R. H. Fitzgerald, and J. E. Rosenblatt.** 1983. Anaerobic osteomyelitis. J. Bone Jt. Surg. Am. Vol. **65**:30–35.
14. **Holdeman, L. V., E. P. Cato, and W. E. C. Moore.** 1977. Anaerobe laboratory manual. Virginia Polytechnic Institute and State University, Blacksburg.
15. **Holdeman, L. V., E. P. Cato, and W. E. C. Moore.** 1984. Taxonomy of anaerobes: present state of the art. Rev. Infect. Dis. **6**(Suppl. 1):3–10.
16. **Huss, V. A. R., H. Festl, and K. H. Schleifer.** 1984. Nucleic acid hybridization studies and deoxyribonucleic acid base compositions of anaerobic, gram-positive cocci. Int. J. Syst. Bacteriol. **34**:95–101.

17. **Kilpper-Bälz, R., and K. H. Schleifer.** 1981. Transfer of *Peptococcus saccharolyticus* Foubert and Douglas to the genus *Staphylococcus: Staphylococcus saccharolyticus* (Foubert and Douglas) comb. nov. Zentralbl. Bakteriol. Parasitenkd. Infektionskr. Hyg. Abt. 1 Orig. Reihe C **2:**324–331.
18. **Lambert, M. A., and C. W. Moss.** 1980. Production of *p*-hydroxyhydrocinnamic acid from tyrosine by *Peptostreptococcus anaerobius*. J. Clin. Microbiol. **12:**291–293.
19. **Marshall, R., V. K. Yasui, R. Prabhala, A. K. Kaufman, and I. Wallace.** 1981. Growth of *Peptococcus* and *Peptostreptococcus*: effect of variations of culture media on efficiency of recovery. Appl. Environ. Microbiol. **42:**493–496.
20. **Rosenblatt, J. E.** 1982. Reevaluation of current methods for incubation of anaerobes, p. 271–274. *In* R. C. Tilton (ed.), Rapid methods and automation in microbiology. American Society for Microbiology, Washington, D.C.
21. **Sutter, V. L.** 1983. Frequency of occurrence and antimicrobial susceptibility of bacterial isolates from the intestinal and female genital tracts. Rev. Infect. Dis. **5**(Suppl.)**:**84–88.
22. **Sutter, V. L., D. M. Citron, and S. M. Finegold.** 1980. Wadsworth anaerobic bacteriology manual, 3rd ed. The C. V. Mosby Co., St. Louis.
23. **Wilkins, T. D., W. E. C. Moore, S. E. H. West, and L. V. Holdeman.** 1975. *Peptococcus niger* (Hall) Kluyver and van Niel 1936: emendation of description and designation of neotype strain. Int. J. Syst. Bacteriol. **25:**47–49.
24. **Wren, M. W., A. W. Baldwin, C. P. Eldon, and P. J. Sanderson.** 1977. The anaerobic culture of clinical specimens: a 14 month study. J. Med. Microbiol. **10:**49–61.

Chapter 40

Gram-Negative, Nonsporeforming Anaerobic Bacilli

SYDNEY M. FINEGOLD AND MARTHA A. C. EDELSTEIN

This chapter covers members of the genera *Bacteroides, Fusobacterium, Wolinella, Leptotrichia, Desulfomonas, Butyrivibrio, Succinivibrio, Succinimonas, Anaerovibrio, Anaerobiospirillum, Mobiluncus,* and *Selenomonas*. These organisms are part of the normal flora of humans and animals and are found in the mouth, upper respiratory tract, intestinal tract, and urogenital tract. Anaerobic spirochetes are considered in Chapter 45. The initial differentiation of these genera is based on motility, flagellar arrangement, cellular morphology, and an analysis of metabolic end products by gas-liquid chromatography (GLC) (Table 1) (8). Species definition is based on biochemical characteristics. In almost all clinical specimens, only the genera *Bacteroides* and *Fusobacterium* need to be considered.

CHARACTERIZATION

This chapter will focus on *Bacteroides* spp. and *Fusobacterium* spp. These species are identified presumptively based on a few observations such as colonial and cellular morphology, fluorescence under long-wave UV light, susceptibility to special-potency antibiotic disks, and certain rapidly determined biochemical characteristics. A definitive identification requires the determination of multiple characteristics with a battery of biochemical tests. A definitive identification is recommended for the confirmation of presumptive identification and for clinical purposes. Additionally, an exact species determination of the isolates from various infections is needed to establish which anaerobes are significant in particular types of infections. See Table 2 for the current taxonomic changes in species nomenclature.

OCCURRENCE IN CLINICAL MATERIAL AND SIGNIFICANCE

Gram-negative anaerobic bacilli are the most commonly encountered anaerobes in clinical infections; they are found in more than half of the specimens yielding anaerobes (3). The *Bacteroides fragilis* group, which is bile resistant (see Table 4), is the most commonly recovered anaerobe of all types found in clinical specimens and is more resistant to antimicrobial agents than are other anaerobes. *B. fragilis* and *B. thetaiotaomicron* are the species in the group of greatest clinical significance. They are recovered from most intraabdominal infections and may be seen in infections at other sites. The *B. fragilis* group constitutes the dominant portion of the normal colonic flora. These organisms may also be found in smaller numbers in the female genital tract, but not usually in the mouth or upper respiratory tract.

The pigmenting bacteroides (formerly *B. asaccharolyticus, B. melaninogenicus* subsp. *intermedius*, and *B. melaninogenicus* subsp. *melaninogenicus*) presently comprise nine species (7), seven of which are found in human clinical material (see Table 6). Three of these newly described species (*B. denticola, B. gingivalis,* and *B. loescheii*) are part of the normal flora of the human mouth, are important pathogens in oral and dental infections, and may produce infections of the head, neck, and lower respiratory tract. The other pigmenting organisms are prevalent in the mouth and the urogenital and intestinal tracts. They are also important in head, neck, oral, dental, bite, and pleuropulmonary infections, as well as in various other infections.

The bile-sensitive saccharolytic bacteroides (see Table 5) are found in the same setting as the pigmenting bacteroides (10). *B. bivius* and *B. disiens* are found particularly in female genital tract and oral infections. *B. oris* and *B. buccae* were formerly classed as *B. ruminicola*, which is presently recognized only as a rumen strain. These species are found in a variety of infections. *B. oralis* is now represented by *B. oralis, B. veroralis*, and *B. buccalis* (19).

The asaccharolytic, formate- and fumarate-requiring, nitrate- or nitrite-reducing organisms include *B. ureolyticus, B. gracilis, Wolinella succinogenes*, and *Wolinella recta* (Table 1). *B. ureolyticus* has been recovered from a variety of infections, including pulmonary, head and neck, intraabdominal, urogenital, bone, and soft tissue infections. *B. gracilis* and the *Wolinella* spp. are primarily oral isolates found in periodontitis and periodontosis (17); further studies are required to determine the extent to which these species are involved in infection.

Another gram-negative anaerobic bacillus commonly encountered in clinical infections is *Fusobacterium nucleatum*. This organism is found in the mouth and the genital, gastrointestinal, and upper respiratory tracts. It is often involved in the same types of infection as the pigmenting bacteroides group. *Fusobacterium necrophorum* is a very virulent anaerobe which may cause widely disseminated infection (commonly originating in a focus of membranous tonsillitis previously known as Vincent's angina). *F. necrophorum* is encountered much less often now than in the era before antimicrobial agents.

Leptotrichia buccalis is a common mouth organism and rarely may be found in the vagina and intestinal tract. *Selenomonas sputigena* is a member of the normal mouth flora, as is *Anaerovibrio lipolytica. Desulfomonas pigra, Succinimonas amylolytica*, and *Butyrivibrio fibrisolvens* are all found as normal colonic flora. *Mobiluncus* spp. are curved rods found in the vagina (15). The sites of normal carriage of *Anaerobiospirillum succiniciproducens* and *Succinivibrio* spp. in humans are unknown at present. Most of these organisms have been encountered on occasion in clinical specimens, either as contaminating normal flora or as causes of infection, usually bacteremia in an immunocompromised patient (4).

COLLECTION AND TRANSPORT OF SPECIMENS

The collection and transport of specimens are discussed in Chapters 8 and 37. The importance of a

TABLE 1. Differentiation of the genera of gram-negative anaerobic bacilli

Characteristic	Genus
I. Nonmotile or peritrichous flagella	
A. Produce butyric acid (without isobutyric and isovaleric acids)	*Fusobacterium*
B. Produce major lactic acid	*Leptotrichia*
C. Produce acetic acid and hydrogen sulfide; reduce sulfate	*Desulfomonas*
D. Not as above (A, B, or C)	*Bacteroides*
II. Polar flagella	
A. Fermentative	
1. Produce butyric acid	*Butyrivibrio*
2. Produce succinic acid	
a. Spiral-shaped cells	*Succinivibrio*
b. Ovoid cells	*Succinimonas*
3. Produce propionic and acetic acids	*Anaerovibrio*
B. Nonfermentative; produce succinic acid from fumarate	*Wolinella*
III. Tufts of flagella on concave side of curved cells	
A. Fermentative	*Selenomonas*
B. Nonfermentative	*Mobiluncus*
IV. Bipolar tufts of flagella	*Anaerobiospirillum*

properly collected specimen cannot be overemphasized. An improperly collected or transported specimen, contaminated with normal flora, means additional work for the laboratory and yields meaningless or misleading results for clinicians.

DIRECT EXAMINATION

Direct examination is of great importance, as it provides immediate semiquantitative information about the types of organisms present in a specimen; culture results will not be available until at least 1 to 4 or more days later. Thus, the initial therapy of a sick patient must be undertaken with information provided by direct examination, along with the clinical data.

Four types of direct examination may be useful for the gram-negative, nonsporeforming anaerobic bacilli: Gram stain, dark-field microscopy, fluorescent-antibody procedures, and GLC. Of these methods, the Gram stain is by far the simplest and the most likely to yield significant information. A Gram stain reveals the type and relative numbers of most microorganisms, anaerobic and otherwise, that are clinically significant. The Gram stain is also important as a quality control procedure for the microbiologist; one should be able to recover all of the different morphological types seen. Pale, irregularly staining gram-negative rods are frequently obligate anaerobes. Coccobacillary forms are suggestive of the pigmenting *Bacteroides* spp. (Fig. 1) or *Haemophilus* spp. Although the *B. fragilis* group of organisms is not unique on Gram stain, pale, pleomorphic or uniform gram-negative rods with irregular staining from an intraabdominal infection are suggestive of this group (Fig. 2). *F. nucleatum* is a thin gram-negative rod with pointed ends; these organisms are often found in pairs end to end (Fig. 3). This morphology is distinctive, being shared only with *Capnocytophaga* spp., a type of microaerophilic organism. *L. buccalis* is a much larger fusiform rod. *Fusobacterium mortiferum* is an extremely pleomorphic rod with filaments containing swollen areas and large round bodies and exhibiting irregular staining (Fig. 4). *F. necrophorum* may have a similar microscopic morphology. Some clostridia are usually gram negative; an example is *Clostridium clostridiiforme*, which appears as a slightly elongated, football-shaped rod and often occurs in pairs.

Dark-field microscopy may be helpful in the detection of small, poorly staining organisms *(B. pneumosintes)*, the direct observation of motility (*Wolinella* spp.), and the noting of spores (*Clostridium* spp.).

Reagents for the direct or indirect fluorescent-antibody testing of clinical material for the detection of

TABLE 2. Recent taxonomic changes

New nomenclature	Previous nomenclature
Bacteroides asaccharolyticus	*B. melaninogenicus* subsp. *asaccharolyticus*
B. gingivalis	*B. melaninogenicus* subsp. *asaccharolyticus*
B. intermedius	*B. melaninogenicus* subsp. *intermedius* (in part)
B. corporis	*B. melaninogenicus* subsp. *intermedius* (in part)
B. levii	*B. melaninogenicus* subsp. *levii*
B. macacae	New species
B. melaninogenicus	*B. melaninogenicus* subsp. *melaninogenicus*
B. loescheii	New species
B. denticola	New species
B. oris	*B. ruminicola*, human strains (in part)
B. buccae	*B. ruminicola*, human strains (in part)
B. veroralis	*B. oralis* (in part)
B. buccalis	*B. oralis* (in part)
B. zoogleoformans	Reinstated
B. gracilis	New species
Fusobacterium periodonticum	New species
Wolinella succinogenes	New species (basonym, *Vibrio succinogenes*)
W. recta	New species
Clostridium symbiosum	*F. symbiosum*
Clostridium bullosum	*F. bullosum*
Eubacterium plauti	*F. plauti*

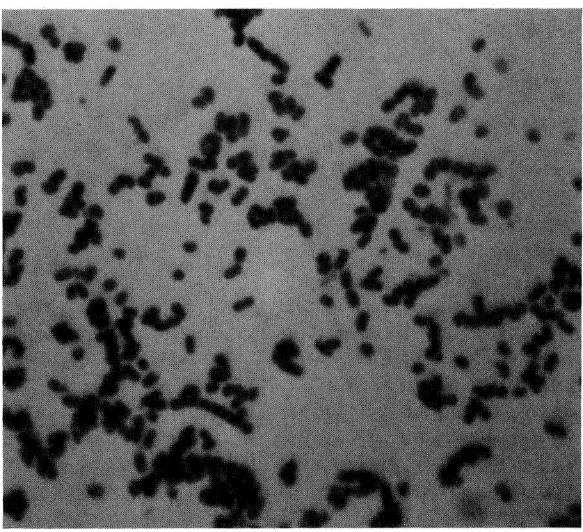

FIG. 1. Coccobacillary cells of *Bacteroides asaccharolyticus*.

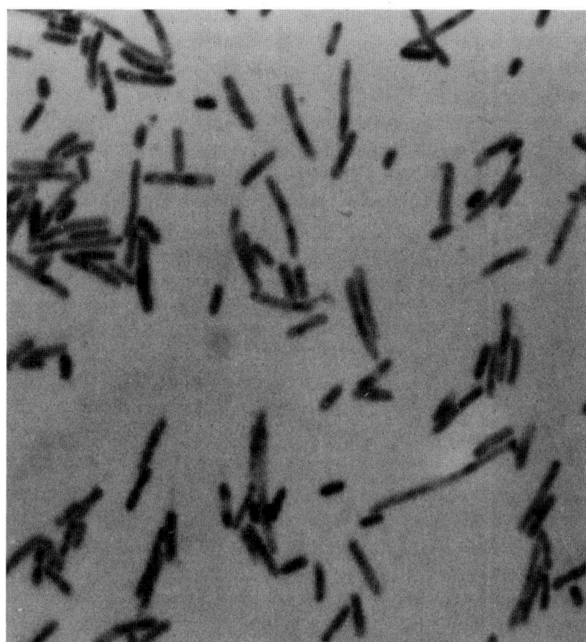

FIG. 2. Pleomorphic, irregularly staining cells of *Bacteroides fragilis*.

the *B. fragilis* group and pigmenting *Bacteroides* spp. are commercially available (General Diagnostics, Morris Plains, N.J.). Studies with these reagents indicate that they reliably detect the presence of the *B. fragilis* and *B. melaninogenicus* groups of organisms in clinical material (21). The *B. melaninogenicus* reagent detects *B. intermedius*, certain other pigmented bacteroides, and nonoral strains of *B. asaccharolyticus* (13). The reagent cross-reacts with *B. bivius* and *B. disiens*, indicating that these organisms share common cell wall antigens with the pigmented bacteroides. Results are obtained in 1 to 2 h.

A few studies indicate that the direct analysis of pus by GLC can provide a rapid presumptive identification of certain anaerobes (5). The presence of gram-negative rods on a Gram stain of such material and a high concentration of butyric acid, in the absence of isobutyric or isovaleric acid, would indicate a *Fusobacterium* sp. All three of these acids plus succinic acid would suggest *Bacteroides* spp. Although *Bacteroides* spp. may produce butyric acid, it is usually produced only in minor quantity.

CULTURE AND ISOLATION

The selection and use of primary isolation media are discussed in Chapter 37. The following procedures are used in the Wadsworth Clinical Anaerobic Bacteriology Research Laboratory (16). Specimens are inoculated onto appropriate aerobic and anaerobic plates, into a tube of thioglycolate broth (BBL-135C) supplemented with hemin, a carbonate chip, and vitamin K_1 or chopped meat-glucose broth, and onto a slide for Gram staining. The anaerobic plates are reduced in an anaerobic atmosphere for at least 24 h before use and consist of (i) a nonselective, enriched, brucella base-sheep blood agar plate supplemented with vitamin K_1 and hemin (BAP); (ii) a kanamycin-vancomycin laked sheep blood agar plate for the selection of *Bacteroides* spp. (16); (iii) a *Bacteroides* bile-esculin agar plate (BBE) for the selection and presumptive identification of the *B. fragilis* group (11); and (iv) a phenylethyl alcohol-sheep blood agar plate for the inhibition of aerobic gram-negative rods, which may overgrow the anaerobes. When fusobacteria are clinically suspected as the cause of infection, a special selective medium may be used (12).

The inoculated anaerobic plates are immediately placed in an anaerobic environment, i.e., an anaerobic jar or chamber or a Marion type A BioBag. Although not optimal, the holding jar method (Chapter 37) may be used until enough plates accumulate to set up a jar. After incubation for 48 h at 36°C, the plates are examined, and the colony types are semiquantitated. In the case of a seriously ill patient, the plates may be

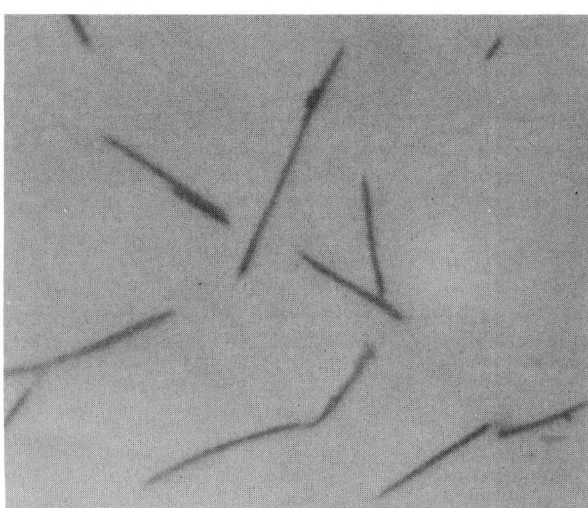

FIG. 3. Cells of *Fusobacterium nucleatum*. Note the slender shape with pointed ends.

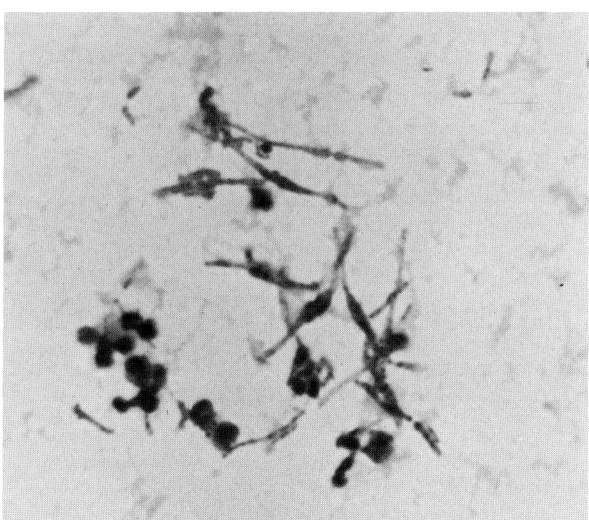

FIG. 4. Microscopic morphology of *Fusobacterium mortiferum*. There is marked pleomorphism and irregularity of staining. Note the filaments with swellings along their course.

examined and processed earlier in an anaerobic chamber. The colonies are described with the aid of an 8× hand lens or, preferably, a stereoscopic dissecting microscope, and they are Gram stained. Different colony types are subcultured to a brucella BAP to which special-potency antibiotic disks (colistin, 10 μg; kanamycin, 1,000 μg; and vancomycin, 5 μg) and a nitrate disk are added (Fig. 5), to a chocolate agar plate which is incubated in 10% CO_2 (for aerotolerance testing), and, if a lipase producer (*F. necrophorum*, *B. intermedius*, or *B. loescheii*) is suspected, to an egg yolk agar plate.

The following characteristics are noted, and tests are made from the BAP: detailed colony morphology, pigment, fluorescence (long-wave UV light), hemolysis, greening, pitting, catalase (15% H_2O_2), spot indole reaction (paradimethylaminocinnamaldehyde reagent), special-potency antibiotic-disk susceptibility, and motility (hanging drop). Since most *Bacteroides* spp. and *Fusobacterium* spp. are nonmotile, motile isolates are checked for suspected *Wolinella* spp. and other motile genera (Table 1). At this time the inoculation of a Lombard-Dowell Presumpto 1 plate (see Chapters 37 and 111) may give additional important information for the presumptive identification of the anaerobic gram-negative rods (2). Good growth is necessary for the proper interpretation of test results on this medium. The primary plates are reinspected after 4 or more days to detect slow growers, new morphotypes, or late pigmenters.

IDENTIFICATION OF SPECIES

Presumptive identification

Most of the clinically significant gram-negative rods can be placed into broad groups with relatively few tests, and some can be presumptively identified with ease (Table 3). The special-potency antibiotic-disk pattern can be used to separate the gram-negative rods into *Bacteroides* spp. or *Fusobacterium* spp. A zone size equal to or greater than 10 mm is considered susceptible. The bacteroides are resistant to vancomycin (as are most gram-negative anaerobes) and kanamycin and variable in susceptibility to colistin, whereas *Fusobacterium* strains are resistant to vancomycin and susceptible to both colistin and kanamycin. Exceptions to this generalization include certain black-pigmented bacteroides and the *B. ureolyticus*-like group. *B. asaccharolyticus* and *B. gingivalis* are usually susceptible to vancomycin and resistant to colistin and kanamycin. *B. ureolyticus*, *B. gracilis*, and *Wolinella* spp. have the same pattern as fusobacteria; however, their colonies are much smaller and more translucent. An anaerobic gram-negative rod which requires formate and fumarate for growth in broth culture or which pits agar may be presumptively identified as a *B. ureolyticus*-like organism. Organisms that fluoresce brick red or produce black colonies are placed in the pigmenting *Bacteroides* spp. (Fig. 6). The *B. fragilis* group can be presumptively identified by their special-potency antibiotic-disk pattern (resistant to all three antibiotics) and growth in a 20% bile tube test, in a bile disk test (20), or on a BBE agar plate (Fig. 7). The term *Bacteroides* spp. is applied to those organisms not fitting the *B. fragilis* group, pigmenting *Bacteroides* spp., *B. ureolyticus*-like group, or *Fusobacterium* spp.

Rapid identification

The anaerobic gram-negative rods that can be rapidly identified are shown in Table 3. *F. nucleatum* is a thin rod with tapering ends and is indole positive. It fluoresces chartreuse and produces greening of the agar. There are three colony morphotypes of *F. nucleatum*, i.e., speckled, breadcrumb, and smooth. The size of the breadcrumb-shaped colonies varies from <0.5 to 1 mm (Fig. 8). A lipase-positive fusobacterium is *F. necrophorum*. It is a pleomorphic rod with round ends and sometimes bizarre forms. It is indole positive, fluoresces chartreuse, and produces greening of the agar. The colonies are umbonate and range in size from 0.5 to 1 mm (Fig. 9). Lipase-positive strains are usually beta-hemolytic. Lipase-negative strains require further biochemical tests for identification. A bile-resistant fusobacterium may be presumptively identified as the *F. mortiferum*-*F. varium* group; however, other *Fusobacterium* spp. (e.g., *F. necrophorum*) may grow in 20% bile; therefore, further testing is required to confirm the presumptive identification.

B. ureolyticus, *B. gracilis*, and *Wolinella* spp. are thin gram-negative rods with rounded ends and the fusobacterium disk pattern. The colonies are small and translucent or transparent and may produce greening of the agar. Three colony morphotypes exist, i.e., smooth and convex, pitting (Fig. 10), and spreading. All three morphotypes can occur in the same culture. These organisms are asaccharolytic and nitrate or nitrite positive, and they require formate and fumarate for growth (Table 3, footnote *c*). The *Wolinella* spp. are motile, *B. ureolyticus* is urease positive, and *B. gracilis* is nonmotile and urease negative (Table 3, footnotes *d*, *e*, and *f*).

An indole- and lipase-positive coccobacillus that forms black-pigmented colonies or fluoresces brick red may be identified as *B. intermedius*. Any lipase-negative strains must be identified by other biochemical tests. A lipase-positive, indole-negative pigmenting bacteroides may be identified as *B. loescheii*.

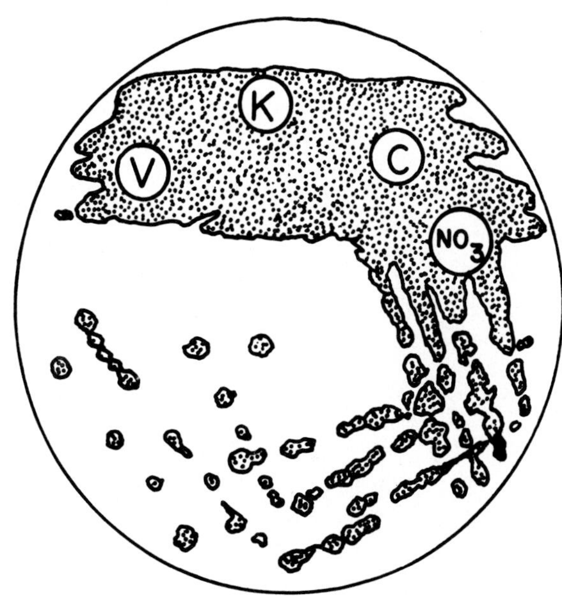

FIG. 5. Placement of special-potency antibiotic and nitrate disks.

TABLE 3. Grouping anaerobic gram-negative rods[a]

Group and species	Kanamycin (1,000 μg)	Vancomycin (5 μg)	Colistin (10 μg)	Growth in 20% bile	Catalase	Indole	Lipase	Slender cells with pointed ends	Growth stimulated by formate-fumarate[b]	Nitrate reduction[c]	Urease[d]	Motility[e]	Pitting of agar	Pigment	Brick-red fluorescence
B. fragilis group	R	R	R	$+^-$	V	V	−								−
Other *Bacteroides* spp.	R	R	V	−	−	V	−			$-^+$					
Pigmenting *Bacteroides* spp.	R	V	V	−	$-^+$	V	V							+	$+^-$
B. intermedius	R	R	S	−	−	+	$+^-$								
B. loescheii	R	R	V	−	−	−	V								
B. ureolyticus-like group[f]	S	R	S	−	−	−	−		+	+	V	V	V		
B. ureolyticus	S	R	S	−	−	−			+	+	+	−	V		
B. gracilis	S	R	S	−	−	−			+	+	−	−	V		
Wolinella sp.	S	R	S	−	−	−			+	+	−	+	V		
Fusobacterium sp.	S	R	S	V	−	V	V	V							−
F. nucleatum	S	R	S	−		+	−	+							
F. necrophorum	S	R	S	$-^+$		+	$+^-$	−							
F. varium-mortiferum	S	R	S	+		V	−	−							

[a] R, Resistant; S, susceptible; V, variable; +, positive reaction for majority of strains; −, negative reaction; $+^-$, most strains positive; $-^+$, most strains negative, some strains positive.

[b] Compare the growth of the organism in an unsupplemented thioglycolate broth with growth in a broth supplemented with formate and fumarate.

[c] Use the spot nitrate disk test (Chapter 112).

[d] Make a heavy suspension of the organism in 0.5 ml of sterile urea broth (Difco Laboratories, Detroit, Mich.). Incubate the suspension aerobically for up to 24 h. A bright pink or red is positive; this color usually appears within 15 to 30 min. If the indicator becomes reduced, add Nessler reagent to determine the ammonia production (ammonia indicates a positive reaction).

[e] Use the hanging-drop method with colonies from the isolation BAP. If negative, check motility with a young broth culture supplemented with formate and fumarate.

[f] Formate and fumarate should be added to test media for this group of organisms.

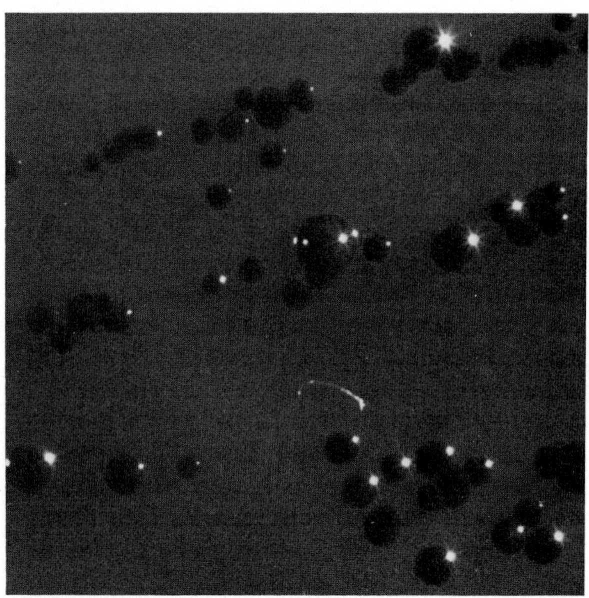

FIG. 6. Black-pigmenting colonies of *Bacteroides asaccharolyticus*.

Definitive identification

The definitive identification of most species requires certain additional biochemical tests and metabolic end product analysis by GLC. Curved rods that are not *Wolinella* spp. should be checked for motility and identified according to Table 1. Organisms with small, translucent, spreading colonies which are not *B. ureolyticus*-like should also be checked for motility. Very large fusiform rods isolated from the mouth or urogenital tract are suggestive of *Leptotrichia* spp. The characteristic GLC pattern should be confirmed. Tables 4 to 7 are based on reactions in prereduced, anaerobically sterilized liquid media (6, 8, 16). Thioglycolate-based liquid and semisolid media with a bromthymol blue indicator give similar results (1) (for the use of these media, see Chapters 37 and 111). Do not interpret the results from other systems with these tables. Gas chromatographic analysis may be performed on any broth which supports good growth of the organism (see Chapter 37 for procedures). Each lot of uninoculated broth is assayed to determine the background amounts of acetic and succinic acids, and an uninoculated chopped-meat broth is additionally assayed for lactic acid. Fermentation end products

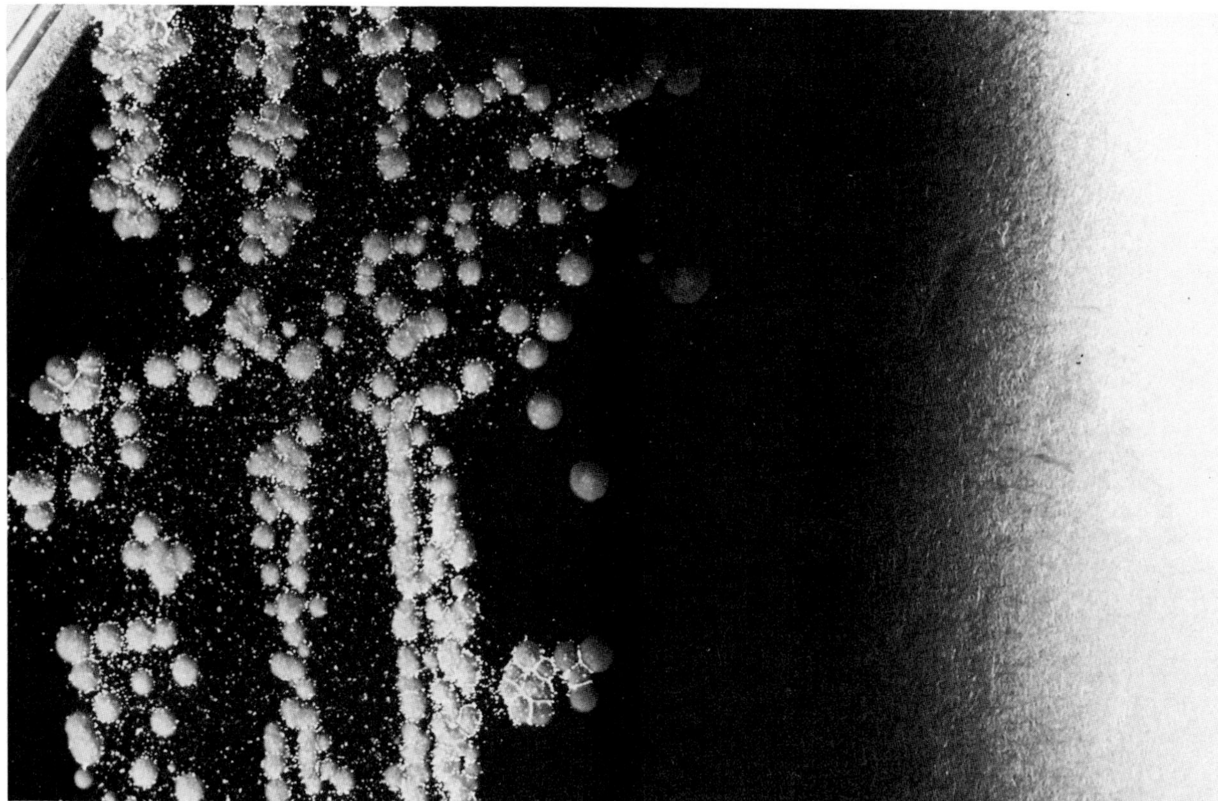

FIG. 7. Colonies of *Bacteroides fragilis* on BBE agar. Note the blackening of the agar and colonies due to esculin hydrolysis and bile precipitation.

vary depending on the substrate available to the organism. This may lead to misinterpretation of the GLC pattern and misidentification of the organism. For instance, saccharolytic organisms will produce greater amounts of isoacids in the absence of a fermentable carbohydrate, and a fermentable carbohydrate is required for the detection of lactic acid.

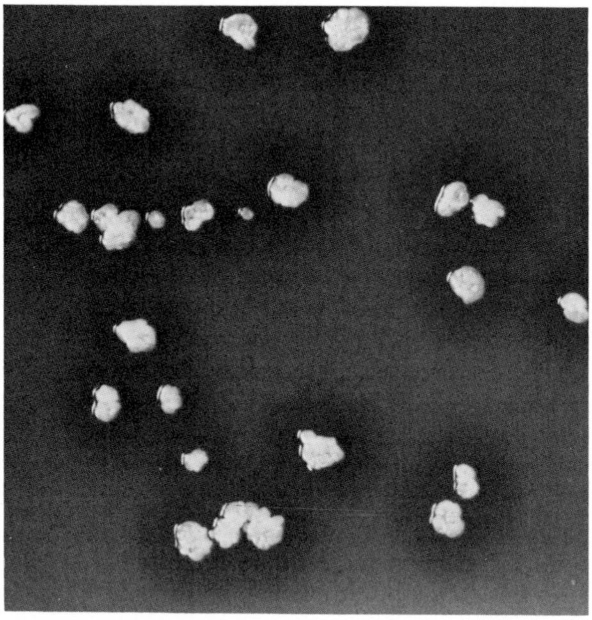

FIG. 8. Breadcrumb-shaped colonies of *Fusobacterium nucleatum*.

Colonies of the *B. fragilis* group on brucella BAP are 2 to 3 mm in diameter, circular, entire, convex, and gray to white. All strains require hemin for growth. The cells may be uniform or pleomorphic (some with vacuoles); this difference is medium and age dependent. Good growth in or stimulation by 20% bile (2%

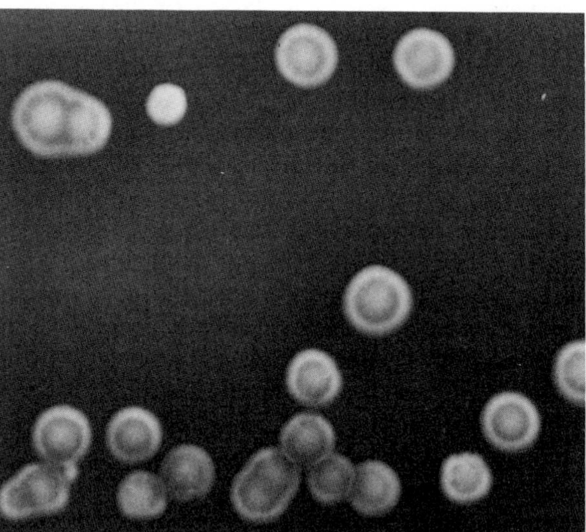

FIG. 9. Umbonate colonies of *Fusobacterium necrophorum*.

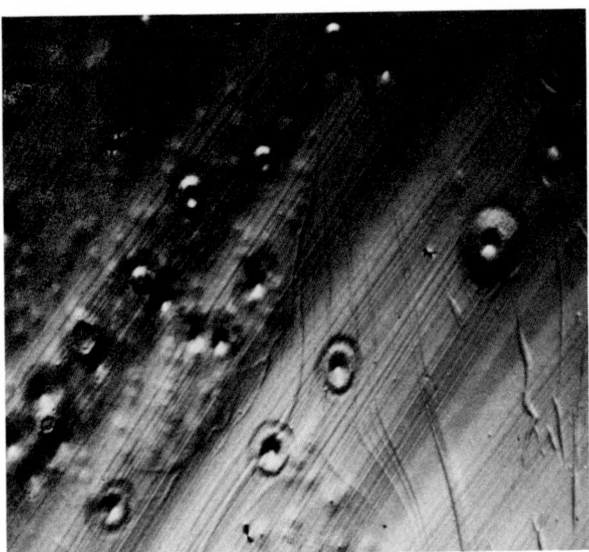

FIG. 10. Pitting colonies of *Bacteroides ureolyticus*.

oxgall) is characteristic of the *B. fragilis* group; however, *B. uniformis* does not grow as well in bile as do other members of the *B. fragilis* group. Some non-*B. fragilis* group organisms are bile resistant; however, *B. splanchnicus* and *B. eggerthii* are sucrose negative. Colonies of the *B. fragilis* group on BBE agar plates are 1 to 3 mm in diameter. Most blacken the agar (esculin positive), except for *B. vulgatus*, which can be esculin negative. Table 4 is a key for the differentiation of the bile-resistant bacteroides. *B. thetaiotaomicron* and *B. ovatus* may be difficult to differentiate; xylan and salicin are useful in separating these two species. *B. ovatus* produces acid from xylan and salicin; *B. thetaiotaomicron* is negative for xylan and is usually negative for salicin (8). In addition, *B. ovatus* has a more ovoid shape on Gram stain than does *B. thetaiotaomicron*.

The unnamed *B. fragilis* group DNA homology species 3452A closely resembles *B. distasonis*. Species 3452A is arabinose positive and catalase negative, whereas *B. distasonis* is catalase positive and usually arabinose negative. Bile-resistant organisms which do not fit any of the described species listed may be reported as *B. fragilis* group.

The bile-sensitive, nonpigmenting bacteroides form three major subgroups: (i) saccharolytic, (ii) saccharolytic and strongly proteolytic, and (iii) asaccharolytic. Table 5 lists the more commonly encountered or important species in this group.

The saccharolytic organisms fall into two categories, i.e., pentose fermenters or pentose nonfermenters (arabinose and xylose are usually tested). *B. oris* and *B. buccae* are pentose fermenters. They are phenotypically very similar, but they can be differentiated by the beta-glucosidase test (or the polyacrylamide gel electrophoretic analysis of cellular proteins). *B. oris* strains are usually more resistant to penicillin (2 U/ml) than are *B. buccae* strains. *B. zoogleoformans* is an oral organism that may ferment pentose, but it is isolated infrequently in the clinical laboratory. This organism forms a highly viscous, tenacious (zoogleal) mass in broth culture. Salicin and xylan are useful in the differentiation of *B. oralis*, *B. buccalis*, and *B. veroralis* (19). Certain strains of the *B. melaninogenicus* group require more than 21 days to develop pigment, and these strains, especially *B. loescheii*, closely resemble the *B. oralis* group. Darker, more opaque colonies and salicin, xylan, and gelatin reactions may aid in differentiating the strains.

TABLE 4. Characteristics of bile-resistant *Bacteroides* spp.[a]

Group and species	Growth in 20% bile	Indole	Catalase	Esculin hydrolyzed	Fermentation of: Glucose	Sucrose	Maltose	Rhamnose	Salicin	Trehalose	Arabinose	Fatty acid from PYG[b]
B. fragilis group												
B. distasonis	+	−	$+^-$	+	+	+	+	V	+	+	$-^+$	A, p, S (pa, ib, iv, l)
3452A homology group	+	−	−	+	+	+	+	$+^-$	$+^-$	+	+	
B. fragilis	+	−	+	+	+	+	+	−	−	−	−	A, p, S, pa (ib, iv, l)
B. vulgatus	+	−	$-^+$	$+^-$	+	+	+	+	−	−	+	A, p, S
B. ovatus	+	+	−	+	+	+	+	+	+	+	+	A, p, S, pa (ib, iv, l)
B. thetaiotaomicron	+	+	+	+	+	+	+	+	$-^+$	+	+	A, p, S, pa (ib, iv, l)
B. uniformis	W^+	+	$-^+$	+	+	+	+	$-^+$	$+^-$	−	+	a, p, l, S (ib, iv)
Other												
B. eggerthii	+	+	−	+	+	−	+	$+^-$	−	−	+	A, p, S (ib, iv, l)
B. splanchnicus	W^+	+	−	+	+	−	−	−	−	−	+	A, P, ib, b, iv, S (l)

[a] +, Positive reaction for the majority of strains (includes weak as well as strong acid production from carbohydrates); −, negative reaction; W, weak reaction; $+^-$, most strains positive; $-^+$, most strains negative, some strains positive. Sugars: +, pH < 5.5; w, pH 5.5 to 5.7; −, pH > 5.7.

[b] Capital letters indicate major metabolic products from peptone-yeast-glucose (PYG), lowercase letters indicate minor products, and parentheses indicate a variable reaction for the following fatty acids: A, acetic; P, propionic; F, formic; IB, isobutyric; B, butyric; IV, isovaleric; V, valeric; L, lactic; S, succinic; PA, phenylacetic. Note that isoacids are primarily from carbohydrate-free media (e.g., peptone-yeast extract) in the case of saccharolytic organisms.

TABLE 5. Characteristics of nonpigmented, bile-sensitive *Bacteroides* spp.[a]

Subgroup and species	Fermentation of: Glucose	Sucrose	Lactose	Arabinose	Xylose	Salicin	Xylan	Esculin hydrolyzed	Beta-glucosidase	Zoogleal mass	Indole	Gelatin	Fatty acids from PYG[b]
Saccharolytic pentose fermenters													
B. oris	+	+	+	+⁻	+	+		+	−	−	−		A, S (p, ib, iv)
B. buccae	+	+	+	+	+	+		+	+	−	−		A, S (p, ib, b, iv, l)
B. zoogleoformans	+	+	+	V	V	V		+		+	+⁻		A, P, S (ib, iv)
Saccharolytic pentose nonfermenters													
B. oralis	+	+	+	−	−	+		+			−		A, f, S (l)
B. buccalis	+	+	+	−	−	−	−	+			−		a, iv, S
B. veroralis	+	+	+	−	−	−	+	+			−		a, S
Saccharolytic and proteolytic													
B. bivius	+	−	+	−	−	−		−			−	+	A, iv, S (f, ib)
B. disiens	+	−	−	−	−	−		−			−	+	A, S (f, p, ib, iv)
Weakly or nonsaccharolytic													
B. capillosus	W⁻	−	−	−	−	−		+			−	−	a, s (f, p, l)
B. praeacutus	−	−	−	−	−	−		−			−	+	A, p, ib, B, IV, s (f, l)
B. putredinis	−	−	−	−	−	−		−			+	+	a, P, ib, b, IV, S (l)

[a] See Table 4, footnote *a*.
[b] See Table 4, footnote *b*.

B. bivius and *B. disiens* are both saccharolytic and strongly proteolytic. Gelatin and milk are usually digested within 2 to 3 days (milk may take longer). Lactose is the key differentiating sugar: *B. bivius* is lactose positive and *B. disiens* is lactose negative. Under long-wave UV light, their colonies may fluoresce a light orange to pink (coral) that should not be confused with the brick-red fluorescence of the pigmenting *Bacteroides* spp. *B. bivius* and *B. disiens* have been called nonpigmenting *B. intermedius-corporis* since they are phenotypically similar. Recent unpublished data of Ueno and of Shaw (personal communication) indicate that *B. bivius* is actually a pigmented bacteroides. Tests for the production of indole and the fermentation of mannose, sucrose, and lactose and an additional check for pigment production on rabbit blood agar are useful.

Asaccharolytic, nonpigmented, bile-sensitive *Bacteroides* spp. are infrequently isolated from clinical specimens. *B. capillosus* may weakly ferment glucose, is esculin positive, coagulates milk, and may grow better with Tween 80-supplemented media. *B. pneumosintes* is a tiny rod and forms minute colonies. *B. praeacutus* and *B. putredinis* are asaccharolytic and proteolytic; an indole test will differentiate them.

The pigmented *Bacteroides* species vary greatly in the degree and rapidity of pigmenting, depending primarily on the type of blood used; laked rabbit blood is the most reliable. A period of 2 to 21 days may be required to detect pigmentation, which ranges from buff to tan to black. A few strains may not pigment within 21 days, and their identity must be established by other biochemical tests. *Actinomyces odontolyticus* may be confused initially with these

TABLE 6. Characteristics of pigmented *Bacteroides* species[a]

Bacteroides sp.	Indole	Lipase	Catalase	Fermentation of: Glucose	Maltose	Lactose	Cellobiose	Esculin	Phenylacetic acid production	Fatty acids from PYG[b]
B. asaccharolyticus[c]	+	−	−	−	−	−	−		−	A, p, ib, B, IV, S
B. gingivalis	+	−	−	−	−	−	−		+	a, p, ib, B, IV, s, pa
B. intermedius	+	+⁻	−	+	+	−	−			A, iv, S, (f, p, ib)
B. corporis	−	−	−	+	+	−	−			A, ib, iv, S, (b)
B. melaninogenicus	−	−	−	+	+	+	−	−		A, S (f, ib, iv, l)
B. denticola	−	−	−	+	+	+	−	+		A, S (f, ib, iv, l)
B. loescheii	−	V	−	+	+	+	+	V		a, S (f, l)

[a] See Table 4, footnote *a*.
[b] See Table 4, footnote *b*.
[c] *B. endodontalis* is a newly described pigmented bacteroides which is phenotypically similar to *B. asaccharolyticus* (18).

TABLE 7. Characteristics of *Fusobacterium* species[a]

Fusobacterium sp.	Distinctive cellular morphology	Indole	Growth in 20% bile	Lipase	Gas in glucose agar	Fermentation of: Glucose	Fructose	Mannose	Esculin hydrolysis	Lactate converted to propionate	Threonine converted to propionate	Fatty acids from PYG[b]
F. nucleatum	Slender pointed ends	+	−[c]	−	$-^{2}$	$-^{w}$	$-^{w}$	−	−	−	+	a, p, B (F, L, s)
F. gonidiaformans	Gonidia forms	+	−	−	4^{2}	−	−	−	−	−	+	A, p, B (f, l, s)
F. necrophorum		+	$-^{+}$	$+^{-}$	4^{2}	$-^{w}$	$-^{w}$	−	−	+	+	a, p, B (l, s)
F. naviforme	Boat shape	+	−	−	$-^{2}$	w^{-}	−	−	−	−	−	a, B, L (f, p, s)
F. varium		$+^{-}$	+	−	4	$+^{w}$	w^{+}	$+^{w}$	−	−	+	a, p, B, L (s)
F. mortiferum	Bizarre; round bodies	−	+	−	4	$+^{w}$	$+^{w}$	$+^{w}$	+	−	+	a, p, B (f, v, l, s)
F. russii		−	−	−	2^{-}	−	−	−	−	−	−	a, B, L (f)

[a] See Table 4, footnote *a*; $+^{w}$, most strains positive, some strains weakly positive; $-^{w}$, most strains negative, some strains weakly positive; w^{-}, most strains weakly positive, some strains negative; w^{+}, most strains weakly positive, some strains positive. Gas in PYG agar deep: −, no gas detected; 2, splits agar horizontally; 4, agar displaced to the top of the tube. Sugars: +, pH < 5.5; W, pH 5.5 to 5.7; −, pH > 5.7.

[b] See Table 4, footnote *b*.

[c] Extremely rarely, strains may be positive.

organisms since it may produce a red pigment. A Gram stain will readily differentiate them. The pigmented bacteroides fluoresce pink, orange, chartreuse, or brick red. Brick red is the only reliable color for presumptive identification, since some nonpigmenting bacteroides fluoresce coral or pink. Fluorescence is masked by pigment production.

Table 6 is a key for the differentiation of the pigmenting *Bacteroides* spp. *B. asaccharolyticus* and *B. gingivalis* are phenotypically very similar. The production of phenylacetic acid, a nonvolatile fatty acid, by *B. gingivalis* will differentiate them (9). Their unusual special-potency antibiotic-disk pattern (susceptible to vancomycin) and nonsaccharolytic properties separate them from the other pigmenters. As noted earlier, *B. melaninogenicus*, *B. loescheii*, and *B. denticola* may require a full 21 days to develop pigment and may, therefore, be confused with *B. oralis*, *B. veroralis*, or *B. buccalis*. *B. oralis* is salicin positive, whereas the pigmenters are salicin negative. *B. macacae* is an indole-positive, catalase-positive, pigmenting bacteroides that has been isolated from the gingivae of monkeys, and we have isolated one strain from a clinical specimen (unpublished data). *B. levii*, which is similar to *B. corporis*, has rarely been isolated from human clinical specimens. Pigmenters not identified by using Table 6 may be designated as pigmenting *Bacteroides* spp.

The name *Bacteroides* species is applied to those organisms that do not fit the species described in Tables 4 through 6.

Fusobacterium spp. have a special-potency antibiotic-disk pattern of resistance to vancomycin and sensitivity to kanamycin and colistin, or they produce butyric acid without isobutyric or isovaleric acid. The fusobacteria are weakly saccharolytic or nonfermentative. Most are nonproteolytic (gelatin negative). Colonial morphology varies greatly. In addition, some strains fluoresce chartreuse under UV light, and other strains produce greening of the agar. The most common clinical isolates, *F. nucleatum* and *F. necrophorum*, are indole positive. These two organisms have been discussed under Rapid Identification.

Table 7 characterizes the more commonly isolated fusobacteria. The conversion of threonine and lactate to propionic acid is important in the differentiation of these species. Bizarre pleomorphic rods with very large coccoid forms (round bodies) are characteristic of *F. mortiferum* (Fig. 4). This organism may grow on BBE agar and turn the agar black. *F. periodonticum* is

TABLE 8. In vitro susceptibility of gram-negative anaerobes to antimicrobial agents[a]

Bacterium	Chloramphenicol	Clindamycin	Erythromycin[d]	Metronidazole	Cefoxitin	Ureido, carboxy, and piperazine penicillins[b]	Penicillin G and ampicillin	Tetracyclines[c]	Vancomycin[d]
B. fragilis group	S	I–S	R–I	S	I–S	I–S	R	R–I	R
Other *Bacteroides* spp.	S	S[e]	I–S	S	S[e]	S[e]	I–S	I	R
F. varium	S	R–I	R	S	S[e]	S[e]	S[e]	I	R
Other *Fusobacterium* spp.	S	S	R	S	S[f]	S[f]	S	S	R

[a] R, Poor or inconsistent activity; I, moderate activity; S, good activity.

[b] Piperacillin, mezlocillin, azlocillin, carbenicillin, and ticarcillin.

[c] Doxycycline and minocycline are more active than tetracycline.

[d] Not approved by the Food and Drug Administration for anaerobic infections.

[e] A few strains are resistant.

[f] Rare strains are resistant.

a newly described oral isolate which is indole positive and bile sensitive, ferments glucose, fructose, and galactose, and converts threonine but not lactate to propionate (14). The term *Fusobacterium* spp. is applied to those organisms that do not fit the species described in Table 7.

Several microbiochemical methods have been developed, including the API 20A (Analytab Products, Inc., Plainview, N.Y.), Minitek (BBL Microbiology Systems, Cockeysville, Md.), Anaerobe-Tek (Flow Laboratories, McLean, Va.), and IDS RapID ANA panel (Innovative Diagnostic Systems, Decatur, Ga.). The API 20A system produces results in 24 to 48 h and has a large, computerized data base to aid in identification. The Minitek system allows for testing a large number of carbohydrates, produces results in 24 to 48 h, and has a computerized data base. The color reaction in both these systems is not always clear-cut; shades of brown (API) and yellow-orange (Minitek) make the interpretation of test results difficult. In this situation, neither system should be relied upon for identification. Neither the API 20A nor the Minitek system is useful for the identification of asaccharolytic organisms. With both systems, supplemental tests, including GLC, are often required for the identification of isolates. In an unpublished study in our laboratory, the Anaerobe-Tek plate did not support the growth of many anaerobes and therefore did not properly identify them. The RapID ANA panel is a 4-h test system based primarily on the detection of preformed enzymes. Gradation in colors may make interpretations moderately difficult. A color chart is provided to aid in the interpretation of reactions read after the developer is added. A computerized data base is available. Preliminary trials indicate that this system, without the use of GLC or supplemental tests, may be an alternative method for the rapid identification of common clinical isolates. Further comparative trials of the biochemical methods with RapID ANA and prereduced, anaerobically sterilized media with fresh clinical isolates will determine the accuracy of the RapID ANA panel system.

Serological procedures are not yet practical for the identification of colonial growth.

Susceptibility testing of anaerobes is discussed in Chapter 103. The usual susceptibility patterns of the most commonly encountered gram-negative rods of clinical significance are noted in Tables 8 through 10. The beta-lactamase-producing (nitrocefin test) bacteroides are the *B. fragilis* group, the pigmented bacteroides, the *B. oralis* group, *B. bivius*, *B. coagulans*, *B. disiens*, *B. oris-buccae*, and *B. splanchnicus*, although not all strains within each species or group produce beta-lactamase. *F. nucleatum* may also produce beta-lactamase.

TABLE 9. Percentage of gram-negative anaerobes susceptible (at breakpoint) to most active drugs presently available

Drug	Breakpoint (μg/ml)	% Susceptible		
		B. fragilis group	*Bacteroides* species other than *B. fragilis* group	*Fusobacterium* species
Metronidazole	16	100	100	100
Chloramphenicol	16	100	100	100
Clindamycin	4	95	97	98
Piperacillin-mezlocillin-azlocillin	128	95	96	99
Carbenicillin-ticarcillin	128	95	96	99
Cefoxitin	32	95	96	99

TABLE 10. Susceptibility of the *B. fragilis* group to other beta-lactam antibiotics

Drug	% Susceptible at breakpoint
Imipenem (thienamycin)[a]	100
Cefotetan	90
Cefoperazone, moxalactam, apalcillin[a]	~70
Cefotaxime, cefmenoxime[a], ceftazidime[a]	~50
Ceftizoxime, ceftriaxone[a], cefpiramide[a]	~30
Cefuroxime	~15

[a] Investigational drugs at present.

LITERATURE CITED

1. **Dowell, V. R., Jr., and T. M. Hawkins.** 1974. Laboratory methods in anaerobic bacteriology, CDC manual. Department of Health, Education, and Welfare publication (CDC) 74-8272. Center for Disease Control, Atlanta.
2. **Dowell, V. R., Jr., and G. L. Lombard.** 1977. Presumptive identification of anaerobic nonsporeforming gram-negative bacilli. Center for Disease Control, Atlanta.
3. **Finegold, S. M.** 1977. Anaerobic bacteria in human disease. Academic Press, Inc., New York.
4. **George, W. L., B. D. Kirby, V. L. Sutter, D. M. Citron, and S. M. Finegold.** 1981. Gram-negative anaerobic bacilli: their role in infection and patterns of susceptibility to antimicrobial agents. II. Little known *Fusobacterium* species and miscellaneous genera. Rev. Infect. Dis. **3**:599–626.
5. **Gorbach, S. L., J. W. Mayhew, J. G. Bartlett, H. Thadepalli, and A. B. Onderdonk.** 1976. Rapid diagnosis of anaerobic infections by direct gas-liquid chromatography of clinical specimens. J. Clin. Invest. **57**:478–484.
6. **Holdeman, L. V., E. P. Cato, and W. E. C. Moore (ed.).** 1977. Anaerobic laboratory manual, 4th ed. Virginia Polytechnic Institute and State University, Blacksburg.
7. **Holdeman, L. V., E. P. Cato, and W. E. C. Moore.** 1984. Taxonomy of anaerobes: present state of the art. Rev. Infect. Dis. **6**(Suppl. 1):S3–S10.
8. **Holdeman, L. V., R. W. Kelley, and W. E. C. Moore.** 1984. Anaerobic gram-negative straight, curved and helical rods. Family 1. *Bacteroidaceae* Pribram 1933, 10[AL], p. 602–662. *In* N. R. Krieg and J. G. Holt (ed.), Bergey's manual of systematic bacteriology, vol. 1. The Williams & Wilkins Co., Baltimore.
9. **Kaczmarek, F. S., and A. L. Coykendall.** 1980. Production of phenylacetic acid by strains of *Bacteroides asaccharolyticus* and *Bacteroides gingivalis* (sp. nov.). J. Clin. Microbiol. **12**:288–290.
10. **Kirby, B. D., W. L. George, V. L. Sutter, D. M. Citron, and S. M. Finegold.** 1980. Gram-negative anaerobic bacilli: their role in infection and patterns of susceptibility to antimicrobial agents. I. Little known *Bacteroides* species. Rev. Infect. Dis. **2**:914–951.
11. **Livingston, S. J., S. D. Kominos, and R. B. Yee.** 1978. New medium for selection and presumptive identification of the *Bacteroides fragilis* group. J. Clin. Microbiol. **7**:448–453.
12. **Morgenstein, A. A., D. M. Citron, and S. M. Finegold.** 1981. New medium selective for *Fusobacterium* species and differential for *Fusobacterium necrophorum*. J. Clin. Microbiol. **13**:666–669.

13. **Mouton, C., P. Hammond, J. Slots, and R. J. Genco.** 1980. Evaluation of Fluoretec-M for detection of oral strains of *Bacteroides asaccharolyticus* and *Bacteroides melaninogenicus.* J. Clin. Microbiol. **11:**682–686.
14. **Slots, J., T. V. Potts, and P. A. Mashimo.** 1983. *Fusobacterium periodonticum*, a new species from the human oral cavity. J. Dent. Res. **62:**960–963.
15. **Spiegel, C. A., and M. Roberts.** 1984. *Mobiluncus* gen. nov., *Mobiluncus curtisii* subsp. *curtisii* sp. nov., *Mobiluncus curtisii* subsp. *holmesii* subsp. nov., and *Mobiluncus mulieris* sp. nov., curved rods from the human vagina. Int. J. Syst. Bacteriol. **34:**177–184.
16. **Sutter, V. L., D. M. Citron, M. A. C. Edelstein, and S. M. Finegold.** 1984. Wadsworth anaerobic bacteriology manual, 4th ed. Star Publishing Co., Belmont, Calif.
17. **Tanner, A. C. R., S. Badger, C.-H. Lai, M. A. Listgarten, R. A. Visconti, and S. S. Socransky.** 1981. *Wolinella* gen. nov., *Wolinella succinogenes (Vibrio succinogenes* Wolin et al.) comb. nov., and description of *Bacteroides gracilis* sp. nov., *Wolinella recta* sp. nov., *Campylobacter concisus* sp. nov., and *Eikenella corrodens* from humans with periodontal disease. Int. J. Syst. Bacteriol. **31:**432–445.
18. **van Steenbergen, T. J. M., A. J. van Winkelhoff, D. Mayrand, D. Grenier, and J. de Graaff.** 1984. *Bacteroides endodontalis* sp. nov., an asaccharolytic black-pigmented *Bacteroides* species from infected dental root canals. Int. J. Syst. Bacteriol. **34:**118–120.
19. **Watabe, J., Y. Benno, and T. Mitsuoka.** 1983. Taxonomic study of *Bacteroides oralis* and related organisms and proposal of *Bacteroides veroralis* sp. nov. Int. J. Syst. Bacteriol. **33:**57–64.
20. **Weinberg, L. G., L. L. Smith, and A. H. McTighe.** 1983. Rapid identification of the *Bacteroides fragilis* group by bile disk and catalase tests. Lab. Med. **14:**785–788.
21. **Weissfeld, A. S., and A. C. Sonnenwirth.** 1981. Rapid detection and identification of *Bacteroides fragilis* and *Bacteroides melaninogenicus* by immunofluorescence. J. Clin. Microbiol. **13:**798–800.

Chapter 41

Gram-Positive, Nonsporeforming Anaerobic Bacilli

STEPHEN D. ALLEN

CHARACTERIZATION

The genera included here are *Actinomyces, Arachnia, Bifidobacterium, Eubacterium, Lachnospira, Lactobacillus,* and *Propionibacterium*. They are all gram positive to gram variable and form rod-shaped cells. However, morphology is highly variable, and they are often pleomorphic. Several of the species discussed in this chapter will grow in a 5 to 10% CO_2-air incubator even though they are classified with the anaerobic bacteria; there is considerable variation in their relationships to oxygen. Most are obligate anaerobes, but some are facultative anaerobes or microaerophiles (capnophiles). None grow rapidly. Most require 2 to 3 days of anaerobic incubation to initiate colonies on anaerobe blood agar. *Actinomyces israelii* is one of the slow-growing anaerobes that sometimes requires more than 2 weeks of incubation to permit primary isolation. Members of each of these genera are part of the normal flora of humans and other animals. Several species are found in other environments, including soil and water, and have not been isolated from humans. All of those that cause disease inhabit the skin or various mucous membrane surfaces or both, e.g., the mouth, intestines, and urogenital tract. Fortunately, many species found in the normal flora appear to be harmless saprophytes. Those that can cause infections are opportunistic pathogens (33).

Morphologic considerations; bacteria that may occur in the form of "diphtheroids"

The gram-positive, nonsporeforming anaerobic bacilli exhibit a distinctively variable morphology that depends on the type of culture medium, length of incubation, and other growth conditions. The finding of gram-positive rods in a smear of thoracic empyema fluid, peritoneal fluid, or the broth of a positive blood culture bottle signifies the possibility of any one of several genera of anaerobic, as well as aerobic (or facultatively anaerobic), bacteria. Among the gram-positive bacilli that should be considered in these settings are the following genera: *Actinomyces, Arachnia, Bacillus, Bifidobacterium, Clostridium, Corynebacterium, Erysipelothrix, Eubacterium, Lactobacillus, Listeria, Nocardia, Propionibacterium,* and *Streptococcus* (24). Isolates of several of the genera of this group may show a pleomorphic cellular morphology that is called diphtheroid ("coryneform" is a synonym which some microbiologists prefer) (Fig. 1). Diphtheroid cells range from coccoid to rod shapes and vary in size. They are often gram variable and tend to stain unevenly, sometimes being mistaken for gram-negative bacilli. Adding to the confusion, *Eubacterium plautii*, discussed below (20), is gram negative by light microscopy. Diphtheroid cells may be arranged in V-shapes, gull wings, birds-in-flight formations, "Chinese letters," "picket fence" rows, clumps, or as single straight or curved rods. The genera *Bacillus, Clostridium, Lactobacillus,* and *Streptococcus* are infrequently diphtheroid in appearance. The presence of bacterial endospores would be consistent with *Bacillus* or *Clostridium* species. Lactobacilli found in clinical specimens usually form chains of uniform rods but are sometimes predominantly diphtheroid. Certain streptococci (e.g., *S. mutans, S. intermedius, S. morbillorum,* and others) may elongate to the extent that they form rod-shaped cells, particularly when grown on solid media, and can be quite pleomorphic. They are more apt to form cocci in chains in broth culture or body fluids. To further complicate matters, some propionibacteria, eubacteria, bifidobacteria, and lactobacilli form short coccoid forms, particularly on solid media. In addition to Gram stains of colonies, broth cultures, which may reveal rod-shaped cells, should be examined microscopically. The remaining genera mentioned above commonly show a diphtheroid morphology. The oxygen tolerance of isolates (Chapter 37) aids in separating the aerobic genus *Nocardia* and the facultatively anaerobic genera *Corynebacterium, Erysipelothrix,* and *Listeria* from the list of gram-positive bacilli currently classified with the anaerobes (Table 1). Rods with clubs or bifurcated ends would be a clue to the genus *Bifidobacterium*, but these shapes may be found in other genera. Branching filaments that are not acid fast would be a major clue to a possible *Actinomyces* species, but such filaments can be seen in other genera (Table 2). Certain clostridia (e.g., *C. perfringens, C. ramosum,* and *C. clostridiiforme*), which may fail to produce spores on the usual lab media or which may appear filamentous, can be a diagnostic problem. The appearance of growth on egg yolk agar may sometimes aid in the separation of the lecithinase-positive clostridia from the nonsporeforming bacilli (34), since none of the latter produce lecithinase (and only a few are lipase positive); for more information on egg yolk reactions of clostridia, see Chapter 38).

Differentiation of the genera of gram-positive, nonsporeforming anaerobic bacilli

Differentiation of the genera of gram-positive, nonsporeforming anaerobic bacilli is based primarily on metabolic product analysis by gas-liquid chromatography (GLC) and on morphology. A catalase test is a particularly useful test to presumptively separate *Propionibacterium* species from less common genera in which organisms are catalase negative (Table 1). Further presumptive and definitive identification of species requires further characterization with batteries of tests (Tables 2 through 5).

Taxonomy and nomenclature

The approved lists of bacterial names, published in 1980 (31), recognized six *Actinomyces* species. Of these, *A. bovis*, the type species, causes infections of animals other than humans, and *A. humiferus*, a soil organism, has not been isolated from humans (3). The remaining species that were recognized in 1980 are isolated from humans; these species include *A. israelii,*

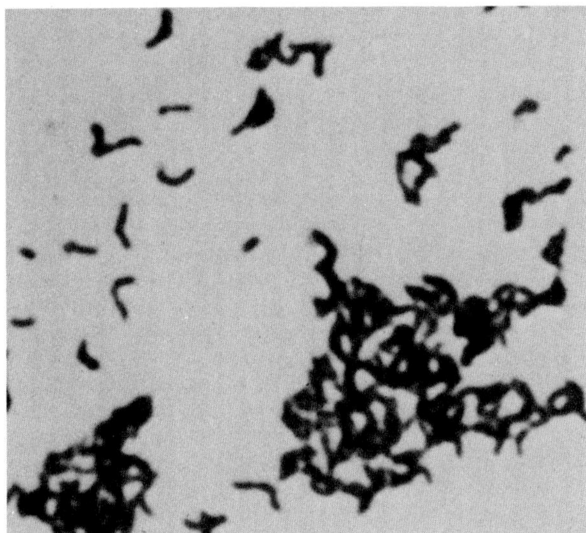

FIG. 1. Microscopic morphology of *P. acnes* from a 2- to 3-day-old anaerobe blood agar plate. Note the Gram-stained diphtheroid cells with V-shapes, irregular bodies, clubbed ends, Chinese letter formations, clumps of cells, coccoid cells, curved forms, and straight rods. ×900. Courtesy of Jean Siders.

A. naeslundii, *A. odontolyticus*, and *A. viscosus*. In 1982, *Corynebacterium pyogenes*, which is occasionally isolated from wounds, septic arthritis, bacteremia, and other conditions of humans, was reclassified as *Actinomyces pyogenes* (5). More recently, two new species have been proposed for two groups of organisms isolated from the dental plaque of cattle, *A. denticolens* (6) and *A. howellii* (7). In addition, *A. meyeri* is described in the manual of Holdeman et al. (22) and in a recent taxonomic proposal (4a). The *Actinomyces* species that are not encountered in infections of humans will not be discussed further in this chapter.

Twenty *Bifidobacterium* species were cited in the approved lists (31). A recent addition, *B. gallinarum*, has been isolated from chickens but not humans (40). The species formerly called *B. eriksonii*, (also previously called *Actinomyces eriksonii*), which has caused significant human illness, did not appear on this list. A recent report by Biavati et al. indicated that *B. eriksonii* is a synonym of *B. dentium* and that the name *B. eriksonii* should not be used (2). Other *Bifidobacterium* species that have been isolated from humans are *B. adolescentis*, *B. catenulatum*, *B. breve*, *B. longum*, *B. globosum*, and *B. infantis* (Table 4). All of these species have been isolated from the normal flora, but their clinical significance is uncertain. One of the difficulties of determining their significance is that fermentation reactions and other commonly used tests for phenotypic characterization do not reliably differentiate between all of the *Bifidobacterium* spp. (2). Polyacrylamide gel electrophoresis of soluble cellular proteins or nucleic acid homology studies, which are necessary for definitive identification and taxonomic work, are not practical for clinical laboratories.

The approved lists (31) cited 29 *Eubacterium* species. In 1980, three new species, *E. timidum*, *E. brachy*, and *E. nodatum*, which were isolated from human periodontal lesions, were described (21). Although it is gram negative by light microscopy, *Fusobacterium plautii* was transferred to the genus *Eubacterium* as *E. plautii*; it has a gram-positive or gram-variable cell wall structure by electron microscopy (20). Among the several *Eubacterium* species that have not been isolated from human clinical sources are *E. budayi*, *E. cellulosolvens*, *E. fissicatena*, *E. ruminantium*, *E. tortuosum*, *E. ventriosum*, and *E. suis* (22, 41). Only those species isolated from human infections will be discussed further.

Lachnospira multiparus, a weakly gram-positive curved rod that is motile, with subterminal monotrichous flagella, has been isolated from the bovine rumen and human intestinal materials but not from properly collected specimens from humans (3, 21). *L. multiparus* produces large amounts of acetic, formic, and lactic acids and resembles *Eubacterium aerofaciens* on metabolic product analysis. Carbohydrate fermentation reactions of these species are similar, but *L. multiparus* differs by being maltose negative (21). It will not be further discussed in this chapter.

Occurrence in clinical materials and nature of diseases

The major pathogens in this group are few in number. They include the species that cause actinomycosis (discussed below), *Propionibacterium* species, and only occasional *Eubacterium* and *Lactobacillus* isolates. In the clinical laboratory, *Propionibacterium* is, by far, the most commonly encountered genus (Chapter 37) (Table 4), and almost all *Propionibacterium* isolates are *P. acnes*, one of the anaerobes that is a rare cause of endocarditis (13, 14). The propionibacteria are usually, though certainly not always, contaminants. The next most common genus is *Eubacterium*. Most *Eubacterium* species are isolated from polymicrobial infections, and it is usually difficult to know the significance of these species. *E. lentum*, *E.*

TABLE 1. Genera of gram-positive, nonsporeforming anaerobic bacilli encountered in human specimens[a]

I. Produce major products of propionic and acetic acids
   - A. Catalase produced; diphtheroid cells common—*Propionibacterium*
   - B. Catalase not produced; diphtheroid cells and spheroplasts common; branching filaments transitory—*Arachnia*

II. Produce acetic and lactic acids (>1-to-1 ratio); diphtheroid and bifid forms common—*Bifidobacterium*

III. Produce lactic acid as sole major product
   - A. Short to long and slender rods; chain formation common; diphtheroid cells uncommon—*Lactobacillus*
   - B. Diphtheroid cells predominant; tend to form branching filaments which are transitory—*Actinomyces*

IV. Produce moderate acetic and (i) major succinic, (ii) major lactic and succinic, or (iii) major lactic acids (see IIIB above); most are facultative anaerobes or aerotolerant—*Actinomyces*

V. Produce mixture of acid metabolic products including (i) butyric acid and others, (ii) acetic and formic acids, or (iii) no major acids; obligate anaerobes; pleomorphic diphtheroid cells common, or uniform—*Eubacterium*

[a] Adapted from references 3 and 22.

TABLE 2. Some key differential tests for presumptive identification of gram-positive, nonsporeforming anaerobic bacilli[a]

Species	Relationship to oxygen	Colonies on blood agar	Red pigment on blood agar	Appearance in enriched thioglycolate broth	Cellular morphology in enriched thioglycolate broth	Production of: Catalase	Production of: Indole	Glucose fermentation	Metabolic products in PYG[b] broth
Actinomyces israelii	M or OA	Rough	−	Granular or diffuse	Branching filaments or diphtheroidal	−	−	+	A, L, S
A. naeslundii	F	Smooth	−	Diffuse	Diphtheroidal, branching	−	−	+	A, L, S
A. odontolyticus	M or OA	Smooth	+	Diffuse	Diphtheroidal, branching	−	−	+	A, L, S
A. viscosus	F	Smooth	−	Diffuse	Diphtheroidal, branching	−	−	+	A, L, S
A. pyogenes	F	Smooth	−	Diffuse	Diphtheroidal, coccoidal	−	−	+	A, L, S
A. meyeri	OA, F	Smooth	−	Diffuse	Diphtheroidal, branching	−	−	+	A, L, S
Arachnia propionica	M or OA	Rough	−	Granular or diffuse	Branching filaments or diphtheroidal	−	−	+	A, P, (L), (S)
Bifidobacterium dentium	OA	Smooth	−	Diffuse	Thin rods, bifid ends, bulbous ends	−	−	+	A, L
Eubacterium alactolyticum	OA	Smooth	−	Diffuse	Thin rods, V-forms, cross-stick arrangements	−	−	+	A, B, C
E. lentum	OA	Smooth	−	Diffuse	Short coccoidal rods, diphtheroidal	−	−	−	(A)
E. limosum	OA	Smooth	−	Diffuse	Plump rods, bulbous and bifid forms	−	−	+	A, B
Lactobacillus catenaforme	OA	Smooth	−	Diffuse (granular)	Short rods in chains or singly	−	−	+	A, L
Propionibacterium avidum	F	Smooth	−	Diffuse	Diphtheroidal	+	−	+	A, P, (L), S
P. acnes	OA[F]	Smooth	−	Diffuse (granular)	Diphtheroidal	+[−]	+[−]	+	A, P, L, S
P. granulosum	F	Smooth	−	Diffuse	Diphtheroidal	+	−	+	A, P, (L), S

[a] Adapted from references 8, 17, 24, and 34. +, Positive reaction for 90 to 100% of strains tested; −, negative reaction for 90 to 100% of strains tested; superscript, reaction shown with 11 to 25% of strains tested; F, facultatively anaerobic; M, microaerophilic; OA, obligately anaerobic.

[b] Metabolic product analysis determined on peptone-yeast extract-glucose (PYG) broth cultures (22). Abbreviations: A, acetic acid; B, butyric acid; C, caproic acid; L, lactic acid; P, propionic acid; S, succinic acid; parentheses, variable or, if produced, usually present only in trace amounts.

aerofaciens, *E. limosum*, *E. cylindroides*, and *E. alactolyticum* are the species that are isolated most often from deep abscesses, wounds, blood, pulmonary infections, and other sources at the Indiana University Medical Center. Certain *Eubacterium* spp. have also been rare causes of endocarditis (30). *Lactobacillus* species are usually of doubtful significance; however, *L. catenaforme* has been isolated from pleuropulmonary infections and miscellaneous other conditions (22, 33). The genera *Actinomyces*, *Arachnia*, and *Bifidobacterium* account for far less than 1% of anaerobe isolates at the Indiana University Medical Center. Nonetheless, these organisms are often of major clinical significance when isolated from appropriate specimens.

Actinomycosis of humans is most commonly caused by *A. israelii*. According to Georg (16), the next most important etiologic agents are *Arachnia propionica* (formerly *Actinomyces propionicus*), *Actinomyces naeslundii*, and *A. viscosus*. *A. odontolyticus*, *A. meyeri*, and *B. dentium* appear to be less common. However, *A. propionica* is seldom encountered at our medical center and seems to be one of the less common species. All of these bacteria are part of the normal flora of the human mouth; *Bifidobacterium* spp. and a few *Actinomyces* spp. have been found in vaginas of healthy individuals, and *Bifidobacterium* spp. inhabit the intestines (33). Classically, actinomycosis has occurred in cervicofacial, thoracic, abdominal, pelvic, and disseminated forms (12, 32). In the cervicofacial form (which is the most common form), the initial lesion may involve the face, neck, mandible, or tongue. The pathogenesis probably involves trauma to underlying tissues of the mouth, for example, dental extraction, and development of a mixed infection, possibly with hypersensitivity to one or more components of the actinomycete that is present (12, 32). The infection is characterized by the slow development of induration, erythema, and swelling, with granulocytic exudation (pus formation with abscesses is typical), necrosis, and chronic draining sinus or fistula tracts. The lesions tend to spread locally, through muscle, fascia, and even bone; older lesions often show considerable fibrosis. Hematogenous dissemination is rare. In the thoracic form, the sequence of events may initially involve the aspiration of oral contents or a foreign body, such as a tooth, on which there are actinomycetes (e.g., from dental plaque), followed by pneumo-

TABLE 3. Additional differential characteristics of gram-positive, nonsporeforming anaerobic bacilli[a]

Species	Hydrolysis of:		Nitrate reduction	Indole production	Action on milk[b]	Urea agar	Fermentation of:								
	Esculin	Gelatin					Glucose	Mannitol	Lactose	Sucrose	Maltose	Salicin	Glycerol	Xylose	Arabinose
Actinomyces israelii	$+^-$	−	V	−	(C)	−	+	V	$+^-$	+	+	V	−	$+^-$	V
A. naeslundii	$+^-$	−	$+^-$	−	(C)	+	+	−	$+^-$	$+^-$	$+^-$	V	V	−	$-^+$
A. odontolyticus	V	−	+	−	(C)	−	+	−	$+^-$	$+^-$	+	$+^-$	$+^-$	V	V
A. viscosus	+	−	$+^-$	−	(C)	+	+	−	$+^-$	+	+	$+^-$	+	−	−
A. pyogenes	−	+	−	−			+	−	$+^-$	V	V	−	V	V	V
A. meyeri	$-^+$	−	−	−	(C)		+	−	+	+	+	−	−	+	V
Arachnia propionica	$-^+$	$-^+$	+	−	(C)	−	+	+	+	+	+	$-^+$	$-^+$	−	−
Bifidobacterium dentium	$+^-$	−	−	−	(CG)		+	V	+	+	+	+	−	+	+
Eubacterium alactolyticum	−	−	−	−	NC		+	$+^-$	−	−	−	−	−	−	−
E. lentum	−	−	V	−	NC		−	−	−	−	−	−	−	−	−
E. limosum	+[c]	−	−	−	C(G)		+	$+^-$	−	−	−	−	−	−	−
Propionibacterium acnes	−	$+^-$	$+^-$	$+^-$	C(G)		+	V	−	−	−	−	+	−	−
P. granulosum	−	V	−	−	(C)		+	−	−	+	+	$-^+$	$+^-$	−	−

[a] Adapted from references 17 and 34. Reactions: +, positive; −, negative; $+^-$, most strains positive, occasional strains negative; $-^+$, most strains negative, occasional strains positive; V, variable reaction; blank, no data available.

[b] C, Coagulated; NC, no coagulation; G, gas. Parentheses indicate variable reaction.

[c] Esculin reaction for *E. limosum* is positive according to the Virginia Polytechnic Institute manual (22) and variable according to the Centers for Disease Control manual (9).

nitis, abscess formation, pleuritis, and development of sinuses that may eventually drain through the thoracic wall. In addition, the thoracic form may follow extension from the head and neck or abdomen; it also may arise from hematogenous spread (12, 32). The abdominal form most commonly occurs in the ileocecal region, often beginning in the appendix or from a site of intestinal perforation. This may lead to peritonitis or the involvement of retroperitoneal areas, with the formation of abscesses, sinus tracts, and flank or groin abscesses. Hematogenous dissemination may occur through the portal venous return, leading to liver abscesses or metastatic suppurative infections in other sites (12, 32).

In 1973, Henderson reported a case of pelvic actinomycosis in the presence of a current modern metallic intrauterine contraceptive device (IUD) (19). By 1979, at least 26 cases of pelvic infection or colonization and one death caused by *A. israelii* in wearers of IUDs had been reported (18). Duguid et al. have reported a significant difference in the prevalence of actinomycete infestation (includes colonization) in cervical materials between users of plastic IUDs (42%) and users of copper IUDs (2%) (10). Although the reasons are not clear, it has been suggested that copper acts as a mild bacteriostatic agent, whereas calcium carbonate deposits on plastic IUDs may form a nidus where actinomycetes are predisposed to grow (10, 11). The difference also may be related to a greater frequency of replacement for the copper devices (10, 39). A recent prospective study of cervical Papanicolaou (PAP) smears from 69,925 women was made for bacterial aggregates consistent with actinomycetes (39). Smears judged to be consistent with actinomycetes were examined at a later time by a direct immunofluorescence antibody procedure for *A. israelii*. *A. israelii*

TABLE 4. Selected characteristics of *Bifidobacterium* species encountered in human normal flora and clinical materials[a]

Species	Fermentation of:					Occurrence in human materials[b]			
	Arabinose	Cellobiose	Glycogen	Melizitose	Sucrose	Clinical	Mouth	Intestines-feces	Cervix-vagina
B. dentium	+	+	+	$+^-$	+	+	+	+	+
B. bifidum	−	−	−	−	−	+		+	+
B. infantis	−	$-^+$	−	−	+			+	+
B. globosum	+	V	+	−	+			+	
B. breve	−	+	$+^-$	V	+	+		+	+
B. longum	+	$-^+$	−	$+^w$	+	+		+	
B. catenulatum	V	+	V	−	+			+	+
B. adolescentis	V	+	+	−	+			+	

[a] Data compiled from references 2 and 22. Fermentation patterns of *Bifidobacterium* spp. overlap considerably; polyacrylamide gel electrophoresis or DNA-DNA homology data or both are required for the definitive identification of some species (2). All *Bifidobacterium* species produce acid from glucose. +, pH below 5.5, or positive; w, pH 5.5 to 5.9, or weak; −, pH above 5.9, or negative; V, variable; superscript, reaction of some strains of a species.

[b] The only documented pathogenic species listed is *B. dentium* (formerly *B. eriksonii*). Few other species have been isolated from clinical specimens; none are commonly encountered.

TABLE 5. Differential characteristics of *Eubacterium* species encountered in human clinical materials[a]

Species	Production of:		Reactions								
	Butyric acid	Caproic acid	Indole	Glucose	Maltose	Lactose	Mannose	Starch hydrolysis	Esculin hydrolysis	Nitrate	Metabolic products in PYG[b]
E. alactolyticum	+	+	−	$+^{w}$	−	−	−	−	−	−	A, B, C
E. rectale	+	−	−	+	$+^{w}$	$+^{w}$	V	+	+	−	B, L
E. limosum	+	−	−	+	$-^{+}$	$-^{w}$	w^{-}	$-^{+}$	+	−	A, B
E. saburreum	+	−	+	$+^{w}$	$+^{w}$	−	w^{-}	−	+	−	a, b, l
E. multiforme	+	−	−	+	−	$+^{w}$	$+^{w}$	−	+	+	A, p, B, L
E. cylindroides	+	−	−	+	−	−	$+^{w}$	$-^{+}$	+	−	a, B, L
E. moniliforme	+	−	−	$+^{w}$	V	−	$+^{w}$	$-^{+}$	−	V	a, B, L
E. combesii	+	−	−	$-^{w}$	$-^{w}$	−	−	−	V	−	A, ib, B, iv
E. nodatum	+	−	−	−	−	−	−	−	−	−	a, B, s
E. tenue	−	−	+	w^{+}	$-^{w}$	−	−	$-^{+}$	−	−	a, f
E. contortum	−	−	−	+	+	+	V	−	+	−	A, f
E. aerofaciens	−	−	−	+	$+^{w}$	$+^{-}$	+	−	$+^{-}$	−	A, F, L
E. lentum	−	−	−	−	−	−	−	−	−	$+^{-}$	a
E. brachy	−	−	−	−	−	−	−	−	−	−	a, IB, IV, IC, s
E. timidum	−	−	−	−	−	−	−	−	−	−	a, s

[a] Data from references 3, 4, 21, and 22. Symbols: +, positive reaction, 90 to 100% of strains, or final pH of 5.5 and lower; w, weak acid, or pH between 5.6 and 6.0; −, negative reaction for 90 to 100% of strains, or pH above 6.0; V, variable; superscript, reaction of some strains.

[b] Metabolic product analysis determined on peptone-yeast extract-glucose (PYG) broth cultures (22). Abbreviations: A, acetic acid; F, formic acid; P, propionic acid; IB, isobutyric acid; B, butyric acid; IV, isovaleric acid; IC, isocaproic acid; C, caproic acid; L, lactic acid; S, succinic acid. Capital letters indicate major products.

was found only in smears from individuals who were wearing IUDs. The prevalence of *A. israelii* among IUD wearers was 1.6% in a general population and 5.3% in a clinic population. Of the 112 women with *A. israelii* infestation, only 2 patients had clinically significant infections; the majority were asymptomatic. It appears that in most instances, when bacterial aggregates suggestive of actinomycetes are seen in routine PAP smears, organisms are present only as a superficial infestation within the endometrial cavity (39). This study did not show any association with the type of IUD but presented evidence that the length of time an IUD is worn is one of the major risk factors for the presence of actinomycetes. IUDs should be removed when patients are symptomatic, and the IUDs should be examined by a pathologist. The presence of granulocytic exudate surrounding the organism would be evidence that the actinomycete may be involved in an infectious process and is not just a colonizer. If there is infection with inflammation, it would be prudent not to insert a new IUD until the infection has cleared (with or without the use of antibiotics) (11). The prevalence of actinomycetes in the vaginal flora of healthy women who are not wearing IUDs seems to be very low but needs further study. IUDs (especially when there are calcium deposits) probably act as foreign bodies, chronically irritating the endometrial lining, and actinomycetes of the normal flora are likely to cause pelvic inflammatory disease by mechanisms that are analogous to cervicofacial forms of actinomycosis (11, 18).

In addition to species involved in actinomycosis, various species of gram-positive, nonsporeforming anaerobic bacilli have been encountered in a variety of clinical settings. These situations include most of the types of infections that involve *Bacteroides* and *Fusobacterium* species, anaerobic cocci, and clostridia, as discussed in Chapter 38. For excellent reviews of diseases caused by gram-positive nonsporeforming anaerobes in general, the reader is referred to Finegold (15) and Smith (33).

COLLECTION AND TRANSPORT OF SPECIMENS

The general principles and discussion of collection and transport procedures in Chapters 8 and 37 apply to this group of organisms. Some additional considerations regarding specimen collection for the laboratory confirmation of actinomycosis follow.

Anaerobic culture requests, when actinomycetes are suspected, should be clearly designated on the laboratory requisition form. The laboratory must know that the physician desires an actinomycete culture, and workers must know the specimen source and means of collection so that appropriate direct examination (e.g., including a search for sulfur granules) and processing for culture are done. Without this kind of appropriate information, the laboratory may not use the indicated culture technique and may not incubate cultures sufficiently long to enrich for *A. israelii* (which may require 2 to 4 weeks of incubation in enriched thioglycolate medium). Examples of appropriate specimens include pus, tissue, body fluids other than urine, and material collected from drainage sites. In terms of sample volume, depending on the site, the following should be collected: 0.5 ml or more of pus collected by needle and syringe aspiration, a small piece of tissue or biopsy, 0.5 to 10 ml of a body fluid, gauze (premoistened with sterile water) which has been collecting wound drainage material (to search for sulfur granules), or other appropriate material as discussed in Chapter 37. Although swabs yield only a limited sample volume and are not as desirable as other means of collection, other methods of collection may not be available at times, and it may be necessary to submit swabs for the laboratory confirmation of actinomycosis. Specimens should be collected aseptically from a site prepared by a surgical scrub technique (see Chapters 8 and 37). Contamination with the normal flora of the skin, mouth, rectum, vaginal contents, urinary tract orifices, and other mucosal surfaces must be avoided. Expectorated sputum is not a desirable specimen for culture because the actinomycetes normally inhabit the oropharynx, tonsils, and dental plaque and may be present in saliva. A protective double-lumen catheter can be used to collect specimens from the endometrial cavity (Chapter 37) (28). In many instances, depending on the anatomic site and clinical circumstances, specimens from lesions suspected of containing *Actinomyces* spp. should also be examined for nocardiae, fungi, and mycobacteria, in addition to the performance of routine bacteriology. The physician must be made aware that the laboratory may need separate requisitions and separate specimens for these other culture requests.

DIRECT MICROSCOPIC EXAMINATION

Microscopic examination provides excellent presumptive evidence that nonsporeforming gram-positive bacilli are present. When blood culture bottles are positive for diphtheroid bacilli, metabolic product analysis of the broth by GLC, as a supplemental direct examination procedure, may aid in rapid diagnosis. A major peak of propionic acid would suggest *P. acnes*. Although *A. propionica* and other *Propionibacterium* species produce propionic acid, they are so rare in blood cultures that the odds would overwhelmingly favor the presence of *P. acnes*. The absence of propionic acid and the presence of a major lactic acid peak in a blood culture in which a relatively small diphtheroid bacillus is seen would raise the possibility of *Listeria monocytogenes* or perhaps a pleomorphic *Streptococcus* sp. or, less likely, one of the coryneform-shaped *Lactobacillus* species.

When actinomycosis is suspected, purulent exudate, bronchial washings, pleural fluid, or other body fluid should be put in a sterile petri dish and examined under a dissecting microscope (×7 to ×15 magnification) for sulfur granules (17, 32). On occasion, a piece of gauze that has been removed from a draining wound will be submitted for examination for sulfur granules. The granules will adhere to the gauze. Also, it is appropriate to examine (with the dissecting microscope) surfaces of tissue removed surgically or at autopsy or IUDs for sulfur granules. When seen in a body fluid or adherent to gauze, sulfur granules usually are white or yellowish, irregularly shaped, hard, grainlike bodies that may range from ≤0.1 to 5 mm in diameter (12, 16, 17, 32). They are composed of masses of filamentous bacteria sometimes surrounded by hyaline clubs (Fig. 2 and 3). Probably they are yellowish because they are surrounded by foamy macrophages which impart a yellowish color (12). Sometimes when sulfur granules are seen in tissue at autopsy, they will appear greenish because they are surrounded by numerous polymorphonuclear leukocytes. When granules are found, they should be placed in a drop of water and gently crushed between two slides. One slide is removed, the preparation is allowed to dry, and then it is heat- or methanol-fixed and examined by Gram stain. Under oil immersion (×1,000 magnification), the presence of thin (1 μm or less in diameter), branched or unbranched filaments within a granule with peripheral clubs is excellent presumptive evidence for one of the actinomycetes (*Actinomyces* spp., *A. propionica*, or *B. dentium*) (Fig. 2 and 3). Of the organisms that produce sulfur granules, *A. israelii* is by far the most common. Sulfur granules are not always seen in actinomycosis (Fig. 4). If granules are absent, but filamentous, branched rods are seen, a modified Kinyoun or Ziehl-Nielson acid-fast stained smear should be examined in addition to a Gram stain. A Fite (or Fite-Ferraco) acid-fast stain is best for tissue. With these stains, *Nocardia asteroides* is usually acid fast. The filamentous, nonsporeforming anaerobic bacilli are non-acid fast. (The Putt acid-fast stain is not useful in this situation and may be misleading, since both *Actinomyces* and *Nocardia* species may stain acid fast with the Putt stain).

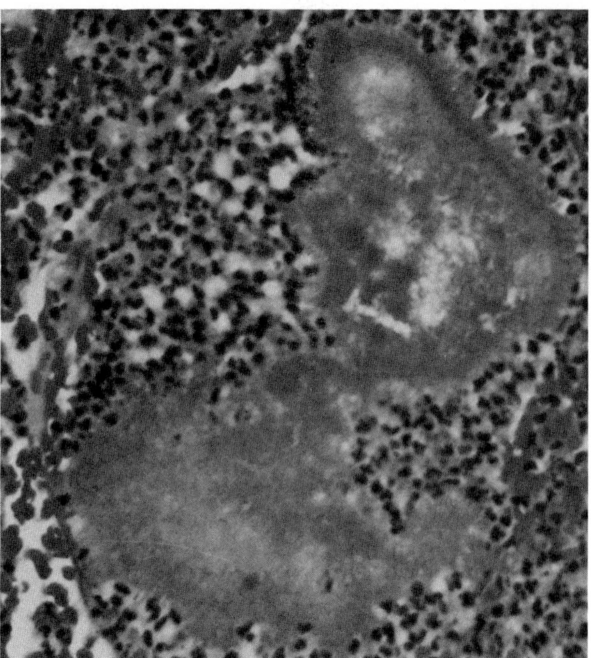

FIG. 2. Sulfur granule in a tissue section stained with hematoxylin and eosin from a patient with an abdominal form of actinomycosis. The granule is an eosinophilic (pinkish), amorphous, irregular mass and is located within an abscess (note myriads of neutrophilic granulocytes surrounding the granule). ×330. Courtesy of William Shasteen.

Staphylococcus aureus and other bacteria can form granules that resemble actinomycotic sulfur granules. However, when Gram stains of *S. aureus* botryomycotic granules are examined at ×1,000 magnification, one should see gram-positive cocci, not branching rods. As indicated previously, sulfur granules can occasionally be found in PAP smears of cervicovaginal materials from women who are wearing IUDs. However, pseudo-sulfur granules, not composed of aggregates of actinomycetes, may be more common in these specimens than true sulfur granules (27). In sections of endometrial curettings stained with hematoxylin and eosin, sulfur granule-like structures have a glassy reddish color and may have internal radiating structures, a fan-shaped appearance, and peripheral clubs. A Gram stain reveals thick, irregular, nonbranching gram-positive structures with peripheral swellings.

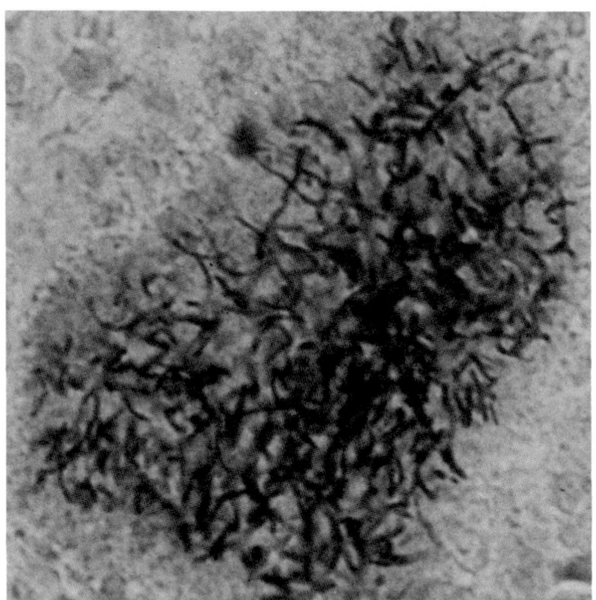

FIG. 3. Gram-stained tissue section of a sulfur granule from the same case as shown in Fig. 2. Note the gram-positive branched filaments and the zone of peripheral clubbing. ×305. Courtesy of William Shasteen.

The origin of these structures is not certain. O'Brien et al. have proposed that some of these granules may be part of the synthetic material of the IUD itself (27).

For demonstrating *A. israelii*, *A. naeslundii*, and *A. propionica* in cervicovaginal smears, Pine et al. have developed a fluorescent-antibody (FA) procedure in which pepsin treatment and a rhodamine conjugate of normal serum are used to reduce background staining (29). The actinomycete cells most often seen in this FA procedure were individual diphtheroid cells or groups of diphtheroid cells. This FA technique is probably a more sensitive and specific way to screen for actinomycetes than PAP smears, since one must see relatively large aggregates or granules of filamentous bacteria in the latter procedure. The clinical significance of finding individual actinomycete cells or cells in small groups in cervicovaginal materials is unknown (29).

CULTURE AND ISOLATION

A variety of solid nonselective media support excellent growth of the gram-positive anaerobic bacilli. These include CDC anaerobe blood agar (discussed in Chapter 37), brain heart infusion agar, brucella agar, and modified Schaedler agar (BBL Microbiology Systems, Cockeysville, Md.) supplemented with 5% sheep or rabbit blood, vitamin K_1, and hemin. These organisms also grow well on phenylethyl alcohol (PEA) blood agar. They are inhibited by vancomycin and should not grow on kanamycin-vancomycin, paromomycin-vancomycin, or gentamicin-vancomycin blood agar. Enriched thioglycolate medium (BBL-135C) and chopped-meat–glucose medium support excellent growth of these bacteria. In addition, a wide variety of blood culture media are satisfactory for their growth. The general directions on the use of primary isolation media, anaerobic systems, incubation, inspection and subculture procedures, and the holding jar technique given in Chapter 37 are relevant for the isolation and cultivation of the gram-positive, nonsporeforming anaerobic bacilli.

There are additional considerations for the primary isolation of the bacteria that can cause actinomycosis. If sulfur granules are found on direct examination, they should be promptly washed in thioglycolate broth within a sterile petri dish, placed in a Ten Broeck grinder (various laboratory supply companies) with a Pasteur pipette, and crushed with the tip of the glass rod within a small volume of thioglycolate medium (e.g., 0.5 ml). With a Pasteur pipette, immediately inoculate one plate of anaerobe blood agar and one plate of PEA blood agar with a drop of inoculum. Streak the inoculum with a platinum or stainless steel loop to achieve well-isolated colonies. Inoculate the liquid medium near the bottom of the tube with 2 or 3 drops of inoculum. Incubate all media in an anaerobic system. Be sure to incubate the thioglycolate medium in an anaerobic chamber or jar with screw cap loosened (to permit the exchange of gas within the tube). If granules were not seen, the same media should be inoculated and processed as above. Additional appropriate media for other anaerobes and nonanaerobic microorganisms should be inoculated, depending on the specimen source, as discussed in Chapter 37. All cultures should be incubated for 48 h, inspected for growth, and then reincubated for at least 5 to 7 days and reinspected for growth. The thioglycolate broth culture is used as an enrichment culture and should be incubated a minimum of 2 weeks, preferably 4 weeks, before it is considered negative. If desired, the anaerobe blood agar and PEA blood agar also may be incubated 2 to 4 weeks.

The isolation of actinomycetes from heavily contaminated specimens is difficult to accomplish, even when sulfur granules have been seen on direct examination. Traynor et al. described the use of a procedure and a selective medium for the isolation of actinomycetes from swab samples taken from the cervical os (38). Their procedure was to soak the swab, or beads of pus from an IUD that had been removed from the uterus, in 5 ml of thioglycolate broth. Tenfold dilu-

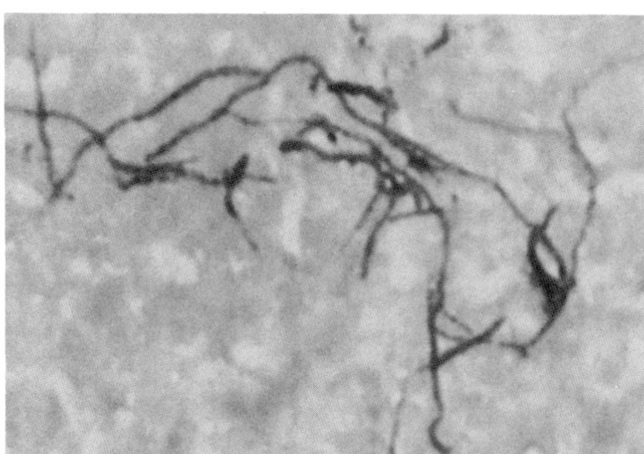

FIG. 4. Microscopic morphology of *A. israelii* in a Gram-stained smear of thoracic empyema fluid. Note the thin (≤ 1 μm in diameter) beaded and branched filaments. The organism was not acid fast with the modified Kinyoun stain. ×670. Courtesy of Linda Marler.

tions (from 10^{-1} through 10^{-4}) were then prepared in more tubes of thioglycolate medium, and 0.1 ml of each dilution was plated confluently on each of two horse blood agar plates (Columbia base; Oxoid). One of each of the duplicate plates contained 2.5 mg of metronidazole per liter to inhibit the growth of certain anaerobic bacteria and permit the actinomycetes to grow. The plates were incubated in an anaerobe jar and inspected after 4, 10, and 14 days for growth. It would seem likely that the use of metronidazole by itself would permit a heavy growth of *Streptococcus* spp., *Lactobacillus* spp., and yeasts which are resistant to metronidazole. Nonetheless, the culture results by this method showed good correlation with the results of direct microscopic examination that were obtained by an FA procedure.

Increasing evidence that *Actinomyces* spp., particularly *A. naeslundii* and *A. viscosus*, are commonly associated with periodontal disease and root caries has stimulated the development of selective media for the culture of these organisms from oral materials (25, 42). Kornman and Loesche added metronidazole (10 μg/ml) and cadmium sulfate (20 μg/ml) to a gelatin-based medium called GMC (25) which was useful for the isolation of *A. naeslundii* and *A. viscosus* from dental plaque. Recently, another selective medium (called CFAT) was developed by Zylber and Jordan for the same purpose (42). The CFAT medium composition was as follows (per liter): Trypticase soy broth (BBL), 30 g; agar, 15 g; glucose, 5 g; cadmium sulfate, 13 mg; sodium fluoride, 85 mg; neutral acriflavin, 1.20 mg; potassium tellurite, 2.50 mg; basic fuchsin, 1.25 mg; and sheep blood, 50 ml. CFAT medium suppressed the growth of most unwanted streptococci, *Rothia* spp., *Bacterionema* spp., *Neisseria* spp., and yeasts and permitted the growth of *A. naeslundii* and *A. viscosus*. No comparative data are available on the performance of GMC and CFAT media.

IDENTIFICATION

Key characteristics which can be used for presumptive identification of the gram-positive anaerobic bacilli include relationship to oxygen, tendency to form smooth versus rough colonies on anaerobe blood agar, red pigment of colonies, diffuse turbidity versus granular appearance of growth in thioglycolate medium, morphology of cells grown in enriched thioglycolate broth, catalase production, glucose fermentation, and metabolic products produced in peptone-yeast extract-glucose broth (Table 2). Directions for the determination of the relationship to oxygen, appearance of growth, and the procedure for metabolic product analysis by GLC are given in Chapter 37. The catalase test can be performed on cultures grown on brain heart infusion agar, Lombard-Dowell agar, or Lombard-Dowell esculin agar, as described in Chapter 111. The recommended procedure for determining the fermentation of glucose and other carbohydrates with CHO-based medium (9) is given in Chapter 111. The use of prereduced, anaerobically sterilized liquid media (22, 35) gives similar fermentation reactions with these bacteria (unpublished data). Definitive identification requires the performance of additional tests. Key differential characteristics of the more commonly encountered gram-positive, nonsporeforming anaerobic bacilli are given in Table 3. All of the tests in Table 3 can be performed according to the directions in Chapter 111 or by using prereduced, anaerobically sterilized liquid media (22). See Dowell and Hawkins (9), Holdeman et al. (22), and Sutter et al. (35) for additional detailed descriptions of differential media and conventional biochemical tests. Do not use these tables to interpret data obtained with commercial packaged micromethod kits (e.g., the Minitek, API-20A, Anaerobe-Tek, or Presumpto system). Tables based on a data base developed specifically for the packaged kit should be used. A discussion of commercial kits and rapid methods is given in Chapter 37 and in reference 24.

Several *Bifidobacterium* and *Eubacterium* species of human origin have been described. Their clinical relevance is uncertain, but there is considerable current interest in the possibility that some of these organisms play a role in the pathogenesis of periodontal disease and other conditions of the oral cavity. Therefore, Tables 4 and 5 provide additional characteristics for the differentiation of the species of these genera. The organisms listed in Tables 4 and 5 are, for the most part, infrequently encountered in clinical materials not derived from the oral cavity or in specimens contaminated with normal flora of other sites. Laboratories with limited facilities and resources can legitimately limit identification to the presumptive level described in Table 2, but they should be prepared to send potentially significant isolates to a reference laboratory for definitive identification.

Serologic test procedures

FA testing has been used at the Centers for Disease Control for several years to identify certain *Actinomyces*, *Arachnia*, *Bifidobacterium*, and *Propionibacterium* species (34). The FA reagents are not available for general use in hospital laboratories, nor are they commercially available.

P. acnes

The majority of the gram-positive, nonsporeforming anaerobic bacilli isolated from human clinical materials will be *P. acnes*. It is a predominant member of the skin flora, inhabiting hair follicles and sebaceous glands. It is also commonly found in the nasopharynx, mouth, intestines, and urogenital tract, including the male urethra. *P. acnes* is a common contaminant of blood cultures, cerebrospinal fluid, and other body fluid specimens that must be obtained by penetrating the skin. It must be differentiated from *L. monocytogenes* and other potential pathogens of similar morphology (see discussion above on morphologic considerations). When two or more sets of blood cultures are positive with *P. acnes*, thought should be given to the possibility of clinically significant bacteremia (e.g., hematogenous infection of a hip or other joint replacement, endocarditis [13–15], etc.). *P. acnes* has also been involved in infections of central nervous system shunts (e.g., for hydrocephalus) and associated with immune complex glomerulonephritis (1). Thus, its presence in cerebrospinal fluid must not always be assumed to represent contamination. *Propionibacterium granulosum* and *Propionibacterium avidum* are rarely encountered in properly collected specimens and are usually of no clinical significance.

In Gram-stained preparations, *P. acnes* varies in size from about 0.3 to 1.3 μm in diameter by about 1 to 10 μm in length. The organisms are highly pleomorphic, usually with a diphtheroid morphology, and may be curved, have clubbed or pointed ends, have elongated forms with beaded uneven staining, or show small coccoid to spherical bodies that can be confused with cocci (Fig. 1). They also may branch and can be confused with *Actinomyces* or *Arachnia* species (33). Gram stains of both colonies and broth culture should be observed because they often show differences. Most fresh clinical isolates of *P. acnes* grow as obligate anaerobes, but some isolates will show sparse growth after incubation in a 5 to 10% CO_2-air incubator, especially on chocolate agar, and are thus facultatively anaerobic (or aerotolerant).

On CDC anaerobe blood agar or PEA blood agar, colonies are 1 to 2 mm, circular, entire, convex, glistening, whitish or tan, and opaque. Some strains may produce narrow zones of hemolysis. *P. acnes* can be presumptively identified without GLC if it produces both indole and catalase. Not all strains produce indole, and these strains might be confused with *A. viscosus*, which is the only other common anaerobic gram-positive rod that is catalase positive. Also, not all strains produce catalase, and these strains might be confused with certain indole-positive *Eubacterium* spp. Those strains that are either indole or catalase negative and rare strains that form branching filaments must be differentiated from the other genera by using GLC and a battery of biochemical tests.

Eubacterium species

The next most commonly isolated gram-positive, nonsporeforming anaerobic bacillus is *E. lentum*. It is usually isolated from miscellaneous wound and abscess cultures from a variety of anatomic sites, and most often it is mixed with other organisms. *Eubacterium* spp. and the remaining anaerobic gram-positive rods discussed in this chapter are not commonly isolated from blood cultures. On Gram stain, *E. lentum* is usually about 0.5 to 0.7 μm in diameter by 1.5 to 3 μm long. It occurs singly, in pairs, and in short chains. In broth culture the cells may form diplococci, and rod shapes may be difficult to ascertain. Colonies are usually 1 mm in diameter, punctate, low convex, entire, and opaque, and there is no hemolysis on sheep blood agar (33). GLC and a battery of biochemical tests are necessary to differentiate *Eubacterium* species from each other. *E. lentum* produces only a minor peak of acetic acid in peptone-yeast extract-glucose broth and is inactive in most biochemical tests (Tables 3 and 5).

E. limosum produces major acetic and butyric acid peaks (with traces of isobutyric acid), hydrolyzes esculin, and ferments glucose, arabinose, mannitol, and other carbohydrates. *E. aerofaciens* was incriminated in a case of subacute bacterial endocarditis (30). For the differentiation of *E. aerofaciens* and other uncommonly encountered *Eubacterium* species, see Table 5.

Actinomyces and *Arachnia* species

A. israelii is rarely isolated from clinical materials, but it is the most common of the *Actinomyces* and *Arachnia* species. The cells are gram-positive rods, approximately 1 μm in diameter, and they vary considerably in length. The organism may be in the form of short diphtheroid rods. The rods may be club shaped, unbranched, or branched and may produce long, thin, beaded filaments. Smooth or rough colonies may be seen. When colonies are smooth (seen with about one-third of *A. israelii* isolates), they are not distinctive, and the organism may be difficult to differentiate from other anaerobic gram-positive rods. Smooth strains may form 1- to 2-mm, white, circular, opaque, smooth, glistening colonies after only 2 to 3 days of incubation. They grow faster than other strains of *A. israelii* (33). Colonies of the more typical rough strains of *A. israelii* take much longer to develop and usually will not be visible to the unaided eye until a full 5 to 7 days, or longer, of incubation. However, if young colonies of *A. israelii* rough strains are viewed with the dissecting microscope, they may appear as colonies of thin radiating filaments called spider colonies (16, 17, 32, 33) (Fig. 5). When the colonies of rough strains are 7 days old, or more, they are usually raised, heaped up, lobate, irregular, opaque, white, and glistening and are called molar tooth colonies (Fig. 6) because they resemble molar teeth. In enriched thioglycolate medium, smooth strains show a diffuse turbidity in the bottom half of the broth tube; rough strains form rough discrete colonies that sink to the bottom. The microscopic and colonial characteristics of *A. israelii* and *A. naeslundii* are very similar. The biochemical characteristics of these species are given in Tables 2 and 3. *A. viscosus* may form young colonies that are filamentous or granular and older colonies which are 0.5 to 2 mm in diameter, entire, convex, translucent, and gray (33). *A. viscosus* produces catalase, is indole negative, and can grow aerobically. It produces acetic, lactic, and succinic acids. *A. odontolyticus* is noteworthy in that it may produce red colonies on anaerobe blood agar after the plates have been held at ambient temperature in room air for several days, or it may produce red colonies after prolonged anaerobic incubation. The colonies of *A. meyeri* and *A. pyogenes* are usually smooth forms, 0.5 to 1 mm in diameter (17). The colonies and microscopic features of *A. propionica* and *A. israelii* are similar. *A. propionica* may form ovoid to spherical bodies that probably are spheroplasts (3).

Bifidobacterium species

The only *Bifidobacterium* species known to be of pathogenic significance in humans is *B. dentium* (formerly *B. eriksonii* and *A. eriksonii*). It normally inhabits the mouth and intestines and has been found in mixed infections of the lower respiratory tract (37). The cells are usually 0.5 to 1 μm in diameter by 2 to 4 μm in length. It commonly forms bifid or clubbed ends and also may branch. Colonies are usually about 1 mm or less in diameter, smooth, convex, entire, translucent, glistening, and white (33). It is an obligate anaerobe that produces major products of acetic and lactic acids in peptone-yeast extract-glucose broth and actively ferments several carbohydrates. A clue to interpreting the GLC results is that *Bifidobacterium* species produce more acetic than lactic acid, which is not typical of the other genera listed in Table 1 that produce both acetic and lactic acids.

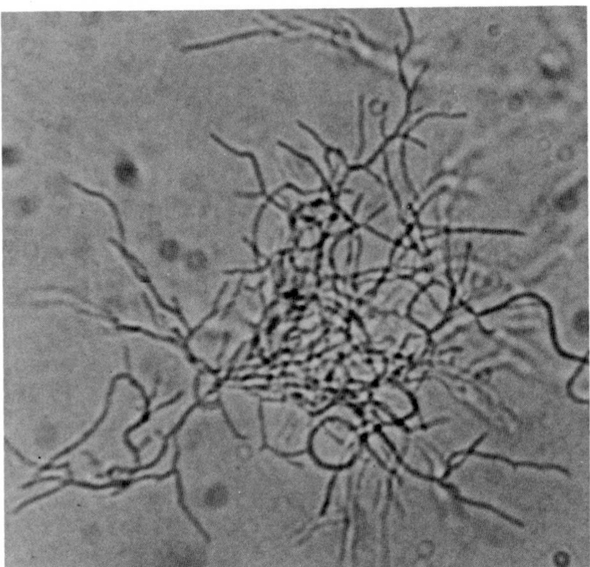

FIG. 5. Dissecting microscope view of early spider-type rough colony of *A. israelii* on brain heart infusion agar. Courtesy of the Center for Disease Control.

Lactobacillus species

Lactobacilli are mentioned here because they are commonly encountered in the clinical laboratory as commensals or contaminants without significance. The majority of *Lactobacillus* species encountered clinically form a large quantity of lactic acid from glucose as the sole major metabolic end product. *L. catenaforme* is found in the oropharynx and intestines and is the anaerobic species that is isolated most often from clinical specimens (33). Its pathogenic significance is uncertain. Colonies are 1 mm or less, smooth, convex, entire, translucent, and grayish to white. The lactobacilli are catalase negative. *L. catenaforme* is indole negative, hydrolyzes esculin and starch, and ferments several carbohydrates. Good presumptive clues for recognizing *Lactobacillus* spp. are that many form slender rods in chains and that they grow on lactobacillus-selective (LBS) agar medium (BBL) (which contains tomato juice and has a low pH); with the exception of some *Bifidobacterium* spp., most other gram-positive bacilli fail to grow on LBS agar. For further information pertaining to the identification of lactobacilli, see Holdeman et al. (22) or Buchanan and Gibbons (3).

SUSCEPTIBILITY TO ANTIMICROBIAL AGENTS; CONSIDERATIONS RELATIVE TO TREATMENT OF ACTINOMYCOSIS

Penicillin is active against most of the gram-positive, nonsporeforming anaerobic bacilli (Table 6) (15, 36). Penicillin, in high doses, is still the antibiotic of choice for the treatment of actinomycosis. Alternatively, tetracycline has been recommended for patients who are allergic to penicillin (23). For patients with cervicofacial, thoracic, and abdominal forms of actinomycosis, intravenous penicillin G, 10×10^6 to 20×10^6 U daily for 4 to 6 weeks, followed by oral phenoxymethylpenicillin for several (6 to 12) months has often been recommended (23). Various clinical circumstances such as the precise anatomic location, evidence of hematogenous dissemination, involvement with other bacteria in a polymicrobial infection, patient compliance (relative to oral therapy), and other factors may alter the approach to treatment. Surgical drainage of abscesses and sinus tracts, removal of necrotic tissue, and debridement of involved areas of suppuration and fibrosis may be a necessary adjunct to anitmicrobial treatment (26). Treatment failures have followed the use of penicillin as the only antibiotic (26). One of five strains of *A. israelii* tested in our laboratory was resistant to 16 μg of penicillin G per ml. It is uncertain whether some strains of *A. israelii* develop resistance to penicillin G during prolonged treatment. Tetracyclines, erythromycin, chloramphenicol, and clindamycin also show good activity in vitro against *Actinomyces* spp. (Table 6) (36). Cefoxitin, moxalactam, and piperacillin show excellent in vitro activity against *A. israelii*, but there is little clinical experience in the treatment of actinomycosis with these agents. Metronidazole shows poor activity against the gram-positive, nonsporeforming anaerobic bacilli, including *Actinomyces, Arachnia, Eubacterium*, and *Propionibacterium* species (Table 6) (36).

The susceptibility of *Eubacterium* and *Propionibacterium* species to penicillin G, chloramphenicol, clindamycin, ticarcillin, and piperacillin is highly predictable. Cefoxitin, moxalactam, other cephalosporins, and other third-generation beta-lactam compounds are usually less active in vitro against *Eubacterium* and *Propionibacterium* species than are the penicillins (Table 6). Antimicrobial susceptibility testing is probably not indicated in most instances for those gram-positive, nonsporeforming anaerobic bacilli that are isolated from mixed wound infections, and such testing is certainly irrelevant for *P. acnes* and other species that are likely to be contaminants. However, antimicrobial susceptibility testing should be done on organisms other than *P. acnes* that are isolated from blood; it may even be relevant at times (clinically) to perform susceptibility tests on *P. acnes* if there are multiple blood isolates (e.g., more than two positive blood cultures), actinomycetes from patients who are diagnosed clinically as having actinomycosis, or gram-positive, nonsporeforming anaerobic bacilli isolated from brain abscess, bone and joint infections, suspected endocarditis, lung abscess, necrotizing pneumonia, thoracic empyema, and other serious conditions that may require prolonged antimicrobial treatment.

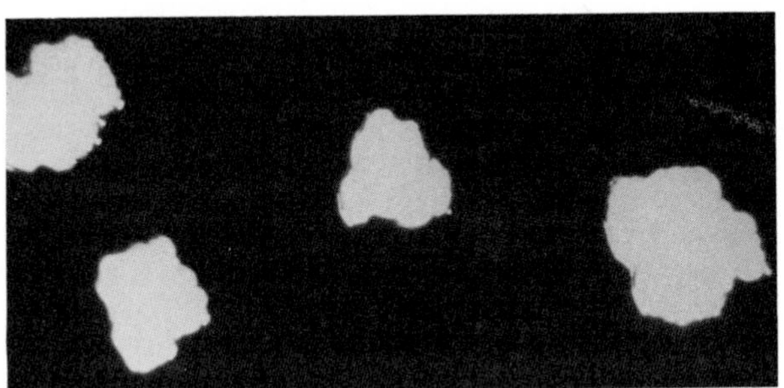

FIG. 6. Ten-day-old molar-tooth rough colony of *A. israelii* on anaerobe blood agar. Courtesy of Jean Siders.

TABLE 6. Percentages of gram-positive, nonsporeforming anaerobic bacilli susceptible to antimicrobial agents[a]

Drug (concn in μg/ml)	% of strains susceptible at breakpoint		
	Actinomyces spp. (5 strains)	*Eubacterium* spp. (9 strains)	*P. acnes* (231 strains)
Penicillin G (4)	80	89	98
Ticarcillin (64)	80	100	100
Piperacillin-azlocillin and mezlocillin (128)	100	100	100
Chloramphenicol (8)	100	100	100
Metronidazole (8)	80	78	3
Clindamycin (4)	100	89	99
Erythromycin (4)	100	89	100
Cefoxitin (32)	100	78	100
Moxalactam (32)	100	44	100

[a] Based on unpublished data of J. A. Siders and S. D. Allen.

LITERATURE CITED

1. **Beeler, B. A., J. G. Crowder, J. W. Smith, and A. White.** 1976. *Propionibacterium acnes*: pathogen in central nervous system shunt infection. Am. J. Med. **61**:935–938.
2. **Biavati, B., V. Scardovi, and W. E. C. Moore.** 1982. Electrophoretic patterns of proteins in the genus *Bifidobacterium* and proposal of four new species. Int. J. Syst. Bacteriol. **32**:358–373.
3. **Buchanan, R. E., and N. E. Gibbons (ed.).** 1974. Bergey's manual of determinative bacteriology, 8th ed. The Williams & Wilkins Co., Baltimore.

3a. **Cato, E. P., W. E. C. Moore, G. Nygaard, and L. V. Holdeman.** 1984. *Actinomyces meyeri* sp. nov., specific epithet rev. Int. J. Syst. Bacteriol. **34**:487–489.

4. **Cato, E. P., C. W. Salmon, and L. V. Holdeman.** 1974. *Eubacterium cylindroides* (Rocchi) Holdeman and Moore: emended description and designation of neotype strain. Int. J. Syst. Bacteriol. **24**:256–259.
5. **Collins, M. D., and D. Jones.** 1982. Reclassification of *Corynebacterium pyogenes* (Glage) in the genus *Actinomyces*, as *Actinomyces pyogenes* comb. nov. J. Gen. Microbiol. **128**:901–903.
6. **Dent, V. E., and R. A. D. Williams.** 1984. *Actinomyces denticolens* Dent & Williams sp. nov: new species from the dental plaque of cattle. J. Appl. Bacteriol. **56**:183–192.
7. **Dent, V. E., and R. A. D. Williams.** 1984. *Actinomyces howellii*, a new species from the dental plaque of dairy cattle. Int. J. Syst. Bacteriol. **34**:316–320.
8. **Dowell, V. R.** 1977. Clinical veterinary anaerobic bacteriology. U.S. Department of Health, Education, and Welfare unnumbered publication. Center for Disease Control, Atlanta.
9. **Dowell, V. R., Jr., and T. M. Hawkins.** 1977. Laboratory methods in anaerobic bacteriology, CDC laboratory manual. U.S. Department of Health, Education, and Welfare publication no. (CDC) 78-8272. Center for Disease Control, Atlanta.
10. **Duguid, H., I. Duncan, D. Parratt, and R. Traynor.** 1982. *Actinomyces* and intrauterine devices. J. Am. Med. Assoc. **248**:1579–1580.
11. **Duguid, H. L. D., D. Parratt, R. Traynor, D. Taylor, I. D. Duncan, J. Elias-Jones, and R. Duguid.** 1982. Studies on uterine tract infections and the IUCD with special reference to actinomycetes. Br. J. Obstet. Gynaecol. **89**(Suppl. 4):32–40.
12. **Emmons, C. W., C. H. Binford, J. P. Utz, and K. J. Kwon-Chung.** 1977. Medical mycology, 34th ed. Lea & Febiger, Philadelphia.
13. **Felner, J. M.** 1974. Infective endocarditis caused by anaerobic bacteria, p. 345–352. *In* A. Balows, R. M. DeHaan, V. R. Dowell, Jr., and L. B. Guze (ed.), Anaerobic bacteria—role in disease. Charles C Thomas, Publisher, Springfield, Ill.
14. **Felner, J. M., and V. R. Dowell, Jr.** 1970. Anaerobic bacterial endocarditis. N. Engl. J. Med. **283**:1188–1192.
15. **Finegold, S. M.** 1977. Anaerobic bacteria in human disease. Academic Press, Inc., New York.
16. **Georg, L. K.** 1974. The agents of human actinomycosis, p. 237–256. *In* A. Balows, R. M. DeHaan, V. R. Dowell, Jr., and L. B. Guze (ed.), Anaerobic bacteria—role in disease. Charles C Thomas, Publisher, Springfield, Ill.
17. **Gerencser, M. A.** 1981. Actinomycosis, p. 143–157. *In* A. Balows and W. J. Hausler, Jr. (ed.), Diagnostic procedures for bacterial, mycotic and parasitic infections, 6th ed. American Public Health Association, Inc., Washington, D.C.
18. **Hager, W. D., and B. Majmudar.** 1979. Pelvic actinomycosis in women using intrauterine contraceptive devices. Am. J. Obstet. Gynecol. **133**:60–63.
19. **Henderson, S. R.** 1973. Pelvic actinomycosis associated with an intrauterine device. Obstet. Gynecol. **41**:726–732.
20. **Hofstad, T., and P. Aasjord.** 1982. *Eubacterium plautii* (Séguin 1928) comb. nov. Int. J. Syst. Bacteriol. **32**:346–349.
21. **Holdeman, L. V., E. P. Cato, J. A. Burmeister, and W. E. C. Moore.** 1980. Descriptions of *Eubacterium timidum* sp. nov., *Eubacterium brachy* sp. nov., and *Eubacterium nodatum* sp. nov. isolated from human periodontitis. Int. J. Syst. Bacteriol. **30**:163–169.
22. **Holdeman, L. V., E. P. Cato, and W. E. C. Moore.** 1977. Anaerobe laboratory manual, 4th ed. Anaerobe Laboratory, Virginia Polytechnic Institute and State University, Blacksburg.
23. **Kinderlehrer, D. A.** 1983. Fungal infections, p. 641–684. *In* R. E. Reese and R. G. Douglas (ed.), A practical approach to infectious diseases. Little, Brown & Co., Boston.
24. **Koneman, E. W., S. D. Allen, V. R. Dowell, Jr., and H. M. Sommers.** 1983. Color atlas and textbook of diagnostic microbiology, 2nd ed. J. B. Lippincott Co., Philadelphia.
25. **Kornman, K. S., and W. J. Loesche.** 1978. New medium for isolation of *Actinomyces viscosus* and *Actinomyces naeslundii* from dental plaque. J. Clin. Microbiol. **7**:514–518.
26. **Lerner, P. I.** 1983. Pneumonia caused by the pathogenic actinomycetes (*Actinomyces, Arachnia, Nocardia*), p. 110–130. *In* L. Weinstein and B. N. Fields (ed.), Seminars in infectious disease, vol. 5. Thieme-Stratton, Inc., New York.
27. **O'Brien, P. K., L. A. Roth-Moyo, and B. A. Davis.** 1981. Pseudo-sulfur granules associated with intrauterine contraceptive devices. Am. J. Clin. Pathol. **75**:822–826.
28. **Pezzlo, M. T., J. W. Hesser, T. Morgan, P. J. Valter, and L. D. Thrupp.** 1979. Improved laboratory efficiency and diagnostic accuracy with new double-lumen-protected swab for endometrial specimens. J. Clin. Microbiol. **9**:56–59.
29. **Pine, L., G. B. Malcolm, E. M. Curtis, and J. M. Brown.** 1981. Demonstration of *Actinomyces* and *Arachnia* species in cervicovaginal smears by direct staining with species-specific fluorescent-antibody conjugate. J. Clin. Microbiol. **13**:15–21.
30. **Sans, M. D., and J. G. Crowder.** 1973. Subacute bacterial endocarditis caused by *Eubacterium aerofaciens*: report of a case. Am. J. Clin. Pathol. **59**:576–580.
31. **Skerman, V. B. D., V. McGowan, and P. H. A. Sneath (ed.).** 1980. Approved lists of bacterial names. Int. J. Syst. Bacteriol. **30**:225–420.
32. **Slack, J. M., and M. A. Gerencser.** 1975. *Actinomyces*, filamentous bacteria: biology and pathogenicity. Burgess Publishing Co., Minneapolis.
33. **Smith, L. D.** 1975. The pathogenic anaerobic bacteria, 2nd ed. Charles C Thomas, Publisher, Springfield, Ill.

34. **Sonnenwirth, A. C., and V. R. Dowell, Jr.** 1980. Gram-positive, nonsporeforming anaerobic bacilli, p. 440–445. *In* E. H. Lennette, A. Balows, W. J. Hausler, Jr., and J. P. Truant (ed.), Manual of clinical microbiology, 3rd ed. American Society for Microbiology, Washington, D.C.
35. **Sutter, V. L., D. M. Citron, and S. M. Finegold.** 1980. Wadsworth anaerobic bacteriology manual, 3rd ed. The C. V. Mosby Co., St. Louis.
36. **Sutter, V. L., M. J. Jones, and A. T. M. Ghoneim.** 1983. Antimicrobial susceptibilities of bacteria associated with periodontal disease. Antimicrob. Agents Chemother. **23:**483–486.
37. **Thomas, A. V., T. H. Sodeman, and R. R. Bentz.** 1974. *Bifidobacterium (Actinomyces) eriksonii* infection. Am. Rev. Respir. Dis. **110:**663–668.
38. **Traynor, R. M., D. Parratt, H. L. D. Duguid, and I. D. Duncan.** 1981. Isolation of actinomycetes from cervical specimens. J. Clin. Pathol. **34:**914–916.
39. **Valicenti, J. F., Jr., A. A. Pappas, C. D. Graber, H. O. Williamson, and N. F. Willis.** 1982. Detection and prevalence of IUD-associated *Actinomyces* colonization and related morbidity: a prospective study of 69,925 cervical smears. J. Am. Med. Assoc. **247:**1149–1152.
40. **Watabe, J., Y. Benno, and T. Mitsuoka.** 1983. *Bifidobacterium gallinarum* sp. nov.: a new species isolated from the ceca of chickens. Int. J. Syst. Bacteriol. **33:**127–132.
41. **Wegienek, J., and C. A. Reddy.** 1982. Taxonomic study of "*Corynebacterium suis*" Soltys and Spratling: proposal of *Eubacterium suis* (nom. rev.) comb. nov. Int. J. Syst. Bacteriol. **32:**218–228.
42. **Zylber, L. J., and H. V. Jordan.** 1982. Development of a selective medium for detection and enumeration of *Actinomyces viscosus* and *Actinomyces naeslundii* in dental plaque. J. Clin. Microbiol. **15:**253–259.

Chapter 42

Section V. Spirochetes

Leptospira

AARON D. ALEXANDER

CHARACTERIZATION

Members of the genus *Leptospira* are serologically heterologous. The basic taxon is the serovar. Two species, *L. interrogans* and *L. biflexa*, are recognized for the so-called pathogenic and saprophytic leptospires, respectively. *L. illini* (syn. *Leptonema illini*), a possible third species or new genus, differs remarkably from other leptospires in DNA base composition (12, 13) and has been designated as *species incertae sedis* in *Bergey's Manual of Systematic Bacteriology*, vol. 1 (14). The saprophytic or water leptospires found predominantly in fresh surface waters and less frequently in seawater are rarely associated with mammalian infections. Pathogenic leptospires occur naturally in a wide variety of wild and domesticated mammals throughout the world and cause acute, febrile, systemic disease of humans and other mammals. These pathogenic leptospires are isolated from clinical specimens, carriers, and natural waters.

Leptospires are helicoidal, flexible organisms, usually 6 to 20 μm long and approximately 0.1 μm in diameter, with semicircular hooked ends (occasionally, the ends are straight) (Fig. 1). They are motile. Electron microscopy reveals a cylindrical body helicoidally wound about two periplasmic flagella which are inserted subterminally at opposite ends of the helicoidal body with their free ends extending toward the middle of the cell (Fig. 2). A common external sheath covers both structures. Leptospires are faintly colored after being stained with aniline dyes. They are invisible by bright-field microscopy but readily seen by dark-field microscopy.

Growth

Leptospires are aerobic and grow in medium containing 10% serum (preferably rabbit) or serum albumin plus long-chain fatty acids at pH 6.8 to 7.8. The optimal temperature is 30°C, and the incubation time for optimal growth ranges from a few days to 4 weeks or longer (usually 6 to 14 days).

Biochemical characteristics

Leptospires within species are not distinguishable on the basis of biochemical characteristics.

Serological identification

Serovars are identified by microscopic agglutination and agglutinin adsorption tests with serovar-specific rabbit antiserum. Serovars with major antigenic affinities as disclosed in cross-agglutination tests are arbitrarily assembled into serogroups (not a taxonomic subdivision).

Animal pathogenicity

Pathogenic leptospires produce lethal to subclinical infections in hamsters, guinea pigs, gerbils, and weanling rabbits after intraperitoneal inoculation.

RECOGNITION OF LEPTOSPIROSIS

The etiological agents of leptospirosis include approximately 180 different serovars (the basic taxon), determined on the basis of their agglutinogenic properties (14). The pathogenic serovars are otherwise indistinguishable by morphology and biochemical activity. Natural reservoirs of infection are rodents and a large variety of other feral and domestic animals. Many serovars occur predominantly in select mammalian hosts, but the distribution of a specific serovar in a select host is not exclusive. The same animal species may be a primary reservoir for several different serovars and may also carry types occurring primarily in other mammals.

The nesting site for leptospires in the natural host is the lumen of nephritic tubules, whence the organisms are shed into the urine. The persistence and intensity of leptospiruria may vary with the host and with the serovar causing the infection. For example, Norway rats infected with serovar *icterohaemorrhagiae* shed profuse numbers of leptospires for the remainder of their natural life, whereas strains of serovar *canicola* apparently persist less efficiently in the kidneys of rats. Shedding by infected dogs, cattle, and swine may be heavy for only a few months after infection and is usually sparse or absent after 6 months (25).

Infections are incurred directly by contact with the urine of carriers or indirectly by contact with streams, ponds, swamps, or wet soils contaminated with the urine of carriers. Pathogenic leptospires can survive for 3 months or longer in neutral or slightly alkaline waters, but they do not persist in brackish or acid waters. Organisms enter their hosts through abrasions of the skin, through mucosal surfaces of the nasopharnyx or esophagus, or through the eye.

Leptospirosis in humans is primarily associated with occupational exposure. Work with animals or in rat-infested surroundings poses infection hazards (e.g., for veterinarians, dairymen, swineherds, abattoir workers, miners, and fish and poultry processors). In various parts of the world, leptospirosis occurs sporadically or in epidemic proportions in agricultural workers engaged in the raising of rice, cane, flax, and vegetables, in rubber plantation workers, and in soldiers exposed to natural environments contaminated by animal carriers. The potential infection hazards of bathing or swimming in ponds or streams around which livestock are pastured have been demonstrated repeatedly (4, 16).

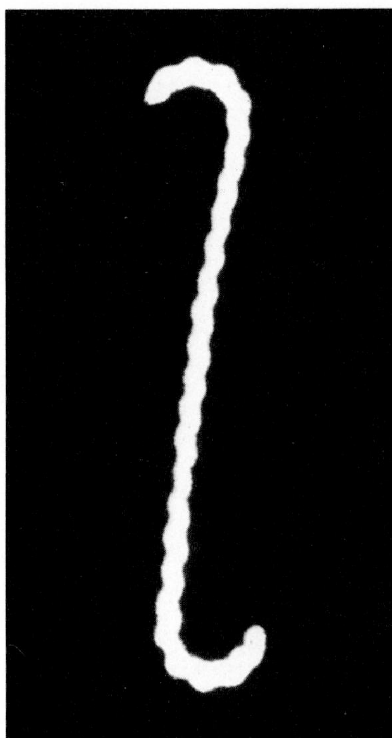

FIG. 1. *Leptospira* sp. Dark-field illumination. ×1,250. Courtesy of C. D. Cox, Department of Microbiology, University of Massachusetts, Amherst.

The clinical manifestations of leptospirosis in humans and animals are variable, ranging from a mild, catarrhlike illness to icteric disease with severe kidney and liver involvement. Diagnosis can only be established in the laboratory by demonstration of the organisms or by serological tests. A variety of procedures are available for laboratory diagnosis. The selection and use of appropriate tests are contingent on an understanding of the course of leptospiral infections.

The incubation period ranges from 3 to 30 days, but is usually 10 to 12 days. Leptospiremia occurs at the time of disease onset and persists for approximately 1 week. During this acute phase, leptospires may also be present in cerebrospinal fluid and in the milk of lactating animals. Antibodies become detectable by days 6 to 10 of disease and reach maximal levels at week 3 or 4 of disease. Thereafter, antibody levels gradually recede but may remain detectable for years (27).

In humans and domestic animals, leptospires may be found in the urine after week 1 of disease. Urinary shedding may persist for 2 to 3 months in a large proportion of cases. In some cases, intermittent shedding may occur for longer periods of time. Infection in pregnant livestock during the latter half of gestation may result in abortion several weeks later, at which time the dam may have detectable antibodies and may be a urinary shedder.

DIRECT EXAMINATION

The concentration of leptospires in the blood and cerebrospinal fluid of patients with naturally occurring infections is low, and the organisms are difficult to demonstrate by direct microscopy. The chances of demonstrating leptospires are increased by the centrifugation of blood treated with sodium oxalate or heparin at low speed to remove cellular elements and then at high speed to concentrate the remaining elements, which are examined microscopically (27). Although this method may be valuable in the establishment of a rapid diagnosis, it frequently results in misdiagnosis by the mistaken identification of fibrils or extrusions from cells as spirochetes. Therefore, direct dark-field examination of blood is not recommended as a single diagnostic procedure.

Direct dark-field examination may be of value for the examination of specimens in which there is a high concentration of leptospires, e.g., blood, peritoneal fluid, or liver suspension from hamsters or guinea pigs infected with clinical material or, frequently, urine or kidney suspensions from wildlife, swine, dogs, and other domestic animals.

The examination of blood or other fluids and tissue suspensions is conducted with a minute drop of fluid distributed in a thin layer between a glass slide and a cover slip. It is important to disperse cellular particles; otherwise, too much light may be reflected, and these reflections would interfere with the detection of leptospires. The typical morphology and motility of leptospires should be evident before a presumptive diagnosis is made. The failure to detect leptospires does not rule out their presence. Diagnosis by microscope examination should be confirmed by cultural or serological tests.

Staining techniques for the demonstration of leptospires in films of blood, urine, and tissue preparations have been described (5, 21). These procedures have the same limitations as dark-field microscopy and are not

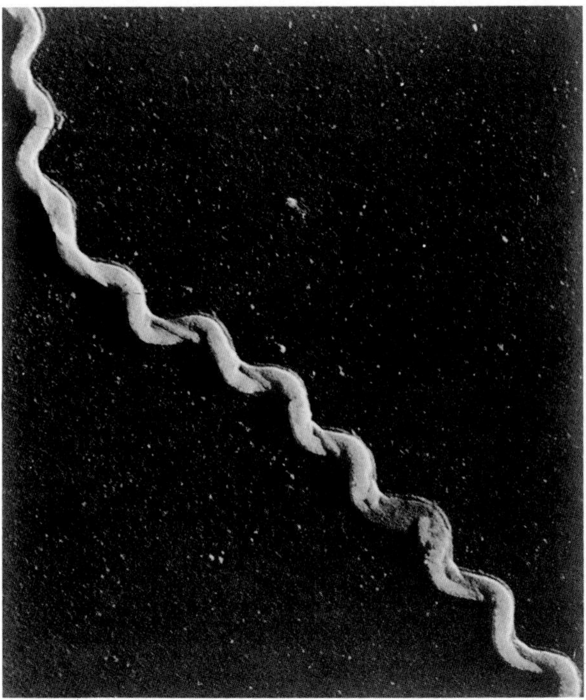

FIG. 2. Electron micrograph of a portion of a leptospire. Courtesy of the Armed Forces Institute of Pathology, Washington, D.C.

routinely used. The use of a silver-deposition technique to demonstrate organisms in a tissue section may be of value when cultural and serological procedures are not possible. Fluorescent-antibody techniques have been used to demonstrate leptospires in urine and tissues from animals (8, 21), but the use of these techniques for routine diagnosis has been limited.

CULTURAL EXAMINATION

During week 1 of disease, the most reliable way to detect leptospires is by the direct culturing of blood on appropriate media (see Chapter 111). Media containing rabbit serum (i.e., Fletcher semisolid and Stuart liquid media) or albumin and fatty acids (i.e., Ellinghausen-McCullough-Johnson-Harris medium) will grow most strains of leptospires (21, 27). A few strains are more readily cultivated in media containing albumin and fatty acids; other strains grow best in rabbit serum media. It is advisable to use at least four tubes of medium from two different types or lots of medium for each sample. Media are dispensed in approximately 5-ml amounts into test tubes (16 by 125 mm, preferably screw capped). Prepared media can be stored for months at room temperature. Repeated daily blood cultures with the use of 1 to 2 drops of blood per 5 ml of medium during week 1 of disease are recommended. The use of minimal inocula, particularly after day 4 of disease, serves to minimize the effect of growth-inhibitory substances that may be present in the blood. If media are not available at the time blood is collected, the blood can be defibrinated or mixed with anticoagulants (heparin or sodium oxalate; citrate solutions may be inhibitory) and subsequently cultured. Alternatively, clotted blood can be triturated and cultured. Spinal fluid obtained during the acute phase of the disease can also be cultured.

After week 1 of disease, blood cultures are rarely successful. However, at this time and for several months thereafter, the urine may contain leptospires, which may be isolated by culturing or animal inoculation. In infected humans, the concentration of leptospires in the urine is low; shedding may be intermittent. Therefore, repeated isolation attempts should be made. Isolation attempts should be made as soon as possible after collection of the specimen. Leptospires may not survive in acid urine for more than a few hours. Urine obtained aseptically, e.g., by perforation of the bladder (a technique used for infected dogs), can be cultured directly (21). Because the undiluted urine may contain growth-inhibiting substances, both undiluted urine and a 10-fold dilution of urine are recommended for culture, with the use of 1 to 2 drops of inoculum. Direct cultural isolation is also possible with midstream urine samples carefully collected from cleansed genitalia. The samples can be cultured directly in medium containing added 5-fluorouracil (5-FU) at a concentration of 200 μg/ml (15). The pyrimidine analog does not inhibit the growth of leptospires but may serve to prevent the growth of contaminating microorganisms. For cultures of aseptically derived material, media containing 5-FU should be used in conjunction with conventional media, as the growth of some leptospires may be slowed in the presence of 5-FU. Pasteur pure-culture techniques—the culturing of serial 10-fold dilutions of voided urine—have been used for the recovery of leptospires from domestic animals (21). The addition to rabbit serum medium of neomycin (5 to 25 mg/liter), either singly or in combination with furazolidone or with sulfathiazole (50 mg/liter) and cycloheximide (0.5 mg/liter), has also been recommended to inhibit the growth of contaminating microorganisms (9, 19). However, the use of these inhibitory substances requires additional study.

In fatal cases, leptospires may be present in various tissues as well as in the blood. Liver and kidney are the tissues of choice for the recovery of organisms. Tissues are triturated by glass grinders or by the use of sand with a mortar and pestle and are suspended in 9 parts of physiological salt solution or medium. The 10% suspension, and 1:10 and 1:100 dilutions thereof, are cultured as in the method for urine. The higher dilutions of tissue suspensions are cultured to limit the effects of growth-inhibitory substances that may be present. Alternatively, especially under field conditions (for example, in culturing the kidneys of trapped small animals), 0.5 to 1.0 g of kidney can be expressed through the barrel of a 2- or 5-ml syringe (without a needle) directly onto medium, which can then be diluted further and subcultured. For culturing the kidneys of large domestic animals, a representative sample can be obtained by scraping the cortex with sterile metal bottle caps with grated surfaces. The grated surface is prepared by punching holes in the cap with a nail. The ground tissue is collected on the undersurface of the bottle cap. Portions of tissue punched out with a capillary pipette can also be cultured directly onto medium, as in blood cultures. The use of media containing 5-FU or antibiotics in the cultivation of tissues may increase the chances of isolation if specimens are contaminated with other microorganisms. If tissues cannot be processed immediately, they can be collected aseptically in sterile bottles, rapidly frozen, and stored at dry-ice temperatures. The addition of a cryoprotective agent such as glycerol (1 part to 20 parts of tissue) before freezing of the specimen may minimize the deleterious effects of freezing on leptospires.

Cultures are incubated in the dark at 30°C or at room temperature. Most pathogenic strains are detected in culture after 6 to 14 days of incubation. In some cases, leptospires may be seen as early as day 3 or as late as 4 to 6 weeks after incubation. There are a few strains, e.g., serovar *hardjo* found in cattle, that are difficult to isolate and maintain in culture media. At the time of optimal growth, the concentration of leptospires may be 1×10^8 to 4×10^8 organisms per ml. Cultures should be examined at 5- to 7-day intervals and discarded if they are negative after 6 weeks of incubation.

Growth in tubes of semisolid medium occurs in the form of a linear disk 1 to 3 cm below the surface. The absence of a disk does not necessarily rule out the presence of leptospires. Fluid media inoculated with leptospires become faintly turbid. For the examination of cultures for growth, a minute drop is obtained from a few centimeters below the surface of fluid cultures or from the ringed area of growth in semisolid media. The drop is placed on a slide, covered with a cover slip, and examined by dark-field microscopy, first at low (×150) and then at high dry (×450) magnification. Leptospires are recognized by their characteristic motility as well as by their morphology. In a fluid medium, leptospires appear to rotate along their longitudinal axis, alternately moving backward and forward without polar differentiation. The motility is characteristic because of the spinning hooked ends. In a semisolid medium or in a more viscous milieu, serpentine as well as boring and flexing movements occur.

Positive cultures should be transferred to fresh medium. The inoculum should make up 5 to 10% of the volume of the subculture. Stock cultures are best maintained in semisolid medium, such as Fletcher medium. After ringed growth occurs at 30°C, cultures can be stored at room temperature. Transfers are usually made at 6- to 8-week intervals. Cultures maintained in fluid medium should be transferred more frequently, at 3- to 6-week intervals. For long-term preservation, cultures of leptospires can be stored by liquid nitrogen refrigeration with the use of glycerol or dimethyl sulfoxide as a cryoprotective agent (3).

Contaminated cultures can be purified by passage in media containing 5-FU or antibiotics with the use of small inocula. If cultures are heavily contaminated, it may be advisable to use three or more serial 10-fold dilutions for the inoculation of subcultures. Cultures can also be purified by filtration through bacteriological filters with average pore sizes ranging from approximately 0.22 to 0.45 μm or by animal inoculation methods. Weanling hamsters or young guinea pigs are inoculated intraperitoneally with 0.5 to 1.0 ml of the contaminated culture. After 10 to 15 min, blood is obtained from the heart and cultured. It may be possible to purify some cultures by subculturing on solid plating medium (26) containing 1% agar, but not all strains readily form colonies in plating medium. Leptospires form subsurface colonies and can be picked for transfer with a sterile capillary pipette.

ANIMAL INOCULATION

Animal inoculation methods are particularly useful for the isolation of strains from tissues or body fluids containing contaminating microorganisms. For material that can be obtained aseptically, animal inoculation techniques provide no greater chances of isolation than direct cultural methods, except for rare instances in which strains are not cultivated easily but can be demonstrated in animals.

The choice laboratory animals for leptospirosis are weanling hamsters and young guinea pigs. *Meriones* species and weanling rabbits have also been used. The laboratory animals selected should be known to be free from natural infections. The course of disease in laboratory animals varies with different serovars or even with different strains of the same serovar and may be inapparent to lethal. Material is inoculated intraperitoneally, preferably in at least three animals. Heart blood for culture and microscope examination is obtained whenever signs of disease are present; otherwise, samples are taken on days 4 and 6 and then at 3- to 4-day intervals up to day 20 after inoculation. Kidney cultures should also be done if animals are alive at the time of the last bleeding.

SEROLOGICAL EXAMINATION

Detection of antibodies

The microscopic agglutination test is the procedure most often used for the detection of antibodies. It is highly sensitive and specific, and it can be used to test both animal and human sera for diagnostic as well as epidemiological purposes. It is highly serovar specific. To ensure the detection of antibodies which may be produced by any of the large number of different serovars, it is necessary to employ a battery of different antigens encompassing cross-reactions with most of the known serovars which may be present. In the continental United States, approximately 23 different serovars have been demonstrated in nonhuman animal hosts. However, relatively few serovars have been associated with infections in humans and domestic animals. At this time, the use of the following eight serovars as antigens would serve to detect all but rare cases of leptospirosis: *copenhageni, grippotyphosa, canicola, wolffi, pomona, djatzi, autumnalis,* and *patoc*. Serovar *patoc*, a biflexa-type leptospire, is used because of the frequent occurrence of cross-reactions with this antigen in sera from human leptospirosis patients. The proposed list of antigens may be modified according to local experience and needs. The substitution of local isolates of the same or a related type could provide a more sensitive test.

Antigens used in microscopic agglutination tests consist of 4- to 7-day-old cultures in fluid medium. The recommended density of cultures is 2×10^8 organisms per ml (28). These organisms may be used live, or they may be treated with Formalin. A detailed description of the conduct of the microscopic agglutination test is given in the *Manual of Clinical Immunology* (1). Generally, test sera are serially diluted two- to fourfold with physiological salt solution to provide serum dilutions from 1:50 to 1:3,200. Amounts of 0.2 ml of each dilution are distributed in a series of agglutination tubes, to each of which is added an equal volume of antigen. The reaction mixtures are shaken, incubated at 30°C for 2 to 3 h, shaken again, and examined microscopically for agglutination by dark-field illumination. The test has been adapted for use with microtitration techniques (7, 21). Titers of 1:100 are considered significant, and they may range as high as 1:25,000 or greater.

The laboriousness of the microscopic agglutination test limits its usefulness for the small diagnostic laboratory. A variety of other serological procedures have been proposed (1, 11, 24). Procedures that have been widely used in lieu of microscopic agglutination tests are the macroscopic (slide or plate) agglutination test with a battery of single or pooled Formalin-fixed antigens (20, 21) and three genus-specific tests in which *L. biflexa* antigens (e.g., strain Patoc) are used in macroscopic agglutination (11, 18), indirect hemagglutination (10), and complement fixation tests (24).

Macroscopic agglutination tests with either single or pooled Formalin-fixed antigens are commercially available from Difco Laboratories, Detroit, Mich., or Fort Dodge Laboratories, Fort Dodge, Iowa. In the genus-specific macroscopic agglutination test (TR/Patoc test), a heat-treated (100°C), concentrated antigen is used. The two types of macroscopic agglutination tests can be used for the serodiagnosis of recent or current infections in humans and animals but have limitations for the detection of antibodies for retrospective studies, e.g., in serological surveys. The hemolytic test entails the use of a 50% ethyl alcohol-soluble, 95% ethyl alcohol-insoluble extract of specific *L. biflexa* strains. This test has been used advantageously in areas of multiple-serovar leptospirosis for the diagnosis of human infections (10). The test has been simplified by the use of sensitized sheep or human type O erythrocytes preserved with glutaraldehyde or pyruvic aldehyde in an indirect hemagglutination procedure (6, 21). A complement fixation test in which a fixed *L. biflexa* antigen is used for the diagnosis of human infections is used in some European countries (24).

Enzyme-linked immunosorbent assay procedures have been advantageously used in several laboratories to detect leptospiral antibodies in humans and animals (11, 22, 23). The technique allows a ready assessment of the relative occurrence of specific immunoglobulins M and G, thereby affording insight on the recency of infection. The procedure of Terpstra et al. (22) entails the use of a genus-specific, heat-extracted, serovar *icterohaemorrhagiae* antigen and appears to have good sensitivity in the detection of leptospiral antibodies, irrespective of the infecting serovar.

Culture typing

Microscopic agglutination techniques are used for culture typing. An isolate employed as antigen is first screened with a select group of 12 or more serovar antisera to determine its serogroup relationship. Isolates are then tested with different serovars of one or more selected serogroups to determine further antigenic relationships. On the basis of observed cross-reactions, representative strains of serovars are chosen for reciprocal agglutinin adsorption tests. "Two strains are considered to belong to different serovars if, after cross-absorption with adequate amounts of heterologous antigen, 10% or more of the homologous titer regularly remains in at least one of the two antisera in repeated tests" (28). Procedures for the conduct of definitive culture typing have been previously described in detail (2, 11). Antigenic-factor analysis has been used for the classification of strains (17), but single-factor sera for such tests are not generally available.

The definitive typing of strains is usually done in leptospirosis reference laboratories. In the United States, such tests are done at the World Health Organization Collaborating Center for the Epidemiology of Leptospirosis, Bacterial Zoonoses Activity, Centers for Disease Control, Atlanta, Ga. Presumptive serogroup or serovar identification can usually be made in diagnostic laboratories by cross-agglutination tests with antisera of strains used for serological diagnosis. Antisera are prepared in rabbits by the injection of successive doses of 0.5, 1.0, 2.0, and 4.0 ml of live cultures into the marginal ear vein at 5- to 7-day intervals. Five- to seven-day-old cultures in rabbit serum medium are commonly used as a source of inoculum. Seven days after the last injection, a blood sample is obtained, and the serum therefrom is tested with homologous antigen. If the titer is 1:6,400 or greater, blood is removed by cardiopuncture. The separated serum is distributed into vials and stored at −20 to −30°C. Alternatively, it can be stored in the freeze-dried state or preserved by the addition of an equal volume of glycerol or by the addition of Merthiolate (concentration, 1:10,000).

DISCUSSION

An unequivocal laboratory diagnosis of current cases of leptospirosis can be established by the isolation of the organism from blood or cerebrospinal fluid or by the demonstration of significant rises in antibody titer in two or more properly timed serum samples.

Direct cultural procedures are relatively simple and within the capabilities of the ordinary diagnostic laboratory. Repeated blood culture attempts during week 1 of disease usually are successful. Preferably, blood cultures should be obtained before antibiotic therapy is started. The use of replicate culture tubes is particularly important. The isolation of leptospires from only one of four or more tubes inoculated is not unusual. The recognition of isolated organisms as leptospires is based on their morphology, motility, and cultural characteristics. The cultural isolation of strains also allows their identification by culture typing tests, which may have epidemiological or forensic importance, as in occupational diseases.

The definitive identification of isolates can be established only by recovery and subsequent typing tests. The microscopic agglutination test, which is highly serogroup and serovar specific, may provide clues to the identity of the infecting serovar. However, the determination of serovars on the basis of the serological response of the patient has the following limitations: the agglutinins may be cross-reacting antibodies initiated by a type not included among the test antigens, higher titers may occur against serologically heterologous but antigenically related strains as well as against unrelated serologically heterologous strains (paradoxical reactions), and complex serological responses may occur in repeated or simultaneous infections.

From the viewpoint of the clinician, the management and treatment of leptospirosis does not depend on the infecting serovar; thus, a laboratory diagnosis of leptospirosis per se serves as confirmation of the clinical diagnosis. In this respect, the macroscopic agglutination and genus-specific test may suffice. Unfortunately, the current serological tests and isolation procedures rarely provide a rapid laboratory confirmation of infection before the end of week 1 of disease. Current methods for the identification of leptospires in blood and other fluids by microscope examination have pitfalls that obviate their usefulness as laboratory confirmatory tests.

LITERATURE CITED

1. **Alexander, A. D.** 1980. Serological diagnosis of leptospirosis, p. 542–546. *In* N. R. Rose and H. Friedman (ed.), Manual of clinical immunology, 2nd ed. American Society for Microbiology, Washington, D.C.
2. **Alexander, A. D., L. B. Evans, A. J. Toussaint, R. Marchwicki, and F. R. McCrumb.** 1957. Leptospirosis in Malaya. II. Antigenic analysis of 110 leptospiral isolates and other serologic studies. Am. J. Trop. Med. Hyg. **6:**871–889.

3. **Alexander, A. D., E. F. Lessel, L. B. Evans, E. Franck, and S. S. Green.** 1972. Preservation of leptospiras by liquid-nitrogen refrigeration. Int. J. Syst. Bacteriol. **22:**165–169.
4. **Alston, J. M., and J. C. Broom.** 1958. Leptospirosis in man and animals. E. and S. Livingstone, Edinburgh.
5. **Babudieri, B.** 1961. Laboratory diagnosis of leptospirosis. Bull. W.H.O. **24:**45–58.
6. **Baker, L. A., and C. D. Cox.** 1973. Quantitative assay for genus-specific leptospiral antigen and antibody. Appl. Microbiol. **25:**697–698.
7. **Cole, J. R., Jr., C. R. Sulzer, and A. R. Pursell.** 1973. Improved microtechnique for the leptospiral microscopic agglutination test. Appl. Microbiol. **25:**976–980.
8. **Cook, J. E., E. H. Coles, F. M. Garner, and L. G. Luna.** 1971. Using scrapings from Formalin-fixed tissues to diagnose leptospirosis by fluorescent-antibody techniques. Stain Technol. **46:**271–274.
9. **Cousineau, J. G., and J. A. McKiel.** 1961. In vitro sensitivity of *Leptospira* to various antimicrobial agents. Can. J. Microbiol. **7:**751–758.
10. **Cox, C. D., A. D. Alexander, and L. C. Murphy.** 1957. Evaluation of the hemolytic test in the serodiagnosis of human leptospirosis. J. Infect. Dis. **101:**210–218.
11. **Faine, S. (ed.).** 1982. Guidelines for the control of leptospirosis. W.H.O. offset publication no. 67. World Health Organization, Geneva.
12. **Hanson, L. E., D. N. Tripathy, L. B. Evans, and A. D. Alexander.** 1974. An unusual leptospira, serovar *illini* (a new serotype). Int. J. Syst. Bacteriol. **24:**355–357.
13. **Hovind-Hougen, K.** 1979. *Leptospiraceae*, a new family to include *Leptospira* Noguchi 1917 and *Leptonema* gen. nov. Int. J. Syst. Bacteriol. **29:**245–251.
14. **Johnson, R. C., and S. Faine.** 1984. Family II. *Leptospiraceae* Hovind-Hougen 1979, 245[AL], p. 62–67. *In* N. R. Krieg and J. G. Holt (ed.), Bergey's manual of systematic bacteriology, vol. 1. Williams & Wilkins, Baltimore.
15. **Johnson, R. C., and P. Rogers.** 1964. 5-Fluorouracil as a selective agent for growth of leptospirae. J. Bacteriol. **87:**422–426.
16. **Kaufman, A. F.** 1976. Epidemiological trends of leptospirosis in the United States, 1965–1974, p. 177–189. *In* R. C. Johnson (ed.), The biology of parasitic spirochetes. Academic Press, Inc., New York.
17. **Kmety, E.** 1967. Factorenanalyse von Leptospiren der Icterohaemorrhagiae und einiger verwandter Serogruppen. Edition of Scientific Committees for General and Special Biology of the Slovak Academy of Science, vol. 13, no. 3. Bratislava, Czechoslovakia.
18. **Mazzonelli, J., G. Dorta de Mazzonelli, and M. Mailloux.** 1974. Possibilité de diagnostic serologique macroscopique des leptospiroses à l'aide d'un antigene unique. Med. Mal. Infect. **4:**253–254.
19. **Myers, D. M.** 1975. Efficacy of combined furazolidone and neomycin in the control of contamination in *Leptospira* cultures. Antimicrob. Agents Chemother. **7:**666–671.
20. **Stoenner, H. G., and E. Davis.** 1967. Further observations on leptospiral plate antigens. Am. J. Vet. Res. **28:**259–266.
21. **Sulzer, C. R., and W. L. Jones.** 1976. Leptospirosis. Methods in laboratory diagnosis (revised edition). Publication no. (CDC) 74-8275. U.S. Department of Health, Education, and Welfare, Washington, D.C.
22. **Terpstra, E. J., G. S. Ligthart, and G. J. Schoone.** 1980. Serodiagnosis of human leptospirosis by enzyme-linked-immunosorbent assay (ELISA). Zentralbl. Bakteriol. Hyg. Abt. 1 Orig. Reihe A **247:**400–405.
23. **Thiermann, A. B., and L. A. Garrett.** 1983. Enzyme-linked-immunosorbent assay for the detection of antibodies to *Leptospira interrogans* serovars *hardjo* and *pomona* in cattle. Am. J. Vet. Res. **44:**884–888.
24. **Turner, L. H.** 1968. Leptospirosis. II. Serology. Trans. R. Soc. Trop. Med. Hyg. **62:**880–899.
25. **Van der Hoeden, J.** 1958. Epizootiology of leptospirosis. Adv. Vet. Sci. **4:**277–339.
26. **Wannon, J. S.** 1958. Isolation of leptospiras from contaminated cultures by plating. Aust. J. Sci. **20:**239.
27. **Wolff, J. W.** 1954. The laboratory diagnosis of leptospirosis. Charles C Thomas, Publisher, Springfield, Ill.
28. **World Health Organization.** 1967. Current problems in leptospirosis. Report of a World Health Organization Expert Group. W.H.O. technical report series, no. 380. World Health Organization, Geneva.

Chapter 43

Borrelia

WILLY BURGDORFER

CHARACTERIZATION

According to *Bergey's Manual of Systematic Bacteriology* (22), spirochetes of the genus *Borrelia* belong to the family *Spirochaetaceae* in the bacterial order *Spirochaetales*. Morphologically, they are helical organisms with 4 to 30 coils and are 5 to 25 μm long and 0.2 to 0.5 μm wide. An outer envelope or membrane encloses the coiled protoplasmic cylinder. The protoplasmic cylinder consists of the peptidoglycan layer and the cytoplasmic membrane enclosing the protoplasmic contents of the cell. Beneath the outer envelope and attached subterminally to opposite ends of the protoplasmic cylinder are the periplasmic flagella (axial filaments). The free ends of the periplasmic flagella extend toward the middle of the cell, where they overlap. Thus, in cross sections, 15 to 22 periplasmic flagella are seen at the terminal portions of the cell, and 30 to 44 periplasmic flagella are present at the middle areas of the cell. Borreliae are highly motile. Multiplication is by binary fission.

Borreliae have an affinity for acid dyes, and they stain with nearly all aniline dyes. They can be demonstrated in tissue sections by silver impregnation techniques. Dark-field microscopy is used for the rapid examination and detection of spirochetes in peripheral blood or in vector tissues.

Borreliae are microaerophilic and require long-chain acids for growth. Glucose is metabolized by the glycolytic pathway, resulting in the accumulation of lactic acid.

The moles percent of guanine plus cytosine (G+C) of the DNA for certain borreliae (*B. hermsii*, *B. turicatae*, and *B. parkeri*) ranges from 28.0 to 30.5. Within the same range are isolates of the recently discovered Lyme disease spirochete for which the name *Borellia burgdorferi* nov. sp. has been proposed (20).

ECOLOGY OF BORRELIAE

All borreliae are arthropod borne; some are pathogenic for humans, rodents, domestic animals, and birds. Those causing relapsing fever in humans are transmitted either by the body louse, *Pediculus humanus humanus*, or by a large variety of soft-shelled ticks of the genus *Ornithodoros*. The etiologic agent of avian borreliosis is transmitted by various species of *Argas* ticks, whereas that of tick spirochetosis in cattle, horses, and sheep is transmitted by ixodid ticks of the genus *Rhipicephalus* and probably by ticks of other genera.

B. burgdorferi, the causative agent of Lyme disease and related disorders, is transmitted by ixodid ticks of the *Ixodes ricinus* complex, namely, *Ixodes dammini* in the northeastern and western United States, *Ixodes pacificus* in the western United States, and *I. ricinus* in Europe (8). Other hematophagous arthropods such as tabanids and mosquitoes may also be involved in the transmission of this agent.

B. recurrentis is the etiologic agent of louse-borne relapsing fever, a disease whose prevalence depends upon ecological and sociological conditions favoring heavy infestations by body lice. Humans are the only reservoir of this agent. Infection in lice is limited primarily to the hemolymph. Thus, transmission does not occur by bite via saliva, but by contamination of the bite wound with the infectious hemolymph of lice that have been smashed or otherwise traumatized by scratching. Borreliae may penetrate normal, unbroken skin.

Borreliae that cause tick-borne relapsing fevers are widely distributed throughout the Eastern and Western Hemispheres. Accordingly, spirochetes causing African, Asiatic, and American relapsing fevers are distinguished from one another (Table 1). There are numerous tick-spirochete associations whose significance in relation to human disease is not known. Vector specificity for *Borrelia* spp. has been reported from several geographical areas. In North America, for instance, *B. hermsii*, *B. turicatae*, and *B. parkeri* are said to be vector specific; i.e., they are maintained and transmitted only by their specific tick vectors, *Ornithodoros hermsi*, *Ornithodoros turicata*, and *Ornithodoros parkeri*, respectively (12). In other areas, such as Iran and Egypt, even local or regional specificity is common. Thus, *Ornithodoros tholozani* from one area in Iran failed to transmit *B. persica* from other parts of that country (3). The mechanism of this vector-spirochete specificity is still conjectural.

The development of borreliae in ticks differs from that in the body louse; after entering the hemocele, spirochetes invade all tissues, including those of the salivary glands and, in many instances, also those of the ovary. The passage of spirochetes via eggs to progeny (transovarial infection) occurs in *Ornithodoros moubata*, *Ornithodoros erraticus* (both varieties), *O. tholozani*, *Ornithodoros tartakowskyi*, *Ornithodoros verrucosus*, *O. turicata*, *O. hermsi*, *Argas persicus*, and *Argas arboreus*, but not in *O. parkeri*, *Ornithodoros talaje*, *Ornithodoros rudis*, *Argas streptopelia*, and *Argas hermanni*.

Transmission of tick-borne spirochetes to a vertebrate host takes place via either infected saliva or infectious body fluids (coxal fluid). Most known *Ornithodoros* and *Argas* vectors are intermittent, mainly nocturnal, fast feeders that live in the soil of rodent burrows (e.g., *O. turicata* and *O. parkeri*), in the crevices of old tree stumps or between the logs of rodent-infested cabins (e.g., *O. hermsi*), and in the dirt floors of native huts (e.g., *O. moubata*) and animal shelters (e.g., *O. tholozani*).

Because *B. burgdorferi* in most of its tick vectors has a distribution limited to the midgut, transmission cannot be via infectious saliva but possibly occurs via infectious midgut suspensions or fecal droplets regurgitated or deposited during the feeding process of a tick.

The main sources for spirochetal infections of arthropods are rodents (rats, mice, and squirrels) and

TABLE 1. Characteristics and distribution of louse- and tick-borne borreliae

Borrelia sp.	Arthropod vector	Animal reservoir	Distribution	Disease
B. recurrentis (syn. *B. obermeyeri, B. novyi*)	*P. humanus humanus*	Humans	Worldwide	Louse-borne, epidemic relapsing fever
B. duttonii	*O. moubata*	Humans	Central, Eastern, and Southern Africa	East African tick-borne, endemic relapsing fever
B. hispanica	*O. erraticus* (large variety)	Rodents	Spain, Portugal, Morocco, Algeria, Tunisia	Hispano-African tick-borne relapsing fever
B. crocidurae, B. merionesi, B. microti, B. dipodilli	*O. erraticus* (small variety)	Rodents	Morocco, Libya, Egypt, Iran, Turkey, Senegal, Kenya	North African tick-borne relapsing fever
B. persica	*O. tholozani* (syn. *O. papillipes, O. crossi?*)	Rodents	From West China and Kashmir to Iraq and Egypt, USSR, India	Asiatic-African tick-borne relapsing fever
B. caucasica	*O. verrucosus*	Rodents	Caucasus to Iraq	Caucasian tick-borne relapsing fever
B. latyschewii	*O. tartakowskyi*	Rodents	Iran, Central Asia	Caucasian tick-borne relapsing fever
B. hermsii	*O. hermsi*	Rodents, chipmunks, and tree squirrels	Western United States	American tick-borne relapsing fever
B. turicatae	*O. turicata*	Rodents	Southwestern United States	American tick-borne relapsing fever
B. parkeri	*O. parkeri*	Rodents	Western United States	American tick-borne relapsing fever
B. mazzottii	*O. talaje (O. dugesi?)*	Rodents	Southern United States, Mexico, Central and South America	American tick-borne relapsing fever
B. venezuelensis	*O. rudis* (syn. *O. venezuelensis*)	Rodents	Central and South America	American tick-borne relapsing fever
B. anserina	*Argas* spp. (mites?)	Fowl	Worldwide	Avian borreliosis
B. theileri	*Rhipicephalus* spp., probably other ixodid ticks	Cattle, horses, sheep	South Africa, Australia	Tick spirochetosis
B. burgdorferi nov. sp.	*I. dammini*	?	Eastern United States	Lyme disease
	I. pacificus	?	Western United States	Lyme disease
	I. ricinus	?	Europe	Lyme disease and related disorders
	?	?	Australia	Lyme disease
B. brasiliensis	*O. brasiliensis*	?	South America (Brazil)	?[a]
B. graingeri	*O. graingeri*	?	East Africa (Kenya)	One laboratory case[a]
B. tillae	*O. zumpti*	Rodents	South Africa	?[a]
B. queenslandica	*O. gurneyi*	Rodents	Australia	?[a]
B. armenica	*O. alactagalis*	Rodents	Armenia	?[a]

[a] Spirochete-tick association of unknown or little human health significance.

lagomorphs. Exceptions include: (i) humans, who appear to be the sole reservoir of *B. recurrentis* and *B. duttonii*, the respective etiologic agents of louse-borne (epidemic) and tick-borne (endemic) relapsing fevers; (ii) birds, the sources of *B. anserina*; and (iii) domestic animals, the reservoirs of *B. theileri*. The sources for the infection of ticks with *B. burgdorferi* are as yet unknown.

CLINICAL SIGNIFICANCE OF BORRELIAE

Relapsing fever in humans

Relapsing fever in humans is a febrile, septicemic disease with sudden onset after an incubation period of 2 to 15 days. Fever persists for 3 to 7 days and is followed by an afebrile interval of several days to several weeks. Thereafter, as many as 10 relapses may occur, especially in untreated patients, as a result of antigenic variations in the causative borreliae. Detailed clinical descriptions of relapsing fever in humans have been presented by Bryceson et al. (6).

According to medical history, more than 50 million persons contracted louse-borne relapsing fever during the first half of this century. Large epidemics have occurred throughout Europe, Africa, Asia, and South America. During World War II, 1 million reported and more than 9 million unreported cases, with a fatality rate of 5%, are said to have occurred.

Since 1967, louse-borne relapsing fever has been reported primarily from African countries, especially Ethiopia (2,278 to 8,700 cases annually) and Sudan, although recent outbreaks have occurred also in South America (World Health Organization Weekly Epidemiological Record, 1967–1978). Information on its prevalence in the USSR and the People's Republic of China is not available.

Because of the sporadic occurrence of tick-borne relapsing fevers, extremely little is known about their incidence worldwide. In Jordan, where tick-borne relapsing fever is caused by *B. persica* transmitted by *O. tholozani*, 723 cases, 4 ending in death, were recorded during 1959–1969 (15). In Rwanda, where *O. moubata* transmits *B. duttonii*, sporadic cases have occurred since 1959. In 1974, however, 103 cases, 2 of them fatal, came to the attention of health authorities and suggested an increase in the incidence in that country (14). Lastly, the microscopic examination of blood during a malarial survey in Iran revealed spirochetes in 13 persons. Most likely, the spirochetes had been transmitted by *O. erraticus* (small form), a vector commonly found in the burrows of gerbils in that area (18).

In the United States during the past 25 years, a total of 382 cases of tick-borne relapsing fever have been recognized (Table 2). Most occurred within the distributional area of *O. hermsi*, and a few occurred also within that of *O. turicata*. Outbreaks usually are sporadic and rarely involve more than two persons. Even so, one outbreak involved 11 Boy Scouts camping near Spokane, Wash. (31), and another outbreak involved 62 employees and tourists at the North Rim of the Grand Canyon in Arizona (5). Tick-borne relapsing fever is underreported in the United States because it is seldom recognized. Most patients are unaware of a tick bite, and unless a history of wilderness exposure, camping, or spending nights in old, rodent-infested cabins is given, the disease is rarely suspected during the initial period of fever (16).

Lyme disease in humans

Lyme disease is an epidemic inflammatory disorder that usually begins in summer with the distinctive skin lesion, erythema chronicum migrans, accompanied by headache, stiff neck, myalgias, arthralgias, malaise, fatigue, or the swelling of lymph nodes. Weeks to months later, some patients may develop meningoencephalitis, myocarditis, or migrating musculoskeletal pain. Still later, patients may develop intermittent attacks of oligoarticular arthritis or chronic arthritis in the large joints, particularly the knees. First described after an outbreak among children in Lyme, Conn., in 1975 (29), the disease appears to be more severe than erythema chronicum migrans, a tick-borne associated syndrome observed as early as 1908 in Europe (1) and for the first time in the United States in 1969 (26). The etiologic agent, *B. burgdorferi*, remained obscure until 1981, when it was discovered in *I. dammini* from New York (10). It is closely related to, if not identical with, isolates of spirochetes from the European tick vector, *I. ricinus* (11). Other clinical syndromes in Europe that appear to be related to the same agent include lymphocytoma (lymphadenosis benigna cutis), acrodermatitis chronica atrophicans, and tick-borne meningoradiculitis (Garin-Bujadoux-Bannwarth's syndrome).

In the United States from 1975 through 1979, about 500 cases of Lyme disease, the majority from the Northeast, were reported. In 1982, as a result of increased surveillance, 487 cases, mostly from the Northeast and Midwest, came to the attention of the Centers for Disease Control in Atlanta, Ga. (25).

Borrelioses in animals

Borrelioses in rodents and lagomorphs may be similar to the diseases observed in humans. There may be one or more relapses accompanied by spirochetemias in various degrees. Not all animals are equally susceptible to the various *Borrelia* species, and even differences in susceptibility to various isolates of a single species of spirochete may be seen. Young animals are generally more susceptible and occasionally may die as the result of infection.

Avian borreliosis, caused by *B. anserina*, affects geese, ducks, turkeys, and chickens in Europe, Siberia, India, Africa, Australia, Indonesia, and South, Central, and North America. In certain countries the disease is of great economic importance because it causes severe losses to the poultry industry. Clinically, the disease begins with a high fever after an incubation period of about 4 days. The birds become cyanotic and have yellowish-green diarrhea. During the early stages of the febrile reaction, spirochetes can be readily detected in the blood. Surviving birds recover after about 2 weeks and have long-lasting immunity.

Tick spirochetosis caused by *B. theileri* in cattle, horses, and sheep in South Africa, and recently also in Australia, is a benign disease characterized by one to two attacks of fever, inappetence, weight loss, weakness, and anemia.

As yet, little is known about the animal reservoirs of *B. burgdorferi* that infect ticks in nature. Spirochetes have so far been recovered from white-tailed deer (*Odocoileus virginianus*), deer mice (*Peromyscus leucopus*), and a raccoon (*Procyon lotor*), suggesting a variety of possible sources (2). In the laboratory, the New Zealand White rabbit is a susceptible experimental host. It responds to the bites of infected ticks with 3 to 8 days of spirochetemias of sufficient concentrations to infect various percentages of normal ticks feeding at that time (9).

DIRECT EXAMINATION

Diagnosis of most borrelioses is based primarily on the detection of spirochetes in the peripheral blood of

TABLE 2. Cases of tick-borne relapsing fever in the United States, 1959–1983[a]

Borrelia sp. and state	No. of cases during: 1959–63	1964–68	1969–73	1974–78	1979–83
B. hermsii					
California	20	13	17	53	58
Arizona	1	0	62	3	4
Oregon	0	14	5	20	20
Washington	0	12	0	4	5
Colorado	0	0	0	4	20
New Mexico	0	0	0	3	3
Idaho	0	0	0	1	3
B. turicatae					
Texas	12	2	1	3	15
Kansas	0	1	0	1	0
Oklahoma	0	0	0	2	0

[a] According to reports from the state health departments.

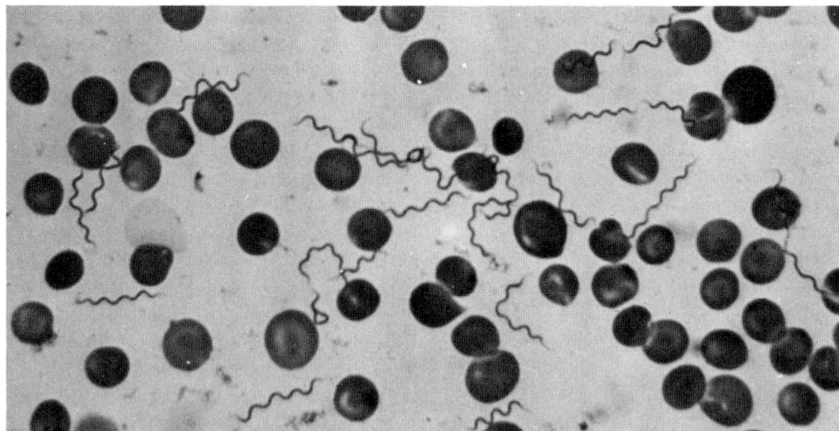

FIG. 1. *B. hermsii* in a thin smear of rodent blood. Giemsa stain. ×1,000.

febrile persons and animals or in the hemolymph and tissues of arthropod vectors (7). In Lyme disease and related disorders, the microscopic detection of spirochetes in the blood of patients and animals has so far not been reported. Direct diagnosis, however, is possible through the demonstration of silver-stained organisms in sections of biopsied skin lesions.

Detection of spirochetes in the blood

During the febrile period, most spirochetes circulate in the blood and can be detected by light or dark-field microscopy of wet preparations made from a drop of blood mixed with a small drop of sodium citrate and covered with a cover slip. Leishman, Giemsa, May-Grünwald, Wright, and other combinations of Romanowsky stains are used to stain thin- and thick-drop films for examination by conventional light microscopy (Fig. 1). For the initial diagnosis, thick films should always be examined, because spirochetemias may be mild and always tend to become milder with each succeeding relapse. A proper thick film is made by placing a drop of blood on a slide and stirring the drop with a toothpick, an applicator stick, or the corner of a microscope slide in a circular motion until it is evenly spread over an area about 1 cm in diameter. This preparation is then air dried for 30 min. During staining, dehemoglobinization renders the smear transparent—a process necessary for the ready detection of blue-stained spirochetes. A microhematocrit concentration technique previously described detects spirochetes more readily in the blood of mildly infected persons (17). The most dependable ways to detect spirochetes in ticks are by the dark-field examination of tissues removed by dissection and by the inoculation of susceptible animals with tissue suspensions.

Detection of spirochetes in arthropod vectors

Spirochetes in the body louse are limited to the hemolymph, which can be readily obtained for microscopic examination by the amputation of the distal portion of one or more legs. The hemolymph test is also useful for the detection of borreliae in large tick vectors, e.g., *O. moubata*, *O. parkeri*, *O. turicata*, *O. erraticus* (large form), and *Argas* spp. The test is not infallible, however, because borreliae are not always abundant in body fluid. This lack of borreliae is particularly true in ticks that have been without a blood meal for a long period. The feeding of infected ticks on experimental animals does not always lead to the transmission of spirochetes. Thus, testing ticks by this method is not recommended unless feeding experiments can be carried out repeatedly. The most dependable ways to detect spirochetes in ticks are by the dark-field examination of tissues removed by dissection and by the inoculation of susceptible animals with tissue suspensions.

ANIMAL INOCULATION

During late relapses of relapsing fever, spirochetes in the blood of a person or an animal may be extremely scarce. Thus, a failure to detect organisms microscopically should never be considered conclusive but should call for the inoculation of susceptible laboratory animals with whole blood or suspensions of triturated blood clots. Suckling Swiss mice and suckling rats are the animals of choice for both louse-borne *B. recurrentis* and tick-borne borreliae. Monkeys, rabbits, and guinea pigs are also useful, depending on the *Borrelia* species involved. The chicken is the animal of choice for *B. anserina*.

Usually, 0.05 to 0.2 ml of a 1:1 dilution of blood in sodium citrate is injected intramuscularly or intraperitoneally into test animals. If the inoculum contains spirochetes, they may be detected in blood preparations as early as 3 days after inoculation. Blood films should be examined daily for at least 14 days before an animal is considered negative.

Microscopically detectable spirochetemias similar to those produced by relapsing fever borreliae in various species of animals have so far not been reported for *B. burgdorferi*.

CULTIVATION OF BORRELIAE

Borreliae can be cultivated readily in their arthropod vectors or in a large variety of vertebrate hosts, particularly small rodents. Cultivation has been successful also in embryonated chicken eggs. A growth medium (Kelly medium) has recently been developed, and so far *B. hermsii*, *B. parkeri*, *B. turicatae*, *B.*

hispanica, and *B. recurrentis* have been cultivated in vitro (21).

Modified versions of Kelly medium have also been successfully used for the cultivation of *B. burgdorferi* from ticks as well as from the blood, skin, and cerebrospinal fluid of Lyme disease patients (4, 27).

SEROLOGICAL DIAGNOSIS OF BORRELIAE

The ability of relapsing fever spirochetes to spontaneously change their antigenic composition makes the serodiagnosis of these diseases difficult and explains the inefficiency and failure of many serological procedures, including agglutination, adhesion, opsonic activity, immobilization, and borreliolysin, in the detection of antibodies. Nevertheless, promising results have been obtained with the indirect immunofluorescence test with cultured spirochetes as antigens. This test may also be used for the characterization of antigenic types of borreliae and for the detection of mixed populations and serotypes of spirochetes (30). *B. burgdorferi*, unlike the relapsing fever spirochetes, is antigenically stable. The indirect immunofluorescence test and the enzyme-linked immunosorbent assay with cultured spirochetes as antigen appear useful in the diagnosis of Lyme disease and related disorders (23). Antigenic cross relationships with other spirochetes, however, may occur, especially in low serum dilutions.

ANTIMICROBIAL AGENT SUSCEPTIBILITY

Although borreliae are susceptible to many antibiotics, tetracyclines are the drugs of choice. They reduce the relapse rate and rid the central nervous system of spirochetes. However, the rapid destruction of organisms may provoke a severe Jarisch-Herxheimer reaction. There is no general agreement about the dosage of the antibiotic and its route of administration. Good results have been obtained with 0.5 g of tetracycline given orally every 6 h for 4 to 5 days or with a single oral dose of 100 mg of doxycycline. To avoid the Jarisch-Herxheimer reaction, a combined penicillin and tetracycline therapy is recommended. It consists of 400,000 U of procaine penicillin administered intramuscularly, to be followed the next day by 500 mg of tetracycline given orally every 6 h for 7 days (24).

Avian borreliosis has been treated successfully with 100,000 U of procaine penicillin given intramuscularly or with oxytetracycline at 2 mg/kg of body weight.

For Lyme disease patients, treatment with 250 mg of oral tetracycline four times a day for at least 10 days and up to 20 days if symptoms persist and recur is recommended. In children, 50 mg of phenoxymethyl penicillin per kg of body weight per day (not less than 1 or more than 2 g/day) in divided doses for the same duration is effective. In case of penicillin allergy, 30 mg of erythromycin per kg of body weight per day in divided doses for 15 or 20 days should be given (28).

IDENTIFICATION OF BORRELIAE

Although DNA base (G+C) content and DNA homology studies provide information about the relation of certain borreliae to other spirochetes of the same genus, these studies are of no value in distinguishing the many relapsing fever spirochetes from one another (19). Similarly, the phenomenon of antigenic phase variations makes a precise serological identification difficult, if not impossible. Therefore, the taxonomic identification of borreliae must take into account the geographical distribution and the natural arthropod vectors.

In certain areas, such as the Western Hemisphere, a specific relationship appears to exist between certain *Borrelia* species and their natural tick vectors. Each species of vector tick maintains and transmits its own spirochete, whose physiological behavior is quite distinct from that of other *Borrelia* species (see above). This phenomenon of specificity has permitted the identification by xenodiagnosis of spirochetes isolated from relapsing fever patients (13). This specificity, however, does not hold true for other areas of the world, and exceptions to it have occurred even in the Western Hemisphere. Thus, for most cases of tick-borne relapsing fevers, taxonomic identification of the etiologic agent is presumptive and is based on a history of exposure to a particular vector.

LITERATURE CITED

1. **Afzelius, A.** 1921. Erythema chronicum migrans. Acta Dermato-Venereol. **2:**120–125.
2. **Anderson, J. F., L. A. Magnarelli, W. Burgdorfer, and A. G. Barbour.** 1983. Spirochetes in *Ixodes dammini* and mammals from Connecticut. Am. J. Trop. Med. Hyg. **32:**818–824.
3. **Baltazard, M., R. Pournaki, and G. Chabaud.** 1954. Sur les fièvres récurrentes à Ornithodores. Bull. Soc. Pathol. Exot. **47:**589–596.
4. **Barbour, A. G., W. Burgdorfer, S. F. Hayes, O. Péter, and A. Aeschlimann.** 1983. Isolation of a cultivable spirochete from *Ixodes ricinus* ticks of Switzerland. Curr. Microbiol. **8:**123–126.
5. **Boyer, K. M., R. S. Munford, G. O. Maupin, C. P. Pattison, M. D. Fox, A. M. Barnes, W. L. Jones, and J. E. Maynard.** 1977. Tick-borne relapsing fever: an interstate outbreak originating at Grand Canyon National Park. Am. J. Epidemiol. **105:**469–479.
6. **Bryceson, A. D. E., E. H. O. Parry, P. L. Perine, D. A. Warrell, D. Vukotich, and C. S. Leithead.** 1970. Louse-borne relapsing fever. A clinical and laboratory study of 62 cases in Ethiopia and a reconsideration of the literature. J. Med. **39:**129–170.
7. **Burgdorfer, W.** 1976. The diagnosis of the relapsing fevers, p. 225–234. *In* R. C. Johnson (ed.), Biology of parasitic spirochetes. Academic Press, Inc., New York.
8. **Burgdorfer, W.** 1984. Discovery of the Lyme disease spirochete and its relationship to tick vectors. Yale J. Biol. Med. **57:**71–76.
9. **Burgdorfer, W.** 1984. The New Zealand White rabbit: an experimental host for infecting ticks with Lyme disease spirochetes. Yale J. Biol. Med. **57:**165–168.
10. **Burgdorfer, W., A. G. Barbour, S. F. Hayes, J. L. Benach, E. Grunwaldt, and J. P. Davis.** 1982. Lyme disease—a tick-borne spirochetosis? Science **216:**1317–1319.
11. **Burgdorfer, W., A. G. Barbour, S. F. Hayes, O. Péter, and A. Aeschlimann.** 1983. Erythema chronicum migrans—a tickborne spirochetosis. Acta Trop. **40:**79–83.
12. **Davis, G. E.** 1942. Species unity or plurality of the relapsing fever spirochetes. Symposium 41. American Association for the Advancement of Science, Washington, D.C.
13. **Davis, G. E.** 1956. The identification of spirochetes from human cases of relapsing fever by xenodiagnosis with comments on local specificity of tick vectors. Exp. Parasitol. **5:**271–275.
14. **deClercq, A. G., A. Z. Meheus, E. de Pierpont, and C.

Nyirashema. 1975. Single-dose doxycycline treatment of tick-borne relapsing fever. East Afr. Med. J. **8:**428–429.

15. **deZulueta, J., S. Nasrallah, J. S. Karam, A. R. Anani, G. K. Sweatman, and D. A. Muir.** 1971. Finding of tick-borne relapsing fever in Jordan by the malaria eradication service. Ann. Trop. Med. Parasitol. **65:**491–495.
16. **Fihn, S., and E. B. Larson.** 1980. Tick-borne relapsing fever in the Pacific Northwest: an underdiagnosed illness? West. J. Med. **133:**203–209.
17. **Goldschmid, J. M., and K. Mahomed.** 1972. The use of the microhematocrit technic for the recovery of *Borrelia duttonii* from the blood. Am. J. Clin. Pathol. **58:**165–169.
18. **Janbakhsh, B., and A. Ardelan.** 1977. The nature of sporadic cases of relapsing fever in Kazeroun area, southern Iran. Bull. Soc. Pathol. Exot. **70:**587–589.
19. **Johnson, R. C., F. W. Hyde, and C. M. Rumpel.** 1984. Taxonomy of the Lyme disease spirochete. Yale J. Biol. Med. **57:**529–537.
20. **Johnson, R. C., G. P. Schmid, F. W. Hyde, A. G. Steigerwalt, and D. J. Brenner.** 1984. *Borrelia burgdorferi* sp. nov.: etiologic agent of Lyme disease. Int. J. Syst. Bacteriol. **34:**496–497.
21. **Kelly, R. T.** 1978. Cultivation and physiology of relapsing fever borreliae, p. 87–94. *In* R. C. Johnson (ed.), Biology of parasitic spirochetes. Academic Press, Inc., New York.
22. **Kelly, R. T.** 1984. Genus IV. *Borrelia* Swellengrebel 1907, 582^{AL}, p. 57–62. *In* N. R. Krieg and J. G. Holt (ed.), Bergey's manual of systematic bacteriology, vol. 1. Williams & Wilkins, Baltimore.
23. **Russell, H., J. S. Sampson, G. P. Smith, H. W. Wilkinson, and B. Plikaytis.** 1984. Enzyme-linked immunosorbent assay and indirect immunofluorescence assay for Lyme disease. J. Infect. Dis. **149:**465–470.
24. **Salih, S. Y., and D. Mustafa.** 1977. Louse-borne relapsing fever. II. Combined penicillin and tetracycline therapy in 160 Sudanese patients. Trans. R. Soc. Trop. Med. Hyg. **71:**49–51.
25. **Schmid, G. P.** 1984. Global distribution of Lyme disease: surveillance in the United States, 1982. Yale J. Biol. Med. **57:**617–618.
26. **Scrimenti, R. J.** 1970. Erythema chronicum migrans. Arch. Dermatol. **102:**104–105.
27. **Steere, A. C., R. L. Grodzicki, A. N. Kornblatt, J. E. Craft, A. G. Barbour, W. Burgdorfer, G. P. Schmid, E. Johnson, and S. E. Malawista.** 1983. The spirochetal etiology of Lyme disease. N. Engl. J. Med. **308:**733–740.
28. **Steere, A. C., G. J. Hutchinson, D. W. Rahn, L. H. Sigal, J. E. Craft, E. T. DeSanna, and S. E. Malawista.** 1983. Treatment of the early manifestations of Lyme disease. Ann. Intern. Med. **99:**22–26.
29. **Steere, A. C., S. E. Malawista, J. A. Hardin, S. Ruddy, P. W. Askenase, and W. A. Andiman.** 1977. Erythema chronicum migrans and Lyme arthritis. The enlarging clinical spectrum. Ann. Intern. Med. **86:**685–698.
30. **Stoenner, H. G., T. Dodd, and C. Larsen.** 1982. Antigenic variation of *Borrelia hermsii*. J. Exp. Med. **156:**1297–1311.
31. **Thompson, R. S., W. Burgdorfer, R. Russell, and B. J. Francis.** 1969. Outbreak of tick-borne relapsing fever in Spokane County, Washington. J. Am. Med. Assoc. **210:**1045–1050.

Chapter 44

Treponema

THOMAS J. FITZGERALD

CHARACTERIZATION

Order *Spirochaetales*, family *Spirochaetaceae*, genus *Treponema*.

The genus *Treponema* contains both pathogenic and nonpathogenic species. At least 10 nonpathogens have been identified as part of the normal flora; they are especially prominent in the oral cavity and are also found in the genital and intestinal tracts. Recently developing evidence suggests that some of these nonpathogenic treponemes are associated with periodontal disease (10). For an extensive reviw of these nonpathogens, see Smibert (13). This chapter focuses on four diseases caused by pathogenic treponemes, collectively termed treponematoses. They are responsible for the human diseases of syphilis (*T. pallidum*), yaws (*T. pertenue*), pinta (*T. carateum*), and endemic syphilis (*T. pallidum* variant?). These treponemal diseases tend to occur in epidemic proportions. Syphilis is found worldwide, yaws is endemic in the tropics, pinta is prevalent in tropical areas of Central and South America, and endemic syphilis occurs in desert regions. Compared to other bacterial infections, treponematoses are very complex. Each exhibits distinct stages of symptomatic manifestations followed by asymptomatic periods. Without antibiotic therapy, these diseases can become chronic and may last for 30 to 40 years.

The newly revised *Bergey's Manual of Systematic Bacteriology* has changed the treponemal nomenclature (14). Treponemes are now divided into the culturable organisms (formerly nonpathogens) and the nonculturable organisms (formerly pathogens). Some species names of the nonculturable treponemes have been altered to reflect their closely similar nature. *T. pallidum* is now termed *T. pallidum* subspecies *pallidum; T. pertenue* is termed *T. pallidum* subspecies *pertenue*; and *T. pallidum* variant (endemic syphilis) is termed *T. pallidum* subspecies *endemicum*. *T. carateum* retains its original name. In this chapter the old terminology of pathogens versus nonpathogens will be used.

The treponemes are gram-negative bacteria that were considered strict anaerobes. It has now been shown that *T. pallidum* takes up oxygen and possesses a functional electron transport system (2); it appears to be microaerophilic (11), as does *T. pertenue* (4). The four species of pathogens are morphologically identical. They are 6 to 15 μm long and 0.1 to 0.2 μm wide (Fig. 1). They exhibit characteristic motility, with rapid rotation about their longitudinal axis and flexing, bending, and snapping about their full length. The ends are usually pointed. Division is by transverse fission. Elongation of organisms and eventual splitting apart account for the variation in length. The ultrastructure of *T. pallidum* and *T. pertenue* has been demonstrated by Hovind-Hougen (5) by transmission electron microscopy. There is an outer membrane (outer envelope) which covers the periplasmic flagella (axial filaments) that wind around the surface of the organism. Three periplasmic flagella arise at each end of the cell and extend halfway down the organism, usually overlapping at midpoint. A cell wall and a cytoplasmic membrane enclose the cytoplasmic contents. Six to eight cytoplasmic tubules are located next to the inner surface of the cytoplasmic membrane. These tubules are attached at each end of the cell and wind around the organism, with the free ends extending toward the midpoint of the cell. The cytoplasm contains ribosomes, mesosomes, and a nuclear region. The susceptibility to penicillin indicates the presence of a peptidoglycan layer. Nonpathogenic treponemes differ in being slightly wider (0.15 to 0.25 μm) and having blunt ends; they also have one to eight periplasmic flagella per cell end and a more structured surface layer.

Studies of DNA homology (8) have demonstrated close similarities between *T. pallidum* and *T. pertenue* and sharp differences between these two pathogens and the nonpathogens. Antigenically, the four pathogens are identical. A species-specific antigen has not yet been identified, although promising results should emerge with the application of newer molecular technology involving polyacrylamide gel electrophoresis. Cross resistance between the pathogens indicates immunologic relatedness (15). In areas where yaws is endemic, the incidence of syphilis is low. After effective campaigns to eradicate yaws, the incidence of syphilis increases dramatically. Similar epidemiologic observations have been made in areas where endemic syphilis is eliminated. Serologic reactions also show immunologic similarities. Both Wasserman and antitreponemal antibodies develop in response to each treponemal disease.

The treponematoses can be differentiated through infections of various laboratory animals. This, however, is a difficult, time-consuming process not readily available to most clinical laboratories. Therefore, the clinical manifestations of the patient are the key to diagnosis. In general, each of the treponematoses can be divided into an early stage (the first 2 years) and a late stage (beyond 2 years). Syphilis is further subdivided into primary, secondary, and tertiary stages. The diagnosis of treponemal disease can be complicated by frequent overlap of the clinical manifestations of early and late stages and also by the widespread occurrence of uncharacteristic lesion appearance. The clinical manifestations of each treponematosis are quite diverse and highly variable. For detailed descriptions of syphilis, yaws, pinta, and endemic syphilis, see Crissey and Denenholz (3), Lomholt (7), and Vegas (16). Comparative generalizations for the treponemal diseases are shown in Table 1, adapted from Vegas (16).

COLLECTION, TRANSPORT, AND STORAGE

Compared with other bacteria, treponemes are very susceptible to environmental influences (15, 17). They are readily inactivated by heat, cold, desiccation,

TABLE 1. Characteristics of the treponematoses[a]

Disease	Agent	Other names	Areas	Predominant age group	Spread by	Congenital infection	Incubation period
Syphilis	*T. pallidum*	Venereal syphilis	Worldwide	Adults	Venereal contact	Yes	10–90 days
Yaws	*T. pertenue*	Frambesia, pian	Tropics	Children	Skin contact	No	14–28 days
Pinta	*T. carateum*	Carate, cute	Tropics, Central and South America	Children, adolescents	Skin contact	No	2–6 mo
Endemic syphilis	*T. pallidum* (variant)	Bjel, dichuchwa	Deserts	Children, adults	Mucous membrane contact	Rarely	?

[a] Common features: highly contagious, generalized infections; regional and general adenopathy; treponemes in the exudate of early lesions; chronic and prolonged; some spontaneous healing; latent stage (except pinta); relatively painless; nonfatal except in some cases of tertiary and congenital syphilis.

most disinfectants, and osmotic changes. *T. pallidum* exhibits relatively narrow optimal ranges of pH (7.2 to 7.4), E_h (−230 to −240 mV), temperature (20 to 35°C), and oxygen concentration (1 to 4%). Because of this delicate nature, it is important to examine lesion material immediately; storage at room or refrigerator temperature even for a few hours is not recommended. Occasionally, tissue samples are required for autopsy (congenital syphilis and tertiary syphilis). These samples should be cut into small fragments (1 mm^3) and stored at −70°C or in liquid nitrogen. Dimethyl sulfoxide (10%) or glycerol (15%) should be used as cryoprotectants. The freezing and subsequent thawing of specimens should be done very rapidly. Repeated freeze-thaw cycles are to be avoided.

Serum or plasma samples for serologic testing are stored at 4°C or at −20°C. A new procedure is especially useful for field surveys which may require prolonged transport to the laboratory (12). A few drops of blood from a finger prick are dropped onto fiber glass disks (the old method used filter paper disks). These disks are dried at room temperature and eluted at the laboratory with phosphate-buffered saline for 2 h. The eluate is then tested for serologic activity. Specimens have been kept for as long as 200 days without significant decreases in antibody titers.

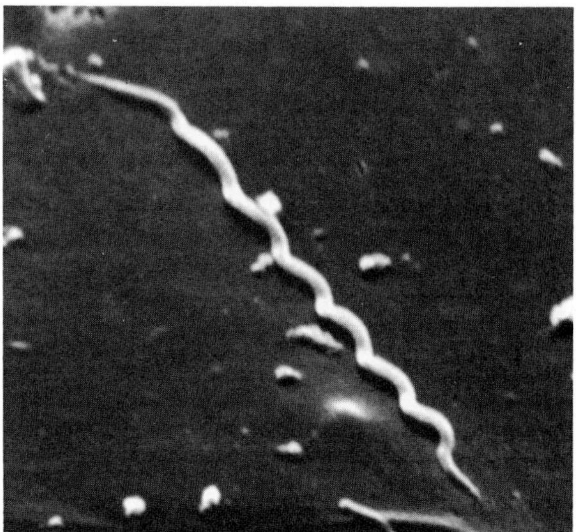

FIG. 1. Scanning electron micrograph of *T. pallidum* (Nichols strain).

DIRECT EXAMINATION

The detection of treponemes within lesion material is a key aid to diagnosis. Although these organisms can be stained, their thinness makes them difficult to see with light microscopy. Dark-field microscopy is recommended. The organisms are white against a black background. Phase-contrast microscopy may also be used, but the organisms are not as readily apparent. Exudates from the lesions of each treponematosis contain both motile and nonmotile treponemes. Depending on the development of the lesion, the numbers may vary from 1 organism per 20 fields to 50 organisms per field. Clinical samples should be obtained before antibiotic therapy since the treponemes are so rapidly cleared.

For most specimens, motile organisms rotate around their longitudinal axis and also bend, snap, and flex along their length. With specimens from well-developed syphilitic lesions, the exudate may contain large amounts of mucoid material. Besides rotating, the organisms in these specimens exhibit a smooth, translational, backward-and-forward movement. This directed motility is also occasionally observed in specimens from yaws lesions.

Samples for dark-field microscopy are obtained in different ways, one of which follows (3). If multiple lesions are present, choose the youngest one available. The chances of visualizing treponemes decrease with increasing lesion development. Clean the surface of the lesion with saline, and blot it dry. Gently remove the crusts if the lesion has already ulcerated. Superficially abrade the lesion until very slight bleeding occurs. The object is to obtain the clear serum exudate from the lesion subsurface. Apply gentle pressure at the base of the lesion and wipe away the first few drops of blood. Touch a glass slide to the clear exudate, place a cover slip on the specimen, and immediately examine the slide by dark-field microscopy. The exudate may also be removed with a capillary pipette and then transferred to a glass slide. If no fluid exudes, a small drop of saline can be added to the lesion.

TABLE 1—*Continued*

Invasiveness	Tissues infected	Predominant cellular infiltrate	Destructive lesions	Granulomas	Gummas	Condyloma lata
High	All	Lymphocytes, plasma cells	Yes	Yes	Yes	Yes
Intermediate	Skin, bones, soft tissues	Mostly plasma cells	Yes	Yes	Yes	Yes
Low	Skin	Mostly lymphocytes	No	No	No	No
Intermediate	Mucous membranes, skin, muscles, bones	Lymphocytes, plasma cells	Yes	Yes	Yes	Yes

Alternatively, lesion material may be aspirated with a 26-gauge needle inserted at the base rather than the center of the lesion. After aspiration, a drop of saline is drawn into the needle, and the material is expressed onto a glass slide for dark-field examination.

Nonpathogenic treponemes may be part of the normal flora of the rectal and oral mucosa. Microscopically, some of these organisms are quite similar to the pathogenic treponemes. Therefore, rectal and oral samples that are dark-field positive must be carefully interpreted.

Accidental infection from laboratory exposure is rare. Nevertheless, the high rates of incidence of the treponematoses indicate their contagious nature. Specimens for dark-field and serologic examination present potential hazards. If the skin of the examiner is exposed, immediate washing with disinfectant is sufficient to prevent infection, and penicillin prophylaxis is not recommended. If the eyes are exposed, immediate, thorough washing with saline followed by penicillin prophylaxis is recommended. Procaine penicillin G, clemizole penicillin, procaine penicillin G with 2% aluminum monostearate, or benzathine penicillin G may be used. In cases of penicillin allergy, tetracycline or erythromycin may be used.

Treponemal lesions frequently exude material that is infectious. For this reason, gloves should always be worn by the examiner. Slides containing dark-field samples, as well as materials used for obtaining these samples, should be placed directly into disinfectant. Most disinfectants rapidly kill treponemes; 70% ethanol is very effective. Freshly isolated serologic specimens may contain infectious organisms. Although these organisms are rapidly inactivated in air, good laboratory procedure dictates sterilization of discarded materials, preferably by autoclaving. After clinical or serologic specimens are handled, the hands should be washed with soap or disinfectant, and the tabletop should be swabbed with disinfectant.

CULTURE AND ISOLATION

Since the four pathogenic treponemes cannot be grown in vitro, little is known about their cultural characteristics, biochemical reactions, metabolic activities, and chemical composition. The organisms can be maintained in vitro for 1 to 6 days after extraction from infected tissues. Maintenance media such as Nelson medium contain rabbit serum, reducing agents (cysteine, thioglycolate, and glutathione), vitamins, cofactors, amino acids, and salts. A variety of cell culture media such as Eagle or McCoy medium supplemented with fetal bovine serum and reducing agents may also be used for maintenance.

Animal inoculation is not routinely performed for patient diagnosis. When necessary, the rabbit is the animal of choice for *T. pallidum*. Primary and secondary infections are quite similar to human infections. The rabbit or the hamster is recommended for *T. pertenue* and *T. pallidum* (endemic syphilis). It is difficult to infect animals with *T. carateum*; minimal success has been achieved with chimpanzees. In places where two treponematoses are prevalent, differentiation can be extremely difficult because of the overlap in clinical manifestations. In this situation, various laboratory animals are inoculated. Comparisons of lesion development then can be used to identify the specific treponemal disease.

Suspected treponemal material for animal inoculation may be contaminated with other microorganisms. To bypass this obstacle, the material is injected intratesticularly. The treponemes rapidly spread to the draining popliteal lymph nodes. Days or weeks later, the nodes can be removed and injected intratesticularly into other animals to produce uncontaminated treponemal preparations. Animal quarters temperature is important to lesion development (15). Optimal infection occurs at 16 to 18°C. Higher temperatures decrease the intensity of the infection; temperatures below 16°C are detrimental (rabbits develop higher rates of respiratory infections, and hamsters tend to hibernate).

IDENTIFICATION

Definitive diagnosis of the treponematoses can be difficult because the four pathogenic *Treponema* species are morphologically identical and immunologically closely related. A species-specific antigen has not been isolated, and serologic reactions for each disease are quite similar. There are no differential biochemical reactions. The four criteria used for diagnosis are geographical area, clinical manifestations, dark-field specimens, and serologic reactions.

The geographical distributions of the four treponematoses are listed in Table 1. The location of a patient within areas in which the disease is endemic is a key initial observation.

Characteristic clinical manifestations of the treponematoses vary greatly. Distinct overlap occurs: late syphilis mimics late yaws; gummas of syphilis, yaws, and endemic syphilis are indistinguishable; condyloma lata of syphilis, yaws, and endemic syphilis are identical; and skin lesions of syphilis, yaws, and endemic syphilis are similar. Despite these potential problems, clinical manifestations usually differ sufficiently to prevent confusion.

The importance of the demonstration of treponemes in lesion material has been stressed. Treponemes are readily observed in lesion exudates in each of the treponemal diseases. It is difficult, however, to observe treponemes in specimens from dry lesions not exuding material.

Serology is a critical aspect in the diagnosis of treponemal disease. Two different types of tests are used. Nontreponemal tests detect Wasserman or reaginic antibodies; these are totally different from the reaginic antibodies associated with immunoglobulin E (IgE) in allergy. A few examples of the nontreponemal tests include the Venereal Disease Research Laboratory, rapid plasma reagin, automated reagin, Kahn, plasmacrit, Hinton, and Kline tests. Treponemal tests detect antibodies specific for treponemal antigens; they include the fluorescent treponemal antibody absorption (FTA-ABS), *T. pallidum* immobilization, and hemagglutination tests. Because *T. pallidum, T. pertenue*, and *T. carateum* are antigenically similar and elicit both nontreponemal and treponemal antibodies, it is not possible to distinguish these organisms or their diseases serologically.

Depending on the stage of the disease, quantitative and qualitative serology varies. Nontreponemal tests usually parallel the infection. Titers are high during clinical infection and then greatly decrease either during subclinical infection (latency) or after effective antibiotic therapy. In contrast, the treponemal tests may not become positive until well after the initial clinical manifestations; titers may remain high in latency and after treatment. Thus, it is recommended that nontreponemal tests be used for diagnosis in the early stages, treponemal tests be used for the diagnosis of latency, and nontreponemal tests be used for the evaluation of cure after antibiotic therapy (6 to 12 months beyond treatment). For an in-depth discussion of serologic testing, see Miller (9).

Two terms relevant to serologic testing are sensitivity and specificity. The perfect test, not yet developed, would detect 100% of the treponemal infections and would be nonreactive in all other diseases. Sensitivity refers to negative serology in patients with treponemal disease. Biological false-negatives may occur early in the infection before antibodies have time to develop. Specificity refers to positive serology in the absence of treponemal disease. Biological false-positives frequently occur in patients with chronic diseases such as leprosy, malaria, collagen disorders, and autoimmune problems.

The most recent recommendation by the World Health Organization (1) is to screen sera with a Venereal Disease Research Laboratory, rapid plasma reagin, or automated reagin test and to confirm positive sera with the FTA-ABS test. A *T. pallidum* immobilization test will definitively determine equivocal reactions. The immobilization test, however, is not recommended except in extreme cases since it requires a viable source of freshly harvested treponemes and the maintenance of a large colony of rabbits.

Two new developments are noteworthy. Microhemagglutination tests have been developed that are rapid and easy to read and have good sensitivity and specificity. Two similar commercial preparations are available. One test uses sensitized sheep erythrocytes (Fujizoki Laboratories, Tokyo, Japan), and the other uses sensitized turkey erythrocytes (Difco Laboratories, Detroit, Mich.). These tests are promising and may eventually supplant the Venereal Disease Research Laboratory, rapid plasma reagin, and automated reagin tests for screening sera.

Another recent development involves the FTA-ABS test (6). This test requires a fluorescence microscope with a dark-field condenser. Suspected serum is added to a commercial preparation of treponemal organisms that have been dried and fixed on a slide. Fluorescein-conjugated antibody to human IgG is then added. Dark-field microscopy is used to locate the treponemes. Fluorescence microscopy is then used to determine whether the treponemes specifically fluoresce. Problems occur with negative serum. The switches between dark-field and fluorescence microscopy frequently require realignment and refocusing of the microscope and thus cause delays during quantitative measurements. A new refinement incorporates the incident illumination of Ploem and a double FTA-ABS reaction. All treponemes are nonspecifically stained with rhodamine-labeled anti-*T. pallidum* globulin. The fluorescein-labeled antibody to human IgG is then added. Selective emission filters differentiate between rhodamine- and fluorescein-stained organisms. This technique eliminates the need for dark-field examination with subsequent errors related to poor focusing.

Quality control is crucial to serologic testing. Controls should be routinely used that include both positive sera of known titers and negative sera that do not react. In addition, the FTA-ABS test requires subjective evaluations of fluorescence which are especially troublesome with borderline sera. Occasionally, reagents are defective, improperly reconstituted, or old. The appropriate controls will detect these problems.

Two final comments are pertinent to treponemal antibody detection. Cerebrospinal fluid should be tested to diagnose symptomatic neurosyphilis or to differentiate between asymptomatic neurosyphilis and latency. Congenital syphilis is difficult to diagnose. Maternal IgG passes the placenta and enters the fetal circulation. At birth, the baby will be serologically positive. Maternal antibodies usually disappear within 2 to 3 months. Quantitative Venereal Disease Research Laboratory or rapid plasma reagin tests should be performed monthly over a period of 6 months. If the titer either increases or stabilizes and does not decrease, congenital syphilis is indicated. Some attempts have been made to detect IgM against *T. pallidum* in a modified FTA-ABS test. A positive reaction at birth would then indicate infection. These tests, however, are still suboptimal and not routinely used.

Serological tests for syphilis are covered in greater detail in Chapter 94. An updated review of syphilis by Crissey and Denenholz (3) has just been published. Many of the concepts discussed in this chapter are also discussed in their excellent review.

LITERATURE CITED

1. **Antal, G. M.** 1979. Present status of therapy and serodiagnosis of syphilis (some selected aspects). W.H.O. document W.H.O./V.D.T./Res. **70**:359.
2. **Cox, C. D., and M. K. Barber.** 1974. Oxygen uptake by *Treponema pallidum*. Infect. Immun. **10**:123–127.
3. **Crissey, J. T., and D. D. Denenholz.** 1984. Syphilis. Clin. Dermatol. **2**:1–166.
4. **Fieldsteel, A. H., J. G. Stout, and F. A. Becker.** 1979. Comparative behavior of virulent strains of *Treponema pallidum* and *Treponema pertenue* in gradient cultures of various mammalian cells. Infect. Immun. **24**:337–345.
5. **Hovind-Hougen, K.** 1976. Determination by means of electron microscopy of morphological criteria of value for classification of some spirochetes, in particular treponemes. Acta Pathol. Microbiol. Scand. Suppl. **255**:1–41.
6. **Hunter, E. G., R. M. McKinney, S. E. Maddison, and D. D. Cruce.** 1979. Double-staining procedure for the fluorescent treponemal antibody absorption (FTA-ABS) test. Br. J. Vener. Dis. **55**:105–108.
7. **Lomholt, G.** 1972. Textbook of dermatology, vol. 1, p. 634–679. Blackwell Scientific Publications, Ltd., Oxford.
8. **Miao, R. M., and A. H. Fieldsteel.** 1980. Genetic relationship between *Treponema pallidum* and *Treponema pertenue*, two noncultivable human pathogens. J. Bacteriol. **141**:427–429.
9. **Miller, J. N.** 1975. Value and limitations of nontreponemal and treponemal tests in the laboratory diagnosis of syphilis. Clin. Obstet. Gynecol. **19**:191–203.
10. **Moore, W. E. C., L. V. Holdeman, R. M. Smibert, D. E. Hash, J. A. Burmeister, and R. R. Ranney.** 1982. Bacteriology of severe periodontitis in young adult humans. Infect. Immun. **38**:1137–1148.
11. **Norris, S. J., J. N. Miller, J. A. Sykes, and T. J. Fitzgerald.** 1978. Influence of oxygen tension, sulfhydryl compounds, and serum on the motility and virulence of *Treponema pallidum* (Nichols strain) in a cell-free system. Infect. Immun. **22**:689–697.
12. **Paris-Hamelin, A., G. Causse, A. Vaisman, S. Fuster-Ibarboure, and N. Tordjman.** 1979. Utilization of fiberglass discs to collect specimens for the serodiagnosis of syphilis by the fluorescent antibody and passive haemagglutination tests. W.H.O. Document W.H.O./V.D.T./Res. **79**:360.
13. **Smibert, R. M.** 1977. CRC handbook of microbiology, 2nd ed., vol. 1, p. 195–228. CRC Press, Inc., Cleveland.
14. **Smibert, R. M.** 1984. Genus III *Treponema* Schaudinn 1905, 1728^{AL}, p. 49–57. *In* N. R. Krieg and J. R. Holt (ed.), Bergey's manual of systematic bacteriology, vol. 1. Williams & Wilkins, Baltimore.
15. **Turner, T. B., and D. H. Hollander.** 1957. Biology of the treponematoses. W.H.O. Monogr. Ser. **35**:1–277.
16. **Vegas, F. K.** 1975. Clinical, tropical dermatology, p. 79–105. Blackwell Scientific Publications, Ltd., Oxford.
17. **Willcox, R. R., and T. Guthe.** 1966. *Treponema pallidum*. A bibliographical review of the morphology, culture and survival of *T. pallidum* and associated organisms. W.H.O. Suppl. **35**:1–169.

Chapter 45

Anaerobic Spirochetes

ROBERT M. SMIBERT

CHARACTERIZATION

The cultivable anaerobic treponemes found in the oral and genital flora of humans and the intestinal and rumen flora of animals will be discussed along with isolation methods and identification procedures.

The anaerobic host-associated treponemes are helically coiled organisms that have one or more periplasmic flagella. The protoplasmic cylinder is covered by an outer membrane. Periplasmic flagella are anchored at each end of the cell and wind around the protoplasmic cylinder within the periplasmic space.

Growth of these organisms requires good anaerobic techniques and the addition of proper nutritional supplements to culture medium. Some treponemes require the medium to be supplemented with 10% inactivated animal serum. Rabbit serum is usually used. Organisms needing serum are *Treponema denticola, T. vincentii, T. phagedenis, T. scoliodontum, T. refringens, T. minutum, T. hyodysenteriae,* and *T. innocens*. In addition to serum, the oral treponemes *T. denticola, T. vincentii,* and *T. scoliodontum* require thiamine pyrophosphate (TPP) (4, 18). The genital organism *T. phagedenis* requires glucose in addition to serum. Other treponemes require both short-chain, volatile fatty acids found in rumen fluid and a fermentable energy source. A medium can be supplemented with 20 to 30% clarified rumen fluid or a fatty acid mixture (5). These organisms requiring short-chain fatty acids are *T. socranskii* (13), *T. pectinovorum* (12), *T. bryantii,* and *T. succinifaciens*. Except for *T. pectinovorum,* these organisms require glucose or other carbohydrate as an energy source. *T. pectinovorum* needs pectin; either glucuronic or galacturonic acid can replace pectin.

The G+C ratio of treponemal DNA ranges from 25 to 52 mol% (11, 13). The organism causing the human treponematoses, *T. pallidum,* is reported to be microaerophilic, whereas the host-associated treponemes found in the normal flora of humans and animals are anaerobes. Most of these organisms, however, have one or more antigens in common with *T. pallidum*.

The host-associated treponemes are found in the normal genital flora of humans and animals, in the rumen of ruminants, and in the intestinal flora of animals. Treponemes are usually not part of the normal intestinal flora of humans. They are also usually not found in the healthy periodontal flora of humans, but they are present in the periodontal flora of people with gingivitis, acute necrotizing ulcerative gingivitis, and periodontal disease.

The anaerobic treponemes are host-associated organisms that belong to the genus *Treponema*. Those organisms belonging to the genus *Spirochaeta* are free living, whereas *Borrelia* species are transmitted by an arthropod vector.

Interest in the anaerobic treponemes has increased in the last few years. The main area of interest is the role of treponemes in periodontal disease in humans. In veterinary microbiology, interest is in *T. hyodysenteriae,* the cause of swine dysentery. To date, this organism is the only known pathogen among the anaerobic treponemes. There are also reports of treponemes in dogs with possible intestinal problems (9) and in humans with intestinal disease (6, 17). A new organism, *Brachyspira aalborgii,* was isolated from rectal biopsies from human patients (6). In rumen microbiology, interest in these organisms involves their role in the physiology of ruminants.

COLLECTION, TRANSPORT, AND STORAGE OF SPECIMENS

Collection

Supragingival periodontal samples can be taken with a sterile dental scaler such as a Morse scaler with a detachable tip or paper points. Subgingival samples from periodontal pockets should be taken with great care to prevent sample contamination with the supragingival flora. Supragingival plaque should be cleaned off of the tooth surface, and the subgingival samples should be taken with either a scaler or paper points. The samples should be placed in anaerobic dilution broth, and the sample should be dispersed. Sonic oscillation should not be used to disperse the organisms in dental samples because treponemes are destroyed by brief sonication. The samples can be dispersed by a few seconds of vortexing in the presence of small glass beads (8).

Intestinal and rumen samples are diluted in anaerobic dilution broth and dispersed by brief vortexing or shaking. Ten-fold serial dilutions of the sample may be made if desired. Each dilution of the sample should be examined by dark-field microscopy for the presence of treponemes.

Samples from pigs with swine dysentery are usually taken from the large colon of dead pigs at necropsy. Intestinal material is gently removed from the intestinal lesion, and the surface of the lesion is washed with sterile saline. The lesion is scraped with a sterile instrument, and the resulting colonic mucosal sample is placed in anaerobic dilution broth or streaked directly onto selective isolation medium. Upon dark-field examination of the sample material, a large treponeme should be seen.

Genital samples can be collected by swabbing the genital region with sterile, moistened swabs. The swabs are placed into anaerobic dilution broth or placed directly onto selective isolation medium.

Transport

Samples in dilution medium or selective broth should be transported at ambient temperature to the laboratory and inoculated onto culture media within 5 to 12 h after collection. Fecal, intestinal, and rumen samples may be placed in a container and transported to the laboratory. They should be cultured within 5 h after collection.

Storage

Intestinal, rumen, fecal, and oral samples may be preserved in liquid nitrogen or at −85°C in a mechanical freezer. Whole sections of intestines from pigs with swine dysentery may be cut up and frozen for future use. Isolated cultures of treponemes are preserved in liquid nitrogen or at −85°C. The culture should contain dimethyl sulfoxide.

DIRECT EXAMINATION

The best method of direct examination of samples for treponemes is by dark-field microscopy. A phase-contrast microscope may also be used; however, dark-field microscopy is the preferred method.

CULTURE AND ISOLATION TECHNIQUES

Two methods can be used for the isolation of treponemes. In one of these methods, antibiotics are used to make the medium selective, and in the other method, membrane filters are used to achieve a mechanical selection of treponemes. Culture and isolation media are listed below. Antibiotic-supplemented selective media for treponemes in oral, intestinal, rumen, and genital samples contain rifampin (7, 15) or rifampin and polymyxin (8, 13).

Antibiotic method

Various dilutions of samples are inoculated into rifampin or rifampin-polymyxin broth or semisolid medium. The cultures are incubated at 37°C for 1 to 2 weeks. Cultures should be examined by dark-field microscopy for treponemes after 1 week of incubation. Supplements in the basic medium will determine which kinds of treponemes are isolated. If only serum or serum and TPP are added, then only serum-requiring treponemes will be selected for and isolated. If only rumen fluid, glucose, and pectin are in the medium, then only treponemes requiring short-chain fatty acids will be selected for and isolated. A general-purpose isolation and culture medium that supports the growth of most described treponemes should contain rumen fluid (fatty acid mixture), serum, TPP, glucose, and pectin.

Cultures containing treponemes are transferred to rifampin-polymyxin broth or semisolid medium. If a culture is contaminated with another kind of bacterium, the treponeme can be separated from the contaminant either by streaking onto a selective agar plate or by using a parabiotic chamber (Bellco Glass, Inc.). The Bellco chamber (no. 1945) with rubber stoppers is put together with a membrane filter (25-mm diameter) with a pore size of either 0.2 or 0.3 μm (4, 8, 10). The contaminated culture is placed into one side of the chamber, and sterile medium is placed into the other side. The chamber should be gassed with oxygen-free nitrogen. After 12 to 18 h of incubation, the motile treponemes migrate through the filter and grow in the medium. The purified culture is inoculated into fresh rifampin-polymyxin broth.

Membrane filter method

Membrane filters (45-mm diameter) with a pore size of 0.2 or 0.3 μm are placed onto the surface of agar isolation medium. The isolation medium may contain antibiotics such as rifampin and polymyxin. A sterile O-ring is placed on top of the filter and sealed to the filter with 3% molten agar in water. The sample, from various dilutions, is placed on the membrane filter, and the plate is quickly placed in an anaerobic jar. The culture is incubated at 37°C for 1 to 2 weeks. After incubation, the filter and the O-ring are carefully removed, and the white, hazy growth in the agar is checked by dark-field microscopy for treponemes. A plug of the hazy growth is removed with a Pasteur pipette and transferred to broth or semisolid medium.

Single-colony isolation of treponemes is necessary for the identification of the organisms. Colonies can be obtained by quickly streaking agar plates with 2- to 4-day-old cultures. The plates are incubated in anaerobic jars at 37°C for 1 to 2 weeks. Two anaerobic jars are used. One jar contains prereduced agar plates, and the other jar is used for the streaked plates. Both jars are gassed with carbon dioxide while they are open. The jar containing the streaked plates should be made anaerobic as quickly as possible with hydrogen and carbon dioxide or with a GasPak envelope (BBL Microbiology Systems). After appropriate incubation (1 to 2 weeks), the white colonies in the agar are plugged with a Pasteur pipette and subcultured in broth or semisolid medium. Cultures in broth containing dimethyl sulfoxide can be preserved in liquid nitrogen or at −85°C. Working cultures should be transferred weekly.

The growth of treponemes on agar plates requires a soft agar medium. A medium with 0.85% Oxoid purified agar (Oxoid USA, Ltd., Columbia, Md.) is recommended. This concentration of agar is the equivalent of 1.2 to 1.3% Bacto-Agar (Difco Laboratories).

To isolate *T. hyodysenteriae*, the colonic mucosal sample is streaked onto TSA-5400 agar plates containing spectinomycin as the selective agent (14). The cultures are incubated in anaerobe jars for 2 to 6 days. The selective plates are examined for areas of beta-hemolysis that show little or no surface growth in the hemolytic zone. A hemolytic zone is scraped with a loop and either streaked onto Trypticase soy agar (BBL) plates containing 5 to 7% citrated bovine blood or inoculated into Trypticase soy broth (BBL) containing 10% serum or bovine fetal calf serum. The plates and tubes are incubated in an anaerobe jar for 3 to 5 days. A weekly transfer of cultures is required. Cultures of *T. hyodysenteriae* in broth containing dimethyl sulfoxide can be preserved in liquid nitrogen.

Other cultural methods

An older isolation method is the well-plate technique. Agar medium is poured into a petri dish to make a thick plate 12 to 14 mm deep. Thicker plates can be made with deeper dishes, and selective agar medium can be used. A well 3 to 5 mm deep is made in the agar with a Pasteur pipette. Any fluid in the well should be removed. The inoculum is placed in the well carefully to avoid contamination of the upper edge of the well. The plates are incubated at 37°C in an anaerobe jar. After 1 to 2 weeks of incubation, the plates are removed from the jar and examined for a circle of white, hazy growth. An agar plug of the hazy growth is removed and inoculated into broth medium (10).

TABLE 1. Some characteristics of serum-requiring treponemes[a]

Species	Cell width (μm) 0.15 to 0.25	>0.36	Fructose	Glucose	Lactose	Maltose	Mannitol	Sucrose	Ribose	H_2S	Indole	Esculin hydrolysis	Phosphatase
T. denticola													
bv. *denticola*	+	−	−	−	−	−	−	−	−	+	+	+	+
bv. *comondonii*	+	−	−	−	−	−	−	−	−	+	−	+	+
T. vincentii	+	−	−	−	−	−	−	−	−	+	+	V	−
T. scoliodontum	+	−	−	−	−	−	−	−	−	−	−	−	−
T. phagedenis													
bv. Reiter	+	−	+	+	+	−	+	+	+	V	+	−	+
bv. Kazan	+	−	+	+	+	−	+	+	+	V	+	+	+
T. refringens													
bv. *refringens*	+	−	−	−	−	−	−	−	−	+	+	+	−
bv. *calligyrum*	+	−	−	−	−	−	−	−	−	+	+	+	−
T. minutum	+	−	−	−	−	−	−	−	−	+	+	+	−
T. hyodysenteriae	−	+	−	V	−	+	−	−	−	−	+	+	
T. innocens	−	+	+	V	−	−	−	−	−	−	−	+	

[a] Reactions: +, positive; −, negative; V, variable; −/w, negative or weak; blank, reaction not known.

In another method, 10-fold dilutions of a sample are inoculated into molten agar deeps (16). The anaerobic medium should contain 0.7% agar. Selective medium can also be used. After incubation, treponeme colonies appear as fluffy white balls in the medium. The soft agar column is removed from the culture tube, and treponeme colonies are picked and subcultured.

Special equipment

The culture and identification of treponemes require good anaerobic methods that can be achieved with the VPI anaerobic system (Bellco) or an anaerobic chamber. The identification of treponemes requires a gas chromatograph to determine the end products of their fermentation.

IDENTIFICATION

Examine a pure culture of treponemes by dark-field microscopy. Estimate the length and diameter of the cells, and determine whether the ends are pointed, blunt, straight, or hooked. The treponemes can be divided into two convenient groups for identification. The first group consists of those species that require serum for growth. The other group is made up of those species that require short-chain fatty acids (rumen fluid) for growth. Table 1 shows the identifying characteristics of serum-requiring treponemes, whereas Table 2 shows the identifying characteristics of fatty acid-requiring organisms. Additional information on the characterization of treponeme species can be found in *Bergey's Manual of Systematic Bacteriology* (11) and other references (6, 12, 13).

Susceptibility to antimicrobial agents

With a broth tube dilution method, *T. phagedenis*, *T. refringens*, *T. denticola*, and *T. vincentii* are inhibited by penicillin (0.1 to 1 U/ml) and by (micrograms per milliliter) ampicillin (0.1 to 1), oxicillin (0.1 to 10), cloxacillin (0.1 to 1), cephalothin (0.1 to 10), vancomycin (0.1 to 10), bacitracin (0.1 to 1), erythromycin (0.1 to 1), novobiocin (1 to 500), tetracycline (1), oxytetra-

TABLE 2. Some characteristics of treponemes requiring short-chain fatty acids[a]

Species	H_2S	Arabinose	Cellobiose	Fructose	Galactose	Glucose	Lactose	Maltose	Mannose	Pectin	Rhamnose
T. socranskii[b]											
subsp. *socranskii*	+	+	−	+	+	+	−	+	+	V	+
subsp. *buccale*	+	+	−	+	V	+	−	+	V	V	V
subsp. *paredis*	+	−	−	+	+	+	−	+	+	V	−
T. pectinovorum	+	−	−	−	−	−	−	−	−	+	−
T. succinifaciens		+	+	−	+	+	+	+	+	−	−
T. bryantii		+	+	−	+	+	+	−	+	−	−

[a] Reactions: +, positive; −, negative; V, variable (less than 70% of the strains are positive); blank, reaction unknown. Positive test with carbohydrates is a pH of less than 6.0 after 5 days of incubation.

[b] *T. socranskii* subsp. *socranskii* and *T. socranskii* subsp. *buccale* can be differentiated only by serologic tests.

TABLE 1—*Continued*

		Source			Major end product									
1% Glycine growth	Beta-hemolysis	Oral	Genital	Intestinal	Acetate	Formate	Propionate	*n*-Butyrate	Lactate	Succinate	Ethanol	Propanol	Butanol	Mol% G+C
V	–	+	–	–	+	–	tr	tr	–	–	–	–	–	37–38
V	–	+	–	–	+	–	tr	tr	–	–	–	–	–	37–38
–	–	+	–	–	+	–	–	+	–	–	–	–	–	44
–	–	+	–	–	+	–	–	tr	–	–	–	–	–	
+	–	–	+	–	+	+	+	+	–	–	+	+	+	38–39
+	–	–	+	–	+	+	+	+	–	–	+	+	+	38–39
–	–	–	+	–	+	–	–	–	–	–	–	–	–	39–43
+	–	–	+	–	+	–	–	–	–	–	–	–	–	39–43
+	–	–	+	–	+	+	–	–	–	–	–	–	–	37
–	+	–	–	+	+	–	–	+	–	–	–	–	–	25–26
–	–/w	–	–	+	+	–	–	+	–	–	–	–	–	25–26

[a] Reactions: +, positive; –, negative; V, variable; –/w, negative or weak; blank, reaction not known.

cycline (1 to 10), doxycycline (0.1 to 1), chloramphenicol (100 to 500), kanomycin (100 to 1,000), gentamicin (1 to 500), and viomycin (10 to 1,000). All are resistant to (micrograms per milliliter) cycloserine (500 to 1,000), polymyxin B (500 to 1,000), nalidixic acid (500 to 1,000), and methenamine mandelate (500 to 1,000). Treponemes that require short-chain fatty acids and are isolated from the human oral cavity and the swine intestinal tract are usually inhibited by higher concentrations of antimicrobial agents than are the serum-requiring organisms. The oral treponemes that require short-chain fatty acids are inhibited by penicillin (10 to 100 U/ml) and by (micrograms per milliliter) ampicillin (10 to 100), oxacillin (100 to 500), cephalothin (0.1 to 1), erythromycin (0.01), vancomycin (1 to 10), tetracycline (1 to 10), chloramphenicol (100), and viomycin (100 to >1,000). The swine intestinal organisms are inhibited by penicillin (1 to 10 U/ml) and by (micrograms per milliliter) ampicillin (0.1 to 100), oxacillin (1 to 10), cephalothin (0.1 to 10), erythromycin (0.1), tetracycline (100 to 500), and chloramphenicol (10 to 100) (1). Bactericidal concentrations of antimicrobial agents are usually 10 to 100 times higher than the inhibitory concentrations (2). Oral and rumen treponemes are resistant to rifampin (1 to 50 μg/ml) (7, 15). A broth-disk method for the studying of the antibiotic sensitivity of treponemes has been reported (3).

Evaluation

Except for *T. hyodysenteriae*, there is no proof that the species listed in this chapter are pathogenic.

MEDIA

Treponeme isolation medium

See references 8, 12, and 13.

Polypeptone (BBL), 5 g; heart infusion broth base (BBL), 5 g; yeast extract (Difco), 5 g; ribose, 0.8 g; pectin, 0.8 g; glucose, 0.8 g; fructose, 0.4 g; starch, 0.8

TABLE 2—*Continued*

					Source			Major end product					
Ribose	Starch	Sucrose	Trehalose	Xylose	Oral	Intestinal	Rumen	Acetate	Formate	Lactate	Propionate	Succinate	Mol% G+C
+	+	+	V	+	+	–	–	+	–	+	–	+	50
+	+	+	+	+	+	–	–	+	–	+	–	+	50
+	+	+	V	V	+	–	–	+	–	+	–	+	50
–	–	–	–	–	+	–	–	+	+	tr	–	–	39
+	+	–	–	+	–	+	–	+	+	+	–	+	36
–	–	+		+	–	–	+	+	+	–	–	+	36

g; sucrose, 0.4 g; maltose, 0.8 g; sodium pyruvate, 0.8 g; K_2HPO_4, 2 g; NaCl, 5 g; $(NH_4)_2SO_4$, 2 g; cysteine hydrochloride, 0.68 g; clarified rumen fluid, 500 ml; distilled water, 500 ml; hemin, 5 mg; vitamin K_1, 1 mg; and resazurin, 2.5 mg. The pH is 7.0. Inactivated rabbit serum (10%) and TPP (0.0075%) are added to the sterile medium. A filter-sterilized yeast autolysate (5%) can also be added.

For a solid medium, add Oxoid Special Agar (0.8%). For a semisolid medium, add Bacto-Agar (Difco) (0.16%).

Note that with the addition of serum and TPP, this medium serves as a general isolation and culture medium for both treponemes requiring serum and those that require short-chain fatty acids.

Rifampin-polymyxin selective medium

See references 8 and 13.

Treponeme isolation medium with rifampin (2 μg/ml) and polymyxin (800 U/ml).

Medium for serum-requiring treponemes

See reference 8.

Trypticase (BBL), 20 g; heart infusion broth base (BBL), 5 g; yeast extract (BBL), 10 g; K_2HPO_4, 2 g; cysteine hydrochloride, 0.68 g; distilled water, 900 ml. The pH is 7.0. Add 100 ml of inactivated rabbit serum and TPP (0.0075%). Solid medium should contain Oxoid special agar (0.8%). The selective agents rifampin and polymyxin can also be added.

Media for the characterization of treponemes

The media described in the *Anaerobe Laboratory Manual* (5) are used for the characterization of treponemes. These media are supplemented with serum and TPP or rumen fluid or both.

TSA-5400 selective medium for *T. hyodysenteriae*

See reference 14.

Trypticase soy agar (BBL); citrated bovine blood, 10%; spectinomycin (The Upjohn Co.), 400 μg/ml.

Isolation medium for *B. aalborgii*

See reference 6.

Tryptose soy agar; calf blood, 10%; spectinomycin, 400 μg/ml; and polymyxin, 5 μg/ml.

For additional media see references 5, 7, 10, 11, and 16.

This work was supported in part by Public Health Service grants DE-05054 and DE-05139 from the National Institute of Dental Research.

LITERATURE CITED

1. **Abramson, I. J., and R. M. Smibert.** 1972. Inhibition of growth of treponemes by antimicrobial agents. Br. J. Vener. Dis. **47:**407–412.
2. **Abramson, I. J., and R. M. Smibert.** 1972. Bactericidal activity of antimicrobial agents for treponemes. Br. J. Vener. Dis. **47:**413–418.
3. **Abramson, I. J., and R. M. Smibert.** 1972. Method of testing antibiotic sensitivity of spirochaetes, using antibiotic discs. Br. J. Vener. Dis. **48:**269–273.
4. **Austin, F. E., and R. M. Smibert.** 1982. Thiamine pyrophosphate, a growth factor for *Treponema vincentii* and *Treponema denticola*. Curr. Microbiol. **7:**147–152.
5. **Holdeman, L. V., E. P. Cato, and W. E. C. Moore (ed.).** 1977. Anaerobe laboratory manual, 4th ed. Virginia Polytechnic Institute and State University, Blacksburg.
6. **Hovind-Hougen, K., A. Birch-Andersen, R. Henrik-Nielsen, M. Orholm, J. O. Pedersen, P. S. Teglbjaerg, and E. H. Thaysen.** 1982. Intestinal spirochetosis: morphological characterization and cultivation of the spirochete *Brachyspira aalborgii* gen. nov., sp. nov. J. Clin. Microbiol. **16:**1127–1136.
7. **Leschine, S. B., and E. Canole-Parola.** 1980. Rifampin as a selective agent for isolation of oral spirochetes. J. Clin. Microbiol. **12:**792–795.
8. **Moore, W. E. C., L. V. Holdeman, R. M. Smibert, I. J. Good, J. A. Burmeister, K. G. Palcanis, and R. R. Ranney.** 1982. Bacteriology of experimental gingivitis in young adult humans. Infect. Immun. **38:**651–667.
9. **Pindak, F. F., W. E. Claper, and J. H. Sherrod.** 1965. Incidence and distribution of spirochetes in the digestive tract of dogs. Am. J. Vet. Res. **26:**1391–1402.
10. **Smibert, R. M.** 1981. The genus *Treponema*, p. 564–577. *In* M. P. Starr, H. Stolp, H. G. Truper, A. Balows, and H. G. Schlegel (ed.), The prokaryotes. A handbook on habitats, isolation, and identification of bacteria. Springer-Verlag, New York.
11. **Smibert, R. M.** 1984. Genus III *Treponema* Schaudin 1905, 1728[AL], p. 49–57. *In* N. R. Krieg and J. G. Holt (ed.), Bergey's manual of systematic bacteriology, vol. 1, Williams & Wilkins, Baltimore.
12. **Smibert, R. M., and J. A. Burmeister.** 1983. *Treponema pectinovorum* sp. nov. isolated from humans with periodontitis. Int. J. Syst. Bacteriol. **33:**852–856.
13. **Smibert, R. M., J. L. Johnson, and R. R. Ranney.** 1984. *Treponema socranskii* sp. nov., *Treponema socranskii* subsp. *socranskii* subsp. nov., *Treponema socranskii* subsp. *buccale* subsp. nov., and *Treponema socranskii* subsp. *paredis* subsp. nov. isolated from the human periodontia. Int. J. Syst. Bacteriol. **34:**457–462.
14. **Songer, J. G., J. M. Kinyon, and D. L. Harris.** 1976. Selective medium for isolation of *Treponema hyodysenteriae*. J. Clin. Microbiol. **4:**57–60.
15. **Stanton, T. B., and E. Canale-Parola.** 1979. Enumeration and selective isolation of rumen spirochetes. Appl. Environ. Microbiol. **38:**965–973.
16. **Stanton, T. B., and E. Canale-Parola.** 1980. *Treponema bryantii* sp. nov., a rumen spirochete that interacts with cellulolytic bacteria. Arch. Microbiol. **127:**145–158.
17. **Tomkins, D. S., M. A. Waugh, and E. M. Cooke.** 1981. Isolation of intestinal spirochetes from homosexuals. J. Clin. Pathol. **34:**1385–1387.
18. **Van Horn, K. G., and R. M. Smibert.** 1983. Albumin requirement of *Treponema denticola* and *Treponema vincentii*. Can. J. Microbiol. **29:**1141–1148.

Chapter 46

Section VI. Fungi

Taxonomy, Classification, and Nomenclature of Fungi

BILLY H. COOPER

Quite often a clinical laboratory worker trained primarily in bacteriology is confused by his or her initial encounter with the fungi because of their unfamiliar names and the different procedures used for their isolation and identification. With careful study, however, the person with an adequate background in the principles and techniques of bacteriology will soon realize that fungi are, in many respects, more easily dealt with than bacteria. For example, most fungi have characteristic conidia that can be observed with a microscope with only high-power (×400) magnification. Fungi grow more slowly than bacteria, and a certain amount of patience is required for their study. However, identification based on their colonial characteristics and microscopic morphology is less difficult than is the identification of bacterial species.

Fungi are ubiquitous, eucaryotic microorganisms, and the majority of species grow well as saprophytes on nonliving organic materials. Only a few of the thousands of known species of fungi cause human diseases. Most species that do infect humans are limited by nutritional requirements and by host defense mechanisms to invasion of the superficial skin and subcutaneous tissues. The species that are capable of invading the deeper tissues to cause serious, life-threatening infections are, fortunately, quite few. Systemic fungus infections are acquired by accidental inhalation of airborne conidia or by traumatic inoculation with contaminated soil or plant materials. They are uncommon in most hospitals except in certain geographical areas where a specific organism may infect almost 100% of the population, usually causing only mild disease. In most hospitals, clinics, and offices of physicians, patients having systemic mycoses are seen infrequently. Consequently, most clinical laboratories are prepared to deal with only the most commonly isolated fungi, i.e., yeasts and dermatophytes.

Two seemingly unrelated factors, the mobility of modern human populations and the widespread therapeutic use of drugs that alter immune defenses, make it imperative that personnel in every sizable clinical laboratory be well trained in techniques for isolation and identification of pathogenic fungi. The availability of modern rapid transportation permits a person to be infected by a fungal agent in an endemic area without manifesting symptoms until some time after returning home, where the disease may be unusual or unexpected. Physicians and laboratory personnel must have a high level of suspicion for the disease in question so as not to delay the correct diagnosis.

Drugs that alter immune defenses predispose patients to infection by several fungi that ordinarily are harmless saprophytes. Similarly, persons with diabetes mellitus, severe burns, hematological malignancies, and other debilitating conditions are susceptible to serious systemic infections by fungi that usually are not pathogenic. In a clinical setting involving an immunocompromised host, any fungus that grows at 37°C should be considered a potential pathogen. Repeated isolation of the same fungus from three or more different specimens from the same patient, isolation of a fungus from a normally sterile body site, or demonstration of tissue invasion by direct microscopic examination of biopsy specimens all strongly suggest pathogenicity of a fungus. When dealing with opportunistic pathogens, it is especially important to maintain quality control checks to ensure that specimens are not contaminated during collection or in the laboratory or plated onto contaminated media. Concurrent use of serological tests that demonstrate a specific rising titer also may help clarify the diagnosis in these clinical settings.

CLASSIFICATION AND NOMENCLATURE

Fungi can be differentiated easily into two types based on the macroscopic appearance of their colonies. Those that produce opaque, creamy, or pasty colonies are called yeasts, and those that produce cottony, woolly, fluffy, or powdery aerial growths above the culture medium are called molds. A third group can be demonstrated to develop as yeasts when cultivated at 37°C and as molds when grown at 25 to 30°C. This environmentally controlled interconversion of morphological phases is called dimorphism. Most systemic human pathogens as well as a few other species of fungi exhibit dimorphism, and it is essential that conversion from one phase to the other be accomplished for exact identification of those species. Upon microscopic examination, the yeasts will be observed to be small, unicellular microorganisms that produce daughter cells from the parent cell by budding. The molds are multicellular microorganisms whose cells are joined together to make up long, tubelike filaments called hyphae. As some hyphae elongate, they form cross walls or septa behind the growing hyphal tip. Other hyphae which do not form cross walls are said to be coenocytic. As hyphae grow, they become intertwined to form a mycelium or mold colony. The portion of the mycelium which penetrates into the substrate from which it absorbs nutrients is called the submerged mycelium. The portion of the mold colony that grows above the substratum in an erect fashion is called the aerial mycelium.

Microorganisms classified as actinomycetes (members of the genera *Nocardia*, *Streptomyces*, and *Actinomyces*) resemble true fungi in many respects; they cause diseases which resemble fungus diseases, and they are encountered in many of the same clinical specimens as fungi. The actinomycetes, however, have distinctive cellular and molecular properties which make them different from fungi, so they are properly classified as gram-positive filamentous bacteria. For these reasons, they have not been included in the chapters dealing with higher fungi.

Identification of fungi is accomplished rather simply by noting the development of their colonies and their gross and microscopic morphology. The surface texture, color, and growth rate of fungal colonies, along with the pigmentation of the reverse side of the colony, are important identifying characteristics. For specific identification, it is necessary to induce a fungus to display its characteristic conidia; although this may require the use of special media or growth conditions, it is not ordinarily a difficult procedure. Biochemical tests are also useful, especially for the identification of yeasts, but the emphasis should be placed on morphological characteristics for correct identification of molds.

In addition to their gross macroscopic properties, fungi are quite distinctive at the cellular and molecular levels. They possess an organized nucleus surrounded by a nuclear membrane; i.e., they are eucaryotic microorganisms. Unlike bacteria, the fungi grow slowly and their cell walls are composed of polysaccharide polymers such as glucan, mannan, cellulose, and chitin. The teichoic and muramic acids found in bacterial cell walls are not present in the cell walls of fungi. Finally, fungi reproduce sexually by the formation of spores and asexually by forming conidia or sporangiospores. Sexual reproduction occurs by the fusion of specialized cells called gametes, and the end result of their fusion is the development of a conspicuous structure in which spores are produced after meiosis. The growth state induced by sexual recombination is referred to as the teleomorph or perfect state and provides the basis for classification of fungi as belonging to the class *Zygomycetes*, *Ascomycetes*, or *Basidiomycetes*. The asexual growth state in which only conidia or sporangiospores are produced is referred to as the anamorph. A fourth form class, *Deuteromycetes* or Fungi Imperfecti, was created to accommodate those species for which the teleomorph is unrecognized.

Recognition of the teleomorph is important taxonomically because the complete morphological potential of a fungal species can only be understood once both the sexual and the asexual states are recognized. Often, two or more different species are described and named which actually just represent different states in the growth cycle of a single species. It is only by understanding the complete reproductive cycles of fungi, i.e., anamorph and teleomorph, that phylogenetic relationships and classification based on these relationships can be established.

The following is a simplified taxonomic scheme illustrating the four major groups in which medically important fungi are classified.

Kingdom *Mycetae* (Fungi)
Division *Amastigomycota*
Subdivision *Zygomycotina* (Division *Zygomycota*)
Class *Zygomycetes*
  Order *Mucorales*
  a. Representative genera: *Rhizopus, Mucor, Absidia, Cunninghamella, Saksenaea, Syncephalastrum*
  b. Human disease: Opportunistic pathogens in patients having diabetes, leukemia, severe burns, or malnutrition
  Order *Entomophthorales*
  a. Representative genera: *Basidiobolus, Conidiobolus*
  b. Human disease: Cause subcutaneous zygomycosis and rhinoenteromophthoromycosis
Subdivision *Ascomycotina* (Division *Ascomycota*)
Class *Ascomycetes*
  Subclass *Hemiascomycetidae*
  a. Representative genera: *Saccharomyces, Pichia*
  b. Human disease: Very rare. Important in baking and brewing industries. Some *Candida* species are related to this subclass.
  Subclass *Plectomycetidae*
    Order *Onygenales*
    Family *Gymnoascaceae*
    a. Representative genera: *Arthroderma*, teleomorph of *Trichophyton; Nannizzia*, teleomorph of *Microsporum; Ajellomyces*, teleomorph of *Blastomyces; Emmonsiella*, teleomorph of *Histoplasma*
    b. Human disease: Teleomorphs of important human pathogens
    Order *Eurotiales*
    Family *Eurotiaceae*
    a. Representative genera: Teleomorphs of some *Aspergillus* and *Penicillium* species
    b. Human disease: Teleomorphs of some human pathogenic *Aspergillus* species
Subdivision *Basidiomycotina* (Division *Basidiomycota*)
Class *Basidiomycetes*
  Subclass *Holobasidiomycetidae*
    Order *Agaricales*
    a. Representative genera: *Amanita, Agaricus*
    b. Human disease: Poisonous and edible mushrooms
  Subclass *Teliomycetidae*
    Order *Ustilagenales*
    Family *Ustilagenaceae*
    a. Representative genus: *Filobasidiella*, teleomorph of *Cryptococcus*
    b. Human disease: *Filobasidiella neoformans*, an important human pathogen
Subdivision *Deuteromycotina* (Division *Deuteromycota*)
Form class *Deuteromycetes*
  Form Subclass *Blastomycetidae*: Imperfect yeasts
  a. Representative genera: *Candida, Cryptococcus*
  b. Human disease: Anamorphs of human pathogenic yeasts
  Form Subclass *Hyphomycetidae*
    Form Family *Moniliaceae*
    a. Representative genera: *Epidermophyton, Coccidioides, Sporothrix, Paracoccidioides*
    b. Human disease: Anamorphs of major human pathogens
    Form Family *Dematiaceae*
    a. Representative genera: *Phialophora, Fonsecaea, Exophiala, Wangiella, Cladosporium*; molds with darkly pigmented hyphae
    b. Human disease: Fungi which cause chromoblastomycosis, phaeohyphomycosis, tinea nigra, and mycetoma

The classification arrangement shown above is patterned after that recommended by Alexopoulos and Mims (2) and is quite similar to the classification system suggested by Ainsworth et al. (1). However, some mycologists prefer to place the higher fungi in four divisions—*Zygomycota, Ascomycota, Basidiomycota*, and *Deuteromycota*—within a taxonomic kingdom which excludes the slime molds and the aquatic molds. Neither of these latter two groups has been associated with human disease.

As fascinating as these subtle taxonomic arguments are, they bear little relevance to the task of laboratory identification of human pathogenic fungi in most instances. It is not essential that the ancestry of a fungal species be traced back into antiquity before a laboratory diagnosis of a fungal disease can be made and treatment for that disease can be prescribed. For that reason, laboratory identification of medically important fungi is most often accomplished by examination of structures produced during the vegetative phase of the life cycle. Demonstration of the sexual phase is sometimes useful for epidemiological purposes and in a few instances is essential for exact identification of particular species.

Asexual reproduction is accomplished by the formation of sporangiospores in members of the class *Zygomycetes* and by the formation of conidia in the other classes. Observation of the size, shape, color, and arrangement of sporangiospores or conidia by use of a microscope is the single most important criterion for laboratory identification of molds. Considerable attention has been focused in recent years on the study of processes of conidium development (conidiogenesis or conidium ontogeny) in a number of research laboratories. These efforts have utilized sophisticated equipment, such as the scanning electron microscope, that ordinarily is not available in clinical microbiology laboratories. However, these research studies do help to point out characteristics of conidia that can sometimes be recognized with a light microscope, and an understanding of conidiogenesis does contribute to an understanding of the unique characteristics of fungal species that allow them to be identified.

Conidia are produced from specialized hyphae called conidiogenous cells (conidiophores). Large, multicellular conidia are called macroconidia, and small, unicellular conidia are called microconidia. Based on studies of conidiogenesis, Barron (4) described 10 different kinds of conidia which have been simplified to the following six types.

1. Aleurioconidia
   a. Conidia produced singly as the blown-out ends of conidiogenous cells
   b. Example: *Blastomyces dermatitidis*
2. Annelloconidia
   a. Conidia produced as blastoconidia in basipetal sequence from a conidiogenous cell (annellide) which elongates after the production of each conidium, leaving a ringlike scar or annulus on its outer surface at the location where each conidium was produced
   b. Example: *Scopulariopsis* spp.
3. Arthroconidia
   a. Conidia produced by fragmentation of hyphae into individual cells
   b. Example: *Geotrichum* spp.
4. Blastoconidia
   a. Conidia arising as small outgrowths (buds) from a portion of a preexisting conidiogenous or parent cell
   b. Example: *Cladosporium* spp.
5. Phialoconidia
   a. Conidia produced endogenously from a conidiogenous cell (phialide) that does not noticeably increase in length
   b. Example: *Phialophora verrucosa*
6. Poroconidia
   a. Conidia produced through pores in the conidiogenous cell
   b. Example: *Drechslera* spp.

Although these are not the only kinds of conidia produced by fungi, an understanding of these basic types of conidia is quite helpful to the laboratory technologist as a starting point for learning to identify molds. Certain types of conidiogenous cells, e.g., annellides and phialides, are not easily distinguished, and certain polymorphic fungi, e. g., *Fonsecaea pedrosoi*, which produce conidia by more than one mechanism are not easily categorized in a system based on conidiogenesis. Most likely, the technologist responsible for identifying fungi in a clinical laboratory will find it helpful to utilize the understanding gained from studies of conidium development in combination with insights derived from simpler approaches to arrive at correct identifications of fungal species. Illustrations of the different conidial types are presented in succeeding chapters.

Conidia and sporangiospores can be demonstrated by removing a small piece of mycelium from a mold colony with a dissecting needle and then gently teasing the mycelium apart in a drop of lactophenol cotton blue. The size, shape, and color of conidia can readily be observed by this technique, but because they are usually dispersed by this procedure, demonstration of the characteristic arrangement of conidia requires that a special microculture called a slide culture be made. For this, a small block of agar is placed on a sterile microscope slide supported on a bent glass rod in a sterile petri dish. The agar block is inoculated on all four sides with the fungus to be studied, and a sterile cover slip is placed over the block of agar. Water is added to the petri dish to prevent dehydration, and the culture is incubated at 25 to 30°C. The slide can be removed from the chamber from time to time, and the fungus can be observed with a microscope as it develops in situ. When the fungus is fully developed, a permanent mount can be made by removing the cover slip from the agar and placing it face downward in a drop of lactophenol cotton blue. The edges of the cover slip are then sealed with nail polish or another mounting medium, creating a permanent mount of the fungus.

The identifying characteristics of the major human pathogens are described in the following chapters. As anyone who is experienced in mycology is well aware, typically sporulating human pathogens are less of a problem to identify than are the numerous fungal contaminants that grow luxuriantly on laboratory culture media. Some of the more confusing contaminants are described in the following chapters. However, it is not within the scope of this manual to provide detailed descriptions of all of the fungi that contaminate laboratory cultures. *Illustrated Genera of Imperfect Fungi*, by Barnett and Hunter (3), and *The Genera of Hyphomycetes from Soil*, by Barron (4), are recommended as useful guides to the identification of most laboratory contaminants.

LABORATORY SAFETY

There seems to be a widespread mystique about the hazards of handling fungi in clinical laboratories. Although there are risks associated with examining the filamentous phases of dimorphic human pathogens, the commonsense application of a few basic safety rules that are commonly followed in clinical microbiology laboratories will suffice to protect laboratory workers against infection. It is a good practice to examine all molds within an enclosure such as a bacteriological glove box or laminar flow hood. This practice not only will protect workers from accidental infection with the systemic mycotic agents but also will reduce contamination of laboratory cultures and help to avoid introducing conidia into a hospital's air-conditioning system. Yeast cultures can be handled somewhat less cautiously, in the same manner that bacterial cultures are routinely handled, but they should never be handled carelessly.

Basically, two types of activities can lead to laboratory infections with fungi: (i) accidental creation of aerosols containing conidia and (ii) accidental inoculation with sharp instruments such as hypodermic needles, dissecting needles, and scalpel blades. Activities such as smoking, drinking, eating, application of cosmetics, and insertion of contact lenses are to be avoided in laboratory work areas. It is a good practice to clean laboratory benches daily with a good disinfectant-containing detergent, not only to prevent potential infections but also to reduce the potential of contaminating laboratory cultures with undesirable molds.

MAINTENANCE OF STOCK CULTURES

It is desirable to maintain stock cultures of typical strains of pathogenic fungi for reference and teaching purposes. Stock cultures may be maintained on Sabouraud agar at room temperature if they are transferred at least every 6 weeks. Cultures stored in a refrigerator at 4°C can be maintained for 3 to 4 months without subculturing. However, some cultures, e.g., *Epidermophyton* spp. and members of the class *Zygomycetes*, will not survive refrigerator temperatures. Agar slant cultures covered with sterile distilled water or sterile mineral oil can be maintained without subculturing for up to 12 months. Stock cultures also can be stored in a freezer at −20°C on agar slants for longer periods of time; lyophilized cultures can be preserved for several years without losing their typical characteristics.

SEROLOGICAL TECHNIQUES

Mycoserological techniques are covered in a separate chapter (Chapter 96) because of the importance attached to development of an accurate understanding of these procedures. Serological testing with fungal antigens presents two major difficulties: (i) the lack of commercial availability of sensitive and specific antigens for all fungus diseases and (ii) the broad cross-reactivity of antigens, which makes interpretation of test results more complex than might be desired. The complexities of mycoserology derive not from the kinds of tests that are employed but from the crudeness and complexity of the antigens that must be used in the tests.

In the past, individual clinical laboratories that wanted to provide serological tests to assist in diagnosis of fungus diseases often had to manufacture their own antigens. Such antigens were carefully standardized by using known positive reference antisera obtained from the Centers for Disease Control, Atlanta, Ga., or from other reference laboratories. Within recent years, however, commercial supply of most of the antigens and control sera needed for mycoserological techniques has become a reality. Suppliers of these reagents are given in Chapter 96, and any laboratory wishing to carry out serological tests for fungus disease should be able to acquire the necessary materials without difficulty. In my experience, the latex test for cryptococcal antigen that is currently available in kit form has been particularly reliable and useful as an aid to detection of cryptococcal infections. The exoantigen test described in Chapter 96 is a recently developed procedure that has proven to be quite useful for serological confirmation of the identification of systemic fungal pathogens.

CONCLUSION

The methods described in the following chapters represent the most up-to-date procedures available at the time of publication. If the recommended procedures are followed, a reasonably well-trained microbiologist should be able to isolate and identify human pathogenic fungi from clinical specimens without much difficulty.

A few selected references that are not directly cited in the text have been listed under Selected References for Further Study to aid those who desire more training in mycology. The Centers for Disease Control and a number of universities in the United States conduct training programs in medical mycology on a regular basis. The *ASM News*, the newsletter of the Medical Mycological Society of the Americas, and the newsletter of the American Mycological Society can be consulted for current announcements concerning these programs. The American Society for Microbiology has made available traveling workshops on a variety of subjects, including dermatophytes, systemic mycoses, yeasts, and saprophytic fungi. These workshops provide access to knowledgable experts and the latest techniques for local groups who wish to schedule them anywhere in the United States and Canada. Requests for these workshops directed to the Manager of Continuing Education at ASM headquarters in Washington, D.C., will be considered. The Centers for Disease Control have made available at nominal cost a set of medical mycology teaching slides, which can be obtained from Color Film Corp., 777 Washington Blvd., Stamford, CT 06901.

LITERATURE CITED

1. **Ainsworth, G. C., F. K. Sparrow, and A. S. Sussmman.** 1973. The fungi. An advanced treatise, vol. 4A and 4B. Academic Press, Inc., New York.
2. **Alexopoulos, C. J., and C. W. Mims.** 1979. Introductory mycology, 3rd ed. John Wiley & Sons, Inc., New York.
3. **Barnett, H. L., and B. B. Hunter.** 1972. Illustrated genera of imperfect fungi, 3rd ed. Burgess Publishing Co., Minneapolis.
4. **Barron, G. L.** 1968. The genera of hyphomycetes from soil. The Williams & Wilkins Co., Baltimore (Robert E. Krieger Publishing Co., Inc., Huntington, N.Y., 1977).

SELECTED REFERENCES FOR FURTHER STUDY

1. **Ainsworth, G. C.** 1971. Ainsworth and Bisby's dictionary of the fungi, 6th ed. Commonwealth Mycological Institute, Kew, England.
2. **Bulmer, G. S.** 1979. Introduction to medical mycology. Yearbook Medical Publishers, Chicago.
3. **Callaway, C. J., and L. D. Haley.** 1978. Laboratory methods in medical mycology, 4th ed. Centers for Disease Control, Atlanta, Ga.

4. **Cole, G. T., and R. A. Samson.** 1979. Pattern of development in conidial fungi. Fearon Pitman Publishers, Inc., Belmont, Calif.
5. **Cooper, B. H.** 1972. The superficial and subcutaneous mycotic agents. *In* R. Clark (ed.), Topics in clinical microbiology. The Williams & Wilkins Co., Baltimore. (Audio tape cassettes with accompanying explanatory manual and Kodachrome slides.)
6. **Cooper, B. H.** 1972. Identification of yeast-like fungi. *In* R. Clark (ed.), Topics in clinical microbiology. The Williams & Wilkins Co., Baltimore.
7. **Cooper, B. H.** 1972. Isolation and identification of the systemic mycotic agents. *In* R. Clark (ed.), Topics in clinical microbiology. The Williams & Wilkins Co., Baltimore.
8. **Dolan, C. T., J. W. Funkhoser, E. W. Koneman, N. Y. Miller, and G. D. Roberts.** 1976. Atlases of medical mycology, vol. 1–6. American Society for Clinical Pathology, Chicago. (Kodachrome slides and explanatory manual.)
9. **Emmons, C. W., C. H. Binford, J. P. Utz, and K. J. Kwon-Chung.** 1977. Medical mycology, 3rd ed. Lea & Febiger, Philadelphia.
10. **Haley, L. D., J. Trandel, and M. B. Coyle.** 1980. Cumitech 11, Practical methods for culture and identification of fungi in the clinical microbiology laboratory. Coordinating ed., J. C. Sherris. American Society for Microbiology, Washington, D.C.
11. **Koneman, E. W., G. D. Roberts, and S. F. Wright.** 1978. Practical laboratory mycology, 2nd ed. The Williams & Wilkins Co., Baltimore.
12. **Krickel, J. H., and L. D. Haley.** 1970. An audio-tutorial kit for training in basic medical mycology. Pan Am. Health Organ. Sci. Publ. **205:**225–227.
13. **LaRone, D. H.** 1976. Medically important fungi. A guide to identification. Harper & Row, Publishers, Hagerstown, Md.
14. **McGinnis, M. R.** 1980. Laboratory handbook of medical mycology. Academic Press, New York.
15. **Rippon, J. W.** 1982. Medical mycology. The pathogenic fungi and the pathogenic actinomycetes, 2nd ed. W. B. Saunders Co., Philadelphia.
16. **Roberts, G. D.** 1977. Laboratory identification of fungi, p. 147–160. *In* E. W. Koneman and M. S. Britt (ed.), Clinical microbiology. Health Education Resources, Inc., Bethesda, Md.

Chapter 47

Detection and Recovery of Fungi in Clinical Specimens

GLENN D. ROBERTS, NORMAN L. GOODMAN, GEOFFREY A. LAND, HOWARD W. LARSH, AND MICHAEL R. McGINNIS

Proper collection, transport, processing, and culturing of clinical specimens are essential for making the diagnosis of mycotic infections. In many instances laboratory workers are unaware of specimen requirements necessary for the optimal detection and recovery of fungal etiologic agents. The emphasis of this chapter is directed toward these criteria, and specific recommendations are presented for individual clinical specimen sources. Recommendations are based on a consensus among the authors and have proven to be effective in providing an accurate and timely diagnosis of fungal infections.

DIRECT EXAMINATION OF CLINICAL SPECIMENS

The direct examination of clinical specimens for fungal elements is certainly not a new concept; however, its usefulness warrants its reiteration. With the increase in the number of immunocompromised patients seen in current medical practice, emphasis has been placed on methods for the rapid diagnosis of infection. Since most fungal serologic tests (except cryptococcal antigen testing) are generally not helpful in making a diagnosis of fungal infections in the immunocompromised patient and most culture results require several days, it is important to use methods that will ensure a rapid diagnosis. With a little experience, most microbiologists should be able to recognize fungal elements in clinical specimens. This, in many instances, can provide the first microbiological proof of fungal infection.

Numerous staining methods have been available for many years to detect fungi in clinical specimens. Often they are underutilized because technologists and microbiologists feel unqualified to examine stained specimens. In addition, there are many instances when fungi are simply not considered when stained specimen is examined, e.g., Gram stain, Wright stain. It should be emphasized that often the tentative or definitive identification of an organism can be made within minutes after receipt of a clinical specimen. The information provided can be very useful to the clinician as a basis for considering chemotherapy. This is particularly important in immunocompromised patients.

Microbiologists and technologists are frequently asked to examine histopathologic sections for the presence of fungi but are reluctant to do so because of a lack of knowledge of host tissue responses. This fear is unfounded since fungi appear the same in histopathologic section as they do in other clinical specimens such as respiratory tract secretions, and one can readily identify them without having extensive knowledge of tissue responses. The use of specific fungal stains, e.g., methenamine silver, has made this task easier; however, these stains must be routinely utilized in laboratory practice. Technologists working in hematology and cytotechnology laboratories can be valuable in detecting fungi in specimens often not suspected of containing them. These individuals should be taught to recognize the fungi, since they can play a definite role in making a rapid diagnosis. All laboratorians are encouraged to rekindle interest in this important area of microbiology to shorten the time required to make the diagnosis of fungal infections.

Table 1 presents commonly used methods for the direct microscopic examination of specimens, the time required for preparing the specimen, and advantages and disadvantages of each method. The decision as to which method to use should be made by each individual laboratory director.

Table 2 is included to assist microbiologists in recognizing the fungi that may be seen in clinical specimens. Their morphologic forms, size ranges, and characteristic features are presented and may be compared with photomicrographs within this chapter. Figures 1 through 26 present representative photomicrographs of organisms commonly encountered.

COLLECTION, TRANSPORT, PROCESSING, AND CULTURING OF CLINICAL SPECIMENS

Prompt handling of freshly collected specimens and the use of enriched culture media (some of which contain blood) and antibiotics are essential for the recovery of pathogenic fungi. A battery of appropriate media should be used since no single culture medium alone is adequate for the recovery of all fungal etiologic agents. It is generally recommended that one medium contain cycloheximide to inhibit the rapidly growing molds that often overgrow the slower-growing dimorphic fungi. However, this compound is known to inhibit the growth of some of the fungi known to be clinically important, e.g., *Cryptococcus neoformans*, and for that reason it is also necessary to use media lacking cycloheximide. It is recommended that at least one medium contain blood enrichment (5 to 10% sheep blood) to ensure the recovery of certain fastidious dimorphic fungi. However, cultures recovered on a blood-containing medium must be subcultured to induce sporulation. In addition, it is necessary that media contain antibacterial antibiotics to reduce the contamination found in most nonsterile clinical specimens. A combination of chloramphenicol (16 μg/ml) and gentamicin (5 μg/ml) or gentamicin alone (50 μg/ml) has been found to be satisfactory. By tradition, it has been suggested that at least one fungal medium be free of antibiotics to ensure the recovery of nocardiae, which are often seen in a mycology laboratory. It is well known that the *Nocardia* spp. are best recovered when specimens are processed and cultured for

TABLE 1. Characteristics of methods available for the direct microscopic detection of fungi in clinical specimens

Method (reference)	Use	Time required	Advantages	Disadvantages
Acid-fast	Detection of mycobacteria and *Nocardia*	12 min	Detects *Nocardia*[a] and *B. dermatitidis.*	Tissue homogenates are difficult to observe due to background staining.
Calcofluor white (4)	Detection of fungi	1 min	Can be mixed with KOH; detects fungi rapidly due to bright fluorescence.	Requires use of a fluorescence microscope; background fluorescence prominent, but fungi exhibit more intense fluorescence. Vaginal secretions are difficult to interpret.
Gram stain	Detection of bacteria	3 min	Is commonly performed on most clinical specimens submitted for bacteriology and will detect most fungi, if present.	Some fungi stain well; however, others, e.g., *Cryptococcus* spp., stain weakly in some instances. Some isolates of *Nocardia* fail to stain or stain weakly.
India ink	Detection of *C. neoformans* in CSF	1 min	When positive in CSF, is diagnostic of meningitis.	Positive in less than 50% of cases of meningitis; not reliable.
Potassium hydroxide (KOH)	Clearing of specimen to make fungi more readily visible	5 min; if clearing is not complete, an additional 5 to 10 min is necessary	Rapid detection of fungal elements.	Experience required since background artifacts are often confusing. Clearing of some specimens may require an extended time.
Methenamine silver stain	Detection of fungi in histologic section	1 h	Best stain to detect fungal elements.	Requires a specialized staining method that is not usually readily available to microbiology laboratories.
Papanicolaou stain	Examination of secretions for presence of malignant cells	30 min	Cytotechnologist can detect fungal elements.	
Periodic acid-Schiff (PAS) stain	Detection of fungi	20 min; 5 min additional if counterstain is employed	Stains fungal elements well; hyphae of molds and yeasts can be readily distinguished.	*Nocardia* spp. do not stain well; *B. dermatitidis* appears pleomorphic.
Wright stain	Examination of bone marrow or peripheral blood smears	7 min	Detects *H. capsulatum.*	Detection is limited to *H. capsulatum.*

[a] Acid-fast bacterium.

mycobacteria. For this reason, an antibiotic-free medium need not be included in the routine battery of fungal culture media. Recommendations for the selection of appropriate culture media are included in this chapter, and laboratory directors may choose those most suitable for their work schema. It should be mentioned that Sabouraud dextrose agar is not recommended for the recovery of fungi from clinical specimens, except for dermatophytes, and then it should contain chloramphenicol and cycloheximide. Sabouraud dextrose agar is a satisfactory subculture medium, but it is a poor primary recovery medium and should not be used as such since better media are available. Table 3 presents a summary of recommended primary culture media and antibiotic supplements.

Procedures designed for the decontamination and concentration of clinical specimens for the recovery of mycobacteria are not recommended for the recovery of fungi. Most fungi will not survive even a 20-min treatment with 2% sodium hydroxide (8). Liquifying agents such as dithiothreitol or *N*-acetyl-L-cysteine have not been shown to enhance the recovery rate of fungi from clinical specimens. The use of such agents, when combined with centrifugation, seems to concentrate not only the pathogens but also other fungi and contaminating bacteria. Careful consideration should be given if this procedure is to be used.

It is recommended that each medium be inoculated with at least 0.5 ml of specimen to ensure the optimal recovery of fungi. The specimen should be streaked across the surface of the agar to provide for the isolation of colonies.

It is generally recommended that cultures be incubated at 30°C since this is the optimal temperature for

TABLE 2. Summary of characteristic features of fungi seen in direct examination of clinical specimens

Morphologic form found in specimens	Organism(s)	Size range (diam) (μm)	Characteristic features
Yeastlike	*Histoplasma capsulatum*	2–5	Small; oval to round budding cells; often found clustered within histiocytes; difficult to detect when present in small numbers.
	Sporothrix schenckii	2–6	Small; oval to round to cigar-shaped; single or multiple buds present; uncommonly seen in clinical specimens.
	Cryptococcus neoformans	2–15	Cells exhibit marked variation in size; usually spherical but may be football shaped; buds single or multiple and "pinched off"; capsule may or may not be evident; occasionally pseudohyphae forms with or without a capsule may be seen in exudates or CSF.
	Blastomyces dermatidis	8–15	Cells are usually large, double refractile when present; buds usually single; however, several may remain attached to parent cells; buds connected by a broad base.
	Paracoccidioides brasiliensis	5–60	Cells are usually large and are surrounded by smaller buds around the periphery ("mariner's wheel appearance"); smaller cells may be present (2 to 5 μm) and resemble *H. capsulatum*; buds have "pinched off" appearance.
Spherules	*Coccidioides immitis*	10–200	Spherules vary in size; some may contain endospores, others may be empty. Adjacent spherules may resemble *B. dermatitidis*; endospores may resemble *H. capsulatum* but show no evidence of budding. Spherules may produce multiple germ tubes if a direct preparation is kept in a moist chamber for ≥24 h.
	Rhinosporidium seeberi	6–300	Large, thick-walled sporangia containing sporangiospores are present; mature sporangia are larger than spherules of *C. immitis*; hyphae may be found in cavitary lesions.
Yeast and pseudohyphae or hyphae	*Candida* spp.	3–4 (yeast) 5–10 (pseudohyphae)	Cells usually exhibit single budding; pseudohyphae, when present, are constricted at the ends and remain attached like links of sausage; hyphae, when present, are septate.
	Malassezia furfur	3 –8 (yeast) 2.5–4 (hyphae)	Short curved hyphal elements are usually present along with round yeast cells that retain their spherical shape in compacted clusters.
Nonseptate hyphae	Zygomycetes: *Mucor*, *Rhizopus*, and other genera	10–30	Hyphae are large, ribbonlike, often fractured or twisted; occasional septa may be present. Smaller hyphae are confused with those of *Aspergillus* spp., particularly *A. flavus*.
Hyaline septate hyphae	Dermatophytes		
	Skin and nails	3–15	Hyaline, septate hyphae are commonly seen. Chains of arthroconidia may be present.
	Hair	3–15	Arthroconidia on periphery of hair shaft producing a sheath are indicative of ectothrix infection. Arthroconidia formed by fragmentation of hyphae within the hair shaft are indicative of endothrix infection.
		3–15	Long hyphal filaments or channels within the hair shaft are indicative of favus hair infection.
	Aspergillus spp.	3–12	Hyphae are septate and exhibit dichotomous, 45° angle branching; larger hyphae, often disturbed, may resemble those of Zygomycetes.
	Geotrichum spp.	4–12	Hyphae and rectangular arthroconidia are present and are sometimes rounded. Irregular forms may be present.
	Trichosporon spp.	2–4 by 8	Hyphae and rectangular arthroconidia are present and are sometimes rounded. Occasionally blastoconidia may be present.
	Pseudallescheria boydii (cases other than mycetoma)		Hyphae are septate and are impossible to distinguish from those of other hyaline molds, e.g., *Aspergillus* spp.

Continued

TABLE 2.—*Continued*

Morphologic form found in specimens	Organism(s)	Size range (diam) (μm)	Characteristic features
Dematiaceous septate hyphae	*Cladosporium* spp. *Curvularia* spp. *Drechslera* spp. *Exophiala* spp. *Phialophora* spp. *Wangiella dermatitidis*	2–6	Dematiaceous polymorphous hyphae are seen; budding cells with single septa and chains of swollen rounded cells are often present. Occasionally aggregates may be present in infection caused by *Phialophora* and *Exophiala* spp.
	Exophiala werneckii	1.5–5	Usually large numbers of frequently branched hyphae are present along with budding cells.
Sclerotic bodies	*Cladosporium carrionii* *Fonsecaea compacta* *Fonsecaea pedrosoi* *Phialophora verrucosa* *Rhinocladiella aquaspersa*	5–20	Brown, round to pleomorphic, thick-walled cells with transverse septations. Commonly, cells contain two fission planes that form a tetrad of cells. Occasionally, branched septate hyphae may be found along with sclerotic bodies.
Granules	*Acremonium* *A. falciforme* *A. kiliense* *A. recifei*	200–300	White, soft granules without a cementlike matrix.
	Curvularia *C. geniculata* *C. lunata*	500–1,000	Black, hard grains with a cementlike matrix at periphery.
	Aspergillus *A. nidulans*	65–160	White, soft granules without a cementlike matrix.
	Exophiala *E. jeanselmei*	200–300	Black, soft granules, vacuolated, without a cementlike matrix, made of dark hyphae and swollen cells.
	Fusarium		
	F. moniliforme	200–500	White, soft granules without a cementlike matrix.
	F. solani	300–600	
	Leptosphaeria		
	L. senegalensis	400–600	Black, hard granules with cementlike matrix present.
	L. tompkinsii	500–1,000	Periphery composed of polygonal swollen cells and center of a hyphal network.
	Madurella		
	M. grisea	350–500	Black, soft granules without a cementlike matrix, periphery composed of polygonal swollen cells and center of a hyphal network.
	M. mycetomatis	200–900	Black to brown, hard granules of two types: (i) rust brown, compact, and filled with cementlike matrix (ii) deep brown, filled with numerous vesicles, 6 to 14 μm in diameter, cementlike matrix in periphery, and central area of light-colored hyphae.
	Neotestudina *N. rosatti*	300–600	White, soft granules with cementlike matrix present at periphery.
	Pseudallescheria *P. boydii*	200–300	White, soft granules composed of hyphae and swollen cells at periphery in a cementlike matrix.
	Pyrenochaeta *P. romeri*	300–600	Black, soft granules composed of polygonal swollen cells at periphery, center is network of hyphae, no cementlike matrix present.

growth. It is not necessary to incubate cultures at 35°C; this additional step is economically and culturally unproductive.

Cultures should be incubated for at least 4 weeks before being reported as negative. Incubation for 6 weeks will perhaps enhance the recovery rate slightly, but is usually unfeasible due to a lack of incubation space in most laboratories.

With extended incubation periods, dehydration of culture media becomes an important problem. This may be prevented if culture dishes contain 40 ml of medium. Placing a pan filled with water in the bottom of the incubator will also reduce dehydration.

Culture dishes provide a greater surface area for the optimal recovery of fungi. Dishes should be taped on both sides to prevent inadvertent opening during handling or incubation. The use of media in screw-capped tubes has the advantages of being safer to handle and reducing the amount of dehydration. However, a lesser number of fungi may be recovered due to the small surface area and the concentration of organisms near the bottom of slants when tubes are

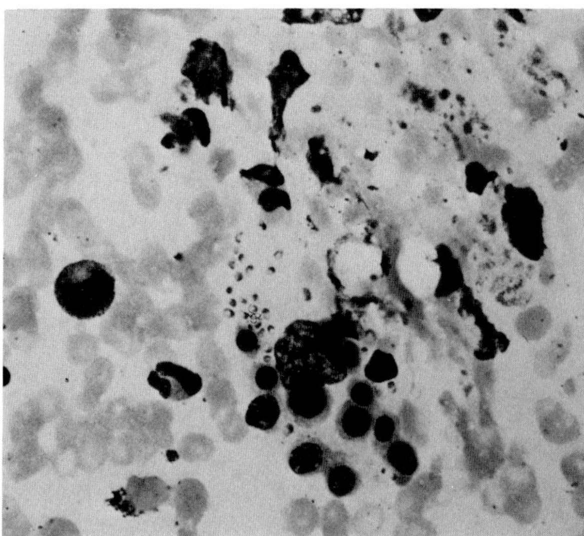

FIG. 1. *Histoplasma capsulatum* in bone marrow. Small intracellular yeast cells are apparent in this Wright-Giemsa preparation. ×2,360.

incubated upright. This problem can be somewhat alleviated by using large tubes and by allowing them to incubate in a horizontal position for 24 h before placing them vertically in a rack for the duration of incubation. The choice of whether to use culture dishes or tubes remains up to the laboratory director; dishes provide for better recovery but are more hazardous to handle. All clinical specimens and cultures should be handled in a certified, laminar flow, biological safety cabinet.

Specific recommendations for the collection, transport, storage, processing, and culturing of individual specimens are presented below. Table 4 presents a guide to the selection of appropriate clinical specimens when a specific fungal infection is suspected.

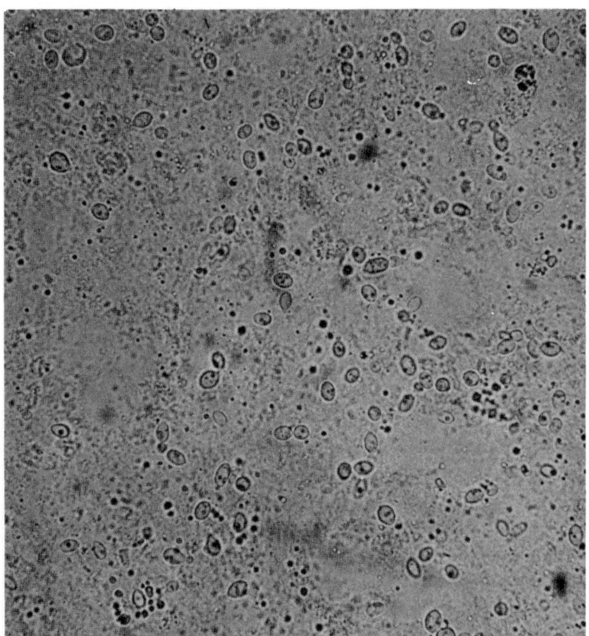

FIG. 2. Bright-field photomicrograph showing numerous small budding yeast cells of *Histoplasma capsulatum* present in sputum. ×1,852.

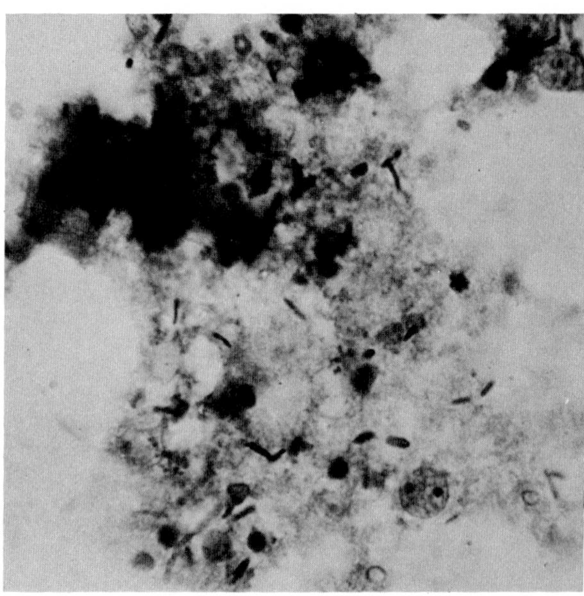

FIG. 3. Periodic acid-Schiff stain of exudate from a cutaneous ulcer, showing numerous elongated yeast cells (cigar bodies) of *Sporothrix schenckii*. ×1,915.

CUTANEOUS SPECIMENS (HAIR, SKIN, AND NAILS)

Collection, transport, and storage

The scalps of patients with suspected tinea capitis may be examined with a Wood's lamp. Fluorescent, distorted, or fractured hairs should be removed with forceps. Infected hairs can easily be removed, but normal hairs are more difficult to dislodge. A comb or brush may be used to collect loose hair and skin squames.

Skin, when involved, should be cleansed with an alcohol wipe before a specimen is collected. Epidermal scales at the active border of a lesion should be removed with a scalpel. Nails should be cleansed with an alcohol wipe, and the outermost layer should then be removed by scraping with a scalpel. Deeper scrapings, debris from under the edges of infected nails, and

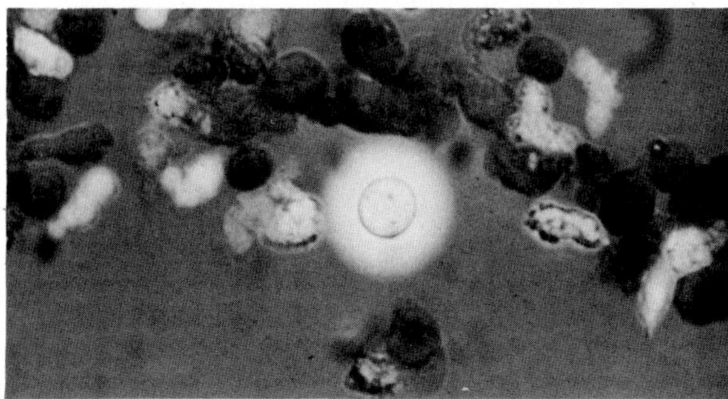

FIG. 4. India ink preparation of cerebrospinal fluid, showing a single encapsulated spherical yeast cell of *Cryptococcus neoformans*. ×2,385.

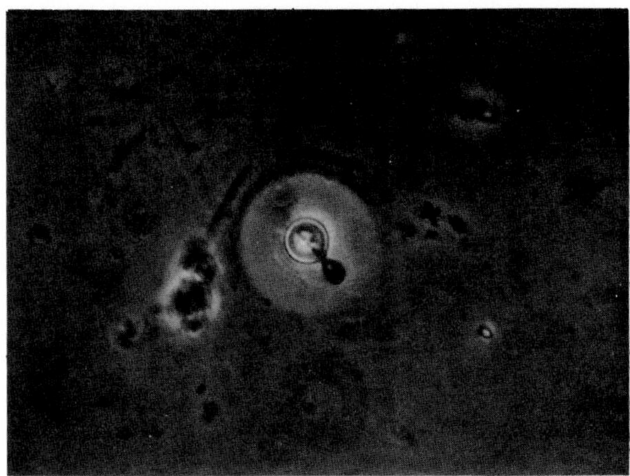

FIG. 5. Phase-contrast photomicrograph of *Cryptococcus neoformans* present in sputum. Note the spherical yeast cell with a "narrow-necked" bud attached. ×2,385.

nail clippings from infected areas are also suitable for culture.

Samples of hair, skin, and nails should be collected and placed in a sterile culture dish or sterile envelope for transport to the laboratory. Storage at 4°C is not recommended since at least one dermatophyte is susceptible to cold temperatures. In addition, storage in closed containers is unsuitable due to overgrowth of contaminating bacteria and saprobic fungi in a moist environment.

Processing and culturing

Nail clippings may be ground in a mortar before being inoculated onto culture media. Skin scrapings and hair may be inoculated directly onto the surface of appropriate culture media.

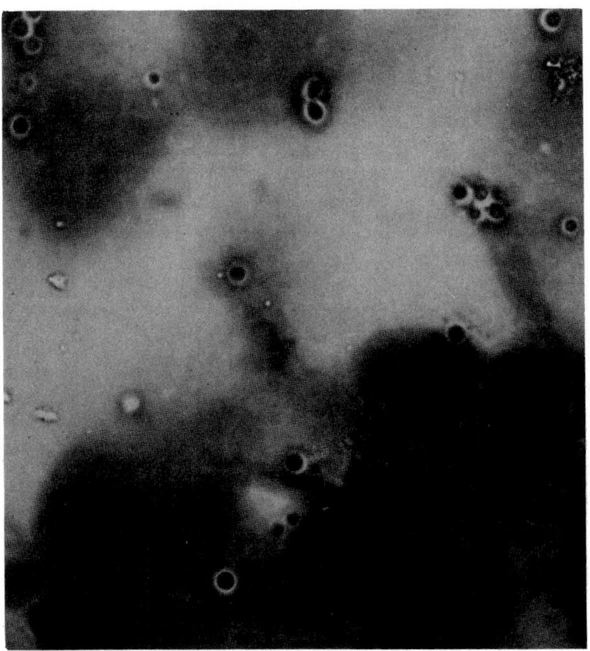

FIG. 6. Periodic acid-Schiff stain of sputum, showing spherical yeast cells of *Cryptococcus neoformans*. Note the presence of a capsule and the variation in size of cells. ×1,590.

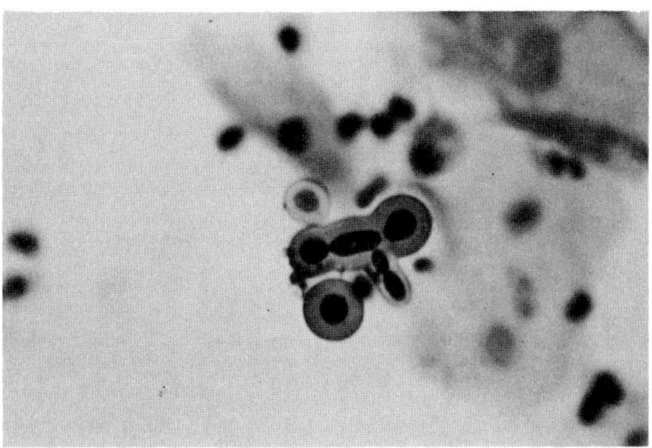

FIG. 7. Papanicolaou stain of sputum showing pseudohyphal form of *Cryptococcus neoformans*. Note the presence of a capsule surrounding all cells. ×2,210.

Sabouraud dextrose agar containing chloramphenicol (50 μg/ml) and cycloheximide (500 μg/ml) is most commonly used for the recovery of dermatophytes. This medium is available commercially as Mycobiotic (Difco Laboratories) or Mycosel (BBL Microbiology Systems) agar. The use of gentamicin may be appropriate for specimens heavily contaminated with bacteria (12). Dermatophyte Test Medium (Clinical Sciences, Whippany, N.J.) is used by some laboratories; however, it is most appropriate for the dermatologist's office and offers no advantages to the clinical laboratory.

Specimens of skin submitted for culture, exclusive of dermatophyte culture, should be inoculated onto at least two enriched media without blood and one blood-enriched medium.

SPECIMENS FROM EYE

Collection, transport, and storage

Material from patients with mycotic keratitis is collected by an ophthalmologist by scraping numer-

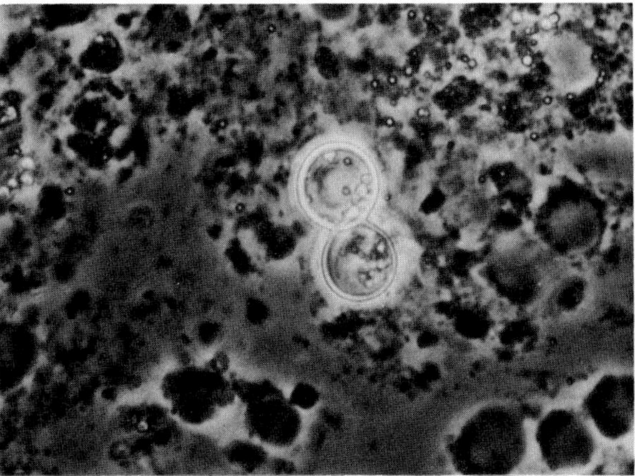

FIG. 8. Phase-contrast photomicrograph of *Blastomyces dermatitidis* in sputum. The presence of a large broad-based budding yeast cell with a "double contoured" wall is characteristic. ×3,040.

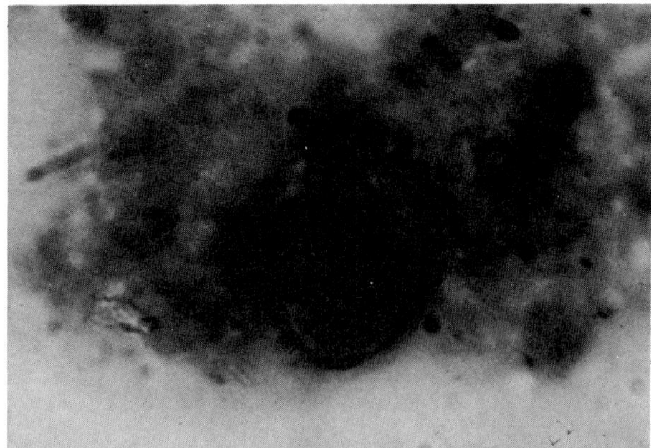

FIG. 9. Gram stain of *Blastomyces dermatitidis* yeast form in sputum. ×1,852.

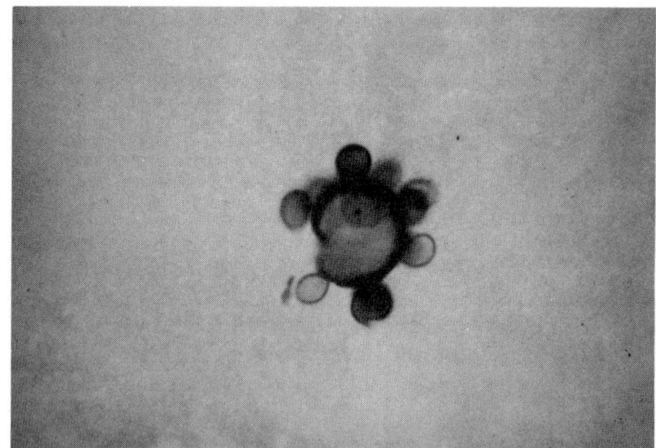

FIG. 11. Methenamine silver stain of *Paracoccidioides brasiliensis* yeast form. ×980.

ous areas of ulceration and suppuration with a Kimura spatula (6).

Processing and culturing

Culture media should be brought to the operating room for direct inoculation by the ophthalmologist. Scrapings of the corneal exudate are inoculated by marking the scrapings slightly into the agar with a platinum spatula in a series of "C"-shaped cuts. The "C" cuts localize the areas of inoculation and provide a crude dilution of the specimen. Growth within the areas of inoculation can be considered significant. After inoculation, cultures should be transported to the laboratory promptly. Ocular aspirates may be inoculated directly onto the surface of appropriate media.

The choice of media for culture is an individual one; however, at least one enriched medium lacking antibiotics and one containing antibiotics except cycloheximide are recommended for culture. Usually bacteriologic media are also inoculated simultaneously; these should be kept for the extended incubation period and examined for fungal growth.

PERITONEAL, PLEURAL, AND SYNOVIAL FLUIDS

Collection, transport, and storage

All body fluids are collected aseptically by needle aspiration and should be sent to the laboratory in a sterile container as quickly as possible. Specimens may be stored at 4°C overnight, if necessary, before culturing.

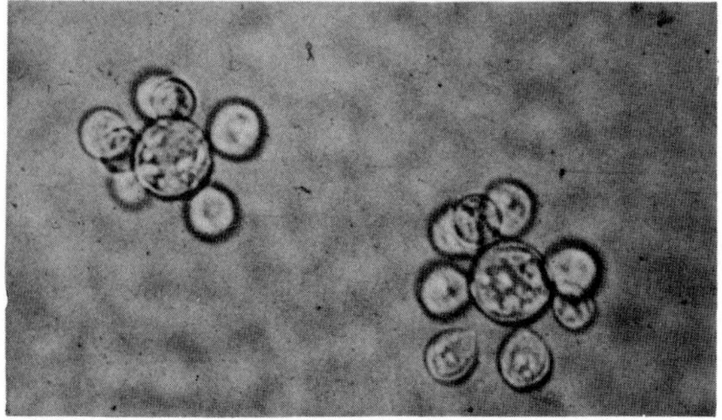

FIG. 10. Bright-field photomicrograph of *Paracoccidioides brasiliensis*, showing multiply budding yeast cells resembling mariner's wheels. ×1,590.

Processing and culturing

All body fluids should be concentrated by centrifugation before culturing. A minimum of 0.5 ml of sediment should be inoculated onto the surface of at least two enriched media, one of which contains blood.

CEREBROSPINAL FLUID

Collection, transport, and storage

As with the collection of other body fluids, careful skin antisepsis is necessary before lumbar puncture. As much cerebrospinal fluid (CSF) as possible (minimum of 2 ml) should be sent to the laboratory in a sterile, tightly sealed tube. CSF should be transported to the laboratory as soon as possible to ensure a rapid diagnosis. However, if a delay is unavoidable, CSF should not be refrigerated, since it is an excellent culture medium and fungi will continue to replicate at 25 to 30°C.

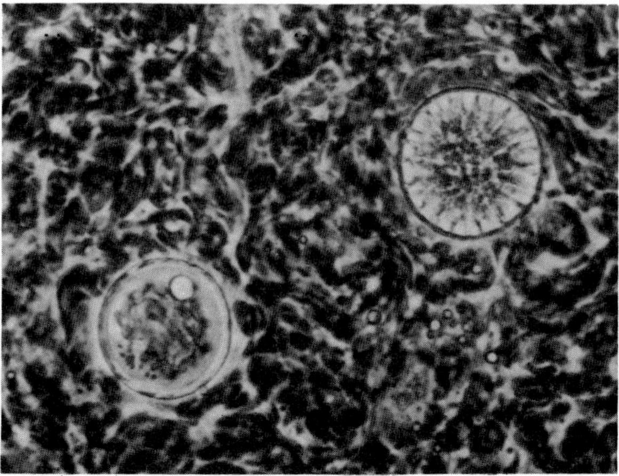

FIG. 12. Phase-contrast photomicrograph of *Coccidioides immitis* spherules in sputum. Note the absence of endospores and the presence of cleavage furrows on one spherule. ×2,400.

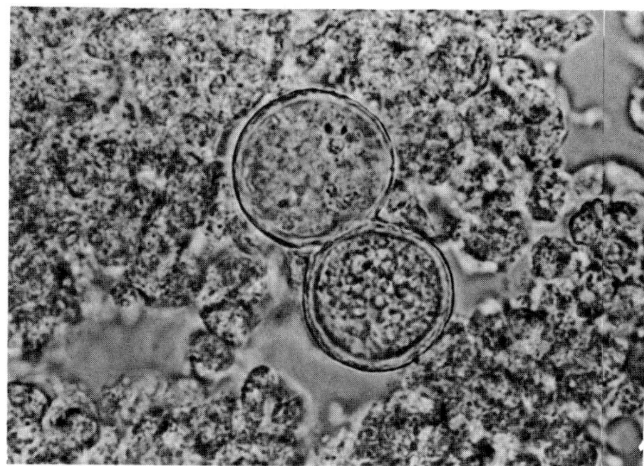

FIG. 13. Bright-field photomicrograph of two adjacent spherules of *Coccidioides immitis* which morphologically resemble *Blastomyces dermatitidis*. Note the presence of endospores in one spherule. ×1,228.

Processing and culturing

Filtration of CSF through a 0.45-μm membrane filter is optimal since it provides concentration of all cellular components onto a localized area. The "organism side" of the filter should be placed on the surface of appropriate culture media and moved to another position every other day for a week. Growth, if present, will be concentrated in the area of filter placement.

Centrifugation for 15 min at 1,000 × *g*, followed by culturing the sediment, is a suitable alternative method.

Since CSF is usually a sterile specimen source, it is not necessary to use culture media containing antibiotics. However, if they are used, cycloheximide should not be included since it inhibits some of the pathogenic fungi, e.g., *C. neoformans*. The membrane filter may be placed on the surface of an enriched medium without blood, or it may be cut with scissors and each half is then placed on a different medium lacking blood enrichment. The sediment from a centrifuged specimen is best cultured by placing several drops on the surface of a medium (media) lacking blood enrichment. This method is preferable to streaking the sediment since it allows the organisms, when present, to slowly acclimate to the environment without rapid desiccation. Small volumes not suitable for centrifugation may be dropped onto the surface of media. Any of the fungal culture media lacking blood enrichment are satisfactory.

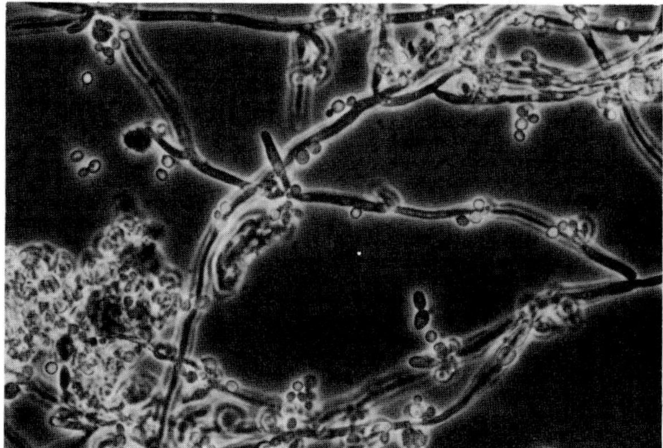

FIG. 14. Phase-contrast photomicrograph of *Candida* sp. of peritoneal fluid. Note the presence of blastoconidia and pseudohyphae charcteristic of the genus. ×1,493.

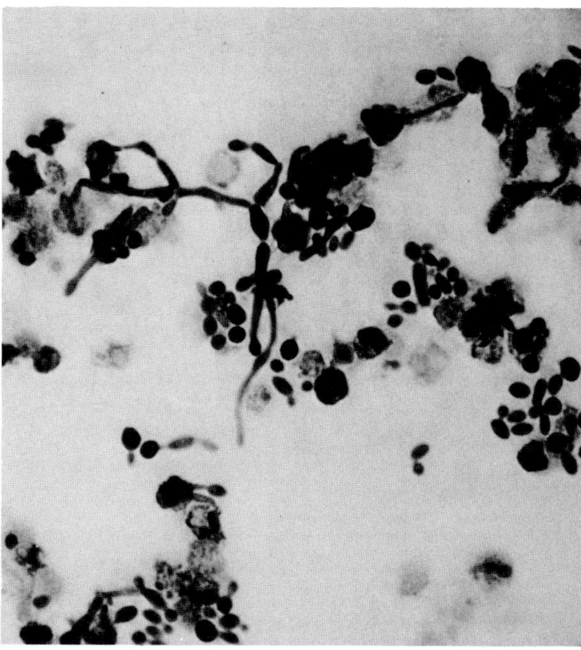

FIG. 15. Periodic acid-Schiff stain of *Candida* sp. ×1,390.

BLOOD

Collection, transport, and storage

As with the collection of other sterile body fluids, good skin antisepsis should be practiced for blood

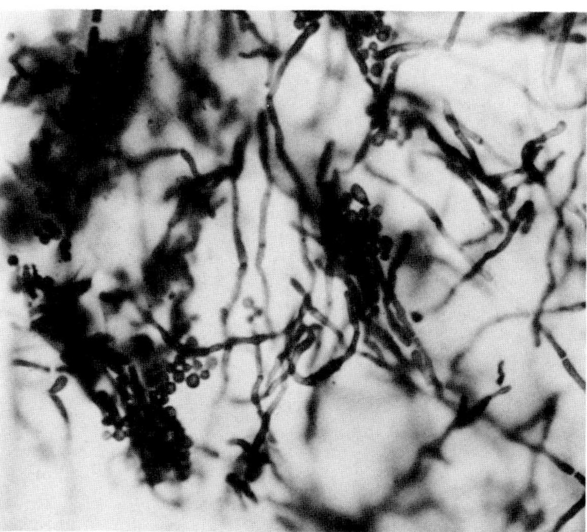

FIG. 16. Periodic acid-Schiff stain of skin showing hyphal and yeast elements characteristic of *Malassezia furfur*. ×1,590.

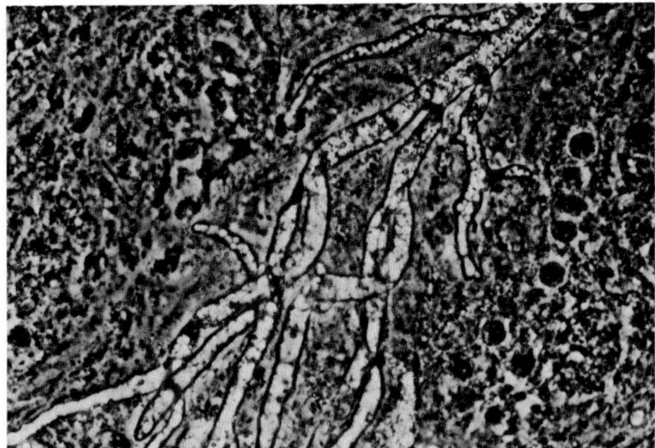

FIG. 17. Phase-contrast photomicrograph of wound exudate, showing large, aseptate, twisted hyphae characteristic of a zygomycete. ×1,493.

sample collection. Ten milliliters of blood should be collected at periodic intervals determined by the physician. If blood culture bottles are used, they should be inoculated at the bedside and incubated at 25 to 30°C until transported to the laboratory.

If a lysis-centrifugation method (1) is utilized, 10-ml tubes should be used and transported to the laboratory as soon as possible. Studies have shown that fungi can still be recovered from these tubes within 9 h of storage at 25°C, but a longer storage time is not recommended (11).

Processing and culturing

Blood culture bottles require no processing but should be examined at daily intervals for visible evidence of growth. In the case of the biphasic brain heart infusion bottle, both the agar slant and broth should be closely examined for growth (9). Chronic venting of bottles is mandatory to ensure optimal recovery.

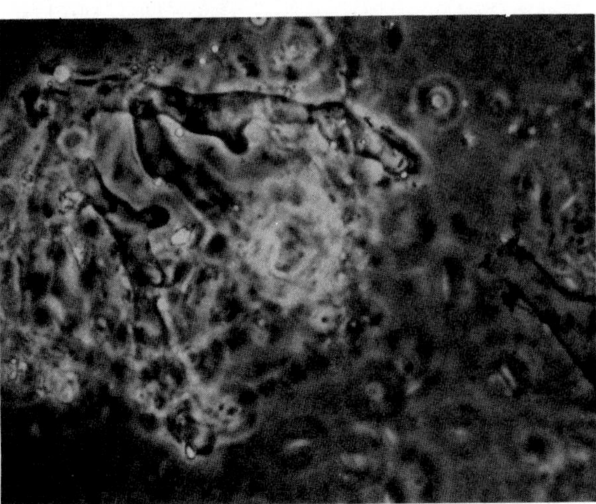

FIG. 18. Phase-contrast photomicrograph showing fractured pieces of large aseptate hyphae characteristic of a zygomycete. ×1,990.

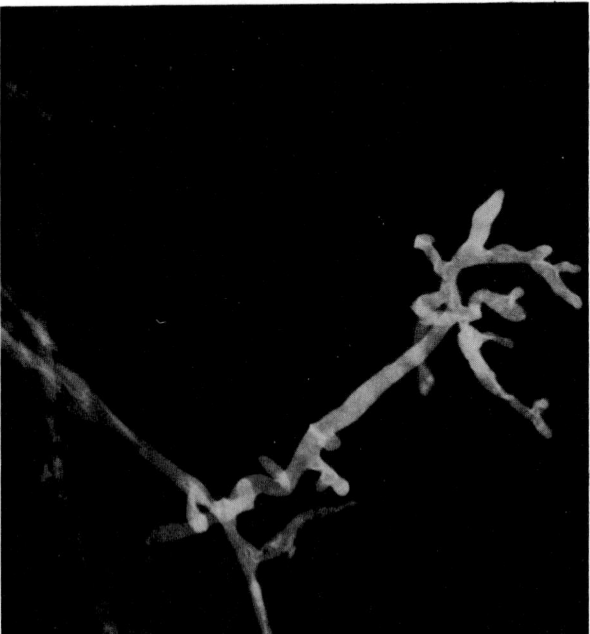

FIG. 19. Calcofluor white-stained sputum showing hyphae of a zygomycete. ×928.

The lysis-centrifugation method requires several procedural steps recommended by the manufacturer. All should be followed except for the selection of media. At least two enriched media without blood and one blood-enriched medium should be used. The blood enrichment ensures the recovery of dimorphic pathogens. In addition, either blood agar or chocolate agar may be used to detect bacteremia along with fungemia. Blood cultures should be incubated for at

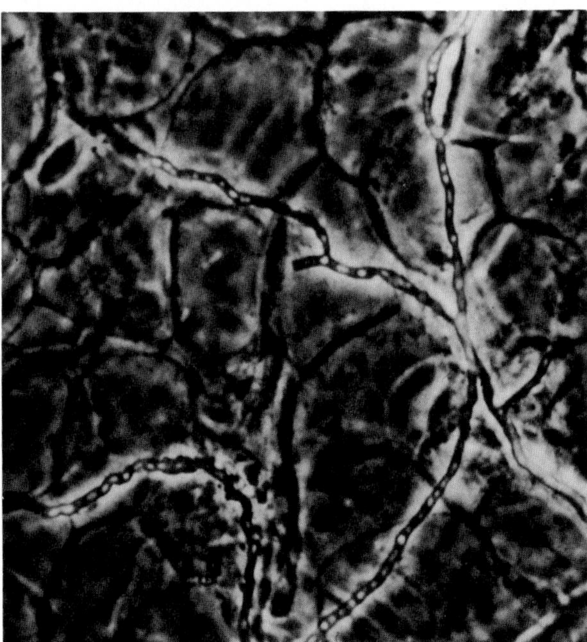

FIG. 20. Phase-contrast photomicrograph of hyphae of a dermatophyte intertwined among squamous cells. ×1,535.

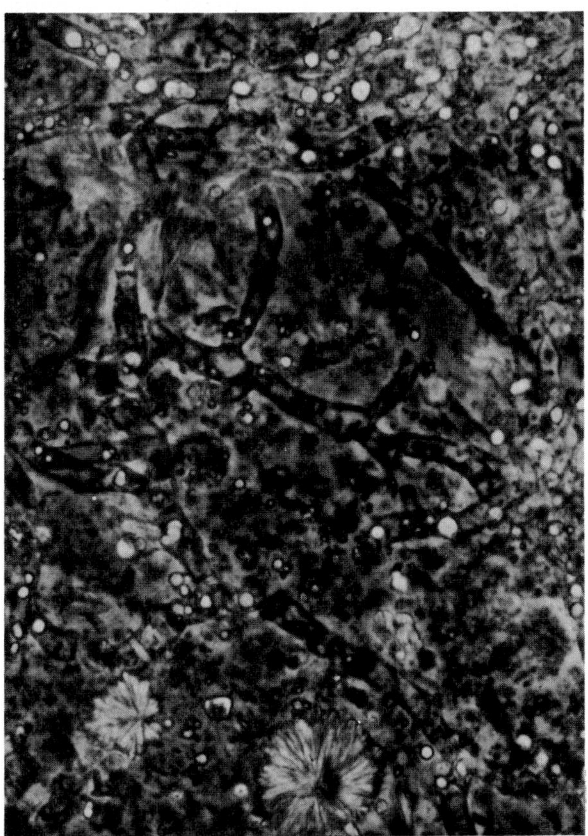

FIG. 21. Phase-contrast photomicrograph of exudate from lung tissue, showing septate hyphae. *Aspergillus* sp. was recovered in culture. ×3,040.

least 30 days before being reported as negative. Most isolates of yeasts are detected during the first 4 days of incubation, while the dimophic fungi, e.g., *Histoplasma capsulatum*, are detected during the first 2 weeks, except for the occasional slower-growing

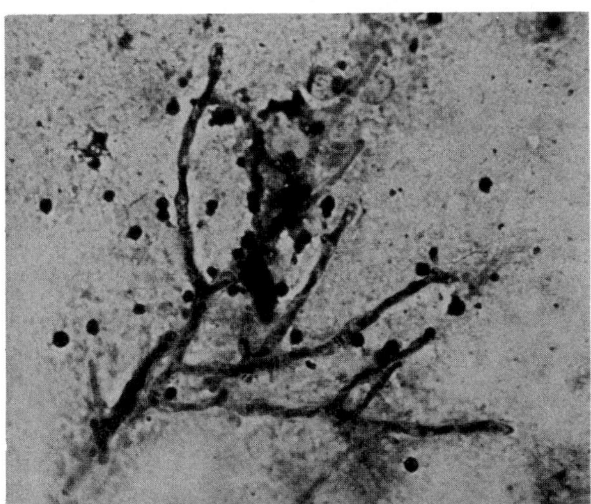

FIG. 22. Gram stain of sputum, showing dichotomously branching septate hyphae consistent with those of *Aspergillus* sp. ×1,860.

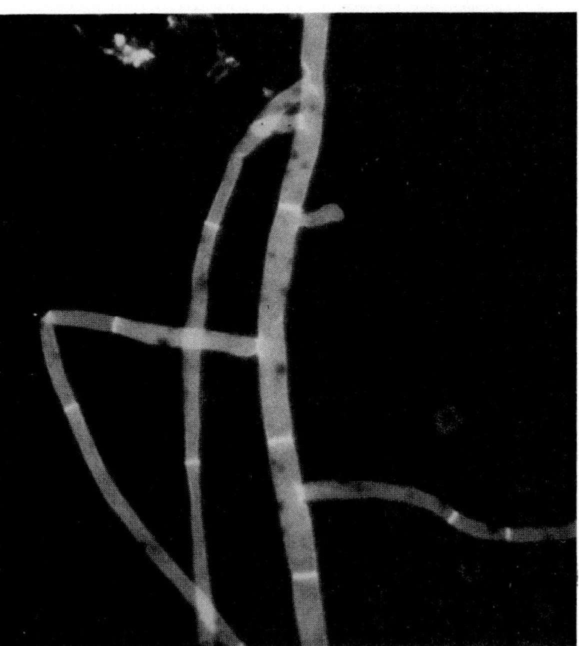

FIG. 23. Calcofluor white-stained hyphae of *Aspergillus* sp. ×1,597.

strains which make the extended incubation period necessary.

The lysis-centrifugation method is recommended since it increases the detection rate of fungemia and decreases the time to recovery. In general the lysis-centrifugation method decreases the time to recovery by one-half as compared to the biphasic bottle system and by one-fourth as compared to conventional broth blood culture systems (1).

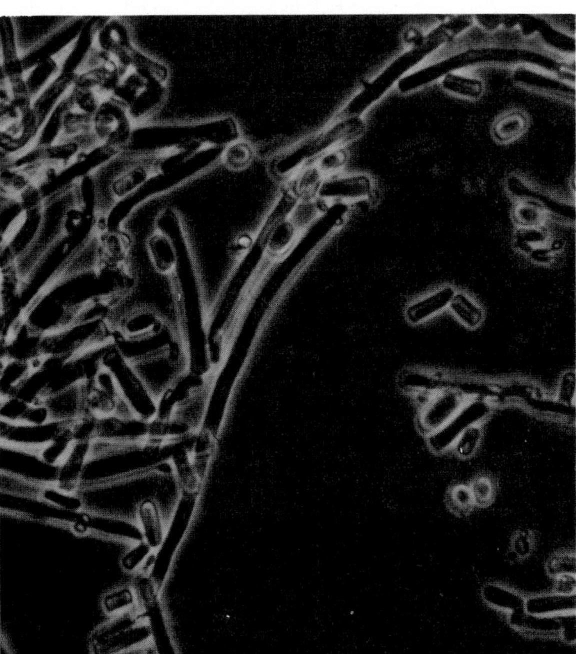

FIG. 24. Phase-contrast photomicrograph showing hyphae and arthroconidia present in sputum. *Geotrichum* sp. was recovered in culture. ×1,597.

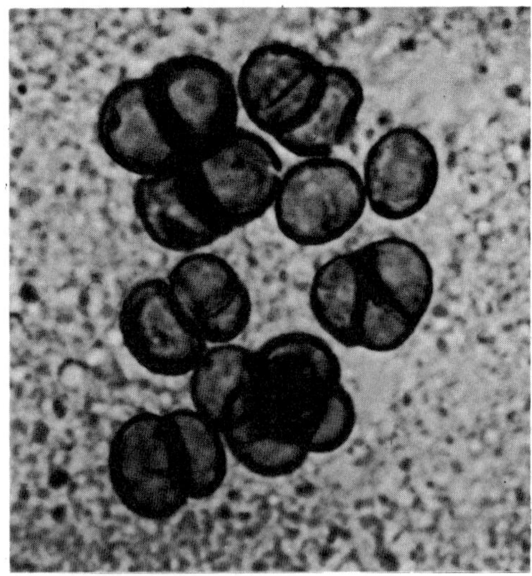

FIG. 25. Sclerotic cells in tissue characteristic of those found in chromoblastomycosis. ×1,535.

TABLE 3. Media and antibiotics recommended for the recovery of fungi from clinical specimens

Media	Antibiotics[a] (concn, μg/ml)
Without blood[b]	Gentamicin (5) and chloramphenicol (16)
Brain heart infusion agar	Penicillin (20) and streptomycin (40)
Inhibitory mold agar	Gentamicin (50)
Sabhi agar	Cycloheximide[c] (0.5)
Yeast extract-phosphate agar[d]	
Blood-enriched[e] (5 to 10% sheep blood)	
Brain heart infusion	
Special	
Caffeic acid or L-Dopa[f]	
Mycobiotic or Mycosel[g]	

[a] Antibiotics may be added to media of choice.
[b] A combination of at least two media in this group is recommended.
[c] Cycloheximide may be added with other antibiotics but at least one medium should be without it.
[d] The concominant use of concentrated NH_4OH is recommended.
[e] Should be included in battery of media used.
[f] Useful for detection of *C. neoformans*.
[g] Used for dermatophyte culture only.

BONE MARROW

Collection, transport, and storage

Bone marrow aspirates and biopsies are collected after good skin antisepsis and are commonly submitted to the laboratory in a sterile syringe or tube containing heparin. The collected specimen should be transported to the laboratory as soon as possible; however, it may be refrigerated, for no longer than 12 h, if immediate culturing is impossible.

Processing and culturing

The amount of bone marrow submitted to the laboratory is variable and precise amounts cannot easily be inoculated onto media. At least two enriched media without blood should be inoculated, and if possible, one enriched medium containing blood should be used. If the amount of specimen is small, nutrient broth or saline (without preservatives) may be added so that more than one medium may be inoculated.

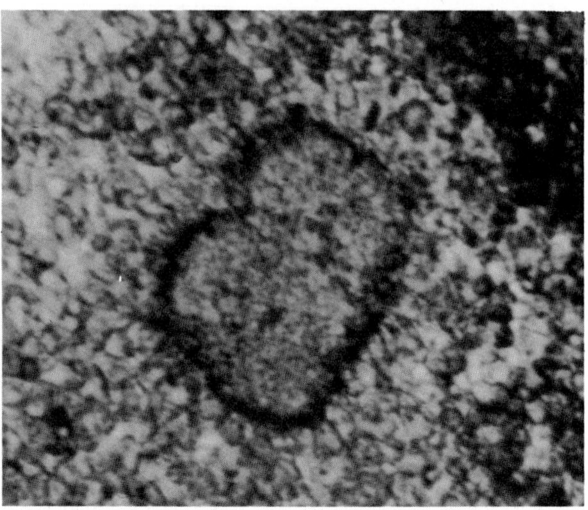

FIG. 26. Sulfur granule in wound exudate characteristically seen in mycetoma. ×500.

URINE

Collection, transport, and storage

The urinary meatus must be adequately cleansed if a clean-catch or catheterized specimen is to be submitted for culture. Suprapubic aspirates are obtained after good skin antisepsis is used in the area of aspiration. Specimens should be cultured promptly since bacteria and yeasts replicate rapidly in specimens kept at room temperature. If specimens cannot be cultured soon after receipt, they should be refrigerated at 4°C for no longer than 12 to 15 h. Twenty-four-hour specimens or those collected from indwelling catheter collection bags are not suitable for culture.

Processing and culturing

The value of quantitation of urine cultures for fungi is equivocal. It has been suggested that candiduria with a colony count of $\leq 10^4$/ml is suggestive of pyelonephritis and that colony counts of $\geq 10^4$/ml represent contamination (3). In a recent review, Fisher et al. (2) suggest that any culture of properly collected urine yielding *Candida* spp. should be regarded as abnormal. The choice of quantitation is up to the individual laboratory. Quantitation may be accomplished by using a quantitative loop and streaking the specimen onto an appropriate medium. To ensure the recovery of fungi other than *Candida*, e.g., *H. capsulatum*, *Blastomyces dermatitidis*, *Coccidioides immitis*, or *Cryptococcus neoformans*, it is necessary to concentrate the specimen by centrifugation before culturing and to inoculate a minimum of 0.5 ml onto culture media. If quantitation is desired, the quantitative culture should be performed before the specimen is centrifuged.

For quantitation of urine cultures, inhibitory mold, brain heart infusion, and Sabhi agars are satisfactory.

TABLE 4. Selection of clinical specimens for fungal culture[a]

Specimen	Suspected infection											
	Blasto-mycosis	Cocci-dioido-mycosis	Histo-plas-mosis	Para-coccidio-idomycosis	Candi-diasis	Crypto-coccosis	Asper-gil-losis	Zygo-mycosis	Dermato-phytes	Chromo-blasto-mycosis	Sporo-tri-chosis	Myce-toma
Lower respiratory tract	1	1	1	2	X	1	1	1			3	
Blood		6	2		1	2					X	
Bone	4	X		X		X	X	X			X	
Bone marrow		X	3	X	X	X						
Brain	X	X		X	X	X	X	3		X		
CSF	X	X	X		X	5					X	
Eye			X		X		X	X				
Nose/nasal sinus	X	X		X	X		2	2			4	
Prostate	X	X			X							
Mucous membranes	3	2	5	3	4	X	X	X			X	
Subcutaneous tissue		X			X			X		2	2	1
Joints	X	X			X							
Urine	5	5	4		2	3	1					
Skin	2	3	X	1	X	X	X	4	1	1	1	2
Hair and nails							X		2			
Multiple systemic sites during severe disseminated infection	6	4	6	4	3	4	3	4			X	

[a] Predominant sites for recovery are ranked in order of importance (based on most common clinical presentations). X indicates other sites from which organisms have been recovered.

VAGINAL SECRETIONS

Collection, transport, and storage

Vaginal and cervical specimens collected by a physician are usually submitted on a swab. Transport to the laboratory should be rapid; however, overnight refrigeration before culturing is satisfactory. Specimens should not be stored at room temperature.

Processing and culturing

Most vaginal and cervical specimens are submitted for culture of yeasts and yeastlike organisms. Any medium shown to recover these organisms is satisfactory. Sabouraud dextrose agar may be used; however, inhibitory mold agar or brain heart infusion agar is recommended.

RESPIRATORY SECRETIONS (Sputum, Bronchial Washings, Transtracheal Aspiration, Tracheal Aspirates, and Gastric Washings)

Collection, transport, and storage

Before a sputum specimen is collected tha patient's teeth must be extensively brushed or his or her dentures must be removed. The mouth should be cleansed by a mouthwash or several rinses of sterile water or saline. Only a specimen expectorated from deep within the lungs is satisfactory. Nasopharyngeal secretions, saliva, or oral secretions are unsatisfactory.

The rationale for collecting a proper specimen should be explained to the patient to ensure the submission of an optimal specimen. In many instances the collection of a specimen must be supervised by personnel aware of its importance. For those patients incapable of expectoration, sputum induction is necessary. Saline or mucolytic agents produce secretions often mistaken for saliva; however, such specimens should not be discarded as unsatisfactory.

All specimens from the respiratory tract should be collected in a sterile wide-mouth bottle or sputum cup. Metal or soft plastic (nonleaking) lids are preferable because they are unbreakable. Containers should not be placed in the patient's room until the approximate time of specimen collection. A first morning expectorated sputum specimen is optimal. The amount of specimen collected need not exceed 10 to 15 ml. The consistency and cellular makeup of sputum are often helpful in determining the quality of a specimen. Those samples containing large numbers of squamous cells often yield predominantly normal flora, whereas those containing alveolar macrophages, granulocytes, or ciliated epithelial cells are more often associated with an infectious process. Fungal pathogens may be recovered from specimens containing squamous cells, and hence those specimens should not be rejected for culture (7). Twenty-four-hour sputum specimens are not acceptable for culture.

Specimens should be transported to the laboratory as soon as possible to ensure maximum recovery of fungi. If culturing is delayed, specimens may be refrigerated at 4°C. Pathogenic fungi have been successfully recovered from specimens cultured after delays of up to 16 days at room temperature (5). It is not recommended that specimens be stored before culturing, but they should not be rejected if a delay in transport or storage has occurred.

Processing and culturing

The processing of clinical specimens should be performed as soon after receipt as possible. Respiratory secretions submitted for fungal culture should not be processed by methods commonly used for mycobacterial culture. Most fungi are susceptible to 15 to 20 min of exposure to 2% NaOH (8). Although it would be convenient to culture specimens for mycobacteria and fungi by the same processing methods, it is not recommended.

The use of mucolytic agents, e.g., *N*-acetyl-L-cysteine or dithiothreitol, is advocated by some microbiologists; however, controlled studies determining their usefulness have not been performed. In addition to concentrating the filamentous fungi present, such procedures likewise concentrate yeasts and bacteria which often overgrow those organisms that are the etiologic agents of infection. The decision to use these compounds must be made by each laboratory director.

The use of a yeast extract-phosphate medium in conjunction with concentrated NH_4OH has proven satisfactory for inhibiting yeasts and bacteria found in respiratory secretions (10). After inoculation of the medium, a drop of concentrated NH_4OH is placed on one side of the culture medium. Diffusion of the NH_4OH inhibits bacteria and yeasts while allowing slower-growing dimorphic fungi to grow. This method is highly recommended for specimens submitted by mail, when a prolonged transit time has occurred.

Bronchoscopy specimens may be centrifuged before culturing since a moderate volume is often collected. Mucus plugs should be separated from the remainder of the specimen and cultured.

At least 0.5 ml of any respiratory tract specimen should be inoculated onto the surface of each appropriate culture medium. Viscous specimens may be liquified as previously mentioned or may be mixed with nutrient broth and shaken before culturing. A wide-bore pipette or glass tubing is very useful for inoculating respiratory secretions onto culture media.

A minimum of two enriched media without blood should be inoculated along with a medium containing blood. Cycloheximide should be included in at least one medium, but it should also be deleted from at least one to ensure the recovery of organisms inhibited by its presence, e.g., *C. neoformans, Pseudallescheria boydii, Aspergillus fumigatus*. If culture tubes are used, they should be allowed to stand in a horizontal position for at least 12 to 24 h to allow the specimen to be absorbed by the medium, thereby preventing growth at the bottom of the slant only.

UPPER RESPIRATORY TRACT (Ear, Nose, Nasopharynx, and Mouth)

Collection, transport, and storage

Specimens from the ear, nose, nasopharynx, and mouth are usually submitted on a sterile swab. Transport to the laboratory should be rapid; however, overnight refrigeration before culturing is satisfactory. Specimens should not be stored at room temperature.

Processing and culturing

Specimens should be streaked onto the surface of two enriched media (excluding yeast extract-phosphate agar) and onto a blood-enriched medium. Cycloheximide should be included in at least one medium; however, it also should be deleted from one medium.

TISSUES

Collection, transport, and storage

In general, tissue should be divided aseptically by the surgeon in the operating room, and material representative of the infectious process should be submitted for culture and direct microscopic examination. All biopsy tissues should be placed in a sterile container containing a small amount of saline without a preservative. When an abscess is drained, a portion of the wall should be submitted for culture.

Granules from eumycotic mycetomas should be washed several times with saline containing antibacterial antibiotics such as penicillin and streptomycin before culturing.

Surgical specimens should be transported to the laboratory as soon as possible after collection. If immediate culturing is impossible, specimens should be stored at 4°C for no longer than 8 to 10 h.

Processing and culturing

All biopsy specimens should be minced or homogenized before being cultured. If a zygomycete infection is suspected, the tissue should be minced and not homogenized since the former reduces the destruction of viable hyphal elements. Tissue may be homogenized with a mechanical homogenizer, or mixed with a sterile abrasive (alundum) in broth and ground with a pestle in a mortar containing enough broth to make a 20% suspension. The most optimal and recently developed system for processing tissue is the Stomacher (Tekmar, Cincinnati, Ohio), which expresses the cytoplasmic contents out of mammalian cells into broth by the action of two rapidly moving metal paddles. Each culture medium should be inoculated with at least 1.0 ml of specimen.

A minimum of two enriched media without blood should be inoculated along with a medium containing blood. Cycloheximide should be included in at least one medium. It is recommended that cultures be kept for 6 weeks before being reported as negative since some of the dimorphic fungi require this extended incubation period.

LITERATURE CITED

1. **Bille, J., L. Stockman, G. D. Roberts, C. D. Horstmeier, and D. M. Ilstrup.** 1983. Evaluation of a lysis-centrifugation system for recovery of yeasts and filamentous fungi from blood. J. Clin. Microbiol. **18:**469–471.
2. **Fisher, J. F., W. H. Chew, S. Shadomy, R. J. Duma, C. G. Mayhall, and W. C. House.** 1982. Urinary tract infections due to *Candida albicans*. Rev. Infect. Dis. **4:**1107–1118.
3. **Goldberg, P. K., P. J. Kozinn, G. J. Wise, N. Nouri, and R. B. Brooks.** 1979. Incidence and significance of candiduria. J. Am. Med. Assoc. **241:**582–584.
4. **Hageage, G. J., and B. J. Harrington.** 1984. Use of calcofluor white in clinical mycology. Lab. Med. **15:**109–112.
5. **Hariri, A. R., H. O. Hempel, C. L. Kimberlin, and N. L. Goodman.** 1982. Effects of time lapse between sputum collection and culturing on isolation of clinically significant fungi. J. Clin. Microbiol. **15:**425–428.
6. **Jones, D. B., T. J. Liesegang, and N. M. Robinson.** 1981. Cumitech 13, Laboratory diagnosis of ocular infections. Coordinating ed., J. A. Washington II. American Society for Microbiology, Washington, D.C.
7. **Murray, P. R., R. E. Van Scoy, and G. D. Roberts.** 1977. Should yeasts in respiratory secretions be identified? Mayo Clin. Proc. **52:**42–45.

8. **Roberts, G. D., A. G. Karlson, and D. R. DeYoung.** 1976. Recovery of pathogenic fungi from clinical specimens submitted for mycobacteriological culture. J. Clin. Microbiol. **3**:47–48.
9. **Roberts, G. D., and J. A. Washington II.** 1975. Detection of fungi in blood cultures. J. Clin. Microbiol. **1**:309–310.
10. **Smith, C. D., and N. L. Goodman.** 1975. Improved culture method for the isolation of *Histoplasma capsulatum* and *Blastomyces dermatitidis* from contaminated specimens. Am. J. Clin. Pathol. **68**:276–280.
11. **Stockman, L., G. D. Roberts, and D. M. Ilstrup.** 1984. Effect of storage of the Dupont lysis-centrifugation system on recovery of bacteria and fungi in a prospective clinical trial. J. Clin. Microbiol. **19**:283–285.
12. **Taplin, D.** 1965. The use of gentamicin in mycology. J. Invest. Dermatol. **45**:549–550.

Chapter 48

Dermatophytes and the Agents of Superficial Mycoses

LIBERO AJELLO AND ARVIND A. PADHYE

When a cutaneous mycotic infection is suspected, two basic procedures must be performed to determine whether the disease is mycotic in nature: (i) direct examination of clinical material with a microscope, and (ii) isolation and subsequent identification of the fungi recovered. Both of these procedures are described in detail in the presentation of the various superficial and cutaneous mycoses.

SUPERFICIAL MYCOSES

All mycotic diseases that affect only the cornified layers of the epidermis and the suprafollicular portion of the hair are classified as superficial mycoses. Neither the stratum granulosum nor the nails have been known to be infected by the agents of the superficial mycoses. Four diseases are classified in this category: black piedra, white piedra, tinea nigra, and tinea versicolor.

Black piedra

Black piedra is a fungus infection of scalp hair but rarely of the axillary and pubic hair of humans. Many genera and species of lower primates and other mammals are also susceptible to this disease (10).

In patients with black piedra, hair filaments above the follicular orifice are overgrown by the mycelium of the dematiaceous ascomycete *Piedraia hortae*. The dark-walled mycelium spreads over and around the hair shaft and forms a cemented mat of hyphae. From this mycelium, nodules eventually are formed which at maturity contain asci and ascospores and may attain a diameter of 0.1 cm and a thickness of 100 μm (Fig. 1). The nodules are hard and gritty, hence the Spanish name of "piedra" (i.e., stone) for this disease.

Black piedra occurs primarily in the tropical areas of the world. Cases have been recorded in Africa, Asia, and Latin America. The etiologic agent, *P. hortae*, is not known to be transmitted from person to person or from infected animals to humans. Presumably this mold occurs as a saprophyte in nature, but it has not yet been isolated from a nonliving source.

Direct examination. Portions of hairs with nodules from patients suspected of having black piedra are examined in wet mounts of 10% KOH. After the wet preparation has been gently heated, the edges of the nodules should be examined under a microscope to reveal septate, dematiaceous mycelium (4 to 8 μm in diameter) on the surface of the hair filament. The nodules themselves are composed of cemented mycelium that forms a pseudoparenchymatous tissue. When mature nodules are crushed, oval asci 44 to 50 μm by 24 to 30 μm are seen. At maturity these asci contain eight aseptate, curved, spindle-shaped ascospores that bear a filament at each pole. The ascospores range from 35 to 55 μm long by 5 to 8 μm wide.

P. hortae usually invades the hair shaft under the cuticle, where its mycelium proliferates. It then breaks out and begins to grow around the hair shaft and causes extensive destruction of the cortex and medulla.

Isolation and culture identification. Sabouraud agar with antibacterial antibiotics is recommended for isolation of *P. hortae*. Media with cycloheximide cannot be used since *P. hortae* is inhibited by that antibiotic. Antibacterial antibiotics such as chloramphenicol, however, are useful to reduce bacterial contamination. Growth of *P. hortae* cultures is slow, but it is stimulated by thiamine (0.01 mg/ml). Colonies are black to dark brown and heaped. They are glabrous or covered with a fine downy mycelium. On Sabouraud agar and most other media, the fungus generally fails to produce asci and ascospores. The hyphae are dematiaceous and of variable diameter.

White piedra

White piedra is an uncommon disease encountered in both tropical and temperate regions of the world. The disease is characterized by the development of soft, yellowish or pale-brown accretions around the shaft of hairs in the axillary, facial, genital, and scalp regions of the body (Fig. 2). Both humans and lower animals are infected.

The fungus responsible for this disease is properly identified as *Trichosporon beigelii* (24). It is frequently, but erroneously, referred to as *Trichosporon cutaneum* in some publications (11).

The accretions are made up of hyaline hyphae that tend to form arthroconidia. The mycelium of *T. beigelii* frequently invades the cortex of hair filaments and thus damages the hair.

Direct examination. In mounts of 10% KOH, the soft nodules of white piedra are readily crushed by covering them with a cover slip and applying light pressure. Hyaline mycelium and arthroconidia will be found in the preparation. The hyphae range in width from 2 to 4 μm, and a cementlike material binds them together.

In some instances, the interior of the hair is found to have been destroyed by the activity of the fungus. In the early stages of hair invasion, the proliferation of mycelium under the cuticle or within the hair shaft causes pilar swelling.

Isolation and culture identification. *T. beigelii* is isolated readily on Sabouraud dextrose agar that contains chloramphenicol or other inhibitors of bacterial growth. Cycloheximide cannot be used because it inhibits the growth of *T. beigelii*. This fungus grows rapidly and in a few days produces a cream-colored yeastlike colony. Microscopically, the growth appears to be composed of hyaline mycelium that produces blastoconidia and that fragments into arthroconidia. Some blastoconidia arise directly from the mycelium, whereas others develop in chains and clusters.

Seven species of *Trichosporon* are currently accepted as valid members of this genus (11). Many others have been reclassified under such genera as *Aciculoconidium*, *Sarcinosporon*, and *Geotrichum*. In *T. beige-*

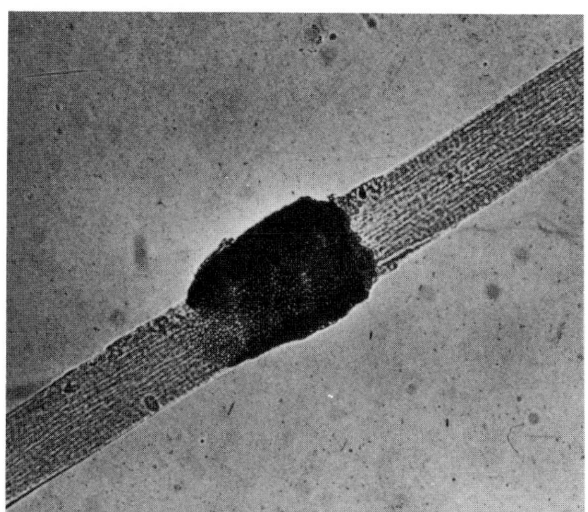

FIG. 1. Nodule of black piedra. ×100.

lii the hyphal septa have dolipores without pore caps. This indicates a relationship to the true basidiomycetes, as does the high guanine-plus-cytosine content of their DNA (24). *T. beigelii* is distinguished from all other members of the genus by its inability to ferment sugars and its ability to assimilate certain compounds (Chapter 49).

Biochemical characteristics. *T. beigelii* assimilates dextrose, lactose, and xylose. The following carbon compounds are assimilated by some isolates and not by others: cellobiose, erythritol, galactose, inositol, maltose, melibiose, melezitose, raffinose, rhamnose, sucrose, and trehalose, Potassium nitrate and sodium nitrite are not assimilated.

Tinea nigra

Tinea nigra is a disease manifested by the development of blackish-brown macular patches on the smooth skin of the body. Most lesions develop on the palms of the hands. Hence the infection is sometimes referred to as tinea nigra palmaris. The etiologic agents of tinea nigra, *Exophiala werneckii* and *Stenella arguata*, are dematiaceous fungi belonging to the form class *Hyphomycetes* of the phylum *Deuteromycota*. For that reason, *E. werneckii* is described in Chapter 52 rather than here. Because of the rarity of *S. arguata*, it is not described in this Manual.

Tinea versicolor

Tinea versicolor, or pityriasis versicolor, is a cosmopolitan disease of the smooth skin of the body. It generally manifests itself by the development of fine, slightly raised scaly patches on the neck and torso. The infection spreads, and the patches of infection enlarge and merge with adjacent ones. On light-skinned subjects, the infected sites tend to be brownish, and on the dark races they are lighter than the normal skin. Sunlight does not tan the infected skin as deeply as normal skin. A rare form of tinea versicolor involves the hair follicles. In such infections, involvement tends to be localized and the lesions become elevated. When irradiated with a Wood's lamp in a dark room, most infected areas fluoresce with a dull reddish to orange color.

The etiologic agent of tinea versicolor is now considered to be best classified in the genus *Malassezia* as *M. furfur*. Its synonymy includes *Pityrosporum furfur*, *Pityrosporum orbiculare*, and *Pityrosporum ovale* (8, 20, 21).

Direct examination. As mentioned above, the use of a Wood's lamp in a dark room is extremely useful in detecting areas infested by *M. furfur* and their distribution on the body.

For microscopic confirmation of infection, the suspected site should be cleaned with 70% alcohol and scraped with a sterile scalpel. The scales so collected are then mounted in a drop of lactophenol cotton blue or Loeffler methylene blue, covered with a cover slip, and examined under a microscope.

Alternatively, infected scales can be stripped from a suspected site by applying a piece of transparent vinyl tape to the area and lifting it off. The tape, with gummed surface down, is placed on a drop of lactophenol cotton blue on a slide. The preparation is then examined under a microscope for the presence of hyphal elements 2.5 to 4 μm in diameter and of variable length, along with unicellular oval or round cells 3 to 7 μm in diameter (Fig. 3).

Isolation and culture identification. Although cultivation of the fungus is not required to establish a diagnosis of pityriasis, at times isolation may be necessary or desired. In such cases, an oil-enriched isolation medium must be used. Sabouraud dextrose agar overlaid with olive oil or Sabouraud dextrose agar with cycloheximide and an olive oil overlay is recommended. The inoculated tubes should be incubated at 37°C and maintained at an angle to keep the agar surface covered with oil.

Growth is slow. In its early stages colonies are cream colored, glossy, and raised, but later they become dull, dry, and beige colored. They are composed of globose to ellipsoidal, hyaline cells ranging from 3 to 7 μm in diameter. Successive budding occurs at a given locus on a mother cell.

Mycelium generally is not produced in culture; however, germ tubes are extruded by some cells, and these resemble short mycelial filaments.

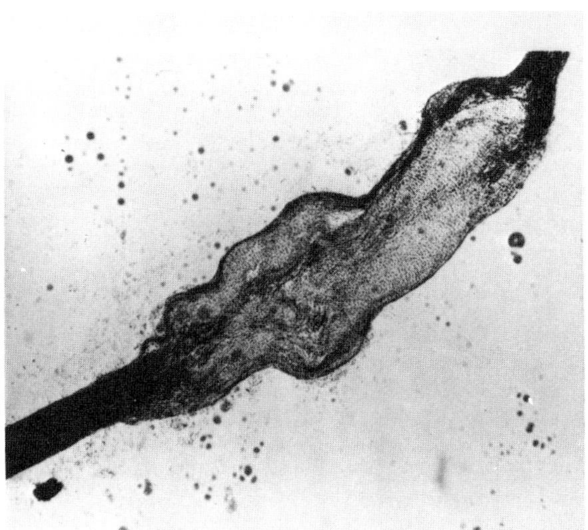

FIG. 2. Nodule of white piedra. ×100.

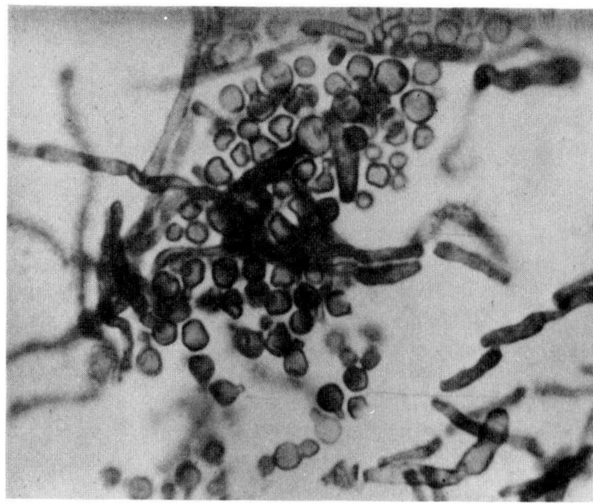

FIG. 3. Tinea versicolor. Skin scraping. Periodic acid-Schiff stain. ×1,475.

CUTANEOUS MYCOSES (RINGWORM)

The cutaneous mycoses are infections of the keratinized epidermal tissues of humans and lower animals. These mycoses are caused by a group of specialized fungi, the dermatophytes, and rarely by other molds. Unlike the agents of the superficial mycoses, these fungi penetrate and parasitize all of the fully keratinized tissues of the body (skin, hair, and nails) and produce infections that give rise to mild to severe symptoms. The dermatophytes are unable, in general, to invade the subcutaneous or deeper tissues of the body, probably because of inhibitory factors in serum and body fluids (6, 12).

Characterization of the dermatophytes

The dermatophytes are mycelial fungi that until recently were classified exclusively in the phylum *Deuteromycota* (Fungi Imperfecti). Some species are now known to produce ascospores. These species are classified in the family *Gymnoascaeae* of the phylum *Ascomycota*, in the genera *Arthroderma* and *Nannizzia* for the *Trichophyton* and *Microsporum* species, respectively, that have teleomorphs (perfect states) (3, 4). All of the dermatophytes posses keratinolytic ability that enables them to parasitize skin, hair, and nails, causing diseases known as the dermatophytoses, ringworm, or tineas. Most of the dermatophytes have a worldwide distribution; however, some species are limited geographically.

The dermatophytes have marked host preferences. Some are basically lower-animal parasites (zoophilic), others almost exclusively infect humans (anthropophilic), and others are essentially soil organisms (geophilic) that only rarely infect humans and animals (1, 2). Many geophilic species, belonging to the genera *Epidermophyton*, *Microsporum*, and *Trichophyton*, which do not infect animals or humans, are erroneously called dermatophytes. Such species cannot be referred to as dermatophytes.

Dermatophyte infections of the skin produce reactions that vary from mild erythema and scaling to severe vesicular, heavily crusted, suppurative, or, rarely, granulomatous lesions. The infections may be asymptomatic or extremely itchy and painful. Ringworm-infected nails are thickened, discolored, and deformed. The distal ends are raised from the nail bed. Usually there is no inflammation of the paronychial tissues and no pain. Infections of the bearded areas of the face and neck of men are frequently suppurative and painful. These resemble bacterial infections. Scalp infections are manifested by scattered loss of hair or by discrete, usually circular lesions with loss of hair, erythema, scaling, vesiculation, and suppuration. In some cases, raised, fluctuant, suppurative lesions known as kerions occur. Favus is another type of clinical manifestation characterized by cup-shaped crusts, "scutulae," which may form heavy confluent masses on the scalp or smooth skin. This type of lesion is usually incited by *Trichophyton schoenleinii*. Depending on the site involved, dermatophyte infections are classified clinically as tinea pedis (athlete's foot), tinea cruris (jock itch), tinea capitis, etc. (7, 19).

Examination of patients

Suspected ringworm of the scalp. Patients are examined first under normal lighting conditions for loss of hair, presence of the stubs of broken hairs, or skin lesions. Patients with suspected tinea capitis are then examined in a darkened room with a Wood's lamp. Hairs infected by certain types of ringworm fluoresce a bright yellow-green and are said to be "Wood's light-positive." Even minimal infections, in which only a few hairs are infected, can be detected by a Wood's lamp. Fluorescence is not observed during the early stages of infections or, in some cases, during treatment. False fluorescence may result from the presence of certain oils or medications on the hair. When hairs are plucked and the roots are examined under the lamp, true fluorescence can be determined.

Suspected ringworm of the skin. A Wood's lamp is also useful in differentiating dermatophyte infections from erythrasma (a bacterial infection of the skin) caused by *Corynebacterium minutissimum*. Ringworm lesions do not fluoresce, whereas the lesions of erythrasma, which are clinically similar to those of ringworm and have a similar distribution over the body, glow with an orange to coral-red fluorescence.

Suspected ringworm of the nails (onychomycosis, tinea unguium). Nail lesions suspected of being mycotic in origin must be differentiated from various other diseases that resemble onychomycosis (such as psoriasis, lichen planus, etc.). This distinction is readily made by directly examining nail scrapings and clippings in 10% KOH. The detection of mycelial elements in nail cells virtually confirms that a fungus is the etiologic agent. The identity of the specific agent involved is established through its isolation from nail material.

Collection of specimens

The following equipment is needed for specimen collection: a disinfectant for cleaning the skin (70% ethyl alcohol or its equivalent), sterile scalpels, epilating forceps, nail clippers or scissors, and clean paper envelopes. Disposable combs or brushes have been recommended for collecting specimens from the scalp or from animals (13). The scales from the active borders of the lesions should be scraped with a scalpel. Hair from the scalp is plucked out with forceps. Infected hairs can be pulled out easily, but normal hairs are harder to remove. Friable material is removed from under the edges of infected nails or clipped from the distal borders.

Specimens should be enclosed in paper packets or envelopes rather than in rubber-stoppered tubes. In closed tubes the specimens retain moisture, and contaminating bacteria and saprophytic fungi may overgrow the material and prevent or make the isolation of any pathogen present more difficult.

Direct examination

Fragments of hair, skin scrapings, or nail clippings are placed in a drop of 10% KOH on a slide, and a cover slip is added. The slide is gently heated, just short of boiling, over a flame for a few seconds. Then it is examined under a microscope. Reheating may sometimes be necessary to clear the specimen sufficiently for examination.

Infected skin and nail scrapings show hyaline, septate, branched hyphae and arthroconidia in chains (Fig. 4). The manner of hair invasion varies with the dermatophyte species involved. In general, there are three types of hair parasitism: ectothrix, endothrix, and favic. The terms ectothrix and endothrix refer to the location of arthroconidia in relation to the hair. All infected hairs show hyphae in the interior of the hair shaft during early stages of invasion. Arthroconidia formed by the fragmentation of the hyphae may be evident later inside the hair shaft (endothrix infections; Fig. 5). In ectothrix infections, the arthroconidia are observed outside the hair shaft, surrounding it in the form of a sheath (Fig. 6). In favic hair invasion, the interior of the hair is filled with long hyphal filaments, with few if any arthroconidia present (Fig. 7).

When septate, branched hyaline hyphae with or without arthroconidia are observed by direct examination of KOH mounts of skin and nail scrapings, a diagnosis of ringworm infection is confirmed, but the etiologic agents cannot be determined in this manner. The agent involved can only be identified by isolation and study of cultures. It should also be remembered that patients with KOH-negative specimens may yield positive cultures. In directly examining hair, the following key may be used as an aid to the preliminary diagnosis of ringworm.

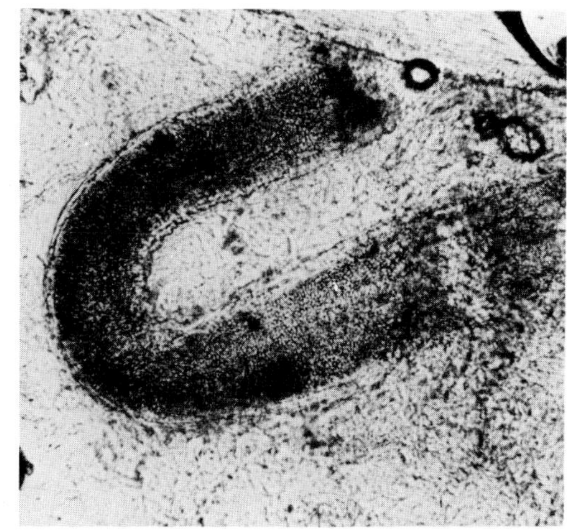

FIG. 5. Endothrix type of hair invasion caused by *Trichophyton tonsurans*. Hair stub filled with arthrospores. ×120.

Key to Direct Examination of Hair

Wood's lamp

1. Bright yellow-green fluorescence of hair shafts: *Microsporum audouinii*, *Microsporum canis*, *Microsporum ferrugineum*, rarely *Trichophyton schoenleinii*.
2. No fluorescence: all the other dermatophyte species.

KOH mounts

1. Ectothrix hairs
   a. Conidia 2 to 3 μm in diameter, forming a mosaic sheath around the hair: *Microsporum*

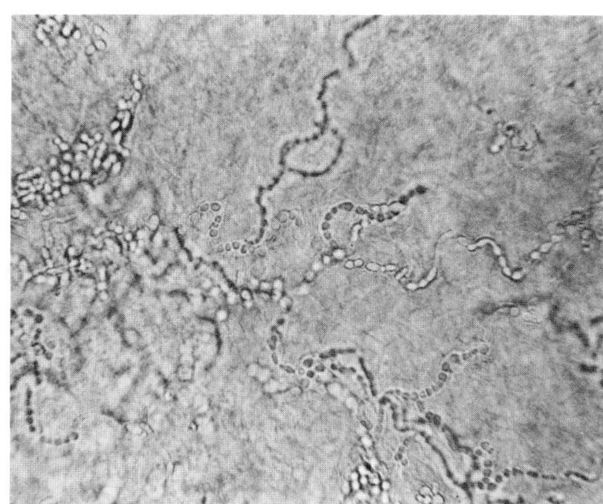

FIG. 4. Dermatophyte mycelium in skin scraping. KOH mount. ×475.

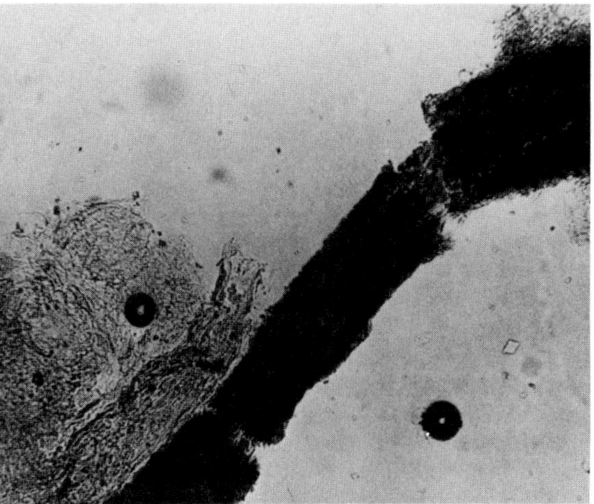

FIG. 6. Ectothrix type of hair invasion caused by *Microsporum audouinii*. Note sheath of small arthrospores surrounding the hair shaft. ×140.

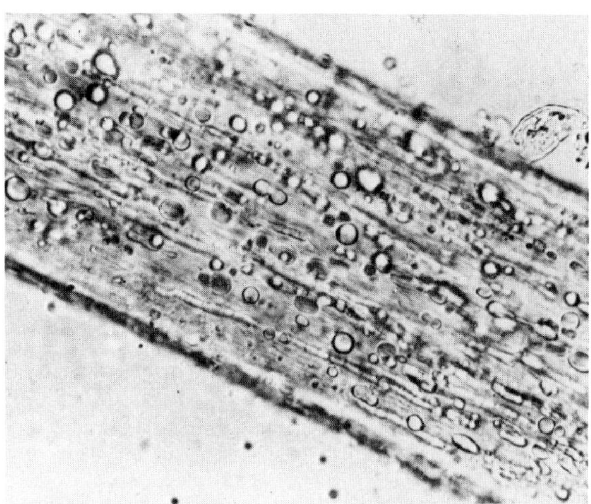

FIG. 7. Favic-type hair invasion caused by *Trichophyton schoenleinii*. Note mycelial filaments and air spaces in the hair shaft. ×620.

audouinii, Microsporum canis, Microsporum ferrugineum (Fig. 6).

b. Conidia 3 to 5 μm in diameter, forming a sheath or in isolated chains on the surface of hairs: *Trichophyton mentagrophytes*.

c. Conidia 5 to 8 μm in diameter, forming a sheath or in isolated chains on the surface of hairs: *Trichophyton equinum*, rarely *Trichophyton rubrum*.

d. Conidia 5 to 8 μm in diameter, in chains or in irregular masses on the hair surface: *Microsporum fulvum, Microsporum gypseum, Microsporum nanum*.

e. Conidia 8 to 10 μm in diameter, forming a sheath or in isolated chains on the surface of hairs: *Trichophyton verrucosum*.

2. Endothrix hairs
Short hair stubs, thick and usually twisted, filled with chains of large spores 4 to 8 μm in diameter: *Trichophyton soudanense, Trichophyton tonsurans, Trichophyton violaceum, Trichophyton yaoundei* (Fig. 5).

3. Favic hairs
Hairs invaded throughout their length by hyphal elements. Empty areas (tunnels) where hyphae have degenerated into fat droplets are commonly seen inside the hair: *Trichophyton schoenleinii* (Fig. 7).

Culture examination

Sabouraud dextrose agar (pH 5.6) is the medium most commonly used for isolating dermatophytes from clinical material. A modification with less dextrose (2%) and a pH between 6.8 and 7.0 is preferable, especially when antibiotics are added to inhibit bacterial and saprophytic fungi, as in cycloheximide medium (see Chapter 111). This medium is available commercially as Mycosel Agar (BBL Microbiology Systems) or Mycobiotic Agar (Difco Laboratories). Use of the antibiotic gentamicin is recommended for specimens heavily contaminated with bacteria (23).

Nail clippings may be ground in a mortar before being inoculated onto media. Animal hairs that are heavily contaminated with bacteria and saprophytic fungi may be teased apart and soaked in an aqueous solution of antibiotics (chloramphenicol and cycloheximide in amounts used in the selective medium) before being inoculated onto selective agar media. Specimens should be pressed gently into the agar surface.

The cultures are incubated at room temperature (25 to 30°C) and examined every 4 to 6 days. When colonies appear, they should be transferred to fresh medium to avoid possible contamination by associated bacteria or saprophytic fungi. Cultures of clinical material from persons suspected of having contracted their infection from cattle should be incubated at 37°C, since *T. verrucosum*, the common cause of cattle ringworm, develops more rapidly at this temperature. Cultures should be held for 4 weeks before being considered negative. The chance of isolating the etiologic agent is increased if four to six tubes of media are inoculated with each clinical specimen.

Morphological study of cultures

Dermatophyte species are identified on the basis of both their gross colony characteristics and their microscopic morphology. The characteristics described are those of colonies grown on Sabouraud agar with or without antibiotics. The characteristics which should be considered are rate of growth, texture, topography of the colony, color of the colony, and production of pigment on the reverse of the colony. Microscopic examination of wet mounts prepared in lactophenol cotton blue is usually adequate for demonstrating the characteristic morphologic structures. Slide cultures, Pablum cereal agar, and special sporulation media such as potato dextrose agar must be used to identify some isolates.

Nutritional tests

In many instances, characteristic conidia are not produced by an isolate or are produced infrequently. With some species, nutritional tests are of utmost importance for their identification. Basal media used for these tests are casein (vitamin-free) agar, to which various vitamin solutions are added, and ammonium nitrate agar, to which amino acids are added. Basal media without additives serve as controls. Inocula may be taken from a culture grown on the usual isolation media. It is important, however, to take only a fragment (about the size of the head of a pin) to avoid carrying over an excess of the medium (9). Nutritional media (Trichophyton agars no. 1 through 7; 26) for dermatophytes may be obtained commercially in desiccated form (Fig. 8).

In vitro hair perforation test

The in vitro hair perforation test, as devised by Ajello and Georg, is most useful in differentiating *T. rubrum* from *T. mentagrophytes* (5) and *Microsporum equinum* from *M. canis* (16, 17). Hair filaments exposed to *T. mentagrophytes* are penetrated radially by organized groups of hyphae that form wedge-shaped perforations. *T. rubrum* grows on the hair and gradu-

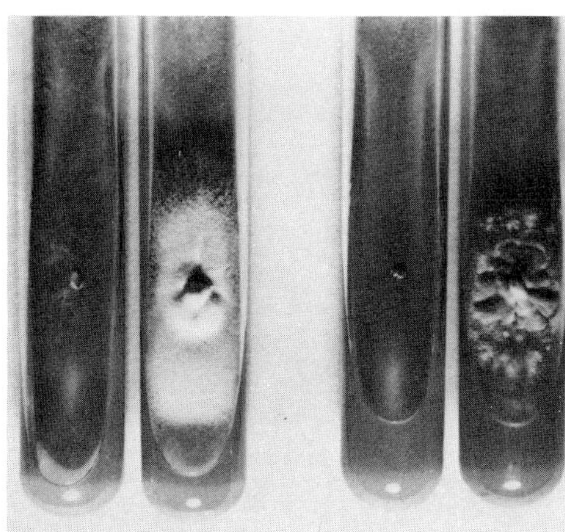

FIG. 8. Dermatophyte nutritional test. Two isolates of *Trichophyton tonsurans* on media with and without thiamine. Note growth stimulation by thiamine.

ally erodes it, but does not form perforations. Under similar conditions, *M. equinum* fails to perforate hair, whereas *M. canis* does.

For the hair penetration test, short strands of human hair or horse hair are placed in petri dishes and sterilized at 120°C for 10 min. Twenty-five milliliters of sterile distilled water and 2 or 3 drops of 10% sterilized yeast extract are added; the strands of hair are then inoculated with several small fragments of the test fungi that have been grown on Sabouraud dextrose agar. The plates are incubated at room temperature and examined at regular intervals over a period of 4 weeks. Hair fragments overgrown with mycelium are removed with sterile forceps and examined in wet mounts of lactophenol cotton blue. Gently heating the slide aids in the detection of perforations (Fig. 9).

The hair perforation test, performed as described above, showed that the ability or inability to perforate hair was a species-specific characteristic that did not vary among the isolates of a given species. (The subvariety *perforans* of *T. tonsurans* var. *sulfureum* is an exception. It perforates hair [14].) Since it was manifested by sporulating as well as nonsporulating isolates of a given species, the test was a valuable tool, especially for those species that do not sporulate readily and thus are difficult to identify by morphologic criteria (16).

Dermatophytes that perforate hair. Dermatophyte species that perforate hair in vitro are: *Microsporum canis*, *M. canis* var. *distortum*, *M. cookei*, *M. fulvum*, *M. gypseum*, *M. nanum*, *M. persicolor*, *M. racemosum*, *M. vanbreuseghemii*, *Trichophyton mentagrophytes*, *T. simii*, and *T. tonsurans* var. *sulfureum* subvar. *perforans*.

Dermatophytes that do not perforate hair. Dermatophyte species that do not perforate hair in vitro are: *Epidermophyton floccosum*, *Microsporum audouinii* (=*M. langeronii*, *M. rivalieri*), *M. ferrugineum*, *M. gallinae*, *M. praecox*, *Trichophyton concentricum*, *T. equinum*, *T. gourvilii*, *T. menginii*, *T. rubrum*, *T. schoenleinii*, *T. soudanense*, *T. tonsurans*, *T. verrucosum*, *T. violaceum*, and *T. yaoundei*.

Generic and species descriptions

Most dermatophyte species produce two types of conidia when grown in culture: small unicellular microconidia and large, septate, thin- or thick-walled, smooth- or rough-walled macroconidia. On the basis of the presence or absence of these types of conidia, the dermatophytes are divided into three genera: *Epidermophyton*, *Microsporum*, and *Trichophyton*. The teleomorphic (perfect) states of the *Microsporum* species that are known to reproduce sexually belong to the genus *Nannizzia*, and those of the *Trichophyton* spp. belong to the genus *Arthroderma* (2, 14). The distinguishing features of the three imperfect genera are as follows.

Epidermophyton. Macroconidia are smooth, large, fusiform to obovate, and multiseptate. They are borne in groups of two or three. Microconidia are not produced.

The genus *Epidermophyton* has two species, *E. floccosum* and *E. stockdaleae* (18). The macroconidia of *E. floccosum* are characteristic of the genus. They measure 20 to 40 μm by 6 to 8 μm with two to four cells. *E. stockdaleae*, known only from one soil isolate, produces club-shaped macroconidia that are 59.5 to 20.0 μm by 12.9 to 3.0 μm with two to nine cells.

Microsporum. Macroconidia are echinulate, multiseptate, variable in shape (fusiform to obovate), and thin or thick walled. They are borne singly on hyphae. They measure 5 to 100 μm in length and 3 to 8 μm in width, with 2 to 15 cells. Macroconidia may be numerous or rare. Microconidia are pyriform to obovate.

Sixteen species are currently classified in the genus *Microsporum*. The type species is *M. audouinii*.

Trichophyton. Macroconidia are smooth and either thin or thick walled and range from clavate to fusiform in shape. They are borne singly or in clusters. Microconidia are spherical, either pyriform or clavate, and are borne singly or in grapelike bunches.

The genus *Trichophyton* has 25 valid species. The type species is *T. tonsurans*.

Characteristics of the common dermatophyte species

The 15 dermatophytes listed below (in alphabetical order) are those species that are commonly isolated

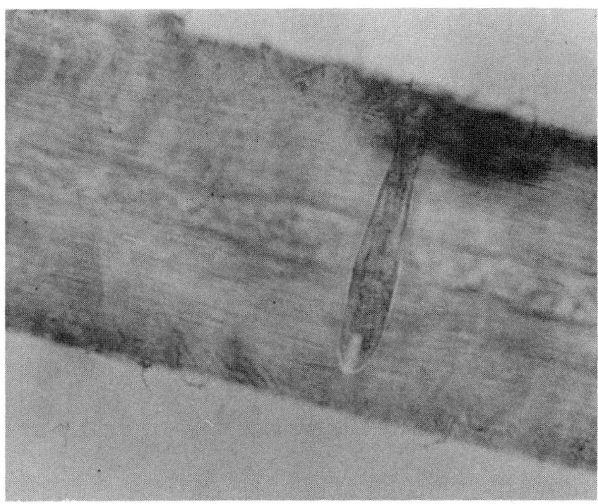

FIG. 9. Positive in vitro hair perforation test. Perforation caused by *Trichophyton mentagrophytes*. ×700.

from humans in the United States. For other species, refer to Rebell and Taplin (19) and Ajello (3, 4).

(i) *Epidermophyton floccosum.* Teleomorph: unknown.

Gross colony characteristics. Growth slow, white and fluffy at first, becoming velvety and powdery, greenish-yellow to tan. Surface flat or radially folded. Reverse tan. Tufts of white sterile growth (pleomorphic) are common.

Microscopic characteristics. Macroconidia numerous in young colonies. At first aseptate and in finger-like groups of two or three, becoming broadly clavate, two- to three-celled, with roundish distal ends and with smooth walls (Fig. 10). Microconidia absent. Spiral hyphae extremely rare. Chlamydospores numerous in old cultures.

Physiologic and pathologic characteristics. No special nutritional requirements. Does not perforate hair in vitro. Common agent of ringworm of the skin, particularly of the feet, groin (tinea curis), and nails of humans. Does not invade hair. Rarely reported in animals. Anthrophophilic.

(ii) *Microsporum audouinii* (=*M. langeronii, M. rivalierii*). Teleomorph: unknown.

Gross colony characteristics. Colony slow growing, flat with short aerial hyphae. Surface gray, cream to tan. Reverse salmon pink to reddish-brown.

Microscopic characteristics. Hyphae usually sterile with occasional chlamydospores. Microconidia usually rare, except on enriched media; clavate, small, borne laterally or terminally on short pedicels, or sessile. Macroconidia usually absent, but a small number are produced by rare isolates; when present, they are large, irregularly spindle shaped, and thick walled, with a smooth or echinulate surface. Abortive or biazarre-shaped macroconidia are more commonly seen.

Physiologic and pathologic characteristics. Grows poorly on rice-grain medium (see Chapter 111). Does not perforate hair in vitro. Formerly common agent of epidemic tinea capitis in young children. Invades skin and hair, very rarely nails. Very rarely infects adults or animals. Anthropophilic.

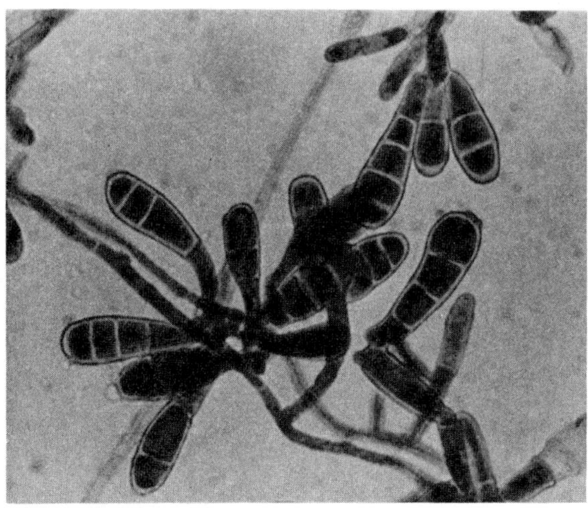

FIG. 10. Clusters of smooth macroconidia of *Epidermophyton floccosum.* ×700.

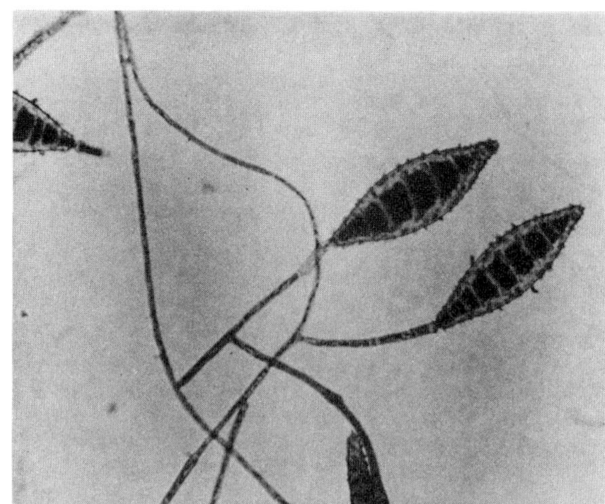

FIG. 11. Echinulate macroconidia of *Microsporum canis.* ×640.

(iii) *Microsporum canis* (*M. distortum, M. felineum, M. lanosum*). Teleomorph: *Nannizzia otae.*

Gross colony characteristics. Colonies are fast growing, white, fluffy at first but becoming silky, with bright-yellow pigment showing through the periphery. After 2 to 4 weeks, aerial mycelium is dense, cottony, and tan, sometimes in irregular tufts of concentric rings. Reverse at first bright yellow, becoming buff orange-brown. Rare isolates show no pigment on reverse.

Microscopic characteristics. Microconidia clavate, small, usually less numerous than macroconidia. Macroconidia usually numerous, large, fusiform, 35 to 110 μm by 12 to 25 μm with up to 14 septa. Thick walled (up to 4 μm) and with verruculose (minutely warty) walls (Fig. 11).

Physiologic and pathologic characteristics. No special nutritional requirements. Perforates hair in vitro. Grows very well and sporulates abundantly on rice-grain medium. Common agent of ringworm in cats and dogs. Most infections in humans are acquired from infected animals rather than from infected humans. Zoophilic.

(iv) *Microsporum gypseum.* Teleomorphs: *Nannizzia gypsea, N. incurvata.*

M. gypseum is a complex of species. Before 1963, *M. fulvum* was also considered to be cospecific with *M. gypseum*. The discovery of the teleomorphs of the *M. gypseum* complex proved that *M. fulvum*, though morphologically similar to *M. gypseum*, is a distinct species. The conidial states of *N. incurvata* and *N. gypsea* are referable to *M. gypseum*, and that of *N. fulva* is referable to *M. fulvum*. The conclusive identification and differentiation of *M. gypseum* and *M. fulvum* is best achieved by mating the conidial isolates with the tester strains of *N. fulva, N. gypsea*, and *N. incurvata*. This generally is not performed in the standard clinical laboratory.

Gross colony characteristics. Growth rapid; colonies flat, coarsely or finely granular, with an irregularly fringed border, buff to rosy buff, with reverse rosy buff to cinnamon.

Microscopic characteristics. Macroconidia numerous, ellipsoid to fusiform, 25 to 60 μm by 8.5 to 15 μm,

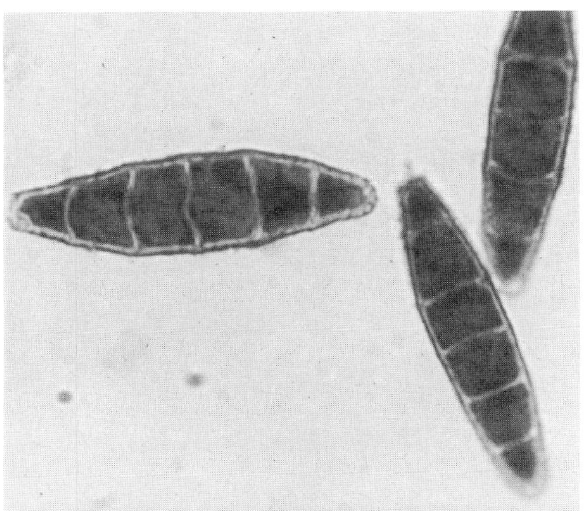

FIG. 12. Echinulate, elliptical macroconidia of *Microsporum gypseum* complex. ×960.

with up to five septa and moderately thick (up to 1.2 μm; Fig. 12), verruculose walls. Microconidia few in number, clavate, sessile or on short pedicels, 3.3 to 8.4 μm by 1.7 to 3.3 μm.

Physiologic and pathologic characteristics. No special nutritional requirements. Perforates hair in vitro. Common in soil. Infection rare in humans and more common in animals, particularly dogs and horses. Geophilic.

(v) *Microsporum fulvum.* Teleomorph: *Nannizzia fulva.*

Gross colony characteristics. Growth rapid; colonies dense, downy to granular, pale buff to rosy buff, usually with white cottony periphery. Reverse rosy buff to amber.

Microscopic characteristics. Macroconidia numerous, predominantly cylindrical, slightly tapering towards each end and with a rounded apex, or clavate, occasionally ellipsoid to fusiform, 25 to 58 μm by 7.5 to 12 μm, with up to five (rarely seven) septa and verruculose walls. Microconidia are clavate, 1.7 to 3.3 μm by 3.3 to 8.3 μm, unicellular, sessile or on short pedicels, and borne on both sides of the hyphae.

Physiologic and pathologic characteristics. No special nutritional requirements. Perforates hair in vitro. Common in soil. Infections are rare in humans. Geophilic.

(vi) *Microsporum nanum.* Teleomorph: *Nannizzia obtusa.*

Gross colony characteristics. Growth rapid; colonies flat, powdery to fluffy, cream to buff to cinnamon colored. Reverse orange-tan, becoming brownish-red. (Colony very similar to that of *M. gypseum.*)

Microscopic characteristics. Macroconidia numerous, small, ovate to elliptical with one to two cells (rarely as many as four cells), echinulate (Fig. 13). Microconidia few, clavate to pyriform, sessile on hyphae.

Physiologic and pathologic characteristics. No special nutritional requirements. Perforates hair in vitro. Grows well and sporulates on rice-grain medium. Geophilic. Commonest agent of ringworm in pigs. Rarely infects humans.

(vii) *Microsporum persicolor.* Teleomorph: *Nannizzia persicolor.*

Gross colony characteristics. Growth moderately rapid; colonies downy, velvety, or rarely powdery; white at first, becoming peach or light buff colored. Reverse reddish-brown. On Pablum cereal agar (see Chapter 111), the majority of the isolates turn peach to rose-violet or even deep vinous red.

Microscopic characteristics. Microconidia numerous, pyriform to spherical, borne along the sides of hyphae or in grapelike bunches. Macroconidia produced by most freshly isolated cultures, evident after 20 to 24 days of incubation at 25°C. Macroconidia clavate, thin walled, and finely echinulate (Fig. 14). The echinulations on the outer walls are evident under oil immersion; they become more pronounced when grown on sterile soil baited with hair. Tightly wound and loose spirals are numerous in fresh isolates.

Physiologic and pathologic characteristics. No special nutritional requirements. Perforates hair in vitro.

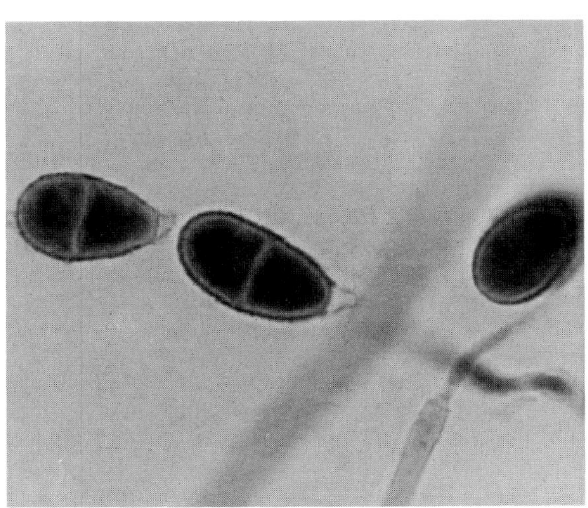

FIG. 13. Two-celled, echinulate macroconidia of *Microsporum nanum.* ×1,210.

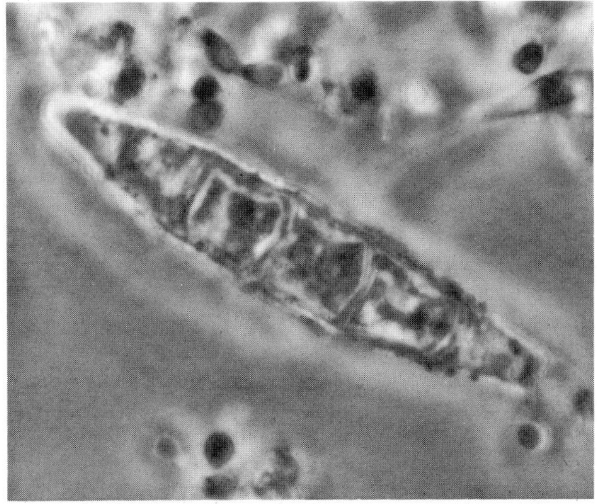

FIG. 14. Echinulate, clavate macroconidium of *Microsporum persicolor.* ×1,650.

M. persicolor is a frequent but mild pathogen of small wild rodents, particularly bank voles and field voles. It infrequently infects humans. Geophilic.

Until recently, *M. persicolor* was believed to be endemic only in western European countries. Recent studies show that it has a wide geographic distribution. It occurs as an infrequent pathogen of humans in the United States, Canada, Africa, and Australia. *M. persicolor* closely resembles *T. mentagrophytes* except for its echinulate macroconidia. It can be differentiated from *T. mentagrophytes* by cultivation on Pablum cereal agar, on which most *M. persicolor* isolates develop characteristics peach to rose-violet colonies. *T. mentagrophytes* shows no such change on Pablum cereal agar.

(viii) *Trichophyton equinum*. Teleomorph: unknown.

Gross colony characteristics. Growth rapid; colonies at first white and fluffy with bright-yellow pigment in peripheral growth. Surface velvety, folded, and cream to tan in older colonies. Reverse bright at first, turning pinkish to deep red-brown.

Microscopic characteristics. Macroconidia numerous, thin, pyriform, occasionally globose, sessile or on short pedicels, and borne along the sides of the hyphae. Macroconidia are very rare; when produced, they are cylindrical with thin, smooth walls.

Physiologic and pathologic characteristics. Most isolates have a complete requirement for nicotinic acid. Recently, isolates not dependent on nicotinic acid have been described from New Zealand and Australia as *T. equinum* var. *autotrophicum*. Occasionally infects humans and other animals. Zoophilic.

(ix) *Trichophyton mentagrophytes* (*T. asteroides*, *T. granulosum*, *T. gypseum*, *T. interdigitale*, *T. quinckeanum*). Teleomorphs: *Arthroderma benhamiae*, *Arthroderma vanbreuseghemii*.

Gross colony characteristics. Colonies fast growing, flat with powdery to granular surface (*T. mentagrophytes* var. *mentagrophytes*), flat with downy to fluffy surface (*T. mentagrophytes* var. *interdigitale*), or heaped and irregularly folded with downy to velvety surface (*T. mentagrophytes* var. *quinckeanum*). Surface usually white to tan or pale yellowish. Reverse usually pale yellowish or rose-brown, occasionally pink to red or orange-red (nodular variety).

Microscopic characteristics. Microconidia numerous, small, spherical or clavate, borne singly along the sides of hyphae, sessile or on short pedicels or in grapelike bunches. Macronconidia usually rare, but abundant in some isolates, two- to five-celled, clavate, thick, and smooth walled (Fig. 15). Spirally coiled hyphae and nodular bodies numerous in some isolates.

Physiologic and pathologic characteristics. Perforates hair in vitro. Common agent of all types of ringworm in humans and animals, especially rodents. Anthropophilic (*T. mentagraphytes* var. *interdigitale*); zoophilic (*T. mentagrophytes* vars. *erinacei*, *mentagrophytes*, and *quinckeanum*).

(x) *Trichophyton rubrum* (*T. purpureum*). Teleomorph: unknown.

Gross colony characteristics. Growth slow, flat or heaped at the center with a white, fluffy surface turning pink-tan with wine-red reverse (downy type),

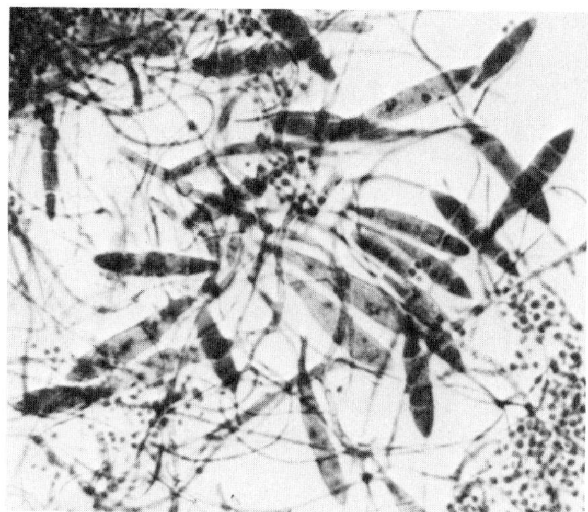

FIG. 15. Micro- and macroconidia of *Trichophyton mentagrophytes*. ×522.

or a powdery suede surface with radial furrows, creamy white, pink, with reverse dark red-tan to wine red (granular type). Some isolates (dysgonic type) are slow growing, deeply pigmented purple with woolly or granular texture and with submerged, feathery peripheral growth. Some downy forms are characterized by the production of diffusible (melanoid type) pigment on reverse. Some isolates (African type) are powdery with compactly heaped, folded center and buff-pink surface, red-tan on reverse. Occasionally, some African isolates produce a diffusible red pigment. The hyperpigmented type produces a violet to red-violet velvety surface with radial furrows with wine-red color on reverse. The pigment in older colonies may diffuse into the medium. Isolates of the *rodhainii* type are slow growing, glabrous, acuminate, folded, and deep purple with a white fringe and white granular center.

Microscopic characteristics. Microconidia thin, clavate, teardrop shaped, borne laterally on undifferentiated hyphae or on short stalks (Fig. 16). Macroconidia typically long, narrow, cylindrical with rounded aspices, three- to eight-celled, with thin, smooth walls (Fig. 17). The production of macroconidia and microconidia varies greatly among the different types of colonies. In the downy type, microconidia are moderate in number and there are no macroconidia. In the granular type, both microconidia and macroconidia are produced in comparatively large numbers. In the dysgonic type, microconidia are produced in varying numbers, and macroconidia are occasionally observed. In the African type, macroconidia are more numerous than microconidia. In hyperpigmented colonies, macroconidia and microconidia are numerous in older colonies of the *rodhainii* type. Many chlamydospores are frequently seen in the dysgonic African, melanoid, hyperpigmented, and *rodhainii* types (27).

Physiologic and pathologic characteristics. No special nutritional requirements. Does not perforate hair in vitro. Common agent of ringworm of the skin and nails. Rarely invades hair. Occasional parasite of animals. Anthropophilic.

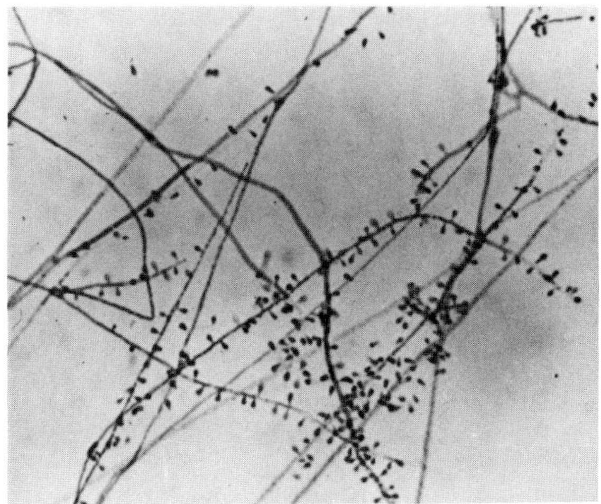

FIG. 16. Microconidia of *Trichophyton rubrum*. ×740.

(xi) *Trichophyton schoenleinii*. Teleomorph: unknown.

Gross colony characteristics. Slow-growing, irregularly heaped and folded colonies. Surface usually glabrous or waxy and cream to yellowish-tan. Old colonies become tough and leathery and may develop a white powdery or downy surface on some areas of the colony. Occasionally, some isolates grow largely submerged in the agar.

Microscopic characteristics. Mycelium highly irregular in diameter. Coarser hyphae become knobby and irregularly branched at ends (favic chandeliers) (Fig. 18). Chlamydospores usually numerous. Microconidia rare. Macroconidia absent to rare.

Physiologic and pathologic characteristics. No special nutritional requirements. Does not have a complete requirement for thiamine, but growth of some isolates is stimulated by thiamine. Growth not stimulated at 37°C. Does not perforate hair in vitro. Commonest cause of favus, a clinical form of ringworm characterized by heavy cup-shaped crusts scutulae and hair invaded throughout its length by hyphae which do not fragment into arthroconidia. Also infects skin and nails. Anthropophilic.

(xii) *Trichophyton soudanense*. Teleomorph: unknown.

Gross colony characteristics. Growth slow, compact, glabrous, velvety with irregularly folded surface and with lanceolate radiations from the central zone. Color varies from yellow or orange-yellow to apricot. Reverse of the colony is deep yellow. On subculture, the apricot color is generally lost. On Lowenstein-Jensen medium, colonies become brown to black.

Microscopic characteristics. Growth on Sabouraud agar at room temperature is generally devoid of microconidia and macroconidia. Occasionally, some isolates produce a few microconidia. On lactrimel agar (Chapter 111), microconidia are moderate to numerous, pyriform, clavate, borne singly on sides of undifferentiated hyphae, either sessile or borne on short conidiophores. Macroconidia are smooth, four- to five-celled, with thin, smooth walls. Characteristic reflexive branching of the hyphae, with a tendency of the lateral branches to grow in the opposite direction to the growth of the main hyphae, is observed when grown on cornmeal agar with 1% Tween 80 (25, 26). In older cultures, numerous chlamydospores are observed.

Physiologic and pathologic characteristics. No special nutrition requirements. In vitro does not perforate hair. Causes large-spored endothrix infections of the scalp. Reported chiefly from Africa, Australia, Brazil, Canada, Israel, several European countries, and occasionally from the United States. Also causes ringworm of the skin and nails. Anthropophilic.

(xiii) *Trichophyton tonsurans* (*T. acuminatum, T. cerebriforme, T. crateriforme, T. sulfureum*). Teleomorph: unknown.

Gross colony characteristics. Growth slow, colonies flat at first with finely powdery surface, later becoming highly heaped, folded with a velvety surface. Center may be acuminate or depressed, or entire surface may be irregularly folded (*T. tonsurans* var. *cerebriforme*). Surface usually cream to tan, rarely rose or bright yellow (*T. tonsurans* var. *sulfureum*). Reverse yellowish to mahogany red.

Microscopic characteristics. Microconidia numerous, delicate, elongate in young colonies, larger and

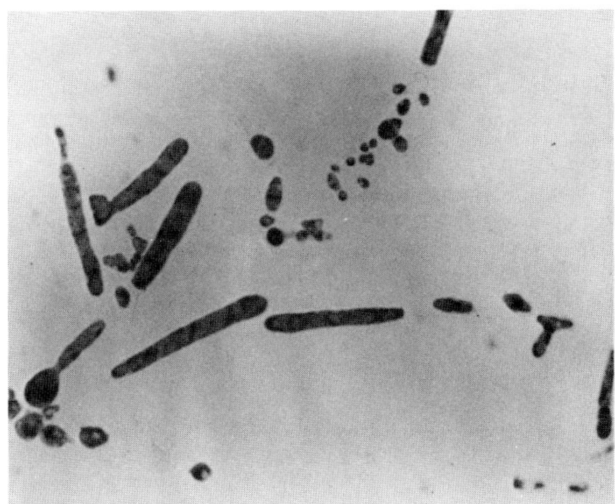

FIG. 17. Smooth, elongated macroconidia of *Trichophyton rubrum*. ×655.

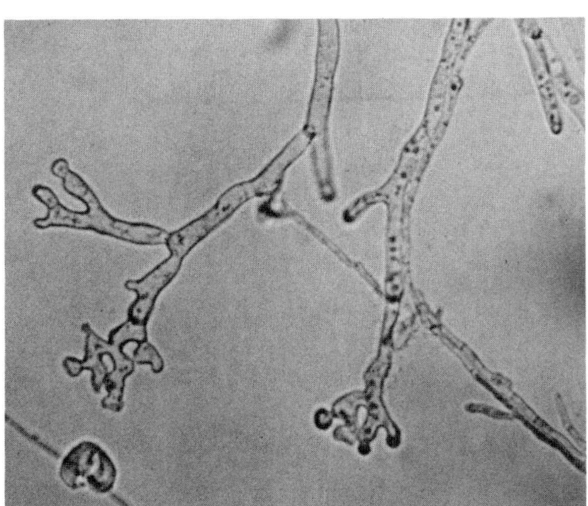

FIG. 18. Favic chandeliers of *Trichophyton schoenleinii*. ×830.

irregular in size and shape in older colonies (Fig. 19). Macroconidia are rare in many isolates; when produced, they are cylindrical to clavate, slightly curved at the tops, two- to three-celled or sometimes five- to seven-celled.

Physiologic and pathologic characteristics. Grows poorly on vitamin-free media. Does not perforate hair in vitro with the exception of *T. tonsurans* var. *sulfureum* subvar. *perforans*. Growth greatly stimulated by thiamine. Common cause of tinea capitis in both children and adults. Causes large-spored endothrix infections. Also commonly infects skin and nails. Infection rare in animals. Anthropophilic.

(xiv) *Trichophyton verrucosum* (*T. album, T. discoides, T. faviforme, T. ochraceum*). Teleomorph: unknown.

Gross colony characteristics. Growth very slow. Colonies at first small, heaped, glabrous, tough, and leathery. May become disk shaped (var. *discoides*) or highly heaped and folded (var. *album*). Most isolates are grayish-white; a few are yellowish-tan (var. *ochraceum*). Old colonies may develop white, powdery to downy, aerial growth.

Microscopic characteristics. On Sabouraud agar at room temperature, growth is thin, consisting of hyphae with many chlamydospores. At 37°C, chlamydospores become numerous, often in chains. On thiamine-enriched media, mycelium is more regular in form, and microconidia may be numerous. Microconidia small, thin, and borne along the sides of the hyphae. Macroconidia extremely rare, three- to five-celled, variable in shape, thin, and smooth walled.

Physiologic and pathologic characteristics. No growth on vitamin-free media. Some isolates require thiamine; others require a combination of thiamine and inositol. Growth more rapid at 37°C than at room temperature. Does not perforate hair in vitro. Common agent of ringworm in cattle. Occasionally infects humans and other animals. Zoophilic.

(xv) *Trichophyton violaceum* (*T. glabrum*). Teleomorph: unknown.

Gross colony characteristics. Growth very slow; heaped or finely folded colonies with a glabrous or waxy surface. At first cream colored, becoming pinkish and then lavender to deep purple. Old cultures may develop color-free sectors with downy grayish aerial hyphae. Some isolates lack pigment when isolated (*T. violaceum* var. *glabrum*).

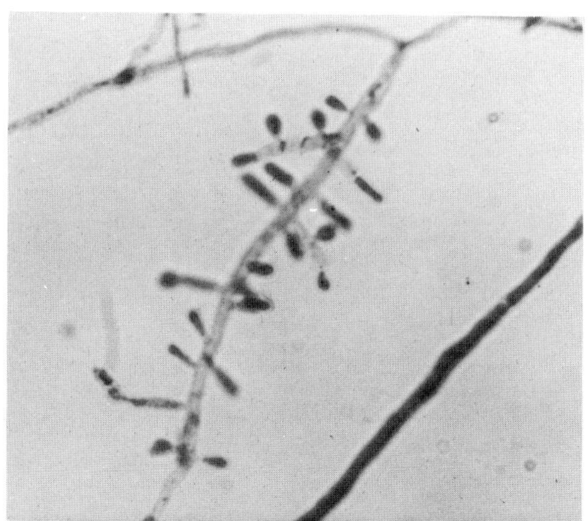

FIG. 19. Microconidia of *Trichophyton tonsurans*. ×1,070.

Microscopic characteristics. Mycelium thin and usually sterile. Occasionally, chlamydospores may be found. Microconidia very rare; macroconidia absent on most media. In some isolates, conidia are produced in old cultures when grown on thiamine-enriched media.

Physiologic and pathologic characteristics. Grows poorly on vitamin-free media. Does not perforate hair in vitro. Growth greatly stimulated by thiamine. Cause of tinea capitis in both children and adults. Also infects skin and nails. Infection very rare in animals. Anthropophilic.

LITERATURE CITED

1. **Ajello, L.** 1962. Present day concepts of the dermatophytes. Mycopathol. Mycol. Appl. **17:**317–324.
2. **Ajello, L.** 1968. A taxonomic review of the dermatophytes and related species. Sabouraudia **6:**147–159.
3. **Ajello, L.** 1977. Taxonomy of the dermatophytes. A review of their imperfect and perfect states, p. 289–297. *In* K. Iwata (ed.), Recent advances in medical and veterinary mycology. University of Tokyo Press, Tokyo.
4. **Ajello, L.** 1978. Gegenwartige Kenntnisse über die imperfekten und perfekten Formen der *Epidermophyton-, Mikrosporum-* und *Trichophytonarten*. Hautarzt **29:**6–9.
5. **Ajello, L., and L. K. Georg.** 1957. In vitro cultures for differentiating between atypical isolates of *Trichophyton mentagrophytes* and *Trichophyton rubrum*. Mycopathol. Mycol. Appl. **8:**3–7.
6. **Blank, H., S. Sagami, C. Boyd, and F. J. Roth, Jr.** 1959. The pathogenesis of superficial fungous infections in cultured human skin. Arch. Dermatol. **79:**524–535.
7. **Conant, N. F., D. T. Smith, R. D. Baker, and J. L. Callaway.** 1971. Manual of clinical mycology, 3rd ed., p. 548–586. The W. B. Saunders Co., Philadelphia.
8. **Emmons, C. W., C. H. Binford, J. P. Utz, and K. J. Kwon-Chung.** 1977. Medical mycology, 3rd ed. Lea & Febiger, Philadelphia.
9. **Georg, L. K., and L. B. Camp.** 1957. Routine nutritional tests for the identification of dermatophytes. J. Bacteriol. **74:**113–121.
10. **Kaplan, W.** 1959. The occurrence of black piedra in primate pelts. Trop. Geogr. Med. **11:**115–126.
11. **Lodder, J.** 1970. The yeasts. North-Holland Publishing Co., Amsterdam.
12. **Lorincz, A. L., J. O. Priestly, and P. J. Jacob.** 1958. Evidence for a humoral mechanism which prevents growth of dermatophytes. J. Invest. Dermatol. **31:**15–17.
13. **Mackenzie, D. W. R.** 1963. "Hairbrush diagnosis" in detection and eradication of nonfluorescent scalp ringworm. Br. Med. J. **2:**263–265.
14. **Matsumoto, T., A. A. Padhye, and L. Ajello.** 1983. *In vitro* hair perforation by a new subvariety of *Trichophyton tonsurans* var. *sulfureum*. Mycotaxon **18:**235–242.
15. **Padhye, A. A., and J. W. Carmichael.** 1971. The genus *Arthroderma* Berkeley. Can. J. Bot. **49:**1525–1540.
16. **Padhye, A. A., I. Weitzman, and L. Ajello.** 1979. Mating behavior of *Microsporum equinum* with *Nannizzia otae*. Mycopathologia **69:**87–90.
17. **Padhye, A. A., C. N. Young, and L. Ajello.** 1980. Hair perforations as a diagnostic criterion in the identification of *Epidermophyton, Microsporum* and *Trichophyton* species, p. 115–120. *In* Superficial, cutaneous, and subcutaneous infections. Scientific publication no. 396. Pan American Health Organization, Washington, D.C.
18. **Prochacki, H., and C. Engelhardt-Zasada.** 1974. *Epider-*

mophyton stockdaleae sp. nov. Mycopathol. Mycol. Appl. **54:**341–345.
19. **Rebell, G., and D. Taplin.** 1970. Dermatophytes, their recognition and identification. University of Miami Press, Coral Gables, Fla.
20. **Rippon, J.** 1982. Medical mycology, 2nd ed. The W. B. Saunders Co., Philadelphia.
21. **Salkin, I. F., and M. A. Gordon.** 1977. Polymorphism of *Malassezia furfur*. Can. J. Microbiol. **23:**471–475.
22. **Takashio, M., and R. Vanbreuseghem.** 1971. Production of ascospores by *Piedraia hortae*. Mycologia **63:**612–618.
23. **Taplin, D.** 1965. The use of gentamicin in mycology. J. Invest. Dermatol. **45:**549–550.
24. **von Arx, J. A., L. Rodrigues de Miranda, M. T. Smith, and D. Yarrow.** 1977. The genera of yeasts and yeast-like fungi, p. 30–31. *In* Studies in mycology, no. 14. Centraalbureau voor Schimmelcultures, Baarn, Netherlands.
25. **Weitzman, I., and S. Rosenthal.** 1984. Studies in the differentiation between *Microsporum ferrugineum* Ota and *Trichophyton soudanense* Joyeux. Mycopathologia **84:**95–101.
26. **Weitzman, I., I. F. Salkin, and S. A. Rosenthal.** 1983. Evaluation of *Trichophyton* agars for identification of *Trichophyton soudanense*. J. Clin. Microbiol. **18:**203–205.
27. **Young, C. N.** 1972. Range of variation among isolates of *Trichophyton rubrum*. Sabouraudia **10:**164–170.

Chapter 49

Yeasts of Medical Importance

BILLY H. COOPER AND MARGARITA SILVA-HUTNER

GROUP CHARACTERIZATION

Summary of distinguishing characteristics

Medically important yeasts are those fungi of primarily unicellular growth habit that are capable of producing or contributing to diseases of humans and animals. Of 361 species in 39 genera of yeasts in the second edition of Lodder's *The Yeasts* (44), and of 434 species recorded by Barnett and Pankhurst (6), only about 24 now meet the above definition (Table 1). Although *Candida lusitaniae* (29) and *Hansenula polymorpha* (51) have recently been added to the list of species of yeasts that are able to cause human disease, the number of pathogens does not appear to be expanding dramatically. The percentage of pathogenic yeasts among the totality of species remains quite small. However, these newcomers to the species reported to cause human disease do require that clinical laboratory personnel better understand the characteristics that delimit yeast species and the methods that can be used to detect those characteristics.

A unicellular growth habit is maintained when these yeasts are actively growing (log phase) under standard temperatures, aeration, pressures, and humidity. Their colonies therefore are glabrous, moist, creamy or membranous in texture, and lack the aerial hyphae that impart a fluffy or velvety texture to the colonies of filamentous fungi (molds). It is true that many yeasts can form filaments as either true or pseudohyphae, but these are produced under diminished O_2 such as may prevail in submerged portions of solid culture media, in an atmosphere of 5 to 10% CO_2, or in the tissues of a parasitized host.

Normal vegetative cells of yeasts are round or oval, 2.5 to 6 μm in diameter, and reproduce asexually by budding, bud fission, or fission. Buds (blastoconidia) can remain attached to the mother cell and continue to bud, producing branching clusters of blastoconidia. Individual blastoconidia adhering to their neighbors in a chain can elongate to produce filaments called pseudohyphae. True septate hyphae result from germination of "transitional cells" (rounded or flattened blastoconidia, or chlamydospores).

According to their method of sexual reproduction, yeasts can be divided into three main groups, Ascomycetes (*Saccharomyces, Endomycopsis, Pichia, Hansenula*, and *Nematospora*), Heterobasidiomycetes (*Leucosporidium, Filobasidiella*, and *Syringospora*), and Deuteromycetes, or Fungi Imperfecti (*Candida, Cryptococcus, Rhodotorula, Torulopsis*, and *Trichosporon*) (34). Ultrastructural and biochemical criteria may be used to suggest affinities of species to higher taxa. DNA hybridization techniques, used increasingly often, may help to clarify relatedness among different isolates.

Unicellular, budding fungi whose cell walls contain melanins are arbitrarily excluded from the species discussed here. A complete discussion of these "black yeasts" (classified in the *Dematiaceae*) can be found in Chapter 52.

Clinical significance

Yeast infections are among the most common fungal infections affecting humans; their incidence has greatly increased with the advent of broad-spectrum antibiotics, immunosuppressive corticosteroids, and antitumor agents. Their severity ranges from benign and localized (either transient or chronic) to disseminated and sometimes fatal. Yeast fungemia occurs frequently in patients with indwelling catheters and can result in endocarditis or pyelonephritis in recipients of organ transplants, artificial heart valves, or other prosthetic devices. *Candida* infections of the eye result from injury to the cornea (keratitis) or can involve the retina (endophthalmitis) as one manifestation of candidiasis spread by hematogenous dissemination. *Candida* infection of the eye is serious and can result in the loss of visual function or enucleation. Yeast endocarditis is also frequent in drug addicts who inject themselves intravenously using nonsterile syringes or needles.

Yeasts exist in nature in a wide variety of substrates, including fruits, vegetables, and homemade fermented beverages. *Candida albicans*, the most frequent pathogen, is a normal inhabitant of the alimentary tract; various surveys show an incidence of 20 to 40% in asymptomatic individuals. Another yeast, *Rhodotorula glutinis*, is a common inhabitant, in the tropics, of moist skin; its role as a pathogen is questionable, although it has been reported to cause fungemia.

Because of the frequent association of yeasts with the internal and external environment of humans, their incidence in clinical specimens is rather high; e.g., yeast colonies grew in cultures from 2,017 of 7,685 clinical specimens received at the mycology laboratory of Columbia-Presbyterian Medical Center during the 1982–1983 fiscal year. During this time, 2,271 cultures were positive for fungi of all types, making yeasts the predominant type of fungi isolated. Most of these yeast isolates were recovered from urine cultures. These numbers may denote increased interest in or awareness of urinary tract infections by the medical staff during this period. Nevertheless, it is estimated that less than 10% of patients examined actually had systemic yeast infections. Hence, collaboration between physicians and laboratory is essential in assessing the possible etiological role of a yeast growing in a clinical culture. Besides clinical and laboratory evaluation of the patients, critical studies should include microscopic examination of fresh specimens and serological tests for specific antigens and antibodies, as well as cultural isolation of such yeasts.

DIRECT EXAMINATION

Identification of yeasts in a clinical specimen begins with the direct microscopic examination of stained or unstained samples of the specimen. Such examination does not permit exact species identification, but does

TABLE 1. Cultural and biochemical characteristics of yeasts frequently isolated from clinical specimens[a]

Species	Growth at 37°C	Pellicle in broth	Pseudo/true hyphae	Chlamydospores	Germ tubes	Capsule, India ink	Assimilation of[b]: Glucose	Maltose	Sucrose	Lactose	Galactose	Melibiose	Cellobiose	Inositol	Xylose	Raffinose	Trehalose	Dulcitol	Fermentation of: Glucose	Maltose	Sucrose	Lactose	Galactose	Trehalose	Cellobiose	Urease	KNO_3 utilization	Phenol oxidase	Ascospores
Candida albicans	+	−	+	+	+	−	+	+	+	−	+	−	−	−	+	−	+	−	F	F	−	−	F	F	−	−	−	−	−
C. guilliermondii	+	−	+	−	−	−	+	+	+	−	+	+	+	−	+	+	+	+	F	−	F	−	F*	F	−	−	−	−	−*
C. krusei	+	+	+	−	−	−	+	−	−	−	−	−	−	−	−	−	−	−	F	−	−	−	−	−	−	+*	−	−	−
C. lipolytica[c]	+	+	+	−	−	−	+	−	−	−	−	−	−	−	−	−	−	−	−	−	−	−	−	−	−	−	−	−	−
C. lusitaniae[d]	+	−	+	−	−	−	+	+	+	−	+	−	+	−	+	−	+	−	F	−	F	−	F	F	F	−	−	−	−
C. parapsilosis	+	−	+	−	−	−	+	+	+	−	+	−	−	−	+	−	+	−	F	−	−	−	−	−	−	−	−	−	−
C. pseudotropicalis (*C. kefyr*)	+	−	+	−	−	−	+	−	+	+	+	−	+*	−	+*	+	−*	−	F	−	F	F*	F	−	−	−	−	−	−
C. rugosa	+	−	+	−	−	−	+	−	−	−	+	−	−	−	+*	−	−	−	−	−	−	−	−	−	−	−	−	−	−
C. stellatoidea	+	−	+	+	+	−	+	+	−	−	+	−	−	−	+	−	+	−	F	F	−	−	−	−	−	−	−	−	−
C. tropicalis	+	+	+	−	−	−	+	+	+	−	+	−	+	−	+	−	+	−	F	F	F	−	F	F	−	−	−	−	−
Torulopsis candida	+	−	−	−	−	−	+	+	+	+*	+	+	+	−	+	+	+	+*	W	−	W	−	−	W	−	−	−	−	−
T. glabrata	+	−	−	−	−	−	+	−	−	−	−	−	−	−	−	−	+	−	F	−	−	−	−	F	−	−	−	−	−
T. pintolopesii	+[e]	−	−	−	−	−	+	−	−	−	−	−	−	−	−	−	−	−	F	−	−	−	−	−	−	−	−	−	−
Cryptococcus neoformans	+	−	R	−	−	+	+	+	+	−	+	−	+	+	+	+*	+	+	−	−	−	−	−	−	−	+	−	+	−
C. albidus	−*	−	−	−	−	+	+	+	+	+*	+*	+	+	+	+	+	+	+*	−	−	−	−	−	−	−	+	+	−	−
C. gastricus	−	−	−	−	−	+	+	+	+*	−	+	−	+	+	+	−	+	−	−	−	−	−	−	−	−	+	−	−	−
C. laurentii	+*	−	−	−	−	+	+	+	+	+	+	+*	+	+	+	+*	+	+	−	−	−	−	−	−	−	+	−	−	−
C. luteolus	−	−	−	−	−	−	+	+	+	−	+	+	+	+	+	+	+	+	−	−	−	−	−	−	−	+	−	−	−
C. terreus	−*	−	−	−	−	+	+	+*	−	+*	+*	−	+	+	+	−	+	−*	−	−	−	−	−	−	−	+	+	−	−
C. uniguttulatus	−	−	−	−	−	+	+	+	+	−	−*	−	−*	+	+	+*	−*	−	−	−	−	−	−	−	−	+	−	−	−
Rhodotorula glutinis	+	−	−	−	−	−*	+	+	+	−	+*	−	+	−	+	+	+	−	−	−	−	−	−	−	−	+	+	−	−
R. rubra	+	−	−	−	−	−*	+	+	+	−	+	−	+*	−	+	+	+	−	−	−	−	−	−	−	−	+	−	−	−
Saccharomyces cerevisiae	+	−	−*	−	−	−	+	+	+	−	+	−	−	−	−	+	+*	−	F	F	F	−	F	F*	−	−	−	−	+
Trichosporon beigelii	+*	+	+	−	−	−	+	+	+	+	+*	+*	+*	+	+	+*	+*	+*	−	−	−	−	−	−	−	+*	−	−	−
T. pullulans	+*	+	+	−	−	−	+	+	+	+*	+	+*	+	+*	+*	+*	+	−	−	−	−	−	−	−	−	+	+	−	−
Geotrichum candidum	−*	+	+	−	−	−	+	−	−	−	+	−	−	−	+	−	−	−	−	−	−	−	−	−	−	−	−	−	−
G. capitatum	+	+	+	−	−	−	+	−	−	−	+	−	−	−	−	−	−	−	−	−	−	−	−	−	−	−	−	−	−
Hansenula anomala	+*	−	−	−	−	−	+	+	+	−	+	−	+	−	+	−	+	−	F	F*	F	−	F	−	−	−	+	−	+

[a] *, Strain variation; R, rare; F, the sugar is fermented (i.e., gas is produced); W, weak fermentation. Based on data from Ahearn and Schlitzer (3) and Kreger-Van Rij (34).

[b] +, Growth greater than that of the negative control.

[c] *C. lipolytica* assimilates erythritol; *C. krusei* does not.

[d] *C. lusitaniae* assimilates rhamnose; *C. tropicalis* does not.

[e] *C. pintolopesii* is a thermophilic yeast capable of growth at 40 to 42°C.

[f] Occasional strains of *C. tropicalis* produce teardrop-shaped chlamydospores.

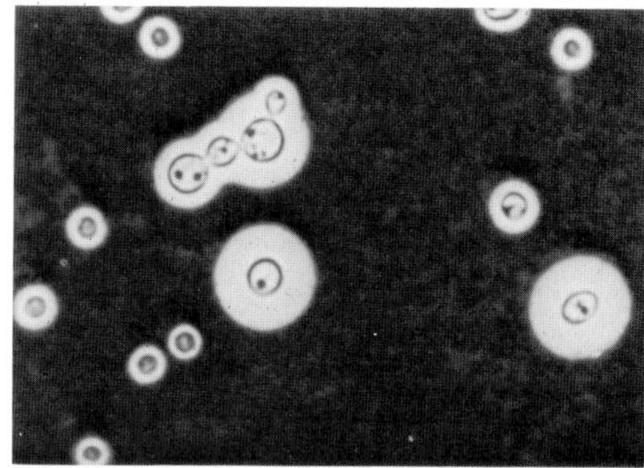

FIG. 1. *Cryptococcus neoformans.* Nigrosin-stained wet preparation of sediment from centrifuged cerebrospinal fluid. Note encapsulated cells, some with wide capsules which probably originated in vivo; narrow capsules are probably on daughter cells "budded out" in vitro. × 1,200.

provide early clues which aid the physician in making a presumptive diagnosis and aids the laboratorian in deciding which further cultural and biochemical tests are needed. The presence of yeasts in properly collected specimens of body fluids which are normally sterile is immediately significant; however, the significance of yeasts in naturally contaminated specimens such as sputum, feces, and urine depends on other considerations. The important findings from direct microscopy include: (i) the presence or absence of encapsulated yeasts (Fig. 1), (ii) the presence or absence of pseudohyphae (Fig. 2), (iii) the presence or absence of true hyphae and arthroconidia (Fig. 3), (iv) the size and shape of the yeast; and (v) the number of buds and nature of their attachment to the mother cell.

COLLECTION, TRANSPORT, AND STORAGE OF SPECIMENS

Specimens to be examined for yeasts of possible etiological significance can be collected in the same way as bacteriological specimens, with certain precautions and exceptions. A minimum of 5 ml of cerebrospinal fluid is essential for adequate sampling when culturing for *Cryptococcus*.

Whenever possible, specimens for yeasts should be examined and cultured at the bedside, or a portion of the specimen should be fixed as a dry smear or preserved in 10% Formalin for microscopic evaluation of the abundance and morphology of the yeast at the moment of collection. For bedside cultures, Sabouraud agar slants in screw-capped tubes or bottles should be available wherever patients are examined. Other culture media are described in Chapter 47.

To facilitate handling, specimens received by the laboratory should be grouped according to type of material submitted.

Smears or scrapings

Smears or scrapings on glass slides can be examined by adding a drop of saline solution or 10% NaOH. The latter should be used with fecal smears, thick exudates, or skin scrapings. Stains are seldom necessary, but yeasts can be detected in smears stained by the Gram, Wright, Ziehl-Neelsen, Giemsa, Papanicolaou, periodic acid-Schiff, or Gomori methenamine silver methods. However, it is needless to duplicate efforts by preparing special smears for yeasts from specimens that will also be examined routinely for bacteria and cytological abnormalities. Gram-stained smears from the vagina are quite acceptable for detecting yeasts, and yeasts can readily be observed in specimens submitted for cytological examination and stained with Papanicolaou stain. Yeasts are also effectively revealed by staining heat-fixed smears with Loeffler methylene blue or with Calcofluor white.

Cerebrospinal fluid or thoracentesis specimens

Cerebrospinal fluid or thoracentesis specimens should be centrifuged or passed through a membrane filter to concentrate solids and increase the probability of finding yeasts. A drop of sediment or a piece of the membrane filter (cut and handled aseptically) can then be examined. India ink or nigrosin may be added by capillarity under the cover slip as a negative "capsule stain" to reveal *Cryptococcus*, if present. Familiarity with the cytological features of *Cryptococcus* is essential to avoid confusing this yeast with blood cells, lipid globules, starch granules, or other artifacts. Examination of India ink stains of living cultures of *Cryptococcus* suspended in body fluids can contribute to this experience. Besides wet preparations, dry smears can be stained as described above, and alcian blue and mucicarmine staining are particularly useful for demonstrating *Cryptococcus neoformans* in smears of sediment from cerebrospinal fluid.

Urine specimens

To assess the significance of *Candida* spp. in urine while handling the specimen appropriately for detecting *C. neoformans* and other fungi, it is necessary to proceed differently than for bacterial cultures of urine. It is generally accepted that 10^4 CFU of *Candida* spp. per ml of urine is suggestive of *Candida* pyelonephritis (25). Although this value is lower than that accepted as being significant for bacterial pyelonephritis (10^5 CFU/ml), numbers of *Candida* colonies lower than 10^4/ml are regarded as being due to contamination from the lower genitourinary tract and not indicative of infection. Hence one must quantitate yeasts in urine by streaking the specimen with a quantitative loop on an appropriate medium (see description of media in Chapter 47) and to examine a sample of the uncentrifuged urine microscopically. On the other hand, a single CFU of *C. neoformans,*

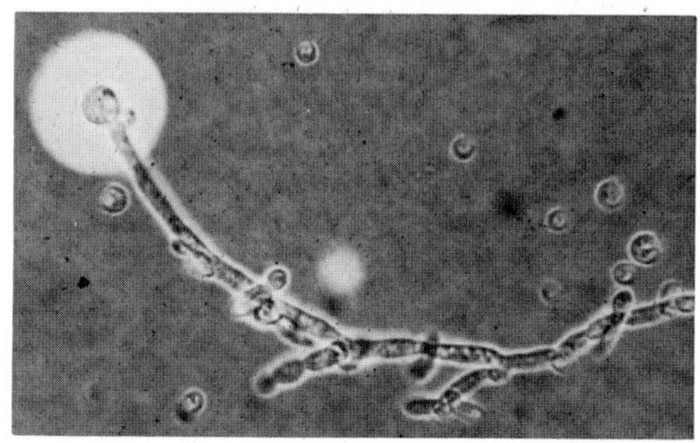

FIG. 2. *Cryptococcus neoformans.* Unusual strain which produced pseudohyphae. Nigrosin-stained wet preparation of sediment from centrifuged cerebrospinal fluid. ×1,200.

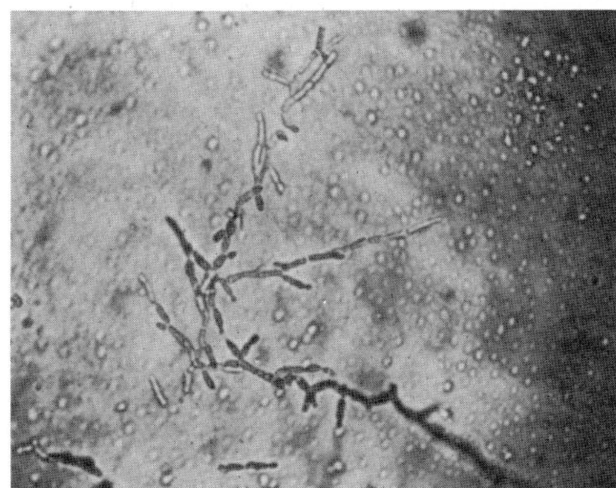

FIG. 3. *Trichosporon beigelii* in fresh unstained preparation of sputum. ×600. Note arthrospores and blastospores characteristic of the genus *Trichosporon*.

Histoplasma sp., or other systemic mycotic agent in the entire urine specimen would suggest infection. This necessitates microscopic examination of centrifuged urinary sediment as well as plating some sediment on isolation medium when a fungus culture is specifically requested, although this practice is contraindicated with bacterial cultures.

Sputum or other mucous secretions

It has been suggested that identification procedures used with yeasts isolated from sputum specimens be limited to those essential for detecting *C. neoformans* (55). However, the isolation of the same species from multiple specimens (sputum, urine, skin lesions, etc.) obtained from an immunocompromised patient suggests infection by that species. For that reason, we recommend that all yeasts isolated from seriously ill patients be identified, whatever the specimen source. Methods for processing respiratory secretions for isolating fungi are discussed further in Chapter 47.

Biopsy or autopsy tissue

Slice tissue with a sterile scalpel and examine all surfaces to detect ulcerations or granulomatous areas which usually contain the fungi.

Because of the possible occurrence of yeasts and yeastlike fungi as part of the transient or resident flora of the body, and because they may proliferate rapidly in vitro, microscopic examination of clinical specimens must be done soon after collection. All specimens should be delivered to the diagnostic laboratory as soon as possible, preferably within minutes of collection.

CULTURE AND ISOLATION

The preparation of specimens for primary culture of yeasts is described in Chapter 47. Selection of media for isolation of yeasts is also discussed in that chapter.

In addition to the media covered in Chapter 47, a diversity of agar media which contain salts of metals such as bismuth and molybdenum, or which contain a tetrazolium dye, are commercially available and provide early clues toward identification of yeast colonies based on color reactions. Precaution: bacteria can produce similar color reactions, and therefore microscopic examination of suspensions made from suspicious colonies is essential. Such media are especially useful for detecting individual colonies in a mixed population of more than one yeast. An excellent differential medium for presumptive identification of *C. neoformans* on primary isolation plates can be prepared by adding to Littman oxgall agar (10) (or to other commonly used media) an extract of seeds from the Indian thistle plant, *Guizotia abyssinica* (a seed common in birdseed mixtures, hence, "birdseed agar"). *C. neoformans* characteristically produces brown colonies on this medium within 1 week when incubated at 30°C, a property not shown by other yeasts including other *Cryptococcus* species. "Birdseed agar" is a general name for the medium first described by Staib (77) and later modified by others to make it more selective (75) or more chemically defined (30, 78). The color reaction of *C. neoformans* colonies on this medium is mediated by a phenol oxidase which leads to the formation of melanin pigment in the yeast cell walls (74). Substitute formulations for birdseed agar have been proposed, and optimal conditions for production of melanin pigment by *C. neoformans* have recently been detailed (30, 57, 74, 78).

Ability to form brown colonies on birdseed agar is one indication of phenol oxidase activity. Various phenols and diaminobenzenes including caffeic acid and dihydroxyphenylalanine are good substrates for the phenol oxidase produced by *C. neoformans*; however, the oxidase does not react with tyrosine (74). A rapid pigmentation test that requires only 6 h has been described (31). A commercial medium containing dihydroxyphenylalanine that is designed to screen yeast isolates for phenol oxidase activity was recently described and has been shown to be reliable for this purpose (15). Most companies dealing in prepared culture media now offer one or another formulation of birdseed agar.

Methods for culturing blood for isolating fungi are discussed in Chapter 47. As stated there, and in our experience, the lysis-centrifugation method for recovering yeasts from blood is recommended.

Whichever isolation medium is employed, human pathogenic yeasts will form pasty, opaque colonies at both 37°C and room temperature. For early presumptive identification of yeasts producing pseudohyphae, specimens positive on direct microscopic examination can be cut directly into cornmeal agar as well as planted onto other isolation media.

IDENTIFICATION

Recommended procedures

Methods for identifying yeasts are presented in Fig. 4; the cultural and biochemical characteristics of the species most often encountered in clinical specimens are summarized in Table 1. Initial steps include: (i) making a wet mount and stained smears for microscopic observation; (ii) making an India ink preparation; (iii) performing a germ tube test; and (iv) inoculating Wolin-Bevis or cornmeal agar to detect pseudohyphae and chlamydospores. While these tests are being done, each distinct colony type should be restreaked on bacteriological medium, such as brain heart infusion agar with or without 10% blood and

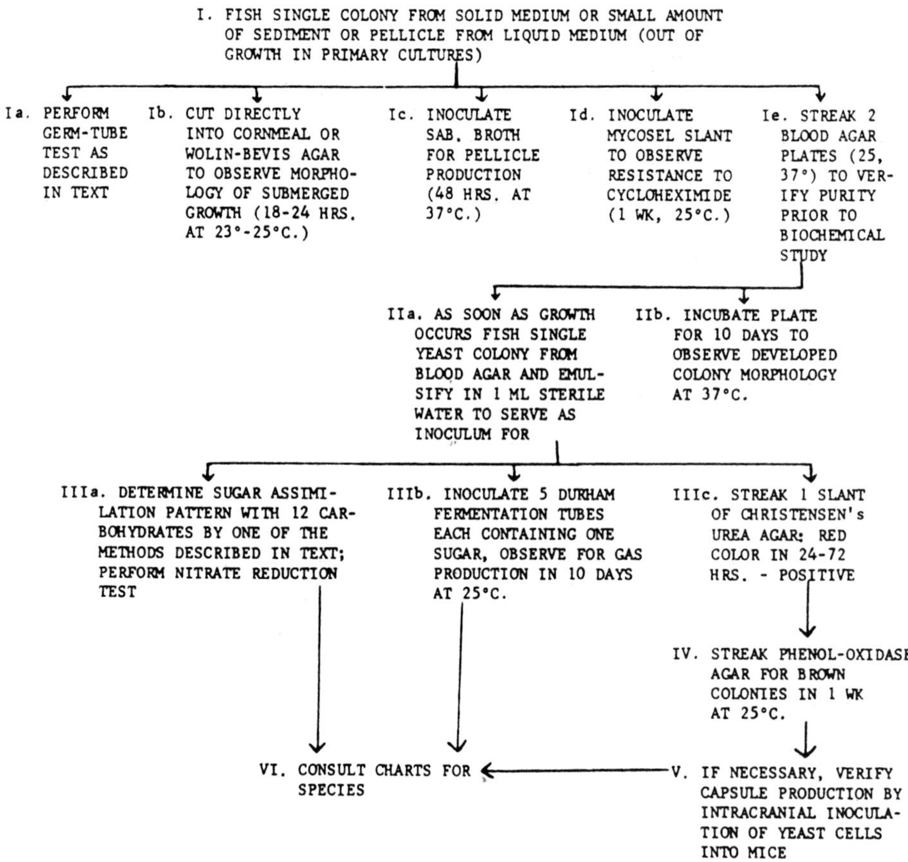

FIG. 4. Schema for a step-by-step procedure that can be followed in the isolation and identification of yeasts from primary cultures of clinical specimens.

incubated at 37°C, or Sabouraud agar incubated at 25°C, to ensure obtaining pure cultures for further biochemical testing. Media for subculturing yeasts before biochemical testing should not contain antibiotics. Inoculation of biochemical test media directly from isolation media is not recommended.

Species identification based on phenotypic characteristics of anamorphic species is often fraught with difficulty. Some species are separable on the basis of a single biochemical test. A case in point is identification of *Candida stellatoidea* by inability to utilize sucrose as a carbon source. Many yeast taxonomists regard *C. stellatoidea* as a variant of *Candida albicans* that lacks an α-glucosidase (34). For simplicity, we retain *C. stellatoidea* as a separate species while recognizing that this species validity is questionable. It does illustrate the dilemma that arises when we try to identify anamorphic form-species.

To complicate matters, mutations affecting phenotypic expression of certain characters in yeasts can be induced by cytotoxic cancer medications. Yeasts isolated from patients receiving such chemotherapy may be suppressed in metabolizing certain carbohydrates, or may require biotin or other growth-factor supplementation to behave appropriately in biochemical tests. Fortunately, correct biochemical reactions often will be obtained upon repeated subculture of the isolate (3).

Devices combining several biochemical tests into a single minaturized plate or plastic strip make the use of a battery of tests very convenient. Such devices encourage more extensive biochemical testing of isolates and, in some instances, provide faster results (11, 13, 16, 41, 76, 84). Automation of biochemical testing of yeasts (17, 27, 56, 59) now provides the capability for carrying out extensive biochemical testing with minimal effort and within as little as 20 to 24 h. The accuracy of identifications achieved with commercial manual and automated devices is, in general, very acceptable. Still, it is unwise to abandon the study of microscopic morphology of yeast isolates grown on Wolin-Bevis agar (89) or a similar medium as part of the identification process. Most commercial devices rely upon the germ tube test as an additional safeguard. Moreover, microscopic morphology on media designed to elicit typical morphology of yeasts will usually aid in interpreting biochemical tests. Some species (e.g., *Trichosporon* and *Geotrichum* spp.) are difficult to distinguish without careful examination of microscopic morphology.

Cultural characteristics

Most yeasts grow well on the common mycological and bacteriological media, producing visible colonies within 48 to 72 h. Most pathogens grow readily at both 25 and 37°C, whereas most saprophytic yeasts encountered in clinical specimens fail to grow at 37°C. The ability to grow at 37°C is an important characteristic for differentiating species. Some species are in-

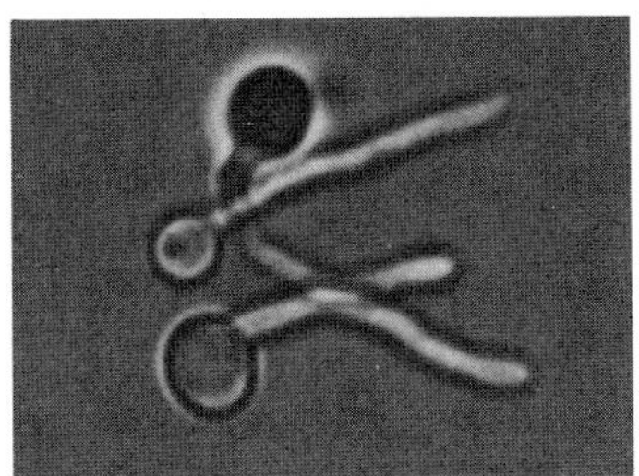

FIG. 5. *Candida albicans.* Germ tubes formed after incubation in serum for 2 h at 37°C. ×2,100.

hibited by cycloheximide or chloramphenicol. In certain species, ascosporulation may be enhanced by the use of special media. The growth and temperature requirements for the several genera and the indications for using special media are summarized later on in this chapter.

Pellicle formation on broth. When yeasts are cultivated in tubes of liquid media such as Sabouraud broth or malt extract broth, growth takes the form of (i) a sediment, (ii) a ring around the circumference of the broth surface, (iii) a film on the surface, or (iv) a surface pellicle. These characteristics depend largely on O_2 requirements and surface tension. In the past, emphasis was placed on the ability to form a surface pellicle on liquid media, but more recent evidence has demonstrated this to be a variable characteristic, and it is no longer stressed as much as formerly. However, the aspect of growth in liquid media can help identify species; e.g., pellicle-forming yeasts often isolated from clinical specimens include *Candida tropicalis, Candida krusei*, and especially *Trichosporon* spp.

Microscope examination of culture growth

Wet preparations of primary cultures in distilled water, India ink, nigrosin, or lactophenol cotton blue, or dried smears stained by the Gram or Ziehl-Neelsen method, provide material for microscopic examination of morphology and ascertaining the presence or absence of capsules, the presence or absence of ascospores, and the purity of cultures. Further morphological study demands special media. Incubation of yeast cells in blood serum or egg white at 37°C for 1 to 4 h permits observation of germ tube production (*C. albicans*; 47). Subculture by cut-streaking cornmeal infusion agar or Wolin-Bevis (89) agar plates, followed by incubation at 23 to 25°C for 18 to 48 h, permits detection of pseudomycelium, true mycelium, and chlamydospores (Table 1). Special sporulation media (Chapter 111) may be necessary to detect production of ascospores, whose presence and characteristic morphology are essential for identification of ascosporogenous genera, such as *Saccharomyces, Pichia, Endomycopsis, Hansenula*, and *Nematospora*, occasionally found in clinical specimens.

Most of these morphological studies can be carried out directly with primary cultures before purification, and even with the original specimen.

Germ tube test

One of the most valuable tests for rapid presumptive identification of *C. albicans* is the germ tube test (47). One need only make a dilute suspension of a yeast colony in 0.5 to 1.0 ml of serum, which is accomplished by touching the tip of a sterile Pasteur pipette to a yeast colony and then gently emulsifying the cells which adhere to the pipette in the serum. The mixture is then incubated at 37°C for 2 to 4 h, after which a drop of the mixture is examined microscopically for germ tubes (as seen in Fig. 5). For convenience, the pipette can be left in the serum during incubation and used to transfer a drop of the serum to a slide for examination. When so incubated, individual cells of *C. albicans* and *C. stellatoidea* produce short lateral hyphal filaments called germ tubes which are not produced by other *Candida* species. We have been unable to verify recent reports of germ tube formation by *C. tropicalis* (49) and regard this species as being germ tube negative under the test conditions described here. *C. stellatoidea* and *C. albicans* are easily differentiated by a sucrose assimilation test and by susceptibility to cycloheximide. *C. albicans* assimilates sucrose and grows on commercial cycloheximide-containing media, i.e., Mycosel (BBL Microbiology Systems); *C. stellatoidea* does neither. Human and bovine sera including commercially available fetal calf serum, bovine serum albumin, ovalbumin, dilute oxgall, or peptone can be used with success in the germ tube procedure. Dehydrated and liquid media for use in the germ tube test are available commercially. Known strains of *C. albicans* and *C. tropicalis* should be included as positive and negative controls, respectively.

Chlamydospores

Another standard technique for identifying yeasts is to test for production of chlamydospores. Various media are available for this test, including chlamydospore agar, Wolin-Bevis (89) agar, cornmeal-polysorbate (Tween 80) agar, and rice-Tween 80 agar. A medium was described recently which permits detection of germ tubes when incubated at 37°C for 3 h, demonstration of typical yeast morphology after incubation for 18 to 24 h at 25°C, and the phenol oxidase reaction of *C. neoformans*, all on the same medium (41). Choice of medium is based largely on individual experience. The technique described here utilizes Wolin-Bevis or cornmeal-Tween 80 agar. Spherical chlamydospores are produced by *C. albicans* and *C. stellatoidea* (Fig. 6), thus requiring further differentiation as described above. *C. tropicalis* occasionally produces distinct oval- to teardrop-shaped chlamydospores on the media mentioned, which are also useful for inducing the formation of pseudohyphae or true hyphae in most species of *Candida* and of true hyphae and arthroconidia in *Trichosporon*.

For the test, cut the yeast colony or original specimen into the agar along two or three parallel lines, no less than 1 cm apart. Alternatively, the Dalmau technique, in which the inoculum is streaked along two shallow scratches in the agar and then covered with a sterile cover slip, can be used. Incubate the plates at room temperature (23 to 25°C) for 18 to 48 h, by which time most strains of *C. albicans* and *C. stellatoidea* will have formed typical chlamydospores. This technique is also useful for demonstrating giant pseudohyphae of *C. parapsilosis*. The experienced eye can also recognize the typical morphological patterns of other *Can-*

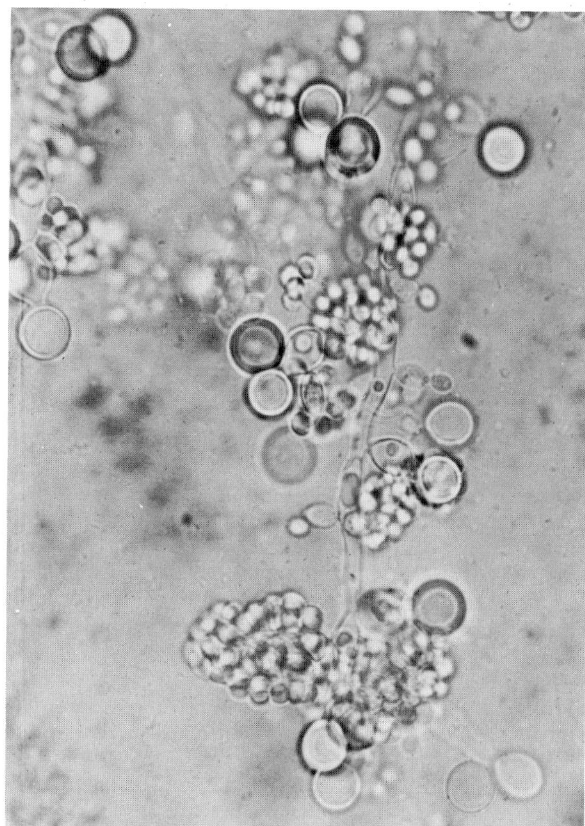

FIG. 6. *Candida albicans.* Submerged mycelium with spherical clusters of blastoconidia and double-walled, spherical chlamydospores on cornmeal agar. ×1,200.

dida species (3). Confirmation by biochemical reactions of purified cultures is recommended.

Biochemical reactions

Before any biochemical reactions are measured, it is essential to purify yeast cultures as described above, even when contamination is not immediately apparent, for a hidden contaminant can significantly affect results. Biochemical tests should not be inoculated from media containing antibiotics because the antibiotics may suppress, but not kill, bacterial contaminants. Once the culture is transferred to biochemical test media, the growth of bacterial contaminants may no longer be inhibited and they will multiply rapidly, confounding interpretation of results.

Biochemical reactions useful in yeast identification include: (i) assimilation (or utilization) of carbohydrates (currently, 12 sugars are used with clinical isolates; some commercial systems employ additional compounds); (ii) assimilation of KNO_3 (equivalent to nitrate reduction); (iii) fermentation reactions (i.e., gas production); (iv) ureas production; (v) ability to produce brown colonies on birdseed agar (or other suitable phenol oxidase detection medium); and (vi) resistance to cycloheximide. Other biological characteristics tested are: (i) ability to grow at 37°C; (ii) ability to form a pellicle on broth; and (iii) animal pathogenicity.

Besides the above routine tests, certain other characteristics are currently exploited by research investigators as taxonomic criteria for yeasts. Some of the following procedures may eventually be adapted for use in clinical laboratories: (i) analysis of cell wall components by gas-liquid chromatography, nuclear magnetic resonance spectra, and methylation techniques; (ii) determination of the guanine-plus-cytosine ratio in DNA of a given strain; (iii) amino acid sequencing of cytochrome *c* or other proteins; (iv) somatic cell or DNA hybridization techniques; (v) determination of optimal temperature, temperature tolerance limits for growth, and other nutritional imbalances; and (vi) detection of polyols (e.g., arabinitol) produced in liquid cultures and quantitated by gas-liquid chromatography.

On the flow sheet for identification procedures (Fig. 4) and in the subsequent discussion, a description is given of the methods we recommend as routine procedures. Alternative methods are also listed.

Urease

Urea is split by yeasts having urease, including all species of *Cryptococcus* and *Rhodotorula* and some species of *Trichosporon*; indeed, this reaction is shown by all heterobasidiomycetes. Occasional strains of *C. krusei* are urease positive. The test is performed by transferring with a sterile loop a portion of a purified yeast colony to a slant of Christensen urea agar. The slant is incubated at 25 to 30°C and examined daily for 4 days. Change in the amber color of the phenol red indicator of the medium to pink or red denotes a positive reaction due to alkalinization of the medium via production of ammonia from the urea. A rapid urease test (15 min) has recently been described for use with yeasts (91).

Carbohydrate assimilation tests

Tests to determine the ability of a yeast species to utilize a carbohydrate as sole source of carbon in a chemically defined medium have long been a mainstay of yeast taxonomists (3, 6, 34) and have become an essential step in yeast identification in the clinical mycology laboratory. Most commercial products now available for identifying yeasts rely heavily on carbohydrate assimilation tests (11, 13, 16, 17, 27, 41, 42, 56, 59, 86).

Assimilation tests provide a clear biochemical basis and shorten the time required for identifying yeasts to the species level. Several methods are available as follows.

(i) Auxanography. In Beijerinck's auxanographic technique (7), small amounts of dry carbohydrates are placed on the surface of a heavily seeded synthetic agar medium. Noticeably greater growth encircles the spot where an assimilated compound was placed as compared with background growth. Lack of enhanced growth indicates a lack of enzymes for utilizing the test carbohydrate; the pattern elicited by utilizable carbohydrates is an auxanogram. Although mutations involving enzymes essential to carbohydrate utilization are well known for yeasts, the auxanogram is a dependable criterion for identification of each species when used in conjunction with other tests. Modification of Beijerinck's method include: (i) streaking the surface, rather than seeding the agar plate; (ii) using filter-paper disks impregnated with the carbohydrates; (iii) placing drops of carbohydrate solutions on the agar; and (iv) placing the carbon sources in wells cut out of the agar in a petri dish.

Auxanography is the method currently employed for assimilation tests in the Mycology Laboratory of Columbia-Presbyterian Medical Center. Large plastic petri dishes (150 mm in diameter) are filled in advance with 30 ml each of reconstituted Wickerham yeast nitrogen base (YNB, Difco Laboratories, Detroit, Mich.) solidified with 2% Noble agar and refrigerated. For the test, the plate is streaked heavily, evenly, and confluently with a cotton swab, transferring the entire growth from a 24-h Sabouraud agar slant culture of the yeast to be tested. Immediately after the plate is streaked, 12 "carbohydrate disks" (glucose, maltose, sucrose, lactose, galactose, melibiose, cellobiose, inositol, xylose, raffinose, trehalose, and dulcitol) are applied to the surface of the plate with a BBL Sensi-Disc dispenser. Cartridges containing small disks of the 12 carbohydrates are obtained from the same company (Fig 7). Additional carbohydrates may be needed for certain species, e.g., rhamnose utilization to separate *Candida lusitaniae* and *C. tropicalis* (Table 1).

A modified auxanographic method makes use of an indicator dye (bromocresol purple) for early, sensitive detection of growth (43). Here a lower concentration of YNB (0.1×) than is usually recommended is employed, and the concentration of each carbohydrate is adjusted for optimum utilization. Correlation between yeast growth with utilization of a test carbohydrate and change in the color of the indicator due to acid production is also demonstrated.

(ii) Wickerham broth technique. The technique originally recommended by Wickerham and Burton (86) uses the same chemically defined medium as discussed above, but in liquid form. The broth tubes containing individual sugars are inoculated either with a depleted suspension of the yeast (grown for 48 h in basal medium without a carbon source or with a drop from a dilute suspension of yeast colony in distilled water). Twelve tubes, each containing one of the sugars listed in Table 1, are inoculated for each isolate, the tubes are incubated at room temperature (25°C), and growth is evaluated turbidimetrically. When a carbon source is assimilated, growth clouds the medium so that black lines or letters will not show through. Ahearn and Schlitzer (3) recommend shaking the tubes during incubation. When screw-capped tubes are used for this technique, the caps should be loosened slightly (one-half turn) to prevent anaerobiosis. A disadvantage of liquid medium is that a false-positive reaction may occur if progeny from a single mutant cell render the entire broth turbid upon prolonged incubation. Such a mutant would appear as a single distinct colony on the surface of an agar plate and thus would not lead to a false-positive reading. By the same token, latent or delayed utilization of carbohydrates can be detected more easily with the broth method than on agar media.

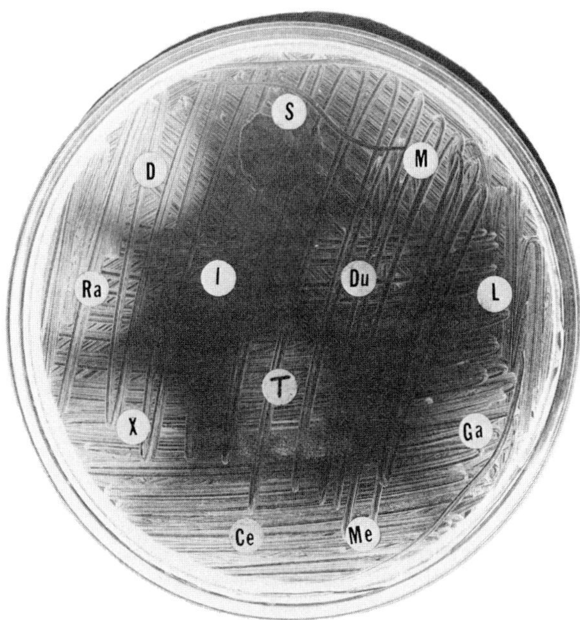

FIG. 7. *Candida guilliermondii*. Auxanographic plate (15 cm in diameter) showing growth around disks of glucose (D), sucrose (S, partly rendered transparent by moisture), maltose (M), galactose (Ga), cellobiose (Ce), xylose (X), raffinose (Ra), dulcitol (Du), and trehalose (T) (two-thirds normal size).

(iii) Assimilation agar slant technique. Individual carbohydrates can also be incorporated in agar slants of Wickerham YNB adjusted to pH 7.0 (1). Addition of an indicator (bromocresol purple) draws attention to early growth by a change in pH to the acid side (indicator turns yellow). For inoculation, a dilute suspension of the yeast to be tested is prepared by suspending a single colony in 9 ml of sterile distilled water. This suspension is inoculated in 0.1-ml volumes onto each assimilation slant and then incubated at 30°C and observed for growth. Assimilations are considered positive when abundant growth appears on a test medium (and the indicator changes to yellow) with negligible or no growth on the control slant (indicator remains purple); results may be reported in 72 h, but slants should be held for 14 days to allow for delayed reactions. Assimilations are considered negative if a carbohydrate medium is not different from the control medium, i.e., the indicator does not change to yellow.

Nitrate assimilation

Tests of ability of a yeast to utilize nitrate as sole nitrogen source can be carried out by any one of these same procedures, except that Wickerham yeast carbon base (Difco) is used as the basal medium and potassium nitrate (final concentration, 0.078 g/100 ml) and peptone (1%) are used as the test nitrogen sources. The peptone serves as a positive growth control.

A rapid (15-min) test for nitrate utilization by yeasts (41) has proved reliable at Columbia-Presbyterian Medical Center. This test detects nitrite, the end product of nitrate reductase, by exposing the yeast to the dehydrated ingredients of a special nitrate broth on a pretreated cotton swab, subsequently dipped in a few drops of an equal mixture of 0.5% α-naphthylamine and 0.8% sulfanilic acid, each in 5 N acetic acid. Standard cotton swabs are pretreated by dipping in a 5× concentration of the special nitrate medium (KNO_3, 2 g; $NaH_2PO_4 \cdot H_2O$, 11.7 g; Na_2HPO_4,

1.14 g; Zephiran chloride, 1.2 ml of a 17% solution; 200 ml of water). The saturated swabs are dried in a desiccator overnight, autoclaved, and stored in a closed container until needed. For the test, the tip of a pretreated swab is swept over several colonies of the yeast in question, and then the swab is inserted in a test tube and pressed against the side of the tube to increase contact between the yeast and the reagents in the swab. The tube and swab are incubated for 10 min at 45°C (or for 2 h at 37°C). At the end of the incubation period, the swab is immersed in 6 drops of the α-naphthylamine–sulfanilic acid mixture (3 drops each), and the color reaction, if any, is compared with that developed by two similarly treated and incubated swabs, one using a yeast known to be nitrate positive and the other with a nitrate-negative yeast. The reaction of the test organism is graded as follows, when compared with the negative control: 0 = no color change; 1+ = faint pink; 2+ = pink; 3+ = dark pink to light red; 4+ = dark red.

In our experience the rapid nitrate and traditional nitrate assimilation tests correlate very well for commonly isolated yeasts. However, some infrequently isolated species, including *Rhodotorula* spp., may give a positive rapid nitrate test and yet be negative for nitrate assimilation. This may be a reflection of a cryptic nitrate reductase enzyme which is detected by the rapid test but poorly expressed in the assimilation test.

Fermentation reactions

Carbohydrate fermentation tests are familiar, useful tests for yeast identification. However, these tests are more prone to variation and are less dependable than carbohydrate assimilation tests. The only reliable evidence for carbohydrate fermentation by yeasts is by production of gas; therefore, Durham tube inserts should be employed for gas detection. The basal medium employed includes peptone, yeast extract, and bromocresol purple indicator; the pH is adjusted to 7.0. Carbohydrates commonly tested include glucose, maltose, sucrose, lactose, galactose, and trehalose. For inoculation, a dilute suspension of yeast cells in sterile distilled water is prepared (to match a McFarland no. 1 standard), and a single drop is added to each tube. After inoculation each tube should be sealed with 1 ml of molten Vaspar (equal parts of paraffin and white petrolatum) placed directly onto the top of the broth. Readings are made after incubation at 25°C for 10 to 14 days. Positive fermentation is indicated by accumulation of gas (CO_2) in the Durham tube or underneath the Vaspar seal (the indicator will also change to yellow). All fermented carbohydrates will also be assimilated; however, certain carbohydrates will be assimilated, but not fermented.

A very few isolates have been reported that were capable of fermenting a particular carbohydrate in the peptone-based fermentation medium, but were unable to grow in the chemically defined YNB used in assimilation tests because biotin or other vitamins were not present in adequate amounts for their growth. Supplementation of YNB with 0.01% yeast extract usually will permit utilization of the carbohydrate in question without yielding false-positive results with other carbohydrates.

Serological methods

Fluorescent-antibody techniques with specific conjugates are extremely useful in those instances in which cultures have failed and yet yeasts are observed in histological sections, body fluids, or exudates (Chapter 96). However, at present these specific conjugates are available only at the Centers for Disease Control and at certain other reference laboratories. These techniques, along with agar gel precipitin tests with soluble antigenic extracts, slide or tube agglutination, or immunoelectrophoresis with soluble extracts, also are useful in specific differentiation of yeasts, provided suitably absorbed, species-specific antisera are employed (76, 79).

A serological identification of *C. neoformans* has recently been described which utilizes a soluble extract of colonies allowed to react with latex particles coated with antibody (54). The same commercial kit used for detecting cryptococcal antigen in patients' body fluids (Chapter 96) can serve for this test and has been shown to be reliable for confirming the identification of a yeast isolate as *C. neoformans*.

Differentiation from related species

Since yeasts, as stated, are a heterogeneous conglomerate of various fungal taxa, this chapter by necessity points out differences among yeast genera and species. It is also important to mention certain other organisms encountered in clinical specimens which produce yeastlike colonies on commonly used isolation media. For the most part, morphological differences between these microorganisms and yeasts will be observed and must be identified by methods other than those used for yeasts. The following are examples of such microorganisms.

Geotrichum. *Geotrichum* is a filamentous fungus whose initial growth may be glabrous and creamy to pasty, but becomes velvety or fuzzy with aging and repeated subculture. *Geotrichum* produces true hyphae and arthroconidia, but neither pseudohyphae nor blastoconidia, thus differing morphologically from the yeast genus *Trichosporon*. Further differentiation can be obtained by physiological and biochemical tests. Fluffy strains of *Geotrichum* superficially resemble the systemic pathogen *Coccidioides immitis* and must be differentiated from it.

Geotrichum and *Trichosporon* spp. have recently been carefully reevaluated by Weijman (85), who analyzed their extractable carbohydrates by gas-liquid chromatography. Species having similarities to ascomycetes were retained in the genus *Geotrichum*, whereas those which contained xylose, a basidiomycetous characteristic, were placed in the genus *Trichosporon*. Weijman thereupon renamed *Trichosporon capitatum* as *Geotrichum capitatum*. Recently Salkin et al. (69) proposed the new genus and species *Blastoschizomyces pseudotrichosporon* as the correct epithet for 13 isolates that had been identified as *T. capitatum*, but formed annelloconidia. It is unclear at this writing whether all isolates of *T. capitatum* should be renamed *B. pseudotrichosporon* or whether some isolates are named *G. capitatum*. It may be necessary to retain all three names! Recent reports (4, 19, 21–23, 28, 60, 88) have documented the pathogenicity of this (these) species. It will be difficult for most laboratories to distinguish between annelloconidia and arthroconidia in these fungi with the microscopes usually available in clinical laboratories. The most practical solution might be to accept *B. pseudotrichosporon* as the correct name for isolates that were formerly called *T. capitatum*. Many mycologists separate *Geotrichum* spp. from *Trichosporon* spp. based on the failure of *Geotrichum* spp. to form blastoconidia.

The colorless alga *Prototheca* (Fig. 8), which is also encountered in clinical specimens (and may be a pathogen), produces colonies resembling those of *Cryptococcus* on Sabouraud agar. Although one species of *Prototheca* produces a capsule (14) resembling that of *Cryptococcus*, other *Prototheca* species differ from *Cryptococcus* in not having a capsule, in having a less refringent cell wall, and especially in producing characteristic endospores by internal division into eight (or more) compartments. *Prototheca* spp. can also be differentiated by staining with specific immunofluorescent conjugates and by carbohydrate assimilation tests (61).

Ustilago. The basidiomycetous genus *Ustilago* is another filamentous fungus which can occasionally grow from sputum specimens and produce yeastlike colonies. One isolate was reported to cause chronic meningitis in a human (53). That identification, however, was based solely on the morphology of the fungus in tissue; it was not isolated in culture. At first unicellular, *Ustilago* spp. soon produce short hyphae with clamp connections, and their colonies become finely powdery or velvety. Even at the unicellular stage, the elongated cells characteristic of this genus differentiate it from yeasts. These elongated cells resemble arthroconidia. Most isolates that we have encountered from clinical specimens are nitrate positive and give physiological reactions similar to those of *Cryptococcus albidus* (Table 1).

Besides the species listed above, occasional isolates of *Sporothrix schenckii* and of the black yeasts *Aureobasidium pullulans* and *Exophiala werneckii* produce white to tan yeastlike colonies on primary isolation media. Careful examination of the microscopic morphology of these isolates growing on cornmeal agar will facilitate their correct identification and avoid the confusion that may result from carrying out biochemical tests with them. Darkly pigmented fungi (dematiaceous fungi) are discussed in Chapter 52.

Animal inoculation

Techniques for isolating *C. neoformans* and other pathogenic yeasts from contaminated specimens by animal inoculation are discussed in Chapter 47. Demonstration of pathogenicity as an aid to differentiation of *C. neoformans* from other cryptococci can be accomplished by intracranial inoculation of 0.05 ml of saline suspension containing a 0.2% concentration of yeast cells (1:500 by volume of packed cells) into each of two 20- to 30-g white mice. Alternatively, 0.5 to 1 ml of the yeast cell suspension can be injected intravenously or intraperitoneally into mice. Most strains of *C. neoformans* kill mice within 4 to 5 days after intracranial injection; examination of brain tissue from infected mice clearly demonstrates encapsulated yeast cells.

The pathogenicity of *C. albicans* can be demonstrated by injecting 0.2 to 0.8 ml of a 1% suspension of packed cells into the marginal ear vein of a 1- to 2-kg rabbit, or a similar volume of a 0.2% suspension into the tail vein of a 20- to 30-g mouse. Death of the animal usually occurs within 1 week; at autopsy miliary abscesses in the kidneys, and sometimes in the spleen and liver, will be observed. Pathogenicity of *C. albicans* can also be demonstrated by chicken embryo inoculation.

CHARACTERISTICS OF INDIVIDUAL MEDICALLY IMPORTANT YEAST GENERA

Genus *Candida*

Candida is a heterogeneous genus presently classified within the family *Cryptococcaceae*, Fungi Imperfecti (Deuteromycetes). *Candida* spp. are frequently present as members of the normal flora of the mouth, throat, large intestine, vagina, and skin and are often contaminants in exudates or other specimens taken from these areas. In patients whose immune defenses have been compromised by disease or by the secondary effects of drugs used to treat their diseases, normal flora organisms may invade deeper tissues, producing severe, life-threatening infections. *C. albicans*, the principal pathogenic species, causes mild to severe or chronic superficial infections of skin, nails, and mucous membranes in individuals with normal immune defenses, as well as serious systemic infections in debilitated patients. *Candida parapsilosis, C. tropicalis*, and *Candida guilliermondii* have become important causes of endocarditis, pyelonephritis, arthritis, and disseminated candidiasis in patients with indwelling intravenous catheters, patients undergoing cardiovascular surgery, and drug addicts.

C. lusitaniae has recently been reported as a cause of human disease (26, 29, 62). It appears to be an important pathogen in immunocompromised patients because it readily develops resistance to amphotericin B (26, 62). It has been confused with *C. tropicalis* and *C. parapsilosis* because of similarities in biochemical reactions. *C. lusitaniae* differs from *C. parapsilosis* by assimilating cellobiose and from *C. tropicalis* by fermenting as well as assimilating cellobiose and by assimilating rhamnose (Table 1).

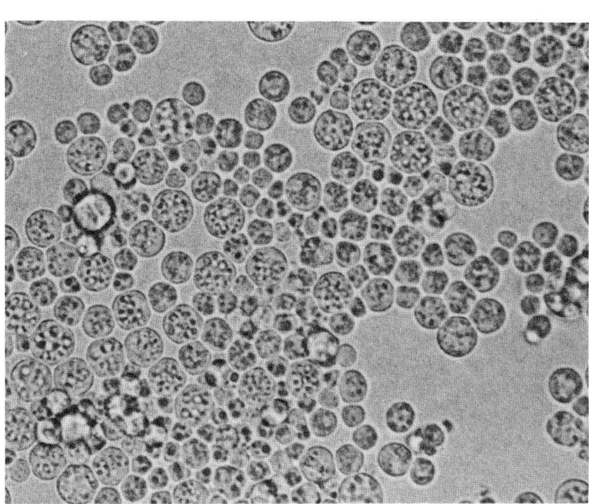

FIG. 8. *Prototheca* sp. grown on cornmeal agar for 72 h at 30°C (ca. ×800).

Candida paratropicalis, described by Baker et al. (5), has been isolated from several different clinical specimens. It varies only slightly in physiological activities from sucrose-negative *C. tropicalis* (2), and based on DNA homology studies, it is considered to be cospecific with *C. tropicalis* (S. A. Meyer, personal communication).

Cultural characteristics

Budding cells (blastoconidia) are round, oval, or oblong, 2.5 μm by 3 to 14 μm, occurring singly or in clusters or chains. Members of the genus *Candida* have been defined on the basis of pseudomycelium formation "by all or most strains of all species and varieties"; this was the main distinction between *Candida* and *Torulopsis*, as the latter was believed not to produce pseudohyphae. However, Yarrow and Meyer (90) reexamined neotype strains of *Torulopsis* and concluded that the distinction between the two genera did not suffice to retain *Torulopsis* as a separate genus. They proposed amending the definition of *Candida* to include the statement "pseudohyphae absent, rudimentary, or well developed," and they renamed all *Torulopsis* species as *Candida* species. This amended definition for *Candida* has been accepted for the third edition of *The Yeasts* (34). However, because *Torulopsis* was described before *Candida* and therefore takes precedence over it, this, as well as other factors, has caused great disagreement among taxonomists (50). The two genera will be treated separately in this chapter until a more final decision on proper taxonomy can be made.

Production of germ tubes (Fig. 5) and spherical chlamydospores (Fig. 6) by *C. albicans* and *C. stellatoidea* is a useful diagnostic characteristic. Ascospores have been reported in certain species later assigned to perfect genera; e.g., ascospore-forming strains of *C. guilliermondii* have been assigned to the genus *Pichia* (34). *Kluyveromyces fragilis* (*Kluyveromyces marxianus*) is the ascospore-forming stage of *C. pseudotropicalis* (*Candida kefyr*) and has been reported by that name as a cause of human disease (45). The teleomorph of *C. krusei* has recently been designated as *Issatchenkia orientalis* (36).

Growth is aerobic. Tiny colonies may be visible as early as 24 to 36 h and reach 1.5 to 2 mm in 5 to 7 days on Sabouraud agar. Colonies are usually stark white, but may become cream colored or tan with age. They are glabrous, creamy, or membranous and may have a fringe of submerged hyphae. Optimal growth temperature is 25 to 37°C.

Biochemical characteristics (Table 1)

Carbohydrates are utilized in oxidative (assimilation) or fermentative patterns, or both, that are helpful in differentiating species. Urease is negative except for occasional strains of *C. krusei*. KNO_3 is not utilized by frequently encountered clinical isolates.

Animal pathogenicity

C. albicans is lethal to mice and rabbits when injected intravenously, causing miliary abscesses in the kidneys and other organs. Other *Candida* spp. may also produce lesions but are seldom lethal for experimental animals that have not been pretreated with corticosteroids or other immunosuppressive drugs.

Genus *Cryptococcus*

Formerly classified in the same family as *Candida*, *Cryptococcus*, like *Candida*, is now believed to be a heterogeneous genus. Its only recognized pathogenic species, *C. neoformans*, and probably also *C. albidus* have teleomorphs that belong to the genus *Filobasidiella* of the Ustilagenales. Other species of *Cryptococcus* have occasionally been reported to cause human disease (9, 18, 32, 35, 46, 52, 87). Evaluation of earlier reports is sometimes difficult because the isolates were not always carefully identified. However, some of the recent reports which are accompanied by demonstration of yeasts that are morphologically compatible with the isolate in tissue suggest that species of *Cryptococcus* besides *C. neoformans* may cause disease in certain debilitated patients.

Five serotypes (A, B, C, D, and AD) of *C. neoformans* have been described, and these have been related to two teleomorphs that produce basidiospores. The demonstration of clamp connections in two isolates by Shadomy (73) provided the first convincing evidence of a taxonomic affinity between *C. neoformans* and basidiomycetes. Subsequent mating experiments by Kwon-Chung (37) with isolates from many different sources demonstrated basidiospores with A and D serotype isolates, and the teleomorph was named *Filobasidiella neoformans*. It was later shown that serotype B and C isolates produced a filobasidiella state that formed rod-shaped basidiospores; that teleomorph was named *Filobasidiella bacillaspora* (38). Subsequently, renaming of the anamorph of B and C serotypes as *C. bacillasporus* was proposed (40). *F. neoformans* then became the teleomorph of *C. neoformans* (serotypes A, D, and AD), and *F. bacillaspora* became the teleomorph of *C. bacillasporus* (serotypes B and C). However, mating studies with appropriate isolates then revealed that *F. neoformans* and *F. bacillaspora* produced a high percentage of fertile basidiospores (39, 71). For that reason, *F. bacillaspora* is now regarded as a variant of *F. neoformans* and the anamorph *C. bacillasporus* is considered a variant of *C. neoformans*. The two varieties differ in serotype, utilization of creatinine and certain dicarboxylic acids, morphology of yeast cells, temperature tolerance, and virulence for mice. DNA hybridization revealed a 60% homology between the two variants. *C. bacillasporus* has also been shown to be identical to the *C. neoformans* subsp. *gatti* described by Vanbreuseghem and Takashio in 1970 (80) using the same criteria given above. It thus becomes a later synonym of *C. neoformans* subsp. *gatti* and is an invalid name. Some regard the two variants as identical and *C. neoformans* (teleomorph *F. neoformans*) as the only valid name necessary to designate all serotypes and variants of the species (71).

The usual biochemical tests (Table 1) do not differentiate the two varieties. To avert confusion, *C. neoformans* is retained here as the sole name to represent both varieties. At present there are no known differences in disease symptoms or severity associated with the two variants; both appear to respond similarly to antifungal agents. There is, however, a distinct difference in their geographic distribution (8). *C. neoformans* subsp. *gatti* (serotype B) is found almost exclusively on the West Coast in the United States, whereas most infections elsewhere in the country are caused by *C. neoformans* subsp. *neoformans* (serotype A).

Infections caused by *C. neoformans* are exogenous; the yeast lives naturally on soil contaminated with bird droppings, notably from pigeons and other seed-eating birds. Meningitis is the most frequently recognized type of cryptococcal infection, followed in frequency by localized abscesses or granulomas (cryptococcoma or toruloma) in lungs, brain, lymph nodes, skin, or bones. Diffuse pulmonary infection is perhaps the most common type of cryptococcal infection, though it is often asymptomatic and unrecognized. The respiratory tract is believed to be the portal of entry for most if not all cryptococcal infections.

Blastoconidia are chiefly spherical and exhibit a wide range of diameters (from 5 to 30 μm). A polysaccharide capsule (Fig. 1) is a constant feature although it is much more prominent in infected tissues or when suspended in immune sera than in cultures. The capsule can be demonstrated in wet preparations by negative staining with India ink or nigrosin or by mounting the cells in a drop of normal serum with a drop of 1% acetic acid, or it can be demonstrated in immune serum. The capsule also is revealed by mucin stains such as mucicarmine or alcian blue. This feature permits the presumptive identification of *C. neoformans* in histopathological sections, since other yeasts which invade human tissues are not revealed by these dyes.

Colonies are often mucoid, becoming dull and drier with age. Colonies on Sabouraud agar initially are pale buff, changing to tan or brown with age. This genus does not ordinarily form mycelium, though occasional isolates have been encountered which produce pseudomycelium (Fig. 2). Hypha-forming strains of *C. neoformans* which produce clamp connections have been observed (73), reflecting the basidiomycetous affinity of the species.

Growth is aerobic. *Cryptococcus* spp. grow well on ordinary bacteriological and mycological media, attaining colony diameters of 1 to 3 mm in less than 1 week. They are inhibited by cycloheximide. Bacteria, serum factors, or tissue inhibitors in clinical specimens may delay the appearance of colonies by as much as 2 to 4 weeks. The optimal temperature varies with species (Table 1).

Biochemical characteristics

Metabolism is strictly oxidative; assimilation of certain sugars and of KNO_3 is useful for differentiating species (Table 1). Utilization of inositol (34) and the usual absence of carotenoid pigments distinguish this genus from *Rhodotorula*; both genera may have capsules, produce starchlike compounds, and are urease positive. *C. neoformans* can be differentiated from other *Cryptococcus* by (i) ability to grow at 37°C, (ii) production of brown colonies on birdseed agar (positive phenol oxidase reaction), (iii) characteristic assimilation pattern, and (iv) pathogenicity for experimental animals. Phenol oxidase may have to be induced in certain isolates (particularly serotype C). Such isolates will not produce brown colonies on birdseed agar when first isolated.

Animal pathogenicity

C. neoformans, but not other species of *Cryptococcus*, is lethal for mice by invasion of the brain, reached by direct intracranial inoculation or by neurotropic extension from the intravenous or intraperitoneal route.

Genus *Hansenula*

Hansenula anomala and *Hansenula polymorpha* have been reported to cause human disease (51, 83). All members of the genus are nitrate positive and produce ascospores that may be hat shaped, hemispheroidal, spherical, or Saturn shaped (34). *H. anomala* ferments as well as assimilates carbohydrates. Colonies vary considerably in macroscopic appearance, from smooth and glistening to white, opaque, and wrinkled, and could easily be confused with colonies of *Cryptococcus* or *Candida* spp. Pseudohyphae may or may not be produced.

Genus *Malassezia* (*Pityrosporum*)

Malassezia is now considered the correct name for the genus of lipophilic yeasts that reproduce by unipolar bud fission and have been associated with tinea versicolor, blepharitis, dacryocystitis, dandruff, and seborrhea. Recent reports of folliculitis (33), isolation from the upper respiratory tract (58) and lungs (65), and isolation of lipophilic yeasts from a case of peritonitis (81) suggest that *Malassezia furfur* can cause systemic human infections. *M. furfur* is now considered the correct name under which both *Pityrosporum ovale* and *P. orbiculare* are combined (68). *M. pachydermatis*, the second species in this genus, has been isolated from dogs and other animals, but rarely from humans (70).

Bottle-shaped, small budding cells are 1 to 2 by 2 to 4 μm. Buds are produced successively from an invaginated end of the cell and are separated from the parent cell by a transverse septum (bud fission). Short hyphae have been observed in special media (68) but are more typically seen in tinea versicolor skin scales. Sexual spores have not been observed.

M. furfur grows well at 37°C when incubated aerobically on media containing olive oil, coconut oil, or other lipid sources. Colonies are creamy and punctiform when young, becoming membranous and confluent when older. *M. pachydermatis* produces similar colonies on complex media without supplementation of lipids, but grows poorly on chemically defined media such as YNB-glucose.

Biochemical characteristics

Malassezia spp. have no fermentative ability. *M. furfur* requires lipids or specific fatty acids, including myristic and palmitic acids, for growth.

Genus *Pichia*

Pichia contains the perfect form of *Candida guilliermondii* and belongs to the subfamily *Saccharomycetoideae* of the family *Saccharomycetaceae* of the Ascomycetes. Of the many strains of *C. guilliermondii* tested, only a few produced ascospores (34). These were hat shaped and ranged from one to four per ascus. Growth characteristics, biochemical reactions, and other laboratory reactions of the teleomorph should be the same as those of the anamorph, which are listed in Table 1.

Genus *Rhodotorula*

Rhodotorula (family *Cryptococcaceae*, subfamily *Rhodotoruloideae*) resembles *Cryptococcus* in rate of growth, colony topography, cell size and shape, occasional rudimentary pseudomycelium, presence of capsule, ability to split urea, and absence of fermentation. Its nonpathogenicity, different serotype, and conspicuous carotenoid pigment have maintained *Rhodotorula* as a distinct genus (34). *Cryptococcus*, furthermore, can use inositol, invade human and animal tissues, and kill (*C. neoformans*). *Rhodotorula* is a normal symbiont of moist skin (due to climate or to the patient's abnormal physiology); *R. rubra* (*mucilaginosa*) and *R. glutinis* have been reported rarely to cause septicemia (63), meningitis (64), and a chronic skin infection (72). Recently, *R. pallida* and *R. marina* were implicated in an invasive infection of a leukemia patient (67). These two species had not been reported to cause human disease before this. Occasionally fungemia associated with the contamination of plastic tubing left in place during prolonged intravenous therapy will resolve without antifungal therapy when the catheter is removed. *Rhodotorula* spp. are also cultured readily from shower curtains, bathtub-wall junctions, and the rubber tips and handles of toothbrushes, if not properly aerated.

Genus *Saccharomyces*

Saccharomyces is a member of the family *Saccharomycetaceae* (ascosporogenous). *S. cerevisiae* is responsible for occasional cases of thrush and vulvovaginitis; it has also been reported from urine specimens in diabetics and has been reported to cause fungemia (20).

Cells are oval to spherical, 3 by 5 μm, and may exist as budding cells as either haploids or diploids (after fusion). Cells may form short chains and elongate as rudimentary pseudohyphae. Ascospores, one to four in number, are in either tetrahedral or linear arrangement (Fig. 9) and are gram negative (vegetative cells are gram positive). An excellent sporulation medium is the potassium acetate-yeast extract-glucose medium (ascospore medium) of McClary modified from Adams (Chapter 111).

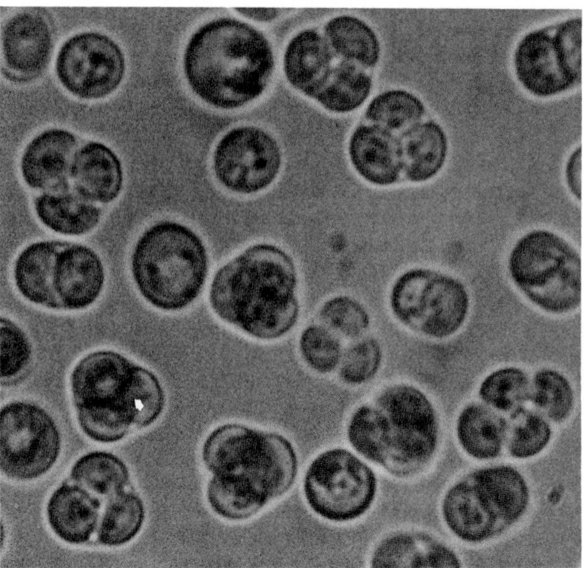

FIG. 9. Ascospores of *Saccharomyces cerevisiae*. Unstained. ×2,500.

Occasional production of rudimentary pseudomycelium occurs; this genus is the teleomorph of *Candida robusta*, a nonpathogen. Growth is rapid on most media, both aerobically and anaerobically; optimal temperature is 25 to 37°C.

Biochemical reactions

Saccharomyces spp. assimilate and ferment sugars in a pattern useful for identification (Table 1). *S. cerevisiae* does not assimilate KNO_3, split urea, or grow on creatine.

Serological examination

A slide agglutination method with specific antisera can be used in identifying species.

Genus *Torulopsis*

Torulopsis glabrata is a common isolate from urine and other specimens. It is considered to be a symbiont of humans, but it has been documented as the cause of pyelonephritis, pneumonia, septicemia, and meningitis in immunocompromised patients. Documentation of pathogenicity can be accomplished only by demonstration by biopsy of tissue invasion. In stained tissues it closely resembles *Histoplasma capsulatum*, from which it must be differentiated. *T. pintolopesii* is a thermophilic yeast capable of growth at 40 to 42°C. It is rarely isolated from humans, but is a symbiont of the intestinal tract of mice. This should be kept in mind when animal pathogenicity studies are performed. *T. candida* is occasionally isolated from specimens as a contaminant, but it has not been shown to produce human disease. It closely resembles *C. guilliermondii* and is not easily differentiated from it with the usual biochemical tests (Table 1).

Genus *Trichosporon*

Trichosporon is a member of the subfamily *Trichosporoideae*, family *Cryptococcaceae*. *T. beigelii* (cutaneum) causes superficial nodules (white piedra) on the distal portion of hair in the scalp, beard, axilla, and pubic area (recently miniepidemics of piedra pubis have been detected in New York City). Occasional opportunistic invasion of the mucous membranes, skin, and deeper tissues by this species has also been reported (12, 24, 28, 48, 66, 82). *T. capitatum*, *T. fermentans*, and *T. penicillatum* have been reclassified in *Geotrichum* (85), and more recently *T. capitatum* has been renamed *Blastoschizomyces pseudotrichosporon* (69).

Trichosporon has well-developed hyphae and pseudohyphae, reproducing also by blastoconidia and arthroconidia; no chlamydospores or sexual spores have been reported.

Growth is aerobic and occurs on all of the usual culture media. Smooth, shiny colonies, 3 to 6 mm, appear within 1 week; later, they become membranous, dry, and cerebriform. The optimal temperature is 25 to 37°C; some species grow equally well at 37°C.

Biochemical characteristics

Most *Trichosporon* spp. are oxidative; an occasional species is also fermentative. A heavy surface pellicle is formed in broth media. Assimilation of KNO_3 is generally absent, but present in *T. pullulans*; *T. beigelii* and *T. pullulans* are urease-positive species (34).

Trichosporon can be distinguished from *Geotrichum* by its production of blastoconidia as well as arthroconidia, its rapid growth at 37°C, and its assimilation of a greater number of the usual sugars (see Table 1).

LITERATURE CITED

1. **Adams, E. D., Jr., and B. H. Cooper.** 1974. Evaluation of a modified Wickerham medium for identifying medically important yeasts. Am. J. Med. Technol. **40**:377–388.
2. **Ahearn, D. G., S. A. Meyer, G. Mitchell, M. A. Nicholson, and A. I. Ibrahim.** 1977. Sucrose-negative variants of *Candida tropicalis*. J. Clin. Microbiol. **5**:494–496.
3. **Ahearn, D. G., and R. L. Schlitzer.** 1981. Yeast infections, p. 991–1012. *In* W. H. Hausler and A. Balows (ed.), Laboratory diagnosis of bacterial, mycotic and parasitic diseases, 6th ed. American Public Health Association, Washington, D.C.
4. **Arnold, A. G., B. Gribbin, M. de Leval, F. Macartney, and M. Slack.** 1981. *Trichosporon capitatum* causing recurrent fungal endocarditis. Thorax **36**:478–480.
5. **Baker, J. G., I. F. Salkin, D. H. Pincus, and R. F. D'Amato.** 1981. *Candida paratropicalis*, a new species of *Candida*. Mycotaxon **13**:115–119.
6. **Barnett, J. A., and R. J. Pankhurst.** 1974. A new key to the yeasts. American Elsevier Publishing Co. Inc., New York.
7. **Beijerinck, M. W.** 1889. L'auxanographie ou le méthode de l'hydro-diffusion dans la gelatine appliquée aux recherches microbiologiques. Arch. Neerl. Sci. Exactes Nat. **23**:367–373.
8. **Bennett, J. E., K. J. Kwon-Chung, and D. H. Howard.** 1977. Epidemiologic differences among serotypes of *Cryptococcus neoformans*. Am. J. Epidemiol. **105**:582–586.
9. **Binder, L., A. Csillag, and G. Toth.** 1956. Diffuse infiltration of the lungs associated with *Cryptococcus luteolus*. Lancet **ii**:1043–1045.
10. **Botard, R. W., and D. C. Kelley.** 1968. Modified Littman Oxgall Agar to isolate *Cryptococcus neoformans*. Appl. Microbiol. **16**:689–690.
11. **Bowman, P. I., and D. G. Ahearn.** 1976. Evaluation of commercial systems for the identification of clinical yeast isolates. J. Clin. Microbiol. **4**:49–53.
12. **Brahn, E., and P. A. Leonard.** 1982. *Trichosporon cutaneum* endocarditis: a sequela of intravenous drug abuse. Am. J. Clin. Pathol. **78**:792–794.
13. **Buesching, W. J., K. Kurek, and G. D. Roberts.** 1979. Evaluation of the modified API 20C system for identification of clinically important yeasts. J. Clin. Microbiol. **9**:565–569.
14. **Camargo, Z. P., and O. Fishman.** 1979. *Prototheca stagnora*, an encapsulated organism. Sabouraudia **17**:197–200.
15. **Cooper, B. H.** 1980. Clinical laboratory evaluation of a screening medium (CN Screen) for *Cryptococcus neoformans*. J. Clin. Microbiol. **11**:672–674.
16. **Cooper, B. H., J. B. Johnson, and E. S. Thaxton.** 1978. Clinical evaluation of the Uni-Yeast-Tek system for rapid presumptive identification of medically important yeasts. J. Clin. Microbiol. **7**:349–355.
17. **Cooper, B. H., S. Prowant, D. Brunson, and B. A. Alexander.** 1984. Collaborative evaluation of the Abbott automated yeast identification system. J. Clin. Microbiol. **11**:853–856.
18. **Cunha, T., and J. Lusins.** 1973. *Cryptococcus albidus* meningitis. South. Med. J. **66**:1230–1243.
19. **Deicke, P., and H. Gemeinhardt.** 1980. Embolisch-metastatische Pilz-Enzephalitis durch *Trichosporon capitatum* nach Infusionstherapie. Dtsch. Gesundheitswes. **35**:673–677.
20. **Eschete, M. L., and B. C. West.** 1980. *Saccharomyces cerevisiae* septicemia. Arch. Intern. Med. **140**:1539.
21. **Gemeinhardt, H.** 1965. Zur Frage der Pathogenitat des Sprosspilzes *Trichosporon capitatum* im Respirationstrakt des Menschen ein Beitrag zur Diagnostik der Lungenmykosen. Z. Tuberk. Erkr. Thoraxorgane **124**:190–197.
22. **Gemeinhardt, H.** 1965. Lungenpathogenitat von *Trichosporon capitatum* beim Menschen. Zentralbl. Bakteriol. Parasitenkd. Infektionskr. Hyg. Abt. 1 **196**:121–133.
23. **Genovez, M. E., E. Porto, and E. M. Heins.** 1979. Isolamento de *Trichosporon capitatum*, de material uterino, de egua puro sangue ingles. Registro de um caso. Biologico Sao Paulo **45**:295–298.
24. **Gold, J. W. M., W. Poston, R. Mertelsmann, M. Lange, T. Kiehn, F. Edwards, E. Bernard, K. Christiansen, and D. Armstrong.** 1981. Systemic infection with *Trichosporon cutaneum* in a patient with acute leukemia: report of a case. Cancer **48**:2163–2167.
25. **Goldbert, P. K., P. J. Kozinn, G. J. Wise, N. Nouri, and R. B. Brooks.** 1979. Incidence and significance of candiduria. J. Am. Med. Assoc. **241**:582–584.
26. **Guinet, R., J. Chanas, A. Goullier, G. Bonnefoy, and P. Ambroise-Thomas.** 1983. Fatal septicemia due to amphotericin B-resistant *Candida lusitaniae*. J. Clin. Microbiol. **18**:443–444.
27. **Hasyn, J. J., and H. R. Buckley.** 1982. Evaluation of the AutoMicrobic System for identification of yeasts. J. Clin. Microbiol. **16**:901–904.
28. **Haupt, H. M., W. G. Merz, W. E. Beschorner, W. P. Vaughan, and R. Saral.** 1983. Colonization and infection with *Trichosporon* species in the immunosuppressed host. J. Infect. Dis. **147**:199–203.
29. **Holzschu, D. L., H. A. Presley, M. Miranda, and H. J. Phaff.** 1979. Identification of *Candida lusitaniae* as an opportunistic yeast in humans. J. Clin. Microbiol. **10**:202–205.
30. **Hopfer, R. L., and F. Blank.** 1975. Caffeic acid-containing medium for identification of *Cryptococcus neoformans*. J. Clin. Microbiol. **2**:115–120.
31. **Hopfer, R. L., and D. Gröschel.** 1975. Six-hour pigmentation test for the identification of *Cryptococcus neoformans*. J. Clin. Microbiol. **2**:96–98.
32. **Kamalam, A., P. Yesudian, and A. S. Thambiah.** 1977. Cutaneous infection by *Cryptococcus laurentii*. Br. J. Dermatol. **97**:221–223.
33. **Klotz, S. A., D. J. Drutz, M. Huppert, and J. E. Johnson.** 1982. *Pityrosporum* folliculitis, its potential for confusion with skin lesions of systemic candidiasis. Arch. Intern. Med. **142**:2126–2129.
34. **Kreger-Van Rij, N. J. W. (ed.).** 1984. The yeasts, a taxonomic study. Elsevier Science Publishing Co., Inc., New York.
35. **Krumholz, R. A.** 1972. Pulmonary cryptococcosis. A case due to *Cryptococcus albidus*. Am. Rev. Respir. Dis. **105**:421–424.
36. **Kurtzman, C. P., M. J. Smiley, and C. J. Johnson.** 1980. Emendation of the genus *Issatchenkia* Kudriavzev and comparison of species by deoxyribonucleic acid reassociation, mating reaction, and ascospore ultrastructure. Int. J. Syst. Bacteriol. **30**:503–513.
37. **Kwon-Chung, K. J.** 1975. A new genus, *Filobasidiella*, the perfect state of *Cryptococcus neoformans*. Mycologia **67**:1197–1200.
38. **Kwon-Chung, K. J.** 1976. A new species of *Filobasidiella*,

the sexual state of *Cryptococcus neoformans* B and C serotypes. Mycologia **68:**942–946.

39. **Kwon-Chung, K. J., J. E. Bennett, and J. C. Rhodes.** 1982. Taxonomic studies on *Filobasidiella* species and their anamorphs. Antonie van Leeuwenhoek J. Microbiol. Serol. **48:**25–38.
40. **Kwon-Chung, K. J., J. E. Bennett, and T. S. Theodore.** 1978. *Cryptococcus bacillisporus* sp. nov.: serotype B-C of *Cryptococcus neoformans*. Int. J. Syst. Bacteriol. **28:**616–620.
41. **Land, G. A., W. H. Fleming III, T. A. Beadles, and J. A. Foxworth.** 1979. Rapid identification of medically important yeasts. Lab. Med. **10:**533–541.
42. **Land, G. A., B. A. Harrison, K. L. Hulme, B. H. Cooper, and J. C. Byrd.** 1979. Evaluation of the new API 20C strip for yeast identification against a conventional method. J. Clin. Microbiol. **10:**357–364.
43. **Land, G. A., E. C. Vinton, G. B. Adcock, and J. M. Hopkins.** 1975. Improved auxanographic method for yeast assimilations: a comparison with other approaches. J. Clin. Microbiol. **2:**206–207.
44. **Lodder, J. (ed.).** 1970. The yeasts. A taxonomic study, 2nd ed. North Holland Publishing Co., Amsterdam.
45. **Lutwick, L. I., H. J. Phaff, and D. A. Stevens.** 1980. *Kluyveromyces fragilis* as an opportunistic fungal pathogen of man. Sabouraudia **18:**69–73.
46. **Lynch, J. P., D. R. Schaberg, D. G. Kissner, and C. A. Kauffman.** 1981. *Cryptococcus laurentii* lung abscess. Am. Rev. Respir. Dis. **123:**135–138.
47. **Mackenzie, D. W. R.** 1962. Serum tube identification of *Candida albicans*. J. Clin. Pathol. **15:**563–565.
48. **Manzella, J. P., I. J. Berman, and M. D. Kukrika.** 1982. *Trichosporon beigelii* fungemia and cutaneous dissemination. Arch. Dermatol. **118:**343–345.
49. **Martin, M. V.** 1979. Germ-tube formation by oral strains of *Candida tropicalis*. J. Med. Microbiol. **12:**187–193.
50. **McGinnis, M. R., L. Ajello, E. S. Beneke, E. Drouhet, N. L. Goodman, C. J. Halde, L. D. Haleg, J. Kane, G. A. Land, A. A. Padhye, D. H. Pincus, M. J. Rinaldi, A. L. Rogers, I. F. Salkin, W. A. Schell, and I. Weitzman.** 1984. Taxonomic and nomenclature evaluation of the genera *Candida* and *Torulopsis*. J. Clin. Microbiol. **20:**813–814.
51. **McGinnis, M. R., D. H. Walker, and J. D. Folds.** 1980. *Hansenula polymorpha* infection in a child with chronic granulomatous disease. Arch. Pathol. Lab. Med. **104:**290–292.
52. **Melo, J. S., S. Srinivasen, M. L. Scott, and M. J. Raff.** 1980. *Cryptococcus albidus* meningitis. J. Infect. **2:**79–82.
53. **Moore, M., W. O. Russell, and E. Sachs.** 1946. Chronic leptomeningitis and ependymitis caused by *Ustilago*, probably *U. zeae* (corn smut). Am. J. Pathol. **22:**761–777.
54. **Muchmore, H. G., F. G. Felton, and E. N. Scott.** 1978. Rapid presumptive identification of *Cryptococcus neoformans*. J. Clin. Microbiol. **8:**166–170.
55. **Murray, P. R., R. E. Van Scoy, and G. D. Roberts.** 1977. Should yeasts in respiratory secretions be identified? Mayo Clin Proc. **52:**42–45.
56. **Ngui Yen, J. H., and J. A. Smith.** 1978. Use of Autobac 1 for rapid assimilation testing of *Candida* and *Torulopsis* species. J. Clin. Microbiol. **7:**118–121.
57. **Nurudeen, T., and D. G. Ahearn.** 1979. Regulation of melanin production by *Cryptococcus neoformans*. J. Clin. Microbiol. **5:**724–729.
58. **Oberle, A. D., M. Fowler, and W. D. Grafton.** 1981. *Pityrosporum* isolate from the upper respiratory tract. Am. J. Clin. Pathol. **76:**112–116.
59. **Oblack, D. L., J. C. Rhodes, and W. J. Martin.** 1981. Clinical evaluation of the AutoMicrobic System Yeast Biochemical Card for rapid identification of medically important yeasts. J. Clin. Microbiol. **13:**351–355.
60. **Oelz, O., A. Schaffner, P. Frick, and G. Schaer.** 1983. *Trichosporon capitatum*: thrush-like oral infection, local invasion, fungemia and metastatic abscess formation in a leukaemic patient. J. Infect. **6:**183–185.
61. **Padhye, A. A., J. G. Baker, and F. G. D'Amato.** 1979. Rapid identification of *Prototheca* species by the API 20C system. J. Clin. Microbiol. **10:**579–582.
62. **Pappagianis, D., M. S. Collins, R. Hector, and J. Remington.** 1979. Development of resistance to amphotericin B in *Candida lusitaniae* infecting a human. Antimicrob. Agents Chemother. **16:**123–126.
63. **Pien, F. D., R. L. Thompson, D. Deye, and G. D. Roberts.** 1980. *Rhodotorula* septicemia: two cases and a review of the literature. Mayo Clin. Proc. **55:**258–260.
64. **Pore, R. S., and J. Chen.** 1976. Meningitis caused by *Rhodotorula*. Sabouraudia **14:**331–335.
65. **Redline, R. W., and B. B. Dahms.** 1981. Malassezia pulmonary vasculitis in an infant on long-term intralipid therapy. N. Engl. J. Med. **305:**1395–1398.
66. **Restrepo, A., and L. De Uribe.** 1976. Isolation of fungi belonging to the genera *Geotrichum* and *Trichosporon* from human dermal lesions. Mycopathologia **59:**3–9.
67. **Rusthoven, J. J., R. Feld, and P. J. Tuffnell.** 1984. Systemic infection by *Rhodotorula* spp. in the immunocompromised host. J. Infect. **8:**244–246.
68. **Salkin, I. F., and M. A. Gordon.** 1977. Polymorphism of *Malassezia furfur*. Can. J. Microbiol. **23:**471–475.
69. **Salkin, I. F., M. A. Gordon, W. A. Samsonoff, and C. L. Rieder.** 1982. *Blastoschizomyces pseudotrichosporon*, gen. et sp. nov. Mycotaxon **14:**497–504.
70. **Sanguinetti, V., M. P. Tampieri, and L. Morganti.** 1984. A survey of 120 isolates of *Malassezia (Pityrosporum) pachydermatis*. Mycopathologia **85:**93–95.
71. **Schmeding, K. A., S. C. Jong, and R. Hugh.** 1981. Sexual compatibility between serotypes of *Filobasidiella neoformans* (*Cryptococcus neoformans*). Curr. Microbiol. **5:**133–138.
72. **Schmidt, U., B. Knopf, and V. Watzig.** 1982. Persistierende umschreibene Hyperkeratosen durch *Rhodotorula rubra*. Dermatol. Monatsschr. **168:**111–115.
73. **Shadomy, H. J.** 1970. Clamp connections in two strains of *Cryptococcus neoformans*, p. 67–72. *In* D. G. Ahearn (ed.), Recent trends in yeast research. Georgia State University, Atlanta.
74. **Shaw, C. E., and L. Kapica.** 1972. Production of diagnostic pigment by phenoloxidase activity of *Cryptococcus neoformans*. Appl. Microbiol. **24:**824–830.
75. **Shields, A. B., and L. Ajello.** 1966. Medium for selective isolation of *Cryptococcus neoformans*. Science **151:**208–209.
76. **Shinoda, T., L. Kaufman, and A. A. Padhye.** 1981. Comparative evaluation of the Iatron serological Candida Check kit and the API 20C kit for identification of medically important *Candida* species. J. Clin. Microbiol. **13:**513–518.
77. **Staib, F.** 1962. *Cryptococcus neoformans* and *Guizotia abyssinica* (syn. *G. oleifera* D.C.) (Farbreaktion für *C. neoformans*). Z. Hyg. Infektionskr. Med. Mikrobiol. Immunol. Virol. **148:**466–475.
78. **Strachan, A. A., R. J. Yu, and F. Blank.** 1971. Pigment production of *Cryptococcus neoformans* grown with extracts of *Guizotia abyssinica*. Appl. Microbiol. **22:**478–479.
79. **Sweet, C. E., and L. Kaufman.** 1979. Application of agglutinins for the rapid and accurate identification of medically important *Candida* species. Appl. Microbiol. **19:**830–836.
80. **Vanbreuseghem, R., and M. Takashio.** 1970. An atypical strain of *Cryptococcus neoformans* (San Felice) Vuillemin 1894. II. *C. neoformans* var. *gattii* var. nov. Ann. Soc. Belge Med. Trop. **50:**695–702.
81. **Wallace, M., H. Bagnall, D. Glen, and S. Averill.** 1979. Isolation of lipophilic yeasts in "sterile" peritonitis. Lancet **ii:**956.
82. **Walsh, T. J., D. H. Orth, C. M. Shapiro, R. A. Levine, and J. L. Keller.** 1982. Metastatic fungal chorioretinitis developing during *Trichosporon* sepsis. Ophthalmology **89:**152–156.

83. **Wang, C. J. K., and J. Schwarz.** 1958. The etiology of interstitial pneumonia: identification as *Hansenula anomala* of a yeast isolated from lungs of infants. Mycopathol. Mycol. Appl. **9:**299–306.

84. **Wehrspann, P.** 1982. Evaluation of the Uni-yeast-tek system for yeast identification against conventional methods. Mykosen **25:**599–605.

85. **Weijman, A. C. M.** 1979. Carbohydrate composition and taxonomy of *Geotrichum, Trichosporon* and allied genera. Antonie van Leeuwenhoek J. Microbiol. Serol. **45:**119–127.

86. **Wickerham, L. H., and K. A. Burton.** 1958. Carbon assimilation tests for the classification of yeasts. J. Bacteriol. **56:**363–371.

87. **Wieser, H. G.** 1973. Zur Frage der Pathogenitat des *Cryptococcus albidus*. Schweiz. Med. Wochenschr. **103:**475–481.

88. **Winston, D. J., G. E. Balsley, J. Rhodes, and S. R. Linne.** 1977. Disseminated *Trichosporon capitatum* infection in an immunosuppressed host. Arch. Intern. Med. **137:**1192–1195.

89. **Wolin, H. L., M. L. Bevis, and N. Laurora.** 1962. An improved synthetic medium for the rapid production of chlamydospores by *Candida albicans*. Sabouraudia **2:**96–99.

90. **Yarrow, D., and S. A. Meyer.** 1978. Proposal for amendment of the diagnosis of the genus *Candida* Berkhout nom. cons. Int. J. Syst. Bacteriol. **28:**611–615.

91. **Zimmer, B. I., and G. D. Roberts.** 1979. Rapid selective urease test for presumptive identification of *Cryptococcus neoformans*. J. Clin. Microbiol. **10:**380–381.

Chapter 50

Fungi of Systemic Mycoses

HOWARD W. LARSH AND NORMAN L. GOODMAN

The classical systemic fungal diseases long have been, and continue to be, a significant health problem. Because of more rapid and accessible transportation, with consequent mass movement of people, the concept of endemicity of fungal diseases is less meaningful now than in the past. It is no longer unusual for residents of the northeastern or eastern United States to have coccidioidomycosis, especially when they have spent their winter vacation in southern Arizona, Texas, or California. It is even less unusual if that person is elderly or suffering from some chronic, debilitating disease.

Numerous factors are changing significantly our views of medical mycology. The increased average age of our population has resulted in more chronic diseases, with their debilitating effects. Medical science has "progressed" to a point where patients are sustained by drugs, chemicals, and mechanical processes that compromise physical barriers to infection, suppress immune mechanisms, or upset the balance of normal flora, rendering hosts more susceptible not only to classically pathogenic fungi, but also to all fungi with which they come in contact. This situation has drastically changed our thinking in terms of which fungi may be pathogenic or nonpathogenic. All fungi isolated from clinical specimens must be considered potentially pathogenic by medical mycologists and reported to physicians in language that is as precise as possible. Considerable judgment is required of the laboratorian in determining the significance of opportunistic fungi isolated from clinical specimens. Serious consideration must be given to the specimen—its source, collection, handling, processing, and culturing—to provide the physician with meaningful results. The ability to identify fungi is only a part of laboratory diagnosis of a fungal disease, and active communication with physicians and nurses is essential if a patient's complete medical situation is to be understood. Active communication between the mycology section of a diagnostic laboratory and those sections dealing with histopathology and mycoserology also is of great benefit in understanding the complete medical picture. Appropriate procedures should be instituted in the diagnostic mycology laboratory to ensure the isolation of all fungi present. Thorough, consistent examination of specimens and of all fungal growth from appropriate media incubated for a sufficient time will result in a more efficient laboratory diagnosis of mycoses.

CHARACTERISTICS OF MICROORGANISMS

The fungi classically referred to as the etiological agents of the systemic mycoses include *Blastomyces dermatitidis, Coccidioides immitis, Cryptococcus neoformans, Histoplasma capsulatum, Histoplasma duboisii*, and *Paracoccidioides brasiliensis*. All of these fungi except *C. neoformans* are diphasic; i.e., they grow as complex mycelial elements in their "natural" state or on media at 30°C, and they assume a yeast or "tissue" form in human or animal tissue or on complex media at 35 to 37°C. These fungi primarily cause pulmonary infections, but they may progress systemically to affect any or all organs of the body and may produce cutaneous and subcutaneous lesions after hematogenous dissemination (6, 19).

Within the past decade, there has been a dramatic increase in systemic infections caused by fungi usually considered to be saprobes. These infections occur in debilitated or compromised patients, and fungi causing these infections are often referred to as "opportunistic" (28). Probably the most common infections are those caused by species of *Candida*, especially *Candida albicans*, and *Cryptococcus* spp. (mostly *C. neoformans*). Discussion of these species can be found in Chapter 49. Other fungi commonly causing opportunistic infections include *Aspergillus* species (primarily *Aspergillus fumigatus*), *Pseudallescheria boydii*, and zygomycetes. Numerous other fungi have been isolated from compromised patients, and medical mycologists should not disregard any fungal isolate as being a contaminant or saprophyte without knowing the clinical history and medical status of the patient.

The epidemiology of systemic mycoses is quite complex. Soil generally is considered the natural habitat of these organisms; however, this assumption is questionable in the case of *B. dermatitidis*. Infection occurs by inhalation of fungal conidia which, in the case of *H. capsulatum, B. dermatitidis*, and *P. brasiliensis*, presumably convert to the yeast state shortly after being inhaled and continue to multiply in the host, eliciting various manifestations of disease, depending on such factors as the size of the inoculum, susceptibility of the host, etc.

The geographical distribution of the fungi causing systemic disease varies with each species. *H. capsulatum* has been isolated throughout the world, but the distribution of *B. dermatitidis* is more limited.

The area having the highest incidence of histoplasmosis includes those states bordered by the valleys of the Ohio, Missouri, and Mississippi Rivers in the United States, although the disease is known to occur throughout the world (2, 25).

H. capsulatum can be isolated regularly from soil in zones in which it is endemic in the United States, especially from soil that has been fertilized by the droppings of chickens, blackbirds, and bats. *B. dermatitidis* has been isolated from soil, but repeated isolation from the same sites could not be accomplished with any regularity. Cases of blastomycosis occurring outside of North America have been reported from Mexico, South America, and North Africa, but the majority of cases occurs in Arkansas, Indiana, Kentucky, Louisiana, Mississippi, North Carolina, and Tennessee in the United States. Focal epidemics occurring as far north as Minnesota have been reported. *C. immitis*, on the other hand, is known to occur only in the arid and semiarid regions of North America, Central America, and South America, although transmission of the disease via fomites beyond these regions has been reported (1, 10, 26). *P. brasiliensis* is found in most countries of South America, in Mexico, and in several of the countries of Central America (15). Thus far, the isolation of *H. duboisii* has been restricted to the continent of Africa.

All of the above species produce mycelia when grown under natural conditions at ambient temperatures; hyphae forming the mycelial mat are septate and vary considerably in size and morphology with environmental conditions. Asexual conidia are produced by all species; however, the tuberculate macroconidium of *H. capsulatum* is the only conidium that is distinctive enough to be diagnostic for an individual species (2, 6, 9, 13, 19).

Taxonomically, the fungi that cause systemic mycoses are placed in the class *Deuteromycetes* of the division *Mycota*, with *H. capsulatum* and *B. dermatitidis* having teleomorphs (perfect stages) in the family *Gymnoascaceae* of the class *Ascomycetes*. The teleomorph of *H. capsulatum* is designated *Emmonsiella capsulata* (12), and that of *B. dermatitidis* is designated *Ajellomyces dermatitidis* (14).

COLLECTION, TRANSPORT, AND STORAGE OF SPECIMENS

The correct laboratory diagnosis of a systemic mycosis is dependent not only upon the mycological expertise of laboratorians, but also upon the quality of the specimen provided for analysis. The proper collection, transport, and processing of specimens, as well as prompt inoculation onto appropriate media, are essential for success in the recovery of pathogenic fungi. Since the collection and transport of specimens usually are not done by laboratory personnel, it is imperative that those responsible for these duties be thoroughly instructed in correct procedures for obtaining appropriate specimens and that they understand the need for the immediate delivery of specimens to the laboratory.

Since the classical systemic mycoses primarily affect the respiratory system, most clinical specimens submitted for examination and culture will be respiratory secretions, i.e., sputum, transtracheal aspirates, endobronchial washings, or tissue taken at biopsy. However, because these diseases may be disseminated beyond the respiratory tract, one must also be prepared to examine fluids of all kinds and tissues from any body system. Collection, transport, and storage of specimens is discussed in greater detail in Chapter 47.

Collection

Sputum. The specimen most frequently submitted to the laboratory for the diagnosis of systemic mycosis is sputum. Because of the probability of contamination by airborne conidia of saprophytic fungi and by yeasts which are residents of the mouth and throat, the collection of expectorated sputum samples from patients with pulmonary mycotic infections presents a serious problem. If successful isolation of the etiological agent is to be achieved, proper procedures must be followed to eliminate excessive contamination by endogenous or saprophytic microorganisms present in the upper respiratory system. The methods used in the proper collection of specimens may be found in Chapter 47.

Exudates and pus. In disseminated fungal diseases, especially blastomycosis, cutaneous and subcutaneous lesions may be present. Also, the prostate usually is infected in disseminated blastomycosis. Specimens obtained from these sites are most productive, since there are usually few contaminating organisms. Swabs are of limited use for the isolation of fungi. When possible, exudates or pus should be aspirated; however, if swabs must be used, small ones with minimal absorbency should be used, and two swabs should be used to collect as much specimen as possible. Swabs should be placed in a sterile tube, and the tube should be sealed and taken immediately to the laboratory. Commercial collection and transportation tubes have proved highly satisfactory because they prevent the drying of specimens.

Tissue. Tissue obtained by biopsy, surgery, or necropsy is an excellent specimen for fungal culture and histopathological study. Formalin-fixed specimens are not suitable for the isolation of fungi; however, these specimens can be used with special stains and fluorescent-antibody (FA) techniques for the demonstration of invading fungi.

Bone marrow. Aspirates and biopsy specimens of bone marrow are particularly useful for the isolation of *H. capsulatum*, although other fungi occasionally can be isolated from this material.

Urine. Urine must be collected in such a manner as to prevent contamination with yeasts from the external urinary tract.

CSF. Cerebrospinal fluid (CSF) generally is collected aseptically in sterile tubes that can be sealed tightly to prevent leakage and contamination. CSF is also used for routine bacteriological cultures; therefore, portions of the specimen should be collected in separate tubes for use in each laboratory (i.e., a specimen for bacteriology, one for mycology, etc.).

Blood. Blood culturing techniques have been extensively studied (3, 11, 17, 20) and are discussed in Chapter 47.

Transport

Systemic fungal pathogens will survive in most clinical specimens for several hours; however, rapidly growing bacteria and fungi present in the material may overgrow and mask slower-growing pathogens.

LABORATORY SAFETY

At this point, it should be made clear that all diagnostic mycological work should be carried out in a laminar-flow hood or a bacteriological glove box. All plating of specimens and examination of cultures or specimens should be done inside a hood. Aerosols of conidia or yeast cells from the organisms are infective and, if inhaled, may cause severe disease.

DIRECT EXAMINATION

The microscopic examination of clinical specimens saves considerable time in the diagnosis of a systemic mycosis. The growth rate of most systemic pathogens dictates that the results of a careful microscopic examination of clinical materials be reported promptly and concisely. Observations can be made on stained or unstained preparations. Good microscopic technique should be followed with all preparations, and even experienced mycologists find it useful to have on hand preserved specimens that are known to contain fungi for quality control of staining and microscopic techniques. When preserved positive specimens are not available, a suspension of yeast cells, inactivated with Formalin, can serve as an acceptable positive control for reference purposes. Phase-contrast microscopy is quite helpful in observing unstained preparations of clinical material, if this type of equipment is available.

Sputum, pus, and exudates

The consistency of the specimen determines what treatment is necessary before microscopic examination. Thick, viscid, or opaque specimens require clearing with 10 to 20% KOH before microscopic observation, or one of the digestion procedures described below can be followed. The simplest procedure is to mix 1 drop of specimen with 20% KOH on a microscope slide, heat the slide gently without boiling, and allow it to clear in a moist chamber for 15 to 20 min before examining it under a microscope with reduced light or with phase-contrast optics. This technique is called a KOH preparation and is useful and simple to carry out. Concentration of the specimen before examination, as described below, often is more productive than examination of sputum or pus that has not been concentrated. The addition of Parker Superchrome ink to KOH permits the staining of fungi and improves the chances of identifying *Aspergillus*, *Blastomyces*, *Coccidioides*, *Cryptococcus*, and *Mucor* species. Concentrated sputum mixed with India ink is quite useful for the detection of the encapsulated yeast cells of *C. neoformans*. Thin smears of the sediment from concentrated sputum should be dried, fixed with gentle heating, and stained with Giemsa or Wright stain for the detection of *H. capsulatum*. Alternatively, the sediment from concentrated sputum can be embedded, sectioned, and stained by standard histological techniques. Paraffin sections of concentrated specimens stained with the Gomori methenamine-silver or periodic acid-Schiff stain frequently reveal pathogenic fungi which cannot be observed in unstained wet mounts; however, these procedures are costly and time consuming.

Cytological studies that are commonly used for the detection of tumor cells are also of great value to medical mycologists. In the hands of experienced observers, sputum smears stained by the Papanicolaou method are excellent for the rapid detection of pathogenic fungi (22). The yeast cells of *H. capsulatum*, *B. dermatitidis*, and *C. neoformans* as well as the spherules of *C. immitis* can be detected by this stain. It is efficient to search for fungi by routine cytological techniques which are more rapid than cultural methods.

Staining with FA techniques, with absorbed antisera, can be used for the rapid and specific detection of fungi in smears of clinical materials, when these techniques are available (see Chapter 96). In many instances, a tentative diagnosis can be made from FA tests several days before cultural confirmation, and with highly contaminated specimens, FA staining may be the only laboratory evidence supporting a clinical diagnosis.

CSF

Systemic fungal pathogens other than *C. neoformans* are not often observed in CSF, but occasionally they may be detected.

The centrifuged (1,000 × *g*, 15 min) sediment of CSF should be examined with the India ink technique for the detection of *C. neoformans*. Alternatively, the sediment can be smeared and stained or collected on a membrane filter for staining and culture (see also Chapter 49).

Urine

Urine specimens should be centrifuged, and a drop of the sediment should be observed as an unstained wet mount; 10% KOH may be added to clear the specimen. Alternatively, a thin smear of the urinary sediment can be stained with a Gram stain (see also Chapter 47).

Bone marrow, buffy-coat smears, and peripheral blood smears

Microscopic examination of bone marrow, buffy-coat smears, and peripheral blood smears stained with Wright, Giemsa, or hematoxylin and eosin stains is of value for the detection of *H. capsulatum* in disseminated infections.

Tissues

Surgically excised tissues are best examined with one of the special histological stains for fungi. Touch preparations, made by touching the cut surface of a tissue specimen onto a microscope slide which is then stained with hematoxylin and eosin, can be quite useful for quickly demonstrating fungi.

Touch preparations of lymph nodes are particularly useful for the demonstration of *H. capsulatum* in disseminated histoplasmosis. Small pieces of minced tissue or a suspension of ground tissue treated with 10% KOH also can reveal fungi; however, this procedure is usually less productive than are standard histological techniques.

H. capsulatum occurs as a single-budding oval yeast (1 to 5 μm in diameter) on microscopic examination; oil immersion is usually required for viewing. It should be noted that other fungi, especially *Torulopsis glabrata*, are morphologically similar to *H. capsulatum*, making it difficult to identify this fungus by microscopic examination.

In wet or stained preparations, *B. dermatitidis* appears as large yeast cells (8 to 15 μm in diameter) that are uniformly spherical with thick, refractile walls (Fig. 1). A single bud is attached to the mother cell by a wide (4- to 5-μm) septum.

C. immitis appears as a mixture of immature and mature spherules. These are nonbudding, spherical, thin-walled structures (30 to 60 μm in diameter) that are filled with numerous endospores (2 to 5 μm wide) at maturity (Fig. 2).

Direct examination of specimens containing *P. brasiliensis* reveals fungal cells varying in size from 19 to 25 μm in diameter. The cells are spherical to ovate or elliptical, with occasional chaining of three or four cells. Budding occurs, with the cell producing one or more buds simultaneously, each bud being attached to a parent cell by a narrow neck.

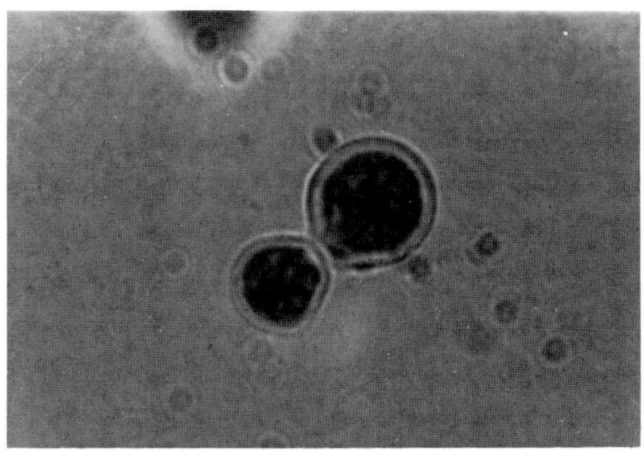

FIG. 1. Yeast cell of *Blastomyces dermatitidis* in wet sputum preparation. ×2,000.

CULTURE AND ISOLATION

Processing of specimens

The processing of specimens should be delayed no longer than 2 h, and immediate processing is preferable. Longer periods between collection and culturing encourage significant increases in the number of rapidly growing organisms. At room temperature, the reproduction of contaminating organisms may lead to excessive overgrowth by these species on various isolation media, preventing the successful isolation of pathogenic fungi. A summary of procedures to be followed in the isolation and identification of fungal pathogens is shown in Fig. 3 and is discussed in Chapter 47.

Because of the frequent contamination of sputum with yeasts, a method of plating concentrated sputum onto plates of yeast extract agar has been developed and shown to be effective for reducing the number of contaminating yeasts while allowing systemic fungal pathogens to proliferate (26). After inoculation of the medium, 1 drop of concentrated NH_4OH is placed on one side of the plate. Diffusion of the NH_4OH into the medium inhibits *Candida* sp., but does not appreciably affect the growth of *Histoplasma, Blastomyces, Paracoccidioides*, and *Coccidiodes* species.

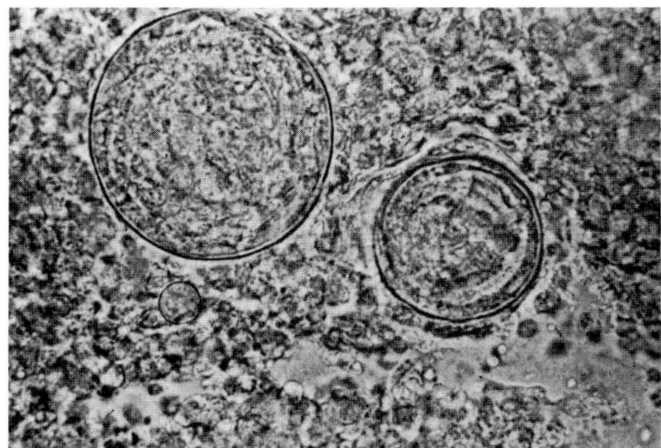

FIG. 2. Spherules of *Coccidioides immitis* seen in an unstained wet preparation of sputum. ×530.

Culturing of specimens

The main purpose of culturing specimens is to grow and isolate the disease-causing fungus in the shortest time possible. Therefore, the best medium, or media, available, incubated under optimal growth conditions, should be used. These media are discussed in Chapter 47. The initial step is to determine if the specimen is likely to contain multiple organisms requiring isolation or the use of selective media. If the specimen is uncontaminated, e.g., CSF, it can be inoculated directly onto an enriched medium, such as blood agar, which will provide an optimal environment for any organisms present to grow most rapidly. However, if the specimen contains multiple organisms, selective media must be used. Logically, since the isolation of colonies is necessary, a large surface area must be provided by using plates or bottles of medium solidified with agar.

There have been numerous studies and discussions on the efficacy of blood cultures in the diagnosis of systemic diseases. Unquestionably, blood cultures are useful in the diagnosis of fungemia, caused by *Candida* species and other yeasts, in the compromised patient. The usefulness of blood cultures in the diagnosis of systemic disease caused by the dimorphic fungi is not so clear. In most cases of disseminated disease, the patient will have gross pulmonary lesions, cutaneous or subcutaneous lesions, or lesions of some other site from which the organism can be demonstrated readily by direct examination and culture. There are, however, cases which are first diagnosed by blood culture. The question of which blood culture system to use has been widely discussed by microbiologists (3, 11, 17, 20).

Regardless of the blood culture system chosen for a particular laboratory, the cultures should be incubated at 30°C under aerobic conditions. Cultures should be examined daily, and broth cultures should be subcultured every week. All cultures should be held at least 6 weeks before being reported as negative.

Fungi causing systemic diseases are aerobic and generally grow optimally at 25 to 30°C.

Many pathogens are inhibited by cycloheximide when they are incubated at 35 to 37°C, but *C. immitis* is an exception. It has been recovered from clinical specimens more rapidly on blood agar containing antibiotics incubated at 35 to 37°C than on media incubated at 25 to 30°C. Mycelial growth appears within 3 to 5 days and usually has a distinctive metallic sheen on the medium.

Special procedures must be used for plated cultures incubated at 30°C. First, the plate should contain 25 to 30 ml of medium. Second, plates should be incubated in an environment with increased humidity. A cabinet containing beakers of water usually is sufficient, but an incubator with controlled humidity is optimal. All cultures should be examined at least weekly for fungal growth and should be held for at least 4 weeks, preferably 6, before being reported as negative. Note again that one should never work with filamentous cultures outside a biological safety hood!

ANIMAL INOCULATION

Although animal inoculation is not mandatory, there are circumstances which make this procedure valuable. It is most useful for the isolation of slowly

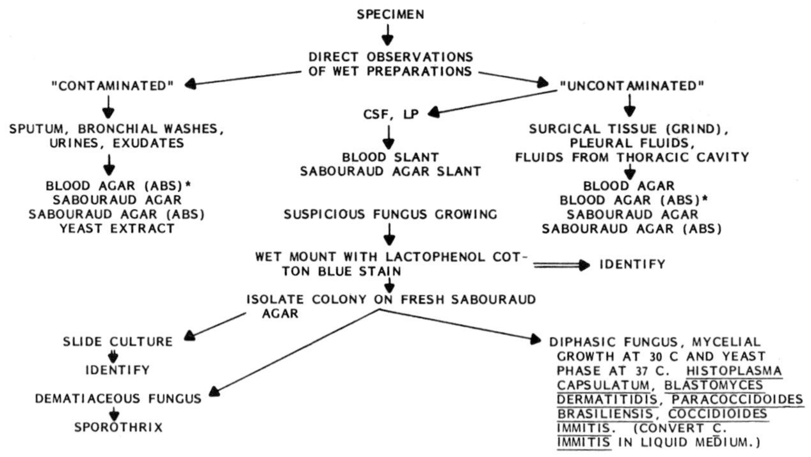

FIG. 3. Procedures used for the identification of systemic fungi in the clinical laboratory.

growing pathogenic fungi from contaminated clinical specimens. It also may be used to determine the pathogenicity of a fungus and to demonstrate the dimorphism of an isolate.

Unfortunately, there is no one experimental animal that is universally susceptible to the fungi pathogenic for humans. White Swiss mice have been used most frequently, but highly pigmented inbred mice are more susceptible to many of the fungi. Rats, hamsters, rabbits, and guinea pigs also have been used for establishing the pathogenesis of various fungi.

For the isolation of fungal pathogens, specimens that are normally sterile can be inoculated directly into the peritoneal cavity of laboratory animals. Antibiotics should be added to contaminated specimens to prevent bacterial sepsis, and it may be necessary to continue treating inoculated animals with antibiotics during the first week after inoculation. The concentration of antibiotics given must be carefully adjusted to prevent toxicity to the animals. It is convenient to inject several animals and to sacrifice some of them each week starting 2 weeks after the initial inoculations. Any animals which succumb to the infection also should be autopsied. The livers and spleens of sacrificed animals are removed and homogenized in a tissue grinder. Each of three plates is inoculated with 1.0 ml of the homogenate from each organ. Sabouraud, blood, and yeast extract agars are the preferred media for plating these tissue homogenates. Plates should be incubated at 30°C and observed for at least 30 days for growth of fungal pathogens.

The use of animal inoculation for determining pathogenicity has been supplanted largely by the exoantigen test (Chapter 96) because of the ease of performing the latter test in comparison with the complexities of housing infected animals. The exoantigen test has proved to be highly specific and reliable for verifying the identities of fungi and provides a reasonable alternative to animal pathogenicity studies (25).

IDENTIFICATION

The laboratory identification of the fungi which cause sytemic mycoses is primarily dependent upon the recognition of morphological characteristics exhibited in all phases of growth, i.e., mycelial, yeast, and tissue phases. In the case of dimorphic fungi, the organisms should be converted from one phase to another to ensure that both phases do indeed exist and thus to strengthen the laboratory diagnosis. A medium that has proven to be useful for the conversion of the mycelial phase of *H. capsulatum* to its yeast phase and for maintenance of the yeast phase was described by Pine and Drouhet (16). The medium is also useful for stimulating yeast phase development of *B. dermatitidis* and *S. schenckii*. A confirmed report should never go to the physician without dimorphism studies of the isolated fungus first being provided.

There is considerable variation in fungi causing systemic disease, not only because of genetic changes, but also because of environmental factors within the host, on culture media, and in incubation methods used in various laboratories. To avoid confusion created by variation, the laboratorian must become familiar with fungi as they appear in laboratory cultures by carefully studying stock cultures including authenticated strains and by keeping up with current literature.

Certain basic procedures have proved successful for the identification of systemic pathogens. In the isolation and identification of dimorphic fungi, it is convenient and simple to isolate the mycelial phase. This isolation is accomplished by incubating the primary isolation plates at 25 to 30°C for at least 4 weeks. Frequent examination of plates will allow the early observation of colony growth and provide a better chance for examining an isolated colony.

Soil extract medium has proved highly efficient in yielding early sporulation of pathogenic fungi (9). This procedure has been used in the laboratory of the Missouri State Chest Hospital for 5 years. Rapid sporulation of pathogenic fungi with diagnostic characteristics being produced within 1 to 5 days has been consistently observed. The medium can be used as an isolation medium, but it excels when early growth of organisms, without diagnostic characteristics, is subcultured onto it. The medium is not useful in determining colony morphology, although extended incu-

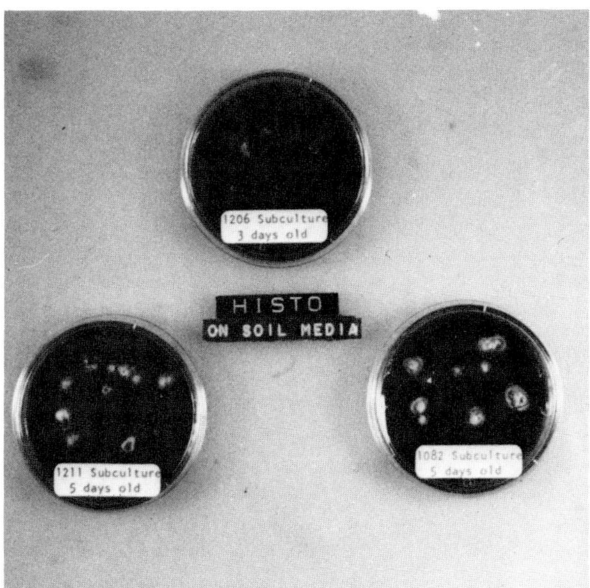

FIG. 4. *Histoplasma capsulatum* on soil extract medium at 3 and 5 days postinoculation.

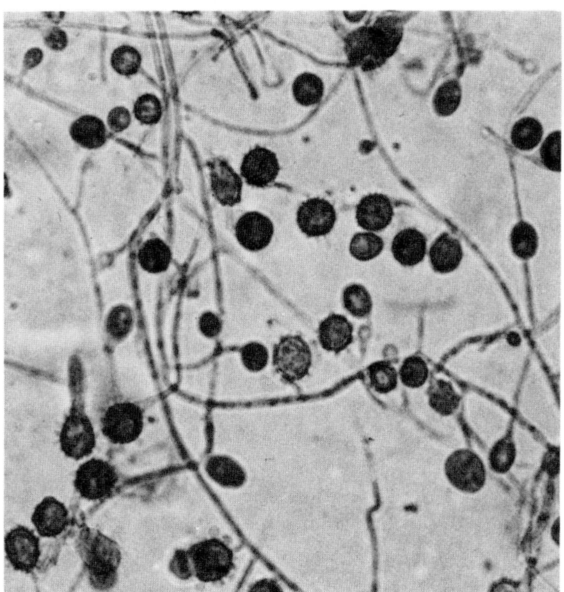

FIG. 6. Wet preparation of *Histoplasma capsulatum* from 3-day-old culture on soil extract medium. ×3,040.

bation usually reveals a well-developed mycelial stage. The sporulation of *H. capsulatum* has been observed within 3 to 5 days (Fig. 4 through 6).

In the early stages of colony growth, a wet mount of the mycelium is prepared inside a biological hood by gently teasing a small piece of the colony apart in lactophenol cotton blue on a microscope slide, and then the slide is examined microscopically for the characteristic mycelium and macroconidia.

At the same time, another small piece of the isolated colony should be transferred to a plate of modified Sabouraud agar and incubated at 25 to 30°C. This culture will serve to demonstrate characteristic colonial morphology and microscopic characteristics.

Also, at this time, yet another small piece of the isolated colony should be transferred to a tube of fresh blood agar, brain heart infusion agar, or other appropriate enriched medium (hereafter referred to as conversion medium). The inoculum is placed firmly onto the medium and smeared over a small area. The culture is incubated at 35°C, and after 48 h, it is examined daily for growth of the yeast form. If conversion does not occur in 4 to 5 days, this culture should be transferred to another fresh tube of conversion medium, and the process should be repeated. Conversion also can be readily accomplished in tissue cultures, with any of the usual cell lines. If *C. immitis* is suspected, conversion to the spherule form may be accomplished by culturing the fungus, at 35°C, in an atmosphere of 5% CO_2 in special liquid medium, e.g., Converse medium (7), 7H10 broth, or 7H11 broth. Dimorphism in *C. immitis* also can be demonstrated by inoculating the fungus into the peritoneal cavity of mice or guinea pigs.

When a pure culture has been obtained on the isolation medium, the fungus can be identified by its gross colonial morphology and microscopic characteristics. Again, one must keep in mind that there are variants of all microorganisms.

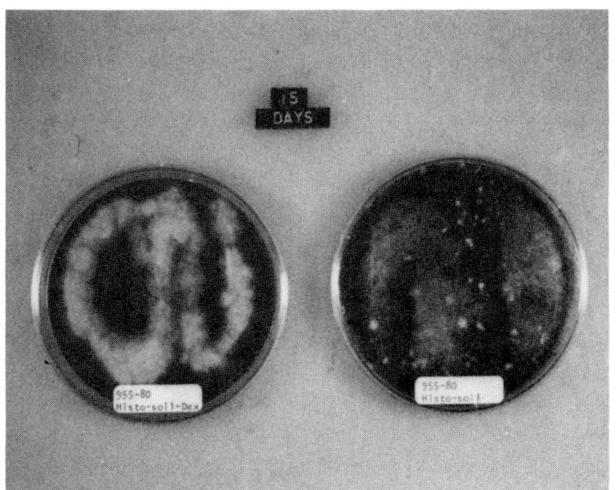

FIG. 5. *Histoplasma capsulatum* on soil extract medium at 15 days postinoculation.

Pathogenesis

Respiratory and lymph gland involvement commonly are observed in systemic mycotic infections of humans and animals. There is a noteworthy host-parasite relationship which depends on many factors, including the size of the inoculum and the immune status of the host. Pathogenic fungi usually enter the host by the airborne route, with the inoculum transmitted from soil or other environmental foci. The host response, which depends on many factors, includes various signs and symptoms characteristic of asymptomatic, mild, moderate, or severe disease. The usual course of the disease is from the inhalation of infectious units through mild infection to the development of some stage of immunity. Most patients convert from a negative to a positive skin test, and precipitins and complement-fixing antibodies are usually pro-

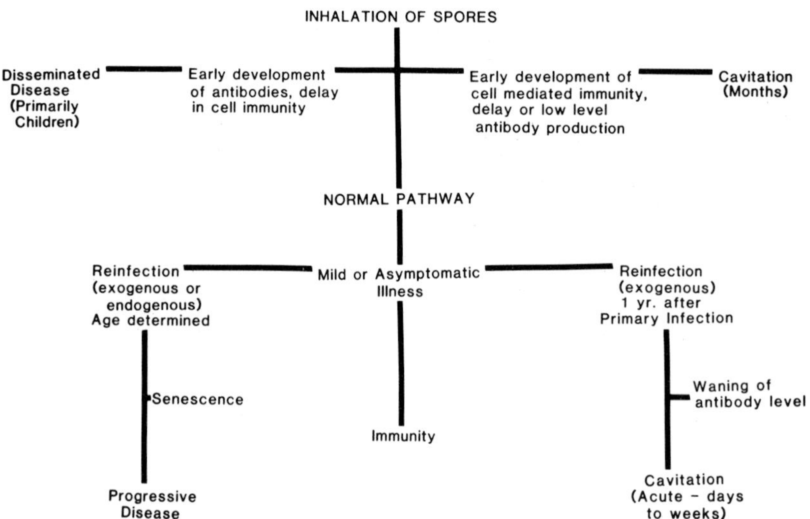

FIG. 7. Generalized natural history of systemic fungal diseases.

duced. Benign disease is the rule, and serious disseminated disease is limited to approximately 1% of those patients who acquire a primary pulmonary infection. The information in Fig. 7 is derived from a limited number of patients seen in recent years and is for study with the hope that further observations will substantiate these findings.

H. capsulatum

When grown on Sabouraud agar at 25 to 30°C, *H. capsulatum* grows slowly and produces a white, cottony, aerial mycelium that usually turns buff to brown with age (Fig. 8). Microscopic examination reveals round to pyriform, smooth, or echinulate microconidia (2 to 5 μm in diameter) which may be sessile on the sides of the hyphae or attached to short, lateral conidiophores. Later, the characteristic and diagnostic large, round, or pyriform tuberculate macroconidia (7 to 15 μm in diameter) appear in the culture (Fig. 9). Similar cultural and microscopic findings may occur on blood agar; however, the colony is usually glabrous, wrinkled, or folded (Fig. 10), and only hyphae can be found on microscopic examination. Such cultures should be transferred to modified Sabouraud agar to demonstrate characteristic morphology. One must be aware that macroconidia may be absent or have only smooth walls when the isolate is from a patient with chronic histoplasmosis. In addition, there are many variants of this fungus, and the mycelial growth is not always typically white and cottonlike, with aerial hyphae. Variants with color ranging from brown to slate gray have been observed. Also, one must be aware of the saprobic fungus *Sepedonium*, which produces spiny tuberculate conidia similar to those produced by *H. capsulatum* (Fig. 11). These two species can be differentiated, however, by the conversion of *H. capsulatum* to its yeast form (*Sepedonium* sp. is not dimorphic), by the failure of most *Sepedonium* isolates to grow on media containing cycloheximide, and by the exoantigen test with *H. capsulatum* antiserum.

The yeast (parasitic) form of *H. capsulatum* may be grown on blood agar or other enriched media at 35°C as a typical oval, budding yeast cell 2 to 3 by 4 to 5 μm in size. The cell reproduces by budding at the small end of the cell, and the pore between the bud and parent cell remains small (Fig. 12).

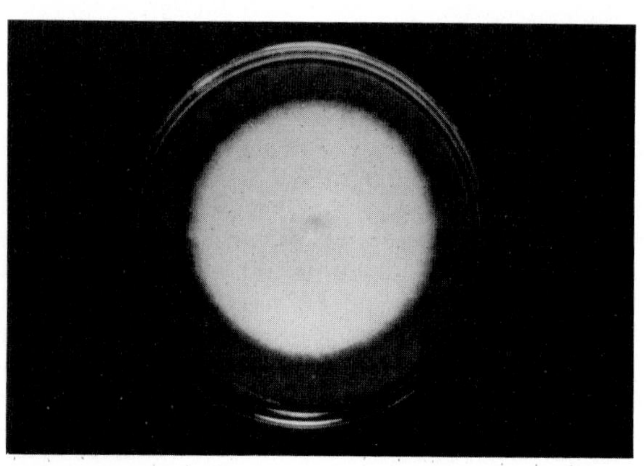

FIG. 8. Colony of *Histoplasma capsulatum* on Sabouraud agar at 30°C.

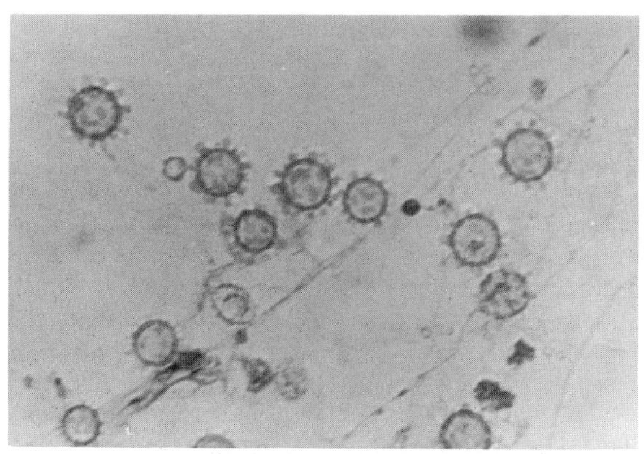

FIG. 9. Microscopic appearance of a wet preparation of *Histoplasma capsulatum* mycelial phase from Sabouraud agar incubated at 30°C. Lactophenol cotton blue stain. ×1,080.

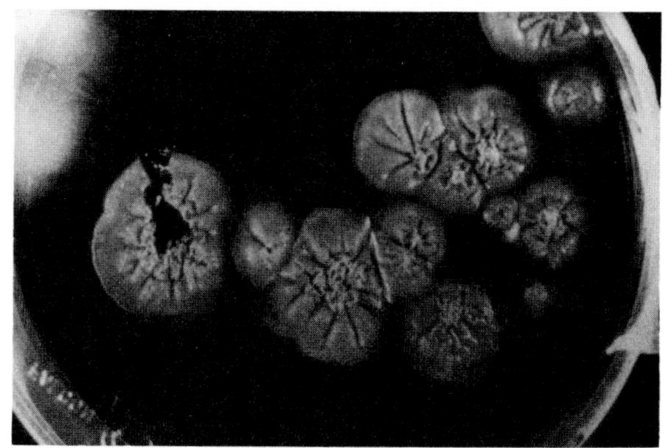

FIG. 10. Colonies of *Histoplasma capsulatum* on blood agar at 30°C.

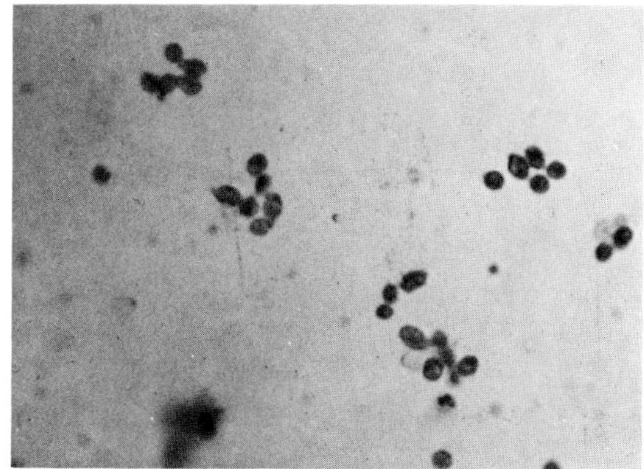

FIG. 12. Wet preparation of *Histoplasma capsulatum* yeast form from blood agar incubated at 37°C. Lactophenol cotton blue stain. ×1,800.

More recent developments of blood culture procedures and techniques have resulted in a greater number of positive isolations of pathogenic fungi. Microbial isolators, biphasic medium, and the radiometric method have been extensively evaluated, and their efficacy in the isolation of pathogenic fungi has been determined (13).

In tissue sections or smears, the fungal cell is seen to be intracellular in macrophages, giant cells, or polymorphonuclear leukocytes. The yeast form may show typical budding, with the cells having the same dimensions as given above (Fig. 13). *H. capsulatum* can be stained in tissue by many techniques, with the Gomori methenamine-silver technique giving the highest contrast between fungal cell and host cell.

B. dermatitidis

The mycelial form of *B. dermatitidis*, when grown on Sabouraud agar at room temperature (25 to 30°C), is a white mold. Seven to 10 days is required to produce a mature colony, which may continue growing to fill the petri plate or agar slant. Occasional isolates may be encountered that grow more slowly and sporulate poorly, and isolates that become brown or produce a dark pigment in the medium (Fig. 14) may be found. On blood agar at 25 to 30°C, *B. dermatitidis* grows as a glabrous or feltlike colony which is usually wrinkled and folded (Fig. 15). Often conidia are not produced in these cultures, and transfer to modified Sabouraud agar is required for demonstration of morphology.

Conidia that are borne on slender, lateral conidiophores or terminally on hyphal branches are smooth walled, spherical to oval, and 2 to 10 μm in size (Fig. 16). In its early stage, the mycelial form of *B. dermatitidis* may very closely resemble, and be confused with, isolates of *P. boydii*. In typical isolates, *B. dermatitidis* becomes tan to brown with age, whereas *P. boydii* turns gray.

The yeast form of *B. dermatitidis* grown at 35°C on an enriched medium such as blood agar is the most easily identified morphological state of this species. The organisms are uniformly spherical with thick, refractile walls and are 8 to 15 μm in diameter. A single bud is characteristically attached to the mother cell by a wide septum 4 to 5 μm wide, which is a

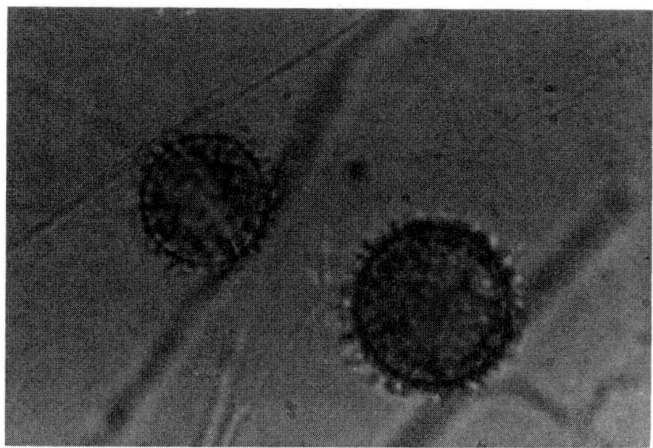

FIG. 11. Tuberculate conidium of *Sepedonium* species. ×2,000.

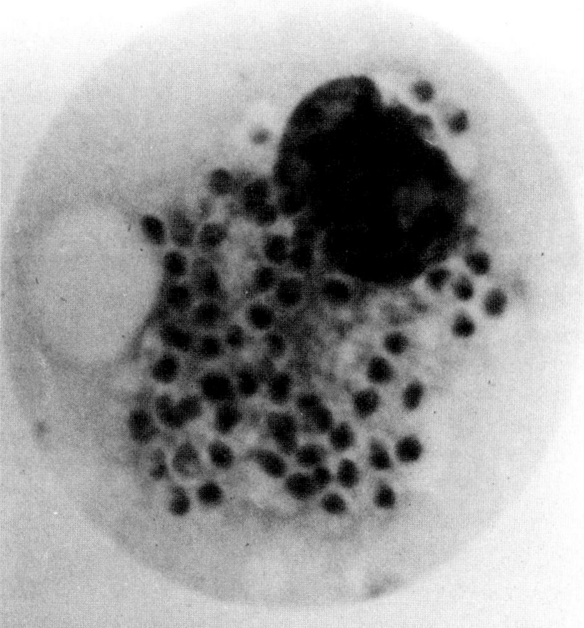

FIG. 13. *Histoplasma capsulatum* in tissue. Hematoxylin and eosin stain. ×1,656.

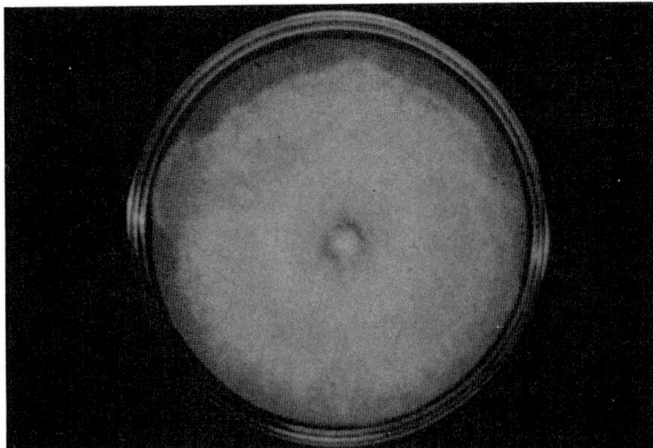

FIG. 14. Colony of *Blastomyces dermatitidis* on Sabouraud agar at 30°C.

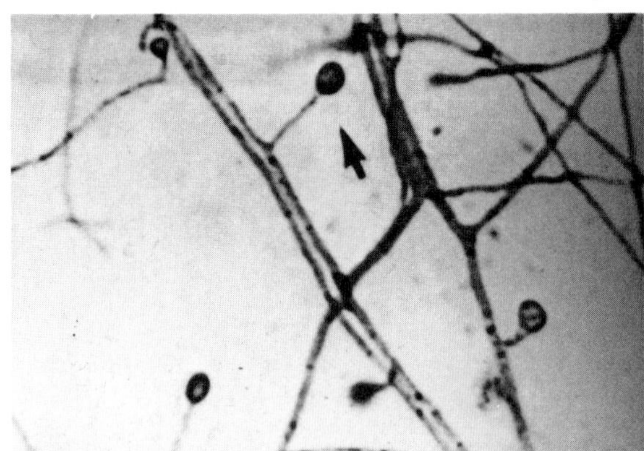

FIG. 16. Wet preparation of mycelial-phase *Blastomyces dermatitidis* grown on Sabouraud agar. Lactophenol cotton blue stain. Arrow, Microaleuriospore. ×590. (Photomicrograph courtesy of L. Ajello.)

diagnostic character (Fig. 17). In a young bud, the wall is much thinner than in the mother cell, and the attachment may be as wide as the bud; the wall gains refractivity with age and is sometimes termed a double-contoured wall. Single budding distinguishes *B. dermatitidis* from its South American counterpart, *P. brasiliensis*, whose pathogenic phase is a multiple-budding yeast.

B. dermatitidis reproduces in tissue by budding, the bud being characterized by its large size, the attachment to the parent cell by a wide septum, and the presence of a thick, refractile wall (Fig. 18).

C. immitis

C. immitis grows rapidly on Sabouraud agar, the mycelium covering the plate or slant within a few days. The young culture is floccose, and with age, the color darkens; some isolates produce a brown or yellow pigment (Fig. 19). Darkening of peptone-containing medium is common. This fungus has many morphological variants (Fig. 20).

Characteristic arthroconidia first appear on the side branches of the hyphae. Septation of hyphae produces arthroconidia that exhibit alternation of the arthroconidia with empty spaces. The arthroconidia are barrel shaped, 2 to 4 by 3 to 6 μm, and are dispersed by fragmentation of the walls of the empty spaces between spores. With age, the entire hyphal mass of a culture may form arthroconidia (Fig. 21).

In cultures on artificial media incubated at room temperature, *C. immitis* must be differentiated from other fungi such as species of *Arthroderma, Auxarthron, Geotrichum, Oidiodendron*, etc. All of these fungi

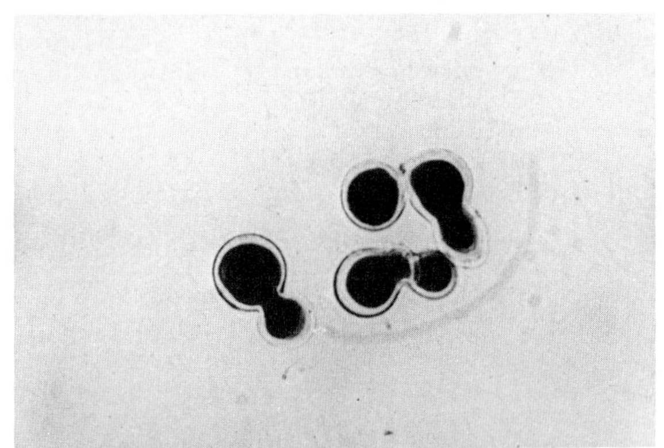

FIG. 17. Wet preparation of *Blastomyces dermatitidis* yeast form. Lactophenol cotton blue stain. ×1,800.

FIG. 15. Colonies of *Blastomyces dermatitidis* on blood agar at 30°C.

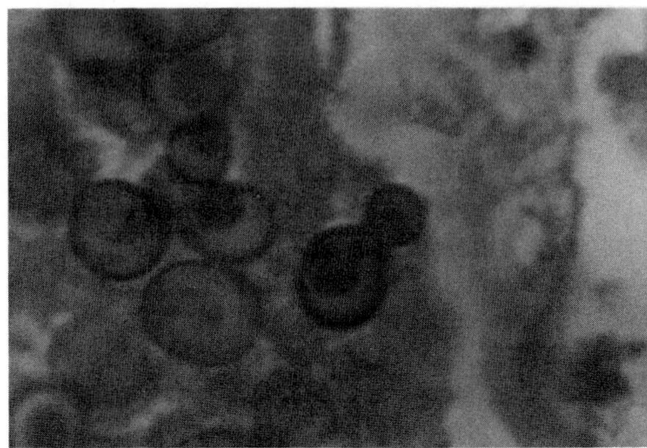

FIG. 18. *Blastomyces dermatitidis* in tissue. Hematoxylin and eosin stain. ×720.

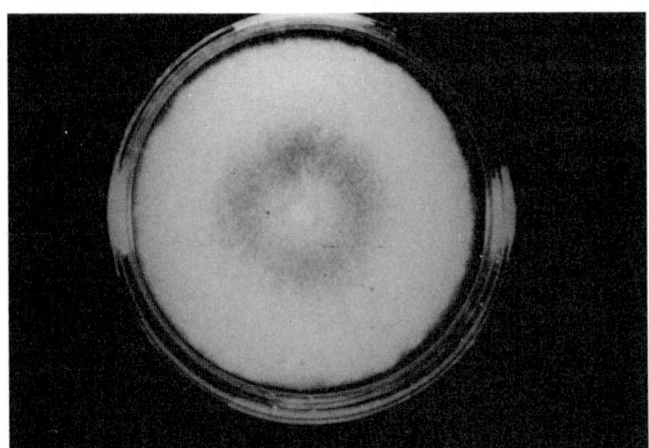

FIG. 19. Colony of *Coccidioides immitis* on Sabouraud agar at 30°C.

produce arthroconidia by hyphal fragmentation, and casual microscopic observation may result in misidentification. Careful studies of growth on Sabouraud agar and of lactophenol cotton blue preparations are valuable for making a correct laboratory diagnosis.

In vitro production of endospores by *C. immitis* has been accomplished in synthetic media (21), but normally this procedure is not done in the clinical laboratory. When living endospore cultures are needed, a liquid synthetic medium modified by Converse can be used (7). Physical-chemical growth factors for *C. immitis* spherules defined by Converse led to improvement in this synthetic medium (8). Agar gel procedures for the cultivation of spherules were introduced by Brosbe (4) and Sun et al. (27). A definitive demonstration of spherules can be made by intraperitoneal inoculation of mice or intratesticular inoculation of guinea pigs.

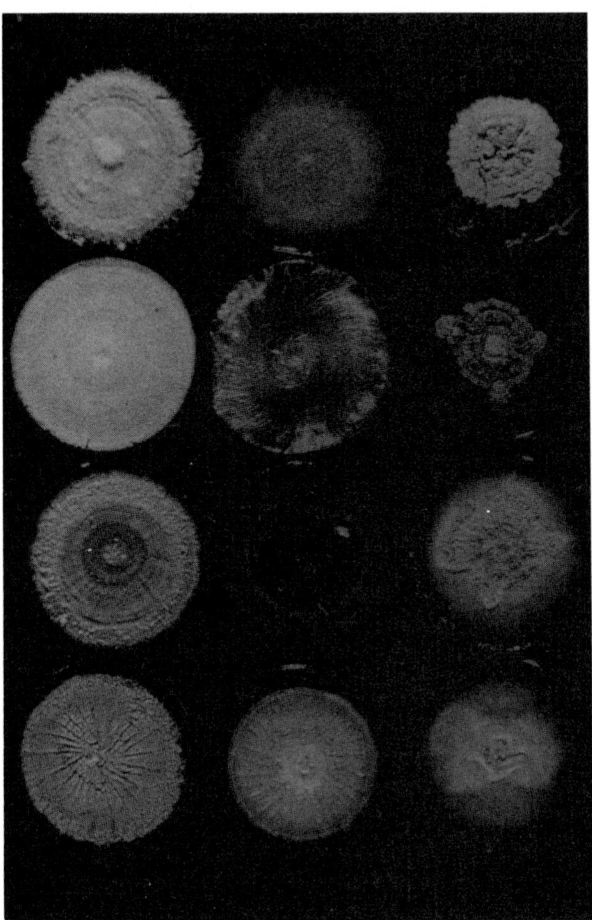

FIG. 20. Variation in colony morphology of *Coccidioides immitis* on Sabouraud agar at 30°C. (Photograph courtesy of M. Huppert.)

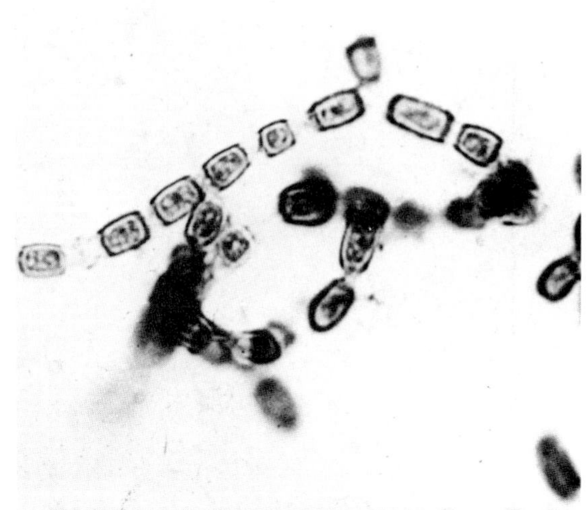

FIG. 21. Wet preparation of *Coccidioides immitis* grown on Sabouraud agar at 30°C. Lactophenol cotton blue stain. ×1,080.

C. immitis appears in tissue as a nonbudding, spherical, thick-walled structure 30 to 60 μm in diameter, which may be filled with numerous small endospores (2 to 5 μm in diameter). The fungus reproduces by means of these endospores, which are released by rupture of the spherule cell wall. These infectious units increase in size and develop into mature spherules. In direct examination of tissue, it is possible to observe only immature forms of *C. immitis*, which do not contain endospores. This stage is easily confused with nonbudding yeasts, especially *B. dermatitidis* and *C. neoformans*, because of variation in the size of

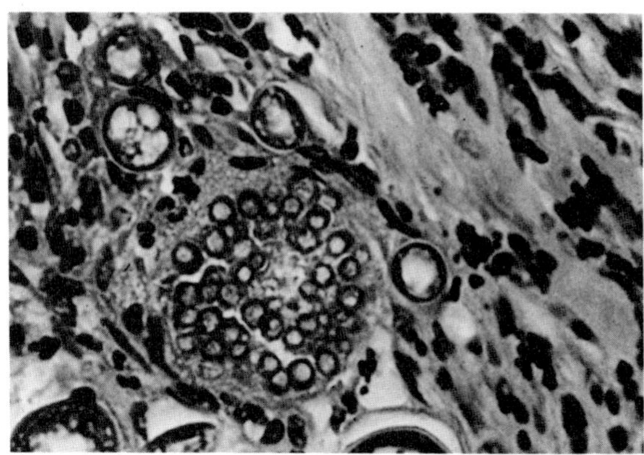

FIG. 22. Spherule of *Coccidioides immitis* in tissue. Hematoxylin and eosin stain. ×900.

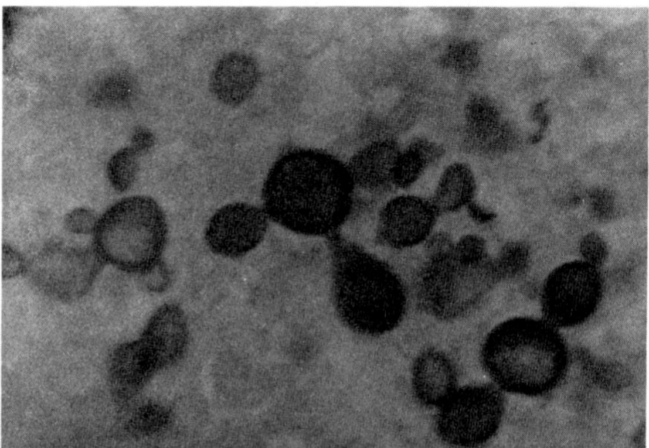

FIG. 23. Yeast form of *Paracoccidioides brasiliensis*. Methenamine-silver stain. ×900.

spherules. Examination of several microscopic preparations of tissue may be required to find the characteristic, endospore-filled, mature spherules (Fig. 22).

P. brasiliensis

The rate of growth of *P. brasiliensis* is slow compared with that of the other systemic fungi. On Sabouraud agar, at 25 to 30°C, the fungus develops as a heaped, glabrous or wrinkled colony with short, white, aerial hyphae, which often turns brown with age. Microscopically, chlamydospores and a few sessile, oval to round conidia may occur. These conidia are indistinguishable from those of *B. dermatitidis*; it is necessary to convert the culture to the yeast form for identification (5, 15).

When *P. brasiliensis* is grown at 37°C on enriched medium, it grows slowly and develops smooth-to-cerebriform yeastlike colonies. Microscopically, the culture is composed of single- and multiple-budding yeasts. The multiple-budding cells are thick walled and 10 to 25 μm in diameter, with buds from 1 to 10 μm in diameter.

In tissue, *P. brasiliensis* appears as single- and multiple-budding yeasts with the dimensions described above; however, care should be taken to look for the multiple-budding cells, as the single-budding cells are indistinguishable from *B. dermatitidis* (Fig. 23 and 24).

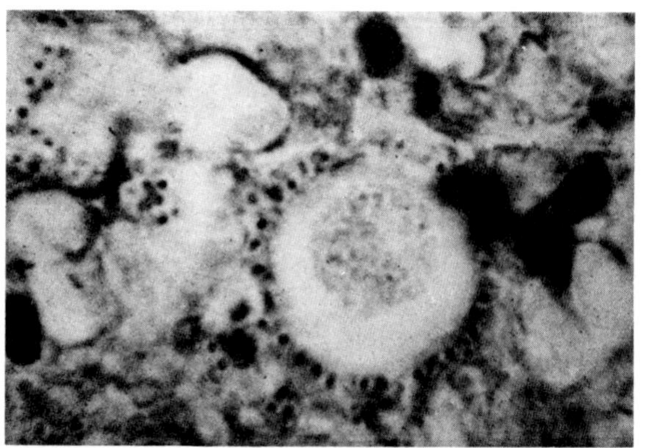

FIG. 24. *Paracoccidioides brasiliensis* in tissue. Note "mariner's wheel" of the multiple-budding yeast. Hematoxylin and eosin stain. ×900.

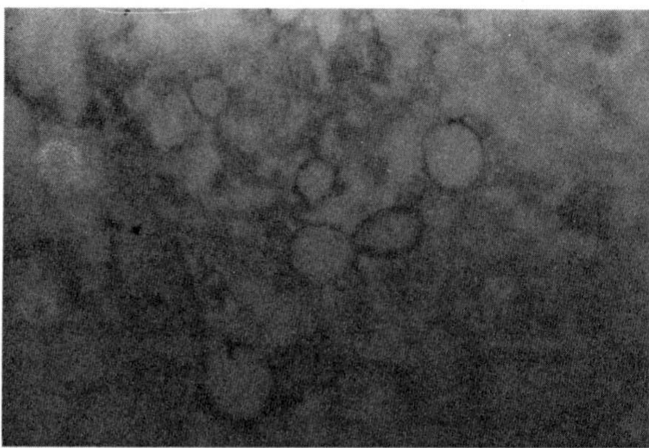

FIG. 25. *Histoplasma duboisii* in tissue. Hematoxylin and eosin stain. ×2,000.

H. duboisii

The mycelial form of *H. duboisii*, grown at 25 to 30°C, is indistinguishable from *H. capsulatum*. The yeast form may be grown on a variety of enriched media at 35°C. Under these conditions, the fungus develops as a white, soft, yeastlike growth, consisting of budding cells 10 to 15 μm in diameter.

In tissue, *H. duboisii* resembles *B. dermatitidis* in size, shape, and thick-walled appearance. The cells are approximately 10 μm in diameter with numerous budding forms that stain well with most histological dyes (Fig. 25).

P. boydii is most noted for causing eumycotic mycetoma; however, it can also produce necrotizing pneumonia (20), meningitis, eye infections, and generalized infections involving multiple organ systems. Infections in immunocompromised patients can be especially severe and are often fatal. Most isolates are resistant to amphotericin B, but most are susceptible to the newer imidazole antifungal agents (see Chapter 104). In infected internal organs, *P. boydii* produces septate hyphae and is difficult to distinguish morphologically from *Aspergillus* species and other fungi which produce septate hyphae in tissues. When isolated from sputum, this species is rarely, if ever, a contaminant. See Chapter 51 for a more detailed description of this species.

Aspergillus species

Aspergillus species grow well on all laboratory media and can be readily identified by their characteristic microscopic morphology. A description of these morphological characteristics can be found in Chapter 54.

Zygomycetes

The zygomycetes commonly causing systemic diseases, *Rhizopus, Mucor, Absidia, Cunninghamella*, and *Saksenaea* species, grow rapidly on all laboratory media not containing cycloheximide. A description of identifying characteristics of these fungi can be found in Chapter 53.

LITERATURE CITED

1. **Ajello, L. (ed.).** 1967. Coccidioidomycosis. University of Arizona Press, Tucson.

2. **Ajello, L., E. W. Chick, and M. L. Furcolow (ed.).** 1971. Histoplasmosis. Proceedings of the Second National Conference. Charles C Thomas, Publisher, Springfield, Ill.
3. **Bille, J., L. Stockman, and G. D. Roberts.** 1982. Detection of yeasts and filamentous fungi in blood cultures during a 10-year period (1972 to 1981). J. Clin. Microbiol. **16:**968–970.
4. **Brosbe, E. A.** 1967. Use of refined agar for the in vitro propagation of the spherule phase of *Coccidioides immitis*. J. Bacteriol. **93:**497–498.
5. **Carbonell, L. M., and J. Rodríguez.** 1965. Transformation of mycelial and yeast forms of *Paracoccidioides brasiliensis* in cultures and in experimental inoculations. J. Bacteriol. **90:**504–510.
6. **Conant, N. F., D. T. Smith, R. D. Baker, and J. L. Galloway.** 1971. Manual of clinical mycology, 3rd ed. The W. B. Saunders Co., Philadelphia.
7. **Converse, J. L.** 1955. Growth of spherules of *Coccidioides immitis* in a chemically defined liquid medium. Proc. Soc. Exp. Biol. Med. **90:**709–711.
8. **Converse, J. L.** 1956. Effect of physico-chemical environment of spherulation of *Coccidioides immitis* in a chemically defined medium. J. Bacteriol. **72:**784–792.
9. **Emmons, C. W., C. H. Binford, M. P. Utz, and K. J. Kwon-Chung.** 1977. Medical mycology, 3rd ed., appendix I, p. 539, no. 12. Lea & Febiger, Philadelphia.
10. **Fiese, M. J.** 1958. Coccidioidomycosis. Charles C Thomas, Publisher, Springfield, Ill.
11. **Kelly, M. T., G. E. Buck, and M. F. Fojtasek.** 1983. Evaluation of a lysis-centrifugation and biphasic bottle blood culture system during routine use. J. Clin. Microbiol. **18:**554–557.
12. **Kwon-Chung, K. J.** 1973. Studies on *Emmonsiella capsulata*. I. Heterothallism and development of the ascocorp. Mycologia **65:**109–121.
13. **Larsh, H. W.** 1970. Isolation and identification media for systemic fungi. Pan American Health Organization Science Publication no. 205, Pan American Health Organization, New York.
14. **McDonough, E. S., and A. L. Lewis.** 1968. The ascigerous stage of *Blastomyces dermatitidis*. Mycologia **60:**76–83.
15. **Pan American Health Organization.** 1972. Paracoccidioidomycosis. Pan American Health Organization Science Publication no. 254, Pan American Health Organization, New York.
16. **Pine, L., and E. Drouhet.** 1963. Sur l'obtention et la conservation de la phase Levure d'*Histoplasma capsulatum* et d'*H. duboisii* en milieu chimiquement definie. Ann. Inst. Pasteur **103:**798–804.
17. **Prevost, E., and E. Bannister.** 1981. Detection of yeast septicemia by biphasic and radiometric methods. J. Clin. Microbiol. **13:**655–660.
18. **Reddy, P. C., C. S. Christianson, D. F. Gorelick, and H. W. Larsh.** 1969. Pulmonary monosporiosis: an uncommon pulmonary mycotic infection. Thorax **24:**722–726.
19. **Rippon, J. W.** 1974. Medical mycology, the pathogenic fungi and the pathogenic actinomycetes. The W. B. Saunders Co., Philadelphia.
20. **Roberts, G. D., and J. A. Washington II.** 1975. Detection of fungi in blood cultures. J. Clin. Microbiol. **1:**309–311.
21. **Roessler, W. G., E. H. Herbert, W. C. McCullough, R. C. Mills, and C. R. Brener.** 1946. Studies with *Coccidioides immitis*. I. Submerged growth in liquid medium. J. Infect. Dis. **79:**12–22.
22. **Sanders, J. S., G. A. Sarosi, D. J. Nollett, and J. L. Thompson.** 1977. Exfoliative cytology in the rapid diagnosis of pulmonary blastomycosis. Chest **72:**193–196.
23. **Schwarz, Jan.** 1981. Histoplasmosis. Praeger Publishers, New York.
24. **Smith, C. D., and N. L. Goodman.** 1975. Improved culture method for the isolation of *Histoplasma capsulatum* and *Blastomyces dermatitidis* from contaminated specimens. Am. J. Clin. Pathol. **62:**276–280.
25. **Standard, P. G., and L. Kaufman.** 1976. Specific immunological test for the rapid identification of members of the genus *Histoplasma*. J. Clin. Microbiol. **3:**191–199.
26. **Stevens, D. A. (ed.).** 1980. Coccidioidomycosis—a text. Plenum Medical Book Co., New York.
27. **Sun, S. H., M. Huppert, and K. R. Vukovich.** 1976. Rapid in vitro conversion and identification of *Coccidioides immitis*. J. Clin. Microbiol. **3:**186–190.
28. **Williams, D. M., J. A. Krick, and J. S. Remington.** 1976. Pulmonary infection in the compromised host. Part I. Am. Rev. Respir. Dis. **114:**359–394.

Chapter 51

Fungi Causing Eumycotic Mycetomas

ARVIND A. PADHYE AND LIBERO AJELLO

A mycetoma is a localized, chronic, granulomatous infection involving cutaneous and subcutaneous tissues and eventually, in some cases, the bones. Mycetomas generally are confined to the feet or hands, but occasionally other parts of the body such as the back, shoulders, and buttocks may be involved. The disease is distributed worldwide and is more commonly seen in humans than in animals. Only a few authentic cases of mycetoma in lower animals such as dogs, horses, and goats have been described in the literature (2, 10, 24). Both filamentous fungi and actinomycetes are known to cause mycetomas. Mycetomas caused by actinomycetes are called actinomycotic mycetomas, and those incited by species of filamentous fungi are referred to as eumycotic mycetomas.

The causal agents of eumycotic mycetomas are saprophytes that live on organic debris in soil. The various causal agents have been isolated from either soil or plant material (1, 4, 15, 21). Segretain and Mariat (37, 40), using specific media and techniques, showed that *Leptosphaeria senegalensis* and *Leptosphaeria tompkinsii* were found on about 50% of the dry thorns of *Acacia* trees that were examined, particularly those that had been stained by mud during the rainy season. *Neotestudina rosatii* was isolated from sandy ground (26), and *Madurella mycetomatis* was recovered from soil and anthills (36, 42). *Acremonium* spp., *Curvularia* spp., *Aspergillus nidulans*, and *Pseudallescheria boydii* frequently have been isolated from soil. Borelli (6) isolated *Madurella grisea* from soil in Venezuela.

Mode of infection and symptoms

A mycetoma develops after an injury by contaminated thorns, splinters from plants, fish scales or fins, snake bites, insect bites, farm implements, knives, etc. The injury allows the entry of potentially pathogenic organisms into the body. The initial lesions are characterized by a feeling of discomfort and pain at the point of inoculation. Weeks or months later, the subcutaneous tissue at the site of inoculation becomes indurated, abscesses develop, and sinuses may drain to the surface. The lesions are characterized by swelling, suppurating abscesses, granulomata, and sinuses from which serosanguineous fluid containing granules oozes out. The mycetoma develops slowly beneath thick fibrosclerous tissue. The subsequent phase of proliferation involves the invasion of muscles and the intramuscular layers. The granulomatous lesions can extend as deep as bony tissue, causing severe destruction of bone. Regardless of the etiologic agent involved, the causal organisms develop in the form of soft or hard, compact mycelial masses, known as granules or grains, within the infected tissue. The hallmark of a mycetoma is the granule composed of nonsporulating mycelium that may or may not be embedded in a cementlike matrix.

Mycetomas develop mainly among people such as field laborers, farmers, sugarcane workers, fishermen, etc., who are in contact with contaminated materials. Even though, because of increased exposure, the prevalence of mycetomas is much higher in males than in females, women and children who walk barefoot also are vulnerable to infection.

Fungus balls, caused by species of *Aspergillus*, *Coccidioides immitis*, or *P. boydii* in preformed lung cavities, are sometimes mistakenly called mycetomas (16). In the absence of well-organized granules, they should be appropriately referred to as fungus balls, aspergillomas, or coccidioidomas. Similarly, mycelial aggregates formed by dermatophytes in deep tissues differ in many respects from the granules of the mycetomas. Such infections caused by dermatophytes are best referred to as pseudomycetomas rather than mycetomas (3).

Geographic distribution of eumycotic mycetomas

Eumycotic mycetomas occur primarily in the tropical and hot temperate zones of the world. They are frequently reported from countries near the Tropic of Cancer, but they also occur beyond this area. Numerous cases have been described from Africa, Asia, and South and Central America. Mycetomas are not as commonly seen in the United States as in tropical countries.

Climatic conditions

Climate has a definite influence on the incidence and distribution of mycetomas. Rivers that flood each year during the wet season in many countries of Africa and Asia influence the distribution of the causal agents. Rainfall also favors the spread of the etiologic agents on organic matter (12).

Collection and examination of the granules

Since all of the agents of eumycotic mycetomas are soil saprophytes and some are encountered as contaminants of clinical specimens, their etiologic role in mycetomas must be established carefully. A definitive diagnosis must be based on the demonstration of granules in tissue and repeated isolation of the causal fungus from granules aspirated from sinuses, preferably unopened.

Pus, exudate, or biopsy material should be examined for the presence of granules, which vary in size from 0.2 to 2 mm or more and are usually detectable with the naked eye. Their color, texture, size, and shape give a fair indication of the identity of the etiologic agent. Actinomycotic and eumycotic mycetomas are differentiated by the examination of crushed granules in KOH or Gram-stained preparations. Actinomycotic granules are composed of gram-positive, interwoven, thin filaments, 0.5 to 1.0 μm in diameter, as well as coccoid and bacillary forms. Granules of the eumycotic agents, on the other hand, are composed of broad, interwoven, septate hyphae, 2 to 5 μm in diameter, with many bizarre-shaped, swollen cells up to 15 μm in diameter, especially at the periphery of the granules. In many species, the granules also contain a cementlike material.

To maximize the chances of obtaining pure cultures of the etiologic agents, granules from the eumycotic mycetomas should be washed several times with saline containing such antibacterial antibiotics as penicillin and streptomycin. The granules thus freed from surface bacteria are cultured in several petri dishes containing Sabouraud agar with chloramphenicol and Sabouraud agar with chloramphenicol and cycloheximide. Several of these should be incubated at 25°C, and some should be incubated at 37°C; they should all be observed at 48-h intervals. Since many of the fungi that cause eumycotic mycetomas grow slowly, culture plates should be incubated for 6 weeks before being discarded as negative. Identification of the isolated fungus is based on the morphology of hyphae and conidia and the mechanism of conidiogenesis. Since certain species (*M. grisea, M. mycetomatis, N. rosatii*, and *Pyrenochaeta romeroi*) do not sporulate readily, additional sporulation media and physiologic tests, such as those for carbohydrate and nitrate utilization, must be used for identification.

The 20 species known to cause eumycotic mycetomas are listed below. Of these species, 5 belong to the phylum *Ascomycota*, and 15 are classified under the phylum *Deuteromycota* (Fungi Imperfecti).

Ascomycota

1. *Emericella nidulans (Aspergillus nidulans)* (23)
2. *Leptosphaeria senegalensis* (26)
3. *Leptosphaeria tompkinsii* (13)
4. *Neotestudina rosatii* (38)
5. *Pseudallescheria boydii* (19, 25, 29)

Deuteromycota

1. *Acremonium falciforme* (17)
2. *Acremonium kiliense* (17)
3. *Acremonium recifei* (17, 24)
4. *Curvularia geniculata* (10)
5. *Curvularia lunata* (23, 24)
6. *Corynespora cassicola* (24)
7. *Exophiala jeanselmei* (34)
8. *Fusarium moniliforme* (24, 26)
9. *Fusarium solani* (24, 26)
10. *Madurella grisea* (6, 20, 22, 23)
11. *Madurella mycetomatis* (6, 20, 26)
12. *Plenodomus avramii* (8)
13. *Pseudochaetosphaeronema larense* (9, 35)
14. *Pyrenochaeta makinnonii* (7, 35)
15. *Pyrenochaeta romeroi* (5, 35, 39)

Depending on the etiologic agent involved, the granules are white to yellow or brown to black and range from 200 μm to 5 mm in diameter. The color gives some clue to the species involved. The gross characteristics of the granules formed by each of the 20 species are summarized in Table 1.

TABLE 1. Gross characteristics of the granules of the eumycotic mycetoma agents

Kind of granule and species	Texture	Size range (mm)	Cementlike matrix
White grains in tissue			
Acremonium falciforme	Soft	0.2–0.5	Absent
Acremonium kiliense	Soft	0.2–0.5	Absent
Acremonium recifei	Soft	0.2–0.5	Absent
Aspergillus nidulans	Soft	1–2	Absent
Fusarium moniliforme	Soft	0.2–0.5	Absent
Fusarium solani	Soft	0.2–0.6	Absent
Neotestudina rosatii	Soft	0.5–1	Present, peripheral
Pseudallescheria boydii	Soft	0.5–1	Absent
Black grains in tissue			
Curvularia geniculata	Hard	0.5–1	Present, peripheral
Curvularia lunata	Hard	0.5–1	Present, peripheral
Corynespora cassicola	Hard	0.2–0.5	Absent
Exophiala jeanselmei	Soft	0.2–0.3	Absent
Leptosphaeria senegalensis	Hard	0.5–2	Present, peripheral
Leptosphaeria tompkinsii	Hard	0.5–2	Present, peripheral
Madurella grisea	Soft	0.3–0.6	Present, peripheral
Madurella mycetomatis	Hard	0.5–5.0	Present, homogeneous
Plenodomus avramii	Soft	0.5–0.8	Absent
Pseudochaetosphaeronema larense	Soft	0.2–0.5	Absent
Pyrenochaeta mackinnonii	Soft	0.2–0.5	Absent
Pyrenochaeta romeroi	Soft	0.2–0.6	Absent

IDENTIFICATION OF SPECIES

The etiologic agents described below represent the species that have been most commonly isolated from human or animal eumycotic mycetomas. The physiologic characteristics of the five species that are known to cause eumycotic mycetomas in the United States are summarized in Table 2.

Aspergillus nidulans (Eidam) Winter, 1884

Teleomorph: *E. nidulans* (Eidam) Vuilleman, 1927.

Cases of mycetoma caused by *A. nidulans* have been described in the literature from Senegal, Sudan, and Tunisia (23).

The granules of *A. nidulans* are white, soft, and 1 to 2 mm in diameter. They are ovoid or lobed, are composed of compact hyphae measuring 1.5 to 2.0 μm in diameter, and have irregular or circular bulbous swellings in the peripheral region.

Colonies of *A. nidulans* are fast growing, dark green-buff to honey-yellow, and downy to powdery. The reverse of the colonies shows various shades of purplish red. Conidial heads are short and columnar, with cinnamon-brown conidiophores that are commonly sinuous with smooth walls and range from 60 to 130 μm in length. Vesicles are hemispherical, bearing phialides in two series. Conidia are globose, finely wrinkled, 3 to 3.5 μm in diameter, and green in mass. When present, cleistothecia are yellowish to cinnamon, globose, and 100 to 200 μm in diameter with eight-spored asci. Ascospores are purple-red, lenticular, and smooth with equatorial crests. Hülle cells usually are formed.

TABLE 2. Physiologic characteristics of some of the causal agents of eumycotic mycetomas

Species	Utilization of:					Starch hydrolysis	Utilization of:			Protease activity	Optimum growth temp (°C)
	Galactose	Glucose	Lactose	Maltose	Sucrose		Asparagine	KNO_3	$(NH_4)_2SO_4$		
Acremonium falciforme	+	+	−	+	+	−	+	+	+	±	37
Exophiala jeanselmei	+	+	−	+	+	+	+	±	±	−	30
Leptosphaeria senegalensis	+	+	±	+	+	+	+	+	?[a]	?[a]	37
Madurella grisea	+	+	−	+	+	+	+	+	+	±	25
Madurella mycetomatis	+	+	+	+	−	+	+	+	+	±	37
Pseudallescheria boydii	±	+	−	−	±	−	+	+	+	+	25

[a] ?, Tests not done.

Leptosphaeria senegalensis Baylet, Camain, and Segretain, 1959, and *Leptosphaeria tompkinsii* El-Ani, 1966

L. senegalensis and *L. tompkinsii* are known to cause mycetomas in the northern tropical zone of west Africa, especially in Senegal and Mauritania. The granules of the two species are indistinguishable from each other. They are black, 0.5 to 2 mm in size, and firm to hard. In tissue sections, the granules are round to polylobulated, with large vesicles. The mycelium is embedded in a black, cementlike substance at the periphery. The central portion of the granules consists of a loose network of hyphae.

In culture, *L. senegalensis* and *L. tompkinsii* grow rapidly and produce gray-brown colonies. On cornmeal agar, both species produce perithecia that are nonostiolate, scattered, immersed or superficial, globose to subglobose, black, and covered with brown, flexuous hyphae. The asci are numerous, eight spored, clavate to cylindrical, and double walled. The major differences between the two species are found in the ascospores, which differ in size, shape, and septation, as well as in the nature of the gelatinous sheath that surrounds them (13, 14).

Neotestudina rosatii Segretain and Destombes, 1961

N. rosatii, described by Segretain and Destombes (38), was found to possess bitunicate (double-walled) asci and was placed by von Arx (43) in the family *Testudinaceae* of the class *Loculoascomycetes*. This transfer, however, was found to be incorrect on the basis of scanning electron microscopic studies of the ascospores of the different genera and species of the family *Testudinaceae* carried out by Hawksworth (18). His studies clearly showed that *Zopfia rosatii* should be treated as a later synonym of *N. rosatii* under the family *Testudinaceae*.

The granules of *N. rosatii* are white to brownish white, 0.5 to 1.0 mm in size, and soft. In tissue sections, they appear to be polyhedral to subregular. The hyphae are embedded in a peripheral cementing material. The central part of the granules consists of more or less degenerative vesicles. Mycetomas due to *N. rosatii* have been described from Australia, Cameroon, Guinea, Senegal, and Somalia (18, 26, 38).

In culture, colonies of *N. rosatii* are slow growing, reaching 25 mm in diameter in 2 weeks, with an aerial mycelium that is grayish black to brownish black. On potato-carrot or cornmeal agar, incubated at 30°C, most of the cleistothecia are submerged (found below the surface of the agar). The cleistothecial walls are smooth and surrounded by a weft of brown to hyaline hyphae. The eight-spored asci, 12 to 35 by 10 to 25 μm, are scattered in the central part of the cleistothecium and are globose to subglobose, thick walled, and bitunicate, becoming evanescent (i.e., disappearing) as the ascospores mature. Ascospores vary in size (9 to 12.5 by 4.5 to 8.0 μm) and shape, ranging from ellipsoid to bicampanulate (campanulate = bell shaped), asymmetrical, or slightly curved, and are constricted at the median transverse septum, with brown, smooth walls.

Pseudallescheria boydii (Shear) McGinnis, Padhye, and Ajello, 1982 (= *Allescheria boydii* Shear, 1922) (= *Petriellidium boydii* (Shear) Malloch, 1970)

Anamorph: *Scedosporium apiospermum* Sacc. ex Castellani and Chalmers, 1919 (= *Monosporium apiospermum* Sacc., 1911).

In 1922, Shear (41) described *Allescheria boydii*, which was isolated by M. F. Boyd from a human case of eumycotic mycetoma. In 1970, while studying the members of the family *Microascaceae*, Malloch (25) found that the binomial *A. boydii* was incorrect for various reasons. Because *A. boydii* could not be accommodated in either the genus *Monoascus* or any other existing genus of cleistothecial ascomycetes, he proposed the genus *Petriellidium* for *A. boydii* in the family *Microascaceae*.

In 1943 and again in 1944 (32, 33), Negroni et al. described the new genus and species *Pseudallescheria shearii* for an ascomycete that they had isolated from a human case of eumycotic mycetoma. Because of the similarity of descriptions of *Petriellidium boydii* and *Pseudallescheria shearii*, McGinnis et al. (29) studied the living cultures derived from the type strains as well as appropriate specimens of these two taxa and concluded that the genera *Pseudallescheria* and *Petriellidium* were congeneric and that *Pseudallescheria shearii* was indistinguishable from *Petriellidium boydii*. Because of priority, they found that the proper binomial for *A. boydii* was *Pseudallescheria boydii*. The correct binomial for the conidial anamorph of *Pseudallescheria boydii* is *Scedosporium apiospermum* Sacc. ex Castellani and Chalmers, 1919.

The granules produced by *P. boydii* are white to yellow and soft to firm; they vary from spherical to subspherical to lobulated and measure from 0.2 to 2.0 mm in diameter. In tissue sections, the granules are composed of hyaline hyphae 1.5 to 5.0 μm in diameter which radiate from the center into terminal, somewhat thick-walled cells 15 to 20 μm in diameter at the periphery of the granules. The granules of *P. boydii* lack a cementlike matrix, and as a result, the central portion of the granules consists of loosely arranged, interwoven hyphae.

The anamorphic state, *S. apiospermum*, produces two types of conidia: (i) broadly clavate conidia that are rounded at the apex and truncate or attenuated at the base and are borne terminally or laterally on annellated conidiogenous cells, with thick walls 6 to 12 by 3.5 to 6.0 μm (Fig. 1); and (ii) hyaline, clavate to cylindrical conidia that are truncate at the base and are borne on short or long conidiophores. They usually are aggregated to form synnemata (Fig. 2).

The ascomycetous state or teleomorph consists of ascocarps (cleistothecia) that develop from conspicuous coiled ascogonia. The ascocarps are spherical, nonostiolate, usually submerged, 140 to 200 μm in diameter, often covered with brown, thick-walled, septate hyphae, 2 to 3 μm wide, with a wall 4 to 6 μm thick composed of two or three layers of interwoven, flattened, dark-brown cells 2 to 6 μm wide. The ascocarp opens at maturity by an irregular rupture of the wall. The asci are ellipsoidal to nearly spherical, 12 to 18 by 9 to 13 μm, and eight spored. The ascospores are ellipsoidal, symmetrical or slightly flattened, and straw colored; they have two germ pores and measure 6 to 7 by 3.5 to 4.0 μm.

Even though *P. boydii* is homothallic, many clinical and soil isolates do not form ascocarps. Their identification, then, is based on the morphology of the conidial state alone. Some may be induced to form ascocarps by cultivation on cornmeal agar. *P. boydii* also is known to cause necrotizing pneumonia and disseminated infections in immunocompromised patients. When isolated from sputum, it should not be regarded casually as a contaminant.

Acremonium falciforme (Carrion) Gams, 1971 (= *Cephalosporium falciforme* Carrion, 1951)

Three species of *Acremonium* (*A. falciforme, A. kiliense,* and *A. recifei*) are known to cause mycetomas. *A. falciforme* was found to be an agent of mycetomas in the San Francisco Bay area of northern California (17).

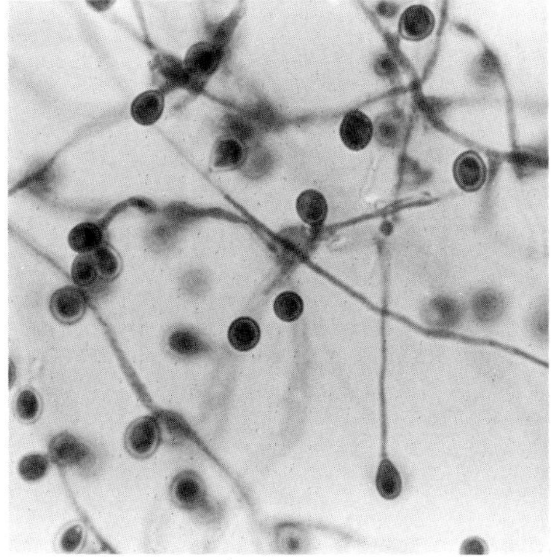

FIG. 1. *Scedosporium apiospermum*. Broadly clavate conidia. ×850.

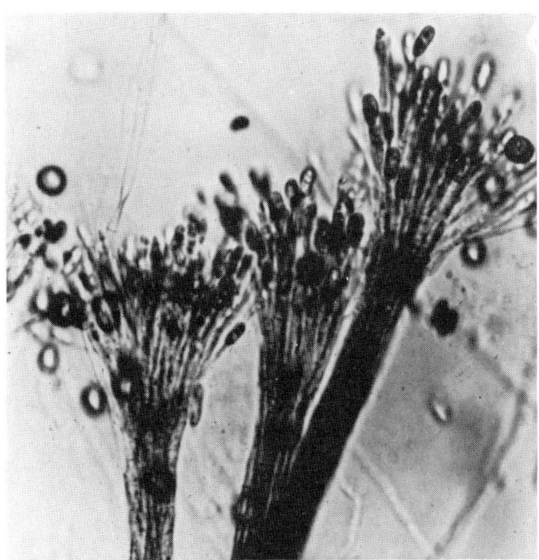

FIG. 2. *Scedosporium apiospermum*. Cylindrical to clavate conidia borne on erect synnemata. ×850.

The granules of *A. falciforme* are white to pale yellow, soft, and 0.2 to 0.5 mm in diameter. They are composed of slender, polymorphic, septate hyphae 1.5 to 2.0 μm in diameter, with irregular bulbous swellings with peripheral cementing material. In tissue sections, the granules resemble those produced by *P. boydii*. The diagnosis, therefore, should not rest on the structure of granules in tissues alone but also on the isolation and identification of the causal fungus.

Colonies of *A. falciforme* on Sabouraud agar are slow growing, reaching 60 to 65 mm in diameter in 2 weeks. They are downy and gray-brown, becoming gray-violet. The reverse of a colony develops a violet-purple pigment. The hyphae are hyaline, septate, smooth, branched, and 1.5 to 2.6 μm in diameter. They bear erect, undifferentiated, unbranched, repeatedly septate conidiophores. The conidia are borne at the tip of the conidiogenous cells in mucoid clusters. The conidia are sausage shaped, slightly curved, nonseptate to one septate, 7 to 8.5 by 2.7 to 3.2 μm (Fig. 3). Intercalary or, rarely, terminal chlamydospores are smooth, thick walled, and 5 to 8 μm in diameter.

The other two species, *A. kiliense* and *A. recifei*, are not known to cause mycetomas as frequently as does *A. falciforme*. The granules produced by *A. kiliense* and *A. recifei* are similar in morphology to those of *A. falciforme*. They cannot be differentiated from each other or from those produced by *P. boydii*. When isolated, *A. kiliense* is differentiated from *A. recifei* and *A. falciforme* by the morphology of conidiophores and conidia. The conidiophores are solitary and rarely branched. The conidia are cylindrical to ellipsoidal with symmetrically rounded ends, usually straight, nonseptate, and very rarely curved when submerged. *A. recifei* is distinguished by its branched conidiophores bearing claviform, nonseptate to one-septate conidia.

Curvularia geniculata (Tracy and Earl) Boedijin, 1933

Teleomorph: *Cochliobolus geniculatus* Nelson, 1964.

C. geniculata has been found to be the etiologic agent of mycetomas in dogs in the United States (10).

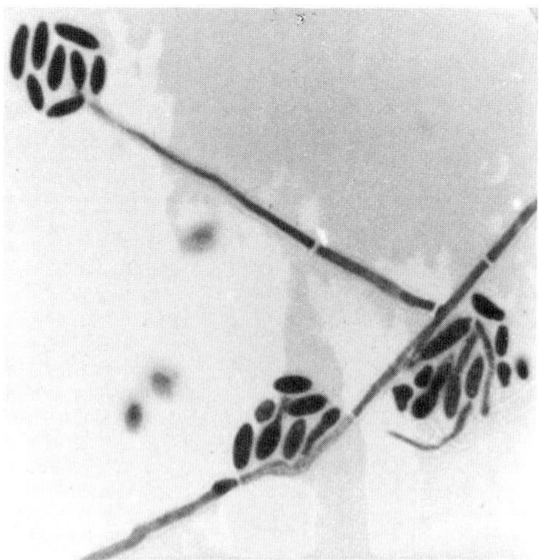

FIG. 3. *Acremonium falciforme*. Erect, multiseptate, unbranched conidiophore and slightly curved conidia. ×850.

Its granules are black to dark brown, firm, and 0.5 to 1.0 mm or more in size. In tissue sections, the granule is spherical, ovoid, or irregularly shaped and often surrounded by a zone of epithelioid cells. The periphery of the granule is composed of a dense, interwoven mass of dematiaceous mycelium and thick-walled, chlamydosporelike cells embedded in a cementlike substance. The interior of the granules is vacuolar and consists of a loose network of septate, hyphal filaments.

In culture, *C. geniculata* develops a cottony to downy, olive-gray to black colony. Microscopically, the dematiaceous, septate mycelium can be seen to bear solitary, geniculate conidiophores that are differentiated from the vegetative hyphae. The conidia are solitary, borne at the tip and on the sides of the conidiophore, simple, ellipsoidal, three to five celled, and curved because of enlargement of the next to the last cell.

C. lunata has been described as a causal agent of mycetomas among humans in Senegal and Sudan (23). The granules resemble those of *C. geniculata* in their morphological characteristics. In culture, *C. lunata* is differentiated from *C. geniculata* by the morphology of the conidia, which are three celled in contrast to the three- to five-celled conidia of *C. geniculata*.

Exophiala jeanselmei (Langeron) McGinnis and Padhye, 1977 (= *Phialophora jeanselmei* (Langeron) Emmons, 1945)

E. jeanselmei, long described as forming phialides and phialoconidia (*P. jeanselmei*), was shown by McGinnis and Padhye (28) to reproduce principally by forming annellophores and annelloconidia. Therefore, they transferred this species from the genus *Phialophora* to the genus *Exophiala* Carmichael.

In the United States, only three cases of mycetoma due to *E. jeanselmei* have been described (34). *E. jeanselmei* produces dark granules in host tissue. They are brown to black, irregular in shape, and fragile, and detached portions or fragments often are found in the lesion within giant cells. When extruded through open sinuses, the granules often look like worm cases because of their elongated shape and irregular surface. In tissue sections, they appear as hollow spheres or as sinuous bands that are worm shaped in appearance. The external surface is composed of brown, thick-walled hyphae and thick-walled chlamydoconidialike cells. The granules are cement-free. Within the hollow granules, smaller, degenerating hyphal fragments with leukocytes and giant cells may be seen.

Initially, the colonies of *E. jeanselmei* may be yeastlike and black, gradually spreading, raised or dome shaped, and 20 to 25 mm in diameter. After 2 weeks on Sabouraud agar, the colonies are covered with short aerial hyphae. At this stage, the colony is mousey gray to olive-gray with an olive-black reverse. The septate mycelium is sometimes toruloid, branched, and pale brown. The conidiophores are partially differentiated from the vegetative hyphae, solitary, branched or unbranched, smooth walled, and pale brown. The conidiogenous cells produce conidia from one or more areas of their cells. The tips of the conidiogenous cells are closely annellated, cylindrical, obclavate, smooth walled, pale brown to black, and elongate due to successive conidial formation. The conidia, which aggregate in masses at the tips of conidiophores, tend to slide down the conidiophore or along hyphae. The conidia are exogenous, nonseptate, subglobose, ellipsoidal to cylindrical, smooth, and 1.5 by 2.8 μm (Fig. 4 and 5).

Madurella grisea Mackinnon, Ferrada, and Montemayer, 1949

M. grisea is one of the etiologic agents of black-grain mycetomas. It is common in Latin America (22). In the United States, three cases due to *M. grisea* have been described (11).

The granules are black, 0.3 to 0.6 mm, and soft to firm. In tissue sections, the granules are oval, lobulated or kidney shaped (reniform), and sometimes elongated to worm shaped (vermiform). They are com-

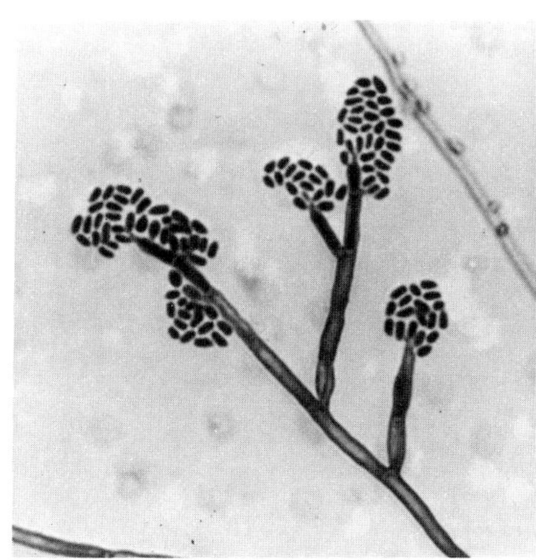

FIG. 4. *Exophiala jeanselmei*. Erect, branched conidiophores and conidia. ×850.

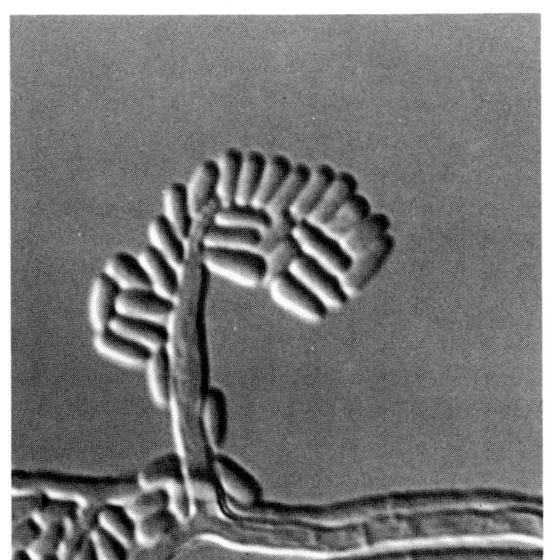

FIG. 5. *Exophiala jeanselmei*. Terminal conidiogenous cell showing annellations and annelloconidia. Nomarski interference contrast microscopy. ×1,600. (Photomicrograph courtesy of M. R. McGinnis.)

posed of a dense network of hyphae, weakly pigmented in the center and brown to blackish brown in the peripheral region because of brown interstitial material.

In culture, *M. grisea* forms slow-growing colonies that are cerebriform, radially furrowed, and dark gray to olive-black. The reverse of a colony is black. In older cultures, a red-brown diffusible pigment is produced by many isolates. Microscopically, the hyphae are septate, dematiaceous, 1 to 3 μm in diameter, and nonsporulating. Chlamydoconidia are rarely observed. Large, moniliform hyphae 3 to 5 μm in diameter are often seen. Some isolates of *M. grisea* have been described as producing abortive pycnidia (27). Such isolates are indistinguishable from *Pyrenochaeta romeroi* (39).

Madurella mycetomatis (Laveran) Brumpt, 1905

The correct binomial for this hyphomycete is *M. mycetomatis*, not *M. mycetomi*. This correction was made in accordance with the Latin requirement that the specific epithet be in the genitive case (30).

The granules produced by *M. mycetomatis* are reddish brown to black. They may reach 5 mm or more in diameter and are firm to hard. In tissue sections, the granules are compact, variable in size and shape, and frequently multilobulated; they are composed of hyphae 1 to 5 μm in diameter that terminate in enlarged hyphal cells at the periphery of the granule and measure 12 to 15 μm in diameter. The cell wall pigment is minimal, but the hyphal cells contain brown particles. The hyphae are embedded in a conspicuous brown matrix that is characteristic of *M. mycetomatis*. Some granules are vesicular and more regular in size and shape. The vesicles are prominently visible in the peripheral zone in a dense, brown, cementlike matrix.

In culture, *M. mycetomatis* shows wide variation. Colonies are slow growing, white at first, becoming olivaceous, yellow, or brown, flat to dome shaped, sometimes powdery or downy, with a characteristic brown diffusible pigment. On nutritionally deficient media, sclerotia 750 μm in diameter develop. These are black and made of undifferentiated polygonal cells. On Sabouraud agar, the mycelium is sterile. On nutritionally poor media, such as soil extract or hay infusion agars, some isolates produce round to pyriform conidia 2 μm in diameter from the tips of phialides.

M. mycetomatis grows better at 37 than at 30°C, whereas *M. grisea* grows better at 30 than at 37°C. *M. mycetomatis* is slowly proteolytic and utilizes glucose, maltose, and galactose, but not sucrose. It utilizes potassium nitrate, ammonium sulfate, asparagine, and urea, and it hydrolyzes starch. *M. grisea*, on the other hand, is weakly proteolytic and assimilates glucose, maltose, and sucrose, but not lactose (20) (Table 2).

Pyrenochaeta romeroi Borelli, 1959

P. romeroi produces soft to firm, black granules that are oval, lobulated, and sometimes vermiform. They resemble those of *M. grisea* (26, 39).

In culture, colonies of *P. romeroi* are fast growing, floccose to velvety, with a gray surface and a whitish margin. The reverse of a colony is black, with no diffusible pigment. The hyphae are hyaline to brown, septate, and branched. On nutritionally poor media, *P. romeroi* produces ostiolate, isolated or aggregated pycnidia, 50 to 160 by 40 to 100 μm, bearing subhyaline, elliptical pycnidiospores, 0.8 to 1.0 by 1.5 to 2.0 μm.

The close resemblance of the granules produced by *P. romeroi* to those produced by *M. grisea* and the morphological similarity of the abortive pycnidia formed by some isolates of *M. grisea* and by *P. romeroi* suggest that these two species are conspecific or very closely related. However, according to Murray and Buckley (31), the two species show serologic differences that allow differentiation between them.

LITERATURE CITED

1. **Ajello, L.** 1962. Epidemiology of human fungus infections, p. 69–83. *In* G. Dalldorf (ed.), Fungi and fungous diseases. Charles C Thomas, Publisher, Springfield, Ill.
2. **Ajello, L.** 1978. Animal mycetomas—a review, p. 270–275. *In* Proceedings: Primer Simposio Internacional de Micetomas. Universidad Centro Occidental, Barquisimeto, Venezuela.
3. **Ajello, L., W. Kaplan, and F. W. Chandler.** 1980. Dermatophyte mycetomas: fact or fiction?, p. 135–140. *In* Superficial, cutaneous and subcutaneous infections. Scientific publication no. 396, Pan American Health Organization, Washington, D.C.
4. **Baylet, R., R. Camain, and M. Rey.** 1961. Champignons de mycétomes isolés des épineux au Sénégal. Bull. Soc. Med. Afr. Noire Lang. Fr. **6**:317–319.
5. **Borelli, D.** 1959. *Pyrenochaeta romeroi* n. sp. Rev. Dermatol. Venez. **1**:325–327.
6. **Borelli, D.** 1962, *Madurella mycetomi* y *Madurella grisea*. Arch. Venez. Med. Trop. Parasitol. Med. **4**:195–211.
7. **Borelli, D.** 1976. *Pyrenochaeta makinnonii* nova species agente de micetoma. Castellania **4**:227–234.
8. **Borelli, D.** 1978. *Plenodomus avramii* nova species agente de micetoma, p. 116–126. *In* Proceedings: Primer Simposio Internacional de Micetomas. Universidad Centro Occidental, Barquisimeto, Venezuela.
9. **Borelli, D., R. Zamora, and G. Senabre.** 1976. *Chaetos-*

phaeronema larense nova species agente de micetoma. Gac. Med. (Caracas) **84**:307–318.
10. **Brodey, R. S., H. S. Schryver, M. J. Deubler, W. Kaplan, and L. Ajello.** 1967. Mycetoma in a dog. J. Am. Vet. Med. Assoc. **151**:442–451.
11. **Butz, W. C., and L. Ajello.** 1971. Black grain mycetoma. Arch. Dermatol. **104**:197–201.
12. **Destombes, P., A. Poirier, and O. Nazimoff.** 1970. Mycoses profondes reconnues en 9 ans de pratique histopathologique à l'Institut Pasteur du Cameroun. Bull. Soc. Pathol. Exot. **63**:310–315.
13. **El-Ani, A. S.** 1966. A new species of *Leptosphaeria*, an etiologic agent of mycetoma. Mycologia **58**:406–411.
14. **El-Ani, A. S., and M. A. Gordon.** 1965. The ascospore sheath and taxonomy of *Leptosphaeria senegalensis*. Mycologia **57**:275–278.
15. **Emmons, C. W.** 1962. Soil reservoirs of pathogenic fungi. J. Wash. Acad. Sci. **52**:3–9.
16. **Fahey, P. J., M. J. Utell, and R. W. Hyde.** 1981. Spontaneous lysis of mycetomas after acute cavitating lung disease. Am. Rev. Respir. Dis. **123**:336–339.
17. **Halde, C., A. A. Padhye, L. D. Haley, M. G. Rinaldi, D. Kay, and R. Leeper.** 1976. *Acremonium falciforme* as a cause of mycetoma in California. Sabouraudia **14**:319–326.
18. **Hawksworth, D. L.** 1979. Ascospore sculpturing and generic concepts in the Testudinaceae (syn. Zopfiaceae). Can. J. Bot. **57**:91–99.
19. **Hughes, S. J.** 1958. Revisiones hyphomycetum aliquot cum appendice de nominibus rejiciendis. Can. J. Bot. **36**:727–836.
20. **Mackinnon, J. E.** 1954. A contribution to the study of the causal organisms of maduromycosis. Trans. R. Soc. Trop. Med. Hyg. **48**:470–480.
21. **Mackinnon, J. E., I. A. Conti-Diaz, E. Gezuele, and E. Civila.** 1971. Datos sobre ecologia de *Allescheria boydii*, Shear. Rev. Urug. Pathol. Clin. Microbiol. **9**:37–43.
22. **Mackinnon, J. E., L. V. Ferrada, and L. Montmayer.** 1949. *Madurella grisea* n. sp., a new species of fungus producing the black variety of maduromycosis in South America. Mycopathol. Mycol. Appl. **4**:385–392.
23. **Mahgoub, E. S.** 1973. Mycetomas caused by *Curvularia lunata, Madurella grisea, Aspergillus nidulans*, and *Nocardia brasiliensis* in Sudan. Sabouraudia **11**:179–182.
24. **Mahgoub, E. S., and I. Murray.** 1973. Mycetoma. William Heinemann Medical Books Ltd., London, England.
25. **Malloch, D.** 1970. New concepts in the Microascaceae illustrated by two species. Mycologia **62**:727–740.
26. **Mariat, F., P. Destombes, and G. Segretain.** 1977. The mycetomas: clinical features, pathology, etiology and epidemiology. Contrib. Microbiol. Immunol. **4**:1–39.
27. **Mayorga, R., and J. E. Close De Leon.** 1966. Sur une souche de *Madurella grisea* sporifere isolée d'un mycétome Guatemaltèque à grains noirs. Sabouraudia **4**:210–214.
28. **McGinnis, M. R., and A. A. Padhye.** 1977. *Exophiala jeanselmei*, a new combination for *Phialophora jeanselmei*. Mycotaxon **5**:341–352.
29. **McGinnis, M. R., A. A. Padhye, and L. Ajello.** 1982. *Pseudallescheria* Negroni et Fischer, 1943, and its later synonym *Petriellidium* Malloch, 1970. Mycotaxon **14**:94–102.
30. **Medical Research Council.** 1977. Nomenclature of fungi pathogenic to man and animals. Medical Research Council Memo 23, 4th ed. Her Majesty's Stationery Office, London, England.
31. **Murray, I. G., and H. R. Buckley.** 1969. Serological differences between *Pyrenochaeta romeroi* and *Madurella grisea*. Sabouraudia **7**:62–63.
32. **Negroni, P., and I. Fischer.** 1944. *Pseudallescheria sheari* n. gen., n. sp. aislada de un paramicetoma de la rodilla. Rev. Inst. Bacteriol. B. Aires **12**:195–204.
33. **Negroni, P., H. Herrmann, and I. Fischer.** 1943. Artritis aguda purulenta producida por el ascomycete *Pseudallescheria sheari* n. g., n. sp. Prensa Med. Argent. **30**:2389–2399.
34. **Neilson, H. S., N. F. Conant, T. Weinberg, and J. F. Reback.** 1968. Report of a mycetoma due to *Phialophora jeanselmei* and undescribed characteristics of the fungus. Sabouraudia **6**:330–333.
35. **Punithalingam, E.** 1979. Sphaeropsidales in culture from humans. Nova Hedwigia **37**:119–158.
36. **Segretain, G.** 1964. Recherches sur l'écologie de *Madurella mycetomi* au Sénégal. Bull. Soc. Fr. Mycol. Med. **3**:121–124.
37. **Segretain, G.** 1972. Epidémiologie des mycétomes. Ann. Soc. Belge Med. Trop. **52**:277–286.
38. **Segretain, G., and P. Destombes.** 1961. Description d'un nouvel agent de maduromycose, *Neotestudina rosatii*, n. gen., n. sp. isolé en Afrique. C. R. Acad. Sci. **253**:2577–2579.
39. **Segretain, G., and P. Destombes.** 1969. Recherche sur les mycétomes à *Madurella grisea* et *Pyrenochaeta romeroi*. Sabouraudia **7**:51–61.
40. **Segretain, G., and F. Mariat.** 1968. Recherches sur la présence d'agents de mycetomes dans le sol et sur les epineux du Senegal et de la Mauritanie. Bull. Soc. Pathol. Exot. **61**:194–202.
41. **Shear, C. L.** 1922. Life history of an undescribed ascomycete isolated from a granular mycetoma of man. Mycologia **14**:239–243.
42. **Thirumalachar, M. J., and A. A. Padhye.** 1968. Isolation of *Madurella mycetomi* from soil in India. Hind. Antibiot. Bull. **10**:314–318.
43. **von Arx, J. A.** 1971. Testudinaceae, a new family of Ascomycetes. Persoonia **6**:365–369.

Chapter 52

Dematiaceous Fungi

MICHAEL R. McGINNIS

CHARACTERIZATION

Dematiaceous fungi are characterized by the development of a brown-to-olive-to-black color in the cell walls of their vegetative cells, conidia, or both. This cell coloring results in colonies that are olive to black. These ubiquitous and cosmopolitan opportunistic pathogens are normally associated with soil and plants, but occasionally they may cause infections in humans and animals. In medical mycology, dematiaceous fungi often are thought of as being exclusively hyphomycetes. This idea is in error because some ascomycetes, basidiomycetes, coelomycetes, and zygomycetes may be dematiaceous.

Mycotic infections caused by dematiaceous fungi include chromoblastomycosis, mycetoma, phaeohyphomycosis, and sporotrichosis. In this chapter, only chromoblastomycosis, phaeohyphomycosis, and sporotrichosis will be considered. Sporotrichosis is treated here because the etiologic agent is dematiaceous in culture, even though the yeast form in tissue is hyaline.

Deciding whether a particular dematiaceous fungus is involved in the disease process can at times be difficult, since these fungi occasionally are recovered from clinical specimens as contaminants. Documentation of a dematiaceous fungus as the etiologic agent of a mycotic infection necessitates sound evidence that the infection is compatible with a mycosis, that the suspected etiologic agent is seen in clinical specimens, that the morphology of the fungus in the clinical specimens is compatible with the suspected etiologic agent, and that the recovered fungus is properly identified. The repeated recovery of a suspected etiologic agent, especially from more than one type of clinical specimen, is highly significant. The recovery of a fungus from body sites that normally are sterile is important.

COLLECTION, TRANSPORT, AND STORAGE OF SPECIMENS

Clinical specimens must be collected aseptically and then promptly transported to the clinical laboratory in a properly labeled sterile container. Specimens collected on swabs or transported to the clinical laboratory in a transport medium are unacceptable for mycological study. An adequate quantity of clinical material is necessary if the information obtained in the laboratory is to be meaningful.

The most frequently submitted specimens for the recovery of dematiaceous fungi include aspirates, biopsy material, scrapings, and tissue specimens. Specimens other than skin scrapings must be protected from dehydration at all times. This protection can be accomplished by ensuring that a few drops of sterile saline or distilled water is added to the specimens at the time of their collection. Biopsy and tissue specimens can be kept moist by placing them between two pieces of sterile gauze moistened with sterile saline or distilled water. Clinical specimens should never be placed on cotton pads, since cotton fibers can be confused with hyphae in the direct microscopic examination. In addition, it is often impossible to recover all of the clinical specimen from among the cotton fibers. Direct microscopic examination of the specimens and subsequent plating must be done promptly.

DIRECT EXAMINATION

Clinical specimens obtained for the recovery of dematiaceous fungi usually do not require extensive processing. If aspirated specimens contain a substantial amount of purulent material, this can be dissolved with *N*-acetyl-L-cysteine without sodium hydroxide. Tissue specimens and biopsy material should be homogenized in a tissue homogenizer after highly suspicious areas consisting of necrotic, purulent, or caseous material are selectively examined microscopically and inoculated onto isolation media.

Specimens are typically examined microscopically in 10% KOH. The clearing process can be accelerated by gently heating the KOH preparation. The dematiaceous nature of fungal elements in clinical specimens should be determined only by bright-field microscopy. Phase-contrast microscopy is an excellent method for examining specimens, but it does not always permit the demonstration of the dematiaceous nature of these fungi. In some instances, the dark color of these fungi can be seen in tissue sections stained with hematoxylin and eosin. However, if a dematiaceous fungus is suspected, an unstained tissue section should be examined microscopically by bright-field microscopy. A drop of immersion oil can be placed directly onto a paraffin section mounted on a microscope slide, and then the section is examined microscopically.

The etiologic agents of chromoblastomycosis may be filamentous at the surface of the skin. In the deeper subcutaneous tissues, they occur as muriform cells (sclerotic bodies). Muriform cells are typically chestnut brown and variable in size, with thick cross walls arranged in a muriform manner. The cells result from vegetative growth without the elongation seen in hyphae. The presence of muriform cells in clinical specimens is diagnostic of chromoblastomycosis. However, the various etiologic agents of this mycosis cannot be identified solely on the basis of their morphology in tissue.

Phaeohyphomycosis (1, 20) is characterized by the presence in tissue of dematiaceous yeastlike cells, hyphae, or both. The hyphae may be regular and uniform in diameter or irregular in shape with many swollen cells, and they can be either short or very long. The name phaeohyphomycosis is not meant to be restricted to hyphomycetes, but it encompasses all dark hyphae causing disease in tissue, regardless of the taxonomic classification of the etiologic agent. As with chromoblastomycosis, the etiologic agents of phaeohyphomycosis cannot be identified in clinical specimens. These fungi must be grown on laboratory

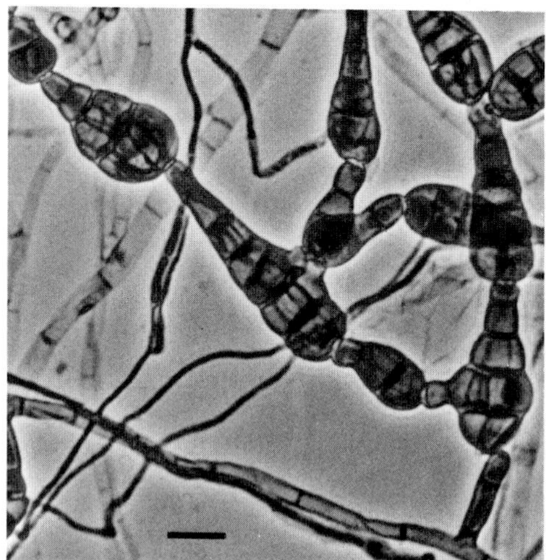

FIG. 1. *Alternaria alternata.* Taken by phase-contrast microscopy after 2 weeks on potato dextrose agar. Bar equals 10 μm.

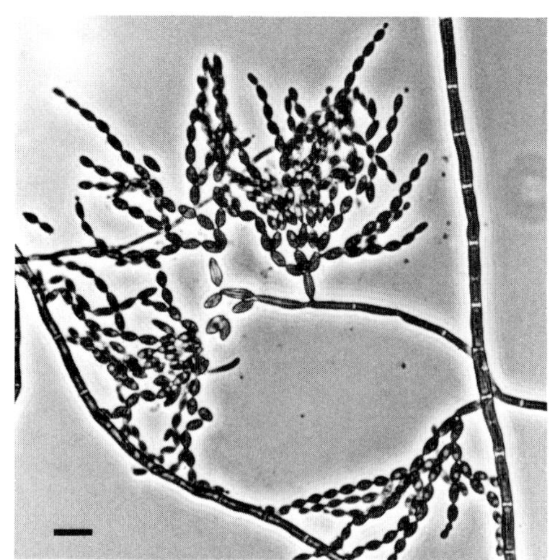

FIG. 3. *Cladosporium carrionii.* Taken by phase-contrast microscopy after 2 weeks on potato dextrose agar. Bar equals 10 μm.

culture medium before they can be identified.

Most mycologists consider it fruitless to directly examine clinical specimens for the yeast form of *Sporothrix schenckii.* The number of yeast cells in the specimen is typically very limited, and their small size and shape are not distinctive. The yeast form of *S. schenckii* is not easily seen in sections of tissue stained with hematoxylin and eosin. The fungus usually can be seen in tissue sections when the sections first are treated with diastase and then stained by the Gomori or periodic acid-Schiff technique. Fluorescent-antibody-specific conjugates are ideal, although they are available only at certain reference laboratories at the present time. Even though the yeast form may be difficult to see in specimens, there is generally no difficulty in recovering this fungus by the cultivation of clinical materials on media that are routinely used for the isolation of fungi.

CULTURE AND ISOLATION

Dematiaceous fungi are easily isolated on most routine media. Some of them are sensitive to cycloheximide, and for that reason, Sabouraud dextrose agar (2% glucose) should be used in conjunction with a medium containing cycloheximide. Most dematiaceous fungi grow well at 30°C; some species grow poorly or not at all at 37°C. The majority of the pathogenic dematiaceous fungi usually are visible on isolation media within a week. However, cultures should not be discarded as negative until 6 weeks. Once a dematiaceous fungus is isolated, it must be determined whether or not the isolate is a pure culture. If the culture is not pure, then the fungus must be purified by the isolation of hyphal tips or individual germinating conidia. This purification is extremely

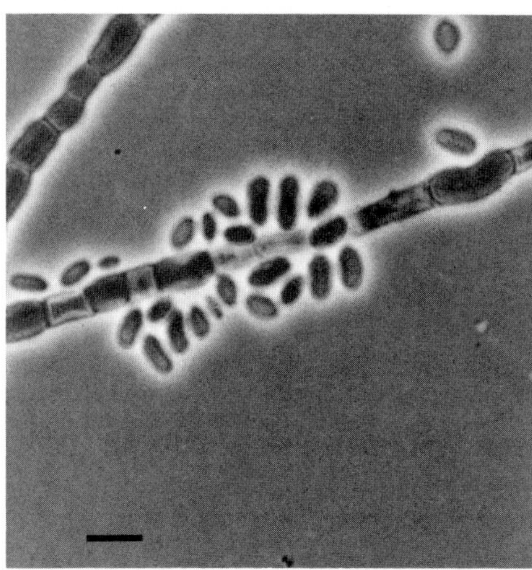

FIG. 2. *Aureobasidium pullulans.* Taken by phase-contrast microscopy after 2 weeks on potato dextrose agar. Bar equals 10 μm.

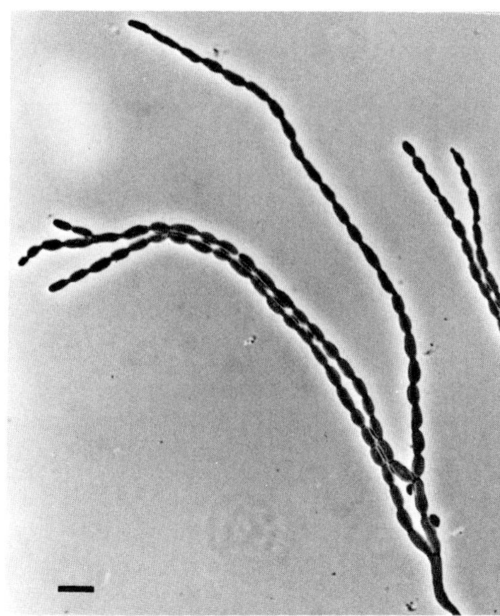

FIG. 4. *Cladosporium bantianum.* Taken by phase-contrast microscopy after 2 weeks on potato dextrose agar. Bar equals 10 μm.

TABLE 1. Diagnostic features for some medically important dematiaceous fungi

Genus	Diagnostic characteristics	Comments	Selected references
Alternaria	Conidiophores dark, septate, simple or branched. Conidia muriform, obclavate, with a beak, darkly pigmented, in simple or branched acropetal chains.	Recognized by the distinctive muriform, obclavate conidia with a beak.	3, 4, 6, 11, 12, 19
Aureobasidium	Conidiophores hyaline to chestnut brown, undifferentiated from hyphae. Conidia borne laterally, hyaline, one celled, often producing secondary blastoconidia. Large, dark, one- or two-celled, thick-walled arthroconidia commonly present.	Differentiated from *Hormonema* spp. by the production of conidia in a synchronous manner. Differentiated from *Phaeococcomyces* spp. by the lack of dematiaceous yeast cells.	10–12, 19
Cladosporium	Conidiophores dark, erect, often septate. Conidia one to several celled in some species, with dark hila, occurring in fragile, branched acropetal chains. Conidia at base of chains usually shield shaped.	*C. bantianum* grows at 42 to 43°C and forms long, sparsely branching chains of conidia from hyphalike conidiophores, with conidia ca. 6.4 μm long. *C. carrionii* grows up to 36°C and forms short, branching chains of conidia from distinct conidiophores, with conidia ca. 5 μm long.	5, 19, 21
Curvularia	Conidiophores dark, erect, geniculate due to sympodial development. Conidia multiseptate, usually curved, with central cell larger and darker than end cells and thickness of septa and outer cell wall approximately the same.	Differentiated from *Drechslera* spp. by possessing conidia which have an enlarged and darker central cell and narrow septa.	11, 12, 19
Drechslera	Conidiophores dark, erect, geniculate due to sympodial development. Conidia multiseptate, cylindrical to oblong, dark, with septal walls thickened.	Differentiated from *Curvularia* spp. by possessing conidia that are oblong to cylindrical with thickened septal walls.	11, 12, 19
Exophiala	Conidiophores hyaline to subhyaline, hyphalike or distinct. Conidiogenous cells annellides that are cylindrical to lageniform. Conidia one to several celled (one species), hyaline to pale brown, accumulating in balls at the apices of the annellides. *Phaeococcomyces* synanamorph often present.	*E. werneckii* has annellides reduced to yeast cells, one to two celled, the latter predominant, tapering towards the end bearing annellations. *E. jeanselmei* has cylindrical-to-lageniform annellides produced from conidiophores, with some annellides intercalary. Growth up to ca. 37°C. *E. jeanselmei* is differentiated from *W. dermatitidis* by lack of the ability to grow at 40°C and by the development of annellides instead of phialides.	10, 17–19, 23
Fonsecaea	Conidiophores pale brown, usually erect, swollen apically due to sympodial development. Conidia one celled, pale brown; primary conidia function as sympodial conidiogenous cells to produce secondary conidia. Tertiary conidia may be formed in the same manner. *Rhinocladiella*, *Cladosporium*, or *Phialophora* synanamorphs often are present.	*F. pedrosoi* is differentiated from *F. compacta* by the formation of conidia that are more elongate and in loose conidial heads.	19
Phaeococcomyces	Conidiophores and hyphae absent. Yeast cells one celled, pale brown to black; pseudohyphae may be formed. May occur as a synanamorph associated with species of *Exophiala*, *Phialophora*, *Wangiella*, and other genera.	Often will produce synanamorphs when grown on either cornmeal agar or potato dextrose agar.	8, 10, 19

Continued

important because many of the opportunistic dematiaceous pathogens are polymorphic; that is, they can produce several different kinds of conidia in the same culture. For an accurate identification, it must be known whether the various types of conidia present in a culture were formed by one fungus or by several fungi.

The colony characteristics and microscopic morphology used for the identification of dematiaceous fungi are based upon cultures that are approximately

TABLE 1—*Continued*

Genus	Diagnostic characteristics	Comments	Selected references
Phialophora	Conidiophores absent or present, pale brown. Conidiogenous cells phialides with distinct collarettes. Conidia one celled, hyaline to pale brown, accumulating as balls at the apices of the phialides.	*P. verrucosa* produces flask-shaped phialides with cup-shaped, dark, often deep collarettes. *P. parasitica* produces phialides of variable length, some isolates forming extremely long phialides; phialides often swollen near base, with prominent encrustations on the cell wall. Conidia elliptical to cylindrical, often curved. *P. repens* produces intercalary phialides without basal septa or phialides cylindrical to slightly lageniform with a delicate collarette. *P. richardsiae* produces phialides of variable size and shape, some phialides long with flaring, flattened collarettes. Conidia of two shapes: globose conidia from phialides with flattened collarettes and cylindrical conidia that are often curved from other phialides.	11–13, 19
Rhinocladiella	Conidiophores pale brown, erect, usually with distinct scars, and sympodial in development. Conidia one celled, fusiform to obovate, pale brown, with a dark basal scar.	Conidia occur along a rachis.	11, 12, 19, 26
Scedosporium	Conidiophores hyaline, short or long. Predominant conidiogenous cells annellides with slight swelling just below the apex. Conidia one celled, obovate, truncate, subhyaline to light black, single or in balls.	Several species of ascomycetes besides *Pseudallescheria boydii* may produce a *S. apiospermum* anamorph.	19, 22
Scytalidium	Hyphae produce one- to two-celled arthroconidia, pale brown to brown, subglobose to ellipsoidal. *H. toruloidea* synanamorph may be produced by some isolates.	May occur as synanamorphs with *Aureobasidium* spp.	19, 27
Sporothrix	Conidiophores erect, hyaline, and sympodial in development. They may be apically swollen or geniculate and gently tapering. Conidia of two types: one celled, hyaline, arising on denticles from sympodial conidiophores, and one celled, thick walled, black, arising laterally from the hyphae in some isolates.	*S. schenckii* is a dimorphic fungus capable of growing as a yeast form at 37°C and as a mold at 25°C. Black aleurioconidia are often no longer produced after repeated subculture.	9, 19
Wangiella	Conidiophores hyphalike, subhyaline to pale brown. Conidiogenous cells phialides without distinct collarettes, intercalary or lateral from hyphae. Phialides cylindrical with rounded apices. Conidia one celled, subglobose, pale brown, occurring as balls that slip down the conidiogenous cells. Annellidic and apical sympodial development also may occur. *Phaeococcomyces* synanamorph often present.	*W. dermatitidis* is differentiated from *E. jeanselmei* and similar fungi by the production of phialides and the ability to grow at 40°C.	16, 19, 24

2 weeks old and have been grown at 25 to 30°C on a medium such as potato dextrose agar or cornmeal agar. These media usually stimulate the formation of conidia. If a suspected pathogen cultivated as described above does not produce conidia, exposure to a naked daylight-type bulb for several days in a 12-h-light, 12-h-dark cycle while the pathogen is growing on a medium such as 2% water agar, sterile wooden sticks, potato dextrose agar, cornmeal agar, or filter paper may stimulate it to form conidia. The culture

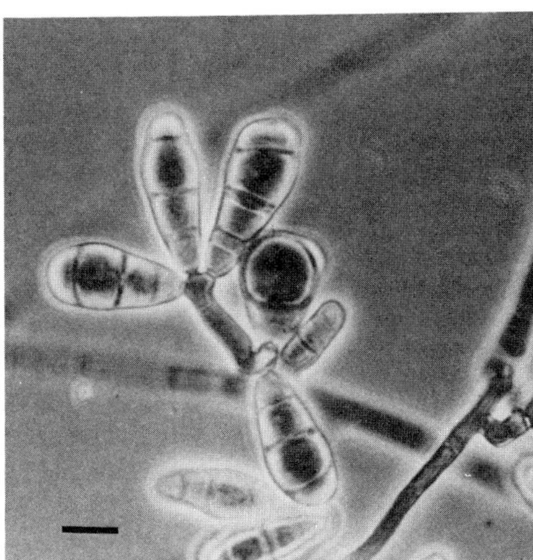

FIG. 5. *Curvularia lunata.* Taken by phase-contrast microscopy after 2 weeks on potato dextrose agar. Bar equals 10 μm.

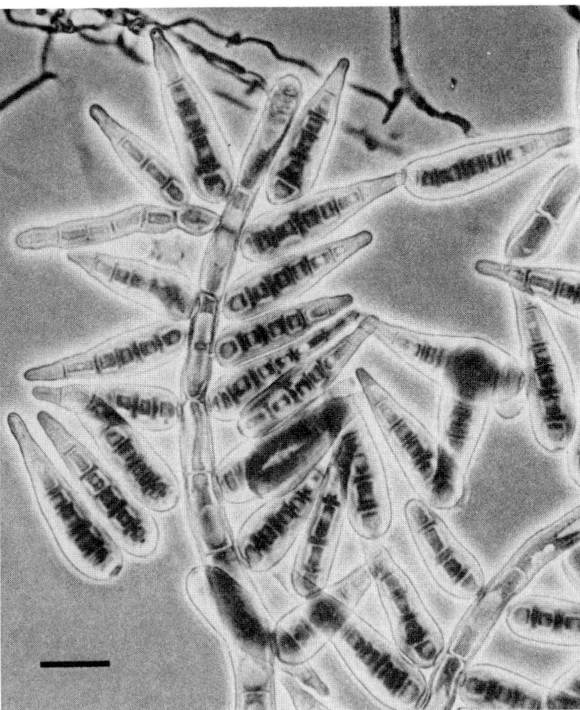

FIG. 7. *Helminthosporium solani.* Taken by phase-contrast microscopy after 2 weeks on potato dextrose agar. Bar equals 10 μm.

also can be lyophilized and then regrown, a procedure which often stimulates the development of conidia. Exposure to UV light often stimulates the development of conidia and other kinds of structures.

It has been suggested that pathogenic and nonpathogenic *Cladosporium* isolates can be distinguished from each other by their inability or ability, respectively, to hydrolyze casein, gelatin, or Loeffler serum medium. Although this test may be of some value for designating a *Cladosporium* species as a saprophyte, it cannot be used in place of morphological studies.

When isolates are believed to be *S. schenckii*, they should be subcultured on an enriched medium such as blood agar to determine whether they are dimorphic; that is, whether they grow vegetatively as hyphae at 25°C and as yeast cells at 35°C. The conversion from the mold form to the yeast form is enhanced by incubation of the inoculated medium in a candle jar. When temperature studies are conducted, it is important to concurrently incubate an additional tube of medium inoculated with the fungus at 25°C to ensure viability of the inoculum. For an isolate to be considered dimorphic, only a few cells of its typical tissue form need to be present; the entire colony does not have to be converted to its corresponding tissue form.

IDENTIFICATION

The identification of dematiaceous fungi (3, 4, 11, 12, 19) ultimately rests upon their microscopic mor-

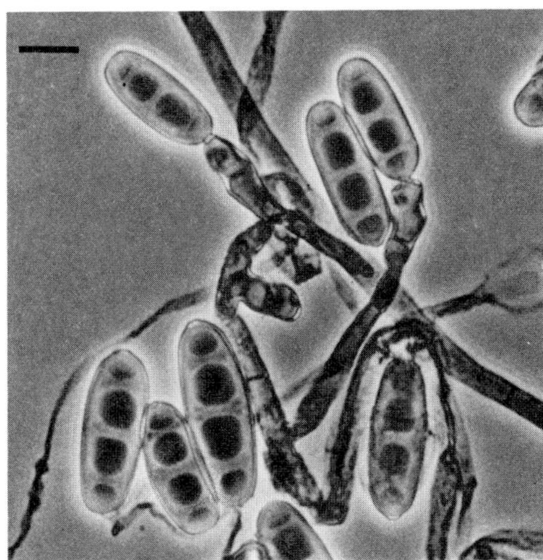

FIG. 6. *Drechslera spicifera.* Taken by phase-contrast microscopy after 2 weeks on potato dextrose agar. Bar equals 10 μm.

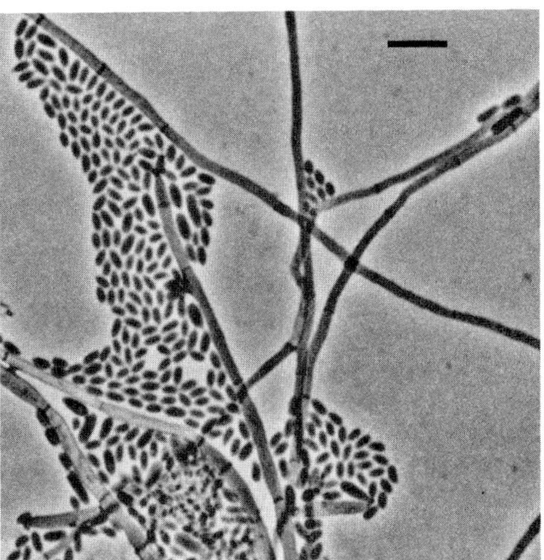

FIG. 8. *Exophiala jeanselmei.* Taken by phase-contrast microscopy after 2 weeks on potato dextrose agar. Bar equals 10 μm.

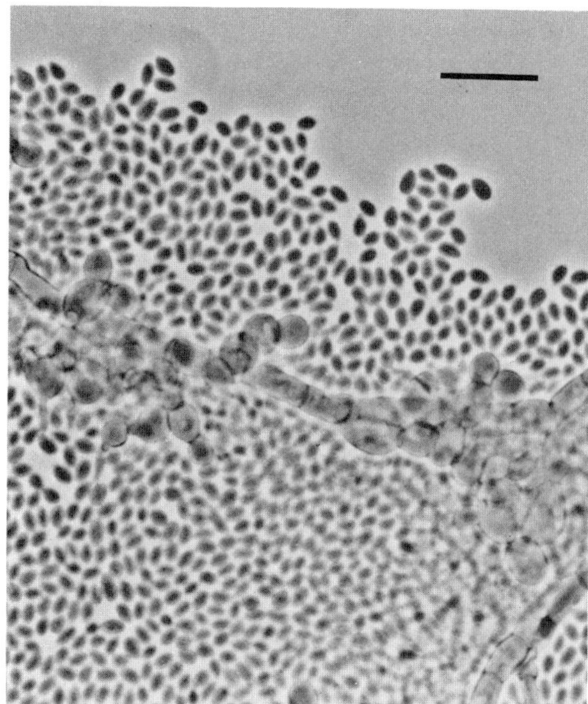

FIG. 9. *Exophiala moniliae.* Taken by phase-contrast microscopy after 2 weeks on potato dextrose agar. Bar equals 10 µm.

phology and, to a lesser extent, upon their gross colonial morphology. The importance of conidium development in defining the numerous genera of dematiaceous fungi makes it essential to determine how a particular fungus forms its conidia. For this reason, slide culture preparations with potato dextrose agar or cornmeal agar are ideal for identification purposes.

Our new understanding of conidium development (7, 15, 19) has resulted in the redefinition of many genera of medically important fungi. Terms such as spore and conidium (plural, conidia) are no longer used interchangeably. Many mycologists consider spores to be propagules that arise either from meiosis (ascospores, basidiospores, oospores, or zygospores) or by mitosis within a sporangium (sporangiospores). All other asexual, nonmotile propagules are considered conidia. Conidia usually occur on specialized hyphae or hyphal branches called conidiophores. The actual cells that give rise to the conidia are referred to as conidiogenous cells. The distinction between the various kinds of conidiogenous cells is important for the identification of species of dematiaceous fungi. Phialides usually are flask shaped to cylindrical and have an apex that neither increases in length nor changes in diameter as the phialoconidia are formed. A cup-shaped structure called a collarette may be present at the apex of the phialide. In contrast to phialides, the apices of annellides increase in length, become narrower, and have apical rings called annellations. The annellations result when the annelloconidia separate from the apex of the annellide (7).

Some fungi produce conidia by a blowing-out process. Such conidia are called blastoconidia. They may occur individually or in chains. The term acropetal is used when the conidia at the apex of the chain are the youngest. The term basipetal refers to the condition of a chain of conidia when the youngest conidium is at the base of the chain. A number of the medically important dematiaceous fungi produce conidiophores that are sympodial. In this type of development, a conidium is formed at the apex of the conidiophore. The conidiophore then increases in length by the formation of a new growing point just below and to one side of the conidium. At the apex of this new growth, a second conidium develops. The entire process is repeated, which often results in a conidiophore that has the appearance of a series of bent knees, which is said to be geniculate.

The term anamorph is used to characterize an asexual reproductive structure or form produced by fungi. Occasionally, some fungi seen in the clinical laboratory produce more than one asexual form, or anamorph. An example of this polymorphic nature is *Fonsecaea pedrosoi*, which may form a sympodial anamorph (*Rhinocladiella* form), a phialide anamorph (*Phialophora form*), and an anamorph consisting of branched chains of blastoconidia (*Cladosporium* form). When a single fungus produces more than one anamorph, the term synanamorph can be used to designate any of these concurrently existing forms. The *Scytalidium* form is often associated with the pycnidial fungus *Hendersonula toruloidae*. This form can be referred to as a synanamorph associated with *H. toruloidae*. The *Rhinocladiella*, *Phialophora*, and *Cladosporium* forms are all synanamorphs of *F. pedrosoi.*

A number of medically important fungi have the ability to produce sexual forms. The sexual form of a fungus is referred to as a teleomorph. *Pseudallescheria boydii* is characterized by the formation of cleistothe-

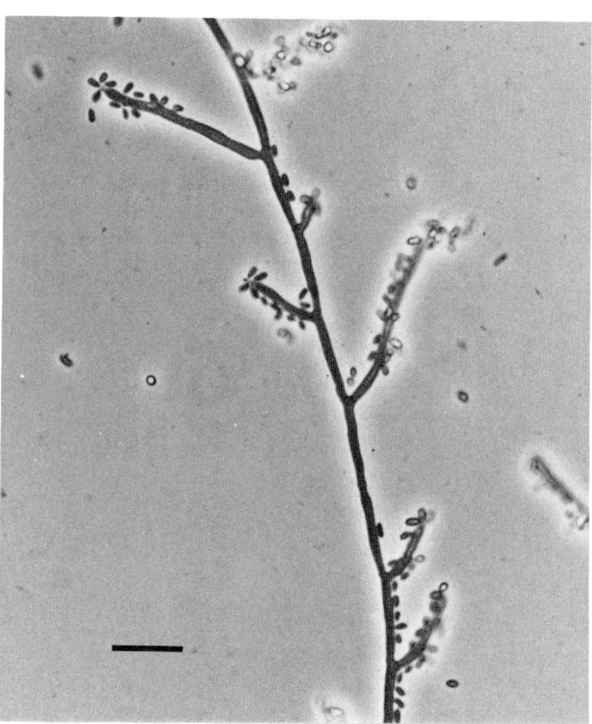

FIG. 10. *Exophiala spinifera.* Taken by phase-contrast microscopy after 2 weeks on potato dextrose agar. Bar equals 10 µm.

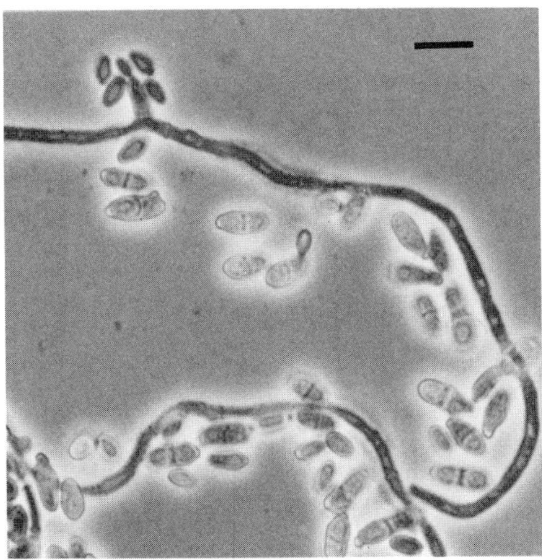

FIG. 11. *Exophiala werneckii.* Taken by phase-contrast microscopy after 2 weeks on potato dextrose agar. Bar equals 10 μm.

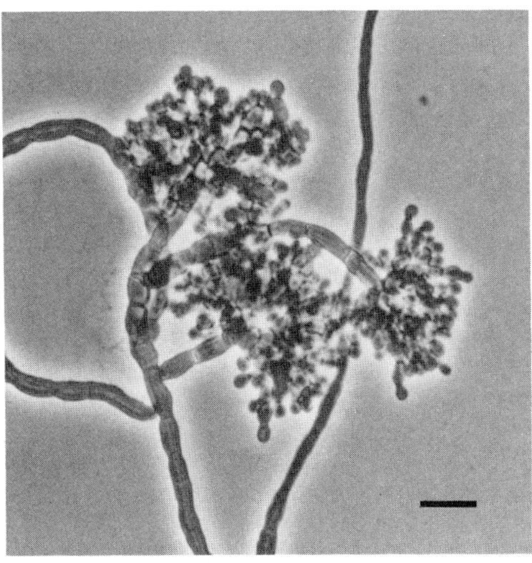

FIG. 13. *Fonsecaea compacta.* Taken by phase-contrast microscopy after 2 weeks on potato dextrose agar. Bar equals 10 μm.

cia; hence, it is a teleomorph. *P. boydii* also may produce two anamorphs, that is, *Scedosporium* and *Graphium* forms. The term holomorph is used to encompass the whole fungus. In this example, the whole fungus consists of the *Pseudallescheria, Scedosporium,* and *Graphium* forms. Because fungi are classified by sexual structures, the name used for the teleomorph also is used for the whole fungus. Problems occasionally arise with anamorph-teleomorph connections because a teleomorph may have more than one anamorph and a single anamorph may be produced by several different teleomorphs. For example, *Scedosporium apiospermum* is one anamorph that is produced by more than one *Pseudallescheria* species.

The black yeasts at times are extremely difficult and frustrating to identify. Black yeasts typically represent one growth form or anamorph of polymorphic fungi. The genus *Phaeococcomyces* was established (8, 10) to accommodate isolates that consisted of black, budding yeasts with occasional short elements of pseudohyphae, or toruloid hyphae. The assumption that a black yeast, regardless of whether the colony is initially dematiaceous or not, should be identified as *Aureobasidium pullulans* is incorrect. One of the most frequently isolated black yeasts in the clinical laboratory is the *Phaeococcomyces* synanamorph of *Exophiala jeanselmei.* When fresh isolates of this fungus are transferred from Sabouraud dextrose agar to potato dextrose agar or cornmeal agar, the typical conidiogenous cells and conidia of *E. jeanselmei* rapidly become evident. Dematiaceous, yeastlike colonies may be formed by such fungi as *A. pullulans, E. jeanselmei,* and many other species as well.

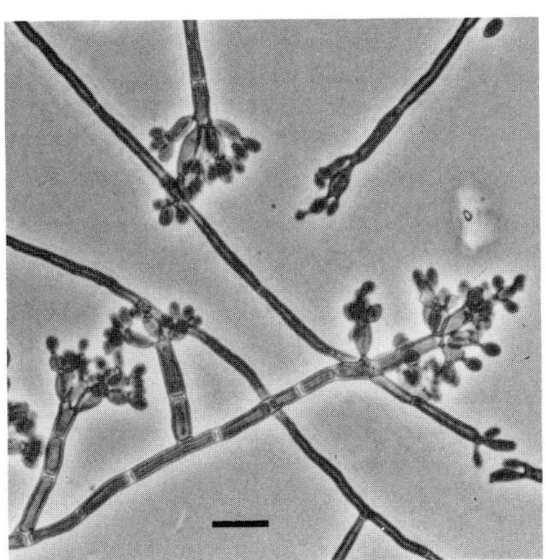

FIG. 12. *Fonsecaea pedrosoi.* Taken by phase-contrast microscopy after 2 weeks on potato dextrose agar. Bar equals 10 μm.

Sterile isolates represent a second group of medically important fungi that are especially difficult to identify. They commonly are referred to as members of the form-order Mycelia Sterilia. These fungi have been shown to cause phaeohyphomycosis and mycetoma. When sterile fungi are isolated, they should be exposed to near-UV radiation from a black light (310 to 410 nm) for several days in a 12-h-light, 12-h-dark cycle, incubated at both low and high temperatures, and subcultured onto media such as 2% water agar, hay infusion agar, soil extract agar, cereal agar, potato dextrose agar, cornmeal agar, and sterile, moist, wooden applicator sticks or filter paper. These media may help stimulate the production of conidia or fruiting bodies. The cultures should be kept for several weeks before they are discarded. With time, some of these fungi may develop structures that produce spores or conidia, such as ascocarps, pycnidia, or synnemata, either in the agar or at the colony surface.

Alternaria spp.

Members of the genus *Alternaria* occasionally are implicated as agents of phaeohyphomycosis. These fungi have been associated with infections involving bone, cutaneous tissue, ears, eyes, and the urinary tract. An *Alternaria* sp. and *Alternaria alternata* (synonym, *A. tenuis*) (14) are the only well-documented human pathogens in this genus. The *Alternaria* ana-

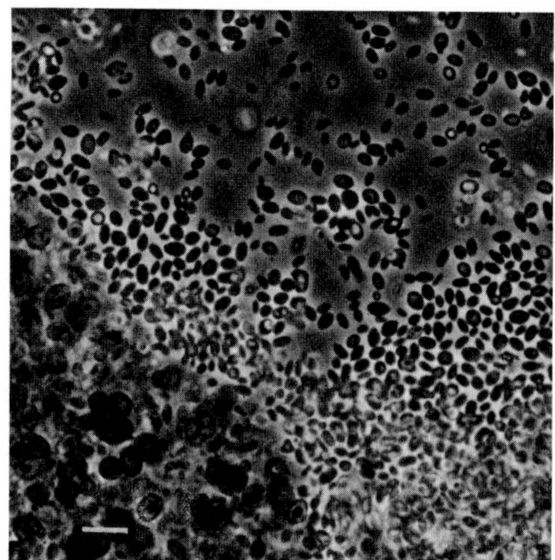

FIG. 14. *Phaeococcomyces exophialae.* Taken by phase-contrast microscopy after 2 weeks on potato dextrose agar. Bar equals 10 μm.

morph of *Pleospora infectoria* has been reported to be a pathogen of humans, but this report has not been convincingly documented.

Alternaria colonies are rapid growing, cottony, and gray to black. The erect conidiophores are dematiaceous, simple or branched, and usually solitary, but they occasionally occur in small groups. The conidia of *Alternaria* spp. develop at the apex of the conidiophore in branching chains, with the youngest conidium at the apex of each chain. The conidia are dematiaceous, muriform, smooth or rough, tapering toward the distal end, and typically with a short cylindrical beak at their apices (Fig. 1).

Alternaria isolates are difficult to identify beyond the generic level. If an isolate is recovered that must be identified to species, it should be sent to a specialist.

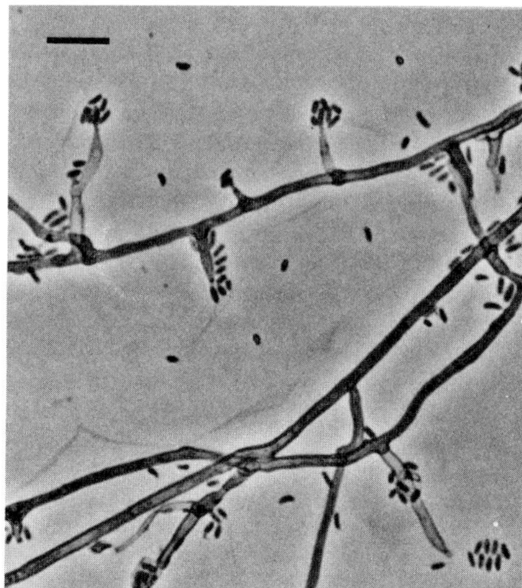

FIG. 15. *Phialophora parasitica.* Taken by phase-contrast microscopy after 2 weeks on potato dextrose agar. Bar equals 10 μm.

Aureobasidium spp.

Aureobasidium pullulans has been implicated as an agent of phaeohyphomycosis in humans and other animals (25). This hyphomycete is capable of causing opportunistic infections and has been reported from skin, nail, subcutaneous, and deeper tissues.

Colonies of *A. pullulans* are smooth, moist, and yellow, white, cream, light pink, or light brown, finally becoming black due to the development of arthroconidia. The conidiogenous cells are undifferentiated from the vegetative hyphae and may be intercalary, terminal, or arising as short lateral branches from the hyphae. The conidia are hyaline, one celled, smooth, ellipsoidal, and variable in shape and size. The conidia develop in a synchronous manner from the conidiogenous cells. Blastoconidia commonly are produced from the conidia that arise from the undifferentiated hyphal cells (Fig. 2). A *Scytalidium* anamorph consisting of dematiaceous arthroconidia typically is present.

Hormonema species occasionally are confused with *Aureobasidium* spp. In the genus *Hormonema*, the conidia arise in a basipetal succession from either hyaline or dematiaceous hyphalike conidiogenous cells. In contrast, *A. pullulans* produces its conidia in a synchronous manner (10). Because of the confusion which has surrounded these two genera, some of the reported cases of infection ascribed to *A. pullulans* may have been caused by misidentified isolates of *Hormonema* spp. This speculation is based upon the fact that several authors have illustrated *Hormonema* spp. under the name *Aureobasidium*.

Cladosporium spp.

Cladosporium bantianum (synonym, *C. trichoides*) and *C. carrionii* are the most important pathogenic members of the genus *Cladosporium* (19). *C. bantianum* is the most frequently reported etiologic agent of cerebral phaeohyphomycosis, whereas *C. carrionii* occasionally is recovered from patients with chromoblastomycosis. *C. cladosporioides* is of some interest since it was the etiologic agent of a pulmonary fungus ball in one patient. This species has been unconvincingly implicated as a pathogen in eye and nail infections. Occasionally, other *Cladosporium* spp. are reported from cutaneous, eye, and nail infections. Because pathogenesis of *C. bantianum* other than cerebral involvement has not been well defined, it would be appropriate to handle this organism in a safety cabinet.

Cladosporium isolates are rapid growing, velvety or cottony, and usually some shade of olive gray to olive brown or black. From the mycelium, erect, tall, dematiaceous conidiophores arise. At the apex of the branching conidiophore, acropetally branching chains consisting of one- to several-celled, smooth or rough, dematiaceous blastoconidia form; the conidia have a dark hilum (basal scar). The conidia at the bottom of the chains tend to have the appearance of and are commonly referred to as shield cells.

C. bantianum and *C. carrionii* are morphologically similar. *C. carrionii* (Fig. 3) can be distinguished from *C. bantianum* (Fig. 4) by its slower growth rate, shorter conidia (2 to 3 by 4 to 5 μm versus 2 to 2.5 by 4 to 7 μm, with some being 3 by 15 to 20 μm), dermo-

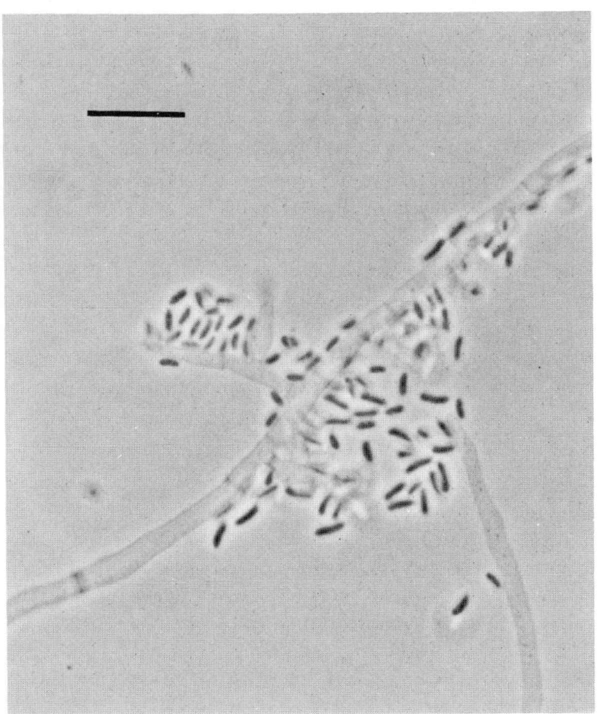

FIG. 16. *Phialophora repens.* Taken by phase-contrast microscopy after 2 weeks on potato dextrose agar. Bar equals 10 μm.

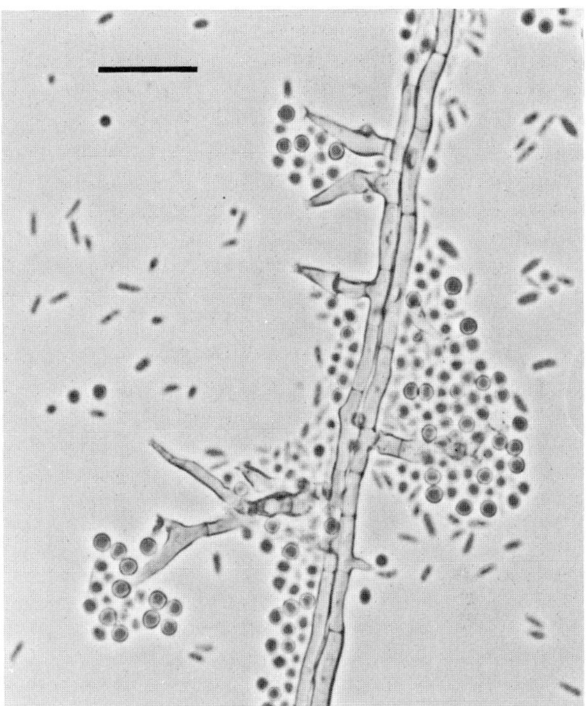

FIG. 17. *Phialophora richardsiae.* Taken by phase-contrast microscopy after 2 weeks on potato dextrose agar. Bar equals 10 μm.

tropic nature in contrast to the neurotropic nature of *C. bantianum*, and maximum growth temperature of 35 to 36°C compared with 42 to 43°C for *C. bantianum* (5, 21). Both of these species may form long chains of blastoconidia (Table 1). *C. carrionii* can hydrolyze casein and starch (though not all isolates can hydrolyze the latter), whereas *C. bantianum* apparently does not have this ability.

Curvularia spp.

Curvularia geniculata, C. lunata, C. pallescens, C. senegalensis, and *C. verruculosa* have been implicated in a number of opportunistic infections (19). Members of this genus have caused endocarditis, eye infections, mycetoma, and pulmonary phaeohyphomycosis.

Curvularia colonies are rapid growing, woolly, and gray to grayish black or brown. The conidiophores are dematiaceous, solitary or in groups, simple or branched, septate, and typically geniculate. The conidia are two to several celled, usually curved, dark with pale ends, solitary, and typically with a dark hilum. The conidia develop from a sympodial conidiophore (Fig. 5). Works by Ellis (11, 12) should be consulted if a *Curvularia* isolate must be identified to species.

Drechslera spp.

Several *Drechslera* species have caused opportunistic infections in humans, including meningitis and cutaneous, eye, nasal, and pulmonary infections. The presently recognized pathogenic members of this genus include an unidentified *Drechslera* sp., *Drechslera hawaiiensis, D. longirostrata, D. rostrata*, and *D. spicifera* (19). Alcorn (2) has suggested that some of these species should be classified in the genera *Bipolaris* and *Exserohilum*. Additional study is necessary before this issue can be adequately resolved.

Drechslera species form rapid-growing, woolly, gray-to-black colonies. The conidiophores are dematiaceous, solitary or in groups, simple or branched, septate, and geniculate. The dematiaceous, oblong-to-cylindrical conidia are multicelled and develop from a sympodial conidiophore (Fig. 6).

Some medical microbiologists have confused *Helminthosporium* spp. with *Drechslera* spp. The conidio-

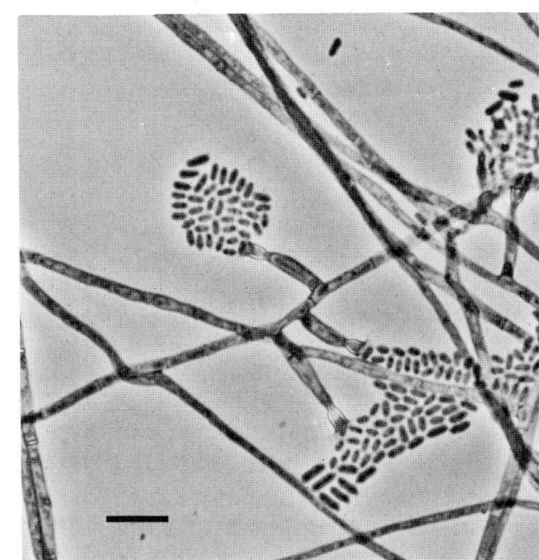

FIG. 18. *Phialophora verrucosa.* Taken by phase-contrast microscopy after 2 weeks on potato dextrose agar. Bar equals 10 μm.

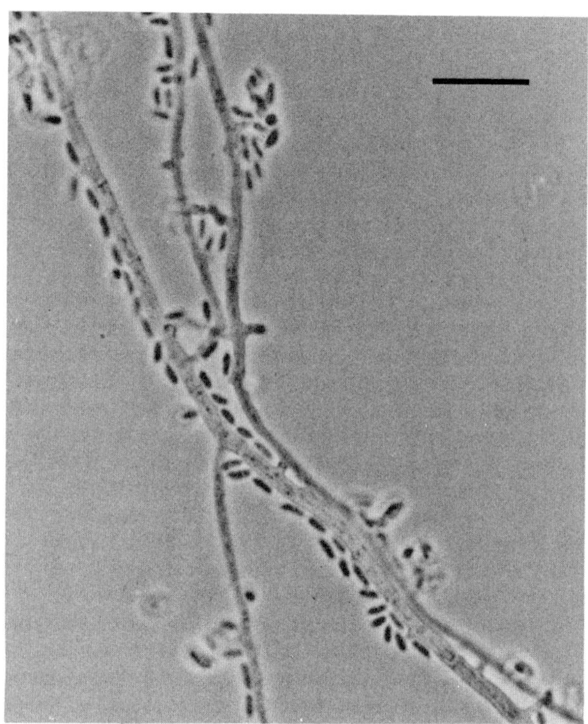

FIG. 19. *Lecythophora mutabilis.* Taken by phase-contrast microscopy after 2 weeks on potato dextrose agar. Bar equals 10 μm.

phores of *Helminthosporium* spp. (Fig. 7) are straight, and they stop lengthening when the terminal conidium is formed. The conidia develop along the conidiophore; hence, the conidiophore is not sympodial. *Helminthosporium* spp. are rarely, if ever, isolated in the clinical laboratory, and members of this genus have not caused phaeohyphomycosis in humans. If it is necessary to identify *Drechslera* isolates, either the works of Ellis (11, 12) or a specialist in this genus should be consulted.

Exophiala spp.

Exophiala jeanselmei, previously known as *Phialophora jeanselmei* or *Phialophora gougerotii*, is a relatively common etiologic agent of mycotic subcutaneous abscesses. This dematiaceous hyphomycete may cause either mycetoma or phaeohyphomycosis. *E. moniliae* and *E. spinifera* also have been reported as agents of phaeohyphomycosis, in which they caused subcutaneous cysts (16). The last member of this genus known to be pathogenic for humans is *E. werneckii* (synonym, *Cladosporium werneckii*), which causes superficial phaeohyphomycosis (synonym, tinea nigra).

The colonial morphology of the members of the genus *Exophiala* is varied. The colonies are slow to rapid growing, often moist and yeastlike at first, becoming woolly with age, and gray to black. Some isolates of *E. werneckii* and the *Phaeococcomyces* synanamorph of *E. jeanselmei* may remain black and yeastlike. Conidiophores are dematiaceous, simple, or hyphalike. The conidiogenous cells are annellides. In *E. jeanselmei* (Fig. 8), the annellides are lageniform to cylindrical, tapering to a narrow apex; in *E. moniliae* (Fig. 9), they are inflated to elliptical, tapering to a very long, narrow apex; in *E. spinifera* (Fig. 10), they are lageniform to cylindrical, tapering to a narrow apex, and they arise from distinct spinelike conidiophores; and in *E. werneckii* (Fig. 11), either they are hyphalike or they consist of two-celled yeast cells that are clavate.

The conidia are one-celled in most species and accumulate in balls at the apices of the annellides. With careful study utilizing the oil immersion objective, annellations (rings) usually can be seen at the apices of the annellides.

E. jeanselmei was incorrectly believed by some to belong to the genus *Phialophora* (17). However, when it was discovered that the conidiogenous cells of *E. jeanselmei* were annellides and not phialides, the fungus was transferred to the genus *Exophiala*. An identical situation occurred when it was discovered that the conidiogenous cells of *E. spinifera* were annellides and not phialides. *E. werneckii* produces one- to two-celled conidia from annellides that exist as either yeast cells or intercalary conidiogenous cells incorporated within the hyphae (18).

The fungus originally described as *Sporotrichum gougerotii* was considered a morphological variant of *Sporothrix schenckii* (17). The name *S. gougerotii* is best considered a nomen dubium. Fungi that are currently identified as either *S. gougerotii* or *P. gougerotii* are typically misidentified isolates of *E. jeanselmei*. At one time, these two supposedly different fungi were erroneously distinguished from each other by their tissue morphology. In the sense of some contemporary medical mycologists, *P. gougerotii* is a misapplied name for isolates of *E. jeanselmei* that do not form granules in tissue.

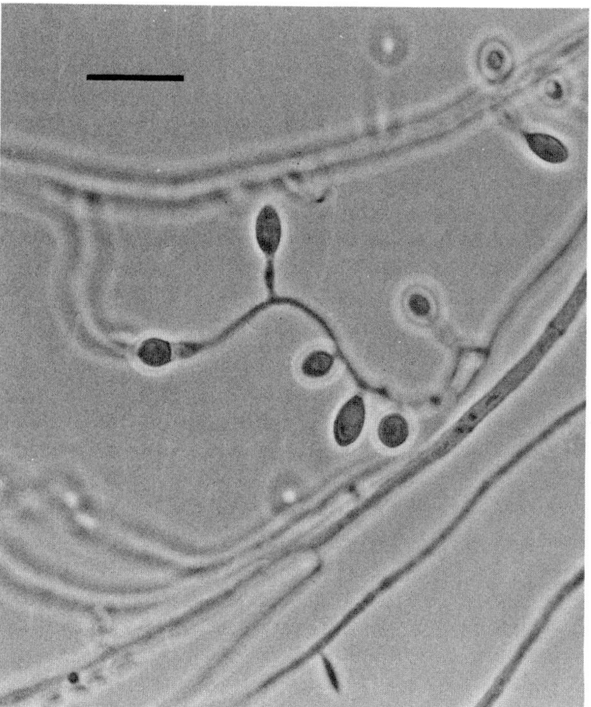

FIG. 20. *Lecythophora mutabilis.* Note the presence of chlamydoconidia. Taken by phase-contrast microscopy after 2 weeks on potato dextrose agar. Bar equals 10 μm.

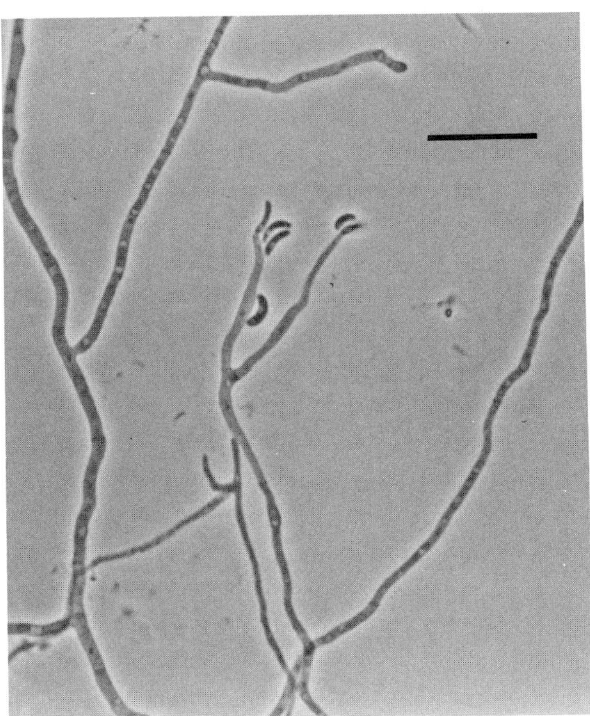

FIG. 21. *Lecythophora hoffmannii.* Taken by phase-contrast microscopy after 2 weeks on potato dextrose agar. Bar equals 10 μm.

Fonsecaea spp.

The genus *Fonsecaea* contains two species, *Fonsecaea compacta* (incorrectly spelled *compactum* by some) and *F. pedrosoi*. Both of these species are agents of chromoblastomycosis.

Fonsecaea colonies are slow growing, velvety to woolly, and olive to black. *Fonsecaea* isolates are extremely polymorphic. They are characterized by the development of one-celled primary conidia that form on erect, dark, sympodial conidiophores. The primary conidia in turn become conidiogenous cells and form secondary one-celled conidia. This form of development has been incorrectly called *Acrotheca*-like by some. It is actually more similar to the form seen in the genus *Rhinocladiella*. Some of the conidia occur as branching chains of blastoconidia, identical to those found in the genus *Cladosporium*. *Fonsecaea* spp. may produce phialides with the collarettes bearing the balls of one-celled conidia that are typical of the genus *Phialophora*.

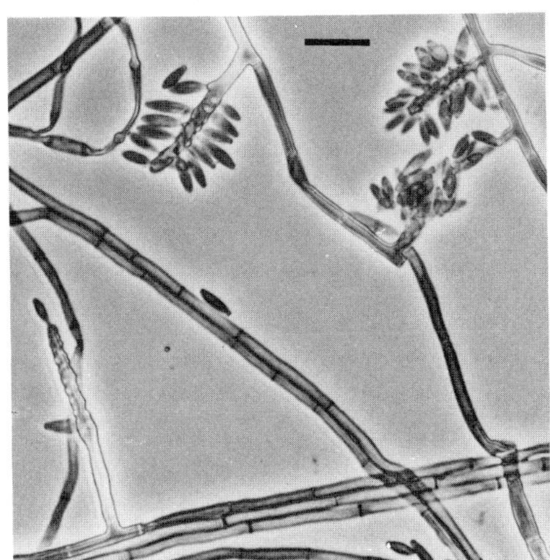

FIG. 22. *Rhinocladiella aquaspersa.* Taken by phase-contrast microscopy after 2 weeks on potato dextrose agar. Bar equals 10 μm.

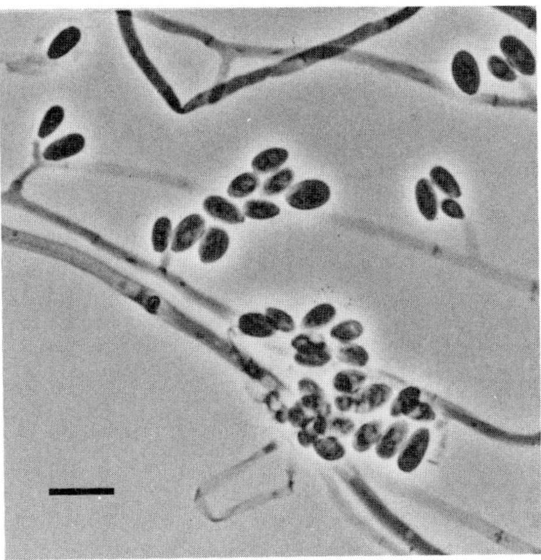

FIG. 23. *Scedosporium apiospermum.* Taken by phase-contrast microscopy after 2 weeks on potato dextrose agar. Bar equals 10 μm.

F. pedrosoi and *F. compacta* are morphologically distinct (19). *F. pedrosoi* is differentiated from *F. compacta* by its elongate conidia that occur in loose heads (Fig. 12), in contrast to the rounded conidia in compact heads produced by *F. compacta* (Fig. 13). As a result of their polymorphic nature, *F. pedrosoi* and *F. compacta* have been placed inappropriately in the genera *Phialophora* and *Rhinocladiella* by some mycologists.

Phaeococcomyces spp.

Phaeococcomyces is a genus that contains black yeasts. Black yeasts are often synanamorphs associated with several of the medically important polymorphic dematiaceous hyphomycetes such as *E. jeanselmei* and *Wangiella dermatitidis*. The genus *Phaeococcomyces*, which was originally named *Phaeococcus* (8, 10), contains four species that are distinguished from each other primarily on morphological criteria.

Phaeococcomyces exophialae forms slimy, mucoid, slow-growing, smooth colonies that are grayish black. Budding yeast cells which are at first subhyaline are abundant. With age, some of the cells become darker, with thickened cell walls (Fig. 14). Some pseudohyphae usually are present. Hyphal development may become dominant in some isolates of this species with subsequent subculture.

Black yeasts occasionally are isolated in the clinical laboratory. They are recognized by their black, mucoid, yeastlike colonies. When grown on cornmeal agar or potato dextrose agar, many black yeasts will rapidly produce the hyphae and conidiogenous cells

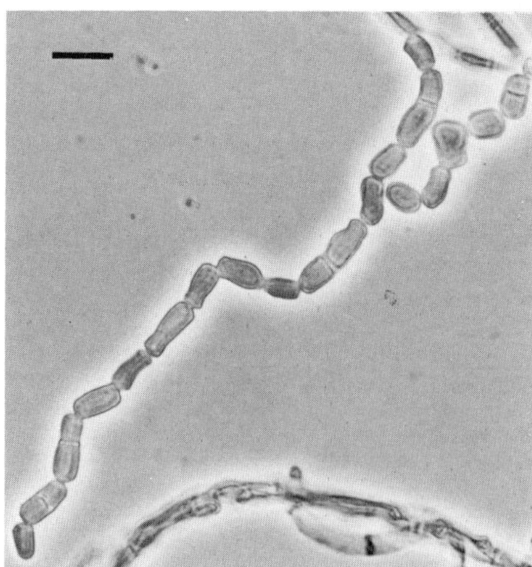

FIG. 24. *Scytalidium lignacola.* Taken by phase-contrast microscopy after 2 weeks on potato dextrose agar. Bar equals 10 μm.

typical of genera such as *Exophiala* and *Wangiella*. Based upon conidiogenesis, other genera of black yeasts probably will be needed in the future to accommodate this group of fungi.

Phialophora spp.

Members of the genus *Phialophora* are well-recognized etiologic agents of phaeohyphomycosis and chromoblastomycosis. In addition to cutaneous and subcutaneous tissue invasion, some species have caused endocarditis and mycotic keratitis. The pathogenic species of *Phialophora* include *Phialophora bubakii, P. parasitica* (Fig. 15), *P. repens* (Fig. 16), *P. richardsiae* (Fig. 17), and *P. verrucosa* (Fig. 18).

Phialophora colonies are rapid growing, cottony to woolly, and usually some shade of olive gray. When conidiophores are present, they usually are short. The conidiogenous cells are hyaline to dematiaceous phialides that are cylindrical to flask shaped. At the apices of the phialides, distinct collarettes are present. The conidia are one celled and usually hyaline, and they occur in balls that may occasionally slip down along the phialides in some species. Some species commonly produce intercalary phialides; a yeast form may occur in some isolates.

P. mutabilis and *P. hoffmannii* have been considered species of the genus *Phialophora* for a number of years. Gams and McGinnis (13) recently have reclassified these species in the genus *Lecythophora* (Fig. 19 through 21). This reclassification was necessary because these fungi produce intercalary phialides with short, lateral, cylindrical necks, which bear the conidia in balls at their apices. Collarettes are present at the tips of the phialides.

Rhinocladiella spp.

Rhinocladiella aquaspersa, previously known as *Acrotheca aquaspersa*, is a rare etiologic agent of chromoblastomycosis (26). Human cases of chromoblastomycosis caused by *R. aquaspersa* have occurred in Brazil and Mexico.

Colonies of *R. aquaspersa* are rapid growing, velvety, slightly elevated, and olive black. The conidiophores are sympodial, usually darker than the vegetative hyphae, unbranched, and erect. Conidia are one celled, rarely two celled, fusiform, elliptical or obovate, smooth, light brown, and with a dark basal scar (Fig. 22). Annellides like those of *Exophiala* spp. and phialides like those of *Wangiella* spp. may be present.

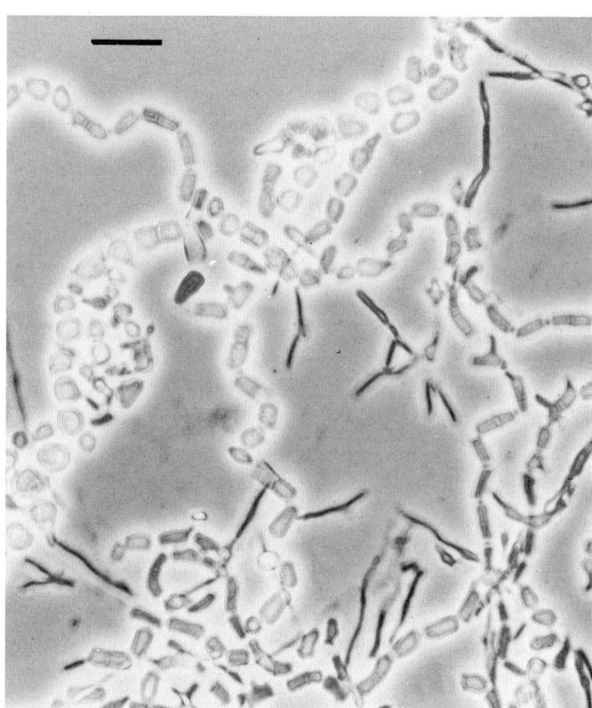

FIG. 25. *Scytalidium hyalinum.* Taken by phase-contrast microscopy after 2 weeks on potato dextrose agar. Bar equals 10 μm.

Scedosporium sp.

Scedosporium apiospermum, previously known as *Monosporium apiospermum*, is an anamorph of *Pseudallescheria boydii* (see Chapter 51 on mycetoma), a fungus once classified as *Petriellidium boydii* and *Allescheria boydii* (22). The fungus may cause mycetoma, as well as infections involving the lungs and brain, where the fungus grows in the form of hyphae that look like those produced by *Aspergillus* spp.

S. apiospermum rapidly produces colonies that are cottony and smoky gray to dark brown. One-celled conidia may occur singly along the hyphae or in clusters at the apices of annellides. The conidia are obovate, truncate, and subhyaline to light black (Fig. 23).

S. apiospermum occasionally has a *Graphium* synanamorph present. Several members of the genera *Pseudallescheria* and *Petriella* may produce a *S. apiospermum* anamorph. Therefore, without having the teleomorph present, it is not possible to determine if an isolate of *S. apiospermum* was produced by *Pseudallescheria boydii*.

Scytalidium spp.

Members of the genus *Scytalidium* (19) have been well documented as opportunistic fungal pathogens of nail, skin, and subcutaneous tissue. The *Scytalidium* synanamorph associated with the pycnidial fungus *H.*

toruloidea, as well as *Scytalidium lignicola* (Fig. 24) and *S. hyalinum* (Fig. 25), all have caused disease. *S. hyalinum*, because of its hyaline nature, would best be classified in a genus other than *Scytalidium*.

Scytalidium spp. produce rapid-growing colonies that are at first white, becoming dark gray with age. In *S. lignicola*, arthroconidia of two types are formed. In the first type, the arthroconidia are cylindrical, one celled, and hyaline. In the second type, they are thick walled, yellowish brown, and one or two celled. The *Scytalidium* synanamorph of *H. toruloidea* forms arthroconidia that are brown, cylindrical at first, becoming rounded, barrel shaped or subglobose, and one or two celled.

The monograph on *Malbranchae* spp. by Sigler and Carmichael (27) should be consulted for the identification of *Scytalidium* species and similar hyphomycetes that produce arthroconidia.

Sporothrix spp.

S. schenckii, a dimorphic fungus, is considered the only pathogenic member of the genus *Sporothrix*. Recently, a new species, *Sporothrix cyanescens*, was added to the genus (9). Some of the isolates upon which the species description was based were isolated from patients with mycosis of human skin. Whether or not *S. cyanescens* is another etiologic agent of sporotrichosis remains to be proven.

Colonies of *S. schenckii* are rapid growing and at first moist, flat, and yeastlike, later developing aerial hyphae. They are initially white, becoming brown to black with age. The conidia are of two kinds in most fresh isolates. Hyaline, one-celled conidia develop solitarily upon denticles along the hyphae; laterally from sympodial, slender, tapering, erect conidiophores; and terminally in clusters at the apices of swollen conidiophores. The second type of conidia are one celled, thick walled, and black. These conidia develop along the hyphae (Fig. 26). At 37°C on enriched media, the mold form of *S. schenckii* converts to a yeast form.

The etiologic agent of sporotrichosis originally was described as *S. schenckii*. Later, the fungus erroneously was transferred to the genus *Sporotrichum*. Members of the genus *Sporotrichum* are characterized by the formation of large hyphae with clamp connections and large, one-celled, thick-walled, golden conidia (19). They are neither dimorphic nor pathogenic for humans and other animals.

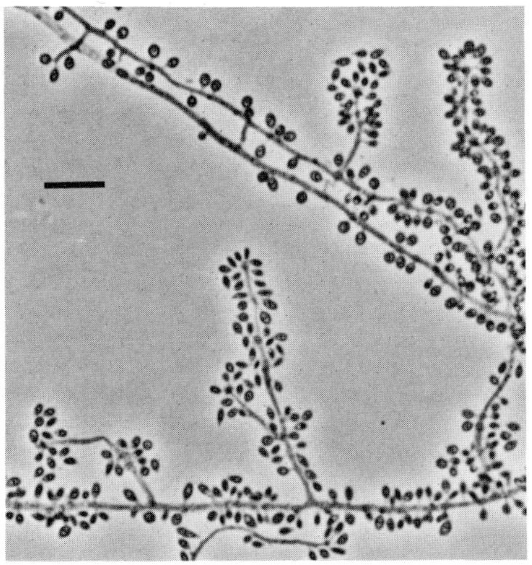

FIG. 26. *Sporothrix schenckii*. Taken by phase-contrast microscopy after 2 weeks on potato dextrose agar. Bar equals 10 μm.

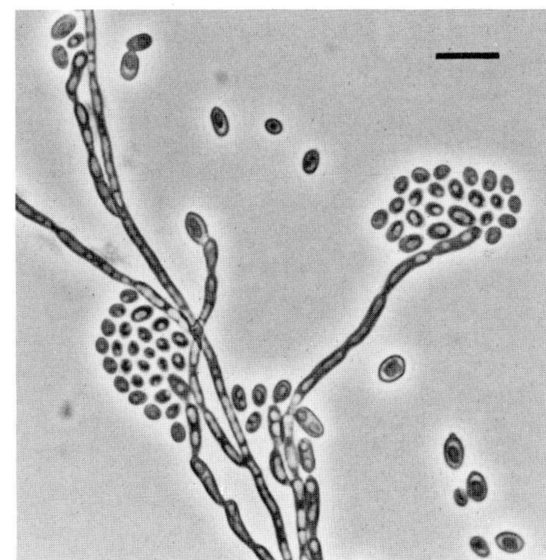

FIG. 27. *Wangiella dermatitidis*. Taken by phase-contrast microscopy after 2 weeks on potato dextrose agar. Bar equals 10 μm.

Wangiella sp.

W. dermatitidis is an agent of phaeohyphomycosis that typically causes infections involving cutaneous and subcutaneous tissue. The fungus most frequently has been seen in patients living in Japan.

The colonies of *W. dermatitidis* are moist and at first yeastlike, developing some aerial hyphae with age. They are olive to black. Distinct conidiophores are absent. The conidiogenous cells are phialides which do not have collarettes. Some conidiogenous cells appear to possess a group of slightly raised, truncate denticles at their apices that occur as a result of sympodial development. Rare annellides may be produced by isolates of this fungus. The one-celled, light-to-dark, smooth phialoconidia form in balls at the apices of the phialides and then slide down their sides (Fig. 27). The phialides develop from conidiophores that are indistinguishable from the hyphae. Most isolates produce an abundant yeast form and large amounts of toruloid hyphae.

The genus *Wangiella* was established to accommodate the fungus known as either *Hormiscium dermatitidis* or *Phialophora dermatitidis* (16). The new genus *Wangiella* was necessary because the phialides without collarettes that are typical of *W. dermatitidis* could not be accommodated in any known genus. *W. dermatitidis* can be recognized by its ability to grow at 40°C, whereas similar dematiaceous hyphomycetes do not grow at that temperature (24).

LITERATURE CITED

1. **Ajello, L.** 1975. Phaeohyphomycosis: definition and etiology. Pan Am. Health Organ. Sci. Publ. **304:**126–133.
2. **Alcorn, J. L.** 1983. Generic concepts in *Drechslera*, *Bipolaris* and *Exserohilum*. Mycotaxon **17:**1–86.

3. **Barnett, H. L., and B. B. Hunter.** 1972. Illustrated genera of imperfect fungi, 3rd ed. Burgess Publishing Co., Minneapolis.
4. **Barron, G. L.** 1968. The genera of Hyphomycetes from soil. The Williams & Wilkins Co., Baltimore.
5. **Borelli, D.** 1960. *Torula bantiana*, agente di un granuloma cerebrale. Riv. Anat. Patol. Oncol. **17:**615–622.
6. **Carmichael, J. W., B. Kendrick, I. L. Conners, and L. Sigler.** 1980. Genera of Hyphomycetes. University of Alberta Press, Edmonton.
7. **Cole, G. T., and R. A. Samson.** 1979. Patterns of development in conidial fungi. Pitman Press, London.
8. **de Hoog, G. S.** 1979. Nomenclatural notes on some black yeast-like hyphomycetes. Taxon **28:**347–348.
9. **de Hoog, G. S., and G. A. deVries.** 1973. Two new species of *Sporothrix* and their relation to *Blastobotrys nivea*. Antonie van Leeuwenhoek J. Microbiol. Serol. **39:**515–520.
10. **de Hoog, G. S., and E. J. Hermanides-Nijhof.** 1977. The black yeasts and allied Hyphomycetes. Studies in mycology no. 15. Centraalbureau voor Schimmelcultures, Baarn, The Netherlands.
11. **Ellis, M. B.** 1971. Dematiaceous Hyphomycetes. Commonwealth Mycological Institute, Kew, England.
12. **Ellis, M. B.** 1976. More dematiaceous Hyphomycetes. Commonwealth Mycological Institute, Kew, England.
13. **Gams, W., and M. R. McGinnis.** 1983. *Phialemonium*, a new anamorph genus intermediate between *Phialophora* and *Acremonium*. Mycologia **75:**977–987.
14. **Goodpasture, H. C., T. Carlson, B. Ellis, and G. Randall.** 1983. *Alternaria* osteomyelitis. Evidence of specific immunologic tolerance. Arch. Pathol. Lab. Med. **107:**528–530.
15. **Kendrick, W. B. (ed.).** 1971. Taxonomy of Fungi Imperfecti. University of Toronto Press, Toronto.
16. **McGinnis, M. R.** 1978. Human pathogenic species of *Exophiala, Phialophora*, and *Wangiella*. Pan Am. Health Organ. Sci. Publ. **356:**37–59.
17. **McGinnis, M. R.** 1979. Taxonomy of *Exophiala jeanselmei* (Langeron) McGinnis and Padhye. Mycopathologia **65:**79–87.
18. **McGinnis, M. R.** 1979. Taxonomy of *Exophiala werneckii* and its relationship to *Microsporum mansonii*. Sabouraudia **17:**145–154.
19. **McGinnis, M. R.** 1980. Laboratory handbook of medical mycology. Academic Press, Inc., New York.
20. **McGinnis, M. R.** 1983. Chromoblastomycosis and phaeohyphomycosis: new concepts, diagnosis, and mycology. J. Am. Acad. Dermatol. **8:**1–16.
21. **McGinnis, M. R., and D. Borelli.** 1981. *Cladosporium bantianum* and its synonym *Cladosporium trichoides*. Mycotaxon **13:**127–136.
22. **McGinnis, M. R., A. A. Padhye, and L. Ajello.** 1982. *Pseudallescheria* Negroni et Fischer, 1943, and its later synonym *Petriellidium* Malloch, 1970. Mycotaxon **14:**94–102.
23. **Padhye, A. A.** 1978. Comparative study of *Phialophora jeanselmei* and *P. gougerotii* by morphological, biochemical, and immunological methods. Pan Am. Health Organ. Sci. Publ. **356:**60–65.
24. **Padhye, A. A., M. R. McGinnis, and L. Ajello.** 1978. Thermotolerance of *Wangiella dermatitidis*. J. Clin. Microbiol. **8:**424–426.
25. **Salkin, I. F., M. A. Gordon, and W. B. Stone.** 1976. Cutaneous infection of a porcupine (*Erethizon dorsatum*) by *Aureobasidium pullulans*. Sabouraudia **14:**47–49.
26. **Schell, W. A., M. R. McGinnis, and D. Borelli.** 1983. *Rhinocladiella aquaspersa*, a new combination for *Acrotheca aquaspersa*. Mycotaxon **17:**341–348.
27. **Sigler, L., and J. W. Carmichael.** 1976. Taxonomy of *Malbranchea* and some other Hyphomycetes with arthroconidia. Mycotaxon **4:**349–488.

Chapter 53

Agents of Zygomycosis (Phycomycosis)

DONALD L. GREER AND ALVIN L. ROGERS

CHARACTERIZATION OF *ZYGOMYCETES*

The heterogeneous group of lower fungi previously included in the class *Phycomycetes* now have been classified into six classes (2). Some species of the class *Zygomycetes* cause disease in humans and lower animals. The obsolete term phycomycosis has been replaced by the more specific term zygomycosis. (Other classes of closely related fungi previously classified as *Phycomycetes* are primarily aquatic forms, not as yet known to cause diseases in humans.) Two orders of the class *Zygomycetes* contain all of the eight genera in which species are presently recognized as the principal zygomycetous pathogens for humans. The genera which contain pathogenic species in the order *Mucorales* include *Rhizopus, Mucor, Absidia, Saksenaea, Cunninghamella*, and *Syncephalastrum* (Fig. 1), and those in the order *Entomophthorales* include *Conidiobolus (Entomophthora)* and *Basidiobolus*. The genus *Rhizopus* contains the agents of zygomycosis most frequently reported in the United States (17). Unidentified species, probably of the genus *Mortierella*, have also been isolated from humans, and *Pythium* species (19) *(Hyphomyces destruens)* and *Absidia corymbifera* have been reported frequently from lower animals (21). *Pythium* is an aquatic species in the class *Oomycetes*; the other genera are members of the order *Mucorales*. The term "mucormycosis" is reserved for infections caused by species within the order *Mucorales*, and the term "entomophthoromycosis" is used for infections caused by species within the order *Entomophthorales*.

Mucormycosis occurs in immunocompromised individuals primarily as an acute, rapidly fatal, necrotic infection of the paranasal sinuses, lungs, and gastrointestinal tract. Recently, serious primary cutaneous infections that have been associated with contaminated bandages have been reported in hospitalized patients (4).

Entomophthoromycosis is a chronic granulomatous infection of the subcutaneous tissues of the limbs of healthy children (subcutaneous phycomycosis) or of adults (rhinoentomophthoromycosis). Unusual cases of deeper organ involvement have been reported (8, 10, 14).

The disease process produced by members in the genera of each order is distinct, both clinically and histologically; therefore, the correct diagnosis of the zygomycoses usually can be made in tissues without cultural confirmation. However, cultures are necessary for the complete identification to species of the etiological agent.

Class *Zygomycetes*

Members of the class *Zygomycetes* are terrestrial fungi with well-developed, broad, usually nonseptate hyphae. Sexual spores (zygospores) are nonmotile (see Fig. 6). All species grow rapidly on routine laboratory media, but most are susceptible to cycloheximide. Temperature range is from 25 to 55°C with an optimal temperature of 28 to 30°C. All of the pathogenic species grow at 37°C (20). All are saprophytic in nature and occur worldwide. No serological or biochemical classification is available. Members of the class *Zygomycetes* are identified primarily on the basis of their morphological characteristics (9).

Order *Mucorales*

The members of the order *Mucorales* are rapidly growing ubiquitous fungi which form loose, gray, woolly colonies filling a petri plate in 2 to 5 days with aseptate mycelium. Asexual spores, sporangiospores, are borne in a closed sac, sporangium, supported on a slender stalk, the sporangiophore. Spores appear to the unaided eye as black pinpoints scattered throughout the mycelial mass of the colony.

Species of the *Mucorales* are opportunistic fungi, i.e., they have a low ability to produce disease and usually can cause disease only in immunologically deficient, metabolically unbalanced, or chronically debilitated individuals (6, 17). Diabetes mellitus is the disease most commonly associated with mucormycosis. In over 50% of affected individuals that have been described in the scientific literature, the diabetes was out of control and acidosis was present. Acidosis can be precipitated by several diseases, including diabetes, leukemia, severe burns, and malnutrition, and appears to be the common predisposing condition for mucormycosis (21).

The incidence of mucormycosis is increasing. New genera of the *Mucorales*, hitherto rarely encountered even as contaminants, are now being reported as causative agents of zygomycosis, namely, *Saksenaea* (1, 7, 23), *Cunninghamella* (16), and *Syncephalastrum* (15). Patients with severe burns, children with acute leukemia, and patients treated with cytotoxic or immunosuppressive drugs and with corticosteroids remain the prime victims (6, 17).

Since species of *Mucorales* are considered common laboratory contaminants, the significance of isolating these fungi from pathological material depends upon the source of the specimen and the condition of the patient. The presence of broad nonseptate hyphal fragments (Fig. 2) in normally sterile body fluids or of invading hyphae in tissue scrapings or biopsy specimens is significant and is often more meaningful than the isolation of the fungus. Conversely, repeated isolation of the same species of *Zygomycetes* from sources that harbor a normal microbial population, e.g., nose, throat, sputum, etc., may be of some diagnostic value, especially if accompanied by clinical signs of infection in a debilitated person.

Order *Entomophthorales*

Members of the order *Entomophthorales* are ubiquitous fungi, forming flat, waxy colonies which may develop a white fuzz on the surface. The hyphae are septate. Asexual spores are either conidia or uninucleate sporangia which are forcibly expelled from the conidiophore at maturity.

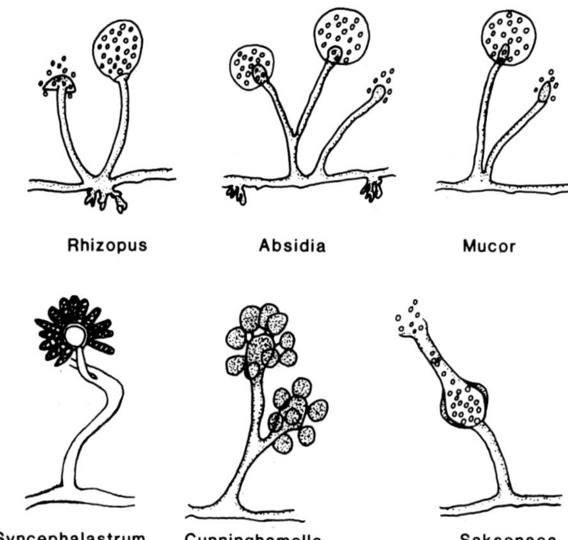

FIG. 1. Diagrammatic representation of the morphological distinction between the genera *Rhizopus, Absidia, Mucor, Saksenaea, Cunninghamella,* and *Syncephalastrum.*

There are no reports of laboratory contamination by the two pathogenic genera, *Basidiobolus* and *Conidiobolus*. Isolation of either of these fungi from clinical material warrants close attention. Factors predisposing to infections caused by these species are not yet known (11).

Entomophthoromycosis caused by *Conidiobolus* has not been reported in humans in the United States, although *C. coronatus* has been isolated from horses in several southern states (9). Only one culturally proven case of *Basidiobolus* infection in a human has been reported in the United States (8). A case of human entomophthoromycosis in a child was diagnosed by histopathology (21). Most *Entomophthorales* infections have been reported from the tropical countries of central Africa and Southeast Asia (12, 18). A few cases are known to have occurred in South America.

COLLECTION, TRANSPORT, AND STORAGE OF SPECIMENS

Care must be taken to obtain an adequate sample, and the pathology of both mucormycosis and entomophthoromycosis must be understood so that this can be accomplished. Light swabs, such as those used for bacteriological specimens, are usually unsatisfactory. Sufficient material for all laboratory examinations can be obtained by aspirating abscesses, by irrigating or scraping lesions on mucous membranes, or by taking multiple biopsies. Zygomycetes regularly invade arteries in affected tissue to produce thrombi and infarcts. For that reason, the fungus may not be found in the center of necrotic tissue resulting from an infarct, but will instead be found at the edge of the infarct or "upstream" from it. This should be taken into account when biopsies of necrotic tissue are performed.

Rhinocerebral infections are acute and often fatal; clinical materials received for examination and culture are frequently tissues from autopsy. The rhinocerebral form of zygomycosis begins in the paranasal sinuses and spreads rapidly by direct invasion to involve the palate, the orbit, the turbinates, and the central nervous system. Antemortem specimens from central nervous system infections usually include exudates from the sinuses, scrapings from nasal or oral mucosa, or cerebrospinal fluid (22). Specimens from pulmonary or abdominal infections may include sputum, bronchial washings, or stools. Biopsy material should be obtained whenever possible. In primary cutaneous infections of healthy individuals or burned patients, the biopsy should be taken at the growing edge of the lesion (4, 5).

The primary responsibility in diagnosing mucormycosis lies with the physician (24). A high level of

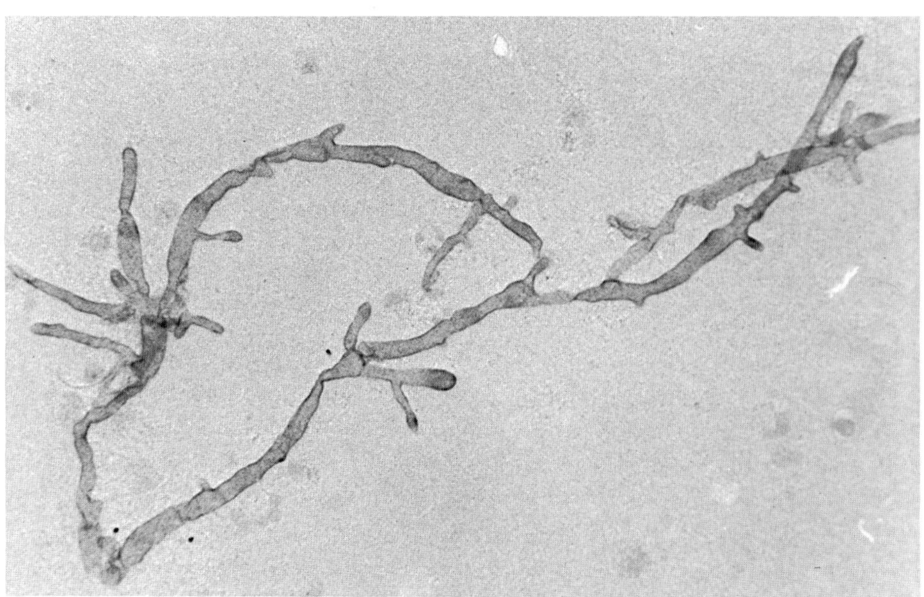

FIG. 2. *Mucor* sp. Nonseptate hyphae in sputum. Potassium hydroxide with ink (Swartz-Lamkins stain). ×300. (Photomicrograph courtesy of B. H. Cooper.)

suspicion is required to make the correct clinical diagnosis, notify the mycology laboratory, and send the appropriate material immediately to the laboratory. A responsive laboratory will rapidly examine the clinical specimens received and make immediate reports, even if there are no positive findings, to the physicians. The examination of clinical material with a direct KOH preparation can provide the physician with enough information to begin treatment. Frequently, the patient is so ill that treatment must be started long before the fungus is cultured and identified.

Clinical specimens should be rapidly transported to the laboratory for examination because of (i) the need for immediate information and (ii) the fragility of the fungal elements. If overnight storage is necessary, specimens should be placed in Stuart bacteriological transport medium and stored at room temperature. Zygomycetes do not survive longer than a few hours at refrigerator temperature.

DIRECT EXAMINATION

A direct wet-mount preparation should be made on all clinical specimens. Necrotic or purulent material may contain only dead or degenerated hyphal elements which, though visible on direct examination of the material, may not grow on culture. The observation of invasion of intact tissue by nonseptate hyphae is good evidence for a zygomycetous infection and confirms the significance of subsequent isolation. Failure to observe fungal elements, however, does not necessarily indicate their absence.

Examine a portion of the necrotic tissue or purulent material by mixing the specimen with several drops of 10 to 20% aqueous potassium hydroxide on a slide. Add a cover slip, warm the preparation slightly to enhance the clearing action of the KOH, and observe under a microscope. Clear body fluids, e.g., cerebrospinal fluid, may be centrifuged, and the sediment is examined directly on a slide with a few drops of KOH.

Study the wet-mount preparations under a microscope for broad, nonseptate hyphal fragments (Fig. 2). In tissues stained with hematoxylin and eosin, abundant large, irregularly branching hyphal elements can be seen.

CULTURE AND ISOLATION

The zygomycetes are not especially fastidious; however, cultures of necrotic tissue frequently fail to yield fungi even though fungi can be observed microscopically in the tissue. Culture media should be heavily inoculated since the number of viable fungal elements in specimens of necrotic tissue usually is small. For that reason, the tissue should be minced instead of being ground in a mortar to avoid killing any viable fungal elements that are present. Sabouraud–2% dextrose agar is an adequate isolation medium, although the use of squares of sterile homemade bread (without preservatives) has also been suggested as a useful isolation medium (these are, after all, bread molds!). Antibacterial antibiotics may be added to the medium, but cycloheximide (Actidione) should not be used, as most zygomycetes are susceptible to this antibiotic. Therefore, media such as Mycosel (BBL Microbiology Systems) or Mycobiotic agar (Difco), which contain cycloheximide and are commonly used for the isolation of pathogenic fungi, cannot be used.

Some specimens received for culture are heavily contaminated with bacteria. To prevent overgrowth by bacteria, specimens such as purulent exudates, mucosal scrapings, biopsies of necrotic tissue, and sputum should be streaked onto petri plates of Sabouraud–2% dextrose agar containing antibacterial agents. Chloramphenicol and polymyxin B are very useful for this purpose. The material should be heavily streaked over the entire plate, with most, if not all, of the specimen being used. Small pieces of minced tissue should be planted directly into the agar medium. Cultures are incubated in duplicate under aerobic conditions at both 25 and 37°C.

Growth of species of *Mucorales* is usually observed in 2 to 5 days. The cultures should be observed for early growth within the area of the inoculum. The fungus grows rapidly and fills the petri plate or tube with an abundance of grayish-white aerial mycelium. The mycelium is tenacious and has a "steel wool" consistency.

Entomophthorales infections are chronic, and virtually all that have been reported were diagnosed from tissue biopsy. The tissue should be minced aseptically into pieces of 1 to 2 mm because the fungi are difficult to isolate when the specimen is ground or macerated too vigorously.

The tissue fragments should be inoculated onto petri plates or slants of Sabouraud–2% dextrose agar containing only antibacterial antibiotics, and the cultures should be incubated at 25 to 30°C.

The typical *Entomophthorales* isolate grows in 3 to 4 days as a thin, flat, waxy colony at the site of inoculation, and the growth adheres tenaciously to the surface of the agar. The colony is gray to pale yellow and may develop radial furrows and a short, white, velvet-like "bloom" composed of conidia. The conidia are forcibly expelled and adhere to the top of the petri plate, giving the plate a "ground glass" appearance.

CULTURAL AND MICROSCOPIC CHARACTERISTICS

The zygomycetes are identified chiefly by their microscopic morphology. This is accomplished by preparing a lactophenol cotton blue wet mount of a small portion of the colony and observing it microscopically for characteristic fungal structures as described below.

There are no special biochemical reactions or nutritional requirements that aid in the identification of the pathogenic zygomycetes. Temperature requirements for optimum growth may suggest the identity of some isolates, particularly those of *Rhizopus*, since the "pathogenic" species often grow better at 37°C than the "saprophytic" ones. The monograph by Zycha et al. (26) can be recommended as a source of definitive information on species of *Mucorales*.

Identification of genera

***Rhizopus* spp. (*R. arrhizus*, *R. oryzae*, *R. microsporus*, *R. rhizopodiformis*).** The colony is rapid growing with voluminous white to gray aerial mycelium (Fig. 3). Mycelium is tenacious, coarse, and woolly. Hyphae are nonseptate and colorless. Sporangio-

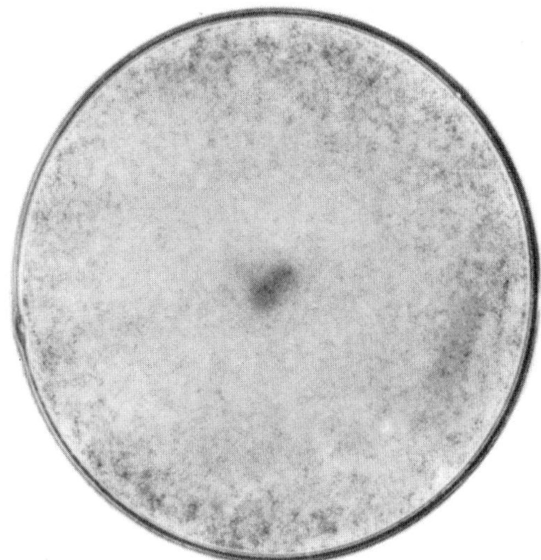

FIG. 3. Colonial morphology of *Rhizopus* sp. grown on Sabouraud agar at room temperature for 4 days. Note the voluminous aerial mycelium and production of sporangia (black dots).

phores are long, unbranched, and clustered at nodes opposite holdfasts (rhizoids) which form along a horizontal runner (stolon). Sporangia are dark walled and spherical and are filled with round hyaline spores. A columella is present (Fig. 4).

***Absidia* spp. (*A. corymbifera*).** Microscopically, rhizoids are present, but the sporangiophores arise between the nodes of the stolon instead of opposite the nodes as in *Rhizopus*. Sporangia are pear shaped, are filled with round to oval spores, and contain a columella (Fig. 5 and 6). Zygospores are similar to those produced by other species of *Mucorales* except that *Absidia* species form appendages which surround the zygospore (Fig. 6), whereas some other species do not. (Zygospores are usually not observed in culture because most species in this order are heterothallic.)

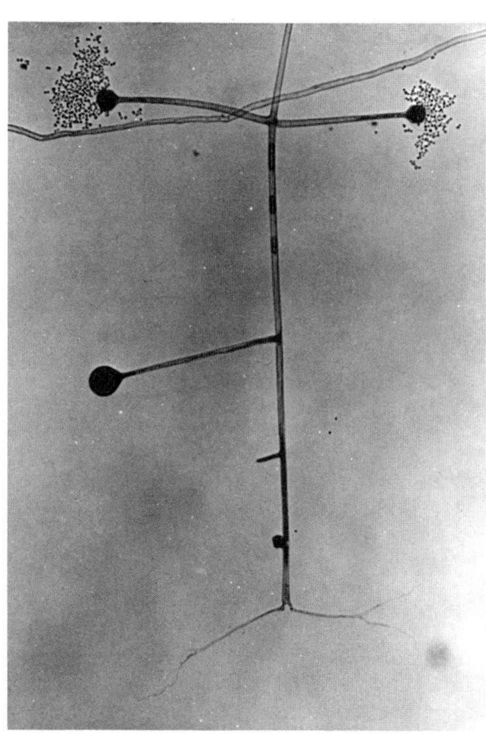

FIG. 5. *Absidia* sp. Large nonseptate hyphae with rhizoids at nodes; sporangiophores arise from stolon other than opposite the rhizoids. The sporangia are pyriform with columellae. ×147.

***Mucor* spp. (*M. pusillus, M. ramosissimus*).** The colony is rapid growing, forming a cottony surface which fills the culture plate in 5 to 7 days. Aerial mycelium is white, later becoming gray to brown. Grossly, the colony resembles that of *Rhizopus*.

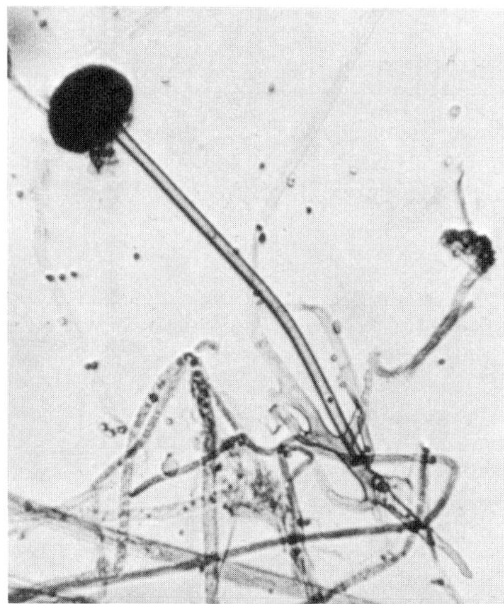

FIG. 4. *Rhizopus* sp. Nonseptate hyphae with rhizoids at the base of sporangiophore. Sporangium is black and spherical and contains a columella. ×400.

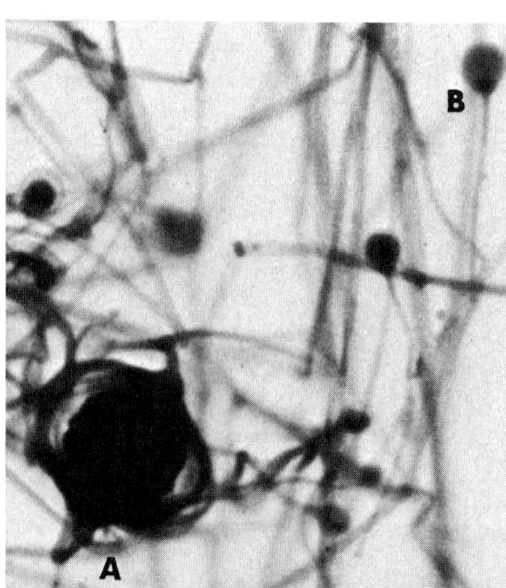

FIG. 6. *Absidia* sp. Note zygospore (A) with accessory appendages and sporangia (B). Lactophenol cotton blue stain. ×400. (Photomicrograph courtesy of B. H. Cooper.)

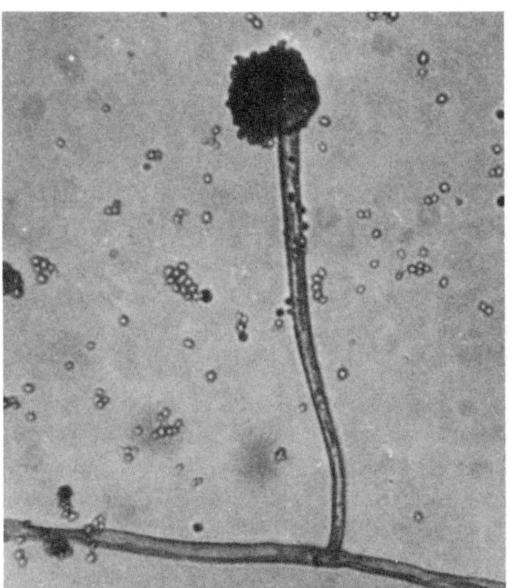

FIG. 7. *Mucor* sp. Nonseptate hyphae forming single or branched sporangiophores. No rhizoids are present. Sporangia contain columellae.

Microscopically, no rhizoids are present. Sporangiophores arise from nonseptate hyphae. Sporangia are spherical and have a columella (Fig. 7). Considerable variation occurs among the different species of *Mucor*. Some mycologists place some of the species in this genus into the genus *Rhizomucor* because rhizoids rarely occur on some species like *M. pusillus* (21).

Saksenaea vasiformis. The colony is rapid growing with white, loosely woolly aerial mycelium. Hyphae are hyaline, broad, nonseptate, and typical of species of *Mucorales*. However, no sporangia are produced on routine laboratory media (Sabouraud-dextrose agar, cornmeal agar, etc.) The characteristic flask-shaped sporangia (Fig. 8) and their rhizoids are produced by cutting a block from a plate of hay infusion agar containing the mycelial growth of the isolate and then floating the block in a plate of sterile distilled water (1). The plate is incubated at 25°C for 1 to 2 weeks.

Cunninghamella bertholletiae. Colonies are rapid growing, gray, and similar to those of *Mucor*. Hyphae are hyaline, broad, and nonseptate and may be filled with oil droplets. Elongated, erect, and highly branching sporangiophores terminate in globose vesicles on which are produced individual sporangiola or "conidia" (Fig. 9). No true sporangia are produced (25).

***Syncephalastrum* spp. (15).** The colonies are fast growing, white at first and turning gray with age. The large nonseptate hyphae (irregular cross walls form with age) are highly branched. The branched sporangiophores are erect and terminate in globose to ovoid vesicles which give rise to fingerlike tubular sporangia called merosporangia, each with a single row of sporangiospores (Fig. 10).

***Basidiobolus* spp. (*B. haptosporus*, synonym *B. meristosporus*).** The fungus grows rapidly as a flat, waxy gray to pale yellow colony which becomes covered with numerous uninucleate sporangia (conidia) and short, white aerial mycelia.

Microscopically, the hyphae are divided by septa into short elements called hyphal bodies. Large numbers of chlamydoconidia and zygospores are usually formed, beginning in the center of the colony. Zygo-

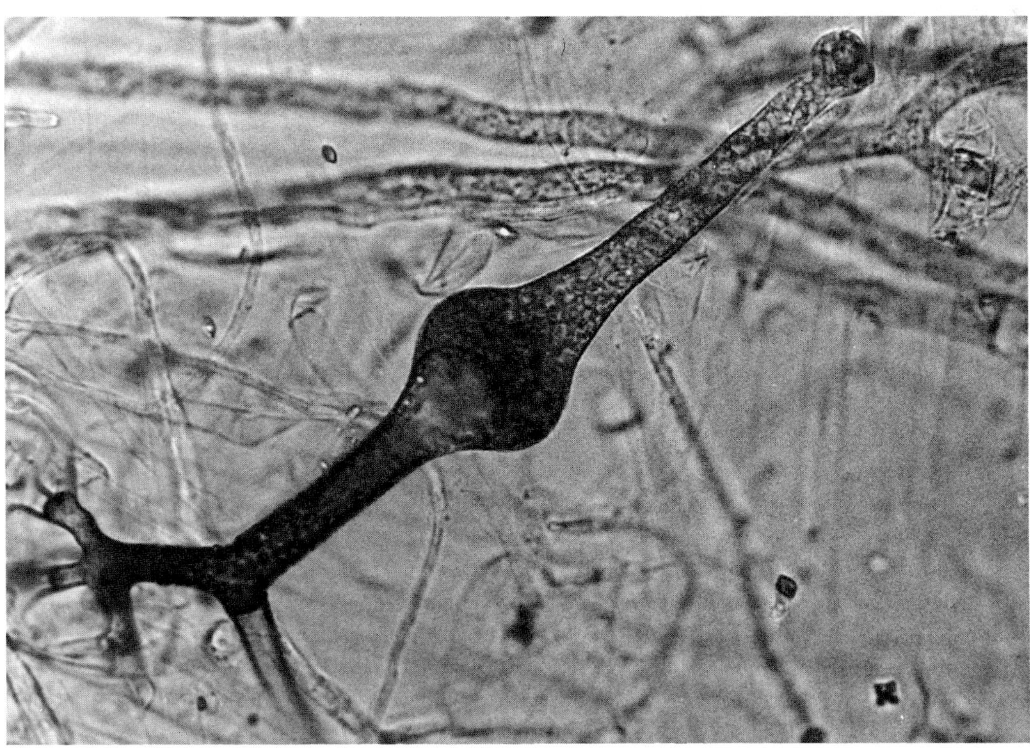

FIG. 8. *Saksenaea vasiformis*. Note flask-shaped sporangium containing sporangiospores. ×480. (Photomicrograph courtesy of B. H. Cooper.)

FIG. 9. *Cunninghamella* sp. Note sporangiophore, vesicles, and sporangiola (conidia). ×400.

spores are the identifying feature and appear as round, thick-walled, smooth structures having two protuberances or "beaks" on their surface (Fig. 11). The beaks are not part of the wall of the zygospore. Numerous globose uninucleate sporangia are borne singly on short club-shaped sporangiophores arising directly from the hyphae. The sporangium and a fragment of the sporangiophore are forcibly ejected at maturity and collect on the surface of the culture container (Fig. 12).

Temperature tolerance, zygospore wall morphology, and odor have been used as criteria to distinguish between strains of *Basidiobolus* isolated from humans and those isolated from nature (11).

Conidiobolus coronatus (Entomophthora coronata). The fungus develops as a flat, gray colony similar to *Basidiobolus*, but sporulation is usually heavier (Fig. 13).

Microscopically, the hyphae are sparingly septate and produce many chlamydoconidia. Zygospores, if present, are not obvious. Numerous globose true conidia are borne singly on short, slender conidiophores. These are forcibly expelled at maturity without an attached hyphal fragment (Fig. 14). The conidia differ from those of *Basidiobolus* by having a prominent papilla and producing multiple secondary conidia in the form of a "corona," from which the species derives its name (Fig. 15). Under unfavorable conditions the conidia may develop short hairlike filaments over their surface.

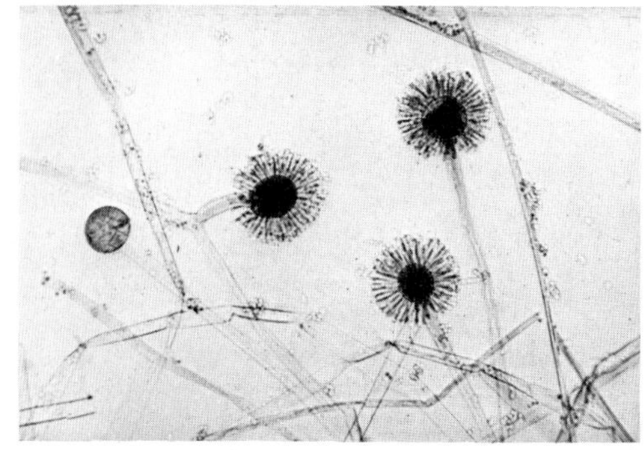

FIG. 10. *Syncephalastrum* sp. Large nonseptate hyphae produce sporangiophores that end in vesicles which give rise to tubular sporangia (merosporangia) with sporangiospores in a single row. ×100.

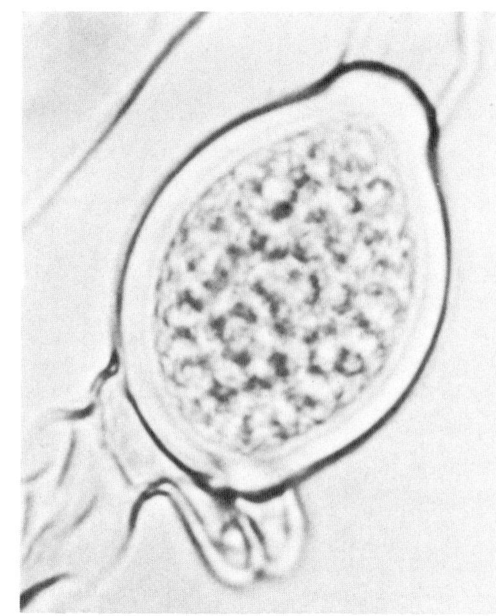

FIG. 11. *Basidiobolus* sp. Smooth-walled intercalary zygospore showing remnants of copulatory tubes (beaks). ×400.

SEROLOGICAL METHODS

No practical serological methods have been developed for the diagnosis of zygomycosis. In patients with proven mucormycosis, serum precipitins against the etiological agents are rare (13). Skin test antigen obtained from the zygomycetes is not specific enough to be of diagnostic value.

HISTOLOGICAL IDENTIFICATION

Since zygomycoses are frequently diagnosed from biopsy sections, it is important for microbiologists

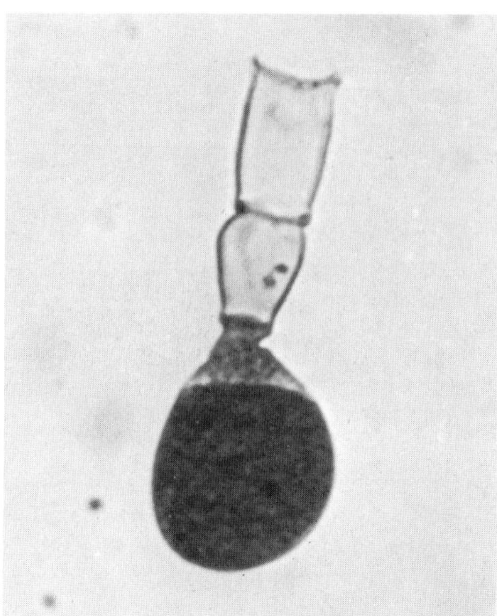

FIG. 12. *Basidiobolus* sp. Smooth, globose, uninucleate sporangium which has been forcibly expelled from the sporangiophore. A fragment of the sporangiophore is attached. ×400.

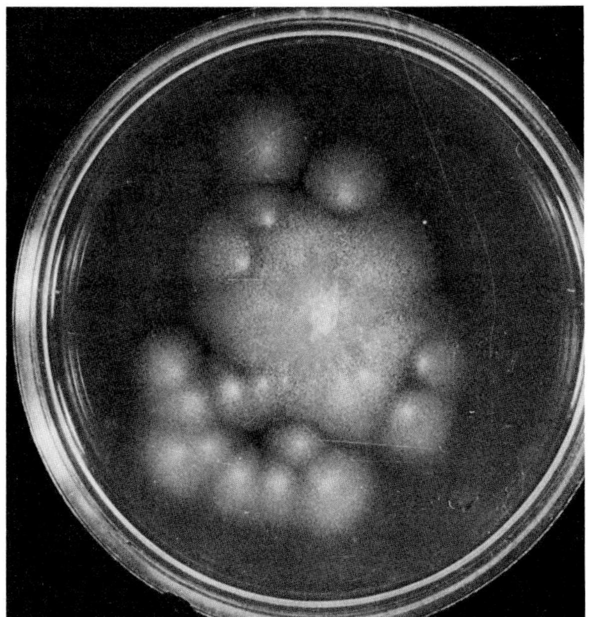

FIG. 13. Colonial morphology of *Conidiobolus (Entomophthora) coronatus* grown on Sabouraud agar at room temperature for 5 days. Note presence of short aerial mycelia, bloom, and ground-glass appearance.

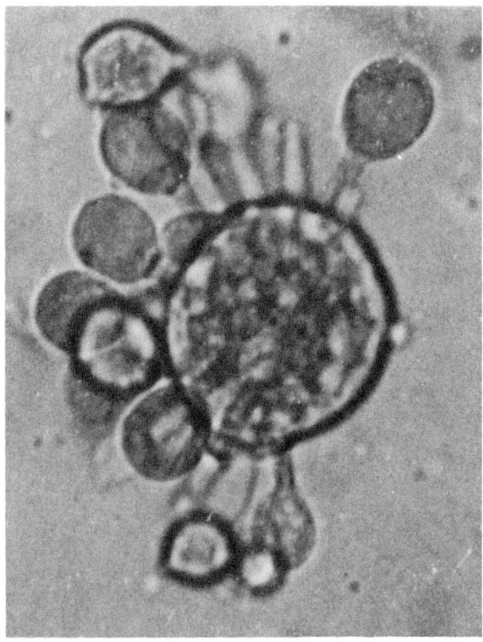

FIG. 15. *Conidiobolus (Entomophthora) coronatus*. Production of secondary globose conidia by multiple replication, showing a corona effect. ×400.

and pathologists to be familiar with their morphology in tissue. The hyphae are easily observed in tissue sections stained with hematoxylin and eosin (3). Special stains for fungi such as the periodic acid-Schiff stain or the Grocott-Gomori methenamine silver stain may be used, if desired, but are not essential for recognition or for identification.

The typical tissue reaction in *Mucorales* infection is different from the tissue reaction produced by the *Entomophthorales*, and this difference is distinctive enough to permit a diagnosis by histopathological methods (3). Species identification, of course, cannot be made from tissue sections.

Mucorales infections in tissues show abundant broad, branched, usually aseptate hyphae 15 to 20 μm in diameter and up to 200 μm in length. Folds in the hyphal walls may resemble septa (Fig. 16).

The tissue reaction usually is acute, showing diffuse infiltration of polymorphonuclear leukocytes and extensive necrosis and edema. The fungus has a marked tendency to invade blood vessels, resulting in vascular thrombosis and perivascular infarction (Fig. 17).

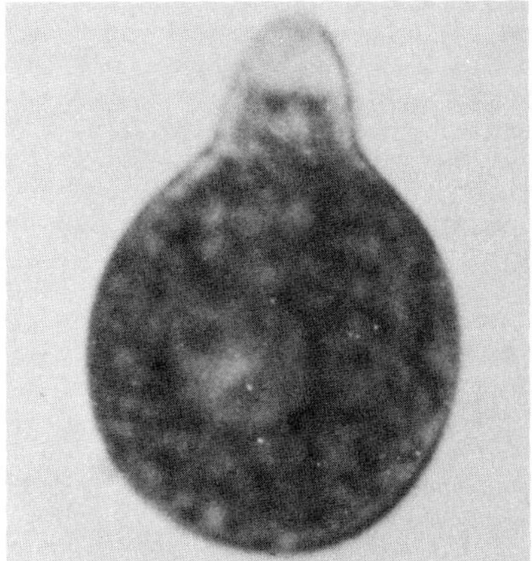

FIG. 14. *Conidiobolus (Entomophthora) coronatus*. Globose, smooth conidium showing prominent papilla. The true conidium has been forcibly expelled without hyphal attachment. ×400.

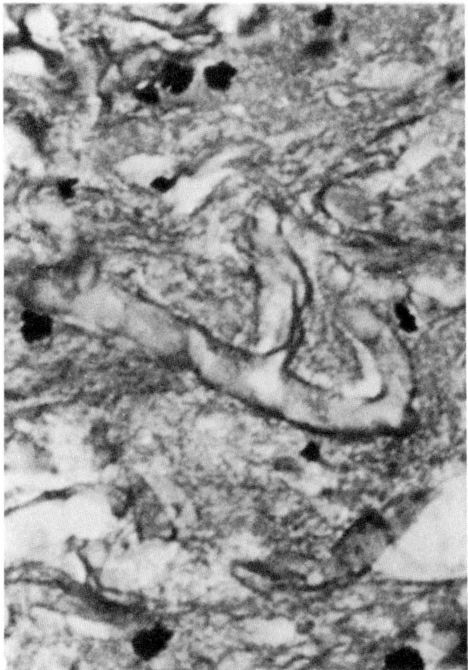

FIG. 16. Mucormycosis. Tissue section (hematoxylin and eosin) showing broad, nonseptate hyphae. Tissue reaction is one of necrosis. ×400.

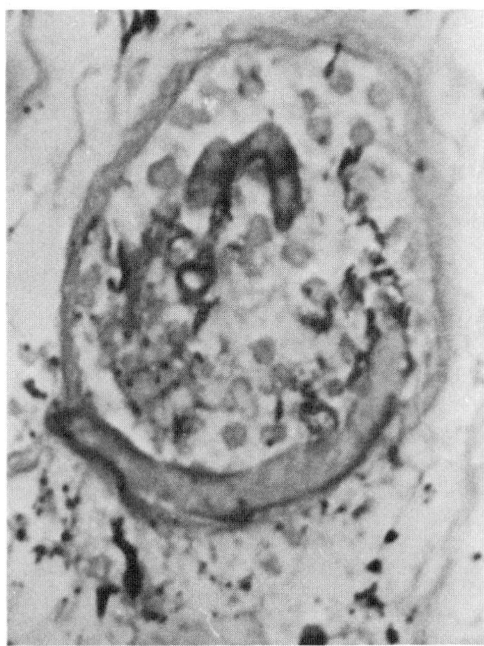

FIG. 17. Mucormycosis. Tissue section (hematoxylin and eosin) showing invasion of blood vessel by broad, nonseptate hyphae. ×400.

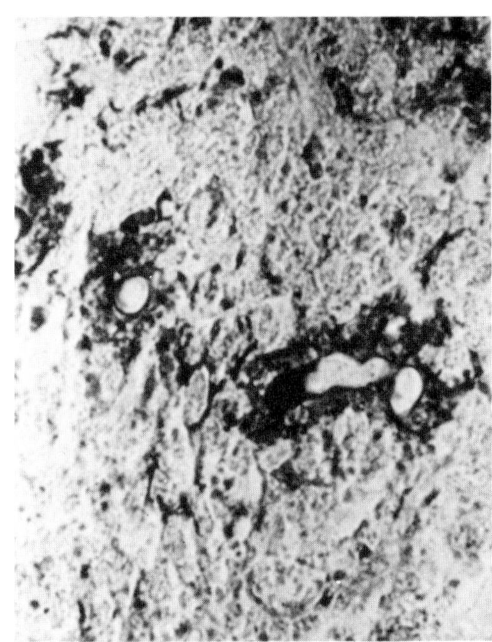

FIG. 18. Entomophthoromycosis. Tissue section (hematoxylin and eosin) showing transverse section of hyphal fragments. Tissue reaction is chronic and granulomatous, with foreign body giant cells and microabscesses. ×400.

Entomophthorales infections in subcutaneous tissue show microabscesses containing fragments of degenerating, thin-walled, broad hyphae surrounded by eosinophilic necrotic precipitate similar in appearance to the Hoeppli phenomenon (3). The tissue reaction is an intense, granulomatous reaction composed of eosinophils, neutrophils, and foreign body giant cells (Fig. 18). The hyphal fragments stain poorly with any stain. Differentiation of *Basidiobolus* from *Conidiobolus* is not possible in tissue sections. Also, infection with the genus *Mortierella* produces a chronic eosinophilic tissue reaction similar to that produced by infection with the *Entomophthorales*.

Rhizopus and *Mucor* must be differentiated in tissue sections from *Aspergillus* (see Chapter 54) and *Candida*. The hyphae of *Aspergillus* and *Candida* do not take the hematoxylin and eosin stain as readily as do the zygomycetes, and usually special histological stains, e.g., the periodic acid-Schiff stain or the Grocott-Gomori methenamine silver stain, are needed to observe them. The hyphae of *Aspergillus* have numerous septa, are narrow (5 μm), and have smooth parallel walls. Usually in tissue sections one sees prominent hyphae branching at acute angles and growing in a radiating pattern (21). *Candida* spp. growing in tissue produce both true septate hyphae and club-shaped pseudohyphae. Often, tissue sections show a mixture of yeast cells and hyphal elements. The hyphae are narrow (4 μm wide).

EVALUATION

Antemortem diagnosis of mucormycosis depends upon close collaboration between the laboratory and the physician (24). Because of the acute nature of mucormycosis, treatment often must be started before there is cultural confirmation of the diagnosis. Therefore, the presence in purulent nasal exudates of broad, nonseptate branching hyphae or of hyphal invasion in biopsies of necrotic tissue is pathognomonic for mucormycosis. Less urgent, but equally important, is the finding in granulomatous inflammation of hyphal fragments surrounded by eosinophilic debris. This suggests a diagnosis of entomophthoromycosis. Whenever possible, the etiological agent should be isolated so that the fungus involved can be identified as to genus and species.

LITERATURE CITED

1. **Ajello, L., D. F. Dean, and R. S. Irwin.** 1976. The zygomycete *Saksenaea vasiformis* as a pathogen of humans with a critical review of the etiology of zygomycosis. Mycologia **68:**52–62.
2. **Alexopoulos, C. J., and C. W. Mims.** 1979. Introductory mycology, 3d ed. John Wiley and Sons, New York.
3. **Baker, R. D.** 1971. Human infection with fungi, actinomycetes and algae. Springer-Verlag, New York.
4. **Bottone, E. J., I. Weitzman, and B. A. Hanna.** 1979. *Rhizopus rhizopodiformis:* emerging etiological agent of mucormycosis. J. Clin. Microbiol. **9:**530–537.
5. **Bruck, H. M., G. Nash, F. D. Foley, and B. A. Pruitt.** 1971. Opportunistic fungal infection of the burn wound with phycomycetes and aspergillus. A clinical pathologic review. Arch. Surg. **102:**476–482.
6. **Chick, E. W., A. Balows, and M. L. Furcolow (ed.).** 1975. Opportunistic fungal infections. Proceedings of the second international conference. Charles C. Thomas Publisher, Springfield, Ill.
7. **Dean, D. F., L. Ajello, R. S. Irwin, W. K. Woelk, and G. J. Sharulis.** 1977. Cranial zygomycosis caused by *Saksenaea vasiformis*. J. Neurosurg. **46:**97–103.
8. **Doworzack, D. L., A. S. Pollock, G. R. Hodges, W. G. Barnes, L. Ajello, and A. Padahye.** 1978. Zygomycosis of the maxillary sinus and palate caused by *Basidiobolus haptosporus*. Arch. Intern. Med. **138:**1274–1276.
9. **Emmons, E. W., C. H. Binford, J. P. Utz, and K. L.**

Kwon-Chung. 1977. Medical mycology, 3rd ed., p. 254–284. Lea & Febiger, Philadelphia.

10. **Gilbert, E. F., G. H. Khoury, and R. S. Pore.** 1970. Histopathological identification of entomophthora-phycomycosis. Deep mycotic infection in an infant. Arch. Pathol. **90:**583–587.
11. **Greer, D. L., and L. Friedman.** 1966. Studies on the genus *Basidiobolus* with reclassification of the species pathogenic for man. Sabouraudia **4:**231–241.
12. **Herstoff, J. K., H. Bogaars, and C. J. McDonald.** 1978. Rhinophycomycosis enthomophthorae. Arch. Dermatol. **114:**1674–1678.
13. **Jones, K. W., and L. Kaufman.** 1978. Development and evaluation of an immunodiffusion test for diagnosis of systemic zygomycosis (mucormycosis): preliminary report. J. Clin. Microbiol. **7:**97–103.
14. **Kamalam, A., and A. S. Thambiah.** 1978. Lymph node invasion by *Conidiobolus coronatus* and its spore formation in vivo. Sabouraudia **16:**175–184.
15. **Kamalam, A., and A. S. Thambiah.** 1980. Cutaneous infection by *Syncephalastrum*. Sabouraudia **18:**19–20.
16. **Kien, T. E., F. Edwards, D. Armstrong, P. P. Rosen, and I. Weitzman.** 1979. Pneumonia caused by *Cunninghamella bertholletiae* complicating chronic lymphatic leukemia. J. Clin. Microbiol. **10:**140–145.
17. **Lehrer, R. I.** 1980. Mucormycosis (review). Ann. Intern. Med. **93:**93–108.
18. **Martinson, F. D.** 1972. Clinical, epidemiological and therapeutic aspects of entomophthoromycosis. Ann. Soc. Belg. Med. Trop. **52:**329–342.
19. **Miller, R. I., D. Wold, W. A. Lindsay, R. E. Beadle, J. J. McClude, J. R. McClude, and D. J. McCoy.** 1983. Complications associated with immunotherapy of equine phycomycosis. J. Am. Vet. Med. Assoc. **182:**1227–1229.
20. **Reinhardt, F. J., W. Kaplan, and L. Ajello.** 1970. Experimental cerebral zygomycosis in alloxan-diabetic rabbits. I. Relationship of temperature tolerance of selected zygomycetes to pathogenicity. Infect. Immun. **2:**404–413.
21. **Rippon, J. W.** 1982. Medical mycology, 2nd ed. The W. B. Saunders Co., Philadelphia.
22. **Schwartz, J. N., E. H. Donnelly, and G. K. Klintworth.** 1977. Ocular and orbital phycomycosis. Surv. Ophthalmol. **22:**3–28.
23. **Torell, J., B. H. Cooper, and N. G. P. Helgeson.** 1981. Disseminated *Saksenaea vasiformis* infection. Am. J. Clin. Pathol. **76:**116–121.
24. **Walker, D. H., and M. R. McGinnis.** 1982. Opportunistic fungal infection: what the clinician, pathologist and mycologist can accomplish if they work together. Clin. Lab. Med. **2:**407–413.
25. **Weitzman, I., and M. Y. Crist.** 1979. Studies with clinical isolates of *Cunninghamella*. I. Mating behavior. Mycologia **71:**250–255.
26. **Zycha, H., R. Siepmann, and G. Linnemann.** 1969. *Mucorales*. Verlag von J. Cramer, Lehre, Federal Republic of Germany.

Chapter 54

Aspergillus Species and Other Opportunistic Saprophytic Hyaline Hyphomycetes

FRANK SWATEK, CARLYN HALDE, MICHAEL J. RINALDI, AND H. JEAN SHADOMY

The occurrence of infection caused by fungi generally considered contaminants has increased to the point that the clinical laboratory now must consider isolates of such fungi with suspicion. Repeated isolation may be of clinical significance. These "opportunistic pathogens" include representatives of every class of the Kingdom *Fungi*. Newer therapy, as well as newly recognized diseases such as AIDS (acquired immune deficiency syndrome), has expanded the incidence of the fungus-infected compromised host.

The establishment of the etiology of disease by opportunistic fungal pathogens requires the laboratory to follow a standard protocol for the culture of fungi as well as to demonstrate fungal elements in tissues or body fluids. Isolation alone is not sufficient to establish that a fungus is a pathogen.

The protocol in the clinical laboratory for the detection, culture, and identification of these agents includes: (i) direct demonstration of mycotic cells within tissue or body fluids; (ii) isolation of the fungus in pure culture from multiple specimens; and (iii) identification by conventional procedures, such as observation of conidial formation.

The clinical conditions caused by the rarer opportunistic fungi are extremely varied; each case must be considered individually. Infection may be the result of a postsurgical event, be associated with medical devices (heart valves, catheters, contact lenses, or intravenous or spinal devices), be caused by chemotherapeutic intervention (e.g., cytotoxic, steroid, or immunosuppressive drugs), or be associated with poorly understood physiological problems. The significance of toxic substances produced by most opportunists is, as yet, unknown in the disease processes.

Opportunistic fungi are widely distributed throughout the Kingdom *Fungi*. Opportunistic fungi may vary with both the site of infection and the geographic location. However, *Aspergillus* spp. must be considered potential pathogens from most specimen sites and in all geographic regions.

The organisms discussed in this chapter do not include opportunistic yeasts (Chapter 49), dematiaceous organisms (Chapter 52), organisms causing eumycotic mycetomas (Chapter 51), or the zygomycetes (Chapter 53). *Trichosporon* and *Geotrichum* species are discussed with the yeasts.

ASPERGILLUS SPECIES

The spectrum of disease caused by pathogenic strains among the approximately 200 *Aspergillus* species is broad. Illnesses vary from allergic reactions after the development of hypersensitivity to aspergilli to different types of clinical problems subsequent to colonization of the surfaces of various body sites, especially colonization of vulnerable areas within the respiratory tract. Other colonized areas include ear canal, burn eschar, nail, and, rarely, other sites. Serious illness results from damage done as these molds invade tissue (24). In addition, strains of some aspergilli produce toxins which, when ingested, cause serious pathologic conditions ranging from acute or chronic toxicoses (e.g., aflatoxicosis) to epidemiologically suspect hepatocarcinoma. Species of aspergilli are found by clinical laboratorians as the second most frequent cause of opportunistic fungal disease (24).

Aspergilli are ubiquitous fungi; they can be found growing on most organic materials and are commonly found on grain and decaying vegetation, as well as in soil. Conidia often become airborne in large numbers; no area open to air is free from aspergilli. Contamination of cutaneous and respiratory tract lesions or of any step during the processing of clinical specimens may occur and increases the difficulty in judging whether an *Aspergillus* isolate is responsible for the disease. Of the many species of *Aspergillus*, only four are commonly encountered as a cause of disease: *A. fumigatus*, *A. flavus*, *A. niger*, and *A. terreus*. Infrequently, other species classified in this genus may incite illness, particularly in an immunocompromised host or a hypersensitive individual.

Nomenclature, morphology, and taxonomy

Molds of the genus *Aspergillus* most often reproduce in the asexual form (anamorph). Strains of some species also have the ability to develop a sexual form (teleomorph).

The anamorph is the most common growth form of *Aspergillus* species seen in clinical laboratories. Typical representations are shown in Fig. 1. As an *Aspergillus* colony matures, typical asexual reproductive structures develop. A specialized hyphal cell, called the foot cell, gives rise to an erect, usually nonseptate, stalklike branch termed the conidiophore (structure bearing conidia). It usually is hyaline, but may be greenish to brown in some species, and varies greatly in length among different species. The tip of the conidiophore expands, developing into a swollen vesicle which assumes a domelike or other shape characteristic of a given species (Fig. 2A through D). The surface of the vesicle may be covered, entirely or partially, with a single layer of flask-shaped cells called phialides. This type of *Aspergillus* species is termed uniseriate. In biseriate aspergilli, the vesicle is similarly covered, but with two layers of cells—the metulae arise first on the surface of the vesicle and serve as supporting cells for the phialides. Both uni- and biseriate forms may be seen in some species. Phialides are the conidiogenous cells which form dry, one-celled, usually globose phialoconidia. These are formed one at a time by a budding-out process from within the phialide but remain attached, resulting in a long chain of conidia emerging from each phialide. Metulae and phialides may be so arranged on the vesicle that conidial chains radiate from the conidio-

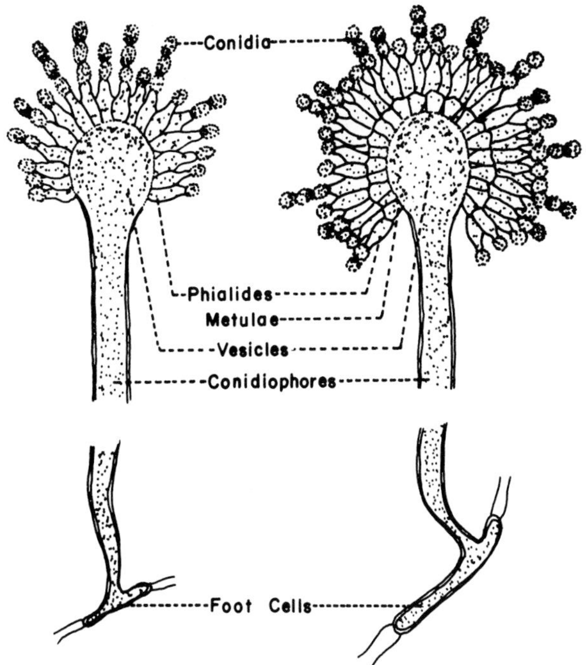

FIG. 1. Morphological aspects of an *Aspergillus* conidiophore. (Left) A uniseriate species. (Right) A biseriate species. Reprinted with permission from The Williams & Wilkins Co. (22) and the University of Chicago Press (24).

phore or develop an upward columnar form. Once disturbed, the conidia separate easily and may become airborne.

Some *Aspergillus* species develop cleistothecia, which consist of spherical structures whose outer walls are composed of interwoven hyphae and whose interiors are filled with asci and ascospores, the sexual structures. Cleistothecia infrequently are observed in the laboratory; when present, they are helpful in identifying an isolate.

Other useful morphological structures are Hülle cells (Fig. 2E). These unusually thick-walled single cells usually are present in clusters and are most often seen in the laboratory associated with cleistothecia of the *A. nidulans* group, but they may be found in colonies of some other groups.

The large number of species within the genus *Aspergillus* makes identification difficult. To overcome this difficulty, the aspergilli have been placed into groups, the members of each group sharing many basic common characteristics. Each group is named after a representative species within it. An exception to this is the Glaucus group, as there is no *A. glaucus*. Some species may be placed in more than one group. Group designations, when determined, may be reported if the laboratorian is unable to identify a specific isolate. Reference laboratories are available to identify isolates when the need for an exact species name is justified.

Collection, transport, and storage of specimens

Collection. Clinical specimens are collected with care by using standard procedures as described in Chapter 47. Biopsy material is not likely to be contaminated when collected and handled carefully. Early-morning sputum samples may be useful in establishing a diagnosis, but they must be collected carefully with regard to exposure to the air. Skin lesions and infected nails should be washed with soap and water before the specimen is collected. Swabs taken from mucous membranes and skin lesions are not recommended and can be useful only if made in duplicate with appropriate labeling of the second specimen. The presence or absence of the same species in the second specimen may aid in judging whether an isolate is a possible pathogen or a contaminant. The isolation of *Aspergillus* species from nail scrapings can be significant if hyphae are observed by direct microscopic examination of the specimen and if a dermatophyte or other recognized pathogen is not isolated after multiple cultures are performed. Blood cultures infrequently are useful in establishing a diagnosis of *Aspergillus* infection, and single isolations from blood cultured in broth should be questioned. The lysis-centrifugation method of culturing blood permits a more sensitive detection of molds causing fungemia and allows for a more confident designation of isolates as contaminants or significant pathogens based on the number of colonies that grow and their location on plates in relation to where the specimen was streaked (Chapter 47).

Transport and storage. All specimens should be covered in sterile containers during transport and processed immediately. It is important to note that refrigeration may prevent the rapid growth of most extraneous fungi and bacteria for 1 to 2 days, but the delay can be detrimental to the aspergilli. The freezing of clinical specimens may kill the hyphae of *Aspergillus* species, but the specimen still may be used for direct microscopic examination. The addition of antibacterial antimicrobial agents will reduce the growth of most bacteria for a few days; however, the addition of these agents may prevent the subsequent growth of aspergilli (but not the growth of all extraneous fungi).

Direct microscopic examination

Clinical materials submitted to the clinical laboratory for microscopic examination are handled by the same procedures as described for all mycoses in Chapter 47. Both a 10% KOH mount and a Gram stain should be performed. Although rapid periodic acid-Schiff and methenamine silver nitrate staining procedures are especially useful for detecting the presence of small fragments of hyphae in a clinical specimen, these methods are time consuming and costly. When present in a specimen, the hyaline hyphae of aspergilli usually are closely septate without constrictions, are 2.5 to 8.0 μm (average, 3.0 μm) wide, and may range from a few fragments up to a small colony (Fig. 2F). In the latter, hyaline hyphae branching at a 45° angle are seen; this characteristic of dichotomously branching parallel hyphae in tissues is shared with many other filamentous fungi causing disease, e.g., *Fusarium*, *Penicillium*, and *Pseudallescheria* species (see Chapter 51).

Hyphal walls usually are parallel and smooth, but occasionally individual cells may be swollen. The cytoplasm of a hyphal tip is refractile, differing from that in older hyphal cells, which are filled with a central vacuole. Occasionally, no septa may be seen,

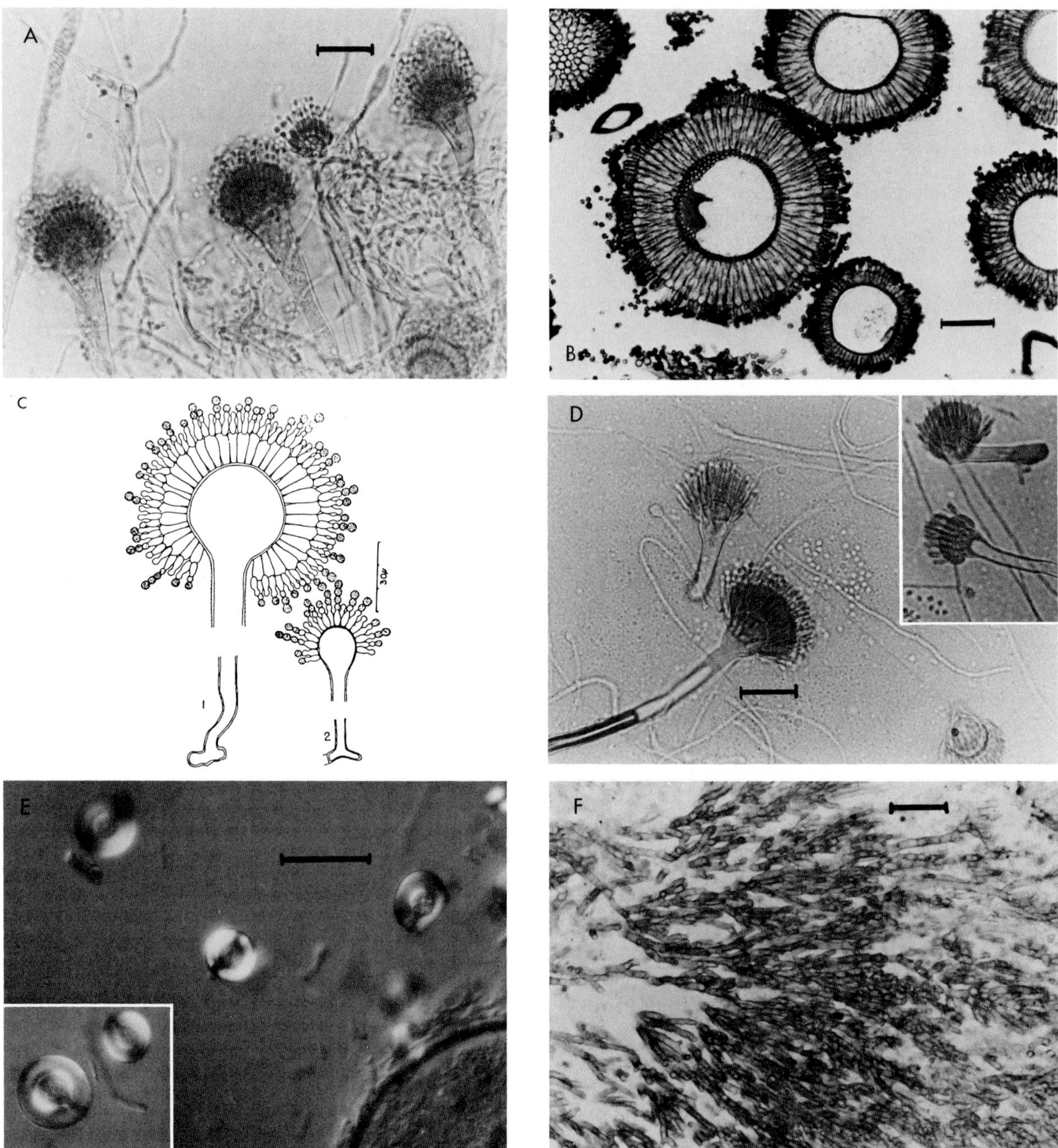

FIG. 2. (A) *Aspergillus fumigatus.* Conidiophores. (B) *Aspergillus niger.* Cross section through conidiophores within lung cavity. (C) *Aspergillus flavus.* Diagram of biseriate (1) and uniseriate (2) conidiophores. (D) *Aspergillus terreus.* Conidiophores. Inset: Two more conidiophores. (E) *Aspergillus nidulans.* Thick-walled Hülle cells near a cleistothecium. Inset: Two more such Hülle cells. (F) *Aspergillus fumigatus.* Hyphae in lung tissue. Hematoxylin-eosin stain. Bars, 15 μm except where noted.

causing confusion with a coenocytic zygomycotic hypha. Whenever *Aspergillus* hyphae grow in vivo in contact with air (cavity in the lung, upper respiratory tract, ear canal, or skin or burn eschar), conidiation is possible, and typical conidiophores and conidia may be present in a clinical specimen (Fig. 2B).

Mycology

The isolation of aspergilli from specimens contaminated by bacteria is accomplished on media containing antimicrobial agents, such as chloramphenicol, gentamicin, or penicillin plus streptomycin. Antifungal agents such as cycloheximide (Actidione; Upjohn Co., Kalamazoo, Mich.) are inhibitory to most aspergilli, as well as to most of the other opportunistic fungi. Media that contain cycloheximide include Mycobiotic agar (Difco Laboratories, Detroit, Mich.) and Mycosel agar (Becton Dickinson and Co., Cockeysville, Md.) and should not be used for this group of organisms. Media that can be used for isolation include Sabouraud agar, brain heart infusion agar, and 2%

malt extract agar. Aspergilli will grow on many other substrates. Some species, not often encountered in a medical setting, require a high osmotic medium for growth.

Standard mycological procedures are used for processing and placing clinical material upon media (see Chapter 47). If possible, duplicate or several different media should be used to allow possible growth from small amounts of fungal material in the clinical specimen. Cultured materials are incubated at both 25 and 35 to 37°C and should not be discarded for lack of growth for at least 4 weeks.

Identification

Table 1 presents characteristics used for the species identification of the most frequently encountered *Aspergillus* pathogens. Other aspergilli which have been recognized less frequently as agents of aspergillosis or which occur commonly in the environment are listed on the table.

The microscopic characteristics that suggest *A. fumigatus* may be found in a culture growing on isolation medium, but other species require transfer to special media to observe their diagnostic features. Transfer of an isolate to Czapek-Dox agar and 2% malt extract agar with incubation at 25°C allows most aspergilli to be identified by the keys found in the standard monograph of Raper and Fennell (22) and in other recent taxonomic treatments of the genus (4, 5; A. A. Musallam, Ph.D. thesis, Rijksuniversiteit Utrecht and Centraalbureau voor Schimmelcultures, Baarn, The Netherlands, 1980). The *Aspergillus* differential medium of Bothast and Fennell (2) is a useful screening medium for the detection of pigment production by *A. flavus* and related species. Temperature studies to determine the ability to grow at 50°C confirm the identification of *A. fumigatus*; no species resembling *A. fumigatus* grows above 45°C. Some infrequently encountered species produce their typical conidia best at 20°C, and others produce conidia best at 30°C; some aspergilli will not grow at 37°C but can colonize the ear canal and burn eschar and skin (including nails).

Exoantigen tests have proved useful for the rapid and specific identification of nonconidiating isolates of *A. fumigatus* (12). This procedure for the identification of *Aspergillus* spp. holds exciting promise. Developing hybridoma technology offers further hope for the rapid immunoidentification of *Aspergillus* species.

Serologic diagnosis

Serologic tests have been developed and may be helpful in the diagnosis of some forms of aspergillosis. Techniques which appear to be most reliable are the immunodiffusion and counterimmunoelectrophoresis tests for detecting antibody and the radioimmunoassays and enzyme immunoassays for detecting circulating antigens in immunocompromised patients with invasive aspergillosis (see Chapter 96).

Identification of aspergilli in tissue by fluorescent antibody. At times, only branching hyphae are seen in tissue sections of a biopsy or autopsy specimen, and no mold is recovered in culture. Such specimens may be tested by *Aspergillus*-specific immunofluorescent antibody when there is a need for fungal identification. Such procedures are accomplished in reference laboratories and may be applied to viable and nonviable fungi in culture as well as in clinical materials or tissue sections. The immunofluorescent-antibody technique, like other mycoserologic procedures, is not entirely specific (12).

Antifungal susceptibility

Information concerning procedures for the determination of in vitro susceptibilities of aspergilli to currently available antifungal antimicrobial agents is given in Chapter 104. However, little information is available that correlates the results of in vitro tests with the outcome of therapy. Correlative problems are amplified in immunocompromised patients with invasive aspergillosis. At present, antifungal susceptibility testing is best accomplished at reference centers.

PENICILLIUM SPECIES

Members of the genus *Penicillium* are among the most prolific and ubiquitous groups of fungi. They commonly are isolated in the clinical laboratory on routine media and at a wide range of incubation temperatures (20 to 45°C). Infections associated with different stages of debilitation and from numerous anatomical sites have been reported. Repeated isolation and the demonstration of fungal elements in tissue are mandatory before *Penicillium* species can be considered opportunistic pathogens, although the literature contains many unsuccessful attempts to ascribe various disease processes to members of this genus. Unlike the aspergilli, which are known to cause invasive disease, the penicilli are much less virulent. Repeated isolation from sputum, skin scrapings, and other specimens, even in large numbers, is not proof of infection. However, the literature does contain bona fide examples of penicilli causing infection, such as in pulmonary granuloma, endocarditis, keratitis, and hypersensitivity pneumonitis.

For example, in one case a pulmonary nodule 2 cm wide detected by a routine chest X ray was removed from an immunocompetent patient and yielded only *Penicillium* sp. in culture. Thick, branching hyphae were seen in partially calcified tissue. No other fungal or bacterial agents were detected. In this case the fungus actually may have been growing saprophytically on necrotic tissue (mere colonization) and not causing an invasive disease (16).

Another case of penicilliosis in an immunocompetent patient was probably acquired after penetration of a sliver of metal deep into the eye. Repeated isolations of *P. chrysogenum* were obtained on blood agar and chocolate agar plates (9). Figure 3a shows the characteristic detail of this organism.

Several bona fide cases of penicilliosis resulted in endocarditis after open-heart surgery for the insertion of mitral or aortic valve prostheses. All patients were immunocompromised; the prosthetic valves may have been contaminated before surgery, or the fungus might have been introduced during the surgical procedure (6, 10).

Hypersensitivity pneumonitis due to *P. casei* (Fig. 3b) has been documented as an occupationally ac-

TABLE 1. Characteristics of a representative species of some *Aspergillus* groups grown on Czapek-Dox agar

Group and organism	Seriation Uni	Seriation Bi	Colony color	Conidiophore	Vesicle
Fumigatus group *A. fumigatus*	+		White to green to gray-green; slate gray with age; reverse side variable in color	Length, up to 300 μm (rarely to 500 μm); width, 5–8 μm; smooth walled; greenish, especially in upper portion	Dome shaped; conidiophore gradually enlarges, merging into vesicle; 20–30 μm in diameter; phialides on upper half only, with axes parallel to that of conidiophore
Flavus group *A. flavus*	+	+	Yellow to yellowish green to green with age	Length, 400–850 μm (rarely to 2.5 mm); uncolored; thick walled; coarsely roughened	Elongate, becoming subglobose to globose; 10–65 (commonly 25–45) μm in diameter
Niger group *A. niger*		+	White, becoming black; reverse side occasionally pale yellow	Length, 1.5–3.0 mm; width, 15–20 μm; smooth walls; colorless with brownish shade on upper half	Globose; most are 45–75 μm in diameter
Terreus group *A. terreus*		+	Cinnamon buff to brown, rarely orange-brown	Length, 100–250 μm; width, 4.5–6 μm; flexous; smooth walled; colorless	Hemispherical or domelike, merging into conidiophore; 10–16 μm in diameter
Versicolor group *A. versicolor*		+	Variable, white to yellow, orange-yellow, tan to yellowish green	Length, up to 500–700 μm; width, 5 μm, merging into funnel shape at vesicle; smooth walled; uncolored or yellowish	Hemispherical or semielliptical; 12–16 μm in diameter
Glaucus group *A. repens*	+		Green to yellow-green to gray-green; orange-yellow areas of cleistothecia with age	Length, 0.5–1.0 mm; smooth walled; uncolored; broadens toward the vesicle	Domelike, hemispherical, 25–40 μm in diameter
Clavatus group *A. clavatus*	+		Blue-green	Length, 1.5–3.0 mm; width, 20–30 μm; smooth walled; uncolored; merging into vesicle	Clavate; fertile over area up to 250 μm long; 40–60 μm in diameter
Nidulans group *A. nidulans*		+	Dark green in conidial forms; buff to yellow in primarily cleistothecial forms; reverse side purplish red	Length, 60–130 (most are 75–100) μm; width, 2.5–3 μm, increasing to 3.5–5 μm near vesicle; smooth walls; brown	Hemispherical; 8–10 μm in diameter
Flavipes group *A. flavipes*		+	White to buff to darker; color due to off-white conidial mass and brownish conidiophores	Length, 500–800 μm (2–3 mm at margin in old colonies); thick walled; yellow to brown; smooth to faintly granular (occasionally disklike concretions on surface)	Subglobose to vertically elongate; up to 18 by 25 μm

quired disease in an entomologist (28). An unusual infection caused by *P. marneffei* (Fig. 3c) was reported in a patient with Hodgkin's disease (8). The organism was isolated from spleen tissue and grew well on Sabouraud medium at 25°C and on blood agar, as well as in thioglycolate broth at 37°C. The primary isolates appeared yeastlike on these media and resembled *Geotrichum candidum* when they were examined microscopically (Fig. 3d). Transfer to Sabouraud agar at 25°C produced membranous colonies. Further subcultures yielded colonies resembling the yeast phase of *Blastomyces dermatitidis*, with yeastlike cells observed microscopically. In infected tissues, the fungus appeared as ovoid cells dividing by fission rather than by true bud formation. The appearance of *P. marneffei* in tissue is unique; occasionally it appears as round cells within macrophages (Fig. 3). This appearance could possibly be confused with kala azar or histoplasmosis.

Mycology

All *Penicillium* species grow readily on standard mycological media. Initial growth usually is rapid, with white growth (occasionally yeastlike) changing with maturation to colors characteristic of each spe-

TABLE 1—*Continued*

Conidial head	Conidia	Cleistothecia	Hülle cells	Comment
Strongly columnar, compact	Globose to subglobose, elliptical in some isolates; echinulate to rarely smooth; most are 2.5–3.0 μm in diameter			Grows well at 50°C; one of the most commonly occurring molds in nature; most common human pathogen
Radiate, splitting into columns with age	Globose; smooth to echinulate; 3–6 (most are 3.5–4.5) μm in diameter			Some strains produce toxins; growth usually enhanced at 37°C; small vesicles uniseriate; dark-brown to black sclerotia; common human pathogen
Globose, then radiate, splitting into columns with age	Globose; thick walls; brown; irregularly roughened; most are 4–5 μm in diameter			Frequent cause of otomycosis; sometimes associated with colonization of cavities
Long, columnar; uniform in diameter throughout length	Globose to slightly elliptical; smooth; 1.8–2.4 μm in diameter			May form single globose conidia on submerged hyphae (aleruioconidia, 6–7 μm)
Radiate	Globose; strongly to delicately echinulate; most are 2–3 μm in diameter		+	Hülle cells, when present, resemble *A. nidulans* type; *A. versicolor* and *A. sydowi* most common pathogens in this group
Radiate to loosely columnar	Ovate to globose; spinulose; most are 5–6.5 μm in diameter	+		Abundant yellow cleistothecia; most of the Glaucus group grow poorly at 37°C; *A. repens*, *A. ruber*, and *A. amstelodami* are most common pathogens in this group
Clavate; splits into 2–3 compact columns with age	Elliptical; thick walled; smooth; 3–4.5 by 2.5–3.5 μm; occasionally larger in some strains			May be involved in allergic conditions
Short; columnar	Globose; rugulose; 3–3.5 μm in diameter	+	+	Cleistothecia small, usually abundant; completely enveloped by or associated with Hülle cells
Loosely columnar to radiate	Globose to subglobose; smooth; 2–3 μm in diameter		+	Hülle cells irregularly swollen, twisted, or branched; *A. niveus* and *A. carneus* included in group

cies. The most commonly isolated species are blue-green or yellow-green (the colors common to many of the species). *P. marneffei* produces a diffusible red pigment and red-stained hyphae, useful in the identification of this species.

On microscopic examination of the colonies of *Penicillium* species, erect, usually branched conidiophores may be seen along the hyphae. These branches, and secondary (and occasionally tertiary) branches, called metulae, may be either symmetrical or asymmetrical in relation to the conidiophore. Flask-shaped, hyaline phialides are borne at the apex of the terminal metulae. Each phialide produces chains of one-celled phialoconidia which may be hyaline to dark and smooth to rough walled. These characteristics, along with either monoverticillate or biverticillate appearance of the structure, are used in the identification of each species (18). The final identification of difficult strains should be confirmed by a reference laboratory experienced with this genus.

PAECILOMYCES SPECIES

Reports of patients with paecilomycosis usually give a history of foreign material coming into contact with, or traumatizing, tissue that later becomes in-

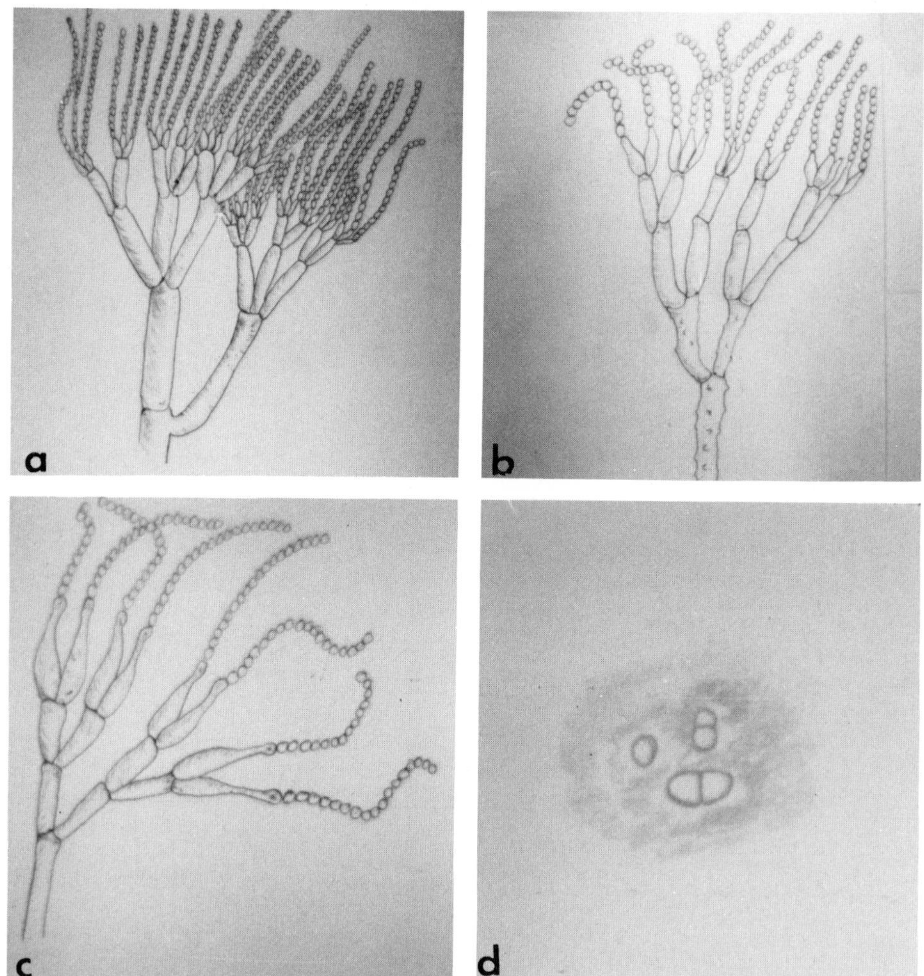

FIG. 3. *Penicillium* species. (a) *P. chrysogenum*. (b) *P. casei*. (c) *P. marneffei*. (d) *P. marneffei* yeastlike cells.

fected. In one reported series of 12 lens transplant patients with eye infections due to *P. lilacinus*, solutions in unopened packages containing similar lens implants were contaminated with this fungus (25). Other reports describe invasion of the eye after some type of trauma, either naturally or surgically induced.

P. lilacinus also has been demonstrated to cause infection of the maxillary sinus (26) and of chronic, cutaneous facial lesions (30).

P. varioti has caused endocarditis in a case involving prosthetic valves (17). Diagnostic samples were obtained from material removed surgically during fur-

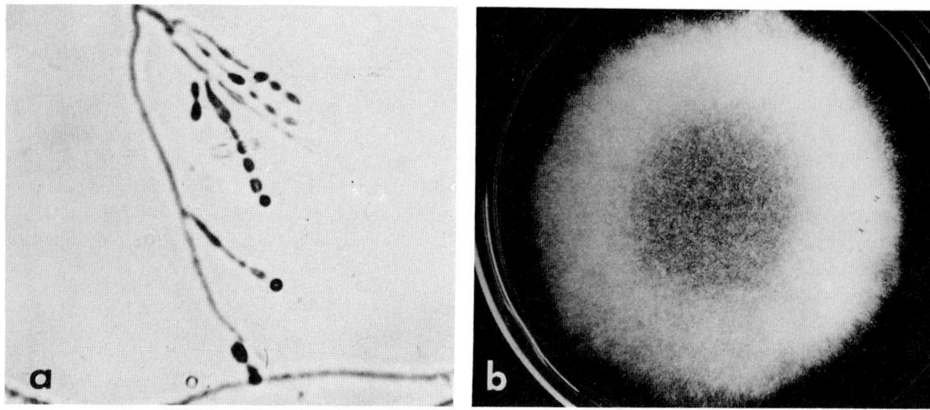

FIG. 4. *Paecilomyces* species. (a) Conidiophore with chains of conidia. (b) Three-week-old colony on Sabouraud medium.

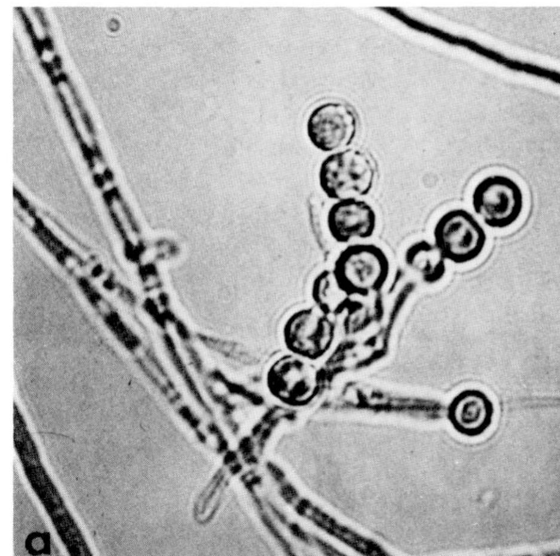

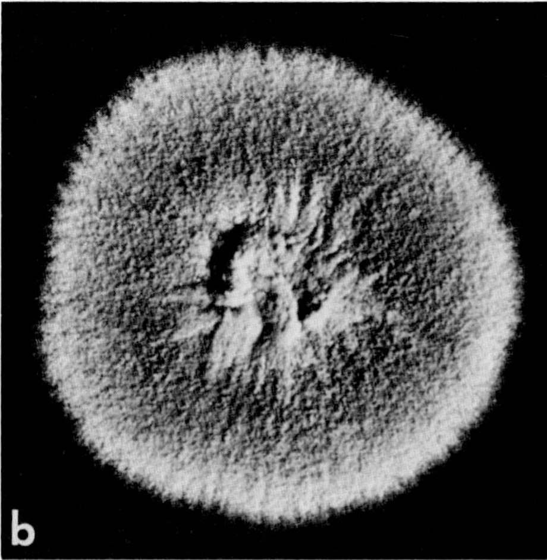

FIG. 5. *Scopulariopsis* species. (a) Conidiophore with chains of conidia. (b) Three-week-old colony on Sabouraud medium.

ther valve replacement. A third species, *P. marquandi*, has caused cellulitis in a renal transplant patient (11).

Mycology

Paecilomyces species resemble *Penicillium* species in growth characteristics and in some elements of microscopic appearance. The primary microscopic characteristic of *Paecilomyces* species is a conidiophore with graceful, elongated, beaklike phialides curving outward, away from each other. Chains of phialoconidia generally are oval and may become enlarged in older chains.

The most commonly isolated species, *P. lilacinus*, produces a characteristic lilac-colored colony on most media, although primary growth from infected tissue may be yellow. The colony growth is rapid, low, and spreading. Figure 4a shows the details of the genus and this species.

P. varioti colonies are fast growing, velvety, and tan to gray, depending on the age of the culture (Fig. 4b).

SCOPULARIOPSIS BREVICAULIS

Scopulariopsis brevicaulis is among the ten most commonly isolated fungal contaminants of humans. It is omnipresent and rarely pathogenic. However, *S. brevicaulis* is recognized as a cause of nail infection. KOH examination of scrapings of infected nails shows chains of conidia typical of the genus, unlike the hyphae of dermatophytes. This fungus grows readily in culture.

The isolation of *S. brevicaulis* from a deep lesion in the soft tissue of an ankle was obtained only after the swollen ankle had been drained for a primary infec-

FIG. 6. *Fusarium* species. (a and b) *Fusarium* macroconidia attached to conidiophore. (c) Branched hypha (arrow) in 10% KOH mount from pus.

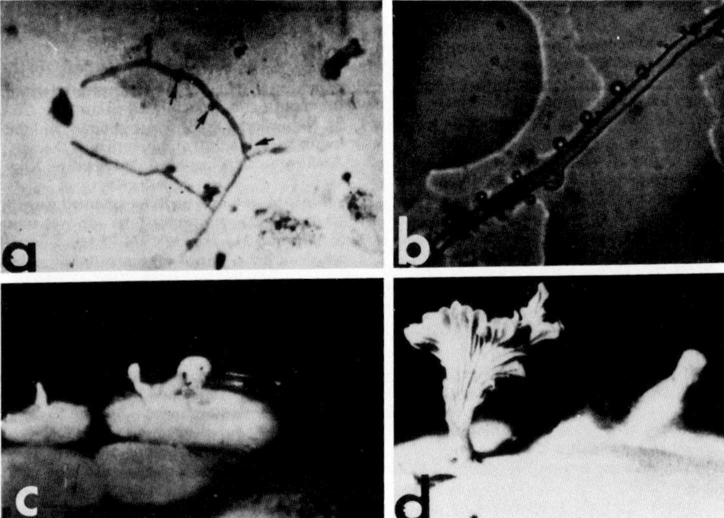

FIG. 7. *Schizophyllum commune.* (a) KOH mount of scrapings from the roof of the mouth (arrows indicate clamp connections). (b) Culture mount (arrows indicate clamp connections) and a series of spicules. (c) Culture from a lesion. (d) Mature basidiocarp culture. Photographs courtesy of A. Restrepo and D. Greer.

tion with *Enterobacter aerogenes*. The secondary infection with *S. brevicaulis* occurred later (27). *Scopulariopsis* sp. has been isolated from a fungus ball in a preexisting pulmonary cavity (14).

Mycology

S. brevicaulis grows rapidly at room temperature on standard mycological media. The yellowish brown of the colony is characteristic. *Scopulariopsis* species produce globose to pyriform, truncate annelloconidia in chains from annellides that are either solitary or in groups on a conidiophore. The conidia of *S. brevicaulis* generally are yellow to brown and often rough walled. These features, along with the "neck" on each conidium and the color of the colony, are the characteristic features of the species. Figure 5 demonstrates the characteristic structures of the genus.

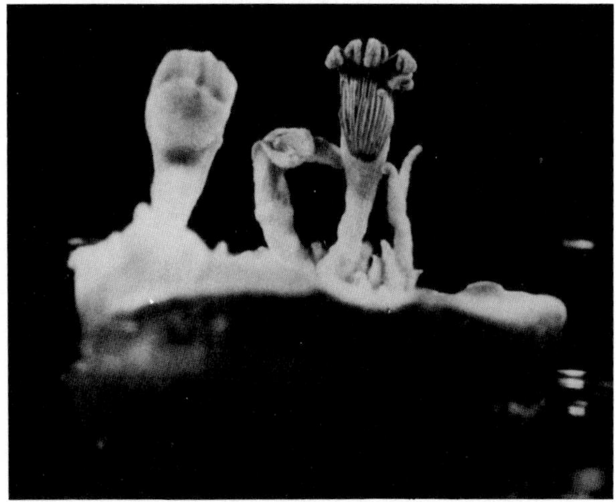

FIG. 8. *Schizophyllum commune* isolated from an inflammatory, swollen lesion of a maxillary sinus. Culture provided by Frank Brancato. Photograph courtesy of B. H. Cooper. Basidiocarp is 2.8 cm in height.

FUSARIUM SPECIES

Several *Fusarium* species have been incriminated as agents causing a range of disease conditions. As with the other organisms discussed here, the ubiquitous nature of this fungus, which grows on plants as a phytopathogen as well as on plant debris, makes it a continuous source of contamination to all surface areas of humans. Best known as a laboratory contaminant, it is now recognized as an opportunistic pathogen causing infection in the eye of both humans and animals (primarily dogs) and in burn patients, recipients of renal transplants, postsurgical patients, and others (15, 20, 21).

F. solani has been found causing infections only in the eye and is third only to the aspergilli and *Candida* spp. as the cause of mycotic keratitis and endophthalmitis. *F. oxysporum and F. moniliforme*, as well as unidentified *Fusarium* species, also have been isolated from eye and other infections (31, 32).

All eye infections with these organisms follow trauma (naturally acquired or post-corneal transplant surgery), with a rapid progression of the infection. Accurate but rapid laboratory identification of infection due to fungi is essential, since steroid therapy (commonly used for treating inflammatory lesions of the eye) disguises the inflammation due to the growth of these (and other) fungi (1, 19, 20).

Mycology

The colonial appearance of the three most commonly isolated *Fusarium* species is greatly influenced by media and cultural conditions. All species grow on Sabouraud agar, but may not sporulate well. A better medium for their study is a carbohydrate-rich medium such as potato agar with double the amount of glucose in the standard formula. On this medium, *F. moniliforme* produces a wine-red pigment that diffuses into the agar and has a gray-white surface. *F. oxysporum* varies in color from brown-white to violet-brown, while the colony of *F. solani* is a "dirty" tan with a hint of pale green. Microscopically, *Fusarium* spp. are characterized by the production of canoe-shaped, two- to multicelled macrophialoconidia and one- or two-celled hyaline microphialoconidia, usually held together in mucus balls. The macrophialoconidia generally are borne in bananalike clusters, which dislodge easily and float free from the hyphae. Not all strains demonstrate two types of conidial production. Difficulty in identification often is due to the absence of macrophialoconidial production. Figure 6 demonstrates the characteristics of the genus.

RARE OPPORTUNISTS: BASIDIOMYCETES

To illustrate the truly unusual nature of fungi that may be found producing opportunistic infection, two examples are given. These examples demonstrate clearly that when a disease process exists and only a fungus, albeit rare, is isolated repeatedly, the identification of the organism by the clinical or reference laboratory is essential.

Coprinus species

Several isolations of *Coprinus* have been reported (7, 13, 29). The most complete account is of the isolation of *C. cinereus* from acute inflammatory endocarditis after the surgical implantation of a mitral valve prosthesis (29). Heavily branched septate hyphae were seen in the aortic valve and within myocardial tissue. The fungus grew readily on Sabouraud agar. The original isolate produced only arthroconidia and was not identifiable; hyphal fusion studies were necessary to induce the typical "mushroom" growth phase.

Schizophyllum species

S. commune has been isolated from a number of different anatomical sites, the first observation being in specimens from infected fingernails (13). Although the fungus grew from nail material, it was so unusual that it was considered a contaminant. The recovery of *S. commune* from cerebrospinal fluid of a patient with meningitis has been reported (3). The patient recovered without specific antifungal treatment. To date there is only one accepted demonstration of this fungus as the agent of disease in a patient with an ulcerative lesion perforating the hard palate (23). Tissue biopsies of the submucosal tissue revealed irregular hyphae with lateral outgrowths. This organism was isolated repeatedly on Sabouraud, brain heart infusion, and blood agar at 25 or 37°C. The basidiocarp (mushroom) develops in 1 to several weeks. Clamp connections were observed on the hyphae in 7 to 10 days. Figures 7 and 8 show the prominent features of this organism.

LITERATURE CITED

1. **Anderson, B., S. S. Roberts, C. Gonzalez, and E. W. Chick.** 1959. Mycotic ulcerative keratitis. Arch. Ophthalmol. **62:**169–179.
2. **Bothast, R. J., and D. I. Fennell.** 1974. A medium for rapid identification and enumeration of *Aspergillus flavus* and related organisms. Mycologia **66:**365–369.
3. **Chavez, B. J., A. Maia, and R. Singer.** 1955. Basidioneuromycosis in man. Soc. Biol. Pernambuco **42:**52–60.
4. **Christensen, C. M., and K. B. Raper.** 1978. Synoptic key to *Aspergillus nidulans* group species and related *Emericella* species. Trans. Br. Mycol. Soc. **71:**177–191.
5. **Christensen, M. A.** 1981. A synoptic key and evaluation of species in the *Aspergillus flavus* group. Mycologia **72:**1056–1084.
6. **DelRossi, A. J., D. Morse, P. M. Spagna, and G. M. Lemole.** 1980. Successful management of Penicillium endocarditis. J. Thorac. Cardiovasc. Surg. **80:**945–947.
7. **DeVries, G. A., R. F. O. Kemp, and D. C. E. Speller.** 1971. Endocarditis caused by *Coprinus delicatulus*, p. 185–186. Compt. Rend. de Communacicones. Fifth Congress, International Society for Human and Animal Mycology, Paris, 1971.
8. **DiSalvo, A. F., A. Fickling, and L. Ajello.** 1972. Infection caused by *Penicillium marneffei*. Am. J. Clin. Pathol. **59:**259–263.
9. **Eschete, M. L., J. W. King, C. West, and A. Aberle.** 1981. *Penicillium chrysogenum* endopthalmitis. Mycopathologia **74:**125–127.
10. **Hall, W. J.** 1974. Penicillium endocarditis following open heart surgery and prosthetic valve insertion. Am. Heart J. **87:**501–506.
11. **Harris, L. F., B. M. Dan, L. B. Lefkowitz, and R. H. Alford.** 1979. Paecilomyces cellulitis in a renal transplant patient: successful treatment with intravenous miconazole. South. Med. J. **72:**897–898.
12. **Kaufman, L.** 1981. Current methods for serodiagnosing systemic fungus infections and identifying their etiologic agents. Estr. L'Ig. Mod. **76:**422–442.
13. **Kligman, A. M.** 1950. A basidiomycete probably causing onychomycosis. J. Invest. Dermatol. **14:**67–70.
14. **Larsh, H. W.** 1977. Opportunistic fungi in chronic disease other than cancer and related problems, p. 221–229. *In* K. Iwata (ed.), Recent advances in medical and veterinary mycology. Proceedings, Sixth Congress, International Society for Human and Animal Mycology. University Park Press, Baltimore.
15. **Lieberman, T. W., P. Perry, and E. J. Bottone.** 1979. *Fusarium solani* endopthalmitis without primary corneal involvement. Am. J. Ophthalmol. **88:**764–767.
16. **Liebler, G. A., G. J. Magovern, P. Sadighi, S. B. Park, and W. J. Cushing.** 1977. Penicillium granuloma of the lung presenting as a solitary pulmonary nodule. J. Am. Med. Assoc. **237:**671.
17. **McClellan, J. R., J. D. Hamilton, J. A. Alexander, W. G. Wolf, and J. B. Reed.** 1976. *Paecilomyces varioti* endocarditis on a prosthetic valve. J. Thorac. Cardiovasc. Surg. **71:**472–475.
18. **McGinnis, M. R.** 1980. Laboratory handbook of medical mycology. Academic Press, Inc., New York.
19. **Moulsdale, M. T., J. M. Harper, and G. N. Thatcher.** 1981. Fungal peritonitis: complication of continuous ambulatory peritoneal dialysis. Med. J. Aust. **1:**88.
20. **Oji, E. O., and D. M. Steele.** 1981. *Fusarium solani* keratitis. East Afr. Med. J. **59:**632–638.
21. **Page, J. C., G. Friedlander, and G. L. Dockery.** 1982. Postoperative Fusarium osteomyelitis. J. Foot Surg. **21:**174–176.
22. **Raper, K. B., and D. I. Fennell.** 1965. The genus *Aspergillus*. The Williams & Wilkins Co., Baltimore.
23. **Restrepo, A., D. L. Greer, M. Robledo, O. Osorio, and H. Mondragon.** 1973. Ulceration of the palate caused by a basidiomycete *Schizophyllum commune*. Sabouraudia **9:**201–204.
24. **Rinaldi, M. G.** 1983. Invasive aspergillosis. Rev. Infect. Dis. **5:**1061–1077.
25. **Rodrigues, M. M., and D. MacLeod.** 1975. Exogenous fungal endophthalmitis caused by Paecilomyces. Am. J. Ophthalmol. **79:**687–690.
26. **Rowley, S. A.** 1982. Paecilomyces fungus infection of the maxillary sinus. Laryngoscope **92:**332–334.
27. **Sekhon, A. S., D. J. Williams, and J. H. Harvey.** 1974. Deep scopulariopsosis: a case report and sensitivity studies. J. Clin. Pathol. **27:**837–843.

28. **Solley, G. O., and R. E. Hyatt.** 1980. Hypersensitivity pneumonitis induced by Penicillium species. J. Allergy Clin. Immunol. **65**:65–70.
29. **Speller, D. C. E., and A. C. MacIves.** 1971. Endocarditis caused by a Coprinus species. A fungus of the toadstool group. J. Med. Microbiol. **4**:370–374.
30. **Takayasu, S. M., M. Akagi, and Y. Shimigu.** 1977. Cutaneous mycosis caused by *Paecilomyces lilacinus*. Arch. Dermatol. **113**:1687–1690.
31. **Young, C. N., and A. M. Meyers.** 1979. Opportunistic fungal infection by *Fusarium oxysporum* in a renal transplant patient. Sabouraudia **17**:219–223.
32. **Young, N. A., K. J. Kwon-Chung, T. T. Kubota, A. E. Jennings, and R. I. Fisher.** 1978. Disseminated infection by *Fusarium moniliforme* during treatment for malignant lymphoma. J. Clin. Microbiol. **7**:589–594.

Chapter 55

Section VII. Parasites

Diagnostic Parasitology: Introduction and Methods

JAMES W. SMITH AND MARILYN S. BARTLETT

Parasitic diseases continue to cause significant morbidity and mortality in the world, particularly in less-developed tropical and subtropical countries. In the United States, indigenous malaria was eradicated long ago, and indigenous nematode infections such as ascariasis, trichuriasis, and hookworm infection have markedly decreased in both incidence and severity. Some other infections are increasing. Giardiasis is a frequent public health problem, with outbreaks related to water supplies and day care centers for children. *Giardia*, ameba, and *Cryptosporidium* infections are increasing in male homosexuals. *Pneumocystis carinii*, *Cryptosporidium* species, *Strongyloides stercoralis*, and *Toxoplasma gondii* are increasingly important causes of serious infections in immunocompromised hosts, especially those with AIDS (acquired immune deficiency syndrome).

The incidence of fecal parasites in specimens submitted to U.S. public health laboratories in 1978 (10) is summarized in Table 1. These figures do not represent prevalence in the U.S. population but show incidence in the fecal specimens submitted for parasite examination. Prevalence for most parasites is probably much lower than these figures indicate.

In addition to infections which are indigenous to the United States, a wide variety of infections may be seen in U.S. citizens who have traveled or worked in foreign countries or in foreign nationals who are visiting or now residing in the United States. Many of these people, such as persons infected with malaria, may be asymptomatic for months or years before disease develops or relapses occur. Some people are recognized as having malaria only when a recipient of their blood develops transfusion-induced malaria or when a baby develops congenital malaria. Other diseases such as echinococcosis may require years before becoming clinically evident.

Efforts to eradicate parasite infections have had variable success. Sanitary fecal disposal, improved water supplies, and improved hygiene in food production and preparation have aided in the control of intestinal parasites. However, much of the earlier enthusiasm for the eradication of malaria has been tempered by the realization that malaria eradication is going to be difficult because parasites are becoming resistant to chemotherapeutic agents, mosquito vectors are becoming resistant to common insecticides, and the use of some insecticides may harm the environment. Human modifications of the environment, such as the building of dams and irrigation systems, have provided an appropriate environment for vectors such as snails and thus allowed diseases such as schistosomiasis to flourish in areas where these diseases had been uncommon. In addition, immunization programs for parasite infections have developed more slowly than was anticipated.

HOST-PARASITE RELATIONS

A knowledge of parasite life cycles is crucial in the understanding of the ways infection is acquired and spread, the pathogenesis of disease, and the ways in which disease might be controlled. Some parasites which infect only humans, such as *Enterobius vermicularis* (pinworm), have a narrow host specificity, whereas others such as *Trichinella spiralis* infect numerous species. When other animals harbor the same parasite stage as humans, these animal species may serve as reservoir hosts. Humans infected with a parasite stage usually seen in other animal species are referred to as accidental hosts.

In the simplest life cycle, the parasite stage from humans is immediately infective for other humans, as in pinworm infection or giardiasis. In other infections such as ascariasis or trichuriasis, a maturation period outside the body is required before the parasite is infective. However, for many parasite infections, a second or even a third host is required for completion of the life cycle. Hosts are defined as intermediate hosts if they do not contain the sexual stage and as definitive hosts if they do contain the sexual stage. Some protozoa, such as the amebae, flagellates and hemoflagellates, do not have a recognized sexual stage. In the intermediate host, there may be a massive proliferation of organisms, as occurs in humans harboring malaria parasites or snails harboring schistosome intermediate stages, or there may be no proliferation, as in mosquitoes which harbor microfilaria undergoing maturation. There may be proliferation in definitive hosts, as in mosquitoes harboring the sexual stage of malaria in which thousands of sporozoites are produced, or there may be no proliferation, as in helminth infections in which one adult is developed from each infective larva. However, in the latter, the adult helminths do produce numerous eggs or larvae.

In some helminth infections, a migration through various body tissues is essential for maturation, as in ascarasis or schistosomiasis, whereas in other infections, the larva leaves the egg and simply matures in the intestinal tract, as in trichuriasis and enterobiasis. Host tissues involved vary depending upon the parasite. In severely immunocompromised patients, sites may be involved that are not involved in normal hosts.

Parasites of humans proliferate tremendously at certain stages, with thousands or even millions of forms being produced for every one that survives to perpetuate the parasite. Parasites may be quite hardy. For example, certain stages, particularly eggs and cysts, may survive for weeks or months in the environment.

Parasites have often developed unique ways of protection from the defense mechanisms of the host.

TABLE 1. Incidence of intestinal parasites in 322,735 fecal specimens examined by state health department laboratories, 1978[a]

Parasite	No. of examinations	% of positive specimens
Protozoa		
Giardia lamblia	12,947	4.0
Entamoeba histolytica	2,409	0.8
Dientamoeba fragilis	1,880	0.6
Balantidium coli	7	
Isospora spp.	1	
Nonpathogenic	21,120	6.5
Nematodes		
Trichuris trichiura	5,481	1.7
Ascaris lumbricoides	4,630	1.4
Enterobius vermicularis	4,344	1.4
Hookworm	2,035	0.6
Strongyloides stercoralis	602	0.2
Trichostrongylus spp.	14	
Trematodes		
Clonorchis-Opisthorchis	205	0.06
Schistosoma mansoni	48	
Fasciola hepatica	1	
Paragonimus westermani	1	
Cestodes		
Hymenolepis nana	1,068	0.3
Taenia spp.	251	0.08
Diphyllobothrium latum	20	
Hymenolepis diminuta	12	
Dipylidium caninum	7	

[a] Adapted from *Center for Disease Control: Intestinal Parasite Surveillance. Annual Summary, 1978.* Issued August 1979. The survey does not include laboratories in Guam, Puerto Rico, or Virgin Islands. One or more parasites were found in 14.7% of specimens. Percentages are not calculated for parasites identified less than 100 times.

These mechanisms include the ability to change antigenic characteristics so that although the host forms antibody, the antibody does not react with the modified parasite, or the parasite may be coated with host immunoglobulins, as in schistosomiasis, so that the host does not recognize the parasite as foreign. Macrophages and both cell-mediated and humoral immunities appear to be important in the host response to infection. Eosinophils are particularly important in the defense against tissue-invading helminths.

DIAGNOSIS OF PARASITIC INFECTIONS

The diagnosis of most parasitic infections is dependent upon the laboratory. For intestinal and blood parasites, morphologic demonstration of diagnostic stage(s) is the principal means of diagnosis, whereas for tissue infections, immunodiagnostic techniques are generally more important. During the early stages before diagnostic forms are produced (prepatent period), patients may be symptomatic. For example, patients may have pulmonary symptoms and eosinophilia due to ascaris larva migration at a time when eggs are not produced. In such patients the physician may suspect parasite infection, but the actual diagnosis must be based on a clinical impression or immunodiagnostic tests, or diagnosis must await the production of diagnostic stages.

In establishing a diagnosis, the clinician places a great deal of trust in the laboratory. This trust can be misplaced if laboratory personnel are not competent to identify or exclude parasites. The literature clearly documents instances in which outbreaks have been overlooked due to incompetent laboratory diagnosis or in which inflammatory cells or other objects have been identified as parasites and outbreaks have been diagnosed when none existed (24). The results of proficiency-testing programs (39) also suggest that laboratories have difficulty with the identification of some parasites, especially intestinal protozoa (Table 2).

Identification may be by gross examination for adult helminths or, more commonly, by microscopic examination for protozoa, helminth eggs, and larvae. The diagnostic forms of some parasites, such as the eggs of *Ascaris* spp., are present on a regular basis. Other forms, such as malaria parasites, *Taenia* eggs, or *Giardia* cysts, vary from day to day.

Most immunodiagnostic tests used today for parasitic infections detect antibody. In recent years, the sensitivity and specificity of many such tests have improved. A number of antigen detection tests have recently been described and show promise, but none of these tests are currently available commercially. Immunodiagnostic tests are more thoroughly discussed in Chapter 97 on parasitic serology and in other chapters under specific infections.

LABORATORY PROCEDURES

Many methods for diagnostic parasitology have been described. There are advantages and disadvantages to each method. Some are particularly valuable for epidemiologic studies or for evaluations of new therapeutic agents, whereas other methods are used primarily for laboratory diagnosis. In this chapter we emphasize the diagnostic procedures. From the numerous methods, we have selected those which are widely used in this country and which are sensitive and relatively easy to perform. These methods should prove adequate for most laboratories. For additional procedures, laboratory manuals (17, 33) or parasitolo-

TABLE 2. Participant performance in College of American Pathologists parasitology survey program, 1973 through 1977[a]

Parasite	No. of specimens	Avg correct identification (%)
Formalin-fixed fecal specimens		
Ascaris lumbricoides eggs	6	90
Hookworm	6	92
Strongyloides stercoralis larvae	4	88
Trichuris trichiura eggs	6	93
Diphyllobothrium latum eggs	6	81
Hymenolepis diminuta eggs	5	91
Taenia sp. eggs	6	87
Paragonimus westermani eggs	5	83
Giardia lamblia cysts	8	65
Entamoeba coli cysts	9	88
No parasite seen	6	92
PVA-fixed specimens		
Entamoeba histolytica	5	73
Entamoeba coli	4	52
Endolimax nana	4	51
Negative for parasites	3	77

[a] See reference 29.

gy books (6, 28) should be consulted. When alternative methods or methods for specific parasites are indicated, references will be given, but the methods will not be described.

PROCEDURES FOR INTESTINAL AND BILIARY PARASITES

Intestinal and biliary parasites are generally diagnosed by finding diagnostic stages in feces or other intestinal material such as duodenal or sigmoidoscopic aspirates. Studies have shown that the eggs of most parasites are uniformly distributed in the fecal mass due to the mixing action of the colon (30), although some, such as schistosome eggs, which originate in the distal colon, may be more numerous on the surface of formed fecal specimens. The distribution of protozoan forms is more variable. There may be fewer protozoan trophozoites in the first part of an evacuation than in the last because they have deteriorated while in the lower colon.

Collection and Handling of Fecal Specimens

The numbers and times of collection for fecal specimens depend somewhat on the diagnosis suspected. As a routine, because some organisms are shed in a variable pattern, it is advisable to examine multiple specimens before excluding parasites. The general recommendation is to collect a specimen every second or third day, for a total of three specimens. From a hospitalized patient, one specimen each day for three days may be more cost effective.

A number of substances may interfere with stool examination. Particulate materials such as barium, antacids, kaolin, and bismuth compounds interfere with morphologic examination, and oily materials such as mineral oil create small, refractile droplets that make examination difficult. Antimicrobial agents, particularly broad-spectrum antimicrobial agents, may suppress amebae. If any of these substances have been used, specimens should not be submitted until the substances have been cleared (generally 5 to 10 days). A fecal specimen may appear satisfactory by gross examination when there is still barium, etc., which can interfere with microscopic examination.

Fecal specimens are best collected into wide-mouthed, water-tight containers with tight-fitting lids such as waxed, pint-sized ice cream cartons or plastic containers. Usually patients can defecate directly into such containers. Urine should not be allowed to contaminate specimens, as it is harmful to some parasites. If specimens are to be collected in a bed pan, the patient should micturate into a separate container before the specimen is collected. Toilet paper should not be included with the specimen. Stool should not be retrieved from toilet bowl water, as various free-living protozoa in water might be confused with the parasites. In addition, water is harmful to some parasites such as schistosome eggs and amebic trophozoites. If the patient is producing formed specimens, stool may be collected by having the patient squat over waxed paper to defecate.

Purgation with sodium sulfate or buffered phosphosoda may be helpful in the diagnosis of amebiasis in some patients (50). Purgation is usually done after a series of fecal specimens have been negative, and it requires the order of a physician. Prior arrangements must be made with the laboratory, and specimens must be collected during regular laboratory hours. The patient is given the appropriate salt solution orally. In approximately 1 to 1.5 h, the patient will begin to pass stool specimens, and each specimen should be promptly transported to the laboratory for examination.

Clinical information such as the suspected diagnosis, travel history of the patient, and clinical findings should be included on the requisition. In addition, the time the specimen was passed and the time it was placed in fixative should be noted. If the specimen is in fixative, the consistency of the original specimen should be stated, or a portion of unfixed specimen should be included with the fixed specimen.

A laboratory may have specimens placed in fixatives in the home or patient care area immediately after passage, may place portions of specimen in fixatives at the time they are received in the laboratory, or may examine the specimen unfixed. Many laboratories use a combination of these methods depending on the location of the patient, consistency of the specimen, time of day, and laboratory work load. Prompt examination or fixation is particularly important for soft, loose, or watery specimens, which are most likely to contain protozoan trophozoites (27).

Formed specimens, which are likely to contain protozoan cysts or helminth eggs or larvae, can remain satisfactory for a number of hours at room temperature or overnight in a refrigerator. Soft and liquid specimens should be examined or placed in fixatives promptly (within 1 h). Specimens which cannot be examined or fixed promptly should be either refrigerated or left at room temperature. They should not be incubated, as incubation speeds the deterioration of the organisms. Feces for parasite examination must not be frozen and thawed.

The fixative system generally used is a two-vial technique with one vial containing 5 to 10% buffered Formalin and the other vial containing polyvinyl alcohol (PVA) fixative. A portion of the specimen is added to the fixative in a ratio of approximately 3 parts fixative to 1 part specimen and thoroughly mixed to ensure adequate fixation. An alternative to Formalin is Merthiolate-iodine-formaldehyde (MIF), which fixes and stains at the same time (38). If unfixed specimens are processed in the laboratory, fecal films may be prepared and immediately fixed in Schaudinn fixative.

Gross Examination of Feces

Specimens should be examined grossly to determine the consistency (hard, formed, loose, or watery), color, and presence of gross abnormalities such as worms, mucus, pus, or blood. It may be profitable to examine flecks of mucus, pus, or blood for parasites. If adult worms or portions of tapeworms are sought, the feces may be carefully washed through a screen. (Small worms may be difficult to see if gauze is used.) The identification characteristics of adult worms are not discussed in this chapter, so parasitology books should be consulted (6, 28).

Procedures for Microscopic Examination

The three principal microscopic examinations performed on stool specimens are direct wet mount, wet mount after concentration, and permanent stain. Although each examination can contribute to diagnosis, the yield of some methods is small with certain kinds of specimens. Procedures to be performed on various types of specimens are outlined in Table 3. As a minimum, formed specimens should be examined by a concentration procedure. Soft specimens should be examined by concentration and permanent stain, and, if submitted fresh, by direct wet mount. Loose and watery specimens should be examined by wet mount and permanent stain. If specimens are received in fixative and the consistency is not known, concentration and permanent stain should be performed. Other examinations may be helpful (Table 3). Special procedures which may assist in the diagnosis of specific parasites are noted below in discussions of the parasites.

Calibration and use of an ocular micrometer

Size is important in the differentiation of parasites and is most accurately determined with a calibrated ocular micrometer; thus, each laboratory performing diagnostic parasitology must have such a micrometer.

An ocular micrometer is a disk on which is etched a scale in units from 0 to 50 or 100. To determine the micrometer value of each unit in a particular eyepiece and at a specific magnification, the unit must be calibrated with a stage micrometer. A stage micrometer has a scale 2 mm long ruled in fine intervals of 0.01 mm (10 μm).

Calibration of the ocular micrometer.

1. Insert the micrometer in the eyepiece so that the micrometer rests on the diaphragm, with the etched scale facing the eye. In many new microscopes, the micrometer can be dropped in and secured with a ring retainer. (It is helpful to have an extra ocular in which the micrometer may be left.)
2. Place the stage micrometer on the microscope stage.
3. Focus on the etched scale. Since the micrometer must be calibrated for each objective, begin with the lowest magnification (e.g., ×10).
4. Align the two scales so that the zero points are superimposed (Fig. 1).
5. Find a point far down the scales at which a line of the stage micrometer coincides with a line of the ocular micrometer. Count the number of ocular units and the number of stage units from zero to these coinciding lines.
6. Multiply the number of stage micrometer units by 1,000 to convert millimeters to micrometers.
7. Divide the product of step 6 by the number of ocular units to determine the value of an ocular unit.

Repeat the calibration for each objective. Keep a record of the unit value for each objective for each microscope used. Calibration must be done separately for each microscope and must be repeated if an ocular or objective is changed.

Use of the micrometer. Insert the eyepiece containing the calibrated ocular micrometer in the microscope. Count the number of ocular units which equal the structure to be measured. Multiply the number by the micrometer value of the ocular unit for the objective being used. If an ocular micrometer is properly used, parasites which are similar in appearance but different in size can be readily differentiated.

TABLE 3. Laboratory examinations for various types of fecal specimens[a]

Type of specimen	Method		
	Direct wet mount	Concentration	Permanent stain
Unpreserved			
Formed	+	++	±
Soft	++	++	++
Loose and watery	++	±[b]	++
Preserved			
Formalin			
Formed or soft	+	++	−
Loose or liquid	++	±[b]	−
PVA fixative			
Formed	−	−	±
Soft, loose, or liquid	−	−	++

[a] ++, Essential for basic examination; +, recommended for basic examination; ±, optional for basic examination.

[b] Concentration is recommended if *Cryptosporidium* sp. is suspected.

Direct wet mount

The direct wet mount made from unconcentrated fresh feces is most useful for the detection of the motile trophozoites of intestinal protozoa and the motile larvae of *Strongyloides* spp. It is also useful for the detection of protozoan cysts and helminth eggs. For fixed feces, the direct wet mount may allow the detection of parasites which do not concentrate well. This method is also useful for the examination of specific portions of feces, such as flecks of blood or mucus.

Direct wet mounts are prepared by placing a small drop of 0.85% saline toward one end of a glass slide (2 by 3 in. [ca. 5 by 7.5 cm]) and a small drop of appropriate iodine solution (see below) toward the other end. With an applicator stick, a small portion of specimen (1 to 2 mg) is thoroughly mixed in each diluent, and a no. 1 cover slip (22 mm) is added. The density of fecal material should be such that newspaper print can be read with difficulty through the smear. The material should not overflow the edges of the cover slip. Grit or debris may prevent the cover slip from seating and may be removed with a corner of the cover slip or an applicator stick. Mounts may be sealed with Vaspar (50% petroleum jelly, 50% paraffin) which is melted on a hot plate (not over an open flame). A cotton applicator or small brush is used to apply small drops of Vaspar to opposite corners to attach the cover slip and then to seal it with even strokes. The amount of Vaspar on top of the cover slip should be minimal. Sealing slows drying and allows oil immersion magnification to be used. Alternatively, drying can be slowed by placing wet gauze or paper toweling in a petri dish, laying portions of applicator sticks or glass rods on the moist material, laying the slide on the sticks or rods, and replacing the lid of the dish.

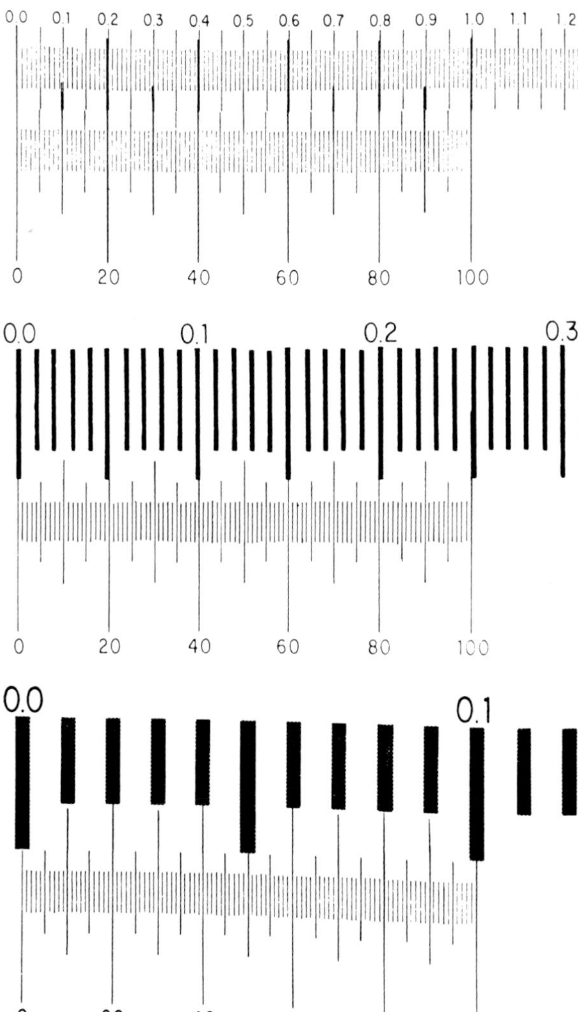

FIG. 1. Calibration of ocular micrometer. The ocular micrometer (lower scale) and stage micrometer (upper scale) appear like this under low (top), high dry (middle), and oil immersion (bottom) magnifications. In these examples, the values of one ocular unit are: low, 10 μm; high dry, 2.5 μm; and oil immersion, 1.0 μm.

The iodine solution should be that of Dobell and O'Connor (1%) or a 1:5 dilution of Lugol iodine. Iodine solution, if too weak, will not stain organisms properly, and if too strong, it will cause clumping of fecal material. Stock iodine solution should be stored in a tightly stoppered brown bottle away from sunlight. Keep the iodine and saline solutions in small dropper bottles, and replace (don't replenish) the solutions weekly. Iodine solution keeps longer if it is refrigerated. Iodine stain solution can be quality controlled by the observation of appropriate staining in positive clinical specimens or Formalin-fixed specimens kept for that purpose.

For the examination of wet mounts, the light of the microscope must be properly adjusted. To achieve optimal resolution, the condenser should be centered and focused for Kohler illumination (racked up). To achieve contrast of the objects in the field, light intensity is diminished with the iris diaphragm of the condenser rather than by lowering the condenser.

The entire saline wet mount cover slip should be systematically scanned with ×100 to ×200 magnification. Suspicious objects are confirmed at higher magnification. In addition, the preparation should be scanned at higher power (×400 to ×500) for a couple of passes across the cover slip to look for protozoan cysts which might be missed with lower power. Screening a slide should take an experienced microscopist about 10 min. If debris is covering a suspicious object, the debris may be removed by pressing or tapping on the cover slip with an applicator stick. This pressure may also help in reorienting an egg, as when one is looking for an operculum. The saline wet mount is best for the detection of helminth eggs and larvae, and it is especially good for protozoan cysts, which appear refractile. The principal usefulness of the iodine mount is to study the morphology of protozoan cysts, as this stain shows nuclear detail and glycogen masses (but does not stain chromatoid material). If suspicious objects are seen, they can be examined under oil immersion (×1,000). If definite or possible protozoan cysts or trophozoites are detected which cannot be identified in wet mounts, permanent stains are required.

A solution of buffered methylene blue (pH 3.6) may be used as a vital stain for the examination of fresh specimens for protozoa (34). The wet mount is prepared as described above, with buffered methylene blue substituted as diluent and 5 to 10 min allowed for the dye to become incorporated in the organisms before examination. Organisms become overstained in 20 to 30 min.

Concentration procedures

Concentration procedures are used to separate parasites from fecal detritus. These procedures are based on differences in the specific gravity of parasite forms and fecal material. In sedimentation, the parasite forms are heavier than the solution and are found in the sediment, whereas in flotation, solutions of high specific gravity are used, and parasite forms float to the surface. An initial washing step removes some of the soluble or finely particulate material, and straining removes larger portions of debris. A wide variety of sedimentation and flotation methods have been described. The Formalin-ethyl acetate modification of the Formalin-ether sedimentation technique and a zinc sulfate flotation technique are widely used and are the only methods described in this chapter. Both methods require that centrifugation be performed in centrifuges with free-swinging carriers. Squeeze bottles for Formalin, saline, or water simplify the processing of large numbers of specimens.

Formalin-ethyl acetate centrifugal sedimentation. The original procedure from which the Formalin-ethyl acetate centrifugal sedimentation technique was adapted was the Formalin-ether concentration method of Ritchie (35). The Formalin-ethyl acetate procedure (51) avoids problems with the flammability and storage of ether. This procedure can be performed on specimens which have been fixed in Formalin for a time or on specimens with Formalin added during the processing. The procedure can also be performed on material fixed in MIF.

The procedure with Formalin-preserved specimens is as follows.

1. Thoroughly mix the formalinized specimens.

2. Depending on the density of the specimen, strain a sufficient quantity through gauze into a 15-ml conical centrifuge tube to give the desired amount of sediment. (Wet gauze in a 4-oz [ca. 120-ml] conical paper cup with the tip cut off can be used for straining.)

3. Add tap water or saline, mix the solution thoroughly, and centrifuge it at 650 × *g* for 1 min. The amount of the resulting sediment should be about 1 ml. The amount of sediment may be adjusted by the addition of more feces and centrifugation again or by the addition of water, suspension again, the removal of an appropriate amount of material, and then recentrifugation.

4. Decant the supernatant, and wash it again with tap water, if desired.

5. To the sediment, add 10% Formalin to the 9-ml mark, and mix the solution thoroughly.

6. Add 4 ml of ethyl acetate, stopper the tube, and shake the tube vigorously in an inverted position for 30 s. Remove the stopper with care.

7. Centrifuge the solution at 450 to 500 × *g* for 1 min. Four layers should result: ethyl acetate, plug of debris, Formalin, and sediment.

8. Free the plug of debris from the sides of the tube by ringing the tube with an applicator stick, and carefully pour the top three layers into a discard container. With the tube still tipped, use a swab to remove debris from the sides of the tube. This step is very important, for lipid droplets which reach the sediment make examination difficult.

9. Mix the remaining sediment with the small amount of fluid that drains back down from the sides of the tube (or add a drop of saline or Formalin). If mounts are to be prepared later, a small amount of Formalin may be added to the sediment and the tube may be stoppered.

10. Prepare wet mounts as described above, and examine them.

The procedure for Formalin-ethyl acetate centrifugal sedimentation with fresh specimens is as follows.

1. Comminute a portion of stool about 1.5 cm in diameter in 10 ml of saline or water.

2. Strain about 10 ml of the fecal suspension into a 15-ml conical centrifuge tube.

3. Centrifuge the suspension at 650 × *g* for 2 min. This step should provide about 1 ml of sediment. If not, adjust the amount of sediment as described above.

4. Wash the sediment again if desired.

5. To the sediment, add 10% buffered Formalin to the 9-ml mark, mix thoroughly, and allow the mixture to stand for 5 min or longer.

6. Proceed as for step 6 of the procedure for fixed specimens.

(Note that either saline or water can be used. Tap water will lyse *Blastocystis hominis*. If schistosomiasis is suspected, the specimen should be preserved in Formalin before concentration to prevent hatching.)

Zinc sulfate centrifugal flotation. The zinc sulfate concentration method originally described by Faust et al. (16) may be performed on unfixed or Formalin-fixed specimens, although the specific gravity of the zinc sulfate solution required differs. The disadvantages of the zinc sulfate concentration are: (i) dense schistosome eggs do not concentrate well; (ii) opercula often pop, and thus operculate eggs may be missed; and (iii) larvae and cysts may collapse. The modified procedure with Formalin-fixed feces (3) slows the collapse of larvae and cysts and largely prevents the popping of opercula. The advantages are that it leaves a rather clean background, has less grit than the sedimentation procedure, and is better for the concentration of some parasites, such as *Giardia* cysts.

The procedure for Formalin-preserved specimens (3) is as follows. The specific gravity of the zinc suflate must be 1.20. Centrifugation must be performed in round-bottomed tubes such as 16- by 100-mm disposable tubes.

1. The Formalin-preserved fecal material is mixed, strained through one layer of cheesecloth into a conical paper cup, poured into the tube to a level about 1 cm from the top, and then centrifuged.

2. The tubes are centrifuged for 3 min at about 650 × *g*. There should be 1 to 1.5 cm of sediment.

3. Decant the supernatant from each tube, and drain the last drop against a clean section of paper towel.

4. To the packed sediment of each tube, add zinc sulfate to within 1 cm of the rim.

5. Insert two applicator sticks, and thoroughly mix the packed sediment.

6. Immediately centrifuge the suspension at 500 rpm for 1.5 min.

7. Very carefully transfer the tubes to a rack of the proper size, so that the tubes remain vertical. Do not disturb the surface films, which now contain the parasites. Allow the tubes to stand for 1 min to compensate for any movement. The countertop must be vibration free.

8. With a loop which is bent at a right angle, transfer to a slide (2 by 3 in.) two loops of surface material beside 1 drop of saline and two loops beside 1 drop of iodine. With the heel of the loop, mix first the saline and then the iodine with the surface material. Cover each mixture with a 22-mm no. 1 cover slip. The slide should be made within 20 min.

9. To retard drying, place each prepared slide on a bent glass rod or portions of applicator sticks in a petri dish containing a damp paper towel. Petri dishes may be placed in the refrigerator if examination will be delayed. Alternatively, cover slips may be sealed with Vaspar.

The procedure with fresh specimens is as follows.

1. Comminute a fecal specimen about 1 cm in diameter in a tube (16 by 100 mm) half filled with tap water. Add additional water to within 1 to 2 cm of the top.

2. Centrifuge the tube at 650 × *g* for 1 min.

3. Discard the supernatant, and add a zinc sulfate solution of specific gravity 1.18 to within 1 cm of the rim.

4. Proceed as from step 5 above.

Sheather sugar flotation. Sheather sugar flotation is recommended for the concentration of *Cryptosporidium* cysts (11). Although these oocysts will concentrate when the Formalin-ethyl acetate or zinc sulfate technique is used, they are more readily detected with the Sheather sugar flotation, for they stand out sharply from the background in this solution of high specific gravity. This procedure may be performed on unfixed or Formalin-fixed feces. The procedure for Sheather sugar flotation is outlined below.

1. (a) Formed stool. Place approximately 0.5 g of stool in a tube (16 by 100 mm) about half full of Sheather sugar solution. Mix the solution thoroughly, and then add more sugar solution to within 1 cm of the rim.

(b) Watery stool. Centrifuge the fecal specimen and mix 0.5 to 1 ml of sediment with Sheather solution as described above.

2. Centrifuge the solution at 400 × *g* for 5 to 10 min.

3. Remove the top portion of the sample with a wire loop bent at a right angle. Place several loopfuls on a glass slide (2 by 3 in.). Cover the specimen with a 22-mm cover slip, and examine the slide with a ×40 objective. Oocysts are found just beneath the cover slip and are refractile. Saline or iodine is not used in the preparation of these mounts.

Baermann concentration. The Baermann concentration technique (47) has greater sensitivity for the detection of strongyloides larvae than do the standard concentration techniques described above. This technique is useful clinically for the diagnosis and monitoring of therapy of strongyloides infections, and it is useful epidemiologically for the examination of soil for the larvae of nematode parasites.

A funnel with a clamped rubber tube on the stem is placed in a ring stand. A circular mesh screen is placed across the funnel approximately one-third from the top, a portion of coarse fabric such as muslin is placed on the screen, and feces is added. Tap water at 37°C is added so that the water just touches the feces. Let the specimen stand 1 h, remove 2 ml of fluid from the stem, and centrifuge the sample at 300 × *g* for 3 min. Prepare a wet mount of sediment, and examine it for larvae.

Hatching technique for detection of viable schistosome eggs. Place a large amount of feces (5 to 10 g) in a large flask (1 to 2 liters), and add water while mixing to break up the feces to a fine suspension. Bring the water level to 2 to 5 cm from the top of the flask. Cover the sides of the flask with foil or other material to shield all but the top of the liquid from light. Allow the flask to stand at room temperature for several hours. With a hand lens, examine the material at the top of the flask neck for swimming miracidia. Remove the miracidia with a Pasteur pipette for examination with a ×10 objective. It is not possible to determine the species of schistosome from the miracidia.

Other concentration procedures. Concentration procedures have been described for feces preserved in MIF (7), sodium acetate-Formalin (49), or PVA fixative (17). MIF– or sodium acetate-Formalin–fixed feces may be used in place of Formalin-fixed feces in the Formalin-ethyl acetate concentration procedure. Some workers feel that organisms do not concentrate as well from material fixed in PVA fixative (9) or from material which has been in MIF for extended periods (6).

If large amounts of specimen are to be concentrated, as when specimens of eggs are prepared for teaching, gravity sedimentation is usually used. The feces is thoroughly mixed in liquid (water, saline, or 10% Formalin) and allowed to settle in a sedimentation jar or funnel for several hours or overnight. Supernatant fluid is discarded, and the sediment is again suspended and allowed to settle. This procedure can be continued if desired until the supernatant is clear.

Permanent stains

Permanent stains of fecal smears are most needed for the detection and identification of protozoan trophozoites, but they are also used for the identification of cysts (18, 26, 46). Wet mounts of fresh feces, even with stains such as methylene blue, are not as sensitive for trophozoites and therefore do not substitute for permanent stains (19). It is sometimes difficult to identify cysts which are detected in wet mounts; thus, for each specimen, regardless of consistency, it may be worthwhile to fix a portion in PVA fixative or to prepare two fecal films fixed in Schaudinn fixative so that permanent stains can be performed if needed. Permanent stains also provide a permanent record and are easily referred to consultants if there are questions on identification.

A number of staining procedures have been described. Some stains, such as chlorazol black (20), require fresh specimens and are not widely used. A variety of stains for fecal smears preserved by Schaudinn or PVA fixative have been described, including various hematoxylin stains (33). The stain most widely used in the United States is the Wheatley trichrome stain (48), which is the only permanent stain described in this chapter. The trichrome staining procedure uses reagents with a relatively long shelf life and is easy to perform. The procedure is outlined below and in Table 4. Note that there are differences in staining times depending on whether the specimen is fixed in Schaudinn or PVA fixative, as penetration is slower in the latter.

Preparation of smears. (i) Unpreserved specimens with Schaudinn fixative.

1. To prepare thin, uniform smears, place a drop of saline on a glass slide (1 by 3 in. [ca. 2.5 by 7.5 cm]). With an applicator stick, transfer a small, representative portion of the specimen to the drop of saline, and mix the two. Spread the solution into a film by rolling the applicator stick along the surface. Remove any lumps.

Before watery specimens are smeared, apply an adhesive such as serum or albumin to the slide. Liquid specimens may be centrifuged, and the sediment may be used for smear preparation.

2. Place fresh smears immediately into Schaudinn fixative. Do not allow the smears to dry at any time before they are stained. Smears should fix for at least 1 h at room temperature or for 5 min at 50°C; however, they can be left in fixative for several days. After fixation, slides may be kept in 70% alcohol indefinitely before they are stained.

(ii) Unpreserved specimens with PVA fixative.

1. On a slide (1 by 3 in.), thoroughly mix 1 drop of unfixed specimen with 1 drop of PVA fixative.

2. Spread the specimen as described below.

3. Allow the smear to dry, preferably overnight, before it is stained.

(iii) PVA fixative-preserved specimens.

1. Preserve 1 part specimen in 3 parts PVA fixative. Mix thoroughly. Fix for at least 1 h. Specimens keep indefinitely.

TABLE 4. Trichrome stain procedure

Step	Reagent	Staining time		Purpose
		PVA fixative	Schaudinn fixative	
1	70% alcohol plus iodine	10 min	1 min	Removal of mercuric chloride, hydration
2	70% alcohol	5 min	1 min	Removal of iodine, hydration
3	70% alcohol	5 min	1 min	Wash, hydration
4	Trichrome stain	6–8 min	2–8 min	Stain
5	90% alcohol acidified	5–10 s	Brief dip	Destain
6	95% alcohol	Rinse	Rinse	Stop destaining
7	95% alcohol	5 min	Rinse	Dehydration
8	Carbol-xylene	10 min	1 min	Clearing and dehydration
9	Xylene	10 min	1–3 min	Clearing

2. Add 1 drop of PVA-fixed specimen to a slide.

(a) If there is little sediment, remove a portion of the sediment with a Pasteur pipette.

(b) If there is abundant sediment, mix the specimen thoroughly, and add 1 drop of specimen to a slide with applicator sticks or a Pasteur pipette.

3. Spread the material over the center third of the slide by rolling the specimen with an applicator stick. Remove any lumps. The film should extend to both the top and bottom edges of the slide, as this helps prevent peeling.

4. Allow the slide to dry overnight at room temperature or 35°C. In an urgent situation, the slide can be dried for 4 h at 35°C and then stained.

Trichrome staining procedure. Table 4 outlines the steps in the trichrome staining procedure.

Permanently stained slides may be mounted with a cover slip or may be air dried and examined after oil is added. Slides should be examined at a magnification of ×400 to ×500 or greater after they are scanned under lower power to find optimal areas. A ×50 oil immersion objective is particularly helpful, as it allows the easy use of a ×100 oil immersion objective for the detailed examination of organisms while allowing more rapid screening with a ×50 objective. Oculars of ×5 or ×6 can provide the same result. A ×20 dry objective may also assist in screening.

Permanently stained slides should be kept for 2 years.

Stain reactions. In an ideal stain, the cytoplasm of cysts and trophozoites is blue-green tinged with purple. *Entamoeba coli* cyst cytoplasm is often more purple than that of other species. Nuclear chromatin, chromatoid bodies, erythrocytes, and bacteria stain red or purplish red. Other ingested particles such as yeasts often stain green. Parasite eggs and larvae usually stain red. Inflammatory cells and tissue cells stain in a fashion similar to that of protozoa. Color reactions may vary from the above.

Incompletely fixed cysts may stain predominantly red, and organisms which have degenerated before fixation often stain pale green. Poor fixation due to an inadequate mixing of the specimen in fixative may result in both of these appearances. In some specimens, degeneration has occurred before the specimen is placed in fixative, either in the patient before the specimen was evacuated or because of delay in fixing the specimen.

Troubleshooting the trichrome stain. Except for problems with delayed or inadequate fixation as noted above, problems with the trichrome stain are usually related to reagents other than the stain. If crystalline material is apparent after the specimen is stained, the crystals are probably mercuric chloride in the fixative which was not adequately removed because the iodine in the alcohol-iodine solution was too weak or because the slide was in this solution too short a time. If crystals are present after treatment with proper-strength iodine-alcohol, they are present in the specimen, which is thus unsatisfactory, and another specimen should be requested.

If the stain appears washed out, it is likely that the slide was destained too much. This washed-out appearance can be either because the specimen was left too long in the acid-alcohol destain or because the alcohol wash after the acid-alcohol destain had become acidic as a result of transfer by previous slides.

The trichrome may become diluted by carry-over alcohol if more than 10 slides per day are stained in one Coplin jar. To restore the stain, the lid may be left off for several hours to allow alcohol to evaporate, and then the volume is replaced with new stain.

Control slides should be used to monitor the staining. Specimens containing protozoa are best for controls; however, feces containing inflammatory cells or added buffy-coat leukocytes also are satisfactory.

Restaining. Should the stain be unsatisfactory, the slide can be destained by placing it in xylene to remove the cover slip or immersion oil and then placing it in 50% alcohol for 10 min to hydrate the slide. Destain the slide in 10% acetic acid in water for several hours, and then wash it thoroughly first in water and then in 50 and 70% alcohols. Place the slide in stain for 8 min, and then complete the stain procedures. It is helpful to eliminate or shorten the destain step.

Acid-fast stain for *Cryptosporidium* sp. Acid-fast staining for *Cryptosporidium* sp. has recently become important because this parasite is now recognized as a cause of severe diarrhea in immunodeficient patients such as those with AIDS, and it can cause transient diarrhea in immunocompetent individuals. The modified acid-fast stain recommended is similar to that used to stain *Nocardia* spp. in that it uses milder acid decolorization. A variety of acid-fast and fluorochrome staining procedures have been described for *Cryptosporidium* spp. (8), and all the procedures appear to work.

The following procedure (21) is useful for staining *Nocardia* species as well as *Cryptosporidium* species. This procedure may be used on fresh or Formalin-fixed material or on material from concentration procedures. If the specimen is liquid, centrifuge it, and use the sediment to prepare a smear.

1. Pick a portion of material with an applicator stick, mix the material in a drop of saline, spread it on a glass slide (1 by 3 in.), and allow it to dry.
2. Fix the dried film in absolute methanol for 1 min, and air dry the slide.
3. Flood the slide with Kinyon carbol-fuchsin, and stain the smear for 5 min.
4. Wash the slide with 50% ethyl alcohol in water, and immediately rinse it with water.
5. Destain the smear with 1% sulfuric acid for 2 min or until no color runs from the slide.
6. Wash the slide with water.
7. Counterstain the smear with Loeffler methylene blue for 1 min.
8. Rinse the slide with water, dry it, and examine the smear with oil immersion.

The results are that *Cryptosporidium* oocysts stain bright red, and background materials stain blue or pale red.

Egg counts

Egg-counting methods are used in clinical studies to assess the intensity of infections (especially infections by intestinal nematodes) and the efficacy of therapeutic agents, and these methods are commonly used for epidemiologic studies. Methods used for scientific studies, such as Kato thick smear (23) or Stoll egg counting (45), require greater accuracy than methods used for patient care. The simplest, most practical method is to use a standard fecal suspension which contains approximately 2 mg of feces mixed in a drop of saline and covered with a cover slip. The entire cover slip is examined at a magnification of 100×, field by field, and the number of eggs is counted. For research work, the density of the smear can be standardized with a light meter (5), but this standardization is not essential for patient care. The number of eggs per cover slip provides a rough index of the severity of the infection.

Duodenal material

The examination of duodenal fluid or duodenal biopsy material may be useful for the diagnosis of giardiasis, strongyloidiasis, or other upper intestinal parasite infections in patients in whom parasites cannot be detected in the feces. In addition, duodenal fluid occasionally can be useful in showing whether helminth eggs are originating in the biliary or intestinal tract. Duodenal material may be obtained by passing a tube through the nose and stomach into the upper small intestine and then aspirating enteric fluid. As an alternative, a string test (Enterotest, Hedeco Corp., Mountain View, Calif.) may be used (4). A weighted gelatin capsule attached to a string is swallowed, and the proximal end of the string is taped to the face of the patient. Over a period of several hours, helped with small sips of water, the string reaches the upper small intestine. After 4 to 5 h, the string is retrieved, and the material on the bile-stained portion is stripped from the string and examined for parasites with direct wet mounts or with permanent stains when wet-mount findings are questionable. Aspirated duodenal fluid is examined in a similar fashion. The material for permanent stains can be fixed in Schaudinn or PVA fixative, although the latter may adhere better to the slide. If questionable organisms are seen in the direct wet mount, the coverslip can be removed, the material can be mixed with a drop of PVA fixative, and a film can be made for later permanent staining.

A duodenal biopsy can be used to demonstrate *Giardia* organisms. A biopsy is usually obtained by a swallowed biopsy capsule. In searching for *Giardia* spp., it is generally preferable to make both impression smears and sections of biopsy tissue. *Giardia* spp. are usually present in mucus or attached to epithelium rather than in tissue. Biopsies occasionally can confirm a diagnosis of strongyloidiasis or cryptosporidiosis.

Sigmoidoscopic material

Materials obtained by sigmoidoscopy may be helpful in the diagnosis or monitoring of amebiases, schistosomiasis, or cryptosporidiosis. Patients suspected of having amebiasis may have ulcerations of the colon which can be visualized by sigmoidoscopy or colonoscopy. Scrapings or aspirates of material from ulcers can be examined by direct wet mounts and permanent stain as described above. The finding of typical, erythrophagocytic, motile trophozoites in direct wet mounts or in permanently stained preparations allows a diagnosis of amebiasis. Material is best aspirated with a pipette or scraped with an instrument. Swabs should not be used, as the parasites may be killed or trapped by swab material.

Biopsy material for amebiasis should be processed for surgical pathology and then examined for ulcers containing amebae. The periodic acid-Schiff stain counterstained with hematoxylin is particularly helpful because amebae stain more positively with periodic acid-Schiff stain than do inflammatory cells, and amebae show typical amebic nuclei. Of course, there are no amebic cysts in tissue.

Biopsy material for schistosomiasis is better examined in teased preparations than in sections, as the entire thickness can be examined at once, and the viability of eggs can be determined by observation of the movement of the larvae within the eggs (25).

In cryptosporidiosis, biopsy material shows organisms at the luminal surface of the epithelial cells, but the organisms are small, and the study of structural detail requires electron microscopy.

Abscess material

Abscesses suspected of being caused by *Entamoeba histolytica* may be aspirated, and the material may be submitted to the laboratory. The last material aspirated is most likely to contain amebae. Material may be examined microscopically in wet mounts and permanent stains, and in addition, it can be cultured for amebae if bacteria are also added to the culture as described below. Abscess material is often thick and difficult to examine. It may be treated with streptokinase and streptodonase enzymes to liquefy the specimen.

1. Reconstitute streptokinase and streptodonase (Varidase; Lederle Laboratories, Pearl River, N.Y.) per the instructions of the manufacturer.
2. Add 1 part enzyme solution to 5 parts aspirated material.
3. Incubate the mixture at 35 to 37°C for 1 h. Shake the mixture at intervals.

4. Centrifuge the mixture at 300 to 400 × *g* for 5 min.

5. The sediment may be used for microscopic examinations for amebae (wet mounts and permanent stains) and for the culture of amebae.

Cellophane tape

Cellophane tape is used for finding the eggs of *Enterobius vermicularis* or *Taenia* species from the perianal area. The tape used must be clear cellophane and not slightly cloudy or opaque. The procedure for obtaining specimens is outlined in Fig. 2. Alternatively, a Vaspar swab may be used (29, 33). Specimens from more than 1 day may be required to diagnose light infections (37).

Examination of cellophane tape.

1. If the specimen is difficult to examine, raise the tape from the front of the glass slide, add a drop of toluene to the slide, and replace the tape smoothly with an applicator stick. (Remember, *Enterobius* spp. and *Taenia solium* eggs are infective!)

2. Examine the entire tape, including the edges, with ×100 magnification (×10 objective).

3. Confirm suspicious objects with high dry objectives (×40 to ×50).

Culture for amebae

Cultures for amebae have improved detection in some studies (15, 31), but they are not widely used. Although *Giardia* spp. have been cultured in research

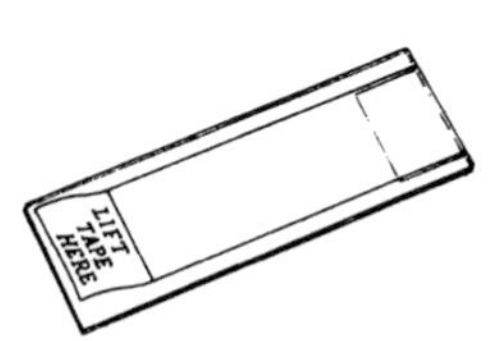

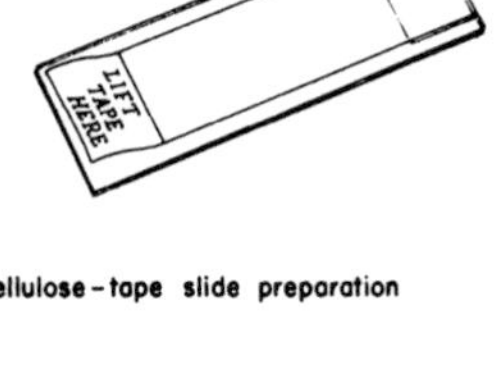

a. Cellulose-tape slide preparation

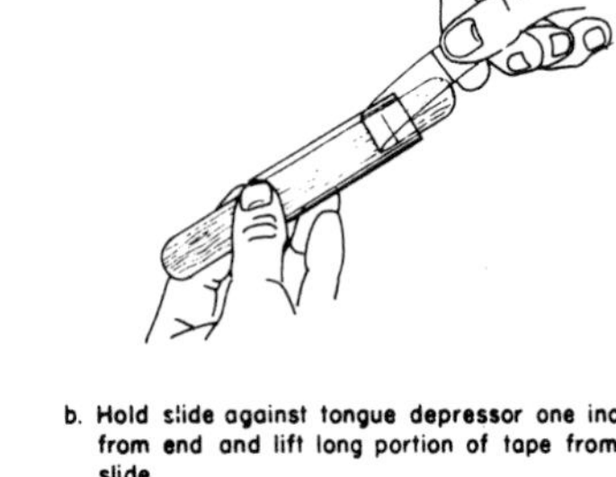

b. Hold slide against tongue depressor one inch from end and lift long portion of tape from slide

c. Loop tape over end of depressor to expose gummed surface

d. Hold tape and slide against tongue depressor

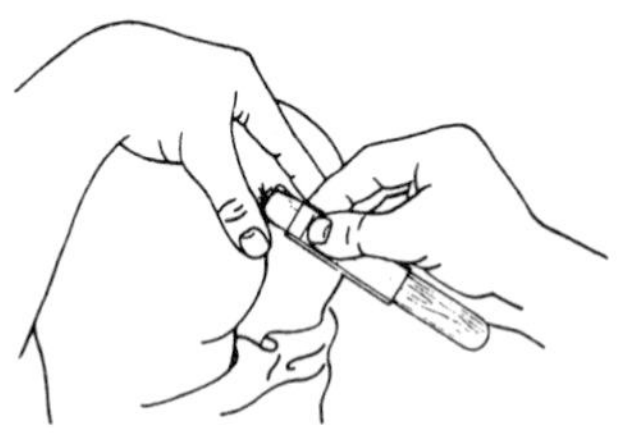

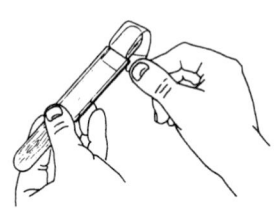

e. Press gummed surfaces against several areas of perianal region

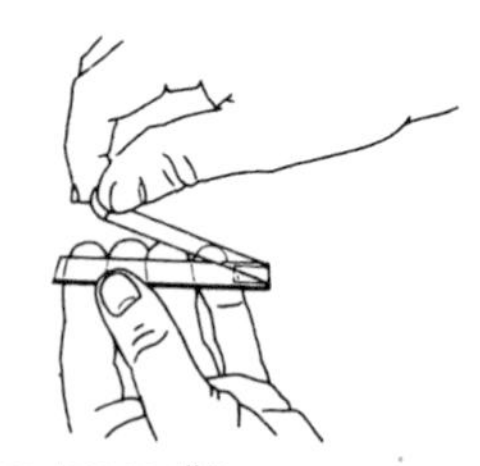

f. Replace tape on slide

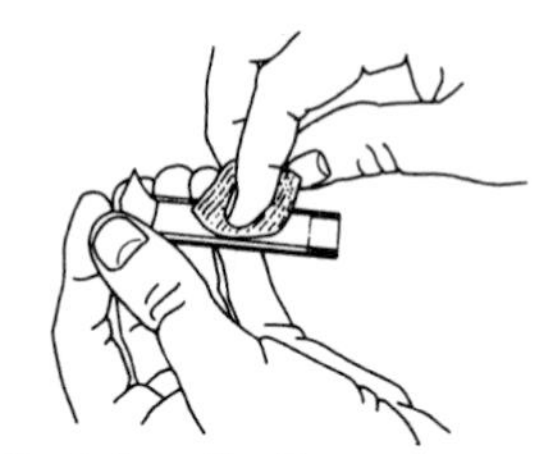

g. Smooth tape with cotton or gauze

Note: Specimens are best obtained a few hours after the person has retired, perhaps at 10 or 11 P.M., or the first thing in the morning before a bowel movement or bath.

FIG. 2. Use of cellulose tape preparation for the diagnosis of pinworm infections. From Melvin and Brooke (33).

laboratories, cultures are not useful for diagnosis.

A variety of culture media for amebae have been described, and some may be purchased from commercial medium manufacturers. The method described here uses the modified charcoal agar slant diphasic medium described by McQuay (31).

1. Place 3 ml of sterile 0.5% phosphate-buffered saline on a charcoal agar slant.

2. Add approximately 30 mg of sterile rice starch to the tube.

3. Warm the medium to 35°C before it is inoculated.

4. (a) Inoculate the medium with fecal specimen (approximately 0.5 ml of liquid specimen or a 0.5-cm sphere of formed specimen) which is mixed with the saline overlay.

(b) If abscess material is cultured, bacteria must be inoculated into the culture in addition to the inoculation with 0.5 ml of specimen. A heavy inoculum with *Clostridium perfringens* or *Escherichia coli* is satisfactory.

5. Incubate the culture at 35°C.

6. At 24 and 48 h, remove 1 drop of liquid from the lowest point of the overlay, and prepare a wet mount.

7. Examine the wet mount for amebae.

8. Permanent stains can be prepared by the fixation of sediment in PVA fixative, with the subsequent preparation of smears and staining.

Larval maturation

Larval maturation studies, sometimes referred to as cultures, can be performed on fecal specimens applied to wet filter paper. Nematode larvae such as *Strongyloides* spp. or hookworm mature to the filariform stages in the culture container and migrate from feces into water, where they are detected microscopically. The procedure can be performed in a petri dish with a square of filter paper or in a large test tube with a strip of filter paper (6, 33).

1. Smear approximately 0.5 g of feces on the filter paper.

2. (a) For the tube method, insert the filter paper strip into the tube so that the bottom of the strip is in 3 ml of water. The feces-smeared portion of the strip need not be immersed in the water.

(b) For the petri dish method, place feces on one half of a piece of filter paper. Lay the feces-bearing end of the filter paper on a glass rod or a portion of an applicator stick in the petri dish. Add approximately 3 ml or sufficient water so that the feces-free end of the filter paper is in the liquid.

3. Leave the tube or dish at room temperature in the dark. Add water as needed to ensure that the filter paper is in contact with the water.

4. Examine the liquid for larvae either by direct microscopic examination with an inverted microscope or by examination of a wet mount of sediment from the liquid. With the petri dish method, the surface of the feces also may be examined with a dissecting microscope.

5. Examine the specimen on days 3, 5, and 7. Strongyloides filariform larvae are found on days 2 and 3, and hookworm larvae are found on days 5 through 7. Larvae are identified by their morphological characteristics.

Adult worms

Adult worms, or objects suspected to be adult worms, may be submitted to the laboratory. The laboratory must determine if these are helminths and, if so, if they are parasites. Identification characteristics are described in standard references (6). Tapeworm proglottids, particularly those of the *Taenia* species, are difficult to differentiate grossly unless they are cleared so that the internal structure can be seen and the number of lateral uterine branches can be counted. One procedure for clearing the proglottids is outlined below.

Clearing *Taenia* proglottids and other helminths. Proglottids are first relaxed by placing them in warm saline (56°C) for 1 h and then clearing them in carbol-xylene while they are kept flat. They may be kept flat in a number of ways. One way is to press the proglottid between two glass slides held together with membrane clips or string. Clearing takes from several hours to overnight.

The proglottid is examined under a dissecting microscope or with a hand lens, and the uterine branching is observed. Glycerine and beechwood creosote can also be used with good results. Cleared proglottids may be mounted or stained if desired.

Small nematodes may also be cleared in carbol-xylene or beechwood creosote and mounted in permount or balsam. This method is particularly good for hookworm adults.

BLOOD AND TISSUE PARASITES

Blood and tissue parasites whose diagnostic forms circulate in the peripheral blood are generally diagnosed by the demonstration of parasites in Giemsa-stained thick or thin films of blood. Special concentration techniques may be helpful for the diagnosis of some diseases such as filarial or trypanosomal infection. Other tissue parasites which do not circulate in the blood may be diagnosed by the detection of parasites in skin snips, lesion scrapings, body fluids, or biopsy material or by the detection of antibody or antigen in serum or other body fluids.

COLLECTION AND HANDLING OF BLOOD SPECIMENS

The timing of the collection of blood specimens depends on the parasite disease suspected. For example, for certain filarial infections, specimens are best obtained between 10:00 p.m. and midnight, whereas for other infections, specimens are best obtained during the day. In malaria, the numbers and stages of parasites in the peripheral blood vary with different parts of the cycle.

Blood films are best made from blood which is not anticoagulated, such as that obtained from finger stick or ear lobe puncture. Anticoagulants may interfere with parasite morphology and staining. Care should be taken that the alcohol disinfectant is allowed to dry before the area is punctured, or there may be fixation of erythrocytes, which will interfere with the preparation and staining of thick films. Both thick and thin films can be prepared from blood obtained by venipuncture, although it is best if the blood remaining in the needle of the venipuncture device is used, because it is anticoagulant free. Thick and thin films can be prepared from blood that is anticoagulated, but the staining characteristics are not as good. EDTA-anticoagulated blood is better for staining than citrate- or heparin-anticoagulated blood.

Both thick and thin blood films are useful. Thick films are more sensitive because the same amount of blood can be examined in a thick film in 5 min as can be examined in a thin film in 30 min. However, thin films allow the study of the effects of parasites on erythrocytes and provide better parasite morphology.

Thick and thin films may be prepared on separate slides or on the same slide, with the thick film at one end and the thin film at the other end.

The thick film is prepared by spreading 1 drop or puddling several small drops of blood into an area approximately 1.5 cm in diameter. A properly prepared thick film should be thin enough so that newspaper print can barely be read through it. If the film is too thick, it will fragment and peel, and if the film is too thin, the increased sensitivity will be lost. Thick films should be allowed to dry overnight and should be stained within 3 days. They must not be heated, and they should be protected from dust. If the erythrocytes are fixed, they will not dehemoglobinize. If prompt examination is required, prepare a slightly thinner thick film, dry it for 1 h, and stain it.

The thin film is prepared in the same manner as a film for a differential leukocyte count. A small drop of blood is placed on one end of a microscope slide. A second slide held at an acute angle of 30 to 45° is backed into the drop of blood, which spreads along the junction of the slides. The spreader slide is then pushed along the slide, and it pulls the drop of blood along behind the angled edge of glass. A properly prepared thin film should have a significant area near the end which is only one erythrocyte thick and in which the erythrocytes show good morphology. The angle and speed of the spreader slide and the size of the drop of blood will influence the thickness and size of the film.

Slides with only a thin film can be fixed by being immersed in absolute methyl or ethyl alcohol for 1 min and allowed to air dry. If the thick film is on one end of the slide and the thin film is on the other end, the thin film is fixed by a brief flooding or by immersion in alcohol and allowed to air dry, while the thick film is protected from alcohol or alcohol fumes. In a well-ventilated area, the slide may be dried vertically with the thick film up or horizontally after the thick film is covered with a dry paper towel.

Thick and thin films are best stained with Giemsa stain, as it provides the most detailed and intense staining of parasites. Wright stain can be used for thin films but not for thick films, as it contains alcohol, which will fix the erythrocytes. Wright stain does not stain parasites as well as Giemsa stain. The staining procedure is outlined below.

Tissue

Biopsy or necropsy tissue may be examined by histology sections or impression smears.

To prepare impression smears (41), tissue should be blotted to remove as much blood or other fluid as possible and then pressed against glass slides (1 by 3 in.) to make a series of impressions. Tissue should stick to the slide slightly and leave an irregular film on the slide. Similar impressions may be made on multiple slides from the same portion of tissue. Portions of biopsy tissue with different gross appearances can be used with the impressions from each portion placed in a longitudinal row. Impressions must be close together, preferably with slight overlapping to make slide scanning easier. Impressions from small fragments may be placed in a small area (1 cm in diameter). After being dried, the area with impressions is circled with a diamond marker to facilitate the location and scanning of the material. Fixatives and stains appropriate for the parasites suspected are used. If amebiasis is suspected, impression smears must be fixed promptly in Schaudinn fixative and not allowed to air dry. For most other parasites, the slides are allowed to dry before fixation in methyl alcohol. Giemsa is the usual stain, but other stains such as Gram-Weigert or hematoxylin may be used depending on the parasite suspected.

Aspirates of bone marrow or spleen

Aspirates of bone marrow or spleen may be useful in the diagnosis of infections such as leishmaniasis, trypanosomiasis, and occasionally malaria. In such instances, Giemsa stains of alcohol-fixed bone marrow films are most useful. Splenic aspiration is rarely performed in the United States because it is dangerous.

Fluids

Fluids such as tissue aspirates, cyst fluid, bronchial washings, cerebrospinal fluid, pleural fluid, and peritoneal fluid can be examined directly, or they can be centrifuged and the sediment examined by wet mounts or stains (or both), depending on the parasite suspected, as described above for abscesses or tissue.

Skin snips

Skin snips may be useful in the diagnosis of microfilarial infections such as onchocerciasis in which the parasites circulate in the skin and not the blood. A small (2-mm) skin snip is taken with a needle and a knife. The needle point is stuck into the skin, and the skin is raised. With a sharp knife or razor blade, the skin is excised just below the needle. Alternatively, a scleral punch may be used. The skin snip is then placed in a small volume (0.2 ml) of saline in a tube or a microtiter well, teased, and allowed to stand for 30 min or more. The microfilariae migrate from the tissue into the saline, which is then examined microscopically to demonstrate the wiggling microfilariae.

Concentration Procedures for Blood

A number of procedures have been described for the concentration of blood specimens. Most of these procedures have been developed to diagnose filarial infections.

The three most widely used methods are membrane filter (13, 14), saponin lysis (32), and Knotts concentration (22). Procedures for the first two methods will be given here, as these methods are the most sensitive. Membrane filter techniques use 5- or 3-μm filters produced by Millipore Corp., Bedford, Mass., or Nuclepore Corp., Pleasanton, Calif. Both filters give satisfactory results, but the procedures with the Nuclepore filters do not require the lysis of erythrocytes (1). Parasites on filters are often not as suitable for morphologic study as are those in thick films.

Membrane filter concentration for filariae

1. Collect approximately 7 ml of blood in EDTA.
2. With a syringe and firm pressure, pass 5 to 7 ml of blood through a 5-μm Nuclepore filter held in a Swinney adapter.
3. Wash the membrane several times with a small amount of distilled water or physiologic saline.
4. The moist filter may be examined directly or fixed and stained in the usual fashion for a thin blood film.

Saponin lysis concentration for filariae

The saponin lysis method (32) can be performed on either EDTA- or citrate-anticoagulated blood. Saponin solution to lyse erythrocytes is available in most laboratories for use with automated hematology instruments.

1. Centrifuge up to 10 ml of blood at 150 × *g* for 10 min.
2. Remove and discard the plasma.
3. Mix the packed erythrocytes with 50 ml of 0.5% saponin solution in 0.85% saline.
4. Mix the solution at intervals for 15 min.
5. Centrifuge the solution at 650 × *g* for 10 min.
6. Decant and discard the supernatant (there should be about 1 ml of sediment).
7. Spread several drops of sediment on a glass slide (1 by 3 in.), and examine two such uncovered wet mounts for motile microfilariae. Allow the wet mounts to dry before they are fixed and stained.
8. Prepare four or five similar wet mounts and examine them as described above. To each slide, immediately add 2 drops of 1% acetic acid solution and mix it well (microfilariae will be killed and straightened). Allow the slide to air dry.
9. Dip the dried slides in buffered methylene blue-phosphate solution.
10. Rinse the slides in distilled water, and let them air dry.
11. Stain the mounts for 10 min in a 1:20 dilution of Giemsa stain in buffered water.
12. Examine the slides microscopically.

Staining Procedures

Giemsa stain procedure

The procedures for staining thick and thin films differ. Staining is usually done in a Coplin jar. The stain must be made fresh each day.

Stain slides with only a thin film as follows.

1. Fix and dry the blood film as described above.
2. Prepare a 1:40 dilution of stock Giemsa stain in neutral buffered water, pH 7.0 to 7.2 (generally, 2 ml of Giemsa stock plus 38 ml of buffered water with 0.01% Triton X-100).
3. Stain the film for approximately 60 min (the time, which will vary slightly with different lots of stock Giemsa stain, can be determined by the staining of leukocytes and erythrocytes).
4. Wash the slide briefly by dipping it in buffered water.
5. Air dry the slide in a vertical position.

Note that, alternatively, a 1:20 dilution for 20 to 30 min may be used.

Stain slides with only a thick film as follows.

1. Do not fix the slide.
2. Prepare a 1:50 dilution of stock Giemsa stain in neutral buffered water (pH 7.0 to 7.2).
3. Stain the film for approximately 50 min (the optimal time may vary with different lots of stain).
4. Wash the slide by placing it in buffered water for 3 to 4 min.
5. Air dry the slide in a vertical position.

For combination thick and thin films, the procedure is as follows.

1. Fix the thin film but not the thick film as described above.
2. Stain the film in a 1:50 dilution of Giemsa stain in neutral buffered water (pH 7.0 to 7.2) for approximately 50 min.
3. Rinse the thin film briefly by dipping it in buffered water. Wash the thick film by immersing it in buffered water for 3 to 5 min.
4. Dry the slide in a vertical position with the thick film down.

Gram-Weigert stain procedure

The Gram-Weigert stain (36) is used to stain the cyst walls of *P. carinii* cysts. It also stains fungi and many bacteria. Impression smears, sediment smears, or sections are fixed in methanol and air dried. For sections, when reagents are added, flood the slide gently from the end opposite the section, and rinse the slide carefully so that the tissue is not washed from the slide.

The stain procedure is as follows.

1. Stain the slide with eosin Y for 5 min.
2. Wash the slide with water.
3. Stain the slide with crystal violet for 5 min.
4. Rinse the crystal violet from the slide with Gram iodine solution.
5. Leave the iodine solution on the slide for 5 min.
6. Rinse the slide with water.
7. Blot the smears carefully (do not blot the sections).
8. Wipe the reverse of the slide.
9. Air dry the slide completely.
10. Decolorize the smear in aniline-xylene, agitating the slide gently until no purple runs from it (the use of a second Coplin jar of aniline-xylene after the majority of blue stain has been removed aids the visual assessment of decolorization).
11. Rinse the slide in xylene.
12. Air dry the slide, add immersion oil to it, and examine it.

P. carinii cysts and fungi stain dark blue and somewhat irregularly. Cell nuclei may stain blue if they are inadequately decolorized, but they are not as dark as *P. carinii* cysts.

Culture Procedures for Blood and Tissue Parasites

Culture procedures have been developed for a number of blood and tissue parasites, but these procedures are used primarily in research. The culturing of *Leishmania* spp. and *Trypanosoma cruzi* may be helpful for diagnosis, and the procedures are easy to use.

Biopsy or blood specimens may be cultured for *Leishmania* spp. or *T. cruzi* with Novy-MacNeal-Nicolle (NNN) medium. Biopsy specimens are ground in a small amount of saline. Biopsies from skin lesions or other tissues which may contain bacteria may have penicillin (0.1 ml of 1,000 U/ml) added to the medium with the inoculum. The inoculum is 1 drop of ground tissue or blood. Incubate the culture at room temperature (22°C), and at days 3 and 7, examine a direct mount of liquid from the bottom of the slant at ×400 magnification. These cultured organisms are potentially infective for humans.

URINE

Urine specimens usually are examined for the eggs of *Schistosoma haemotobium* or the trophozoites of *Trichomonas vaginalis*, although occasionally the larvae of *Strongyloides stercoralis* may be found in patients with hyperinfection syndrome. Urine is the usual specimen for the diagnosis of *Trichomonas* infection in males. See below (Vaginal Material) for culture method. Urine is centrifuged, and the sediment is examined microscopically.

SPUTUM

Sputum may be examined to diagnose *Paragonimus* infection or hyperinfection due to *Strongyloides stercoralis*. Occasionally an amebic abscess or hydatid cyst may rupture, and amebic trophozoites or hydatid sand, respectively, may be found in sputum. *Entamoeba histolytica* must be differentiated from *Entamoeba gingivalis*, which may be found in the oral cavity of over 30% of people (12). Occasionally, the migrating larvae of ascarids, strongyloides, or hookworm can be found. Sputum may be examined directly by wet mount or treated with a mucolytic agent such as *N*-acetyl-cysteine and then concentrated by simple centrifugation, with subsequent examination of the sediment.

VAGINAL MATERIAL

T. vaginalis frequently infects the vagina, and *Enterobius vermicularis* adults or eggs occasionally may be found. Direct wet mounts of vaginal material for typical, tumbling *T. vaginalis* organisms are widely used and generally allow the diagnosis of symptomatic infection, but wet mounts are not as sensitive as culture methods.

Vaginal material is best submitted as liquid in a tube, although swabs submitted in a small amount of saline may be used. A drop of the material is covered with a cover slip and examined with reduced light. To culture, 1 or 2 drops of urine sediment or vaginal exudate are inoculated into tubes of warmed, modified Diamond medium. If vaginal swabs are submitted, the swab is immersed in the medium and pressed against the side of the tube to express material. Tubes are incubated at 35°C, and drops of culture are examined by wet mount at 48 and 72 h for motile trophozoites.

REFERRAL OF MATERIALS

Few laboratories perform complete parasitological examination, whereas many perform limited studies, and some perform none. Referral laboratories may provide services not available in the individual laboratory and can provide consultation on specimens with questionable laboratory findings. Referral laboratories with a special interest and competence in parasitology may be found in major cities, university medical centers, and state public health laboratories. The major national resource is the Centers for Disease Control in Atlanta, Ga. Specimens for the Centers must be sent via the state health laboratory, and appropriate clinical information must be provided. Guidelines for handling specimens to be sent to a referral laboratory are outlined in Table 5. Of course, the recommendations of the specific referral laboratory should supersede these guidelines.

SAFETY

The parasitology laboratory has infection hazards for personnel. Blood, feces, and other body materials as well as parasite cultures may be infective. Eggs of *Ascaris* spp. can survive and embryonate even in Formalin, and *Cryptosporidium* oocysts are hardy. In fresh fecal specimens, the cysts of *Entamoeba histolytica* and *Giardia* spp., the oocysts of *Cryptosporidium* spp., the eggs of *Enterobius vermicularis, Taenia solium*, and *Hymenolepis nana*, and the larvae of *Strongyloides stercoralis* may be infective. In addition, feces may contain other infectious agents such as hepatitis A, rotavirus, *Salmonella* spp., *Shigella* spp., and *Campylobacter* spp. Blood and tissue specimens can be infectious for trypanosomes, *Leishmania* spp., malaria, and *Babesia* spp., as well as for non-A, non-B hepatitis, hepatitis B, and possibly AIDS.

Reagents such as mercury-containing fixatives (Schaudinn and PVA fixatives) may be toxic, and solvents such as ether may be flammable. These materials must be handled and discarded properly.

QUALITY ASSURANCE

The parasitology laboratory must have an up-to-date procedure manual and appropriate reference materials which might include color atlases (2, 44) or 35-mm slide collections (40, 42, 43), permanently stained glass slides, wet fecal material containing parasites, and one or more standard reference books on laboratory methods (17, 33) or general medical parasitology (6, 28). The persons performing parasitic examinations must be competent in the identification of parasites which might be found in patients from whom they receive specimens. Methods should allow the ready use of outside consultants, if there is a question of diagnosis. Personnel may maintain proficiency through participation in formal courses or workshops, review of self-study sets, and periodic review of known positive materials. Participation in external survey programs is particularly valuable, as the performance of the laboratory in the identification of unknown specimens can be compared with the performance of other laboratories.

If a laboratory is unable to do accurate parasitology because of either the types of procedures offered or the

TABLE 5. Handling of specimens for referral

Specimen	Handling
Feces, for	
Helminths	Fix in 10% buffered Formalin.
Protozoa	Fix a portion in 10% buffered Formalin and either fix a portion in PVA fixative or prepare three Schaudinn-fixed fecal films.
Cryptosporidium spp.	Fix a portion in 10% buffered Formalin.
Material from suspected amebic abscess	Place the last material aspirated in a sterile tube and send it on ice for culture (do not freeze). Prepare Schaudinn-fixed fecal films, or fix a portion in PVA fixative. Obtain serum for serology.
Duodenal aspirate	Centrifuge, and remove the supernatant. Prepare two films from sediment. Fix in Schaudinn or PVA fixative. Preserve the remainder of sediment in 10% Formalin.
Urine, for	
Trichomoniasis	Centrifuge. Cover the sediment with sterile saline and send it on ice (not frozen) for direct mounts and culture.
Schistosomiasis	Centrifuge entire midday urine. Add an equal volume of 10% buffered Formalin to the sediment.
Sputum, for	
Nematode larvae or *Paragonimus* eggs	Break up mechanically or digest 1 part sputum plus 5 parts 3% NaOH for 1 h. Centrifuge, and preserve the sediment in an equal volume of 10% buffered Formalin.
Amebae	Prepare films fixed in Schaudinn fixative, or fix a portion in PVA fixative.
Blood	
Malaria and babesiasis	Send unstained and, if available, Giemsa-stained thick and thin films. Fix thin film (but not thick) in alcohol before it is sent.
Filariasis	Send 5 ml of citrate- or EDTA-anticoagulated blood on ice (not frozen). Unfixed thick films may be sent in addition. Send serum for serologic tests.
Trypanosomiasis	Send 5 ml of anticoagulated blood as for filariasis (above). Send fixed thin films.
Cerebrospinal fluid	
Trypanosomes, toxoplasma, leishmania, trichinella	Send on ice (not frozen).
Free-living amebae	Send in a sterile container without refrigeration.
Sigmoidoscopic material	Fix films in Schaudinn fixative or mix material with PVA fixative.
Tissue	For impression smears when *E. histolytica* is suspected, fix in Schaudinn or PVA fixative. When toxoplasma, leishmania, *Pneumcoystis* spp., or *Trypanosoma cruzi* is suspected, prepare multiple impression smears and fix in methyl alcohol. For surgical pathology, fix the tissue in buffered Formalin.
Whole worms or proglottids	Wash debris from the specimen and send it in saline. If there are multiple worms or proglottids, some may be fixed in Formalin.

quality of personnel available, it should arrange to have specimens appropriately prepared and submitted to a reference laboratory.

LITERATURE CITED

1. **Abaru, D. E., and D. A. Denham.** 1976. A comparison of the efficacy of the Nuclepore and Millipore filtration systems for detecting microfilariae. Southeast Asian J. Trop. Med. Public Health **7**:367–369.
2. **Ash, L. R., and T. C. Orihel.** 1984. Atlas of human parasitology, 2nd ed. American Society of Clinical Pathologists Press, Chicago.
3. **Bartlett, M. S., K. Harper, N. Smith, P. Verbanac, and J. W. Smith.** 1978. Comparative evaluation of a modified zinc sulfate flotation technique. J. Clin. Microbiol. **7**:524–528.
4. **Beal, C. B., P. Viens, R. G. L. Grant, and J. M. Hughes.** 1970. A new technique for sampling duodenal contents. Am. J. Trop. Med. Hyg. **19**:349–352.
5. **Beaver, P. C.** 1950. The standardization of fecal smears for estimating egg production and worm burden. J. Parasitol. **36**:451–455.
6. **Beaver, P. C., R. C. Jung, and E. W. Cupp.** 1984. Clinical parasitology, 9th ed. Lea & Febiger, Philadelphia.
7. **Blagg, W., E. L. Schlaegel, N. S. Mansour, and G. I. Khalaf.** 1955. A new concentration technic for the demonstration of protozoa and helminth eggs in feces. Am. J. Trop. Med. Hyg. **4**:23–28.
8. **Bronsdon, M. A.** 1984. Rapid dimethyl sulfoxide-modi-

fied acid-fast stain of *Cryptosporidium* oocysts in stool specimens. J. Clin. Microbiol. **19**:952–953.

9. **Carroll, M. J., J. Cook, and J. A. Turner.** 1983. Comparison of polyvinyl alcohol- and Formalin-preserved fecal specimens in the Formalin-ether sedimentation technique for parasitological examination. J. Clin. Microbiol. **18**:1070–1072.
10. **Center for Disease Control.** 1979. Intestinal parasite surveillance annual summary 1978. Center for Disease Control, Atlanta.
11. **Current, W. L., N. C. Reese, J. V. Ernst, W. S. Bailey, M. B. Heyman, and W. M. Weinstein.** 1983. Human cryptosporidiosis in immunocompetent and immunodeficient persons. N. Engl. J. Med. **308**:1252–1257.
12. **Dao, A. H., D. P. Robinson, and S. W. Wong.** 1983. Frequency of *Entamoeba gingivalis* in human gingival scrapings. Am. J. Clin. Pathol. **80**:380–383.
13. **Dennis, D. T., E. McConnell, and G. B. White.** 1976. Bancroftian filariasis and membrane filters: are night surveys necessary? Am. J. Trop. Med. Hyg. **25**:257–262.
14. **Desowitz, R. S., and J. C. Hitchcock.** 1974. Hyperendemic bancroftian filariasis in the Kingdom of Tonga: the application of the membrane filter technique to an age-stratified blood survey. Am. J. Trop. Med. Hyg. **23**:877–879.
15. **Edelman, M. H., and C. L. Spingarn.** 1977. Cultivation of *Entamoeba histolytica* as a diagnostic procedure: a brief review. Mt. Sinai J. of Med. **43**:27–32.
16. **Faust, E. C., J. S. D'Antoni, V. Odom, M. J. Miller, C. Peres, W. Sawitz, L. F. Thomen, J. E. Tobie, and J. H. Walker.** 1938. A critical study of clinical laboratory technics for the diagnosis of protozoan cysts and helminth eggs in feces. Am. J. Trop. Med. **18**:169–183.
17. **Garcia, L. S., and L. R. Ash.** 1979. Diagnostic parasitology: clinical laboratory manual, 2nd ed. The C. V. Mosby Co., St. Louis.
18. **Garcia, L. S., T. C. Brewer, and D. A. Bruckner.** 1979. A comparison of the formalin-ether concentration and trichrome stained smear methods for the recovery and identification of intestinal protozoa. Am. J. Med. Technol. **45**:932–935.
19. **Gardner, B. B., D. J. Del Junco, J. Fenn, and J. H. Hengesbaugh.** 1980. Comparison of direct wet mount and trichrome staining techniques for detecting *Entamoeba* species trophozoites in stools. J. Clin. Microbiol. **12**:656–658.
20. **Gleason, N. N., and G. R. Healy.** 1965. Modification and evaluation of Kohn's one-step staining technic for intestinal protozoa in feces or tissue. Am. J. Clin. Pathol. **43**:494–496.
21. **Haley, L. D., and P. G. Standard.** 1973. Kenyoun's acid-fast stain (modified). Laboratory methods in medical mycology, 3rd ed. Center for Disease Control, Atlanta.
22. **Knott, J. I.** 1939. A method for making microfilarial surveys on day blood. Trans. R. Soc. Trop. Med. Hyg. **33**:191–196.
23. **Komiya, Y., and A. Kobayashi.** 1966. Evaluation of Kato's thick smear technic with a cellophane cover for helminth eggs in feces. Jpn. J. Parasitol. **19**:59–64.
24. **Krogstad, D. J., H. C. Spencer, Jr., G. R. Healy, N. N. Gleason, D. J. Sexton, and C. A. Herron.** 1978. Amebiasis: epidemiologic studies in the United States, 1971–1974. Ann. Intern. Med. **88**:89–97.
25. **Lichtenberg, F., and C. Valladares.** 1955. Compression examination of fresh tissue for ova of *Schistosoma mansoni*. Am. J. Clin. Pathol. **25**:1099–1102.
26. **Markell, E. K., L. R. Ash, D. M. Melvin, D. V. Moore, F. Sogandares-Bernal, and M. Voge.** 1977. Procedures suggested for use in examination of clinical specimens for parasitic infection. A statement by the Subcommittee on Laboratory Standards, Committee on Education, American Society of Parasitology. J. Parasitol. **63**:959–960.
27. **Markell, E. K., and P. M. Quinn.** 1977. Comparison of immediate polyvinyl alcohol (PVA) fixation with delayed Schaudinn's fixation for the demonstration of protozoa in stool specimens. Am. J. Trop. Med. Hyg. **26**:1139–1142.
28. **Markell, E. K., and M. Voge.** 1981. Medical parasitology, 5th ed. The W. B. Saunders Co., Philadelphia.
29. **Markey, R. L.** 1950. A vaseline swab for the diagnosis of *Enterobius* eggs. Am. J. Clin. Pathol. **20**:493.
30. **Martin, L. K.** 1965. Randomness of particle distribution in human feces and the resulting influence on helminth egg counting. Am. J. Trop. Med. Hyg. **14**:747–759.
31. **McQuay, R. M.** 1956. Charcoal medium for growth and maintenance of large and small races of *Entamoeba histolytica*. Am. J. Clin. Pathol. **26**:1137–1141.
32. **McQuay, R. M.** 1970. Citrate-saponin-acid method for the recovery of microfilariae from blood. Am. J. Clin. Pathol. **54**:743–746.
33. **Melvin, D. M., and M. M. Brooke.** 1982. Laboratory procedures for the diagnosis of intestinal parasites, 3rd ed. U.S. Department of Health, Education and Welfare publication no. (CDC) 82-8282. Centers for Disease Control, Atlanta.
34. **Nair, C. P.** 1953. Rapid staining of intestinal amoebae on wet mounts. Nature (London) **172**:1051.
35. **Ritchie, L. S.** 1948. An ether sedimentation technique for routine stool examinations. Bull. U.S. Army Med. Dept. **8**:326.
36. **Rosen, P. P., N. Martini, and D. Armstrong.** 1975. *Pneumocystis carinii* pneumonia: diagnosis by lung biopsy. Am. J. Med. **58**:795–801.
37. **Sadun, E. H., and D. M. Melvin.** 1956. The probability of detecting infections with *Enterobius vermicularis* by successive examinations. J. Pediatr. **48**:431–438.
38. **Sapero, J. J., and D. K. Lawless.** 1953. The "MIF" stain preservation technique for the identification of intestinal protozoa. Am. J. Trop. Med. Hyg. **2**:613–619.
39. **Smith, J. W.** 1979. Identification of fecal parasites in the special parasitology survey of the College of American Pathologists. Am. J. Clin. Pathol. **72**(Suppl.):371–373.
40. **Smith, J. W., L. R. Ash, J. H. Thompson, R. M. McQuay, D. M. Melvin, and T. C. Orihel.** 1976. Diagnostic parasitology—intestinal helminths. American Society of Clinical Pathologists, Chicago.
41. **Smith, J. W., and M. S. Bartlett.** 1982. Laboratory diagnosis of *Pneumocystis carinii* infection, p. 393–406. *In* W. C. Winn (ed.), Clinics in laboratory medicine. The W. B. Saunders Co., Philadelphia.
42. **Smith, J. W., R. M. McQuay, L. R. Ash, D. M. Melvin, T. C. Orihel, and J. H. Thompson.** 1976. Diagnostic parasitology—intestinal protozoa. American Society of Clinical Pathologists, Chicago.
43. **Smith, J. W., D. M. Melvin, T. C. Orihel, L. R. Ash, R. M. McQuay, and J. H. Thompson.** 1976. Diagnostic parasitology—blood and tissue parasites. American Society of Clinical Pathologists, Chicago.
44. **Spencer, F. M., and L. S. Monroe.** 1982. The color atlas of intestinal parasites, 2nd ed. Charles C Thomas, Publisher, Springfield, Ill.
45. **Stoll, N. R.** 1923. Investigations on the control of hookworm disease. XV. An effective method of counting hookworm eggs in feces. Am. J. Hyg. **3**:59–70.
46. **Thornton, S. A., B. S. Hanna West, H. L. Dupont, and L. K. Pickering.** 1983. Comparison of methods for identification of *Giardia lamblia*. Am. J. Clin. Pathol. **80**:858–860.
47. **Watson, J. M., and R. Al-Hafidh.** 1957. A modification of the Baermann funnel technique and its use in establishing the infection potential of human hookworm carriers. Ann. Trop. Med. Parasitol. **51**:15–16.
48. **Wheatley, W. B.** 1951. A rapid staining procedure for intestinal amoebae and flagellates. Am. J. Clin. Pathol. **21**:990–991.
49. **Yang, J., and T. Scholten.** 1977. A fixative for intestinal

parasites permitting the use of concentration and permanent staining procedures. Am. J. Clin. Pathol. **67**:300–304.

50. **Yarinsky, A., and S. D. Sternberg.** 1963. A study of paired purged stool specimens for the recovery of *Entamoeba histolytica*. Am. J. Clin. Pathol. **40**:598–600.
51. **Young, K. H., S. L. Bullock, D. M. Melvin, and C. L. Spruill.** 1979. Ethyl acetate as a substitute for diethyl ether in the Formalin-ether sedimentation technique. J. Clin. Microbiol. **10**:852–853.

Chapter 56

Blood and Tissue Protozoa

DONALD J. KROGSTAD, GOVINDA S. VISVESVARA, KENNETH W. WALLS, AND JAMES W. SMITH

In addition to the protozoa which infect the gastrointestinal tract, other protozoa may infect the blood and tissues. Protozoan infections readily demonstrable in blood or tissue accessible to biopsy (malaria, cutaneous leishmaniasis) are usually diagnosed morphologically. Infections of the central nervous system and other deep sites (toxoplasmosis) are usually diagnosed by serology.

MALARIA

Malaria is of overwhelming importance in the developing world with 150 to 200 million cases causing over 1 million deaths each year. Although malaria is distinctly less common in industrialized countries, 300 to 400 cases are diagnosed each year in the United States and reported to the Centers for Disease Control (CDC) (4). Infections caused by *Plasmodium falciparum* are more fulminant than those caused by other plasmodia and may produce coma and renal failure within 2 to 3 days in nonimmune patients. In addition, *P. falciparum* infections are often resistant to chloroquine. Therefore, the laboratory must provide a rapid diagnosis of the infecting species so the clinician can choose appropriate antimalarial agents and anticipate the likely complications.

Four plasmodia infect humans: *P. falciparum, Plasmodium vivax, Plasmodium ovale,* and *Plasmodium malariae.* For clinical purposes, *P. ovale* (which is uncommon) is very similar to *P. vivax*. In contrast, *P. malariae* rarely causes acute illness. Therefore, the differential diagnosis of malaria in the acutely ill patient can usually be considered as the distinction between *P. falciparum* and *P. vivax*.

Malaria is transmitted to humans through the inoculation of infectious sporozoites by female anopheline mosquitoes. Those sporozoites then travel via the bloodstream to the liver, where they infect hepatocytes. After a delay of 8 to 25 days (depending on the plasmodial species), the sporozoites mature to tissue schizonts and release merozoites which enter the bloodstream and infect erythrocytes (Fig. 1). An asexual replication cycle then recurs in the bloodstream at regular intervals, depending on the species (48 h for *P. falciparum*, 48 h for *P. vivax* and *P. ovale*, or 72 h for *P. malariae*), until chemotherapy, acquired immunity, or death supervenes. In relapsing malarias (*P. vivax* and *P. ovale*), some of the sporozoites entering hepatocytes become dormant. These hypnozoites can develop to mature tissue schizonts 6 to 24 or more months later and provide a morphologic correlate to explain the phenomenon of relapsing malaria. Primaquine, which is used to prevent relapses, eradicates hypnozoites (25, 26). Although gametocytes do not produce disease in humans, they are essential to complete the life cycle of the parasite. In the mosquito, macro- and microgametocytes (derived from infected erythrocytes) fuse to form an ookinete, which eventually produces infectious sporozoites (Fig. 1) (24).

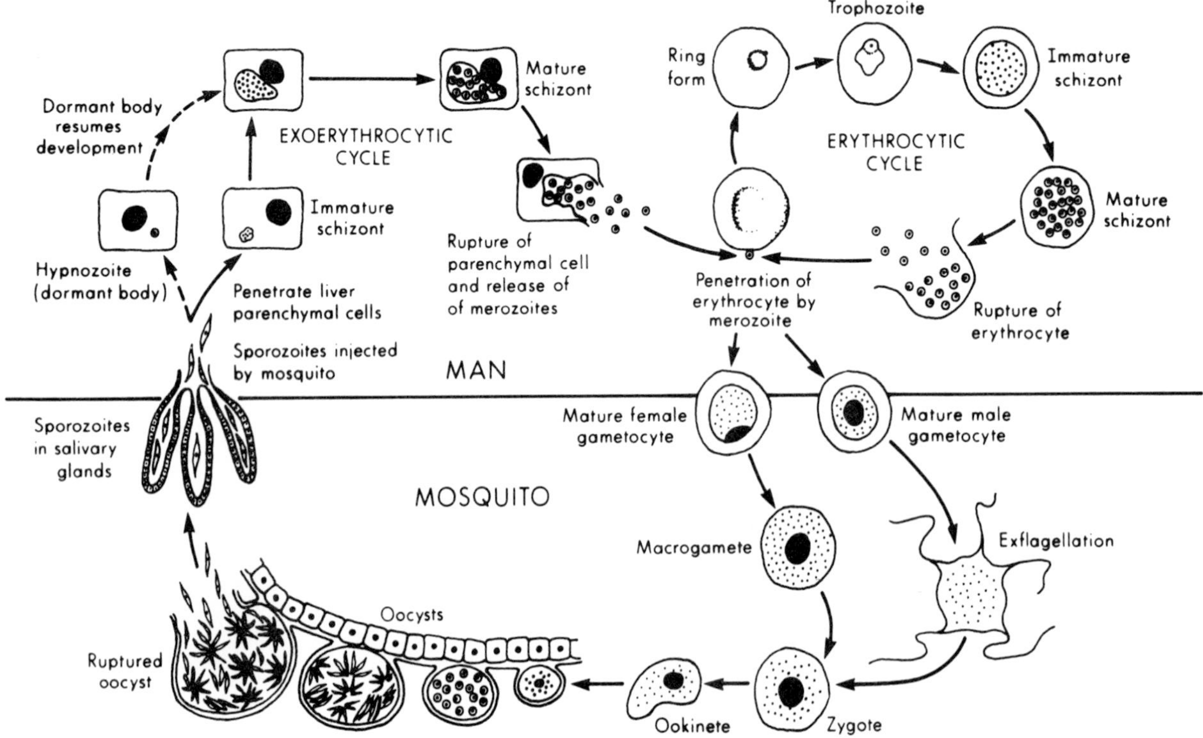

FIG. 1. Life cycle of the malaria parasite (reproduced from reference 24 with permission of McGraw-Hill Book Co.).

TABLE 1. Simplified identification of plasmodia[a]

Characteristic	*P. vivax* (and *P. ovale*)	*P. falciparum*	*P. malariae*
Enlarged infected cells	+	0	0
Schuffner stippling	+	0	0
Multiple erythrocytic stages on smear	+	0	+
Multiply infected erythrocytes	±	+	0

[a] The most reliable criteria are erythrocyte enlargement and Schuffner stippling. Other criteria (in the bottom two lines of the table) are relative and should not be taken as absolute (see the text for details).

For practical purposes, malaria is endemic only in the Third World. However, the traveler acutely ill with malaria is often misdiagnosed because the typical symptoms of fever, chills, and myalgia are nonspecific and may not develop until after one's return to an area in which malaria is not endemic. Malaria may also be transmitted by the transfusion of infected blood or blood products, including the sharing of syringes among drug addicts, and on occasion from mother to child (congenital malaria) (24).

TABLE 2. Plasmodial infection and erythrocyte age[a]

Plasmodium species	Erythrocyte age	Maximal parasitemia (per μl)
P. malariae	Old	10,000
P. vivax (and *P. ovale*)	Young	25,000
P. falciparum	Any age	≥1,000,000

[a] Adapted from Neva (33).

Diagnosis of the infecting species

Characters for identification of malaria species are outlined in Tables 1 through 3 and illustrated in Fig. 2

TABLE 3. Morphology of *Plasmodium* species affecting humans[a]

Species	Appearance of erythrocyte: Size	Appearance of erythrocyte: Schuffner stippling	Appearance of parasite: Cytoplasm	Appearance of parasite: Pigment	Appearance of parasite: Number of merozoites	Stages found in circulating blood
P. vivax	Enlarged; maximum size (attained with mature trophozoites and schizonts) may be 1.5 to 2 times normal erythrocyte diameter	+ With all stages except early ring forms	Irregular, ameboid in trophozoites; has "spread-out" appearance	Golden brown, inconspicuous	12–24; avg, 16	All stages; wide range of stages may be seen on given film
P. malariae	Normal	– (Ziemann's dots rarely seen)	Rounded, compact trophozoites with dense cytoplasm; band-form trophozoites occasionally seen	Dark brown, coarse, conspicuous	6–12; avg, 8; "rosette" schizonts occasionally seen	All stages; wide variety of stages usually not seen; relatively few rings or gametocytes generally present
P. ovale	Enlarged; maximum size may be 1¼ to 1½ times normal erythrocyte diameter; ca. 20% or more of infected erythrocytes are oval or fimbriated or both (border has irregular projections)	+ With all stages except early ring forms	Rounded, compact trophozoites; occasionally slightly ameboid; growing trophozoites have large chromatin mass	Dark brown, conspicuous	6–14; avg, 8	All stages
P. falciparum	Normal; multiply infected erythrocytes are common	– (Maurer's dots occasionally seen)	Young rings are small, delicate, often with double chromatin dots; gametocytes are crescent or elongate	Black; coarse and conspicuous in gametocytes	6–32; avg, 20–24	Rings or gametocytes or both; other stages develop in blood vessels of internal organs but are not seen in peripheral blood except in severe infections

[a] See reference 50.

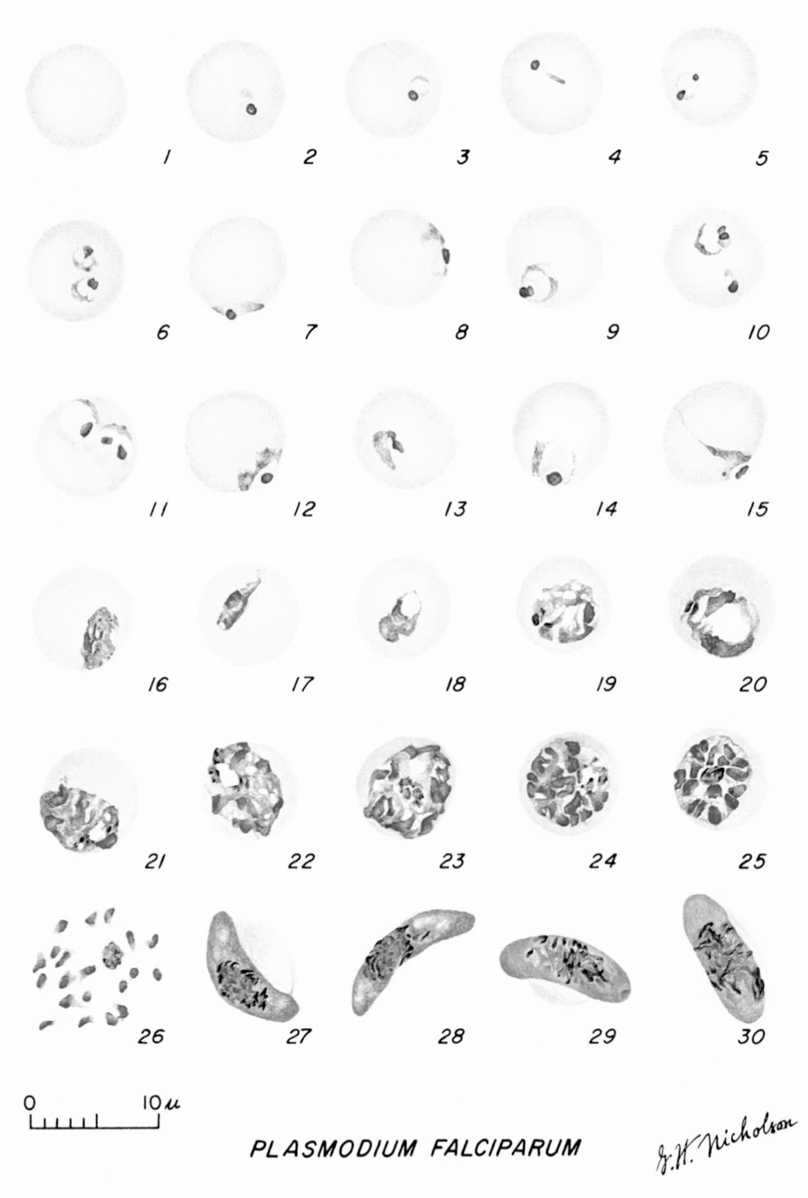

FIG. 2. *Plasmodium falciparum* growth stages. Successive developmental stages as they appear in stained blood films. Rings and gametocytes will be found in peripheral blood films; other stages occur in erythrocytes but are sequestered in the capillaries of internal organs and hence are rarely found in peripheral blood films. Drawings: 1, normal erythrocyte; 2–11, young trophozoites; 12–15, growing trophozoites; 16–18, mature trophozoites; 19–22, immature schizonts; 23–26, nearly mature and mature schizonts; 27 and 28, mature macrogametocytes; 29 and 30, mature microgametocytes.

through 8. Morphologic diagnosis is the only practical means of diagnosing malaria within a clinically relevant time frame. Because *P. falciparum* infections are often chloroquine resistant (4a, 24), rapid diagnosis of the infecting species is essential for appropriate therapy. Thin and thick films of blood are prepared and stained with Giemsa as described in Chapter 55. Serologic tests are not useful for the diagnosis of acute infections because most patients require 3 or more weeks to produce a diagnostic rise in antibody titer.

The more sensitive thick blood films (thick smears; Fig. 6 through 8) permit one to examine volumes of blood approximately 10-fold greater than on thin smears (Fig. 2 through 5). The sensitivity is greater because the erythrocytes on the thick smear are lysed by exposure to hypotonic Giemsa stain without prior methanol fixation. The result is that only parasites and leukocytes remain after staining. One disadvantage of thick smears is that one cannot determine whether a given plasmodium increased the size of its host erythrocyte as it matured, as do both *P. vivax* and *P. ovale*. Although most malariologists prefer thick

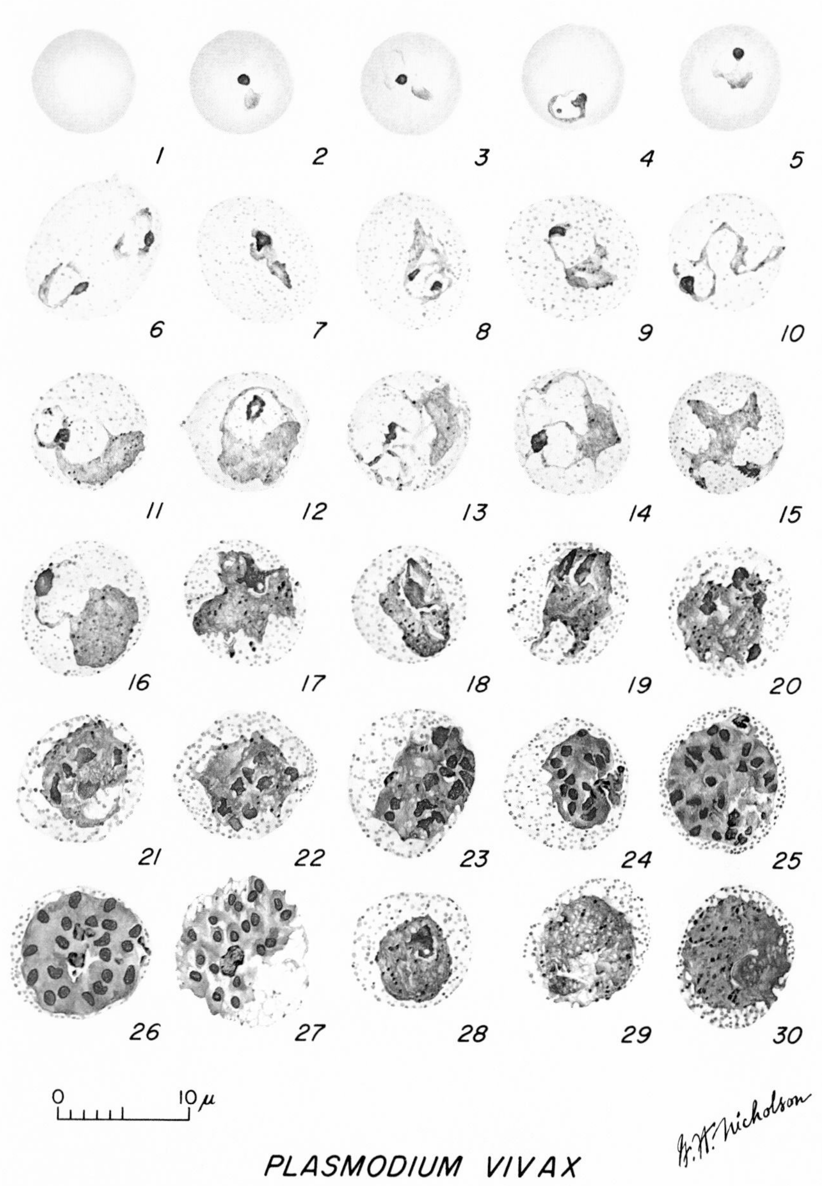

FIG. 3. *Plasmodium vivax* growth stages. Successive developmental stages as they appear in stained blood films. Drawings: 1, normal erythrocyte; 2–5, young trophozoites; 6–16, growing trophozoites; 17 and 18, mature trophozoites; 19–21, young (early) immature schizonts; 22 and 23, older immature schizonts; 24–27, nearly mature and mature schizonts; 28 and 29, nearly mature and mature macrogametocytes; 30, mature microgrametocyte.

smears, investigators who infrequently examine positive blood films usually find that thick smears are more difficult to read than thin smears. Thus, despite the theoretical advantage of thick smears (greater sensitivity), we believe they are not optimal for laboratories that rarely examine positive smears.

The most important pitfall in Giemsa staining is failure to control the pH of the phosphate buffer, which should be between 7.0 and 7.2. The staining of Schuffner stippling (which is present in erythrocytes infected by *P. vivax* or *P. ovale*) is particularly pH dependent, and incorrectly buffered stain may cause *P. vivax* parasites without visible Schuffner stippling to be mistaken for *P. falciparum*.

In microscopic examination of the stained blood film, inexperienced observers commonly make two mistakes.

1. Inadequate magnification. Ring-stage malaria parasites are often ≤2 μm in diameter. For this reason, oil immersion magnification (≥×1,000) is essential. Standard high-power magnification without oil (×440) is inadequate to distinguish malaria parasites from platelets, precipitated stain, and nonspecific debris.

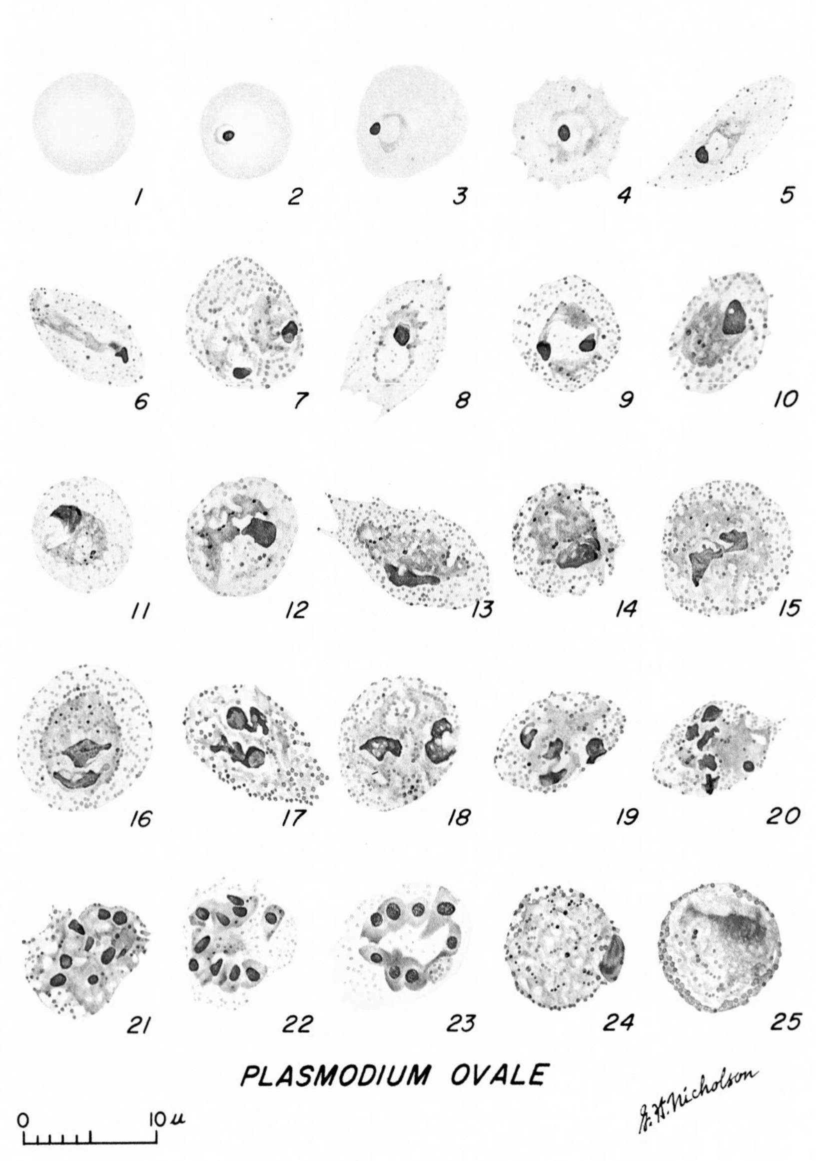

FIG. 4. *Plasmodium ovale* growth stages. Successive developmental stages as they appear in stained blood films. Drawings: 1, normal erythrocyte; 2–5, young trophozoites; 6–12, growing trophozoites; 13–15, mature trophozoites; 16–22, immature schizonts; 23, mature schizont; 24, mature macrogametocyte; 25, mature microgametocyte.

2. Confusion of parasites with platelets. Platelets are similar in size to malaria parasites and are often mistaken for plasmodia when they are on top of an erythrocyte in the blood film. This confusion can usually be resolved by identifying other platelets which are not within erythrocytes by their similar morphology (on thin smears) and by determining that no chromatin dots, signet rings, or pigment are present.

Identification of the infecting species

For acutely ill patients, the infecting species is usually either *P. falciparum* or *P. vivax* and can often be identified microscopically by using the relatively simple criteria of cell size, Schuffner stippling, variety of parasite stages, and multiply infected erythrocytes (Table 1). However, if Schuffner stippling is not present and if most parasites are ring stages, it may be difficult or impossible to distinguish between *P. falciparum* and *P. vivax* (or *P. ovale*). In such situations and in patients who may be infected by more than one malaria species, more subtle criteria are used, such as the number of parasites per cell and the central versus peripheral location of the parasite within the erythrocyte. However, these criteria are less reliable and can be misleading (e.g., *P. vivax* infections may produce

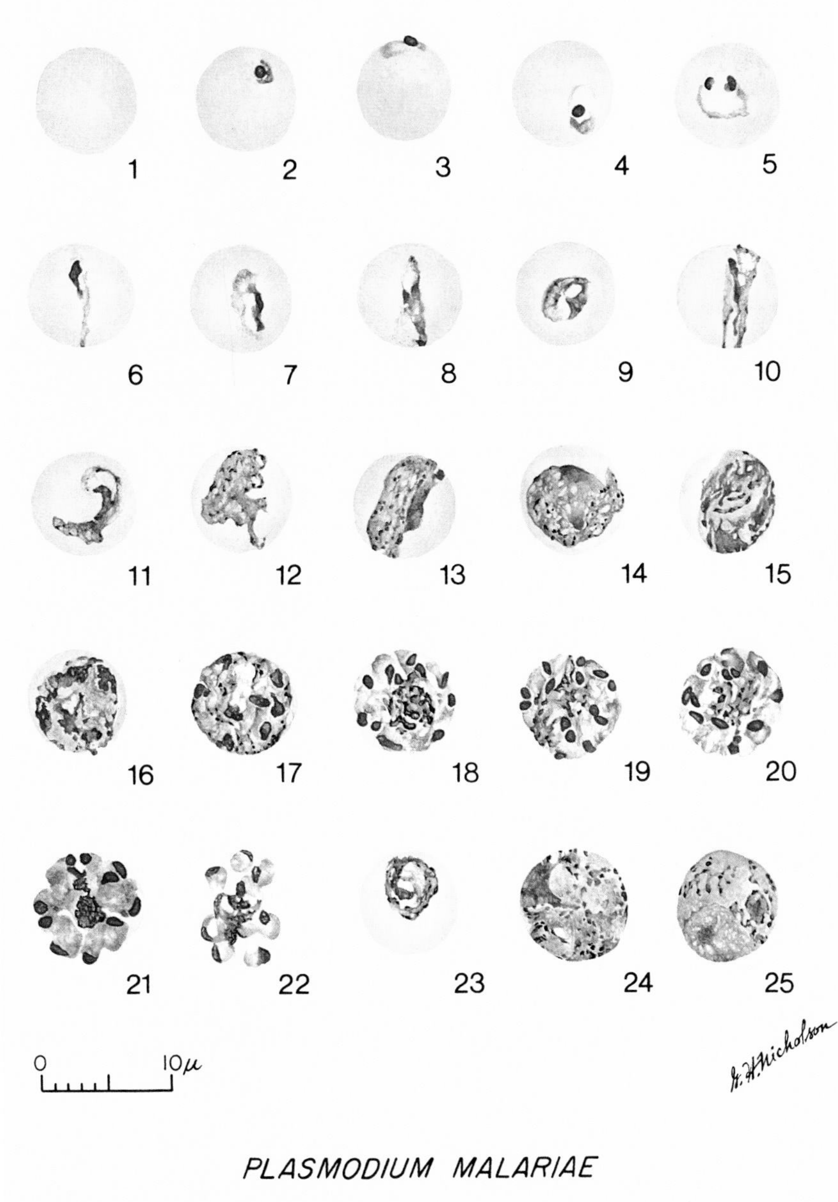

FIG. 5. *Plasmodium malariae* growth stages. Successive developmental stages of *P. malariae* as they appear in stained blood films. Drawings: 1, normal erythrocyte; 2–5, young trophozoites; 6–11, growing trophozoites; 12, 13, nearly mature and mature trophozoites; 14–20, immature schizonts; 21, 22, mature schizonts; 23, developing gametocyte; 24, mature macrogametocyte; 25, mature microgametocyte.

more than one parasite per erythrocyte, have double chromatin dots, and have parasites at the periphery of the erythrocyte). Other ancillary criteria include the intensity of the parasitemia (*P. falciparum* produces the highest parasitemias because it can invade erythrocytes of all ages) (Table 2) (33), the presence of various stages including trophozoites or schizonts on the blood smear (only ring forms and gametocytes circulate in *P. falciparum* infection), and the number of merozoites in a mature schizont. Characteristics for all four species are summarized in Table 3.

Although gametocytes of *P. falciparum* are diagnostic, their absence does not permit one to exclude that diagnosis. *P. falciparum* gametocytes take longer to mature than the asexual stages of the parasite (8 to 10 days versus 48 h), and nonimmune travelers often become severely ill before sufficient time has elapsed to permit the maturation of gametocytes in vivo. Thus, despite their prominence in textbooks, gametocytes are rarely present on blood smears obtained from acutely ill nonimmune patients.

Examination of the blood smear after treatment

Treatment with antimalarial agents may radically

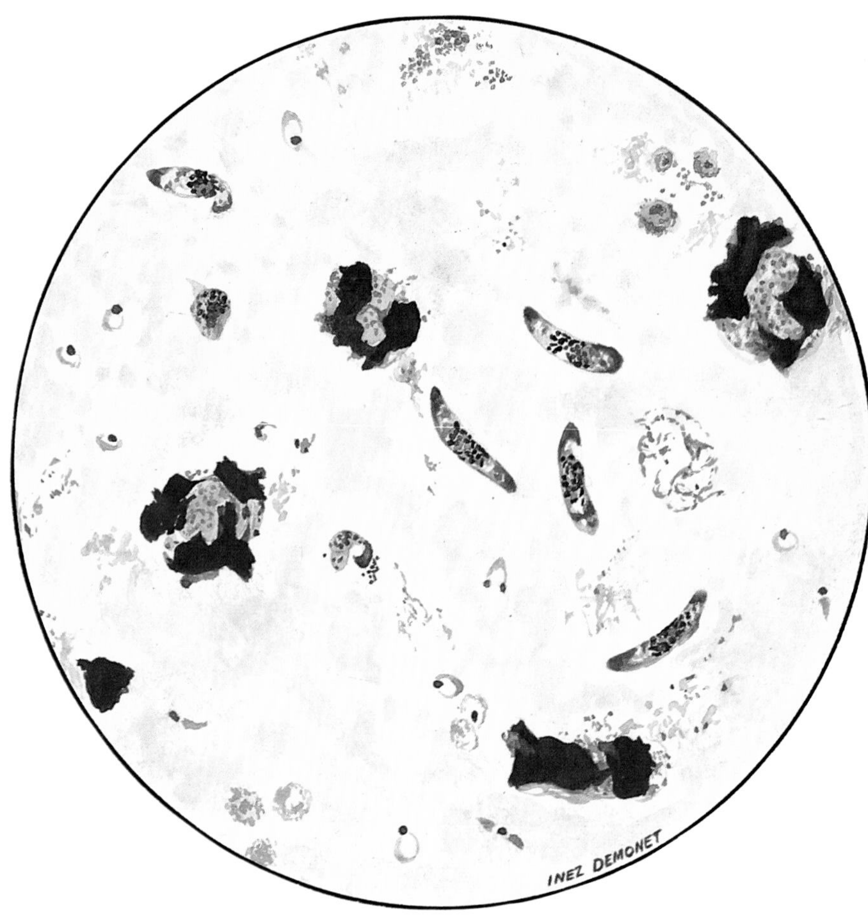

FIG. 6. Morphology of *Plasmodium falciparum* in thick blood film. There are multiple small (ring form) trophozoites and sausage-shaped gametocytes. In addition, there are two gametocytes toward 10 o'clock which have other shapes. (Reproduced from reference 59.)

change parasite morphology within hours. In addition, the synchronous nature of these infections means that the apparent parasitemia may drop precipitously during day 2 of the cycle in infections such as *P. falciparum* (which are associated with peripheral sequestration of trophozoite- and schizont-containing erythrocytes) even if therapy has been ineffective. For these reasons, apparent changes in viability and numbers of parasites should be interpreted with caution.

Serologic diagnosis

Because serologic testing for malaria requires several weeks for a change in antibody titer to occur and another delay while the specimen is sent to the CDC or another reference laboratory, it is not useful for the diagnosis of acutely ill patients. However, if the morphologic diagnosis is not clear (e.g., because smears were obtained only after treatment, because the patient was treated empirically, or because of a very low parasitemia), serologic testing may be helpful. For example, patients with indirect immunofluorescence (IIF) patterns suggestive of recent *P. vivax* infection (51) should receive a 14-day course of primaquine (to prevent a late relapse due to persistent hypnozoites in the liver) even if their original illness responded to treatment with chloroquine alone. Antigens currently tested at CDC are *P. falciparum*, *P. vivax*, and *P. malariae*. No commercial reagents or kits are available. Serologic testing is often helpful in making retrospective diagnoses in tourists or others with single exposures because the titers observed correlate with the infecting species in acute infections. Serologic testing is also helpful in the identification of infected blood donors associated with transfusion malaria. It is not particularly useful for individual patients in endemic areas because of cross-reactivity and because titers may remain elevated for years, although the testing may be helpful in epidemiologic studies. In persons with a defined single exposure (e.g., U.S. citizens), titers of $<$1:64 probably indicate no clinical involvement, 1:64 indicates a probable recent exposure, and higher titers indicate recent clinical involvement.

BABESIOSIS

Babesia species that may infect humans include *B. microti*, a rodent parasite, and *B. bovis*, a parasite of cattle which has been associated with disease in splenectomized patients. Like plasmodia, babesia live

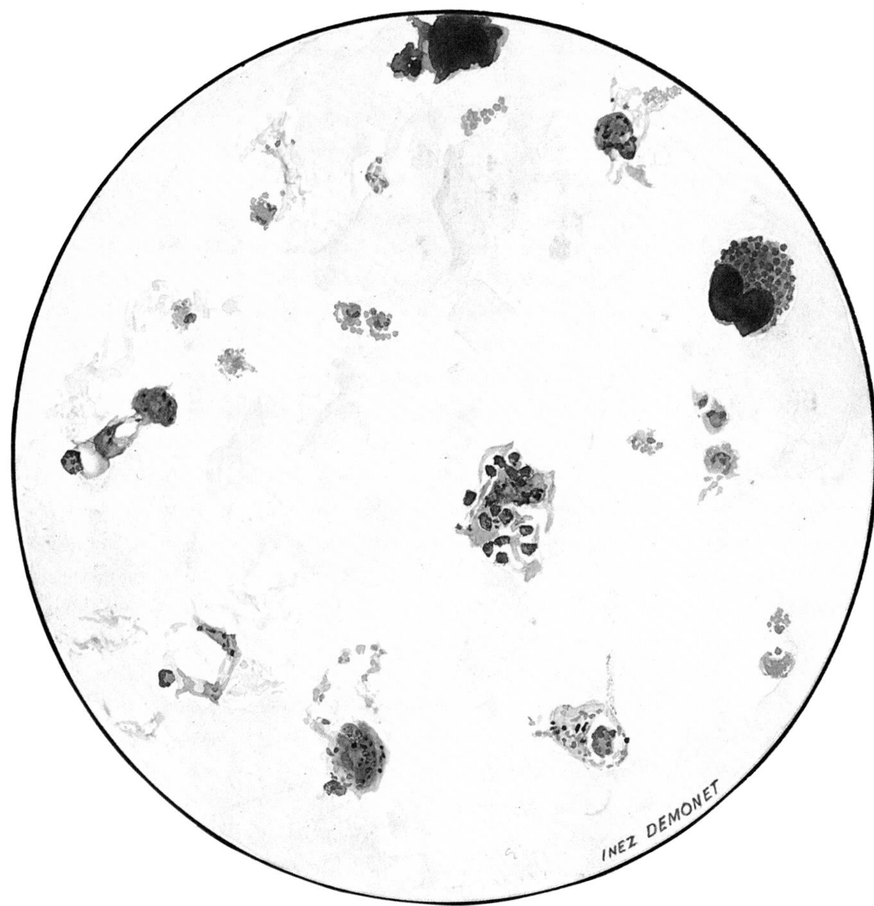

FIG. 7. Morphology of *Plasmodium vivax* in thick blood film. Ameboid trophozoites at 2, 8, and 9 o'clock. Immature schizont with two chromatin masses at 7 o'clock. Macrogametocyte at 5 o'clock. Mature schizont near center. (Reproduced from reference 59.)

within erythrocytes and (in severe infections) may produce massive hemolysis. In contrast to plasmodia, babesia have no known exoerythrocytic stages. Babesiosis was first described in humans as a fatal infection of splenectomized patients and other immunosuppressed hosts. However, subsequent serologic studies have shown that undiagnosed babesia infection is relatively common among residents of areas such as Martha's Vineyard in which the disease is known to be endemic, and that most patients recover uneventfully (41).

Babesiosis is a zoonosis that involves humans only accidentally. Its reservoirs are wild mammals such as deer and mice and the tick vector, which can perpetuate the infection by transovarial transmission. The distribution of infected wildlife and the insect vector explains the prevalence of babesiosis on the east coast of the United States, including Nantucket Island, Martha's Vineyard, and Shelter Island.

Diagnosis

Recent studies by Wittner and his colleagues indicate that the combination of clindamycin plus quinine is effective for the treatment of babesiosis in humans (61). Because chloroquine is ineffective against the parasite despite its antipyretic activity (32), the laboratory must be able to recognize babesiosis and to distinguish it from malaria, for which chloroquine is usually the treatment of choice.

In contrast to patients with malaria, most patients with babesiosis are not critically ill. For patients with subacute babesia infection, the delays involved in serologic diagnosis are acceptable, and initial serologic titers are often diagnostic. However, morphologic diagnosis is essential for critically ill patients who must be diagnosed and treated rapidly. Diagnosis by the inoculation of susceptible animal hosts is more sensitive than diagnosis by morphology and is often useful in chronically infected patients. However, this technique is available only in research laboratories.

Babesiosis is usually diagnosed by using Giemsa-stained thin smears. Babesia parasites are typically smaller than plasmodia, but there is overlap. Erythrocytes may contain multiple parasites which are usually close together. Pigment is not seen, even in erythrocytes with multiple parasites, whereas pigment would be present in malaria schizonts. If only ring forms are seen and there are neither pigment nor gametocytes, it may be impossible to differentiate babesiosis from malaria (especially *P. falciparum*) on a single smear. Except in splenectomized patients, babesia parasite-

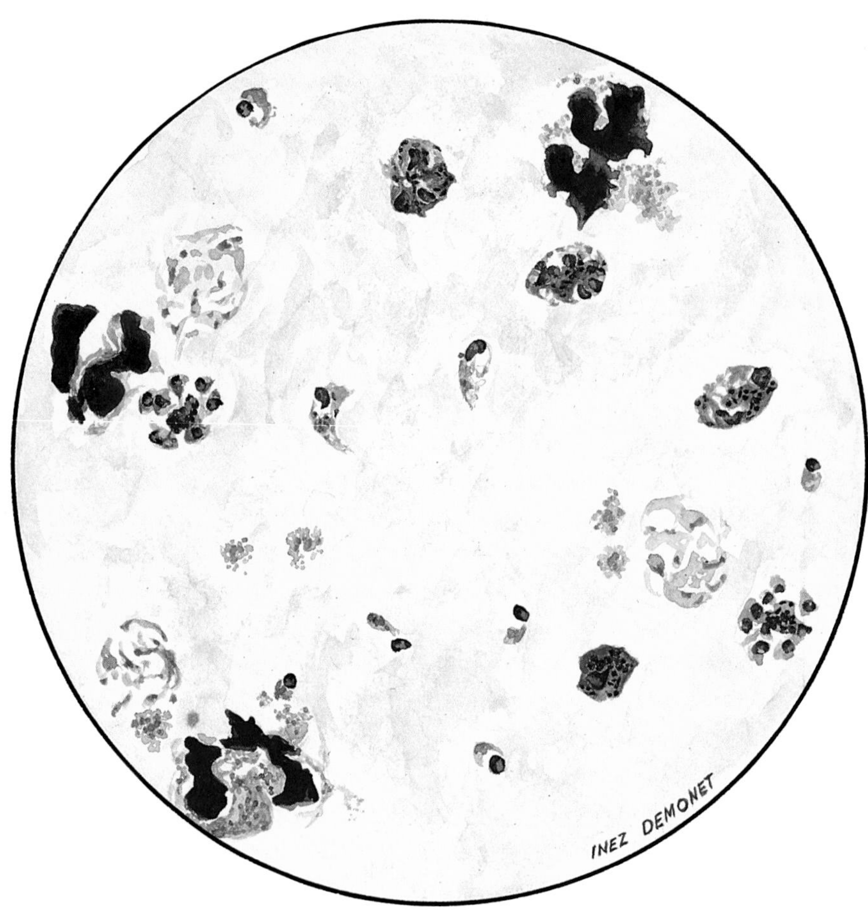

FIG. 8. Morphology of *Plasmodium malariae* in thick blood film. Small trophozoites toward 6 o'clock. Growing trophozoites slightly above center. Mature trophozoite at 5 o'clock. Immature schizonts at 12, 1, and 3 o'clock. Mature schizonts at 4 and 9 o'clock. (Reproduced from reference 59.)

mias are typically low. The small ring forms are frequently mistaken for platelets. The most readily recognizable stage is the tetrad, which has four pyriform organisms grouped together with no pigment (Fig. 9). Trophozoites occasionally may also be seen free (outside erythrocytes) in the peripheral smear (17).

Serologic testing is most frequently employed for epidemiologic studies of patient populations potentially exposed to babesia. However, it may also be useful for the evaluation of individual patients with subacute symptoms who do not have identifiable parasites in their blood smears, since less severely ill patients may have parasitemias of <0.1%, which are difficult or impossible to diagnose by morphology alone. The value of serology is unfortunately compromised by cross-reactivity with malaria. Therefore, because there are ambiguities of both morphology and serology, a careful history of overseas travel and exposure is essential before interpreting a positive result as diagnostic of babesiosis. The test is available at CDC (no commercial reagents or kits are available) and uses IIF.

LEISHMANIASIS

Leishmania spp. cause a variety of clinical illnesses in humans depending on the ability of the organism to proliferate in deep tissues (at 37°C) or near the skin surface at lower temperatures (e.g., 25°C). Although molecular studies with restriction endonucleases and isozyme patterns should ultimately provide a sound biochemical basis for species identification of the leishmania (22, 60), laboratory diagnoses are generally based on smears or histopathology (only occasionally on cultures at this time) and additional clinical information to characterize the species (i.e., the geographic area of exposure, anatomic sites involved, and overall clinical picture).

Human infections may be produced by *Leishmania donovani* (kala-azar), which typically involves liver, spleen, and bone marrow, or *Leishmania tropica*, *Leishmania mexicana*, and *Leishmania braziliensis*, which produce cutaneous or mucocutaneous infections.

Each of the leishmania requires a phlebotomus sandfly to complete its life cycle. The epidemiology of these infections also requires a reservoir of infected mammalian hosts (dogs, jackals, and rats for *L. dono-*

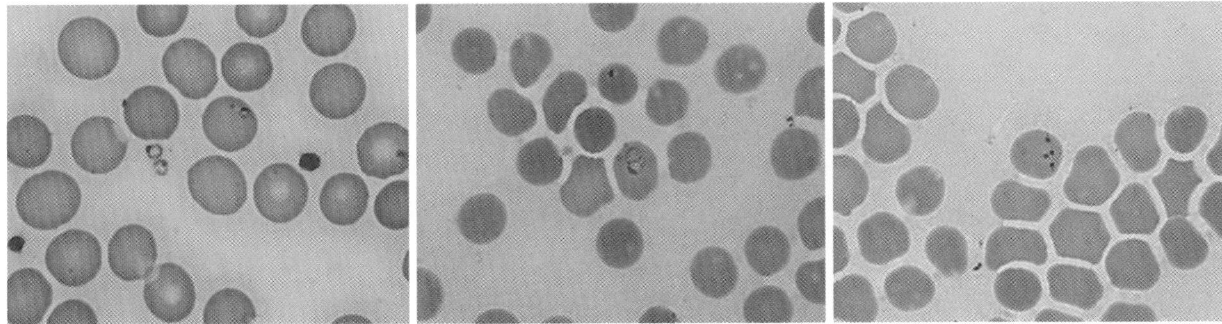

FIG. 9. Morphology of *Babesia microti* in thin blood films, Giemsa stain. (A) Erythrocyte containing ring form plus two extraerythrocytic babesia organisms. (B) Erythrocyte containing small ring and erythrocyte containing more mature trophozoite without pigment. (C) Erythrocyte containing tetrad. (Courtesy of George Healy, CDC.)

vani; gerbils and other rodents for *L. tropica*, *L. mexicana*, and *L. braziliensis*). Because of these constraints, leishmaniasis is an imported disease in countries such as the United States.

Diagnosis

Morphologic diagnosis is the most accepted method for the identification of these intracellular parasites which are typically found in the vacuoles of mononuclear cells or macrophages. In tissue sections or impression smears stained with Giemsa, the amastigote form of the parasite (Fig. 10) is identified by the presence of both the darkly staining kinetoplast and a lighter-staining nucleus (34). *L. donovani* is usually diagnosed in specimens from liver, spleen, bone marrow, or lymph nodes. When scrapings or biopsies of cutaneous or mucocutaneous lesions are obtained for other leishmania, care must be taken to sample the active margin of the lesion and to avoid confusing gram-positive cocci (which are normal skin flora) with leishmania. Gram-positive cocci typically resemble the kinetoplast alone, without a nucleus or a surrounding mononuclear phagocytic cell.

Culture of the blood (or buffy coat) in kala-azar or of aspirates or skin scrapings in the cutaneous forms of the disease is definitive (Chapter 55). However, such culture may provide false-negative results if antibiotics are not added to the cultures to suppress bacterial growth or if the biopsy or scraping is inadequate.

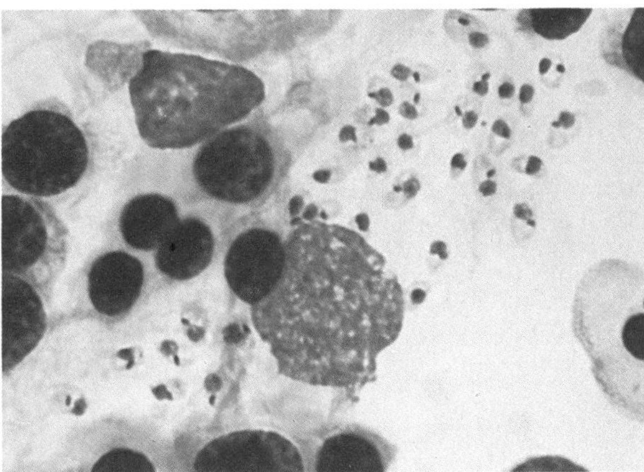

FIG. 10. Macrophage containing intracellular amastigotes of *Leishmania donovani* (reproduced with permission from reference 50).

Serologic diagnosis. Serologic tests may be of value in visceral leishmaniasis but are of more limited usefulness in the cutaneous form of the disease. Positive serologic results are especially helpful in residents of the developed world who have had defined exposures. However, because the test may remain positive for years, it is less useful for the study of persons who have resided for years in endemic areas. The most frequently used serologic test is the IIF. Although a promastigote antigen obtained from cultured parasites may be used to diagnose visceral disease, an amastigote antigen is necessary to test for the cutaneous form (57). The older complement fixation (CF) test using a mycobacterial antigen is also useful (30). Specimens should be tested with both IIF and CF. Titers of ≥1:16 (IIF) or ≥1:8 (CF) are significant. Identification of the infecting species is not possible with serology alone, and cross-reactions occur in patients with Chagas' disease. No commercial reagents or kits are available.

TRYPANOSOMIASIS

Neither American trypanosomiasis (Chagas' disease) nor African trypanosomiasis (sleeping sickness) is a major health problem in the United States or other developed countries. Although virtually all human cases of these diseases are imported, occasional endogenous human cases of Chagas' disease may occur in the South, Southeast, and Southwest United States.

Trypanosomes known to infect humans include *Trypanosoma cruzi* (the cause of Chagas' disease) and *Trypanosoma brucei gambiense* and *Trypanosoma brucei rhodesiense* (which cause sleeping sickness).

Chagas' disease

The life cycle of *T. cruzi* requires both an animal reservoir (usually rodents) and an insect vector. Both the reduviid (triatomid) vector and an animal reservoir exist in an area extending from Georgia to California and presumably account for the few reported cases of Chagas' disease in lifelong residents of the United States (62). Infectious parasite forms are present in feces of the vector and are rubbed into the bite

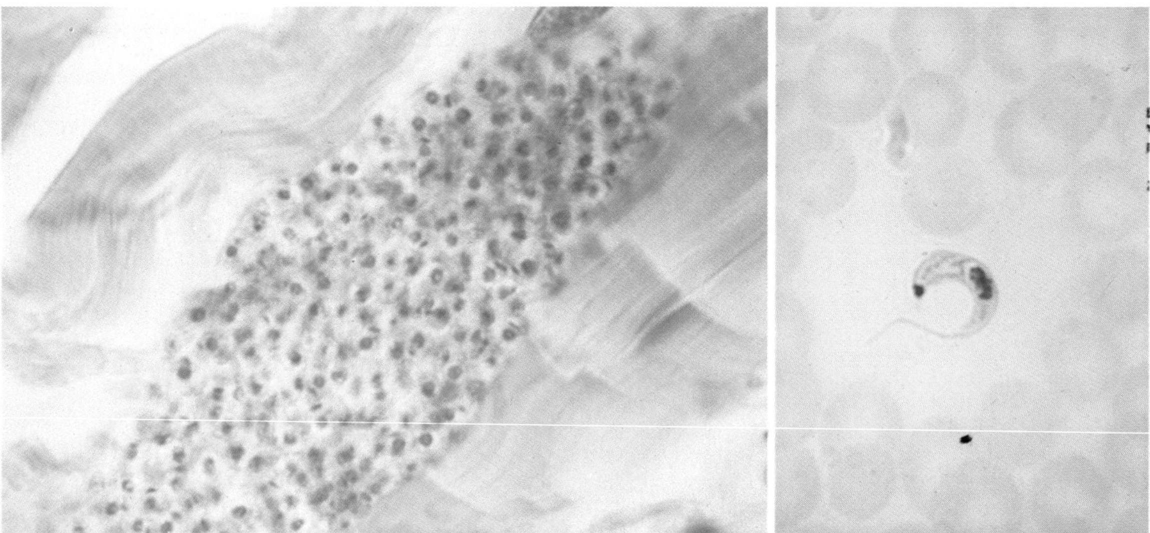

FIG. 11. *Trypanosoma cruzi* amastigotes in skeletal muscle, hematoxylin and eosin stain, and trypomastigote in peripheral blood, Giemsa stain (reproduced with permission from reference 50).

wound, causing a local lesion (the chagoma). Proliferation of the amastigote stage occurs in organs such as the heart. Trypomastigotes may be present in peripheral blood early in the disease, but they do not proliferate. Acute disease usually resolves without treatment. More chronic sequelae of the disease such as myocardial conduction defects, megacolon, and megaesophagus develop years after the initial infection.

African trypanosomiasis

The transmission of African trypanosomiasis to humans requires tsetse flies. These insects are not present in the United States, which helps to limit this disease to Africa. The reservoir of African trypanosomiasis is primarily in humans in West Africa (*T. brucei gambiense*) and in wildlife such as impala in East Africa (*T. brucei rhodesiense*).

Infection with *T. brucei rhodesiense* classically produces septicemic disease with generalized lymphadenopathy and may also cause fatal encephalitis within a few months. In contrast, *T. brucei gambiense* causes a more indolent disease which rarely produces encephalitis (sleeping sickness) in less than 2 to 3 years.

Diagnosis

In Chagas' disease, trypomastigote forms of the parasite may be found in the peripheral blood early in the disease (within the first several months). Later, the organism can be seen (in the amastigote form) on histopathologic examination of involved organs such as the heart (Fig. 11), although the organism is often difficult to find. Culture of peripheral blood on NNN media (Chapter 111), when positive, is also diagnostic. Trypomastigotes of *T. cruzi* are typically less sinuous than those of the African trypanosomes and have a larger kinetoplast (Fig. 11). However, the diagnosis is usually established by serologic testing.

A number of tests have been evaluated, with indirect hemagglutination (5) and CF (28) persisting as the tests of choice. Both tests are sensitive but have major cross-reactivity with leishmaniasis. The CF test, less sensitive than the indirect hemagglutination test but more specific, is most useful in the diagnosis of acute infection. Rising CF titers are highly suggestive of active acute disease. Realizing that serologic results are not absolute, a CF titer of $\geq$1:8 is considered indicative of acute infection, whereas an indirect hemagglutination result of $\geq$1:128 indicates chronic disease, especially if the CF test is negative. Usually both tests are performed on each specimen to provide the maximum information. Recently, an enzyme-linked immunosorbent assay (ELISA) test has been evaluated. The assay is promising, but has not been adopted for clinical use. There are no commercial reagents in the United States.

In contrast to *T. cruzi*, the organisms of African trypanosomiasis remain extracellular as trypomastigotes and proliferate in that stage. They are more sinuous and have a smaller kinetoplast than that of *T. cruzi* (Fig. 12). They circulate in the peripheral blood acutely and may be seen in biopsies or aspirates of involved lymph nodes, as well as in bone marrow. Late in the disease, trypanosomes may be found in cerebrospinal fluid by Giemsa stain of smears of the cell pellet after centrifugation. Increased levels of immunoglobulin M (IgM) in spinal fluid are characteristic, although not diagnostic.

Because single exposures tend to produce titers that remain positive, serologic testing is most valuable for the study of tourists who have had only a single defined exposure. No kits or reagents are commercially available.

TOXOPLASMOSIS

The high prevalence and protean manifestations of toxoplasmosis make diagnosis extremely challenging. Since *T. gondii* is an obligate intracellular parasite, culture attempts are frequently negative even in

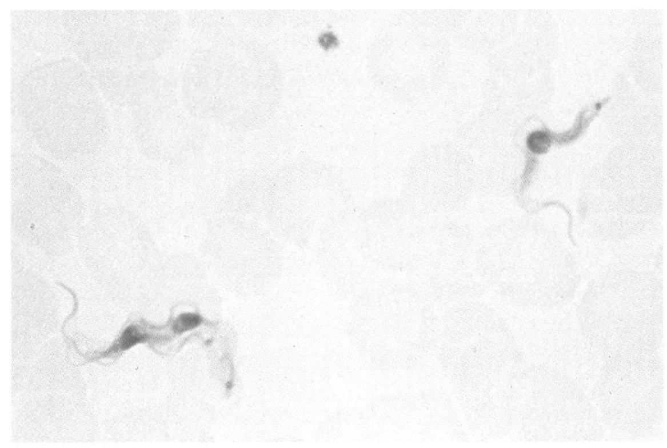

FIG. 12. Trypomastigotes of *Trypanosoma brucei rhodesiense* in mouse; thin film, Giemsa stain. One trypomastigote is in the process of cell division (reproduced with permission from reference 50).

known cases. Conversely, a large number of healthy individuals have asymptomatic chronic infections with pseudocysts (Fig. 13), and the organism can be isolated from biopsied tissue. Thus, isolation of the parasite from tissue by culture or mouse inoculation, even when successful, is only an aid to diagnosis and is not diagnostic of acute infection. Trophic forms (tachyzoites) are rarely seen in tissue sections or body fluids, but when present, they are diagnostic of acute infection (Fig. 13). Infection may cause a variety of manifestations. Acute infection in healthy individuals may resemble infectious mononucleosis. Congenital infection, acquired in utero from mothers with acute infection, may cause a devastating syndrome with severe central nervous system and ocular abnormalities which may be fatal. However, symptomatic ocular toxoplasmosis may first develop in adults years after the primary infection. Reactivation of infection in immunocompromised patients, such as those with AIDS (acquired immunodeficiency syndrome) or those receiving immunosuppression for transplantation, also typically develops years after the primary infection. Characteristic features of reactivated infections in immunosuppressed patients include encephalitis, pneumonitis, and myocarditis (23, 42).

Humans and a variety of other animals serve as intermediate hosts for toxoplasma. The sexual stage, which occurs in cats, causes an intestinal infection resembling isosporiasis (14). Humans may acquire the infection by the ingestion either of oocysts in material contaminated with cat feces or of inadequately cooked meat containing cysts from other intermediate hosts.

In most industrialized nations, approximately 30% of the population have IIF or dye test titers of 1:16 or greater, although the prevalence of antibodies varies among the populations being studied. In certain areas in France, prevalence ranges as high as 80%, and in some areas of the United States it is as low as 1%. The overall prevalence of antibodies in the adult population in the United States is assumed to be 25 to 30%. New infections in the United States are estimated to be 1% per year.

In addition to the large number of asymptomatic cases, there are four major forms of clinical toxoplasmosis, each of which requires a different interpretation of serologic test results: congenital infection, maternal infection acquired during pregnancy, ocular infection, and adult disseminated cerebral infection. More than with any other form of disease, the serodiagnosis of congenital infection depends upon the accurate measurement of specific IgM antibodies. IgG antibodies are efficiently transported from maternal to fetal circulation and in some cases may be slightly higher (twofold) in the fetus. IgM antibodies, on the other hand, are not shared by mother and fetus, except when placental damage and leakage occur. This situation occurs in only a small percentage of cases. IgM antibody in the newborn circulation is generally considered of fetal origin and is diagnostic of congenital infection. However, the detection and measurement of IgM antibody are fraught with problems. Rheumatoid factor and antinuclear antibodies give false-positive reactions in the IIF test (2), IgM antibodies are frequently blocked by IgG antibodies (38), and until recently, IgM-specific serologic tests were insensitive (39). Consequently, serologic results must be considered carefully. The diagnosis of toxoplasmosis in a newborn infant requires a positive toxoplasma IgM test and an IgG titer equivalent to maternal IgG. Infant sera must always be tested in parallel with a maternal sample.

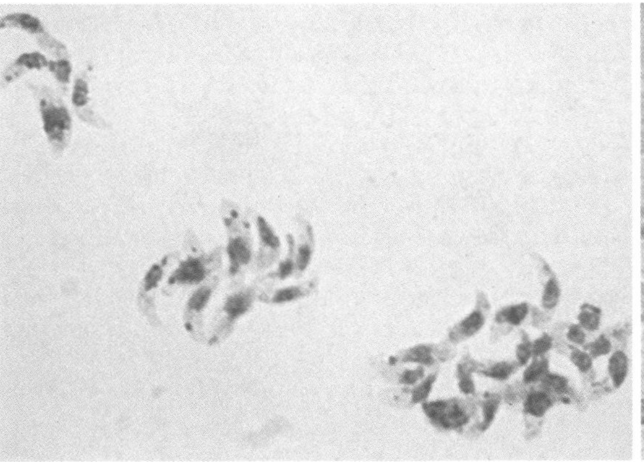

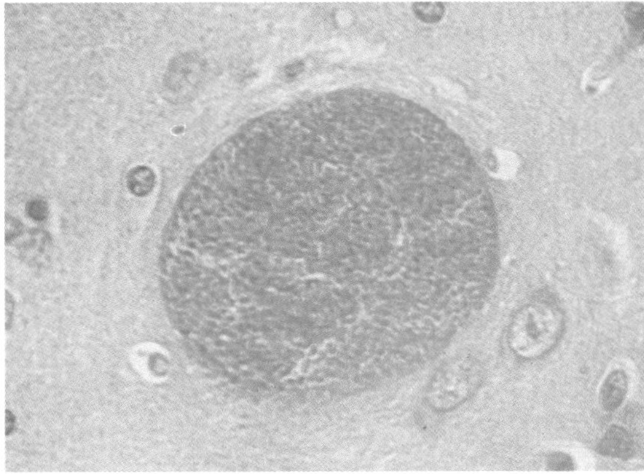

FIG. 13. Trophozoites (tachyzoites) and pseudocyst of *Toxoplasma gondii* (reproduced with permission from reference 50).

Most tests on obstetric patients are performed for screening rather than diagnosis. The presence of maternal antibody before conception indicates immunity, although a few exceptions have been documented. In general, serologic procedures for screening are used qualitatively to test for the presence of antibody, and the absolute titers are less important. Infections acquired during pregnancy are associated with a risk of transmission to the fetus. To determine whether a positive titer in a pregnant woman represents acute infection, a prepregnancy or very early pregnancy sample is extremely valuable. However, in most cases these samples are not available, and clinical decisions must be based on more limited serologic information. Since approximately one-third of pregnant women have IgG antibody without evidence of active disease, IgG levels must be unusually high to be of significance. IgM antibodies, on the other hand, do not persist, and their presence indicates a relatively recent infection. For the serodiagnosis of acute infection, titers of 1:64 are usually accepted as minimal for IgM, and titers of 1:1,024 are minimal for IgG in adults. Interpretations based on these serologic guidelines are, of course, influenced by the clinical presentation. Significant levels of IgM antibody are assumed to indicate that the infection was acquired within the past 6 months, although newer tests can measure IgM in some patients for up to 1 year. The absence of IgM antibody, particularly by newer procedures, mitigates strongly against recent infection.

Ocular toxoplasmosis is usually diagnosed clinically, and serology serves only for confirmation. These cases are often confusing because the antibody response may be extremely poor when the major, or only, clinical manifestation is ocular. These patients commonly have titers of ≤1:8—far below those accepted as diagnostic in other adult infections. Although the importance of these low levels is questionable, the absence of IgG antibody indicates that the likelihood of ocular toxoplasmosis is nil. Since virtually all ocular cases are considered a reactivation of old infection, IgM antibody testing rarely plays a role in the serodiagnosis of the ocular disease.

Adult disseminated cerebral toxoplasmosis is a clinical entity which was extremely rare until the appearance of AIDS in the past 5 years. Serologic study of these cases suggests that they represent the reactivation of chronic toxoplasmosis, rather than recent infection. Most cases of cerebral toxoplasmosis in these patients have been characterized by high levels of IgG antibody and an absent IgM response, even early in the disease (16). Because AIDS is relatively new and few cases have been studied, the interpretation of serologic results is still uncertain and serology can serve only as an adjunct to clinical and pathological diagnosis.

A variety of serologic procedures are available both commercially and at specialty laboratories. In general, these procedures can be divided into three types: screening tests, quantitative tests, and tests for measuring IgM antibody. Qualitative screening tests include latex agglutination, direct agglutination, indirect hemagglutination, and ELISA. These tests have all been evaluated and shown to be reliable. For diagnosis, screening tests should not be used alone. Rather, positive results and questionable negative results should be confirmed by quantitative testing. Quantitative procedures include IIF tests (20), methylene blue dye test (43), and ELISA (55). ELISA procedures are relatively new, and more experience will be necessary before ELISA can be accepted as the test of choice. Numerous reports have demonstrated the comparability of IIF and dye tests (56). Because of its ease of performance and the availability of reagents, the IIF test is the test of choice for the clinical laboratory. It is both specific and sensitive.

IIF and ELISA procedures have been developed to detect IgM antibody. Until recently, IIF was the only acceptable test, and it was recognized as having limitations. Rheumatoid reactions could be removed only by adsorption, and false-negative reactions were not uncommon, especially in newborns. Antitoxoplasma IgG, which blocked the detection of antitoxoplasma IgM, could be removed only with the use of columns or ultracentrifugation to separate IgG from IgM. However, ELISA systems have now been developed which use a capture method (12, 45). IgM antibody is captured by anti-μ antibody on the solid phase, and conventional ELISA techniques are then used to complete the test. By this method, IgM is isolated from IgG, and both rheumatoid reactions and blocking by IgG can be eliminated. Although the sensitivity of this test is greater than that of the IIF test, false-negative results may occur, especially in newborns. When it becomes commercially available, the capture system will undoubtedly be the test of choice for the detection of IgM antibody.

Commercially, a variety of reagents and kits are available. All have been evaluated and found adequate, although quality control problems continue to plague the industry. Latex, direct agglutination, indirect hemagglutination, ELISA, and IIF kits are available. Consumers should restrict products precisely to their intended use. Screening tests or semiquantitative procedures should not be used for clinical diagnosis.

PNEUMOCYSTIS INFECTION

Pneumocystis infection should be considered in patients who develop bilateral diffuse pulmonary infiltrates, particularly patients on immunosuppressive therapy or with immune deficiencies such as AIDS (61a). Infection appears to be acquired by the airborne route, and subclinical infection of humans is quite common. Infection in compromised persons probably arises from the activation of dormant cysts in their lungs or from the inhalation of cysts from the environment. Organisms proliferate within cysts and in the free trophozoite stage.

At present, morphologic demonstration of organisms is the most reliable way to diagnose *Pneumocystis* infection. Although antibody and antigen detection methods have been described, the usefulness of the currently available methods is unclear (58). For instance, many patients do not form antibody because of their underlying diseases, and many normal persons have antibody. Circulating antigen has also been detected in the serum of patients without apparent *Pneumocystis* infection.

Specimen collection and handling

Specimens submitted for the detection of *Pneumocystis carinii* must be material from the lung. They

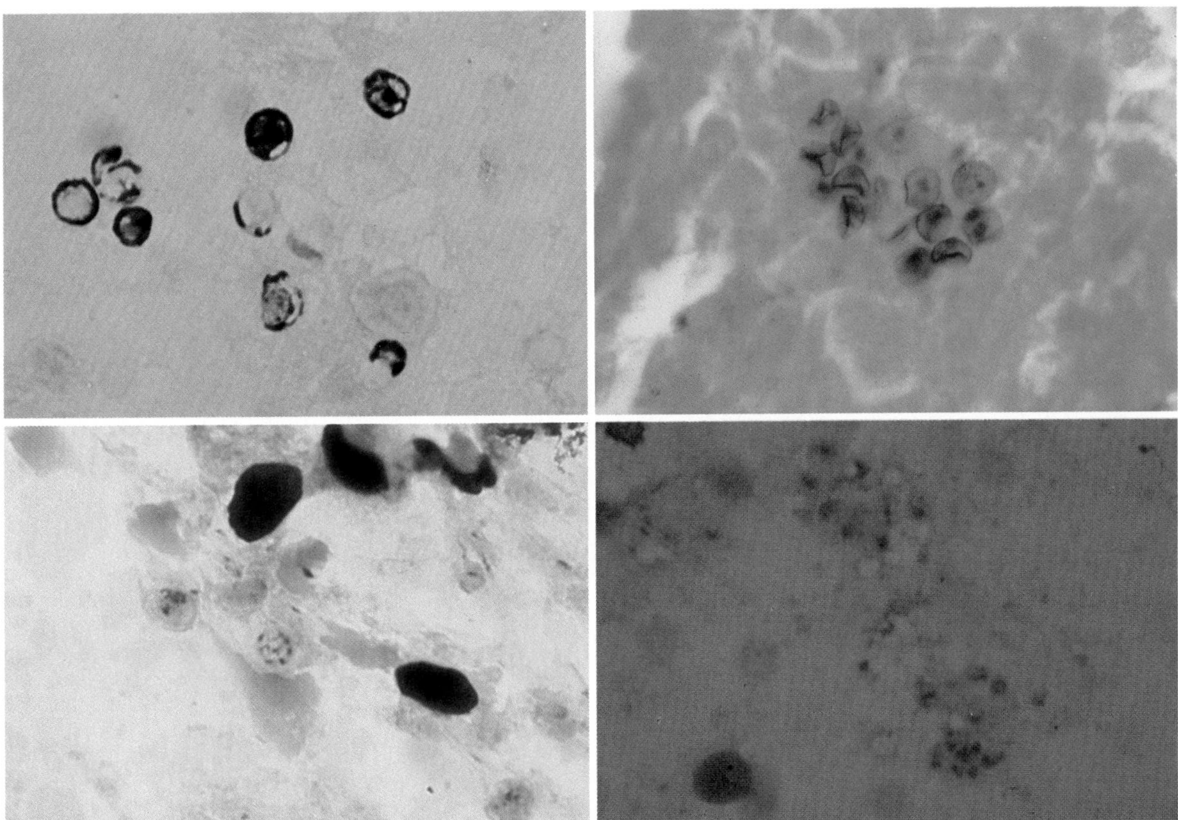

FIG. 14. *Pneumocystis carinii* organisms. (Top, left) Cysts in impression smears, Gram-Weigert stain. (Top, right) Cysts in tissue section, methenamine silver stain. (Bottom, left) Cyst in impression smear showing intracystic bodies, Giemsa stain. (Bottom, right) Trophozoites in impression smear, Giemsa stain.

may be tissue obtained from biopsy (either open lung or transbronchial biopsy), transthoracic aspirates, bronchial brushings, bronchoalveolar lavage fluid, or bronchial washings (49). Lung tissue is the best specimen for diagnosis. Open lung biopsy has the greatest sensitivity and may also allow the diagnosis of other infectious and noninfectious processes. Transbronchial biopsy often allows diagnosis, although multiple specimens should be examined. Transthoracic needle aspirates have been useful, particularly in children. In our experience, bronchial brushings have had a poor yield. Bronchoalveolar lavage may be particularly useful in AIDS patients, who generally have large numbers of organisms. Expectorated sputum usually does not contain sufficient organisms for detection with stains and often contains numerous yeast cells which must be differentiated from *P. carinii*. Thus, sputum is not an appropriate specimen. Most laboratories refuse to examine expectorated sputum for *Pneumocystis* spp. except perhaps in AIDS patients. The usefulness of these various procedures depends not only on the procedure selected, but also on the skill of the individuals collecting the specimens, the competence of the persons examining the material, and the patient population.

Tissue obtained from open lung biopsy should be initially examined macroscopically for areas of consolidation. From a consolidated area, a portion of tissue should be selected for impression smears and frozen sections. A portion of the tissue should also be fixed in 10% Formalin for permanent sections. Fragments of tissue obtained from transbronchial biopsy are too small to be used for both sections and impression smears. At least two portions should be used for impression smears. Additional portions should be submitted for surgical pathology.

Stains

Stains used to demonstrate *P. carinii* are Giemsa or Wright stains, which stain the trophozoites and intracystic organisms, and several stains which stain the cyst wall but not trophozoites or intracystic organisms, including toluidine blue O (6), methenamine silver nitrate, cresyl echt violet (1), or Gram-Weigert (40). There are several modifications of the methenamine silver nitrate stain (29, 37). A laboratory should be able to do a Giemsa stain and one of the cyst wall stains. Any of the cyst wall stains is satisfactory in the hands of experienced personnel. We prefer the Gram-Weigert stain because it employs reagents with long shelf life and is easy to perform.

Giemsa and Wright stains demonstrate trophozoites and *Pneumocystis* organisms within cysts. The cyst wall does not stain but may be evident as a negative-staining halo around a group of intracystic organisms. Organisms have red nuclei and pale blue cytoplasm. Mature cysts measure 4 to 7 μm and contain eight intracystic organisms, often arranged in a clock face pattern (Fig. 14). Finding typical cysts such as this is pathognomonic of *Pneumocystis* infection. Trophozoites measure 2 to 6 μm in diameter and also have blue cytoplasm and red nuclei (Fig. 14). The trophozoites are difficult to differentiate from cell fragments, particularly from cells containing granules. The finding of objects suspiciously like trophozoites should stimulate careful searching for typical cysts. Both cysts and trophozoites often occur in clumps.

Intracystic organisms and free trophozoites are not seen with cyst wall stains, although a more intensely stained dot may be evident in methenamine silver (Fig. 14) or toluidine blue O stains. In Gram-Weigert stains there is usually some irregularity of staining of the cyst wall (Fig. 14). Cysts are usually round, although cup-shaped cysts are common. Some workers feel that the latter represent collapsed cysts which have lost their intracystic bodies. More cysts are found with cyst wall stains than with Giemsa stains, probably because empty cysts stain with cyst wall stains, and immature cysts cannot be differentiated from trophozoites with Giemsa stain. If only occasional organisms are seen, it may not be possible to differentiate *Pneumocystis* cysts from fungi which are of a similar size. The presence of budding allows yeast cells to be recognized, and cup-shaped organisms are suggestive of *Pneumocystis* sp.

Although immunospecific stains such as immunofluorescence and immunoperoxidase have been described, they have not been widely used for diagnosis, and neither reagents nor kits are commercially available (27).

FREE-LIVING PATHOGENIC AMEBAE

Small free-living amebae belonging to the genera *Naegleria*, *Hartmanella*, and *Acanthamoeba* are commonly found in soil, fresh water, and even sewage and sludge (47). Several *Acanthamoeba* species have also been isolated from brackish water and seawater (44). Organisms of the genera *Naegleria* and *Acanthamoeba* may cause a fatal disease of the central nervous system in humans.

Only one *Naegleria* species, *N. fowleri* (*N. aerobia* and *N. invades* are not valid synonyms), is known to cause human disease. *Naegleria* amebae cause an acute and fulminating primary amebic meningoencephalitis (PAM) which generally produces death within 5 to 7 days after the onset of symptoms (3, 13). Of more than 120 cases of PAM reported worldwide, only a few patients have survived.

Several *Acanthamoeba* species (*A. culbertsoni*, *A. castellanii*, *A. polyphaga*, and *A. astronyxis*) are considered pathogenic to humans (15). *Acanthamoeba* spp. typically cause a chronic granulomatous amebic encephalitis, which may last for more than a week and sometimes for even months before causing death (13, 31). A total of 28 cases of granulomatous amebic encephalitis have been recorded worldwide. In addition to causing granulomatous amebic encephalitis, *Acanthamoeba* spp. may infect the eye, often leading to the loss of the eye (17a). Over 30 such eye infections, principally keratitis, have been recorded at the CDC, and many more are probably not diagnosed. *Acanthamoeba* spp. have also been isolated from ear discharge, pulmonary secretions, nasopharyngeal swabs, mandibular autografts, and stool samples (15, 54). Excellent reviews (3, 7, 9, 13, 15, 18) and one book (47) summarize the biology, pathogenicity, and epidemiology of these amebae and the diseases they cause.

Culbertson and his colleagues (10) first demonstrated that free-living amebae caused brain lesions in mice and monkeys, which died within 1 week after intracerebral inoculation. These researchers suggested that the free-living amebae could be human pathogens. Fowler and Carter in 1965 described the first fatal infection due to free-living amebae in an Australian patient with meningoencephalitis (3). Although that infection was thought to be caused by *Acanthamoeba* sp., it is now believed to have been due to *N. fowleri* (15). Butt et al. in 1966 described the first case in the United States and coined the term PAM (3).

Although most human infections have been caused by *N. fowleri*, at least 28 fatal infections were probably due to *Acanthamoeba* spp., based on the morphology of the cysts, immunofluorescence staining of the amebae in tissues, and the epidemiologic and clinical picture (31).

In most instances, *Naegleria* organisms enter via the nasal passage when persons swim in lakes and other bodies of water that harbor these amebae. The organisms then cross the cribriform plate and make their way into the olfactory lobes of the brain. In addition, some persons, especially those who are immunologically compromised, may inhale *Acanthamoeba* cysts when passing through areas where soil has been freshly turned over. The amebae may then excyst and invade the nasal mucosa.

Pathogenic *N. fowleri* are susceptible to amphotericin B (3, 13), and at least three PAM patients have recovered after receiving intrathecal and intravenous injections of this drug alone or in combination with miconazole (3, 18, 46). Culbertson found sulfadiazine to be active against experimental *Acanthamoeba* infections in mice (9). Jones et al. (19) found that paromomycin, clotrimazole, and hydroxystilbamidine isethionate were active against *A. polyphaga* in vitro. Recent studies suggest that oral ketoconazole, together with topical miconazole, another antifungal imidazole, may prove to be useful for the treatment of *Acanthamoeba* keratitis (17a).

Morphology

N. fowleri trophozoites measure 10 to 35 μm in diameter, exhibit eruptive locomotion by producing smooth hemispherical bulges, and have a sticky posterior end (uroid) which often has several trailing filaments. *N. fowleri* may transiently have a pear-shaped biflagellate stage in altered environmental conditions, can also produce smooth-walled cysts which measure 7 to 15 μm across (35) (Fig. 15), and may have one or more pores plugged with a mucoid material (18).

Acanthamoeba organisms are slightly larger than *Naegleria* organisms (15 to 45 μm), produce fine, tapering hyaline projections called acanthopodia (Fig. 15), and have no flagellate stage but produce a double-walled cyst (10 to 25 μm) with a wrinkled outer wall (ectocyst) and a stellate, polygonal, or even round inner wall (endocyst) (36, 48).

Both *Naegleria* and *Acanthamoeba* organisms are uninucleate and have a nucleus characterized by a large, dense, centrally located nucleolus. *Naegleria* spp. exhibit the promitotic pattern of cell division, in which the nucleolus and the nuclear membrane per-

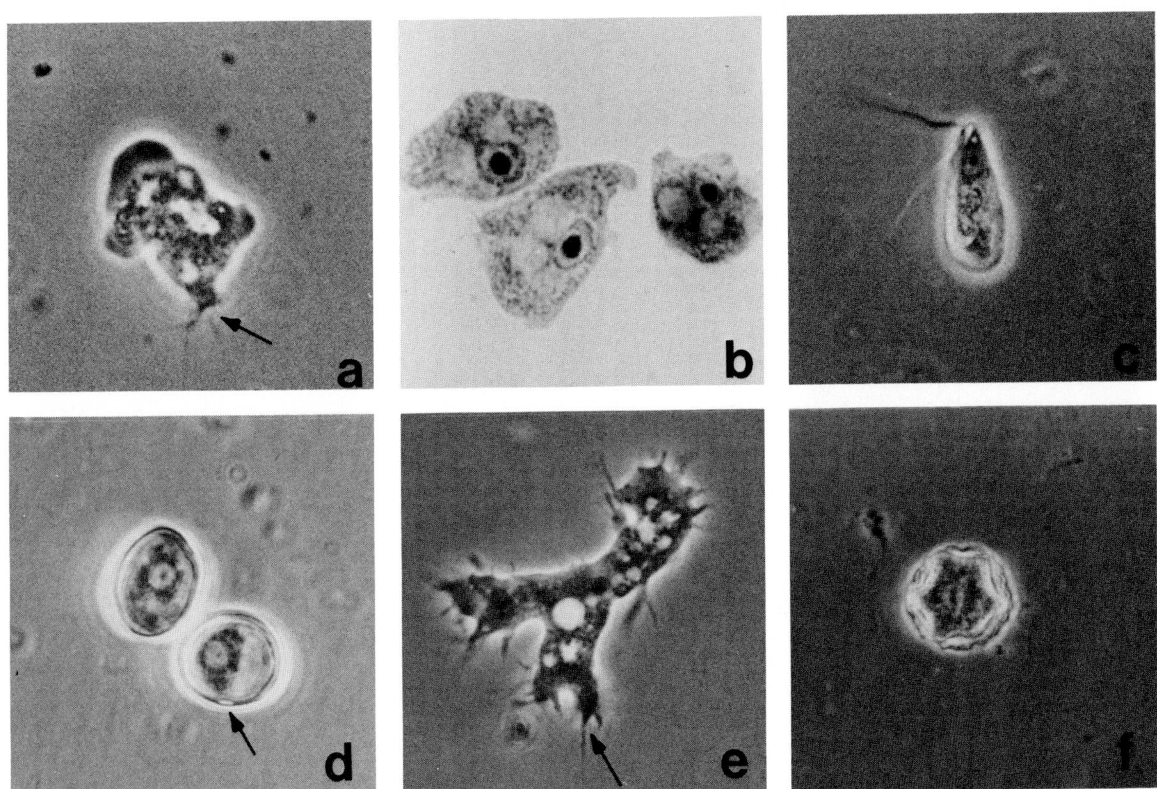

FIG. 15. Free-living pathogenic amebae. (a through d) *Naegleria fowleri*: (a) trophozoite, phase contrast (note the uroid and filaments at arrow); (b) trophozoite, trichrome stain; (c) biflagellate, phase contrast; (d) smooth-walled cyst, phase contrast (note the pore at arrow). All are ×1,100. (e and f) *Acanthamoeba castellanii*: (e) trophozoite, phase contrast (note the acanthopodia at arrow); (f) double-walled cyst, phase contrast. All are ×1,100.

sist during division. *Acanthamoeba* spp., however, divide by conventional mitosis; i.e., the nucleolus and the nuclear membrane disappear during division.

Collection, handling, and storage of specimens

Cerebrospinal fluid or small pieces of tissue (brain, lungs, corneal biopsy material, etc.) from the affected area must be obtained aseptically to isolate the etiologic agent. Specimens should be kept at room temperature (24 to 28°C) and should not be frozen or refrigerated. Personnel working with these specimens should take appropriate precautions, such as using surgical masks and gloves and working in a biological safety cabinet. The remaining tissue (which is not being used for culture) should be preserved for histopathology in 10% neutral buffered Formalin or another appropriate fixative.

Methods of examination

Direct examination. Direct examination of the sample as a wet-mount preparation is the principal means of diagnosing PAM and other diseases caused by these amebae. Since the amebae tend to attach to the surface of the specimen container, the container should be shaken gently before removing a small drop of fluid to examine. Cerebrospinal fluid may be centrifuged at 250 × *g* for 5 min to concentrate the amebae. Most of the supernatant should be aspirated carefully, and the sediment should be gently suspended in the remaining fluid before 1 drop of this suspension is aspirated for microscopic examination on a clean microscope slide with a no. 1 cover slip.

The slide preparation should be examined with ×10 and ×40 objectives. Phase-contrast optics are preferable, but bright-field illumination may be used with diminished light. The slide may be warmed to 35°C (optional). Amebae, if present, are detected by their active directional movement.

Permanently stained preparations. A small drop of the sedimented cerebrospinal fluid or other sample is placed in the middle of a slide and allowed to stand in a moist chamber for 5 to 10 min at 37°C. This treatment will allow the amebae to attach to the surface of the slide. Several drops of warm (37°C) Schaudinn fixative are dropped directly onto the sample and allowed to stand for 1 min. The slide is then transferred to a Coplin jar containing Schaudinn fixative and left there for 1 h. The slide may be stained in Wheatley trichrome (Chapter 112) or Heidenhain iron hematoxylin stain.

Culture

The following is a recommended procedure for the isolation of free-living pathogenic amebae from biological specimens.

Preparation of agar plates

1. Remove plates of nonnutrient agar with Page's ameba saline (35) from the refrigerator and place them in a 37°C incubator for 30 min.
2. Add 0.5 ml of Page's ameba saline to an 18- to 24-

h slant culture of *Escherichia coli* K-12, *E. coli* U5-41, or *Enterobacter aerogenes*. Gently scrape the surface of the slant with a sterile bacteriologic loop (do not break the agar surface). With a sterile Pasteur pipette, gently and uniformly suspend the bacteria. Add 2 to 3 drops of this suspension to the middle of a warmed (37°C) agar plate, and spread the bacteria over the surface with a bacteriologic loop.

Inoculation of plates with specimens

1. Cerebrospinal fluid. Centrifuge the cerebrospinal fluid at 250 × *g* for 5 to 8 min. With a sterile serologic pipette, carefully transfer all but 0.5 ml of the supernatant to a sterile tube, and store the tube at 4°C for possible future use. Mix the sediment with the remaining fluid, use a sterile Pasteur pipette to place 2 to 3 drops in the center of the agar plate, precoated with bacteria, and incubate the culture at 37°C.
2. Tissue. Triturate a small piece of the tissue in a small quantity of ameba saline. With a sterile Pasteur pipette, place 2 to 3 drops in the center of the agar plate. Incubate the plate at 37°C.
3. Water and soil samples. Handle in the same manner as cerebrospinal fluid and tissue specimens, respectively.

Examination of plates

1. With the low-power (×10) objective of the microscope, observe the plates daily for 7 days for evidence of amebae.
2. If amebae are seen, circle that area with a wax pencil. With a fine spatula, cut a small piece of agar from the circled area, place it face down on the surface of a fresh agar plate precoated with bacteria, and incubate as described above. Both *Naegleria* spp. and *Acanthamoeba* spp. can easily be cultivated in this way and can be maintained indefinitely with periodic transfers. When the plate is examined under a microscope, the amebae will look like small blotches, and if they are observed carefully, their movement can be discerned. After 2 to 3 days of incubation, the amebae will start to encyst. If a plate is examined after 4 to 5 days of incubation, both trophozoites and cysts should be visible.

Identification

Identification of amebae to the generic level is based on characteristic patterns of locomotion, morphologic features of the trophic and cyst forms, and enflagellation experiments.

Immunofluorescence or immunoperoxidase tests with monoclonal or polyclonal antibodies are used by reference laboratories for the identification of the species, especially *Acanthamoeba* spp. in culture or in fixed tissue. These reagents are not available commercially. *Naegleria* species can also be differentiated by isoenzyme electrophoretic patterns (11).

Enflagellation experiment

1. Mix 1 drop of the sedimented cerebrospinal fluid containing amebae with about 1 ml of sterile distilled water in a sterile tube, or scrape the surface of a plate that is positive for amebae with a bacteriologic loop and transfer a loopful of the scraping to a sterile tube containing approximately 1 ml of distilled water.
2. Gently shake the tube, and transfer a drop of the suspension to the center of a cover slip, the edges of which have been coated thinly with petroleum jelly. Place a microscope slide over the cover slip and invert the slide. Seal the edges of the cover slip with Vaspar (a 1:1 mixture of petrolatum and paraffin). Place the slide in a moist chamber, and incubate the slide and tube at 35°C for 24 h.
3. Periodically examine the tube and the slide preparation microscopically for free-swimming flagellates. *Naegleria* spp. have a flagellated stage; *Acanthamoeba* spp. do not. If the sample contains *N. fowleri*, about 30 to 50% of the amebae will transform into pear-shaped biflagellated organisms within 2 h.

Laboratories unfamiliar with these amebae should have their identifications confirmed by reference laboratories. Either specimens or cultures should be sent at ambient temperature (not frozen) and delivered within 24 h.

Other culture methods

A variety of other culture methods including axenic culture and culture on mammalian cells have been described, but these methods are used primarily in research laboratories. Animal inoculation has generally been used for laboratory investigation and not for diagnosis. Mice develop brain infection after intranasal inoculation of amebae.

Serology

Serologic tests are not useful for the diagnosis of *Naegleria* infections because most patients die too rapidly to produce antibody. CF antibody to *Acanthamoeba* spp. has been demonstrated in the serum of patients suffering from upper respiratory tract distress and those with optic neuritis and macular disease (15, 52). Kenney (21) demonstrated increasing CF antibody titers to *Acanthamoeba* spp. in three successive samples of serum from each of two patients. Precipitin antibody has also been demonstrated in the serum of a patient suffering from *A. polyphaga* keratitis (53). Recently a Nigerian patient, who made a partial recovery from *A. rhysodes*-induced central nervous system disease, was shown to have an increase in his ameba immobilization antibody titer over a 16-month period (54). Serologic tests for diagnosis are not readily available.

LITERATURE CITED

1. **Bowling, M. C., I. M. Smith, and S. L. Wescott.** 1973. A rapid staining procedure for *Pneumocystis carinii*. Am. J. Med. Technol. **39:**267–268.
2. **Camargo, M. G., P. G. Leser, and A. Rocca.** 1972. Rheumatoid factors as a cause for false-positive IgM antitoxoplasma fluorescent tests: a technique for specific results. Rev. Inst. Med. Trop. Sao Paulo **14:**310–313.
3. **Carter, R. F.** 1972. Primary amoebic meningo-encephalitis: an appraisal of present knowledge. Trans. R. Soc. Trop. Med. Hyg. **66:**193–208.
4. **Centers for Disease Control.** 1984. Malaria surveillance report. U.S. Public Health Service, Atlanta.

4a. **Centers for Disease Control.** 1984. Prevention of malaria in travelers—1984. Morbid. Mortal. Weekly Rep. **33**(Suppl. 2)**:**75S–103S.

5. **Cerisola, J. A.** 1970. Immunodiagnosis of Chagas' dis-

ease: hemagglutination and immunofluorescence tests. J. Parasitol. **56**:409–410.

6. **Chalvardjian, A. M., and L. A. Grawe.** 1963. A new procedure for the identification of *Pneumocystis carinii* cysts in tissue sections and smears. J. Clin. Pathol. **16**:383–384.
7. **Chang, S. L.** 1971. Small, free-living amebas: cultivation, quantitation, identification, classification, pathogenesis, and resistance. Curr. Top. Comp. Pathobiol. **1**:201–254.
8. **Coatney, G. R., W. E. Collins, M. Warren, and P. G. Contacos.** 1971. The primate malarias. U.S. Department of Health, Education and Welfare, Washington, D.C.
9. **Culbertson, C. G.** 1971. The pathogenicity of soil amebas. Annu. Rev. Microbiol. **25**:231–254.
10. **Culbertson, C. G., J. W. Smith, and J. R. Minner.** 1958. *Acanthamoeba*: observations on animal pathogenicity. Science **127**:1506.
11. **DeJonckheere, J. F.** 1982. Isoenzyme patterns of pathogenic and non-pathogenic *Naegleria* spp. using agarose isoelectric focusing. Ann. Microbiol. **133A**:319–342.
12. **Duermeyer, W., F. Wielaard, and J. Van der Veen.** 1979. A new principle for the detection of specific IgM antibodies applied in an ELISA for hepatitis. Am. J. Med. Virol. **4**:25–32.
13. **Duma, R. J.** 1972. Primary amoebic meningoencephalitis. Crit. Rev. Clin. Lab. Sci. **3**:163–192.
14. **Frenkel, J. K., J. P. Dubey, and N. L. Miller.** 1970. *Toxoplasma gondii* in cats: fecal stages identified as coccidian oocysts. Science **167**:893–896.
15. **Griffin, J. L.** 1978. Pathogenic free-living amebae, p. 507–549. *In* J. P. Kreier (ed.), Parasitic protozoa, vol. 5. Academic Press, Inc., New York.
16. **Hauser, W. E., B. J. Loft, F. K. Conley, and J. S. Remington.** 1982. Central nervous system toxoplasmosis in homosexual and heterosexual adults. N. Engl. J. Med. **307**:498–499.
17. **Healy, G. R., and T. K. Ruebush II.** 1980. Morphology of *Babesia microti* in human blood smears. Am. J. Clin. Pathol. **73**:107–109.

17a. **Hirst, L. W., R. W. Green, W. Merz, C. Kaufmann, G. S. Visvesvara, A. Jensen, and M. Howard.** 1984. Management of *Acanthamoeba* keratitis: a case report and a review of the literature. Ophthalmology **91**:1105–1111.

18. **John, D. T.** 1982. Primary amoebic meningoencephalitis and the biology of *Naegleria fowleri*. Annu. Rev. Microbiol. **36**:101–123.
19. **Jones, D. B., G. S. Visvesvara, and N. M. Robinson.** 1975. *Acanthamoeba polyphaga* keratitis and *Acanthamoeba* uveitis associated with fatal meningoencephalitis. Trans. Ophthalmol. Soc. U.K. **95**:221–232.
20. **Kelen, A. E., L. Ayllon-Leindl, and N. A. Labzoffsky.** 1962. Indirect fluorescent antibody method in serodiagnosis of toxoplasmosis. Can. J. Microbiol. **8**:545–554.
21. **Kenney, M.** 1971. The micro-Kolmer complement fixation test in routine screening for soil ameba infection. Health Lab. Sci. **8**:5–10.
22. **Kreutzer, R. D., M. E. Semko, L. D. Hendricks, and N. Wright.** 1983. Identification of *Leishmania* sp. by multiple isozyme analysis. Am. J. Trop. Med. Hyg. **32**:703–715.
23. **Krick, J. A., and J. S. Remington.** 1978. Toxoplasmosis in the adult—an overview. N. Engl. J. Med. **298**:550–553.
24. **Krogstad, D. J., and M. A. Pfaller.** 1983. Prophylaxis and treatment of malaria. Curr. Clin. Top. Infect. Dis. **3**:56–73.
25. **Krotoski, W. A., W. E. Collins, R. S. Bray, P. C. C. Garnham, F. B. Cogswell, R. W. Gwadz, R. Killick-Kendrick, R. Wolf, R. Sinden, L. C. Koontz, and P. S. Stanfill.** 1982. Demonstration of hypnozoites in sporozoite-transmitted *Plasmodium vivax* infection. Am. J. Trop. Med. Hyg. **31**:1291–1293.
26. **Krotoski, W. A., D. M. Krotoski, P. C. C. Garnham, R. S. Bray, R. Killick-Kendrick, C. C. Draper, G. A. T. Targett, and M. W. Guy.** 1980. Relapses in primate malaria: discovery of two populations of exoerythrocytic stages—preliminary note. Br. Med. J. **1**:153–154.
27. **Lim, S. K., W. C. Eveland, and R. J. Porter.** 1974. Direct fluorescent-antibody method for the diagnosis of *Pneumocystis carinii* pneumonitis from sputa or tracheal aspirates from humans. Appl. Microbiol. **27**:144–149.
28. **Mackett, G. A.** 1960. Die Komplementbindungsreaktian der Chagaskrankheit. Z. Tropenmed. Parasitol. **11**:152–186.
29. **Mahan, C. T., and G. E. Sale.** 1978. Rapid methenamine silver stain for pneumocystis and fungi. Arch. Pathol. Lab. Med. **102**:351–352.
30. **Mansueto, S., G. Migneco, and C. LaCascia.** 1975. The complement fixation test with BCG in the diagnosis of leishmaniasis. Boll. Ist. Sieroter. Milan **54**:140–144.
31. **Martinez, J. A., C. Sotelo-Avila, J. Garcia-Tamayo, J. T. Moron, E. Willaert, and W. P. Stamm.** 1977. Meningoencephalitis due to *Acanthamoeba* sp.: pathogenesis and clinico-pathological study. Acta Neuropathol. **37**:183–191.
32. **Miller, L. H., F. A. Neva, and F. Gill.** 1978. Failure of chloroquine in human babesiosis (*Babesia microti*): case report and chemotherapeutic trials in hamsters. Ann. Intern. Med. **88**:200–202.
33. **Neva, F. A.** 1977. Looking back for a view of the future: observations on immunity to induced malaria. Am. J. Trop. Med. Hyg. **26**(Suppl.):211–215.
34. **Neva, F. A.** 1982. Diagnosis and treatment of cutaneous leishmaniasis. Curr. Clin. Top. Infect. Dis. *3*:364–380.
35. **Page, F. C.** 1967. Taxonomic criteria for limax amoebae, with descriptions of 3 new species of *Hartmanella* and 3 of *Vahlkampfia*. J. Protozool. **14**:499–521.
36. **Page, F. C.** 1967. Redefinition of the genus *Acanthamoeba* with descriptions of three species. J. Protozool. **14**:709–724.
37. **Pintozzi, R. L.** 1978. Technical methods—modified Grocott's methenamine silver nitrate method for quick staining of *Pneumocystis carinii*. J. Clin. Pathol. **31**:803–805.
38. **Pyndiah, N., U. Krech, P. Price, and J. Wilhelm.** 1979. Simplified chromatographic separation of immunoglobulin M from G and its application to toxoplasma indirect immunofluorescence. J. Clin. Microbiol. **9**:170–174.
39. **Remington, J. S., and G. Desmonts.** 1973. Congenital toxoplasmosis: variability in the IgM fluorescent antibody response and some pitfalls in diagnosis. J. Pediatr. **83**:27–30.
40. **Rosen, P. P., N. Martini, and D. Armstrong.** 1975. *Pneumocystis carinii* pneumonia: diagnosis by lung biopsy. Am. J. Med. **58**:794–801.
41. **Ruebush, T. K., II, D. D. Juranek, E. S. Chisholm, P. C. Snow, G. R. Healy, and A. J. Sulzer.** 1977. Human babesiosis on Nantucket Island: evidence for self-limited and subclinical infections. N. Engl. J. Med. **297**:825–827.
42. **Ruskin, J., and J. S. Remington.** 1976. Toxoplasmosis in the compromised host. Ann. Intern. Med. **84**:193–199.
43. **Sabin, A. B., and H. A. Feldman.** 1948. Dyes as microchemical indicators of a new immunity phenomenon affecting a protozoan parasite (toxoplasma). Science **108**:660–663.
44. **Sawyer, T. K., G. S. Visvesvara, and B. A. Harke.** 1976. Pathogenic amoebas from brackish and ocean sediments, with a description of *Acanthamoeba hatchetti*, n. sp. Science **196**:1324–1325.
45. **Schmitz, H., V. von Diemling, and B. Flehmig.** 1980. Detection of IgM antibodies to cytomegalovirus (CMV) using enzyme-labelled antigen. J. Gen. Virol. **50**:59–68.
46. **Seidel, J. S., P. Harmatz, G. S. Visvesvara, A. Cohen, J. Edwards, and J. Turner.** 1982. Successful treatment of primary amebic meningoencephalitis. N. Engl. J. Med. **306**:346–348.
47. **Singh, B. N.** 1975. Pathogenic and non-pathogenic amebae. John Wiley & Sons, Inc., New York.
48. **Singh, B. N., and S. R. Das.** 1970. Studies on pathogenic

and non-pathogenic small free-living amebae and the bearing of nuclear division on the classification of the order Amebida. Philos. Trans. R. Soc. London Ser. B **259**:435–476.

49. **Smith, J. W., and M. S. Bartlett.** 1982. Laboratory diagnosis of *Pneumocystis carinii* infection, p. 393–406. *In* W. C. Winn (ed.), Clinics in laboratory medicine. The W. B. Saunders Co., Philadelphia.
50. **Smith, J. W., D. M. Melvin, T. C. Orihel, L. R. Ash, R. M. McQuay, and J. H. Thompson, Jr.** 1976. Atlas of diagnostic medical parasitology: blood and tissue parasites. American Society of Clinical Pathologists, Chicago.
51. **Sulzer, A. J., and M. Wilson.** 1971. The fluorescent antibody test for malaria. Crit. Rev. Clin. Lab. Sci. **2**:601–609.
52. **Visvesvara, G. S., and W. Balamuth.** 1975. Comparative studies on related free-living and pathogenic amebae with special reference to *Acanthamoeba*. J. Protozool. **22**:245–256.
53. **Visvesvara, G. S., D. B. Jones, and N. M. Robinson.** 1975. Isolation, identification, and biological characterization of *Acanthamoeba polyphaga* from a human eye. Am. J. Trop. Med. Hyg. **24**:784–790.
54. **Visvesvara, G. S., S. S. Mirra, F. H. Brandt, D. M. Moss, H. M. Mathews, and A. J. Martinez.** 1983. Isolation of two strains of *Acanthamoeba castellanii* from human tissue and their pathogenicity and isoenzyme profiles. J. Clin. Microbiol. **18**:1405–1412.
55. **Walls, K. W., S. L. Bullock, and D. K. English.** 1977. Use of the enzyme-linked immunosorbent assay (ELISA) and its microadaptation for the serodiagnosis of toxoplasmosis. J. Clin. Microbiol. **5**:273-277.
56. **Walton, B. C., B. M. Benchoff, and W. H. Brooks.** 1966. Comparison of the indirect fluorescent antibody test and the methylene blue dye test for the detection of antibodies to *Toxoplasma gondii*. Am. J. Trop. Med. Hyg. **15**:149–152.
57. **Walton, B. C., W. H. Brooks, and I. Arjona.** 1972. Serodiagnosis of American leishmaniasis by indirect fluorescent antibody test. Am. J. Trop. Med. Hyg. **21**:296–299.
58. **Walzer, P. D., and L. S. Young.** 1984. Clinical relevance of animal models of *Pneumocystis carinii* pneumonia. Diagn. Microbiol. Infect. Dis. **2**:1–6.
59. **Wilcox, A.** 1960. Manual for the microscopical diagnosis of malaria in man. U.S. Department of Health, Education and Welfare, Washington, D.C.
60. **Wirth, D. F., and D. M. Pratt.** 1982. Rapid identification of *Leishmania* species by specific hybridization of kinetoplast DNA in cutaneous lesions. Proc. Natl. Acad. Sci. U.S.A. **79**:6999–7003.
61. **Wittner, M., K. S. Rowin, H. B. Tanowitz, J. F. Hobbs, S. Saltzman, B. Wenz, R. Hirsch, E. Chisholm, and G. R. Healy.** 1982. Successful chemotherapy of transfusion babesiosis. Ann. Intern. Med. **96**:601–604.

61a. **Young, L. S.** 1984. *Pneumocystis carinii* pneumonia: pathogenesis, diagnosis, treatment. *In* C. Enfant (ed.), Lung biology in health and disease, vol. 22. Marcel Dekker, Inc., New York.

62. **Zeledon, R.** 1974. Epidemiology, modes of transmission and reservoir hosts of Chagas' disease, p. 51–85. *In* K. Elliott, M. O'Connor, and G. E. W. Wolstenholme (ed.), Trypanosomiasis and leishmaniasis: a Ciba Foundation symposium. Elsevier North-Holland Publishing Co., Amsterdam.

Chapter 57

Intestinal and Urogenital Protozoa

DOROTHY M. MELVIN AND GEORGE R. HEALY

The protozoa that parasitize the intestinal and urogenital systems of humans belong to four groups: amebae, flagellates, ciliates, and coccidia (Table 1). In addition, *Blastocystis hominis*, once considered a yeast, has been identified as a protozoan and placed in a separate group in the subphylum Sporozoa (the same subphylum to which the coccidia belong). With the exception of *Trichomonas vaginalis*, a flagellate, all of the organisms live in the intestinal tract.

The species of intestinal protozoa vary in prevalence and in pathogenicity (Table 1). Some species are rarely encountered in patients in the United States but may be found in Americans who travel to areas in which the organisms are endemic and in persons from those areas who visit or emigrate to the United States. Therefore, clinicians and laboratory personnel should be aware of both common and uncommon parasite species that might be found in their patients.

In addition to the protozoan species generally considered human parasites, some species parasitic in animals may also infect humans. For example, *Cryptosporidium* species, long recognized as pathogens in calves, lambs, and other animals, have recently been found in humans and have caused severe infections in patients with AIDS (acquired immune deficiency syndrome).

Most of the intestinal protozoa (except *Sarcocystis* spp., which are acquired by the ingestion of the infective stages in raw or poorly cooked beef or pork) are transmitted through fecally contaminated food, water, or other materials. Prevalence of intestinal protozoa is correlated with socioeconomic conditions, and higher rates of infection occur in people who have poor personal hygiene or who live in areas with poor sanitation.

Some of the intestinal protozoa are commensals or nonpathogenic organisms that produce no evidence of disease; however, microscopists must be able to distinguish pathogenic from nonpathogenic species. Several species are capable of causing mild to severe gastrointestinal symptoms, and one species, *Entamoeba histolytica*, may produce extraintestinal lesions in various areas of the body. However, pathogenic or potentially pathogenic protozoa do not always produce symptoms in infected people. Such asymptomatic persons may serve as reservoirs for the infection. In addition, finding a potentially pathogenic protozoan does not necessarily prove that it is causing the illness of the patient. The pathogenicity of some species (*Sarcocystis* spp., for example) has been questioned but not definitely proven. *T. vaginalis*, a urogenital protozoan, is also considered pathogenic and may cause mild to severe vaginitis and other urogenital problems.

This chapter covers information on the morphologic identification of organisms (presented in tabular form and diagrams), recommended procedures for laboratory diagnosis, and clinical aspects of important pathogens.

In the descriptions of diseases, the clinical manifestations noted refer to findings in patients with symptomatic disease and do not necessarily refer to findings in every person infected with the parasite species.

Laboratory diagnosis

Because the symptoms produced by pathogenic intestinal protozoa are usually nonspecific, diagnosis requires laboratory detection of the parasite by the microscopic examination of feces or other body material. Immunodiagnostic methods are useful for the diagnosis of extraintestinal amebiasis, but they are of limited usefulness for intestinal diseases.

Although not all intestinal protozoa are pathogenic, microscopists must be capable of identifying both pathogenic and nonpathogenic species, with the possible exception of species which are rarely found in patients in the United States. Morphology, especially that of amebae, varies, and species characteristics often overlap so that individual nonpathogenic organisms may have characteristics which resemble those of pathogens and vice versa. Thus, for reliable identification, microscopists must be able to differentiate all species regardless of their potential for causing disease. Special attention should be given to the recognition and identification of the clinically significant pathogens, especially *E. histolytica* and *Giardia lamblia*.

The identification of protozoan species is based on the morphology of the diagnostic stages. The particular features or characteristics used for identification vary with the group of organisms (for example, amebae or flagellates), the species, and the stage(s) of parasite present. The diagnostic stages are trophozoites or cysts for the amebae, flagellates, and ciliates; oocysts or sporocysts for the coccidia; and vacuolated forms for *B. hominis*.

The type of material to be examined depends on the parasite and its location in the body. For the intestinal protozoa, feces are commonly submitted for examination, although other materials are occasionally obtained.

Four types of procedures are used to recover and demonstrate intestinal protozoa: direct wet mount examinations, concentration techniques, permanently stained preparations, and cultivation (21) (procedures are discussed in detail in Chapter 55). All of these methods may not be needed in every case. The selection of appropriate techniques depends on the species of parasite suspected and the stage(s) of parasite likely to be found in the specimen. For example, trophozoite stages of amebae and flagellates are more likely to be present in soft or diarrheic stools, and cysts are more likely in formed feces. Thus, the techniques used to examine diarrheic fecal specimens may differ from those used for formed specimens.

For accurate and reliable identification, specimens must be properly collected and handled before examination. Protozoa, especially trophozoites, may develop atypical morphology or die in old or poorly collect-

TABLE 1. Intestinal and urogenital protozoa that may be found in specimens from patients in the United States

Type and species	Relative prevalence[a]	Pathogenicity
Amebae		
Entamoeba histolytica	+	+
Entamoeba hartmanni	+	−
Entamoeba coli	++	−
Entamoeba polecki	R	−
Endolimax nana	++	−
Iodamoeba bütschlii	+	−
Ciliate		
Balantidium coli	R	+
Flagellates		
Dientamoeba fragilis	+	+
Giardia lamblia	++	+
Trichomonas vaginalis	++	+
Trichomonas hominis	+	−
Chilomastix mesnili	+	−
Enteromonas hominis	R	−
Retortamonas intestinalis	R	−
Coccidia		
Isospora belli	R	+
Sarcocystis spp.	R	?
Cryptosporidium sp.	+	++
Blastocystis hominis	+	±

[a] R, Rare.

ed specimens. Ideally, fecal specimens should reach the laboratory within 1 to 2 h after passage; other materials (urine, duodenal material, or aspirates from lesions) should be sent to the laboratory immediately after collection. If transportation is delayed, specimens should be appropriately preserved to maintain the diagnostic characteristics of organisms that might be present.

AMEBAE

Five species of intestinal amebae may live in the cecum and colon of humans. They are *E. histolytica, Entamoeba hartmanni, Entamoeba coli, Endolimax nana*, and *Iodamoeba bütschlii*. Infection is acquired by the ingestion of cysts, which then excyst in the intestine. The cysts are quite hardy and can survive for days or weeks in water or the environment. Infection is usually diagnosed by the identification of organisms in feces, although other materials may be examined in symptomatic cases. Immunodiagnostic tests are useful for the diagnosis of extraintestinal amebiasis.

Entamoeba histolytica

E. histolytica causes amebiasis and is the only ameba pathogenic for humans. Infections with *E. histolytica* are classified as amebiasis irrespective of whether the person exhibits symptoms. A number of outbreaks have occurred in the United States (15). Strains which cause clinical disease can be distinguished from commensal strains by isoenzyme analysis (25). The incubation period is variable, from as short as a few days to weeks or even months (7). Clinical amebiasis, i.e., infection with symptoms produced presumably by an amebic invasion of colonic tissue, may present with several manifestations. These include amebic dysentery, amebic colitis, and ameboma. Amebic dysentery is an acute diarrhea with ulcerations of the colonic mucosa. Symptoms include crampy, lower abdominal pain, with bloody mucoid diarrhea in severe cases. In some people, an increased frequency of bowel movements with or without blood and mucus may occur. A chronic form, amebic colitis, produces symptoms similar to those of ulcerative colitis or other forms of inflammatory bowel disease, with diarrhea, sometimes bloody, occurring over a long period. Some patients with amebiasis have been misdiagnosed as having ulcerative colitis. Another less common form of intestinal disease, ameboma, is produced by the growth of granulomatous tissue in response to the infecting amebae, resulting in a large local lesion of the bowel which symptomatically and radiologically resembles colon cancer.

Infections with *E. histolytica*, with or without a history of antecedent, gastrointestinal symptoms, may result in hematogenous spread of the organisms to the liver via the portal system, resulting in amebic abscess or abscesses of the liver. This occurs in up to 5% of patients with symptomatic intestinal amebiasis. Approximately 40% of patients with amebic liver abscess do not give a history of prior bowel symptoms, and in many patients, *E. histolytica* is not present in stool at the time liver disease becomes manifest. Amebic abscesses occasionally occur in the lung, brain, or other organs.

Intestinal infection is usually diagnosed by the detection of organisms microscopically in feces or in sigmoidoscopic material from ulcerations. Depending on the type of fecal specimen, morphologic examinations by direct wet mount, concentration, and permanent stain may be useful. Purged stool specimens occasionally show parasites when they are not detected by routine examinations. Cultures for amebae may be helpful. Abscesses are generally diagnosed by serologic tests (see below), although organisms may sometimes be demonstrated in abscess material by morphology or culture.

Immunodiagnosis of amebiasis

In general, serology is unnecessary in intestinal amebiasis, although it may be positive in 70 to 80% of intestinal cases. In extraintestinal amebiasis, serology is extremely useful, being positive in over 90% of cases. Titers are generally higher in patients with extraintestinal diseases.

Five tests are presently accepted as useful in clinical diagnosis: indirect hemagglutination (IHA) (14) and enzyme-linked immunosorbent assay (ELISA) (2) are quantitative tests, and double diffusion (DD) (18), countercurrent immunoelectrophoresis (CIE) (16), and latex agglutination are qualitative tests. The IHA test has been the most widely used and evaluated and is considered the reference test (10). It is rapid, simple, and inexpensive and requires no special equipment or training. The test is sensitive and reacts in over 90% of extraintestinal cases. The IHA test has the nonspecificities of all tests with erythrocytes, primarily autoagglutinins and heterophilelike antibody, but these rarely occur at diagnostic levels. Clinical correlation is excellent; if a value of 1:256 or greater is considered significant, the test is virtually 100% specific for amebiasis.

TABLE 2. Characteristics of intestinal amebae visible in different types of fecal preparations[a]

Characteristic	Unstained		Temporary stain		Permanent stain
	Saline	Formalin	Iodine (cysts)	Buffered methylene blue (trophozoites)[b]	
Trophozoite					
Motility	+	−		−	−
Cytoplasm					
Appearance	+	+		+	+
Inclusions (erythrocytes, bacteria)	+	+		+	+
Nucleus	−	+[c]		+	+
Cyst					
Nucleus	−	+	+		+
Chromatoid bodies	+	+	+[d]		+
Glycogen	−	−	+		− (vacuole present)

[a] Table from Brooke and Melvin (4).
[b] Quensel stain may be substituted for buffered methylene blue.
[c] Nuclei of trophozoites are visible in Formalin-fixed material but are usually not sufficiently distinctive for species identification.
[d] Chromatoid bodies are more easily seen in unstained wet mounts than in iodine preparations.

The ELISA procedure is newer and less well evaluated. Published data indicate results comparable to those obtained by the IHA test (17). Even though ELISA is not plagued by the nonspecificities of the IHA test, it is more difficult to perform, even with automated instruments to alleviate some of the problems. Because it is more adaptable to clinical situations, further evaluation may prove ELISA to be the test of choice.

The DD and CIE tests have been available for some time, but because of the inability to quantitate reactivity, they have not been widely used (13). Each has its disadvantage: the DD test requires relatively large quantities of reagents to obtain an appropriate precipitate; the CIE test requires the use of specialized equipment. Both, however, are simple, yield acceptable qualitative data, and serve as useful screen tests.

Because of the difficulty in clinically differentiating echinococcosis from amebiasis, all cases of suspected hydatid disease should be tested for amebiasis as well. The significance of the results of amebic serologic tests, as designated by the Centers for Disease Control, are as follows.

IHA test.
- ≤1:128, probably no extraintestinal involvement
- 1:256 to 1:512, possible extraintestinal involvement
- ≥1:1,024, likely extraintestinal involvement

DD and CIE tests. One or more bands are equivalent to IHA ≥1:256. There is no correlation between the number of bands and the titer.

Latex agglutination test. Highly reactive; false-positives occur. Should be used for screen test purposes only.

Amebic antibodies may persist for long periods of time, a fact which should be considered in any interpretation of serologic results.

Individual reagents of antigen and antisera are commercially available. Kits standardized for reactivity can be purchased for the IHA, ELISA, DD, and CIE tests. All appear to have acceptable performance characteristics.

Morphologic identification of amebae

Both trophozoites and cysts are diagnostic stages of the amebae, and either or both stages can be detected in feces. Microscopists must be familiar with the morphologic characteristics used for the differentiation of species and must be able to distinguish trophozoites from epithelial cells and macrophages, as well as cysts from pus cells, yeast cells, molds, and other objects that may be present in feces.

Characteristics used to distinguish species of amebae are as follows.

- Trophozoites
  - Motility (progressive or nonprogressive)
  - Cytoplasm
    - Appearance (finely granular or coarse)
    - Inclusions (erythrocytes, yeast cells, molds, or bacteria)
  - Nucleus
    - Peripheral chromatin (present or absent; if present, the arrangement and size of the granules are important)
  - Karyosome (size and position)
- Cysts
  - Nucleus
    - Number
    - Peripheral chromatin (present or absent)
    - Karyosome (size and position)
  - Cytoplasmic inclusions (chromatoid bodies or glycogen; these are more often seen in young cysts)

Size is not a reliable feature for species differentiation of either trophozoites or cysts except in separating *E. histolytica* and *E. hartmanni*.

Not all of the characteristics listed can be seen in a single type of preparation; stained and unstained wet mounts and permanent stained smears are necessary to demonstrate all of the features (Table 2). Unstained wet mounts may reveal trophozoites and cysts. The motility of trophozoites (in viable saline mounts) and the cytoplasmic inclusions such as erythrocytes in trophozoites and chromatoid bodies in cysts can be observed. However, stained preparations are usually needed for reliable species identification. Buffered methylene blue solution (Nair stain) can be used for

TABLE 3. Differential morphology of ameba trophozoites found in human stool specimens[a]

Species	Size (diam or length)	Motility	Nucleus			Cytoplasm	
			No. and characteristics	Peripheral chromatin	Karyosomal chromatin	Appearance	Inclusions
Entamoeba histolytica	10–60 μm; usual range, 15–20 μm, commensal form[b]; over 20 μm invasive form[c]	Progressive, with hyaline, fingerlike pseudopods	1, not visible in unstained preparations	Fine granules; usually evenly distributed and uniform in size	Small, discrete; usually centrally located, but occasionally eccentric	Finely granular	Erythrocytes occasionally; noninvasive organisms may contain bacteria
Entamoeba hartmanni	5–12 μm; usual range, 8–10 μm	Usually nonprogressive, but may be progressive occasionally	1, not visible in unstained preparations	Similar to *E. histolytica*	Small, discrete, often eccentric	Finely granular	Bacteria
Entamoeba coli	15–50 μm; usual range, 20–25 μm	Sluggish, nonprogressive, with blunt pseudopods	1, often visible in unstained preparations	Coarse granules, irregular in size and distribution	Large, discrete, usually eccentric	Coarse, often vacuolated	Bacteria, yeast cells, or other material
Entamoeba polecki	10–25 μm; usual range, 15–20 μm	Usually sluggish, similar to *E. coli*; occasionally in diarrheic specimens, motility may be progressive	1, may be slightly visible in unstained preparations; occasionally distorted by pressure from vacuoles in cytoplasm	Usually fine granules evenly distributed; occasionally, granules may be irregularly arranged; chromatin sometimes in plaques or crescents	Small, discrete, eccentric; occasionally large, diffuse, or irregular	Coarsely granular, may resemble *E. coli*; contains numerous vacuoles	Bacteria, yeast cells
Endolimax nana	6–12 μm; usual range, 8–10 μm	Sluggish, usually nonprogressive with blunt pseudopods	1, visible occasionally in unstained preparations	None	Large, irregularly shaped, blotlike	Granular, vacuolated	Bacteria
Iodamoeba bütschlii	8–20 μm; usual range, 12–15 μm	Sluggish, usually nonprogressive	1, not usually visible in unstained preparations	None	Large, usually central; surrounded by refractile, achromatic granules often not distinct even in stained slides	Coarsely granular, vacuolated	Bacteria, yeast cells, or other material
Dientamoeba fragilis[d]	5–15 μm; usual range, 9–12 μm	Pseudopods are angular, serrated, or broad lobed; hyaline, almost transparent	2 (in ca. 20% of organisms, only 1 nucleus is present), nuclei invisible in unstained preparations	None	Large cluster of 4–8 granules	Finely granular	Bacteria; occasionally erythrocytes

[a] Table from Brooke and Melvin (4).
[b] Commensal form usually found in asymptomatic or chronic cases; may contain bacteria.
[c] Invasive form usually found in acute cases; often contains erythrocytes.
[d] Flagellate included with amebae for diagnostic purposes.

TABLE 4. Differential morphology of ameba cysts found in human stool specimens[a]

Species	Size	Shape	Nucleus			Cytoplasm	
			No. and characteristics	Peripheral chromatin	Karyosomal chromatin	Chromatoid bodies	Glycogen
Entamoeba histolytica	10–20 μm; usual range, 12–15 μm	Usually spherical	4 in mature cyst; immature cysts with 1 or 2 occasionally seen	Present; fine, uniform granules, evenly distributed	Small, discrete, usually centrally located	Present; elongated bars with bluntly rounded ends	Usually diffuse; concentrated mass often present in young cysts; stains reddish brown with iodine
Entamoeba hartmanni	5–10 μm; usual range, 6–8 μm	Usually spherical	4 in mature cyst; immature cysts with 1 or 2 often seen	Similar to *E. histolytica*	Similar to *E. histolytica*	Present; elongated bars with bluntly rounded ends	Similar to *E. histolytica*
Entamoeba coli	10–35 μm; usual range, 15–25 μm	Usually spherical; occasionally oval, triangular, or other shapes	8 in mature cyst; occasionally supernucleated cysts with 16 or more are seen; immature cysts with 2 or more occasionally seen	Present; coarse granules irregular in size and distribution, but often more uniform than in trophozoites	Large, discrete, usually eccentric, but occasionally centrally located	Present, but less frequently seen than in *E. histolytica*; usually splinterlike with pointed ends	Usually diffuse, but occasionally a well-defined mass in immature cysts; stains reddish brown with iodine
Entamoeba polecki	9–18 μm; usual range, 11–15 μm	Spherical or oval	1, rarely 2; occasionally visible in unstained preparations	Usually fine granules evenly distributed	Usually small and eccentric	Present; many small bodies with angular or pointed ends or few large bodies; may be oval, rodlike, or irregular	Usually small, diffuse masses; stains reddish brown with iodine, dark area called inclusion mass (possibly concentrated cytoplasm) often also present; mass does not stain with iodine
Endolimax nana	5–10 μm; usual range, 6–8 μm	Spherical, ovoidal, or ellipsoidal	4 in mature cysts; immature cysts with less than 4 rarely seen	None	Large (blotlike), usually central	Occasionally granules or small oval masses, but bodies as seen in *Entamoeba* spp. are not present	Usually diffuse; concentrated mass seen occasionally in young cysts; stains reddish brown with iodine
Iodamoeba bütschlii	5–20 μm; usual range, 10–12 μm	Ovoidal, ellipsoidal, triangular, or other shapes	1 in mature cyst	None	Large, usually eccentric; refractile, achromatic granules on one side of karyosome; indistinct in iodine preparations	Occasionally granules present, but chromatoid bodies as seen in *Entamoeba* spp. are not present	Compact, well-defined mass; stains dark brown with iodine

[a] Table from Brooke and Melvin (4).

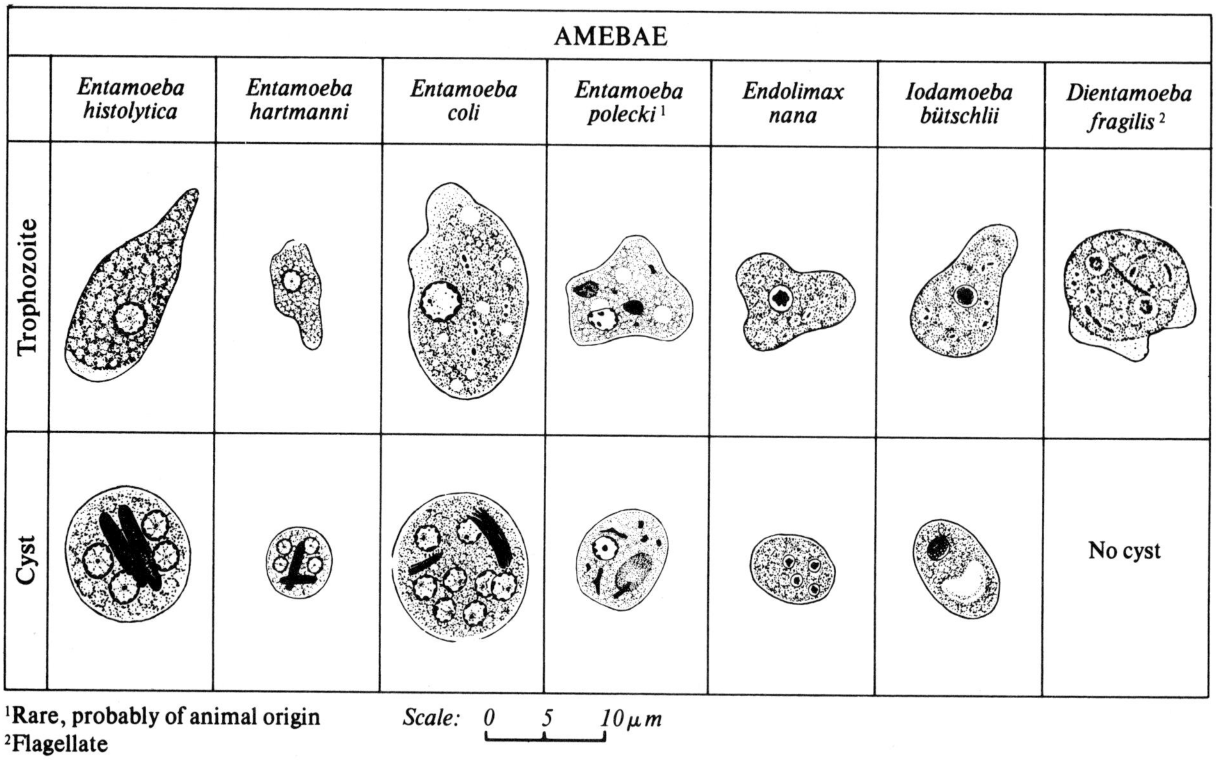

AMEBAE
Entamoeba histolytica
Entamoeba hartmanni
Entamoeba coli
Entamoeba polecki 1
Endolimax nana
Iodamoeba bütschlii
Dientamoeba fragilis 2
Trophozoite
Cyst
No cyst
1Rare, probably of animal origin
2Flagellate
Scale: 0 5 10 μm

FIG. 1. Amebae found in human stool specimens. From Brooke et al. (5).

temporary stains of trophozoites in fresh specimens and will permit the microscopist to distinguish host cells from amebae and *Entamoeba* sp. trophozoites from those of other genera, but permanent stains are necessary for accurate species identification. Iodine solutions are used for temporary cyst stains in fresh or fixed specimens. Characteristics of cysts are less variable than those of trophozoites, and species of cysts can frequently be identified in iodine-stained wet mounts, especially if the organisms are examined with oil-immersion magnification.

Regardless of the types of materials examined or the methods used to demonstrate organisms, species identifications are based on microscopic observations of

Entamoeba histolytica
Entamoeba hartmanni
Entamoeba coli
Endolimax nana
Dientamoeba fragilis*
Iodamoeba bütschlii**

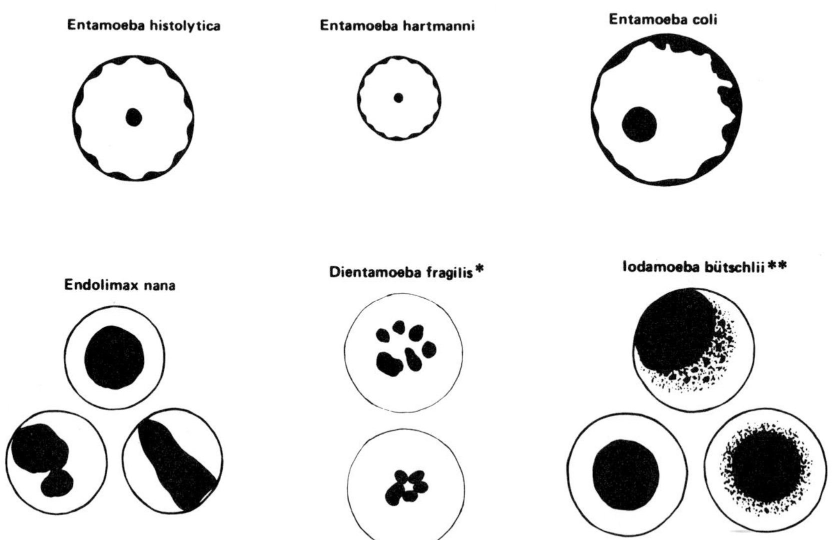

FIG. 2. Nuclei of amebae. *Flagellate. ***Iodamoeba* cysts often have eccentric karyosomes against the nuclear membrane, as in the upper *Iodamoeba* nucleus, whereas the trophozoite nuclei are usually not against the nuclear membrane. Achromatic granules are not always visible (lower left) in *Iodamoeba* nuclei. From Smith et al. (26).

morphologic characteristics. The typical characteristics of trophozoites and cysts of the ameba species are listed in Tables 3 and 4; diagrams are presented in Fig. 1. Diagrams of nuclei are shown in Fig. 2; photomicrographs are presented in Fig. 3 through 5. Although *Dientamoeba fragilis* is taxonomically a flagellate, it is included in these tables and figures because it resembles and must be differentiated from the amebae.

Morphologic characteristics of species overlap, and some organisms may be atypical, thus making identification difficult. For example, distinguishing trophozoites of *E. histolytica* from those of *E. coli* is often difficult because of morphologic variations. Rarely can species of trophozoites be identified from a single feature, such as karyosome location, or from a single organism. The microscopist must observe both the cytoplasmic and nuclear characteristics of several organisms before making a species identification. Although cysts are more easily identified than trophozoites, several cysts (particularly if they are immature) should be observed to ensure that the identification is reliable. If two species are identified, there should be distinct populations of each.

Sometimes, although organisms are recognized, species cannot be identified. In these instances, the laboratory should report "unidentified ameba trophozoites (or cysts)," or if the genus can be determined but the species cannot, "unidentified *Entamoeba* trophozoites or cysts" should be reported, and another specimen should be requested.

FLAGELLATES

The flagellates inhabit the intestinal and atrial areas. Intestinal infections by flagellates with cyst stages are acquired by the ingestion of cysts. Infections are usually diagnosed by morphologic examinations of feces or other body materials. For *G. lamblia*, immunodiagnostic tests have been reported for the detection of antibody in serum (31) and parasite antigens in stool specimens (30). Species identifications, however, are generally based on microscopic observations of morphologic features of trophozoites or cysts. Two species, *G. lamblia* and *D. fragilis*, cause clinically significant intestinal disease, and *T. vaginalis* is a frequent cause of vaginitis.

Giardia lamblia

Giardiasis is acquired by the ingestion of the hardy cysts of *G. lamblia*. Outbreaks related to contaminated water are common in the United States, and infections are frequent in day-care centers and among campers and male homosexuals (27, 31). Trophozoites infect the upper small intestine but do not invade the tissues to produce ulcers. Infection may elicit a variety of symptoms (33), or, in some patients, it may be asymptomatic. The incubation period is variable, ranging from a few days to several weeks, with an average of about 9 days. In acute giardiasis, symptoms include nausea, upper intestinal cramping or pain, and malaise. There is often explosive, watery diarrhea, characterized by foul-smelling stools. These symptoms are accompanied by flatulence and abdominal distention. The acute stage of clinical giardiasis may be followed by a chronic stage, or the chronic type of infection may be the first indication of infection. In such infections, there are flatulence, mushy, foul-smelling stools, upper intestinal cramping, and abdominal distention. A number of patients also exhibit belching, nausea, anorexia, vomiting, and symptoms of heartburn. Fever and chills may be present, but to a lesser degree than the aforementioned symptoms. Symptoms may mimic peptic ulcer or gallbladder disease. In some patients, the cysts may be present in stools in a variable pattern, although the reasons for this are not clear (6, 23). This variable presence of cysts may occur even in the presence of classic symptoms of disease and numerous trophozoites in the upper small intestine.

Diagnosis is usually established by the demonstration of cysts or, occasionally, trophozoites in feces or of trophozoites in duodenal material (Fig. 5 through 7). Because of the variable shedding of organisms, several specimens should be examined before ruling out the infection. It is best to examine a total of three specimens collected 2 to 3 days apart, although daily specimens for a total of three can be used. Feces should be examined by direct wet mounts and a concentration method. If viable trophozoites are present, they can be readily identified by the characteristic falling-leaf or tumbling motion in saline mounts of fresh feces. The large, vertical sucking disk can be seen as the organism turns. In a lateral view, the trophozoite appears spoon shaped. Permanent stains may be needed if organisms cannot be identified in wet mounts or concentration (29). Small plant cells can resemble *Giardia* cysts, so organisms should be carefully examined for the fibrils and nuclei characteristic of *Giardia* species. Permanent stains may help.

When *Giardia* organisms are not found in stool specimens, duodenal aspirates, string test mucus, or biopsied mucosal tissue can be examined. The string test (1) is used to collect mucus from the duodenal area and may be less traumatic for the patient than other methods would be. Materials obtained by drainage, aspiration, or the string test can be examined by simple, direct wet mounts. Biopsy tissue may be processed and stained by the usual histopathologic methods. Initially, an imprint smear of the mucosal surface on a slide can be made and stained with trichrome or Giemsa stain.

Chilomastix mesnili

C. mesnili is a nonpathogenic flagellate which inhabits the cecum and colon. The diagnostic stages, trophozoites and cysts, are passed in feces. The trophozoites have a characteristic rotating, wobbling motion which is readily recognized in saline mounts of fresh material. The spiral groove, which extends along the body, is sometimes visible as the organism turns. In permanent stained preparations, trophozoites are usually lightly stained and sometimes distorted, and they may be overlooked. The most prominent feature is the long cytostome which extends about one-third to one-half the length of the body. The nucleus also may have a collection of chromatin along one side, giving it a lopsided look (Fig. 6 and 7).

The presence of the cytostome and spiral groove, the location of the nucleus at one end with tapering of the opposite end, and nuclear characteristics aid in the

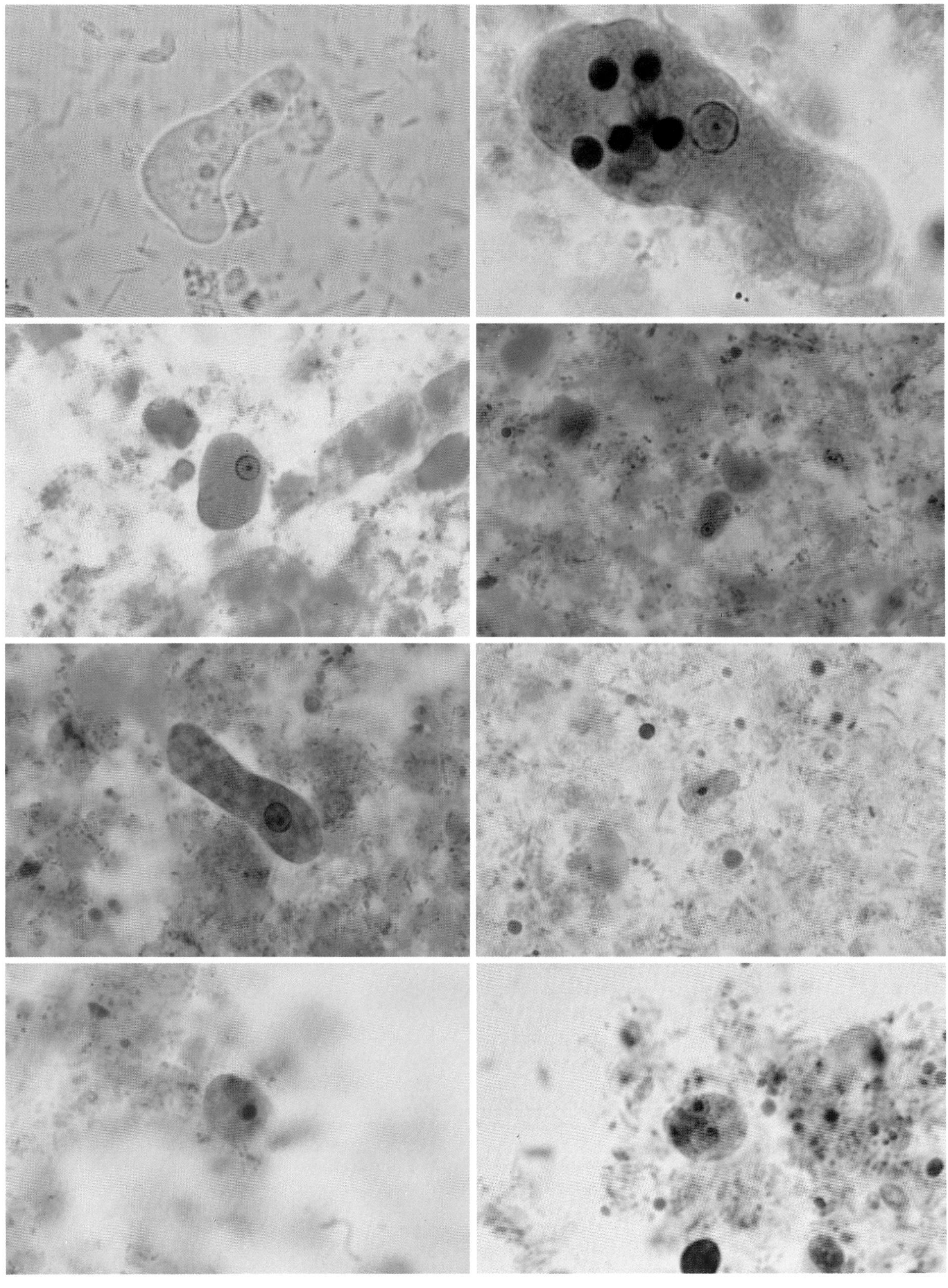

differentiation of *C. mesnili* from other intestinal protozoa.

C. mesnili cysts are usually identified by their lemon shape, single large nucleus, and fibrils (Fig. 6). Not all cysts are lemon shaped; they may be rounded or, if viewed on end, may appear rounded.

Dientamoeba fragilis

D. fragilis was originally thought to be an ameba but is now recognized as a flagellate with no flagella; thus, the disease dientamebiasis is discussed in the flagellate section, but the organism is included in Fig. 1 through 3 and Tables 2 and 3, which concern amebae, because it must be differentiated from the amebae. The organism does not have a cyst stage, and the exact means of spread is not clear. *D. fragilis* infection is commonly associated with enterobiasis, and it has been suggested that *D. fragilis* may infect *Enterobius* eggs and thus bypass gastric acidity. Although clinical infections with *D. fragilis* occur, they are infrequently reported. This low incidence of reporting may be due to the self-limited nature of the infection or, more probably, to the difficulty of identification of the organism in stool specimens from individuals with suggestive symptoms, so that many infections are not diagnosed. The incubation period for clinical dientamebiasis is not known with certainty. Symptoms have been reported more frequently in children than in adults and are predominately diarrhea and abdominal distention (28). Nausea, vomiting, and weight loss have been recorded in from one-third to one-fifth of the cases reported in the literature.

Permanent stains are required to diagnose most infections. The delicately staining trophozoites are usually (60 to 80%) binucleate, though the nuclei may be in different planes of focus. They must be differentiated from the trophozoites of *Endolimax nana*, *I. bütschlii*, and *E. hartmanni*. Nuclear characteristics, the presence of binucleate forms, and the absence of cysts aid in identification of this organism.

Trichomonas hominis

T. hominis is a nonpathogenic protozoan inhabitant of the colon which has only a trophozoite stage. The motile trophozoites have a characteristic nervous, jerky motion and can be readily identified in saline mounts. They possess an undulating membrane which extends most of their length (Fig. 6 and 7) and often can be seen in wet mounts, especially if the organisms have slowed down. Iodine stains are of little value for the identification of *T. hominis* because the organisms tend to become distorted. Permanent stains are also of limited value; although organisms can be seen, they are often distorted and difficult to recognize.

Trichomonas vaginalis

T. vaginalis inhabits the urogenital systems of both males and females and is considered a pathogen. The trophozoites (the only stage) are found in the urine of both sexes, in material from the vagina, and in prostatic secretions. It is estimated that approximately 5 million women in the United States have trichomoniasis, and roughly 1 million men harbor the parasite. Infection is usually, but not always, acquired by sexual contact. The infection in males is generally asymptomatic, but 25 to 50% of infected women exhibit symptoms (32) which include dysuria, vaginal itching and burning, and in severe infections, a foamy, yellowish-green discharge with a foul odor. In many women the infection becomes symptomatic and chronic, with periods of relief in response to therapy. Recurrences of infection and disease may be caused by reinfection from an asymptomatic sexual partner, in the true sense of a sexually transmitted disease, or by failure of the drug metronidazole to eliminate the parasite completely. Symptomatic infections in males, although rarely reported, have included prostatitis, urethritis, epididymitis, and urethral stricture. Rarely, *T. vaginalis* infections occur in ectopic sites, and parasites may be recovered from areas of the body other than the urogenital system (20).

Most cases of *T. vaginalis* are detected by finding the motile trophozoites in wet mounts of vaginal fluid, prostatic fluid, or sediments of freshly passed urine. Like *T. hominis*, the trophozoites move with a nervous, jerky motion and possess an undulating membrane, but it extends only half the length of the organism. In old urine specimens, the organisms may be dead or badly distorted and thus cannot be identified or may be confused with host cells. In addition, old urine specimens may be contaminated with *Bodo*

FIG. 3. Ameba trophozoites (trichrome stain, except upper left; oil immersion magnification). Row 1: (left) *E. histolytica*, unstained. The clear pseudopod is evident at the bottom of the organism. Several ingested erythrocytes are present in the cytoplasm. The nucleus is not visible. (Right) *E. histolytica*. This large trophozoite contains numerous ingested erythrocytes to the left of the nucleus. The nucleus has a small, dotlike central karyosome and peripheral chromatin which although granular is distributed in fairly uniform fashion. This large erythrophagocytic trophozoite is from a patient with amebic dysentery. Row 2: (left) *E. histolytica*. The trophozoite is typical of the commensal form of *E. histolytica*. It is smaller than the invasive form seen in Row 1, right, and does not contain ingested erythrocytes. The nucleus is typical with a dotlike central karyosome and evenly distributed peripheral chromatin. (Right) *E. hartmanni*. This organism has characteristics similar to those of *E. histolytica* but is smaller. The nucleus is at the lower portion of the trophozoite. Row 3: (left) *E. coli*. This elongated trophozoite of *E. coli* has a nucleus with a large eccentric karyosome. There is some unevenness in distribution of peripheral chromatin. Cytoplasm contains numerous vacuoles. (Right) *E. nana*. This trophozoite has a nucleus with a large karyosome and no peripheral chromatin on the nuclear membrane. A clear halo surrounds the karyosome and extends to the nuclear membrane. It is most evident beneath the karyosome. Cytoplasm is delicate and vacuolated. Row 4: (left) *I. bütschlii*. This organism has a large karyosome and no chromatin on the nuclear membrane. Distinct achromatic granules are not evident in the nucleus, but the karyolymph is "muddy" in contrast to the clearer karyolymph noted in the *E. nana* above. The cytoplasm of this organism is rather homogeneous, although the cytoplasm of *I. bütschlii* trophozoites often contains numerous ingested bacteria. (Right) *D. fragilis* trophozoite. This flagellate is included here because it must be differentiated from the amebae. There are two nuclei, one above the other. The granular karyosome of the lower nucleus is evident. There is no chromatin on the nuclear membrane. Cytoplasm is vacuolated.

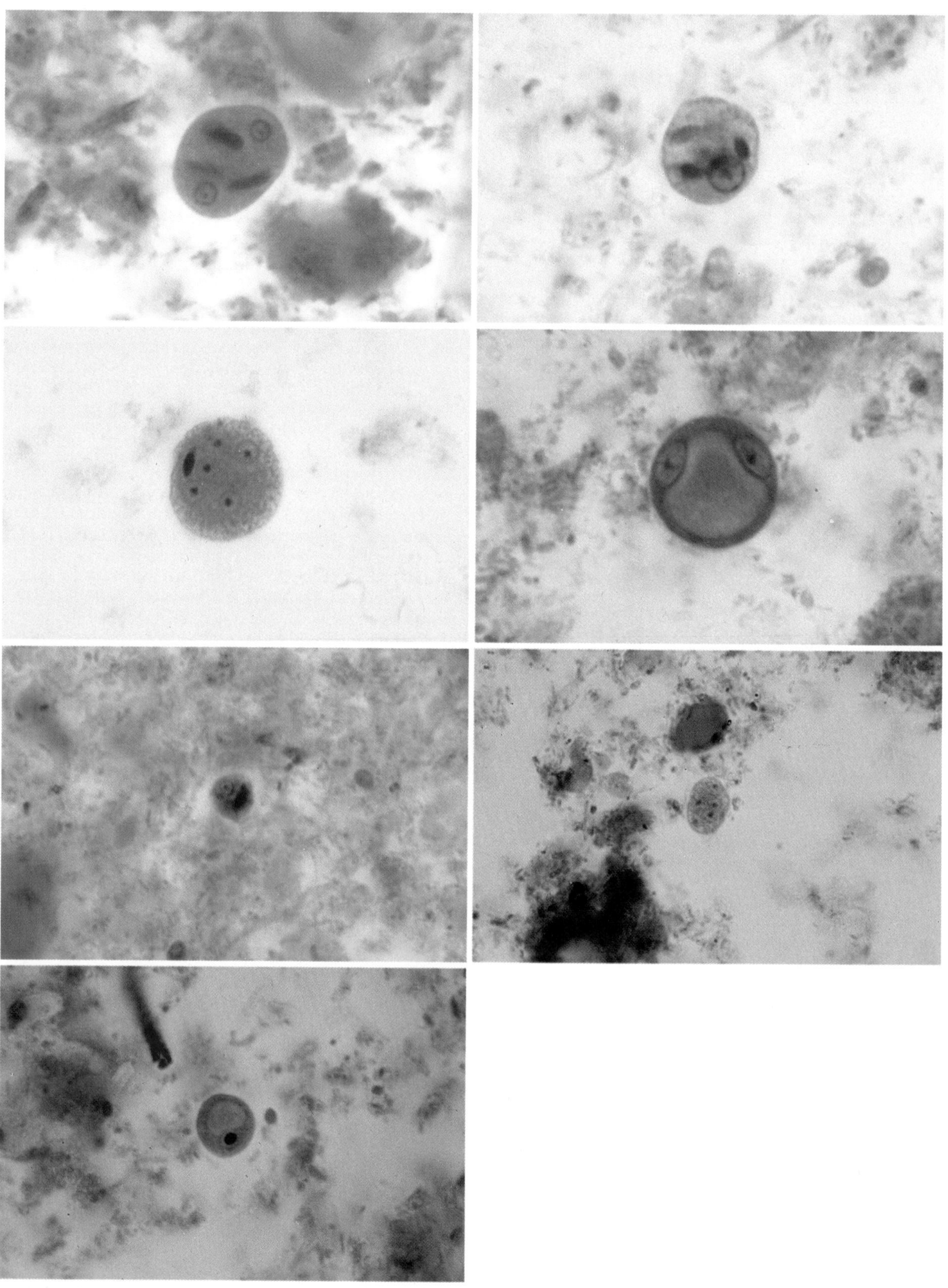

TABLE 5. Characteristics of intestinal flagellates, a ciliate, and coccidia visible in different types of fecal preparations[a]

Group characteristics	Unstained		Temporary stain		Permanent stain
	Saline	Formalin	Iodine (cysts)	Neutral red[b] (trophozoites)	
Flagellates					
Trophozoite					
Motility	+	−		+	−
Shape	+	+		+	+ (may be distorted)
Nucleus	−	+		+	+
Flagella	±	−		+	±
Other features[c]	+	+		+	+
Cyst					
Shape	+	+	+		+
Nucleus	−	+	+		+
Fibrils	±	+	+		+
Ciliate *(Balantidium coli)*					
Trophozoite					
Motility	+	−		+	−
Macronucleus	+	+		+	+
Cilia	+	+		+	+
Cyst					
Macronucleus	+	+	±		+
Coccidia					
Oocyst and sporocyst	+	+	+		+[d]

[a] Table adapted from Brooke and Melvin (4).
[b] Neutral red dye in methocel solutions.
[c] The undulating membrane of *Trichomonas* and the spiral groove of *Chilomastix* may not be visible in all cases.
[d] *Cryptosporidium* oocysts can be demonstrated in acid-fast stains.

species or other free-living flagellates, especially if the urine collection vessel is open to the air and not sterile.

Vaginal materials commonly used for the diagnosis of *T. vaginalis* infections are vaginal fluid, scrapings, or washings. These samples may be examined morphologically in a saline wet mount or as a stained smear, or the material can be cultured. Because of the pronounced distortion in stained vaginal smears, these preparations are not recommended. If organisms are not seen in saline wet mounts, material should be cultured. Although some workers feel that wet mount examinations are as efficient as cultures in revealing infections, current evidence suggests that cultivation methods are superior (9).

Morphologic identification of flagellates

The characteristics of trophozoites and cysts which aid in identification are outlined below. The flagellates are a more diverse group than the amebae.

Trophozoites
- Motility (in saline mounts, the type of trophozoite movement is characteristic and species specific)
- Shape
- Number of nuclei (the character of the nucleus is not generally used for species identification)
- Other features (undulating membrane, sucking disk, prominent cytostome, spiral groove)
- Flagella (number and location, but since flagella are difficult to see and count, they are not practical diagnostic features for species identification; however, their presence distinguishes the organism as a flagellate trophozoite)

Cysts
- Shape
- Size
- Number of nuclei
- Fibrils (arrangement or pattern within the cyst; their presence distinguishes the cyst as a flagellate rather than an ameba cyst)

Not all the features listed can be seen in a single type of preparation (Table 5). In many cases, species can be determined by the examination of either direct or concentrated wet mounts, without resorting to permanent stains. If viable trophozoites are present,

FIG. 4. Ameba cysts (trichrome stain; oil immersion magnification). Row 1: (left) *E. histolytica.* Two of the four nuclei are seen in this plane of focus. Three chromatoid bodies are evident which stain dark blue and have rounded contours. (Right) *E. histolytica.* This uninucleate cyst has numerous rounded chromatoid bodies which in this organism stain red. The pale staining areas represent glycogen masses. Row 2: (left) *E. coli.* Five nuclei are evident in this plane of focus, and there is a red chromatoid body to the left. Nuclear karyosomes are large and central. Although karyosomes are typically eccentric, they may be centered as in nuclei of this cyst. Cytoplasm is granular. (Right) *E. coli.* This binucleate cyst of *E. coli* contains a large glycogen vacuole. Immature cysts such as this are typical of *E. coli.* Row 3: (left) *E. hartmanni.* This small cyst has one nucleus in this plane of focus. A large chromatoid body is present on the right. (Right) *E. nana.* All four dotlike nuclei are evident at this focal plane. Halos are evident around some of them. Row 4: *I. bütschlii.* This cyst has a large glycogen vacuole with the nucleus below it. Achromatic granules are not evident, and the karyosome is large and rounded.

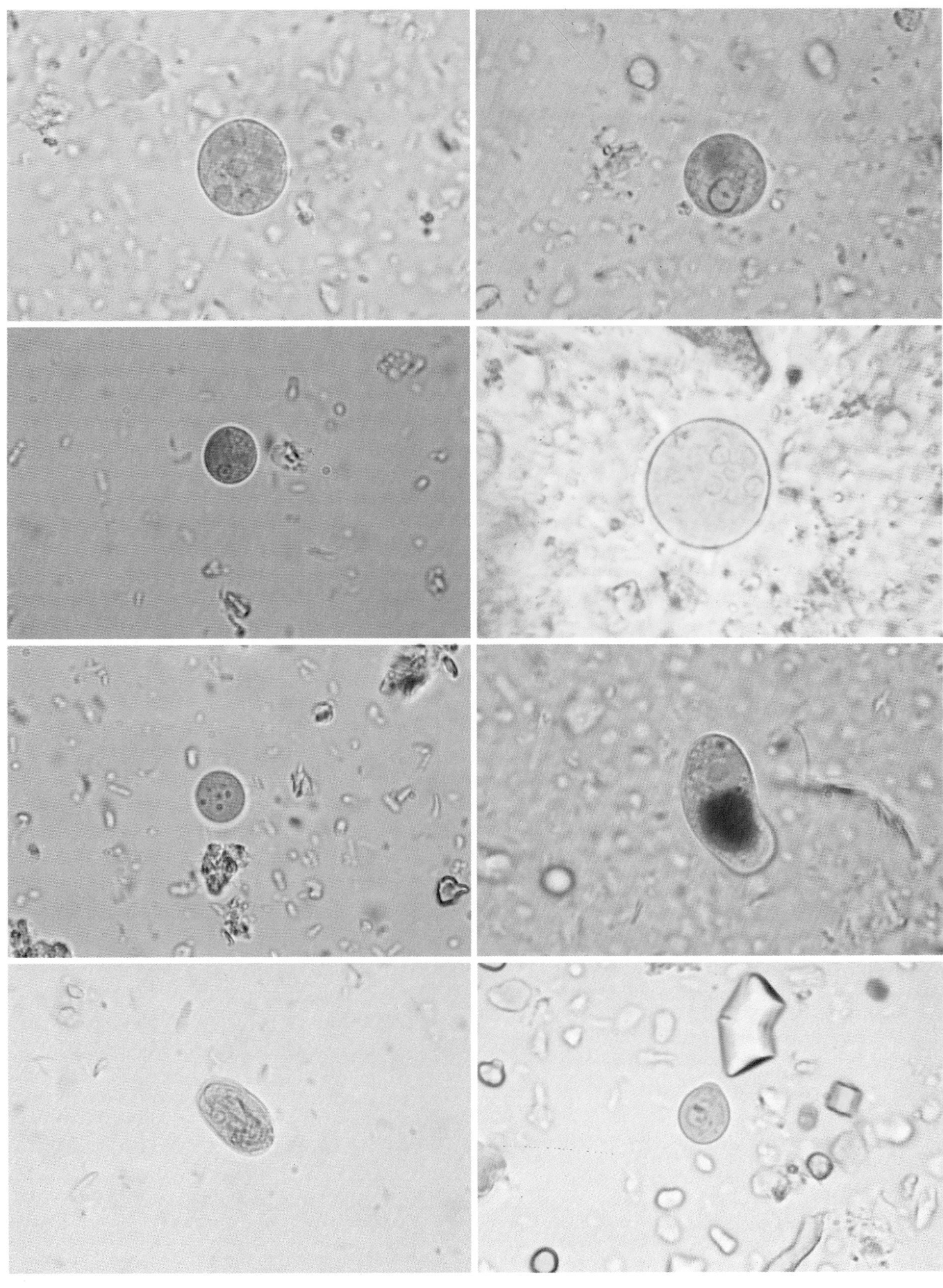

TABLE 6. Differential morphology of flagellate trophozoites found in human stool specimens[a]

Species	Length	Shape	Motility	No. and characteristics of nuclei	No. and location of flagella[b]	Other features
Trichomonas hominis	8–20 μm; usual range, 11–12 μm	Pear shaped	Nervous, jerky	1, not visible in unstained mounts	3–5 anterior 1 posterior	Undulating membrane extending length of body
Chilomastix mesnili	6–24 μm; usual range, 10–15 μm	Pear shaped	Stiff, rotary	1, not visible in unstained mounts	3 anterior 1 in cytostome	Prominent cytostome extending 1/3–1/2 length of body; spiral groove across ventral surface
Giardia lamblia	10–20 μm; usual range, 12–15 μm	Pear shaped	Falling leaf	2, not visible in unstained mounts	4 lateral 2 ventral 2 caudal	Sucking disk occupying 1/2–3/4 of ventral surface; median bodies lying horizontally or obliquely in lower part of body
Enteromonas hominis	4–10 μm; usual range, 8–9 μm	Oval	Jerky	1, not visible in unstained mounts	3 anterior 1 posterior	One side of body flattened; posterior flagellum extends free posteriorly or laterally
Retortamonas intestinalis	4–9 μm; usual range, 6–7 μm	Pear shaped or oval	Jerky	1, not visible in unstained mounts	1 anterior 1 posterior	Prominent cytostome extending approx 1/2 length of body

[a] Table from Brooke and Melvin (4).
[b] Not a practical feature for the identification of species in routine fecal examinations.

identification can readily be made by the type of motility in direct saline mounts; no further observations are necessary. Species of cysts can usually be identified in iodine-stained mounts. However, permanent stains are necessary if organisms are atypical or degenerate or cannot be positively identified in wet mounts. The diagnostic features of flagellate trophozoites and cysts are described in Tables 6 and 7, respectively; diagrams are shown in Fig. 6, and photomicrographs are presented in Fig. 5 and 7. *D. fragilis* is not included in Table 6; because it closely resembles the amebae, it has been included with the ameba trophozoites for diagnostic purposes.

CILIATE

Balantidium coli, a pathogenic ciliate inhabiting the colon, is the only ciliate and the largest protozoan parasitizing humans. Both trophozoites and cysts may be found in the feces.

Balantidiasis in humans is rarely reported in the United States. The disease is more prevalent where there is a close association of humans with pigs, the natural hosts from which humans contract the infection. The organism also infects nonhuman primates, especially the great apes. The symptoms of infection with *B. coli* are referable to the large bowel and similar to those of amebiasis: lower abdominal pain, nausea, vomiting, and tenesmus. Chronic infections may present with cramps, frequent episodes of watery, mucoid diarrhea, and rarely with bloody diarrhea. Chronic infections have been known to last for several months. In tropical areas in which the parasite is endemic, the infection often is severe in patients who also have other parasitic, bacterial, and viral infections and are undernourished. *B. coli* causes colonic ulcers similar to those caused by *E. histolytica*, but it does not cause lesions in other organs.

In human feces, trophozoites are readily recognized by their large size, their shape, and their rapid, rotating motion. Cysts are less easily identified, but they usually cause few diagnostic problems. The morphology of trophozoites and cysts is described in Table 8; diagrams are shown in Fig. 8. Characteristics that are visible in different types of preparations are listed in Table 5.

The examination of direct saline mounts is the most practical method of detecting infections. Cysts can be recovered by concentration, but in human cases, trophozoites are usually seen more frequently than are cysts. Iodine-stained mounts and permanent stains are of little value because the organisms tend to overstain.

COCCIDIA

The intestinal coccidia that parasitize humans belong to the subphylum Sporozoa and are obligatory

FIG. 5. Ameba and flagellate cysts (iodine-stained wet mounts; oil immersion magnification). Row 1: (left) *E. histolytica*. Three nuclei are evident in this focal plane. (Right) *E. histolytica*. This immature uninucleate cyst has a reddish-staining glycogen mass above the nucleus. Row 2: (left) *E. hartmanni*. There is one nucleus in this focal plane. An irregular reddish glycogen mass is evident above the nucleus. (Right) *E. coli*. In this focal plane a cluster of six nuclei may be recognized toward the right of the cyst. Row 3: (left) *E. nana*. Four nuclei are evident, though the one in the center is out of focus. (Right) *I. bütschlii*. The large, reddish glycogen mass is prominent. Above it is the nucleus, which has a large pale karyosome surrounded by a pale irregular karyolymph space. Row 4: (left) *G. lamblia*. Two nuclei are evident toward the upper left, and multiple fibrils are present. (Right) *C. mesnili*. This small, lemon-shaped cyst has the nucleus on the left and faint fibrils on the right.

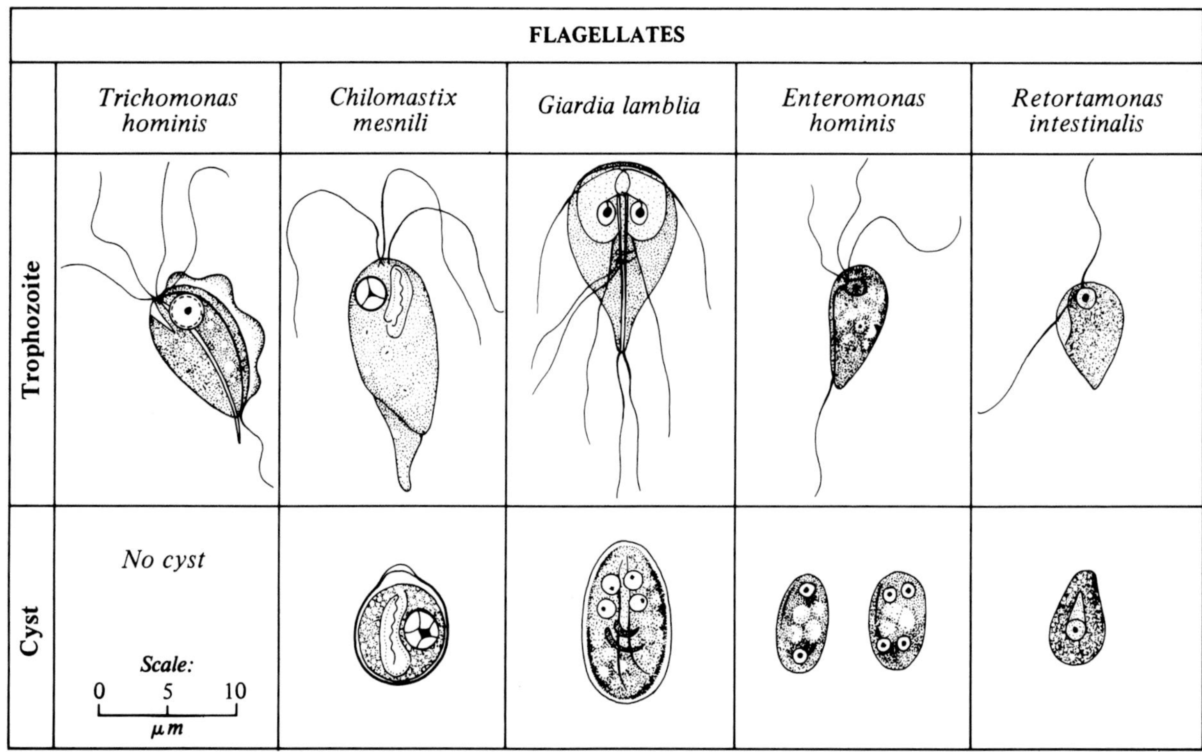

FIG. 6. Flagellates found in human stool specimens. From Brooke and Melvin (4).

tissue parasites that inhabit the mucosa of the small intestine. Species of three genera (*Isospora, Sarcocystis*, and *Cryptosporidium*) parasitize humans. The intestinal phase of toxoplasmosis which occurs in cats is similar to the intestinal infections of *Isospora* and *Sarcocystis* species in humans. The growth stages resemble those of malaria (also a sporozoan) and involve asexual and sexual generations. Therefore, the diagnostic stages, which are passed in feces, are unlike those of other intestinal protozoa. For both *Isospora* and *Cryptosporidium* species, oocysts, either unsporulated (immature) as in *Isospora belli* or sporulated (mature) as in *Cryptosporidium*, are diagnostic stages. The diagnostic stages for *Sarcocystis* species are free sporocysts and mature oocysts. Oocysts and sporocysts are almost transparent and are difficult to see in unstained preparations unless the microscope light is reduced and carefully regulated. Descriptions of the

TABLE 7. Differential morphology of flagellate cysts found in human stool specimens[a]

Species	Size	Shape	No. and characteristics of nuclei	Other features
Trichomonas hominis	No cyst			
Chilomastix mesnili	6–10 μm; usual range, 8–9 μm	Lemon shaped with anterior hyaline knob	1, not visible in unstained preparations	Cytostome with supporting fibrils; usually visible in stained preparations
Giardia lamblia	8–19 μm; usual range, 11–12 μm	Oval or ellipsoidal	Usually 4, not distinct in unstained preparations; usually located at one end	Fibrils or flagella longitudinally in unstained cysts; deep-staining fibers or fibrils lying laterally or obliquely across fibrils in lower part of cyst; cytoplasm often retracts from a portion of cell wall
Enteromonas hominis	4–10 μm; usual range, 6–8 μm	Elongated or oval	1–4, usually 2 lying at opposite ends of cyst; not visible in unstained mounts	Resembles *E. nana* cyst; fibrils or flagella usually not seen
Retortamonas intestinalis	4–9 μm; usual range, 4–7 μm	Pear shaped or slightly lemon shaped	1, not visible in unstained mounts	Resembles *Chilomastix* cyst; shadow outline of cytostome with supporting fibrils extending above nucleus

[a] Table from Brooke and Melvin (4).

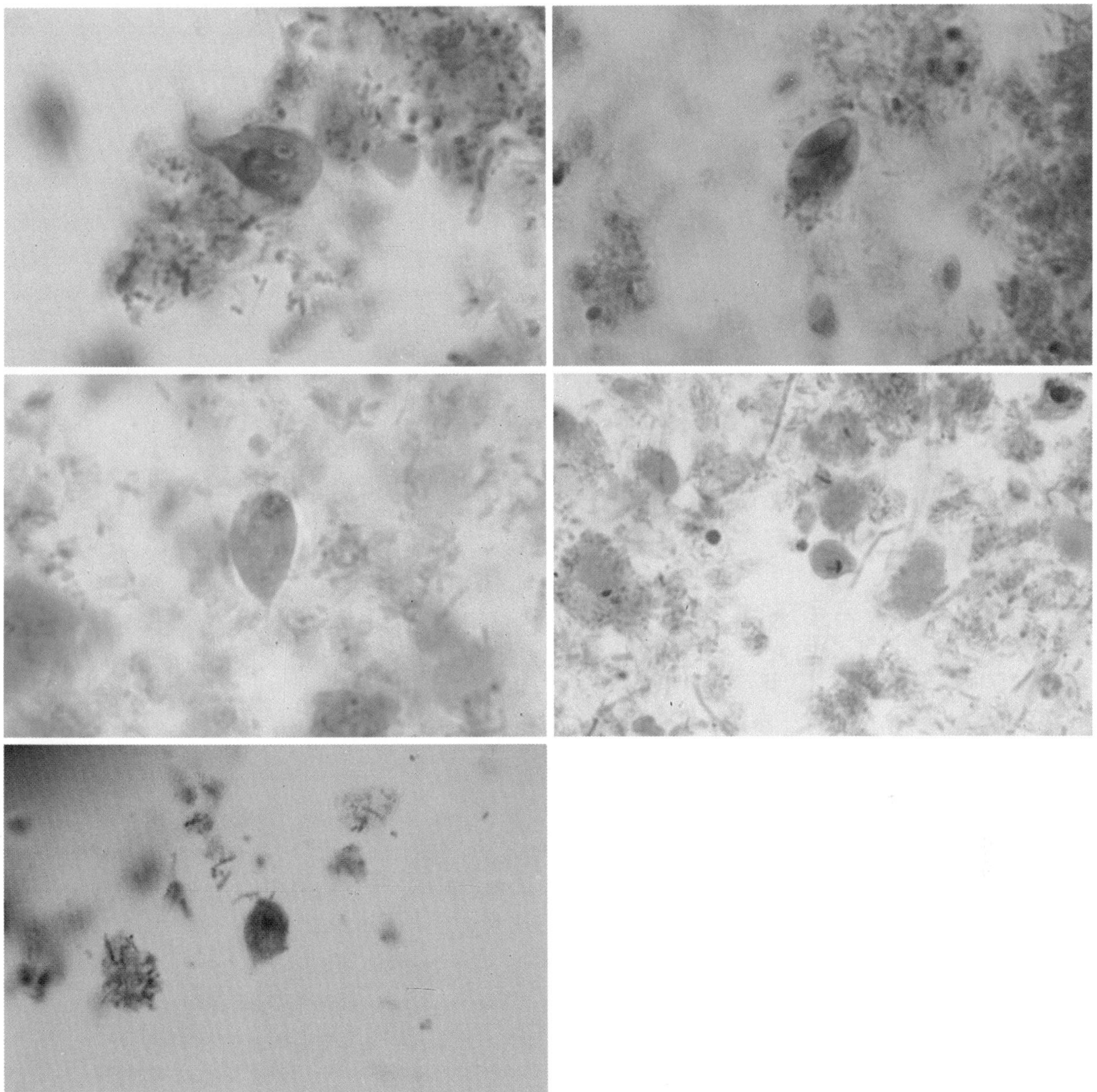

FIG. 7. Flagellates (trichrome stain; oil immersion magnification). Row 1: (left) *G. lamblia* trophozoite. Two nuclei and prominent median body are evident in this pyriform organism. (Right) *G. lamblia* cyst. There are two nuclei toward the bottom in this plane of focus. Fibrils are evident in the cytoplasm. Row 2: (left) *C. mesnili* trophozoite. The nucleus is at the upper end. The pale cytostome is evident to the left of the nucleus. The posterior end is tapered. (Right) *C. mesnili* cyst. The lemon-shaped cyst is in the center of this field. The nucleus is in the lower portion of the cyst. Fibrils are faintly visible. Row 3: (left) *T. hominis* trophozoite. The nucleus is toward the top of the organism. A portion of the undulating membrane is evident to the right of the nucleus. The axostyle is evident at the bottom of the organism.

diagnostic stages are presented in Table 8; diagrams are shown in Fig. 8. Photomicrographs are presented in Fig. 9. Characteristics that are visible in different types of mounts are listed in Table 5.

Isosporiasis is caused by *I. belli* and, although infrequently recognized, can produce severe intestinal disease (3). Deaths from overwhelming infections have been reported, especially in immunocompromised patients. Organisms infect the overall intestine, and symptoms, apparently caused by the schizogony of the developing forms in the epithelial cells and perhaps by the toxins produced, include diarrhea, nausea, fever, steatorrhea, headache, and weight loss. The disease may persist for months or years.

Sarcocystis species

Sarcocystis infection occurs in a variety of hosts, rarely including humans. The sexual stage of *Sarcocystis* species, similar to that of *Isospora* species,

TABLE 8. Differential morphology of ciliate, coccidia, and *Blastocystis* species found in human stool specimens[a]

Type, species, and form	Size (length)	Shape	Motility	No. and characteristics of nuclei	Other features
CILIATE					
Balantidium coli					
Trophozoite	40–70 μm or more; usual range, 40–50 μm	Ovoid with tapering anterior end	Rotary, boring	1 large, kidney-shaped macronucleus; 1 small subspherical micronucleus immediately adjacent to macronucleus; macronucleus occasionally visible in unstained preparations as hyaline mass	Body surface covered by spiral, longitudinal rows of cilia; contractile vacuoles present
Cyst	45–65 μm; usual range, 50–55 μm	Spherical or oval		1 large macronucleus visible in unstained preparations as hyaline mass	Macronucleus and contractile vacuole are visible in young cysts; in older cysts, internal structure appears granular
COCCIDIA					
Isospora belli	Oocyst: 25–30 μm; usual range, 28–30 μm	Ellipsoidal	Nonmotile		Usual diagnostic stage is immature oocyst with single granular mass (zygote) within; mature oocyst contains 2 sporocysts with 4 sporozoites each
Sarcocystis spp.	Sporocyst[b]	Oval	Nonmotile		Mature oocysts with thin wall collapsed around 2 sporocysts or free, fully mature sporocysts with 4 sporozoites inside are usually seen in feces
S. hominis	13–17 μm; usual range, 14–16 μm				
S. suihominis	11–15 μm; usual range, 12–13 μm				
Cryptosporidium spp.	Oocyst: 3–6 μm; usual range, 4–5 μm	Spherical or oval	Nonmotile		Mature oocyst contains 4 naked sporozoites; no sporocysts
BLASTOCYSTIS					
Blastocystis hominis[c]					
Vacuolated form	5–30 μm; usual range, 8–10 μm	Spherical, oval, or ellipsoidal	Nonmotile	1, usually, but 2–4 may be present; located in rim of cytoplasm; in binucleated organisms, nuclei may be at opposite poles; in quadrinucleated forms, nuclei are evenly spaced around the periphery of the cell	Cell contains large central body, or vacuole, with a thin band, or rim, of cytoplasm around the periphery; occasionally a ring of granules may be seen in cytoplasm, and the cell appears to have a beaded rim

[a] Table adapted from Brooke and Melvin (4).
[b] Sizes are based on information from Rommel and Heydorn (24) and Heydorn et al. (12).
[c] Description based on information from Zierdt (34) and McClure et al. (19).

occurs in the intestine of the carnivorous host, and the asexual stage occurs in the muscles of a prey animal which ingests infective sporozoites from the feces of the carnivore. Humans may accidentally serve as either a definitive or an intermediate host for various *Sarcocystis* species (24), but they do not usually have clinical disease.

Isospora and *Sarcocystis* intestinal infections are diagnosed by the identification of the organisms in direct or concentrated wet mounts of feces. Iodine stains the oocysts and sporocysts and makes them more readily visible. Permanent stains, however, are of little or no value in the demonstration of organisms. In addition to being difficult to detect microscopically, oocysts of *I. belli* are sometimes not passed in feces until the symptoms of the infection have subsided. The duration of oocyst passage varies considerably, from a few days to a few weeks. Therefore, in the establishment of diagnoses, several stool specimens should be collected and examined. This pattern of passage of diagnostic stages of coccidia in feces may also be true of *Sarcocystis* species (8). The diagnostic

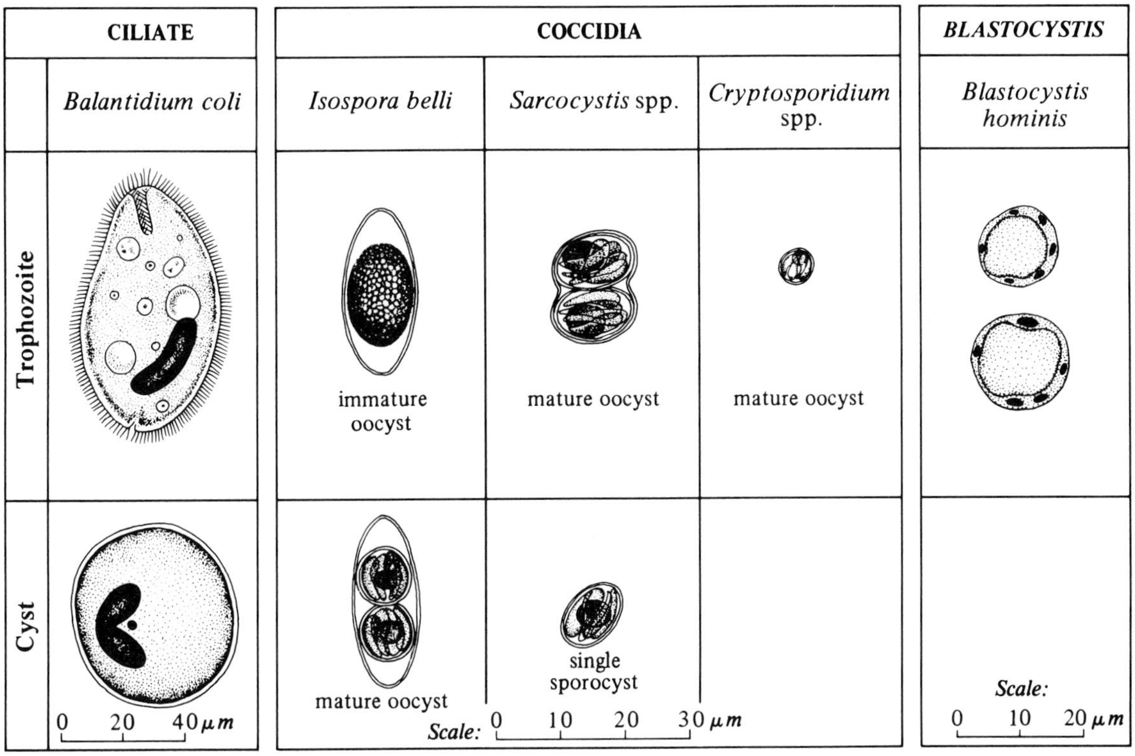

	CILIATE	COCCIDIA			*BLASTOCYSTIS*
	Balantidium coli	*Isospora belli*	*Sarcocystis* spp.	*Cryptosporidium* spp.	*Blastocystis hominis*
Trophozoite		immature oocyst	mature oocyst	mature oocyst	
Cyst	0 20 40 μm	mature oocyst	single sporocyst		
		Scale: 0 10 20 30 μm			Scale: 0 10 20 μm

FIG. 8. Ciliate, coccidia, and *B. hominis* found in human stool specimens. From Brooke and Melvin (4).

stage of *I. belli* is an unsporulated or immature oocyst which will mature in several days at room temperature to an oocyst containing two sporocysts which in turn contain four sporozoites each. *Sarcocystis* species already have mature sporocysts, sometimes in oocysts, in fresh species.

Cryptosporidium species

Cryptosporidium species infect the brush border of intestinal epithelial cells and cause cryptosporidiosis (22). They may occasionally infect the cells of other organs in immunocompromised hosts. Clinically apparent infections with *Cryptosporidium* species are separable into two categories. Patients with intact immune function develop a profuse, watery diarrhea accompanied by mild epigastric cramping pain, nausea, and anorexia which is generally self-limited, lasting for 10 to 15 days. Immunocompromised patients, such as those having AIDS or receiving immunosuppressive therapy, develop a more severe, long-lasting infection. Symptoms are as noted above, but the disease is prolonged, with profuse, watery diarrhea persisting from several weeks to months or years. There is no therapy.

Specimens from patients suspected of having cryptosporidiosis should be preserved in Formalin before being sent to the laboratory or immediately upon receipt. *Cryptosporidium* oocysts are very small (4 to 5 μm) and can easily be overlooked in fecal preparations or confused with yeast cells. Both immature and mature oocysts may be present, although usually those present are mature and contain four naked sporozoites; sporocysts are not present (Table 8, Fig. 8, Fig. 9). However, identification is often difficult. Direct wet mounts can be used to examine feces, but in light infections, organisms may not be found. Specimens can also be concentrated by the Sheather sugar flotation method or the Formalin-ethyl acetate method. Both unstained and iodine-stained mounts should be prepared. In unstained mounts containing mature oocysts, the refractile residual body can be detected, but the sporozoites may not be distinct. Also, oocysts have a low specific gravity and are usually found in the upper levels of the mount, just below the cover slip. Oocysts do not stain with iodine (unless they are exposed to it for long periods), but yeast cells do stain, thus helping to distinguish oocysts from yeast cells. Various acid-fast stains may be used to detect oocysts in feces or concentrates or to confirm the identification of organisms seen in those preparations. Oocysts stain intensely acid-fast, whereas yeast cells and fecal material do not. In many mature oocysts, sporozoites can be seen. Recently, a carbol-fuchsin negative stain has been developed (11) which is rapid and easily performed. However, the preparation is temporary and the results are variable. Organisms stand out as unstained oval structures measuring 4 to 6 μm, whereas other fecal material is stained red by carbol-fuchsin. Oocysts may take up the stain and seem to disappear in about 10 min. Although the negative stain may be useful for those studying the disease or frequently seeing infected patients, permanent, acid-fast stained smears are more reliable and easier to perform and interpret.

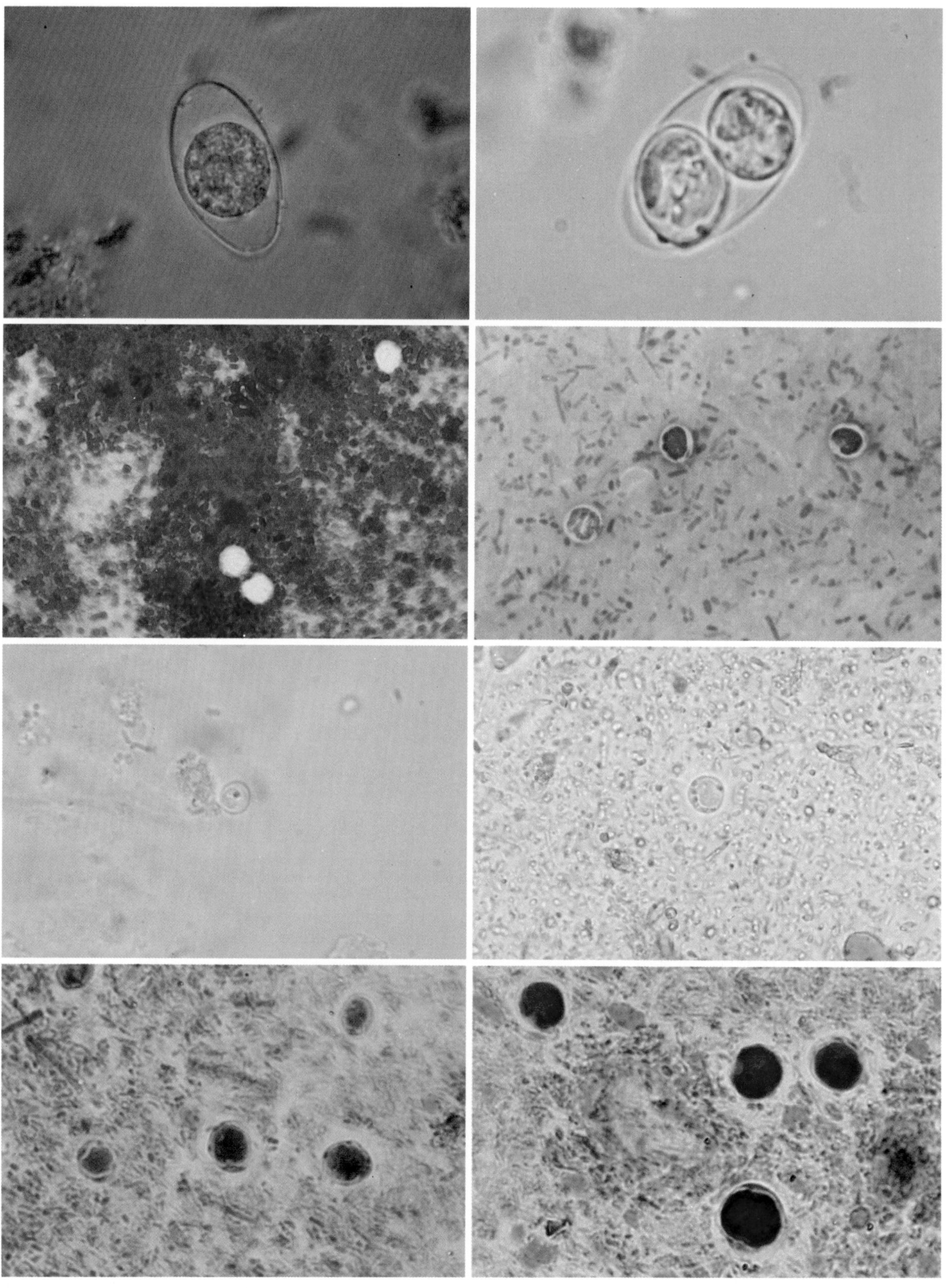

FIG. 9. Coccidia, stained and unstained. Row 1: (left) *I. belli*, immature oocyst, one celled, unstained; (right) *I. belli* mature oocyst, unstained. Row 2: (left) *Cryptosporidium* oocysts with negative stain; (right) *Cryptosporidium* oocysts with acid-fast stain. Row 3: (left) *Cryptosporidium* oocyst, unstained; (right) *B. hominis*, vacuolated forms, iodine stain. Row 4: (left) *B. hominis*, vacuolated forms, trichrome stain; (right) *B. hominis*, vacuolated forms, trichrome stain.

BLASTOCYSTIS HOMINIS

B. hominis inhabits the large intestine, and organisms are passed in feces. Three morphologic forms have been described: the ameba, granular, and vacuolated forms (34). The vacuolated form is the form most commonly seen in fecal specimens (19).

Although *B. hominis* may be found in up to 25% of stool specimens examined, only occasional patients develop clinical symptoms attributable to this organism. Blastocystosis may be suspected when the complete battery of parasitologic, bacterial, and viral tests on stools have failed to disclose any agent other than *B. hominis* and *Blastocystis* organisms are numerous in the specimen. The predominant and virtually only symptom has been persistent, mild diarrhea.

Infection is diagnosed by finding the familiar spherical or ovoid form with a large vacuole or area in the center and granules arranged around the periphery. The form is described in Table 8 and shown in Fig. 8 and 9. *Blastocystis* organisms can be demonstrated by any of the methods usually used for the diagnosis of intestinal parasite infections, although exposing unfixed feces to water in the performance of concentration procedures causes lysis of *Blastocystis* organisms.

LITERATURE CITED

1. **Beal, C. B., P. Viens, R. G. L. Grant, and J. M. Hughes.** 1970. A new technique for sampling duodenal contents. Am. J. Trop. Med. Hyg. **19:**349–352.
2. **Bos, H. J., A. A. Van Den Eijk, and P. A. Steerenberg.** 1975. Application of ELISA in the serodiagnosis of amoebiasis. Trans. R. Soc. Trop. Med. Hyg. **69:**440.
3. **Brandborg, L. L., S. B. Goldberg, and W. C. Breidenbach.** 1970. Human coccidiosis—a possible cause of malabsorption. N. Engl. J. Med. **283:**1306–1313.
4. **Brooke, M. M., and D. M. Melvin.** 1984. Morphology of diagnostic stages of intestinal parasites of humans, 2nd ed. U.S. Department of Health and Human Services publication no. (CDC) 84-8116. Centers for Disease Control, Atlanta.
5. **Brooke, M. M., D. M. Melvin, and G. R. Healy.** 1983. Common intestinal protozoa of humans—life cycle charts, 2nd ed. U.S. Department of Health and Human Services, Centers for Disease Control, Atlanta.
6. **Danciger, M., and M. Lopez.** 1975. Numbers of *Giardia* in the feces of infected children. Am. J. Trop. Med. Hyg. **24:**237–242.
7. **Elsdon-Dew, R.** 1968. The epidemiology of amoebiasis. Adv. Parasitol. **6:**1–62.
8. **Fayer, R.** 1982. Other protozoa: *Eimeria, Isospora, Cystoisospora, Besnoitia, Hammondia, Frenkelia, Sarcocystis, Encephalitozoo,* and *Nosema*, p. 187–196. *In* L. Jacobs (ed.), Handbook series in zoonoses, section C: parasitic zoonoses, vol. 1. CRC Press, Inc., Boca Raton, Fla.
9. **Fouts, A. C., and S. J. Kraus.** 1980. *Trichomonas vaginalis*: reevaluation of its clinical presentation and laboratory diagnosis. J. Infect. Dis. **141:**137–143.
10. **Healy, G. R.** 1968. The use of and limitations to the indirect hemagglutination test in the diagnosis of intestinal amebiasis. Health Lab. Sci. **5:**174–179.
11. **Heine, J.** 1982. Eine einfache Nachweismethod für Kryptosporidien in Kot. Zentralbl. Veterinaermed. Reihe B **29:**324–327.
12. **Heydorn, A. O., R. Gestrich, H. Melhorn, and M. Rommel.** 1975. Proposal for a new nomenclature of the *Sarcosporidia*. Z. Parasitenkd. **48:**73–82.
13. **Kagan, I. G.** 1980. Serodiagnosis of parasitic diseases, p. 573–604. *In* N. R. Rose and H. Friedman (ed.), Manual of clinical immunology, 2nd ed. American Society for Microbiology, Washington, D.C.
14. **Kessel, J. F., W. P. Lewis, C. M. Pasquel, and J. A. Turner.** 1965. Indirect hemagglutination and complement fixation tests in amebiasis. Am. J. Trop. Med. Hyg. **14:**540–550.
15. **Krogstad, D. J., H. C. Spencer, G. R. Healy, N. N. Gleason, D. J. Sexton, and C. A. Herron.** 1978. Amebiasis: epidemiologic studies in the United States, 1971–1974. Ann. Intern. Med. **88:**89–97.
16. **Krupp, I. M.** 1974. Comparison of counterimmunoelectrophoresis with other serologic tests in the diagnosis of amebiasis. Am. J. Trop. Med. Hyg. **23:**27–30.
17. **Lin, T. M., S. P. Halbert, C. T. Chiu, and R. Zarco.** 1981. Simple standardized enzyme-linked immunosorbent assay for human antibodies to *Entamoeba histolytica*. J. Clin. Microbiol. **13:**646–651.
18. **Maddison, S. E.** 1965. Characterization of *Entamoeba histolytica* antigen-antibody reaction by gel diffusion. Exp. Parasitol. **16:**224–235.
19. **McClure, H. M., E. A. Strobert, and G. R. Healy.** 1980. *Blastocystis hominis* in a pig-tailed macaque: a potential enteric pathogen for nonhuman primates. Lab. Anim. Sci. **30:**890–894.
20. **McLaren, L., L. Davis, G. Healy, and G. James.** 1983. Isolation of *Trichomonas vaginalis* from the respiratory tract of infants with respiratory diseases. Pediatrics **71:**888–890.
21. **Melvin, D. M., and M. M. Brooke.** 1982. Laboratory procedures for the diagnosis of intestinal parasites, 3rd ed. U.S. Department of Health and Human Services publication no. (CDC) 82-8282. Centers for Disease Control, Atlanta.
22. **Navin, T. R., and D. D. Juranek.** 1984. Cryptosporidiosis: clinical, epidemiologic and parasitologic review. Rev. Infect. Dis. **6:**313–327.
23. **Rendtorff, R. C.** 1954. The experimental transmission of human intestinal protozoan parasites. Am. J. Hyg. **59:**209–220.
24. **Rommel, M., and A. O. Heydorn.** 1972. Beiträge zum Lebenszykles der Sarkosporidien. III. *Isospora hominis* (Raillet und Lucet, 1891) Wenyon, 1923, eine Dauerform der Sarkosporidien des Rindes und des Schweins. Berl. Muench. Tiertaerztl. Wochenschr. **85:**143–145.
25. **Sargeaunt, P. G., and J. E. Williams.** 1980. A comparative study of *Entamoeba histolytica* (NIH:200,HK9, etc.), "*E. histolytica*-like" and other morphologically identical amoebae using isoenzyme electrophoresis. Trans. R. Soc. Trop. Med. Hyg. **74:**469–474.
26. **Smith, J. W., R. M. McQuay, L. R. Ash, D. M. Melvin, T. C. Orihel, and J. H. Thompson, Jr.** 1976. Atlas of diagnostic parasitology. II. Intestinal protozoa. American Society for Clinical Pathology, Chicago.
27. **Smith, J. W., and M. W. Wolfe.** 1980. Giardiasis. Annu. Rev. Med. **31:**373–383.
28. **Spencer, M. J., L. S. Garcia, and M. R. Chapin.** 1979. *Dientamoeba fragilis*: an intestinal pathogen in children. Am. J. Dis. Child. **133:**329–393.
29. **Thornton, S. A., A. H. West, H. L. DuPont, and L. K. Pickering.** 1983. Comparison of methods for identification of *Giardia lamblia*. Am. J. Clin. Pathol. **80:**858–860.

30. **Ungar, B. L., R. H. Yoken, T. E. Nash, and T. C. Quinn.** 1984. Antigenicity of *Giardia lamblia* in fecal specimens. J. Infect. Dis. **149:**90–97.
31. **Visvesvara, G. S., and G. R. Healy.** 1984. Antigenicity of *Giardia lamblia* and the current status of serologic diagnosis of giardiasis, p. 219–221. *In* S. L. Erlandsen and E. A. Meyer (ed.), Giardia and giardiasis. Plenum Publishing Corp., New York.
32. **Walsh, J.** 1982. Human helminthic and protozoan infections in the north, p. 45–62. *In* K. W. Warren and J. Z. Bowers (ed.), Parasitology, a global perspective. Springer-Verlag, New York.
33. **Wolfe, M. S.** 1984. Symptomatology, diagnosis, and treatment, p. 147–161. *In* S. L. Erlandsen and E. A. Meyer (ed.), Giardia and giardiasis. Plenum Publishing Corp., New York.
34. **Zierdt, C. H.** 1973. Studies of *Blastocystis hominis*. J. Protozool. **20:**114–121.

Chapter 58

Tissue Helminths

THOMAS C. ORIHEL AND LAWRENCE R. ASH

There are a large number of helminth parasite species, including nematodes, flukes, and tapeworms, which, as adults or larvae, live in the tissues of humans. These include, among the nematodes, the filariae, *Trichinella*, and *Capillaria hepatica*, as well as zoonotic species of ascarids, metastrongyles, and hookworms. The cestodes are represented by larval tapeworms which produce cysticercosis, coenurosis, sparganosis, and hydatid disease. Schistosomes (blood flukes) inhabit blood vessels of the abdomen but are discussed in the chapter on intestinal helminths (Chapter 59) because their eggs are passed in feces and urine.

NEMATODES

Filariae

The filarial worms are arthropod-transmitted parasites of the lymphatic, subcutaneous, and cutaneous tissues of humans. All share a unique characteristic: the adult female worm produces a primitive larva called a microfilaria, which is found in the peripheral blood or in the skin and cutaneous tissues. Certain species of microfilariae circulate in the blood with a well-defined circadian rhythm or "periodicity" which may be nocturnal or diurnal; other species lack periodicity and circulate in the blood at all hours of the day and night. When absent from the peripheral blood, microfilariae are to be found in the deeper visceral capillaries. Because adult worms are typically sequestered in the tissues, diagnosis of filarial infections depends on finding microfilariae in the blood or skin. The microfilaria is simple in its structure. It is vermiform in shape and, in stained preparations, appears to be composed of a column of nuclei which is interrupted along its length by spaces or special cells which are the precursors of the body organs. Some species of microfilariae are enveloped in a sheath, whereas in others the sheath is absent (Fig. 1).

All filariae are transmitted by species of bloodsucking arthropods such as mosquitoes, midges, blackflies, and tabanid flies, in which the microfilaria develops to the infective stage. Development of the infective larva to the gravid, adult stage in the vertebrate host requires several months, in some cases as long as a year or more. Although these parasites are not endemic in humans in the United States, they are often seen in immigrants or in individuals who have resided or traveled in endemic areas. There are several species of filariae which infect humans.

Wuchereria bancrofti, causing an infection often referred to as bancroftian filariasis, is the most common and widespread species of filaria infecting humans. It has cosmopolitan distribution throughout tropical and subtropical areas of the world. Adult worms live in the host's lymphatic system and produce lymphangitis, lymphadenitis, and obstructive fibrosis which restricts the flow of lymph, with a resultant lymphedema. Long-term, chronic infection may result in elephantiasis of the extremities and genitalia. The microfilaria circulates in the peripheral blood with a nocturnal periodicity in most regions of the world; however, in the South Pacific region, the microfilaria is essentially without any periodicity. The microfilaria is sheathed, lies in smooth curves, and measures about 298 μm in length by 7.5 to 10 μm in diameter. Column nuclei are dispersed; there is a short head space, and the pointed tail is devoid of nuclei (Fig. 1A). In Giemsa stain, the sheath stains faintly or not at all. The microfilaria must be distinguished from other sheathed microfilariae. This is done most easily on the basis of the arrangement of nuclei, particularly in the tail (Fig. 2). Since microfilariae may be present in the blood only in small numbers, sensitive procedures such as thick blood films, Saponin lysis, Knott concentration, or membrane filter concentration are used routinely to detect infections.

Brugia malayi is another mosquito-borne filaria which inhabits the lymphatic system of humans. It is restricted in its geographical distribution to Asia and the Indian subcontinent. In some regions it is coendemic with *W. bancrofti*. Lymphatic pathology similar to that produced in bancroftian filariasis is seen in chronic infections with this parasite. The microfilaria, which circulates in the blood, may be periodic or subperiodic. It is similar in its structure to that of *W. bancrofti*, being sheathed but somewhat smaller in size (270 μm by 5 to 6 μm). It can be differentiated from the *W. bancrofti* microfilaria by the presence of subterminal and terminal nuclei in the tail (Fig. 1B and 2b). These nuclei may be smaller than other nuclei. With Giemsa stain, the sheath stains a bright pink color, whereas that of *W. bancrofti* does not.

Another species of *Brugia*, *B. timori*, infects humans in the Lesser Sunda Islands in the Indonesian archipelago. It too is a lymphatic dweller and produces a microfilaria very similar to that of *B. malayi*. The two can be most easily differentiated on the basis of size. *B. timori* is larger, measuring more than 300 μm in length; also, its sheath tends not to stain with Giemsa stain.

Loa loa, a common filarial parasite of humans, is endemic only in West and Central Africa. It is often called the eye worm because the adult worms, which live in the subcutaneous tissues, often migrate into the conjunctiva. The adult worms move freely through the tissues, often producing transient inflammatory reactions referred to as "Calabar" swellings. The microfilariae circulate in the blood, often in large numbers, with a diurnal periodicity. They are sheathed and measure up to 300 μm in length. In contrast to the other sheathed microfilariae, nuclei extend to the end of the tapered tail; however, they are somewhat irregularly arranged along the length of the tail (Fig. 1C and 2d). When adult worms enter the conjunctiva, they can be surgically extracted. Diagnosis of infection depends on identifying the microfilaria in diurnal blood films or removal of adult worms from the conjunctiva.

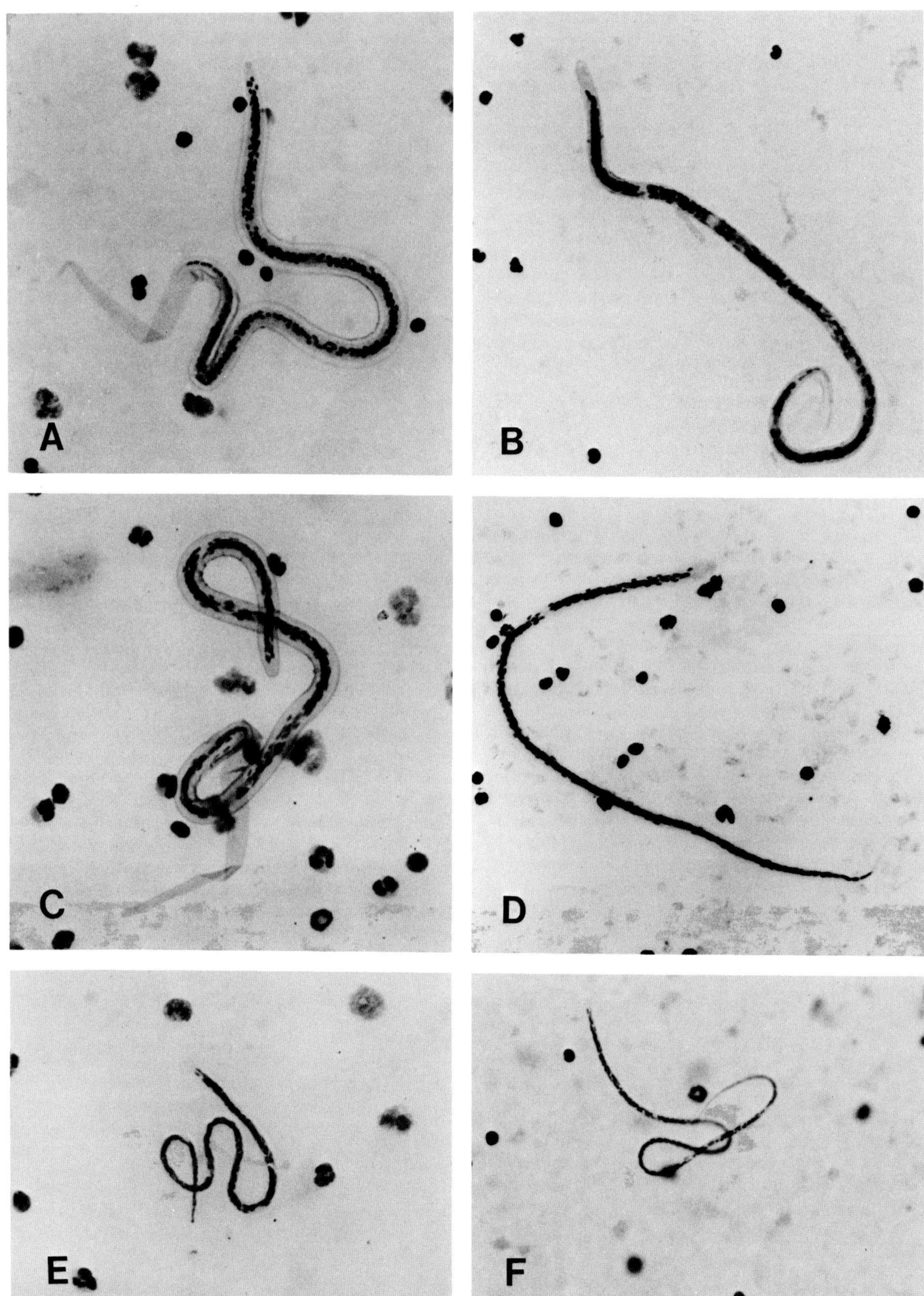

FIG. 1. Common microfilariae found in humans. Hematoxylin stain, ×400. (A) *Wuchereria bancrofti*, (B) *Brugia malayi*, (C) *Loa loa*, (D) *Onchocerca volvulus*, (E) *Mansonella perstans*, and (F) *Mansonella ozzardi*.

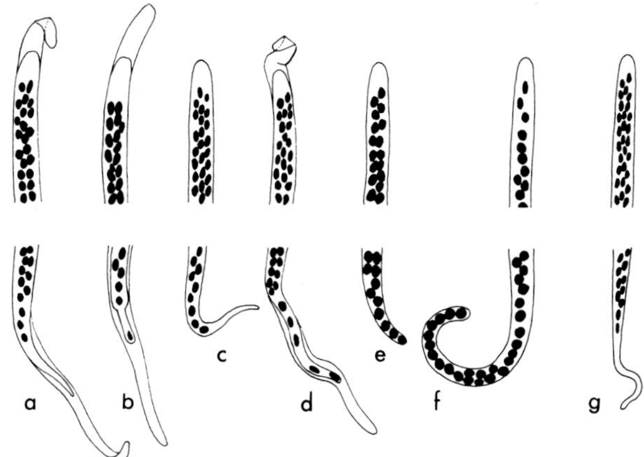

FIG. 2. Diagrammatic representation of the anterior and posterior extremities of the common microfilariae found in humans. (a) *Wuchereria bancrofti*, (b) *Brugia malayi*, (c) *Onchocerca volvulus*, (d) *Loa loa*, (e) *Mansonella perstans*, (f) *Mansonella streptocerca*, and (g) *Mansonella ozzardi*.

Onchocerca volvulus is an important human parasite in both hemispheres. It occurs in people across Central Africa, in a small area in the Middle East (Yemen), in Mexico, in portions of Central America, and in several countries in the northern part of South America. Adult worms are found embedded in fibrous nodules in the subcutaneous tissues and sometimes in deeper tissues. These nodules or "onchocercomata" may be found on the head, trunk, and extremities; their anatomical location frequently correlates with the geographical strain of the parasite. The microfilaria is found in the skin. It lacks a sheath and measures approximately 309 μm in length by 5 to 9 μm in diameter. The tail is tapered, usually bent or flexed, and without nuclei (Fig. 1D and 2c). The microfilaria is rarely found in the blood but may appear in the urine, particularly after treatment. Diagnosis is made by finding the typical microfilaria in skin snips teased in water or saline, in the fluid expressed from scarified skin, or in aspirates from nodules. Also, adult worms may be demonstrated in excised nodules which have been sectioned and stained. Microfilariae may also be seen in the cornea and in the anterior chamber of the eye viewed with the aid of a slit lamp.

Skin snips are best obtained using a biopsy punch. Samples should be taken from the region of the scapula or the iliac crest although other sites (near nodules) are acceptable. Teasing the skin snip tends to liberate the microfilariae from the tissues.

Mansonella streptocerca is another skin-dwelling filaria found in humans in the rainforest belt of Africa. This filaria has been known by a variety of generic names, including *Acanthocheilonema, Dipetalonema*, and *Tetrapetalonema*, but has recently been placed in *Mansonella* (9). The adult worms are found in the dermal layers of the skin, as are the microfilariae. The microfilaria has no sheath, is long and slender, and measures approximately 210 μm by 5 to 6 μm. Its most characteristic feature is its "crooked" tail (Fig. 2f); in addition, the column of nuclei extends to the end of the tail. In areas where this species overlaps with *O. volvulus*, great care must be given to proper identification of the microfilariae found in skin snips.

Two other species of *Mansonella* are parasites of humans. One of these, *Mansonella ozzardi*, is restricted in its geographical distribution to the Western Hemisphere. It is endemic in portions of Mexico, Panama, and northern South America, especially in the Amazon basin. It is also found in several Caribbean islands including Hispaniola. The adult worms inhabit the subcutaneous tissues, and the microfilariae circulate in the blood. The microfilaria is small, measuring about 224 μm by 4 to 5 μm, and has a long attenuated tail devoid of nuclei (Fig. 1F and 2g). It circulates in the blood at all hours of the day and night. It has been observed in experimental animals that an infection requires approximately 5 months to reach the patent (microfilaremic) state. This filaria, like *L. loa*, readily infects visitors to endemic areas and is frequently found in missionaries and others residing temporarily in endemic areas.

Mansonella perstans (formerly *Dipetalonema perstans*), a filaria with wide geographical distribution in tropical Africa but less widely seen in South America, also is seen frequently in individuals who have lived temporarily in endemic areas. The location of the adults of this filaria is not certain, but it is believed that they inhabit the peritoneal cavity and mesenteries of humans. The microfilaria has no sheath, is small (measuring approximately 203 μm by 4 to 5 μm), and circulates in the peripheral blood without any periodicity. Its most characteristic feature is its blunt tail filled with nuclei (Fig. 1E and 2e). In the Western Hemisphere, *M. perstans* often is found in association with *M. ozzardi*. In areas where the two species coexist, special care must be taken to make accurate identifications.

Laboratory diagnosis of filariasis. Microfilariae may be detected in samples of blood by a variety of techniques. In thick, wet blood films prepared during either the day or night, microfilariae may be quickly recognized by their size and rapid movement in the blood. However, individual species are not likely to be identified in this manner, so that stained blood films are generally required. Thin blood films usually are not adequate because of the small volume of blood used. In contrast, thick blood films of approximately 20 μl, dehemoglobinized and stained with Giemsa or hematoxylin stains, bring out the diagnostic morphological features of each species.

Often, the numbers of microfilariae in the blood may be too scanty to find even in thick blood films so that it is necessary to examine a larger volume. Procedures for this are given in Chapter 55. Microfilariae are readily identified on the basis of their size, the presence or absence of a sheath and its staining characteristics with Giemsa, and the structure of the tail.

Skin snips must be examined when searching for microfilariae of *O. volvulus* and *M. streptocerca*. Care should be taken to obtain "bloodless" snips so as not to contaminate the sample with microfilariae that may be present in the blood. Techniques have been described in which one simply abrades a small area of skin and collects the exuded tissue juices on a slide which can be either examined immediately for living microfilariae or allowed to dry and then stained and examined. Serological tests have poor sensitivity and specificity and generally are not useful for diagnosis.

The patient's history of travel and residencies is particularly helpful when considering a diagnosis of filariasis.

Trichinosis

Trichinella spiralis, the agent of human trichinosis, is a parasite of carnivores with very little evidence of host specificity. Infection in humans results from the ingestion of insufficiently cooked or raw pork and pork products containing the encysted larvae. Bear meat is also a frequent source of human infection. Adult worms live in the mucosa of the small intestine. The female produces and discharges larvae which enter the bloodstream and invade the skeletal musculature, where they undergo further development and encapsulation. These larvae remain viable for many years. Initially there may be a nonspecific gastroenteritis, fever, eosinophilia, myositis, and circumorbital edema. The adult worms survive in the intestine for up to 6 to 8 weeks and then are expelled from the host.

The definitive diagnosis is made by demonstration of encapsulated larvae in biopsy specimens from skeletal muscle, particularly deltoid and gastrocnemius muscles (Fig. 3A). The digestion of muscle tissue in artificial gastric juice followed by examination of the sediment for larvae is more sensitive (Fig. 3B). Serological tests are widely used with good results. A highly antigenic parasite, *Trichinella* stimulates a very strong antibody response that can be measured by a variety of procedures. For many years the bentonite flocculation procedure has been the test of choice (8). Although somewhat difficult to perform, the test is highly specific and a reliable indicator of infection. There is some correlation between infective dose and level of antibody activity, and in some cases very light infections are serologically negative. Measurable antibody does not appear until 3 to 4 weeks postinfection, so that sera negative during the acute attack should be confirmed by testing a second specimen 2 to 4 weeks later. Clinically active cases usually respond with relatively high titers, but because of individual variations, titers as low as 1:5 are considered significant. Although the specificity is excellent, titers persist for years, not becoming negative for 2 to 3 years. A variety of other tests including indirect hemagglutination, double diffusion, latex agglutination, and counterimmunoelectrophoresis have been evaluated, but none has replaced the bentonite flocculation test.

Recently, an enzyme-linked immunosorbent assay (ELISA) has been described which is far more sensitive than the bentonite flocculation test and becomes positive in 3 to 5 days after infection (12). It has been effective for screening swine intended for market but has not been adequately tested in human cases to determine its effectiveness and to interpret titers clinically. There are no commercial reagents available.

Capillariasis

Though normally a parasite of rodents, *Capillaria hepatica* can produce infection in humans. Human and animal infections are acquired by ingestion of infective eggs in soil; adult worms mature and lay eggs directly into the liver parenchyma, where they remain in an undeveloped state until the liver is eaten and the eggs are digested free to pass in the feces of the predator animal. If, as in the human host, the liver is not eaten, the eggs are never liberated into the external environment. Diagnosis is established by liver biopsy or necropsy. In genuine human cases, eggs are not passed in feces (1a). When humans are found passing eggs of *C. hepatica* it is indicative of spurious infection and represents an instance where the livers of infected animals have been eaten (e.g., squirrel pie). Eggs will be passed for several days and will then disappear from the stool.

The eggs of *C. hepatica* must be distinguished from the eggs of *Trichuris* or other species of human *Capillaria*. Eggs are 51 to 67 μm long by 30 to 35 μm wide and have thick, striated walls and inconspicuous "plugs" at both ends (Fig. 3C). These eggs are unembryonated when seen in feces. Eggs in liver biopsy specimens can be readily recognized on the basis of their typical morphological features (Fig. 3D).

Other nematode infections

Larva migrans. Several species of nematode parasites which are natural parasites of lower animals may gain entry into the human host and undergo partial development. The severity of subsequent disease manifestations may vary from asymptomatic to serious in the form of skin rash, pneumonitis, and involvement of the central nervous system.

(i) Cutaneous larva migrans. Cutaneous larva migrans, also known as creeping eruption or ground itch, refers to the production of serpiginous, inflamed trails in the skin (Fig. 3E) resulting in intense pruritus, and may be caused by various species of *Strongyloides*, *Ancylostoma*, or other animal hookworm species. It occurs as a result of skin contact with sandy-loam types of soil that contain filariform larvae of hookworms and *Strongyloides* spp. deposited in the feces of dogs, cats, and other animals. In the southeastern part of the United States this may be an occupational hazard of electricians, plumbers, and construction workers who inadvertently come in contact with infected soil when working under or near houses where animals have defecated. In addition, species of *Strongyloides* occurring in wild animals such as nutria and raccoons have been incriminated in causing dermatitis in oil field workers and trappers in Louisiana. *Ancylostoma braziliense*, a hookworm of dogs and cats, has been the species usually identified with classical creeping eruption along sandy beaches, but other species of the genus may be involved too. Invasion of human skin by the dog hookworm, *Ancylostoma caninum*, can produce creeping eruption; these larvae may also invade other tissues and organs of the body, including the eye, where they produce granulomatous, visceral larva migrans-like lesions.

Diagnosis of cutaneous larva migrans is based almost entirely on clinical findings and the demonstration of the classical serpiginous trails (Fig. 3E).

(ii) Visceral larva migrans. Visceral larva migrans is a syndrome originally associated with the larval migration in humans, especially children, of dog and cat ascarids of the genus *Toxocara*. Although many other helminths of animals may infect humans and migrate through organs and deeper tissues of the body, *Toxocara canis* is still the classical example of this kind of infection.

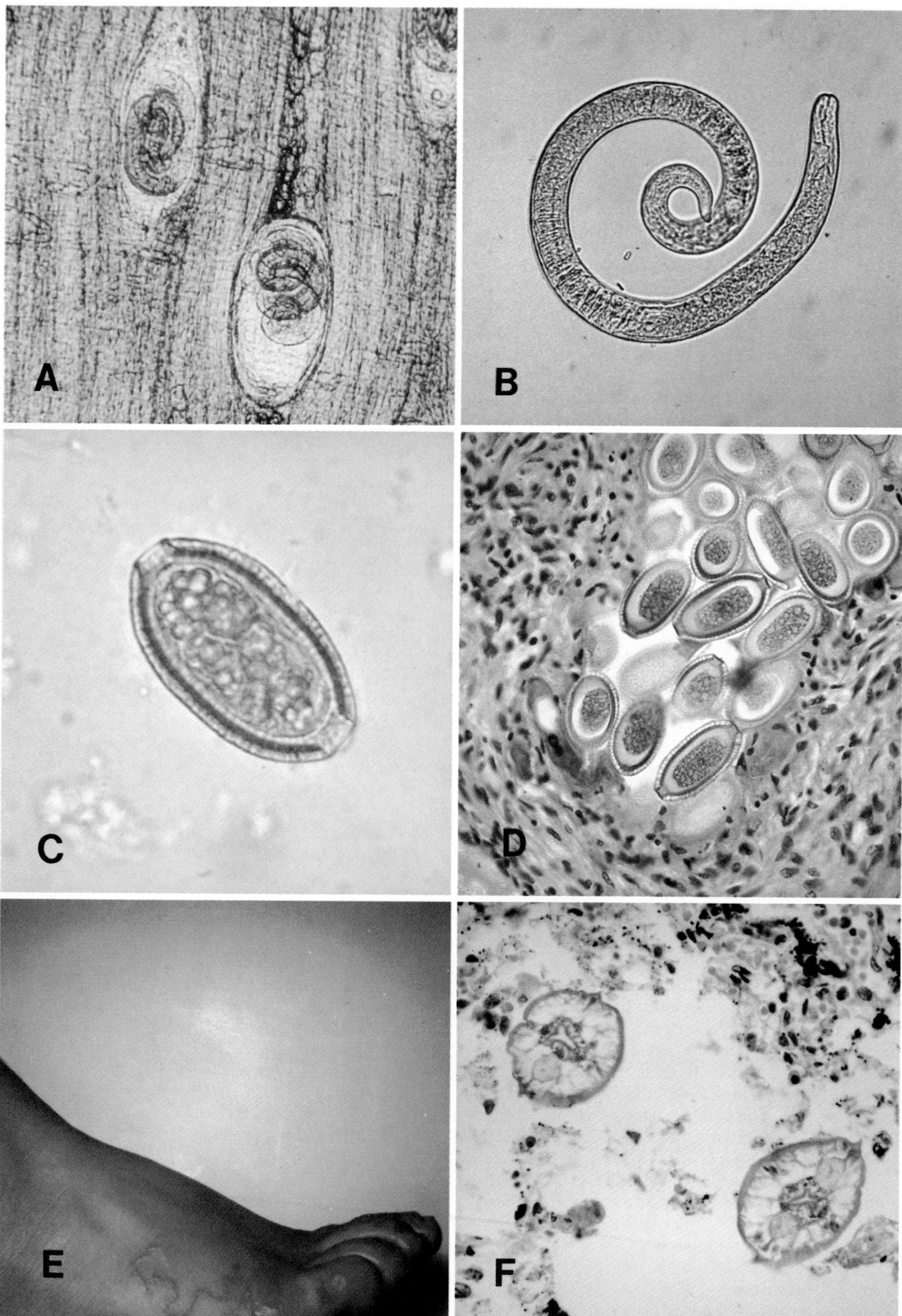

FIG. 3. Nematode parasites in tissues. (A) *Trichinella spiralis* infective-stage larvae. In this press preparation of diaphragm muscle one can see the encapsulated, infective larvae (×100). (B) *Trichinella spiralis* infective-stage larva. This larva has been digested free from muscle tissue (×175). (From reference 14, used by permission.) (C) *Capillaria hepatica* egg. Note that this unembryonated egg resembles that of *Trichuris* spp. However, the plugs at either end are less prominent and the shell has a striated appearance (×750). (D) *Capillaria hepatica* eggs. Unembryonated eggs are expelled by the female into the burrows produced by the adult. Characteristic features of eggs are seen even in tissue sections (×290). (From reference 1, used by permission.) (E) Hookworm, cutaneous larva migrans. Infective-stage larvae of dog or cat hookworms wander through the layers of skin, producing serpiginous trails; this condition is often referred to as creeping eruption. (F) *Ascaris* larva in section of human lung; hematoxylin-eosin stain (×900).

T. canis, the dog ascarid, is more important as a human parasite than is *T. cati*, the species occurring in felines. Toxocariasis results from the ingestion of infective eggs from soil and is characterized by hypereosinophilia, hepatomegaly, fever, pneumonitis, and sometimes death. The visceral larva migrans syndrome may persist for many years and may result in severe complications involving the eye and central nervous system. Ocular larva migrans may mimic retinoblastoma, a malignant tumor of the eye. Other nematode larvae producing the syndrome include *A. caninum* (the dog hookworm), *Gnathostoma spinigerum* (a spirurid parasite of dogs and cats in Southeast Asia), and some trematode and cestode larval parasites of wild animals that also invade human tissues.

Diagnosis of visceral larva migrans is difficult and is generally based on clinical findings and immunodiagnostic procedures. The indirect hemagglutination test (6) has poor sensitivity and has been superseded by an ELISA procedure utilizing embryonated egg antigens of *Ascaris* and *Toxocara* which improve both sensitivity and specificity (3, 10). The ELISA was further improved by using antigen prepared from excretions and secretions of *Toxocara* larvae maintained in cell culture medium (5). It is now accepted that ELISA values of ≥1:8 are significant in ocular larva migrans and ELISA values of ≥1:32 are significant in visceral larva migrans when appropriate clinical symptoms are present. No commercial reagents are available.

Occasionally diagnosis is established by finding focal inflammatory lesions containing larvae in tissue specimens such as liver. Sections of nematode larvae can sometimes be identified in human tissues after biopsy or necropsy, and these can be identified by personnel expert in helminth microanatomy (Fig. 3F).

Anisakiasis. The growing popularity of eating raw fish dishes—sushi, sashimi, ceviche, "Tahitian salad," and others—has led to increased human infection with various larval ascaridoid nematodes. A number of genera, including *Anisakis*, *Phocanema*, and *Terranova*, are parasites, as adults, of marine mammals, utilizing shrimplike crustaceans and fish and squid as first and second intermediate hosts, respectively, in their life cycle. Larval stages in the mesenteries or flesh of fish have the ability to pass from one fish to another when they are ingested without maturing to adult worms. When humans accidentally ingest the larval stages of these nematodes, the larvae may partially penetrate the wall of the stomach or the intestine and produce an eosinophilic granuloma. Many human infections present with acute abdomen or other signs suggestive of intestinal obstruction. In some instances the long, whitish worms, up to several centimeters long, may be coughed up by the patient or be removed by the throat.

Diagnosis of the nematodes producing anisakiasis can be accomplished by study of the distinctive morphological features of the whole worms or by examination of the microanatomy of the parasites in histologic section.

***Angiostrongylus cantonensis* and *A. costaricensis*.** Metastrongylid nematodes of the genus *Angiostrongylus* have been found to be important human parasites. *A. cantonensis*, a lungworm that lives in the pulmonary artery of rats (*Rattus* spp.), is a cause of human eosinophilic meningitis or eosinophilic meningoencephalitis in many Pacific islands, including Hawaii, in Thailand, Indonesia, and other parts of Southeast Asia, and in Cuba. Though human cases have not been reported, the parasite has been found in rats in Puerto Rico. In the life cycle, rats excrete first-stage larvae in feces, and these penetrate directly into, or are ingested by, many species of terrestrial snails or slugs. Larvae reach the infective, third stage in the molluscs, and when the molluscs are eaten by rats, the larvae make a prolonged migration through the brain before they become adults in the pulmonary artery. Here the worms mate and the females lay eggs which will develop and hatch in the lung tissue.

Infective larvae in terrestrial molluscs can be eaten by a wide range of vertebrate and invertebrate animals including planarians, shrimp, crabs, fish, amphibians, and reptiles. In these hosts the larvae remain in various tissues in the infective stage. Human infections are rarely derived from eating ordinary terrestrial molluscs; instead, infection is acquired by eating raw or poorly cooked, infected shrimp, crabs, or large edible snails such as *Pila*; accidental infection also may occur from eating infected planarians or small slugs on improperly washed vegetables, such as lettuce, or fruits, such as strawberries.

In the human host, larvae migrate to the brain, spinal cord, or eye, where they become immature adults. Rarely can the worms complete the migration to the pulmonary artery; instead, they usually die in the meninges or the parenchyma of the brain, giving rise to meningeal symptoms. Diagnosis is usually based on a history of being in an area where the parasite occurs, combined with the development of appropriate clinical symptoms such as severe and prolonged headaches, nerve involvement, and the presence of eosinophils in cerebrospinal fluid. Occasionally, immature worms may be found in aspirated cerebrospinal fluid. Serological diagnosis has not been reliable.

A closely related parasite, *Angiostrongylus costaricensis*, causes human abdominal angiostrongyliasis in most countries of Central and South America. This parasite normally lives in small arteries and arterioles of the ileocecal region of various rodents. Eggs are produced and develop in the intestinal wall, and first-stage larvae then migrate into the intestinal lumen to pass in feces. Various terrestrial slugs serve as intermediate hosts; most likely, as for *A. cantonensis*, most terrestrial snails and slugs will support development of the parasite to the infective third stage. Infection is acquired by accidental ingestion of slugs or other hosts. Transport hosts probably play an important role in human infection. Most human infections occur in children, but all age groups may be infected. The distal small intestine is the usual site, where eggs stimulate a granulomatous inflammation which leads to symptoms of acute abdomen. Surgery is often performed and the affected tissues may be resected; infection is subsequently diagnosed by finding eggs in tissue. Larvae have not been found in human feces. Though serological tests have been described, their reliability is still in question.

Subcutaneous and pulmonary dirofilariasis. Species of *Dirofilaria*, which are common filarial parasites of animals and are transmitted by mosquitoes, occasionally infect humans. Those species which ordinarily inhabit the subcutaneous tissues of their animal hosts (e.g., *Dirofilaria repens, D. tenuis,* etc.) usually are found in human subcutaneous tissues (Fig. 4C). *Dirofilaria immitis,* the common heartworm of dogs and other canids, characteristically lodges in human pulmonary arteries, where granulomatous nodules are produced (Fig. 4D). Typically, microfilariae are not produced or do not circulate in the blood of the human host. Diagnosis is usually based on identification of the worm extracted from the nodule or in stained sections of the lesion.

CESTODES

There are a number of tapeworm species that, as adults, live in the intestine of humans or animals and whose larval stages often are found in human tissues. They can be recognized in most cases on the basis of their morphological features. These are described below.

Sparganosis

Human sparganosis is caused by larval cestodes of the genus *Spirometra,* which are parasites of cats, dogs, and various wild canids and felids. These parasites are closely related to *Diphyllobothrium* and have similar life cycles involving copepods as first intermediate hosts and fish as second intermediate hosts. *Spirometra mansoni* is found in many parts of Asia, in particular in China, Japan, Korea, Vietnam, and other areas, and *Spirometra mansonoides* is the common species in the United States. Characteristically, in sparganosis, the plerocercoid larva (also called a sparganum) migrates in the tissues of the human host after ingestion of these larvae in poorly cooked fish, ingestion of infected copepods containing procercoid larvae, or the use of infected hosts such as frogs and snakes as poultices, a practice followed in many Asian countries. In the United States, typical cases of sparganosis present as subcutaneous swellings which may be migratory and sometimes tender; usually there is little or no peripheral eosinophilia. Diagnosis is usually made after surgical removal of the mass and histopathological study.

Coenurosis

Coenurosis is produced by the larval stage of the tapeworm *Taenia multiceps,* which in the adult stage is a common intestinal parasite of canid and felid species, e.g., dogs, foxes, and cats. The adult worms live in the small intestine of their host. They produce eggs indistinguishable from those of other species of *Taenia.* Herbivores and omnivores, as well as some rodents, are the usual intermediate hosts. The hexacanth embryo that hatches from the egg migrates to extraintestinal sites such as the brain, subcutaneous tissues, or musculature and develops to the coenurus stage. Typically, the coenurus, which may vary from a few millimeters to several centimeters in diameter, has multiple scoleces which develop from the germinal layer lining the bladderlike larva (Fig. 4A). Humans usually acquire the infection by ingesting the eggs from dog feces. Diagnosis is generally made from material surgically removed from the human host.

Cysticercosis

Human cysticercosis is produced by the ingestion of the eggs of the pork tapeworm, *Taenia solium.* It can be found throughout the world, although it has its greatest prevalence in Mexico, other areas of Latin America, India, China, Africa, and Europe. Acquisition of cysticercosis in the United States is uncommon, but immigrants from endemic areas are frequently diagnosed with the infection.

The cysticercus or bladder worm is round to oval and translucent and measures about 5 mm or more in diameter (Fig. 4B). It has an invaginated scolex bearing four suckers and a circle of hooks. This cysticercus will develop in any organ or tissue of the body; they are most serious when they occur in the central nervous system and the eye. Diagnosis of the infection may be difficult. It may be established by the recovery of whole cysticerci or the histologic demonstration of cysticerci in surgically removed tissues. Diagnosis may also be accomplished by detection of calcifying cysticerci by X ray and the use of computerized tomography. The indirect hemagglutination procedure is the most reliable serological test (11), but suffers from both poor sensitivity and strong cross-reactivity with echinococcosis (13).

Hydatid disease

Various forms of hydatid disease are caused by a number of species of the tapeworm *Echinococcus. E. granulosus* is a minute tapeworm, measuring 3 to 6 mm in length, found in domestic and wild canids. It occurs extensively in sheep- and cattle-raising areas of the world. Unilocular hydatid infections results from humans ingesting the taeniid-like eggs of this parasite. The normal life cycle of the parasite involves canids as definitive hosts, who pass eggs in feces which are ingested by sheep, cattle, and other animals. In these animals, the embryo released from the egg migrates to various organs, principally the liver and lungs, where it develops into a hydatid cyst. Within this cyst scoleces develop and proliferate from a germinal membrane so that ultimately the cysts may become very large and contain many hundreds, and even thousands, of scoleces, usually referred to as "hydatid sand" (Fig. 4E and F); each of these is a potentially infective organism should the cyst rupture. When canids ingest cysts in the tissues of sheep, each of the scoleces ingested will mature in the intestine to produce an adult tapeworm. Thus, infected dogs may harbor heavy worm burdens.

Only the larval stage, the hydatid cyst, develops in humans as a result of accidental ingestion of *E. granulosus* eggs, and this may take many years. Hydatid cysts may grow in almost all organs and tissues of the body, as they do in normal intermediate hosts, but they are most common in the liver and lungs. Involvement of the central nervous system and bone is not uncommon. These cysts continue to grow, and their presence frequently goes undetected until they reach sufficient size to cause clinical symptoms. Infections of the central nervous system usually become apparent much earlier than those of other organs.

Diagnosis of hydatid infection may be difficult but can be accomplished by several means. X rays, ultrasonic scanning, and computerized tomography can

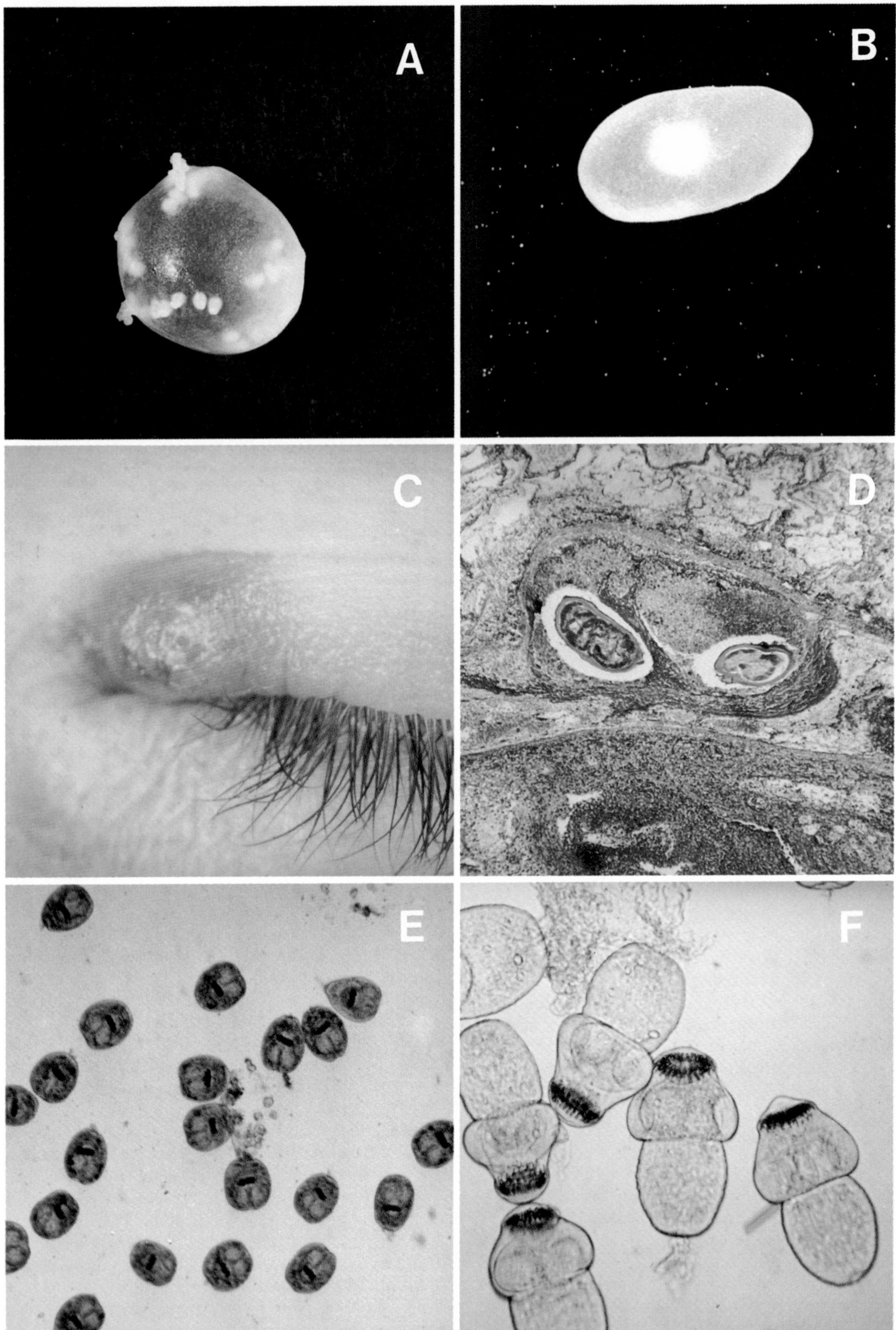

FIG. 4. Helminth parasites in tissues. (A) Coenurus of *Taenia multiceps* from the eye of a man (×6). (B) Cysticercus of *Taenia solium*, also called *Cysticercus cellulosae* (×10). (C) A stylelike lesion from the outer canthus of the eye, which contained an immature female filaria identified as a species of *Dirofilaria*. This represents a typical clinical presentation. (D) Sections of an immature filaria identified as a species of *Dirofilaria* occluding a pulmonary artery of a human in the United States (×45). (E) *Echinococcus granulosus*, hydatid sand. Aspirated fluid from a hydatid cyst usually contains numerous scoleces. Characteristically, the scoleces are invaginated in their own bodies as shown here. (F) *Echinococcus granulosus*, hydatid sand. Here the scoleces are everted and the hooks are clearly evident. (Panels A, B, E, and F from reference 14, used by permission.)

detect cysts in tissues. When combined with an appropriate residential or travel history, or an occupation such as sheep raising in endemic areas, observation of tissue cysts provides reasonable grounds for suspicion of infection. Many immunodiagnostic tests have been utilized, but there is wide variability in their sensitivity and specificity (1a). Indirect hemagglutination is the principal serological test used at present. Sensitivity and specificity depend upon the location of the cyst in the patient. Cysts in the liver or peritoneum stimulate a vigorous antibody response, and 88% of these patients develop titers of ≥1:256 with <5% false-positives. Less than 10% of patients with lung cysts or calcified cysts have positive serology, and even those have titers of ≤1:256. Cross-reactivity with cysticercosis is very strong, and the indirect hemagglutination test will not differentiate between echinococcosis and cysticercosis (13). A specific band, "Arc 5," has been found by counterimmunoelectrophoresis in 40% of patients with hydatid disease; if present, this aids in differentiation (2). An intradermal test using an antigen from hydatid fluid has been available for some time, but is of limited value. The sensitivity of this test is low, but its positivity is lifelong. Aspiration of material from a hydatid cyst in situ is dangerous and not a recommended procedure, since accidental spillage of cyst material in the body may result in further dissemination of infection, an anaphylactic reaction, or both.

Multilocular hydatid (alveolar hydatid) infection

A second form of hydatidosis occurring in humans is alveolar hydatid disease, caused by the larval stage of *Echinococcus multilocularis*. The life cycle of this parasite involves foxes, wolves, and dogs as definitive hosts for the adult tapeworms and small rodents and voles as intermediate hosts. Human ingestion of the typical taeniid eggs may result in the development of this invasive cyst in the liver. These cysts usually lack scoleces, and they may ramify throughout the tissue. The geographical distribution of this infection is usually the northern parts of Europe, Japan, and North America, but the infection is also present in Australia, New Zealand, and South America. Diagnosis is difficult and frequently is accomplished only by histological examination of tissues removed surgically. Serological tests are of some value, although their sensitivity and specificity, as mentioned, are a problem.

Polycystic hydatid infection

A third type of hydatid disease, caused by *Echinococcus vogeli* in Latin America, has been reported (4). The normal life cycle of this parasite involves bush dogs and large rodents, or pacas. A polycystic hydatid develops primarily in the liver where, in humans, it is an invasive type of cyst. Diagnosis has been by examination of pathological specimens removed by surgery.

LITERATURE CITED

1. **Ash, L. R., and T. C. Orihel.** 1984. Atlas of human parasitology, 2nd ed. American Society of Clinical Pathologists, Chicago.

1a. **Beaver, P. C., R. C. Jung, and E. W. Cupp.** 1984. Clinical parasitology, 9th ed. Lea & Febiger, Philadelphia.

2. **Capron, A., L. A. Yarzabal, A. Vernes, and J. Fruit.** 1970. Le diagnostic immunologique de l'echinococcose humaine. Pathol.-Biol. **18:**357–365.

3. **Cypess, R. H., M. H. Karol, J. L. Zidian, L. T. Glickman, and D. Gitlin.** 1977. Larvae-specific antibodies in patients with visceral larva migrans. J. Infect. Dis. **135:**633–640.

4. **D'Alessandro, A., R. L. Rausch, C. Cuello, and N. Aristizabal.** 1979. *Echinococcus vogeli* in man with a review on polycystic hydatid disease in Colombia and neighboring countries. Am. J. Trop. Med. Hyg. **28:**303–317.

5. **deSavigny, D. H.** 1975. In vitro maintenance of *Toxocara canis* larvae and a simple method for the production of Toxocara ES antigen for use in serodiagnostic tests for visceral larva migrans. J. Parasitol. **61:**781–782.

6. **Kagan, I. G.** 1968. Serologic diagnosis of visceral larval migrans. Clin. Pediatr. (Philadelphia) **7:**508–509.

7. **Kagan, I. G.** 1976. Serodiagnosis of hydatid disease, p. 130–142. *In* S. Cohen and E. Sadun (ed.), Immunoserology of parasitic diseases. Blackwell Scientific Publications, London.

8. **Norman, L., and I. G. Kagan.** 1963. Bentonite, latex and cholesterol flocculation tests for the diagnosis of trichinosis. Public Health Rep. **78:**227–232.

9. **Orihel, T. C., and M. L. Eberhard.** 1982. *Mansonella ozzardi*: a redescription with comments on its taxonomic relationships. Am. J. Trop. Med. Hyg. **31:**1142–1147.

10. **Pollard, Z. F., W. H. Jarrett, W. S. Hagler, D. S. Allain, and P. M. Schantz.** 1979. ELISA for diagnosis of ocular toxocariasis. J. Ophthalmol. **86:**743–749.

11. **Proctor, E. M., S. J. Powell, and R. Elsdon-Dew.** 1966. The serological diagnosis of cysticercosis. Ann. Trop. Med. Parasitol. **60:**146–151.

12. **Ruitenberg, E. S., P. A. Slenberg, B. T. M. Brosi, and J. Buys.** 1976. Reliability of the enzyme-linked immunosorbent assay (ELISA) for the serodiagnosis of *Trichinella spiralis* infections in conventionally raised pigs. J. Immunol. Methods **10:**67–83.

13. **Schantz, P. M., D. Shanks, and M. Wilson.** 1980. Serologic cross reactions with sera from patients with echinococcosis and cysticercosis. Am. J. Trop. Med. Hyg. **29:**609–612.

14. **Smith, J. W., L. R. Ash, R. M. McQuay, D. M. Melvin, and T. C. Orihel.** 1976. Atlas of diagnostic medical parasitology: intestinal helminths. American Society of Clinical Pathologists, Chicago.

Chapter 59

Intestinal Helminths

LAWRENCE R. ASH AND THOMAS C. ORIHEL

Intestinal helminths are usually diagnosed by detection of eggs or larvae in fecal specimens. Characteristics used in identifying eggs (Fig. 1) include size, shape, thickness of shell, special structures of shell (mammillated covering, operculum, knob, spine), and development stage of egg contents (undeveloped, developing, embryonated).

Objects which might be confused with helminth eggs include pollen grains, portions of vegetable material, and mite eggs. Plant hairs may be confused with nematode larvae. Confusing objects do not show the proper structure and content and may not be the proper size.

In the period after infection is acquired, but before helminths have matured to produce eggs (prepatent period), infections are difficult to diagnose. Occasionally larvae may be found in sputum, but in most instances diagnosis during the prepatent periods is established on clinical grounds.

NEMATODES

The nematodes, or roundworms, are small to large, elongate, cylindrical parasites living primarily as adult worms in the intestinal tract. Nematodes mainly have life cycles that are direct and require no intermediate hosts, but some utilize one or more intermediate hosts. All of the nematodes of humans, with the exception of *Strongyloides*, are dioecious (have two sexes) and characteristically have four larval stages and a fifth, adult stage. The stage of infectivity for the human host varies from the first- to the third-stage larva depending on the species of parasite involved. The infective stages of nematodes are usually within eggs, but in a few species the larvae hatch from the eggs and become infective in the soil. Diagnosis of most of the human intestinal nematode parasites depends on finding characteristic eggs in feces. An important exception to this is *Strongyloides stercoralis* infection, in which first-stage larvae are excreted in fresh feces.

The parasites which commonly infect humans are described below.

Enterobius vermicularis

Infection with *Enterobius vermicularis*, the human pinworm parasite, is cosmopolitan in its distribution, and its true prevalence is probably considerably underestimated. The organism is primarily a parasite of young children; the rapid development of its eggs to the infective stage and their ability to persist for extended periods on fomites lead to rapid dissemination of the infection from child to child and to adults.

Pinworm adults are small, with females being 8 to 13 mm long and barely visible to the naked eye as white, motile worms on the surface of stool specimens or on the perianal skin. Males are only 2 to 3 mm long and usually are not seen. The adult female has a characteristic long and pointed tail, whereas in males the posterior end is blunt. Adult worms typically live in the cecum, colon, appendix, and rectum. Usually, adult females migrate out of the anal orifice at night and lay their eggs in the perianal area, where they adhere to the skin, hair, or bed clothing and bed linen that come in contact with these eggs. Females may lay their eggs on top of a fecal mass, where they are not well mixed into the fecal stream and often are not detected in routine stool examinations.

The eggs of *Enterobius* are elongate and flattened on one side, with a thick, colorless shell. They are 50 to 60 μm long by 20 to 40 μm wide and are partially embryonated when laid. The eggs develop rapidly and become infective within 4 to 6 h, at which time the egg contains a tadpolelike larva (Fig. 2A). Adherence of these eggs to fingers and fomites results in their ready transfer to the mouth and further infection.

The infection is especially troublesome in young children; large worm burdens may cause extreme pruritus, loss of sleep, and irritability. Adult females may enter the vagina and subsequently the uterus or fallopian tubes, where they die. Disintegration of the dead worms and liberation of the eggs contained in utero result in an inflammatory response and granuloma formation about the eggs in these sites.

Diagnosis of the infection is best effected by the use of a cellophane tape procedure (see Chapter 55). Examinations on multiple days may be required to diagnose infection. The proportion of specimens positive correlates with severity of infection (7).

Other methods have been utilized to demonstrate this infection, in particular the use of Vaspar anal swabs; however, the cellophane tape method is most widely used.

Ascaris lumbricoides

The largest and probably the most prevalent of the human intestinal roundworms, *Ascaris*, has a worldwide distribution. The flesh-colored adult worms are large, with adult females usually ranging from 20 to 35 cm long by 3 to 6 mm wide and males from 15 to 31 cm by 2 to 4 mm. Males, with their ventrally curved tails, can be readily distinguished from females, which have straight tails. Females produce large numbers of eggs, perhaps up to 200,000 eggs per female worm per day. These eggs must undergo a developmental period in soil of approximately 3 weeks before they are infective.

In the human host, infective eggs hatch in the intestine and the second-stage larvae undergo an obligatory migration through the liver to the lungs, where they grow and develop for 8 to 9 days to a length of approximately 1 mm before returning to the small intestine to complete maturity. The prepatent period is approximately 2 months.

Pathology in humans can be caused by both larval and adult *Ascaris*. The larval migration phase in infections with large numbers of eggs, and in particular with repeated infections, may result in *Ascaris* pneumonitis (Loeffler's syndrome) consisting of dys-

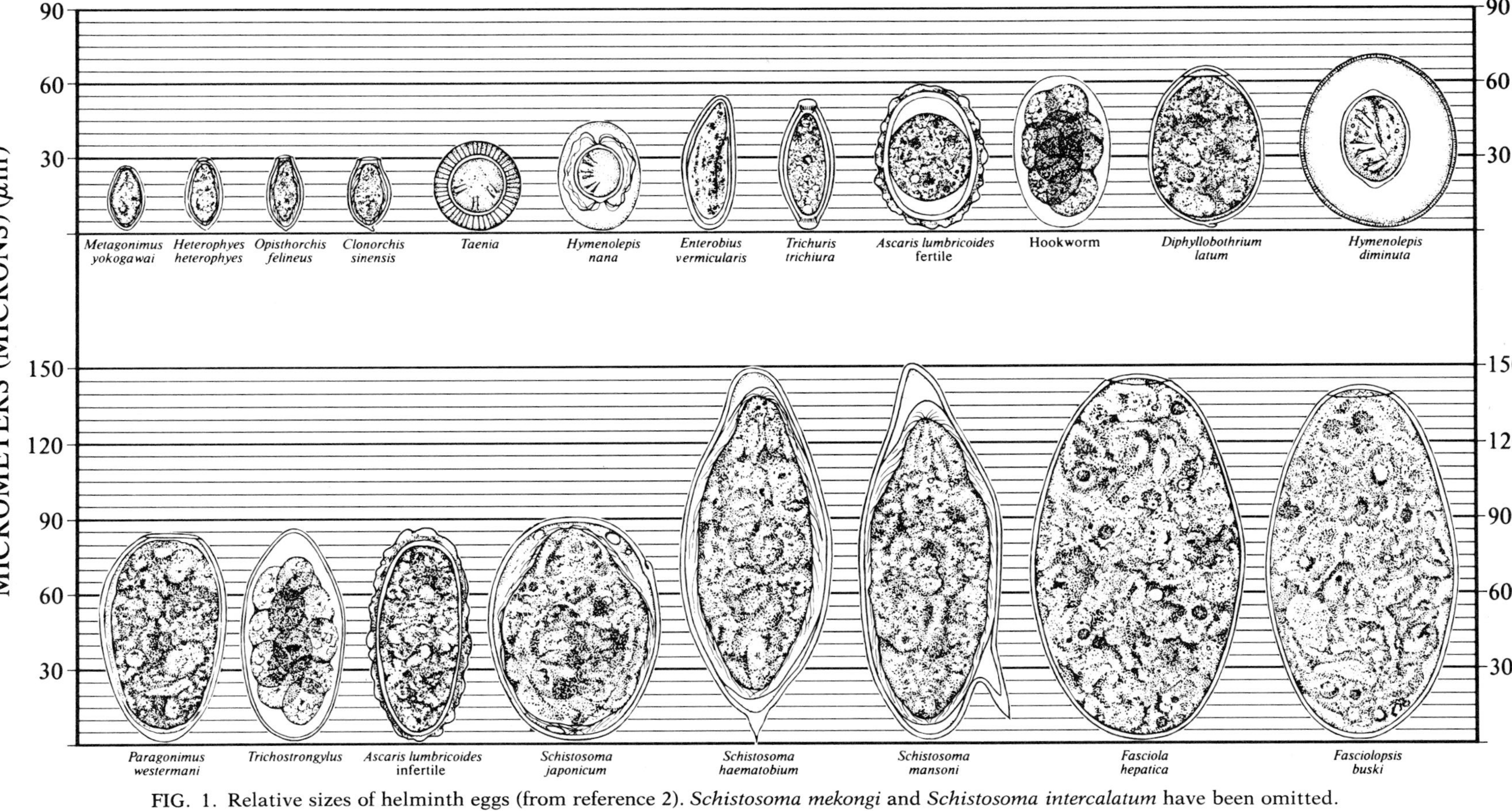

FIG. 1. Relative sizes of helminth eggs (from reference 2). *Schistosoma mekongi* and *Schistosoma intercalatum* have been omitted.

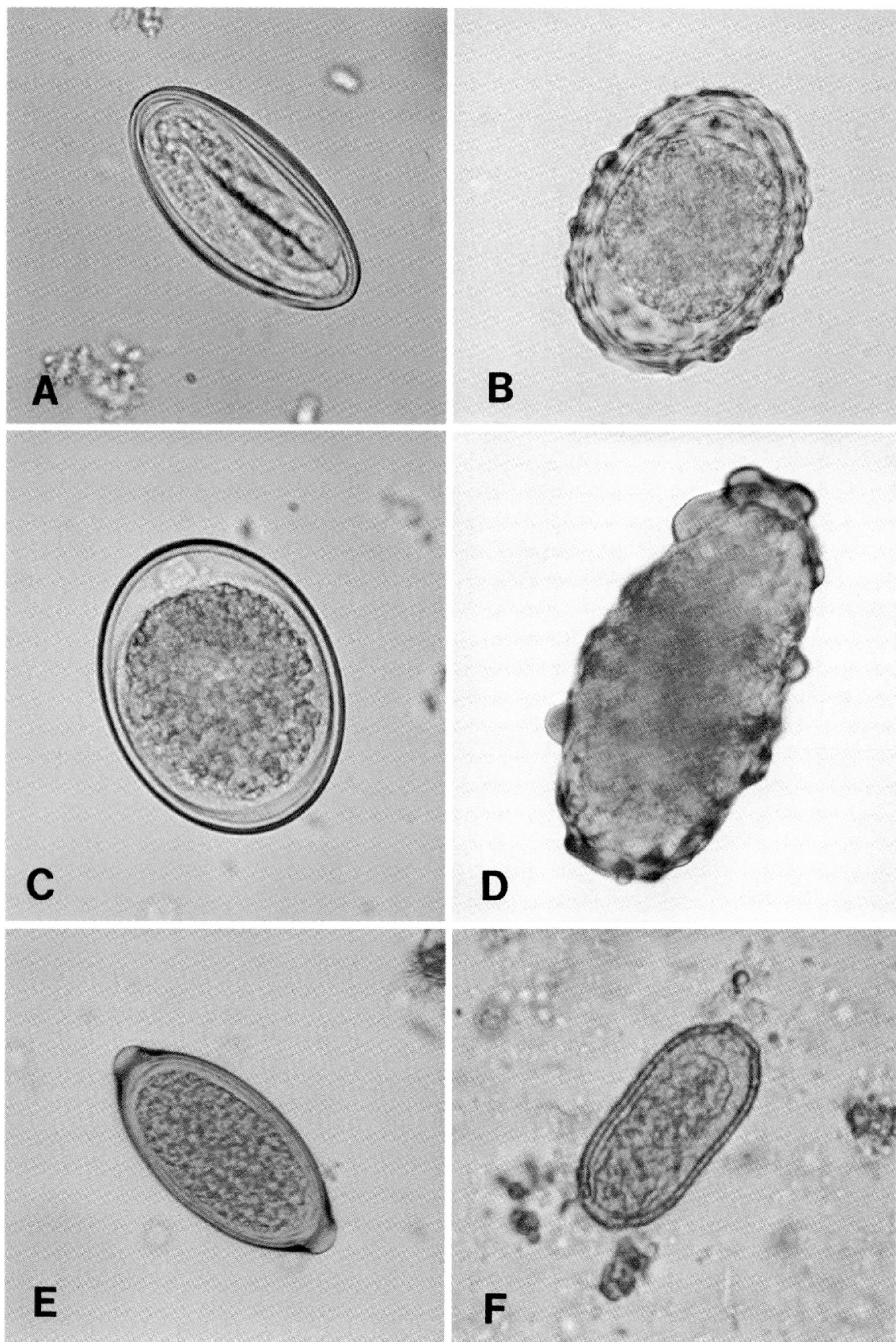

FIG. 2. Eggs of intestinal nematode parasites (×850). (A) Embryonated, infective egg of *Enterobius vermicularis*. (B) Fertile egg of *Ascaris lumbricoides*. (C) Decorticated, fertile egg of *A. lumbricoides*. (D) Infertile egg of *A. lumbricoides*. (E) *Trichuris trichiura*. (F) *Capillaria philippinensis*. (Panels A, C, and D from reference 1, used by permission.)

pnea, cough, rales, eosinophilia, and transient, shifting lung infiltrates as seen by X ray. Adult worms may cause intestinal blockage when present in large numbers. However, the presence of small numbers or even one adult worm is potentially dangerous because of their tendency to migrate to ectopic sites, particularly the liver, when febrile illness occurs. The normal life span for adult worms is approximately 1 year, although female worms may persist for 16 to 20 months.

Diagnosis of infection is usually made by demonstration of typical fertile eggs in feces. Fertile eggs are ovoid, contain a single-celled ovum, and measure 55 to 75 μm long by 35 to 50 μm wide. Eggs are bile stained and have a thick, transparent shell that is covered by a mammillated albuminoid outer layer (Fig. 2B). Occasionally the outer mammillated layer is absent; these eggs are called decorticated eggs (Fig. 2C). Female worms that have never been fertilized, or have exhausted their supply of sperm, will produce infertile eggs. These are elongate, 85 to 90 μm by 43 to 47 μm, and have a thin shell which may lack mammillations entirely or may have grossly irregular mammillations scattered unevenly over the surface of the shell. The contents of these eggs are a mass of disorganized, highly refractive granules and fat globules of varying sizes (Fig. 2D). Because of the large numbers of eggs produced by the female worms, they can usually be found in direct fecal smears and are readily detected by either flotation or sedimentation concentration procedures. Infertile eggs do not float in the standard zinc sulfate solution (sp. gr. 1.18), and they may be missed if only this flotation concentration is used. Adult worms may be spontaneously passed in feces or emerge from the anus, mouth, or nares; young developing worms, especially in heavy infections, may be found in feces. The characteristic, prominent three lips at the anterior end of the worm aid in the diagnosis of these immature or adult parasites (1a).

Trichuris trichiura

Infection with *Trichuris*, known as whipworm infection, is cosmopolitan in its geographic distribution and is found in the warmer, moist regions of the world. Adult worms live attached to the wall of the cecum and, less commonly, to the wall of the large intestine, appendix, and lower part of the ileum. Males and females are of similar size, ranging from 30 to 50 mm long, and have a long attenuated anterior portion and a thicker, short posterior end. The long, slender anterior end is threaded into the mucosal epithelium, and the posterior portion hangs free in the lumen. Adult worms are long lived, commonly up to 10 years and often longer. Though frequently found in association with *Ascaris lumbricoides* infection, because soil requirements for the development of infective eggs are similar, *Trichuris* infections are more frequently seen in older children because of the longer life span of the adult parasites. Egg production by *Trichuris* females probably does not exceed several thousand eggs per day.

Eggs are barrel shaped and brown in color, have thick shells, and measure 50 to 55 μm long by 22 to 24 μm wide. At both ends of the egg are prominent, clear, mucoid "plugs" (Fig. 2E). The eggs contain an unsegmented ovum when passed in feces, and once in the soil it takes 2 to 3 weeks for the infective, first-stage larva to develop. When infective eggs are ingested, the prepatent period in humans is approximately 3 months.

Light infections with *Trichuris* are usually not troublesome, but when large numbers of parasites are present there may be diarrhea or dysentery with abdominal cramping, and occasionally rectal prolapse may occur. Heavy infections may result in dehydration, weight loss, and anemia.

Diagnosis, in particular with heavy infections, is readily made by finding the characteristic eggs in direct wet mounts of feces. In light infections, when eggs are few in number, the use of concentration procedures is required to find eggs. In some instances, larger than normal eggs (70 to 83 μm long by 33 to 35 μm wide) are produced by female worms for reasons not well understood. In addition, the dog whipworm (*Trichuris vulpis*) can reach maturity in humans; the eggs of this parasite are normally larger than those of *T. trichiura*, being 72 to 90 μm long by 32 to 40 μm wide and more barrel shaped.

Capillaria philippinensis

Capillaria philippinensis has been recognized as a human pathogen only since the mid-1960s. Its geographic distribution is restricted primarily to the Philippines, Thailand, and areas bordering on the South China Sea, but one case has been reported from Japan, and ultimately the parasite may be found in other parts of Southeast Asia. *C. philippinensis* appears to be normally a parasite of fish-eating birds, and various fish serve as the necessary intermediate host. In endemic areas human infection is acquired by the ingestion of raw or poorly cooked fish harboring infective larvae in their tissues. Patent infections develop in approximately 1 month. Females produce thin-shelled eggs with developed larvae or free larvae and thick-shelled eggs which are unembryonated (3). Internal autoinfection from the larvae is a normal feature of the life cycle in mammalian hosts, resulting in large worm burdens that cause diarrhea, wasting, dehydration, and, if untreated, death of the host.

Diagnosis of the infection depends on finding the characteristic unembryonated eggs in feces. The eggs of *C. philippinensis* are 36 to 45 μm long by 21 μm wide, have a moderately thick, striated shell, and possess inconspicuous mucoid "plugs" at both ends of the egg (Fig. 2F). It is not uncommon in individuals with chronic diarrhea for eggs, larvae, and even adult worms to be passed simultaneously in feces.

Hookworm infection

The two principal human hookworm parasites are *Necator americanus* and *Ancylostoma duodenale*, but other species of the genus *Ancylostoma* may also produce infections in humans in various parts of the world. *Necator* is found in the United States as well as other parts of the world, but *A. duodenale* is not present in the United States, although its geographic distribution elsewhere frequently overlaps that of *Necator*. Hookworm infection is widely distributed in the tropics and subtropics and also extends into moist, temperate climates.

Adult hookworms are characterized by an anterior end modified into a buccal capsule that contains either teeth or cutting plates with which they can anchor themselves to the wall of the small intestine. Male hookworms are further characterized by having posterior ends modified to form an umbrellalike structure referred to as a bursa; this structure does not occur in female worms, which have straight, pointed tails. *Necator* adults are 7 to 11 mm long by 0.3 mm wide, and they have a buccal capsule provided with cutting plates. *A. duodenale* adults are somewhat larger, 8 to 13 mm long by 0.4 mm wide, and they have a buccal capsule containing two pairs of teeth.

The thin-shelled eggs of *Necator* and *A. duodenale* are unembryonated when passed in feces, and they are essentially indistinguishable from each other (Fig. 3A); their size ranges from 55 to 75 μm long by 36 to 40 μm wide. In the soil, the eggs embryonate and hatch within 1 to 2 days as first-stage, rhabditoid larvae measuring 250 to 350 μm long by 17 μm wide.

In the hookworm life cycle, the first-stage larvae that hatch in the soil develop into infective, third-stage, filariform larvae in approximately 1 week, and these larvae initiate human infection by direct penetration of the skin. *A. duodenale* infective larvae can also infect via the mouth, but *Necator* cannot; additionally, *Necator* requires an obligatory lung migration, whereas if *A. duodenale* infection is acquired orally there is direct maturation in the intestine to the adult stage. Patent infections develop in 5 to 6 weeks, and the life span of the adults of both species is usually only 1 to 2 years but may be as long as 10 years or more.

The pathogenesis of hookworm infections is directly related to the worm burden. Light infections are well tolerated and exhibit few symptoms. Acute, heavy infections may result in fatigue, weakness, abdominal pain, and diarrhea with blood loss, the latter being more severe in *A. duodenale* infections. Chronic hookworm infection results in an iron-deficient anemia, listlessness, pallor, and general retardation of development in afflicted children.

Diagnosis is accomplished by demonstration of characteristic eggs in feces (Fig. 3A). Counts of fewer than five eggs per cover slip in wet mounts denote light infections unlikely to cause anemia. Counts of over 25 eggs per cover slip suggest heavy infection. If there has been prolonged delay in examination of the feces, larvae may develop and hatch, and it is then necessary to differentiate hookworm first-stage larvae from those of *Strongyloides stercoralis*, the stage typically passed in the feces in human strongyloidiasis. Hookworm rhabditoid larvae have a long, narrow buccal chamber (Fig. 3E) and an inconspicuous genital primordium which is an elongate cluster of cells located midway along the intestine, between it and the ventral body wall.

Accurate identification of the hookworm species causing infection depends on recovering and examining adult worms or on culturing larval stages from eggs to the infective, filariform stage with subsequent morphological study of these larvae to distinguish between *Necator* and *Ancylostoma* (1a).

Trichostrongylus species

Human infections with species of the genus *Trichostrongylus* are found throughout the world, in particular in rural areas where herbivorous animals are raised. Although these infections are rarely troublesome in terms of human disease, they do cause some diagnostic difficulties, since trichostrongyle eggs resemble hookworm eggs in shape, although they tend to be much larger in size.

Adult worms are rarely seen in feces. They are small, slender worms, usually measuring less than a centimeter in length, and males have a large prominent bursa. Many different species of trichostrongyles may infect humans, depending on the area of the world in which the infections are acquired.

Trichostrongyle eggs are larger than hookworm eggs, from 75 to 95 μm long by 40 to 50 μm wide. They have a colorless, thin shell, and the egg is tapered slightly at one end. The inner vitelline membrane is frequently wrinkled at the tapered end of the egg, and the germinal mass does not fill the shell (Fig. 3B).

Strongyloides stercoralis

Strongyloidiasis is widely distributed in the tropics and subtropics and also extends into moist, temperate regions. Even in endemic areas its distribution is extremely focal because the existence of the parasite is dependent upon a high groundwater table. Adult parasitic females are parthenogenetic, and parasitic males do not occur. These minute females, only 2 to 3 mm long, live within the mucosal epithelium of the small intestine, where they produce thin-shelled eggs which embryonate and hatch in the mucosa of the intestine. The first-stage larvae migrate into the intestinal lumen and are the usual stage passed in feces (Fig. 3C).

In the soil, first-stage larvae may follow a direct or indirect course of development. In the direct cycle the larvae develop into filariform, third-stage infective larvae which can initiate human infection by direct penetration of the skin. Alternatively, the first-stage larvae may have an indirect cycle in the external environment in which the larvae develop into a free-living generation of adult male and female worms. When these free-living adults mate, the female lays eggs which embryonate and hatch in the soil as first-stage larvae, and these then develop into filariform larvae which may initiate infection by skin penetration. Although multiple free-living generations have been suggested to occur, it appears that this is not common.

Infection may persist for decades as a result of autoinfection either in the colon or perianal area (6). Such infection may be asymptomatic or produce minimal symptoms. However, individuals with these latent infections and who develop malnutrition, frequently in association with alcoholism, and patients who are immunocompromised or receive immunosuppressive therapy, are at great risk for developing hyperinfection with the parasite (8). Hyperinfection results in rapid multiplication of these parasites within the intestinal tract and subsequent reinvasion of the bowel wall, with migration of third-stage larvae through the viscera and back to the intestine to develop into adult females. This greatly accelerated life cycle can result in overwhelming infection, with massive numbers of adult parasites in the intestinal epithelium and migrating larvae invading all organs of the body. Unless detected rapidly, these fulminating infections may result in death.

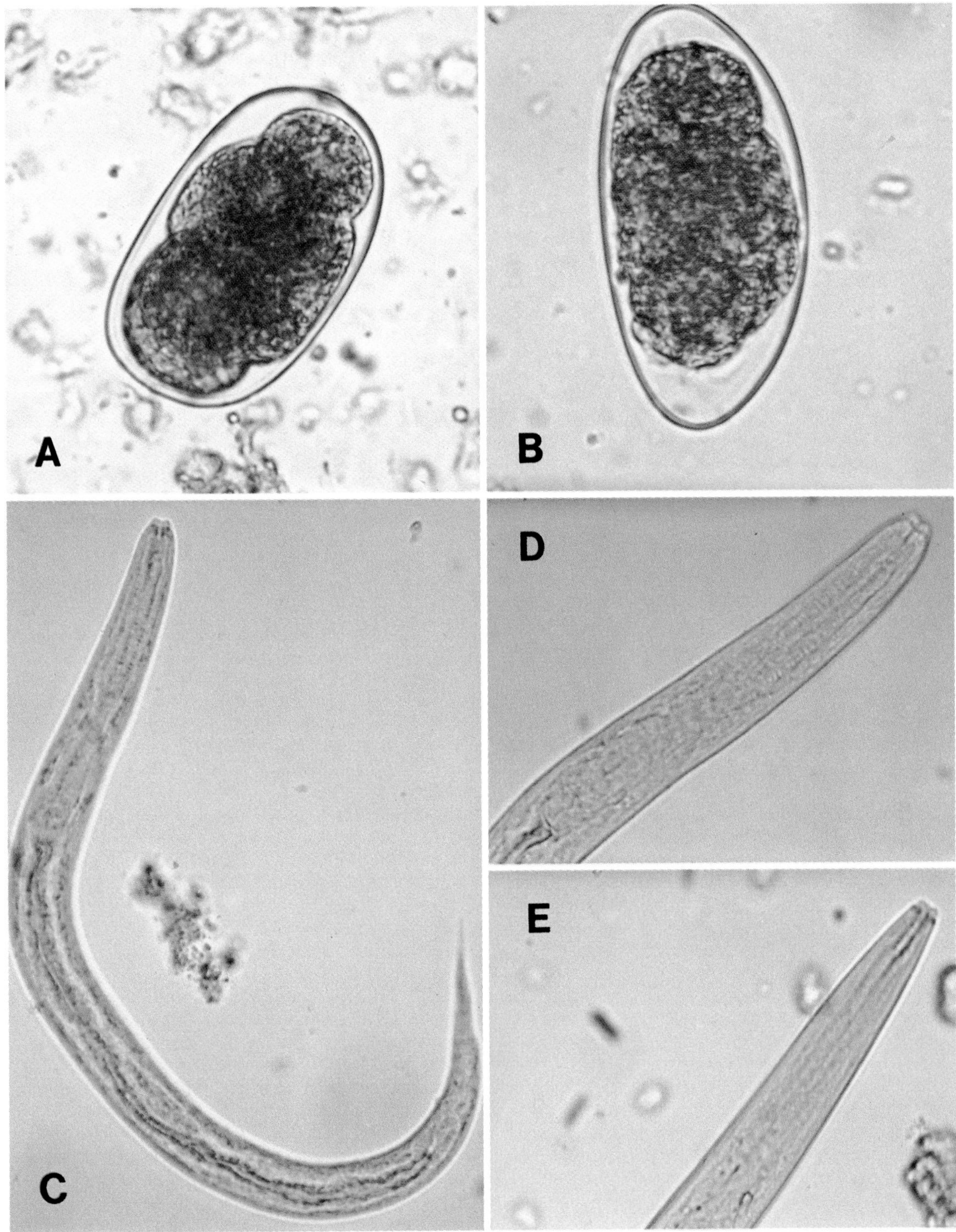

FIG. 3. Eggs and larvae of intestinal nematode parasites (×850). (A) Hookworm egg. (B) *Trichostrongylus* sp. (C) First-stage larva of *Strongyloides stercoralis*. (D) Anterior end of first-stage larva of *S. stercoralis* to show short buccal cavity. (E) Anterior end of first-stage hookworm larva to show long buccal cavity. (Panels A, C, and E from reference 1, used by permission.)

Diagnosis may be difficult. In those patients with no symptoms and in whom few parasites are present, direct wet mount and standard concentration procedures may fail to reveal the first-stage larvae. Individuals who are candidates for immunosuppressive therapy, and who are from known geographic areas where strongyloidiasis is endemic, must be carefully screened for the possible presence of this infection. In latent infections it is common for only small numbers of larvae to be passed in feces, and in many instances the larvae will be passed on an irregular basis. Thus, multiple stool examinations, several days apart, should be performed. In addition to normal concentration procedures, it is recommended that whole fecal specimens should be examined by the Baermann procedure (see Chapter 55). Multiple examinations performed in this manner are likely to detect light infections, if present. Duodenal aspiration techniques, including the Entero-test, have also resulted in improved diagnosis of this infection.

First-stage larvae of *Strongyloides* are the diagnostic stage found in feces, and they measure from 180 to 380 μm in length by 14 to 24 μm in width. They have a short buccal chamber and a prominent cluster of cells, the genital primordium, located midway along the intestine between it and the ventral body wall (Fig. 3C and D). If stool specimens have been allowed to sit at warm room temperatures for more than 24 h before examination, it may be necessary to distinguish between *Strongyloides* first-stage larvae and the first-stage larvae of hookworm that may have embryonated and hatched from eggs present in the feces. Hookworm first-stage larvae are of the same size as *Strongyloides* larvae, but they have a long buccal chamber (Fig. 3E) and the genital primordium is inconspicuous and usually cannot be seen.

Occasionally, third-stage larvae of *Strongyloides* may be seen in feces, particularly in cases of hyperinfection. These larvae are approximately 500 μm long; they have a ratio of esophagus length to intestine length of 1:1, and the tail of the larva is notched.

TREMATODES

Adult trematode parasites of humans live in the intestine, liver, lung, or blood vessels. The flukes, as they are frequently called, all have complex life cycles that involve snails as first intermediate hosts, and many must utilize a second intermediate host, most frequently fish, in which the infective stage for humans will develop. In the life cycles of the various flukes, specific freshwater molluscs are used as first intermediate hosts by each species of trematode. The molluscs are infected by a ciliated larva, the miracidium, which emerges from the trematode eggs. Within the snail tissues a complex developmental process involving several different stages results finally in the production of a tailed larva called a cercaria. Cercariae are released into water, and some (as in the schistosomes) may infect humans directly, although in most species the cercariae invade the tissues of a second intermediate aquatic host within which the cercaria develops into the infective, metacercaria stage. Some metacercariae occur on various types of aquatic vegetation rather than in the flesh of second intermediate hosts. Human infection with trematodes is usually acquired by ingestion of infective stages, although schistosomiasis is produced by direct skin invasion by cercariae. Diagnosis of trematode infections usually is accomplished by identification of eggs in feces or, more rarely, in sputum and urine. The eggs of all human trematodes, with the exception of the schistosomes, have an operculum through which the miracidium escapes. Small trematode eggs usually contain a fully developed miracidium when passed in feces, as do the schistosome eggs; larger trematode eggs are undeveloped when passed and must undergo a period of development in water for several weeks before the miracidium is produced.

Intestinal flukes

Fasciolopsis buski. *Fasciolopsis buski* is the largest and most pathogenic of the human intestinal flukes. It occurs in many parts of Asia, including China, India, Indonesia, Taiwan, Thailand, and Vietnam. Pigs and humans are the primary hosts for this parasite, and infection is acquired by ingestion of various types of aquatic vegetation upon which the metacercarial stage is encysted. Such plants as "water chestnuts" and the water caltrop are important sources of infection. It takes approximately 3 months from ingestion of metacercariae until eggs are found in feces. The eggs are large, 130 to 140 μm by 80 to 85 μm, ovoid, and unembryonated when passed in feces. The operculum is not conspicuous, and at the abopercular end the relatively thin shell often appears to be blemished (Fig. 4A). The eggs of *Fasciolopsis* are essentially indistinguishable from those of *Fasciola hepatica*, the sheep, cattle, and human liver fluke. Since *Fasciolopsis* is restricted to the Orient, whereas *Fasciola* is of cosmopolitan distribution, it is important to establish the geographic history of the patient with the infection as well as the clinical symptomatology, if any, to establish the correct diagnosis.

Heterophyid infections. There are a large number of genera and species of minute intestinal flukes that parasitize humans in many parts of the world. Though of little medical significance, heterophyid eggs resemble those of the liver flukes *Clonorchis* and *Opisthorchis*, which may pose diagnostic problems in the laboratory. The most important species are *Heterophyes heterophyes* and *Metagonimus yokogawai*, both of which occur principally in Asia, although the former also is found in Egypt and Turkey and the latter extends into the Balkan states. All of the heterophyids show little host specificity, and dogs, cats, and various wild animals are the usual definitive hosts in nature. Adult flukes live in the crypts of the small intestine and are only a few millimeters long. Patent infections develop within 2 to 3 weeks of ingestion of infected, poorly cooked fish harboring metacercariae under their scales. When large numbers of worms are present there may be a mild diarrhea and abdominal cramping, but ordinarily there is little or no symptomatology associated with the infections.

All heterophyid eggs are small, from 25 to 30 μm long by 15 to 17 μm wide, and ovoid and have an inconspicuous operculum. The eggs contain a miracidium when laid (Fig. 4B). Though somewhat similar in morphology to the eggs of *Clonorchis*, heterophyid eggs lack the seated operculum characteristic of *Clonorchis* and *Opisthorchis* eggs. These eggs are usually passed in small numbers, and the infections are self-limiting, rarely lasting for more than several months.

Liver flukes

Fasciola hepatica. *Fasciola hepatica* has a cosmopolitan distribution in sheep- and cattle-raising areas of the world, where it is a not infrequent human infection. Infection is acquired by ingestion of aquatic vegetation, such as watercress in salads, on which metacercariae have encysted. Metacercariae migrate from the intestine to the bile ducts of the liver by passing through the intestinal wall into the abdominal cavity and entering the liver through its outer surface. As a consequence of this migration, migrating

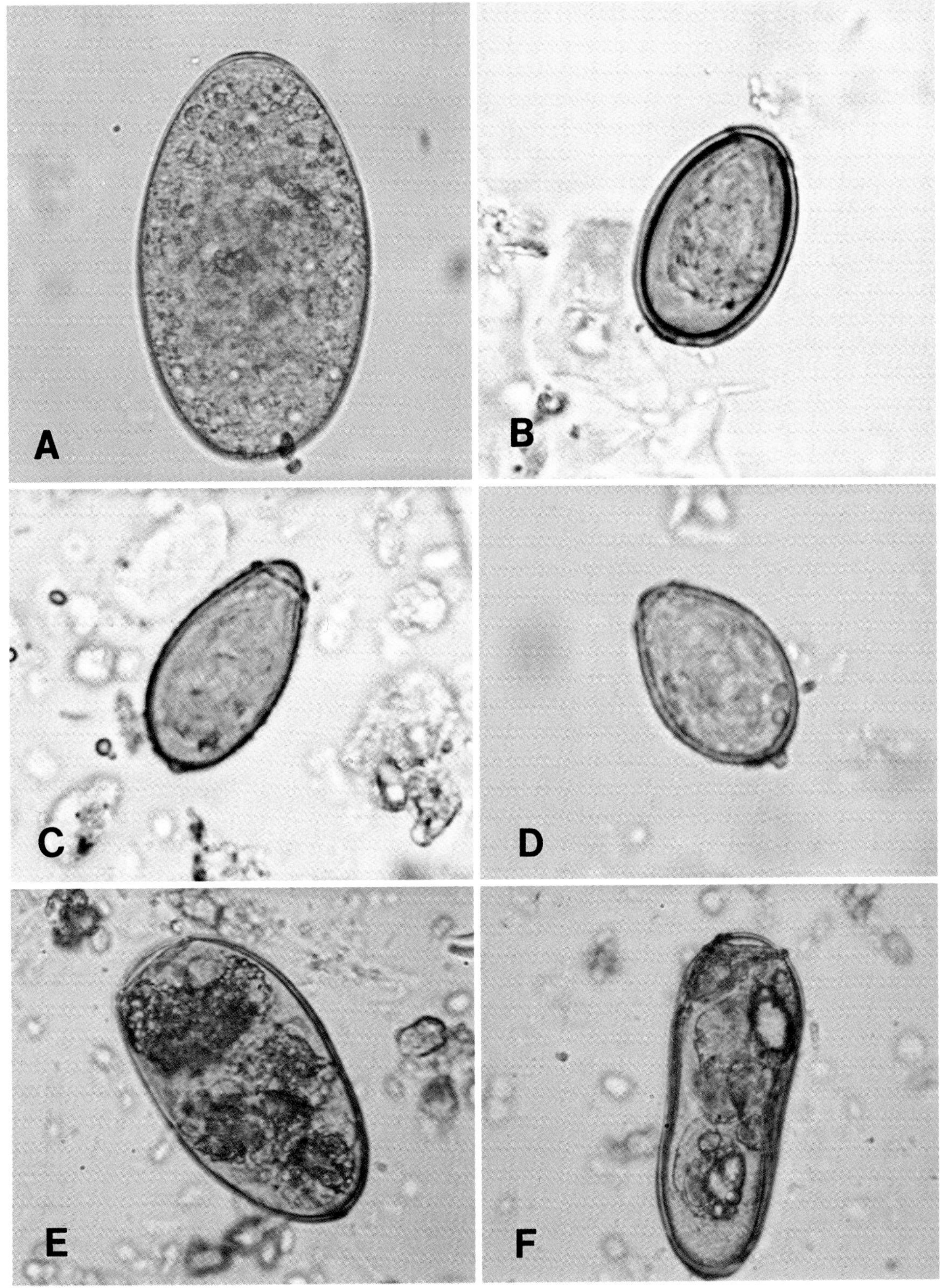

FIG. 4. Eggs of trematode parasites. (A) Egg of *Fasciolopsis buski* (×500). (B) Egg of *Heterophyes heterophyes* (×1,500). (C) Egg of *Clonorchis sinensis* (×1,500). (D) Egg of *Opisthorchis viverrini* (×1,500). (E) Egg of *Paragonimus westermani* (×600). (F) Somewhat atypical egg of *P. westermani* (×600). (Panels B through F from reference 1, used by permission.)

worms may end up in ectopic locations such as the body wall and the cutaneous tissues. The prepatent period is approximately 3 months. Adult worms are large, fleshy flukes that may cause severe liver damage, in particular when present in large numbers.

Eggs are large, 130 to 150 μm long by 63 to 90 μm wide, and ovoid and have a yellowish-brown shell. There is an inconspicuous operculum, and the eggs are unembryonated when laid. These eggs are morphologically similar to *Fasciolopsis buski* eggs. Geographic history and symptomatology aid in diagnosis. Since the ingestion of eggs in infected cattle or sheep liver will result in the passage of these eggs in feces, it is necessary to rule out spurious infection by examination of feces several days after individuals have stopped eating liver.

Clonorchis sinensis. The so-called Oriental liver fluke is commonly seen in the United States now, in particular in immigrants from Southeast Asia. *Clonorchis* and the closely related species, *Opisthorchis viverrini*, live as adults in the bile ducts of humans and reservoir host animals, including cats and dogs, throughout the Orient. Infections are acquired by ingestion of metacercariae encysted under the scales of fish that have been insufficiently cooked. Infections may persist for 20 years or longer. Pickled fish imported from the Orient have been an occasional source of human infection in the United States.

Diagnosis of infection depends on finding the characteristic eggs in feces. Embryonated *Clonorchis* eggs measure 22 to 30 μm long by 12 to 19 μm wide, are ovoid, and have a yellowish-brown, moderately thick shell with a seated operculum (Fig. 4C). At the end opposite the operculum, there is usually a small knob or short, commalike extension of the shell. The eggs of *Clonorchis* and *Opisthorchis* (Fig. 4D) are similar in size and appearance to heterophyid eggs, except that the latter lack the knob at the abopercular end and they do not have a seated operculum.

Lung flukes

Paragonimus westermani. Human lung fluke infections are caused by a number of species of *Paragonimus* in various parts of the world. *P. westermani*, found in the Orient, is the most important species, but other species in Africa, Central and South America, and other parts of Asia are also responsible for human disease. Dogs, cats, and wild animals serve as reservoirs of infection. Adult flukes usually live in pairs in fibrous capsules in the lung parenchyma of their hosts. Eggs that are produced pass up the bronchial tree and may be found in sputum or in feces after they have been swallowed. Crabs and crayfish are second intermediate hosts, and human infections derive from eating raw or poorly cooked infected crustaceans.

In the human host, metacercariae migrate from the intestine into the body cavity, through the diaphragm, and invade the lungs, where the worms mature and lay eggs after 5 to 6 weeks. These infections may be long lived, frequently 10 to 20 years. In the course of larval migration, the flukes may take aberrant migratory routes and end up in ectopic locations such as the body wall, the rib cage, and the brain. In the third location the infection is frequently fatal.

Diagnosis depends on finding eggs in feces or, more rarely, in sputum. The broadly ovoid eggs are thick shelled, yellowish-brown, and unembryonated and have a distinct operculum. They measure 80 to 120 μm long by 45 to 70 μm wide. At the abopercular end the shell is distinctly thickened but does not have a knob (Fig. 4E). Though the eggs of the other species of *Paragonimus* may differ somewhat in size, shape, and morphology, they are usually sufficiently similar that the diagnosis of paragonimiasis can be made (Fig. 4E and F).

Blood flukes (schistosomes)

Schistosomiasis (bilharziasis) afflicts in excess of 250 million people in the world and as such is, along with malaria, one of the most important of all human parasitic diseases. The etiologic agents are the blood flukes, which are markedly different from the other human trematodes. The schistosomes have separate sexes and live in blood vessels of the abdominal cavity. The three most important human schistosomes are *Schistosoma mansoni*, *S. japonicum*, and *S. haematobium*, and there are a number of other species of lesser importance which infect humans in Africa and Asia. Though each of the schistosomes utilizes specific and different snail intermediate hosts, their life cycles are similar. Each produces thin-shelled eggs which lack an operculum and contain a miracidium when excreted in feces or urine. Infected snails produce fork-tailed cercariae which directly penetrate the skin and establish infection in human and animal hosts. In the mammalian host the larval blood flukes migrate through the lungs and become established in venous blood vessels of the mesenteries or the bladder, where they mature and females deposit eggs. Through a mechanism not well understood, the eggs make their way through the wall of the intestine or bladder and pass in feces or urine. Pathology due to schistosome infection is primarily attributable to egg deposition in tissue at the site of adult worms or the areas of venous drainage (liver, lung).

S. mansoni. *S. mansoni* has the widest geographic distribution of the schistosomes, including Africa, the Arabian peninsula, Brazil, Surinam, Venezuela, Puerto Rico, and a number of Caribbean islands. Adult worms live in the portal system of the liver and the small venules of the lower ileum and colon. Eggs are laid in the blood vessels, make their way through the wall of the intestine, and are passed in feces. *S. mansoni* eggs measure 114 to 175 μm long by 45 to 70 μm wide; they contain a miracidium, and the shell has a prominent lateral spine (Fig. 5A and B). In acute schistosomiasis, blood and mucus appear in feces along with the lateral-spined eggs. The prepatent period for the infection is approximately 6 weeks. In chronic schistosomiasis, eggs will accumulate in the walls of the intestine and rectum and also the liver (Fig. 5F); correspondingly, fewer eggs will be found in feces, and concentration procedures will be required to diagnose infection reliably. Rectal biopsies may also be useful in diagnosis (see Chapter 55).

S. japonicum. *S. japonicum* is found in China, the Philippines, and other countries of Southeast Asia. A zoophilic strain of the parasite infects animals but not humans in Taiwan, and the infection has been virtually eliminated from Japan, where occasional infections may be found in cattle. Animal reservoirs of infection are important and include water buffaloes, pigs, dogs,

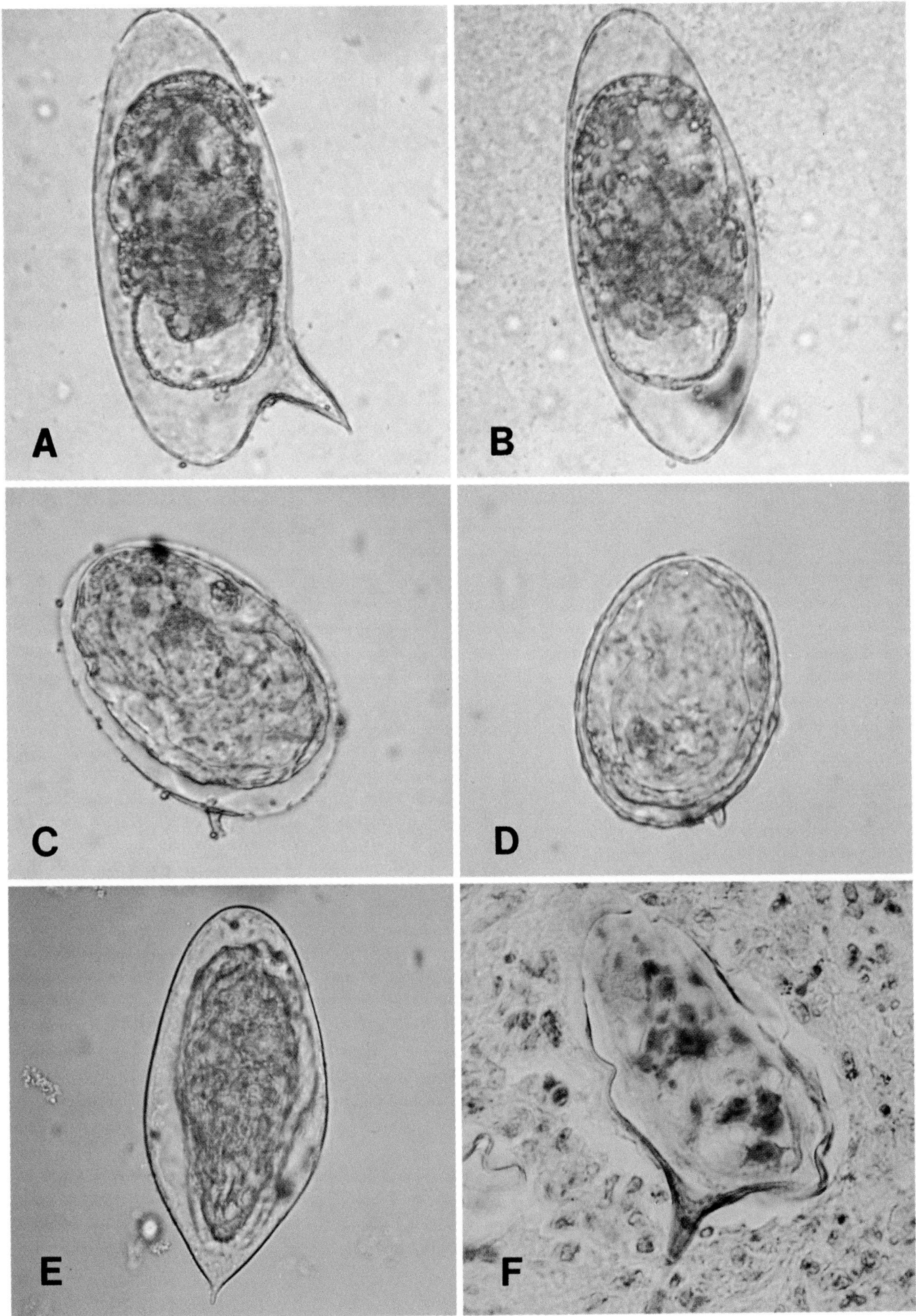

FIG. 5. Eggs of schistosome species (×600). (A) Egg of *Schistosoma mansoni*. (B) *S. mansoni* egg with typical lateral spine not in view. (C) *S. japonicum*. (D) *S. mekongi*. (E) *S. haematobium*. (F) *S. mansoni* egg showing lateral spine in tissue section. (Panels C through E from reference 1, used by permission.)

cats, and wild rodents. Adult worms live in mesenteric veins, and the prepatent period is 5 to 6 weeks. The embryonated eggs found in feces are round to ovoid, lack an operculum, and measure 70 to 100 μm long by 55 to 65 μm wide. The thin shell has a small, inconspicuous spine which frequently is not seen (Fig. 5C). In addition, the surface of the egg frequently has fecal debris adhering to it, which can make the eggs difficult to see or recognize. Rectal biopsy may be an important diagnostic tool when fecal examinations are negative.

S. mekongi. Closely related to *S. japonicum*, but described as a separate species, is *S. mekongi*, which is a parasite of humans and dogs in countries bordering on the Mekong River, specifically Laos and Cambodia. The egg is similar in morphology to that of *S. japonicum*, but smaller (9), ranging from 51 to 78 μm long by 39 to 66 μm wide. Eggs from dogs are usually smaller than those from human cases. There is a small, knoblike spine on the shell that may be difficult to see (Fig. 5D).

S. haematobium. *S. haematobium* occurs in Africa, Lebanon, Syria, Iran, the Arabian peninsula, and Malagasy, where it causes urinary schistosomiasis. Adult worms reside in the venous plexuses of the bladder, and eggs that are laid move through the wall of the bladder and pass in urine. In chronic infections, accumulation of eggs in the bladder wall can lead to bladder and ureter pathology. The thin-shelled eggs of this parasite measure 112 to 170 μm long by 40 to 70 μm wide, are embryonated, and have a terminal spine (Fig. 5E). Hematuria is frequently present with this infection, and diagnosis is made by finding eggs in urine sediment. Eggs can sometimes be found in feces and in the wall of the rectum as well as the bladder.

Serodiagnosis. Although schistosomiasis is perhaps immunologically the most thoroughly studied parasitic infection, serodiagnosis is quite inadequate. Indirect immunofluorescence, using cryostat sections of adult worms as antigen (10), is the test of choice. Sensitivity is 85%, and cross-reactions are limited to patients with trichinosis or filariasis. Titers are of little value, and tests are reported as positive or negative. Laboratories in endemic areas frequently perform the circumoval precipitin test (5), which is less sensitive; the requirement for schistosome eggs makes the test impractical except in very few laboratories. Serology is most useful for ruling out schistosomiasis in persons who have been in endemic areas for short periods and for epidemiologic studies. Recent studies using enzyme-linked immunosorbent assays with cercarial and adult antigens and deriving a cercaria/adult ratio show promise in differentiating acute and chronic schistosomiasis (4). Reagents for serodiagnosis of schistosomiasis are not available commercially.

CESTODES

The four most common adult tapeworm parasites of the human small intestine are *Diphyllobothrium latum*, *Taenia saginata*, *Taenia solium*, and *Hymenolepis nana*. Two other tapeworms, primarily animal parasites, reach maturity in humans and may cause infrequent infections; they are *Hymenolepis diminuta* and *Dipylidium caninum*. In addition to the adult tapeworms, a number of larval cestodes can produce serious human disease including cysticercosis (*Taenia solium*), hydatid disease (*Echinococcus* spp.), and coenurosis (*Taenia multiceps*). The large adult tapeworms, *D. latum*, *T. saginata*, and *T. solium*, usually occur singly in the intestine and may live up to 20 years or longer. Diagnosis of tapeworm infections is achieved by finding characteristic eggs, proglottids, or both in feces, depending on the species involved. All of the adult tapeworms in humans require an intermediate host with the exception of *H. nana*, which may or may not utilize an intermediate host. Morphologically, adult tapeworms consist of the anterior head end, or scolex; a short undifferentiated neck region that gives rise to the proglottids; and the main body (strobila) of the tapeworm, consisting of immature, mature, and gravid proglottids. As tapeworms increase in age and size, the most posterior gravid proglottids may break off or disintegrate and pass out in feces. Some tapeworms lay eggs (*D. latum*, *H. nana*, and *H. diminuta*) which pass in feces; others (*T. saginata* and *T. solium*) typically have proglottids which break off and pass in feces or actively migrate out of the anus. Sometimes *Taenia* proglottids will rupture in the intestine, and eggs will then appear in feces. The eggs of all species of *Taenia* and the related genus *Echinococcus* are morphologically identical and cannot be distinguished from one another. The eggs of *T. solium*, *T. multiceps*, and *Echinococcus* spp. can infect humans directly and cause disease.

Diphyllobothrium latum

Known as the broad fish tapeworm, *D. latum* differs from other adult tapeworms infecting humans in its morphology, biology, and epidemiology. Its geographic distribution includes areas with cold, clear lakes as in Scandinavia, other areas of northern Europe, the USSR, northern Japan, and North America, principally the upper Midwest, Canada, and Alaska. The adult parasite can attain a length of 10 to 15 m, is ivory in color, and has a scolex that is provided with shallow grooves on its dorsal and ventral aspects.

In its life cycle, unembryonated, operculate eggs, resembling those of trematodes, are passed in feces and must undergo embryonation in water for several weeks. Ciliated, six-hooked embryos (coracidia) hatch from these eggs and must be ingested by appropriate species of freshwater copepods. Within the copepod a solid-bodied larval stage, the procercoid, develops and becomes infective to the second intermediate host, fish. In fish, the procercoids migrate into the flesh and develop into the plerocercoid (sparganum) stage, which is then infective to the human or animal hosts. After ingestion of the sparganum stage, it takes 3 to 5 weeks for the adult tapeworm to attain maturity and begin to lay eggs. The parasite may produce no clinical symptoms in some people, but when it reaches large size it may cause mechanical obstruction of the bowel, may cause diarrhea and abdominal pain, and in some individuals, particularly in northern European countries, may be responsible for a vitamin B_{12} deficiency resulting in pernicious anemia.

Diagnosis is made by finding the characteristic operculated eggs in feces. They measure 58 to 75 μm in length by 40 to 50 μm in width. At the abopercular end there is frequently a small knoblike protrusion (Fig. 6F). Individuals with long-standing infection are

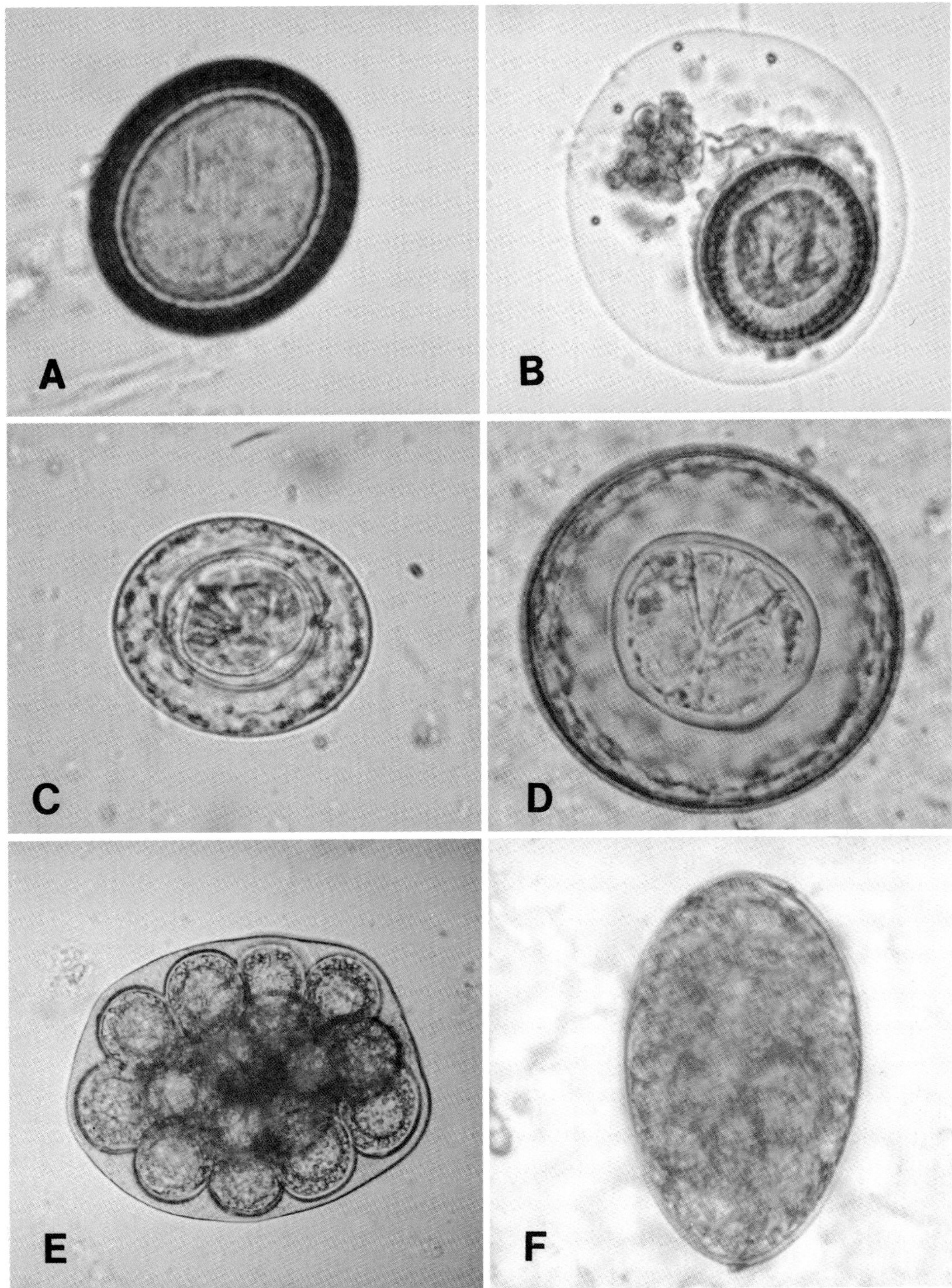

FIG. 6. Eggs of cestodes. (A) Egg of *Taenia* spp. Eggs of all species of *Taenia* and *Echinococcus* are identical (×1,000). (B) *Taenia* egg surrounded by the primary membrane frequently seen around eggs directly liberated from gravid proglottids (×800). (C) *Hymenolepis nana* (×900). (D) *Hymenolepis diminuta* (×900). (E) Egg packet of *Dipylidium caninum*. (F) *Diphyllobothrium latum* (×800). (Panels B through F from reference 1, used by permission.)

likely to have large numbers of eggs in feces. As infections grow older, one or a small chain of proglottids may break off and be passed in feces. These proglottids are wider than long (3 by 1 mm), and the genital pore is situated on the midventral surface rather than laterally as in the other human tapeworms. In freshly passed proglottids the coiled uterus appears yellowish in the center of the proglottid.

Taenia saginata

T. saginata is the beef tapeworm, which has a cosmopolitan distribution but is particularly prevalent in Mexico, South America, and eastern and western Africa. Cattle serve as the intermediate host, and ingestion of eggs on contaminated pastureland by grazing cattle results in the development in their tissues of the infective cysticercus stage (*Cysticercus bovis*).

The cysticercus is 0.5 to 2.0 mm in diameter and has a pearllike appearance in tissues, where the unarmed scolex is invaginated into a fluid-filled bladder. After ingestion of the cysticercus in raw or poorly cooked beef, it takes approximately 2 to 3 months for the infection to become patent. Adult tapeworms attain lengths of 4 to 8 m and have a scolex with four suckers and an unarmed rostellum; gravid proglottids are longer than they are wide (18 to 20 mm by 5 to 7 mm). Each proglottid has a genital pore at the midlateral margin. Gravid proglottids, which are highly muscular and active, will break off from the strobila and can actively migrate out of the anus. Although patients may exhibit no symptomatology with this infection, the mature worm may cause abdominal discomfort, diarrhea, and occasionally intestinal obstruction due to its large size.

Diagnosis of species usually is made by identification of proglottids that have passed in feces or actively migrated out of the anus. Identification of the proglottids is based on morphology of the uterus which has either been injected via the genital pore with India ink or stained with carmine or hematoxylin stains, or after the specimen has been cleared (see Chapter 55). In *T. saginata* there are 15 to 20 lateral branches on each side of the central uterine stem. If the proglottids rupture in the intestine, the typical *Taenia* eggs can be seen in feces although specific identification on the basis of these eggs is impossible. Taeniid eggs are spherical and have a thick, yellow-brown, prismatic shell (Fig. 6A). Within the egg is a six-hooked embryo, the oncosphere. Occasionally, especially when eggs are liberated directly from proglottids, there is a thin outer membrane around the eggs (Fig. 6B). Eggs measure 31 to 43 μm in diameter.

It is generally accepted that the eggs of *T. saginata* are not directly infective to humans, but caution should be exercised in the handling of all proglottids and taeniid eggs since the eggs of *T. solium*, *T. multiceps*, and *Echinococcus* spp. are directly infective and can cause cysticercosis, coenurosis, and hydatidosis, respectively.

Taenia solium

Known as the pork tapeworm, *T. solium* has an extensive geographic distribution throughout Europe, Mexico and other Latin American countries, China, and India. It is no longer commonly found in the United States. Human infection is acquired by ingestion of infective cysticerci (*Cysticercus cellulosae*) in poorly cooked pork or pork products. The adult worms may reach lengths of 2 to 7 m; they have a scolex with four suckers and a rostellum armed with two rows of hooklets. Gravid proglottids have 7 to 13 lateral uterine branches off the central uterine stem. Since the eggs of *T. solium* are infective to humans and can cause cysticercosis, extreme caution in the handling of these proglottids is recommended.

Hymenolepis nana

H. nana is the smallest of the adult tapeworms infecting humans, attaining lengths of 2.5 to 4.0 mm, and is the most common tapeworm infection of humans in the United States. Though it is normally a parasite of mice, in which the life cycle characteristically involves various beetles as intermediate hosts, transmission to humans, by direct ingestion of eggs containing six-hooked embryos, is common. When eggs are ingested by humans, a solid-bodied larva, the cysticercoid, first develops in the wall of the small intestine and subsequently migrates back into the intestinal lumen, where it reaches maturity as an adult tapeworm in 2 to 3 weeks. In beetles that ingest eggs of *H. nana*, the cysticercoids develop in the body cavity and have a thick protective wall about them. Although humans may acquire infection by accidental ingestion of infected beetles, direct infection is far more common and is the primary reason why *H. nana* usually occurs in institutional and familial settings where hygiene is substandard. A feature of human *H. nana* infection is the opportunity for internal hyperinfection, which may result in the presence of large numbers of adult tapeworms.

Diagnosis of *H. nana* infection rests on finding the spherical to subspherical, embryonated eggs in feces. The eggs are 30 to 47 μm in diameter and thin shelled and contain an oncosphere (Fig. 6C). There are two thickenings at opposite poles of the membrane around the embryo, and from these arise four to six polar filaments which extend into the space between the embryo and the outer shell.

Hymenolepis diminuta

An occasional human parasite, *H. diminuta* is primarily a parasite of rats. Beetles and other arthropods serve as obligatory intermediate hosts, with humans generally acquiring infection accidentally by ingesting infected meal beetles present in various grains and cereals. Cysticercoids, the infective stage, develop in the hemocoel of beetles after ingestion of eggs. As an adult, *H. diminuta* may be from 20 to 60 cm long, and multiple specimens may be present in the same host. In humans, hyperinfection or direct infection by ingestion of eggs, as occurs in *H. nana*, does not occur.

Diagnosis of *H. diminuta* infection is by demonstration of eggs in feces. The eggs are spherical and large, 70 to 85 μm by 60 to 80 μm, and have a yellowish-brown, moderately thick shell (Fig. 6D). The six-hooked embryo in the egg is considerably separated from the outer membrane. There are no polar filaments, so that the combination of size and lack of these filaments allows for ready differentiation of these eggs from those of *H. nana*.

Dipylidium caninum

D. caninum is the most common and widespread adult tapeworm of dogs and cats. Human infection has been reported in many parts of the world. Children are more frequently infected as a result of more intimate contact with dogs and their fleas, which serve as obligatory intermediate hosts. Because the infection is not a troublesome one and is self-limiting, there are probably many more cases of it than are reported in the literature.

Adult tapeworms may be present in considerable numbers and may vary in length from 10 to 70 cm. The scolex is conical in shape with four prominent suckers and a small retractile rostellum that bears multiple rows of small spines. Gravid proglottids are elongate (23 by 8 mm), have a genital pore on each lateral margin (hence the name double-pored tapeworm), and are divided into small compartments, each of which contains 8 to 15 six-hooked onchospheres that are enclosed in a thin, embryonic membrane. In dogs, proglottids are frequently passed in small chains; when they undergo dehydration on carpets and floors, they resemble grains of rice. In the life cycle, larval fleas ingest the eggs and cysticercoids develop in the hemocoel; the cysticercoids remain viable when larval fleas undergo metamorphosis to the adult flea, and human and animal infections usually are acquired by ingestion of the adult fleas. The tapeworms reach maturity in the small intestine in approximately 1 month.

Diagnosis of infection is usually made by finding the typical double-pored, compartmented proglottids in feces, or by finding packets of oncospheres liberated by disintegration of the proglottids (Fig. 6E).

LITERATURE CITED

1. **Ash, L. R., and T. C. Orihel.** 1984. Atlas of human parasitology, 2nd ed. American Society of Clinical Pathologists, Chicago.

1a. **Beaver, P. C., R. C., Jung, and E. W. Cupp.** 1984. Clinical parasitology, 9th ed. Lea & Febiger, Philadelphia.

2. **Brooke, M. M., and D. M. Melvin.** 1984. Morphology of diagnostic stages of intestinal parasites of humans. CDC 84-8116. Centers for Disease Control, Atlanta, Ga.

3. **Cross, J. H., T. Banzon, M. D. Clarke, V. Basaca-Servilla, R. H. Watten, and J. J. Dizon.** 1972. Studies on the experimental transmission of *Capillaria philippinensis* in monkeys. Trans. R. Soc. Trop. Med. Hyg. **66:**819–827.

4. **Lunde, M. N., E. A. Ottesen, and A. W. Cheever.** 1979. Serologic differences between acute and chronic *Schistosomiasis mansoni* detected by enzyme-linked immunosorbent assay (ELISA). Am. J. Trop. Med. Hyg. **28:**87–91.

5. **Oliver-Gonzales, J.** 1956. Anti-egg precipitins in the serum of humans infected with *Schistosoma mansoni*. J. Infect. Dis. **95:**86–91.

6. **Pelletier, L. L., Jr.** 1984. Chronic strongyloidiasis in World War II Far East ex-prisoners of war. Am. J. Trop. Med. Hyg. **33:**55–61.

7. **Sadun, E. H., and D. M. Melvin.** 1956. The probability of detecting infections with *Enterobius vermicularis* by successive examination. J. Pediatr. **48:**438–441.

8. **Scowden, E. B., W. Schaffner, and W. J. Stone.** 1978. Overwhelming strongyloidiasis: an unappreciated opportunistic infection. Medicine **57:**527–544.

9. **Voge, M., D. Bruckner, and J. I. Bruce.** 1978. *Schistosoma mekongi* sp. n. from man and animals, compared with four strains of *Schistosoma japonicum*. J. Parasitol. **64:**577–584.

10. **Wilson, M., A. J. Sulzer, and K. W. Walls.** 1974. Modified antigens in the indirect immunofluorescent test for schistosomiasis. Am. J. Trop. Med. Hyg. **23:**1072–1076.

Chapter 60

Arthropods of Medical Importance

HARRY D. PRATT AND JAMES W. SMITH

Arthropods or arthropod larvae may affect human health in several ways.

Arthropods may transmit disease-causing pathogens mechanically, as when flies carry bacteria causing diarrhea and dysentery from filth to human food. Other arthropods are biological vectors in the transmission of viruses, bacteria, protozoa, and metazoa that cause human disease. For example, mosquitoes may transmit malaria parasites, or fleas may transmit bacteria causing plague (4, 11).

Arthropods also may be parasites on a patient, such as lice, scabies mites, and tissue-invading maggots. Others may attack a patient: mosquitoes, certain flies, fleas, bed bugs, some spiders, ticks, and mites, which bite with their mouthparts, or bees, wasps, and scorpions, which sting with an apparatus at their posterior end. Such bites or stings usually cause only temporary local swelling and itching which are treated with calamine or other soothing lotions, although secondary infection may occur. A few persons become sensitized to protein in the insect bite or sting, and if they are bitten or stung at a later date, these people develop larger, more severe welts around each lesion. For example, a person repeatedly infested with scabies mites may develop a generalized rash and intense itching over much of the body, not just in the areas of scabies mite infestation. A few people who are sensitized to the stings of bees, wasps, or hornets, if stung again may have a severe anaphylactic reaction and may die. Such reactions usually occur within an hour after the sting.

Arthropods are very widespread in the environment, and excrement and fragments of arthropods are present in soil and dust. Persons may become hypersensitive to these materials and develop allergic manifestations such as asthma and hay fever (4, 10, 11).

Parasitism by arthropod parasites which are attached to the skin or temporarily invade superficial tissues is generally called infestation. In contrast, parasitism of body tissues, intestine, or atria by protozoan or helminth parasites is generally called infection.

The arthropods are the largest of the animal phyla, with over a million species. It is beyond the scope of this chapter to deal with all the arthropods of medical importance or the therapy and control of arthropod infestations. The most common ectoparasites submitted to laboratories for identification in the United States are discussed and pictorial keys are presented which show the principles of identification. Microbiologists can recognize the major groups of arthropods sent to laboratories and, in some cases as with head or crab lice, can identify a specimen to genus and species with reasonable confidence. Arthropods may be brought to the laboratory because a person wishes to know whether they are of concern and what might be done to control them. Some, such as cockroaches, clothes moths, termites, and stored-food insects, are widespread and require special literature on household pest control (9, 20, 28). More detailed information on arthropods of medical importance and their identification, treatment, and control is available in various references (4, 9, 11, 14, 20, 26, 28).

PHYLUM ARTHROPODA

The phylum Arthropoda comprises invertebrate animals with a segmented body, several pairs of jointed appendages, and a rigid chitinous exoskeleton which is molted periodically and renewed as the animal grows. Two classes are of importance to the clinical microbiologist.

Arthropods develop from egg to adult through a developmental process known as metamorphosis, literally a change in form (3). Insects with gradual metamorphosis pass through three stages during their life: egg, nymphs, and adult. Examples are sucking lice, bed bugs, and cockroaches. Nymphs are smaller but resemble adults. They change (metamorphose) gradually through a succession of molts until they become adults. In some with gradual metamorphosis, such as cockroaches and kissing bugs, the nymphs have wing pads while adults have wings. Nymphs are always sexually immature, while adults are sexually mature, ready to mate and reproduce. Nymphs have the same type of mouthparts as adults and live in the same environment, for example, head lice nymphs and adults on the scalp.

Insects with complete metamorphosis pass through four stages during their life: egg, larva, pupa, and adult. Examples are flies, fleas, mosquitoes, and bees. Insects with this type of metamorphosis differ greatly in the immature and adult stages. Typical larvae are wigglers of mosquitoes, maggots of flies, and caterpillars of moths and butterflies. The pupa is a nonfeeding stage during which the larva undergoes profound external and internal changes to become the adult. Typically larvae have different mouth parts and live in a different environment than the adult.

Ticks and mites have a still different type of metamorphosis, typically with four stages: egg, larva, nymph (or several nymphal stages), and adult. Tick and mite larvae have three pairs of legs, while nymphs and adults have four pairs of legs.

Class Insecta

Representatives. Lice, bugs, fleas, flies, cockroaches, and others.

Characterization. As adults with three body regions (head, thorax, and abdomen), one pair of antennae, three pairs of legs, and wings in many species.

Medical importance. May be venomous or parasitic or act as vectors of disease.

Class Insecta includes the following important species:

- Order Hemiptera, family Cimicidae (bed bugs)
- Order Anoplura, family Pediculidae (human lice)
- Order Anoplura, family Pthiridae (crab lice)
- Order Siphonaptera, family Pulicidae (fleas)
- Order Diptera (flies, maggots)

Class Arachnida

Representatives. Ticks, mites, spiders, and scorpions.

Characterization. As adults with one or two body regions (cephalothorax and abdomen, or one body region in ticks and mites), no antennae, four pairs of legs, and no wings. Larvae of ticks and mites have only three pairs of legs.

Medical importance. May be venomous or parasitic or act as vectors of disease.

Class Arachnida includes the following important species:

- Order Araneida, family Theridiidae (black widow spider)
- Order Araneida, family Loxoscelidae (brown recluse spider)
- Order Acarina, family Ixodidae (hard ticks)
- Order Acarina, family Argasidae (soft ticks)
- Order Acarina, family Sarcoptidae (scabies mites)
- Order Acarina, family Demodecidae (follicle mites)
- Order Acarina, family Trombiculidae (chiggers)
- Order Acarina, family Dermanyssidae (rodent and bird mites)
- Order Acarina, family Pyroglyphidae (house dust mites)

SPECIMEN HANDLING AND EXAMINATION

Clinical microbiologists are often asked to identify arthropods or arthropod larvae. Specimens of ectoparasites should be preserved and shipped for identification in 70% ethyl alcohol (Formalin is irritating to the eyes of the identifier). Fly larvae should be washed in water to remove debris, particularly those collected from fecal samples or wounds, then killed in hot (not boiling) water (about 80°C) for a few minutes so that they do not become dark when placed in 70% alcohol.

Most ectoparasites can be identified with a dissecting microscope with magnifications of ×25 to ×75 power. However, some may have to be mounted on microscope slides and examined with a compound microscope. Some fleas, fly larvae, lice, and mites may have to be treated overnight with cold 10% potassium or sodium hydroxide to remove the internal flesh and permit better visibility of key characters. Such specimens can then be washed in water, dehydrated serially in 70% ethyl alcohol, Cellosolve, and clove oil, and then mounted in Canada balsam as permanent slides. Quick nonpermanent mounts can be made with commercially available clearing and mounting medium, such as Hoyer mounting medium, or by mounting the specimen directly in 85% lactic acid in water.

Assistance in identifying arthropods may be obtained from entomologists in public health laboratories, educational institutions, natural history museums, or agriculture extension agencies.

BED BUG, ORDER HEMIPTERA (5, 9, 11, 20, 28)

The bed bug (*Cimex lectularius*) is a reddish-brown insect 4 to 5 mm long, with a pair of four-segmented antennae and a three-segmented, blood-sucking proboscis which lies in a groove on the underside of the head and thorax (Fig. 1). It has short wing pads but cannot fly. The pronotum is broad and concave on the anterior margin. The abdomen is flattened and somewhat heart-shaped.

Normally bed bugs feed at night and hide during the day in cracks and crevices in bedsteads, in seams or tufts of mattresses, or under loose wallpaper in bedrooms. However, infestations also occur in overstuffed chairs, sofas, and furniture in theaters and clubs. Sometimes the first signs of these insects are tiny spots of blood on bedding or the dead insects themselves, killed as people crushed them rolling in their sleep. Some people are very sensitive to bed bug bites and have reddish wheals as big as a quarter or half dollar, while other people are hardly aware of them. In addition to local lesions, bed bugs can cause nervous disorders and sleeplessness in children and adults. Although bed bugs feed repeatedly on humans, they have not been incriminated in the transmission of any disease.

HUMAN LICE, ORDER ANOPLURA (5, 9, 11, 16, 17, 19, 20, 23–25, 28)

The sucking lice are small, wingless, dorso-ventrally flattened insects belonging to the order Anoplura. They differ from other ectoparasites by their specialized legs, each ending in a single (usually curved) claw. This claw and the opposing thumblike process enable lice to cling to hair or fibers of clothing or bedding. Three types of sucking lice are specific parasites of humans, usually, but not always, confined to a certain part of the body. They are named according to the region of the body that they infest or their general appearance: head louse, body louse, and pubic or crab louse (Fig. 2).

The head louse (*Pediculus humanus capitis*) is 1 to 2 mm long, with three pairs of legs of approximately equal size and an elongate abdomen with slightly darkened margins but without lateral hair tufts. The adults and immatures, called nymphs, are found on the head and neck, particularly behind the ears and on the back of the neck. The eggs, called "nits," are glued to the hairs.

The body louse (*Pediculus humanus humanus*) is usually 2 to 4 mm long with three pairs of legs of approximately equal size and an elongate abdomen with pale margins but without lateral hairy tufts. The adults and immatures are found on the hairy parts of the body below the neck and frequently rest on clothing when they are not feeding. Typically the eggs are laid on clothing, particularly along the seams.

The pubic or crab louse (*Pthirus pubis*) is 0.8 to 1.8 mm long, with the first pair of legs much smaller and more slender than the second and third pairs of legs, and having a short crablike abdomen with lateral hairy tufts. The adults and immatures are typically found on the pubic (hence the species name "pubic louse") and anal areas of the body or on other parts of the body with widely spaced hairs, such as the chest, armpits, mustache, beard, eyebrows, or eyelashes. The eggs are glued to hairs.

These three species can be identified by referring to Fig. 2.

Body lice are the vectors of pathogens that cause epidemic typhus, epidemic relapsing fever, and trench fever. Fortunately, none of these diseases occurs in the United States today. Infestations with body lice are not common in this country, possibly as a result of the

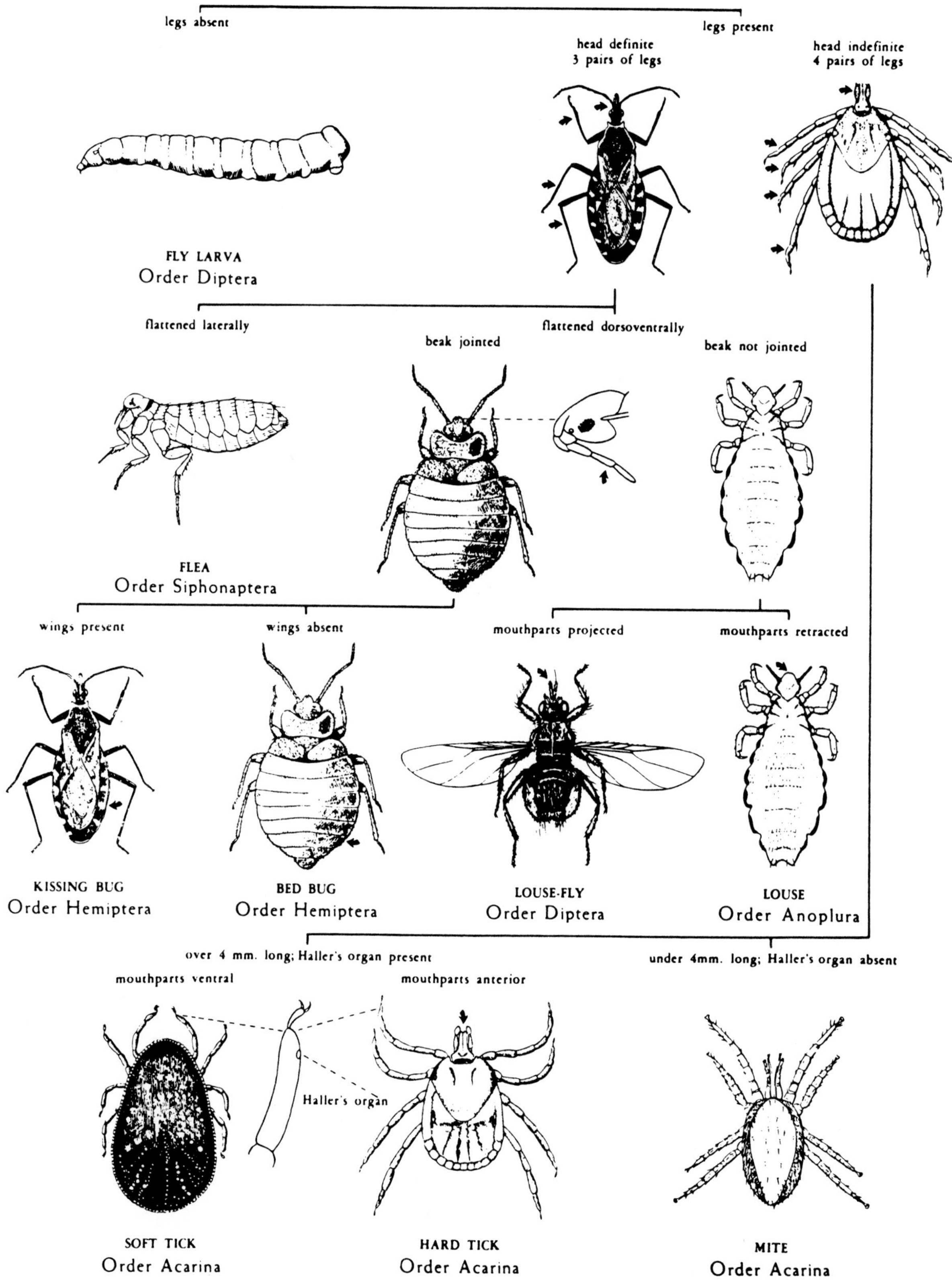

FIG. 1. Pictorial key to common groups of human ectoparasites (C. J. Stojanovich and H. G. Scott, Centers for Disease Control, Atlanta, Ga.).

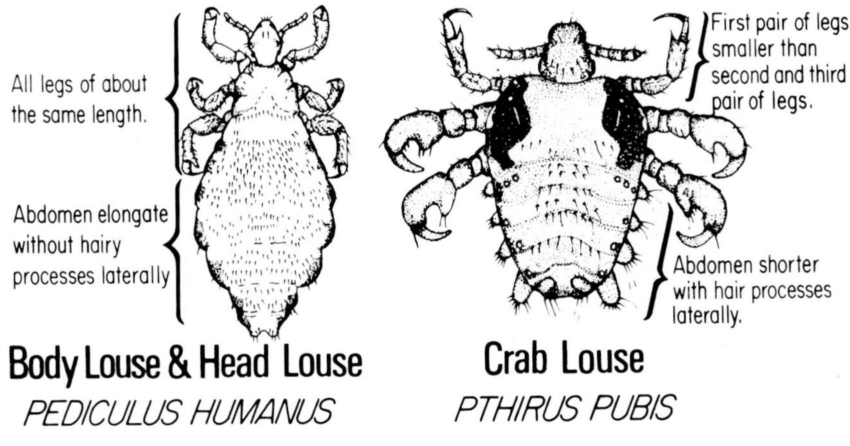

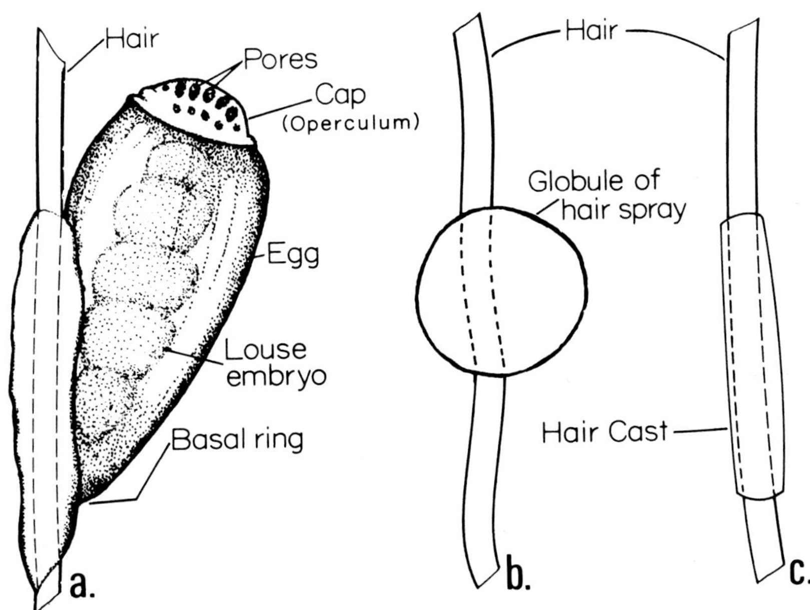

FIG. 2. Identification characteristics of lice and nits commonly found on humans. The upper figures show the important identifying characteristics of adult lice. In the lower figure are shown (a) an egg (nit) and (b) a globule of hair spray and (c) a hair cast, which must be differentiated from the nit (H. D. Pratt and K. S. Littig, Centers for Disease Control, Atlanta, Ga.).

widespread use of automatic washing machines and hot-air dryers to launder clothing regularly. However, infestations of crab lice are reported frequently, possibly as a result of a more permissive society in recent years. Crab lice usually are transmitted from person to person during sexual intercourse; rarely are they acquired from infested toilet seats.

Epidemics of head lice are reported frequently in the United States and in many other parts of the world, particularly among school children who wear infested garments such as caps, scarves, and other clothing left in crowded cloakrooms, use infested combs or brushes, or lie on infested carpets, beds, or other upholstered furniture previously used by an infested person. In examining a person for head lice, check especially the area behind the ears and on the back of the neck looking for adults and nymphs on the scalp and nits on the head hair. Head lice attach their eggs to hairs very close to the scalp, normally within 0.25 in. (ca. 0.6 cm) of the scalp surface. Hatched eggs will be 0.5 in (ca. 1.3 cm) or more from the scalp—the so-called "half-inch rule" for determining whether nits are live or not. The best method for determining whether or not the eggs have hatched is to examine them with a hand lens or under a dissecting microscope. Live eggs have an embryo inside and a cap (or operculum); hatched eggs have no cap. It is important to differentiate nits from hair casts and globs of hair spray (Fig. 2).

FLEAS, ORDER SIPHONAPTERA (5, 9, 11, 12, 20, 24, 28)

Fleas are small, wingless insects with bodies compressed from side to side and long legs adapted for jumping. The name of the flea order, Siphonaptera, refers to their blood-sucking or "siphoning" mouthparts and to their lack of wings. They vary from 1 to 8.5 mm in length, averaging about 2 to 4 mm. Fleas develop through complete metamorphosis with four stages in their life cycle: egg, larva, pupa, and adult. Fleas often cause loss of blood, irritation, and extreme discomfort. Public health workers are concerned with fleas as vectors of plague and flea-borne or murine typhus from rats to humans and as vectors of rural or sylvatic plague among wild rodents and occasionally to humans. In addition, fleas are intermediate hosts for the double-pored dog tapeworm (*Dipylidium caninum*) and two rodent tapeworms (*Hymenolepis diminuta* and *Hymenolepis nana*) that occasionally infect humans (4, 11). In the United States, most problems with fleas occur when persons are bitten by dog or cat fleas. This is most frequent when the preferred host (dog or cat) is no longer available, as when the animal dies or leaves home.

Causative organisms

The important structures used in identifying fleas are shown in Fig. 3. The presence or absence of the genal or pronotal combs, the shape of the head and spermatheca in the female, the length of the labial palpi, and the number and position of bristles on the head, abdomen, and tarsi offer important characters for identification.

The Oriental rat flea (*Xenopsylla cheopis*) is the chief vector of the rickettsiae causing flea-borne or murine typhus from rats to humans. It was the most important species transmitting urban plague in the United States during the period 1900 to 1925. The Oriental rat flea normally parasitizes Norway and roof rats, but bites humans readily. It has neither genal nor pronotal combs, the front margin of the head is rounded and the thorax is of normal length, the mesopleuron is divided by a vertical thickening, the ocular bristle is inserted in front of the eye, and the female has a large, dark, C-shaped spermatheca (Fig. 3).

The cat flea (*Ctenocephalides felis*) and the dog flea (*Ctenocephalides canis*) both have genal and pronotal combs, each with seven or eight pointed black teeth on each side of the body (Fig. 3). The head of the cat flea is about twice as long as it is high and the first two spines of the genal comb are almost the same length, whereas the head of the dog flea is less than twice as long as it is high and the first spine of the genal comb is definitely shorter than the second. It is usually not necessary to distinguish between these two species because they have similar habits and both species attack cats, dogs, rats, and humans.

FLY MAGGOTS, ORDER DIPTERA (4, 9, 11, 15, 20, 26, 28)

Flies belong to the insect order Diptera, or two-winged insects. Flies develop by complete metamorphosis with four stages in their life cycle: egg, larva, pupa, and adult. Fly larvae are often called maggots. Clinical microbiologists may receive specimens of fly larvae submitted from wounds or sinuses, the umbilical area of newborn babies, or stool or urine samples.

A typical muscoid fly larva is legless and somewhat cone-shaped, with a narrow anterior end bearing the mouthparts and anterior spiracles and a broader posterior end with two prominent posterior spiracles. The structure of the mouthparts and the anterior and posterior spiracles provide characters used in identification. Most muscoid fly larvae are rather similar and are identified by characters shown in Fig. 4. Two types easily recognized with the naked eye are the lesser house fly and latrine fly (*Fannia* spp.), which have prominent lateral processes, and the rat-tailed maggot (*Eristalis tenax*), which has a long, telescopic respiratory tube.

Clinical manifestations and epidemiology

Infestation with fly maggots causes a condition known as "myiasis," in which the fly larvae feed on living or dead tissues of humans or on food in the human tract. A number of terms have been used to describe the various types of myiasis, depending on the location of the fly larvae: enteric or gastrointestinal (digestive tract); dermal, subdermal, or cutaneous (skin); auricular (ear); ocular (eye); nasopharyngeal (nose); and urinary or urogenital (urogenital tract).

Fly larvae in the alimentary canal cause nausea, pain in the abdomen, diarrhea, dysentery (with actual discharge of blood as a result of injury to the intestinal mucosa), and nervousness. Fifty species of fly larvae have been reported, either positively or questionably, from cases of enteric myiasis in humans (11). Many of these cases involved flies which lay their eggs or larvae on fish, cold meat, cheese, ripe fruit, and other foods. Normally such eggs or larvae are destroyed by the digestive juices in the human alimentary tract. However, there are apparently reliable records of living larvae found in the stool or vomit or both. These cases include children who drank "dirty water" from a ditch or container containing rat-tailed maggots (*Eristalis*) and people who ate meat or fish containing larvae (*Sarcophaga*).

Laboratory workers should be very careful in reporting enteric myiasis. Stool samples can easily be contaminated in the laboratory, particularly by flesh flies (*Sarcophaga* spp.), which are strongly attracted by the smell of feces. Flesh flies lay larvae rather than eggs, and the first two larval stages are often completed in a day, so that it is possible to find third-stage larvae in stool samples only a day old. In cases of "questionable enteric myiasis," a second stool sample should be passed in a fly-free room and the material should be examined at once.

Although myiasis due to maggots feeding on dead tissue may be found in wounds when dressings or casts are removed, invasive disease with maggots invading living tissue is uncommon in the United States.

SPIDERS, ORDER ARANEIDA (5, 9, 11, 18, 20, 28)

Spiders are readily identified by a number of characters: eight legs, no antennae, body divided into two regions (cephalothorax and abdomen) joined by a narrow pedicel, and unsegmented abdomen with spinnerets at the tip. All spiders kill their prey by biting and injecting venom. Therefore, it is not surprising that there are reports of some 50 species of spiders biting humans, usually in self-defense (9). In most cases the bite is no worse than the sting of a bee

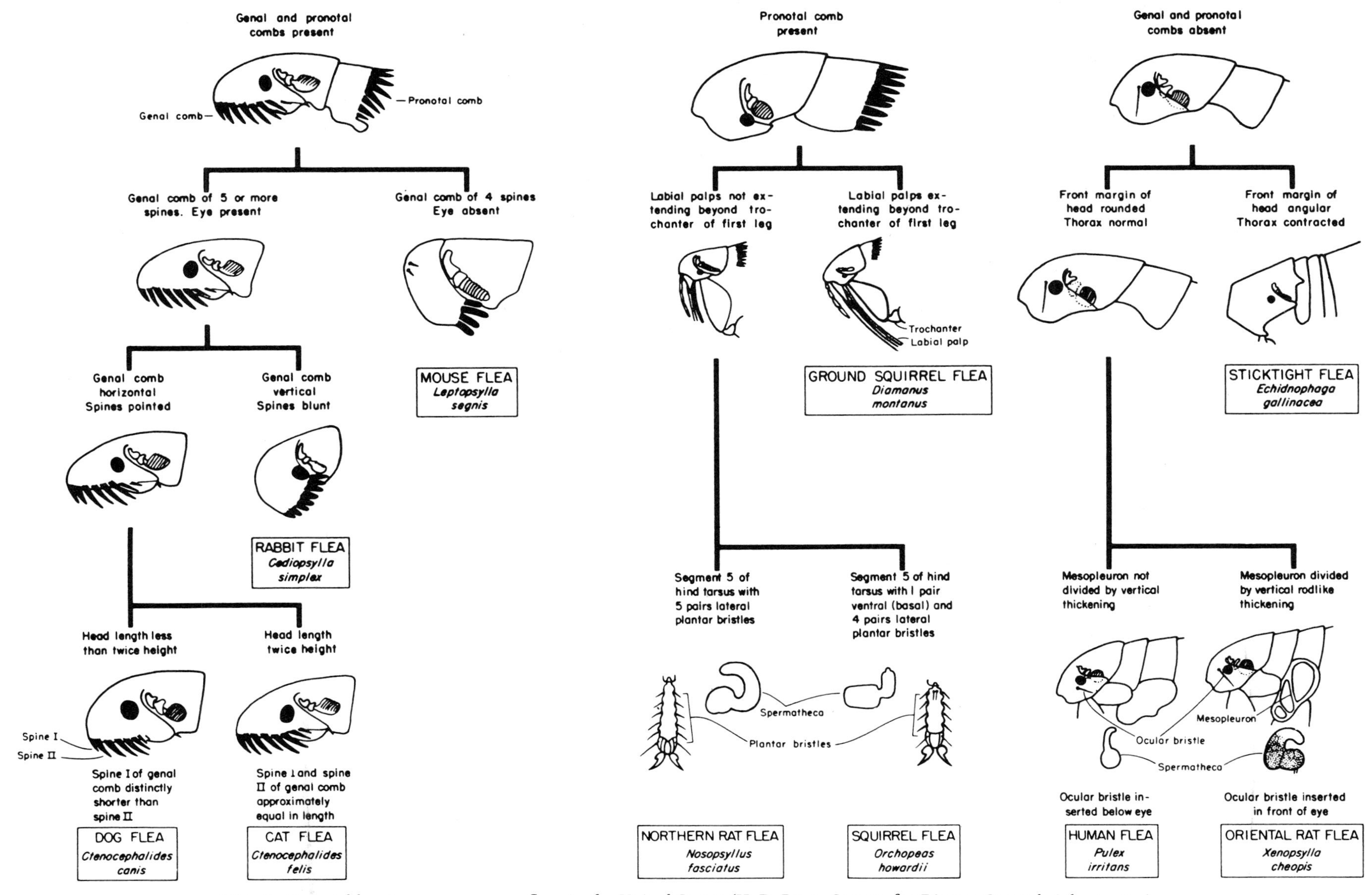

FIG. 3. Pictorial key to some common fleas in the United States (H. D. Pratt, Centers for Disease Control, Atlanta, Ga.).

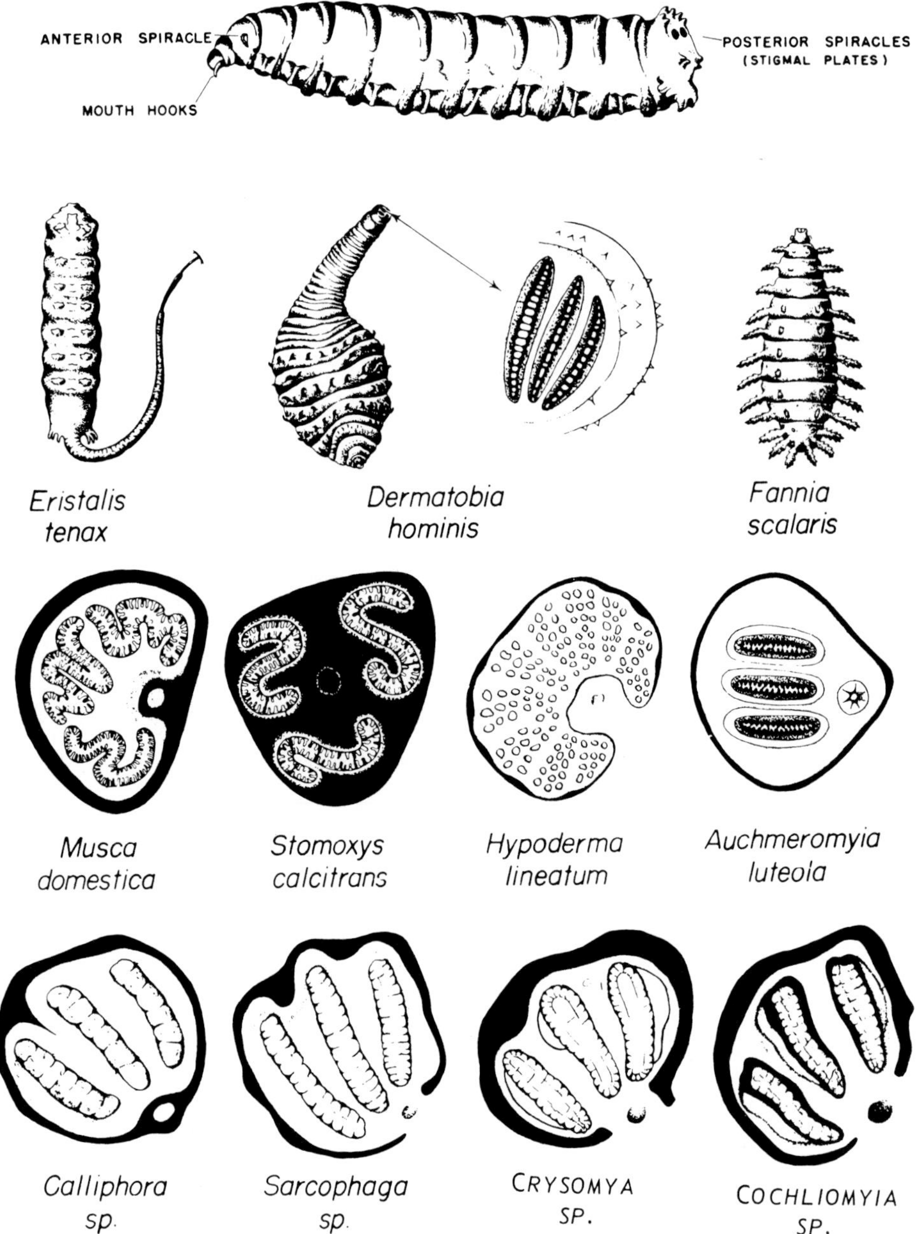

FIG. 4. Key characters of myiasis-producing fly larvae (Centers for Disease Control, Atlanta, Ga.). Top, Mature larva of a muscoid fly (from R. Hegner et al., © 1938 by D. Appleton-Century Co., Inc., New York; used by permission).

or wasp, with some redness, swelling, and pain for a few hours, but it can occasionally be more severe. However, the bite of two spiders, the black widow spider and the brown recluse spider (and their relatives), can cause serious illness or death as discussed below.

The black widow spider (*Latrodectus mactans*) is easily recognized in all stages by the orange or reddish hourglass marking on the ventral side of the abdomen (Fig. 5). Females vary from 5 to 13.5 mm in body length, with a globose, shiny black abdomen with one or more reddish dots on the dorsal side and all-black legs. Males and immatures are smaller than the females and have additional reddish or pale spots on the abdomen and banded legs. The venom injected by the female is a neurotoxin and can cause severe pain within an hour, a "board-like abdomen," severe illness for a day or more, or even death. In the United States, the venom of four other species of *Latrodectus, L. bishopi, L. geometricus, L. hesperus*, and *L. variolus*, also can cause illness.

The brown recluse or violin spider (*Loxosceles reclusa*) is a brownish species with two characteristics in combination: a dark fiddle or violin marking on the tan or brownish cephalothorax, and six eyes arranged in three pairs forming a semicircle (see Fig. 5). The body is 8 to 9 mm long. The venom injected by the female or male is a necrotoxin which causes necrotic lesions with deep tissue damage that extend for several days after the initial bite, heal very slowly (often 6

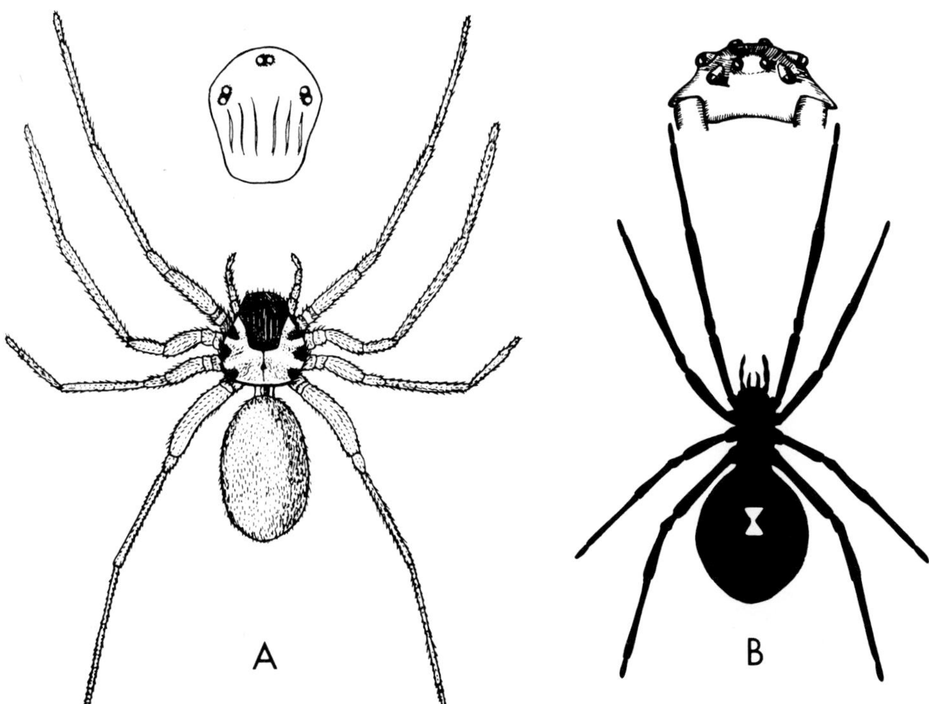

FIG. 5. Key characteristics of *Loxosceles reclusa* (brown recluse spider) and *Latrodectus mactans* (black widow spider). The brown recluse (A) has a fiddle-shaped marking on the cephalothorax and has six eyes in three pairs. The black widow (B) has a red hourglass on the underside of the abdomen and has eight eyes (Centers for Disease Control, Atlanta, Ga.).

to 8 weeks or longer), and leave disfiguring scars. In the United States the venom of three other species of *Loxosceles, L. arizonensis, L. deserta,* and *L. laeta,* can also cause serious illness and leave ugly scars.

TICKS, ORDER ACARINA (2, 5–9, 11, 13, 20, 28)

Ticks are blood-sucking arachnids that are ectoparasites of many vertebrates such as reptiles, birds, mammals, and humans. Ticks have a four-stage life cycle: egg; six-legged larva; eight-legged, sexually immature nymph; and eight-legged, sexually mature adult. Usually a blood meal is necessary for the larva to molt to become a nymph, and another blood meal is needed for the nymph to molt before becoming an adult. Hard ticks have only one nymphal stage, whereas soft ticks may have more than one.

There are two families of ticks (Fig. 6): hard ticks in the family Ixodidae and soft ticks in the family Argasidae. Hard ticks have a hard dorsal plate or scutum, and the mouth parts are clearly visible from above. Soft ticks have a leathery body but no hard plate on the dorsal part of the body, and the mouthparts are located ventrally and are not visible from above. Both hard and soft ticks can bite humans and can cause painful, itching lesions.

Causative organisms

The four most important species of hard ticks in the United States are the American dog tick (*Dermacentor variabilis*), the Rocky Mountain wood tick (*Dermacentor andersoni*), the lone star tick (*Amblyomma americanum*), and the brown dog tick (*Rhipicephalus sanguineus*). In addition, ticks in the genus *Ixodes* are becoming more important. Ticks in the genera *Dermacentor* and *Amblyomma* are often called "ornate" ticks because they have whitish markings on the scutum easily seen with the naked eye, hand lens, or stereoscopic microscope. The ticks in the other genera of North American hard ticks do not have these whitish markings and are called "inornate" ticks.

The American dog tick (*D. variabilis*) is found in most of the eastern United States and in limited areas on the Pacific Coast, in northern Idaho, and in eastern Washington. Small males of these brownish ticks may be only 3 mm long, whereas engorged females may grow from 5 to 13 mm or more in length. As shown in the pictorial key (Fig. 6), the mouthparts (palpi and hypostome) are about as long as the basis capituli, and the sides of the basis capituli are parallel. The scutum has diffuse whitish markings that may be faint or well defined. The fine punctuation, called "goblets," on the spiracular plates on the underside of the abdomen, behind and lateral to the fourth pair of coxae, is finer in *D. variabilis* than in *D. andersoni*. The American dog tick is the important vector of Rocky Mountain spotted fever in the eastern United States. It may also transmit pathogens causing tularemia and Q fever.

The American dog tick, Rocky Mountain wood tick, and lone star tick all have been reported to cause tick paralysis. After these ticks feed for several days, they inject a toxin which causes an ascending flaccid paralysis which can progress to death. The paralysis can be reversed by removal of the tick. Use forceps to remove the tick by a slow steady pull that will not break off the mouthparts and leave them in the wound. Some-

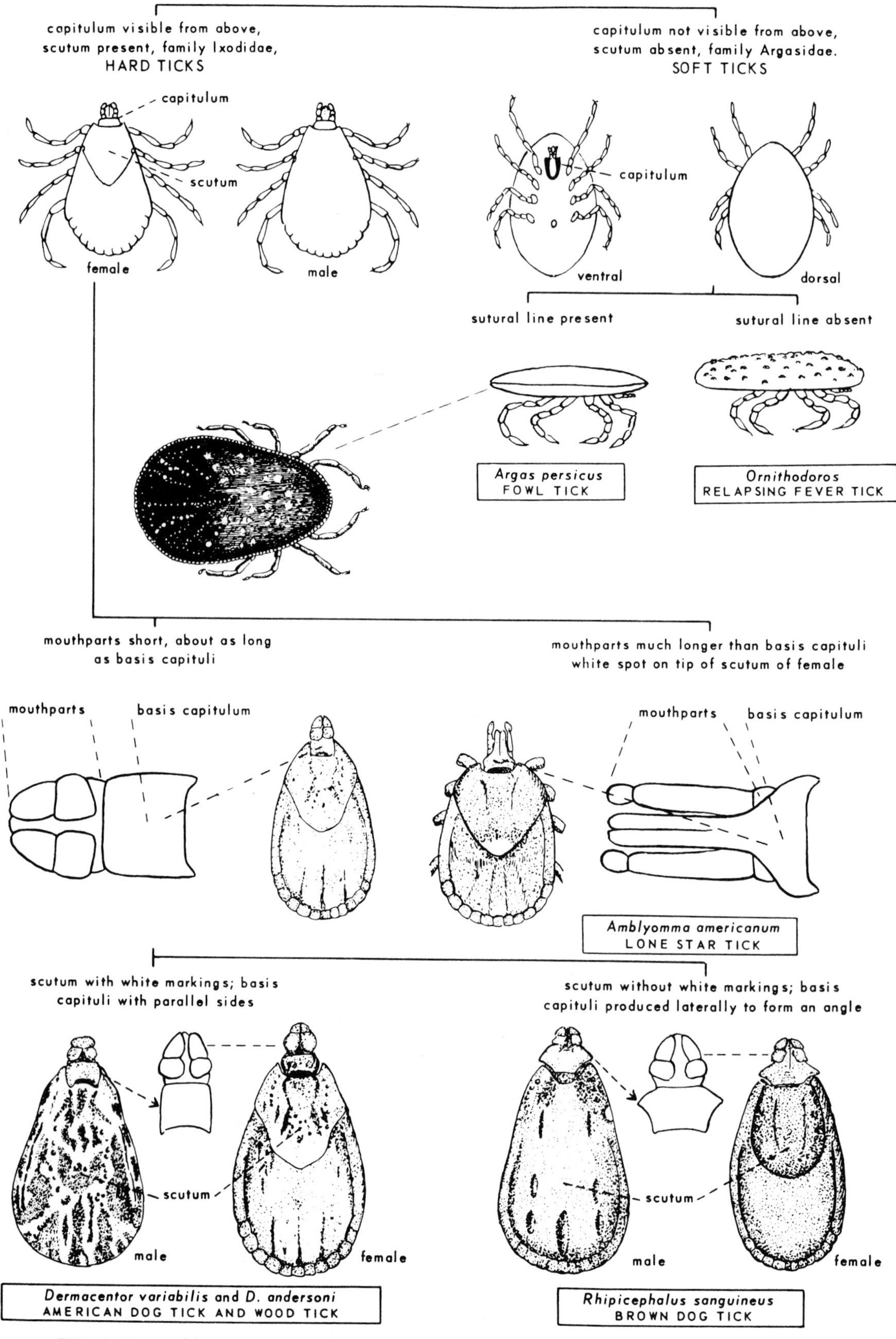
capitulum visible from above,
scutum present, family Ixodidae,
HARD TICKS
capitulum not visible from above,
scutum absent, family Argasidae.
SOFT TICKS
capitulum
scutum
female
male
capitulum
ventral
dorsal
sutural line present
sutural line absent
Argas persicus
FOWL TICK
Ornithodoros
RELAPSING FEVER TICK
mouthparts short, about as long
as basis capituli
mouthparts much longer than basis capituli
white spot on tip of scutum of female
mouthparts
basis capitulum
mouthparts
basis capitulum
Amblyomma americanum
LONE STAR TICK
scutum with white markings; basis
capituli with parallel sides
scutum without white markings; basis
capituli produced laterally to form an angle
scutum
male
female
scutum
male
female
Dermacentor variabilis and D. andersoni
AMERICAN DOG TICK AND WOOD TICK
Rhipicephalus sanguineus
BROWN DOG TICK

FIG. 6. Pictorial key to some common ticks (H. D. Pratt, Centers for Disease Control, Atlanta, Ga.).

times a drop of chloroform, ether, alcohol, Vaseline, or fingernail polish will help make the tick detach its mouthparts. After removal of the tick, the wound should be disinfected.

The Rocky Mountain wood tick (*D. andersoni*) is similar to the American dog tick. However, the whitish markings on the scutum are generally more evident and the "goblets" on the spiracular plate are larger and less numerous than in the American dog tick. *D. andersoni* is the major vector of Rocky Mountain spotted fever and Colorado tick fever in the Rocky Mountain region. It may transmit tularemia, Q fever, and Powassan encephalitis and may cause tick paralysis.

The lone star tick (*A. americanum*) has mouthparts much longer than the basis capituli. The female has a conspicuous whitish marking on the scutum, from which is derived the name "lone star tick," for the Lone Star State of Texas. This tick may serve as a vector for Rocky Mountain spotted fever and tularemia and may cause tick paralysis. Unlike the American dog and Rocky Mountain wood ticks, whose larvae and nymphs do not normally feed on humans, the larvae, nymphs, and adults of the lone star tick all feed on persons. There are many reports of people who were bitten by lone star tick larvae and nymphs and developed severe itching and redness comparable to attacks of chiggers.

The brown dog tick (*R. sanguineus*) rarely bites humans in North America. However, it is commonly found on dogs and in buildings where both dogs and humans live. It is an inornate brown species which varies from 3 mm long when unengorged to 13 mm or more in engorged females. The sides of the basis capituli are angled. It is not uncommon to find larvae, nymphs, and adults of the brown dog tick in homes, kennels, veterinary hospitals, or laboratory animal houses, because all stages obtain their blood meals from dogs.

The northern deer tick (*Ixodes dammini*), in the nymphal stage, occasionally feeds on humans from June to October and may transmit the protozoan parasite, *Babesia microti*, that causes human babesiosis in northeastern United States and Canada and spirochetes (*Borrelia* spp.) that cause Lyme disease (13, 27). Several other species of *Ixodes* are of considerable interest because of their role in the transmission of Powassan encephalitis and other diseases. *Ixodes* ticks are easily recognized because they have the anal groove curved in a U-shape in front of the anus, whereas other genera of hard ticks have the anal groove behind the anus or absent. Species identification of ticks in the genus *Ixodes* is difficult, and the services of a specialist may be required (8).

Soft ticks in the genus *Ornithodoros*, family Argasidae, transmit relapsing fever spirochetes (*Borrelia* spp.) in limited areas in 13 western states. These ticks are dull-colored, leathery species with the mouthparts on the underside of the body, not visible from above. In the United States, at least four species of *Ornithodoros* (*O. hermsi*, *O. parkeri*, *O. talaje*, and *O. turicata*) are proven vectors of *Borrelia* spp., which cause tick-borne relapsing fever. Their identification requires specialized literature (7). Many of these cases of relapsing fever were contracted in rural cabins inhabited by small rodents, such as chipmunks. As they slept, people were bitten by infected relapsing-fever ticks which came from rodent nests in the cabins (11, 13).

MITES (ORDER ACARINA) (1, 2, 5, 9, 16, 20–24, 28)

Mites are tiny arthropods with eight legs in the adult stage, a saclike body, and no antennae. Clinical laboratories may be requested to identify mites which parasitize humans. Because of their small size, mites are generally difficult for the nonspecialist to identify. However, well-mounted specimens of the more common and important parasitic mites may be identified by reference to pictorial keys (Fig. 7) or to illustrations in specialized literature (1, 2). Some of the most important mites are discussed below.

The scabies, itch, or mange mite (*Sarcoptes scabiei*) causes an infectious skin disease called scabies, mange, or "the itch." Diagnosis is made by locating the mite, usually a female with an egg, in a papule or vesicle in the epidermis or in a tiny burrow in the superficial skin. In young children the mites may be found on any part of the body. In adults, up to 80% of the mites occur on some part of the hands or arms, particularly the webbing between the fingers or the folds of the wrists, on the external genitalia of males, and on the breasts of women. Mellanby (21) recommends teasing the mite from its burrow or papule with a needle (or cutting a tiny bit of skin with a razor blade), mounting the specimen on a slide with a cover glass, and then examining it under a compound microscope. Muller et al. (22) prefer scraping the skin with a sharp scalpel with a drop of mineral oil and then examining this liquid with a microscope having a magnification of at least 50 diameters. The mites average 0.3 to 0.4 mm in length, with an oval saclike body in which the first and second pairs of legs are widely separated from the third and fourth pairs (Fig. 7). The mouthparts at the anterior end have chelicerae and palpi, and the anus is at the posterior end. The skin has fine wrinkles on the dorsal side with a number of prominent blunt spines and many backward-projecting triangular points.

The scabies mites cause intense itching and rash, particularly at night. In a previously unexposed individual, a month or more may elapse between initial infestation and onset of symptoms; in reinfested persons, rash and itching may develop quickly after penetration of even a single mite, suggesting that both the itching and rash are due to hypersensitivity to the mite. Although the majority of the papules or burrows with mites are found on the hands or arms, the "scabies rash" has a characteristic distribution in the armpits, waist, buttocks, inner thigh, and ankles, areas that do not correspond with the location of the mites.

Scabies is often considered to be a "family disease," with mites transferred from husband to wife, or from one child to another, by simple contact such as touching or shaking hands. The disease is also acquired through sexual contact. Outbreaks may occur in hospitals or institutions. If one member of a family or group of persons has scabies, it is advisable to treat all close contacts (16, 17, 21).

The follicle mite (*Demodex folliculorum*) is a microscopic, cigar-shaped creature with eight stumpy legs

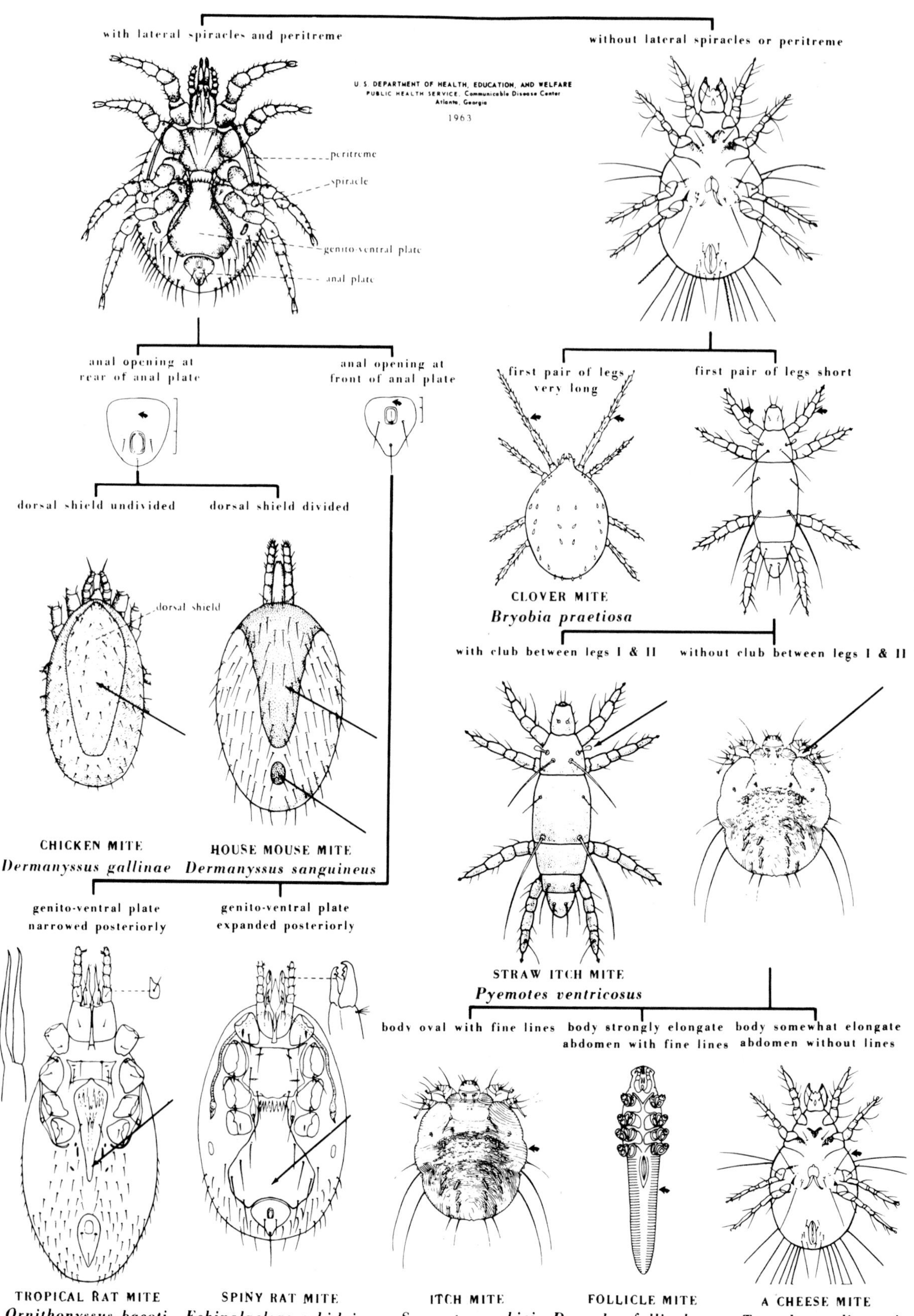

FIG. 7. Pictorial key to some mites of public health importance (H. G. Scott and C. J. Stojanovich, Centers for Disease Control, Atlanta, Ga.).

and an annulate abdomen. They are found in hair follicles and sebaceous glands, particularly on the nose and face, and probably infest over half of middle-aged adults. The infestation is usually asymptomatic but may be associated with blackheads. The mites may be squeezed out of hair follicles, often by pressing a slide against the nose, or they may be incidental findings in histological sections of facial skin.

Chiggers are larvae of mites in the family Trombiculidae. In Asia, these mites are vectors of scrub typhus. In other parts of the world, trombiculid mites infest grass and bushes, and their six-legged larvae, i.e., chiggers (red bugs, harvest mites), may attack humans. Mite larvae appear as tiny reddish dots attached to skin, usually in areas where clothing is tight such as the ankles at the top of the socks, or skin touched by belts or elastic bands. Sensitive individuals react to the secretions of the larvae with swollen itching areas at the sites of attachment which persist for days. Scratching may lead to secondary infections. The six-legged larvae have characteristic branched, feather-like hairs. The eight-legged nymphs and adults are nonparasitic vegetarians.

The straw or hay itch mite (*Pyemotes ventricosus*) normally is a parasite of insect larvae which infest straw, hay, grains, wood, and other plant products. These mites may attack people who handle these products, such as people sleeping on hay or crawling under houses infested with wood borers, and may cause severe itching and rash. Adult mites have a characteristic club-shaped hair between the first and second pairs of legs. In gravid females, the abdomen becomes enormously enlarged, resembling a small pearl.

Mites of the family Dermanyssidae are normally parasites of rodents or birds but may attack people, causing dermatitis. People raising domestic fowl may suffer from attacks of the chicken mite (*Dermanyssus gallinae*), northern fowl mite (*Ornithonyssus sylviarum*), or tropical bird mite (*Ornithonyssus bursa*). These same species are often parasites of birds nesting in buildings, such as pigeons, starlings, or sparrows, and may leave the bird nest and attack people.

The tropical rat mite (*Ornithonyssus bacoti*) is a common parasite of rats, particularly in the southern United States. There are many reports of this species attacking humans when rats are killed in buildings. The mites leave the dead animal and seek a blood meal from the nearest warm-blooded animals, people.

The house mouse mite (*Liponyssoides sanguineus*), often reported as belonging to the genera *Allodermanyssus* or *Dermanyssus*, is a parasite of house mice. It is sometimes abundant in buildings in the northeastern United States and may transmit the rickettsiae causing rickettsialpox from mice to people.

Grain and cheese mites are frequently found in tremendous numbers in flour, grain, cheese, and dried fruits, particularly when humidity and temperatures are high. Some of these mites may cause dermatitis among persons who handle these infested foods.

In recent years there have been many reports of allergic reactions to mites (entire or in fragments) and their excreta that produce conditions similar to asthma in some persons. The chief offenders seem to be the house dust mites (*Dermatophagoides farinae* and *Dermatophagoides pteronyssoides*), which occur in great numbers in dust from mattresses and bedroom floors and in smaller numbers in other parts of houses.

LITERATURE CITED

1. **Baker, E. W., T. M. Evans, D. J. Gould, W. B. Mull, and H. L. Keegan.** 1956. A manual of parasitic mites of medical or economic importance. National Pest Control Association, Inc., New York.
2. **Baker, E. W., and G. W. Wharton.** 1952. An introduction to acarology. Macmillan Publishing Co., New York.
3. **Borror, D. J., D. M. DeLong, and C. A. Triplehorn.** 1981. An introduction to the study of insects, 5th ed. Saunders College Publishing Co., Philadelphia.
4. **Brown, M. D., and F. A. Neva.** 1983. Basic clinical parasitology, 5th ed. Appleton-Century-Crofts, Norwalk, Conn.
5. **Center for Disease Control.** 1967. Pictorial keys: arthropods, reptiles, birds, and mammals of public health importance. Center for Disease Control, Atlanta, Ga.
6. **Cooley, R. A.** 1938. The genera *Dermacentor* and *Otocentor* (Ixodidae) in the United States with studies in variation. Natl. Inst. Health Bull. no. 171.
7. **Cooley, R. A., and G. M. Kohls.** 1944. The Argasidae of North America, Central America, and Cuba. Am. Midl. Nat. Monogr. 1.
8. **Cooley, R. A., and G. M. Kohls.** 1945. The genus *Ixodes* in North America. Natl. Inst. Health Bull. no. 184.
9. **Ebeling, W.** 1975. Urban entomology. Division of Agriculture Sciences, University of California, Richmond.
10. **Frazier, C. A., and F. K. Brown.** 1980. Insects and allergy and what to do about them. University of Oklahoma Press, Norman.
11. **Harwood, R. E., and M. T. James.** 1979. Entomology in human and animal health, 7th ed. Macmillan Publishing Co., New York.
12. **Holland, G. P.** 1949. The Siphonaptera of Canada. Canadian Department of Agriculture Technical Bulletin no. 70.
13. **Hoogstraal, H.** 1981. Changing patterns of tickborne diseases in modern society. Annu. Rev. Entomol. **26:**75–99.
14. **Hunter, G. W., III, J. C. Schwartzwelder, and D. F. Clyde.** 1976. Tropical medicine, 5th ed. W. B. Saunders, Philadelphia.
15. **James, M. T.** 1948. The flies that cause myiasis. USDA Misc. Pub. 631.
16. **Juranek, D. D.** 1976. The nuisance diseases: pediculosis and scabies. Association for Practitioners in Infection Control **4:**1–5.
17. **Juranek, D. D.** 1984. *Pediculus capitis* in school children: epidemiologic trends, risk factors, and recommendations for control. *In* M. Orkin and M. L. Maibach (ed.), Cutaneous infestations and insect bites. Marcel Dekker, Inc., New York.
18. **Kaston, B. J.** 1978. How to know the spiders, 3rd ed. W. C. Brown Co., Dubuque, Iowa.
19. **Kim, K. C., and H. L. Ludwig.** 1978. The family classification of the Anoplura. Syst. Entomol. **3:**249–284.
20. **Mallis, A. (ed.).** 1982. Handbook of pest control, 7th ed. Franzak and Foster Co., Cleveland.
21. **Mellanby, K.** 1972. Scabies. E. W. Classey, Ltd., Hampton, Middlesex, England.
22. **Muller, G., P. H. Jacobs, and N. E. Moore.** 1973. Scraping for human scabies. Arch. Dermatol. **107:**70.
23. **Orkin, M., H. L. Maibach, L. C. Parish, and R. M. Schwartzman.** 1977. Scabies and pediculosis. J. B. Lippincott, Philadelphia.
24. **Parish, L. C., W. B. Nutting, and R. M. Schwartzman.** 1983. Cutaneous infestation of man and animal. Praeger, New York.
25. **Pratt, H. D., and K. S. Littig.** 1973. Lice of public health

importance and their control. Center for Disease Control, Atlanta, Ga.

26. **Smith, F. G. V.** 1975. Insects and other arthropods of medical importance. British Museum (Natural History), London.

27. **Spielman, A., C. M. Clifford, J. Piesman, and M. D. Corwin.** 1979. Human babesiosis on Nantucket Island, USA: description of the vector, *Ixodes* (*Ixodes*) *dammini* n. sp. (Acarina, Ixodidae). J. Med. Entomol. **15:**218–234

28. **Truman, L. C., G. W. Bennett, and W. L. Butts.** 1976. Scientific guide to pest control operations. Harvest Publishing Co., Cleveland.

Chapter 61

Section VIII. Viruses, Rickettsiae, and Chlamydiae

Collection and Preparation of Specimens for Virological Examination

DAVID A. LENNETTE

The laboratory diagnosis of viral infections is based upon three general approaches: (i) the direct detection of viral antigens or structures, either in cells derived from infected tissues or free in fluid specimens; (ii) isolation and identification of viruses, usually accomplished in cell cultures; and (iii) demonstration of a significant increase in serum antibodies to an etiologically plausible virus during the course of an illness.

Specimens for virus isolation and direct detection, as well as acute-phase blood samples, must be collected within the first few days of an illness if adequate sensitivity of testing is to be expected. A brief clinical résumé should accompany the initial specimen from each patient, specifying date of onset, type of disease suspected, and the major clinical findings. The optimum time for collection of specimens is shown in Fig. 1, and the type of specimen to be collected for virus isolation in various forms of viral infection is shown in Table 1.

Additional information may be obtained from a number of books on viral and rickettsial diseases (1, 3, 7–9, 12). Details on collecting and processing highly infectious agents (e.g., Lassa virus, Marburg virus, Ebola virus, etc.) are outlined in Chapter 79.

SPECIMENS FOR VIRUS ISOLATION ATTEMPTS

Collect specimens promptly, preferably within 3 days and not longer than 7 days after the onset of illness. Collect postmortem specimens as soon as possible after death, using aseptic techniques. Label each specimen container with the patient's name, the site of the specimen, and the date of collection. Specimens which must be held for long intervals before testing should be promptly frozen to −70°C or below. Otherwise, specimens should be refrigerated promptly after collection. Most viruses are better recovered from specimens held at 2 to 6°C for up to several days before testing than from specimens that have been frozen, with few exceptions. Do not freeze specimens at −20°C, as the infectivity of many viruses is rapidly lost at this temperature (2). Fluid specimens (urine, cerebrospinal fluid [CSF]) do not require any transport medium and should not be diluted. Suitable holding media for use with swabs and washings are shown in Table 2; others are available in commercial formulations. There is no evidence that any one medium is greatly advantageous over the others. Although any type of swab may be used satisfactorily with most specimens, calcium alginate fiber tips may inactivate herpes simplex virus and chlamydiae (2, 11) and probably should be avoided when possible. Some laboratories prefer metal or plastic swab shafts over wooden shafts.

Nasal and pharyngeal swabs

A dry swab (either cotton or synthetic fiber) may be used to swab each nostril, and the swab should be allowed to remain in the nose for a few seconds to absorb secretions. Throat swabs are best collected by rubbing the tonsils and posterior pharynx with a cotton or synthetic fiber swab, either dry or wetted with viral transport medium.

Both nasal and pharyngeal swabs should be broken off just above the tip into screw-cap vials containing a few milliliters of an appropriate transport medium (see Table 2).

Nasal washings

Nasal washings can be obtained by instilling several milliliters of sterile saline into each nostril while the patient's head is tilted back slightly; the head is then brought forward and the saline is allowed to flow into a small container held beneath the nose. In infants, a small catheter with a suction trap may be employed. Gelatin or bovine serum albumin (1%) may be added to the washing to stabilize any virus that may be recovered.

Throat washings

Adult patients should gargle with the smallest convenient volume (10 to 20 ml) of cell culture medium or general purpose bacteriological broth and then expectorate into a paper cup. The cup contents are then poured into a screw-cap vial. Pediatric patients may collect a specimen in the same manner, if able to cooperate; otherwise throat swabs will suffice. Throat washings may give a somewhat higher yield of virus than swabs, but are not as convenient to collect.

Oral swabs

Swabs may be collected from oral lesions by rubbing a dry cotton swab over the lesions and transferring the swab immediately to a vial of virus transport medium.

Eye swabs

If any exudate or pus is present in the eye, it should first be removed with a sterile swab. Then a second swab, moistened with transport medium or saline, should be used to rub the affected conjunctiva. The swab tip should be immediately clipped off into a vial of transport medium to retain any cells trapped in the

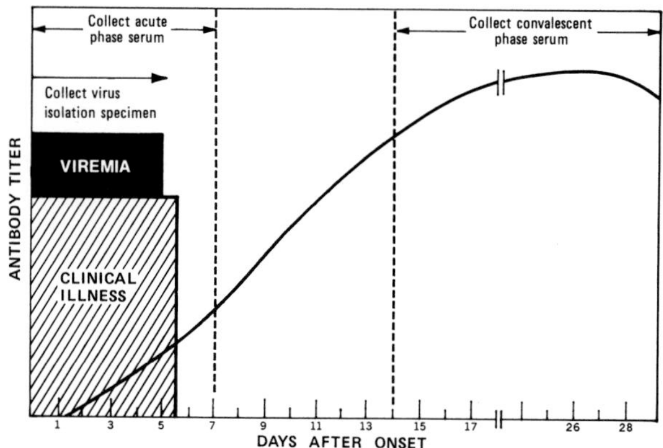

FIG. 1. Optimal time for collecting specimens with regard to illness. Modified from reference 12.

fibers. Corneal specimens should be collected by an ophthalmologist or other adequately trained physician, using a spatula.

Cervical swabs

If more than one swab is used to obtain a cervical specimen, more infected cells will be recovered and better results may be obtained. One swab is used first to clean the cervix of mucus and is discarded; another swab is then inserted about 1 cm into the cervical canal and rotated. If any lesions are seen, they should be swabbed, and the swab then should be removed to a vial of transport medium. One study in which four swabs were collected showed improved recovery of chlamydiae from the last swabs collected (4). Some physicians have even resorted to using scrapers to obtain increased numbers of cells, but this procedure

TABLE 1. Specimens for virus isolation and types of serological tests employed for diagnosis

Clinical manifestations and common etiological agents	Source of specimen for virus isolation		Serological tests[a]	
	Clinical	Postmortem	Usual	(Special)
Upper respiratory tract infections				
Rhinovirus	Throat swab or nasal secretions	—	NA	(Nt)
Mycoplasma			CF	
Parainfluenza			CF, HI	
Epstein-Barr virus			FA	
Adenovirus	Throat swab and feces	—	CF, HI, Nt	(Nt, HI)
Enterovirus			NA	
Reovirus			HI, Nt	
Lower respiratory tract infections				
Influenza	Throat swab and sputum	Lung, bronchus, trachea	CF, HI	
Adenovirus			CF, HI, Nt	
Parainfluenza			CF	
Mycoplasma			CF	
Chlamydiae			CF, FA	
Pleurodynia				
Coxsackievirus	Feces and throat swab	—	NA	(Nt)
Cutaneous and mucous membrane diseases				
Vesicular				
Small pox and vaccinia	Vesicle fluid and scrapings	Lung, liver, spleen, brain	CF, HI	(FA)
Herpes simplex			CF, Nt, ELISA, FA	
Varicella-zoster			CF, ELISA, FA	
Enterovirus	Vesicle fluid, feces, and throat swab	—	NA	(Nt, HI)
Exanthematous				
Measles	Throat swab	—	CF, HI	(Nt, FA)
Rubella	Throat swab	—	HI, ELISA	(Nt)
Enterovirus	Feces and throat swab	—	NA	(Nt, HI)
Central nervous system infections				
Enterovirus	Feces and CSF	Brain, tissue, intestinal contents	NA	(Nt, HI)
Herpes simplex	Brain biopsy and CSF	Brain tissue	CF, ELISA	(Nt, FA)
Mumps	Throat swab, CSF, and urine		CF, HI	(Nt)
Lymphocytic choriomeningitis	Blood and CSF	Brain tissue	CF	(FA)
Arbovirus				
Western equine encephalitis	Blood and CSF	Brain tissue	CF, HI	(Nt)
Eastern equine encephalitis			CF	(Nt)
Venezuelan equine encephalitis			CF	(Nt)

Continued

TABLE 1. Continued

Clinical manifestations and common etiological agents	Source of specimen for virus isolation		Serological tests[a]	
	Clinical	Postmortem	Usual	(Special)
Arbovirus continued				
California encephalitis	Usually not possible to isolate virus from clinical specimens	Brain tissue	CF, CIEP	(Nt)
St. Louis encephalitis			CF, HI	(Nt)
Japanese B encephalitis			CF, IAHA	(Nt)
Rabies	Saliva	Brain tissue	Nt, FA	
Parotitis				
Mumps	Throat swab and urine	—	CF, HI	
Parainfluenza	Throat swab		CF, HI	
Severe undifferentiated febrile illnesses				
Colorado tick fever	Blood	Liver, spleen, lung, brain	CF	
Yellow fever			CF	
Dengue			CF	
Congenital anomalies				
Cytomegalovirus	Urine and throat swab	Kidney, lung, other tissues	CF, ELISA	(FA)
Rubella	Throat swab, CSF, and urine	Lymph nodes, lung, spleen, other tissues	HI, ELISA	(Nt, FA)
Hepatitis				
Virus B (HB Ag)	Agent not recoverable	Agent not recoverable	ELISA, RIA	
Virus A (HAV)				
Enteritis				
Rotavirus	Agent not recoverable	Agent not recoverable	ELISA, RIA	(IEM)
Norwalk agent				(RIA, IEM)
Hemorrhagic fevers				
Lassa	Blood, urine, and throat swab	Liver	FA	
Machupo				(Nt, CF)
Junin				(Nt, CF)
Marburg[b]				(Nt, FA)
Ebola[b]				(Nt, FA)
Hantaan				(Nt, FA)

[a] Usual indicates types of serological tests commonly performed; (Special) indicates serological tests which may be used for special studies, not feasible for routine diagnosis. NA, Tests either not available or generally not feasible as routine diagnostic procedure; Nt, neutralization; CF, complement fixation; HI, hemagglutination inhibition; FA, fluorescent antibody; ELISA, enzyme-linked immunosorbent assay; CIEP, counterimmunoelectrophoresis; RIA, radioimmunoassay; IEM, immunoelectron microscopy; IAHA, immune adherence hemagglutination.

[b] Occasional isolations reported from seminal fluid.

TABLE 2. Holding media for specimens for virus isolation

Type of specimen[a]	Method of medium prepn
Swabs, exudates, cellular scrapings, or washings, e.g.: Nasal Nasopharyngeal Throat Mouth Skin or mucous membrane lesions Eye exudate Rectal	*Buffered tryptose phosphate broth with gelatin* Tryptose phosphate broth .... 2.95 g $Na_2HPO_4 \cdot 7H_2O$ .... 2.06 g $NaH_2PO_4 \cdot H_2O$ .... 0.08 g Gelatin .... 0.5 g Phenol red (optional) .... 0.002 g Distilled water .... 100.00 ml Dispense in 3.5-ml amounts in screw-cap vials. Autoclave at 15 lb/in^2 for 15 min.
	Hanks balanced salt solution with gelatin Hanks balanced salt solution .... 100.00 ml Gelatin .... 0.5 g Dispense in 3.5-ml amounts in screw-cap vials. Autoclave at 15 lb/in^2 for 15 min; pH approximately 7.0.

[a] Holding media are not used for feces (other than swab), CSF, or organ samples such as brain, lung, kidney, etc. These should be refrigerated or frozen without preservative or holding medium; see instructions for each specimen.

has not been demonstrated to be necessary to obtain satisfactory results.

Vesicle fluids and skin scrapings

Collect specimens of vesicle fluids and cellular material from the base of lesions during the first 3 days of an eruption, as the recovery rate from specimens collected later drops sharply. Prior preparation of the site with disinfectants (e.g., alcohol or iodophors) may inactivate the viruses; if possible, it is preferable to use local disinfection after specimens have been collected. In the case of primary infections with herpes simplex virus, however, the virus may be recovered for up to 7 to 10 days after onset. Aspirate vesicle fluids with a 26- or 27-gauge needle attached to a tuberculin syringe or with a capillary pipette. The fluids obtained with either method should be rinsed promptly into a small volume of transport medium to prevent loss of the specimen by clotting. Swab open lesions to obtain both fluid and cells from the lesion base. Immediately clip off the swab tip into a vial of transport medium to retain any cells trapped in the fibers.

Stools and rectal swabs

A suitable stool sample is obtained by transferring a small (1- to 4-g) portion of stool (either formed or liquid) into a small leakproof container, such as a 1-oz. (ca. 29.6-ml) screw-cap jar. Cardboard or waxed containers are unsuitable, as they are not leakproof and allow desiccation of the sample. No transport medium is required. A rectal swab should not be regarded as an expedient substitute for a stool specimen, but rather as a specimen appropriate for the recovery of agents which cause proctitis. A dry swab should be inserted 3 to 5 cm past the anal sphincter, rotated, and then withdrawn. The swab should immediately be placed in a vial of transport medium and refrigerated. Rectal swabs are generally inadequate specimens for the detection of rotavirus or the cytotoxin produced by *Clostridium difficile* (often assayed in the virus laboratory with available cell cultures).

Urine

Clean-voided specimens collected in conventional containers are quite satisfactory for isolation of viruses; special collection methods are not required. Provided that the specimen is refrigerated at 2 to 6°C soon after collection, even viruses often regarded as "labile," e.g., cytomegalovirus, may be recovered for from several days to as much as a week after collection. Addition of antibiotics to the specimen may be useful in suppressing bacterial overgrowth, but this should not be required if the specimen is kept cold. Urine specimens are not adequate substitutes for urethral swabs for the recovery of chlamydiae. Recovery of cytomegalovirus is improved by processing several specimens when possible, as shedding may be intermittent.

CSF

Because the concentration of infectious virus is seldom very high in CSF, it is important to obtain an adequate sample volume. It is desirable to obtain at least 2 ml for virological work, collected in a sterile screw-cap tube or vial. Samples of at least 1 ml in volume should be obtained from infants; volumes of less than 0.5 ml are of less value, considering the low recovery rate to be expected. The specimen should not be diluted in any manner and should be refrigerated as soon as possible until processed by the laboratory. If the specimen cannot be processed within 24 h, the specimen may be frozen to below −70°C to preserve the infectivity of any virus that is present; the specimen should not be frozen at −20°C, as many viruses lose infectivity rapidly at this temperature.

Serum and blood

Serum is rarely used for the recovery of viruses; it is, however, reported to be a suitable specimen for isolation of enteroviruses from infected infants (11a). The buffy coat cells from heparinized blood are also occasionally useful for detection of viremia, primarily for patients with cytomegalovirus infections (6). Blood clots are used for the recovery of Colorado tick fever virus (an agent found in erythrocytes); the virus may also be detected by immunofluorescent staining of blood smears (5).

Autopsy and biopsy specimens

Collect fresh tissue from any affected site or obvious lesion, using separate sterile instruments for each site sampled. Autopsy samples need not be larger than 1 or 2 g. Each specimen should be placed in a separate sterile container and clearly labeled. Frequently sampled tissues for cases of suspected viral etiology include brain, lung, heart muscle, lymph node, and kidney. Liver tissue is often collected, but is frequently toxic to cell cultures; tracheal/bronchial tissue is often overlooked, but is often superior to lung tissue for recovery of respiratory viruses. Samples should be kept refrigerated in a small volume of viral transport medium or saline, but should not be fixed or placed in any sort of preservative solution. This renders them useless for virus isolation and often for immunofluorescent staining tests as well. If the specimens cannot be processed within 1 or 2 days, it may be preferable to freeze them to −70°C or below.

BLOOD SPECIMENS FOR SEROLOGICAL TESTS

Blood specimens are usually collected to obtain serum for serological tests to measure antibodies. Only rarely are they useful for virus isolation. Acute- and convalescent-phase sera must be tested together to determine that antibodies have appeared or increased in titer during the course of the illness. Collect an acute-phase specimen as soon as possible, not later than 5 to 7 days after onset of the illness. Collect a convalescent-phase specimen 14 to 21 days after onset, or 7 to 14 days after the acute-phase specimen. Useful results may sometimes be obtained by testing a single serum specimen. Circumstances vary for different agents according to the test methods available, so that the laboratory should be consulted for advice and limitations of such testing.

Blood specimens should be collected without anticoagulants or preservatives, which may affect the results of serological tests. The usual volume of blood collected is 8 to 10 ml, although 3- to 4-ml specimens (normally collected from pediatric patients) usually provide enough serum to complete all necessary tests. Allow the specimen to clot at room temperature, and then separate the serum by centrifugation and remove it to a separate vial. Serum should not be shipped in its collection tube to a remote laboratory, as the clot tends to disintegrate and hemolyse during transit. Many laboratories are willing to do limited testing with blood samples obtained with capillary collection techniques, which yield less than 0.5 ml of whole blood—very helpful with newborn infants. The serum may be stored at 4 to 6°C for up to several weeks, pending the completion of tests. For longer storage, serum is usually frozen to −20°C or below. Do not freeze whole blood; this causes severe hemolysis and may render the specimen unusable for serological testing. Either submit each specimen when collected, carefully identifying the specimen as "acute" or "convalescent," or store the acute-phase specimen to be submitted for testing with the convalescent-phase specimen. If the date of onset of a patient's illness is obtained, it will usually be evident whether a serum is acute or convalescent phase, even if not otherwise indicated. Paired acute- and convalescent-phase sera from a patient should always be tested simultaneously in one laboratory, as results obtained from two laboratories cannot be accurately compared for changes in antibody titer. If the specimen is a random sample for determination of immunity, it should be identified as "for immunity status."

PROCESSING OF SPECIMENS FOR VIRUS ISOLATION

Throat, nasal, oral, and genital swabs

Frozen specimens may be thawed in a beaker of cold tap water. Dislodge any cells that are trapped in swab fibers by agitating the specimen vial on a Vortex mixer for a few seconds or by vigorous shaking. Add antimicrobial agents to the specimen fluid and mix thoroughly. After part of the specimen has been used for inoculating hosts, store the remaining material at or below −70°C for use in any later studies. The antibiotics that have been most widely used for specimen treatment are mixtures of penicillin and streptomycin with nystatin or amphotericin B (Fungizone); more recent experience indicates that a combination of gentamicin (GRS) and amphotericin B is quite effective. Gentamicin may be preferred for its greater stability and broad spectrum against common specimen contaminants. Gentamicin sulfate is sometimes incorporated into the viral transport medium at concentrations of 10 to 50 μg/ml.

Throat and nasal washes

The antimicrobial treatment described above for swabs is also used for throat and nasal washes. If much particulate matter or mucus is present, transfer the specimen to a centrifuge tube containing a few glass beads and shake or mix it vigorously. Then centrifuge the tube for 30 min at 3,000 × *g* in a refrigerated centrifuge. If a high-speed centrifuge capable of 8,000 × *g* or greater is available, centrifugation times may be reduced in proportion to the force used. For centrifugation times of 10 min or less, refrigeration of the specimen tube may not be required. Remove as much of the supernatant fluid for use as is feasible without disturbing the pellet and beads. Add antibiotics to the fluid after the centrifugation step.

Rectal swabs and stools

Rectal swabs are processed in the same manner as throat and nasal washes, described above. Stool specimens are usually preferable to rectal swabs, as the swabs are rarely adequate for some tests commonly performed with stools, such as detection of rotaviruses or *C. difficile* cytotoxin. About 1 g of stool is added to a screw-cap centrifuge tube containing a few glass beads and diluted to a consistency suitable for extraction by the addition of 5 to 10 ml of a buffered diluent such as Hanks balanced salt solution or phosphate-buffered saline with added antibiotics. After the tube contents are thoroughly dispersed by violent agitation, the tube is centrifuged as described above for throat and nasal washes.

Body fluids

Most body fluids, such as CSF and pleural fluid, need not be treated with antibiotics before inoculation unless bacterial contamination of the specimen is suspected, in which case the specimen should be treated as described for throat swabs. Because routinely collected urine specimens contain bacteria, urine specimens not collected by catheter should be treated with antibiotics. Excessively acidic urine samples may be partially neutralized by addition of a few drops of sodium bicarbonate solution (7.5%).

Autopsy and biopsy specimens

For larger pieces of tissue (several grams), extracts may be prepared from tissue suspensions. The suspensions can be made by use of a mortar and pestle or a tissue homogenizer. First, place the specimen in a sterile petri dish and mince it into small pieces, using sterile instruments (forceps, scalpel, and scissors). The minced tissue may then be transferred to a mortar containing a little sterile sand and ground to a paste with a minimum volume of a diluent such as Hanks balanced salt solution with antibiotics. Gradually add enough diluent to produce a homogeneous suspension of approximately 10% (wt/vol), then pipette the suspension into a screw-capped centrifuge tube and centrifuge it at 1,000 × *g* for 30 min at 4°C. Remove the supernatant extract and dispense it into sterile vials for inoculation of host systems and storage at −70°C.

For small biopsy specimens, the tissue may simply be minced into fragments with dimensions of <0.5 mm, using sterile scalpel and scissors. Place the tissue in a sterile petri dish with enough cell culture medium to prevent the specimen from drying. Using the minced tissue to initiate a cell culture (explant) or cocultivating it with various cultured cell lines may permit more efficient recovery of any virus present in the sample than would the inoculation of an extract of homogenized tissue.

Specimens for direct detection of virus and viral antigens

Many common viruses may be detected by immunofluorescent or immunoenzyme staining of an appropriately prepared specimen. The more commonly used immunofluorescence procedures are rapid and quite reliable when good-quality reagents are available; their use should be considered if the number of agents of interest is small and a suitable specimen (one containing infected cells) is available. For some agents, direct detection is equal to isolation in sensitivity (10). Selection of an appropriate specimen is often the most important means of obtaining good results with direct detection methods, as often only a limited amount of a sample can be examined. At present, immunofluorescence tests are most useful for the detection of respiratory infections and infections causing lesions of the skin or mucous membranes (5, 10), but there are a number of cases where rapid direct detection of other infections has already been shown to be practical, and wider use of these methods should be expected in the future.

All specimens for immunofluorescent staining consist of cells deposited on some type of microscope slide; plain, precleaned 1-mm-thick glass slides are suitable for any type of specimen. For examination of cell suspensions, some laboratories prefer slides which have a printed overlay with wells for the cell spots. For immunofluorescence work, all slides should be air-dried as rapidly as possible and fixed in acetone for 5 to 10 min.

Direct smears from vesicular lesions or nasopharyngeal swabs may be employed as described below; it is also possible, however, to examine the cellular sediment from these sites, which is sometimes available after initial processing for cultures by centrifugation. The procedure is the same as described for cell sediments.

Vesicular lesion smears

Exudate or vesicular fluids should first be removed by gentle blotting of open lesions; closed lesions must first be incised around the edge with a suitable instrument. Cells are obtained by rubbing the base of the lesion with the tip of a swab, which is then rolled firmly over a small area on one or more microscope slides. Rubbing motions are ineffective in transferring cells to the slide and should be avoided. Alternatively, scrapings may be obtained from accessible lesions with a spatula or similar instrument, held at a right angle to the plane of the lesion. Good collection technique should avoid causing the lesion to bleed. Cells obtained with a spatula should be thinly spread over one or two 5- to 10-mm-diameter areas on a slide, which is then allowed to air dry.

Nasopharyngeal smears

For detection of viruses such as respiratory syncytial virus, influenza viruses, parainfluenza viruses, and adenoviruses, cells from the nasopharynx are most satisfactory and are readily obtained. Cells may be recovered from nasal washes by centrifugation of the wash sample; cells are more conveniently obtained, however, by use of nasopharyngeal swabs. Good smears are prepared by using two swabs, the first to remove excess mucous secretion and a second to collect the cells needed for examination. Make smears by rolling the swab tip firmly on the slide surface, as described above for vesicular lesions; if properly made, these smears are entirely satisfactory. Prepare at least one smear for each agent to be tested. Additional smears are often needed for controls.

Tissue specimens

Appropriate specimens for direct examination by immunofluorescence may be prepared from tissue obtained at biopsy or autopsy. Frozen sections, cut thinly (<7 μm thick), are best for almost any tissue. If cut sections are not available, impression smears may be adequate, although they do not allow as many cells to be examined. Tissue that has been fixed in Formalin is often not suitable for immunofluorescent staining, due to denaturation of the viral antigens. Some success has been obtained by careful digestion with trypsin to "reexpose" viral antigens, but the procedure is not routine in most virology laboratories. Touch preparations can be made with a freshly cut piece of tissue. Results are improved by partially drying the wet surface by "blotting" it against the surface of a petri dish. The cut surface is then pressed firmly, with a sliding motion, against a microscope slide.

Smears from cell sediments

Cells can be obtained by centrifugation of liquid specimens such as urine, amniotic fluid, and peritoneal fluid. In some cases it may be worthwhile attempting to rescue cells from swabs collected for isolation attempts, as follows. The specimen (swab in viral transport medium) is agitated thoroughly with a Vortex mixer to dislodge cells from the swab fibers, and the swabs are removed. (This procedure is not effective with calcium alginate swabs, which tend to dissolve partially and gel in the transport medium.) The remaining specimen is centrifuged for a few minutes at >1,000 × *g*, after which the supernatant fluid can be used to inoculate cell cultures. Any pellet is dispersed in a small amount (<0.1 ml) of buffered saline containing 0.1% bovine serum albumin and spotted onto slides for immunofluorescent examination. This technique may allow the detection of virus in specimens that have been too long in transit for recovery of the virus by isolation.

LITERATURE CITED

1. **Behbehani, A. M.** 1972. Human viral and rickettsial diseases. A handbook of laboratory diagnosis for practicing physicians. The University of Kansas School of Medicine, Kansas City.
2. **Bettoli, E. J., P. M. Brewer, M. J. Oxtoby, A. A. Zaidi, and M. E. Guinan.** 1982. The role of temperature and swab materials in the recovery of herpes simplex virus from lesions. J. Infect. Dis. **145:**339.
3. **Chernesky, M. A., C. G. Ray, and T. F. Smith.** 1982. Cumitech 15, Laboratory diagnosis of viral infections. Coordinating ed., W. L. Drew. American Society for Microbiology, Washington, D.C.
4. **Embil, J. A., H. J. Thiebaux, F. R. Manuel, L. H. Pereira, and S. W. MacDonald.** 1983. Sequential cervical specimens and the isolation of *Chlamydia trachomatis:* factors affecting detection. Sex. Transm. Dis. **10:**62–66.
5. **Emmons, R. W., and J. L. Riggs.** 1977. Application of immunofluorescence to diagnosis of viral infections. Methods Virol. **6:**1–28.
6. **Howell, C. L., M. J. Miller, and W. J. Martin.** 1979.

Comparison of rates of virus isolation from leukocyte populations separated from blood by conventional and Ficoll-Paque/Macrodex methods. J. Clin. Microbiol. **10**:533–537.

7. **Hsiung, G. D.** 1973. Diagnostic virology, revised ed. Yale University Press, New Haven, Conn.
8. **Lennette, D. A., and E. T. Lennette.** 1981. A user's guide to the diagnostic virology laboratory. University Park Press, Baltimore.
9. **Lennette, E. H., and N. J. Schmidt (ed.).** 1979. Diagnostic procedures for viral, rickettsial, and chlamydial infections, 5th ed. American Public Health Association, Inc., Washington, D.C.
10. **Lyerla, H. C.** 1979. Diagnostic applications of immunofluorescence tests in the virology laboratory, p. 103–113. *In* D. A. Lennette, S. Specter, and K. D. Thompson (ed.), Diagnosis of viral infections: the role of the clinical laboratory. University Park Press, Baltimore.
11. **Mardh, P. A., L. Westrom, S. Colleen, and P. Wolner-Hanssen.** 1981. Sampling, specimen handling, and isolation techniques in the diagnosis of chlamydial and other genital infections. Sex. Transm. Dis. **8**:280–285.

11a. **Prather, S. L., J. A. Jenista, and M. A. Menegus.** 1984. The isolation of nonpolio enteroviruses from serum. Diagn. Microbiol. Infect. Dis. **2**:353–357.

12. **Virus Laboratory.** 1972. Viral and rickettsial diseases. Physicians handbook, 4th ed. Virus Laboratory, Ontario Department of Health, Toronto.

Chapter 62

Taxonomy of Viruses

JOSEPH L. MELNICK

Viruses are separated into families on the basis of the type and form of the nucleic acid genome and the size, shape, substructure, and mode of replication of the virus particle. Within each family, classifications of genera and species are based on antigenicity in addition to other properties.

Significant developments in classification and nomenclature of viruses are documented in the reports of the International Committee on Taxonomy of Viruses, formerly the International Committee on Nomenclature of Viruses. These reports, published in 1971 (6), 1976 (2), 1979 (3), and 1982 (4), have dealt with viruses of humans, lower animals, insects, plants, and bacteria and have included summaries of the properties of those groups of viruses as related to their taxonomic placement. It seems probable that most of the major groups of viruses have been recognized, particularly with regard to those infecting humans and the vertebrate animals of direct importance to humans. Many of them now have been officially placed in families, genera, and species; within some families, subfamilies or subgenera or both also have been established. Although the focus of this chapter is on these families, progress also has been made with respect to the taxonomy of the viruses of other host groups. For more detailed discussion of virus taxonomy, references 1 through 6 are recommended to the reader.

In Table 1, properties of the major families of RNA-containing viruses of humans and other vertebrate animals are summarized; in Table 2, vertebrate viruses with a DNA genome are similarly treated. In Table 3, these families are listed along with the genera and the individual members that may be of special concern for the viral diagnostic laboratory. Where subfamilies have been designated, these also are included if they represent agents having direct or indirect bearing upon human diseases. Since most of these virus agents are dealt with more fully in the other chapters of this volume, the brief text which follows is confined to explanation and commentary on some examples from the tables, together with some notes about recent developments.

Picornaviridae. Of particular importance is the recent classification of hepatitis A virus as enterovirus 72, within the family Picornaviridae. This virus has been shown to have the physicochemical properties of a member of the genus *Enterovirus*. These properties include a nonenveloped icosahedral (cubic) virion about 27 nm in diameter, a buoyant density in CsCl of ca. 1.33 to 1.34 g/cm^3, and four major polypeptides with molecular weights of about 33,000, 27,000, 23,000, and 6,000. The genome consists of a single piece of single-stranded RNA of molecular weight ca. 2.5×10^6. Like the other enteroviruses, hepatitis A virus is stable to acid pH and resistant to ether. In its resistance to thermal inactivation this serotype differs somewhat from other enteroviruses. In comparative studies, 50% of the particles in a poliovirus type 2 preparation disintegrate during heating at pH 7 for 10 min at 43°C, whereas under the same conditions 61°C is required to produce disintegration of 50% of the hepatitis A virus (enterovirus 72) particles. However, enterovirus 72, like all other enteroviruses, is stabilized by $MgCl_2$ against thermal inactivation. In the presence of 1 M $MgCl_2$, the temperatures required to produce 50% destruction of the particles are 61°C for poliovirus type 2 and 81°C for enterovirus 72. Stabilization of infectivity by $MgCl_2$ is a well-established characteristic of enteroviruses. Enterovirus 72 shows unusual thermal resistance without $MgCl_2$ (loss of only 1.3 logs of infectivity after heating at 65°C for 10 min), but in the presence of $MgCl_2$ it is even more resistant. There is no reduction after 20 min at 65°C, and only 1.5 log units of infectivity is lost after 10 min at 80°C.

Reoviridae. For all of the virus families listed in Table 1, the RNA genome is single stranded except in the case of the family Reoviridae, whose RNA is double stranded. The genus *Reovirus* differs somewhat from the other genera in its possession of an outer protein shell and the larger molecular weight of its genome (15×10^6 versus 12×10^6). There are three serotypes within the *Reovirus* genus that infect humans, monkeys, dogs, and cattle; in addition, at least five avian reoviruses are known. In the genus *Orbivirus*, important human pathogens include Colorado tick fever virus and the Kemerovo viruses; other important orbiviruses are the bluetongue viruses of sheep and the viruses of African horse sickness. The most recently established genus is *Rotavirus*, which includes several viruses that infect humans as well as viruses of many other mammalian species, typified by simian virus SA-11 and Nebraska calf diarrhea virus. The human rotaviruses are increasingly being recognized as the cause of a large share of the serious episodes of nonbacterial infantile diarrhea. Rotavirus gastroenteritis is one of the most common childhood illnesses throughout the world and is a leading cause of infant deaths in developing countries. These viruses also infect adults, particularly those in close contact with infants and children, but infected adults usually experience no symptoms or may have only minor illness.

Caliciviridae. Other recent additions to the taxonomic roll of RNA-containing viruses are the Caliciviridae and the Bunyaviridae. The Caliciviridae include a number of viruses of pigs, cats, and sea lions and may include agents that infect humans. Calicivirus-like particles have been observed in human feces in association with gastroenteric disease; preliminary results have failed to show relationship to the feline calicivirus. The possible relationship of these agents to the virus of Norwalk gastroenteritis also remains to be resolved. The Norwalk virus, a widespread human agent causing acute epidemic gastroenteritis, has a virion protein structure similar to that of the caliciviruses; it also resembles caliciviruses in several other characteristics. Because these agents have not yet been successfully adapted to tissue culture, it has been

TABLE 1. Current classification of RNA-containing viruses of vertebrates

Characteristic	Classification										
Nucleic acid core	RNA	RNA	RNA	RNA	RNA	RNA	RNA	RNA	RNA	RNA	RNA
Capsid symmetry	Cubic	Cubic	Cubic	Cubic	Helical	Helical	Helical	Helical	Helical	Uncertain	Uncertain
Virion: naked or enveloped	Naked	Naked	Naked	Enveloped	Enveloped	Enveloped	Enveloped	Enveloped	Enveloped	Enveloped	Enveloped
Site of capsid assembly	Cytoplasm	Cytoplasm	Cytoplasm	Cytoplasm	Cytoplasm	Cytoplasm	Cytoplasm	Cytoplasm	Cytoplasm	Cytoplasm	Cytoplasm
Site of nucleocapsid envelopment				Surface membrane[a]	Surface membrane	Surface membrane	Surface membrane	Intracytoplasmic membranes	Intracytoplasmic membranes	Surface membrane	Surface membrane
Reaction to ether treatment	Resistant	Resistant	Resistant	Sensitive	Sensitive	Sensitive	Sensitive	Sensitive	Sensitive	Sensitive	Sensitive
Number of capsomeres	32	32	32	32							
Diameter of helix (nm)					9–15	18[b]	18	11–13	10–12		
Diameter of virion (nm)[c]	24–30	35–39	60–80	40–70[a]	80–120	150–300	60 × 180	80–130	80–110	About 100	50–300
Mol wt of nucleic acid (×10⁶)	2.3–2.8	2.6	12–15	3–4	4–5	5–8	3.5–4.6	5–6	6–7	6–7	3–5
Virus family	Picornaviridae	Caliciviridae	Reoviridae	Togaviridae	Orthomyxoviridae	Paramyxoviridae	Rhabdoviridae	Coronaviridae	Bunyaviridae	Retroviridae	Arenaviridae

[a] All of the properties shown for Togaviridae apply to the genera *Alphavirus* and *Rubivirus*; however, for the genera *Pestivirus* and *Flavivirus*, envelopment takes place at intracytoplasmic membranes, the number of capsomeres is not clearly established, and the virions are somewhat smaller.

[b] All of the properties shown for Paramyxoviridae apply to the genera *Paramyxovirus* and *Morbillivirus*; however, for members of the genus *Pneumovirus*, the diameter of the helix is 12 to 15 nm.

[c] Diameter or diameter × length.

difficult to study their properties.

Bunyaviridae. The Bunyaviridae form a family of more than 200 viruses, at least 145 of them belonging to the Bunyamwera supergroup of serologically interrelated arboviruses. With the taxonomic placement of this large group, the vast majority of the viruses of the classical arbovirus groupings, initially based on ecological properties and subdivided by serological interrelationships, have been assigned to families on the basis of biophysical and biochemical characteristics. Human illness caused by Hantaan virus has been recognized in the Far East as Korean hemorrhagic fever, and a variant is known in Scandinavian and eastern European countries as epidemic nephropathy. The illness has been variously named "hemorrhagic fever with renal syndrome" or "muroid virus nephropathy." Recent studies show the causative agent to be a member of the Bunyaviridae. The virus has a labile membrane and a tripartite single-stranded RNA genome. The most common natural hosts are mice (in Korea) and voles (in Europe). There have been several instances of infection of staff members handling laboratory rats infected with the virus, both in the Far East and more recently in Europe. In Belgium there also have been sporadic cases with no apparent link to an outbreak or to each other.

Retroviridae. The family Retroviridae has been divided into subfamilies (see Table 3). The best-known retroviruses belong to the subfamily Oncovirinae, the RNA tumor virus group, which has been the focus of special interest because its members, long recognized as causing leukemia and sarcoma in animals, serve as valuable animal models of oncogenic viruses.

The Retroviridae are enveloped viruses whose genome contains a single-stranded RNA of the same polarity as viral mRNA. The virion contains a reverse transcriptase enzyme. Replication proceeds off an integrated "provirus" DNA copy in infected cells. Study of retroviruses, and oncoviruses in particular, has permitted the identification of cellular "oncogenes." Normal cells of several animal species contain integrated copies of the genes of the endogenous

TABLE 2. Current classification of DNA-containing viruses of vertebrates

Characteristic	Classification						
Nucleic acid core	DNA						
Capsid symmetry	Cubic						Complex
Virion: naked or enveloped	Naked			Enveloped			Complex coats
Site of capsid assembly[a]	Nucleus			Nucleus	Nucleus	Cytoplasm	Cytoplasm
Site of nucleocapsid envelopment				Cytoplasm	Nuclear membrane	Cytoplasmic membrane	
Reaction to ether (or other lipid solvents)	Resistant				Sensitive	Sensitive	Resistant
Number of capsomeres	32	72	252		162	1,500	
Diameter of virion (nm)[b]	18–26	45–55	70–90	40–50	100[c]	130–300	230 × 300
Mol wt of nucleic acid ($\times 10^6$)	1.5–2.0	3.0–5.0	20–30	2.1	80–150	100–250	160
Virus family	Parvoviridae	Papovaviridae	Adenoviridae	Hepadnaviridae	Herpesviridae	Iridoviridae	Poxviridae

[a] For the DNA-containing viruses whose capsid assembly takes place in the nucleus, a phase of replication occurs in the cytoplasm, as evidenced by the detection of viral mRNA associated with polyribosomes.
[b] Diameter or diameter × length.
[c] The naked virus is 100 nm in diameter; however, the enveloped virions range up to 150 nm in diameter.

oncovirus. The oncovirus genes may not be expressed but can be activated by physical and chemical agents, by superinfection with other oncoviruses, and even by herpesviruses. Recently, members of the subfamily Oncovirinae have been shown to cause human disease. These include the agents generally known as the human T-cell leukemia viruses and the agent known as the lymphadenopathy-associated virus. There are at least three types of human T-cell leukemia viruses, and type 3 apparently is identical with the lymphadenopathy-associated virus; this serotype seems to be the etiological agent of AIDS (acquired immune deficiency syndrome).

Parvoviridae. All of the virus families shown in Table 2 have their DNA genome in double-stranded form except the Parvoviridae, whose DNA is single stranded within the virion. As indicated in Table 2, members of Parvoviridae are very small viruses. The molecular weight of the nucleic acid in the virion is relatively very low, 1.5×10^6 to 2.0×10^6 (as compared, for example, with 160×10^6 for the DNA of poxviruses). Some members display resistance to high temperatures (60°C, 30 min).

The family Parvoviridae encompasses viruses of numerous species of vertebrates, including humans. Two members of the genus *Parvovirus*, the members of which are able to replicate independently, have been found to be associated with disease problems of human beings. Parvovirus B19 has been shown to cause a transient shutdown of erythrocyte production by killing the late erythroid progenitor cells. This shutdown presents particular problems for individuals already suffering from hemolytic anemias such as sickle cell anemia, causing aplastic crises. A virus named RA-1, which is associated with rheumatoid arthritis, is another newly identified member of the genus.

A host range mutant of feline panleukopenia parvovirus, known as canine parvovirus, induces acute enteritis with leukopenia in young and adult dogs as well as myocarditis in puppies. Infections with this virus have reached enzootic proportions around the world.

Several serotypes of adeno-associated viruses, belonging to the *Dependovirus* genus, are known to infect humans, but they have not been shown to be associated with any human disease. Members of this genus cannot multiply in the absence of a replicating adenovirus which serves as a "helper virus." The single-stranded DNA is present within the virion as either plus or minus complementary strands in separate particles. Upon extraction, the plus and minus DNA strands unite to form a double-stranded helix.

Papovaviridae. Members of the family Papovaviridae have DNA in double-stranded, circular form. The human representatives are the papilloma or wart viruses and the JC and BK viruses; these latter were isolated, respectively, from the brain tissue of patients with progressive multifocal leukoencephalopathy and from the urine of immunosuppressed renal transplant recipients. In addition, several isolates which appear to be identical to simian virus 40 of monkeys have also

TABLE 3. Members of virus families, with emphasis on viruses that infect humans

Family[a]	Genus	Common species	No. of members
Picornaviridae	*Enterovirus*	Polioviruses	3
		Coxsackieviruses, group A	23
		Coxsackieviruses, group B	6
		Echoviruses	31
		Enteroviruses 68 through 71	4
		Enterovirus 72 (hepatitis A virus)	1
		Viruses of other vertebrates	>34
	Cardiovirus	Encephalomyocarditis virus and mengovirus, mouse encephalomyelitis virus	3
	Rhinovirus	Virus types infecting humans	>115
		Viruses of cattle	2
	Aphthovirus	Foot-and-mouth disease viruses of cattle and other cloven-hoofed animals	7
Caliciviridae	*Calicivirus*	Vesicular exanthema of swine virus	13
		Viruses of cats, sea lions	Many
		(Possible member: Norwalk gastroenteritis virus of humans)	?
Reoviridae	*Reovirus*	Viruses of humans, monkeys, and lower vertebrates	3
		Viruses of birds	>5
	Orbivirus	Seventeen subgroups, including Colorado tick fever and Kemerovo viruses of humans and also bluetongue virus of sheep and African horse sickness virus	>90
	Rotavirus	Human rotaviruses	>4
		Rotaviruses of many mammals, including SA-11 virus of monkeys and Nebraska calf diarrhea virus	Many
Togaviridae	*Alphavirus* (arbovirus group A)	Sindbis virus and many other mosquito-borne viruses, including the viruses of eastern equine, Venezuelan, and western equine encephalitis and Semliki Forest virus	23
	Flavivirus (arbovirus group B)	Yellow fever virus and other mosquito-borne viruses, including the viruses of dengue, of Japanese, Murray Valley, and St. Louis encephalitis, and of West Nile fever	26
		Tick-borne viruses, including the viruses of Kyasanur Forest disease, Omsk hemorrhagic fever, European and Far Eastern tick-borne encephalitis of humans, and louping ill of sheep	11
		Viruses whose vectors are unknown	17
	Rubivirus	Rubella virus	1
	Pestivirus	Viruses of cattle and pigs	>3
Orthomyxoviridae	*Influenzavirus*	Influenza virus type A	Many
		Influenza virus type B	Several
	(Probably a separate genus)	Influenza virus type C	1
Paramyxoviridae	*Paramyxovirus*	Human parainfluenza viruses, including Sendai virus	4
		Mumps virus	1
		Newcastle disease virus of fowl and viruses of other diseases of birds and mammals	>6
	Morbillivirus	Measles virus	1
		Rinderpest virus of cattle	1
		Distemper virus of dogs	1
		Peste-des-petits-ruminants virus of sheep and goats	1
	Pneumovirus	Human respiratory syncytial virus	1
		Respiratory disease viruses of cattle and of mice	?
Rhabdoviridae	*Vesiculovirus*	Vesicular stomatitis virus of horses, cattle, and pigs	Several
	Lyssavirus	Rabies virus	1
		Lagos bat virus and others	>5

Continued

TABLE 3—*Continued*

Family[a]	Genus	Common species	No. of members
Coronaviridae	*Coronavirus*	Human coronavirus	1
		Mouse hepatitis virus, infectious bronchitis virus of fowl, and other agents infecting pigs and other vertebrates	>4
Bunyaviridae	*Bunyavirus*	Bunyamwera virus California encephalitis viruses LaCrosse virus, other serologically cross-related groups, and several ungrouped viruses	>145
	Phlebovirus	Sandfly fever viruses Other viruses of humans and animals, including Rift Valley fever virus of sheep and other ruminants, which may cause human disease	>30
	Nairovirus	Crimean-Congo hemorrhagic fever Viruses of five other serogroups, including the virus of Nairobi sheep disease	>27
	Uukuvirus	Uukuniemi virus and six other agents, all belonging to the same serogroup (infect rodents and ticks)	7
	Hantaanvirus	Hantaan virus of Korean hemorrhagic fever (hemorrhagic fever with renal syndrome)	1
Retroviridae			
Oncovirinae[a] (RNA tumor virus group)	Type C oncovirus group	Sarcoma and leukemia viruses of mice, cats, cattle, birds, snakes, and primates, including HTLV and LAV[b] of humans	>15
	Type B oncovirus group	Mammary tumor virus of mice (and humans?)	?
	(Proposed genus: type D retrovirus group)	Monkey (mammary tumor?) virus (Mason-Pfizer monkey virus)	?
Spumavirinae (foamy virus group)		Syncytial and foamy viruses of humans, monkeys, cattle, and cats	>4
Lentivirinae (maedi-visna virus group: "slow viruses")		Visna virus of sheep, maedi, progressive pneumonia viruses of sheep	?
Arenaviridae	*Arenavirus*	Lymphocytic choriomeningitis virus of mice	1
		Lassa fever virus	1
		Viruses of the Tacaribe complex, including Junin and Machupo viruses of South American hemorrhagic fevers	>8
Parvoviridae	*Parvovirus*	Human parvoviruses: B19, RA-1	>2
		Aleutian mink disease virus and viruses of rodents, pigs, cattle, cats, and dogs	Many
	Dependovirus	Adeno-associated virus (adeno-satellite virus): human (types 1–3); monkey (type 4); also of cattle, dogs, birds	>8
Papovaviridae	*Papillomavirus*	Human papilloma (warts) viruses	Many
		Rabbit (Shope) papilloma virus	1
		Papilloma viruses of other mammals	Many
	Polyomavirus	Polyoma virus of mice	1
		JC and BK viruses of humans	2
		Simian virus 40 of rhesus monkey	1
		Lymphotropic virus of African green monkey	1
		Viruses of mouse, rabbit, hamster, and baboon	4
Adenoviridae	*Mastadenovirus*	Human adenoviruses	36
		Viruses of other mammals	>45
	Aviadenovirus	Viruses of birds	>13
Herpesviridae			
Alphaherpesvirinae[a]	*Simplexvirus*	Human herpes simplex virus types 1 and 2	2
		Bovine mammillitis virus	1
		Herpes B virus of monkeys	1

Continued

TABLE 3—*Continued*

Family[a]	Genus	Common species	No. of members
	Poikilovirus	Pseudorabies virus Equine rhinopneumonitis virus	2
	Varicellavirus	Varicella-zoster virus	1
Betaherpesvirinae	*Cytomegalovirus* *Muromegalovirus*	Human cytomegaloviruses Mouse cytomegaloviruses	1 1
Gammaherpesvirinae	*Lymphocryptovirus*	Epstein-Barr virus	1
	Thetalymphocryptovirus	Marek's disease herpesvirus of fowl	1
	Rhadinovirus	Herpesvirus saimiri and others	>2
Iridoviridae (icosahedral cytoplasmic deoxyriboviruses)	African swine fever virus group	African swine fever virus	Several
	Ranavirus	Frog virus 3 and other viruses of amphibians	>30
Poxviridae			
Chordopoxvirinae[a] (poxviruses of vertebrates)	*Orthopoxvirus*	Vaccinia virus Smallpox virus (variola) Poxviruses of lower animals	1 1 >6
	Parapoxvirus	Orf virus and other viruses of ungulates Virus of milker's nodule	? 1
	Avipoxvirus	Fowlpox virus and other viruses of birds	8
	Capripoxvirus	Viruses of sheep and goats	3
	Leporipoxvirus	Myxoma virus of hares Fibroma viruses of rabbits and squirrels	4
	Suipoxvirus	Swinepox virus	1
Entomopoxvirinae		Poxviruses of insects	>24
Hepadnaviridae	*Hepadnavirus*	Human hepatitis B virus Hepatitis viruses of woodchuck, ground squirrel, and duck	1 >3

[a] Where subfamilies have been designated, they are listed in this column, indented below the family name (see Retroviridae, Herpesviridae, and Poxviridae).

[b] HTLV, Human T-cell leukemia viruses; LAV, lymphadenopathy-associated virus.

been isolated from patients with progressive multifocal leukoencephalopathy. Papovaviruses produce latent and chronic infections in their natural hosts. Many of them produce tumors, particularly in experimentally infected rodents, thus serving as models for studying viral carcinogenesis. The viral DNA integrates into cellular chromosomes of transformed cells.

When simian virus 40 and adenoviruses replicate together, they may interact to form "hybrid" virus particles, in which a defective simian virus 40 genome is covalently linked to adenovirus DNA and is carried within an adenovirus capsid.

Hepadnaviridae. Ample evidence has accumulated for the formation of a new virus family. The name, Hepadnaviridae, reflects the DNA-containing genomes of its members and their replication within hepatocytes. These viruses have a circular DNA genome that is double stranded except for a region of variable length that is single stranded. In the presence of appropriate substrates, DNA polymerase within the virion can complete the single-stranded region to its full length of 3,200 nucleotides.

Hepatitis B virus of humans and three similar viruses found in woodchucks, Beechey ground squirrels, and Pekin ducks share many basic features. All members of the family share antigens as well as similar morphology and behavior in the infected host. Large amounts of excess viral coat protein are produced in the form of small 22-nm spherical and tubular particles (in the human virus, the antigen is known as hepatitis B surface antigen). The viruses replicate in the liver and are associated with acute and chronic hepatitis. More than 200 million persons are persistent carriers of the human virus and are at very high risk of developing liver cancer. The woodchuck hepatitis B virus also causes liver cancer in its natural host. Fragments of viral DNA may be found in the liver cancer cells of both species.

LITERATURE CITED

1. **Andrewes, D., H. G. Pereira, and P. Wildy.** 1978. Viruses of vertebrates, 4th ed. Macmillan Publishing Co., New York.
2. **Fenner, F.** 1976. Classification and nomenclature of viruses: second report of the International Committee on Taxonomy of Viruses. Intervirology **7:**1–116.
3. **Matthews, R. E. F.** 1979. Classification and nomenclature of viruses: third report of the International Committee on

Taxonomy of Viruses. Intervirology **12:**129–296.
4. **Matthews, R. E. F.** 1982. Classification and nomenclature of viruses: fourth report of the International Committee on Taxonomy of Viruses. Intervirology **17:**1–199.
5. **Melnick, J. L.** 1966–1982. Summaries on viral taxonomy (published annually). *In* Progress in medical virology. S. Karger, Basel.
6. **Wildy, P.** 1971. Classification and nomenclature of viruses: first report of the International Committee on Nomenclature of Viruses. Monogr. Virol. **5:**1–81.

Chapter 63

Adenoviruses

MARION K. COONEY

CLINICAL BACKGROUND

Adenovirus infections in humans are predominantly associated with respiratory and ocular disease symptoms. Except in military recruits, adenoviruses cause little significant respiratory disease in adults, but in children under 6 years of age they make a relatively important contribution to febrile upper respiratory illness such as pharyngitis, pharyngoconjunctival fever, and influenza-like disease, as well as croup, bronchiolitis, and pneumonia. Adenoviruses were first discovered by Rowe et al. (13) as latent viruses in adenoidal tissue, removed surgically from asymptomatic children, which had been used to establish cell lines for virus isolation. In the same year, adenoviruses were firmly established as respiratory pathogens in military recruits, causing acute respiratory disease in an outbreak studied by Hilleman and Werner (8). Epidemic keratoconjunctivitis is the most serious ocular disease associated with adenoviruses. It typically begins with redness of the conjunctiva, followed by involvement of the cornea. Healing proceeds without sequelae, and recurrent infections have not been reported. Adenovirus type 8 was the principal cause of epidemic keratoconjunctivitis in the United States for many years after this disease was recognized, but it has apparently been replaced by adenovirus types 19 and 37 (10).

Longitudinal surveillance studies of family groups, in which both fecal and respiratory specimens were examined, defined the characteristics of adenovirus infections (2). Adenovirus infections are predominantly enteric, may be abortive or invasive, and are followed by persistent intermittent excretion for periods as long as 18 months after initial infection. This excretion pattern is most characteristic of types 1, 2, 3, and 5, which may explain why these types are usually endemic. Worldwide, adenovirus types 1, 2, 3, 5, and 7 are isolated most frequently, and adenovirus types 4, 6, and 8 are encountered less often. All other species were found much less frequently (15). After initial infection, the viruses may persist as a latent infection in the lymphoid tissues of the nasopharynx and enteric tract and be reactivated by some mechanism such as infection with another agent, e.g., *Bordetella pertussis* (1). Children are usually infected with the endemic types before the age of 2 years. Approximately half of those infected are ill, with symptoms of pharyngitis, tonsillitis, pharyngoconjunctival fever, or influenza-like disease. Types 3 and 7 have also been associated with "swimming pool conjunctivitis" (3) and with lower respiratory disease in children under 6. Types 4, 7, and 21 are the major cause of epidemics of acute respiratory disease in military recruits and cause sporadic cases of pneumonia in children. Type 7 has been reported to cause community outbreaks of lower respiratory disease. Adenovirus types 40 and 41, not cultivable in cell systems usually employed for adenovirus isolation, have been established as significant pathogens associated with diarrheal disease in hospitalized pediatric patients (4).

Syndromes of intussusception, meningitis, and hemorrhagic cystitis are sometimes associated with adenovirus infection.

Adenoviruses, like other latent viruses, are a problem in immunosuppressed patients, as exemplified by the isolation of adenovirus types 34 and 35 from kidney transplant patients (16).

DESCRIPTION OF AGENT

Because of their many unique characteristics, in addition to their role as disease agents, adenoviruses have served as models for studies of viral ultrastructure, latency, oncogenesis, the biochemistry of virus-cell interaction, and genomic interactions between adenoviruses and two groups of unrelated small DNA viruses, the adeno-associated viruses and simian virus 40. The virions are nonenveloped perfect icosahedrons, 70 to 90 nm in diameter. The nucleoprotein core consists of double-stranded DNA, associated with several species of arginine-rich polypeptides. The capsid has 252 capsomers, 240 of which are hexons, hollow polygonal-shaped structures surrounded by six neighbors. At each of the 12 vertices are plate-like capsomers (penton base) which have an attached projection called a fiber which ends in a terminal knob. Each vertex capsomer is surrounded by five neighbors, hence the term "penton." Adenoviruses multiply in the nucleus of the cell. They are chloroform stable and acid stable and are inactivated at pH above 8 and by heating at 56°C for 30 min (6, 9). Over 90 serotypes are included in the family Adenoviridae. Adenoviruses isolated from mammalian species share a common complement-fixing (CF) antigen and constitute the genus *Mastadenovirus*. Adenoviruses isolated from fowl lack the common CF antigen and constitute the genus *Aviadenovirus* (20). Since adenoviruses, like most virus groups, have a strict host specificity, only the 41 human serotypes are of interest here. Capsid substructures have been identified with antigenic specificity or functions. Hexons are responsible for group CF antibody and type-specific neutralizing antibody. Pentons have cell-detaching activity and type-specific as well as subgroup-reactive antigen. Fibers are the organ of attachment to cells and are responsible for hemagglutinating activity. Fiber antigen, although type specific, does not induce neutralizing antibody. Human adenoviruses have been assigned to groups according to their hemagglutination of rat or rhesus monkey erythrocytes (12). They have also been grouped according to oncogenic potential, and this grouping has been correlated with DNA-DNA hybridization and guanine-cytosine content (Table 1). Although several adenoviruses have been shown to be oncogenic (5), no association with human tumors has been shown to date. Endonuclease restriction enzyme analysis has been employed to differentiate genome types within species, as well as to classify adenoviruses within subgenera (19). Genome typing has been useful in epidemiological studies (18).

TABLE 1. Subgenera of adenovirus species defined by DNA homology and hemagglutination patterns[a]

Subgenus	Species	% DNA homology[b]	Rosen hemagglutination group	Hemagglutination	
				Rhesus	Rat
A	12, 18, 31	48–69 (8–20)	IV	0	0
B	3, 7, 11, 14, 16 21, 34, 35	89–94 (9–20)	I	+	0
C	1, 2, 5, 6	99–100 (10–16)	III	0	Partial
D	8, 9, 10, 13, 15 17, 19, 20, 22, 23, 24, 25 26, 27, 28, 29, 30, 32 33, 36, 37, 38, 39	95–99 (4–17)	II	+ or 0	+
E	4	(4–23)	III	0	Partial

[a] Adapted from Wigand et al. (20) and Rosen (12).
[b] Percent homology within group. Parentheses indicate percent homology with other types.

COLLECTION AND STORAGE OF SPECIMENS

Adenoviruses may be isolated from nasopharyngeal secretions, eye exudates, fecal specimens, and urine specimens. Adenoviruses may also be isolated from body fluids, urethral and cervical specimens, and tissues obtained at biopsy or autopsy.

The techniques for collection, shipment, and storage of specimens for virus isolation described in Chapter 61 are generally applicable to the adenoviruses. Adenoviruses are hardier than many viruses sought in clinical specimens, and they survive well under the conditions recommended.

DIRECT EXAMINATION

Examination of stool specimens by electron microscopy for agents associated with gastroenteritis revealed viruses that have typical adenovirus conformation and possess adenovirus capsid antigen (adenovirus types 40 and 41), but which will not replicate in cells usually susceptible to adenovirus infection (4). Electron microscopy and immunoperoxidase and immunofluorescent techniques using antihexon antibody (to the common antigen) are useful for direct diagnosis for examination of stool specimens from gastroenteritis patients. Direct examination of conjunctival scraping specimens may be used when rapid diagnosis is important (7, 11). A method for direct detection of adenovirus DNA has recently been described (17). Giemsa-stained smears will reveal cells with intranuclear inclusions in infected cell cultures and provide useful information in some instances. Direct examination is not usually used for respiratory specimens, since results are difficult to interpret in relation to disease symptoms under investigation.

ISOLATION OF VIRUS

Adenoviruses grow only in cell cultures. Primary, diploid, and continuous human cell lines are susceptible to infection with adenoviruses. Primary human embryonic kidney (HEK) cells are most susceptible and are the system of choice. However, HEK cells are expensive and sometimes difficult to obtain regularly, so most laboratories depend on heteroploid lines, such as HeLa, HEp-2, or KB cells. Diploid lines, particularly fetal tonsil (FT), are also suitable for adenovirus isolation. Some of the higher types, such as adenovirus 19, are very slow to show cytopathic effect (CPE), but can be isolated in FT cells when not detectable in HeLa or KB cells. In heteroploid cells the types most often seen (1 through 7) show CPE, rounding and swelling of cells, refractile appearance, and clumping in grapelike clusters 4 to 6 days after inoculation with specimens, depending on virus content. If heteroploid cells degenerate within 14 days, at least one blind passage should be made. The pH of adenovirus-infected cultures usually drops dramatically. In FT cells, the first indication of adenovirus infection may be the appearance of foci of four or five greatly enlarged, rounded cells, not unlike foci produced by cytomegalovirus. Unlike cytomegalovirus, however, adenovirus infection progresses rapidly to involve the entire monolayer and produces typical clumps of cells. Appearance of CPE is usually later (6 to 10 days) in diploid than in heteroploid cells. Confirmatory passage of infected cell culture fluid usually produces typical CPE within 24 to 48 h in either diploid or heteroploid cell systems.

IDENTIFICATION OF ADENOVIRUS

Adenovirus CPE is usually recognizable as such, and if confirmatory passage shows typical CPE the infected cell culture fluid is suitable for typing. For rapid identification of an isolate as an adenovirus, antihexon antibody may be used in immunofluorescent or enzyme-linked immunosorbent assay procedures. Reliable conjugates of antihexon antibody are commercially available and are useful for group-specific identification of isolates.

Neutralization with type-specific antiserum is the method of choice for type identification of the virus. Type-specific antisera may be produced by injection of rabbits with adenovirus immunogens produced in cell cultures. Antisera for types 1 through 7 are commercially available, and reference antisera for the higher types may be obtained from the American Type Culture Collection, Rockville, Md.

Cross relationships have been shown by neutralization or hemagglutination inhibition, or both, between many of the serotypes (9). Since types 1, 2, and 3 are endemic in most populations, and types 4 through 7 account for most of the residual isolations, the most practical approach is to attempt typing with antisera to the most common types. Equal volumes of infected cell culture fluid (such fluid, undiluted or diluted 1:10

at subsequent passage levels, usually contains approximately 100 TCD_{50} [50% tissue culture doses] per 0.1 ml) and type-specific immune serum diluted to contain at least 4 antibody units per 0.1 ml are mixed and allowed to stand for 1 h at room temperature. Two tubes of the appropriate cell culture are inoculated with 0.2 ml of each serum-virus mixture. A virus control, diluted to the same virus concentration but without serum, is included in each set of typings. Incubation is at 37°C. If virus is neutralized, no CPE will be seen for at least 4 days after the virus control shows 4+ CPE. If the isolate is not identified as type 1 through 7, it can be identified as an adenovirus by the fluorescent-antibody test or can be tested for adenovirus group CF antigen against reference human antiserum with high-titer adenovirus CF antibody (commercially available). Human serum used for identification should be tested against a battery of viral antigens so that titers to other viral agents are known. Heteroploid cells infected with the virus usually have the highest CF antigen titer (at least 1:8) at the second or third passage level. It is important to heat the infected cell culture fluid at 56°C for 30 min before testing. Determination of the hemagglutination group (rat or monkey erythrocytes) (Table 1) is also very useful, since this narrows the possible number of types, and hemagglutination inhibition can be used for type identification.

SEROLOGICAL DIAGNOSIS

Neutralization, hemagglutination inhibition, and CF tests may be used for serological diagnosis (14). The microtiter technique is usually used. For detection of current infections with adenoviruses, the CF test is the test of choice. Since the antigen is shared by all members of the group, any adenovirus type can be used for the CF antigen for serological testing, but adenovirus types 3 and 5 give the best results with sera from both children and adults. A fourfold or greater rise in CF antibody titer is indicative of current infection with an adenovirus, although no indication of type is possible. The use of the CF test for detection of rises to group antigen lends itself to "battery testing" in a diagnostic laboratory, in which all paired specimens submitted from suspected cases of respiratory disease are tested against a large number of antigens known to be associated with respiratory disease. Unfortunately, young children often do not respond with CF antibody after infection. If information on response to specific types is required, hemagglutination inhibition tests with monkey or rat erythrocytes can be used (Table 1). Microneutralization tests for type-specific antibodies are more sensitive and less subject to error. Serial dilutions of serum, 0.025 ml in volume, are made in tissue culture flat-bottom microtiter plates. The diluent used is cell culture growth medium with serum content reduced 50%. A 0.025-ml amount of the appropriate adenovirus type, containing 100 TCD_{50} of virus, is added to each serum dilution, and plates are allowed to stand at room temperature for 1 h. HeLa cells, suspended in growth medium (with serum content reduced to 5% fetal calf serum), are added to each cup (20,000 cells per cup). Plates are covered with Lucite covers and incubated at 37°C in a CO_2 incubator. A virus titration is included with the test. When the virus control titration shows the presence of 100 TCD_{50} of virus, 2 drops of a mixture of equal parts of Formalin and 1% crystal violet in ethanol is added to each well to inactivate virus and stain uninfected monolayers. After 30 min, plates are flushed with a stream of water and allowed to dry. Deeply stained monolayers of cells remain in wells in which virus was completely neutralized, and all infected cells are detached and removed by the washing. Neutralizing antibody titers are recorded as the highest dilution of serum which completely neutralizes the virus.

EVALUATION AND INTERPRETATION OF RESULTS

Since adenoviruses were identified as latent viruses by the circumstances of their initial demonstration and subsequently proved to be capable of latency and recrudescence in human infections, and since recrudescent shedding of virus can be accompanied by a significant rise in antibody titer, a certain amount of caution must be applied to interpretation of the significance of either virus isolation or antibody response as associated with specific symptoms. Virus isolation from fecal specimens only, with no recovery from throat specimens, during a respiratory illness would suggest lack of correlation with illness. Isolation of adenovirus from throat secretions during an episode of pharyngitis or pharyngoconjunctivitis would be expected to have some relevance to the clinical symptoms. Isolation of adenovirus from conjunctivae in an episode of eye disease is unequivocal evidence of infection with that agent. Adenovirus isolation from other body fluids and tissues must be interpreted with caution. Gastroenteritis appears to be associated with uncultivable adenoviruses in fecal specimens detected by electron microscopy or enzyme-linked immunosorbent assay. If an adenovirus antibody response occurs in the absence of involvement with other respiratory agents or in conjunction with virus isolation from the throat or eyes, one can have confidence in the etiological significance of these findings. Somewhat different criteria may be applied to the significance of findings in investigating a particular outbreak, such as pneumonia in a recruit population or pharyngoconjunctivitis and conjunctivitis in a group using the same swimming pool.

LITERATURE CITED

1. **Baraff, L. J., J. Wilkins, and P. F. Wehrle.** 1978. The role of antibiotics, immunizations and adenoviruses in pertussis. Pediatrics **61**:224–230.
2. **Fox, J. P., C. E. Hall, and M. K. Cooney.** 1977. The Seattle Virus Watch. VII. Observations of adenovirus infections. Am. J. Epidemiol. **105**:362–368.
3. **Foy, H. M., M. K. Cooney, and J. B. Hatlen.** 1968. Adenovirus type 3 associated with irregular chlorination of a swimming pool. Arch. Environ. Health **17**:795–802.
4. **Gary, G. W., Jr., J. C. Hierholzer, and R. E. Black.** 1979. Characteristics of noncultivable adenoviruses associated with diarrhea in infants: a new subgroup of human adenoviruses. J. Clin. Microbiol. **10**:96–110.
5. **Gilead, Z., and H. S. Ginsberg.** 1965. Characterization of a tumorlike antigen in type 12 and type 18 adenovirus-infected cells. J. Bacteriol. **90**:120–125.
6. **Ginsberg, H. S.** 1972. Adenoviruses. Am. J. Clin. Pathol. **57**:771–776.
7. **Hawkes, R. A.** 1979. Detection of virus or viral antigen

without isolation, p. 29–42. *In* E. H. Lennette and N. J. Schmidt (ed.), Diagnostic procedures for viral, rickettsial and chlamydial diseases, 5th ed. American Public Health Association, Washington, D.C.

8. **Hilleman, M. R., and J. H. Werner.** 1954. Recovery of new agents from patients with acute respiratory illness. Proc. Soc. Exp. Biol. Med. **85:**183–188.
9. **Kasel, J. A.** 1979. Adenoviruses, p. 229–255. *In* E. H. Lennette and N. J. Schmidt (ed.), Diagnostic procedures for viral, rickettsial and chlamydial diseases, 5th ed. American Public Health Association, Washington, D.C.
10. **Keenlyside, R. A., J. C. Hierholzer, and L. J. D'Angelo.** 1983. Keratoconjunctivitis associated with adenovirus type 37: an extended outbreak in an ophthalmologist's office. J. Infect. Dis. **147:**191–198.
11. **Lyerla, H. C.** 1979. Diagnostic applications of immunofluorescence tests in the virology laboratory, p. 103–113. *In* D. A. Lennette, S. Specter, and K. D. Thompson (ed.), Diagnosis of viral infections: the role of the clinical laboratory. University Park Press, Baltimore.
12. **Rosen, L.** 1958. Hemagglutination by adenoviruses. Virology **5:**574–577.
13. **Rowe, W. P., R. J. Huebner, L. K. Gilmore, R. H. Parrott, and T. G. Ward.** 1953. Isolation of a cytopathogenic agent from human adenoids undergoing spontaneous degeneration in tissue culture. Proc. Soc. Exp. Biol. Med. **84:**570–573.
14. **Schmidt, N. J., E. H. Lennette, and C. J. King.** 1966. Neutralizing, hemagglutination-inhibiting and group complement-fixing antibody responses in human adenovirus infections. J. Immunol. **97:**64–74.
15. **Schmitz, H., R. Wigand, and W. Heinrich.** 1983. Worldwide epidemiology of human adenovirus infection. Am. J. Epidemiol. **117:**455–466.
16. **Stalder, H., J. C. Hierholzer, and M. N. Oxman.** 1977. New human adenovirus (candidate type 35) causing fatal disseminated infection in a renal transplant recipient. J. Clin. Microbiol. **6:**257–265.
17. **Virtanen, M., M. Laaksonen, H. Soderlund, A. Palva, P. Halonen, and M. Ranki.** 1983. Novel test for rapid viral diagnosis: detection of adenovirus in nasopharyngeal mucus aspirates by means of nucleic-acid sandwich hybridization. Lancet **i:**381–383.
18. **Wadell, G., J. C. De Jong, and S. Wolontis.** 1981. Molecular epidemiology of adenoviruses: alternating appearance of two different genome types of adenovirus 7 during epidemic outbreaks in Europe from 1958 to 1980. Infect. Immun. **34:**368–372.
19. **Wadell, G., M. L. Hammarskjold, G. Winberg, T. M. Varsanyi, and G. Sundell.** 1980. Genetic variability of adenoviruses. Ann. N.Y. Acad. Sci. **354:**16–42.
20. **Wigand, R., A. Bartha, R. S. Dreizin, H. Esche, H. S. Ginsberg, M. Green, J. C. Hierholzer, S. S. Kalter, J. B. McFerran, U. Pettersson, W. C. Russell, and G. Wadell.** 1982. *Adenoviridae*: second report. Intervirology **18:**169–176.

Chapter 64

Herpes Simplex Viruses

W. LAWRENCE DREW AND WILLIAM E. RAWLS

CLINICAL BACKGROUND

There are two types of herpes simplex virus, types 1 and 2 (HSV-1 and HSV-2); both are common human pathogens (33). They establish latent infections, which may be associated with recurrent episodes of virus excretion. These episodes may or may not be associated with lesions, but in either case they provide a source for infection of susceptible individuals. A primary infection in either children or adults who have no HSV antibody is commonly mild or inapparent but may be quite severe. The common sites of lesion development, depending upon the type involved, include the genitalia, oral cavity (gingivostomatitis), pharynx (18, 43), esophagus (47), eye (27), and skin of the fingers (whitlow). Pain, fever, and lymphadenopathy are often associated with primary infections; dysuria may occur when the genitalia are the site of virus replication. Abraded or traumatized skin is more susceptible to infection than is intact skin, and patients with eczema may acquire a serious generalized infection (Kaposi's varicelliform eruption). Serious infections of the central nervous system include acute necrotizing encephalitis, usually caused by HSV-2 in neonates and HSV-1 thereafter, and meningitis, usually caused by HSV-2 (12). The viruses may also produce fatal illnesses in patients with deficiencies in immune functions. Despite HSV antibody, a portion of the population can experience recurrent lesions around the mouth, genitalia, fingers, and eyes.

HSV-1 and HSV-2 are closely related, yet they are biologically and serologically distinguishable (33). HSV-1 is usually associated with skin lesions above the waist, encephalitis, stomatitis, eye infections, and some cases of generalized herpes simplex. HSV-2 is found primarily in and on the genitalia and surrounding areas, is usually venereally transmitted, and causes most cases of generalized infections of the newborn. The relationship between the location of lesions and the HSV type involved is not, however, absolute. Up to 25% of HSV recovered from genital sites may be type 1, and a similar percentage of oral isolates are type 2. The reported association of HSV-2 with cervical carcinoma has led to substantial interest in HSV as a possible cause of human cancer.

HSV disease can be confused with herpangina (caused by type A coxsackieviruses), aphthous stomatitis, varicella-zoster, and other diseases involving skin eruptions. Encephalitis due to HSV-1 in postneonates is often indistinguishable from that due to other viruses, or even other types of microorganisms. In such patients, the virus is rarely recovered from cerebrospinal fluid (CSF), and a brain biopsy is mandatory to confirm the diagnosis. The accurate diagnosis of HSV infections is important, since antiviral agents may alter the outcome of these illnesses.

DESCRIPTION OF AGENT

HSV consists of a central core surrounded by an envelope derived from host cell membranes. The genome is a linear, double-stranded, single molecule of DNA with a molecular weight of just less than 10^8 (16). The DNA is contained in a protein core surrounded by 162 capsomeres arranged with icosahedral symmetry. During maturation, the nucleocapsid is enveloped by a membrane when the virus buds out of the nucleus. Virus-specified glycoproteins accumulate in the membranes through which the virus buds. Between the envelope membrane and the nucleocapsid is a space filled with an amorphous material. The complete virus particles have a diameter of between 110 and 220 nm and a buoyant density of 1.271 g/cm^3 in cesium chloride. As would be expected from the size of the genome, HSV induces the synthesis of multiple viral proteins; 49 and 57 polypeptides have been identified in cells infected with HSV-1 and HSV-2, respectively (11, 23, 38). The lipids of the virion are exclusively associated with the enveloping membrane and are similar to those found in the nuclear membrane of the host cell.

Since the envelope contains essential phospholipid, the virus is inactivated by lipid solvents. As with other viruses sensitive to lipid solvents, HSV is also inactivated by a pH lower than 4, as well as by temperatures of 56°C for 0.5 h. HSV replication can be blocked by a variety of inhibitors of DNA synthesis.

The two serotypes of HSV share common antigens, and both are at least partially neutralized by antisera produced to the heterologous type. HSV-1 and HSV-2 also have specific antigens and differ from each other in many biological characteristics, mostly of a quantitative nature (33). Only those differences that seem pertinent to diagnostic procedures will be mentioned in this chapter.

COLLECTION AND STORAGE OF SPECIMENS

Specimens for HSV culture can be readily obtained by swabbing open skin lesions or mucous membrane ulcers with a cotton-tipped applicator. Calcium alginate swabs should not be used (13). Fluid from fresh vesicles provides a rich source of virus. It can be aspirated with a tuberculin syringe equipped with a 25-gauge needle or absorbed onto a cotton swab after the vesicle is aseptically opened with a sterile needle. Swabs of pharyngeal or vaginal secretions may yield virus in cases where no apparent oral or perineal lesions are present. Dry swabs produce fewer viral isolations than do swabs that are moist.

In addition to vesicle fluids and debris from ulcers, CSF, urine, blood leukocytes, and specimens from diseased organs may yield virus. Samples from these sources should be collected in sterile containers and sent immediately to the laboratory for processing. A common practice is to homogenize biopsy or necropsy brain tissue from suspected cases of HSV encephalitis, producing a 10% tissue suspension in sterile broth. HSV can be isolated even more efficiently by teasing or trypsinizing the tissues before inoculation into cell cultures (4, 45). HSV-1 is rarely isolated from CSF;

however, HSV-2 can be isolated from CSF samples from patients with meningitis (12). HSV-2 can also be isolated from urine samples taken from patients with primary genital herpes complicated by cystitis. Demonstrable viremia may also occur in primary HSV infections.

Specimens for culture can be directly inoculated into tubes of cell culture if possible. There are antibiotics in the fluid media of the cell culture tubes which will inhibit the few bacteria that may contaminate the specimen. If such direct inoculation is not feasible, then transportation in a broth medium is necessary. For short-term transportation, i.e., 2 to 4 h, buffered salt solution such as in the convenient "Culturette" (Marion Scientific Corp., Rockford, Ill.) device may be satisfactory. For longer transportation, the use of a broth which contains protein (for example, TSY broth, skim milk, veal infusion broth) is probably superior.

HSV is hardier than respiratory syncytial virus but certainly less hardy than the picornaviruses. Highest isolation rates are generally thought to occur if a swab extract is inoculated into cell cultures on the day the specimen is taken, but if a delay in inoculating cell culture must occur, specimens should be refrigerated but *not* frozen (54). The cycle of freezing and thawing is extremely deleterious to HSV. HSV can be stored for prolonged periods at low temperatures (−60°C) after addition of 20% heat-inactivated serum (56°C for 0.5 h), but freezing and thawing rates should be rapid (1).

DIRECT EXAMINATION

Cells scraped from the base of herpes-like lesions may show cytological changes characteristic of HSV infection. In the Tzanck test, the cells are smeared onto a slide, fixed, and stained with Wright or Giemsa preparations (7). Syncytial giant cells (polykaryocytes) and "ballooning" cytoplasm are observed in lesions produced by either HSV or varicella-zoster virus; thus the test does not distinguish between lesions caused by these two viruses. Cowdry type A intranuclear inclusions may be found in infected cells which are stained with hematoxylin and eosin; however, these inclusions may also be produced by either HSV or varicella-zoster virus. Since cytological examination is a relatively insensitive technique, intranuclear inclusions are often not detected in tissues which are virus positive by culture.

Vesicle fluids usually contain sufficient particles to be detected with ease by electron microscopy (46). The electron microscopist cannot distinguish HSV from varicella-zoster virus but can readily distinguish herpesvirus from poxvirus. A rapid (1.5 to 2 h), definitive determination of HSV involvement can also be made by demonstrating virus antigen in cells by immunofluorescence (IF) (22, 29, 41, 49) or by immunoperoxidase (5) techniques. In the direct fluorescent-antibody test, the globulin is prepared and conjugated with fluorescein isothiocyanate. The specificity and titer of the labeled antibody should be determined by using cells infected with known strains of HSV and uninfected cells. Cells scraped directly from the base of herpetic lesions can be smeared onto slides, fixed in acetone for 10 min at room temperature, and stained with 1 drop of the conjugate. After incubation at 37°C for 30 min, the excess conjugate is removed by three washes in buffered saline and one wash in distilled water. The air-dried slide is then mounted in 25% buffered glycerol with a cover slip and examined with a fluorescence microscope. Although they are more rapid than culture, cytological and antigen detection techniques are, respectively, only 50% and 70 to 80% as sensitive as viral isolation (29, 41). Conversely, there definitely are instances in which the IF test is positive but the culture is negative. If appropriate controls, including more than one antibody, have been used, these are true-positive results attributable to a loss of viral viability (41). False-positive IF tests are less of a problem with the direct IF procedure than with the indirect test (41). Since culture and the IF test may complement each other, both procedures should be performed in critical situations (see Interpretation of Results, below).

VIRUS ISOLATION

Recovery of HSV in culture is at least partially related to the type of lesion; in one study, virus was recovered from 94% of vesicular lesions, 87% of pustular lesions, 70% of ulcers, and 27% of crusted lesions (29). HSV-1 and HSV-2 replicate readily in a number of primary and established cell culture systems, and a variety of these have been used for virus isolation. Most workers have found that primary or established cell lines of human or primate origin are satisfactory. Primary or passaged cultures of cells from human embryonic tissues are convenient since they are commercially available and are sensitive to a wide range of other viruses isolated from clinical specimens. Of the established cell lines, Vero cells and human diploid lines are recommended. They multiply well in Eagle basal medium containing 10% fetal bovine serum and maintain their morphology for prolonged periods in the same medium supplemented with 20 mM HEPES (*N*-2-hydroxyethylpiperazine-*N*′-2-ethanesulfonic acid) to achieve a higher buffer capacity. Fetal bovine serum, which can be reduced to 2% in maintenance medium, has no suppressive effect on the isolation of HSV-1 or HSV-2.

HSV will produce cytopathic effects (CPE) in heteroploid cells such as HeLa and HEp-2, but these cells are generally considered to be inferior to Vero cells or human embryonic fibroblasts, as are rhesus monkey kidney cells (20). Primary rabbit kidney cells are popular for research and are suggested by some as the most sensitive cell for the isolation of HSV. Unfortunately, this cell type has not been shown to be useful for detecting the many other human viruses that are routinely isolated in human fibroblasts in a comprehensive viral diagnostic program. Moreover, they are more expensive to utilize since they are only used as primary cells.

With specimens containing high concentrations of virus, CPE may be seen within 24 h, but with lesser virus inocula more time is required to destroy the cell monolayer (19). In general, the mean time to detection of CPE should be 2 to 3 days. Blind passage of cultures which do not develop visible CPE after 7 days of incubation is rarely fruitful. Two types of HSV CPE can be observed. The usual form begins with cytoplasmic granulation, after which the cells become enlarged and appear ballooned. The enlarged cells become rounded and take on a refractile appearance. Less commonly, the virus induces fusion of neighboring cells and the nuclei of the cells aggregate, giving rise to multinucleated giant cells. Although not absolute, the CPE induced by HSV-2 tends to be focal whereas HSV-1 produces CPE scattered throughout the monolayer. The distinctive features of the CPE produced by HSV may be lost when all the cells in the monolayer become involved. In rare cases, with a vesicle fluid rich in varicella-zoster virus, human embryonic fibroblasts may show a generalized and advanced CPE which may be confused with that caused by HSV (20, 21). Such confusion may be resolved by the subpassage of cell-free culture medium to fresh cultures. Varicella-zoster virus will not subpassage from cell-free medium.

IDENTIFICATION OF HSV

In a comprehensive viral diagnostic program, frequent isolations of HSV provide the necessary familiarity with viral CPE to permit tentative identification of the isolates without definitive testing. It has been shown that 95% of HSV isolated can be accurately reported before serotyping (21). However, in critical situations, e.g., brain or other visceral isolates and isolates from newborns, the virus should be definitively identified.

Isolates of HSV-1 may be distinguished from isolates of HSV-2 by biochemical, biological, or immunological methods. The restriction endonuclease cleavage patterns of the DNA of HSV-1 and HSV-2 are unique and allow unequivocal typing of isolates. In addition, minor variations in patterns between isolates of either type of virus exist and may be of epidemiological interest (9). The expertise required to examine isolates by this method is not available in most routine diagnostic laboratories, and interested readers should refer to details published elsewhere (9). Of the biological differences noted between HSV-1 and HSV-2, replication in chicken embryo cells represents one of the most reliable methods of differentiation. Newly isolated strains of HSV-1 do not replicate well in chicken embryo cells, whereas HSV-2 strains do. Titers of the unknown isolate are determined simultaneously in chicken embryo cells and Vero or human embryo cells; this can be efficiently done in microtiter plates (34, 53). Similar titers in the two cell systems will be found for HSV-2 isolates, whereas HSV-1 isolates will have titers 2 $\log_{10}$ lower in chicken embryo cells than in the primate cells.

The immunological methods used to differentiate HSV-1 from HSV-2 isolates have included kinetics of neutralization (42), endpoint neutralization (37), IF (30), solid-phase radioimmunoassay (14), inhibition of passive hemagglutination (3), and immunoperoxidase assays (5). Of the described methods, IF has been most widely used. For this test, cells are harvested from infected cultures exhibiting 3+ to 4+ CPE and are removed from the culture medium by centrifugation at 200 × *g* for 10 min. The cells are then suspended to 10^6 cells per ml in phosphate-buffered saline containing 2% fetal bovine serum. About 0.05-ml amounts of cell suspension are spotted onto microscope slides, and after the spots are air dried the cells are fixed for 10 min in cold acetone. Since antiserum to either virus will contain substantial cross-reacting antibody, it is necessary to assay the antisera against known HSV-1 and HSV-2 strains to determine the dilution of the antisera that must be used to differentiate isolates. Enhanced specificity can be obtained by adsorbing antisera with heterologous virus antigens. Monoclonal antibodies to HSV-1 and -2 are now commercially available; if they are used, titration and absorption are unnecessary but the reagents should be tested for the presence of heterologous antibody. The test is performed by covering the cell spot with a drop of appropriately diluted typing antiserum. After incubation at 37°C for 30 min, the slides are washed three times in phosphate-buffered saline, and a drop of fluorescein-conjugated antibody directed against the immunoglobulin of the typing antiserum is added. The slides are incubated for an additional 30 min and washed three times in phosphate-buffered saline, cover slips are mounted over the spots, and the cells are examined under a fluorescence microscope. Each assay should include cells infected with known HSV-1 and HSV-2 as well as uninfected cells for control purposes.

SEROLOGICAL PROCEDURES

Seroconversion in paired sera may provide evidence of a primary HSV infection. However, a significant rise in antibody titers does not always accompany recurrent disease; accordingly, serological tests should not be used in an attempt to diagnose recurrent herpes infection. To demonstrate the development of antibody to HSV, the complement fixation test may be used but is being supplanted by more convenient procedures. All complement fixation reagents, including HSV antigen and positive control sera, are readily obtained from commercial sources. Complement fixation antigen can also be readily prepared by infecting monolayers of cells and scraping the cells into the culture medium when 4+ CPE is evident. The cells are resuspended to about 10^8/ml in Veronal buffer and disrupted by three cycles of freeze-thawing. The preparation is then clarified by centrifugation and stored frozen. An alternative is to extract the infected cells with saline buffered with 0.05 M glycine-NaOH (35). The procedure for the complement fixation test is as described by Palmer et al. (34a). The adequacy of the HSV antigen should first be determined in a box titration using both known negative and positive sera. The test proper would, of course, require at least two sera, one taken as early as possible in the course of the disease and a second taken no less than 2 weeks later. Both sera must be heated at 56°C for 0.5 h to destroy any complement activity.

Other serological procedures can be used to quantitate antibodies to HSV-1 and HSV-2. The more sensitive methods have been modified to quantitate type-specific antibodies to HSV-1 and HSV-2 by adsorbing sera with heterologous antigens. These include indirect IF (17), indirect hemagglutination (2, 26), solid-phase radioimmunoassay (15, 36), and enzyme immunoassay procedures. Readers interested in these methods are referred to the published descriptions.

The neutralization test is still another method that may permit diagnosis of primary infections with either HSV-1 or HSV-2. This can be satisfactorily performed in microtiter plates (37, 39, 48). The test is performed by diluting the serum 1:10 to 1:640 in Eagle medium containing 2% fetal bovine serum. Each serum dilution is divided into 0.45-ml amounts. Stocks of HSV-1 and HSV-2 are diluted to contain 5×10^4 infectious units per ml. To 0.45 ml of each serum dilution is added 0.05 ml of either HSV-1 or HSV-2, and the virus-serum mixtures are held for 1 h at 4°C. The cells to be used for substrate, i.e., Vero cells, are monodispersed with trypsin and suspended at a concentration of 3×10^5/ml in Eagle medium containing 10% fetal bovine serum. Each well of the microtiter plate receives 0.05 ml of virus-serum mixture and 0.05 ml of cell suspension; the first six wells of each row receive a serum dilution mixed with HSV-1, and the second six wells receive the same serum dilution mixed with HSV-2. Cells and the lowest serum dilution are placed in wells of the last row in the plate as controls. After 3 days of incubation at 37°C, the plates are immersed for 1 min in a solution of 5% Formalin in phosphate-buffered saline (pH 7.2) and then immersed in a solution of crystal violet. The plates are then washed in tap water and examined. The highest dilution of the serum which still protects 50% of the monolayer from viral CPE is recorded as the antibody titer. Neutralizing titers against HSV-2 equal to or greater than the titers against HSV-1 are commonly observed in patients infected with HSV-2 (26, 39). However, relatively high titers to HSV-2 are not often observed among patients infected with both HSV-1 and HSV-2 (28).

Virus isolation and the demonstration of viral antigen in cells are the methods of choice in confirming the diagnosis of herpetic lesions. It is clear that neither IF nor culture is 100% sensitive, and therefore both tests should be performed in critical situations, e.g., in pregnant women near term when a genital lesion is observed. Determination that the lesion is herpetic may influence the decision to perform a cesarean section if the patient is about to deliver. If such a patient has a history of herpes and there are clinical features which support this diagnosis, many experts feel that the lesions should be considered as herpetic, irrespective of the result of culture or IF. On the other hand, IF may well be able to immediately confirm the clinical impression, and if time permits, culture confirmation may be obtained in as little as in 1 or 2 days. Direct fluorescent-antibody examination should not be performed on blind specimens from the cervix or vagina since cellular material in these secretions may interfere with the reading of the test. HSV infections of the central nervous system represent another critical situation. Although a definitive diagnosis of meningitis can be made by isolating HSV-2 from CSF, in postneonates with encephalitis it is usually necessary to perform a brain biopsy to demonstrate virus. These examinations should include culture, antigen detection, and electron microscopy if available.

INTERPRETATION OF RESULTS

Chemotherapeutic agents may be of value in altering the clinical course of certain HSV infections, and thus rapid, accurate diagnosis is important for optimal patient management. Inhibitors of DNA synthesis such as adenine arabinoside have demonstrable therapeutic value for eye and brain infections (27, 52). Recent studies with another inhibitor, acyclovir, suggest that this drug may be more effective than earlier agents since it is much less toxic (40, 44).

In less critical, ordinary herpetic infections, HSV can be readily isolated and promptly reported to the physician, and it should rarely be necessary to perform direct IF testing. A positive viral culture from the throat must be interpreted with some caution because herpes simplex may reactivate during any febrile episode. Therefore, the pertinence of virus isolation with respect to disease rests primarily upon the clinical circumstances associated with the particular patient.

Serological diagnosis of HSV infections is more limited in value. Significant increases in antibody titers are normally found only with primary infections, and this method is of little value in diagnosing recurrent disease or reinfections. In addition, patients with varicella-zoster infections may have a rise in titer to HSV antigens. Serological diagnosis may be helpful in differentiating between primary infection with HSV-1 versus HSV-2. However, among patients with a prior HSV-1 infection, a subsequent HSV-2 infection may not produce sufficient amounts of HSV-2 antibodies to be detected by most available methods.

B VIRUS

B virus (*Herpesvirus simiae*), which is indigenous to Asian monkeys (24), is important because it can cause a highly lethal central nervous system infection in humans when transmitted by monkey bites, saliva, or even tissues and cells so widely used in virus laboratories. B virus is the simian counterpart of HSV in humans and causes subclinical infections as well as dermal, oral, or eye lesions in monkeys. Once infected, a human may show localized redness and vesicles and may have pain at the site of virus entrance. Vesicles on the mucous membranes, pneumonia, diarrhea, abdominal pain, and pharyngitis have been reported, but in all cases an encephalopathy occurs which is frequently fatal; most who survive are left with serious brain damage (8).

B virus is similar to HSV in morphology and size and is also inhibited by agents that suppress DNA synthesis. It is assumed that this is a typical enveloped herpesvirus and hence sensitive to lipid solvents and an acid pH. There is an antigenic relationship between HSV and B virus in that antisera to B virus will neutralize both viruses equally well whereas antisera to HSV neutralize homologous virus at much higher titers than B virus (17, 25, 50). The magnitude of the cross-neutralization is enhanced by complement.

B virus is reported to be somewhat hardier than HSV; it will survive for a few weeks at 4°C, loses little titer when stored at −70°C, and is inactivated at 37°C within 7 days. Like HSV, B virus can be isolated in a number of cell and animal systems. An interesting difference is the ability of rhesus monkey kidney cells to readily support the replication of B virus, whereas HSV isolates grow poorly in these cells. The CPE produced by B virus in cultured cells is similar to that produced by HSV.

Because of the potential hazard of B virus to laboratory workers, caution should be exercised in handling specimens from patients suspected of suffering from B virus-induced disease. Investigations of specimens from these patients might best be referred to those expert in handling this virus. Inquiries can be directed to the Centers for Disease Control or similar agencies. The diagnosis of B virus infections can be established by virus isolation or by serological means. Specimens of vesicular fluids or, more commonly, biopsy specimens from diseased tissues can be inoculated onto cultures of Vero cells or primary rhesus monkey kidney cells. HSV-like CPE will become evident within 7 to 10 days in cultures inoculated with positive specimens (51). The virus isolate can be identified by neutralization tests using antisera to B virus and HSV. An isolate neutralized well by antiserum to B virus but poorly by anti-HSV serum is B virus, whereas an isolate neutralized by both antisera is HSV. Both specific sera and virus are obtainable from the Research Resources Branch of the National Institutes of Health, but use of this virus requires a Public Health Service permit. The serological diagnosis of B virus infection requires the demonstration of a fourfold or greater rise in titers to the virus in paired serum samples. Patients with preexisting antibodies to HSV may have sufficient titers to cross-react with B virus (10). In infected patients a rise in titers to HSV is observed along with an increase in titers to B virus. Increases in antibodies are observed about 2 weeks after the onset of illness, and with time the ratio of antibody titers for B virus to titers for HSV approaches unity. Antibodies may be measured by a plaque reduction test (50), by microneutralization assay (17), or by complement fixation (17). It is also possible to make a presumptive diagnosis of B virus encephalitis by quantitating antibody activity to B virus and HSV in serially collected samples of CSF (8).

LITERATURE CITED

1. **Ash, R. J., and E. R. Barnhart.** 1975. Optimal cooling and warming rates in the preservation of herpes simplex virus (type 2). J. Clin. Microbiol. **2**:270–271.
2. **Back, A. F., and N. J. Schmidt.** 1974. Indirect hemagglutinating antibody response to *Herpesvirus hominis* types 1 and 2 in immunized laboratory animals and in natural infections of man. Appl. Microbiol. **28**:392–399.
3. **Back, A. F., and N. J. Schmidt.** 1974. Typing *Herpesvirus hominis* antibodies and isolates by inhibition of the direct hemagglutination reaction. Appl. Microbiol. **28**:400–405.
4. **Baringer, J. R., and P. Swoveland.** 1973. Recovery of herpes simplex virus from trigeminal ganglions. N. Engl. J. Med. **288**:648–650.
5. **Benjamin, D. R.** 1974. Rapid typing of herpes simplex virus strains using the indirect immunoperoxidase method. Appl. Microbiol. **28**:568–571.
6. **Bishai, F. R., and N. A. Labzoffsky.** 1974. Stability of different viruses in a newly developed transport medium. Can. J. Microbiol. **20**:75–80.
7. **Blank, H., C. F. Burgoon, G. D. Baldridge, P. L. McCarthy, and F. Urbach.** 1951. Cytologic smears in diagnosis of herpes simplex, herpes zoster and varicella. J. Am. Med. Assoc. **146**:1410–1412.
8. **Bryan, B. L., C. D. Espana, R. W. Emmons, N. Vijayan, and P. D. Hoeprich.** 1975. Recovery from encephalomyelitis caused by Herpesvirus simiae. Report of a case. Arch. Intern. Med. **135**:868–870.
9. **Buchman, T. G., B. Roizman, G. Adams, and B. H. Stover.** 1978. Restriction endonuclease fingerprinting of herpes simplex virus DNA: a novel epidemiological tool applied to a nosocomial outbreak. J. Infect. Dis. **138**:488–498.
10. **Cabasso, V. J., W. A. Chappell, J. E. Avampato, and J. L. Bittle.** 1978. Correlation of B virus and herpes simplex virus antibodies in human sera. J. Lab. Clin. Med. **70**:170–178.
11. **Cassai, E. N., M. Sarmiento, and P. G. Spear.** 1975. Comparison of the virion proteins specified by herpes simplex virus types 1 and 2. Virology **16**:1327–1331.
12. **Craig, C. P., and A. J. Nahmias.** 1973. Different patterns of neurologic involvement with herpes simplex virus types 1 and 2: isolation of herpes simplex virus type 2 from the buffy coat of two adults with meningitis. J. Infect. Dis. **127**:365–372.
13. **Crane, L. R., P. A. Gutterman, T. Chapel, and A. M. Lerner.** 1980. Incubation of swab materials with herpes simplex virus. J. Infect. Dis. **141**:531.
14. **Forghani, B., N. J. Schmidt, and E. H. Lennette.** 1974. Solid phase radioimmunoassay for identification of *Herpesvirus hominis* types 1 and 2 from clinical materials. Appl. Microbiol. **28**:661–667.
15. **Forghani, B., N. J. Schmidt, and E. H. Lennette.** 1975. Solid phase radioimmunoassay for typing herpes simplex viral antibodies in human sera. J. Clin. Microbiol. **2**:410–418.
16. **Frenkel, N., and B. Roizman.** 1972. Separation of the herpesvirus deoxyribonucleic acid duplex into unique fragments and intact strand on sedimentation in alkaline gradients. J. Virol. **10**:565–572.
17. **Gary, G. W., Jr., and E. L. Palmer.** 1977. Comparative complement fixation and serum neutralization antibody titers to herpes simplex virus type 1 and *Herpesvirus simiae* in *Macaca mulatta* and humans. J. Clin. Microbiol. **5**:465–470.
18. **Glezen, W. P., G. W. Fernold, and J. B. Lohr.** 1975. Acute respiratory disease of university students with special reference to the etiologic role of *Herpesvirus hominis*. Am. J. Epidemiol. **101**:111–121.
19. **Hanna, L. H., H. Keshishyan, F. Jawetz, and V. R. Coleman.** 1975. Diagnosis of *Herpesvirus hominis* infections in a general hospital laboratory. J. Clin. Microbiol. **1**:318–323.
20. **Herrmann, E. C., Jr.** 1967. Experiences in laboratory diagnosis of herpes simplex, varicella-zoster, and vaccinia virus infections in routine medical practice. Mayo Clin. Proc. **42**:744–753.
21. **Herrmann, E. C., Jr.** 1971. Efforts toward a more useful viral diagnostic laboratory. Am. J. Clin. Pathol. **56**:681–686.
22. **Hitchcock, G. P., P. L. Randell, and M. M. Wishart.** 1974. Herpes simplex lesions of the skin diagnosed by the immunofluorescence technique. Med. J. Aust. **2**:280–284.
23. **Honess, R. W., and B. Roizman.** 1974. Regulation of herpesvirus macromolecular synthesis. I. Cascade regulation of the synthesis of three groups of viral proteins. J. Virol. **14**:8–19.
24. **Hull, R. N.** 1968. The simian viruses. Virol. Monogr. **2**:22–25.
25. **Hull, R. N., F. B. Peck, T. G. Ward, and L. C. Nash.** 1962. Immunization against B virus infection. II. Further laboratory and clinical studies with an experimental vaccine. Am. J. Hyg. **76**:239–251.
26. **Johnson, L. D., D. A. Fuccillo, H. Stadler, M. A. Oxman, C. E. O. Fraser, and D. L. Madden.** 1979. Comparison of indirect hemagglutination and indirect immunofluorescence tests with microneutralization test for detection of type-specific *Herpesvirus hominis* antibodies. J. Clin. Microbiol. **9**:384–390.
27. **Kaufman, H. E.** 1978. Herpetic keratitis. Invest. Ophthalmol. Visual Sci. **17**:941–957.
28. **McClung, H. P., P. Seth, and W. E. Rawls.** 1976. Relative concentrations in human sera of antibodies to cross-reacting and specific antigens of herpes simplex virus types 1 and 2. Am. J. Epidemiol. **104**:192–201.

29. **Moseley, R. C., L. Corey, D. Benjamin, C. Winter, and M. L. Remington.** 1981. Comparison of viral isolation, direct immunofluorescence, and indirect immunoperoxidase techniques for detection of genital herpes simplex virus infection. J. Clin. Microbiol. **13:**913–918.
30. **Nahmias, A., I. DelBuono, J. Pipkin, K. Hutton, and C. Wickliffe.** 1971. Rapid identification and typing of herpes simplex virus types 1 and 2 by a direct immunofluorescence technique. Appl. Microbiol. **22:**455–458.
31. **Nahmias, A., C. Wickliffe, J. Pipkin, A. Leibovitz, and R. Hutton.** 1971. Transport media for herpes simplex virus types 1 and 2. Appl. Microbiol. **22:**451–454.
32. **Nahmias, A. J., W. R. Dowdle, Z. M. Naib, A. Highsmith, R. W. Harwell, and W. E. Josey.** 1968. Relation of pock size on chorioallantoic membrane to antigenic type of herpesvirus hominis. Proc. Soc. Exp. Biol. Med. **127:**1022–1028.
33. **Nahmias, A. J., and B. Roizman.** 1973. Infection with herpes simplex 1 and 2. N. Engl. J. Med. **289:**667–674, 719–725, 781–789.
34. **Nordlund, J. J., C. Anderson, G. D. Hsiung, and R. B. Tenser.** 1977. The use of temperature sensitivity and selective cell culture systems for differentiation of herpes simplex virus types 1 and 2 in a clinical laboratory. Proc. Soc. Exp. Biol. Med. **155:**118–123.
34a. **Palmer, D. F., L. Kaufman, W. Kaplan, and J. J. Cavallaro.** 1977. Serodiagnosis of mycotic diseases. Charles C Thomas, Publisher, Springfield, Ill.
35. **Palmer, E. L., M. L. Martin, and D. T. Warfield.** 1971. Preparation of type 2 herpes simplex virus complement-fixing antigen. Appl. Microbiol. **22:**925–927.
36. **Patterson, W. R., W. L. Rawls, and K. O. Smith.** 1978. Differentiation of serum antibodies to herpesvirus types 1 and 2 by radioimmunoassay. Proc. Soc. Exp. Biol. Med. **157:**273–277.
37. **Pauls, F. P., and W. R. Dowdle.** 1967. A serologic study of *Herpesvirus hominis* strains by microneutralization tests. J. Immunol. **98:**941–947.
38. **Powell, K. L., and R. Courtney.** 1975. Polypeptides synthesized in herpes simplex virus type 2 infected cells. Virology **66:**217–228.
39. **Rawls, W. E., K. Iwamoto, E. Adam, and J. L. Melnick.** 1970. Measurement of antibodies to herpesvirus types 1 and 2 in human sera. J. Immunol. **104:**599–606.
40. **Schaeffer, H. J., L. Beauchamp, P. deMiranda, G. B. Elian, D. J. Bauer, and P. Collins.** 1978. 9-(2-Hydroxyethoxy methyl) guanine activity against viruses of the herpes group. Nature (London) **272:**583–585.
41. **Schmidt, N. J., J. Dennis, V. Devlin, D. Gallo, and J. Mills.** 1983. Comparison of direct immunofluorescence and direct immunoperoxidase procedures for detection of herpes simplex virus antigen in lesion specimens. J. Clin. Microbiol. **18:**445–448.
42. **Schneweis, K. E., and H. Brandis.** 1961. Typendifferenzen beim Herpes Simplex Virus. Zentralbl. Bakteriol. Parasitenkd. Infektionskr. **183:**556–558.
43. **Sheridan, P. J., and E. C. Herrmann, Jr.** 1971. Intraoral lesions of adults associated with herpes simplex virus. Oral Surg. **32:**390–397.
44. **Shiota, H., S. Inoue, and S. Yamane.** 1979. Efficacy of acycloguanosine against herpetic ulcers in rabbit cornea. Br. J. Ophthalmol. **63:**425–428.
45. **Shope, T. C., J. Klein-Robbenhaar, and G. Miller.** 1972. Fatal encephalitis due to *Herpesvirus hominis*: use of intact brain cells for isolation of virus. J. Infect. Dis. **125:**542–544.
46. **Smith, K. O., and J. L. Melnick.** 1962. Recognition and quantitation of herpesvirus particles in human vesicular lesions. Science **137:**543–544.
47. **Springer, D. J., L. R. Da Costa, and I. T. Beck.** 1979. A syndrome of acute self-limiting ulcerative esophagitis in young adults probably due to herpes simplex virus. Am. J. Dig. Dis. **24:**535–539.
48. **Stalder, H., M. N. Oxman, and K. L. Herrmann.** 1975. Herpes simplex virus microneutralization: a simplification of the test. J. Infect. Dis. **131:**423–430.
49. **Tomlinson, A. H., I. J. Chinn, and F. O. MacCallum.** 1974. Immunofluorescence staining for the diagnosis of herpes encephalitis. J. Clin. Pathol. **27:**495–499.
50. **Veda, Y., I. Tagaya, and K. Shiroki.** 1968. Immunological relationship between herpes simplex virus and B virus. Arch. Gesamte Virusforsch. **24:**231–244.
51. **Vigoso, A. D.** 1975. Recovery of herpes simiae (B virus) from both primary and latent infections in rhesus monkeys. Br. J. Exp. Pathol. **56:**485–488.
52. **Whitley, R. J., S.-J. Soong, R. Dolin, G. J. Galasso, L. T. Ch'ien, C. A. Alford, and the Collaborative Study Group.** 1977. Adenine arabinoside therapy of biopsy-proved herpes simplex encephalitis: NIAID collaborative antiviral study. N. Engl. J. Med. **297:**289–294.
53. **Yang, J. P. S., W.-T. Chiang, J. L. Gale, and N. S. T. Chen.** 1975. A chick embryo cell microtest for typing of herpes virus hominis. Proc. Soc. Exp. Biol. Med. **148:**324–328.
54. **Yeager, A. S., B. A. Morris, and C. G. Prober.** 1979. Storage and transport of cultures for herpes simplex virus, type 2. Am. J. Clin. Pathol. **72:**977–979.

Chapter 65

Human Cytomegalovirus

STUART E. STARR AND HARVEY M. FRIEDMAN

CLINICAL BACKGROUND

Cytomegalovirus (CMV) infections are common and usually asymptomatic; however, the incidence and spectrum of disease in newborns and in immunocompromised hosts establish this virus as an important human pathogen (34). CMV infections can be classified as those acquired before birth (congenital), at the time of delivery (perinatal), or later in life (postnatal).

CMV infection has been detected in 0.5 to 2.5% of newborn infants and is the most common identified cause of congenital infection. Fewer than 5% of congenitally infected infants develop symptoms during the newborn period; possible manifestations range from severe disease with jaundice, hepatosplenomegaly, petechiae, and central nervous system abnormalities to more limited involvement. Symptomatic infants may die of complications within the first months of life; more commonly, they survive but are neurologically damaged. In recent years, it has been recognized that even those congenitally infected infants who are asymptomatic early in life may develop hearing defects.

Newborns can also acquire infection at the time of delivery by contact with virus in the birth canal. Such infants begin to excrete virus at 3 to 12 weeks of age but usually remain asymptomatic. Thus far, it appears that such perinatally infected infants do not develop late neurological sequelae of infection.

Most postnatal infections are acquired by close contact with individuals who are shedding virus. Since CMV has been detected in several body fluids, including saliva, urine, breast milk, cervical secretions, and semen, transmission can occur in a variety of ways. Prolonged shedding of virus after congenital or acquired CMV infection contributes to the ease of virus spread. In addition, CMV can be transmitted by blood transfusion and organ transplantation.

The vast majority of children and adults who acquire CMV infection postnatally remain asymptomatic. Hepatosplenomegaly and thrombocytopenia have been described in high-risk premature newborns infected as a result of blood transfusions. In young adults a mononucleosis-like illness with fever, lethargy, and atypical lymphocytosis can occur. A similar syndrome has been described after transfusion of large amounts of blood for open-heart surgery.

CMV infections are frequent and occasionally severe in children or adults with congenital or acquired defects of cellular immunity; these include patients with the acquired immunodeficiency syndrome (AIDS), cancer patients (particularly those with leukemia and lymphoma), and recipients of organ transplants. Infections in these patients may be due to primary infection, reactivation of latent virus, or reinfection with exogenous virus. Possible sources of virus include blood transfusions and, in the case of renal transplantation, the grafted kidney. After renal transplantation virtually all seropositive recipients shed virus, but most remain asymptomatic. Recipients undergoing primary infections are more likely to develop symptoms which may include fever, leukopenia, thrombocytopenia, pneumonitis, hepatitis, chorioretinitis, and encephalitis. Death may occur as a result of various complications, including bacterial and fungal superinfections. CMV infection, particularly when associated with pneumonitis, is an important cause of morbidity and mortality after bone marrow and cardiac transplantation.

Description of the agent

CMV is a member of the family Herpesviridae, which includes Epstein-Barr virus, herpes simplex virus, and varicella-zoster virus. Complete CMV particles consist of a core containing double-stranded DNA, an icosahedral capsid, and a surrounding envelope. Electron microscopic features of CMV include virions morphologically indistinguishable from those of other herpesviruses, a high ratio of defective viral particles, and the presence of spherical particles called dense bodies.

Molecular virological techniques have been used to study variation among CMV strains. By DNA-DNA reassociation kinetics analysis, various CMV strains have been found to share considerable homology with AD-169, a standard laboratory strain, and by restriction endonuclease analysis, DNAs from various strains have been shown to have similar but distinctive fragment migration patterns (18). These studies suggest that CMV strains are closely related, more so than are herpes simplex virus types 1 and 2. Antigenic heterogeneity among CMV strains has been detected in cross-neutralization and other serological assays, but evidence for distinct serotypes is limited (36).

CMV is inactivated by a number of physical and chemical treatments, including heat (56°C for 30 min), low pH, ether, and cycles of freezing and thawing.

COLLECTION AND STORAGE OF SPECIMENS

Specimens for virus isolation

CMV can be isolated from a variety of body fluids; however, urine and throat washings are the most useful for diagnostic purposes. In the evaluation of immunocompromised patients, buffy coat cultures are particularly useful since in some instances detection of CMV in leukocytes is a better indicator of symptomatic CMV infection than is shedding of virus in urine or throat. Biopsy and autopsy specimens can also be processed for virus isolation. The details of specimen collection and processing are given in Chapter 8. Since CMV loses titer when subjected to freezing and thawing, specimens should be kept at 4°C in an ice-water bath or in a refrigerator until they can be used to inoculate cultures, preferably within a few hours after collection. When prolonged transport times are unavoidable, infectivity is reasonably well preserved for at least 48 h at 4°C. If storage in the frozen state is necessary, an equal volume of 0.4 M sucrose-phosphate added to the specimen helps to preserve viral infectivity (16).

Specimens for serology

Single serum samples are useful to screen for evidence of past infection with CMV. This is especially helpful when testing sera from organ transplant donors and from donors of blood products which are to be administered to premature infants. For the diagnosis of recent CMV infection, paired sera should be obtained at least 2 weeks apart. If congenital infection is suspected, both maternal and infant sera should be submitted.

DIRECT EXAMINATION OF SPECIMENS

Histopathology

Characteristic large cells with intranuclear and, on occasion, cytoplasmic inclusions can be seen in routine sections of biopsy or autopsy material. Wright-Giemsa–stained touch imprints of lung or other biopsy specimens may demonstrate such cells (30). Although the presence of characteristic cytological changes suggests CMV infection, virological or serological confirmation is necessary. Since CMV can infect tissues without producing morphological changes, failure to find typical cytomegalic cells does not exclude CMV infection.

Exfoliative cytology

With exfoliative cytological techniques, a presumptive diagnosis of CMV can be made in 25 to 50% of cases of symptomatic congenital infection. Specimens from older infected individuals are rarely positive. Several fresh urine specimens should be submitted since exfoliated cells disintegrate rapidly and may be shed only intermittently. For best results the urine should be passed through a membrane filter and the trapped cells should be stained with hematoxylin-eosin or Papanicolaou stain. Characteristic large cells with prominent inclusions are seen in positive preparations. Rarely, similar-appearing cells are seen in herpes simplex virus infection. Exfoliative cytology is most useful when virus isolation techniques are not available.

Immunofluorescence

Monoclonal antibodies to CMV can be used for immunofluorescence tests on lung tissue obtained at the time of open-lung biopsy or autopsy (11, 35). The sensitivity of the assay is improved when mixtures of monoclonal antibodies are used (35). The main advantage of immunofluorescent staining is its rapidity; results are available within several hours after tissue is obtained. The sensitivity of this technique as compared to virus isolation remains to be determined.

Electron microscopy

The pseudoreplica method of electron microscopy can be used to detect CMV in urine and oral specimens of congenitally infected infants (25). Positive results can be obtained with almost all specimens which have infectivity titers of $\geq 10^4$ ml. An advantage of electron microscopy is its rapidity; also, stored or contaminated specimens, unsuitable for isolation attempts, can be examined. The main disadvantage of the technique is its relative insensitivity.

Hybridization

Molecular hybridization techniques have recently been described for detection of the CMV genome in urine (3, 31) and peripheral blood leukocytes (26, 31). Biotin-labeled DNA probes are currently being investigated as a substitute for radioactive probes. The relative sensitivity of these techniques needs to be further defined.

ISOLATION OF AGENT

Processing of specimens

Adjustment of urine specimens to pH 7.0 with 0.1 N NaOH or 0.1 N HCl is recommended to reduce toxicity to cell cultures. Centrifuging urine specimens to obtain sediment-enriched samples has been advocated, but is usually unnecessary. Sediment-enriched urines of renal transplant recipients may produce toxicity more frequently than do uncentrifuged urines (23).

A number of procedures have been described for obtaining buffy coat cells. We have found the following method to be suitable for a clinical laboratory. Dextran (6%) is added to a syringe containing 2 to 4 ml of heparinized peripheral blood, at a ratio of 1 ml of dextran per 2 ml of blood. The syringe is placed in an inverted position in a 37°C incubator for 30 to 60 min, and the upper leukocyte-rich fraction is expelled through a bent 21-gauge needle into a centrifuge tube. The cells are pelleted by low-speed centrifugation (300 × *g*), washed twice with phosphate-buffered saline (PBS), and resuspended in 1 ml of PBS. Alternatively, heparinized peripheral blood can be centrifuged on Ficoll-Hypaque gradients (17). Mononuclear cells are obtained from the resultant band, and polymorphonuclear cells are obtained by allowing the pellet to sediment with 6% dextran. A greater number of isolates may be obtained by the Ficoll-Hypaque method than by the conventional method (17).

Throat, biopsy, autopsy, and other specimens are processed as described in Chapter 8. All specimens are treated with antibiotics (as described in Chapter 8) before inoculation of cell cultures. Portions of specimens not used to inoculate cell cultures should be mixed with equal volumes of 0.4 M sucrose-phosphate and frozen at −70°C; some infectivity may be preserved if further isolation attempts are necessary.

Cell cultures

Human fibroblast cells best support the growth of CMV and are therefore used for diagnostic purposes. CMV will not replicate in standard laboratory cell cultures such as HeLa, HEp-2, or monkey kidney cells. Acceptable fibroblast cultures include those prepared from human embryonic tissues or foreskins and serially passaged diploid human fetal lung strains such as WI-38, MRC-5, or IMR-90 (9). Several of these fibroblast cell cultures are commercially available.

Inoculation and observation of cultures

Specimens to be tested are added in a volume of 0.2 ml to duplicate tubes of confluent fibroblasts maintained on Eagle minimal essential medium with 2% fetal bovine serum. Alternatively, the tubes are drained of medium, the inocula are allowed to absorb for 1 h, and then fresh medium is added. After inoculation, the tubes can be rolled or kept stationary at 36°C. Twenty-four hours later, the medium is changed for tubes inoculated with urine or buffy coat specimens. Thereafter, and for other types of specimens, medium changes are performed once a week, or more frequently if toxicity appears. When excessive toxicity necessitates passage of the culture, cells are removed by addition of 0.25% trypsin–0.1% EDTA to monolayers and incubation at 36°C for 5 or 10 min. When the cells detach, minimal essential medium with 2% fetal bovine serum is added, and the cells are used to inoculate fresh tubes. Tubes are examined for cytopathic effect (CPE) daily for the first 5 days and then twice a week for at least 4 weeks for most specimens and for 6 weeks for buffy coat specimens. Control, uninoculated cultures are handled in the same manner as those inoculated with clinical specimens.

The time of appearance and the extent of CPE depend on the amounts of virus present in specimens. In cultures inoculated with urine from a congenitally infected newborn, CPE may develop by 24 h and progress rapidly to involve most of the monolayer, if the virus titer of the urine is extremely high. More commonly, foci of CPE, consisting of enlarged, rounded, refractile cells, appear during the first week, and progression of CPE to surrounding cells proceeds slowly (Fig. 1). In cultures inoculated with urine or throat specimens from older individuals, CPE usually appears within 2 weeks. Buffy coat cultures may not become positive until after 3 to 6 weeks. The usual slow progression of CPE in cultures inoculated with clinical specimens is due, at least in part, to limited release of virus into extracellular fluid. With strains of CMV that have been serially passaged, including laboratory-adapted strains, greater amounts of extracellular virus are released, and CPE progresses more rapidly.

For storage of fresh isolates, monolayers exhibiting CPE are treated with trypsin-EDTA, and the cells obtained are suspended in Eagle minimal essential medium with 10% fetal bovine serum and 10% dimethyl sulfoxide and are then frozen at −70°C. Infectivity can be better maintained for long periods of time by storage in liquid nitrogen.

IDENTIFICATION OF ISOLATES

In many laboratories CMV isolates are identified solely on the basis of characteristic cytopathology and host cell range. Alternatively, the anticomplement immunofluorescence (ACIF) assay or the indirect immunofluorescent-antibody assay (IFA) can be performed using human serum known to contain antibodies to CMV but not to herpes simplex or varicella-zoster virus (see below for details of methods). In both immunofluorescence assays, the presence of typical nuclear fluorescence of infected cells indicates the presence of CMV. The ACIF test has also been used for rapid identification of CMV in cultures before the appearance of advanced CPE (32). Monoclonal antibodies, particularly to early antigens of CMV, can also be used for this purpose and may permit a more rapid diagnosis of CMV infection. Identification of isolates can also be done by means of complement-requiring neutralizing assays in which animal sera to CMV are used (38), but this is not practical for routine diagnosis.

SEROLOGICAL DIAGNOSIS

A variety of tests are available for serodiagnosis of CMV infection.

CF test

The complement fixation (CF) test is a commonly used assay in many clinical laboratories. All of the necessary reagents are commercially available. Generally, the broadly reactive AD-169 antigen is used. It can be prepared by glycine extraction (22) or by freeze-thawing (15). The CF test with glycine-extracted antigen is more sensitive for distinguishing seropositive from seronegative specimens (22), whereas

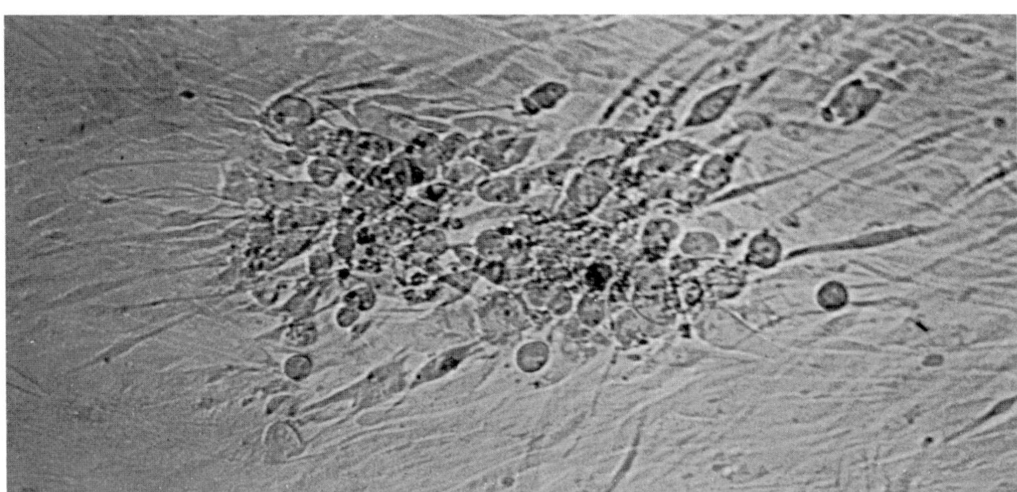

FIG. 1. CPE produced by a CMV isolate in human skin fibroblasts 10 days postinoculation. Unstained preparation; ×100. Courtesy of Sergio Stagno.

the freeze-thawed antigen is slightly better for detection of fourfold or greater titer rises (5). Choice of antigen preparation should in part depend on the purpose of the serological test. For example, in our laboratory we place considerable emphasis on a sensitive assay to distinguish seronegative from seropositive renal transplant donors and recipients, and therefore we use the glycine-extracted CMV antigen.

Before the test is performed, titrations of antigen, hemolysin, and complement are made so that 2 U of each reagent can be used in the assay (see reference 26a for details of the CF test). For performance of the test, the patient's serum is diluted 1:8 and is heat inactivated for 30 min at 56°C. Each serum is tested at dilutions of 1:8 to 1:1,024 against CMV antigen, at dilutions of 1:8 and 1:16 against control antigen (prepared from uninfected cells), and against buffer alone (to detect anticomplementary activity in the serum). Two units of complement are used, and complement controls contain 2.0, 1.5, 1.0, and 0.5 U of complement that has reacted with 2 U of CMV antigen or 2 U of control antigen. After overnight incubation at 4°C, a 1.4% suspension of hemolysin-sensitized sheep erythrocytes is added to each well, and plates are incubated at 37°C until the 2.0-, 1.5-, and 1.0-U complement control wells clear (15 to 30 min). Results of the antibody testing can then be read immediately, but generally endpoints are easier to interpret if the plates are incubated at 4°C for 1 to 2 h. A 3+ or greater cell button (on a scale of 0 to 4) is considered a positive reaction.

The antibody titer is the highest dilution of serum which gives a 3+ or greater reaction against CMV antigen while giving no reaction against control antigen or against buffer alone. For those sera which react against control antigen or buffer, an endpoint can still be determined if the reaction against CMV is at least two dilutions higher than the reaction in the control wells.

We find it unnecessary to carry out hemolysin and complement titrations each time the CF test is performed. Hemolysin is very stable at 4°C and can be titrated every 3 to 6 months. Complement, once titrated, can be dispensed into volumes appropriate for a single day's tests and frozen at −70°C. Complement should be purchased in large enough quantities so that each lot lasts approximately 6 months. To ensure adequate control, we include two negative, one low-positive, and one high-positive serum in each run. These samples must yield consistent results; otherwise, the test results are discarded.

ACIF test

The ACIF test is an immunofluorescence assay which detects CMV antibody (21, 29). An important advantage of this assay compared with CF testing is its rapidity, since results can be available within 2 to 3 h. In addition, sera which are anticomplementary by the CF procedure can be tested by the ACIF test. Also, nonspecific cytoplasmic staining caused by Fc receptors is avoided. Disadvantages of the ACIF test compared to the CF test are that fewer specimens can be handled daily, the test requires cell culture facilities and an immunofluorescence microscope, and examination of slides requires considerable time and experience.

For performance of the ACIF test, human embryonic lung fibroblasts are infected with CMV (generally the AD-169 strain) and are harvested with trypsin-EDTA treatment when the fibroblasts show advanced CPE. The cells are sedimented at $300 \times g$ for 5 min and are suspended to approximately 5×10^6 cells per ml. One drop is placed in each well of an eight-well Teflon-coated slide. One 75-cm^2 flask provides enough cells for 16 slides. The slides are air dried, fixed in cold acetone for 10 min, and stored at −20 or −70°C for subsequent use. Cells can be stored for several months without loss of CMV antigens. Before staining, each well of the slide is circled with embroidery ink to prevent serum from running into the adjacent wells. Heat-inactivated (56°C for 30 min) serum is added to each well in dilutions of 1:8 to 1:1,024. The slides are incubated for 30 min at 37°C in a moist chamber, washed twice for 5 min in PBS, and allowed to react for 45 min at 37°C with an optimal dilution of cold guinea pig complement. Slides are again washed twice in PBS, allowed to react with an optimal dilution of anti-guinea pig C3 fluorescent conjugate (commercially available) for 30 min at 37°C, washed twice in PBS for 5 min each time, and dipped in distilled water. Slides are air dried and mounted in 25% glycerol in PBS for reading at ×250 or ×400 magnification.

The optimal dilutions for complement and the fluorescent conjugate must be determined before testing by allowing CMV antibody-positive serum and serial dilutions of complement and conjugate to react with CMV-infected and uninfected cells. The optimal dilutions of complement and conjugate are the highest dilutions of reagents producing 3 to 4+ nuclear fluorescence of infected cells but little or no (0 to 1+) background fluorescence of uninfected cells.

As controls for the test, known positive and negative samples are included in each run. In addition, a single dilution of serum (1:8) is allowed to react with uninfected cells to detect antinuclear antibodies, which can produce false-positive results. The antibody titer is the highest dilution of serum producing 3+ or greater (on a scale of 0 to 4) nuclear fluorescence against CMV-infected fibroblasts while giving no reaction against uninfected control cells.

IFA

As an alternative to the ACIF test, the IFA can be performed (29). This test is slightly faster than the ACIF test but has the disadvantage that the assay should be performed on monolayers of infected cells; these are more difficult to prepare than the cell suspensions used for the ACIF test. In addition, specific nuclear fluorescence, which indicates a positive result, is sometimes difficult to distinguish from nonspecific cytoplasmic fluorescence. The latter occurs because CMV infection of fibroblasts induces a cytoplasmic Fc receptor which binds viral and nonviral immunoglobulin G (IgG) antibody (20).

For performance of the test, fibroblasts are grown in cell culture chamber slides and infected with CMV (generally AD-169 strain). When CPE is 3 to 4+ (approximately 3 days after inoculation), cells are fixed in cold acetone and stored (−20 or −70°C) or are used directly for antibody determination. Human sera are serially diluted from 1:8 to 1:1,024 and allowed to react with infected fibroblasts. As a control for antinuclear antibodies, serum diluted 1:8 is allowed to react with uninfected cells. Slides are incubated at 37°C for 30 min in a moist chamber, washed as described for the ACIF test, and stained with an optimal dilution of fluorescein-conjugated anti-human IgG. After a 30-min incubation at 37°C, slides are washed and mounted for microscopic reading as described in the ACIF procedure. The antibody titer is the highest serum dilution producing 3+ or greater (on a scale of 0 to 4) nuclear fluorescence when allowed to react with CMV-infected fibroblasts, while giving no nuclear staining of uninfected control cells.

The optimal dilution of fluorescent conjugate must be determined before the test is performed, by allowing CMV antibody-positive serum and serial dilutions of conjugate to react with CMV-infected and uninfected fibroblasts. The optimal dilution is the highest dilution of conjugate producing 3 to 4+ nuclear fluorescence of infected cells but little or no (0 to 1+) staining of uninfected cells.

Several immunofluorescent assay systems for CMV antibody detection are commercially available. These include a solid-phase immunofluorescent assay (FIAX) in which CMV antigen and control antigen are absorbed onto solid surfaces. A single dilution of the patient's serum is applied to the solid surfaces, an anti-human IgG fluorescein conjugate is added, and the intensity of fluorescence is measured in a fluorometer (10). CMV-infected culture cells on glass slides are also available commercially for measuring CMV antibodies.

Immune adherence hemagglutination assay

The immune adherence hemagglutination assay measures antibodies which react with CMV antigen and complement to form complexes (7, 24). The complexes bind to C3b receptors on human type O erythrocytes, causing the erythrocytes to agglutinate. Many of the necessary reagents are commercially available, including viral antigen and control antigen, complement, and Veronal-buffered saline (VBS). The immune adherence hemagglutination assay is faster than the CF test because it requires no overnight incubation. A large number of samples can be handled in a single day, and the assay is sensitive for detecting seropositivity and fourfold changes in titer (24). Disadvantages of the immune adherence hemagglutination assay include the need for human donors of type O erythrocytes (not all type O donors cells react equally well) and a lack of experience with this test, since relatively few laboratories have performed the assay.

The optimal concentrations of CMV antigen and complement are determined by box titration before the test is performed. Both antigen and complement are used in mild excess (slightly more than 1 U of each). Control antigen is used in the same dilution as CMV antigen. Human blood type O is collected in 2 volumes of sterile Alsever solution and stored at 4°C for up to 5 weeks. Before use, erythrocytes are washed twice in VBS and once in gelatin-Veronal buffer (prepared by adding autoclaved gelatin to a concentration of 0.125%) and are resuspended to 0.75% concentration in gelatin-Veronal buffer. Each time the test is performed, known negative and positive serum samples are included as controls.

For performance of the test, sera are diluted 1:4 in VBS and heat inactivated at 56°C for 30 min. All reagents are added in 0.025-ml volumes. Microtiter V-well plates are washed with gelatin-Veronal buffer and tapped dry against an absorbent surface before use (some lots of microtiter plates are unsatisfactory). Bovine serum albumin-VBS (0.1% fraction V bovine serum albumin) is distributed to all microtiter wells. Serial twofold dilutions of sera are prepared using microdiluters prewet with gelatin-Veronal buffer. CMV antigen and control antigen are diluted in bovine serum albumin-VBS. CMV antigen is distributed to nine wells containing serum dilutions from 1:8 to 1:2,048, and control antigen is added to the wells containing serum dilutions from 1:8 to 1:32. Plates are vibrated briefly, covered, and incubated at 37°C for 30 min. Complement, optimally diluted in bovine serum albumin-VBS, is added, and the plates are vibrated and then incubated at 37°C for 45 min. Dithiothreitol-EDTA-VBS in a volume of 0.025 ml is then added to stop the reaction, followed by the addition of the type O erythrocyte suspension. (Dithiothreitol-EDTA-VBS is prepared by mixing 2 volumes of 0.1 M disodium EDTA with 3 volumes of VBS and adding 3 mg of dithiothreitol per ml.) The plates are vibrated and left at room temperature for 1 to 3 h until a hemagglutination pattern develops. Agglutination of 3+ or more is considered positive. The antibody titer is the highest dilution of serum producing 3+ or greater (on a scale of 0 to 4+) hemagglutination with CMV antigen while giving weak or no agglutination with control antigen. Prozone reactions may occur at high excesses of antibody or antigen, and thus it is important to test sera at dilutions past the expected antibody endpoint, rather than screening at one or a few low dilutions.

Enzyme immunoassays

Kits that detect CMV IgG are available from a number of commercial sources. The kits are easy to use, and the manufacturers provide detailed instructions. All of the necessary materials are included to perform the assay. Some companies also provide a spectrophotometer, which otherwise must be purchased separately at considerable expense.

The tests are performed as follows. Microtiter wells are precoated with CMV antigen. A small volume (for example, 10 μl) of patient serum is added to a diluent which is then incubated with CMV antigen or control antigen in microtiter wells. Unbound antibody is removed by washing, and the conjugate (such as alkaline phosphatase-conjugated anti-human IgG) is added. After incubation, unbound conjugate is removed by washing, and enzyme substrate is added. The reaction is stopped by addition of sodium hydroxide, and the intensity of the color reaction is quantitated in a spectrophotometer. As controls, known positive and negative samples are included in each run, and each serum is reacted with microtiter wells coated with control antigen.

As described below, a number of studies have examined the specificity and sensitivity of CMV enzyme immunoassays. Some manufacturers provide kits which are intended only to discriminate between seronegative and seropositive samples, while others provide assays which will also detect titer rises. However, to date little information is available on the accuracy of commercially available assays for detecting significant titer rises. Therefore, the commercial kits may have limited utility in those laboratories which use CMV serology to determine recent infection.

Indirect hemagglutination

In the indirect hemagglutination assay, human type O erythrocytes are sensitized with CMV antigen. Human serum or plasma is added to antigen-coated erythrocytes which agglutinate if CMV antibody is present. The erythrocytes can be prepared as described by Yeager (39), using glutaraldehyde fixation and optimal concentrations of tannic acid and CMV antigen. The glutaraldehyde permits long-term storage of fixed cells, and use of optimal dilutions of tannic acid for each cell batch greatly reduces nonspecific hemagglutination and improves the sensitivity of the assay. Addition of 0.4 M lysine to the PBS used to dilute the sera and to adjust the erythrocytes to the working concentration also reduces nonspecific hemagglutination.

Recently, an indirect hemagglutination assay to determine the presence of CMV antibodies in human serum or plasma has become commercially available (28). As described below, the sensitivity and specificity of this assay compare favorably with those of several well-established techniques for measuring CMV antibody. However, this assay is not approved for detecting rises in antibody titers and therefore has some limitations for use in a clinical virology laboratory.

CMV IgM antibodies

Several methods for measuring CMV IgM antibodies have been described. A recognized pitfall of CMV IgM assays is false-positive results because of serum rheumatoid factor (6). This is an immunoglobulin, usually of the IgM class, which reacts with IgG. Rheumatoid factor is produced in some rheumatological, vasculitic, and viral diseases, including CMV infection (32). IgM rheumatoid factor forms a complex with IgG which may contain CMV-specific IgG. The CMV IgG binds to CMV antigen, carrying nonviral IgM with it. A test designed to detect IgM will produce a false-positive result. Therefore, testing for rheumatoid factor and removing it if present is important when measuring IgM antibodies.

Yolken and Leister described an enzyme immunoassay for CMV IgM antibody (40). Polyvinyl, round-bottom microtiter plates are coated with goat anti-human IgM at a concentration of 1 μg/ml in 0.06 M carbonate buffer (pH 9.6). The plates are incubated for 14 h at 4°C and stored until used. Plates are washed five times with PBS containing 0.05% Tween 20 (PBS-Tween), and serum is added at an initial dilution of 1:160. The serum is diluted in PBS-Tween containing 1% normal goat serum and 0.5% gelatin. Serum is incubated for 2 h at 37°C, plates are washed, and CMV antigen (commercially available CF antigen) or control antigen is added. The optimal concentration of antigen must first be determined by checkerboard titration. Antigen is added for 1 h at 37°C, plates are washed, and peroxidase-labeled goat anti-CMV, diluted 1:200 in PBS-Tween with 1% normal goat serum and 0.5% gelatin, is added for 1 h at 37°C. Plates are washed, and *O*-phenylenediamine substrate is added for 30 min at 37°C. H_2SO_4 (2 M) is added to stop the reaction, and the color generated is measured by spectrophotometry at 492 nm. The specific activity of a sample is calculated by subtracting the mean optical density in the wells containing control antigen from that in those containing CMV antigen. A dilution is considered positive for CMV IgM antibody if its specific activity is 2 standard deviations greater than the mean of three negative control serum samples.

Kangro and Griffiths et al. described a solid-phase radioimmunoassay test for measuring CMV-specific IgM antibodies (14, 19). Glycine-extracted CMV antigen is diluted to a concentration corresponding to 1 U of CF antigen in 50 μl. The viral antigen, or an equivalent dilution of control antigen, is desiccated onto polyvinyl chloride wells of microtiter plates and fixed with 10% Formalin (pH 7.3). Plates are then incubated overnight at 4°C with PBS containing 2% normal rabbit serum. Reciprocal serum dilutions of 100, 400, 1,600, and 6,400 are prepared in PBS containing 5% normal rabbit serum. Serum is added to wells containing CMV antigen or control antigen and incubated for 3 h at 37°C. Wells are washed with PBS containing 0.2% gelatin. Affinity-purified rabbit antiserum to human IgM is labeled with $Na^{125}I$ by the iodogen method and added at a dilution giving approximately 30,000 cpm per well. After a 2-h incubation at 37°C, the wells are washed with water and bound radioactivity is determined in a gamma scintillation counter. Specific binding of radiolabel is expressed by dividing the binding to CMV antigen by the binding to control antigen. A value of 2 or greater is considered as positive. Serum samples giving a positive result at any dilution are absorbed for 1 h at 37°C with 500 μl of a latex bead suspension containing human IgG to remove rheumatoid factor if present. After centrifugation at 4,000 × *g* for 30 min the supernatants are titrated in twofold dilutions and retested for CMV IgM antibody.

Comparing the two IgM assays described above, that of Yolken and Leister has the advantage of showing little or no interference by rheumatoid factor (40), and it uses alkaline phosphatase, which has a longer shelf-life than ^{125}I. However, at present there is no commercial source for goat anti-CMV antibody, nor are other equivalent anti-CMV antibodies available. The method described by Kangro utilizes commercially available reagents. The need to retest positive sera after removal of rheumatoid factor is inconvenient, as is the use of ^{125}I. Perhaps substituting an enzyme for radioactive iodine would make this latter method more convenient to use in a clinical laboratory.

Comparison of serological assays

Griffiths et al. compared CF (using glycine-extracted antigen), IFA, and ACIF for discriminating between CMV-seropositive and seronegative sera (12). Generally, the titers obtained by IFA and ACIF were higher than those measured by CF. However, there was excellent concordance of all three assays for detecting seropositive and seronegative samples.

Pass et al. compared CF, ACIF, and IFA for antibody to late antigens, IFA for antibody to early antigens, and neutralizing antibody assays for detecting changes in antibody levels after renal transplantation (27). Changes in titers roughly paralleled each other in all assays, although the antibody response to late antigens was generally higher using ACIF and IFA.

Brandt et al. compared CF (using glycine-extracted antigen), IFA, indirect hemagglutination (IHA), and enzyme immunoassay (EIA) for detecting CMV IgG antibody (2). The EIA, IFA, and IHA tests were in close agreement on all samples tested; the CF test gave results which differed from the other methods on 4 of 43 samples. In addition, antibody titers were considerably higher by EIA, IFA, and IHA than by CF.

Booth et al. compared EIA, radioimmunoassay, ACIF, CF, and passive hemagglutination for distinguishing negative from positive antibody levels (1). Passive hemagglutination gave the most discordant results, whereas agreement among the other assays was excellent. Generally, the levels of antibody detected by EIA and radioimmunoassay were 10- to 100-fold higher than by ACIF, CF, and passive hemagglutination.

Phipps et al. compared commercially available CF, IFA, FIAX, EIA, and IHA kits for screening blood donors (28). When ranked for accuracy, sensitivity, specificity, false-positives, and false-negatives, the authors judged the tests in the following order: IHA, EIA, FIAX, CF, and IFA. They also considered the IHA to be an easy test to perform and to have a rapid turnaround time.

Dylewski et al. compared EIA, CF, and radioimmunoassay for CMV antibody screening in homosexual males (8). The correlation between EIA and radioimmunoassay was 99%, and that between EIA and CF, using glycine-extracted antigen, was 96%. Cremer et al. compared freeze-thawed CF antigen, glycine-extracted CF antigen, and IHA for detecting rises in antibody titers (5). The freeze-thawed CF antigen detected more rises than did glycine-extracted antigen. IHA detected the least number. The apparent reason for fewer rises with the latter two assays was the detection (by these methods) of high antibody levels in the first serum sample. Friedman et al. compared CF (using glycine-extracted antigen) with FIAX for measuring immunity to CMV and detecting titer rises (10). The specificity and sensitivity of the two assays were similar, as was the ability to detect significant titer rises.

The above findings indicate that many techniques are sensitive and specific for measuring CMV antibodies. In deciding which test to perform, such factors as cost, turnaround time, equipment needs, and ease of performance should be considered. Which method is chosen will depend on which best meets the needs of individual laboratories.

Other serological tests

Neutralizing antibody to CMV can be measured by a microtiter method (38). However, the test is tedious and impractical for examining large numbers of clinical specimens.

INTERPRETATION OF RESULTS

Recovery of CMV from urine, throat, or other body fluids within the first week of life is the preferred method for diagnosis of congenital infection. Serological tests are less useful because of transplacental passage of maternal antibody and because of technical problems with detection of CMV-specific IgM. Recently described immunoassays for CMV-specific IgM may alleviate this latter problem (13, 19, 40). Most symptomatic congenitally infected infants have relatively high initial CF antibody titers, which persist at stable levels for many months (33). Congenital infection should therefore be suspected in infants with typical symptoms whose CMV titers remain high 2 to 3 months after birth.

Infants not previously tested and found to be excreting virus after 1 week of age may have either congenital or acquired infection. Standard serological tests do not differentiate between these possibilities; however, results of serial determinations of immunofluorescent antibody to early CMV antigens may be helpful (33).

Interpretation of serological tests performed on sera of patients 6 months of age or older is facilitated by the absence of passively acquired maternal antibody. If the initial serum is negative for CMV antibodies and the second serum is positive, a diagnosis of primary infection can be made, provided that the type of assay employed is a sensitive one for distinguishing seronegative from seropositive titers. If a serum sample from early in an illness contains CMV antibodies and a second sample which is taken several weeks later demonstrates a fourfold or greater rise in titer, a diagnosis of recent infection, due to reactivation or reinfection, can be made. Fourfold falls in titer are seldom observed early enough to be useful for laboratory diagnosis, since antibody levels tend to decline slowly over several months after infection. If acute- and convalescent-phase sera are both positive for CMV antibody but the antibody titer is unchanged, a diagnosis can be made of CMV infection at some time in the past. If the titers of the positive sera are high and the first specimen was obtained late in the patient's illness, the results may indicate active infection, and appropriate specimens for viral isolation should be obtained. Whenever possible, serological diagnoses of CMV infection should be confirmed by virus isolation, particularly since fourfold fluctuations in CMV antibody titers have been noted in some apparently healthy seropositive individuals (37).

Virological or serological detection of CMV indicates active infection, but does not establish whether such infection is responsible for symptomatic illness. To implicate CMV as a cause of an illness, laboratory confirmation of active infection in an appropriate clinical setting is required. When CMV is isolated from buffy coat cells, the likelihood that the infection is symptomatic increases, but asymptomatic viremia has also been described (4). Isolation of CMV from liver and particularly lung biopsy specimens must be interpreted with caution since other pathogens (*Chlamydia trachomatis, Pneumocystis carinii*, bacteria, or fungi) may also be present; when dual infection is documented, the relative importance of each pathogen in producing clinical illness may be difficult to determine.

LITERATURE CITED

1. **Booth, J. C., G. Hannington, T. M. F. Bakir, H. Stern, H. Kangro, P. D. Griffiths, and R. B. Heath.** 1982. Compari-

son of enzyme-linked immunosorbent assay, radioimmunoassay, complement fixation, anticomplement immunofluorescence and passive haemagglutination techniques for detecting cytomegalovirus IgG antibody. J. Clin. Pathol. **35**:1345–1348.

2. **Brandt, J. A., J. D. Kettering, and J. E. Lewis.** 1984. Immunity to human cytomegalovirus measured and compared by complement fixation, indirect fluorescent-antibody, indirect hemagglutination, and enzyme-linked immunosorbent assays. J. Clin. Microbiol. **19**:147–152.
3. **Chou, S., and T. C. Merigan.** 1983. Rapid detection and quantitation of human cytomegalovirus in urine through DNA hybridization. N. Engl. J. Med. **308**:921–925.
4. **Cox, F., and W. T. Hughes.** 1975. Cytomegaloviremia in children with acute lymphatic leukemia. J. Pediatr. **87**:190–194.
5. **Cremer, N. E., M. Hoffman, and E. H. Lennette.** 1978. Analysis of antibody assay methods and classes of viral antibodies in serodiagnosis of cytomegalovirus infection. J. Clin. Microbiol. **8**:152–159.
6. **Cremer, N. E., M. Hoffman, and E. H. Lennette.** 1978. Role of rheumatoid factor in complement fixation and indirect hemagglutination tests for immunoglobulin M antibody to cytomegalovirus. J. Clin. Microbiol. **8**:160–165.
7. **Dienstag, J. L., W. L. Cline, and R. H. Purcell.** 1976. Detection of cytomegalovirus antibody by immune adherence hemagglutination. Proc. Soc. Exp. Biol. Med. **153**:543–548.
8. **Dylewski, J. S., L. Rasmussen, J. Mills, and T. C. Merigan.** 1984. Large-scale serological screening for cytomegalovirus antibodies in homosexual males by enzyme-linked immunosorbent assay. J. Clin. Microbiol. **19**:200–203.
9. **Friedman, H. M., and C. Koropchak.** 1978. Comparison of WI-38, MRC-5, and IMR-90 cell strains for isolation of viruses from clinical specimens. J. Clin. Microbiol. **7**:368–371.
10. **Friedman, H. M., N. B. Tustin, M. M. Hitchings, and S. A. Plotkin.** 1981. Comparison of complement fixation and fluorescent immunoassay (FIAX) for measuring antibodies to cytomegalovirus and herpes simplex virus. Am. J. Clin. Pathol. **76**:305–307.
11. **Goldstein, L. C., J. McDougall, R. Hackman, J. D. Meyers, E. D. Thomas, and R. C. Nowinski.** 1982. Monoclonal antibodies to cytomegalovirus: rapid identification of clinical isolates and preliminary use in diagnosis of cytomegalovirus pneumonia. Infect. Immun. **38**:273–281.
12. **Griffiths, P. D., K. J. Buie, and R. B. Heath.** 1978. A comparison of complement fixation, indirect immunofluorescence for viral late antigens, and anti-complement immunofluorescence tests for the detection of cytomegalovirus specific serum antibodies. J. Clin. Pathol. **31**:827–831.
13. **Griffiths, P. D., S. Stagno, R. F. Pass, R. J. Smith, and C. A. Alford.** 1982. Congenital cytomegalovirus infection: diagnostic and prognostic significance of the detection of specific immunoglobulin M antibodies in cord serum. Pediatrics **69**:544–549.
14. **Griffiths, P. D., S. Stagno, R. F. Pass, R. J. Smith, and C. A. Alford.** 1982. Infection with cytomegalovirus during pregnancy: specific IgM antibodies as a marker of recent primary infection. J. Infect. Dis. **145**:647–653.
15. **Hanshaw, J. B.** 1966. Cytomegalovirus complement fixing antibody in microencephaly. N. Engl. J. Med. **275**:476–479.
16. **Howell, C. L., and M. J. Miller.** 1983. Effect of sucrose phosphate and sorbitol on infectivity of enveloped viruses during storage. J. Clin. Microbiol. **18**:658–662.
17. **Howell, C. L., M. J. Miller, and W. J. Martin.** 1979. Comparison of rates of virus isolation from leukocyte populations separated from blood by conventional and Ficoll-Paque/macrodex methods. J. Clin. Microbiol. **10**:533–537.
18. **Huang, E.-S., H. A. Kilpatrick, Y.-T. Huang, and J. S. Pagano.** 1976. Detection of human cytomegalovirus and analysis of strain variation. Yale J. Biol. Med. **49**:29–43.
19. **Kangro, H. O.** 1980. Evaluation of a radioimmunoassay for IgM-class antibodies against cytomegalovirus. Br. J. Exp. Pathol. **61**:512–520.
20. **Keller, R., R. Peitchel, J. N. Goldman, and M. Goldman.** 1976. An IgG-Fc receptor induced in cytomegalovirus-infected human fibroblasts. J. Immunol. **116**:772–777.
21. **Kettering, J. D., N. J. Schmidt, D. Gallo, and E. H. Lennette.** 1977. Anti-complement immunofluorescence test for antibodies to human cytomegalovirus. J. Clin. Microbiol. **6**:627–632.
22. **Kettering, J. D., N. J. Schmidt, and E. H. Lennette.** 1977. Improved glycine-extracted complement-fixing antigen for human cytomegalovirus. J. Clin. Microbiol. **6**:647–649.
23. **Lee, S. L., and H. H. Balfour.** 1977. Optimal method for recovery of cytomegalovirus from urine of renal transplant patients. Transplantation **24**:228–230.
24. **Lennette, E. H., and D. A. Lennette.** 1978. Immune adherence hemagglutination: alternative to complement fixation serology. J. Clin. Microbiol. **7**:282–285.
25. **Macris, M. P., A. J. Nahmias, P. D. Bailey, F. K. Lee, A. M. Visintine, and A. W. Braun.** 1981. Electron microscopy in the routine screening of newborns with congenital cytomegalovirus infection. J. Virol. Methods **2**:315–320.
26. **Martin, D. C., D. A. Katzenstein, G. S. M. Yu, and M. C. Jordan.** 1984. Cytomegalovirus viremia detected by molecular hybridization and electron microscopy. Ann. Intern. Med. **100**:222–225.

26a. **Palmer, D. F., L. Kaufman, W. Kaplan, and J. J. Cavallaro.** 1977. Serodiagnosis of mycotic diseases. Charles C Thomas, Publisher, Springfield, Ill.

27. **Pass, R. F., P. D. Griffiths, and A. M. August.** 1983. Antibody response to cytomegalovirus after renal transplantation: comparison of patients with primary and recurrent infections. J. Infect. Dis. **147**:40–46.
28. **Phipps, P. H., L. Gregoire, E. Rossier, and E. Perry.** 1983. Comparison of five methods of cytomegalovirus antibody screening of blood donors. J. Clin. Microbiol. **18**:1296–1300.
29. **Rao, N., D. T. Waruszewski, J. A. Armstrong, R. W. Atchison, and M. Ho.** 1977. Evaluation of anti-complementary immunofluorescence test in cytomegalovirus infection. J. Clin. Microbiol. **6**:633–638.
30. **Shulman, H. M., R. C. Hackman, G. E. Sale, and J. D. Meyers.** 1982. Rapid cytological diagnosis of cytomegalovirus interstitial pneumonia on touch imprints from open-lung biopsy. Am. J. Clin. Pathol. **77**:90–94.
31. **Spector, S. A., J. A. Rua, D. H. Spector, and R. McMillan.** 1984. Detection of human cytomegalovirus in clinical specimens by DNA-DNA hybridization. J. Infect. Dis. **150**:121–126.
32. **Stagno, S., R. F. Pass, D. W. Reynolds, M. A. Moore, A. J. Nahmias, and C. A. Alford.** 1980. Comparative study of diagnostic procedures for congenital cytomegalovirus infection. Pediatrics **65**:251–257.
33. **Stagno, S., D. W. Reynolds, A. Tsiantos, D. A. Fuccillo, W. Long, and C. A. Alford.** 1975. Comparative serial virologic and serologic studies of symptomatic and subclinical congenitally and natally acquired cytomegalovirus infections. J. Infect. Dis. **132**:568–577.
34. **Starr, S. E.** 1979. Cytomegalovirus. Pediatr. Clin. North Am. **26**:283–293.
35. **Volpi, A., R. J. Whitley, R. Ceballos, S. Stagno, and L. Pereira.** 1983. Rapid diagnosis of pneumonia due to cytomegalovirus with specific monoclonal antibodies. J. Infect. Dis. **147**:1119–1120.
36. **Waner, J. L., and T. H. Weller.** 1978. Analysis of antigenic diversity among human cytomegaloviruses by kinetic

neutralization tests with high-titered rabbit antisera. Infect. Immun. **21**:151–157.

37. **Waner, J. L., T. H. Weller, and S. V. Kevy.** 1973. Patterns of cytomegaloviral complement-fixing antibody activity: a longitudinal study of blood donors. J. Infect. Dis. **127**:538–543.
38. **Waner, J. L., T. H. Weller, and J. A. Stewart.** 1976. Cytomegalovirus, p. 423–427. *In* N. E. Rose and H. Friedman (ed.), Manual of clinical immunology. American Society for Microbiology, Washington, D.C.
39. **Yeager, A. S.** 1979. Improved indirect hemagglutination test for cytomegalovirus using human O erythrocytes in lysine. J. Clin. Microbiol. **10**:64–68.
40. **Yolken, R. H., and F. J. Leister.** 1981. Enzyme immunoassays for measurement of cytomegalovirus immunoglobulin M antibody. J. Clin. Microbiol. **14**:427–432.

Chapter 66

Varicella-Zoster Virus

NATHALIE J. SCHMIDT

CLINICAL BACKGROUND

Varicella (chickenpox) and zoster represent different clinical manifestations of infection with the same virus. Varicella occurs most frequently in children and is characterized by fever and a generalized vesicular exanthem. Zoster (shingles) generally occurs in adults and consists of a painful, circumscribed eruption of vesicular lesions with accompanying inflammation of associated dorsal root or cranial sensory nerve ganglia. Varicella constitutes the primary infection, whereas zoster occurs in individuals with partial immunity resulting from a prior varicella infection. Although there may be rare exceptions, it is generally considered that zoster results from reactivation of the virus harbored in a latent state, rather than from reintroduction of the virus into the host. This is supported by the fact that zoster does not exhibit the seasonal prevalence seen with varicella (late winter and spring). Often there are no apparent inciting factors for episodes of zoster, but they frequently occur in persons undergoing trauma or certain types of drug therapy, and particularly in patients receiving immunosuppressive therapy. Results of certain serological studies (1, 8) suggest that reinfection or reactivation of varicella-zoster virus (VZV) may occur in the absence of clinical symptoms.

There are several situations in which providing a specific laboratory diagnosis of VZV infection is crucial. VZV infection may cause severe or fatal disease in individuals who are on immunosuppressive therapy or who have genetic defects in their immune responses. Progressive, generalized varicella occurs in as many as 30% of children who acquire chickenpox while on cancer therapy, and mortality in these cases has ranged from 7 to 28%. In older immunodeficient patients, there is an increased risk of disseminated zoster, and mortality rates in these infections range from 3 to 5% (21). Providing a specific diagnosis of VZV infection in immunosuppressed patients or their contacts may guide in the administration of antiviral agents or immune globulin. Determining the immunity status (presence or absence of antibody) in high-risk immunocompromised individuals exposed to VZV infection also guides in the management of these patients. It is important to provide a specific diagnosis of some of the less common manifestations of VZV infection, such as varicella pneumonia and encephalitic complications, particularly as antiviral agents become available. It is sometimes necessary to differentiate VZV infection from the vesicular eruptions caused by certain enteroviruses, bacterial agents, or hypersensitivity reactions, and also from generalized vesicular eruptions caused by herpes simplex virus (HSV). The need for providing a differential diagnosis of varicella and variola-vaccinia viruses rarely occurs at present, but it is still important for key viral diagnostic laboratories to have this capability.

DESCRIPTION OF THE AGENT

VZV has the typical morphology of members of the family Herpesviridae. It has a DNA genome with a molecular weight of approximately 80×10^6. The DNA exists in two isomeric forms that differ by inversion of one short terminal genome segment. The short genome segment consists of a terminal sequence of about 3.4×10^6 daltons that is separated from an internal inverted repeat of itself by a 5.8×10^6-dalton unique DNA segment. The virion has a diameter of 150 to 200 nm and consists of an inner icosahedral capsid composed of 162 capsomeres and surrounded by an envelope composed of two or more membranes. The envelope contains essential lipid, and thus infectivity of the virus is inactivated by lipid solvents (17). To date, antigenic variation has not been demonstrated, but strain differences can be demonstrated by restriction enzyme analysis of the viral genome (16). The virus shares major antigens with herpesviruses which produce varicella-like disease in various simian species (3), and it appears to share minor antigens with HSV (11, 15).

COLLECTION, SHIPMENT, AND STORAGE OF SPECIMENS

Specimens usually examined for VZV are smears of vesicular lesions, vesicular fluid and scrapings, and tissues obtained at autopsy. Rarely, virus has been demonstrated in buffy coat or cerebrospinal fluid of immunosuppressed patients. Lung is the autopsy tissue from which VZV is most frequently recovered. Smears of cellular material from vesicular lesions should be collected for immunofluorescence (IF) staining wherever feasible, since this is more sensitive than isolation in cell culture for virus detection (2, 12). An acute-phase blood specimen should be taken as soon as possible after the onset of symptoms, to be tested in parallel with a convalescent-phase specimen, collected at least 14 days later, in an effort to demonstrate a diagnostically significant increase in VZV antibody titer. VZV immunity status is determined by testing a single blood specimen collected before, or as soon as possible after, exposure to VZV infection.

Methods for collection of vesicular fluids and cellular material from the base of lesions are described in Chapter 61.

Smears of cellular material from lesions

Smears of cellular material from the base of fresh lesions are used for direct examination by IF or histological staining. Since differentiation of VZV and HSV by IF staining is accomplished by testing against conjugates to each of these viruses and demonstrating staining with only a single conjugate, it is essential to prepare at least two smears of cellular material for IF examination. In special cases it may be necessary to prepare a third smear for examination with a vaccinia conjugate. If insufficient material is available for IF examination, what is available should be submitted for electron microscopy and virus isolation only. Cellular material should be placed in holding medium (see Chapter 61) to be used for virus isolation.

Vesicular fluids

Vesicular fluids collected into capillary pipettes or syringes can be used for virus isolation attempts and electron microscopy, but fluids are unsatisfactory for IF staining. Fluids collected onto swabs and placed in holding medium (see Chapter 61) may also be used for virus isolation. Lesions more than 4 days old rarely yield infectious virus, but the fluids may give positive results by electron microscopy.

Crusts from lesions

Crusts do not yield infectious virus and are not suitable for IF staining, but if nothing else is available they may be used for examination by electron microscopy.

Lung, liver, or skin tissue

Smears of autopsy tissue for IF staining are prepared by excising three or four pieces of autopsy tissue about 10 to 15 mm in size, holding the tissue with forceps, and gently pressing the cut surface to the clean surface of a slide. A series of impressions is made over an area 30 to 40 mm in length. Three or four slides should be prepared from each specimen. The slides are allowed to dry at room temperature for 15 to 20 min and then are fixed in acetone for 10 min at room temperature. Suspensions of tissue are prepared for virus isolation as described in Chapter 61.

Storage and transport of specimens

Specimens for virus isolation attempts should be inoculated into suitable cell cultures as soon as possible after collection. If inoculation is to be delayed for no longer than 24 h, specimens may be transported or held on wet ice or in a refrigerator. For longer periods of holding, dry-ice temperatures are required. Refrigeration is not necessary for smears to be examined for viral antigen or intranuclear inclusions. It should be recognized that smears from vesicular lesions may be infectious, particularly in the case of vaccinia or smallpox materials, and suitable precautions should be taken in packing and storage to protect postal and laboratory personnel.

DIRECT EXAMINATION OF LESION MATERIAL

Examination of vesicular material for virus or viral antigen provides the most rapid diagnosis of VZV infection. A differential diagnosis between herpesviruses and poxviruses may be made by cytological examination of smears from the base of vesicular lesions or by electron microscopy of vesicular materials. These procedures do not, however, distinguish between VZV and HSV. Specific identification of VZV in lesion material can be accomplished by IF staining.

Cytological examination of smears from vesicular lesions

Smears of cellular material collected from the base of vesicular lesions are prepared as described in Chapter 61. They are fixed with methanol and stained with buffered Giemsa stain at pH 7.0 to 7.2. Microscopic examination with an oil immersion lens reveals the presence of multinucleated, giant epithelial cells with altered chromatin patterns (see Fig. 1); these are

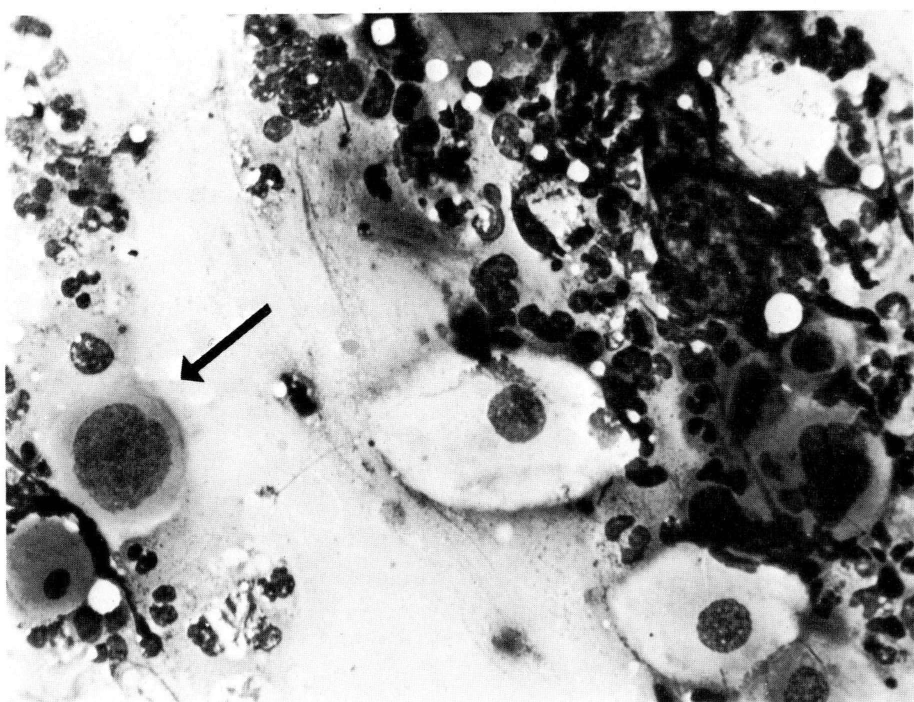

FIG. 1. Giemsa-stained preparation of material from the base of a vesicular lesion. Magnification, ×250. Arrow shows a giant cell with folded nucleus characteristic of VZV or HSV infection.

characteristic of infection with either VZV or HSV and tend to rule out the possibility of infection with variola-vaccinia viruses.

Electron microscopy

Vesicular fluids are examined directly by electron microscopy. Crusts from lesions are ground in 1 or 2 drops of distilled water. When transportation of specimens to the laboratory presents a problem, heavy smears of fluids and crusts may be prepared on glass microscope slides and air dried; material from these smears is reconstituted in a drop of water for examination by electron microscopy.

A drop of the specimen is placed on a grid and blotted with filter paper. A drop of 3% phosphotungstic acid (prepared in distilled water buffered to neutrality with 1 N KOH) is then added and blotted. The specimen should be examined in an electron microscope as soon as possible. The demonstration of virus particles with the typical morphology of herpesviruses (see Fig. 2) identifies the etiological agent as a member of this group and distinguishes it from viruses of the variola-vaccinia group, but it does not provide a specific diagnosis of VZV infection.

IF staining

Smears properly collected from the base of fresh vesicular lesions may be used for IF staining for the specific demonstration of VZV antigen in the infected epithelial cells. It is essential that the smears contain a large number of infected cells, and two or three smears should be made to permit staining with conjugates to VZV, HSV, and in special situations vaccinia virus. The direct method of staining, with the use of fluorescein-conjugated immune globulins from hyperimmune animal sera or ascitic fluid, is the most specific procedure. Direct IF staining is described below.

IF staining on frozen sections of punch biopsies from prevesicular skin lesions has also been reported for the rapid and early diagnosis of VZV infections (10).

ISOLATION AND IDENTIFICATION OF VZV

Host systems

Diploid human cell lines or primary human cell cultures are the most sensitive host system for isolation of VZV from clinical materials. In the rare event that smallpox is suspected, the material should also be inoculated onto the chorioallantoic membranes of 11- to 13-day-old embryonated hen eggs; this should be done in a reference laboratory.

In this laboratory, and others, human fetal diploid kidney (HFDK) or human fetal diploid lung (HFDL) cells have been found to be highly satisfactory for primary isolation of VZV; these cells are available from commercial sources. Some laboratories have used human foreskin fibroblasts for isolation of VZV; sensitivity as compared to that of human fetal diploid cells is uncertain. Cell cultures are initiated with Eagle minimal essential medium prepared in a Hanks balanced salt solution base and supplemented with 10% fetal bovine serum. Inoculated cell cultures are maintained on 98% Eagle minimal essential medium prepared in Earle balanced salt solution (or Leibovitz medium no. 15) and supplemented with 2% fetal bovine serum. Either of these media will maintain the cultures for 14 days without a fluid change, a period usually sufficient for the development of a specific viral cytopathic effect (CPE). Primary monkey kidney cells are less sensitive than human fetal diploid cells for virus isolation, but isolates are sometimes recovered in these cells.

Recent studies have shown virus isolation, even in optimal cell culture systems, to be less sensitive than IF staining on lesion materials for diagnosis of VZV infection (2, 12). The lability of viral infectivity in vesicular lesions and the strongly cell-associated nature of the virus probably both contribute to the difficulty of virus isolation.

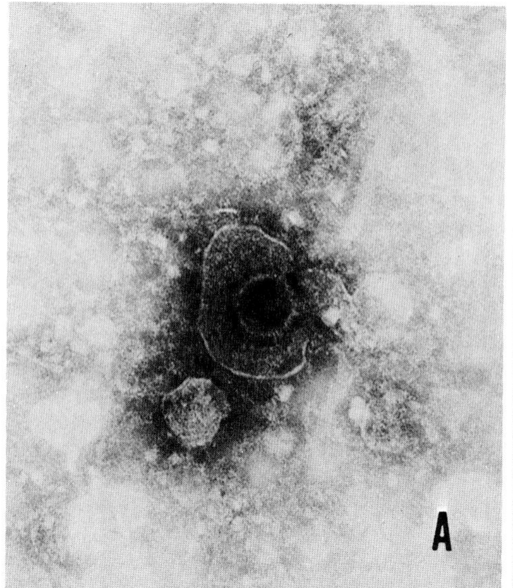

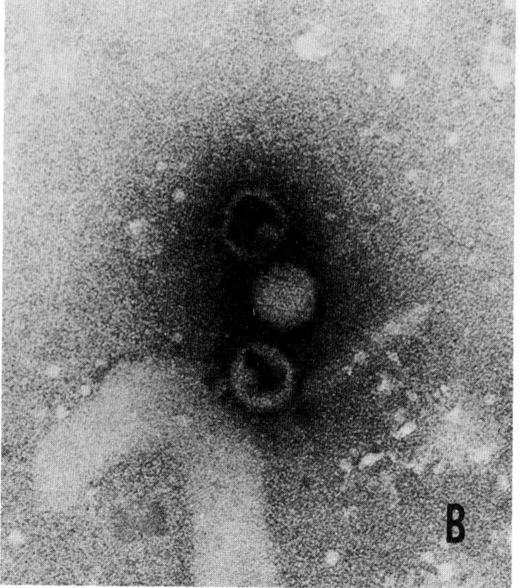

FIG. 2. Electron micrograph of VZV in a preparation from a lesion scab. Magnification, ×80,000. (A) Enveloped virus. (B) Naked nucleocapsids.

VZV cannot be isolated in mouse or embryonated egg host systems.

Evidence of infection in cell cultures

Figure 3 shows the characteristic CPE of VZV in HFDK cells. Initially the CPE consists of small, discrete foci of rounded and swollen refractile cells; in HFDK and HFDL cells these may appear from 3 to 14 days after inoculation of the cultures with clinical materials, but in most instances CPE is first apparent at 4 to 7 days. The foci of infected cells enlarge and may slowly involve most of the monolayer. The spread of infection can be accelerated by dispersing cells in the infected culture with trypsin (see below) and replanting them in growth medium in the same culture vessel.

Subpassage and storage of virus

Infectious VZV remains in close association with the host cell, and therefore it is necessary to use virus-infected cells, rather than culture fluids, as an inoculum for serial propagation of the virus. Trypsin-dispersed infected cells are most suitable for subpassage of VZV. The medium is removed from cell cultures in which CPE involves approximately 50% of the cell sheet, and 2 ml of 0.25% trypsin solution is added to each tube culture. After 30 s at room temperature, most of the trypsin is removed, and the small amount of residual trypsin is flooded over the monolayer; as soon as the cells detach from the glass surface they are dispersed in the original volume of maintenance medium and inoculated into fresh cell cultures.

Virus in infected cells can be stored in the frozen state if the viability of the cells is maintained through the use of a cryoprotective agent such as dimethyl sulfoxide or glycerol in the freezing medium. Infected cells are dispersed with trypsin as described above and then suspended in Eagle minimal essential medium containing 10% fetal bovine serum and 10% dimethyl sulfoxide or glycerol. Infectious virus can be recovered after 18 months or more of storage at −70°C, but infectivity is preserved more effectively by storage of the infected cells in liquid nitrogen. High-titered cell-free virus prepared by sonic treatment of infected cells (14) is stable at −70°C in minimal essential medium with 10% sorbitol and 10% fetal bovine serum for at least 5 years.

Identification of VZV isolates

Presumptive identification. Presumptive identification of VZV may be made on the basis of a typical CPE, which is more focal in nature and progresses more slowly than that of HSV. VZV may also be differentiated from HSV by its failure to produce a CPE in rabbit or hamster kidney cell cultures, whereas HSV rapidly produces a CPE. The source of the specimen and the clinical manifestations of the illness should usually prevent confusion between VZV and human cytomegalovirus (CMV). However, if isolations are made from tissue specimens, there may be a need to distinguish between these herpesviruses. VZV grows well in epithelial cells, whereas CMV generally produces a CPE only in human fibroblast cells. Production of a CPE by CMV generally is slower and more focal in nature than that produced by VZV. Furthermore, human CMV strains fail to produce a CPE in primary monkey kidney cells, but after initial isolation VZV strains will generally do so.

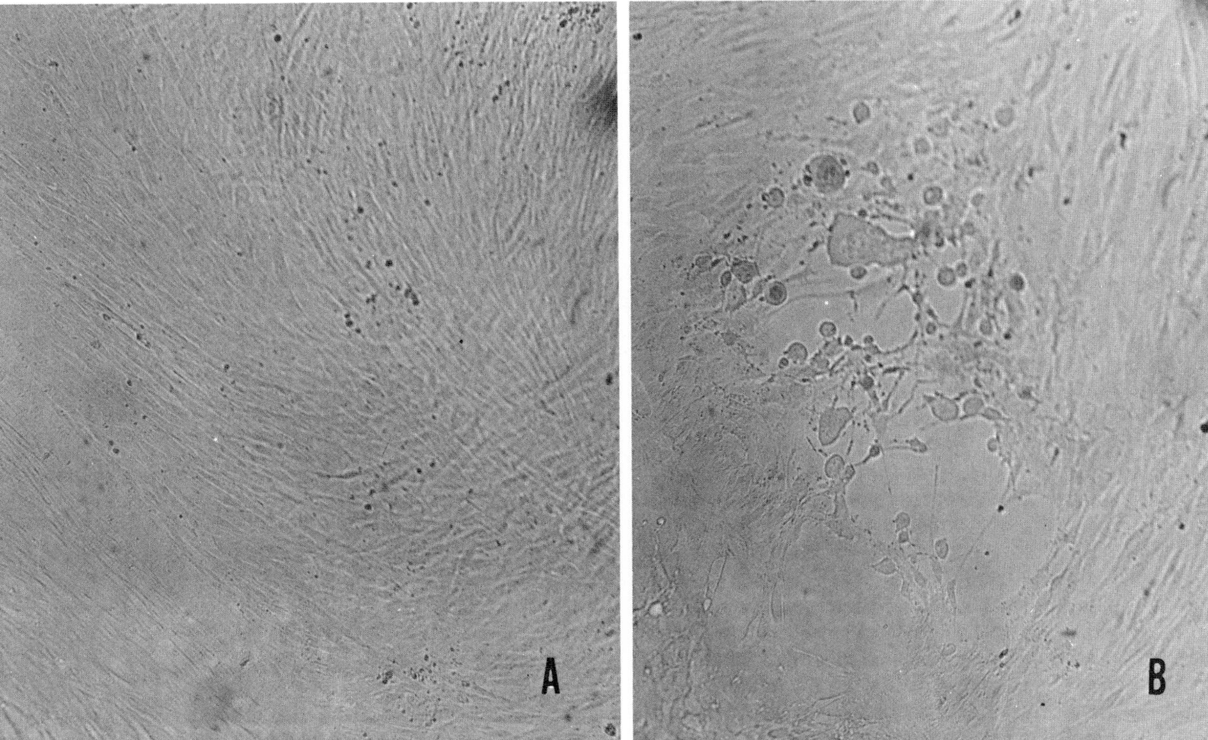

FIG. 3. CPE of VZV in HFDK cells. Magnification, ×100. (A) Uninfected cells. (B) Virus-infected cells.

Specific identification. Specific identification of VZV isolates is made by demonstrating their ability to react with a known positive antiserum. Animal immune sera to VZV have not been consistently available from commercial sources, but monoclonal antibodies to VZV are beginning to become available commercially, and these should soon overcome problems previously encountered with antisera of inadequate potency and specificity and with inconsistencies in supply of VZV immune reagents. In the absence of satisfactory immune animal sera to VZV, acute- and convalescent-phase sera from a known case of varicella have been employed. Identification is based upon the demonstration of a greater degree of reactivity (IF or complement fixation [CF]) of the isolate with the convalescent-phase serum than with the acute-phase serum. However, the use of human sera of uncertain antibody content for identification of viral isolates is not as reliable as is the use of specific immune animal sera. Human sera employed for identification of VZV should be free from antibodies to other human herpesviruses from which it is important to distinguish VZV, namely, HSV and CMV.

Specific identification of VZV isolates is accomplished most readily by IF staining by the direct method. Alternative methods are indirect IF staining (18) and CF.

Antisera to VZV

Problems encountered in production of VZV antisera have related in large part to the strongly cell-associated nature of the virus, and various approaches have been used to produce adequately potent VZV antisera free from unwanted antibodies to host proteins. Immunization of monkeys with virus grown in monkey kidney cell cultures provided this laboratory with a satisfactory VZV antiserum for virus detection and identification by direct IF staining. Other approaches have included the use of density gradient-purified virus for immunization of rabbits, guinea pigs, mice, or rats, immunization of rabbits made tolerant to human immunoglobulin G with antigen immunoprecipitated with human immunoglobulin, and the use of hybridoma technology to produce monoclonal antibodies of high potency and specificity (5, 18). Methods for preparation of antisera to VZV are given in the original publications (see references 5, 17, and 20). However, most clinical virology laboratories will depend upon commercially available immune reagents for virus detection and identification. The commercial reagents, monoclonal antibodies for the most part, may be obtained as fluorescein-labeled immunoglobulins for use in direct IF staining or unlabeled for use in indirect IF staining.

IF STAINING FOR VZV

Preparation of VZV slides for IF staining

Smears of VZV-infected cells are required for standardization of IF reagents, as positive controls when IF staining is used for virus detection in clinical materials or for identification of viral isolates, and as an antigen substrate for antibody assays.

The stock virus is subpassaged two to three times to increase infectivity by inoculating trypsinized infected cells from one culture into two cultures of uninfected cells. When the cultures show a 3+ CPE at 24 h after inoculation, they are used for slide preparation. The infected cells are dispersed with trypsin and mixed with trypsin-dispersed cells from two uninfected cultures. The inclusion of uninfected cells in the infected cell smears serves as one control on the specificity of IF staining; only roughly one-third of the cells in the smear should show staining. In addition, uninfected cell controls are prepared from trypsinized uninfected cells of the same lot as those infected with VZV. The dispersed cells are suspended in phosphate-buffered saline (PBS) with 2% fetal bovine serum and sedimented by centrifugation at 2,000 rpm for 5 min. The supernatant fluids are removed, and the packed cells are resuspended in PBS with 2% fetal bovine serum, using a volume of 0.01 ml per tube culture or 0.4 ml per 32-oz (ca. 0.95-liter) bottle culture. Smears approximately 5 mm in diameter are made from this cell suspension by placing small drops (approximately 0.005 ml) on microscope slides. The smears are dried at room temperature, fixed with acetone at room temperature for 10 min, and then dried at room temperature. Smears can be stored at −70°C until they are used. Before staining, the smears are ringed with a liquid embroidery pen to retain the conjugate.

Standardization of immune reagents

Fluorescein-conjugated VZV immune conjugates should be titrated against smears of VZV-infected cells to determine the appropriate "working dilution" for use in virus identification. This should be determined in the user's laboratory, even though a working dilution may be specified by the manufacturer. If unlabeled VZV antibodies are to be used in indirect IF staining, optimal working dilutions of both the viral immune reagent and the conjugated antispecies immune reagent must be determined by preliminary block titrations.

IF staining is done by either the direct or indirect method (see below), using dilutions of the reagent against smears of VZV-infected and uninfected cells, and the degree of specific immunofluorescence is graded as 1+, 2+, 3+, and 4+. A reading of 4+ indicates glaring yellow-green fluorescence; 3+ indicates bright green, but not glaring, fluorescence; 2+ is dull-green fluorescence, and 1+ and ± designate faint and questionable fluorescence, respectively. The working dilution of the reagent is the highest dilution giving specific staining of 4+ and showing no reactivity with uninfected control cells.

Direct IF staining for VZV detection and identification

Direct IF staining of vesicular lesion or tissue specimens should be done whenever suitable specimens are available, since this has been found to be more sensitive than isolation in cell culture for detection of VZV (2, 12).

Smears of epithelial cells from the base of vesicular lesions or impression smears of autopsy or biopsy tissue are prepared as described above. It is essential to make enough smears to permit staining with VZV and HSV immune reagents and, in rare situations, with a vaccinia reagent as well.

Working dilutions of the immune conjugates are

prepared in 0.1 M PBS (pH 7.3). Each of the conjugates is added to the specimen smear in a volume sufficient to cover the ringed area (ca. 0.05 ml). Positive control slides for VZV and HSV are also stained with each conjugate in every run as a control on the specificity and reactivity of the conjugates. After addition of the conjugates, tests are incubated in a humidified atmosphere at 35 to 37°C for 20 min, and the slides are then rinsed twice in 0.01 M PBS (pH 7.2 to 7.5) (10 min for each rinsing), followed by a rinse in distilled water. They are allowed to dry and are then mounted in buffered glycerol solution (1 part glycerol and 4 parts PBS [pH 7.2 to 7.5]).

Examination with a fluorescence microscope should reveal specific staining of 3+ to 4+ intensity, associated with both the cytoplasm and nucleus of the epithelial cells stained with the VZV conjugate and little or no staining in cells treated with the HSV conjugate. Vesicular lesion specimens containing too few epithelial cells for definitive examination are reported as unsatisfactory rather than negative.

Identification of VZV isolated in cell cultures is performed by direct IF staining when the inoculated cultures show a 2+ viral CPE. Trypsin-dispersed cells from an infected tube culture are mixed with those from two uninfected cultures of the same lot and suspended in 3 ml of PBS with 2% fetal bovine serum, and cells are sedimented by centrifugation at 2,000 rpm for 5 min. After the supernatant fluid is removed, the packed cells are suspended in approximately 0.03 ml of PBS with 2% fetal bovine serum and smears are prepared on microscope slides as described above. Four smears of each suspension are made to permit staining with both VZV and HSV conjugates. Smears from uninfected cultures are prepared similarly. The smears are fixed and stained as described above. Positive VZV and HSV control slides are included in each run. Again, positive results are based upon staining of 3+ or 4+ intensity with the VZV conjugate, little or no staining with the HSV conjugate, and no staining with the uninfected control cells.

Indirect IF staining for VZV detection and identification

Unconjugated monoclonal antibodies can be used for direct detection of the virus in clinical specimens or for identification of virus isolated in cell culture (18). It is essential to use a conjugate against mouse immunoglobulins which is free from nonspecific reactivity with host tissues and host serum components in clinical specimens. This is determined by testing dilutions of the conjugate against infected and uninfected cell smears in the absence of intermediate mouse antibodies. The appropriate working dilutions of the VZV reagent and the anti-mouse globulin conjugate are determined by preliminary block titration. The working dilution of the VZV immune reagent is applied to the specimen smear, and the specimen is also stained with HSV antibodies as a control. Slides are incubated at 35 to 37°C in a humidified atmosphere for 20 min and then are washed twice in PBS (pH 7.2 to 7.5). The working dilution of the conjugate is then added, and tests are incubated at 35 to 37°C for 20 min. The slides are washed twice in PBS and mounted in buffered glycerol-saline solution. Positive results are based upon 3+ to 4+ staining of the specimen with the VZV reagent and little or no staining with the HSV reagent.

Identification of VZV isolates by CF

In laboratories lacking equipment for IF examinations, specific identification of VZV isolates may be accomplished by preparing a complement-fixing antigen from the isolate and testing it against known negative and positive VZV sera, either pre- and postimmunization animal sera or acute- and convalescent-phase human sera. Again, the human sera must be free from antibodies to other human herpesviruses.

The isolate may require one or more subpassages to increase its infectivity. When it produces 2+ CPE within 2 to 3 days after inoculation, antigen is prepared in bottle cultures of HFDL or HFDK cells. The antigen is prepared as described below, with the exception that the infected cells are dislodged into 1/50 of the original culture volume. Control antigen is prepared in the same manner from uninoculated cultures of the same lot in which the VZV was propagated.

The antigen prepared from the isolate is run in a box titration (see CF procedure in reference 10a) against known positive and negative VZV sera and for anticomplementary activity. Identification is based upon CF activity with the positive serum but not the negative serum.

SEROLOGICAL METHODS

Serological procedures are used for laboratory diagnosis of varicella or zoster infections and also for determining immunity status (presence or absence of VZV antibody) in high-risk individuals exposed to infection.

Serodiagnosis

The value of serological procedures for diagnosis of VZV infection is limited to some extent by the fact that heterotypic antibody titer rises to VZV may occur in certain patients with HSV infection who have experienced a prior infection with VZV. This would appear to be a heterotypic antibody response to common antigens in the two viruses (11, 15).

In the past the CF test has been the most widely applicable procedure for serodiagnosis of VZV infections. However, as viral diagnostic laboratories increasingly are adopting enzyme immunoassays (EIAs) as their principal serological tool for viral diagnosis, it can be expected that the CF test for VZV will be displaced to a large extent by EIA. In general, EIA methods use the same type of VZV antigens as those used in CF tests.

VZV antigens are available from commercial sources, or they can be prepared from infected HFDL or HFDK cultures. Cultures are inoculated with a high concentration of trypsin-dispersed, infected cells and harvested 4 to 5 days later, when the cultures show an advanced CPE. Infected cultures are not trypsinized, since this tends to make antigens anticomplementary; instead, cells are dislodged into 1/20 of the original culture volume of fresh maintenance medium by shaking with glass beads. Cells are then disrupted by sonic treatment for 1 min, and the antigen is clarified by centrifugation at 1,500 rpm for 15 min. Uninfected control antigen is prepared in the same manner.

The technique for the CF test is described in reference 10a.

For serodiagnosis of infection, acute- and convalescent-phase sera must be tested in parallel in the same run. Twofold dilutions of inactivated (56°C, 30 min) serum are tested against a dilution of VZV antigen containing 2 CF antigenic units, and the lower serum dilutions are tested against a comparable dilution of uninfected control antigen and for possible anticomplementary activity. A fourfold or greater increase in CF antibody titer betwen the acute- and convalescent-phase sera is considered diagnostically significant. In varicella, CF antibodies are usually demonstrable by 7 to 10 days after appearance of the exanthem, and they reach peak titers at 2 to 3 weeks after onset of the illness. The antibody response is usually more rapid in zoster, and CF antibodies may be present on day 1 or 2 of the eruption. VZV antibodies may be demonstrable earlier in the course of infection by EIA than by CF, but this does not preclude demonstration of significant titer increases, as convalescent-phase titers demonstrable by EIA are markedly higher than acute-phase titers (4).

EIA tests for VZV antibody (4) are performed in microtiter plates marketed specifically for EIA procedures. The wells are coated with an optimal dilution of VZV antigen or control antigen in 0.06 M bicarbonate buffer (pH 9.5) in a volume of 0.2 ml. After overnight incubation at 4°C, the unadsorbed antigen is removed by vacuum suction, the wells are washed with 0.01 M PBS (pH 7.3) containing 0.05% Tween 20, and protein adsorption sites are saturated by adding 0.35 ml of a 5% bovine albumin solution in PBS and incubating for 4 h at room temperature. The fluids are then aspirated, and the wells are washed with PBS. Serial twofold dilutions of serum are prepared with 0.05-ml microdiluters in wells coated with VZV antigen and with control antigen. PBS in a volume of 0.05 ml is added to each well, and tests are incubated overnight at room temperature. Controls include known positive and negative sera and wells incubated with diluent instead of serum. The next day the contents of the wells are aspirated, and the wells are washed three times with PBS. An optimal dilution of alkaline phosphatase-labeled antibodies to human immunoglobulin G, prepared in PBS with 5% bovine albumin, is added in a volume of 0.1 ml, and tests are incubated for 2 h at room temperature. The conjugate is then aspirated from the wells and they are washed three times with PBS. The enzyme substrate, *p*-nitrophenylphosphate (1 mg/ml) in diethanolamine (10%) buffer (pH 9.8) with 10^{-3} M $MgCl_2$, is added in a volume of 0.1 ml, and after incubation at room temperature for 30 min the reaction is stopped by the addition of 0.05 ml of 3 N NaOH per well. Results can be read by visual inspection against a white background. Antibody endpoints are taken as the highest serum dilution showing a visible yellow color, and reactions are considered to be specific only if the corresponding wells containing uninfected control antigen show no color.

Alternative methods which are applicable to the routine serodiagnosis of VZV infections are the immune adherence hemagglutination test (see reference 4), which can be performed with commercially available CF antigens and control antigens, and anticomplement immunofluorescence (ACIF) staining (see below). Neutralizing antibody to VZV can be assayed by plaque reduction techniques with cell-free virus obtained by sonic treatment of infected human fetal diploid cells (13); however, this is primarily a research, rather than a diagnostic, tool.

Determination of immunity status

A rapid and sensitive serological method is required for determining past infection, and presumed immunity, to VZV. The CF test is not adequately sensitive for detection of antibody elicited by past infection, giving positive results in only 50% or fewer of individuals with VZV antibody demonstrable by more sensitive methods. Although the presence of CF antibody can be taken as evidence of past infection, a negative CF result leaves the immunity status in doubt. Fluorescent-antibody staining of membrane antigen in VZV-infected cells (FAMA) has been used as a method for determining immunity status (19), and results have correlated well with resistance or susceptibility to infection in clinical settings. However, the FAMA method is somewhat cumbersome and is not generally available in most laboratories.

A method which gives results comparable to FAMA assay is the ACIF test (6). This method avoids nonspecific reactivity which may be encountered in indirect IF staining and permits examination of serum at dilutions as low as 1:2 and 1:4. Smears of VZV-infected and uninfected cells are prepared as described above for IF staining. Optimal dilutions of complement and conjugate are determined by box titrations of various dilutions of each reagent (diluted in Veronal-buffered saline, pH 7.2, 0.1 ionic strength) against VZV-infected cells treated with appropriate dilutions of known positive and negative sera. Smears of VZV-infected and uninfected cells are treated with dilutions of inactivated (56°C, 30 min) test serum prepared in 0.01 M PBS (pH 7.2) for 20 min at 37°C in a humidified chamber. After rinsing for 5 min in PBS, an optimal dilution of complement is added, and incubation is conducted for 45 min at 37°C in a moist chamber. After a 5-min rinse in PBS, an optimal dilution of fluorescein-conjugated anti-guinea pig complement is added, and incubation is conducted for 20 min at 37°C in a moist chamber. After a 5-min rinse, the slides are air dried, mounted in 25% glycerol in PBS, and examined under UV illumination.

EIA has also shown good sensitivity for detection of VZV antibody from past infection and is a good alternative method for determination of immunity status (4, 7). Immune adherence hemagglutination has shown greater sensitivity than CF, but less than that of EIA, for detection of VZV antibody of long duration.

EVALUATION AND INTERPRETATION OF RESULTS

Isolation of VZV or demonstration of viral antigen in lesion material or autopsy tissue is diagnostic of a current infection.

The most rapid presumptive diagnosis of VZV or HSV infection can be made by electron microscopy of lesion materials or by Giemsa staining of cellular material from the lesions. This only identifies members of the herpesvirus group and does not distinguish between VZV and HSV. Specific identification of VZV antigen can be made directly by IF staining of lesion material.

Isolation of the virus is a relatively slow method and is less sensitive than electron microscopy and IF staining, since infectious virus persists for a shorter length of time in vesicles and is more labile than the viral particles and antigens which can be detected by electron microscopy and IF staining. Presumptive identification of isolates can be made on the basis of a typical VZV CPE, failure of the virus to replicate in rabit or hamster kidney cells, which distinguishes VZV from HSV, and ability of the virus to propagate in primary monkey kidney cells and human epithelial cells, which distinguishes it from human CMV. With the increasing availability of VZV-specific immune reagents from commercial sources, specific identification of isolates readily can be made by IF staining. Pitfalls in the use of human sera for specific identification have been indicated.

A fourfold or greater increase in antibody titer to VZV antigen, in the absence of a fourfold rise to HSV antigen, is diagnostic of a current VZV infection. However, a high proportion (up to one-third) of individuals with primary HSV infections who have experienced a prior VZV infection show a heterotypic CF antibody response to the VZV antigen, making a differential diagnosis between VZV and HSV infection difficult in the absence of clear-cut clinical findings. Frequently, a differential diagnosis can be made on the basis of the fact that antibody to the infecting virus type is absent or at a very low titer in the acute-phase specimen, whereas antibody to the viral heterotype is already present.

Although negative results by FAMA, ACIF, and EIA are reliable indicators of susceptibility to VZV infection, the occurrence of clinical VZV infections in a few individuals with low titers of antibody demonstrable by FAMA (9) and by ACIF indicates that low levels of antibody demonstrable by these sensitive methods are not always a reliable indication of protection against clinical illness. However, the detection of serum antibody to VZV has correlated with protection in up to 96% of individuals in one study (9).

LITERATURE CITED

1. **Arvin, A. M., C. M. Koropchak, and A. C. Wittek.** 1983. Immunologic evidence of reinfection with varicella-zoster virus. J. Infect. Dis. **148**:200–205.
2. **Drew, W. L., and L. Mintz.** 1980. Rapid diagnosis of varicella-zoster virus infection by direct immunofluorescence. Am. J. Clin. Pathol. **73**:699–701.
3. **Felsenfeld, A. M., and N. J. Schmidt.** 1977. Antigenic relationships among several simian varicella-like viruses and varicella-zoster virus. Infect. Immun. **15**:807–812.
4. **Forghani, B., N. J. Schmidt, and J. Dennis.** 1978. Antibody assays for varicella-zoster virus: comparison of enzyme immunoassay with neutralization, immune adherence hemagglutination, and complement fixation. J. Clin. Microbiol. **8**:545–552.
5. **Forghani, B., N. J. Schmidt, C. K. Myoraku, and D. Gallo.** 1982. Serological reactivity of some monoclonal antibodies to varicella-zoster virus. Arch. Virol. **73**:311–317.
6. **Gallo, D., and N. J. Schmidt.** 1981. Comparison of anticomplement immunofluorescence and fluorescent antibody-to-membrane antigen test for determination of immunity status to varicella-zoster virus and for serodifferentiation of varicella-zoster and herpes simplex virus infections. J. Clin. Microbiol. **14**:539–543.
7. **Gershon, A. A., H. M. Frey, S. P. Steinberg, M. D. Seeman, D. Bidwell, and A. Voller.** 1981. Determination of immunity to varicella using an enzyme-linked-immunosorbent-assay. Arch. Virol. **70**:169–172.
8. **Gershon, A. A., S. P. Steinberg, W. Borkowsky, D. Lennette, and E. Lennette.** 1982. IgM to varicella-zoster virus: demonstration in patients with and without clinical zoster. Ped. Infect. Dis. **1**:164–167.
9. **Gershon, A. A., S. Steinberg, and L. Gelb.** 1984. Clinical reinfection with varicella-zoster virus. J. Infect. Dis. **149**:137–142.
10. **Olding-Stenkvist, E., and M. Grandien.** 1976. Early diagnosis of virus-caused vesicular rashes by immunofluorescence on skin biopsies. I. Varicella, zoster and herpes simplex. Scand. J. Infect. Dis. **8**:27–35.

10a. **Palmer, D. F., L. Kaufman, W. Kaplan, and J. J. Cavallaro.** 1977. Serodiagnosis of mycotic diseases. Charles C Thomas, Publisher, Springfield, Ill.

11. **Schmidt, N. J.** 1982. Further evidence for common antigens in herpes simplex and varicella-zoster virus. J. Med. Virol. **9**:27–36.
12. **Schmidt, N. J., D. Gallo, V. Devlin, J. D. Woodie, and R. W. Emmons.** 1980. Direct immunofluorescence staining for detection of herpes simplex and varicella-zoster virus antigens in vesicular lesions and certain tissue specimens. J. Clin. Microbiol. **12**:651–655.
13. **Schmidt, N. J., and E. H. Lennette.** 1975. Neutralizing antibody responses to varicella-zoster virus. Infect. Immun. **12**:606–613.
14. **Schmidt, N. J., and E. H. Lennette.** 1976. Improved yields of cell-free varicella-zoster virus. Infect. Immun. **14**:709–715.
15. **Shiraki, K., T. Okuno, K. Yamanishi, and M. Takahashi.** 1982. Polypeptides of varicella-zoster virus (VZV) and immunological relationship of VZV and herpes simplex virus (HSV). J. Gen. Virol. **61**:255–269.
16. **Straus, S. E., J. Hay, H. Smith, and J. Owens.** 1983. Genome differences among varicella-zoster virus isolates. J. Gen. Virol. **64**:1031–1041.
17. **Takahashi, M.** 1983. Chickenpox virus. Adv. Virus Res. **28**:285–356.
18. **Weigle, K. A., and C. Grose.** 1983. Common expression of varicella-zoster viral glycoprotein antigens in vitro and in chickenpox and zoster vesicles. J. Infect. Dis. **148**:630–638.
19. **Williams, V., A. A. Gershon, and P. A. Brunell.** 1974. Serological response to varicella-zoster membrane antigens measured by indirect immunofluorescence. J. Infect. Dis. **130**:669–672.
20. **Wroblewska, Z., M. Devlin, K. Reilly, H. van Trieste, M. Wellish, and D. H. Gilden.** 1982. The production of varicella zoster virus antiserum in laboratory animals. Arch. Virol. **74**:233–238.
21. **Zaia, J. A.** 1981. Clinical spectrum of varicella-zoster virus infections, p. 10–19. *In* A. J. Nahmias, W. R. Dowdle, and R. F. Schinazi (ed.), The human herpesviruses: an interdisciplinary perspective. Elsevier-North Holland, Inc., New York.

Chapter 67

Epstein-Barr Virus

EVELYNE T. LENNETTE

CLINICAL BACKGROUND

Epstein-Barr virus (EBV) is the principal etiological agent of infectious mononucleosis (IM) (3), as well as a contributory factor in the etiology of Burkitt's lymphoma (BL) and nasopharyngeal carcinoma (NPC) (5). It has a worldwide distribution, such that 80 to 90% of all adults have been infected. Primary infections occur during the first decade of life in areas with crowded living conditions and poor hygiene. Childhood infections are mostly asymptomatic, but infrequently may be associated with classical IM. In contrast, 50 to 75% of young adults undergoing primary EBV infections are symptomatic, with illness ranging from mild to severe. As with other herpesviruses, EBV causes a persistent, latent infection which can be reactivated under immunosuppression. Reactivated infections are generally asymptomatic (7).

Transmission of EBV requires salivary contact; airborne transmission is not an important route of infection. The virus is excreted orally by IM patients and by asymptomatic individuals with reactivated infections. The virus can be recovered in the saliva of 10 to 20% of healthy adults sampled.

In most cases of IM, clinical diagnosis can be made from the characteristic triad of fever, pharyngitis, and cervical lymphadenopathy lasting 1 to 4 weeks. Normally a self-limiting illness, IM may be complicated by splenomegaly, hepatitis, pericarditis, or central nervous system involvement. Hematological features include lymphocytosis with prominent atypical lymphocytes. In 85 to 90% of IM patients, Paul-Bunnell heterophile tests are positive; false-positives may occur in 2 to 3% of patients and can be excluded only by EBV-specific serology. Specific laboratory diagnosis is also needed to differentiate the 10 to 15% of heterophile-negative EBV infections from mononucleosis illnesses induced by other agents such as cytomegalovirus, adenovirus, and *Toxoplasma gondii*. In primary infections of young children and for adults with clinically atypical disease, EBV-specific laboratory diagnosis may also be helpful.

BL is primarily a tumor of children in Africa and New Guinea. Elsewhere in the world, EBV-associated BL has been reported mostly in adults and, more recently, among severely immunosuppressed male homosexuals. EBV has also been consistently associated with undifferentiated squamous cell carcinoma of the nasopharynx, with particularly high incidence among southern Chinese. In both tumors, EBV-induced antigens and genomes can be detected in malignant tissues (16, 20). Antibodies to several EBV antigens are present in unusually high titers in the tumor patients, and their level can be correlated with the patient's tumor burden. Serology can be helpful in the management of patients with either of these malignancies and to monitor the effectiveness of their therapies (6, 10).

Description of agent

EBV is a member of the family Herpesviridae; it contains DNA and has the characteristic herpetic 120-nm enveloped morphology, with 162 capsomeres in an icosahedral arrangement. In vivo, the virus can be found both in lymphoid tissues and in epithelial-derived tissues of the nasopharynx. In vitro, EBV has only been propagated in B lymphocytes from humans and subhuman primates. Antigenic variations have not been demonstrated among human isolates of EBV. Although it shares common antigenic determinants with other EBV-like subhuman primate viruses, EBV is antigenically distinct from other human herpesviruses.

Infection of lymphocytes by EBV leads to their transformation into lymphoblastoid cell lines, capable of continuous growth in culture. Once infected, only rarely will a cell produce infectious virus in vitro. Most transformed cell lines are nonproducers, although EBV-induced antigens can be reliably detected in the cells.

Collection and storage of specimens

Generally, only a single acute-phase serum sample (1 to 5 ml) is needed for diagnosis by serological testing. Convalescent-phase serum collected 1 to 2 months after onset is occasionally needed for confirmation and interpretations. If collected aseptically, serum can be stored at 5°C for a period of several months. For longer term storage, freezing at −20°C is recommended.

For isolation of excreted virus, a throat gargle can be collected using 5 to 10 ml of serum-free cell culture medium, although Hanks balanced salt solution is also satisfactory. Fetal bovine serum (2 to 5%) can then be added as a stabilizer, as well as antibiotics to suppress microbial growth. Specimens should be refrigerated promptly. For long-term storage, they should be frozen at −70°C with added bovine serum.

Tissues to be examined for viral antigens are collected aseptically and refrigerated in cell culture medium. Tissues suitable for this purpose are lymph nodes, spleen, and biopsies of tumors. Thin frozen sections, but not Formalin-fixed tissues, are also suitable. Biopsies may be examined for the presence of virus either by selection of EBV-transformed cells or by direct examination of EBV antigens by specific immunostaining.

For cultivation of EBV-infected peripheral blood lymphocytes, 10 ml of heparinized (5 to 10 U/ml) blood from IM patients is sufficient. Blood specimens should be processed as soon as possible, although refrigeration is adequate for up to 24 h.

Direct examination

EBV is generally excreted in the saliva in concentrations insufficient for direct detection by either electron microscopy or immunospecific staining techniques. Similarly, viremia in IM patients does not produce enough infected peripheral leukocytes for direct detection.

Neither saliva nor peripheral blood is a satisfactory specimen for direct detection assays; tissues, however, do contain enough infected cells for examination by two alternative methods. The first (and more practical) method uses anticomplement indirect immunofluorescence staining (ACIF) to demonstrate the presence of EBV-associated nuclear antigen (EBNA) in lymphoid tissues of IM patients; EBNA can also be detected in tumor biopsies of NPC and BL patients (11, 13). The second approach uses nucleic acid hybridization techniques to demonstrate the presence of EBV genomes in the tissues (16); though this method has been very useful in research studies, it is not practicable in most clinical laboratories.

Isolation of virus

Only B lymphocytes have been found to be susceptible to infection by EBV. Preparations of fetal B lymphocytes are most commonly used, due to their good susceptibility of infection. Lymphoid tissues and leukocytes from seronegative adults are, for practical purposes, not useful for EBV isolation due to suppressive interference from T cells usually found in partially fractionated lymphocyte preparations.

Indicator cells. A 10-ml sample of human umbilical-cord blood should be collected aseptically with 5 to 10 U of preservative-free heparin per ml. For optimal recovery of lymphocytes, the blood sample should be fractionated through polysaccharide density gradients (Hypaque-Ficoll, Histopaque, etc.). Depending on the formulation of the gradient material, heparinized blood is diluted with phosphate-buffered saline to the supplier's recommended concentration. This diluted blood sample is layered on the gradient material and centrifuged in clear centrifuge tubes at 400 × *g* for 30 min at room temperature. During centrifugation, erythrocytes and granulocytes are aggregated and sedimented to the bottom. Lymphocytes can be recovered at the plasma-gradient interphase, washed once with saline, and suspended in RPMI 1640 medium supplemented with 10% fetal bovine serum to a density of 5×10^6 cells per ml. After incubation at 37°C in 5% CO_2 overnight to check sterility, the lymphocytes are ready for inoculation and should be used as soon as possible. In our laboratory, uninfected lymphocytes thus prepared can be frozen at liquid N_2 temperature in 10% dimethyl sulfoxide. After thawing and a brief washing to remove the preservative, they are usable for EBV isolation, though not without some loss in viability.

Inoculation. Throat garglings must be centrifuged at 1,500 × *g* for 10 min. The supernatant fluid is filtered through a 0.45-μm filter to remove remaining cell debris and microorganisms. The specimen can be frozen at −70°C at this stage for future inoculations. When possible, prompt inoculation is recommended. Just before inoculation, 1 ml of the fractionated leukocyte suspension is centrifuged at 1,500 × *g* for 10 min to remove the culture fluid. A 0.5-ml sample of the filtered throat specimen is added to the leukocyte pellet. After an adsorption period of 1 to 2 h at 37°C, 1 ml of culture medium with 10% fetal bovine serum is added. An uninfected cell control and an EBV-infected control should be included in parallel, as rare cord lymphocyte preparations have shown spontaneous transformation. All cultures are incubated for 4 weeks with weekly replacements of growth medium. Necrosis of uninfected cultures is usually observable after 2 weeks, whereas virus-positive cultures should show clusters of large proliferating lymphoblastoid cells. The identification of the transforming agent can then be made by the detection of EBNA in these cells by using the ACIF procedure.

Identification of virus

The viral expression in transformed cell lines varies widely, from almost silent, with only a few detectable viral antigens, to a fully productive cycle with late proteins, such as viral capsid antigen (VCA) and infectious particles. Depending on the exact "block" on the viral genome expression, varying spectra of EBV antigens are detectable in the transformed cells. The only antigen present in all EBV-infected cells, however, is EBNA. It can be reliably used as an EBV marker, assayable by the ACIF procedure (18). Currently, there are no available hyperimmune animal anti-EBNA sera, hence human sera provide the only source of reagents suitable for EBNA detection. For typing reagents, sera from EBV-infected and susceptible individuals should be used in parallel. Care should be taken in the selection of negative sera to exclude those with nonspecific antinuclear antibodies.

To detect EBNA in transformed cells in culture, the medium is discarded after centrifugation at 1,500 × *g* for 10 min. The cell pellet is suspended in phosphate-buffered saline containing 0.5% bovine serum albumin. The cell density is adjusted to 5×10^6 cells per ml, and the suspension is dropped on glass slides. The air-dried slides are fixed for 1 min with an ice-cold acetone-methanol (1:1) mixture. Fixed smears can be held at −20°C until tested. Several slides should be prepared in this manner to allow for necessary controls.

Serological diagnosis

EBV isolation is usually not diagnostically useful due to the narrow host range of EBV in vitro, the long period needed for isolation of cell-free virus, and the ubiquity of EBV in healthy individuals. Serological testing is the method of choice for the diagnosis of primary infections (8). In patients with symptoms compatible with IM, a positive Paul-Bunnell antibody result is diagnostic and no further testing is necessary. Rapid qualitative agglutination test kits are widely available and are effective for 80 to 85% of IM patients. Quantitative testing procedures include several variations on the Paul-Bunnell differential test (1, 17), the ox cell hemolysis assay (15), and the immune adherence hemagglutination assay (12). Moderate to high levels of Paul-Bunnell antibody are seen during the first month of illness and decrease rapidly after week 4. The false-positive rate for the Paul-Bunnell antibodies is approximately 3%, mostly from individuals who maintain a low but persistent level of these antibodies long after their primary illness (9). The false-negative rate is 10 to 15%; it is more frequent among children than adults. For these patients, EBV-specific serological testing is needed.

Humoral responses to primary EBV infections appear to be quite rapid. Eighty percent of the patients usually already have peak titers by the time they consult their physicians (8). Hence, testing of paired sera is not useful in demonstrating significant antibody changes in the majority of the cases. Effective laboratory diagnosis can be made, on the other hand, on a single acute-phase serum by the testing of antibodies to several EBV-associated antigens measured simultaneously. The level and spectrum of antibodies are sufficiently distinct in most cases to allow classification as to whether the patient: (i) is still susceptible, (ii) has a current primary infection, (iii) has had a recent primary infection (within 2 to 3 months), (iv) had a past infection, or (v) may be having reactivated EBV infection (see Table 1).

Antibodies to four antigens may be measured, including VCA, early antigen (EA) diffuse component (EA/D), EA restricted component (EA/R), and EBNA. In addition, differentiation of immunoglobulin G (IgG), IgM, and IgA subclasses to VCA can often be helpful for confirmatory purposes.

Anti-VCA. During the acute phase, both IgG-VCA and IgM-VCA are detectable. Whereas IgM-VCA disappears after 4 weeks, IgG-VCA declines to a lower level but persists for life. Neutralizing antibodies can also be detected early after onset and persist for life. In practice, the complex neutralizing assays are rarely used to test immunity. As all patients with IgG-VCA also have neutralizing antibodies, anti-VCA titers are an accurate indicator of immunity. In patients with BL and NPC, the levels of IgG-VCA are maintained at very high levels, usually 8 to 10 times the geometric mean titers of healthy adults. NPC patients have an additional high IgA-VCA titer as an outstanding EBV serological feature.

Anti-EA/D and -EA/R. In the majority of patients, antibodies to EA/D show a transient rise in the acute phase. They are generally undetectable 3 to 6 months after onset. Anti-EA/R antibodies are sometimes present in children younger than 2 years of age and in asymptomatically infected patients. Antibodies to either or both early antigen components at moderate titers can reappear during EBV reactivation. With BL patients, anti-EA/R antibodies are present at moderate to high levels, whereas with NPC patients, high anti-EA/D titers are common. In the latter patients, both IgG and IgA subclasses of anti-EA/D are present in high titers.

Anti-EBNA. Antibodies to EBNA are rarely present in acute-phase serum. A gradual increase in EBNA occurs during convalescence, and near-peak titers are maintained for life.

The levels of each of the antibodies mentioned are usually lower in young patients. However, the profile does not differ with age. The exact titers to each antigen and the time needed to develop a full spectrum of antibodies vary widely with individuals. Also, many individuals may maintain EBV antibodies at high levels due to reactivations. For these reasons, diagnosis based on a single "screening" titer is not feasible. In addition, diagnosis of primary infection should not rely on the detection of IgM-VCA alone. Both false-positive and false-negative results occur in IgM-VCA testing. The former is due to the presence of rheumatoid factor (4), and the latter results from late collection of serum samples.

Fluorescent-antibody tests

Listed below are the most commonly used cell line combinations for measuring antibodies to VCA, EA/D, EA/R, and EBNA (9). All cell lines grow well in RPMI 1640 supplemented with 10% fetal bovine serum.

(i) P3-HR1 is a virus-producer cell line. From 10 to 15% of the cells in the culture express VCA at any given time. This degree of viral expression can be maintained indefinitely. To prepare smears, centrifuged cells from a 1-week-old culture are suspended in phosphate-buffered saline containing bovine serum albumin to a final density of 5×10^6 to 1×10^7 cells per ml. The cell suspension (10 to 20 μl) is dropped on glass slides with a Pasteur pipette. Air-dried slides are fixed for 3 min in acetone. Fixed slides can be stored at −20°C indefinitely.

(ii) The Raji line, in contrast, is positive only for EBNA. VCA and EA are rarely detectable if it is grown under normal culture conditions. Hence, Raji cells are suitable for the detection of anti-EBNA. Air-dried smears from 5-day-old cultures are prepared as described for P3-HR1 cells. Fixation, however, is with an ice-cold mixture of equal volumes of acetone and methanol for 1 to 2 min.

(iii) EA-positive cells can be prepared several ways. The conventional method involves superinfection of Raji cells with infectious virus concentrated from a P3-HR1 culture with centrifugation of culture fluid for 1 h at $10{,}000 \times g$. EA smears of consistent quality can be prepared with pretitered virus stock frozen at −70°C. The advantage of this method lies in the ability

TABLE 1. Serological profiles of EBV-associated syndromes

Antibodies-antigen	Nonimmune	Current primary infection	Recent primary infection	Past infection	Reactivation	BL	NPC
IgM-VCA	−	+	−	−	−	−	−
IgG-VCA	−	+	+	+	+	++	++
IgA-VCA	−	+/−	−	−	+/−	−	++
IgG-EA/D	−	+	+	−	−	−	++
IgA-EA/D	−	*	−	−	+/−	−	++
IgG-EA/R	−	+/−	+/−	−	+/−	++	−
Anti-EBNA	−	−	−	+	+	+	+

[a] −, Negative (<1:10); +, positive (>1:10); +/−, either positive or negative; *, not known.

to control the number of EA-positive cells desired. The alternative method involves EA induction of Raji cells by treatment with various chemicals, including tumor-promoting agents (19), idodeoxyuridine, and sodium butyrate (2, 14). By controlling the concentration and duration of the treatment with the various chemicals, varying degrees of EA expression can be achieved.

(iv) An EBV-negative cell line is necessary as negative control to exclude nonspecific antinuclear antibodies in some sera. The most appropriate cell line is BJAB, a B-lymphocyte EBV-negative line, although MOLT-4, a T-lymphoid line, is sometimes used as an alternative.

For the indirect immunofluorescent-antibody assay, fourfold dilutions of the patient's serum are incubated with the fixed cells, followed by fluorescein isothiocyanate-conjugated antiserum to the appropriate anti-human immunoglobulin subclasses. Incubations are for 30 min at 37°C in moist chambers. The acetone-fixed P3-HR1 line can be used for measurements of IgG, IgM, and IgA antibodies to VCA. IgM-VCA testing is sometimes prolonged to 3 h in the first incubation. This may be necessary to enhance the intensity of the staining.

Acetone-fixed EA smears are suitable for indirect immunofluorescent-antibody assays of antibodies to both EA/D and EA/R, as both antigens are present in the EA-producing cells. The corresponding antigens can be differentiated by their characteristic staining morphologies. EA/D-positive cells appear as cells with diffusely distributed speckled staining in the cytoplasm. Frequently, these cells have a halo of fine granular staining where the antigens have leaked out during fixation. EA/R, in contrast, is "restricted" to the cytoplasmic regions within the cells. Methanol fixation of the EA smears preferentially removes EA/R only and hence can be used to prepare smears for the differentiation of the two antigens.

The concentration of EBNA in EBV-transformed cells is usually low, and the more sensitive ACIF assay must be used to detect anti-EBNA. Raji smears are incubated with the patient's serum. This is followed by successive 30-min incubations with EBV-negative complement (either pretested human or guinea pig) and appropriately fluorescein isothiocyanate-conjugated anti-C3 antiserum. If human complement is used, fluorescein isothiocyanate–anti-B_1C/B_1A is preferable. The staining pattern is nuclear and should be present in all the cells.

Commercial smears are now available for all of the above antigens, though some are not yet licensed for diagnostic use.

Interpretations

From the titers and profile of antibodies to VCA, EA, and EBNA in the acute-phase serum, the patient can be classified as: (i) susceptible if anti-VCA is absent (<1:10); (ii) with primary EBV infection if anti-VCA is present and anti-EBNA is absent; or (iii) immune with past infection if both anti-VCA and EBNA are present. Eighty percent of patients with active EBV infections produce anti-EA/D titers. These antibodies can be very useful as indicators of current or reactivated infections. In the absence of anti-EBNA, anti-EA confirms a primary infection. In the presence of anti-EBNA, anti-EA suggests a reactivated past infection.

IgM-VCA is present in approximately 85 to 90% of sera from IM patients submitted to our laboratory for EBV testing. Most of the 10 to 15% IgM-VCA–negative sera had low or undetectable levels of anti-EBNA, indicative of a primary infection within the past 6 to 8 weeks. A second serum tested 4 to 6 weeks later should show a significant rise in anti-EBNA titers in all of the patients with primary disease. Hence, reliance on anti-EBNA instead of IgM-VCA effectively "extends" the acute phase of the illness.

LITERATURE CITED

1. **Davidsohn, I., and C. L. Lee.** 1969. The clinical serology of infectious mononucleosis, p. 177–200. *In* R. L. Carter and H. G. Penman (ed.), Infectious mononucleosis. Blackwell Scientific Publications, Oxford.
2. **Gerber, P., and S. Lucas.** 1972. Epstein-Barr virus associated antigens activated in human cells by 5-bromodeoxyuridine. Proc. Soc. Exp. Biol. Med. **141**:431–435.
3. **Henle, G., and W. Henle.** 1979. The virus as the etiologic agent of infectious mononucleosis, p. 297–307. *In* M. A. Epstein and B. G. Achong (ed.), The Epstein-Barr virus. Springer Verlag, Berlin.
4. **Henle, G., E. T. Lennette, M. A. Alspaugh, and W. Henle.** 1979. Rheumatoid factor as a cause of positive reactions in tests for Epstein-Barr virus specific IgM antibodies. Clin. Exp. Immunol. **36**:415–422.
5. **Henle, W., and G. Henle.** 1974. Epstein-Barr virus and human malignancies. Cancer **34**:1368–1374.
6. **Henle, W., and G. Henle.** 1979. Seroepidemiology of the virus, p. 61–102. *In* M. A. Epstein and B. G. Achong (ed.), The Epstein-Barr virus. Springer-Verlag, Berlin.
7. **Henle, W., and G. Henle.** 1980. Consequences of persistent Epstein-Barr virus infections: viruses in naturally occurring cancers. Cold Spring Harbor Conf. Cell Proliferation **7**:3–9.
8. **Henle, W., G. Henle, and C. A. Horwitz.** 1974. Epstein-Barr virus specific diagnostic tests in infectious mononucleosis. Hum. Pathol. **5**:551–564.
9. **Henle, W., G. Henle, and C. A. Horwitz.** 1979. Infectious mononucleosis and Epstein-Barr virus associated malignancies, p. 441–470. *In* E. H. Lennette and N. J. Schmidt (ed.), Diagnostic procedures for viral, rickettsial and chlamydial infections, 5th ed. American Public Health Association, Washington, D.C.
10. **Henle, W., J. H. C. Ho, G. Henle, J. C. W. Chan, and H. C. Kwan.** 1977. Nasopharyngeal carcinoma: significance of changes in Epstein-Barr virus related antibody patterns following therapy. Int. J. Cancer **20**:663–672.
11. **Huang, D. P., J. H. C. Ho, W. Henle, and G. Henle.** 1974. Demonstration of Epstein-Barr virus associated nuclear antigen in nasopharyngeal carcinoma cells from fresh biopsies. Int. J. Cancer **14**:580–588.
12. **Lennette, E. T., G. Henle, W. Henle, and C. A. Horwitz.** 1978. Heterophil antigen in bovine sera detectable by immune adherence hemagglutination with infectious mononucleosis sera. Infect. Immun. **19**:923–927.
13. **Lindahl, T., G. Klein, B. Johansson, and S. Singh.** 1974. Relationship between Epstein-Barr virus (EBV) DNA and the determined nuclear antigen (EBNA) in Burkitt lymphoma biopsies and other lymphoproliferative diseases. Int. J. Cancer **13**:764–772.
14. **Luka, J., B. Kallin, and G. Klein.** 1979. Induction of the Epstein-Barr virus (EBV) cycle in latently infected cells by N-butyrate. Virology **94**:228–231.
15. **Mikkelsen, W., C. J. Tupper, and J. Murray.** 1958. The ox cell hemolysin test as a diagnostic procedure in infectious mononucleosis. J. Lab. Clin. Med. **52**:648–652.
16. **Nonoyama, M., C. H. Huang, J. S. Pagano, G. Klein, and**

S. Singh. 1973. DNA of Epstein-Barr virus detected in tissue of Burkitt's lymphoma and nasopharyngeal carcinoma. Proc. Natl. Acad. Sci. U.S.A. **70**:3265–3268.

17. **Paul, J. R., and W. W. Bunnell.** 1932. The presence of heterophile antibodies in infectious mononucleosis. Am. J. Med. Sci. **183**:80–104.
18. **Reedman, B. M., and G. Klein.** 1973. Cellular localization of an Epstein-Barr virus (EBV)-associated complement fixing antigen in producer and nonproducer lymphoblastoid cell lines. Int. J. Cancer **11**:499–520.
19. **zur Hausen, H., F. J. O'Neill, and U. K. Freese.** 1978. Persisting oncogenic herpesvirus induced by the tumour promoter TPA. Nature (London) **272**:373–375.
20. **zur Hausen, H., H. Schulte-Holthausen, G. Klein, W. Henle, G. Henle, P. Clifford, and L. Santesson.** 1970. EBV DNA in biopsies of Burkitt's tumours and anaplastic carcinomas of the nasopharynx. Nature (London) **228**:1056–1058.

Chapter 68

Poxviruses

JAMES H. NAKANO

CLINICAL BACKGROUND

Smallpox (variola)

Smallpox, perpetuated only in humans, was differentiated into two types by the degree of severity and proportions of fatal cases it produced during outbreaks. Variola major, the severe type with a case-fatality ratio of 15 to 40%, prevailed in the "modern days" in the Asiatic subcontinent. Variola minor, also known as alastrim, amaas, or Kaffir pox, was the mild type, with fatalities of less than 1%. It first appeared in South Africa and spread to the West Indies, Brazil, North America, and England (6), and in the 1970s, before the world-wide eradication in 1977, it was prevalent in Ethiopia and Somalia.

The virus enters through the mucosa of the upper respiratory tract, migrates to the lymphatic glands, and is carried by the bloodstream to the internal organs. During an incubation period of 7 to 17 days, the virus multiplies in the internal organs and overflows into the bloodstream. During this prodromal period the disease is manifested by fever, headache, and backache. A widespread infection of the skin and mucous membranes follows, and skin lesions become apparent in 2 to 3 more days. As the lesions break down, virus is liberated and the patient becomes highly infective. The immune response that follows determines to some extent the severity of the disease.

Monkeypox

Monkeypox was first recognized as an exanthematous disease in 1958 in captured cynomolgus monkeys (11). The human disease caused by monkeypox virus infection, which evolved to be the most important poxvirus disease since the eradication of smallpox, was first recognized in Zaire (Congo), Africa, in 1970. The clinical features of monkeypox virus infection in humans, designated human monkeypox, resemble those of smallpox so much that the final differentiation between human monkeypox and smallpox usually depends on the isolation and identification of monkeypox virus. Because of this close resemblance, human monkeypox was not recognized until smallpox was eradicated. Since 1970 and through 1983, 150 cases of human monkeypox have been identified, mostly in Zaire but also in Liberia, the Ivory Coast, Sierra Leone, Nigeria, Benin (the patient traveled from Nigeria), and Cameroon. The fatality rate of human monkeypox is 15%, but the transmission rate to unprotected individuals is much lower than that of smallpox. Tertiary transmission was seen for the first time in 1983. Although outbreaks have occurred in monkeys in captivity, the natural reservoir of the virus is still unknown, and consequently the means of its initial transmission to humans is unknown.

Vaccinia

Rare but possibly serious complications following vaccination or contact with a recent vaccinee include progressive vaccinia (vaccinia gangrenosa, vaccinia necrosum), postvaccinal encephalitis, eczema vaccinatum, and congenital vaccinia. Less serious complications include erythema multiforme and generalized vaccinia. The number of these complications will decrease in the United States since smallpox vaccine is no longer available for the civilian population as of 1983. However, occasional contact vaccinia continues to occur in family members and close contacts of recently vaccinated members of the U.S. Armed Forces.

Because of recent enthusiasm in using vaccinia virus as a carrier of genome segments of other infectious agents for immunization, the Advisory Committee for Immunization Practice recommended in 1983 that laboratory investigators using orthopoxviruses, mainly those working with vaccinia virus but also those handling monkeypox virus and smallpox virus, be vaccinated. Vaccine for laboratory investigators is available from the Centers for Disease Control, Atlanta, GA 30333.

Cowpox

Cowpox was always believed to be transmitted to humans by direct contact with infected cows, but infected rats and other rodents are now implicated in transmitting the disease to humans. The lesions in humans are found on the fingers, with reddening and swelling. The lesions become papular and then, in 4 to 5 days, vesiculate and heal in 2 to 4 weeks.

Whitepox

Six isolates of whitepox virus have been found since 1964. Two were isolated from two healthy monkeys in captivity, and four were isolated from wild animals (chimpanzee, sala monkey, and two rodents) captured in Zaire in the same localities where human monkeypox cases had occurred. These isolates cannot be differentiated from human variola virus by biological genetic marker tests or genome characterization. A recent investigation, however, showed that at least two of the six whitepox virus isolates are products of laboratory cross-contamination (3).

Tanapox

Tanapox in humans, first described by Downie et al. (2), occurred in Kenya along the Tana River. A World Health Organization surveillance of human monkeypox in Zaire revealed that the disease is also prevalent there. Lesions found on the skin of the upper arms, face, neck, or trunk start as papules, then become vesicles (from which fluid is difficult to extract). The lesions umbilicate without pustulations and heal in 2 to 4 weeks (some to 7 weeks).

Milker's nodule

Milker's nodule, known as pseudocowpox in cattle, is transmitted to humans by direct contact. The infection in humans starts at the site of abraded skin on the hands and fingers which have been in direct contact with an animal's infected udder and teats.

Orf

Contagious ecthyma, contagious pustular dermatitis, contagious pustular stomatitis, or sore mouth, all naming the same disease in sheep and goats, is transmitted to humans by direct contact. The infection in humans is usually found on the fingers, hands, and arms but may be found on the face and neck.

Molluscum contagiosum

Two forms of molluscum contagiosum occur in humans. In one form found in children, the lesions are found on the face, trunk, and limbs, and the infection is transmitted by skin-to-skin or fomites-to-skin contact. In the other form, found in young adults, the lesions are mostly in the lower abdominal wall, pubis, inner thighs, and genitalia, and the infection is transmitted by sexual contact.

DESCRIPTION OF AGENTS

All poxviruses described in this chapter belong to the family Poxviridae. The viruses of smallpox, monkeypox, vaccinia, cowpox, and whitepox are in the genus *Orthopoxvirus*. The viruses of milker's nodule and orf are in the genus *Parapoxvirus*. The viruses of tanapox and molluscum contagiosum are still unclassified in the family Poxviridae. The genomes of these viruses consist of a single molecule of double-stranded DNA.

The orthopoxviruses (smallpox, monkeypox, vaccinia, cowpox, and whitepox viruses) cannot be distinguished from each other on the basis of morphology when visualized by electron microscopy (EM). The parapoxviruses (milker's nodule and orf) cannot be distinguished from each other by morphology but can be distinguished from orthopoxviruses, tanapox virus, and molluscum contagiosum virus. The tanapox and molluscum contagiosum viruses can at times be distinguished from orthopoxviruses and can always be distinguished from parapoxvirus by morphology (see section on EM, below).

The viruses of the orthopoxvirus group are serologically so closely related that each virus can be identified as belonging to this group by a routine serological test but cannot be distinguished from the others. There is no serological relationship among orthopoxviruses, parapoxviruses, tanapox virus, and molluscum contagiosum virus.

Orthopoxviruses produce a soluble hemagglutinin, but parapoxvirus, tanapox virus, and molluscum contagiosum virus do not. Orthopoxviruses grow on embryonated chicken egg chorioallantoic membranes (CAMs), but parapoxviruses, tanapox virus, and molluscum contagiosum virus do not.

COLLECTION AND STORAGE OF SPECIMENS

In the United States and its territories, a suspected case of smallpox is to be reported immediately upon discovery, by telephone, to the respective state or territorial health department. After the state or territorial health department reviews the case and if it still appears to be a suspected case, it should be immediately reported to the International Health Program Office [(404) 329-3111] or the Poxvirus Laboratory [(404) 329-3667], Viral Exanthems and Herpesvirus Branch [(404) 329-3532], Division of Viral Diseases, Center for Infectious Diseases, Centers for Disease Control (CDC). The Poxvirus Laboratory also serves as one of the World Health Organization Collaborating Centers for Smallpox and Other Poxvirus Infections (the other center is in Moscow, USSR). Both Collaborating Centers serve other countries.

Collection of specimens

The basic procedures for collecting specimens can be found in Chapter 61. In addition, however, it is important to collect an amount of specimen sufficient to permit effective testing; use of an inadequate amount of a specimen decreases the dependability of laboratory tests for diagnosing smallpox and other poxvirus diseases.

Suitable specimens from patients with smallpox, vaccinia, or cowpox collected for virological tests are vesicular fluid (including the cells at the base of the vesicule) and scabs. If these specimens are not available, such as in cases of milker's nodule and orf, a biopsy from the lesion's outer area is useful.

Collect vesicular fluid in capillary tubes or on glass slides as a thick smear (without spreading the smear) or on swabs. Collect at least two or three capillary tubes of vesicular fluid, or at least four glass slides with thick smear, or at least three to four swabs and at least four to six scabs. Send the collected specimens in a dry condition in a screw-capped container. Do not add "transport fluid" to the specimens. Suitable autopsy specimens include sections of the lungs, liver, spleen, and kidneys. Specimens collected from cases of any poxvirus infection can be stored at −20°C, but −70°C is preferred for long-term storage.

DIRECT EXAMINATION

To be effective, laboratory diagnostic tests (direct examination) for smallpox should produce accurate, easily interpreted results. They should be relatively quickly performed, simple, and direct. Seven methods that may be used for the laboratory diagnosis of smallpox and other orthopoxvirus diseases are listed in Table 1. The first three methods, namely, EM, chicken CAM culture, and cell culture, have been the combination of choice (at CDC) which gives the most dependable results with the least confusion. The other four methods, agar gel precipitation (AGP), stained smear, fluorescent antibody, and complement fixation (CF), do not provide additional advantages to the efficacy of laboratory diagnosis.

I shall discuss the use of these tests to identify the etiological agents for milker's nodule, orf, tanapox, and molluscum contagiosum when these tests are discussed later.

EM

Because EM has been well established as a standard rapid laboratory diagnostic method, I shall not describe the procedures for grid preparation of specimens and grid examination by EM, but they can be found in the manual *Diagnostic Procedures for Viral, Rickettsial and Chlamydial Infections* (8).

TABLE 1. Accepted direct examination methods for the laboratory diagnosis of smallpox, human monkeypox, vaccinia, and cowpox

Method	Accomplishment
Preferred methods	
EM	Direct visualization of virus
Chicken CAM culture	Growth of smallpox, human monkeypox, vaccinia, cowpox, whitepox, and herpes simplex viruses with definitive pock characteristics
Cell culture	Growth of smallpox, human monkeypox, vaccinia, whitepox, and herpes simplex viruses with definitive CPE characteristics
Other methods	
AGP	Antigenic identification
Stained smear	Visualization of elementary bodies
Fluorescent antibody	Visualization of virus-antibody complex; antigenic identification
CF	Visualization of reaction dependent on virus-antibody complex; antigenic identification

Vesicular fluid collected in capillary tubes, lesion fluid (pustular) collected as smears, and ground scab suspensions are all excellent sources of virus to be examined by EM.

In examining a prepared grid, one must note that a grid too dense and showing no transparency indicates that too much material has been put onto the grid, and a grid not dense enough and showing too much transparency indicates that too little material has been used. Either condition greatly diminishes the reliability of the test.

Figures 1-1 and 1-5 illustrate the "M" form of smallpox virus described by Harris and Westwood (5), most prevalent in vesicular fluid collected from patients with smallpox. The M form is also prevalent in specimens of other orthopoxvirus diseases when a specimen is vesicular fluid or material from a wet lesion. It is also prevalent in poxvirus populations growing in cell cultures. Figure 1-7 illustrates the "C" form of Harris and Westwood, most prevalent in scabs from patients with smallpox as well as with other poxvirus diseases. Figures 1-9 and 1-11 show the typical enveloped virion, and Fig. 1-15 shows the typical naked capsids of varicella virus. The enveloped virions are found more often in vesicular fluid, and the naked capsids are found more often in scabs collected from patients with chickenpox. Figures 1-12, 1-13, 1-14, and 1-16 illustrate the less typical forms of varicella virions (note that Fig. 1-16 is shown at a magnification twice that of the others).

In examining a grid, not only must one know that a sufficient amount of material is on the grid, but one must also learn to differentiate nonviral particles which resemble poxvirus or herpesvirus from the real viral particles. Nonviral particles that resemble the "M" and "C" forms of poxviruses are illustrated in Fig. 1-2, 1-4, 1-6, and 1-8, and a particle that resembles a varicella virion is shown in Fig. 1-10.

The reliability of EM for detecting viruses in specimens collected from cases of smallpox and human monkeypox at CDC has been over 98%. An even higher percentage of positive specimens could have been detected by EM if the quantity of vesicular, pustular, or crust materials in all specimens had been adequate.

Compared with the reliability of EM for positive diagnosis of smallpox and human monkeypox, the reliability for diagnosis of vaccinia infection was only 67%. This low percentage can be partially explained by the fact that inadequate amounts of specimens were obtained or, more likely, that specimens were obtained when the numbers of virus present in the lesion were not at peak.

To determine the reliability of EM for the detection of varicella virus is difficult. Of the specimens from 6,919 patients with suspected smallpox, 1,936 were identified as having varicella viruses. EM examination identified all of these. Varicella virus is considerably labile as compared with poxviruses, and since many specimens received were scabs or vesicular fluid which were not fresh, virus isolation was impossible. Therefore, EM remains the only dependable method for the detection of this virus. The reliability of EM for detecting varicella virus in vesicular fluid collected in adequate amounts (which includes the basal cell layer of the vesicle) and in scabs collected in the very early stage of their formation is very high. However, the reliability decreases in scabs collected at later stages and in scabs which are ready to fall off.

Because varicella is clinically similar to and sometimes confused with smallpox and human monkeypox, the use of EM is the fastest method by which to distinguish among these. As seen in Fig. 1, varicella virus can easily be differentiated from poxvirus by EM. EM cannot, however, differentiate the orthopoxviruses (smallpox, human monkeypox, vaccinia, cowpox, and whitepox).

Although there are no data to measure the reliability of EM in detecting viruses of milker's nodule, orf, tanapox, and molluscum contagiosum, I found it most useful. Parapoxviruses (milker's nodule and orf) are smaller, and their morphology is distinctly different from that of orthopoxvirus and the viruses of tanapox and molluscum contagiosum. Parapoxviruses are elongated (oblong) and the tubules on the viral surface are arranged in parallel and form a criss-cross pattern, whereas the orthopoxviruses and those of tanapox and molluscum contagiosum are more brick shaped and the tubules are without the parallel arrangement. Tanapox virus morphology is similar to that of orthopoxviruses, but in about 80% of the specimens the viral particles of tanapox visualized by EM possess an envelope similar to that seen on the vaccinia M-form virus illustrated in Fig. 1-3. Although vaccinia, smallpox, and human monkeypox viruses in lesion materials are seen usually without an envelope, vaccinia virus, especially when grown in cell culture, shows an envelope (about 10% of the viral population). In contrast, tanapox virus visualized by EM in lesion materials shows an envelope surrounding each virion. Molluscum contagiosum virus seen in lesion material morphologically resembles orthopoxviruses, but the surface tubules of the M-form virions are more pronounced than those usually seen on orthopoxviruses. Because of the unique characteristic of the

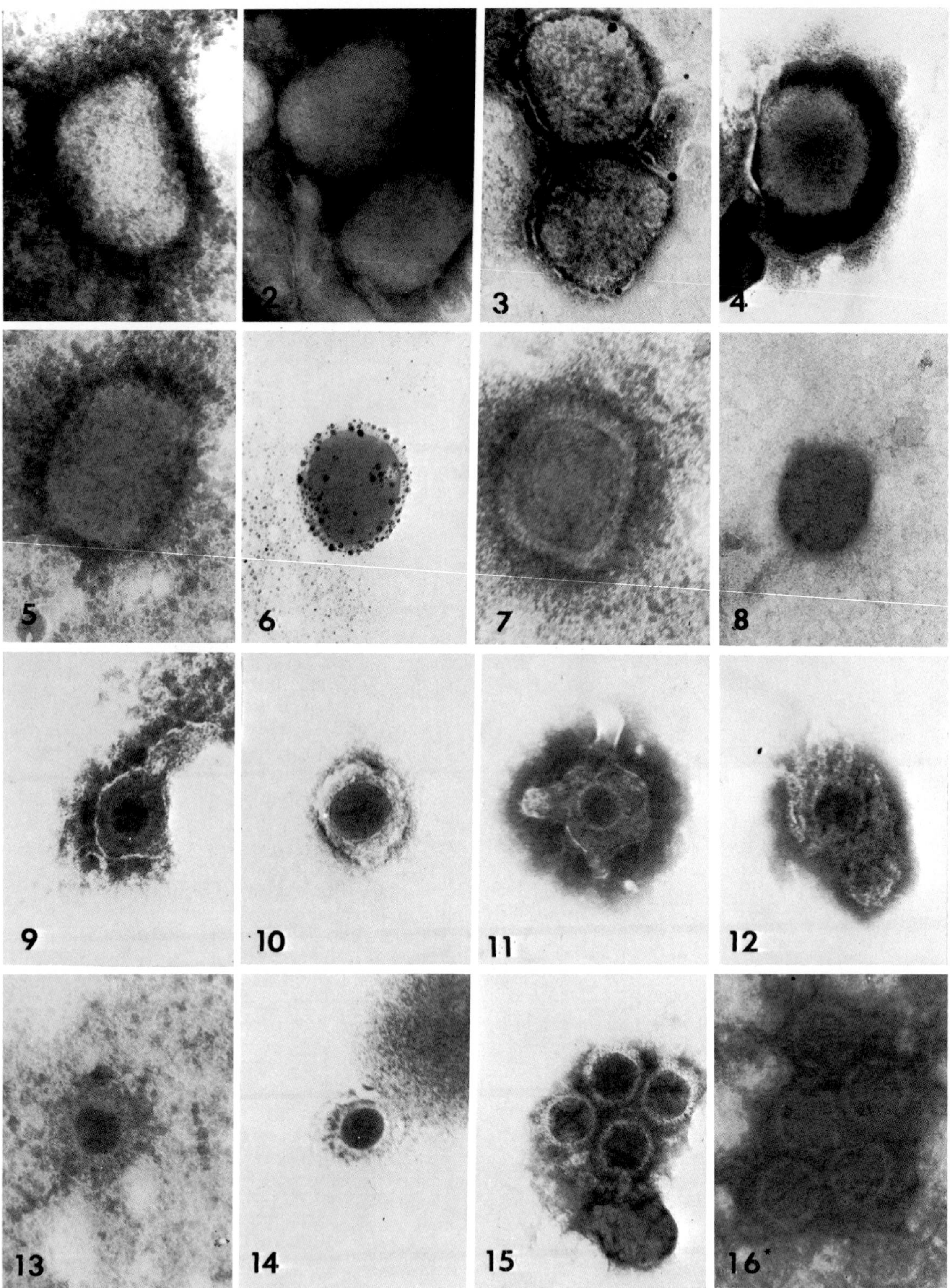

FIG. 1. Electron micrographs at the original magnification on negative plates of ca. ×41,000. 1. An M form of variola virus from crust, with indistinguishable surface filaments. 2. Two nonviral particles resembling the M form of variola virus (from crust); note the absence of the surface filamentous structure. 3. Two particles of vaccinia virus M form (from cell culture), each with an outer envelope; note the clearly distinguishable surface filaments. 4. A nonviral particle which can be mistaken for an M form of poxvirus. 5. An M form of variola virus (from crust) showing a more distinct filamentous structure than that in Fig. 1-1. 6. A nonviral particle (a stain spot). 7. A C form of variola virus (from crust) with a capsule; visible internal nucleoid. 8. A nonviral particle resembling the C form of variola virus; note that it is much smaller than a poxvirus particle. 9. A typical enveloped varicella-zoster virus found in diagnostic materials; note the capsomeres on the capsid surface. 10. A nonviral particle resembling an enveloped varicella-zoster virus particle; note that it is almost twice as large as varicella-zoster virus and has no capsomeres. 11. An enveloped varicella-zoster virus particle with three pseudopod-like structures (from vesicular fluid). 12. Disintegrating envelope of a varicella-zoster virus particle (from crust). 13. A varicella-zoster virus particle (from crust) with disintegrating envelope and an unevenly stained capsid. 14. A varicella-zoster virus particle (from vesicular fluid). 15. Typical capsids with capsomeres of varicella-zoster virus (from crust). 16. Varicella-zoster virus capsids with staining characteristics uncommon in viral particles found in crusts or vesicular fluid; these particles are twice the magnification of particles in the other micrographs.

lesions found in molluscum contagiosum, the visualization of poxviruses is usually adequate for the laboratory diagnosis of this disease.

AGP

Methods for the preparation of vaccinia antigen in CAM and hyperimmune antivaccinia serum in rabbits are described in the Appendix. An end-frosted glass slide (75 by 25 mm), precoated with 0.2% purified agar (Difco Laboratories, Detroit, Mich.) in distilled water, is prepared in the following manner for use as an agar gel slide. A line is drawn between the frosted and the clear area with a marking pen containing a fast-drying, oil-base paint. The slightly raised line produced prevents the melted agar from running into the frosted area. A 1.5-ml amount of melted, 1.0% purified agar in distilled water containing a 1:10,000 dilution of thimerosal is carefully delivered and spread onto the entire clear area of the slide. This forms a layer of agar about 2 mm thick. After the agar hardens, wells are made with a plastic template which cuts a pattern of wells so that a centrally located well is surrounded by six wells. The wells are 4 mm in diameter, and they are separated from each other by a distance of 5 mm from the center of one well to another. Agar cores are removed by suction with a Pasteur pipette attached to a vacuum line.

A 10% suspension of specimens is used for the AGP test. The limit of dilution of the 10% crust suspension is about 1:60 (wt/vol) (4). The specimen suspension to be tested is placed in the wells at the 12 and 6 o'clock positions. The positive control vaccinia antigen is placed in the well at 2 o'clock, and the normal rabbit serum is placed in the well at the 4 o'clock position. Rabbit antivaccinia serum is placed in the central well. The slide is then placed in a humid chamber and incubated at 35°C. Lines of precipitation (positive reaction) will occur within 2 to 4 h between the wells containing the specimen and the antivaccinia serum if the specimen is from lesions of smallpox, human monkeypox, vaccinia, or cowpox. The line(s) of precipitation formed between the wells containing the specimen and the antivaccinia serum must fuse or join (form a line of identity) with at least one of the lines between the wells of the positive control vaccinia antigen and the rabbit antivaccinia serum. Specimens are not considered negative unless diagnostic lines fail to appear by 24 h of incubation.

A specimen negative on first testing may be retested with rabbit antivaccinia serum diluted 1:2, 1:4, and 1:8. The use of the diluted reagent antiserum in the test sometimes results in an optimal antigen-antibody proportion and consequently gives a positive result.

The reliability of the AGP test at CDC was over 78% in detecting viral antigens of smallpox and human monkeypox. The failure to detect was due to an inadequate amount of specimens received for testing (especially when vesicular fluid was received) or the probable degeneration of soluble precipitating antigens, a consequence caused by prolonged exposure of some of the specimens to high ambient temperatures during shipment. Heating crust suspension at 60°C for 15 min greatly weakens AGP reactions (4).

Human convalescent-phase smallpox serum or vaccinia serum should not be substituted for hyperimmune antivaccinia rabbit serum as a testing reagent. In human serum, antibodies other than those specific for smallpox, human monkeypox, vaccinia, or cowpox may be present, in which case a precipitation line other than that against vaccinia may be observed and may confuse the diagnosis.

The AGP can identify the viral antigens of smallpox, human monkeypox, vaccinia, cowpox, and whitepox, but because of the close similarity of these antigens, one antigen cannot be differentiated from another. Despite the test's disadvantage of giving false-negative results, the results can be obtained in relatively short time; therefore, it is useful when EM is not available.

Although the AGP test has been used to identify the viral antigens of milker's nodule, orf, tanapox, and molluscum contagiosum, adequate data are not yet available to evaluate the test.

Stained smears

Stained smears made with lesion materials of smallpox, human monkeypox, vaccinia, and cowpox can be examined by light microscopy when EM is not available. This method, however, does not equal EM in reliability. For details consult Downie and Kempe (1), who described and evaluated Gutstein's method and Gispen's modification of Morosow's silver stain method. Very limited information is available for using this method for field specimens of milker's nodule, orf, tanapox, and molluscum contagiosum.

Fluorescent antibody

Although the fluorescent-antibody test can be used for the identification of antigens of smallpox, human monkeypox, vaccinia, cowpox, milker's nodule, orf, tanapox, and molluscum contagiosum, CDC has found that it can give false-positive results under certain conditions; therefore, it should be used with caution.

CF

The CF test is more sensitive than the AGP test for identifying poxvirus antigens in lesions caused by poxvirus infections. Lesion materials must be "cleaned up" before they can be tested (8), however, because these materials unless cleaned can cause anticomplementary reaction.

ISOLATION OF VIRUS

Chicken embryo CAM culture

A detailed technique to prepare eggs so that the CAM can be used is described in reference 8. Fertile chicken eggs must be incubated at 38 to 39°C for 11 to 13 days to be useful for isolating and identifying the viruses of smallpox, human monkeypox, vaccinia, cowpox, and whitepox. Lower incubation temperatures render the CAM less susceptible or totally unsusceptible to poxviruses at the recommended time of 12 days.

As a routine procedure, the inoculated eggs are incubated at 35°C instead of 37°C to avoid the complicating effects which result from supraoptimal temperatures. Some eggs are opened for examination at 48 h, if necessary, but the usual incubation time is 72 h. Negative CAMs are passed again in eggs (blind passage).

It has been observed since 1966 at CDC that at times CAMs do not support the growth of viruses of smallpox, human monkeypox, vaccinia, or herpes simplex virus. Possible reasons are: (i) eggs from physiologically different flocks of hens; (ii) use of unusual antibiotics in the flock; (iii) viral infection in the flock, causing infection of the embryo and interference; (iv) incubation of the eggs at a temperature lower than 38°C, resulting in physiologically less developed embryos than the normal 12-day-old embryos; (v) insufficient humidity during incubation; (vi) use of a buffer solution too highly concentrated to dilute the test specimens; and (vii) incubation of inoculated eggs at a supraoptimal temperature. Although the exact cause of the variable susceptibility is uncertain, the effects are manifested in several ways: (i) CAMs support no growth of control virus strains of smallpox, human monkeypox, vaccinia, and herpes simplex; (ii) CAMs support the growth of a control strain of vaccinia virus, but the pocks are atypically small; or (iii) CAMs support the growth of a "house standard" vaccinia, but the pock titers may be 0.5 to 1.0 $\log_{10}$ lower than usual, although the pock morphology is characteristic of vaccinia virus.

Differentiation of virus on the basis of pock morphology

The viruses of smallpox, human monkeypox, vaccinia, cowpox, and types 1 and 2 herpes simplex are differentiated primarily by the morphology of the pocks they form on the CAM. Herpes simplex virus is included in the discussion because it is sometimes isolated in our investigation of smallpox and human monkeypox. Varicella-zoster virus does not grow on chicken CAMs.

Smallpox virus pocks at 72 h of incubation are about 1 mm in diameter, grayish-white to white, opaque, convex or dome shaped, raised above the CAM, round, regular, smooth on the surface, and not hemorrhagic, and all are of nearly the same size. They generally resemble "sunny-side-up" fried eggs when examined with a magnification of ×10. Whitepox virus pocks cannot be differentiated from those of variola virus.

Human monkeypox virus pocks at 72 h are about the same size as those of variola virus, but they are not as raised; many pocks have a pinpoint hole in the center, and they are mostly hemorrhagic when incubated at 35°C (more hemorrhagic at 34°C).

Vaccinia virus pocks at 72 h are 3 to 4 mm in diameter, flattened with central necrosis and ulceration, and sometimes slightly hemorrhagic. Some strains of vaccinia virus are quite hemorrhagic.

Cowpox virus pocks at 72 h are 2 to 4 mm in diameter, flattened, and rather round and have a bright red central area (hemorrhagic). When the pocks are examined at ×10, the erythrocytes are in the pock proper.

Herpes simplex virus type 1 pocks at 72 h are pinpoint size, not raised, not opaque, and not regular shaped. When many pocks are present, they are in a lattice-work arrangement.

Herpes simplex virus type 2 pocks at 72 h are about 1 mm in diameter, white, flat, and irregular in shape and size. They are large and appear mucoid when initially isolated on the CAM.

Viruses of tanapox, milker's nodule, orf, and molluscum contagiosum do not grow on CAMs.

The reliability for detecting smallpox and human monkeypox has been over 91% by the CAM culture method at CDC. This might have been closer to 100% if the specimens had been fresh when received. Most of the specimens received at CDC had been in transit from 2 to 4 weeks.

When the eggs are candled, examine the pointed end of the egg to determine whether the CAM is adequately developed into the area; eggs with underdeveloped CAMs should not be used because they may be less sensitive to poxviruses. Eggs with albumen sac encroachment on the dropped area of the CAM can give only about one-tenth of the vaccinia pock count of the control (12).

In examining the pocks on the CAM, one must be careful not to mistake nonspecific lesions for true pocks. Of the several causes for the appearance of nonspecific lesions, the most common is mechanical trauma. In addition, isotonic sodium phosphate buffer (0.122 M) used for inoculation can induce a high incidence of nonspecific lesions (12).

A large dose of variola virus inactivated by heat or UV irradiation can cause a general thickening of the CAM and obscure the effect of a small amount of viable virus which may be present. A similar effect has been observed at CDC when CAMs are inoculated with diagnostic specimens containing a very high virus titer; this effect could be avoided by diluting the specimens 10^{-3} or 10^{-4} before inoculation. CAMs showing the thickening effect should always be put through a second passage at several 10-fold dilutions in attempts to obtain definitive evidence of pock growth.

Cell cultures

The use of cell culture has been a necessary alternative virus isolation method because of the periodic unpredictable insensitivity of the CAM for smallpox and human monkeypox viruses.

Orthopoxviruses (viruses of variola, human monkeypox, vaccinia, and cowpox) can be isolated in human and nonhuman primate cells (e.g., embryonic human diploid cells, LLC-MK_2, and Vero).

Variola virus in clinical specimens may produce cytopathic effect (CPE) within 1 to 3 days, with a rounding-up of the cells and the presence of hyperplastic foci, followed by the formation of small plaques (1 to 3 mm). The CPE spreads rapidly when a high-titered inoculum is used, and the cells eventually slough off. Human monkeypox, vaccinia, and cowpox viruses may also cause CPE in 1 to 3 days, as characterized by fused cells and the formation of foci, followed in 2 to 3 days by the formation of plaques which measure 2 to 6 mm in diameter. The plaques usually show cytoplasmic bridging. Again, when a high-titered inoculum is used, the entire cell sheet becomes involved, and the cells will eventually slough off. Therefore, when cell cultures are used to produce viral plaques that can be characterized morphologically, one must inoculate several dilutions of the stock viruses (or specimens) so that the plaques formed are properly separated on the cell monolayers.

In some instances, orf virus can be isolated in primary rhesus monkey kidney cells or embryonic human fibroblasts, and in other instances it may require ovine cells for the isolation. However, after it is isolated, orf virus can grow in embryonic human fibroblast cells and LLC-MK_2 or other nonhuman primate cells.

Milker's nodule virus is more demanding and usually requires bovine cells for initial isolation, but once isolated it can also be propagated in embryonic human fibroblasts or LLC-MK_2.

Molluscum contagiosum virus has been reported to produce CPE in primary human amnion cells or primary rhesus monkey kidney cells, but the virus has never been successfully passed continuously in a cell culture.

SEROLOGICAL DIAGNOSIS

Orthopoxvirus

Serological methods of choice to assay antibodies evoked by variola, human monkeypox, vaccinia, and cowpox are hemagglutination inhibition (HI) (8), neutralization (18), indirect fluorescent antibody (IFA) (9), enzyme-linked immunosorbent assay (ELISA) (10), radioimmunoassay (14), and radioimmunoassay adsorption (7). For each serological test, a fourfold rise in titer between acute- and convalescent-phase serum specimens is considered diagnostic, but often only a single serum specimen taken at one stage or another of the illness is available. Therefore, it is important to know how to interpret the serological result obtained with such a serum.

The HI, neutralization, IFA, ELISA, and radioimmunoassay procedures are not effective in differentiating the antibodies produced against viruses of variola, human monkeypox, vaccinia, and cowpox, but radioimmunoassay adsorption can identify specific antibodies of variola, human monkeypox, and vaccinia.

HI test

Pretested chicken erythrocytes that can be agglutinated by vaccinia hemagglutinin are required for the HI test, and vaccinia virus grown in BHK-21 cells may be used. (See Appendix for the preparation of chicken erythrocytes and vaccinia antigen.) The HI test probably detects antibody earliest after an infection by virus of variola, human monkeypox, vaccinia, or cowpox. The HI antibody titers in patients with smallpox and human monkeypox are generally greater than 1:80 and may be greater than 1:1,000. Sera with extremely high HI titers should be suspected of containing nonspecifically reactive substances, especially if the blood specimens are old and improperly stored or if they were collected during autopsy. (See Appendix for periodate treatment to remove nonspecific HI titers.) Contrary to the belief that HI antibody is short-lived, CDC found HI titers of 1:10 to 1:20 in cases of human monkeypox at 3 to 4 years after onset.

Neutralization test

The antigen used at CDC for the routine neutralizing test is monkeypox virus.

Neutralizing antibody for variola, monkeypox, or vaccinia viruses is usually detected in the latter part of the first week and during the second week after onset; the antibody may persist for a number of years. When only one serum specimen is taken from a patient sometime after the onset of rash, and if it shows only a moderate titer (less than 1:500), the diagnosis is difficult, especially if the patient had been vaccinated previously. However, a neutralizing titer of greater than 1:1,000 is probably significant for the diagnosis of an infection.

IFA

The antigen used for routine IFA test at CDC is cells infected with monkeypox virus. The cells are not fixed with methanol, acetone, or other similar chemical, but simply dried onto the slides. A titer of ≥1:32 for a patient without previous vaccination indicates recent infection. IFA titers begin to decrease at about 6 months after the onset of illness.

ELISA

The antigen used for routine ELISA test at CDC is also monkeypox virus. In patients with an appropriate history, a titer of ≥1:360 signifies a recent infection by monkeypox virus or recent vaccination.

Radioimmunoassay

Radioimmunoassay for orthopoxvirus antibodies is probably the most sensitive test and is useful to detect a long-existing antibody. Titers found in human monkeypox cases with recent infection by monkeypox virus range from 1:3,000 to 1:20,000 or more. At titers of <1:50 the specificity of this test is questionable.

Radioimmunoassay adsorption

Radioimmunoassay adsorption is presently the only method by which specific antibodies to human monkeypox, vaccinia, and variola viruses can be differentiated.

Milker's nodule, orf, and tanapox

The serological methods of choice for milker's nodule and orf diagnosis are IFA and ELISA; for tanapox they are IFA, ELISA, and neutralization. Perhaps because of the relatively less extensive nature of the infections found in these diseases (usually only one to a few lesions on the skin), the serum should be collected at 3 to 5 weeks after the onset. Any positive results can be considered significant for diagnosis. The HI test is useless for these diseases because the viruses do not produce hemagglutinin.

Molluscum contagiosum

Although tests such as AGP, IFA, and CF have been used for antibody assay of molluscum contagiosum, they are not practical for routine use because the virus cannot be propagated easily.

Other methods

AGP and CF tests can still be used when other tests are not available. However, with AGP, precipitating antibody is not detected in every clinically positive case of variola, human monkeypox, or vaccinia. It is rarely positive in individuals recently vaccinated or revaccinated. With the CF test, the antibody in smallpox patients may not appear until the second week after infection. Serum specimens from unvaccinated smallpox patients tested for CF antibody even after day 8 of onset can be negative; therefore, the test may have limited diagnostic value because only the positive results are useful. After primary vaccination or revaccination for smallpox, the CF antibody titer may not be detected. For parapoxvirus diseases (orf and milker's nodule), the antibody is also not always detected. For these diseases, negative results are not useful.

EVALUATION OF VIROLOGICAL DIAGNOSTIC METHODS

Orthopoxvirus infections, especially smallpox and human monkeypox, can be confidently diagnosed in the laboratory by the "three-test combination" method, namely, the combined use of EM, CAM culture, and cell culture techniques for each specimen. It must be emphasized, however, that an adequate amount of specimen is needed. Strongly positive specimens pose no problem; the problem is in making a negative diagnosis with confidence, and this cannot be done if the quantity of specimen received is inadequate.

The diagnostic confidence for orf, milker's nodule, and tanapox is relatively less than that for orthopoxvirus infections, because the viruses of these diseases are more difficult to isolate. Furthermore, although visualization of virus particles by EM for tanapox specimens is good, it can be difficult for orf and milker's nodule specimens because the chances of receiving unsuitable specimens are greater.

APPENDIX

Vaccinia antigen produced on CAMs, positive control for AGP

1. Harvest 20 CAMs confluently infected with the Wyeth smallpox vaccine strain of vaccinia virus.
2. Place the 20 infected CAMs in a 250-ml Sorvall Omnimixer cup and homogenize them for 3 min at full speed with the cup immersed in an ice-water bath.
3. Add 20 ml of sterile phosphate-buffered saline (PBS) and homogenize the mixture for 2 min with the cup immersed in an ice-water bath.
4. Add 1 part of Genetron 113 to 3 parts of the homogenate. Homogenize for 3 min with the cup immersed in an ice-water bath.
5. Centrifuge the mixture at 600 × *g* for 10 min. Draw off and save the supernatant fluid. Add 1 part of 0.01% trypsin to 9 parts of the supernatant fluid, bringing the trypsin concentration to 0.001%. Mix and place in an incubator at 36°C for 1 h.
6. Dialyze the mixture against polyethylene glycol (20 M) to a final volume of 20 ml. Distribute in 1-ml portions and store at −20°C.

For normal CAM control, 20 uninfected CAMs are treated in the same manner.

Chicken erythrocytes for hemagglutination

Because only about 50% of the chicken population has erythrocytes that can be agglutinated by vaccinia hemagglutinin, samples of erythrocytes from several chickens must be pretested to select those that do hemagglutinate.

Selection of chicken

1. Obtain 5 ml of blood from several 7- to 14-month-old chickens.
2. Wash the erythrocytes from each of the chickens three times with PBS (pH 7.2) by centrifugation.
3. Prepare a 0.5% erythrocyte suspension from each chicken in PBS.
4. Set up the standard hemagglutinin titration for each chicken erythrocyte suspension, using a known positive hemagglutinin vaccinia virus antigen.
5. Keep only the chickens whose erythrocytes are agglutinated by the hemagglutinin antigen at a dilution equal to or greater than 1:64.

Preparation of erythrocytes for the HI test

1. Collect 10 ml of blood from an approved chicken and mix the blood in a bottle at a ratio of 1 part of blood to 4 parts of Alsever solution. (The blood-Alsever solution mixture can be stored in a refrigerator for 2 weeks.)
2. Wash erythrocytes three times with PBS (pH 7.2) by centrifugation in 15-ml graduated conical centrifuge tubes.
3. Remove and discard the buffy coat after each wash.
4. After the third wash, prepare a 5% erythrocyte suspension in PBS.
5. From the 5% suspension, prepare daily a 0.5% erythrocyte suspension for the HI test.

Preparation of vaccinia antigen for HI and CF tests

1. Inoculate each of eight 1-liter bottles containing confluent BHK-21 cell monolayers with 2 ml of vaccinia-infected cell culture with a titer of $10^{6.0}$ pock-forming units per 0.1 ml.
2. Allow the inoculated virus to adsorb for 1 h at 35°C, and then add 50 ml of Eagle minimal essential medium with 0.4% bovine albumin.
3. Harvest the cells by scraping them from the bottle surface when they show 3+ to 4+ CPE.
4. Centrifuge the harvested cells at 1,500 rpm for 15 min and discard the supernatant fluid.
5. Wash the cells three times with 30 ml of reticulocyte swelling buffer (RSB = 0.01 M Tris hydrochloride–0.01 M NaCl–0.0015 M $MgCl_2$, pH 7.8; this solution is made each time by mixing 20 ml of 0.5 M Tris, pH 7.8, 2 ml of 5 M NaCl, and 1.5 ml of 1.0 M $MgCl_2$ and adding distilled water to a final volume of 1 liter).
6. After the third washing, resuspend the cells in 8 ml of RSB and leave the suspension overnight at 4°C.
7. Rupture the cells in the suspension by 20 to 30 strokes with a Dounce homogenizer.
8. Centrifuge at 600 rpm for 15 min to pellet the nuclei and cell debris.
9. Save the supernatant fluid (labeled no. 1).
10. Repeat the treatment of the sediment with a Dounce homogenizer, if many intact cells remain, by resuspending the sediment in 8 ml of RSB and subjecting it to another 20 to 30 strokes.
11. Centrifuge at 600 rpm for 15 min.
12. Combine the 8 ml of supernatant fluid and the previously saved 8 ml of supernatant fluid (no. 1) and discard the sediment.
13. Centrifuge the 16 ml of supernatant fluid at 25,000 rpm for 45 min in an SW25 or SW27 Beckman rotor.
14. Resuspend the sediment in 20 ml of RSB and add sufficient thimerosal to give a final concentration of 1:10,000.

15. Evaluate by performing hemagglutination, HI, and CF tests.

Preparation of antivaccinia rabbit serum

1. Inoculate one 1-liter bottle containing a confluent primary rabbit kidney cell monolayer (grown with Eagle minimal essential medium containing 10% inactivated normal rabbit serum) with 2 ml of vaccinia cell culture with a titer of $10^{6.0}$ pock-forming units per 0.1 ml.

2. Allow the virus to adsorb at 35°C for 1 h and add 50 ml of Eagle minimal essential medium with 2% inactivated normal rabbit serum.

3. Incubate at 36°C until 3+ to 4+ CPE can be observed (2 to 3 days).

4. Discard the culture medium, scrape off the cells, and suspend them in 10 ml of McIlvaine buffer (pH 7.4).

5. Freeze and thaw through three cycles. Titrate the virus on CAMs. A titer of 10^6 to 10^7 pock-forming units per 0.1 ml is acceptable.

6. Mix the suspension with an equal volume of Freund complete adjuvant.

7. Inoculate each prebled rabbit (older than 6 months) with 0.3 ml of the mixture in each front footpad.

8. Inoculate 1 ml subcutaneously in the right hindquarter.

9. Inoculate a 1-ml booster of viral immunogen intramuscularly in the same hindquarter 20 days later.

10. Exsanguinate the rabbit 14 days after the booster.

11. Store serum at −20°C.

Periodate and erythrocyte treatment of serum to remove nonspecific hemagglutination inhibitors and nonspecific hemagglutinin (13)

1. Prepare fresh 0.011 M solution of KIO_4 (do not substitute $NaIO_4$) by dissolving 0.256 g of KIO_4 in 100 ml of PBS. Stir constantly by use of a magnetic stirrer (do not use heat).

2. Prepare 3% glycerol in PBS (may be stored in a refrigerator for a few weeks).

3. To 0.1 ml of serum in a test tube, add 0.3 ml of the 0.011 M KIO_4 solution and leave the mixture at room temperature for exactly 15 min.

4. Add 0.1 ml of the 3% glycerol-PBS solution and leave at room temperature for at least 15 min (to stop the oxidation reaction).

5. Add 0.05 ml of the 50% chicken erythrocytes (prepared with equal volumes of packed cells and PBS) and leave at 4°C for 1 h (for removal of nonspecific hemagglutinin).

6. Centrifuge at 600 × *g* for 10 min and transfer the supernatant fluid to another tube. This adsorbed serum product is considered to be a 1:5 dilution.

7. Inactivate the 1:5-diluted serum at 56°C for 30 min; proceed with further dilution of the serum and test for HI antibody. This treatment removes the nonspecific hemagglutination inhibitors and also the nonspecific hemagglutinin in the serum.

LITERATURE CITED

1. **Downie, A. W., and C. H. Kempe.** 1969. Poxviruses, p. 281–320. *In* E. H. Lennette and N. J. Schmidt (ed.), Diagnostic procedures for viral and rickettsial diseases, 4th ed. American Public Health Association, Inc., New York.
2. **Downie, A. W., C. H. Taylor-Robinson, A. E. Caunt, G. S. Nelson, P. E. C. Manson-Bahr, and T. C. H. Matthews.** 1971. Tanapox: a new disease caused by a poxvirus. Br. Med. J. **1**:363–368.
3. **Dumbell, K. R., and J. G. Kapsenberg.** 1983. Laboratory investigation of "whitepox" viruses and comparison with two variola strains from Southern India. Bull. W.H.O. **60**:381–387.
4. **Dumbell, K. R., and M. Nizamuddin.** 1959. An agar gel precipitation test for the laboratory diagnosis of smallpox. Lancet **i**:916–917.
5. **Harris, W. J., and J. C. W. Westwood.** 1964. Phosphotungstate staining of vaccinia virus. J. Gen. Microbiol. **34**:491–495.
6. **Hopkins, D. R.** 1983. Princes and peasants. University of Chicago Press, Chicago.
7. **Hutchinson, H. D., D. W. Ziegler, D. E. Wells, and J. H. Nakano.** 1977. Differentiation of variola, monkeypox, and vaccinia antisera by radioimmunoassay. Bull. W.H.O. **55**:613–623.
8. **Nakano, J. H.** 1979. Poxviruses, p. 257–308. *In* E. H. Lennette and N. J. Schmidt (ed.), Diagnostic procedures for viral, rickettsial and chlamydial infections, 5th ed. American Public Health Association, Inc., Washington, D.C.
9. **Riggs, J. L.** 1979. Immunofluorescent staining, p. 141–151. *In* E. H. Lennette and N. J. Schmidt (ed.), Diagnostic procedures for viral, rickettsial and chlamydial infections, 5th ed. American Public Health Association, Inc., Washington, D.C.
10. **Volles, A., D. Bidwell, and A. Bartlett.** 1980. Enzyme-linked immunosorbent assay, p. 359–371. *In* N. R. Rose and H. Friedman (ed.), Manual of clinical immunology, 2nd ed. American Society for Microbiology, Washington, D.C.
11. **Von Magnus, P., E. K. Andersen, K. B. Petersen, and A. Birch-Andersen.** 1959. A pox-like disease in cynomolgus monkeys. Acta Pathol. Microbiol. Scand. **46**:156–176.
12. **Westwood, J. C. N., P. H. Phipps, and E. A. Boulter.** 1957. The titration of vaccinia virus on the chorioallantoic membrane of the developing chick embryo. J. Hyg. **52**:123–139.
13. **World Health Organization.** 1959. Expert Committee on Respiratory Virus Disease, First Report. WHO Tech. Rep. Ser. **170**:40–41.
14. **Ziegler, D. W., H. D. Hutchinson, J. P. Koplan, and J. H. Nakano.** 1975. Detection by radioimmunoassay of antibody in human smallpox patients and vaccines. J. Clin. Microbiol. **1**:311–317.

Chapter 69

Reoviruses

MARILYN A. MENEGUS

CLINICAL BACKGROUND

Reoviruses infect humans and lower animals and are found worldwide. Through serosurveys we know that infection of humans is common and occurs at an early age. Although reoviruses have been isolated from individuals with a variety of illnesses, no firm association between infection and human disease has been established. Apparently most infections are subclinical (2–4).

DESCRIPTION OF THE AGENT

The family Reoviridae is composed of three genera: *Reovirus*, *Orbivirus*, and *Rotavirus*. The genus *Reovirus* consists of three virus serotypes designated 1, 2, and 3. Reoviruses have a naked, double capsid, 70 nm in diameter, with icosahedral symmetry. The capsid contains a genome consisting of 10 segments of double-stranded RNA. Reoviruses are resistant to lipid solvents and are acid stable. Virus synthesis and maturation occurs in the cytoplasm of the cell and is associated with the formation of large, eosinophilic, intracytoplasmic inclusions (1).

COLLECTION AND STORAGE OF SPECIMENS

Reoviruses can be recovered from throat swabs, nasal secretions, and feces. Since the viruses are quite stable, specimens can be collected and processed as described in Chapter 61.

VIRUS CULTIVATION

Reoviruses replicate and produce cytopathic effect in a wide variety of cell cultures, including primary monkey and human embryonic kidney, HeLa, and human diploid fibroblasts. However, the recognition of reovirus-induced cytopathology can be challenging even for experienced personnel. Infected cells become granular, remain loosely fastened to the glass, and when agitated, flutter in the medium. These changes are easily mistaken for nonspecific cell degeneration. The presence of reovirus in cell cultures can be confirmed by the demonstration of cytoplasmic inclusions in stained preparations of the infected cells and hemagglutinating activity (human O erythrocytes) in the cell culture media. The viruses can be definitively identified by a hemagglutination inhibition test with type-specific antisera (2–4). Despite their prevalence, reoviruses are seldom isolated from clinical specimens.

IDENTIFICATION OF THE AGENT

Since all three reovirus serotypes hemagglutinate human erythrocytes, the method of choice for the identification of specific serotypes is the hemagglutination inhibition test. The test is carried out as follows. Typing sera are adsorbed with kaolin by mixing a 1:5 dilution of serum in 0.85% saline with an equal volume of a 25% suspension of acid-washed kaolin in 0.85% NaCl (25 g of kaolin plus 100 ml of saline solution), and the mixture is allowed to stand for 20 min at room temperature. The mixture is then centrifuged briefly to sediment the kaolin, and the decanted supernatant fluid is considered a 1:10 dilution of serum. A 0.2-ml amount of hemagglutinin diluted in 0.85% NaCl to contain 20 U/ml (4 U/0.2 ml) is added to 0.2-ml amounts of serial twofold dilutions of serum in 0.85% NaCl. Each serum is used from a dilution of 1:10 to, or beyond, its endpoint. Mixtures are shaken briefly and then allowed to stand for 1 h at room temperature before the addition of 0.2 ml of the standard (0.75%) erythrocyte suspension. Erythrocytes are allowed to settle at room temperature, and the titer of the serum is then taken as that dilution which completely inhibits agglutination. The lowest dilution of each serum used is tested for the presence of nonspecific erythrocyte agglutinins by substituting 0.2 ml of saline solution for the antigen. An antigen titration is also included in the test.

An isolate is considered typed if it is inhibited at a titer of at least 1:40 by one of the typing antisera and not at a dilution of 1:10 by the other antisera (2, 3).

SEROLOGICAL METHODS

Practically all naturally occurring and experimental infections with reoviruses are accompanied by a fourfold or greater rise in homologous hemagglutination inhibition antibody. Antibody can be detected for at least 1 year after natural infection and probably persists much longer (2–4). However, tests for reovirus antibody are seldom requested because no clinical situation demands them.

EVALUATION AND INTERPRETATION OF RESULTS

The ubiquity of reoviruses and their lack of association with human disease make it impossible in the individual patient to attach any significance to the isolation of virus from nonsterile body sites. If there is some compelling clinical reason to build a case for a causal relationship with the illness of the patient, additional diagnostic methods (e.g., histopathology and serology) must be employed.

LITERATURE CITED

1. **Joklik, W. K.** 1983. The reovirus particle, p. 9–78. *In* W. K. Joklik (ed.), The reoviridae. Plenum Publishing Corp., New York.
2. **Rosen, L.** 1968. Reoviruses, p. 73–107. *In* H. A. Wenner and A. M. Behbehani (ed.), Echoviruses and reoviruses. Springer-Verlag, New York.
3. **Rosen, L.** 1979. Reoviruses, p. 577–584. *In* E. H. Lennette and N. J. Schmidt (ed.), Diagnostic procedures for viral, rickettsial and chlamydial infections, 5th ed. American Public Health Association, Inc., New York.
4. **Stanley, N. F.** 1977. Diagnosis of reovirus infections: comparative aspects, p. 385–421. *In* E. Kurstak and C. Kurstak (ed.), Comparative diagnosis of viral diseases, vol. 1. Human and related viruses. Academic Press, Inc., New York.

Chapter 70

Enteroviruses

MARILYN A. MENEGUS

CLINICAL BACKGROUND

Enteroviruses commonly infect humans, and the consequences of infection are either asymptomatic virus shedding or a broad spectrum of acute diseases, including undifferentiated febrile illness, aseptic meningitis, encephalitis, paralysis, a sepsislike picture in neonates, myopericarditis, pleurodynia, conjunctivitis, enanthems, exanthems, pharyngitis, and pneumonia. Clinical illness is most frequently seen in infants and young children. In addition, a role has also been hypothesized for enteroviruses in the etiology of diabetes, chronic cardiomyopathy, and fetal malformations (2, 7, 11).

In temperate climates, enteroviruses are prevalent in the summer and fall; outbreaks occur each year and are associated with significant morbidity. A more endemic prevalence pattern is seen in tropical and semitropical areas. By 2 years of age, regardless of where they reside, most children have already experienced several asymptomatic or mildly symptomatic enterovirus infections. The viruses are highly transmissible and are spread for the most part via the fecal-oral route. In a family setting and in closed populations, the rate of infection among nonimmune individuals can reach as high as 80%, with children more likely to be infected than adults. Antibody appears to offer significant protection against disease and also reduces the likelihood of, but does not prevent, infection (7, 11).

The incubation period for enteroviruses ranges from 1 day to 3 weeks but is generally 3 to 5 days. The viruses enter the alimentary tract via the mouth and begin replicating in the lymphoid tissue of the pharynx and gut. In some cases, viremia then occurs, leading to the involvement of target organs such as the spinal cord, heart, and skin. Virus can be found in the pharynx for 1 to 2 weeks postinfection and may be excreted in feces for even longer periods (2 to 6 weeks). Viremia, on the other hand, is present for only a short time (several days) and is found only early in infection (7, 11).

DESCRIPTION OF THE AGENTS

Viruses belonging to two genera of the family Picornaviridae, *Enterovirus* and *Rhinovirus*, commonly infect humans. They share the following composition: a naked, ether-resistant, 20- to 30-nm icosohedral capsid made up of four proteins and a single-stranded, unsegmented, plus-sense RNA genome with a molecular weight of 2.5×10^6 to 3.0×10^6. The main characteristics which distinguish enteroviruses from rhinoviruses are the stability at acid pH (3.0) and the lower buoyant density in CsCl of the enteroviruses. In addition, enteroviruses, unlike rhinoviruses, replicate better at 36 than at 33°C and can be isolated from feces as well as from the respiratory tract (9).

The viruses belonging to the genus *Enterovirus* were initially divided into four groups (poliovirus, echovirus, coxsackievirus A, and coxsackievirus B) based on their pathogenicity for laboratory animals. However, later it was realized that with this system, the lines of demarcation between groups were not sufficiently clear. Therefore, beginning in 1970, new members of the genus were merely called enteroviruses and designated by sequential numbers beginning with 68. There are presently 68 distinct enterovirus serotypes: polioviruses 1 through 3, echoviruses 1 through 34, group A coxsackieviruses 1 through 24, group B coxsackieviruses 1 through 6, and enteroviruses 68 through 72. The numbers are not additive because four of the echovirus serotypes have been reclassified (echovirus 10 as reovirus 1; echovirus 28 as rhinovirus 1A; echovirus 34 as a prime strain of coxsackievirus A-24; and echovirus 9 and coxsackievirus A-23 were found to be identical) (9). The latest addition to the genus, hepatitis A virus, was officially designated enterovirus 72 by the International Committee on Taxonomy of Viruses in 1983, but to avoid unnecessary confusion, the Committee also recommended retaining the name hepatitis A virus.

COLLECTION AND STORAGE OF SPECIMENS

Specimens for enterovirus isolation should be collected as soon after the onset of clinical symptoms as possible. Depending on the clinical picture, one or more of the following specimens may be appropriate: throat swab, feces or rectal swab, cerebrospinal fluid, blood, vesicular fluid, and tissue and conjunctival swab (for enterovirus 70 only). Methods for specimen collection are described in Chapter 61. Occasionally enteroviruses are also isolated from urine, but the significance of such isolates is unknown, and urine is not a recommended specimen. The sampling of multiple sites does result in maximal enterovirus recovery, but before this approach is accepted, some consideration should be given to the clinical usefulness of the information provided relative to its cost.

Enteroviruses are hardier than most other viruses sought in clinical specimens, and they survive well under the conditions recommended in Chapter 61 for specimen shipment and storage. Unlike many other viruses, enteroviruses can be held at 4°C for weeks or frozen at either −20 or −70°C indefinitely without loss of infectivity. A measurable loss of infectivity does occur, however, if specimens are kept at room temperature for many hours or allowed to dry (9).

Paired sera for antibody studies can also be useful in the diagnosis of enterovirus infections. Blood specimens should be collected and processed as described in Chapter 61; however, particular emphasis should be placed on the collection of acute-phase blood as soon as possible after the onset of illness. The antibody responses to enteroviruses are often brisk, and titer rises can be missed by even a few days of delay (9, 11).

DIRECT EXAMINATION OF SPECIMENS

Enteroviruses can be demonstrated in clinical specimens by a number of direct diagnostic techniques, including electron microscopy, enzyme-linked immunoassay, and immunofluorescence staining. However, thus far, none of these methods is in widespread use, and the sensitivity and specificity of each method relative to cell culture is yet to be defined. The one attempt made to detect enteroviruses in clinical specimens by DNA hybridization techniques was unsuccessful. Nevertheless, further work in the area is anticipated because hybridization could provide a generic diagnosis (5, 9).

CULTIVATION OF VIRUSES

Isolation in cell culture

With the exception of most group A coxsackieviruses, isolation in cell culture is the method of choice for the diagnosis of enterovirus infection. For virus recovery from clinical specimens, primary monkey (rhesus, African green, or cynomolgus) kidney (PMK) cells and human diploid fibroblast (e.g., WI-38) cells are the most efficient and widely used dual culture system (1, 3, 9, 13). Because primary kidney cell cultures derived from other species of monkeys and other animals may not be as susceptible as PMK, these alternative cell cultures should not be used as substitutes for PMK cell cultures (4). Unfortunately, no specific fibroblast strain can be recommended, since data are limited regarding the relative sensitivity of fibroblasts for primary enterovirus isolation. The use of two continuous cell lines, RD and BGM, in addition to the PMK and WI-38 cell lines, can improve enterovirus isolation in terms of speed, sensitivity, and the spectrum of serotypes detected (1, 10, 14; R. Dagan, J. A. Jenista, and M. A. Menegus, *in* L. M. de la Maza and E. M. Peterson, ed., *Medical Virology IV*, in press), but many laboratories may find the cost of this approach prohibitive. In addition to the cell types already mentioned, various other commonly used cells, including HeLa, human embryonic kidney, and Hep-2, also support the growth of a variety of enterovirus serotypes, but their sensitivity for primary isolation is not as well defined as that of PMK and WI-38 cells (9). For the most part, all of the cultures referred to above are now readily available from commercial sources.

Inoculated cultures should be incubated at 37°C on a roller drum. They can also be held in a stationary position, but this may delay the development of viral cytopathic effect (CPE) in some cases. The CPE produced by enteroviruses in cell culture is quite distinctive and can be recognized with greater than 90% accuracy by experienced technologists (3). Enterovirus CPE is often evident quite early; up to 69% of all positive cultures can be detected within 4 days of specimen inoculation (Dagan et al., in press). Therefore, daily culture readings through day 4 or 5 are recommended to laboratories that place a high priority on providing clinically useful results. Inoculated cultures should be examined for CPE for 10 to 14 days before being discarded as negative. Blind passage of cultures yields only a small number of additional positives.

The growth characteristics of enteroviruses in vitro vary considerably. Polioviruses, both wild and vaccine strain, produce CPE in a wide range of cell types, including PMK, human embryonic kidney and fibroblasts, HeLa, BGM, and RD. The CPE develops quickly and often destroys the monolayer within 3 days. Group B coxsackieviruses, on the other hand, do not replicate well in RD cells and human fibroblasts, but do replicate in the other cell systems mentioned. The growth characteristics of echoviruses and the group A coxsackieviruses that do replicate in cell culture vary from type to type, but in general, these viruses can be detected with equal frequency in PMK, RD, and human fibroblast cells, but seldom in BGM cells. For the most part, the CPE caused by these viruses does not progress as rapidly as that caused by the polioviruses (1, 3, 9, 10, 13, 14; Dagan et al., in press).

Isolation in suckling mice

Inoculation into suckling mice (24 to 48 h old) is the method of choice for the primary isolation of most of the group A coxsackieviruses; once isolation is accomplished, the viruses can then be adapted to growth in cell culture. Group B coxsackieviruses and some of the newer strains of enterovirus are also pathogenic for suckling mice, but because cell culture is so much simpler, it remains the preferred method for the isolation of these viruses. Only rare strains of echovirus replicate and cause pathology in suckling mice.

Specimens should be inoculated by three routes: 0.02 ml intracerebrally, 0.05 ml intraperitoneally, and 0.03 ml subcutaneously; after inoculation, the mice should be observed for at least a 14-day period. Signs of illness and then death generally occur in infected animals within 4 to 7 days if virus is present. Group A coxsackievirus infections are characterized by progressive, flaccid paralysis without signs of encephalitis and, histologically, by a generalized myositis, whereas group B coxsackievirus infections cause spastic paralysis, frequently accompanied by encephalitis. Marked panniculitis is also a common feature of group B coxsackievirus infection, but myositis is generally absent or limited and focal (9).

IDENTIFICATION OF THE AGENTS

It is possible to identify an isolate as enterovirus with reasonable certainty based on the clinical history of the patient, the time of year when the specimen was obtained, the type of specimen, the cell culture system(s) which supports its growth, and its CPE. Reporting a presumptive diagnosis of enterovirus infection without waiting for the specific typing of the isolate can be of real value in the clinical management of the patient (Dagan et al., in press).

A specific identification of the serotype is most readily accomplished by virus neutralization, with intersecting pools of hyperimmune serum. These pools may be constructed in many ways, but the most widely used system is that of Lim and Benyesh-Melnick (6). The Research Resources Branch of the National Institute of Allergy and Infectious Diseases once supplied equine antisera combined into eight pools (A through H) to identify 42 enteroviruses that grow readily in cell culture. Detailed directions for the use of these convenient pools are supplied with the reagents. The responsibility for the preparation and distribution of the pools as described by Lim and Benyesh-Melnick is now in the hands of the World Health Organization. It is important to conserve these valuable reagents, because the supply is limited. If, therefore, an epidemic due to a single serotype is in progress, the use of a single hyperimmune serum to identify isolates is clearly indicated. It should also be added that it is not always necessary to identify the serotype of each isolate. In most cases, it is probably sufficient for the physician to know simply that a nonpolio enterovirus was isolated. However, it is important to exclude poliovirus as a possibility, because generally such isolates represent clinically irrelevant, long-term shedding of vaccine strain virus. Poliovirus can be excluded if antisera to the three poliovirus serotypes are pooled and a simple neutralization test is performed.

Mouse neutralization tests are used to identify isolates that do not replicate in cell culture systems. Again, intersecting serum pools (J through P) are available for this purpose (8); these pools are distributed by the National Institute of Allergy and Infectious Diseases.

A number of other methods, including hemagglutination inhibition (for those strains which hemagglutinate), complement fixation, immunofluorescence, counterimmunoelectrophoresis, and virus agglutination, can be used for virus identification, but none is as specific or widely used as neutralization (9, 11).

If a virus isolate cannot be identified, it may represent a new enterovirus, a mixture of viruses, or some other kind of virus. If there is uncertainty about the identity of an isolate, the determination of size, nucleic acid content, ether stability, and acid lability should show whether or not it is an enterovirus.

SEROLOGICAL PROCEDURES

Neutralization test

No rapid or inexpensive serological screening procedure is presently available to detect antibody to all enteroviruses. The neutralization test is accurate, but it requires the use of over 60 virus serotypes in combined cell culture-mouse systems to screen completely for rises in antibody titer. A virus neutralization test is really feasible for serological diagnosis only when a virus isolate from the patient is available, when an epidemic due to a single serotype is present, or when a clinical picture such as pleurodynia or carditis clearly implicates a small number of agents, such as the group B coxsackieviruses.

Paired sera are collected from the patient as soon as possible after the onset of symptoms (acute serum) and 2 to 3 weeks later (convalescent serum) and are diluted at 1:2, 1:8, 1:32, 1:128, and 1:512. Twofold dilutions (1:2, 1:4, 1:8, etc.) are preferred by some laboratories. A 0.5-ml amount of each serum dilution is then mixed with an equal amount of virus diluted to contain approximately 100 50% cell culture infective doses per 0.1 ml, and the mixture is incubated for 1 h at 37°C. Virus-serum mixtures in 0.2-ml amounts are then inoculated into the appropriate cell culture system. Two cell culture tubes are usually used for each dilution. A control virus titration to determine the 50% cell culture infective dose is always included. Virus is diluted in 10-fold amounts (1:10, 1:100, 1:1,000, etc.), and 0.1 ml of each dilution is inoculated per tube of cell culture. Tubes are observed for 6 to 7 days, but results are usually apparent by day 3 or 4. Serum titers are calculated and expressed as the 50% endpoint. A fourfold or greater rise in antibody titer is considered significant. Neutralization tests can also be performed in vitro by the plaque reduction method or in microtiter and in vivo in newborn mice (9).

Other methods

A number of other techniques, including complement fixation, immunoprecipitation, indirect fluorescent antibody, passive hemagglutination, and, more recently, enzyme-linked immunoassay, have been used to detect antibody to enteroviruses. In addition, several of these methods have also been adapted to demonstrate virus-specific immunoglobulin M (9, 11, 12).

EVALUATION AND INTERPRETATION OF RESULTS

In the individual patient, it is difficult to establish that an enterovirus infection is causally related to disease because these viruses are so prevalent in the general population, particularly during epidemic periods. A causal relationship is most likely if virus is isolated from a normally sterile body site (e.g., blood, cerebrospinal fluid, brain, or liver), but the causal relationship is presumptive if virus is isolated only from pharyngeal or fecal specimens. The latter can be positive in up to 50% of well individuals. A significant rise in antibody titer adds weight to a postulated etiological role for a pharyngeal or fecal isolate, but single high titers are not useful, because neutralizing antibodies can remain elevated for years after an acute infection. Serological tests alone are not useful for the determination of the specific enterovirus serotype causing disease, because heterotypic antibody responses are common and unpredictable in both magnitude and diversity.

LITERATURE CITED

1. **Bell, E. J, and B. P. Cosgrove.** 1980. Routine enterovirus diagnosis in a human rhabdomyosarcoma cell line. Bull. W.H.O. **58:**423–428.
2. **Cherry, J. D.** 1981. Nonpolio enteroviruses, coxsackieviruses and enteroviruses, p. 1316–1365. *In* R. D. Feigin and J. D. Cherry (ed.), Textbook of pediatric infectious diseases. The W. B. Saunders Co., Philadelphia.
3. **Herrmann, E. C., Jr., D. A. Person, and T. F. Smith.** 1972. Experience in laboratory diagnosis of enterovirus infections in routine medical practice. Mayo Clin. Proc. **47:**577–586.
4. **Hsuing, G. D., and J. L. Melnick.** 1957. Comparative susceptibility of kidney cells from different monkey species to enteric viruses (poliomyelitis, coxsackie, and ECHO groups). J. Immunol. **78:**137–146.
5. **Hyypiä, T., P. Stålhandske, R. Vainionpää, and U. Pettersson.** 1984. Detection of enteroviruses by spot hybridization. J. Clin. Microbiol. **19:**436–438.
6. **Lim, K. A., and M. Benyesh-Melnick.** 1960. Typing of viruses by combinations of antiserum pools. Application to typing of enteroviruses (coxsackie and echo). J. Immunol. **84:**309–317.
7. **Melnick, J. L.** 1982. Enteroviruses, p. 187–251. *In* A. S.

Evans (ed.), Virus infections of humans: epidemiology and control. Plenum Publishing Corp., New York.

8. **Melnick, J. L., N. J. Schmidt, B. Hampil, and H. H. Ho.** 1977. Lyophilized combination pools of enterovirus equine antisera: preparation and test procedures for the identification of field strains of 19 group A coxsackievirus serotypes. Intervirology **8:**172–181.
9. **Melnick, J. L., H. A. Wenner, and C. A. Phillips.** 1980. Enteroviruses, p. 471–534. *In* E. H. Lennette and N. J. Schmidt (ed.), Diagnostic procedures for viral, rickettsial and chlamydial infections, 5th ed. American Public Health Association, Inc., New York.
10. **Menegus, M. A., and G. E. Hollick.** 1982. Increased efficiency of group B coxsackievirus isolation from clinical specimens by use of BGM cells. J. Clin. Microbiol. **15:**945–948.
11. **Moore, M., and D. M. Morens.** 1984. Enteroviruses, including polioviruses, p. 407–483. *In* R. B. Belshe (ed.), Textbook of human virology. PSG Publishing Co., Inc., Littleton, Mass.
12. **Pattison, J. R.** 1983. Tests for coxsackie B virus-specific IgM. J. Hyg. **90:**327–332.
13. **Schmidt, N. J.** 1972. Tissue culture in the laboratory diagnosis of viral infections. Am. J. Clin. Pathol. **57:**820–828.
14. **Schmidt, N. J., H. H. Ho, and E. H. Lennette.** 1975. Propagation and isolation of group A coxsackieviruses in RD cells. J. Clin. Microbiol. **2:**183–185.

Chapter 71

Rhinoviruses and Coronaviruses

JACK H. SCHIEBLE

CLINICAL BACKGROUND

Rhinoviruses

Rhinoviruses are a subgroup of the viral family Picornaviridae and as such are now generally recognized as the principal etiological agents of mild upper respiratory tract infections in humans. The association of rhinoviruses with the lower respiratory tract appears to be infrequent and therefore less important than their relationship with the upper respiratory tract, where the pathological effects of the virus result in the syndrome of the "common cold."

Characteristically, the cold begins with a prodromal irritation and stuffiness in the nose, which is followed rapidly by a sometimes profuse watery discharge. There is frequently a sore throat with a dry cough, and occasionally a slight fever may be noted. The duration of symptoms in naturally acquired infection usually ranges from 4 to 24 days, with a mean of 7 to 10 days. In natural infections the incubation period varies between 1 and 5 days with a mean of 2 days. Virus shedding appears to begin with the onset of symptoms and may continue for as long as 1 week. In a recent study (22) of human volunteers, rhinovirus was found in ciliated epithelial cells from 16 of 17 experimentally infected individuals and from none of 10 uninfected control subjects. The pattern of shedding of ciliated epithelial cells containing virus was similar to the development of nasal symptoms in infected volunteers but was not correlated with the severity of nasal symptoms. Interestingly, in view of the widespread use of aspirin for symptomatic relief of the common cold, there is reason to believe that the use of aspirin may prolong virus shedding. Transmission of infectious rhinovirus from contaminated surfaces to susceptible human volunteers has been shown to be a more efficient means of transmission than either large- or small-particle aerosols. More recent studies (5) extend and confirm earlier findings that human rhinoviruses remain infectious on hands and environmental surfaces and are capable of being transmitted to susceptible individuals. Additional studies further indicate that disinfectant treatment of such surfaces may reduce the risk of transmission by this route.

Rhinovirus infections are prevalent throughout the year, with peak incidence often occurring during the fall and spring. The large number of immunotypes accounts for repeated episodes of naturally acquired infections. Long-term studies indicate that several immunotypes may be present and contribute concurrently to an outbreak within a limited geographical area. Such studies also suggest that there may be a slow progressive change with time in the prevalence of particular rhinovirus immunotypes. This could be accounted for by either the very large number of immunologically distinct rhinovirus types or some mechanism of antigenic variation causing the emergence of new immunotypes. For more detailed information on the clinical, pathological, and epidemiological aspects of rhinovirus infection the reader is referred to a recent review by Gwaltney (4).

Coronaviruses

It would seem reasonable to assume that the rhinoviruses are the etiological agents in 50 to 60% of common colds. The etiology of the remaining infections is attributable to several different viral groups, and of these the human coronavirus (HCV) group is now generally recognized as the second most important viral group responsible for the common cold. The HCVs were first isolated independently by two groups of investigators at approximately the same time (1965–1966). One of the groups recovered an agent in organ cultures of human nasal epithelium inoculated with nasal secretions from human volunteers. The virus isolated was called B814, and subsequent studies demonstrated it to be morphologically similar to infectious bronchitis virus of chickens. The second group of investigators successfully isolated five viruses from medical students with symptoms of mild upper respiratory tract illness. The prototype strain was designated 229E. These virus strains grew slowly and with some difficulty on initial isolation. Adaptation was quickly achieved on subsequent cell culture passages. The prototype strain was partially characterized and shown to be sensitive to ether, to be ca. 89 nm in diameter, and to contain RNA.

The HCVs as a group are very fastidious in their growth requirements. Approximately one third of them were isolated in organ cultures of human nasal epithelium. The remaining HCVs were isolated in human primary or diploid cell cultures. All of the viruses isolated in monolayer cell cultures have been of a single serological type, similar to the prototype 229E strain.

Coronavirus infection is clinically manifested by a general malaise, headache, rhinorrhea, and sore throat in most cases. A cough is observed less frequently, being present in less than a third of infected individuals. The incubation period, as determined in human volunteer studies, appears to be 2 to 5 days, with symptoms lasting up to a week.

DESCRIPTION OF AGENT

Rhinovirus

Early studies defined the optimal cultural conditions which achieve the isolation of rhinoviruses from a high proportion of patients with mild upper respiratory infections. For best results human fetal lung fibroblast cell cultures should be maintained in medium of near neutral pH and incubated at 33°C in a roller apparatus. The effect of higher incubation temperature tends to reduce rhinovirus growth, and this is especially evident as the incubation temperature approaches 37°C (11). Use of these cultural conditions in conjunction with methods used for the propagation of human fetal diploid cells resulted in the isolation and characterization of numerous new rhinovirus immunotypes. In 1971 the numbering system for rhinoviruses was extended from types 1A through 55 to types 1A through 89. Subsequently, approximately 20 additional rhinoviruses have been shown to be serologically distinct from types 1A through 89 (6).

Rhinoviruses, being a genus within the family Picornaviridae, are small single-stranded RNA viruses with icosahedral symmetry. Although rhinoviruses share many of the biochemical and biophysical properties of enteroviruses, they are readily distinguishable from enteroviruses because of their acid lability. This is fortunate because certain enteroviruses are frequently encountered in clinical specimens from respiratory disease and it is often necessary to distinguish between these two virus groups. The infectivity of rhinoviruses is significantly reduced (95% or more) by exposure to pH 3.0 to 5.0 for 3 h at 25°C; such treatment has no effect on the infectivity of enteroviruses. Early studies of virus size were based on graded membrane filtration and showed a size range of 15 to 30 nm. A more precise size range of 20 to 27 nm was obtained by negative staining techniques (9).

Rhinoviruses have a buoyant density of 1.38 to 1.41 g/cm^3 in CsCl and are resistant to ether and chloroform, thus indicating that lipids are not an essential component of their structure.

Studies of rhinovirus type 14 reveal that the virion contains approximately 30% RNA; the calculated molecular weights of the virion and RNA are 7×10^6 to 8×10^6 and 2×10^6 to 2.5×10^6, respectively (15, 17). Polyacrylamide gel analysis of rhinoviruses indicates that the capsid is composed of 60 copies each of four distinct polypeptides, VP 1 through 4 (17). This composition is similar to that of other picornaviruses.

The original classification describing the first 89 rhinovirus serotypes noted some antigenic overlap. Within type 1 there are two subtypes, A and B, which give low-level reciprocal cross-reactions. Subsequent studies have shown additional antigenic relationships. Within type 22 a prime strain which possesses broader antigenic reactivity than the prototype strain has been described (20). In addition, low-level reciprocal cross-neutralization was demonstrated between types 9 and 32, 13 and 14, and 2 and 4. A recent study involving 90 rhinovirus serotypes has revealed extensive antigenic variation among new isolates identified as types 12 and 78 and types 36 and 58, which are known to be reciprocally related pairs (2).

Rhinoviruses are inactivated at 56°C. Heat treatment results in the conversion of complete virus particles to empty capsids. As for polioviruses, heat treatment also results in rhinovirus with C-type serological reactivity. The effect of trypsin on virus infectivity appears to be variable and type dependent. Some viral types are readily inactivated, some are inactivated only partially, and others appear to be quite resistant. Although only a few rhinovirus serotypes have been studied, it appears that they are readily inactivated by brief UV exposure. The half-lives for types 17 and 40 were found to be approximately 3 and 5 s, respectively (6).

Coronavirus

The coronaviruses appear as round or elliptical particles with moderate pleomorphism in negative-stained electron micrographs (1). Characteristically the particles have widely spaced, club-shaped projections, which present an appearance similar to a solar corona. The diameter of the virus including the projections ranges from 80 to 160 nm. The internal structure of the virus as observed in thin sections consists of an outer double membrane and a dense inner core. The projections are not readily discernible in thin-section electron micrographs.

Sensitivity to a variety of physical and chemical agents was determined for two strains of HCV. Infectivity of these strains was readily lost on exposure to ether or chloroform, acid, trypsin, and UV light. The genome of coronavirus 229E is a single-stranded RNA with a molecular weight of 5.6×10^6 (14). The International Committee on the Taxonomy of Viruses has suggested the genus *Coronavirus* in the family Coronaviridae. The genus comprises seven species, one infecting humans and six infecting a variety of animal species. Within the HCV species only one serotype, 229E, was isolated in monolayer cell culture. The remaining human strains were isolated in organ cultures of human nasal or tracheal epithelium. Two of the organ culture viruses, OC-38 and OC-43, serologically identical virus strains, were adapted to grow in the brains of suckling mice and subsequently in a cell monolayer culture. It has been shown (10) that OC-43-infected mouse brain suspension would directly agglutinate erythrocytes from rats, mice, and chickens. Monolayer cell cultures infected with OC-43 virus also exhibit the phenomenon of hemadsorption. The mechanism of hemadsorption by OC-43 virus differs from that produced by myxoviruses. In the late stages of infection large numbers of virus particles are frequently observed adhering to the outer surfaces of the plasma membranes of cells undergoing lysis. It is the location of these newly formed viruses that results in the hemadsorption observed in infected monolayer cell cultures, which has been referred to as a pseudohemadsorption.

The presence of cross-reacting antigens in both HCVs and animal coronaviruses is well established. A recent study (7) has shown that the ribonucleoprotein components of both HCV 229E and a murine coronavirus (mouse hepatitis virus) share a cross-reacting antigen. There appear to be other cross-reacting antigens or related antigens among the HCVs (16).

All of the HCVs that have been recovered in monolayer cell cultures are antigenically similar and comprise a single serotype. Among the other HCVs isolated in organ culture of human respiratory epithelial cells, there are at least two and probably several distinct serotypes. Except for the serologically similar HCV strains OC-38 and OC-43, the remaining HCVs isolated in organ culture are not antigenically well characterized because of their limited in vitro growth capacity.

COLLECTION AND STORAGE OF SPECIMENS

Specimens

Rhinovirus and coronavirus isolations are usually made from nasal and pharyngeal swabs or from nasal washings. Although washings are reported to yield the greatest number of isolates, collection of such specimens is cumbersome, especially under field conditions. Generally, satisfactory rates of isolation are obtained by using a combination of nasal and pharyngeal swabs.

Collection and storage

Specimens should be collected within 3 days, and not later than 4 to 5 days, after the onset of symptoms. If possible, the specimen should be inoculated onto cell cultures within 2 to 3 h of collection, and during the interval between collection and inoculation it is best to keep the specimen at approximately 4°C. If storage is necessary, the specimens can be frozen at −70°C in a mechanical freezer; if dry ice is used, the specimen must be stored in a flame-sealed glass ampoule, or other provisions must be made to prevent absorption of CO_2 by the specimen. This is especially important for storage of rhinoviruses and coronaviruses which are inactivated at low pH. The recommended procedures for collection of nasal washings and nasal and pharyngeal swabs are described in Chapter 61.

DIRECT EXAMINATIONS

Rhinoviruses

Studies on the direct detection of rhinoviruses or rhinovirus antigens in clinical specimens have been very limited. The indirect immunofluorescence test was used successfully for antigen detection in respiratory specimens; however, the correlation with viral isolation was poor (3). More recently the immunoperoxidase technique was used for direct detection of viral antigen in ciliated epithelial cells from infected volunteers (22). Although direct detection of viral antigen in clinical material is certainly feasible, the large number of rhinovirus serotypes essentially precludes the use of either of these techniques as a practical means of early diagnosis.

Coronaviruses

In general the direct examination of clinical material for the presence of HCV or HCV antigen is not practical for many laboratories. However, because of the distinctive morphology and size of HCV, it is recognizable by electron microscopic examination of clinical specimens. A recent report (13) has demonstrated the usefulness of the enzyme-linked immunosorbent assay (ELISA) for direct detection of HCV antigen in clinical specimens from children. The application of immunological methods for direct detection of HCV antigens is essentially limited to two viral serotypes, 229E and OC-43.

ISOLATION OF VIRUS: HOST SYSTEMS

Rhinovirus

Because of their sensitivity to rhinoviruses, availability, and ease of handling, human diploid cell strains similar to WI-38 are the cell systems of choice and are widely used for the isolation of rhinoviruses. Diploid cell cultures can also be readily prepared in the laboratory. Diploid cells are grown in Eagle minimal essential medium containing 10% fetal bovine serum. For maintenance, Leibovitz medium no. 15 with 2% inactivated (56°C, 30 min) fetal bovine serum will maintain these cells for 2 to 3 weeks without a change of medium. Primary cultures of human embryonic kidney or lung have also been used successfully. Many of these cell cultures can be obtained from commercial sources. A heteroploid cell line designated HeLa M, grown in the presence of 30 nM $MgCl_2$, has been used successfully for the isolation of rhinoviruses from clinical material. One disadvantage of the HeLa M cell system, however, is that it is difficult to maintain the cells for more than 5 or 6 days without a change of medium.

Coronavirus

Because of the fastidious growth requirements of HCV, successful isolation of most of its serotypes is dependent on the utilization of organ cultures of human ciliated respiratory epithelial tissues. This technique is essentially a research method, requiring a variety of sophisticated procedures for detection and identification of viruses. It is therefore not a technique to be applied as a routine isolation method. Details of organ culture methods for HCV isolation have been described (19). The only HCV strain to be isolated in monolayer cell culture of human tissues is the 229E virus strain. Successful isolation of this virus serotype has been accomplished in primary culture human embryonic kidney cells and several different human diploid cell strains derived from embryonic kidney and lung tissues. For the isolation of HCV strain 229E the procedure described for rhinovirus isolation described above is recommended.

INOCULATION OF HOST CELL SYSTEM

For virus isolation attempts, respiratory tract specimens should be treated with antibiotics to suppress the growth of contaminating microorganisms. In this laboratory, a final concentration of 1,000 U of penicillin, 5,000 μg of streptomycin, and 10 μg of amphotericin B per ml has been found to be satisfactory for this purpose. The treated specimens are inoculated in a volume of 0.2 ml into two tube cultures of human fetal diploid cells. Inoculated cultures are incubated at 33°C in a roller apparatus revolving at approximately 12 revolutions per h. The cultures are observed daily, or on alternate days, for cytopathic effect (CPE) for approximately 2 weeks. If no evidence of infection is observed, one blind passage may be carried out by inoculating 0.1 ml of the tissue culture fluids from the inoculated cultures onto fresh cell cultures and observing the new cultures for about 7 days.

EVIDENCE OF INFECTION

Rhinovirus

The growth of rhinoviruses in human fetal diploid cell cultures produces a CPE which can be easily seen under the low-power objective of a light microscope. Typical CPE induced by rhinoviruses is shown in Fig. 1. The first cellular changes are in the appearance of the fibroblast. The typically long slender cell becomes rounded or oval in appearance and may swell in size, become refractile, and eventually lyse. These changes initially occur in discrete foci throughout the cell sheet but eventually involve all, or nearly all, of the cell sheet.

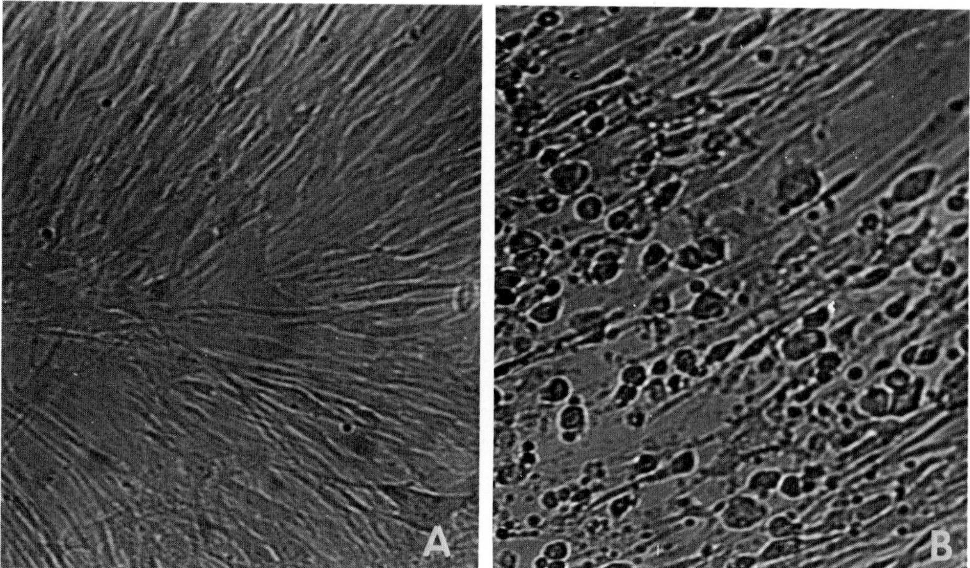

FIG. 1. Human fetal diploid lung cells, (A) uninfected and (B) infected with rhinovirus type 2. Magnification, ×930.

On occasion, adverse conditions in the cell culture may cause nonspecific changes in cellular morphology which may be confused with specific viral CPE. If this is suspected, a subculture should be carried out as described above. The additional passage not only serves as a check on the specificity of the viral CPE, but also serves to increase the virus titer. Some rhinoviruses require several passages before they become adapted to in vitro growth.

Coronavirus

HCV strain 229E on initial isolation in human diploid cell cultures does produce degenerative cell changes, but they are frequently minimal, and an additional passage may be required before the CPE is readily visible. Small focal lesions may develop on initial isolation; these sometimes heal and the cell monolayer may resemble the uninfected cell control. Eventually, with additional passage, the CPE becomes more granular in appearance, the cells detach from the glass surface, and the degenerative changes eventually involve the entire cell monolayer.

IDENTIFICATION OF AGENT

Rhinovirus

A flow scheme for the identification of rhinoviruses is given in Fig. 2. Viruses causing a picornavirus-like CPE can be presumptively identified as either an enterovirus or a rhinovirus by determining their acid stability.

For acid lability testing, dilute the test virus 1:10 in (i) Eagle minimal essential medium prepared without sodium bicarbonate (pH 3.0) and also in (ii) the same medium buffered at pH 7.0 with Tris. Both mixtures are incubated at room temperature for 3 to 4 h. Infectivity titers of the treated (pH 3.0) and control (pH 7.0) preparations are then determined utilizing serial 10-fold dilutions prepared in Hanks balanced salt solution. Inoculate 0.1 ml of each virus dilution into each of four tubes of human fetal diploid cells and observe for 7 to 10 days for evidence of CPE. A virus is considered to be acid labile if a 100-fold or greater reduction in infectivity titer occurs in the acid-treated preparation as compared to that of the virus control.

An alternative acid lability test involves mixing equal volumes of virus-infected cell culture fluid and 0.1 M sodium citrate-citric acid buffer at pH 3; for control purposes, equal volumes of virus suspension and a similar buffer system at pH 7.0 are mixed. The conditions of incubations and assay are the same as those described above.

Acid-labile viruses which cause picornavirus-like CPE can be presumptively identified as rhinoviruses. Definitive identification, however, is based upon neutralization of the isolate by a type-specific rhinovirus immune serum. Typing sera can be easily prepared in guinea pigs if animal facilities are available. Rhinovirus antigen pools are best prepared with virus adapted to a sensitive human heteroploid cell line (HeLa M) because higher titered antigen pools are more easily obtained in these cells than in human diploid cell strains.

Antigen pools with a minimal titer of 10^5 50% tissue culture infective doses ($TCID_{50}$) per 0.1 ml are partially purified by mixing equal volumes of antigen and fluorocarbon (Freon 13) and blending for three 1-min intervals in a blender (Sorval Omni-mixer or similar blender) at approximately 3/4 maximal speed. An ice bath should be used to prevent heating of the antigen pool. The aqueous phase containing the virus is removed after centrifugation at 1,000 × *g* for 10 min and then homogenized with an equal volume of incomplete Freund adjuvant. Guinea pigs are injected intramuscularly with 1 to 1.5 ml of the homogenized virus-adjuvant mixture in the thigh of each hind leg. A second intramuscular injection is given 6 weeks later, and the animals are bled 2 weeks after the second injection. It is important to verify both the homotypic and heterotypic reactivity against known reference rhinoviruses and to determine antibody titer against a standard challenge of 100 $TCID_{50}$ of homologous virus. Rhinovirus typing sera, in small reference vol-

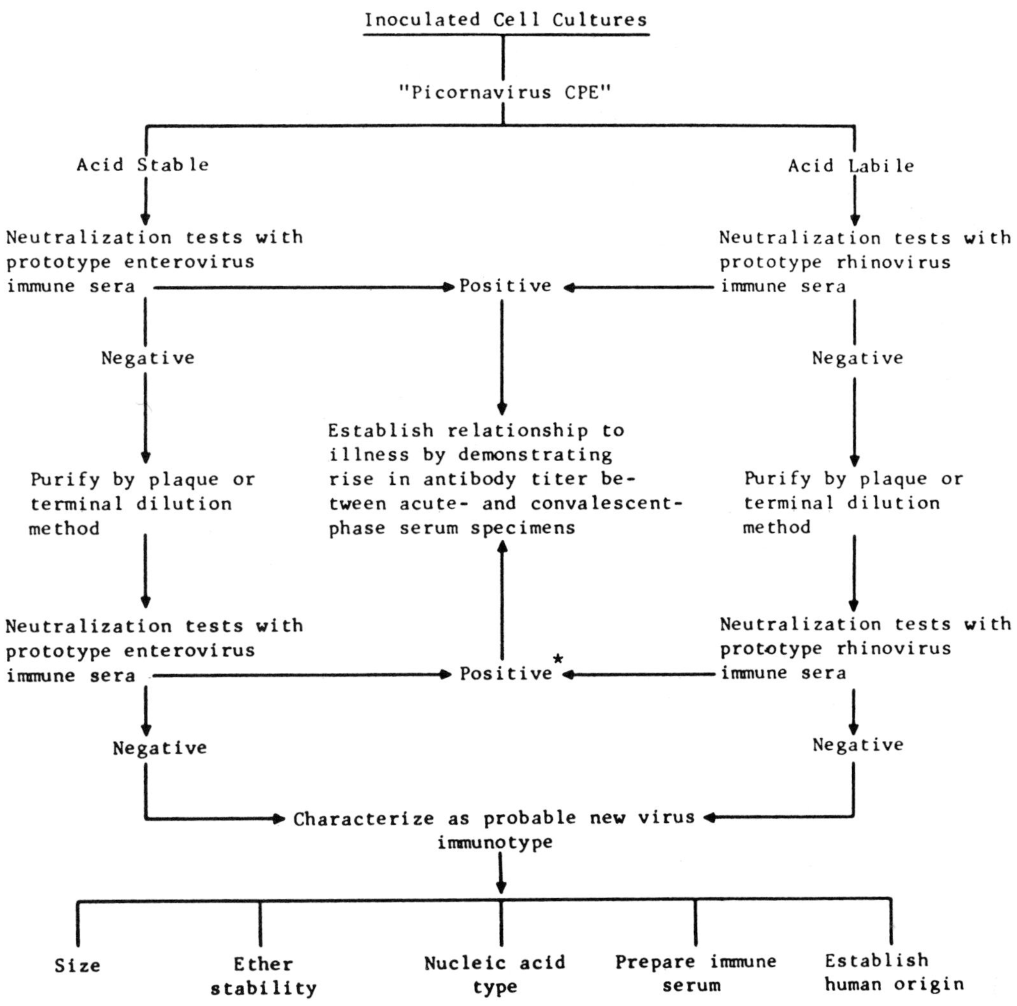

* Indicates a possible mixture of two or more viruses in the unpurified isolate; attempt to identify the other virus(es) by use of appropriate immune serum.

FIG. 2. Identification of rhinoviruses.

umes, can be obtained from the American Type Culture Collection, Rockville, Md.

Identification of specific rhinovirus immunotypes by serum neutralization tests is greatly facilitated by the use of combination or intersecting serum pools. In the combination serum pools a virus is identified by demonstrating neutralization in one, two, or three serum pools; in the intersecting serum scheme, identification is accomplished by demonstrating neutralization with two serum pools sharing a common virus type.

If the test virus is not neutralized by type-specific immune sera for the currently recognized rhinovirus serotypes, it is possible that the isolate may consist of two or more viruses, and therefore the isolate must be purified by either the plaque or the terminal dilution procedure. For the latter procedure, half-log dilutions of the isolate are prepared in Hanks balanced salt solution, and each dilution is inoculated in a volume of 0.1 ml into 10 human fetal diploid cell cultures. The cultures are maintained for 14 to 18 days and observed for CPE. Three successive passages are made from the single positive culture at the limiting (highest) dilution or from the second of two cultures to exhibit CPE at the limiting dilution.

Viral aggregation may be a cause for failure to neutralize a test virus; although it is not a common problem with rhinoviruses, it does occur on occasion. Diluting 1 volume of the test virus with 9 volumes of 1% deoxycholate is effective in disaggregating rhinoviruses.

The purified or disaggregated virus should be retested with immune serum for each of the prototype rhinoviruses as previously described. In the event that the test virus is not neutralized, the isolate may possibly represent a new immunological type. However, to establish the isolate as a candidate prototype virus, one must show it to be serologically distinct from the prototype rhinoviruses by reciprocal serum neutralization tests; i.e., in addition to showing that the purified unknown virus is not neutralized by immune sera to each of the prototype rhinoviruses, immune serum prepared with the unknown virus must show no neutralizing activity for any of the known prototype viruses.

To be considered as a new rhinovirus serotype, serologically distinct candidate viruses must also be characterized in terms of their biophysical and biochemical properties (see Fig. 2). Details of these methods have been published (6).

Coronavirus

Presumptive identification of an isolate as a coronavirus can be accomplished by determining that the isolate possesses the essential characteristics of the Coronaviridae family: ether and acid lability, virus growth in presence of DNA inhibitors, and typical size and morphology in the electron microscope. Details of these methods have been published (6). Definitive serological identification of HCV strain 229E is made by the neutralization test (19) in human diploid cells.

Coronavirus antisera are not available commercially but are easily prepared as follows. Infected cell culture fluid containing 10^3 $TCID_{50}$ of virus is first lightly clarified by centrifugation at 2,000 rpm for 10 min. The infectious cell culture fluid is then emulsified with an equal volume of Freund adjuvant; 2.0 ml of the mixture is inoculated intramuscularly in the hind legs of several guinea pigs. A second intramuscular injection is given 6 weeks later, and the animals are trial bled 10 to 14 days after second injection. The trial bleeding is titrated by viral neutralization against 100 $TCID_{50}$ of reference 229E virus. A titer of 1:250 or greater is sufficient for routine serological identification. If needed, a booster injection can be given, and the animal is then bled 7 to 10 days later.

SEROLOGICAL DIAGNOSIS

Rhinovirus

Serological diagnosis of rhinovirus infection is impractical at present because of the large number of distinct virus types and the lack of a simple and reliable serological method for detecting rhinovirus antibody.

Hemagglutination and hemagglutination inhibition have been reported for certain rhinovirus serotypes (21). The development of hemagglutination inhibition antibody closely parallels that of neutralizing antibodies in sera from natural infections. However, hemagglutination has been demonstrable for only 16 of 55 serotypes examined. Thus, until hemagglutination can be demonstrated for a significant number of additional serotypes, this test is presently not practical for the serological diagnosis of rhinovirus infection.

Complement-fixing antigens have been prepared by concentrating virus from infected cell culture fluids. Complement fixation reactions are relatively insensitive for detection of rhinovirus antibody in human infection and when positive are nearly always accompanied by a significant rise of antibody to heterotypic rhinoviruses. Thus, the complement fixation test is of little value in the serological diagnosis of natural rhinovirus infection.

Certain rhinoviruses have been used in immunodiffusion tests, and the native antigens produce a single, highly specific precipitin line with homotypic animal sera. However, when rhinovirus antigens are diffused against human sera containing rhinovirus neutralizing antibody, two precipitin lines are frequently found. Subsequent testing has indicated that rhinoviruses probably share a common antigen with other members of the human picornavirus family. This suggests that immunodiffusion may be a useful qualitative test or screening procedure for neutralizing rhinovirus antibody. However, the test so far has not been used successfully for quantitative measurements of neutralizing antibody in paired human sera.

The neutralization test remains the one serological method available at this time which can be used for the quantitative measurement of rhinovirus antibody responses in humans. Antibody responses to natural infection are frequently minimal, and the detection of small amounts of antibody requires that the neutralization test be carefully performed with a small test dose of virus (10 to 30 $TCID_{50}$). Acute-phase serum specimens should be collected as soon as possible after the onset of symptoms, and convalescent-phase specimens should be obtained approximately 3 weeks later. Sera are separated from the clot and stored at −20°C until needed for testing. The 50% endpoint neutralization test is performed as follows. All sera are inactivated at 56°C for 30 min. Equal volumes of the test virus dilution and twofold dilutions of serum in Hanks balanced salt solution are incubated at room temperature for 2 h, and 0.2-ml amounts of each serum-virus mixture are inoculated into two or four tubes of human fetal diploid cells. These tubes are examined for CPE after 2 or 3 days, when a simultaneous virus titration indicates that 10 to 30 $TCID_{50}$ doses of virus are present in the test. Serum neutralization endpoints are calculated by the method of Reed and Muench and are expressed in terms of the initial serum dilutions. The test is considered positive or significant if there is a fourfold or greater rise in antibody titer to the homologous or infecting virus between the acute- and convalescent-phase serum specimens.

Several microneutralization tests have been described for the quantitative assay of rhinovirus antibody. One of these (18) employs disposable microtiter plates in which serum-virus neutralization mixtures are added to wells containing HeLa M cells. The microtiter plates are examined daily for CPE, and antibody titers based on a 50% endpoint are calculated. The microneutralization tests are particularly suitable for those laboratories conducting large-scale serological surveys or epidemiological studies.

Coronavirus

The serological diagnosis of HCV infections is at present practical for two serotypes, 229E and OC-43. The diagnosis is dependent upon the demonstration of a fourfold or greater increase in antibody. Both the neutralization and complement fixation tests have been widely used for both virus types. The hemagglutination inhibition test has had broad acceptance for the OC-43 HCV serotype. More recently the ELISA has been reported to be more sensitive than neutralization (12).

The neutralization test is performed in human diploid cells, and the method is essentially that described above for rhinoviruses. However, the virus challenge dose should be increased to 32 to 100 $TCID_{50}$.

Complement-fixing antigens are readily prepared in human diploid cells for 229E and OC-43 serotypes. In addition, infected suckling mouse brain is an excellent source of complement-fixing antigen for OC-43 virus. The general procedure for the complement fixation test is as described by Palmer et al. (18a).

The hemagglutination inhibition test has been most useful in seroepidemiological studies of OC-43 virus; however, it is important to test, and if necessary treat, the test sera for presence of nonspecific inhibitors (8).

EVALUATION AND INTERPRETATION OF RESULTS

The laboratory diagnosis of rhinovirus infection is dependent, in the absence of a simple serological method, upon the isolation of the virus. Rhinoviruses are recovered from respiratory tract specimens preferentially in human fetal diploid cells or primary culture of human embryonic kidney cell strains. Once isolated, the viruses can be adapted to sensitive heteroploid cells such as HeLa M. Adaption to heteroploid cells usually results in higher virus titers, and infectious fluids from these cells yield potent antigens for use in immune serum production.

Of the many factors contributing to the successful isolation of rhinoviruses, the sensitivity or susceptibility of the host cells is of paramount importance. This may vary depending on the particular fetus from which the cultures were derived or even change with passage levels of the same cell strain.

The physiological condition of cells in culture can be influenced by the buffer capacity of the maintenance medium or the method of cultivation. Whole serum incorporated into the medium may contain nonspecific inhibitors or even specific antibodies to animal viruses which may share antigens with the viruses under study.

Isolates which produce a picornavirus CPE are readily identified as either enteroviruses or rhinoviruses, depending on their acid lability. The test is reliable and simple to perform. It is important, however, that the isolate have an infectivity titer of at least $10^{3.0}$ $TCID_{50}$/ml before the acid lability test is attempted. This will not be a problem with most isolates, but in those few instances where virus titers are inadequate one or two rapid passages in cell cultures are usually sufficient to raise the titer to the minimal level.

In this laboratory, for those isolates which induce picornavirus-like CPE and are acid labile, a preliminary report is made identifying the isolate as "rhinovirus, type undetermined." A final report giving the specific type is made when serotyping tests have been successfully completed. Although type-specific identification is of epidemiological interest and is useful for public health virological surveillance purposes, it may not be feasible, nor generally warranted, as a diagnostic procedure for the clinical virus laboratory.

The fastidious nature of the HCV group has been a serious impediment in developing a better understanding of the natural history of these viruses. In particular it has limited our capacity to apply routine isolation and identification procedures. The HCVs isolated in human diploid cell culture comprise an antigenically homogeneous group that do not cross-react with viruses of the OC-43 serotype. There are, however, numerous antigenic relationships between various HCV serotypes and several animal coronaviruses.

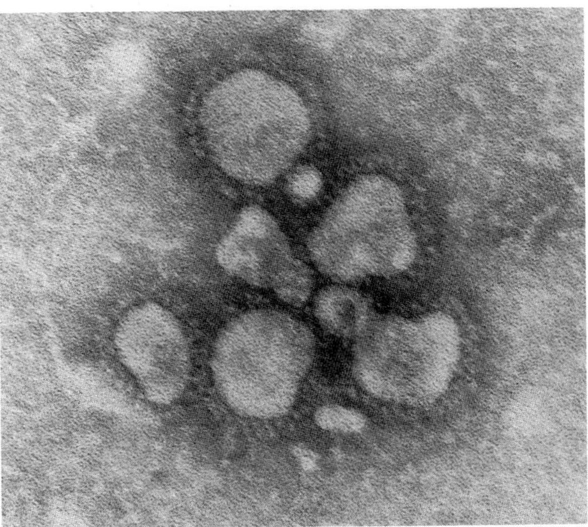

FIG. 3. Coronavirus strain 229E. Phosphotungstic acid stain. Magnification, ×127,000.

Members of the coronavirus group can be presumptively identified by demonstration of their unique morphology and size in electron micrographs (Fig. 3). Adaptation for growth in monolayer cell culture will also permit additional characterization of the biophysical and biochemical characteristics of the Coronaviridae family.

LITERATURE CITED

1. **Almeida, J. D., and D. A. J. Tyrrell.** 1967. The morphology of three previously uncharacterized human respiratory viruses that grow in organ cultures. J. Gen. Virol. **1**:175–178.
2. **Cooney, M. K., J. P. Fox, and G. E. Kenny.** 1982. Antigenic groupings of 90 rhinovirus serotypes. Infect. Immun. **37**:642–647.
3. **Dreizin, R. S., N. M. Borokova, T. I. Ponomareva, E. M. Vikhnovich, A. A. Kheinites, and T. V. Leichinskaya.** 1975. Diagnosis of rhinovirus infections by virological and immunofluorescent methods. Acta Virol. **19**:413–418.
4. **Gwaltney, J. M., Jr.** 1982. Rhinoviruses, p. 491–517. *In* A. S. Evans (ed.), Viral infections of humans. Plenum Publishing Corp., New York.
5. **Gwaltney, J. M., Jr., and J. O. Hendley.** 1982. Transmission of experimental rhinovirus infection by contaminated surfaces. Am. J. Epidemiol. **116**:828–833.
6. **Hamparian, V. V.** 1979. Rhinoviruses, p. 535–575. *In* E. H. Lennette and N. J. Schmidt (ed.), Diagnostic procedures for viral and rickettsial infections. American Public Health Association, Inc., New York.
7. **Hasony, H. J., and M. R. Macnaughton.** 1982. Serological relationships of human coronavirus strain 229E and mouse hepatitis virus strains. J. Gen. Virol. **58**:449–452.
8. **Hovi, T. H., H. Kainulainen, B. Ziola, and A. Salmi.** 1979. OC-43 strain related coronavirus antibodies in different age groups. J. Med. Virol. **3**:313–320.
9. **Kapikian, A. Z., J. D. Almeida, and E. J. Stott.** 1972. Immune electron microscopy of rhinoviruses. J. Virol. **10**:142–146.

10. **Kaye, H. S., and W. R. Dowdle.** 1969. Some characteristics of hemagglutination of certain strains of IBV-like viruses. J. Infect. Dis. **120:**576–581.
11. **Killington, R. A., E. J. Stott, and D. Lee.** 1977. The effect of temperature on the synthesis of rhinovirus type 2. J. Gen. Virol. **36:**403–411.
12. **Kraaijeveld, C. A., S. E. Reed, and M. R. Macnaughton.** 1980. Enzyme-linked immunosorbent assay for the detection of antibody in volunteers experimentally infected with human coronavirus strain 229E. J. Clin. Microbiol. **12:**493–497.
13. **Macnaughton, M. R., D. Flowers, and D. Isaacs.** 1983. Diagnosis of human coronavirus infections in children using enzyme-linked immunoabsorbent assay. J. Med. Virol. **11:**319–325.
14. **Macnaughton, M. R., and M. H. Madge.** 1978. The genome of human coronavirus 229E. J. Gen. Virol. **39:**497–504.
15. **McGregor, S., and H. D. Mayor.** 1971. Biophysical and biochemical studies on rhinoviruses and poliovirus. J. Virol. **44:**259–270.
16. **McIntosh, K., A. Z. Kapikian, K. A. Hardison, J. W. Hartley, and R. M. Chanock.** 1969. Antigenic relationship among coronaviruses of man and between human and animal coronaviruses. J. Immunol. **102:**1109–1118.
17. **Medeppa, K. C., C. McClean, and R. R. Ruckert.** 1971. On the structure of rhinovirus 1A. Virology **44:**259–270.
18. **Monto, A. S., and E. R. Bryan.** 1974. Macro-neutralization test for detection of rhinovirus antibodies. Proc. Soc. Exp. Biol. Med. **145:**690–694.
18a. **Palmer, D. F., L. Kaufman, W. Kaplan, and J. J. Cavallaro.** 1977. Serodiagnosis of mycotic diseases. Charles C Thomas, Publisher, Springfield, Ill.
19. **Schieble, J. H., and A. Z. Kapikian.** 1979. Coronaviruses, p. 709–723. *In* E. H. Lennette and N. J. Schmidt (ed.), Diagnostic procedures for viral and rickettsial infections. American Public Health Association, Inc., New York.
20. **Schieble, J. H., E. H. Lennette, and V. L. Fox.** 1970. Antigenic variation of rhinovirus type 22. Proc. Soc. Exp. Biol. Med. **133:**329–333.
21. **Stott, E. J., and R. A. Killington.** 1972. Hemagglutination of rhinoviruses. Lancet **i:**1369–1370.
22. **Turner, R. B., J. O. Hendley, and J. M. Gwaltney, Jr.** 1982. Shedding of infected ciliated epithelial cells in rhinovirus colds. J. Infect. Dis. **145:**849–853.

Chapter 72

Influenza Viruses

ALAN P. KENDAL, WALTER R. DOWDLE, AND GARY R. NOBLE

CLINICAL BACKGROUND

Influenza is an acute respiratory disease which characteristically comes to attention through dramatically sudden increases in visits to clinics, physicians' offices, or hospital emergency rooms or through focal outbreaks of respiratory illness with high attack rates in schools or residential institutions, such as nursing homes. After these first indications of activity, rapid, community-wide spread normally, but not always, occurs over the next month before the epidemic ends. The disease usually begins abruptly with chilliness followed by fatigue, headache, and myalgia. Fever appears early and may persist for 2 to 4 days. Constitutional symptoms are more striking with influenza than with other respiratory diseases. A nonproductive cough is frequently present after the first several days of illness, but coryza and pharyngitis are less common. In infants, croup and bronchiolitis are frequent, and gastrointestinal symptoms have been reported in up to about 25% of school children. Physical findings are consistent with those of an acute febrile respiratory illness. Recovery from uncomplicated influenza begins 3 to 4 days after onset, although complaints of weakness, fatigue, and cough may persist for 1 week or longer. Pneumonia of viral or secondary bacterial origin may occur, and such complications can be fatal, particularly in the elderly or in patients of any age with chronic debilitating illnesses. Compromised pulmonary or cardiac function is a particular indication of risk of severe influenza infection. Reye's syndrome, characterized by noninflammatory encephalopathy and fatty infiltration of the liver, occurs rarely in children and young adults after infection with influenza A and B viruses.

Epidemics of influenza type A may occur in about 2 of every 3 years. Outbreaks of type B are less frequent and usually occur every 3 to 5 years. Influenza type C is associated with subclinical or mild common cold-like illness and does not cause recognizable epidemics.

Influenza may not always follow the classic clinical course or occur in epidemic patterns. Sporadic cases, particularly among children, may be difficult to recognize when viewed against a background of other febrile and respiratory diseases in the community, yet are often of great value in providing advance warning that influenza virus is spreading in the community before it causes an epidemic. For this reason, laboratory diagnosis is essential, particularly in the fall and early winter months. Although there is no absolute rule, strains which appear in the United States late in one season, and are the minority of isolates, often reappear in epidemic form the following year. Thus, continuation of influenza virus isolation efforts through almost the entire year contributes to national surveillance efforts and improves the likelihood that the following year's strain will be correctly anticipated.

For additional information on influenza, several monographs are available (1, 5, 15).

DESCRIPTION OF AGENT

Influenza viruses contain a segmented, single-stranded RNA genome enclosed within a virus-modified host cell membrane. Influenza A and B virion RNA contains eight segments. Three RNA segments code for the polypeptides believed to be required for virus-specific, RNA-dependent RNA polymerase activity. Two code for the surface polypeptides, the hemagglutinin and neuraminidase. The three remaining RNA segments code for an internal structural virion protein (M or matrix protein), the nucleoprotein, and one or more nonstructural polypeptides found only in virus-infected cells. Influenza C viruses are believed to contain seven polypeptides, but precise information on their coding arrangements is lacking. Hemagglutinating and receptor-destroying activities of influenza C virus are associated with a single type of surface glycoprotein (like parainfluenza viruses) but do not have the chemical specificity for sialic acid-containing substrates that those of influenza A and B viruses (or parainfluenza viruses) have.

Influenza viruses may be spherical, with an approximate diameter of 100 nm, or filamentous, with a total length of several micrometers. Newly isolated viruses grown in eggs may show a high degree of pleomorphism and aggregation. Surface projections about 8 to 10 nm in length and spaced at 8-nm intervals may usually be observed in influenza viruses examined by negative-staining techniques. Influenza C viruses may, in addition, possess a reticular structure on the surface, although this may not be reliably observed on newly isolated virions. Influenza A and B nucleocapsids appear to be packaged as a large coil about 50 nm in diameter and of variable length.

Influenza virus infectivity may be destroyed by heating to 56°C or treatment with lipid solvents, acid, formaldehyde, β-propiolactone, UV light, or gamma radiation. Infectivity is lost on repeated freezing and thawing or after storage in standard −20°C freezers. Infectious virus may be preserved by lyophilization with 0.5% gelatin and maintenance at 4°C or by freezing liquid suspensions to −60°C or colder in the presence of at least 1% protein stabilizer.

The family Orthomyxoviridae contains the genus *Influenzavirus* (which includes influenza species A and B) and a probable genus (with no approved name) for the type species, influenza C (3). The World Health Organization provides rules for nomenclature and periodically updates them on the basis of new information; this was done most recently in 1981 (23). Influenza virus nomenclature is based on antigenic type and epidemiological information, which includes host of origin and place and year of isolation. For influenza A viruses, an antigenic description follows the strain designation and indicates the antigenic character of the hemagglutinin and neuraminidase subtypes. In the recent revision of nomenclature, the use of species designation for the hemagglutinin and the neuraminidase subtypes found only in animals was abandoned, and several former subtypes were merged. In particular, as a result of genetic data that supported earlier antigenic results, the hemagglutinins previously designated HSW1, H0, and H1 were placed in a single H1 subtype. Prototype strains for the three hemagglutinin subtypes thus far found in humans are A/Puerto Rico/8/34(H1N1), A/Japan/305/57(H2N2), and A/Hong Kong/1/68(H3N2). A total of 12 different hemagglutinin subtypes (a 13th has been proposed) and 9 different neuraminidase subtypes were recognized among type A viruses, including those isolated from horses, pigs, and birds. These additional subtypes are now designated H4 through H13 and N3 through N9. Certain hemagglutinin and neuraminidase subtypes are shared among strains isolated from different species. (For example, viruses with H7 hemagglutinins have been isolated from birds, horses, and seals.) Each hemagglutinin and neuraminidase subtype may encompass strains exhibiting a considerable degree of antigenic heterogeneity (this is most evident in viruses from humans), which may be readily differentiated by hemagglutinin inhibition (HI) and neuraminidase inhibition tests. Thus, the provision for full names for each strain is required so that the relatedness of viruses to various reference strains within each subtype can be made.

The nucleoprotein or "soluble" antigen of influenza viruses is type specific for influenza A, B, or C viruses. The M protein associated with the inner structure of virion envelopes is also type specific.

Antigenic variation in influenza viruses occurs primarily in the surface antigens of virions and is mediated by multiple determinants in the hemagglutinin and neuraminidase antigens. Abrupt change in the antigenic composition of influenza A viruses is called antigenic shift and, in the case of the hemagglutinin, may be associated with worldwide epidemics such as those occurring in 1957 with the Asian (H2) virus and in 1968 with the Hong Kong (H3) virus. The appearance in 1977 of the "Russian" influenza virus (H1N1) constituted an antigenic shift (although the virus was antigenically indistinguishable from strains previously isolated in 1950), but the ensuing worldwide epidemic was limited, affecting primarily persons born after about 1955. Persons born earlier have been largely protected by virtue of previous infection with strains of the H1N1 subtype which circulated from at least 1918 to 1957.

More gradual changes in the antigens within a subtype are described as antigenic drift, which may or may not be associated with epidemics.

Contemporary influenza B strains diverge considerably from the prototype strain (B/Lee/40), but strains from intermediate years cross-react with earlier and later isolates. All influenza C viruses isolated thus far cross-react to a considerable extent in HI tests.

Influenza viruses grown in eggs contain host carbohydrate antigens (12). In some circumstances antibodies to egg host antigens can result in cross-reactivity between unrelated strains. The nonstructural protein synthesized in influenza-infected cells and the viral RNA polymerase have not been studied for type or strain specificity.

COLLECTION AND STORAGE OF SPECIMENS

Two types of specimens are routinely collected for the diagnosis of influenza: throat and nasal specimens (individually or combined) for virus isolation and paired blood specimens for serological examination. Specimens for virus isolation should be collected within the first 3 days after onset of disease. Although throat garglings and nasal washings may be used successfully, throat and nasal swabs are generally more convenient for the physician and less objectionable to the patient. Throat swabs are obtained by vigorously rubbing the tonsils, soft palate, and the back wall of the lower pharynx with dry cotton applicators. The cotton tips of the applicators are then broken off into screw-capped vials containing 2 to 5 ml of tryptose phosphate broth with 0.5% gelatin. Secretions may also be collected from the nasal mucosa by use of a thin wire nasopharyngeal swab with a calcium alginate tip, which is eluted in the same vial containing the throat swabs. Specimens for viral isolation should be kept cold (4°C) at all times. If they are not to be inoculated within 48 h after collection, specimens should be sealed and stored in dry ice or under mechanical refrigeration (−70°C).

Blood specimens for serological diagnosis are drawn by venipuncture during the acute stage of the disease, usually at the time a throat swab is collected, and again 2 to 3 weeks later during the convalescent stage. Sera collected aseptically from the blood clots need not be refrigerated during transport to the laboratory, but should be stored at 4°C or −20°C.

In fatal cases of suspected influenza pneumonia, virus isolation should be attempted from lung tissue and tracheal mucosa. Tissue specimens, about 1 g, should be collected as soon as possible on autopsy and frozen at −70°C in sterile screw-capped containers if there is to be a delay in processing.

DIRECT EXAMINATION

As early as 1956, Liu (18) showed that 71% of influenza A and 38% of influenza B infections could be diagnosed by direct fluorescent-antibody staining of nasal epithelial cells. Other workers have also reported successful use of the technique (9, 20), confirming that fluorescent-antibody staining may be useful for rapid diagnosis of influenza virus infections. Experience with direct antigen detection in immunoassay procedures indicates these methods can detect antigen with a frequency comparable to fluorescence microscopy. The quality of specimens in many cases may be the limiting factor in direct antigen detection methods, with nasal aspirates collected from children being the best for testing. Supplies of reagents for antigen detection are not generally available, and this practical problem probably contributes to the relatively low rate of use of rapid diagnostic tests.

ISOLATION OF VIRUS

Influenza virus types A, B, and C are isolated in embryonated hen eggs. Influenza A and B viruses are isolated also in primary rhesus or cynomolgus monkey kidney cell cultures. The host range for production of infectious influenza virus may be increased in the presence of certain proteolytic enzymes, particularly trypsin, so that influenza A and B can also be isolated in the MDCK cell line (8, 21, 22). Influenza A viruses can be expected to grow with approximately the same success in each of these hosts (when several years' experiences are taken into account), whereas the cell culture systems are often superior for growth of influenza B virus. Differences between growth properties of isolates from year to year, however, are common. Because of variations in growth properties and because of possibilities for variability in batches of tissue culture, inoculation of both embryonated eggs and primary monkey kidney or MDCK cells is recommended to ensure the timely isolation and identification of all influenza types and subtypes that may cocirculate.

Use of eggs can have the additional advantage of shortening the time required to diagnose an influenza infection; the virus frequently grows to detectable titers within 2 to 3 days of inoculation, and influenza viruses are the only human viral pathogens which commonly produce hemagglutinating activity in eggs. Mumps virus is rarely isolated with the conditions recommended for influenza virus.

Antibiotics are added from a 10-fold concentrate to throat and nose specimens to yield a final concentration of 800 U of penicillin and 400 μg of streptomycin sulfate per ml; specimens are incubated for 30 min at 37°C. Alternatively, 0.1 ml of gentamicin (10 μg/ml) is added per ml of specimen. Antibiotic-treated specimens are centrifuged at approximately 2,000 × *g* for 15 min to remove particulate matter, including fibers from swabs that may prove toxic to cell cultures. Lung and trachea autopsy tissues are ground in a sterile mortar with small amounts of Alundum (60 mesh) and made up to an approximately 10% suspension in tryptose phosphate broth with 0.5% gelatin and antibiotics. Samples of specimens lacking antibiotics should be maintained if isolation of nonviral respiratory pathogens (e.g., *Legionella pneumophila*, *Mycoplasma pneumoniae*, rickettsiae, or chlamydiae) may be attempted.

Embryonated eggs

Clinical specimens are prepared as above. After initial isolation, isolates being grown in eggs for reagent production may be diluted to contain about 10^4 50% egg infectious doses per ml in tryptose phosphate broth before inoculation, to limit production of defective-interfering virus particles.

Ten- to 11-day-old embryonated eggs are swabbed with 70% alcohol directly over the air sac, and a small hole is punched in the shell in the center of the area. The egg is placed on a candler, with the air sac up, and rotated to locate the embryo. A 23-gauge, 3.8-cm needle attached to a 1-ml syringe is inserted into the amniotic sac with a jabbing motion. The needle is in the correct position when the embryo can be moved with the needle. A 0.2-ml amount of inoculum is injected into the amniotic cavity. The needle is withdrawn slightly, and an additional 0.2 ml is injected into the allantoic cavity. The hole in the shell is sealed with "model airplane cement" or wax. Two or three eggs are inoculated per specimen. Eggs are incubated at 33°C for 2 to 3 days. Eggs as young as 7 or 8 days may be used to permit longer periods of virus growth while allowing amniotic fluid to be recovered in reasonable yields (e.g., for poorly growing influenza C viruses). Incubator temperatures should not exceed 35°C because many influenza viruses have an inherent temperature sensitivity.

Eggs are chilled overnight at 4°C to minimize bleeding during harvest (more rapid harvesting can be done if individual fiberboard trays of eggs are placed at −70°C for 10 min and then transferred to 4°C for 1 h). The area over the air sac is swabbed with 70% alcohol. The eggshell is broken away to the level of the allantoic membrane, and the membrane is pulled away with sterile forceps. The allantoic fluid is harvested with a pipette or syringe, and the amniotic fluid is harvested with a short 20- to 22-gauge needle attached to a 1-ml syringe. Fluids are clarified by centrifugation at about 800 × *g* for 15 min. If volumes of amniotic fluid are very small, the amnions can be washed with 0.5 to 1 ml of allantoic fluid to obtain a reasonable working volume. Pools of amniotic fluid are prepared from all eggs inoculated with the same specimen, but allantoic fluids are maintained separately. Undiluted samples of both types of fluids are tested in tubes (0.5-ml volumes) or in microtiter plates (0.05-ml volumes) for the presence of virus by adding equal volumes of guinea pig, chicken, or human O erythrocytes. (Erythrocytes should be collected at a ratio of 1 volume of whole blood to 4 volumes of Alsever solution. Suspensions in Alsever solution may be held as long as 1 week under refrigeration. Just before use, the erythrocytes are washed three times with phosphate-buffered saline [PBS], pH 7.2, centrifuged at 1,000 rpm [about 250 × *g*] for 10 min, and resuspended to a concentration of 0.4% for guinea pig or human O cells and 0.5% for chicken cells.) When microtiter plates are used, the tests should be done in U plates with guinea pig or human O cells. The fluid-erythrocyte mixtures are incubated at room temperature for 30 to 60 min or at 4°C for several hours, or until control cells have settled. Some viruses, particularly influenza C, may rapidly elute from erythrocytes at ambient temperatures, and influenza C does not agglutinate guinea pig erythrocytes. Guinea pig or human O cells may be more sensitive for detecting some strains of influenza A virus in early passage. Titration with chicken erythrocytes at ambient temperature is the most rapid test for detecting virus hemagglutinin in egg fluids.

Two or three blind egg passages, via both allantoic and amniotic routes, should be undertaken before a specimen is considered not to contain virus that can grow in eggs. For these passages, mixtures should be prepared containing equal amounts of pooled amniotic and allantoic fluids obtained with each specimen and inoculated as above into another series of eggs. Once influenza A and B viruses have been adapted to growth in the allantoic cavity, amniotic inoculation is no longer necessary. Influenza C, however, must normally continue to be propagated in the amnion. Presence of hemagglutinating activity in allantoic fluid can occur, probably by leakage from the amnion, and is not a reliable indicator of allantoic growth.

Because of the high infectious titers that may be attained with egg harvests of influenza virus (e.g., 10^8 to 10^9 50% egg infectious doses per ml) and the relatively large number of manipulations required when processing eggs, cross-contamination may occur, jeopardizing the interpretation of results. Precautions to prevent contamination include: prohibiting the growth of laboratory virus strains in areas where diagnostic specimens are handled; maintenance of independent supplies of media, antibiotics, and other supplies in the separate areas where diagnostic and reference positive specimens are processed; use of disposable pipettes, syringes, and needles when possible; and working with separate specimens and processing batches from different sources independently of each other. High virus titers in positive egg fluids also pose a safety hazard for members of the laboratory staff. Workers should minimize the generation of aerosols when transferring fluids with pipettes and syringes. Centrifuges should be kept out of working areas to prevent exposure to aerosols that might be generated if tubes break or leak during centrifugation steps.

Cell culture

Inhibition of influenza viruses by nonspecific substances in cell culture media containing serum may be avoided by washing the monolayers twice with Hanks balanced salt solution before replacing the growth medium with 1 ml of serum-free Eagle minimal essential medium. Two or three tubes are inoculated with 0.3 ml of the antibiotic-treated specimen in a total volume of about 1.5 ml. Highly purified trypsin at a concentration of about 2 μg/ml is included for cultures other than primary kidney cells; maximum trypsin concentration depends on the resistance of the particular cells. Cultures are incubated at 33 to 35°C, observed microscopically each day to detect viral cytopathic effect (CPE), and tested every other day to detect hemadsorption. The appearance of CPE and production of hemagglutinins can be hastened by placing the tubes on a roller drum. In the absence of serum, frequent adding or changing of medium is required to maintain the cell cultures.

Tubes should be harvested as soon as CPE is detected. In the absence of CPE, tubes are harvested between days 7 and 14. Medium from one tube is transferred to the remaining one or two tubes and replaced with 1 ml of Hanks balanced salt solution. A 0.2-ml amount of fresh 0.4% guinea pig erythrocytes is added, and the tube is returned to a slanted position. The cell sheet is examined for hemadsorption after 10 min and again after 30 min of incubation at room temperature; alternatively, it may be incubated at 4°C for 30 to 60 min (as with parainfluenza viruses), and then examined. Uninfected control tubes should be tested to rule out nonspecific adsorption or the presence of simian paramyxoviruses such as simian virus 5 (6).

If a culture is positive for hemadsorption, a hemagglutinin titration should be performed on the pooled medium with 0.4% guinea pig erythrocytes at 4°C. If the specimen is negative for hemadsorption or if the hemagglutinin titer is less than 8, the harvested material should be repassaged. When viruses produce low hemagglutinin titers in cell culture, passage into eggs may speed up the process of amplifying the titer to a level adequate for identification.

In monkey kidney cells, most influenza B isolates are detected during day 3 to 7 of the first passage by hemadsorption or hemagglutination. CPE may also be evident. Cells become progressively granular, swollen, and round; later, they become pyknotic and fragmented, and the cell sheet is eventually destroyed. Influenza A virus may grow less vigorously in cell culture. The CPE, consisting of slowly vacuolating or lacy cells which degenerate and detach from the glass, is often difficult to detect and may not become apparent until late in the second passage. Hemadsorption of varying degree is generally detected on the first passage, well in advance of an obvious CPE. Findings vary for different strains.

As is the case with virus isolation in eggs, cross-contamination with laboratory reference strains is always possible when these are used in the same facility (2).

IDENTIFICATION OF VIRUS

HI test

Isolates are best typed by the HI test, using sera prepared with currently prevalent virus strains. Because influenza viruses exhibit characteristics of continual antigenic change, commercial sera prepared with prevalent strains may not always be available. However, if antisera are available to recent strains of the same subtype as those in circulation, sufficient cross-reactivity should occur for identification of the isolates as influenza viruses. Each year the Centers for Disease Control makes available antisera to current strains through state health laboratories.

Typing sera for diagnostic use are best prepared in chickens by intravenous injections of about 500 to 1,000 hemagglutination units (HAU) of virus present in infected allantoic fluid, diluted as necessary to about 5 ml. The animals are exsanguinated 12 to 14 days later. Such sera have the advantage of not containing antibodies to certain host components that may be present in egg-grown virus, and they usually (but not always) have very low activities of nonspecific inhibitors. Sera with greater strain specificity may be prepared by infecting ferrets intranasally with 1 ml of allantoic fluid, diluted 1- to 100-fold, depending on strain and infectious virus titer. Ferrets are bled after 14 days. Because ferrets may be infected by exposure during human epidemics or by cross-infection from other infected ferrets, control preimmunization sera should be tested for antibodies to prevalent strains, and animals should be strictly quarantined before and after infection. Usually minimum HI titers of 160 are required for sera to be useful in diagnostic tests, and sera should not react with strains outside the type or subtype of the immunizing virus. (Occasionally, heterotypic cross-reactivity is observed which cannot be explained on the basis of known antigenic properties of viruses, e.g., chickens immunized with A/Texas/1/77[H3N2] usually produce antibodies that react with A/PR/8/34[H1N1].)

Hemagglutinin titration. The following procedure is described for the use of microtiter equipment. (Titration may be done in tubes [e.g., 12 by 75 mm], using 10-fold greater volumes of all reagents. Because dilutions can then be made with pipettes, test reproducibility can be improved compared to that obtained by use of microtiter equipment. The ratio of all reagents is similar regardless of the total test volume so that expression of titer is independent of the test volumes.) Beginning with undiluted virus, prepare twofold serial dilutions of the virus in 0.1 M PBS (pH 7.2) in 0.05-ml volumes. Add 0.05 ml of chicken or guinea pig erythrocytes to each well in the series, using the erythrocytes which give the highest titer with prevalent strains. Include one well for the erythrocyte suspension as a cell control (diluent plus erythrocytes). Mix and incubate at room temperature until the cells settle (about 30 min with chicken erythrocytes at room temperature; up to several hours with guinea pig cells at 4°C). Consider the highest dilution of virus causing agglutination as the titration endpoint; this dilution produces a solution containing 1 HAU per unit volume of 0.05 ml (or 0.5 ml). For influenza C, the test must be done at 4°C.

Preparation of test antigen. The HI test is performed with the most sensitive erythrocyte system (chicken or guinea pig) and a test antigen preparation containing 4 HAU in 0.025 ml. To determine the antigen dilution factor, divide its hemagglutinin titer by 8. For example, if the hemagglutinin titer is 160, then a 1:20 dilution would contain 8 HAU in 0.05 ml or 4 HAU in 0.025 ml.

To control possible errors in dilution and to confirm the hemagglutinin titers, retitrate the test or working virus dilution. Prepare a row of six wells, each containing 0.05 ml of diluent. Add 0.05 ml of the working dilution of virus (8 HAU) to the first well, and make the twofold dilution series through five wells. To the sixth well, add 0.5 ml of diluent in place of virus. This will serve as the cell control.

Add 0.05 ml of the appropriate erythrocyte suspension to all wells and mix; allow the contents to settle. The cell control and the last two wells should show compact buttons of normal settling. Agglutination in the first three wells of the series confirms that the working dilution contains 8 HAU of virus per 0.05 ml, or 4 HAU per 0.025-ml volume as required for use in the HI test. Adjust the virus concentration of the working dilution, if necessary, by adding PBS or virus as appropriate. If adjustment of working antigen dilution is required, confirm the hemagglutinin titer of the final sample by retitration as just described, in parallel with the HI test.

Serum treatment. Many influenza isolates are highly sensitive to serum factors which may nonspecifically inhibit agglutination. Such inhibitors in human, chicken, and most rabbit sera can often be successfully removed by treatment with the receptor-destroying enzyme of *Vibrio cholerae* (14). Successful destruction of inhibitors, however, varies with the virus to be tested and may require treatment with potassium periodate (sometimes in conjunction with trypsin) (14) or kaolin (10). Receptor-destroying enzyme treatment is performed as follows. Add 4 volumes of receptor-destroying enzyme (100 U/ml) to each volume of serum. Incubate the mixture overnight in a water bath at 37°C. Add 5 volumes of 1.5% sodium citrate and incubate the mixture at 56°C for 30 min. If the treated serum contains nonspecific agglutinins when serum controls are tested (see below), adsorb the treated serum at a rate of 0.1 ml of 50% erythrocytes to 1 ml of the 1:10 serum. Allow adsorption to proceed for 1 h at 4°C, and remove the erythrocytes by centrifugation.

HI tests. Prepare twofold dilutions of treated reference antiserum from 1:10 through 1:2,560 in 0.025-ml volumes. Add 0.025 ml of the test virus suspension containing 4 HAU to each well. To test for erythrocyte agglutinins in the serum, add diluent instead of antigen to a well containing the lowest dilution of serum. Also prepare cell controls (PBS only) and antigen controls (PBS and antigen) for each test. Shake and incubate at room temperature for 30 min.

Add 0.05 ml of erythrocytes to each well, shake, and incubate at room temperature (4°C for influenza C) until the cell control shows the button of normal settling. The HI titer is defined as the dilution factor of the highest dilution of serum which completely inhibits agglutination. Complete inhibition is determined by tilting the plates and observing the "tear-shaped" streaming of cells which flow at the same rate as cell controls.

For maximum test reproducibility, volumes of each series of serum dilutions adequate for testing each antigen should be prepared in tubes, using pipettes ("master dilution" technique). Droppers are then used to add 0.025 ml of each serum dilution to the appropriate wells of microtiter plates for each test antigen.

Hemadsorption inhibition test

Since many myxoviruses may be isolated in cell culture, and all exhibit the property of hemadsorption, it is impractical to attempt to identify the virus by hemadsorption inhibition unless there is sufficient evidence to narrow the possibility to only one or two suspected types. Such evidence may consist of the type of CPE, its rate of appearance, hemadsorption patterns, or epidemiological information. In general, specific antisera are employed to best advantage in the HI test. Hemadsorption inhibition is less sensitive than HI; a serum with an HI titer of less than 80 may fail to inhibit hemadsorption of recent isolates which show slight changes in antigenic composition. The hemadsorption inhibition test is useful, however, in some instances, particularly when there is a need to rapidly determine the cause of an outbreak of influenza-like illness but the hemagglutinin titer of virus growing in cell culture is too low to perform the HI test. The hemadsorption inhibition test is performed as follows.

Treat antisera with receptor-destroying enzyme in the same manner as described for the HI test. Wash infected monkey kidney cell cultures twice with Hanks balanced salt solution. Add 0.2 ml of the receptor-destroying enzyme-treated serum diluted 1:10, and then add 0.6 ml of Hanks balanced salt solution. Incubate the cultures for 30 min at room temperature with the entire cell sheet covered; add 0.2 ml of 0.4% guinea pig erythrocyte suspensions to each tube. Reincubate the cultures at room temperature for 20 min and examine microscopically for the presence of hemadsorption. The isolate is identified by the serum that prevents hemadsorption.

Fluorescent-antibody stain

Demonstration that viruses growing in cell culture are influenza A or B (or C) strains may be done by scraping cells and fixing multiple aliquots on slides which are then stained with fluorescent-antibody preparations by direct or indirect procedures. The greatest value of the test is that rapid discrimination between commonly encountered respiratory viruses may be obtained by staining replicate samples from a single tube of cells with an appropriate battery of typing sera. Generally, influenza typing sera used for this purpose are reactive with ribonucleoprotein (RNP), M protein, or both, so that the test is type specific. Monoclonal antibody pools are being developed at the Centers for Disease Control to provide reproducible, specific reagents for this purpose. Use of sera specific for hemagglutinin can permit the test to distinguish subtypes. If it is important to distinguish between strains within one subtype, however, virus antigen will still require identification by HI, neutralization, or strain-specific complement fixation (CF) testing.

Immunoassays

In the same way that immunoassays are equivalent to fluorescent-antibody procedures for direct detection of antigen in clinical species, they may also be used to type influenza isolates. Various types of immunoassay procedures have been found suitable for this purpose (see the chapter by Kendal in reference 1), and it is likely that, in due course, virus isolates will be conveniently typed by this method, as reagents become available.

Neutralization test

Influenza viruses may be identified by neutralization tests performed in either eggs or cell cultures. Mix equal volumes of 100 to 300 50% infectious doses of virus and serial twofold dilutions of reference antisera, incubate for 1 h at room temperature, and inoculate into the appropriate test system. Allantoic harvests from eggs should be tested for the presence of hemagglutinins, and cell cultures should be tested for hemadsorption. Viruses are identified by inhibition of growth in the presence of one serum. Neutralization test results generally parallel those obtained by the HI test, but the procedures are more complex than the HI test.

Other tests

Neuraminidase inhibition. A complete description of the influenza A viruses requires characterization of the neuraminidase as well as the hemagglutinin antigens. This may be done by the neuraminidase-inhibition test (14).

CF test. Influenza viruses may be typed as A, B, or C through identification of their RNP antigen by the CF test. This test is most useful in the event of a major change in the composition of the surface (hemagglutinin and neuraminidase) antigens of a new variant or if strains are encountered which are poorly reactive in the HI test (17). Diagnostic laboratories should, however, anticipate replacing such a method with immunoassay or fluorescence microscopy procedures.

Double immunodiffusion or counterimmunoelectrophoresis tests. Double immunodiffusion and counterimmunoelectrophoresis tests have the potential to rapidly identify the type and subtype of influenza virus isolates (4, 16). Because many future reagents are likely to be prepared by hybridoma technology and monoclonal antibodies are often poorly reactive in precipitin reactions, such tests may only be possible in specialty laboratories with access to hyperimmune sera specific for viral components.

SEROLOGICAL DIAGNOSIS

The serological diagnosis of influenza is based upon demonstration of a fourfold or greater increase in antibody titer between acute- and convalescent-phase sera. This may be measured by the type-specific CF test with RNP antigens or by the HI test with antigens carefully selected as similar to currently prevalent strains. Immunoassay procedures may detect the greatest number of infections, possibly because as usually performed they have a broad specificity for both type- and subtype-specific antibodies (11), or possibly because they are amenable to statistical analysis of results, which can effectively narrow the standard error of the test so that small differences in antibody titer achieve experimental significance (19). The CF test is often used because of its wide applicability in the virus laboratory. Unlike the HI test, the type-specific CF test is influenced neither by antigenic variability of prevalent strains nor by nonspecific inhibitors. In many years, the HI test has been a more sensitive test than CF for serological diagnosis of influenza A infection, but considerable difficulties have been observed in obtaining adequate sensitivity with the HI test for influenza A(H1N1) and influenza B viruses circulating since about 1977. Although it is possible to increase the sensitivity of the HI test by treating virus with ether, some loss of specificity may result (13). The HI test for serological diagnosis is performed in the same manner as described previously for virus identification, except that acute- and convalescent-phase sera are titrated simultaneously against known reference virus antigens. Another useful method of serological diagnosis is the single radial hemolysis test for assay of hemagglutinin and neuraminidase antibodies (7).

The CF test may be performed as described by Palmer et al. (20a). Influenza antigens may be readily standardized in the procedure used for the diagnosis of other virus diseases. RNP antigens and control sera for type A and type B influenza viruses may be obtained from commercial sources or prepared in the laboratory. Sera for control of the CF test with RNP antigens are prepared by intranasal inoculation of guinea pigs or ferrets (17). Adaptation of the CF test to a single radial system performed in immunodiffusion plates has been described (24).

EVALUATION, INTERPRETATION, AND REPORTING OF RESULTS

The laboratory diagnosis of influenza is based upon the isolation and identification of the virus, the demonstration of a fourfold or greater increase in antibody titer, usually by the CF or HI test (Fig. 1), or both. The identity of the isolate should be reported only as to type, unless tests have been performed with hemagglutinin- or neuraminidase-specific antisera to characterize the antigens fully. If the virus was identified with influenza A hemagglutinin-specific antisera, then the corresponding subtype of the neuraminidase is usually inferred, because thus far epidemic isolates of human influenza A have almost never been found to have exchanged HA and NA antigens even when two subtypes (H1N1 and H3N2) cocirculate.

In the absence of virus isolation, caution should be exercised in the interpretation of serological results. A

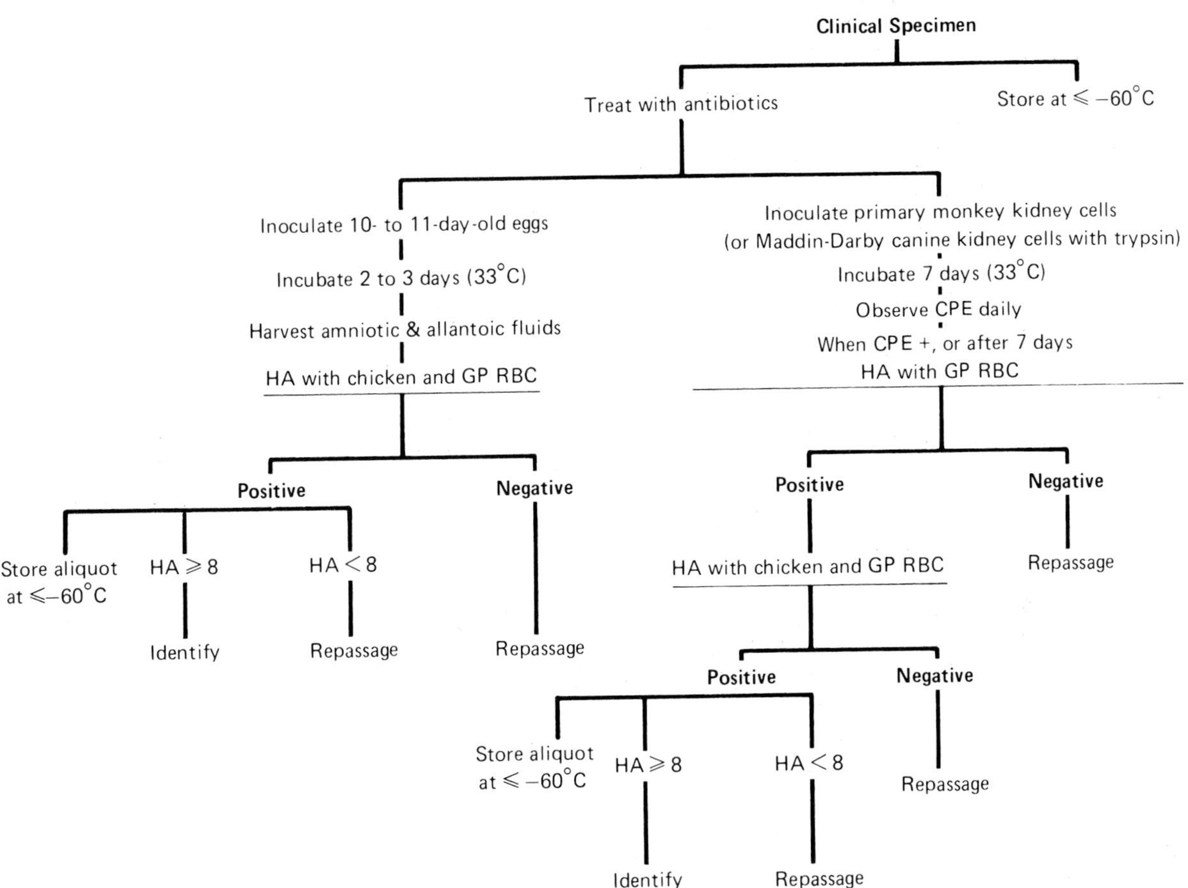

FIG. 1. Scheme for influenza virus isolation. When virus isolates in cell culture have a consistently low hemagglutinin titer, it may be advantageous to pass them into eggs to boost the titer for identification in the HI test.

fourfold or greater increase in antibody titer measured in the CF test with RNP antigen is interpreted only to mean infection, or vaccination, with type A or type B influenza virus. Since the RNP antigen is identical for all viruses of a given type, this technique does not provide information identifying the subtype causing the disease.

Results of HI tests on patients' sera should also be interpreted with caution. The specific antigen used for diagnostic serology does not necessarily identify the infecting virus. Anamnestic responses frequently occur, depending on the previous immunological experience of the patient. Antibody response to an earlier antigen may be greater than to the current infecting virus. For maximal diagnostic efficiency, antigens that closely resemble currently prevalent strains and antigens of recently prevalent strains may both be included in the HI test.

At present, influenza A viruses of subtypes H1N1 and H3N2 and influenza B viruses are in circulation. How long this will continue cannot be predicted, but presumably antigens representing current strains of all these viruses will be required for serodiagnosis for some time to come.

In epidemic situations when influenza is suspected, a rapid presumptive diagnosis can often be made by examining single serum specimens from a group of selected individuals. Sera are collected from 10 or more patients who are in the acute stage of the disease and from the same number of age-matched cohorts who experienced the same symptoms 10 or more days earlier. All sera are tested simultaneously for influenza A and B hemagglutinin or RNP antibody titers. If the epidemic was caused by influenza, the geometric mean antibody titer for type A or B should be significantly higher (by the *t* test) in the latter group than in the former (14). If the rise in antibody titer is fourfold or greater, the difference may be considered significant without resorting to statistical analysis. A diagnosis made on this basis should be confirmed by conventional methods of virus isolation and serological diagnosis with paired sera.

LITERATURE CITED

1. **Beare, A. S. (ed.).** 1982. Basic and applied influenza research. CRC Press, Inc., Boca Raton, Fla.
2. **Budnick, L. D., M. E. Moll, H. F. Hull, J. M. Mann, and A. P. Kendal.** 1984. A pseudo-outbreak of influenza A associated with the use of a laboratory stock strain. Am. J. Public Health **74:**607–609.
3. **Dowdle, W. R., F. M. Davenport, H. Fukumi, G. C. Schild, B. Tumova, R. G. Webster, and L. Y. Zakstelskaja.** 1975. Orthomyxoviridae. Intervirology **5:**245–251.
4. **Dowdle, W. R., J. C. Galphin, M. T. Coleman, and G. C. Schild.** 1974. A simple double immunodiffusion test for typing influenza viruses. Bull. W.H.O. **51:**213–218.
5. **Dowdle, W. R., G. R. Noble, and A. P. Kendal.** 1977. Orthomyxovirus–influenza: comparative diagnosis unifying concept, p. 448–501. *In* E. Kurstak and C. Kurstak (ed.), Comparative diagnosis of viral diseases, vol. 1, part

A, Human and related viruses. Academic Press, Inc., New York.

6. **Dowdle, W. R., and R. Q. Robinson.** 1966. Non-specific hemadsorption by rhesus monkey kidney cells. Proc. Soc. Exp. Biol. Med. **121:**193–198.
7. **Farrohi, K., F. K. Farrohi, G. R. Noble, H. S. Kaye, and A. P. Kendal.** 1977. Evaluation of the single radial hemolysis test for measuring hemagglutinin- and neuraminidase-specific antibodies to H3N2 influenza strains and antibodies to influenza B. J. Clin. Microbiol. **5:**353–360.
8. **Frank, A. L., R. B. Couch, C. A. Griffis, and B. D. Baxter.** 1979. Comparison of different tissue cultures for isolation and quantitation of influenza and parainfluenza viruses. J. Clin. Microbiol. **10:**32–36.
9. **Gardner, P. S., and J. McQuillin.** 1974. Rapid virus diagnosis—application of immunofluorescence. Butterworth, London.
10. **Hammon, W. M., and G. E. Sather.** 1969. Arboviruses, p. 227–280. *In* E. H. Lennette and N. J. Schmidt (ed.), Diagnostic procedures for viral and rickettsial infections, 4th ed. American Public Health Association, Inc., New York.
11. **Hammond, G. W., S. J. Smith, and G. R. Noble.** 1980. Sensitivity and specificity of enzyme immunoassay (EIA) for diagnosis of influenza A virus infections. J. Infect. Dis. **141:**644–651.
12. **Harboe, A.** 1963. The influenza virus hemagglutination inhibition by antibody to host material. Acta Pathol. Microbiol. Scand. **57:**317–330.
13. **Kendal, A. P., and T. R. Cate.** 1983. Increased sensitivity and reduced specificity of hemagglutination inhibition tests with ether-treated influenza B/Singapore/222/79. J. Clin. Microbiol. **18:**930–934.
14. **Kendal, A. P., M. S. Pereira, and J. J. Skehel.** 1982. Concepts and procedures for laboratory-based influenza surveillance. Centers for Disease Control, Atlanta, Ga.
15. **Kilbourne, E. D. (ed.).** 1975. The influenza viruses and influenza. Academic Press, Inc., New York.
16. **Lecomte, J., L. Berthiaume, and A. Boudreault.** 1979. Influenza viruses in birds: rapid identification by counterimmunoelectrophoresis. J. Clin. Microbiol. **9:**128–133.
17. **Lief, F. S.** 1963. Antigenic analysis of influenza viruses by complement fixation. VII. Further studies on production of pure anti-S serum and on specificity of type A S antigens. J. Immunol. **90:**172–177.
18. **Liu, C.** 1956. Rapid diagnosis of human influenza infection from nasal smears by fluorescein-labeled antibody. Proc. Soc. Exp. Biol. Med. **92:**883–887.
19. **Madore, H. P., R. C. Reichman, and R. Dolin.** 1983. Serum antibody responses in naturally occurring influenza A virus infections determined by enzyme-linked immunosorbent assay, hemagglutination inhibition, and complement fixation. J. Clin. Microbiol. **18:**1345–1350.
20. **Minnich, L., and G. Ray.** 1980. Comparison of direct immunofluorescent staining of clinical specimens for respiratory virus antigens with conventional isolation techniques. J. Clin. Microbiol. **12:**391–394.

20a. **Palmer, D. F., L. Kaufman, W. Kaplan, and J. J. Cavallaro.** 1977. Serodiagnosis of mycotic diseases. Charles C Thomas, Publisher, Springfield, Ill.

21. **Tobita, K.** 1975. Permanent canine kidney (MDCK) cells for isolation and plaque assay of influenza B viruses. Med. Microbiol. Immunol. **162:**23–27.
22. **Tobita, K., A. Sugiura, C. Enomuto, and M. Furuyama.** 1975. Plaque assay and primary isolation of influenza A viruses in an established line of canine kidney cells (MDCK) in the presence of trypsin. Med. Microbiol. Immunol. **162:**9–14.
23. **World Health Organization Committee.** 1980. A revision of nomenclature for influenza viruses: a WHO memorandum. Bull. W.H.O. **58:**585–591.
24. **Yamane, N., M. Yuki, and Y. Nakamura.** 1983. Single radial complement fixation test for assaying antibody to influenza virus type-specific antigens. J. Clin. Microbiol. **18:**837–843.

Chapter 73

Parainfluenza and Respiratory Syncytial Viruses

KENNETH McINTOSH AND JULIA C. CLARK

CLINICAL BACKGROUND

The parainfluenza viruses (PIVs) and respiratory syncytial virus (RSV) constitute the most important respiratory viral pathogens for infants and children (5, 7, 8, 19). As with the other respiratory viruses, all members of the group cause a wide range of respiratory illness, but particular viruses tend to be associated with particular diseases. PIV type 1 (PIV 1) is the most important cause of viral croup and tends to be epidemic in the autumn months, usually in an every-other-year pattern. PIV 2 is the second most common cause of croup and has a similar seasonality. Both of these viruses tend to produce the most severe illness in children in years 2 to 4 of life. PIV 3 is an important cause of pneumonia and bronchiolitis. Severe disease caused by PIV 3 tends to occur during the first year of life; there is little seasonality and little tendency to occur in large outbreaks. RSV is the most important pathogen of the group, being the major cause of bronchiolitis and pneumonia in infants and children. As with PIV 3, the most severe disease occurs in the first year of life. RSV infections appear in large outbreaks every winter.

All of these viruses reinfect with great frequency, RSV and PIV 3 perhaps more often than the others. RSV has recently been shown to cause significant respiratory illness in normal and elderly adults (16). PIV 3 has been recovered frequently from adults with chronic respiratory disease, often in the absence of acute symptoms and sometimes for prolonged periods of time (14).

In addition to the severe, occasionally fatal, diseases mentioned above, all members of this group commonly cause less severe illness—the common cold and bronchitis. Infection is extremely widespread: most children have experienced infection with RSV and PIV 3 by the end of year 2 of life and with PIV 1 and PIV 2 by the end of year 5. Reinfections occur throughout life. In general, disease associated with reinfection tends to be less severe than with primary infection, but this is not always so.

DESCRIPTION OF THE AGENTS

Parainfluenza viruses

The PIVs were first isolated in the late 1950s and were rapidly associated with respiratory disease in children (5, 21). There are four serotypes, numbered 1 through 4, the last of which contains two closely related subtypes, A and B (3). The nucleic acid type is RNA, which is single stranded. The RNA-protein capsid has helical symmetry and by electron microscopy has a diameter of 18 nm and a length of about 1 μm. It is coiled inside a lipid-containing membrane. The membranous coat has short glycoprotein spikelike projections of two sorts. One has both hemagglutinating and neuraminidase activity. The second (the F protein) has cell-fusing activity which develops only after it has been cleaved by proteolytic enzymes (28). Other proteins of major importance in PIV structure are the M or membrane protein and the P or polymerase protein, an RNA-dependent RNA polymerase. The entire virus particle is highly pleomorphic, is 150 to 300 nm in diameter, and buds from the cell surface during maturation.

The human PIVs are antigenically distinct from influenza viruses and RSV but share common antigens among themselves and with mumps virus, Newcastle disease virus, Sendai virus, shipping fever virus, and simian virus type 5 (SV5). Despite these relationships, however, the human PIVs can be antigenically distinguished from one another with the use of commercially available or laboratory-manufactured animal antisera. On the other hand, cross-reacting antibodies often lead to confusion when paired sera are used for diagnosis.

On isolation from clinical specimens, PIVs have a limited host range. The most useful cell culture types are primary monkey or human kidney cells. Cynomolgus and rhesus monkey cells are of roughly equal sensitivity. The continuous cell line LLC-MK2, maintained in medium with 2 μg of trypsin per ml, has also been reported to be sensitive for primary isolation of parainfluenza viruses (12). Types 1 and 3 are easily adapted to growth in other cell culture types, including diploid fibroblast strains and multiple continuous lines. Moreover, a number of laboratory animal species can be infected by the respiratory route. PIVs have been adapted to growth in embryonated eggs.

PIVs are heat and acid labile and are destroyed by ether. They are also labile to freezing and thawing, but considerably less so than RSV (17).

Respiratory syncytial virus

RSV is morphologically similar to the PIVs with the exception that the diameter of its helical nucleocapsid is smaller, 13 to 14 nm rather than 18 nm. Biochemically, it is also distinct: there is no neuraminidase, and to date hemagglutinating activity has not been described. Its polypeptides have a distinct pattern in polyacrylamide gels (4). It has been set off in a genus separate from that of the PIVs, the "pneumoviruses," still part of the larger family of paramyxoviruses.

RSV is an antigenically heterogeneous species, but the differences between strains are small and probably of little or no practical importance (11). A closely related bovine virus (bovine RSV) causes respiratory disease in calves. The biophysically related virus, pneumonia virus of mice, is antigenically distinct.

RSV is very sensitive to freezing and thawing (17). This is of great importance in the handling of clinical specimens, since recovery of RSV from a specimen which has been frozen is uncommon.

The host range of RSV is narrow on primary isolation, although it is somewhat broader than that of PIVs. It grows best in continuous cell lines of human origin, particularly HEp-2 and some strains of HeLa cells ("Bristol" HeLa), but can also be recovered in human diploid cell strains (1) and primary monkey kidney (MK) cells. Laboratory strains have a somewhat broader host range. Several simian and small animal species can be infected by the respiratory route. The virus has not been adapted to growth in embryonated eggs.

COLLECTION AND STORAGE OF SPECIMENS

PIVs and RSV are recovered almost exclusively from the respiratory tract. The specimens containing the most abundant virus are secretions obtained early in the course of the illness. Secretions are ideally obtained as such, either by aspiration through a catheter (13) or by suction into a soft rubber bulb (15). Swabs, either nasopharyngeal or throat, may also be used. Swabs must be immediately mixed with a suitable volume of holding medium at 4°C. In all instances, specimens should be inoculated onto tissue cultures as soon as possible.

In infants and small children we obtain secretions as follows. A no. 8 French soft plastic feeding tube is attached through a valve-containing trap to an electric suction apparatus. The sterile catheter tip is introduced through the nares to the back of the nose, and suction is intermittently applied by means of the thumb valve while the catheter is slowly withdrawn. This process may in most infants be repeated once so that 0.2 to 0.8 ml of secretion is obtained in the trap. The trap is placed in wet ice at 4°C and transported immediately to the laboratory, where mixture with holding medium and cell culture inoculation are performed.

Hall and Douglas (15) have described the use of a rubber suction bulb for obtaining secretions from infants and young children. In this method 3 to 7 ml of phosphate-buffered saline (PBS) is aspirated into the bulb. The point of the bulb is then placed in the child's nose so as to completely occlude one side. The PBS is then squeezed into the nose and rapidly aspirated. The secretion thus obtained is expelled into holding medium.

Swabs are more convenient in adults or older children and are adequate but inferior to secretions. Throat and nasopharyngeal swabs may both be used and combined in a single vial of holding medium.

Holding medium consists of a buffered salt solution containing protein and antibiotics. Veal infusion broth, PBS, or Hanks balanced salt solution with 0.5% bovine albumin or gelatin may be used. We add 250 U of penicillin, 20 μg of gentamicin, and 5 μg of amphotericin B per ml. An amount of 3 ml per specimen is adequate.

Rapid transport to the laboratory is desirable. Although Hall and Douglas recommend bedside inoculation of specimens into cell culture (15), we feel this is unnecessary, and a delay of minutes to a few hours probably has little effect on the success of virus recovery as long as specimens are kept cold. In any case, materials should be held at 4°C in wet ice and *not* frozen unless the delay before cell culture inoculation is over 72 h. If freezing is necessary, it should be accomplished rapidly in a bath of dry ice and acetone or alcohol. Slow freezing (for example, placing the vial on the shelf of a −20°C freezer) virtually always destroys the infectivity of RSV in clinical specimens.

DIRECT EXAMINATION

Direct examination of secretions has great attraction for several reasons. It is rapid, and the information obtained is therefore more clinically useful. Problems of virus lability are rarely encountered since infectivity is no longer used to detect and identify viruses. Finally, with increasing availability of serological reagents, particularly for fluorescence tests, the methods are widely applicable.

A number of methods have been described including immunofluorescence (IF) staining, electron microscopy, enzyme-linked immunosorbent assay (ELISA), and immunoperoxidase. The last method, which uses specimens similar to those useful for IF, has not thus far been of proven value. The ELISA has recently been described for examination of secretions for RSV (10, 25), but is not yet widely available. Electron microscopy has been described for detection of PIVs (22) and has the advantages of being rapid and requiring no serological reagents. Disadvantages include relatively poor sensitivity (22), the inability to identify viruses as to PIV type, and the necessity for an expensive instrument.

IF is the method most widely and successfully used for direct examination of respiratory secretions for PIVs and RSV. The technique is described in detail in Gardner and McQuillin's book (13). Four elements are essential to success in this venture: (i) meticulous attention to preparation of specimens, (ii) high-quality antisera and conjugates, (iii) a good fluorescence microscope, and (iv) practice. Despite this list, however, an increasing number of laboratories are using the method and finding it helpful. In experienced hands, it is as sensitive as or slightly more sensitive than cell culture for identification of RSV and PIV infection.

The best specimens for IF tests are obtained by aspiration of secretions. About 0.2 to 0.4 ml of secretion is mixed with 2 ml of cold PBS by using a Pasteur pipette and gently suctioning and blowing out with a rubber bulb. This is then made up to about 10 ml with PBS from a squeeze bottle and centrifuged at 800 to 1,000 rpm for 10 min in a clinical centrifuge. The supernatant wash fluid is pipetted off with a Pasteur pipette, leaving the loosely packed cell button at the bottom. The same process is repeated until the cells are free from mucus (usually two or three times in all). Thorough washing of cells is essential to minimize nonspecific fluorescence.

The cells are then suspended in several drops of PBS (experience dictates the volume of PBS to be used at this stage), and about 20 μl is placed in each of six to nine spots on two or three Teflon-coated slides. We use slides with three 14-mm spots on each and find that they are clean enough for use when taken directly from the box. Alternatively, 10-mm squares can be etched on plain slides with a glass-marking pencil.

The drops of suspended cells are spread to cover each spot, allowed to dry, and then fixed for 10 min in acetone at 4°C. At this point they are either stained or stored in airtight boxes at 4°C for up to 1 week and −20°C for longer.

Indirect reagents of high quality for staining RSV, PIV 1, and PIV 3 in clinical specimens are available at this writing. In addition, monoclonal antibodies for RSV have been found to be excellent tools for rapid diagnosis and should be commercially available soon, either as indirect or direct (fluorescein-conjugated) reagents (2). Staining procedures are simple: 30 min at 37°C in a humidified atmosphere for the antiviral serum, followed by three 10-min washings in a bath of PBS at room temperature, and 30 min at 37°C for the conjugate, followed by another three 10-min washes in PBS. Finally, the slide should be rinsed for 2 min in water and dried.

Fluorescence microscopes of high quality are widely available, albeit expensive. Epi-illumination is preferred. An oil- or glycerol-immersion ×40 objective is essential for examination of specimens. Filter systems must be individualized for instruments from each manufacturer.

Experience is gained through the examination of specimens which have been also inoculated into tissue culture. This point must be emphasized: each worker must obtain confidence in his technique and reading ability. This can be done only by comparing readings with virus recovery. IF does not become a substitute for cell culture until such confidence has been achieved through extensive practice.

CULTIVATION OF AGENTS

Parainfluenza viruses

The most sensitive system for the cultivation of PIVs is primary human embryonic, or rhesus or cynomolgus monkey, kidney cell culture (6, 8). Because of the difficulties in obtaining suitable human cells, most workers now depend on MK cells. Such cultures are grown with 10% fetal calf serum and 0.2% antiserum to one or several simian viruses (SV5 and SV40). However, because of the sensitivity of influenza viruses to serum inhibitors, we maintain all our MK cultures in 1.0 ml of Eagle minimal essential medium or an equivalent medium (e.g., CMRL 1969, Connaught Laboratories, Willowdale, Ontario) with antibiotics and no serum. We inoculate 0.2 to 0.4 ml of specimen (in holding medium) into each of two MK tubes and incubate them either stationary or on a roller drum at 33 to 36°C. Specimens are also inoculated onto monolayers of human fetal diploid fibroblasts and of HEp-2 cells at the same time.

PIVs do not always produce a significant cytopathic effect (CPE), but with practice changes can often be seen. An increase in the number of small rounded cells in the monolayer is typical of PIV 1, syncytial cells similar to those produced by SV40 or foamy virus are seen with PIV 2, and a stretching of cells at the edge of the cell sheet ("stringy edges") is produced by PIV 3. PIV 4 may show more extensive cell degeneration. Final detection of viral growth is accomplished by hemadsorption with guinea pig erythrocytes (GP-RBC) (6).

The cell cultures are examined for hemadsorption every 5 to 7 days or when CPE is detected. For use in the hemadsorption test, GP-RBC are prepared as a 10% suspension in Alsever solution. These can be obtained commercially and stored in Alsever solution for 14 days. Once a week we wash them three times in normal saline and store them in saline as a 10% suspension. A suitable volume of this suspension is then diluted to a packed concentration of 0.4% immediately before each use. To each culture tube, 0.2 ml of the 0.4% suspension is added to the 1.0 ml of medium already present. The tubes are then placed at 4°C for 30 min in a horizontal position to permit contact of the GP-RBC with the cell monolayer. PIV 4 and mumps virus both require incubation at room temperature for optimal hemadsorption. The tubes are then read individually with a low-power (×6) objective. Inverted microscopes are not suitable for this procedure. Adsorption of the GP-RBC to the cells is sought. Strong hemadsorption is easily read. Rotating the tube gently often dislodges nonspecifically adsorbed cells and assists in the reading. Weak reactions may be difficult to interpret, and nonspecific hemadsorption is common. If there is doubt about the specificity, the cell monolayer can be washed free from erythrocytes; new maintenance medium is then added, and the tube is incubated for 3 or 4 days longer and retested. If the reaction is still questionable, passage to fresh MK tubes is advisable. Nonspecificity is easier to interpret if uninoculated control tubes are tested or if a large number of inoculated tubes from the same lot of MK cells are hemadsorbed at the same time. If all monolayers show the same pattern of hemadsorption, nonspecificity is likely. Bacterial contamination may also cause hemadsorption or hemagglutination. Medium and cells from all positive tubes should be passaged to fresh MK tubes and retested by hemadsorption.

If growth of PIV is strong, enough hemagglutinin will be released to cause agglutination of the GP-RBC in the culture medium. Hemadsorption may be inhibited if there is sufficient free virus present to coat erythrocytes.

Respiratory syncytial virus

Different strains of HEp-2 or HeLa cells, as well as different passages of the same strain, vary greatly in their sensitivity to RSV. Most diagnostic laboratories use certain precautionary measures to assure that sensitivity is optimal. HEp-2 cells may be removed from a frozen source at intervals. Alternatively, a known source of RSV may be periodically titrated in the cells in use. HEp-2 cells will, at certain passages, permit growth of virus but cease to demonstrate a syncytial CPE. At other passages sensitivity may be markedly reduced. It is important to be aware of such pitfalls. Variation in the sensitivity of diploid strains and primary cultures is not so evident.

HEp-2 cells should be seeded lightly (5×10^4 to 6×10^4 cells per tube) and inoculated at a time when islands of actively growing cells are still evident. Maintenance medium is Eagle minimal essential medium or its equivalent (e.g., CMRL 1969), supplemented with glutamine (which is necessary for syncytium formation [20]), 2% fetal calf serum, and antibiotics. Bovine sera may contain antibody to RSV and should be avoided except in the form of precolostral fetal calf serum. Cultures should be incubated at 33 to 36°C in a stationary rack.

Virus growth is detected by the characteristic syncytium formation in HEp-2 cells. Adjacent cells fuse into irregular, refractile "blobs" or into large sheets of cellular material with indistinct borders and multiple nuclei. Fibroblast strains show areas of destruction which follow the lines of cell growth and occasional syncytia. Primary MK cells show clusters of enlarged cells.

A change of medium at 5 to 7 days is often very helpful in accelerating the CPE of RSV. Cell cultures which are read as negative at 1 week may develop typical changes soon after the medium is changed.

IDENTIFICATION OF ISOLATES

Parainfluenza viruses

Because of serological cross-reactions between PIV types, identification is sometimes difficult. For most laboratories, the most reliable typing system is neutralization, with the use of type-specific, commercially obtained antisera. We have recently found that the IF test is easier, less expensive, and more rapid, and we now use this method almost exclusively. Isolates may also be identified by hemadsorption inhibition (6), by hemagglutination inhibition (HI), and by complement fixation (CF).

Neutralization is performed with approximately 100 50% tissue culture infective doses of virus, with the use of primary MK cell monolayers and positive hemadsorption of GP-RBC at 5 to 7 days as an indication of virus growth. Since frozen PIVs usually retain infectivity titer, the correct estimate of virus dose is usually accomplished without difficulty.

The major problem with identification by IF is availability of antisera. As mentioned above, reagents for PIV 1 and PIV 3 can be purchased, but none at present are available for PIV 2. In our laboratory we use a rabbit serum of our own making for this serotype. As soon as a specimen is considered to show definite hemadsorption, and before CPE or cell degeneration is extensive, we prepare it for IF staining as follows. The monolayer is washed three times with PBS to remove GP-RBC. The cells are scraped with the tip of a pipette into a few drops of fresh PBS and gently broken up by pipetting up and down. The cells are then transferred to at least three, and usually six, etched spots on one or two microscope slides and allowed to dry. Fixation is accomplished by immersing the slides in acetone at 4°C for 5 min. The spots are then stained as described above with antisera to the three major PIV types and, if appropriate, mumps and influenza A and B viruses.

Respiratory syncytial virus

The usual syncytial CPE in HEp-2 cells is sufficiently characteristic to permit presumptive identification of RSV. Many laboratories will end their identification efforts at this point. With the possible exceptions of PIV 3 and mumps virus, no other virus from the human respiratory tract produces a similar CPE in human heteroploid cells. Cell culture-adapted strains of measles virus (including vaccine strains) also do so but are unlikely to present problems. However, we also believe that passage of isolates and production of a characteristic CPE a second time is an important identifying feature. Moreover, specific identification is always desirable. We primarily use IF for identification (26). CF may also be used, but achieving an adequate titer of antigen is sometimes difficult. Neutralization is problematic because of the unpredictability of infectivity titers in samples frozen for storage.

The IF test is essentially the same as that used for PIVs. Antisera may be prepared specifically for the test or obtained commercially.

SEROLOGICAL METHODS

In general, it is more satisfactory to make a specific diagnosis of PIV or RSV infection by recovery of the virus (or identification by rapid methods) from a properly obtained and handled secretion specimen than by serological methods. This is true for several reasons. First, virus excretion rarely is found in asymptomatic children, and therefore recovery during illness is fairly convincing evidence of the etiological involvement of the virus in the illness. Second, cross-reactions between PIV types make the interpretation of serological data difficult: rises in antibody to multiple types are often seen, and assignment of the infection to a single type with confidence is thus not possible. Third, rises in serum antibody often fail to occur in very young children with RSV infections. And finally, the obtaining of 2- to 3-week convalescent sera in a clinical setting is frequently not possible, particularly in young infants. Therefore, serological methods are often of secondary importance, although in large studies they may give valuable information, and in individual instances in which cultures were not obtained they may be well worth performing. If, however, a virus has been recovered, the presence or absence of a homologous serum antibody titer rise adds little useful information, from a clinical standpoint.

As with infection by other viruses, a fourfold antibody rise is considered evidence of infection. Neither a single high titer nor a fourfold fall in titer can be interpreted with any confidence as evidence of recent infection.

Parainfluenza viruses

Three antibody tests are available: the HI antibody test, the neutralizing antibody test, and the CF antibody test. Because of its sensitivity and relative ease, the HI test is preferred. On the other hand, except for PIV 1 infections, the CF test is probably equivalent (9). The HI test will be described in detail.

Antigens for PIV types 1, 2, and 3 can be prepared by harvesting MK cell culture at the point of near-maximal hemadsorption. Pools of supernatant fluids are treated with Tween 80 and ether to increase the hemagglutinin titer (18). Dissolve 0.125 ml of Tween 80 in 2 ml of maintenance medium, and then add the dissolved Tween 80 to the virus suspension (0.1 ml for each 5 ml of virus). Mix by shaking at room temperature for 5 min, and then add ether to a final concentration of 33.3%. Mix well by shaking at 4°C for 15 min. Centrifuge this mixture at 2,000 rpm for 30 min. Remove the aqueous phase and place it in a freshly washed open container to allow the ether to evaporate. Treated antigen can be stored at 4°C. Untreated antigens are stored in sealed ampoules in small quantities at −70°C.

Serum must be treated before testing to remove nonspecific inhibitors and agglutinins. Incubate equal volumes (0.1 ml) of undiluted receptor-destroying enzyme (RDE) and serum overnight at 37°C, then heat inactivate the sera at 56°C for 30 min. Add 0.2 ml of 15% GP-RBC and leave at 4°C for 1 h. Centrifuge (2,000 rpm for 10 min), pour the supernatant into a labeled tube, and add 0.5 ml of PBS (0.01 M, pH 7.2). This is a 1:8 dilution.

One can use a microtechnique in microtiter plates or a macrotechnique in Kahn tubes. In the microtechnique, serial twofold dilutions of the virus in 0.05-ml amounts are prepared in a microtiter "V" plate to which are added equal volumes of 0.4% GP-RBC or human erythrocytes to determine the hemagglutinin titer. The endpoint is the highest dilution of antigen which produces partial agglutination of the erythrocytes. Partial agglutination is a small button of unagglutinated cells surrounded by agglutinated cells. The dilution of virus needed to provide 4 U of hemagglutinin in 0.025 ml, for use in the HI test, is determined by dividing the titer obtained in this titration by 8.

The actual performance of the HI microtiter technique is as follows.

1. Perform the antigen control in triplicate by adding 0.05 ml of PBS to three rows of five wells and the three erythrocyte control wells, using the microtiter "V" plates. Add 0.05 ml of the test antigen diluted to contain 8 hemagglutinating units per 0.05 ml to the first well in each row of five wells. Make serial dilutions using 0.05-ml microdiluters. The five wells then contain 4, 2, 1, 0.5, and 0.25 U of antigen. Add 0.05 ml of 0.4% erythrocyte suspension to the wells containing diluted antigen and the erythrocyte control wells. The 4- and 2-U wells should show complete agglutination, the 1-U well should show partial agglutination, and the 0.5- and 0.25-U wells should show no agglutination.

2. Include a standard reference serum of known titer in each test to verify specificity and sensitivity of the test.

3. Include an erythrocyte control, consisting of 0.05-ml volumes of PBS and 0.4% erythrocytes, in each test.

4. Include a serum control for each serum tested; add the 1:8 dilution of RDE-treated serum to an equal volume of PBS (0.025 ml of each).

5. Make the dilutions of the absorbed and RDE-treated sera by adding 0.025 ml of PBS to all dilution wells except those for the initial (1:8) serum dilution. Add 0.05 ml of the 1:8 dilution of treated serum to the first well and then make serial twofold dilutions, using 0.025-ml microdiluters. Add the antigen, diluted to contain 4 hemagglutinating units, to the serum dilutions in 0.025-ml amounts and mix. Incubate the virus-serum mixtures at room temperature for 1 h. Add GP-RBC (0.4%) to each well with an 0.05-ml dropper or by pipette in 0.05-ml amounts. Shake the resultant mixture and incubate it at 37°C (some prefer room temperature; both should be tried initially, and the temperature which gives the clearest endpoint should be selected). We use the serum controls and erythrocyte control cups to determine the appropriate time to read the test—usually 60 min is required, but occasionally a longer interval is necessary. Titers are read as the highest dilution of serum which completely inhibits hemagglutination (button of unagglutinated erythrocytes in the bottom of the cup).

Respiratory syncytial virus

A number of tests for serum antibody to RSV have been developed, including CF, tube neutralization, plaque reduction, and ELISA. At the present time the CF test is the most practical. A plaque reduction neutralization test (11), which includes fresh guinea pig serum in the reaction mixture as a source of complement, is a very sensitive test for serum antibody and offers some advantage in the detection of antibody rises during infection. Likewise, the ELISA offers precision and sensitivity not found in most other serological tests, and significant changes in titer may be found when they are not detectable by more traditional methods (23). Nevertheless, both of the last two methods involve technology and instrumentation which for most purposes are impractically complex.

The CF test is performed by standard methods (27). The antigen is available commercially or may be made from infected HEp-2 cells harvested by freezing when the CPE has been 4+ for several days. When paired sera from infants are being tested, there is some advantage to using the antigen at two to four times normal concentrations (8 to 16 U). Even with this modification, however, the CF test remains insensitive for infants under 6 months of age undergoing what is assumed to be primary infection (24).

EVALUATION AND REPORTING

Viral cultures

Often, the piece of information most valuable to the clinician is a report of whether or not a particular clinical specimen contains a virus, rather than specific information on what the precise viral species is. For this reason, a preliminary report (often by telephone) should be made as soon as a tentative recognition of CPE or hemadsorption is made. This is, of course, carefully labeled a tentative report. Often the label "RSV" or "hemadsorbing agent" can be attached at this point, with more definitive information to follow.

Since RSV and the PIVs are rarely present in normal healthy individuals, it is likely that recovery of the virus during disease means both an infection and the etiological association of the virus with the disease in question.

A negative report, on the other hand, does not prove the absence of infection. There are many possible reasons for a "false-negative" result, the most likely of which is faulty specimen collection and transport.

Serological results

The problems of interpretation of serological data in RSV and PIV infections are mentioned above. On the other hand, antibody rises may not be detected during bona fide infection (particularly with RSV in young infants, but also in reinfections with PIVs). Moreover, assignment of an infection to a particular PIV type on the basis of serological data alone is hazardous because of heterologous responses.

LITERATURE CITED

1. **Anderson, J. M., and M. O. Beem.** 1966. Use of human diploid cell cultures for primary isolation of respiratory syncytial virus. Proc. Soc. Exp. Biol. Med. **121:**205–209.
2. **Bell, D. M., E. E. Walsh, J. F. Hruska, K. C. Schnabel, and C. B. Hall.** 1983. Rapid detection of respiratory syncytial virus with a monoclonal antibody. J. Clin. Microbiol. **17:**1099–1101.
3. **Canchola, J., A. J. Vargosko, H. W. Kim, R. H. Parrott,**

E. Christmas, B. Jeffries, and R. M. Chanock. 1964. Antigenic variation among newly isolated strains of parainfluenza type 4 virus. Am. J. Hyg. **79**:357–364.

4. **Cash, P., W. H. Wunner, and C. R. Pringle.** 1977. A comparison of the polypeptides of human and bovine respiratory syncytial viruses and murine pneumonia virus. Virology **82**:369–379.
5. **Chanock, R. M., and L. Finberg.** 1957. Recovery from infants with respiratory illness of a virus related to chimpanzee coryza agent (CCA). II. Epidemiologic aspects of infection in infants and young children. Am. J. Hyg. **66**:291–300.
6. **Chanock, R. M., K. M. Johnson, M. Cook, D. C. Wong, and A. Vargosko.** 1961. The hemadsorption technique, with special reference to the problem of naturally occurring simian parainfluenza virus. Am. Rev. Respir. Dis. **83**:125–129.
7. **Chanock, R. M., H. W. Kim, A. J. Vargosko, A. Deleva, K. M. Johnson, C. Cumming, and R. H. Parrott.** 1961. Respiratory syncytial virus. I. Virus recovery and other observations during 1960 outbreak of bronchiolitis, pneumonia, and minor respiratory diseases in children. J. Am. Med. Assoc. **176**:647–653.
8. **Chanock, R. M., R. H. Parrott, K. M. Johnson, A. Z. Kapikian, and J. A. Bell.** 1963. Myxoviruses: parainfluenza. Am. Rev. Respir. Dis. **88**:152–166.
9. **Chanock, R. M., D. Wong, R. J. Huebner, and J. A. Bell.** 1960. Serologic response of individuals infected with parainfluenza viruses. Am. J. Public Health **50**:1858–1865.
10. **Chao, R. K., F. M. Fishaut, J. D. Schwartzman, and K. McIntosh.** 1979. Detection of respiratory syncytial virus in nasal secretions from infants by enzyme-linked immunosorbent assay. J. Infect. Dis. **139**:483–486.
11. **Coates, H. V., D. W. Alling, and R. M. Chanock.** 1966. An antigenic analysis of respiratory syncytial virus isolates by a plaque reduction neutralization test. Am. J. Epidemiol. **83**:299–313.
12. **Frank, A. L., R. B. Couch, C. A. Griffis, and B. D. Baxter.** 1979. Comparison of different tissue cultures for isolation and quantitation of influenza and parainfluenza viruses. J. Clin. Microbiol. **10**:32–36.
13. **Gardner, P. S., and J. McQuillin.** 1974. Rapid virus diagnosis: application of immunofluorescence. Butterworth & Co. Ltd., London.
14. **Gross, P. A., R. H. Green, and M. G. M. Curnen.** 1973. Persistent infection with parainfluenza type 3 virus in man. Am. Rev. Respir. Dis. **108**:894–898.
15. **Hall, C. B., and R. G. Douglas, Jr.** 1975. Clinically useful method for the isolation of respiratory syncytial virus. J. Infect. Dis. **131**:1–5.
16. **Hall, W. J., C. B. Hall, and D. M. Spears.** 1978. Respiratory syncytial virus infection in adults. Ann. Intern. Med. **88**:203–205.
17. **Hambling, M. H.** 1964. Survival of the respiratory syncytial virus during storage under various conditions. Br. J. Exp. Pathol. **45**:647–655.
18. **John, T. J., and V. A. Fulginiti.** 1966. Parainfluenza 2 virus: increase in hemagglutinin titer on treatment with tween-80 and ether. Proc. Soc. Exp. Biol. Med. **121**:109–111.
19. **Kim, H. W., J. O. Arrobio, C. D. Brandt, B. C. Jeffries, G. Pyles, J. L. Reid, R. M. Chanock, and R. H. Parrott.** 1973. Epidemiology of respiratory syncytial virus infection in Washington, D.C. Am. J. Epidemiol. **98**:216–225.
20. **Marquez, A., and G. D. Hsiung.** 1967. Influence of glutamine on multiplication and cytopathic effect of respiratory syncytial virus. Proc. Soc. Exp. Biol. Med. **124**:95–99.
21. **McLean, D. M., R. D. Bach, R. P. B. Larke, and G. A. McNaughton.** 1963. Myxoviruses associated with acute laryngotracheobronchitis in Toronto 1962–63. Can. Med. Assoc. J. **89**:1257–1259.
22. **Pavilanis, V., J. H. Joncas, R. Skvorc, L. Berthiaume, M. F. Blanc, B. Martineau, and M. O. Podoski.** 1971. Microscopie électronique et diagnostic de laboratoire des maladies à virus. Union Med. Can. **100**:2195–2202.
23. **Richardson, L. S., R. H. Yolken, R. B. Belshe, E. Camargo, H. W. Kim, and R. M. Chanock.** 1978. Enzyme-linked immunosorbent assay for measurement of serological response to respiratory syncytial virus infection. Infect. Immun. **20**:660–664.
24. **Ross, C. A. C., E. J. Stott, S. McMichael, and I. A. Crowther.** 1964. Problems of laboratory diagnosis of respiratory syncytial virus infection in childhood. Arch. Gesamte Virusforsch. **14**:553–562.
25. **Sarkkinen, H., P. E. Halonen, P. P. Arstila, and A. A. Salmi.** 1981. Detection of respiratory syncytial, parainfluenza type 2, and adenovirus antigens by radioimmunoassay and enzyme immunoassay on nasopharyngeal specimens from children with acute respiratory disease. J. Clin. Microbiol. **13**:258–265.
26. **Schieble, J. H., E. H. Lennette, and A. Kase.** 1965. An immunofluorescent staining method for rapid identification of respiratory syncytial virus. Proc. Soc. Exp. Biol. Med. **120**:203–208.
27. **Sever, J. L.** 1962. Application of a microtechnique to viral serological investigations. J. Immunol. **88**:320–329.
28. **Silver, S. M., A. Scheid, and P. W. Choppin.** 1978. Loss on serial passage of rhesus monkey kidney cells of proteolytic activity required for Sendai virus activation. Infect. Immun. **20**:235–241.

Chapter 74

Measles Virus

ERLING NORRBY

Measles is a community-related, acute, febrile disease of variable severity (2, 8). The readily distinguishable features of its clinical symptoms, e.g., the characteristic generalized rash, make diagnosis of most cases possible without laboratory assistance. However, in thoroughly vaccinated populations the frequency of occurrence of the disease will be very low, and even pediatricians may become unfamiliar with the disease. Further, persons with low-grade immunity may contract a mitigated form of the disease. In this situation, the laboratory may be called on to assist in giving a diagnosis.

The acute disease may be complicated by bacterial infections in the respiratory tract or middle ear and by encephalitis, usually of a postinfectious type. In patients with an inherited or acquired defective cell-mediated immunity, a progressive form of measles without a rash can develop. This form of the disease involves many organs, particularly lung tissue (leading to the appearance of giant-cell pneumonia), kidneys, and frequently brain tissue (resulting in encephalitis).

An uncommon but severe late complication of measles is subacute sclerosing panencephalitis (SSPE). This disease is caused by a disseminating infection in the brain with a defective, cell-associated measles virus variant, which most likely has remained in this organ since the time of the primary infection.

Certain diseases, such as systemic lupus erythematosus, active chronic hepatitis, and multiple sclerosis, are associated with an increase in measles antibody titers. The significance of this phenomenon remains to be defined (11).

DESCRIPTION OF THE AGENT

Measles virus is a member of the family Paramyxoviridae, genus *Morbillivirus* (4, 12). The center of the virion is composed of a helical nucleocapsid which contains a single piece of RNA enclosed in one species of protein and associated with an RNA polymerase. The coiled nucleocapsid is surrounded by an envelope, and the overall diameter of the particle is 120 to 200 nm. The envelope is constructed from three viral components: the matrix protein, the hemagglutinin protein, and the fusion (hemolysin) protein. The measles virus envelope has the capacity to agglutinate and lyse erythrocytes from Old World monkeys and to fuse nucleated cells. No neuraminidase activity has been detected, and the hemagglutinin does not elute at 37°C.

Despite the high contagiousness of measles, the virus is not readily isolated from patients with the disease. However, several strains adapted to grow effectively in cell cultures are available. The rate of growth of virus varies, depending on the degree of adaptation of virus, the occurrence of autointerference phenomena, and the cell system used. In particular, under conditions of slow growth, the infection tends to cause an accumulation of smooth nucleocapsids in the cell nucleus. Persistent infections are readily established.

The characterization of measles virus strains by polyclonal hyperimmune or convalescent sera does not reveal any antigenic heterogeneity. However, variations in the occurrence of antigenic epitopes on both surface and internal components has been demonstrated by the use of monoclonal antibodies (19). Measles virus does not show immunological cross-reactions with other human viruses, but there is a close relationship with canine distemper and rinderpest viruses.

Virion infectivity is a labile property (half-life of 2 h at 37°C in protein-containing medium), but the virus can readily be stored at −20°C or preferably −60°C or lower. The relative sensitivity of different biological activities to inactivating agents is infectivity > hemolytic activity > hemagglutinating activity > antigen activities.

COLLECTION AND STORAGE OF SPECIMENS

Samples for virus isolation or for the detection of viral antigens include blood (leukocytes), nasopharyngeal and conjunctival secretions, urine, and in special circumstances, skin and brain biopsies. Blood should be collected during the prodromal stage until 1 to 2 days after the rash appears. Urine (preferably sediment) may be infectious up to 7 days after onset of the rash. Material to be used for virus isolation should not be frozen, but should be held at 4°C. Except for blood and blood fractions, samples should be mixed with balanced salt solution or medium containing 2 to 10% serum. Transport time should be minimal, especially in the case of brain biopsies for the establishment of explant cultures. Specimens for rapid diagnosis by immunofluorescence require special consideration. Preferably, smears should be fixed as early as possible in cold acetone (10 min at room temperature), transported at 4°C, and stored in the laboratory at −20°C.

Paired serum samples can be used for serological analysis, and the initial specimen should be obtained as early as possible. A rise of antibody titers starts at the appearance of the rash. Single specimens collected within 10 to 20 days after the debut of the rash may be useful for the identification of virus-specific immunoglobulin M (IgM). Matched serum and cerebrospinal fluid samples can be utilized for the demonstration of a local synthesis of virus antibodies within the central nervous system. Serological samples should be stored at −20°C, but antibody activities are stable for extended periods at 4°C.

DIRECT EXAMINATION

In acute measles, cytological examination can be made of cells in nasopharyngeal secretions and urine. Occurrence of giant cells and characteristic intranuclear inclusions may aid in the diagnosis. However, the fluorescent-antibody (FA) technique allows a much more specific identification of the presence of virus. This technique can be applied to cells in the above-mentioned kinds of specimens and also to skin biopsies. Either the direct or indirect FA technique can be used. The indirect technique has the advantage that the same conjugated anti-immunoglobulin can be used in many different systems. Paired human sera may be used, but preferably, appropriate animal hyperimmune sera or especially murine monoclonal antibodies are used. The most distinct fluorescence is obtained with monoclonal antibodies against antigen epitopes on the nucleocapsid or polymerase protein (13). Both of these structural components have the advantage of occurring in abundance in both acutely and persistently infected cells (16). Also, their epitope characteristics do not vary with virus strains (19). In special cases, it may be of interest to identify separately other structural components by the performance of FA analysis with selected monoclonal antibodies.

Direct examination is of particular importance in SSPE since the virus cannot be isolated directly (20). Intranuclear Cowdry type A inclusions can be seen in various cells, especially oligodendroglia cells. The FA technique may directly identify measles antigen. However, since antibody from the serum of the patient is complexed to measles virus antigen in the diseased brain, hyperimmune or monoclonal measles IgG from an animal species should be used to avoid misleading results.

In the future, the enzyme-linked immunosorbent assay (ELISA; see below) will likely be used for the specific identification of viral antigen in clinical specimens.

ISOLATION OF VIRUS

Because of the time required and the technical complexity of the procedures, attempts to isolate the virus should be made only under special circumstances. Primary human or simian cells should be used for the isolation of virus from patients with measles. Primary human kidney cells have been most extensively exploited, but primary human amnion cells have also been found useful. Primary simian cells, e.g., rhesus, cynomolgus, or *Cercopithecus* monkey kidney cells, have the disadvantage of possible contamination with indigenous viruses. Human diploid cells and monkey cell lines such as Vero or BS-C-1 may offer a satisfactory alternative, but data on comparative susceptibility are not available.

Only rarely has virus been isolated from brain or cerebrospinal fluid material of patients with measles encephalitis. Explant cultures of brain tissue from patients with SSPE may carry a persistent, nonproductive infection demonstrable by the FA technique or by hemadsorption. The production of extracellular, infectious virus can be induced in a certain proportion of cases by cocultivation with susceptible cells, e.g., Vero or BS-C-1 cells (20).

IDENTIFICATION OF VIRUS

Syncytial lesions are the most common manifestation of measles virus during primary isolation. Nuclei frequently form a ring around a granular mass of cell organelles in the center of a giant cell. The occurrence of intranuclear inclusions is characteristic of measles virus and may assist in distinguishing this virus from other paramyxoviruses. Hemadsorption properties can also be used for this purpose since measles-infected cells interact only with simian erythrocytes, whereas other paramyxoviruses may adsorb nonsimian cells also. However, measles virus isolates may fail to demonstrate hemadsorption at early passage levels. Upon passage, some measles virus strains induce a different cytopathology, which includes the occurrence of long, narrow cells with a branched cytoplasm (spindle cells). This change often is connected with an increased production of hemagglutinin.

Measles virus strains may be identified by FA staining (7) or by neutralization (see below), preferably by use of animal hyperimmune sera or murine ascites containing monoclonal antibodies. In the future, it may be possible to distinguish different measles virus strains by their capacity to react with a panel of monoclonal antibodies. Low-passage virus rarely produces sufficient hemagglutinin or complement-fixing antigen to permit the use of these procedures for identification. However, antigen identification by ELISA may be performed in the future.

SEROLOGICAL DIAGNOSIS

During recent years there has been a gradual exchange of traditional antibody tests such as complement fixation (CF) and hemagglutination inhibition (HI), which has been used extensively in measles serology, for the new generation of tests, i.e., radioimmunoassay and particularly ELISA. The latter tests provide a higher sensitivity and give better opportunities for the separate identification of IgG and IgM antibodies. ELISA is to be preferred before radioimmunoassay, since it avoids the need to use radioisotopes.

ELISA

ELISA is generally performed in microtiter plates in which antigen is passively adsorbed to the polystyrene material.

Antigens. The yield of viral antigen varies considerably in different virus systems. Various primate cell culture lines such as Vero or HeLa provide useful systems for the propagation of virus. A suitable strain of virus should be selected. Development of a strand-forming rather than an extensive syncytium-forming cytopathic effect usually is a good indicator of efficient antigen production. Cultures are harvested when the cell monolayer displays an advanced cytopathic effect. Alternatively, the cells may be scraped off into the medium or harvested separately into a 10% (vol/vol) suspension in an appropriate buffer. Disruption of cells can be achieved by freezing and thawing, treatment with ultrasonics, or both. In some cases, detergent-containing buffers may be used, provided they do not interfere with the test. The material should be clarified at 2,000 to 3,000 rpm before use to remove particulate material. For the purpose of standardization, the protein content should be determined by some method, such as the Lowry test.

Test procedure. Other reagents required for the test are commercially available. The general procedure for the ELISA is described in Chapter 98. In short, the following steps should be followed. Antigen in an appropriate dilution in 0.05 M bicarbonate buffer (pH 9.5) is added in a volume of 0.1 to 0.3 ml to wells in polystyrene plates. After incubation at room temperature for 2 h or at 4°C overnight, the plates are washed three times in phosphate-buffered saline (PBS) containing 0.05% Tween 20. To saturate protein adsorption sites, 5% bovine serum albumin may be added. Serial dilutions of sera are added to individual wells, and the plates are incubated for 2 h at room temperature (or 1 h at 37°C). The plates are then washed again and incubated with enzyme-labeled anti-immunoglobulin. After an additional round of incubation and washing, substrate is added. The test allows a separate identification of IgG and IgM by the use of anti-immunoglobulins specific for each immunoglobulin class. However, if the proportion of IgM to IgG is low, competition for antigen sites will reduce the possibility of detecting IgM. The so-called μ-capture technique (see below) therefore is to be preferred for IgM detection.

HI test

The HI test has specificity and sensitivity comparable to those of the neutralization test and higher sensitivity than the CF test. However, it is at least 20 times less sensitive than the ELISA (3, 5). Since the HI test is simple to perform and rapid, it is a useful test for serological diagnosis of measles. One limitation of the test is the fact that it requires availability of monkey erythrocytes.

Treatment of sera. Nonspecific agglutinins are removed by absorption of sera with an equal volume of 10% green monkey erythrocytes for 30 min at room temperature. Erythrocytes are sedimented at 1,000 × *g* and discarded. This absorption is mandatory when sera are tested at dilutions of 1:10 or lower. Nonspecific inhibitors cause relatively little difficulty in measles HI tests as compared with HI tests with other paramyxoviruses. When sera are being tested at low dilutions, the following treatment may be used. Mix 1 volume of serum with 9 volumes of 14% (wt/vol) kaolin in isotonic PBS (pH 7.4), and let it stand for 20 min at room temperature. Remove the kaolin by sedimentation at 2,000 rpm.

If serum is to be used for IgM antibody analysis, methods other than kaolin treatment should be used for the removal of nonspecific inhibitors. The immunoglobulins may be precipitated by mixing the serum with 0.5 volume of saturated $(NH_4)_2SO_4$ and letting the mixture stand at 4°C for 1 h. The mixture is then centrifuged at 2,000 rpm for 2 min at 4°C, and the pellet is suspended in PBS.

Antigens. Viral antigen is prepared as described above for the ELISA. The hemagglutinin titer may be increased 4- to 10-fold by treatment with Tween 80 and ether (10). To do this, add 0.1 volume of a 1.25% solution of Tween 80 in PBS. Mix carefully, and incubate for 3 to 5 min at room temperature. Then add an equal volume of anesthetic ether, and mix carefully for 10 min in an ice bath. Remove the ether phase by centrifugation at 3,000 rpm for 10 min, and remove ether dissolved in the aqueous phase by bubbling nitrogen gas through the preparation. The small-size hemagglutinin obtained by this treatment is more readily inhibited by antibodies than is whole virus. Therefore, higher serum HI values are recorded.

Cells. Erythrocytes from Old World monkeys must be used. Variations in sensitivity within the Cercopithecoidea (catarrhines) are small, but cells of *Cercopithecus aethiops* (green monkey) are preferred. Erythrocytes are collected in Alsever or acid-citrate-glucose solution, stored at 4°C, and washed three to four times by centrifugation in PBS before use.

Test procedure. Prepare serial twofold dilutions of sera in a volume of 0.025 ml in microtiter wells. Add an equal volume of antigen diluted in PBS to contain 4 hemagglutinating units, and incubate for 1 h at 37°C. Then add 0.05 ml of a 1% erythrocyte suspension, shake, and allow the cells to settle at 37°C. The erythrocyte suspension should contain 0.1% normal serum to allow rapid sedimentation of erythrocytes and the formation of distinct bottom patterns. The highest dilution of serum which gives complete inhibition of hemagglutination is taken as the endpoint.

CF test

The CF test is less sensitive than the HI or neutralization test, but few immune persons have serum titers less than 1:4, and this test may be more convenient than the other in laboratories in which the technique is routine. Use the standard procedure described previously (17). Antigen is commercially available or may consist of selected, untreated or heat-inactivated, measles-infected cell culture (HEp-2, Vero, etc.) materials.

Hemolysis inhibition test

The hemolytic activity of measles virus is carried by a population of peplomers separate from the hemagglutinin peplomers. However, for the hemolytic activity to become expressed, virus particles have to be anchored to erythrocytes via the hemagglutinin. Hemolysis can therefore be blocked in two different ways: either directly via antibodies reacting with the hemolysin or indirectly via HI antibodies which prevent the attachment of particles carrying hemolysin to erythrocytes (15). The absolute titer of antibodies reacting with hemolysin in the presence of an excess amount of HI antibodies can be determined only after removal of the latter antibodies. This removal can be achieved by absorption with an antigen treated with Tween 80 and ether (14).

Antigen. Different virus-cell systems give variable yields of hemolytically active material. After the selection of a suitable system, the following kinds of preparations can be used: 10% cell pack material or extracellular material concentrated 10- to 20-fold by differential centrifugation (15,000 × *g* for 60 min) or by forced dialysis against polyethylene glycol (Carbowax 6000). Virus material should be frozen and thawed five times or more. Hemolytic activity is determined by preparing serial twofold dilutions in 0.4 ml of PBS, adding 0.1 ml of a recently washed 10% suspension of green monkey erythrocytes, and incubating the mixture for 3 h at 37°C in a water bath. After the cells are removed by centrifugation at 1,500 rpm for 10 min, optical extinction values are determined on supernatant fluids at 540 nm.

Test procedure. Prepare serial twofold dilutions of serum in 0.2 ml of PBS. Mix with an equal volume of antigen diluted to give an extinction value at 540 nm of 0.4 to 0.5. After incubation for 1 h at room temperature, add 0.1 ml of a 10% suspension of green monkey erythrocytes. Incubate for 3 h at 37°C in a water bath, and then remove the erythrocytes by centrifugation at 400 × *g* for 10 min. Read the optical extinction of supernatant fluids at 540 nm. The highest dilution of serum which gives at least 50% reduction of hemolysis is considered to represent the antibody endpoint.

Other tests

Various additional tests are available for the determination of measles antibodies, but usually the information provided by the tests described above is sufficient.

Neutralization tests may be performed as endpoint assays based either on inhibition of cytopathology in cell culture tubes or on plaque reduction. In the selection of an antigen to be used in the tests, it is advantageous to take a 2-h harvest from a culture with advanced cytopathology. Such a preparation contains a relatively high proportion of infectious over noninfectious antigen, which increases the sensitivity of the test. Plaque reduction is more sensitive than a method based on cytopathology. The sensitivity of neutralization tests may be increased by the addition of anti-immunoglobulin to the reagent mixture.

Another test for antibody determination is the indirect FA technique. By the use of conjugated anti-immunoglobulins with different class specificities, IgG and IgM may be separately determined.

Tests for IgM

Analysis of measles virus-specific IgM antibodies provides an opportunity for the identification of recent and, occasionally, persistent virus infections. The duration of the IgM antibody response after an acute infection is 1 to 3 months, depending on the sensitivity of the assay employed. As already mentioned, it is possible in ELISA and immunofluorescence tests to select reagents which separately identify IgG and IgM. Since, however, the antibodies compete for the same antigenic sites, optimal detection of IgM requires separation of the two classes of antibodies. This separation can be achieved by rate zonal centrifugation in sucrose gradients, ion-exchange chromatography (9), and the ELISA μ-capture technique. To confirm the class nature of the isolated IgM, the sensitivity of the immunoglobulin to reduction by 2-mercaptoethanol in a final concentration of 0.1 M for 30 min at 37°C is determined. Titrate treated and untreated serum fractions in parallel by HI or ELISA. IgM antibody is inactivated by 2-mercaptoethanol, whereas IgG antibody is not. The μ-capture technique includes as a first step appropriately diluted polyclonal (18) or monoclonal (1) anti-IgM for the coating of polystyrene plates. After a careful washing and supplementary coating with albumin as described above, serially diluted whole serum is added. Subsequently, the plates are washed, incubated with viral antigen, and washed again, and finally a conjugated nonhuman antiserum against measles virus is added. Alternatively, the latter antiserum may not be conjugated, and the addition of a conjugated anti-immunoglobulin specific for the nonhuman hyperimmune serum can be used.

In cases in which the human sera contain rheumatoid factor in addition to measles-specific IgM and IgG antibodies, false-positive reactions may occur. This possibility is inherent in all the tests discussed. There are two ways of circumventing this problem. One possibility is to remove anti-IgG from test sera by absorption with aggregated or latex particle-attached IgG. The other possibility may be to use $F(ab_2)$ fragments for enzyme labeling.

EVALUATION

The acute form of measles shows such distinct clinical features that usually there is no demand for laboratory diagnosis. The identification of virus or viral antigen in body sites other than the central nervous system is indicative of an acute infection. The pattern of antibody development is highly characteristic. HI, CF, and neutralizing antibodies start to appear with development of the rash and reach peak titers about 10 days later. Thus, the first specimen in a pair should be collected no later than a few days after the rash appears. IgM antibodies appear at the same time as IgG antibodies but decline to undetectable levels within 30 to 90 days. Identification of IgM may therefore assist in defining a primary infection. After reaching peak titers, IgG antibodies decline gradually over 6 months and remain relatively stable thereafter. Persons who live where measles is prevalent display somewhat higher antibody titers than do persons who live in a virgin milieu, presumably due to the occurrence of subclinical reinfections.

An atypical form of measles has been encountered in individuals who have received their primary immunization by inactivated measles vaccine. This vaccine immunized against the measles hemagglutinin, but not against the hemolysin (14). Because of this partial immunity, replication of virus may cause immunopathological complications. The selective antibody response, antihemagglutinin but no antihemolysin, can be identified serologically after vaccination and frequently also after boosting caused by virus replication.

Measles infection in the central nervous system may be connected with distinct serological responses. Postinfectious encephalitis by definition does not allow the isolation of virus or the identification of an enhanced measles antibody response. However, certain forms of acute encephalitis occur in immune-suppressed individuals, from whom virus can be recovered. In some cases, these patients display a local production of measles antibodies in the central nervous system, identifiable by comparative serological analysis of serum and cerebrospinal fluid samples, but in other cases the immune response is poor, due to extensive effects of the immune suppressive state.

In SSPE, virus cytopathology and antigen can be identified in brain biopsies, and in a certain proportion of cases virus can be recovered after cocultivation of explanted brain tissue with cells susceptible to measles virus. Serological analysis reveals increased antibody titers in both serum and cerebrospinal fluid. Hemolysis-inhibiting and CF antibody titers frequently are more elevated than are HI antibody titers (14). A pronounced local production of virus-specific antibodies inside the central nervous system can be identified. In some SSPE patients, IgM antibodies can be demonstrated. However, this is not found consistently enough to be useful as a marker for the persisting virus infection in these patients (21).

Measles antibody titers in body fluids may also increase in certain conditions, which as yet have not been correlated with persistent measles virus infections. Increasing titers of serum antibodies against measles virus and concomitantly against certain other viruses, e.g., rubella, have been associated with active chronic hepatitis, systemic lupus erythematosus, and infectious mononucleosis (6). Finally, increased measles antibody titers have been found in serum and particularly in cerebrospinal fluid samples from patients with multiple sclerosis (11). In 60% of patients with this disease, local production of measles virus antibodies in the central nervous system has been demonstrated. However, titers of antibodies against other enveloped viruses also have been found to be increased in a considerable proportion of patients with multiple sclerosis. The significance of these serological changes remains to be identified.

LITERATURE CITED

1. **Forghani, B., C. K. Myoraku, and N. J. Schmidt.** 1983. Use of monoclonal antibodies to human immunoglobulin M in "capture" assays for measles and rubella immunoglobulin M. J. Clin. Microbiol. **18:**652–657.
2. **Fraser, K. B., and S. J. Martin.** 1978. Measles virus and its biology. Academic Press, Inc., New York.
3. **Kahane, S., V. Goldstein, and I. Sarov.** 1979. Detection of IgG antibodies specific for measles virus by enzyme-linked immunosorbent assay (ELISA). Intervirology **12:**39–46.
4. **Kingsbury, D. W., M. A. Bratt, P. W. Choppin, E. P. Hansen, Y. Hosaka, V. ter Meulen, E. Norrby, W. Plowright, R. Rott, and W. H. Wunner.** 1978. Paramyxoviridae. Intervirology **10:**137–152.
5. **Kleiman, M. B., C. K. L. Blackburn, S. E. Zimmerman, and M. L. V. French.** 1981. Comparison of enzyme-linked immunosorbent assay for acute measles with hemagglutination inhibition, complement fixation, and fluorescent-antibody methods. J. Clin. Microbiol. **14:**147–152.
6. **Laitinen, O., and A. Vaheri.** 1974. Very high measles and rubella virus antibody titres associated with hepatitis, systemic lupus erythematosus and infectious mononucleosis. Lancet **i:**194–198.
7. **McQuillin, J., T. M. Bell, P. S. Gardner, and M. A. P. S. Dounham.** 1976. Application of immunofluorescence to a study of measles. Arch. Dis. Child. **51:**411–419.
8. **Morgan, E. M., and F. Rapp.** 1977. Measles virus and its associated diseases. Bacteriol. Rev. **41:**636–666.
9. **Nagy, G., S. Kósa, S. Takátsy, and M. Koller.** 1984. The use of IgM tests for analysis of the causes of measles vaccine failures. J. Med. Virol. **13:**93–103.
10. **Norrby, E.** 1962. Hemagglutination by measles virus. 4. A simple procedure for production of high potency antigen for hemagglutination-inhibition (HI) tests. Proc. Soc. Exp. Biol. Med. **111:**814–818.
11. **Norrby, E.** 1978. Viral antibodies in multiple sclerosis. Prog. Med. Virol. **24:**1–39.
12. **Norrby, E.** 1978. Myxoviridae: pseudomyxovirus-measles virus, p. 387–399. *In* G.-D. Hsiung and R. H. Green (ed.), CRC handbook series in clinical laboratory science. Section H: virology and rickettsiology, vol. 1, part 1. CRC Press, Inc., West Palm Beach, Fla.
13. **Norrby, E., C.-N. Chen, T. Togashi, H. Sheshberadaran, and R. P. Johnson.** 1982. Five measles virus antigens demonstrated by use of mouse hybridoma antibodies in productively infected tissue culture cells. Arch. Virol. **71:**1–11.
14. **Norrby, E., G. Enders-Ruckle, and V. ter Meulen.** 1975. Differences in the appearance of antibodies to structural components of measles virus after immunization with inactivated and live virus. J. Infect. Dis. **132:**262–269.
15. **Norrby, E., and Y. Gollmar.** 1975. Identification of measles virus-specific hemolysis-inhibiting antibodies separate from hemagglutination-inhibiting antibodies. Infect. Immun. **11:**231–239.
16. **Norrby, E., T. A. Haase, K. P. Johnson, and C. Örvell.** 1982. Persistent infections with paramyxoviruses, p. 217–236. *In* L. M. de la Maza and E. M. Peterson (ed.), Medical virology. Elsevier Biomedical, New York.
17. **Palmer, D. F., L. Kaufman, W. Kaplan, and J. J. Cavallaro.** 1977. Serodiagnosis of mycotic diseases. Charles C Thomas, Publisher, Springfield, Ill.
18. **Pedersen, I. R., A. Antonsdottir, T. Evald, and C. H. Mordhorst.** 1982. Detection of measles IgM antibodies by enzyme linked immunosorbent assay (ELISA). Acta Pathol. Microbiol. Scand. Sect. C **90:**153–160.
19. **Sheshberadaran, H., S.-H. Chen, and E. Norrby.** 1983. Monoclonal antibodies against five structural components of measles virus. I. Characterization of antigenic determinants on nine strains of measles virus. Virology **128:**341–353.
20. **Wechsler, S. L., and H. C. Meissner.** 1982. Measles and SSPE viruses: similarities and differences. Prog. Med. Virol. **28:**65–95.
21. **Ziola, B., A. Salmi, M. Panelius, P. Halonen, and B. Friis.** 1979. Measles virus-specific IgM antibodies and IgM class rheumatoid factor in serum and cerebrospinal fluid of subacute sclerosing panencephalitis patients. Clin. Immunol. Immunopathol. **13:**462–474.

Chapter 75

Mumps Virus

ERLING NORRBY

CLINICAL BACKGROUND

Mumps is an acute, generally self-limiting, contagious disease with moderate fever of short duration. Bilateral or unilateral parotitis is the most common clinical feature. In patients with this symptom, a laboratory confirmation of the diagnosis usually is not required. Secondary involvement concerns the testes, ovaries, central nervous system, and, more rarely, pancreas, peripheral nerves, eye, inner ear, and other organs. The incubation period in most cases ranges between 18 and 21 days. Infections are spread generally by droplets via the upper respiratory route. Virus is excreted in saliva and urine from 6 days before to 9 or more days after the appearance of glandular enlargement. Between 25 and 50% of all infections are silent. Immunity after infection appears to be lifelong. Silent reinfections may occur. Attenuated live virus vaccine is available. The Formalin-inactivated vaccine does not induce a complete antibody response (8).

DESCRIPTION OF AGENT

Mumps virus is a member of the family Paramyxoviridae, genus *Paramyxovirus* (2, 4). It contains one piece of RNA associated with a capsid protein and an RNA-dependent RNA polymerase to form a helical structure. The coiled nucleocapsid is surrounded by an envelope, giving the virion a diameter of 120 to 200 nm. The envelope is composed of a matrix protein and two surface glycoproteins, the hemagglutinin-neuraminidase and the fusion factor (hemolysin) (10). The two biological activities of the former glycoprotein are carried by topographically distinct parts of the polypeptide, as demonstrated by the use of monoclonal antibodies (11, 16).

Infectivity is destroyed by ether, heating at 56°C for 20 min, UV irradiation, and treatment with 0.1% Formalin. The virus envelope can agglutinate erythrocytes of fowls, humans, and other species, and it can cause cell fusion or hemolysis. These activities are carried out by two immunologically distinct peplomers (9). For a long time, viral (V) and soluble (S) complement-fixing (CF) antigens (3) have been recognized. Only one distinct antigenic type is known, but some antigenic cross-reactivity exists with other paramyxoviruses, particularly parainfluenza type 1 (Sendai) and Newcastle disease viruses. By the use of monoclonal antibodies, differences in the occurrence of antigen epitopes on the hemagglutinin-neuraminidase glycoprotein of mumps virus strains have been demonstrated (11, 16). Mumps virus can be isolated and grown in various cell cultures and in embryonated hen eggs, but hemagglutinin and CF antigens are most readily produced by virus adapted to grow in the latter system.

COLLECTION AND STORAGE OF SPECIMENS

The specimens to be examined include swabs obtained from the area around Stensen's duct, saliva, and urine. Cerebrospinal fluid (CSF) is of special interest in cases with central nervous system symptoms. Specimens should be collected at the onset of disease or within 5 days of illness by the use of procedures described in Chapter 61. Note that mumps virus is relatively thermolabile. Bovine serum should not be added to the buffered salt solution, since it may contain inhibitors. The frequency of isolation of virus from urine and CSF can be increased by the concentration of virus by sedimentation. The specimen should be centrifuged at 20,000 rpm for 60 min, and the pellet is suspended in one-tenth the original volume in Hanks balanced salt solution (BSS) containing 2% serum (19).

Paired serum samples should be used for serological analysis, and the initial specimen should be obtained as early as possible after the debut of symptoms. Single specimens collected within 10 to 15 days after the debut are useful for the identification of virus-specific immunoglobulin M (IgM) (17). Matched serum and CSF samples can be utilized for the demonstration of a possible local synthesis of viral antibodies within the central nervous system (18, 21). Serological samples should be stored at −20°C, but antibody activities are stable for extended periods at 4°C.

DIRECT EXAMINATION

The direct or indirect fluorescent-antibody (FA) technique (see below) should be useful for the identification of viral antigen in infected tissues, but only limited studies of this kind have been performed. By the use of monoclonal antibodies, it should be possible to selectively identify the presence of different structural components, as illustrated by findings in animal experimental systems (5).

At the present stage of development of the enzyme-linked immunosorbent assay (ELISA) for demonstration of mumps virus antibodies (see below), it should be possible to apply this test also for the identification of mumps virus antigen in tissues or secretions. So far, no application of the ELISA for this purpose has been described.

ISOLATION OF VIRUS

Cell cultures of primate origin, chicken embryo fibroblast cell cultures, and embryonated hen eggs can be used for the isolation of mumps virus. The best results are obtained with primary monkey kidney or human embryonic kidney cell cultures.

Primary monkey or human kidney cells (150,000 to 200,000 cells per ml) or cells of continuous lines (40,000 to 80,000 cells per ml) are planted in standard culture tubes in 1-ml amounts (or correspondingly smaller volumes in plastic microplates) of growth medium containing 10 parts inactivated fetal calf serum in Eagle basal medium (improved, containing nonessential amino acids and pyruvate) in Hanks BSS, with a final addition of 3 ml of 7.5% $NaHCO_3$ per

100 ml, 100 U of penicillin per ml, 50 μg of streptomycin per ml, and 5 μg of amphotericin B per ml. When large islands of actively growing cells have formed, the growth medium is removed and replaced with a corresponding volume of maintenance medium containing 3% inactivated fetal calf serum in Eagle basal medium in Earle BSS, with a final addition of 3.0 ml of 7.5% $NaHCO_3$ per 100 ml, 100 U of penicillin per ml, 50 μg of streptomycin per ml, and 5 μg of amphotericin B per ml. Ingredients for these culture media or media ready for use are available commercially. Other cell culture media in common use may be substituted; however, calf serum should not be used. Cultures are inoculated with 0.2 to 0.4 ml of specimen and incubated in a stationary or rotating position at 36°C for 6 to 7 days.

Cytopathic effects such as giant-cell formation and rounding of cells do not occur regularly enough to be of diagnostic value. A hemadsorption test is the simplest and most reliable assay for the presence of mumps virus in cell cultures. (Caution: as cells, particularly those of monkey kidney, may contain other latent hemadsorbing viruses, several uninoculated control cultures should always be tested for hemadsorption.)

The hemadsorption test is performed as follows. Cell culture medium is removed, and 0.05 to 0.2 ml of a precooled 1% chick erythrocyte, guinea pig erythrocyte, or human group O erythrocyte suspension in Veronal-buffered saline (VBS) is added. Incubation is at 4°C for 45 min with the erythrocytes in contact with the cell monolayer. The cells are then washed two or three times with 1- to 2-ml volumes of cold buffered saline or Hanks BSS, and the cultures are examined microscopically for hemadsorption of erythrocytes to the infected cells. Adherence of erythrocytes to the infected cells indicates the presence of virus.

If, however, more rapid detection of virus is desired, the FA technique can be used up to ca. 3 days postinfection. This technique allows a direct specific identification of mumps virus, particularly if appropriately selected monoclonal antibodies are used. The details of the FA procedure are given below under Serological Diagnosis.

Alternative procedure

The amniotic cavity of 7- to 8-day-old embryonated hen eggs is inoculated with 0.2 ml of the specimen. Inoculated eggs are incubated for 6 days at 36°C and then harvested. The amniotic fluids from each of the eggs are tested individually.

The presence of virus in the amniotic fluids is determined by performing a hemagglutination test as follows. Volumes of 0.05 ml of egg fluid are placed in plastic hemagglutination plates, 0.05 ml of a 1% suspension of chick, guinea pig, or human O erythrocytes in VBS is added, and the tubes or plates are shaken well. An erythrocyte control (erythrocytes and VBS only) is included. The tubes or plates are refrigerated at 4°C for 45 min or longer to allow the erythrocytes to settle, and then they are read for hemagglutination.

IDENTIFICATION OF VIRUS

Virus isolated in cell cultures is best identified by a hemadsorption inhibition or a neutralization test or by FA techniques. Virus isolated in embryonated hen eggs is most rapidly identified by a hemagglutination inhibition (HI) test or by the demonstration of mumps virus CF antigens in the infected egg fluids. Infected egg materials may also be used for a neutralization test in cell cultures. In the future, antigen ELISA tests may be employed. Methodological procedures for the characterization of isolates by antibodies are described below under Serological Diagnosis. The virus-containing material, cells plus medium frozen and thawed twice or amniotic fluid, should be centrifuged at 1,000 × *g* for 10 min to eliminate cell debris and then stored at −20°C or, if possible, at −65°C. Polyclonal typing antisera are commercially available.

SEROLOGICAL DIAGNOSIS

A large number of different tests for the determination of antibodies to mumps virus have been described. The continued evolution of serological methods reflects the need for increased sensitivity, specificity, and simplicity of tests. The traditional assays, the neutralization, HI, and CF tests, all have their drawbacks. The neutralization test is too cumbersome for routine serological work (1). In contrast, the HI and CF tests are simple to perform, but they have shortcomings in regard to sensitivity and reliability. Both of them have a relatively low sensitivity, and in addition, cross-reacting antibodies against other paramyxoviruses may pose a problem. The presence of nonspecific inhibitors causes a special problem in the determination of low titers of antibodies by the HI test. The hemolysis-in-gel (HIG) test is not influenced by nonspecific factors to the same extent. Thus, no pretreatment of sera is needed in this test, which, however, has a moderate sensitivity. Further, plates for the HIG test have a limited storage time. The immunofluorescence test is not practically suited for large-scale antibody determinations, but it has the advantage of allowing the separate identification of IgG and IgM. This separate identification is also possible with the recently introduced ELISA test. Since this test has a high sensitivity, it appears to be the current method of choice (6, 14, 17).

ELISA

Antigens. A virus strain adapted to growth in the allantoic fluid of embryonated hen eggs preferably should be used in the ELISA, since the yield in this system markedly exceeds that of cell culture systems. However, harvested chorioallantoic fluid should not be used directly (14, 17). Purify and concentrate virus by sedimentation at 60,000 × *g* for 1 h at 4°C. Suspend the pellet in 1/10 to 1/100 of the original volume. Determine the protein content (e.g., by the Lowry test).

Test procedure. Dilute antigen in 0.05 M carbonate-bicarbonate buffer (pH 9.6) to give a protein content of 10 to 20 μg/ml, and incubate polystyrene plates with 0.05 ml per well. Incubate at room temperature for 2 h or at 4°C overnight. Wash two to three times with phosphate-buffered saline with 0.05% Tween 20. This buffer may have 1% bovine serum albumin (or heterologous serum) added to cover sites in the plates which have not been saturated by antigen. Add serial dilutions of serum (four- or fivefold) in the corresponding buffer. Incubate at 37°C for 1 to 2 h. Wash again three times in the same buffer. Add either conjugated anti-human immunoglobulin or heavy-chain-specific anti-human IgG or IgM (commercially available). The latter reagents allow a separate identification of immunoglobulin classes of specific antibodies. Wash and add appropriate substrate, following the procedure in Chapter 95.

Test for IgM

The determination of mumps virus-specific IgM antibody allows the identification of a recent infection in a single serum sample (6, 14, 17). Cross-reactions involving antibodies against other paramyxoviruses do not appear to interfere with this test. The test procedure is described above. Perhaps in the future, the first layer in the test could be a human, heavy-chain-specific heterospecies antiserum to avoid competition with mumps virus-specific IgG antibodies.

HIG test

Gels for the hemolysis-in-gel (HIG) test are commercially available (20). The principle of the test is as follows. Mumps viral antigen is coupled to erythrocytes with $CrCl_3$, and these are suspended in an agarose gel containing diluted guinea pig serum. Sera are applied in wells in the gel, and rapidly diffusing antibodies form a zone of hemolysis in the presence of complement. The diameter of the zone is proportional to the antibody titer of the serum.

HI test

Twofold serial dilutions of inactivated (56°C, 30 min) serum in VBS (pH 7.2) are prepared in a volume of 0.025 ml in microplates. Sera often contain nonspecific inhibitors, which may be removed by treatment with $NaIO_4$. (This is done by adding 0.15 ml of freshly prepared 0.1 M $NaIO_4$ to 0.5 ml of serum. The mixture is allowed to stand for 30 min at 37°C, and then 0.15 ml of 40% glucose solution is added. The volume of serum is then brought to 1 ml by the addition of BSS.) Nonspecific agglutinins may be removed by absorption with packed 10% erythrocytes of the kind used in the test for 30 min at room temperature. Pretitrated mumps virus antigen diluted to contain 4 hemagglutinating units is added in a volume of 0.025 ml; plates are shaken well. The antigen preparation preferably should be treated with Tween 80 and ether before use (1). (This is done by adding 0.1 volume of a 1.25% solution of Tween 80, mixing carefully, and incubating the mixture for 3 to 5 min at room temperature. An equal volume of anesthetic ether is added, and the mixture is shaken intermittently for 10 min on an ice bath. It is then centrifuged at 3,000 rpm for 5 min. The aqueous phase is recovered, and the ether is removed by bubbling nitrogen gas through the preparation.) Virus-serum mixtures are incubated for 1 h at 37°C or overnight at 4°C. The addition of appropriately diluted anti-human immunoglobulin is a means of further increasing the sensitivity of the HI test. Serum, virus, and erythrocyte controls are included. A 1% suspension of chick, guinea pig, or human O erythrocytes in VBS is added in a volume of 0.05 ml, and tests are shaken well. Addition of a small quantity of protein, e.g., 0.1% normal calf serum, improves the sedimentation of erythrocytes. Erythrocytes are allowed to settle for 60 to 90 min at 4°C or at room temperature. The highest serum dilution that inhibits hemagglutination is considered the antibody endpoint. To ascertain results with low serum dilutions, the test plate may be tilted. Nonagglutinated erythrocytes will then stream towards the low side of the cups, whereas agglutinated cells will retain the pattern.

CF test

Mumps V and S antigens for the determination of CF antibodies can be prepared as described by Henle et al. (3). Whole virus antigen prepared by three cycles of freezing and thawing of infected cells may also be used. The CF test is performed as described previously (13).

Neutralization test

Blocking of cytopathic changes or hemadsorption. Twofold serial dilutions of inactivated sera (56°C, 30 min) are prepared in Hanks BSS, and to each is added an equal volume of a mumps virus material pretitrated in serial 10-fold dilutions to contain 30 to 100 50% infective doses per 0.1 ml. The 50% infective dose is defined as the highest dilution which infects 50% of the infected cultures in the pretitration. The virus-antibody mixtures are shaken well and incubated for 1 h at 37°C or for 1 h at room temperature plus overnight at 4°C. Cell cultures on maintenance medium are inoculated with 0.2 ml of the mixtures (proportionally smaller volumes in microcultures) and incubated for 4 to 7 days at 37 or 32°C. Primary monkey or human kidney cells, chicken embryo fibroblasts, HEp-2, HeLa, or other diploid or heteroploid cell cultures may be used. Cytopathic changes are read or a hemadsorption test is performed (1). The highest dilution of serum that prevents the multiplication of mumps virus is considered the endpoint. The test may also be performed as a microtiter assay.

Inhibition of plaque formation. Twofold serial dilutions of inactivated sera (56°C, 30 min) are prepared in Hanks BSS, and to each is added an equal volume of virus pretitrated to give 50 to 60 or 20 to 30 PFU, depending upon whether a macro- or microassay is used. After careful mixing and incubation for 1 h at 37°C, serum-virus mixtures are dispersed onto pretested susceptible cells, e.g., Vero cells. After adsorption for 1 h at 35°C, overlay medium consisting of Eagle minimal essential medium containing 3 to 5% fetal calf serum, 0.03% glutamine, and 0.5% agarose is added. These preparations are incubated for 5 days at 35°C in a CO_2 atmosphere, and then an additional 0.5 volume of overlay medium containing neutral red (1:10,000) is added. Plaques are counted on the next day. The highest dilution of serum giving a reduction in plaque number exceeding 50% is taken as the endpoint. The addition of anti-immunoglobulin to the virus-antibody mixture may markedly increase the sensitivity of the test (15).

FA test

As mentioned, the immunofluorescence technique is too cumbersome for routine serological work, but it allows a possibility of selective determination of IgG and IgM. Since, however, the test has a potential usage for the identification of virus isolates, it is briefly described here.

Cells may be trypsinized off from any kind of culture, washed, and used for the preparation of smears on microscope slides. However, it is preferable to grow the cells directly on such slides, either in Leighton tubes or in petri dishes in a CO_2 incubator. The cells are air dried and fixed in cold acetone (−20°C) for 10 min at room temperature. The indirect FA is carried out by the use of paired human sera, animal hyperimmune sera, or murine ascites-containing monoclonal antibodies. This last kind of reagent is to be preferred because of its specificity and potency (12). Selection of monoclonal antibodies against the nucleocapsid or polymerase proteins may be of value, since these antigens dominate in infected cells.

Determination of antibodies by other techniques

Other techniques for the determination of antibodies include (i) immunodiffusion, which allows the separate identification of different antigen-antibody systems; (ii) hemolysis inhibition, which makes it possible to separately identify antibodies to the fusion (hemolyzing) component on the virion surface, if HI antibodies are removed by adsorption with Tween 80–ether-treated antigen (8); (iii) single radial immunodiffusion with immobilized virions and mixed hemadsorption, which detect antibodies to virion surface components (7); (iv) imprint immunofixation, which allows the identification of mumps virus-specific antibody in fractionated immunoglobulin (21); and (v) radioimmune precipitation assay, which makes it possible to identify the structural component(s) participating in specific antigen-antibody reactions (11).

EVALUATION AND INTERPRETATION OF RESULTS

Isolation of mumps virus is diagnostic for a current infection, since persistent infection with a prolonged excretion of infectious virus has not been seen. There are indications of the occasional persistence of mumps virus infections in the central nervous system of humans, but judging from comparative data in experimental animals (5), it seems likely that such infection may be caused by defective virus. The nature of such infections may be elucidated by the use of antigen-ELISA, which most likely also will have an increasing application for direct examination in acute infections, and by immunofluorescence techniques employing monoclonal antibodies, selectively identifying individual structural components.

A serological diagnosis can be made by the demonstration of a fourfold or greater antibody rise by any of the described methods. Their advantages and shortcomings have been briefly discussed. It is an old observation that different kinetics of the antibody response are seen when V and S antigens are used in a serological test, e.g., the CF test (3). Antibodies to S antigen appear earlier than do antibodies to V antigen. This asynchrony is confirmed by the application of modern techniques. Presumably, S-antigen preparations predominantly contain nucleocapsids, and this is also the case for antigen preparations conventionally used in CF, HIG, and ELISA tests. Maximal antibody titers in these tests are reached about 3 weeks after the onset of symptoms (17). In contrast, antibodies to virion surface antigens, presumably dominating in V-antigen preparations, show peak titers 1 to 2 weeks later. This is evidenced by the results of HI and neutralization tests. Thus, when the first serum sample is collected relatively late, it may be an advantage to employ the latter tests. In tests for both virion surface and virion internal components there is, however, a certain risk in the identification of cross-reacting antibodies deriving from infection with other paramyxoviruses (6, 14, 17). This risk appears eliminated when IgM antibody determinations are performed instead. The IgM antibody response reaches maximum titers about 1 week after the onset of symptoms, and although it is transient, it appears to last for at least 1 month and in many cases probably even a good bit longer. IgM serology, therefore, is very attractive, since it combines specificity with simplicity, and only a single serum sample is required. Obviously, the IgM ELISA is not useful for the determination of immunity, but in this case the IgG ELISA appears to be the method of choice. Cross-reactions with other paramyxoviruses pose less of a problem in this context, since they do not seem to be durable.

Involvement of the central nervous system in mumps infections motivates the search for virus and viral antigen and also the performance of comparative serology with serum and CSF samples. A local production of antibodies is more frequently encountered by IgG ELISA than IgM ELISA tests (18). Such a production of IgG and IgM is seen in ca. 80 and 65%, respectively, of all cases of meningitis, and maximal antibody titers in CSF occur 1 to 2 weeks after the onset of symptoms of meningitis. A similar high frequency of local antibody production is observed by the use of imprint immunofixation (21). The same technique also allows the demonstration of an occasional parallel mobilization of an antibody response to other viruses, such as measles, rubella, and herpes simplex viruses. This response was interpreted to represent a nonspecific activation of antibody-producing cells. The local immune response in the central nervous system after mumps meningitis in some cases lasts for more than 2 years. It remains to be determined whether this reflects some tendency for occasional viral persistence.

LITERATURE CITED

1. **Buynak, E. B., J. E. Whitman, Jr., R. R. Roehm, D. H. Morton, G. P. Lampson, and M. R. Hilleman.** 1967. Comparison of neutralization and hemagglutination-inhibition techniques for measuring mumps antibody. Proc. Soc. Exp. Biol. Med. **125:**1068–1071.
2. **Compans, R. W., and K. Nakamura.** 1978. Myxoviridae: paramyxoviruses—parainfluenza, mumps and Newcastle disease, p. 361–385. *In* G.-D. Hsiung and R. H. Green (ed.), CRC handbook series in clinical laboratory science. Section H: virology and rickettsiology. CRC Press, Inc., West Palm Beach, Fla.
3. **Henle, G., S. Harris, and W. Henle.** 1948. The reactivity of various human sera with mumps complement-fixation antigens. J. Exp. Med. **88:**133–147.
4. **Kingsbury, D. W., M. A. Bratt, P. W. Choppin, E. P. Hansen, Y. Hosaka, V. ter Meulen, E. Norrby, W. Plowright, R. Rott, and W. H. Wunner.** 1978. Paramyxoviridae. Intervirology **10:**137–152.
5. **Kristensson, K., C. Örvell, G. Malm, and E. Norrby.**

1984. Mumps virus infection of the developing mouse brain: appearance of structural virus proteins demonstrated with monoclonal antibodies. J. Neuropathol. Exp. Neurol. **43:**131–138.

6. **Nicolai-Scholten, M. E., R. Ziegelmaier, F. Behrens, and W. Höpken.** 1980. The enzyme-linked immunosorbent assay (ELISA) for determination of IgG and IgM antibodies after infection with mumps virus. Med. Microbiol. Immunol. **168:**81–90.
7. **Norrby, E., M. Grandien, and C. Örvell.** 1977. New tests for characterization of mumps virus antibodies: hemolysis-inhibition, single radial immunodiffusion with immobilized virions, and mixed hemadsorption. J. Clin. Microbiol. **5:**346–352.
8. **Norrby, E., and K. Penttinen.** 1978. Differences in antibodies to surface components of mumps virus after immunization with formalin-inactivated and live virus vaccines. J. Infect. Dis. **138:**672–676.
9. **Örvell, C.** 1978. Immunological properties of mumps virus glycoproteins. J. Gen. Virol. **41:**517–526.
10. **Örvell, C.** 1978. Structural polypeptides of mumps virus. J. Gen. Virol. **41:**527–539.
11. **Örvell, C.** 1984. The reactions of monoclonal antibodies with structural proteins of mumps virus. J. Immunol. **132:**2622–2631.
12. **Örvell, C., and E. Norrby.** 1984. Antigenic structure of paramyxoviruses. *In* A. R. Neurath and M. H. V. van Regenmortel (ed.), Immunochemistry of viruses—the basis for serodiagnosis and vaccines. Elsevier Press, Amsterdam.
13. **Palmer, D. F., L. Kaufman, W. Kaplan, and J. J. Cavallaro.** 1977. Serodiagnosis of mycotic diseases. Charles C Thomas, Publisher, Springfield, Ill.
14. **Popow-Kraupp, T.** 1981. Enzyme-linked immunosorbent assay (ELISA) for mumps virus antibodies. J. Med. Virol. **8:**79–88.
15. **Sato, H., P. Albrecht, J. T. Hicks, B. C. Meyer, and F. A. Ennis.** 1978. Sensitive neutralization test for virus antibody. I. Mumps antibody. Arch. Virol. **58:**301–311.
16. **Server, C. A., C. D. Merz, N. M. Waxham, and S. J. Wolinsky.** 1982. Differentiation of mumps virus strains with monoclonal antibody to the HN glycoprotein. Infect. Immun. **35:**179–186.
17. **Ukkonen, P., M.-L. Granström, and K. Penttinen.** 1981. Mumps-specific immunoglobulin M and G antibodies in natural mumps infection as measured by enzyme-linked immunosorbent assay. J. Med. Virol. **8:**131–142.
18. **Ukkonen, P., M.-L. Granström, J. Räsänen, M. E. Salonen, and K. Penttinen.** 1981. Local production of mumps IgG and IgM antibodies in the cerebrospinal fluid of meningitis patients. J. Med. Virol. **8:**257–265.
19. **Utz, J. P., J. A. Kasel, H. G. Cramblett, C. F. Szwed, and R. H. Parrot.** 1957. Clinical and laboratory studies of mumps. I. Laboratory diagnosis by tissue-culture techniques. N. Engl. J. Med. **257:**497–502.
20. **Väänänen, P., T. Hovi, E.-P. Helle, and K. Penttinen.** 1976. Determination of mumps and influenza antibodies by haemolysis-in-gel. Arch. Virol. **52:**91–99.
21. **Vandvik, B., R. E. Nilsen, F. Vartdal, and E. Norrby.** 1982. Mumps meningitis: specific and non-specific antibody responses in the central nervous system. Acta Neurol. Scand. **65:**468–487.

Chapter 76

Rubella Virus

KENNETH L. HERRMANN

CLINICAL BACKGROUND

Rubella (German or 3-day measles) is a mild, contagious rash illness, primarily of children and young adults. Although the disease was first described as a clinical entity in the 18th century, the causal relationship between maternal infection with rubella virus during pregnancy and congenital malformations was not appreciated until 1941, when Sir Norman Gregg, an Australian ophthalmologist, reported an epidemic of congenital cataracts occurring in the wake of a large outbreak of rubella (3). This observation established a much more serious significance to what previously had been considered a mild and rather innocuous rash illness.

The viral etiology of rubella was not confirmed until 1962, with the successful isolation of the virus in cell cultures by Weller and Neva (16) at Harvard and by Parkman and colleagues (9) at Walter Reed Army Institute of Research. This discovery led to the development of tests for accurate laboratory diagnosis and subsequently to the development of live attenuated rubella virus vaccine. The incorporation of rubella vaccination into routine pediatric and public health practice has now resulted in rubella rapidly becoming an uncommon illness in the United States. During 1983 fewer than 1,000 cases of rubella were reported to the Centers for Disease Control.

Acute rubella virus infection in a child or adult is usually a self-limited, benign disease, characterized by a low-grade fever, mild upper respiratory symptoms, an erythematous maculopapular rash, and suboccipital lymphadenopathy. In young adults, the illness may be complicated by transient arthralgias or arthritis, but severe complications, such as encephalitis or thrombocytopenia purpura, occur only rarely. During the 14- to 18-day incubation period, prodromal symptoms are notably absent in most cases. Although the rash is the most conspicuous feature of this disease, it is of such variable character that it may be confused with that produced by other infectious diseases and even by certain drugs. Diagnosis of rubella on clinical grounds alone, except possibly during epidemics, may therefore be grossly inaccurate.

The consequences of rubella infection in pregnant women are varied and unpredictable. Infection of the fetus during the first trimester of pregnancy, and to a much lesser degree during the second trimester, may result in congenital defects. The triad of anatomic abnormalities including cataracts, neurosensory deafness, and congenital heart disease has classically been referred to as the rubella syndrome. However, after extensive studies following the 1964 rubella epidemic, the definition of the syndrome had to be greatly expanded to include thrombocytopenia, hepatitis, microcephaly, intrauterine growth retardation, long-bone lesions, retinitis, and encephalitis.

Frequently, rubella defects may not be apparent until weeks or months after birth. In one study, no abnormal findings were detected in 68% of rubella-infected newborns during the neonatal period, but among those who were followed, 71% developed clinically apparent disease during the first 5 years of life (12). Obviously, many important rubella defects can be overlooked during the first few months of life. Careful ophthalmological examination, X-ray examinations of long bones, neurological testing, and detailed cardiac evaluations should be performed on all suspect neonates, whether symptomatic or not.

Antibodies to the virus appear as the rash fades, and initially, both immunoglobulin G (IgG) and IgM antibodies can be detected. Antibodies of the IgM class generally do not persist beyond 4 to 5 weeks after onset of illness, but IgG antibodies usually persist throughout the patient's life. Reinfection with the virus can occur, but it is almost always asymptomatic and can be detected by a rise in IgG antibodies. The risk of fetal damage resulting from rubella reinfection during pregnancy is negligible. The attenuated virus vaccines induce the production of IgM and IgG antibodies similar to that observed with natural infections except that the titers are somewhat lower. Reinfection rates with wild virus are greater among vaccinees than among persons previously infected under natural conditions.

DESCRIPTION OF THE AGENT

Rubella virus has been classified as a member of the family Togaviridae. Studies of rubella virus RNA and its structural proteins, however, have differentiated rubella virus from all other members of the togavirus family. Only one immunologically distinct type of rubella virus has been described, and no serological relationship is known to exist between rubella and any other known virus group. Rubella virus contains three major structural proteins: two glycoproteins, E1 (molecular weight, 58,000) and E2 (molecular weight, 42,000 to 47,000), associated with the viral envelope, and an internal nucleocapsid protein, C (molecular weight, 33,000) (7). Each of these polypeptides is immunogenically active; however, no single distinct antigenic site has yet been associated with virus infectivity.

Morphologically, rubella virus is a spherical particle with a diameter of 60 to 70 nm. The virion contains a dense central core surrounded by a lipid bilayer. Replication occurs in the cytoplasm of the cell, and the virus matures by budding into cytoplasmic vesicles or from the marginal plasma membrane. The virus is destroyed by trypsin and lipid solvents, but is relatively stable to repeated freezing and thawing or to ultrasound treatments.

COLLECTION AND STORAGE OF SPECIMENS

For details of procedures for the collection and storage of specimens, see Chapter 95.

ISOLATION OF VIRUS

Virus isolation has only limited applicability for the routine diagnosis of rubella. Most clinical situations requiring laboratory diagnosis are better investigated serologically. The few situations in which rubella virus isolation is indicated include suspected rubella with severe complications, fatal cases for which serological confirmation of the etiology would not be possible, and cases where strain characterization of the infecting agent (i.e., vaccine-like versus wild-like) may be required for epidemiological purposes.

A wide variety of cell types are susceptible to infection by rubella virus. For primary isolation of rubella virus from clinical specimens, however, primary African green monkey kidney (AGMK), Vero, or RK-13 cell cultures are recommended. Isolation of rubella virus in primary cultures of AGMK cells has been considered the standard method since 1962 (9). Rubella virus is detected in this cell type by interference with the cytopathic effects (CPE) of a challenge virus.

For virus isolation in AGMK, four tubes containing cell monolayers are drained of medium; 0.2 ml of the specimen is inoculated into each tube and allowed to adsorb for 1 h at 35 to 36°C. A 1.5-ml amount of maintenance medium (Eagle basal medium with 2% inactivated fetal bovine serum) is then added to each tube, and the tubes are incubated in stationary racks at 35 to 36°C for 10 days. Uninoculated AGMK cell cultures serve as controls. At the end of the incubation period the tubes are examined for any CPE. If none is found, two inoculated and two control tubes are challenged with 100 to 1,000 50% tissue culture infectious doses of a challenge virus. Those viruses most commonly used as the challenge agent for rubella isolation in AGMK cells are echovirus type 11 or coxsackievirus A9. Challenged tubes are read 3 to 4 days after challenge. The presence of rubella virus is indicated by complete destruction of the control cells but little or no CPE in the inoculated tubes. Absolute identification of the isolate requires specific neutralization of the interference with rubella antibody.

Specimens with low concentrations of rubella virus may produce little or partial interference in the culture tubes initially inoculated. The culture fluids may have to be passaged to demonstrate the virus. Fluid from the cultures which were not challenged should be inoculated into an additional four tubes of AGMK cells. Additional control tubes are also included. These tubes should be incubated for an additional 7 to 8 days. Two inoculated tubes and two control tubes are then challenged with echovirus 11 and observed for CPE. The absence of interference with the echovirus cytopathogenicity on passage of fluid from tubes which did not demonstrate interference initially confirms the absence of an interfering agent. If virus is not found after such a passage, it is rarely found by further passages.

Some laboratories use RK-13 or Vero cells for isolation of rubella virus. In these cell systems, rubella virus produces CPE; however, the CPE is not always clear on primary isolation, and cell culture fluids may need to be passaged several times for full detection of virus. These cell systems, however, do offer the advantage of direct neutralization for identification of an isolate. Furthermore, an indirect immunofluorescence staining method has been shown to be specific and sensitive for identifying rubella virus isolates in these cells (13).

Fresh unfrozen tissue specimens may be processed for the propagation of explant cell cultures (4). This approach is of particular value in attempts to isolate virus from fetal tissues and organs. A convenient method is to explant minced tissue fragments with growth medium and allow sufficient time for the outgrowth of cells. When the cells have formed monolayers, the extracellular fluids can be harvested and tested for the presence of an interfering agent, as described above. This is a more sensitive method for rubella virus isolation than the method in which tissue extracts or homogenates of ground tissue are used.

IDENTIFICATION OF VIRUS

Rubella virus can be specifically neutralized with rubella antiserum prepared in rabbits. Such antisera are available from several commercial sources. Immune rabbit serum is diluted to contain 4 U of neutralizing antibody. A normal preimmune (rubella antibody-free) rabbit serum is diluted similarly for the control titration. The media from the two companion, unchallenged cultures containing the interfering agent are pooled, and serial 10-fold dilutions are made in maintenance medium (undiluted to 10^{-6}); 0.1-ml samples of each dilution are inoculated into three culture tubes. The 10^{-1}, 10^{-2}, and 10^{-3} dilutions are also combined with an equal volume of the prediluted rubella antiserum and of the prediluted normal rabbit serum. After 1 h of incubation at 35°C, 0.2 ml of each of the mixtures is inoculated into three AGMK tubes. The tubes are incubated at 35 to 36°C for 7 to 8 days, challenged with echovirus 11, and observed for the development of enterovirus CPE. Destruction of the AGMK monolayers inoculated with the isolate dilution containing between 10 and 100 50% tissue culture infectious doses plus the immune rabbit serum, but not of those with the normal rabbit serum, indicates that the isolate is rubella virus.

SEROLOGICAL DIAGNOSIS

Serological techniques for the detection of antibodies to rubella virus provide the approach of choice for laboratory diagnosis of acute and congenital rubella infections and for the determination of rubella immune status. Methods currently available include hemagglutination inhibition (HI), passive hemagglutination (PHA), hemolysis in gel, latex agglutination, enzyme immunoassay (EIA), fluorescent immunoassay, radioimmunoassay, complement fixation (CF), and a variety of rubella-specific IgM antibody assays.

Over 95% of the rubella antibody testing in the United States is currently being performed with commercially prepared test kits. The Centers for Disease Control evaluate these kits and monitor the performance of these products on a lot-to-lot basis. Although the rubella HI test remains the benchmark for rubella serology, the numerous more recently developed test systems are now much more widely used than HI for routine rubella antibody testing.

HI

Rubella HI antibody appears very soon after the onset of rubella symptoms and rises rapidly, often reaching peak levels within 7 to 10 days. Thereafter, rubella HI antibodies are usually detectable throughout life, and thus the presence of HI antibody is a reliable indicator of past rubella virus infection.

HI was the first widely used rubella test. In 1970, in an effort to improve the reliability and reproducibility of this test, a standardized reference HI protocol was developed. This standardized protocol has since been adopted by the National Committee for Clinical Laboratory Standards and the World Health Organization as the recommended rubella reference test. The precision of other rubella serodiagnostic test methods should now be measured against that of the standard reference HI method before being accepted for diagnostic use. The reader is referred to the referenced procedural guide for a more detailed description (4).

Satisfactory rubella hemagglutinin (HA) antigen can be prepared in cultures of BHK-21 or Vero cells or obtained commercially. The minimal acceptable titer of HA antigen for this test is 1:64. Erythrocytes to be used in the standard HA and HI tests are obtained from 1- to 3-day-old unfed baby chicks. These cells may be stored for up to 2 weeks in Alsever solution without loss of sensitivity. After the cells have been washed three times in dextrose-gelatin-Veronal buffer, they are suspended in HEPES (*N*-2-hydroxyethylpiperazine-*N'*-2-ethanosulfonic acid)-saline-albumin-gelatin (HSAG) diluent. Cell suspensions of 0.25% concentration in HSAG are used as the indicator cells in the test proper; 50% cell suspensions are used for serum adsorption to remove natural agglutinins.

The HA and HI tests are conveniently performed in disposable microtitration plates. The rubella HA antigen is pretitrated by making serial twofold dilutions of the antigen in 0.025 ml of HSAG, to which 0.025 ml of the same diluent is subsequently added. To each well of the titration series, 0.05 ml of the 0.25% suspension of chick erythrocytes is then added. The plate is incubated at 4°C for 90 min and then read for HA. The highest dilution that produces a pattern of complete HA is considered 1 HA unit.

In performing the HI test, the serum must first be treated with heparin and manganous chloride to remove the nonspecific beta-lipoprotein inhibitors and then adsorbed with chick erythrocytes to remove natural cell agglutinins. Serum must not be heat inactivated before this treatment. One-tenth milliliter of serum is mixed with 0.15 ml of HSAG diluent. Then 0.1 ml of a 1:1 mixture of heparin (5,000 U/ml) and 1 M $MnCl_2$ is added and mixed gently. The mixture is incubated at 4°C for 15 min; 0.2 ml of 50% chick erythrocytes is then added and mixed. After incubation for an additional 1 h, 0.4 ml of HSAG is added, and the mixture is centrifuged. The supernatant fluid, which represents a 1:8 dilution of serum, is removed for testing. Serial twofold dilutions of the treated serum are made in 0.025-ml amounts of cold HSAG diluent. To these dilutions are added 4 HA units of antigen in volumes of 0.025 ml. The serum-antigen mixtures are incubated for 1 h at 4°C, after which 0.05 ml of the 0.25% suspension of chick erythrocytes is added to each well. The HA patterns are read after 90 min at 4°C, and that dilution of serum which completely inhibits HA is taken as the endpoint. Each test must be controlled by including HA antigen back-titration, serum agglutination controls for each serum tested, and known positive and negative serum titer controls.

HI has the advantage that considerable experience has been gained over the past 15 years regarding the clinical significance of HI titer results. It is well established that the presence of rubella antibody detectable by the HI test accurately correlates with clinical protection of the individual. The HI test does, however, have several disadvantages. It is time consuming and highly technique-dependent for accuracy, and it requires pretreatment of serum. One of the most critical steps in rubella HI antibody testing is the removal of nonspecific inhibitors before the serum is assayed for antibody. Failure to completely remove the beta-lipoprotein inhibitors may result in false-positive test results.

PHA

The PHA test is most useful in detecting serological evidence of immunity from past rubella infection. Human erythrocytes for the test are coated with a soluble rubella virus antigen and used in a one-stage PHA system. Standard microtitration techniques are used. Test and control sera are diluted with phosphate-buffered saline in wells of a V-type microtitration plate. Sensitized erythrocytes are added to each well and thoroughly mixed. The test is incubated at room temperature until the cells have settled. These patterns are read and compared with the controls. The presence of antibody is detected by an agglutination pattern in the well.

The rubella PHA test is currently available commercially in kit form. These kits are designed to be used for rubella immunity screening; test results are recorded as "positive" or "negative" for antibody. Since sera in the PHA test do not have to be treated to remove nonspecific reactants before testing, specimens can be tested rapidly. Immunity screening results obtained with the commercial PHA kits correlate over 98% with HI test results.

Latex agglutination

A passive latex agglutination card test, recently developed commercially for detecting rubella antibody in human serum, offers another alternative system for rubella diagnosis and immunity screening. The test requires no serum pretreatment and no elaborate equipment and can be performed in a matter of minutes. Rubella antibody detectable by this latex agglutination test appears within a few days after onset of acute illness and thereafter persists indefinitely. This test method appears to be most useful for the clinical laboratory performing tests in which immediate results are required for patient management.

EIA

Solid-phase EIA methods have been successfully applied for the detection and quantitation of rubella antibodies (2, 15). A variety of rubella EIA test kits are available commercially. One distinct advantage of the solid-phase "sandwich" immunoassay techniques, such as EIA, over the standard HI technique is that serum can be tested without pretreatment because no natural serum inhibitors of the reaction are known. These tests are simple to perform and practical for use in clinical laboratories, given the required reagents and equipment.

There is, at present, no standard method for reporting results of EIA tests. Without an accepted standard reference serum, comparison of results from one commercial EIA method to another is not possible. When comparing results of paired sera from patients suspected to have rubella, it is important that both sera be run simultaneously (preferably, in duplicate) in the same test and that results be interpreted as described by the specific test protocol.

Hemolysis in gel

The rubella hemolysis-in-gel, or radial hemolysis, test is widely used in Europe as a simple and sensitive method for rubella antibody testing (5). Commercial development of this test in the United States has been hampered by the relatively short 2- to 3-week shelf life of the reagents. This test has the advantage of simplicity; it requires no special or expensive equipment and can be performed on untreated serum. At present it is recommended that a control reference serum with a minimum potency, such as that widely used in England (10), be included in each plate to define the breakpoint differentiating immunes from susceptibles. Because such a standard reference antibody control is not currently available in the United States, laboratories using this method for rubella immunity screening or diagnosis must establish their own criteria for interpreting of their test results.

Fluorescent immunoassay

An indirect fluorescent-antibody (IFA) test for rubella antibody was first described in 1964 (1); chronically infected LLC-MK_2 cell cultures were used as the solid-phase antigen. Other acutely infected cell systems or even purified rubella antigens have since been used for rubella IFA. For the classic IFA test, acutely infected cells are grown either on Leighton tube cover slips or in culture flasks which are then trypsinized and deposited on slides to form smears. The cytoplasm of cells infected with rubella virus contains rubella antigens. These antigens are used to detect specific antibodies by the indirect method in which an anti-human globulin conjugated with fluorescein isothiocyanate is employed. The technique is rapid and relatively inexpensive and allows quantitation of IgG and IgM antibodies; however, IFA results are read visually with a fluorescence microscope and are open to subjective interpretation.

More recently, a soluble rubella antigen immobilized on an opaque plastic surface has been used in the indirect fluorescence immunoassay test. In this test system, marketed commercially as the FIAX test, the antigen-sensitized surface is allowed to react in a two-step procedure with the serum and the fluorescein-labeled conjugate, and the resulting fluorescence signal is measured objectively with a fluorometer. The intensity of the fluorescence signal correlates with the titer of rubella antibody. The sensitivity and specificity of this assay method correlate well with those of the HI test.

Specific rubella IgM assays

The presence of specific IgM antibodies to rubella virus has special diagnostic significance. Specific rubella antibodies in the IgM class can be detected by various methods, including IFA, EIA, radioimmunoassay, and adsorption of serum IgG with staphylococcal protein A, or by physical separation of IgM from IgG by density gradient ultracentrifugation or column chromatography followed by HI assay of the fractions. Sucrose gradient ultracentrifugation is considered at present to be the most reliable and specific method for specific rubella IgM antibody assay (14); however, only a few diagnostic laboratories have the equipment necessary to perform this test.

Some of the more recently developed methods for detecting rubella IgM antibody, such as the IgM-radioimmunoassay (6) and the IgM-EIA (11), offer simpler and more rapid approaches, but questions have been raised regarding the specificity of such indirect assay methods. The presence of rheumatoid factor (IgM anti-IgG antibodies) in serum may cause false-positive IgM results in these systems. Effective methods for avoiding these nonspecific reactions, including preadsorption of serum to remove IgG and IgG-rheumatoid factor complexes or IgM capture assays utilizing anti-IgM capture antibody on the solid-phase carrier, are being incorporated into these procedures to make them acceptable for specific IgM antibody assay in routine diagnosis. Acceptable commercial rubella IgM assays are now available in the United States.

A list of currently acceptable commercial rubella antibody test kits is presented in Table 1.

EVALUATION AND INTERPRETATION OF RESULTS

Rubella virus infection produces essentially two clinical entities, congenital rubella and rubella acquired after birth. Establishing the diagnosis of rubella in the first trimester of pregnancy is of importance and may influence the decision to terminate or continue the pregnancy. Because of the time required for isolating and identifying rubella virus, the serological approach to diagnosis is preferred. Rubella HI antibodies appear promptly as the rash fades and reach peak titers in 10 to 20 days (Fig. 1). When feasible, an acute-phase serum should be obtained as early as possible after the onset of disease, and a convalescent-phase specimen should be obtained as early as 5 to 7 days later. For diagnosis, acute- and convalescent-phase sera together with appropriate negative and positive controls must be run in the same test. A fourfold or greater titer rise between acute- and convalescent-phase serum specimens is considered diagnostic of acquired rubella. High, stable HI titers (>256) should not be interpreted as evidence of recent rubella infection. Rubella CF antibodies appear somewhat later, rise more slowly, attain a relatively lower maximal titer, and decline more rapidly than antibodies measured by the HI method. Thus, the CF test may occasionally be useful for diagnosis in cases in which the acute-phase serum has been collected too late for the HI titer rise to be detected. CF antibodies, however, often persist for several years after a rubella infection, so the mere presence of an elevated CF

TABLE 1. Commercial rubella serodiagnostic test kits

Test	Kit name (manufacturer)
HI	RubaTect (Abbott Laboratories, North Chicago, Ill.)
	Rubella HI (American Dade, Miami, Fla.)
	Rubasure (Calbiochem-Behring, La Jolla, Calif.)
	Rubella HI (Flow Laboratories, Inc., McLean, Va.)
	Rubella HI (GIBCO Laboratories, Grand Island, N.Y.)
	Rubindex (Ortho Diagnostics, Raritan, N.J.)
EIA	Rubazyme (Abbott Laboratories)
	Rubazyme-M (Abbott Laboratories)
	Rubella Enzygnost (Calbiochem-Behring)
	Cordia-R (Cordis Laboratories, Miami, Fla.)
	Rubella EIA-Nidas (Fisher Scientific Co., Pittsburgh, Pa.)
	Rubella BioEnzaBead (Litton Bionetics, Charleston, S.C.)
	Rubelisa (MA Bioproducts, Walkersville, Md.)
	RubaSTAT (MA Bioproducts)
	Rubelisa-M (MA Bioproducts)
	Rubella EIA (Ortho Diagnostics)
Immunofluorescence	Rubella IFA (Electro-Nucleonics, Inc., Bethesda, Md.)
	Rubella FIAX (International Diagnostic Technologies, Santa Clara, Calif.)
	Rubella IFA (Microbiological Research Corp., Bountiful, Utah)
PHA	Rubacell II (Abbott Laboratories)
	Rubella PHAST (Calbiochem-Behring)
	Rubindex Direct (Ortho Diagnostics)
Other methods	Rubella GAMMACOAT RIA (Clinical Assays, Cambridge, Mass.)
	Rubascan Latex Aggl (Hynson, Westcott, & Dunning, Baltimore, Md.)
	Orivir Rubella HIG (Orion Diagnostics, Helsinki, Finland)

antibody titer during convalescence from a rash illness does not confirm recent rubella in the patient.

In patients with histories compatible with rubella, who are not seen until 2 to 4 weeks after the illness, the diagnosis can sometimes be established retrospectively by demonstrating specific rubella IgM antibodies. This approach has proved to be very useful in my laboratory. However, IgM antibodies have a short 5-day half-life and wane rapidly after recovery from the acute illness. Many rubella patients may lose all detectable specific IgM antibody long before the end of the 4-week period; thus, only the presence of rubella IgM has diagnostic value.

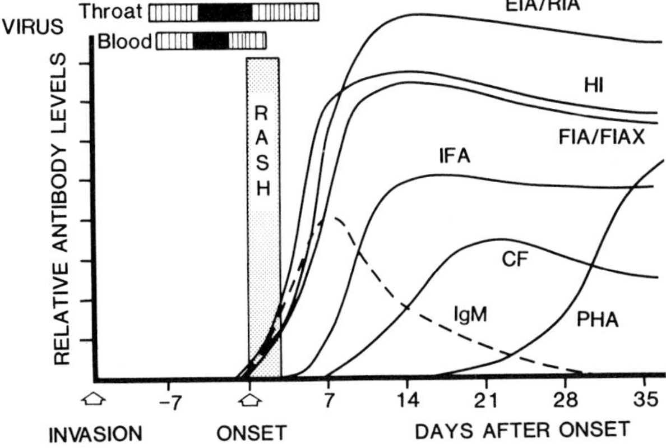

FIG. 1. Schema of immune response in acute rubella infection. RIA, Radioimmunoassay; FIA/FIAX, fluorescence immunoassay.

Reinfection with rubella virus may occur in some persons who have had rubella or rubella vaccination and are reexposed to the disease. Such reinfections are almost always subclinical, and the risk of congenital infection resulting from reinfection during pregnancy appears to be minimal. The rise in antibody titer during reinfection is due to an increase in IgG antibody only. Therefore, a primary infection can be distinguished from a reinfection by evaluating the serum for the presence of IgM antibodies to rubella; these are absent in reinfection.

Congenital rubella infection can be confirmed virologically by isolating the virus from the infant. It can be confirmed serologically by demonstrating the persistence of rubella antibody at a higher level and for a longer period than expected of antibody passively transferred from the mother, and also by detecting specific rubella IgM antibody in the newborn. Although congenitally infected infants may excrete the virus for many months, repeated isolation attempts may be necessary to confirm the diagnosis. The serological approach is therefore recommended for routine diagnosis. Since IgM antibodies are not acquired transplacentally, the presence of rubella IgM antibody in the newborn infant implies active production of antibody by the infant and likelihood of congenital infection. Furthermore, the persistence of rubella antibody beyond 6 months of age, after the loss of passively acquired maternal antibody, is usually considered diagnostic of congenital rubella. In children above the age of 12 months, however, the chance that the antibody resulted from rubella immunization or natural postnatal rubella must be weighed against the likelihood that it resulted from congenital infection.

Although the rubella HI test remains as the benchmark for rubella serology, numerous new test systems are now available or are being developed for use in confirming rubella illness or assessing immune status. It is apparent that a number of these newer tests are capable of greater sensitivity than the HI test. The clinical significance of low levels of rubella antibody, below those detectable by the standard HI test, has not as yet been clearly defined. Recent reports of viremic reinfection in persons with such very low antibody levels suggest that such individuals should not be considered immune to all rubella infection problems (8; G. M. Schiff, Proc. Int. Symp. Prevention of Congenital Rubella Infection, 1984).

LITERATURE CITED

1. **Brown, G. C., H. F. Maassab, J. A. Veronelli, and T. J. Francis.** 1964. Rubella antibodies in human serum: detection by the indirect fluorescent-antibody technique. Science **145:**943–945.
2. **Gravell, M., P. H. Dorsett, O. Gutenson, and A. C. Ley.** 1977. Detection of antibody to rubella virus by enzyme-linked immuno-sorbent assay. J. Infect. Dis. **136:**S300–S303.
3. **Gregg, N. M.** 1941. Congenital cataract following German Measles in the mother. Trans. Ophthalmol. Soc. Aust. **3:**35–46.
4. **Herrmann, K. L.** 1979. Rubella virus, p. 725–766. *In* E. H. Lennette and N. J. Schmidt (ed.), Diagnostic procedures for viral, rickettsial and chlamydial infections, 5th ed. American Public Health Association, Washington, D.C.
5. **Kurtz, J. B., P. P. Mortimer, P. R. Mortimer, P. Morgan-Capner, M. S. Shaf, and G. B. B. White.** 1980. Rubella antibody measured by radial haemolysis: characteristics and performance of a simple screening method for use in diagnostic laboratories. J. Hyg. **84:**213–222.
6. **Meurman, O. H., M. K. Viljanen, and K. Granfors.** 1977. Solid-phase radioimmunoassay of rubella virus immunoglobulin M antibodies: comparison with sucrose density gradient centrifugation test. J. Clin. Microbiol. **5:**257–262.
7. **Oker-Blom, C., N. Kalkkinen, L. Kaariainen, and R. F. Pettersson.** 1983. Rubella virus contains one capsid protein and three envelope glycoproteins, E1, E2a, and E2b. J. Virol. **46:**964–973.
8. **O'Shea, S., J. M. Best, and J. E. Banatvala.** 1983. Viremia, virus excretion, and antibody responses after challenge in volunteers with low levels of antibody to rubella virus. J. Infect. Dis. **148:**639–647.
9. **Parkman, P. D., E. L. Buescher, and M. S. Artenstein.** 1962. Recovery of rubella virus from army recruits. Proc. Soc. Exp. Biol. Med. **111:**225–230.
10. **Pattison, J. R. (ed.).** 1982. Laboratory investigation of rubella. Public Health Laboratory Service Monograph Series no. 16. Her Majesty's Stationery Office, London.
11. **Prevot, J., and J. L. Guesdon.** 1977. Titrage immunoenzymatique des anticorps IgG et IgM specifiques de la rubeole. Ann. Microbiol. (Paris) **128:**531–540.
12. **Schiff, G. M., J. Sutherland, and I. Light.** 1971. Congenital rubella, p. 31–36. *In* O. Thalhammer (ed.), Prenatal infections. International Symposium in Vienna, September 2–3, 1970. Georg Thieme Verlag, Stuttgart.
13. **Schmidt, N. J., E. H. Lennette, J. D. Woodie, and H. H. Ho.** 1966. Identification of rubella virus isolates by immunofluorescent staining, and a comparison of the sensitivity of three cell culture systems for recovery of virus. J. Lab. Clin. Med. **68:**502–509.
14. **Vesikari, T., and A. Vaheri.** 1968. Rubella: a method for rapid diagnosis of a recent infection by demonstration of the IgM antibodies. Br. Med. J. **1:**221–223.
15. **Voller, A., and D. E. Bidwell.** 1975. A simple method for detecting antibodies to rubella. Br. J. Exp. Pathol. **56:**338–339.
16. **Weller, T. H., and F. A. Neva.** 1962. Propagation in tissue cultures of cytopathic agents from patients with rubella-like illness. Proc. Soc. Exp. Biol. Med. **111:**215–225.

Chapter 77

Arboviruses

ROBERT E. SHOPE

CLINICAL BACKGROUND

The arboviruses are a heterogeneous group of animal viruses, usually biologically transmitted by hematophagous arthropods (mosquitoes, ticks, *Phlebotomus* spp., and *Culicoides* spp.). One or more of the following syndromes may be associated with human infection: fever, encephalitis, aseptic meningitis, hemorrhagic fever, rash, acute arthritis, hepatitis, or retinitis. Subclinical infection is common. The virus may be recovered from blood and occasionally from cerebrospinal fluid or throat washings. At autopsy, it may be recovered from the central nervous system, liver, and spleen. However, isolation attempts are frequently unsuccessful, and diagnosis must depend on serological tests.

Hantaan virus, a rodent-associated virus which causes hemorrhagic fever with renal syndrome in Asia and Europe, will also be considered in this chapter.

DESCRIPTION OF AGENTS

Important New World human pathogens include the viruses of eastern, western, Venezuelan, St. Louis, Powassan, and California encephalitis, yellow fever, dengue, Oropouche, Caraparu, Guaroa, Chagres, and Colorado tick fever. These viruses contain RNA and are usually spherical; most have an essential lipid envelope.

The laboratory host of choice is the baby mouse. Other, less susceptible hosts are wet chicks, hamsters, primary cell cultures (chicken embryo or hamster kidney), and cell culture lines (Vero, BHK-21, or HeLa). Mosquitoes and insect cell cultures are sometimes highly susceptible.

Arboviruses pass 0.22-μm, and in many cases 0.1-μm, membrane filters or the equivalent. They are readily heat inactivated, and most of them agglutinate chick and goose erythrocytes. Most are inactivated by sodium deoxycholate (SDC), diethyl ether, and chloroform.

Serological grouping is done by use of polyvalent sera in complement fixation (CF), hemagglutination inhibition (HI), immunofluorescence, enzyme-linked immunosorbent assay (ELISA), and neutralization tests. Typing is done by use of type-specific sera in these same tests. Arboviruses embrace five genera in four families: Togaviridae (*Alphavirus* and *Flavivirus*), Rhabdoviridae, Bunyaviridae, and Reoviridae (*Orbivirus*), defined by physical characteristics including appearance by electron microscopy, but these groupings are not generally useful to the diagnostic laboratory.

COLLECTION AND STORAGE OF SPECIMENS

Virus isolation is usually successful only if specimens are obtained within the first few days of illness. For virus isolation, collect blood, cerebrospinal fluid, and brain or other organs aseptically; refrigerate the specimens at 4°C. Separate the serum from the clot. Heparinized (0.16 mg/ml) whole blood or homogenized clots are satisfactory for virus isolation and are preferable to serum for some viruses (Colorado tick fever). If immediate inoculation of the specimen is not possible, store the material at −60°C or colder. Store it in sealed ampoules if dry ice is used. Collect throat washings in 0.75% bovine albumin in Hanks balanced salt solution or phosphate buffers (BAP) prepared from 0.12 M NaCl, 0.02 M sodium phosphate buffer, and bovine albumin (fraction V) powder (final pH, 7.2).

For antibody study, collect acute-phase and 3-week or later convalescent-phase sera, and store them at 4°C or colder. Collect acute-phase cerebrospinal fluid from encephalitis cases. Specimens may be shipped without refrigeration.

For long-term virus storage, add 50% normal serum or 7.5% BAP, and store the virus at −60°C or colder. Freeze-drying is excellent for the preservation of both virus and antibody.

ISOLATION OF VIRUS

Preparation of specimen

Mince the tissue with scissors and grind in a mortar or tissue grinder, or use a homogenizer. For safety, work in a vertical laminar-flow containment cabinet. Make a 10% suspension in 0.75% BAP, and centrifuge it at 480 × *g* for 10 min. If bacterial contamination is suspected, centrifuge the suspension for 1 h at 12,000 × *g* in the cold (4°C), or add penicillin (1,200 U/ml) and streptomycin (10 mg/ml). Serum and cerebrospinal fluid are inoculated without preparation or dilution.

Mouse inoculation

The mouse is susceptible to most arboviruses. Inoculate 0.015 ml intracerebrally into one or two litters of 1- to 4-day-old Swiss mice from a colony as free as possible from contaminating murine viruses. Observe the animals for 21 days for illness (lethargy, nonfeeding, tremor, loss of equilibrium, alopecia, paralysis, apnea, cyanosis, hyperexcitability, clonus, circling) or death. Kill such mice by exsanguination (ether and chloroform inactivate arboviruses), and passage a pool of liver and brain intracerebrally until mice uniformly sicken. Test harvested tissues from each passage for bacterial sterility.

Alternative laboratory hosts

If wet chicks are used, inoculate them subcutaneously (5); inoculate baby hamsters intracerebrally. Monolayer cell cultures in tubes, i.e., chicken embryo, hamster kidney, Vero, BHK-21, HeLa, LLC-MK_2, and others, should be inoculated with 0.1 ml and observed for cytopathic effect (18); cell cultures under agar should also be inoculated with 0.1 ml and observed for plaques (12, 13). If dengue virus is suspected, inoculate mosquitoes and test the heads for antigen at 7 days by the fluorescent-antibody test (15); alternatively, inoculate an *Aedes* cell culture (14). If Hantaan-related virus is suspected, inoculate Vero E6 or A549 cells.

Reisolation of the virus is important to rule out laboratory contamination or a murine virus. Alternatively, it is necessary to show antibody rise in an infected natural host to prove the validity of the isolation.

IDENTIFICATION OF VIRUS

Characterization of the isolate

Pathogenicity for laboratory hosts, incubation periods, infectivity titers by various routes of inoculation, filterability, biochemical tests for RNA, and sensitivity to chemicals may aid in identification. A comprehensive listing of these properties for 388 arboviruses is available elsewhere (2, 11).

The World Health Organization centers for arbovirus research and reference at the Vector-Borne Virus Diseases Division of the Centers for Disease Control, P.O. Box 2087, Fort Collins, CO 80521, and the Yale Arbovirus Research Unit, Box 3333, New Haven, CT 06510, do special diagnostic tests, accept presumed arboviral isolates for identification, and offer consultation and technical training.

SDC test

Inactivation by SDC is a reliable method of ruling out enteroviruses such as encephalomyocarditis virus, mouse encephalomyelitis virus, and coxsackievirus, but not myxoviruses, poxviruses, rabies virus, lymphocytic choriomeningitis virus, and herpes simplex virus. For the SDC test, prepare a 10% virus (usually brain) suspension in 0.75% BAP, and centrifuge the suspension at 12,000 × *g* for 1 h at 4°C. Mix the supernatant fluid with an equal volume of 1:500 SDC in 0.75% BAP, and mix equal volumes of virus and BAP as a control. After incubating both mixtures for 1 h at 37°C, prepare 10-fold dilutions of SDC and control mixtures, and inoculate them intracerebrally into baby mice. A reduction of titer of 1 $\log_{10}$ or more is significant. Include SDC tests of a known arbovirus and an enterovirus as a control. Enteroviruses may give a higher titer in SDC than in the control. Arboviruses of the *Orbivirus* genus (Colorado tick fever) are relatively insensitive to SDC.

Inactivation of arboviruses by ether or chloroform is equally satisfactory in distinguishing them from enteroviruses.

SEROLOGICAL IDENTIFICATION

Principle

Compare the new virus by the neutralization, CF, immunofluorescence, ELISA, and HI tests with viruses known, or suspected to be, in the geographical region of isolation (2, 4). Initially, test the new virus against polyvalent sera (prepared with agents known to exist in the geographical area). (Polyvalent sera are available from the Research Resources Branch, National Institute of Allergy and Infectious Diseases, Bethesda, MD 20205.) Once grouping is accomplished, test the virus with type-specific sera against viruses in the group. For definitive identification, the new virus and homologous immune serum must be compared in reciprocal cross-tests with type strain reagents.

Mouse-pathogenic viruses which might be isolated in the United States and for which sera should be prepared are listed in Table 1.

Preparation of immune serum

The animal of choice for the preparation of immune serum is the mouse. Alternatively, guinea pigs, rabbits, monkeys, or horses are used. Mice are preferred because they are susceptible to most arboviruses, they do not usually produce mouse tissue antibody, and the serum is useful in CF, HI, immunofluorescence, ELISA, and neutralization tests. Immunize mice with uncentrifuged, infected, 10% mouse brain suspension in saline according to the following schedule: 0.2 ml intraperitoneally on days 1 and 30; bleed on day 37 (alternative schedule: 0.2 ml intraperitoneally on days 1, 3, 10, 15, and 20; bleed on day 27). Freund complete adjuvant may be used to increase the titer (3). For polyvalent (grouping) sera, immunize the mice with a mixture of viruses (polyvalent inoculum). For type-specific sera, use a single-virus inoculum. If the virus is lethal intraperitoneally, inactivate a 10% brain suspension by incubation for 1 h at 37°C in 0.1% β-propiolactone in normal saline.

Neutralization test

The constant virus-serum dilution method, with inoculation of baby mice intracerebrally, is applicable to most arboviruses. Other types of neutralization tests vary as to the age of the animal, route of inoculation, and temperature and time of incubation (8, 17). (The constant virus-serum dilution test is commonly used in cell culture studies.) Use 1- to 4-day-old Swiss mice and randomize the babies. Inoculate them intracerebrally with 0.015 ml per mouse. Use 0.75% BAP as the diluent. Prepare the virus stock as a 10 or 20% brain (liver and serum used in some cases) suspension from sick mice. Centrifuge the stock virus at 12,000 × *g* for 1 h at 4°C.

Titrate part of the supernatant fluid under test conditions, and store the remainder at −60°C.

Make serial 10-fold virus dilutions, changing pipettes with each dilution. The number of dilutions to be used is determined from pretitration of the stock virus. In the run, include negative- and positive-control sera with the test sera. Each serum is tested undiluted against several 10-fold dilutions of the virus. Shake the virus-serum mixtures in an ice bath during inoculation, and inoculate at least six mice intracerebrally per dilution. Record deaths over a 2-week period, or at least until 3 days after the last death. Calculate the log 50% lethal dose for each test and control serum by the method of Reed and Muench (14a). The log neutralization index is the difference between the log 50% lethal dose of the negative-control serum and that of the test serum.

Alternatively, the plaque reduction neutralization test (16) is used to assay antibody and is usually quite specific.

Sera stored for long periods may lose neutralizing capacity. This is restored by performing the test intraperitoneally or by inactivating the test serum

TABLE 1. Mouse-pathogenic viruses occurring in the United States (including Puerto Rico and the Virgin Islands)

Group	Virus
Group A (alphavirus)	Venezuelan encephalitis Eastern encephalitis Western encephalitis Mayaro
Group B (flavivirus)	Cowbone Ridge Modoc St. Louis encephalitis Powassan Dengue types 1, 2, 3, and 4 U.S. bat salivary gland Montana *Myotis* leukoencephalitis Yellow fever[a]
Bunyavirus	Tensaw Northway Cache Valley La Crosse (and several closely related viruses) Trivittatus Buttonwillow Mermet Patois Pahayokee Shark River Mahogany Hammock Gumbo Limbo Lokern Main Drain Hantaan-related viruses[b]
Group vesicular stomatitis (VS)	VS-Indiana VS-New Jersey
Group Turlock	Turlock
Other groups and ungrouped	Flanders Tacaiuma New Minto Rio Grande Sunday Canyon Oceanside Tillamook Hart Park Colorado tick fever Hughes Silverwater Bluetongue Epizootic hemorrhagic disease of deer Farallon Kern Canyon Lone Star Mono Lake Yaquina Head Sawgrass
Nonarbovirus, SDC sensitive	Tamiami Lymphocytic chroiomeningitis Herpes simplex Poxviruses Newcastle disease Rabies

[a] May be recovered from individuals inoculated with live, attenuated virus vaccine.
[b] May not always be mouse pathogenic.

and adding accessory factor (normal rhesus or human serum freshly collected) in a volume equal to that of the test serum.

CF test

See reference 13a for details of the CF test.

Antigen preparation. Sucrose-acetone-extracted mouse brain (or liver) is the preferred antigen because it is useful in both CF and HI tests (6). Alternatively, 10% brain in Veronal buffer can be used as a CF antigen.

Use a hood for safety. Prepare brains (20%, wt/vol) in 8.5% sucrose by use of a blender, and express the mixture through an 18-gauge needle into 20 volumes (20 times the volume of the sucrose suspension) of chilled acetone. Shake the mixture thoroughly, and then decant the acetone. Immediately add 20 volumes of acetone and shake again; the sediment should appear as a finely dispersed particulate. Let the mixture stand for 1 h at 4°C without shaking. Carefully decant the acetone and dry the sediment (vacuum pump for 1 h). Rehydrate the powder with normal saline in a volume twice the original brain weight, and centrifuge at 12,000 × *g* for 30 min. The supernatant fluid is the antigen.

For the identification of isolates, antigen made from the new isolate is used in two dilutions: the optimal dilution, as determined in grid titration with homologous serum (usually 1:32 to 1:64), and the lowest dilution (most concentrated antigen) which is not significantly anticomplementary (usually 1:4). The antigen is tested with polyvalent sera to determine the group and then with type-specific sera to determine the type. Definitive identification involves reciprocal cross-CF tests, with the use of antigen and serum of the new isolate and antigen and serum of the type virus in a checkerboard scheme.

Fluorescent-antibody test

The indirect fluorescent-antibody test to assay antibody is used for Colorado tick fever (7) and may be adapted for other arboviruses. BHK-21 cell cultures are inoculated with virus; after 24 h, cells are removed from the culture vessel with 0.25% trypsin in phosphate-buffered saline (PBS), pH 7.5, and concentrated to give 2.0×10^7 cells per ml. Smears of 0.05 ml of cells are placed on microscope slides, air dried, fixed in acetone at room temperature, air dried again, and stored at −65°C.

Test serum and known positive- and negative-control sera are inactivated at 56°C for 30 min, diluted serially in a 20% suspension of beef brain in PBS, pH 7.3, and then added to virus-infected cell smears. Test serum is added to uninfected cells as a control. Slides are incubated at 36°C for 20 min, rinsed briefly in PBS, washed twice for 10 min in PBS, pH 7.3, and rinsed in distilled water. Rabbit anti-species-specific serum conjugated to fluorescein isothiocyanate is added to the slide and incubated for 20 min at 36°C; the slide is washed in PBS and distilled water and then dried and examined for specific staining.

This test has been adapted to rapid, specific typing of dengue infections by using monoclonal antibodies developed at the Walter Reed Army Institute of Research (9) and distributed by the Vector-Borne Virus Diseases Division, Centers for Disease Control.

ELISA

The ELISA is comparable in sensitivity to the plaque reduction neutralization test and is simple and rapid. Ammonium sulfate precipitate of hyperimmune mouse or rabbit serum is used to coat wells of polystyrene microplates. This antibody captures antigen from infected mouse tissue or cell culture or from patient tissues. Positive serum of a different species from the coat antibody is detected by antispecies antibody conjugated to peroxidase (or other enzyme). A color change is detected when substrate is added; the reaction is assayed in a spectrophotometer. The test is used for antigen detection or antibody assay.

The immunoglobulin M capture ELISA has special application to the detection of recent infection by using a single serum or cerebrospinal fluid of patients with encephalitis. Immunoglobulin M in patient material is captured by anti-mu chain antibody attached to the solid phase. Antigen is added and the plate is washed; peroxidase-conjugated antiviral antibody and substrate lead to a color change in a positive test (10).

Hemagglutination and HI tests

Antigen preparation. The sucrose-acetone antigen described in the section on the CF test is used. Alternatives are (i) acetone extraction of mouse or hamster serum (6); (ii) 10% brain in borate-saline buffer, pH 9.0; or (iii) sonically treated sucrose-acetone antigens (1). Sonic treatment utilizes uncentrifuged sucrose-acetone antigen and a Sonifier with a probe microtip in a glass tube (16 by 160 mm) immersed in an ice bath at −10°C (salt water-ice mixture stirred magnetically). Sonic treatment is conducted at a 20-W output, with the use of four cycles of 2 min each and 2 min of rest between cycles. Freezing of the antigen during this procedure should be avoided.

Chemical reagents. Borate-saline buffer (0.05 M borate–0.12 M NaCl), pH 9.0, plus 0.4% bovine albumin (Armour fraction V) is used as the diluent of antigens and sera; 0.15 M NaCl–0.2 M Na_2HPO_4 mixed with 0.15 M NaCl–0.2 M NaH_2PO_4 in the proportions shown in Table 2 is used as an adjusting diluent for the addition of cell suspensions. For bunyaviruses (Table 1), 0.4 M NaCl–0.2 M Na_2HPO_4 mixed with 0.4 M NaCl–0.2 M NaH_2PO_4 as an adjusting diluent may give higher antigen titers.

Treatment of sera. See reference 6. Test sera are extracted with acetone for removal of nonspecific inhibitors of viral hemagglutinins. Dilute the serum 1:10 in 1% bovine albumin in demineralized water; then express the diluted serum through a 23-gauge needle into 12 volumes (referred to the volume of the diluted serum) of chilled acetone, and shake the mixture. After 5 min, centrifuge lightly (bring the mixture to 270 × *g* on an International PR-2 centrifuge or the equivalent, and then turn the machine off). Excessive centrifugation packs the sediment and hinders subsequent extraction. Decant the acetone, replace it immediately with 12 volumes of chilled acetone, and shake the mixture. Centrifuge it for 5 min in the cold at 750 × *g*. Decant the acetone. Spread the sediment over the surface of the tube and dry it for 1 h (vacuum pump). Several sera may be dried simultaneously in a vacuum jar. Rehydrate to 10 times the original serum volume with borate-saline buffer, pH 9.0, and let the mixture stand overnight at 4°C.

Remove erythrocyte agglutinins (if present) by mixing diluted serum 1:50 with the packed cells (1 part cells, 49 parts 1:10 serum); keep the mixture at 4°C for 20 min with occasional shaking. Centrifuge it at 400 × *g* for 10 min at 4°C. The supernatant serum is used in testing.

Preparation of erythrocytes. See reference 6. Goose or chick erythrocytes are washed and stored not longer than 2 weeks in a solution of dextrose-gelatin-Veronal (recrystallized Veronal, 0.58 g; gelatin, 0.60 g; sodium Veronal, 0.38 g; $CaCl_2$, 0.02 g; $MgSO_4 \cdot 7H_2O$, 0.12 g; NaCl, 8.5 g; dextrose, 10.0 g; made up to 1 liter and autoclaved) as an 8% suspension and are diluted 1:24 in an adjusting diluent (Table 2) just before use.

Hemagglutinin titrations. Plastic microplates (nondisposable or disposable) and dropping pipettes with 18-gauge needles (as described for CF tests in reference 13a) are used in hemagglutinin and HI tests. Serial twofold antigen dilutions made in tubes are transferred to plates, 1 drop per well. Rows of antigen dilutions are made to correspond to different pH values (for instance, pH 5.75, 6.0, 6.2, 6.4, etc.). For each pH used, 1 drop of diluent (borate-saline buffer with 0.4% bovine albumin) is added to each well, and 2 drops are added to the control well. Cells are suspended in adjusting diluents corresponding to the desired final pH values, and 2 drops of cell suspension are added to each well. The plates are shaken by tapping their corners on a table top, and then the cells are allowed to settle (about 30 min) at room temperature or 37°C. Results are recorded in four grades: complete, nearly complete, partial, and no agglutination.

One unit is contained in the antigen dilution which gives complete or nearly complete agglutination; 4 to 8 units are used in the test. That pH which gives an adequate titer and is on the alkaline side of the optimal pH (to increase the sensitivity of the HI test) is chosen for the HI test.

HI test. Serial twofold dilutions of serum, 1 drop per well, are added to plates, followed by 1 drop of antigen per well. Serum controls (test serum plus diluent) and positive-control serum are included in each test. The antigen titration is repeated with each test. The test is incubated overnight (16 h) at 24°C (the HI test is more sensitive with incubation at 24°C than at 2°C). Erythrocytes in appropriate pH-adjusting diluent are added, 2 drops per well, and the test is read

TABLE 2. Composition of adjusting diluents used for the addition of cell suspensions[a]

Final pH[b]	0.15 M NaCl–0.2 M Na_2HPO_4 (%)	0.15 M NaCl–0.2 M NaH_2PO_4 (%)
5.75	3.0	97.0
6.0	12.5	87.5
6.2	22.0	78.0
6.4	32.0	68.0
6.6	45.0	55.0
6.8	55.0	45.0
7.0	64.0	36.0

[a] Adapted from Clarke and Casals (6).

[b] Final pH is that pH realized by mixing equal volumes of the adjusting diluent and borate-saline buffer, pH 9.0.

as described above. Complete inhibition is considered positive. Serum controls must show no agglutination. For identification of a new isolate, reciprocal cross-HI testing is needed to establish identity with the type virus.

SEROLOGICAL DIAGNOSIS

Neutralization test

Test undiluted sera in the mouse neutralization test with 10-fold virus dilutions. A difference of 1.7 log 50% lethal dose between the acute- and convalescent-phase sera is of diagnostic significance. Alternatively, test twofold serial dilutions of serum in a plaque reduction neutralization test in cell culture. A fourfold rise in antibody titer between the acute- and convalescent-phase sera is of diagnostic significance.

CF test

Use antigen in two dilutions, namely, optimal, as determined in a grid titration with homologous serum, and the most concentrated which at the same time is not significantly anticomplementary (for example, antigen dilutions of 1:64 and 1:4). A fourfold rise in antibody titer between the acute- and convalescent-phase sera is of diagnostic significance.

HI test

Two to eight antigen units are used. Test the sera in twofold dilutions starting at 1:10. A fourfold rise in antibody titer between the acute- and convalescent-phase sera is of diagnostic significance.

Fluorescent-antibody and ELISA tests

For diagnosis of illness, rabbit (or other species) anti-human globulin conjugate is used. A fourfold rise in antibody titer between the acute- and convalescent-phase sera is of diagnostic significance.

INTERPRETATION OF LABORATORY RESULTS

Definitive diagnosis can be made only with the isolation and identification of virus. A presumptive diagnosis can be made on a diagnostically significant rise in antibody titer between acute- and convalescent-phase blood specimens; however, antibody may be heterologous (related to a virus of the same group). A high titer of CF (≥1:32) or HI (≥1:320) antibody in a single convalescent-phase specimen or a fall in antibody titer during convalescence is indicative of recent infection, but it must be interpreted cautiously. The detection of immunoglobulin M usually indicates a recent infection. The detection of specific viral immunoglobulin M in the cerebrospinal fluid is excellent evidence of viral encephalitis. Isolation of the virus and serological diagnosis may indicate association of the virus with clinical illness, but they are not proof of causation.

LITERATURE CITED

1. **Ardoin, P., and D. H. Clarke.** 1967. The use of sonication and of calcium-phosphate chromatography for preparation of group C arbovirus hemagglutinins. Am. J. Trop. Med. Hyg. **16:**357–363.
2. **Berge, T. O. (ed.).** 1975. International catalogue of arboviruses. Department of Health, Education, and Welfare publication no. (CDC) 75:8301. U.S. Government Printing Office, Washington, D.C.
3. **Brandt, W. E., E. L. Buescher, and F. M. Hetrick.** 1967. Production and characterization of arbovirus antibody in mouse ascitic fluid. Am. J. Trop. Med. Hyg. **16:**339–347.
4. **Casals, J.** 1961. Procedures for identification of arthropod-borne viruses. Bull. W.H.O. **24:**723–734.
5. **Chamberlain, R. W., R. K. Sikes, and R. E. Kissling.** 1954. Use of chicks in eastern and western encephalitis studies. J. Immunol. **73:**106–144.
6. **Clarke, D. H., and J. Casals.** 1958. Techniques for hemagglutination and hemagglutination-inhibition with arthropod-borne viruses. Am. J. Trop. Med. Hyg. **7:**561–573.
7. **Emmons, R. W., D. V. Dondero, V. Devlin, and E. H. Lennette.** 1969. Serologic diagnosis of Colorado tick fever. A comparison of complement-fixation, immunofluorescence, and plaque-reduction methods. Am. J. Trop. Med. Hyg. **18:**796–802.
8. **Hammon, W. M., and T. H. Work.** 1964. Arbovirus infection in man, p. 268–311. *In* E. H. Lennette and N. J. Schmidt (ed.), Diagnostic procedures for viral and rickettsial diseases, 3rd ed. American Public Health Association, Inc., New York.
9. **Henchal, E. A., J. M. McCown, M. C. Seguin, M. K. Gentry, and W. E. Brandt.** 1983. Rapid identification of dengue virus isolates by using monoclonal antibodies in an indirect immunofluorescence assay. Am. J. Trop. Med. Hyg. **32:**164–169.
10. **Jamnback, T. L., B. J. Beaty, S. W. Hildreth, K. L. Brown, and C. B. Gundersen.** 1982. Capture immunoglobulin M system for rapid diagnosis of La Crosse (California encephalitis) virus infections. J. Clin. Microbiol. **16:**577–580.
11. **Karabatsos, N.** 1978. Supplement to international catalogue of arboviruses including certain other viruses of vertebrates. Am. J. Trop. Med. Hyg. **27:**372–440.
12. **Karabatsos, N., and S. M. Buckley.** 1967. Susceptibility of the baby hamster kidney-cell line (BHK-21) to infection with arboviruses. Am. J. Trop. Med. Hyg. **16:**99–105.
13. **Lennette, E. H., M. I. Ota, H. Ho, and N. J. Schmidt.** 1961. Comparative sensitivity of four host systems for the isolation of certain arthropod-borne viruses from mosquitoes. Am. J. Trop. Med. Hyg. **10:**897–904.

13a. **Palmer, D. F., L. Kaufman, W. Kaplan, and J. J. Cavallero.** 1977. Serodiagnosis of mycotic diseases. Charles C Thomas, Publisher, Springfield, Ill.

14. **Race, M. W., R. A. J. Fortune, C. Agostini, and M. G. R. Varma.** 1978. Isolation of dengue viruses in mosquito cell cultures under field conditions. Lancet **i:**48–49.

14a. **Reed, L. J., and H. Muench.** 1938. A simple method for estimating fifty per cent endpoints. Am. J. Hyg. **27:**493–497.

15. **Rosen, L., and D. Gubler.** 1974. The use of mosquitoes to detect and propagate dengue viruses. Am. J. Trop. Med. Hyg. **23:**1153–1160.
16. **Russell, P. K., A. Nisalak, P. Sukhavachana, and S. Vivona.** 1967. A plaque reduction test for dengue virus neutralizing antibody. J. Immunol. **99:**285–290.
17. **Schmidt, N. J., and E. H. Lennette.** 1965. Appendix. Basic techniques for virology, p. 1189–1231. *In* F. L. Horsfall, Jr., and I. Tamm (ed.), Viral and rickettsial infections of man, 4th ed. J. B. Lippincott Co., Philadelphia.
18. **Webb, P. A., K. M. Johnson, R. B. Mackenzie, and M. L. Kuns.** 1967. Some characteristics of Machupo virus, causative agent of Bolivian hemorrhagic fever. Am. J. Trop. Med. Hyg. **16:**531–538.

Chapter 78

Rabies Virus

GEORGE M. BAER AND JEAN S. SMITH

Rabies is an acute viral encephalitis that follows the bite of rabid animals. In the United States, Canada, and western Europe, the disease is spread by wild animals, especially skunks, raccoons, foxes, and insectivorous bats (13, 40). Few human deaths occur in these regions despite the large number of rabid animals reported every year. The relatively limited problem that now exists there resulted from the successful efforts to control dog rabies after World War II. Between 1946 and 1957, 113 of the 131 (86.2%) cases of human rabies in the United States were caused by rabid dogs, but since then the few human rabies cases in this country (one or two per year) usually develop after wildlife exposure (13). The human disease in most developing countries, however, is the result of uncontrolled dog rabies; in those countries the epidemiology continues unchanged, with many more human deaths occurring, and up to 25 times as many human vaccinations as in the United States (Table 1) (49). Another example of that increased incidence is a report from Cali, Colombia, where a careful survey of the brain stems of all people dying between 1962 and 1966 showed that 27 of 1,596, an astounding 1.7%, died of rabies (C. Sanmartín, P. Correa, A. Dueñas, and N. Muñoz, Mem. del Primer Seminario Nacional Sobre Rabia, Medellín, Colombia, 1967). The authors concluded that: "This alarming proportion of rabies cases in autopsies carried out . . . gives a clear idea of the situation in our community and which really cannot be very different from that in the rest of the country. A careful study of pathological material from other teaching centers and an increased awareness to the clinical and pathological aspects of rabies could show that the situation in Cali is really not peculiar." There is no doubt that rabies continues as a major public health problem in most of the world.

To improve postexposure treatment in these areas, many studies have recently been carried out on the efficacy of rabies vaccination and on the pathogenesis of the virus. It is now known that virus remains at or near the bite site for weeks or months after its entry (8); it then replicates in myocytes (32, 33) and finally advances up the peripheral nerves to the central nervous system (17). Treatment procedures are still effective during the first period when the virus is at or near the bite area (8, 31), that is, during most of the usual incubation period of 1 to 2 months (14). The prophylaxis recommended in all exposed persons is a combination of local wound treatment, antiserum injection, and vaccination (10, 51).

The disease in nature continues to spread with ease because of the long incubation periods, which permit infected animals to wander over large areas, and the subsequent lengthy period of virus excretion in the saliva. There is also a period before the appearance of signs, during which a clinically normal animal can excrete virus in the saliva. This may be many weeks in foxes (43), skunks (35), and insectivorous bats (7). In dogs the critical presymptomatic period appears to be a maximum of 14 days (20). In wild animals and stray dogs the laboratory diagnosis (by immunofluorescence) is made by killing the animal and examining the brain. When the brain is positive, virus may or may not be present in the salivary glands, but virus has never been found in salivary glands when it is absent from the brain (25). Before the advent of the fluorescent-antibody technique no reliable method was available to separate the negative brain impressions from the positive ones, and all biting dogs were held for a 10-day observation period to allow the clinician to observe the animal for the clinical signs of rabies.

Although there have been occasional reports of animals living for prolonged periods, even months, while still excreting the virus in the saliva (19), these cases should be viewed as extreme exceptions and not pertinent to the public health importance of the disease.

A separate cycle of the disease is found in vampire bats. These animals produce rabies in hundreds of thousands of cows annually in Latin America (1). An occasional person is also bitten; upwards of 150 human deaths from vampire bats have been reported from Trinidad, Brazil, Mexico, and Suriname (6).

Surveillance systems in North America (13) and Europe (40) are published periodically to give the location of rabid animals, by species. A map of skunk rabies in the United States is shown in Fig. 1. Any decision on rabies treatment and diagnosis should be made only after such surveillance information is reviewed. Any additional information needed can readily be obtained from state health departments.

RABIES DIAGNOSIS

There are two separate procedures for the laboratory diagnosis of rabies, as follows. (i) The postmortem diagnosis of the brains of infected animals, most reliably done by immunofluorescent rabies antibody staining, is important for protecting persons exposed to possibly rabid animals. (ii) Antemortem diagnosis of rabies encephalitis in humans may be performed by detection of rabies antigen in central nervous system (CNS) biopsy tissue, parafollicular nerve endings, or corneal epithelium or by isolation of virus from saliva or nasal secretions. The presence of rabies antibody in cerebrospinal fluid (CSF) or a significant serum antibody rise in patients who have never received rabies vaccine or passive antibody is also used for antemortem diagnosis.

Postmortem Rabies Diagnosis

Techniques for the diagnosis of rabies in infected animals evolved from Pasteur's 1881 report (36) that the disease-producing organisms of rabies were located in the CNS of infected animals and that "submeningeal" inoculation of infected CNS material into rabbits caused their death. The possibility of a rapid diagnostic procedure led pathologists to look for histological lesions in the CNS of humans and animals

TABLE 1. Antirabies vaccination rates for the Americas[a]

Area	Population	No. vaccinated	Rate per 100,000
North	224,803,608	31,600	14.1
Central	62,623,159	59,711	95.3
South	179,702,190	178,192	99.2

[a] From Turner (49).

dying of rabies. Some of the first pathological lesions to be described were the perivascular accumulation of mononuclear cells in neurophagic areas (Babes' nodules) (5) and the ganglionic lesions observed by Van Gehuchten and Nelis (50). These lesions were not specific for rabies, however, and were often found in other viral diseases and toxic conditions.

The true beginning of rabies diagnosis was Adelchi Negri's 1903 description (34) of inclusion bodies in the cytoplasm of certain nerve cells. Although at first mistakenly identified as protozoa and as the etiological agent of rabies, Negri bodies represent a specific and diagnostic lesion of the disease. In the half-century preceding the introduction of the immunofluorescent-antibody (IFA) technique by Goldwasser and Kissling in 1958 (22), the diagnosis of rabies was made by Negri body detection. Although specific, the recognized low sensitivity of the Negri body technique (50 to 80%) has made IFA the preferred method of rabies diagnosis.

IFA technique

Although rabies virus may be detected in all parts of the CNS of infected animals, its distribution is frequently uneven (30), and reliable diagnoses can be made only when sufficient material is examined. This should include the medulla (brain stem), the cerebellum, and the hippocampus. Thin-tissue impressions are made by lightly touching cut sections from each of the three areas of the brain to clean microscope slides. Duplicate impressions are also prepared.

Before staining, impressions are fixed in acetone at −20°C for 4 h or overnight. In emergency situations, examination of slides fixed for only 10 min gives satisfactory results, although a second test of slides fixed for the longer intervals should be used for confirmation of negative results.

After removal from the acetone, the slides are allowed to air dry for 10 to 15 min (acetone-fixed slides should be handled as containing potentially infectious material [21]). The tissue impressions are then ringed with heavy ink or nail polish to contain the antirabies fluorescent-antibody conjugate.

The IFA test must have proper controls to insure specificity and reliability. Slides made from known positive and known negative animals should be included with each test. Two impressions are made on each slide. The specificity of the reaction is ensured by absorbing the antirabies conjugate with either normal brain tissue (on one impression) or with rabies-infected brain tissue (on the other). If the slide contains

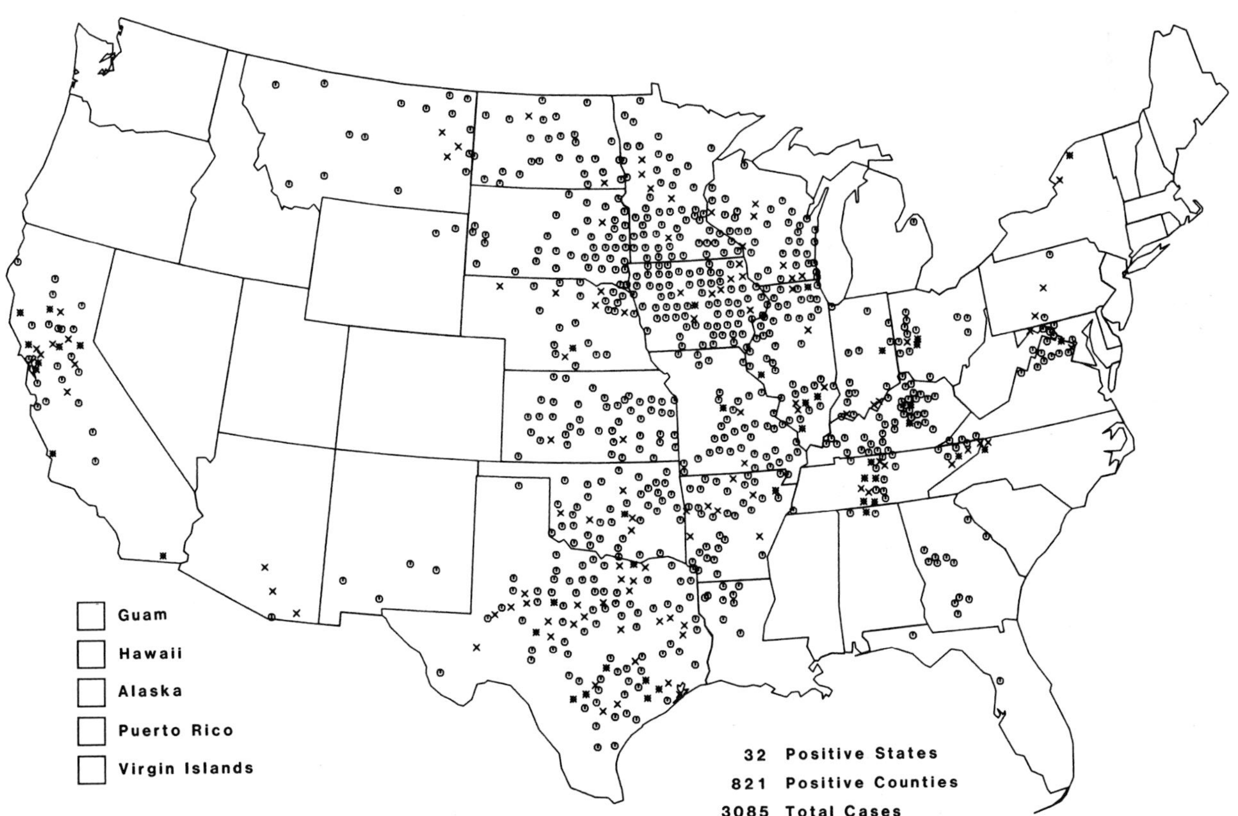

FIG. 1. Counties reporting skunk rabies, 1982. Symbols: ①, 1 to 5 cases; ×, 6 to 10 cases; *, over 10 cases. Symbols are plotted at the population center of each county.

rabies antigen, rabies antibody remaining in the conjugate absorbed with normal brain tissue should react with the antigen on test slides and the positive control slide to produce fluorescence. If the reaction is specific for rabies, tissue stained with the conjugate absorbed with rabies antigen should not fluoresce. The negative control slide is used to detect nonspecific staining which may be mistaken for rabies antigen. The rabies-positive slide serves as a sensitivity control for the conjugate.

Each impression is covered with absorbed conjugate, and the slides are incubated for 30 min at 37°C in a humidified chamber. Unbound conjugate is then removed by rinsing in phosphate-buffered saline (pH 7.6) and one or two 10-min soaks in phosphate-buffered saline. Salts are removed by a final rinse in distilled water. Phosphate-buffered glycerol mounting medium (pH 8.5) is dropped on each impression, and a cover slip is added. A standard microscope equipped with a UV light source and appropriate filters is used to observe fluorescing rabies antigen in infected tissues. Fluorescent material may vary from dustlike particles of <1 μm to large masses or threads 2 to 10 μm in diameter.

Comparative studies of IFA and virus isolation techniques have shown a 97 to 98% correlation of diagnoses (25). Since more tissue is sampled by virus isolation, that technique may be the more sensitive procedure when a very small amount of antigen is present. On the other hand, IFA may detect rabies antigen in tissue which may no longer contain infectious virus because of tissue decay or the presence of antibody.

Although the IFA technique has been thoroughly evaluated in the 20 years since its introduction, some laboratories still confirm the absence of rabies virus in IFA-negative clinical samples by the inoculation of brain tissue suspensions into susceptible animals, especially when human exposure to the submitted animals has occurred.

Histological examination for Negri bodies

If facilities for IFA detection of rabies antigen are unavailable, staining techniques for the demonstration of Negri bodies may be used as a method of diagnosis. The simplest, fastest, and most economical procedure uses Sellers stain, a 1% solution of basic fuchsin and methylene blue in absolute methanol (42). Tissue impressions or smears are prepared from the medulla, cerebellum, and hippocampus. While the tissue impressions or smears are still wet (no separate fixation is needed), the slides are immersed in Sellers stain for 1 to 5 s and then immediately rinsed in tap water or phosphate-buffered water (pH 7.0).

When examined under high magnification, these stained tissues appear reddish violet to purplish blue. The Negri body takes up the fuchsin and stains magenta or cherry red, appearing as a sharply defined, spherical, oval, or elongated body from 0.25 to 27 μm in diameter. For the inclusion to be confirmed as rabies specific, basophilic dark-blue-staining granules (0.2 to 0.5 μm in diameter) must appear in it.

Negri body development appears to be related to the length of time an animal lives after it develops rabies. Although the observation of Negri bodies confirms the diagnosis of rabies, the inclusions are found in only 75 to 80% of specimens found positive by other, more sensitive diagnostic methods. In some animal species the sensitivity of this diagnostic method may even drop to 50% or lower. In addition, there may be nonspecific inclusions in some animals, especially in cats and rodents, giving false-positive rabies diagnoses. Because of the low sensitivity and possibility of false-positives with the Sellers test, results should always be confirmed by mouse inoculation (as described below) when IFA is not available.

Mouse inoculation test for rabies virus isolation

The absence of infectious virus in Negri bodies or IFA-negative material may be confirmed by the inoculation of suckling or weanling mice (26).

A 20% suspension of tissue from medulla, cerebellum, and hippocampus is prepared in phosphate-buffered saline containing a protein stabilizer (such as 0.75% bovine albumin fraction V). Penicillin (500 U) and streptomycin (2 mg/ml) should also be added to prevent animal death from bacterial contamination of the tissue suspension. Five weanling mice or two families of suckling mice are then inoculated intracerebrally with a 1:2 dilution of centrifuged suspension (500 × *g* for 5 min) and observed daily for 30 days for signs of rabies (trembling, humping, paralysis, or prostration). A 7- to 20-day incubation period is expected in mice inoculated with street virus preparations, but incubation periods of 30 days or more are sometimes observed.

The presence of rabies virus antigen in the brains of dying mice is usually confirmed by IFA.

Collecting and shipping of specimens for diagnosis

Any animal suspected of having rabies should be killed immediately, and specimens should be sent to the laboratory for examination. A domestic or pet animal which has bitten a human or another animal but otherwise has no signs of illness, nor had contact with rabid animals, should be confined and observed for 14 days. Both wild and stray animals which have bitten a human or domestic animal are always considered as possibly rabid and must be killed immediately and sent to the laboratory for examination (51). Circumstances may dictate what method of euthanasia is used, but an intact head is necessary since rabies diagnosis is made by examination of the brain. Large animals should be decapitated, and the head is then submitted for testing. Specimens should be placed in a suitable watertight metal container which is sealed tightly and then placed in a large, watertight, insulated container. If transit time to state or regional diagnostic laboratories is less than 2 days, specimens may be refrigerated with wet ice or cold packs. For longer transit times, freezing of the brain sample on dry ice is recommended. An alternative method of preservation (suspension of 0.5-in [ca. 1.3-mm]-square pieces of brain samples in 50% glycerine-saline) can also be used.

Each specimen must be accompanied by a detailed history of the date and nature of human exposures and the names and addresses of the contacts and owners. Also included should be the species or breed of animal, whether the animal died or was killed, and its vaccination status.

Because rabies diagnosis requires handling of infected tissues, care must be taken to protect those involved in every step of collection and processing. Pre-exposure rabies immunization is recommended for veterinarians, animal control personnel, and diagnosticians. Those assisting in removal of the brain from a rabies-suspected animal should wear heavy rubber gloves, a face shield, and protective clothing.

Antemortem Diagnosis

The recently developed methods for the antemortem diagnosis of rabies are based on previous studies of rabies pathogenesis (17, 33, 37). These studies showed that, after multiplication in the CNS, rabies virus may be transmitted centrifugally along nerves to peripheral organs. Peripheral nerves, cornea, salivary glands, and other tissues near the CNS may be infected early in the disease. The detection of viral antigen in IFA-stained impressions of corneal epithelium (39) and frozen sections of skin biopsy (46) and isolation of virus from saliva and tracheal aspirates are reliable indicators of rabies infection of the CNS. Other satisfactory methods of diagnosis are the demonstration of a significant rise in titer of antibodies to rabies virus in serum, in the absence of passive or active immunization, and especially the appearance of antibody in CSF (18, 24, 38) during infection.

Laboratory techniques used in these procedures are described below.

Nuchal biopsy

Biopsies of the skin, 5 to 6 mm in diameter and of the same depth, should be taken from the posterior region of the neck just above the hairline. Specimens should be placed in a small sterile sealed container such as a screw-cap tube or plastic bag with a small amount of moist tissue paper to prevent desiccation. This material should then be enclosed in a sealed mailing can and shipped on dry ice. No preservative or fluid should be used. Acetone-fixed frozen sections 4 to 6 μm thick are then examined by IFA for the presence of rabies antigen in cutaneous nerves surrounding the hair follicles. The percentage of rabies-infected individuals with positive nuchal biopsy appears to be lower than 50% (9, 11, 12).

Corneal impression

Cleaned microscope slides (prepared by rinsing with alcohol, flame drying, and then cooling) should be pressed firmly onto the cornea and released (one impression per slide). Two slides should be prepared from each eye. A frosted end should be used to identify the side of the slide containing the impression. A local anesthetic can be used when the impressions are made, if necessary, but its use should be noted on the information sent to the laboratory. Slides should be air dried, placed in an airtight slide container in a sealed mailing can, and shipped on dry ice. Acetone fixation of the tissue is performed in the diagnostic laboratory. Clinical specimens are then examined by the IFA procedure for the presence of rabies antigen in corneal epithelial cells. This technique appears to be less sensitive than is the examination of nuchal biopsy, and false-positive results have been reported (9).

Isolation of virus from saliva

Virus may be present in oral secretions collected during the course of the disease. Screw-cap vials containing 1 to 2 ml of a buffered cell culture medium, such as Eagle minimal essential medium, may be used for saliva or oral swab samples. Saliva may be collected by using a sterile eyedropper pipette, which is then rinsed in the medium. If the patient does not salivate, a swab of the oral cavity is rotated in the medium, and fluid is expressed from the swab by rolling it against the side of the tube.

Isolation of rabies virus may be attempted by mouse inoculation, as described previously (26), or by inoculation of susceptible cell culture lines with subsequent detection of viral antigen by IFA techniques (22). Cell lines successfully used to grow street rabies virus isolates are murine neuroblastoma cells, chicken embryo-related cells, and baby hamster kidney cells (44, 47).

All conventional methods can be employed for rabies virus propagation in these cell cultures. Monolayers or suspension cultures are susceptible when inoculated with tissue suspension or saliva from infected individuals. Seeding of cell culture chamber slides or culture tubes containing glass cover slips allows examination of cultures at daily intervals for the presence of infected cells. As observed in our laboratory and others (44), murine neuroblastoma cell lines are the most sensitive cells to street rabies virus strains.

Measurement of antibody in serum and CSF samples

Also of use in the intravitam diagnosis of rabies is the measurement of antirabies antibody in serum and CSF samples. Antibodies induced by viral membrane antigens (the glycoprotein and possibly the M or matrix proteins) have the ability to neutralize rabies virus and are measured by neutralization tests in mice (3) or cell culture (15, 28, 41, 45). Antibody induced by internal viral antigens (ribonucleoproteins) may be measured by indirect immunofluorescence (48), enzyme-linked immunosorbent assay (4), and some other antibody-binding tests (16, 23, 27).

Antibody does not usually appear in the serum until day 8 to 10 of clinical illness (18, 24, 38), which may be several months after the patient's rabies exposure. Specimens collected before day 8 are not usually helpful except as the first of paired samples. As with other viral infections, a rise in serum antibody titers is indicative of current infection.

In rabies-infected individuals, antibody also appears in the CSF a day to a week (or longer) after antibody is detected in the serum. No significant amount of rabies antibody has been detected in the CSF of rabies-immunized subjects; therefore, rabies antibody in the CSF, regardless of the rabies immunization history, confirms rabies infection (2, 18, 24, 38).

Although any method for the detection of rabies-specific antibody may be used for diagnostic purposes, antibody induced by glycoprotein surface antigens of the virus can prevent infection by rabies virus, and tests measuring this ability (neutralization tests) are employed to test the efficacy of pre- or postexposure antirabies prophylaxis. The Centers for Disease Control has adopted the rapid fluorescent focus inhibition test (RFFIT) (45) because of its sensitivity, rapidity, economy, and reproducibility. Briefly, the test is performed as described below.

Dilutions of test serum in cell culture medium are mixed with a constant amount of rabies virus in a multichambered cell culture slide. Neutralization is achieved by incubation of serum and virus for 90 min at 37°C. Suspensions of BHK-21/13S cells (29) are then added to each well, and incubation at 37°C is continued for 24 h. Cultures are then rinsed, fixed in acetone, and stained by IFA. A reduction of 50% or more in the number of cells infected with virus indicates that the serum contains neutralizing antibody. An estimation of the amount of antibody in a particular serum can be made by determining the reduction of challenge virus by successive dilutions of serum. The results of this assay may be expressed as a serum titer or as international units of antibody if a reference standard serum has been included in the test.

Interpretation of antemortem diagnostic results

Antemortem diagnosis of rabies is one of the most difficult procedures attempted by the reference laboratory. The psychological impact on a patient's family and on the hospital staff when rabies infection is considered as a diagnosis should not be underestimated. To make patient handling even more difficult, negative findings on all of the above-mentioned tests cannot rule out rabies infection, although the possibility that rabies virus is the cause of illness is certainly reduced. In addition, no single diagnostic test has been positive in every case of human rabies, although a positive result on any of the currently used tests is an indication of rabies infection. The antibody response to infection, usually found on days 8 to 10 of clinical illness, may on rare occasion be absent (11), even 33 days after clinical illness. Virus isolation or detection of antigen may be negative because of sampling time (rabies virus may not have spread peripherally from the CNS) or insensitivity of the assay methods used (12). The distribution of antigen in nuchal biopsy and corneal impressions may be extremely irregular (for instance, only one of six hair follicles in a recent diagnostic test contained rabies-specific antigen) (11; John Shaddock, personal communication), making specificity controls difficult to interpret. In some instances false-positive diagnoses have been obtained in IFA examination of corneal epithelium (9).

For these reasons three important points concerning antemortem diagnosis should be emphasized: (i) antemortem diagnosis should be attempted only by experienced laboratories; (ii) if rabies is suspected, samples should be collected for testing by all currently used diagnostic procedures; (iii) repeated samples may be necessary because the absence of rabies antigen or antibody in early samples does not rule out infection.

LITERATURE CITED

1. **Acha, P. N.** 1967. Epidemiology of paralytic bovine rabies and bat rabies. Bull. Off. Int. Epizoot. **67**:343–382.
2. **Arko, R. J., L. G. Schneider, and G. M. Baer.** 1973. Nonfatal canine rabies. Am. J. Vet. Res. **34**:937–938.
3. **Atanasiu, P.** 1966. Quantitative assay and potency test of antirabies serum, p. 167–172. *In* Laboratory techniques in rabies, 2nd ed. Monograph series no. 23. World Health Organization, Geneva.
4. **Atanasiu, P., V. Savy, and P. Perrin.** 1977. Epreuve immunoenzymatique pour la detection rapide des anticorps antirabiques. Ann. Microbiol. (Inst. Pasteur) **128A**:489–498.
5. **Babes, V.** 1892. Sur certains-caractères des lesions histologiques de la rage. Ann. Inst. Pasteur Paris **6**:209–223.
6. **Baer, G. M.** 1975. Bovine paralytic rabies and rabies in the vampire bat, p. 155–175. *In* G. M. Baer (ed.), The natural history of rabies, vol. 2. Academic Press, Inc., New York.
7. **Baer, G. M., and G. L. Bales.** 1967. Experimental rabies infection in the Mexican freetail bat. J. Infect. Dis. **117**:82–90.
8. **Baer, G. M., and W. F. Cleary.** 1972. A model in mice for the pathogenesis and treatment of rabies. J. Infect. Dis. **125**:520–527.
9. **Centers for Disease Control.** 1979. Human rabies—U.S. Morbid. Mortal. Weekly Rep. **28**:315.
10. **Centers for Disease Control.** 1980. Recommendation of the Immunization Practices Advisory Committee (ACIP)—rabies prevention. Morbid. Mortal. Weekly Rep. **29**:265–272, 277–280.
11. **Centers for Disease Control.** 1981. Human rabies acquired outside the U.S. from a dog bite. Morbid. Mortal. Weekly Rep. **30**:537.
12. **Centers for Disease Control.** 1983. Human rabies—Michigan. Morbid. Mortal. Weekly Rep. **32**:159.
13. **Centers for Disease Control.** 1983. Rabies surveillance annual summary 1980–1982. Centers for Disease Control, Atlanta, Ga.
14. **Dean, D. J.** 1963. The pathogenesis of rabies. N.Y. State J. Med. **63**:3507–3513.
15. **Debbie, J. G., J. A. Andrulonis, and M. K. Abelseth.** 1972. Rabies antibody determination by immunofluorescence in tissue culture. Infect. Immun. **5**:902–904.
16. **Dierks, R. E., and P. M. Gough.** 1973. Passive hemagglutination test for rabies antibodies, p. 147–150. *In* M. M. Kaplan and H. Koprowski (ed.), Laboratory techniques in rabies, 3rd ed. World Health Organization, Geneva.
17. **DiVestea, A., and G. Zagari.** 1889. La transmission de la rage par voie nerveouse. Ann. Inst. Pasteur Paris **3**:237–248.
18. **Fekadu, M., and G. M. Baer.** 1980. Recovery from clinical rabies of two dogs inoculated with a rabies virus strain from Ethiopia. Am. J. Vet. Res. **41**:1632–1634.
19. **Fekadu, M., J. H. Shaddock, and G. M. Baer.** 1981. Intermittent excretion of rabies virus in the saliva of a dog two and six months after it had recovered from experimental rabies. Am. J. Trop. Med. Hyg. **30**:1113–1115.
20. **Fekadu, M., J. H. Shaddock, and G. M. Baer.** 1982. Excretion of rabies virus in the saliva of dogs. J. Infect. Dis. **145**:715–719.
21. **Fischman, H. R., and F. E. Ward.** 1969. Infectivity of fixed impression smears prepared from rabies virus-infected brain. Am. J. Vet. Res. **30**:2205–2208.
22. **Goldwasser, R. A., and R. E. Kissling.** 1958. Fluorescent antibody staining of street and fixed rabies virus antigen. Proc. Soc. Exp. Biol. Med. **98**:219–223.
23. **Halonen, P., F. A. Murphy, B. N. Field, and D. R. Reese.** 1968. Hemagglutinin of rabies and some other bullet-shaped viruses. Proc. Soc. Exp. Biol. Med. **123**:1037–1042.
24. **Hattwick, M. A. W., T. T. Weiss, C. J. Stechschulte, G. M. Baer, and M. G. Gregg.** 1972. Recovery from rabies. Ann. Intern. Med. **76**:931–942.
25. **Kissling, R. E.** 1975. The fluorescent antibody test in rabies, p. 401–416. *In* G. M. Baer (ed.), The natural history of rabies, vol. 1. Academic Press, Inc., New York.
26. **Koprowski, H.** 1973. The mouse inoculation test, p. 85–93. *In* M. M. Kaplan and H. Koprowski (ed.), Laboratory

techniques in rabies, 3rd ed. World Health Organization, Geneva.

27. **Kuwert, E.** 1973. The complement fixation test, p. 124–134. *In* M. M. Kaplan and H. Koprowski (ed.), Laboratory techniques in rabies, 3rd ed. World Health Organization, Geneva.
28. **Lennette, E. H., and R. W. Emmons.** 1971. The laboratory diagnosis of rabies: review and perspective, p. 77–90. *In* Y. Nagano and F. Davenport (ed.), Rabies (proceedings of the Conference on Rabies sponsored by the Japan-United States Cooperative Medical Science Program, Univ. of Tokyo). University Park Press, Baltimore.
29. **MacPherson, I., and M. Stoker.** 1962. Polyoma transformation of hamster cell clones. An investigation of genetic factors affecting cell competence. Virology **16**:147–151.
30. **Maserang, D. L., and L. Leffingwell.** 1981. Single-site localization of rabies virus: impact on laboratory reporting policy. Am. J. Public Health **71**:428–429.
31. **Moreno, J. A., S. D. Baughcum, H. B. Levy, and G. M. Baer.** 1979. Further studies on rabies postexposure prophylaxis in mice: a comparison of vaccine with interferon and vaccine. J. Gen. Virol. **42**:219–222.
32. **Murphy, F. A., S. P. Bauer, A. K. Harrison, and W. C. Winn.** 1973. Comparative pathogenesis of rabies and rabies-like viruses. Viral infection and transit from inoculation site to the central nervous system. Lab. Invest. **28**:361–376.
33. **Murphy, F. A., A. K. Harrison, W. C. Winn, and S. P. Bauer.** 1973. Comparative pathogenesis of rabies and rabies-like viruses. Infection of the CNS and centrifugal spread of virus to peripheral tissues. Lab. Invest. **29**:1–16.
34. **Negri, A.** 1903. Beitrag zum Studium der Aetiologie der Tollwut. Z. Hyg. Infektionskr. **43**:507–528.
35. **Parker, R. L., and R. E. Wilsnack.** 1966. Pathogenesis of skunk rabies virus: quantitation in skunks and foxes. Am. J. Vet. Res. **27**:33–38.
36. **Pasteur, L., S. Chamberland, E. Roux, and E. Thuillier.** 1881. Sur la rage. C. R. Acad. Sci. **92**:1259–1260.
37. **Perl, D. P.** 1975. The pathology of rabies in the central nervous system, p. 236–242. *In* G. M. Baer (ed.), The natural history of rabies, vol. 1. Academic Press, Inc., New York.
38. **Porras, C. J., T. Barboza, E. Fuenzalida, H. Lopez Adaros, A. M. Oviedo de Diaz, and T. Furst.** 1976. Recovery from rabies in man. Ann. Intern. Med. **85**:44–48.
39. **Schneider, L. G.** 1969. The cornea test: a new method for the intra-vitam diagnosis of rabies. Zentralbl. Veterinaermed. Reihe B **16**:24–31.
40. **Schneider, L. G.** 1983. Rabies bulletin Europe/information surveillance research, vol. 7, no. 3. WHO Collaborating Centre for Rabies Surveillance and Research, Federal Research Institute for Animal Virus Diseases, Tübingen, Federal Republic of Germany.
41. **Sedwick, W. D., and T. J. Wiktor.** 1967. Reproducible plaquing system for rabies, lymphocytic choriomeningitis, and other ribonucleic acid viruses in BHK-21/13S agarose suspensions. J. Virol. **1**:1224–1226.
42. **Sellers, T. F.** 1927. A new method for staining Negri bodies of rabies. J. Publ. Health **17**:1080–1081.
43. **Sikes, R. K.** 1962. Pathogenesis of rabies in wildlife. I. Comparative effect of varying doses of rabies virus inoculated into foxes and skunks. Am. J. Vet. Res. **23**:1041–1047.
44. **Smith, A. L., G. H. Tignor, R. W. Emmons, and J. D. Woodie.** 1978. Isolation of field rabies virus strains in CER and murine neuroblastoma cell cultures. Intervirology **9**:359–361.
45. **Smith, J. S., P. A. Yager, and G. M. Baer.** 1973. A rapid reproducible test for determining rabies neutralizing antibody. Bull. W.H.O. **48**:535–541.
46. **Smith, W. B., D. C. Blenden, F. Tsu-Huei, and L. Hiler.** 1972. Diagnosis of rabies by immunofluorescent staining of frozen sections of skin. J. Am. Vet. Med. Assoc. **161**:1495–1501.
47. **Sokol, F., E. Kuwert, T. Wiktor, K. Hummeler, and H. Koprowski.** 1968. Purification of rabies virus grown in tissue culture. J. Virol. **2**:836–849.
48. **Thomas, J. B., R. K. Sikes, and A. S. Ricker.** 1963. Evaluation of indirect fluorescent antibody techniques for detection of rabies antibody in human sera. J. Immunol. **91**:721–723.
49. **Turner, G. S.** 1976. A review of the world epidemiology of rabies. Trans. R. Soc. Trop. Med. Hyg. **70**:175–178.
50. **Van Gehuchten, A., and C. Nelis.** 1900. Les lesions histologiques de rage chez les animaux et chez l'homme. Bull. Acad. R. Med. Belg. **14**:31–36.
51. **World Health Organization.** 1973. Expert Committee on Rabies, sixth report. World Health Organization, Geneva.

Chapter 79

Marburg Virus, Ebola Virus, and the Arenaviruses

PETER B. JAHRLING

CLINICAL BACKGROUND

Viruses from two taxa, the family Arenaviridae (11) and the proposed family Filoviridae (15), are important human pathogens causing severe, frequently fatal viral hemorrhagic fevers. Although exotic to North America, these viruses are important public health problems in Africa and South America and may be introduced elsewhere by travelers returning from these areas. The procedures for the initial isolation, clinical management, and virological diagnosis of patients with suspected arenavirus or filovirus infections are similar, and it is appropriate to group these viruses together in this chapter. Patients with these infections frequently present with similar, nonspecific clinical manifestations. A detailed travel history may be helpful in early diagnosis, which is critical to timely intervention with specific immune globulin and appropriate antiviral drugs.

Among the 11 recognized members of the family Arenaviridae (11), 4 are human pathogens. Lassa virus is the etiological agent of Lassa fever, a West African disease. The original, isolated outbreaks described were hospital-associated, occurring first in Nigeria in 1969 and later in Liberia and Sierra Leone. High case-fatality rates ranging from 20 to 40% were reported among hospitalized cases. Lassa virus is now known to persist in endemic foci in Liberia and Sierra Leone, and the true incidence of human Lassa fever cases is believed to be thousands to tens of thousands per year. In hospitals located near endemic foci in Sierra Leone, 30% of all medical deaths are attributed to Lassa fever. Among hospitalized Lassa fever cases, the mortality rate is currently estimated to be 15 to 20%. Overall mortality, including subclinical cases identified serologically, is substantially less, perhaps less than 1%. Deaths occur all year, with some increase during the dry season.

Junin virus (causing Argentine hemorrhagic fever [AHF]) was first isolated in 1959, although the disease had been recognized since 1943 in Argentina, where it has been associated with annual outbreaks of disease, primarily among agricultural workers. Annual numbers of AHF cases from 1958 to the present have fluctuated from 100 to 3,500, with several hundred cases occurring most years. The mortality rate for patients with laboratory-confirmed AHF who do not receive specific immune plasma therapy is reported to be 14 to 17%.

Machupo virus, the etiological agent of Bolivian hemorrhagic fever, was first isolated in 1963, although reports of sporadic outbreaks began in 1959 and a series of devastating epidemics occurred from 1962 to 1964; these involved more than 1,000 patients and had an 18% mortality rate. Another severe outbreak occurred in Cochabamba, Bolivia, in 1971, and was associated entirely with nosocomial spread from an index case apparently infected in the endemic region. Lassa, Machupo, and Junin viruses are associated with specific rodent hosts; chronically infected rodents may serve as the primary reservoirs for transmission of the viruses to humans who come into contact with materials or food contaminated with rodent excreta. Geographic distributions of the viruses are limited by the restricted ranges of the rodent hosts.

In contrast to the viruses described above, lymphocytic choriomenigitis (LCM) virus may have a worldwide distribution due to its association with the house mouse, *Mus musculus*. The presence of LCM virus is well documented in North America, South America, and Europe, except for Scandinavia. LCM virus has also not been isolated in Australia or Africa. Recently, human infections with LCM virus have been traced to contact with pet Syrian hamsters (*Mesocricetus auratus*).

Marburg (MBG) and Ebola (EBO) viruses are tentatively classified as Filoviridae (15). MBG virus was first recognized in 1967 when 25 persons in Germany and Yugoslavia became infected after contact with monkey kidneys or cell cultures derived from monkeys imported from Uganda. Seven of these patients died. In addition, six secondary human cases occurred among persons having direct contact with the primary cases. Since then, sporadic cases have occurred in Zimbabwe, South Africa, and Kenya. EBO virus first emerged in two major disease outbreaks, which occurred simultaneously in Zaire and Sudan in 1976. Over 500 cases were reported, with case-fatality rates of 88% in Zaire and 53% in Sudan. Serological studies to date suggest that infections with EBO or related viruses have occurred in Zaire, Sudan, Central African Republic, Gabon, and Kenya. The geographic range of EBO viruses may extend to other African countries for which adequate serosurveys are lacking. Subclinical infections may occur. The natural reservoirs for human infection with MBG and EBO viruses have not been identified.

Clinically, LCM virus infection is the least severe of the group, presenting most commonly as an influenza-like syndrome and less frequently as meningoencephalitis or aseptic meningitis (17). Even when the disease progresses to the neurological state, it is rarely fatal, and complete recovery usually ensues. A large proportion of LCM virus infections are thought to be subclinical. Lassa, Junin, and Machupo virus infections are more severe. For Lassa virus infection, fever and myalgia develop insidiously between 3 and 16 days after exposure and increase in severity during the following week. There are few specific clinical manifestations. Lassa fever patients usually come to the hospital within 5 to 7 days of onset, with sore throat, severe lower back pain, and conjunctivitis. Pneumonitis and pleural and pericardial effusions with friction rub frequently occur. A maculopapular rash may develop, but although Lassa fever is grouped with the hemorrhagic fevers, frank hemorrhage is seen only in a proportion of the more severe cases. Death, due to sudden cardiovascular collapse as a consequence of hepatic, pulmonary, and myocardial damage, occurs in the second or third week in ca. 15 to 20% of hospitalized patients. Few Lassa fever patients develop central nervous system signs, although tinnitis or deafness may develop as recovery begins. Lassa fever is a particularly severe disease among pregnant women, for whom mortality rates are somewhat higher.

The clinical pictures for AHF (Junin virus) and Bolivian hemorrhagic fever (Machupo virus) are similar. Incubation periods range from 7 to 14 days, and very few subclinical cases are thought to occur. After a gradual onset of fever, anorexia, and malaise over several days, constitutional signs involving the gastrointestinal, cardiovascular, and central nervous systems become apparent by the time patients present to the hospital. On initial examination, AHF and Bolivian hemorrhagic fever patients are febrile, acutely ill, and mildly hypotensive. They frequently complain of back pain, epigastric pain, headache, retroorbital pain, photophobia, dizziness, constipation or diarrhea, and coughing. Vascular phenomena, including flushing of the face, neck, and chest and bleeding from the gums, are common. Enanthem is almost invariably present; petechiae or tiny vesicles spread over erythematous palate and fauces. Neurological involvement, ranging from mild irritability and lethargy to abnormalities in gait, tremors of the upper extremities, and, in severely ill patients, coma, delirium, and convulsions, occurs in more than half the patients. During the second week of illness, clinical improvement may begin, or complications may develop. These include extensive petechial hemorrhages, oozing from puncture wounds, melena, and hematemesis. These manifestations of capillary damage and thrombocytopenia do not result in life-threatening blood loss. However, hypotension and shock, complicated by gross hemorrhaging, may develop, often in combination with serious neurological signs in the 15% of patients who die. Survivors begin to show improvement by the third week. Recovery is slow; weakness, fatigue, and mental difficulties may last for weeks.

MBG (19) and EBO virus infections are clinically indistinguishable. After incubation periods of 4 to 16 days, fever develops either suddenly or insidiously with headache, myalgia, and conjunctival effusions, followed by nausea, vomiting, and diarrhea. Biochemical evidence of hepatitis without clinical jaundice is usual, and multiple organ systems become involved. A characteristic maculopapular rash over the trunk, petechiae, and mucous membrane hemorrhages appear. Protracted vomiting and intense epigastric pain is accompanied by uncontrollable bleeding from venipuncture sites. Development of shock occurs soon before death, often 7 to 10 days after the onset of illness. Fibrin split products and abnormalities in coagulation parameters suggest that disseminated intravascular coagulation is a terminal event, particularly in MBG disease.

DESCRIPTION OF THE AGENTS

The family Arenaviridae comprises 11 viruses which share unique morphological and physiochemical characteristics. Antigenic relationships are established mainly on the basis of broadly reactive complement fixation (CF) and indirect fluorescent-antibody (IFA) tests, whereas fine discriminations among virus strains are made by neutralization tests. Two taxonomic complexes are generally accepted (3, 29). The LCM, or Old World, complex includes LCM and Lassa viruses and a number of Lassa-like strains recently isolated from rodents in Mozambique, Zimbabwe, and Central African Republic. These strains may be unique viruses or substrains of Lassa. All have been isolated from rodents of the family Muridae. The Tacaribe, or New World, complex includes Tacaribe, Junin, Machupo, Amapari, Parana, Latino, Pichinde, Tamiami, and Flexal viruses. All New World complex viruses have been isolated from rodents of the family Cricetidae or, in the case of Tacaribe, from bats. Old World and New World complex viruses are distantly related, and cross-reactions by CF or IFA test are obtained only when high-titered antisera are used.

All arenaviruses share a unique morphology when observed in thin-section preparations by electron microscopy (22). This morphology is so distinctive that it was the basis for first associating these viruses into the present taxon. Individual virions are pleomorphic and range in size from 60 to 280 nm (mean, 110 to 130 nm). A unit membrane envelopes the structure and is covered with club-shaped, 10-nm projections. No symmetry has been discerned. The most prominent and distinctive feature of these virions is the presence of several electron-dense particles (usually 2 to 10), which may be connnected to each other by fine filaments. These particles, 20 to 25 nm in diameter, have been shown to be identical with host cell ribosomes by biochemical and oligonucleotide analysis. Three major virion structural proteins are usually found (24). Two are glycosylated; G1 (50,000 to 72,000 daltons) and G2 (31,000 to 41,000 daltons) form the virion envelope and spikes. The N protein (63,000 to 72,000 daltons) is clearly associated with the virion RNA and is considered the nucleocapsid protein. The N protein in intact cells or virions is not accessible to antibody but can readily be detected in acetone-fixed cells by immunofluorescence. Nucleocapsids can be isolated by treatment of intact virions with detergent. Liberated nucleocapsids are 10 nm in diameter and range to 450 nm in length. Two size classes of closed or circular nucleocapsid have been identified, 640 and 1,300 nm. Small, 3- to 4-nm, beaded strands can also be resolved. Four RNA species with distinct oligonucleotide fingerprints can be isolated from intact virions. The large (31S) and small (22S) RNA species are virus specific; the 28S and 18S RNA species, isolated in various proportions, are ribosomal. Arenaviruses are negatively stranded; for some, an RNA-dependent RNA polymerase activity has been demonstrated. The molecular weights of large and small viral RNAs have been estimated at 2.6×10^6 and 1.4×10^6, respectively. The group-reactive arenavirus antigen measured in both the CF and IFA tests is associated with the nucleocapsid and is probably N protein. A soluble (non-virion-associated) CF antigen is also elaborated by infected cells in culture and, in the case of Pichinde virus, is antigenically related to the N protein. The glycoproteins G1 and G2 are elaborated on the envelope and in surface projections, and they probably serve as neutralization targets which are highly type specific (2). No hemagglutinins have been found. Arenaviruses mature by budding at the cytoplasmic membrane, and host proteins are incorporated into the virion envelope. Buoyant densities of intact arenavirus particles in cesium chloride range from 1.17 to 1.18. All arenaviruses are readily inactivated by ethyl ether, chloroform, sodium deoxycholate, and acid media (pH less than 5). Beta-propiolactone (27) and gamma irradiation (8) are both reported to inactivate arenavirus infectivity while preserving reactivity in standard serological tests.

MBG and EBO viruses were originally classified as members of the family Rhabdoviridae; however, based on biochemical, ultrastructural, and serological criteria, these viruses are now considered members of a new taxonomic group, the proposed family Filoviridae (15). MBG and EBO virus particles are very large, typically 800 to 900 nm long and consistently 80 nm in diameter. Bizarre structures of widely different lengths up to 14,000 nm, as well as branching, circular, or "6" shapes, are frequently found in negatively stained preparations (21). All MBG and EBO virus particles, regardless of length, contain a nucleocapsid consisting of a 20- to 30-nm central axis surrounded by a helical capsid, 40 to 50 nm in diameter, with cross-striations at 3- to 5-nm intervals. This nucleocapsid is surrounded by a host cell-derived envelope bearing 7- to 10-nm projections in a regular array. MBG and EBO virion RNA is single stranded, 4.0×10^6 to 4.5×10^6 daltons, of negative polarity, and noninfectious (7). Both viruses have at least four major structural proteins, designated VP-1 to -4. VP-1 and VP-3 are associated with the RNA and make up the nucleocapsid; VP-2 is glycosylated and is associated with the envelope. Tryptic peptide maps comparing MBG and EBO virus structural proteins are entirely dissimilar (1). EBO virus isolated from Zaire (EBO-Z virus) and Sudan (EBO-S virus) share a similar VP-2, whereas the VP-1 glycoprotein and the VP-3 nucleoprotein show major polypeptide differences. Oligonucleotide maps of MBG virus RNA suggest that 50 to 60% of its oligonucleotides are distinct from those of EBO virus. Between EBO-Z and EBO-S viruses, 30% of the oligonucleotides are significantly different. EBO-Z and EBO-S viruses are clearly related by broadly cross-reactive serological techniques, such as IFA tests, but are readily distinguished by radioimmunoassay techniques (25). Neutralization tests for MBG and EBO viruses are not sufficiently reliable to determine taxonomic relationships. Despite their unusual morphological properties, MBG and EBO viruses resemble the arenaviruses and other lipid-enveloped viruses in being susceptible to lipid solvents, beta-propiolactone (27), formaldehyde, UV light, and gamma radiation (8). These viruses are stable at room temperature for several hours, but they are inactivated by incubation at 60°C for 1 h.

COLLECTION AND STORAGE OF SPECIMENS

Serum, heparinized plasma, or, less ideally, whole blood for virus isolation should be collected during the acute, febrile stages of these illnesses and frozen on dry ice or in liquid nitrogen vapor. Throat wash and urine specimens should also be collected and mixed with an equal volume of buffered diluent containing serum proteins to stabilize viral infectivity before the specimens are frozen. Storage at higher temperatures may be unavoidable but will lead to rapid losses in infectivity. Lassa virus is frequently isolated from acute-phase sera and throat washings for several weeks after the onset of clinical signs, but it is only occasionally isolated from urine. Sera obtained as late as 15 days after onset have yielded Lassa virus. Junin virus is usually recoverable from serum for a period of 3 to 10 days after onset and often from throat washings for a similar period, but rarely from urine. Machupo virus is recovered from only one of five acute-phase sera and even less frequently from throat washings or urine. LCM virus may be recovered from acute-phase sera obtained within the first week after onset, but rarely, if ever, from throat washings or urine specimens. LCM virus may also be isolated from the cerebrospinal fluid during the period of meningeal involvement and from the brain at autopsy. MBG and EBO viruses are usually recoverable from acute-phase sera; various specimens, including throat wash, urine, soft tissue effusates, semen, and anterior eye fluid, have yielded these viruses, even when the specimens are obtained late in convalescence. Lassa, Machupo, Junin, MBG, and EBO viruses are all readily isolated from spleen, lymph nodes, liver, and kidney obtained at autopsy, but rarely, if ever, from brain or other central nervous system tissues. Notably, Lassa virus is usually isolated from the placentas of infected pregnant women.

Blood obtained in early convalescence for serodiagnosis may be infectious despite the presence of antibodies and should be handled accordingly (5, 6). Maintenance of such samples at −20°C or below is sufficient to preserve antibody titers, but lower temperatures are required to preserve infectivity. Certain anticoagulants should be avoided: citrate interferes with the IFA test, both citrate and oxalate cause nonspecific cytopathic effect in Vero cells used for virus isolation, and EDTA interferes with enzyme-linked immunosorbent assay (ELISA) techniques for both antigen and antibody determinations (for Lassa fever). Heparinized plasma or serum samples are preferable.

Manipulation of these specimens and tissues, including sera obtained from convalescent patients, may pose a serious biohazard, and their handling should be minimized outside a maximum containment laboratory. Current recommendations in the United States are that such samples be manipulated only at the P-4 containment level (6). Since P-4 containment facilities are rarely available where primary cases are occurring, it is recommended that the handling of patient samples be minimized in the field. Personnel caring for the patient and obtaining diagnostic specimens should wear disposable caps, gowns, shoe covers, surgical gloves, and full-face respirators equipped with high-efficiency particulate air filters. Gloves should be disinfected immediately if they come in direct contact with infected blood or secretions. All equipment (preferably disposable) should be placed directly in a suitable disinfectant solution, such as sodium hypochlorite, phenolic detergent, or a solution of quaternary ammonium compounds. Reusable equipment should be similarly disinfected before sterilization. To minimize the risk of autoinoculation, needle sheaths should not be replaced on used needles; needles should be immediately disinfected. Glass equipment poses a significant hazard also (e.g., microscope slides, microhematocrit tubes, and syringes), and its use should be minimized. One exception is the use of Vacutainer tubes for the collection of blood. This method is considered safer than the use of syringes (which must be emptied into another tube). Procedures which generate aerosols (e.g., trituration and centrifugation) should be minimized and conducted only if additional protective equipment, such as a flexible film plastic isolator capable of maintaining negative pressure and a high-efficiency particulate air-filtered exhaust, is available. For specialized procedures, samples may be inactivated by the addition of beta-propiolactone (27) and may then be safely tested in the field. Heating to 60°C for 1 h will render diagnostic specimens noninfectious and is feasible for the measurement of heat-stable substances such as electrolytes, blood urea nitrogen, and creatinine.

For all testing of infectious material, samples should be packaged in accordance with current recommendations (4) and forwarded, after consultation, to one of the following laboratories.

Center for Disease Control
Special Pathogens Branch
Virology Division, Bureau of Laboratories
Atlanta, GA 30333 U.S.A.

U.S. Army Medical Research Institute of Infectious Diseases
Medical Division
Fort Detrick
Frederick, MD 21701 U.S.A.

Center for Applied Microbiology and Research
Special Pathogens Unit
Salisbury
Porton Down, Wiltshire, SP4 OJG, England

Institut de Medicine Tropicado Prince Leopold
Nationalestraat 155
B-2000 Antwerp, Belgium

National Institute for Virology
Private Bag X4
Sandringham, Johannesburg, South Africa

National Institute of Health of Japan
Gakuen 4-7-1
Masashimurayama-shi
Tokyo, Japan 190-12

Instituto Nacional de Estudios Sobre Virosis Hemorragicas
Casilla Correo 195
2700 Pergamino, Argentina

DIRECT EXAMINATION

MBG and EBO viruses have been successfully visualized directly by electron microscopy of heparinized blood drawn during the febrile period and immediately fixed with glutaraldehyde before separation of the erythrocytes by low-speed centrifugation. Virions are then sedimented at 20,000 × *g* directly from plasma onto coated electron microscopy grids, negatively stained, and examined. The combination of size and shape of the virions is sufficiently characteristic to allow a morphologic diagnosis (21, 26). Differentiation of these virions and identification as MBG and EBO virus are accomplished by immune electron microscopic techniques. Application of similar techniques to the diagnosis of arenavirus infections has not been reported. Retrospective examination of ultrathin sections of Formalin-fixed tissues obtained from patients at autopsy have occasionally revealed typical MBG, EBO, or arenavirus particles, most often in the liver, spleen, or kidney, but only when infectious virus concentrations are very high. Direct fluorescent-antibody (DFA) and IFA staining of impression smears or air-dried suspensions from liver, spleen, or kidney have been used successfully to detect cytoplasmic inclusion bodies associated with MBG virus infection; clumps of MBG virus antigen have also been observed by DFA examination of infected, dried, citrated blood smears (26). This approach has not been successfully applied to the diagnosis of EBO virus or arenavirus infections. Early reports on the utility of DFA techniques to detect Lassa virus-infected cells in conjunctival scrapings have not been borne out. A similar approach for the detection of Junin virus-infected cells in urinary sediment is more promising. Examination of frozen sections of infected tissues by DFA techniques should be feasible but has not been reported, a consequence of the biohazard associated with this procedure and the difficulty in obtaining fresh, well-preserved tissues. Experience with DFA staining of tissues from monkeys experimentally infected with Lassa virus suggests that antigens for this virus can be detected in diverse tissues when infectious virus concentrations exceed 6 $\log_{10}$ PFU/g. Development of immunohistochemical techniques for the detection of arenavirus antigens in Formalin-fixed tissues has been attempted without convincing success. Presently, direct detection methods are of little practical value in the diagnosis of these infections, since tissues which contain sufficient virus to be detected directly will yield infectious virus within 1 to 2 days by conventional isolation procedures, at far less risk, or in even less time by ELISA-based antigen detection systems.

ISOLATION OF VIRUS

The best general method currently available for the isolation of MBG and EBO viruses and the pathogenic arenaviruses is the inoculation of Vero cells, followed by the examination of inoculated cells at intervals for the presence of viral antigens by using an immunologically specific staining method such as DFA or immunoperoxidase. This approach has been successfully adopted for routine isolations of Lassa, Machupo, Junin, MBG, and EBO-Z viruses from materials collected in the field. Vero cell inoculation should work for LCM virus as well, but intracranial (i.c.) inoculation of weanling mice is still regarded as the most sensitive established indicator of LCM virus (10). EBO-S virus has also been isolated by Vero cell inoculation, but less reliably since several blind passages are usually required (20). Other cell lines, including human diploid lung and BHK-21 cells, have been substituted for Vero cells with reasonable success. Although historically Machupo and Junin viruses were isolated by i.c. inoculation of newborn hamsters and mice, respectively, Vero cells are approximately as sensitive and far less cumbersome to manage in a P-4 containment system. Furthermore, Vero cells permit isolation and identification usually within 1 to 5 days, a significant advantage over animals, which require a 7- to 20-day incubation period for illness to develop.

Clinical specimens and clarified tissue homogenates (usually 10% [wt/vol]) are diluted in a suitable maintenance medium, such as Eagle minimal essential medium with Earle salts and 2% heat-inactivated calf serum, and adsorbed in small volumes to monolayers of Vero cells grown in suitable vessels, such as cell culture T-25 flasks or 60-mm petri dishes. It is important to test higher dilutions of these specimens as well as more concentrated material, since autointerference (thought to be due to defective interfering particles present in lower dilutions) may totally inhibit the development of infectious virus, cytopathic effect, or antigen detectable by the CF or immunofluorescence test. After adsorption, sufficient maintenance medium is added to maintain the cells for 7 days. Inoculation of replicate vessels permits the examination of cells at frequent intervals. Vero cells inoculated with high-titer samples contain detectable antigen within 1 to 3 days. Cells inoculated with low-titer material may require up to 7 days to accumulate detectable antigen. If no antigen is detected after 7 days, the sample is considered negative, but supernatant fluids may be blind-passaged to confirm the absence of virus. For Junin virus detection, immunoperoxidase procedures have been substituted for DFA procedures, with similar success (16). When EBO-S virus is suspected, the requirement for blind passage is anticipated, and supernatant fluids are harvested and passed at 3- to 5-day intervals. To confirm the presence and identity of the virus, supernatant fluids collected from infected cells may be tested for infectious virus by plaquing (requires 4 to 8 more days) and for antigen detection by the CF test (4 h) or by ELISA techniques when available.

LCM virus is still conventionally isolated in mice. Since LCM virus does not require P-4 containment, the problems associated with maintenance of animals in maximum containment laboratory are not factors in the selection of an isolation system. Mice, 3 to 4 weeks old, are inoculated i.c. with undiluted samples. Since the high-dose phenomenon of viral interference is occasionally a problem, a dilution of each sample (perhaps 1:100) is ideally also inoculated into a second group of mice. Many LCM virus isolates produce a characteristic convulsive disease within 5 to 7 days, but some isolates require a longer time or may not produce death at all. To circumvent the problem of variable responses to different LCM virus strains, endotoxin obtained from *Escherichia coli* (100 μg per 0.2 ml) is inoculated intraperitoneally into mice 4 to 7 days after virus inoculation. All LCM virus-infected mice die with a typical LCM disease within 24 h of endotoxin inoculation, regardless of the virus strain, whereas uninfected mice are unaffected (10). Brains from dead mice may be used to prepare CF antigens or may be stained by DFA to obtain presumptive identification, which may be confirmed by neutralization with clarified mouse brain homogenate.

For primary isolation of other arenaviruses and of MBG and EBO viruses, animal inoculations are still used when cell cultures are not available. Newborn mice (1 to 3 days old) are highly susceptible to Junin virus inoculated i.c.; newborn hamsters are the animal of choice for Machupo virus, although newborn mice are sufficiently sensitive to be useful. Newborn hamsters and mice die 7 to 20 days after i.c. inoculation, usually exhibiting a characteristic "tailspin" reaction. For the South American hemorrhagic fever viruses, particularly Junin virus, young adult guinea pigs inoculated either i.c. or peripherally have been used; the option to inoculate large volumes of potentially infectious material compensates for the lower sensitivity of guinea pigs compared with that of newborn mice and hamsters. Guinea pigs die 7 to 18 days after Junin virus inoculation. Most LCM virus strains are lethal for guinea pigs also. For Lassa virus, animals have never been used routinely for virus isolation, although inbred strain 13 guinea pigs are exquisitely sensitive to certain Lassa virus strains and uniformly die 12 to 18 days after inoculation; outbred Hartley strain guinea pigs are somewhat less susceptible. Pathogenicity of virulent Lassa virus strains for mice inoculated i.c. varies with different sources of outbred Swiss albino mice and with inbred mouse strains. MBG, EBO-Z, and EBO-S viruses produce febrile responses in guinea pigs 4 to 10 days after inoculation; however, none of these viruses kills guinea pigs consistently on primary inoculation, and only EBO-Z virus has been adapted to uniform lethality by sequential guinea pig passages. EBO-Z virus is usually pathogenic for newborn mice inoculated i.c., but EBO-S and MBG viruses are not.

IDENTIFICATION OF VIRUS

Typing antisera

Detection of viral antigens in infected cells (usually Vero) in culture constitutes a presumptive diagnosis, provided the serological reagents have been tested against the prototype arena- and filoviruses expected in a given laboratory. Virus isolates in cell culture supernatant fluids or tissue homogenates are presumptively or specifically identified by their reactivity with diagnostic antisera in various serological tests. Immune sera are prepared in adult guinea pigs, hamsters, or mice inoculated intraperitoneally with infectious virus. Rhesus and cynomolgus monkeys, convalescent from experimental infections, are also reasonable sources for larger quantities of immune sera. To compensate for any expected mortality, additional animals should be inoculated, with the virus dose adjusted to ensure uniform infection with minimum mortality. Mortality may also be reduced by treatment of the animals with appropriate antiviral drugs, such as ribavirin or specific immune plasma (13). Diagnostic antisera produced in this way are less cross-reactive and usually higher titered than those produced by multiple injections of inactivated antigens. To further reduce the induction of extraneous antibodies, inoculum virus should be derived from tissues or cells homologous to the species being immunized; likewise, the virus suspension should be stabilized with homologous serum or serum proteins. Sera produced for use in the CF, DFA, and IFA tests should be collected 30 to 60 days after inoculation; sera for neutralization tests should be collected later. All sera should be rigorously tested for the presence of live virus before they are removed from a maximum containment system. Recently, monoclonal antibodies have been produced with specificity for individual nucleocapsid and viral glycoproteins for Lassa and LCM viruses, other arenaviruses (2), and MBG and EBO virus strains. These reagents promise to be useful in refining taxonomic relationships among these strains and in rapid diagnostic procedures. Reference reagents for Lassa, Machupo, MBG, and EBO viruses are not generally available; hyperimmune mouse ascitic fluids for LCM and Junin viruses are available from the National Institutes of Health, Research Resources Branch.

Immunofluorescence procedures

To process infected cells for DFA examination and presumptive identification, inoculated cell monolayers are dispersed by trypsinization (0.05% trypsin with 0.02% EDTA for 10 min at 37°C), diluted in phosphate-buffered saline (PBS) containing 10% calf serum, and centrifuged at 400 × *g* for 10 min. The cell pellet is washed by resuspension in PBS, centrifugation, and resuspension in PBS to a final concentration of 10^6 cells per ml. Small drops (10 to 20 μl) of cell suspension are placed onto circular areas of specially prepared epoxy-coated slides, cleaned previously by immersion in ethanol followed by polishing to remove residual oily deposits. These "spotslides" are air dried, fixed in acetone at room temperature for 10 min, and either stained immediately or stored frozen at −70°C until used. Although acetone fixation greatly reduces infectious intracellular virus, spotslides prepared in this manner should still be considered infectious and handled accordingly. Recently, spotslides have been rendered noninfectious by gamma radiation (8), with no diminution in fluorescent antigen intensity reported. Alternatively, infected cells may be biologically inactivated with beta-propiolactone (27). Gamma radiation is recommended if the appropriate equipment is available.

For DFA tests, specific immune globulin, prepared by ethanol or ammonium sulfate precipitation of immune sera followed by conjugation with fluorescein, is diluted to a working concentration predetermined by box titration and flooded onto the infected cells, which are incubated at room temperature for 30 min in a moist chamber. After incubation, the slides are washed by immersion in PBS for two 10-min periods, dipped in water to remove salts, dried by evaporation, and mounted under cover slips in PBS-glycerol (pH 7.8). Specific viral fluorescence is characterized as intense, punctate-to-granular aggregates confined to the cytoplasm of infected cells. In addition, specific MBG or EBO virus fluorescence may include large, bizarre-shaped aggregates up to 10 μm across. Nonspecific fluorescence is rarely a problem in DFA procedures for these viruses. Detection of MBG, EBO, Lassa, and LCM virus antigens by DFA is usually considered sufficient for a definitive diagnosis, although Lassa and LCM virus antigens cross-react at low levels in this test. Detection of Junin or Machupo virus antigens by DFA constitutes a presumptive diagnosis since these viruses can be reliably distinguished from each other only by neutralization tests.

CF test

Until recently, the CF test was routinely employed for the detection and presumptive identification of the arenaviruses and MBG virus. However, the CF test is rarely used now, since the development of reliable and simplified immunofluorescence procedures described above. Detailed instructions for performing the CF test for arenaviruses are available (3), and the method can be applied to MBG and EBO viruses as well. In brief, CF antigens are prepared from infected Vero cell culture supernatant fluids or from suckling mouse or suckling hamster brains extracted by sucrose-acetone. For reasons of safety, CF antigens are frequently inactivated by the addition of beta-propiolactone and may be stored frozen at −70°C indefinitely or preserved by lyophilization.

ELISA

An ELISA for the detection and presumptive identification of Lassa virus antigens in viremic sera was recently developed (23). Although not yet field tested, it promises to detect antigen in sera with viremia titers as low as 2.1 $\log_{10}$ PFU/ml and may be sufficiently sensitive to detect antigen in throat wash and urine samples as well. The test reliably detects Lassa virus antigen in beta-propiolactone–inactivated samples and could be safely conducted without elaborate containment facilities. This experimental test in its present form is a triple antibody (sandwich) ELISA utilizing highly avid, Lassa virus-specific, affinity-purified globulin preparations in a conventional ELISA format. Briefly, test sample dilutions are added to wells of polystyrene microtiter plates preincubated with monkey anti-Lassa virus immunoglobulin. After incubation and washing of the reaction wells, guinea pig anti-Lassa virus immunoglobulin is added as a detector antibody. The plate is reincubated and washed, and rabbit anti-guinea pig immunoglobulin G (IgG) is then added, incubated, and washed. Finally, alkaline phosphatase-labeled swine anti-rabbit IgG is added, incubated, and washed, and *p*-nitrophenylphosphate diluted in 1 M diethanolamine buffer is added. The reaction is read spectrophotometrically after 20 min at room temperature. A sample is considered positive if the optical density is more than the mean background (mean plus 2 standard deviations for 30 negative samples). All samples are tested in duplicate. To detect nonspecific binding (i.e., "sticky" sera), each serum sample is also added to one well treated as described above, but with normal (nonimmune) guinea pig immunoglobulin substituted for Lassa virus-immune guinea pig immunoglobulin. Presently, sticky sera, for which the optical density of the control well appears positive, cannot be assayed reliably. Various laboratories are currently developing similar ELISA techniques for Junin, Machupo, LCM, MBG, and EBO virus antigen detection. Substitution of monoclonal antibodies of high avidity and appropriate specificities for the polyclonal sera presently used may increase the sensitivities of these antigen detection ELISA tests.

Neutralization tests

Neutralization tests for these viruses take many forms; depending on the virus, neutralization tests range from extremely sensitive and reliable (e.g., with Junin and Machupo viruses) to moderately insensitive but reliable (with Lassa and LCM viruses) to totally unreliable (with MBG and EBO viruses). The common denominator in all neutralization tests is measurement of an inhibition of viral replication by reaction with immune serum. Thus, the form of the test is determined in part by the availability of tools to measure infectious virus or viral antigens (e.g., DFA, CF, or ELISA procedures, plaques, or animal infectivity) as well as the kinetics of the virus-antibody interactions. The unreliability of the MBG and EBO virus neutralization tests may be partially a function of the primitive and cumbersome viral quantitation procedures available, as well as low-avidity antibody. For the New World arenaviruses (Junin and Machupo), the most generally applied neutralization test is a plaque reduction test with Vero cells and the serum dilution-constant virus format. The serum dilution calculated (by probit analysis) to reduce the control number of plaques by 50% is usually taken as the endpoint, although in some laboratories, the highest serum dilution producing 80% reduction is used. The 80% plaque reduction test is commonly used to distinguish Junin from Machupo virus. For Lassa and LCM viruses, neutralizing antibody activity is rapidly lost on dilution; for this reason, the constant serum dilution-varying virus format is preferred. Neutralization of Old World arenaviruses is also markedly enhanced by the addition of complement; thus, 10% fresh guinea pig serum is routinely added to the diluent. Plaque reduction is further enhanced by the addition of either anti-immunoglobulin or protein A. Plaque reduction tests in various formats have largely replaced animal protection tests, which were notoriously imprecise for the measurement of arenavirus neutralizing antibody.

SEROLOGICAL DIAGNOSIS

IFA test

The IFA test is clearly the method of choice for documenting recent infections with MBG, EBO, and arenaviruses. The preparation of spotslides with infected Vero cells is identical to the procedure described above. Uninfected cells are often admixed with the virus-infected cells to aid discrimination between specific and nonspecific fluorescence. Although monovalent spotslides are usually desired and are prepared with cells optimally infected with a single virus, polyvalent spotslides can also be prepared by mixing cells infected with different viruses selected from these or other taxonomic groups which share similar geographic distributions (14). Test sera are diluted serially, usually in twofold increments, starting at 1:4 or 1:10. Prozones may occur in low dilutions. Thus, for screening procedures, sera are commonly tested at both 1:10 and 1:80 dilutions. Infected cells (and uninfected control cells) are incubated with serum dilutions, washed, incubated with appropriate fluorescein-conjugated anti-globulin (or specific anti-IgM or -IgG), washed, mounted, and observed. Endpoint determination is very subjective. Most experienced observers consider the endpoint to be the highest dilution producing typical cytoplasmic fluorescence clearly positive relative to uninfected cells. Although it is possible to obtain reproducible endpoints within individual laboratories, discrepancies in titers determined by different laboratories are common and probably relate to variations in interpretation, epiillumination intensity, filtration systems, and fluorescein conjugates. Antibodies measured by the IFA test are usually the first to appear, often becoming detectable within the first few days of hospitalization for Lassa virus, within 10 days of onset for MBG and EBO viruses, and somewhat later for Junin and Machupo viruses. The presence of specific IgM antibodies or a rising IFA titer comprises a presumptive diagnosis of acute infection. IgM antibodies measured by IFA decline to undetectable titers within several months, whereas IgG antibodies persist at least several years.

CF test

The CF test was used initially to classify the arenaviruses and detect seroconversions. However, the CF test is rarely used now because it is inferior to the IFA test for both arenavirus and filovirus infections. CF antibody titers evolve more slowly (3 to 4 weeks after onset) and recede rapidly to undetectable titers (usually within 1 year). The CF test is less specific than the IFA test, and anticomplementary sera are frequently encountered. Since the CF test is often used to screen aseptic meningitis specimens against a battery of antigens, a CF antigen for LCM virus is often appropriately included in this battery.

Neutralization tests

As described above, reliable tests for measuring neutralizing antibody to MBG and EBO viruses are not available. For the arenaviruses, plaque reduction tests with Vero cells are generally used. For measuring neutralizing antibody to Lassa and LCM viruses, which are difficult to neutralize and are poor inducers of this antibody, test sera are diluted, usually 1:10, in medium containing 10% guinea pig serum as a complement source and mixed with serial dilutions of challenge virus. Titers are expressed as a $\log_{10}$ neutralization index, defined as $\log_{10}$ PFU in control serum $-$ $\log_{10}$ PFU in test serum. For Junin and Machupo viruses, the more conventional serum dilution-constant virus format is usually used, although the constant serum-virus dilution format is equally useful for distinguishing among strains. Neutralizing antibody responses require weeks to months to evolve, but persist for years. Performance of these tests is retricted to laboratories equipped to handle the infectious viruses.

Other serological tests

ELISA procedures for Lassa virus-specific IgG have been developed (23) and are ready for field testing. The success of these ELISA procedures depends on the availability of highly avid and specific antiviral globulin to capture antigen on the microtiter plates. Development of ELISA procedures for the remaining viruses considered in this chapter has not advanced to the field testing stage. Another test developed for Lassa virus (and antibodies) is a reversed passive hemagglutination (and inhibition) test with Lassa virus antibody-coated erythrocytes which agglutinate in the presence of viral antigen (9). The test has not gained widespread acceptance, perhaps because meticulous care is required to obtain satisfactory antibody-erythrocyte conjugates. For EBO virus, a radioimmune assay with ^{125}I-labeled staphylococcal protein A was successfully applied to a wide range of human and animal sera and to the discrimination of EBO-Z from EBO-S virus strains (25). The eventual places these experimental procedures will assume in the routine diagnosis of MBG, EBO, and arenavirus infections remain to be determined.

EVALUATION AND INTERPRETATION OF RESULTS

Early diagnosis of arenavirus and filovirus infections is desirable since specific immune plasma and appropriately selected antiviral drugs are often effective when treatment is initiated soon after onset. Early recognition of these infections should also trigger strict isolation procedures to prevent spread of the disease to patient contacts. In areas where specific viruses are endemic, the index of suspicion is often high, and experienced clinicians may be remarkably accurate in rendering an accurate diagnosis of fully developed cases on clinical grounds alone. Yet even in these areas, specific virological and serological tests are required to confirm clinical impressions, since many other diseases, including malaria, typhoid, rickettsiosis, idiopathic thrombocytopenia, and viral hepatitis, may masquerade as an arena- or filovirus infection. Although the availability of inactivated antigen spotslides for IFA tests in field hospitals has facilitated diagnosis, based on seroconversion by IFA testing, timely diagnosis requires a means of detecting infectious virus or antigen in the field. The ELISA for Lassa virus antigen detection holds promise, since it will detect clinically relevant viremia in beta-propiolactone–inactivated sera. Similar detection systems are urgently required for the other viruses.

Inoculation of cell cultures for DFA examination and isolation of these viruses should not be conducted outside a maximum containment laboratory, with the exception of LCM virus, which may be handled at a lower containment level (6). The detection of viral antigens by DFA in Vero cells inoculated with patient specimens constitutes a definitive diagnosis, except for Junin and Machupo viruses, which are reliably discriminated from each other only by the neutralization test. Other virus strains that cross-react by DFA, such as Lassa with LCM, and EBO-Z with EBO-S, can be distinguished in quantitative cross-testing by DFA. For virus isolates originating from areas where the geographic distributions of related viruses overlap, it is essential that quantitative DFA testing be done and desirable that identification be confirmed by neutralization tests when available.

Although the interpretation of serological data is usually facilitated by the generally restricted geographic ranges of these viruses, ranges do overlap, and occasionally IFA and CF data are ambiguous. In one documented outbreak, patients with clinical AHF developed CF antibody rises to both Junin and LCM viruses (18). The geographic distributions of Junin and Machupo viruses certainly overlap that of LCM virus in South America. In Africa the distribution of Lassa may overlap the distributions of newly isolated virus strains from rodents in Zimbabwe, Mozambique, and Central Africa Republic which cross-react strongly by IFA with Lassa virus and, to a lesser extent, with LCM virus. Although these virus strains are not known to be associated with human disease, their presence may confuse the interpretation of serological data from African surveys. The extent to which heterologous arenavirus infection or reinfection broadens antibody specificity has not been systematically evaluated for any of the available serological tests. For EBO virus, antibodies reacting in the IFA test have been detected in populations such as Panamanian Indians, who have never experienced clinical EBO virus infections, thus casting doubt on the validity of this test for any population, including any in which EBO virus infections have been documented virologically. Despite these potential problems, the experience to date has been that in the midst of outbreaks caused by MBG, EBO, or arenaviruses, identifications of the etiological agents by DFA and IFA tests have been clear and unambiguous, especially when the diagnoses were confirmed by neutralization tests.

Because of the biohazard, virus isolation data for these viruses are usually available only retrospectively. MBG and EBO-Z viruses are usually isolated from acute-phase sera, whereas EBO-S virus is isolated less often, perhaps because of the need for blind passage. Lassa virus is usually recovered from acute-phase sera of hospitalized patients soon after admission, frequently in the presence of specific IgM antibody. Junin, Machupo, and LCM viruses are recovered less frequently, and diagnosis is usually based on seroconversion. The IFA response is the earliest for all these viruses, detectable 7 to 10 days after onset for Lassa and LCM viruses, 10 to 14 days after onset for MBG and EBO viruses, 12 to 17 days after onset for Junin virus, and 17 to 30 days after onset for Machupo virus. The CF antibodies evolve several days after the IFA response. The presence of specific IgM antibodies detected by IFA is indicative of recent infection, since IFA-IgM titers persist for less than 3 months. The presence of specific IFA-IgM in the cerebrospinal fluid of LCM patients constitutes a definitive diagnosis. For all the arenavirus and filovirus pathogens, a rising IFA-IgM or -IgG titer constitutes a strong presumptive diagnosis. Since IFA-IgM titers, as well as CF titers, do not persist long, a decreasing titer suggests a recent infection which occurred perhaps several months previously.

For Lassa fever patients, a detectable IFA response does not necessarily signal imminent recovery; viremia frequently persists, and patients die after an IFA response (28). For Junin, LCM, MBG, and EBO virus infections, the appearance of antibodies detectable by IFA coincides with disappearance of viremia and with recovery. In Machupo virus infection, IFA titers appear even later, 1 week or more after the crisis has passed. For the arenaviruses, neutralizing antibodies appear much later in convalescence than do IFA or CF antibodies. Reliable neutralizing antibody data for MBG and EBO virus infections are not currently available. Neutralizing antibodies against arenaviruses persist for long periods, perhaps for life, and thus provide the most reliable basis for determining the minimum resistance of a population to reinfection.

The role of neutralizing antibody in acute recovery is less clear. The protective efficacy of passively administered immune plasma is believed to be a function of neutralizing antibody titers, and selection of plasma should be on this basis, especially for Lassa fever, since protective efficacy is predicted by neutralizing antibody titers and not by IFA (12). A serological test to predict the efficacy of MBG and EBO virus immune plasma is urgently needed.

The highest priority for future development is refinement of the available diagnostic tools to permit definitive virus identifications in the field. A reasonable approach is adaptation of the available Lassa virus antigen ELISA to the other viruses, with biologically inactivated samples. For broad screening procedures, a highly avid, cross-reactive monoclonal antibody might be employed, while a battery of type-specific monoclonal antibodies could be used for definitive identification. This investment in rapid diagnosis will permit more timely intervention with effective treatment regimens and, through implementation of appropriate public health measures, may reduce dissemination of these highly virulent viral pathogens.

LITERATURE CITED

1. **Buchmeier, M. J., R. U. DeFries, J. B. McCormick, and M. P. Kiley.** 1983. Comparative analysis of the structural polypeptides of Ebola viruses from Sudan and Zaire. J. Infect. Dis. **147**:276–281.
2. **Buchmeier, M. J., H. A. Lewicki, O. Tomori, and M. B. A. Oldstone.** 1981. Monoclonal antibodies to lymphocytic choriomeningitis and Pichinde viruses: generation, characterization, and cross-reactivity with other arenaviruses. Virology **113**:73–85.
3. **Casals, J.** 1977. Serologic reactions with arenaviruses. Medicina (Buenos Aires) **37**(Suppl. 3):59–68.
4. **Centers for Disease Control.** 1980. Interstate shipment of etiologic agents. Fed. Reg. **45**:48626–48629.
5. **Centers for Disease Control.** 1983. Viral hemorrhagic fever: initial management of suspected and confirmed cases. Morbid. Mortal. Weekly Rep. **32**(Suppl.):27–40.
6. **Centers for Disease Control and National Institutes of Health.** 1983. Biosafety in microbiological and biomedical laboratories. Centers for Disease Control, Atlanta.
7. **Cox, N. J., J. B. McCormick, K. M. Johnson, and M. P. Kiley.** 1983. Evidence for two subtypes of Ebola virus based on oligonucleotide mapping of RNA. J. Infect. Dis. **147**:272–275.
8. **Elliott, L. H., J. B. McCormick, and K. M. Johnson.** 1982. Inactivation of Lassa, Marburg, and Ebola viruses by gamma irradiation. J. Clin. Microbiol. **16**:704–708.
9. **Goldwasser, R. A., L. H. Elliott, and K. M. Johnson.** 1980. Preparation and use of erythrocyte-globulin conjugates to Lassa virus in reversed passive hemagglutination and inhibition. J. Clin. Microbiol. **11**:593–599.
10. **Hotchin, J., and E. Sikora.** 1975. Laboratory diagnosis of lymphocytic choriomeningitis. Bull. W.H.O. **52**:555–558.
11. **International Committee on Taxonomy of Viruses.** 1982. Arenaviruses. Intervirology **17**:119–122.
12. **Jahrling, P. B., and C. J. Peters.** 1984. Passive antibody therapy of Lassa fever in cynomolgus monkeys: importance of neutralizing antibody and Lassa virus strain. Infect. Immun. **44**:528–533.
13. **Jahrling, P. B., C. J. Peters, and E. L. Stephen.** 1984. Enhanced treatment of Lassa fever by immune plasma combined with ribavirin in cynomolgus monkeys. J. Infect. Dis. **149**:420–427.
14. **Johnson, K. M., L. H. Elliott, and D. L. Heymann.** 1981. Preparation of polyvalent viral immunofluorescent intracellular antigens and use in human serosurveys. J. Clin. Microbiol. **14**:527–529.
15. **Kiley, M. P., E. T. W. Bowen, G. A. Eddy, M. Isaacson, K. M. Johnson, J. B. McCormik, F. A. Murphy, S. R. Pattyn, D. Peters, O. W. Prozesky, R. L. Regnery, D. I. H. Simpson, W. Slenczka, P. Sureau, G. van der Groen, P. A. Webb, and H. Wulff.** 1982. Filoviridae: taxonomic home for Marburg and Ebola viruses? Intervirology **18**:24–32.
16. **Lascano, E. F., M. I. Berria, and N. A. Candurra.** 1981. Diagnosis of Junin virus in cell cultures by immunoperoxidase staining. Arch. Virol. **70**:79–82.
17. **Lehmann-Grube, F.** 1971. Lymphocytic choriomeningitis virus. Virol. Monogr. **10**:1–173.
18. **Maiztegui, J. I., G. M. Aguirre, M. S. Sabattini, and J. G. B. Oro.** 1971. Actividad de dos "arenavirus" en seres humanos y roedores en un mismo lugar de la zona endemica de fiebre hemorragica Argentina. Medicine (Buenos Aires) **31**:509–510.
19. **Martini, G. A.** 1971. Marburg virus disease. Clinical syndrome, p. 1–9. *In* G. A. Martini and R. Siegert (ed.), Marburg virus disease. Springer-Verlag, New York.
20. **McCormick, J. B., S. P. Bauer, L H. Elliott, P. A. Webb, and K. M. Johnson.** 1983. Biologic differences between strains of Ebola virus for Zaire and Sudan. J. Infect. Dis. **147**:264–267.
21. **Murphy, F. A., G. Van der Groen, S. G. Whitfield, and J. V. Lange.** 1978. Ebola and Marburg virus morphology and taxonomy, p. 61–84. *In* S. R. Pattyn (ed.), Ebola virus haemorrhagic fever. Elsevier/North Holland Biomedical Press, Amsterdam.
22. **Murphy, F. A., and S. G. Whitfield.** 1975. Morphology and morphogenesis of arenaviruses. Bull. W.H.O. **52**:409–419.
23. **Niklasson, B. S., P. B. Jahrling, and C. J. Peters.** 1984. Detection of Lassa virus antigens and Lassa virus-specific immunoglobulins G and M by enzyme-linked immunosorbent assay. J. Clin. Microbiol. **20**:239–244.
24. **Pedersen, I. R.** 1979. Structural components and replication of arenaviruses. Adv. Virus Res. **24**:277–330.
25. **Richman, D. D., P. H. Cleveland, J. B. McCormick, and K. M. Johnson.** 1983. Antigenic analysis of strains of Ebola virus: identification of two Ebola virus serotypes. J. Infect. Dis. **147**:268–271.
26. **Siegert, R., and W. Slencyka.** 1971. Laboratory diagnosis and pathogenesis, p. 157–160. *In* G. A. Martini and R. Siegert (ed.), Marburg virus disease. Springer-Verlag, New York.
27. **Van der Groen, G., and L. H. Elliott.** 1982. Use of betapropiolactone-inactivated Ebola, Marburg, and Lassa intracellular antigens in immunofluorescent antibody assay. Ann. Soc. Belge Med. Trop. **62**:49–54.
28. **Wulff, H., and K. M. Johnson.** 1979. Immunoglobulin M and G responses measured by immunofluorescence in patients with Lassa or Marburg virus infections. Bull. W.H.O. **57**:631–635.
29. **Wulff, H., J. V. Lange, and P. A. Webb.** 1978. Interrelationships among arenaviruses measured by indirect immunofluorescence. Intervirology **9**:344–350.

Chapter 80

Viral Gastroenteritis Agents

NEIL R. BLACKLOW AND GEORGE CUKOR

CLINICAL BACKGROUND

Viral gastroenteritis is a common disease that affects all age groups and occurs in both epidemic and endemic forms (3, 5). Although the disease is usually self-limited, it can be lethal in the very young, the very old, the debilitated, or the malnourished patient.

Several viruslike particles have been associated with gastroenteritis by electron microscopic examination of stool specimens; however, this chapter deals with the two major viral agents which have been demonstrated conclusively to be of epidemiological and medical importance to date. These two viral gastroenteritis agents are rotavirus and Norwalk virus. Other agents associated with gastroenteritis include enteric adenovirus, calicivirus, astrovirus, small round viruses, and coronaviruslike particles.

Rotavirus

Rotavirus is known to cause both sporadic and epidemic cases of gastroenteritis in infants and young children throughout the world (1, 3, 8, 10). Approximately one-half of infants and young children hospitalized with diarrhea are infected with rotavirus. In pediatric patients, this virus is also the most common cause of gastroenteritis not requiring hospitalization. In temperate climates, the peak prevalence of rotaviral disease is during the winter months, and infection is uncommon in the summer. Transmission of the virus is thought to be by the fecal-oral route, and nosocomial spread on pediatric hospital wards has been well documented. The estimated incubation period is 24 to 48 h. The maximum amount of virus is found in the feces from day 3 to 5 after disease onset. Shedding of virus is greatly diminished or absent after day 8. Mild to asymptomatic infections are common among adult contacts of ill children, although more severe epidemics of gastroenteritis among adults have also been described. Asymptomatic infection has been noted during the newborn period and is currently unexplained. The indication for hospitalization of rotavirus-infected infants and young children is usually severe dehydration requiring fluid replacement. Diarrhea of 5 to 8 days' duration is the main feature of rotaviral disease, and it is frequently accompanied by vomiting and fever. Death due to rotavirus, although rare in developed areas, has been reported.

At least four serotypes and two or three subgroups of human rotavirus are described, all of which produce indistinguishable forms of clinical illness in humans (8, 20, 22). Serum antibodies, reactive with a common antigen shared by all rotaviruses, are acquired rapidly by most children between the ages of 6 and 24 months, indicating the early age for infection with rotavirus. A heteroserotypic and heterosubgroup serum antibody response occurs in both naturally infected young children and adult volunteers after infection with one rotaviral strain (8). Although young children rarely experience severe rotavirus gastroenteritis more than once, sequential infections by different serotypes or subgroups do occur (2, 8). The heteroserotypic and heterosubgroup nature of the immune response to rotavirus is now being exploited in early approaches towards vaccine development.

Norwalk virus

Norwalk virus has been associated with epidemics of gastrointestinal illness occurring in families, schools, and communities (14, 15). Settings for outbreaks include recreational camps, cruise ships, areas with contaminated water, and nursing homes, and have involved eating raw shellfish or other uncooked foods handled in an unsanitary manner. Nearly one-half of the outbreaks of epidemic nonbacterial gastroenteritis in the United States are caused by Norwalk virus (15). Serum antibody to the virus is regularly found in populations from various parts of the world. In contrast to rotavirus, antibody to Norwalk virus is not commonly acquired in temperate climates until the second or third decade of life. The prevalence of Norwalk virus antibody has been found to be 50 to 75% among American adults (4, 11). The Norwalk virus is therefore a common pathogen of older children and adults in the United States.

The clinical and pathogenic features of Norwalk gastroenteritis have been studied by disease production in healthy volunteers who orally ingested filtrates of feces containing the virus (3–5, 11). Experimentally induced illness is indistinguishable from the naturally occurring disease. The incubation period is 18 to 48 h, and symptoms last for 24 to 48 h. Symptoms vary among volunteers receiving identical inocula, and the illness attack rate is approximately 60 to 70%. Some subjects experience mild to severe vomiting without diarrhea, others develop mild to severe diarrhea without vomiting, and still others experience both diarrhea and vomiting. Abdominal cramps, headache, malaise, myalgia, low-grade fever, and nausea are commonly present. A characteristic histological lesion of the mucosa of the proximal small intestine develops during the course of the illness, accompanied by transient malabsorption of xylose. Half of ill volunteers excrete detectable virus in their stools during the first 72 h after onset of symptoms. Virus is not shed before the onset of symptoms and has been found in fewer than 20% of stool specimens collected 72 h or more after disease onset.

Rechallenge studies in volunteers (3) indicated the presence of two cohorts of individuals with different clinical forms of immunity to Norwalk virus. One cohort possessed long-term immunity as evidenced by resistance to illness on viral challenge and late homologous rechallenge. The other cohort became ill on first challenge and again when rechallenged several years later, but remained well when a third challenge was given 1 to 2 months after the second, indicating short-term immunity. Most of the first cohort persistently lacked serum antibody to Norwalk virus, while paradoxically many of the volunteers who developed illness possessed preexisting antibody. Although the presence of serum antibody was not protective, all volunteers who developed Norwalk illness showed at least a fourfold rise in serum antibody titer to the virus. A virus-specific immunoglobulin M response was associated with each bout of illness in repeatedly infected volunteers.

Other agents

There are several enteric viruses that clearly produce gastroenteritis, but their medical importance has yet to be conclusively shown through definitive epidemiological studies (8). A newly recognized subgroup of adenoviruses, termed enteric or fastidious adenoviruses, is a cause of diarrheal illness in infants and young children. These viruses fail to grow in standard cell cultures, unlike the well-characterized 39 conventional serotypes of adenovirus. The enteric adenoviruses, currently termed serotypes 40 and 41, do replicate in adenovirus-transformed 293 cells and are clearly associated with some cases of pediatric gastroenteritis, unlike the conventional serotypes. Although serotypes 40 and 41 have been found by electron microscopy in the stools of asymptomatic children (which is commonly the case for the conventional adenoviruses), they have also been noted six to nine times more frequently in children with acute gastroenteritis in early epidemiological studies. Simplified and specific immunological techniques to identify the enteric adenoviruses are being developed and should permit a definitive assessment of their medical importance.

Similarly, a human calicivirus-like particle, recognized by electron microscopy in stools, is a cause of gastroenteritis that usually occurs in young children. This 31- to 35-nm virus has yet to be cultivated in vitro and was first recognized by its characteristic morphology. It is not known whether it may be related to Norwalk virus. Recent development of an immunoassay for the detection of the calicivirus-like particle in stools should provide data regarding its importance in human disease.

Additional agents may be associated with medically important gastroenteritis, such as astrovirus, certain small 27-nm-sized round viruses (e.g., Hawaii, W-Ditchling, Snow Mountain), and enteric coronavirus-like particles; however, there is less clinical and epidemiological evidence for these to date than exists for enteric adenoviruses and caliciviruses.

DESCRIPTION OF THE AGENTS

Rotavirus

Rotavirus is classified as a member of the Reoviridae family of animal viruses which contain segmented double-stranded RNA as their genetic material (3, 10). The rotavirus is approximately 70 nm in diameter and has a double-shelled capsid. The human reoviruses are not related antigenically to rotaviruses. Rotaviruses have been observed in diarrheal feces from the young of numerous mammalian species (e.g., mice, calves, piglets, foals, monkeys, lambs, rabbits, deer, antelope, and goats). Cross-infection among species may occur experimentally, but there is no evidence currently that this occurs in nature.

Considerable information is available about the molecular virology of rotavirus, which has aided in its classification and immunological characterization (8). The virus contains 11 segments of RNA, and the functions associated with several of these genes are defined. Antigenicity of an outer capsid glycoprotein, the product of gene 8 or 9, determines the rotavirus serotype, of which at least 4 are known for human strains (20, 22). Serotypes are recognized diagnostically by neutralization tests such as plaque reduction, cytopathic effect neutralization, and fluorescent focus reduction. Antigenicity of the major inner capsid protein, the product of gene 6, determines the rotavirus subgroup, of which there are two and possibly three. Subgroups are recognized diagnostically by procedures such as enzyme immunoassay, complement fixation, counterimmunoelectrophoresis, and immune adherence hemagglutination (8). Inasmuch as a common rotaviral antigen shared by all mammalian rotaviruses is also determined by the product of gene 6, diagnostic procedures such as enzyme immunoassay suffice to identify rotaviruses of all serotypes in human stools. Some animal and human rotaviruses share subgroup specificities, and others are also related serotypically. Rotaviruses can also be classified by the electrophoretic migration patterns of their RNA gene segments; however, these patterns are so varied that electropherotypic classification has only a limited epidemiological value.

Approximately 50 to 70% of field strains of human rotavirus in stools can be cultivated inefficiently in roller-tube cell cultures that are treated with low concentrations of trypsin (8, 19, 20, 22). Thus, viral cultivation, which often requires several blind passages, is neither a practical nor a sensitive diagnostic method for human rotavirus. In contrast, many animal rotaviruses are readily cultivated in vitro and can serve as clinical reagents in diagnostic procedures for humans because of their immunological relatedness to human rotaviruses (7, 13). Rotavirus remains antigenically stable for years when stored at −20°C.

Norwalk virus

Immune electron microscopy has revealed the Norwalk virus to be small (27 nm in diameter) and nonenveloped (5, 14). The buoyant density of the virus in cesium chloride is approximately 1.38 g/cm^3. Volunteer studies have demonstrated that Norwalk virus remains infectious after exposure to 20% ether, acid (pH 2.7), and heat (60°C for 30 min). The virus has not been propagated in cell culture.

Norwalk virus does contain a single protein of a molecular weight characteristic of the RNA-containing calicivirus group, but its final classification must await analysis of its nucleic acid (8).

COLLECTION AND STORAGE OF SPECIMENS

The laboratory diagnosis of viral gastroenteritis is dependent on either the direct detection of virus in a stool specimen or the demonstration of a fourfold rise in antibody titer from acute- to convalescent-phase serum samples. Only diarrheal stools from the acute phase of illness are likely to contain virus. Stool samples should be stored frozen at −70°C and shipped on dry ice. Stool specimens are processed as for routine virus isolation. No serum should be used in the diluent, as any serum is suspected of containing antibody to rotavirus. In our laboratory, 2 to 5% suspensions of the stool are prepared in veal infusion broth supplemented with 0.5% bovine serum albumin (BSA) mixed on a Vortex apparatus at high speed for 1 min, and clarified by centrifugation at 2,000 × *g* for 15 min. Filtration of the suspension is not required. Convalescent-phase serum should be collected 2 to 4 weeks after the onset of illness.

DIRECT EXAMINATION FOR ROTAVIRUS

Direct examination of stool specimens is the primary method for diagnosis of rotavirus infection, inasmuch as the virus is cultivated inefficiently in vitro. Rotavirus was originally discovered by electron microscopic examination of feces (1), and use of an electron microscope was initially a prerequisite for the diagnosis of rotavirus infection. Techniques for detection of rotavirus by electron microscopy are well described elsewhere (1, 10). Relatively little processing of stool samples is necessary for specimens to be negatively stained with phosphotungstic acid and rendered suitable for study by electron microscopy. However, to remove the need for an electron microscope, immunofluorescence, counterimmunoelectrophoresis, and complement fixation techniques were developed for direct examination. These latter techniques lack the convenience, sensitivity, and large-scale applicability of solid-phase immunoassays, which have been developed for detection of rotavirus in stools. Solid-phase radioimmunoassay (RIA), (7, 12) and enzyme-linked immunosorbent assay (ELISA) (6) procedures are now available as diagnostic tools for rotavirus and seem to be the preferred diagnostic tests for use in most laboratories.

For the clinical diagnosis of rotavirus infection, it is not necessary to determine the specific viral serotype or subgroup since all strains produce indistinguishable forms of clinical illness (8). Therefore, this chapter will not cover diagnostic procedures for serotypes or subgroups, which are of value predominantly for research and epidemiological purposes. Rather, diagnostic immunoassays that detect a common rotaviral antigen shared by all mammalian strains will be described.

Features of solid-phase immunoassays

Solid-phase immunoassays are technically simple to perform and are relatively easy to establish for use on a single occasion. However, considerable effort is required in empirically determining and maintaining optimal assay conditions for maximum specificity and sensitivity. The assay for virus detection should be initially validated by the use of both positive and negative stool samples which have been independently examined by another method such as electron microscopy. At this time, there are no standard reagents available for the detection of gastroenteritis viruses, nor is there an accepted standardized procedure for the rotavirus RIA and ELISA tests. Published reports, although all using the same type of assays in principle, vary widely in critical details such as the choice of the solid phase, incubation time and temperature, composition of the washing solution, and the method used for calculation and interpretation of results. Therefore, until standard reagents and procedures are agreed upon, each laboratory should validate its test for sensitivity and specificity.

We present in this chapter details of reagent preparation and assay procedures which we have found to be valid, practical, and convenient in our laboratory. All dilutions specified are, of course, for our own reagents and are provided as guidelines only. Optimal dilutions of reagents must always be determined by a preliminary checkerboard titration.

ELISA offers major advantages over RIA in that it does not require the use of radioactive materials and gamma-radiation counting equipment. A spectrophotometer capable of rapidly handling small sample volumes is, however, required for ELISA. All reagents for ELISA are stable during storage, unlike the labeled antibody used in RIA. Furthermore, many of the detection antibody-enzyme conjugates used in ELISA are commercially available. In our laboratory, for the rotavirus ELISA test to achieve the sensitivity of RIA, an indirect test including an extra amplification step is required. This is accomplished by the use of two antirotavirus sera derived from different animal species. Our RIA test, therefore, has the advantage that a direct test may be performed, requiring hyperimmune serum from only one animal species.

A commercially available rotavirus ELISA test, Rotazyme (Abbott Laboratories, North Chicago, Ill.) (18), is of the direct test design; although it appears to be somewhat less sensitive in detecting rotavirus than most electron microscopic or indirect ELISA assays, it seems to be satisfactory for the detection of rotavirus in the stools of pediatric patients (who excrete higher titers of virus than do infected adults).

Preparation and selection of hyperimmune serum

The most important step in establishing the RIA and ELISA procedures is the proper preparation and selection of immune sera. For detection of rotavirus in stools, some assays have employed as critical reagents sera which have been prepared against rotavirus purified either from human stools or from the stools of gnotobiotic calves experimentally infected with human rotavirus (12, 23). These sera are difficult to prepare in most viral diagnostic laboratories because it is necessary to hyperimmunize animals either with a carefully selected preparation of human stool containing a high titer of rotavirus or with stool collected from experimentally infected gnotobiotic animals, which are expensive and not readily available. Another potential problem with these assays is that the hyperimmune sera occasionally possess nonspecific reactivity, presumably against extraneous stool components. Our laboratory has developed an RIA technique (7) for detection of human rotavirus in stool specimens which uses an easily prepared antiserum against a cell culture-grown simian rotavirus that is immunologically very closely related to human rotavirus. The technique is specific, sensitive, and practical and can be readily established in viral diagnostic laboratories.

SA-11 virus, a cytopathic simian rotavirus originally isolated by H. Malherbe and M. Strickland-Cholmley, University of Texas Health Science Center at San Antonio (16), is grown in low-passage cultures of a continuous line of rhesus embryonic kidney cells (MA 104). Cells are grown to confluence at 36°C in an equal mixture of Eagle minimal essential medium and medium 199 supplemented with 10% heat-inactivated fetal calf serum and 25 mM HEPES (*N*-2-hydroxyethylpiperazine-*N*′-2-ethanesulfonic acid) buffer. Cells are then washed twice in maintenance medium consisting of Earle balanced salt solution with 0.5% lactalbumin hydrolysate and 0.1% Yeastolate, renewed with the same medium, and then infected with SA-11 virus at a multiplicity of 25 $TCID_{50}$ (50% tissue culture infectious doses) per cell. Over 90% cytopathic effect is normally noted 48 h later, at which time the cultures are scraped with a rubber policeman, frozen and thawed three times, and then clarified by centrifugation at 2,000 × *g* for 15 min. The supernatant fluid (1.5 liters containing 10^7 viral $TCID_{50}$/0.1 ml) is concentrated by ultrafiltration and chromatographed at 4°C on a Sephadex G-100 column with phosphate-buffered saline (PBS) as the eluant. The volume of the first peak of UV light-absorbing material to emerge from the column is adjusted to 30 ml. This material usually contains 10^8 $TCID_{50}$/0.1 ml. Five-milliliter portions of this virus pool are applied to six 30-ml 20 to 40% sucrose density gradients and centrifuged for 75 min at 82,000 × *g* in a swinging-bucket rotor. Gradients are then fractionated into 1-ml portions. A distinct peak of UV light-absorbing material appearing between 10 and 15 ml from the bottom of the gradient is pooled and pelleted at 100,000 × *g* for 1 h. The pellet is suspended in 1.5 ml of PBS. This preparation, which contains both single- and double-shelled rotavirus particles, has a viral titer of approximately 10^9 $TCID_{50}$/0.1 ml and a 260/280-nm absorption ratio of 1.3. The preparation is mixed with an equal volume of Freund incomplete adjuvant and used as the inoculum for the immunization of guinea pigs.

We have been able to purchase rotavirus antibody-free guinea pigs in Massachusetts (7), but such animals may not always be available from other suppliers. Preimmunization serum should be obtained from each guinea pig. Five guinea pigs are then each immunized by the injection of a total of 0.4 ml of the virus-adjuvant mixture into the rear footpads and thighs. Three weeks later, a booster inoculum, prepared in the same manner, is administered in the front footpads and thighs. Four weeks later, the animals are exsanguinated, and serum is separated and stored frozen. Goats are immunized with 1.5 ml of the same virus-adjuvant mixture by the intramuscular route, with a booster immunization and bleeding schedule similar to that used for guinea pigs.

Not all of these hyperimmune animal sera give equally satisfactory results in the solid-phase immunoassay. Therefore, selection of the most suitable serum for the detection antibody (see below) is a critical step. Sera from all five guinea pigs should be iodinated and used as detection antibody in RIA tests or, in the case of ELISA tests, used as the first detection antibody (see below). Five diarrheal stools positive for rotavirus by electron microscopy, five control rotavirus-negative stools, and a 1:100 dilution of purified SA-11 virus should be used as samples in this preliminary test. The detection antibody which gives the highest positive/negative (P/N) ratio (see details of test below) for positive samples and low P/N ratios for negative samples is selected for all subsequent tests.

RIA

For performance of the RIA test, a stool sample is applied to the solid phase, which has previously been coated with hyperimmune serum prepared against rotavirus (coating antibody). A second hyperimmune serum, which can be from the same animal species, is radiolabeled (detection antibody) and allowed to react with virus from stool that may be bound to the solid phase. Presence of virus in the stool sample is indicated by the amount of radioactivity that remains bound to the solid phase.

The detection antibody for use in the RIA test is iodinated by the following procedure (7, 12, 17). First, the immunoglobulin fraction of the hyperimmune serum is prepared as follows. To 0.5 or 1 ml of serum an equal volume of saturated (4.1 M) $(NH_4)_2SO_4$ is added at room temperature. The precipitate is pelleted by centrifugation at 2,000 × *g* for 15 min, washed once with half-saturated $(NH_4)_2SO_4$, and suspended to the original volume of serum in 0.005 M sodium phosphate buffer (pH 8.0), followed by exhaustive dialysis at 4°C against the same buffer containing 0.1% azide. A column of DE52 cellulose ion-exchange resin is prepared by the manufacturer's directions. It is critical that the fines be removed and that the pH and ionic strength of the column be properly adjusted. The dialyzed material is then chromatographed, using the same 0.005 M phosphate buffer for the eluant. The first peak of UV light-absorbing material eluted contains the immunoglobulin. The protein concentration of the peak is adjusted to 1 mg/ml, and portions are stored at −70°C.

Second, the immunoglobulin fraction is iodinated. The following materials are needed for this iodination procedure:

Sodium phosphate buffer, pH 7.4, 0.25 M

High-specific-activity $Na^{125}I$ (200 μCi/2μl)

Immunoglobulin to be labeled (1 mg/ml)

Freshly made chloramine T (3.5 μg/μl). This, as well as the sodium metabisulfite, is weighed out in advance and stored in small screw-capped vials. They are reconstituted in 0.25 M sodium phosphate buffer (pH 7.4) just before use.

Freshly made sodium metabisulfite (4.8 μg/μl)

Solution containing sucrose (22.5%), KI (2 mg/ml), and aqueous phenol red (0.025%). This is stored at −20°C in small volumes.

Sephadex G-100 column, 0.9 by 15 cm, equilibrated with PBS containing 0.1% azide (PBS-Z). The Sephadex G-100 is packed in a disposable column. PBS-Z with 1% BSA and phenol red is applied to the column until the entire column turns red. One hour later, the column is washed with PBS-Z until all the red color is removed. The collection tubes are similarly pretreated with BSA to diminish the nonspecific adsorption of the globulin to the walls of the vessels.

Fetal calf serum negative for rotavirus antibody. This is best obtained from a supplier who sells serum from individual calves, rather than serum pooled from several animals (12).

For the iodination procedure, the following reagents are added in sequence to a small conical tube: 20 μl of phosphate buffer, 2 μl of ^{125}I, 10 μl of immunoglobulin, and 10 μl of chloramine T. After the addition of the last solution, a timed interval of 10 s is allowed to elapse, and the reaction is stopped by the addition of 20 μl of sodium metabisulfite and 20 μl of the solution containing sucrose, KI, and aqueous phenol red. The mixture is immediately applied to the top of the Sephadex G-100 column and allowed to penetrate completely into the column. The column is eluted with PBS-Z, and 1-ml fractions are collected. The first peak of radioactivity to emerge from the column contains the labeled immunoglobulin, which is usually contained in one or two fractions. This material is immediately diluted with an equal volume of fetal calf serum negative for rotavirus antibody. This labeled detection antibody is stored at 4°C and should be used in the RIA test as soon as possible, certainly within 4 weeks of its preparation.

The steps for testing a stool specimen by RIA are now outlined. Half the wells of a 96-well polyvinyl microtiter plate are coated with 75 μl of hyperimmune guinea pig serum diluted 1:10,000 in PBS-Z (coating antibody). The remaining wells are coated with a 1:10,000 dilution of preimmunization serum. After absorption of coating serum in a moist chamber at room temperature for 4 h, the plates are washed twice with PBS-Z, the wells are filled with 1% BSA in PBS-Z, and the plates are stored at 4°C until use. They should be stored for at least 12 h.

Stool suspension (50 μl per well) is inoculated into two wells coated with preimmunization sera and two wells coated with postimmunization sera. A 25-μl amount of a 2% solution of fetal calf serum negative for rotavirus antibody is added to each well. The plates are placed in a moist chamber for incubation at room temperature overnight and are then washed five times with PBS-Z.

The iodinated immunoglobulin (detection antibody) is diluted to contain 2×10^5 cpm/25 μl in PBS-Z, and 25 μl is added to each well. After 4 h of incubation at 37°C, the wells are again washed five times and are either punched out with a commercially available device or cut apart.

The individual wells are placed into test tubes, and the amount of radioactivity bound to each well is determined in a gamma counter.

A P/N ratio is calculated for each sample by dividing the mean counts per minute of the postimmunization serum-coated wells by the mean counts per minute of the preimmunization serum-coated wells. Controls for each test should include PBS-Z alone and a 1:100 dilution of purified SA-11 virus. A sample having a P/N ratio of 2 or greater is considered positive for rotavirus. The amount of nonspecific reactivity with guinea pig serum, as seen by the counts per minute in the preimmunization serum-coated wells, will vary with individual stool samples. Very rarely, a stool sample may give high counts on both pre- and postimmunization serum-coated wells and therefore must be reported as "nontestable." Nonspecific reactions have been a more serious problem in some configurations of the ELISA (see below). Various measures which have been employed in attempts to circumvent nonspecific reactions include addition of normal serum, reducing agent, chelating agent, and protease inhibitor to stool specimens as well as adjustment of the stool to neutral pH. None of these measures can completely eliminate false-positive reactions, making the inclusion of a confirmatory (preimmunization serum-coated well) test highly desirable (6, 8).

ELISA

For performance of the ELISA test (6, 21), a stool sample is applied to the solid phase which has previously been coated with hyperimmune serum prepared against rotavirus (coating antibody). Another hyperimmune serum (first detection antibody), derived from a different animal species than the coating serum, is allowed to react with virus from stool that may be bound to the solid phase. A second detection antibody, conjugated with the enzyme alkaline phosphatase, is directed against the immunoglobulin of the animal species which was the source of the first detection antibody. The amount of virus present in the stool sample is proportional to the amount of alkaline phosphatase bound to the solid phase by the second detection antibody. Quantitation of the bound alkaline phosphatase is then accomplished by the addition of a substrate which changes color in the presence of this enzyme. Details of the ELISA procedure used in our laboratory are as follows.

It has been suggested that only the inner 60 wells of a 96-well round-bottomed polyvinyl microtiter plate be used for ELISA tests, since it has been noted that the ELISA test may be sensitive to defects in the manufacturing process of microtiter plates, resulting in variable antibody binding. The outer wells should be filled with PBS-Z at each step to provide even heat distribution. Half of the 60 inner wells are coated with 75 μl of hyperimmune goat serum (coating antibody) diluted 1:20,000 in 0.05 M carbonate-bicarbonate buffer (pH 9.6). The remaining 30 inner wells are coated with a 1:20,000 dilution of preimmunizaton serum. After absorption of coating antibody in a moist chamber at room temperature for 4 h, the plates may be sealed and stored at 4°C until used.

For testing of a stool specimen, the coated plates are first washed five times with PBS-Z containing 0.05% Tween 20. A commercially available washer-aspirator apparatus is used in some laboratories for performing the multiple washing steps. Stool suspension (50 μl per well) is then inoculated into two wells coated with preimmunization goat sera and two wells coated with postimmunization goat sera. A 25-μl amount of a 2% solution of fetal calf serum negative for rotavirus antibody is added to each well. The plates are incubated for 1 h at 37°C and washed five times with PBS-Z–Tween.

A 50-μl amount of a 1:1,000 dilution of guinea pig antiserum to SA-11 (first detection antibody) prepared in PBS-Z–Tween is added to each well, and the plate is incubated for 1 h at 37°C.

The plate is again washed five times with PBS-Z–Tween, and 100 μl of a commercially available alkaline phosphatase-conjugated anti-guinea pig serum (second detection antibody) at a 1:500 dilution in PBS-Z–Tween is added.

The plate is again washed five times with PBS-Z–Tween, and 200 μl of substrate solution prepared in 10% diethanolamine buffer (described below) is added. The plate is then incubated at room temperature until a yellow color develops and a weakly positive standard (e.g., 1:1,000 dilution of purified SA-11 virus) becomes positive. In practice, this usually requires about 15 to 30 min. The reaction is stopped by the addition of 50 μl of 3 N NaOH, and the optical density at 405 nm is determined for each well in a rapid sampling spectrophotometer.

P/N calculations and controls are the same as described above for the RIA test.

The substrate solution is prepared as follows. Diethanolamine buffer (21) containing 97 ml of diethanolamine, 800 ml of water, 0.2 g of NaN_3, and 100 mg of $MgCl_2 \cdot 6H_2O$ is prepared, and 1 M HCl is added until the pH is 9.8. The total volume is made up to 1 liter with distilled water, and the buffer is stored at 4°C in the dark. The buffer is warmed to room temperature before use, and the substrate *p*-nitrophenyl phosphate is added to a concentration of 1 mg/ml. The substrate solution must be used on the same day that it is prepared.

Commercially available ELISA

As mentioned above, the Rotazyme assay is of the direct test design (18). The test employs polystyrene beads coated with guinea pig antibody to SA-11 rotavirus, instead of antibody-coated polyvinyl microtiter plates. The detection antibody is peroxidase-conjugated rabbit antibody to SA-11 rotavirus. Rotazyme does not provide a built-in confirmatory test; that is, results are not determined by comparison of reactivities of test stool specimens with both rotavirus-positive (postimmunization) serum and rotavirus-negative (preimmunization) serum. Such a confirmatory test, when included in an indirect ELISA procedure, has been shown to reduce problems of false-positive results by 73% (6). However, Rotazyme has not encountered many problems with nonspecificity, apparently because of the direct test design and special formulations of sample diluent and wash solutions. A large number of specimens have been reported to be tested with the Rotazyme assay in several laboratories, with favorable results (18). However, one report, requiring confirmation, does indicate that Rotazyme has encountered frequent false-positive reactions in stool specimens from neonates (8).

Monoclonal antibody test

Recent experience in our laboratory indicates that use of a rotavirus-specific monoclonal antibody reagent improves the sensitivity and specificity of rotavirus detection (9). A monoclonal antibody which reacts with the common rotavirus antigen on the sixth viral gene product was prepared. Mice were immunized with murine rotavirus, and cell fusions were subsequently performed by standard techniques. Murine rotavirus cannot be classified in either of the two common rotaviral subgroups, and therefore the monoclonal antibody that has been prepared apparently lacks rotavirus subgroup prejudice and reacts only with the group-common antigen. When the monoclonal antibody was used as a capture reagent in our direct RIA procedure, a concordance of 96% with Rotazyme was observed in 177 tested stool samples. Six of seven discrepant specimens were shown to be false-negative by Rotazyme, and the seventh was shown to be false-positive by Rotazyme, by use of electron microscopy and confirmatory (blocking) immunoassay procedures. This direct monoclonal antibody immunoassay can be performed in less than 4 h and merits additional evaluation as a rapid, convenient, and sensitive test that can reduce currently encountered problems associated with diagnosing rotavirus by immunoassay.

DIRECT EXAMINATION FOR NORWALK VIRUS

Direct examination of stool specimens is the primary method for diagnosis of Norwalk virus infection, inasmuch as the virus has not been cultivated in vitro. Norwalk virus was originally visualized by immune electron microscopic examination of diarrheal stools derived from human volunteer studies (14). All tests for Norwalk virus of necessity rely for critical reagents upon clinical materials from volunteers, because it has not been possible to purify Norwalk antigen from stool sufficiently to permit preparation of useful hyperimmune animal sera. Furthermore, unlike the situation with rotavirus, there are no known animal viruses that cross-react serologically with Norwalk virus and can be used as substitute antigens. Thus, the diagnostic capability for Norwalk virus is currently limited to a very few research laboratories that have access to human clinical material from volunteer studies.

Immune electron microscopy was first used for detection of Norwalk virus and is described in detail elsewhere (14). Briefly, the procedure involves the incubation of a virus particle-containing fecal filtrate with a selected convalescent serum from an individual known to have had recent infection with Norwalk virus. Antibody in the convalescent-phase serum aggregates the viral particles, which can be recovered by centrifugation and visualized by electron microscopy after negative staining with phosphotungstic acid. Various technical manipulations, such as the concentration of stool filtrates and alterations of antigen-antibody ratios, may be required to achieve optimal conditions for Norwalk virus particle visualization by immune electron microscopy.

Clearly, the RIA test for Norwalk virus (4, 11) is the most sensitive, practical, and quantitative assay currently available and is therefore the procedure of choice. An ELISA test has yet to be developed for detection of Norwalk virus, as immune serum is readily available only from one species, humans.

The RIA test for Norwalk virus is similar in principle to the one described above for rotavirus. The assay will not be outlined in detail since, as indicated above, it is currently limited to those laboratories possessing precious human reagents from volunteer studies. In brief, wells of microtiter plates are coated with acute- and convalescent-phase sera from a selected volunteer who has experienced experimentally induced Norwalk illness. The stool samples to be tested are added, and convalescent-phase serum from a different volunteer is used as the source of radioiodine-labeled detection antibody. P/N ratios are calculated as described for rotavirus, and stool samples having P/N values greater than 2 are considered positive.

ISOLATION OF GASTROENTERITIS VIRUSES

To date, Norwalk virus has not been propagated in cell culture. Although some stools containing rotavirus can be grown in roller-tube cell cultures of human or simian origin, viral growth is inefficient and achievable only with 50 to 70% of virus-positive specimens (8, 19, 20, 22). In addition, specimens and cells must be treated with trypsin. Thus, although in vitro cultivation is an invaluable research tool, it is not practical as a diagnostic procedure for rotavirus. The sensitivity, convenience, rapidity, and large-scale applicability of immunoassays clearly make them the clinical diagnostic procedures of choice.

SEROLOGICAL DIAGNOSIS FOR ROTAVIRUS

The diagnosis of rotavirus infection is primarily dependent on the demonstration of viral antigen in diarrheal stool specimens. Children, in particular, excrete copious amounts of virus over a period of several days, making antigen detection practical. Demonstration of a virus-specific antibody rise in acute- and convalescent-phase serum specimens has a number of major disadvantages for rotavirus diagnosis in a clinical setting. Approximately 90% of the population have various levels of preexisting rotavirus-specific serum antibody (8). Infants have rotavirus antibody of maternal origin. Seroconversions in all age groups are associated with both symptomatic and asymptomatic rotavirus infections (2, 8). In addition, there are the inherent problems of obtaining two serum specimens and of not being able to make the diagnosis until after the clinical illness is over. Serological diagnosis of rotavirus infection is, however, useful for certain epidemiological studies (8).

Several serological assays for rotavirus exist, including complement fixation, immunofluorescence, neutralization, and solid-phase immunoassay (RIA and ELISA) tests (8, 13, 18, 23). Serological assays rely either upon human rotavirus from stool as antigen or upon use of an animal rotavirus as a convenient substitute antigen for the human agent. As is the case for direct examination, RIA and ELISA provide the most sensitive methods for detecting seroconversion to rotavirus.

The procedure used in our laboratory is as follows. A purified SA-11 virus preparation is titrated by the antigen test and diluted to contain 4 U of antigen in 50 μl. Mock-infected cells are used to prepare control antigen, which is diluted to the same extent as the virus. Microtiter plates are coated with hyperimmune antirotavirus serum and postcoated with BSA as described above. To half the wells of the microtiter plate 50 μl of virus is added, and an equivalent amount of control antigen is added to the remaining wells. After a 4-h incubation the plates are washed with PBS-Z, and fourfold dilutions of acute- and convalescent-phase test sera (which are prepared in PBS-Z–Tween) are added in duplicate to both virus-containing and control wells. After a 1-h incubation, the plates are washed five times with PBS-Z–Tween. A commercially available antibody to human immunoglobulin, which is either radioiodinated or enzyme labeled, is used as a detection antibody. A serum dilution is considered to be positive if it shows at least twice the reactivity on the virus-coated well as on the control antigen-coated well. A fourfold rise in antibody titer is considered diagnostic of rotavirus infection.

SEROLOGICAL DIAGNOSIS FOR NORWALK VIRUS

In contrast to rotavirus, the serological diagnosis of Norwalk virus is of more practical importance than antigen detection. Norwalk antigen is shed in feces in small amounts for a brief time period (3, 8). Whereas seroconversions could be demonstrated in all volunteers experiencing experimentally induced Norwalk illness, antigen shedding could only be shown for half the subjects. In naturally occurring outbreaks the experience has been similar. Almost all affected individuals seroconvert, but antigen can be demonstrated usually in only a third or fewer of the cases (8, 15). Unlike the situation with rotavirus, the major interest in diagnosing Norwalk infection is epidemiological. Public health authorities commonly are not involved until the peak of the outbreak has passed, making it easier to obtain paired serum samples rather than illness-phase stools.

Immune electron microscopy, immune adherence hemagglutination assay, and RIA have been used for serological diagnosis of Norwalk virus infection (4, 8, 14). As noted above, laboratory studies with Norwalk virus are currently restricted to a very few research laboratories that possess the required reagents derived from human volunteer studies. RIA is the preferred serological assay because it possesses greater sensitivity than the other assays, and it is more convenient and quantitative than immune electron microscopy.

The serological test for antibody to Norwalk virus is a blocking test (4, 11) based upon a modification of the RIA procedure for detecting Norwalk antigen (see above). In brief, wells are coated with convalescent-phase serum from a volunteer who developed Norwalk illness. A stool specimen containing Norwalk antigen and obtained from another volunteer is then added. After the wells are washed, test serum specimens are added, starting at an initial dilution of 1:50. Labeled detection antibody, prepared from a convalescent volunteer's serum, is added next. A given dilution of test serum is considered to be positive for Norwalk antibody if it reduces the number of counts bound by at least 50% as compared to a buffer control.

LITERATURE CITED

1. **Bishop, R. F., G. P. Davidson, I. H. Holmes, and B. J. Ruck.** 1974. Detection of a new virus by electron microscopy of faecal extracts from children with acute gastroenteritis. Lancet **i:**149–151.
2. **Bishop, R. F., G. L. Barnes, E. Cipriani, and J. S. Lund.** 1983. Clinical immunity after neonatal rotavirus infection. A prospective longitudinal study in young children. N. Engl. J. Med. **309:**72–76.
3. **Blacklow, N. R., and G. Cukor.** 1981. Viral gastroenteritis. N. Engl. J. Med. **304:**397–406.
4. **Blacklow, N. R., G. Cukor, M. K. Bedigian, P. Echeverria, H. B. Greenberg, D. S. Schreiber, and J. S. Trier.** 1979. Immune response and prevalence of antibody to Norwalk enteritis virus as determined by radioimmunoassay. J. Clin. Microbiol. **10:**903–909.
5. **Blacklow, N. R., R. Dolin, D. S. Fedson, H. DuPont, R. S. Northrop, R. B. Hornick, and R. M. Chanock.** 1972.

Acute infectious nonbacterial gastroenteritis: etiology and pathogenesis. Ann. Intern. Med. **76:**993–1008.

6. **Brandt, C. D., H. W. Kim, W. J. Rodriguez, L. Thomas, R. H. Yolken, J. O. Arobio, A. Z. Kapikian, R. H. Parrott, and R. M. Chanock.** 1981. Comparison of direct electron microscopy, immune electron microscopy, and rotavirus enzyme-linked immunosorbent assay for detection of gastroenteritis viruses in children. J. Clin. Microbiol. **13:**976–981.
7. **Cukor, G., M. K. Berry, and N. R. Blacklow.** 1978. Simplified radioimmunoassay for detection of human rotavirus in stools. J. Infect. Dis. **138:**906–910.
8. **Cukor, G., and N. R. Blacklow.** 1984. Human viral gastroenteritis. Microbiol. Rev. **48:**157–179.
9. **Cukor, G., D. M. Perron, R. Hudson, and N. R. Blacklow.** Detection of rotavirus in human stools by using monoclonal antibody. J. Clin. Microbiol. **19:**888–892.
10. **Flewett, T. H., and G. N. Woode.** 1978. The rotaviruses. Arch. Virol. **57:**1–23.
11. **Greenberg, H. B., R. G. Wyatt, J. Valdesuso, A. R. Kalica, W. T. London, R. M. Chanock, and A. Z. Kapikian.** 1978. Solid-phase microtiter radioimmunoassay for detection of the Norwalk strain of acute nonbacterial, epidemic gastroenteritis virus and its antibodies. J. Med. Virol. **2:**97–108.
12. **Kalica, A. R., R. H. Purcell, M. M. Sereno, R. G. Wyatt, H. W. Kim, R. M. Chanock, and A. Z. Kapikian.** 1977. A microtiter solid phase radioimmunoassay for detection of the human reovirus-like agent in stools. J. Immunol. **118:**1275–1279.
13. **Kapikian, A. Z., W. L. Cline, H. W. Kim, A. R. Kalica, R. G. Wyatt, D. H. VanKirk, R. M. Chanock, H. D. James, and A. L. Vaughan.** 1976. Antigenic relationships among five reovirus-like (RVL) agents by complement fixation (CF) and development of new substitute CF antigens for the human RVL agent of infantile gastroenteritis. Proc. Soc. Exp. Biol. Med. **152:**535–539.
14. **Kapikian, A. Z., R. G. Wyatt, R. Dolin, T. S. Thornhill, A. R. Kalica, and R. M. Chanock.** 1972. Visualization by immune electron microscopy of a 27-nm particle associated with acute infectious nonbacterial gastroenteritis. J. Virol. **10:**1075–1081.
15. **Kaplan, J. E., G. W. Gary, R. C. Baron, N. Singh, L. B. Schonberger, R. Feldman, and H. B. Greenberg.** 1982. Epidemiology of Norwalk gastroenteritis and the role of Norwalk virus in outbreaks of acute nonbacterial gastroenteritis. Ann. Intern. Med. **96:**756–761.
16. **Malherbe, H. H., and M. Strickland-Cholmley.** 1967. Simian virus SA-11 and the related O agent. Arch. Gesamte Virusforsch. **22:**235–245.
17. **Purcell, R. H., D. C. Wong, H. J. Alter, and P. V. Holland.** 1973. Microtiter solid-phase radioimmunoassay for hepatitis B antigen. Appl. Microbiol. **26:**478–484.
18. **Rubenstein, A. S., and M. F. Miller.** 1982. Comparison of an enzyme immunoassay with electron microscopic procedures for detecting rotavirus. J. Clin. Microbiol. **5:**938–944.
19. **Sato, K., Y. Inaba, T. Shinozaki, R. Fujii, and M. Matumoto.** 1981. Isolation of human rotavirus in cell cultures: brief report. Arch. Virol. **69:**155–160.
20. **Urasawa, S., T. Urasawa, and K. Taniguchi.** 1982. Three human rotavirus serotypes demonstrated by plaque neutralization of isolated strains. Infect. Immun. **38:**781–784.
21. **Voller, A., D. Bidwell, and A. Bartlett.** 1980. Enzyme-linked immunosorbent assay, p. 359–371. *In* N. R. Rose and H. Friedman (ed.), Manual of clinical immunology, 2nd ed. American Society for Microbiology, Washington, D.C.
22. **Wyatt, R. G., H. D. James, A. L. Pittman, Y. Hoshino, H. B. Greenberg, A. R. Kalica, J. Flores, and A. Z. Kapikian.** 1983. Direct isolation in cell culture of human rotaviruses and their characterization into four serotypes. J. Clin. Microbiol. **18:**310–317.
23. **Yolken, R. H., R. G. Wyatt, G. Zissis, C. D. Brandt, W. J. Rodriguez, H. W. Kim, R. H. Parrott, J. J. Urrutia, L. Mata, H. B. Greenberg, A. Z. Kapikian, and R. M. Chanock.** 1978. Epidemiology of human rotavirus types 1 and 2 as studied by enzyme linked immunosorbent assay. N. Engl. J. Med. **299:**1156–1161.

Chapter 81

Hepatitis Viruses

F. BLAINE HOLLINGER AND JULES L. DIENSTAG

CLINICAL AND EPIDEMIOLOGICAL FEATURES

Viral hepatitis is a systemic disease primarily involving the liver. Most cases of acute viral hepatitis seen in children and adults are caused by one of the following agents (Table 1): hepatitis A virus (HAV), the etiological agent of viral hepatitis type A (infectious hepatitis or short-incubation hepatitis); hepatitis B virus (HBV), which is associated with viral hepatitis type B (serum hepatitis or long-incubation hepatitis); and the more recently recognized non-A, non-B hepatitis viruses. The disease caused by these last viruses has been tentatively designated non-A, non-B hepatitis because no specific serological assays are currently available for their identification, and the hepatitis cannot be ascribed to either HAV, HBV, cytomegalovirus, or the Epstein-Barr virus. Non-A, non-B hepatitis accounts for most of the transfusion-associated hepatitis cases seen in the United States and a sizable portion of sporadic hepatitis. The most recently discovered hepatitis agent, the delta agent, now designated hepatitis D virus (HDV), is a defective virus which coinfects with, and requires, HBV for its own expression. Other well-characterized viruses which infrequently cause sporadic hepatitis, such as yellow fever virus, cytomegalovirus, Epstein-Barr virus (infectious mononucleosis), herpes simplex virus, rubella virus, and the enteroviruses, are discussed in other chapters of this Manual.

The pertinent clinical and epidemiological features of viral hepatitis types A, B, and non-A, non-B are presented in Table 2. All of these agents produce characteristic but generally indistinguishable histopathological lesions in the liver. In individual cases, differentiation among the various types of hepatitis on clinical grounds alone is not feasible. Moreover, clinical expression of disease is extremely variable, ranging from asymptomatic (inapparent infection) to anicteric or icteric hepatitis to fulminant hepatitis and even death. The prodromal or preicteric phase is characterized frequently by fever (<39.5°C), fatigability, malaise, myalgia, blunting of olfactory and gustatory senses, anorexia, nausea, vomiting, and, especially in patients with HBV infection, maculopapular and urticarial rash, arthralgias, or polyarthritis (seen in 10 to 15% of cases). Symptoms are followed by right upper quadrant discomfort or pain associated with hepatomegaly and the appearance of dark urine and clinical jaundice. Patients with acute viral hepatitis usually recover completely. The frequency of fulminant hepatitis among icteric hepatitis B patients is probably less than 2% now that posttransfusion hepatitis B has been significantly reduced by the screening of donor units for hepatitis B surface antigen. Mortality data for hepatitis A suggest that the frequency of fulminant disease in icteric cases is less than 0.5%. The importance of the delta agent is its ability to convert an asymptomatic or mild, acute or chronic HBV infection into fulminant or severe, progressive disease.

Distinguishing between viral hepatitis type A and type B assumes practical importance in light of demonstrations that hepatitis A, an infection spread by close contact via the fecal-oral route, can, on rare occasions, be transmitted parenterally, and hepatitis B, which occurs sporadically after parenteral inoculation of virus-contaminated blood or blood products, can also be spread by close intimate contact. The discovery of a unique hepatitis B antigen by Blumberg and his associates (3) and its specific relationship to HBV infection allow serological identification of type B hepatitis cases. Now that serological markers for HAV infection are also available, all cases of acute viral hepatitis can be categorized serologically as type A or B or, by exclusion, as non-A, non-B.

DESCRIPTION OF AGENTS (5, 21, 22, 53)

HAV is a nonenveloped icosahedral 27- to 32-nm particle (Fig. 1) which may appear "full" or "empty" under an electron microscope. Its peak buoyant density in cesium chloride is 1.33 to 1.34 g/cm^3, but both heavier and lighter particles exist; the virus has a sedimentation coefficient ($s_{20,w}$) of 160S in a neutral sucrose solution. Studies of its nucleic acid composition indicate that HAV is an RNA virus with a linear, single-stranded genome of approximately 2.3×10^6 daltons. HAV has four major, structural polypeptides similar in molecular weight to those of poliovirus, and it localizes exclusively in the cytoplasm of hepatocytes. Lipid is not an integral component of HAV, which is stable to treatment with 20% ether, acid, and heat (60°C for at least 1 h); its infectivity can be preserved for years at −20°C and for at least 1 month after being dried and stored at 25°C. Viral infectivity is destroyed by autoclaving (121°C for 20 min), by boiling in water for 1 min, by dry heat (180°C for 1 h), by UV irradiation (1 min at 1.1 W), by treatment with Formalin (1:4,000 [wt/vol] for 3 days at 37°C), or by treatment with chlorine (10 to 15 ppm for 30 min) or chlorine-containing compounds (e.g., sodium hypochlorite, 10 mg/liter for 15 min). Only one serotype of HAV has been defined. It does not cross-react with HBV, and its host range appears to be limited to humans, marmosets, owl monkeys, Malaysian cynomolgous monkeys, and chimpanzees. Its properties most closely resemble those of the enterovirus subgroup of the family Picornaviridae, and it has been classified as enterovirus type 72.

Sera from patients with hepatitis B infections reveal three distinct morphological entities in varying proportions (Fig. 2). The more numerous forms (by a factor of 10^3 to 10^4) are the small pleomorphic spherical particles measuring 17 to 25 nm in diameter (mean of 22 nm). Particle counts of 10^{13} or higher have been detected in some sera. Tubular or filamentous forms of various lengths, but with a diameter similar to that of the smaller particles, are also observed. HBV is a complex, double-shelled particle with a diameter of 42 nm. Originally designated the Dane

TABLE 1. Terminology of viral hepatitis

Agent	Preferred terminology	Equivalent terminology
HAV	Viral hepatitis type A	Infectious hepatitis Epidemic jaundice Short-incubation hepatitis
HBV	Viral hepatitis type B	Serum or transfusion hepatitis Homologous serum jaundice Long-incubation hepatitis
Non-A, non-B hepatitis viruses	Viral hepatitis type non-A, non-B	Non-A, non-B hepatitis
HDV	Viral hepatitis type D	Delta agent hepatitis

particle, it consists of a 27-nm core surrounded by a 7- to 8-nm viral protein coat.

The nomenclature used in viral hepatitis B is shown in Table 3. HBV is the prototype agent for a new family of viruses designated Hepadnaviridae. The complex antigen determinant found on the surface of HBV is called hepatitis B surface antigen (HBsAg). Previous designations include Australia or Au antigen and hepatitis-associated antigen. HBsAg on the hepatitis B virion is biochemically identical to that detected on the smaller particles and the filamentous forms and is composed of proteins, carbohydrates, and lipids. Seven or more polypeptides ranging in molecular weight from 25,000 to 100,000 have been found in purified preparations of HBsAg. Two of the polypeptides, with molecular weights of 25,000 and 30,000, comprise approximately 55% by weight of the whole particle. The larger polypeptide (p30) and one or two of the larger minor components appear to be glycosylated. The 22-nm particles have an average buoyant density of 1.20 g/cm^3 in CsCl and 1.17 g/cm^3 in sucrose, a molecular weight of 3.7×10^6 to 4.6×10^6, and a sedimentation coefficient that ranges from 39 to 54S. Analysis has revealed that one antigenic specificity, designated *a*, is common to all HBsAg preparations. In addition, there are two sets of mutually exclusive determinants, *d* or *y* and *w* or *r*. This results in four principal subtypes of HBsAg: *adw*, *ayw*, *adr*, and *ayr*. Because of antigenic heterogeneity of the *w* determinant, there are 10 major serotypes of HBV. The predominant subtype found in North America is *adw*, followed by *ayw*. HBV subtypes do not change after infection. This fact is useful when attempting to trace an infection from one source to another. Complete

TABLE 2. Epidemiological and clinical features of viral hepatitis types A, B, and non-A, non-B

Feature	Type A	Type B	Non-A, non-B
Incubation period	2–7 weeks (avg, 4 ± 1)	4–20 weeks (avg, 11 ± 4)[a]	2–20 weeks (avg, 7 ± 2)
Principal age distribution	Children,[b] young adults	Adults[c]	?
Seasonal incidence	Throughout the year but tends to peak in autumn	Throughout the year	Throughout the year
Route of infection	Predominantly fecal-oral	Parenteral, sexual contact	Predominantly parenteral
Occurrence of virus			
Blood	2 weeks before to <1 week after jaundice	Months to years	Months to years
Stool	2 weeks before to 2 weeks after jaundice	Absent	Probably absent
Urine	Rare	Absent	Probably absent
Saliva, semen	Rare (saliva)	Frequently present	Unknown
Clinical and laboratory features			
Onset	Usually abrupt	Usually insidious	Insidious
Fever > 38°C	Common early	Less common	Less common
Duration of transaminase elevation	2–6 weeks	2–6+ months	2–6+ months
Immunoglobulins (IgM levels)	Elevated	Normal to slightly elevated	Normal to slightly elevated
Complications	Uncommon, no chronicity	Chronicity in 5 to 10%	Chronicity in 30 to 50%
Mortality rate (icteric cases)	<0.5%	<1–2%	0.5–1%
HBsAg	Absent	Present	Absent
Immunity			
Homologous	Yes	Yes	?
Heterologous	No	No	No
Duration	Probably lifetime	Probably lifetime	?
Gamma globulin (immune globulin USP) prophylaxis	Regularly prevents jaundice	Prevents jaundice only if gamma globulin is of sufficient potency against HBV	?

[a] Longer incubation periods, up to 9 months, have been observed, but the delay has usually been due to the administration of specific antibody (anti-HBs) after infection with HBV.

[b] Nonicteric hepatitis A is common in children.

[c] Among the 15- to 29-year age group, hepatitis B is often associated with drug abuse or promiscuous sexual behavior. Patients with transfusion-associated hepatitis B are generally over age 29.

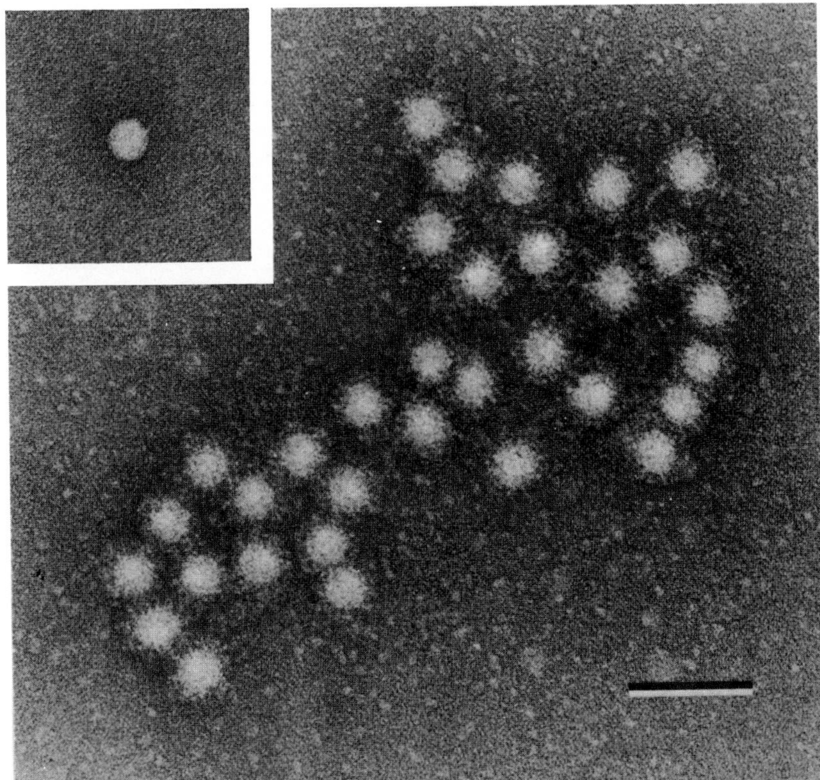

FIG. 1. Electron micrograph of 27-nm HAV particles purified by a combination of isopycnic banding in CsCl, column chromatography, and rate-zonal separation in sucrose. A single isolated HAV particle (upper left) is compared to the same preparation after incubation with convalescent-phase serum from a patient with type A hepatitis. ×165,000. The bar is equivalent to 100 nm (from reference 44).

(DNA-containing) HBV has a buoyant density of 1.28 g/cm^3 in CsCl. The 27-nm internal core of HBV contains the hepatitis B core antigen (HBcAg), a small, circular, partially double-stranded DNA molecule, and specific DNA polymerase activity. The presence of a single-stranded region of variable length in the circular DNA molecules results in genetically heterogeneous particles with buoyant densities in CsCl that range from 1.28 to 1.38 g/cm^3. The core particle contains several unique polypeptides ranging in molecular weight from 17,000 to 80,000. The hepatitis B e antigen (HBeAg) has not been completely elucidated but is postulated to be an integral component of the HBV core particle, presumably existing in a cryptic form. When core particles are disrupted, a 19,000-dalton polypeptide is released which appears to contain epitopes for both HBcAg and HBeAg.

The stability of HBV does not always coincide with that of HBsAg. Immunogenicity and antigenicity are retained after exposure to ether, acid (pH 2.4 for at least 6 h), heat (98°C for 1 min; 60°C for 10 h), and up to 40 cycles of freeze-thawing. HBV is stable at 37°C for 60 min but not at 60°C for 10 h. However, inactivation may be incomplete under these conditions if the concentration of virus is excessively high. Exposure of HBsAg to 0.25% sodium hypochlorite for 3 min destroys antigenicity (and presumably infectivity). Infectivity in serum is lost after direct boiling for 2 min, autoclaving at 121°C for 20 min, or dry heat at 160°C for 1 h. Recent studies have shown that HBV is inactivated by exposure to sodium hypochlorite (500 mg of free chlorine per liter) for 10 min, 0.1 to 2% aqueous glutaraldehyde, Sporicidin (pH 7.9), 70% isopropyl alcohol, 80% ethyl alcohol at 11°C for 2 min, Wescodyne diluted 1:213, or combined β-propiolactone and UV irradiation (4, 39, 46). HBV has been shown to retain infectivity when stored at 30 to 32°C for at least 6 months, when frozen at −20°C for 15 years, and after being dried and stored at 25°C for at least 1 week.

The delta agent (HDV) is a 35- to 37-nm virus composed of a delta antigen (HDAg)-expressing core encapsidated by HBsAg and requiring the helper function of HBV to support its replication. As anticipated, HDV assumes the HBsAg subtype of the HBV infection present in the host. HDV has a small RNA genome (5.5 × 10^5 daltons) that is nonhomologous with HBV DNA; HDAg is a 68,000-dalton protein. The virus particle has a buoyant density of 1.25 g/cm^3 and a sedimentation coefficient intermediate between those of HBsAg and intact HBV particles. It is inactivated by Formalin under conditions similar to those which inactivate HBV (50). Both immunoglobulin M (IgM) and IgG antibodies to HDAg (anti-HD) can be detected during the course of HDV infection. HDV can infect only persons simultaneously or already infected with HBV.

PROCESSING OF SPECIMENS AND ENVIRONMENTAL CONTROL

The stability of the various serological markers for HBV and HAV eliminates the need for extraordinary

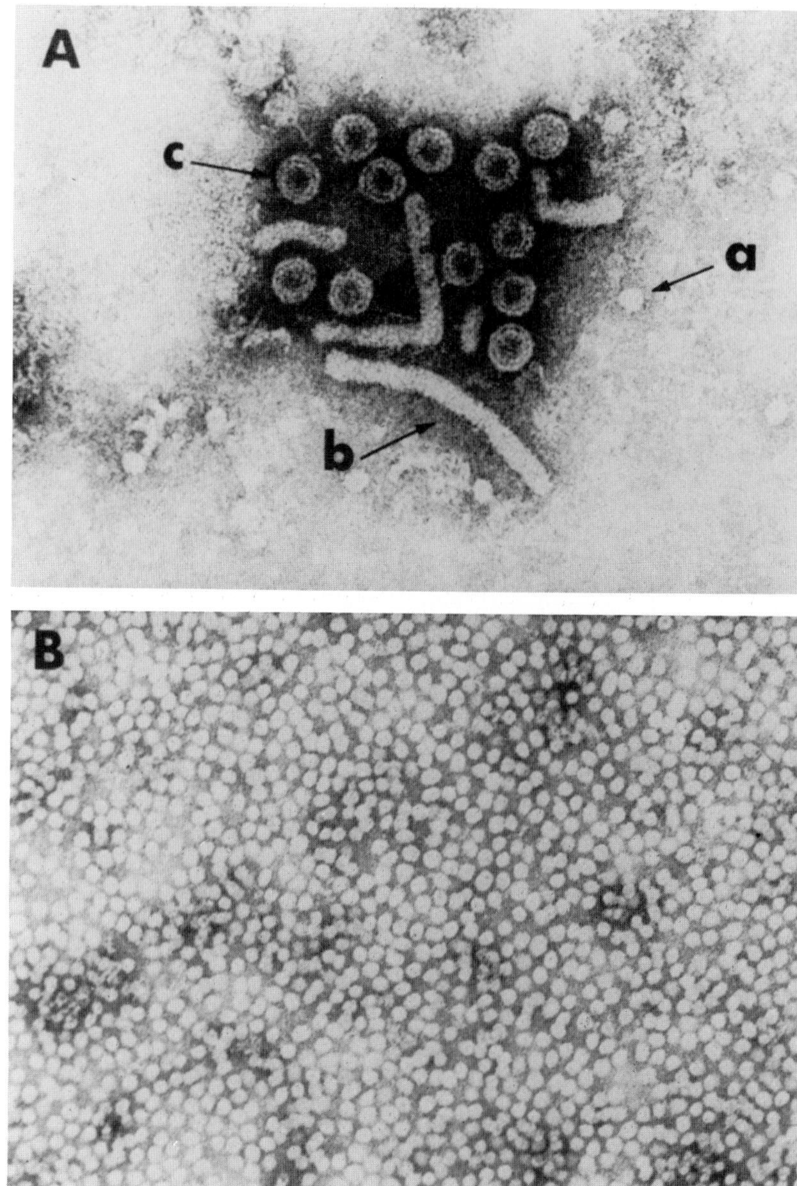

FIG. 2. (A) Electron micrograph of serum showing the presence of three distinct morphological entities: (a) 20-nm pleomorphic spherical particles; (b) tubular or filamentous forms with a diameter of 20 nm; and (c) 42-nm spherical particles now considered to be HBV (Dane particle). ×132,000. (B) Electron micrograph of a purified preparation of 17- to 25-nm pleomorphic spherical particles containing HBsAg. ×77,000. Similar preparations currently are being used in human vaccine trials.

collection and storage procedures if bacterial contamination is minimized. Samples can be stored at 4 to 8°C if testing is to take place within 5 to 7 days. Should longer delays be anticipated, freezing is desirable. The addition of a bacteriostatic agent to the samples is rarely indicated. Should this become necessary, a final concentration of 0.01% thimerosal, 0.1% sodium azide, or gentamicin sulfate at 25 to 50 μg/ml is preferred. The use of anticoagulants or the presence of severely hemolyzed blood may occasionally cause false-positive immunoassay responses. Serum samples can be shipped at ambient temperatures if delivery is expected to be made within 48 h. Unseparated clotted blood sent in the original collection tube and received in the laboratory within 24 h does not adversely affect assay results. However, safety is best achieved by shipping samples frozen in dry ice and in doubly sealed containers, as stipulated by the Interstate Quarantine Regulations (Code of Federal Regulations, Title 42, Part 72.25, Etiologic Agents). Although antibodies to HAV and HBV antigens are stable for years at −20°C and for at least 10 days at 37°C, repetitive freezing and thawing may lead to substantial losses in titer.

Laboratory personnel should regard all specimens collected from hepatitis patients as potentially dangerous, especially stool samples from hepatitis A candidates and blood or body fluids collected from patients with hepatitis B or non-A, non-B hepatitis. Mouth pipetting and smoking, eating, or drinking in

TABLE 3. Nomenclature for viral hepatitis type B

Term	Abbreviation	Description
Hepatitis B virus	HBV	The 42-nm double-shelled particle that consists of a 7-nm outer shell and a 27-nm inner core. The core contains a small, circular, partially double-stranded DNA molecule and DNA polymerase activity. Originally called the Dane particle. This is the prototype for the family Hepadnaviridae.
Hepatitis B surface antigen	HBsAg	The complex antigenic determinant that is found on the surface of HBV and on the 22-nm particles and tubular forms. Formerly designated Australia (Au) antigen or hepatitis-associated antigen (HAA).
Hepatitis B core antigen	HBcAg	The antigenic specificity associated with the 27-nm core of HBV.
Hepatitis B e antigen	HBeAg	An antigenic determinant that is closely associated with the nucleocapsid of HBV.
Antibody to HBsAg, HBcAg, and HBeAg	Anti-HBs, anti-HBc, and anti-HBe	Specific antibodies produced in response to their respective antigenic determinants.

the laboratory should be strictly forbidden. In our laboratories, needles are placed in glass bottles or metal cans and autoclaved before being discarded. All other materials are placed in discard pans and autoclaved at 121°C for 40 min. Work areas are decontaminated with 0.5% sodium hypochlorite, e.g., a 1:10 dilution of Clorox, that is prepared fresh each month. It is used for physical cleaning of environmental surfaces, for washing hands that have been inadvertently contaminated with virus, and for decontaminating metal instruments or rotors and most labile plastics. Because hypochlorites may be corrosive to many metals, the solutions are not allowed to remain in contact with such materials for more than 5 min. Disposable gloves and gowns are worn when personnel are working with known infectious blood products or stool specimens, and hand-washing procedures are strictly enforced. Additional safety recommendations can be found in the cited literature (5, 61).

DIRECT EXAMINATION OF HAV

Immunofluorescence and immunoperoxidase staining techniques as well as thin-section electron microscopic techniques have been applied to the demonstration of HAV in hepatocytes from chimpanzees and marmosets infected with HAV. However, methods for detecting intrahepatic HAV are not applicable to clinically available specimens, for liver biopsy is rarely, if ever, performed on patients with hepatitis A. On the other hand, immune electron microscopy (IEM), which was the first method used successfully to visualize HAV (21), remains an important reference technique. By allowing visualization of specific immune aggregates, which are more easily detected than are monodispersed virions, this procedure enhances the sensitivity of conventional electron microscopy approximately 1,000-fold (sensitivity threshold of 10^5 to 10^6 particles per ml for IEM, compared to 10^8 to 10^9 particles per ml for electron microscopy).

For detecting HAV in stools, a 2% fecal extract is prepared from an early-acute-phase stool specimen by mixing 0.2 g of stool in 10 ml of veal infusion broth supplemented with 0.5% bovine serum albumin (BSA). The suspension is homogenized by shaking it vigorously for 5 to 10 min in a securely sealed tube containing glass beads and then is clarified by centrifugation at 4°C for 1 h at 1,000 × *g*. Subsequently, the supernatant fluid is passed through a series of 1.2- and 0.45-μm microfiltration membranes premoistened with 0.5% BSA to minimize virus adsorption. To reduce the hazard of infection in the laboratory, preparation of stool filtrates and of electron microscope grids must be done in an appropriate containment facility. A 0.9-ml volume of stool filtrate is incubated with 0.1 ml of a 1:10 dilution of convalescent-phase serum known to contain antibody to HAV (anti-HAV). Important controls include stool filtrates incubated with diluent alone (phosphate-buffered saline [PBS], pH 7.4) and with preillness serum from a patient with hepatitis A. (Reference chimpanzee preinoculation and convalescent-phase serum samples are available from the Research Resources Branch, National Institute of Allergy and Infectious Diseases, Bethesda, MD 20205.)

After incubation for 1 h at room temperature or, alternatively, overnight at 4°C, the suspension is centrifuged at 47,000 × *g* for 90 min at 4°C to pellet antigen-antibody complexes. The resulting pellet is suspended in 50 μl (1 to 2 drops) of distilled water and then mixed with an equal volume of 2% phosphotungstic acid (pH 7.2). One drop of this preparation is applied to a 400-mesh Formvar carbon-coated copper grid. After 1 min, excess fluid is absorbed with filter paper, and the grid is air dried. Grids are examined by electron microscopy at a magnification of ×40,000 to 60,000, and antibody-coated aggregates or single antibody-coated particles are counted (Fig. 1). Quantitation is achieved if a reference antiserum is used and the particles in a fixed number of grid squares are enumerated.

Because antibody-coated viruslike particles other than HAV may appear in fecal extracts, a comparison between stool filtrates incubated with preillness serum and convalescent-phase serum is mandatory and preferably is done under code. This technique for visualizing HAV can be applied with minor modifications to detection of HAV in homogenates of liver (from experimental animals) and to HAV purified from liver or stool. IEM can also be modified to quantitate anti-HAV in serum (see below).

Finally, in choosing fecal specimens for visualization of HAV, an attempt should be made to collect the earliest possible sample, for the bulk of fecal HAV shedding precedes the onset of jaundice (Fig. 3). Because such early specimens are rarely available, detection of fecal HAV is usually not a practical clinical technique. Instead, clinical diagnosis relies on the demonstration of a serological response.

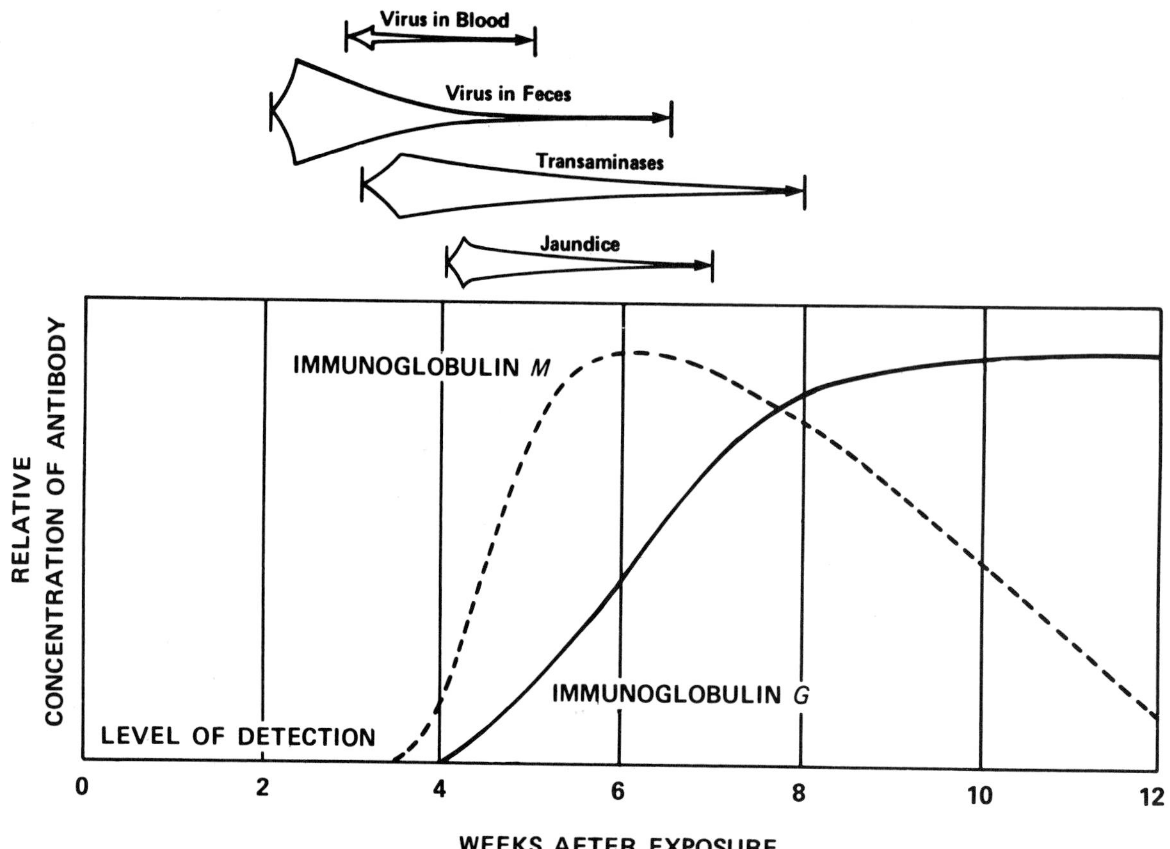

FIG. 3. Immunological and biological events associated with viral hepatitis type A.

BIOLOGICAL INVESTIGATIONS OF HAV

Nonhuman primates, primarily chimpanzees (14, 42) and marmosets (specifically genus *Saguinus*) (10, 48), have been experimentally infected with fecal specimens and blood samples and have been found to develop viremia and to excrete the virus in bile and stool. Both naturally and experimentally infected animals can transmit HAV to animal caretakers. Attempts to isolate and propagate HAV in cell or organ cultures were unsuccessful until 1979, when marmoset-adapted HAV was cultivated serially in primary explant cultures of adult *Saguinus labiatus* marmoset livers and in a normal fetal rhesus monkey kidney cell line (FRhK6) (47). A noncytopathic infection occurs with this marmoset-adapted virus.

Recently, HAV has been cultivated in vitro in a large variety of cell lines (African green monkey kidney, additional strains of fetal rhesus monkey kidney, Vero cells, the Alexander human hepatoma cell line, and human diploid lung cells). Although direct inoculation of feces and serum containing HAV has led to successful virus cultivation in vitro, isolation of HAV from clinical specimens is not sufficiently sensitive for routine diagnostic purposes. Moreover, inoculation is usually followed by a long eclipse period, rendering rapid viral diagnosis impractical. Because HAV is not cytopathic in cell culture, detection of virus in vitro requires demonstration of HAV antigen by immunoassay or immunofluorescence microscopy. A by-product of in vitro cultivation of HAV is a limitless supply of HAV antigen as a diagnostic reagent. In addition, cell culture provides a simple, sensitive method for testing the infectivity of clinical specimens. Still, cultivation of HAV remains limited to a select group of research laboratories. Practical, rapid viral diagnosis is achieved most readily by serological testing.

SEROLOGICAL METHODS TO IDENTIFY HAV AND ITS ANTIBODY

After the visualization of HAV in human stools and marmoset livers, a variety of in vitro assays were developed to detect HAV and its antibody, anti-HAV (Table 4). Of these methods, IEM is valuable as a research and reference tool but impractical for routine diagnostic purposes. Similarly, techniques for localizing HAV in tissue have limited clinical applications. Although complement fixation is a potentially quantitative, rapid assay, its usefulness is limited by its low sensitivity, its nonspecificity, and the high frequency of anticomplementarity in acute-phase hepatitis serum. The immune adherence hemagglutination assay (IAHA) was a popular approach to hepatitis A diagnostics, but, eclipsed by more sensitive, practical assays, this method is rarely used. In contrast, radioimmunoassay (RIA) and enzyme-linked immunosorbent assay (ELISA) have been applied widely and will be described in detail below.

Preparation of reagents

The availability of commercial assays for the diagnosis of HAV infection has diminished the need for

TABLE 4. In vitro techniques for detecting HAV and anti-HAV

Technique	HAV	Anti-HAV	Remarks
Immune electron microscopy	+	+	
Complement fixation	±	+	Less useful for detecting HAV in unpurified clinical specimens
Immune adherence hemagglutination	±	+	
RIA	+	+	
Immunofluorescence/immunoperoxidase	+	−	Detection of HAV in liver cells
ELISA	+	+	

preparing reagents except in research laboratories. A combination of preparative techniques (including differential centrifugation, isopycnic ultracentrifugation, isopycnic banding in cesium chloride, rate-zonal separation in sucrose, preparative electrophoresis, gel filtration, ion-exchange chromatography, affinity chromatography, and organic solvent extraction) can be employed to purify HAV from human or nonhuman primate stool, from homogenized marmoset liver, or from cell culture for use in serological assays. If stool specimens are to be used, they should be collected as early during acute illness as possible, preferably *before* the onset of jaundice, after which virus excretion declines rapidly (Fig. 3). Thus, most patients presenting with clinically obvious hepatitis A are poor sources of virus-rich stool. Success is much more likely if stool is collected from close contacts of the index case, even in the absence of clinical signs of hepatitis. Although generally impractical, this approach has been rewarding in the setting of large common-source outbreaks (15, 23). Now that HAV has been grown in cell culture, an inexhaustible supply of virus antigen for diagnostic purposes has become available.

Details of these purification techniques can be found in the 3rd edition of this Manual.

Little difficulty is encountered in obtaining a source of anti-HAV-positive serum, for the antibody response after natural infection is brisk, reaching peak titers—on the order of 1:100,000—approximately 2 to 3 months after acute illness. Furthermore, relatively high titers of serum anti-HAV are maintained years after infection (56). If necessary, anti-HAV can be raised in animals by immunization with purified HAV (16). Depending on the purity of the HAV preparation used for immunization, antisera raised in animals may have antibodies to normal liver proteins, to stool antigens, or to cell culture components. Therefore, under certain circumstances, appropriate control antigens and antisera must be used, as discussed below.

IEM

IEM can be used to detect either HAV (see Direct Examination of HAV, above) or anti-HAV. The presence of anti-HAV in serum is determined by incubating the serum with a source of HAV containing at least 50 HAV particles per electron microscope grid square. Traditionally, a 2% stool filtrate containing HAV has been used for this purpose, although other sources may be substituted. Paired serum samples (acute phase, or preillness and convalescent phase) are tested under code to allow comparison between them and to avoid observer bias. Serum diluted 1:10 in PBS is incubated with the HAV-positive sample, and IEM is performed as described above. Particles visualized are rated on a scale of 0 to 4+, with 0 signifying no antibody and 4+ representing such heavy antibody coating that the particle surfaces are almost obscured. A rise in antibody rating of 1+, which corresponds roughly to a 10-fold increase in titer by IAHA, is considered a serological response indicative of acute infection. Preillness samples should be anti-HAV negative; acute-phase samples will be negative or barely positive (1 to 2+), and late-convalescent-phase samples, taken several weeks to months after illness, will have high antibody ratings (3 to 4+).

Difficulties in interpretation may arise when serum contains antibodies to other viruslike particles that may be prevalent in stools. Generally, these particles are smaller than the 27-nm HAV particles, and an increase in antibody coating is not observed when paired sera are examined. The presence of these ubiquitous particles emphasizes the importance of testing paired serum samples simultaneously for anti-HAV and of using HAV purified from cell culture. Correspondingly, tests for HAV should include anti-HAV-negative (preinoculation or preillness) control sera.

IAHA

IAHA is a complement-dependent assay 10 to 100 times more sensitive than complement fixation and easily adapted for quantitative determinations on large numbers of samples (43, 44). Although it can be used to detect HAV during virus purification, it is poorly suited to detection of HAV in unpurified liver or stool homogenates. Its widest application, therefore, has been as a serological test to detect anti-HAV. At present, IAHA is used more frequently in areas of the world other than the United States.

IAHA takes advantage of the fact that primate erythrocytes bear surface receptors for the third component of complement (C3). When a finite concentration of partially purified HAV is added to a heat-inactivated serum sample containing anti-HAV, immune complexes form. These immune complexes are incubated with a fresh complement source, followed by exposure to dithiothreitol, which stabilizes the immune complex containing C1, C4, C2, and C3 by protecting it from attack by C3b inactivator. After the addition of human group O erythrocytes, a stable complex forms between C3 and C3 receptors on the erythrocytes, resulting in hemagglutination. Hemagglutination is evaluated on a scale of 0 to 4+. All wells with hemagglutination patterns of 3+ and 4+ are considered positive if hemagglutination is absent in buffer control wells. Details of the assay can be found in the 3rd edition of this Manual.

Because a prozone effect may be encountered, serum samples to be tested are screened at dilutions of 1:10, 1:100, and 1:1,000, and endpoint titers are determined by serial twofold dilutions within or beyond these limits. It is also important to realize that not all erythrocytes are suitable. A large number of donors

TABLE 5. Principles of the procedures for various hepatitis assays

Serological assay	Vol required (μl)	Principle of procedure (RIA or ELISA)	Type of support system	Adsorbed reagent	Label or conjugate	Positive result (cpm or absorbance value)	
						Low	High
HBsAg	200	Sandwich	Bead, tube, well, pellet, disk, glass	Anti-HBs	Anti-HBs		X
Anti-HBc	100	Competitive binding	Bead	HBcAg	Anti-HBc	X	
IgM Anti-HAV	10	Sandwich[a]	Bead	Anti-IgM	Anti-HAV		X
Anti-HAV	10	Competitive binding	Bead	HAV Ag	Anti-HAV	X	
Anti-HBs	200	Sandwich	Bead, tube	HBsAg	HBsAg		X
HBeAg	200	Sandwich	Bead	Anti-HBe	Anti-HBe		X
Anti-HBe	50/100[b]	Competitive binding[c]	Bead	Anti-HBe	Anti-HBe	X	
IgM anti-HBc	10	Sandwich[a]	Bead	Anti-IgM	Anti-HBc		X

[a] Modified sandwich: IgM-specific antibody (anti-HAV or anti-HBc) that is immunologically bound to the bead is detected by adding the corresponding antigen (HAV antigen or HBcAg), followed by the appropriate antibody label or conjugate.

[b] RIA requires 100 μl; ELISA requires 50 μl.

[c] Anti-HBe in the test sample and adsorbed anti-HBe compete for a standardized amount of HBeAg added to the reaction well before the addition of anti-HBe label or conjugate.

must be screened; only about 20% of group O donors show marked hemagglutination activity. Once appropriate donors are found, their erythrocytes, stored in Alsever solution, can be used for 2 to 4 weeks. In addition to the difficulties of maintaining a steady source of erythrocytes, IAHA has other limitations. Stool- and liver-derived HAV for IAHA must be purified by one or more biophysical techniques, and only high-titer preparations work satisfactorily. Furthermore, nonionic detergents and CsCl, both used in purification of HAV, interfere with IAHA and must be removed by dialysis. Unlike other techniques for identifying anti-HAV, IAHA reactions cannot be tested for immunological specificity by blocking with purified antigen; instead, specificity must be inferred from a panel of positive and negative control sera. In addition, development of anti-HAV, as measured by IAHA, is delayed from 1 to 4 weeks compared to the earlier detection of antibody possible with other techniques. Finally, nonspecific hemagglutination is encountered when IAHA is used to measure anti-HAV in lots of immune serum globulin; treatment with kaolin has been used to minimize this problem (43).

RIA (Table 5)

Commercial RIA kits are available for diagnosing acute HAV infection (HĀVAB-M, Abbott Laboratories, North Chicago, Ill.), for determining the immune status of an individual to HAV after exposure, or for assessing risk in a person traveling to an endemic area or working in a high-risk environment (HĀVAB, Abbott Laboratories). Advantages of RIA over other assay systems include excellent precision, ease of performance, and unexcelled specificity and sensitivity. Limiting factors include the need for a gamma counter and the relatively short shelf life of the radiolabeled reagent.

Because anti-HAV is present in the serum of about 40% of normal urban adults, its detection cannot, by itself, establish a diagnosis of acute disease without further analysis. In some patients, seroconversion or a rise in anti-HAV titer may eventually result in a correct diagnosis. However, the antibody titer is already high in most patients at the time medical attention is solicited, thereby limiting the value of this approach. On the other hand, a diagnosis of acute HAV infection can readily be achieved if distinction can be made between anti-HAV of the IgM class, which appears during acute illness, and IgG antibody, which becomes pronounced during convalescence and remains elevated for years.

The HĀVAB-M assay is a specific and sensitive solid-phase RIA which measures IgM-specific anti-HAV in sera (9, 54, 59). Conceptually, anti-HAV of the IgM class is removed from the serum by anti-human IgM (mu-chain specific) bound to polystyrene beads. The subsequent addition of HAV results in the attachment of the antigen to IgM anti-HAV that has combined with the solid-phase anti-human IgM. The "sandwich" is completed by adding radiolabeled anti-HAV. Specifically, serum is diluted 1:200 in PBS, and 10 μl is added to a well containing 0.2 ml of PBS (final serum concentration is 1:4,000). A goat anti-human IgM-coated bead is then added to each well, and the trays are covered with an adhesive-backed cardboard and incubated for 2 h at 24 ± 2°C. Two positive and three negative control samples are included in every test run. The beads are washed three times with distilled water, followed by the addition of 0.2 ml of HAV reagent. The covered trays are incubated for 18 to 24 h at 24 ± 2°C, after which the beads are washed again. Next, 0.2 ml of ^{125}I-labeled human anti-HAV is added to each well, the trays are covered, and incubation is continued for 4 h at 45°C. Finally, the beads are washed and counted. A cutoff value is calculated based on the positive and negative control samples and on the potency of the HAV and ^{125}I-labeled anti-HAV reagents used. This value is determined by dividing the mean positive control counts per minute (cpm) by 10 and adding this to the mean negative control cpm. Samples with counts above the cutoff are considered positive.

Other approaches have been developed to detect IgM-specific anti-HAV. These employ either concanavalin A to preferentially bind IgM (unpublished data) or staphylococcal protein A to adsorb IgM from serum samples (6). Details of these procedures can be found in the 3rd edition of this Manual.

The HĀVAB test measures total anti-HAV antibody (IgG and IgM). It is based on the principle of competition between anti-HAV in test serum or plasma and ^{125}I-labeled anti-HAV for binding to HAV coated on a polystyrene bead. Ten microliters of each specimen is added to a microtiter well containing 0.2 ml of ^{125}I-labeled human anti-HAV. (The HĀVAB protocol suggests adding the label to the 10-μl specimen, but unless the label is added promptly, the specimen may dry out or adsorb to the well.) Three negative and two positive controls are incorporated into every test run. A Formalin- and heat-inactivated HAV (primate)-coated bead is added to each well, and an adhesive cover-sealer is applied to the plate. Plates are incubated for 4 h in a 45°C water bath or for 18 to 24 h at room temperature. During this time, anti-HAV, if present in the test serum, competes with the labeled antibody for HAV binding sites on the bead. After the incubation period, the fluid contents of the wells are aspirated, and the beads are washed three times with distilled or deionized water and then transferred to counting tubes. Radioactivity bound to the beads is measured in a gamma scintillation counter, and the net (minus gamma counter background) count rate of the test sample is compared to a calculated cutoff value (one-half of the combined negative control mean and the positive control mean). Specimens with counts greater than the cutoff value are negative (no anti-HAV was available to interfere with binding of ^{125}I-labeled anti-HAV to HAV), and specimens with a count rate lower than the cutoff value are considered positive for anti-HAV. If necessary, an anti-HAV titer can be derived by testing serial serum dilutions from 1:50 to 1:3,200 (1:1,050 to 1:67,200 final concentration when the original dilution of 1:21 is taken into consideration). Dilutions are made in PBS with 0.5% BSA. The diluent is substituted for negative control serum in this situation. The anti-HAV titer is considered to be that dilution with mean cpm most nearly equivalent to, but not greater than, the cutoff value. Validity of these tests is verified by checking the N/P ratio of the negative (N) to positive (P) kit control cpm, which should be ≥5.

For clinical investigators interested in preparing their own microtiter solid-phase and competitive-inhibition RIAs for detecting HAV antigen in stool filtrates or cell culture homogenates or during the various stages of purification, a simple microtiter solid-phase RIA is the most direct method available. Both methods have been used for anti-HAV.

The microtiter solid-phase RIA utilizes the sandwich principle, an arrangement created when antibody adsorbed to an inert substance binds an antigen that is subsequently detected by the addition of specific antibody tagged with ^{125}I (59). To test for HAV antigen, 100 to 150 μl of a predetermined optimal concentration of high-titer anti-HAV is added to each well of a polyvinyl microtiter "U" plate. In most instances, the optimal antibody dilution lies between 1:100 and 1:10,000. The antibody is diluted in 0.05 M glycine buffer (pH 7.2) or in PBS (pH 7.4). An immune globulin preparation may also be used for coating the wells, but purified IgG is usually unnecessary. The plates are incubated overnight at 25°C (or for 2 h at 37°C). Desiccation is minimized by covering the plates with Parafilm "M" (Dixie/Marathon, Greenwich, Conn.) held in place by a close-fitting plastic lid (no. 3041; Falcon Plastics, Oxnard, Calif.). The plates are washed four times with 0.15 M NaCl containing 0.02% sodium azide and then once with PBS containing 2% fetal bovine serum. The PBS-fetal bovine serum is allowed to remain in contact with the wells at 25°C for at least 30 min to saturate any remaining binding sites. One percent BSA or gelatin works equally well. After the wash procedure, residual fluid is removed to avoid subsequent dilution of samples or reagents if the plates are to be used immediately. Conversely, the plates can be stored at 4°C for at least 6 months without significant loss of reactivity. Duplicate 25- to 75-μl samples of the material suspected of containing HAV are added to the coated wells, and incubation is continued for 2 h at 45°C by floating the Parafilm-covered plates in a water bath. Alternatively, the incubation period may be extended for 1 to 2 days at 4°C.

Appropriate controls should also be evaluated. These include HAV-positive reference material, PBS, and HAV-negative stool or gradient material, e.g., CsCl. After incubation, the wells are aspirated and washed four times with saline or PBS containing 2% fetal bovine serum or 0.05% Tween 20, and the residual fluid is removed before the addition of ^{125}I-labeled IgG containing high-titer anti-HAV. Specific activity of the antibody preparation should range from 15 to 25 μCi/μg, and a total of 200,000 to 300,000 cpm per well is employed. In general, the volume of labeled IgG added to each well should not exceed the sample volume used. To reduce background counts significantly without altering the specific binding, the label is diluted in PBS containing 50% fetal bovine serum and 10% normal human serum. Each lot of fetal bovine serum and normal human serum should be evaluated separately before use. After an additional incubation period of 1 h at 45°C (or 4 h at 25°C) and a corresponding wash step, clear adhesive tape (no. 1-220-30; Cooke Engineering Co.) is applied to the plate, and the wells are numbered. Each well is cut out, and residual radioactivity is measured in a gamma scintillation counter. Mean cpm of duplicate samples are compared with mean cpm for at least five negative control samples. The test for HAV is considered positive if the ratio of cpm in the sample to the mean cpm in negative control wells is ≥2.1 (sample/negative, S/N ratio). Specificity is demonstrated by blocking the binding of ^{125}I-labeled anti-HAV (IgG) with convalescent-phase, but not with preinoculation, serum from persons or experimental animals previously infected with viral hepatitis type A; by comparing results with appropriate antigen-negative controls, such as HAV-negative stool filtrates or gradient fractions; or by both methods. Similarly, hepatitis B antigens and antibodies should yield negative results.

The microtiter solid-phase RIA can also be modified in a number of ways to detect anti-HAV. Perhaps the most sensitive method is to perform an indirect blocking or inhibition assay. Briefly, 0.1 ml of a 1:10 serum dilution is preincubated for 30 min at 25°C with an equal volume of a partially purified standard preparation of HAV. Specimens containing anti-HAV will combine with HAV antigen, thus reducing the amount of HAV available for binding to an antibody-coated microtiter well. Duplicate samples (75 μl) of the reaction mixture are placed in microtiter wells coated with anti-HAV, and incubation is carried out as described above. After the addition of labeled anti-HAV, the wells are counted. A reduction in residual radioactivity of ≥40% as compared with an anti-HAV-negative control sample is indicative of anti-HAV in the test serum.

A second approach substitutes serum dilutions of the sample to be tested for the precoat first antibody. If the serum contains anti-HAV, specific binding will result when HAV antigen is added to the microtiter well. Bound antigen will subsequently be detected after the addition of labeled anti-HAV to the wells. The test is considered positive if the P/N ratio is ≥2.1. A shortcoming of this procedure, however, is a reduction in sensitivity that occurs at serum dilutions of <1:200 (i.e., at higher protein concentrations).

ELISA (Table 5)

ELISA was first utilized in hepatitis research to detect HBsAg and anti-HBs, using the sandwich principle employed in the solid-phase RIA procedure. In this technique, an enzyme-labeled antibody is substituted for a radionuclide-labeled preparation, its presence being reflected by hydrolysis of a subsequently added enzyme substrate. The color produced can be quantitated colorimetrically. The development of ELISA technology has spawned an interesting rivalry between the proponents of RIA and ELISA. Major objections to RIA include the expense of a gamma counter and the presence of a radiation hazard. However, in many situations the large, expensive, general-purpose gamma counters with automatic sample changers are not necessary, and hand-operated gamma counters or ^{125}I-spectrometers are similar in price to colorimeters. In addition, the radiation hazard is greatly overstated. Ten microcuries of ^{125}I (commercial kits usually contain a total of 7 to 15 μCi of ^{125}I) held at a distance of 10 cm for 1 h yields an exposure rate of only 0.06 mrem (milliroentgen equivalent, man). The exposure rate is inversely proportional to the square root of the distance in centimeters. For comparison, background radiation exposure rates are about 150 mrem per year, a chest X-ray provides 10 mrem of exposure, and a gastrointestinal series provides about 8,000 mrem. Nevertheless, ELISA has a number of advantages which could recommend it over RIA, particularly in those countries where restrictive laws govern the handling of radioisotopes and the disposal of radioactive wastes. A major factor is the stability of the enzyme-antibody conjugates. These have a shelf life of months to years compared to radionuclide-labeled antibody conjugates which must be freshly prepared every 3 to 5 weeks. On the other hand, an additional manipulation is required during ELISA in that a specific substrate must be added to detect fixation of the conjugated probe. In addition, some of the enzyme substrates used are potentially carcinogenic; therefore, they should be handled with caution. Both tests are objective, quantitative, reliable, and reproducible.

Simplified, standardized commercial ELISA test kits are now available to measure anti-HAV and IgM-specific anti-HAV (HĀVAB EIA and HĀVAB-M EIA, Abbott Laboratories). These commercial tests are analogous to the commercial RIA procedures produced by the same manufacturer (see preceding section). In the ELISA for anti-HAV, 10 μl of each serum or plasma specimen is added with 0.2 ml of horseradish peroxidase-conjugated human anti-HAV to a microtiter well. Three negative and two positive control wells are incorporated into every test run. A HAV (human)-coated polystyrene bead is added to each well, and the microtiter plates are sealed and incubated for 3 h in a 40°C water bath or for 18 to 24 h at 24 ± 2°C. After the incubation period, the fluid contents of the wells are aspirated, and the beads are washed three times with distilled or deionized water and transferred to counting tubes. Then 0.3 ml of freshly prepared *o*-phenylenediamine substrate solution containing H_2O_2 is added to each bead-containing tube as well as to two empty tubes to serve as substrate blanks. After another 30 min of incubation, shielded from light and at 24 ± 2°C, the reaction is stopped by adding 1.0 ml of 1 N sulfuric acid to each tube. Absorbance is determined (within 2 h after the addition of sulfuric acid) in a spectrophotometer at 492 nm, and the presence of anti-HAV is determined by comparing the absorbance value of the specimen to a cutoff value (one-half the sum of the negative control mean and the positive control mean absorbance). Specimens with absorbance levels higher than the cutoff value are negative, and those with absorbance values equal to or below the cutoff are considered reactive for anti-HAV. Because of low confidence in values near the cutoff, samples with absorbances within 10% of the cutoff should be retested to confirm the initial result. Validity of the test procedure is determined by calculating the difference between the negative (N) and positive (P) control mean absorbances (N minus P), which should be ≥0.400. If desired, the titer of anti-HAV can be derived by testing serial dilutions of serum from 1:50 to 1:3,200 in PBS containing 0.5% BSA. The diluent should be used as the negative control instead of the negative control serum. Remember that the final concentration of the sample is 21 times higher, based on the initial dilution of sample in the test well. The titer of anti-HAV is considered to be the sample dilution whose absorbance is most nearly equivalent to, but not greater than, the cutoff value.

In the ELISA for IgM-specific anti-HAV, 10 μl of a serum or plasma sample is added to 2.0 ml of normal saline; 10 μl of the diluted sample or undiluted positive and negative control samples, as described above, is then added to a microtiter well containing 0.2 ml of specimen diluent (final serum concentration of 1:4,221). Polystrene beads coated with goat antibody to human IgM (μ-chain specific) are added to each well, and the plates are incubated for 1 h at 40°C, after which the fluid contents of the well are aspirated and the beads are washed three times in distilled or deionized water. Next, 0.2 ml of a solution containing HAV (human) is added to each well, and the plates are incubated for 20 ± 2 h at 24 ± 2°C, after which the wells are aspirated and the beads are washed three more times. Then, 0.2 ml of horseradish peroxidase-conjugated human anti-HAV is added to each well. The plates are incubated at 40°C for 2 h. After the fluid is aspirated from the wells, the beads are washed and transferred to counting tubes. Freshly prepared *o*-phenylenediamine solution (0.3 ml) containing H_2O_2 is added to each tube as well as to two substrate blank tubes, and the tubes are shielded from light and incubated for 30 min at 24 ± 2°C. The reaction is terminated by adding 1.0 ml of 1 N sulfuric acid to each tube, and absorbance is read as described above. Samples with absorbance levels equal to or above the cutoff value (the sum of the negative control mean absorbance plus one-tenth of the positive control mean absorbance) are considered reactive for IgM-specific anti-HAV, while samples with absorbance values below the cutoff level are considered negative. As noted above, samples with absorbance levels within 10% of the cutoff should be retested to confirm the original result. Validity of the test requires that the difference in the mean absorbance between the positive (P) and negative (N) control samples (P minus N) is ≥0.400.

Commercial ELISA tests for HAV antigen are not available. To conduct such an assay (41), 75 μl of a predetermined optimal dilution of anti-HAV (diluted in PBS) is added to wells of polyvinyl microtiter U plates, which are incubated at 4°C for 4 h and then washed three times with PBS containing 0.05% Tween 20. The antibody-coated wells are incubated overnight at 4°C with PBS containing 1% BSA and then are aspirated and washed once before the 25-μl samples being evaluated for HAV antigen are added. The plates are covered and incubated for 15 to 18 h at 4°C; they are then washed once, and 25 μl of PBS-Tween is added and left for 15 min at room temperature. Concurrently, several wells should be incubated with preillness and convalescent-phase sera instead of PBS-Tween; these will serve to demonstrate specificity of the assay. To each well is added 25 μl of anti-HAV conjugated to horseradish peroxidase (type VI, RZ = approximately 3.0; Sigma Chemical Co., St. Louis, Mo.) by the method of Nakane and Kawaoi (45). The enzyme conjugate is diluted in PBS containing 50% BSA. The plates are incubated at room temperature for 2 h and washed three times with PBS-Tween; 100 μl of freshly prepared *o*-phenylenediamine-H_2O_2 substrate is then added. After an additional 30-min incubation period in the dark, 50 μl of 2 M H_2SO_4 is added to each well to stop the reaction, and the optical density is measured at 492 nm in a spectrophotometer. Alternatively, color can be quantitated visually on a scale of 0 to 3+ (40). Before a sample is considered positive for HAV, specificity of the reaction should be checked by demonstrating that binding of enzyme-labeled anti-HAV is blocked with convalescent-phase but not with preillness serum from a patient with previously documented viral hepatitis type A.

INTERPRETATION OF TEST RESULTS IN HEPATITIS A

Because viremia is limited (28, 30) and because cell culture techniques are cumbersome and insensitive, demonstration of circulating HAV is difficult, impractical, and probably unwarranted for diagnostic purposes. Furthermore, because fecal shedding of virus occurs so early and because liver and bile are not routinely available as diagnostic specimens, a diagnosis of HAV infection can be made most practically by demonstrating a specific antibody response. Classically, this is done by measuring an increase in serum anti-HAV between the acute phase and convalescence; however, the interval required to demonstrate a perceptible change in antibody titer may be as long as 4 to 6 weeks. In many situations appropriately spaced serum specimens are not available, and in any event, the ability to make a diagnosis during acute illness, without having to wait for a convalescent-phase sample, is preferable. This can be achieved by demonstrating the presence of anti-HAV of the IgM class, and this approach has become the method of choice for making a diagnosis of acute viral hepatitis type A (see Fig. 3).

When testing for the presence of HAV in stool or when using HAV prepared from stool to detect anti-HAV in serum, it is important to realize that a false-positive test may result from binding of antibodies in serum to non-HAV antigens in stool. Therefore, to ensure specificity of positive results, testing with preinoculation and convalescent-phase reference reagents or appropriate blocking of the specific reaction between HAV antigen and anti-HAV conjugate with convalescent-phase, but not preinoculation, serum is necessary. Many of the difficulties encountered with false-positive results can be eliminated or minimized by using highly purified antigen, by using high-titered hyperimmune animal sera, or by using antigen and antibody reagents derived from heterologous sources. For example, nonspecific binding of antibodies in human serum to human stool components can be eliminated by using antigen derived from marmoset liver or cell culture.

Selection of the best test method in any laboratory depends on the facilities and expertise available. For detection of HAV in stool, IEM, solid-phase RIA, and ELISA are of comparable sensitivity, although the ELISA may be less specific, especially in competitive binding assays. Although RIA and ELISA are the most sensitive tests for anti-HAV, levels of anti-HAV achieved during illness and persisting thereafter are so substantial that any of the available methods is quite satisfactory. Prevalence of anti-HAV in populations determined with all assays is virtually the same.

DIRECT EXAMINATION OF HBV

Immunofluorescence, immunoperoxidase staining, and electron microscopy have been used extensively to examine pathological specimens and serum samples for the presence of HBV-associated antigens or particles (36, 49). These procedures are not applicable to rapid, large-scale screening of HBV infections by clinical laboratories, but they have been invaluable in elucidating the biosynthetic origin of the various antigens within infected cells and for detecting the presence of immune complexes. Within the liver, HBcAg-containing particles are found predominantly in the nuclei of hepatocytes, whereas HBsAg reactivity is observed exclusively in the cytoplasm. Detection of complete virions is uncommon. The techniques employed for immunoperoxidase or immunofluorescence are well known. As a general rule, immunofluorescence staining of cryostat sections is superior to the use of paraffin sections. However, cryostat sections are not always available, and cellular outlines are better resolved in uniformly cut paraffin sections than in frozen sections. Reduction of nonspecific background staining of Formalin-fixed sections can be achieved by treating tissue for 2 to 6 h at 37°C with 0.1% trypsin before staining. Muscle tissue is more resistant to digestion than liver and kidney tissue. In situations in which autologous antibody might bind to cell-associated antigen, cryostat sections can be treated for 5 min with 0.1 M glycine HCl buffer (pH 1.2) before specific antibody (labeled or unlabeled) is added.

Recently, biotin-avidin systems have been introduced for detecting viral antigens on cells or in tissues. This system offers a number of advantages over other immunodiagnostic clinical techniques: (i) the binding of avidin to biotin is extremely rapid and essentially irreversible because of the high affinity constant ($>10^{15}$ M^{-1}) that exists between these two proteins (affinity constants of antibody for most antigens are generally 10^6-fold lower); (ii) sensitivity is enhanced, even when highly diluted primary antibody is used, because the four binding sites on avidin permit amplification of the response; (iii) Formalin-fixed, paraffin-embedded histological sections, smears, and frozen sections can be examined by conventional light microscopy; and (iv) background stain is greatly reduced. Many types of immunoperoxidase staining procedures are available. The "ABC" (avidin-biotin–labeled horseradish peroxidase complex) procedure developed by Su-Ming Hsu and his associates (34, 35) has been found to be highly sensitive and specific. Briefly, Formalin-fixed tissue sections are deparaffinized and hydrated through xylene and a graded alcohol series. The fixation process should employ a buffered Formalin, not exceeding 4% formaldehyde, sufficient to maintain integrity of the tissue without destroying the antigenic determinants being evaluated. Sections are incubated for 30 min with primary antibody raised against the antigen of interest. The antiserum is diluted in buffer (10 mM PBS, pH 7.6). Slides are washed for 10 min in buffer, after which diluted, biotin-labeled secondary antibody is added and incubation is continued for 30 min. This antibody, directed against the immunoglobulin class of the species used as primary antiserum, introduces biotin-labeled residues into the section at the location of the primary antibody. After washing, an avidin-biotin–labeled horseradish peroxidase complex (Vectastain ABC reagent; Vector Laboratories, Burlingame, Calif.) is added which binds to the biotin-labeled secondary antibody during a 30- to 60-min incubation period. The slides are rewashed for 10 min in buffer, and tissue antigen is localized by incubating the sections for 5 min in freshly prepared peroxidase substrate solution, i.e., equal volumes of 0.1% diaminobenzidene tetrachloride prepared in 0.1 M Tris buffer (pH 7.2) and 0.02% hydrogen peroxide prepared in distilled water. The hydrogen peroxide should be prepared fresh from a concentrated stock solution. Sections are washed for 5 min in tap water, counterstained with hematoxylin, eosin, or both, and mounted. Sections should not be allowed to dry out during the staining procedure; therefore, a humidified chamber is recommended for incubations. If antigen concentration is low, the peroxidase substrate incubation step may be lengthened to achieve maximal staining. Many mammalian tissues contain endogenous peroxidase. If this problem exists, sections may be incubated for 30 min in 0.3% hydrogen peroxide in methanol either before primary antibody is added or before the ABC reagent step. The latter point is desirable in cases where the antigenic determinant may be destroyed by hydrogen peroxide. If nonspecific staining occurs, it can be reduced by incubating the sections for 20 min with diluted normal serum prepared from the species in which the secondary antibody is made. Excess serum is blotted from the sections before the primary antibody is added.

CELL CULTURES AND ANIMAL MODELS OF HEPATITIS B

Despite strong evidence for the growth of HBV in some cell and organ cultures, serial propagation over a prolonged period of time has not been accomplished. Chimpanzees and other high-order primates are highly susceptible to experimental induction of hepatitis B, but are not routinely available (2). The pattern of infection is similar to that observed in humans except that the disease is milder; e.g., jaundice is rare. Nevertheless, the chimpanzee system continues to play an essential role in inactivation studies (vaccine safety, disinfection kinetics), infectivity determinations, and immunopathology.

SEROLOGICAL IDENTIFICATION OF HEPATITIS B ANTIGENS AND ANTIBODIES

A number of serological techniques with varying degrees of sensitivity have been developed for the detection of HBV antigens and their specific antibodies (Table 6). These methods include agarose gel diffusion, counterimmunoelectrophoresis, rheophoresis, complement fixation, latex agglutination, hemagglutination, IEM, ELISA, and RIA. Each method offers certain advantages and disadvantages and differs markedly in sensitivity, specificity, simplicity, and expense from the others. A description of the ELISA, RIA, and DNA polymerase methods follows. Readers interested in detailed information concerning the purification of hepatitis B-associated antigens, production of antibodies to hepatitis B, agarose gel diffusion, discontinuous counterimmunoelectrophoresis, rheophoresis, reverse passive hemagglutination, and hemagglutination are referred to the 3rd edition of this Manual. A complete list of licensed manufacturers for hepatitis B products can be obtained by writing to the Director, Bureau of Biologics, Food and Drug Administration, Bethesda, MD 20205.

Purification of hepatitis B-associated antigens (18, 32)

The widespread availability of sensitive and specific commercial kits for the detection of hepatitis B anti-

TABLE 6. Relative sensitivity of methods used to detect HBV antigens and antibodies

Relative sensitivity (HBsAg)	Assay method	Time to complete test (h)
Least (≥10 μg/ml)	Agarose gel diffusion	24–72
Intermediate (1–5 μg/ml)	Counterimmunoelectrophoresis	1–2
	Rheophoresis	24–72
	Complement fixation	18
Most (0.1–40 ng/ml)	Latex agglutination (third generation)	1
	Hemagglutination	3–6
	Immune adherence	4–24
	ELISA	4–24
	RIA	3–24

gens and antibodies has markedly diminished the laboratory's need to prepare highly purified antigens or antibody of high specificity, affinity, and avidity. However, situations may exist, e.g., subtyping, in which reagent preparation may be desirable. For those who are interested, methods for purifying HBsAg or HBcAg are described in the 3rd edition of this Manual. The HBsAg procedure can be completed in 1 week and incorporates an acid pH step to help remove antibody specific for HBsAg and other proteins nonspecifically associated with the particles. The low pH also results in partial disruption of the 42-nm HBV particles and the tubular forms. This may have additional benefits since 27-nm particles with their higher densities (1.32 to 1.36 g/cm^3 in CsCl versus 1.24 to 1.28 g/cm^3 for HBV particles) are more easily separated from the lighter 22-nm HBsAg particles, thereby producing a final product that is less likely to be contaminated with HBcAg or to be infectious. Finally, immunogenicity and radioisotopic labeling of these purified HBsAg preparations seem to be superior to those found with other methods of purification that lack a low-pH step (31).

Production of antibodies to hepatitis B (18)

Antisera from multiply transfused (or immunized) individuals have provided investigators with a rich source of anti-HBs of moderate titer. The major advantage of human (or chimpanzee) serum is its freedom from anti-normal human serum contaminants. Satisfactory anti-HBs for screening sera for HBsAg also has been prepared in a variety of animals, including horses, goats, guinea pigs, rabbits, mice, rhesus monkeys, baboons, and chimpanzees. However, virtually all of these preparations have detectable levels of antibody to human serum proteins, particularly albumin. These contaminating antibodies can be effectively removed by adsorption with glutaraldehyde cross-linked preparations of normal human serum.

Techniques for immunizing goats, guinea pigs, and rabbits can be found in the 3rd edition of this Manual. Antibody with excellent antigen-precipitating capacity can be prepared in goats. In general, the predominant antibody is directed against the group-specific *a* antigenic determinant unless specimens are obtained early in the course of immunization. Therefore, goat antiserum is better for screening than for subtyping. Antiserum prepared in guinea pigs is an excellent reagent for subtyping and for screening tests. Specificity of antisera produced against subtype *ay* is better than that produced against subtype *ad*, which suggests that the *a* subdeterminant in the *ad* preparation may be more immunogenic, resulting in a more broadly reacting antiserum. Rabbits are also useful for preparing subtyping reagents (especially against the *w* and *r* determinants) and for preparing antiserum to HBcAg. For subtyping, monospecific IgG containing anti-*d* or anti-*y* can be obtained by affinity chromatographic techniques in which the HBV-specific antiserum is passed through a column containing the heterologous HBsAg subtype. Anti-*a* (and presumably anti-*w*) can subsequently be recovered from the column by elution with acid or potassium thiocyanate. More recently, commercial firms have produced a number of monoclonal antibodies of precise, quantifiable specificity and antiserum homogeneity directed against the group *a* determinant and subtypes *ad* and *ay*.

Agarose gel diffusion

Agarose gel diffusion is the least sensitive method available for detecting HBsAg, but it permits direct comparisons to be made between positive specimens regarding their identity, partial relatedness, or non-identity. Reinforcement of weak precipitin lines is accomplished by placing reference HBsAg-positive specimens in peripheral wells between the unknown specimens. Because HBsAg is a large molecule and diffuses slowly compared with IgG antibody, preincubation of the slides for 2 h at room temperature in a humidified chamber before the addition of anti-HBs in the center well permits precipitin lines to develop more distal to the peripheral well. The slides are observed for the development of precipitin lines at 24, 48, and 72 h.

For subtyping antigens, the same reinforcement pattern can be used. Monospecific anti-HBs containing anti-*d* or anti-*y* antibody only can be substituted for the reference antiserum to enhance specificity. Slides are observed for lines of identity or for the formation of a "spur" signifying different antigenic determinants.

Discontinuous counterimmunoelectrophoresis

Discontinuous counterimmunoelectrophoresis or counterelectrophoresis was the most widely utilized method for detecting HBsAg before the introduction of more sensitive and specific assays such as RIA. It is still used for subtyping. The method is based on the principle that, in a relatively alkaline environment, HBsAg, which has an isoelectric pH between 4.4 and 5.2, is negatively charged and migrates in an electrophoretic field toward the anode. Conversely, IgG, being closer to its isoelectric pH, travels by electroendosmosis toward the cathode. This condition is caused by charged groups within the agarose that promote the movement of buffer through the gel toward the cathode, thereby drawing the gamma globulin with it. In this regard, agarose powders with high relative mobility (m_r) are available, making these gels especially suitable for counterimmunoelectrophoresis. A precipitin line forms when optimal concentrations of the antigen and antibody meet. By reducing the ionic strength of the agarose buffer, as compared to the buffer used in the electrophoresis chambers, a discontinuous buffer system can be prepared which enhances the movement of acidic proteins toward the anode and globulins toward the cathode. This results in increased sensitivity and speed of reaction and in sharper, more easily read precipitin lines (58). Plates must be examined carefully for weak precipitin lines; a magnifying lens, a darkened room, and a good oblique viewing light are helpful. Artifacts may be encountered in this system which must not be confused with a true positive reaction. These include a halo of precipitation around the well and movement of lipid over the surface of the agarose adjacent to the sample well. The latter artifact can easily be distinguished from a true precipitin reaction within the gel by wiping the area gently with a cotton swab.

Both specificity and sensitivity depend on the use of potent precipitating antibody rendered free from anti-normal human serum by prior adsorption. False-positive results are uncommon. An imbalance between reactants, either excess HBsAg or anti-HBs, can lead to the establishment of a prozone and a false-negative result. The prozone phenomenon, which occurs in the region of HBsAg excess, can be minimized by diluting the reagent antibody in normal homologous whole serum or its globulin fraction rather than in a buffer (19). Two-dimensional box titrations are essential to determine the optimal concentration of reagent antibody needed in subsequent tests. Specifics of the test can be found in the previous edition of this Manual.

Rheophoresis

The bidimensional immunorheophoresis (rheophoresis) method relies on continuous evaporation of water through a central hole placed directly over an antigen or antibody well (37). Protein solutions placed in peripheral wells are transported by hydrodynamic forces to the central area of dehydration through the flow of low-ionic-strength buffer (0.01 M Tris, pH 7.6) placed external to the agarose. Sensitivity is equivalent to that of counterimmunoelectrophoresis.

Reverse passive latex agglutination

The major advantages of the reverse passive latex agglutination test (38) are speed (1 h) and simplicity, coupled with a reagent shelf life of 5 months. This makes it useful for emergency situations that do not allow ample time to evaluate a sample by standard testing procedures, provided the person performing the assay has experience in reading agglutination reactions. Unfortunately, a relatively high number of false-negative (about 5%) and false-positive (about 20%) reactions occur. The causes of these unwanted reactions include the presence of rheumatoid factor, autoimmune antibodies, heterophile antibodies, lipemic serum, albumin/globulin imbalance, electrolyte abnormalities, pH imbalance, and various drug metabolites. The false-positive rate can be cut in half by heat inactivation and by adsorbing out the rheumatoid factor. All other agglutination reactions should be confirmed by a blocking test. Confirmation of weak positive reactions by another method of equivalent or greater sensitivity and specificity is essential.

Hemagglutination (32, 33)

Erythrocytes (human group O, turkey, sheep) are coated with purified HBsAg (passive hemagglutination) or with anti-HBs (reversed passive hemagglutination). Anti-HBs antisera are prepared in guinea pigs, sheep, chimpanzees, or horses. Detailed methodology for preparing erythrocytes and for conjugating them with IgG or HBsAg can be found in the 2nd edition of this Manual. Hemagglutination tests have been used extensively for the detection of anti-HBs or HBsAg in human serum or recalcified plasma. Their major advantages are conservation of sera (less than 10 μl is required), the absence of a requirement for expensive equipment, rapid completion (1 to 3 h), good proficiency, and easy quantification. Disadvantages include a relatively large number of nonspecific reactions and the need for personnel experienced in hemagglutination techniques. False-positive reactions frequently occur at dilutions below 1:8, thereby reducing the sensitivity accordingly. Antibodies against ruminant IgG and Forssman antigen may be removed after heat inactivation by adsorbing the sera with uncoated erythrocytes. However, the use of control erythrocytes coated with normal immunoglobulin from the same species is preferred. False-positive reactions may also occur if the serum or buffer is contaminated with certain microorganisms or if it contains rheumatoid factor. Low dilution of high-titered antibody or HBsAg may result in a false-negative reaction because of the formation of a prozone. The use of a vibration-free surface is also required in the reversed passive hemagglutination test to avoid false-negative results.

ELISA (Table 5)

Currently, two ELISA procedures are licensed in the United States for the detection of HBsAg in human serum or plasma. These are Auszyme II (Abbott Laboratories) and Cordia H (Cordis Corp., Miami, Fla.); Organon's Hepanostika has been used extensively in Europe. Specific details of the assays can be obtained from the package insert accompanying each kit. In the Auszyme II assay, HBsAg is detected by incubating 200 μl of human serum or plasma with a polystyrene bead coated with guinea pig anti-HBs. Trays are sealed with an adhesive cover, and incubation proceeds overnight (usually 20 ± 2 h) at 24 ± 2°C. Contents of the wells are aspirated, the beads are washed, and goat anti-HBs–horseradish peroxidase conjugate is added. Incubation continues at 40 ± 1°C for 1 h. During this incubation period, reactive sites on the bound HBsAg combine with the enzyme-labeled antibody to form an antibody-antigen–labeled antibody sandwich. The subsequent wash period is considered extremely critical for reducing the number of nonrepeatable false-positive reactions. The colorless enzyme substrate *o*-phenylenediamine, containing hydrogen peroxide, is added to the beads, resulting in hydrolysis and the production of a yellow-orange color in tubes containing beads with adsorbed HBsAg and substrate. After a 30-min incubation period, the enzyme reaction is stopped by adding 1 ml of 1 N HCl or sulfuric acid. Absorbance is read at 492 nm in a spectrophotometer. Specimens giving absorbance values equal to or greater than the absorbance value of the negative control mean plus the factor 0.050 are considered positive for HBsAg. Runs are not valid if the difference between the positive and negative controls is less than 0.400 absorbance units. The Cordia H assay (24) uses an inert disk coated with horse anti-HBs. The conjugate is horse anti-HBs labeled with alkaline phosphatase, and the enzyme substrate is *p*-nitrophenyl phosphate. NaOH is used to stop the reaction, which is read colorimetrically at 405 nm. The Hepanostika assay (60) uses polystyrene microtiter wells coated with sheep anti-HBs. The conjugate is sheep anti-HBs IgG labeled with horseradish peroxidase. Hydrolysis of the substrate (*o*-phenylenediamine-urea peroxide) is stopped with H_2SO_4, and this timing is rather critical. The reaction is read colorimetrically at 492 nm or visually using a viewbox.

In each of these assays, appropriate positive and negative controls are added to ensure specificity of the test. The total incubation time is 20 ± 2 h for Auszyme II (although shorter times of 1.5 to 3.5 h can be employed), 2.5 h for Cordia H, and 4 h for Hepanostika. The incubation temperatures are 24, 40, and 24°C for Auszyme II (routine overnight procedure), 43°C for Cordia H, and 37 and 25°C for Hepanostika. In each assay, final readings should be made within 2 h after the addition of acid (or base for the Cordia H test). The blank should be repeated whenever prolonged interruptions occur. Care should be taken to avoid splashing specimens or reagents outside of wells or high up on the rim of the well, as such splashes may not be removed in subsequent washing and could be transferred to tubes, causing test interference. Sodium azide will poison the enzyme substrate, so care should be taken that this reagent is not present in the wash reagent.

Commercial ELISA kits are available from Abbott Laboratories for the detection of IgM-specific anti-HBc (Corzyme-M) and HBeAg and anti-HBe (Abbott HBe-EIA). The Corzyme-M test uses a modified sandwich technique to measure IgM-specific anti-HBc in serum or plasma. In this assay, an appropriately diluted specimen (10 μl in 0.5 ml of PBS) is added to a polystyrene bead on which antibody specific for human IgM (mu-chain specific) is adsorbed. This removes IgM from the patient samples. The wells are sealed with an adhesive cover, and incubation proceeds at 40 ± 1°C for 1 h. Liquid is then aspirated, and the beads are washed three times with distilled or deionized water. HBcAg is added to the bead to which IgM-specific anti-HBc may be immunologically bound, and the plates are sealed. After another incubation period at 24 ± 2°C for 20 ± 2 h, the beads are rewashed three times as previously described. In the third incubation period (40°C for 2 h), human anti-HBc, conjugated with horseradish peroxidase, is added to each well to react with any HBcAg retained on a bead by the patient's IgM-specific anti-HBc. After an incubation period of 2 h at 40°C, the liquid is aspirated from the wells. The beads are washed and transferred to tubes, and *o*-phenylenediamine solution containing hydrogen peroxide is added. After incubation at 24 ± 2°C for 30 min, 1 ml of 1 N sulfuric acid is added to each tube to stop the reaction. A yellow-orange color develops in proportion to the amount of anti-HBc–horseradish peroxidase which bound to the bead during the previous incubation. The intensity of the color is measured with a spectrophotometer at 492 nm. The absorbance is proportional to the quantity of IgM-specific anti-HBc present in the patient's serum. A cutoff value of 0.25 times the positive control mean plus the negative control mean is determined. Specimens giving absorbance values equal to or greater than the cutoff are considered reactive for IgM antibodies to HBcAg. If rapid results are essential, the second incubation period can be reduced to 3 h at 37°C so that the test can be completed the same day.

The tests for HBeAg and anti-HBe are performed with the same commercial kit (Abbott HBe-EIA) but employ different assay principles. The HBeAg test uses a sandwich principle to measure HBeAg in serum or plasma. Plastic beads coated with human anti-HBe are added to test samples (200 μl). After the trays are sealed, the samples are incubated at 24 ± 2°C for 20 ± 2 h. The fluid is aspirated, the beads are washed, and anti-HBe (human) conjugated to horseradish peroxidase is then added. After an additional incubation period of 2 h at 40°C, the beads are washed again and transferred to tubes. *o*-Phenylenediamine solution (with hydrogen peroxide) is added, and the reactants are allowed to incubate for another 30 min at 24 ± 2°C. The reaction is stopped with 1 N sulfuric acid, and absorbancy values are determined at 492 nm. Specimens giving absorbancy values equal to or greater than the absorbancy value of the negative control mean plus a factor of 0.060 are considered reactive for HBeAg.

The ELISA is about as sensitive as RIA but is superior to reversed passive hemagglutination and reverse passive latex agglutination. Improvements in the test have reduced the number of repeatable false-positive reactions, but specificity is still lower than that observed by RIA, primarily due to technical errors. However, confirmatory testing with known negative and positive sera gives reliable final results.

RIA (Tables 5 and 7)

The RIA technique continues to be the most sensitive and specific method available for detecting the various serological markers of hepatitis B (HBsAg, HBeAg, anti-HBs, anti-HBc, anti-HBe). The methods most commonly employed utilize the solid-phase sandwich RIA technique. The double-antibody procedure, a research tool which measures the primary interaction between antigen and antibody, can be used for studying the kinetics of this reaction (31, 32). It is highly sensitive, specific, and reproducible. Complete details of this technique, including labeling of purified HBsAg, can be found in the 2nd edition of this Manual.

The solid-phase sandwich method (59) for HBsAg detection is comparable in sensitivity to the double-antibody RIA method, is less cumbersome, and currently is the principal system used by all commercially licensed U.S. manufacturers of hepatitis B RIA kits (Table 7). These assays are capable of detecting HBsAg to a level of 0.1 to 0.5 ng/ml. The concept is identical to that described in the hepatitis A RIA section. Briefly, 200 μl of serum or plasma is added to a solid-phase system to which anti-HBs is adsorbed. The solid-phase support systems used in HBsAg assays include Sepharose, polystyrene or polyethylene beads (pearls, pellets), polystyrene tubes, or controlled-pore glass. During incubation (usually 2 h at 40 to 45°C or 18 ± 2 h at 24 ± 2°C), HBsAg forms an immunological complex with the anti-HBs at the liquid-surface interface. After the beads are washed, anti-HBs tagged with ^{125}I is added and incubation is continued (1 h at 40 to 45°C). The labeled antibody binds to any HBsAg on the support system, creating an antibody–antigen–^{125}I-antibody sandwich. The washed beads are counted in a gamma counter. By dividing cpm of the test sample (S) by the mean cpm of the negative control samples (N), an S/N ratio is calculated. Values of 2.1 and above are generally considered to be reactive (positive). Within limits, there is a direct correlation between the final count rate and the concentration of HBsAg in the specimen. However, size of the HBsAg and surface

TABLE 7. Comparison of commercial solid-phase RIAs for the detection of HBsAg

Assay	Source[a]	Support system	Adsorbed antibody	Sample vol (μl)	Incubation: Total time (h)	Incubation: Temp (°C)	Wash solution	Labeled antibody
AusRia II	Abbott	Polystyrene beads	Guinea pig	200	3/18	45	DW[b]	Human
Clinical Assays	Connaught	Polystyrene tube	Guinea pig	200	3/18	45	NaCl	Chimpanzee
RIAUSURE II	Electro-Nucleonics	Controlled-pore glass	Goat	200	2/18	25	PBS	Goat
NML HBsAg	Nuclear-Medical/ Centocor	Polystyrene beads	Mouse monoclonal IgG	100/200	4/18	45	DW	Mouse monoclonal IgG
Hepria B	RIA International	Polystyrene beads	Goat	200	3/18	45	DW	Human

[a] Abbott, Abbott Laboratories, North Chicago, Ill.; Connaught, Connaught Laboratories, Willowdale, Ontario, Canada (distributed in the United States by Travenol Laboratories, Inc., Deerfield, Ill.); Electro-Nucleonics, Electro-Nucleonics Laboratories, Inc., Bethesda, Md.; Nuclear-Medical, Nuclear-Medical Laboratories, Inc., Dallas, Tex.; RIA International, RIA International, Inc. (NABI), Miami, Fla.

[b] DW, Distilled water.

area of the support system restrict the working range to between 0.1 ng/ml and 1 μg/ml, above which level saturation (a plateau) is reached. Sera from some carriers have HBsAg concentrations above 100 μg/ml. In these situations, further dilutions are required to permit quantitation. Repeatably positive reactions that cannot be confirmed as positive for HBsAg are highly unusual (<0.1%). Nevertheless, the seriousness of the diagnosis, with its attendant personal, social, and economic repercussions, mandates that a confirmatory test be attempted on all repeatably reactive specimens. This can be approached in one of two ways: (i) specimens may be tested with another licensed HBsAg assay, or (ii) a blocking or inhibition assay can be performed to see whether unlabeled anti-HBs will specifically inhibit the reaction. The detection of anti-HBc in the absence of anti-HBs also corroborates a positive HBsAg result.

To avoid an erroneous interpretation, the negative control samples must be comparable to the test specimen. Unfortunately, the kit "negative control" values are usually 10 to 40% higher than values obtained using fresh normal (nonreactive) sera. Thus, S/N values between 1.5 and 2.1 should be viewed with suspicion whenever the kit negative control is used. Correspondingly, tests for HBsAg in cerebrospinal fluid must use "normal" cerebrospinal fluid as the control. In general, protein-deficient specimens and recalcified plasma result in higher background levels.

In an emergency, the solid-phase RIA may be modified to provide an answer within 1 h. Sensitivity is slightly less than that of the regular assay, but the number of positive specimens likely to be missed should be relatively small. To further reduce the number of false-negative values, at the expense of increasing the number of false-positive reactions, the cutoff value can be reduced to ≥1.5 times the negative control. Ultimately, verification of a positive reaction by the regular procedure is essential.

Solid-phase RIA for the detection of anti-HBs is similar in principle to HBsAg detection except that the specimen is incubated with polystyrene beads coated with HBsAg. Radiolabeled HBsAg is used to detect anti-HBs bound to the fixed HBsAg. Specimens with a count rate equal to or greater than 2.1 times the negative control mean cpm are reactive. Unfortunately, inter- and intralot variations in commercial anti-HBs kits and diminishing sensitivity as the kits approach their expiration date limit the usefulness of the S/N ratio for comparative purposes. With the advent of the HBsAg vaccine, the need for precision and accuracy in the assay has become more important. To permit results that can be expressed in milli-International Units per milliliter (mIU/ml), Hollinger et al. (26) have modified the RIA test. Anti-HBs concentrations are based on the First International World Health Organization (WHO) Reference Preparation for HBIG (lot 26.1.77) provided by the International Laboratory for Biological Standards, Central Laboratory of The Netherlands Red Cross Transfusion Services. An arbitrary value of 50 IU of anti-HBs has been assigned to this product. To determine mIU/ml, an S/R ratio is computed as follows:

$$\frac{\text{Sample cpm} - \text{negative control mean cpm}}{\text{Reference control cpm} - \text{negative control mean cpm}}$$

The WHO anti-HBs reference standard is diluted to contain 125 mIU/ml. Regression of the S/R ratio on the WHO reference anti-HBs concentration (0.1 to 500 mIU/ml) (Fig. 4) using the computer nonlinear regression program BMDP3R (17) yielded the following formula:

$$\text{mIU/ml} = 130.75\ (e^{0.66765(\text{S/R})} - 1)$$
$$\text{(reciprocal of dilution)}$$

The lower limit of detection for anti-HBs with the current commercially available solid-phase RIA (Ausab, Abbott Laboratories) is 0.7 mIU/ml. Dilutions are usually required when anti-HBs concentrations exceed 200 mIU/ml. To conserve the WHO reference reagent, the laboratory can prepare a large batch of reference anti-HBs and determine its relationship to the WHO reference standard (diluted to 125 mIU/ml) by using the S/R ratio. The exponential value in the formula must be adjusted proportionally to agree with the new laboratory standard. For example, if the S/R relationship between the laboratory standard and the WHO reference standard is 0.500, the exponential value in the formula (0.66765) must be reduced by a factor of 0.5. Conversely, an S/R ratio of 1.5 would

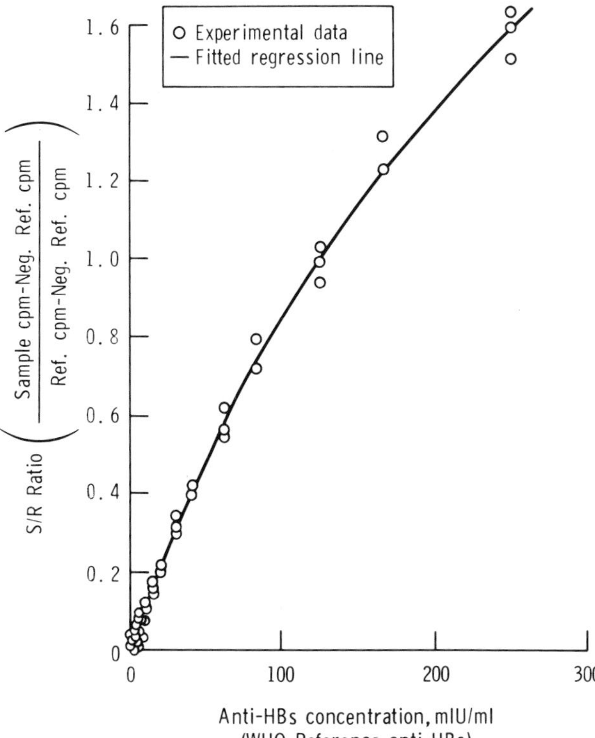

FIG. 4. RIA standard curve for anti-HBs. Regression of the S/R ratio on the anti-HBs concentration (mIU/ml) (26). For the regression equation, see the text.

increase it by a factor of 1.5. Once this adjustment is made, the laboratory reference standard can be substituted for the WHO reference standard. Periodic reevaluation against the WHO standard should be performed to verify the accuracy of the laboratory standard, which is stored in small aliquots at −30°C.

The anti-HBc test (Corab, Abbott Laboratories) is a competitive binding assay in which anti-HBc from the test sample (100 μl) competes with a constant amount of human ^{125}I-tagged anti-HBc for a limited number of binding sites found on beads coated with HBcAg. Within limits, the proportion of radioactive anti-HBc bound to the bead is inversely proportional to the concentration of anti-HBc in the test specimen. Incubation is carried out at 24 ± 2°C for 20 ± 2 h. A specimen is considered to be reactive for anti-HBc if the count rate of the unknown is less than the cutoff value (one-half the sum of the negative control mean and the positive control mean).

Other RIA tests include those for HBeAg and anti-HBe (Abbott-HBe, Abbott Laboratories). The HBeAg test uses the sandwich principle to measure HBeAg in serum or plasma (200 μl). Polystyrene beads coated with anti-HBe are added to the test samples, and incubation proceeds at 24 ± 2°C for 20 ± 2 h. After the beads are washed, anti-HBe tagged with ^{125}I is added and incubation is continued at 45°C for 1 h. Count rates are increased when HBeAg is present in the serum. A positive reaction is one which is equal to or greater than the cutoff value of 2.1 times the negative control mean count rate. The test for anti-HBe is a modified competitive binding assay in which anti-HBe in the test serum (100 μl) competes with anti-HBe-coated beads for a standardized amount of added HBeAg. Trays containing the beads are incubated for 20 ± 2 h at 24 ± 2°C, at the end of which time the beads are washed. If anti-HBe is present in the test sample, less HBeAg would be coupled to the bead and therefore less ^{125}I-labeled anti-HBe would bind to the bead to complete the sandwich. Within limits, the greater the amount of anti-HBe in the specimen, the lower the count rate. The count rate is compared to the cutoff value, which is one-half of the sum of the count rates of the negative control mean plus the positive control mean. Specimens with count rates equal to or less than the cutoff are considered reactive for anti-HBe, similar to the other competitive binding assays.

Tests for HBV-associated DNA polymerase (52)

Samples are diluted 3- to 30-fold in Tris-saline buffer (pH 7.4; 0.01 M Tris hydrochloride and 0.15 M NaCl) and are clarified at 10,000 × *g* for 10 min; 3.0 ml is then layered over 2.5 ml of 30% (wt/vol) sucrose containing Tris-saline buffer, 0.001 M EDTA, 0.1% 2-mercaptoethanol, and 1 mg of BSA per ml which has been precentrifuged for 10 min at 10,000 × *g* to remove precipitated BSA. After centrifugation for 4 h at 250,000 × *g* (50,000 rpm in an SW65 rotor), the supernatant fluid is removed, residual fluid is adsorbed with paper, and the pellet is resuspended in 50 μl of Tris-saline buffer containing 0.1% Nonidet P-40 and 0.1% 2-mercaptoethanol. For evaluation of a larger number of specimens, a Beckman type 25 rotor that holds 100 1-ml tubes can be used. A 200-μl sample of a precentrifuged undiluted serum is layered over 0.5 ml of 30% (wt/vol) sucrose. The tubes are centrifuged at 24,000 rpm for 15 h at 4°C, and the pellets are recovered as described above. The removal of serum proteins from the pelleted material is essential because their presence will result in high background counts that cannot be washed away. To the resuspended pellet is added 25 μl of the reaction mixture (0.2 M Tris, pH 7.4, 0.08 M $MgCl_2$, 0.24 M NH_4Cl, 1.0 mM dATP, 1.0 mM TTP, and 0.025 nM each of [^{3}H]dCTP and [^{3}H]dGTP both at 21 Ci/mmol). The reactants are incubated at 37°C for 3 h, and then 50 μl is placed on a Whatman 3-mm paper disk, washed, and assayed for acid-precipitable ^{3}H. Confirmation that the reactivity is associated with HBV is accomplished by its immunoprecipitation with anti-HBs before Nonidet P-40 detergent treatment and with anti-HBc after the addition of Nonidet P-40.

Hybridization techniques (55)

While hybridization procedures are beyond the scope of most clinical laboratories, an awareness of their availability is desirable. Human serum can be analyzed for HBV DNA sequences using molecular hybridization techniques. To conduct these studies, 10 to 15 μl of serum is applied to a 0.45-μm nitrocellulose filter sheet, using 5 μl for each application. The paper should be dried between applications. Known positive and negative control samples are included in each assay. The paper is treated with 0.5 N NaOH neutralized with 1 M Tris hydrochloride (pH 7.4) containing 0.5 M NaCl and treated with proteinase K at 200 U/ml for 1 h at 37°C. After this treatment, the paper is washed twice with 0.3 M NaCl containing 0.03 M sodium citrate and then baked in vacuum at 80°C for 2 h, prehybridized, and hybridized for 24 to 36 h in a solution containing Denhardt's solution (0.1% BSA, 0.1% Ficoll, 0.1% polyvinylpyrrolidone), 0.6 M NaCl, 0.06 M sodium citrate, 0.1% sodium dodecyl sulfate, 0.025 M sodium phosphate buffer (pH 6.5), 200 μg of denatured calf thymus DNA per ml, and 10^7 cpm of repurified recombinant cloned HBV DNA labeled with ^{32}P (specific activity, 2×10^8 to 4×10^8 cpm/μg of DNA). After hybridization, the nitrocellulose filter is washed, dried, and autoradiographed. This method is capable of detecting visually quantities from 0.2 to 0.5 pg of HBV DNA sequences within 24 h with a 2- to 10-fold increase after 5 days of autoradiography. Similar hybridization studies can be performed using DNA extracted from small portions of frozen biopsy specimens (55).

INTERPRETATION OF TEST RESULTS IN HEPATITIS B

Successful detection of HBV serological markers depends not only on the relative sensitivity of the test procedures but also on the availability of experienced personnel who comprehend the procedure used and its idiosyncrasies and who are meticulous in their performance of the test. Provided that these conditions are met, the final evaluation and interpretation of any positive test result will be determined by the specificity of the reagents used. It is essential for the diagnostic virologist to appreciate the difficulties encountered in preparing from human sources quality reagents that are free from contaminating human proteins. These contaminants result in the production of low concentrations of unwanted antibodies during immunization. The preparation of monospecific antibody or the prior adsorption of antisera with an insoluble immunoadsorbent prepared from normal human serum has eliminated most of these problems as has the production of HBsAg by using recombinant DNA methodology. Specificity testing with a reference antigen or antiserum also provides confirmation of the laboratory result. Additional verification of an HBsAg reaction is generated when anti-HBc is also detected.

The marked increase in sensitivity of the RIA, ELISA, or hemagglutination tests as compared to the discontinuous counterimmunoelectrophoresis test should not imply that an equivalent increase in the number of HBsAg-positive persons will be detected. Experience indicates that 70% of the positive carriers in the United States have HBsAg concentrations that are detectable by second-generation assays, leaving 30% undetectable except by the more sensitive procedures. HBsAg levels between 0.1 and 0.5 ng/ml (the lower level of sensitivity for most third-generation tests) are found in less than 5% of the HBV carriers, representing about 0.01% of the donor population. Conversely, 10 to 15% of the HBV carrier population will have HBsAg concentrations above 100 μg/ml.

As shown in Fig. 5 and summarized in Table 8, the presence of HBsAg in a serum is indicative of an active

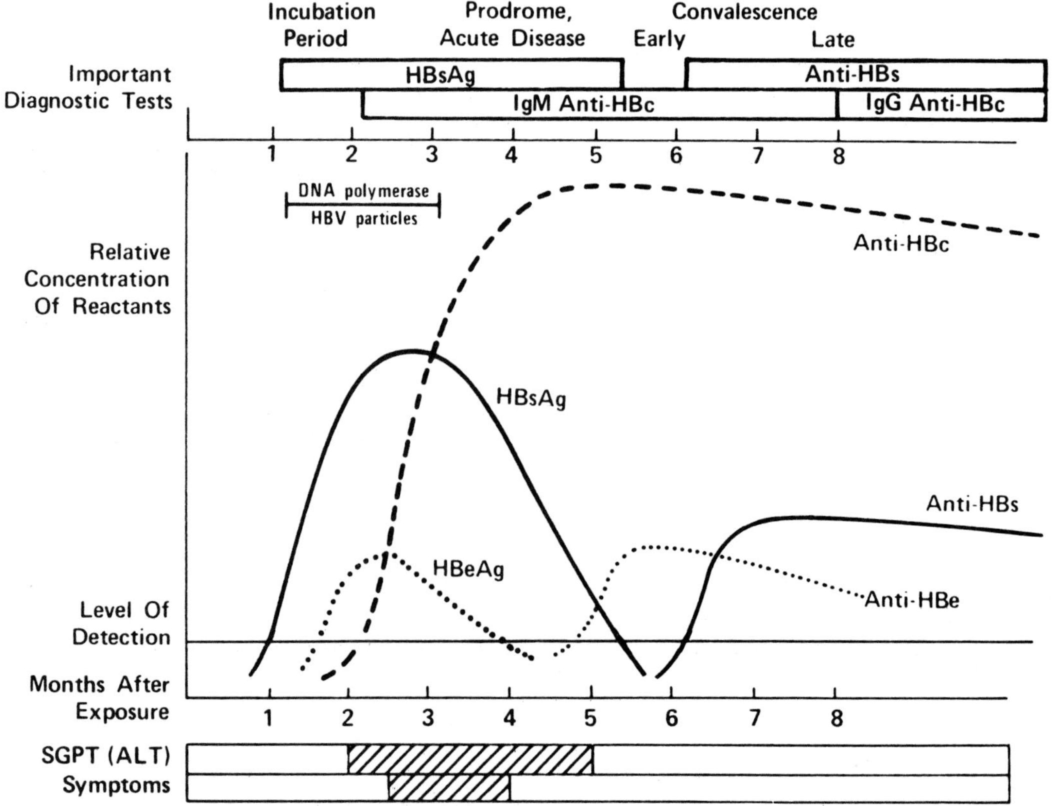

FIG. 5. Serological and clinical patterns observed during acute HBV infection. SGPT, Serum glutamic pyruvic transaminase; ALT, alanine aminotransferase.

TABLE 8. Interpretation of viral hepatitis screening profile

Assay results			Interpretation
IgM-specific anti-HAV	HBsAg	Anti-HBc	
Positive	Negative	Negative	Recent acute HAV infection.
Negative	Positive	Negative	Early acute hepatitis B infection. Confirmation is required to exclude nonrepeatable or nonspecific reactivity.
Negative	Positive	Positive	HBV infection, either acute or chronic.[a] Symptoms may be unrelated to hepatitis B.
Negative	Negative	Positive	Active HBV infection cannot be excluded. Test for anti-HBs and anti-HBe. A positive anti-HBs test indicates a previous HBV infection and confirms immunity to hepatitis B.[b] A negative anti-HBs and a positive anti-HBe indicate recent HBV infection (<5% of cases). Confirm by demonstrating anti-HBs seroconversion in 2 to 6 weeks or by examining the sample for high-titer IgM-specific anti-HBc.
Negative	Negative	Negative	Possible non-A, non-B hepatitis, other viral infections, or toxic (drug-induced) liver disease.
Positive	Positive	Positive	Recent acute HAV infection in an HBV carrier.

[a] Differentiate acute from chronic hepatitis by examining the sample for high-titer IgM-specific anti-HBc. Use HBsAg quantitation of paired specimens to predict the potential for chronicity or to verify existing chronic hepatitis B. The presence of HBeAg indicates those specimens which exhibit the potential for enhanced infectivity. HBeAg is found only in the presence of HBsAg. Other HBV serological markers that may be present at the same time include HBV (Dane) particles observable by electron microscopy. By disrupting the virion, HBcAg and viral DNA polymerase can be measured.

[b] Exclude recent transfusions, immune globulin administration, or a maternal antibody source within the previous 6 months.

hepatitis B infection, either acute or chronic. In a typical HBV infection, HBsAg will be detected 2 to 4 weeks before the transaminase level becomes abnormal and 3 to 5 weeks before the patient develops symptoms or becomes jaundiced. Anti-HBc, primarily of the IgM immunoglobulin class, usually appears when the transaminase levels begin to increase. The presence of IgM-specific anti-HBc in high titer is evidence of an acute infection. These elevated titers decline regardless of whether the disease is resolved or becomes chronic. HBsAg that persists for more than 4 to 6 months after the onset of clinical illness specifies those persons who are likely to become carriers of HBV. The HBsAg level rises and falls steadily in acute hepatitis B, whereas the HBsAg concentration is maintained within a very narrow range in the untreated chronic carrier. Thus, it is possible to predict the potential for chronicity in patients with acute HBV disease, e.g., those who are positive for IgM-specific anti-HBc, by performing a quantitative RIA test on paired sera collected 2 to 3 weeks apart. Most comparisons can be evaluated at a dilution of 1:10,000 (prepared in PBS containing 0.5% BSA), although some samples may require a lower dilution to obtain an S/N ratio which is on the descending slope of the RIA dose-response curve. Extreme care must be employed in preparing the dilutions. A decline in HBsAg concentration predicts eventual recovery, whereas no change in the concentration indicates that the patient has developed a persistent infection.

Because the anti-HBc test is invariably positive when HBsAg is present in a clinically ill patient (see Fig. 5 and Table 8 for rare exceptions to this statement), its immunodiagnostic value is to validate the HBsAg reaction. Tests with discordant results always should be repeated. However, in perhaps 5 to 10% of the acute cases of hepatitis B (especially those with fulminant disease), and more frequently during early convalescence, serum levels of HBsAg may be undetectable (Fig. 5). Examination of these sera for IgM-specific anti-HBc may help in establishing the correct diagnosis. In the absence of anti-HBc and HBsAg, active hepatitis B disease can be excluded. In contrast, the presence of anti-HBc alone is presumptive evidence for an active HBV infection. However, this relationship is not infallible, as many patients who have recovered from hepatitis B with the development of anti-HBs and anti-HBc eventually lose one or the other.

Specimens which are HBsAg positive may be evaluated for HBeAg, anti-HBe, specific DNA polymerase reactivity, or HBV particles to determine their potential for enhanced infectivity. HBeAg-positive specimens contain high concentrations of HBV particles and are more likely to transmit hepatitis B, in contrast to anti-HBe-positive samples in which the number of HBV particles is markedly reduced.

Antibody to HBsAg usually becomes detectable 1 to 2 months after the disappearance of HBsAg. It is assumed that anti-HBs production occurs much earlier, but is not observed as a result of the formation of immune complexes with excess HBsAg. Antibody to HBsAg, with or without anti-HBc, specifies immunity against reinfection. However, RIA S/N ratios lower than 10 should be viewed with caution in the absence of anti-HBc. It has been proposed that anti-HBc, not anti-HBs, be used to determine susceptibility or immunity to HBV or to decide whether to recommend vaccination with HBsAg. Conversely, only anti-HBs develops in persons who receive the HBsAg vaccine. Depending on the circumstances, passive transfer of anti-HBs or anti-HBc is often observed in patients receiving these antibodies during blood transfusions, after hepatitis B immune globulin administration, or in neonates of mothers with recent or past hepatitis B. Recognition of these possibilities will avoid an erroneous diagnosis of HBV infection, since passive antibodies gradually disappear over a 2- to 4-month observation period in contrast to actively produced antibodies, which are remarkably stable over many years.

Subtyping of specimens by the clinical laboratory provides additional information to the clinician or hospital epidemiologist, since the mutually exclusive *d* and *y* subdeterminants are virus specific and not host determined. This can be helpful in determining the source of infection or can provide epidemiological evidence for the relatedness among cases.

DIRECT EXAMINATION OF SAMPLES FOR THE DELTA VIRUS

Immunofluorescence and immunoperoxidase staining of liver tissue from patients or experimental animals (chimpanzees, woodchucks with woodchuck hepatitis virus) are reliable approaches for the identification of HDAg; however, these methodologies are restricted to specialized pathology laboratories and are not practical for adoption as rapid screening tests by clinical laboratories. The methodologies for such immunohistochemical staining are routine for pathology and immunology laboratories. The anti-HD probe, to which fluorescein or peroxidase is conjugated, is usually derived from serum or plasma which contains a higher titer of anti-HD and a low level of antibody to HBcAg. Thus, the anti-HBc reactivity can be reduced substantially or eliminated entirely by dilution. However, an anti-HD–negative, anti-HBc–positive control probe should be used in parallel with the anti-HD probe to substantiate the anti-HD specificity of staining. Controls for other sources of nonspecificity, including autoantibodies to nuclear antigens, should be incorporated into procedures for immunohistochemical staining. With these techniques, investigators have shown that the HDAg localizes to the liver cell nucleus primarily, but rarely can be detected in the cytoplasm. Although frozen cryostat sections are preferable, HDAg is stable to Formalin fixation and paraffin embedding. Therefore, HDAg can be studied in stored paraffin-embedded tissue after digestion of the section with trypsin or pronase (8).

SEROLOGICAL IDENTIFICATION OF DELTA ANTIGEN (HDAg) AND ANTIBODY (ANTI-HD)

Antibodies to HDAg have been identified and quantitated primarily by RIA or ELISA (7, 51). The major limitation to wider availability of these diagnostic techniques has been a shortage of HDAg-containing clinical material for use in immunoassays. As discussed below, however, the ability to perform these tests is increasing among diagnostic laboratories.

Purification of HDAg and preparation of antibodies

Tests for anti-HD require a source of HDAg. Previously the HDAg employed for these purposes had been obtained from HDAg-positive human (postmortem) or chimpanzee liver or serum. More recently, a more reliable source, HDAg-positive woodchuck liver, has become available. HDAg can be extracted from human liver with strong dissociating agents such as 6 M guanidine hydrochloride or 8 M urea. When liver tissue is obtained from experimentally infected chimpanzees or woodchucks at the peak of intrahepatic HDAg expression, i.e., before the appearance of anti-HD, HDAg can be harvested by simple aqueous extraction. Although HDAg is rarely detectable in patients with acute delta infection, occasionally sufficient HDAg is present in serum to serve as a source of antigen for diagnostic testing.

Serum containing anti-HD can be obtained from patients or experimental animals with acute or chronic delta infection. As mentioned above, anti-HBc reactivity, invariably present in anti-HD-positive serum, is often low in serum samples with a high level of anti-HD activity. Residual anti-HBc activity can be diluted out, and HBsAg is removed by ultracentrifugation.

RIA

Because the availability of HDAg remains limited, the only practical approach to routine laboratory immunodiagnosis is commercial immunoassays in which HDAg is derived from the livers of woodchucks with experimental woodchuck hepatitis virus and delta infection. A commercial RIA for anti-HD has recently become available for research purposes (Abbott Anti-Delta, Abbott Laboratories). The test is a competitive-binding RIA in which anti-HD in the test serum competes with ^{125}I-labeled human anti-HD for woodchuck liver-derived HDAg coating a solid-phase polystyrene bead. In this assay, 100 μl of ^{125}I-labeled anti-HD and 100 μl of serum or plasma to be tested are delivered to a microtiter well and an HDAg-coated bead is added. Three negative control and two positive control wells are incorporated into each test run. Microtiter plates are incubated for 20 ± 2 h at 24 ± 2°C. After incubation, liquid is aspirated from the wells. The beads are washed three times with distilled or deionized water and transferred to counting tubes, and the tubes are assayed in a gamma scintillation counter. The net cpm (minus background) for the test sample is compared with a calculated cutoff (0.4 of the mean negative control counts plus 0.6 of the mean positive control counts). Samples with a count rate equal to or below the cutoff are considered reactive for anti-HD; counts above the cutoff range are considered negative. Validity of the test requires that, for each run, the ratio of the mean negative (N) control counts to the mean positive (P) control counts should be ≥ 4.0. The titer of anti-HD can be determined by testing serial dilutions of serum or plasma and selecting the dilution which yields counts closest to, but not greater than, the cutoff value as the anti-HD titer.

Delta antigen can be detected by solid-phase sandwich RIA, as described by Rizzetto et al. (51). The test is based on binding of HDAg in serum to anti-HD adherent to a solid phase, followed by incubation with an ^{125}I-labeled purified IgG anti-HD probe. Because HDAg always circulates within an HBsAg-encapsidated virion, detection of HDAg requires detergent (0.5% Tween-80, Nonidet P-40, or deoxycholate) disruption of the virion to expose the internal, otherwise sequestered antigen. Adding to the difficulty of HDAg detection is the transience of delta antigenemia, occurring during early infection. This simple solid-phase RIA for HDAg can be modified for detection of anti-HD in an RIA blocking assay. A standardized quantity of HDAg is added to beads to which anti-HD is adsorbed. Nonreactive anti-HD from serum or plasma competes with a constant amount of ^{125}I-labeled anti-HD for the HDAg immunologically bound to the beads. A reduction in the count rate of ≥50% is evidence for the presence of anti-HD in the sample. Measurement of anti-HD titers is achieved by diluting serum and determining that dilution which inhibits 50% of binding as compared to negative control sera (51).

ELISA

A commercial enzyme immunoassay for anti-HD, based on the configuration described above for the commercial RIA, is being developed and will be available shortly.

New diagnostic approaches

IgM anti-HD. Acute delta hepatitis infection (simultaneous with acute HBV infection or acute delta hepatitis superinfection of an HBV carrier) is accompanied by an early anti-HD response predominantly of the IgM class. An antibody class-capture solid-phase RIA for IgM anti-HD has been described in which antibody to human IgM (mu-chain specific) is bound to a solid phase and test serum is added. If the test serum contains IgM anti-HD, it will bind to the solid phase. When HDAg is added subsequently, it binds to the IgM anti-HD. The sandwich is completed when the labeled probe, ^{125}I-conjugated IgG anti-HD, binds to the HDAg (57). The methodology for this technique is analogous to that for the detection of IgM-specific anti-HBc and IgM-specific anti-HAV. Interested readers are referred to the sections above in which details of these assays are provided.

cDNA probe. Because HDAg is rarely detectable in serum, even when the serum is known to be infectious, and because anti-HD is present indefinitely in patients with chronic delta hepatitis infection, liver biopsy with immunohistochemical staining for intrahepatic HDAg is the only reliable way to demonstrate ongoing delta hepatitis replication. A noninvasive approach for detecting HDV RNA in serum has been made possible with the recent availability of cloned cDNA probes (57a). The cDNA probe for HDV-associated RNA is incubated with serum, and HDV RNA is identified by dot-blot hybridization analysis. This technique remains limited to a small number of research laboratories but holds promise as a simple, noninvasive test for the diagnosis of chronic delta hepatitis infection.

INTERPRETATION OF TEST RESULTS IN DELTA HEPATITIS

Infection with HDV can occur in the presence of acute or chronic HBV infection. The duration of HBV infection determines the duration of delta infection. When acute delta and HBV infection occur simultaneously, clinical and biochemical features may be indistinguishable from those of HBV infection alone. The presence of delta hepatitis infection can be identified by demonstrating intrahepatic HDAg or, more practically, an anti-HD seroconversion (a rise in titer of anti-HD or de novo appearance of IgM-specific anti-HD, which is briefly detectable, if at all). Because IgM-specific anti-HD is transient and IgG anti-HD is often undetectable once HBsAg disappears, retrospective serodiagnosis of acute, self-limited, simultaneous HBV and HDV infection is difficult.

In contrast to patients with acute HBV infection, patients with chronic HBV infection can support HDV replication indefinitely. This can happen when acute HDV infection occurs in the presence of a nonresolving acute HBV infection. More commonly, acute HDV infection becomes chronic when it is superimposed on an underlying chronic HBV infection. In such cases, the delta superinfection appears as a clinical exacerbation or an episode resembling acute viral hepatitis in someone already chronically infected with HBV. In the past, events resembling acute hepatitis in an HBV carrier or a patient with chronic hepatitis B were attributed to superimposed non-A, non-B hepatitis or to the natural history of the disease. A proportion of such episodes, however, represents acute superinfection with HDV.

When a patient presents with acute hepatitis and has HBsAg and anti-HD in his serum, determination of the class of anti-HBc is helpful in establishing the relationship between infection with HBV and HDV. Although IgM-specific anti-HBc does not distinguish absolutely between acute and chronic HBV infection, its presence is a reliable indicator of recent infection and its absence is a reliable indicator of infection in the remote past. In simultaneous acute HBV and HDV infections, IgM-specific anti-HBc will be detectable, whereas in acute delta hepatitis infection superimposed upon chronic HBV infection, anti-HBc will primarily be of the IgG class.

As noted above, cDNA tests for the presence of HDV-associated RNA will be useful in the future for determining the presence of ongoing HDV replication and relative infectivity. Currently, probes for this marker are restricted to a limited number of research laboratories.

NON-A, NON-B HEPATITIS (11, 12, 27)

Non-A, non-B hepatitis accounts for approximately 90% of the cases of transfusion-associated hepatitis seen in the United States and may be responsible for up to 20% of endemic and sporadic hepatitis (1, 13). Transmission mechanisms are similar to those for hepatitis B. The incubation period ranges from 5 to 10 weeks, although longer and shorter periods have been observed. Clinical characteristics are similar to those observed after hepatitis B except that chronic liver disease may be more prominent. Experimental transmission of non-A, non-B hepatitis agents has been accomplished in chimpanzees (27, 29). Patients with or without biochemical evidence of liver disease can transmit this disease, although the risk is significantly greater when the donor has alanine aminotransferase abnormalities (25). Viremia appears to precede the liver disease by at least 12 days. Evidence for at least two distinct non-A, non-B hepatitis agents has been obtained (27).

Diagnosis of this entity is by exclusion of HAV, HBV, cytomegalovirus, and Epstein-Barr virus in a patient who presents biochemical evidence of hepatitis. Several investigators have reported the detection of new antigens or antibodies, and even HBV-like serological markers, in the sera and liver biopsies of these patients, but none of these putative virus markers has yet been proven to be specific for non-A, non-B hepatitis (11, 12, 20). Recently, Seto et al. (54a) reported the detection of particle-associated reverse transcriptase activity in two plasma-derived products and four human sera which previously had been shown to transmit non-A, non-B hepatitis to humans and chimpanzees. In addition, sera collected from 12 patients with acute or chronic non-A, non-B hepatitis also were positive, whereas reverse transcriptase activity was found in only 2 of 48 controls. Unexpectedly, neutralization of infectivity with presumably convalescent-phase serum or immune globulin has been unsuccessful. In another recent development, Prince et al. (45a) reported the successful isolation of infectious membrane-coated virus particles from chimpanzee liver cell cultures inoculated with sera previously implicated in transmission of non-A, non-B hepatitis. The particles were 85 to 90 nm in diameter with a core of 40 to 45 nm. Cell homogenates from infected cultures were capable of causing hepatitis in chimpanzees. Chloroform extraction of this material appeared to inactivate the virus. If each of the above findings can be confirmed, progress toward the eventual control of non-A, non-B hepatitis will have been advanced.

LITERATURE CITED

1. **Aach, R. D., W. Szmuness, J. W. Mosley, F. B. Hollinger, R. A. Kahn, C. E. Stevens, V. M. Edwards, and J. Werch.** 1981. Serum alanine aminotransferase of donors in relation to the risk of non-A, non-B hepatitis in recipients. N. Engl. J. Med. **304:**989–994.
2. **Barker, L. F., J. E. Maynard, R. H. Purcell, J. H. Hoofnagle, K. R. Berquist, and W. T. London.** 1975. Viral hepatitis, type B, in experimental animals. Am. J. Med. Sci. **270:**189–195.
3. **Blumberg, B. S., A. I. Sutnick, W. T. London, and L. Millman.** 1971. The discovery of Australian antigen and its relation to viral hepatitis. Perspect. Virol. **7:**223–240.
4. **Bond, W. W., M. S. Favero, N. J. Petersen, and J. W. Ebert.** 1983. Inactivation of hepatitis B virus by intermediate-to-high-level disinfectant chemicals. J. Clin. Microbiol. **18:**535–538.
5. **Bond, W. W., N. J. Petersen, and M. S. Favero.** 1977. Viral hepatitis B: aspects of environmental control. Health Lab. Sci. **14:**235–252.
6. **Bradley, D. W., H. A. Fields, K. A. McCaustland, J. E. Maynard, R. H. Decker, R. Whittington, and L. R. Overby.** 1979. Serodiagnosis of viral hepatitis A by a modified competitive binding radioimmunoassay for immunoglobulin M anti-hepatitis A virus. J. Clin. Microbiol. **9:**120–127.
7. **Crivelli, O., M. Rizzetto, C. Lavarini, A. Smedile, and J. L. Gerin.** 1981. Enzyme-linked immunosorbent assay for detection of antibody to the hepatitis B surface antigen-associated delta antigen. J. Clin. Microbiol. **14:**173–177.
8. **Crivelli, O., J. W. K. Shih, and M. Rizzetto.** 1983. Methods for detection of the delta antigen and antibody in liver and serum, p. 121–126. *In* G. Verme, F. Bonino, and M. Rizzetto (ed.), Viral hepatitis and delta infection. Alan R. Liss, New York.
9. **Decker, R. H., S. M. Kosakowski, A. S. Vanderbilt, C.-M. Ling, R. Chairez, and L. R. Overby.** 1981. Diagnosis of acute hepatitis A by HAVAB-M, a direct radioimmunoassay for IgM anti-HAV. Am. J. Clin. Pathol. **76:**140–147.
10. **Deinhardt, F., D. Peterson, G. Cross, L. Wolfe, and A. W. Holmes.** 1975. Hepatitis in marmosets. Am. J. Med. Sci. **270:**73–80.
11. **Dienstag, J. L.** 1983. Non-A, non-B hepatitis. I. Recognition, epidemiology, and clinical features. Gastroenterology **85:**439–462.
12. **Dienstag, J. L.** 1983. Non-A, non-B hepatitis. II. Experimental transmission, putative virus agents and markers, and prevention. Gastroenterology **85:**743–768.
13. **Dienstag, J. L., A. Alaama, J. W. Mosley, A. G. Redeker, and R. H. Purcell.** 1977. Etiology of sporadic hepatitis B surface antigen-negative hepatitis. Ann. Intern. Med. **87:**1–6.
14. **Dienstag, J. L., S. M. Feinstone, R. H. Purcell, J. H. Hoofnagle, L. F. Barker, W. T. London, H. Popper, J. M. Peterson, and A. Z. Kapikian.** 1975. Experimental infection of chimpanzees with hepatitis A virus. J. Infect Dis. **132:**532–545.
15. **Dienstag, J. L., J. A. Routenberg, R. H. Purcell, R. R. Hooper, and W. O. Harrison.** 1975. Foodhandler-associated outbreak of hepatitis type A. An immune electron microscopic study. Ann. Intern. Med. **83:**647–650.
16. **Dienstag, J. L., A. N. Schulman, R. J. Gerety, J. H. Hoofnagle, D. E. Lorenz, R. H. Purcell, and L. F. Barker.** 1976. Hepatitis A antigen isolated from liver and stool: immunologic comparison of antisera prepared in guinea pigs. J. Immunol. **117:**876–881.
17. **Dixon, W. J., and M. B. Brown.** 1979. BMDP-79. University of California, Berkeley.
18. **Dressman, G. R., F. B. Hollinger, R. M. McCombs, and J. L. Melnick.** 1972. Production of potent anti-Australia antigen sera of high specificity and sensitivity in goats. Infect. Immun. **5:**213–221.
19. **Dreesman, G. R., F. B. Hollinger, and J. L. Melnick.** 1972. Detection of hepatitis B antigen by counter-immunoelectrophoresis enhancing role of homologous serum diluents. Appl. Microbiol. **24:**1001–1002.
20. **Feinstone, S. M., and J. H. Hoofnagle.** 1984. Non-A, maybe-B hepatitis. N. Engl. J. Med. **311:**185–189.
21. **Feinstone, S. M., Y. Moritsugu, J. W.-K. Shih, J. L. Gerin, and R. H. Purcell.** 1978. Characterization of HAV, p. 41–48. *In* G. N. Vyas, S. N. Cohen, and R. Schmid (ed.), Viral hepatitis. The Franklin Institute Press, Philadelphia.
22. **Gerin, J. L., and J. W.-K. Shih.** 1978. Structure of HBsAg and HBcAg, p. 147–158. *In* G. N. Vyas, S. N. Cohen, and R. Schmid (ed.), Viral hepatitis. The Franklin Institute Press, Philadelphia.
23. **Gravelle, C. R., C. L. Hornbeck, J. E. Maynard, C. A. Schable, E. H. Cook, and D. W. Bradley.** 1975. Hepatitis A: report of a common source outbreak with recovery of a possible etiologic agent. II. Laboratory studies. J. Infect. Dis. **131:**167–171.
24. **Halbert, S. P., and M. Anken.** 1977. Detection of hepatitis B surface antigen (HBsAg) with use of alkaline phosphatase-labeled antibody to HBsAg. J. Infect. Dis. **136**(Suppl.)**:**318–323.
25. **Hollinger, F. B.** 1984. Prevention of posttransfusion hepatitis, p. 319–337. *In* G. N. Vyas, J. L. Dienstag, and J. H. Hoofnagle (ed.), Viral hepatitis and liver disease. Grune & Stratton, Inc., Orlando, Fla.
26. **Hollinger, F. B., E. Adam, D. Heiberg, and J. L. Melnick.** 1981. Response to hepatitis B vaccine in a young adult population, p. 451–466. *In* W. Szmuness, H. J. Alter, and J. E. Maynard (ed.), Viral hepatitis, 1981 International Symposium. The Franklin Institute Press, Philadelphia.
27. **Hollinger, F. B., H. J. Alter, P. V. Holland, and R. D. Aach.** 1981. Non-A, non-B posttransfusion hepatitis in the United States, p. 49–70. *In* R. J. Gerety (ed.), Non-A, non-B hepatitis. Academic Press, Inc., New York.
28. **Hollinger, F. B., D. W. Bradley, J. E. Maynard, G. R. Dreesman, and J. L. Melnick.** 1975. Detection of hepatitis A viral antigen by radioimmunoassay. J. Immunol. **115:**1464–1466.
29. **Hollinger, F. B., G. L. Gitnick, R. D. Aach, W. Szmuness, J. W. Mosley, C. E. Stevens, R. L. Peters, J. M. Weiner, J. B. Werch, and J. J. Lander.** 1978. Non-A, non-B hepatitis transmission in chimpanzees: a project of the transfusion-transmitted viruses study group. Intervirology **10:**60–68.
30. **Hollinger, F. B., N. C. Khan, P. E. Oefinger, D. H. Yawn, A. C. Schmulen, G. R. Dreesman, and J. L. Melnick.** 1983. Posttransfusion hepatitis type A. J. Am. Med. Assoc. **250:**2313–2317.

31. **Hollinger, F. B., M. Morrison, R. Chairez, and G. R. Dreesman.** 1975. Immunological and biophysical properties of hepatitis B antigen labeled by the chloramine-T and by the lactoperoxidase methods. J. Immunol. Methods **8:**67–84.
32. **Hollinger, F. B., V. Vorndam, and G. R. Dreesman.** 1971. Assay of Australia antigen and antibody employing double-antibody and solid-phase radioimmunoassay techniques and comparison with the passive hemagglutination methods. J. Immunol. **107:**1099–1111.
33. **Hollinger, F. B., C. Wasi, G. R. Dreesman, and J. L. Melnick.** 1973. Subtyping hepatitis B antigen using monospecific antibody-coated cells. J. Infect. Dis. **128:**753–760.
34. **Hsu, S.-M., L. Raine, and H. Fanger.** 1981. A comparative study of the peroxidase-antiperoxidase method and an avidin-biotin complex method for studying polypeptide hormones with radioimmunoassay antibodies. Am. J. Clin. Pathol. **75:**734–738.
35. **Hsu, S.-M., L. Raine, and H. Fanger.** 1981. Use of avidin-biotin-peroxidase complex (ABC) in immunoperoxidase techniques: a comparison between ABC and unlabeled antibody (PAP) procedures. J. Histochem. Cytochem. **29:**577–580.
36. **Huang, S., H. Minassian, and J. D. More.** 1976. Application of immunofluorescent staining on paraffin sections improved by trypsin digestion. Lab. Invest. **35:**383–390.
37. **Jambazian, A., and J. C. Holper.** 1972. Rheophoresis: a sensitive immuno-diffusion method for detection of hepatitis associated antigen. Proc. Soc. Exp. Biol. Med. **140:**560–564.
38. **Kachani, Z. F., and D. J. Gocke.** 1973. An agglutination-flocculation test for rapid detection of hepatitis B antigen. J. Immunol. **111:**1564–1570.
39. **Kobayashi, H., M. Tsuzuki, K. Koshimizu, H. Toyama, N. Yoshihara, T. Shikata, K. Abe, K. Mizuno, N. Otomo, and T. Oda.** 1984. Susceptibility of hepatitis B virus to disinfectants or heat. J. Clin. Microbiol. **20:**214–216.
40. **Locarnini, S. A., S. M. Garland, N. I. Lehmann, R. C. Pringle, and I. D. Gust.** 1978. Solid-phase enzyme-linked immunosorbent assay for detection of hepatitis A virus. J. Clin. Microbiol. **8:**277–282.
41. **Mathiesen, L. R., S. M. Feinstone, D. C. Wong, P. Skinhøj, and R. H. Purcell.** 1978. Enzyme-linked immunosorbent assay for detection of hepatitis A antigen in stool and antibody to hepatitis A antigen in sera: comparison with solid-phase radioimmunoassay, immune electron microscopy, and immune adherence hemagglutination assay. J. Clin. Microbiol. **7:**184–193.
42. **Maynard, J. E., D. Lorenz, D. W. Bradley, S. M. Feinstone, D. H. Krushak, L. F. Barker, and R. H. Purcell.** 1975. Review of infectivity studies in nonhuman primates with virus-like particles associated with MS-1 hepatitis. Am. J. Med. Sci. **270:**81–85.
43. **Miller, W. J., P. J. Provost, W. J. McAleer, O. L. Ittensohn, V. M. Villarejos, and M. R. Hilleman.** 1975. Specific immune adherence assay for human hepatitis A antibody, application to diagnostic and epidemiologic investigations. Proc. Soc. Exp. Biol. Med. **149:**254–261.
44. **Moritsugu, Y., J. L. Dienstag, J. Valdesuso, D. C. Wong, J. Wagner, J. A. Routenberg, and R. H. Purcell.** 1976. Purification of hepatitis A antigen from feces and detection of antigen and antibody by immune adherence hemagglutination. Infect. Immun. **13:**898–908.
45. **Nakane, P. K., and A. Kawaoi.** 1974. Peroxidase-labeled antibody. A new method of conjugation. J. Histochem. Cytochem. **22:**1084–1091.
45a. **Prince, A. M., T. Huima, B. A. A. Williams, L. Bardina, and B. Brotman.** 1984. Preliminary communication: isolation of a virus from chimpanzee liver cell cultures inoculated with sera containing the agent of non-A, non-B hepatitis. Lancet **ii:**1071–1075.
46. **Prince, A. M., W. Stephan, and B. Brotman.** 1983. β-Propiolactone/ultraviolet irradiation: a review of its effectiveness for inactivation of virus in blood derivatives. Rev. Infect. Dis. **5:**92–107.
47. **Provost, P. J., and M. R. Hilleman.** 1979. Propagation of human hepatitis A virus in cell culture *in vitro*. Proc. Soc. Exp. Biol. Med. **160:**213–221.
48. **Provost, P. J., W. J. Miller, B. S. Wolanski, O. L. Ittensohn, V. M. Villarejos, W. J. McAleer, and M. R. Hilleman.** 1978. Studies of human HAV strain CR326, p. 49–64. *In* G. N. Vyas, S. N. Cohen, and R. Schmid (ed.), Viral hepatitis. The Franklin Institute Press, Philadelphia.
49. **Ray, M. B.** 1979. Hepatitis B virus antigens in tissue. University Park Press, Baltimore.
50. **Rizzetto, M.** 1983. The delta agent. Hepatology **3:**729–737.
51. **Rizzetto, M., J. W. Shih, and J. L. Gerin.** 1980. The hepatitis B virus-associated δ antigen: isolation from liver, development of solid-phase radioimmunoassays for δ antigen and anti-δ and partial characterization of δ antigen. J. Immunol. **125:**318–324.
52. **Robinson, W. S.** 1975. DNA and DNA polymerase in the core of the Dane particle of hepatitis B. Am. J. Med. Sci. **270:**151–159.
53. **Robinson, W. S.** 1978. Hepatitis B Dane particle DNA structure and the mechanism of the endogenous DNA polymerase reaction, p. 139–145. *In* G. N. Vyas, S. N. Cohen, and R. Schmid (ed.), Viral hepatitis. The Franklin Institute Press, Philadelphia.
54. **Roggendorf, M., G. G. Frosner, F. Deinhardt, and R. Scheid.** 1980. Comparison of solid phase test systems for demonstrating antibodies against hepatitis A virus (anti-HAV) of the IgM-class. J. Med. Virol. **5:**47–62.
54a. **Seto, B., W. G. Coleman, Jr., S. Iwarson, and R. G. Gerety.** 1984. Detection of reverse transcriptase activity in association with the non-A, non-B hepatitis agent(s). Lancet **ii:**941–943.
55. **Shafritz, D. A., and S. J. Hadziyannis.** 1984. Hepatitis B virus DNA in liver and serum, viral antigens and antibodies, virus replication, and liver disease activity in patients with persistent hepatitis B virus infection, p. 80–90. *In* F. V. Chisari (ed.), Advances in hepatitis research. Masson Publishing USA, Inc., New York.
56. **Skinhøj, P., F. Mikkelsen, and F. B. Hollinger.** 1977. Hepatitis A in Greenland: importance of specific antibody testing in epidemiologic surveillance. Am. J. Epidemiol. **105:**140–147.
57. **Smedile, A., C. Lavarini, O. Crivelli, G. Raimondo, M. Fassone, and M. Rizzetto.** 1982. Radioimmunoassay detection of IgM antibodies to the HBV-associated delta (δ) antigen: clinical significance in δ infection. J. Med. Virol. **9:**131–138.
57a. **Smedile, A., M. Rizzetto, F. Bonino, J. Gerin, and B. Hoyer.** 1984. Serum delta-associated RNA (DAR) in chronic HBV carriers infected with the delta agent, p. 613–614. *In* G. N. Vyas, J. L. Dienstag, and J. H. Hoofnagle (ed.), Viral hepatitis and liver disease. Grune and Stratton, Inc., Orlando, Fla.
58. **Wallis, C., and J. L. Melnick.** 1971. Enhanced detection of Australia antigen in serum hepatitis patients by discontinuous counter-immunoelectrophoresis. Appl. Microbiol. **21:**867–869.
59. **Wide, L.** 1970. Solid phase antigen-antibody systems, p. 199–206. *In* K. E. Kirkham and W. M. Hunter (ed.), Radioimmunoassay methods. E. & S Livingstone, Edinburgh.
60. **Wolters, G., L. P. C. Kuijpers, J. Kacaki, and A. H. W. M. Shuurs.** 1977. Enzyme-linked immunosorbent assay for hepatitis B surface antigen. J. Infect. Dis. **136**(Suppl.)**:**311–317.
61. **World Health Organization Scientific Group.** 1973. Viral hepatitis. W.H.O. Tech. Rep. Ser. no. 512.

Chapter 82

Human Parvoviruses and Papovaviruses

LINDA L. MINNICH AND C. GEORGE RAY

At present, human parvoviruses and papovaviruses are rarely sought in the clinical laboratory; however, the increasing recognition of their roles in human diseases and the advancing technology used in their detection mandate at least a brief discussion of their salient features. Since they represent two disparate families, they will be considered separately in this chapter.

PARVOVIRUSES

Clinical Background

The Parvoviridae family includes two genera known to infect humans. The adenosatelloviruses, serotypes 2 and 3, have been shown to naturally infect humans but to date have not been associated with disease (5). In 1975, Cossart et al. described a serum parvovirus-like virus (SPLV) (4), also known as B19, which has subsequently been associated with aplastic crises in patients with chronic hemolytic anemias such as sickle-cell disease, pyruvate kinase deficiency, and hereditary spherocytosis (1, 11, 16). It has also been associated with acute febrile illness and erythema infectiosum (2).

Patients with chronic hemolytic anemia attempt to compensate for the continual lysis of their erythrocytes (average life span of 10 to 14 days versus 120 days in normal individuals) by increased production of reticulocytes. The bone marrow in these patients is hypercellular with respect to erythroid precursors. During aplastic crisis, the bone marrow becomes hypocellular for erythroid precursors, and this, combined with the chronic lysis of peripheral erythrocytes, results in a marked decrease in peripheral hemoglobin.

Evidence for the association of SPLV with aplastic crisis in chronic hemolytic anemia patients is threefold. First, SPLV has been demonstrated by electron microscopy (EM) in the serum of patients in aplastic crisis. Disappearance of the virus particles from serum is noted during recovery from the transient aplastic crisis. Evidence for the SPLV being a single virus serotype includes agglutination with specific antiserum and demonstration of a precipitin line by counterimmunoelectrophoresis (CIE) (4, 11). Second, the production of specific antibody of the immunoglobulin M (IgM) class to SPLV indicates acute infection by SPLV in these patients (4, 11). Third, serum containing SPLV has been shown by Mortimer et al. to inhibit erythropoietic colony formation in vitro. Inhibition disappears with clearance of virus from the serum (13).

The incubation period is estimated as approximately 17 days, with peak cycling of SPLV infection in populations occurring every 3 years. Antibody to SPLV can be demonstrated in 30 to 50% of healthy adults (1). Although transmission via blood products has been described (12), the route of transmission for natural infection is still unknown.

Description of the Agent

The SPLVs are naked icosahedral viruses 18 to 28 nm in diameter. They contain single-stranded DNA and replicate in the nucleus of actively dividing cells. Parvoviruses are stable at 56°C for >1 h and survive for years at room temperature. They are also stable at pH 3 to 9 and resistant to ether and chloroform.

Collection and Storage of Specimens

Standard collection and serum separation methods are adequate for detection of SPLV particles by EM and detection of SPLV antigen by CIE.

Direct Examination

SPLVs are detected in serum by EM to visualize virus particles or by CIE with SPLV-specific antiserum. SPLV-containing serum has been shown to inhibit erythroid colony formation in vitro. The effect is neutralized with specific antiserum.

Isolation of Virus

No isolation system is presently available for SPLV. SPLV has not been shown to agglutinate rat, mouse, hamster, fowl, rhesus monkey, or human erythrocytes.

Identification of Parvovirus

Identification of SPLV particles is performed by immune EM (IEM) or detection of antigen by CIE. Antisera are prepared by the investigator against virus purified from infected patients and previously identified as SPLV.

Serological Diagnosis

Antibody to SPLV is detected by capture radioimmunoassay, IEM, and CIE. IgM-specific anti-SPLV is detected by capture radioimmunoassay. These procedures are currently available on a research-investigational basis only.

Evaluation and Interpretation of Results

The presence of SPLV in serum, determined by EM, or the presence of IgM-specific anti-SPLV, determined by capture radioimmunoassay, indicates active infection with SPLV. Aplastic crisis associated with infection has been reported only in patients with chronic hemolytic anemia. Incidental findings of virus or IgM-specific anti-SPLV in normal individuals is an indication of acute, but usually asymptomatic, infection. Between 30 and 50% of normal individuals have antibody to SPLV (1).

PAPOVAVIRUSES

Clinical Background

Papillomaviruses

The human papillomaviruses (HPVs) are the etiological agents for a variety of benign cutaneous and mucosal lesions, including verruca plantaris (plantar warts), verruca plana (juvenile or flatwarts), verruca vulgaris (common warts), condylomata (anal and genital warts), epidermodysplasia verruciformis, and oral and laryngeal papillomas (5). Although the various papillomatous lesions may be distinguished histologically, they have common features, including a proliferation of the epidermis with hypertrophy or hyperkeratinization of the superficial layers of the cutaneous lesions. Basophilic inclusions may be seen in the upper layers of the epidermis. Virus particles have been seen in warts and condylomata by EM, and antisera to the genus-specific capsid proteins have been used to identify HPV as the etiological agent for juvenile laryngeal papilloma and cervical flat warts (9).

Although models for non-HPV-induced malignant tumors have been well studied, the carcinomas associated with HPV, especially HPV-5, also appear to involve environmental factors such as exposure to UV light or other irradiation. The specific role of HPV in human malignancies remains to be established. Future investigations will be facilitated by the use of antisera to genus-specific cross-reactive antigens and HPV DNA probes to locate the viral DNA within the cells of cutaneous and mucosal neoplasms (9, 10).

Polyomaviruses

Three polyomaviruses, JC, BK, and PML-SV40, have been associated with disease in humans. Both PML-SV40 and JC viruses have been isolated from the brains of patients with progressive multifocal leukoencephalopathy (14, 15, 18). JC virus and BK virus have been isolated from renal transplant recipients (6, 8). Both JC and BK viruses, as well as another polyomavirus designated AS, were also isolated from the urine of 12 of 1,235 pregnant women studied prospectively (3). No evidence of transmission of BK virus to the fetus was found. Self-limiting acute hemorrhagic cystitis of childhood has also been associated with the presence of papovavirus particles in the urine (7).

Progressive multifocal leukoencephalopathy (PML) is a demyelinating disease of the brain usually associated with an underlying disease of the reticuloendothelial system (e.g., systemic lupus erythematosus, leukemia, lymphoma), immunodeficiency, or immunosuppression after renal transplantation. It is characterized clinically by multifocal neurological deficits, with death usually following within 3 to 6 months. Survival up to 2 years has been documented. From 10^7 to 10^9 virus particles per g of brain are present in cells surrounding the demyelinated lesions. These cells have also been shown to contain large amounts of JC virus antigen (by immunofluorescence) and numerous viral particles (by EM).

The PML-SV40 virus, which is a variant of simian virus 40, has been isolated from two PML patients (18). It is antigenically distinct from polyomavirus JC. Thus far, BK virus has not been associated with PML. The JC virus isolates from various cases are antigenically distinct from BK virus but indistinguishable from the prototype JC virus and each other (15).

Disease associated with polyomavirus BK is seen primarily in the renal tract. Viruria with BK virus was documented in a 6-year-old boy with hypergammaglobulinemia M immunodeficiency who subsequently died of tubular-interstitial nephritis. Marked viruria and extensive distribution of viral antigen implicated BK virus as the etiological agent of the severe renal injury (17). Urinary excretion of BK and JC viruses has been documented in renal transplant recipients in association with complications such as diabetes mellitus, arterial occlusive disease, and ureteral stricture with loss of renal function. Antibody rises to BK virus were associated with rising serum creatinine levels and need for transplant biopsy. Infection with JC virus appeared to be primary, whereas BK viruria was associated with reactivation (8).

As many as 40 to 70% of healthy adults are positive for antibody to polyomaviruses, suggesting that these viruses are indeed ubiquitous, but do not usually cause disease unless the patient is immunocompromised.

Description of the Agents

Papillomaviruses

These members of the Papovaviridae family are naked icosahedral viruses, 43 to 55 nm in diameter, which contain double-stranded, circular DNA. They replicate in the nucleus of the host cell. Papillomaviruses have a restricted host range, cause persistent infections, and are potentially oncogenic. To date, 26 different HPVs have been identified by genotyping. Each genotype has <50% DNA homology with previously described HPVs when analyzed by DNA hybridization under strictly controlled conditions. Investigations of bovine papillomavirus 1 indicate that the DNA exists as a stable, multicopy plasmid in transformed cells, and integration into host cell chromosomes is not necessary for transformation. Cell culture systems for either growth of or transformation by HPV have yet to be developed (18).

Polyomaviruses

Polyomaviruses are also members of the Papovaviridae family, and their general characteristics are described under papillomaviruses.

At least three polyomaviruses (JC, BK, PML-SV40) are associated with disease in humans. Additional ones will undoubtedly be characterized in the future. In addition to the family characteristics, they have been shown to produce tumors in newborn hamsters, with JC virus being the more highly oncogenic. No hemagglutinin has been identified for PML-SV40, but it has been grown in Vero and primary simian cell lines. BK virus was originally isolated in Vero cell cultures and the human fibroblast line WI-38; primary human fetal glial cell cultures have been subsequently used for primary isolation. The BK virus will agglutinate human O and guinea pig erythrocytes at 4°C. Thus far, JC virus has only been grown in human fetal glial cell cultures and will agglutinate human O, guinea pig, and chicken erythrocytes at 4°C (5).

Collection and Storage of Specimens

Papillomaviruses

Specimens for histological examination are collected and processed by standard methods. Specimens for HPV antigen detection or DNA hybridization should be processed according to the investigator's specifications.

Polyomaviruses

Specimens for the isolation of polyomaviruses should be collected and stored at 4°C for <24 h or <−70°C if longer storage is required. Specimens for histological, cytological, or EM examination should be collected and processed by standard methods.

Direct Examination

Papillomaviruses

HPV may be visualized by EM in some lesions, e.g., common warts. In other lesions, such as flat cervical warts, genus-specific viral antigens or genotype-specific DNA localization is required. These procedures are not routinely available in clinical laboratories.

Polyomaviruses

Histopathology of brain tissue from PML patients, processed by standard methods, is considered diagnostic for the disease. Virus visualized by EM in the lesions may be identified as papovavirus.

Polyomaviruses may be detected in the urine of patients. Active infection may also be documented with cytological examination of urine sediment. Stained with hematoxylin and eosin, the enlarged transitional epithelial cells contain a single, large, homogeneous, basophilic inclusion which fills the swollen nucleus (14).

Antigens of the JC and BK viruses may be identified in tissue (brain and renal) by immunofluorescence and immunoenzymatic methods. Antisera for this have been prepared by individual investigators, but are not commercially available.

Isolation of the Viruses

Papillomaviruses

No in vitro systems for isolation of HPV are currently available.

Polyomaviruses

The polyomaviruses BK and JC may be initially isolated in human fetal glial cell cultures, producing vacuolization of the cells within 10 to 12 days. Further biological identification can be made by subculture. The JC virus will replicate only in brain-derived cell cultures, whereas BK virus will replicate on subculture to human embryonic kidney and human diploid fibroblasts. The PML-SV40 virus grows preferentially in simian-derived cell lines. Further identification can be made by hemagglutination. The PML-SV40 virus has no known hemagglutinin. Both JC and BK viruses agglutinate human O and guinea pig erythrocytes at 4°C, and BK virus will agglutinate chicken erythrocytes at 4°C.

Identification of the Viruses

Papillomaviruses

Definitive identification of HPV is accomplished by genotyping, using DNA hybridization.

Polyomaviruses

Further identification of polyomaviruses BK, JC, and PML-SV40 is made by IEM, hemagglutination-inhibition, or neutralization testing with specific antisera.

Serological Diagnosis

Papillomaviruses

Routine serological procedures are not available. Antibodies to HPV antigens in common warts and genital warts have been detected by immunofluorescence and immunoenzymatic methods on an investigational basis only.

Polyomaviruses

Because antibodies to JC, BK, and SV40 are so prevalent in the normal population, serology is of limited value. Changes in antibody titers to polyomaviruses in immunocompromised patients may reflect activation of latent virus infection. As always, serological tests must be correlated with the patient's clinical presentation. Hemagglutination inhibition is the test of choice. Indirect immunofluorescence, IEM, complement fixation, and neutralization have also been used; however, these serological tests are not routinely available.

Evaluation and Interpretation of Results

Papillomaviruses

Diagnosis of HPV lesions is done primarily by clinical appearance and histopathology. Genotyping of virus is an investigational procedure and not available on a routine clinical basis.

Polyomaviruses

The diagnosis of PML is usually made by clinical presentation and confirmed by histological examination of brain tissue. Since no therapy is available and the prognosis is not affected by the etiological agent (JC versus PML-SV40), further definition of the etiology remains at the investigational level. Etiological studies serve to enhance our understanding of the natural history of the disease and substantiate JC and PML-SV40 as the etiological agents. Serological studies are of little clinical value in the diagnosis of PML, as these patients are immunocompromised and therefore subject to activation of latent virus.

Although no therapy is available for BK virus infection, documentation of polyomavirus excretion may aid the clinician in management of post-renal transplant patients. Increases in antibody titers to BK virus have been associated with increased post-renal transplant complications (8).

LITERATURE CITED

1. **Anderson, M. J.** 1982. The emerging story of a human parvovirus like agent. J. Hyg. **89:**1–8.
2. **Clewley, J. P.** 1984. Biochemical characterization of a human parvovirus. J. Gen. Virol. **65:**241–245.
3. **Coleman, D. V., M. R. Wolfendale, R. A. Daniel, N. K. Dhanjal, S. D. Gardner, P. E. Gibson, and A. M. Field.** 1980. A prospective study of human polyomavirus infec-

tion in pregnancy. J. Infect. Dis. **142:**1–8.

4. **Cossart, Y. E., A. M. Field, B. Cant, and D. Widdows.** 1975. Parvovirus-like particles in human serum. Lancet **i:**72–73.
5. **Gardner, S.** 1977. Implication of papovaviruses in human disease, p. 41–84. *In* E. Kurstak and C. Kurstak (ed.), Comparative diagnosis of viral diseases, vol. 1. Academic Press, Inc., New York.
6. **Gardner, S. D., A. M. Field, D. V. Coleman, and B. Hulme.** 1971. New human papovavirus (B.K.) isolated from urine after renal transplantation. Lancet **i:**1253–1257.
7. **Hashida, Y., P. C. Gaffney, and E. J. Yunis.** 1976. Acute hemorrhagic cystitis of childhood and papovavirus-like particles. J. Pediatr. **89:**85–87.
8. **Hogan, T. F., E. C. Borden, J. A. McBain, B. L. Padgett, and D. L. Walker.** 1980. Human polyomavirus infections with JC virus and BK virus in renal transplant patients. Ann. Intern. Med. **92:**373–378.
9. **Howley, P. M.** 1982. The human papillomavirus. Arch. Pathol. Lab. Med. **106:**429–432.
10. **Howley, P. M.** 1983. The molecular biology of papillomavirus transformation. Am. J. Pathol. **113:**414–420.
11. **Kelleher, J. E., M. L. C. Luban, P. P. Mortimer, and F. Kamimura.** 1983. Human serum "parvovirus." A specific cause of aplastic crisis in children with hereditary spherocytosis. J. Pediatr. **102:**720–721.
12. **Mortimer, P. P.** 1983. Transmission of serum parvovirus-like virus by clotting-factor concentrates. Lancet **ii:**482–484.
13. **Mortimer, P. P., R. K. Humphries, J. C. Moore, R. H. Purcell, and N. S. Young.** 1983. A human parvovirus-like virus inhibits haematopoietic colony formation *in vitro.* Nature (London) **302:**426–429.
14. **Padgett, B. L., D. L. Walker, G. M. ZuRhein, and R. J. Eckroade.** 1971. Cultivation of papovavirus-like virus from human brain with progressive multifocal leucoencephalopathy. Lancet **i:**1257–1260.
15. **Padgett, B. L., D. L. Walker, G. M. ZuRhein, A. E. Hodach, and S. M. Chou.** 1976. JC papovavirus in progressive multifocal encephalopathy. J. Infect. Dis. **133:**686–690.
16. **Rao, K. R. P., A. P. Patel, M. J. Anderson, J. Hodgson, S. E. Jones, and J. R. Pattison.** 1983. Infection with parvovirus-like virus and aplastic crisis in chronic hemolytic anemia. Ann. Intern. Med. **98:**930–932.
17. **Rosen, S., W. Harmon, A. Krensky, P. J. Edelson, B. L. Padgett, B. W. Grinnell, M. J. Rubino, and D. L. Walker.** 1983. Tubulo-interstitial nephritis associated with polyomavirus (BK type) infection. N. Engl. J. Med. **308:**1192–1196.
18. **Weiner, L. P., R. M. Herndon, O. Narayan, R. T. Johnson, K. Shah, L. J. Rubinstein, T. J. Preziosi, and F. K. Conley.** 1972. Isolation of virus related to SV40 from patients with progressive multifocal leukoencephalopathy. N. Engl. J. Med. **286:**385–389.

Chapter 83

Human T-Cell Leukemia Viruses

M. ESSEX

CLINICAL BACKGROUND

Adult T-cell leukemia

The existence of a class of retroviruses that are etiologically linked to the occurrence of a particular family of T-cell leukemias was first reported by Gallo and his colleagues in 1980 (13, 17). The first virus was isolated from a case diagnosed as a malignant variant of mycosis fungoides. The diagnosis of mycosis fungoides, Sezary's syndrome, and related cutaneous lymphoid neoplasms has been somewhat variable and vague, especially before the availability of monoclonal antibodies specifically directed to T-cell subsets. The T-cell leukemias that are related to exposure to the human T-cell leukemia virus (HTLV) are regularly of the OKT 4 or Leu 3a subset of "helper" cells. In southwestern Japan, primarily on the major islands of Kyushu and Shikoku, a leukemia of the same phenotype has been substantially more frequent. This particular leukemia, which is highly lethal and insensitive to existing modalities of therapy, was initially classified as a distinct entity by Takatsuki and his colleagues in about 1976 (18). He called this tumor adult T-cell leukemia/lymphoma (ATL). This tumor occurs at substantially elevated rates in those regions where HTLV is present. The strains or isolates of HTLV that are regularly linked to this disease have been called type I (HTLV-I).

At about the same time a disease of the same general type was recognized in blacks in the Caribbean Basin. Designated a lymphosarcoma-cell leukemia, it is also a tumor of T-helper cells in which the same general features as the ATL-type disease are present, suggesting that the two are in fact caused by the same agent, presumably HTLV-I. In all these diseases, several common features are observed. These include, in addition to the presence of the T-helper cell as the malignant clone that develops, cutaneous infiltration, hypercalcemia, and a rapidly progressing clinical course. The tumor cell itself appears to be a very mature T-cell, terminal transferase negative, and with somewhat distinctive polylobated nuclei. While phenotypically characterized as a helper cell, the tumor cell itself actually functions in a suppressor capacity when analyzed in in vitro tests.

A second type of HTLV (designated type II) was isolated from a T-cell form of hairy-cell leukemia (10). This appeared to be closely related to but distinct from the prototype HTLV-I isolates that had been described in the United States, southern Japan, and the Caribbean, which all seemed to be either very closely related to each other or essentially the same agent. Although numerous isolations have been made of the type I strain, only a very small number of HTLV-II isolates have thus far been identified. Thus, it is too early to say whether or not the type II isolate will be regularly associated with any particular disease syndrome as are HTLV-I and HTLV-III (see below).

It is generally accepted that the most practical and reliable method for assessing virus infection with HTLV-I or -II is by the presence of antibodies, especially antibodies to the virion *env* glycoproteins (11). While essentially all cases of ATL clearly have circulating antibodies, such antibodies are also found in 3 to 30% of selected healthy adults from regions of the world where these viruses are endemic (1, 8). The presence of antibodies apparently reflects a persistent lifetime infection with the virus, as would be anticipated for agents such as the Epstein-Barr virus. The estimated lifetime risk for development of ATL for carriers in endemic regions is about 1 in 100. However, infection with HTLV-I also appears to be clinically associated with the development of certain infectious diseases at higher-than-expected rates. The diseases observed would not be classified as AIDS (acquired immunodeficiency syndrome), but represent the same types of infections (bacterial pneumonias, hepatitis, septicemias) that would occur in the absence of persistent HTLV infections (4). The degree to which HTLV-I and -II cause clinical problems by predisposing the host to other infectious diseases has not yet been fully analyzed.

AIDS

AIDS was first recognized in Los Angeles in 1981 (6), although most epidemiologists believe that a restricted number of cases had appeared for a 1- to 3-year period before the first recognition and definition of this disease as a distinct clinical entity. The disease has clearly occurred at an increasing frequency in selected categories of people at risk. As of this writing, about 5,000 cases have been reported to the Centers for Disease Control within the last 3-year period. Although the rate of increase may be tapering off, it is clear that this disease will continue to be a major threat for individuals within the high-risk category for the foreseeable future. The disease has been observed at dramatically increasing rates in cities in Western Europe as well as in selected areas of Africa. Smaller numbers of cases have occurred in other geographical areas. The disease is categorized by the cardinal sign of irreversible damage to the immune system. This is illustrated most vividly by the observed depletion of T-helper cells which, in advanced disease, are essentially absent. As a result, immune responses to otherwise controllable infectious agents are almost totally obliterated, and the disease is ultimately fatal. Those agents that are most responsible for opportunistic infections that result in the clinical manifestations are *Pneumocystis carinii*, *Mycobacterium avium*, *Cryptosporidium*, *Toxoplasma*, cytomegalovirus, and others that produce either pneumonias, encephalitis, or bowel infections that eventually become untreatable. AIDS patients also develop selected cancers, with Kaposi's sarcoma and B-cell lymphomas being the most common, but these tumors are generally not as lethal by themselves as are the infectious agents. Both the tumors and the infectious agents appear to grow as a result of the inadequate T-helper cell function which would normally control the disease process.

Several population groups are at particularly high risk for development of AIDS. These include homosexual men with multiple partners, intravenous drug abusers, hemophiliacs, Haitians, Africans from selected areas in the west-central regions, infants born in environments of extensive exposure to those in other risk groups, and, to a lesser degree, recipients of multiple blood transfusions. The incubation or induction period for development of AIDS is apparently both prolonged and variable. Whereas some cases appear about a year after exposure to a carrier individual (for example, by blood transfusion), most cases do not appear until about 18 to 24 months after such exposure, and the induction period can apparently be at least 4 years in duration. It now appears quite likely that many healthy homosexuals are carriers of the virus for prolonged periods during which they can transmit the agent, but it is not clear how many of the individuals that appear infected within the high-risk groups will ultimately develop the disease as opposed to remaining healthy in a subclinical status. It appears that up to half of the male homosexual population in selected major cities may have been exposed to the virus. As yet it is too early to tell what proportion of these will ultimately develop the disease. In the case of male homosexuals, generalized lymphadenopathy has also been observed in association with AIDS, and this lymphadenopathy or "AIDS-related complex" has actually occurred at higher rates than AIDS. Yet by most estimates, only 10 to 20% of the male homosexuals with lymphadenopathy progress to the development of AIDS within a 2-year period. Many appear to undergo regression or resolution of the lymphadenopathic state. Although some risk groups have opportunistic infections that appear at least proportionally different from those of other high-risk groups, it seems very likely that the cause of the immune suppression is the same in all instances.

The agent that apparently causes AIDS is HTLV-III, a distinct but distantly related member of the HTLV family (5, 14–16). The observation that an HTLV-type agent is apparently involved in AIDS came initially because antibodies that were cross-reactive with other types of HTLV were detected at higher than expected rates in patients with AIDS and AIDS-related complex (3). This was apparently due to distantly shared antigens of the virus envelope gene products.

With the possible exception of asymptomatic hemophiliacs, it is believed that all antibody-positive individuals from the high-risk groups, with or without AIDS or AIDS-related complex, are persistent carriers of virus. Thus, as in the case of HTLV-I and -II, the presence of antibodies generally reflects a latent infection. It is as yet unclear whether or not this is true for most asymptomatic hemophiliacs. While most have antibodies to HTLV-III, and a minority have antibodies to HTLV-I, it is possible that some of the types of exposures in this population occurred as a result of inactivated virus antigens that were administered with clotting-factor preparations. Further studies on virus isolations will be necessary to address this issue.

Both HTLV-I and HTLV-III are believed to be transmitted by blood transfusions (9). Both have also been postulated to be transmitted by sexual intercourse and by exposure of newborn infants to body fluids of infected mothers. These exposures cannot, however, account for the vast majority of persistent infections observed in areas such as southern Japan. How the largest population of healthy carriers become infected in such regions remains unknown. Furthermore, recent evidence from central Africa suggests that AIDS occurs there at similar rates in both males and females. This suggests that additional routes of transmission for HTLV-III would also have to be considered for such regions. While transmission by sexual intercourse would not seem to be unusual, the propensity of the agent for homosexual males does seem unusual. However, even for the HTLV-I agent in Japan it has been reported that wives of husbands with ATL are themselves more often infected with HTLV-I than are husbands of wives with ATL (7). Other possible explanations for transmission of these agents include the presence of an arthropod vector that could be present in geographically appropriate areas. No concrete evidence to support this hypothesis has yet been presented.

DESCRIPTION OF THE AGENT

HTLV-I and -II are typical retroviruses in morphologic appearance. They are about 100 nm in diameter, have a buoyant density of 1.16 to 1.18 g/cm^3, and would be indistinguishable from most other C-type retroviruses that were grown in lymphoid cells in culture. The HTLV-III agents have a similar size but a somewhat different morphologic appearance. The electron-dense core within the particle itself is more likely to appear distorted (i.e., in a cylindrical form) or shrunken. All HTLVs appear to bud from the cytoplasmic membrane of infected cells. The HTLVs have three major structural genes, as do retroviruses of lower animals. The entire genome is about 9 kilobases long and flanked by two long terminal repeat sequences at each end. Within the 9 kilobases are three genes that encode for proteins common to all retroviruses, the *gag*, *pol*, and *env* genes. The *gag* gene encodes virus core antigens, and the *env* gene encodes those glycoproteins that appear at the surface of virions as well as at the cytoplasmic membranes of infected cells. HTLV-I and -II are only weakly cytopathic and usually exist in infected cells in a steady-state manner without causing cell death. Formation of syncytia can occur in sensitive cells. HTLV-III, on the other hand, is more cytopathic and will kill most of the cells that it infects. All HTLV types are highly tropic for the particular subpopulation of T lymphocytes designated OKT 4 or Leu 3a. HTLVs are also different from most other type C retroviruses in that their polymerase uses Mg^{2+} rather than Mn^{2+} as the divalent cation. The HTLVs bud from the cytoplasmic membrane in the shape of an inverted C, as do most other retroviruses.

In the case of HTLV-I and -II, free circulating virus is rarely found in vivo, and most cell cultures are poor producers. Conversely, with HTLV-III, although the in vivo status has not yet been fully evaluated, cultures of T lymphocytes that are able to survive at all in vitro, such as the H-9/HTLV-III line developed in the Gallo laboratory, make large amounts of virus.

The HTLV-I and -II viruses are closely related by genomic hybridization as well as by antigenic cross-reactivity of the major *gag* and *env* antigens. The most immunogenic *gag* antigen, p24, is highly cross-reactive, as is the envelope precursor polyprotein, gp61 (11). By all these criteria, HTLV-III and lymphadenopathy-associated virus agents cross-react with the other HTLV types, but the degree of cross-reactivity between type III and either type I or type II is significantly less than that seen between types I and II when these two are compared (17). The HTLV types are distantly related to bovine leukemia virus. Again, this is reflected in antigenic cross-reactivities, but only under conditions designed to detect distant relationships.

The HTLVs also have another region of the genome adjacent to the 3′ end that is believed to be important in relation to their oncogenic potential. One antigen, designated p42, has been identified as encoded from this region. This marker may eventually prove to be important for evaluating the state of the virus-infected cell in relation to immortalization and transformation.

COLLECTION AND STORAGE OF SPECIMENS

For virus isolation, the major material to use is freshly frozen (or freshly maintained) T lymphocytes. In the case of ATL, this can be either tumor cells, circulating lymphocytes, or bone marrow. In any case, the cells should be maintained in a viable state, if at all possible, for cocultivation with sensitive cord blood lymphocytes. Sera or other body fluids are not known to contain significant levels of free infectious virus, at least in the case of HTLV-I and -II. In the case of HTLV-III, although not fully established, it appears likely that higher levels of free infectious virus are available in body fluids, including plasma and perhaps secretory fluids. However, T lymphocytes should also be collected whenever possible. This will be difficult for patients with AIDS because a hallmark of the disease is the depletion of T-helper cells. Greater success may be possible with asymptomatic carriers or individuals with AIDS-related complex that have not developed AIDS. Specimens should be stored frozen until examined. Viable cells should be stored in liquid nitrogen, using conditions for the long-term maintenance of lymphoblastoid lines.

DIRECT EXAMINATION

The direct examination of freshly biopsied tissues from either healthy carriers or patients with ATL has not been found to be very efficient for detection of HTLV-I and -II. It appears that fresh tumor tissues do not express mature virus particles or even virus structural antigens that can be a target for immunodiagnosis. When tissues are available, it is clearly more efficient to try cultivation of the T cells with T-cell growth factor before examination by either immunodiagnostic or electron microscopic techniques. Particles can occasionally be seen by electron microscopy, perhaps especially in cutaneous sites, but this approach is inefficient. Immunodiagnosis, using antiserum directed to the p24 or gp61 antigens, may be more efficient after only a brief period in culture (e.g., 1 to 2 days), but this approach is still much less efficient than examination for antibodies in serum.

Perhaps the most logical procedure for direct examination when using tumor tissue is nucleic acid hybridization by Southern blotting. In essentially all tumors, one can find either one appropriate band or a limited number of bands that hybridize specifically with an HTLV provirus probe.

Little or no information is available about the approaches one could use for the direct examination of AIDS tissues to detect HTLV-III. Direct examination by electron microscopy and the serologic examination of fresh tissues for antigen have not been recorded, but it appears unlikely that these techniques would be productive because of the substantial loss of target cells that occurs by the time the disease has been identified. Additionally, studies using nucleic acid hybridization with AIDS tissues are less likely to be successful than those using fresh ATL tissues because the T-helper cells that harbor the virus have largely been destroyed.

Thus far, no tests have been established for the detection of soluble circulating HTLV antigens in either AIDS or ATL patients. If such tests are developed, they may be useful. It seems possible that many patients that lack free circulating virus would yield sufficient amounts of viral proteins to allow some degree of diagnosis or prognosis. It is unlikely, however, that such an approach would be as efficient as examination of serum or plasma for HTLV-specific antibodies (see below).

ISOLATION OF VIRUS

Attempts to isolate HTLV from cases of either ATL or AIDS can be highly successful, but require careful and prolonged attention to cell cultivation. In the case of HTLV-I and ATL, the most appropriate procedure is to cocultivate patient tumor cells, blood or bone marrow lymphocytes, or all of these with susceptible human cord blood T-cells. This procedure should be conducted in the manner required to cultivate new T-cell lines. Of greatest importance is the presence of human T-cell growth factor or IL-2 (12). After several weeks of cultivation, the examination of survivor T cells for the presence of HTLV is done using immunodiagnostic procedures, nucleic acid hybridization, electron microscopic examination, or analysis for reverse transcriptase activity in the medium. The presence of HTLV-I type antigens or sequences in T-cell tissues that grow, whether they are donor or recipient cells, will be an indication of the presence of virus in the patient.

HTLV-III is more cytolytic, and apparently the agent is present in larger amounts in the donor as free virus. For this reason the direct addition of blood lymphocytes or lymph node cells to fresh cord blood or bone marrow should result in a cytolytic effect within a few weeks. At this time, it will still be necessary to examine the culture for particles by electron microscopy, reverse transcriptase activity, or viral antigen analysis. An alternative approach is to cocultivate patient cells with the H-9 cell line, a recently established cell line of the T-helper type that is both resistant to lysis by HTLV-III and sensitive to infection (15). In this instance, the AIDS agent will infect the target culture without lysis, and the medium or cells of the culture can subsequently be examined by the procedures mentioned.

IDENTIFICATION OF VIRUS

Certification that the agent detected is HTLV can be conducted with reference antiserum to the p24 proteins of the virus. This can be done readily by using hyperimmune rabbit or goat antiserum to the p24 antigen and procedures such as Western blotting, radioimmunoprecipitation with sodium dodecyl sulfate-polyacrylamide gel electrophoresis, or competition radioimmunoassays. These procedures can utilize either infected cells or ultracentrifuged virus appropriately labeled with radioactive isotopes as necessary (11, 16). In the same manner, nucleic acid hybridization can be used with reference cloned probes. In any case, positive reference infected cell lines, which are available from the American Type Culture Collection, should be used as controls. Differentiation between HTLV-I and -II can only be done either by carefully controlled nucleic acid hybridization experiments or by the use of human type-specific antiserum to neutralize pseudotypes of HTLV-I or -II made with vesicular stomatitis virus in plaque assays (2). By most criteria, there is significant cross-reactivity between HTLV-I and -II, and a separation of types can only be conducted in a small number of research laboratories at this time.

The identification of HTLV-III is possible by immunodiagnosis of target H-9 cells infected with a newly isolated agent. Antiserum of human origin that is type specific for the envelope glycoproteins of type III are available, and the profile of glycoproteins is distinctively different for type III from that for type II or I. Rabbit or goat antisera to HTLV-III core proteins (p24) would also allow easy distinction between HTLV-III and HTLV-I or -II, as will nucleic acid hybridization using probes specific for type III.

SEROLOGIC DIAGNOSIS

Clearly the method of choice for rapid detection of infection in individuals is examination of serum or plasma for the presence of antibodies. The relative prevalence of antibodies in infected individuals in the case of either ATL patients (with HTLV-I) or AIDS patients (HTLV-III) is at least 90 to 95%. Some procedures, however, are significantly more efficient than others. Procedures based on the detection of antibodies directed to virus surface glycoproteins are considerably more efficient because these proteins appear much more immunogenic in infected people (especially disease cases) than the virus core proteins. Antibodies directed to virus envelope glycoproteins can be readily detected in plasma by using cell-surface immunofluorescence, fixed-cell (cytoplasmic) immunofluorescence, immunoprecipitation combined with sodium dodecyl sulfate-polyacrylamide gel electrophoresis, Western blotting, or enzyme-linked immunoassay (ELISA). Thus far, however, HTLV glycoproteins (either type I or III) have not been available in purified form except under very expensive conditions for laboratory use. Thus, most preparations to be used as purified virion proteins for commercial kits are likely to be deficient in envelope glycoprotein and thus insensitive for the detection of antibodies in many people who only have significant titers of HTLV-specific antibodies to the envelope proteins. Similarly, the use of p24-directed radioimmunoassays has been relatively inefficient (as compared to other procedures), especially when groups such as AIDS patients, who are likely to be harboring only low titers of antibodies to the virus or protein, are examined. Hopefully, kits using ELISA or similar rapid tests that employ envelope glycoprotein made by recombinant DNA techniques will be available in the near future. Such proteins have been produced for experimental use. Some procedures such as ELISA have been employed using crudely purified virus. Nonspecific reactions are quite possible in these tests, due to contamination of the antigen with cellular debris.

The most appropriate procedure currently available to detect different types of HTLV (including I and III) based on cross-reactivities is indirect membrane immunofluorescence (3, 9). This procedure, however, will only detect a limited number of infected individuals when a target of the opposite type is used.

In summary, the detection of antibodies in diseased individuals and healthy carriers is clearly the method of choice for diagnosis of infection. This is true for individuals infected with either HTLV-I or HTLV-III, with the possible exception of some asymptomatic hemophiliacs, who may have been exposed to inactivated antigen. The presence of antibodies is taken as a sign of latent infection, even in healthy individuals. Some kits for rapid detection of antibodies are currently available. They are not ideal, but it seems very likely that efficient techniques will be widely available in the very near future.

EVALUATION AND INTERPRETATION OF RESULTS

In the case of either HTLV-I or HTLV-III, when antibodies are found this is probably an indication of a permanent rather than transitory infection. As indicated above, exceptions such as hemophiliacs who may only have been exposed to blood products containing largely inactivated antigen are at least a theoretical exception to this general rule. However, many healthy carriers infected with HTLV-I have no greatly increased risk for development of ATL (perhaps 1 in 100 lifetime risk). Although infection with HTLV-I has also been indirectly associated with development of common infectious diseases, there is no evidence that the common infections are any less sensitive to usual therapeutic procedures and resolution. The detection of evidence for HTLV-I infection is of little value to the therapy of individuals with ATL; this disease unfortunately has a poor prognosis.

The presence of antibodies to HTLV-III, although probably indicative of a persistent infection (again with the possible exception of some hemophiliacs), only indicates that the patient is infected with the virus. Although most AIDS patients and a significant proportion of healthy male homosexuals with multiple partners will have antibodies, it is certainly not yet clear that all healthy individuals with antibodies are at risk for development of AIDS. It appears likely that, although the incubation period for the disease is very long, many antibody-positive healthy homosexuals will not develop AIDS. This, of course, suggests that some other set of secondary circumstances comes into play to precipitate the disease in infected individuals. Once AIDS has been diagnosed, however, it is likely to be terminal.

The major value for serologic tests to determine antibody-positive carriers of the HTLV-III or HTLV-I viruses is probably for screening blood to be used for transfusion or the preparation of blood products. Both agents can be transmitted efficiently by transfusion, especially when cellular materials are present in the preparation, and increasing proportions of asymptomatic healthy carriers capable of transmitting the agent are present throughout the world.

LITERATURE CITED

1. **Blattner, W. A., V. S. Kalyanaraman, M. Robert-Guroff, T. A. Lister, D. A. G. Galton, P. S. Sarin, M. H. Crawford, D. Catovsky, M. Greaves, and R. C. Gallo.** 1982. The human type C retrovirus, HTLV, in blacks from the Caribbean region, and relationship to adult T-cell leukemia/lymphoma. Int. J. Cancer **30:**257–264.
2. **Clapham, P., K. Nagy, and R. A. Weiss.** 1984. Pseudotypes of human T-cell leukemia virus types 1 and 2: neutralization by patients' sera. Proc. Natl. Acad. Sci. U.S.A. **81:**2886–2889.
3. **Essex, M., M. F. McLane, T. H. Lee, L. Falk, C. W. S. Howe, J. Mullins, C. Cabradilla, and D. P. Francis.** 1983. Antibodies to cell membrane antigens associated with human T-cell leukemia virus in patients with AIDS. Science **220:**859–862.
4. **Essex, M., M. F. McLane, N. Tachibana, D. P. Francis, and T. H. Lee.** 1984. Seroepidemiology of HTLV in relation to immunosuppression and the acquired immunodeficiency syndrome, p. 355–362. *In* R. C. Gallo, M. Essex, and L. Gross (ed.), Human T-cell leukemia viruses. Cold Spring Harbor Press, Cold Spring Harbor, N.Y.
5. **Gallo, R. C., S. Z. Salahuddin, M. Popovic, G. M. Shearer, M. Kaplan, B. F. Haynes, T. J. Palker, R. Redfield, J. Oleske, B. Safai, G. White, P. Foster, and P. D. Markham.** 1984. Frequent detection and isolation of cytopathic retroviruses (HTLV-III) from patients with AIDS and at risk for AIDS. Science **224:**500–503.
6. **Gottlieb, M. S., R. Schroff, H. M. Schranker, J. D. Weisman, P. T. Fan, R. C. Wolf, and A. Saxon.** 1981. *Pneumocystis carinii* pneumonia and mucosal candidiasis in previously healthy homosexual men. N. Engl. J. Med. **305:**1425–1429.
7. **Hinuma, Y.** 1982. Association of a retrovirus (ATLV) with adult T-cell leukemia: review of serologic studies. Gann **28:**211–221.
8. **Hinuma, Y., H. Komoda, T. Chosa, T. Kondo, M. Kohakura, T. Takenaka, M. Kikuchi, M. Ichimaru, K. Yunoki, M. Sato, R. Matsuo, Y. Takiuchi, H. Uchino, and M. Hanaoka.** 1982. Antibodies to adult T-cell leukemia virus associated antigen (ATLA) in sera from patients with ATL and controls in Japan: a nation-wide sero-epidemiologic study. Int. J. Cancer **29:**631–635.
9. **Jaffe, H. W., D. P. Francis, M. F. McLane, C. Cabradilla, J. W. Curran, B. W. Kilbourne, D. N. Lawrence, H. W. Haverkos, T. J. Spira, R. Y. Dodd, J. Gold, D. Armstrong, A. Ley, J. Groopman, J. Mullins, T. H. Lee, and M. Essex.** 1984. Transfusion-associated AIDS: serologic evidence of human T-cell leukemia virus infection of donors. Science **223:**1309–1312.
10. **Kalyanaraman, V. S., M. G. Sarngadharan, M. Robert-Guroff, I. Miyoshi, D. Blayney, D. Golde, and R. C. Gallo.** 1982. A new subtype of human T-cell leukemia virus (HTLV-II) associated with a T-cell variant of hairy cell leukemia. Science **218:**571–573.
11. **Lee, T. H., J. E. Coligan, T. Homma, M. F. McLane, N. Tachibana, and M. Essex.** 1984. Human T cell leukemia virus associated membrane antigens (HTLV-MA): identity of the major antigens recognized following virus infection. Proc. Natl. Acad. Sci. U.S.A. **81:**3856–3860.
12. **Morgan, D. A., F. W. Ruscetti, and R. C. Gallo.** 1976. Selective *in vitro* growth of T-lymphocytes from normal human bone marrow. Science **193:**1007–1009.
13. **Poiesz, B. J., F. W. Ruscetti, A. F. Gazdar, P. A. Bunn, J. D. Minna, and R. C. Gallo.** 1980. Detection and isolation of type C retrovirus particles from fresh and cultured lymphocytes of a patient with cutaneous T cell lymphoma. Proc. Natl. Acad. Sci. U.S.A. **77:**7415–7419.
14. **Popovic, M., M. G. Sarngadharan, E. Read, and R. C. Gallo.** 1984. Detection, isolation, and continuous production of cytopathic retroviruses (HTLV-III) from patients with AIDS and pre-AIDS. Science **224:**497–500.
15. **Sarngadharan, M. G., M. Popovic, L. Bruch, J. Schupbach, and R. C. Gallo.** 1984. Antibodies reactive with human T-lymphotropic retroviruses (HTLV-III) in the serum of patients with AIDS. Science **224:**506–508.
16. **Schupbach, J., M. Popovic, R. V. Gilden, M. A. Gonda, M. G. Sarngadharan, and R. C. Gallo.** 1984. Serological analysis of a subgroup of human T-lymphotropic retroviruses (HTLV-III) associated with AIDS. Science **224:**503–505.
17. **Shaw, G. M., S. Broder, M. Essex, and R. C. Gallo.** 1984. Human T-cell leukemia virus: its discovery and role in leukemogenesis and immunosuppression. Adv. Intern. Med. **30:**1–27.
18. **Uchiyama, T., J. Yodoi, K. Sagawa, K. Takatsuki, and H. Uchino.** 1977. Adult T-cell leukemia: clinical and hematological features of 16 cases. Blood **50:**481–491.

Chapter 84

Rickettsiae

RICHARD A. ORMSBEE

The important, naturally occurring rickettsial diseases in the United States are Rocky Mountain spotted fever (with a recorded incidence of approximately 1,000 cases per year during the period 1979–1984), Q fever, and murine typhus. Consequently, diagnosis of these diseases is emphasized in this chapter.

Rickettsialpox occurs sporadically in cities but is considered to be a mild disease. The rapidity of international air travel and the lack of specific vaccines may permit the occasional, unwitting entry into the United States of individuals with scrub typhus, particularly from the Far East, or with spotted fever group disease, particularly from the Indian subcontinent and some parts of Africa.

Rickettsial diseases of humans can be classified conveniently into four major groups, namely, typhus fevers, spotted fevers, Q fever, and trench fever. They are caused by various members of the family *Rickettsiaceae*. Rickettsiae are characterized by their ability to multiply in one or more arthropods and by their existence in natural reservoirs of one or several warm-blooded animal hosts, including humans. Effective therapy of human infections is usually provided by chloramphenicol and the tetracycline group of broad-spectrum antibiotics (questionably effective in Q fever).

The difficulties and dangers of working with live pathogenic rickettsiae are great enough to preclude such efforts in most clinical laboratories. Consequently, the reader is referred to the appropriate chapters in *Viral and Rickettsial Diseases of Man* (26) and *Diagnostic Procedures for Viral, Rickettsial and Chlamydial Infections* (28) for the details of these procedures and for clinical details also.

DESCRIPTION OF AGENTS

Rickettsiae essentially are highly fastidious bacteria. They are obligate intracellular parasites (except *Rochalimaea quintana*), pleomorphic but typically rodlike in form, and often found in pairs in smears of infected cells. They possess double- or triple-layered limiting membranes which are susceptible to lysozyme digestion, but not to tryptic digestion, and they contain muramic acid and diaminopimelic acid, in common with most bacteria except the gram-positive cocci. They possess both synthetic and energy-yielding enzyme systems, contain RNA and DNA, and multiply by binary fission. Characteristically, they vary in diameter; members of the spotted fever group are the largest, *Coxiella burnetii* is the smallest, and members of the typhus group are intermediate. They also vary in their resistance to physical and chemical agents. *C. burnetii* is more resistant than any other pathogenic rickettsia, indeed more resistant than most nonsporogenic microorganisms. Effective disinfection can be achieved with 2% Formalin USP, 1% Lysol (a mixture of saponified aryl and alkyl derivatives of phenol), 100% ethyl ether, and 5% H_2O_2. Other common antiseptics should not be relied upon.

The newest member of the genus, *Rickettsia canada*, was isolated from the tick *Haemaphysalis leporispalustris* taken from a domestic rabbit in eastern Canada (31), and later isolated from the same tick species taken from a jack rabbit, *Lepus californicus*, in northern California (39). This rickettsia is related antigenically to the typhus group and possesses a DNA with a guanine-plus-cytosine composition similar to that of members of the typhus group (47). Koch's postulates, incriminating this organism unequivocally as a cause of human disease, have not yet been fulfilled. Only scanty serological data (7) suggest that this rickettsia is a significant factor in tick-borne human disease.

Rickettsia prowazekii recently has been isolated in the United States from flying squirrels, *Glaucomys volans volans* Linn., and from their ectoparasites, including the flea *Orchopeas howardi* and the louse *Neohaematopinus sciuroptera* (8). The public health significance of these enzootic infections has not been fully assessed. However, eight cases have been identified in which the serological evidence suggests that the etiological agent was *R. prowazekii* (30).

Table 1 lists the common rickettsial diseases of humans, their etiological agents, and other information descriptive of ecological characteristics of the rickettsiae and epidemiological features of rickettsial diseases.

COLLECTION, STORAGE, AND SHIPMENT OF SPECIMENS

For serological examination

In general, the only specimens pertinent to laboratory diagnosis are tissues obtained under aseptic conditions at autopsy, or blood samples. Laboratory diagnosis in most cases depends upon demonstration of specific antibodies in serum specimens and an increase in antibody titer as the disease progresses. Three samples of blood should be taken from the patient: one as soon as possible after onset of illness, one during week 2, and one during week 4 after onset of illness.

For attempted isolation

If material suspected of containing rickettsiae is to be injected into animals, the specimens should be injected within 30 min after the suspensions are made; alternatively, the material can be placed in a glass or plastic container, quickly frozen in an alcohol-dry ice bath, and stored at −70°C until used. Whole blood should be shell-frozen in an alcohol-dry ice bath. If possible, it is preferable to take a blood sample of 10 to 12 ml, allow it to clot, centrifuge it immediately, and freeze the clot and serum separately. The clot can then be injected as a 10% suspension in skim milk or brain heart infusion broth (100 g of Difco skim milk or 37 g of Difco brain heart infusion dissolved in 1,000 ml of cold, distilled water, distributed in tubes or flasks, and sterilized in an autoclave for 15 min at 15 lb/in^2) (49). The use of other diluents may result in a great loss of infectivity of rickettsiae present. The serum can be used for subsequent serological examination. If facilities permit, a noncoagulated blood specimen may be taken with a syringe wet with USP heparin and injected immediately into susceptible animals. Blood specimens taken later than week 1 of illness should routinely be separated into cells and serum (or plasma), which may by then contain antibodies. The cells should be suspended in an equal volume of brain heart infusion or skim milk for injection into animals.

It is imperative that properly prepared frozen tissue and blood specimens be shipped to the reference laboratory by air in a properly insulated container with enough dry ice to maintain the inside temperature at −70°C for at least 24 h longer than the estimated length of time required for shipment. It is not sufficient that the sample merely be kept frozen. Once properly shell-frozen in an alcohol-dry ice bath, the specimen must be held at a constant −70°C until quickly thawed, by immersing the container in tepid water, immediately before injection into experimental animals. In addition, at the time of shipment the reference laboratory should be notified by telephone or telegram of time of departure of the shipment, airline and flight number, and adequate identification marks to permit the parcel to be individually spotted and retrieved at the receiving end. Neglect of any of these precautions may cause failure of the considerable efforts required to isolate and identify the rickettsiae.

If ticks or other arthropods collected from the bodies or clothing of patients suspected of rickettsial disease are to be examined directly for detection and identification of rickettsial infection (9), they should

TABLE 1. Rickettsial diseases of humans

Disease	Etiological agent	Geographical distribution	Natural cycle		Mode of transmission to humans	Environmental associations
			Arthropod	Mammal		
Spotted fever group						
Rocky Mountain spotted fever	*R. rickettsii*	Western Hemisphere	Ixodid ticks	Rodents, dogs, foxes	Tick bite	Tick-infested terrain, houses, dogs
Boutonneuse fever	*R. conorii*	Mediterranean, Black Sea, and Caspian Sea littorals, Middle East, India, Africa	Ixodid ticks	Dogs, rodents	Tick bite	Tick-infested terrain, houses, dogs
Rickettsialpox	*R. akari*	North America, USSR, Southern Africa, Korea	Mites	House mice, other commensal rodents	Mite bite	Rodent- and mite-infested urban premises
Siberian tick typhus	*R. sibirica*	Armenia, Central Asia, Siberia, Mongolia, Central Europe	Ixodid ticks	Rodents	Tick bite	Tick-infested terrain, houses, dogs
Queensland tick typhus	*R. australis*	Australia	Ixodid ticks	Marsupials	Tick bite	Tick-infested terrain; rainy season
Typhus group						
Epidemic typhus	*R. prowazekii*	Worldwide	Human body louse, squirrel ectoparasites	Humans, flying squirrels	Infected louse feces into skin	Crowded, filthy conditions, lousy population
Murine typhus	*R. typhi (mooseri)*	Worldwide	Flea	Rodents	Infected flea feces into skin	Rat- and flea-infested premises
Brill-Zinsser disease	*R. prowazekii*	North America, Europe, Africa	Recrudescences of latent epidemic typhus infection few to many years after primary attack. Stress may precipitate attack.			
Q fever	*Coxiella burnetii*	Worldwide	Ixodid ticks	Rodents, cattle, sheep, goats	Inhalation of infectious aerosol, tick bite (?)	Animal husbandry, dairies, lambing pens, abattoirs
Trench fever	*Rochalimaea quintana*	Europe, Africa, North America	Human body louse	Humans	Infected louse feces into skin	Crowded, filthy conditions, lousy population
Scrub typhus	*Rickettsia tsutsugamushi*	Asia, Indian subcontinent, Australia, Pacific Islands	Trombiculid mites	Rodents	Mite bite	Chigger-infested terrain, secondary scrub, grass airfields, golf courses

be shipped to the reference laboratory in a tightly stoppered receptacle which contains a pledget of moist cotton. Ticks are easily killed by desiccation.

DIRECT MICROSCOPIC OBSERVATION OF SPECIMENS

Smears of infected tissue may be made and examined directly for the presence of rickettsiae after appropriate staining. The stains commonly used are Gimenez (19), Macchiavello (28), and Giemsa. Gimenez stain is recommended. Rickettsiae (except for *Rickettsia tsutsugamushi*) stain brilliant red with Gimenez stain, red with Macchiavello stain, purple with Giemsa stain, and gram negative (but very poorly) with Gram stain. *R. tsutsugamushi* stains reddish-black with a modification of Gimenez stain. None of these stains will reliably differentiate rickettsiae from bacteria.

Rickettsiae appear as pleomorphic coccobacillary forms and may vary in length from 0.25 to ≥2.0 μm within individual preparations. They are sometimes observed in chains or pairs. *C. burnetii* is the smallest, with average dimensions of 0.25 by 1.0 μm. It is seen usually as a bipolar rod (as are the other rickettsiae) near cells or in the cytoplasm. Typhus-group rickettsiae are a bit larger, averaging 0.3 by 1.2 μm, and also are typically found near cells or in the cytoplasm. The spotted fever rickettsiae are the largest and average 0.6 by 1.2 μm. They often may be found within the nuclei of infected cells, as well as in the cytoplasm, and often are seen in pairs surrounded by a halo, as if encapsulated. Electron microscope studies of most of the pathogenic rickettsiae have been published (4).

The technique of the Gimenez stain and preparation of the reagents are described in Chapter 112.

Infected tissues or cells may be directly examined by fluorescence microscopy. Specific antisera tagged with fluorescein isothiocyanate (FITC) have been used successfully to detect and identify pathogenic rickettsiae, including those located in biopsy material from the early skin lesions produced in spotted fever (51). (See below, section on immunofluorescence tests.)

ISOLATION AND IDENTIFICATION OF RICKETTSIAE

As previously mentioned, the difficulties and dangers of working with most rickettsiae are such as to preclude attempts at isolation in animals or embryonated eggs except in Class III containment facilities. If isolation and identification of a suspected rickettsial agent are deemed important, blood samples taken during the febrile period may be sent to a reference laboratory through the office of the local state board of health.

Tissue culture techniques have been successfully used for the primary isolation of spotted fever rickettsiae from infected ticks (48) and presumably could be used effectively in the primary isolation of rickettsiae from blood and other tissues. The appropriate use of FITC-tagged specific antiserum on infected monolayers of mouse L cells or primary chicken embryo cells should provide direct identification of the agent, regardless of whether or not conditions were optimal for plaque formation. *Rickettsia rickettsii*, the agent of spotted fever, forms characteristic plaques in monolayers of primary chicken embryo cells or Vero monkey cells within 6 to 8 days after inoculation. *Rickettsia typhi* and *R. prowazekii* plaques appear at 8 to 11 days, and *C. burnetii* plaques appear at 12 to 14 days after inoculation.

SEROLOGICAL DIAGNOSIS

Rocky Mountain spotted fever and Q fever are the two rickettsial diseases most likely to be encountered in the continental United States. Others, however, do occur, either because of indigenous infections or, as a result of rapid air transport, because of infections acquired in other parts of the world. Thus, murine typhus, Brill-Zinsser disease, and rickettsialpox are endemic in the United States, and scrub typhus and boutonneuse fever are known to have occurred in the United States, although the infections were acquired in Asia and Africa, respectively. *R. prowazekii* appears to be enzootic in flying squirrels (*Glaucomys volans*) from Virginia to Florida in the eastern United States (8). The clinical laboratory should be aware of these possibilities and should keep itself equipped to perform the initial serological tests which can classify the disease as belonging to the typhus fever, spotted fever, or Q fever group. More definitive identification can then be secured by sending the appropriate serum samples to a reference laboratory.

Rickettsial diseases are acute illnesses of sudden onset. They are characterized by fever, severe headache, myalgia, chills, and pneumonitis. Except in Q fever, they are accompanied by a rash and disseminated focal peripheral vasculitis. Q fever infections may cause hepatitis, and chronic infection with *C. burnetii* may result in fatal subacute endocarditis. Furthermore, *C. burnetii* may heavily infect the placenta many months after recovery of the mother from Q fever. These aspects of rickettsial disease should be kept in mind by the laboratory diagnostician as well as by the clinician.

A serological test for Q fever phase I antibodies should be done routinely in cases of subacute endocarditis or granulomatous hepatitis in which blood cultures are bacteriologically negative (37).

CF test

See reference 35a for the complement fixation (CF) test procedure.

IF tests

Immunofluorescence (IF) tests have been developed for the direct identification of rickettsiae in infected tissues and fluids (9, 22, 29, 38, 51), as well as for the indirect detection and measurement of specific rickettsial antibodies (6, 10, 14, 20, 29, 35, 40). These tests, in terms of specific information which can be derived from them, comprehensive detection of antibody, relative freedom from artifacts, economy of antigen, sensitivity, reproducibility, and convenience, represent a truly major advance in serology of rickettsial diseases.

In the direct method, specific antiserum is conjugated with FITC (36). This conjugate is then used to stain the suspected organisms directly. Direct IF staining of infected tissue has been found useful for identification of pathogenic rickettsiae. This procedure has been effective in identifying *R. rickettsii* in infected ticks (9) and in skin lesions (51) and other human tissues (23). Group-specific fluorescent-antibody reagents for identification of typhus, Q fever, and spotted fever group infections have been produced for this purpose (23).

In the indirect method, the specific antibody is allowed to fix to antigen. This complex in turn is allowed to react with FITC-conjugated antibody against the specific antiserum. This technique is most useful for the detection of rickettsial antibody but also can be employed for the detection of antigen in infected tissues (38).

The indirect method of Weller and Coons as modified by Goldwasser and Shepard (20) has been used effectively by Elisberg and Bozeman for the diagnosis of scrub typhus and spotted fever (6, 14). In this procedure, test serum is allowed to react with a smear of yolk sac tissue heavily infected with the appropriate rickettsiae. After 30 min, excess serum is thoroughly rinsed away, and a drop of horse anti-human immunoglobulin serum conjugated with FITC is applied to the slide. After 30 min it is rinsed off, and the slide is air dried, mounted, and examined the same day under a microscope with UV light optics. A positive reaction results in the fluorescence of the rickettsiae in UV light. With this technique (as with the CF reaction), one can differentiate between sera from patients with murine and epidemic typhus fever.

The IF technique of Elisberg and Bozeman gives a broader group reaction than does the CF test with strains of *R. tsutsugamushi*. This is an advantage, because in the CF test, strain specificity is so marked that tests with antigens from several strains must be used to ensure adequate testing of suspect sera. The technique does require the availability of fresh, infected yolk sacs and preservation of the smears at −70°C.

The micro-IF dot technique described by Philip et al. (40), the development of stable, Formalin-fixed standard suspensions of rickettsial antigens (35), and the use of completely heavy-chain-specific anti-human immunoglobulin G (IgG) and anti-human IgM have permitted the indirect micro-IF test to become a routine, reliable, and sensitive test for the measurement of humoral antibodies against rickettsiae. With the indirect micro-IF technique, it is possible routinely to test 10 dilutions of serum plus positive and negative serum controls against 9 different rickettsial antigens on a single microscope slide. The microscope slide must be given a final rinse in acetone and air dried before being used in this procedure. Rickettsial antigens are applied by dip-pen point (C. Howard Hunt Pen Co, Camden, N.J.) to microscope slides positioned over marked templates. After application to the slides, antigens are dried for 30 min and fixed in acetone for 10 min at room temperature. Serial two-fold dilutions of serum in phosphate-buffered saline containing 5% normal yolk sac to reduce background fluorescence are then placed on sets of antigen dots with a 3-mm bacteriological loop. The slides are placed in a moist chamber for 30 min at 37°C and then washed with two changes of phosphate-buffered saline for 5 min each. After air drying, each antigen set is overlaid with loopfuls of FITC-conjugated anti-human IgG or anti-human IgM and again incubated for 30 min in a moist chamber at 37°C, washed as before, mounted in buffered glycerol (or simply air dried), and examined under a UV microscope with an oil immersion lens. It is important to use low-fluorescence immersion oil. Refer to the review of Liu (29) and the manual of Cherry et al. (10) for further information on IF tests.

Antigen quality influences the sensitivity and specificity of any given test. Antigen preparations made from smears of infected tissue for use in IF tests tend to deteriorate unless stored at −70°C until used. However, it is possible to make suspensions of purified rickettsiae which are stable, sensitive to antibody, and fluoresce as brilliantly as live rickettsiae in infected yolk sac suspensions stored at −70°C until used. This method (35) includes initial inactivation with formaldehyde followed by differential centrifugation in 2 M NaCl. Such purified preparations have been stored as aqueous suspensions for 2 to 5 years at 4°C without detectable deterioration. It is particularly difficult to produce purified suspensions of phase II *C. burnetii* which fluoresce brilliantly in the IF test. Purified and stabilized by this method, phase II organisms fluoresce as brilliantly as phase I organisms in IF tests.

Treatment of rickettsial antigens with ether, a procedure commonly employed in making purified suspensions of rickettsiae, should be avoided because it reduces the brilliance of fluorescent staining. The effects of ether treatment in terms of reduction in specificity and sensitivity of the antigen have not been systematically examined.

Immunoperoxidase test

A variant of the usual indirect IF test was developed as a field test for diagnosis of scrub typhus (12). This test, the paper enzyme-linked immunosorbent assay (PELISA), employs a 20% suspension of infected yolk sac as antigen. The test is performed on a microscope slide as in the indirect IF test. The novel feature of this test is the use of a strip of filter paper saturated with an aqueous solution of the solution of the substrate (5-aminosalicylic acid plus water). This strip of paper is laid on the slide, which bears spots of the antigen-antibody-peroxidase–tagged anti-human serum sandwich. The purple-brown reaction product forms in the filter paper at sites where human antibody was bound to the antigen. The strip of paper is then removed and dried, and the test is read with the naked eye. The piece of filter paper constitutes a permanent record.

A related test in which antigen consisted of peritoneal tissue smears from infected and cyclophosphamide-treated mice also has been described (52). In this test peroxidase also has been simply substituted for FITC as a label on the indicator antiserum. Tests are read with an ordinary microscope, the slide preparations are permanent, and endpoints are easily determined.

Serum antibody titers and specificity in both tests equal corresponding values obtained by indirect IF test.

Agglutination tests

Rickettsial agglutination. A variety of rickettsial agglutination tests have been developed for diagnostic use in epidemic and murine typhus (28, 45) and in Q fever (17, 34). Agglutination tests with highly purified antigen are as sensitive as the CF test in detecting antibody and often are more specific. In addition, they have the virtue of being simple to perform. Unfortunately, purified antigens suitable for use in agglutination tests are not yet commercially available in the United States.

Rickettsial agglutinating antigens are ordinarily prepared from infected chicken yolk sac tissue, although satisfactory suspensions of *R. prowazekii* have been prepared from mouse lungs. Excellent antigen preparations of *C. burnetii*, *R. rickettsii*, *R. prowazekii*, and others have been made from various types of cell cultures.

The capillary agglutination test of Luoto (34) was developed for the detection of Q fever antibodies in milk and other opaque liquids. It is equally effective in detecting antibodies in sera of humans and animals. This test depends for its unique usefulness on purified suspensions of phase I *C. burnetii* which have been stained with hematoxylin. The technique has been used extensively for detection of antibodies in animals' sera with unstained purified suspensions of *R. prowazekii*, *Rickettsia mooseri*, *Rickettsia conorii*, and *C. burnetii*. In this test, capillary tubes, 9 by 0.4 mm (i.d.), are one-third filled with antigen by capillary attraction. Similar volumes of test sera are next drawn in; the tubes are inverted and stuck into modeling clay in a vertical position with the antigen on the bottom. After incubation for 2 h at 37°C or 4 h at room temperature, clumping of the rickettsiae is visible macroscopically in tubes containing antibody. Results obtained are specific and reproducible. This is a particularly useful technique for field use. The antigen preparations are stable for long periods when stored at 4°C.

By far the most useful and practical rickettsial agglutination test that has been developed is the microagglutination test (17), which employs the microtiter equipment used for the CF test (35a). The microagglutination test uses a 0.025-ml volume (containing 8 μg) of antigen and a similar volume of test serum. The plate is incubated overnight at room temperature, after which the agglutination pattern is read from the bottom like a hemagglutination test. Results indicate good reproducibility, titers as high as or higher than corresponding CF titers, and great economy of time and laboratory glassware. The purified antigens necessary for this test can be used interchangeably in the CF test. Differentiation of phase I and phase II antibodies against *C. burnetii* is as sharp as in the CF test. The microagglutination test also has been used successfully with antigens of *R. prowazekii*, *R. typhi*, *R. conorii*, and *R. rickettsii*. This test is relatively more sensitive than the CF test in detecting early (IgM) antibody because of the relatively greater efficiency of IgM antibodies in producing agglutination than in causing CF.

C. burnetii in phase I is relatively easy to purify, and antigen preparations made from it are ordinarily stained with Harris hematoxylin to facilitate the reading of microagglutination tests. Suspensions of *C. burnetii* in phase II (egg adapted) tend to agglutinate spontaneously, particularly if stained with hematoxylin (Harris). However, one can remove the phase I antigen from phase I rickettsiae by extraction with trichloroacetic acid (17) or sodium periodate (42), thus converting the organisms to the phase II condition. Such "synthetic" phase II antigen does not spontaneously agglutinate, even when stained with hematoxylin, and constitutes a stable and sensitive antigen for the detection of phase II agglutinins. It is entirely comparable to the natural phase II antigen when used in the CF test.

Addition of a drop of aqueous acridine orange (1:5,000) to each microtiter well after the agglutination reaction is complete facilitates reading of tests in which *R. rickettsii*, *R. prowazekii*, or *R. typhi* antigens are employed.

Latex agglutination test. The latex agglutination test has been developed and tested particularly for serodiagnosis of spotted fever (24). It employs antigens (1) absorbed to latex particles which then can be agglutinated by specific serum antibodies. It is a simple slide test which can be performed in 15 min. It requires no elaborate instrumentation or other equipment and is read simply by naked eye.

In a 2-year evaluation of this test in 11 separate laboratories (25), reproducibility of the test proved excellent and specificity was approximately equal to that of the indirect IF test. In terms of early antibody detection, both latex agglutination and IF tests detected specific antibody 4 to 6 days after onset of illness. Antibody titers by latex agglutination averaged two- to fourfold lower than corresponding IF titers. In terms of diagnostic efficiency, i.e., percentage of patients correctly classified, the latex agglutination test achieved an average score of >93%.

Hemagglutination tests. Various attempts have been made to use human erythrocytes coated with soluble rickettsial antigen as an indicator system for rickettsial antibodies. An adaptation of this test to the use of sheep erythrocytes in place of human erythrocytes has been made by Anacker et al. (2). They demonstrated that fresh sheep erythrocytes coated with soluble antigen from *R. rickettsii* were agglutinated by specific antibodies formed in humans, rabbits, mice, and guinea pigs as a result of infection or immunization. Glutaraldehyde-treated erythrocytes, as contrasted to fresh erythrocytes, prevented detection of antibodies in guinea pig sera but were as effective as fresh erythrocytes in detecting antibodies in the other three species. This indirect hemagglutination test appears to detect IgM antibodies more efficiently than IgG antibodies.

Glutaraldehyde will stabilize sensitized sheep erythrocytes for 3 to 6 months (43). Indirect hemagglutination tests with this reagent give results with human sera comparable to those obtained by IF and better than those obtained by CF, but the test generally did not detect antibodies in guinea pig nor in rabbit sera.

ELISA

The enzyme-lined immunosorbent assay (ELISA) is an indirect test in which the antigen is applied to interior surfaces of the wells of microtiter plates. Indicator antiserum is linked to alkaline phosphatase, as the indicator antiserum is linked to FITC in indirect IF tests. The enzyme substrate is *p*-nitrophenyl phosphate. Reaction with enzyme produces a colored soluble product whose concentration may be measured at 400 nm in a spectrophotometer.

This test has been widely applied in systems in which the antigen is a soluble protein. It has been adapted for use in rickettsial antibody detection with those rickettsiae that possess soluble antigens (13, 21) or can be broken up effectively by sonication (11). The ELISA procedure is relatively labor-intensive, but does have the advantage that it can be read by machine, thus eliminating reader bias.

An ELISA has been used recently to measure class-specific immunoglobulins against *R. rickettsii* in volunteers who had received spotted fever vaccine (11). Results indicated that the ELISA was as specific and as sensitive as the indirect IF test in monitoring antibody increases, and superior to it in detecting seroconversion in vaccinees. This finding probably was due to better detection of IgM antibodies by the ELISA. In an admittedly small series of volunteers, ELISA was superior to the IF test in detecting low levels of antibodies appearing after vaccination and for as long as a year after an attack of spotted fever. For this reason the ELISA may be particularly suited for use in retrospective seroepidemiology.

Weil-Felix test

The Weil-Felix test (15) depends upon the agglutination of *Proteus vulgaris* strains OX-19, OX-2, and OX-K by antibodies produced by rickettsiae of the typhus and spotted fever groups. It is included here because the antigens and standard antisera are often more readily available than are rickettsial antigens, notwithstanding the fact that its use has probably produced more erroneous and misleading results than any other serological test employed for the detection of rickettsial antibodies. Nevertheless, if the proper precautions are taken, the Weil-Felix test is capable of establishing useful presumptive diagnoses in diseases caused by the typhus and spotted fever groups of rickettsiae. Table 2 summarizes the usual reactions encountered in cases of rickettsial disease. Unfortunately, Weil-Felix reactions may vary widely from case to case of spotted fever and therefore may be of little help in either detecting the disease or differentiating it from murine typhus.

Reagents. Commercial antigens and antisera are available, but certain precautions with their use must be observed. Sometimes commercial antigens are standardized against rabbit antisera from animals which have been immunized against the homologous strain of *P. vulgaris*. Such rabbit sera will agglutinate *Proteus* spp. regardless of whether or not the organisms constituting the antigen are sensitive to rickettsial antibodies. Such antigens have not been acceptably tested and should not be used in the Weil-Felix test. Such rabbit sera cannot be used as positive control sera in the Weil-Felix test. The only acceptable commercial antigens are those which have been standardized against the appropriate human convalescent-phase serum, and the only acceptable positive controls are human convalescent-phase sera. If these conditions cannot be met, do not use the Weil-Felix test. The results will mean nothing at best and cause confusion at the worst.

TABLE 2. Weil-Felix reactions in rickettsioses

Disease	OX-19	OX-2	OX-K
Epidemic typhus	++++	+	0
Murine typhus	++++	+	0
Brill-Zinsser disease	Variable, often negative		0
Scrub typhus	0	0	+++
Spotted fever group	++++	+	0
	+	+++	0
Q fever	0	0	0
Rickettsialpox	0	0	0

Test proper. The macroscopic agglutination test is performed by mixing 0.5-ml volumes of antigen suspension and test serum. Serum dilutions of 1:10 to 1:640 (final dilutions, 1:20 to 1:1,280) should be tested (41). A control tube containing 0.5 ml of antigen in 0.85% NaCl and one containing positive serum should be included. The tubes are incubated in a water bath at 37°C for 2 h, followed by incubation overnight at 4°C. Complete agglutination is shown by complete clearing of the supernatant fluid and the presence of large white masses in the bottom of the tube (4+). Partial agglutination is indicated by partial clearing of the supernatant fluid and the formation of smaller masses of agglutinated antigen in the bottom of the tube. Shaking the tubes does not resuspend the agglutinated antigen to give the even turbidity seen in the negative control. Reading the tubes can be facilitated by shielded illumination and viewing at a critical angle.

This entire test can be more conveniently run in microtiter plates with hematoxylin-stained antigen (17). The agglutination patterns obtained are sharp, reproducible, and easy to read. An entire test can be run in a fraction of the time required to set up and read the comparable macroscopic agglutination test.

Other tests

There are a number of other laboratory tests which are useful in research but which are of little practical use in the routine diagnosis of human disease. These include toxin neutralization and other neutralization tests, antiglobulin sensitization, radioisotope precipitation, and precipitin tests.

Toxin neutralization test. Injection into mice of suspensions of chicken embryo yolk sac heavily infected with various members of the typhus and spotted fever groups of rickettsiae causes death within a few hours (45, 50). Death can be prevented by mixing the rickettsial suspension with antiserum before injection. This phenomenon has been employed to demonstrate antigenic relationships among members of the spotted fever group and to differentiate between the antibodies caused by murine and epidemic typhus rickettsiae: it is still used in the standard assay of epidemic typhus vaccine.

Other neutralization tests. Other neutralization tests have been used with many of the rickettsiae. Such tests are similar to the toxin neutralization test in principle and differ only in that the aim is to test the ability of a given serum to prevent typical rickettsial disease in laboratory animals rather than to prevent an acute toxic reaction. The techniques are costly in animals, laboratory space, and time and therefore are employed only for special purposes.

Antiglobulin sensitization test. The antiglobulin sensitization test (28) essentially is a hyperagglutination test. It has been used for the detection and measurement of Q fever antibodies in human serum and for the detection of small amounts of antibody in bulk pools of cows' milk. Purified antigen is sensitized with the test serum and then treated with rabbit antispecies globulin to produce agglutination. The test is specific and results in much higher titers than those achieved by direct agglutination. It is of limited value as a diagnostic aid because of the relatively large amounts of purified antigen and time necessary for its performance.

Radioisotope precipitation test. The radioisotope precipitation test (18) is essentially a modification of the antiglobulin sensitization test, but differs in the method of detecting and measuring agglutination of the sensitized antigen. It has been used for the detection of Q fever antibodies (phase I) in human and animal sera (27).

The heart of the technique is the antigen tagged with a radioisotope. *C. burnetii* may be tagged with ^{32}P while growing in chicken embryo yolk sacs or with ^{131}I after being purified. In either case, a purified suspension of phase I *C. burnetii* is sensitized with the suspect serum and then treated with the appropriate rabbit antiglobulin. The suspension is then centrifuged in a capillary tube lightly enough so that the antigen in a negative control suspension is not sedimented. The upper and lower portions of the suspensions are then placed in planchets, and the radioactivity in both portions is measured in a Q-gas counter. A positive reaction is one in which most of the radioactivity is found in the lower portion of the capillary tube after centrifugation.

The radioisotope precipitation test gives specific results and is extremely sensitive. Titers against *C. burnetii* as high as 1:32,000 have been recorded. Antibody levels which are not detectable by CF and direct agglutination techniques will give significant titers in the radioisotope precipitation test.

This test is not easily adapted to the requirements of a routine diagnostic tool, and highly purified and radiolabeled phase I antigen is required for its performance.

Precipitin test. The precipitin reaction has been used in studies on soluble antigens of rickettsiae (26, 28). Again, there is the usual limitation in the requirement for large amounts of relatively pure antigen. Several workers have demonstrated the presence of specific, serologically active substances in the blood and urine of patients in acute stages of epidemic typhus, and this phenomenon was adapted to a slide test for early rapid diagnosis of the disease (44). The method has not been widely used and is not now routinely employed as a diagnostic test in the USSR, where it was originally developed.

INTERPRETATION OF LABORATORY RESULTS

The single most important diagnostic aid in the identification of rickettsial disease is the demonstration of a rise in serum antibody when paired acute- and convalescent-phase sera are titrated simultaneously in the same test. Absolute antibody titers usually are of little diagnostic significance, whereas a rise in titer as the disease progresses is of the greatest importance in establishing a diagnosis. A significant rise in titer is defined as a fourfold increase. Sometimes only single serum specimens will be available for laboratory test. In such circumstances, although it is necessary to perform the appropriate tests for antibody, it is well to be conservative in assigning much significance to a positive finding. Certain rickettsial diseases, including Rocky Mountain spotted fever, Q fever, and murine typhus, are widespread in the United States, and in certain areas there may be ecological "islands of infection" (5) in which a particular disease will have a much higher incidence than in the surrounding territory. The offending rickettsiae often produce inapparent infections or very mild disease from which the individual recovers without benefit of diagnosis or medical care. This is particularly, perhaps typically, true of *C. burnetii* infections. To complicate the picture, vaccines against Rocky Mountain spotted fever have been used widely in the civilian population, and typhus vaccine has been employed routinely in the military population. Also, in the immigrant population, particularly in cities along the Atlantic seaboard, there are many individuals who have come from persisting endemic areas of typhus in Europe and Asia. Such individuals may still possess antibodies resulting from prior epidemic typhus disease, or they may become ill with Brill-Zinsser disease (recrudescent epidemic typhus) many years after recovery from the initial infection. Although the public health significance of *R. prowazekii* infections of flying squirrels is still under investigation, recent data (30) make it highly probable that cases of epidemic typhus due to contact with flying squirrels have occurred in the southeastern United States. These factors make it difficult to assess the significance of results from tests on a single serum specimen, particularly if the titer is low.

The time of appearance of humoral antibodies produced in response to rickettsial infections is partially summarized in Table 3. Appearance of antibodies (particularly CF antibodies) may be delayed by broad-spectrum antibiotic therapy begun early in the course of disease. Negative CF findings in such circumstances therefore must be interpreted cautiously. It should be mentioned that antibodies arising in response to Rocky Mountain spotted fever have also been detected by indirect IF and latex agglutination within 4 to 7 days after onset of disease (25).

The initial antibody response to rickettsial disease typically is IgM or a mixed IgM-IgG response. Thus for early diagnosis a test which can readily detect IgM antibody is the one of choice. IF, latex agglutination, and ELISA tests all are excellent detectors of IgM, as are most agglutination tests, provided the proper indicator antiserum is employed. The CF test, however, is relatively unresponsive to early IgM antibodies in rickettsial disease.

CF test

The lowest serum titer usually considered significant is 1:8 (original dilution), particularly if washed antigen is used. During convalescence from rickettsial disease, CF titers of 1:128 to 1:256 are often found. CF antibodies usually appear within 7 to 14 days after onset of disease. After convalescence, the titer slowly falls, but low levels may persist for several years.

The CF test is diagnostically useful in all rickettsioses except scrub typhus, in which the multiplicity of strains, which are mutually exclusive with respect

TABLE 3. Time of appearance of antibodies in serum from human patients

Disease	CF	Weil-Felix	Agglutination
Spotted fever group[a]	8–10 days	5–12 days[a]	Unknown
Q fever	8–14 days	None	5–8 days
Epidemic and murine typhus	7–9 days	7–14 days	5–7 days
Scrub typhus	Unknown	10–14 days	Unknown

[a] Except for rickettsialpox, in which Weil-Felix antibodies are not found.

to CF antigens, makes the test impractical except for a very few reference laboratories. Phase II Q fever antigen and group antigens for the typhus and spotted fever groups are commercially available. The available group antigens are actually partially group specific and partially species specific and are composed of a mixture of soluble and particulate antigen. Caution must be used in interpreting results of tests with such antigens. Positive results by themselves indicate only infection with an anonymous member of the group, not with any specific member. Results should be so reported.

Commercial spotted fever group antigen is often made from *Rickettsia akari* rather than *R. rickettsii*. In most cases, but not invariably, the use of *R. akari* antigen will detect antibodies produced by *R. rickettsii*. At least two cases have been noted in which sera from patients with Rocky Mountain spotted fever gave good CF titers with *R. rickettsii* antigen but low or negative titers with *R. akari* antigen.

Q fever infections in humans produce two distinguishable CF antibodies (16). Phase II CF antibodies are produced initially; phase I CF antibodies (not phase I agglutinins) are produced irregularly several weeks or months later and remain at low levels, except in cases of subacute rickettsial endocarditis. In such cases, the invariable finding has been one of very high phase I CF antibody titers (1:256 to 1:2,048) and similarly high phase II CF titers (37, 46). Patients with a diagnosis of subacute bacterial endocarditis with negative blood cultures should be examined routinely for Q fever antibodies. A similar stipulation applies to patients with hepatitis of unknown etiology.

Murine typhus in individuals previously given epidemic typhus vaccine may present an antibody pattern which is difficult or impossible to decipher by CF test, even by a reference laboratory with specific antigens available. Because of the presence of group antigens, the patient may show an anamnestic response to the group antigen and a primary response to the murine-specific antigen. The result may be a higher CF titer with specific epidemic typhus antigen than with specific murine typhus antigen.

Brill-Zinsser disease, which results in the production primarily of IgG (7S) antibodies, can be distinguished from primary epidemic typhus, in which the major antibody present up to 21 days after onset is IgM (19S), by selective destruction of the 19S CF antibodies with the aid of heat and ethanethiol (33). This results in a drop in the CF titer of sera from patients with primary typhus but little change in the CF titer of sera from patients with Brill-Zinsser disease.

Although the Weil-Felix is probably the most widely used test for detection of rickettsial antibodies, the CF test may be the most useful test now widely used for serodiagnosis of rickettsial disease. The recent availability from the Centers for Disease Control (Atlanta, Ga.) of IF test kits for serodiagnosis of spotted fever and typhus suggests that this technique may, in the future, at least partially supplant the CF test. For the diagnosis of endocarditis or granulomatous hepatitis caused by *C. burnetii*, however, the CF test is still the most widely available.

It must be emphasized that quite erroneous and misleading information can be produced by the CF test. It is essential that the details of the test be rigorously and faithfully followed if reliable and reproducible results are to be obtained.

IF tests

The use of IF as a means of identifying and measuring rickettsial antibody, as well as identifying rickettsiae in infected tissues, is a most promising development in the diagnosis of rickettsial disease. The discovery that Formalin-fixed antigens for IF use will remain stable and reactive for 2 to 5 years as aqueous suspensions stored at 4°C (35, 40), the development of the micro-IF dot technique by Philip et al. (40), and the great economy of antigen and antisera that this affords make the micro-IF a practical, reliable, and sensitive test for any laboratory equipped with a UV microscope and the requisite reagents.

The use of specific anti-IgM and anti-IgG sera in the micro-IF test makes the diagnosis of Brill-Zinsser disease much easier and more certain than before, because the immune response to this recrudescent form of epidemic typhus is a pure IgG antibody response (35). The ability to differentiate between IgG- and IgM-specific antibody in a single serum sample makes it easier to identify the low residual titers of antibodies due to previous vaccination or disease as such, because such antibodies are ordinarily IgG in character. The use of the indirect IF test also has improved the serodiagnosis of rickettsial endocarditis and granulomatous hepatitis caused by *C. burnetii* and the serological differentiation of these two diseases from each other.

The direct IF test for specific identification of rickettsiae in ticks, other arthropod carriers, tissue cultures, or infected tissues has already shown its usefulness.

The present lack of commercially available rickettsial antigens suitable for the indirect micro-IF test poses the only obstacle to widespread routine use of this technique.

Agglutination tests

Rickettsial agglutinin titers of 1:4 (phase I) and 1:8 (phase II) have been judged to be diagnostically significant in Q fever. Although phase II CF antibodies as well as phase I and II agglutinins appear early in the course of disease, phase I CF antibodies appear, if at all, only after several weeks and never achieve high titers. (The known exception to this is in rickettsial endocarditis caused by *C. burnetii*.) Diagnosis may be based, therefore, on a rise in phase I or phase II agglutinins or phase II CF antibodies. Agglutinins are demonstrable in 50% of patients in 5 to 7 days, in 92% within 14 days, and in 100% of patients within 30 days. In contrast, CF antibody is rarely detected within 7 days after onset, although 65% of patients have significant CF antibody levels by 14 days after onset and 90% of patients are positive by 30 days. Both types of antibody may persist in high titer for many months after recovery from the acute disease. In general, agglutinin titers produced during the course of rickettsial disease appear earlier, rise earlier to higher titers, and decline on a time course comparable to that of CF antibodies. Thus, in patients with Q fever or in vaccinees, the agglutination test has been found to be more sensitive than the CF test in detecting early rises in specific antibody. Increased sensitivity of microagglutination over the CF test in these circumstances is due to the early appearance of IgM antibody, which is a more effective agglutinin than is IgG antibody.

Antibodies to *C. burnetii* do not cross-react with other pathogenic rickettsiae, and a rise in agglutinin titer is considered to be convincing evidence of Q fever infection (except in vaccinees).

Agglutinin titers of 1:32 have been considered to be diagnostically significant in persons ill with murine or epidemic typhus. As in Q fever, agglutinins usually appear 5 to 8 days after onset of disease and reach titers comparable to or higher than CF antibody titers.

The rickettsial agglutination test appears to be consistently more species specific than the CF test, even if the same antigen is used in both tests. Thus, it has been found possible to identify by the agglutination test murine typhus infection in patients previously given epidemic typhus vaccine, although it was impossible to do this by the CF test. Parallel tests by toxin neutralization and microagglutination of sera from individuals with laboratory-acquired infections of murine and epidemic typhus have been in close agreement. Both CF and agglutination tests can be used to distinguish Brill-Zinsser disease from primary epidemic typhus by the different susceptibilities of IgM and IgG antibodies to heat and ethanethiol (33) or 3-mercaptoethanol.

Agglutination tests for spotted fever antibodies have been little used, primarily because *R. rickettsii* and closely related species grow so poorly in embryonated eggs that the cost of producing diagnostic antigen from infected chicken embryo yolk sacs for agglutinin measurements is prohibitive. The exploitation of modern cell culture techniques may change this. In any case, when agglutinating antigens made from *R. rickettsii* or other members of the spotted fever group have been used to measure agglutinins in sera from experimentally infected guinea pigs, the tests have been found to be quite species specific and sensitive.

In addition to antigens or antigenic determinants unique to the typhus or spotted fever groups, there are other antigenic determinants which these two groups of rickettsiae share in common. This intergroup relationship has been revealed by simultaneous antibody rises to both *R. prowazekii* and *R. rickettsii* in the sera of persons who were convalescent from epidemic typhus fever. It is probable that the microagglutination test is as effective as, if not more effective than, the CF test in detecting antibodies to common determinants. The possibility of concurrent rises in antibody titer to more than one species of rickettsiae, due to the existence of interspecific antigens, should be kept in mind when the significance of serological findings is being assessed.

The simplicity of the microagglutination test (17) and the generally good reproducibility of results are strong recommendations for its general use when appropriate antigens become commercially available. No antigens for the microagglutination test are available from the Centers for Disease Control or from any commercial source at this time.

Weil-Felix test

The Weil-Felix test is traditionally the convenient test which can be performed in any laboratory and often gives positive results which are later proven to be nonspecific. It is worth emphasizing again that reliable results depend upon the proper reagents and the proper controls, and that even so the test has limitations which must be kept in mind in attempting to determine the significance of the results.

The lowest titer that may be considered significant is 1:160 (final dilution), and a rise in titer must be at least fourfold to be considered significant. Although Weil-Felix agglutinins never appear in the blood of many scrub typhus patients, they may be detected during week 2 after onset of the disease. Weil-Felix agglutinins may appear somewhat earlier in spotted fever group disease, epidemic typhus, and murine typhus. Peak titers are usually found during week 3 after onset.

Q fever and rickettsialpox do not give a positive Weil-Felix test. Brill-Zinsser disease gives variable and often negative results. In most cases, it is impossible to differentiate between the typhus and spotted fever groups on the basis of the Weil-Felix reaction, and it is impossible to differentiate between murine and epidemic typhus with this test. Infections with *Proteus* spp. and relapsing fever caused by spirochetes of the genus *Borrelia* also cause a rise in Weil-Felix agglutinins.

A fourfold rise in OX-2 and OX-19 agglutinins with a negative OX-K result is presumptive evidence of typhus group or spotted fever group disease. Similarly, a rise in OX-K agglutinins only is presumptive evidence of scrub typhus infection.

Future possibilities

The identification and isolation from human sera of specifically protective antibodies (3) make it possible, in principle, to design a subunit vaccine comprised of constituents specifically able to elicit protective antibodies against a particular microbial agent. With the possibility of successful autocatalytic replication of recombinant RNA (32), it now seems possible in principle to synthesize in vitro diagnostic antigens able to identify with great precision the etiological cause of any given disease. Such an accomplishment could make it possible to routinely produce the necessary antigens for diagnosis of rickettsial and viral disease without the present necessity of growing and purifying highly dangerous agents. Such a development may finally make reliable reagents for diagnosis of rickettsial disease widely available.

LITERATURE CITED

1. **Anacker, R. L., R. K. Gerloff, L. A. Thomas, R. E. Mann, and W. D. Bickel.** 1975. Immunological properties of *Rickettsia rickettsii* purified by zonal centrifugation. Infect. Immun. **11**:1203–1209.
2. **Anacker, R. L., R. N. Philip, L. A. Thomas, and E. A. Casper.** 1979. Indirect hemagglutination test for detection of antibody to *R. rickettsii* in sera from humans and laboratory animals. J. Clin. Microbiol. **10**:677–684.
3. **Anacker, R. L., R. N. Philip, C. M. Wilfert, K. T. Kleeman, L. Turner, J. N. MacCormack, and K. E. Hechemy.** 1983. Detection of *Rickettsia rickettsii* antibodies in human sera by crossed immunoelectrophoresis. J. Clin. Microbiol. **18**:569–577.
4. **Anderson, D. R., H. E. Hopps, M. F. Barile, and B. C. Bernheim.** 1965. Comparison of the ultrastructure of several rickettsiae, ornithosis virus, and *Mycoplasma* in tissue culture. J. Bacteriol. **90**:1387–1404.
5. **Audy, J. R., and J. L. Harrison.** 1951. A review of investigations on mite typhus in Burma and Malaya, 1945–1950. Trans. R. Soc. Trop. Med. Hyg. **44**:371–395.
6. **Bozeman, F. M., and B. L. Elisberg.** 1963. Serological diagnosis of scrub typhus by indirect immunofluorescence. Proc. Soc. Exp. Biol. Med. **112**:568–573.
7. **Bozeman, F. M., B. L. Elisberg, J. W. Humphries, K. Runcik, and D. B. Palmer, Jr.** 1970. Serologic evidence of *Rickettsia canada* infection of man. J. Infect. Dis. **121**:367–371.
8. **Bozeman, F. M., N. S. Williams, N. I. Stocks, D. P. Chadwick, B. L. Elisberg, D. E. Sonenshine, and D. M. Lauer.** 1978. Ecologic studies on epidemic typhus infection in the eastern flying squirrel, p. 493–501. *In* J. Kazar, R. A. Ormsbee, and I. N. Tarasevich (ed.), Rickettsiae and rickettsial diseases. Proceedings of the 2nd International Symposium on Rickettsiae and Rickettsial Diseases, Smolenice, 21–25 June 1976. VEDA, Bratislava.
9. **Burgdorfer, W.** 1970. Hemolymph test. A technique for detection of rickettsiae in ticks. Am. J. Trop. Med. Hyg. **19**:1010–1014.
10. **Cherry, W. B., M. Goldman, T. R. Corski, and M. D. Moody.** 1960. Fluorescent antibody techniques in the diagnosis of communicable diseases. U.S. Public Health Service Publication No. 729. Government Printing Office, Washington, D.C.
11. **Clements, M. L., J. S. Dumler, P. Fiset, C. L. Wisseman, Jr., M. J. Snyder, and M. M. Levine.** 1983. Serodiagnosis of Rocky Mountain spotted fever: comparison of IgM and IgG enzyme-linked immunoabsorbent assays and indirect fluorescent antibody test. J. Infect. Dis. **148**:876–880.
12. **Crum, J. W., S. Hanchalay, and C. Eamsila.** 1980. New paper enzyme-linked immunoabsorbent technique compared with microimmunofluorescence for detection of human serum antibodies to *Rickettsia tsutsugamushi*. J. Clin. Microbiol. **11**:584–588.
13. **Dasch, G. A., S. Halle, and A. L. Bourgeois.** 1979. Sensitive microplate enzyme-linked immunosorbent assay for detection of antibodies against the scrub typhus rickettsia, *Rickettsia tsutsugamushi*. J. Clin. Microbiol. **9**:38–48.
14. **Elisberg, B. L., and F. L. Bozeman.** 1966. Serological diagnosis of rickettsial diseases by indirect immunofluorescence. Arch. Inst. Pasteur (Tunis) **43**:193–204.
15. **Felix, A.** 1933. Serological types of typhus virus and corresponding types of *Proteus*. Trans. R. Soc. Trop. Med. Hyg. **27**:147–172.
16. **Fiset, P.** 1959. Serological diagnosis, strain variation and antigenic variation, p. 28–38. *In* J. E. Smadel (ed.), Symposium on Q fever, Medical Science Publication No. 6. Walter Reed Army Medical Center, Washington, D.C.
17. **Fiset, P., R. A. Ormsbee, R. Silberman, M. Peacock, and S. H. Spielman.** 1969. A microagglutination technique for the detection and measurement of rickettsial antibodies. Acta Virol. **13**:60–66.
18. **Gerloff, R. K., B. H. Hoyer, and L. C. McLaren.** 1962. Precipitation of radio-labeled poliovirus with specific antibody and antiglobulin. J. Immunol. **89**:559–570.
19. **Gimenez, D. F.** 1964. Staining rickettsiae in yolk-sac cultures. Stain Technol. **39**:135–140.
20. **Goldwasser, R. A., and C. C. Shepard.** 1959. Fluorescent antibody methods in the differentiation of murine and epidemic typhus sera; specificity changes resulting from previous immunization. J. Immunol. **82**:373–380.
21. **Halle, S., G. A. Dasch, and E. Weiss.** 1977. Sensitive enzyme-linked immunosorbent assay for detection of antibodies against typhus rickettsiae, *Rickettsia prowazekii*, and *Rickettsia typhi*. J. Clin. Microbiol. **6**:101–110.
22. **Hanon, N., and K. O. Cooke.** 1966. Assay of *Coxiella burnetii* by enumeration of immunofluorescent infected cells. J. Immunol. **97**:492–497.
23. **Hébert, G. A., T. Tzianabos, W. C. Gamble, and W. A. Chappell.** 1980. Development and characterization of high-titered, group-specific, fluorescent-antibody reagents for the direct identification of rickettsiae in clinical specimens. J. Clin. Microbiol. **11**:503–507.
24. **Hechemy, K. E., R. L. Anacker, R. N. Philip, K. T. Kleeman, J. N. MacCormack, S. J. Sasowski, and E. E. Michaelson.** 1980. Detection of Rocky Mountain spotted fever antibody by a latex agglutination test. J. Clin. Microbiol. **12**:144–150.
25. **Hechemy, K. E., E. E. Michaelson, R. L. Anacker, M. Zdeb, S. S. Sasowski, K. T. Kleeman, J. M. Joseph, J. Patel, J. Kudlac, L. B. Elliott, J. Rawlings, C. E. Crump, J. D. Folds, H. Dowda, Jr., J. H. Barrick, J. R. Hindman, G. E. Kilgore, D. Young, and R. H. Altieri.** 1983. Evaluation of latex-*Rickettsia rickettsii* test for Rocky Mountain spotted fever in 11 laboratories. J. Clin. Microbiol. **18**:938–946.
26. **Horsfall, F. L., Jr., and I. Tamm (ed.).** 1965. Viral and rickettsial diseases of man, 4th ed. J. B. Lippincott Co., Philadelphia.
27. **Lackman, D. B., G. Gilda, and R. N. Philip.** 1964. Application of the radioisotope precipitation test to the study of Q fever in man. Health Lab. Sci. **1**:21–28.
28. **Lennette, E. H., and N. J. Schmidt (ed.).** 1979. Diagnostic procedures for viral, rickettsial and chlamydial infections, 5th ed. American Public Health Association, Inc., Washington, D.C.
29. **Liu, C.** 1969. Fluorescent antibody techniques, p. 179–204. *In* E. H. Lennette and N. J. Schmidt (ed.), Diagnostic procedures for viral and rickettsial diseases, 4th ed. American Public Health Association, Inc., New York.
30. **McDade, J. E., C. C. Shepard, M. A. Redus, V. F. Newhouse, and J. D. Smith.** 1980. Evidence of *Rickettsia prowazekii* infections in the United States. Am. J. Trop. Med. Hyg. **29**:277–284.
31. **McKiel, J. A., E. J. Bell, and D. B. Lackman.** 1966. *Rickettsia canada*; a new member of the typhus group of Rickettsiae isolated from *Haemaphysalis leporispalustris* ticks in Canada. Can. J. Microbiol. **13**:503–510.
32. **Miele, E. A., D. R. Mills, and F. R. Kramer.** 1983. Autocatalytic replication of a recombinant RNA. J. Mol. Biol. **171**:281–295.
33. **Murray, E. S., J. M. O'Connor, and J. A. Gaon.** 1965. Differentiation of 19S and 7S complement fixing antibodies in primary versus recrudescent typhus by either ethanethiol or heat. Proc. Soc. Exp. Biol. Med. **119**:291–297.
34. **Ormsbee, R. A.** 1965. Q fever rickettsiae, p. 1144–1160. *In* F. L. Horsfall, Jr., and I. Tamm (ed.), Viral and rickettsial infections of man. J. B. Lippincott Co., Philadelphia.
35. **Ormsbee, R., M. Peacock, R. Philip, E. Casper, J. Plorde, T. Gabre-Kidan, and L. Wright.** 1977. Serologic diagnosis of typhus fever. Am. J. Epidemiol. **105**:261–271.

35a. **Palmer, D. F., L. Kaufman, W. Kaplan, and J. J. Cavallaro.** 1977. Serodiagnosis of mycotic diseases. Charles C Thomas, Publisher, Springfield, Ill.

36. **Peacock, M., W. Burgdorfer, and R. A. Ormsbee.** 1970. Rapid fluorescent-antibody conjugation procedure. Infect. Immun. **3**:355–357.

37. **Peacock, M. G., R. N. Philip, J. C. Williams, and R. S. Faulkner.** 1983. Serological evaluation of Q fever in humans: enhanced phase I titers of immunoglobulins G and A are diagnostic for Q fever endocarditis. Infect. Immun. **41**:1089–1098.

38. **Pedersen, C. E., Jr., L. R. Bagley, R. H. Kenyon, L. S. Sammons, and G. T. Burger.** 1975. Demonstration of *Rickettsia rickettsii* in the rhesus monkey by immune fluorescence microscopy. J. Clin. Microbiol. **2**:121–125.

39. **Philip, R. N., E. A. Casper, R. L. Anacker, M. G. Peacock, S. F. Hayes, and R. S. Lane.** 1982. Identification of an isolate of *Rickettsia canada* from California. Am. J. Trop. Med. Hyg. **31**:1216–1221.

40. **Philip, R. N., E. A. Casper, R. A. Ormsbee, M. G. Peacock, and W. Burgdorfer.** 1976. Microimmunofluorescence test for the serological study of Rocky Mountain spotted fever and typhus. J. Clin. Microbiol. **3**:51–61.

41. **Plotz, H.** 1944. The rickettsiae, p. 559–578. *In* J. S. Simmons and C. J. Gentzkow (ed.), Laboratory methods of the United States Army, 5th ed. Lea & Febiger, Philadelphia.

42. **Schramek, S., R. Brezina, and J. Urvolgyi.** 1972. A new method of preparing diagnostic Q fever antigen. Acta Virol. **16**:487–492.

43. **Shirai, A., J. W. Dietel, and J. V. Osterman.** 1975. Indirect hemagglutination test for human antibody to typhus and spotted fever group rickettsiae. J. Clin. Microbiol. **2**:430–437.

44. **Smorodintzeff, A. A., and R. V. Fradkina.** 1944. Slide agglutination test for rapid diagnosis of pre-eruptive typhus fever. Proc. Soc. Exp. Biol. Med. **56**:93–94.

45. **Snyder, J. C.** 1965. Typhus fever rickettsiae, p. 1059–1094. *In* F. L. Horsfall, Jr., and I. Tamm (ed.), Viral and rickettsial infections of man. J. B. Lippincott Co., Philadelphia.

46. **Turck, W. P. G., G. Howitt, L. A. Turnberg, H. Fox, M. Longson, M. B. Mathews, and R. DasGupta.** 1976. Chronic Q fever. Q. J. Med. **45**:193–217.

47. **Tyeryar, F. J., E. Weiss, D. B. Millar, F. M. Bozeman, and R. A. Ormsbee.** 1973. DNA base composition of rickettsiae. Science **180**:415–417.

48. **Wike, D. A., and W. Burgdorfer.** 1972. Plaque formation in tissue cultures by *Rickettsia rickettsii* isolated directly from whole blood and tick hemolymph. Infect. Immun. **6**:736–738.

49. **Wike, D. A., R. A. Ormsbee, G. Tallent, and M. G. Peacock.** 1972. Effect of various suspending media on plaque formation by rickettsiae in tissue culture. Infect. Immun. **6**:550–556.

50. **Woodward, T. E., and E. B. Jackson.** 1965. Spotted fever rickettsiae, p. 1095–1129. *In* F. L. Horsfall, Jr., and I. Tamm (ed.), Viral and rickettsial infections of man, 4th ed. J. B. Lippincott Co., Philadelphia.

51. **Woodward, T. E., C. E. Pedersen, Jr., C. H. Oster, L. R. Bagley, J. Romberger, and M. J. Snyder.** 1976. Prompt confirmation of Rocky Mountain spotted fever: identification of rickettsiae in skin tissue. J. Infect. Dis. **134**:297–301.

52. **Yamamoto, S., and Y. Minamishima.** 1982. Serodiagnosis of tsutsugamushi (scrub typhus) disease by the indirect immunoperoxidase technique. J. Clin. Microbiol. **15**:1128–1132.

Chapter 85

Chlamydiae (Psittacosis-Lymphogranuloma Venereum-Trachoma Group)

JULIUS SCHACHTER

CHARACTERIZATION OF THE ORGANISM

The chlamydiae are among the more common pathogens throughout the animal kingdom (12, 17, 18, 22). They are nonmotile, gram-negative, obligate intracellular bacteria. Their unique developmental cycle differentiates them from all other microorganisms. They replicate within the cytoplasm of host cells, forming characteristic intracellular inclusions which can be seen by light microscopy. They differ from the viruses by possessing both RNA and DNA and cell walls quite analogous in structure to those of gram-negative bacteria. They are susceptible to many broad-spectrum antibiotics, possess a number of enzymes, and have a restricted metabolic capacity. None of these metabolic reactions results in the production of energy. Thus, they have been considered as energy parasites that use the ATP produced by the host cell for their own requirements (10).

Growth cycle

Chlamydiae are phagocytized by susceptible host cells (4). The phagocytic process is directly influenced by the chlamydiae and is specifically enhanced. After attachment at specific sites on the surface of the cell, the elementary body enters the cell in a phagosome, where the entire growth cycle is completed. The chlamydiae prevent phagolysosomal fusion. Once the elementary body (diameter, 0.25 to 0.35 nm) has entered the cell, it reorganizes into a reticulate particle (initial body) which is larger (0.5 to 1 nm) and richer in RNA. After approximately 8 h, the initial body begins dividing by binary fission. Approximately 18 to 24 h after infection, these initial bodies become elementary bodies by a poorly understood reorganization or condensation. The elementary bodies are then released to initiate another cycle of infection. The elementary bodies are specifically adapted for extracellular survival and are the infectious form, whereas the intracellular metabolically active and replicating form, the initial body, does not survive well outside the host cell and seems adapted for an intracellular milieu.

Taxonomy

Chlamydiae are presently placed in their own order, the *Chlamydiales*, family *Chlamydiaceae*, with one genus, *Chlamydia* (13). There are two species, *C. trachomatis* and *C. psittaci*. *C. trachomatis* includes the organisms causing trachoma, inclusion conjunctivitis, lymphogranuloma venereum (LGV), and the other sexually transmitted infections, and some rodent pneumonia strains (17, 18). *C. trachomatis* strains are sensitive to the action of sulfonamides and produce an iodine-staining glycogen-like material within the inclusion vacuole. *C. psittaci* strains infect many avian species and mammals, producing the diseases psittacosis, ornithosis, feline pneumonitis, bovine abortion, etc. (12, 22). They are resistant to the action of sulfonamides and produce inclusions which do not stain with iodine.

Culture

All chlamydiae appear to grow well in the yolk sac of embryonated hen eggs. Most *C. psittaci* strains grow well in cell culture systems, but *C. trachomatis* strains are more difficult to grow. LGV strains of *C. trachomatis* have the broadest host spectrum in vitro. The other *C. trachomatis* strains generally require some mechanical assistance to facilitate their growth in cell culture, particularly centrifugation of the inoculum to enhance the *Chlamydia*-host cell contact. Chemical treatment of host cells (to change surface charge or render them nonreplicating or metabolically inhibited) will also enhance the growth of *C. trachomatis* strains.

Antigenic relationships

The chlamydiae possess group (genus)-specific, species-specific, and type-specific antigens. Most of these are apparently located within the cell wall, but precise structural relationships are not known. The group antigen, shared by all members of the genus, appears to be a lipopolysaccharide complex with a ketodeoxyoctanoic acid as the reactive moiety (7). It may be quite similar to the lipopolysaccharide of certain gram-negative bacteria (11). Species-specific protein antigens have been identified but have not been characterized (5). Some type-specific antigens of *C. trachomatis* have been identified and appear to be proteins with an approximate molecular weight of 30,000 (16). Specific antigens of *C. psittaci* strains can be demonstrated by neutralization tests (1). The specific antigens of *C. trachomatis* are best recognized by a microimmunofluorescence (micro-IF) technique (25), although these antigens are also associated with a toxic factor (large numbers of viable chlamydiae may kill mice in less than 24 h after intravenous inoculation). One-way cross-reactions have been reported between chlamydiae and some bacteria, but these do not appear to influence serodiagnosis.

CLINICAL SIGNIFICANCE

C. psittaci is a common pathogen in many avian and mammalian species. Specific strains may be associated with specific disease patterns in these hosts. Thus, the same serotype seems to cause arthritis in both sheep and cattle. Although this group of agents can be found throughout the animal kingdom, *C. psittaci* infections in humans almost always result from exposure to infected avian species. Psittacosis is the most common human infection with *C. psittaci*, and infections derived from mammals, although known, are relatively rare (18).

C. trachomatis, on the other hand, is almost exclusively a human pathogen (9, 17). Serotypes within this species cause trachoma (serotypes A, B, Ba, and C have been associated with endemic trachoma, the most common preventable form of blindness), inclusion conjunctivitis, and LGV (serotypes L1, L2, and L3). When sexual transmission of *C. trachomatis* strains other than LGV has been studied, serotypes D through K have been found to be the major identifiable cause of nongonococcal urethritis in men and may also cause epididymitis. Proctitis may occur in either sex. In women, cervicitis is a common result of chlamydial infection, and acute salpingitis may occur. The agent in the cervix may be transmitted to the neonate by passing through the infected birth canal, and eye disease (inclusion conjunctivitis) and a characteristic chlamydial pneumonia of infants may develop (2). Vaginal infection and enteric infection in neonates are also recognized.

COLLECTION, TRANSPORT, AND STORAGE OF SPECIMENS

Methods of collection

For cytological studies, impression smears of involved tissues or scrapings of involved epithelial cell sites should be appropriately fixed (cold acetone for immunofluorescence, heat for Gimenez or Macchiavello stains, and methanol for Giemsa stain). As is also true for the isolation attempts, it is imperative that samples be collected from the involved epithelial cell sites by vigorous swabbing or scraping. Purulent discharges are inadequate and should be cleaned from the site before sampling.

For agent isolation, different specimens are collected depending upon the disease, the chlamydial agent, and the host involved. Thus, for chlamydial infections in mammals, tissue samples of involved sites, secretions, or fecal material may be collected. From birds, fecal material or blood may often yield the agent without sacrificing the bird. At necropsy, air sacs (when involved), liver, spleen, and pericardium are sites to be sampled.

With human psittacosis, sputum, blood, or biopsy material (spleen, liver) may yield the organism. With LGV, bubo aspirates of fluctuant nodes are the most likely source of agent. For the other *C. trachomatis* infections of humans, the involved mucous membranes should be vigorously swabbed or sampled by scraping. Thus, the conjunctiva for trachoma-inclusion conjunctivitis, the anterior urethra (several centimeters into the urethra), or the cervix at the endocervical canal would be tested. Since these strains appear to infect only columnar cells, cervical specimens must be collected at the transitional zone or within the os. Since the organism also can infect the urethra of the female, recovery rates may be improved if another sample is collected from the urethra and sent to the laboratory for testing in the same tube with the cervical sample. For women with salpingitis the samples may be collected by needle aspiration of the involved fallopian tube, or endometrial specimens may yield the agent. Rectal mucosa, nasopharynx, and throat may also be sampled. For infants with pneumonia, swabs may be collected from the posterior nasopharynx or the throat, although aspirates collected by intubation appear to be a superior source of agent.

Transport

Swabs, scrapings, and small tissue samples should be collected in a special transport medium. Because *Chlamydia* spp. are bacteria, the selection of antibiotics to prevent other bacterial contamination is restricted. Broad-spectrum antibiotics such as tetracyclines, macrolides, or penicillin must be excluded. Aminoglycosides and fungicides are the mainstays. The chlamydial specimens should be refrigerated if they can be processed within 24 h after collection. If not, they should be frozen at −60°C. *C. psittaci* strains are much more stable than *C. trachomatis* strains.

For isolation in cell culture, one suspending medium (2SP) consists of 0.2 M sucrose in 0.02 M phosphate buffer (pH 7.0 to 7.2) with 5% fetal calf serum and added antibiotics. A sucrose-phosphate-glutamate (SPG) medium has also been commonly used. (See Chapter 111 for media formulations.) It may be simpler to place the clinical specimen directly into standard cell culture growth medium containing streptomycin (200 μg/ml) or gentamicin (10 μg/ml) together with vancomycin (100 μg/ml) and amphotericin B (4 μg/ml).

DIRECT EXAMINATION

Tissues from infected avian species or lower mammals may be examined by use of impression smears stained with Giemsa, Gimenez, or Macchiavello stains (see Chapter 112). The latter two are preferred. Inclusions are relatively difficult to demonstrate in sections. Direct microscopy is not a useful diagnostic tool for human psittacosis.

LGV is rarely diagnosed by microscopy, and this technique is not recommended. *C. trachomatis* infections of the conjunctiva, urethra, or cervix can be diagnosed by demonstrating typical intracytoplasmic inclusions, but cytological procedures are usually less sensitive than isolation in cell culture. The Giemsa stain is the historic method, but immunofluorescence procedures are more sensitive. These procedures are particularly useful in diagnosing acute, severe inclusion conjunctivitis of the newborn and are less effective in diagnosing adult conjunctival and genital tract infections. With infants the ability to detect intracellular diplococci, if the child has gonococcal ophthalmia neonatorium, is another benefit, and obviously, direct microscopy is much faster than the isolation procedures. If isolation techniques are available, the use of cytology for genital tract disease is not recommended. However, where diagnostic facilities are not available, direct microscopy of epithelial cell scrapings may be useful in diagnosing the chlamydial infections.

Fluorescent-antibody technique

Most of the published experience with immunofluorescent procedures reflects the use of polyclonal antibody in either direct or indirect fluorescent-antibody procedures. These procedures represented efforts to detect typical chlamydial inclusions within epithelial cells. The fluorescent-antibody procedures were more sensitive than other cytological methods (Giemsa or iodine) but required an adequate source of appropriate antisera. There were no commercial sources, and laboratories had to prepare their own reagents. The antisera were usually standardized against infected cell culture monolayers to determine appropriate working dilutions. Trained microscopists were required to identify chlamydial typical intracytoplasmic inclusions. The procedure is less sensitive than isolation of the agent in cell culture. Interested readers are referred to previous editions of this manual for the technique.

More recently, fluorescein-conjugated monoclonal antibodies have been made available (21). Although they are routinely used in some laboratories to identify *C. trachomatis* inclusions in cell culture (20), the test may be applied directly to clinical specimens. There is inadequate experience with this procedure to recommend it at this writing. Anecdotal experience indicates that it will require trained microscopists and rigorous criteria for interpretation of the microscopic analysis of the smears. The procedure seems somewhat less sensitive than culture but may represent an alternative method of diagnosing chlamydial infections in settings where culture is not available.

Iodine-staining technique

Scrapings are air dried, fixed in absolute methanol, and stained with Lugol iodine or 5% iodine in 10% potassium iodide for 3 to 5 min. Slides are examined as wet mounts. The matrix of inclusions may appear as a reddish-brown mass recognizable under low magnification. The slides may be decolorized with methanol and restained with Giemsa stain. This technique is the least sensitive cytological procedure. It is not recommended for use with clinical specimens. Its speed and simplicity have made it the popular test for examining *C. trachomatis*-infected cell cultures.

CULTURE AND ISOLATION

Recommended procedure for primary culture

The recommended procedures for primary isolation of chlamydiae are cell culture techniques. The most popular technique involves inoculation of clinical specimens onto cycloheximide-treated McCoy cells (15). The basic principle involves centrifugation of the inoculum onto the cell monolayer at approximately 2,800 × *g* for 1 h, incubation of monolayers for 48 to 72 h, and then staining with iodine to detect the glycogen-positive inclusions (with *C. psittaci* strains, Giemsa stains must be used, or alternatively, acridine orange staining can be used). Fluorescent-antibody staining may allow earlier detection of the inclusions (23). Use of the above-mentioned fluorescein-conjugated monoclonal antibodies represents the most sensitive method for detecting *C. trachomatis* inclusions in cell culture. The procedure requires more attention to staining than the iodine technique and is more costly.

McCoy cells are planted onto 13-mm cover slips contained in 15-mm-diameter (1 dram) disposable glass vials. Cell concentration (approximately 1×10^5 to 2×10^5 per vial in minimum essential medium with 10% fetal calf serum) is selected to give a light, confluent monolayer after 24 to 48 h of incubation at 37°C. The cells should, for optimal results, be used within 24 to 72 h after reaching confluency. If the laboratory is only passing cells on a sporadic basis, they may then be held at room temperature or in a low (2%) serum medium for at least 2 weeks before inoculation.

The clinical specimen is inoculated onto the cells, and isolation medium containing 1 to 2 μg of cycloheximide per ml (this too must be titrated for each batch) is placed onto the cells 2 h after centrifugation. Although it is usual to wash the inoculated cells before incubation, for greater convenience it is possible to collect the clinical specimen into a cycloheximide-containing medium and simply incubate the cells at 37°C immediately after centrifugation. More debris from the inoculum is found on the cells if they are not washed, but this procedure offers considerable time saving.

Standard inoculation procedure involves removing medium from the cell monolayer and replacing it with the inoculum in a volume of 0.1 to 1 ml. The clinical specimens should be shaken with glass beads before inoculation. This is safer and more convenient than sonication. The specimen is then centrifuged onto the cell monolayer at approximately 3,000 × *g* at 35°C for 1 h. The vials should be held at 37°C for 2 h before the cells are washed or the medium is changed. The cells are then incubated at 37°C for 48 to 72 h, after which one cover slip is examined for inclusions by use of iodine, Giemsa, or immunofluorescence staining. The use of immunofluorescence can speed up the process, as inclusions can clearly be seen (although smaller) at 24 h postinfection, but this requires availability of immunological reagents and uses more difficult microscopic procedures. Giemsa stain is more sensitive than the iodine stain, but the microscopic evaluation is more difficult. Slide reading can be facilitated by examining the Giemsa-stained cover slip by dark-field, rather than bright-field, microscopy (6). The iodine stain is the simplest procedure and the one most commonly used, although it is less sensitive than either of the other two. If passage of positive material or blind passage of negative material is desired, the material should be passed at 72 to 96 h postinoculation. The cell monolayer is disrupted by shaking with glass beads on a Vortex mixer; the material is treated by low-speed centrifugation to remove cell debris, and the supernatant is inoculated as above. Ninety percent of positive specimens are inclusion positive in the first passage.

With trachoma, inclusion conjunctivitis, and the genital tract infections, the technique is exactly as described above. In LGV, the aspirated bubo pus is diluted (10^{-1} and 10^{-2}) and treated as described above. Second passages are always made because detritus from the inoculum may make it difficult to read the slides. In an alternative procedure which may be used for LGV and as a routine for *C. psittaci* isolation attempts, the inoculation monolayers are incubated for 10 days, with routine examination at 5 and 10 days or whenever a cytopathic effect is noted. The cells are stained by the Giemsa method and examined for inclusions.

For laboratories processing large numbers of specimens, it may be convenient to use flat-bottomed 96-well microtiter plates rather than vials for the specimens (27). Cells are plated onto cover slips or can be placed directly onto the plastic. Processing and incubation will be as described above, but microscopy will be modified to use either long working distance objectives or inverted microscopes. This procedure is less sensitive than the vial technique but offers considerable savings in terms of reagents and time and may be suitable for settings where mostly symptomatic patients are being screened. Such patients would yield higher amounts of agent and thus minimize the impact of the decreased sensitivity of the test.

Note. Laboratory workers should be warned that isolation of psittacosis agents should not be undertaken except in laboratories equipped with appropriate isolation facilities. These agents have been among the most common sources of laboratory infections.

Animal, egg inoculation

Most strains of *C. psittaci* of avian origin will grow well in the brains of mice after intracranial inoculation, and many grow well in liver or spleen after intraperitoneal inoculation. These procedures have been useful for primary isolation of many strains. Since host range and pathogenicity by different routes of inoculation have been used as a crude typing procedure for chlamydiae in the past, it is obvious that no single host system of living animal is routinely useful for all chlamydiae. LGV strains will also grow in the brains of mice after intracerebral inoculation. Most *C. trachomatis* strains readily infect only primates.

The yolk sac technique is one which is capable of supporting the growth of all *Chlamydia* isolates. It is usually less sensitive than cell culture procedures and requires much more time. The yolk sac technique is still a useful one for preparation of large quantities of agent.

Yolk sac isolation technique

Clinical specimens are collected in an appropriate antibiotic broth. A suitable one contains streptomycin, neomycin, and nystatin (2.5 mg/ml, 0.5 mg/ml, and 100 U/ml, respectively). Other antibiotics may also be used (vancomycin, ristocetin, gentamicin, and amphotericin). The specimen is held for 1 h at room temperature before inoculation of 0.25 ml into the yolk sac, using a 3.2-cm 22-gauge needle. Before inoculation, the fertile eggs are incubated at 38.5 to 39°C in a moist atmosphere. The eggs to be used must be obtained from a flock fed an antibiotic-free diet. They should be free from *Mycoplasma*. When 7 days old, embryonated hen eggs are candled for viability, the location of air sacs is painted with tincture of iodine, and a hole is gently punched. The specimen is inoculated at a slight angle away from the embryo; three or four eggs are labeled with a pencil or marking pen. After inoculation, the shell is again swabbed with iodine and the hole is sealed (with glue or tape). The eggs are then incubated in a moist environment at 35°C and candled daily for 13 days. Eggs that die in the first 3 days after inoculation are discarded.

The yolk sacs of eggs dying thereafter are harvested. This procedure entails painting the shell with iodine, cracking and removing the shell over the air sac, dissecting the shell and chorioallantoic membranes away, and removing the yolk sac with forceps. Excess yolk material may be stripped away. It is important that all instruments be sterile and that fresh instruments be used for each specimen. Impression smears are made and stained (Gimenez or the modified Macchiavello method). Sterility tests are performed on yolk sac with thioglycolate broth. If the embryos are still viable 13 days postinoculation, the eggs are chilled for several hours and yolk sacs are harvested, ground in nutrient broth, and centrifuged lightly. The supernatant is passed to another group of four 7-day-old embryonated hen eggs (1 ml of 50% yolk sac per egg). After two blind passages, attempts are terminated as negative.

The generally acceptable criteria for positive isolation are the finding of elementary bodies in the impression smears, serially transmissible egg mortality, the presence of group antigen in the yolk sac, and the absence of contaminating bacteria.

IDENTIFICATION

Since most laboratories will be using cell culture isolation systems, the basic procedure for identification of chlamydiae involves demonstration of typical intracytoplasmic inclusions by appropriate (Giemsa or iodine) staining procedures. However, in laboratories initiating work with chlamydiae, it must be considered prudent to use at least one other parameter for identification for chlamydiae. Fluorescent-antibody staining provides both a morphological and an immunological identification.

The two species may be differentiated on the basis of sulfonamide sensitivity and iodine staining ability. Serotyping *C. psittaci* strains is not currently feasible. Although specific antigens have been demonstrated, the techniques of plaque reduction assay or neutralization tests are arduous and provide only crude grouping. *C. trachomatis* strains may be serotyped by the micro-IF technique of Wang and Grayston (25). In this procedure antisera are produced by intravenous inoculation of mice at day 0, a booster at day 7, and exsanguination at day 11. The mouse antiserum is then tested in a titration against all serotypes, as well as the immunizing agent, and the serotype is identified presumptively by the pattern of reactivity and finally by box titration with the appropriate prototype serotype.

Serological methods

The most widely used serological test for diagnosing chlamydial infections is the complement fixation (CF) test. This is useful in diagnosing psittacosis, in which paired sera often show fourfold or greater increases in titer. It may also be useful in diagnosing LGV, in which single-point titers greater than 1:64 are highly supportive of this clinical diagnosis. With LGV it is difficult to demonstrate rising titers since the nature of the disease results in the patient being seen by the physician after the acute stage. Any titer above 1:16 is considered significant evidence of exposure to chlamydiae. The CF test is not particularly useful in diagnosing trachoma-inclusion conjunctivitis or the related genital tract infections, and it plays no role in diagnosing neonatal chlamydial infections.

The micro-IF method is a much more sensitive procedure for measuring antichlamydial antibodies. It may be used in diagnosing psittacosis, in which paired sera will show rising immunoglobulin G (IgG) titers (and often IgM antibody). With LGV it is again difficult to demonstrate rising titers, but single-point titers in active cases usually have relatively high levels of IgM antibody (>1:32) and IgG levels of ≥1:2,000. Trachoma, inclusion conjunctivitis, and the genital tract infections may be diagnosed by the micro-IF technique if appropriately timed paired acute- and convalescent-phase sera can be obtained. However, it is often difficult to demonstrate rising antibody titers, particularly in highly sexually active populations. Many of these individuals will be seen for chronic or repeat infections. The background rate of seroreactors in venereal disease clinics is ≥60%, making it particularly difficult to demonstrate seroconversion. In general, first attacks of chlamydial urethritis have been regularly associated with seroconversion (3). Individuals with systemic infection (epididymitis, salpingitis) usually have much higher antibody levels than those with superficial infections, and women tend to have higher antibody levels than men.

Serology is particularly useful in diagnosing chlamydial pneumonia of neonates. In this case, high levels of IgM antibody are regularly found in association with disease (19). IgG antibodies are less useful because the infants are being seen at a time when they have considerable levels of circulating maternal IgG (since all these infections are acquired from the infected mother, who is always seropositive). It takes between 6 and 9 months for maternal antichlamydial antibodies to disappear. Infants older than that age may be tested for determination of prevalence of chlamydial infection without fear of confounding effects of maternal antibody. Infants with inclusion conjunctivitis or respiratory tract carriage of *Chlamydia* spp. without pneumonia usually have very low levels of IgM antibodies. Thus, a single titer of ≥1:128 may support the diagnosis of chlamydial pneumonia.

The micro-IF technique uses many serotypes of chlamydiae, and the procedure as simplified by Wang et al. is recommended (26). Since serology is particularly useful in diagnosing neonatal infection and the IgM antibody responses tend to be markedly specific, the use of single broadly reacting antigens will miss at least 15 to 25% of the infections that can be proven to be due to *Chlamydia* spp. by other procedures, or that would be positive by a multiple-antigen micro-IF. The single-antigen tests may involve either yolk sac suspensions of agent or identification of fluorescent inclusions in cell monolayers (2, 14, 24). DEL serotype antigens are commonly chosen for this purpose.

Research workers should be warned that monotypic A seroreactions, at least in the United States, are liable to be spurious. Unpublished long-term longitudinal studies on infants that I am performing suggest that the appearance of antibodies against serotype A (and to a lesser extent the cross-reacting CJI complex) may appear in response to a nonchlamydial antigenic stimulus. These antibodies are usually transient and do not result in the persistent high levels of IgG antibodies that usually follow chlamydial infections.

Enzyme immunoassay techniques have been described which measure antichlamydial antibodies (8). These procedures are usually less sensitive than the micro-IF test. They have not been successful in measuring IgM antibody. One test is commercially available, but as yet there is no published experience with it. On the basis of published findings, it is likely that it will be less sensitive than the micro-IF, will miss some C-complex reactors, and cannot be readily applied to IgM antibody determination. The procedure may be of some use in selected instances and for serological surveys in laboratories where micro-IF techniques are not available.

Complement fixation. The CF test may be performed in either the tube system or the microtiter system. Reagents should be standardized in the tube system, regardless of which system is being used for test. The microtiter systems are most useful in screening large numbers of sera, but it is preferable to retest all positive sera in the tube system. Occasionally, sera giving titers in the 1:4 to 1:8 range in the microtiter system are positive at 1:16 (the significant level) in the tube system. The microtiter system uses standard plates and volumes one-tenth of those used in the tube test. The CF test is performed on serum specimens heated at 56°C for 30 min (preferably acute- and convalescent-phase paired sera tested together). In each test a positive control serum of high titer is included with a known negative serum. The reagents for the CF test are standardized by the Kolmer technique and include special buffered saline, group antigen, antigen (normal yolk sac) control, the positive serum, the negative serum, guinea pig complement, rabbit anti-sheep hemolysin, and sheep erythrocytes. (The guinea pig complement should be carefully tested for chlamydial antibodies since many herds are enzootically infected with a chlamydial agent, guinea pig inclusion conjunctivitis.) The hemolytic system is titrated, and the complement unitage is determined. The standard units used in the test are 4 U of antigen and 2 exact units of complement. The test may be performed by either the water bath technique or the overnight (icebox) technique, the former being preferable. Doubling dilutions of the serum (from 1:2) are made in a 0.25-ml volume of saline. The antigen is added at 4 U (0.25 ml), and 2 exact units of complement (0.5 ml) is added. Standard reagent controls are always included. The normal yolk sac control is used at the same dilution as the group antigen. The tubes are shaken well and incubated in a water bath at 37°C for 2 h. Then 0.5 ml of sensitized sheep erythrocytes is added and the tubes are placed in a water bath for 1 h, after which they are read for hemolysis on a 1+ to 4+ scale (roughly equivalent to 25 to 100% inhibition of erythrocyte lysis). The endpoint of the serum is considered the highest dilution inhibiting at least 50% (2+) hemolysis after a complete inhibition of hemolysis has been observed. It is good practice to shake the tubes to resuspend the settled cells and then to refrigerate them overnight and recheck the results the next morning.

All reagents are available commercially, except for high-titer group antigen. This may be prepared as follows. Yolk sacs of 7-day embryonated eggs are inoculated with *Chlamydia* sp. (e.g., psittacosis isolate 6BC) at a dose estimated to result in death of about 50% of inoculated eggs in 5 to 7 days. Eggs are candled daily, and those dying early are discarded. When the

50% death endpoint is approached, the remaining eggs (recently dead or live) are refrigerated for 3 to 24 h. The yolk sacs are then harvested. If examination of random samples shows large numbers of particles, the yolk sacs are pooled. This preparation may be stored at −20°C until further processing is required. The yolk sacs are ground in a mortar with sterile sand. Beef heart broth (pH 7.0) is added to make a 20% suspension, and the material is cultured for bacteriological sterility. The suspension is placed in a flask containing sterile glass beads and stored at 4°C for 3 to 6 weeks with daily shaking. It is then centrifuged at ca. 500 × *g* to remove coarse particles, transferred to a heavy sterile flask, and steamed at 100°C or immersed in boiling water for 30 min. After it has cooled, liquefied phenol is added to 0.5%. The antigen should then be refrigerated for at least 1 week before being used. It is stable for at least 1 year if not contaminated and should have an antigen titer of 1:256 or greater. A similar preparation from uninfected yolk sacs must be included as one of the controls.

Micro-IF. The micro-IF test is performed against chlamydial organisms grown in yolk sac. The individual yolk sacs are selected for elementary body richness and pretitrated to give an even distribution of particles. It is generally found that a 1 to 3% yolk sac suspension (phosphate-buffered saline, pH 7.0) is satisfactory. The antigens may be stored as frozen aliquots; after thawing, they are well mixed on a Vortex mixer before use. Antigen dots are placed on a slide in a specific pattern, with separate pen points used for each antigen. Each cluster of dots includes all the antigenic types to be tested. The antigen dots are air dried and fixed on slides with acetone (15 min at room temperature). Slides may be stored frozen. When thawed for use, they may "sweat," but they can be conveniently dried (as can the original antigen dots) with the cool air flow of a hair dryer. The slides have serial dilutions of serum (or tears or exudate) placed on the different antigen clusters. The clusters of dots are placed sufficiently separated to avoid the running of the serum from cluster to cluster. After the serum dilutions have been added, the slides are incubated for 0.5 to 1 h in a moist chamber at 37°C. They are then placed in a buffered saline wash for 5 min, followed by a second 5-min wash. The slides are then dried and stained with fluorescein-conjugated anti-human globulin. Conjugates are pretitrated in a known positive system to determine appropriate working dilutions. This reagent may be prepared against any class of globulin being considered (IgA or secretory piece for secretions, IgG, or IgM). Counterstains such as bovine serum albumin conjugated with rhodamine may be included. The slides are then washed twice again, dried, and examined by standard fluorescence microscopy. Use of a monocular tube is recommended to allow greater precision in determining fluorescence of individual elementary body particles. The endpoints are read as the dilution giving bright fluorescence clearly associated with the well-distributed elementary bodies throughout the antigen dot. Identification of the type-specific response is based upon dilution differences reflected in the endpoints for different prototype antigens.

For each run of either CF or micro-IF, known positive and negative sera should always be included. These sera should always duplicate their titers as previously observed within the experimental (±1 dilution) error of the system.

LITERATURE CITED

1. **Banks, J., B. Eddie, M. Sung, N. Sugg, J. Schachter, and K. F. Meyer.** 1970. Plaque reduction technique for demonstrating neutralizing antibodies for *Chlamydia*. Infect. Immun. **2:**443–447.
2. **Beem, M. O., and E. M. Saxon.** 1977. Respiratory-tract colonization and a distinctive pneumonia syndrome in infants infected with *Chlamydia trachomatis*. N. Engl. J. Med. **296:**306–310.
3. **Bowie, W. R., S.-P. Wang, E. R. Alexander, J. Floyd, P. Forsyth, H. Pollock, J.-S. Tin, T. Buchanan, and K. K. Holmes.** 1977. Etiology of nongonococcal urethritis: evidence for *Chlamydia trachomatis* and *Ureaplasma urealyticum*. J. Clin. Invest. **59:**735–742.
4. **Byrne, G. I., and J. W. Moulder.** 1978. Parasite-specified phagocytosis of *Chlamydia psittaci* and *Chlamydia trachomatis* by L and HeLa cells. Infect. Immun. **19:**598–606.
5. **Caldwell, H. D., and C. C. Kuo.** 1977. Serologic diagnosis of lymphogranuloma venereum by counter immunoelectrophoresis with a *Chlamydia trachomatis* protein antigen. J. Immunol. **118:**442–445.
6. **Darougar, S., J. R. Kinnison, and B. R. Jones.** 1971. Simplified irradiated McCoy cell culture for isolation of chlamydiae, p. 63–70. *In* R. L. Nichols (ed.), Trachoma and related disorders caused by chlamydial agents. Excerpta Medica, Amsterdam.
7. **Dhir, S. P., G. E. Kenny, and J. T. Grayston.** 1971. Characterization of the group antigen of *Chlamydia trachomatis*. Infect. Immun. **4:**725–730.
8. **Finn, M. P., A. Ohlin, and J. Schachter.** 1983. Enzyme-linked immunosorbent assay for immunoglobulin G and M antibodies to *Chlamydia trachomatis* in human sera. J. Clin. Microbiol. **17:**848–852.
9. **Grayston, J. T., and S.-P. Wang.** 1975. New knowledge of chlamydiae and the diseases they cause. J. Infect. Dis. **132:**87–105.
10. **Moulder, J. W.** 1966. The relation of the psittacosis group (chlamydiae) to bacteria and viruses. Annu. Rev. Microbiol. **20:**107–130.
11. **Nurminen, M., M. Leinonen, P. Saikku, and P. H. Makela.** 1983. The genus-specific antigen of *Chlamydia*: resemblance to the lipopolysaccharide of enteric bacteria. Science **220:**1279–1281.
12. **Page, L. A.** 1972. Chlamydiosis (ornithosis), p. 414–417. *In* M. S. Hofstad (ed.), Diseases of poultry, 6th ed. Iowa State University Press, Ames.
13. **Page, L. A.** 1974. Order II. *Chlamydiales* Storz and Page 1971, 334, p. 914–928. *In* R. E. Buchanan and N. E. Gibbons (ed.), Bergey's manual of determinative bacteriology, 8th ed. The Williams & Wilkins Co., Baltimore.
14. **Richmond, S. J., and E. O. Caul.** 1977. Single-antigen indirect immunofluorescence test for screening venereal disease clinical populations for chlamydial antibodies, p. 259–265. *In* K. K. Holmes and D. Hobson (ed.), Nongonococcal urethritis and related infections. American Society for Microbiology, Washington, D.C.
15. **Ripa, K. T., and P.-A. Mårdh.** 1977. New simplified culture technique for *Chlamydia trachomatis*, p. 323–327. *In* K. K. Holmes and D. Hobson (ed.), Nongonococcal urethritis and related infections. American Society for Microbiology, Washington, D.C.
16. **Sacks, D. L., and A. B. MacDonald.** 1979. Isolation of type-specific antigen from *Chlamydia trachomatis* by sodium dodecyl sulfate-polyacrylamide gel electrophoresis. J. Immunol. **122:**136–139.
17. **Schachter, J.** 1978. Chlamydial infections. N. Engl. J. Med. **298:**428–435, 490–495, 540–549.

18. **Schachter, J., and C. R. Dawson.** 1978. Human chlamydial infections, p. 273. Publishing Sciences Group, Littleton, Mass.
19. **Schachter. J., M. Grossman, and P. H. Azimi.** 1982. Serology of *Chlamydia trachomatis* in infants. J. Infect. Dis. **146:**530–535.
20. **Stamm, W. E., M. Tam, M. Koester, and L. Cles.** 1983. Detection of *Chlamydia trachomatis* inclusions in McCoy cell cultures with fluorescein-conjugated monoclonal antibodies. J. Clin. Microbiol. **17:**666–668.
21. **Stephens, R. S., M. R. Tam, C.-C. Kuo, and R. C. Nowinski.** 1982. Monoclonal antibodies to *Chlamydia trachomatis*: antibody specificities and antigen characterization. J. Immunol. **128:**1083–1089.
22. **Storz, J.** 1971. *Chlamydia* and *Chlamydia*-induced diseases. Charles C Thomas, Publisher, Springfield, Ill.
23. **Thomas, B. J., R. T. Evans, G. R. Hutchinson, and D. Taylor-Robinson.** 1977. Early detection of chlamydial inclusions combining the use of cycloheximide-treated McCoy cells and immunofluorescence staining. J. Clin. Microbiol. **6:**285–292.
24. **Thomas, B. J., P. Reeve, and J. D. Oriel.** 1976. Simplified serological test for antibodies to *Chlamydia trachomatis*. J. Clin. Microbiol. **4:**6–10.
25. **Wang, S.-P., and J. T. Grayston.** 1970. Immunologic relationship between genital TRIC, lymphogranuloma venereum and related organisms in a new microtiter indirect immunofluorescence test. Am. J. Ophthalmol. **70:**367–374.
26. **Wang, S.-P., J. T. Grayston, E. R. Alexander, and K. K. Holmes.** 1975. Simplified microimmunofluorescence test with trachoma-lymphogranuloma venereum (*Chlamydia trachomatis*) antigens for use as a screening test for antibody. J. Clin. Microbiol. **1:**250–255.
27. **Yoder, B. L., W. E. Stamm, M. C. Koester, and E. R. Alexander.** 1981. Microtest procedure for isolation of *Chlamydia trachomatis*. J. Clin. Microbiol. **13:**1036–1039.

Chapter 86

Section IX. Sexually Transmitted Diseases

Sexually Transmitted Diseases

STEPHEN A. MORSE AND SAMUEL K. SARAFIAN

Before World War II, laboratories had to contend primarily with the five classical venereal diseases: gonorrhea, syphilis, chancroid, lymphogranuloma venereum, and granuloma inguinale. With the introduction of penicillin and other antibiotics, the annual incidence of gonorrhea and syphilis declined, and the remaining three classical venereal diseases all but disappeared from developed countries.

The introduction of oral contraceptives and intrauterine devices in the 1960s was followed by changes in human behavior as well as a marked decrease in the use of condoms, diaphragms, and spermicidal preparations for contraception. In addition, there was a significant increase in the proportion of the population between the ages of 15 and 29 years. These factors were responsible, in part, for the tremendous increase in the incidence of gonorrhea and other sexually transmitted diseases in the United States. Although chancroid, lymphogranuloma venereum, and granuloma inguinale are relatively uncommon in the United States, a number of other diseases have emerged that are considered sexually transmissible. The etiological agents known or thought to be transmitted sexually are presented in Table 1. Most of these agents are discussed in other chapters of this Manual as indicated.

The incidence of specific sexually transmitted diseases varies markedly in the United States (Table 2). Recent advances in microbiological and serological tests have greatly facilitated the identification of their causative agents (see Table 1). It has become apparent that many pathogens can be transmitted via sexual contact. In heterosexuals and particularly in homosexual men, oral-genital and oral-rectal sexual practices have resulted in nonvenereal infections with viral and gastrointestinal pathogens (Table 1).

In women, lower genital tract infections generally involve the vulva, vagina, or cervix. The collection of specimens from the genital tract of women is summarized in Table 3. Usually, the causative organisms of these infections are acquired by sexual or direct contact. However, some organisms that are part of the flora and are normally present in very low numbers may increase sufficiently in number to cause symptoms. Infections may extend to the upper genital tract, involving the uterus, fallopian tubes, ovaries, or abdominal cavity (pelvic inflammatory disease). In men, genital tract infections are invariably caused by orga-

TABLE 1. Sexually transmitted agents and diseases they cause[a]

Etiological agent	Disease or syndrome	Refer to:
Bacterium		
Neisseria gonorrhoeae	Urethritis, epididymitis, cervicitis, proctitis, pharyngitis, conjunctivitis, endometritis, salpingitis, perihepatitis, bartholinitis, amniotic infection syndrome, disseminated gonococcal infection (arthritis, dermatitis, tenosynovitis), prepubertal vaginitis, prostatitis(?), accessory gland infection, chorioamnionitis, premature rupture of membranes, premature delivery	Chapter 17
Treponema pallidum	Syphilis	Chapter 44
Haemophilus ducreyi	Chancroid	Chapter 87
Calymmatobacterium granulomatis	Donovanosis (granuloma inguinale)	Chapter 87
Chlamydia trachomatis	Urethritis, epididymitis, proctitis, cervicitis, endometritis, salpingitis, inclusion conjunctivitis, infant pneumonia, trachoma, lymphogranuloma venereum, perihepatitis, bartholinitis, prepubertal vaginitis, otitis media in infants, chorioamnionitis(?), premature rupture of membranes(?), premature delivery(?)	Chapter 85
Ureaplasma urealyticum	Nongonococcal urethritis(?), acute urethral syndrome in women(?), chorioamnionitis(?), premature delivery(?)	Chapter 36
Mycoplasma hominis	Salpingitis, amnionitis, nonspecific vaginitis(?), postpartum fever	Chapter 36
Gardnerella vaginalis	*Gardnerella*-associated nonspecific vaginosis	Chapter 88
Mobiluncus spp.	Nonspecific vaginosis(?)	
Campylobacter spp. and *Campylobacter*-like organisms	Enteritis, proctitis, and proctocolitis in homosexual men	Chapter 27
Group B beta-hemolytic streptococci	Neonatal sepsis, neonatal meningitis	Chapter 16
Shigella spp. and other *Enterobacteriaceae*	Shigellosis, salmonellosis, enteropathogenic *Escherichia coli*, and other enteric infections in homosexual men	Chapter 24

Continued

TABLE 1—*Continued*

Etiological agent	Disease or syndrome	Refer to:
Virus		
Herpes simplex virus (types 1 and 2)	Primary and recurrent genital herpes, aseptic meningitis, neonatal herpes, cervical dysplasia and carcinoma(?), carcinoma in situ of the vulva	Chapter 64
Cytomegalovirus	Heterophile-negative infectious mononucleosis, congenital infection, gross birth defects and infant mortality, cognitive impairment, cervicitis(?), protean manifestations in the immunosuppressed host	Chapter 65
Human papillomavirus	Genital warts, condyloma acuminatum, laryngeal papilloma(?), cervical and rectal carcinoma(?)	Chapter 82
Molluscum contagiosum virus	Genital molluscum contagiosum	Chapter 68
Hepatitis A virus	Classic hepatitis	Chapter 81
Hepatitis B virus	Classic hepatitis, fulminant hepatitis, chronic active hepatitis, submassive hepatic necrosis, persistent (unresolved) hepatitis, polyarteritis nodosa, chronic membranous glomerulonephritis	Chapter 81
Protozoan		
Trichomonas vaginalis	Vaginitis	Chapter 57
Giardia lamblia	Enteritis in homosexual men (giardiasis)	Chapter 57
Cryptosporidium	Enteritis in immunosuppressed homosexual men (cryptosporidiosis)	Chapter 57
Entamoeba histolytica	Amebiasis in homosexual men	Chapter 57
Fungus		
Candida albicans	Vulvovaginitis, penile candidiasis	Chapter 49
Arthropod		
Phthrius pubis	Pubic lice infestation	Chapter 60
Sarcoptes scabiei	Scabies, Norwegian scabies	Chapter 60

[a] Evidence for sexual transmission of some of these agents is circumstantial or is limited to homosexual men.

nisms which are acquired by sexual contact, since the urethra is a relatively noncolonized environment (Table 4).

Sexually transmitted diseases may involve other anatomical sites in both men and women, depending upon their sexual practices and preferences. Consequently, appropriate specimens must be obtained depending on the site(s) of infection and the organisms suspected. Moreover, it has been shown that a significant proportion of infected individuals harbor two or more sexually transmitted disease agents (5, 7). Therefore, it becomes imperative to obtain the proper specimens and to ensure that all the suspected organisms are sought (refer to Tables 3 and 4).

IDENTIFICATION OF SEXUALLY TRANSMITTED DISEASE AGENTS

Most of the causative agents of sexually transmitted diseases are discussed in detail in individual chapters of this Manual (Table 1). Many infections are caused by agents which can also be acquired by nonsexual contact.

Swabs are often used to obtain specimens for the culture of many of the agents listed in Table 1. The composition of both the tip and shaft of the swab is important. Calcium alginate-tipped swabs are not recommended for obtaining specimens used to inoculate cell cultures for *Chlamydia trachomatis* and herpes simplex virus (1, 6, 10). Apparently, the calcium alginate binds directly to these agents, rendering them noninfectious in cell cultures (1, 10). Instead, swabs with tips composed of rayon, Dacron, cotton (preferably carbon treated), or polyester may be used. For the collection of urethral specimens, swabs should preferably have a narrow diameter that allows them to be inserted approximately 3 to 4 cm into the urethra without causing pain. Swabs with wooden shafts should not be placed in transport media. The wooden shaft contains substances (probably fatty acids) which are inhibitory to *Neisseria gonorrhoeae* and *Ureaplasma urealyticum* (8). The appropriate swabs for

TABLE 2. Incidence of various sexually transmitted diseases in the United States

Disease	No. of cases[a]	No. of visits to office-based private physicians
Gonorrhea	1,003,958	
Syphilis	67,049	
Trichomoniasis		967,900[b]
Genital herpes		200,000–500,000[c]
Genital warts		900,000[d]
Nongonococcal urethritis		1,249,300[b]
Scabies		746,900[b]
Pubic lice		58,500[b]
Chancroid	840	
Granuloma inguinale	76	
Lymphogranuloma venereum	250	

[a] Number of cases reported in 1979.
[b] Figures from the National Ambulatory Medical Care Survey, National Center for Health Statistics; 2-year average, 1975–1976.
[c] Estimated annual incidence.
[d] Data for 1979 from reference 4.

TABLE 3. Collection of specimens from the genital tract of women[a]

Site	Organism	Specimen type and preparation	Collection technique
Vulva	*T. pallidum*	Lesion exudate; clean surface with saline and remove crust if present	Abrade lesion until serous fluid emerges. Wipe away. Try to avoid bleeding. Press base of lesion until clear fluid is expressed. Touch slide to fluid, and then cover with cover slip. Examine immediately by dark-field microscopy.
	Herpes simplex virus	Vesicle fluid (culture) and lesion scrapings (culture and direct examination); gently clean surface with saline	1. Aspirate vesicular fluid with a 26- to 27-gauge needle on a tuberculin syringe, and immediately place it in carrier medium or 2. Unroof vesicle and collect fluid with a swab (not calcium alginate). 3. Lesion scrapings: scrape the base of an open vesicle with a sterile scalpel blade or rub the base vigorously with a swab.
	H. ducreyi	Aspirate or swab; clean lesion with saline	1. Irrigate with 0.5 ml of saline in portions. Aspirate with flame-smoothed Pasteur pipette. 2. Moisten swab with saline. Swab lesion base.
	C. trachomatis (types L_1, L_2, L_3:LGV)	Lymph node aspirate; skin decontamination	1. Fluctuant lymph node: aspirate pus. 2. Lymph node not fluctuant: inject a small amount of sterile saline and aspirate.
	C. granulomatis	Scrapings or biopsy; clean lesion with saline	Scrape beneath extending border of ulcer. Punch biopsy
	Molluscum contagiosum virus	Biopsy	Punch biopsy
	Human papillomavirus	Biopsy	Punch biopsy
Bartholin's gland	*N. gonorrhoeae, C. trachomatis*, aerobic and anaerobic bacteria	Aspirate; skin decontamination with povidone-iodine, not alcohol	Express exudate from duct. Aspirate abscesses with needle and syringe.
Vagina	*T. vaginalis*	Vaginal secretions; speculum without lubricant (cultures of the external vaginal orifice not useful)	1. Collect secretions with a pipette or a swab of mucosa high in vaginal canal. 2. Prepare wet mount with saline for *T. vaginalis* and clue cells. 3. Gram stain for nonspecific vaginosis.
	C. albicans		
	G. vaginalis, Mobiluncus spp. (nonspecific vaginosis)		
	Group B beta-hemolytic streptococci (*Streptococcus agalactiae*)	Swab	Mucosal swab
Cervix	*N. gonorrhoeae*	Swab; wipe cervix clean of vaginal secretions and mucus, and use a speculum with no lubricant	Gently compress cervix between speculum blades to express any endocervical exudate. Collect on swab with a ringing motion.

Continued

TABLE 3—*Continued*

Site	Organism	Specimen type and preparation	Collection technique
Cervix *continued*			
	C. trachomatis (serotypes A through K)	As for *N. gonorrhoeae*	Insert swab a few millimeters past cervical os; rotate firmly to obtain cervical cells.
	Herpes simplex virus	See under Vulva	
Urethra	*N. gonorrhoeae*	Swab; wipe urethra clean; specimen should be collected at least 1 h after urination	"Milk" urethra and collect discharge. If no discharge, collect as for *C. trachomatis*.
	C. trachomatis	As for *N. gonorrhoeae*	Insert thin urogenital swab ca. 2 cm into the urethra. Gently rotate and remove.
Rectum	*N. gonorrhoeae*	Rectal swab	Insert swab past anal sphincter, move swab from side to side, allow 10 to 30 s for absorption of organisms, and withdraw.
	C. trachomatis	Rectal swab	As for *N. gonorrhoeae*
Endometrium	Aerobic and anaerobic bacteria, *C. trachomatis*	Aspirate; swab	1. Transabdominal aspirate 2. Protected swab
Cul-de-sac	*N. gonorrhoeae*, aerobic and anaerobic bacteria	Aspirate of peritoneal fluid; skin decontamination	Aspirate through posterior vaginal vault or directly during surgery.
Fallopian tubes	*N. gonorrhoeae, C. trachomatis*, aerobic and anaerobic bacteria	Aspirate or swab at surgery	Surgery is necessary. Bronchoscopy cytology brushes may be used if exudate is not expressed.
Amniotic fluid	Aerobic and anaerobic bacteria	Fluid	Pressure catheter, amniocentesis, aspiration with syringe at cesarean section

[a] Table adapted from *Cumitech 17* (2).

TABLE 4. Collection of specimens from the urogenital tract of men

Site	Organism	Specimen type and preparation	Collection technique
Urethra	*N. gonorrhoeae*	Urethral exudate	Use a narrow-diameter swab inserted 3 to 4 cm into the urethra.
	C. trachomatis	Urethral exudate	As for *N. gonorrhoeae*
	U. urealyticum	Urethral exudate	As for *N. gonorrhoeae*
Rectum	*N. gonorrhoeae*	Anoscopy followed by collection of specimen from areas containing pus	With a swab, rub areas containing pus.
	C. trachomatis	Rectal swab	As for *N. gonorrhoeae*
	T. pallidum	Lesion exudate; clean surface with saline and remove crust if present	Abrade lesion until serous fluid emerges. Wipe away. Transfer serous fluid to slide. If specimen is scant, add a small amount of saline and prepare a wet mount. Examine immediately by dark-field microscopy.
	Herpes simplex virus	Vesicle fluid (culture) and lesion scrapings (culture and direct examination); gently clean surface with saline	1. Unroof vesicle and collect fluid with a swab. 2. Lesion scrapings: rub base of open vesicle vigorously with a swab.
Penis	*T. pallidum*	As for rectum	As for rectum
	H. ducreyi (chancroid)	Clean lesion thoroughly with sterile saline	Moisten cotton-tipped swab with saline, and swab the lesion.
	Herpes simplex virus	As for rectum	As for rectum

TABLE 5. Diagnostic procedures for identification of common agents of sexually transmitted diseases[a]

Etiological agent	Direct observation	Transport medium[b]	Culture medium[b]	Serological diagnosis
Bacterium				
N. gonorrhoeae	1. Gram stain of exudate 2. Direct FA on exudate	1. Nonnutritive: Stuart or Amies 2. Nutritive: JEMBEC	Modified Thayer-Martin, Martin-Lewis, or New York City	Antigen detection by EIA; antibody tests not adequate
T. pallidum	1. Dark-field microscopic examination of lesion exudate (not for oral lesions) 2. Direct FA on lesion exudate	None	Cannot be cultivated on laboratory media	Screening tests (reaginic antibodies): VDRL, RPR, ART, etc. Treponeme-specific antibody tests: FTA-ABS, TPI, hemagglutination
H. ducreyi	Stained smears of lesion exudate (Giemsa, Wright, Gram, Unna-Pappenheim, or Sellers stain)	Direct inoculation	Enriched chocolate agar	Not recommended
C. granulomatis	Giemsa-, Dieterle-, or Wright-stained tissue smears	None	Routine culture not recommended	Not routinely available
C. trachomatis	Direct FA on exudate (LGV rarely diagnosed by microscopy)	0.2 M sucrose in 0.02 M phosphate buffer (pH 7.0 to 7.2) with 5% fetal calf serum and antibiotics (aminoglycosides and fungicides); refrigerate if specimen can be processed within 24 h after collection; if not, store at −60°C	Cycloheximide-treated McCoy cells	1. Antigen detection by EIA 2. CF test for LGV; micro-IF test for other serotypes on acute- and convalescent-stage sera
U. urealyticum	Giemsa-stained smears; requires experienced observer	1. Tryptic soy broth with 0.5% bovine serum albumin 2. Sucrose phosphate buffer transport medium (as for *C. trachomatis*) without gentamicin	E agar, MES agar, urea broth, or New York City medium	Not routinely available
M. hominis	Giemsa-stained smears; requires experienced observer	Same as *U. urealyticum*	E agar, MES agar, or New York City medium	CF on acute- and convalescent-stage sera
G. vaginalis	Presence of clue cells in Gram-stained smear		HBT medium	No
Mobiluncus spp.	Kopeloff modification of Gram method with basic fuchsin counterstain (3)		Brain heart infusion medium containing 2% agar, 5% defibrinated sheep blood, and 2% Difco Proteose Peptone no. 3 (11); incubate anaerobically	No
Campylobacter spp. and *Campylobacter*-like organisms	No	Campy-thio, Cary-Blair, alkaline peptone water, and buffered glycerol-saline; all effective	Campy-BAP for direct isolation, Campy-thio for indirect isolation	
Virus				
Herpesvirus hominis (types 1 and 2)	1. Tzanck test 2. Electron microscopy 3. Direct immunofluorescence or direct immunoperoxidase 4. Pap smear in women	Viral transport medium (Hanks balanced salt solution, 2% fetal calf serum, and antibiotics); store refrigerated, but do not freeze	Cell culture; confirmation and typing by immunofluorescence	Limited value
Human papillomavirus	1. Clinical diagnosis 2. Pap smear in women 3. Immunofluorescent or peroxidase-antiperoxidase staining of tissue sections		No	No

Continued

TABLE 5—*Continued*

Etiological agent	Direct observation	Transport medium[b]	Culture medium[b]	Serological diagnosis
Virus *continued*				
Molluscum contagiosum virus	Clinical diagnosis		No	No
Protozoan				
T. vaginalis	Wet mount		Cysteine-peptone-liver-maltose (9)	No
Cryptosporidium	Auramine staining or DMSO-modified, acid-fast stain of fecal smears		Not routinely available	No
Fungus				
C. albicans	1. Gram-stained smear 2. Wet mount 3. KOH preparation		Sabouraud medium; results should be interpreted in conjunction with clinical presentation	No
Arthropod				
P. pubis	Direct observation of adults or nits			No
S. scabiei	Direct observation of adults or ova			No

[a] Abbreviations: FA, fluorescent-antibody test; EIA, enzyme immunoassay; VDRL, Venereal Disease Research Laboratory; RPR, rapid plasma reagin-card test; ART, automatic reagin test; FTA-ABS, fluorescent treponemal antibody-absorption test; TPI, *Treponema pallidum* immobilization; LGV, lymphogranuloma venereum; CF, complement fixation test; micro-IF, micro-immunofluorescence test; DMSO, dimethyl sulfoxide.

[b] See Chapter 111 in this Manual for descriptions of media.

each agent are mentioned in the individual chapters of this Manual.

Ideally, specimens should be inoculated onto primary isolation media at the time of collection. However, from a practical point of view, this is not always feasible. Various transport media are available. Appropriate transport media and the conditions of transport are indicated in other chapters of this Manual. For instance, transport medium for *C. trachomatis* should be placed in plastic tubes. Transport medium stored in glass tubes can cause cytopathic effects when inoculated onto cell cultures (6).

A summary of diagnostic procedures for the identification of sexually transmitted disease agents is presented in Table 5. Most of these procedures are detailed in the appropriate chapters of this Manual.

LITERATURE CITED

1. **Crane, L. R., P. A. Gutterman, T. Chapel, and A. M. Lerner.** 1980. Incubation of swab materials with herpes simplex virus. J. Infect. Dis. **141**:531.
2. **Eschenbach, D., H. M. Pollock, and J. Schachter.** 1983. Cumitech 17, Laboratory diagnosis of female genital tract infections. Coordinating ed., S. J. Rubin. American Society for Microbiology, Washington, D.C.
3. **Holdeman, L. V., E. P. Cato, and W. E. C. Moore (ed.).** 1977. Anaerobe laboratory manual, 4th ed. Virginia Polytechnic Institute and State University, Blacksburg.
4. **Holmes, K. K., T. A. Bell, and R. E. Berger.** 1984. Epidemiology of sexually transmitted diseases. Urol. Clin. North Am. **11**:3–13.
5. **Johannisson, G., G.-B. Löwhagen, and E. Lycke.** 1980. Genital *Chlamydia trachomatis* infection in women. Obstet. Gynecol. **56**:671–675.
6. **Mardh, P.-A., and B. Zeeberg.** 1981. Toxic effect of sampling swabs and transportation test tubes on the formation of intracytoplasmic inclusions of *Chlamydia trachomatis* in McCoy cell cultures. Br. J. Vener. Dis. **57**:268–272.
7. **Oriel, J. D., P. Reeve, B. J. Thomas, and C. S. Nicol.** 1975. Infection with *Chlamydia* group A in men with urethritis due to *Neisseria gonorrhoeae*. J. Infect. Dis. **131**:376–382.
8. **Poulin, S. A., R. B. Kundsin, and H. W. Horne, Jr.** 1979. Survival of *Ureaplasma urealyticum* on different kinds of swabs. J. Clin. Microbiol. **10**:601–603.
9. **Rayner, C. F. A.** 1968. Comparison of culture media for the growth of *Trichomonas vaginalis*. Br. J. Vener. Dis. **44**:63–66.
10. **Smith, T. F., and L. A. Weed.** 1983. Evaluation of calcium alginate-tipped aluminum swabs transported in Culturettes® containing ampules of 2-sucrose phosphate medium for recovery of *Chlamydia trachomatis*. Am. J. Clin. Pathol. **80**:213–215.
11. **Spiegel, C. A., and M. Roberts.** 1984. *Mobiluncus* gen. nov., *Mobiluncus curtisii* subsp. *curtisii* sp.nov., *Mobiluncus curtisii* subsp. *holmesii* subsp. nov., and *Mobiluncus mulieris* sp. nov., curved rods from the human vagina. Int. J. Syst. Bacteriol. **34**:177–184.

Chapter 87

Haemophilus ducreyi and *Calymmatobacterium granulomatis*

WILLIAM L. ALBRITTON, FRANK A. PLUMMER, FRANCES O. SOTTNEK, AND STEPHEN J. KRAUS

CHARACTERIZATION

Haemophilus ducreyi

The causative agent of chancroid, *Haemophilus ducreyi*, was first recognized in stained smears of genital ulcer material by Auguste Ducrey in 1889 (15). Initial isolation of the organism has been credited (27, 43) to Petersen (1895), Istoamanov and Akopiantz (1897), Lenglet (1898), and Bezancon, Griffin, and Lesourd (1900). The organism was described as a nonmotile, pleomorphic, gram-negative bacillus occurring in streptobacillary chains and parallel chains referred to as "schools of fish." This bacillus of Ducrey was included in the genus *Haemophilus* when it was proposed by Winslow et al. (45) and subsequently was shown by Lwoff and Pirosky (33) to require hemin (X factor). Kilian and Theilade (30) confirmed the gram-negative character of the cell wall by electron microscopy. Kilian determined the guanine plus cytosine (G+C) content to be 0.38 mole fraction and confirmed the hemin requirement (29). These characteristics are sufficient for inclusion of the organism in the genus *Haemophilus*. Although *H. ducreyi* shares some antigenic relationships with other members of the genus *Haemophilus*, DNA-DNA hybridization studies suggest that it is unrelated, or only distantly related, to other species in the genus (1).

Calymmatobacterium granulomatis

The causative agent of granuloma inguinale (donovanosis), *Calymmatobacterium granulomatis*, was first recognized as inclusions in mononuclear cells in smears from genital ulcers by C. Donovan in 1905 (14). The initial isolation has been credited to Anderson (2). The organism has been described as a nonmotile, pleomorphic, gram-negative, encapsulated bacillus, occasionally showing a bipolar condensation of chromatin described as safety-pin forms. Referred to as *Donovania granulomatis* by Anderson, DeMonbreun, and Goodpasture (3), current nomenclature has reverted to *C. granulomatis*, the earlier designation of Donovan bodies by Aragao and Vianna (4). Although there are many descriptions in the literature between 1945 and 1965 of the direct staining and growth characteristics of *C. granulomatis* (12, 16, 18, 21), there are no organisms currently available in known culture collections, and it has not been possible to study relationships between strains of *C. granulomatis* or between *C. granulomatis* and other species by modern techniques such as G+C content and DNA-DNA hybridization. For this reason, a systematic description of the genus is not available, and the relationship of *C. granulomatis* to Donovan bodies and the syndrome granuloma inguinale is unclear.

CLINICAL SIGNIFICANCE

Chancroid

Chancroid, caused by *H. ducreyi*, was first differentiated from syphilis by Bassereau in 1842. The clinical features associated with chancroid include painful genital ulcers, single or multiple, accompanied by painful and occasionally suppurative inguinal buboes. In sporadic cases, it is not possible to clinically distinguish chancroid from other genital ulcer diseases such as syphilis, herpes genitalis, lymphogranuloma venereum, granuloma inguinale, and traumatic ulcer. Reported cases of chancroid have declined in North America since 1960, but major urban epidemics have occurred (25, 26). Outbreaks of chancroid have usually been prostitute associated, and disease is reported 5 to 10 times more frequently in males. In tropical areas, the incidence of chancroid is frequently second only to gonorrhea as a sexually transmitted disease syndrome (37). The causative agent has been isolated only from humans, and reports of asymptomatic carriage have not been regularly confirmed.

The chancre begins as a tender papule after a 3- to 10-day incubation period, becoming gradually eroded and ulcerated over a 24- to 48-h period. The ulcer is characteristically painful, ragged, and undermined with an elevated erythematous border. In men, ulcers occur most frequently on the prepuce, frenulum, or coronal sulcus (Fig. 1). In women, the vaginal fourchette and labia are commonly involved. Secondary lesions due to autoinoculation may occur. Inguinal adenitis occurs in about 30% of patients, and in one-third of patients, it progresses to fluctuant bubo formation.

Granuloma inguinale

Granuloma inguinale, caused by *C. granulomatis*, was first recognized by McLeod in India in 1882 (35). There have been few studies of biopsy- or culture-proven cases in recent years, and the clinical and epidemiologic features are less clear than for chancroid (19, 32). The onset of clinical features is gradual, with incubation periods estimated from days to months. Lesions begin as small nonpainful papules or nodules which eventually erode, leaving beefy red, indurated, hypertrophic granulation tissue having a velvety appearance. Lesions show an anatomical distribution similar to those of chancroid. Several studies have reported nongenital lesions, and nonsexual transmission is supported by the reported occurrence of the disease in young children and sexually inactive individuals, its rarity in prostitutes, and its rarity in sexual partners. Unlike chancroid, bubo formation is rare, but involvement of subcutaneous tissue in the

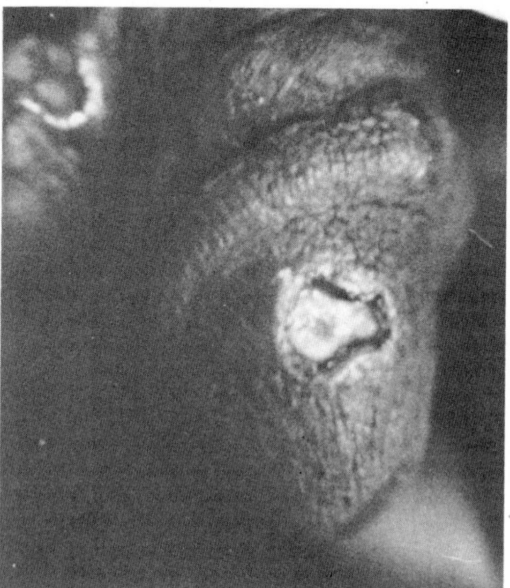

FIG. 1. Typical chancroid ulcer.

inguinal region can result in pseudobubo formation, resembling the lymphadenopathy seen in lymphogranuloma venereum. Untreated lesions are usually progressive and are deforming (17). The relationship of chancroid and granuloma inguinale is not clear, but the clinical presentation of chancroid may resemble granuloma inguinale (31).

COLLECTION, TRANSPORT, AND STORAGE OF SPECIMENS

Specimens for the diagnosis of chancroid or granuloma inguinale should, in general, be obtained as part of the diagnostic workup for genital ulcer disease, with or without inguinal adenopathy. As such, in addition to the specimens necessary for the diagnosis of chancroid or granuloma inguinale, specimens should be obtained for the diagnosis of syphilis, herpes genitalis, occasionally lymphogranuloma venereum, and rarely other noninfectious conditions such as malignancy.

Chancroid

Specimens for the isolation of *H. ducreyi* should be obtained from the base of the ulcer or from the inguinal bubo. Isolation from inguinal bubo aspirates has been less successful than from genital ulcer material. Specimens should be obtained from the base of the ulcer with a swab and inoculated directly onto selective media. Precleaning the ulcer is not required. Comparative studies of swab material are not available, but cotton swabs have been used successfully in several studies. In general, the transport and storage of primary specimens have not been successful; however, there is a single report of a study in which Amies medium was successfully used (8). Heavy growth of 24- to 48-h subcultures of presumptively positive isolates may be scraped with a sterile swab and transported as subsurface stabs in chocolate agar. Similar growth from pure subcultures may be suspended in skim milk or lysed horse blood and frozen at −70°C or lyophilized in similar media or serum-inositol for long-term storage.

Granuloma inguinale

There is no literature on the collection, transport, and storage of specimens for the isolation of *C. granulomatis*. The most suitable material for histologic diagnosis, and presumably for culture, is a biopsy of subsurface tissue from an area of active granulation. Formalin-fixed tissue is less satisfactory for examination by staining; thus, tissue biopsy material should be transported to the laboratory in a sterile, dry container or a container containing preservative-free saline. The absence of strains of *C. granulomatis* in existing culture collections precludes data on the long-term storage of isolates.

DIRECT EXAMINATION

For the diagnosis of chancroid, direct examination of genital ulcer material by Gram stain is not recommended. In Nairobi, Kenya, where 70% of genital ulcers were confirmed by culture to be due to *H. ducreyi*, the sensitivity of the Gram stain was less than 50%, with similarly low specificity and positive predictive value (37). Similar findings have been reported in other studies (8). Direct examinations of genital ulcer material by an indirect fluorescent-antibody technique (11) and electron microscopy (34) have been reported to be useful for the diagnosis of chancroid. Nevertheless, direct examination may be misleading due to the polymicrobial flora of genital ulcers (7). Until sufficient information is available to assess sensitivity, specificity, and predictive value, routine use of direct examination methods cannot be recommended.

For the diagnosis of granuloma inguinale, direct examination of stained smears or tissue sections is the usual method (9, 10, 13). Freshly prepared smears from a crushed piece of biopsy material are stained with Giemsa, Dieterle, or other suitable stain, and the diagnosis is based on finding typical, intracellular, encapsulated *C. granulomatis* organisms within histiocytes (Fig. 2).

CULTURE AND ISOLATION

H. ducreyi

The recommended procedure for the isolation of *H. ducreyi* is the direct inoculation of specimens onto one or more selective solid media. The use of more than one type of medium generally improves isolation rates. Agar media supplemented with IsoVitaleX (BBL Microbiology Systems) or CVA enrichment (GIBCO Laboratories) and incorporating vancomycin (3 mg/liter) as a selective antibiotic have been most successful for isolation. Various base formulations have been used, including gonococcal agar base with bovine hemoglobin (24), Mueller-Hinton agar base with chocolatized horse blood (5) or sheep blood (38), and heart infusion agar base with 5% defibrinated rabbit blood (42) or sheep blood (31). Improved isolation rates have been obtained with most media by the incorporation of fetal bovine serum. There are currently no commercial sources of selective media for the isolation of *H. ducreyi*. The ability of nonselective chocolate agar to support the growth of *H. ducreyi* varies from lot to lot, but Hannah and Greenwood (26) were successful in isolating *H. ducreyi* during an

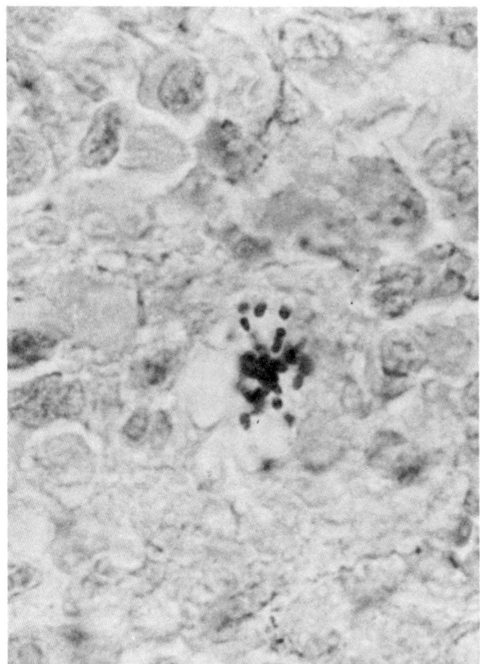

FIG. 2. Histologic appearance of Donovan bodies in dermal granulation tissue stained with Dieterle silver impregnation stain. ×1,120.

epidemic of chancroid with a commercial source of chocolate agar. Inoculated media should be incubated in a water-saturated atmosphere containing 5 to 10% CO_2. Candle extinction jars in which a moistened towel has been placed are satisfactory for this purpose. A reduced incubation temperature of 33 to 35°C is important for the isolation of *H. ducreyi*. Small, nonmucoid, yellow-grey, semiopaque or translucent colonies that can be pushed intact across the agar surface usually appear in 24 to 48 h but may appear as late as 96 h after inoculation. Recovery of viable organisms declines after 48 h; therefore, early subculturing of suspicious colonies is recommended.

Older cultural methods, such as autoinoculation and growth in human serum or other liquid blood media, offer no advantages over new solid media (24, 38). Considering negative-culture clinical chancroid as a false negative, the culture sensitivity for a single medium was found to be ca. 70% and was over 80% when two media were used (37a). Since the etiology of negative-culture clinical chancroid may not necessarily be *H. ducreyi*, culture sensitivity with available media may actually be higher. It is not uncommon for laboratories to experience initial difficulty in the isolation of *H. ducreyi*. Every effort should be made, however, to establish a cultural diagnosis in patients with genital ulcers determined not to be herpes genitalis or syphilis.

C. granulomatis

Initial isolations of *C. granulomatis* were reported after the inoculation of embryonated hen eggs. The best in vitro media were various liquid or biphasic media which contained yolk material and showed a reduced oxidation-reduction potential, such as the yolk-thioglycolate medium described by Goldberg (20). Recovery of the organism has been largely unsuccessful. Reported procedures for the culture and isolation of *C. granulomatis* cannot be recommended until additional studies are available which demonstrate acceptable sensitivity of the isolation procedure when compared with direct examination for Donovan bodies.

IDENTIFICATION

H. ducreyi

Cultural characteristics. *H. ducreyi* has a characteristic colonial morphology (Fig. 3). Colonies are yellow-grey and can be moved intact across the surface of the agar. Pitting of the agar surface is not observed. On rabbit blood agar, zones of alpha-hemolysis may be seen, especially in areas of subsurface inoculation.

Microscopic examination. On Gram stain, organisms appear as gram-negative, pleomorphic coccobacilli, with many in single chains (streptobacillary chaining), parallel chains (railroad track or school-of-fish patterns), and clumps or whorls (fingerprint patterns) (Fig. 4). Organisms with the colonial and microscopic appearance of *H. ducreyi* should be picked for further identification.

Biochemical reactions. The minimal biochemical characteristics required to identify an organism as *H. ducreyi* are given in Table 1. Although the hemin (X factor) requirement may be determined by satellite growth around factor disks on nutritionally deficient media, as for other *Haemophilus* species, many clinical isolates of *H. ducreyi* fail to grow on such media. The preferred test for the hemin requirement, therefore, is the porphyrin test described by Kilian (22, 28). Secondary tests of differential value in the identification of *H. ducreyi* are also given in Table 1. The oxidase reaction is method dependent (36) and re-

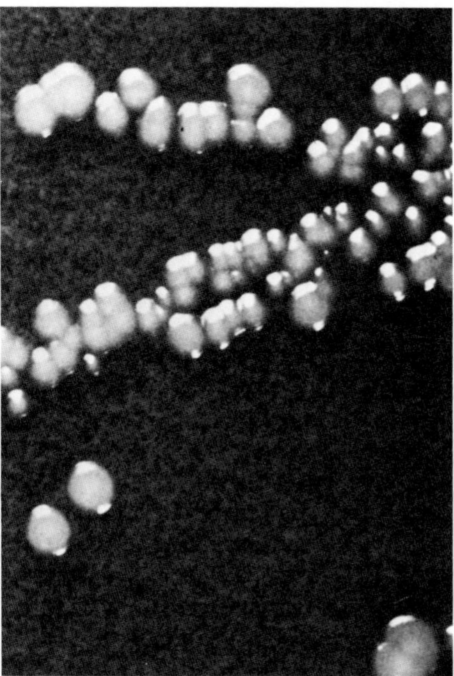

FIG. 3. Colonial morphology of *H. ducreyi*.

quires *N,N,N,N*-tetramethyl-*p*-phenylenediamine as substrate. This test is generally negative for *H. ducreyi* with *N,N*-dimethyl-*p*-phenylenediamine as substrate. Many strains of *H. ducreyi* will produce a change in the color of the indicator during incubation in enteric base medium containing various carbohydrates. In general, most of these strains will cause a decrease in the pH of the medium when they are incubated in the absence of carbohydrate, so that acid production alone should not be used as a differential characteristic. Strains of *H. ducreyi* show a wide spectrum of peptidase activity, and the enzymatic profile may be useful for the recognition of strains within the species or for identification (6). The characteristics which differentiate *H. ducreyi* from related species are presented in Chapter 33. Ducreyi-like bacteria described by Ursi et al. (44) and isolated from genital ulcers may be differentiated from *H. ducreyi* on the basis of factor requirements and the absence of alkaline phosphatase.

Other methods of identification. Strain- and species-specific protein profiles have been observed with isolates of *H. ducreyi* (39). Such patterns may be useful for epidemiologic studies, but they have not yet been developed for use in identification. Serologic tests have been developed for the diagnosis of chancroid, but these have not been evaluated in studies with the more sensitive isolation procedures. Animal studies have demonstrated the ability of isolated strains to produce ulcers on rabbit skin (23), but such pathogenicity tests are not required to confirm a diagnosis of chancroid or to identify *H. ducreyi*.

C. granulomatis

Other than the characteristics presented for the direct examination of tissue for *C. granulomatis*, insufficient information is available to provide procedures for the identification of gram-negative, pleomorphic organisms such as *C. granulomatis*. Serologic tests have shown relationships of *C. granulomatis* to the *Enterobacteriaceae*, specifically *Klebsiella* spp. (40), but the tests require supporting data and are not useful at present for the diagnosis of granuloma inguinale.

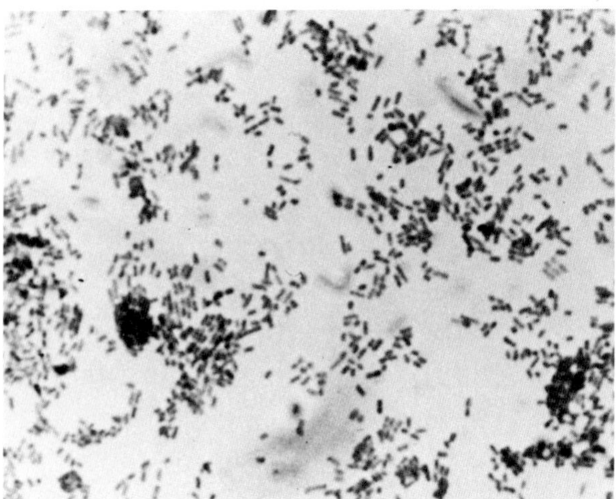

FIG. 4. Gram stain of agar-grown *H. ducreyi* showing typical pleomorphic, gram-negative bacilli in clumps, single chains, and parallel chains.

TABLE 1. Major differential biochemical characteristics of *H. ducreyi*

Characteristic	Reaction[a]
Factor requirement[b]:	
X	+
V	−
Nitrate reduction[b]	+
Alkaline phosphatase[b]	+
Catalase	−
Oxidase	+[c]
Indole	−
Urease	−
Ornithine decarboxylase	−
γ-Glutamyl aminopeptidase	−
Acid production from:	
Glucose	−[d]
Sucrose	−
Lactose	−

[a] +, >90% of strains positive; −, >90% of strains negative.
[b] Minimum biochemical characteristics required for identification.
[c] With *N,N,N,N*-tetramethyl-*p*-phenylenediamine indicator.
[d] Some strains may show acid production in enteric base medium containing glucose but, in general, they also show acid production in the same basal medium without glucose or other carbohydrate source.

LITERATURE CITED

1. **Albritton, W. L, J. K. Setlow, M. Thomas, F. Sottnek, and A. G. Steigerwalt.** 1984. Heterospecific transformation in the genus *Haemophilus*. Mol. Gen. Genet. **193**:358–363.
2. **Anderson, K.** 1943. The cultivation from granuloma inguinale of a microorganism having the characteristics of Donovan bodies in the yolk sac of chick embryos. Science **97**:560–561.
3. **Anderson, K., W. A. DeMonbreun, and E. W. Goodpasture.** 1945. An etiologic consideration of *Donovania granulomatis* cultivated from granuloma inguinale (three cases) in embryonic yolk. J. Exp. Med. **81**:25–39.
4. **Aragao, H., and G. Vianna.** 1913. Pesquizas sobre o granuloma venereo. Mem. Inst. Oswaldo Cruz **5**:211–238.
5. **Bilgeri, Y. R., R. C. Ballard, M. O. Duncan, A. C. Mauff, and H. J. Koornhoff.** 1982. Antimicrobial susceptibility of 103 strains of *Haemophilus ducreyi* isolated in Johannesburg. Antimicrob. Agents Chemother. **22**:686–688.
6. **Casin, I. M., M. J. Sanson-Le Pors, M. F. Gorce, M. Ortenberg, and Y. Perol.** 1982. The enzymatic profile of *Haemophilus ducreyi*. Ann. Microbiol. **133**:379–388.
7. **Chapel, T. A., W. J. Brown, C. Jeffries, and J. A. Stewart.** 1977. How reliable is the morphologic diagnosis of penile ulcerations? Sex. Transm. Dis. **4**:150–154.
8. **Choudhary, B. P., S. Kumari, R. Bhatia, and D. S. Agarwal.** 1982. Bacteriologic study of chancroid. Indian J. Med. Res. **76**:379–385.
9. **Davis, C. M.** 1970. Granuloma inguinale. A clinical, histological, and ultrastructural study. J. Am. Med. Assoc. **211**:632–636.
10. **Davis, C. M., and C. Collins.** 1969. Granuloma inguinale: an ultrastructural study of *Calymmatobacterium granulomatis*. J. Invest. Dermatol. **53**:315–321.
11. **Denys, G. A., T. A. Chapel, and C. D. Jeffries.** 1978. An indirect fluorescent antibody technique for *Haemophilus ducreyi*. Health Lab. Sci. **15**:128–132.
12. **Dienst, R. B., R. B. Greenblatt, and C. H. Chen.** 1948. Laboratory diagnosis of granuloma inguinale and studies on the cultivation of the Donovan body. Am. J. Syph. Gon. Vener. Dis. **32**:301–306.
13. **Dodson, R. F., G. S. Fritz, W. R. Hubler, A. H. Rudolph,**

J. M. Knox, and L. W.-F. Chu. 1974. Donovanosis: a morphologic study. J. Invest. Dermatol. **62:**611–614.

14. **Donovan, C.** 1905. Medical cases from Madras General Hospital. Indian Med. Gaz. **40:**411–414.
15. **Ducrey, A.** 1889. Experimentelle Untersuchungen uber den Ansteckungsstoff des weichen Schankers und uber die Bubonen. Montasch Prakt. Dermatol. **9:**387–405.
16. **Dunham, W., and G. Rake.** 1948. Cultural and serologic studies on granuloma inguinale. Am. J. Syph. Gon. Vener. Dis. **32:**145–149.
17. **Fritz, G. S., W. R. Hubler, R. F. Dobson, and A. Rudolph.** 1975. Mutilating granuloma inguinale. Arch. Dermatol. **111:**1464–1465.
18. **Goldberg, J.** 1962. Studies on granuloma inguinale. V. Isolation of a bacterium resembling *Donovania granulomatis* from the faeces of a patient with granuloma inguinale. Br. J. Vener. Dis. **38:**99–102.
19. **Goldberg, J.** 1964. Studies on granuloma inguinale. VII. Some epidemiological considerations of the disease. Br. J. Vener. Dis. **40:**140–145.
20. **Goldberg, J., R. H. Weaver, and H. Packer.** 1953. Studies on granuloma inguinale. Bacteriologic behavior of *Donovania granulomatis*. Am. J. Syph. Gon. Vener. Dis. **37:**60–70.
21. **Greenblatt, R. B., R. B. Dienst, and E. R. Pund.** 1939. Experimental and clinical granuloma inguinale. J. Am. Med. Assoc. **113:**1109–1116.
22. **Hammond, G. W., C.-J. Lian, J. C. Wilt, W. L. Albritton, and A. R. Ronald.** 1978. Determination of the hemin requirement of *Haemophilus ducreyi*: evaluation of the porphyrin test and media used in the satellite growth test. J. Clin. Microbiol. **7:**243–246.
23. **Hammond, G. W., C. J. Lian, J. C. Wilt, and A. R. Ronald.** 1978. Antimicrobial susceptibility of *Haemophilus ducreyi*. Antimicrob. Agents Chemother. **13:**608–612.
24. **Hammond, G. W., C. J. Lian, J. C. Wilt, and A. R. Ronald.** 1978. Comparison of specimen collection and laboratory techniques for isolation of *Haemophilus ducreyi*. J. Clin. Microbiol. **7:**39–43.
25. **Hammond, G. W., M. Slutchuk, J. Scatliff, E. Sherman, J. C. Wilt, and A. R. Ronald.** 1980. Epidemiologic, clinical, laboratory and therapeutic features of an urban outbreak of chancroid in North America. Rev. Infect. Dis. **2:**867–879.
26. **Hannah, P., and J. R. Greenwood.** 1982. Isolation and rapid identification of *Haemophilus ducreyi*. J. Clin. Microbiol. **16:**861–864.
27. **Himmel, J.** 1901. Contribution a l'etude de l'immunite des animaux vis-a-vis du bacille du chancre mou. Ann. Inst. Pasteur **15:**928–940.
28. **Kilian, M.** 1974. A rapid method for the differentiation of *Haemophilus* strains. The porphyrin test. Acta Pathol. Microbiol. Scand Sect. B **82:**835–842.
29. **Kilian, M.** 1976. A taxonomic study of the genus *Haemophilus* with the proposal of a new species. J. Gen. Microbiol. **93:**9–62.
30. **Kilian, M., and J. Theilade.** 1975. Cell wall ultrastructure of strains of *Haemophilus ducreyi* and *Haemophilus piscium*. Int. J. Syst. Bacteriol. **25:**351–356.
31. **Kraus, S. J., B. S. Werman, J. W. Biddle, F. O. Sottnek, and E. P. Ewing.** 1982. Pseudogranuloma inguinale caused by *Haemophilus ducreyi*. Arch. Dermatol. **118:**494–497.
32. **Kuberski, T.** 1980. Granuloma inguinale (donovanosis). Sex. Transm. Dis. **7:**29–36.
33. **Lwoff, A., and I. Pirosky.** 1937. Determination du facteur de croissance pour *Haemophilus ducreyi*. C. R. Seances Soc. Biol. **124:**1169–1171.
34. **Marsh, W. C., N. Haas, and G. Stuttgen.** 1978. Ultrastructural detection of *Haemophilus ducreyi* in biopsies of chancroid. Arch. Dermatol. Res. **263:**153–157.
35. **McLeod, K.** 1882. Precis of operations performed in the wards of the first surgeon Medical College Hospital during the year 1881. Indian Med. Gaz. **17:**121.
36. **Nobre, G. N.** 1982. Identification of *Haemophilus ducreyi* in the clinical laboratory. J. Med. Microbiol. **15:**243–245.
37. **Nsanze, H., M. V. Fast, L. J. D'Costa, P. Tukei, J. Curran, and A. R. Ronald.** 1981. Genital ulcers in Kenya: a clinical and laboratory study. Br. J. Vener. Dis. **59:**378–381.

37a. **Nsanze, H., F. A. Plummer, A. B. N. Maggwa, G. Maitha, J. Dylewski, P. Piot, and A. R. Ronald.** 1984. Comparison of media for the primary isolation of *Haemophilus ducreyi*. Sex. Transm. Dis. **11:**6–9.

38. **Oberhofer, T. R., and A. E. Back.** 1982. Isolation and cultivation of *Haemophilus ducreyi*. J. Clin. Microbiol. **15:**625–629.
39. **Odumeru, J. A., A. R. Ronald, and W. L. Albritton.** 1983. Characterization of cell proteins of *Haemophilus ducreyi* by polyacrylamide gel electrophoresis. J. Infect. Dis. **148:**710–714.
40. **Packer, H., and J. Goldberg.** 1950. Studies of the antigenic relationship of *D. granulomatis* to members of the tribe *Eschericheae*. Am. J. Syph. Gon. Vener. Dis. **34:**342–350.
41. **Sng, E. H., A. L. Lim, V. S. Rajan, and A. J. Goh.** 1982. Characteristics of *Haemophilus ducreyi*: a study. Br. J. Vener. Dis. **58:**239–242.
42. **Sottnek, F. O., J. W. Biddle, S. J. Kraus, R. E. Weaver, and J. A. Stewart.** 1980. Isolation and identification of *Haemophilus ducreyi* in a clinical study. J. Clin. Microbiol. **12:**170–174.
43. **Sullivan, M.** 1940. Chancroid. Am. J. Syph. **24:**482–521.
44. **Ursi, J. P., E. VanDyck, R. C. Ballard, W. Jacob, P. Piot, and A. Z. Meheus.** 1982. Characterization of an unusual bacterium isolated from genital ulcers. J. Med. Microbiol. **15:**97–103.
45. **Winslow, C.-E. A., J. Broadhurst, R. E. Buchanan, C. Krumwiede, Jr., L. A. Rogers, and G. H. Smith.** 1917. The families and genera of the bacteria. Preliminary report of the Committee of the Society of American Bacteriologists on characterization and classification of bacterial types. J. Bacteriol. **2:**505–566.

Chapter 88

Gardnerella vaginalis

PETER PIOT

CHARACTERIZATION

Organisms formerly classified as *Haemophilus vaginalis* or *Corynebacterium vaginale* have been transferred recently to the genus *Gardnerella*, which contains only one species, *Gardnerella vaginalis* (8, 16). The type strain is 594 Gardner and Dukes (ATCC 14018; NCTC 10287). Based on DNA hybridization studies and on the unique trilayered laminar cell wall of the organism, this genus is not related to any other taxon. Although the cell wall morphologically resembles that of a gram-negative organism (8), in chemical composition (i.e., amino acids and carbohydrates) it is more typical of that of a gram-positive bacterium (10). The cells are nonmotile, pleomorphic, nonencapsulated, gram-negative to gram-variable rods which average 0.5 by 1.5 μm.

G. vaginalis is facultatively anaerobic and fermentative, with acetic acid as the major end product. Catalase and oxidase are not produced. Most strains are fastidious in their growth requirements, but there is no need for either X or V factor or coenzyme-like substances (4). Colonies on human blood bilayer agar medium are 0.3 to 0.5 mm in diameter after incubation for 48 h at 35 to 37°C in a humidified atmosphere with 5% CO_2. The optimal pH for growth ranges between 6 and 7. A distinct beta-hemolysis is produced on human and rabbit but not on sheep blood agar media. Acid is produced from a wide variety of carbohydrates. Nitrate is not reduced, and indole and urease reactions are negative.

CLINICAL SIGNIFICANCE

G. vaginalis is associated with bacterial vaginosis (nonspecific vaginitis) (5, 15). Bacterial vaginosis is a defined syndrome, the hallmark of which is an excessive, malodorous vaginal discharge associated with a significant increase in the number of *G. vaginalis* and various obligate anaerobes, mainly *Bacteroides* spp. and *Peptococcus* spp. (20), and a decrease in the number of vaginal lactobacilli. A minimum diagnostic requirement for bacterial vaginosis is the presence of at least three of the following signs: excessive vaginal discharge, vaginal pH of >4.5, "clue" cells (vaginal epithelial cells covered by small gram-negative rods), and a fishy, amine-like odor in the KOH test (1). Other features of bacterial vaginosis include the presence and elevated concentration of certain vaginal diamines (3) and a distinct gas-liquid chromatographic pattern of fatty acids in vaginal secretions (20). The succinate/lactate ratio is significantly increased (>0.4), and volatile organic acids, including acetic, propionic, and butyric acids, are found in the majority of patients. Although *G. vaginalis* is found consistently and in high numbers in women with bacterial vaginosis, the organism can also be isolated from as many as 20 to 40% of healthy women depending on the sensitivity of the culture technique. The pathogenesis of bacterial vaginosis is still unclear; hypothesis favors some as yet unidentified factor(s) leading to conditions which disrupt the interaction between *G. vaginalis* and certain anaerobes, resulting in an increase in the numbers of *G. vaginalis*.

G. vaginalis is a common blood isolate in postpartum and postabortal fever (24, 26; L. G. Reimer and L. B. Reller, Obstet. Gynecol., in press) and has been isolated from amniotic fluid (6). This organism is frequently isolated from the endometrium of women with postpartum endometritis (D. A. Eschenbach, M. G. Gravett, K. C. S. Chen, Scand. J. Urol. Nephrol., in press). It can cause fatal and nonfatal septicemia and soft tissue infection in neonates (11, 18). In two studies on urinary tract infections during pregnancy, *G. vaginalis* was isolated from suprapubic aspirates of urine (13, 14). *G. vaginalis* has also been isolated from vaginal abscesses, bartholinitis, abdominal fluid, and the oropharynx (12).

COLLECTION OF SPECIMENS

Vaginal discharge can be collected with calcium alginate or cotton-tipped swabs. One swab is either placed into a transport medium such as Amies or Stuart or, preferably, used to inoculate the isolation medium. A second swab is used to prepare a Gram-stained smear and a wet mount. The vaginal pH should be measured with pH paper strips, but care should be taken not to touch the exocervix, as the pH of cervical mucus is different from that of the vaginal fluid. Isolation media should be inoculated as soon as possible after specimen collection, at most within 24 h when a transport medium is used.

DIRECT EXAMINATION

Direct examination of vaginal secretions is more relevant for the diagnosis of bacterial vaginosis than is culture of *G. vaginalis* from these specimens, because *G. vaginalis* can be part of the normal vaginal flora. The presence of clue cells correlates well with a diagnosis of bacterial vaginosis and with *G. vaginalis* in vaginal fluid. The classic description of clue cells was based on a wet-mount preparation (5). However, the observation of clue cells in a Gram-stained smear is preferable. This smear allows not only for a better evaluation of the vaginal flora but also for preservation of the preparation for later examination. In typical smears from patients with bacterial vaginosis, clue cells are accompanied by a mixed flora consisting of very large numbers of small gram-negative and gram-variable rods and coccobacilli in the absence of larger gram-positive rods (lactobacilli). Gram-negative curved rods may be found as well. When the clinical examination is normal, *Lactobacillus* spp. morphotypes are found alone or in the presence of small numbers of *G. vaginalis* morphotypes. In general, there is a strong inverse relationship between the numbers of *G. vaginalis* and *Lactobacillus* spp. morphotypes (21). When 10% KOH is added to vaginal discharge from women with bacterial vaginosis, a fishy, amine-like odor is immediately apparent (15).

TABLE 1. Selected features of *G. vaginalis* (175 strains) (from reference 17)

Test	Test result[a]	% Positive
Beta-hemolysis (human blood-bilayer)	+	99
Hippurate hydrolysis	+	90
Starch hydrolysis	+	100
α-Glucosidase	+	100
β-Glucosidase	−	0
β-Galactosidase	V	45
Lipase	V	64
Acid from:		
Arabinose	V	38
Glucose	+	100
Maltose	+	100
Mannitol	−	0
Starch	+	100
Sucrose	+	85
Trehalose	V	72
Xylose	V	44
Zone of inhibition with:		
Metronidazole (50 μg)	+	92
Trimethoprim (5 μg)	+	100
Sulfonamide (1 mg)	−	0
Bile (10%)	+	100

[a] V, Variability between strains; +, positive; −, negative.

CULTURE AND ISOLATION

Semiselective human blood bilayer agar (HBT agar) (23) is inoculated by rolling a swab across a sector of the plate. This inoculum is streaked with a loop to allow a semiquantitative estimate of the growth of the isolate. Plates are incubated in a candle jar or a humid atmosphere containing 5% CO_2. Obligately anaerobic strains of *G. vaginalis* occasionally occur (12), but routine anaerobic incubation is not recommended. After 48 h of incubation, the plates are examined for colonies exhibiting diffuse beta-hemolysis. Alternatively, vaginalis agar (nonselective single layer human blood agar) (9) may be used. Isolates should be subcultured at least every 72 h, since older cultures are often no longer viable.

IDENTIFICATION

A presumptive identification of *G. vaginalis* may be based on the typical cell morphology in the Gram stain, a beta-hemolysis with diffuse edges on human blood bilayer agar, and a negative catalase test (7, 17, 19, 22). Laboratory reports should include a quantitative estimate of the growth of *G. vaginalis*.

When a more accurate identification is required, α- and β-glucosidase, hippurate hydrolysis, and starch hydrolysis tests yield a maximum discriminative value (Table 1). The presence of α- and β-glucosidase is determined with a solution of 0.1% (wt/vol) 4-nitrophenyl-α-D-glucopyranoside or 4-nitrophenyl-β-D-glucopyranoside, respectively, in 0.067 M Sorensen phosphate buffer (pH 8.0). Tubes containing 0.5 ml of the substrate solution are inoculated with a loopful of bacteria from an overnight culture. The tubes are incubated at 35°C in a water bath and examined after 4 h for the appearance of a yellow color, which for the hippurate hydrolysis test is composed of 1% (wt/vol) sodium hippurate in 0.067 M Sorensen phosphate buffer (pH 6.4). Tubes containing 0.5 ml of this solution are inoculated as for the glucosidase tests above and incubated at 35°C for 2 h, after which 0.2 ml of a solution of 3.5% (wt/vol) ninhydrin dissolved in equal parts of acetone and butanol is added. Development of a deep blue-purple color within 5 min indicates a positive hippurate hydrolysis test. Starch hydrolysis can be determined on Mueller-Hinton agar enriched with 5% (vol/vol) sterile horse serum and detected by the addition of Lugol iodine.

Inhibition tests with disks containing 50 μg of metronidazole (zone of inhibition present), 5 μg of trimethoprim (zone of inhibition present), 1 mg of sulfonamide (no zone of inhibition), and 10% bile (zone of inhibition present) are also useful in the differentiation of *G. vaginalis*, vaginal lactobacilli, and unclassified catalase-negative coryneforms. The latter two groups of organisms are resistant to metronidazole, trimethoprim, and bile and are variably susceptible to sulfonamide (2, 17, 22). Traditional identification schemes for *G. vaginalis* were based on sugar fermentation reactions, but these are of little value, as other organisms from the vagina frequently give the same reactions as *G. vaginalis*. Moreover, fermentation tests with *G. vaginalis* are often inconsistent. However, when extragenital isolates must be fully characterized, acid production from the carbohydrates listed in Table 1 are determined. Acid production may be determined by the rapid method of Greenwood et al. (9). To obtain reproducible results, tests should always be performed with large inocula from cultures not older than 24 h. Gas-liquid chromatography of organic acids may also provide additional information. Most *G. vaginalis* strains produce acetic and lactic acids. However, these acids are also produced by bifidobacteria and some unclassified vaginal coryneforms. Antimicrobial susceptibility testing of genital isolates of *G. vaginalis* is not recommended.

Fluorescent-antibody tests for the identification of *G. vaginalis* have been reported, but the reagents are not routinely available (18, 25). Isolates can be subdivided into six biovars based on the production of lipase and β-galactosidase and the hydrolysis of hippurate.

LITERATURE CITED

1. **Amsel, R., P. A. Totten, C. A. Spiegel, K. C. S. Chen, D. Eschenbach, and K. K. Holmes.** 1983. Nonspecific vaginitis: diagnostic criteria and microbial and epidemiologic associations. Am. J. Med. **74:**14–22.
2. **Bailey, R. K., J. L. Voss, and R. F. Smith.** 1979. Factors affecting isolation and identification of *Haemophilus vaginalis (Corynebacterium vaginale)*. J. Clin. Microbiol. **9:**65–71.
3. **Chen, K. C. S., P. S. Forsyth, T. M. Buchanan, and K. K. Holmes.** 1979. Amine content of vaginal fluid from untreated and treated patients with nonspecific vaginitis. J. Clin. Invest. **63:**828–835.
4. **Dunkelberg, W. E., and I. McVeigh.** 1969. Growth requirements of *Haemophilus vaginalis*. Antonie van Leeuwenhoek J. Microbiol. Serol. **35:**129–145.
5. **Gardner, H. L., and C. D. Dukes.** 1955. *Haemophilus vaginalis* vaginitis. A newly defined specific infection previously classified "nonspecific" vaginitis. Am. J. Obstet. Gynecol. **69:**962–976.
6. **Gibbs, R. S., J. D. Blanco, P. J. St. Clair, and Y. S. Castaneda.** 1982. Quantitative bacteriology of amniotic fluid from women with clinical intraamniotic infection at term. J. Infect. Dis. **145:**1–8.

7. **Greenwood, J. R., and M. J. Pickett.** 1979. Salient features of *Haemophilus vaginalis*. J. Clin. Microbiol. **9:**200–204.
8. **Greenwood, J. R., and M. J. Pickett.** 1980. Transfer of *Haemophilus vaginalis* Gardner and Dukes to a new genus, *Gardnerella, G. vaginalis* (Gardner and Dukes) comb. nov. Int. J. Syst. Bacteriol. **30:**170–178.
9. **Greenwood, J. R., M. J. Pickett, W. J. Martin, and E. G. Mack.** 1977. *Haemophilus vaginalis (Corynebacterium vaginale)*: method for isolation and rapid biochemical identification. Health Lab. Sci. **14:**102–106.
10. **Harper, J. J., and G. H. G. Davis.** 1982. Cell wall analysis of *Gardnerella vaginalis (Haemophilus vaginalis)*. Int. J. Syst. Bacteriol. **32:**48–50.
11. **Leighton, P. M., B. Bulleid, and R. Taylor.** 1982. Neonatal cellulitis due to *Gardnerella vaginalis*. Pediatr. Infect. Dis. **1:**339–340.
12. **Malone, B. H., M. Schreiber, N. J. Schneider, and L. V. Holdeman.** 1975. Obligately anaerobic strains of *Corynebacterium vaginale (Haemophilus vaginalis)*. J. Clin. Microbiol. **2:**272–275.
13. **McDowall, D. R. M., J. D. Buchanan, K. F. Fairley, and G. L. Gilbert.** 1982. Anaerobic and other fastidious microorganisms in asymptomatic bacteriuria in pregnant women. J. Infect. Dis. **144:**114–122.
14. **McFadyn, I. D., M. B. Glasy, and S. J. Eykin.** 1968. Suprapubic aspiration of urine in pregnancy. Lancet **i:**1112–1114.
15. **Pheifer, T. A., P. S. Forsyth, M. A. Durfee, H. M. Pollock, and K. K. Holmes.** 1978. Nonspecific vaginitis: role of *Haemophilus vaginalis* and treatment with metronidazole. N. Engl. J. Med. **298:**1429–1434.
16. **Piot, P., E. Van Dyck, M. Goodfellow, and S. Falkow.** 1980. A taxonomic study of *Gardnerella vaginalis (Haemophilus vaginalis)* Gardner and Dukes 1955. J. Gen. Microbiol. **119:**373–396.
17. **Piot, P., E. Van Dyck, P. A. Totten, and K. K. Holmes.** 1982. Identification of *Gardnerella (Haemophilus) vaginalis*. J. Clin. Microbiol. **15:**19–24.
18. **Platt, M. S.** 1971. Neonatal *Haemophilus vaginalis (Corynebacterium vaginale)* infection. Clin. Pediatr. **10:**513–516.
19. **Shaw, C. E., M. E. Forsyth, W. R. Bowie, and W. A. Black.** 1981. Rapid presumptive identification of *Gardnerella vaginalis (Haemophilus vaginalis)* from human blood agar media. J. Clin. Microbiol. **14:**108–110.
20. **Spiegel, C. A., R. Amsel, D. Eschenbach, F. Schoenknecht, and K. K. Holmes.** 1980. Anaerobic bacteria in nonspecific vaginitis. N. Engl. J. Med. **303:**601–607.
21. **Spiegel, C. A., R. Amsel, and K. K. Holmes.** 1983. Diagnosis of bacterial vaginosis by direct Gram stain of vaginal fluid. J. Clin. Microbiol. **18:**170–177.
22. **Taylor, E., and I. Phillips.** 1983. The identification of *Gardnerella vaginalis*. J. Med. Microbiol. **16:**83–92.
23. **Totten, P. A., R. Amsel, J. Hale, P. Piot, and K. K. Holmes.** 1982. Selective differential human blood bilayer media for isolation of *Gardnerella (Haemophilus) vaginalis*. J. Clin. Microbiol. **15:**141–147.
24. **Venkataramani, T. K., and H. K. Rathbun.** 1976. *Corynebacterium vaginale (Haemophilus vaginalis)* bacteremia: clinical study of 29 cases. Johns Hopkins Med. J. **139:**93–97.
25. **Vice, J. L., and M. F. Smaron.** 1973. Indirect fluorescent-antibody method for the identification of *Corynebacterium vaginale*. Appl. Microbiol. **25:**908–916.
26. **Vontver, L. A., and D. A. Eschenbach.** 1981. The role of *Gardnerella vaginalis* in nonspecific vaginitis. Clin. Obstet. Gynecol. **24:**439–460.

Chapter 89

Section XI. Immunodiagnostic Tests

Specific Immunoglobulin Detection

RAYMOND W. RYAN AND IRENE KWASNIK

The isolation and identification of infectious agents is generally the preferred method of diagnosis of an infection. However, for many years, serological procedures have been widely used as a diagnostic tool. This indirect approach relies on host response to the infectious agent during the course of the disease. Such serological tests to detect the presence of or significant rises in specific antibody include complement fixation, passive hemagglutination, hemagglutination inhibition, indirect fluorescent antibody (IFA), and recently, enzyme immunoassay (EIA). In a primary infection, the absence of specific immunoglobulin G (IgG) in the acute phase and a significant titer (fourfold or greater) in the convalescent phase can be used to confirm the diagnosis. In recently acquired or reactivated infection, a fourfold or greater rise in antibody titer must be clearly demonstrated. There are several disadvantages in measuring only IgG. They are: (i) antibodies may have reached peak titers before the collection of the acute specimen, so that clear-cut seroconversion is absent; (ii) an assay of low sensitivity may produce a false seroconversion due to fluctuations between the negative and low-positive titers; (iii) IgG persists for long periods, often for life, and may vacillate independently of active infection; (iv) cases may arise in which a fourfold or greater rise is not adequate to determine clinical significance; (v) IgG freely crosses the placenta from mother to fetus, and virtually all the IgG present in the newborn has been passively transferred from the mother. On the other hand, IgM is produced in initial or primary infection and is detectable only for a limited time after primary infection (Fig. 1). In secondary infections, the IgM response follows essentially the same time course as in primary infection, but the peaks may be higher. IgM does not cross the placenta from mother to fetus, but is synthesized by the fetus in utero after the end of the first trimester. The presence of specific IgM antibody in the neonate or the adult is usually a clear-cut indication of current or recent infection and obviates the need for paired serum specimens. The period during which specific IgM can be detected after primary infection varies with the type of infection, course of the disease, and method used to detect the antibody. In general, IgM reaches a maximum within 1 to 2 weeks and declines to undetectable levels within 2 to 3 months. There are, however, several problems associated with the detection of specific IgM antibody. Competitive immunoglobulin binding between IgG and IgM may produce false-negative results. High levels of specific IgG present in serum can produce a blocking effect on the binding of the less avid IgM antibody to the available antigen binding sites during the serum-substrate incubation period. False-positive results in the presence of specific IgG antibody can occur when rheumatoid factor produced against maternal IgG is present (20, 23).

IMMUNOGLOBULIN DETECTION TECHNIQUES

For several years, the detection of IgM, as well as other specific immunoglobulins, has been performed by IFA. Within the past 10 years, EIA has been developed, a new technique which possesses high sensitivity and specificity and is technically simple to perform. Although the techniques involved in doing IFA and EIA are simple, rapid, and relatively inexpensive, there are some advantages and disadvantages to each assay. For example, IFA suffers from subjectivity in the quantitative interpretation of endpoints. However, an experienced reader can visually differentiate between specific and nonspecific reactions by the appearance and localization of the fluorescence. Initially, the sensitivity of specific IgM detection by IFA was not very high, presumably because high levels of specific IgG competed with the IgM for antigenic binding sites during the staining procedure. This problem was overcome by simply increasing the binding time (4), thus eliminating false-negative reactions. Sera containing rheumatoid factor, on the other hand, may produce false-positive reactions. This result usually occurs in infants with congenital infections. If the passively acquired maternal IgG complexes with the infecting organism, the infant usually produces specific IgM directed against the maternal IgG. In the IFA procedure, the specific maternal IgG reacts with the antigen. The IgM anti-IgG of the infant complexes with the maternal IgG. This complex is then detected by the fluorescein-conjugated anti-IgM, thus producing false-positive results (Fig. 2). The nonspecific binding of IgM to various antigens has been shown to occur, and this binding can interfere with specific IgM detection in both IFA and EIA systems. For example, cytomegalovirus (CMV)-infected human fibroblasts have induced Fc receptors, which can bind human IgM molecules with no antiviral activity (8). This binding results in false-positive assays by both methodologies. The use of receptor-free antigens in the EIA systems resolves this problem. In the IFA test, only nuclear fluorescence, not cytoplasmic, should be interpreted as positive.

SEPARATION OF SPECIFIC IMMUNOGLOBULINS

The use of whole serum for IgM detection has been criticized because of the problems cited above. An early approach to the dilemma was a comparison of titers before and after the selective destruction of IgM. Serum is exposed to 2-mercaptoethanol, which breaks the bonds between the structural units of the IgM

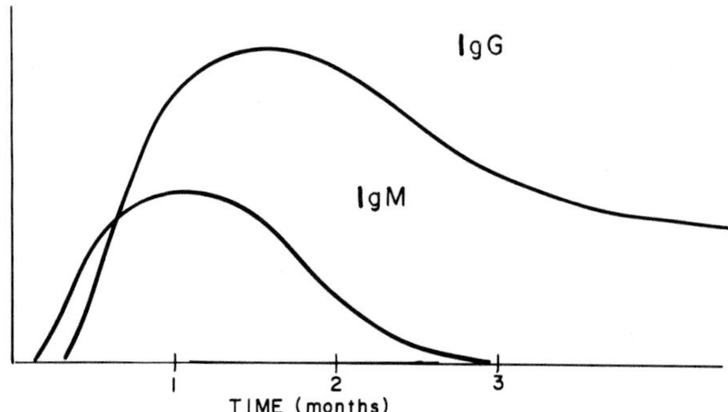

FIG. 1. Immunoglobulin response in primary infections.

molecule. This breaking of bonds destroys immunological activity. This method was insensitive, unreliable, and dependent on an IgM titer higher than that of IgG (4).

Separation of the IgM fraction before specific antibody testing by either EIA or IFA can be achieved by several methods, each with inherent problems. These methods include protein separation techniques and adsorption of IgG.

Sucrose gradient centrifugation

The sucrose gradient centrifugation method consists of layering a sample on a density gradient made up of several layers of sucrose solutions of different concentrations. The gradient is subjected to high-speed centrifugation, after which the bottom of the tube is pierced, and several fractions are collected. The presence of separate immunoglobulin classes in the fractions is detected by any one of several methods. Peak fractions containing IgM are pooled and assayed for specific activity.

Gel filtration

IgM can be easily separated from other immunoglobulins by gel filtration with a Sephadex G-200 column (7). This technique separates molecules according to size. The serum can be eluted from the column by Veronal buffer (pH 7.0) to which 1% glucose and 0.01% sodium azide have been added. The optical density of each fraction collected is measured at 280 nm to identify the major protein peaks. The fractions composing each peak are pooled and assayed.

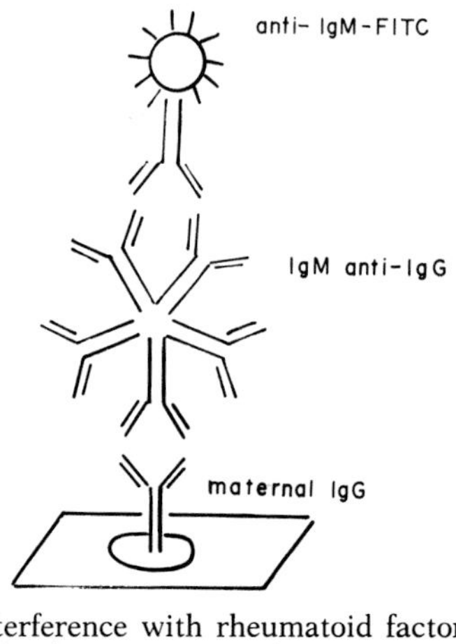

FIG. 2. Interference with rheumatoid factor in congenitally acquired infections. FITC, Fluorescein isothiocyanate.

Ion exchange

One simple, efficient method for the isolation of IgM utilizes the principle of ion-exchange chromatography. The Quik-Sep kit (Isolab, Inc., Akron, Ohio) consists of IgM isolation columns, wash buffer, IgM elution buffer, and Tris buffer. The column is primed with wash buffer, and a known quantity of serum is added. The column is then washed, thus removing IgG1, IgG2, IgG3, and most of the lipoproteins. The IgM elution buffer is added, and the eluate is collected; this fraction contains IgM, IgG4, and all other serum proteins, including IgA. This eluate will have an acidic pH and may be neutralized by the Tris buffer.

Separation by either sucrose gradient centrifugation or gel filtration is time consuming and impractical for the routine clinical laboratory. The requirement of specialized equipment is also a major drawback. Ion exchange with the commercially available product is quick and simple to perform and easily adaptable to the routine clinical laboratory. It also provides the final IgM in a convenient and known dilution. Recovery of IgM is greater than 80% of the original.

Adsorption of IgG

There are several adsorption methods for the removal of IgG from whole serum. They include the use of heat-aggregated IgG (23), glutaraldehyde-aggregated IgG (11), human IgG linked to Sepharose (11), and protein A coupled to Sepharose gel (4).

The heat-aggregated IgG and glutaraldehyde-aggregated IgG methods are the same except for the aggregation step. The lyophilized human gamma globulin is aggregated either by heating for 10 min at 73°C or by the addition of 2.5% glutaraldehyde. The aggregates are washed with phosphate-buffered saline, and then 110 μl of a 20% suspension is mixed with 10 μl of undiluted serum, incubated, and centrifuged.

The human IgG-Sepharose and the protein A-Sepharose methods are similar. A 110-μl amount of 20% suspension in phosphate-buffered saline of either compound is mixed with 10 μl of undiluted serum. The human IgG-Sepharose-serum mixture is stirred at room temperature and incubated overnight at 4°C. The suspension is centrifuged, and the supernatant is used in the assay. The protein A-Sepharose-serum mixture is shaken and centrifuged at 700 × *g* for 10 min at room temperature, and the supernatant is used in the assay. The principle involved in the use of protein A is the nonspecific reaction of this protein with the Fc fragment of the gamma globulin molecule. This method does not eliminate IgG3, as protein A reacts with IgG subclasses 1, 2, and 4 only, leaving IgM, IgA, and IgD antibodies intact (11). The reported loss of original IgM has been as low as 15% and as high as 40% (4).

DETECTION OF CIRCULATING IMMUNE COMPLEXES

The determination of circulating immune complexes has been used as a diagnostic tool for estimating the severity of a disease and for investigating the role of these complexes in the pathogenesis of a disease. However, the direct applicability of the detection of immune complexes in the diagnosis and management of certain infectious diseases is limited by the sensitivity, specificity, and reproducibility of the currently available methods. These methods are based on various biologic or chemical properties of the immune complexes, such as interaction with the C1q component of complement, rheumatoid factor, or cell receptors. Each of these techniques has different sensitivities and different pitfalls in the detection of immune complexes of specific diseases. Physical methods such as analytic ultracentrifugation and gel filtration to demonstrate aggregated IgG or IgM are too slow and too complex for clinical use. Also, they separate rather than detect the immune complexes, and they lack specificity, as does the cryoprecipitation assay. This assay also lacks sensitivity in that large amounts of immune complexes are required before a positive result is obtained.

Several assays utilizing the C1q-binding ability of immune complexes have been devised. The initial tests were precipitin reactions, but these have been replaced by more sensitive methods, such as the radiolabeled-C1q binding assay, solid-phase C1q assay, and C1q deviation tests (24). These all have moderate to high sensitivity.

The Raji cell assay uses a B-cell tissue line which contains surface receptors for C3 breakdown products as well as receptors for the Fc region of IgG and some receptors for C1q. When complement is activated by the immune complexes, the C3 cleavage products bind to the immune complexes. Any immunoglobulin bound to these cells via these receptors can be detected by radiolabeled anti-immunoglobulin antibody.

Monoclonal or polyclonal rheumatoid factors can be used in assays because they react with antigenic determinants on IgG. The monoclonal rheumatoid factors have increased affinity for aggregated IgG and can be used in solid-phase or fluid-phase assays.

Hepatitis B infection is one disease in which the detection of immune complexes may play a diagnostic role. Circulating immune complexes have been demonstrated during various stages of the disease by both the C1q binding assay and the Raji cell test. Recently, a radioimmunoassay based on the selective absorption of IgM on solid phase has been described (25). The hepatitis B surface antigen-IgM complexes were found in patients with acute hepatitis B on admission to the hospital and subsequently decreased and disappeared in patients who recovered. In patients who developed chronic hepatitis, the immune complexes persisted, thereby signaling an unfavorable outcome of hepatitis B infection.

ANTIBODY CAPTURE ASSAY

Within the past few years, an EIA capture assay (double-sandwich, reverse enzyme-linked immunosorbent assay) for specific IgM has been developed. It is especially useful in the detection of small amounts of IgM in specimens such as cerebrospinal fluid, due to its increased sensitivity. This method consists of coating a solid phase with antisera (goat, rabbit, or monoclonal) of human mu chain, the capture antibody.

When monoclonal antibodies are used as the capture antibodies, many of the former problems of low specificity are overcome. Test samples are added, and the IgM is captured and bound to the solid phase. An appropriate dilution of the specific antigen is added and incubated. A hyperimmune globulin, directed against the antigen and conjugated to the indicator enzyme, is then added, and the addition of substrate follows. The products of the enzyme-substrate reaction are then measured at a predetermined incubation time. This assay is said to eliminate the immunoglobulin class competition for antigen binding sites and the requirement for specimen pretreatment, although preabsorption of test sera with aggregated human IgG to remove rheumatoid factor has been reported (2).

SPECIFIC IgM DETECTION IN INFECTIOUS DISEASE

There are several instances in which the rapid diagnosis of certain infectious diseases can be made by specific IgM detection, particularly in congenital or persistent infections. Some indications for serological testing are the inability to culture suspected pathogens such as Epstein-Barr virus (EBV) and hepatitis virus, too much time required for isolation and identification, inadequate local facilities for specimen processing, and an element of risk to the patient in the procedure necessary to obtain an adequate specimen.

The ability to provide an accurate and timely diagnosis allows the physician to institute preventive measures, health precautions, and chemotherapy when appropriate. With the advent of specific antiviral drugs, this capability will become exceedingly important and may obviate the need for antibacterial therapy and extended hospitalization, as well as provide important information regarding the prognosis.

Viral infections are an important cause of neonatal morbidity and mortality. Most of these infections are asymptomatic; however, in the symptomatic newborn, the clinical manifestations are frequently overlapping. Therefore, specific etiologic diagnosis usually depends on laboratory results. Research efforts have been directed toward rapid cultural and serological tests in this area.

The definitive and most sensitive means for the diagnosis of CMV is the isolation of the virus. The presence of specific IgM antibody in the neonate is virtually diagnostic of infection with this agent. Serial serum specimens must be obtained to differentiate between congenital and postnatal infections. In adults, CMV IgG antibodies are not helpful, due to significant fluctuations in titers even in the absence of illness (6). In overt infections, both primary and reactivated, the presence of specific IgM is a reliable diagnostic tool (19).

Specific IgM antibodies can usually be detected in serum 1 to 2 weeks after the onset of herpes simplex virus infection. However, herpes simplex virus-specific IgM may or may not be detected during recurrent or reactivated infections. In most instances, isolation of the virus can be accomplished within 24 to 48 h, usually before any IgM is detectable. Rapid, more sensitive techniques of antigen detection by EIA and the use of DNA probes will play a major role in the timely diagnosis of this disease.

Rubella is a proven teratogenic agent, and the time during gestation at which rubella infection occurs is the most critical factor determining the frequency of fetal infection and the severity of the disease in the neonate. If a pregnant woman is exposed to a suspected case of rubella and serum samples are not available until several weeks later, diagnosis may be difficult. Detectable antibody is present in the blood within a few days after the onset of a rash and reaches peak titers within 2 to 3 weeks. A significant rise in titer probably cannot be demonstrated. Rubella-specific IgM peaks 3 to 6 weeks after infection and remains detectable for several months. In the case of congenital rubella, virus isolation can be attempted, although it may be difficult. Specific IgM can be detected at birth, generally persists for about 6 months, and is diagnostic in a single serum sample. Attempts at antenatal diagnosis by examination of the amniotic fluid have not proved successful. In rubella infections acquired postnatally, IgM can be detected quite readily during the first days of illness and remains easily detectable for 1 to 2 months.

Infection with *Toxoplasma gondii* is often clinically inapparent. However, in utero transmission of the parasite can occur as a complication of first infection in a pregnant woman. Since about 50% of the adult population in the United States has detectable antibodies to this parasite, the measurement of specific IgM is extremely important. The presence of IgM in the neonate is considered indicative of congenital infection; however, IgM is present only 20% of the time. A repeat test 2 to 3 weeks later must be done. It has been reported that the detection of specific IgM to *T. gondii* by EIA is more sensitive than detection by IFA (17).

The detection of antibody against hepatitis A virus antigen can be demonstrated during the acute illness by various types of immunoassay. It has been reported that at least 30% of the general population have antibody to hepatitis A virus. Antibody of the IgG type appears after the acute illness and usually is detectable for a lifetime. Thus, a rapid diagnosis of acute hepatitis A or differentiation between recent and past infections can be achieved by the detection of hepatitis A virus IgM by either EIA or radioimmunoassay.

Specific IgG and IgM to the viral capsid antigen of EBV are present in the sera of essentially all patients with current EBV infections. The rise of IgG-viral capsid antigen is so rapid that a fourfold rise in titer may not be demonstrated. The presence of IgM in the absence of antibody to EBV nuclear antigen is a reliable indicator of past infection. The major drawbacks to the performance of these tests in routine clinical laboratories are the problems inherent in the production of reliable antigens. Gallo et al. (3) have recently reported new and improved antigen preparations for use in the measurement of viral capsid antigen-IgM and EBV nuclear antigen-antibody levels in patients suspected of having EBV infections. Both preparations proved to be sensitive and reliable in the IFA test. Recently, Luka at al. (13) reported an enzyme-linked immunosorbent assay technique for the detection of the antibody responses to the major EBV-associated antigens—viral capsid, EBV nuclear, early, and membrane antigens—in patients with EBV-associated diseases. The enzyme-linked immunosorbent assay titers correlated well with those obtained by immunofluorescent testing, with significantly greater sensitivity in some instances.

It is apparent from the literature that methods for the detection of specific IgM, whether IFA, EIA, or radioimmunoassay, should not be used without a carefully controlled system. With the knowledge of the potential pitfalls which can occur with any one of these methods, the detection of specific IgM can be a sensitive and reliable diagnostic tool.

IgA

IgA, although present in serum, is the predominant immunoglobulin in secretions from mucous membranes. It exists in both secretory and serum forms. The origin of serum IgA is unknown. The secretory IgA arises from plasma cells within the mucosal surfaces of the body which are adjacent to the external environment (1). Secretory IgA is a dimer of IgA molecules with a sedimentation coefficient of 11S. This dimer is held together by a polypeptide J chain bound near the end of the heavy chain on each monomer. Once secreted by the plasma cell, the IgA dimer complexes with a glycoprotein or secretory component which becomes covalently attached by disulfide bonds to one of the two IgA monomeric constituents (26). There is evidence to suggest that this secretory component, which is produced by epithelial cells, functions as a receptor protein that transports the IgA dimer through the epithelial cell onto the mucous membrane. Human secretory IgA is a much larger molecule than serum IgA, the major difference being the addition to secretory IgA of J chain, secretory component, and another IgA monomer.

Secretory IgA may contain antibody activity against any of several infectious agents or their toxins. The exact mechanism by which these antigens are inactivated is still unclear, since IgA does not appear to activate the complement system (28). Secretory IgA, as well as other immunoglobulins, appears to bind to epithelial cell surfaces, thereby preventing the adherence of bacteria to susceptible tissues (27). Since the adherence of bacteria may be required to establish colonization or initiate the disease process, the prevention of adherence may be the means by which secretory IgA defends against infection.

IgA protease

The two subclasses of human IgA are IgA1, which predominates in serum, and IgA2, which is largely found in secretions. IgA protease activity was first reported in 1973 (15, 27). This enzyme degrades human IgA1 into Fab and Fc fragments. The only structural difference between the two IgA subclasses is that IgA2, which is not susceptible to the enzymatic action of IgA protease, has a deletion of 13 amino acid residues in the hinge region between the Fc and Fab portions of the molecule. Other classes of human immunoglobulins, as well as immunoglobulins from several subhuman primates, are not affected by this protease activity. Since the discovery of IgA protease, only four of several bacteria genera examined have species which possess this activity. It was first reported in *Streptococcus sanguis* (5), a member of the oral microflora which colonizes the surface of teeth and is an important cause of subacute bacterial endocarditis. Of all the other streptococci tested, only strains of *Streptococcus pneumoniae* were found to produce IgA protease (14). All strains of *Neisseria gonorrhoeae*, both virulent (piliated) and avirulent (nonpiliated), are enzyme positive (18). Strains of *Neisseria meningitidis*, regardless of serotypes, are IgA protease positive (16). Isolates of *Haemophilus influenzae* from middle ear infections, cerebrospinal fluid, and pneumonia are all positive (10). Most recently, IgA protease activity has been reported from strains of *Ureaplasma urealyticum* (21). A comprehensive list of the numerous organisms screened for the presence of IgA protease activity, including viruses and fungi, can be found in a recent review article by Kornfeld and Plaut (10).

Organisms which produce IgA protease do so during the logarithmic phase of growth, with production ceasing as they enter the stationary phase. IgA protease is mainly an extracellular protein, with only 5% present in the cell. There is some evidence to suggest that most IgA proteases isolated are metal dependent, since in the presence of EDTA, about 50% of the enzymatic activity is lost (9, 14, 18). After dialysis to remove the EDTA, protease activity can be restored by the addition of any one of several divalent cations.

The role of IgA protease activity has yet to be clarified. However, as one gains a better understanding of both antibody function on mucosal surfaces and the significance of IgA1 and IgA2, the function of IgA protease in the pathogenesis of human infection may be better understood.

Detection of specific IgA antibodies

Several different methodologies have been used to measure specific IgA antibodies in human sera. Specific IgA to human CMV was found to correlate well with the presence of IgM in eight of nine patients with primary CMV infection (12). Likewise, in kidney transplant patients with recurrent CMV infections, radioimmunoassay, enzyme-linked immunosorbent assay, and immunoperoxidase were used to measure CMV IgA, which was detected at about the same time the rise in IgG occurred (22).

The measurement of specific IgA may prove to be a valuable tool in the diagnosis of certain infectious diseases.

LITERATURE CITED

1. **Crabbé, P. A., and J. F. Heremans.** 1966. The distribution of immunoglobulin containing cells along the human gastrointestinal tract. Gastroenterology **51:**305–316.
2. **Forghani, B., C. K. Myoraku, and N. J. Schmidt.** 1983. Use of monoclonal antibodies to human immunoglobulin M in "capture" assays for measles and rubella immunoglobulin M. J. Clin. Microbiol. **18:**652–657.
3. **Gallo, D., K. H. Walen, and J. L. Riggs.** 1982. Improved immunofluorescence antigens for detection of immunoglobulin M antibodies to Epstein-Barr viral capsid antigen and antibodies to Epstein-Barr virus nuclear antigen. J. Clin. Microbiol. **15:**243–248.
4. **Gardner, P. S., and J. McQuillin.** 1980. Rapid virus diagnosis, 2nd ed., p. 259–287. Butterworth and Co., London.
5. **Genco, R. J., A. G. Plaut, and R. C. Moellering, Jr.** 1975. Evaluation of human oral organisms and pathogenic streptococcus for production of IgA protease. J. Infect. Dis. **131**(Suppl):17–21.
6. **Gershon, A. A.** 1979. Problems of infancy TORCH infections and diagnostic virology, p. 205–214. *In* D. Lennette, S. Specter, and K. D. Thompson (ed.), Diagnosis of viral infections. University Park Press, Baltimore.
7. **Gupta, J. D., V. Peterson, M. Stout, and A. M. Murphy.** 1971. Single sample diagnosis of recent rubella by fractionation of antibody on Sephadex G-200 column. J. Clin. Pathol. **24:**547–550.
8. **Keller, R., R. Peitchel, J. N. Goldman, and M. Goldman.** 1976. An IgG-Fc receptor induced in cytomegalovirus-infected human fibroblasts. J. Immunol. **116:**772–777.
9. **Kilian, M., J. Mestecky, R. Kulhavy, M. Tomana, and W. T. Butler.** 1980. IgA1 proteases from *Haemophilus influenzae, Streptococcus pneumoniae, Neisseria meningitidis*, and *Streptococcus sanguis*: comparative immunochemical studies. J. Immunol. **124:**2596–2600.
10. **Kornfeld, S. J., and A. G. Plaut.** 1981. Secretory immunity and the bacterial IgA proteases. Rev. Infect. Dis. **3:**521–534.
11. **Leinikki, P. O., I. Shekarchi, P. Dorsett, and J. L. Sever.** 1978. Determination of virus-specific IgM antibodies by using ELISA: elimination of false positive results with protein A-sepharose absorption and subsequent IgM antibody assay. J. Lab. Clin. Med. **92:**849–857.
12. **Levy, E., and I. Sarov.** 1980. Determination of IgA antibodies to human cytomegalovirus by enzyme-linked immunosorbent assay. J. Med. Virol. **6:**249–257.
13. **Luka, J., R. C. Chase, and G. R. Pearson.** 1984. A sensitive enzyme-linked immunosorbent assay (ELISA) against the major EBV-associated antigens. I. Correlation between ELISA and immunofluorescence titers using purified antigens. J. Immunol. Methods **67:**145–156.
14. **Male, C. J.** 1979. Immunoglobulin A1 protease production by *Haemophilus influenzae* and *Streptococcus pneumoniae*. Infect. Immun. **26:**254–261.
15. **Mehta, S. K., A. G. Plaut, N. J. Calvanico, and T. B. Tomasi.** 1973. Human immunoglobulin A: production of an Fc fragment by an enteric microbial proteolytic enzyme. J. Immunol. **111:**1274–1276.
16. **Mulks, M. H., and A. G. Plaut.** 1978. IgA protease production as a characteristic distinguishing pathogenic from harmless Neisseriaceae. N. Engl. J. Med. **299:**973–976.
17. **Naot, Y., E. V. Barnett, and J. S. Remington.** 1981. Method for avoiding false-positive results occurring in immunoglobulin M enzyme-linked immunosorbent assays due to presence of both rheumatoid factor and antinuclear antibodies. J. Clin. Microbiol. **14:**73–78.
18. **Plaut, A. G., J. V. Gilbert, M. S. Artenstein, and J. D. Capra.** 1975. *Neisseria gonorrhoeae* and *Neisseria meningitidis*: extracellular enzyme cleaves human immunoglobulin A. Science **190:**1103–1105.
19. **Rasmussen, L., D. Kelsall, R. Nelson, W. Carney, M. Hirsch, D. Winston, J. Preiksaitis, and T. C. Merigan.** 1982. Virus-specific IgG and IgM antibodies in normal and immunocompromised subjects infected with cytomegalovirus. J. Infect. Dis. **145:**191–199.
20. **Reimer, C. B., C. M. Black, D. J. Phillips, L. C. Logan, E. F. Hunter, B. J. Pender, and B. E. McGrew.** 1975. The specificity of fetal IgM: antibody or anti-antibody? Ann. N.Y. Acad. Sci. **254:**77–93.
21. **Robertson, J. A., M. E. Stemler, and G. W. Stemke.** 1984. Immunoglobulin A protease activity of *Ureaplasma urealyticum*. J. Clin. Microbiol. **19:**255–258.
22. **Sarov, I., E. Levy, M. Aymard, Y. Chardonnet, S. Bosshard, J. P. Revillard, M. Frideman, E. Nord, M. Greiff, and H. Haikin.** 1982. Detection of virus-specific IgA antibodies in serum of kidney transplant patients with recurrent cytomegalovirus infections by enzymeimmuno and radioimmunoassay techniques. Clin. Exp. Immunol. **48:**321–328.

23. **Shirodaria, P. V., K. B. Fraser, and F. Stanford.** 1973. Secondary fluorescent staining of virus antigens by rheumatoid factor and fluorescein-conjugated anti-IgM. Ann. Rheum. Dis. **32:**53–57.
24. **Stites, D. P.** 1982. Clinical laboratory methods for detection of antigens and antibodies, p. 325–365. *In* D. P. Stites, J. D. Stobo, H. H. Fudenberg, and J. V. Wells (ed.), Basic and clinical immunology, 4th ed. Lange Medical Publications, Los Altos, Calif.
25. **Toti, M., R. Rizzi, P. Almi, M. Palla, and F. Bonino.** 1983. Complexes between HBsAg and IgM in serum of patients with acute hepatitis. J. Med. Virol. **11:**139–145.
26. **Underdown, B. J., J. DeRose, and A. Plaut.** 1977. Disulfide bonding of secretory component to a single monomer subunit in human secretory IgA. J. Immunol. **118:**1816–1821.
27. **Williams, R. C., and R. J. Gibbons.** 1972. Inhibition of bacterial adherence by secretory IgA: a mechanism of antigen disposal. Science **177:**697–699.
28. **Wilson, I. D.** 1972. Studies on the opsonic activity of human secretory IgA using an in vitro phagocytosis system. J. Immunol. **108:**726–730.

Chapter 90

Detection of Bacterial Antigens by Counterimmunoelectrophoresis, Coagglutination, and Latex Agglutination

JOAN C. FUNG AND RICHARD C. TILTON

Bacterial antigen detection has become an important adjunct to the laboratory diagnosis of infectious diseases. It is especially useful when applied to body fluid specimens, because the rapid results can aid in directing immediate, appropriate antimicrobial therapy. Moreover, in patients previously treated with antibiotics, the detection of bacterial antigen may be the only laboratory test available to establish the etiologic agent of infection. However, antigen detection cannot completely replace culture techniques since it does not predict antibiotic susceptibility patterns of the infecting organism. Several investigators have observed a correlation between the quantity of bacterial antigen in body fluid and patient morbidity and mortality, so the detection of bacterial antigen in body fluids may also have prognostic implications.

Methods in bacterial antigen detection have also been used in the identification of clinical isolates. The culture broths of Gram stain-positive blood cultures have been examined for specific bacterial antigens, as the rapid presumptive identifications of these cultures can aid in clinical management. In addition, immunoserologic methods of serogrouping and serotyping bacteria can provide important epidemiologic information.

The results of immunologic detection of bacterial antigens must be interpreted with care. As with all such assays, several factors may affect the sensitivity and specificity of the test, and since the results of these assays may determine clinical management, it is essential that they be used with a knowledge of their applications and limitations.

The three methods of antigen detection covered in this chapter are counterimmunoelectrophoresis (CIE), coagglutination, and latex agglutination.

CIE

Principle

CIE is based on the principle of immunodiffusion modified by electrophoresis to drive the antigen and antibody toward each other. The specimen to be tested for antigen is placed in the cathodic well, and the antiserum is placed in the anodic well. At neutral or alkaline pH, bacterial polysaccharides are negatively charged and will migrate toward the anode in electrophoresis. Most bacterial antigens detected by CIE are polysaccharides. Immunoglobulins, on the other hand, are less negatively charged and will be swept to the cathode by the streaming of buffer ions, an effect known as endosmosis. A precipitin line is formed at the region of the field containing the optimum proportion of antigen and antibody. Two reviews of CIE by Anhalt et al. (2) and Tilton (21) provide specific clinical and laboratory details.

Preparation of CIE plates

Prepare 30 ml of 1% agarose in barbital buffer (0.3 g of agarose and 30 ml of barbital buffer, pH 8.2, 0.05 M). Carefully bring the solution to a boil until there are no suspended solids left in solution. Avoid overheating. On a level surface, apply the molten agarose with a 10-ml pipette onto a 5- by 7-in. (13- by 18-cm) Mylar sheet or glass plate. When the agarose has gelled, it may be stored in a moist chamber at 4°C for up to 4 days until ready for use. Pairs of wells, with each well 3 mm in diameter, are punched from the agarose plate and removed by vacuum aspiration. Antigen and antibody wells should be 4 mm apart, and pairs of wells should be 6 mm apart.

Specimen preparation

Virtually any body fluid may be examined for bacterial antigens by CIE. Table 1 lists the body fluids from which the most common bacterial antigens detected by CIE have been found. In suspected cases of bacterial meningitis, cerebrospinal fluid (CSF), urine, and serum are the specimens most frequently submitted. Blood and any other body fluids which may clot are allowed to clot, and then they are centrifuged and the supernatant is tested by CIE. All bloody or cloudy body fluids should be centrifuged before testing. Gram stain-positive CSF for bacteria should also be diluted 1:10 with sterile saline or water. The diluted and undiluted CSF specimens are tested by CIE. This avoids a false-negative CIE result due to a prozone effect, that is, antigen in excess. Viscous body fluids (such as sputum) may have to be solubilized before testing. Solubilization can be accomplished by the addition of an equal volume of 10% dithiothreitol (DTT) solution to the specimen. Then the specimen is vortexed and kept at room temperature for 15 min. Alternatively, an equal volume of 20% *N*-acetyl cysteine can be added to the specimen, which is then incubated at 35°C for 30 min. Urine specimens should be centrifuged if there is any visible debris and then concentrated 20- to 50-fold. The method of concentration should not concentrate ions, which may inhibit precipitin formation. Urine samples are concentrated either by ethanol precipitation or by ultrafiltration. In the ethanol precipitation method, 15 ml of 95% ethanol is added to 5 ml of urine and held at 4°C for 1 h. The mixture is then centrifuged, and the pellet is suspended with 0.25 ml of saline for CIE. This resuspended pellet represents a 20-fold concentration. Ultrafiltration with the Minicon-15 (Amicon Corp., Lexington, Mass.) can be used to concentrate the urine up to 50-fold. All specimens should be kept at 4°C until used or kept frozen if stored for more than 24 h.

TABLE 1. Body fluids in which common bacterial antigens have been detected

Antigen detected	Body fluids
S. pneumoniae	Serum, sputum, CSF, urine, bullous fluid, pleural fluid, synovial fluid
H. influenzae type b	Serum, sputum, CSF, urine, pericardial fluid, joint fluid, pleural fluid
N. meningitidis	Serum, CSF, urine, synovial fluid, pericardial fluid
Group B streptococcus	Serum, CSF, urine, gastric aspirate, amniotic fluid, peritoneal fluid

Electrophoresis

The antigen and antibody wells are filled with disposable capillary pipettes. The gel is placed in the CIE chamber with the antigen well near the cathode, and the electrophoretic chambers are filled with the barbital buffer. Strips of Whatman no. 1 filter paper, cut 7 by 7 in. (17.8 by 17.8 cm), are folded, wetted with buffer, and used as connecting wicks between the CIE plate and the buffer chambers. At least one positive control antigen should be included in each CIE run. The control antigen chosen may be that from the most highly suspected etiologic agent. Electrophoresis is usually performed under constant voltage at 8 to 10 V/cm. For a 5- by 7-in. gel plate, this is approximately 100 to 130 V. Electrophoresis is carried out for about 30 to 60 min.

With the same buffer, the polarity of the power supply should be changed after each run to prevent the accumulation of salts on either side of the chamber. Buffer in the chambers should be changed biweekly or when there is visible microbial growth in the buffer.

Examination of gel

CIE plates are examined in bright oblique light. A true precipitin between the antigen and antibody wells generally forms a sharp line, either straight or slightly arced. Its location between the wells depends on the migrational properties and concentration of the antigen and the potency of the antisera. The visibility of weak precipitin reactions may be enhanced by cooling the slides to 4°C for 15 min or up to 24 h. Protein staining of the CIE plate may also enhance the detection of precipitin reactions.

In some cases, a hazy zone around the antigen or antibody well may be observed. This type of nonspecific precipitation may be distinguished from true precipitin by washing the CIE plate or by further incubation of the gel plate in a moist chamber overnight at room temperature. Nonspecific precipitation may be removed by washing the CIE plate, whereas true antigen-antibody precipitin will remain. Alternatively, further overnight incubation of the gel plate at room temperature will result in a longer precipitin arc, due to further antigen and antibody diffusion. This further incubation will not affect the nature of the nonspecific precipitation.

A second type of nonspecific precipitation can occur when the specimen precipitates with a number of different antisera. This nonspecific reaction is the result of rheumatoid factor (RF) and other heat-labile factors in the specimen, such as complement (C1q). RF is an antibody against immunoglobulin G; thus, binding of immunoglobulin G by RF will form a nonspecific precipitate. C1q can bind several Fc portions, probably by way of its six globular carboxy-terminal areas. C1q will bind with aggregated immunoglobulins or immune complexes, thereby forming nonspecific precipitation or agglutination (2). This nonspecific reaction cannot be eliminated by washing the gel after electrophoresis, but it may be eliminated by treatment of the specimen before electrophoresis with reducing agents to destroy RF or with heating to destroy C1q.

Treatments have included heating the specimen to 100°C for 3 to 5 min or treating it with reducing agents such as DTT. In the latter method, an equal volume of specimen is mixed with filter-sterilized, freshly prepared DTT (0.003 M), and the solution is allowed to stand at room temperature for 1 h (12). The specimen should be tested within 3 h after DTT treatment, since it will not remain in this reduced state. Serum and other protein-rich specimens can be treated by heating to 56°C for 30 min or by the addition of 3 volumes of 0.1 M EDTA to 1 volume of specimen, followed by heating at 100°C for 3 min. EDTA aids in the extraction of the antigen from the heat-denatured protein after the centrifugation. The resulting supernatant is tested for heat-stable bacterial polysaccharide antigen. In the absence of EDTA, the serum will form a gel when boiled, thereby entrapping the antigen (8). Another means of eliminating nonspecific interference is by treatment of the specimen with 5 mg of pronase per ml at 56°C for 15 min, followed by exposure to 100°C for 5 min (20). Pronase, DTT, and heat treatment cannot be used to remove nonspecific reactions if the antigen of interest (e.g., a protein) can be denatured by the treatment. Heating at 56°C inactivates C1q but not RF activity.

Washing and staining

Immerse the agarose sheet in the deproteinizing wash solution overnight and then in distilled water for 1 h. Place the sheet between layers of filter paper until the agarose gel is dried. Stain the sheet in Coomassie brilliant blue R, wash it with distilled water, place it in destaining solution for at least 10 min, rinse it with distilled water, and blot it dry.

Reagents include: (i) a deproteinizing solution of sodium phosphate dibasic, heptahydrate, 1.88 g; NaCl, 8.0 g; sodium phosphate monobasic, 0.08 g; and sodium azide, 0.1 g, dissolved in 1 liter of distilled water; (ii) Coomassie brilliant blue R stain, made of Coomassie brilliant blue R, 6 g; absolute methanol, 1,350 ml; glacial acetic acid, 300 ml; and distilled water, 1,350 ml; and (iii) a destaining solution of ethanol (95%), 475 ml; glacial acetic acid, 100 ml; and distilled water, 425 ml.

Quality control

Each lot of antiserum should ideally be tested for sensitivity and specificity before use and then rechecked monthly. Table 2 lists the minimum detectable concentrations of bacterial antigens observed with CIE. The wide range of sensitivity reported by different investigators can be the result of several variables, including the antisera. The minimum detectable concentration of antigen and the percentage of positive clinical specimens detected by CIE have been shown to vary with the source of the antiserum, the type or group specificity of the bacterial strain, and the difference between monovalent and polyvalent antisera. The lot-to-lot variability of antisera can be tested against a preparation of the bacterial antigen. An overnight broth culture of the organism which has been centrifuged and passed through a 0.45-μm membrane filter can be used. Samples of this prepared antigen control can be kept frozen until needed.

Other variables affecting CIE performance are the agarose, the ionic strength and pH of the buffer, and electrophoretic conditions (1, 21).

Specificity of the antiserum may be harder to determine. Table 3 lists some of the known cross-reactions observed in CIE. These cross-reactions have been observed more often with culture isolates and have been found infrequently with clinical specimens (<0.1 to 3.6%). This difference may be due to the larger amount of cross-reacting antigen in culture, compared with that in clinical specimens.

Reporting

CIE results should not be reported as positive or negative for the etiologic agents tested. False-negative CIE results could occur in culture-positive clinical specimens as a result of a low antigen level in the specimen (for example, early in infection) or a low sensitivity of detection by the CIE system used. False-positive CIE results can occur in culture-negative clinical specimens due to cross-reacting antigens. Instead, positive CIE results should be reported as reactive or as demonstrating a positive precipitin reaction against the specific antiserum tested. Conversely, negative CIE results should be reported as nonreactive or as demonstrating a negative precipitin reaction against the specific antiserum used. In specimens demonstrating nonspecific precipitation even after the specimen has been treated to prevent nonspecificity, the result should be reported as equivocal or as demonstrating nonspecific reaction with the antisera used.

Interpretation

Many variables already discussed which affect CIE sensitivity must be kept in mind when interpreting CIE results, including the potency of the antiserum used. High titers of precipitating antibody affect the reliability of CIE in the detection of bacterial antigens, especially those from urine specimens. Bacterial polysaccharides in urine have fewer binding sites than their respective polysaccharides in other body fluids, as was observed with the type-specific pneumococcal polysaccharide (5, 7). Urine polysaccharide is a degradation product of the native bacterial polysaccharide. It has a lower molecular weight than the native polysaccharide and partial immunologic identity.

Antiserum may also vary in its sensitivity of detection of different type- or group-specific strains. For example, CIE can detect a lower limit of 0.024 to 0.075 μg of group A or C meningococcal polysaccharide per ml, but only 100 μg of group B *Neisseria meningitidis* polysaccharide per ml. Though certain antiserum preparations to *N. meningitidis* group B can detect group B antigen in CSF, there is currently no reliable, commercially available CIE product. The group B capsular polysaccharide of *N. meningitidis* is a poor immunogen.

TABLE 2. Minimum concentration of bacterial antigen detected by CIE, coagglutination, and latex agglutination

Antigen	Minimum antigen (ng/ml) detected by:		
	CIE	Coagglutination	Latex agglutination
S. pneumoniae capsular polysaccharide	24–1,000	6	0.2–50
Group B streptococcus capsular polysaccharide	500–14,000		62
H. influenzae type b polyribophosphate polysaccharide	1–25	2–25	0.1–5
N. meningitidis groups A, C, D, and Y	24–75	1.5	0.75–50

Possible cross-reaction should also be taken into account in the interpretation of CIE results.

These problems of unpredictable immunological response, antiserum titer, and cross-reactivity may not be solved with the use of monoclonal technology. Monoclonal antibodies recognize only a single antigenic site on a molecule, and with most antigens they cannot form the lattice necessary for precipitation. This problem may be solved by using a mixture of two to three different monoclonal antibodies to the antigen in CIE (i.e., against different epitopes). Antigens composed of repeating units may, on the other hand, have several antigen sites capable of precipitating with one type of monoclonal antibody.

Migrational characteristics of the antigen in the CIE system must also be taken into consideration. The failure of reactivity in CIE by an antigen may be caused by its lack of anodal migration in the electric field. This has been observed with pneumococcal capsular antigens types 7 and 14, which have a neutral or slightly positive charge. In the barbital buffer system, these polysaccharides migrate toward the cathode. Since both types are common causes of pneumococcal infections, detection of their polysaccharide requires the use of a barbital buffer containing a boric acid derivative. This derivative consists of either a sulfonated derivative of phenylboronic acid or *m*-carboxyphenylboronic acid (3, 6, 14). Since boronic acid-containing buffer systems have decreased sensitivities in the detection of other antigens, they should be used only for the detection of neutrally charged antigens.

Migrational characteristics of the antigen are dependent on its composition. For example, the group A carbohydrate of *Streptococcus pyogenes* is a heterogeneous molecule with varying amounts of *N*-acetylglucosamine attached to its polyrhamnose backbone. In CIE, extracts of group A carbohydrate contain both cathodic and anodic migrating material. The heterogeneity in migration results from the relative extent of *N*-acetylglucosamine attachment (11).

The sensitivity of CIE is also affected by the amount of antigen in the specimen. Increased levels of antigen

TABLE 3. Antigenic cross-reactions observed in CIE, coagglutination, and latex agglutination

Species and antiserum	Method[a]	Cross-reacting strain
S. pneumoniae		
Omniserum	CIE, Coagglut.	*Klebsiella* spp.
	CIE	*N. meningitidis* group B
	CIE	Alpha streptococcus (certain strains)
	CIE	Group B streptococcus
	CIE	*H. influenzae* type b
S. pneumoniae		
Type 3	CIE	*S. pneumoniae* types 7 and 12
Type 2	CIE	*S. pneumoniae* types 7, 12, 13, 22, and 23
Type 19	CIE	*H. influenzae* type a
Group B streptococcus	Coagglut.	Groups A, C, D, and G streptococci
	CIE	Groups A, C, and G streptococci
	Coagglut., latex	*S. pneumoniae*
	Latex	*Proteus mirabilis*
Group A streptococcus	Coagglut.	Groups B, C, D, and G streptococci
Group C streptococcus	Coagglut.	Groups A, B, D, and G streptococci
	CIE	Group B streptococci
	Coagglut.	*S. pneumoniae*
	Coagglut.	*Klebsiella pneumoniae*
Group G streptococcus	Coagglut.	Groups A, B, C, and D streptococci
Group F streptococcus	Coagglut.	*Streptomyces albus* enzyme
N. meningitidis		
Group B	Latex	*N. meningitidis* groups A and C
	CIE	*N. meningitidis* group C
	Coagglut.	*H. influenzae*
	Coagglut.	*S. pneumoniae*
	Coagglut.	*E. coli*
Group A	CIE	*N. meningitidis* groups B and C
Group C	CIE	*E. coli*
Group D	CIE	*N. meningitidis* group B
Group Z	CIE	*N. meningitidis* groups B and C
H. influenzae		
Type b	Latex	*H. influenzae* types e and f
	Latex	*N. meningitidis* group B
	CIE	*S. aureus*
	CIE	Group B streptococcus
	CIE, latex	*S. pneumoniae*
	CIE, latex	*E. coli*
Type c	CIE	*H. influenzae* type a
Type d	CIE	*H. influenzae* type f
Type e	CIE	*N. meningitidis* groups B and C
Type f	CIE	*N. meningitidis* groups B and C
	CIE	*E. coli*

[a] Coagglut., Coagglutination assay; Latex, latex agglutination assay.

detection by CIE are observed with concentrated versus nonconcentrated urine specimens. Disadvantages in the use of concentrated urine in antigen detection include the possible false-positive reactions in patients with urinary tract infections or urine specimens contaminated with bacteria and the time-consuming (1- to 2-h) process of urine concentration. Precipitation of urine polysaccharides in particular can be enhanced by cooling the agarose plate after electrophoresis.

COAGGLUTINATION AND LATEX AGGLUTINATION

Coagglutination and latex agglutination assays have several advantages over CIE. A laboratory should not attempt CIE unless it is committed to quality control of the variables delineated in the previous section. Although quality control is still essential, the coagglutination and latex agglutination assays are simpler to perform. No special equipment is required; less time is required (about 3 min each, compared with a minimum of 30 min for CIE); and no staining, washing, or cooling is required, as may at times be necessary in CIE. This simplicity makes the coagglutination and latex agglutination assays more convenient and practical to perform in the routine laboratory when individual specimens arrive at different times. There is also no dependence on the migrational property of the antigen. Less antiserum is required in the agglutination assays than with CIE, making these assays more cost effective. More importantly, the agglutination assays are more sensitive than CIE in the detection of bacterial antigens. For these reasons and due to the increasing commercial availability of reliable kits for coagglutination and latex agglutination, CIE has become transitional technology and is rapidly being replaced by the agglutination tests. In the rest of this chapter, unless specifically stated, the term agglutination will refer to both coagglutination and latex agglutination.

Principle

Staphylococcus aureus strains containing protein A on their cell surface bind immunoglobulin G at its Fc portion, thus leaving the Fab portion free to bind antigen. In coagglutination, the antibody-coated *S. aureus* agglutinates in the presence of homologous antigen.

In latex agglutination, antibodies attached to polystyrene latex particles agglutinate in the presence of homologous antigen.

Preparation of agglutination reagents

S. aureus Cowan (ATCC 12598) for the coagglutination assay is grown in tryptic soy broth for 18 h at 35°C, harvested, and washed three times with phosphate-buffered saline (PBS), pH 7.3. The bacteria are suspended in 0.5% formaldehyde in PBS for 3 h at room temperature. The cells are then washed three times in buffer and adjusted to a final 10% suspension in PBS. The suspension is heated for 1 h at 80°C and cooled. Antiserum (0.1 ml) is added to 1 ml of *S. aureus* suspension, and the mixture is left to stand at room temperature for 1 h. The suspension is brought to 10 ml with PBS and stored at 4°C until used.

The latex agglutination is prepared by the addition of 1 part polystyrene latex particles (0.81 μm) to 1 part antiserum diluted to a concentration yielding optimum reactivity. After the suspension is incubated at 37°C for 2 h, 2 volumes of glycine-buffered saline (pH 8.2) containing 0.1% bovine serum albumin is added. Glycine-buffered saline consists of 7.3 g of glycine and 10 g of sodium chloride in 1 liter of distilled water, adjusted to pH 8.2 with sodium hydroxide. The sensitized latex reagent is stored at 4°C until used. Nonimmune serum may be substituted to prepare control latex or coagglutination reagent.

Also see references 15 and 18 for information on the preparation of agglutination reagents.

Test procedure

In coagglutination, 1 drop (approximately 50 μl) of clinical specimen and 1 drop of antibody-coated *S. aureus* are added to a glass slide, mixed thoroughly with an applicator stick, and rotated manually (15 to 30 times per min) or on a rotating apparatus for up to 2 min.

In latex agglutination, 2 drops of clinical specimen is mixed with 1 drop of latex reagent. The subsequent process is identical to coagglutination.

With both assays, macroscopic agglutination usually occurs within 30 s to 3 min. The slides should be read immediately since drying out of the reagents by evaporation could be misinterpreted as a positive reaction.

Protocols for coagglutination and latex agglutination assays performed with commercial kits should follow the recommendations of the manufacturer.

A positive control antigen should be included with each run or be performed daily.

Examination of agglutination

Agglutination reactions should be viewed with indirect light against a dark background. Agglutination appears as visible clumping, and no agglutination produces a milky-looking mixture. Agglutination can be graded from 1+ to 4+ as follows: 1+, fine granularity, milky background; 2+, small clumps, milky background; 3+, small and large clumps, clear background; and 4+, large clumps, clear background.

Agglutination demonstrating the presence of an antigen should ideally occur only with homologous antibody-coated reagent and not with heterologous antibody-coated reagent or with nonimmune control reagent. However, cross-reactivity may cause agglutination in heterologous as well as homologous antibody-coated reagents. In these cases, agglutination usually appears first in the homologous reagents and with larger clumps than in the heterologous reagents. In this case, the specimen can be considered to have a positive agglutination reaction mainly with the single antibody-coated reagent.

If the degree of agglutination is equal with all the antibody-coated reagents and the nonimmune control reagent, the specimen should be treated to remove nonspecific interference by methods outlined above, in the section on CIE. Nonspecific agglutination may occur more frequently with coagglutination than with latex agglutination and more frequently with serum and urine than with CSF specimens.

Quality control

As with CIE, each lot of agglutination reagents should be tested for sensitivity and specificity before use and then rechecked monthly. Table 2 lists the minimum detectable levels of bacterial antigens observed with coagglutination and latex agglutination. As with CIE, specificity of the agglutination assays is determined by the antiserum. Table 3 lists some known cross-reactivities observed with coagglutination and latex agglutination.

If the specimen is tested against several antibody-coated test reagents, the heterologous reagents may serve as negative controls. In this case, it is not necessary to use a nonimmune control reagent. On the other hand, the nonimmune control reagent is needed to exclude the possibility of a nonspecific reaction if the specimen is tested against a single antibody-coated reagent. In all cases, a positive antigen control should be tested at least once daily in a laboratory in which agglutination assays are performed throughout the day.

Reporting

As with CIE, agglutination results should not be reported as positive or negative for the etiologic agents tested. Instead, positive agglutination results should be reported as reactive or as demonstrating a positive agglutination reaction against the specific antibody-coated reagent tested. Likewise, negative

TABLE 4. Detection sensitivity of bacterial antigen in body fluids

Infection	Body fluid	% Positive reactions in:		
		CIE	Coagglutination	Latex agglutination
S. pneumoniae				
Empyema	Pleural fluid	100		
Meningitidis	CSF	44–100	67–100	61–88
	Serum	35–75		
	Urine	50–100		
Pneumonia	Sputum	74–99	84	
	Serum	4–41		22
	Urine	19–47		
Group B streptococcus				
Meningitis	CSF	63–100	83	79–92
	Serum	27–63		27
	Urine	80–100		93–100
Bacteremia	CSF	0		
	Serum	88		
	Urine	89–100		89
N. meningitidis meningitis	CSF	30–92	33–49	71–88
	Serum	14–67		
	Urine	83		
H. influenzae type b meningitis	CSF	63–100	57–100	78–100
	Serum	70–83		
	Urine	94		

agglutination should be reported as nonreactive or as demonstrating a negative agglutination reaction, and nonspecific interference should be reported as equivocal or as demonstrating nonspecific reactions with the antibody-coated reagents used.

Interpretation

The sensitivity and specificity of the assays must be considered in the interpretation of agglutination results. The agglutination assays are at least as sensitive as, if not more sensitive than, CIE. In some reports, agglutination assays are 10- to 100-fold more sensitive than CIE in the detection of bacterial antigens. Similarly, latex agglutination may be slightly more sensitive than coagglutination.

APPLICATIONS OF BACTERIAL ANTIGEN DETECTION

Diagnostic indications of infectious agents in body fluids

Coagglutination, latex agglutination, and CIE assay have been incorporated in many laboratories because they are simple, rapid, and relatively sensitive and specific, especially in the diagnosis of suspected cases of meningitis. Table 4 lists the frequencies of positive antigen detection in various body fluids from patients with meningitis, bacteremia, pneumonia, or empyema. The highest rate of antigen detection usually occurs in body fluids taken from closed-spaces infections, such as CSF in meningitis, sputum in pneumonia, and pleural fluid in empyema. Up to 92 to 100% of patients with pneumococcal, group B streptoccal, meningococcal, and *Haemophilus influenzae* type b meningitis have detectable bacterial antigen in their CSF, compared with reported figures of 63 to 88% gram-positive smears of CSF from patients with meningitis. Thus, antigen detection is more sensitive than Gram stain.

Several investigators reported that concentrated urine is the single best source for antigen detection in group B streptococcal infections in newborns (4). Urine specimens for antigen detection may be especially useful in pediatric patients, since other specimens such as serum are more difficult to obtain.

There is good correlation (74 to 99%) between the presence of pneumococcal capsular polysaccharide in sputum and pneumococcal pneumonia. In some series, sputum antigen detection was more sensitive than sputum culture (55%) or sputum Gram stain (62%) in the identification of patients with pneumococcal pneumonia. However, 5 to 60% of the normal population may be carriers of pneumococci in the upper respiratory tract, and pneumococcal polysaccharide can be detected in up to 48% of sputum specimens from patients with nonpneumococcal pneumonia.

Bacterial antigen detection may be the only means of establishing an etiologic agent of infection in a previously treated patient. Since antigen detection results are available at least 18 h earlier than culture results, they may dictate appropriate antimicrobial therapy and appropriate infection control measures. Bacterial antigen detection may help to determine the appropriate chemoprophylaxis of household contacts of patients with *N. meningitidis* meningitis (17). Since 91% of groups A and C isolates of *N. meningitidis* are sulfonamide resistant compared with 1 to 4% of groups B and Y isolates, the detection of group A or C antigen would dictate chemoprophylaxis with rifampin instead of sulfonamide.

Recently, a number of latex agglutination test kits have become available for the detection of group A streptococci directly from throat swab specimens. An extraction procedure such as nitrous acid or enzyme extraction treatment is required to release the streptococcal group antigen before testing. The decision on whether these rapid tests can completely replace culture techniques awaits the completion of many clinical trials.

Prognostic indication of antigen detection in body fluids

Antigen concentration in body fluid may be determined by testing serial dilutions of the fluid. High concentrations of bacterial antigen in body fluids have been associated with increased patient morbidity and mortality. Also, the duration of positive antigen detection in these specimens was associated with a complicated clinical course. The persistence of the antigen may reflect the antigen concentration in the initial specimen. Thus, antigen quantitation from the initial body fluid may be a useful predictor of clinical outcome.

Such associations have been observed with bacterial meningitis. In infants with type III, group B streptococcal meningitis, a high concentration of type III antigen in CSF and the persistence of antigen in CSF and concentrated urine correlated strongly with mortality or neurologic sequelae (4). In children with *H. influenzae* type b meningitis, the presence of greater than 1.28 μg of *H. influenzae* type b capsular polysaccharide per ml was significantly related to the development of early and permanent neurologic sequelae (13). The CSF of neonates with *Escherichia coli* K1 meningitis was examined with antiserum against *N. meningitidis* group B. This was possible because *E. coli* K1 capsular antigen and meningococcal group B polysaccharide are immunochemically identical. The amount and persistence of K1 antigen in the CSF and serum were directly related to clinical outcome (16).

Several other associations between bacterial concentration, antigen concentration, and patient morbidity have been noted in bacterial meningitis (9). (i) The pretreatment concentration of bacteria was significantly related to the pretreatment concentration of bacterial antigen. (ii) The pre- and posttreatment concentrations of antigen were related. (iii) Patients with $>10^7$ CFU per ml in pretreatment CSF were more likely to have complications or more severe disease, demonstrated by the presence of convulsions, subdural effusion, bacteremia, and duration of fever.

Identification of clinical isolates

Bacterial polysaccharides are released into the growth medium during bacterial growth. Antigen detection of positive blood cultures is useful in the presumptive identification of the infecting agent, and bacteremia caused by *Streptococcus pneumoniae, H. influenzae, Neisseria* spp., *S. aureus*, streptococci, or *Klebsiella* spp. can be identified by the examination of the supernatants of blood cultures. The bacterial antigens to be tested are predetermined by Gram staining the blood culture.

As with body fluids, similar precautions with antigen detection should be applied to clinical isolates. Nonspecific reactions may occur and must be eliminated by heating or DTT treatment. In the event that antigen detection is performed on broths or on broth-grown clinical isolates, uninoculated broth (medium) control should also be tested. Precipitation of uninoculated growth medium with antiserum has been observed in CIE (10).

The detection of bacterial antigens in clinical isolates is also dependent on the concentration of bacteria in the culture, the location of the antigen within the cell, and the excretion of the antigen into the growth medium. If the bacterial antigen has to be extracted from the cell, the method of extraction may also influence the sensitivity of antigen detection.

Earlier identification of blood culture isolates is important for clinical management and infection control. During a hospital outbreak of a multiply resistant *Klebsiella aerogenes*, blood cultures containing gram-negative rods were tested for the K2 capsular antigen of the organism. The detection of this antigen aided in directing antimicrobial therapy in patients with gram-negative bacteremia (19).

Apart from the organisms already mentioned, antigen detection by CIE, coagglutination, and latex agglutination has been used for identifying, serotyping, or serogrouping *Legionella* spp., *Pseudomonas* spp., *Peptostreptococcus* spp., *Salmonella* spp., and *Clostridium difficile* cytotoxin.

LITERATURE CITED

1. **Agnello, V., R. J. Winchester, and H. G. Kunkel.** 1970. Precipitin reactions of the Clq component of complement with aggregated γ-globulin and immune complexes in gel diffusion. Immunology **19:**909–919.
2. **Anhalt, J. P., G. E. Kenny, and M. W. Rytel.** 1978. Cumitech 8, Detection of microbial antigens by counterimmunoelectrophoresis. Coordinating ed., T. L. Gavan. American Society for Microbiology, Washington, D.C.
3. **Anhalt, J. P., and P. K. W. Yu.** 1975. Counterimmunoelectrophoresis of pneumococcal antigens: improved sensitivity for the detection of types VII and XIV. J. Clin. Microbiol. **2:**510–515.
4. **Baker, C. J., B. J. Webb, C. V. Jackson, and M. S. Edwards.** 1980. Countercurrent immunoelectrophoresis in the evaluation of infants with group B streptococcal disease. Pediatrics **65:**1110–1114.
5. **Coonrod, J. D.** 1974. Physical and immunologic properties of pneumococcal capsular polysaccharide produced during human infection. J. Immunol. **112:**2193–2201.
6. **Coonrod, J. D.** 1983. Urine as an antigen reservoir for diagnosis of infectious diseases. Am. J. Med. Suppl., p. 85–92.
7. **Coonrod, J. D., and M. W. Rytel.** 1973. Detection of type-specific pneumococcal antigens by counterimmunoelectrophoresis. I. Methodology and immunologic properties of pneumococcal antigens. J. Lab. Clin. Med. **81:**770–777.
8. **Doskeland, S. O., and B. P. Berdal.** 1980. Bacterial antigen detection in body fluids: methods for rapid antigen concentration and reduction of nonspecific reactions. J. Clin. Microbiol. **11:**380–384.
9. **Feldman, W. E.** 1977. Relation of concentrations of bacteria and bacterial antigen in cerebrospinal fluid to prognosis in patients with bacterial meningitis. N. Engl. J. Med. **296:**433–435.
10. **Finch, C. A., and H. W. Wilkinson.** 1979. Practical considerations in using counterimmunoelectrophoresis to identify the principal causative agents of bacterial meningitis. J. Clin. Microbiol. **10:**519–524.
11. **Fung, J. C., K. Wicher, and M. McCarty.** 1982. Immunochemical analysis of streptococcal group A, B, and C carbohydrates, with emphasis on group A. Infect. Immun. **37:**209–215.

12. **Gordon, M. A., and E. W. Lapa.** 1974. Elimination of rheumatoid factor in the latex test for cryptococcosis. Am. J. Clin. Pathol. **61**:488–494.
13. **Kaplan, S. L., and R. D. Feigin.** 1980. Rapid identification of the invading microorganisms. Pediatr. Clin. North Am. **27**:783–803.
14. **Kelsey, M. C., and C. S. Reed.** 1979. Countercurrent immunoelectrophoresis: improved detection of pneumococcal capsular antigens in sputum by incorporation of a carboxylated derivative of phenyl boronic acid. J. Clin. Pathol. **32**:960–962.
15. **Kronvall, G.** 1973. A rapid slide-agglutination method for typing pneumococci by means of specific antibody adsorbed to protein A-containing staphylococci. J. Med. Microbiol. **6**:187–190.
16. **McCracken, G. H., Jr., M. P. Glode, L. D. Sarff, S. G. Mize, M. S. Schiffer, J. B. Robbins, E. C. Gotschlich, I. Orskov, and F. Orskov.** 1974. Relation between *Escherichia coli* K1 capsular polysaccharide antigen and clinical outcome in neonatal meningitis. Lancet **i**:246–250.
17. **Pickering, L. K.** 1976. Chemoprophylaxis against *Neisseria meningitidis*. The role of countercurrent immunoelectrophoresis. J. Am. Med. Assoc. **236**:1882–1883.
18. **Severin, W. P. J.** 1972. Latex agglutination in the diagnosis of meningococcal meningitis. J. Clin. Pathol. **25**:1079–1082.
19. **Simpson, R. A., and D. C. E. Speller.** 1977. Detection of bacteraemia by countercurrent immunoelectrophoresis. Lancet **i**:1206.
20. **Stockman, L., and G. D. Roberts.** 1982. Specificity of the latex test for cryptococcal antigen: a rapid, simple method for eliminating interference factors. J. Clin. Microbiol. **16**:965–967.
21. **Tilton, R. C.** 1978. Counterimmunoelectrophoresis in biology and medicine. Crit. Rev. Clin. Lab. Sci. **9**:347–365.

Chapter 91

Immunofluorescence Microscopy

ROGER M. McKINNEY AND WILLIAM B. CHERRY

This chapter is designed to provide a practical discussion of immunofluorescence (IF) equipment and reagents. The principles involved in the use of various types of fluorescence optical and lighting equipment are reviewed, and the roles of important components of the system are analyzed. Techniques for the preparation and evaluation of fluorescent-antibody (FA) reagents that are discussed include serum production, fractionation, and labeling and evaluation. For detailed information on all aspects of IF techniques and their applications, the monograph of Nairn (19) should be consulted. Goldman (4) and Kawamura (10) also have published useful books on the subject.

FA MICROSCOPY

The success of FA microscopy depends on the specific binding of a fluorochrome-labeled antibody to antigenic determinants on the surface of the organism that is being tested for. The specimen is illuminated with excitation light in the wavelength range that produces maximum emission of the particular fluorescent dye that is covalently bound to the antibody. The excitation light must then be very efficiently filtered from the emitted fluorescent light so that essentially all that is seen by the microscopist is the fluorescing antibody, which coats and outlines the microorganism. The intensity of emission or brightness of the image depends on the number of antigenic sites that are available, the quality and concentration of the fluorochrome-labeled antibody, the intensity of the excitation light of appropriate wavelength that reaches the specimen, and the quality and arrangement of the filters and optics of the fluorescence microscope.

The essential components of fluorescence lighting and optical equipment for transmitted- and incident-light fluorescence are shown in Fig. 1 and 2. Lighting sources are usually mercury arc or the newer quartz-iodine-tungsten (halogen) lamps, although xenon lamps also are used for photography for high-intensity excitation at the longer visible wavelengths and as a stable direct-current light source for quantitative IF measurements with incident light. Xenon lamps require expensive direct-current power supplies. The HBO-100 mercury arc lamp may also be operated from a direct-current power supply. The characteristics of the most useful systems will be discussed briefly.

Survey of microscopes and illuminators

The choice of equipment is determined by (i) how it will be used—that is, whether for research, routine testing, or diagnostic work, etc., (ii) the systems to be studied, (iii) personal preference, and (iv) financial considerations. The best arrangement is that which gives the desired results for the biological systems being investigated.

In recent years, two important developments for IF microscopy have occurred. One was the introduction of the halogen lamp with the interference filter for the excitation of fluorochromes whose maximum absorption peaks are at the longer visible wavelengths, e.g., fluoroscein isothiocyanate (FITC). These low-wattage (50- or 100-W) lamps are inexpensive and do not require expensive transformers as do mercury or xenon lamps. Because their emission intensity is low in the shorter-wavelength portion of the visible spectrum, they could not be used until band-pass interference filters with very high transmittance became available and allowed excitation of the fluorochromes at their absorption maxima.

Figure 3 shows a comparison of the relative intensity of various excitation sources. Note that the most commonly used halogen lamp (100 W) has only a fraction of the intensity of the HBO-200 mercury arc lamp at the absorption maxima of FITC (490 nm) and tetramethylrhodamine isothiocyanate (TMRI; 555 nm). Nevertheless, halogen lamps are suitable for many routine applications of IF procedures, provided that suitable interference filters are used for excitation and that these are matched with appropriate barrier filters. These lamps are, however, unsatisfactory for the excitation of rhodamine.

The second important development for IF microscopy was the introduction of incident light or epi-illumination by Ploem (20). Excitation radiation of the desired wavelength is deflected downward into the objective by a beam-splitting or dichroic mirror (Fig. 2) (11). The full aperture of the objective is used both for excitation and for the transmission of fluorescence to the observer; the light beam is automatically centered because the objective also acts as condenser. For incident lighting, objectives of the largest possible numerical aperture should be used to obtain high intensity. Stimulation of fluorescence in the surface layers of the specimen ensures minimal loss of emission intensity by absorption within the specimen itself. The success of incident-light fluorescence depends upon the selection of appropriate dichroic beam-splitting mirrors and suppression filters supplemented by excitation filters selected for their ability to further narrow and separate excitation radiation from fluorescence emission.

The plethora of equipment manufacturers and illuminating systems makes it impracticable to discuss them individually. The most commonly used excitation sources are the mercury arc and the halogen lamp, and the two basic illumination systems are transmitted- and incident-light fluorescence. Excitation sources and illumination systems may be combined in many ways, including the simultaneous use of incident and transmitted excitation. For example, the latter is well adapted to the demonstration of two different antigens in the same smear when one is labeled with FITC and the other with rhodamine. Since the absorption maxima of the two fluorochromes are quite different, they cannot be excited to maximum fluorescence by a single narrow band-pass filter. In the dual instrument, the excitation and

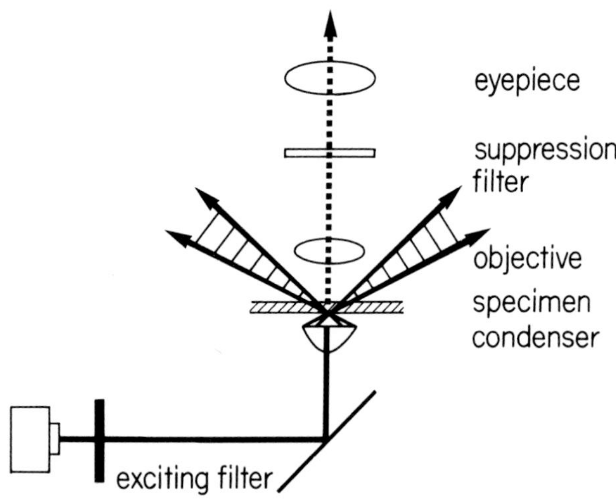

FIG. 1. Transmitted-light fluorescence. Exciting light, ——; fluorescent light, – – – –. Reproduced from Koch (11) with permission of E. Leitz GmbH.

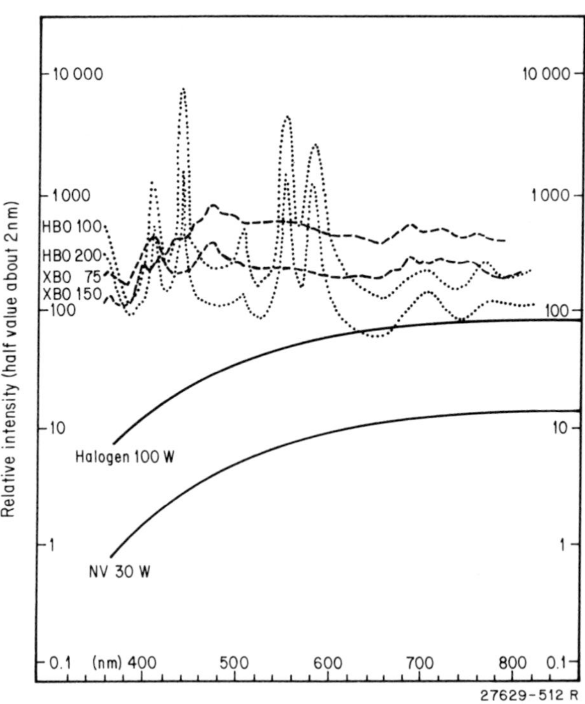

FIG. 3. Intensity comparison of various light sources suitable for fluorescence microscopy (referred to 6-V 30-W lamp). Reproduced from Koch (11) with permission of E. Leitz GmbH.

emission filter system most favorable for each fluorochrome may be used.

For research purposes, workers should attempt to purchase fluorescence equipment that has the versatility required for a variety of uses. Expenditures for combined incident- and transmitted-light systems, including interchangeable mercury arc and halogen lamp housings, will undoubtedly prove to be a good investment. This equipment probably will not become obsolete for many years.

Objectives

The four major types of objectives that are of interest to the fluorescence microscopist are the achromats, the apochromats, the fluorite, and the plano. All objectives are characterized by certain numerical apertures, magnifications, and working distances.

Achromatic objectives are the least corrected and least expensive lenses. They are satisfactory for most fluorescence microscopy when used with either transmitted (dark-field) or incident illumination. Fluorite objectives are also useful under the same conditions. If maximum resolution and brightness are required, the highly corrected but expensive apochromats may be desired. Plano objectives are designed primarily for photomicroscopy in which flatness of field is of great importance. The improved performance is gained at the price of higher sensitivity to the thickness of the slide and cover glass and substantially increased costs.

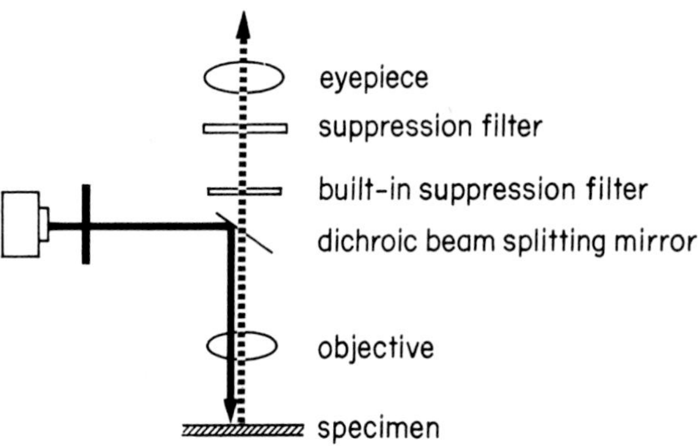

FIG. 2. Incident-light fluorescence. Exciting light, ——; fluorescent light, – – – –. Reproduced from Koch (11) with permission of E. Leitz GmbH.

Condensers

Dark-field condensers are required for IF work with transmitted light. The bispheric or cardioid type of doubly reflecting condenser is preferred. The maximum numerical aperture of the oil immersion objectives which can be used successfully with such condensers is ca. 0.05 less than the numerical aperture of the condenser. If the relationship is not maintained, light enters directly into the objective and destroys the dark-field effect. Control is achieved preferably by an iris diaphragm, which permits variable adjustment of the numerical aperture of the objective. All condensers must be oiled to the under surface of the slide on which the preparation is mounted. For dark-field fluorescence microscopy, the importance of maintaining a homogeneous optical system from condenser to oil immersion objective cannot be overemphasized.

Slides

For some years, we have used successfully, with a variety of fluorescence microscopes, special slides whose thickness specification is 1.1 to 1.2 mm and on which are etched two circles, each 14 mm in diameter (Trident Fluoro Slides no. V58958; A. S. Aloe Co., St.

Louis, Mo.). Cel-Line Associates (Newfield, N.J.) markets standard slides printed with various numbers and sizes of wells, and the company will custom prepare slides of any design. If slides are used with acetone, specifications should call for an acetone-resistant coating. These multiwell slides expedite the performance and decrease the cost of routine IF testing.

Cover glasses

The thickness of the cover glasses used for IF microscopy is critical. Commercial cover slips are rather uniform in refractive index but vary widely in thickness; the latter condition is frequently responsible for the hazy, bleached appearance of the image (Technical Information Bulletin, vol. 1, no. 4, 1961, E. Leitz, Inc., New York, N.Y.). If the refractive index of the mounting medium differs from that of the cover glass, it may have considerable influence upon the image quality. The microscopist should strive to achieve as homogeneous a dispersive system as possible.

Binocular versus monocular heads

The choice of binocular versus monocular heads for microscopes used in IF work is purely a practical matter. The binocular head is more comfortable. It can be used with mercury arc and xenon sources but usually not with halogen lamps. The intensity of illumination from any light source increases with the square of the numerical aperture and decreases with the square of the magnification. Thus, combining an objective of high numerical aperture with an ocular of low magnification produces the highest field illumination.

Immersion oil

For fluorescence microscopy, a nonfluorescent oil, such as Cargille type A low-fluorescence, nondrying immersion oil for microscopy, must be used (Fisher Scientific Co., Pittsburgh, Pa.). It has a refractive index of 1.515 at 23°C.

Mounting fluid

Buffered glycerol (9 parts of glycerol to 1 part [vol/vol] of 0.5 M $NaHCO_3$-Na_2CO_3 buffer [pH 9.0]) has been used extensively as a mounting medium. Cover slips may be sealed with clear fingernail polish, and some preparations may be preserved at 4°C for several weeks or months. Results are not consistent, however, from one antigen or one specimen to another.

Mounting media may be adjusted to any pH compatible with the fluorescence of the fluorochrome used, but for FITC it should not be lower than 7.0 because fluorescence decreases rapidly below this pH. Because oxidation and absorption of CO_2 decrease the pH of glycerol, a pH of at least 8.0 should be maintained, and this pH should be checked at least once a month.

Filters

Selection of an appropriate combination of exciter (primary) and barrier or suppression (secondary) filters is essential for successful IF work. Starting from the light source, the following filters are required in the optical path: (i) a heat-absorbing filter to prevent damage to the exciter filter; (ii) a red-absorbing filter to stop the transmission of visible red light that is secondarily transmitted by all blue filters used for excitation of fluorescein; (iii) an exciter filter to transmit a wave band of light appropriate for the excitation of the fluorochrome; and (iv) a suppression filter to eliminate most, if not all, of the short-wavelength excitation light while transmitting the longer-wavelength fluorescent light. The complementary manner in which the exciter and suppression filters work together is illustrated in Fig. 4. This arrangement maximizes the darkness of the background and enhances contrast. Some IF experts also prefer near-ultraviolet (365-nm) excitation because the various colors produced by autofluorescence of tissue components aid in the histological localization of the specific fluorescence.

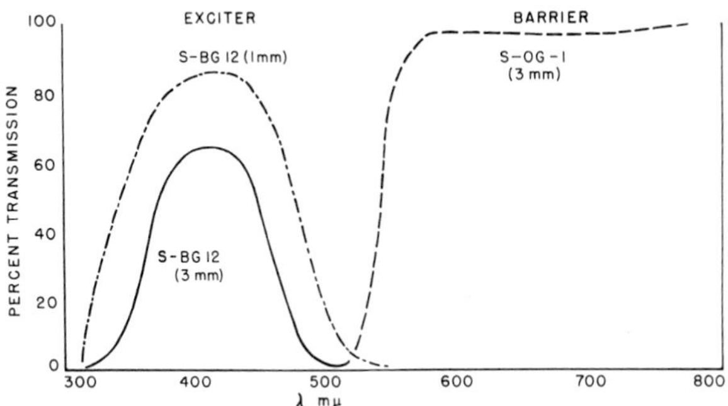

FIG. 4. Principle of complementary filters.

Until the introduction of interference filters, which made possible the use of weaker light sources, mercury arc lamps were used almost exclusively for excitation. When these lamps are used with glass filters, the fluorescence of FITC-stained specimens usually is excited with light in the wavelength band 350 to 450 nm, although fluorescein absorbs maximally at 490 nm (Fig. 5.). When mercury arc lamps are used with interference filters for FITC, excitation occurs at the absorbance maximum for FITC but at reduced lamp output because the intensity spectrum of the mercury arc is relatively low at 490 nm (Fig. 3).

Fluorescence microscopy with incident light required the introduction of a dichroic (beam-splitting)

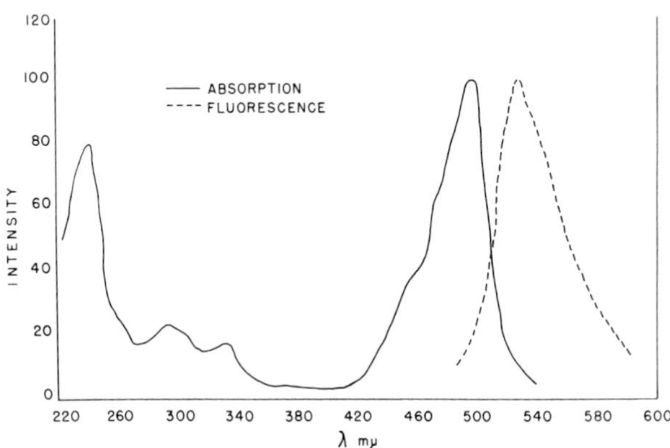

FIG. 5. Absorption and fluorescence of fluorescein in alkaline solution.

mirror whose function is similar to that of the primary interference filter except that it is spatially positioned to deflect the excitation beam downward onto the specimen at a 90° angle (Fig. 2). In the case of FITC, the mirror reflects the blue excitation beam onto the specimen but transmits the green fluorescence emission vertically to the observer. TMRI is excited to fluorescence by green light at a wavelength of 555 nm, and the orange-red fluorescence emission occurs at a wavelength of ca. 575 nm. Interference filters combined with appropriate dichroic mirrors and suppression filters make it possible to utilize halogen lamps for the excitation of FITC, but not TMRI, by either transmitted- or incident-light microscopy.

Suppliers of fluorescence equipment often provide different combinations of filters for use with a single fluorochrome. Furthermore, the same filters may be given different designations by various suppliers. Most companies, however, can be depended upon to furnish the purchaser a rather well-standardized set of filters suitable for use with a given fluorochrome. E. Leitz, Inc. (Rockleigh, N.J.) and Carl Zeiss, Inc. (New York, N.Y.) supply informative literature on filter systems for both transmitted- and incident-light fluorescence of a variety of fluorochromes used in biology and medicine.

IF REAGENTS

Reagents are not commercially available for many specilized applications of IF. For this reason we briefly describe general methods for the preparation of reagents and indicate reference sources for more detailed information.

Antiserum production

Rowe (21) described methods for purification of proteins of the various immunoglobulin classes for use as immunogens. Recommended schedules, dosage, and route of adminstration of the immunogen vary considerably among investigators. With soluble protein antigens, it appears to be standard procedure to administer the first injection as an emulsion in complete Freund adjuvant by the intramuscular (i.m.) route. Booster injections of the antigen in incomplete Freund adjuvant are then given by the i.m. route at intervals. Rowe recommended 3 mg of protein antigen per dose to immunize rabbits and 2 mg of protein antigen per dose for sheep by the i.m. route. Antibody levels should be determined by test bleeding and evaluating the serum by indirect FA (IFA) testing before large volumes of serum are harvested. Particulate antigens such as bacteria can be injected either as a saline suspension or as an emulsion in Freund adjuvant. Experience with rabbits in our laboratory with *Legionella* species seems to indicate that Freund complete adjuvant is beneficial with the first injection. After a 30-day rest period, this injection is followed by 1 i.m. dose in incomplete Freund adjuvant and then biweekly by intravenous injections of a saline suspension of killed bacteria. The dosage in each case is ca. 4×10^9 cells. Test bleeding is done 4 or 5 days after each booster injection, and serum is harvested when the titer appears to plateau. The dosage and schedule for producing the best antiserum for IF applications may vary with the particular bacterial species. Cells of some bacterial species are too toxic to be given intravenously, in which case the i.m. route can be used for booster injections.

Antiserum fractionation

Practical procedures for fractionating serum to obtain antibody are ammonium sulfate precipitation, column separations on DEAE–Sephadex G-50 or DEAE-cellulose, and chromatography on protein A-Sepharose. The simplest and least expensive of these is that of salting out the globulins with $(NH_4)_2SO_4$. Generally, a satisfactory globulin fraction for fluorochrome labeling can be obtained by precipitating three times with $(NH_4)_2SO_4$ at 40% saturation. Critical studies have been done to optimize concentrations for precipitating immunoglobulin from various animal species (7). Fractionation on protein A-Sepharose CL4B (Pharmacia Fine Chemicals, Inc., Piscataway, N.J.) columns by affinity chromatography is an efficient method for obtaining immunoglobulin G (IgG) from human, rabbit, and guinea pig sera. Precise details of the technique are given by Goding (3). The method is not equally applicable to sera of all animal species. Fractionation of human and rabbit serum proteins on DEAE–Sephadex A-50 (Pharmacia) is an efficient way of obtaining pure IgG (22). As with protein A, the method is not equally effective with sera of all animal species.

Labeling of immunoglobulins

The conjugation of protein with fluorochromes is a chemical reaction. If the serum contains the desired antibody at a reasonable titer, careful control of production usually results in a specific reagent. Conjugation procedures may be designed to produce reagents with almost any degree of fluorochrome-protein labeling desired (15). The choice of the fluorochrome/protein ratio for a conjugate may vary with different antigen-antibody systems, with the nature of the fluorochrome, and with the specific diagnostic application (8, 9). Therefore, the degree of nonspecific staining which can be tolerated is a primary factor in determining the optimal fluorochrome/protein ratio for any antigen-antibody system. A relatively low fluorochrome/protein ratio must be used for rhodamine conjugates to avoid extensive denaturing of the labeled antibodies (14).

The fluorochromes most often used for antibody labeling are FITC and TMRI. When one antibody species is labeled with FITC and another with TMRI and the two conjugates are combined, two different antigens may be recognized simultaneously in a mixed antigen smear. This double-staining procedure simplifies diagnostic IF tests in some applications.

Titration, sorption, and standardization of conjugates

Titers of conjugates should be determined against the antigens they are designed to detect, and their ability to stain related antigens should be determined. Usually a series of twofold dilutions of the conjugate in saline is prepared and used to stain appropriate smears; positive and negative control conjugates and antigens are used. Frequently, a diagnostic dilution can be selected which is specific and makes it unnecessary to absorb the reagent. Important aspects of the standardization of IF reagents are discussed in two papers (2, 18). Nonspecific staining may be a problem when FAs are used to specifically detect pathogenic organisms in clinical specimens. This may be particularly troublesome if the conjugate must be used at a relatively low dilution. Frequently this nonspecific staining can be dramatically reduced by using rhodamine-labeled bovine serum albumin as a counterstain (23). Alternatively, in our laboratory, when double staining is not an objective, we routinely use rhodamine-labeled whole serum from nonimmunized animals of the homologous species as diluent for the fluorescein conjugates. Nonspecific staining by rhodamine-conjugated antibody can likewise be greatly reduced by using unlabeled 2% bovine serum albumin as diluent for the conjugate.

Anti-immunoglobulin reagents for the indirect IF tests should be examined by means of a box or checkerboard titration in which series of doubling dilutions of both the unlabeled antibody (test serum) and the labeled antiglobulin are tested against each other (1, 6). In these tests the plateau titer is defined as the highest dilution of a positive serum which gives maximum fluorescence with the highest dilution of conjugate. This dilution of the conjugate is referred to as the plateau endpoint of the conjugate.

The most commonly used diagnostic applications based on indirect IF are probably the fluorescent treponemal-antibody absorption and the antinuclear-factor tests and tests for antibody to *Toxoplasma gondii* and to *Legionella* spp. Appropriate positive and negative control conjugates constitute a vital part of these tests. In the indirect IF procedure, controls should consist primarily of both homologous and heterologous species antisera known to be either positive or negative for the antigen. The indicator reagent (labeled homologous antiglobulin) control should consist of labeled normal globulin or labeled heterologous antiglobulin.

Commercial FA diagnostic reagents

Following is a list of bacterial, fungal, parasitic, and viral pathogens for which FA diagnostic reagents are currently available commercially. The list is not necessarily complete, does not include FA reagents for detection of autoimmune diseases, and does not include monoclonal antibody reagents which are just beginning to become available commercially. Etiological agents for which direct FA (DFA) conjugates are available include: *Bordetella pertussis*, *Bordetella parapertussis*, *Escherichia coli* (K and O antigens), *Francisella tularensis*, *Haemophilus influenzae*, *Klebsiella* spp., *Legionella* spp. (monovalent and polyvalent), *Leptospira* spp., *Listeria* spp., *Neisseria gonorrhoeae*, *Neisseria meningitidis*, *Pseudomonas pseudomallei*, *Salmonella* spp. (O, H polyvalent), *Shigella* spp., *Staphylococcus* spp., *Streptococcus pneumoniae*, *Streptococcus pyogenes* (groups A and B), *Treponema pallidum*, *Chlamydia trachomatis*, *Candida albicans*, adenovirus, cytomegalovirus, Epstein-Barr virus (infectious mononucleosis), herpes simplex virus types 1 and 2 (HSV-1 and HSV-2), influenza, measles (rubella), mumps, parainfluenza, and rabies.

Some IFA reagents are available only as kits, which include the standard antigen and appropriate serum controls. Others are available as the individual components. Commercial IFA reagents for serological testing are currently available for the following etiological agents: *Legionella* spp. (monovalent and polyvalent), *Treponema pallidum*, *C. trachomatis*, *Toxoplasma gondii*, cytomegalovirus, Epstein-Barr virus, HSV-1 and -2, influenza, mumps, measles (rubella), and varicella-zoster virus.

All commercial FA diagnostic reagents should be critically evaluated by the user with appropriate positive and negative controls before they are used in clinical diagnostic testing.

Monoclonal antibodies

Improvement in the specificity of IF reagents depends upon finding methods for making labeled antibodies monospecific for their respective antigens. This goal is now readily achievable through hybridoma technology resulting from the pioneering work of Kohler and Milstein (12, 13). Monoclonal antibodies can be used in IFA tests, either directly or diluted, as the cell-free, spent-culture fluids of the hybridoma cell line. They may also be used in IFA tests as diluted, fibrin-free ascites fluids without further purification. If conjugates are to be made for DFA testing, then purification is necessary. If the monoclonal antibody is of an IgG subclass that has affinity for staphylococcal protein A, then it can most conveniently be isolated on a staphylococcal protein A column. If the monoclonal antibody is IgM or an IgG subclass that does not have affinity for staphylococcal protein A, then it can be purified by 50% saturated ammonium sulfate precipitation. If the monoclonal antibody is a euglobulin, it can be further purified from other proteins of ascites fluid through precipitation by dialysis against deionized or distilled water. Fluorescein or rhodamine conjugates can be made with monoclonal antibodies in the same manner as with polyclonal antibodies. However, the solubility and stability properties of conjugates of monoclonal antibodies may vary greatly from one monoclonal antibody to another.

IF monoclonal antibody reagents should soon become available commercially for specific applications in diagnostic microbiology. Nahm et al. (17) developed a monoclonal antibody that was reactive in DFA tests with all group A streptococcal strains that were tested and gave no false-positive reactions. Stephens et al. (24) produced a monoclonal antibody against membrane protein of *C. trachomatis* that appeared to be active in DFA tests with all strains of *C. trachomatis* and did not cross-react with the closely related *Chlamydia psittaci*. Fluorescein-labeled monoclonal antibody to *C. trachomatis* is now commercially available from Syva Co., Palo Alto, Calif. Tam et al. (25) prepared monoclonal antibodies to outer membrane protein of *N. gonorrhoeae* that reacted in DFA tests only with specific subsets of *N. gonorrhoeae* strains. They found that monoclonal antibodies that reacted broadly with *N. gonorrhoeae* also cross-reacted with other *Neisseria* species. To prepare a DFA diagnostic reagent that was specific to and reactive with all *N. gonorrhoeae* reference strains, they found it necessary to pool two monoclonal antibodies with reactivity to subsets within the species. Goldstein et al. (5) developed a panel of four monoclonal antibodies to react broadly with HSV and to distinguish unambiguously between HSV-1 and HSV-2. Investigations in our laboratory with monoclonal antibodies to *Legionella pneumophila* serogroup 1 led to the development of a panel of nine monoclonal antibodies which could be used as epidemiological markers to identify serological subsets of *L. pneumophila* serogroup 1 by IFA and to relate clinical cases of legionellosis to possible sources of infection in the environment through subgrouping of relevant isolates (16). A combination of two monoclonal antibodies from this panel gave a positive IFA reaction with all of the several hundred *L. pneumophila* serogroup 1 strains that were tested in our laboratory.

Monoclonal antibodies are relatively in their infancy as diagnostic reagents in IF techniques. The development of an array of hybridoma cell lines and monoclonal antibodies for all IF applications in diagnostic microbiology presents an appreciable challenge. The exquisite specificity of monoclonal antibodies may in some instances be a disadvantage, since it greatly limits the range of diagnostic testing. In the usual fusion experiment, many cell lines result which produce antibodies having properties that make them undesirable or impractical as diagnostic reagents. Cell lines are selected for the production of monoclonal antibodies that have the desired avidity and specificity and are of the desired class. The cell line must be stable, so as not to lose the ability to synthesize the desired antibody. This can only be developed and demonstrated by repeated passaging, routine cloning, and testing the antibody that is produced. Within the present state of development of the technology, it appears to be quite feasible, through judicious use of purified immunogens and appropriate selection of hybridoma cell lines, to tailor monoclonal antibodies for reactivities that are specific to any serological variant or subset within a serogroup or that are broadly specific to a given serogroup or species of microorganisms. The potential for the development of useful monoclonal antibodies for bacterial, viral, and fungal diagnostic reagents appears to be great indeed.

CLINICAL APPLICATIONS OF FA MICROSCOPY

FA microscopy is a means of rapidly diagnosing illness when appropriate specimens are available. Results with FA testing can be obtained within several hours after receipt of the specimen, whereas the culture and characterization of an etiological agent may require from 1 to 2 days to weeks. FA testing can be used to advantage when mixed infections are present or in situations in which the specimen was not handled aseptically. The method can also be used with Formalin-fixed specimens, when culturing is not possible.

The specificity of testing varies depending on the quality of the reagents that are used, the quality and timeliness of the specimen, and the ease of identification of the particular microbial pathogen that is tested for. In the case of bacterial pathogens, the bacterial cells are visualized by fluorescence microscopy, so the microscopist can use cellular morphology as well as fluorescence in interpreting the test. The location of the antigens within the tissue architecture or the type of fluorescence observed can be informative. For example, the location of FA-stained cells in macrophages in lung tissue specimens is highly diagnostic in tests for legionellosis. Likewise, the capsular fluorescence observed in FA staining of tissues infected with *Bacillus anthracis* is highly diagnostic.

It has been estimated that ca. 10^5 stained organisms per ml or per g of sample are required to ensure the detection of FA-stained *Legionella* spp. within a reasonable period of searching by fluorescence microscopy. This approximate figure is probably applicable to most other systems in which the DFA test is used to detect pathogenic bacteria in clinical specimens. Diagnosis on the basis of positive DFA findings is presumptive and should be interpreted on the basis of the recognized specificity of the test for the particular microbial infection. Negative DFA findings by tissue specimen examination should never be considered as an absolute determination of absence of infection by the pathogen in question, since the specimen may not have been collected at the appropriate stage of the disease or the concentration of organisms may be too low to permit reliable detection.

FA microscopy should be considered as but one method in the armament of the clinical diagnostic laboratory. Under favorable conditions it can provide an early, rapid, presumptive diagnosis at the time when maximum benefit can be gained by guiding the clinician to the proper choice of therapy. When possible, FA microscopy should be used in conjunction with other supportive or confirmative tests such as tests for soluble antigens of microbial pathogens, tests for serological conversion to antigenic determinants of microbial pathogens, and ultimately, the culture and characterization of the etiological agent.

LITERATURE CITED

1. **Beutner, E. H., M. R. Sepulveda, and E. V. Barnett.** 1968. Quantitative studies of immunofluorescent staining. Bull. W.H.O. **39:**588–606.
2. **Cherry, W. B., and C. B. Reimer.** 1974. The standardization of diagnostic immunofluorescence. Bull. W.H.O. **48:**737–746.
3. **Goding, J. W.** 1978. Use of staphylococcal protein A as an immunological reagent. J. Immunol. Methods **20:**241–253.
4. **Goldman, M.** 1968. Fluorescent antibody methods. Academic Press, Inc., New York.
5. **Goldstein, L. C., L. Corey, J. K. McDougall, E. Tolentino, and R. C. Nowinski.** 1983. Monoclonal antibodies to herpes simplex viruses: use in antigen typing and rapid diagnosis. J. Infect. Dis. **147:**829–837.
6. **Hardy, P. H., and E. E. Nell.** 1971. Characteristics of fluorescein-labelled anti-globulin preparations that may affect the fluorescent treponemal antibody-absorption test. Am. J. Clin. Pathol. **56:**181–186.
7. **Hébert, G. A.** 1976. Improved salt fractionation of animal serums for immunofluorescence studies. J. Dent. Res. **55**(Special Issue A):A33–A37.
8. **Hébert, G. A., B. Pittman, and W. B. Cherry.** 1967. Factors affecting the degree of nonspecific staining given by fluorescein isothiocyanate labeled globulins. J. Immunol. **98:**1204–1212.
9. **Jones, G. L., G. A. Hébert, and W. B. Cherry.** 1979. Fluorescent antibody techniques and bacterial applications. Center for Disease Control, Atlanta.
10. **Kawamura, A., Jr.** 1977. Fluorescent antibody techniques and their applications. University Park Press, Baltimore.
11. **Koch, K. F.** 1972. Fluorescence microscopy. Instru-

ments, methods, applications. Ernst Leitz GmbH, Wetzlar, Federal Republic of Germany.

12. **Kohler, G., and C. Milstein.** 1975. Continuous cultures of fused cells secreting antibody of predefined specificity. Nature (London) **256:**495–497.
13. **Kohler, G., and C. Milstein.** 1976. Derivatization of specific antibody producing tissue culture and tumor cell lines by cell fusion. Eur. J. Immunol. **6:**511–519.
14. **McKinney, R. M., and J. T. Spillane.** 1975. An approach to quantitation in rhodamine isothiocyanate labeling. Ann. N.Y. Acad. Sci. **254:**55–64.
15. **McKinney, R. M., J. T. Spillane, and G. W. Pearce.** 1964. Factors affecting the rate of reaction of fluorescein isothiocyanate with serum proteins. J. Immunol. **93:**232–242.
16. **McKinney, R. M., L. Thacker, D. E. Wells, M. C. Wong, W. L. Jones, and W. F. Bibb.** 1983. Monoclonal antibodies to *Legionella pneumophila* serogroup 1: possible applications in diagnostic tests and epidemiologic studies. Zentralbl. Bakteriol. Mikrobiol. Hyg. Abt. 1 Orig. A **255:**91–95.
17. **Nahm, M. H., P. R. Murray, B. L. Clevinger, and J. M. Davie.** 1980. Improved diagnostic accuracy using monoclonal antibody to group A streptococcal carbohydrate. J. Clin. Microbiol. **12:**506–508.
18. **Nairn, R. C.** 1968. Standardization in immunofluorescence. Clin. Exp. Immunol. **3:**465–476.
19. **Nairn, R. C.** 1976. Fluorescent protein tracing, 4th ed. Churchill Livingstone, New York.
20. **Ploem, J. S.** 1967. The use of a vertical illuminator with interchangeable dichroic mirrors for fluorescence microscope with incident light. Z. Wiss. Mikrosk. Mikrosk. Tech. **68:**129–142.
21. **Rowe, D. S.** 1970. Production of specific antisera, p. 27–38. *In* E. J. Holborough (ed.), Standardization in immunofluorescence. Blackwell Scientific Publications, Oxford.
22. **Sela, M., D. Givol, and E. Mozes.** 1963. Resolution of rabbit γglobulin into two fractions by chromatography on diethylaminoethylsephadex. Biochim. Biophys. Acta **78:**649–657.
23. **Smith, C. W., J. D. Marshall, and W. C. Eveland.** 1959. Use of contrasting fluorescent dye as a counterstain in fixed tissue preparations. Proc. Soc. Exp. Biol. Med. **102:**179–181.
24. **Stephens, R. S., M. R. Tam, C.-C. Kuo, and R. C. Nowinski.** 1982. Monoclonal antibodies to *Chlamydia trachomatis*: antibody specificities and antigen characterization. J. Immunol. **128:**1083–1089.
25. **Tam, M. R., T. M. Buchanan, E. G. Sandström, K. K. Holmes, J. S. Knapp, A. W. Siadak, and R. C. Nowinski.** 1982. Serological classification of *Neisseria gonorrhoeae* with monoclonal antibodies. Infect. Immun. **36:**1042–1053.

Chapter 92

Immunology of Bacterial Infections

SILAS G. FARMER

Serological testing can be an important addition to our armamentarium in the diagnosis of bacterial diseases. However, care must be exercised in the use of this indirect approach. The qualitative and quantitative presence of various antibodies in the absence of overt disease may vary significantly among individuals. Furthermore, serological tests vary in their sensitivity and specificity, antigens vary in their immunogenicity, and patients vary in their ability to respond to antigenic stimuli. One aspect of the serological approach to the diagnosis of bacterial disease that has been poorly understood, or even ignored, is the prevalence of the disease for which the serological test is being used. A knowledge of disease prevalence is very important in the determination of the predictive value (PV) of the positive (PV-pos) or negative (PV-neg) result. In other words, the value of the test lies in its ability to enable the clinician to predict the presence of the disease from a positive result or to rule out the disease from a negative result. It is the belief of the author that all serological tests should be accompanied with the PV of the result. The serologist is referred to the excellent discussion by Galen and Gambino (4) on the value of test results and how to calculate their PV.

Timing is important in serum collection, and in almost every instance, it is necessary to collect two samples. The first sample should be collected soon after the onset of illness (acute stage) and the second should be collected 1 or more weeks later (convalescent stage), depending on the antibody response. Technical procedures are frequently difficult to duplicate exactly. For this reason, it is preferable to assay both samples at the same time. This procedure eliminates day-to-day, person-to-person, and batch-to-batch variabilities. Simultaneous testing would require the freezing and saving of the acute-stage specimen until the convalescent-stage sample is tested. Occasionally, it might be useful to find a very high titer in an individual who has been ill for several weeks. Although such a high titer might suggest an infection at some undetermined time in the past, it is rare that a serological diagnosis can be made in these instances. An increase in titer must be two dilution increments (usually fourfold) or greater to be significant. The serologist should evaluate the serological response within the framework of guidelines by the author of the procedure, providing the antibody titers have been correlated with the intended infectious disease.

STREPTOCOCCAL INFECTIONS

Group A beta-hemolytic streptococci continue to be important in the etiology of human streptococcal disease. They are important not only for primary inflammatory components, but also for secondary toxicosis and immunological sequelae to infections. It is the sequelae that are of interest to the serologist. Many intracellular and extracellular antigenic components of these streptococci stimulate the production of antibodies in the infected patient. The two antibody determinations that are most useful in the evaluation of patients with recent streptococcal disease are anti-streptolysin O (ASO) and anti-DNase B (ADN-B). The ASO determination is the best known and most frequently performed of the two. However, ASO titers are unreliable for the determination of streptococcal pyoderma and its complications. The ADN-B determination is very useful in the detection of streptococcal infections of the latter type.

ASO titer

The procedure for the determination of the ASO titer will not be given here, but the recommendation of the author is to use the microstreptolysin O test of Klein et al. (15).

Interpretation and significance. The ASO titer is expressed in Todd units or international units, depending on whether the potency of the ASO standard was adjusted against the Todd standard or the international (World Health Organization) standard. The number of units will be the same as the titer when streptolysin O produced in the United States is used.

A rise in titer of two or more dilution increments (0.3 log) between the acute- and convalescent-stage specimens is usually considered significant regardless of the magnitude of the titer. The upper limit of the normal range varies with the age of the patient, season, and geographical area. Therefore, each laboratory should establish its own upper limits of normal range. There is no clear-cut distinction between a normal and a low elevated titer in the absence of a demonstrable increase in the antibody titer.

Limitations of the ASO titer and sources of error. Serum beta-lipoprotein in liver disease and growth products of some bacterial species can neutralize the hemolytic property of streptolysin O and give false-positive titers. The oxidation of streptolysin O from its reduced state may result in erroneously high titers. Antibiotic treatment during the streptococcal infection may abort or depress the antibody response. Only about 80 to 85% of patients with acute rheumatic fever have significant ASO titer response.

ADN-B test

The serologist is referred to the method of Klein et al. (14) for the performance of the micro-ADN-B test.

Clinical indications and interpretation. The ADN-B test is thought by many to be the best single test for the serological detection of streptococcal infection. In addition to its usefulness for the confirmation of a recent streptococcal infection when acute rheumatic fever or acute glomerulonephritis is suspected, it is also useful for the detection of Sydenham's chorea. It surpasses the ASO test in usefulness for patients with acute glomerulonephritis after streptococcal pyoderma. The ADN-B test does not suffer the shortcomings of the ASO. False-positives are not caused by an increase in beta-lipoproteins in liver disease, by bacterial growth in serum, or by oxidation of the antigen. The rise in ADN-B titer is expressed later than the rise in ASO titer, peaking in 4 to 6 weeks, but it remains longer. A normal titer varies with the age of the patient, season, and geographical area, in a manner similar to that of the ASO titer. A rise in titer of two or more increments (0.3 log) between the acute- and convalescent-stage specimens is considered significant.

Streptozyme test

A commercial slide test for the detection of antibodies to five streptococcal exoenzymes (streptolysin, streptokinase, hyaluronidase, DNase, and NADase) was marketed some years ago by Wampole Laboratories of Stamford, Conn., and was purported to be more sensitive and less subject to error than existing procedures. Although the initial reported impressions were favorable, this test has not performed in the laboratory of the author in the manner claimed by the manufacturer. In a more comprehensive analysis, Golubjatnikov et al. (7) demonstrated the unreliability of this multienzyme test. They found both false-positive and false-negative results compared with results of the ASO and ADN-B tests. The best correlation was found with young children; false-positive Streptozyme test results increased with the age of the study subjects.

Latex test for screening

The standard ASO and ADN-B procedures are rather labor intensive. A simple, reliable screening test for these antibodies would be a welcome addition to most serological laboratories. A recent latex test, LEAP-STREP (liposome-enhanced agglutination procedure), has been introduced by Cooper Biomedical Diagnostics Div., Malvern, Pa. This rapid slide test requires only a few minutes to perform. From the data of Ploussard and Sloyer (20), it appears to have a sensitivity of 100% and a specificity of 99%. However, inasmuch as a false-positive of 4% is observed and there is a low incidence of streptococcal disease, the predictive value of a positive test is only 0.004%. It is obvious that quantitative ASO and ADN-B procedures must be performed in the event of a positive LEAP-STREP test. The only comparison published to date was with the Streptozyme test, which has been shown to be an unreliable procedure. Future studies will determine the true usefulness of this procedure in the clinical laboratory (20).

Febrile agglutinins or Widal test

The traditional purpose of the Widal test has been to screen suspected cases or carriers of enteric fevers, i.e., pyrexia of unknown origins. The test reached its greatest popularity decades ago when cultural isolation and identification methodologies were poorly developed. The agglutination test by itself can never afford more than a presumptive diagnosis of infection, and it must always be interpreted in relation to the clinical condition of the patient and to previous experience. There are so many limitations to the test that it should always be subordinated to the direct cultural demonstration of the causative agent. Many variables plague this test: (i) the frequency of distribution of agglutinins in the population, (ii) the effect of previous immunizations or inoculations, (iii) the stage of the disease, and (iv) the effect of antimicrobial therapy in aborting the immune response. In addition, (v) agglutinins may result from the colonization of the gut with other members of the family *Enterobacteriaceae*, and (vi) due to the smooth to rough dissociation of members of this family upon cultivation, spontaneous agglutination may occur in saline suspensions.

If the test must be performed, it should be done with the original Widal tube test, in which serial twofold dilutions are made in 0.9% saline and incubated at 50 to 52°C for 2 h. Antigens that reflect the local distribution of members of the family should be used. Flagellar H agglutinins exhibit large floccular aggregations that are easily broken up by agitation. Somatic O agglutinins exhibit small granular aggregates that are better seen with a magnifying or hand lens. A microscope ocular held against the eye also works well.

Agglutinins appear at the beginning of week 2, reach a maximum during week 3, and may persist for weeks or months after convalescence. No single titer can be accepted as significant. A fourfold or greater rise in titer must be demonstrated.

The slide test, although commercially available and popular, has not been correlated with human disease and thus has very limited clinical usefulness.

YERSINIA SPP.

Yersinia enterocolitica

Clinical indications. Attempts to make a serological diagnosis of *Y. enterocolitica* infection may be indicated in small children with fever and symptoms of acute gastroenteritis of short duration. Older children and youth will have symptoms of appendicitis. In reality, however, they probably will have acute terminal ileitis, mesenteric lymphadenitis, or pseudoappendicitis. Older individuals present with symptoms of acute enteritis, with or without erythema nodosum, fever, and arthritis, or they may present as terminal ileitis.

It is best to combine serology and culture for the evaluation of these cases. The Widal-type agglutination is considered the best procedure for evaluation of the patient. In the United States, serotype O8 is the antigen of choice. In other parts of the world, serotypes O3 and O9 may be considered. The agglutinins have immunological specificity and seldom cross-react with other gram-negative bacilli. However, it has been reported that serotype O9 will cross-react with *Brucella* spp.

Interpretation. A titer of 160 or greater is significant, indicating an actual infection. Titers of 1,280 or greater are diagnostic of acute, actual infection with *Y. enterocolitica*. One may encounter low titers of 40 to 80 for months or years after infection.

Yersinia pestis

Immunological diagnoses of *Y. pestis* should be attempted in those parts of the world in which the disease is endemic or in travelers who are returning from those areas. The passive agglutination test with sheep erythrocytes sensitized with fraction I antigen is the method of choice (21). A titer of 256 or greater is considered evidence of a specific immunological response to plague infection.

Yersinia pseudotuberculosis

Infections by *Y. pseudotuberculosis* may present as mesenteric lymphadenitis in children and young people. Although the complement fixation (CF) test and Widal-type agglutinins have all been applied to the study of the immunological response to this organism, the Widal agglutinins test is considered the method of choice. One should employ antigens from serotypes 1a and 1b. A titer of 160 or greater indicates an actual infection. Titers may be very high in the acute symptomatic stage (25).

MYCOPLASMA SPP.

Mycoplasma pneumoniae

Mycoplasma spp. lack a cell wall; the antigens most important in the immune response are either membrane components or membrane associated. These organisms may also be inhibited or killed by specific antiserum directed against antigenic determinants on the membrane.

The serological approach for the evaluation of respiratory infection with *M. pneumoniae* is reasonable because culture procedures are time consuming and not readily available in many clinical laboratories. Two general procedures applied toward the determination of antibody response are CF and metabolic inhibition (MI). Both of these procedures appear to measure antibody to lipid determinants, and there is reasonable correlation between them.

CF test

The CF test is identical to the test commonly used in most laboratories. The antigen is commercially available or can be prepared by the serologist (12). It is preferable to use the lipid antigen rather than the whole organism, inasmuch as the lipid antigen is less anticomplementary, and greater differences are found between acute- and convalescent-stage sera (13).

Interpretation. Titers of 256 or greater suggest recent infection. A fourfold or greater rise in titer indicates recent infection. However, the CF test is not a sensitive procedure; only 58% of pneumonia patients with *M. pneumoniae* isolates showed a fourfold rise in titer, and only 65% of those with a fourfold rise yielded *M. pneumoniae* isolates (9).

Limitations. *M. pneumoniae* glycolipids cross-react with glycolipids from *Mycoplasma neurolyticum* and glycerolipids from spinach. In addition, patients with pancreatitis have high immunoglobulin M (IgM) antibodies that cross-react with the lipids of *M. pneumoniae*. However, the CF test is acceptable for the diagnosis of respiratory infections.

MI test

The MI test may be a viable alternative to the CF test in laboratories that routinely cultivate *Mycoplasma* spp. (12). The MI test also measures antibodies against the membrane-related lipid determinants.

Interpretation. A fourfold rise in titers to *M. pneumoniae* can readily be demonstrated with the MI test. However, inasmuch as the endpoint may vary from day to day, it is necessary to test the paired sera at the same time. In addition, antibiotic carry-over may inhibit growth. The latter problem may be overcome by using an erythromycin-resistant strain in the assay procedure (18).

IAHA test

In an attempt to overcome the obvious shortcomings of the CF test in sensitivity, Lennette and Lennette (16) modified the immune adherence hemagglutination (IAHA) procedure of Gershon et al. (6) and applied this modified procedure to viral, chlamydial, and mycoplasmal antigens. Antigen titers were higher for the IAHA procedure, and the method was 4 to 20 times more sensitive than the CF test. The IAHA procedure required only 4 h, compared with 20 h for the CF test. No sera found to be positive by CF were found to be negative by the IAHA test, and all sera found to be negative by the IAHA test were also found to be negative by the CF test. Although good specificity appears to exist among these two tests, it will be necessary to apply the IAHA test to the general population and determine the upper limits of normal for the general population and IAHA titer responses in disease before the test can be incorporated into the laboratory routine. It does deserve our attention for further development and standardization.

LEGIONELLOSIS

Patients infected with *Legionella* spp. respond with the production of antibodies in the IgG, IgA, and IgM classes. Because of the slowness of the antibody rise, requiring 3 to 6 weeks for a peak antibody response, serological diagnoses are made retrospectively only. A plethora of serological procedures have been used in attempts to diagnose infections with these organisms, e.g., microagglutination, enzyme-linked immunosorbent assay (ELISA), micro-ELISA, kinetic ELISA, immunodiffusion, counterimmunoelectrophoresis, indirect immunofluorescence assay (IFA), and solid-phase immunofluorescence assay.

IFA

The IFA is the serological test most often used for legionellosis, and accordingly, more data have been accumulated on the clinical significance of antibody responses with the IFA than with experimental or recently introduced procedures. The IFA is considered the reference procedure (24). However, significant variability is encountered with this procedure, depending on the source of the antigen, temperature of incubation, formulation of the medium on which the antigen is cultivated, method of killing and fixing the organisms, and whether or not the conjugates react with antibodies to IgG, IgM, and IgA (3). It has been shown that maximum reactivity is obtained when antigen is prepared from growth on buffered charcoal-yeast extract agar grown at 25°C and heat killed (2). However, even though 1% Formalin- or 37% formaldehyde vapor-fixed cells have 20% less relative antigenicity than cells from the buffered charcoal-yeast extract agar, the cell walls stain more clearly and have fewer aberrations. Polyvalent antigen containing serogroups 1 through 4 is recommended for the IFA.

An acute-stage specimen should be collected within 1 week of the onset of symptoms, and a convalescent-stage specimen should be collected 22 days after onset. If the titer is on the rise but has not reached a fourfold increase to 128 or greater, another serum should be obtained 42 days after onset. The IFA is a specific test with a sensitivity, at 3 weeks from onset, of about 75 to 85%. Another specimen taken later increases the sensitivity to about 90%.

The immune response to legionellosis is diverse in terms of antigenic specificity and class of immunoglobulins. Significant titers may result from responses to serogroup-specific antigens, *Legionella* common antigens, and nonspecific antigens. The antibodies may be exhibited in IgG, IgM, and IgA, singly or in various combinations (24). No one immunoglobulin class appears to be more specific than other classes for making a serological diagnosis (3).

IFA titers are not affected by cross-reaction with Q fever, psittacosis, *Chlamydia trachomatis*, tularemia, or *Mycoplasma* spp., by the presence of rheumatoid factor, or by heat-labile serum factors.

Interpretation. It is considered diagnostic if a fourfold increase in titer is demonstrated and maximum titer is 128 or greater. A single titer of 256 or greater is inconclusive.

Limitations. In some communities, the prevalence of titers of 128 or greater is 25%. In areas in which legionellosis is hyperendemic, titers may exceed 256. Accordingly, one must exercise great care and restraint in the interpretation of titers of any magnitude on a single specimen.

ELISA

In theory, the ELISA and IFA are very similar. The advantages of the ELISA are the potentials for automation, use of purified antigenic preparations and conjugates from several immunoglobulin classes, spectrophotometric readout, and computer analysis of test results (22). If this test is used, it must react with antibodies in the IgG, IgM, and IgA classes.

A kinetic ELISA, which measures the rate of substrate conversion by the enzyme when the concentration of antibody is rate limiting, has been compared to IFA (22). The sensitivity, specificity, or PV could not be ascertained because of the paucity of serum from culture or epidemiologically confirmed cases. However, the procedure was reproducible with coefficients of variation of 4.8 to 5.8% and demonstrated good correlation with the IFA. This procedure holds much promise for the future if the correlation can be made with clinical disease.

Solid-phase immunofluorescence technique

The semiautomated solid-phase immunofluorescence technique demonstrated an agreement with the IFA for 91.8% of the serum pairs tested but gave evidence of recent *Legionella* infection significantly fewer times than did the IFA. Further study is required before this procedure is considered for diagnostic use (23).

LEPTOSPIROSIS

The antibodies associated with human infection caused by *Leptospira interrogans* are demonstrable at the end of 1 week after onset of symptoms, reach maximum titer at about 3 to 4 weeks, and slowly decline over the course of months or years. These pathogenic leptospires have distinct antigenic properties and thus are important for serological diagnoses. Over 180 serovars have been arbitrarily assembled into 18 serogroups based on common cross-reacting agglutinogens (1).

Because of the high serological specificity, a battery of serovars must be used in diagnostic work. Eight serovars, which represent those associated with human infection, are currently recommended for this battery. They are copenhageni, canicola, pomona, autumnalis, grippotyphosa, wolffii, dzatzi, and patoc. If the serum is negative to this battery and leptospirosis is still highly suspected, the serum should be retested with a larger battery of antigens (1). Extended testing is available at the Leptospirosis Reference Laboratory, Center for Infectious Diseases, Centers for Disease Control, Atlanta, Ga.

The two most commonly used tests for the serological diagnosis of leptospirosis are the microscopic agglutination test and the macroscopic (slide) agglutination test.

Microscopic agglutination test

The microscopic agglutination test is the standard reference test. It has excellent sensitivity for the diagnosis of recent or past infection. Because of the high specificity, a battery of serovars must be used. Antigens must be cultivated in the laboratory; no commercial ones are available. However, antisera are available from Difco Laboratories, Detroit, Mich. The test may be conducted with either live or Formalin-fixed antigens. The live-antigen test is read by dark-field microscopy; agglutination and lysis are observed in the presence of homologous antibody. No lysis is observed with the Formalin-fixed antigens. In addition, with Formalin-fixed antigens, the sensitivity of the test is lower, and prozone reactions may occur in high-titered serum. The endpoint is 50% agglutination of the cells.

However, the microagglutination test is labor intensive and time consuming, and it involves the handling of live cultures, with risk to laboratory personnel. To give broad coverage, it is necessary to maintain a large number of serovars.

Macroscopic (slide) agglutination test

Although the titers obtained with the macroscopic (slide) agglutination test are somewhat lower than those obtained with the microscopic agglutination test, it is still considered the best assay for laboratory diagnosis per se. It has good sensitivity and specificity, it is simple to perform, and commercial antigens are available (Difco; Ft. Dodge Laboratories, Ft. Dodge, Iowa). The test reacts more broadly than the microscopic test, but serological diagnoses can be made if paired (acute- and convalescent-stage) sera are tested.

Significance. Titers may be very high, even in excess of 25,600. Since leptospiral agglutinins persist for extended periods, one must evaluate the titer of any single serum specimen with great caution. Titers in excess of [illegible]600 are strong presumptive evidence of recent infection; a fourfold increase in titer on paired acute- and convalescent-stage sera is diagnostically significant (1).

RICKETTSIA SPP.

Procedures that may be used to evaluate the immunological response to rickettsial diseases are CF, microagglutination, indirect hemagglutination, IFA, Weil-Felix (WF), ELISA, and latex agglutination (LA).

IFA

The IFA is the most sensitive and specific procedure for the demonstration of immunological response from epidemic, endemic (murine), and scrub typhus, as well as Rocky Mountain spotted fever. The specificity of the IFA can be enhanced by adsorption of group-specific antibodies before performance of the IFA. The antigen is readily prepared from infected chicken yolk sacs (19).

Interpretation. A fourfold rise in paired serum titers, a single titer of 128 or greater, or an IgM titer of any level (suggestive of recent infection) is diagnostically significant.

WF test

Members of the spotted fever group, typhus, and scrub typhus share some minor antigens with members of the genus *Proteus* and serve as the basis of the WF test. *Proteus vulgaris* OX19 is agglutinated by sera of patients convalescing from epidemic and murine typhus, but rarely from patients with Brill-Zinsser disease. *P. vulgaris* OX19 and OX2 are agglutinated by the immune sera of patients recovering from most of the spotted fever group. Scrub typhus may elicit antibodies that react to *Proteus mirabilis* OXK. Human response is variable, and only ca. 50% of these patients exhibit a demonstrable rise in titer against this strain.

Rickettsia akari of rickettsialpox and *Coxiella burnetii* of Q fever do not share antigenic determinants with members of the genus *Proteus*.

The WF test is not specific for rickettsial infections, and agglutination may occur as the result of *Proteus* organisms. If the WF test is to be used, great care must be exercised in the interpretation of the results. The WF tests are the least sensitive of all procedures used in rickettsial serology. However, a fourfold rise in titer of paired sera, along with proper history and symptoms, is considered of diagnostic importance.

LA test for Rocky Mountain spotted fever

A latex test for Rocky Mountain spotted fever has been developed and extensively evaluated (10). The erythrocyte-sensitizing substance from *Rickettsia rickettsii* was adsorbed onto latex particles. In a collaborative evaluation of this test in 11 laboratories, it was judged to have an efficiency (true positives plus true negatives per 100,000) of 93.3%, compared with the IFA. Both the LA test and the IFA detected antibodies within 7 to 9 days of onset. A high LA titer of 128 or greater from a single serum was diagnostic of active Rocky Mountain spotted fever. The LA test detects rickettsial antibodies in patients with active infection, but usually does not detect antibodies in patients with past infections. The test reactivity is not uniquely linked to any particular immunoglobulin class (11).

Q fever

C. burnetii rickettsiae, the causative agents of Q fever, are antigenically distinct from the genus *Rickettsia*. They are known to undergo phase variation, which is a host-dependent change in antigenic structure. Newly isolated strains from animals and ticks are characteristically of phase 1; after passages in the embryonated egg, phase 2 is expressed.

Individuals convalescing from Q fever are generally reactive in the CF test with phase 2 antigen. If high antibody titers are found to phase 1 antigens, it is suggestive of chronic Q fever, such as subacute Q fever endocarditis.

Antibodies against *C. burnetii* are highly specific regardless of the diagnostic techniques employed. A fourfold or greater rise in titer has diagnostic significance.

STAPHYLOCOCCI

Numerous attempts have been made to categorize the seriousness of staphylococcal disease and the duration of antibody therapy based on the detection of antigens, antibodies, and immune complexes in the sera of patients with suspected staphylococcal disease. Most efforts have been directed toward the teichoic acid cell wall component. Recent attention has also been directed toward the peptidoglycan. The spectrum of methodologies has varied from the relatively insensitive double immunodiffusion to the more sensitive counterimmunoelectrophoresis, ELISA, and solid-phase radioimmunoassay.

It has been abundantly shown that people with staphylococcal disease develop antibodies to cell wall components. In addition, titers tend to be higher in patients with deep sepsis or endocarditis than in those with benign bacteremia.

Unfortunately, almost every group of investigators uses an antigen either from a different staphylococcal strain or prepared dissimilarly. In the hands of a few groups, it appeared that the data supported the categorization of patients; in others, it was not possible to discern significant differences among the groups studied.

One group reported that "the enzyme-linked immunosorbent assay for teichoic acid antibodies was found to be a sensitive and specific method for diagnosing staphylococcal endocarditis and septicemia" (8). Their data showed a sensitivity of only 74% and a specificity of 95.8%. However, given the low prevalence of staphylococcal sepsis and endocarditis and the 4.2% false-positive rate, the PV-pos was only 0.22%. In other words, if this test is not applied to a well-defined population, more than 99 of every 100 positives will be with individuals who do not have staphylococcal septicemia or endocarditis. Conversely, their data, as with most other investigators, have a PV-neg in excess of 99%. These results would best serve to rule out staphylococcal endocarditis or septicemia, rather than confirm it.

Lentino and Rytel (17) tried to circumvent some of the problems inherent in traditional methods of antigen-antibody detection by testing not only for free antigenemia but for complexed antigens as well as ELISA. They demonstrated a very high mean titer of 32 in the staphylococcal endocarditic group, which was significantly higher than in the group with staphylococcal bacteremia or nonbacteremic staphylococcal disease. Despite this, the PV-pos was still only 0.17; the PV-neg was 98.9%. Although much remains to be discovered before the detection of staphylococcal antigens, antibodies, or immune complexes can serve to confirm or categorize staphylococcal disease, present technologies do permit the exclusion of significant staphylococcal disease.

ACB

Numerous investigators have studied the role of antibody-coated bacteria (ACB) in the differentiation of upper from lower urinary tract infection, in infection from colonizing bacteria in lower respiratory tract secretions, and in the identification of patients who are at high risk of treatment failures.

Urinary tract infections

Although initially it was reported that the presence of ACB from the urinary tract indicated kidney invasion, current thought is that it indicates that the problem is more than a simple lower urinary tract infection. Significant information can be obtained only if the conditions are standardized and correlated with diagnosis or treatment. Gargan et al. (5) used 1% ACB as positive and found that the cure rate was significantly lower in those patients with greater than 1% ACB in midstream catch specimens. They found no correlation of ACB and symptoms of upper urinary tract infection. Questions which must still be answered before this test realizes its full potential are: (i) what is the source of the antibody? and (ii) where does the coating occur?

ACB in pneumonia

Winterbauer et al. (26) combined quantitative culture and ACB to diagnose bacterial pneumonia by fiberoptic bronchoscopy. Their quantitative culture was reported to be 100% specific with 88% sensitivity. This result is certainly better than that obtained with expectorated sputum or transtracheal aspirate cultures. Other investigators have obtained lower sensitivity and specificity with quantitative cultures. The sensitivity of the ACB as applied to lower respiratory tract secretions was 73% with a specificity of 98%. However, given the prevalence of pneumonia, the PV-pos was still only 10.98%; the PV-neg was 99.9%. Further studies are indicated before this procedure finds clinical application in the laboratory.

LITERATURE CITED

1. **Alexander, A. D.** 1980. Serological diagnosis of leptospirosis, p. 542–546. *In* N. R. Rose and H. Friedman (ed.), Manual of clinical immunology, 2nd ed. American Society for Microbiology, Washington, D.C.
2. **Benson, R. F., G. B. Malcolm, L. Pine, and W. K. Harrell.** 1983. Factors influencing the reactivity of *Legionella* antigens in immunofluorescence tests. J. Clin. Microbiol. **17:**909–917.
3. **Blackman, J. A., F. W. Chandler, W. B. Cherry, A. C. England III, J. C. Feeley, M. D. Hicklin, R. M. McKinney, and H. W. Wilkinson.** 1981. Legionellosis. Am. J. Pathol. **103:**429–465.
4. **Galen, P. S., and S. R. Gambino (ed.).** 1975. Beyond normality: the predictive value and efficiency of medical diagnosis. John Wiley & Sons, Inc., New York.
5. **Gargan, R. A., W. Brumfitt, and J. M. T. Hamilton-Miller.** 1983. Antibody-coated bacteria in urine: criterion for a positive test and its value in defining a higher risk of treatment failure. Lancet **ii:**704–706.
6. **Gershon, A. A., Z. G. Kalter, and S. Steinberg.** 1976. Detection of antibody to varicella-zoster virus by immune adherence hemagglutination. Proc. Soc. Exp. Biol. Med. **151:**762–765.
7. **Golubjatnikov, R., J. E. Koehler, and J. Buccowitch.** 1977. Comparative study of antistreptolysin O, antideoxyribonuclease B and multienzyme test in streptococcal infections. Health Lab. Sci. **14:**284–290.
8. **Granström, M., I. G. Julander, S.-Å. Hedström, and R. Möllby.** 1983. Enzyme-linked immunosorbent assay for antibodies against teichoic acid in patients with staphylococcal infections. J. Clin. Microbiol. **17:**640–646.
9. **Grayston, J. T., H. M. Foy, and G. E. Kenny.** 1969. The epidemiology of mycoplasma infections of the human respiratory tract, p. 651–682. *In* L. Hayflick (ed.), The mycoplasmatales and L-phase of bacteria. Appleton-Century-Crofts, New York.
10. **Hechemy, K. E., R. L. Anacker, R. N. Philip, K. T. Kleeman, J. N. MacCormack, S. J. Sasowski, and E. E. Michaelson.** 1980. Detection of Rocky Mountain spotted fever antibodies by a latex agglutination test. J. Clin. Microbiol. **12:**144–150.
11. **Hechemy, K. E., E. E. Michaelson, R. L. Anacker, M. Zdeb, S. J. Sasowski, K. T. Kleeman, J. M. Joseph, J. Patel, J. Kudlac, L. B. Elliott, J. Rawlings, C. E. Crump, J. D. Folds, H. Dowda, Jr., J. H. Barrick, J. R. Hindman, G. E. Killgore, D. Young, and R. H. Altieri.** 1983. Evaluation of latex-*Rickettsia rickettsii* test for Rocky Mountain spotted fever in 11 laboratories. J. Clin. Microbiol. **18:**938–946.
12. **Kenny, G. E.** 1980. Serology of mycoplasmic infections, p. 547–552. *In* N. R. Rose and H. Friedman (ed.), Manual of clinical immunology, 2nd ed. American Society for Microbiology, Washington, D.C.
13. **Kenny, G. E., and J. T. Grayston.** 1965. Eaton PPLO (*Mycoplasma pneumoniae*) complement fixing antigen: extraction with organic solvents. J. Immunol. **95:**19–25.
14. **Klein, G. C., C. N. Baker, B. V. Addison, and M. D. Moody.** 1969. Micro test for streptococcal anti-deoxyribonuclease B. Appl. Microbiol. **18:**204–206.
15. **Klein, G. C., M. D. Moody, C. N. Baker, and B. V. Addison.** 1968. Micro antistreptolysin O test. Appl. Microbiol. **16:**184.
16. **Lennette, E. T., and D. A. Lennette.** 1978. Immune adherence hemagglutination: alternative to complement-fixation serology. J. Clin. Microbiol. **7:**282–285.
17. **Lentino, J. R., and M. W. Rytel.** 1982. Detection of circulating free and complexed staphylococcal antigens by enzyme-linked immunosorbent assay. J. Clin. Microbiol. **16:**1019–1024.
18. **Niitu, Y., S. Hasegawa, and H. Kubota.** 1974. Usefulness of an erythromycin-resistant strain of *Mycoplasma pneumoniae* for the fermentation-inhibition test. Antimicrob. Agents Chemother. **5:**111–113.
19. **Osterman, J. V., and C. S. Eisemann.** 1980. Rickettsiae, p. 707–713. *In* N. R. Rose and H. Friedman (ed.), Manual of clinical immunology, 2nd ed. American Society for Microbiology, Washington, D.C.
20. **Ploussard, J. H., and J. L. Sloyer, Jr.** 1984. Antibody screening for Group A streptococcal antibodies: an overview and comments on the LEAP-STREP test. Adv. Ther. **1:**129–136.
21. **Rust, J. H., Jr., S. Berman, W. H. Habig, J. D. Marshall, Jr., and D. C. Cavanaugh.** 1972. Stable reagent for the detection of antibody to the specific fraction I antigen of *Yersinia pestis*. Appl. Microbiol. **23:**721–724.
22. **Sampson, J. S., H. W. Wilkinson, V. C. W. Tsang, and B. J. Brake.** 1983. Kinetic-dependent enzyme-linked immunosorbent assay for detection of antibodies to *Legion-*

ella pneumophila. J. Clin. Microbiol. **18:**1340–1344.

23. **Thompson, T. A., and H. W. Wilkinson.** 1982. Evaluation of a solid-phase immunofluorescence assay for detection of antibodies to *Legionella pneumophila*. J. Clin. Microbiol. **16:**202–204.
24. **Wilkinson, H. W., B. J. Fikes, and D. D. Cruce.** 1979. Indirect immunofluorescence test for serodiagnosis of Legionnaires disease: evidence for serogroup diversity of Legionnaires disease bacterial antigens and for multiple specificity of human antibodies. J. Clin. Microbiol. **9:**379–383.
25. **Winblad, S.** 1980. Immune response to *Yersinia* and *Pasteurella*, p. 474–478. *In* N. R. Rose and H. Friedman (eds.), Manual of clinical immunology, 2nd ed. American Society for Microbiology, Washington, D.C.
26. **Winterbauer, R. H., J. F. Hutchinson, G. N. Reinhardt, S. E. Sumidi, B. Dearden, C. A. Thomas, P. W. Schneider, N. E. Pardee, E. H. Morgan, and J. W. Little.** 1983. The use of quantitative cultures and antibody coating of bacteria to diagnose bacterial pneumonia by fiberoptic bronchoscopy. Am. Rev. Respir. Dis. **128:**98–103.

Chapter 93

Monoclonal Antibodies in Clinical Microbiology

MILTON R. TAM, LYNN C. GOLDSTEIN, AND DALE E. YELTON

The diagnosis of infectious disease is performed with a variety of techniques, including microscopic examination of tissues and exudates, measurement of microbial antigens in patient specimens or antibodies in patient sera, and culture methods to amplify the numbers of organisms (5). Clinical microbiology laboratories use one or a combination of these methods, depending on the pathogen to be identified. Rarely does one method prove optimal for all situations. For certain organisms, diagnosis by direct microscopic identification is adequate. Many other organisms, however, require culture, which is the most sensitive method. Unfortunately, culturing is labor intensive, time consuming, and impossible for some pathogens.

Historically, the clinical laboratory has relied upon immunological methods for the diagnosis of a variety of infectious agents. Immunological techniques, however, have always depended upon the quality of the antisera used to perform the tests. To produce useful antisera, animals are usually inoculated several times with antigen. The best responders are then identified, and their sera are pooled. However, animals inoculated in a uniform manner with apparently pure antigens will frequently make antisera of various titers, avidities, and specificities. Because of the dynamic nature of the immune response, the antibodies usually vary from one bleeding to the next, even in a single animal. In addition, antisera may often cross-react between pathogenic and nonpathogenic species, since many related microorganisms share antigens with one another. When cross-reactive and nonspecific antibodies are present in an antiserum, they must be absorbed to produce a useful diagnostic reagent. As a result, antisera of good quality and specificity are difficult to prepare in large quantities and are therefore valuable commodities.

Although many excellent diagnostic tests have been formulated with polyclonal antisera, difficulties in the preparation of highly specific antibodies have limited more extensive development and utilization of immunoassays, especially those involving direct identification of microorganisms in patient specimens. Recently, however, monoclonal antibody techniques have afforded researchers an opportunity to reevaluate the use of immunological methods for the diagnosis of infectious diseases. Not only can the problems of heterologous antisera be overcome with the use of monoclonal antibodies of defined specificities, avidities, and titers, but these antibodies can be produced in unlimited quantities, assuring a constant, uniform supply.

HYBRIDOMA TECHNOLOGY

Immunologists have attempted to overcome the difficulties associated with antiserum production by searching for methods to reproducibly generate large quantities of antibodies (8). A model recognized as potentially valuable was that of multiple myeloma, a B-cell malignancy of mice, humans, and some other mammalian species. The tumor is a result of the transformation of a single B cell, which proliferates and secretes large amounts of the immunoglobulin produced before malignant transformation. Such myelomas were useful for the study of immunoglobulins since they could be easily cultured, cloned, and manipulated genetically to produce variants in immunoglobulin structure and production (1, 17). With the introduction of drug resistance markers and the development of techniques to fuse different myeloma lines, examination of the somatic cell genetics of immunoglobulin production was possible (11, 13). Hybridoma technology is a direct extension of these studies.

A major advance was provided by Kohler and Milstein (7), who fused cultured myeloma cells with spleen cells from mice immunized with sheep erythrocytes. They obtained hybrid cell lines secreting antibodies which lysed sheep erythrocytes in the presence of complement. Each cell line produced only one antibody derived from a single B cell and hence was termed a monoclonal antibody. The hybrid cell lines (hybridomas) exhibited antibody specificity, contributed by the spleen cell, and the abilities to grow in continuous cell culture and produce tumors in mice, contributed by the myeloma cell. In a short time, many laboratories had generated hybridomas secreting monoclonal antibodies to a variety of antigens (6, 12). The technology has become more widespread, but the basic procedure has not changed appreciably from the one originally described by Kohler and Milstein.

PRODUCTION OF MONOCLONAL ANTIBODIES

To appreciate the complexity of and problems in the production of monoclonal antibodies (Fig. 1), an understanding of the hybridoma procedure is necessary. Mice or rats are immunized one or more times with the antigen of interest. Two to 4 days after the last immunization, the spleen (and sometimes additional lymphoid organs) is removed, and a cell suspension is prepared. Spleen cells are mixed with myeloma cells adapted for growth in culture, and polyethylene glycol is added to promote the fusion of cell membranes. The method results in a mixture of unfused cells and fused cells of various combinations, including spleen-spleen, myeloma-myeloma, and spleen-myeloma. Spleen-myeloma cell hybrids, however, occur with only low frequency and must be placed in a selective medium to favor their growth. To provide for the survival of only the spleen-myeloma hybrids, the classic hypoxanthine-aminopterin-thymidine system popularized by Littlefield (9) is usually employed. Myeloma cells are engineered with a defect in production of the enzyme hypoxanthine-guanine phosphoribosyltransferase. Cells lacking this enzyme cannot utilize hypoxanthine to synthesize purines, and they die in the presence of aminopterin, a drug which blocks de nova synthesis of nucleic acid precursors. In spleen-myeloma hybrids, this genetic defect is complemented by the introduction of the normal gene

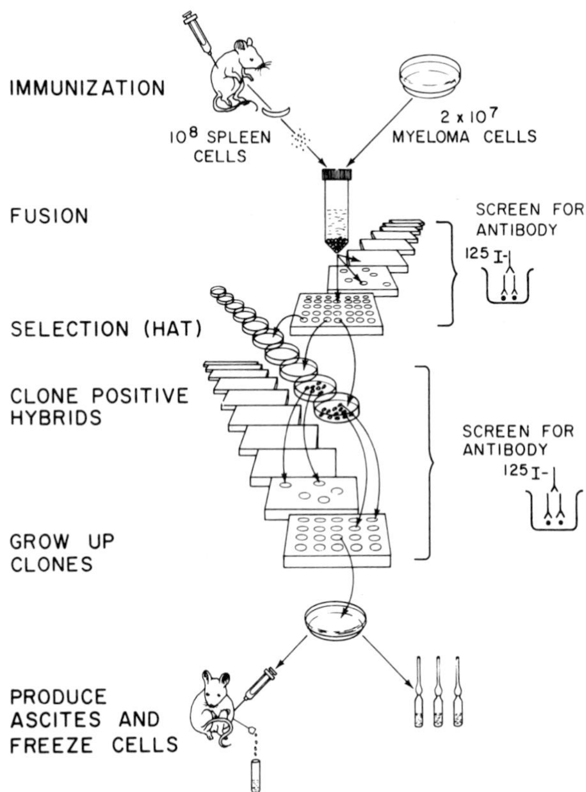

FIG. 1. Fusion of mouse myeloma cells and immune spleen cells. Mouse myeloma cells that no longer secrete their own immunoglobulin and lack the enzyme hypoxanthine-guanine phosphoribosyltransferase are fused in polyethylene glycol to spleen cells from an immunized mouse. The hybrid cells are selected in a medium containing hypoxanthine-aminopterin-thymidine (HAT), which kills the unfused myeloma cells, whereas the unfused spleen cells do not grow in tissue culture. (Reprinted with the permission of Lab. Manage. **19**:19–24, 1981.)

from the spleen cell. Spleen cells, though not killed by hypoxanthine-aminopterin-thymidine selection, are capable of living for only a short time in culture, so that after a few days, the only vigorously dividing cells are hybrids.

After a week or two, hybrid cells multiply to sufficient numbers so that culture fluids may be assayed for the presence of monoclonal antibodies. At this stage an assay should be rapid, simple, sensitive, and reproducible. Often many more hybrids are generated than can be conveniently handled beyond the initial culture vessels, so a quick and accurate assessment of the antibodies produced is necessary to select those of desired specificities.

To obtain maximum information about the antigenic specificity of monoclonal antibodies in the early screening, we have adapted replica-plating techniques from bacterial genetics (10). Hybrid cells are grown in 96-well microtest plates. When the hybrids have achieved a sufficient density, samples of culture fluid are removed and assayed on as many as six to eight replicate antigens. Each of the replica plates contains a different microbial or control antigen adsorbed onto the surfaces of the wells. Immune reactions are then detected by means of a radioimmune or enzyme-mediated assay. Figure 2 shows an example of a radioimmune assay of culture fluids containing antibodies to *Neisseria gonorrhoeae* (23). The culture fluids from a 96-well culture plate were tested by replica plating onto outer membrane extracts of three gonococcal strains. Several antibodies with differing specificities were detected. In this manner, rapid screening of individual antibodies made possible the selection of those desired for continued development.

Selected antibodies should also be evaluated in an assay similar to the one for which the antibody will ultimately be used. This evaluation is necessary because immunoassays have differing requirements for antibody affinity and specificity and for biological properties such as the abilities to agglutinate cells and fix complement. For example, monoclonal antibodies identified by the enzyme-linked immunosorbent assay, a sensitive technique, will not always react with antigen in fluorescent-antibody assays. Other enzyme-linked assays, such as those used in the immunohis-

Replicate Test of Culture Fluids from
Plate GC3/1 on GC Strains

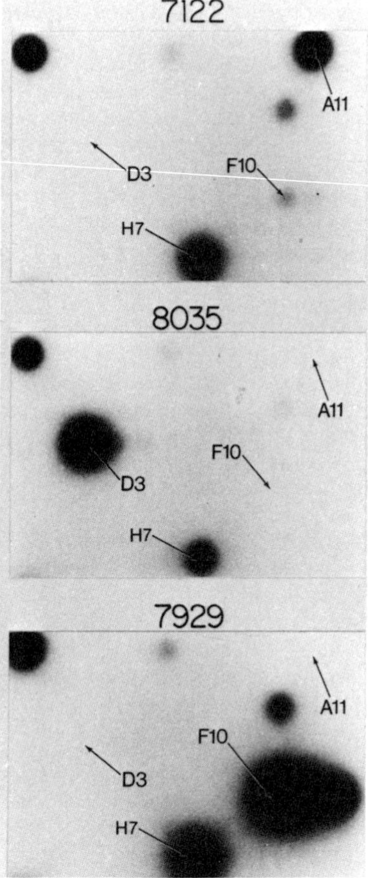

FIG. 2. Replica plate method for detection of monoclonal antibodies against *N. gonorrhoeae*. Medium from hybridomas growing in a 96-well culture plate was replica plated onto microtest plates on which three different *N. gonorrhoeae* antigen extracts were adsorbed. Immune reactions were detected by the addition of ^{125}I-labeled protein A and subsequent autoradiography.

tochemical staining of tissue sections, are not necessarily more sensitive than fluorescent-antibody techniques. Not all of the parameters that contribute to the relative sensitivity of assays are currently understood, so it may be useful to empirically test the monoclonal antibodies as soon as possible in their intended assay system.

After the useful hybridomas have been identified, they are cloned as quickly as possible to avoid overgrowth by other hybridomas, especially nonproducers. These cells frequently arise as variants, probably as the result of random chromosome loss which occurs during the initial culture period. Once the hybridomas have been cloned and reassayed until phenotypically stable for antibody production, they are grown to large numbers, frozen for future recovery, and inoculated into syngenic mice for the production of high-titered ascites fluids. When sufficient antibody has been produced, the specificity of a monoclonal antibody may be examined in greater detail by any of several biochemical techniques. Antibody can be used to precipitate a radiolabeled antigen for examination on one- or two-dimensional gel systems (10). Alternatively, antigens may be electrophoretically separated on gels, transferred to nitrocellulose paper, and detected by the monoclonal antibody with an immunoblot (26). When multiple antibodies to the same antigen are generated, the fine specificity of the antibodies can be assessed by competition assays (22).

Although there has been considerable success in generating monoclonal antibodies, problems persist with the hybridoma technique that make it impractical or difficult to generate monoclonal antibodies to every antigen of interest. Successful production of monoclonal antibodies is contingent upon: (i) the nature and origin of the myeloma cell line used for fusion, since, to date, only rat or mouse hybrids can be generated with consistent success; (ii) the inherent immunogenicity of the antigen of interest in these species; (iii) the relatively low, often unpredictable frequency of hybrid cell generation; (iv) difficulties in initial propagation, cloning, and stabilizing of hybridomas; and (v) the development of rapid, sensitive methods to detect and select useful specific hybridomas.

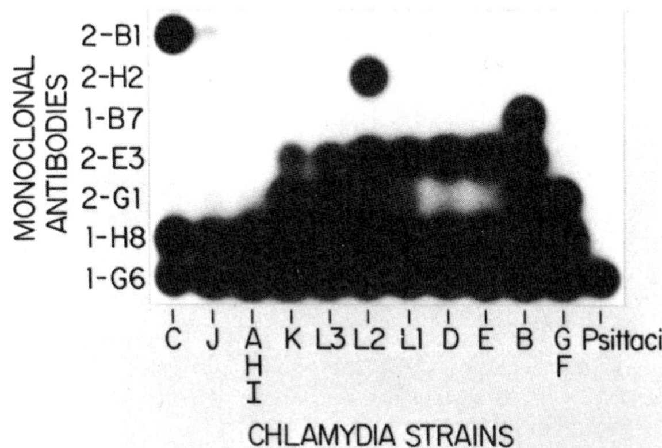

FIG. 3. Classification of monoclonal antibodies against *C. trachomatis*. *C. trachomatis* immunotypes A through K, L_1-L_3, and *Chlamydia psittaci* strain Mn were adsorbed to wells of a microtest plate and reacted with monoclonal antibodies. Immune reactions were detected with ^{125}I-labeled protein A and subsequent autoradiography. Reaction patterns of monoclonal antibodies 2-B1, 2-H2, and 1-B7 were type specific; patterns of 2-E3 and 2-G1 were subspecies specific; the pattern of 1-H8 was species specific; and the pattern of 1-G6 was genus specific.

APPLICATIONS OF MONOCLONAL ANTIBODIES

One of the most important applications of monoclonal antibody technology is in the identification of microorganisms. Panels of monoclonal antibodies to a wide range of viruses, bacteria, and parasites have been prepared in many research laboratories. By the selection of monoclonal antibodies with comparatively broad or narrow specificity, as we have done for *Chlamydia trachomatis* (Fig. 3), immunological assays can be created for different types of analyses. For example, monoclonal antibodies of cross-species (or genus) reactivity will have applications in the phylogenetic classification and taxonomy of microorganisms (2). Antibodies which define species-specific antigens or have broad subspecies reactivity will be especially valuable in the development of diagnostic reagents (15). Antibodies which exhibit more restricted specificity have applications in the characterization of strains or the determination of biotypes within a species, studies of antigenic drift, strain analysis for strain matching in epidemiological studies, and the mapping of individual antigenic determinants on proteins or other molecules (3, 21, 23).

In the area of diagnostic microbiology, monoclonal antibodies to bacterial or viral antigens are starting to replace polyclonal antibodies in the marketplace for use in culture confirmation when the accurate definition of microbial species or types is of primary concern. With the use of sensitive techniques such as fluorescent-antibody assays, it has been possible to perform culture confirmatory assays with improved accuracy (14) or early in culture when fewer organisms are present (18, 20). In addition, monoclonal antibodies provide researchers the opportunity to develop direct assays which were not possible with use of conventional antisera (24). The formulation of new direct assays has been possible with monoclonal antibody reagents which contain no contaminating antibodies and produce a minimum of artifacts. The highly defined and reproducible properties of monoclonal antibodies also invite their incorporation into immunoassays for use with the next generation of instruments for the automated analysis of blood, blood cells, tissues, and serum antibodies and for the detection of microbial antigens.

Beyond applications in microbial taxonomy and clinical use as diagnostic reagents, monoclonal antibodies have been valuable in the study of: (i) normal or altered proteins, hormones, or nucleic acids, (ii) cell surface markers, (iii) tumor markers, and (iv) autoimmune diseases. Monoclonal antibodies have found application in industry, as adsorption to solid-phase supports provides a highly efficient, one-step elution of compounds from complex or very dilute mixtures. Monoclonal antibodies have also been useful as second antibodies in immunoassays in which they are fluorescein- or enzyme-conjugated or radiolabeled. In addition, a promising area of study involves the application of monoclonal antibodies to immunotherapy of microbial or neoplastic disease.

HUMAN MONOCLONAL ANTIBODIES

There has been considerable interest in the generation of human monoclonal antibodies, primarily for immunotherapy of human disease. Although there have been isolated reports of human-human hybridomas secreting predefined monoclonal antibodies (16), formidable technical problems remain. First, a human myeloma cell line capable of serving as a consistent, high-frequency fusion partner must be derived. Human-mouse fusions have in the past proven to be unsatisfactory because the hybrids almost always lose the capacity to produce human antibody. Second, it is necessary either to have a supply of preimmunized human B cells or to devise efficient methods for their in vitro antigenic stimulation. An alternative approach to producing human monoclonal antibodies is the generation of antibody-secreting cells by mouse-human heteromyeloma cell construction (25) or by a process of cellular transformation which results from exposure to Epstein-Barr virus in culture (4, 19). Once produced, these human antibodies should be less immunogenic in patients than are murine antibodies and therefore should produce fewer allergic reactions upon repeated inoculation. Passively administered human antibodies should also be valuable for the treatment of acute, life-threatening allergies, drug overdoses or other toxemias, infectious diseases, or neoplasms.

ADVANTAGES AND DISADVANTAGES OF MONOCLONAL ANTIBODIES

Until the advent of monoclonal antibody technology, the specific composition of antibodies in antisera could never be controlled. Antibodies to minor or less immunodominant antigens could not be separated from more predominant species of antibodies. The monoclonal antibody technique, on the other hand, provides an opportunity to scan the large number of individual antibodies which make up the total immune response and to select those which will be useful. Monoclonal antibodies can be generated against either major serological antigens or minor, previously unrecognized or undetected antigenic determinants which may be very useful in diagnostic assays for the differentiation or identification of microorganisms. The composition and specificity of antibody reactants in immunodiagnostic reagents can be precisely controlled. Once hybrid cell lines are established, a constant supply of monoclonal antibodies for use in standardized reagents can be assured. Ascites fluid from a single mouse can produce enough monoclonal antibody for between 10,000 and 50,000 tests.

Despite their wide use and enormous potential as analytical and diagnostic tools, monoclonal antibodies will probably never solve all of the problems in conventional serology. Although monoclonal antibodies are displacing conventional antisera for the bulk production of commercial reagents, in basic research in which small amounts of antibodies are required, antisera may still be preferable. Despite many technical advances, the hybridoma technique is still relatively labor intensive and complex. It requires a comparatively long time (usually 4 to 6 months, at best) to stabilize and characterize a typical hybrid cell line. The expense of equipment and materials for the maintenance of a complete cell culture laboratory is also a significant consideration.

Some problems often encountered in hybridoma technology are associated with the clonal derivation of antibody-secreting cell lines. Since each monoclonal antibody represents one of a large number of antibodies which make up the total immune response, each will retain only a subset of the properties of a conventional antiserum. It is, therefore, unsafe to assume that a monoclonal antibody and an antiserum can be automatically interchanged. Each monoclonal antibody must be characterized for biological properties such as complement fixation, precipitation, and agglutination of antigen. The antibodies need to be evaluated for their physical or biochemical properties such as retention of activity after labeling with radioisotopes or fluorochromes, solubility as influenced by ionic strength or pH of the buffer, and stability as influenced by temperature or freeze-drying. These properties affect the suitability of a monoclonal antibody for specific types of assays, as well as its quality as a commercial reagent.

Other concerns for monoclonal antibodies include: (i) their more restricted specificity compared with antisera, (ii) their avidity, sometimes lower than hyperimmune antisera, and (iii) the stability of long hybridoma lines in long-term culture. Since monoclonal antibodies are building blocks of antisera, we can assume that if the correct monoclonal antibody is selected for use, it would match the desired specificity or other properties of the antiserum. If a single monoclonal antibody does not provide the appropriate specificity, it may be necessary to mix and match different antibodies to achieve the desired result. The monoclonal antibody technique allows screening of a great number of antibodies. With the proper selection techniques, it is possible to choose antibodies of the desired avidity. Alternatively, two or more antibodies can be mixed; often this mixing will have an additive or even synergistic effect, producing more stable immune complexes or better precipitating antibodies. Further, we have found that most cell lines are stable in long-term culture. Although a few lines are problematic and lose the capacity to produce immunoglobulin over long periods in cell culture or during serial passage in animals, these lines can be maintained through careful attention, frequent recloning, and freezing of cells for regeneration of stocks.

CONCLUSIONS

The future development of hybridoma technology will take advantage of the defined specificity, avidity, high specific activity, and selected isotype of monoclonal antibodies to produce homogeneous reagents for use in novel immunodiagnostic functions and automated instruments. An individual antibody can be selected for its unique reactivity with particular antigenic determinants and may react with antigens which had not been previously identified. Monoclonal antibodies will also provide unique opportunities for the dissection and understanding of complex biological processes and for the diagnosis and treatment of infectious diseases.

LITERATURE CITED

1. **Cohn, M.** 1967. The natural history of the myeloma. Cold Spring Harbor Symp. Quant. Biol. **32:**211–221.

2. **de Macario, E. C., and A. J. L. Macario.** 1983. Monoclonal antibodies for bacterial identification and taxonomy. ASM News **49:**1–7.
3. **Gerhard, W., C. M. Croce, D. Lopes, and H. Koprowski.** 1978. Repertoire of antiviral antibodies expressed by somatic cell hybrids. Proc. Natl. Acad. Sci. U.S.A. **75:**1510–1514.
4. **Irie, R. F., L. L. Sze, and R. E. Saxton.** 1982. Human antibody to OFA-1, a tumor antigen, produced *in vitro* by Epstein-Barr virus-transformed human B-lymphoid cell lines. Proc. Natl. Acad. Sci. U.S.A. **79:**5666–5670.
5. **Isenberg, H. D., J. A. Washington II, A. Balows, and A. C. Sonnenwirth.** 1980. Collection, handling, and processing of specimens, p. 52–82. *In* E. H. Lennette, A. Balows, W. J. Hausler, Jr., and J. P. Truant (ed.), Manual of clinical microbiology, 3rd ed. American Society for Microbiology, Washington D.C.
6. **Kennett, R., and T. McKearn (ed.).** 1980. Monoclonal antibodies. Plenum Publishing Corp., New York.
7. **Kohler, G., and C. Milstein.** 1975. Continuous cultures of fused cells secreting antibody of predefined specificity. Nature (London) **256:**495–497.
8. **Krause, R. M.** 1970. The search for antibodies with molecular uniformity. Adv. Immunol. **12:**1–56.
9. **Littlefield, J. W.** 1964. Selection of hybrids from matings of fibroblasts *in vitro* and their presumed recombinants. Science **145:**709–710.
10. **Lostrom, M. E., M. R. Stone, M. R. Tam, W. N. Burnette, A. Pinter, and R. C. Nowinski.** 1979. Monoclonal antibodies against murine leukemia viruses: identification of six antigenic determinants on the p15(E) and gp70 envelope proteins. Virology **98:**336–350.
11. **Margulies, D. H., W. Cieplinski, B. Dharmgrongartama, M. Gefter, S. L. Morrison, T. Kelly, and M. D. Scharff.** 1977. Regulation of immunoglobulin expression in mouse myeloma cells. Cold Spring Harbor Symp. Quant. Biol. **41:**781–791.
12. **Melchers, F., M. Potter, and N. Warner (ed.).** 1978. Lymphocyte hybridomas. Curr. Top. Microbiol. Immunol. **81**.
13. **Milstein, C., K. Adetugbo, J. J. Cowan, G. Kohler, and D. S. Secher.** 1977. Somatic cell genetics of antibody-secreting cells: studies of clonal diversification and analysis by cell fusion. Cold Spring Harbor Symp. Quant. Biol. **41:**793–803.
14. **Nahm, M. H., P. R. Murray, B. L. Clevinger, and J. M. Davie.** 1980. Improved diagnostic accuracy using monoclonal antibody to group A streptococcal carbohydrate. J. Clin. Microbiol. **12:**506–508.
15. **Nowinski, R. C., M. R. Tam, L. C. Goldstein, L. Stong, C. C. Kuo, L. Corey, W. E. Stamm, H. H. Handsfield, J. S. Knapp, and K. K. Holmes.** 1983. Monoclonal antibodies for diagnosis of infectious diseases in humans. Science **219:**637–644.
16. **Olsson, L., and H. S. Kaplan.** 1980. Human-human hybridomas producing monoclonal antibodies of predefined antigenic specificity. Proc. Natl. Acad. Sci. U.S.A. **77:**5429–5431.
17. **Potter, M.** 1972. Immunoglobulin-producing tumors and myeloma proteins of mice. Physiol. Rev. **52:**631–719.
18. **Stamm, W. E., M. Tam, M. Koester, and L. Cles.** 1983. Detection of *Chlamydia trachomatis* inclusions in McCoy cell cultures with fluorescein-conjugated monoclonal antibodies. J. Clin. Microbiol. **17:**666–668.
19. **Steinitz, M., G. Klein, S. Koskimies, and O. Makela.** 1977. EB virus-induced B lymphocyte cell lines producing specific antibody. Nature (London) **269:**420–422.
20. **Stephens, R. S., C.-C. Kuo, and M. R. Tam.** 1982. Sensitivity of immunofluorescence with monoclonal antibodies for detection of *Chlamydia trachomatis* inclusions in cell culture. J. Clin. Microbiol. **16:**4–7.
21. **Stephens, R. S., M. R. Tam, C. C. Kuo, and R. C. Nowinski.** 1982. Monoclonal antibodies to *Chylamydia trachomatis*: antibody specificities and antigen characterization. J. Immunol. **128:**1083–1089.
22. **Stone, M. R., and R. C. Nowinski.** 1980. Topological mapping of murine leukemia virus proteins by competition-binding assays with monoclonal antibodies. Virology **100:**370–381.
23. **Tam, M. R., T. M. Buchanan, E. G. Sandström, K. K. Holmes, J. S. Knapp, A. W. Siadak, and R. C. Nowinski.** 1982. Serological classification of *Neisseria gonorrhoeae* with monoclonal antibodies. Infect. Immun. **36:**1042–1053.
24. **Tam, M. R., W. E. Stamm, H. H. Handsfield, R. Stephens, C. C. Kuo, K. K. Holmes, K. Ditzenberger, M. Krieger, and R. C. Nowinski.** 1984. Culture independent diagnosis of *Chlamydia trachomatis* using monoclonal antibodies. N. Engl. J. Med. **310:**1146–1150.
25. **Teng, N. N. H., K. S. Lam, F. C. Riera, and H. S. Kaplan.** 1983. Construction and testing of mouse-human heteromyelomas for human monoclonal antibody production. Proc. Natl. Acad. Sci. U.S.A. **80:**7308–7312.
26. **Towbin, H., T. Staehelin, and J. Gordon.** 1979. Electrophoretic transfer of proteins from polyacrylamide gels to nitrocellulose sheets: procedure and some applications. Proc. Natl. Acad. Sci. U.S.A. **76:**4350–4354.

Chapter 94

Serologic Tests for Syphilis

LYNDA L. BRADFORD AND SANDRA A. LARSEN

Syphilis is a sexually transmitted (venereal) disease caused by *Treponema pallidum*. Endemic syphilis (bejel), yaws, and pinta are nonvenereal diseases caused by pathogenic *Treponema* spp. which are morphologically and antigenically similar to *T. pallidum*. Differentiation of these diseases is based on clinical manifestations and epidemiology (see Chapter 44).

Tests for the detection of antibodies resulting from infection with *T. pallidum* are of unique diagnostic importance. In few other conditions are serologic tests as critical for diagnosis, the opportunity for misdiagnosis so great, and the potential consequences to the patient so drastic. It is paradoxical that the first-generation, and still first-line, tests for syphilis are biologically nonspecific. In these tests, cardiolipin antigens, derived from beef heart, are used to detect antilipid antibodies, traditionally termed reagin. Antilipid antibodies, immunoglobulin G (IgG) and IgM, are formed by the host in response to lipoidal material released from damaged host cells early in the infection as well as to lipid present on the cell surface of the treponeme (23). In the past, these tests were referred to as cardiolipin or reagin tests, but they are currently termed nontreponemal, in contrast to second-generation tests with *T. pallidum* as a source of antigen.

The history of the nontreponemal tests is marked by efforts to standardize and simplify test procedures so that reliable tests for syphilis can be performed in any competent laboratory. The nontreponemal tests most frequently used in the United States today are the Venereal Disease Research Laboratory (VDRL) test (11) and the Rapid Plasma Reagin (RPR) 18-mm circle card test (27). Both are simple, rapid, reproducible, and inexpensive; both may be used either qualitatively or quantitatively. The RPR card test has several advantages over the VDRL: (i) it is a kit test, containing all needed reagents and controls, including pre-prepared antigen suspension; (ii) unheated serum is used; (iii) the reaction is read macroscopically; and (iv) most materials are disposable. One disadvantage is that the RPR test cannot be used with cerebrospinal fluid specimens. The VDRL slide test procedure is given in the 1969 *Manual of Tests for Syphilis* (36) and in the third edition of this Manual (3). The RPR 18-mm circle card test procedure is described in this chapter.

The nontreponemal tests are of greatest value when used as screening procedures and for following therapy. Quantitation is necessary to determine the response to treatment and to detect reinfection. The level of reactivity of the nontreponemal tests is a compromise between sensitivity and specificity. The VDRL test, for example, will detect ca. 75% of the cases of primary syphilis and is somewhat less reactive than the treponemal tests in late latent and late symptomatic syphilis (4, 7). False-positive reactions occur not infrequently and have been attributed to a variety of acute and chronic conditions (30).

Treponemal tests are most commonly used to determine if a reactive nontreponemal test is the result of syphilis or some other condition. They may also be used to detect syphilis in patients with negative nontreponemal tests but with clinical evidence of late syphilis. However, they should not be used indiscriminately to test reagin-negative sera, since false-positive treponemal reactions may be a problem in this population (6). The treponemal tests do not indicate the response of the patient to treatment, and quantitative tests are of no value in diagnosis or prognosis. There are no standard procedures for testing cerebrospinal fluid with the commonly used treponemal tests. A recent publication (16) confirms previous reports that treponemal tests are of doubtful value in the diagnosis of active neurosyphilis. At present, the quantitative VDRL spinal fluid test is the test of choice for cerebrospinal fluid.

The *T. pallidum* Immobilization (TPI) test, developed by Nelson and Mayer in 1949, was the first well-evaluated treponemal test and became the standard with which other tests were compared. Because the TPI test is complex, technically exacting, and difficult to control, research efforts have focused on the development of simpler treponemal procedures. Unfortunately, these efforts have been hindered by the continuing inability to cultivate *T. pallidum* in vitro. Because the organism must be grown intratesticularly in rabbits, treponemal antigens are difficult to produce and are contaminated with rabbit tissue components. Major improvements in treponemal antigens and tests are unlikely until in vitro cultivation of *T. pallidum* is achieved. Treponemal tests most commonly used in the United States today are the Fluorescent Treponemal Antibody-Absorption (FTA-ABS) test (7), the FTA-ABS Double-Staining (DS) test (14), the Microhemagglutination-*T. pallidum* (MHA-TP) test (5, 32), and the Hemagglutination Treponemal Test for Syphilis (HATTS) test (37).

The FTA-ABS test is more sensitive than the TPI test in all stages of syphilis, especially in the very early and very late stages. Extensive evaluation by several investigators (7, 33, 38) indicates good agreement between the FTA-ABS and TPI tests in patients presenting diagnostic problems. Although the specificity of the FTA-ABS test is considered good, the possibility of false-positive reactions should be recognized. False-positive reactions have been documented in patients with diseases associated with increased or abnormal globulins (2, 22) and in patients with lupus erythematosus and antinuclear antibodies (17, 18). The FTA-ABS test procedure is given in the 1969 *Manual of Tests for Syphilis* (36) and in the third edition of this Manual (3). The FTA-ABS DS technique is a modification of the standard FTA-ABS test. It was developed specifically for use with microscopes with epi- or incident illumination and is described in this chapter.

The use of monospecific IgM conjugates in the FTA-ABS test to differentiate fetal from maternal antibody is still under investigation. The FTA-ABS IgM test is subject to a number of technical problems and is not satisfactory for diagnostic use at this stage of development (1). Quantitative VDRL tests on serial specimens yield more information and are currently the test of choice in suspected congenital syphilis.

The microhemagglutination tests are simple, rapid, reproducible, passive hemagglutination tests employing microtiter equipment. Reagents for the MHA-TP test are manufactured by Fujizoki Pharmaceutical Co., Tokyo, Japan, and are distributed in kit form in the United States only by the Ames Co., Elkhart, Ind. The MHA-TP test procedure is described in this chapter. Reagents for the HATTS test kit are manufactured by Difco Laboratories, Detroit, Mich. The test procedure is given in the package insert. The major difference between these two tests is the source of erythrocyte carriers. The MHA-TP test uses sheep cells; the HATTS test uses turkey cells.

Several studies have shown that MHA-TP and FTA-ABS test results are comparable in all categories of syphilis except in the primary stage, in which the MHA-TP test is less reactive than either the FTA-ABS or VDRL tests (29). The MHA-TP test appears to have a high degree of specificity and compares favorably with the FTA-ABS test on selected normal sera (15). False-positive reactions in the MHA-TP test have been reported in some patients with connective tissue disease, in lepromatous leprosy, and in infectious mononucleosis (29). Studies in the laboratory of one of us (L.L.B.) indicate that the MHA-TP test is at least as specific as the FTA-ABS test on routine diagnostic sera. Agreement between the two treponemal tests on 4,100 VDRL-reactive sera ranged from 92 to 98%, depending on the lots of MHA-TP and FTA-ABS reagents used. Average agreement was 95%, with 4% reactive in the FTA-ABS test only and 1% reactive in the MHA-TP test only. On 671 sera that presented diagnostic problems and were submitted for TPI testing, the MHA-TP test agreed slightly better with the TPI test than did the FTA-ABS test. Other investigators have found the HATTS test to have sensitivity and specificity comparable to that of the MHA-TP and FTA-ABS tests (9, 19, 37).

The MHA-TP and HATTS tests both have standard test status and are considered satisfactory substitutes for the FTA-ABS test. The chief advantages are technical simplicity and economy. Reagent costs for the hemagglutination and FTA-ABS tests are comparable, but the FTA-ABS test requires three to five times as much staff time for testing. The reading of the FTA-ABS test is more subjective, and quality control is appreciably more difficult. Potential disadvantages of the microhemagglutination tests are their lack of sensitivity in primary syphilis and their unfamiliarity to clinicians.

All of the treponemal tests require good quality control at the user level. A laboratory should not attempt treponemal testing if it lacks resources for the quality control described under Test Procedures. The activities of the venereal disease laboratories of the Centers for Disease Control have been a major influence in the United States in the improvement of test performance. This improvement has been accomplished by the development of new test procedures, notably the VDRL and FTA-ABS tests, by checking commercial reagents for these and other tests, by the publication of standard procedures (36), and by the provision and promotion of training and proficiency-testing programs. These activities provide an excellent foundation upon which the individual laboratory can build to produce reliable test results within the limitations of the state of the art in syphilis serology.

CLINICAL INDICATIONS

Syphilis is a complex, acute, chronic infectious disease frequently referred to as the great imitator. Clinical manifestations, when present, are diverse; they vary depending upon the stage of infection and the individual response. Infection with *T. pallidum* may affect any organ or tissue and can be transferred via the placenta to the fetus, causing infection in utero. Depression of the cellular immune function, including skin test anergy and decreased lymphocyte responsiveness to mitogens and antigens, occurs during the first weeks of infection (20, 21, 24, 25). Humoral antibodies are not usually detected by the standard serologic tests for syphilis until 1 to 4 weeks after the formation of the primary lesion (chancre). Approximately one-third of the patients with untreated syphilis have a spontaneous cure, one-third remain in the latent stage, and one-third develop the destructive lesions of late syphilis. A diagnosis of syphilis may be based on one or more of the following: clinical manifestations, historical or epidemiological evidence, and laboratory findings (35).

The incubation period for syphilis ranges from 10 days to 3 months. During this period there is no clinical or serologic evidence of infection, but *T. pallidum* is multiplying at the site of inoculation and is invading the lymphatics and blood stream.

The primary stage is characterized by the appearance of a painless lesion (chancre) at the site of inoculation. Chancres usually appear ca. 3 to 4 weeks (with an extreme range of 10 to 90 days) after infection, and they usually heal spontaneously within a few weeks. Chancres most often occur on the external genitalia but may be present in the mouth or cervix or on the anal mucosa, where they may never be seen.

The manifestations of secondary syphilis usually follow healing of the chancre, 6 weeks to 6 months after infection. Generalized skin and mucous membrane lesions are most common, but invasion of hair follicles, bones, eyes, viscera, and the central nervous system may also occur. Systemic reactions to secondary infection are not uncommon. Antigenic changes within *T. pallidum* apparently do not occur in secondary syphilis (8), and as in primary syphilis, an effective host immune response develops, with secondary lesions healing over a period of 2 to 6 weeks.

The early infectious stages are followed by a period of clinical latency which may last from a few weeks to a lifetime. The early latent period (1 to 5 years after infection) may be interrupted by the recurrence of infectious lesions of the skin and mucous membranes or by the development of lesions of the eye and central nervous system.

The manifestations of late syphilis usually occur 5 to 20 years after infection. The most frequent complications of late syphilis involve destructive lesions in the central nervous and cardiovascular systems and gummatous lesions of skin, bone, and viscera. Congenital syphilis may present with secondary lesions of the skin or mucous membranes or with lesions of the bone, cartilage, eye, or central nervous system.

The detection of *T. pallidum* in lesions by conventional dark-field or fluorescence dark-field microscopy (3) provides the earliest and most reliable evidence for infection. With properly taken specimens, treponemes can be demonstrated in over 95% of the primary lesions (35). Finding treponemes in primary lesions is especially helpful because antibody may not yet be detectable. Treponemes may also be demonstrated in a large proportion of lesions associated with secondary and congenital syphilis.

Although the immunological response to infection with *T. pallidum* is complex and poorly understood, serologic tests provide a strong basis, and in many cases the only basis, for the diagnosis of syphilis and for evaluation of the response of the patient to treatment.

Nontreponemal tests usually become reactive within 4 to 6 weeks after infection (or 1 to 2 weeks after the chancre first appears). Titers peak during the secondary stage and then decline slowly. Approximately half of the untreated infections will continue to show some level of reactivity, some of them for life. Treatment in the early stages of infection may completely suppress antibody formation, and the tests may remain nonreactive. Treatment in the primary or secondary stage usually results in a rapid decline in titer, frequently to nonreactive in 6 to 18 months. Treatment given in the latent or late stage has less effect on antibody levels, and tests may remain reactive at low titers indefinitely.

Nontreponemal tests lack specificity and are reactive in a variety of other conditions. In the absence of supporting clinical, historical, or epidemiological evidence, reactive results must be confirmed with the more specific treponemal tests.

Treponemal tests vary in their ability to react in early syphilis (Fig. 1). The FTA-ABS test is reactive as early as the nontreponemal tests, but the TPI and MHA-TP tests become reactive more slowly. Once reactive, all treponemal tests tend to remain so for many years. In fact, in some patients with late syphilis, the treponemal tests are the only reactive test. Treponemal tests do not indicate response to therapy and for this reason are not suitable for monitoring treatment. None of the treponemal tests distinguishes between syphilis and other treponematoses such as yaws, pinta, and bejel (endemic syphilis).

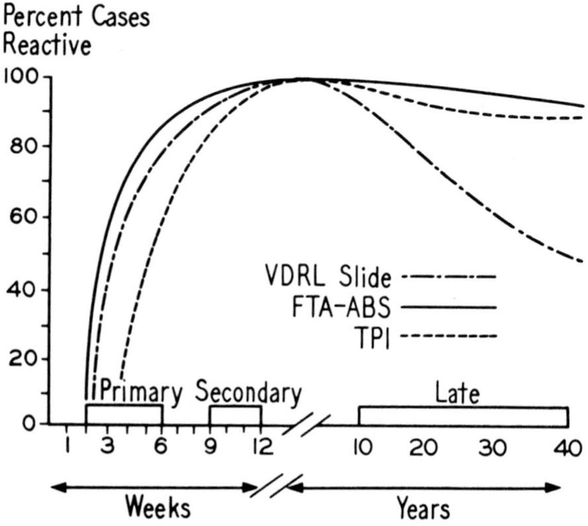

FIG. 1. Reactivity of nontreponemal and treponemal tests in untreated syphilis. MHA-TP curve approximates TPI in early syphilis and FTA-ABS in late syphilis. (I. Davidsohn and J. B. Henry, ed., *Todd-Sanford Clinical Diagnosis by Laboratory Methods*, 15th ed., 1974, W. B. Saunders Co., Philadelphia.)

TEST PROCEDURES

RPR 18-mm Circle Card Test

See references 10, 26, 28, and 36.

Principle

A stabilized suspension of standardized VDRL Antigen (36) and sized charcoal particles is used as the RPR card antigen. This antigen is added to the serum of the patient, mixed on a mechanical rotator, and examined macroscopically for degree of flocculation. Any degree of reactivity obtained in the undiluted serum indicates a need to perform the quantitative test to determine the endpoint titer.

Kit

The RPR 18-mm circle card kit contains test antigen, a disposable 20-gauge needle, an antigen-dispensing bottle, plastic-coated test cards, disposable specimen-dispensing devices, and stirrers. Some kits also contain sera with reactivities specified as reactive, minimally reactive, and nonreactive.

Preliminary testing of antigen suspension

Before the test is begun, the RPR card antigen is transferred to the dispensing bottle, and the accuracy of the dispensing needle is checked. The needle should deliver 0.017 ml or 60 drops ± 2 drops per ml of antigen. To check the accuracy of the needle, place the needle on a 1-ml pipette, fill the pipette with RPR antigen suspension, hold it in a vertical position, and count the number of drops delivered in 0.5 ml. The needle is considered satisfactory if 30 drops ± 1 drop are obtained in 0.5 ml. A needle that does not meet this specification should be replaced with another needle that does.

Control sera are used to test the antigen suspension to confirm optimal reactivity before routine testing is performed. Control sera for nontreponemal tests are not used as reading standards. One must test the control sera of graded reactivity (reactive, minimally reactive, and nonreactive) each day, as described in the next section. Only RPR card antigen suspensions that reproduce the established reactivity pattern of the control sera should be used.

Equipment

Mechanical rotator. A rotator with a fixed speed or adjustable to 100 ± 2 rpm that circumscribes a circle 2 cm (3/4 in.) in diameter on a horizontal plane.

Humidity cover. Any convenient cover containing a moistened blotter or sponge to cover cards during rotation.

Safety pipetting device. Must deliver 0.05 ml (50 μl) of reagent for use in the quantitative test; may also be used for the qualitative test.

Light source for the reading of test results. A high-intensity lamp can be used.

Reagents

Saline, 0.9%. A 900-mg volume of dry sodium chloride (A.C.S.) in each 100 ml of distilled water.

Diluent. A 1:50 dilution of serum nonreactive for syphilis prepared in 0.9% saline. This diluent should be used in the quantitative procedure for making 1:32 and higher dilutions of serum specimens.

Antigen. Store antigen suspension in ampoules or in the plastic dispensing bottle at 2 to 8°C. An unopened ampoule of antigen is stable until the expiration date. Antigen suspension in the plastic dispensing bottle (stored at 2 to 8°C) is stable for 3 months or until the expiration date, whichever comes first. Do not use suspension beyond the expiration date. A new lot of antigen suspension should be tested in parallel with a reference reagent to verify that it is of standard reactivity before it is placed in routine use.

Collection, preparation, and storage of specimens

Collect blood by venipuncture into dry tubes without anticoagulant, and allow the blood to clot. If necessary, centrifuge the specimen with sufficient force to sediment cellular elements from the serum. Generally, 1,500 to 2,000 rpm for 5 min is sufficient. Serum specimens may be retained in the original collection tube if the test is to be performed immediately. Store clear serum removed from the clot at refrigerator temperature (2 to 8°C) or frozen (−20°C) if a delay of more than 48 h is anticipated before testing. Avoid repeated freezing and thawing of specimens. Sera are tested without heating and must be at 23 to 29°C (73 to 85°F) at the time of testing.

Precaution. Human serum is a potential source of hepatitis B virus infection and acquired immunodeficiency syndrome. Handle all specimens as though they were capable of transmitting these diseases. Do not pipette by mouth. Decontaminate all materials before disposal.

RPR circle card test on serum: qualitative test

Note that slide flocculation tests for syphilis are affected by room temperature. For reliable and reproducible results, control sera, RPR card test antigen suspension, and test specimens must be at room temperature, 23 to 29°C (73 to 85°F), when tests are performed.

1. Place 50 μl of unheated serum onto an 18-mm circle of the RPR test card with a dispensing or an automatic pipetting device that delivers 0.05 ml (50 μl).
2. With the reverse end of the dispensing device or a toothpick, spread the specimen to fill the entire circle. Care should be taken not to allow the serum to spread beyond the confines of the circle. Use a clean spreader for each sample.
3. Gently suspend the RPR antigen suspension. Hold the dispensing bottle in a vertical position and add exactly 1 free-falling drop (1/60 ml) of suspension to each test area containing serum.
4. Place the card on a mechanical rotator under a humidity cover.
5. Rotate for 8 min at 100 rpm.
6. Read the test reactions in the wet state under a high-intensity light source immediately after removing the card from the rotator. Read the test without magnification. To better differentiate minimally reactive serum from nonreactive, immediately after rotation tilt the card to about 30° from horizontal, and briefly rotate the card manually.
7. Report results as follows: characteristic clumping of the RPR card antigen slight but definite, minimally reactive; clumping marked and intense, reactive; no clumping or only slight roughness, nonreactive.

Note that results are reported as reactive or nonreactive regardless of the degree of reactivity. Slight but definite flocculation should always be reported as reactive.

Specimens giving any degree of reactivity should be quantitated. Rough nonreactive results should also be quantitated to identify possible prozones.

8. Upon completion of the daily tests, remove the needle, rinse it with distilled or deionized water, and air-dry it. (Avoid wiping the needle because this removes the silicone coating.) Recap the dispensing bottle and store it in the refrigerator.

RPR card test on serum: quantitative test

1. For each specimen to be tested, place 50 μl of 0.9% saline onto circles 2 through 5. Do not spread the saline.
2. With a safety pipetting device, place 50 μl of serum specimen onto circles 1 and 2.
3. Mix the saline and the specimen in circle 2 by drawing the mixture up and down in the safety pipettor five or six times. Avoid the formation of bubbles.
4. Transfer 50 μl from circle 2 (1:2) to circle 3 (1:4), and mix.
5. Transfer 50 μl from circle 3 (1:4) to circle 4 (1:8), and mix.
6. Transfer 50 μl from circle 4 (1:8) to circle 5 (1:16), mix, and discard 0.05 ml.
7. With a clean stirrer for each specimen, start at the highest dilution, 1:16 (circle 5), and with the broad end of the stirrer, spread the serum dilution within the confines of the circle. Repeat this action with the same sampling stirrer in circles 4, 3, 2, and 1.
8. Gently suspend the RPR card antigen suspension in the dispensing bottle.
9. Holding the antigen suspension bottle in a vertical position, dispense 1 or 2 drops to clear the needle of air, and then place exactly 1 free-falling drop (1/60 ml) of antigen suspension onto each test area.
10. Place the card on the rotator under a humidity cover.
11. Rotate for 8 min at 100 rpm.
12. Read the test reactions as described in Step 6 of the qualitative test.
13. Report results in terms of the highest dilution, giving any reactivity in accordance with the examples in Table 1.

TABLE 1. Method for reporting RPR quantitative test results[a]

Undiluted serum (1:1)	Serum dilutions				Report
	1:2	1:4	1:8	1:16	
R_m	N	N	N	N	Reactive, 1:1 dilution or undiluted only
R	R	N	N	N	Reactive, 1:2 dilution
R	R	R_m	N	N	Reactive, 1:4 dilution
R	R	R	R	N	Reactive, 1:8 dilution

[a] N, Nonreactive; R, reactive; R_m, minimally reactive.

14. If the highest dilution tested (1:16) is reactive, proceed as follows.
    a. Prepare a 1:50 dilution (2.0%) of nonreactive serum in 0.9% saline. (This dilution is to be used for making 1:32 and higher dilutions of specimens to be quantitated.)
    b. Prepare a 1:16 dilution of the test specimen by adding 100 μl of the specimen to 1.5 ml of 0.9% saline. Mix thoroughly.
    c. Place 50 μl of the 1:50 nonreactive serum in circles 2, 3, 4, and 5 of an RPR card.
    d. Measure 50 μl of the 1:16 dilution of the test specimen to circles 1 and 2.
    e. With the same pipettor, make serial twofold dilutions and complete the test as described in steps 2 through 13 of the RPR card quantitative test. Higher dilutions may be prepared, if necessary, in a similar manner.

Note that all dilutions may be made on the card if the 1:50 nonreactive serum is used as the diluent from circle 6 (1:32) on.

FTA-ABS DS Test

See references 13, 14, and 36.

Principle

The FTA-ABS DS test is a modification of the standard FTA-ABS test (36) developed specifically for use with microscopes with incident illumination. As with the standard test, the FTA-ABS DS is an indirect fluorescent-antibody procedure with *T. pallidum* as the antigen and sorbent prepared from nonpathogenic Reiter treponemes to increase specificity. The DS test employs a tetramethyl-rhodamine isothiocyanate (TMRITC)-labeled anti-human IgG globulin and a counterstain fluorescein (FITC)-labeled anti-*T. pallidum* conjugate which enables one to locate the organisms in a nonreactive test.

Equipment

Incubator. Adjustable to 35 to 37°C.
Bibulous paper.
Slide board or holder.
Moist chamber. Place moistened paper inside a convenient cover fitting the slide board.
Loop. Bacteriological, standard 2 mm, 26 gauge, platinum.
Oil. Immersion, low fluorescence, nondrying, type A, Cargille code 1248.
Safety pipetting devices.
Water bath adjustable to 56°C.

Microscope assembly

Lamp. HBO-50, HBO-100, or HBO-200.
Oculars. 6×, 8×, or 10×.
Objectives. 40×/1.30 oil, 63× oil, or 100×/1.25 oil.
Filters. FITC: BG38, K480, 2KP490, TK510, K515. TMRITC: BG38, BG36, KP560, K530, TK580, K590.

Glassware

Microscope slides. 1 by 3 in. (2.54 by 7.62 cm), frosted end, ca. 1 mm thick, with two etched circles, 1-cm inside diameter.
Cover slips. No. 1, 22 mm square.
Dish. Staining, glass or plastic, with removable slide carriers.
Test tubes. 12 by 75 mm.

Reagents

T. pallidum antigen. The antigen for the FTA-ABS test is a lyophilized suspension of *T. pallidum* (Nichols strain) extracted from rabbit testicular tissue and washed in phosphate-buffered saline (PBS) to remove rabbit globulin. The rehydrated suspension should adhere to the slide and yield sufficient treponemes for the easy interpretation of staining. The antigen may be stored at 2 to 8°C for 1 week or placed on microscope slides, fixed in acetone, and frozen at −20°C or below. Discard the antigen suspension if it becomes bacterially contaminated or does not demonstrate the proper reactivity with control sera.

FTA-ABS test sorbent. Sorbent is a product prepared from cultures of Reiter treponemes. This sorbent may be purchased in lyophilized or liquid state. The reagent is standardized by titration with a known serum containing high-titered nonspecific staining. Store the sorbent and rehydrate it if it is lyophilized, according to accompanying directions.

TMRITC-labeled anti-human globulin (conjugate). The FTA-ABS anti-human conjugate recognizes heavy-chain IgG. The conjugate should be evaluated in the FTA-ABS test to determine a satisfactory working titer and to verify that it meets the criteria concerning nonspecific staining and standard reactivity. Follow the directions of the manufacturer when commercial reagents are used. Store lyophilized conjugate at 2 to 8°C. Dispense rehydrated conjugate in not less than 0.3-ml quantities and store it undiluted at −20°C or lower. A conjugate with a working titer of 1:1,000 or higher may be diluted 1:10 with sterile PBS containing 0.5% bovine serum albumin and 0.1% NaN_3 before storage. When the conjugate is thawed for use, do not refreeze it, but store it at 2 to 8°C. It may be used as long as satisfactory reactivity is obtained with test controls. Discard the conjugate if it becomes bacterially contaminated. Centrifuge before use if the reagent is cloudy when thawed or becomes cloudy at 2 to 8°C. If a change in FTA-ABS test reactivity is noted in routine laboratory testing, the conjugate should be retitered to determine if it is the contributing factor.

FITC-labeled antitreponemal conjugate (counterstain). The conjugate should be of proven quality for the FTA-ABS DS test. Test each new lot of conjugate to determine its working dilution and to verify that it meets the criteria concerning standard reactivity for a counterstain. Store as described above for the TMRITC-labeled conjugate.

PBS, pH 7.2 ± 0.1. Formula per liter: NaCl, 7.65 g; Na_2HPO_4, 0.724 g; KH_2PO_4, 0.21 g. Several liters may be prepared and stored in a large Pyrex (or equivalent) or polyethylene bottle. Determine the pH of each lot of PBS prepared for the FTA-ABS DS test. PBS outside the range of pH 7.2 ± 0.1 should be adjusted with 1 N NaOH. Discard bacterially contaminated PBS.

2.0% Tween 80 (polysorbate 80). To prepare PBS containing 2% Tween 80, heat the two reagents in a 56°C water bath. To 49 ml of sterile PBS, add 1 ml of Tween 80 by measuring from the bottom of a pipette, and rinse out the pipette. The 2% Tween 80 solution should be pH 7.0 to 7.2. Check the pH periodically because the solution may become acid. Discard the solution if a precipitate develops or if the pH changes.

Mounting medium consists of 1 part PBS (pH 7.2) plus 9 parts glycerol (reagent quality).

Acetone (A.C.S.).

Preparation of *T. pallidum* antigen smears

Mix the antigen suspension thoroughly by mixing it on a Vortex Jr. for 10 s or by drawing the suspension into a syringe with a 25-gauge needle and expelling it vigorously. Determine by dark-field examination that treponemes are adequately dispersed before making slides for the FTA test. Additional mixing may be required. Wipe slides with pre-etched circles with clean gauze to remove dust particles. Note that unclean slides and slides of poor quality can affect the reactivity of the test. Slides should be cleaned by sonic vibration or alcohol wiping if treponemes are not clearly observed after staining. Prepare very thin *T. pallidum* antigen smears within each circle by using a standard 2-mm, 26-gauge, platinum wire loop. Generally, one-half of a loop is sufficient antigen for two circles of 1-cm inside diameter each. Allow the circles to air dry at least 15 min. Two-circle slides are preferred for FTA-ABS DS tests to prevent serum runover when the slides are processed. (Multispot antigen slides are available commercially from several manufacturers. These must be sufficiently hydrophobic to prevent serum and conjugate runover during slide processing. Follow the recommendations of the manufacturers to be sure that serum and conjugate do not run between smears. If cross-contamination occurs, the test must be repeated. Reactivity from cross-contamination may occur in 30 s [12].) Fix the smears in acetone for 10 min and allow them to air dry thoroughly. Not more than 60 slides should be fixed with 200 ml of acetone. Store acetone-fixed smears at −20°C or below. Fixed, frozen smears are usable indefinitely, provided that satisfactory results are obtained with the control. Do not thaw and refreeze antigen smears.

Criteria of acceptability for antigen, sorbent, conjugate, and counterstain

An antigen is considered acceptable if: (i) a sufficient number of organisms remain on the slides after staining so that tests may be read without difficulty; (ii) the antigen does not contain background material that stains to the extent that it interferes with the test reading; (iii) the number of organisms remains stable on antigen smears stored at −20°C; (iv) the antigen does not stain nonspecifically with a standard conjugate at its working titer; and (v) reportable test results on controls and individual sera are comparable with those obtained with an antigen previously found to be satisfactory. A sorbent is acceptable if it removes nonspecific reactivity of the nonspecific control serum and does not reduce the intensity of fluorescence of the reactive (4+) control serum to less than 3+ and if test results are comparable with those obtained with a previously satisfactory sorbent. An acceptable conjugate does not stain reference antigen nonspecifically at 3 doubling dilutions below the working titer of the conjugate, and test results on controls and individual sera are comparable with those obtained with a conjugate previously found to be satisfactory.

The counterstain is considered acceptable if it stains *T. pallidum* with 3+ to 4+ intensity when used at the working dilution recommended by the manufacturer. The minimal acceptable working dilution is 1:10.

Preparation of sera

Bacterial contamination or excessive hemolysis may render specimens unsatisfactory for testing. Heat the test and control sera at 56°C for 30 min before testing. Reheat previously heated test sera for 10 min at 56°C on the day of testing.

Controls

Store and use control sera from commercial sources according to accompanying directions. Include the following controls in each test run.

1. Reactive (4+) control. Reactive serum or a dilution of reactive serum demonstrating strong (4+) fluorescence when diluted 1:5 in PBS and only slightly reduced fluorescence (a reduction of no more than 1+ fluorescence; i.e., 4+ changing to 3+) when diluted 1:5 in sorbent. Add 50 μl of reactive control serum to a tube containing 200 μl of PBS. Mix well at least eight times. (Tips may be left in the tube and used to dispense material to the slide.) Add 50 μl of reactive control serum to a tube containing 200 μl of sorbent. Mix well at least eight times.
2. Minimally reactive (1+) control. The dilution of reactive serum demonstrating the minimal degree of fluorescence reported as reactive for use as a reading standard. The reactive (4+) control serum may be used for this control when diluted in PBS according to directions furnished by the manufacturer. This control is not further diluted in sorbent.
3. Nonspecific serum control. A serum from a nonsyphilitic individual known to demonstrate equal or greater than 2+ nonspecific reactivity in the FTA test at a 1:5 PBS dilution with essentially no staining when diluted 1:5 in sorbent. Add 50 μl of nonspecific control serum to a tube containing 200 μl of PBS. Mix well at least eight times. Add 50 μl of nonspecific control serum to a tube containing 200 μl of sorbent. Mix well at least eight times.
4. Nonspecific staining controls. Antigen smear treated with 30 μl of PBS. Antigen smear treated with 30 μl of sorbent.

Note that controls 1, 3, and 4 are included for the purpose of controlling reagents and test conditions. Control 2 (minimally reactive [1+] control serum) is included as a reading standard and equipment control. If a gradual decrease of fluorescence in the reactive controls is observed over a period of time, this decrease may indicate deterioration of control sera, reagents, or the light source.

Control pattern illustration

R, Reactive; N, nonreactive. Pluses indicate intensity of fluorescence.

Reactive controls.	
a. 1:5 PBS dilution	R4+
b. 1:5 sorbent dilution	R(4+–3+)
Minimally reactive (1+) control.	R1+
Nonspecific serum controls.	
a. 1:5 PBS dilution	R(2+–4+)
b. 1:5 sorbent dilution	N–±
Nonspecific staining controls.	
a. Antigen, PBS, and conjugate	N
b. Antigen, sorbent, and conjugate	N

Test runs in which these control results are not obtained are considered unsatisfactory and should not be reported.

Test procedure

1. Identify previously prepared slides by numbering the frosted end with a lead pencil.
2. Number the tubes to correspond to the sera and control sera being tested, and place the tubes in racks.
3. Prepare reactive (4+), minimally reactive (1+), and nonspecific control sera dilutions in sorbent, PBS, or both, according to the directions.
4. Pipette 200 μl of sorbent into a test tube for each test serum.
5. Add 50 μl of the heated test serum into the appropriate tube and mix eight times.
6. Cover the appropriate antigen smears with 30 μl of the reactive (4+), minimally reactive (1+), and nonspecific control serum dilutions.
7. Cover the appropriate antigen smears with 30 μl of PBS and 30 μl of the sorbent for nonspecific staining controls.
8. Cover the appropriate antigen smears with 30 μl of the test serum dilutions.
9. Prevent evaporation by placing the slides within a moist chamber.
10. Place the slides in an incubator at 35 to 37°C for 30 min.
11. Rinsing procedure.
    a. Place the slides in slide carriers and rinse the slides with running PBS for ca. 5 s.
    b. Place the slides for 5 min in a staining dish containing PBS.
    c. Agitate the slides by dipping them in and out of the PBS at least 30 times.
    d. Using fresh PBS, repeat steps b and c.
    e. Rinse the slides in running distilled water for ca. 5 s.
12. Gently blot the slides with bibulous paper to remove water drops.
13. Dilute TMRITC-labeled anti-human IgG globulin to its working titer in PBS containing 2% Tween 80.
14. Place ca. 30 μl of diluted conjugate on each smear.
15. Repeat steps 9 through 12.
16. Dilute FITC-labeled antitreponemal globulin to its working titer in PBS containing 2% Tween 80.
17. Place ca. 30 μl of the diluted conjugate on each smear.
18. Repeat step 9.
19. Place the slides in an incubator at 35 to 37°C for 20 min.
20. Repeat steps 11 and 12.
21. Mount the slides immediately by placing a small drop of mounting medium on each smear and applying a cover glass.
22. Examine the slides as soon as possible. If a delay in reading is necessary, place the slides in a darkened room and read them within 4 h.
23. Locate and focus treponemes with the FITC filter system.
24. After the treponemes have been located, dial in the TMRITC filters to read specific fluorescence.
25. With the minimally reactive (1+) control slide as the reading standard, record the intensity of fluorescence of the treponemes as follows: 2+ to 4+, moderate to strong; 1+, equivalent to minimally reactive (1+) control; ± to <1+, visible staining but less than 1+; −, none or vaguely visible but without distinct fluorescence.

MHA-TP Test

See references 5, 31, and 36.

Principle

The MHA-TP is a hemagglutination test which uses sheep erythrocytes as carrier for *T. pallidum* antigen. The serum of the patient is first absorbed with an absorbing diluent made from nonpathogenic Reiter treponemes, and then the serum is reacted with sensitized sheep erythrocytes in a microtiter plate. Serum containing antibodies will react with the sensitized cells to form a smooth mat of agglutinated cells. Unsensitized sheep erythrocytes are used as a control for heterophile reactions.

Equipment

Safety pipette devices calibrated to deliver 20, 25, and 50 μl.
Pipette dropper calibrated to deliver 25 or 75 μl.
Disposable, clear plastic trays with 8 rows of 15 cups each with round, U-shaped bottoms. Trays should be free from dust and lint.
Two-milliliter serologic pipettes graduated in 1/100 ml or safety pipette device calibrated to deliver 0.38 ml.
Test tubes. 12 by 75 mm.
Tray viewer.

Reagents

Rehydrate the lyophilized reagents with the rehydrating fluid supplied, according to the instructions of the manufacturer. (Note that rehydrated reagents should stand for 1 h before being used.) Store lyophilized and rehydrated reagents at 2 to 8°C. These reagents should be discarded if they become contam-

TABLE 2. Reporting system for the FTA-ABS DS test

Initial test reading	Repeat test reading	Report
4+		Reactive
3+		Reactive
2+		Reactive
1+	>1+	Reactive
	1+	Reactive minimal[a]
	<1+	Nonreactive
<1+	<1+/N[b]	Nonreactive
N–±		Nonreactive

[a] In the absence of historical or clinical evidence of treponemal infection, this test result should be considered equivocal. A second specimen should be submitted for serologic testing.

[b] N, No fluorescence.

inated or do not demonstrate the proper reactivity with control sera.

Absorbing diluent (liquid). Sonicated cell membranes from sheep and ox erythrocytes, normal rabbit testicular extract, sonicated Reiter treponemes, normal rabbit serum, and stabilizers in PBS (pH 7.2). This reagent is used to preabsorb and dilute sera and to prepare the working dilutions of the sensitized and unsensitized cell suspensions.

Antigen. *T. pallidum*-sensitized sheep cells (lyophilized). The rehydrated antigen is a 2.5% suspension of formalinized, tanned sheep erythrocytes which have been sensitized with sonicated *T. pallidum* (Nichols strain). The working dilution of antigen for 1 test day is prepared by adding 1 part of the rehydrated suspension to 5.5 parts of absorbing diluent (1:6.5 dilution). The quantity of working dilution needed is 75 μl for each serum dilution tested, plus a slight excess.

Unsensitized sheep cells (lyophilized). When rehydrated, this is a 2.5% suspension of formalinized, tanned sheep erythrocytes (not sensitized with *T. pallidum* antigen). Reconstituted reagents should be used within 5 days. Prepare the working dilution (1:6.5) for 1 test day, allowing 75 μl for each serum tested, plus a slight excess.

Controls

Control sera (lyophilized). Store lyophilized control sera at 2 to 8°C. When rehydrated, the control sera may be stored in aliquots at −20°C for up to 4 weeks. These sera are supplied unabsorbed.

a. The reactive control serum should not vary more than ±1 doubling dilution from the endpoint titer established for that serum. (The dilutions are expressed in terms of the final serum dilution obtained after the addition of all reagents.)

b. The nonreactive control serum should be nonreactive at the 1:80 serum dilution.

Unsensitized erythrocyte-serum control. Each test and control serum tested with sensitized sheep erythrocytes (antigen) is also tested with unsensitized erythrocytes. This serum control should be nonreactive.

Reagent controls. Both sensitized erythrocytes (antigen) and unsensitized erythrocytes mixed with absorbing diluent should give nonreactive test results.

Preparation of sera

Add 20 μl of unheated test or control serum to 0.38 ml of absorbing diluent in test tubes (1:20 dilutions). Mix at least eight times and incubate at room temperature (25°C ± 5°C) for 30 min. The absorbed test and control sera (1:20 dilutions) are now ready to be tested. The residual of the absorbed sera may be stored at 2 to 8°C and can be retested on the same day. Absorbed sera should be at room temperature when tested.

Test procedure

1. Record the control and serum numbers on the daily worksheets to correspond to their respective tray and cup numbers.
2. Set up reactive and nonreactive control sera.
   a. In cups 1 through 10 of tray row A, prepare twofold dilutions of the absorbed reactive control as follows. Place 50 μl of absorbed reactive control serum in cup 1. Place 25 μl of absorbing diluent in cups 2 through 10 of that row. Ten cups are used for the dilution to ensure sufficient twofold dilution to exceed the endpoint titer for the reactive control serum. Prepare the serial dilution with either 25-μl microdiluters or 25-μl safety pipettes. Discard 25 μl from cup 10. Measure 25 μl of the absorbed reactive control serum (1:20 dilution) in row B, cup 1, for the unsensitized erythrocyte-serum control.
   b. Measure 25 μl of the absorbed nonreactive control serum (1:20 dilution) into each of two adjacent cups, row A, cup 14, and row B, cup 14.
   c. Add 75 μl of the sensitized cells to all cups (row A, cups 1 through 10, and row A, cup 14) containing reactive control serum dilutions and the nonreactive control serum dilution, either with 3 drops from a 25-μl dropper or with a safety pipette device or dropper set to deliver 75 μl.
   d. With a pipette dropper, as in step c, add 75 μl of the unsensitized cells to row B, cup 1, and row B, cup 14, for the reactive and nonreactive unsensitized erythrocyte-serum controls.
3. Set up reagent controls.
   a. Add 75 μl of the sensitized erythrocytes to 25 μl of absorbing diluent in one cup (row A, cup 15).
   b. Add 75 μl of the unsensitized erythrocytes to 25 μl of absorbing diluent in another cup (row B, cup 15).
4. Place 25 μl of each absorbed test serum (1:20 dilution) into two adjacent cups in the tray, e.g., cup 1 in rows C and D, cup 2 in rows C and D, etc.
5. Add 75 μl of the working dilution of sensitized cells (antigen) to each cup in rows C, E, and G (absorbed serum test).
6. Add 75 μl of the working dilution of unsensi-

tized cells to each cup in rows D, F, and H (absorbed serum control).

7. The final serum dilution in each test and control cup is 1:80.

Note that other methods of adding the 75-μl quantities of sensitized cells and unsensitized cells may be used if delivery is checked for accuracy.

8. Shake the trays gently, and stack and cover them with an empty tray.

9. Incubate the trays at room temperature (25°C ± 5°C) for at least 4 h. The incubation period may be extended overnight.

10. Read the settling patterns of the erythrocytes with an angled mirror (tray viewer) to visualize the patterns from below.

11. Readings are scored on a scale of − to 4+ hemagglutination; the degree of hemagglutination is judged according to the following criteria: smooth mat of cells covering the entire bottom of the cup, with edges sometimes folded, 4+, reactive; smooth mat of cells covering less area of the cup, 3+, reactive; smooth mat of cells surrounded by a red circle, 2+, reactive; smooth mat of cells surrounded by a smaller red circle with hemagglutination outside the circle, 1+ reactive; button of cells with a small hole in the center, ±, nonreactive; definite compact button in the center of the cup, with or without a very small hole in the center, −, nonreactive.

Note that specimens giving ± hemagglutination should be repeated before a report is made. Specimens repeating at ± or − should be reported as nonreactive, whereas specimens repeating as 1+ should be reported as reactive.

12. The results of the controls included for each test day should conform to the criteria outlined in the Controls section.

13. Reporting scheme for qualitative test.
   a. Report as reactive a serum showing hemagglutination of 1+ or higher with sensitized cells (antigen), provided there is no hemagglutination with unsensitized cells.
   b. Report as nonreactive a serum showing no hemagglutination (− or ±) with sensitized cells and unsensitized cells.

14. If a serum specimen gives nonspecific hemagglutination with the unsensitized erythrocytes (serum control), retest the serum in the following manner.
   a. Prepare dilutions of the absorbed serum in two rows of cups.
   b. Add 75 μl of sensitized cells to each cup in one row.
   c. Add 75 μl of unsensitized cells to each cup in the other row.
   d. Report as reactive, without reference to titer, if the hemagglutination with sensitized cells is at least 2 doubling dilutions (four times) greater than with unsensitized cells and if the first dilution showing no hemagglutination with unsensitized cells has a 3+ or 4+ reaction with sensitized cells.

QUALITY CONTROL

The reliability and accuracy of tests for syphilis is dependent upon the quality of reagents used and tight control of technical factors in the test procedure.

The quality of reagents can be controlled by adequate testing before use. New lots of reagents must always be tested in parallel with reference reagents or with other lots of the reagent known to be satisfactory.

A number of technical factors are known to affect test sensitivity: specimen collection, condition, and treatment; glassware condition; time of incubation and time and speed of rotation; microscope condenser adjustment and lighting; temperature; reagent volumes, pH, and electrolyte concentration; antigen preparation and handling; and timing in the performance and reading of tests. Control of these factors is essential to obtaining reliable test results.

Plasma and cord blood specimens may cause problems in reading card test results, and these specimens should not be used except in special situations. Reactivity of nontreponemal tests may be masked by prozone reactions, and specimens showing rough or other atypical reactions in qualitative tests should be quantitated to help avoid this problem.

INTERPRETATION OF SEROLOGIC TESTS

Complex and poorly defined immunological phenomena complicate the interpretation of serologic tests for syphilis. As pointed out by Turner (34), there is inherent conflict between the desire by the clinician for simplicity in the message conveyed by these tests and the underlying complexity of the tests.

Results of serologic tests for syphilis must be interpreted in relation to the stage of syphilis suspected, the presence of other underlying diseases or conditions, a history of previous infection or treatment, and epidemiological findings.

The nontreponemal tests are useful as screening tests, in monitoring therapy, and in detecting reinfection. They are biologically nonspecific and are known to react in a variety of diseases and conditions other than syphilis. The more specific treponemal tests are helpful in confirming the specificity of reactive nontreponemal tests.

Once reactive, treponemal tests tend to remain reactive. They are helpful in the detection of late infections and the confirmation of reactive nontreponemal test results. They are not useful in following therapy or in the detection of reinfection.

Currently used serologic tests lack sensitivity in the early stages of infection. Since *T. pallidum* is present in lesions before the appearance of detectable serum antibody, conventional and fluorescent-antibody dark-field tests are frequently helpful in the detection of early infection. Negative results do not necessarily exclude syphilis, since the ability to demonstrate *T. pallidum* can be affected by the age and condition of the lesion and the adequacy of the specimen.

Currently used serologic tests also detect antibody produced as a result of other treponemal infections (bejel, yaws, and pinta). Differentiation between these diseases and syphilis must be based on clinical and epidemiological findings.

When treponemal test results and clinical opinion disagree, the tests should be repeated, and the clinician should endeavor to obtain as much additional clinical, historical, and epidemiological information as possible. If the disagreement persists, it may be helpful to send the specimen to a reference laboratory for both fluorescent-antibody and microhemagglutination tests.

The clinician must interpret treponemal test results with an understanding of the limitations of sensitivity and specificity of each test and with as much clinical information as possible. In the final analysis, the diagnosis will depend on clinical judgment.

LITERATURE CITED

1. **Balows, A., J. C. Feeley, and H. W. Jaffe.** 1976. Laboratory diagnosis of treponematoses, p. 201–208. *In* R. C. Johnson (ed.), The biology of parasitic spirochetes. Academic Press, Inc., New York.
2. **Bradford, L. L., D. L. Tuffanelli, J. Puffer, M. L. Bissett, H. L. Bodily, and R. M. Wood.** 1967. Fluorescent treponemal absorption and *Treponema pallidum* immobilization tests in syphilitic patients and biologic false-positive reactors. Am. J. Clin. Pathol. **47:**525–532.
3. **Coffey, E., L. Bradford, and S. A. Larsen.** 1980. Serological tests for syphilis, p. 509–520. *In* E. H. Lennette, A. Balows, W. J. Hausler, Jr., and J. P. Truant (ed.), Manual of clinical microbiology, 3rd ed. American Society for Microbiology, Washington, D.C.
4. **Coffey, E. M., L. L. Bradford, L. S. Naritomi, and R. M. Wood.** 1972. Evaluation of the qualitative and automated quantitative microhemagglutination assay for antibodies to *Treponema pallidum*. Appl. Microbiol. **24:**26–30.
5. **Cox, P. M., L. C. Logan, and L. C. Norins.** 1969. Automated, quantitative microhemagglutination assay for *Treponema pallidum* antibodies. Appl. Microbiol. **18:**485–489.
6. **Dans, P. E., F. N. Judson, S. A. Larsen, and M. A. Lantz.** 1977. The FTA-ABS test: a diagnostic help or hindrance. South. Med. J. **70:**312–315.
7. **Deacon, W. E., J. B. Lucas, and E. V. Price.** 1966. Fluorescent treponemal antibody-absorption (FTA-ABS) test for syphilis. J. Am. Med. Assoc. **198:**624–628.
8. **Fitzgerald, T. J.** 1981. Pathogenesis and immunology of *Treponema pallidum*. Annu. Rev. Microbiol. **35:**29–54.
9. **Friedly, G., M. V. Zartarian, J. C. Wood, C. M. Floyd, E. M. Peterson, and L. M. De La Maza.** 1983. Hemagglutination treponemal test for syphilis. J. Clin. Microbiol. **18:**775–778.
10. **Hambie, E. A., S. A. Larsen, M. W. Perryman, D. E. Pettit, R. L. Mullally, and W. Whittington.** 1983. Comparison of a new rapid plasma reagin card test with the standard rapid plasma reagin 18-mm circle card test and the Venereal Disease Research Laboratory slide test for serodiagnosis of syphilis. J. Clin. Microbiol. **17:**249–254.
11. **Harris, A., A. A. Rosenberg, and L. M. Riedel.** 1946. A microflocculation test for syphilis using cardiolipin antigen. Preliminary report. J. Vener. Dis. Infect. **27:**169–174.
12. **Hunter, E. F., M. R. Adams, L. H. Orrison, B. J. Pender, and S. A. Larsen.** 1979. Problems affecting performance of the fluorescent treponemal antibody-absorption test for syphilis. J. Clin. Microbiol. **9:**163–166.
13. **Hunter, E. F., W. E. Deacon, and P. E. Meyer.** 1964. An improved test for syphilis—the absorption procedure (FTA-ABS). Public Health Rep. **79:**410–412.
14. **Hunter, E. F., R. M. McKinney, S. E. Maddison, and D. D. Cruce.** 1979. Double-staining procedure for the fluorescent treponemal antibody absorption (FTA-ABS) test. Br. J. Vener. Dis. **55:**105–108.
15. **Jaffe, H. W., S. A. Larsen, O. G. Jones, and P. E. Dans.** 1978. Hemagglutination tests for syphilis antibody. Am. J. Clin. Pathol. **70:**230–233.
16. **Jaffe, H. W., S. A. Larsen, M. Peters, D. F. Jove, B. Lopez, and A. L. Schroeter.** 1978. Tests for treponemal antibody in CSF. Arch. Intern. Med. **138:**252–255.
17. **Jokinen, E. F., A. Lassus, and E. Linder.** 1969. Fluorescent treponemal antibody (FTA) reaction in sera with antinuclear factors. Ann. Clin. Res. **1:**77–80.
18. **Kraus, S. J., J. R. Haserick, and M. A. Lantz.** 1970. Fluorescent treponemal antibody-absorption test reactions in lupus erythematosus: atypical beading pattern and probable false-positive reactions. N. Engl. J. Med. **282:**1287–1290.
19. **Larsen, S. A., E. A. Hambie, D. E. Pettit, M. W. Perryman, and S. J. Kraus.** 1981. Specificity, sensitivity, and reproducibility among the fluorescent treponemal antibody-absorption test, the microhemagglutination assay for *Treponema pallidum* antibodies, and the hemagglutination treponemal test for syphilis. J. Clin. Microbiol. **14:**441–445.
20. **Levene, G. M., J. L. Turk, D. J. M. Wright, and A. G. S. Grimble.** 1969. Reduced lymphocyte transformation due to a plasma factor in patients with active syphilis. Lancet **ii:**246–247.
21. **Levene, G. M., D. J. M. Wright, and J. L. Turk.** 1971. Cell-mediated immunity and lymphocyte transformation in syphilis. Proc. R. Soc. Med. **64:**426–428.
22. **Mackey, D. M., E. V. Price, J. M. Knox, and A. Scotti.** 1969. Specificity of the FTA-ABS test for syphilis: an evaluation. J. Am. Med. Assoc. **207:**1683–1685.
23. **Matthews, H. M., T.-K. Yang, and H. M. Jenkin.** 1979. Unique lipid composition of *Treponema pallidum* (Nichols virulent strain). Infect. Immun. **24:**713–719.
24. **Musher, D. M., R. F. Schell, R. H. Jones, and A. M. Jones.** 1975. Lymphocyte transformation in syphilis: an in vitro correlate of immune suppression in vivo? Infect. Immun. **11:**1261–1264.
25. **Pavia, C. S., J. D. Folds, and J. B. Baseman.** 1978. Cell-mediated immunity during syphilis: a review. Br. J. Vener. Dis. **54:**144–150.
26. **Perryman, M. W., S. A. Larsen, E. A. Hambie, D. E. Pettit, R. L. Mullally, and W. Whittington.** 1982. Evaluation of a new rapid plasma reagin card test as a screening test for syphilis. J. Clin. Microbiol. **16:**286–290.
27. **Portnoy, J.** 1963. Modifications of the rapid plasma reagin (RPR) card test for syphilis for use in large scale testing. Am. J. Clin. Pathol. **40:**473–479.
28. **Portnoy, J., J. H. Brewer, and A. Harris.** 1962. Rapid plasma reagin card test for syphilis and other treponematoses. Public Health Rep. **77:**645–652.
29. **Shore, R. N.** 1974. Hemagglutination tests and related advances in serodiagnosis of syphilis. Arch. Dermatol. **109:**854–857.
30. **Sparling, P. F.** 1971. Medical progress. Diagnosis and treatment of syphilis. N. Engl. J. Med. **284:**642–653.
31. **Tomizawa, T., and S. Kasamatsu.** 1966. Hemagglutination tests for diagnosis of syphilis. A preliminary report. Jpn. J. Med. Sci. Biol. **19:**305–308.
32. **Tomizawa, T., S. Kasamatsu, and S. Yamaya.** 1969. Usefulness of hemagglutination test using *Treponema pallidum* antigen (TPHA) for the serodiagnosis of syphilis. Jpn. J. Med. Sci. Biol. **22:**341–350.
33. **Tuffanelli, D. L., K. D. Wuepper, L. L. Bradford, and R. M. Wood.** 1967. Fluorescent treponemal-antibody absorption tests: studies of false-positive reactions to tests for syphilis. N. Engl. J. Med. **276:**258–262.
34. **Turner, T. B.** 1970. Syphilis and the treponematoses, p. 346–390. *In* S. Mudd (ed.), Infectious agents and host reactions. The W. B. Saunders Co., Philadelphia.
35. **Varmus, H. E.** 1975. Syphilis, p. 830–835. *In* M. A. Krupp and M. J. Chatton (ed.), Current medical diagnosis and treatment. Lange Medical Publications, Los Altos, Calif.
36. **Venereal Disease Program, National Communicable Disease Center, U.S. Department of Health, Education, and Welfare.** 1969. Manual of tests for syphilis. U.S. Government Printing Office, Washington, D.C.
37. **Wentworth, B. B., M. A. Thompson, C. R. Peter, R. E. Bawdon, and D. L. Wilson.** 1978. Comparison of a hemagglutination treponemal test for syphilis (HATTS) with

other serologic methods for the diagnosis of syphilis. Sex. Transm. Dis. **5**:103–111.

38. **Wood, R. M., Y. Inouye, W. Argonza, L. Bradford, R. Jue, Y. Jeong, J. Puffer, and H. L. Bodily.** 1967. Comparison of the fluorescent treponemal antibody absorption and *Treponema pallidum* immobilization tests on serums from 1182 diagnostic problem cases. Am. J. Clin. Pathol. **47**:521–524.

Chapter 95

Viral Serology

KENNETH L. HERRMANN

Immunological investigations have played an important role in the identification and study of viruses as causes of human illness. Viral serology provides both clinicians and epidemiologists with a powerful tool for the diagnosis of infection in individual patients, the determination of susceptibility to a specific virus, and the assessment of the prevalence of a particular virus infection in a community. Technological advances during the past several years have led to more rapid, sensitive, and accurate serodiagnostic tests for viral diseases, and the development of automated technology has made these tests easier to perform and has encouraged more clinical laboratories to perform them.

Immunological responses of the human host to viral infections involve both the B cells, with production of antiviral antibodies, and the T cells, with induction of cell-mediated immunity. Circulating antiviral antibodies, particularly neutralizing antibodies, are widely accepted as prima facie evidence of past infection with the particular virus. Such humoral antibodies, however, may be of limited importance in determining resistance to reinfection with the given virus. There is growing evidence that cell-mediated immune factors are of primary importance in protecting a person against viral infections. This chapter, however, will focus only on the humoral immune responses to viral infections.

SPECIMENS

Serologic confirmation of a suspected viral etiology of an illness optimally requires two serum specimens, the first obtained as soon as possible after the onset of the illness and the other obtained 1 to 2 weeks later. Whole blood, collected aseptically and allowed to clot, is the specimen of choice for most viral antibody studies (5). A 10-ml volume of venous blood is generally requested for most viral serodiagnostic studies. However, lesser volumes are sufficient for the performance of many antibody tests. The serum should be separated from the clot within 24 h after collection and then stored at 4 to 8°C or frozen until tested. Whole blood should not be frozen because the resulting lysis of erythrocytes will make the specimen unsuitable for most serologic tests. Capillary blood specimens, collected either in capillary tubes or on filter paper, may be suitable for certain specific test procedures.

Specimens for serology should be considered potentially infectious and handled accordingly in all phases of processing and testing. In particular, mouth pipetting of specimens and reagents for serologic tests should be prohibited. Laboratory workers handling large numbers of blood and serum specimens should be considered candidates for hepatitis B virus vaccination.

ANTIBODY RESPONSES

Most viruses induce detectable levels of specific antibodies after a primary infection and may stimulate a boost in titer after reinfection or reactivation of latent infection (1). Serodiagnosis of a viral infection generally requires paired acute- and convalescent-phase sera. Confirmation of a suspected viral etiology of the illness usually requires the demonstration of a significant rise of specific antibody, and not merely the presence of antibody to the viral agent. Acute- and convalescent-phase sera must be tested together in the same test run for differences in antibody levels to be meaningfully interpreted. In tests that use doubling dilutions of the serum, a fourfold or greater rise in antibody titer is considered significant, whereas a twofold difference is within the technical error of the technique. In tests by other methods of quantitating antibody, significance limits of diagnostic titer changes must be established independently. Frequently, a critical ratio of convalescent- to acute-phase antibody levels is used to define a significant antibody rise.

The recognition of temporal differences in antibody responses to the infecting agent may also be of diagnostic value. For example, antibodies detected by one assay method may appear very early in the course of an illness, whereas antibodies measured by another assay may not be detectable until a week or more after illness onset. For this reason, the performance of more than one type of antibody assay is recommended for the serodiagnosis of certain virus infections such as hepatitis A and B.

When antibody development cannot be demonstrated because an acute-phase service is unavailable, the presence of specific antiviral antibody in the immunoglobulin M (IgM) fraction of the patient's convalescent-phase serum is often indicative of recent or current infection with the virus (2). The clinical interpretation of antiviral IgM tests, however, should be made with caution. For antiviral IgM tests to be useful in diagnosis, the IgM antibody response must be specific, measurable with adequate reliability and sensitivity, and transient (i.e., present only with recent active infection with the specific virus).

The transient nature of the IgM antibody response appears to hold true for most primary viral infections. Specific IgM generally does not persist longer than 1 to 2 months after the acute viral infection. Some investigators, however, have reported the persistence of IgM antibody for several months or even years after the onset of primary infection. Such reports are more common for viruses, such as cytomegalovirus, which characteristically produce chronic or persistent infections in the patient. Other investigators have reported the appearance of specific IgM antibody to agents other than the agent responsible for the current infection. Such true heterologous IgM responses are probably rare but should be considered when the results of IgM tests are interpreted for clinical management.

On the other hand, the absence of specific IgM antibody in a serum sample containing IgG antibody may be misleading. Such a finding may indicate a past infection with the virus in question, may be the result of inappropriate timing in obtaining the sample (i.e., serum collected after the IgM has disappeared), or may reflect a current secondary infection or reinfection with the virus. Clinical interpretation of negative IgM antibody assays, therefore, may also be quite difficult.

The sensitivity and specificity of IgM antibody assays determine the reliability of these methods for diagnosis and thus directly influence the level of confidence placed in these tests by the clinician. A potential source of error in the identification of viral IgM antibody is the presence of rheumatoid factor (RF) in the patient's serum. IgM-RF may be present in the serum of patients with nonrheumatoid conditions, including infections with a number of viruses such as cytomegalovirus, herpes simplex, measles, and influenza. IgM-RF is also commonly found in the serum of neonates, presumably elicited by the allotypic antigens (Gm antigens) of the maternal IgG. If specific viral antibody of the IgG class is present in serum that also contains IgM-RF, the RF may react with the viral IgG and be erroneously identified as specific antiviral IgM antibody by any of a number of the indirect IgM immunoassays. This pitfall can be avoided by first removing the IgM-RF.

A second potential problem in the detection of specific antiviral IgM is the inhibitory effect of specific IgG antibody when both immunoglobulins are present in the same serum sample. Competition for available antigen may significantly reduce the sensitivity of the IgM assay. This inhibitory effect can be reduced substantially either by a physical separation of the immunoglobulin classes by gel filtration or ultracentrifugation before the assay for specific IgM or by the preabsorption of the serum with staphylococcal protein A. This protein binds to the Fc portion of the IgG molecule, allowing the IgG to be removed by centrifugation.

Specific IgM assays are not available for many viral infections. When the lack of an acute-phase serum precludes the demonstration of a rising antibody titer and when specific IgM testing is not available, it is often tempting to place clinical significance on highly elevated antiviral titers in a patient's serum during the convalescent phase of illness. Occasionally, the presence of very high levels of complement-fixing antibody in convalescent-phase serum may be diagnostic of recent infection. As a rule, however, single serum antiviral IgG titers, regardless of the degree of elevation, should not be given diagnostic significance.

TABLE 1. Immunodiagnosis of common viral diseases of humans[a]

Disease agent	Available tests	Comments
Adenovirus	CF, EIA	Group specific; good for diagnosis.
	HI, NT	Type specific; limited diagnostic value.
Cytomegalovirus	CF, IHA, EIA, IFA	Serologic tests of limited diagnostic value.
Enteroviruses		
Poliomyelitis	CF, NT	For confirmation of diagnosis.
Non-polio	NT, EIA	Heterotypic responses often make serologic tests difficult to interpret.
Hepatitis A virus	RIA, EIA	IgM anti-HAV for diagnosis.
Hepatitis B virus	RIA, EIA	IgM anti-HBc diagnostic; antigen detection tests preferred for diagnosis.
Herpes simplex virus	CF, IHA, EIA, IFA	Serologic tests of limited diagnostic value.
Influenza virus	CF, HI	Need antigen of current strain for optimal sensitivity; for diagnosis.
Measles (rubeola) virus	HI, IFA, EIA	For either diagnosis or immune screening.
	CF	Good only for diagnosis; antibody is short lived.
Mumps virus	HI, IFA, EIA	Heterologous responses common with other paramyxovirus infections; for diagnosis or immune screening.
	CF	For diagnosis only; antibody is short lived.
Parainfluenza virus	HI, CF, EIA	Heterotypic responses common with other paramyxovirus infections.
Respiratory syncytial virus	HI, CF, EIA	EIA more sensitive than CF or HI for diagnosis.
Rabies virus	RFFIT, IFA	For determining immune status following rabies vaccination, and for confirming diagnosis.
Rotavirus	EIA	Confirmation of diagnosis; antigen detection preferred for diagnosis.
Rubella virus	HI, EIA, IFA, LA	For diagnosis or immune screening.
	PHA	For immune screening only.
Infectious mononucleosis virus (Epstein-Barr virus)	Heterophil tests	Preferred for diagnosis; may be negative in some children with infectious mononucleosis.
	IFA for anti-VCA, anti-EBNA, and anti-EA	Tests are of limited diagnostic value.
Varicella-zoster virus (chickenpox)	CF	For diagnosis only.
	IFA, FAMA	For assessment of immune status; FAMA generally acknowledged to be most sensitive.

[a] Abbreviations: CF, complement fixation; EA, early antigen; EBNA, Epstein-Barr nuclear antigen; EIA, enzyme immunoassay; FAMA, fluorescent antibody to membrane antigen assay; HI, hemagglutination inhibition; IFA, indirect fluorescent antibody assay; IHA, indirect hemagglutination; LA, latex agglutination; NT, neutralization; PHA, passive hemagglutination; RFFIT, rabies fluorescent focus inhibition test; RIA, radioimmune assay; VCA, viral capsid antigen.

For certain viruses, however, multiple antibodies that develop in response to infection are detectable at various times postinfection, making it possible to determine whether the infection is current or recent rather than long past by using only a single serum. Hepatitis B virus and Epstein-Barr virus infections have been assessed in this manner.

SEROLOGIC TESTS

It is beyond the scope of this chapter to attempt to describe all the serologic techniques currently used in diagnostic virology. Immunodiagnostic tests for selected viral diseases, however, are presented in Table 1. Specific test methods will be described in detail in subsequent chapters. Several factors, however, should be considered when a serologic test is selected for a particular situation. Viruses generally possess a number of antigens, i.e., some which are specific for a particular strain of virus and others which are shared by related viruses. Tests with these antigens can therefore be either strain specific or group specific. For general diagnostic purposes, a single group-specific test is preferable to a battery of type-specific tests. The temporal pattern of antibody response should also be considered when a test is selected. A test which easily detects an antibody rise after illness onset but reverts to negative within a few months or years after the infection would be suitable for diagnosis, but useless for the serologic assessment of immune status. On the other hand, a test which detects antibodies that persist for life would be a better indicator of prior infection in assessments of immunity.

GENERAL CONSIDERATIONS

The use of serology has several important advantages in diagnostic virology (3, 4). First, serology often enables a diagnosis to be established when virus isolation either is negative or cannot be attempted. Second, serology may be used to confirm the temporal relationship of a virus isolate to the clinical illness. This relationship is particularly important with viruses such as cytomegalovirus or herpes simplex, which produce chronic infections lasting for months or years after onset. Third, serodiagnosis is often more rapid than virus isolation, especially for the slow- or difficult-to-grow viruses. Fourth, the use of serology is often the least expensive way to establish a laboratory diagnosis. And finally, serologic techniques can be very useful in ruling out a particular viral diagnosis. Negative viral serologic test results can often exclude a specific virus from the etiologic possibilities, whereas failure to isolate virus or detect specific viral antigen in tissues cannot.

LITERATURE CITED

1. **Drew, W. L. (ed.).** 1976. Viral infections. F. A. Davis Co., Philadelphia.
2. **Lennette, E. H., and N. J. Schmidt (ed.).** 1979. Diagnostic procedures for viral, rickettsial and chlamydial infections. American Public Health Association, Washington, D.C.
3. **McLean, D. M.** 1980. Virology in health care. The Williams & Wilkins Co., Baltimore.
4. **Milgrom, F., C. J. Abeyounis, and K. Kano (ed.).** 1981. Principles of immunologic diagnosis in medicine. Lea & Febiger, Philadelphia.
5. **National Committee for Clinical Laboratory Standards.** 1984. Procedures for the collection of diagnostic blood specimens by venipuncture: approved standard, 2nd ed. National Committee for Clinical Laboratory Standards publication H3-A2. National Committee for Clinical Laboratory Standards, Villanova, Pa.

Chapter 96

Serodiagnosis of Fungal Diseases

LEO KAUFMAN AND ERROL REISS

Meticulous consideration of signs, symptoms, and epidemiologic data may result in an accurate clinical diagnosis of a mycotic disease; however, such presumptive diagnoses should be confirmed by standard culture and histologic laboratory procedures. Unfortunately, diagnosis of a mycotic infection cannot always be proven by culture or histology, despite repeated efforts. In such situations, immunologic procedures can be used to provide rapid and presumptive evidence of infection, and immunologic reactions often give the first clues to the existence of a fungal infection. Serologic tests can also yield information on the effects of chemotherapy and, in many cases, lead to increased efforts to isolate and identify the etiologic agent.

A positive serologic reaction, particularly at a titer above 1:32, even though on a single specimen, can be diagnostically significant. Titers of 1:32 or less may reflect early infection, a cross-reaction, or residual antibody from a previous infection. Many of the diagnostic antigens used in medical mycology are unpurified mixtures of multiple antigenic factors, some of which are shared by fungi of different genera and by other microorganisms. When a serum titer is low or when cross-reactions are encountered, a prudent diagnosis rests on (i) the results of a variety of serologic tests performed with a battery of antigens (including those representing antigenically related species), (ii) the examination of serial serum specimens for titer changes, (iii) tests for specific precipitins, and (iv) information on the acquisition of hypersensitivity. Some persons suffering with a systemic mycosis are immunologically unresponsive. Others may not demonstrate antibody levels against a fungus because serum is taken in the latent period of the primary immune response or because the patient is immunologically impaired. Consequently, negative immunological results do not exclude a diagnosis of mycotic infection. Knowledge of the clinical history and therapy of the patient is helpful in the interpretation of serologic data.

DETECTION OF HYPERSENSITIVITY

Dual immediate (15-min) and Arthus-type (4- to 6-h) skin test reactions, which develop after the intradermal injection of *Aspergillus* species antigens, are important criteria in the diagnosis of allergic bronchopulmonary aspergillosis.

Most patients with coccidioidomycosis and histoplasmosis develop a hypersensitive state that is readily and reliably demonstrated by coccidioidin or spherulin and histoplasmin skin tests, respectively. The hypersensitive state has been demonstrated in persons suffering from blastomycosis, candidiasis, cryptococcosis, and paracoccidioidomycosis, but not with sufficient frequency or specificity to warrant the widespread use of corresponding skin test antigens. *Candida* skin test results show no significant difference in reactivity between acutely ill and normal subjects and are perhaps best used to assess immunocompetence. Delayed-type skin test reactions are frequently negative in patients with chronic mucocutaneous candidiasis.

The skin test is generally administered by an intradermal injection of 0.1 ml of an appropriate dilution of antigen into the volar surface of the forearm. In sensitized persons, an area of induration and erythema develops at the injection site. Ideally, the coccidioidin and spherulin tests should be read at 24- to 48-h intervals for maximal reaction, and the histoplasmin test should be read at 48- to 72-h intervals. The largest diameter of induration, not erythema, is recorded. An induration of 5 mm or more in diameter is considered a positive reaction.

The skin test is most useful in the definition of areas in which a disease is endemic. The test has limited value as a diagnostic tool, since it does not distinguish between past and present infections. In general, a positive reaction is of diagnostic value only if a negative reaction was obtained before the onset of clinical symptoms. Except in infants, a positive reaction with no patient history available has little diagnostic value. A negative reaction is of greater significance because it shows a definite absence of the disease, except when the patient is in the very early or terminal stages of disseminated infection or has a defective cellular immune system. In these cases, the test is negative even though disease may be present. In an ill person, a negative skin test that reverts to positive is a good prognostic sign.

For certain diseases, especially coccidioidomycosis, physicians rely on the coccidioidin skin test to assess the prognosis. A reversion to a negative result indicates a state of anergy and a poor prognosis. The prognostic value of the spherulin skin test antigen in patients with disseminated coccidioidomycosis has yet to be determined. In a healthy person, a positive coccidioidin skin test implies resistance to the infection. Whether a positive histoplasmin skin test in a healthy subject implies such a state is not known.

Because fungi share common antigens and symptoms resemble each other, skin tests with several antigens, such as coccidioidin or spherulin and histoplasmin, should be performed simultaneously. Interpretation is thus facilitated when only one of the antigens produces a positive reaction or possibly when one antigen elicits a larger area of induration than the other. The magnitude of the reaction, however, does not necessarily indicate the homologous reaction. The proper interpretation of tests with multiple antigens must be balanced with the clinical picture and other laboratory data.

COLLECTION, PRESERVATION, AND SHIPMENT OF SPECIMENS

Specimens for serologic tests must be taken aseptically, and 10 ml of blood should be drawn. After the blood has clotted and the serum has separated, the serum is removed aseptically and preserved by the addition of Merthiolate (ethylmercurithiosalicylic acid, sodium salt) to make a final concentration of 1:10,000.

It is convenient to maintain a 1% stock solution of Merthiolate in the laboratory. Dissolve 1.4 g of sodium borate in distilled water, add 1.0 g of Merthiolate, and adjust the solution to a final volume of 100 ml with distilled water. Use 10 μl of the stock solution per 1.0 ml of serum or other clinical specimen. Specimens so treated do not require refrigeration during shipment.

Spinal fluid specimens should be aseptically taken when meningeal involvement is suspected. No preservative should be added if the material is to be cultured before serologic examination, but preservative may be added if the specimen is not to be cultured. Preferably, the specimen should be frozen or refrigerated if it is to be shipped to a laboratory for culture. This treatment will curtail bacterial growth should the specimen become contaminated during transit.

All specimens should be enclosed in heavy-walled glass or plastic tubes that are secured with a tight-fitting screw cap with a rubber liner. Specimens should be properly insulated against breakage and sent by airmail or air express to ensure prompt arrival.

DETECTION OF CIRCULATING ANTIBODIES AND ANTIGENS

Antibody responses are useful indices in the determination of the cause and the prognosis of a mycosis. Dependable diagnoses can frequently be made from one serologic test. Low titers or cross-reactions in the complement fixation (CF) test are, however, difficult to interpret. In such cases, serum samples taken 3 weeks apart should be studied. Ideally, when the course of an infection is monitored, serum samples should be taken early in the course of illness, at its height, during convalescence, and several weeks after recovery. Fourfold or greater rises in titer are usually acceptable diagnostic signs of disease, although cross-reactions with heterologous antigens may appear early in an illness and cause confusion. Usually, as the disease progresses, cross-reacting antibody titers remain stable or increase at a lower rate than the homologous titer. False serodiagnoses may be avoided by the use of reference precipitates in immunodiffusion (ID) tests.

Because antibody responses do not always reflect active disease and because the responses are variable and may be absent in certain compromised patients, many investigators have focused on the development of sensitive tests for the detection of fungal antigens in body fluids. The detection and quantitation of circulating antigens have proven important for the diagnosis and monitoring of patients with cryptococcosis. The diagnostic value of enzyme immunoassays (EIA) and radioimmunoassays (RIA) in the detection of antigenemia in patients with invasive aspergillosis and candidiasis has been the subject of several investigations (5).

Aspergillosis

CF, counterimmunoelectrophoresis (CIE), and ID tests for aspergillosis have been developed and evaluated. Many researchers have shown the ID test to be an effective and specific method for the diagnosis of aspergillosis in persons whose immune systems are intact (9). The sensitivity of the CIE test has been reported to equal or exceed that of the ID test, whereas the specificity of the CIE test is reported to range from the same to less than that of the ID test (9). The CIE test may be useful for screening purposes.

The value of the CF test for the diagnosis of pulmonary aspergillosis is open to question. Some researchers have found the CF test to demonstrate a high degree of sensitivity and specificity, comparable to that of the ID test. Others, however, reported the CF test to be less specific than the ID test but useful for the detection of active or very recent aspergillosis (9). Studies in our laboratory indicate that the CF test is less sensitive than the ID test, but it does demonstrate good specificity. Sera analyzed by CF should be tested with a battery of *Aspergillus* antigens.

Clinical indications

Patients with allergic bronchopulmonary disease, patients with suspected pulmonary aspergilloma (fungus ball), immunologically intact patients with obscure pulmonary or meningeal infections, and patients receiving immunosuppressive therapy who develop an unexplained fever or whose disease follows an unusual course should be tested for *Aspergillus* precipitins. Allergic bronchopulmonary aspergillosis should be considered a possibility in patients with asthma, transient pulmonary infiltrates, or peripheral eosinophilia. This last disease represents a hypersensitive state characterized by both immediate and Arthus-type skin test reactions at the site of aspergillin injection and by the formation of precipitins. Pulmonary aspergilloma or fungus ball occurs when *Aspergillus fumigatus* or other *Aspergillus* species colonize open-healed cavities of tuberculosis, sarcoidosis, or carcinoma. Invasive aspergillosis includes those cases in which aspergilli have been shown to actually penetrate tissue.

The greatest number of aspergillosis cases may be detected by the use of *A. fumigatus*, *Aspergillus flavus*, and *Aspergillus niger* precipitinogens (9) in separate ID tests performed at the same time. Recent experience indicates that the *Aspergillus terreus* antigens should also be used. Precipitins can be found in over 90% of aspergillomas and in 70% of the cases of allergic bronchopulmonary aspergillosis. They are found less frequently in patients with invasive disease. The ID test, however, may be helpful in the diagnosis of systemic aspergillosis (10) and should be applied in suspected invasive aspergillosis. Some *Aspergillus* antigens contain C substance. This substance is capable of reaction with the C-reactive protein frequently found in patients with inflammatory diseases. This complex forms a precipitate which may be erroneously interpreted as being due to *Aspergillus* antibodies, but this false-positive reaction can easily be eliminated if the test plates are soaked in 5% sodium citrate for 45 min before a final reading is taken. In addition, C substance produces lines of nonidentity with reference antisera.

For those patients suspected of having invasive aspergillosis whose serum samples are negative for antibody, antigenemia tests should be done to try to obtain a specific antemortem diagnosis. Antigenemia in aspergillosis has been detected by CIE and RIA in experimentally infected animals and in human patients. A modified Farr-type RIA for antigenemia was developed by Weiner and Coats-Stephen (25). When the RIA was performed, immune complexes were dissociated by heating them at an acidic pH. The sensitivity of the assay for an unequivocally positive test is 6.86 ng of *A. fumigatus* antigen per ml of serum. Studies with two series of patients with acute leukemia have demonstrated the ability of the RIA to detect a carbohydrate-containing antigen in serum during invasive aspergillosis (26). The test is highly specific in that no antigen has been found in acute leukemics without aspergillosis. The most recent expression of the sensitivity of the test is that antigenemia was detected in five of seven patients (71%) histologically shown to have invasive aspergillosis (26).

ID test for antibody

The micro-ID test (18) is recommended. In addition to reference *Aspergillus* antigens and antisera, buffered phenolized agar is required.

Stock barbital buffer. pH 8.6, 0.1 ionic strength made up to 500 ml.

Sodium diethylbarbiturate	5.16 g
Diethylbarbituric acid	0.92 g
Sodium acetate	2.05 g

Phenolized medium

Noble agar (or equivalent)	1.0 g
Phenol, liquified	0.25 ml
Barbital buffer, 0.1 ionic strength, pH 8.6	25.0 ml
Distilled water	to 100 ml

Heat to boiling until the agar is completely dissolved.

Equipment

1. Plexiglas matrix (3 mm) with 17 patterns of seven wells each (L. L. Pellet Co., Dallas, Tex.)
2. Plastic petri dish, 15 by 100 mm
3. Spatula, 1-mm tip (flattened)
4. Pasteur pipette
5. Viewing box for reading plates

Procedure

1. Pipette 6.5 ml of agar into a petri dish (15 by 100 mm), and allow the agar to harden.
2. Overlay the first agar layer with 3.5 ml of hot agar, and immediately place the matrix in the liquid agar.
3. Use the plates 30 min after the gel has formed and stabilized, or store them in a moist chamber at 4°C for up to 1 week.
4. Number each pattern on the bottom of the dish (not the template).
5. Remove the excess agar from the wells down to layer 1 with a spatula.
6. Place the reference serum in the top and bottom wells of each pattern, and place the unknown or test sera in the four lateral wells. Place the reference antigen in the center well of each pattern. All wells must be checked for air bubbles. If bubbles are found, they should be broken by gently piercing them with toothpicks. Separate toothpicks should be used for the different antigen and antibody solutions. The reactants are incubated in a moist chamber for 48 h at 25°C.
7. Remove the matrix by gently pressing the sides of the petri dish against the matrix, and remove the agar overlay by gently sweeping the surface with a cotton swab. Wash the agar with distilled water to remove excess reactants. Cover the agar with distilled water and examine it for lines of identity with the reference sera.

Controls. Positive control sera must be included in each test. Three or more distinct precipitin lines should be formed when *A. fumigatus* reference antiserum is allowed to react with *A. fumigatus* antigen. One or more distinct precipitin lines should be formed when *A. flavus*, *A. niger*, or *A. terreus* reference antiserum is allowed to react with the homologous antigen.

Reagents. Standardized and reproducible *A. fumigatus*, *A. flavus*, and *A. niger* antigens with either no or minimal C substance can be prepared from 5-week-old stationary Sabouraud broth cultures grown at 31°C. The culture filtrates are precipitated with cold acetone and concentrated to one-eighth the original volume. The carbohydrate content of these antigens is determined by the anthrone test and adjusted with distilled water to contain 1,000 to 1,500 μg/ml (9). After standardization, all *Aspergillus* antigens should be checked for the presence of C substance with serum known to contain C-reactive protein.

Aspergillosis test reagents may be obtained from the following commercial sources: Greer Laboratories, Inc., Lenoir, N.C.; Hollister-Stier Laboratories, Spokane, Wash.; Immuno-Mycologics, Inc., Norman, Okla.; Meridian Diagnostics, Inc., Cincinnati, Ohio; M. A. Bioproducts, Walkersville, Md.; and Nolan-Scott Biological Laboratories, Inc., Tucker, Ga.

Interpretation. In the ID test, only serum samples that produce a line or lines of identity with a reference serum from a proven case of human aspergillosis are considered positive. The demonstration of one or more precipitating antibodies indicates infection, aspergilloma, or allergy due to an *Aspergillus* species. Although one or two precipitins can occur with any clinical form of aspergillosis, the presence of three or more precipitins is invariably associated with either an aspergilloma or an invasive disease. Physicians must be aware that fungus balls can be produced by *Pseudallescheria boydii* and other fungi and that some cysts or other abnormalities may be misconstrued on an X ray as aspergillomas. In such situations, tests with *Aspergillus* antigens are negative, and thus one can rule out colonization by an *Aspergillus* sp. (10).

When used with reference sera, the ID test is 100% specific. Occasionally, bands of nonidentity are detected that are associated with aspergillosis cases. These bands should make one suspect aspergillosis, but a specific diagnosis cannot be made in the absence of reference bands. One cause of nonspecific bands is C-reactive protein. Such precipitates do not produce lines of identity and will disappear after treatment with sodium citrate. Serum samples that produce lines that are not identical with the specific reference lines and which remain after citrate treatment warrant further study.

Serum samples from most patients with aspergillomas and allergic bronchopulmonary aspergillosis do not have to be concentrated to demonstrate precipitin bands. Some (nonconcentrated) serum samples from patients with invasive disease are precipitin positive. Precipitin-negative serum samples from patients with suspected invasive aspergillosis should be retested after the samples have been concentrated to one-fourth the original volume. This retesting is especially necessary with serum from immunosuppressed patients.

Blastomycosis

Two tests, the CF and ID tests, are widely used in the serodiagnosis of blastomycosis. The CF test is less sensitive and specific than the ID test, but because of its widespread use, the CF test will be discussed in this section.

Clinical indications

Serologic tests for blastomycosis should be sought when a patient shows signs of a respiratory infection which progresses gradually, with fever, loss of weight, cough, and purulent sputum. The test should also be performed when skin lesions are present, since the disease may have spread to localized subcutaneous or cutaneous sites.

Blastomycosis has no pathognomonic symptoms or specific radiologic features. Diagnosis by histologic or culture studies, although ideal, may take time or be negative. The ID test for blastomycosis with a yeast-form culture filtrate containing the A antigen is specific. Positive reactions can be the basis for immediate treatment of the patient without the need for parallel tests with coccidioidin and histoplasmin. The test has a sensitivity of approximately 80% and detects more blastomycosis cases than the CF test with mechanically disrupted, whole-cell, yeast-form antigens. Negative tests, however, do not exclude a diagnosis of blastomycosis (9).

The blastomycosis CF test with broken yeast cells as antigen suffers from the current lack of specific antigens and from insensitivity. Studies with sera from patients with culturally or histologically proven cases of blastomycosis indicate that only about 50% of the serum samples from these cases react in the CF test with yeast-type antigens. In addition, positive reactions also occur with serum samples from patients with diseases other than blastomycosis, such as coccidioidomycosis, histoplasmosis, and paracoccidioidomycosis. In documented, serologically positive cases of blastomycosis, the CF test may, however, have prognostic value.

More recently, the A antigen, identified in yeast-form culture filtrates as specific for *Blastomyces dermatitidis*, was purified by DEAE-cellulose chromatography with a discontinuous salt gradient. Purified A antigen is stable for several months at 5°C in phosphate buffer (6). In CF tests, the A antigen reacted with 70% of the serum samples from patients with proven blastomycosis and was negative with a battery of serum samples from patients with other systemic fungal infections. Significant increases in sensitivity (90%) occurred with EIA; however, some cross-reactions were evident with sera from histoplasmosis cases (10).

The ID test is of value in the interpretation of CF results with sera from patients with suspected blastomycosis that react solely with the *B. dermatitidis* yeast-form antigen or with the *B. dermatitidis* antigen and heterologous antigens. A positive ID test with such a serum indicates *B. dermatitidis* infection.

Test 1. ID test

The micro-ID procedure (18) is recommended. In addition to reference *B. dermatitidis* antigens and antisera, phenolized agar is required.

Phenolized medium

Sodium chloride	0.9 g
Noble agar (or equivalent)	1.0 g
Sodium citrate ($Na_3C_6H_5O_7 \cdot 2H_2O$)	0.4 g
Phenol, liquified	0.25 ml
Glycine	7.5 g
Distilled water	to 100 ml

Autoclave the mixture at a pressure of 15 lb/in^2 for 10 min. The final pH of the medium should be 6.3 to 6.4.

Equipment. Use the same equipment as that for the aspergillosis ID test.

Procedure. Follow the same procedure as that for the aspergillosis ID test except that serum samples are preincubated for 45 min at 37°C before antigen is added. Reactants are incubated for 48 h at 37°C.

Controls. Positive control sera must be included in each test. The *B. dermatitidis* antiserum must react with the homologous reference antigen to form the specific A precipitin band. The *B. dermatitidis* antigen must react with the homologous reference antiserum to yield the precipitin band designated A.

Reagents. Suitable reproducible antigens are prepared from 1-week-old brain heart infusion (Difco Laboratories, Detroit, Mich.) broth cultures of yeast-form *B. dermatitidis* cells shaken at 150 rpm and maintained at 37°C. The culture filtrates are acetone precipitated. The precipitate is then dissolved in a volume of phosphate-buffered saline (PBS), pH 7.2, equal to 1/10 the original filtrate volume (9).

B. dermatitidis CF and ID antigens and antisera may be obtained from Immuno-Mycologics, Meridian Diagnostics, M. A. Bioproducts, and Nolan-Scott.

Interpretation. Sera from blastomycosis cases that react with yeast-form filtrate antigen(s) frequently give rise to what has been designated the A-precipitin line. Serum samples containing the A precipitin from patients or animals with proven blastomycosis are used as references. Only serum samples that produce lines of identity with the A reference band are considered positive for blastomycosis. A positive reaction denotes a recent or current infection by *B. dermatitidis*. In a study with sera from 113 proven cases of blastomycosis, the test permitted the serodiagnosis of 80% of the subjects. However, the sera of some blastomycosis cases are not easily diagnosed by currently available serologic procedures. These sera may be CF and ID negative or CF positive and ID negative. Patients with sera in those categories should be studied intensively for culture or histologic evidence of blastomycosis. In addition, several sera should be drawn at 3-week intervals and examined by CF and ID tests with *B. dermatitidis*, *Coccidioides immitis*, and *Histoplasma capsulatum* antigens. This testing will detect the appearance of CF antibodies, significant changes in homologous titer levels, or the development of precipitin bands diagnostic for blastomycosis, coccidioidomycosis, or histoplasmosis. In established cases of blastomycosis, a decline in the number or the disappearance of precipitin lines is evidence of a favorable prognosis. The serological response, however, is often not as rapid as the clinical response.

Test 2. CF test

The standardized Centers for Disease Control (CDC) Laboratory Branch Complement Fixation (LBCF) test (9) is recommended when serum samples are titrated for complement-fixing antibodies. Either the macro- or the micro-CF test, with an optimal dilution of a suspension of homogenized yeast-form antigen of *B. dermatitidis* (9), may be used. Five 50% units of complement are used in the LBCF test with the optimal concentration of antigen and test serum. The antigen-antibody-complement mixture is incubated for 15 to 18 h at 4°C. Sensitized sheep erythrocytes are added, and the mixture is incubated for 30 min at 37°C. The percentages of hemolysis in the controls and the tests are read. When the controls are satisfactory, sera that demonstrate 30% or less hemolysis at a particular dilution are recorded as positive. Anticomplementary serum samples are those showing less than 75% hemolysis in the serum control without antigen. In the microtest, the initial 1:8 dilutions of heat-inactivated sera are prepared with conventional pipettes and transferred to the microplates. Ensuing dilutions are made with microloops. Detailed directions for the performance of the LBCF test have been published previously (18).

As controls, serum samples from human blastomycosis cases that demonstrate a homologous CF titer of 1:16 or greater should be tested each time the blastomycosis CF test is performed.

Reagents. Yeast-form *B. dermatitidis* soluble antigens for use in the CF test are prepared from 6- to 8-day-old cultures grown on brain heart infusion agar (9). The harvested cells are disrupted in a Braun cell homogenizer, the suspension is centrifuged, and the supernatant is retained. The cells are extracted again with Merthiolate-containing saline, and the supernatants are pooled and adjusted to an absorbance at 540 nm (A_{540}) of between 0.3 and 0.7 with a Coleman Junior spectrophotometer. An LBCF box titration of this antigen with one or more known positive human serum samples should be performed. To be acceptable, the antigen should have a CF titer of 1:32 or higher.

Interpretation. Titers of 1:8 or greater with *B. dermatitidis* yeast-form, homogenate-supernate antigen are considered positive. When reactions occur solely with the *B. dermatitidis* antigen, one is inclined to suspect blastomycosis. This antigen, however, frequently reacts in low titers with sera from patients who show no evidence of blastomycosis and also with sera from persons with confirmed coccidioidomycosis and histoplasmosis. With such sera, a serologic diagnosis would be based on the comparative reactions of several serum specimens taken 3 to 4 weeks apart. High or rising titers indicate that the patient probably has blastomycosis. If precipitins are demonstrated in the blastomycosis ID test, the testing of several serum samples by CF is unnecessary. Because fewer than 50% of the serum samples from persons with proven blastomycosis react in the CF test, a negative CF reaction has little value and does not exclude the existence of active blastomycosis. Although the CF test for blastomycosis may have a limited diagnostic value, it is frequently of prognostic value in the study of culturally proven, serologically positive cases. CF tests with purified antigen A yield results which are very specific (10).

Candidiasis

Latex agglutination (LA), ID, and CIE tests are valuable in the diagnosis of systemic candidiasis in the immunologically intact host (9). In contrast, the serodiagnosis of candidiasis by agglutination and CF tests has proven to be of little reliability because of positive responses in healthy subjects and in persons with superficial candidiasis without systemic involvement (9). Negative results by these tests may be of value in the exclusion of systemic candidiasis as a diagnosis. The quantitative LA, the ID, and the CIE tests appear to give the most reliable results for antibody detection of systemic candidiasis in immunologically intact hosts. The ID and CIE procedures yield results that are apparently comparable (9).

Immunosuppressed patients often fail to produce antibodies, so a negative antibody test does not necessarily rule out the disease. Such patients may be in a state of antigen excess, so that tests for antigen will be positive. Encouraging reports from many laboratories indicate that tests for *Candida* antigenemia are proving useful for the diagnosis of invasive candidiasis in immunologically compromised hosts (5, 10). EIA tests for *Candida* antigenemia have proven useful for the diagnosis of invasive candidiasis in immunologically compromised patients. Antigenemia occurs when mannan polysaccharide is sloughed off *Candida* cell walls and persists for a few days in the plasma. Mannanemia has also been observed in laboratories in which diverse methods of detection are used in patients with chronic mucocutaneous candidiasis (5).

The treatment of sera with alpha-mannosidase resulted in a loss of mannan and a fourfold rise in antibodies to mannan. These results provided evidence that mannan can occur as soluble immune complexes. Procedures that make no provision to detect or dissociate such complexes can be expected to have greatly reduced sensitivity. To provide effective coverage of the clinical levels of mannanemia, a test should be capable of the detection of a low level (nanograms per milliliter) of mannan; such conditions are satisfied by either the RIA or the EIA (5).

Dissociation of mannan-serum complexes was accomplished by heat and alkali digestion, followed by dialysis to neutralize and desalt the digested serum. A prototype indirect EIA inhibition test was developed that required enzyme-labeled anti-human immunoglobulin G (IgG) as the indicator antibody (5). The variables in the cumbersome prototype EIA have been analyzed (5), and it was concluded that an EIA inhibition test has no advantages over the double-antibody sandwich EIA. When the dissociation of complexes is accomplished by boiling sera in the presence of disodium EDTA and the mannan is detected by sandwich EIA, the time of performance is reduced to 3 h. Mannanemia was evident in the 2-week period preceding dissemination and rose in proportion to disease severity (17).

The results obtained with the double-antibody sandwich EIA are exemplified by the studies of de Repentigny et al. (L. de Repentigny, L. D. Marr, J. W.

Keller, A. W. Carter, R. J. Kuykendall, L. Kaufman, and E. Reiss, submitted for publication). The mean mannan concentration in normal blood donors was 0.04 ng/ml, and in leukemia patients without candidiasis, the mean concentration was 0.08 ng/ml. The mean plus 2 standard deviations for mannan in the serum of leukemia patients without candidiasis of 0.5 ng/ml was used as the upper limit of normal. With this concentration as test positive, 16 of 23 patients with invasive candidiasis had elevated mannan in serum. The sensitivity was thus 70%, and the specificity, defined as the percentage of patients without invasive candidiasis whose mannan in serum did not exceed 0.5 ng/ml, was 87%. If an upper limit of normal is taken as 1 ng/ml, the sensitivity of detection was 65%, and the specificity was 100%.

Clinical indications

The LA, ID, or CIE test for antibodies to *Candida* species should be applied to sera from patients with persistent candidemia, pneumonitis, endocarditis, wound or intraabdominal abscess, and indwelling urinary or intravenous catheters. Debilitated patients—those receiving immunosuppressive agents, prolonged courses of antibacterial antibiotics, or cytotoxic anticancer therapy—who are granulocytopenic and develop an unexplained fever should be tested for antibodies to *Candida* species and for antigenemia.

Serologic tests are frequently used to ascertain the clinical significance of *Candida* isolates. The detection of precipitins or the recognition of fourfold changes in agglutinin titers is considered presumptive evidence of systemic candidiasis. These findings can also indicate colonization or transient candidemia. The ID, CIE, and LA tests have a sensitivity of about 90% for proven candidiasis cases in immunologically intact hosts. The CIE and ID tests are the most specific. Extrageneric cross-reactions occur only with *Torulopsis glabrata* antisera. It should be noted that, in a controversial move, this taxon was transferred to the genus *Candida* as *C. glabrata*. In contrast, the LA test shows more nonspecific reactions. Sera from patients with cryptococcosis, torulopsosis, and tuberculosis have reacted with this test. The LA test is quantitative and appears to have prognostic value. When candidiasis is suspected and ID reactions are negative with *C. albicans* antigens, a test should be performed with *Candida krusei* antigen to rule out infection with this species (9).

A decision to treat a patient must not be based on serologic data alone, but rather on a consideration of all the available clinical and laboratory data.

The CIE test yields results essentially similar to those of the ID test, but in less time. The antigen used for the ID test is also used for the CIE test. However, the serum of each patient must be tested with dilutions of antigen to determine the optimum antigen concentration for testing with various dilutions of the patient serum. A fourfold rise in titer or a titer of 1:8 or greater is considered highly suggestive of invasive candidiasis (9).

Test 1. ID test

Medium (0.9% agar)

Sodium chloride	0.9 g
Sodium citrate ($Na_3C_6H_5O_7 \cdot 2H_2O$)	0.4 g
Phenol, liquified	0.25 ml
Glycine	7.5 g
Noble agar (or equivalent)	0.9 g
Distilled water	to 100 ml

Autoclave at 15 lb/in^2 for 10 min

Equipment

1. ID template for cutting wells with outer diameters of 6 and 8 mm. The pattern consists of three serum wells (8 mm) and two antigen wells (6 mm) placed laterally; the reactant reservoirs are placed at a distance of 10 mm.
2. Petri dishes, 140 mm in diameter, with covers
3. Vacuum sidearm flask with rubber tubing and Pasteur pipette
4. Glass slides, 50 by 75 mm
5. Humid chamber (petri dishes containing water-soaked filter paper may be used)

Procedure

1. Pipette 7.0 ml of hot molten agar onto each slide.
2. Cover the slides with petri dish halves, and allow the agar to solidify for approximately 20 min at room temperature.
3. With a template, cut out five wells.
4. Remove the agar plugs from the wells by suction with a Pasteur pipette connected to a vacuum source.
5. With separate Pasteur pipettes, add specimens to the wells as follows.
   a. Fill the center 8-mm well with the positive reference anti-*C. albicans* serum.
   b. Fill the upper 8-mm well with serum from a patient, and fill the lower 8-mm well with serum from another patient.
   c. Finally, fill the two 6-mm wells with *C. albicans* antigen.
6. After adding all the specimens to slides, incubate the specimens in a humid chamber at room temperature for 72 h.

Controls. A positive control serum containing at least three precipitins should be included in the test each time it is performed.

Reagents. Whole-cell antigens are prepared from 48-h glucose-peptone-yeast extract broth cultures grown at 37°C. A 1:4 suspension of the yeast cells in saline buffered with 2-amino-2-(hydroxymethyl)-1,3-propanediol (Tris hydrochloride, 0.05 M, pH 7.6) is disrupted in a Braun MSK homogenizer for 3 min and centrifuged for 30 min at 3,000 × *g*. The homogenate-supernatant antigens are collected and concentrated to yield a total protein concentration of 1.0 g/100 ml. Merthiolate is added to the antigen to a final concentration of 0.01% (wt/vol). The optimal antigen dilution is determined by testing various dilutions (1:1 to 1:8) prepared with PBS, pH 7.2, against a positive reference *C. albicans* antiserum. The optimal antigen dilution is the highest dilution that demonstrates distinct A, B, and C precipitin bands (9).

Satisfactory ID antigens may be obtained from Hollister-Stier Laboratories as a 1:10 extract of *C. albicans* in 50% glycerol-saline solution.

Interpretation. Sera from candidiasis cases that react with homogenate antigens of *C. albicans* in the ID test may produce between one and seven precipitates. The production of one or more lines by a serum interacting with antigen constitutes a positive reaction whether the antigen to which the antibody is directed is mannan or protein. Systemic candidiasis should be strongly suspected when serial serum specimens demonstrate serologic conversion (i.e., when negative antibody tests become positive) or show increases in the number of precipitins. In addition to the diagnosis of systemic *Candida* spp. infections, positive reactions may reflect colonization or infection due to *T. glabrata*.

Test 2. LA test

See reference 9 for further information.

Equipment and reagents

1. Glycine-buffered saline (GBS), pH 8.4
2. GBS–0.1% bovine serum albumin (BSA)
3. Polystyrene latex particle suspension, 0.81 μm (Difco and Dow Chemical Co., Indianapolis, Ind.)
4. *C. albicans* homogenate antigen
5. Rotary shaker
6. Glass slides, 50 by 75 mm (marked with 12 circles, each with a 1.5-cm diameter)

Procedure

1. Prepare a standardized suspension of sensitized latex particles.
2. Inactivate sera at 56°C for 30 min.
3. Prepare 1:4 dilutions of serum in GBS-BSA.
4. Include a negative control serum and a positive control serum that is known to give a 2+ reaction.
5. With a 0.1-ml pipette, add 0.02 ml of the optimally sensitized latex suspension to each of the circles on a slide.
6. With separate 0.1-ml pipettes, add 0.04 ml of each 1:4 dilution of the serum of the patient, the 1:4 dilution of the negative control serum, and the positive reference serum to the latex on the slide.
7. Rotate the slide at 150 rpm for 5 min on a rotary shaker.
8. Read macroscopically: the positive control must show 2+ agglutination (small but definite clumps with a slightly cloudy background); the negative control must show no agglutination. Inspect the reactions of the serum diluted 1:4 and record as positive all specimens showing agglutination equal to or greater than the 2+ positive reference serum.
9. All specimens positive in the screening test should be further diluted 1:8 to 1:64 and tested. If an endpoint is not reached, dilute the specimen to 1:1,024.
10. Record the endpoint of each specimen as the highest dilution of serum that gives a 2+ agglutination.

Controls. A positive control serum sample showing 2+ agglutination (small but definite clumps with a slightly cloudy background) and a negative control serum sample must be included each time the test is performed.

Reagents. See test 1 for candidiasis above. Perform a box titration by diluting the 1.0 g of protein homogenate antigen per 100 ml in GBS, pH 8.4, and adsorbing the diluted antigens to 0.81-μm latex particles. The sensitized particles are then tested against positive human serum controls. Select the highest dilution of antigen showing 2+ agglutination with the highest dilution of positive serum (diluted in GBS-BSA) and no reaction with normal human serum.

The antigen may be purchased from Hollister-Stier Laboratories as a 1:10 extract of *C. albicans*. This antigen should be dialyzed against GBS to remove glycerol, adjusted to a final protein concentration of 1.0 g/100 ml, and box titrated to obtain the optimal dilution for the sensitization of latex particles.

Interpretation. A serum titer of 1:8 or greater is considered presumptive evidence of systemic candidiasis. Patients whose sera are LA positive at 1:4 and demonstrate precipitins in the ID test are regarded as having possible early infections or as being colonized; a patient whose serum shows only a 1:4 LA titer may have early disease, be colonized, or show a nonspecific reaction. A serologic conversion from negative to positive (1:4) for agglutinins or a fourfold or greater increase in titer between serum specimens is considered presumptive evidence of infection. The sera of patients colonized by *Candida* spp. or *T. glabrata* may frequently show positive titers, but the recognition of heavy colonization can be important, since it sometimes precedes invasion (9). Fourfold declines in titer may denote the success of antifungal therapy or the elimination of colonization due to the removal of contaminated intravascular catheters or prosthetic valves.

Test 3. Double-antibody sandwich EIA to detect mannan antigenemia

See reference 21 for further information.

Antisera, conjugates, and mannan standards. Antiserum is produced in rabbits by weekly injections for 5 weeks in the marginal ear vein with the cell walls of *C. albicans* serotype A (22), 0.5 mg/0.5 ml of 0.85% NaCl. The concentration of anti-mannan IgG is assessed by indirect EIA (14), with mannan polysaccharide extracted and purified from whole *C. albicans* blastoconidia as the antigen adsorbed to the solid phase (15). The IgG fraction of the antiserum is conjugated to horseradish peroxidase and stored in vaccine bottles at 2.5 mg/ml at or below 40°C, as a 50% (vol/vol) PBS solution in glycerol. Another portion of the IgG is adjusted to 10 mg/ml to use as the capture antibody. These reagents are not commercially available. Inquiries about the acquisition of reference reagents should be directed to the Division of Mycotic Diseases, Center for Infectious Diseases, CDC. Known mannan standards in normal human serum covering the range 1.6 to 50 ng/ml are produced by the appropriate dilution of a stock solution of 0.1 mg of mannan per ml of 0.85% NaCl. The serum samples are immediately dissociated (see below) and stored at or below −40°C. Positive control sera can be obtained from rabbits immunosuppressed with cortisone and infected with *C. albicans* (5).

Dissociation of mannan-serum complexes by boiling the reagents. (i) Disodium EDTA solution, 0.1 M, pH 7.2. To prepare 100 ml of solution, add 0.01 mol of disodium EDTA (= 3.72 g; Baker #8993-1 or equivalent) to about 50 ml of distilled (or deionized) water (dH_2O) with a pH electrode immersed in the beaker, and stir the mixture with a magnetic spin bar. Add 1 M NaOH dropwise to dissolve the disodium EDTA and to adjust the pH to 7.2. Adjust the final volume to 100 ml. Store at 4°C and discard after 6 months.

(ii) Procedure. Combine 1 ml of serum and 1 ml of disodium EDTA solution, pH 7.2, in a 10-ml polycarbonate Oak Ridge tube (Nalgene no. 3118-0010). Boil the mixture for 3 min in a water bath. Cool it.

Centrifuge it at 13,000 rpm, 30 min, in a Sorvall RC-2 centrifuge or equivalent. Aspirate the supernatant, transfer it to a labeled 1-dram vial, and store it at −40°C.

Coating plates with capture antibody for sandwich EIA. (i) Materials.

Microtiter plates. Immulon II MicroELISA, flat bottom, 96 wells, no. 011-010-3450 (Dynatech Laboratories, Inc., Alexandria, Va.); plastic plate sealers (no. 1-200-30, Dynatech).

Carbonate buffer. 0.06 M, pH 9.6 (3.81 g of $NaHCO_3$, 1.93 g of Na_2CO_3, 500 ml of H_2O). Adjust the pH to 9.6 with 1 M NaOH, and dilute to 1 liter.

Stock solution of anti-*C. albicans*–cell walls–IgG (stored at or below −40°C at an IgG concentration of 10 mg/ml).

(ii) Procedure

1. Dilute 25 μl of IgG stock solution in 100 ml of 0.06 M carbonate buffer (1/4,000 dilution). This solution will be sufficient for five plates. Add 0.2 ml to each well. Wells on the outer edges will, however, not be used for unknowns or standards because of possible edge effects.

2. Incubate the coated plates for 3 h at 37°C, and then store them in the cold at 4°C (label and date them and seal them with plastic sealer). The plates will retain activity for 1 month. The optimal dilution for each IgG preparation is determined separately; in our experience, a 1/1,000 to 1/4,000 dilution of the IgG has proved optimal.

Performance of the EIA. (i) Equipment. The contents of the wells are aspirated with an eight-channel stainless steel manifold (Bellco Glass, Inc., Vineland, N.J.) into a 2-liter sidearm vacuum flask.

Washing device. The washing device is composed of a 2-ml Cornwall repeating, dispensing syringe fitted with an identical eight-channel stainless steel manifold. Rinse the device thoroughly after use with first tap and then distilled water. Store the pipettor filled with distilled water. Prime the pipettor with PBS plus Tween 20 (PBS-T) before each use.

Filling aid. An eight-channel manifold pipettor (Flow Laboratories, Inc., McLean, Va.) with volumes adjustable from 50 to 200 μl is almost indispensable. It is fitted with disposable plastic tips for each solution. When set at about 25 μl, it is also used to dispense the 4 M H_2SO_4.

EIA reader. MicroELISA MR600 (Dynatech) or equivalent with the 490-nm filter in place. If the reader is interfaced with a Hewlett-Packard 86 computer, a software diskette with complete data reduction capability can be provided.

(ii) Chemicals and reagents. *o*-Phenylenediamine (Aldrich Chemical Co., Inc., Milwaukee, Wis.). Stock solution: 10 mg/ml of methanol. Store the solution cold in a dark bottle, i.e., a 5-dram (18-ml), screw cap, amber glass vial. Discard the solution when it becomes yellow tinged, and make fresh stock solution twice weekly or before each use.

Hydrogen peroxide (H_2O_2), 3% aqueous (Mallinckrodt, Inc., St. Louis, Mo.). Store at 4°C. Shelf-life, 1 year.

Sulfuric acid, 4 M.

Plates coated with IgG (see above).

Conjugate. Peroxidase-labeled anti-*C. albicans* IgG (conjugate): 50 ml of PBS-T plus 10 μl of conjugate = 1/5,000 optimal dilution. The optimal dilution is determined for each batch of conjugate.

PBS-T. 0.01 M, pH 7.2. 0.14 M NaCl prepared from a 10× stock solution made as follows. A total of 10.96 g of Na_2HPO_4, 3.15 g of $NaH_2PO_4 \cdot H_2O$, and 85 g of NaCl are dissolved in 1 liter of deionized water. The stock solution is sterilized and stored at ambient temperature in screw cap bottles. For use, a 10-fold dilution is made in deionized water (2-megohm resistance), and Tween 20 (polyoxyethylene sorbitan monolaurate; Fisher Chemical Co., Fair Lawn, N.J.) is added to give a final 0.05% concentration. The diluted buffer is stored at 4°C and made fresh weekly.

(iii) Controls. For controls, mannan standards (1.6 to 50 ng/ml) are added to normal human serum samples and dissociated (5). Dissociated serum samples from mannanemic rabbits (see below) are used. PBS-T serves as an additional blank. Human antigenemic standards are not usually available in sufficient quantity for routine use.

(iv) Procedure

1. Remove the IgG-coated plate from 4°C storage and wash it three times with 0.25 ml of PBS-T per well; for the fourth wash, let the buffer stand in the wells for 5 min. Aspirate the contents. Strike the plate vigorously on an absorbent paper towel after each wash. Consult Fig. 1 for the layout of samples, standards, and controls.

2. Add dissociated serum at 0.2 ml per well in triplicate. Include triplicate mannan–dissociated-serum standards, normal human dissociated serum (negative control), and dissociated serum from a mannanemic rabbit (positive control). Cover the plate with another microtiter plate and incubate it for 30 min at 4°C. Aspirate the contents of the wells to avoid aerosols.

3. Wash the plate three times with PBS-T, and the fourth time, let the plate stand for 5 min as above.

4. Add appropriately diluted, peroxidase-labeled anti-*C. albicans* IgG-conjugate, 0.2 ml per well, and incubate the plate at 4°C for 30 min.

5. Wash the plate as described above.

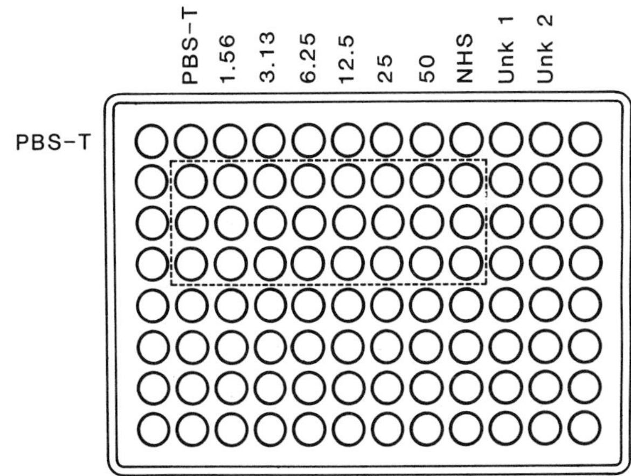

FIG. 1. Microtitration plate layout for samples and mannan standards in the EIA for mannan antigenemia. NHS, Normal human serum; Unk, unknown; PBS-T, phosphate buffered saline plus Tween 20. Dashed line indicates the boundary of the wells occupied by triplicate standards.

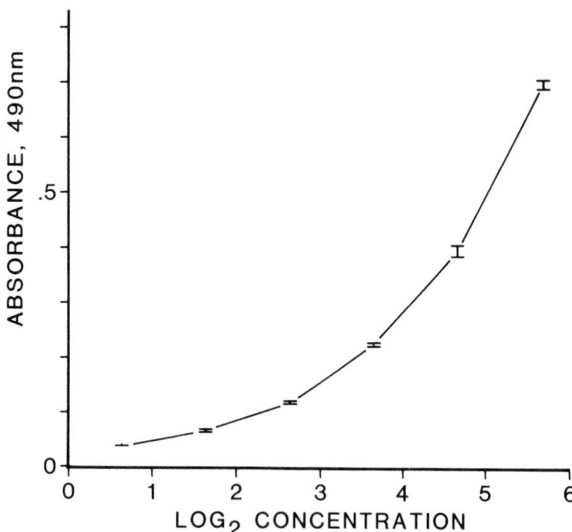

FIG. 2. Standard curve (semilog$_2$) of mannan in serum (1.56 to 50 ng/ml) in the EIA for antigenemia. Triplicates at each mannan concentration.

6. Prepare the substrate solution by mixing the stock solution of OPD, *o*-phenylenediamine–3% H_2O_2–dH_2O, in the proportions 0.5 ml:0.1 ml:50 ml. Add 200 μl of this solution to each well, and incubate the plate for 1 h at room temperature in the dark.

7. Add 25 μl of 4 M H_2SO_4 per well to stop the reaction. (Protect the eyes.)

8. Scan the plate on an ELISA reader at 490 nm.

9. Plot a graph of absorbance (*y* axis, log scale) versus log concentration (*x* axis) (Fig. 2 and 3).

10. Calculate nanograms of mannan per milliliter with respect to the standard curve.

Calculating mannan concentrations

1. Obtain the mean of triplicate samples for each mannan concentration for the normal human serum blanks and for the unknowns. Subtract the A_{490}, if any, of the blank from each mannan-containing sample and from the unknown.

2. Plot the standard curve, A_{490} versus mannan concentration, on graph paper.

3. Interpolate the value of the unknown with respect to the standard curve.

4. The significance of pertinent positive and negative controls for the EIA are reviewed as follows.

The wells receiving the normal human serum serve as blank controls for the background. Typically, these wells develop no color. The absorbance of mannan concentrations spiked into serum allows the construction of a standard curve and controls for run-to-run variation. The in vivo mannanemia sample, obtained from infected rabbits, more closely represents mannan in the sera of patients and provides another benchmark for run-to-run variations.

A computer program that prints the standard curve, calculates the concentration of mannan in unknown samples, and prints the results, including mean A_{490}, mean mannan concentration (in nanograms per milliliter), and the coefficient of variation, is available in a diskette (5 1/4 in. [ca. 13.3 cm]) for use with the Hewlett-Packard 86 computer, through the Division of Mycotic Diseases, CDC. A sample mannan determination is shown in Table 1, giving the A_{490} in each well. Table 2 shows how the computer program reduces the data and computes variation in the standard curve. Table 3 shows how the absorbance values of the standards and unknowns are converted into antigen concentrations in nanograms per milliliter with respect to the best-fit line through all points of the standard curve. These computations are determined by the Hewlett-Packard 86 computer program. Note that small deviations occur between the nominal mannan concentrations of standards spiked into sera and the values recorded with respect to the standard curve.

Interpretation. Heat-stable mannan-polysaccharide antigens have been detected by EIA in the sera of immunosuppressed rabbits infected with *C. albicans* (4) and in humans (de Repentigny et al., submitted for publication). The sensitivity of detection is 65 to 70% in human cancer patients. Specificity is 100%. Since mannan is not a normal serum constituent, concentrations greater than 2 ng/ml are presumptive evidence of infection. The sensitivity limit of the mannanemia EIA is 1 ng/ml of serum. Mannans of *C. albicans* A and *Candida tropicalis* are detected. Mannanemia may be a transient event, and close monitoring of high-risk patients is warranted. Patients receiving systemic immunosuppressive drugs who are profoundly granulocytopenic and have a fever of unknown origin despite broad-spectrum antibacterial therapy should be monitored twice weekly. At the same time, blood cultures should be drawn (de Repentigny et al., submitted for publication).

Coccidioidomycosis

The CF and tube precipitin (TP) tests are valuable aids in the determination of the diagnosis and prognosis of coccidioidomycosis. The two tests measure at least two different antigen-antibody systems. The antigen responsible for the evocation of IgM in the TP

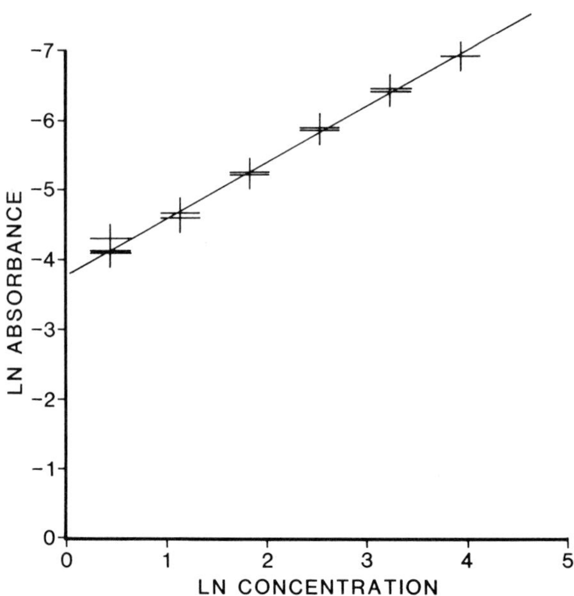

FIG. 3. Logarithmic transformation (ln-ln) of the standard curve shown in Fig. 2 generated by the Hewlett-Packard 86.

TABLE 1. A_{490} for mannan standards, normal human serum control, and unknowns[a] in the double-antibody sandwich enzyme immunoassay for antigenemia

Well		Mannan standard (ng/ml)[b]						NHS[c]	Control[d]			
		1.56	3.13	6.25	12.5	25	50					
A	0.000	0.000	0.000	0.000	0.000	0.000	0.000	0.000	0.000	0.000	0.000	0.000
B	0.000	0.076	0.125	0.264	0.454	0.691	1.041	0.000	0.124	0.001	0.000	0.002
C	0.000	0.073	0.126	0.245	0.424	0.710	1.059	0.000	0.141	0.000	0.000	0.003
D	0.000	0.069	0.117	0.243	0.424	0.640	1.050	0.000	0.114	0.000	0.000	0.010
E	0.000	0.000	0.013	0.560	0.029	0.005	0.522	0.030	0.000	0.271	0.000	0.000
F	0.000	0.000	0.021	0.614	0.042	0.004	0.498	0.028	0.000	0.269	0.000	0.000
G	0.000	0.000	0.000	0.383	0.003	0.000	0.508	0.023	0.000	0.251	0.000	0.000
H	0.000	0.000	0.000	0.000	0.000	0.000	0.000	0.000	0.000	0.000	0.000	0.000

[a] Unknowns, wells B, C, and D in columns 4, 7, 8, and 10. Columns 1, 11, and 12 not used.
[b] Wells B, C, and D.
[c] NHS, Normal human serum. Wells B, C, and D.
[d] Positive mannanemic control (rabbit). Wells B, C, and D.

test has been identified as factor 2 of coccidioidin, a protein that is stable to heat (60°C, 30 min), alkali, and trypsin (3). The TP test is most effective in the detection of an early primary infection or an exacerbation of existing disease. It is most frequently used in areas in which the disease is endemic. The CF procedure is the most widely used serologic test for coccidioidomycosis, and its reacting antibodies persist for longer periods than do those reactive in the TP test. Smith et al. (23) found that the combination of CF and TP tests yielded positive results in over 90% of the primary symptomatic cases of coccidioidomycosis. Screening tests (9), such as the latex particle agglutination (LPA) and ID tests, which yield results comparable to the TP and CF tests, respectively, can be used by those laboratories not in a position to perform the TP or CF test. A variation of the ID test, with a heated, toluene-induced lysate of mycelium as antigen, has also been shown to yield qualitative results that correlate with those obtained with the TP test. This test, referred to as the IDTP procedure, is recommended as a screen for the detection of sera that may yield a positive TP test (9). The parallel use of heated and unheated coccidioidin in ID tests readily permits the detection and distinction of heat-labile precipitins and complement-fixing antibodies reactive with the F antigen. In addition, an 8- to 10-fold concentration of clinical specimens appears to enhance the detection of these antibodies (18).

Clinical indications

Serologic tests for coccidioidomycosis should be considered whenever patients display symptoms of pulmonary or meningeal infection and have lived or traveled in areas in which *C. immitis* is endemic. These tests should be used particularly when such patients demonstrate sensitivity to a coccidioidin or spherulin skin test.

Serum precipitins may be detected within 1 to 3 weeks after the onset of primary infections in a large percentage of cases in which CF tests have not yet become positive. Precipitins are diagnostic but not prognostic. They are rarely detected 6 months after infection, but they can reappear if the infection spreads or relapse occurs. Precipitins may persist in disseminated cases. They are rarely found in the cerebrospinal fluid (CSF) of patients with coccidioidal meningitis. The TP test is of little value in the analysis of CSF specimens. The CF test becomes positive later than the precipitin test. It is most effective in the determination of disseminated disease. The CF titer results parallel the severity of the infection (9); titers rise as the disease progresses and decline as the patient improves.

Qualitative data similar to those obtained in the TP and CF tests may be obtained from the screening LPA and ID procedures, respectively.

TABLE 2. Statistical analysis of the standard curve for the determination of mannan concentrations in serum[a]

Mannan standard (ng/ml of serum)	Mean A_{490}	SD	CV (%)[b]
1.56	0.073	0.004	4.8
3.12	0.123	0.005	4.0
6.25	0.251	0.012	4.6
12.5	0.434	0.017	4.0
25.0	0.680	0.036	5.3
50.0	1.050	0.009	0.9

[a] Slope, 0.79; y intercept, −2.92; coefficient of determination, 0.9944.
[b] CV, Coefficient of variation.

TABLE 3. Mannanemia EIA interpolations of standards and unknowns with respect to the standard curve of mannan in serum

Specimen	Antigenemia (ng/ml)	Mean A_{490}	SD	CV (%)[a]
Mannan standard (ng/ml)				
1.56	1.46	0.073	0.004	4.8
3.13	2.86	0.123	0.005	4.0
6.25	7.1	0.251	0.012	4.6
12.5	14.3	0.434	0.017	3.9
25.0	25.3	0.680	0.036	5.3
50.0	44.1	1.050	0.009	0.8
Positive mannanemic control	2.97	0.126	0.014	10.8
Unknowns[b]				
1	17.9	0.519	0.121	23.2
2	17.5	0.509	0.012	2.3
3	0.4	0.027	0.004	13.3
4	7.5	0.264	0.011	4.1

[a] CV, Coefficient of variation.
[b] Infected rabbits.

Test 1. TP test

The TP test is performed with two dilutions of coccidioidin and constant amounts of serum (9). The antigen dilutions are used to obviate the possible occurrence of a false-negative result due to inhibition of the precipitin reaction by excess antigen.

Reagents and equipment

1. Undiluted serum containing Merthiolate diluted 1:10,000
2. *C. immitis* precipitinogen (coccidioidin), both undiluted and diluted 1:10, with a final concentration of 1:10,000 Merthiolate added as a preservative
3. Saline buffered at pH 7.0 with 0.067 M phosphate buffer for use as a diluent and a control containing Merthiolate at a concentration of 1:10,000
4. Culture tubes, 7 by 70 to 75 mm

Procedure

1. Add 0.2 ml of test serum to each of three tubes.
2. Add 0.2 ml of undiluted antigen to serum in test tube 1; add a 1:10 antigen dilution to serum in test tube 2.
3. Add 0.2 ml of control saline to the test specimen in test tube 3.
4. Thoroughly mix the contents of each test tube.
5. Incubate the test tubes at 37°C, and read them daily for 5 days by sharply flicking the bottom of each tube while holding the top of the tube between thumb and forefinger.
6. A button or a flake of precipitate in test tube 1 or 2 is a positive test. The TP test may be applied to sera and pleural fluids.

Reagents. The antigens for both the TP and CF tests are filtrates of mycelial cultures of multiple or single isolates of *C. immitis*. Coccidioidin is prepared by a variety of procedures. In the most widely known procedure, filtrates are produced from cultures grown in a synthetic asparagine-glycerol-salts medium originally devised for tuberculin production. The preparation of coccidioidin in this medium usually requires incubation for several weeks at room temperature. Coccidioidin antigens can be prepared within 1 week by a toluene lysis technique (9).

Heating coccidioidin at 60°C for 30 min destroys the F antigen responsible for the CF activity, but the precipitinogen associated with the TP reaction is retained (9).

The LPA test uses latex particles sensitized with coccidioidin heated at 60°C for 30 min. LPA kits may be obtained from Meridian Diagnostics. The LPA test should not be applied to CSF or to diluted sera, because one is likely to encounter false-positive reactions. It is not known whether such difficulties would occur with pleural, joint, or ascitic fluid (9).

Interpretation. Early coccidioidomycosis is usually detected by the TP, IDTP, and LPA tests. A positive TP test, indicated by the appearance of a precipitated button or flake in any dilution, is considered diagnostic. In about 80% of all infections, the TP test becomes positive within 2 weeks of the onset of symptoms. It is diagnostic but not prognostic. Precipitins are infrequently detected 6 months after infection.

The IDTP test is more sensitive than the TP test but less sensitive than the LPA test. The LPA test, however, is not as specific as the IDTP and TP tests; approximately 6 to 10% false-positive reactions may occur with the LPA test. The LPA test, however, may become positive before the TP test. A positive LPA reaction must be confirmed by a TP or CF test. The LPA results can be obtained in 4 min (9).

Test 2. CF test

The standardized LBCF test with coccidioidin is recommended for the titration of sera from patients with suspected coccidioidomycosis. Details for the performance of the test have been published previously (18). In addition, see the blastomycosis tests described above. The microadaptation of the CF test for coccidioidomycosis gives results comparable to those of the macrotest (9). The CF tests may be performed on serum, CSF, plasma, and pleural and joint fluids.

Coccidioidin is prepared as described above for the TP test. Since the CF antigen is destroyed by heating at 60°C for 30 min, this antigen should not be heated. The CF antigen may be purchased from Immuno-Mycologics, M. A. Bioproducts, or Meridian Diagnostics.

The qualitative ID test with unheated coccidioidin yields results comparable to those obtained with the CF test. The concentration of sera before ID and CF tests are performed can be useful in the detection of specimens of chronic cases which might ordinarily be missed in such tests (19). Recent studies suggest that titers similar to those obtained in CF tests may also be obtained in quantitative ID tests (27).

Spherulin, an extract of in vitro-produced spherules of *C. immitis*, appears to be as sensitive as coccidioidin in the CF test but less specific (9).

Interpretation. Any CF titer with coccidioidin should be considered presumptive evidence for *C. immitis* infection. The ID test with a filtrate antigen (9) gives results that correlate with those observed with the CF test. The ID test with reference reagents is highly specific. Titers of 1:2 and 1:4 in the CF test usually indicate early, residual, or meningeal coccidioidomycosis. However, sera that demonstrate such titers have also been obtained from patients not known to have coccidioidomycosis. The parallel use of CF and ID tests is an effective means for the specific diagnosis of coccidioidomycosis in patients with low levels of complement-fixing antibodies. Studies indicate that sera positive in the CF test at the 1:2 to 1:8 range and also positive in the ID test reflect active or recent *C. immitis* infections (9). Obviously, when low titers are obtained, a diagnosis of coccidioidomycosis must be based on subsequent serologic tests and preferably on clinical and mycologic studies. Generally, CF titers greater than 1:16 indicate disseminating disease. Negative serologic test results do not exclude a diagnosis of coccidioidomycosis. About 5% of all spinal fluid specimens from patients with coccidioidal meningitis are negative in the CF test, and serum samples from many patients with chronic cavitary coccidioidomycosis are negative.

The coccidioidin skin test is considered a valuable screen for serologic testing. Conversion from a negative to a positive skin test reaction is pathognomonic and is usually the earliest immunologic response to infection. Studies have indicated a strong positive correlation in patients with primary coccidioidomycosis (without impending or concomitant disseminating disease) between serologic positivity and a positive coccidioidin skin test. Unlike the serologic reactions sometimes noted after the histoplasmin skin test, coccidioidin and spherulin skin tests do not elicit an antibody response to coccidioidin in previously sensitized persons (9).

Cryptococcosis

Conventional methodology for the diagnosis of cryptococcosis is time consuming and, in many cases, inadequate. Until recently, persons suffering from cryptococcosis were considered to be essentially immunologically inert. Previous immunologic tests had only limited applications (9), and even then results were difficult to interpret. During the last two decades, work on serologic procedures for cryptococcosis has resulted in the development of diagnostically and prognostically useful tests. These procedures are an indirect fluorescent-antibody (IFA) technique, an EIA, and a tube agglutination (TA) test for cryptococcal antibodies and a LA test and an EIA for cryptococcal antigen (9, 10). The antibody tests are of value in the detection of early or localized cryptococcosis and in the determination of a prognosis. These tests are, however, less specific than the LA test. More recently, an EIA with *Cryptococcus neoformans* galactoxylomannan antigen was evaluated for its ability to detect IgM antibody. Such antibodies were found in approximately 22% of 55 cryptococcosis patients examined (20). Cross-reactions similar to those noted with IgG antibodies with the IFA test, although markedly reduced, were still evident.

The EIA potentially can detect cryptococcal antigen earlier and at lower concentrations than the LA test can. The EIA is also not subject to prozone reactions. Preliminary EIA tests for cryptococcal antigens detected 6 ng of the major viscous capsular polysaccharide per ml, in contrast to 35 ng/ml detectable by the LA test. The EIA, however, requires a few hours to perform, in contrast to the few minutes needed for the simple LA test (10). Because the LA test is very specific, has both diagnostic and prognostic value, and is simple to perform and widely used, this procedure is herein described in detail.

Clinical indications

Serologic tests for *C. neoformans* antigens or antibodies or both should be considered with patients who have symptoms of pulmonary or meningeal infection. Cutaneous, skeletal, and visceral involvement occurs as the result of dissemination. The disease may be primary, but many cases are associated with various debilitating diseases, such as Hodgkin's disease, AIDS (acquired immune deficiency syndrome), leukemia, or diabetes.

It is interesting to note that the CDC AIDS Task Force has documented 5,091 cases of AIDS as of July 1984. Of these, 333 cases (6.5%) demonstrated complications from cryptococcal infection, mainly of the meningeal form (personal communication, Richard Selik).

The IFA and TA antibody tests are reactive with about 50% of the sera from patients with active cases of extrameningeal cryptococcosis. The IFA test has a specificity of about 77%, whereas the TA test has a specificity of about 89% (9). Although the IFA test is not entirely specific, it is valuable in the detection of those cases of cryptococcosis that are negative for *C. neoformans* agglutinins and antigens. The TA test for cryptococcal antibodies has been found to be diagnostically reliable. Agglutinins were detected in the early stages of central nervous system infection and in infections with no central nervous system involvement (9).

The LA test has been successfully used for the specific detection of cryptococcal antigen in sera and CSF from persons with proven cryptococcosis. The test is valuable in the diagnosis of active nonmeningeal and meningeal cryptococcosis, particularly the latter. Of 330 patients with proven meningeal cryptococcosis who were recently studied, 328 (99%) had spinal fluids positive for cryptococcal antigen by the LA test (9). The test is also more sensitive in the diagnosis of cryptococcal meningitis than is the India ink stain for *C. neoformans* yeast cells in the spinal fluid.

LA titers of 1:8 or greater are considered strong evidence of active infection. Titers of 1:4 or less, although diagnostic in many instances, have been demonstrated in sera and CSF specimens of symptomatic patients without corroborating culture or histologic evidence of cryptococcosis (9). False-positive serum reactions are uncommon and occur mainly with sera from some patients with severe rheumatoid arthritis. These specimens may be recognized by their reactivity with latex sensitized with normal globulins or by their inactivation after treatment with 0.003 M dithiothreitol (9). Nonspecific agglutination has also been noted in CSF (9) and sera from patients without rheumatoid arthritis. False-positives due to rheumatoid factor or other interfering proteins in sera may be eliminated by boiling the specimens with disodium EDTA (10) or by treatment of the specimens with a final concentration of 5 mg of pronase per ml (24).

Cryptococcosis is best diagnosed serologically through the concurrent use of three tests: the LA test for antigen and the IFA and TA tests for *C. neoformans* antibodies (9).

Test 1. LA test for cryptococcal antigen

Equipment and reagents

1. GBS–0.1% BSA, pH 8.4
2. Polystyrene latex particle suspension, polystyrene spheres (0.81-μm diameter; Difco and Dow)
3. Latex particles optimally sensitized with rabbit anti-*C. neoformans* globulin (LI)
4. Latex sensitized with rabbit normal (preimmune) globulin (LN)
5. Sera
   a. Positive human cryptococcosis case reference serum with a known antigen titer (expressed as the highest dilution that demonstrates 2+ agglutination)
   b. Negative control human serum
   c. Human serum positive for rheumatoid factor (negative for *C. neoformans* antigen)
   d. Serum, CSF, or urine specimen from the patient
6. Water bath, 56°C
7. Rotary shaker
8. Glass slides, 50 by 75 mm (marked with 12 circles, each 1.5 cm in diameter)

9. Microtitration droppers or microliter pipettes, 0.025 and 0.05 ml

Procedure

1. Inactivate sera and CSF specimens at 56°C for 30 min. Urine specimens should be inactivated by being heated in boiling water for 10 min. CSF specimens may be similarly treated, if necessary, and then retested.

2. Place positive and negative control sera and the specimen from the patient in a test tube rack in the order to be tested.

3. Add a 0.025-ml drop of LI reagent to each of the circles on a slide.

4. Add 0.05-ml drops of positive and negative control sera and up to 10 undiluted specimens from patients to the drops of the LI reagent in separate circles. Mix the drops.

5. Place the slide on a rotary shaker and rotate it at 125 ± 25 rpm for 5 min.

6. Read the test immediately for macroscopic agglutination by visual inspection over a dark background: the positive control must show 2+ agglutination (small but definite clumps with a slightly cloudy background); the negative control must show no agglutination. Check the reactions of the specimens from the patient, and record as positive all reactions showing agglutinations equal to or greater than the 2+ positive control serum.

7. Test all serum and CSF specimens that are positive in the screening test with the LN control reagent to rule out false-positive reactions due to rheumatoid factor or other interfering proteins. Include the rheumatoid factor-positive serum as a control.

8. Serially dilute all specimens positive with the LI reagent and negative with the LN reagent to make 1:2, 1:4, and 1:8 dilutions, etc. Prepare the dilutions in GBS-BSA.

9. Record the titer of each specimen as the highest dilution that gives a 2+ agglutination.

10. Consider the results of tests in which sera react with both the LI and LN reagents as equivocal. Then proceed to treat the specimen to eliminate any interfering protein, and retest for the presence of antigen.

Controls. A positive control serum showing 2+ agglutination (small but definite clumps with a slightly cloudy background) and a negative control serum must be included each time the test is performed. Rheumatoid factor in serum may interfere with the test. To avoid false-positive results due to rheumatoid factor or other interfering proteins, sera positive with the LI reagent should always be tested with the LN reagent.

Reagents. A properly standardized suspension of latex particles, with an absorbance of 0.30 ± 0.02 when diluted 1:100, is sensitized with an optimal dilution of 4% rabbit anti-*C. neoformans* globulins. Similarly, such a standardized suspension is sensitized with a dilution (the same as for LI) of preimmune 4% rabbit immunoglobulin obtained from the rabbit(s) later used to produce the anti-*C. neoformans* globulin.

Kits for the detection of *C. neoformans* antigen in clinical specimens are available from American Scientific Products, McGaw Park, Ill., M. A. Bioproducts, Meridian Diagnostics, and Wampole Laboratories, Inc., Cranbury, N.J.

Interpretation. The LA test for *C. neoformans* antigen has both diagnostic and prognostic value. A positive reaction in the serum or CSF of an untreated patient at titers of 1:4 or less is highly suggestive of cryptococcal infection. LA titers of 1:8 or greater usually indicate active cryptococcosis. The antigen titer is usually proportional to the extent of infection, with increasing titers reflecting progressive infection and a poor prognosis and declining titers indicating a response to chemotherapy and progressive recovery. Failure of the titer to fall during therapy suggests inadequate treatment (9).

LA tests in which sera react with both LI and LN reagents should be considered equivocal. Since cryptococcosis and arthritic conditions may occur concomitantly, tests with both LI and LN reagents should be performed. A fourfold or greater titer with the LI reagent suggests cryptococcosis, but additional specimens taken subsequently should be examined for titer change.

The controlled LA test appears to be highly specific. Some researchers, however, have reported occasional false-positive reactions at low dilutions, particularly in CSF (9). A negative reaction should not exclude a diagnosis of cryptococcosis, especially when only one specimen has been tested and the patient shows symptoms consistent with those of cryptococcosis. False-negative reactions are uncommon. The few that have occurred were associated with a prozone in specimens with high antigen titers, specimens from patients infected with nonencapsulated or dry variants of *C. neoformans*, and specimens obtained in the very early stages of infection. Weakly reactive undiluted sera (±, +) should be checked for prozone reactivity by testing higher dilutions of sera. Specimens may also be negative due to immune complex formation. Dissociation of such complexes with pronase will free antigen so that it can be detected by antibody-coated latex particles (Glenn Roberts, personal communication).

Test 2. IFA test for *C. neoformans* antibody

Heat-killed *C. neoformans* serotype A cells are heat fixed to a slide and covered with a 1:16 dilution of the heat-inactivated serum. After incubation, the preparation is washed, air dried, and treated with anti-human immunoglobulin conjugated to fluorescein isothiocyanate. A positive reaction is indicated by the cells staining to an intensity of 2+ or greater.

Test 3. TA test for *C. neoformans* antibody

Formalin-killed whole yeast cells heated at 56°C for 30 min are used in the agglutination test (9). The cells are adjusted to a concentration of 7.5×10^6 cells per ml, and 0.25-ml volumes of serial twofold dilutions of serum (inactivated at 56°C for 30 min) are mixed with equal volumes of yeast cells. The mixtures are placed on a rotary shaker for 2 min, incubated at 37°C for 2 h, and then refrigerated at 4°C for 72 h, during which time readings are taken at 24-h intervals. The serum titer is the highest dilution that shows any degree of agglutination. The antibody tests are performed only on serum. Positive and negative controls must always be included in each run.

Reagents for the antibody test are not commercially available.

Interpretation. A positive antibody test is suggestive of infection by *C. neoformans* and could also reflect a past infection or a cross-reaction. Antibodies may be detected in the early course of the disease and in localized infections. As the disease progresses, abundant antigens may be produced and detected, with concurrent exclusion of antibody. The antibody test may have prognostic value. With effective chemotherapy, the antigen titer declines, and antibody may become demonstrable by one or both tests.

Histoplasmosis

Serologic evidence is often the prime factor responsible for a definitive diagnosis of histoplasmosis. Such evidence can be obtained through CF, ID, and LA tests, used singly or in some combination. Of these procedures, the most widely used is the CF test. Properly performed, as either a tube or a microtitration procedure (9), it can yield information of diagnostic and prognostic value. Over 90% of the culturally proven cases of histoplasmosis may be positive by the CF test if the illness of the patient is monitored by testing sera collected at 2- to 3-week intervals (9). Unfortunately, CF tests are relatively complex and expensive and should be performed only by highly trained technicians.

Clinical indications

Serologic tests for histoplasmosis should be applied to clinical specimens (serum, plasma, peritoneal fluid, or CSF) from patients with respiratory illness, hepatosplenomegaly, signs of extrapulmonary systemic infection, or meningeal involvement. The history of the residence, travel, and occupation of the patient may also be used as a guide for the application of these tests. The CF test is very sensitive; however, with currently available antigens, the test is not entirely specific. Cross-reactions may occur with sera from patients with blastomycosis, coccidioidomycosis, and other fungal infections. Sera from patients with leishmaniasis may cross-react in the CF test when *H. capsulatum* yeast forms are used as the antigen. In addition, positive reactions cannot be obtained with anticomplementary specimens. The histoplasmosis ID and CF tests, with histoplasmin as antigen, will react with about 85% of histoplasmosis sera. The CF test with yeast-form antigen has greater sensitivity (9). This test with the yeast-form antigen should be used in the diagnostic laboratory, and where possible, it should be supplemented with either the ID or CIE test with histoplasmin (9). There is a greater than 90% agreement between results obtained with the ID and CIE tests. The latter tests are very useful for the examination of anticomplementary sera. Because of their greater specificity, they provide a more accurate diagnosis with those sera that cross-react in CF tests.

The histoplasmin LA test is satisfactory for the detection of acute primary infections, but it may be negative with sera from persons with chronic histoplasmosis (9). It is particularly valuable in the detection of early disease. Because of the transitory nature of these agglutinins, the LA test cannot be considered a replacement for the CF test, especially with the intact yeast-form antigens.

Test 1. CF test

The standardized LBCF test with *H. capsulatum* yeast-form cells and histoplasmin antigens is recommended for the titration of sera from persons with suspected cases of histoplasmosis. Details for the performance of the test have been previously published (18). In addition, see the CF test procedure described above for blastomycosis. The microadaptation of the CF test for histoplasmosis gives results comparable to those of the macrotest (9).

Reagents. Two antigens are used in the CDC LBCF test. One is a suspension of Merthiolate-treated, intact, yeast-form cells of *H. capsulatum*, and the other is a soluble, mycelial-form filtrate antigen, histoplasmin, harvested after growth of the fungus for approximately 6 months in Smith asparagine medium (9). The optimal dilution for each antigen is determined by a block titration with low- and high-titer positive human sera. These antigens may be purchased from American Scientific Products, Immuno-Mycologics, M. A. Bioproducts, and Meridian Diagnostics.

Interpretation. CF tests are valuable in the diagnosis of acute, chronic, disseminated, and meningeal histoplasmosis. Antibodies in primary pulmonary infections are generally demonstrable within 4 weeks after exposure to the fungus or, frequently, by the time symptoms appear. These antibodies are usually antibodies to the yeast form of *H. capsulatum*. Antibodies to histoplasmin usually develop later in primary pulmonary cases, but titers are considerably lower than those with the yeast-form antigen. Histoplasmin titers are usually higher in sera from chronic cases. CF test results can be difficult to interpret, because cross-reactions or nonspecific reactions with the yeast or histoplasmin antigens are often encountered. In such instances, titers usually range between 1:8 and 1:32 and occur mainly with the yeast-form antigen. Many serum samples from culturally proven cases of histoplasmosis, however, give titers in the same range. Consequently, titers of 1:8 and greater with either antigen are generally considered presumptive evidence of histoplasmosis. Titers above 1:32 or rising titers offer strong presumptive evidence of histoplasmosis. The probability of infection increases in proportion to the height of the CF titer.

Nonetheless, one cannot rely solely on CF titers above 1:32 as a means of diagnosis, since false-positive reactions of that magnitude may occur in patients with other diseases. Titer changes are often of great assistance in the diagnosis of histoplasmosis. Fourfold changes in titer in either direction are significant indicators of disease progression or regression. However, cultural, clinical, and other laboratory data should also be considered in the determination of the prognosis or in deciding whether or not to treat the patient. Occasionally, in some patients, positive titers that slowly decline are noted for a long time after the patient has been cured. Reactions with heterologous antigens may complicate the interpretation of a result when only a single serum sample has been tested. For example, in some situations, the first serologic response noted in a person suffering from histoplasmosis may be obtained only with the nonpurified *B. dermatitidis* antigen. Some patients with histoplasmosis responses may even show antibody to the antigens of *H. capsulatum*, *B. dermatitidis*, and *C. immitis*, to only some of them, or to none. Furthermore, a lack of immunologic response does not exclude histoplasmosis, particularly when only one specimen has been tested and when the clinical picture strongly suggests pulmonary mycotic disease. In disseminated or terminal histoplasmosis, humoral antibody responses may or may not be positive. The CF test is usually positive with CSF specimens from cases of chronic meningitis. CF titers may range from 1:8 to 1:128 (10).

Attempts to replace the CF test with a primary binding assay, such as an EIA, have been frustrated, even with column-purified M antigen, by the presence of a galactomannan that cross-reacts with *B. dermatitidis*, *C. immitis*, and *P. brasiliensis*. Recently, Brock et al. (2) showed how specificity could be increased in the EIA to detect antibodies against the M antigen of histoplasmin. Column-purified M antigen was subjected to periodate oxidation, which inactivated the contaminating polysaccharide. The EIA was further modified by a competitive-binding format with enzyme-labeled rabbit anti-M IgG as the indicator antibody.

As indicated above, the test antigens may cross-react in cases of blastomycosis and coccidioidomycosis, other fungal diseases, and leishmaniasis. If cross-reactions are observed or suspected, the laboratorian should base the interpretation of results on the study of serial specimens in CF and ID tests used in combination, the clinical picture, and other laboratory tests.

Test 2. ID and CIE tests

ID test. The micro-ID procedure is recommended for the detection of *H. capsulatum* precipitins against the H and M protein antigens of histoplasmin. Diagnostic precipitins can frequently be detected in CSF specimens from patients with meningeal histoplasmosis before isolates of *H. capsulatum* are obtained and when culture attempts are negative. The ID procedure used is the same as that described for blastomycosis, except that antigen, unknown sera, and control sera with H and M precipitins are added immediately. The reactants are then allowed to diffuse while the gel is incubating in a moist chamber for 24 h at 25°C.

CIE test. The histoplasmosis CIE procedure (9) is performed as follows. Ten milliliters of an equal mixture of 0.85% agarose and 0.85% Ionagar no. 2 dissolved in 0.01 M Veronal buffer, pH 7.2, is applied to a projector slide cover glass (3.25 by 4 in. [8.2 by 10.2 cm]), and 5-mm wells are cut into the agar. Each antigen well is 3 mm from each of two serum wells. Sera are placed in the anodic wells of each pair, and histoplasmin is placed in the cathodic wells. A control serum containing H and M antibodies is placed in the well adjacent to the serum to be tested. Electrophoresis is performed at room temperature with 0.05 M Veronal buffer, pH 7.2, in each chamber. A constant current of 25 mA is applied across the narrow dimension of the slide for 90 min. After electrophoresis, the slides are removed and read for lines of identity. ID and CIE test results are valid only when control reference sera showing H and M bands are positive.

Reagents. Histoplasmin is made as described for the CF test. The mycelial-form filtrate antigen is concentrated 5 to 10 times and titrated to determine the optimal dilution that demonstrates well-defined H and M bands when the antigen is allowed to react with serum from a proven human histoplasmosis case. Control antisera containing H and M precipitins may be prepared in animals by using precipitin arcs as vaccines (9). Histoplasmosis ID reagents or kits may be purchased from American Scientific Products, Immuno-Mycologics, M. A. Bioproducts, Meridian Diagnostics, and Nolan-Scott.

Interpretation. The ID or CIE test is a useful screening procedure or adjunct in the serologic diagnosis of histoplasmosis. The results usually obtained are qualitative. The ID test was first applied to the diagnosis of histoplasmosis in 1958 by Heiner (9). He demonstrated six precipitin bands when concentrated histoplasmin antigen interacted with serum from patients with histoplasmosis. Two of these bands had diagnostic value. One, designated H, was found in the serum of patients with active histoplasmosis. The second, designated M, was found in acute and chronic histoplasmosis and also appeared after normal sensitized persons had been skin tested with histoplasmin. Although the H band is usually associated with the M band, the M band frequently is the first to appear and frequently occurs without the H band. The M band has been considered presumptive evidence of infection with *H. capsulatum*. The presence of only M antibodies in serum may be attributed to active disease, inactive disease, or skin testing (9).

The serum from about 70% of patients with proven histoplasmosis contains M precipitins, whereas only 10% of the sera demonstrate both the H and M precipitins. The detection of the H precipitin is increased by CIE.

To interpret the ID and CIE reactions properly, laboratory workers must know whether the patient whose serum sample is being analyzed was recently skin tested. If the patient has not had a recent histoplasmin skin test, the detection of an M band may serve as an indicator of early disease, since antibody to M appears before the H precipitin and disappears more slowly. The demonstration of both the M and H bands is highly suggestive of active histoplasmosis, regardless of other serologic results. An additional precipitin, y, has been reported to indicate acute histoplasmosis, particularly when the serum sample is devoid of M and H precipitins (9). The detection of M and H precipitins in CSF specimens indicates meningeal histoplasmosis (10).

Test 3. LA test

The LA test is useful for the early detection of acute histoplasmosis. Commercially prepared antigen in the form of histoplasmin-sensitized latex particles is available from Spectrum Diagnostic Inc., Glenwood, Ill. When the LA test is performed, serial twofold dilutions of serum samples ranging from 1:4 to 1:512 are prepared in tubes, and optimally diluted antigen is added. The centrifuged reactants are then examined for strong agglutination (9).

Interpretation. The LA test yields results in 24 h and may even be used with anticomplementary sera. Although the test may be negative with sera from persons with chronic histoplasmosis, it is an excellent aid in the diagnosis of acute histoplasmosis (9).

Some workers consider an LA titer of 1:16 or greater to be significant, whereas others consider titers of 1:32 or greater strong evidence for active or very recent disease. Although a positive LA test can be demonstrated as early as 2 to 3 weeks after exposure to infection by *H. capsulatum* (9), such a reaction should be confirmed by an ID test or other laboratory data. False-positives can occur with the LA test, and results should be interpreted with caution, particularly if only one specimen has been examined and the titer is low.

Caution. Levels of CF antibodies, precipitins, and agglutinins to *H. capsulatum* antigens may be significantly increased in histoplasmin-sensitized persons after one histoplasmin skin test (9). This boosting of antibodies makes subsequent changes in serologic titers uninterpretable. For this reason, patients with suspected active histoplasmosis should not be skin tested. In the CF test, these antibody responses were detected in sera drawn 15 days after skin testing. Blood should be drawn for serologic studies before skin testing, but obviously, the specimen can be taken within 2 to 3 days after the skin test, since antibodies do not develop that soon. Furthermore, it is the serum reaction with the histoplasmin antigen that is affected, although effects on the yeast cell titer have also been reported (9). One histoplasmin skin test does not produce a serologic response in unsensitized persons.

Actinomycotic Hypersensitivity Pneumonitis

The ID test is widely used for screening or for the confirmation of a diagnosis of hypersensitivity pneumonitis (HP) or extrinsic allergic alveolitis resulting from sensitization to a thermophilic actinomycete. The CIE test, although more rapid and sensitive than the ID test, is not as specific or reproducible (9).

This section will be devoted to farmer's lung, but when the appropriate antigens are used (9), the tests described can also be applied to the diagnosis of other hypersensitivity diseases, such as bagassosis, mushroom worker's lung, and bird breeder's lung. Since the eliciting antigen is usually not known at the time of initial testing, it is a common practice to screen sera with a battery of antigens, including those from *Aspergillus* spp., *Aureobasidium pullulans*, and avian proteins, as well as antigens from thermophilic actinomycetes.

Clinical indications

The demonstration of precipitating antibodies to antigens derived from thermophilic actinomycetes or other offending antigens is an important aid in the diagnosis of farmer's lung and related HPs. The ID tests should be performed on patients who show respiratory disease that appears to be environmentally related, i.e., related to occupation (farmer's lung), hobby (pigeon breeder's disease), or home or office (forced-air-system disease). The typical symptoms of these diseases are chills, fever, cough, and dyspnea, usually occurring 4 to 6 h after exposure to the antigen. Patients usually demonstrate crepitant rales in the lower lung fields. Chest X rays may show infiltrates with a pattern indistinguishable from the patterns of other interstitial pneumonias. The demonstration of precipitating antibodies in the sera of patients with suspected HP is frequently used to determine the etiologic agent of the disease and to confirm the tentative diagnosis (9). Immediate hypersensitivity associated with elevated levels of IgE does not appear to play a role in HP. Quantitative IgE tests carried out at the same time as the ID tests would be expected to yield normal values in patients with HP.

Test

The ID test with antigens from *Faenia rectivirugula (Micropolyspora faeni), Thermoactinomyces candidus, Thermoactinomyces vulgaris*, and reference homologous rabbit antiserum is used to detect precipitating antibodies in the sera of patients with clinical and radiologic evidence of farmer's lung. Many types of ID tests can be used, but the test performed at CDC is the same as that used for aspergillosis.

Reagents. Suitable *T. candidus, Thermoactinomyces sacchari, Saccharomonospora viridis*, and *T. vulgaris* antigens may be prepared in tryptic soy broth (9). *F. rectivirugula* antigen can be prepared in an AOAC (Difco) synthetic broth containing 1.0% lactose (13).

HP test reagents may be obtained from Greer or Hollister-Stier.

Interpretation. The demonstration of precipitins to a particular antigen in the serum of a patient is highly suggestive of a diagnosis of HP. The presence of precipitins per se is not diagnostic, since some healthy persons have precipitins. A proper diagnosis of hypersensitivity lung disease in a precipitin-positive patient must also be based on proper clinical, radiologic, and preferably biopsy or inhalation data. Precipitins provide information about the lung disease and the type of sensitizing antigen. A negative result does not rule out a diagnosis of HP (9).

Paracoccidioidomycosis

CF, ID, and CIE tests are useful in the diagnosis of paracoccidioidomycosis and in the monitoring of response to treatment (9).

Clinical indications

Serologic tests for paracoccidioidomycosis should be performed on patients displaying symptoms of chronic disease with lung involvement or ulcerative lesions of the mucosa (oral, nasal, or intestinal) and the skin. In addition, patients with paracoccidioidomycosis often have lymphadenopathy. A history of travel or residence in Latin America also suggests the possibility of paracoccidioidomycosis.

The CF test will detect antibodies in 80 to 96% of patients with paracoccidioidomycosis (9). Complement-fixing antibodies are diagnostic. However, the CF test results with pooled filtrate antigens of the yeast form of *Paracoccidioides brasiliensis* are not always specific, and cross-reactions may be obtained with sera from patients with other diseases. These cross-reactions, however, are infrequent and occur mainly at the 1:8 level. The ID test (9), with concentrated yeast-form filtrate antigens, has a sensitivity of 94% with sera from patients with paracoccidioidomycosis. A 79% correlation was reported betwen ID and CF test results. The ID test used with reference sera is entirely specific (9). An initial serodiagnosis of paracoccidioidomycosis can be obtained in over 98% of cases with the concomitant use of the ID and CF tests (9). Specific *P. brasiliensis* antigen may also be prepared from the mycelial form of the fungus grown in

SABHI broth with inocula derived from potato-glucose agar slants. This antigen in ID tests detected 103 (90%) of 114 proven cases of paracoccidioidomycosis but did not appear to be suitable for use in CF tests (1).

Test

The standardized LBCF test with *P. brasiliensis* yeast-form filtrate antigens is recommended for the titration of sera from suspected cases of paracoccidioidomycosis. Details for performing the test have been previously published (18). In addition, see the description of the CF test under blastomycosis.

Reagents. Paracoccidioidin antigens for CF and ID tests are produced from yeast-form shake cultures of three CDC stock cultures of *P. brasiliensis* (B339, B341, and B1183) grown singly at 35°C in a tryptic soy broth dialysate medium supplemented with glucose, ammonium sulfate, and vitamins (9). The 4-week culture filtrates of each isolate are dialyzed, concentrated 10 times, and mixed in equal volumes. The optimal dilution for each antigen pool is determined by titration with low- and high-titer CF-positive sera from human paracoccidioidomycosis cases or by ID tests with precipitin-positive sera. This antigen is not commercially available, but reference antigen can be obtained from CDC.

Interpretation. CF titers of 1:8 or greater are considered presumptive evidence of paracoccidioidomycosis. Titers may range from 1:8 to 1:16,384, depending on the severity and extent of infection. Serum samples from 85 to 95% of the patients with active disease demonstrate CF titers of 1:32 or greater (9). Low CF titers are usually associated with localized disease or with patients having reticuloendothelial involvement, whereas high CF titers are found in patients with pulmonary lesions or disseminating disease. Young children with disseminated paracoccidioidomycosis are an exception and ordinarily show low or negative CF reactions. Serial CF determinations are of prognostic value. Declines in titer generally indicate effective therapy, whereas clinical relapses are accompanied by increases in humoral antibodies. High and fluctuating titers suggest a poor prognosis. Complement-fixing antibodies at low levels may persist long after the patient is cured.

When they are examined in ID tests with paracoccidioidin, the sera of patients with paracoccidioidomycosis may contain one to three precipitins to *P. brasiliensis*. The ID test is excellent for the diagnosis of progressive pulmonary and disseminated paracoccidioidomycoses (16). Band 1 is found close to the antigen well, and band 3 is found near the serum well. Precipitin to antigen 1, the antibody most frequently encountered, is found in 95 to 98% of the seroactive cases of paracoccidioidomycosis (9). ID and CF tests for paracoccidioidomycosis are available at CDC.

A diagnostic precipitin has been consistently found in the sera of patients with paracoccidioidomycosis. This precipitin reacts with a specific soluble antigen (designated E) which demonstrates cationic electrophoretic mobility. The E arc appears to be identical to band 1 (9). High-titer antigen, equivalent to E and 1, may be produced in commercially prepared medium in 2 weeks or less (1). The highest number of precipitin bands is usually found with sera from patients with lung involvement or disseminated disease. Precipitating antibodies, like those that react in the CF tests, are long lasting. At least one of the three precipitins that might occur in blood could, however, disappear after successful treatment (9).

Sporotrichosis

Serologic tests can be used in the diagnosis of sporotrichosis. These tests are especially helpful in the diagnosis of the extracutaneous or systemic form of sporotrichosis when distinct clinical features are lacking. Two tests, the TA and LA tests, are reliable and sensitive. The antigen that is the basis for the agglutinin test is the peptido-L-rhamno-D-mannan that is the outer layer of the cell well. Comparable sensitivity is not obtained with the CF and ID tests with *Sporothrix schenckii* antigens. The slide LA and the TA tests are preferred because they are both highly sensitive and specific. The LA tests provides results in minutes, but the TA test must be incubated overnight (18). Both tests are performed at CDC.

Clinical indications

Serologic tests for sporotrichosis may be applied to sera from patients with skin lesions, subcutaneous nodules, bone lesions, lymphadenopathy, or pulmonary disease and to CSF from patients with undiagnosed chronic meningitis. The disease should be suspected in patients who handle thorny plants, timber, or sphagnum moss.

Because of its sensitivity (94%), high specificity, and ability to provide results in 5 min, the LA test is highly recommended for routine use in the clinical laboratory. The TA test has a comparable sensitivity, but sera being tested for sporotrichosis may show false-positive reactions with 1:8 and 1:16 dilutions of sera from patients with leishmaniasis (9).

Test. LA test for *S. schenckii* antibody

Equipment and reagents

1. GBS (pH 8.4)–0.1% BSA
2. A spectrophotometrically standardized suspension of 0.81-μm polystyrene latex particles sensitized with an optimal dilution of *S. schenckii* (yeast-form) culture filtrate antigen
3. Sera
   a. Positive reference human anti-*S. schenckii* serum with a known titer
   b. Negative control human serum
   c. Sera from patients
4. Water bath, 56°C
5. Rotary shaker
6. Test tubes, 12 by 75 mm
7. Glass slides, 50 by 75 mm
8. Serologic pipettes, 0.1, 0.5, and 1.0 ml

Procedure

1. Inactivate all sera at 56°C for 30 min.
2. Prepare enough slides (50 by 75 mm; 12 circles per slide) to accommodate the specimens to be tested. Ten sera plus negative and positive control specimens can be screened on each slide.
3. Place the positive and negative control sera and sera from the patients in a test tube rack in the order to be tested. These specimens should be diluted with GBS-BSA.

4. With a 0.1-ml pipette, add 0.02 ml of the optimally sensitized latex suspension to each of the circles on a slide.

5. With separate 0.1-ml pipettes, add 0.04 ml of each 1:4 dilution of sera from the patients and of the positive and negative controls to the circles on the slide. Mix the drops with applicator sticks.

6. Place the slide on a rotating shaker, and rotate it at 150 rpm for 5 min.

7. Immediately read the test macroscopically over a dark background for agglutination. The positive control serum must show 2+ agglutination (small but definite clumps with a slightly cloudy background); the negative control must show no agglutination. Check the reactions of the sera from patients and record as positive all specimens showing agglutination equal to or greater than the 2+ positive reference serum.

8. All specimens positive in the screening test should be diluted serially with GBS-BSA to make 1:8, 1:16, 1:32, and 1:64 dilutions of each serum for titration.

9. The test is performed as described in steps 2 through 7. One slide will accommodate the four dilutions of each of two serum samples plus the positive and negative control serum samples.

10. Record as positive all dilutions that show agglutination equal to or greater than the 2+ positive reference serum.

If the reaction of a 1:64 dilution is greater than 2+, the specimen must be diluted through four more serial dilutions (1:128 to 1:1,024), and the testing procedure must be repeated until an endpoint is reached. The titer of each specimen is the highest serum dilution that gives a 2+ agglutination.

Controls. Positive and negative control sera must be included each time the test is performed.

Reagents. A properly standardized suspension of latex particles with an absorbance of 0.30 ± 0.02 when diluted 1:100 is sensitized with an equal volume of an optimal dilution of yeast-form *S. schenckii* culture filtrate antigens. The optimal quantity of filtrate is the highest dilution that produces a clear 2+ agglutination with the highest reactive dilution of rabbit *S. schenckii* reference antiserum or human sporotrichosis case serum (18). *S. schenckii* antibody test reagents are not commercially available.

Interpretation. Slide LA titers of 1:4 or greater are considered presumptive evidence of sporotrichosis. False-positive reactions have, however, been noted in the 1:4 to 1:8 range with sera from patients with nonfungal infections. Sera from patients with localized cutaneous, subcutaneous, disseminated subcutaneous, or systemic sporotrichosis may show titers ranging from 1:4 to 1:128. An increasing titer or a sustained high titer is helpful in the diagnosis of pulmonary sporotrichosis. The test has limited prognostic value, since antibody levels may show little change during and after convalescence. Slide latex agglutinin titers of 1:32 or greater with CSF are considered evidence of meningeal sporotrichosis (10).

Zygomycosis

An ID test in which homogenate antigens of *Absidia corymbifera, Rhizomucor pusillus, Rhizopus arrhizus,* and *Rhizopus oryzae* are used has been developed for the diagnosis of human and animal zygomycosis caused by these etiologic agents. Preliminary studies indicate that the test demonstrates a sensitivity of about 70%. The present state of research suggests that peptido-L-fuco-D-mannans are antigens that are common to zygomycetes and that differentiate them from other fungi. Cross-reactions occur among the zygomycete genera. The specificity of the test, however, appears good when reference precipitates are used (9). Studies, however, are needed to develop more sensitive and reliable methods for the immunologic diagnosis of zygomycosis.

Sera from patients with diabetic ketoacidosis who present evidence of rhinocerebral disease, from other compromised persons with renal disease or acute leukemia, and from debilitated patients who have signs of pulmonary or systemic infection should be tested for zygomycosis.

IN VITRO IDENTIFICATION OF MYCELIAL-FORM PATHOGENIC FUNGAL CULTURES BY EXOANTIGEN TECHNIQUES

Isolates of *B. dermatitidis, C. immitis, H. capsulatum,* and *P. brasiliensis* often vary in their gross and microscopic features and are frequently difficult or impossible to convert to their tissue forms in vitro or in vivo. A simple diagnostic procedure, the exoantigen technique, can be used to rapidly identify these four pathogens regardless of whether they are sporulating or nonsporulating (12). Studies have revealed that *B. dermatitidis, C. immitis, H. capsulatum,* and *P. brasiliensis* mycelial-form cultures readily produce cell-free antigens (exoantigens) that are homologous to the diagnostic precipitins discussed in the appropriate sections of this chapter.

The detection of specific exoantigens homologous to *B. dermatitidis* precipitin A, *C. immitis* precipitins HS, HL, or F, *H. capsulatum* precipitins H or M, or *P. brasiliensis* precipitins reactive with antigen 1 (E) or 2 in ID tests permits the early and accurate identification of these pathogens. A 10-day or older Sabouraud glucose agar slant culture with a growth of at least 15 by 30 mm is extracted with 8 to 10 ml of 1:5,000 aqueous Merthiolate solution for 24 to 48 h at 25°C. Fungi that form arthroconidia are tested for *C. immitis* antigens, those that form tuberculate macroconidia are tested for *H. capsulatum* antigens, and those devoid of characteristic conidia are tested for antigens to all four of the pathogens. The cellular extract is concentrated by ultrafiltration and tested in the ID test for diagnostic exoantigen(s). Technical details and information on the value of the method for the identification of other hyaline species (*Aspergillus* spp., *H. capsulatum* var. *duboisii, Histoplasma farciminosum, Penicillium marneffei, Pseudallescheria boydii,* and *S. schenckii*) and dematiceous fungi (*Cladosporium bantianum, Cladosporium carrionii, Exophiala jeanselmei,* and *Wangiella dermatitidis*) have been previously published (11, 12).

Reagents

Fungal reagents for use in ID tests for serodiagnosis are available from numerous commercial sources. These reagents are designed to detect antibodies and are not always reliable for use in exoantigen tests.

They may, however, with proper concentration or dilution and after verification of the presence of specific antibodies and antigens, be used in exoantigen tests. At present, Immuno-Mycologics and Nolan-Scott sell kits specifically designed for the immunoidentification of *B. dermatitidis, C. immitis*, and *H. capsulatum* cultures.

IN VITRO AND IN VIVO IDENTIFICATION OF PATHOGENIC FUNGI BY FA TECHNIQUES

Fluorescent-antibody (FA) procedures provide mycologists with a valuable adjunct to conventional diagnostic tests. FA procedures permit the rapid acquisition of presumptive diagnostic data and the rapid screening of clinical material for pathogenic fungi. Above all, with fungus cultures, these procedures permit rapid identification, whereas conventional procedures might take 2 weeks or longer. FA procedures can be applied to viable as well as nonviable fungi in culture and in clinical materials or tissue sections. At the present time, these tests are available at CDC and at certain other reference laboratories in the United States.

With transmitted-light fluorescence microscopes, a combination of a 5113 Corning glass primary filter (3 mm thick) and a Wratten 2A secondary filter (2 mm thick) is satisfactory for use with the fungi. Some researchers use a BG-12 primary filter (3 mm thick) with a Schott GG-9 secondary filter. Recent studies indicate that an American Optical interference exciter filter used with a Schott GG-9 ocular filter gives excellent results with FA-stained fungi (9). Incident-light fluorescence microscopes are excellent for FA studies with fungi. They provide a greater intensity of fluorescence, and they offer the possibility of working with a completely dry system. The following information pertains to those FA procedures that have been developed to a practical level.

Actinomyces species and related organisms

Immunofluorescence procedures readily permit the detection and identification of the principal etiologic agents of actinomycosis. FA reagents have been produced for the specific staining of the serotypes of *Actinomyces israelii, Actinomyces naeslundii, Actinomyces viscosus*, and *Arachnia (Actinomyces) propionica* either in smears of tissues and exudates or in culture (8, 9).

B. dermatitidis

Specific FA preparations for *B. dermatitidis* have been developed (9) by adsorbing rabbit anti-yeast-form, *B. dermatitidis*-labeled antiglobulins with yeast-form cells of *H. capsulatum* and *Geotrichum candidum*. These yeast-form-specific conjugates make possible the rapid and accurate detection of *B. dermatitidis* in culture and in clinical materials. FA techniques cannot identify the mycelial form of this fungus.

Candida species

Numerous investigators have studied the application of the FA technique to the detection and identification of *Candida albicans* and other *Candida* species. All found that *Candida* species are closely related antigenically. Attempts to isolate species-specific FA preparations useful for a definitive identification of *C. albicans* in clinical materials and cultures have failed. As a result, single specific reagents are not available for use in the clinical laboratory. Some researchers have reported the successful use of a combination of reagents to identify *Candida* sp. isolates (9). Despite the lack of species-specific FA reagents, some reagents that demonstrate broad intrageneric cross-staining properties can be used for screening clinical specimens for the presence of *Candida* spp. (8).

Coccidioides immitis

FA reagents specific for the tissue form of *C. immitis* have been developed (8). These reagents have been produced from the antisera of rabbits infected with viable *C. immitis* cultures. They generally stain the walls of endospores and the contents of spherules. Cross-staining of heterologous fungal antigens by the conjugates is eliminated by dilution or adsorption with yeast-form cells of *H. capsulatum*. Alternatively, specific conjugates can be prepared from the antisera of rabbits immunized with suspensions of the arthroconidia of *C. immitis* killed with Formalin. Cross-staining is eliminated by adsorption with yeast-form cells of *H. capsulatum*.

The specific conjugates are used to detect *C. immitis* in a variety of specimens from humans and animals with coccidioidomycosis. The conjugates can also be used in the diagnostic laboratory for the rapid and specific demonstration of the tissue form of *C. immitis* in clinical materials from laboratory animals injected with suspected *C. immitis* cultures.

Cryptococcus neoformans

Unadsorbed *C. neoformans* conjugates have been used successfully to study the distribution of *C. neoformans* and its polysaccharide products in Formalin-fixed tissue. One group of researchers (9) stained histopathological sections from patients with cryptococcosis with Mayer mucicarmine stain and FA preparations. They found that the conjugate, although nonspecific for *C. neoformans*, stained the yeast cells more intensely and rapidly than did mucicarmine. An effective and practical diagnostic FA reagent of higher specificity is produced by the adsorption of *C. neoformans* conjugates with cells of *Cryptococcus diffluens* and *Candida krusei* (9).

H. capsulatum

A specific FA reagent for *H. capsulatum* is produced by the adsorption of conjugated homologous antiglobulins with the yeast-form cells of *B. dermatitidis* (8). This conjugate is used to identify the yeast-form cells of *H. capsulatum* in culture and in tissue impression smears from humans with histoplasmosis and from experimentally infected mice. Several workers have investigated the applicability of FA reagents to the rapid detection of *H. capsulatum* in human clinical specimens (9). The direct FA procedure has been recommended as a rapid screening procedure for *H. capsulatum* and for the staining of sputum smears as an adjunct to conventional cultural methods.

Serologic studies demonstrated the existence of five *H. capsulatum* serotypes (8). Of these, only one, the 1,4 type, consistently failed to react with the FA reagents then available. Recent studies indicate that this serotype reacts with *H. capsulatum* var. *duboisii* and *B. dermatitidis* isolates.

A diagnostically useful polyvalent reagent has been developed (9) for the detection and identification of *H. capsulatum* var. *capsulatum*, regardless of serotype. The adsorption of labeled antibodies produced against the most complete *H. capsulatum* serotype (1,2,3,4) with cells of *C. albicans* yielded a reagent that intensely stained only *H. capsulatum* var. *capsulatum*, *H. capsulatum* var. *duboisii*, and *B. dermatitidis*. Despite this cross-staining, the *C. albicans*-adsorbed reagent can be used diagnostically by using the *B. dermatitidis*-specific FA reagent along with the polyvalent conjugate. Both of these FA reagents stain isolates of *B. dermatitidis*, whereas only the polyvalent conjugate stains *H. capsulatum*. The two varieties of *H. capsulatum* cannot be differentiated from each other with the available FA reagents.

Paracoccidioides brasiliensis

FA reagents for the diagnosis of paracoccidioidomycosis have been developed (8). Tissue-form-specific reagents are produced from the antisera of rabbits immunized with suspensions of yeast-form cells of *P. brasiliensis* killed with Formalin. Cross-reactions are eliminated by multiple adsorptions with selected heterologous fungi. These reagents are used in the direct FA procedure to demonstrate *P. brasiliensis* cells in smears of clinical materials. They are especially useful in clinical materials that contain few *P. brasiliensis* cells or no morphologically typical cells.

Pseudallescheria boydii

Histopathology is not sufficiently definitive to permit a precise diagnosis of pseudallescheriasis. A FA reagent for the detection and identification of *P. boydii* in tissue has been developed (7). Labeled rabbit antisera to mycelial and conidial antigens of *P. boydii* brightly stain all *P. boydii* isolates, but they cross-react with *A. flavus*, *A. fumigatus*, *Aspergillus nidulans*, *A. niger*, *A. terreus*, *Fusarium oxysporum*, *Fusarium solani*, and a *Scopulariopsis* sp. Cross-staining antibodies can be removed by adsorption with *A. fumigatus* and *F. oxysporum* cellular elements. The adsorbed conjugate specifically stains *P. boydii* in Formalin-fixed tissues from humans and animals with pseudallescheriasis.

S. schenckii

Good-quality reagents for the detection of the tissue form of *S. schenckii*, both in clinical materials and in culture, have been developed (8). Such reagents are produced from antiserum obtained by the immunization of rabbits with suspensions of whole yeast cells of *S. schenckii* killed with Formalin.

Cross-reactions are readily eliminated by dilution, which does not compromise the staining qualities. Although, as a rule, few *S. schenckii* cells are found in the exudates of lesions, the cells are readily detected with the FA tests. In a few cases, particularly if the patient is already under therapy, a number of fields may have to be searched before the fungal cells are found.

Application of FA reagents to clinical materials

The FA technique is most effective for the detection of fungal antigens in cultures, pus, exudates, blood, tissue impression smears, and spinal fluid specimens. The technique is, however, more difficult to use with sputum and tissue sections. Not only does one have to cope with tissue elements that autofluoresce, but for unknown reasons, the staining capacity of the conjugate is impaired in sputum and tissue sections. This impairment may result from a lack of surface interaction between antigen and antibody, because either enzymatic or chemical digestion of sputum specimens results in a more effective staining of fungus elements (8). Such treatment may also be applied to tissue sections. FA work with tissue sections has shown that fungus elements are stained more intensely in thin sections (4 μm) than in thicker sections. Glass slides 1 mm or less in thickness are recommended.

Direct FA staining permits the rapid detection of fungi in paraffin sections of Formalin-fixed tissue. In addition, one can also identify fungi in tissue sections that were previously stained with hematoxylin and eosin, Brown and Brenn, and Giemsa stains. The conjugates, however, will not stain fungi in tissues stained previously by the Gomori methenamine-silver nitrate, the periodic acid-Schiff, or the Gridley procedure.

Prolonged storage of Formalin-fixed tissues, either wet or in paraffin blocks, does not appear to have adverse effects on fungal antigens. Therefore, the FA procedure can well be used to make retrospective immunohistologic diagnoses. Although fluorescein-labeled *H. capsulatum* antiglobulins regularly stain *H. capsulatum* in sections of fixed tissue from patients with active histoplasmosis, they do not regularly stain *H. capsulatum* in healed, calcified lesions (8).

LITERATURE CITED

1. **Blumer, S. O., M. Jalbert, and L. Kaufman.** 1984. Rapid and reliable method for production of a specific *Paracoccidioides brasiliensis* immunodiffusion test antigen. J. Clin. Microbiol. **19:**404–407.
2. **Brock, E. G., E. Reiss, L. Pine, and L. Kaufman.** 1984. Effect of periodate oxidation on the detection of antibodies against the M-antigen of histoplasmin by enzyme immunoassay (EIA) inhibition. Curr. Microbiol. **10:**177–180.
3. **Cox, R. A., M. Huppert, P. Starr, and L. A. Britt.** 1984. Reactivity of alkali-soluble, water-soluble cell wall antigen of *Coccidioides immitis* with anti-*Coccidioides* immunoglobulin M precipitin antibody. Infect. Immun. **43:**502–507.
4. **de Repentigny, L., R. J. Kuykendall, F. W. Chandler, J. R. Broderson, and E. Reiss.** 1984. Comparison of serum mannan, arabinitol, and mannose in experimental disseminated candidiasis. J. Clin. Microbiol. **19:**804–812.
5. **de Repentigny, L., and E. Reiss.** 1984. Current trends in immunodiagnosis of candidiasis and aspergillosis. Rev. Infect. Dis. **6:**301–312.
6. **Green, J. H., W. K. Harrell, J. E. Johnson, and R. Benson.** 1980. Isolation of an antigen from *Blastomyces dermatitidis* that is specific for the diagnosis of blastomycosis. Curr. Microbiol. **4:**293–296.
7. **Jackson, J. A., W. Kaplan, L. Kaufman, and P. Standard.** 1983. Development of fluorescent-antibody reagents for demonstration of *Pseudallescheria boydii* in tissues. J. Clin. Microbiol. **18:**668–673.
8. **Kaplan, W.** 1975. Practical application of fluorescent

antibody procedures in medical mycology, p. 178–186. *In* Mycoses. Proceedings of the Third International Symposium on Mycoses. Scientific publication no. 304. Pan American Health Organization, Washington, D.C.

9. **Kaufman, L.** 1980. Serodiagnosis of fungal diseases, p. 553–572. *In* N. R. Rose and H. Friedman (ed.), Manual of clinical immunology, 2nd ed. American Society for Microbiology, Washington, D.C.
10. **Kaufman, L.** 1983. Mycoserology: its vital role in diagnosing systemic mycotic infections. Jpn. J. Med. Mycol. **24:**1–8.
11. **Kaufman, L.** 1984. Antigen detection: its role in the diagnosis of mycotic disease and the identification of fungi, p. 203–212. *In* A. Sanna and G. Morace (ed.), New horizons in microbiology. Elsevier, Amsterdam.
12. **Kaufman, L., P. Standard, and A. A. Padhye.** 1983. Exoantigen tests for the immunoidentification of fungal cultures. Mycopathologia **82:**3–12.
13. **Kurup, V. P., and J. N. Fink.** 1977. Extracellular antigens of *Micropolyspora faeni* grown in synthetic medium. Infect. Immun. **15:**608–613.
14. **Lehmann, P. F., and E. Reiss.** 1980. Detection of *Candida albicans* mannan by immunodiffusion, counterimmunoelectrophoresis, and enzyme-linked immunoassay. Mycopathologia **70:**83–88.
15. **Lloyd, K. O.** 1970. Isolation, characterization, and partial structure of peptido-galactomannans from the yeast form of *Cladosporium werneckii*. Biochemistry **9:**3446–3470.
16. **Londero, A. T., J. O. S. Lopes, C. D. Ramos, and L. C. Severo.** 1981. A prova da dupla difusão em gel de ágar no diagnóstico da paracoccidioidomicose. Rev. Assoc. Med. Rio Grande do Sul **25:**272–275.
17. **Meckstroth, K. L., E. Reiss, J. W. Keller, and L. Kaufman.** 1981. ELISA detection of antibodies and antigenemia in leukemia patients with candidiasis. J. Infect. Dis. **144:**24–32.
18. **Palmer, D. F., L. Kaufman, W. Kaplan, and J. J. Cavallaro.** 1977. Serodiagnosis of mycotic diseases. Charles C Thomas, Publisher, Springfield, Ill.
19. **Pappagianis, D.** 1980. Serology and serodiagnosis of coccidioidomycosis, p. 97–112. *In* D. A. Stevens (ed.), Coccidioidomycosis. Plenum Publishing Corp., New York.
20. **Reiss, E., R. Cherniak, R. Eby, and L. Kaufman.** 1984. Enzyme immunoassay detection of IgM to galactoxylomannan of *Cryptococcus neoformans*. Diagn. Immunol. **2:**109–115.
21. **Reiss, E., L. Stockman, R. J. Kuykendall, and S. J. Smith.** 1982. Dissociation of mannan-serum complexes and detection of *Candida albicans* mannan by enzyme immunoassay variations. Clin. Chem. **28:**306–310.
22. **Reiss, E., S. H. Stone, and H. F. Hasenclever.** 1974. Serological and cellular immune activity of peptidoglucomannan fractions of *Candida albicans* cell walls. Infect. Immun. **9:**881–890.
23. **Smith, C. E., M. T. Saito, R. R. Beard, R. M. Kepp, R. W. Clark, and B. U. Eddie.** 1950. Serological tests in the diagnosis and prognosis of coccidioidomycosis. Am. J. Hyg. **52:**1–21.
24. **Stockman, L., and G. D. Roberts.** 1983. Corrected Version. Specificity of the latex test for cryptococcal antigen: a rapid, simple method for eliminating interference factors. J. Clin. Microbiol. **17:**945–947.
25. **Weiner, M. H., and M. Coats-Stephen.** 1970. Immunodiagnosis of systemic aspergillosis. Antigenemia detected by radioimmunoassay in experimental infection. J. Lab. Clin. Med. **93:**111–119.
26. **Weiner, M. H., G. H. Talbot, S. L. Gerson, G. Felice, and P. A. Cassileth.** 1983. Antigen detection in the diagnosis of invasive aspergillosis. Utility in controlled, blinded trials. Ann. Intern. Med. **99:**777–782.
27. **Wieden, M. A., J. N. Galgiani, and D. Pappagianis.** 1983. Comparison of immunodiffusion techniques with standard complement fixation assay for quantitation of coccidioidal antibodies. J. Clin. Microbiol. **18:**529–534.

Chapter 97

Serodiagnostic Tests for Parasitic Diseases

KENNETH W. WALLS

Despite the fact that many parasitologic infections are diagnosed by blood or stool examinations, clinical evaluation, or exposure or travel history, serology plays an important role in parasitic diseases. In some diseases, the diagnostic stage may be rare or missing, while in other diseases, the diagnostic stage may not occur or requires invasive procedures to obtain. In diseases such as ascariasis, filariasis, malaria, and trichinellosis, serology only occasionally is diagnostic but serves mainly as confirmation or simply as a guide to the effectiveness of therapy. On the other hand, the serology of diseases such as extraintestinal amebiasis, toxocariasis, and toxoplasmosis is frequently the only method of diagnosis. In Chagas' disease and schistosomiasis, in which the parasite often is rare or absent, or in echinococcosis and cysticercosis, in which hazardous invasive procedures may be necessary for morphologic diagnosis, serology is the alternative diagnostic method.

Virtually every seroimmunologic procedure has been evaluated for use in the diagnosis of parasitic infections. Most of these procedures are of very limited use or have proven to be ineffective or of no advantage over other diagnostic procedures. Acceptance, in general, has depended upon the ability of the test to detect a particular stage of the life cycle of the parasite, the ability to more clearly identify the infective agent, the ease and economy of performance, or the adaptability to large-scale epidemiological studies. Relatively few seroimmunologic procedures meet any or all of these criteria.

In this section, the types of serodiagnostic tests used in parasitology are discussed along with their advantages and disadvantages. The specific use and interpretation of each test in the diagnosis of parasitic diseases are included in the discussion of the appropriate disease in the parasitology portion of this Manual. Detailed procedures for performing the tests are found in the references listed in the Literature Cited, the *Manual for Clinical Microbiology* (4), or the package inserts of the suggested manufacturers.

A variety of serologic procedures have been evaluated for use in the diagnosis of most parasitic infections with various degrees of success. Some of these procedures have been sufficiently successful to merit commercial production; some are successful but of limited demand, and the reagents used in the procedures must be prepared in individual laboratories performing the tests; and many procedures have been unsuccessful due to technical difficulties or the inability of laboratory workers to prepare quality reagents in sufficient quantity to support the procedure. Table 1 lists the tests used at the Centers for Disease Control (which, in general, are the most commonly accepted procedures throughout the world), the reactivity levels of each, and a commercial source of reagents when one is available. In some cases, other tests are fully acceptable, but for a variety of reasons, they are not the tests of choice at the Centers for Disease Control. Where applicable, these tests are noted. In addition, it should be noted that there is no effort to list all manufacturers, and inclusion or exclusion in Table 1, which is only a summary, does not constitute endorsement or disapproval of any manufacturer or product. The text presented in Chapter 55 in this Manual should be referred to for specific details or interpretation and for the cross-reactions which occur.

A number of diseases have been omitted from the table because serology is of no value, e.g., ascariasis; because the disease is rarely seen and does not justify performance of the tests available, e.g., babesiosis; or because there are no tests acceptable at present, e.g., pneumocystis. Only a few tests have been produced commercially, and for most tests, reagents must be prepared in the users' own laboratories. This method of preparation results in a lack of standardization in both reagents and test procedures and makes test interpretation difficult.

Criteria used by the Centers for Disease Control for the interpretation of test results have been published (9) and serve as the guidelines for the discussions in this Manual. Changing titers are always more helpful than a single result, but in many parasitic infections, the patient is from an area in which the disease is endemic, where there are multiple exposures and chronic infections, or the disease has a prolonged prepatent period. In either case, antibody response is stable by the time testing is initiated. As a result, interpretation is generally made on a single specimen and is sometimes confirmed by two or more procedures.

Although new tests are continually being developed, diagnosis primarily depends upon complement fixation (CF), indirect hemagglutination (IHA), indirect immunofluorescence (IIF), enzyme-linked immunosorbent assay (ELISA), immunoelectrophoresis (IEP), and double diffusion (DD) tests. References concerning the techniques for the performance of each test can be found in the Literature Cited.

The CF test was one of the earliest successful tests used in parasitology. In the United States, the Laboratory Branch CF test described by Casey (2) is the most common micromodification used. Although the test is still used in some infections, technical problems have decreased its popularity. Complement is labile and nonspecifically inactivated by a variety of substances and conditions leading to anticomplementary activity. In general, CF titers are lower than those in other procedures, but specificity is usually very good. Although a variety of antigens are used, care must be taken to prevent the incorporation of nonspecific components and anticomplementary activity. CF is still used for Chagas' disease, paragonimiasis, and leishmaniasis.

Perhaps the most commonly used procedure is the IHA test (3). The test is simple to perform and requires no special training or equipment. A variety of antigens can be adsorbed to the tanned erythrocytes, making the test very versatile and adaptable to testing a variety of diseases. More reactive than the CF test, the

TABLE 1. Significance of and some commercial sources for serologic tests of choice for parasitic diseases

Disease	Test choice[a]	Alternate test	Sensitivity (%)	Specificity	Significant titer	Commercial source
Amebiasis	IHA		95 (Extraintestinal) 70 (Intestinal)	Excellent	1:256	Calbiochem-Behring VIRATEK, Inc.
		CEP				Cordis Laboratories, Inc.
		DD				Cordis Laboratories, Inc.
		ELISA				Cordis Laboratories, Inc. Litton Bionetics
Chagas'	CF		29 (Acute disease)	Strong cross-reactions with leishmaniasis	1:8	None in the United States
	IHA		96 (Chronic disease)		1:28	
		ELISA				Litton Bionetics
Echinococcosis	IHA		88 (Liver cyst) 10 (Lung cyst)	Cross-reaction with cysticercosis	1:128	None in the United States
		IEP	Very high	Arc 5 indicates echinococcosis	Arc 5	None in the United States
Cysticercosis	IHA		50 (Seizures) 70 (Meningitis) 98 (Intracranial pressure)	Cross-reactions with echinococcosis	1:128	None in the United States
Malaria	IIF		100	Excellent (not species specific)	1:64	None
		ELISA	100	Excellent	1:32	None
Schistosomiasis	IIF		85 (Not correlated with clinical disease)	Excellent	Positive	None
		ELISA	?	May be species specific	1:64	None
Toxocariasis	ELISA		90 (Ocular) 78 (Visceral)	Excellent	1:8 1:32	None
Toxoplasmosis	IIF (IgG)		100	Excellent	1:256	Clinical Sciences, Inc. Microbiological Research Corp. Electro-Nucleonics, Inc.
		ELISA	100	Excellent	Each kit is different	Microbiological Associates Litton Bionetics Abbott Laboratories Cordis Laboratories, Inc.
		DA (IgG)	80	Good	Positive	Hynson, Westcott & Dunning
		LA (IgG)	80	False positives with rheumatoid; may miss infants	1:100	Synkit, Inc.
		IHA (IgG)	80	Some false positives; may miss infants	1:64	Wampole Laboratories
	IIF (IgM)		70	False positives with rheumatoid and antinuclear antibody	1:64 (Adults)	Microbiological Associates
				False negatives occur	Any measurable in newborns	Electro-Nucleonics, Inc. Clinical Science, Inc.
		ELISA	90	Excellent		Litton Bionetics
Trichinosis	BFT		75 (May miss light infections)	Excellent	1:5	None

[a] Tests: IHA, Indirect hemagglutination; CEP, counterelectrophoresis; DD, agar gel double diffusion; ELISA, enzyme-linked immunosorbent assay; CF, complement fixation; IEP, immunoelectrophoresis; IIF, indirect immunofluorescence; DA, direct agglutination; LA, latex agglutination; BFT, bentonite flocculation test; IgG, immunoglobulin G.

IHA test suffers the problems of tests with erythrocytes. Nonspecific activity is common, due to heterophilelike reactions and the presence of some therapeutic agents. Nonsensitized cell controls must be included for each serum, and the sera should be heat inactivated to remove any complement activity. The IHA test is the test of choice for amebiasis, cysticercosis, echinococcosis, filariasis, and strongyloidiasis.

The IIF test (6) continues to gain in popularity. Because the IIF test requires particulate antigens to permit visualization, the variety of antigens available for the IHA test cannot be used in IIF tests, but in most

TABLE 2. Parasitic diseases in which soluble antigens have been detected

Disease	Source or body fluid in which detected	Test method[a]
Protozoan		
Chagas' disease	Serum	EIA
Giardiasis	Duodenal aspirate	DD
Primary amebic meningoencephalitis	Serum	DD
Amebiasis	Liver abscess, liver biopsy, serum, feces	CIE, EIA
Malaria	Serum, urine	DD, CIE
	Kidney biopsy	IF
Pneumocystis	Serum	CIE
Toxoplasmosis	Serum, spinal fluid	EIA
Helminth		
Onchocerciasis	Serum	DD, RIA, RIP
Bancroftian filariasis	Serum	CIE
Schistosomiasis	Urine	DD
	Serum	DD, RIA
	Breast milk	DD
Trichinosis	Urine	Precipitation

[a] Test method: CIE, Counterimmunoelectrophoresis; DD, agar gel double diffusion; EIA, capture enzyme immunoassay; IF, direct immunofluorescence; Precipitation, tube precipitation; RIA, capture radioimmunoassay; RIP, radioimmunoprecipitation.

cases the IIF test is more specific than the IHA test, and in some cases it is more sensitive. In contrast to the IHA test, the IIF test is somewhat difficult to perform and requires specialized equipment. It is not well suited to epidemiological studies and is primarily used for diagnosis. Antigens are usually whole organisms, fragments of organisms, or sections of adult worms. By the use of specific fluorescein-labeled antiglobulins, reactivity can be identified as specifically immunoglobulin G, M, or A. In many cases, such identification of reactivity is useful in identifying the stage of disease. One disadvantage of the IIF test is the need for the appropriate antiglobulin conjugate. In epidemiological investigations for which animal sera must be evaluated, the specific antispecies conjugate must be used for each animal species tested. The IIF test is the test of choice for leishmaniasis, malaria, schistosomiasis, and toxoplasmosis.

Perhaps the most promising of the new procedures to be introduced is the ELISA (8). In principle, the procedure is the same as that of the IIF test except that an enzyme label is used in place of fluorescein, and the readout is color instead of fluorescence. More difficult to perform than the previously described procedures, ELISA approaches the "universal" test. Almost any antigen can be used to produce a highly sensitive and specific test. The procedure can be performed qualitatively by testing a single serum dilution and reading the results visually; it can be performed semiquantitatively by testing a single serum dilution but reading the results with a special colorimeter and interpolating the concentration by the quantitative color determination; or it can be performed quantitatively by performing serial dilutions, reading the results colorimetrically, and precisely determining the reactivity. As with the IIF test, specific immunoglobulin activity can be determined by use of the appropriate labeled antiglobulin. Ideally suited for epidemiological studies, ELISA can be used to test large numbers of sera either in a single-dilution screen test or a quantitative test. Because ELISA is much more susceptible to technical variations and nonspecific binding of reagents than are other tests, strict quality control must be maintained, and training must include a thorough understanding of the principles of the test. For proper quantitative performance, a specialized colorimeter is required. As improved antigens are developed, the ELISA procedure will undoubtedly become one of the more prominent procedures in serodiagnosis. Currently it is being used as an ancillary test in amebiasis, Chagas' disease, leishmaniasis, malaria, schistosomiasis, and toxoplasmosis; it is the primary test of toxocariasis and for the detection of immunoglobulin M antibody in toxoplasmosis.

The IEP (1) and DD (5) tests are gel tests widely used for confirmation and rapid screening. Although they are not quantitative, both tests are recognized for their ability to identify specific precipitin bands useful in differentiating reactivities to antigenically related organisms. The counterelectrophoresis test is most commonly used to identify a specific "arc 5," which identifies echinococcosis. Recently, another band, unfortunately called "band 5," has been identified as being specific for Chagas' disease. The IEP procedure is rapid, the bands appearing in as little as 90 min, but it requires electrophoresis equipment. DD, although requiring 24 to 48 h to obtain results, requires no special equipment. Both tests require relatively large amounts of antigen and antisera to perform. While not as precise as other serologic procedures, test parameters should be adhered to closely. False-positive and false-negative results are obtained. Counterelectrophoresis and DD tests are used primarily as ancillary tests in amebiasis, cysticercosis, echinococcosis, and trichinellosis.

Direct agglutination tests have been used in several parasitic diseases with varying degrees of success (7). In these tests, whole organisms are reacted with patients' serum, and in the presence of antibody, agglutination occurs. In most cases it was found that "natural" antibodies occurred which resulted in nonspecific agglutination. To overcome this agglutination, the organisms were treated with trypsin, and the patient serum was inactivated with 2-mercaptoethanol to destroy its nonspecific activity. Some laboratories have reported success by such methods—other laboratories have not proven these methods to be as successful. The test is used as an ancillary test in Chagas' disease, leishmaniasis, and toxoplasmosis.

Other indirect agglutination procedures with carrier particles such as latex, bentonite, and cholesterol have been described and are modifications of the IHA test. These other procedures now have limited use and are being replaced by newer procedures. Particle sensitization and stability continue to be a problem. The bentonite flocculation test is still used for the serodiagnosis of trichinellosis, but its use is restricted to only a few laboratories.

Since serologic results are as dependent upon patient responses as upon the test and its reagents, there is no perfect test, and all results must be interpreted in light of the clinical information. In some of the diseases, however, serology is the only reliable diagnostic method, and decisions on patient management must be based on these data. Quality control cannot be overemphasized. Whether commercially or locally prepared reagents are used, the same controls and subjective evaluation must be used. Paired specimens are always better than one specimen and should be tested in the same test. Since in parasitic infections, unlike many other diseases, considerable information can be gleaned from the initial specimen, this specimen should be tested upon receipt and then tested again with the second specimen when it becomes available.

The detection of antigen is felt to be an important adjunct to the diagnosis of parasitic infections. Table 2 lists a variety of soluble antigens which have been detected in protozoal and helminthic diseases. To date, relatively few tests for antigen have been pursued as to their usefulness on a routine basis, and none of these procedures has yet been successfully adopted. For a short time, a commercial ELISA capture method was available for detecting ameba antigen in stool and showed great promise of being equal or superior to standard coprologic methods. This procedure is no longer being offered.

Although a number of articles have been published on the detection of *Toxoplasma gondii* antigen in serum, the correlation of the antigen with clinical disease was insufficient to justify the use of the test. Recently, enzyme immunoassay methods for detecting antigen in tissue have suggested this assay as a method for supporting histopathological diagnosis. However, the presence of antigen in tissue during chronic infections interferes with interpretation in clinical cases.

One of the primary tests for the diagnosis of *Pneumocystis carinii* pneumonia was the detection of circulating antigen by counter IEP. When counter IEP was used in conjunction with the detection of increased levels of antibody, diagnosis was presumed. Recently it has been shown that there is no correlation between clinical disease and the detection of antigen or antibody or both by present techniques. Consequently, until improved methods are developed, neither the antigen- nor the antibody-detecting test appears to have clinical value.

Many of the other systems described in Table 2 have been shown to correlate with disease, but in most cases, the technique for identifying antigen is more difficult or time consuming than present methods for detecting the parasite. The promise of more sensitive and specific procedures utilizing enzyme immunoassay methods and monoclonal antibodies holds out hope for a future use of these procedures in diagnosing many parasitic diseases.

LITERATURE CITED

1. **Capron, A., A. Vernes, and J. Biguet.** 1967. Le diagnostic immunoelectrophoretique de l'hydatidose, p. 27. *In* J. Coudert (ed.), Le kyste hydatique der foie. Journees Lyonnaises d'Hydatidologie SIMEP, Lyon, France.
2. **Casey, H.** 1965. Standardized diagnostic complement fixation method and adaptation to microtest. Public Health Monogr. 74. U.S. Government Printing Office, Washington, D.C.
3. **Healy, G. R.** 1968. The use of and limitations to the indirect hemagglutination test in the diagnosis of intestinal amebiasis. Health Lab. Sci. **5**:174–179.
4. **Kagan, I. G.** 1980. Serodiagnosis of parasitic diseases, p. 724–750. *In* E. H. Lennette, A. Balows, W. J. Hausler, Jr., and J. P. Truant (ed.), Manual of clinical microbiology, 3rd ed. American Society for Microbiology, Washington, D.C.
5. **Maddison, S. E.** 1965. Characterization of *Entamoeba histolytica* antigen-antibody reaction by gel diffusion. Exp. Parasitol. **16**:224–235.
6. **Palmer, D. F., J. J. Cavallaro, K. W. Walls, A. J. Sulzer, and M. Wilson.** 1976. Serology of toxoplasmosis. Immunology Series no. 1, procedural guide. U.S. Department of Health, Education and Welfare, Public Health Service, Center for Disease Control, Atlanta.
7. **Vattsione, N. H., and J. F. Yanovsky.** 1971. *Trypanosoma cruzi*: agglutination activity of enzyme-treated epimastigotes. Exp. Parasitol. **30**:349–355.
8. **Walls, K. W., S. L. Bullock, and D. K. English.** 1977. Use of the enzyme-linked immunosorbent assay (ELISA) and its microadaptation for the serodiagnosis of toxoplasmosis. J. Clin. Microbiol. **5**:273–277.
9. **Walls, K. W., and M. Wilson.** 1983. Immunoserology of parasitic infections, p. 191–214. *In* R. M. Aloisi and J. Hyum (ed.), Immunodiagnostics. Arlan R. Liss, Inc., New York.

Chapter 98

Solid-Phase Enzyme Immunoassays for the Detection of Microbial Antigens in Body Fluids

ROBERT H. YOLKEN

Traditionally, the diagnosis of infectious diseases has been accomplished by the growth, isolation, and identification of the infecting microorganism in pure culture, utilizing an in vitro cultivation system or a laboratory animal. However, there are situations in which the direct cultivation of a microorganism from a clinical specimen can be difficult to accomplish quickly enough to be useful for patient management. In addition, the cultivation of some microorganisms requires techniques which are available only in central or reference laboratories. These constraints are particularly evident in the case of viruses such as those which cause respiratory diseases and exanthems. With some of these viruses, careful collection and immediate inoculation onto multiple cell culture cell lines is required for successful cultivation (8, 23). With fastidious viruses such as those which cause viral gastroenteritis and hepatitis, cultivation is difficult to achieve in a consistent fashion in any tissue cell lines which are generally available to the diagnostic microbiologist (8, 16). In addition, there are bacterial and fungal organisms which, although they can be isolated in pure culture, require extended periods of time for accurate identification. This is particularly true with some strains of myobacteria, anaerobes, rickettsiae, and pathogenic fungi (23).

For this reason, there has been interest in the development of assay systems capable of direct detection of microbial antigens in body fluids without the requirement for cultivation. Although there are a number of ways in which microorganisms can be detected in body fluids, the specificity and reproducibility of antigen-antibody reactions has led to the widespread application of immunoassays for this purpose. There are a number of formats in which immunoassays can be performed. These formats differ both in the way in which antigens and antibodies are reacted and in the indicator labels which are utilized to quantitate the immune reaction. However, regardless of format, an assay system must meet several requirements to be useful for microbial diagnosis. For example, the system must be sensitive, since many microorganisms are present in body fluids at concentrations less than 1 ng/ml. In addition, the assay must be relatively unaffected by interfering materials such as rheumatoid factors and other immunoglobulin-binding proteins which might be present in body fluids (28). In addition, it would be preferable that the assay system be quantitative, since this will allow for the objective determination of assay endpoints as well as for the performance of quantitative control reactions. Although numerous immunoassay systems meet these specifications, one assay system which has attained widespread usage for the direct detection of microbial antigens in body fluids is that of enzyme immunoassay (EIA), also known as enzyme-linked immunosorbent assay. As in the case of other solid-phase systems, such as solid-phase radioimmunoassays, EIAs involve a series of reaction steps in which the immunoreactants are bound to a solid-phase surface by a series of antigen-antibody reactions. The use of solid-phase supports to bind immunoreactants allows for the detection of antigens in a wide range of body fluid specimens, since many potentially interfering materials in specimens can be removed by washing before the performance of subsequent antigen-antibody reactions. However, EIAs differ from other solid-phase immunoassays in that the terminal immunoreactants are labeled with molecules with enzymatic activity. The EIA reaction is quantitated by the measurement of the enzyme-substrate reaction which results from the interaction of enzyme-labeled immunoreactants bound to the solid phase with added substrate.

The utilization of enzyme-substrate reactions offers the inherent advantage that enzymes are biological magnifiers. Thus, a small number of enzyme molecules can convert a large number of substrate molecules to a measurable product. This magnification allows for the detection of small amounts of enzyme-labeled immunoreactants by using relatively low-energy markers such as those which generate visibly colored products. This capability allows the formulation of an assay system which has a high degree of sensitivity without the need for marker molecules which generate large amounts of energy. Thus, EIAs can utilize stable reagents with relatively long shelf life and can be performed in situations which are not suitable for the use of radioactive isotopes (16, 20, 22).

To be useful for EIA systems, an enzyme must have a high turnover rate. In addition, it is necessary that the enzyme be capable of linkage to immunoreactants in a consistent fashion to formulate stable conjugates which retain a high proportion of the original immunological and enzymatic activity of the marker molecules. Although a large number of enzymes meet these requirements, most of the assays to date have utilized either horseradish peroxidase, alkaline phosphatase, or beta-galactosidase as the enzyme marker. These enzymes can be efficiently linked to immunoreactants by a number of means and detected at concentrations as low as 10^{-12} mol/liter (approximately 0.1 ng/ml) by the use of colorimetric instrumentation and at somewhat higher concentrations by the naked eye by colorimetric means (1, 12). It should be noted, however, that there are additional enzymes which might prove useful in EIA systems. The enzyme beta-lactamase, for example, has the advantage of being present in large concentrations in a wide range of microorganisms but is generally not found in uninfected human body fluids (17). The measurement of this enzyme thus offers the advantage of avoiding the possibility of background enzymatic activity due to the presence of mammalian enzymes. In addition, beta-lactamase can be utilized with a starch-iodine-penicillin G substrate which yields a very distinct color reaction. This reaction is based on the conversion of a deep blue starch-iodine complex to a colorless iodine-starch complex by the penicillinoic acid generated by the beta-lactamase reaction. This reaction allows for sensitive visual readings as well as for the recording of immunoassay results with standard office photocopying instrumentation (29; R. H. Yolken, M. Van Regenmortel, and S. B. Wee, J. Immunol. Methods, in press). In addition, the substrates utilized for beta-lactamase, namely the penicillin and starch-iodine compounds, are inexpensive, widely available, and free of known carcinogenic or mutagenic potential in the concentrations at which they are utilized.

As discussed above, most EIAs utilize enzyme-substrate reactions which generate visible color upon enzymatic reaction. However, EIAs can also be formulated with substrates which generate higher-energy substrates, such as those with fluorescent, chemiluminescent, or radioactive properties. Since these substrates can be detected in lower concentrations than the ones which generate visible color, they detect lower concentrations of enzyme-labeled immunoreactants (7, 14, 16). Some of the practical high-energy substrates which might be utilized in EIA systems are listed in Table 1. However, it should be noted that the principal limit of sensitivity of immunoassay systems is determined by the binding characteristics of the immunoreactants which are utilized in the system. Thus, while high-energy substrates will generally result in a decrease of reaction volumes and time required for the generation of measurable substrate, the use of such substrates will not always result in an absolute increase in assay sensitivity, especially if prolonged incubation times are utilized (23).

REAGENTS

The principal determinants of the sensitivity and specificity of EIA are the immunoreagents which are utilized in the assay system. Of particular importance is the use of immune reagents which will specifically react with small quantities of antigen from the microbial antigens of interest but will not react with antigens from other microbes or from mammalian cells. Two methods of antibody preparation have proven useful in solid-phase immunoassay systems. One method involves the classical techniques for the preparation of hyperimmune antisera from immunized animals. In such cases, it is necessary that the antigen be purified as much as possible to remove extraneous antigens which might result in the generation of antibody to heterologous antibody. In addition, it is necessary that the animals be immunized in such a way that the production of the high-affinity antibody is ensured. We have found that, since protocols for immunization are difficult to standardize, it is advisable that the immunized animals be bled at regular intervals after immunization and that the resulting antisera be tested in solid-phase systems for sensitivity and specificity. It should be noted that reaction conditions of solid-phase immunoassay systems are somewhat different than conditions in other assay systems utilized for the measurement of antimicrobial

TABLE 1. Comparison of enzymes available for EIA[a]

Enzyme	Source	Uses	pH optimum	Sp act (U/mg)	Mol wt (approx)	Practical methods of conjugation	Practical substrates available		
							Visual	Fluorescent	Radioactive
Alkaline phosphatase	Calf intestine	EIA	8–10	400	100,000	Glutaraldehyde, 1 step	NP-PO_4	MU-PO_4	[^{3}H]AMP
Peroxidase	Horseradish	EIA, HIST	5–7	900	40,000	Glutaraldehyde, 1 step	H_2O_2 + 5-AS	NADH	NA
						Glutaraldehyde, 2 step	H_2O_2 + OPD	NPA	
						NaIO	H_2O_2 + ABTS		
						Biotin			
β-Galactosidase	*Escherichia coli*	EIA	6–8	400	540,000	Glutaraldehyde, 1 step	NP-Gal	MU-Gal	^{3}H-GalP
						p-Benzoquinine			
						SDPD			
Glucose oxidase	*Aspergillus niger*	EIA, HIST	4–7	200	160,000	Glutaraldehyde, 1 step	Glu + 5-AS Glu + NBT	NADH	^{3}H-Glu
β-Lactamase	Bacteria	EIA	5–8	2,400	45,000	Glutaraldehyde, 1 step	PenG + Starch-I_2	NA	^{14}C-PenG
						Biotin	CPD		
Catalase	Calf liver	EIA	6–8	40,000	250,000	Glutaraldehyde, 1 step	H_2O_2 + I_2	NA	NA

[a] Abbreviations: NP-PO_4, *p*-nitrophenyl phosphate; MU-PO_4, methylumbelliferyl phosphate; HIST, enzyme-mediated histochemical procedure; 5-AS, 5-aminosalicylic acid; NADH, nicotinamide adenine dinucleotide phosphate, reduced form; NA, not available; OPD, *o*-phenylenediamine; HPA, *p*-hydroxyphenylacetic acid; ABTS, 2,2-azino-di-(*d*-ethylbenzothiazoline sulfone-b) diammonium salt; NP-Gal, nitrophenyl galactose; MU-Gal, methylumbelliferyl galactose; GalP, galactose 6-phosphate; SDPD, *N*-succinimidyl-3(2-pyridyldithio)propionate; Glu, glucose; NBT, *p*-nitro blue tetrazolium chloride; PenG, penicillin G; CPD, chromagenic penicillin derivatives.

antibodies. Thus, the antisera should be evaluated by solid-phase methods to determine their suitability for use in EIA systems. Once antibody with suitable properties is documented by the evaluation of the test bleeds, larger quantities of serum should be obtained from the animals, divided into aliquots, and stored until used. Since only a small portion of immunoglobulin from an animal which has been immunized with purified antigens is actually directed at the immunizing antigen, reaction parameters can be improved by increasing the concentration of specific antibody in the immune serum. This increase can be most readily accomplished by means of immunoprecipitation or immunochromatography techniques (15). However, the optimal utilization of these techniques is often limited by the availability of purified antigens in quantities sufficient for the performance of the techniques.

An alternative method for the production of antibody for solid-phase immunoassay systems involves the production of monoclonal antibodies in in vitro hybridoma systems after fusion with effector cells from immune animals (4). An advantage of these systems is that clones can be produced which generate antibodies with only a defined range of antigenic reactivity. Since each hybridoma clone produces a single antibody, highly purified reagents are not necessary for the immunization of the animal before the isolation of the antibody-producing clones. This situation can be advantageous when it is difficult to obtain such purified antigens. In addition, the monoclonal nature of the immunoglobulins produced by the clones decreases the likelihood of nonspecific reactions due to extraneous antigens in the clinical specimen. Furthermore, virtually all of the antibody produced by the clones is directed at an epitope on the immunizing antigen. Hence, reaction kinetics are more favorable in the solid-phase immunoassay. In addition, the replicative nature of hybridoma systems ensures that there will be a large supply of the desired immunoglobulin, since the antibody-producing cells have a virtually unlimited capacity for cellular division. This supply obviates the need for constant preparation and restandardization of immunoreagents.

Despite these advantages, a number of problems are inherent in the use of monoclonal antibodies for solid-phase EIA systems. The most important theoretical limitation is that the range of reactivity of a monoclonal antibody might be too narrow to ensure the detection of antigen from a microbial pathogen in all forms which might be present in clinical specimens. For this reason, it is important that monoclonal antibodies be evaluated not only for reactivity to standard antigens, but also for reactivity to a large range of clinical specimens containing the antigen. In addition, many monoclonal antibodies do not have sufficient affinity to allow for the detection of small concentrations of antigens in solid-phase immunoassay systems. However, when high-affinity monoclonal antibodies can be obtained, their properties often favor their usage in solid-phase immunoassay systems for microbial detection.

All EIAs involve the linkage of enzymes to immunoreactants. In some cases this is accomplished by the direct, covalent linkage of enzymes to immunoglobulin molecules. This covalent linkage can be accomplished by the cross-linking of ϵ-amino groups of lysine residues of the enzyme and immunoglobulin molecules by means of bifunctional reagents such as gluteraldehyde (3). In addition, in the case of glycosylated enzymes such as horseradish peroxidase, the carbohydrate can be oxidized by mild oxidizing agents, and linkage can be accomplished by the formation of a covalent bond between aldehyde groups formed on the carbohydrate portion of the enzyme and free amino groups on the immunoglobulin (10). In addition, conjugation can be accomplished by utilizing linkage between sulfhydryl groups of cystine residues on the immunoglobulin or enzymes. For example, the enzyme beta-galactosidase contains available sulfhydryl residues and thus can be linked to immunoglobulin by sulfhydryl-reacting molecules such as *N*-succinimidyl 3-(2-pyridyldithio)propionate (11). With other enzymes, sulfhydryl groups can be introduced into the molecule by chemical methods to allow for sulfhydryl linkage. In addition, the sulfhydryl groups at the hinge region of Fab fragments of the immunoglobulin molecule can be used for linkage to enzymes (6). Since the hinge area is located at a defined position on immunoglobulin molecules, at a site removed from the antigen-binding site of the immunoglobulin, this method results in the reproducible generation of a conjugate with virtually complete retention of immunological activity.

Alternatively, noncovalent methods can be utilized to link enzyme to immunoglobulin. One such method involves the use of enzyme-labeled protein A conjugated with peroxidase to form a noncovalent bond to immunoglobulin molecules (24). Such conjugates can be prepared by the simple mixing of enzyme-labeled protein A and immunoglobulin subclasses with protein A-binding capability. Since protein A binds reproducibly to the Fc portion of the immunoglobulin molecule at a site removed from the Fab antigen-binding site, antibody activity is consistently preserved. However, one problem with protein A-binding reactions is that dissociation can occur during the subsequent antigen-antibody reaction. This dissociation can result in nonspecific binding to other immunoreactants in the assay systems, with the subsequent generation of nonspecific reactions. Another means of noncovalent linkage involves the use of antibodies directed at enzymes to accomplish the linkage between immunoglobulin and enzyme. Such methods are widely used in histology, since efficient conjugates with favorable reaction kinetics (25) can be formed. In solid-phase EIAs, enzyme-antienzyme linkage can be accomplished either by separate incubations with antibody to immunoglobulins and to enzymes such as peroxidase (peroxidase-antiperoxidase method) or by the covalent formation of hybrid antibodies prepared from immunoglobulins directed at the antigen and other immunoglobulins directed at the enzyme (9). In this case, the reaction is completed by the subsequent binding of enzyme to the antiantigen-antienzyme hybrid and the reaction with substrate. This format has the advantage of obviating the need for large amounts of highly purified enzyme, since the antienzyme will specifically bind enzyme from a crude mixture.

All of the above conjugation methods can be utilized to generate stable enzyme-immunoglobulin conjugates. However, the direct conjugation of enzyme to immunoglobulin can result in a macromolecular conjugate with diminished immunological and enzymatic activities compared with those of the parent molecules. This diminished activity is particularly problematic in situations in which large polymers are formed, since such polymers can display markedly decreased activity due to the steric blockage of active sites within the polymer (2). Many of the problems inherent in such conjugation can be overcome by the linkage of immunoglobulins to a low-molecular-weight marker. The reaction can then be quantitated by the subsequent reaction of this marker to the enzyme. The use of a low-molecular-weight marker has the advantage that polymerization is unlikely to occur. In addition, the small size of the marker relative to immunoglobulin ensures that unreacted material can be removed by simple separation techniques such as dialysis or gel exclusion chromatography. This is useful when a large number of different immunoreactants must be labeled in a single run. One low-molecular-weight marker which is particularly suited for microbial detection is biotin (5). Biotin is a low-molecular-weight vitamin with a high affinity for avidin, a protein found in abundance in egg white and some species of fungi. Biotin can be linked covalently with immunoglobulin molecules through amino, carboxyl, or sulfhydryl residues on the immunoglobulin. Linkage to immunoglobulin is most easily accomplished through amino groups by reaction with nucleophilic biotin derivatives such as *N*-hydroxy-biotin succinimide ester. This reaction can be accomplished by the simple mixing of reagents in an alkaline buffer, with the subsequent removal of unreacted biotin by dialysis. Alternatively, the amino groups of proteins can be reacted with larger, active biotin molecules such as biotin ϵ-amino caproic acid esters. Linkage with these molecules results in the location of the active site of the biotin at a greater intramolecular distance from the immunoglobulin compound, resulting in the facilitated interaction between biotin-labeled immunoglobulin and labeled avidin due to decreased steric hindrance. After reaction with the biotin-labeled immunoreactant, the amount of biotin-labeled antibody which binds to the antigen is quantitated by subsequent reaction with enzyme-linked avidin. The avidin can be linked to the enzyme by covalent attachment in a manner similar to the methods described. Alternatively, a complex of avidin and biotin-labeled enzyme can be formed by the simple mixing of these reagents under conditions which result in the production of free biotin-binding sites on the avidin (27). It should be noted that unmodified avidin purified from egg white displays a significant amount of nonspecific binding to solid-phase surfaces. This problem can be overcome by the use of succinylated avidin or other avidin preparations with decreased rates of nonspecific binding. Avidin from *Streptomyces* spp. can also be utilized in place of avidin from egg white. If these avidin preparations are utilized, efficient avidin-biotin systems can be formulated with minimal degrees of nonspecific interactions.

PREPARATION OF SPECIMENS

The optimal methods for obtaining the specimen and ensuring maximum reactivity vary, depending upon the nature of the clinical specimen and the antigen to be detected. One important parameter in assay sensitivity is the accessibility of the antigen in the specimen. To be detectable in a solid-phase immunoassay, the antigen in the clinical specimen must be in a form which is capable of reacting with solid- and liquid-phase antibodies. Antigens which are inaccessible to antibodies can thus be difficult to detect in solid-phase immunoassay formats. In practice, such inaccessible sites include intracellular locations as well as antigens which have been complexed with endogenous antibody in the form of immune complexes (21). Antigens which are very stable, such as bacterial polysaccharides, can be exposed in both intracellular and immune-complexed sites by simple treatments directed at denaturing the lipid and protein components of such complexes (18). This can be accomplished by the use of lipases and proteases or by heat denaturation such as incubation for 30 min at 78°C. With the more labile protein antigens, such harsh treatments cannot be utilized since they can also destroy the antigen to be measured. In such cases, detergents such as polysorbates (Tween) and Triton X can be utilized to liberate protein antigens from membrane surfaces. Immune complexes can be dissociated with an agent such as urea or sodium thiocyanate (13). However, in such cases it is important that the dissociating agent be removed before the testing to prevent interference with subsequent antigen-antibody reactions. Since such procedures can be laborious and time consuming, it is often more practical to simply test the specimens after incubations for extended periods with high-affinity, solid-phase immunoreactants to allow for the binding of the immunoreactants to free antigenic sites which are exposed on the complexes. Although such procedures do not allow for the detection of all complexed antigens, there is generally sufficient free antigen circulating early in the course of infection to allow for the diagnosis of most infectious diseases during that period.

An additional important consideration in the collection of specimens concerns the possible presence of interfering materials in the clinical specimen. Although endogenous molecules with enzymatic activity or molecules which inhibit enzymatic activity are frequently present in clinical specimens, these materials generally do not result in problems in the performance of solid-phase immunoassays since they are removed by washing before the enzyme-substrate reaction. However, a more serious problem involves the presence in clinical specimens of materials which are capable of desorbing immunoreactants from the solid-phase surface (19). Some body fluid specimens, especially stools and intestinal fluids, contain proteases and other molecules capable of desorbing reactants from solid-phase surfaces. The presence of such desorbing materials in the specimen can lead to a markedly decreased reactivity in the solid-phase immunoassay systems. Such desorbing materials in clinical specimens can be readily detected by measurement of the desorption of enzyme- or biotin-labeled immunoglobulins which have been bound to the solid-phase surface. In addition, the effect of the desorbing material can be neutralized by the dilution of the specimen in materials which interfere with protease activity. Although numerous protease inhibitors can be utilized, we have found that the most practical are concentrated (50%) solutions of animal sera such as fetal calf serum or solutions containing 5% albumin in an acid buffer (0.1 M citrate, pH 5.0). We have found that such treatments will result in virtually complete restoration of activity. In most cases, it is important to determine that the serum or albumin does not contain antibody directed at the antigen to be measured, since such antibodies will decrease the specific reactivity of the system.

An additional problem related to the testing of clinical specimens is that the specimens can contain materials which will react nonspecifically with the immunoglobulins which are utilized as immunoreactants in the assay system. Such nonspecifically reacting materials include rheumatoid-like factors and bacterial products with immunoglobulin-binding capacity (28). Such materials can cause false-negative reactions by sterically inhibiting specific binding, or they can cause false-positive reactions by nonspecifically cross-linking labeled reactants with the solid-phase antibodies in the absence of specific antigen. There are several potential approaches to the reduction or elimination of the effect of such nonspecific antiglobulins. In the case of heat-stable antigens, such as bacterial or fungal polysaccharides, the effect can be eliminated by heating or chemical denaturation, which destroys the interfering proteins but will not alter the antigenic activity of the infecting microorganism. In the case of more labile antigens, the nonspecific binding can be decreased by the addition of reducing agents such as dithiothreitol, which will denature immunoglobulin M reactants. The reactivity of the immunoglobulin G or bacterial antiglobulins for solid-phase immunoreactants can be decreased by the addition of an excess of immunoglobulins or Fc fragments to the reaction mixture. For example, nonimmune goat or guinea pig immunoglobulin can be added to the reaction mixture when goat or guinea pig immunoglobulin is utilized as the solid phase. This will result in the binding of the nonspecific antiglobulins to the added immunoglobulin rather than to the solid-phase or labeled immunoreagent. Since most of the antiglobulin factors react with the Fc portion of the immunoglobulin molecule, the extent of nonspecific binding can be decreased by the use of the Fab fragments as the solid-phase immunoreactant. The use of Fab fragments is thus recommended in situations in which nonspecific reactions can occur and in which the Fab fragment of the immunoglobulin can be prepared in sufficient quantity for use in the immunoassay systems.

IMMUNOASSAY FORMATS

Solid-phase immunoassays can be performed in a number of different formats. In each case, the preferred format will be determined by the needs of the laboratory, the nature of the antigen to be measured, and the available immunoreactants. In addition, opti-

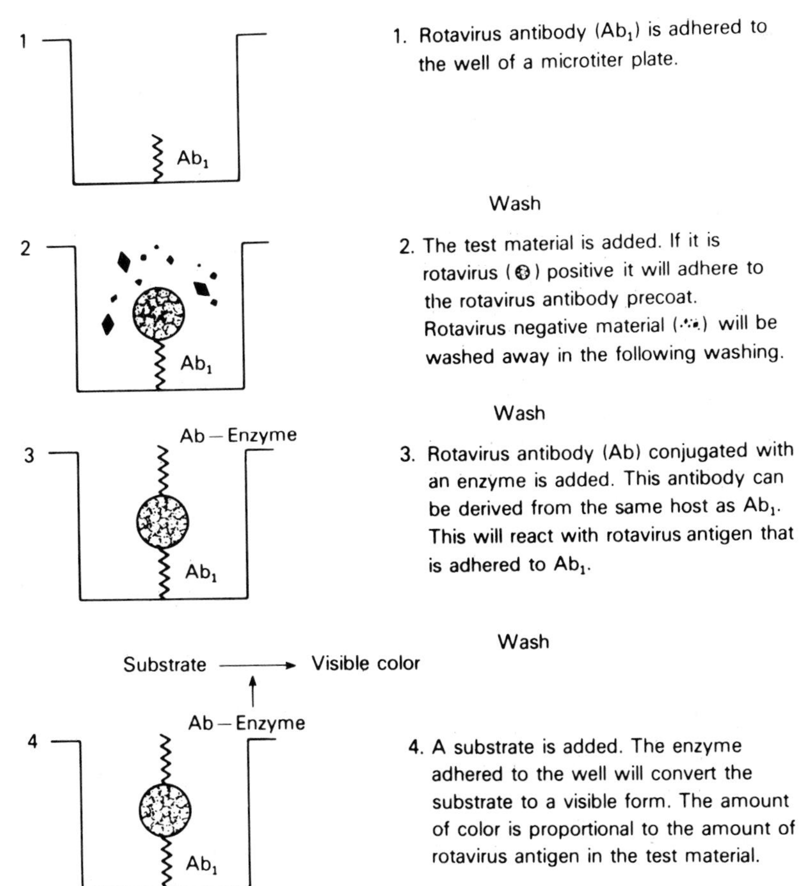

1
Ab1
1. Rotavirus antibody (Ab1) is adhered to the well of a microtiter plate.
Wash
2
Ab1
2. The test material is added. If it is rotavirus positive it will adhere to the rotavirus antibody precoat. Rotavirus negative material will be washed away in the following washing.
Wash
3
Ab – Enzyme
Ab1
3. Rotavirus antibody (Ab) conjugated with an enzyme is added. This antibody can be derived from the same host as Ab1. This will react with rotavirus antigen that is adhered to Ab1.
Wash
Substrate → Visible color
4
Ab – Enzyme
Ab1
4. A substrate is added. The enzyme adhered to the well will convert the substrate to a visible form. The amount of color is proportional to the amount of rotavirus antigen in the test material.

FIG. 1. Direct enzyme-linked immunosorbent assay for antigen measurement. From reference 8; used by permission.

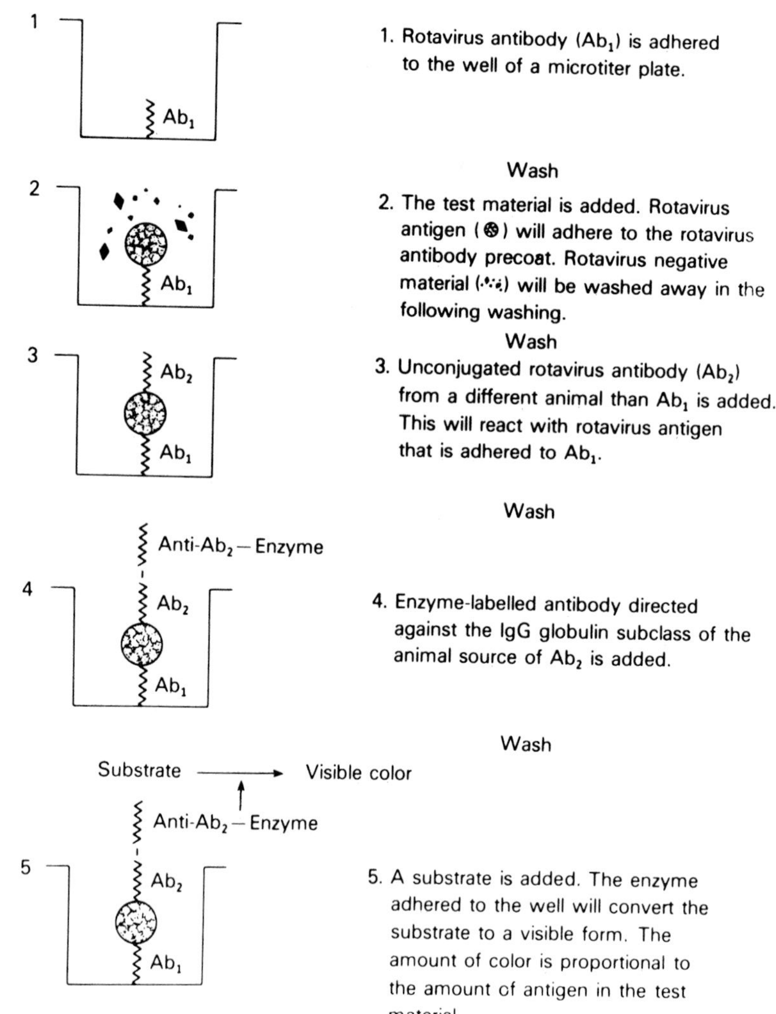
1
Ab₁
1. Rotavirus antibody (Ab₁) is adhered to the well of a microtiter plate.
Wash
2
Ab₁
2. The test material is added. Rotavirus antigen ( ) will adhere to the rotavirus antibody precoat. Rotavirus negative material ( ) will be washed away in the following washing.
Wash
3
Ab₂
Ab₁
3. Unconjugated rotavirus antibody (Ab₂) from a different animal than Ab₁ is added. This will react with rotavirus antigen that is adhered to Ab₁.
Wash
Anti-Ab₂—Enzyme
4
Ab₂
Ab₁
4. Enzyme-labelled antibody directed against the IgG globulin subclass of the animal source of Ab₂ is added.
Wash
Substrate
Visible color
Anti-Ab₂—Enzyme
5
Ab₂
Ab₁
5. A substrate is added. The enzyme adhered to the well will convert the substrate to a visible form. The amount of color is proportional to the amount of antigen in the test material.

FIG. 2. Indirect enzyme-linked immunosorbent assay for rotavirus antigen measurement. From reference 8; used by permission.

mal reagent dilutions and reaction conditions should be determined for each assay system by appropriate titrations.

One practical EIA format is the direct immunoassay (Fig. 1). In this type of immunoassay, antibody directed at the antigen to be measured or control antibody is bound to a solid-phase surface. Although numerous surfaces have been devised for use in EIA systems, the most widely used surfaces are plastics which are capable of binding immunoglobulins by means of hydrophobic interactions or covalent attachment. The plastic can be used in a variety of forms; however, the most convenient solid phases generally consist of beads or microtiter plates.

After incubation of the specimen with solid-phase antibody (generally for 2 h at 37°C or overnight at room temperature or 4°C), the unreacted material is removed by washing, and enzyme-labeled antibody directed at the antigen is added. After another period of incubation (generally 1 or 2 h at 37°C), unreacted antibody is removed by washing, and substrate is added. The presence of antigen in the original specimen will be manifested by the binding of enzyme-labeled antibody to the solid phase and the subsequent enzymatic conversion of substrate to detectable product. Alternatively, haptene-labeled reagents prepared as described above can be utilized in place of the enzyme-linked reagents. In such cases, the reaction is completed by the addition of an enzyme-linked reagent capable of binding with the haptene and substrate. In each case, specific activity can be determined by a comparison of optical densities generated by reaction with immune and nonimmune reagents as described above, and the positivity of an individual specimen can be determined by comparison of its specific activity with known negative-control specimens.

EIAs can also be performed in an indirect format in which the reaction is completed by the addition of enzyme-labeled antiglobulin directed at the species of the antibody utilized to bind the antigen (Fig. 2). An advantage of such systems is that they can utilize widely available enzyme-labeled antispecies immunoglobulins and thus do not require the preparation of labeled reagents directed at specific antigens. However, care must be taken to avoid nonspecific reactions between the labeled antiglobulin and the solid-phase or endogenous immunoglobulins in the specimen

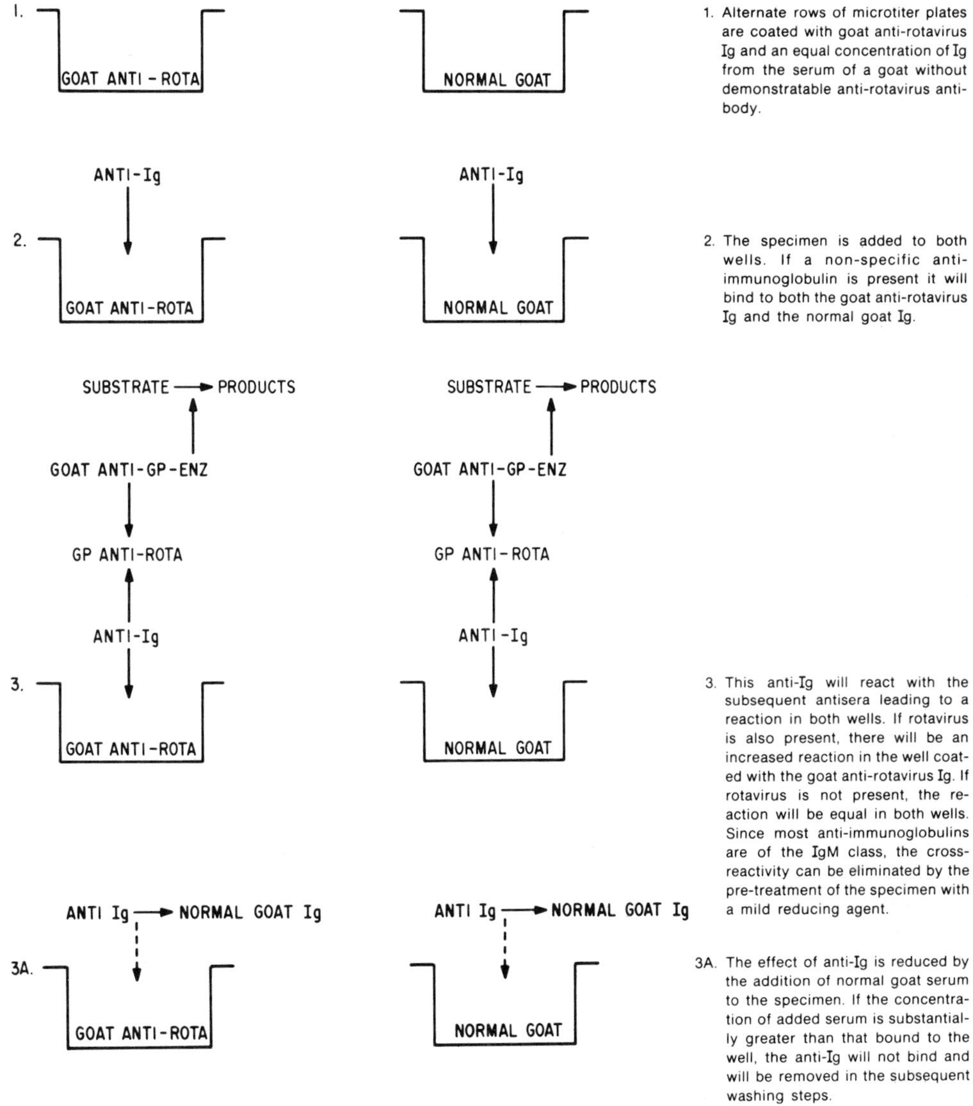
1.
GOAT ANTI - ROTA
NORMAL GOAT
1. Alternate rows of microtiter plates are coated with goat anti-rotavirus Ig and an equal concentration of Ig from the serum of a goat without demonstratable anti-rotavirus antibody.
ANTI-Ig
ANTI-Ig
2.
GOAT ANTI-ROTA
NORMAL GOAT
2. The specimen is added to both wells. If a non-specific anti-immunoglobulin is present it will bind to both the goat anti-rotavirus Ig and the normal goat Ig.
SUBSTRATE → PRODUCTS
SUBSTRATE → PRODUCTS
GOAT ANTI-GP-ENZ
GOAT ANTI-GP-ENZ
GP ANTI-ROTA
GP ANTI-ROTA
ANTI-Ig
ANTI-Ig
3.
GOAT ANTI-ROTA
NORMAL GOAT
3. This anti-Ig will react with the subsequent antisera leading to a reaction in both wells. If rotavirus is also present, there will be an increased reaction in the well coated with the goat anti-rotavirus Ig. If rotavirus is not present, the reaction will be equal in both wells. Since most anti-immunoglobulins are of the IgM class, the cross-reactivity can be eliminated by the pre-treatment of the specimen with a mild reducing agent.
ANTI Ig → NORMAL GOAT Ig
ANTI Ig → NORMAL GOAT Ig
3A.
GOAT ANTI-ROTA
NORMAL GOAT
3A. The effect of anti-Ig is reduced by the addition of normal goat serum to the specimen. If the concentration of added serum is substantially greater than that bound to the well, the anti-Ig will not bind and will be removed in the subsequent washing steps.

FIG. 3. Cross-reactivity due to anti-immunoglobulin. From reference 8; used by permission.

which have bound nonspecifically to the solid-phase surface. It is thus preferable to utilize antiglobulins which have been extensively purified to ensure a minimum of cross-reaction. High-affinity monoclonal antibodies directed at a species of immunoglobulins can also be utilized for this purpose.

One advantage of quantitative solid-phase immunoassays is that control reactions to determine the specificity of the assay result can easily be performed. One such format utilizes nonimmune serum bound to the solid phase in place of specific antiserum (Fig. 3). In such cases, specific reactions should be manifested by a marked decrease or absence of activity in wells in which nonimmune serum is utilized in place of specific antisera directed at the antigen. On the other hand, nonspecific reactivity is noted by the generation of signal in wells coated with nonimmune serum. Alternatively, specificity can be documented by the performance of blocking assays, in which the ability of immune and nonimmune sera to inhibit the reaction of antigen in the clinical specimen can be determined. In such blocking assays, specificity is manifested by a decrease in activity after reaction with specific antiserum compared with reaction with nonimmune serum. It should be noted, however, that the antigen must be reacted at a concentration which is not in excess of the added antibody so that the inhibition reaction can be easily measured. One advantage of the blocking formats is that, since the blocking reagent can be added in excess, low-affinity, postinfection serum or monoclonal antibodies can be utilized as the confirmatory reagents. However, in practice, one disadvantage of the blocking assays is that they generally require the performance of a separate confirmatory assay to determine specificity, whereas in the case of the control format described above, the detection and the confirmatory assays can be performed in the same test run.

Antigen can also be detected by competitive formats, which can be performed with either labeled antibody or labeled antigen. In one practical competitive format (Fig. 4), antigen is bound to the solid-phase surface. The unknown specimen and labeled

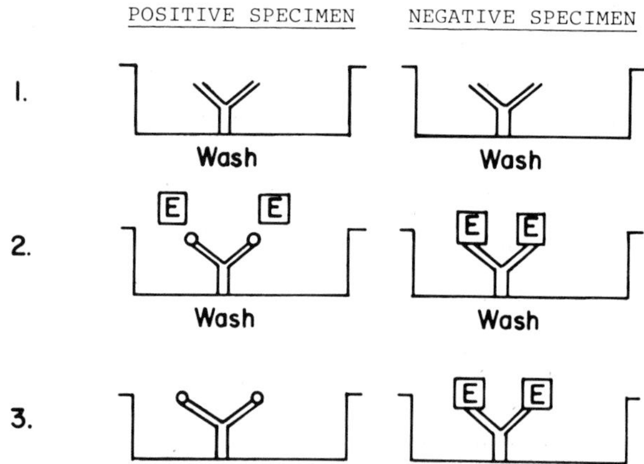

FIG. 4. Competitive format for antigen detection. From reference 8; used by permission. (1) Antibody (Y) is bound to the solid phase. Unbound antibody is removed in the washing step. (2) The specimen is added. If it contains antigen (○), it will bind to the solid-phase antibody. Enzyme-labeled antigen (E) is then added. This will react with antibody sites not occupied by antigen from the specimen. (3) Unbound enzyme-labeled antigen is removed by washing, and the amount of bound enzyme-labeled antigen is quantitated by the addition of the appropriate substrate. The amount of substrate product is inversely proportional to the amount of antigen in the test specimen.

antibody are incubated together and added to the solid-phase antigen. If antigen is present in the specimen, it will compete with the solid-phase antigen for binding sites on the labeled antibody. The presence of antigen in the specimen will thus be manifested by a decreased amount of labeled antibody bound to the solid phase and a decreased amount of signal generated by the subsequent enzyme-substrate reaction. Alternatively, the reaction can be completed with antispecies immunoglobulin, avidin-biotin, or haptene-antihaptene reactions as described above. Although such assays are somewhat less sensitive than the noncompetitive assays discussed above, they offer a number of advantages for the rapid diagnosis of infectious diseases. The most important advantage is that they require fewer incubations and reaction procedures and thus allow for more rapid diagnosis of the infectious disease. In addition, the specificity of competitive reactions is determined only by the portion of the antibody which is capable of reacting with the solid-phase antigen. This is advantageous when it is difficult to get large amounts of highly purified high-affinity antibody. However, a disadvantage of the competitive format is that large amounts of antigens suitable for binding to the solid phase must be available. This can be problematic with some viruses for which it is difficult to obtain such antigens. In addition, the nature of the solid-phase reactions makes them highly susceptible to interfering effects from desorbing material which might be present in clinical specimens. As described above, such desorbing material can cause the elution of reactants from the solid-phase surface. In the case of the competitive formats, this elution will result in a decrease of the signal and thus a false-positive reaction. It is thus advisable that specimens which are likely to contain such desorbing material, such as stools, be processed with protease inhibitors such as animal sera to prevent desorption and false-positive reactions.

One disadvantage of the above sequential-reaction formats is that they require a number of separate incubation and washing steps for assay performance. However, when the immunoreactants are directed at distinct antigenic determinants, the reaction steps consisting of the binding to solid- and liquid-phase antibodies can be performed simultaneously (26). In one simultaneous-reaction format (Fig. 5), the antibodies utilized as the solid- and liquid-phase immunoreactants are reacted simultaneously with the antigen. Since the antibodies react with distinct determinants, these simultaneous interactions will not result in significantly decreased reactivity due to steric hindrance (23). After these reactions, the solid phase is washed, and the amount of bound label is determined by the addition of substrate. This reaction format can be formulated most easily by the use of distinct monoclonal antibodies. However, such assays can also be formulated with hyperimmune animal sera prepared in different species if care is taken in the selection of the reagents in terms of antigenic reactivity. We have found that under appropriate conditions

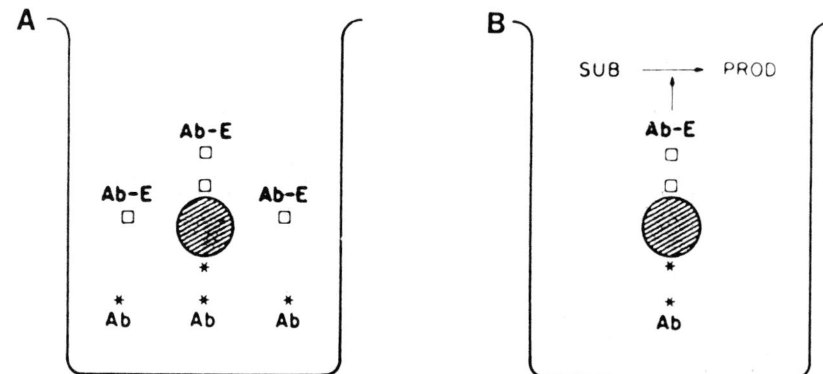

FIG. 5. Double-determinant enzyme immunoassay. From reference 8; used by permission. (A) Antigen is added to a solid phase coated with antibody directed against one antigenic site (⋆Ab). Enzyme-labeled antibody directed against a different site on the antigen (□Ab-E) is then added. This will react with unbound sites on the antigen. (B) After a washing step to remove unreacted □ Ab-E, substrate is added. This will be converted by bound □Ab-E to a measurable product. The amount of product formed will be proportional to the concentration of antigen in the specimen.

this format can be utilized to detect antigen in specimens utilizing reaction times as short as 30 min. Although this shortened format can be somewhat less sensitive than sequential-reaction formats, the availability of such rapid-detection systems can be extremely useful, not only for the rapid diagnosis of infectious diseases, but also for the control of the transmission of infectious agents within a population.

This work was supported by the National Institute of Allergy and Infectious Diseases contract NO1-AI-22680.

LITERATURE CITED

1. **Avrameas, S., and T. Ternynck.** 1971. Peroxidase-labelled antibody and Fab conjugates with enhanced intracellular penetration (letter). Immunochemistry **8:**1175–1179.
2. **Boorsma, D. M., J. G. Streefkerk, and N. Kors.** 1976. Peroxidase and fluorescein isothiocyanate as antibody markers: a quantitative comparison of two peroxidase conjugates prepared with glutaraldehyde or periodate and a fluorescein conjugate. J. Histochem. Cytochem. **24:**1017–1025.
3. **Engvall, E., and P. Perlmann.** 1972. Enzyme linked immunosorbent assay. ELISA. III. Quantitation of specific antibodies by enzyme-linked anti-immunoglobulin in antigen coated tubes. J. Immunol. **109:**129–135.
4. **Gerhard, W., C. M. Croce, D. Lopes, and H. Koprowski.** 1978. Repertoire of antiviral antibodies expressed by somatic cell hybrids. Proc. Natl. Acad. Sci. U.S.A. **75:**1510–1514.
5. **Guesdon, J. L., T. Ternynck, and S. Avrameas.** 1979. The use of avidin-biotin interaction in immunoenzymatic techniques. J. Histochem. Cytochem. **27:**1131–1139.
6. **Ishikawa, E., M. Imagawa, and S. Hashida.** 1983. Ultrasensitive enzyme immunoassay using fluorogenic, luminogenic, radioactive and related substrates, p. 219–232. *In* S. Avrameas, P. Druet, H. R. Massayeff, and G. Feldman (ed.), Immunoenzymatic techniques, 2nd ed. Elsevier, Amsterdam.
7. **Iwasha, S., H. Veno, T. Miya, M. Wakimasu, K. Kondo, and A. Ohneda.** 1979. Enzyme immunoassay of pancreatic glucagon at the picogram level using β-D-galactosidase as a label. J. Biochem. **86:**943–949.
8. **Kapikian, A. Z., R. H. Yolken, H. B. Greenberg, R. G. Wyatt, A. R. Kalica, R. M. Chanock, and H. W. Kim.** 1979. The gastroenteritis viruses, p. 927–995. *In* E. H. Lennette and N. Schmidt (ed.), Diagnostic procedures for viral, rickettsial, and chlamydial infections, 5th ed. American Public Health Association, Inc., New York.
9. **Mandache, E., E. Moldoveanu, G. Mota, I. Moraru, and V. Ghetie.** 1980. Multivalent hybrid antibody with double specificity as a tool for locating cell surface antigens by electron microscopy. J. Immunol. Methods **36:**33–41.
10. **Nakane, P. K., and A. Kawaoi.** 1974. Peroxidase-labelled antibody—a new method of conjugation. J. Histochem. Cytochem. **22:**1084–1091.
11. **Nilsson, P., N. R. Bergquist, and M. S. Grundy.** 1981. A technique for preparing defined conjugates of horseradish peroxidase and immunoglobulin. J. Immunol. Methods **41:**81–93.
12. **Pesce, A. J., R. R. Modesto, D. J. Ford, K. Sethi, D. N. Clyne, and V. E. Pollak.** 1976. Preparation and analysis of peroxidase antibody and alkaline phosphatase antibody conjugates, p. 7–23. *In* G. Feldman, P. Druett, J. Bignon, and S. Avrameas (ed.), Immunoenzymatic techniques. North-Holland, Amsterdam.
13. **Popova, O. Y., and L. S. Kositskaya.** 1977. Dissociation of immune complexes and inactivation of bound antibodies by reducing agents. Immunochemistry **14:**633–635.
14. **Pronovost, A. D.** 1980. A highly sensitive enzyme immunoassay using chemiluminescence. *In* Workshop on new and useful techniques in rapid viral diagnosis, National Institutes of Health, Bethesda, Md., May, 1980. Summarized in J. Infect. Dis. **142:**793–802, 1980.
15. **Robbins, J. B., J. Haimovich, and M. Sela.** 1967. Purification of antibodies with immunoadsorbents prepared using bromoacetyl cellulose. Immunochemistry **4:**11–22.
16. **Rosenthal, J. D., K. Hayashi, and A. L. Notkins.** 1973. Comparison of direct and indirect solid-phase microradioimmunoassays for the detection of viral antigens and antiviral antibody. Appl. Microbiol. **25:**567–573.
17. **Sykes, R. B., and M. Matthew.** 1978. The β-lactamase of gram-negative bacteria and their role in resistance to β-lactam antibiotics. J. Antimicrob. Chemother. **2:**115–157.
18. **Tabbarah, Z. A., L. J. Wheat, R. B. Kohler, and A. White.** 1980. Thermodissociation of staphylococcal immune complexes and detection of staphylococcal antigen in serum from patients with *Staphylococcus aureus* bacteremia. J. Clin. Microbiol. **11:**703–709.
19. **Visicidi, R., B. E. Laughon, M. Hanvanich, J. G. Bartlett, and R. H. Yolken.** 1984. Improved enzyme immunoassays for the detection of antigens in fecal specimens. Investigation and correction of interfering factors. J. Immunol. Methods **67:**129–143.
20. **Voller, A., D. Bidwell, and A. Bartlett.** 1980. Enzyme-linked immunosorbent assay, p. 359–371. *In* N. R. Rose and H. Friedman (ed.), Manual of clinical immunology, 2nd ed. American Society for Microbiology, Washington, D.C.
21. **Wanatabe, H., I. D. Gust, and I. H. Holmes.** 1978. Human rotavirus and its antibody: their coexistence in feces of infants. J. Clin. Microbiol. **7:**405–409.
22. **Wisdom, G. B.** 1976. Enzyme immunoassay. Clin. Chem. **22:**1243–1255.
23. **Yolken, R. H.** 1982. Enzyme immunoassays for the detection of infectious antigens in body fluids: current limitations and future prospects. Rev. Infect. Dis. **4:**35–68.
24. **Yolken, R. H., and F. J. Leister.** 1981. Investigation of enzyme immunoassay time courses: development of rapid assay systems. J. Clin. Microbiol. **13:**738–741.
25. **Yolken, R. H., and F. J. Leister.** 1982. Comparison of fluorescent and colorigenic substrates for enzyme immunoassays. J. Clin. Microbiol. **15:**757–760.
26. **Yolken, R. H., and F. J. Leister.** 1982. Rapid multiple-determinant enzyme immunoassay for the detection of human rotavirus. J. Infect. Dis. **146:**43–46.
27. **Yolken, R. H., F. J. Leister, L. S. Whitcomb, and M. Santosham.** 1983. Enzyme immunoassays for the detection of bacterial antigens utilizing biotin-labelled antibody and peroxidase biotin-avidin complex. J. Immunol. Methods **56:**319–327.
28. **Yolken, R. H., and P. J. Stopa.** 1979. Analysis of nonspecific reactions in enzyme-linked immunosorbent assay testing for human rotavirus. J. Clin. Microbiol. **10:**703–707.
29. **Yolken, R. H., and S. B. Wee.** 1984. Enzyme immunoassays in which biotinillated β-lactamase is used for the detection of microbial antigens. J. Clin. Microbiol. **19:**356–360.

Chapter 99

Section XI. Laboratory Tests in Chemotherapy

General Considerations

CLYDE THORNSBERRY AND JOHN C. SHERRIS

A number of considerations are involved in selecting an appropriate antimicrobial agent to treat an infection. These include: (i) knowledge of the inherent in vitro susceptibility of the infecting organism to appropriate antimicrobial agents; (ii) the relationship of the susceptibility of the strain to that of other members of the same species; (iii) pharmacological properties, including toxicity, protein binding, distribution, absorption, and excretion, particularly under circumstances of existing or developing hepatic or renal failure; (iv) previous clinical experience of efficacy in the treatment of infections due to the same species; (v) the nature of the underlying pathological process, its natural history, and its influence on chemotherapy; and (vi) the immune status of the host.

Of these factors, the concentrations of antimicrobial agents required to inhibit or kill organisms in vitro and those attained in body fluids during treatment are subject to direct measurement in the clinical laboratory. The purpose of this section is to provide detailed descriptions of appropriate procedures for these purposes. The susceptibility methods described are for use with bacteria other than mycobacteria, which are considered in Chapter 22.

The role of the laboratory in the selection and monitoring of chemotherapy was succinctly expressed by Theodore G. Anderson in the first edition of this Manual: "When selecting an antimicrobial agent for therapy, it is the physician's responsibility to take into consideration the pharmacological characteristics of the several drugs as well as their relative antimicrobial effectiveness. The responsibility of the laboratory is to provide information, through standardized in vitro tests, of the activity of appropriate antimicrobial agents against the organism in question." The methods given in subsequent chapters constitute accepted approaches among the authors providing this information. Different procedures have been developed by others in a number of countries, and the reader is referred to more detailed reviews for further information and a broader consideration of the theory of the subject (1, 2, 4, 7, 10, 14–16, 18, 19).

ANTIMICROBIAL SUSCEPTIBILITY TESTING

Influence of technical variation on susceptibility test results

The results of both dilution and diffusion susceptibility tests may be influenced markedly by the reagents and conditions of the tests, and these variables have been the source of considerable confusion in the past. Inoculum density, incubation time and temperature, pH, atmosphere, and stability of antimicrobial agents may all influence the endpoints obtained. Differences in constituents or ionic content of the medium, even between batches, may influence test results, particularly with the sulfonamides, tetracyclines, polymyxins, and aminoglycosides. In addition, diffusion tests are influenced by the growth rate of the organism and by the type, depth, and concentration of the agar used. For these reasons, special emphasis has been placed on reference procedures and methodological standardization (4–6, 8, 11–13, 20, 22, 23), because only in this way can adequate reproducibility be obtained in investigative and clinical work.

In each of the susceptibility tests described, the inoculum is derived from several colonies. This procedure is designed to reduce the chance of selecting variants derived from loss mutations (e.g., loss of penicillinase production in staphylococci) or segregants from R-factor resistance markers. It also increases the chance of including representatives of the more resistant organism if more than one strain is represented by colonies that cannot be distinguished morphologically. The final inocula are reasonably heavy, which increases the chance of detecting high-frequency mutations to resistance and heteroresistant strains. The media selected show generally good buffering qualities and reproducibility and are of physiological pH. A central criterion of the conditions to be used in effective diffusion, dilution, or automated tests is that the tests must be able to detect strains which carry clinically important resistance determinants.

Selection of susceptibility test methods

Diffusion test. The most widely used procedure is still the disk diffusion method, although nearly half of the clinical laboratories now do dilution tests (8) as a supplemental procedure or the routine method. The disk diffusion procedure, described in Chapter 102, has been accepted by the Food and Drug Administration (FDA) (5, 6) and as a standard by the National Committee for Clinical Laboratory Standards (NCCLS) (12). This procedure, as normally used, is essentially a qualitative test which allocates organisms to the susceptible, intermediate (or moderately susceptible), or resistant categories discussed below. The procedure is flexible in regard to the antimicrobial agents that can be tested, and it is easy to set up individual tests at different times. It is technically simple, although it requires careful attention to detail. It is generally applicable to organisms whose growth rates approximate those of the members of the *Enterobacteriaceae* family, *Staphylococcus* species, and enterococci, i.e., the so-called rapid growers, and the procedure has now also been adapted to test other streptococci and to detect penicillinase-producing strains of *Haemophilus influenzae* and *Neisseria gonorrhoeae* and strains of pneumococci that have developed increased resistance to penicillin and some other antimicrobial agent

TABLE 1. Suggested groupings of antimicrobial agents that should be considered for routine testing and reporting by clinical microbiology laboratories[a] (modified from Table 1 of NCCLS standard M2-T3 [12])

Antimicrobial agents against:				
Enterobacteriaceae[b]	*Pseudomonas* spp.[b]	Staphylococci	Enterococci[c]	Nonenterococcal streptococci
Amikacin	Amikacin	Amikacin or	Penicillin G or	Cephalothin
Ampicillin	Azlocillin[d] or	gentamicin,[d]	ampicillin[e]	Chloramphenicol[d,h]
Carbenicillin[d] or	piperacillin	kanamycin,[d]	Vancomycin[g]	Clindamycin[d]
ticarcillin[d]	Carbenicillin,	netilmicin,[d] or	Erythromycin[i]	Erythromycin[d]
Cefamandole,[d]	mezlocillin, or	tobramycin[d]	Nitrofurantoin[i]	Penicillin G
cefonicid,[d] or	ticarcillin	Cephalothin[f]	Tetracycline[i]	Tetracycline[d]
cefuroxime[d]	Cefoperazone[d] or	Chloramphenicol[d,h]		Vancomycin[d]
Cefazolin	ceftazidime[d]	Clindamycin		Nitrofurantoin[i]
Cefotaxime,[d]	Cefotaxime,[d]	Erythromycin		
ceftazidime,[d]	ceftizoxime, or	Methicillin, nafcillin,		
ceftizoxime,[d] or	moxalactam[d]	or oxacillin		
moxalactam[d]	Chloramphenicol[d,h,j]	Penicillin G		
Cefoperazone[d]	Gentamicin	Tetracycline[k]		
Cefoxitin[d]	Netilmicin[d]	Trimethoprim-		
Cephalothin	Tobramycin	sulfamethoxazole[d]		
Chloramphenicol[h]	Trimethoprim-	Vancomycin[d]		
Gentamicin	sulfamethoxazole[d,j]	Nitrofurantoin[i]		
Kanamycin[d]	Sulfisoxazole[i,j]	Sulfisoxazole[i]		
Mezlocillin[d] or	Tetracycline[i,j,k]	Trimethoprim[i]		
piperacillin[d]				
Netilmicin[d]				
Tetracycline[k]				
Tobramycin				
Trimethoprim-				
sulfamethoxazole				
Cinoxacin[i] or				
nalidixic acid[i]				
Nitrofurantoin[i]				
Sulfisoxazole[i]				
Trimethoprim[i]				

[a] Selection of the most appropriate antimicrobial agents to test is a decision best made by each clinical laboratory in consultation with its medical staff, infectious disease practitioners, and pharmacy. The lists present antimicrobial agents with proven clinical efficacy for each organism group and, additionally, with acceptable in vitro test performance.

[b] Also applies to *Acinetobacter* spp.

[c] Antimicrobial agents not listed in this column, such as the cephalosporins, clindamycin, and the aminoglycosides, should not be tested or reported against the enterococci because the reporting of their results can be dangerously misleading.

[d] Secondary antimicrobial agent that may require testing as a primary agent in those institutions harboring endemic or epidemic resistance to one or more of the primary drugs (especially in the same family; e.g., β-lactam or aminoglycoside drugs), as an epidemiological aid for the treatment of patients allergic to primary drugs (see nonenterococcal streptococci). This designation was applied to some drugs in the *Pseudomonas* spp. list due to lack of long-term clinical experience with that antimicrobial agent or, in the case of the cephalosporins, after their first clinical application.

[e] Penicillin susceptibility may be used to predict susceptibility to ampicillin, ampicillin analogs, amoxicillin, and acylamino penicillins, to which enterococci are also moderately susceptible. However, combination therapy of penicillin G or ampicillin, plus an aminoglycoside, is usually indicated for serious enterococcal infections such as endocarditis.

[f] The cephalothin disk test cannot be relied upon to detect cephalosporin resistance in methicillin-resistant staphylococci.

[g] Often used for serious enterococcal infections in patients with significant penicillin allergy.

[h] Not routinely applied to organisms isolated from the urinary tract.

[i] Tested only with organisms recovered from urinary tract infections.

[j] May be indicated for testing some *Pseudomonas* spp. other than *P. aeruginosa* or for *Acinetobacter* spp.

[k] Doxycycline or minocycline may be tested in place of or in addition to tetracycline for some isolates of *S. aureus* and nonfermentative, gram-negative bacilli, e.g., *Acinetobacter* spp.

(see Chapter 102). Unless it is specifically indicated in standard publications (e.g., this Manual and the NCCLS disk diffusion standard) that an organism can be tested by this method, efficacy of the method must be determined by appropriate studies. In cases of clinical urgency, clinical material may serve as the inoculum for the test if the precautions indicated in Chapter 102 are followed, but this is not generally recommended. More experience has been gained over the years with this diffusion procedure than with any other test.

The deficiencies of the diffusion test are its nonquantitative interpretation, its inapplicability to many slowly growing organisms and anaerobes, and its inaccuracy in predicting susceptibility (as opposed to resistance) with some antimicrobial agents, e.g., the polymyxins, that diffuse poorly. Overall, however, it is an effective procedure for most routine testing, but it should be supplemented with a dilution test when it is inapplicable or when more quantitative results are needed.

Dilution test. The most quantitative method for

antimicrobial susceptibility testing is one of the dilution tests (considered in Chapters 100, 101, and 103), which are basically derived from the International Collaborative Study recommendations (4) or from NCCLS standards (13). These tests yield direct quantitative results, are essentially uninfluenced by the growth rate of the organism, and avoid some of the complexities of diffusion properties of antimicrobial agents. Dilution tests do not have the flexibility of the diffusion test, generally cannot be used for direct tests of clinical material because of the difficulty in detecting contamination, and if reported quantitatively, require that the clinician be able to interpret the result or be helped in doing so.

If dilution tests are not the routine method of testing, the primary indication for them is to obtain quantitative susceptibility data when these are important or necessary for proper management of antimicrobial therapy. Although qualitative data are usually adequate for guiding the therapy of most infections, quantitative information may be needed when drug dosage schedules and levels in serum must be closely monitored or under the conditions in which disk test results are inapplicable, equivocal, or unreliable. These conditions include tests on slowly growing organisms, confirmation of susceptibility (as opposed to resistance) to the polymyxins, confirmation of resistance to the aminoglycosides (particularly gentamicin, tobramycin, and amikacin), and tests with potentially toxic but clinically useful antimicrobial agents which yield intermediately susceptible, or equivocal, results by the disk diffusion test. Infections due to microorganisms which are categorized by the disk test as resistant or intermediate to the relatively nontoxic penicillins and cephalosporins may occasionally be treated preferentially and safely with large doses of one of these agents. Some urinary tract infections may respond to ordinary dosages of some antimicrobial agents because of the high levels which these agents attain in the urine. In these cases, the precise degree of susceptibility of an organism may influence the choice of antimicrobial agent, its dosage, and its route of administration. Other indications for dilution methods are for testing the susceptibilities of anaerobes by the methods described in Chapter 103 and for the determination of bactericidal activity or evidence of synergism or antagonism between antimicrobial agents against particular microorganisms. These procedures are considered in Chapter 105. Finally, dilution tests have been found to be practical and economical for routine purposes through the use of semiautomated microdilution techniques (see Chapter 101) or by replica-plating and agar dilution methods (see Chapter 100). The tests yield results that are both accurate and reproducible, and as indicated above, almost 50% of the laboratories participating in the College of American Pathologists surveys now perform dilution tests (although not necessarily as the routine test) (8).

Any laboratory that intends to use the dilution test routinely and to prepare its own reagents and antibiotic dilutions must have the ability to obtain, prepare, and maintain fully potent stock solutions of antimicrobial agents and to produce working dilutions on a regular basis. As with all susceptibility tests, the laboratory must control inoculum size and endpoint reading and must develop or use a quality control system, with standard reference strains, that will give endpoints within the range of each series of dilutions of antimicrobial agent. For example, if seven concentrations were being tested, the ideal reference strain would have as its endpoint the fourth concentration, but an endpoint at the third or fifth concentration would be very acceptable. Recent data from proficiency testing surveys show that intralaboratory and interlaboratory reproducibility are better when well-controlled, prediluted commercial systems are used, but similar results can be expected if the protocols used in Chapters 100 and 101 are followed.

The routine use of dilution procedures in which preprepared commercial plates are not used is probably most appropriate for larger laboratories, in which many tests are performed daily. Preprepared systems, when adequately controlled, are probably more applicable to smaller laboratories.

Tests for antimicrobial agent-inactivating enzymes. In some instances, resistant strains of bacteria among originally susceptible species owe their resistance exclusively to their ability to destroy or inactivate particular antimicrobial agents. This is the case with penicillin and ampicillin resistance of strains of staphylococci, *H. influenzae*, *N. gonorrhoeae*, and *Bacteroides* species that produce β-lactamase. These enzymes can be detected rapidly and accurately with simple chemical procedures (21), and results are available much more rapidly that those of orthodox susceptibility tests. When used for decision making for therapy, generally only the organisms listed above should be tested. Many microbiologists and clinicians feel that since the incidence of β-lactamase is so high in *Staphylococcus aureus*, it is not cost effective to test these organisms, and all should be considered β-lactamase positive and thus penicillin resistant. Procedures for this purpose are described in Chapter 105.

Automated tests. A variety of mechanized or automated procedures have been developed for susceptibility testing, and these are considered in Chapter 107. Some facilitate the performance and reading of traditional overnight susceptibility tests. Others are designed to yield qualitative or quantitative results on the same day that the test is set up. Several of these procedures have already been evaluated by collaborative studies and have been reported to have a high degree of reproducibility. However, comparability of procedures providing rapid results with overnight dilution or diffusion tests has been more difficult to achieve with a few organism-antimicrobial agent combinations because the extent of inhibition of growth in the first few hours of contact may differ from that seen after overnight incubation. Most of these difficulties are being overcome by the use of heavier inocula or computer analyses of growth patterns in the presence of one or more concentrations of antimicrobial agent. These approaches are discussed in more detail in Chapter 107. Suffice it to say that automated methods have already been shown to have a place in routine work of some larger laboratories, but their limitations must be recognized and avoided by using a traditional test on organism-antimicrobial agent combinations for which a given automated test is inappropriate.

Quality control

The development of quality control parameters by testing standard reference strains on a daily basis has been an important factor in the high level of performance that most laboratories have attained for antimicrobial susceptibility testing (8). There are, however, some concerns. Because of the cost of daily quality control and because of the high level of performance of most laboratories, there has been a move toward weekly, rather than daily, quality control tests. The high level of performance must be demonstrated, however, before a laboratory can make this change. Another problem is the difficulty in quality controlling those tests with one or two concentrations of antimicrobial agent to derive a category result. Such tests are being commonly used in the broth-disk tests with anaerobes (Chapter 103), in automated tests (Chapter 107), and for some drugs in the commercial microdilution systems (Chapter 101). In the latter systems, one or two concentrations are used to test more of the newer antimicrobial agents, particularly the β-lactam drugs. Although the suitability of some elution disks can be controlled by diffusion tests (Chapter 102), this quality control problem is mostly unsolved.

Interpretation of susceptibility tests: susceptibility and resistance

The interpretation of the quantitative MIC-susceptibility test result and, indirectly, the disk diffusion test, has three major components.

The first component is the relationship of the MIC or MBC of the antimicrobial agent for the organism to the concentration of antimicrobial agent in the blood, or in some cases in the urine or other fluid, obtained with the dosage given. Generally, this approach has proved to be clinically useful, but it is inevitably an incomplete model of the in vivo situation because the levels of antimicrobial agents in tissue are not usually known, the degrees of protein binding varies by drug, the interaction of host defense mechanisms exerts an unmeasurable effect, and some aspects of the selection of test conditions are purely arbitrary.

The second component is the relationship of the susceptibility of the strain under test to that of other members of the same species. This knowledge is useful because the selection of resistant mutants or strains with extrachromosomal determinants of resistance has led to the appearance of populations of strains of some species well separated from the wild types that were previously uniformly susceptible to the antimicrobial agent. The resulting bimodal distribution of susceptibilities correlates well with clinical responsiveness. Thus, a strain falling in the more resistant population is considered, a priori, a resistant member of that species. In some cases, the emerging resistance is more difficult to categorize, e.g., the relatively penicillin-resistant pneumococci. A strain with an MIC of 0.5 μg/ml may be resistant to penicillin if the organism is causing meningitis, but susceptible if it is causing pneumonia.

The third component is clinical experience with the treatment of the particular type of infection involved. An ideal interpretation of susceptibility test results takes into account these factors independently. From a practical point of view, however, organisms are frequently allocated to predetermined susceptible, resistant, and one or more intermediate categories. This approach was considered by the International Collaborative Study, and more recently by the NCCLS, to be still useful and sometimes necessary, in the light of the available technical methods and general understanding of the principles of chemotherapy (4). The three categories recommended for the diffusion test given in Chapter 102 have been based on the synthesis of the first two criteria given above. The categories have been defined (11) as: (i) susceptible, implying that an infection due to the strain tested may be appropriately treated with the antimicrobial agents and dosages recommended for that type of infection and infecting species, unless otherwise contraindicated; (ii) resistant, containing strains not completely inhibited within the usual therapeutic dosage range; and (iii) intermediate, comprising a buffer zone which prevents major interpretive discrepancies that might result from small, uncontrolled technical factors. The last category also includes strains which may respond to concentrations attainable by the maximum safe dosage and strains which may respond because they are causing infections in areas, such as portions of the urinary tract, in which the antimicrobial agent is concentrated. More recently, those organisms that would likely respond to maximal dosages of the relatively nontoxic β-lactam antimicrobial agents have been designated moderately susceptible (12, 13), and those that would likely respond because of concentration of the drug (as in the urinary tract) have been designated conditionally susceptible (13). The three category systems of the disk diffusion method do not take into consideration the higher levels in the urinary tract for most drugs or the relatively low levels in blood achieved with oral as opposed to parenteral dosages of some antimicrobial agents. The clinical extrapolation of these categories is, of course, subject to the considerations given in the first paragraph of this chapter.

Indications for susceptibility tests in the clinical laboratory

Tests are indicated for those organisms contributing to the infectious process whose susceptibility cannot be predicted from knowledge of their identity. This applies, in particular, to *S. aureus*, to gram-negative fermentative and nonfermentative organisms, to some anaerobes, and to unusual and opportunistic species playing a pathogenic role. Specific patterns of susceptibility (antibiograms) may be characteristic of a species and may, therefore, assist in identification of the species. Antibiograms may also be determined for epidemiological reasons because the occurrence of an unusual antibiogram for a given species often assists in the recognition of common-source outbreaks and patterns of cross-infection.

Routine susceptibility tests are not needed when resistance to the antimicrobial agent of choice, e.g., that of *Streptococcus pyogenes* and *Neisseria meningitidis* to penicillin, has not been described. However, this situation needs to be reviewed constantly, because the recent emergence of β-lactamase-producing strains of *H. influenzae* and *N. gonorrhoeae* (3) and of pneumococci resistant to penicillin and other antibiotics (9) has changed the requirements for testing for these species. Such emerging resistance is most often recognized due to failure of an infected patient to respond to drugs of choice, e.g., lack of response of *H. influenzae* infections to ampicillin. One should remember, however, that these failures are not always due to resistance of the organism to the antimicrobial agent. Susceptibility testing should be avoided on members of the normal flora in their normal habitat and on organisms that are known not to be playing a pathogenic role. To make such tests is both wasteful and misleading. The routine diffusion susceptibility test described in Chapter 102 should never be made on organisms for which its interpretive criteria are not applicable. We have classified such organisms as fastidious or unusual. For the susceptibility tests, the fastidious organisms are those that do not grow on or in Mueller-Hinton medium unless it is supplemented and that may require an atmosphere other than air. Blood is a common supplement and for most organisms does not significantly alter results (i.e., does not alter the categories of susceptibility), but there are exceptions, e.g., some enterococci with some cephalosporins (17). The unusual organisms are those that have not been tested enough to determine if they can be correctly categorized with the standard interpretive breakpoints.

Selection of antimicrobial agents for testing

Because of the recent rapid proliferation of antimicrobial agents, selection of a panel for routine testing has become more difficult. To attempt to test them all is not only unfeasible for economic reasons, it is also unnecessary because of similarities in spectra of activity and pharmacokinetics. In the past, the selection process was made easier because the FDA recognized specific class disks for use in the diffusion test (see Chapter 102). The designation of class disks has become increasingly more difficult because many of the newer antimicrobial agents are sufficiently different to demand separate disks.

It is likely that some of the older class disks will remain, but there will be some changes. For example, a disk for cefazolin has been approved since it is clear that cephalothin (the class disk) will not correctly categorize the activity of cefazolin for some gram-negative bacilli (12).

Most of the antimicrobial agents approved since the mid-1970s have had a disk approved by the FDA, and it is likely that all future antimicrobial agents will also have disks approved if they are requested.

As a result of these developments, each laboratory needs to develop its own policy for the selection of a basic set of drugs for routine testing. Factors which should be considered in making these selections include the epidemiology for the individual medical center (what organisms are causing infections and what antimicrobial agents they are susceptible to), hospital antimicrobial agent usage patterns including those that are in the formulary, and for the drugs themselves, their toxicity, cost, and similarities of activity and pharmacokinetics.

One way of approaching this problem is to have a committee composed of the microbiologist, hospital epidemiologist, and pharmacist and representatives from pathology, infectious disease, and other medical and surgical services.

A guide to making these selections for drugs presently available is shown in Table 1 (12). We have grouped a number of agents that have similar activity, and it is suggested that only one of them be selected for routine tests (if the group is to be tested at all). For example, if testing of a third-generation cephalosporin is desired for enteric organisms, select from cefotaxime, ceftizoxime, and moxalactam (this group will enlarge if some others are approved). Note that cefoperazone is not included since it may be affected differently by some β-lactamases. To keep these lists current, the NCCLS standards are recommended, since frequent supplements are used to update the standards.

The selection of antimicrobial agents to be tested should also be limited to those that are clinically useful and appropriate for the site of infection, except when the procedure is being used to determine antibiograms for epidemiological purposes or when particular antimicrobial agents yield taxonomically useful information. For example, tests with nitrofurantoin and nalidixic acid should be limited to bacteria isolated from the urinary tract. Tests with methenamine mandelate should not be performed at all because its in vivo activity depends on urinary acidification to a pH of 5.0 or less, and this condition is not reproduced in the ordinary test systems.

To avoid confusion, results and tests on antimicrobial agents that are inappropriate in therapy and that are tested for epidemiological or taxonomic purposes should not be reported to the physician unless such tests are part of a collaborative research effort.

In recent years, there has been a concerted effort by various individuals and groups to establish interpretive breakpoints and quality control parameters for new antimicrobial agents at the time the agents are approved for use. The FDA, of course, has the legal authority and charge to set these standards, but the NCCLS also makes recommendations. For most drugs, the FDA and NCCLS agree on their recommendations, but sometimes they do not, thus creating some confusion. When this occurs, there is no answer to the problem, and the user must decide which to use. Usually these differences are not great and can be reconciled.

SPECIAL TESTS AND ASSAYS

Susceptibility tests make up the bulk of the clinical laboratory tests that are ordered to assist the clinician in the choice of chemotherapeutic agents. They may need to be supplemented with other procedures in certain complex clinical situations, especially in subacute bacterial endocarditis, severe infections in the immunologically compromised patient, and situations where tolerant organisms may be suspected. In these cases, a determination of bactericidal concentrations or of the effect of combinations of antimicrobial agents may need to be made. Direct tests of the ability of the antimicrobial agent in the serum of the patient to inhibit or kill the infecting organism may also be helpful in monitoring the adequacy of dosage schedules (see Chapters 105 and 106). So far, there is still no general agreement on methods for the determination of MBCs, study of antimicrobial agent combinations, or determination of serum inhibitory or bactericidal activity against the infecting organisms.

The procedures given in this Manual can continue to serve as a basis for further studies towards methodological standardization of these important tests, but clinicians should recognize that the usefulness of data collected by the present methods may be limited by the problems inherent in these methods.

It has become increasingly necessary to determine the amount of antimicrobial agent present in serum, urine, other fluids, or tissues. In clinical practice, this applies particularly to agents such as gentamicin, tobramycin, and amikacin, for which potentially toxic levels and therapeutic levels are very close. More recently it also includes vancomycin because of its increased use. Serum assays are thus required to ensure that antimicrobial agent concentrations in the blood are within a safe but effective range. This is particularly true in patients with renal deficiency, in whom levels of antimicrobial agent in serum may be less predictable. A number of newer methods have been developed for performing these tests, and most of them are available commercially. Older methods as well as some of the newer ones are described in Chapter 106.

FUTURE NEEDS

Since the last edition of this Manual, the NCCLS standard for diffusion susceptibility tests has been so extensively revised that it was returned to a tentative status (M2-T3). It is anticipated that it will soon be returned to an approved status. The standards for anaerobe susceptibility testing (12) and for dilution tests (13) also will soon reach approved status. These standards should be used to provide reference points for clinical, epidemiological, and research studies and, where applicable or desirable, could also be used as routine methods. Quality control values for standard strains have been developed and updated to include new antimicrobial agents, and this practice should continue. Data collected through major proficiency testing programs (by the Centers for Disease Control and the College of American Pathologists) appear to be a valuable tool for measuring the improvement of susceptibility test procedures, their accuracy, and their interlaboratory reproducibility. We should continue to study these kinds of data to confirm their value in the prediction of the overall performance of susceptibility testing.

It is especially important that mechanisms be maintained for the regular updating of reference procedures, interpretive standards for new antimicrobial agents, recommendations for basic routine sets of antimicrobial agents for testing, and data on performance of quality control strains. The NCCLS is now doing this for the disk diffusion standard by publishing supplements whenever new information is available. These supplements usually consist of tables that can be inserted into the standard. The use of such supplements should be extended to the other NCCLS standards.

Developments in mechanization, automation, and commercially available test kits have now brought the capacity for quantitative susceptibility testing to small laboratories, and this trend is likely to continue. In a recent College of American Pathologists survey, which involved thousands of laboratories, almost 50% were determining MICs either as a routine or as a supplemental procedure; the majority of these MICs were determined by microdilution. Automated procedures that yield rapid results and can be directly interfaced with computer reporting systems are already available, and more will probably be developed, but the trend toward these systems seems to have slowed. There is a clear need for agreement on evaluative protocols by which the relationship between the performance of new systems and reference procedures can be assessed and for definitions of acceptable performance. Potential purchasers need access to such data before they decide whether to incorporate a system into their routine work (see Chapter 107).

Media for susceptibility testing remain a problem. For instance, there can be variations in the performance of *Pseudomonas aeruginosa* with the newer aminoglycosides when it is tested with different methods and different batches of media from the same or different manufacturers, resulting in MICs ranging over several dilution steps if the cation content is not controlled. Another well-known example is the inhibition of sulfonamide and trimethoprim activity by thymidine-containing media. In more recent studies, it has been shown that with some enterococci and some cephalosporins, not only the brand of Mueller-Hinton agar but also blood supplementation can grossly affect the susceptibility results (17). Although manufacturers have controlled some of the problems, a method of standardizing Mueller-Hinton agar is needed. A subcommittee of the NCCLS has actively pursued this problem and a reference standard lot has been selected from which manufacturers have chosen secondary standard lots for their own use. Further studies will determine the success of this approach in the elimination of these variations. If successful, this approach should be used with other susceptibility test media.

With the wider acceptance of reference procedures, there is a need for review of the qualitative interpretive categories used in diffusion and in some dilution and automated procedures. In the International Collaborative Study report, four categories were recommended that were essentially based on dosages and pharmacokinetics (4). In the NCCLS dilution standard, similar kinds of categories were recommended (13). In the first dilution standard, M7-P, the recommended categories of susceptibility were very susceptible, moderately susceptible, moderately resistant, and resistant, but in the latest document, M7-T, the recommended categories are susceptible, moderately susceptible, resistant, and conditionally susceptible. Moderately susceptible indicates the need for the maximum safe dosage, and conditionally susceptible indicates that these organisms would be susceptible if the infection were in a site in which the antimicrobial agent was concentrated, e.g., the urinary tract. The boundaries of interpretive categories involve some best-judgment decisions and subjective clinical experience. Validation of these categories by detailed clinical studies is needed but has been made difficult in the past by inadequate methodological standardization. This standardization should now be possible and should at least be done with new antimicrobial agents.

Methodological agreement is needed on techniques for the determination of bactericidal endpoints, measurement of the effects of combinations, and performance of serum inhibitory or bactericidal tests. In the absence of standardized methods or reference procedures, results from different laboratories cannot be compared with confidence, and an adequate base of experience has not been developed for fully satisfactory interpretation of the results. For all of these procedures, the kinetics of microbial killing by antimicrobial agents make it essential that statistical endpoints be accepted, such as those recommended in Chapter 105. Studies are needed to develop standard methods that take into consideration the problems of ensuring contact between antimicrobial agent and organism and of transferring an adequate sample to determine numbers of unkilled organisms without carrying over inhibitory levels of the antimicrobial agent.

When standard methods for the performance of MBCs are developed, further studies of "tolerance" should be done. At present, most procedures used are poorly reproducible because some organisms escape exposure to antimicrobial agents. With a well-developed standard method, the true incidence of tolerance can be established for each institution, and the marked interlaboratory discrepancies in the designation of tolerance can be eliminated. More importantly, the therapeutic significance of tolerance can be established.

The present mood of cost consciousness and the use of diagnosis-related groups for reimbursement will probably have an effect on antimicrobial susceptibility testing in clinical laboratories. The cost of quality control is already an issue that has resulted in the proposal to test reference strains weekly, rather than daily, in the disk diffusion test (12). It is likely that these developments will require a more creative approach in the selection of organisms and drugs to test and in providing data that will permit the clinician to develop a therapeutic scheme that will be the most effective at the least cost. This approach might include the development of computer programs to integrate MICs, MBCs, and levels of drugs in serum to determine the best and most economical dosage schedules. Studies in these areas should be considered.

Finally, the clinical microbiology laboratory is still plagued with the problem of excessive time lapse between the receipt of a specimen and the reporting of results. Although we have called some tests rapid, they are, in fact, rapid only after isolation of the organism(s). One of the greatest needs is to find ways to identify the pathogen and perform susceptibility testing within 1 to 2 h of collecting the specimen.

In summary, we may reasonably look forward to improved performance of media and of procedures for orthodox susceptibility tests and to improvements in the selection and dissemination of interpretive recommendations. Weekly, rather than daily, quality control tests may be used in some laboratories that have demonstrated high levels of proficiency, but quality control of nondiffusion category tests is likely to remain a problem. There has been a large increase in the number of laboratories in which MIC tests, mostly microdilution, are done, and this trend will likely continue (8). It is likely that, in the near future, all laboratories will have the capability of performing MIC tests in conjunction with diffusion or other routine tests. Commercial systems for microdilution MICs will generally result in better reproducibility through the elimination of many sources of technical error, but tests prepared in-house should be acceptable if the standard recommendations are followed. There continues to be a need for reference procedures for the determination of bactericidal endpoints, interactions of combinations of antimicrobial agents, and bactericidal activities of serum so that interpretations of the results of these tests can be refined through cumulative experience. Recent developments have increased the usefulness of laboratory procedures in the selection and monitoring of chemotherapy, and this trend will likely continue.

LITERATURE CITED

1. **Balows, A. (ed.).** 1974. Current techniques for antibiotic susceptibility testing. Charles C Thomas, Publisher, Springfield, Ill.
2. **Barry, A. L.** 1976. The antimicrobic susceptibility test: principles and practices. Lea & Febiger, Philadelphia.
3. **Elwell, L. P., M. Roberts, and S. Falkow.** 1978. Common β-lactamase-specifying R plasmid isolated from the genera *Haemophilus* and *Neisseria*, p. 255–256. *In* D. Schlessinger (ed.), Microbiology—1978. American Society for Microbiology, Washington, D.C.
4. **Ericsson, H. M., and J. C. Sherris.** 1971. Antibiotic sensitivity testing. Report of an international collaborative study. Acta Pathol. Microbiol. Scand. Sect. B Suppl. 217.
5. **Federal Register.** 1972. Rules and regulations. Antibiotic susceptibility disks. Fed. Regist. **37:**20525–20529.
6. **Federal Register.** 1973. Rules and regulations. Antibiotic susceptibility disks: correction. Fed. Regist. **38:**2576.
7. **Garrod, L. P., H. P. Lambert, and F. O'Grady, with a chapter on laboratory methods by P. M. Waterworth.** 1981. Antibiotic and chemotherapy, 5th ed. Churchill Livingstone Ltd., Edinburgh.
8. **Jones, R. N.** 1983. Antimicrobial susceptibility testing (AST): a review of changing trends, quality control guidelines, test accuracy, and recommendations for testing of β-lactam drugs. Diagn. Microbiol. Infect. Dis. **1:**1–24.
9. **Koornhof, H. J., M. R. Jacobs, J. I. Ward, P. C. Appelbaum, and F. A. Hallett.** 1979. Therapy and control of antibiotic-resistant pneumococcal disease, p. 286–289. *In* D. Schlessinger (ed.), Microbiology—1979. American Society for Microbiology, Washington, D.C.
10. **Lorian, V. (ed.).** 1979. Antibiotics in laboratory medicine. The Williams & Wilkins Co., Baltimore.
11. **National Committee for Clinical Laboratory Standards.** 1982. Tentative reference dilution procedure for antimicrobic susceptibility testing of anaerobic bacteria. M11-T. National Committee for Clinical Laboratory Standards, Villanova, Pa.
12. **National Committee for Clinical Laboratory Standards.** 1983. Performance standards for antimicrobial disk susceptibility tests. M2-T3. National Committee for Clinical Laboratory Standards, Villanova, Pa.
13. **National Committee for Clinical Laboratory Standards.** 1983. Standard methods for dilution antimicrobial susceptibility tests for bacteria that grow aerobically. M7-T. National Committee for Clinical Laboratory Standards, Villanova, Pa.
14. **Petersdorf, R. G., and J. J. Plorde.** 1963. The usefulness of in vitro sensitivity tests in antibiotic therapy. Annu. Rev. Med. **14:**41–56.
15. **Petersdorf, R. G., and J. C. Sherris.** 1965. Methods and significance of in vitro testing of bacterial sensitivity to drugs. Am. J. Med. **39:**766–779.
16. **Reeves, D. S., I. Phillips, J. D. Williams, and R. Wise (ed.).** 1978. Laboratory methods in antimicrobial chemo-

therapy. Churchill Livingstone Ltd., New York.

17. **Sahm, D. F., C. N. Baker, R. N. Jones, and C. Thornsberry.** 1983. Medium-dependent zone size discrepancies associated with susceptibility testing of group D streptococci against various cephalosporins. J. Clin. Microbiol. **18:**858–865.
18. **Schoenknecht, F. D., and J. C. Sherris.** 1980. Recent trends in antimicrobial susceptibility testing. Lab. Med. **11:**824–832.
19. **Sherris, J. C.** 1977. The antibiotic sensitivity test. Variability. Interpretation. Rapid versus overnight. Agar versus broth, p. 170–185. *In* V. Lorian (ed.), Significance of medical microbiology in the care of patients. The Williams & Wilkins Co., Baltimore.
20. **Thornsberry, C., and C. N. Baker.** 1981. Antimicrobial susceptibility tests for bacteria, p. 747–766. *In* A. Balows and W. J. Hausler (ed.), Bacterial, mycotic, and parasitic infections, 6th ed. American Public Health Association, Washington, D.C.
21. **Thornsberry, C., T. L. Gavan, and E. H. Gerlach.** 1977. Cumitech 6, New developments in antimicrobial agent susceptibility testing. Coordinating ed., J. C. Sherris. American Society for Microbiology, Washington, D.C.
22. **World Health Organization.** 1961. Standardization of methods for conducting microbic sensitivity tests. Second Report of the Expert Committee on Antibiotics. W.H.O. Tech. Rep. Serv. **210:**1–24.
23. **World Health Organization Expert Committee on Biological Standardization Technical Report Service.** 1977. Requirements for antibiotic susceptibility tests. 1. Agar diffusion tests using antibiotic susceptibility disks. Rep. no. 610. World Health Organization, Geneva.

Chapter 100

Susceptibility Tests: Agar Dilution

JOHN A. WASHINGTON II

Dilution tests are used to determine the minimal concentration of an antimicrobial agent required to inhibit or kill a microorganism. Serial dilutions of the antimicrobial agent are inoculated with the organism and incubated. The MIC is the lowest concentration without visible growth. The terms broth (or tube) and agar (or plate) are added to the term dilution test to indicate tests performed in liquid and agar media, respectively. Both terms are actually misnomers because it is the antimicrobial agent that is being diluted rather than the broth or agar.

CONCENTRATIONS OF ANTIMICROBIAL AGENTS

For most purposes, a concentration of 128 μg/ml is a satisfactory upper limit for routine testing with any antimicrobial agent. Important exceptions are tests of carbenicillin and certain expanded-spectrum cephalosporins, for which strains requiring MICs of 256 μg/ml may be considered susceptible if infections occur in tissues or body fluids where attainable concentrations greatly exceed those in blood (11). Under these conditions, higher concentrations will need to be tested. In other cases, upper limits of 32 μg/ml are suitable. The lowest concentration selected for routine dilution testing will vary according to the antimicrobial agent. In general, however, this concentration should be below the upper limit of a high degree of susceptibility; inhibition by such a concentration makes in vivo response probable when mild to moderately severe systemic infections are treated with the usual dosage (susceptible category as defined by the National Committee for Clinical Laboratory Standards [NCCLS] [11]). The range of concentrations should include the endpoint for appropriate standard strains to permit adequate control.

AGAR DILUTION METHOD

The agar dilution method is described first because of its convenience for testing a number of strains simultaneously, its ability to detect microbial heterogeneity or contamination, and its slightly better reproducibility than the broth dilution method (6). Each step in the procedure for performing the agar dilution susceptibility test is described in detail in a tentative standard published by the NCCLS (11).

Preparation of dilutions of antimicrobial agents

Dilutions of antimicrobial agents are prepared from stock solutions (see Appendix 1) at 10 times the concentrations required in the final test. Log_2 (twofold) dilutions are normally used for determining MICs and may be prepared for reference work according to the volumetric schedule shown in Table 1. Dilution schedules should be selected to include a concentration of 1 μg/ml or 1 IU/ml to permit comparison of results from different laboratories and their easy expression as log_2, which facilitates statistical manipulations (6). The dilution method shown in Table 1 is convenient and economical in pipettes because only one need be used for each block of dilutions. This method is not subject to the cumulative error inherent in traditional serial dilution methods.

On a routine basis, it is possible to reduce the number of concentrations tested to a few that correspond with levels readily attainable in serum or urine after administration of various dosage regimens of each antimicrobial agent. An example of the concentrations tested in such an approach is shown in Table 2.

Selection and preparation of medium

For rapidly growing aerobic and facultatively anaerobic bacteria, Mueller-Hinton agar is recommended. Although this medium supports the growth of most bacterial pathogens, supplementation with 5% defibrinated sheep, horse, or other animal blood may be necessary to ensure growth of some more fastidious organisms. Blood supplementation of the Mueller-Hinton agar has little effect on antibiotic susceptibility test endpoints except in the case of highly protein-bound agents such as novobiocin (3, 16) or with nafcillin. The activity of the sulfonamides and trimethoprim is partly antagonized by components of all bloods except lysed horse blood (7). For routine purposes, supplementation of Mueller-Hinton agar with blood is usually unnecessary; however, defibrinated blood is useful for testing streptococci, and hemoglobin and IsoVitaleX (BBL Microbiology Systems, Cockeysville, Md.) or supplement VX (Difco Laboratories, Detroit, Mich.), each in a final concentration of 1%, may be used for testing *Haemophilus* spp. (19). Unsupplemented Mueller-Hinton agar has been reported as satisfactory for testing of *Neisseria meningitidis* (2); however, GC medium base (Difco) (14) supplemented according to the manufacturer's recommendations or with 5% chocolatized sheep blood is suggested when *Neisseria gonorrhoeae* is tested for epidemiologic or research purposes. The reader is referred to the proposed international reference procedure of Reyn et al. (13).

The appropriate amounts of medium (100-mm plates require 25 ml of agar) are bottled in a screw-capped container and autoclaved; the medium is then allowed to equilibrate in a constant-temperature water bath to 50°C. Addition of the antimicrobial agent to the agar at higher temperatures may lead to deterioration; its addition to the agar at lower temperatures will preclude adequate mixing. Defibrinated blood may be added to the agar after the antimicrobial agent has been added and thoroughly mixed. Molten agar may be dispensed into smaller, sterile containers, which are sealed for purposes of storage, and then allowed to equilibrate to 48 to 50°C before addition of

TABLE 1. System for preparing dilutions for the agar dilution method[a]

Antimicrobial solution		+	Sterile water (volumes[b])	=	Intermediate concn (μg/ml or IU/ml)	=	Final concn at 1:10 in agar plates	
Volumes[b]	Concn[c] (μg/ml or IU/ml)						μg/ml or IU/ml	Log_2 concn
6.4 ml	2,000		3.6		1,280		128	7
2 vol	1,280		2		640		64	6
1 vol	1,280		3		320		32	5
1 vol	1,280		7		160		16	4
2 vol	160		2		80		8	3
1 vol	160		3		40		4	2
1 vol	160		7		20		2	1
2 vol	20		2		10		1	0
1 vol	20		3		5		0.5	−1
1 vol	20		7		2.5		0.25	−2

[a] Modified from Ericsson and Sherris (6).
[b] The volume size is determined by number of tests.
[c] Intermediate concentration, from column 4.

antimicrobial solutions (and, when necessary, growth supplements) and before pouring the plates. The pH of each batch of medium must be checked to ensure that it is in the range of 7.2 to 7.4.

For reference work, 1 volume of each dilution of antimicrobial agent is added to each 9 volumes of agar. For example, a final concentration in agar of 128 μg/ml is attained by adding 10 ml of the 1,280-μg/ml solution (Table 1) to 90 ml of agar. For routine work, the ratio of the volume of antimicrobial dilution to volume of agar can be 1:100 or 1:1,000 (17). It is essential to mix the contents of the container thoroughly and to pour the agar into the plates (25 ml per plate for reference work and 10 ml per plate for routine work) as quickly thereafter as possible to prevent cooling and partial solidification in the container. Although 100-mm round plates may be used, 100-mm square plates with a 13-mm grid embossed on the dish bottom provide a convenient means of identifying the location of each organism. The agar is permitted to solidify in the plates on a level surface. Control plates, with and without defibrinated sheep blood, containing no antimicrobial agent also should be prepared.

Once the agar has solidified, the plates are stored at 4°C. Ryan and co-workers (15) have shown that there is no significant loss of activity of a wide range of antimicrobial agents in agar stored at 4°C in Mylar bags for 1 week. We have confirmed this finding with other antimicrobial agents, including carbenicillin. For routine purposes, plates containing antimicrobial agents should be used within this period. For reference work, it is desirable to use agar plates that have not been stored for longer than 24 h.

TABLE 2. Concentrations of antimicrobial agents tested against bacterial isolates by the agar dilution method at Mayo Clinic

Antimicrobial agent	Concn (μg/ml)									
	0.1	1	2	4	8	16	32	64	128	256
Amikacin					X	X				
Ampicillin			X		X	X			X[a]	
Cefuroxime					X	X	X			
Cefoxitin					X	X			X[a]	
Cefazolin					X	X			X[a]	
Chloramphenicol					X	X				
Clindamycin		X		X						
Erythromycin	X	X		X						
Gentamicin		X	X	X						
Mezlocillin						X	X	X	X	
Moxalactam					X	X	X		X[a]	
Nalidixic acid[a]						X				
Nitrofurantoin[a]							X			
Oxacillin		X	X							
Penicillin	X	X			X					
Tetracycline		X		X	X					
Trimethoprim[a]					X					
Trimethoprim-sulfamethoxazole[b]										
Vancomycin[c]	X		X							

[a] Urinary isolates only.
[b] Trimethoprim-sulfamethoxazole, 0.5 to 9.5 and 2 to 38 μg/ml, respectively.
[c] Special request.

Preparation of inoculum

Portions of four or five discrete colonies representative of the organisms to be tested are inoculated into 4 to 5 ml of a suitable broth medium, such as soybean-casein digest (Trypticase soy broth [BBL] or tryptic soy broth [Difco]), and adjusted to the turbidity of the barium sulfate standard described in Chapter 102 by the methods discussed there. This turbidity is equivalent to approximately 5×10^7 CFU per ml for members of the family *Enterobacteriaceae* and 1×10^8 to 5×10^8 CFU/ml for *Pseudomonas aeruginosa*. A 1:20 dilution is then prepared in saline or Mueller-Hinton broth for inoculation on the agar containing antimicrobial agents. Other methods of adjusting the inoculum size of log-phase cultures to approximately this concentration are acceptable. Inoculation of media should be made within 30 min of adjusting the inoculum.

The importance of testing pure cultures at properly standardized concentrations cannot be overemphasized. Primary or direct susceptibility testing should be restricted to normally sterile body fluids or to broth cultures of such fluids when microscopic examination suggests the presence of a single organism and when it is possible to standardize the inoculum. If the culture is mixed, susceptibility tests should be repeated by the standard method.

Inoculation of medium

The agar surface of the plates containing the dilutions of antimicrobial agents and the control plate containing no antimicrobial agent are spot inoculated (without spreading) with a loop calibrated to deliver 0.001 to 0.002 ml (1 to 2 μl) or with an inoculum-replicating apparatus such as that described by Steers et al. (18). In each case, about 10^4 CFU is delivered to a spot 5 to 8 mm in diameter. The Steers replicator may be purchased with an aluminum head carrying 32 (for 100-mm round petri dishes) or 36 (for 100-mm square dishes) equally spaced inoculating rods (Craft Machine, Inc., Chester, Pa.). Stainless-steel heads are easier to clean and less subject to corrosion. Identical results have been obtained by the loop method and with the replicator (6). If other types of inoculum-replicating apparatus are used, the inoculum should be adjusted so that the equipment delivers the same inoculum and volume to the surface of the plate or yields results identical to those obtained by the methods described above.

When the Steers replicator is used, a portion of the adjusted broth suspension is pipetted to the appropriate well in the seed plate, and then the inocula are picked up and gently transferred onto the agar surface with the replicator to avoid splashing. The plates containing the lowest concentration of antimicrobial agent should be seeded first, although transfer of significant amounts of agent back to the wells does not appear to occur. In routine practice, up to six antimicrobial agents can be tested without changing the replicator head. Control plates should be seeded last to ensure that viable organisms were present throughout the procedure.

Incubation

Inoculated agar plates are allowed to stand undisturbed until the inoculum spots are completely absorbed and are then incubated at 35°C for 16 to 20 h. Incubation in an atmosphere with CO_2 is not recommended because of the influence of surface pH on various antimicrobial agents (6).

Controls

Staphylococcus aureus ATCC 29213, *Escherichia coli* ATCC 25922, *Streptococcus faecalis* ATCC 29212, and *P. aeruginosa* ATCC 27853 are inoculated daily onto each set of agar plates. The MIC ranges of various antimicrobial agents for each of the reference strains are listed in Table 3 (11). At least 95% of MICs should fall within these MIC ranges, usually close to the middle of the pertinent range. Significant deviation (±1 twofold dilution) from the pertinent range requires a careful search for possible errors in the procedure or contamination of the control organism. The agar plates without antimicrobial agents and with and without added blood are inoculated at the end of the procedure to determine whether each organism was able to grow on agar alone or required blood supplementation.

Although the presence of contamination or a mixture of organisms in one inoculum site on the agar is usually readily detectable on close scrutiny of the control plates, it is recommended that a loopful of the broth culture remaining in each well in the seed plate be streaked onto a properly labeled quadrant of a standard blood agar plate upon completion of the replicating process. In this manner, detection of mixtures is facilitated and isolated colonies are made available for retesting in pure culture.

Results

The MIC represents the lowest concentration of antimicrobial agent at which complete inhibition occurs; a very fine, barely visible haze or a single colony is disregarded.

Results of testing sulfonamides and trimethoprim with the agar dilution procedure have proved satisfactory, but the endpoint must be taken as the plate showing sudden sharp (80 to 90%) diminution of growth (1, 2, 10). Some growth may be seen up to the highest concentration tested because of delay in inhibition due to carry-over of sulfonamide antagonists in the inoculum. Endpoints are sharper on media containing no thymidine, 5 to 10% horse blood lysed by freezing and thawing, or thymidine phosphorylase (7). However, results with unmodified Mueller-Hinton medium obtained in the United States have been generally satisfactory (1, 2).

The total divalent cation content of most Mueller-Hinton agar preparations brings them to concentrations approximating the physiological range; however, substantial lot-to-lot variations in MICs of aminoglycosides occur when they are tested in agar against *P. aeruginosa* (20). These variations appear to be due to the levels of ionized calcium and soluble magnesium in the medium (4), as well as the temperature and duration of autoclaving (21).

Methicillin

Heteroresistant strains of *S. aureus* express their resistance to methicillin, oxacillin, and nafcillin readily with the agar dilution procedure at 35°C, but

TABLE 3. Ranges of MIC[a] for standard strains[b]

Antimicrobial agent	MIC (μg/ml) for:			
	S. aureus ATCC 29213	*S. faecalis* ATCC 29212	*E. coli* ATCC 25922	*P. aeruginosa* ATCC 27853
Ampicillin	0.25–1.0	0.5–2.0	2.0–8.0	—
Azlocillin	2.0–8.0	1.0–4.0	8.0–32	2.0–8.0
Carbenicillin	2.0–8.0	16–64	4.0–16	16–64
Methicillin	0.5–2.0	>16	—	—
Mezlocillin	1.0–4.0	1.0–4.0	2.0–8.0	8–32
Nafcillin	0.12–0.5	2.0–8.0	—	—
Oxacillin	0.12–0.5	8.0–32	—	—
Penicillin	0.25–1.0	1.0–4.0	—	—
Piperacillin	1.0–4.0	1.0–4.0	1.0–4.0	2.0–8.0
Ticarcillin	2.0–8.0	16–64	2.0–8.0	8.0–32
Cefamandole	0.5–2.0	16–64	0.25–1.0	—
Cefoperazone	1.0–4.0	8.0–32	0.12–0.5	2.0–8.0
Cefotaxime	1.0–4.0	>32	0.06–0.25	8.0–32
Cefoxitin	1.0–4.0	>128	1.0–4.0	—
Cephalothin	0.12–0.5	16–64	4.0–16	—
Moxalactam	4.0–16	>128	0.12–0.5	8.0–32
Amikacin	1.0–4.0	64–256	0.5–4.0	2.0–8.0
Gentamicin	0.25–1.0	4.0–16	0.25–1.0	1.0–4.0
Kanamycin	1.0–4.0	16–64	1.0–4.0	—
Netilmicin	≤0.25	4.0–16	≤0.5–1.0	2.0–8.0
Tobramycin	0.25–1.0	8.0–32	0.25–1.0	0.5–2.0
Chloramphenicol	2.0–8.0	4.0–16	2.0–8.0	—
Clindamycin	0.06–0.25	4.0–16	—	—
Erythromycin	0.12–0.5	1.0–4.0	—	—
Tetracycline	0.25–1.0	8.0–32	1.0–4.0	8.0–32
Vancomycin	0.5–2.0	1.0–4.0	—	—
Nalidixic acid	—	—	1.0–4.0	—
Nitrofurantoin	8.0–32	4.0–16	4.0–16	—
Sulfisoxazole[c]	32–128	32–128	8.0–32	—
Trimethoprim[c]	1.0–4.0	≤1.0	0.5–2.0	>64
Trimethoprim-sulfamethoxazole (1:19)[c]	≤0.5-9.5	≤0.5-9.5	≤0.5-9.5	8.0-152–32-608

[a] These MICs were obtained in several reference laboratories by agar dilution or by broth microdilution with cation-supplemented broth.

[b] From the NCCLS Tentative Standard (11).

[c] Very medium dependent, especially with enterococci.

less rapidly and clearly at 37°C or above unless incubated for at least 48 h. Careful control of incubator temperatures at 35°C is thus important (5).

INTERPRETATION OF RESULTS OBTAINED FROM DILUTION TESTS

Interpretation for clinical purposes involves the factors discussed in Chapter 99. In general, in the treatment of systemic infections the dosage employed should yield a peak concentration in the blood substantially higher than the MIC (12). Factors of three- to fivefold have been suggested, but it must be realized that such recommendations have been made without standardized dilution procedures and on the basis of few well-controlled laboratory-clinical correlative studies. Many uncomplicated urinary tract infections respond to levels sufficient to inhibit the infecting organism in urine rather than in blood, and in such cases account may be taken of the urine levels.

As indicated in Chapter 99, interpretation is best made not only on the basis of data correlating blood level and MIC, but also by comparing the susceptibility of the strain under examination with that of other strains of the same species and by considering clinical experience in therapy with the agent being used. Appendix B of this section summarizes blood level data on different dose schedules for many antibiotics. For further information on available blood levels and usual MIC data, the reader is referred to articles by Goodman and Gilman (8), Hewitt and McHenry (9), and Wise (22). For a detailed description of a dilution testing procedure utilizing selected, clinically relevant concentrations of antimicrobial agents and providing precise interpretative categories, the reader is referred to the NCCLS Tentative Standard (11).

LITERATURE CITED

1. **Bauer, A. W., and J. C. Sherris.** 1964. The determination of sulfonamide susceptibility of bacteria. Chemotherapia **9:**1–19.
2. **Bennett, J. V., H. M. Camp, and T. C. Eickhoff.** 1968. Rapid sulfonamide disc sensitivity test for meningococci. Appl. Microbiol. **16:**1056–1060.
3. **Brenner, V. C., and J. C. Sherris.** 1972. Influence of

different media and bloods on the results of diffusion antibiotic susceptibility tests. Antimicrob. Agents Chemother. **1**:116–122.

4. **Casillas, E., M. A. Kenny, B. H. Minshew, and F. D. Schoenknecht.** 1981. Effect of ionized calcium and soluble magnesium on the predictability of the performance of Mueller-Hinton agar susceptibility testing of *Pseudomonas aeruginosa* with gentamicin. Antimicrob. Agents Chemother. **19**:987–992.
5. **Drew, W. L., A. L. Barry, R. O'Toole, and J. C. Sherris.** 1972. Reliability of the Kirby-Bauer disc diffusion method for detecting methicillin-resistant strains of *Staphylococcus aureus*. Appl. Microbiol. **24**:240–247.
6. **Ericsson, H. M., and J. C. Sherris.** 1971. Antibiotic sensitivity testing. Report of an international collaborative study. Acta Pathol. Microbiol. Scand. Sect. B Suppl. **217**:1–90.
7. **Ferone, R., S. R. M. Bushby, J. J. Burchall, W. D. Moore, and D. Smith.** 1975. Identification of Harper-Cawston factor as thymidine phosphorylase and removal from media of substances interfering with susceptibility testing to sulfonamides and diaminopyrimidines. Antimicrob. Agents Chemother. **7**:91–98.
8. **Goodman, L. S., and A. Gilman.** 1980. The pharmacological basis of therapeutics, 6th ed. The Macmillan Co., New York.
9. **Hewitt, W. L., and M. C. McHenry.** 1978. Blood level determinations of antimicrobial drugs. Med. Clin. N. Am. **62**:1119–1140.
10. **Kirven, L. A., and C. Thornsberry.** 1978. Minimum bactericidal concentration of sulfamethoxazole-trimethoprim for *Haemophilus influenzae*: correlation with prophylaxis. Antimicrob. Agents Chemother. **14**:731–736.
11. **National Committee for Clinical Laboratory Standards.** 1983. Methods for dilution antimicrobial susceptibility tests for bacteria that grow aerobically. Tentative standard. M7-T. National Committee for Clinical Laboratory Standards, Villanova, Pa.
12. **Petersdorf, R. G., and J. J. Plorde.** 1963. The usefulness of *in vitro* sensitivity tests in antibiotic therapy. Annu. Rev. Med. **14**:41–56.
13. **Reyn, A., M. W. Bentzon, J. D. Thayer, and A. E. Wilkinson.** 1965. Results of comparative experiments using different methods for determining the sensitivity of *Neisseria gonorrhoeae* to penicillin G. Bull. W.H.O. **32**:477–495.
14. **Ronald, A. R., J. Eby, and J. C. Sherris.** 1969. Susceptibility of *Neisseria gonorrhoeae* to penicillin and tetracycline, p. 431–434. Antimicrob. Agents Chemother. 1968.
15. **Ryan, K. J., G. M. Needham, C. L. Dunsmoor, and J. C. Sherris.** 1970. Stability of antibiotics and chemotherapeutics in agar plates. Appl. Microbiol. **20**:447–451.
16. **Sherris, J. C., A. L. Rashad, and G. A. Lighthart.** 1967. Laboratory determination of antibiotic susceptibility to ampicillin and cephalothin. Ann. N.Y. Acad. Sci. **145**:248–265.
17. **Snyder, R. J., P. C. Kohner, D. M. Ilstrup, and J. A. Washington II.** 1976. Analysis of certain variables in the agar dilution susceptibility test. Antimicrob. Agents Chemother. **9**:74–76.
18. **Steers, E., E. L. Foltz, B. S. Graves, and J. Riden.** 1959. An inocula replicating apparatus for routine testing of bacterial susceptibility to antibiotics. Antibiot. Chemother. (Basel) **9**:307–311.
19. **Thornsberry, C., T. L. Gavan, and E. H. Gerlach.** 1977. Cumitech 6, New developments in antimicrobial agent susceptibility testing. Coordinating ed., J. C. Sherris. American Society for Microbiology, Washington, D.C.
20. **Washington, J. A., II, R. J. Snyder, P. C. Kohner, C. G. Wiltsie, D. M. Ilstrup, and J. T. McCall.** 1978. Effect of cation of agar on the activity of gentamicin, tobramycin, and amikacin against *Pseudomonas aeruginosa*. J. Infect. Dis. **137**:103–111.
21. **Waterworth, P. M.** 1978. The aminoglycosides, p. 85–87. *In* D. S. Reeves, I. Phillips, J. D. Williams, and R. Wise (ed.), Laboratory methods in antimicrobial chemotherapy. Churchill Livingstone, Edinburgh.
22. **Wise, R.** 1978. Table of expected concentrations of antibiotic, p. 151–156. *In* D. S. Reeves, I. Phillips, J. D. Williams, and R. Wise (ed.), Laboratory methods in antimicrobial chemotherapy. Churchill Livingstone, Edinburgh.

Chapter 101

Susceptibility Tests: Microdilution and Macrodilution Broth Procedures

RONALD N. JONES, ARTHUR L. BARRY, THOMAS L. GAVAN, AND JOHN A. WASHINGTON II

Dilution antimicrobial susceptibility tests in a broth medium have been used for decades and represent only a minor modification of the agar-based method (see Chapter 100). These dilution procedures are so similar that the previously cited general statements about stock solutions, medium quality or performance, and suggested testing of antimicrobial agent concentrations will not be repeated. However, some characteristics are unique to the broth tests, and these will be discussed with regard to their variability from other methods or their contribution to test results.

The broth dilution susceptibility test was first utilized in large test tubes (at least 13 by 100 mm) with broth volumes of at least 1.0 ml. This method was not standardized by regulatory agencies or voluntary groups until late 1977, although the report from the International Collaborative Study described in some detail procedures for the agar and broth dilution methods (4). The National Committee for Clinical Laboratory Standards (NCCLS) published a proposed and later a tentative standard for dilution methods that generally followed those guidelines suggested by the International Collaborative Study report (4, 20). For routine use in clinical microbiology laboratories, the popularity of the broth dilution procedure in volumes of ≥1.0 ml was and has remained very low (4, 12). The broth dilution method received a great boost in popularity in the late 1960s when clinical laboratorians used serologic dispensing and diluting devices to perform the antimicrobial susceptibility test (8, 9, 18, 24). This simple miniaturization of the older broth tube dilution procedure became known as the microdilution broth method. A number of mechanical devices are currently available which enable laboratories to prepare large numbers of plastic trays with antimicrobial agent-containing broths in volumes of 0.05 to 0.2 ml. Broth dilution tests have now become an economical, technically accurate susceptibility procedure for routine use, especially for those clinical microbiology facilities that can prepare their own reagents. In 1975, commercial microdilution trays became available in a frozen form. Since the introduction of this first frozen product, several other frozen and dry-form systems have been marketed, differing only subtly in tray design, antimicrobial agents tested, inoculum methods, test performance or accuracy, and the volume of the test medium (2, 7, 15, 16).

Among the currently available products, several have been evaluated by well-developed protocols in multilaboratory collaborative trials (2, 7, 10, 11, 15, 16). The results from these products show comparable accuracy (Table 1), and reports from the College of American Pathologists Microbiology Surveys confirm that excellent levels of test performance and accuracy are being attained in numerous laboratories (12). Two tray forms are available, i.e., the frozen antimicrobial agents in broth (Micro-Media Systems, Potomac, Md.; Pasco Laboratories, Wheatridge, Colo.; and American MicroScan, Campbell, Calif.) and the dry-form trays requiring rehydration of the antimicrobial agents with a broth or broth-inoculum solution (Sensititre, GIBCO Diagnostics, Lawrence, Mass.; Sceptor, BBL Microbiology Systems, Cockeysville, Md.; MICUR, Boehringer Mannheim Diagnostics, Houston, Tex.; Microbial Profile System, 3M Co., St. Paul, Minn.; and API Uniscept, Analytab Products, Inc., Plainview, N.Y.). The dried type has the advantage of simple and prolonged storage, e.g., at room temperature for as long as 2 years. These products have been compared with the NCCLS (20) and the International Collaborative Study (4) methods and they yield MICs comparable to the reference procedures (2, 7, 10, 11, 15, 16). The systems are also very reproducible and accurate when used in the manner specified by the manufacturer.

There is a trend in clinical microbiology laboratories toward the use of the microdilution broth test and a decreased use of the standardized disk diffusion method (1, 12, 21). This trend is shown by the data in Table 2 from the College of American Pathologists Surveys, which show a remarkable increase in microdilution broth and automated test procedures since 1979. In 1983, 40% of the reporting laboratories used methods other than, but not to the exclusion of, the disk test. The greatest number (74%) of those laboratories used frozen trays, made either commercially (68%) or in house (6%).

For those laboratories wishing to produce their own broth dilution test reagents, the following sections are offered as a description of contemporary practice in the use of the broth methods (20).

STOCK SOLUTIONS, DILUTION SCHEDULES, AND MEDIA

The antimicrobial agent stock solutions are prepared and stored as described in Appendix 1 of this Manual. Additional information can be found in the NCCLS M7-T standard (20). The dilutions recommended in Chapter 100 can also be applied to all dilution methods. The flexibility of the dilution schedule selection and the ability not to be limited to serial twofold schedules from unity (1.0 μg/ml) or to choose half-dilution intervals are advantages of the preparation of one's own reagent trays.

The broth recommended for testing the majority of rapidly growing facultative anaerobes is Mueller-Hinton broth supplemented with the divalent cations Mg^{2+} and Ca^{2+}, although a more nutritive broth may be needed for some microorganisms. The divalent-cation supplementation has the most effect on the MICs of aminoglycosides, tetracyclines, and polymyxins, especially when *Pseudomonas aeruginosa* is tested (3, 6, 22, 25). The broth should contain 50 mg of calcium and 25 mg of magnesium per liter (22). These

TABLE 1. Results of several comparative trials of commercially produced microdilution broth antimicrobial susceptibility testing systems

Test system (reference)	Precision of reproducibility[a] (%)		Accuracy (% ±1 dilution) compared with reference result
	Intra-laboratory	Inter-laboratory	
Micro-Media (2)	96.0	96.0	90.6
MicroScan and Pasco	ND[b]	ND	ND
MICUR (11)	98.4	95.0	92.5–96.6
Sceptor (16)	96.8–97.6	97.2	92.9–98.3
Sensititre (7, 15)	80.4[c]	64.9[c]	87.6–93.8
Microbial Profile (10)	97.7	96.2	94.7–97.3

[a] Percentage of total comparisons. Some values are expressed as ranges from more than one protocol or published evaluation.

[b] ND, No data available from a collaborative trial.

[c] Reproducibility analyzed by statistical methods differing significantly from other studies. Please refer to published report (7). The reproducibility of the Sensititre system was equal or superior to that of the reference microdilution procedure.

concentrations can be achieved by the addition of the appropriate amounts of filter-sterilized $CaCl_2$ and $MgCl_2$ stock solutions to cold, sterile broth (20). The divalent-ion content of Mueller-Hinton broth is usually so low that it can be ignored (22). Since some medium manufacturers might supplement Mueller-Hinton broth, the laboratorian must refer to the dry-powder package insert for details before considering further supplementation with divalent cations. This supplementation will produce aminoglycoside (gentamicin, tobramycin, and amikacin) MICs for *P. aeruginosa* comparable to those obtained on Mueller-Hinton agar. Achieving acceptable results for *P. aeruginosa* ATCC 27853 with an aminoglycoside (usually gentamicin) should be the minimal criterion for acceptable cation-supplemented medium quality (20).

MICRODILUTION BROTH METHOD

Preparation and storage of plastic trays

For reference work, microdilution trays with antimicrobial agents should ideally be prepared each day they are to be used. However, with the availability of semiautomated dispensing devices for the preparation of dilutions and/or the dispensing of prediluted antimicrobial agents, it is possible, and convenient, to prepare large numbers of trays at one time to be frozen until used. The following procedure has been found to be satisfactory for the preparation of these trays.

As trays are filled, they can be stacked in groups of 5 to 10 trays and covered with an empty tray on the top of the stack, or the top tray can be covered with tape. In this way, each tray fits on top of the other tightly enough to provide a cover that minimizes evaporation and contamination. The stacks are then sealed in plastic bags and frozen at −20°C or colder. Household freezers have been found to be satisfactory, but self-defrosting units must be avoided because there may be a fluctuation in temperature during the defrost cycle significant enough to thaw and refreeze the antimicrobial agents, thus contributing to their rapid deterioration. If the equipment is available, freshly filled trays should be quick frozen at −70°C and then, if necessary, transferred to −20°C for longer-term storage. This treatment ensures maximum stability of the antimicrobial solutions. Once a stack of trays is removed from the freezer, the trays should be allowed to warm to room temperature before use. Unused thawed trays should be discarded, never refrozen. When stored as described, trays have a useful shelf life of at least 6 weeks. Storage at −70°C can significantly increase the shelf life to at least 3 months.

Inoculation of the trays

Actively growing broth cultures may be diluted to match the turbidity of a McFarland 0.5 standard (ca. 10^8 CFU/ml) and then further diluted to achieve the desired inoculum. Alternatively, a 4- to 8-h, 0.5-ml brain heart infusion broth culture (ca. 10^9 CFU/ml) may simply be diluted according to a pre-established dilution scheme to the desired concentration. The exact dilution scheme will vary, depending upon the volume of inoculum to be delivered to each well in the system being utilized. For those systems in which 50 μl of inoculum is added to each well containing 50 μl of diluted antimicrobial agent, the inoculum is standardized and prepared in the same way as for the macrodilution broth method described below. The final inoculum should be approximately 5×10^5 CFU/ml. Because the antimicrobial agent is diluted 1:2 with the inoculum, the final concentration will be one-half that which was originally dispensed (e.g., if the original concentration is 64 μg/ml, then after the addition of the inoculum, the concentration will be 32 μg/ml).

Multipoint plastic or metal inoculum replicators are used in several commercial and semiautomated systems. In these systems, tray wells are usually filled with 100 μl of antimicrobial agent and inoculated with approximately 1 to 5 μl of a standardized inoculum. With such small volumes being added, the dilution of the antimicrobial agent is not enough to require correction of the concentration. To obtain an inoculum density of approximately 5×10^5 CFU/ml (5×10^4 CFU per well, since the volume used per well is 0.1 ml), the culture is standardized to contain 5×10^7 CFU/ml if each well receives 1 μl. Since each commercial inoculation device delivers a slightly different volume, an appropriate adjustment of the initial inoculum must be made to fit each system. The operat-

TABLE 2. Trends in dilution susceptibility testing as monitored by the College of American Pathologists Microbiology Surveys, 1979–1983, Bacteriology Survey Group (800 to 1,200 laboratories)

Susceptibility method	% of laboratories reporting dilution results		
	1979	1981	1983
Agar dilution	9	4	1
Macrodilution broth	7	1	2
Microdilution broth			
In-laboratory preparation	18	9	6
Commercial preparation			
Frozen	65	65	68
Dry form	1	9	11
Automated		6	10
Total % of all tests reported	18	31	40

ing instructions from the manufacturer should be followed for each device. It is important to check the actual inoculum density delivered to the wells of a test tray by using conventional colony count techniques as part of a regular quality control program. Immediately after inoculation, each well should contain approximately 5 × 10^5 CFU/ml (5 × 10^4 CFU per well). Variations in inoculum concentrations may profoundly affect the test results. Sanders and others have reported low-frequency mutant subpopulations of antimicrobial-resistant organisms that are best identified by broth dilution systems at an inoculum concentration of $\geq 10^5$ CFU/ml (19, 23). Also, higher concentrations of inocula may be used in some emergent clinical settings to speed up the susceptibility results without significantly differing from routine standardized test results (17).

Tray incubation

After the microdilution trays are inoculated, they should be covered with sealing tape to minimize evaporation. Alternatively, when large numbers of tests are to be performed, trays can be stacked five high and covered with an empty tray, or the top tray can be taped. The stacks are then placed in plastic refrigerator bins or other acceptable containers with tightly sealed tops. The base of each bin is covered with wet paper towels. This arrangement forms a small humidity chamber which can be placed in an incubator at 35°C for 16 to 20 h. The humidity within the chamber is sufficient to prevent evaporation, and test results are equivalent to those obtained with sealed trays incubated without the humidity chamber. Excessive humidity should be avoided so that moisture does not condense on the surfaces of the trays, possibly contributing to contamination.

Interpretation of test results

After 16 to 20 h of incubation, the microdilution trays may be examined from below with a reflective viewer, and MICs are then determined. The trays may also be read visually from the top (remove the tape if it was used). Semiautomated and automated devices are also available for reading the plates. The endpoint (MIC) is taken as the lowest concentration of drug at which the microorganism tested does not show visible growth. The criterion of growth is a definite turbidity, a single, sedimented button of organisms 2 mm or greater in diameter, or more than one button (even if each is less than 2 mm in diameter). In judging the endpoint, it is important to consider the growth (or lack of growth) in the test well in comparison with growth characteristics of the microorganism in the growth control well (with no antimicrobial agent) used in each tray. For most endpoints, the distinction is an easily read turbid well versus a clear well. Some antimicrobial agents such as the sulfonamides and trimethoprim may have endpoints that "trail." The endpoint for these drugs should be an 80 to 90% decrease in growth compared with the growth of the microorganism in the control well. To achieve readable endpoints for these drugs, the broth medium lot used should be free of significant sulfonamide-interfering substances such as thymidine and *p*-aminobenzoic acid (5). Criteria for the determination of acceptable medium lots can be found in the NCCLS susceptibility test standards (20, 21) and include performance tests with *Streptococcus faecalis* ATCC 29212.

MACRODILUTION BROTH METHOD

Reagent and dilution preparations

A working antimicrobial solution is prepared by diluting the drug in Mueller-Hinton broth to the highest final concentration desired. Perform the test on sterile covered (screw-capped, cotton-plugged, etc.) test tubes (13 by 100 mm). For a small number of tests, prepare twofold dilutions directly in test tubes as follows. Add 2 ml of the working solution of drug to test tube 1 of the dilution series. To each remaining tube, add 1.0 ml of the Mueller-Hinton test broth. With a sterile pipette, transfer 1.0 ml from tube 1 to tube 2. After a thorough mixing, transfer 1.0 ml (with a separate pipette for this and each succeeding transfer) to tube 3. This process is continued through the next-to-last tube, from which 1.0 ml is removed and discarded. The last tube receives no antimicrobial agent and serves as a growth control. The final concentrations of antimicrobial agent in this test will be half those of the initial dilution series because of the addition of an equal concentration of inoculum in broth (see below).

A volumetric method or a bulk method of preparation of dilutions similar to that described for the agar dilution procedure may also be used (4, 20).

Inoculation and incubation of tubes

The inoculum is prepared to contain 10^5 to 10^6 CFU/ml by adjusting the turbidity of a broth culture to match the turbidity standard (see Chapter 102) and then further diluting it 1:200 in broth. Add 1.0 ml of the adjusted inoculum to each test tube. Incubate the tubes at 35°C for 16 to 20 h. Incubation in CO_2 is not recommended unless it is essential for organism growth.

Test result interpretations

The lowest concentration of antimicrobial agent that results in complete inhibition of visible growth represents the MIC. A very faint haziness or a small button of possible growth is generally disregarded, whereas a large button of growth or definite turbidity is considered evidence that the drug has failed to inhibit growth completely at that concentration.

QUALITY CONTROL

Control of MIC results

The goals of a quality control procedure are to ensure the precision and accuracy of the test procedure and to monitor the performance of the reagents and the individuals who perform and read the results. These aims are best realized through the use of reference strains selected for their genetic stability and their usefulness in the particular method being monitored.

The ideal reference strain for quality control of dilution susceptibility methods should have MIC endpoints near the middle of the range of the concentra-

TABLE 3. MICs for commonly used quality control strains by the broth microdilution procedure

Antimicrobial agent	MIC (μg/ml)			
	E. coli ATCC 25922	*Streptococcus faecalis* ATCC 29212	*Staphylococcus aureus* ATCC 29213	*P. aeruginosa* ATCC 27853
Penicillins				
Penicillin	VR[a]	2	0.25	VR
Ampicillin	2	1	0.5	VR
Methicillin	VR	VR	1–2	VR
Carbenicillin	4–8	32	4	32
Ticarcillin	2–4	32	4	16
Mezlocillin	4	2	4	16
Piperacillin	1	2	2	2
Cephalosporins				
Cephalothin	8	16–32	0.12–0.25	VR
Cefazolin	1	16	0.5–1	VR
Cefamandole	0.5	32	1	VR
Cefoxitin	2–4	VR	2–4	VR
Cefuroxime	2	VR	0.5	VR
Cefotaxime	<0.12	VR	2	8
Cefoperazone	0.25	16–32	1–2	2–4
Moxalactam	0.25	VR	4–8	4–8
Aminoglycosides				
Kanamycin	2–4	32	1	VR
Gentamicin	0.5	8	0.5–1	2–4
Tobramycin	0.5	16	1	1
Amikacin	1–2	128	2	4–8
Others				
Chloramphenicol	4	8	4	VR
Clindamycin	VR	8–16	0.06–0.12	VR
Erythromycin	VR	1	0.12–0.25	VR
Tetracycline	2	VR	0.5	ND[b]
Sulfamethoxazole-trimethoprim (19/1)	≤9.5/0.5	X[c]	≤9.5/0.5	ND[b]
Vancomycin	VR	2	1	VR
Nalidixic acid	2	VR	VR	VR
Nitrofurantoin	8	8	16	VR

[a] VR, Strain is very resistant to the antimicrobial agent.
[b] ND, No data available.
[c] X, This strain can be used to detect interfering substances (thymidine, etc.) in broth media. A value of ≤9.5/0.5 implies an acceptable broth lot.

tions being tested. For example, if seven concentrations are being tested, the ideal strain would have an MIC at dilution 4, but strains with endpoints at dilution 3 or 5 would be acceptable. Before a strain is accepted as a control strain, it should be tested for a period of time adequate to demonstrate that its antimicrobial susceptibility pattern is genetically stable.

The following reference strains are recommended for the control of dilution tests:

Escherichia coli ATCC 25922
Pseudomonas aeruginosa ATCC 27853
Staphylococcus aureus ATCC 29213
Streptococcus faecalis ATCC 29212

The expected MICs of various antimicrobial agents for these strains are shown in Table 3. For these antimicrobial agents, 95% of the MICs should be within ±1 $\log_2$ dilution of the MIC shown in Table 3, and most of the tests should be at the modal MIC listed.

Representative trays from each new batch or lot of microdilution trays should be tested with the reference strains to determine the acceptability of the batch. The MICs obtained should be no more than one dilution interval above or below the expected MIC; if the difference is greater, either the batch should be rejected or the results with the affected antimicrobial agent should not be recorded or reported. Representative uninoculated trays from various parts of the production lot should be incubated overnight to assess the sterility of the medium. The performance of each new lot of cation-supplemented Mueller-Hinton broth should be determined by testing dilutions of gentamicin with *P. aeruginosa* ATCC 27853. The MIC obtained in the supplemented broth should be within ±1 doubling dilution of the expected MIC (Table 3).

Appropriate reference strains should be included whenever tests are performed or on a weekly basis after reproducibility and accuracy have been documented by daily quality control practices (13, 14; also see reference 21). These quality control procedures provide a continuous review of the variables in dilution susceptibility methodology such as antimicrobial potency and stability, instrument function, and technical proficiency. The careful application of these quality control procedures will ensure that accurate and reproducible results are being obtained. Record the MICs obtained with each reference strain in an ongoing and current record. Check the MIC to see if it is within one dilution of the expected MIC (Table 3). If the MIC is not within the control limits, consider this difference as tentative evidence that a technical systematic error exists which may lead to a significant misinterpretation of test results. Conduct an immediate investigation to determine the source of the error(s), and then take corrective action to eliminate the error (21).

Other control procedures

Each microdilution broth tray used in a test should include a growth control well consisting of the basal medium without an antimicrobial agent to assess the viability of the test organism and to serve as a turbidity control for some bacteria-drug combinations for which growth comparisons are required for interpretation.

A sample of each inoculum should be streaked on a suitable agar plate (purity control) and incubated overnight to detect mixed cultures and to provide freshly isolated colonies should retesting be necessary. This step is particularly important for broth dilution methods, in which mixed cultures are more likely to go unrecognized from the routine test results.

Periodically, the laboratory should perform plate counts on representative inocula (inoculum control) to ensure that the inoculum standardization and dilution procedures remain accurate. Samples should be removed from the growth control well of randomly selected microdilution trays, and after appropriate dilution, measured volumes should be plated for colony counts.

To minimize variability in the interpretation of MIC endpoints among observers, periodic monitoring of personnel should be performed. A selected set of dilution tests should be read independently by all laboratory personnel who do these tests, and the results should be recorded and compared with the results obtained by a selected "standard" observer.

INTERPRETATION OF RESULTS

The category interpretation of broth dilution test results (MICs) should take into consideration all those criteria discussed in Chapter 99. The reader is also referred to the NCCLS M7-T standard (20), which contains interpretative tables that will be amended on a regular short-term basis by the NCCLS subcommittees as test factors change or as new antimicrobial agents are released by the Food and Drug Administration.

LITERATURE CITED

1. **Balows, A., C. T. Hall, and T. L. Gavan.** 1977. Standardization and quality control of disc susceptibility testing in the United States, p. 19–43. *In* A. Bondi, J. T. Bartola, and J. E. Prier (ed.), The clinical laboratory as an aid in chemotherapy of infectious disease. University Park Press, Baltimore.
2. **Barry, A. L., R. N. Jones, and T. L. Gavan.** 1978. Evaluation of the Micro-Media System for quantitative antimicrobial drug susceptibility testing: a collaborative study. Antimicrob. Agents Chemother. **13**:61–69.
3. **D'Amato, R. F., C. Thornsberry, C. N. Baker, and L. A. Kirven.** 1975. Effect of calcium and magnesium ions on the susceptibility of *Pseudomonas* species to tetracycline, gentamicin, polymyxin B, and carbenicillin. Antimicrob. Agents Chemother. **7**:596–600.
4. **Ericsson, H. M., and J. C. Sherris.** 1971. Antibiotic sensitivity testing. Report of an International Collaborative Study. Acta Pathol. Microbiol. Scand. Sect. B Suppl. **217**:1–90.
5. **Ferone, R., S. R. M. Bushby, J. J. Burchall, W. D. Moore, and D. Smith.** 1975. Identification of Harper-Cawston factor as thymidine phosphorylase and removal from media of substances interfering with susceptibility testing to sulfonamides and diaminopyrimidines. Antimicrob. Agents Chemother. **7**:91–98.
6. **Garrod, L. P., and P. M. Waterworth.** 1969. Effect of medium composition and the apparent sensitivity of *Pseudomonas aeruginosa* to gentamicin. J. Clin. Pathol. **22**:534–538.
7. **Gavan, T. J., R. N. Jones, and A. L. Barry.** 1980. Evaluation of the Sensititre system for quantitative antimicrobial drug susceptibility testing: a collaborative study. Antimicrob. Agents Chemother. **17**:464–469.
8. **Gavan, T. L., and M. A. Town.** 1970. A microdilution method for antibiotic susceptibility testing: an evaluation. Am. J. Clin. Pathol. **53**:880–885.
9. **Gerlach, E. H.** 1973. Microdilution: a comparative study, p. 63–76. *In* A. Balows (ed.), Current techniques for antibiotic susceptibility testing. Charles C Thomas, Publisher, Springfield, Ill.
10. **Gerlach, E. H., R. N. Jones, and A. L. Barry.** 1983. Collaborative evaluation of the Microbial Profile System for quantitative antimicrobial susceptibility testing. J. Clin. Microbiol. **17**:436–444.
11. **Jones, R. N., A. L. Barry, J. Bigelow, T. L. Gavan, and C. Thornsberry.** 1982. Evaluation of the MICUR system for quantitative antimicrobial susceptibility testing: a multiphasic comparison with reference methods. J. Clin. Microbiol. **16**:153–163.
12. **Jones, R. N., and D. C. Edson.** 1982. Interlaboratory performance of disk agar diffusion and dilution antimicrobial susceptibility tests, 1979–1981. Am. J. Clin. Pathol. **78**(Suppl.):651–658.
13. **Jones, R. N., D. C. Edson, and B. F. Gilmore.** 1983. Contemporary quality control practices for antimicrobial susceptibility tests: a report from the microbiology portion of the College of American Pathologists (CAP) Surveys Program. Am. J. Clin. Pathol. **80**(Suppl.):622–625.
14. **Jones, R. N., D. C. Edson, and J. V. Marymont.** 1982. Evaluation of antimicrobial susceptibility test proficiency by the College of American Pathologists Survey Program, a clarification of quality control recommendations. Am. J. Clin. Pathol. **78**:168–172.
15. **Jones, R. N., T. L. Gavan, and A. L. Barry.** 1980. Evaluation of the Sensititre microdilution antibiotic susceptibility system against recent clinical isolates: three-laboratory collaborative study. J. Clin. Microbiol. **11**:426–429.
16. **Jones, R. N., C. Thornsberry, A. L. Barry, and T. L. Gavan.** 1981. Evaluation of the Sceptor microdilution antibiotic susceptibility testing system: a collaborative investigation. J. Clin. Microbiol. **13**:184–194.
17. **Lampe, M. F., C. L. Aitken, P. G. Dennis, P. S. Forsythe, K. E. Patrick, F. D. Schoenknecht, and J. C. Sherris.** 1975. Relationship of early readings of minimal inhibitory concentrations to the results of overnight tests. Antimicrob. Agents Chemother. **8**:429–433.
18. **MacLowry, J. D., and H. H. Marsh.** 1968. Semiautomatic microtechnique for serial dilution-antibiotic sensitivity testing in the clinical laboratory. J. Lab. Clin. Med. **72**:685–687.
19. **Minami, S., A. Yotsuji, M. Inoue, and S. Mitsuhashi.** 1980. Induction of β-lactamase by various β-lactam antibiotics in *Enterobacter cloacae*. Antimicrob. Agents Chemother. **18**:382–385.
20. **National Committee for Clinical Laboratory Standards.** 1982. Tentative standard: M7-T. Standard methods for dilution antimicrobial susceptibility tests for bacteria

which grow aerobically. National Committee for Clinical Laboratory Standards, Villanova, Pa.

21. **National Committee for Clinical Laboratory Standards.** 1983. Tentative standard: M2-T3. Performance standard for antimicrobial disk susceptibility tests. National Committee for Clinical Laboratory Standards, Villanova, Pa.

22. **Reller, L. B., F. D. Schoenknecht, and M. A. Kenny.** 1974. Antibiotic susceptibility testing of *P. aeruginosa*: selection of a control strain and criteria for magnesium and calcium content in media. J. Infect. Dis. **130:**454–463.

23. **Sanders, C. C., and W. E. Sanders.** 1983. Emergence of resistance during therapy with newer beta-lactam antibiotics: role of inducible beta-lactamases and implications for the future. Rev. Infect. Dis. **5:**639–648.

24. **Takatsy, G.** 1956. The use of spiral loops in serological and virological micro methods. Acta Microbiol. Hung. **3:**191–202.

25. **Washington, J. A., II, R. J. Snyder, P. C. Kohner, C. G. Wiltsie, D. M. Ilstrup, and J. T. McCall.** 1978. Effect of cation of agar on the activity of gentamicin, tobramycin and amikacin against *Pseudomonas aeruginosa*. J. Infect. Dis. **137:**103–111.

Chapter 102

Susceptibility Tests: Diffusion Test Procedures

ARTHUR L. BARRY AND CLYDE THORNSBERRY

Bacterial susceptibility to antimicrobial agents may be measured in vitro by utilizing the principles of agar diffusion. Reasonably accurate and precise results can be obtained with agar diffusion techniques, provided that all procedural details are carefully standardized and controlled. Diffusion techniques can be utilized as qualitative, semiquantitative, or quantitative procedures, but generally microorganisms are simply categorized as being resistant, intermediate, or susceptible to each antimicrobial agent.

Antimicrobial agents are commonly applied to the test plates in the form of dried filter paper disks. When a disk is applied to the inoculated surface of the test medium, several events progress simultaneously. First, the dried disks absorb water from the agar medium, and thus the drug is dissolved. The antimicrobial agent is then free to diffuse through the adjacent agar medium, according to the physical laws that govern the diffusion of molecules through an agar gel (2, 4). The result is a gradually changing gradient of drug concentrations in the agar surrounding each disk. As the diffusion of the antimicrobial agent progresses, microbial multiplication also proceeds. After an initial lag phase, a logarithmic growth phase is initiated. At that time, bacterial multiplication proceeds more rapidly than the drug can diffuse, and bacterial cells which are not inhibited by the antimicrobial agent will continue to multiply until a lawn of growth can be visualized. No growth will appear in the area where the drug is present in inhibitory concentrations, and the more susceptible the test organism, the larger the zone of inhibition will be. The position of the edge of the zone of inhibition for most antimicrobial agents and microorganisms is determined during the first few hours of incubation (lag phase plus two or three generations). Obviously, microorganisms with prolonged generation times will appear to be more susceptible to each antimicrobial agent because the drugs will have more time to diffuse before the position of the edge of the zone is determined. Diffusion procedures have been primarily standardized for testing commonly isolated, rapidly growing bacterial pathogens such as members of the *Enterobacteriaceae* family and *Staphylococcus aureus*, which demonstrate fairly consistent, predictable growth rates when tested under standardized conditions. Microorganisms demonstrating marked strain-to-strain variability in growth rates cannot be tested reliably by the standardized diffusion procedures.

The size of the zone of inhibition is also affected by the rate of diffusion of the drug through the agar gel, and different drugs diffuse at different rates. Consequently, zones observed with one drug cannot be compared with those obtained with another antimicrobial agent. The diameter of the zone of inhibition, however, is indirectly proportional to the MIC, as measured by a dilution procedure. Table 1 includes the currently recommended zone standards that can be used to interpret disk tests performed by the procedures outlined in this chapter. The approximate MIC correlates which correspond to the susceptible and resistant categories are also listed in Table 1. In the following discussion, the principles that may be followed to establish such zone size interpretive standards and to estimate the MIC correlates are briefly outlined.

INTERPRETIVE ZONE STANDARDS

According to the guidelines outlined in Chapters 100 and 101, one must first select the MIC breakpoints which define the resistant and susceptible categories for each antimicrobial agent. Many antimicrobial agents include an intermediate category between the two MIC breakpoints. This category includes strains which are truly intermediate in susceptibility and are not properly categorized as resistant or susceptible. With some of the newer beta-lactam antibiotics, a broader, moderately susceptible category has been defined to identify those strains which would require maximal safe dosage of the drug for successful treatment. The moderately susceptible category is not the same as the intermediate category.

To establish zone diameters which correspond to

TABLE 1. Zone diameter interpretive standards and approximate MIC correlates used to define the interpretive categories[a]

Antimicrobial agent (amt per disk) and organism	Zone diam (nearest whole millimeter) for each interpretive category[b]:				Approximate MIC correlates[c] (μg/ml) for:	
	R	I	MS	S	R	S
Amikacin (30 μg)	≤14	15–16		≥17	≥32	≤12
Ampicillin (10 μg)						
Enterobacteriaceae	≤11	12–13		≥14	≥32	≤8
Staphylococcus spp.	≤28			≥29	β-Lactamase[d]	≤0.25
Haemophilus spp.	≤19			≥20	≥4	≤2
Enterococci[e]	≤16		≥17[e]		≥16	
Other streptococci	≤21		22–29	≥30	≥4	≤0.12
Augmentin[f] (20/10 μg)						
Haemophilus spp.	≤19			≥20		≤4/2
Staphylococcus spp.	≤19			≥20		≤4/2
Other organisms	≤13	14–17		≥18	≥32/16	≤8/4

Continued

TABLE 1—*Continued*

Antimicrobial agent (amt per disk) and organism	Zone diam (nearest whole millimeter) for each interpretive category[b]:				Approximate MIC correlates[c] (μg/ml) for:	
	R	I	MS	S	R	S
Azlocillin (75 μg)						
Pseudomonas spp.	≤14	15–17		≥18	≥256	≤64
Carbenicillin (100 μg)						
Enterobacteriaceae	≤17	18–22		≥23	≥32	≤16
Pseudomonas spp.	≤13	14–16		≥17	≥512	≤128
Cefamandole (30 μg)	≤14	15–17		≥18	≥32	≤8
Cefazolin (30 μg)	≤14	15–17		≥18	≥32	≤8
Cefonicid (30 μg)	≤14	15–17		≥18	≥32	≤8
Cefoperazone (75 μg)	≤15		16–20	≥21	≥64	≤16
Cefotaxime (30 μg)	≤14		15–22	≥23	≥64	≤8
Cefoxitin (30 μg)	≤14	15–17		≥18	≥32	≤8
Ceftazidime (30 μg)	≤14	15–17		≥18		≤8
Cefuroxime (30 μg)	≤14	15–17		≥18	≥32	≤8
Cephalothin (30 μg)	≤14	15–17		≥18	≥32	≤8
Chloramphenicol (30 μg)	≤12	13–17		≥18	≥25	≤12.5
Clindamycin (2 μg)	≤14	15–16		≥17	≥2	≤1
Doxycycline (30 μg)	≤12	13–15		≥16	≥16	≤4
Erythromycin (15 μg)	≤13	14–17		≥18	≥8	≤2
Gentamicin (10 μg)	≤12	13–14		≥15	≥8	≤6
Kanamycin (30 μg)	≤13	14–17		≥18	≥25	≤6
Methicillin[g] (5 μg)	≤9	10–13		≥14		≤3
Mezlocillin (75 μg)	≤12	13–15		≥16	≥256	≤64
Minocycline (30 μg)	≤14	15–18		≥19	≥16	≤4
Moxalactam (30 μg)	≤14		15–22	≥23	≥64	≤8
Nafcillin[g] (1 μg)	≤10	11–12		≥13		≤1
Nalidixic acid (30 μg)	≤13	14–18		≥19	≥32	≤12
Netilmicin (30 μg)	≤12	13–14		≥15	≥32	≤8
Nitrofurantoin (300 μg)	≤14	15–16		≥17	≥100	≤25
Oxacillin (1 μg)						
Staphylococcus spp.[g]	≤10	11–12		≥13		≤1
Pneumococci for penicillin	≤19			≥20		≤0.06
Penicillin G (10 U)						
Staphylococcus spp.	≤28			≥29	β-Lactamase	≤0.1
N. gonorrhoeae	≤19			≥20	β-Lactamase	≤0.1
Enterococci[e]	≤14		≥15[e]		≥16	
Other streptococci	≤19		20–27	≥28	≥4	≤0.12
Piperacillin (100 μg)	≤14	15–17		≥18	≥256	≤64
Streptomycin (10 μg)	≤11	12–14		≥15		
Sulfonamides[h] (250 or 300 μg)	≤12	13–16		≥17	≥350	≤100
Tetracycline (30 μg)	≤14	15–18		≥19	≥16	≤4
Ticarcillin (75 μg)	≤11	12–14		≥15	≥128	≤64
Trimethoprim[h] (5 μg)	≤10	11–15		≥16	≥16	≤4
Trimethoprim-sulfamethoxazole[h] (1.25/23.75 μg)	≤10	11–15		≥16	≥8/152	≤2/38
Tobramycin (10 μg)	≤12	13–14		≥15	≥8	≤6
Vancomycin (30 μg)	≤9	10–11		≥12		≤5

[a] Adapted from the October 1983 document (M2-T3) of the NCCLS. Refer to the most current NCCLS documents for updates and changes.

[b] R, Resistant; I, intermediate; MS, moderately susceptible; S, susceptible. An I result should be reported since it indicates an equivocal test result that may require further testing. When designated in the table, an MS result should be reported to indicate a level of susceptibility that should require the maximal safe dosage for therapy. Strains in the MS category are susceptible and not intermediate.

[c] Approximate MIC correlates used for the definition of the resistant and susceptible categories. These correlates should not be used for the interpretation of antimicrobial dilution test results.

[d] Resistant staphylococci are those that produce a β-lactamase, although their MICs may be low. Tests with a 10-U penicillin disk are preferred. Penicillin-resistant strains are also resistant to other penicillinase-sensitive penicillins, i.e., ampicillin, amoxicillin, azlocillin, carbenicillin, mezlocillin, piperacillin, ticarcillin, etc.

[e] Enterococci (*Streptococcus faecalis*, *Streptococcus faecium*, and *Streptococcus durans*) with zones of ≥17 mm (ampicillin) or ≥15 mm (penicillin) should be reported as moderately susceptible since they require a high dosage of penicillin or ampicillin usually combined with an aminoglycoside.

[f] Comprises 2 parts amoxicillin and 1 part clavulanic acid.

[g] Oxacillin disks are the preferred representative of the penicillinase-resistant penicillins for the testing of staphylococci and for the prediction of penicillin resistance among pneumococci. Staphylococci that are resistant to the penicillinase-resistant penicillins should be considered resistant to the cephalosporins.

[h] Sulfisoxazole disks may be used as class representatives of the sulfonamides. For the testing of sulfonamides or trimethoprim, the Mueller-Hinton agar should be shown to be relatively free of thymidine and thymine by screening with *S. faecalis* (ATCC 29212 or ATCC 33186). The latter should produce relatively clear zones of ≥24 mm in diameter around trimethoprim-sulfamethoxazole disks.

these MIC breakpoints, studies must be performed to establish the correlation between zone diameters and MICs for each antimicrobial agent. At least 100 to 150 microbial strains should be tested. These strains should include representatives of the common species for which the antimicrobial agent might be used but not strains with delayed or variable growth rates. Scattergrams are first prepared by plotting each MIC (log_2 scale) against the corresponding zone diameter (arithmetic scale). By convention, the MICs are plotted as the dependent variable (y axis), and the zone diameters are plotted as the independent variable (x axis).

The linear relationship between zone diameters and log_2 of the corresponding MICs can be expressed mathematically by applying the statistical method of least squares. This method provides a mathematical formula which permits an MIC correlate to be calculated for any given zone diameter. A straight regression line can be superimposed on the scattergram to display the theoretical line of best fit. An examination of the scattergram will then reveal whether one particular type of microorganism consistently deviates from linearity. For regression analysis, all tests showing no zone of inhibition and all tests with MICs outside the range of concentrations actually tested should be excluded from the calculations but included in the scattergram. In practice, the MICs should include concentrations of at least two doubling dilutions above and below the MIC breakpoints. With some antimicrobial agents, the regression line becomes parabolic with strains showing very low MICs. Because the method of least squares assumes a linear relationship, this method would be inappropriate for such parabolic relationships. In such situations, one may simply exclude data on the extremely susceptible strains from the calculations; however, these data should be included in the scattergram. Strains with endpoints near the MIC breakpoints are most important because the analysis is performed primarily to define the MIC-zone size relationship at that level to establish zone-size breakpoints.

Regression analysis is affected by the distribution of endpoints. The slope and intercept of the line can be distorted if a disproportionately large number of strains is clustered at either end of the spectrum. Ideally, there should be fairly uniform distribution of endpoints along the range of concentrations tested. When the distribution of endpoints does not permit a valid regression analysis, the data are best expressed as a simple scattergram. This situation often occurs with antimicrobial agents for which there is clearly a bimodal distribution of MICs with very few intermediate strains. By a simple examination of such scattergrams, interpretive zone standards can be selected as zone sizes which best separate the two populations. The error rate-bounded method of Metzler and DeHaan (17) has been suggested to formalize this type of analysis. That approach needs to be modified if an intermediate or a moderately susceptible category is to be defined. Also, acceptable error rates need to be modified for each antimicrobial agent; i.e., the criteria defined by Metzler and DeHaan (17), which were recommended for anaerobes, are only guidelines that may or may not be applicable to tests with aerobes and other antimicrobial agents.

Gentamicin disk tests were improved by the establishment of an intermediate zone size category (18). However, some strains which are susceptible by the disk test are resistant to 4 μg/ml but inhibited by 8 μg/ml; most of those strains are actually inhibited by 6 μg/ml, and thus the susceptible breakpoint may be redefined as a MIC of ≤6.0 μg/ml rather than ≤4.0 μg/ml (9). Similar half-dilution interval MIC correlates are also appropriate for tests with other aminoglycosides (9). Because MICs are frequently determined by testing twofold dilutions of antimicrobial agents (even log_2 dilution schedule), a significant number of strains may appear to be discrepant. This statistical artifact, which occurs because of the discontinuous nature of most MIC determinations, can be resolved by testing each isolate three or more times and then plotting the geometric mean of the MICs against the arithmetic mean of the zone diameters.

SELECTION OF DISK POTENCY

For agar diffusion susceptibility tests, the amount of antimicrobial agent in the disk must be standardized. In the United States, only one disk potency is recommended for each antimicrobial agent. If the disk content is changed and the interpretive zone standards are adjusted appropriately, the result will not be affected a great deal. Small changes in disk content produce rather minor changes in zone diameters; e.g., a twofold increase in disk potency generally increases zone sizes by only 2 or 3 mm (2). The standardized potency of the agent in the disk is selected after the examination of scattergrams and regression lines generated by testing disks with various concentrations of the drug. The most appropriate disk contains just enough antimicrobial agent to produce zones at least 10 mm in diameter with all strains which are inhibited by the greatest concentration that could be of clinical interest. On the other hand, the disk should not be so potent that susceptible strains yield unusually large zones of inhibition (in most cases <30 mm, rarely >40 mm).

CLASS CONCEPT OF DISK TESTING

For practical reasons, the number of antimicrobial agents which are tested routinely must be limited. The general principles outlined in Chapter 99 serve as guides for the selection of the most appropriate battery of drugs to be tested for a given patient population. Antimicrobial agents with a similar chemical structure can be classified into a family. In turn, it is possible to categorize the members of each chemically related family into drug classes based on similar spectra of activity. A single representative of each class of drugs may often be used to predict the susceptibility or resistance of an organism to the other drugs in that class. Some families of related drugs may be represented by a single class representative; other families may require two or three class representatives. The class concept of disk testing is applicable only when members of the class demonstrate almost identical spectra of activity against those microorganisms that are ordinarily subjected to disk diffusion tests.

When a new antimicrobial agent is introduced, its

spectrum of activity should be compared with those of chemically related antimicrobial agents, especially the established class representatives, by using a large collection of representative species. If the number of qualitative interpretive discrepancies (susceptible to one drug but resistant to another) is relatively small and judged to be of little clinical significance, the class disk may continue to be applicable. Routine disk tests with a new drug are necessary only when the drug demonstrates a uniquely different spectrum of activity. The drug may have other advantages, but the spectrum of activity is the only aspect to be considered in the selection of a drug for testing. If the class concept is judged to be applicable, the class representative can be used to predict the susceptibility or resistance of an organism to the new drug, with a relatively small number of interpretive errors. The class representative should be the least active drug within each class so that any differences that do occur will be in the direction of false resistance rather than false susceptibility. In this way, all isolates found to be susceptible to the class representative can be assumed to be susceptible to the other related drugs. However, a few strains which are resistant to the class representative might be susceptible to one or more drugs within that class, and if testing is clinically indicated and no disk is available, the resistant strains may require testing against those drugs by dilution procedures.

When several chemically related drugs have somewhat dissimilar spectra of activity, the class concept is not necessarily applicable, but the expense involved in routine testing of all drugs in the spectrum-related group would not be justified. In that situation, the clinical laboratory should test the representative that is most frequently used in the institution being served. Other drugs in the spectrum-related group may or may not be so similar that the class concept is applicable.

Methicillin is the least active member of the β-lactamase-resistant penicillins, but oxacillin disks are preferred as the class representative because of their stability and because they may detect heteroresistant staphylococci more efficiently (16). Nafcillin disks may also be used, but they should not be tested on blood-containing media because the activity of nafcillin is diminished by the blood. Cephalothin is the class representative for most first-generation cephalosporins, but the second- and third-generation cephalosporins have somewhat broader spectra of activity. Tests with three cephalosporins and a cephamycin might be judged necessary in dealing with certain patient populations. The tetracyclines are represented by disks containing tetracycline hydrochloride, but doxycycline and minocycline may be active against tetracycline-resistant strains of *S. aureus* or *Acinetobacter* spp. Testing selected isolates with doxycycline or minocycline disks might be appropriate if the isolates are resistant to tetracycline disks.

RECOMMENDED DISK DIFFUSION TECHNIQUES

The disk diffusion test currently recommended by the U.S. Food and Drug Administration (15) and by the National Committee for Clinical Laboratory Standards (NCCLS) (18) is a slight modification of that described by Bauer et al. (10). This method should be followed exactly as outlined below if accurate, reproducible results are to be obtained. Only one other method has been adequately studied and shown to give comparable zone sizes, similar precision, and satisfactory correlation with MICs. That method is the agar overlay method of Barry et al. (6), which has been recognized formally (15, 18) as the acceptable alternative method when testing the commonly isolated, rapidly growing bacterial pathogens such as *S. aureus*, members of the *Enterobacteriaceae* family, and *Pseudomonas aeruginosa* (3).

Agar medium

Both disk diffusion methods have been standardized with Mueller-Hinton agar. The unsupplemented medium supports the growth of most of those microorganisms for which susceptibility tests are most relevant. Some microorganisms may require the addition of 5% defibrinated sheep, horse, or other animal blood, and in most cases this addition has little effect on the results. Nafcillin and novobiocin, however, should not be tested on media supplemented with blood. The agar overlay method has not been evaluated for testing those microorganisms which require the addition of defibrinated blood or other supplements to the base medium.

Because there is some lot-to-lot variation in the performance of Mueller-Hinton agars (14, 19), each new lot should be tested with the recommended control strains (19, 20) before it is released for use in testing clinical isolates. Media which fail to perform satisfactorily should not be used. When the control strains yield zones within the control limits in Table 2, performance is satisfactory.

The pH of each batch of Mueller-Hinton agar should be checked at the time the medium is poured for use. The pH should be 7.2 to 7.4 after equilibration at room temperature and may be measured by allowing the agar to solidify around the electrodes of a pH meter, by macerating the medium in neutral distilled water, or by using a surface electrode. The freshly prepared and cooled medium is poured into petri plates on a level horizontal surface to give a uniform depth of approximately 4 mm; this depth requires approximately 60 to 70 ml of medium in 150-mm plates and approximately 25 to 30 ml of medium in 100-mm plates. After the medium has been allowed to cool to room temperature, it should be stored in a refrigerator (2 to 8°C). The plates should be sealed in plastic to minimize evaporation, especially if they are to be stored for more than 7 days. Just before use, the plates should be placed in an incubator (35°C) with their lids ajar until excess surface moisture is lost by evaporation (usually about 10 to 20 min). There should be no droplets of moisture on the surface of the medium or on the petri plate cover. With the agar overlay method of inoculation, the plates must be warmed to room temperature, but the surface need not be dried before inoculation.

Storage of antimicrobial disks

Filter paper disks containing antimicrobial agents specifically certified for susceptibility testing are generally supplied in separate containers, each with a desiccant. The disks should be stored under refrigera-

TABLE 2. Zone diameter limits for quality control of antimicrobial disk susceptibility tests on Mueller-Hinton agar without blood or other supplements

Antimicrobial agent (disk content)	Zone diam limits (mm)			
	E. coli (ATCC 25922)	*S. aureus* (ATCC 25923)	*P. aeruginosa* (ATCC 27853)	*E. coli* (ATCC 35218)
Amikacin (30 μg)	19–26	20–26	18–26	
Ampicillin (10 μg)	16–22	27–35		
Augmentin[a] (20/10 μg)	19–25	28–36		18–22
Azlocillin (75 μg)			24–30	
Carbenicillin (100 μg)	23–29		18–24	
Cefamandole (30 μg)	24–30	26–34		
Cefazolin (30 μg)	23–29	29–35		
Cefonicid (30 μg)	25–29	22–28		
Cefoperazone (75 μg)	28–34	24–33	23–29	
Cefotaxime (30 μg)	29–35	25–31	18–22	
Cefoxitin (30 μg)	23–29	23–29		
Cefuroxime (30 μg)	20–26	27–35		
Ceftazidime (30 μg)	25–32	16–20	22–29	
Ceftizoxime (30 μg)	30–36	27–35	12–17	
Cephalothin (30 μg)	17–22	29–37		
Chloramphenicol (30 μg)	21–27	19–26		
Cinoxacin (100 μg)	26–32			
Clindamycin (2 μg)		24–30		
Doxycycline (30 μg)	18–24	23–29		
Erythromycin (15 μg)		22–30		
Gentamicin (10 μg)	19–26	19–27	16–21	
Kanamycin (30 μg)	17–25	19–26		
Methicillin (5 μg)		17–22		
Mezlocillin (75 μg)	23–29		19–25	
Minocycline (30 μg)	19–25	25–30		
Moxalactam (30 μg)	28–35	18–24	17–25	
Nafcillin (1 μg)		16–22		
Nalidixic acid (30 μg)	22–28			
Netilmicin (30 μg)	22–30	22–31	17–23	
Nitrofurantoin (300 μg)	20–25	18–22		
Oxacillin (1 μg)		18–24		
Penicillin G (10 U)		26–37		
Piperacillin (100 μg)	24–30		25–33	
Streptomycin (10 μg)	12–20	14–22		
Sulfisoxazole (250 or 300 μg)[b]	18–26	24–34		
Tetracycline (30 μg)	18–25	19–28		
Ticarcillin (75 μg)	24–30		21–27	
Tobramycin (10 μg)	18–26	19–29	19–25	
Trimethoprim (5 μg)[b]	21–28	21–28[c]		
Trimethoprim-sulfamethoxazole (1.25/23.75 μg)[b]	24–32	24–32[c]		
Vancomycin (30 μg)		15–19		

[a] Comprises 2 parts amoxicillin and 1 part clavulanic acid.

[b] For the testing of sulfonamides or trimethoprim, the media should be shown to be relatively free of thymidine and thymine by screening tests with *S. faecalis* (ATCC 29212 or ATCC 33186). Trimethoprim-sulfamethoxazole disks should produce zones essentially free of fine colonies and ≥24 mm in diameter.

[c] Several laboratories have reported difficulty with these quality control parameters.

tion (2 to 8°C) or frozen at −14°C or colder until needed. Disks containing drugs which belong to the penicillin or cephalosporin family should always be kept frozen to ensure maintenance of their potency (10, 14), but a small working supply may be held in a refrigerator at 2 to 8°C for as long as 1 week. Unopened containers should be removed from the refrigerator or freezer 1 or 2 h before the disks are to be used and allowed to equilibrate with room temperature before being opened. This step is done to minimize the amount of condensation that occurs when warm room air reaches the cold containers. If a disk-dispensing apparatus is used, it should be fitted with a tight cover and supplied with an adequate indicating desiccant. Also, it should be allowed to warm to room temperature before being opened. When not in use, the dispensing apparatus should always be kept covered and refrigerated. Only those disks that have not reached the stated expiration date of the manufacturer should be used.

Inoculation of test plates

Standard method. By the standard method of inoculation (10, 18), an inoculating needle or loop is touched to each of four or five well-isolated colonies of the same morphological type, and the inoculum is inoculated into 4 or 5 ml of a suitable broth medium such as soybean-casein digest broth. The broth cultures are then allowed to incubate at 35°C until a slightly visible turbidity appears (usually 2 to 5 h). The turbidity of actively growing broth cultures is then adjusted with saline or broth to obtain a turbidity visually comparable to that of a turbidity standard prepared by adding 0.5 ml of 0.048 M $BaCl_2$ (1.75% [wt/vol] $BaCl_2 \cdot 2H_2O$) to 99.5 ml of 0.36 N H_2SO_4 (1%, vol/vol). This turbidity is half the density of a McFarland no. 1 standard and is often referred to as a McFarland 0.5 standard. This turbidity standard is agitated on a Vortex mixer immediately before use. Sealed tubes should be stored in the dark at room temperature and replaced if the inocula are shown (by count) to be aberrant. For proper turbidity adjustment, it is helpful to use a white background and a contrasting black line(s) in combination with an adequate light source. A device such as the modified Rh-typing view box described by Stemper and Matsen (21) facilitates the standardization of cultures. When time does not permit the development of a turbid broth culture, isolated colonies from an overnight plate can be suspended directly into a small volume of saline which is then further diluted until the turbidity matches that of the McFarland standard (1, 7, 13). This method is preferred for the detection of methicillin-resistant strains (16, 24) of *S. aureus* and for the testing of species that will not grow well in the broth medium, i.e., *Haemophilus influenzae, Neisseria gonorrhoeae*, and *Streptococcus pneumoniae* (23). The inoculum suspension should not be allowed to stand longer than 15 to 20 min before the plates are inoculated. A device that permits the direct standardization of the inoculum without an adjustment of turbidity and without preincubation in a broth medium has been found to be satisfactory (1).

To inoculate the agar medium, a sterile, nontoxic swab on an applicator stick is dipped into the standardized suspension, and excess broth is expressed by pressing and rotating the swab firmly against the inside of the tube above the fluid level. The swab is then streaked evenly in three directions over the entire surface of the agar plate to obtain a uniform inoculum. A final sweep is made of the agar rim with the cotton swab. This plate is then allowed to dry for 3 to 5 min, but no longer than 15 min, before the disks are applied. The inoculum should yield confluent or almost confluent growth.

Alternative agar overlay method. For the alternative agar overlay method (6), four or five isolated colonies of the same morphological type are selected, and a visibly turbid suspension is prepared in 0.5 ml of brain heart infusion broth in tubes (13 by 100 mm). Changes that may occur as a result of evaporation during storage of this small volume of broth are avoided by aseptic transfer of the broth into sterile tubes on the day it is to be used. The small-volume broth cultures are then allowed to incubate in a 35 to 37°C water bath or heating block for 4 to 8 h. By this time, maximal growth has occurred. A 0.001-ml calibrated loopful of a well-mixed broth culture is transferred to 9.0 ml of a 1.5% aqueous solution of agar which has been melted and cooled to 45 to 50°C. Screw-capped tubes of agar may be held at this temperature for up to 8 h in a heating block before inoculation. After inoculation, the seeded agar is quickly mixed by gentle inversion and spread evenly over the surface of a 150-mm petri plate containing Mueller-Hinton agar (4 mm deep). This procedure is facilitated by bringing the plates to room temperature before attempting to spread the thin layer of seeded agar. The inoculated plates are allowed to stand for 3 to 5 min undisturbed on a flat, level surface before susceptibility disks are applied.

Test procedure: either method of inoculation

Within 15 min after the plates are inoculated, antibiotic-impregnated disks are applied to the surface of the inoculated plates either with a mechanical dispenser or by hand with sterile forceps. All disks must be gently pressed down onto the agar with forceps or an inoculating needle to ensure complete contact with the agar surface (if the dispenser does not do this automatically). The spatial arrangement of the disks should be such that they are no closer than 15 mm to the edges of the plate and far enough apart to prevent overlapping of zones of inhibition. Generally, this arrangement limits the number of disks which can be placed on a single plate to 12 or 13 on a 150-mm plate or only 4 or 5 on a 100-mm plate. Within 15 min after the disks are applied, the plates are inverted and placed in an incubator at 35°C. Any longer delay before incubation will allow excess prediffusion of the antimicrobial agents. Incubation in an environment of increased CO_2 is to be avoided, because the CO_2 may alter the surface pH enough to affect the antimicrobial activity of some agents.

Reading and interpretation

After 16 to 18 h of incubation, the plates are examined, and the diameters of the zones of complete inhibition are measured to the nearest whole millimeter by sliding calipers, a ruler, or a template prepared for this purpose. When unsupplemented medium is used, the measuring device is held on the back of the petri plate, which is illuminated with reflected light from a source at an angle of approximately 45° against a black nonreflecting background (5). Zones on blood-containing media are measured at the agar surface. The endpoint by all reading systems is complete inhibition of growth, disregarding tiny colonies which can be detected only by very close scrutiny or by using transmitted light or mechanical enlargers. When staphylococci are tested against the penicillinase-resistant penicillins, the plates should be examined with transmitted light to visualize the very small colonies or light growth that may occur with methicillin-resistant strains (16). Large colonies growing within the clear zone of inhibition may represent resistant variants or a mixed inoculum and may require reidentification and retesting. In the case of sulfonamides or sulfonamide-trimethoprim mixtures, the microorganisms may grow through several generations before inhibition occurs. In this instance, slight growth (80% inhibition) is disregarded, and the margin of heavy growth is measured (18). A veil of swarming *Proteus* sp. is also disregarded, and the margin of heavy growth is measured. In clinically urgent situations, preliminary readings can often be obtained within 5 or 6 h after inoculation, but the plates should always be reincubated, and a final report should be withheld until a full 16 to 18 h has elapsed (7).

The zone diameters for individual antimicrobial agents are translated into susceptible, intermediate, moderately susceptible, or resistant categories by referring to an interpretive table. The interpretations for the antibiotics in Table 1 are those presently recommended by the Food and Drug Administration (15) or by the NCCLS (18). For netilmicin, the interpretive standards recommended by Barry et al. (8) are currently being considered. Approximate MIC correlates of the breakpoints are also given in Table 1 (18). For the aminoglycosides, MIC correlates are probably more appropriately located at half-dilution intervals (9).

The MIC breakpoints listed in Table 1 are related to the levels of drug in blood usually expected with frequently used dose schedules or, in the case of nitrofurantoin or nalidixic acid, to levels of drug in urine. The breakpoints were tested against the distribution of zone sizes and MICs among a variety of species of known clinical responsiveness or lack of responsiveness to check their appropriateness and were modified, when considered necessary, for adequate discrimination. The resistant and susceptible categories for most drugs were developed to apply to systemic infections and appropriate dose schedules. Where applicable, a moderately susceptible category has been identified to separate those susceptible strains that are likely to need the maximal safe dosage from those that are likely to respond to a reduced dosage. In situations in which a high dosage of nontoxic agent may be given, the levels of drug in blood may greatly exceed those considered in the establishment of the interpretive values for the resistant category. Similarly, the concentration of certain antibiotics by the kidneys may result in drug levels in urine manyfold higher than the levels considered in the development of breakpoints for systemic infections. In such situations, organisms which are resistant by the disk method might be treated successfully, but antimicrobial dilution tests may need to be done before such therapy is considered.

LIMITATIONS OF THE METHODS AND SPECIAL PRECAUTIONS

Slowly growing organisms, obligate anaerobes, and capnophiles should not be tested with the disk diffusion method, which has been standardized for testing rapidly growing aerobic or facultative organisms, but rather with dilution methods if susceptibility tests are needed. Special precautions must be taken and special interpretative standards must be used to test *Neisseria meningitidis* against the sulfonamides (11). The method has not been standardized for testing other antimicrobial agents against *N. meningitidis*. Disk diffusion susceptibility testing is not done with either methenamine mandelate or methenamine hippurate, as there is no corollary between the in vivo and in vitro conditions (26).

Special problems are posed by heteroresistant, "methicillin-resistant" *S. aureus*. These strains appear to have an increased clinical resistance to the penicillins and cephalosporins, although they may appear to be susceptible in in vitro tests. Although some resistant strains may be very difficult to recognize, they can usually be detected with oxacillin or methicillin disks. Tests should be done at 35°C because resistance is often not seen at 37°C (22). If incubators cannot be controlled at 35°C, separate tests with only one of these agents should be made on segments of Mueller-Hinton agar plates incubated at 30°C. Resistant strains often produce a haze or film of small colonies growing within an otherwise definite zone of inhibition; that inner haze of growth is easily overlooked unless the zones are carefully examined with transmitted and reflected light (16). Diffusion tests with these strains often fail to indicate resistance to cloxacillin and cephalothin, although dilution tests often show them to be resistant. Thus, strains proven to be resistant to methicillin, oxacillin, or nafcillin should be considered potentially resistant to the whole group of penicillinase-resistant penicillins and to the cephalosporins, and the clinician should be alerted to this possible resistance (16, 18, 24).

Many false reports of methicillin resistance have resulted from the deterioration of methicillin disks during refrigeration. Attention to the recommendations given for disk storage and quality control should avoid this difficulty. Use of the more stable oxacillin disk is preferred for the detection of most methicillin-resistant strains; some strains will be detected with methicillin disks but not with oxacillin disks, but many more strains will be missed if only methicillin disks are tested.

As discussed in Chapters 100 and 101, the results of tests for gentamicin susceptibility of *P. aeruginosa* are highly dependent on the amount of soluble magnesium and calcium in the medium. Most batches of Mueller-Hinton agar are satisfactory for routine testing and interpretation by the criteria given in Table 1. However, it is important to use a control strain of *P. aeruginosa* to detect errors from this source (18–20). Other control strains are less sensitive to minor changes in the agar medium.

When trimethoprim or trimethoprim-sulfamethoxazole disks are tested, each lot of Mueller-Hinton agar should be first screened with *Streptococcus faecalis* ATCC 29212 or ATCC 33186. Satisfactory media should produce clear zones of inhibition >24 mm in diameter around trimethoprim-sulfamethoxazole disks. Trace amounts of thymidine or thymine will be antagonistic to this control strain, thus producing a haze of growth inside the zone of inhibition.

The accuracy of this disk test is dependent on the adequate diffusion of antimicrobial agents. Polymyxin B and E (colistin) both diffuse very poorly through an agar gel. Although resistance to polymyxins in the disk test is significant, it is important to confirm susceptibility by a dilution test if the use of these agents for systemic therapy is proposed.

MODIFICATIONS OF THE DIFFUSION TEST FOR CERTAIN FASTIDIOUS ORGANISMS

The routine methods described above are generally unacceptable when the more fastidious microorganisms are tested. If disk tests are to be performed with such microorganisms, the method must be modified to fit each microorganism. Disk tests for *H. influenzae* (23), *N. gonorrhoeae* (23), and *S. pneumoniae* (25) have been standardized for certain antimicrobial agents and are described below.

***H. influenzae*.** Ampicillin-resistant, β-lactamase-producing strains of *H. influenzae* (18, 23) are not uncommon in the United States. In critical infections caused by *H. influenzae*, the isolate should be tested for β-lactamase production (Chapter 105) to rapidly detect ampicillin resistance as soon as colonies are observed on the primary isolation media. Disk tests may then be performed with ampicillin, chloramphenicol, or tetracycline.

Mueller-Hinton agar is supplemented by adding 1% hemoglobin (or 5% horse blood) and 1% IsoVitaleX (BBL Microbiology Systems, Cockeysville, Md.), Supplement VX (Difco Laboratories, Detroit, Mich.), or any equivalent synthetic supplement. Growth from an overnight agar plate is emulsified in a small volume of Mueller-Hinton broth. This suspension is then adjusted to match the turbidity of a McFarland 0.5 standard. Disk tests are then performed as described for the standard disk diffusion technique. Incubation in CO_2 is not required.

With ampicillin disks, susceptible strains produce zones of ≥20 mm, and resistant, β-lactamase-producing strains yield zones of ≤19 mm. Strains which are resistant to ampicillin are also resistant to other penicillins. For the other antimicrobial agents, the interpretive zone standards listed in Table 1 may be tentatively applied to tests with *H. influenzae*.

***N. gonorrhoeae*.** Some isolates of *N. gonorrhoeae* (23) demonstrate decreased susceptibility or moderate resistance to penicillin that is not associated with β-lactamase production. Dilution tests are the preferred method for the measurement of the susceptibility of such isolates. Although disk diffusion tests have been used as screening tests, the method has not been standardized for these organisms. However, some strains have been found to be highly resistant as a result of their ability to produce β-lactamase. Such microorganisms can be detected by directly testing colonies for β-lactamase activity (Chapter 105). A disk test can also be used for the detection of β-lactamase-producing strains.

The agar medium for the disk technique is GC Agar Base (Difco) without hemoglobin but supplemented with 1% IsoVitaleX (BBL), Supplement VX (Difco), or an equivalent supplement. Colonies from an overnight agar plate are emulsified in a small volume of Mueller-Hinton broth. The suspension is then diluted to match the turbidity of a McFarland 0.5 standard. Disk tests are then performed as described for the standard disk diffusion technique except for incubation of the test plates under an atmosphere of increased CO_2.

β-Lactamase-producing strains will produce zones of ≤19 mm around 10-U penicillin disks. A few β-lactamase-negative strains which are relatively resistant to penicillin will also produce zones of ≤19 mm. Strains with zones of ≤19 mm may be considered resistant to penicillin. Susceptible strains usually produce 30- to 60-mm zones.

It should be noted that as a standardized procedure, disk diffusion is applicable only to the detection of a high-level penicillin resistance in gonococci, but it has been used widely to detect spectinomycin resistance, using a 100-μg disk. Resistant strains have zones of ≤14 mm, and susceptible strains have zones of ≥18 mm.

***S. pneumoniae*.** Some strains of *S. pneumoniae* (25) have been found to be resistant to penicillin (MIC of ≥2 μg/ml), and other strains are relatively resistant to penicillin (MIC of 0.12 to 1.0 μg/ml). The latter may not respond to penicillin therapy of meningitis or other serious, life-threatening diseases. Consequently, pneumococci recovered from spinal fluid, blood, or other body fluids should be tested for resistance or relative resistance to penicillin. Penicillin-resistant or relatively resistant strains can be detected by the use of a disk diffusion test, with a 1-μg oxacillin disk to screen for penicillin resistance. A 10-U penicillin disk should not be used for the detection of penicillin resistance. Nafcillin disks should not be used on blood-containing media.

Mueller-Hinton agar with 5% defibrinated sheep blood is the preferred agar medium. The inoculum is standardized by the selection of bacterial colonies from an overnight agar plate and the preparation of an emulsion in Mueller-Hinton broth. The broth suspension is then diluted until it matches the turbidity of a McFarland 0.5 standard. A test plate is inoculated as described previously for the standard Bauer-Kirby method, and then a 1-μg oxacillin disk is applied. The tests are allowed to incubated for 18 to 24 h at 35°C without increased CO_2.

Penicillin-susceptible strains produce oxacillin zones of ≥20 mm in diameter. Penicillin-resistant or relatively resistant strains produce oxacillin zones of ≤19 mm in diameter. Chloramphenicol disks may also be tested to detect chloramphenicol-resistant strains. Zone standards listed in Table 1 can be applied to interpret the results.

The most recent NCCLS standard should be consulted for revised or updated data on testing or breakpoints.

QUALITY CONTROL OF DIFFUSION TESTS

Standard control strains of *Escherichia coli* (ATCC 25922), *S. aureus* (ATCC 25923), and *P. aeruginosa* (ATCC 27853) have been designated for monitoring the accuracy and precision of disk diffusion tests. Another strain of *E. coli* (ATCC 35218) has been designated for monitoring the performance of tests with disks containing a β-lactam antibiotic combined with a β-lactamase inhibitor, e.g., augmentin (amoxicillin plus clavulanic acid). This strain produces a β-lactamase which should be inactivated by the inhibitor. When used in conjunction with *E. coli* ATCC 25922, both components of the disk can be monitored.

The first three control strains mentioned above and *S. faecalis* (ATCC 29212 or ATCC 33186) should be used to test the performance of every new batch of Mueller-Hinton agar before the medium is released for use with clinical specimens. In addition, the appropriate control strains should be tested every day that a batch of susceptibility tests is performed. However, in those laboratories that have documented a record of satisfactory performance for 30 consecutive days, the frequency of control tests may be reduced to once a week (18).

Stock cultures of all control strains should be obtained from a reliable source and maintained in a manner which will ensure continued viability with minimal opportunity for the selection of resistant variants (12). Working cultures can be maintained at 4 to 8°C on soybean-casein digest agar, with weekly subcultures. The working cultures should be replaced at least once a month from frozen, lyophilized, or commercial cultures, or sooner if results are questionable. This replacement is especially important with *P. aeruginosa* ATCC 27853, which tends to lose susceptibility to the ureidopenicillins if it is repeatedly transferred on agar slants. Cultures may be maintained frozen at −20°C or lower (in either a freezer or a nitrogen chest) in a suitable stabilizer such as a broth medium with 15% glycerol defibrinated sheep or rabbit blood, 50% fetal calf serum in broth, etc. Cultures may also be maintained in a lyophilized state. With either method, the cultures can be stored without significant risk of altering their antimicrobial susceptibility.

Before being tested, the cultures should be transferred to a nutrient broth, incubated for 4 to 18 h, and then streaked onto an agar plate to obtain isolated colonies. Control tests should be performed only with 18- to 24-h isolated colonies and never initiated from stored cultures. If the results of control tests suggest contamination of the stock cultures or changes in the susceptibility of the organism, fresh cultures should be obtained. Such a problem might be suspected if there is a sudden, dramatic change in test results which cannot be explained by methodology.

Table 2 lists the minimum and maximum zone sizes that should be observed with a single control test. These limits generally represent 95% confidence limits, i.e., 1 of every 20 tests might be outside the stated limits. Zones should never be more than four standard deviations above or below the midpoint of the stated limits, i.e., midpoint ± the stated range (maximum zone diameter minus minimum zone diameter). The mean of a series of tests should approach the midpoint value between the upper and lower limits defined in Table 2.

Several laboratories have reported difficulties with certain control limits, possibly as a result of minor changes in the Mueller-Hinton agars that have been introduced since the control limits were established. These parameters are being reevaluated, and if necessary, revised standards will be published as supplements to the current standards of the NCCLS (18). The most recent NCCLS document should be consulted for the most current quality control guidelines.

COMMON SOURCES OF ERROR

Although the disk diffusion method is a fairly forgiving procedure, technical errors can compromise accuracy and reliability. The following are some of the more common sources of error encountered in clinical microbiology laboratories.

1. Improper preparation of Mueller-Hinton agar, especially failure to measure pH at the time of preparation
2. Use of outdated medium or unsatisfactorily stored plates
3. Variability in Mueller-Hinton agars. Each new lot should be checked by testing appropriate control strains before it is used
4. Improper storage of disks
5. Inadequate standardization of broth culture density. More often than not, the culture is too heavy, but too-light inocula are not uncommon
6. Inaccurate preparation or maintenance of turbidity reference standard
7. Failure to express surplus fluid from the swab before the plates are inoculated
8. Excessive delay between culture standardization and plate inoculation
9. Excessive delay in the application of the disk after the inoculation of the plates
10. Excessive delay in the incubation of the plates after the application of the disks
11. Incubation temperature deviating from 35°C or incubation in an increased CO_2 atmosphere
12. Premature reading of test results before the full 16 to 18 h of incubation
13. Failure to measure zone borders carefully and with a standardized angle and source of illumination
14. Attempts to test mixed cultures
15. Application of the procedure to slow growers and anaerobes
16. Failure to include quality control strains or to record the results of control tests at appropriate intervals
17. Transcription error in recording the results of individual tests

INDICATIONS FOR DIRECT SUSCEPTIBILITY TESTING ON CLINICAL MATERIAL

The direct inoculation of susceptibility plates can sometimes provide invaluable preliminary information on urgent clinical infection problems. For example, direct tests may be made on plates seeded with emergency specimens, such as cerebrospinal fluid, other body fluids, or purulent specimens, if direct Gram-stained smears indicate that a large number of bacteria of a single species may be expected to grow. However, routine direct susceptibility tests on clinical material are to be avoided. Mixtures of organisms, common in many specimens, frequently produce inaccurate interpretations (7). Furthermore, it is very difficult to standardize the density of an inoculum from direct clinical material. The use of a purity check plate will be of great assistance in these emergency situations, as will an assessment of the nature of the lawn of inoculum on the susceptibility test plate. The results of emergency tests should be reported as preliminary or tentative and should be repeated and confirmed by one of the recommended methods. When directly inoculated test plates are unsatisfactory, valuable preliminary information can be obtained by making preliminary readings of the regular test after 5 to 6 h of incubation at 35°C. When this reading is done, the plate must be reincubated, and a final report is issued after overnight incubation.

LITERATURE CITED

1. **Baker, C. N., C. Thornsberry, and R. W. Hawkinson.** 1983. Inoculum standardization in antimicrobial susceptibility testing: evaluation of overnight agar cultures and the Rapid Inoculum Standardization System. J. Clin. Microbiol. **17:**450–457.

2. **Barry, A. L.** 1976. The antimicrobic susceptibility test: principles and practices. Lea & Febiger, Philadelphia.
3. **Barry, A. L.** 1976. Standardized disk diffusion methods for antimicrobic susceptibility testing of common rapid-growing bacterial pathogens, p. 9–20. *In* A. Bondi, J. T. Tartola, and J. E. Prier (ed.), The clinical laboratory as an aid in chemotherapy of infectious diseases. University Park Press, Baltimore.
4. **Barry, A. L.** 1980. Procedures for testing antimicrobial agents in agar media; theoretical considerations, p. 1–23. *In* V. Lorian (ed.), Antibiotics in laboratory medicine. The Williams & Wilkins Co., Baltimore.
5. **Barry, A. L., M. B. Coyle, C. Thornsberry, E. H. Gerlach, and R. W. Hawkinson.** 1979. Methods of measuring zones of inhibition with the Bauer-Kirby disk susceptibility test. J. Clin. Microbiol. **10:**885–889.
6. **Barry, A. L., F. Garcia, and L. D. Thrupp.** 1970. An improved single disk method for testing the antibiotic susceptibility of rapidly growing pathogens. Am. J. Clin. Pathol. **53:**149–158.
7. **Barry, A. L., L. J. Joyce, A. P. Adams, and E. J. Benner.** 1973. Rapid determination of antimicrobial susceptibility for urgent clinical situations. Am. J. Clin. Pathol. **59:**693–699.
8. **Barry, A. L., G. H. Miller, R. S. Hare, R. N. Jones, and C. Thornsberry.** 1984. Modification of interpretive breakpoints for netilmicin disk susceptibility tests with *Pseudomonas aeruginosa*. J. Clin. Microbiol. **19:**311–314.
9. **Barry, A. L., C. Thornsberry, and R. N. Jones.** 1981. Gentamicin and amikacin disk susceptibility tests with *Pseudomonas aeruginosa*: definition of minimal inhibitory concentration correlates for susceptible and resistant categories. J. Clin. Microbiol. **13:**1000–1003.
10. **Bauer, A. W., W. M. M. Kirby, J. C. Sherris, and M. Turck.** 1966. Antibiotic susceptibility testing by a standardized single disk method. Am. J. Clin. Pathol. **45:**493–496.
11. **Bennett, J. V., H. M. Camp, and T. C. Eickhoff.** 1968. Rapid sulfonamide disc sensitivity test for meningococci. Appl. Microbiol. **16:**1056–1060.
12. **Coyle, M. B., M. F. Lampe, C. L. Aitkin, P. Feigl, and J. C. Sherris.** 1976. Reproducibility of control strains for antibiotic susceptibility testing. Antimicrob. Agents Chemother. **10:**436–440.
13. **D'Amato, R. F., and L. Hochstein.** 1982. Evaluation of a rapid inoculum preparation method for agar disk diffusion susceptibility testing. J. Clin. Microbiol. **15:**282–285.
14. **Ericsson, H. M., and J. C. Sherris.** 1971. Antibiotic sensitivity testing. Report of an international collaborative study. Acta Pathol. Microbiol. Scand. Suppl. **217:**1–90.
15. **Federal Register.** 1972. Rules and regulations: antibiotic susceptibility disks. Fed. Regist. **37:**20525–20529.
16. **McDougal, L. K., and C. Thornsberry.** 1984. New recommendations for disk diffusion antimicrobial susceptibility tests for methicillin-resistant (heteroresistant) staphylococci. J. Clin. Microbiol. **19:**482–488.
17. **Metzler, C. M., and R. M. DeHaan.** 1974. Susceptibility of anaerobic bacteria: statistical and clinical considerations. J. Infect. Dis. **130:**588–594.
18. **National Committee for Clinical Laboratory Standards.** 1984. (Approved standard M2-A3). Performance standards for antimicrobial disk susceptibility tests. National Committee for Clinical Laboratory Standards, Villanova, Pa.
19. **Pollock, H. M., B. H. Minshew, M. A. Kenny, and F. D. Schoenknecht.** 1978. Effect of different lots of Mueller-Hinton agar on the interpretation of the gentamicin susceptibility of *Pseudomonas aeruginosa*. Antimicrob. Agents Chemother. **14:**360–367.
20. **Reller, L. B., F. D. Schoenknecht, M. A. Kenny, and J. C. Sherris.** 1974. Antibiotic susceptibility testing of *Pseudomonas aeruginosa*: selection of a control strain and criteria for magnesium and calcium content in media. J. Infect. Dis. **130:**454–463.
21. **Stemper, J. E., and J. M. Matsen.** 1970. Device for turbidity standardizing of cultures for antibiotic sensitivity testing. Appl. Microbiol. **19:**1015–1016.
22. **Thornsberry, C., J. Q. Caruthers, and C. N. Baker.** 1973. Effect of temperature on the in vitro susceptibility of *Staphylococcus aureus* to penicillinase-resistant penicillins. Antimicrob. Agents Chemother. **4:**263–269.
23. **Thornsberry, C., T. L. Gavan, and E. H. Gerlach.** 1977. Cumitech 6, New developments in antimicrobial agent susceptibility testing. Coordinating ed., J. C. Sherris. American Society for Microbiology, Washington, D.C.
24. **Thornsberry, C., and L. K. McDougal.** 1983. Successful use of broth microdilution in susceptibility tests for methicillin-resistant (heteroresistant) staphylococci. J. Clin. Microbiol. **18:**1084–1091.
25. **Thronsberry, C., and J. M. Swenson.** 1980. Antimicrobial susceptibility tests for *Streptococcus pneumoniae*. Lab. Med. **11:**83–86.
26. **Waterworth, P. M.** 1962. A misapplication of the sensitivity test: mandelamine disks. J. Med. Lab. Technol. **19:**163–168.

Chapter 103

Susceptibility Testing of Anaerobes

VERA L. SUTTER

Routine susceptibility testing of all anaerobic isolates appears to be unnecessary since many of the isolates have predictable susceptibility patterns to commonly recommended antimicrobial agents. In some instances, test results may be misleading, since most anaerobic infections are polymicrobial and the nature of the infectious process may change in the interval between the collection of the specimen and the reporting of results. However, susceptibility tests are often needed for patients with serious infections such as endocarditis or brain abscess, with infections that require prolonged therapy such as osteomyelitis, or with infections that persist or recur despite appropriate, empirical antimicrobial therapy. Tests are also needed to monitor the susceptibility of commonly isolated species so that changing patterns of resistance can be detected and the empirical basis of therapy can be changed to reflect changes in the antibiogram.

Methods similar to those used for aerobic and facultative bacteria are useful for estimating the susceptibility of anaerobes to various antimicrobial agents. The working group on the standardization of susceptibility tests for anaerobic bacteria has published a tentative reference agar dilution procedure (4) which has been shown to give reproducible results (5). Broth dilution tests, both macro- and microdilution, are convenient procedures in some laboratories. However, many laboratories are not properly staffed or equipped to do these tests and require simpler procedures. Broth-disk procedures are gaining acceptance in many laboratories. They have been found to be reproducible and are recommended for those who cannot use the more complex procedures. Agar diffusion tests are not recommended because of the complexities and variation introduced by the slow and varied growth rates of anaerobic bacteria.

ANTIMICROBIAL AGENTS TO BE TESTED

Antimicrobial agents to be tested are those commonly recommended for therapy of anaerobic infections, i.e., cefoxitin, chloramphenicol, clindamycin, and metronidazole. If drugs such as penicillins, cephalosporins, erythromycin, tetracyclines, or the new combinations of penicillins or cephalosporins with β-lactamase inhibitors are being considered for use, they should also be tested. Aminoglycosides (gentamicin, kanamycin, etc.) are generally inactive against anaerobes; therefore, tests with these agents are of little or no value.

AGAR DILUTION TEST

The test described below is the tentative agar dilution reference method (4). Modifications of the incubation time or medium will be necessary with unusually slow-growing or nutritionally fastidious strains. Modified procedures should be tested with the recommended control strains and should give results consistent with those given in Table 1.

Preparation of inoculum

For each culture to be tested, inoculate portions of five or more colonies (to provide a 3-mm loopful of growth) into thioglycolate medium without indicator. This medium should be supplemented with hemin (5 μg/ml) and vitamin K_1 (0.1 μg/ml) before the medium is autoclaved and with sodium bicarbonate (1 mg/ml) just before the medium is used. Incubate the medium 18 to 20 h, or longer if required, to obtain sufficient growth of the test organism. Just before the test, adjust the turbidity to match that of the barium sulfate standard described in Chapter 102. An alternative preparation of the inoculum may be made by suspending the growth from an agar plate (not more than 72 h old) into a clear medium such as brucella broth and adjusting the turbidity to the above standard (3, 6).

Preparation of test plates

On the day of the test, prepare dilutions of antimicrobial agents according to the procedure described in Chapter 100 and incorporate the dilutions into Wilkins-Chalgren agar (7).

Performance of test

Inoculate the adjusted inoculum onto the surface of the agar of each plate with a 0.001-ml calibrated loop or with a Steers replicator. Allow the plates to dry; then incubate the culture at 35°C in anaerobic jars for 42 to 48 h. The MIC of each strain is the lowest concentration of drug yielding no growth, one discrete colony, or a fine, barely visible haze.

Control procedures

Before and after the inoculation of each series of antimicrobial agent-containing plates, inoculate two plates of the Wilkins-Chalgren agar without antimicrobial agents. Incubate one set of plates in the anaerobic jars to serve as growth controls, and incubate one set of plates in air to determine whether contamination with aerobic or facultatively anaerobic bacteria has occurred. Each time tests are performed, include one or more of the recommended control strains (Table 1) which have been selected to yield at least one on-scale endpoint with each antimicrobial agent tested. Results obtained with these strains should be consistent with those given in Table 1.

BROTH DILUTION TESTS

Macrodilution test

Conventional broth dilution tests may be performed in a manner similar to that described in Chapter 101, with modifications necessary for the growth of anaerobic bacteria. Brucella broth, containing vitamin K_1 (0.1 μg/ml) and either hemin at a final concentration of 5 μg/ml or peptic digest of sheep blood at a final concentration of 5%, may be used. Other broths which

TABLE 1. Acceptable MIC ranges for control strains[a]

Antimicrobial agent	MIC range[b] for:		
	Bacteroides fragilis ATCC 25285	*Bacteroides thetaiotaomicron* ATCC 29741	*Clostridium perfringens* ATCC 13124
Carbenicillin	16–64	16–64	0.25–1
Cefamandole	32–128	32–128	0.063–0.25
Cefoperazone	32–64	32–128	NR[c]
Cefotaxime	8–32	16–64	0.063–0.25
Cefoxitin	4–16	8–32	0.25–1
Chloramphenicol	2–8	4–16	2–8
Clindamycin	0.5–2	2–8	0.031–0.125
Metronidazole	0.25–1	0.5–2	0.125–0.5
Mezlocillin	16–64	8–32	0.063–0.25
Moxalactam	0.25–1	4–16	0.031–0.125
Penicillin	16–64	16–64	0.063–0.25
Tetracycline	0.125–0.5	8–32	0.031–0.125
Ticarcillin	16–64	16–64	0.25–1

[a] Data from Sutter et al. (5), Zabransky et al. (9), and R. Zabransky, R. Birk, A. Helstad, P. Murray, J. Emmerman, and V. Sutter (Program Abstr. Intersci. Conf. Antimicrob. Agents Chemother. 23rd, Las Vegas, Nev., abstr. no. 168, 1983).

[b] MIC given in micrograms per milliliter, except for penicillin, which is given in units per milliliter.

[c] NR, No range has been recommended with this combination of organism and antibiotic.

can be used are Schaedler, brain heart infusion, or a broth formulation of the Wilkins-Chalgren medium.

Prepare the inoculum as described above for the agar dilution tests; then dilute it 1:200 in the broth medium with minimal aeration. Add an equal volume of inoculum to the antimicrobial agent-containing tubes, and incubate the cultures in anerobic jars for approximately 48 h. An inoculated broth containing no antimicrobial agent is included as a growth control for each strain tested. A tube of uninoculated broth is also included with the tests each day since a slight turbidity often develops in broth during incubation, without any actual growth of microorganisms. This uninoculated tube is included to help distinguish between this type of turbidity and actual growth. The MIC is the lowest concentration of drug with no visible growth.

Bactericidal endpoints can be determined as described in Chapter 105.

Microdilution test

Microdilution tests may be performed in a manner similar to that described in Chapter 101, with modifications to accommodate the special requirements of anaerobic bacteria. Prepare antimicrobial agent dilutions in one of the broths listed above for the macrodilution test. Schaedler broth should not be used if trays are prepared in advance and frozen at −20°C, because penicillins have been shown to be unstable in this broth (1).

On the day of the test, precondition the trays by placing them in an anerobic environment 3 to 4 h before the test. Prepare the inoculum from an overnight broth culture, and adjust it so that the final concentration in each well is between 10^5 and 10^6 CFU/ml. Inoculate the trays, and incubate them in anaerobic jars or a similar anaerobic environment for 48 h. The MIC is the lowest concentration of drug that allows no visible growth.

Quality control of broth dilution tests should follow the principles outlined in Chapter 101.

BROTH-DISK ELUTION TESTS

In broth-disk elution tests, antimicrobial agents are eluted into broth from paper disks marketed for use in the diffusion test to provide the specific concentrations needed.

Broth-disk test of Wilkins and Thiel

See reference 8 for more information on the Wilkins and Thiel broth-disk test.

Preparation of tubes. Tubes should have occlusive, rubber-stoppered seals. In an atmosphere of oxygen-free CO_2 (see Chapter 37), add one or more disks of the appropriate antimicrobial agent (Table 2) to 5 ml of prereduced brain heart infusion broth supplemented with cysteine (0.05%), hemin (0.0005%), menadione (0.002%), and yeast extract (0.5%). Examples of the concentrations obtained are given in Table 2. Other concentrations can be achieved by an increase in the volume of medium in the tubes or by the addition of more disks.

Inoculum. Use an 18- to 24-h culture in prereduced chopped-meat medium.

Performance of test. In an atmosphere of oxygen-free CO_2, add 1 drop of inoculum to the broth in each tube containing antimicrobial agent and in one tube without antimicrobial agent. Reseal the tubes and incubate them for 18 to 24 h at 35°C. Compare the turbidity of the broth in the tubes containing drug with that of the growth control. Susceptibility to the drug is defined as either the absence of turbidity or turbidity less than 50% of that of the growth control.

Report the results as either susceptible or resistant to the concentration of each antimicrobial agent tested.

Broth-disk elution test of Kurzynski et al.

See reference 2 for more information on the broth-disk elution test of Kurzynski et al.

TABLE 2. Preparation of broth-disk tubes

Antimicrobial agent	Disk content	Disks per 5 ml of medium	Concn (per ml)
Ampicillin	10 μg	2	4 μg
Carbenicillin	100 μg	6	120 μg
Cefoperazone	75 μg	4	60 μg
Other cephalosporins[a]	30 μg	3	18 μg
Chloramphenicol	30 μg	3	18 μg
Clindamycin[b]	2 μg	8	3.2 μg
Metronidazole	80 μg	1	16 μg
Mezlocillin	75 μg	4	60 μg
Penicillin G	10 U	1	2 U
Tetracycline	30 μg	1	6 μg
Ticarcillin	75 μg	4	60 μg

[a] Includes cephalothin, cefoxitin, cefamandole, cefotaxime, and moxalactam.

[b] A 10-μg disk may become available and would be preferable. Two 10-μg disks added to 5 ml of medium will give a concentration of 4 μg/ml. One 10-μg disk added to 5 ml of medium will give a concentration of 2 μg/ml. If oral therapy is to be used, the lower concentration should be tested.

A simple modification of the test described above does not require manipulation in oxygen-free CO_2. Tests are performed in thioglycolate medium incubated aerobically.

Add the appropriate number of disks to boiled and cooled thioglycolate medium without indicator. Leave the tubes at room temperature for 2 h to allow elution of the antimicrobial agent into the medium, because the medium contains 0.07% agar. Inoculate the tubes with 0.1 ml of an overnight culture or with a suspension prepared as described above for the agar dilution procedure, and tighten the caps. Invert the tubes twice to ensure a uniform distribution of drug and inoculum; then incubate the cultures at 35°C overnight, or for 48 h for fastidious bacteria. A tube of thioglycolate medium without antimicrobial agent should also be inoculated with each organism tested. Read the results and report them as described above for the Wilkins-Thiel procedure. Strains listed in Table 1 may be used for quality control.

Contamination introduced with the disks has not been a problem in practice, but it is wise to subculture tubes that show growth to segments of a blood agar plate and to incubate the subcultures aerobically overnight to exclude aerobic contamination.

LITERATURE CITED

1. **Jones, R. N., R. P. Packer, P. C. Fuchs, A. L. Barry, and K. Borchardt.** 1978. Stability of antimicrobials in Schaedler's anaerobic and brain heart infusion broths stored at −20°C. J. Antibiot. **31:**226–228.
2. **Kurzynski, T. A., J. W. Yrios, A. G. Helstad, and C. R. Field.** 1976. Aerobically incubated thioglycolate broth disk method for antibiotic susceptibility testing of anaerobes. Antimicrob. Agents Chemother. **10:**727–732.
3. **Murray, P. R., and A. C. Niles.** 1983. Inoculum preparation for anaerobic susceptibility tests. J. Clin. Microbiol. **18:**733–734.
4. **National Committee for Clinical Laboratory Standards.** 1982. Tentative standard reference agar dilution procedure for antimicrobial susceptibility testing of anaerobic bacteria, vol. 2, p. 70–101. National Committee for Clinical Laboratory Standards, Villanova, Pa.
5. **Sutter, V. L., A. L. Barry, T. D. Wilkins, and R. J. Zabransky.** 1979. Collaborative evaluation of a proposed reference dilution method of susceptibility testing of anaerobic bacteria. Antimicrob. Agents Chemother. **16:**495–502.
6. **Swenson, J. M., and C. Thornsberry.** 1984. Preparing inoculum for susceptibility testing of anaerobes. J. Clin. Microbiol. **19:**321–325.
7. **Wilkins, T. D., and S. Chalgren.** 1976. Medium for use in antibiotic susceptibility testing of anaerobic bacteria. Antimicrob. Agents Chemother. **10:**926–928.
8. **Wilkins, T. D., and T. Thiel.** 1973. Modified broth-disk method for testing the antibiotic susceptibility of anaerobic bacteria. Antimicrob. Agents Chemother. **3:**350–356.
9. **Zabransky, R. J., E. Randall, V. L. Sutter, R. J. Birk, G. Westenfelder, J. Emmerman, and A. T. M. Ghoneim.** 1983. Establishment of minimum inhibitory concentrations of cefoperazone for control and reference anaerobic organisms. J. Clin. Microbiol. **17:**711–714.

Chapter 104

Laboratory Studies with Antifungal Agents: Susceptibility Tests and Bioassays

SMITH SHADOMY, ANA ESPINEL-INGROFF, AND RODNEY Y. CARTWRIGHT

Infections caused by the pathogenic and saprophytic fungi have increased in terms of annual morbidity and mortality (15). A major part of this increase has been attributed to the increased use of cytotoxic agents as well as to newer and more effective antibacterial agents (5). Paralleling the increase in the occurrence of fungal infections over the last decade has been the introduction of a number of new antifungal agents, both topical and systemic. As a result of both developments, the clinical laboratory now is assuming a greater role in the selection and monitoring of antifungal chemotherapy, and efforts now are being made to standardize laboratory tests with antifungal agents (16).

SUSCEPTIBILITY TESTS WITH ANTIFUNGAL AGENTS

Susceptibility tests with antifungal agents are performed for the same reason that tests with antibacterial agents are performed: to provide reliable data which will permit selection of the most suitable agents for use in the treatment of human infections. Such data may be either quantitative, i.e., the determination of MICs, or qualitative, i.e., the prediction of probable clinical response. Other reasons for the performance of such tests include our inability to predict probable clinical or in vitro responses, the need to explain apparent therapeutic failures, and the problems associated with primary as well as selected or secondary resistance.

In vitro test procedures with antifungal agents are similar in design to those employed with antibacterial agents (22). However, there are certain differences which make testing with antifungal agents somewhat more difficult. Of primary importance are differences in the basic properties of some of the compounds to be tested, as well as unique differences in the growth characteristics of many of the organisms involved.

A variety of antifungal agents are now available for the treatment of human and animal mycotic infections. Five such agents are discussed in this chapter. They include the antifungal antibiotics amphotericin B and nystatin, the synthetic imidazoles miconazole and ketoconazole, and the synthetic antimetabolite 5-fluorocytosine (5-FC; flucytosine).

Amphotericin B and nystatin are polyene macrolide antibiotics. Amphotericin B is used primarily in the treatment of systemic and life-threatening fungal infections. It is usually administered by intravenous infusion, although other routes may be employed. Nystatin is used only in the treatment of nonsystemic or noninvasive *Candida* infections such as vaginal, mucocutaneous, and intestinal candidiasis.

The chemical properties of amphotericin B and nystatin present certain problems relative to in vitro susceptibility testing. Both are light sensitive and subject to thermal decay upon incubation. Both are insoluble in water and unstable in the presence of acid. Accordingly, certain requirements exist for the preparation of solutions of these drugs and of media for testing them, as well as for the conditions under which the drugs are tested. The mode of action of the polyenes relates to their ability to bind to sterols contained in cell membranes. The specificity of this binding determines the utility of the drug; amphotericin B, for example, binds more avidly to ergosterol in fungal membranes than to cholesterol in mammalian cell membranes. Changes in membrane sterols can be correlated with the development of apparent resistance (20). Such resistance has now assumed clinical importance (19).

The imidazole derivatives represent the most fertile field of the current development of new antifungal agents. As many as eight such agents are now used clinically, and even more are under development (15a). However, only ketoconazole and miconazole need be discussed at this time. Both are broad-spectrum agents active against a variety of fungal pathogens including yeasts, dimorphic organisms, dermatophytes, and opportunistic pathogens. Ketoconazole is used either topically or orally in the treatment of a variety of different fungal infections. Miconazole is available for the topical treatment of smooth skin and vaginal fungal infections. In addition, it is available as an intravenous infusion for the treatment of systemic fungal infections including histoplasmosis, blastomycosis, coccidioidomycosis, and cryptococcosis.

The imidazoles share certain properties which affect in vitro testing. These properties include broad spectra, poor solubility in water, and relatively good stability. The property of broad spectra infers the potential need for testing against a variety of fungal pathogens. Poor solubility in water implies the use of other solvents such as dimethyl sulfoxide, dimethylformamide, or polyethylene glycols. Stability suggests the possible use of diffusion disk tests with these substances.

Several problems exist with the imidazoles in general and with miconazole and ketoconazole in particular which make laboratory studies with these agents difficult. First, MICs are subject to considerable variation. The sources of such variation include inoculum size, method of testing, and growth phase of the inoculum (17) as well as the rate of growth and the time of incubation (7). Second, the pharmacokinetics of oral ketoconazole are not fully predictable. Factors which can modify the bioavailability of ketoconazole include cytochrome P-450 inducers (25) and the concomitant administration of cimetidine (6). In one clinical study, as many as 25% of serum samples assayed failed to contain the anticipated amounts of the drug (13). Third, although resistance to ketoconazole and other imidazoles has been demonstrated in *Candida albicans*, such resistance is not measurable by all in vitro test procedures (29).

Flucytosine, or 5-FC, is a water-soluble, stable compound employed orally in the treatment of systemic infections caused by susceptible pathogenic or opportunistic yeasts and fungi. It acts as a competitive antimetabolite for uracil in the synthesis of yeast RNA, and it also interferes with thymidylate synthetase (28, 40). These activities can be antagonized in vitro by a variety of purine and pyrimidine bases and nucleosides (27). Because of this antagonism, the antifungal activity of 5-FC can be demonstrated only in synthetic media free of such substances.

In vitro testing with 5-FC is more important clinically than testing with either the polyene compounds or the imidazoles because of the repeated demonstration of de novo resistance to the drug as well as the emergence of resistant strains of yeasts and fungi after therapeutic exposure to the drug (26). At least two metabolic sites are responsible for this resistance (28): one involves the enzyme cytosine permease, which is responsible for the uptake of 5-FC into fungal cells, whereas the second site involves the enzyme cytosine deaminase, which is responsible for the deamination of 5-FC to 5-fluorouracil, the metabolically active form of the drug.

Susceptibility tests with 5-FC should be performed on all isolates of pathogenic yeasts or fungi isolated from patients destined to receive the drug. Such tests also should be performed on all isolates recovered during therapy. However, the results of such tests must be interpreted with caution, as they cannot always be regarded as fully predictive of clinical responses (34).

In vitro testing with antifungal agents is further complicated by certain characteristics of the fungi themselves. These characteristics include dimorphism, requirements for prolonged growth times, and requirements for specific growth temperatures. Dimorphic fungi require incubation at 37°C for conversion to the yeast phase. However, such conversions are difficult to perform and not always consistent. Also, this temperature may be detrimental to the polyene compounds. Further, although growth at 37°C is a specific characteristic of *Cryptococcus neoformans*, not all strains grow readily at this temperature. Thus, a compromise is required, and 30°C has been selected for the incubation temperature in most of the following test procedures.

Three methods for performing in vitro susceptibility tests with antifungal agents will be described. These include broth dilution, agar dilution, and diffusion disk tests.

Broth Dilution Method

The broth dilution method provides a quantitative in vitro assessment of the MIC of a given antifungal agent for a specific microorganism. In addition, the test may also be used to provide an assessment of the minimal fungicidal concentration of the compound. This procedure is employed primarily when yeasts or yeastlike fungi are tested, but it can be adapted to the testing of filamentous fungi. Unfortunately, the broth dilution method may not be applicable to tests with ketoconazole and isolates of yeasts without some modification, as it has been shown that ketoconazole is inactive against yeasts under anaerobic conditions (35).

Media

The selection of a medium depends upon the drug being tested. Unbuffered yeast-nitrogen base (YNB; 0392, Difco Laboratories, Detroit, Mich.) supplemented with glucose (1%) and asparagine (0.15%) is used in tests with 5-FC. This medium is most conveniently prepared as a 10× concentrate which is sterilized by filtration and stored at 4°C. A 1× solution is prepared by making a 1:10 dilution of the 10× concentrate in sterile distilled water. Tests with amphotericin B and nystatin are performed in Antibiotic Medium 3 FDA (Penassay broth, 0243, Difco; Antibiotic Assay Broth, 10932, BBL Microbiology Systems, Cockeysville, Md.) (38). Unbuffered YNB is not suitable for testing with polyenes because of the low pH (<5.0) which develops as a result of the fermentation or assimilation of glucose. Miconazole and ketoconazole are antagonized by rich media such as Sabouraud and brain heart infusion broth. Suitable media for these substances include YNB and casein-yeast extract-glucose broth (39). A single broth medium suitable for all three groups of compounds would be buffered YNB prepared by dilution of the 10× YNB concentrate in 9 volumes of sterile 0.01 M phosphate buffer, pH 7.0 (22).

Drug solutions

A solution of standard 5-FC powder (10,000 μg/ml; Hoffmann-LaRoche, Inc., Nutley, N.J.) is prepared in distilled water and sterilized by filtration. This solution may be used indefinitely if uncontaminated and stored at −80°C. When testing with amphotericin B or nystatin (E. R. Squibb & Sons, Princeton, N.J.), sufficient standard drug is weighed to prepare a solution of 5,000 μg/ml. The actual amount to be weighed must be adjusted according to the specific biological activity of the standard. If standard amphotericin B is not available, Fungizone as either the pharmaceutical preparation (Fungizone for injection, Squibb) or as a laboratory reagent (Fungizone for laboratory use, Squibb) may be substituted. A preparation of nystatin designed for use in viral cell culture media is available and may be used in lieu of standard material (Mycostatin suspension, 10,000 U/ml; GIBCO Laboratories, Grand Island, N.Y.). Standard substances of both amphotericin B and nystatin may be solubilized in dimethyl sulfoxide or dimethylformamide. These solutions must be protected from light and should be allowed to stand for 30 min before use to permit autosterilization. The solutions of 5,000 μg/ml may be used for 1 week if stored in the dark at 4°C.

A variety of solvent systems are used with the imidazoles. These systems include 10% stock solutions in polyethylene glycol, dimethyl sulfoxide, and dimethylformamide. Ketoconazole is best dissolved in 0.2 N HCl. None of these substances gives a solution which will remain completely solubilized upon dilution into aqueous media. Thus, at higher concentrations, a certain amount of turbidity will be encountered, which may interfere with the interpretation of in vitro susceptibility tests.

Preparation of inocula

Inocula are prepared from 24- to 48-h-old cultures grown either on Sabouraud agar (Sabouraud dextrose agar Emmons, 11589, BBL; Sabouraud agar modified, 0747, Difco) or in YNB. Sabouraud broth cultures should not be used in tests with 5-FC because of possible antagonism. Suspensions are prepared in sterile saline and adjusted to a transmission of 95% as measured at 530 nm. An alternative method of adjusting inocula employs the Wickerham card technique. In this procedure, the suspension to be adjusted is diluted into a tube of sterile saline which is then placed against a white card bearing several sharply ruled lines. The desired endpoint is reached when there is obvious turbidity but the lines are still sharply defined when viewed through the suspension (this turbidity is approximately the same as one-half of a McFarland no. 1 standard). A similar suspension should also be prepared for a control organism. Suitable susceptible control organisms include *Saccharomyces cerevisiae* ATCC 36375, *S. cerevisiae* ATCC 9763, *S. cerevisiae* ATCC 2601, *Candida albicans* ATCC 10231, *Candida tropicalis* ATCC 13803, and *Candida pseudotropicalis* ATCC 28838.

Drug dilutions and performance of the test

The following procedure gives sufficient material to test (in duplicate) one isolate and the appropriate control organism. With 1× YNB, Antibiotic Medium 3 FDA, casein-yeast extract-glucose broth, or buffered 1× YNB as required, prepare 10 ml of a solution of the drug or drugs to be tested (128 μg/ml); amphotericin B requires a solution of 32 μg/ml.

Place 12 sterile, disposable tubes (16 by 125 mm) in a rack, and add 5.0 ml of the appropriate broth medium to tubes 2 through 12. Add 5.0 ml of the solution (128 μg/ml) in broth, or a solution of 32 μg/ml in the case of amphotericin B, to tubes 1 and 2. Mix the contents of tube 2 and then serially dilute the drug, using 5.0-ml volumes and fresh pipettes for each dilution, through the remaining 10 tubes, discarding 5 ml from the last tube. This procedure will give a dilution series ranging in concentration from 128 to 0.06 μg/ml, or from 32 to 0.02 μg/ml in the case of amphotericin B. Remove 4.0 ml from each dilution, and divide the 4 ml equally among four sterile, disposable tubes. The remaining 1.0-ml volume for each dilution may be kept as controls for contamination of the serial dilution. Add 1 ml of drug-free broth to each of four additional sterile tubes for use as growth controls. Inoculate each of two tubes of each concentration of the drug with 0.05 ml of the standardized suspensions of the test and control organisms. Also inoculate two tubes of drug-free media with each suspension for the growth controls. Incubate the cultures at 30°C for 48 h or until growth is visible in the growth control tubes. Tests with 5-FC should not be read in less than 48 h, as investigations have shown that both the time of incubation and the size of the inoculum are highly critical in susceptibility testing with 5-FC and *Candida* species (33) or *C. neoformans* (4). Tests with nystatin and amphotericin B should be read as soon as the growth control becomes positive (i.e., at 24 to 48 h), as should tests with the imidazole compounds.

After incubation, the tubes are examined, and the MIC is recorded. The MIC is defined as the lowest concentration of drug which inhibits clearly visible growth, with a faint haze or slight turbidity being ignored. The minimal fungicidal concentration may be determined by subculturing approximately 0.01 ml from each negative tube and from the positive growth control tubes onto drug-free Sabouraud agar, with subsequent incubation at 30°C for 48 h or until growth of the subcultures from the growth control tubes is apparent. The minimal fungicidal concentration is defined as the lowest concentration of drug from which subcultures were negative or which yielded fewer than three colonies.

Expected results

Most isolates of susceptible yeasts will be inhibited by 8 μg or less of 5-FC per ml and killed by 16 μg or less per ml (Table 1). Isolates of *C. neoformans* with intermediate susceptibilities (MIC of 16 to 64 μg/ml) as well as totally resistant isolates (MIC of >1,000 μg/ml) may be recovered from patients during treatment with 5-FC. Amphotericin B is both inhibitory and fungicidal for most yeasts at a concentration of 0.05 μg or less per ml. An MIC of 2 to 4 μg/ml for amphotericin B suggests probable intermediate susceptibility, as this MIC only approximates the concentrations routinely achievable in serum and exceeds the concentrations achievable in cerebrospinal fluid (CSF). Nystatin is inhibitory and fungicidal for most *Candida* isolates at concentrations of 4 μg or less per ml. Prolonged incubation or incubation at 37°C may result in erroneously high MICs because of inactivation of the drug. Miconazole will be fungistatic for most isolates of *Candida* species, *C. neoformans*, and dermatophytic fungi at concentrations in the range of 0.5 to 8 μg/ml, and it will be fungicidal at higher concentrations; ketoconazole will compare favorably with miconazole in vitro but may differ from it depending on the organism being tested as well as the medium and test system being used (14).

Agar Dilution Test Methods

As with dilution test procedures with antibacterial agents, broth dilution tests with antifungal agents can be adapted to agar dilution tests. However, the results of agar dilution tests with both imidazoles and 5-FC are highly dependent upon the inoculum size and the time of incubation, and the results of such tests and broth dilution tests tend to be divergent (7). In addition, agar dilution tests may not detect resistance to the imidazoles in isolates of *C. albicans* (29). Despite these problems, agar dilution testing provides a cost-effective alternative to broth dilution tests when the occasion or work load warrants.

Drug solutions and media

Stock drug solutions are prepared as described above. Dilutions of drug are prepared in appropriate broth at 10 times the final desired concentrations (1,280 to 0.63 or 1,000 to 0.5 μg/ml for 5-FC and the imidazoles; 128 to 0.063 or 100 to 0.05 μg/ml for the polyenes). The dilutions are then added in ratios of 1:10 to sterile blanks of molten agar. Appropriate agar media include antibiotic medium 12 or buffered yeast morphology agar (YMA) for the polyenes, unbuffered YMA for 5-FC, and casein-yeast extract-glucose agar for the imidazoles. One medium which is suitable for all drugs except 5-FC is Kimmig agar (Fungus Agar

TABLE 1. In vitro antifungal activities of four antifungal agents against pathogenic fungi[a]

Organism	Amphotericin B		Flucytosine (5-FC)		Miconazole		Ketoconazole
	MIC (μg/ml)	MFC (μg/ml)	MIC (μg/ml)	MFC (μg/ml)	MIC (μg/ml)	MFC (μg/ml)	MIC (μg/ml)
Pathogenic yeasts							
Cryptococcus neoformans	0.05–0.78[b]	0.1–12.5	0.10–100[c]	0.39–>100	0.05–3.13	0.05–25	0.1–32
Candida albicans	0.2–0.78[d]	0.39–0.78	0.05–12.5[c]	0.10–>100	0.1–2.0[d]	0.1–10	<0.1–80
Candida spp. not *C. albicans*	0.2–1.56[d]	0.39–6.25	0.10–50[c]	0.20–>100	<0.1–2.0	0.1–>10	<0.1–64
Torulopsis glabrata	0.1–0.4	0.2–0.78	0.05–1.56	0.4–>100	0.5–10	2–10	1–64
Trichosporon sp.	0.78–3.13	1.56–3.13	25–100	>100	0.2–25	0.2–>100	
Geotrichum sp.	0.4–1.56	0.78–3.13	1.56–12.5	25–>100	0.1–2	0.5–>10	
Filamentous fungi							
Pseudallescheria (Petrillidium) boydii	1.56–>100[c]	>100	Resistant		0.05[e]	0.05	0.1–4[c]
Aspergillus spp. including *A. fumigatus*	0.05–8	6.25–>100	0.2–1.56[c]	>100	0.4–>100	0.8–>100	0.1–100
Blastomyces dermatidis	0.05–0.2	0.1–0.4	Resistant		≤0.25	ND	0.1–2
Cladosporium trichoides	3.13–>100	3.13–>100	3.13–12.5[c]	12.5–>100	0.5–>64	ND	0.1–64
Coccidioides immitis	0.1–0.78	0.78–1.56	Resistant		0.25–1.0	ND	0.1–0.8
Histoplasma capsulatum	0.05–1.0	0.05–0.2	Resistant		≤0.25	ND	0.1–0.5
Phialophora spp. and other dematiaceous fungi	0.05–>128	6.25–>128	Variable susceptibility	Resistant	0.05–32	ND	0.1–64
Sporothrix schenckii	1.56–12.5	3.13–>100	Resistant		1–2	ND	0.1–16
Zygomycetes	0.78–1.56	1.56–>100	Variable susceptibility	Resistant			
Dermatophytic fungi							
Epidermophyton floccosum					0.5–10	2–10	0.1–8
Microsporum spp.					0.5–10	0.5–>10	0.1–64
Trichophyton spp.					0.5–10	0.5–10	<0.05–128
Control organisms							
S. cerevisiae ATCC 36375, etc.	0.1	0.2	0.05	0.10	0.20	0.39	0.20
C. pseudotropicalis ATCC 28838			0.05	0.10	0.10	0.20	0.05

[a] Based upon both data obtained at the Medical College of Virginia, Virginia Commonwealth University, Richmond, and a review of the literature. In vitro data for nystatin is not included because of the narrow clinical spectrum of this agent; however, most isolates of *Candida* species and *Torulopsis* species should be clinically susceptible (MIC of ≤10 μg/ml) to nystatin. MFC, Minimal fungicidal concentration; ND, not determined.

[b] Expected ranges of MICs and MFCs.

[c] Resistance not uncommon.

[d] Resistance reported but rare.

[e] Only limited data available.

[f] In vitro susceptibility of *Aspergillus* spp. to ketoconazole is highly species dependent.

according to Kimmig, 5414, E. Merck; available from E. M. Industries, Hawthorne, N.Y.). Buffered YMA has been recommended by some workers as being suitable for agar dilution tests with all antifungal agents (9, 22).

Inocula and performance of test

Inocula for agar dilution tests should be prepared as described above for broth dilution tests. They should contain at least 10^6 CFU/ml, as the volume delivered by most mechanical replicators is approximately 0.001 to 0.003 ml. Inoculum size is critical with both 5-FC and the imidazole compounds. Agar dilution tests should include a pair of drug-free plates, one to be inoculated before and one to be inoculated after inoculation of the drug dilution plates. Positive growth responses must be obtained on both control plates for the results to be valid with any given organism. Appropriate control organisms of known susceptibility or resistance should be included in all agar dilution tests.

In tests with yeasts and rapidly growing filamentous organisms, all prongs of a mechanical replicator may be employed. In tests with slow-growing organisms, only every other prong should be used. Incubation should be at 30°C. The time of incubation will be controlled by the appearance of growth on the drug-free control plates. Results should be read only when maturing colonies are clearly visible on these plates. The MIC will be defined as the lowest concentration of drug preventing growth of macroscopically visible colonies. Hazy responses and pinpoint colonies should be regarded as negative.

Expected results

In vitro MICs obtained by agar dilution for the polyenes should agree with similar data obtained by broth dilution tests (Table 1). The relationship between agar dilution and broth dilution results for 5-FC and the imidazoles is less clear. This lack of a clear-cut relationship reflects problems associated both with inoculum size and with the selection of media for the performance of such tests. Inoculum size is most critical in tests with 5-FC; overly small inocula may result in false reports of in vitro susceptibility. The selection of media is most critical in tests with the imidazoles. Overly rich media may result in false reports of in vitro resistance. Both problems demonstrate the need for standardization of both broth and agar dilution test procedures with the antifungal agents (16).

Diffusion Disk Tests

Diffusion disk tests with antifungal agents have been described but not yet fully standardized (2, 9, 31, 37). There are reports regarding a variety of experimentally prepared disks (2, 37), and several commercial disks and tablets are now available in the United Kingdom and Europe (9, 31).

The application of diffusion disk tests with antifungal agents is limited. Polyene compounds lack stability (21), and both the various solubilities and partial stabilities of the imidazoles (18) may restrict the application of the test to these agents.

5-FC is one substance readily adaptable to diffusion disk testing. A 1-μg disk has been available for some time in Europe. This disk is used in a diffusion disk test standardized by a special Working Group of the British Society for Mycopathology which will be described here (8).

Flucytosine diffusion disk test for yeasts

Dispense molten YMA agar in 15-ml volumes into petri dishes (90-mm diameter) placed on a level surface, and allow the agar to solidify. Once the agar is hardened, invert the dishes and draw lines on the reverse sides so as to besect them.

Inoculate 2 ml of sterile distilled water or saline with *C. pseudotropicalis* ATCC 28838 or *C. pseudotropicalis* NCPF 3234 (National Collection of Pathogenic Fungi, Mycological Reference Laboratory, London School of Hygiene and Tropical Medicine, London, England). Adjust the turbidity of this preparation to that of a 0.5 McFarland tube, or use a Wickerham card to adjust the turbidity. Now prepare suspensions of the patient isolates to be tested in a similar fashion.

With sterile cotton swabs, inoculate one half of each plate with the suspension of the susceptible *C. pseudotropicalis* indicator strain. Then inoculate the other half of each test plate with the test strains. Once the plates are dry, place a 1-μg 5-FC disk on the surface in the center of each plate. Incubate the plates at 35°C.

The plates should be first examined after 24 h of incubation to determine obvious resistance. Final readings are made after 48 h of incubation. Measure the radius of each zone of inhibition of the test and indicator strains. Calculate the percentage of difference for each of the test strains by using the radius obtained for the indicator strain on the same plates as follows: test strain zone radius/indicator strain zone radius × 100. Isolates with zones of inhibition whose radii are not less than 80% of the indicator strain radius are regarded as susceptible. Highly resistant isolates will have no zones or very small zones of inhibition. Intermediate zones (less than 80% of the indicator zone) may imply susceptibility to 5-FC when it is used in combination with amphotericin B. Intermediate zones may also suggest the likelihood of the emergence of more highly resistant strains during treatment. Development of resistance will also be suggested by the presence of persisting colonies in otherwise clear zones of inhibition. In such situations, the resistant organism should be retested, and the susceptibility of the isolate should be monitored during treatment to ensure that a resistant mutant does not become dominant.

Alternative Diffusion Tests

One commercial system employing tablets rather than disks is available for testing with most commonly used antifungal agents (9). In one comparative study limited to pathogenic yeasts, an overall correlation of 90.9% between results obtained with amphotericin B, flucytosine, and miconazole tablets and with paired broth dilution tests was obtained with this system (12). The correlation, 87.8%, with tests employing a 1-μg 5-FC diffusion disk test was nearly as good.

The diffusion tablet test is performed in a manner analogous to that of the standard Food and Drug Administration disk test. Plates of buffered YMA are prepared (provided by the manufacturer of the tablets in 50-ml volumes of Shadomy Agar, i.e., YNB agar at a pH of 7.0 containing 0.33 g of NaH_2PO_4 and 0.92 g of K_2HPO_4 per liter with 1% glucose and 0.15% L-asparagine). One 50-ml bottle of agar is sufficient for the preparation of two or three 90-mm plates or one 140-mm plate. Inocula are adjusted to a density of 10^4 to 10^5 CFU/ml and the plates are inoculated with a cotton swab, after which the tablets are placed on the surfaces and the plates are incubated at 37°C for 24 to 48 h. Zones of inhibition are then measured. In measuring zones around tablets containing imidazoles, the measurements are based upon colonies of normal size, and smaller, partially inhibited colonies are ignored. The results obtained with the diffusion tablets are interpreted as shown in Table 2 (10). However, it must be noted that these criteria have not been validated.

Synergism

One of the more important advances in antifungal chemotherapy has been the use of synergistic combinations of antifungal agents. One such combination, amphotericin B and flucytosine (3), is advocated as the regimen of choice in the treatment of cryptococcal meningitis. Although therapeutically superior to either drug alone, this regimen is not without disadvantages, the most important being the toxicity associated with sustained flucytosine levels in blood in excess of 100 to 125 μg/ml (A. Stamm, R. Diasio, W. Dismukes, S. Shadomy, G. Cloud, and the NIH-NIAID Mycoses Study Group, Program Abstr. Intersci. Conf. Antimicrob. Agents Chemother. 24th, Washington, D.C., abstr. no. 1102, 1984).

The advent of combination antifungal chemotherapy introduces the possibility of a requirement for the performance of tests for antifungal synergy in the clinical laboratory. Such tests are performed by using procedures similar to those employed with combinations of antibacterial agents.

Synergism can be measured by testing a multiple series of dilutions of two drugs in a so-called checker-

TABLE 2. Interpretation of results obtained with diffusion tablets

Interpretation	Diameter (mm) of zone of inhibition for		
	Flucytosine	Imidazoles	Polyenes
Susceptible	≥30	≥20	≥15
Intermediate	23–29	12–19	10–14
Resistant	≤22	≤11	≤9

board titration (see Chapter 105). Both inhibitory and fungicidal endpoints for each drug alone and in combination are determined and then plotted as an isobologram (24). The resulting graph will be interpreted in terms of possible synergism, antagonism, or indifference. Specific considerations for the application of such a test to combinations of amphotericin B and 5-FC include the selection of media and the conditions for incubation. In this case, specific recommendations include the use of buffered YNB and incubation at 30°C for 48 h.

BIOASSAYS

Bioassays for antifungal agents can be performed by using standard microbiological bioassay procedures (see Chapter 106). The classical radial agar diffusion method is readily adaptable for use with such drugs as amphotericin B, 5-FC, and ketoconazole. Bioassays for levels of topical antifungal agents in serum are not warranted. Specific considerations include the selection of media and bioassay indicator organisms as well as design of the standard or dose-response curve (Table 3). Specimens for bioassay may include such body fluids as sera, CSF, and synovial fluids, as well as other specimens such as dialysis fluids. Assays of antifungal agents in serum by high-pressure liquid chromatography have been described (1). Such assays are rapid and convenient but require special equipment. In addition, approximately 10% of the antifungal agent may be lost in extracting the drug from plasma, so that artificially low values may be obtained.

The bioassay of 5-FC levels in sera is the most critical of such determinations and may be required for all patients with impaired renal function and in whom 5-FC-associated toxicity is suspected (neutropenia, leukopenia, and thrombocytopenia). Levels of 5-FC in serum in excess of 100 to 125 μg/ml should be regarded as potentially toxic (3). Less essential are determinations of amphotericin B serum levels. In adult patients receiving 1 mg/kg daily or every other day, levels of amphotericin B in serum will range from 4 μg/ml or more to 0.2 μg/ml or less for 24 to 48 h after infusion; CSF levels rarely exceed 1 μg/ml. Pharmacokinetic studies (36) with intravenous miconazole have shown that this drug will produce maximum serum concentrations ranging from 2 to 8 μg/ml depending on dosage (400 to 1,000 mg/day). Levels of miconazole in CSF rarely exceed 0.1 to 0.3 μg/ml except when the drug is administered intrathecally. Peak levels of ketoconazole in serum, usually attained 2 to 4 h after the ingestion of a 400-mg dose, will be in the range of 4 to 6 μg/ml (23). The poorly predictable bioavailability of this drug makes the determination of levels of ketoconazole in serum the second most critical of such determinations.

Bioassay for Flucytosine

Several bioassays for flucytosine have been reported. The one to be described here is modified from a bioassay procedure again evaluated by the Working Group of the British Society for Mycopathology (8).

Preparation of standards

Standard solutions of flucytosine at concentrations of 6.25, 12, 25, 50, and 100 μg/ml will be required. These solutions can be prepared from a concentrated standard stock solution in sterile distilled or deionized water or in sterile, pooled, normal human serum. The standards can be dispensed in 0.5- or 1-ml volumes and frozen at −20°C for up to 12 months until used.

Indicator organism

Inoculate 5 ml of standard 1× YNB with *C. pseudotropicalis* ATCC 28838 or *S. cerevisiae* ATCC 36375. Incubate the culture at 37°C overnight. Adjust the culture to a density of 10^7 CFU/ml.

TABLE 3. Recommended test conditions for bioassay of other antifungal agents in body fluids by a radial agar diffusion bioassay technique

Drug	Bioassay indicator organism	Medium	Standard dose-response curve (μg/ml)[a]	Maximum therapeutic levels (μg/ml)
Amphotericin B (in the absence of flucytosine)	*Paecilomyces varioti* ATCC 36257	Antibiotic medium 12 FDA (nystatin assay agar)	0.03, 0.06, 0.125, 0.25, 0.5, and 1.0	In serum: 1.0–4.0 In CSF: 0.02–1.0
Amphotericin B (in the presence of flucytosine)	*Chrysosporium pruinosum* ATCC 36374	As above but supplemented with 10 μg of cytosine per ml	As above	As above
Miconazole	*Candida stellatoidea* ATCC 36232 or *C. pseudotropicalis* ATCC 28838	Emmons Sabouraud dextrose agar or Kimmig agar	0.5, 1, 2, 4, 8, and 16	In serum: 2.0–8.0 (depending on dosage) In CSF: 0.1–0.3

[a] Standard dose-response curves may contain as few as three concentrations spanning the therapeutic range. A five- or six-point curve permits use of the equations $H = (3a + 2b + c - e)/5$ and $L = (3e + 2d + c - a)/5$, where H and L are the high and low plotting points for drawing of the standard curve, and *a* through *e* represent the diameters of the zones of inhibition produced by the highest through the lowest of five standard drug concentrations in a twofold increment serial. The standard curve drug concentrations should be prepared in the same type of fluid(s) as is being assayed. Both standards and specimens may be applied to the seeded agar plates with cylinders, paper disks, or cut holes.

Preparation of test plates

Melt 90 ml of YMA agar, cool it to 55°C, and inoculate it with 5 ml of the suspension of the indicator organism. Pour the inoculated agar into an assay plate (25 cm^2; A/S Nunc Bio Plate, no. 1015), and allow it to cool on a level surface. Once the medium has solidified, place the plate in a 37°C incubator and leave it there until the surface of the agar has dried.

Cut 30 holes, 4 mm each in diameter, in the agar layer (six rows of five holes each). Introduce 10 μl of a standard or patient specimen into each hole. A control sample of a known concentration may also be incorporated. The samples and standards should be placed on the plate in a randomized distribution. Each specimen or standard should be present in triplicate in this assay. The design of this assay approximates a "13 × 4 (3 + 1) incomplete block assay," which has a 95% confidence limit of ±11%, which is adequate for the assay of antibiotics in clinical materials.

The plates are incubated without prediffusion overnight at 35 to 37°C. After incubation, the diameters of the zones of inhibition are measured. Mean diameters are then calculated for each standard and specimen.

Several methods are available for the calculation of serum concentrations in the clinical specimens. The British Society for Mycopathology Working Group protocol calls for plotting the mean diameters of the standard concentration against the drug concentration on semilogarithmic paper, with drug concentrations on the logarithmic ordinate. The resulting graph is then used to estimate the concentrations of the drug in the clinical specimens. This method is subject to error. Fortunately, two other procedures are available for the analysis of bioassay data (32).

Bioassay data are best analyzed statistically, and a number of minicomputers are available which have the necessary programs or functions for such analyses. This type of analysis is best performed by first using a regression analysis function to analyze the data for the standard curve. Such an analysis yields three statistics which describe the curve: r, or the regression coefficient, which is a measurement of the linearity of the curve; α, which describes the intercept of the curve (the value for y when $x = 0$); and β, which describes the slope of the curve. The second step is the application of the estimating equation, $y = \alpha + \beta \times \log x$, to calculate drug levels in serum from zone diameter values. This equation uses the values for α and β derived from the regression analysis of the data for the standard curve.

A second procedure for the analysis of bioassay data combines both mathematical and graphic procedures. First, the data for the standard curve are used to calculate "high" and "low" plotting points. A minimum of five concentrations is required for these calculations. The plotting points are calculated as follows: $H = (3a + 2b + c - e)/5$ and $L = (3e + 2d + c - a)/5$, where H and L are the high and low plotting points of the standard curve, and a through e are the diameters of the zones of inhibition produced by the highest to lowest of five standard drug concentrations prepared in a twofold increment serial. The two plotting points now are plotted graphically as above, and the resulting graph is used to estimate concentrations of drug in the clinical specimens by plotting the diameters of zones of inhibition against the standard curve.

Peak levels of 5-FC in serum should be in the range of 60 to 80 μg/ml in patients with normal renal function, with steady-state troughs of not less than 20 μg/ml. Paired CSF levels should be in the range of 40 to 60 μg/ml. Lower peaks are desirable in patients with renal impairment, as sustained 5-FC levels will be prolonged in such patients (30).

Serum specimens from patients receiving both 5-FC and amphotericin B may be heated (90°C, 30 min) or subjected to ultrafiltration before assay to eliminate the latter drug. Serum specimens containing more than 100 μg of 5-FC per ml should be reassayed after being diluted 1:1 in pooled, normal human serum.

Bioassay for Ketoconazole

As already noted, the bioavailability of ketoconazole is subject to considerable variation and is not fully predictable. Thus, it may be necessary to determine levels of this drug in serum in patients who are not responding therapeutically or in whom there is reason to suspect poor absorption of the drug. The bioassay to be described here was developed during an exhaustive clinical trial with ketoconazole (13) and has been validated by comparing it with results obtained by a high-pressure liquid chromatography procedure.

Preparation of standard

Standard solutions of ketoconazole at concentrations of 0.25, 0.5, 1, 2, 4, 8, and 16 μg/ml are required. These solutions should be prepared in pooled human serum. The standards can be dispensed in 1.0-ml volumes and frozen for up to 6 months at −20°C until used.

Indicator organism

Inoculate 5 ml of Sabouraud broth with *C. pseudotropicalis* ATCC 28838, and incubate the culture overnight at 35 to 37°C. Adjust the culture to a density of 10^6 CFU/ml.

Preparation of test plates

Melt 100 ml of Kimmig agar, cool it to 55°C, and inoculate it with 5 ml of the suspension of *C. pseudotropicalis*. Pour the inoculated agar into an assay plate (25 cm^2), and allow it to cool on a level surface. Once the medium has solidified, place it in a 37°C incubator, and leave it until the surface has dried.

Place 12 or 16 stainless steel cylinders on the surface of the plate. If cylinders are not available, sterile paper disks or cut wells may be employed. However, the use of disks or wells will reduce the sensitivity of the bioassay. Each standard and clinical specimen is tested in duplicate. Using a random pattern of distribution, introduce 0.1 ml of a standard solution or a specimen into each cylinder. Be sure not to spill the sample or tip the cylinder. Samples of pooled human serum used to prepare the standard curve should also be tested, as should samples containing known amounts of ketoconazole.

The plates are incubated without prediffusion overnight at 30°C. After incubation, the diameters of the resulting zones of inhibition are measured. A mean diameter is then calculated for each standard and specimen.

Data from the ketoconazole bioassay are analyzed in the same fashion as described above for 5-FC. The peak concentration of approximately 4 to 21 μg/ml will be measurable at 1.5 to 3 h, depending upon dosage and concomitant factors which may modulate absorption (11).

Other Antifungal Bioassays

Bioassays for amphotericin B are not warranted under most circumstances. Exceptions would include unusual clinical situations in which the penetration of the drug into specific sites or the pharmacokinetics of the drug are unknown. Two bioassays for amphotericin B can be used in such situations. These assays use the same design as described above for 5-FC and ketoconazole and differ only in media and indicator organisms. These changes are detailed in Table 3.

There are few reasons for requiring bioassays for miconazole. The role of this drug in the intravenous preparation in clinical medicine is uncertain. However, should the need exist for such a bioassay, it can be performed using the same design as described above and with the media and indicator organism detailed in Table 3. One further modification is that the assay plates need to prediffuse overnight at 4°C before incubation.

LITERATURE CITED

1. **Alton, K. B.** 1980. Determination of the antifungal agent, ketoconazole, in human plasma by high-performance liquid chromatography. J. Chromatogr. **221**:337–344.
2. **Baier, R., and U. Puppel.** 1978. Antimykotika-Empfindlichkeit von Hefen aus klinischem Untersuchungsmaterial im Blättchendiffusionstest. Dtsch. Med. Wochenschr. **103**:1113–1116.
3. **Bennett, J. E., W. E. Dismukes, R. J. Duma, G. Medoff, M. A. Sande, H. Gallis, J. Leonard, B. T. Fields, M. Bradshaw, H. Haywood, Z. A. McGee, T. R. Cate, C. G. Cobbs, J. F. Warner, and D. W. Alling.** 1979. A comparison of amphotericin B alone and combined with flucytosine in the treatment of cryptococcal meningitis. N. Engl. J. Med. **301**:126–131.
4. **Block, E. R., A. E. Jennings, and J. E. Bennett.** 1973. Variables influencing susceptibility testing of *Cryptococcus neoformans* to 5-fluorocytosine. Antimicrob. Agents Chemother. **4**:392–395.
5. **Brass, C.** 1984. Antimycotic therapy, a critical appraisal. Antibiot. Newsl. **1**:36–39.
6. **Brass, C., J. N. Galgiani, T. F. Blaschke, R. Defelice, R. A. O'Reilly, and D. A. Stevens.** 1982. Disposition of ketoconazole, an oral antifungal, in humans. Antimicrob. Agents Chemother. **21**:151–158.
7. **Brass, C., J. Z. Shainhouse, and D. A. Stevens.** 1979. Variability of agar dilution-replicator method of yeast susceptibility testing. Antimicrob. Agents Chemother. **15**:763–768.
8. **British Society for Mycopathology.** 1984. Report of a Working Group. Laboratory methods for flucytosine (5-fluorocytosine). J. Antimicrob. Chemother. **14**:1–8.
9. **Casals, J. B.** 1979. Tablet sensitivity testing of pathogenic fungi. J. Clin. Pathol. **32**:719–722.
10. **Casals, J. B., and O. G. Pederson.** 1981. Antimicrobial sensitivity testing using Neo-Sensitabs. Technical manual. A/S Rosco, Taastrap, Denmark.
11. **Daneshmend, T. K., D. W. Warnock, M. D. Ene, E. M. Johnson, M. R. Potten, M. D. Richardson, and P. J. Williamson.** 1984. Influence of food on the pharmacokinetics of ketoconazole. Antimicrob. Agents Chemother. **25**:1–3.
12. **Delhalle, E., and M. Carpentier.** 1981. Sensitivity of yeast-like organisms to amphotericin B, 5-fluorocytosine and miconazole, p. 121–127. *In* A. Adam, P. Ers, and J. Pijck (ed.), Current concepts in microbiology. Elsevier, Amsterdam.
13. **Dismukes, W. E., A. M. Stamm, J. R. Graybill, P. C. Craven, D. A. Stevens, R. L. Stiller, G. A. Sarosi, G. Medoff, C. R. Gregg, H. A. Gallis, B. T. Fields, Jr., R. L. Marier, T. A. Kerkering, L. G. Kaplowitz, G. Cloud, C. Bowles, and S. Shadomy.** 1983. Treatment of systemic mycoses with ketoconazole: emphasis on toxicity and clinical response in 52 patients. Ann. Intern. Med. **98**:13–20.
14. **Dixon, D., S. Shadomy, H. J. Shadomy, A. Espinel-Ingroff, and T. M. Kerkering.** 1978. Comparison of the in vitro antifungal activities of miconazole and a new imidazole, R41,400. J. Infect. Dis. **138**:245–248.
15. **Fraser, D. W., J. I. Ward, L. Ajello, and B. D. Plikaytis.** 1979. Aspergillosis and other systemic mycoses: the growing problem. J. Am. Med. Assoc. **242**:1631–1635.
15a. **Fromtling, R. A.** 1984. Imidazoles as medically important antifungal agents: an overview. Drugs Today **20**:325-349.
16. **Galgiani, J. N.** 1984. Why not standardize antifungal susceptibility testing. Antimicrob. Newsl. **1**:40.
17. **Galgiani, J. N., and D. A. Stevens.** 1978. Turbidimetric studies of growth inhibition of yeasts with three drugs: inquiry into inoculum-dependent susceptibility testing, time of onset of drug effect, and implications for current and newer methods. Antimicrob. Agents Chemother. **13**:249–254.
18. **Grendahl, J. G., and J. P. Sung.** 1978. Quantitation of imidazoles by agar-disk diffusion. Antimicrob. Agents Chemother. **14**:509–513.
19. **Guinet, R., J. Chanas, A. Goullier, G. Bonnefoy, and P. Ambroise-Thomas.** 1983. Fatal septicemia due to amphotericin B-resistant *Candida lusitaniae*. J. Clin. Microbiol. **18**:443–444.
20. **Hamilton-Miller, J. M. T.** 1973. Chemistry and biology of the polyene macrolide antibiotics. Bacteriol. Rev. **37**:166–196.
21. **Hoeprich, P. D., and A. C. Houston.** 1978. Stability of four antifungal antimicrobics in vitro. J. Infect. Dis. **137**:87–90.
22. **Holt, R. J.** 1975. Laboratory tests of antifungal agents. J. Clin. Pathol. **28**:767–774.
23. **Hume, A. L., and T. M. Kerkering.** 1983. Ketoconazole (Nizoral, Janssen Pharmaceutica, Inc.). Drug Intell. Clin. Pharm. **17**:169–174.
24. **Jawetz, E.** 1968. Combined antibiotic action: some definitions and correlations between laboratory and clinical results, p. 203–209. Antimicrob. Agents Chemother. 1967.
25. **Loose, D. S., P. B. Kan, M. A. Hirst, R. A. Marcus, and D. Feldman.** 1983. Ketoconazole blocks adrenal steroidogenesis by inhibiting cytochrome P-450-dependent enzymes. J. Clin. Invest. **71**:1495–1499.
26. **Normark, S., and J. Schönebeck.** 1972. In vitro studies of 5-fluorocytosine resistance in *Candida albicans* and *Torulopsis glabrata*. Antimicrob. Agents Chemother. **2**:114–121.
27. **Polak, A., and H. J. Scholer.** 1973. Fungistatic activity, uptake and incorporation of 5-fluorocytosine in *Candida albicans* as influenced by pyrimidines and purines. I. Reversal experiments. Pathol. Microbiol. **39**:148–159.
28. **Polak, A., and H. J. Scholer.** 1975. Mode of action of 5-fluorocytosine and mechanisms of resistance. Chemotherapy **21**:113–130.
29. **Ryley, J. F., R. G. Wilson, and J. Barrett-Bee.** 1984. Azole resistance in *Candida albicans*. Sabouraudia **22**:53–63.
30. **Scholer, H. J.** 1980. Flucytosine, p. 36–100. *In* D. C. E. Speller (ed.), Antifungal chemotherapy. John Wiley, Chichester, England.
31. **Scholer, H. J., and A. Polak.** 1973. Fungistatic and fungicidal properties of 5-fluorocytosine; methods for routine

sensitivity tests, p. 162-163. *In* Symposium International de Mycologie Medicale, Bukarest, 20–22 September 1973.

32. **Shadomy, S., and A. Espinel-Ingroff.** 1984. Methods for bioassay of antifungal agents in biologic fluids. *In* A. Laskin and H. Lechevalier (ed.), CRC handbook of microbiology, vol. 6. CRC Press, Boca Raton, Fla.
33. **Shadomy, S., C. B. Kirchoff, and A. E. Ingroff.** 1973. In vitro activity of 5-fluorocytosine against *Candida* and *Torulopsis* species. Antimicrob. Agents Chemother. **3:**9–14.
34. **Stiller, R. L., J. E. Bennett, H. J. Scholer, M. Wall, A. Polak, and D. A. Stevens.** 1983. Correlation of in vitro susceptibility test results with in vivo response: flucytosine therapy in a systemic candidiasis model. J. Infect. Dis. **147:**1070–1077.
35. **Sud, I. J., and D. S. Feingold.** 1981. Heterogeneity of action mechanisms among antimycotic imidazoles. Antimicrob. Agents Chemother. **20:**71–74.
36. **Sung, J. P., and J. G. Grendahl.** 1977. Clinical experimental therapy with miconazole for human disseminated coccidioidomycosis, p. 293–309. *In* L. Ajello (ed.), Coccidioidomycosis, current clinical and diagnostic status. Symposia Specialists, Miami.
37. **Utz, C. J., and S. Shadomy.** 1977. Antifungal activity of 5-fluorocytosine as measured by disc diffusion susceptibility testing. J. Infect. Dis. **135:**970–974.
38. **Utz, C. J., S. White, and S. Shadomy.** 1976. New medium for in vitro susceptibility studies with amphotericin B. Antimicrob. Agents Chemother. **10:**776–777.
39. **Van den Bossche, H., G. Willemsens, and J. M. Van Cutsem.** 1975. The action of miconazole on the growth of *Candida albicans*. Sabouraudia **13:**63–73.
40. **Wagner, G. E., and S. Shadomy.** 1979. Studies on the mode of action of 5-fluorocytosine in *Aspergillus* species. Chemotherapy **25:**61–69.

Chapter 105

Susceptibility Tests: Special Tests

F. D. SCHOENKNECHT, L. D. SABATH, AND C. THORNSBERRY

The following tests may be requested of the microbiology laboratory because they may aid in establishing optimal therapeutic regimens for certain infections. None of these tests should be used for all microorganisms, and the situations for which they are used and the frequency with which they are done will vary from one institution to another. Most would agree, however, that bactericidal and synergy tests are essential in cases of infective endocarditis and that β-lactamase tests should be done on *Haemophilus influenzae*, probably on *Neisseria gonorrhoeae*, and maybe on staphylococci. The screening test for methicillin-resistant (heteroresistant) staphylococci has been included in this chapter because of the increasing problem with these organisms in the United States and because the test appears to be very accurate (13, 47).

TESTS FOR BACTERICIDAL ACTIVITY

Routine susceptibility tests, including MIC determinations, estimate the bacteriostatic or inhibitory action of antimicrobial agents. In practice, this is quite sufficient for guiding chemotherapy in most cases, since the success of in vivo antimicrobial action depends very much on the host's own defense mechanisms which will ultimately kill and eliminate microorganisms that have been reduced by bactericidal or bacteriostatic antimicrobial agents. For antibiotics with bactericidal action (mainly aminoglycosides and beta-lactams), it may occasionally be necessary to make an additional quantitative assessment of this killing effect on a given microorganism. This can be done in several ways: (i) by determining the MBC of an antibiotic for an infecting organism; (ii) by determining the number of surviving bacteria at a fixed concentration of the drug, using the average obtainable blood level, at defined time intervals (killing curve); or (iii) by estimating the titer at which the serum of a patient receiving antimicrobial chemotherapy kills the offending organism (serum bactericidal titer or test). Surprisingly, no standardized or reference methods have been established or accepted, although such tests have been conducted and were the subject of many reports for over 40 years. Of these procedures, the MBC test is probably the one most frequently requested in clinical settings and used in the evaluation of drugs with antimicrobial activity. It will be dealt with and described below in more detail. The principles employed, and particularly the influence of technical factors, apply to all three.

The MBC test is based on the common broth dilution method of susceptibility testing, with the MIC as the endpoint (48). It determines the proportion of survivors after 20 to 24 h of incubation in broth containing antibiotic. The MBC then is defined as that lowest concentration of antibiotic which kills at least 99.9% of the original inoculum.

Problems of poor reproducibility have beset this test from early on, and this has not been helped by the considerable differences in MBC test methods employed in routine practice (28). The ratio of MBC to MIC has become a matter of considerable practical and scientific interest, especially so when MBC values are obtained which are 32 or more times the MIC with a given organism-bacterial combination. For that reason, questions of procedural variation and method-related discrepancies are of great concern and need to be resolved. This section deals with the test as it applies to bacteria only. It will address briefly the indications for such a test, discuss the influence and significance of certain technical factors, and outline a step-by-step procedure for the execution of such tests.

Definitions

When the MIC of a given antimicrobial agent for a susceptible microorganism is determined by a broth dilution method, the endpoint is usually taken as the lowest concentration of the drug showing no growth by naked-eye inspection. However, subculture to antibiotic-free medium will often show that this inhibition is a reversible process and that viable organisms (survivors) can be recovered from the tube or well at the MIC level and sometimes at levels several dilution steps higher. While this is to be expected with bacteriostatic drugs, with bactericidal antimicrobics there will be a concentration at which no survivors can be recovered, as a result of the killing effect of the agent. This effect can be measured in the laboratory and is conventionally defined as that lowest concentration of a given antimicrobial agent which reduces the population of a test organism to 0.1% or less of the number of cells present in the original inoculum. This 1,000-fold reduction of the original inoculum is an arbitrary requirement, and there is not much evidence that 99.0 or 98.0% killing would be a less useful definition. However, most work in the past has been based on this requirement for 99.9% killing, and it will be used in the procedure described below. A broader term, minimal lethal concentration (MLC), is preferred by many workers. It is applicable to bactericidal and fungicidal tests.

Several terms and phenomena of resistance have been the subject of study, controversy, and confusion and should be briefly defined, as follows.

(i) Paradoxical effect. The "paradoxical effect" (8–10, 20) also called the Eagle phenomenon, has been recognized since the early days of penicillin and originally referred to the puzzling appearance of increasing numbers of survivors at concentrations higher than the MBC, sometimes separated from this concentration by one or more tubes of no growth. It has been described for several species of bacteria (8) and for other beta-lactam antibiotics (8) and is generally thought to be the result of interference with protein synthesis by higher concentrations of the beta-lactam. No therapeutic implications appear to be known.

(ii) Persisters. Another term referring to incomplete killing at or above bactericidal levels is "persisters." By numbers, the survivors are usually well below the 0.01% level. They do not differ in their susceptibility from the parent strain when subcultured and retested. Most workers consider them metabolically inactive forms that were not growing at the time of the test and consequently were not killed by the beta-lactam, the type of antibiotic with which their occurrence has been mostly observed (5, 6).

(iii) Tolerance. The third phenomenon of resistance is referred to as tolerance. One form of tolerance involves the decreased susceptibility of certain bacterial species to the killing effect of beta-lactam antibiotics, mainly, and has been unequivocally demonstrated under experimental conditions for strains of *Streptococcus pneumoniae* with cycloserine, penicillin, and phosphonomycin (48, 49). The mechanism for this form of tolerance involves the absence or suppression of the physiological autolytic enzyme system (48). Less well understood is another possibly similar form of tolerance with potential clinical significance, reported by several authors for clinical isolates of *Staphylococcus aureus* (25, 26, 30, 36, 38). This form has been defined by Sabath as being associated with an MBC-MIC ratio of 32 or greater after 24 h of incubation (37). Its expression appears to decrease with prolonged incubation, it is dependent on the phase of growth of the inoculum, and during storage the trait may be lost (25, 37). This phenomenon has been ascribed to interference with normal bacterial autolytic enzyme mechanisms. The conditions employed for detecting its presence in clinical isolates vary considerably (16, 38).

It is the contention of an increasing number of authors that careful attention to a variety of technical factors while determining MBCs in the laboratory will significantly reduce or eliminate most of the phenomena described above (12, 15, 23, 44).

Indications

The role that host defenses play in successful chemotherapy cannot be stressed enough. Where such defenses are seriously compromised, clinicians often prefer bactericidal or fungicidal drugs and may ask for a determination of the MBC in addition to the MIC. Endocarditis was and still is one of the major indications for such tests. Infections in seriously immunosuppressed patients and compromised hosts have made the determination of MBCs a frequently requested test for transplant and cancer chemotherapy patients. Some authorities would also include osteomyelitis and systemic infections of patients with prosthetic devices as indications for the MBC test.

Methodological variables

Recent studies have emphasized and documented the effect of several technical factors on the outcome of MBC tests and their reproducibility, the most important of which are listed in Table 1. Two major problems involve insufficient contact between test organism and antibiotic (12, 16, 23, 26, 44) and the phase of growth of the inoculum (16, 23, 25, 44). The first problem is the result of transfer of viable organisms to the inside of test tube walls above the meniscus, either during inoculation or through subsequent handling and shaking of the tubes. Such viable cells will remain adherent to the test tube wall for the time of the test and in this way escape the action of the antibiotic; they may then be resuspended into the broth during the time of sampling. Such adherence is most marked to polypropylene plastic and least marked to acid-washed borosilicate glassware. Gwynn and others actually saw and stained material of bacterial regrowth at the liquid-solid-air interface at the meniscus when studying this phenomenon with *Pseudomonas aeruginosa* and carbenicillin (12, 23). The procedure outlined below specifies both the subsurface inoculation of a small-volume inoculum (0.1 ml) into acid-washed borosilicate glass tubes and the avoidance of shaking.

The second factor involves the phase of growth of the inoculum. The most effective killing occurs with organisms in midlogarithmic phase (ca. 5 h) (44). The least effective killing takes place with organisms in stationary phase (16, 19, 25). Even organisms from late logarithmic phase have been shown to be less effectively killed than those from midlogarithmic-phase cultures (44). Horne and Tomasz (15) have shown convincingly for penicillin and group B streptococci that growth-phase-dependent tolerance was caused by the gradual increase in acidity of the cultures as the cell concentration increased. Further work is necessary to show whether the same is true for oxacillin and *S. aureus* (50). pH-dependent oxacillin tolerance may just be acidity, associated with stationary-phase growth and its well-known diminished response to the bactericidal action of beta-lactams.

TABLE 1. Summary of technical factors which influence MBC tests[a]

Variable factor	Effect
Phase of growth	Increased survivors in stationary phase; exaggeration of paradoxical effect in late-logarithmic phase
Type of tube or glassware	Adhesion to inside of test tube varies with material and may falsely elevate counts of survivors, give erratic results, or both
Mode of inoculation	Eliminate adhesion above meniscus by using a small-volume inoculum below the meniscus and avoid shaking
Mixing at 20 h	Vortexing is needed to resuspend all cells
Reincubation for 4 h	Allows all cells resuspended at 20 h from above meniscus to be killed before sampling
Antibiotic carry-over	Gives falsely low counts of survivors at higher antibiotic concentrations
Reincubation of recovery media	Total of 48 h for staphylococci (72 h for fastidious organisms) may be necessary for final results

[a] Adapted from reference 44.

Although these two critical factors have been known and stressed repeatedly for many years, they have never been consistently applied in actual laboratory execution of MBC testing. The interested reader may find recommendations towards preventing contamination of the insides of test tube walls and specific advice for mixing test suspensions before sampling already in the official methods of analysis of the Association of Official Agricultural Chemists (3) and in a description of the Rideal-Walker test for disinfectants (35). The need for young growing cultures for MBC tests was stressed by many early workers (5, 10, 14, 20) and emphasized more recently (16, 19, 25, 44).

Volume and inoculum density

Since the determination of bactericidal endpoints is done as a follow-up to MIC testing, the volume used is determined by the broth dilution MIC method. For the purposes of MBC testing, it is recommended to use the broth macrodilution procedure for MIC testing with a 2-ml final volume in test tubes, described in an earlier chapter of this Manual. Although smaller volumes and microtitration techniques have been successfully employed for MIC testing, most of the critical work with MBC tests has been done using 2-ml or larger volumes. The estimation of survivors requires a certain relation between the inoculum density, the volume cultured, and the volume transferred for the count of survivors. Keeping in mind that actual cell counts and suspensions are standardized by turbidity standards and may vary by a factor of 10, the initial inoculum and the transfer volume for MBC testing should be such that at least 10 colonies are available for endpoint determination. For a 99.9% killing endpoint, an initial inoculum of 10^6 CFU/ml should result in ca. 100 colonies growing on subculture if 0.1 ml is transferred. Larger transfer volumes may result in drug carry-over, a problem that occurs mainly at higher concentrations and is usually eliminated by spreading the subculture onto agar plates. Much smaller transfer volumes (less than 0.01 ml) can result in too few colonies and make the test unreliable.

Media

The media recommended for the broth macrodilution MIC test should be used, and that means, in most cases, Mueller-Hinton broth (MHB) (29). When *P. aeruginosa* is tested against aminoglycosides, the broth needs to be supplemented with calcium and magnesium as described for the broth dilution MIC test. For fastidious organisms Trypticase soy broth (BBL Microbiology Systems), tryptic soy broth (Difco), or Levinthal broth may all be used. For subculture, any solid agar supporting the organism is suitable, e.g., Mueller-Hinton agar, brain heart infusion agar, or Trypticase soy agar with 5% defribinated sheep blood.

Incubaton and sampling

All inoculated tubes should be incubated at 35°C for 20 h in a static incubator. After this time all tubes are mixed for 15 s by holding the tubes vertically in the cup of a Vortex mixer, being sure to avoid aerosolization by applying pressure on the caps of the tubes. After 4 h of additional incubation, the tubes are mixed again for 15 s with the Vortex mixer and then sampled. Volumes of 0.1 ml are transferred with a pipette from any tube showing no growth by naked-eye inspection to the surface of a suitable agar plate and spread over the entire plate with a sterile, bent glass rod. Colony counts should be recorded after 24 h of incubation at 35°C. It is recommended that plates be incubated for an additional 24 h (48 h for fastidious organisms) before final results are reported.

Recording of MBC endpoints

The MBC is considered the lowest concentration of the antibiotic which prevented growth and reduced the inoculum by 99.9% within 24 h. Survivors at higher concentrations are ignored as long as they number below 0.1% of the count from the original inoculum. Growth at antibiotic concentrations higher than this cut-off point should be recorded and listed as the number of survivors at concentrations higher than the MBC. Such strains show the paradoxical phenomenon.

Procedure

The first part of the MBC susceptibility testing procedure (Table 2) follows the general outline of the broth macrodilution MIC test as described in an earlier chapter of this Manual. However, the steps are listed again in Table 2 for the sake of convenience in a form suitable for follow-up with MBC testing. As described, this method will yield the MIC in micrograms per milliter, after about 24 h of incubation, and the MBC in micrograms per milliliter. The minimum time for the total test then will be approximately 48 h.

Killing Curve and Serum Bactericidal Titer

Two additional methods to assess the killing effect of an antimicrobial agent against a given microorganism should be briefly discussed. These are the killing curve and the serum bactericidal titer. There is no generally accepted method for either, and the technical variables and factors discussed above for the MBC test apply as well to these two tests. An additional variable will have to be considered if serum is used as diluent as recommended by several workers (32, 34, 42, 43). Various levels of complement (7) and even substances with antimicrobial activity or antimicrobic-neutralizing effects may influence the outcome of the test between the patient's organism and the drug. The main reason for adding serum is that a more realistic test situation is created for highly protein-bound antimicrobial agents such as the semisynthetic penicillins. Consequently, the greatest effect is seen when these drugs are tested. For other drugs the addition of serum does not make much difference.

Killing curve

The determination of the killing rate or establishment of a killing curve has been widely applied to the evaluation and comparison of new drugs and the study of differences and changes in the antimicrobial susceptibility of clinically important bacterial isolates. Such determinations are used less frequently for guiding chemotherapy in an individual case. Only one concentration of the antimicrobic is tested, usually

TABLE 2. MBC test procedure

Procedure steps	Comments
1. Subculture organisms onto appropriate medium (usually a blood agar plate) and incubate overnight at 35°C.	Blood agar will facilitate the detection of contaminants and variants.
2. Inoculate a tube containing 3 ml of saline or MHB with five or more colonies from the plate incubated overnight (step 1) to achieve a turbidity equivalent to a no. 1 McFarland standard (ca. 10^8 organisms per ml).	
3. Transfer 0.1 ml of turbid inoculum into 10 ml of MHB or other appropriate broth. Incubate in shaking water bath or equivalent at 35°C until turbid. This corresponds to an endpoint between a no. 1 McFarland standard and an overnight suspension and will take between 5 and 6 h for rapid growers.	Trypticase soy or other broths may be substituted when testing fastidious organisms. This represents a midlogarithmic-phase inoculum.
4. The standard control organism (*Escherichia coli, S. aureus,* etc.) should be inoculated into 3 ml of broth and incubated (without shaking) at 35°C until turbid.	The first three steps of the procedure apply only to the patient's sample.
5. Prepare twofold serial dilutions of the antibiotic in 2 ml of MHB (total volume per acid-washed borosilicate glass tube).	Glass tubes (16 by 100 mm) with loose-fitting metal caps are preferred by many workers.
6. Standardize inoculum (patient's organism and control organism) to equal a 0.5 McFarland turbidity standard (ca. 5×10^7 organisms per ml) in 3 ml of saline or broth.	The phase of growth of the control organism is not critical if only the MIC determination is intended.
7. Dilute adjusted inocula 1:10 (0.2 ml in 1.8 ml of MHB or appropriate substitute). This equals about 5×10^6 organisms per ml.	
8. Using an Eppendorf or equivalent pipette, dispense 100 μl (0.1 ml) of diluted inoculum into tubes containing serial dilutions of the antibiotic. To inoculate, insert pipette tip well under the surface of antibiotic-containing broth. Avoid any contact between tip and walls of the tube. Rinse tip five times in solution. The same tip may be used throughout the test if inoculating from lowest to highest concentration of antibiotic. The final inoculum size is approximately 2.5×10^5 organisms per ml.	It is critical to avoid transfer of organisms to the inside of test tubes above the meniscus by shaking or splashing, etc.
9. Incubate for 20 h at 35°C.	
10. From the 1:10 dilution of the 0.5 McFarland adjusted inoculum, which should be about 5×10^6 organisms per ml (step 7), dilute serially 1:10 in MHB four times to achieve a final inoculum of 5×10^2 organisms per ml.	
11. Allocate 0.1 ml of this suspension and dispense either into a tube of melted agar for the preparation of pour plates or onto an appropriate agar plate (e.g., blood agar) and distribute evenly using sterile, bent glass rods. This procedure should be done in duplicate. Incubate overnight at 35°C.	Pour plates are not advised for fastidious organisms.
12. Read and record MIC of control organisms.	
13. For patient's sample only: vortex tubes without visible growth vigorously for 15 s and reincubate for an additional 4 h.	This resuspends, and exposes to the antibiotic, organisms adhering to the test tube wall above the meniscus. CAUTION: AVOID AEROSOLS!
14. Mix again on a Vortex mixer and sample tubes for MBC determination; spread 100-μl samples across the surface of dried Trypticase soy agar plates with sterile, bent glass rods.	This facilitates reading of survivor colonies and dilutes out any antibiotic carry-over.
15. Record patient's MIC.	
16. Incubate plates overnight at 35°C for the MBC test.	For fastidious organisms longer incubation may be necessary.
17. After 1 day (or 2) count the number of colonies per plate from the original inoculum plates or pour plates and average. Determine a colony count that will represent 0.1% of the original inoculum (i.e., 99.9% reduction).	
18. Count colonies from MBC plates. Any number equal to or less than the determined colony count from step 17 will be considered as a 99.9% kill or bactericidal result.	Erratic colony counts above this cut-off point at higher concentrations of antibiotic should be reported as representing the paradoxical phenomenon as described above.

the one representing the average obtainable blood level during therapy. At periodic intervals, at least at 0, 4, and 24 h of incubation, colony counts are performed and charted on semilog paper with the survivor colony count on the ordinate (*y* axis) in logarithmic scale and the time on the abscissa (*x* axis) in arithmetic scale. Performed with various organisms against one drug or with several drugs against the same organism, such killing rate determinations then allow one to compare the rates of decline in the survivor count and thus to find the most rapidly bactericidal agent, or to detect the organism most susceptible to the killing action of the drug.

Procedure. In actual practice the killing curve test is done in glass tubes, each containing 10 ml of MHB or similar suitable broth and the chosen concentration of the antimicrobial agent to be tested. If survivors are to be counted after 4, 8, and 24 h of incubation, the procedure should be set up as follows (duplicates or multiple samples are recommended).

1. Prepare and label at least one antibiotic-containing tube each for sampling at 0, 4 (8), and 24 h.
2. Set up one tube without antibiotic as a growth control.
3. Prepare inoculum, approximately 5×10^7 CFU/ml, according to steps 1 to 3, 5, and 6 of the MBC test procedure.
4. Inoculate tubes according to step 8 of the MBC test procedure, avoiding any splashing on inside of test tubes above the meniscus. The final concentration should now be approximately 5×10^5 CFU/ml of broth in each tube.
5. Immediately vortex for 15 s the 0 growth tube and dispense 0.1 ml into a tube of melted agar for the preparation of a pour plate. Incubate at 35°C for 18 to 24 h.
6. In the same fashion prepare pour plates after 4 and 8 h of incubation from the next tube.
7. For testing at 24 h, vortex the last tube at 20 h and reincubate for an additional 4 h.
8. At 24 h, vortex again and prepare pour plate(s).
9. Plot colony counts on semilog graphing paper.

Comment. Depending on the actual organism-antibiotic combination and sampling time, there may be a final increase in CFU after an initial decrease, due to the selection of resistant mutants. In some instances, total incubation may have to be extended to 48 or 72 h. After such prolonged incubation, inactivation of the antimicrobial agent may account for increased CFU.

Determination of bactericidal activity in serum

Directly estimating the antimicrobial activity in the serum of a patient during treatment, using the infecting organism as a test strain, theoretically would appear to be the simplest and most logical approach to monitoring chemotherapy in complicated cases. Antimicrobial agents singly or in combination are being evaluated as to their efficacy in the patient's "own system." However, in practice there are several problems to contend with. Many variations of this procedure exist. As originally described by Schlichter (39, 40), determination of antibiotic activity in the patient's serum was used only as a bacteriostatic test. The bactericidal part was added later, possibly because of difficulties in reading endpoints in serum-containing tubes. Washington has listed in a table all the differences in methodology established over three decades: timing, inoculum, diluent, subculture volume, titer, endpoint percentage, and lethal or static endpoint (51). Partly because of the variability of methods, proof for the clinical value of the test was not clear cut. Originally it served mainly to monitor penicillin therapy of subacute bacterial endocarditis; it was generally agreed that if titers were kept at 1:16 or higher, clinical outcome was better. Recent papers provide more evidence of the test's clinical worth (43), particularly as applied to monitoring antimicrobial therapy in neutropenic and cancer patients (21, 41).

Several reasonably standardized methods have been published from laboratories with considerable experience (2, 34, 51). They differ mainly in the choice of the diluent, namely, whether pooled human serum is used or not. The procedure outlined below generally follows the published versions of the method used at Mayo Clinic, with attention to growth phase of the inoculum and its delivery without splashing as specified for the MBC test.

Procedure. Use the same type of tubes and care in preparing and delivering the inoculum as specified in the MBC test.

1. Serially dilute the patient's serum twofold to a titer of at least 1:64, using MHB as diluent. Final volume per tube should be 1 ml.
2. Add the patient's organism in the form of 0.1 ml of a carefully prepared inoculum (steps 10 and 11 of MBC procedure).
3. Prepare pour plate for establishing original inoculum count of CFU (steps 10 and 11 of MBC procedure).
4. Include a growth control tube containing MHB and inoculum but no serum.
5. Incubate for 18 to 24 h.
6. Subculture for 99.9% bactericidal endpoint as described for MBC procedure.

Comment. Interpretation should be individualized. A dilution titer of 1:8 or 1:16 is reportedly adequate for most infections. For endocarditis it should be higher.

EFFECT OF ANTIMICROBIAL AGENTS IN COMBINATION

Combinations of antimicrobial agents are often used in the treatment of serious infections in an attempt to increase therapeutic effect. The value of this approach has been clearly documented in the treatment of tuberculosis, enterococcal endocarditis, and some infections due to gram-negative bacilli (1, 27). A combination can be considered synergistic when the effect observed with the combination is greater than the sum of the effects observed with the two drugs independently. The interaction of drugs is considered additive or indifferent when the combined effect is equal to the sum of the effects produced by the drugs independently, or equal to that of the most active drug in the combination. The interaction of drugs can also be antagonistic when the combination is clearly less effective than the more active drug alone. One important reason for determining the in vitro effect of antibiotic combinations is to avoid antagonism. The effect may be measured in terms of MIC, MBC, rate of killing, suppression of resistance, or other factors (22), but methods which determine the

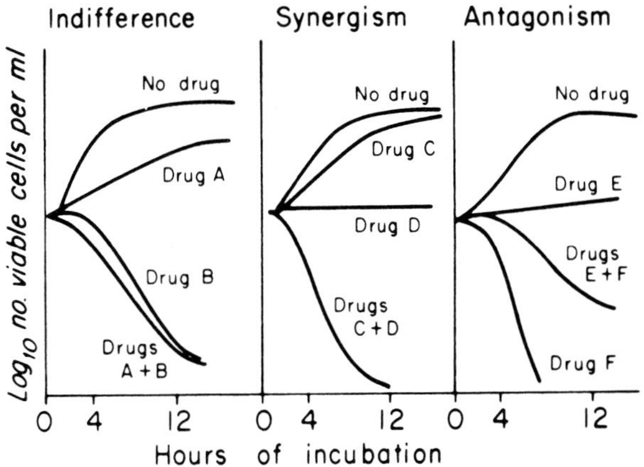

FIG. 1. Rate of killing with antimicrobial agents singly and in combination. Schematic representation showing three different types of results.

lethal activity of combinations are generally preferred. The methods described for determination of MBC and killing rate can be modified to include testing antimicrobial combinations.

The results from studies of the effect of a combination on rate of killing are presented in Fig. 1. In this example the interpretation appears obvious, but not all studies are so simple. Often, both antimicrobial agents will show killing. The definition of synergy requires that the combined effect be greater than the sum of the independent effects. Each effect could be calculated as a survival rate and summed, but this is rarely done. Usually, a combination is defined as synergistic simply when the number of viable cells at a given time is 10-fold lower than with the more effective antibiotic alone.

An alternative method for evaluating the activity of drugs in combination is to test multiple combinations and express the results as an isobologram (4, 24). This procedure requires that a specific endpoint be determined for each drug in a combination. Those combinations that result in the same endpoint are then plotted with the concentrations expressed on an arithmetic scale (Fig. 2). Since the coordinates of an isobologram are in units of drug concentration rather than effect, the definition for synergy based on effect must be changed (24). The task of defining synergy based on drug concentration is simplified when concentrations are expressed as fractions of the concentration of each drug which gave the selected endpoint when tested alone. A drug combination, then, is defined as synergistic when the sum of the fractions for an effective combination is less than unity. Determination of drug interactions can be simplified by measuring the results for only a few selected concentrations. A useful approach is to test a concentration one-fourth the MBC for one drug in combination with concentrations of the second drug ranging from the MBC to one-eighth or less of the MBC. The choice of which drug is tested at constant concentration should be made by considering achievable serum levels and toxicity.

Both of the above procedures require multiple colony counts and are time-consuming. Bacteriostatic interactions are far easier to demonstrate and can be shown simply by placing paper disks or strips impregnated with antimicrobial agents on an agar plate that has been inoculated with a bacterial suspension. Characteristic zones of inhibition are obtained for the different types of interaction. Modifications of this procedure have been used to measure bactericidal activities (33).

DETECTION OF β-LACTAMASE

Direct tests on colonies for β-lactamase production are rapid and effective methods of detecting penicillin- and ampicillin-resistant β-lactamase-producing strains of *S. aureus*, *H. influenzae*, and *N. gonorrhoeae*. These strains comprise the great majority of penicillin- and ampicillin-resistant members of their species.

Several methods are available for detecting β-lactamase (33, 46). The most common and rapid methods depend on: (i) using pH indicators to detect increased acidity resulting from cleavage of the β-lactam ring of a penicillin to yield a penicilloic acid; (ii) decolorization of a starch-iodine mixture as a result of the ability of penicilloic acid to reduce iodine; or (iii) a color change resulting from hydrolysis of the β-lactam ring of a chromogenic cephalosporin. The first three procedures which follow are identical or similar to those described in *Cumitech 6* (46) for testing *H. influenzae* and *N. gonorrhoeae*, but other variations of these methods have been used. Longer reaction times than those indicated may be necessary when *S. aureus* is being tested or when the bacterial suspension is too dilute. β-Lactamase production in staphylococci, especially coagulase-negative species, may be inducible, and results are enhanced by taking growth from the zone margins around a methicillin, oxacillin, or nafcillin disk. Control strains must be included with each test. A β-lactamase-producing *S. aureus* strain (ATCC 29213) and a β-lactamase-negative *S. aureus* strain (ATCC 25923) are satisfactory positive and negative controls, as are known β-lactamase-positive and -negative strains of *H. influenzae* and *N. gonorrhoeae*.

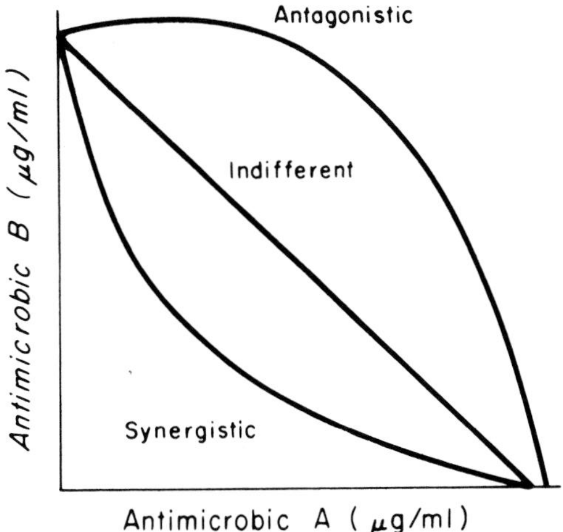

FIG. 2. Isobologram portraying three possible results when two antimicrobial agents (A and B) are tested singly and in various combinations; either MIC or MBC endpoints may be plotted. A straight line joining the values obtained with each drug separately represents an isobol which indicates indifference. Antagonism is indicated by an isobol which bows upward away from the coordinates, and a bowing toward the coordinates indicates synergism.

Rapid acidimetric method

Add 2 ml of a 0.5% solution of phenol red to 16.6 ml of sterile distilled water and inject the mixture into an ampoule containing 20×10^6 U of penicillin G which is designated for parenteral use and contains a citrate buffer. Transfer the mixture to a test tube and add 1 M NaOH by drops until the pH reaches approximately 8.5, as shown by the development of a violet color or as measured by a pH meter. Solution that is not used immediately should be divided into portions and frozen at −60°C or colder. Discard the solution when it has turned yellow.

Place 0.05 to 0.1 ml of the above substrate in a well of a microdilution plate or in a small tube. Prepare a heavy suspension with colonies from a culture grown overnight, and mix it into the substrate to form a suspension more turbid than a no. 4 McFarland standard. If the culture produces β-lactamase, the solution will turn yellow within 15 min (in most cases within 1 min).

Rapid iodometric method

To prepare the substrate, dissolve penicillin G in 0.1 M phosphate buffer (pH 6.0) to a concentration of 6,000 μg/ml. To prepare the starch reagent, add 1 g of soluble starch to 100 ml of distilled water and heat the container in a boiling-water bath until the starch dissolves. Prepare the iodine reagent by dissolving 2.03 g of iodine and 53.2 g of potassium iodide in a small volume of distilled water, and then dilute to 100 ml. Store in a brown glass bottle at room temperature.

The penicillin and starch solutions should be freshly prepared. The iodine solution should be replaced when excessive precipitate develops.

Dispense 0.1 ml of the penicillin G solution into a well of a microdilution plate or a small test tube. Prepare a heavy suspension of the test organism and mix it with the substrate as described above. Let the mixture stand at room temperature for 30 min. Add 2 drops of the starch solution and mix. Add 1 drop of the iodine reagent. A dark blue or purplish color will develop immediately as a result of the reaction of iodine and starch. Shake or stir the mixture for 1 min at room temperature. Decolorization (to a whitish, starchy color) in less than 10 min indicates the production of β-lactamase.

Rapid chromogenic cephalosporin method

The rapid chromogenic cephalosporin method can be used with either nitrocefin (Glaxo Ltd.) or PADAC (pyridine-2-azo-*p*-dimethylaniline cephalosporin; Hoechst-Roussel Pharmaceuticals). With nitrocefin, dissolve 10 mg in 1 ml of dimethyl sulfoxide. Dilute with 0.1 M phosphate buffer (pH 7.0) to a concentration of 500 μg/ml. This solution will be yellow to light orange. It is stable for many weeks at 4 to 10°C. With PADAC, dissolve 10 mg in 1 ml of dimethyl sulfoxide. Dilute with 0.1 M phosphate buffer (pH 7.0) to contain 256 μg/ml. This solution will be purple.

Add 0.1 ml of the cephalosporin substrate to the well of a microdilution plate or to a small tube. Prepare a heavy suspension of organisms and mix with the substrate as described above. The nitrocefin substrate will turn red within 10 min at room temperature (usually 1 to 2 min), whereas the PADAC substrate turns yellow in 60 min if the culture produces β-lactamase. The PADAC reaction occurs in two steps and the intermediate compound is colorless, so a positive reaction goes from purple to colorless to yellow.

Paper strip or disk methods

The acidimetric, iodometric, and chromogenic cephalosporin methods can be modified to be performed on paper. The acidimetric and chromogenic cephalosporin methods done on paper or plastic are available commercially, but the iodimetric technique on paper must be prepared in the laboratory. Below is such a method.

Saturate strips of Whatman no. 3 filter paper with a 1% penicillin G solution prepared in water. Allow the strips to dry at room temperature. The dried strips can be stored at −20°C or colder for at least 1 year.

Place a dried strip in a disposable petri dish and thoroughly moisten with either Gram or Lugol iodine solution. Pour off excess liquid. Remove approximately 10 colonies from an agar plate with an inoculating loop and apply to the strip. The organisms should be spread over an area about 5 mm in diameter. The production of β-lactamase is evidenced by a change from the original dark purple or brownish color to white within 1 min (18).

Methods available commercially

Several commercial methods for detecting β-lactamase are now available. As indicated previously, these tests are modifications of either the acidimetric or chromogenic cephalosporin methods described above. The commercial acidimetric methods consist of a paper carrier (usually a paper disk or strip) containing penicillin and a pH indicator, usually bromcresol purple. The paper is moistened, and growth is rubbed onto it. A positive reaction, i.e., acid production, is indicated by a change from purple to yellow. Nitrocefin and PADAC are tested similarly. Nitrocefin is contained in a paper disk, and PADAC comes on a plastic strip. A positive test for β-lactamase with the nitrocefin disk will be a change from yellow to red, and that for PADAC would be from purple to yellow, as described previously.

All of these methods are accurate with *Haemophilus* or *Neisseria* spp., but with other organisms methods should be chosen carefully. The PADAC method should not be used to test staphylococci. Only the chromogenic cephalosporin methods should be used to test *Bacteroides* spp. β-Lactamase tests for therapeutic purposes are not recommended for *Enterobacteriaceae*, nonfermentative gram-negative bacilli, or anaerobes other than *Bacteroides* spp.

Other methods

Methods based on principles other than those discussed above can be used but are generally not convenient, rapid methods. The hydrolysis of some β-lactams can be detected and followed by spectrophotometry, gas-liquid chromatography, and mass spectrometry (11, 17, 31, 45). The "cross-streak" assay is a very sensitive and simple test but requires overnight incubation (45).

SCREENING TEST FOR METHICILLIN-RESISTANT (HETERORESISTANT) STAPHYLOCOCCI

The incidence of methicillin-resistant *S. aureus* in hospitals in the United States has increased, particularly in the large teaching hospitals affiliated with medical schools; community-acquired infections have also been documented (13). These organisms may be identified by the susceptibility tests described in other chapters of this section. The addition of salt to the medium is advantageous, particularly in the broth microdilution test (47). In the agar screening test described below, the Mueller-Hinton agar is supplemented with NaCl and either methicillin, oxacillin, or nafcillin. It is reported to be a very sensitive test for detecting these staphylococci (47).

For the test medium, add 4% NaCl to Mueller-Hinton agar and prepare as directed by the manufacturer. After sterilization, cool the agar to approximately 50°C, add (per ml) either 10 μg of methicillin, 6 μg of oxacillin, or 6 μg of nafcillin, and mix. Pour the appropriate amount into a petri plate (e.g., 25 ml for a 100-mm round plate), let cool, seal in plastic bags, and store at 4 to 8°C until needed. The precautions for storing antimicrobic-containing agar listed in Chapter 100 should be observed. Known heteroresistant and nonheteroresistant *S. aureus* strains can be used as controls.

Prepare inocula by suspending colonies from a 24-h agar plate in MHB (or other similar broths or saline) and adjusting to a no. 0.5 McFarland standard (Chapter 102). Inoculate the screen plate with a "spot" containing approximately 10^4 CFU as described for the agar dilution test (Chapter 100). Alternatively, spot inoculate the plate with a swab as described in the disk diffusion procedure (Chapter 102). Incubate the plates at 35°C (no higher) for 24 h. Observe the plate for growth. If colonies are present, the strain can be assumed to be resistant; if no growth occurred, it can be assumed to be susceptible to these drugs.

Oxacillin is generally the antimicrobial agent of choice to use in these tests because it is more stable and because it usually detects heteroresistance more efficiently. The choice may depend, however, on the characteristics of the strains within the individual hospital.

LITERATURE CITED

1. **Anderson, E. T., L. S. Young, and W. L. Hewitt.** 1978. Antimicrobial synergism in the therapy of gram-negative rod bacteremia. Chemotherapy **24:**45–54.
2. **Anhalt, J. P., L. D. Sabath, and A. L. Barry.** 1980. Special tests: bactericidal activity, activity of antimicrobics in combination, and detection of β-lactamase production, p. 478–484. *In* E. H. Lennette, A. Balows, W. H. Hausler, Jr., and J. P. Truant (ed.), Manual of clinical microbiology, 3rd ed. American Society for Microbiology, Washington, D.C.
3. **Association of Official Agricultural Chemists.** 1945. Official methods of analysis, 6th ed. Association of Official Agricultural Chemists, Washington, D.C.
4. **Berenbaum, M. C.** 1978. A method for testing synergy with any number of agents. J. Infect. Dis. **137:**122–130.
5. **Bigger, J. W.** 1944. The bactericidal action of penicillin on *S. aureus*. Ir. J. Med. Sci. Part 1 **227:**553–568.
6. **Bigger, J. W.** 1944. Treatment of staphylococcal infections with penicillin by intermittent sterilization. Lancet **ii:**497–500.
7. **Bryan, C. S., S. R. Marney, Jr., R. H. Alford, and R. E. Bryant.** 1975. Gram-negative bacillary endocarditis. Interpretation of the serum bactericidal test. Am. J. Med. **58:**209–215.
8. **Eagle, H., and A. D. Musselman.** 1948. The rate of bactericidal action of penicillin in vitro as a function of its concentration, and its paradoxically reduced activity at high concentrations against certain organisms. J. Exp. Med. **88:**99–131.
9. **Erikson, K. R.** 1946. Some studies on the lytic action of penicillin on staphylococci and pneumococci. Acta Pathol. Microbiol. Scand. **23:**221–228.
10. **Garrod, L. P.** 1945. The action of penicillin on bacteria. Br. Med. J. **i:**107–110.
11. **Gibbs, D. L., and C. Thornsberry.** 1979. Susceptibility of gram-negative bacteria to β-lactam antibiotics and rapid characterization of β-lactamase activity. Curr. Microbiol. **2:**239–242.
12. **Gwynn, M. N., L. T. Webb, and G. N. Rolinson.** 1981. Regrowth of *Pseudomonas aeruginosa* and other bacteria after the bactericidal action of carbenicillin and other beta-lactam antibiotics. J. Infect. Dis. **144:**263–269.
13. **Haley, R. W., A. W. Hightower, R. F. Khabbaz, C. Thornsberry, W. J. Martone, J. R. Allen, and J. M. Hughes.** 1982. The emergence of methicillin-resistant *Staphylococcus aureus* infections in United States hospitals. Ann. Intern. Med. **97:**297–308.
14. **Hobby, G. L., K. Meyer, and E. Chaffee.** 1942. Observations on the mechanism of action of penicillin. Proc. Soc. Exp. Biol. **50:**281–285.
15. **Horne, D., and A. Tomasz.** 1981. pH-dependent penicillin tolerance of group B streptococci. Antimicrob. Agents Chemother. **20:**125–128.
16. **Ishida, K. P., A. Guze, G. M. Kalmason, K. Albrandt, and L. B. Guze.** 1982. Variables in demonstrating methicillin tolerance in *Staphylococcus aureus* strains. Antimicrob. Agents Chemother. **21:**688–690.
17. **Jones, R. N., H. W. Wilson, and W. J. Novick.** 1982. In vitro evaluation of pyridine-2-azo-*p*-dimethylanaline cephalosporin, a new diagnostic chromogenic reagent, and comparison with nitrocefin, cefacetrile, and other β-lactam compounds. J. Clin. Microbiol. **15:**677–683.
18. **Jorgensen, J. H., J. C. Lee, and G. A. Alexander.** 1977. Rapid penicillinase paper strip test for detection of beta-lactamase-producing *Haemophilus influenzae* and *Neisseria gonorrhoeae*. Antimicrob. Agents Chemother. **11:**1087–1088.
19. **Kim, K. S., and B. F. Anthony.** 1981. Importance of bacterial growth phase in determining minimal bactericidal concentration of penicillin and methicillin. Antimicrob. Agents Chemother. **19:**1075–1077.
20. **Kirby, W. M. M.** 1945. Bacteriostatic and bacteriolytic actions of penicillin on sensitive and resistant staphylococci. J. Clin. Invest. **24:**165–169.
21. **Klastersky, J., D. Daneau, G. Swings, and D. Weerts.** 1974. Antibacterial activity in serum and urine as a therapeutic guide in bacterial infections. J. Infect. Dis. **129:**187–193.
22. **Lacey, B. W.** 1958. Mechanisms of chemotherapeutic synergy. Symp. Soc. Gen. Microbiol. **8:**247–288.
23. **Layte, S., P. Harris, and G. N. Rolinson.** 1983. Factors affecting the apparent regrowth of *Pseudomonas aeruginosa* following exposure to bactericidal concentrations of carbenicillin. Chemotherapy **30:**26–30.
24. **Loewe, S.** 1953. The problem of synergism and antagonism of combined drugs. Arzneimittelforsch. **3:**285–290.
25. **Mayhall, C. G., and E. Apollo.** 1980. Effect of storage and changes in bacterial growth phase and antibiotic concentrations on antimicrobial tolerance in *Staphylococcus aureus*. Antimicrob. Agents Chemother. **18:**784–788.
26. **Mayhall, C. G., G. G. Medhoff, and J. J. Mar.** 1976. Variation in the susceptibility of strains of *Staphylococcus aureus* to oxacillin, cephalothin, and gentamicin. Antimicrob. Agents Chemother. **10:**707–712.

27. **McCabe, W. R.** 1968. Clinical use of combinations of antimicrobial agents, p. 225–233. Antimicrob. Agents Chemother. 1967.
28. **Murray, P. R., and J. H. Jorgensen.** 1981. Quantitative susceptibility test methods in major United States medical centers. Antimicrob. Agents Chemother. **20:**66–70.
29. **National Committee for Clinical Laboratory Standards.** 1983. Tentative standard. Methods for dilution antimicrobial susceptibility tests for bacteria that grow aerobically, M7-T, vol. 3, no. 2. National Committee for Clinical Laboratory Standards, Villanova, Pa.
30. **Nelson, R. E., and J. A. Washington II.** 1981. Paradoxic and tolerant effects of moxalactam on *Staphylococcus aureus*. J. Infect. Dis. **144:**178.
31. **O'Callaghan, C. H., A. Morris, S. M. Kirby, and A. H. Shingler.** 1972. Novel method for detection of β-lactamases by using a chromogenic cephalosporin substrate. Antimicrob. Agents Chemother. **1:**283–288.
32. **Pien, F. D., R. D. Williams, and K. L. Vosti.** 1975. Comparison of broth and human serum as the diluent in the serum bactericidal test. Antimicrob. Agents Chemother. **7:**113–114.
33. **Reeves, D. S., I. Phillips, J. D. Williams, and R. Wise (ed.).** 1978. Laboratory methods in antimicrobial chemotherapy, p. 64–69. Churchill Livingstone, Edinburgh.
34. **Reller, L. B., and C. W. Stratton.** 1977. Serum dilution test for bactericidal activity. II. Standardization and correlation with antimicrobial assays and susceptibility tests. J. Infect. Dis. **136:**196–204.
35. **Report.** 1909. The standardization of disinfectants. Lancet **ii:**1516–1531.
36. **Rozenberg-Arska, M., G. T. J. Fabius, M. A. A. J. Beens-Dekkers, S. A. Duursma, L. D. Sabath, and J. Verhoef.** 1979. Antibiotic sensitivity and synergism of penicillin-tolerant *Staphylococcus aureus*. Chemotherapy **25:**352–355.
37. **Sabath, L. D.** 1979. Staphylococcal tolerance to penicillins and cephalosporins, p. 299–303. *In* D. Schlessinger (ed.), Microbiology—1979. American Society for Microbiology, Washington, D.C.
38. **Sabath, L. D., N. Wheeler, M. Laverdiere, D. Blazevic, and B. J. Wilkinson.** 1977. A new type of penicillin resistance of *S. aureus*. Lancet **i:**443–447.
39. **Schlichter, J. G., and H. MacLean.** 1947. A method of determining the effective therapeutic level in the treatment of subacute bacterial endocarditis with penicillin. Am. Heart J. **34:**209–211.
40. **Schlichter, J. G., H. MacLean, and A. Malzer.** 1949. Effective penicillin therapy in subacute bacterial endocarditis and other chronic infections. Am. J. Med. Sci. **217:**600–608.
41. **Sculier, J. P., and J. Klastersky.** 1984. Significance of serum bactericidal activity in gram-negative bacillary bacteremia in patients with and without granulocytopenia. Am. J. Med. **76:**429–435.
42. **Stratton, C. W., and L. B. Reller.** 1977. Serum dilution test for bactericidal activity. I. Selection of a physiologic diluent. J. Infect. Dis. **136:**187–195.
43. **Stratton, C. W., M. P. Weinstein, and L. B. Reller.** 1982. Correlation of serum bactericidal activity with antimicrobial agent level and minimal bactericidal concentration. J. Infect. Dis. **145:**160–168.
44. **Taylor, P. C., F. D. Schoenknecht, J. C. Sherris, and E. C. Linner.** 1983. Determination of minimum bactericidal concentrations of oxacillin for *Staphylococcus aureus*: influence and significance of technical factors. Antimicrob. Agents Chemother. **23:**142–150.
45. **Thornsberry, C., J. W. Biddle, L. A. Kirven, and C. W. Moss.** 1977. Penicillin resistance in *Neisseria gonorrhoeae* due to β-lactamase production. Microbios **20:**39–46.
46. **Thornsberry, C., T. L. Gavan, and E. H. Gerlach.** 1977. Cumitech 6, New developments in antimicrobial agent susceptibility testing, p. 1–2. Coordinating ed., J. C. Sherris. American Society for Microbiology, Washington, D.C.
47. **Thornsberry, C., and L. K. McDougal.** 1983. Successful use of broth microdilution in susceptibility tests for methicillin-resistant (heteroresistant) staphylococci. J. Clin. Microbiol. **18:**1084–1091.
48. **Tomasz, A.** 1979. From penicillin-binding proteins to the lysis and death of bacteria: a 1979 view. Rev. Infect. Dis. **1:**434–467.
49. **Tomasz, A.** 1981. Penicillin tolerance and the control of murein hydrolases, p. 227–247. *In* M. Salton and G. D. Shockman (ed.), Beta-lactam antibiotics. Academic Press, Inc., New York.
50. **Venglarick, J., III, L. Blair, and L. Dunkle.** 1983. pH-dependent oxacillin tolerance of *Staphylococcus aureus*. Antimicrob. Agents Chemother. **23:**232–235.
51. **Washington, J. A., II.** 1981. Bactericidal tests, p. 715–723. *In* Laboratory procedures in clinical microbiology. Springer-Verlag, New York.

Chapter 106

Assays for Antimicrobial Agents in Body Fluids

JOHN P. ANHALT

The determination of the concentration of an antimicrobial agent in serum or other fluids is justified only when that concentration can be correlated with benchmarks of efficacy or excess and cannot be predicted adequately from dosage. For most antimicrobial agents, the range between therapeutic and toxic levels is large, and the tendency is to administer doses that essentially assure levels in serum severalfold the assumed minimal effective level. Assays are not required for these drugs unless there is a reasonable suspicion that the level in serum may be markedly different than usual, such as in patients with severe renal disease. The effects of dialysis on the elimination of many antimicrobial agents is sufficiently predictable (2), however, that this factor alone does not mandate the use of assays for dosage adjustment. A relatively small number of antimicrobial agents—the most common being aminoglycosides, chloramphenicol, and vancomycin—exhibit a narrow range between therapeutic and toxic concentrations. Calculated dosages are often inadequate to ensure appropriate levels in serum and are particularly inaccurate for critically ill patients in whom the rapid attainment of maximum levels is needed to treat a life-threatening infection. Assays for these drugs are needed frequently for optimal patient care.

SPECIMENS

The most appropriate specimen for the assay of common antimicrobial agents is serum. Plasma can be used in some assay procedures; however, anticoagulants may interfere. For example, heparin used to anticoagulate plasma binds gentamicin and can adversely affect a bioassay. Levels of antimicrobial agents in urine are rarely indicated, and the results are difficult to interpret. Data concerning optimal storage conditions for clinical specimens are generally lacking. However, there are some drugs and drug combinations that are known to be particularly unstable. The combination of an aminoglycoside and a β-lactam drug, usually carbenicillin or ticarcillin, is particularly unstable, and significant losses may occur rapidly even at −20°C. This reaction can be almost eliminated by the addition of a β-lactamase. The cephalosporins are very unstable in serum, presumably due to the presence of lipoproteins (4). Addition of sodium dodecyl sulfate to a final concentration of 1% (wt/vol) at pH 6 or below will stabilize cephalosporins. Specimens should be analyzed as soon as practical after receipt. However, most clinical specimens can be stored at −20°C for up to 2 days without greatly affecting results and without the need for added stabilizers. Some drugs, such as thienamycin, show decreased stability upon freezing, presumably due to high local concentrations that occur as the water crystallizes.

Timing

Most antimicrobial agents are administered at intervals that are greater than the half-lives for excretion. A steady state is reached after only two or three doses (i.e., after five half-lives) and is characterized by a large difference between the maximal concentration (peak) and the minimal concentration (trough) attained in serum between doses. These large changes mandate that times for administering doses and obtaining blood for assays be carefully planned and accurately recorded. Samples drawn when the times are inappropriate or uncertain will result in values that are misleading or uninterpretable. For only a few antibiotics, namely, gentamicin, amikacin, and other aminoglycosides, is the usefulness of both peak and trough measurements established. For most other antibiotics, a determination of either the peak or the trough concentration will suffice for dosage adjustments and will be more cost effective. Of the two choices, trough levels provide a more sensitive indication of decreased clearance and drug accumulation. Peak levels in a stable patient will tend to vary more because of uncontrolled variables, such as rate of drug infusion, rate of absorption, and differences in sampling time.

The exact time that a peak level occurs varies depending on the drug, route of administration, and clinical status of the patient. It is impossible to predict accurately the time that the peak level will occur. As a general practice, blood for peak levels should be obtained 30 min after the completion of an intravenous infusion, 1 h after an intramuscular dose, and 1 to 2 h after an oral dose. Absorption of an oral medicine is particularly variable. Therefore, if the level in serum after an oral dose is unexpectedly low, the possibility of slow absorption should be tested by obtaining a sample of blood 3 h after a dose. The tetracyclines and erythromycin typically are absorbed slowly, and for these drugs, expected peak concentrations occur 2 to 4 h and 2 to 6 h, respectively, after oral doses. Blood for trough levels should be obtained immediately before a dose.

Information needed

A test request for antimicrobial assay should include the names of any other antimicrobial agents the patient received within the previous 48 h. This information is essential, particularly when a bioassay is used, if the laboratory expects to provide a result that does not represent a combined effect of the agents. Equally important is that the dose, exact time of dose, and time of collection be given. The absence of this information can cause additional work for the laboratory using bioassays, because of the difficulty in estimating the proper dilutions of serum to be tested. More important, the assay result cannot be interpreted without this information, and it is unlikely that the facts can be reconstructed later from notes made in a patient's record. Some assay results will indicate medical emergencies (e.g., unusually high or low values). The laboratory should report these values immediately to a physician or other responsible party listed on the test request.

TABLE 1. Therapeutic ranges and pharmacokinetics for selected antimicrobial agents[a]

Antimicrobial agent	Usual dose	Dosage interval (h)	Maximum dose (per 24 h)	Normal serum half-life (h)	Major route of elimination	Removed by:		Therapeutic range (μg/ml)		Recommended basis for dosage adjustment[b]
						Hemodialysis	Peritoneal dialysis	Peak	Trough	
Amikacin	5–7.5 mg/kg	8–12	15 mg/kg	2–3	Renal	Yes	Yes	20–25	5–10	P,T
Gentamicin	1.7 mg/kg	8	5 mg/kg	2–3	Renal	Yes	Yes	4–8	1–2	P,T
Kanamycin	5–7.5 mg/kg	8–12	15 mg/kg	2–3	Renal	Yes	Yes	20–25	5–10	P,T
Netilmicin	2–2.5 mg/kg	8	7.5 mg/kg	2–3	Renal	Yes	Yes	6–10	0.5–2	P,T
Streptomycin	0.5–1 g	8–12	2 g	2–3	Renal	Yes		5–20	<5	P,T
Tobramycin	1.7 mg/kg	8	5 mg/kg	2–3	Renal	Yes	Yes	4–8	1–2	P,T
Chloramphenicol	0.5–1 g	6	4 g	4[c]	Hepatic-renal	Yes	No	15–25	8–10	T
Flucytosine	37.5 mg/kg	6	150 mg/kg	4	Renal	Yes	Yes	100	50	T
TMP-SMX[d]										
TMP	5 mg/kg	6	20 mg/kg	11	Renal	Yes		≥5		P
SMX	25 mg/kg	6	100 mg/kg	13	Renal	Yes		≥100		P
Vancomycin	0.5–1 g	8–12	2 g	6	Renal	No	No	20–40	5–10	T

[a] From Hermans et al. (9). Used with permission.

[b] P, Peak level; T, trough level. These recommendations are valid when the dosage interval is no greater than twice the usual dosage interval. Blood for peak levels should be drawn 30 min after completion of an intravenous infusion, 1 h after an intramuscular dose, and 1 to 2 h after an oral dose.

[c] Half-life in children less than 4 weeks old can be prolonged greatly. Half-life is affected only slightly in renal failure, but it can be greatly prolonged with liver disease.

[d] TMP, Trimethoprim; SMX, sulfamethoxazole. Serum half-life is shortened in adolescents and children. Measurement of either TMP or SMX alone is sufficient for dosage adjustment.

INTERPRETATION OF RESULTS

In many cases, a specific therapeutic range for antimicrobial agents has not been defined. Instead, levels attained with doses that have been shown to be effective in clinical trials are considered desirable (Appendix 2). For a small number of relatively toxic agents, therapeutic ranges have been defined more narrowly (Table 1). A more extensive listing of agents has been compiled by Hermans et al. (9).

METHODS

The discovery of each new antimicrobial agent has been accompanied by the development of an assay procedure. Since these discoveries are made in many different laboratories, it is no surprise that assays exist in great profusion and serve several functions. To select an appropriate method, one must consider: (i) availability and cost of specialized equipment; (ii) availability of personnel with special expertise; (iii) reagent cost; (iv) work load; (v) versatility; and (vi) needs for specificity, precision, speed, and sensitivity. Work load does not mean simply the total number of tests. One should also consider whether the work occurs in a steady stream or in spurts and whether work can be batched. The need for specificity will be influenced by the degree of control a laboratory has over ordering and the availability of accurate patient information. For example, a referral laboratory might require greater specificity than a hospital laboratory. Pharmacologic studies require sensitive and accurate assays, but rapidity and specificity are less important. The need for specificity, however, is easily underestimated. Not only is the presence of unreported antimicrobial agents a concern (16), but microbiologically active and inactive metabolites may be present. For example, the metabolites of sulfonamides and chloramphenicol are inactive; the metabolites of cefotaxime, cephalothin, cephapirin, and rifampin have only a portion of the activity of the parent drugs; and degradation of carbenicillin to benzylpenicillin gives a product with a different spectrum of activity. Accuracy is less important in clinical assays than in pharmacologic studies. Reeves and Wise (18) concluded that for clinical assays of drugs such as aminoglycosides, a 95% confidence limit of ±25 to 30% (a coefficient of variation of about 10 to 15%) was adequate. For drugs with a low potential for toxicity, a 95% confidence limit of ±50% was considered adequate. These limits are within the capabilities of most assay procedures. Speed is often emphasized for clinical assays. What is actually needed in most cases is a result within the time limit of a dosing interval.

Bioassay

Bioassays are based on a comparison of the response of a susceptible organism to an unknown concentration of antibiotic with the response of the same organism, under identical test conditions, to a known concentration of antibiotic. In essence, a bioassay is the converse of a susceptibility test of a standard organism, and the response is usually measured by either broth dilution or agar diffusion techniques. To avoid the inherent inaccuracies of a broth dilution method with serial twofold dilutions, various continuous measures of growth in broth have been used. These methods include turbidimetric assays, potentiometric (15) and titrimetric assays (3), radiometric assays (8), and assays based on measurement of intracellular bacterial ATP concentration (14). Comprehensive reviews of the theory and practice of bioassays have been prepared by Grove and Randall (7), Kavanaugh (10, 11), and Reeves and associates (17).

Because clinical specimens often contain more than one antimicrobial agent, an indicator organism that is multiply resistant should be used whenever practical to improve specificity. Specificity can also be improved by antagonism or inactivation of other antimicrobial agents present or by separation of the antimicrobial agents in a mixture before assay. The most common of these methods are the addition of a β-lactamase to inactivate penicillins and cephalosporins, addition of 50 mg of *p*-aminobenzoic acid and 5 mg of thymidine per liter to antagonize the activity of sulfonamides and trimethoprim, and addition of calcium salts or cation-exchange resin (24) to antagonize or remove aminoglycoside activity. Before a bioassay is used to measure antimicrobial agents in mixtures, the specific mixtures of interest should be tested to rule out the possibility of synergistic or antagonistic activity. Antineoplastic agents may have antimicrobial activity and should not be neglected as a potential source of interference in bioassays.

Chemical assay

Classical chemical methods are seldom used today for therapeutic monitoring. The principal problems are lack of specificity and lack of versatility of individual procedures. For example, the Bratton-Marshall procedure for sulfonamides is not specific. Similarly, the colorimetric procedure for chloramphenicol, which involves reduction of the nitro group followed by the Bratton-Marshall reaction, responds to several metabolites as well as the prodrug, chloramphenicol succinate. Although relatively specific chemical methods are available for selected antimicrobial agents, e.g., some of the penicillins and cephalosporins, these methods are probably not practical for most clinical laboratories, because these drugs are measured too infrequently.

Enzymatic assay

Enzymatic assays have been developed for several antimicrobial agents and can now be classified into two types. In one type, the drug being measured is the substrate for an enzyme which is capable of modifying the drug to render it microbiologically inactive. Such enzymes are isolated from resistant bacteria, and enzymes that have found clinical use either acetylate or adenylylate the antimicrobial agent (21, 22). In the usual procedure, the enzyme cofactor is labeled with a radioisotope that is transferred to the antimicrobial agent. The reaction components are adjusted so that the antimicrobial agent is the only limiting factor. After reaction, the modified antimicrobial agent is separated from unreacted cofactor and quantitated by its radioactivity. A modified acetylation procedure for aminoglycosides avoids the use of radioisotopes. In this procedure, the amount of coenzyme A released from acetyl coenzyme A is measured spectrophotometrically (25). In the other type of enzymatic assay, the antimicrobial agent functions to inhibit enzymatic activity, which is measured by some indicator reaction. This approach has been used to measure β-lactam antibiotics, with the inhibited enzyme being either DD-carboxypeptidase or β-lactamase. In the β-lactamase reaction, a chromogenic cephalosporin is also present (20). Because the β-lactam antibiotic inhibits or competes for the β-lactamase, the rate of color formation is inversely proportional to the concentration of antibiotic. The DD-carboxypeptidase-based method uses a polypeptide substrate with D-alanyl-D-alanine as the carboxy terminus. Cleavage of D-alanine is detected by using D-amino acid oxidase coupled with peroxidase and a chromogenic substrate (5).

Enzymatic assays offer a unique degree of specificity, which is best illustrated by the assays for aminoglycosides and β-lactam antibiotics. In these assays, high specificity is attained for the particular class of antibiotics (e.g., any β-lactam antibiotic) that will function as the enzyme substrate; however, members within a class may not be distinguished.

Immunoassay

Immunoassays have been developed for several antimicrobial agents, and they have been used most successfully and widely for the aminoglycosides. Immunoassays do have limitations, and although they are useful for aminoglycosides and vancomycin, they have not been successfully applied to β-lactam antibiotics or chloramphenicol. Immunoassays for antimicrobial agents are based upon competition between the drug in the specimen with a labeled form of the drug for a limited number of antibody-binding sites. Because the antibody and the labeled drug are present in fixed amounts, a standard curve can be developed that relates the proportion of labeled drug that is displaced from antibody to the concentration of drug in the sample. Assays are termed heterogeneous when there is an actual physical separation of antibody-bound drug from free labeled drug. Radioimmunoassay (12) and the Stratus (American Dade) enzyme immunoassay are examples of heterogeneous immunoassays. An assay is termed homogeneous when there is not a physical separation of antibody-bound from free labeled drug (19). In these assays, the proportion of labeled drug that is bound to antibody is determined by measuring some property of the label (e.g., fluorescence intensity, fluorescence polarization, or enzymatic activity) that changes as the result of antibody binding and can be measured in the presence of free label. Examples of homogeneous immunoassays are EMIT (Syva Co.), TDx (Abbott Diagnostics Div.), and TDA (Miles Laboratories, Inc.). Nephelometric immunoassays (e.g., ICS, Beckman Instruments, Inc.) and agglutination inhibition immunoassays (e.g., MacroVue, Becton Dickinson and Co.) share some of the characteristics of both homogeneous and heterogeneous immunoassays.

Immunoassays offer good precision, and the coefficients of variation for national proficiency surveys are generally 8 to 15%. Such high precision may be related to conformity of methods because of kits. Immunoassay kits are available for many of the aminoglycosides (gentamicin, tobramycin, amikacin, netilmicin, kanamycin, sisomicin, and streptomycin) and for vancomycin. Some of the assays for gentamicin can be extended to netilmicin by using reagents for gentamicin with netilmicin standards. The principal advantages of immunoassays are the availability of kits and automated or semiautomated instruments. The reagents are expensive, and the methods are limited in scope. The manufacturers of some automated instruments have addressed the problem of cost by greatly reducing the frequency with which standards must be measured. With these instruments, a standard curve is determined and used for subsequent analyses as long as control results are within acceptable limits.

Chromatographic assay

Chromatographic assays have been developed for almost all antimicrobial agents and offer a versatility that immunoassays and enzymatic assays lack (6, 26). Chromatographic assays can be highly specific and more economical when applied to small work loads of only a few specimens each day. Fully automated systems are not available, but some advances have been made. Most useful of the advances are automatic injectors and data systems. Sample preparation is still largely manual, and vendors have not been active in the production of standardized procedures and kits. With a largely do-it-yourself procedure like chromatography, one problem, which is shared by bioassays and enzymatic assays, is the wide variety of methods in use. The precision and accuracy of chromatographic methods, as reflected by national surveys, appear slightly poorer than those of immunoassays. This weakness probably results more from a lack of standardization than from an inherent property of chromatography. In considering chromatography, the laboratory must be prepared to do developmental work. There are no set criteria by which to evaluate various published methods, and the laboratorian must be acutely aware that many published methods, although adequately documented for some purposes, are often poorly documented for routine clinical specimens. Specific applications of liquid chromatography to antibiotic assays are described in several reviews, and selected, specific procedures have been described by Gerson and Anhalt (6).

Theory. Liquid chromatography is basically a method for the separation of complex mixtures. Separation occurs as a result of differences in the interactions of the molecules of each component in the sample with a mobile phase and a stationary phase. The mobile phase is usually a simple mixture of methanol or acetonitrile with water or an aqueous buffer. The stationary phase is typically a C_8 or C_{18} hydrocarbon chemically bonded to a microparticulate silica support. The support particles average 5 to 10 μm in diameter and are densely packed into a stainless steel column with an internal bore of 2 to 5 mm and a length of 10 to 25 cm. The combination of a polar mobile phase with a relatively nonpolar stationary phase is termed reversed-phase chromatography. The mobile phase is forced through the column at a flow rate of 1 to 3 ml/min, which requires a pressure in excess of 1,000 lb/in^2 (6.9 MPa). A solution to be analyzed (20 to 100 μl) is introduced into the mobile phase from an injector valve at one end of the column. The mobile phase carries the sample components through the column and into a detector. The effluent from the detector is usually directed to waste, but a device for collecting the separated components can be used. Components of a mixture that are retarded by the column to different extents will be separated and detected at different times. The retention time of a particular compound is characteristic and highly reproducible, and it forms the basis for chromatographic identification of compounds in a mixture. Empirically, the more nonpolar a compound, the longer it will be retained by a reversed-phase column. Conversely, when the mobile phase is made less polar, as by the addition of methanol or acetonitrile, the retention of a compound will be reduced. By manipulation of the mobile-phase composition, one is often able to adjust retention times so the compounds of interest are completely separated from other substances in the sample and elute in a relatively short time (e.g., 4 to 15 min). Once established, the composition of the mobile phase for a particular analysis rarely needs to be changed significantly. (On occasion, one may have to add a little more or less methanol or acetonitrile to adjust retention times after replacing a column.)

Quantitation. Quantitation is achieved by analysis of the separated moieties of a mixture as they emerge from the chromatography column. This analysis can take many forms, including bioassay and immunoassay; however, in clinical laboratories, quantitation is usually based on the response of a spectrophotometer. By an appropriate selection of wavelength, sensitivity for the compounds of interest can be optimized while response to potential interferences is minimized. The height or area of the chromatographic peak can be related directly to the concentration of drug in the sample if the dilution and response factors are known. Because these factors may vary, the most accurate results are usually obtained by the addition of an internal standard to the specimen before analysis (23). The internal standard should be chemically similar to the analyte so that changes affecting dilution or response factors will affect each compound to the same degree. By knowing only the amount of standard added and the relative responses of the standard and the analyte from calibration standards, the analyte can be quantitated.

Sample preparation. Accurate results are obtained only when the analyte and the internal standard are completely separated from potential interferences. The adjustment of chromatographic conditions and the use of more selective detectors, such as fluorescence detectors, can decrease interferences. These approaches affect the entire chromatographic system and may decrease its utility for other drugs. For example, the use of an ion-pairing reagent in the mobile phase to avoid an interference with one assay may render the system unsuitable for analysis of another drug without first changing the column. Alternatively, various methods of sample preparation can be used to remove interferences. Because these procedures are done separately from the chromatography, they do not affect the utility of the whole system.

Sample preparation for many drugs consists simply of precipitating protein or filtering the specimen. These methods primarily prevent the analytical column from becoming plugged by particulate matter or by proteins that would precipitate when mixed with the mobile phase. Potential interferences are not removed.

Solvent extraction usually involves dilution of the sample with a buffer and addition of an internal standard, which is often contained in the dilution buffer to minimize pipetting. An organic solvent such as ethyl acetate, diethyl ether, or methylene chloride is then mixed with diluted sample, and the phases are allowed to separate. The organic phase, which now contains the drug of interest and the internal standard, is either evaporated to a residue that is dissolved in a suitable solvent or back extracted under different conditions into an aqueous solution. For example, a weak acid can often be extracted into ethyl acetate from an acidic solution (pH $\leq$ [pK_a of the analyte $-$ 1.5]) and then back extracted into an alkaline buffer. In either case, the solvent eventually used for the sample must be compatible with the mobile phase and the detector. The conditions for solvent extraction can be manipulated to remove many interferences from a sample. Because retention time in reversed-phase chromatography is largely controlled by nonpolar differences between compounds, sample preparation based on acid-base reactions complements rather than simply extends the analytical separation. The disadvantages of solvent extraction are that more time is required, an internal standard is essential, and some of the solvents pose health risks unless handled with appropriate care.

Selective adsorption and elution from small chromatography columns before analytical chromatography is a third method of sample preparation that can be used when solvent extraction cannot. For example, the aminoglycosides are too polar to be extracted into an organic solvent, but they can be extracted from serum by adsorption to an ion-exchange resin (1a) or to silica gel (13) followed by elution. The use of extraction involving a different mode of separation than the analytical column (i.e., ion-exchange or adsorption versus reversed phase) may be the best way to remove a difficult interference. However, the technique is not as widely applied and suffers from some of the same disadvantages as solvent extraction. Both methods of extraction can be automated, as illustrated by the Technicon FAST-LC for solvent extraction and the DuPont Prep I for column extraction.

Summary. The advantages of liquid chromatography over alternative biologic, immunologic, and chemical methods for antibiotic assays are high specificity and the capability for an analysis of several drugs with only minor modification, if any, in methodology. The principal disadvantages are the lengthy procedures sometimes needed for sample preparation and the fact that samples are processed sequentially through the time-consuming chromatography step. Immunoassay is the method of choice for aminoglycosides because of the large volume of tests. For small numbers of tests and when metabolites may interfere with other methodologies, liquid chromatography is the method of choice. Further development is needed in assays of β-lactam drugs to provide methods applicable in a clinical laboratory.

REPRESENTATIVE DISK-PLATE BIOASSAY FOR PENICILLIN

Several variations of the disk-plate bioassay exist, but all are characterized by the use of absorbent paper disks to contain specimens and standards. The disks are placed on the surface of agar containing a dispersion of an indicator organism. Antibiotic diffuses from the disks and inhibits growth of the organism. After a suitable incubation period, zones of inhibition around each disk are measured. The inhibition zones produced by known concentrations are plotted against the concentrations to form a standard curve. The drug concentration in the sample corresponding to the zone of inhibition produced is determined from the standard curve.

Preparation of assay plates

Several assay plates may be prepared at a time and stored at 4°C for up to 7 days before use. Each plate consists of a flat-bottomed petri dish, 100 by 15 mm (Falcon Optilux), into which a base layer and a layer of agar containing inoculum have been cast. Antibiotic medium no. 5 (Difco Laboratories), with 10% $CaCl_2$ per liter added and the pH adjusted to 7.0, is used for both layers. The base layer is prepared by adding 9 ml of agar to each dish. The dishes are tilted back and forth to distribute the agar evenly and then placed on a level surface while the agar solidifies. The inoculum is prepared by adding 2 ml of an overnight culture of *Micrococcus luteus* (ATCC 9341) incubated at 30°C in Trypticase soy broth (BBL Microbiology Systems) to a bottle containing 100 ml of melted agar maintained at 50°C in a water bath. (The temperature of the agar should be measured with an alcohol-flamed thermometer to ensure that it does not exceed 50°C.) The inoculum should be carefully mixed by inverting the bottle several times and will contain about 10^7 CFU/ml. The base layer is then overlaid with 4 ml of the seeded agar. The plates are left on a level surface to harden and then stored at 4°C until used.

Assay procedure

Prepare several dilutions of specimen so that the concentration of antimicrobial agent in at least one dilution will be approximately 0.3 μg of penicillin per ml or, for other antimicrobial agents, within the range of the standard curve. Specimen dilutions of 1:4, 1:10, 1:20, and 1:40 will usually suffice. For trough levels, an undiluted specimen is tested, and for peak levels, a 1:100 dilution may be necessary. Serum specimens are diluted in sterile human serum that has been shown to lack both antimicrobial and antimicrobial-inactivating activity and to be free of hepatitis B surface antigen. Urine and cerebrospinal fluid are diluted in 0.1 M phosphate buffer, pH 7.0. Standards are prepared in either buffer or serum, as appropriate. Usually, five concentrations containing 0.1, 0.2, 0.3, 0.4, and 0.5 μg of penicillin per ml are prepared. For other antimicrobial agents, the concentrations are adjusted to give appropriate zones of inhibition. Spread 30 blank paper disks (6.35-mm diameter; Schleicher & Schuell, Inc.) in a single layer in a petri dish. Pipette 20 μl of the appropriate standards (triplicate sets plus three additional sets at 0.3 μg of penicillin per ml) and test fluid dilutions (triplicate sets) onto separate disks. Apply each disk to the surface of the seeded agar plates, and gently tap the disk with forceps to ensure complete contact with the agar. Do not pipette fluid onto more than six disks before transferring them to the assay plate. All manipulations should be done by sterile technique with sterile equipment. On each of three plates, place disks with each concentration of standard. On each of three additional plates, place disks with each specimen dilution and a disk with the reference concentration of standard (0.3 μg/ml for penicillin). This reference concentration should be chosen as approximately the midpoint of the standard curve. The disks should be evenly spaced and about 2 cm from the center of the plate. Incubate the plates, without stacking, at 30°C for 18 h.

Interpretation and calculations

Measure the diameters of the zones of inhibition for each calibration concentration, reference concentration, and specimen dilution. Calculate the means of the zone diameters for each calibration concentration on the three plates with standards only. Construct a standard curve by plotting the mean zone diameter on the abscissa against logarithms of the concentration on the ordinate. Calculate the means of the zone diameters of the reference concentration and of each specimen dilution. In the mean for the reference concentration, do not include zone diameters from the plates containing standards. Correct the mean zone diameter of each specimen dilution by adding algebraically the difference between the mean zone of the reference concentration on the plates containing specimen and the zone corresponding to the reference concentration as determined from the standard curve. Because the best line for the standard curve usually does not go through each point, the zone corresponding to the reference concentration will be slightly different from the mean zone diameter of the reference concentration. Locate the corrected zone diameter for each specimen dilution on the standard curve, and read the corresponding concentration. Multiply this concentration by the dilution factor to determine the antibiotic concentration in the original specimen. Use the result from the specimen dilution that gives a zone diameter nearest the reference zone diameter. Results from other dilutions may be compared with this result to check for consistency and to protect against gross errors, but exact calculations do not need to be made with them. When no dilution falls within the standard curve, the test must be repeated with different specimen dilutions. The standard curve should never be extrapolated to accommodate zones larger or smaller than the largest or smallest zone observed.

Specifications for the bioassay of other antimicrobial agents by this procedure are published elsewhere (1).

LITERATURE CITED

1. **Anhalt, J. P.** 1985. Antimicrobial assays. *In* J. A. Washington II (ed.), Laboratory procedures in clinical microbiology, 2nd ed. Sprinter-Verlag, New York.

1a. **Anhalt, J. P., and S. D. Brown.** 1978. High-performance liquid chromatographic assay of aminoglycoside antibiotics in serum. Clin. Chem. **24:**1940–1947.

2. **Bennett, W. M., R. S. Muther, R. A. Parker, P. Feig, G. Morrison, T. A. Golper, and I. Singer.** 1980. Drug therapy in renal failure: dosing guidelines for adults. I. Antimicrobial agents, analgesics. Ann. Intern. Med. **93:**62–89.

3. **Bourne, P. R., I. Phillips, and S. E. Smith.** 1974. Modification of the urease method for gentamicin assays. J. Clin. Pathol. **27:**168–169.

4. **Broughall, J. M., M. J. Bywater, H. A. Holt, and D. S. Reeves.** 1979. Stabilization of cephalosporins in serum and plasma. J. Antimicrob. Chemother. **5:**471–472.

5. **Frère, J.-M., D. Klein, and J.-M. Ghuysen.** 1980. Enzymatic method for rapid and sensitive determination of β-lactam antibiotics. Antimicrob. Agents Chemother. **18:**506–510.

6. **Gerson, B., and J. P. Anhalt.** 1980. High-pressure liquid chromatography and therapeutic drug monitoring. American Society of Clinical Pathologists, Chicago.

7. **Grove, D. C., and W. A. Randall.** 1955. Assay methods of antibiotics: a laboratory manual (antibiotics monograph 2). Medical Encyclopedia, New York.

8. **Gunn, B. A., S. L. Brown, C. S. Otey, C. A. Gaydos, J. F. Keiser, F. A. Meeks, and R. G. Trahan.** 1980. Serum gentamicin assay by a radiometric procedure. Am. J. Clin. Pathol. **73:**259–262.

9. **Hermans, P. E., J. P. Anhalt, and J. A. Washington II.** 1984. Pocket guide to antimicrobial agents 1984. Centrum, Philadelphia.

10. **Kavanagh, F. (ed.).** 1963. Analytical microbiology. Academic Press, Inc., New York.

11. **Kavanagh, F. (ed.).** 1972. Analytical microbiology, vol. 2. Academic Press, Inc., New York.

12. **Lewis, J. E., J. C. Nelson, and H. A. Elder.** 1972. Radioimmunoassay of an antibiotic: gentamicin. Nature (London) **239:**214–216.

13. **Maitra, S. K., T. T. Yoshikawa, J. L. Hansen, I. Nilsson-Ehle, W. J. Palin, M. C. Schulz, and L. B. Guze.** 1977. Serum gentamicin assay by high-performance liquid chromatography. Clin. Chem. **23:**2275–2278.

14. **Nilsson, L., H. Höjer, S. Ansehn, and A. A. Thore.** 1977. A rapid semiautomated bioassay of gentamicin based on luciferase assay of bacterial adenosine triphosphate. Scand. J. Infect. Dis. **9:**232–236.

15. **Noone, P., J. R. Pattison, and D. Samson.** 1971. Simple, rapid method for assay of aminoglycoside antibiotics. Lancet **ii:**16–19.

16. **Reeves, D. S., and H. A. Holt.** 1979. Resolution of antibiotic mixtures in serum samples by high-voltage electrophoresis. J. Clin. Pathol. **28:**435–442.

17. **Reeves, D. S., I. Phillips, J. D. Williams, and R. Wise (ed.).** 1978. Laboratory methods in antimicrobial chemotherapy. Churchill Livingstone, Ltd., Edinburgh.

18. **Reeves, D. S., and R. Wise.** 1978. Antibiotic assays in clinical microbiology, p. 137–143. *In* D. S. Reeves, I. Phillips, J. D. Williams, and R. Wise, (ed.), Laboratory methods in antimicrobial chemotherapy. Churchill Livingstone, Ltd., Edinburgh.

19. **Rubenstein, K. E., R. S. Schneider, and E. F. Ullman.** 1972. Homogeneous enzyme immunoassay—a new immunochemical technique. Biochem. Biophys. Res. Commun. **47:**846–851.

20. **Schindler, P., and G. Huber.** 1980. Use of PADAC, a novel chromogenic β-lactamase substrate, for the detection of β-lactamase-producing organisms and assay of β-lactamase inhibitors/inactivators. *In* U. Brodbeck (ed.), Enzyme inhibitors. Verlag Chemie, Weinheim.

21. **Shaw, W. V., J. Carter, and J. Sachs.** 1972. Enzymatic assay of gentamicin and kanamycin in body fluids. J. Clin. Res. **20:**83.

22. **Smith, D. H., B. Van Otto, and A. L. Smith.** 1972. A rapid chemical assay for gentamicin. N. Engl. J. Med. **286:**583–586.

23. **Snyder, L. R., and Sj. van der Wal.** 1981. Precision of assays based on liquid chromatography with prior solvent extraction of the sample. Anal. Chem. **53:**877–884.

24. **Stevens, P., and L. S. Young.** 1977. Simple method for elimination of aminoglycosides from serum to permit bioassay of other antimicrobial agents. Antimicrob. Agents Chemother. **12:**286–287.

25. **Williams, J. W., J. S. Langer, and D. B. Northrop.** 1975. A spectrophotometric assay for gentamicin. J. Antibiot. **28:**982–987.

26. **Yoshikawa, T. T., S. K. Maitra, M. C. Schotz, and L. B. Guze.** 1980. High-pressure liquid chromatography for quantitation of antibacterial agents. Rev. Infect. Dis. **2:**169–181.

Chapter 107

Automated Procedures for Antimicrobial Susceptibility Tests

CLYDE THORNSBERRY

Automation in the clinical microbiology laboratory has not developed as rapidly or to the same extent as it has in clinical chemistry and hematology laboratories. However, in recent years, much greater effort has been expended by microbiologists and industry to develop automated or mechanized procedures that identify bacteria and test for antimicrobial susceptibility.

Several novel methods, such as radiometry (7), microcalorimetry (2), bioluminescence (10), and electrical impedance (6), have been considered for automation. Some of these approaches have looked promising, but systems using these methods have not yet been studied extensively by collaborative groups or submitted to the Food and Drug Administration (FDA) for approval. All the procedures that have been fully evaluated or approved are based on optical determinations of bacterial growth by light transmission or light-scattering photometry.

Automated or mechanized systems for antimicrobial susceptibility testing yield either a category result (i.e., susceptible, intermediate, or resistant) or an MIC. In some cases the same systems can yield either result, depending on the selection by the user of the appropriate test unit (cards, cuvettes, etc.). In one system, the MS-2 (Abbott Diagnostics, Irving, Tex.), a computer program determines whether the susceptible or resistant category or an intermediate range of MIC results is reported.

Some systems have automated or mechanized traditional susceptibility testing methods (22). In most of these, some part (e.g., dispensing or inoculating) of the broth microdilution method has been automated, and the reading and recording phases of both the broth microdilution and the agar dilution methods have been automated. Some systems were designed to give readings after only a few hours of incubation and involved more complex automation.

The first automated short-incubation system that was extensively studied was the TAAS system (Technicon Instruments Corp., Tarrytown, N.Y.) (12). In this system, growth (number of bacteria) after 3 h in the presence of one concentration of an antimicrobial agent was compared with growth after 3 h in a control broth that contained no drug. Category judgments (susceptible or resistant) were made on the basis of an index calculated by comparing the number of bacteria in the presence of the drug with the number of bacteria in the control. The TAAS system was totally automated except for inoculum preparation and contained its own computer system. Although this system is not available, some of the systems that were subsequently developed were based on similar principles.

Among the systems that were developed after the TAAS, the Autobac I and Autobac MIC systems (General Diagnostics, Morris Plains, N.J.) and the MS-2 system and AutoMicrobic System (Vitek Systems, Inc., St. Louis, Mo.) have been extensively evaluated by collaborative groups of investigators (13, 17, 21, 22) and have been approved by the FDA.

Most of these systems also contain a data management component which, for the most part, provides epidemiological data and summaries. They also contain components for identifying gram-negative and gram-positive microorganisms.

In this chapter I extend the descriptions given in Chapters 100, 101, and 102 of systems that have been used to automate or mechanize classical methods and that may not require FDA approval, as well as of automated systems that have been approved by the FDA.

AUTOMATION IN AGAR AND BROTH DILUTION TESTS

Agar dilution

The Steers replicating device (19) (Chapters 100 and 103) has been used for more than 20 years to mechanize the inoculation of the agar plates used in agar dilution tests. With this device, plates can be rapidly inoculated with as many as 36 cultures. In the Repliscan System (Diagnostic Equipment, Inc., St. Paul, Minn.), identification and susceptibility plates can be inoculated in the same manner, and after overnight incubation, results can be semiautomatically read and recorded (3). The susceptibility results are based on growth or lack of growth in a single concentration of antimicrobial agent. Organisms are identified by the data base and computer programs of the system itself. The identification and susceptibility data can be viewed on a screen and are printed on a report form.

An agar dilution system (Mastascan) containing automated components has been developed in the Mast Laboratories in the United Kingdom. This system has been used in Europe but has not been approved for use in the United States.

Broth dilution

The availability and use of frozen or lyophilized preprepared microdilution plates is considered in Chapter 101. In some commercially available systems (MicroScan, API, Sensititre, and Dynatech), an automatic photometric device is available to read the results of microdilution tests (8, 24).

For those who wish to prepare their own plates, automated dispensers are available that allow simultaneous dispensing of predetermined volumes of broth and antimicrobial agent into the cups of microdilution plates from bulk dilutions of the drugs. The concentrations used are, therefore, not restricted to twofold dilutions, as was the case with the earlier mechanized dilution systems. Three automated dispensers available commercially are the MIC-2000 system (Dynatech Laboratories, Inc., Alexandria, Va.), the Anderson-Pasco system (Pasco, Inc., Wheatridge, Colo.), and the Quick Spense II system (Bellco Glass, Inc., Vineland, N.J.). The MIC-2000 and Anderson-Pasco systems also provide automated or mechanized inoculators; the Quick Spense inoculator is not automated but does inoculate an entire plate at one time.

AUTOBAC

The Autobac I system is an FDA-approved system in which susceptibility tests yield category results (susceptible, intermediate, or resistant) (23). These results are based on comparisons of the amount of growth after ca. 3 h of incubation in the presence of the antimicrobial agent with the amount of growth after the same time in the absence of the antimicrobial agent. The amount of growth is determined by forward light scattering, and calculations are made with a built-in computer. The category of susceptibility is based on a light scatter index, which is determined by comparison of the light scatter in the presence of antimicrobial agent with light scatter in the growth control (no antimicrobial agent).

The Autobac I system was extensively evaluated in a collaborative investigation by seven laboratories located in various areas of the country (23). The results obtained by Autobac I, by the standardized agar disk diffusion test (1, 16), and by the agar dilution test (Chapter 100) were compared. The overall agreement of the three tests was similar and exceeded 90%. Most of the discrepancies occurred with certain drug-organism combinations: *Enterobacter* spp. versus ampicillin and cephalothin; enterococci versus gentamicin, kanamycin, and tetracycline; indole-positive *Proteus* spp. versus nitrofurantoin and tetracycline; *Pseudomonas aeruginosa* versus chloramphenicol, gentamicin, and kanamycin; *Serratia* spp. versus polymyxins; and *Staphylococcus epidermidis* versus penicillin and tetracycline. In some cases the cause of these discrepancies could be determined and remedied; e.g., in the collaborative study, a large number of discrepancies between nitrofurantoin and *Proteus* spp. were due to inappropriate concentrations of the drug in the elution disk (4). However, some drug-organism combinations have been contraindicated, e.g., cephalothin-enterococcus and kanamycin-*P. aeruginosa*. Other published reports of Autobac I trials in clinical laboratories have shown general agreement with disk diffusion tests, but some additional problems have been demonstrated with certain drug-organism combinations, e.g., *Escherichia coli* and ampicillin (20, 25), *Staphylococcus aureus* and erythromycin (15), and methicillin-resistant *S. aureus* and methicillin (5).

The Autobac I system and method have been modified to permit determination of MICs, and the method has been approved by the FDA for this purpose. The Autobac MIC system was first developed to use three concentrations of antimicrobial agent, each contained in a disk, as in the Autobac I test. Growth was measured by light scattering, and the MIC was reported as the lowest concentration that inhibited growth.

This method was extensively evaluated in a collaborative study in which the results were compared with those of conventional microdilution MICs (17). During this study, it was determined that a more precise MIC could be derived from two results (rather than three) by use of a statistical program in the computer of the system. MICs obtained by the statistical calculations were generally equivalent to those obtained with three concentrations and with microdilution methods.

MS-2 SYSTEM

In the MS-2 system, susceptibility results are determined by comparing the growth curve generated in the presence of one concentration of antimicrobial agent with that generated in a control broth with no antimicrobial agent (21). These curves are constructed by computer through optical density readings made at short intervals (usually 5 min). When adequate growth has occurred, the appropriate portions of the two curves are compared, and if the organism is resistant to the drug, the slope of the curve will be essentially the same as the slope of the control curve. If the organism is susceptible, the slope of the curve will be significantly different. The susceptibility result is printed out either as an MIC or as a category (susceptible or resistant). The MIC is printed only when results are within the intermediate range, and the category results are reported as susceptible and less than the lower limit of the intermediate range or resistant and greater than the upper limit of the intermediate range, e.g., susceptible, <4 μg/ml; resistant, >12 μg/ml; or MIC = 9 μg/ml.

The MS-2 system has been evaluated in a collaborative study by four laboratories in different areas of the United States (21). In addition to evaluation with current clinical isolates, a set of selected challenge organisms was also tested to ensure that the system was challenged with as many drug-organism resistance patterns as possible. For example, the challenge set contained inducible erythromycin-resistant *S. aureus* strains, which are seldom seen in this country but are common in some parts of the world. In general, there was good correlation between MS-2, disk diffusion, and microdilution results. Discrepancies were usually associated with certain drug-organism combinations: enterococci versus cephalothin, penicillin, gentamicin, and kanamycin; *Enterobacter* spp. versus β-lactam antibiotics; and *Serratia* spp. versus colistin.

AUTOMICROBIC SYSTEM

The AutoMicrobic System is an outgrowth of systems that were developed for the National Aeronautics and Space Administration. Its initial purpose was to identify microorganisms, but its functions have been expanded to include antimicrobial susceptibility tests.

The tests are performed in small plastic cards that contain small wells, or microcuvettes, which are closed except for capillaries through which a saline suspension of the test bacteria passes for rehydration and inoculation. A carousel of inoculated cards is placed in a reader-incubator, and subsequent manipulations are done automatically. An initial reading of light transmission serves as the basis for reading future changes, i.e., the amount of growth that occurs. Readings are made at intervals by a series of light-emitting diodes. Results are usually available in 3 to 6 h.

The card for susceptibility tests contains 30 wells; one is a growth control, and the others contain one or more concentrations of an antimicrobial agent. Category results, i.e., susceptible or resistant, are based on two concentrations of drugs, and MICs are based on more than two concentrations or, in some cases, statistical inference.

The system has been tested in collaborative studies and approved by the FDA. Recent reports indicate that agreement with the standard disk diffusion method generally varies from 85 to >95%, depending on the drug-organism combination. Overall, the agreement with MIC tests is less, but is still >80% for most drug-organism combinations (11, 13, 14).

FACTORS AFFECTING RESULTS FROM AUTOMATED SYSTEMS

The methods and systems we have described are major technical advances in antimicrobial susceptibility testing, but they are not without problems. Many of the factors that affect the results of other susceptibility methods—type and pH of media, inoculum, time and temperature of incubation, and end-point determination (9)—also can affect automated susceptibility tests. The two factors that appear to cause the greatest problems in these systems are the time of incubation and the inoculum. Most of the automated systems, in contrast to the microdilution systems, have been developed as rapid tests, with tests and results being completed the same day and growth time being as little as 3 h. In some cases, factors that affect an overnight result have not been manifested at the time of the rapid read. For example, a β-lactamase-positive staphylococcal strain may not produce adequate enzyme during a short incubation period to inactivate the penicillin and thus would appear susceptible, although after longer incubation it would inactivate the penicillin and be correctly interpreted as resistant. This problem can be partially resolved by manipulation of the drug content or modification of the interpretive breakpoint. Short incubation times may also contribute to problems of detecting methicillin-resistant staphylococci (5). Unfortunately, the literature tends to vary as to the efficacy of these systems for testing these organisms, so the extent of the problem is difficult to ascertain. Similar problems exist for drug-organism combinations in which the organism has high mutation rates.

There may also be a more general disagreement between rapid-read results and overnight results. Lampe et al. (15) found that 28% of the tests showed fourfold or greater increases in MICs after 18 h of incubation compared with 3-h readings. The correlation improved considerably when the inoculum for the 3-h test was 10^7 CFU/ml and the inoculum for the overnight test was 10^5 CFU/ml. They concluded that inoculum manipulation should be considered as a way to improve the correlation between rapid and overnight tests.

In the collaborative investigation of the Autobac I (23), discrepancies were described as very major if the Autobac I system indicated that the organism was susceptible and the reference method indicated resistance, major if the reverse discrepancy occurred, and minor if an intermediate result was obtained in one system and not the other. Obviously, the very major discrepancies are the most serious and should occur in only a very low percentage of cases. An arbitrary rule of thumb has been that complete correlation should be over 90% and that the combined major and very major discrepancies attributable to the new system should be less than 5%. Sherris and Ryan have recommended that the overall percentage of errors attributable to the new system should not exceed 5% and that very major errors should not exceed 1.5% (18). Sometimes these levels of accuracy have not been achievable because of variability or inaccuracy in the reference method, but combinations with values less than 90% should always be examined carefully to determine whether the procedure could be changed to alleviate the problem or whether the drug-organism combination should be excluded. Again, in a comparison study, no system can be better than the reference method, since with any discrepancy the reference method is assumed to be correct.

A correlation of 90% or better is not sufficient if the population tested did not contain an adequate number of strains that might show a disproportionate number of discrepancies. This situation is likely to occur if only sequential clinical isolates are used in the evaluation. Therefore, any new system should be tested with a set of selected strains to ensure that the system is challenged with all clinically relevant organism-antimicrobial agent combinations (21). Special attention should be paid to strains with resistance that has been difficult to detect by other methods in other studies.

Some problem drug-organism combinations have not had therapeutic significance, e.g., the kanamycin-*P. aeruginosa* problem in the Autobac I system. Tests with combinations such as these can be contraindicated, and this has been done for certain combinations in both the Autobac I and MS-2 systems. With some combinations it may be necessary to develop override systems in which the automated system is supplemented by another procedure known to yield the correct result.

QUALITY CONTROL

Quality control of automated susceptibility tests is usually difficult because testing reference strains allows little useful information in the form of on-scale quantitative results. Reference strains are generally useful only to monitor gross changes that lead to changes in categories of susceptibility. The antimicrobial content of some elution disks can be monitored by disk diffusion (Chapter 102).

CONSIDERATIONS FOR USERS

If an automated system is being considered for routine clinical use, the prospective user should determine: (i) whether the system has been tested against an approved standard method in a collaborative study with an adequate sample of clinical isolates and a set of challenge strains containing selected resistance patterns and whether the results of the study are available (preferably in a refereed publication); (ii) whether problematic drug-organism combinations are excluded from the system to be tested by another method; (iii) whether the system is reproducible; (iv) whether the system is mechanically reliable; (v) whether the system is flexible enough to be modified for newer developments or other uses; (vi) whether the number of tests that can be performed compensate for the cost of the instrument and the cost per test; (vii) whether laboratory reports can be obtained within the normal working day; and (viii) whether the system is approved by the FDA or whether it needs approval.

Users should also read publications in which the systems are compared not only with standard methods, but also with each other (11, 13, 14). Prospective users should discuss a system with other users, if possible. Furthermore, users should do their own comparative studies before adopting a new system. They should determine how the results compare with those for the current method, how the system fits into the work schedule, whether it saves labor or requires more than does the current method, and whether it is cost efficient. (See reference 18 for further discussion of this subject.) If the results do not compare favorably with those of the method already in use, the problem may be technical or due to an unusually large number of isolates that are difficult to test accurately by the automated system.

Other automated or mechanized systems are usually being evaluated or are in early phases of development, and other systems could be approved by the FDA (including some developed outside the United States) before another edition of this Manual is published. Most of the systems now being used and those in development will combine antimicrobial susceptibility testing with identification, and the prospective user will have to determine the suitability of both the identification system and the susceptibility testing system. It is likely, however, that the manufacturers of some of these systems will offer the opportunity to buy either the entire system or only one of the methods.

LITERATURE CITED

1. **Bauer, A. W., W. M. M. Kirby, J. C. Sherris, and M. Turck.** 1966. Antibiotic susceptibility by a standardized single disk method. Am. J. Clin. Pathol. **45:**493–496.
2. **Binford, J. S., L. F. Binford, and P. Adler.** 1973. A semi-automated microcalorimetric method of antibiotic sensitivity testing. Am. J. Clin. Pathol. **59:**86–94.
3. **Brown, S. D., and J. A. Washington II.** 1978. Evaluation of the Repliscan system for identification of *Enterobacteriaceae*. J. Clin. Microbiol. **8:**695–699.
4. **Butler, D. A., and T. L. Gavan.** 1977. Autobac 1: a rapid semi-automated antimicrobial susceptibility test system, p. 53–77. *In* A. Bondi, J. T. Bartola, and J. E. Prier (ed.), The clinical laboratory as an aid in chemotherapy of infectious disease. University Park Press, Baltimore.
5. **Cleary, T. J., and D. Maurer.** 1978. Methicillin-resistant *Staphylococcus aureus* susceptibility testing by an automated system, Autobac I. Antimicrob. Agents Chemother. **13:**837–841.
6. **Colvin, H. J., and J. C. Sherris.** 1977. Electrical impedance measurements in the reading and monitoring of broth dilution susceptibility tests. Antimicrob. Agents Chemother. **12:**61–66.
7. **DeBlanc, H. J., Jr., P. Charache, and H. N. Wagner, Jr.** 1972. Automatic radiometric measurement of antibiotic effect on bacterial growth. Antimicrob. Agents Chemother. **2:**360–366.
8. **DeGirolami, P. C., K. A. Eichelberger, L. C. Salfity, and M. F. Rizzo.** 1983. Evaluation of the AutoSCAN-3, a device for reading microdilution trays. J. Clin. Microbiol. **18:**1292–1295
9. **Ericsson, H. M., and J. C. Sherris.** 1971. Antibiotic sensitivity testing. Report of an international collaborative study. Acta Pathol. Microbiol. Scand. Sect. B Suppl. 217.
10. **Gutekunst, R. R.** 1977. Rapid procedures under development and evaluation: bioluminescence and impedance measurement, p. 85–100. *In* A. Bondi, J. T. Bartola, and J. E. Prier (ed.), The clinical laboratory as an aid in chemotherapy of infectious disease. University Park Press, Baltimore.
11. **Hansen, S. L., and P. K. Freedy.** 1983. Concurrent comparability of automated systems and commercially prepared microdilution trays for susceptibility testing. J. Clin. Microbiol. **17:**878–886.
12. **Isenberg, H. D., A. Reichler, and D. Wiseman.** 1971. Prototype of a fully automated device for determination of bacterial antibiotic susceptibility in the clinical laboratory. Appl. Microbiol. **22:**980–986.
13. **Johnson, J. E., J. H. Jorgensen, S. A. Crawford, J. S. Redding, and R. C. Pruneda.** 1983. Comparison of two automated instrument systems for rapid susceptibility testing of gram-negative bacilli. J. Clin. Microbiol. **18:**1301–1309.
14. **Kelly, M. T., J. M. Latimer, and L. C. Balfour.** 1982. Comparison of three automated systems for antimicrobial susceptibility testing of gram-negative bacilli. J. Clin. Microbiol. **15:**902–905.
15. **Lampe, M. F., C. L. Aitken, P. G. Dennis, P. S. Forsythe, K. E. Patrick, F. D. Schoenknecht, and J. C. Sherris.** 1975. Relationship of early readings of minimal inhibitory concentrations to the results of overnight tests. Antimicrob. Agents Chemother. **8:**429–433.
16. **National Committee for Clinical Laboratory Standards.** 1983. M2-T3. Performance standards for antimicrobial disk susceptibility tests. National Committee for Clinical Laboratory Standards, Villanova, Pa.
17. **Schoenknecht, F. D., J. A. Washington II, T. L. Gavan, and C. Thornsberry.** 1980. Rapid determination of minimum inhibitory concentrations of antimicrobial agents by the Autobac method: a collaborative study. Antimicrob. Agents Chemother. **17:**824–833.
18. **Sherris, J. C., and K. J. Ryan.** 1982. Evaluation of automated and rapid methods, p. 1–5. *In* R. C. Tilton (ed.) Rapid methods and automation in microbiology. American Society for Microbiology, Washington, D.C.
19. **Steers, E., F. Foltz, B. S. Graves, and J. Riden.** 1959. An inocula replicating apparatus for routine testing of bacterial susceptibility to antibiotics. Antibiot. Chemother. **9:**307–311.
20. **Stubbs, K. G., and K. Wicher.** 1977. Laboratory evaluation of an automated antimicrobial susceptibility system. Am. J. Clin. Pathol. **68:**769–777.
21. **Thornsberry, C., J. P. Anhalt, J. A. Washington II, L. R. McCarthy, F. D. Schoenknecht, J. C. Sherris, and H. J. Spencer.** 1980. Clinical laboratory evaluation of the Abbott MS-2 automated antimicrobial susceptibility testing system: report of a collaborative study. J. Clin. Microbiol. **12:**375–390.
22. **Thornsberry, C., and C. N. Baker.** 1981. Antimicrobial susceptibility tests for bacteria, p. 747–766. *In* A. Balows and W. J. Hausler (ed.), Bacterial, mycotic, and parasitic infections, 6th ed. American Public Health Association, Washington, D.C.
23. **Thornsberry, C., T. L. Gavan, J. C. Sherris, A. Balows, J. M. Matsen, L. D. Sabath, F. Schoenknecht, L. D. Thrupp, and J. A. Washington II.** 1975. Laboratory evaluation of a rapid, automated susceptibility testing system: report of a collaborative study. Antimicrob. Agents Chemother. **7:**466–480.
24. **Turner, A., P. M. Hawkey, and S. J. Pedler.** 1983. Computer-assisted determination of antibiotic susceptibility in photometer-read microtiter plates. J. Clin. Microbiol. **18:**996–998.
25. **Waterworth, P. M.** 1976. Automated sensitivity tests. J. Antimicrob. Chemother. **2:**104–106.

Appendix 1

Preparation and Storage of Antimicrobial Solutions

JOHN P. ANHALT AND JOHN A. WASHINGTON II

Standard or reference preparations of antimicrobial agents should be obtained directly from the manufacturers, from any laboratory supply house which makes available powders specifically for susceptibility tests, or from the U.S. Pharmacopeia Convention, Inc. (12601 Twinbrook Parkway, Rockville, MD 20852). Clinical preparations should not be used because they are less precisely standardized and because some develop full activity only after hydrolysis to the active substance in vivo (e.g., chloramphenicol sodium succinate). Containers of powder should bear a label stating activity, expressed in micrograms or in international units per milligram, and an expiration date.

Store the powders according to the directions of the manufacturer. As a class, the aminoglycosides are hygroscopic and should be stored at room temperature in a desiccator. Storage of other antimicrobial agents at room temperature is preferred whenever stability permits. Many antimicrobial agents require refrigerated storage in a desiccator or in sealed ampoules. Be sure to allow the desiccator and contents to equilibrate to room temperature before the desiccator is opened to avoid condensation of water. After the desiccator is opened, the air inside is humid and should be removed by evacuation or allowed to dry over the desiccant at room temperature before the desiccator is replaced in the refrigerator. Routine storage of antimicrobial agents at −20°C is inadvisable and increases the risk of water condensation on the powders. All antimicrobial agents should be protected from direct sunlight. Some are very sensitive to light (e.g., rifampin and amphotericin B) and must be stored in the dark.

Powders are generally used as received, and special drying procedures are not necessary for the clinical laboratory. Published procedures for drying antimicrobial agents are available (1–3). The recommended

TABLE 1. Solvents and diluents for stock solutions of antimicrobial agents[a]

Antimicrobial agent	Solvent	Diluent
Amoxicillin[b]	Phosphate buffer, 0.1 M, pH 8	Phosphate buffer, 0.1 M, pH 6; water
Amphotericin B	Dimethylformamide	Water
Ampicillin[b]	Phosphate buffer, 0.1 M, pH 8	Phosphate buffer, 0.1 M, pH 6; water
Ceftazidime[c]	Na_2CO_3 in water	Water
Cephalothin and other cephalosporins[d]	Phosphate buffer, 0.1 M, pH 6	Water
Chloramphenicol	Methanol	Water
Erythromycin[e]	Methanol	Phosphate buffer, 0.1 M, pH 8
Imidazoles	Dimethylformamide	Water
Moxalactam[f]	HCl, 0.04 or 0.08 M	Water
Nalidixic acid[g]	NaOH, 0.1 M	Water
Oxolinic acid[g]	NaOH, 0.1 M	Water
Rifampin	Methanol	Water
Sulfonamides[g]	NaOH, 0.1 M	Water
Trimethoprim[h]	HCl, 0.05 M	Water

[a] Water should be used as both solvent and diluent for the following antimicrobial agents: aminoglycosides (amikacin, gentamicin, kanamycin, neomycin, netilmicin, streptomycin, and tobramycin), azlocillin, bacitracin, carbenicillin, clindamycin, cloxacillin, colistin (polymyxin E), cycloserine, dicloxacillin, doxycycline, ethambutol, flucytosine, isoniazid, lincomycin, methicillin, metronidazole, mezlocillin, minocycline, nafcillin, nitrofurantoin (sodium salt), oxacillin, *p*-aminosalicylic acid, penicillin G, piperacillin, polymyxin B or E, spectinomycin, tetracycline, ticarcillin, trimethoprim lactate, and vancomycin.

[b] Dilute with water after first 1:10 dilution with phosphate buffer at pH 6.0.

[c] The weight of Na_2CO_3 should be 10% of the weight of the ceftazidime pentahydrate used.

[d] Except as noted, phosphate buffer at pH 6 is used to dissolve cephalosporins. For many cephalosporins (cefamandole, cefazolin, cefotaxime, ceftizoxime, cefuroxime, and cephalexin), water may be used as the solvent. Should cephradine fail to dissolve in water, add a small amount (e.g., 20 μl/ml) of 5 to 10% (wt/vol) $NaHCO_3$ solution.

[e] Alcohol solutions of erythromycin are unstable because of ester formation and should be diluted immediately. Solutions in acetone are more stable and can be stored. Water can also be used as the diluent.

[f] To prepare a solution containing 1,000 μg/ml, dissolve the diammonium salt at 10,000 μg/ml in 0.04 M HCl. Let the solution stand for 1.5 to 2 h at room temperature to allow R and S isomers to equilibrate. Dilute the solution with 9 volumes of 0.1 M phosphate buffer at pH 6. To prepare a solution containing 10,000 μg/ml, dissolve the diammonium salt at 20,000 μg/ml in 0.08 M HCl, and let the solution stand for 1.5 to 2 h. Dilute with 0.1 M phosphate buffer at pH 8 to 10,000 μg/ml. Swirl the solution while adding the buffer to avoid exposing the drug to local, high concentrations of alkaline buffer.

[g] Use 1.0 ml of 0.1 M NaOH per 10 mg of antimicrobial agent to dissolve this agent.

[h] Use 1.0 ml of 0.05 M HCl per 10 mg of trimethoprim to dissolve this agent. Use water to dissolve trimethoprim lactate.

procedure for aminoglycosides suggests heating at 60°C for 3 h at a pressure not in excess of 5 mm of Hg (0.67 kPa).

Powders are weighed on an analytical balance and dissolved to yield the required concentration based on labeled activity or potency. The following formulas may be used to determine the amount of powder or diluent necessary for a stock solution:

$$\text{weight (mg)} = \frac{\text{volume (ml)} \times \text{concentration } (\mu\text{g/ml})}{\text{potency } (\mu\text{g/mg})}$$

or

$$\text{volume (ml)} = \frac{\text{weight (mg)} \times \text{potency } (\mu\text{g/mg})}{\text{concentration } (\mu\text{g/ml})}$$

In most cases, it is advisable to weigh a portion of antimicrobial agent roughly in excess of that needed and to calculate the volume of solvent required. This procedure avoids multiple adjustments of the powder during weighing and minimizes time and exposure to atmospheric moisture. The alternative of weighing a predetermined amount of powder is advisable when an organic solvent is used to dissolve the antimicrobial agent, and the solution is diluted to a fixed volume in a volumetric flask.

Solvents and diluents for common antimicrobial agents are listed in Table 1. These recommendations are based on an effort to use a minimal number of solvents or buffers and may not give optimal stability. Solvents are used to dissolve dry powder to give stock solutions containing high concentrations of antimicrobial agent (i.e., greater than 1,000 μg/ml). Diluents are used to prepare working dilutions from stock solutions. In general, ethanol can be substituted for methanol, dimethyl sulfoxide can be substituted for dimethylformamide, and phosphate buffer (0.1 M, pH 6) can be used instead of water to dissolve β-lactam antibiotics to improve stability (water should be used as the diluent).

Stock solutions containing high concentrations of antimicrobial agents do not usually need to be sterilized; however, sterile water or buffers should be used in their preparation. Should it be necessary to sterilize a solution of antimicrobial agent, membrane filtration should be used to minimize loss due to adsorption. Rifampin and amphotericin B solution should not be filtered. Antimicrobial agents can also adsorb to the walls of glass vessels, and polypropylene or polyethylene vials and laboratory ware are recommended for the preparation and storage of solutions containing low concentrations. Adsorption to surfaces is decreased in the presence of protein or broth media.

The stability of antimicrobial solutions varies greatly, and data regarding maximum storage times at −20 or −60°C are generally lacking. Stock solutions of most antimicrobial agents will remain stable for at least 6 months at −20°C and longer at −60°C in concentrations of 1,000 μg/ml or greater. Ampicillin and amoxicillin are unstable, particularly in concentrations greater than 1,000 μg/ml, and should be stored for no longer than 6 weeks at −20°C or 6 months at −60°C. Cephaloglycin, cefaclor, and rifampin are even less stable and should be freshly prepared with each use. Antimicrobial agents are often more unstable in serum than in buffer. The β-lactam antibiotics should not be stored in serum even in the frozen state.

LITERATURE CITED

1. **Code of Federal Regulations.** 1979. Title 21, parts 300 to 499. U.S. Government Printing Office, Washington, D.C.
2. **Kavanagh, F. (ed.).** 1963. Analytical microbiology. Academic Press, Inc., New York.
3. **Kavanagh, F. (ed.).** 1972. Analytical microbiology, vol. 2. Academic Press, Inc., New York.

Appendix 2

Approximate Concentration of Antimicrobial Agents Achieved in Blood

CLYDE THORNSBERRY AND L. D. SABATH

The concentrations of antimicrobial agents listed below are approximations taken from various reported studies. Several factors can influence the level of antimicrobial agent achieved in individual patients, including inherent differences in the patients themselves, their physical condition, the dosages, and the routes of administration. The values can also be influenced by the assay methods used to obtain them. These concentrations should therefore be used only as approximate values, and clinicians should use their knowledge of the patient and the drugs, the recommendations from FDA-approved inserts, or other reputable sources in planning therapeutic regimens.

Antimicrobial agent[a]	Unit dose	Avg peak blood level[b] (μg/ml)		
		PO	IM	IV
Amdinocillin	700 mg		26	28
Amikacin	7.5 mg/kg		30	30
Amphotericin B	0.37 mg/kg			1–4
Amoxicillin	500 mg	8		
Ampicillin	500 mg	4		17
Augmentin	(See Amoxicillin)			
Azlocillin	2 g			250
Aztreonam	1 g		65–85	150
Carbenicillin	4 g			150
Cefaclor	500 mg	12		
Cefadroxil	500 mg	16		
Cefamandole	0.5–1 g		28	60
Cefazolin	1–2 g		64	121
Cefmenoxime	1 g			42
Cefonicid	1 g		83	
Cefoperazone	1–2 g		65	100–200
Ceforanide	1–2 g		72	136
Cefotaxime	1 g		18–23	47–70
Cefoxitin	1 g		35	70
Cefsulodin	1 g		20	
Ceftazidime	1 g		20	80
Ceftizoxime	1 g		30	60
Ceftriaxone	1 g		60	150
Cefuroxime	1 g		40	43–98
Cephalexin	500 mg	18–25		
Cephalothin	1 g			70
Cephapirin	1 g			37
Cephradine	0.5–1 g	12–25	10	86
Chloramphenicol	1 g	8–14		8–14
Cinoxacin	250 mg	8		
Clindamycin	150–600 mg	2.5	5	12
Cloxacillin	500 mg	2.5	3–15	
Cycloserine	0.5–1 g	15–25		
Dicloxacillin	250–500 mg	4	12–20	
Doxycycline	100 mg	3		
Erythromycin	500 mg	0.5–1.7		8
Flucytosine	2.5 g	78		
Gentamicin	1.3 mg/kg		5–7	5–7
Isoniazid	10 mg/kg	1–5		
Kanamycin	500 mg		15	15–20

Continued

Continued

Antimicrobial agent[a]	Unit dose	Avg peak blood level[b] (μg/ml)		
		PO	IM	IV
Ketoconazole	200 mg	3.5		
Methicillin	500 mg			16
Metronidazole	500 mg	11		21
Mezlocillin	2 g			166–255
Miconazole	200–400 mg			1–4
Minocycline	200 mg	3		
Moxalactam (latamoxef)	1 g		39–59	76
Nafcillin	0.5–1 g	5		10
Netilmicin	1.3 mg/kg		8	
Nitrofurantoin	300 mg	1.8		
Oxacillin	500 mg	5	10	
Benzathine penicillin G	1.2×10^6 U		0.1	
Procaine penicillin G	0.6×10^6 U		1	
Penicillin G	500 mg			16
Penicillin V	500 mg	4		
Piperacillin	2 g			200
Rifampin	8 mg/kg	10		
Spectinomycin	2 g		30–50	
Streptomycin	1 g		25–50	
Sulfamethoxazole	1 g	92		
Sulfisoxazole	2 g	110		
Tetracycline	250 mg	3		
Ticarcillin	3 g			250
Tobramycin	1–2 mg/kg		4–8	4–8
Trimethoprim	100 mg	1		
Trimethoprim/sulfamethoxazole	80/400–160/800 mg	1/20		3.4/46
Vancomycin	0.5–1 g			20–40

[a] Not all of these agents were approved for use at the time this table was prepared.
[b] PO, Oral; IM, intramuscular; IV, intravenous.

Chapter 108

Section XII. Molecular Methods

DNA Methods in Clinical Microbiology

LUCY S. TOMPKINS

By applying new molecular technology, it is now possible to analyze both chromosomal and plasmid DNA in the clinical laboratory. While the procedures of DNA cloning and labeling cannot be carried out in diagnostic laboratories yet, other techniques have been simplified to the extent that they can be used to solve clinical problems. The analysis of plasmid DNA, in particular, is an extremely useful epidemiological tool. In addition, plasmid and chromosomal genetic sequences have been cloned and can now be employed to detect microorganisms or identify them to species levels.

Like the chromosome, plasmids are double-stranded, covalently closed circular molecules of DNA which are found in the cytoplasm of procaryotes. The organization of plasmid DNA is also like that of the bacterial chromosome, with genes which ensure replication, maintenance, and distribution to daughter cells during division of the host cell. Some plasmids, usually those of lower molecular weight, can be found in multiple copies in a single bacterial cell, while larger plasmids usually have restricted replication and are found in fewer copies. Nonessential, plasmid-encoded genes which may also be present are those which promote conjugation and transfer of plasmid DNA to a recipient cell. Larger plasmids are more likely to contain conjugation determinants than smaller ones. Small plasmids may be transferred, or mobilized, by a conjugative plasmid, however. For example, the plasmids encoding for beta-lactamase production in *Neisseria gonorrhoeae* range from 3.2 to 4.4 megadaltons and are nonconjugative, but when present in cells containing a large (26-megadalton) conjugative plasmid, transfer of penicillin resistance to susceptible recipient cells can occur (18).

Plasmid DNA can also mediate physiologic properties which tend to confound some laboratory identification schemes. The fermentation of lactose by *Salmonella* isolates, urease production in *Providencia stuartii*, and utilization of citrate and production of H_2S and urease by some strains of *Escherichia coli* are examples.

Properties which determine pathogenicity have also been located on plasmids; these properties include the enterotoxins of *E. coli*, *E. coli* attachment proteins, and determinants of invasiveness in *E. coli* and *Shigella* spp. Residence of such genes on plasmids has made cloning a somewhat simpler task; indeed, the first diagnostic molecular probes were devised to detect enterotoxigenic *E. coli* (15).

Plasmids containing sequences which encode for antibiotic resistance are known as R factors or R plasmids. Recently, it has been discovered that plasmid-mediated antibiotic resistance is frequently carried on transposons which can insert into and excise from replicons by a process of "illegitimate" recombination (6). Transposons, designated by the letters Tn, can mediate resistance to virtually all classes of antibiotics. Therefore, many R plasmids are made up of essential genes needed for maintenance, replication, etc., in addition to one or more resistance transposons. These so-called "jumping genes" can move between plasmids or between plasmids and chromosomes and may be major contributors to the rapid evolution of resistance in many clinical isolates.

Resistance plasmids are distributed through nearly all genera of medically important bacteria, with the notable exceptions of *Neisseria meningitidis* and *Streptococcus pneumoniae*, and can be transferred via conjugation even between members of different species or genera. While the phenomenon of plasmid epidemics was first noted in multiresistant gram-negative bacilli (19, 22), gram-positive bacteria, including staphylococci, also owe much of their resistance to plasmids. Whereas gram-negative enteric bacteria may contain a single, large plasmid encoding for resistance to several antibiotics, staphylococci often have resistance determinants distributed on several small plasmids. Instability of these small plasmids apparently accounts for the variable antibiograms observed in isolates derived from a single colony or from cultures obtained at different intervals from a single patient (P. A. Mickelsen, J. J. Plorde, J. McClure, F. D. Schoenknecht, F. C. Tenover, K. Gordon, and L. S. Tompkins, Abstr. Annu. Meet. Am. Soc. Microbiol. 1984, L22, p. 311).

The majority of plasmids found in clinical isolates are deemed "cryptic" because their gene products have not yet been determined. The distribution of cryptic plasmids in nearly all genera of clinically important bacteria serves as a means of marking strains of the same species. The technique of plasmid fingerprinting exploits this property of many bacterial species of carrying "excess baggage" and has become a powerful tool for investigating epidemics.

The principle of plasmid fingerprinting is that isolates which are of the same strain contain the same number of plasmids with the same molecular weights, and tend to have similar phenotypes, whereas those isolates which are phenotypically distinct also have distinct plasmid fingerprints. Plasmid fingerprinting of a large number of *Enterobacteriaceae* and *Pseudomonas* species has demonstrated that this method is often more accurate than other phenotyping methods, including biotyping, antibiotic resistance patterns, phage typing, and serotyping (9, 22). This method of strain differentiation was first applied to gram-negative enteric bacilli causing nosocomial infections (20), but it has now been employed in solving community-acquired epidemics, including those due to *Salmonella* (9, 20) and *Campylobacter* (2) spp.

Staphylococci, including *Staphylococcus aureus* and *S. epidermidis*, may also be typed in this fashion (16; Mickelsen et al., Abstr. Annu. Meet. Am. Soc. Microbiol. 1984, L22, p. 311). The method is especially helpful in differentiating coagulase-negative species, since phage typing is not widely available or even applicable to many strains. Strains of *Acinetobacter* and *Citrobacter*, and other species for which other phenotyping schemes have not been developed, may also be satisfactorily differentiated by this method. Although in some investigations more than one phenotyping method must ultimately be employed, plasmid fingerprinting can be performed more rapidly and on a routine basis if necessary.

The technique of plasmid fingerprinting involves relatively few steps, as noted below: strains are grown in broth or on agar plates; the cells are lysed by exposure to detergent; the plasmid DNA is separated from the chromosomal DNA and other cellular components in an extraction procedure which takes advantage of the covalently closed supercoiled structure of plasmid DNA; and finally the DNA is applied to agarose gels and electrophoretically separated. The gel is then stained with ethidium bromide, which intercalates into DNA (and RNA), causing it to fluoresce under UV light. Since the rate of migration of plasmid DNA in agarose is inversely proportional to molecular weight, different-sized plasmids appear as distinct bands in the stained gel. Small plasmids migrate more rapidly through the agarose matrix than large molecules. During the extraction procedure, the chromosome is reduced to linear fragments and is seen as a more diffuse band which migrates at a position equivalent to a 12-megadalton plasmid. The molecular weight of each plasmid species can then be extrapolated from a curve obtained by plotting the distance migrated from the origin versus the logarithms of the molecular weights of plasmids of known size which have been electrophoresed simultaneously in the same gel (Fig. 1) (12).

Because the rate of migration of a plasmid is inversely proportional to its frictional coefficient, which depends on shape as well as size, a variety of conformations of one plasmid can be observed. The original covalently closed, supercoiled molecules, which migrate farthest from the cathode, can be converted to nicked, open circles and linear molecules during the extraction procedure. The supercoiled form migrates more rapidly than the relaxed circle, but the relationship that all three species have to one another depends on the electrical current, the buffer system, and the gel pore size, as well as on the size of the DNA molecules. Thus, the appearance of multiple bands may be erroneously interpreted as showing additional plasmids (12).

Since the agarose gel electrophoretic mobility of two plasmids which are not actually homologous may be the same, it may be necessary to examine the sequence of nucleotide bases to determine whether plasmids of equivalent molecular weight are genetically identical. Such an analysis can be performed by digesting plasmids with restriction endonuclease enzymes which cleave double-stranded DNA at specific restriction sites determined by the location of short nucleotide sequences. Restriction sites generally have twofold rotational (dyad) symmetry, meaning that the nucleotide sequence of one DNA strand is the same as the sequence of the complementary strand when each strand is read in the 5′-to-3′ direction. As an example, the enzyme *Eco*RI recognizes the sequence GAATTC (reading from 5′ to 3′) on one strand and contains the sequence CTTAAG (reading from 3′ to 5′) on the complementary strand. The enzyme cleaves each strand between the G and A to give overlapping, cohesive termini with a protruding 5′ phosphate residue. Each molecularly distinct plasmid (or chromosome) contains a unique number of restriction endonuclease sites which are placed in distinct locations, and therefore plasmids which are cleaved in identical loci are considered to be homologous. However, changes involving a few base pair substitutions at other sites may be present and will not necessarily be detected by this method. The number and distribution of restriction sites of any DNA molecule can be determined by electrophoresing the digestion fragments in agarose. The electrophoretic mobility of fragments is inversely related to length so that each linear frag-

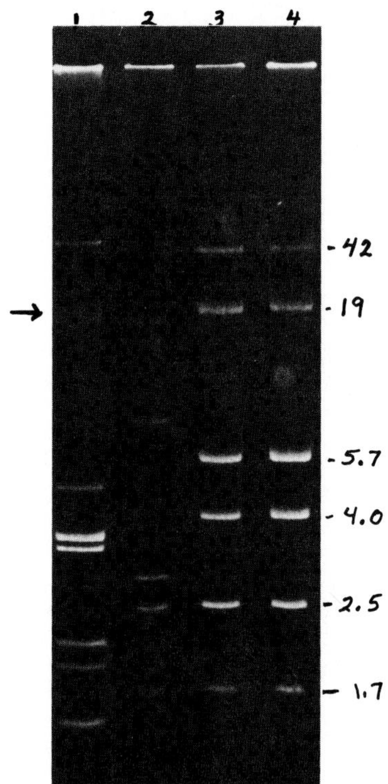

FIG. 1. Polaroid photograph of agarose gel electrophoresis of *S. epidermidis* plasmids. Lane 1 contains plasmids of known molecular weight. Lane 2 contains five plasmids from a strain of *S. epidermidis* which was nontypable by phage typing. The faint band at the position of the arrow is composed of residual chromosomal DNA. Lanes 3 and 4 contain two isolates of the same phage type, derived from two different blood cultures from the same patient. Each contains six plasmids, estimated to be 42, 19, 5.7, 4.0, 2.5, and 1.7 megadaltons, respectively. Five of the plasmids are of different molecular weights than those from the isolate in lane 2, indicating that the isolates in lanes 3 and 4 are the same, but distinct from the isolate in lane 2.

ment appears as a separate band in the gel, providing another kind of molecular fingerprint. Each different endonuclease recognizes a different nucleotide sequence and is active under special conditions of incubation temperature, buffer composition, and salt concentration.

Restriction endonuclease fingerprinting can also be applied to the chromosomal DNA of bacteria as well as to plasmids. Chromosomal digestion fingerprints have been used to differentiate strains of some bacteria, including *Vibrio cholerae* (10) and *Campylobacter jejuni* (3), although this method has not yet been widely applied or confirmed in other species. Restriction endonuclease fingerprinting has also been applied to viral DNA and is now the basis for specifically typing herpes simplex viruses (4).

A new development in clinical microbiology is the application of DNA as a diagnostic reagent for the detection or identification of microorganisms. The use of cloned DNA as a probe is based upon the tendency for single-stranded DNA to anneal with a complementary strand, forming a double-stranded DNA hybrid. Thus, a single-stranded sequence derived from one microorganism (the probe) is used to search for others containing the same gene. The hybridization reaction may be applied to purified DNA preparations, to colonies, and even to clinical materials, including tissues, serum, and pus, and is usually carried out on a solid matrix such as a nitrocellulose filter. The DNA in the sample is rendered single stranded by treatment with NaOH and is then exposed to labeled probe DNA which has been heat treated to denature it to single strands. Double-stranded duplexes which form, composed of labeled and unlabeled DNA, remain bound to the nitrocellulose after unreacted single-stranded DNA is removed by washing the filter. At present, probe DNA is most often radiolabeled by reacting it with a mixture of ^{32}P-labeled nucleotides in the presence of DNA polymerase I and DNase. The result of this enzymatic process, termed nick translation, is the formation of double-stranded sequences into which labeled nucleotides have been substituted for "cold" residues, forming radiolabeled double-stranded DNA (17).

The inherent specificity of DNA is the basis for using it as a diagnostic reagent. If one can select and clone a sequence which is found in only one strain, species, or genus, then this DNA should hybridize only with the DNA extracted from specific microorganisms and not others. This feature of specificity has been exploited in the development of probes which can detect enterotoxigenic *E. coli* strains and differentiate them from commensals (13, 15). In this instance, probe DNA consists of portions of cloned sequences encoding for three *E. coli* enterotoxins, i.e., LT, ST I, and ST II. Only microorganisms containing one or more of these genes will react with these probes. Because of the extensive homology between the *E. coli* enterotoxin and *V. cholerae* enterotoxin (14), the cloned LT *E. coli* sequence can be employed to probe *V. cholerae* isolates as well as *E. coli*. This is an example in which cloned virulence genes have been used to detect pathogenic strains or determine toxigenicity. Theoretically, any sequence of DNA specific for a particular group can serve as a probe. For example, the cryptic plasmid found in most strains of *N. gonorrhoeae* specifically hybridizes only with gonococci, not with other species of *Neisseria* or other bacteria, and can detect *N. gonorrhoeae* in urethral pus samples (23). Currently, DNA probes have been developed to detect several viruses (5, 7), parasites, and many bacteria. Although not commercially available yet, these probes will undoubtedly become important diagnostic reagents soon.

While the technique of colony hybridization, first developed by Grunstein and Hogness (8), is relatively straightforward and can be performed in laboratories having access to probe DNA (see below), the process of developing molecular probes is too complex to be carried out in diagnostic laboratories. However, cloning methods and the procedures by which probes are derived should be understood by clinical microbiologists.

The initial step in constructing a probe is to isolate the specific gene and remove it. This is achieved by digesting DNA with restriction endonucleases, thus reducing the plasmid or chromosome containing the unique gene to linear fragments with "sticky" ends, one of which will contain the gene of interest. To amplify this sequence, the fragment containing it must be inserted into a replicon, such as a plasmid, a virus, or a cosmid (a new type of genetically engineered plasmid), which serves as a vector to carry the foreign insert DNA into a bacterial cell which will produce many copies of the DNA probe. The vector plasmid not only serves to carry the inserted DNA but also contains a selectable marker, usually resistance to an antibiotic such as tetracycline. To insert the foreign DNA into the cloning vector, both DNA molecules must be digested with the same endonuclease, creating complementary single-stranded, overlapping ends at each restriction site. The DNA molecules are then mixed together in the presence of DNA ligase, an enzyme which forms covalent bonds between inserted DNA and vector DNA. Newly formed, covalently closed circular recombinant plasmids are then introduced into *E. coli* cells by transformation. Cells which have been transformed are selected by plating onto medium containing antibiotics to which the transformants are resistant. The probe sequence has now been cloned.

Often the most difficult step in probe development is the last one, that of identifying an appropriate gene or gene fragment which will be specific. One strategy is to select a gene encoding for virulence which is present in virulent strains but found in no other strains or species. A second approach is to find a sequence which does not encode for virulence properties per se which is found in all isolates of one particular species. The cryptic plasmid of *N. gonorrhoeae* is such an example. A third method of developing a species-specific probe would be to hybridize its DNA with DNA from organisms with which it shares significant homology; this would "absorb out" all homologous sequences, producing nonhomologous, unique DNA.

Once the appropriate fragment has been cloned it must be removed from the cloning vector by digestion with the same endonuclease which was used to insert it and purified. The two fragments formed by digestion can be separated by electrophoresing the DNA in agarose and removing the probe band from the gel.

Lastly, the probe DNA is tagged, either with radiolabeled nucleotides or with nucleotides which have been linked to another molecule, such as biotin. Radiolabeled DNA is detected by exposing the nitrocellulose filter to X-ray film, creating an autoradiograph in which black spots signify the location of cells or material containing the gene which has been sought. Biotinylated DNA can be detected by overlaying the filter with avidin, a protein which has an extremely high affinity constant for biotin, coupled with a biotin-polymeric enzyme complex. The addition of substrate brings about a color reaction which can be detected visually.

The specificity of a probe can be ascertained by reacting it with isolates of the same species as well as with a wide variety of other microorganisms which might be expected to share some DNA homology or would be found in the same clinical samples. Sensitivity can be determined by seeding uninfected clinical material with various concentrations of the microorganism. These samples are then reacted with the probe to ensure that the selected sequence can detect a concentration of organisms which is relevant to the clinical situation.

PLASMID FINGERPRINTING

Reagents

1. Solution 1 (1), made up freshly

Lysozyme (Sigma Chemical Co.)	0.02 g
EDTA (0.5 M, pH 8.0)	0.20 ml
Tris (1 M Tris hydrochloride, pH 8.0)	0.25 ml
Glucose (20%, wt/vol)	0.45 ml
Distilled water	9.5 ml

2. Solution 2

NaOH (10 N)	0.20 ml
Distilled water	9.3 ml
Sodium dodecyl sulfate (20%, wt/vol, in water)	0.5 ml

3. Solution 3
   Sodium acetate, 3 M, pH 4.8. For 100 ml, use 0.3 mol of sodium acetate (rehydrated) and 50 ml of distilled water. Adjust the pH to 4.8 with glacial acetic acid. Bring the final volume to 100 ml with distilled water.
4. TES buffer
   30 mM Tris, pH 8
   5 mM EDTA
   50 mM NaCl
5. Stop mix

Glycerol	50 ml
EDTA (0.5 M)	7 ml
Bromphenol blue	5 mg

   Bring to 100 ml with distilled water.
6. Ethanol, 95%, −20°C
7. Agarose, 0.7% (wt/vol) in Tris borate buffer, boiled and cooled to 60°C.
8. Tris borate buffer (1×, made from a 10× stock solution)
   89 mM Tris, pH 8.2
   2.5 mM Na_2EDTA, pH 8
   89 mM boric acid
9. Ethidium bromide
   Stock solution, 10 mg/ml; final concentration, approximately 1 to 5 μg/ml in distilled water
10. L broth
11. Lysostaphin (Sigma), 1-mg/ml stock solution
12. Molecular weight standards (12)

Equipment

Vertical slab gel electrophoresis apparatus (Fig. 2) with power source

Two glass plates, one notched (Fig. 1)

Spacers and comb (Fig. 1)

Large binder clips (Fig. 1)

UV light box (UV Light Products Inc., San Gabriel, Calif.)

Polaroid MD4 Land Camera with no. 9 Wratten filter and Polaroid type 55 film

Microcentrifuge, nonrefrigerated

Micropipettes

Procedure for preparing plasmid DNA (modified from the method of Birnboim and Doly [1])

1. Grow the culture to stationary phase of cell growth in 1 ml of L broth. Transfer the culture to a microcentrifuge tube and spin it for 2.5 min in the microcentrifuge. Pour off the supernatant and remove the last drops with paper toweling.

2. Suspend the cell pellet, or 1 or 2 colonies scraped from an agar plate, in 100 μl (0.1 ml) of solution 1. Mix by inversion or by vortexing, and let stand for 30 min on ice. When staphylococci are lysed, lysostaphin must be substituted for lysozyme, in a final concentration of 130 μg/ml.

3. Add 200 μl of solution 2, gently mix by inversion only, and let stand on ice for 5 min. The solution will become clear and viscous.

4. Add 150 μl of solution 3, mix by inversion, and let stand for 60 min on ice. Sodium dodecyl sulfate and denatured chromosomal DNA will precipitate out of solution at this stage.

5. Spin tubes in the microcentrifuge for 5 min. Decant the supernatant into clean microcentrifuge tubes. Add 1.0 ml of 95% ethanol chilled to −20°C. Mix by inversion. Let stand at −20°C (or in a bath of dry ice and ethanol) for 10 min.

6. Spin the frozen tubes for 5 min in the microcentrifuge, with the tubes oriented in the centrifuge so that the pellet will be deposited in the same location in each tube. Pour off the supernatant, being careful not to dislodge the tiny amount of precipitated DNA on the side of the tube. Wash the pellet by gently adding 1 to 2 ml of 95% ethanol to each tube; add the ethanol so as not to dislodge the pellet. Invert the

FIG. 2. Typical vertical gel electrophoresis chamber, a flask containing molten agarose, and the proper arrangement of the components (glass plates, spacers, comb, and clips) needed to make the gel.

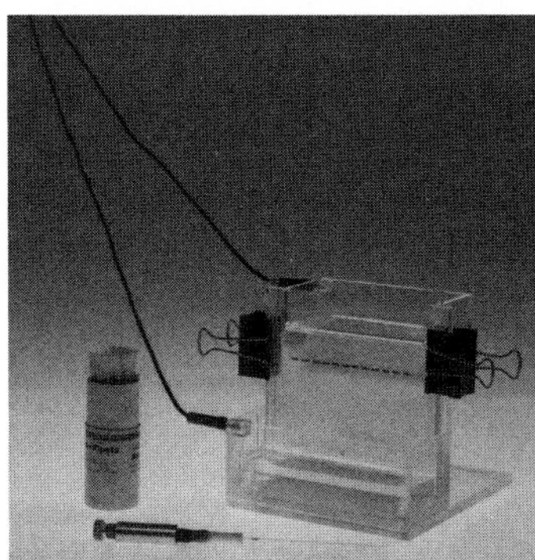

FIG. 3. Demonstration of a vertical gel apparatus with the gel in place. The wells of the gel have been loaded with samples containing DNA and stop mix containing blue dye. The micropipetter and capillary pipettes used to load the gel are also shown. Note clips securing the glass plates onto the face of the gel box. Enough buffer has been added to the upper chamber to cover the agarose gel by 1 to 2 cm.

tubes to drain immediately without allowing the precipitate to resuspend.

7. Vacuum dry the tubes to remove residual ethanol, using a dessicator under vacuum. Tubes may be air dried at room temperature for several hours if a vacuum apparatus is not available.

8. Suspend the pellet in 50 μl of TES.

9. Remove 10 μl of the DNA preparation to a fresh tube. Add 5 μl of stop mix. Apply to gel as described below.

Agarose gel electrophoresis

1. Assemble glass plates with spacers. Hold them together firmly with clips as shown in Fig. 2. Pipette 2 ml of molten agarose between the plates to seal the bottom of the gel mold. Allow the agarose to solidify at room temperature for 10 min.

2. Pour the remaining 50 ml of cooling, molten agar into the mold up to top of the glass plates. Insert the comb at an angle to avoid trapping air bubbles. Allow the gel to solidify at room temperature.

3. Remove the clips and bottom spacer, being careful not to fracture the gel. Carefully remove the comb. Remove any loose agarose bits with a cotton-tipped swab dipped in buffer.

4. Position the sponge in the bottom chamber to prevent gel from slipping out from between the glass plates. After applying petroleum jelly to the back of the notched glass plate, position the plates so that the glass plates are resting on the raised pedestals of the bottom chamber with the notched plate to the back. Secure the plates and gel with clips attached to the sides of the apparatus, as shown in Fig. 2. Fill both chambers of apparatus with Tris borate buffer.

5. Load DNA samples, including molecular weight standards, by using thin pipettes or capillary tubes.

6. Attach electrodes so that the positive pole (cathode, black) is at the top and the negative pole (anode, red) is at the bottom of the vertical apparatus (Fig. 3). Run vertical slabs at 100 V (25 to 30 mA) for 120 min. The extent of migration can be followed by observing the tracking dye in the stop mix. Turn off the power when the dye reaches the bottom of the gel.

7. Transfer the gel to a staining tray containing 200 ml of water mixed with ethidium bromide. Wear gloves at all times when handling gels stained with this potent mutagen.

8. After staining for 5 min, transfer the gel to a second tray of water for 5 min of destaining. The gel is now ready to be viewed and photographed.

9. With a Polaroid camera and UV transilluminator, the stained gel can be photographed through a Wratten no. 9 filter onto Polaroid type 55 film. Always wear goggles or glasses when viewing gels on the UV transilluminator to prevent corneal burns.

LITERATURE CITED

1. **Birnhoim, H. C., and J. Doly.** 1979. A rapid alkaline extraction procedure for screening recombinant plasmid DNA. Nucleic Acids. Res. **7:**1513–1523.
2. **Bradbury, W. C., M. A. Marko, J. N. Hennessy, and J. L. Penner.** 1983. Occurrence of plasmid DNA in serologically defined strains of *Campylobacter jejuni* and *Campylobacter coli*. Infect. Immun. **40:**460–463.
3. **Bradbury, W. C., A. D. Pearson, M. A. Marko, R. V. Congi, and J. L. Penner.** 1984. Investigation of a *Campylobacter jejuni* outbreak by serotyping and chromosomal restriction endonuclease analysis. J. Clin. Microbiol. **19:**342–346.
4. **Buchman, T. G., B. Roizman, G. Adams, and B. H. Stover.** 1978. Restriction endonuclease fingerprinting of *Herpes simplex* virus DNA: a novel epidemiological tool applied to a nosocomial outbreak. J. Infect. Dis. **138:**488–498.
5. **Chou, S., and T. C. Merigan.** 1983. Rapid detection and quantitation of human cytomegalovirus in urine through DNA hybridization. N. Engl. J. Med. **308:**921–925.
6. **Cohen, S. N.** 1976. Transposable genetic elements and plasmid evolution. Nature (London) **263:**731–737.
7. **Gissman, L., L. Wolnik, H. Ikenberg, U. Koldovsky, H. G. Schnurch, and H. zur Hausen.** 1983. Human papillomavirus types 6 and 11 DNA sequences in genital and laryngeal papillomas and in some cervical cancers. Proc. Natl. Acad. Sci. U.S.A. **80:**560–563.
8. **Grunstein, M., and D. S. Hogness.** 1975. Colony hybridization: a method for the isolation of cloned DNAs that contain a specific gene. Proc. Natl. Acad. Sci. U.S.A. **72:**3961–3965.
9. **Holmberg, S. D., I. K. Wachsmuth, F. W. Hickman-Brenner, and M. L. Cohen.** 1984. Comparison of plasmid profile analysis, phage typing, and antimicrobial susceptibility testing in characterizing *Salmonella typhimurium* isolates from outbreaks. J. Clin. Microbiol. **19:**100–104.
10. **Kaper, J. B., H. B. Bradford, N. C. Roberts, and S. Falkow.** 1982. Molecular epidemiology of *Vibrio cholerae* in the U.S. Gulf Coast. J. Clin. Microbiol. **16:**129–134.
11. **Langer, P. R., A. A. Waldrop, and D. C. Ward.** 1981. Enzymatic synthesis of biotin-labeled polynucleotides: novel nucleic acid affinity probes. Proc. Natl. Acad. Sci. U.S.A. **78:**6633–6637.
12. **Macrina, F. L., D. J. Kopecko, K. R. Jones, D. J. Ayers, and S. R. McCowen.** 1978. A multiple plasmid-containing *Escherichia coli* strain: convenient source of size reference plasmid molecules. Plasmid **1:**417–420.
13. **Moseley, S. L., P. Echeverria, J. Seriwatana, C. Tirapat, W. Chaicumpa, T. Sakuldaipeara, and S. Falkow.** 1982. Identification of enterotoxigenic *Escherichia coli* by colony hybridization using three enterotoxin gene probes. J. Infect. Dis. **145:**863–869.
14. **Moseley, S. L., and S. Falkow.** 1980. Nucleotide sequence homology between the heat-labile enterotoxin gene of *Escherichia coli* and *Vibrio cholerae* deoxyribonucleic acid. J. Bacteriol. **144:**444–446.

15. **Moseley, S. L., I. Huq, A. R. M. A. Alim, M. So., M. Samadpour-Motalebi, and S. Falkow.** 1980. Detection of enterotoxigenic *E. coli* by DNA colony hybridization. J. Infect. Dis. **142:**892–898.

16. **Parisi, J. T., and D. W. Hecht.** 1980. Plasmid profiles in epidemiologic studies of infections by *Staphylococcus epidermidis*. J. Infect. Dis. **141:**636–643.

17. **Rigby, P. W. J., M. Dieckmann, C. Rhodes, and P. Berg.** 1977. Labeling deoxyribonucleic acid to high specific activity in vitro by nick translation with DNA polymerase I. J. Mol. Biol. **113:**237–251.

18. **Roberts, M., L. P. Elwell, and S. Falkow.** 1977. Molecular characterization of two beta-lactamase-specifying plasmids isolated from *Neisseria gonorrhoeae*. J. Bacteriol. **131:**557–563.

19. **Sadowski, P. L., B. C. Peterson, D. N. Gerding, and P. P. Cleary.** 1979. Physical characterization of ten R plasmids obtained from an outbreak of nosocomial *Klebsiella pneumoniae* infections. Antimicrob. Agents Chemother. **15:**616–624.

20. **Schaberg, D. R., L. S. Tompkins, and S. Falkow.** 1981. Use of agarose gel electrophoresis of plasmid deoxyribonucleic acid to fingerprint gram-negative bacilli. J. Clin. Microbiol. **13:**1105–1108.

21. **Taylor, D. N., I. K. Wachsmuth, Y.-H. Shangkuan, E. V. Schmidt, T. J. Barrett, J. S. Schrader, C. S. Scherach, H. B. McGee, R. A. Feldman, and D. J. Brenner.** 1982. Salmonellosis associated with marijuana. A multi-state outbreak traced by plasmid fingerprinting. N. Engl. J. Med. **306:**1249–1253.

22. **Tompkins, L. S., J. J. Plorde, and S. Falkow.** 1980. Molecular analysis of R-factors from multiresistant nosocomial isolates. J. Infect. Dis. **141:**625–636.

23. **Totten, P. A., K. K. Holmes, H. H. Handsfield, J. S. Knapp, P. Perine, and S. Falkow.** 1983. DNA hybridization technique for the detection of *Neisseria gonorrhoeae* in men with urethritis. J. Infect. Dis. **148:**462–471.

Chapter 109

Uses of Gas-Liquid Chromatography and High-Pressure Liquid Chromatography in Clinical Microbiology

C. WAYNE MOSS

Chromatography is the science of separation. Since the introduction of chromatography in 1905, microbiologists have used chromatographic procedures for the isolation, purification, and separation of microbial components, by-products, and products from metabolism. The concept of using gas-liquid chromatography (GLC) as an analytical tool for the identification of microorganisms was advanced in the middle 1960s. The results from early studies of organic acids and other microbial metabolites, cellular fatty acids, and products from the pyrolysis of whole bacterial cells demonstrated the feasibility of this approach and provided a basis for subsequent detailed investigations of a large number of microorganisms. The reliability, precision, and accuracy of recent studies have been improved significantly by continued advancements in GLC instrumentation and columns, by the introduction of accurate and reliable columns and instruments for high-pressure liquid chromatography (HPLC), and by the use of mass spectrometry (MS), nuclear magnetic resonance (NMR) spectroscopy, and associated analytical techniques for the identification of components separated by the chromatographic processes. These combined techniques have also been used for the direct analysis of body fluids to discover specific chemical markers of the etiological agents of infection.

This chapter provides an overview of chromatographic instrumentation and techniques and a brief review of major applications in microbiology. Specific methods and procedures for the chromatographic analysis of cellular fatty acids, isoprenoid quinones, and microbial metabolites are described in detail, as these data have proved to be of major taxonomic value. For detailed information on the use of chemical data for taxonomy and the diagnosis of disease, the reader should consult the original references.

BASIC PRINCIPLES, INSTRUMENTATION, AND TECHNIQUES

All chromatography systems consist basically of two phases, the stationary and the mobile phases. In conventional column and thin-layer chromatography (TLC), the stationary phase is a solid adsorbent and the mobile phase is a liquid. In gas chromatography, the stationary phase may be a solid adsorbent (gas-solid chromatography) or an involatile liquid adsorbed on an inert support to render it stationary (GLC). In the field of separation science, GLC has had much greater application than gas-solid chromatography.

The basic components of a GLC system are the injector, column, and detector, each of which has separate temperature controls. Samples are introduced onto the column and into the mobile (carrier gas) phase through the injector, where the sample components are heated and thereby converted to the vapor state. Since all compounds analyzed by GLC must pass through the column in the vapor state, analysis is restricted to relatively low-molecular-weight compounds which are volatile or to compounds that can be converted to a derivative which can be volatilized (without destruction) upon heating in the injector port. The volatilized components of the mixture are distributed between the mobile and stationary phases, and the separation is effected by the relative affinity of the components for the liquid phase coated on the inert support. The compound with the least affinity for the stationary phase will elute first from the column, followed by other components which have progressively higher affinities. Ideally, all components of the sample mixture will emerge from the column as separate bands within a reasonable analysis time. Located at the end of the column is a detector which senses the separated components which are simultaneously recorded as a series of peaks to form the chromatogram. The time required for the component to emerge from the column is referred to as the retention time (RT). The RT is the basic parameter for the identification of separated components, as it is constant for a given set of chromatographic conditions (temperature, flow, stationary phase, etc.). Therefore, a tentative identification of the separated components is made by RT comparisons with known standards; accurate quantitative measurements are established by determining peak heights or peak areas compared with known concentrations of standards.

The detection system most widely used in GLC is the flame ionization detector (FID). The advantages of the FID are high sensitivity (10^{-12} g/s) for essentially all organic compounds, ease of operation and maintenance, and linear response over a wide range of concentrations. There are modifications of the FID which provide selectivity and increased sensitivity for sulfur- and nitrogen-bearing compounds. Electron capture detectors are extremely sensitive (10^{-15} g/s) for compounds such as halogens, which have a high affinity for capturing electrons. The electron capture detector has been used effectively by the conversion of compounds to halogenated derivatives, with subsequent GLC analysis to obtain high sensitivity and selectivity. The thermal conductivity detector is less sensitive but more versatile than the electron capture detector; its major use in microbiology has been for the analysis of short-chain acid products from anaerobic bacteria. The analytical column is the "heart" of GLC, since this is where all the separation is accomplished. Packed columns, which have been used extensively, consist of an inert supporting surface such as silica or Celite on which the liquid phase (carbowax, methyl silicone, etc.) is coated. The type of liquid stationary phase chosen is based on the affinity of the liquid for the class of compounds to be separated, its stability, and its temperature range. The recent use of open tubular and fused-silica capillary columns (0.1 to 0.3 mm inside diameter) has improved and extended GLC applications because of the increased resolution and separating efficiency of these columns, which are now available with both polar and nonpolar stationary phases.

High-pressure ("high-performance") liquid chromatography (HPLC) is an advanced form of liquid chromatography. As noted above, in liquid chromatography the mobile phase is a liquid solvent which carries the mixture to be separated over an immobilized stationary phase where the separation takes place. In HPLC, the mobile phase is pumped rapidly and at high pressure through a highly efficient microparticulate (particles, 5 to 10 μm) stationary phase contained in a relatively narrow-bore stainless steel column (0.4 by 30 cm). The separated individual compounds are detected as they elute from the column, tentatively identified by comparing their RT with known standards, and quantitated by measuring the area or heights of peaks from a strip chart recorder. The major advantages of HPLC over gravity-fed column chromatography are better resolution, shorter analysis time, and superior precision, reproducibility, and sensitivity. Since HPLC can be used in a wide variety of separation modes (reversed phase, size exclusion, ion-exchange, etc.) with various detection modes (UV absorption, fluorescence, electrochemical, postcolumn reactions), it is a powerful analytical technique for solving a variety of separation problems. Reverse-phase HPLC (RPHPLC) utilizes a nonpolar stationary phase (usually a C-8 or C-18 silane functional group chemically bonded to silica) and a polar mobile phase. This column is particularly useful for the analysis of biological samples, since aqueous samples can be introduced directly into the HPLC, and a wide range of compounds of various molecular weights and polarities can be separated by a modification of the composition of the mobile phase. A major application of HPLC in microbiology is the measurement of antibiotics in body fluids (62); other major uses are in the analysis of isoprenoid quinones (7, 42, 56) and of proteins, peptides, and polynucleotides (51).

Paper chromatography, reversed-phase thin-layer chromatography (RPTLC), conventional column chromatography, and TLC continue as useful analytical procedures. Often these techniques are used simultaneously or in combination with other chromatographic procedures and physicochemical methods; they are relatively simple, and none requires expensive instrumentation. Their major use is to obtain qualitative data by comparing R_f values of separated components with those of known standards. Qualitative measurements are limited unless a device such as a densitometer is used to determine the intensity of the separated bands or spots. Both TLC and RPTLC have been used effectively in chemotaxonomic studies of mycobacteria (32, 34).

Associated analytical procedures often used in combination with chromatographic techniques include the computer, NMR and infrared spectroscopy, and MS. The last three techniques are most often used for the identification of compounds which are detected and separated with the chromatographic processes, whereas the computer is used for the collection, storage, and management of the large amount of data generated with these procedures. Combined GLC-MS is an established technique, and efficient combined HPLC-MS procedures (i.e., thermospray) will be available in the near future. These combined techniques provide powerful analytical tools for the detection, separation, and identification of important marker compounds in microorganisms and body fluids. In some instances, the mass spectrometer has been used as a GLC detector for the identification and quantitation of compounds which are specific for a particular infectious agent (46, 52). Routine applications of these latter techniques are not practical at this time because of the expense and the knowledge and experience required to effectively utilize the procedures. However, the trend of manufacturers of analytical instruments is toward cheaper, simpler, and more versatile instruments, with special emphasis on applications in the biomedical field.

IDENTIFICATION AND TAXONOMY

Microbial metabolites

A major use of GLC in microbiology laboratories has been in the analysis of bacterial metabolites. Systematic studies have demonstrated the value of short-chain acids in the identification of anaerobic bacteria (17, 20, 29, 43, 55, 58). These compounds are easily removed from growth media by extraction with an organic solvent and are most often analyzed with an isothermal GLC system equipped with a thermal conductivity detector (20, 55, 58). A typical procedure consists of growing the culture in a prereduced medium (i.e., peptone-yeast extract-glucose) for 24 to 40 h under anaerobic conditions. After incubation, the medium is adjusted to pH 2.0 or below with H_2SO_4, and the acids are extracted by shaking with diethyl ether. The ether layer is removed, and a sample is injected into the GLC system. With this procedure, both volatile acids (formic, acetic, propionic, isobutyric, butyric, isovaleric, valeric, isocaproic, and caproic) and nonvolatile acids (lactic, pyruvic, succinic, and other dicarboxylic acids and phenylacetic and other aromatic acids) will be present in the ether layer, but only the volatile acids are detected under isothermal operating conditions. To determine nonvolatile acids, a second sample is processed by adding methanol to the broth culture and heating it overnight to convert these acids to methyl esters which are sufficiently volatile for subsequent isothermal GLC analysis (20, 55, 58). In our laboratory, short-chain acids are determined with a one-step procedure in which all the acids in the ether layer (from extraction) are converted to volatile derivatives (butyl-trifluoroacetyl) with subsequent analysis by a temperature-programmed GLC system equipped with a packed column and an FID (29, 43, 45). With this procedure, pyruvic, lactic, dicarboxylic, volatile, and aromatic acids are detected and resolved in a single chromatographic run (29, 43). Significantly improved resolution of these acids and the discovery of other acids not detected with packed columns and the thermal conductivity detector have been demonstrated recently with the fused-silica capillary column (45). These data indicate that further systematic GLC studies of the acid products from anaerobic bacteria with capillary columns and sensitive detectors (FID and electron capture detector) may provide additional valuable information for the rapid identification and classification of these organisms. Methods have also been developed for GLC determination of short-chain acids from organisms grown on agar media (10, 11, 40). *Pseudomonas* spp. were found to produce propionic, isobutyric, isovaleric, and glutaric acids after growth on Trypticase soy agar. *Achromobacter*-like organisms produced large amounts of 2-ketoisocaproic acid, and *Flavobacterium meningosepticum* produced large amounts of 2-ketoisovaleric acid. The presence and relative amounts of these acids were found to be useful for distinguishing these organisms (10, 40). Acid products from *Pseudomonas* spp. grown on chemically defined media are different from the products of organisms grown on Trypticase soy agar (47); these differences emphasize the effects of media and culture conditions on short-chain acids.

Bacterial metabolites other than short-chain acids are very amenable to analysis by GLC and other chromatographic procedures such as HPLC and headspace GLC (9, 19, 31). Relatively simple procedures have been developed for amines, alcohols, carbonyl, and sulfur-containing compounds (4, 6, 45). Amines are readily removed from growth media by adjustment of the pH to 11 or above and extraction with an organic solvent. The extracted amines are converted to volatile derivatives by reaction with an anhydride and analyzed by GLC. Both aliphatic and aromatic amines have been detected and shown to be of taxonomic value in the identification of *Proteus* spp. and *Clostridia* spp. (6, 9). Headspace GLC techniques have been used to detect acids, amines, methyl mercaptan, and dimethyl sulfide, but these studies have been limited to only a few microorganisms (9, 19, 31). Additional investigations with state-of-the-art chromatographic instrumentation, detectors, columns, and techniques are needed to further determine the value of these metabolites for the identification of microorganisms.

Cellular components: cellular fatty acids

The value of data on cellular fatty acids for the identification and classification of microorganisms is now firmly established. The current use of these data in both research and clinical laboratories is a result of detailed studies on the physiology and metabolism of lipids and fatty acids of various microorganisms and the development of practical methods and procedures for their determination. The lipids of microorganisms are found in the cell wall-cell membrane fraction, in which the unit fatty acids are chemically bonded to other cellular constituents. Most but not all lipids are removed by extraction with an organic solvent, but to determine total cellular fatty acids, the cells must be hydrolyzed with acid or base. The hydrolysis procedure is critical to the final results, as acid hydrolysis destroys cyclopropane acids, and base hydrolysis fails to liberate all the amide-linked hydroxy acids (30). In this laboratory, we have developed a simple and rapid base hydrolysis procedure which gives accurate and reproducible results with either lyophilized or fresh whole cells, as follows (30). Growth from one agar slant or an agar plate is removed by the addition of 0.5 to 1.0 ml of distilled water and a gentle scraping. The cells are transferred to a screw-capped test tube (20 by 150 mm) with a Teflon-lined cap and saponified by the addition of 2 to 4 ml of 15 NaOH in 50% aqueous methanol and heating at 100°C for 30 min. The sample is cooled to ambient temperature, 3 to 5 ml of 25% hydrochloric acid-methanol reagent is added, and the mixture is heated to 85°C and held for 15 min. The resulting fatty acid methyl esters (FAME) are extracted twice with two 5-ml aliquots of a 1:1 mixture of diethyl ether-hexane, evaporated to 0.3 ml with nitrogen, washed with 0.5 ml of 0.1 M phosphate buffer, dried with 0.5 g of anhydrous Na_2SO_4, and stored at −20°C until analysis by GLC.

The FAME can be determined by GLC on glass columns (6 or 8 ft [183 or 244 cm] by 4.0 mm [inside diameter]) packed with a nonpolar (i.e., SE-30, OV-1, or OV-101) phase followed by analysis on a polar phase (i.e., OV-17). However, much superior resolution and separating efficiency is obtained with the newer fused-silica capillary columns (28, 38). In this laboratory, FAME are now routinely analyzed on a fused-silica capillary column (50 m by 0.2 mm inside diameter) with cross-linked methyl silicone (OV-1) as the stationary phase. The column is installed in a gas chromatograph equipped with an FID and a column oven which can be temperature programmed. For analysis, the column oven is temperature programmed from 180 to 260°C at 8°C/min and maintained at 260°C for 8 min before recycling to the initial temperature of 180°C. The injector temperature is maintained at 250°C, and the detector is maintained at 280°C. Hydrogen is used as carrier gas with a flow rate of approximately 0.7 ml/min. The sample size is 0.8 to 1.0 μl with a split ratio of approximately 50:1. The separated FAME of the sample are recorded with a computing integrator or a computer which accurately measures RT and peak areas. The peaks in the chromatogram are identified by the comparison of RT with known FAME standards which are available from commercial companies. Some fatty acids reported in bacteria are not available commercially; the RT of these relatively unusual acids can be established by consulting the literature and then analyzing microorganisms which contain these acids. Most of the fatty acids known to occur in bacteria can be resolved on a 50-m OV-1 capillary column, and the accuracy of RT measurements with the newer gas chromatographs is routinely greater than 0.005 min. Thus, precise RT matches of FAME peaks from the analytical sample with those of known standards under the above GLC conditions give a very high probability of identification. The accuracy of RT measurements is increased by the hydrogenation of unsaturated acids or the acylation of hydroxy acids or both, with subsequent GLC analysis under conditions identical to those of the original sample. Thus, peaks of unsaturated acids will disappear, as these have been converted to their standard homolog (i.e., $16{:}1 \xrightarrow{H_2} 16{:}0$); the peak for the saturated homolog increases in size compared with the peak for the original sample. With acylation, hydroxy acid methyl esters are converted to a more volatile diester derivative which elutes faster from the OV-1 column than the original methyl ester.

With these combinations of GLC analyses, essentially all the fatty acids of microorganisms can be identified by RT alone. Other analytical techniques (MS and NMR and infrared spectroscopy) are useful for the confirmation and identification of any component not identified by RT measurements. The fatty acid compositions of a number of genera, species, and closely related groups of microorganisms have been determined. The specific uses and applications of these data for identification and taxonomy are discussed in detail in the original publications (2, 10, 21, 23, 28, 32, 34, 37–41, 44, 53, 60).

Cellular components: isoprenoid quinones

Isoprenoid quinones are another class of compounds which have been shown to be of value for the identification and classification of microorganisms (8, 24, 61). These compounds are widely distributed in nature and are divided into two major structural groups: naphthoquinones, which include menaquinones and demethylmenaquinone, and benzoquinones, which include ubiquinones. Menaquinones (vitamin K_2) and ubiquinones (coenzyme Q) occur most frequently in microorganisms and are located in the plasma membrane, where they function in electron transport. Menaquinones have a 1,4-naphthoquinone ring nucleus, whereas ubiquinones have a 1,4-benzoquinone ring; both compounds have polyprenyl side chains which vary in the number of isoprenoid units as well as in the degree of unsaturation. This structural variation in the polyprenyl side chain is the basis for classification.

Since isoprenoid quinones are soluble in lipid solvents, most procedures specify the direct extraction of lyophilized cells with ether, hexane, chloroform, methanol, acetone, or combinations of these solvents. In this laboratory, we found that isoprenoid quinones are also effectively extracted from wet cells, and thus we have eliminated the lyophilization step from our procedure. We use approximately 0.5 g (wet weight) of cells harvested from agar plates or a broth culture. The cells are transferred to a test tube (20 by 150 mm) with a Teflon cap, 3 ml of 1% (wt/vol) pyrogallol and 0.2 ml of 50% aqueous KOH are added, and the mixture is heated at 100°C for 10 min. The mixture is immediately cooled to room temperature under running tap water. A 1-ml volume of distilled water and 5 ml of diethyl ether-hexane (1:1, vol/vol) are added, and the mixture is shaken vigorously for 5 min on a wrist-action shaker and centrifuged at 1,200 × *g* for 10 min at room temperature. The upper organic layer is recovered, and the lower aqueous layer is extracted three additional times. The combined organic layers are evaporated to dryness under subdued light with a gentle stream of nitrogen and resuspended in 0.5 ml of methanol for subsequent chromatographic analysis.

The differentiation of menaquinones from ubiquinones is accomplished by TLC with commercial silica gel thin-layer plates which are activated by heating at 100°C for 30 min before spotting. The sample and known standards are applied to the same TLC plate, dried, and developed by ascending chromatography with a solvent system of hexane-diethyl ether (85:15, vol/vol). Development is completed in approximately 60 min, after which the plates are removed from the solvent chamber, dried briefly, and examined under 254-nm UV light. Under these conditions, menaquinones and ubiquinones develop as discrete spots with R_f values (×100) of 75 and 35, respectively. After determination of the class of isoprenoid quinone, individual components are separated by RPTLC or RPHPLC. RPTLC is done by spotting the sample and known standards (menaquinone, MK-3 to MK-10, or ubiquinones, Q-6 to Q-10) on a commercially prepared $KC_{18}F$ RPTLC plate (activated at 100°C for 30 min) and developing the plate in a solvent system of methanol-acetone (50/50, vol/vol). After approximately 35 min for development, the plate is removed from the solvent, dried, and examined under 254-nm UV light. With RPTLC, R_f values decrease linearly as a function of increasing isoprene units in the side chain. Thus, under the above conditions, the R_f values (×100) for quinone standards are as follows: MK-3, 64.7; MK-4, 58.5; MK-5, 51.7; MK-6, 45.1; MK-7, 38.4; MK-8, 32.0; MK-9, 26.3; MK-10, 21.1; Q-6, 54.9; Q-7, 38.4; Q-9, 35.5; and Q-10, 29.2. Recently, our laboratory and others have developed RPHPLC techniques for the determination of quinones (7, 42, 56). The major advantages of these RPHPLC procedures are a better resolution of individual quinones and the capability of accurate quantitative measurements. In this laboratory, samples are analyzed by HPLC on a μ-Bondapak C_{18} reverse-phase column (300 by 3.9 mm, packed with 10-μm particle size stationary phase) with a gradient liquid-phase system composed of methanol-isopropanol-water. Initially, the solvent system is 75:20:5, and this ratio is changed linearly to 35:65:0 at 25 min with a flow rate of 1 ml/min. Separated components are monitored at 248 or 275 nm with a variable-wavelength UV detector and tentatively identified by comparing RTs (or retention volumes) with known standards. Since neither RPTLC or RPHPLC techniques alone will completely resolve all known isoprenoid quinones, the accurate identification of these compounds requires the combination of chromatographic procedures and ancillary techniques such as NMR spectroscopy, MS, and UV spectrophotometry. A listing of the isoprenoid quinones of a variety of microorganisms and a discussion of their taxonomic significance have been detailed in an excellent review (8). Recent reports of major amounts of ubiquinones with more than 10 isoprene units in *Legionella* species and the presence of a novel methyl-substituted menaquinone-6 in *Campylobacter* species provide additional support to the taxonomic significance of quinones (7, 24, 39, 44).

Carbohydrates

Compared with cellular fatty acids and quinones, there have been relatively few studies to determine the taxonomic value of the carbohydrate composition of microorganisms. Reasons for these limited studies are due to one or more of the following technical considerations: (i) difficulties in the development of rapid and simple hydrolysis procedures with little or no destruction or conversion of sugars during the hydrolysis procedure; (ii) lengthy and detailed derivatization procedures before analysis by GLC; (iii) the formation of multiple derivatives from a single sugar during derivatization, leading to multiple peaks upon subsequent GLC analysis; and (iv) difficulty in the identification and quantitation of individual sugar components due to multiple peaks and peaks which often overlap with other sugars in a complex biological mixture. Despite these difficulties, data from several studies have demonstrated the value of carbohydrate analysis in distinguishing between various streptococci (1, 49), nocardiae (33), mycobacteria (2), and other microorganisms (3, 57). However, the extensive application of carbohydrate data for taxonomic purposes will require the development of much simpler sample preparation and derivatization techniques as well as improved chromatographic procedures.

Proteins, peptides, and amino acids

Chromatographic processes have been used extensively for the isolation, purification, and identification of proteins, peptides, and amino acids of microorganisms. The major application for these compounds in chemotaxonomy has been in comparisons of the amino acid composition of isolated cell walls (3, 57). Recently, however, there has been considerable interest in the evaluation of the chemical relatedness of microorganisms by comparisons of their soluble cellular protein patterns after electrophoresis on polyacrylamide gels. These electrophoretic protein profiles have provided additional valuable information in a number of taxonomic studies (15, 18). With the continuing advancements in HPLC instrumentation, detectors, columns, and procedures, it is possible that HPLC techniques will be used routinely for protein and peptide mapping and for the rapid analysis of nucleic acids, DNA and its fragments, and polynucleotides (51).

Cellular components: PGLC

Pyrolysis GLC (PGLC) is a technique in which an organic sample is decomposed by rapid heating that produces decomposition products which are separated by the GLC column. With the major exception of the introduction of the sample, the technique is identical to other forms of GLC requiring an analytical column, a detector, and a strip chart recorder or other device to record the peaks of decomposition products. As with other GLC techniques, only those components which are volatile under the conditions of analysis will be separated and detected. The degradation of macromolecules such as whole microorganisms produces a wide range of compounds (aldehydes, ketones, acids, and nitrites) and many yet-unidentified chemical types. Some of the earliest applications of GLC in microbiology were PGLC studies. Data from these early studies were not convincing, as the chromatograms (or pyrograms) of different bacterial species were essentially identical, with purportedly small quantitative differences which were difficult to discern by visual comparison of the chromatograms. Also, many of the early studies failed to consider strain variability, effects of growth media, and other environmental parameters, and essentially all studies failed to test a sufficient number of strains within a species for meaningful comparisons. Over the years, there have been major technical improvements in PGLC instrumentation, the use of capillary columns for increased resolution, and application of the computer for the collection, storage, and processing of the data. In addition, PGLC has been interfaced with a mass spectrometer to obtain chemical information on the type of products resulting from pyrolysis. This PGLC-MS technique is much more complex and expensive than PGLC alone, but it has the advantage of speed, sensitivity, and reproducibility, and the data are much more amenable to computer analysis. Published results with PGLC-MS are encouraging (27), but insufficient data are available for a critical evaluation of the technique or for assessing any significant advantages over PGLC alone. Thus, in view of these limited data, it is appropriate for the microbiologist to view PGLC and PGLC-MS as potentially useful (but expensive and complex) analytical techniques of the future. An excellent review of the state of the art and a future prospectus for the identification of microorganisms by pyrolysis techniques has been published and should be consulted for additional information (50).

DIRECT DIAGNOSIS OF INFECTION

After GLC was applied for the identification and taxonomy of pure cultures of microorganisms, studies were initiated to apply the technique to the direct examination of body fluids in the hope of discovering unique or characteristic compounds which would serve as markers of specific disease. If the technique was successful, this information would be available to the clinician within a short time, and appropriate treatment could be initiated. Although this approach was new to microbiology, it had been applied extensively in the clinical laboratory to determine the short-chain acids in the urine of patients with various metabolic disorders. Continuation of these studies and the addition of computer techniques and MS have resulted in the ability to study about half of the approximately 200 currently recognized metabolic disorders (22). The early applications of GLC to infectious diseases by Mitruka and co-workers (35, 36) and Brooks and co-workers (4, 5, 13) were designed to detect relatively small compounds such as short-chain acids, alcohols, and amines in body fluids which might serve as diagnostic markers. The appearance of unique compounds as well as large quantitative differences between compounds in body fluid from a healthy person may be correlated with specific infectious diseases.

The application of GLC to anaerobic infections has shown strong correlation between the detection of isobutyric, butyric, isovaleric, and succinic acids in pus and wound drainage fluids and the subsequent culture of anaerobes such as *Bacteroides* spp. and *Fusobacterium* spp. (16). In later studies, isobutyric, butyric, and isovaleric acids were confirmed as markers of anaerobic infection, as these were not detected in pus specimens from which aerobic organisms were isolated (48). However, the value of succinic acid as a possible marker of *Bacteroides fragilis* was limited by the finding of relatively large amounts of this acid from aerobic and facultative anaerobic bacteria. The presence of isobutyric, butyric, and isovaleric acids has also been shown to be a reliable indicator of the presence of anaerobes in blood cultures (54). Thus, it appears that the analysis of short-chain acids by GLC is a simple, rapid, and reliable test for anaerobes versus aerobes in clinical specimens; however, the acid profiles are not sufficiently distinct to predict the genera or species.

More sophisticated GLC and combined GLC-MS techniques have been applied to studies of sera for possible changes that might provide diagnostic markers. In an examination of sera of patients with tuberculosis, pneumococcal and *Klebsiella* pneumonia, and streptococcal pharyngitis, Mitruka et al. (36) found a correlation between "characteristic" GLC peaks in serum with peaks produced in vitro by each of the causative organisms. High levels of arabinitol and mannose in serum have been correlated with *Candida* infections (25, 52); other studies have shown that these compounds are also elevated in sera of patients with renal insufficiency (without mycotic infection) and in sera of rabbits which were immunosuppressed with cortisone and infected with *Candida albicans* (12, 14, 52). However, recent studies have shown that arabinitol can be used as a marker of candidiasis by correcting for changes in renal function by the accurate determination and application of arabinitol/creatinine ratios (59). In this laboratory, Brooks and coworkers found differences in the GLC patterns of short-chain acids in sera from human controls and from patients with Rocky Mountain spotted fever, varicella infections, *Neisseria meningitidis* infections, and infections due to the viruses of measles and rubella. Although most of the peaks were unidentified and only a relatively small number of specimens were tested, consistent and reproducible qualitative and large quantitative differences were observed among the chromatograms from these specimens (4, 13).

Frequency-pulsed electron capture GLC (FPEC-GLC) techniques have been used by Brooks and coworkers (4, 5) to study cerebrospinal fluid (CSF) from patients with lymphocytic meningitis. Acids, alcohols, amines, and hydroxy acids were extracted from CSF, converted to electron-capturing derivatives (i.e., esters and amides), and analyzed by FPEC-GLC. The data were recorded, standardized, sorted, and managed with a computer (4, 5). These techniques provide a high level of sensitivity and selectivity in that essentially only derivatized compounds with an affinity for free electrons are detected. Using FPEC-GLC, Brooks et al. (4, 13) found differences in the amine profiles of CSF from patients with viral and cryptococcal meningitis which were each different from a control group. In subsequent studies (4, 13), 3(2'-ketohexyl)indoline was found in CSF specimens from patients with acute tubercular meningitis. Initially this compound was thought to be unique for tubercular meningitis, but later studies indicated that it was present in only 50% of those cases and was also present in some CSF with acute bacterial infections; however, the compound has not been detected in cryptococcal, viral, or parasitic infections (4, 13). Elevated lactate levels have been observed in CSF from patients with meningitis, but elevated levels have also been detected in patients without infection (26).

GLC techniques have been used to examine other body fluids for diagnosis of disease. A wide variety of amines and nitrosamines (i.e., *N*-nitrosodimethylamine) were detected in urine from a patient with cystitis due to *Proteus mirabilis* (4). GLC techniques have been used for the rapid detection of *Escherichia coli* and *Proteus* spp. in urine by measuring for ethanol and dimethyl sulfide by headspace analysis (9, 19). GLC and combined GLC-MS techniques have been used to measure tuberculostearic acid (10-methyloctadecanoate) in sputum; this acid was detected in sputum from each of eight patients with pulmonary tuberculosis but not from six patients with other pneumonias (46). Recently, Brooks and co-workers have applied FPEC-GLC techniques in the study of stool specimens from patients with various types of diarrheal disease. Preliminary data have shown marked differences in FPEC-GLC profiles among stools from confirmed cases of antimicrobial-associated pseudomembranous colitis due to *Clostridium difficile*, enterotoxigenic *E. coli* (producing heat-labile or heat-stable enterotoxin or both), rotavirus, and adenovirus. Presently, additional specimens are being studied to expand the data base, and other efforts are being made to identify some compounds which appear to be useful chemical markers of these diseases.

The present status of GLC and associated techniques for the direct diagnosis of disease must be viewed as in the active research stage. The complexity of body fluid and factors which control or influence its composition, specimen collection and preparation, extraction and derivatization of clinical samples, analytical analysis, and interpretation of complex data are major factors for consideration in all studies. Sufficient numbers of well-documented clinical and normal control specimens must be examined to critically evaluate biological variability, and special emphasis must be given to the identification and documentation of the specificity of marker compounds. The use of NMR spectroscopy, HPLC, and combined HPLC-MS techniques in studies of body fluids have been limited, but these techniques may prove to be important analytical tools for the rapid diagnosis of infectious diseases. The potential of direct diagnosis without resort to culture warrants continued research with these techniques in various areas of medical microbiology.

LITERATURE CITED

1. **Abuyi, H. S., and D. B. Drucker.** 1983. Trimethylsilyl-sugar profiles of *Streptococcus milleri* and *Streptococcus mitis*. J. Appl. Bacteriol. **54:**391–397.
2. **Alvin, C., L. Larsson, M. Magnusson, P. A. Mardh, G. Odham, and G. Westerdahl.** 1983. Determination of fatty acids and carbohydrate monomers in microorganisms by glass capillary gas chromatography: analysis of *Mycobacterium gordonae* and *Mycobacterium scrofulaceum*. J. Gen. Microbiol. **129:**401–405.
3. **Berd, D.** 1973. Laboratory identification of clinically important aerobic actinomycetes. Appl. Microbiol. **25:**665–681.
4. **Brooks, J. B.** 1983. Gas-liquid chromatography as an aid in rapid diagnosis by selective detection of chemical changes in body fluids, p. 313–334. *In* J. D. Coonrod, L. J. Kunz, and M. J. Ferrano (ed.), The direct detection of microorganisms in clinical samples. Academic Press, Inc., New York.
5. **Brooks, J. B., D. S. Kellogg, Jr., M. E. Shepherd, and C. C. Alley.** 1980. Rapid differentiation of the major causative agents of bacterial meningitis by use of frequency-pulsed electron capture gas-liquid chromatography: analysis of amines. J. Clin. Microbiol. **11:**52–58.
6. **Brooks, J. B., and W. E. C. Moore.** 1969. Gas chromato-

graphic analysis of amines and other compounds produced by several species of *Clostridium*. Can. J. Microbiol. **15**:1433–1437.

7. **Carlone, G. M., and F. A. L. Anet.** 1983. Detection of menaquinone-6 and a novel methyl-substituted menaquinone-6 in *Campylobacter jejuni* and *Campylobacter fetus* subsp. *fetus*. J. Gen. Microbiol. **129**:3385–3393.
8. **Collins, M. D., and D. Jones.** 1981. Distribution of isoprenoid quinone structural types in bacteria and their taxonomic implications. Microbiol. Rev. **45**:316–354.
9. **Coloe, P. J.** 1978. Headspace gas liquid chromatography for rapid detection of *Escherichia coli* and *Proteus mirabilis* in urine. J. Clin. Pathol. **31**:365–369.
10. **Dees, S. B., and C. W. Moss.** 1978. Identification of *Achromobacter* species by cellular fatty acids and by production of keto acids. J. Clin. Microbiol. **8**:61–66.
11. **Dees, S. B., and C. W. Moss.** 1979. Analysis of short chain acids by gas-liquid chromatography on SP-1220. J. Chromatogr. **171**:466–468.
12. **de Repentigny, L., R. J. Kuykendall, and E. Reiss.** 1983. Simultaneous determination of arabinitol and mannose by gas-liquid chromatography in experimental candidiasis. J. Clin. Microbiol. **17**:1166–1169.
13. **Edman, D. C., and J. B. Brooks.** 1983. Gas-liquid chromatography-frequency pulsed-modulated electron-capture detection in the diagnosis of infectious diseases. J. Chromatogr. **274**:1–25.
14. **Eng, R. H. K., H. Chmel, and M. Buse.** 1981. Serum levels of arabinitol in the detection of invasive candidiasis in animals and humans. J. Infect. Dis. **143**:677–683.
15. **Ferragut, C., D. Izard, F. Gavini, K. Kersters, J. De Ley, and H. Leclerc.** 1983. *Klebsiella trevisanii*: a new species from water and soil. Int. J. Syst. Bacteriol. **33**:133–142.
16. **Gorbach, S. L., J. W. Mayhew, J. G. Bartlett, H. Thadepalli, and A. B. Onderdonk.** 1976. Rapid diagnosis of anaerobic infections by direct gas-liquid chromatography of clinical specimens. J. Clin. Invest. **57**:478–484.
17. **Guerrant, G. O., M. A. Lambert, and C. W. Moss.** 1982. Analysis of short-chain acids from anaerobic bacteria by high-performance liquid chromatography. J. Clin. Microbiol. **16**:355–360.
18. **Hanna, J., S. D. Neill, J. J. O'Brien, and W. A. Ellis.** 1983. Comparison of aerotolerant and reference strains of *Campylobacter* species by polyacrylamide gel electrophoresis. Int. J. Syst. Bacteriol. **33**:143–146.
19. **Hayward, N. J., T. H. Jeavons, A. J. C. Nicholson, and A. G. Thornton.** 1977. Methyl mercaptan and dimethyl sulfide production from methionine by *Proteus* species detected by head-space gas-liquid chromatography. J. Clin. Microbiol. **6**:187–194.
20. **Holdeman, K. V., E. P. Cato, and W. E. C. Moore (ed.).** 1977. Anaerobic laboratory manual, 4th ed. Virginia Polytechnic Institute and State University, Blacksburg.
21. **Jantzen, E., B. P. Berdal, and T. Omland.** 1979. Cellular fatty acid composition of *Francisella tularensis*. J. Clin. Microbiol. **10**:928–930.
22. **Jellum, E.** 1977. Profiling of human body fluids in healthy and diseased states using gas chromatography and mass spectrometry with special reference to organic acids. J. Chromatogr. **143**:427–462.
23. **Kaneda, T.** 1967. Fatty acids in the genus *Bacillus*. I. Iso- and anteiso-fatty acids as characteristic constituents of lipids in 10 species. J. Bacteriol. **93**:894–903.
24. **Karr, D. E., W. F. Bibb, and C. W. Moss.** 1982. Isoprenoid quinones of the genus *Legionella*. J. Clin. Microbiol. **15**:1044–1048.
25. **Kiehn, T. E., E. M. Bernard, J. W. M. Gold, and D. Armstrong.** 1979. Candidiasis: detection by gas-liquid chromatography of D-arabinitol, a fungal metabolite, in human serum. Science **206**:577–580.
26. **Komorowski, R. A., S. G. Farmer, G. A. Hanson, and L. L. Hause.** 1978. Cerebrospinal fluid lactic acid in diagnosis of meningitis. J. Clin. Microbiol. **8**:89–92.
27. **Kristemaker, P. G., H. L. Meuzelaar, and M. A. Posthumus.** 1975. Rapid and automated identification by curie-point pyrolysis techniques. II. Fast identification of microbiological sample by curie-point pyrolysis mass spectrometry, p. 179–191. *In* C. G. Haden and T. Illeni (ed.), New approaches to the identification of microorganisms. John Wiley & Sons, London.
28. **Lambert, M. A., F. W. Hickman-Brenner, J. J. Farmer III, and C. W. Moss.** 1983. Differentiation of *Vibrionaceae* species by their cellular fatty acid composition. Int. J. Syst. Bacteriol. **33**:777–792.
29. **Lambert, M. A., and C. W. Moss.** 1980. Production of *p*-hydroxyhydrocinnamic acid from tyrosine by *Peptostreptococcus anaerobius*. J. Clin. Microbiol. **12**:291–293.
30. **Lambert, M. A., and C. W. Moss.** 1983. Comparison of the effects of acid and base hydrolyses on hydroxy and cyclopropane fatty acids in bacteria. J. Clin. Microbiol. **18**:1370–1377.
31. **Larsson, L., P. A. Mardh, and G. Odham.** 1978. Analysis of amines and other bacterial products by head space gas chromatography. Acta Pathol. Microbiol. Scand. Sect. B **86**:207–213.
32. **Lechevalier, M. D.** 1982. Lipids in bacterial taxonomy, p. 436–508. *In* A. I. Laskin and H. A. Lechevalier (ed.), CRC handbook of microbiology. CRC Press, Inc., Boca Raton, Fla.
33. **Lechevalier, M. P., H. Lechevalier, and A. C. Horan.** 1973. Chemical characteristics and classification of nocardiae. Can. J. Microbiol. **19**:965–972.
34. **Minnikin, D. E., and M. Goodfellow.** 1980. Lipid composition in the classification and identification of acid-fast bacteria, p. 189–210. *In* M. Goodfellow and R. G. Board (ed.), Microbiological classification and identification. Academic Press, Ltd., London.
35. **Mitruka, B. M., A. M. Jonas, M. Alexander, and R. S. Kundargi.** 1973. Rapid differentiation of certain bacteria in mixed populations by gas-liquid chromatography. Yale J. Biol. Med. **46**:104–112.
36. **Mitruka, B. M., R. S. Kundargi, and A. M. Jonas.** 1972. Gas chromatography for rapid differentiation of bacterial infections in man. Med. Res. Eng. **11**:7–11.
37. **Moss, C. W.** 1978. New methodology for identification of non-fermenters: gas-liquid chromatographic chemotaxonomy, p. 188–195. *In* G. L. Gilardi (ed.), Glucose nonfermenting gram-negative bacteria in clinical microbiology. CRC Press, Inc. West Palm Beach, Fla.
38. **Moss, C. W.** 1981. Gas-liquid chromatography as an analytical tool in microbiology. J. Chromatogr. **203**:337–347.
39. **Moss, C. W., W. F. Bibb, D. E. Karr, and G. O. Guerrant.** 1983. Chemical analysis of the genus *Legionella*: fatty acids and isoprenoid quinones. Inst. Natl. Sante Rech. Med. Symp. **114**:375–381.
40. **Moss, C. W., and S. B. Dees.** 1976. Cellular fatty acids and metabolic products of *Pseudomonas* species obtained from clinical specimens. J. Clin. Microbiol. **4**:492–502.
41. **Moss, C. W., V. R. Dowell, Jr., D. Farshtchi, L. J. Raines, and W. B. Cherry.** 1969. Culture characteristics and fatty acid composition of propionibacteria. J. Bacteriol. **97**:561–570.
42. **Moss, C. W., and G. O. Guerrant.** 1983. Separation of bacterial ubiquinones by reverse-phase high-pressure liquid chromatography. J. Clin. Microbiol. **18**:15–17.
43. **Moss, C. W., C. L. Hatheway, M. A. Lambert, and L. M. McCroskey.** 1980. Production of phenylacetic and hydroxyphenylacetic acids by *Clostridium botulinum* type G. J. Clin. Microbiol. **11**:743–745.
44. **Moss, C. W., A. Kai, M. A. Lambert, and C. Patton.** 1984. Isoprenoid quinone content and cellular fatty acid composition of *Campylobacter* species. J. Clin. Microbiol. **19**:772–776.
45. **Moss, C. W., and O. L. Nunez-Montiel.** 1982. Analysis of

short-chain acids from bacteria by gas-liquid chromatography with a fused-silica capillary column. J. Clin. Microbiol. **15:**308–311.

46. **Odham, G., L. Larsson, and P. A. Mardh.** 1979. Demonstration of tuberculostearic acid in sputum from patients with pulmonary tuberculosis by selected ion monitoring. J. Clin. Invest. **63:**813–819.
47. **Peladan, F., and H. Monteil.** 1983. Apports de la chromatographie en phase gazeuse a la chimiotaxonomie des *Pseudomonas*. Inst. Natl. Sante Rech. Med. Symp. **114:**301–310.
48. **Phillips, K. O., P. V. Tearle, and A. T. Willis.** 1976. Rapid diagnosis of anaerobic infections by gas-liquid chromatography of clinical materials. J. Clin. Pathol. **29:**428–432.
49. **Pritchard, D. G., J. E. Coligan, S. E. Speed, and B. M. Gray.** 1981. Carbohydrate fingerprints of streptococcal cells. J. Clin. Microbiol. **13:**89–92.
50. **Quinn, P. A.** 1976. Identification of microorganisms by pyrolysis: the state of the art, p. 178–186. *In* Proceedings of the Second International Symposium on Rapid Methods and Automation. Learned Information, Inc., Oxford.
51. **Regnier, F. E.** 1983. HPLC of proteins, peptides, and polynucleotides. Anal. Chem. **55:**1299A–1306A.
52. **Roboz, J., R. Suzuki, and J. F. Holland.** 1980. Quantification of arabinitol in serum by selected ion monitoring as a diagnostic technique in invasive candidaisis. J. Clin. Microbiol. **12:**594–602.
53. **Shaw, N.** 1974. Lipid composition as a guide to the classification of bacteria. Adv. Appl. Microbiol. **17:**63–108.
54. **Sondag, J. E., M. Ali, and P. R. Murray.** 1980. Rapid presumptive identification of anaerobes in blood cultures by gas-liquid chromatography. J. Clin. Microbiol. **11:**274–277.
55. **Sutter, V. L., D. M. Citron, and S. M. Finegold.** 1980. Wadsworth anaerobic bacteriology manual, 3rd ed. The C. V. Mosby Co., St. Louis.
56. **Tamooka, J., Y. Katayome-Fijimura, and H. Kuraishi.** 1983. Analysis of bacterial menaquinone mixtures by high performance liquid chromatography. J. Appl. Bacteriol. **54:**31–36.
57. **Wilkinson, S. G., and K. A. Carby.** 1971. Amino sugars in the cell walls of *Pseudomonas* species. J. Gen. Microbiol. **66:**221–227.
58. **Willis, A. T.** 1977. Anaerobic bacteriology, clinical and laboratory practice, 3rd ed., p. 96–105. Butterworths, London.
59. **Wong, B., E. M. Bernard, J. W. M. Gold, D. Fong, and D. Armstrong.** 1982. The arabinitol appearance rate in laboratory animals and humans: estimation from the arabinitol/creatinine ratio and relevance to the diagnosis of candidiasis. J. Infect. Dis. **146:**353–359.
60. **Yabuuchi, E., and C. W. Moss.** 1982. Cellular fatty acid composition of strains of three species of *Sphingobacterium* gen. nov. and *Cytophaga johnsonae*. FEMS Microbiol. Lett. **13:**87–91.
61. **Yamada, Y., G. Inouye, Y. Tahara, and K. Kondo.** 1976. The menaquinone system in the classification of coryneform and nocardiaform bacteria and related organisms. J. Gen. Appl. Microbiol. **22:**203–214.
62. **Yashikawa, T. T., S. K. Maitra, M. C. Schotz, and L. B. Guze.** 1980. High-pressure liquid chromatography for quantitation of antimicrobial agents. Rev. Infect. Dis. **2:**169–181.

Chapter 110

Section XIII. Media, Reagents, and Stains

Quality Control of Culture Media

JOHN E. FORNEY AND J. MICHAEL MILLER

Culture media are used for a variety of purposes in the laboratory. Their performance must therefore be judged acceptable to ensure that they serve the purpose for which they are prepared. Culture media are used to propagate bacteria, fungi, viruses, and, in some cases, animal parasites. Not only must a medium support the growth of an organism, but that organism should exhibit typical colonial and microscopic morphology on the medium. Variations in composition of the medium may alter these characteristics. The typical appearance of bacteria and fungi on culture media is described in this Manual as well as in other publications. Media are also used to demonstrate many other characteristics of organisms, e.g., production of acid and gas in carbohydrate fermentation media or hemolysis of erythrocytes in blood agar media.

Most media used in the clinical bacteriology laboratory belong to one of the following types: enriched, differential, selective, or one-purpose media.

Enriched media

Enriched media include 5% sheep blood agar, enriched chocolate agar, and a number of infusion-based media, all of which may be further enriched by additives. These media must be carefully tested for sterility before they are used, since even fastidious organisms are likely to grow.

Differential media

Differential media include, among others, MacConkey and eosin methylene blue agars, which contain dyes, sugars, and indicators designed to elicit a characteristic biochemical response (usually a color) and are used to differentiate groups of organisms, such as lactose fermenters from non-lactose fermenters.

Selective media

Selective media contain, in addition to components found in differential media, agents designed to further inhibit most *Enterobacteriaceae* and to select from the specimen those strains resistant to the higher concentrations of the selective components. Some enteric media combine both selective and differential characteristics. Hektoen enteric agar and xylose-lysine-desoxycholate agar may be classed as selective because they are somewhat more inhibitory than MacConkey and eosin methylene blue agars. Many selective media also are capable of detecting H_2S production.

One-purpose media

One-purpose media are highly selective and are usually designed to isolate a specific organism. Brilliant green and bismuth sulfite agars are designed specifically for isolating *Salmonella* species from feces, whereas *Campylobacter* media contain a series of antimicrobial agents that allow growth of *Campylobacter jejuni* and inhibit most other fecal flora. Media selective for *Neisseria gonorrhoeae* also inhibit most other flora.

Most cell culture media used in virology contain antibacterial agents to prevent overgrowth from the bacterial flora found in a specimen. Mycoplasma is often a problem, and media must be carefully tested before they are used to ensure their sterility.

Though the frequency with which commercially prepared media are tested may change, there must be assurance in the laboratory using it that an enriched medium is indeed sterile and that it will support fastidious isolates. Differential and selective media must exhibit the appropriate biochemical response as well as support the growth of a light inoculum of the organism used in the test. Heavy inocula may give misleading results. Similarly, a one-purpose medium must support a light inoculum of the organism for which that medium was designed. Maintenance media may not provide all the essential elements for growth and reproduction of organisms and yet will maintain them in a viable state over time. Such media are more widely used in virology than in other areas of microbiology.

SOURCES OF MEDIA

Before the 1950s, most laboratories manufactured their own media, frequently starting with basic raw ingredients. Laboratories no longer prepare media in this way, with the possible exception of a few laboratories that produce some media for use in their tuberculosis laboratory. The most common sources of media today are as follows.

Dehydrated media

Dehydrated media have been available for over 50 years, but beginning in the post-World War II era, many more became available. They merely require the addition of water to be reconstituted for use. With the introduction of manufactured dehydrated media, partial responsibility for quality control rested with the manufacturer; few laboratories attempted to control the quality of the rehydrated final product.

Dehydrated media with additives

For some media, the basic formulation can be prepared in dehydrated form, but to make the media supportive of more fastidious organisms, certain additives need to be used when the media are prepared in the laboratory. The additives usually are unstable materials such as blood, serum, or other growth factors.

Commercially prepared media

In the early 1960s, companies began preparing and selling plated and tubed media. Early offerings included sheep blood agar and certain tuberculosis media such as Lowenstein-Jensen. The burden for quality control in preparation of the media falls upon the manufacturer up to the point of release to the consumer. However, manufacturer quality control does not relieve the laboratory of responsibility for the product's performance in the laboratory.

SOURCES OF ERROR

Inappropriate medium

Since dehydrated media are usually arranged alphabetically on a shelf, one may select the wrong bottle inadvertently, or an improper additive might be selected, making the medium unsuitable for use. It is always important to read the label, particularly when a new lot of medium has been received in the laboratory.

Water

Measure carefully the amount of water that is added when reconstituting media. Since impurities render tap water unsuitable for the preparation of most biological media, laboratories should use either distilled water, deionized water, or water that has been treated in both ways.

Weighing

Accurate balances should be used for weighing dry materials. Weighing errors significantly alter the composition of the final product.

Dispensing

Media should be dispensed accurately and aseptically into plates and tubes. Failure to measure the amount accurately may result, for example, in too shallow or too deep agar medium, either of which may make the medium unsuitable for use.

Proper sterilization

A common error in media preparation is sterilizing media at too high a temperature or for too long a period, or both. This may result in deterioration or decomposition of some constituents of the media, which will render the media useless for the intended purpose.

Glassware

Care should be taken to use clean glassware, since residues on glass may be inhibitory to some fastidious microorganisms, particularly viruses grown in cell culture, or to the cells themselves.

QUALITY CONTROL

Any quality control program for culture media must, in the final analysis, assure that a medium will support the growth of the organisms likely to be in the specimen. It must, if specified, inhibit the growth of commensal organisms, exhibit a typical biochemical response, be stable, and have a reasonable shelf life.

Sterility

A few media are used without terminal sterilization, but they are the exceptions; most media must be sterile when they are inoculated. Each batch of medium, whether prepared in the laboratory or received from a commercial source, should be sampled for sterility. This is best done by removing 1 to 5% of the batch and placing it in a bacteriologic incubator at 35°C for 48 h. If contaminants appear in the medium as a result of inadequate sterilization, a new lot should be obtained. Those containers that are used for sterility testing should be discarded at the completion of the test, since they are unsuitable for inoculation because of the dehydration that occurs after up to 48 h in the incubator.

Growth

Determine the ability of the medium to support the growth of suspected organisms by inoculating the medium with a typical stock culture isolate. A frequent quality control error is the use of a heavy inoculum for this purpose. For most media, inoculating with a stock culture that is too heavy may result in misleading growth. In a specimen, the organism may be much more fastidious or present in very small numbers; therefore, the medium may not support its growth. When testing for ability to support growth, it is well to prepare a dilute suspension to use as the inoculum. This suspension will give greater assurance that the medium is adequate for the growth of a small number of organisms in a patient's specimen. In selecting an organism for testing, one should select from among the more fastidious species of organisms that one may be looking for in specimens received from patients.

Biochemical response

When inoculating media used to identify a specific reaction, such as fermentation or H_2S production, it is necessary to use only a species or strain of organism that will produce the desired reaction.

Selective medium

Since selective media are designed not only to support the growth of organisms but to inhibit the growth of others, it is necessary to inoculate the medium with representatives of both groups of organisms. To demonstrate the inhibitory effect, one can challenge the medium with a heavy inoculum, since if the medium will prevent the growth of a large inoculum it will inhibit the small number of organisms that may be present in the primary specimen. The medium must also support the growth of the selected organisms.

A long list of media has been published with recommended organisms to be used for determining the quality of media in the laboratory (3). It is not necessary to repeat these lists in this publication but only to establish the proper techniques for maintaining quality control on culture medium. There are several sources of cultures that can be used for quality con-

TABLE 1. Media for aerobic bacteriology[a]

Medium	Quality control procedures: Sterility	Growth	Inhibition	Biochemical response	Organism(s)	Expected reaction(s)	Comment
Arginine dihydrolase	X			X	*Salmonella typhimurium*	Positive = red-violet or purple within 48 h	Liquid medium must be overlaid with mineral oil or petrolatum.
					Proteus vulgaris	Negative = yellow	
Azide blood agar	X	X	X		*Streptococcus pyogenes*	Small colony with beta-hemolysis	Azide is added to media to inhibit gram-negative organisms and allow gram positives (streptococci) to grow. Principle also applies to blood agar with trimethoprim-sulfamethoxazole, neomycin, etc.
					Escherichia coli	Growth inhibited	
Bile-esculin agar	X			X	*Streptococcus faecalis* (group D)	Growth; black-colored medium within 48 h	Used for presumptive identification of group D *Streptococcus*. Enterococci also may be identified by their growth in NaCl medium.
					Streptococcus pyogenes	Medium not blackened	
Bismuth sulfite agar			X	X	*Salmonella typhimurium* (some strains inhibited)	Growth; colonies black with metallic silver sheen in 48 h	Medium used to isolate *Salmonella* species, including *S. typhi*, from fecal specimens. Inhibits other organisms. Use on day of preparation. Should be a gray color, not green, when fresh.
					Escherichia coli	No growth or inhibited growth; colorless to pale-green colonies in 48 h	
Blood agar	X	X			Group A beta-hemolytic *Streptococcus*	Small, clear colonies exhibiting beta (complete)-hemolysis in 48 h	Blood agar should allow these typical isolates to exhibit characteristic hemolysis. The CAMP test can be performed to ensure proper sheep cell activity.
					Streptococcus viridans or *Streptococcus pneumoniae*	Small to medium colonies exhibiting alpha (green)-hemolysis	
Brain heart infusion agar or broth	X	X			*Staphylococcus aureus*	Growth in 24 h	Highly nutritious medium for support of bacterial growth.
Brilliant green agar	X		X	X	*Salmonella typhimurium*	Red or pink colonies in 48 h	Selective for *Salmonella* spp. other than *S. typhi*. Sulfadiazine may be added to inhibit *Pseudomonas* sp.
					Proteus mirabilis	Yellow or yellow-green colonies with a yellow-green halo in agar	
					Escherichia coli	Growth inhibited	
Chocolate agar	X	X			*Haemophilus influenzae*	Growth in 24 h	Should support growth of fastidious pathogens.
Desoxycholate agar			X	X	*Escherichia coli*	Red colonies in 24 h	Differential medium for the demonstration of lactose fermentation by gram-negative rods. H_2S indicator is not present in the medium.
					Proteus mirabilis	Colorless colonies	
					Streptococcus sp.	Growth inhibited	

Continued

TABLE 1—*Continued*

Medium	Quality control procedures				Organism(s)	Expected reaction(s)	Comment
	Sterility	Growth	Inhibition	Biochemical response			
DNase test agar	X			X	*Staphylococcus aureus* or *Serratia marcescens*	Clear zone (in acid reaction with HCl) or pink zone (toluidine blue-containing media) around colony in 24 h	Plates may be spot inoculated rather than streaked for isolation.
					Streptococcus faecalis	No zone around colonies	
Eosin methylene blue	X		X	X	*Escherichia coli*	Colonies blue-purple, usually with green metallic sheen in 24 h	Differential medium for the demonstration of lactose fermentation by gram-negative rods. Other lactose fermenters may produce green sheen. H_2S indicator is not present in the medium.
					Shigella sonnei	Colorless colonies	
					Staphylococcus aureus	Growth inhibited	
Glucose utilization oxidative-fermentative medium	X			X	*Pseudomonas aeruginosa* (oxidizer)	Overlaid tube: no color change Open tube: yellow (acid) at surface	*E. coli* (a fermenter) should cause both tubes to turn yellow (acid).
					Acinetobacter lwoffi (non-utilizer)	Overlaid tube: no change Open tube: no change	
GN broth	X	X	X		*Shigella sonnei*	Growth when subcultured to EMB, MAC, XLD, or HE[b] in 8 to 24 h	This medium was designed for subculture after 6 h of incubation.
					Escherichia coli	Growth inhibited for up to 6 to 8 h; light growth on subculture	
Heart infusion agar	X	X			*Streptococcus* sp.	Growth in 24 h	
Hektoen agar	X		X	X	*Escherichia coli*	Orange colonies (may be inhibited) in 24 h	Black center indicates H_2S production. *Shigella sonnei* should exhibit green, transparent colonies.
					Salmonella typhimurium	Green colonies with black centers in 24 h	
					Streptococcus sp.	Growth inhibited	
Indole (peptone water)	X			X	*Escherichia coli*	Positive in 24 h = red ring	Add a few drops of Kovac's reagent to broth and shake tube gently.
					Enterobacter aerogenes	Negative = yellow ring	
Kligler iron agar	X			X	*Escherichia coli*	Acid slant/acid butt with gas; no H_2S in 18 to 24 h	Both lactose and glucose are fermented.
					Proteus mirabilis	Alkaline slant/acid butt with H_2S and perhaps gas in 18 to 24 h	Glucose but not lactose is fermented.

Continued

TABLE 1—*Continued*

Medium	Quality control procedures: Sterility	Growth	Inhibition	Biochemical response	Organism(s)	Expected reaction(s)	Comment
Kligler iron agar					*Pseudomonas aeruginosa*	Alkaline slant/alkaline butt; no gas or H_2S in 18 to 24 h	Neither glucose nor lactose is fermented.
Lysine decarboxylase agar or broth	X			X	*Klebsiella pneumoniae*	Positive in 24 h = red-violet to purple	Liquid medium is overlaid with mineral oil or petrolatum.
					Proteus mirabilis	Negative = yellow	
Lysine iron agar	X			X	*Enterobacter aerogenes*	Positive in 18 to 24 h = alkaline slant and butt (purple)	Decarboxylase activity is read in the butt of the tube: positive, purple; negative, yellow.
					Providencia sp.	Negative = red slant (deamination), yellow butt; H_2S negative	
MacConkey agar	X		X	X	*Escherichia coli*	Pink colonies (lactose fermenter) in 24 h	Differential medium for the demonstration of lactose fermentation by gram-negative rods. H_2S indicator is not present in medium.
					Proteus mirabilis	Colorless colonies (lactose nonfermenter) in 24 h	
					Streptococcus sp.	Growth inhibited	
Malonate broth	X			X	*Klebsiella pneumoniae*	Positive in 24 h = blue	Used to separate *Salmonella* (negative) and *Arizona* spp. (positive).
					Escherichia coli	Negative = green (no color change)	
Mannitol salt agar	X		X	X	*Staphylococcus aureus*	Growth, yellow colonies (acid) in 24 h	Mannitol fermentation alone is *not* used for the presumptive identification of *S. aureus* or *S. epidermidis*.
					Staphylococcus epidermidis	Growth, no color change in 24 h	
					Escherichia coli	Growth inhibited	
Motility medium	X	X			*Escherichia coli*	Motile; medium becomes cloudy and stab line disappears in 24 h (no color involved)	Motility medium also may be prepared with triphenyltetrazolium chloride, which will turn red where organisms grow.
					Klebsiella pneumoniae	Nonmotile; growth only at stab line (no color involved)	
Motility-indole-ornithine medium	X	X		X	*Escherichia coli*	Motility positive, indole positive, ornithine positive at 24 h	Motility: cloudy medium. Indole (after addition of Kovac's reagent): positive, red ring; negative, yellow ring. Ornithine: positive, purple; negative, yellow.
					Klebsiella pneumoniae	Motility negative, indole negative, ornithine negative at 24 h	
Nutrient agar or broth	X	X			*Staphylococcus aureus*	Growth in 18 to 24 h	
Ornithine decarboxylase	X			X	*Escherichia coli*	Positive in 24 h = purple broth	Liquid medium must be overlaid with mineral oil or petrolatum.
					Proteus vulgaris	Negative = yellow broth	

Continued

TABLE 1—*Continued*

Medium	Quality control procedures				Organism(s)	Expected reaction(s)	Comment
	Sterility	Growth	Inhibition	Biochemical response			
Pfizer selective enterococcus agar	X		X	X	Group D *Streptococcus* (enterococci)	Growth, black colonies in 24 h	
					Escherichia coli	Growth inhibited or light growth; clear colonies	
Phenylalanine deaminase agar	X			X	*Proteus vulgaris*	Positive in 24 h = green slant after reagent	Add 2 to 3 drops of 10% $FeCl_3$ to growth on the slant.
					Escherichia coli	Negative = no color change after reagent	
Salmonella-shigella agar			X	X	*Enterobacter aerogenes*	Pink to red colonies in 24 h	Lactose used as a differential sugar. H_2S indicator is present in the medium. Not recommended for *Shigella sonnei* (does not grow well).
					Escherichia coli	Little or no growth in 24 h	
					Salmonella typhimurium	Colorless colonies, black centers in 24 h	
Phenylethyl alcohol agar	X	X	X		*Staphylococcus aureus* or *Streptococcus* sp.	Growth in 24 h	Used especially for specimens likely to be contaminated with aerobic gram-negative organisms, particularly swarming *Proteus* sp.
					Escherichia coli or *Proteus vulgaris*	Growth inhibited or little growth	
Selenite F broth		X	X		*Salmonella typhimurium*	Growth when subcultured to MAC, EMB, XLD, HE, or SS[b] in 8 to 24 h	Should be a straw color with no orange precipitate. Overheating causes precipitate. Cystine-selenite medium acceptable for stool but not good for isolation of *S. typhi.*
					Escherichia coli	Light growth when subcultured	
Simmons citrate	X	X		X	*Enterobacter aerogenes*	Positive = growth, slant turns blue in 24 h	
					Escherichia coli	Negative = no growth, no color change (green)	
Sodium azide agar	X	X	X		*Staphylococcus aureus* or *Streptococcus pyogenes*	Growth in 24 h	Inhibits gram-negative organisms.
					Escherichia coli	Growth inhibited	
Streptococcus faecalis broth or 6.5% NaCl	X	X	X	X	*Streptococcus faecalis* (enterococci)	Growth, color change to yellow, or both in 48 h	Growth (cloudy) without color change is considered positive.
					Streptococcus sp. (not enterococci)	Growth inhibited	
Sulfide-indole motility medium	X			X	*Proteus vulgaris*	H_2S positive (black color), indole positive (add Kovac's reagent), motility (cloudy medium) in 24 h	

Continued

TABLE 1—*Continued*

Medium	Quality control procedures				Organism(s)	Expected reaction(s)	Comment
	Sterility	Growth	Inhibition	Biochemical response			
Sulfide-indole motility medium					*Klebsiella pneumoniae*	H_2S negative, indole negative, nonmotile in 24 h	
Tetrathionate broth		X	X		*Salmonella typhimurium* *Escherichia coli*	Growth when subcultured to MAC, EMB, XLD, HE, or SS[b] in 18 h Light growth when subcultured	
Thayer-Martin agar	X	X	X		*Neisseria gonorrhoeae* *Escherichia coli*	Growth in 24 h No growth	Same organisms used for modified Thayer-Martin, Martin-Lewis, etc. Use fresh strains of *N. gonorrhoeae* since laboratory-adapted strains may be less fastidious.
Thioglycolate broth	X	X			*Bacteroides* sp. (if anaerobes are cultured in lab) *Staphylococcus aureus*	Growth in 24 h; culture in anaerobic atmosphere Growth in 24 h; aerobic incubation	Thioglycolate should be carefully inoculated at the bottom of the tube.
Transgrow	X	X	X		*Neisseria gonorrhoeae* *Escherichia coli*	Growth in 24 h No growth	If bottle is tilted during inoculation, the CO_2 atmosphere may be lost.
Triple sugar iron agar	X			X	*Escherichia coli* *Proteus mirabilis* *Pseudomonas aeruginosa*	Acid slant, acid butt, gas, no H_2S in 18 to 24 h Alkaline slant, acid butt, gas, H_2S in 18 to 24 h Alkaline slant, alkaline butt, no gas, no H_2S in 18 to 24 h	Glucose, lactose, and/or sucrose fermented. Only glucose fermented. Nonfermenter.
Tryptic soy agar	X	X			*Streptococcus pyogenes*	Growth in 24 h	
Tryptic soy broth	X	X			*Streptococcus pyogenes*	Growth in 24 h	
Urea agar or broth	X			X	*Proteus vulgaris* *Escherichia coli*	Positive in 4 to 18 h = pink Negative = yellow or no change	Pink color should appear within 4 to 6 h.
Voges-Proskauer	X			X	*Enterobacter aerogenes* *Escherichia coli*	Positive in 48 h = pink color after reagents Negative = yellow	Broth medium is methyl red-Voges-Proskauer. Formulation A includes α-naphthol; formulation B includes 40% KOH (creatine).

Continued

TABLE 1—*Continued*

Medium	Quality control procedures				Organism(s)	Expected reaction(s)	Comment
	Sterility	Growth	Inhibition	Biochemical response			
Xylose-lysine-desoxycholate agar			X	X	*Escherichia coli* *Salmonella typhimurium* *Shigella sonnei* *Streptococcus* sp.	Yellow colonies in 24 h Pink-red colonies, black center in 24 h Transparent colonies in 24 h Growth inhibited	(see below)

	E. coli	*Salmonella*	*Shigella*
Xyl	+	+	−
Lac	+	−	−
Suc	+	−	−
Lys	+/−	+	−

[a] From references 1 and 2.
[b] Abbreviations: EMB, eosin methylene blue agar; MAC, MacConkey agar; XLD, xylose-lysine-desoxycholate agar; HE, Hektoen enteric agar; SS, salmonella-shigella agar.

TABLE 2. Media for anaerobic bacteriology[a]

Medium	Quality control procedures				Organism(s)	Expected reaction(s)	Comment
	Sterility	Growth	Inhibition	Biochemical response			
Anaerobic transport medium	X	X			*Bacteroides fragilis*	Recovery when plated on isolation medium after 1 h	
Blood agar (anaerobe formula)	X	X			*Fusobacterium necrophorum*	Growth in 24 h	*F. necrophorum* will grow if cysteine is in the medium. Beta-hemolysis on rabbit blood. No hemolysis on sheep blood.
Carbohydrate fermentation base medium (bromthymol blue indicator)	X	X		X	*Bacteroides vulgatus* and *Clostridium tetani*	Growth; no color change	Base without carbohydrate is inoculated as a growth control. No color change expected. pH 7.0, blue-green; pH 6.0, yellow.
Arabinose (0.6%)	X			X	*Bacteroides vulgatus* *Clostridium tetani*	Positive = yellow Negative = blue-green	Other pH indicators may give alternative color patterns for positive and negative reactions.
Dextrose (glucose) (0.6%)	X			X	*Bacteroides vulgatus* *Clostridium tetani*	Positive = yellow Negative = blue-green	
Glycerol (0.6%)	X			X	*Propionibacterium acnes* *Clostridium tetani*	Positive = yellow Negative = blue-green	
Lactose (0.6%)	X			X	*Bacteroides vulgatus* *Clostridium tetani*	Positive = yellow Negative = blue-green	
Maltose (0.6%)	X			X	*Bacteroides vulgatus* *Clostridium tetani*	Positive = yellow Negative = blue-green	

Continued

TABLE 2—*Continued*

Medium	Quality control procedures				Organism(s)	Expected reaction(s)	Comment
	Sterility	Growth	Inhibition	Biochemical response			
Mannitol (0.6%)	X			X	*Propionibacterium acnes* *Clostridium tetani*	Positive = yellow Negative = blue-green	
Mannose (0.6%)	X			X	*Bacteroides vulgatus* *Clostridium tetani*	Positive = yellow Negative = blue-green	
Rhamnose (0.6%)	X			X	*Bacteroides vulgatus* *Clostridium tetani*	Positive = yellow Negative = blue-green	
Salicin (0.6%)	X			X	*Clostridium tertium* *Clostridium tetani*	Positive = yellow Negative = blue-green	
Starch hydrolysis (0.25%) agar broth	X			X	*Bacteroides vulgatus* *Clostridium tetani*	Positive, agar = clear zone after addition of iodine Positive, broth = remains clear after addition of iodine Negative, agar = no zone after addition of iodine Negative, broth = turns black after addition of iodine	
Sucrose (0.6%)	X			X	*Bacteroides vulgatus* *Clostridium tetani*	Positive = yellow Negative = blue-green	
Trehalose (0.6%)	X			X	*Clostridium tertium* *Clostridium tetani*	Positive = yellow Negative = blue-green	
Xylose (0.6%)	X			X	*Bacteroides vulgatus* *Clostridium tetani*	Positive = yellow Negative = blue-green	
Chopped-meat broth	X	X			*Clostridium sporogenes*	Good growth from a small inoculum (0.01 ml of a 24- to 48-h LD broth culture diluted to 10^{-3}) in 24 h	Proteolysis occurs with *C. sporogenes*. Useful as a holding medium and for sporulation of clostridium and toxin production of some clostridia. Glucose (0.3%) may be added as enrichment with same quality control.
Egg yolk agar	X	X		X	*Clostridium perfringens* *Fusobacterium necrophorum* *Clostridium sporogenes*	Lecithinase positive; no lipase or proteolysis Lipase positive; no lecithinase or proteolysis Lipase and proteolysis positive; no lecithinase	Read up to 48 h.

Continued

TABLE 2—*Continued*

Medium	Quality control procedures				Organism(s)	Expected reaction(s)	Comment
	Sterility	Growth	Inhibition	Biochemical response			
Esculin agar and broth	X			X	*Bacteroides fragilis*	Agar: moderate growth, hydrolysis (deep brown); broth: brown after indicator added	Unhydrolyzed esculin fluoresces under a Wood's lamp. Positive hydrolysis = no fluorescence; negative hydrolysis = fluorescence.
					Clostridium tetani	Agar: moderate growth, no hydrolysis (no color change); broth: no change after indicator added	
Gelatin (Thiogel)	X			X	*Clostridium perfringens*	Abundant growth, liquefaction	
					Bacteroides fragilis	Abundant growth, no liquefaction	
H_2S semisolid medium (with lead acetate)	X			X	*Clostridium sordellii*	H_2S produced (blackening); motility	
					Bacteroides fragilis	No H_2S; no motility	
Indole-nitrate medium	X			X	*Clostridium perfringens*	Growth abundant; NO_3 produced; no indole	Fastidious anaerobes may require hemin, normal serum, or both as supplements for growth. Read at 48 h.
					Propionibacterium acnes	Growth moderate; indole and NO_3 produced	
					Bacteroides fragilis	Growth moderate; no indole or NO_3	
Iron-milk medium	X			X	*Clostridium perfringens*	Positive coagulation, gas; negative digestion, blackening	
					Clostridium sporogenes	Negative coagulation; positive gas, digestion, blackening	
					Fusobacterium necrophorum	Negative coagulation, gas digestion, blackening	
Milk digestion (agar)	X			X	*Clostridium sporogenes*	Milk digested (clear zone)	
					Clostridium perfringens	No change (no zone); cloudy	
Motility	X	X			*Bacteroides fragilis*	Growth moderate; no motility	Read at 48 h.
					Clostridium sordellii	Growth abundant; motility	
Peptone yeast broth	X	X			Uninoculated	See comments	Uninoculated broth should show only trace amounts of volatile and nonvolatile fatty acid when tested with gas-liquid chromatography. Addition of glucose (1%) should allow characteristic metabolic by-products on testing with gas-liquid chromatography.

Continued

TABLE 2—*Continued*

Medium	Quality control procedures: Sterility	Growth	Inhibition	Biochemical response	Organism(s)	Expected reaction(s)	Comment
Phenylethyl alcohol agar	X	X	X		*Clostridium perfringens* *Escherichia coli*	Growth in 24 h No growth	Only inhibits facultative gram-negative rods. Facultative gram-positives and all obligate anaerobic organisms should grow.
Thioglycolate (enriched)	X	X			*Bacteroides fragilis*	Growth in 24 h	Also use as control for 20% bile broth.
Urea	X			X	*Clostridium sordellii* *Clostridium bifermentans*	Positive = pink Negative = no change	

[a] From references 1 and 2.

trol. The American Type Culture Collection can provide representative strains of essentially any microorganism that the clinical laboratory may want to use for a quality control test. Organisms are also available from several commercial sources. Many laboratories participate in proficiency testing programs sponsored by governmental agencies or professional organizations. Cultures distributed by these agencies can be saved and used for quality control testing. Finally, wild strains isolated in the laboratory may be saved and used for quality control if they are properly characterized and reactions are well verified. The test organisms should be checked for purity on agar plates. Incubation should be under the same conditions as the incubation of patient materials. In certain instances, such as the use of primary isolation media for *Mycobacterium tuberculosis*, it may be necessary to conduct quality control testing simultaneously with

TABLE 3. Media and reagents for mycobacteriology[a]

Medium or reagent	Quality control procedures: Sterility	Growth	Inhibition	Biochemical response	Organism(s)	Expected reaction(s)	Comment
Arylsulfatase	X			X	Uninoculated (reagent control) *Mycobacterium intracellulare* TMC 1403 (3-day test) *M. tuberculosis* TMC 201 (14-day test) *M. fortuitum* TMC 1529 (3-day test) *M. intracellulare* TMC 1403 (14-day test)	Negative (no color change) Positive (pink to red)	1. Compare positive with color standards 2. Sodium carbonate (2 N) is used for the 3-day and the 2-week tests. 3. Three-day test is for rapid growers. 4. Inoculate with 7- to 10-day 7H9 broth culture.
Catalase (semiquantitative) (68°C)				X	*M. tuberculosis* TMC 201 or *M. bovis* BCG, TMC 1011 (68°C test only) *M. intracellulare* TMC 1403 (semiquantitative only) *M. kansasii* TMC 1201 (semiquantitative and 68°C)	Negative (no bubbles) within 10 min Positive (bubbles) (<45 mm) Positive (bubbles) (≥45 mm)	1. Positive may take 5 min after reagent (30% H_2O_2 + 10% Tween 80). 2. The room temperature and 68°C tests can be done using the same slant or plate. Scrape colonies for 68°C test before room temperature test.
Hydroxylamine		X			*M. bovis* BCG, TMC 1011	Negative (growth on reagent slant <1%)	1. Reagent slant and plain L-J are inoculated.

Continued

TABLE 3—*Continued*

Medium or reagent	Quality control procedures				Organism(s)	Expected reaction(s)	Comment
	Sterility	Growth	Inhibition	Biochemical response			
Hydroxylamine					*M. kansasii* TMC 1201	Positive (growth on reagent slant ≥1%)	2. Incubate for 4 weeks. 3. Estimate proportion of growth on reagent slant relative to control (plain L-J).
Iron uptake (ferric ammonium citrate incorporated into L-J before inspissation)		X		X	*M. chelonae* TMC 1542 or *M. bovis* BCG, TMC 1011 *M. fortuitum* TMC 1529	Negative (no color change) Positive (rusty-brown colony on tan medium)	1. Final concn of ferric ammonium citrate in L-J is 2.5%. 2. Slants are incubated for 4 weeks. 3. Compare color of colonies on reagent slant with those on plain L-J.
Isoniazid		X			*M. tuberculosis* TMC 201 Not done	Negative (growth on reagent slant <1%) Positive (growth on reagent slant ≥1%)	1. Reagent slant and plain L-J are inoculated. 2. Estimate proportion (%) of growth in reagent slant relative to control slant.
MacConkey agar (use medium without crystal violet)	X	X			Uninoculated plate *M. kansasii* TMC 1201 or *M. bovis* BCG, TMC 1011 *M. fortuitum* TMC 1529	Negative (no growth) Negative (no growth) Positive (growth; with or without color change)	1. Use 3-mm loop of 7-day liquid culture. 2. Inoculate in a spiral from center outward. 3. If turntable unavailable, streak for isolation.
Niacin	X	X		X	*M. intracellulare* TMC 1403 or *M. bovis* BCG, TMC 1011 *M. tuberculosis* TMC 201	Negative (no color change) Positive (yellow color) (If 7H10 is used, must use warm [37°C] extracting fluid and incubate medium at 37°C for 2 h to get good niacin extraction)	1. Test only rough nonchromogenic strains from L-J. 2. Must have 50 to 100 colonies with growth 3 to 4 weeks old. 3. Niacin is extracted from the medium, not the organism. 4. Reagents require special safety precautions.
Nitrate reduction	X	X		X	Uninoculated and *M. bovis* BCG, TMC 1011 or *M. intracellulare* TMC 1403 *M. kansasii* TMC 1201 or *M. tuberculosis* TMC 201	Negative (no color change after reagents, but red after zinc addition) Positive (pink to deep red after reagents or no color change after zinc addition)	1. Test 3- to 4-week-old colonies. 2. Compare results with color standard.

Continued

TABLE 3—*Continued*

Medium or reagent	Quality control procedures				Organism(s)	Expected reaction(s)	Comment
	Sterility	Growth	Inhibition	Biochemical response			
Photochromogenicity				X	*M. kansasii* TMC 1201	Positive (colony change from buff to yellow)	1. Use lightly inoculated L-J slants (isolated colonies) incubated in the dark for about 2 weeks. 2. Expose to light source for 1 to 2 h. Loosen cap. 3. Observe after 18 h for pigment. May need to recheck in 1 week.
Pyrazinamidase		X		X	*M. bovis* BCG, TMC 1011 *M. tuberculosis* TMC 201	Negative (no pink band in agar) Positive (pink band in agar)	1. Use heavy inoculum in two tubes; test one at 4 days and one at 7 days. 2. Read after 4 h. 3. If the 4-day test is positive, do not perform the 7-day test.
7H10	X	X			Uninoculated *M. tuberculosis* TMC 201	Negative (no growth) Positive (growth; nonpigmented)	1. Six stock solutions can be prepared in advance. 2. Solutions remain stable for at least 1 month. 3. If precipitation or flocculation occurs in any stock solution, discard. 4. Do not boil basal commercial medium.
Sodium chloride (5%)		X			*M. tuberculosis* TMC 201 or *M. kansasii* TMC 1201 *M. fortuitum* TMC 1529	Negative (growth on plain L-J but not reagent slant) Positive (any growth on reagent slant)	1. Reagent slant is 5% NaCl in L-J. 2. Make 10-fold dilution of culture and inoculate with a full loop (0.02 ml). 3. Incubate for 4 weeks.
Tellurite reduction		X		X	Uninoculated medium and reagent *M. intracellulare* TMC 1403	Negative (no black precipitate) Positive (black metallic precipitate of tellurium	1. Incubate test culture for 7 days. 2. Must have very heavy growth at 7 days; if not, start over.
Thiophene-2-carboxylic acid hydrazide		X			*M. bovis* BCG, TMC 1011 *M. tuberculosis* TMC 201	Negative (growth on reagent slant <1%) Positive (growth on reagent slant ≥1%)	1. Thiophene-2-carboxylic acid hydrazide at 0.15 mg/ml, aqueous stock. Final concn, 1 μg/ml. 2. Inoculate reagent slant and plain L-J.

Continued

TABLE 3—*Continued*

Medium or reagent	Quality control procedures: Sterility	Growth	Inhibition	Biochemical response	Organism(s)	Expected reaction(s)	Comment
Tween 80 hydrolysis (degradation)		X		X	Uninoculated and *M. intracellulare* TMC 1403 or *M. bovis* BCG, TMC 1011	Negative (amber fluid)	1. Commercial reagent is available. 2. Disregard color of cells.
					M. kansasii TMC 1201	Positive (pink to red fluid)	
Urease broth	X	X		X	Uninoculated	Negative (no color change)	Inoculate liquid medium and incubate for 3 days.
					M. fortuitum TMC 1529	Positive (pink to red)	
Urease disk				X	Uninoculated and *M. intracellulare* TMC 1403	Negative (no color change)	1. Heavy inoculum in 0.5 ml of sterile distilled water. 2. Add 1 reagent disk (Difco). 3. Incubate for 72 h.
					M. tuberculosis TMC 201 or *M. fortuitum* TMC 1529	Positive (cherry-red color)	

[a] From references 1 and 2. The test procedures for reagents should be checked when each new vial or container is prepared or opened, and on each day of use, with at least one acid-fast organism that produces a positive reaction. Negative controls with these reagents are optional and are prepared at the discretion of the laboratory. Reagents used for the iron uptake test should be checked when each new vial or container is prepared or opened, and on each day of use, with at least one acid-fast organism that produces a positive reaction and with an acid-fast or non-acid-fast organism that produces a negative reaction. The other controls listed are strongly recommended as good laboratory practice. L-J, Lowenstein-Jensen medium.

culture because of the slow growth of the organisms. When batches of media have been proven to support the growth of organisms and provide the essential growth characteristics, the batches of media can be released for use.

SPECTRUM OF QUALITY CONTROL

The frequency of performing quality control procedures needs to be determined from the experience of the laboratory. Restrictions, however, are placed upon laboratories that are involved in certification programs. To meet their licensure certification requirements, laboratories need to perform quality control procedures according to a prescribed pattern. Careful records of quality control procedures should be made and maintained, and the records should be reviewed to determine which media are rarely out of control as opposed to the media that, for reasons of instability or other problems, may frequently not be up to standards of quality. Quality control of culture media should not be a blind procedure, but should be approached in a rational and disciplined manner.

Tables 1 through 3 offer suggestions on organisms to be used with selected bacteriologic media and their expected reactions (1, 2). Generally, two organisms will suffice to check the growth characteristics, selective or inhibitory characteristics, and biochemical response. Occasionally, more organisms may be required to confirm all of the reactions.

Note. The media quality control practices for aerobic and anaerobic commercial test systems are identical to those used for conventional media.

Note. A "batch" of medium refers to all of the tubes, plates, or containers of medium prepared at the same time in the laboratory or all of the plates, tubes, or containers having the same lot number that are received in a single shipment from an outside supplier.

LITERATURE CITED

1. **Miller, J. M.** 1983. Quality control in microbiology. Centers for Disease Control, Atlanta, Ga.
2. **Miller, J. M., and B. B. Wentworth (ed.).** 1985. Methods for quality control in diagnostic microbiology. American Public Health Association, Washington, D.C.
3. **Vera, H. D., and D. A. Power.** 1980. Culture media, p. 965–999. In E. H. Lennette, A. Balows, W. J. Hausler, Jr., and J. P. Truant (ed.), Manual of clinical microbiology, 3rd ed. American Society for Microbiology, Washington, D.C.

Chapter 111

Culture Media

ELIZABETH PHILLIPS AND PETER NASH

The microorganisms listed in this Manual vary in their growth requirements. All organisms require sources of nitrogen, carbon, and trace elements. Some organisms can utilize a chemically simplified medium such as ammonia or nitrate, whereas others can fix free nitrogen. Other organisms require protein hydrolysates in the form of peptones, which provide nitrogenous compounds in a more available form. These peptones are water-soluble materials derived from proteins by means of acid, alkali, or added or intrinsic enzymes. Most microorganisms listed have limited ability to synthesize compounds, so they require complex compounds to grow on. The user should be very careful in the preparation of all media (60).

PREPARATION

Media prepared from dehydrated materials should be prepared in accordance with directions given by the manufacturer. The use of chemically clean equipment and glassware is essential. Distilled or demineralized water should always be used unless specified otherwise. Dry materials must be accurately weighed, and water must be accurately measured. Heating for solution of ingredients and for sterilization of media may be accomplished by the use of direct heat, a boiling-water bath, or an autoclave and should be used for the shortest time possible. Excessive heating should be avoided.

Since the microorganisms found in clinical specimens grow best at a pH near neutrality, it is essential that the pH after autoclaving be checked regularly. It is especially important to check the final pH of media prepared from dehydrated materials which have been opened, to detect changes which may have occurred.

Sterilization is usually accomplished by autoclaving or by filtration. Volumes of up to 500 ml of most media are autoclaved at 121°C for 15 min. Larger volumes may need 20 to 30 min or more.

Temperatures from 116 to 118°C should be used for media containing heat-stable carbohydrates. The pH of these media should be checked before sterilization. For media containing heat-labile carbohydrates, the carbohydrates should be sterilized by filtration and added to cooled, autoclaved medium by aseptic technique. Carbohydrates may also be added by using impregnated paper disks. There are various filters which can be used to filter sterilize carbohydrates. All filtration equipment must be sterile, and care should be taken to use aseptic technique. The dispensing of sterile media and the addition of sterile substances should be carried out in a laminar-flow hood by strict aseptic technique.

Enrichments such as blood should be added aseptically to cooled base media, and care should be taken to avoid forming froth or bubbles during mixing. Media enriched with blood should not be incubated before use. Sterility should be checked by incubating several tubes or plates from each batch of medium at 20 to 25°C or at 30 to 35°C for up to 1 week. The remainder of each batch should be refrigerated as soon as possible.

Blood or serum used for enrichment or for determination of hemolysis should be from animals which are free of antimicrobial agents. Defibrinated sheep blood is the blood of choice for routine work since the hemolytic reactions of streptococci on sheep blood are "true" and the growth of *Haemophilus haemolyticus*, which may be mistaken for hemolytic streptococci, is inhibited.

Blood from other animal species, including humans, may be used to demonstrate hemolysis. However, hemolytic reactions on these bloods may differ from those on sheep blood, as follows.

1. Hemolytic reactions on defibrinated rabbit blood correlate with those of sheep blood; however, both *H. haemolyticus* and *Haemophilus influenzae* will grow on media containing rabbit blood.
2. Hemolytic reactions on defibrinated horse blood are not dependable, and some streptococci, e.g., group D, may give the hemolytic reactions of group A.

Blood from the blood bank should be used with caution since this blood contains citrates and glucose. Citrates are inhibitory to some organisms, and glucose may cause alpha- rather than beta-hemolysis. Antibodies and antimicrobial agents in human blood may also cause inhibition of growth or false hemolytic reactions or both.

QUALITY CONTROL

As stated in the previous chapter, quality control of the media used is very important. The performance characteristics of raw materials of biological origin are determined by the nature of the starting material and the methods of processing. Variations may occur between lots or batches of ingredients. It may be necessary to adjust or supplement media formulations to meet performance criteria. Media should conform to accepted performance criteria (60). Examples of media for which performance standards have been prepared are given in Chapter 110.

Sterility and performance of media

The sterility and performance of all media must be verified before use, whether the media are prepared by an individual laboratory from "scratch" or from dehydrated materials or whether they are received ready to use from commercial companies. Commercially prepared media include systems in kit form which are used for the identification of microorganisms. Although commercially prepared media must meet standards set by the Food and Drug Administration, these products should be subjected to tests for sterility and performance to detect changes that may have occurred during shipment.

Materials used in the preparation of media should be dated when received, and the instructions of the manufacturer for storage and expiration dates should be followed carefully. Complete records should be kept for each batch of medium prepared in the laboratory. A 5 to 10% sample of each batch of medium should be tested for sterility, especially with those media to which one or more components are added after the basal media are sterilized. In addition, as each plate is used, it should be examined for random surface contamination.

Special care should be taken in testing the sterility of selective media, since contaminating organisms may be inhibited on these media but appear later when transferred to nonselective media. For this reason, the sterility of selective media should be tested by inoculating samples of each batch onto noninhibitory medium.

Media should also be examined for color, clarity, and evidence of dehydration. Precipitates may appear in some media during storage. If such precipitates do not disappear on heating, the medium should be discarded unless the medium normally contains some insoluble component.

The pH of each batch of medium should be checked electrometrically after the completed medium or medium base has cooled to room temperature. The pH should be in a range of ±0.2 of the value indicated by the manufacturer.

All media should be subjected to performance testing unless the reliability of a certain medium has been ensured by extensive prior testing. The laboratory should maintain a collection of stock cultures with known typical culture characteristics adequate for testing all media used in the laboratory. Performance testing should be based on the use for which the medium being tested is intended and should include organisms which give both positive and negative results. For multipurpose media such as blood agar plates which are used for the isolation of fastidious organisms and the demonstration of hemolysis, it may be necessary to use more than one test organism.

FORMULAS AND MEDIA

The formulas are listed according to requirements determined by the authors of individual chapters in this Manual. The generic names for components such as peptones are given, and in some cases, trade names are noted. Many of the media listed are available from commercial sources. Note that different terminologies and spellings may be given. Contact the manufacturers if questions arise. Note that media or components designated by an asterisk (*) are available commercially in dehydrated or prepared form. The composition of the commercial preparation may be either identical or comparable to the formula submitted by the author, and the similarity may be determined by comparing the formulas.

When commercial media are used, preparation should be in accordance with the directions of the manufacturers. The water of hydration for individual preparation is included if submitted by the author. The formulas used here, although not traditional or always chemically correct, are reproduced essentially as received from individual authors, except for changes made in the interest of uniform presentation. It should be noted that there is no effort to list all manufacturers of media and inclusion or exclusion does not constitute endorsement or disapproval of any manufacturer or product.

Full names and addresses of suppliers will appear at the end of this chapter.

Acetate agar*

For differentiation of species of Shigella and Escherichia (56)

Sodium acetate	2.00 g
Sodium chloride	5.00 g
Magnesium sulfate	0.20 g
Monoammonium phosphate	1.00 g
Dipotassium phosphate	1.00 g
Bromothymol blue	0.08 g
Agar, dried	17.00 g
(*or* Not dried	20.00 g)
Distilled water	1.00 liter
Final pH 6.7 ± 0.2	

Dissolve and dispense into tubes for 2.5-cm butts and 3.8-cm slants. Autoclave at 121°C for 15 min. Inoculate and incubate in the same manner as Simmons citrate agar (see citrate agar).

For nonfermenting gram-negative bacteria

A. Method of Tatum et al. (54)

Add a 0.2% final concentration of sodium acetate to commercial Simmons agar base. Adjust pH to 6.8. Dispense into tubes (4-ml, 13 by 100 mm, screw capped), autoclave at 121°C for 15 min, and slant. Inoculate and incubate in the same manner as Simmons citrate agar (see citrate agar).

B. Method of Gilardi (19)

See carbon assimilation medium.

Aeromonas differential agar (45)*

Casein peptone	10.00 g
Meat extract	3.00 g
NaCl	5.00 g
Dextrin	15.00 g
Sodium sulfite	1.60 g
Fuchsin	0.25 g
Na_2HPO_4	7.75 g
Agar	13.00 g
Distilled water	1.00 liter
Final pH 7.5 ± 0.2	

To dissolve the fuchsin, one may add 50 ml of 5% aqueous Dioxan (Merck). Boil to solution. Autoclave.

A redness of the medium can be removed by adding a few drops of sodium sulfite solution.

Albumin-fatty acid broth. Ellinghausen and McCullough, modified* (25–27)

Leptospira medium and enrichment EMJH*

Basal medium

Disodium phosphate (anhydrous)	1.0 g
Monopotassium phosphate (anhydrous)	0.3 g

Sodium chloride	1.0 g
Ammonium chloride, 25% aqueous	1.0 g
Thiamine hydrochloride, 0.5% aqueous	1.0 ml
Sodium pyruvate, 10% aqueous	1.0 ml
Glycerol, 10% aqueous	1.0 ml
Distilled water	996.0 ml

Dissolve salts in 996 ml of distilled water and then add stock solutions and adjust to pH 7.4. Autoclave at 121°C for 20 min.

Albumin-fatty acid supplement

Bovine albumin fraction V*	20.0 g
Calcium chloride · $2H_2O$ and magnesium chloride · $6H_2O$; aqueous solutions, 1.5% each salt	2.0 ml
Zinc sulfate · $7H_2O$, 0.4% aqueous	2.0 ml
Copper sulfate · $5H_2O$, 0.3% aqueous	0.2 ml
Ferrous sulfate · $7H_2O$, 0.5% aqueous	20.0 ml
Vitamin B_{12}, 0.2% aqueous	2.0 ml
Polysorbate (Tween) 80, 10% aqueous	25.0 ml
Distilled water, to	200.0 ml

1. Add the bovine albumin slowly, with careful stirring (to avoid foaming), to 100 ml of water.
2. Slowly add the remaining ingredients (prepared as stock solutions) to the albumin solution, with constant stirring.
3. Adjust the pH to 7.4, and add water to a 200-ml final volume.
4. Sterilize by filtration through a membrane or Seitz filter (porosity, 0.2 to 0.3 µm).
5. Store at 4 or −20°C.

Aseptically combine 1 part supplement with 9 parts basal medium, and dispense into sterile containers.

Note. If deionized water is used, preheat it to 56°C for 30 min to destroy naturally occurring water leptospires which are not retained by membrane and asbestos filters (0.2 µm).

Albumin-fatty acid semisolid medium. Ellinghausen and McCullough, modified* (25–27)

1. To 900 ml of basal broth medium (see above), add 2.0 g of agar (quality tested).
2. Heat to dissolve the agar.
3. Dispense.
4. Autoclave at 121°C for 20 min.
5. Cool to 50°C.
6. Add 1 volume of prewarmed (45 to 50°C) albumin supplement (see above) to 9 volumes of the agar basal medium.
7. Media may be stored at room temperature if dispensed into screw-capped tubes.

Alkaline peptone water

Peptone (Difco)	10.0 g
NaCl	5.0 g
Distilled water	1.0 liter

Final pH 8.4

Autoclave at 121°C for 20 min.

Amies transport medium without charcoal*†

Sodium thioglycolate	1.00 g
Sodium chloride	3.00 g
Potassium chloride	0.20 g
Calcium chloride	0.10 g
Magnesium chloride	0.10 g
Potassium phosphate, monobasic	0.20 g
Sodium phosphate, dibasic	1.15 g
Agar	4.00 g
Demineralized water	1 liter

Final pH 7.4 ± 0.2 at 25°C

† Amies transport medium with charcoal contains, additionally, 10 g of charcoal per liter.

Ammonium nitrate agar

Trichophyton agars 6 and 7*

Ammonium nitrate	1.5 g
Glucose	40.0 g
Magnesium sulfate	0.1 g
Monopotassium phosphate	1.8 g
Agar	20.0 g
Distilled water	1.0 liter

Adjust pH to 6.8, and dispense in 100-ml amounts into flasks.

For histidine ammonium nitrate agar, add 2 ml of histidine solution at 150 mg/100 ml of water. Tube, sterilize at 121°C for 15 min, and slant.

Antibiotic medium 3 FDA* (23)

Antibiotic assay broth* (BBL); Penassay broth* (Difco)

For susceptibility testing with polyene antifungal antibiotics

Peptone or Gelysate (BBL)	5.00 g
Yeast extract	1.50 g
Beef extract	1.50 g
Sodium chloride	3.50 g
Glucose	1.00 g
Dipotassium phosphate	3.68 g
Potassium dihydrogen phosphate	1.32 g
Distilled water	1.00 liter

Final pH 7.0±

1. Heat to boiling to dissolve.
2. Dispense, and autoclave at 121°C for 15 min.

Antibiotic medium 12 FDA* (23)

Nystatin assay agar*

For agar dilution susceptibility tests with polyene antifungal antibiotics

Peptone or Gelysate (BBL)	9.40 g
Yeast extract	4.70 g
Beef extract	2.40 g
Sodium chloride	10.00 g
Glucose	10.00 g
Agar	23.50 g
Distilled water	1.00 liter

Final pH 6.1±

1. Heat to boiling to dissolve.
2. Dispense, and autoclave at 121°C for 15 min.

Arenavirus plaquing medium

Eagle basal medium (pH 7.3) without phenol red, 2× concentration
40 mM HEPES (*N*-2-hydroxyethylpiperazine-*N'*-2-ethanesulfonic acid) buffer
26 mM $NaHCO_3$
Add NaOH; adjust the pH to 7.0

Add an equal volume of the above solution to a 2% (wt/vol) solution of agarose in distilled water, at 50°C. Add fetal calf serum or heat-inactivated (56°C, 30 min) newborn calf serum to a final concentration of 5%.

Arginine hydrolysis (Moeller decarboxylase media)

Peptone (Orthana special) .... 5 g
Beef extract .... 5 g
Bromocresol purple .... 0.625 ml
Cresol red (0.2%; prepared by grinding 0.5 g of cresol red powder to a fine powder, adding 26.2 ml of 0.01 N NaOH, and diluting to 250 ml with distilled water) .... 2.5 ml
Pyridoxal .... 5 mg
L-Arginine (if DL-arginine is used, add 20 g) .... 10 g
Glucose .... 0.5 g
Distilled water .... 1,000 ml

1. Adjust the pH to 6.0 to 6.5. Dispense in 3-ml amounts into 13- by 100-mm screw-capped tubes.
2. Sterilize in an autoclave for 10 min at 121°C.
3. Immediately after inoculation, add a layer (about 10 mm) of sterile mineral oil. Reactions: a positive reaction is recorded when the indicators turn violet to reddish violet. Yellow does not indicate a positive reaction; it indicates an acid reaction rather than deamination.

Auxanographic agar media

For carbohydrate assimilation tests

1. Prepare a 10× concentration of yeast nitrogen base* (YNB). Sterilize by filtration.
2. Autoclave a 2% aqueous solution of washed agar. Autoclave at 121°C for 15 min, and cool to 50°C.
3. Pipette 2.0 ml of the YNB into each petri plate. Mix 20 ml of sterile agar; allow the medium to solidify.
4. Use a sterile swab to streak an actively growing yeast culture from a Sabouraud slant for confluent growth. Allow the surface to dry.
5. Place carbohydrate disks near the periphery of each plate. Use glucose, maltose, sucrose, lactose, galactose, cellobiose, melibiose, xylose, raffinose, inositol, trehalose, and dulcitol.
6. Incubate at 25°C, and examine for growth around disks.

For nitrate assimilation tests

1. Use 10× yeast carbon base* (YCB), and proceed as described above.
2. Add disks containing potassium nitrate or peptone (or amino acids).

Bacteroides bile esculin agar (BBE) (37)

Trypticase soy agar* (BBL) .... 40.0 g
Oxgall .... 20.0 g
Esculin .... 1.0 g
Ferric ammonium citrate .... 0.5 g
Hemin solution (5 mg/ml) .... 2.0 ml
Gentamicin solution (40 mg/ml) .... 2.5 ml
Distilled water .... 1,000 ml

1. Adjust pH to 7.0.
2. Heat to dissolve. Autoclave at 121°C for 15 min. Cool to 50°C, and pour plates.

Beef infusion agar*

Ground defatted beef, infusion from .... 453.6 g
Peptone .... 10.0 g
Sodium chloride .... 5.0 g
Agar .... 20.0 g
Distilled water .... 1.0 liter

1. Allow meat to infuse in water overnight at 4 to 5°C. Cook for 1 h at 80 to 90°C. Allow to stand for 2 h, and filter through muslin.
2. Add peptone and salt, and adjust to pH 7.6 with 4% NaOH. Filter.
3. Add agar, and autoclave. Filter through cotton or several layers of milk filter disks. Autoclave again.

Beef infusion broth*

Prepare as described above, but omit the agar.

Bennett agar

Pancreatic digest of casein USP* .... 2.0 g
Yeast extract* .... 1.0 g
Beef extract* .... 1.0 g
Glucose .... 10.0 g
Agar .... 15.0 g
Demineralized water .... 1 liter

Heat with agitation to dissolve. Dispense and autoclave.
Recommended for production of aerial hyphae and spores by *Nocardia* and *Streptomyces* species.

Bicarbonate agar

Trypticase soy agar* (BBL; sterile fluid at 50°C) .... 90 ml
Sodium bicarbonate (7% aqueous, filter sterilized) .... 10 ml

Bile esculin agar*

Nutrient agar .... 23 g
or
(Beef extract .... 3.0 g)
(Peptone .... 5.0 g)
(Agar .... 15.0 g)
Oxgall .... 40.0 g
Ferric citrate .... 0.5 g
Water .... 1 liter

1. Dissolve the nutrient agar, or beef extract, peptone, and agar, in 400 ml of water; heat until colloidal. Mix oxgall with 400 ml of water and heat into solution. Mix the ferric citrate with 100 ml of water and heat into solution.

2. Combine the solutions and mix well. Heat to 100°C for 10 min.

3. Autoclave at 121°C for 15 min. Cool to 50°C. (This is the base medium.)

4. Aseptically add 100 ml of esculin solution (100 ml of water plus 1 g of esculin, heated gently to obtain solution, and filter sterilized).

5. Dispense into sterile, 16- by 125-mm screw-capped tubes. Tighten the caps and cool in slanted position.

6. Alternatively, use dehydrated medium of this formula and prepare according to directions.

7. Optional: add 50 ml of horse serum to the sterile cool base.

Authors' note. The original medium contained horse serum (50 ml), added to the base medium. In a controlled study we found, however, that this addition was not necessary (13a). At least one commercial source (Difco) adds the esculin to the base medium. The dehydrated medium sold by Difco can be resuspended, tubed, autoclaved, slanted, and used with excellent results.

Bile esculin agar can be used to identify group D streptococci. All group D streptococci (including all enterococci) will blacken the slant, usually within 48 h. Most non-group D streptococci do not blacken the medium.

Birdseed agar

See Staib agar.

Blood agar base*
Heart infusion agar

Beef heart muscle, infusion from	375.0 g
Tryptose or Thiotone peptic digest of animal tissue USP	10.0 g
Sodium chloride	5.0 g
Agar	15.0 g
Distilled or demineralized water	1 liter
pH 7.4	

Autoclave at 121°C for 15 min. Cool to 50°C, and add 5% sterile defibrinated rabbit blood, if desired.

For aerobic actinomycetes, add defibrinated horse blood, 50 ml/900 ml of base.

Blood agar, diphasic, for trypanosomes and leishmanias, NIH method (25)

Lean beef, desiccated*	25.0 g
Neopeptone* or other peptone	10.0 g
Agar	10.0 g
Sodium chloride	2.5 g
Distilled water	500.0 ml

1. Infuse the beef and distilled water in a water bath for 1 h. Heat for 5 min at 80°C (176°F) to coagulate a portion of the protein.

2. Filter with ordinary-grade filter paper.

3. Add the rest of the above ingredients, and adjust the pH to 7.2 to 7.4 with NaOH.

4. Autoclave at 120°C for 20 min.

5. Cool to 45°C, and aseptically add 10% defibrinated rabbit blood.

6. Dispense 5-ml quantities into sterile tubes. Slant and cool.

7. Just before inoculation, overlay with 2 ml of sterile Locke solution.

Locke solution

Sodium chloride	8.0 g
Potassium chloride	0.2 g
Calcium chloride	0.2 g
Monopotassium phosphate	0.3 g
Glucose	2.5 g
Distilled water	1.0 liter

Editors' note. The author recommends Difco ingredients.

Bordet-Gengou (BG) agar without peptone*
For isolation of Bordetella pertussis and Bordetella parapertussis.

Peeled, diced potatoes	187.5 g
Distilled water	1,500.0 ml
Glycerol	15.0 ml
NaCl, reagent grade	10.0 g
Agar	30.0 g
Sterile defibrinated sheep blood	300.0 ml

1. Boil the potatoes in water for about 30 min, until soft, and filter through several layers of gauze.

2. Add glycerol, NaCl, and agar to 1,000 ml of the filtrate, adjust the pH to 7.0, and bring to a boil.

3. Dispense 100-ml amounts into screw-capped bottles or 20-ml amounts into screw-capped tubes.

4. Autoclave, cool, seal, and store refrigerated (for up to 1 year).

5. Just before use, melt the base and hold it at 50 to 60°C. Add 30 ml of sterile defibrinated sheep blood per 100 ml, mix well, and pour five or six plates. (Or add 6 ml of blood to a 20-ml pour.)

6. For selective medium, add 3.25 ml of freshly prepared stock methicillin (at 100 μg/ml) per bottle just before adding blood. (Or add 0.65 ml of stock methicillin per 26-ml pour, or add Cephalex to a final concentration of 40 μg/ml.) Mix methicillin well with agar. (Methicillin concentration is 2.5 μg/ml.)

7. Properly prepared plates should be cherry red, moist, and free from bubbles. Use immediately if possible, but plates may be maintained for about 1 week at 5°C if tightly sealed.

Bordet-Gengou agar base containing peptone should not be used.

Comments. Homemade BG base is superior to commercially prepared dehydrated medium. If commercial BG base is used, it should not contain added peptone, as this formulation is inhibitory to many clinical isolates. BG plates may be stored sealed at 4°C for 1 week for subculture work, but they should be freshly poured for isolation of *B. pertussis* from clinical specimens.

Brain heart infusion agar*
For primary recovery of fungi from clinical specimens

Calf brains, infusion from .... 200.0 g
Beef heart, infusion from .... 250.0 g
Proteose or Gelysate (BBL) pancreatic
digest of gelatin .... 10.0 g
Glucose .... 2.0 g
Sodium chloride .... 5.0 g
Disodium phosphate .... 2.5 g
Agar .... 15.0 g
Distilled or demineralized water .... 1.0 liter
Final pH 7.4

Dispense and autoclave at 121°C for 15 min. For *Actinomyces* spp., the use of freshly poured plates is recommended.

Bromocresol purple (BCP) milk*

Purple milk .... 100.0 g
Demineralized water .... 1.0 liter

Mix to obtain a homogeneous suspension, and dispense 2- to 2.5-ml amounts into 11- by 75-mm tubes. Autoclave at 115 to 118°C for 15 min.

This medium is recommended for determining proteolytic properties of the aerobic actinomycetes. Observe for peptonization. *Nocardia asteroides* does not peptonize, but usually turns this medium alkaline. Incubate an uninoculated tube with every set of tests.

Brucella agar*

Pancreatic digest of casein USP .... 10.0 g
Peptic digest of animal tissues USP ...... 10.0 g
Yeast autolysate .... 2.0 g
Glucose .... 1.0 g
Sodium chloride .... 5.0 g
Sodium bisulfite .... 0.1 g
Agar .... 15.0 g
Distilled water .... 1.0 liter
Final pH 7.0 ± 0.2

Heat with agitation until dissolved. Dispense, and autoclave at 121°C for 15 min. Cool to 50°C.

Additives for anaerobes (53)

For nonselective media

1. Supplement before autoclaving with hemin (5 μg/ml) and, after autoclaving, with 50 ml of sterile defibrinated sheep blood and 1 ml of vitamin K_1 solution per liter.
2. Use 50 ml of laked blood and 1 ml of vitamin K_1 per liter.

Note. Prepare and use vitamin K_1 (44) solution as follows.

a. Weigh 0.2 g on a previously flamed aluminum foil square.
b. Aseptically place 20 ml of absolute alcohol in a sterile tube.
c. Add foil with sterile forceps.
d. When solution has occurred, remove foil. Concentration is 0.01 g/ml.
e. Use 1 ml/liter in agar media (concentration, 10 μg/ml).
f. For fluid media, dilute stock solution 1:100 in sterile water. Use 0.01 ml/liter (concentration, 0.1 μg/ml).

Prepare and use hemin solution as follows.

a. Dissolve 0.5 g of hemin in 10 ml of commercial ammonia water (or 1 N NaOH).
b. Bring the volume to 100 ml with distilled water.
c. Autoclave at 121°C for 15 min. Stock solution is 5 mg/ml. Use as a medium supplement in a final concentration of 5 μg/ml.

For selective media

1. For *Bacteroides* spp. (53), add kanamycin, 100 μg/ml, to brucella agar before autoclaving. Cool, and add blood and vitamin K_1 solution. Also add filter (0.45-μm pore size)-sterilized aqueous vancomycin to a concentration of 7.5 μg/ml.
2. For *B. melaninogenicus*, add kanamycin as for *Bacteroides* spp., but first lyse the blood by freezing and thawing.
3. For certain *Fusobacterium*, *Eubacterium*, and *Clostridium* species, add rifampin (50 μg/ml) and blood just before pouring plates.
4. For *Fusobacterium* and *Veillonella* species, add neomycin (100 μg/ml) before autoclaving; add blood, vitamin K_1, and 7.5 μg of vancomycin per ml just before pouring plates.
5. For *Clostridium* spp. and anaerobic cocci, add neomycin (100 μg/ml) before autoclaving. Add blood and vitamin K_1 just before pouring plates.

Brucella broth*

See brucella agar.

Leave out agar base. Heat with agitation until dissolved.

Dispense, and autoclave at 121°C for 15 min.

Buffered charcoal-yeast extract medium supplemented with α-ketoglutarate (BCYEα) (11)

ACES [*N*-(2-acetamido)-2-
aminoethanesulfonic acid] buffer
(Research Organics) .... 10.00 g
KOH pellets (85% KOH, reagent grade) ..... 2.80 g
Activated charcoal (Norit A) .... 1.50 g
Yeast extract (Difco) .... 10.00 g
Agar (Difco) .... 17.00 g
α-Ketoglutarate, monopotassium salt
(Sigma) .... 1.00 g
L-Cysteine hydrochloride in 10 ml of
distilled water (ICN) .... 0.40 g
Ferric pyrophosphate in 10 ml of distilled
water (Sigma) .... 0.25 g

1. Add ACES buffer to 900 ml of distilled water in a flask. Heat in a water bath at 50°C for about 1.5 h or until buffer has dissolved.
2. Add KOH pellets to buffer solution.
3. Add dry Norit A, yeast extract, agar, and α-

ketoglutarate to solution. Mix well. Use 80 ml of distilled water to rinse into flask.
4. Autoclave for 15 min at 121°C.
5. Cool to 50°C in water bath.
6. Add separately filter-sterilized (0.2-μm filter) ferric pyrophosphate and L-cysteine hydrochloride.
7. Check pH. Should be between 6.85 and 6.95 at 40°C. Correct with 1 N HCl or KOH pellets, if needed.
8. Cool to 40°C, and pour.
9. Preincubate the entire batch for 1 day at 35°C to detect contaminants.
10. Cool to room temperature, and then store noncontaminated plates at 5°C in plastic sleeves. Shelf life is 2 months at 5°C.

Buffered charcoal-yeast extract medium with cefamandole, polymyxin B, anisomycin, and α-ketoglutarate (BMPAα medium) (11)

Yeast extract (Difco)	10.00 g
Agar (Difco)	17.00 g
Activated charcoal (Norit A)	1.50 g
ACES buffer (Research Organics)	10.00 g
KOH pellets (reagent grade, 85% KOH)	2.80 g
α-Ketoglutarate, monopotassium (Sigma)	1.00 g
L-Cysteine hydrochloride (ICN)	0.40 g
Ferric pyrophosphate (Sigma)	0.25 g
Distilled water	1 liter
Polymyxin B sulfate (Pfizer or Sigma) (as base)	80,000 U
Cefamandole lithium (Lilly) (as base)	4.00 μg
Anisomycin (General Diagnostics)	80.00 g

1. Prepare BCYEα as detailed above, and cool to 50°C in a water bath. Do not check pH until after antimicrobial agents are added.
2. Add filter-sterilized antimicrobial agents, and then mix by swirling. Check pH, which should be 6.85 to 6.95 at 40°C.
3. Pour into petri dishes, with about 24 ml per plate.
4. Do not preincubate the entire batch to check for sterility. Instead, preincubate one plate to check for contaminants. Preincubation degrades the cefamandole.
5. Refrigerate at 5°C in plastic sleeves. Shelf life is 3 to 4 weeks at 5°C.

Butzler medium (4)

Thioglycolate medium	1 liter
Sheep blood	50 to 70 ml
Bacitracin	25,000 IU
Novobiocin	5 mg
Actidione (cycloheximide)	50 mg
Cefazolin	15 mg
Colistin	10,000 U
Agar	10.5 g

1. Mix thioglycolate medium and agar.
2. Sterilize the mixture, and cool to 50°C.
3. Add blood and other ingredients to the sterile molten agar.

Caffeic acid (niger seed) agar
For primary recovery of Cryptococcus neoformans

Pulverized *Guizottia abyssinica* seed	50 g
Agar	15 g

Add seed to 100 ml of distilled water, and grind in a beaker. Boil for 0.5 h in 1 liter of water. Strain through cloth to remove the water extract from the seed. Adjust volume of extract to 1 liter with distilled water. Add agar, and heat until dissolved. Autoclave at 121°C for 15 min. Final pH should be 5.5.

Or

Glucose	5 g
Ammonium sulfate	5 g
Yeast extract	2 g
Potassium phosphate	0.8 g
Magnesium sulfate	0.7 g
Caffeic acid	0.18 g
Ferric citrate solution	4.0 ml
Agar	20 g
Distilled water	1 liter

Prepare ferric citrate solution (10 mg plus 20 ml of water). Mix reagents in 1 liter of distilled water, autoclave at 121°C for 12 min, and dispense.
A comparable medium (CN-Screen) may be obtained from Flow Laboratories, catalog no. 3710045.

***Calymmatobacterium granulomatis* semidefined medium (20)**

Lactalbumin hydrolysate or papaic digest of soy meal USP*	20.0 g
Sodium chloride	2.5 g
Dipotassium phosphate	1.5 g
Sodium thioglycolate	0.6 g
L-Cystine	0.4 g
Distilled water	1.0 liter

1. Dissolve with heat, cool, and adjust to pH 7.2.
2. Dispense in 20- to 22-ml amounts into screw-capped tubes.
3. Autoclave at 121°C for 15 min, and tighten caps to maintain reduced conditions.

Campy-BAP (1)

Brucella agar base (Oxoid, CM 169)	1 liter
Sheep blood	100 ml
Vancomycin	10 mg
Trimethroprim	5 mg
Polymyxin B	2,500 IU
Amphotericin B	2 mg
Cephalothin	15 mg

1. Sterilize brucella agar base and cool to 50°C.
2. Add blood and other ingredients to sterile molten agar.

Carbohydrate fermentation broths
For streptococci identification (arabinose, inulin, lactose, mannitol, raffinose, sorbitol, sorbose, trehalose)

Heart infusion broth, 22.5 g in 900 ml of distilled water
Carbohydrate, 10 g in 200 ml of distilled water

Indicator, 1 ml (1.6 g of bromocresol purple in 100 ml of 95% ethanol)

1. Add ingredients together, and dispense in 3-ml amounts into 13- by 100-mm screw-capped tubes.
2. Sterilize in an autoclave for 10 min at 121°C.
3. A positive reaction is recorded when the indicator changes from purple to yellow.

For anaerobes, see fermentation broth (CHO) base.

Carbohydrate-peptone broth

Peptone	10.0 g
Sodium chloride	5.0 g
Test carbohydrate	10.0 g
Distilled water	1.0 liter
Andrade indicator	10.0 ml

1. Adjust the pH to 7.4 to 7.5.
2. Autoclave at 121°C for 15 min.

Andrade indicator

Acid fuchsin (EM Science)	0.15 g
NaCl, 1 N	16.00 ml
Distilled water	100.0 ml

Carbon assimilation medium

Mineral base medium (MBM)

Magnesium sulfate, anhydrous	0.1 g
Sodium chloride	5.0 g
Ammonium phosphate, monobasic	5.0 g
Potassium hydroxide, dibasic anhydrous	1.0 g
Distilled water	1.0 liter
Final pH 6.5 ± 0.1	

1. Prepare MBM at double concentration, and sterilize by filtration.
2. Add carbon sources to a final concentration of 0.03 M pelargonate or DL-norleucine or 0.015 M sodium acetate.

Agar solution

Agar	32.0 g
Distilled water	1.0 liter

1. Dissolve agar in water, bring to boil, and autoclave at 121°C for 15 min.
2. Add an equal volume of sterile, melted agar solution to MBM containing the desired carbon source.
3. Dispense into sterile tubes, and slant.
4. Test for utilization of the carbon source by inoculating the slant with 1 drop of an overnight broth culture. Incubate at 30°C and observe for growth.

Cary and Blair transport (holding) medium*

Sodium thioglycolate	1.5 g
Disodium phosphate	1.1 g
Sodium chloride	5.0 g
Agar	5.0 g
Distilled or demineralized water	991.0 ml

1. Prepare in chemically clean glassware rinsed with Sorensen 0.067 M buffer (pH 8.1).
2. Heat with agitation until the solution just becomes clear.
3. Cool to 50°C, add 9 ml of freshly prepared aqueous 1% $CaCl_2$, and adjust the pH to about 8.4.
4. Dispense 7 ml into previously rinsed and sterilized 9-ml screw-capped vials.
5. Steam vials for 15 min, cool, and tighten the caps.

Cary and Blair transport medium, modified (PRAS)

Sodium thioglycolate	1.5 g
Calcium chloride	0.1 g
Disodium phosphate	0.1 g
Sodium chloride	5.0 g
Sodium bisulfite	0.1 g
Agar	5.0 g
Resazurin solution	4.0 g
Distilled water	1.0 liter

1. Mix, and boil ingredients to dissolve.
2. Gas out with carbon dioxide.
3. Add 0.5 g of L-cysteine hydrochloride.
4. Adjust so that final pH is 8.4.
5. Tube in roll tubes gassed out with nitrogen.
6. Stopper with butyl stoppers.
7. Steam for 15 min on 3 successive days.

Note. Prepare resazurin solution (42) by dissolving 0.05 g of resazurin in 200 ml of 95% ethanol and adding 180 ml of distilled water.

Casein agar
Trichophyton agars 1–5*

A. *With casein peptone for fungi*

Casein, 10% acid hydrolyzed, vitamin free	25.0 ml
(*or* Acid hydrolysate of casein	2.5 g)
Glucose	40.0 g
Magnesium sulfate	0.1 g
Monopotassium phosphate	1.8 g
Agar	20.0 g
Distilled water, to	1.0 liter

1. Adjust pH to 6.8.
2. Dissolve by heating, distribute 100-ml quantities into flasks, and autoclave at 121°C for 15 min.
3. Contents of several flasks may be distributed into test tubes and slanted for use as vitamin-free controls.
4. To 100 ml of sterile, melted casein agar, add autoclaved vitamin solutions as follows: thiamine, 2 ml of a stock solution containing 10 mg/liter of water; inositol, 2 ml of a stock solution containing 250 mg/100 ml of water; thiamine-inositol, 2 ml of each of the above stock solutions; nicotinic acid, 2 ml of a stock solution containing 10 mg/100 ml of water.
5. Tube and slant.

B. *With skim milk for aerobic actinomycetes*

Skim milk (dried)	75.0 g
Agar	20.0 g
Demineralized water	1.0 liter

1. Add the milk to 500 ml of water, a little at a time, stirring constantly; do not leave lumps.
2. Autoclave at 113 to 115°C for 15 min.
3. Dissolve the agar in 500 ml of water, and autoclave.
4. Cool the solutions to 60 to 65°C, and pour the agar solution into the skim milk suspension.
5. Mix gently, and pour 20 ml per plate while hot.

Recommended for differentiation of species of aerobic actinomycetes. Streak a pure culture heavily on duplicate plates, and incubate at room temperature or 37°C for up to 2 weeks. Observe for growth and clearing of medium.

Casein hydrolysate (CAS)
For brief transport and fluorescent-antibody procedures for Bordetella pertussis

Casamino Acids* .... 1 g
Distilled water .... 100 ml

1. Dissolve the casein hydrolysate in the water in an acid-cleaned Pyrex beaker, and adjust the pH to 7.2.
2. Dispense 0.5-ml amounts into 16- by 125-mm acid-cleaned Pyrex test tubes with screw caps.
3. Autoclave, cool, seal, and store refrigerated for up to 1 year.
4. For use, immerse calcium alginate nasopharyngeal swabs in liquid, snip off the end of the wire, seal the tube, and keep it cool. Streak plates or prepare smears within 2 h.

Casein-yeast extract-glucose agar (CYG agar) (57)
For agar dilution susceptibility tests with imidazole antifungal agents

Casein hydrolysate .... 5.00 g
Yeast extract .... 5.00 g
Glucose .... 5.00 g
Agar .... 20.00 g
Distilled water .... 1.00 liter
Final pH 7.0±

1. Heat to boiling to dissolve.
2. Dispense, and autoclave at 121°C for 15 min.

Casein-yeast extract-glucose broth (CYG broth) (57)
For susceptibility testing with imidazole antifungal agents

Casein hydrolysate .... 5.00 g
Yeast extract .... 5.00 g
Glucose .... 5.00 g
Distilled water .... 1.00 liter
Final pH 7.0±

1. Heat to dissolve.
2. Dispense, and autoclave at 121°C for 15 min.

CDC anaerobe blood agar
For anaerobes

Trypticase soy agar* (BBL) .... 40 g
Agar .... 5 g
Yeast extract .... 5 g
**Hemin .... 5 mg
**L-Cystine .... 400 mg
***Vitamin K_1 stock solution .... 1 ml
Distilled water .... 1,000 ml
Final pH 7.5

** Dissolve the hemin and L-cystine in 5 ml of 1 N NaOH before adding the other ingredients.
*** Vitamin K_1 (3-phytylmenadione; ICN) stock solution is made by adding 1 g of vitamin K_1 to 99 ml of absolute ethanol.

Dissolve the ingredients by heating, adjust the pH, and autoclave at 121°C for 15 min. Cool to 50°C, add 50 ml of sterile defibrinated sheep blood, mix, and dispense 20 ml into sterile 15- by 100-mm plastic petri plates. After the medium has solidified, place the plates in cellophane bags and store in a refrigerator (4°C).

CDC modified McClung-Toabe egg yolk agar (EYA) (9)
For anaerobes

Pancreatic digest of casein* .... 40.0 g
$NaHPO_4$ .... 5.0 g
NaCl .... 2.0 g
$MgSO_4$ (5% aqueous solution) .... 0.2 ml
Yeast extract* .... 5.0 g
D-Glucose .... 2.0 g
Agar .... 25.0 g
Distilled water .... 900.0 ml
Egg yolk suspension* .... 100.0 ml
Final pH 7.4

Except for the egg yolk suspension, dissolve the ingredients by heating, autoclave at 121°C for 15 min, and cool to 60°C. Bring the temperature of the egg yolk suspension up to 60°C, and then add to the basal medium. Mix well and dispense 20 ml per 15- by 100-mm plastic petri plate.

The development of an insoluble, opaque precipitate within the agar indicates lecithinase activity. An iridescent sheen or "oil on water" appearance (pearly layer) on the surface growth indicates lipase activity. Proteolysis is indicated by a zone of translucent clearing in the medium around the colonies. These same reactions can be determined on Lombard-Dowell egg yolk agar (Presumpto quadrant plate).

Cell wall-defective forms
Media for recovery

Mycoplasma agar base (PPLO agar base)*

Beef heart, infusion from .... 50 g
Peptone or Biosate (BBL) .... 10 g
Sodium chloride .... 5 g
Agar .... 14 g

Mycoplasma broth base (PPLO broth base)*

Beef heart, infusion from .... 50 g
Peptone or Biosate (BBL) .... 10 g
Sodium chloride .... 5 g

Agar and broth media can be prepared by using PPLO agar and broth with supplementation just before use. Unsupplemented media are prepared according to the manufacturer's directions, with the addition of 20% sucrose and 0.2% $MgSO_4 \cdot 4H_2O$ and adjustment of pH to 7.6 to 7.8. The unsupplemented media may be distributed in appropriate aliquots, autoclaved, and stored at 4 to 5°C for up to several months. The agar is melted in boiling water and cooled to 50°C for final preparation of media.

To 7 parts of the above add 2 parts horse serum and 1 part yeast extract for preparation of supplemented media. The horse serum is not inactivated. The serum should at least be shown to support growth of mycoplasmas (pretested horse serum is available commercially) if there is not an opportunity to test it with cell wall-defective organisms. Agamma horse serum can help in reduction of pseudocolony formation. The yeast extract may also be purchased or prepared as follows. Add 1 part active dry bakers' yeast (Fleischman 20-40) to 4 parts (wt/vol) distilled water, and boil. Filter by gravity through medium-porosity paper. Adjust the pH to 8.0 with sodium hydroxide. Distribute in appropriate aliquots, and autoclave. The yeast extract may be stored at −20°C for up to a few months, but avoid using the precipitate which usually forms.

Some of the variations in the media include: using an unsupplemented base of beef heart infusion broth rather than the PPLO broth, reducing the agar concentration to 1.0 to 1.2% for solid medium, reducing the sucrose to 10% in agar media, substituting 3 to 5% NaCl for the sucrose, heat inactivating (56°C, 30 min) the horse serum, and reducing the yeast extract supplement to 1.0%.

Hypertonic media in bottles under CO_2 and vacuum and intended for blood cultures are also available commercially.

Cetrimide agar, non-USP (54)

Heart infusion agar .... 40.0 g
Cetrimide (hexadecyltrimethylammonium bromide) .... 0.9 g
Distilled water........................... 1.0 liter
Final pH 7.2

Heat to dissolve. Dispense in 5-ml amounts into 15- by 125-mm test tubes, shaking flask to ensure even distribution of sediment.

Note. Lots of cetrimide vary, and the amount needed must be determined with known cultures.

Cetrimide agar, USP (19)
Pseudocel agar*
For nonfermenting gram-negative bacteria

Gelysate pancreatic digest of gelatin ..... 20.0 g
Magnesium chloride .... 1.4 g
Potassium sulfate .... 10.0 g
Agar (dried) .... 13.6 g
Cetrimide* .... 0.3 g
Distilled water........................... 1.0 liter

Suspend ingredients in the water, and add 10 ml of glycerol. Heat with frequent agitation, and boil for 1 min. Dispense, and autoclave at 118 to 121°C for 15 min.

Charcoal agar slant: diphasic medium for amoeba culture

A. Buffered saline overlay

1. Solution A (0.067 M KH_2PO_4)

Potassium dihydrogen phosphate (KH_2PO_4), anhydrous .... 9.07 g
Distilled water, to .... 1,000.0 ml

Dissolve the KH_2PO_4 in a small amount of water in a clean 1-liter volumetric flask. Add water to the 1-liter mark. Mix, and store in a glass-stoppered bottle.

2. Solution B (0.67 M Na_2HPO_4)

Disodium phosphate (Na_2HPO_4), anhydrous .... 9.46 g
Distilled water, to .... 1,000.0 ml

Prepare as directed for solution A.

Note. $Na_2HPO_4 \cdot 12H_2O$ may be substituted for anhydrous Na_2HPO_4 by using 23.88 g of crystalline compound per liter of solution.

3. Buffered saline (0.5%, pH 7.4)

Sodium chloride (NaCl).................... 5.0 g
Solution A.............................. 190.0 ml
Solution B.............................. 810.0 ml

Combine solutions A and B and add the NaCl. Stir thoroughly to dissolve. Sterilize at 15 lb/in^2 pressure for 15 min. Store in the refrigerator.

B. Preparation of agar slants

Disodium phosphate ($Na_2HPO_4 \cdot 12H_2O$) .... 3.0 g
Potassium phosphate (KH_2PO_4) .... 4.0 g
Sodium citrate crystals .... 1.0 g
Magnesium sulfate crystals .... 0.1 g
Ferric ammonium citrate .... 0.1 g
Asparagin (Difco)......................... 2.0 g
Tryptone (Difco).......................... 5.0 g
Glycerol (reagent grade) .... 10.0 ml
Distilled water .... 1,000.0 ml

Add the various ingredients, in the order listed, to the distilled water, and heat to dissolve. Do not boil. Stir thoroughly. Then add, stirring thoroughly after each addition:

Bacto-Agar (Difco)....................... 10.0 g
Norit A (charcoal; Pfanstiehl) .... 10.0 g
Cholesterol in acetone, 1% solution (0.25 g of cholesterol in 25 ml of acetone) (cholesterol, C.P., ash free, for Kline Test; Pfanstiehl) .... 25.0 ml

Note: Be sure to keep the flask away from flames when adding the cholesterol-acetone.

1. Heat the entire mixture to boiling to dissolve the agar.
2. Stir frequently to keep the charcoal in suspension.
3. Dispense the hot solution in 3-ml amounts into tubes, and plug or cap.
4. Autoclave for 15 min at 15 lb/in^2 pressure.
5. Resuspend the charcoal, and allow the tubes to cool in a slanted position to form short butts or no butts.
6. Add 3 ml of sterile 0.5% buffered saline overlay to the slants before specimen inoculation.

Commercial media are available. Dehydrated charcoal agar modified may be purchased as Hirsh charcoal agar (BBL) and used for slants to be overlaid with sterile Locke solution or 0.5% buffered saline. The medium is excellent for maintenance of amoeba cultures.

Chlamydial media

A. Growth medium

Eagle minimum essential medium in Earle salts (10×)	50 ml
Fetal calf serum	50 ml
L-Glutamine, 200 mM solution	5 ml
Sterile distilled water, to	500 ml

Adjust the pH to 7.4 with 7.5% sodium bicarbonate.

B. Isolation medium

Growth medium (see above) containing added:

Vancomycin	50 μg/ml
Gentamicin	10 μg/ml
Amphotericin B	2 μg/ml
Glucose	0.594 mg/ml
Clycoheximide	1 to 2 μg/ml

This medium may be used as a collecting medium by doubling the concentrations of vancomycin and amphotericin B.

C. Sucrose phosphate transport medium (2SP)

Sucrose	68.46 g
K_2HPO_4	2.088 g
KH_2PO_4	1.088 g
Distilled water	1,000 ml

Adjust the pH to 7.0 and autoclave; add:

Bovine serum, to	5%
Streptomycin	50 μg/ml
Vancomycin	100 μg/ml
Nystatin	25 U/ml

D. Sucrose-phosphate-glutamate transport medium (SPG)

Sucrose	75.00 g
KH_2PO_4	0.52 g
Na_2HPO_4	1.22 g
Glutamic acid	0.72 g
Distilled water, to	1,000 ml

Adjust the pH to 7.4 to 7.6, and autoclave; add antibiotics as described above (2SP).

Chocolate agars

Use one of the following as preferred, or as required to obtain satisfactory growth.

Beef infusion agar*
Blood agar base*
Casein peptone agar*
Eugonic agar*
Mueller-Hinton agar*

In addition, GC Agar Base* or GC Medium Base* may be employed.

Pancreatic digest of casein USP	7.5 g
Peptic digest of animal tissue USP	7.5 g
(*or* Other peptone	15.0 g)
Cornstarch	1.0 g
Dipotassium phosphate	4.0 g
Monopotassium phosphate	1.0 g
Sodium chloride	5.0 g
Agar	10.0 g
Distilled water	1.0 liter

Prepare sterile base. Add sterile 5 to 10% defibrinated blood, and heat at about 80°C for 15 min or until the color is chocolate brown.

Variations

1. Add chemical supplement. See Thayer-Martin agar.
2. Add yeast supplements.
3. Prepare double-strength medium and add an equal volume of sterile 2% hemoglobin. Also add chemical supplement, especially for *Neisseria* and *Haemophilus* species.

Chopped meat (CM) medium (9)
For anaerobes

Ground beef, lean (connective tissue trimmed off before putting meat through grinder)	500 g
Distilled water	1,000 ml
Sodium hydroxide (1 N solution)	25 ml

Mix the above ingredients, heat to boiling, and cool at 4°C overnight. Skim remaining fat from surface. Filter through several layers of gauze. Save the liquid filtrate. Wash the meat particles with distilled water to remove excess NaOH, and spread on a clean towel to partially dry. Add enough distilled water to the liquid filtrate to give a final volume of 1,000 ml.

Trypticase (BBL)	30.0 g
Yeast extract (Difco)	5.0 g
K_2HPO_4	5.0 g
L-Cysteine	0.5 g

**Hemin (1% solution) .... 0.5 ml
**Vitamin K_1 (1% solution) .... 0.1 ml

** See Lombard-Dowell broth for directions to make 1% hemin and vitamin K_1 solutions.

Add the ingredients listed above, except for the L-cysteine, heat to dissolve, and cool to 50°C. Add the L-cysteine and mix to dissolve. Dispense 0.5 g of meat into 15- by 90-mm screw-capped tubes. Add 7 ml of enriched broth, autoclave at 121°C for 15 min, and cool.

See note on CHO fermentation broth for anaerobes. Place tubes in glove box, tighten caps, remove from glove box, and store at 4°C or at ambient temperature.

Chopped meat medium is used to detect digestion (D) of meat particles, gas (G), or blackening (B) of the meat. Plain chopped meat broth is also an excellent holding medium for anaerobe stock cultures (18).

Chopped meat-glucose (CMG) medium
For anaerobes

Follow the formula for chopped meat medium. In addition, add 3.0 g of D-glucose to the broth before autoclaving.

Citrate agar*
Simmons citrate agar

Sodium citrate .... 2.00 g
Sodium chloride .... 5.00 g
Magnesium sulfate .... 0.20 g
Monoammonium phosphate .... 1.00 g
Dipotassium phosphate .... 1.00 g
Agar, dried .... 15.00 g
(*or* Not dried .... 20.00 g)
Bromothymol blue .... 0.08 g
Distilled water .... 1.00 liter
Final pH 6.9

Dispense into tubes, autoclave at 121°C for 15 min, and cool in the slanted position for 2.5-cm butt and 3.8-cm slant.

Inoculate the slant with a straight wire from a saline suspension of a young agar culture. Incubate at 37°C for 4 days. If equivocal results are obtained, as sometimes happens with members of the *Providencia* genus, for example, the test should be repeated and incubated at room temperature for 7 days.

Columbia agar base*

Polypeptone (BBL) or Pantone (Difco) .... 10.0 g
Biosate (BBL) or Bitone .... 10.0 g
Myosate (BBL) or tryptic digest of beef heart .... 3.0 g
Cornstarch .... 1.0 g
Sodium chloride .... 5.0 g
Agar .... 13.5 g
Distilled or demineralized water .... 1.0 liter
Final pH 7.3 ± 0.2

Heat with agitation until the medium boils. Dispense and autoclave at 121°C for 15 min.

Columbia CNA agar*

Polypeptone peptone .... 10.0 g
Biosate peptone .... 10.0 g
Myosate peptone .... 3.0 g
Cornstarch .... 1.0 g
Sodium chloride .... 5.0 g
Agar .... 13.5 g
Final pH 7.3

Suspend 42.5 g of Columbia agar base in 1 liter of distilled water, and heat with frequent agitation. Boil for 1 min, dispense, and autoclave for 15 min at 121°C.

Cool to 45°C, and then add 50 ml of defibrinated sheep blood and 5 ml of a solution containing 10 mg of colistin and 10 to 15 mg of nalidixic acid.

Note. The above is BBL formation. Different suppliers have other trade names for their peptones.

Converse liquid medium (Levine modification) for *Coccidioides* species (6)
For induction of spherules of Coccidioides immitis

Ammonium acetate .... 1.23 g
Glucose .... 4.0 g
Dipotassium phosphate .... 0.52 g
Potassium phosphate .... 0.4 g
Magnesium sulfate .... 0.4 g
Zinc sulfate .... 0.002 g
Sodium chloride .... 0.014 g
Sodium carbonate .... 0.012 g
Tamol .... 0.5 g
Calcium chloride .... 0.002 g
Agar (Ionagar No. 2, Agarose, or purified agar) .... 10.0 g
Distilled water .... 1 liter

1. Mix reagents.
2. Bring to a boil.
3. Autoclave for 15 min at 15 lb/in^2.
4. Dispense 15.0-ml aliquots into sterile petri dishes.
5. Allow to harden.

Inoculate with a suspension of arthroconidia. Incubate at 40°C in a candle jar or anaerobic incubator. Examine tease mounts (prepared in a bacteriological glove box or laminar flow hood) microscopically for spherules. Endosporulating spherules should be apparent after 4 to 5 days of incubation.

Cornmeal agar, modified (medium B)

Cornmeal agar* .... 17 g
Glucose .... 2 g
Sucrose .... 3 g
Yeast extract .... 1 g
Water .... 1 liter

1. Melt to suspend the ingredients.
2. Autoclave at 121°C for 15 min.
3. Cool to 45 to 50°C, and dispense into sterile petri dishes.

Cornmeal-polysorbate 80 agar*

Cornmeal . . . . 40 g
Polysorbate (Tween) 80 . . . . 3 ml
Water . . . . 1 liter
Agar (not dried) . . . . 20 g

1. Simmer the cornmeal and water for 1 h.
2. Filter through gauze. Restore filtrate to 1 liter.
3. Add agar. Mix, and heat to melt the agar. Filter again if necessary.
4. Add polysorbate 80. Add glycerol (20 to 30 ml) if desired.
5. Autoclave at 121°C for 15 min.
6. Cool to 45 to 50°C, and dispense into sterile petri dishes.

CTA medium (57)
Cystine Trypticase agar medium (BBL)

Cystine . . . . 0.500 g
Pancreatic digest of casein USP . . . . 20.000 g
Agar . . . . 3.500 g
Sodium chloride . . . . 5.000 g
Sodium sulfite . . . . 0.500 g
Phenol red . . . . 0.017 g
Distilled water . . . . 1.000 liter
Final pH 7.3

Mix, and heat with agitation until solution occurs. Dispense, and autoclave at 115 to 118°C for 15 min. Add filter-sterilized carbohydrates to a final concentration of 1%.

Cycloheximide-chloramphenicol agar* (18)
Mycosel agar (Difco), Mycobiotic agar (BBL), etc.

Sabouraud dextrose agar . . . . 1.00 liter
Agar . . . . 5.00 g
Cycloheximide, in 10 ml of acetone . . . . . . 0.50 g
Chloramphenicol, in 10 ml of 95% alcohol . . . . 0.05 g

Heat to dissolve the agar and the Sabouraud dextrose agar. Dispense into tubes, and autoclave at 121°C for 10 min. Cool in a slanted position.

Recommended for the routine isolation of most pathogenic fungi. For species susceptible to cycloheximide, the medium can be prepared without it.

Czapek agar*
Czapek Dox medium

Sucrose . . . . 30.00 g
Sodium nitrate . . . . 3.00 g
Dipotassium phosphate . . . . 1.00 g
Magnesium sulfate . . . . 0.50 g
Potassium chloride . . . . 0.50 g
Ferrous sulfate . . . . 0.01 g
Agar . . . . 15.00 g
Demineralized water . . . . 1.00 liter
Final pH 7.3

Suspend 50 g of the dehydrated material in a liter of distilled water. Mix thoroughly. When a uniform mixture has been obtained, heat with frequent agitation and boil for about 1 min to dissolve the agar. Sterilize by autoclaving at 121°C for 15 min. Aseptically dispense into plates before the agar solidifies.

Recommended for species identification of aspergilli.

Decarboxylase broths, Moeller*

Peptic digest of animal tissue USP . . . . . . 5.000 g
Beef extract . . . . 5.000 g
Bromocresol purple . . . . 0.10 g
Cresol red . . . . 0.005 g
Glucose . . . . 0.500 g
Pyridoxal . . . . 0.005 g
Distilled water . . . . 1.000 liter
Final pH 6.0

1. Divide into 4 parts, with tube 1 as a control, and add 1% L-arginine monohydrochloride, L-lysine dihydrochloride, and L-ornithine dihydrochloride to the remaining portions. Alternatively, add 2% DL-amino acids.
2. Readjust the pH of the ornithine portion.
3. Dispense in 3- to 4-ml amounts into 13- by 100-mm screw-capped tubes, and autoclave at 121°C for 10 min.
4. A small amount of floccular precipitate in the ornithine does not interfere with its use.

Inoculate test and control portions lightly from a young agar slant culture. Cover with a 4- to 5-mm layer of sterile mineral (paraffin) oil. Incubate at 37°C, and examine daily for 4 days. Positive reactions are indicated by alkaline (purple) reaction. The media turn yellow first, due to acid production from glucose.

Dermatophyte test medium (DTM)
For primary recovery of dermatophytes

Phytone . . . . 10 g
Glucose . . . . 10 g
Agar . . . . 20 g
Phenol red solution . . . . 40 ml
HCl, 0.8 N . . . . 6 ml
Cycloheximide . . . . 0.5 g
Gentamicin . . . . 0.1 g
Chlortetracycline hydrochloride . . . . 0.1 g
Distilled water, to . . . . 1 liter

Add phytone, glucose, and agar to 1 liter of distilled water, and boil. Add phenol red solution (0.5 g in 15 ml of 0.1 N NaOH made up to 100 ml with water). Dissolve cycloheximide in 2 ml of acetone and add while the medium is hot. Dissolve gentamicin in 2 ml of distilled water before adding to medium. Autoclave at 121°C for 10 min. Add chlortetracycline dissolved in a 25-ml volume of sterile distilled water.

Diphasic BCYEα blood culture medium (11)

Agar slant

Agar (Difco) . . . . 7.00 g
Yeast extract (Difco) . . . . 3.50 g
Activated charcoal (Norit A) . . . . 1.40 g
KOH pellets (reagent grade, 85% KOH) . . . 0.95 g
ACES buffer (Research Organics) . . . . 3.50 g
α-Ketoglutarate monopotassium salt (Sigma) . . . . 0.35 g
Distilled water . . . . 350.0 ml

1. Add ACES buffer to 350 ml of distilled water in flask. Heat in water bath at 50°C for about 0.5 h or until the buffer has dissolved.
2. Add KOH pellets to buffer solution.
3. Add dry Norit A, yeast extract, agar, and α-ketoglutarate. Mix well, and boil to dissolve the agar. Check the pH (should be 6.85 to 6.95).
4. Dispense 30 ml each into 100-ml bottles, and autoclave (121°C for 15 min) the bottles with the stoppers vented. Remove vents immediately after removing the bottles from the autoclave.
5. Let slant solidify with the bottle flat. This usually takes several hours, as the glass retains heat.

Broth

ACES buffer	6.000 g
KOH pellets	1.470 g
α-Ketoglutarate monopotassium salt	0.600 g
L-Cysteine hydrochloride in 10 ml of distilled water (ICN)	0.398 g
Ferric pyrophosphate in 10 ml of distilled water (Sigma)	0.1537 g
Yeast extract (Difco)	6.200 g
Sodium polyanetholesulfonate (SPS) (Sigma)	0.183 g
Distilled water	600 ml

1. Add ACES buffer to 600 ml of distilled water in flask. Heat in water bath at 50°C for about 0.5 h or until buffer has dissolved.
2. Add KOH pellets to buffer solution.
3. Add yeast extract, SPS, and α-ketoglutarate. Autoclave at 121°C for 15 min.
4. Filter sterilize the L-cysteine and ferric pyrophosphate, and add them to the cooled (50°C), autoclaved distilled water mixture.
5. Check the pH (should be between 6.85 and 6.95).
6. Add 50 ml each to 100-ml blood culture bottles with a medium syringe and needle.
7. Preincubate for 24 h at 35°C to check for contaminants. Shelf life is 2 months at 5°C.

DNase test agar* (54)
For nonfermenting gram-negative bacteria

DNA	2.0 g
Pancreatic digest of casein USP	15.0 g
Papaic digest of soy meal USP	5.0 g
Sodium chloride	5.0 g
Agar	15.0 g
Distilled water	1.0 liter

Final pH 7.3

Inoculate culture heavily over a 1-cm^2 area of the plate. Several cultures may be tested on the same plate. Incubate for 18 to 24 h, and flood with 1 N HCl. A zone of clearing around the colony indicates a positive DNase test.

Method of Gilardi (19)

DNase test agar	1.0 liter
Toluidine blue	0.1 g

Inoculate as described above, and observe for a change in the indicator to pink around the growth. Strains producing equivocal results are tested as described above.

Dulaney slants (10)

Aseptically remove yolks from 5- to 8-day hen egg embryos and place in an equal volume of sterile Locke solution containing glass beads. Homogenize. Dispense the homogenate into slanted tubes, and coagulate with steam at 80°C for 15 min.

Prepare Locke solution as follows:

Sodium chloride	0.900 g
Calcium chloride	0.024 g
Potassium chloride	0.042 g
Sodium carbonate	0.020 g
Glucose	0.250 g
Distilled water	100.000 ml

Egg yolk agars

A. Blood agar base
B. Brain heart infusion agar
C. Soybean casein digest agar USP or other

Prepare egg yolk emulsion as described below. Alternatively, use commercial suspension as directed.

D. McClung and Toabe agar
For Bacillus anthracis

Pancreatic digest of casein USP	40.0 g
Disodium phosphate	5.0 g
Monopotassium phosphate	1.0 g
Sodium chloride	2.0 g
Magnesium sulfate	0.1 g
Glucose	2.0 g
Agar	25.0 g
Distilled water	1.0 liter

1. Dissolve, and adjust the pH to 7.6.
2. Autoclave at 121°C for 15 min.
3. Cool to 50 to 55°C.
4. Meanwhile, scrub, and then soak, an antibiotic-free hen egg in 95% ethanol for 1 h. Aseptically aspirate or separate the egg yolk.
5. Add one egg yolk to 500 ml of agar base, and stir to a smooth suspension with a sterile pipette. Alternatively, add commercial egg yolk emulsion* as directed.

E. McClung and Toabe agar, modified

For lecithinase and lipase tests

Proteose no. 2 peptone* or Polypeptone* (BBL)	40.0 g
Disodium phosphate	5.0 g
Monopotassium phosphate	1.0 g
Sodium chloride	2.0 g
Magnesium sulfate	0.1 g
Glucose	2.0 g
Hemin solution, 5 mg/ml	1.0 ml
Agar	20.0 g
Water	1.0 liter

1. Suspend ingredients, and adjust the pH to 7.6.
2. Mix, and boil to dissolve.

3. Dispense 20 ml per tube, and autoclave at 118°C for 15 min.
4. Cool to 50°C, and add 2 ml of egg yolk emulsion* or 1 ml of laboratory-prepared egg yolk emulsion per tube. Mix, and pour the plates. Use an emulsion of equal volumes of egg yolk in sterile saline. See D above.

For Clostridium species (53)

1. Add neomycin (100 μg/ml) to medium base before autoclaving.
2. Cool, and add the egg yolk emulsion.

For Clostridium botulinum

Yeast extract	5.0 g
Pancreatic digest of casein	5.0 g
Proteose no. 2 peptone* or Polypeptone* (BBL)	20.0 g
Sodium chloride	5.0 g
Sodium thioglycolate	1.0 g
Agar	20.0 g
Distilled water	1.0 liter

1. Sterilize, and cool the medium base to 45 to 50°C.
2. Add 80 ml of sterile egg yolk suspension.
3. Mix, and pour the plates immediately.
4. Dry the plates, and store in a refrigerator.

Elek medium

See toxigenicity test agar.

Enriched thioglycolate (THIO) medium (9)

Thioglycolate medium without indicator*	30.0 g
**Hemin (1% solution)	0.5 ml
**Vitamin K_1 (1% solution)	0.1 ml
Distilled water	1,000.0 ml

** See Lombard-Dowell broth for directions to make 1% hemin and vitamin K_1 solutions.

Add all the ingredients and heat to dissolve. Dispense 7 ml into 15- by 90-mm screw-capped tubes, autoclave at 121°C for 15 min, and cool.

See note on fermentation broth for anaerobes. Place tubes in anaerobic glove box, tighten caps, remove from glove box, and store at 4°C or ambient temperature.

Enriched thioglycolate medium with 20% bile (THIO + bile) (9)

Prepare as described above, but add 20 g of oxgall* per liter of enriched thioglycolate medium before autoclaving (2% oxgall = 20% bile).

Compare growth in bile medium with growth in enriched thioglycolate medium. Record S for stimulated, N for no change, or I for inhibited.

Enrichment broth for *Aeromonas hydrophila* (28)

Maltose	3.500 g
L-Cysteine hydrochloride	0.300 g
Bile salts no. 3 (Difco)	1.000 g
Novobiocin	0.005 g
Yeast extract	3.000 g
Sodium chloride	5.000 g
Bromothymol blue	0.030 g
Distilled water	1.000 liter
Final pH 7.0	

Enteric base for *Listeria* species (38)
Phenol red broth base at pH 7.4

Peptone	10.000 g
Beef extract (optional)	1.000 g
Sodium chloride	5.000 g
Phenol red	0.018 g
Distilled water	1,000.000 ml

Enteric fermentation base (13)

Peptone	10.0 g
Meat extract	3.0 g
Sodium chloride	5.0 g
Andrade indicator	10.0 g
Distilled water	1.0 liter

Dispense medium in 3-ml amounts into 15- by 125-mm tubes containing inverted Durham vials. Autoclave at 121°C for 15 min. After cooling, 0.3 ml of a 10% filter-sterilized carbohydrate solution is added aseptically to each tube to give a final concentration of 1%.

Esculin agar, modified

Esculin	1.0 g
Ferric citrate	0.5 g
Heart infusion agar* (blood agar base)	40.0 g
Distilled water	1.0 liter

Heat to dissolve. Cool to 55°C; adjust the pH to 7.0. Dispense in 5-ml amounts into 16- by 125-mm cotton-plugged tubes. Autoclave at 121°C for 15 min. Cool in slanted position.

Esculin hydrolysis is indicated when the medium turns black.

Esculin broth (7)
For anaerobes

Heart infusion broth*	25.0 g
Esculin	1.0 g
Agar	1.0 g
Distilled water	1,000.0 ml
Final pH 7.0	

Dissolve the ingredients by boiling, adjust the pH, dispense 7 ml into 15- by 90-mm screw-capped tubes, autoclave at 121°C for 15 min, and cool.

See note on fermentation broth for anaerobes. Place tubes in glove box, tighten caps, remove tubes from glove box, and store at 4°C or ambient temperature.

Editor's note. The author recommends the use of Difco heart infusion broth.

If esculin broth is used, read the reaction after 48 h of incubation. Add 1 dropperful of 1% ferric ammonium citrate, and read the immediate reaction. Interpret brownish-black color as positive and no color change as negative (9).

Esculin hydrolysis

Heart infusion agar .... 40 g
Esculin .... 1 g
Ferric chloride .... 0.5 g
Distilled water .... 1,000 ml

Heat to dissolve ingredients, dispense into 13- by 100-mm screw-capped tubes, and sterilize in an autoclave for 15 min at 121°C. Slant the tubes for the cooling period.

Esculin hydrolysis is indicated when the medium turns black.

Farrell selective medium

Add the following antibiotics to 1 liter of serum dextrose agar after autoclaving and cooling to 50°C.

1. Bacitracin (Burroughs-Wellcome). Dissolve in distilled water to a concentration of 2,000 U/ml. Add 12.5 ml of stock; final concentration is 25 U/ml of medium.
2. Polymyxin B (Burroughs-Wellcome). Dissolve in distilled water to a concentration of 5,000 U/ml. Add 1 ml of stock; final concentration is 5 U/ml.
3. Actidione (Upjohn). Dissolve 1 g in 5 ml of acetone, and dilute 1:20 in distilled water (10,000 μg/ml). Add 10 ml of stock; final concentration is 100 μg/ml of medium.
4. Vancomycin (Lilly). Dissolve 50 mg in 1 ml of distilled water. Add 0.4 ml of stock; final concentration is 20 μg/ml of medium.
5. Nalidixic acid (Winthrop). A 5% (wt/vol) stock solution is prepared in 0.5 M NaOH. Immediately before use, dilute 1:10 in distilled water. Add 1 ml to medium; final concentration is 5 μg/ml of medium.
6. Nystatin (Squibb). Dissolve in distilled water to a concentration of 50,000 U/ml. Add 2 ml to medium; final concentration is 100 U/ml of medium.

Fermentation basal medium (51)

Diammonium phosphate .... 1.0 g
Potassium chloride....................... 0.2 g
Magnesium sulfate .... 0.2 g
Yeast extract .... 0.2 g
Agar .... 15.0 g
Distilled water........................... 1.0 liter
Bromocresol purple, 0.04% .... 20.0 ml

Tube and sterilize. Filter sterilize the appropriate sugar and add 10 to 15% after the basal medium has cooled.

Fermentation broth (CHO base) (9)
For anaerobes

CHO medium base* .... 26.0 g
Distilled water.......................... 900.0 ml

Mix well and heat to dissolve. Autoclave at 121°C for 15 min. Cool to 50°C. Prepare fermentation base control medium by adding 100 ml of sterile distilled water. To prepare carbohydrate fermentation medium, add 100 ml of filter-sterilized aqueous carbohydrate stock solution to 900 ml of the sterile basal medium. Except for starch, each carbohydrate is prepared as a 6% stock solution, and the final concentration in the prepared CHO fermentation medium is 0.6%. The concentration of the starch stock solution is 2.5%; the final concentration of starch in the CHO fermentation medium is 0.25%. Mix well and adjust the final pH to 7.0 ± 0.1 at 25°C. Dispense 7 ml into 15- by 90-mm screw-capped tubes.

Note. Pass the cooled tubes (with caps loose) into an anaerobic chamber containing an atmosphere of 85% N_2–10% H_2–5% CO_2. Fasten the caps securely, remove from the chamber, and store at ambient or refrigerator (4°C) temperature.

If an anaerobic chamber is not available, an alternative on the day of use would be to boil or steam the medium (with tube caps loose) for 10 min, cool, and inoculate immediately.

Editor's note. The CHO medium base is available from Difco.

Inoculate the tubes of liquid media near the bottom (with a capillary pipette) with a few drops of culture per tube. Be sure to expel air from the pipette before placing it in the medium. One pipette may be used to inoculate multiple tubes. Fermentation media should be incubated under anaerobic conditions (glove box or jar) with tube caps loosened. Acid production as an indication of fermentation is determined by inspecting the tubes on days 1, 2, and 7. The bromothymol blue indicator in the tubes will turn yellow at pH 6.0 or lower. This change indicates a positive reaction. If the medium is blue or blue-green, the color indicates no change or a negative reaction. Some clostridia will reduce the indicator. When this occurs, remove 2 to 3 drops from each tube with a sterile capillary pipette to a spot plate, and add 2 to 3 drops of dilute bromothymol blue (prepared by adding 2 to 3 drops of 1% aqueous bromothymol blue to 30 ml of water in a 30-ml dropper bottle). Each tube with an acid reaction can be discarded. Reincubate all negative tubes. A final reading should be made at 7 days. Note that the majority of clinical isolates can be read after good growth is observed (after 24 to 48 h) and need not be held for 7 days.

Flagella broth

Tryptose (Difco) or Biosate (BBL)........ 10.0 g
Dipotassium phosphate .... 1.0 g
Sodium chloride .... 2.5 g
Distilled water........................... 1.0 liter

1. Dissolve ingredients, and adjust the pH to 7.0.
2. Dispense in 5-ml volumes into 15- by 125-mm test tubes.
3. Autoclave at 121°C for 15 min.

Fletcher semisolid medium* (17)

Pancreatic digest of casein USP or
 pancreatic digest of gelatin .... 0.3 g
Beef extract............................... 0.2 g
Sodium chloride .... 0.5 g
Agar...................................... 1.5 g
Water .... 920.0 ml
 Final pH 7.4 to 8.0

1. Suspend ingredients in cold water, and heat to boiling to dissolve. Autoclave at 121°C for 15 min.
2. Cool to 56°C, and add filter-sterilized rabbit serum.
3. Dispense into screw-capped tubes in 5- to 7-ml amounts.
4. Inactivate the medium by placing the tubes in a water bath at 56°C for 1 h on 2 successive days.
5. Medium can be stored at room temperature if dispensed into screw-capped tubes.

Fluorouracil leptospira medium

1. Dissolve 10 g of 5-fluorouracil in approximately 50 ml of distilled water, add 1.0 to 2.0 ml of 2 N NaOH, and heat gently (less than 56°C) for 1 to 2 h or until soluble. Adjust the pH to 7.4 to 7.6 with 1 N HCl, and bring the volume to 100 ml with distilled water.
2. Sterilize by filtration, and store in a refrigerator.
3. Add aseptically, just before use, 0.1 ml of 5-fluorouracil solution (Roche) for each 5 ml of albumin fatty acid broth, Fletcher medium, or Stuart leptospira broth.

Formate-fumarate additive

Sodium formate . . . . 3.0 g
Fumaric acid . . . . 3.0 g
Distilled water . . . . 50.0 ml

To adjust the pH, add 20 pellets of NaOH, stirring until pellets are dissolved and the fumaric acid is in solution. Bring the final pH to 7.0 with 4 N NaOH. Sterilize by filtration.

Fungus agar according to Kimmig (32)

Standard II nutrient broth . . . . 15.0 g
Peptone (trypsin digest) . . . . 5.0 g
D-Glucose . . . . 10.0 g
NaCl . . . . 5.0 g
Agar . . . . 15.0 g

Suspend the above 50 g in 1 liter of freshly distilled or completely demineralized water in which 5 ml of glycerol has previously been dissolved. Allow to stand for 15 min, and then boil to dissolve completely with frequent shaking. After it is dispensed, sterilize in the autoclave (15 min at 121°C). Use for cultivation and strain preservation of fungi as well as for the assay of fungistatic agents.

Gelatinase test medium (11)

ACES buffer (Research Organics) . . . . 1.000 g
KOH pellets in 1.0 ml of distilled water (reagent grade, 85% KOH) . . . . 0.280 g
Activated charcoal (Norit A) . . . . 0.150 g
Yeast extract (Difco) . . . . 1.000 g
Gelatin (Difco) . . . . 3.000 g
α-Ketoglutarate monopotassium salt (Sigma) . . . . 0.100 g
L-Cysteine hydrochloride in 1.0 ml of distilled water (ICN) . . . . 0.040 g
Ferric pyrophosphate in 1.0 ml of distilled water (Sigma) . . . . 0.015 g
Distilled water . . . . 100 ml

1. This medium is prepared in the same way BCYEα medium is made except that gelatin is used in place of agar.
2. Dispense in 2-ml volumes into small screw-capped tubes.
3. Preincubate for 72 h at 35°C before using.
4. Store at 4°C. Shelf life is 2 months at 5°C.

H agar for mycoplasma media (22, 30, 31)

Papaic digest of soy meal USP (Hysoy [soy peptone], Sheffield) . . . . 20.0 g
Sodium chloride . . . . 5.0 g
Purified water . . . . 1.0 liter
Phenol red, 2% aqueous . . . . 2.0 ml
Agarose . . . . 10.0 g

1. Adjust the pH to 7.3 with 1 N NaOH before adding agarose.
2. Heat with agitation to obtain solution.
3. Dispense, and autoclave at 121°C for 15 min. Cool to 50°C.
4. To 70 ml of solution, add 10 ml of yeast dialysate, 20 ml of horse serum, 2 ml of penicillin (10,000 U/ml), and 1 ml of 3.3% aqueous thallium acetate.
5. Dispense in 5-ml amounts into 10- by 35-mm petri dishes, and incubate overnight at room temperature.

Note. Prepare yeast dialysate as follows. Suspend 450 g of active dried yeast (e.g., Fleischman) in 1,250 ml of water at 40°C. Heat in an autoclave at 121°C for 5 min or long enough to kill yeast cells. Place in dialysis casing and dialyze against 1 liter of water at 4°C for 2 days. Discard casing and contents. Autoclave dialysate at 121°C for 15 min. Store frozen.

H broth for mycoplasma (22, 30, 31)

Papaic digest of soy meal USP . . . . 20 g
Sodium chloride . . . . 5 g
Glucose . . . . 10 g
Purified water . . . . 1 liter
Phenol red, 2% aqueous . . . . 2 ml

1. Dissolve, and adjust the pH to 7.3.
2. Dispense, and autoclave at 121°C for 15 min. Cool.
3. To 65 ml of broth, aseptically add the same solutions as for H agar.
4. For *Mycoplasma hominis*, glucose may be omitted, and 5 ml of 1 M arginine is then added per 100 ml.

H medium, diphasic, for *Mycoplasma pneumoniae*

Dispense H agar in 3-ml amounts into 16- by 125-mm screw-capped tubes, allow to solidify, and overlay with 3 ml of H broth. Store at 4°C for not more than 2 weeks. Tubes must have good closures to avoid excessive elevation of pH upon loss of CO_2 from horse serum.

HBT medium (human blood bilayer with Tween 80) for *Gardnerella vaginalis*; selective (55)

Bottom layer

Colistin-nalidixic acid-Columbia agar base (BBL) . . . . 21.25 g

Proteose Peptone no. 3 (Difco)............. 5.0 g
Tween 80 (10% vol/vol)................... 1.00 ml
Distilled water......................... 500 ml
Final pH 7.3

Heat to dissolve, and autoclave at 121°C for 15 min. Add 1 mg of amphotericin B.

Upper layer

Medium is the same as above except for the addition of 37 ml of sterile human blood after autoclaving.

Heart infusion agar

Same as blood agar base, but no blood added.

Heart infusion agar for isolation of *Brucella* species

Heart infusion agar .... 40.0 g
Gelatin .... 1.0 g
Glucose.................................. 2.5 g
Distilled water .... 1,000.0 ml

Mix ingredients, autoclave for 15 min at 121°C, allow to cool to 50°C, and add 10 ml of sterile sheep blood and the following antibiotics:

Actidione (Upjohn) 0.2 ml of a 10-mg/ml stock solution; final concentration, 100 μg/ml of agar
Bacitracin (Upjohn), 1 ml of a 5,000-μg/ml stock solution; final concentration, 25 μg/ml of agar
Circulin (Upjohn), 0.3 ml of a 10-mg/ml stock solution; final concentration, 15 μg/ml of agar
Polymyxin B (Burroughs-Wellcome), 1.2 ml of a 1,000-μg/ml stock solution; final concentration, 6 μg/ml of agar

Heart infusion broth
Infusion broth

Same as blood agar base or heart infusion agar, but without agar.

Heart infusion-tyrosine agar
For browning of blood-free media to identify Bordetella parapertussis

Heart infusion agar .... 40 g
L-Tyrosine .... 1 g
Distilled water.......................... 1,000 ml

1. Dissolve agar and tyrosine in water by boiling, and dispense into 15- by 125-mm tubes for slants.
2. Autoclave, slant the tubes, and allow to cool. Seal. Refrigerate the slants.
3. Shelf life is not known, so a positive control must be used when tests are done.
4. Inoculate a slant heavily with suspected *B. parapertussis*; incubate for 24 to 48 h at 35°C, with the cap loose to allow access of air. Development of a brown pigment in the medium is characteristic of *B. parapertussis*.

Editors' note. The author recommends the use of Difco heart infusion agar.

Hickey and Tresner agar (21)

Dextrin .... 10 g
Pancreatic digest of casein USP* .... 2 g
Beef extract* .... 1 g
Yeast extract* .... 1 g
Agar .... 20 g
Cobalt chloride · $6H_2O$ .... 20 mg
Demineralized water...................... 1 liter

Dissolve and dispense for agar deeps. Autoclave at 121°C for 30 min. For pour slides, melt in a water bath and pipette 0.5 to 1.0 ml per slide.

Recommended for production of aerial hyphae, arthrospores, and pigment by *Nocardia* and *Streptomyces* species. For best results, allow surface to dry well before inoculation.

***Histoplasma capsulatum* agar media**

For conversion to or maintenance of the yeast phase (42)

Required solutions and chemicals

Solution 1 (basal salts, in grams)

KH_2PO_4.................................. 8.00
$(NH_4)_2SO_4$ 8.00
$MgSO_4 \cdot 7H_2O$ 0.86
$CaCl_2$ (anhydrous) 0.08
$ZnSO_4 \cdot 7H_2O$ 0.05

Dissolve the above in 500 ml and make to 1 liter with distilled water; store at 5°C.

Solution 2 (trace elements, in grams)

$FeSO_4 \cdot 7H_2O$......................... 5.70
$MnCl_2 \cdot 6H_2O$......................... 0.80
$NaMoO_4 \cdot 2H_2O$........................ 0.15

Add 1.0 ml of concentrated HCl to 100 ml of distilled water in a 1-liter volumetric flask. Dissolve each component completely in the sequence given, and make to 1 liter.

Store at 5°C; good indefinitely; discard if red color or red precipitate appears.

Solution 3 (amino acids)

Casein hydrolysate, 10% acid-hydrolyzed, vitamin-free solution. (*Note.* Do not use "enzymatically digested" casein hydrolysate.)

Solution 4 (vitamins, in milligrams)

Inositol 200
Thiamine hydrochloride..................... 200
Calcium pantothenate....................... 200
Riboflavin................................. 200
Nicotinamide............................... 100
Biotin..................................... 10

Suspend and make to 1 liter in distilled water. Store frozen; if stored at 5°C, suspension is good for 1 year unless microbial growth occurs.

Solution 5 (hemin)

Suspend 200 mg of hemin in 10 to 20 ml of distilled water, and bring into solution by the addition of a few drops of concentrated ammonium hydroxide; bring final volume to 100 ml; store at 5°C.

Solution 6 (thioctic acid)

Dissolve 10 mg of DL-thioctic acid in 10 ml of 95% alcohol; store frozen.

Solution 7 (coenzyme A)

Dissolve 10 mg of coenzyme A in 10 ml of distilled water; add 2 drops of 0.05% $Na_2S \cdot 5H_2O$ solution made in freshly boiled distilled water. Store frozen.

Solution 8 (oleic acid)

Suspend 100 mg of oleic acid in 50 ml of distilled water and neutralize with NaOH to pH 9.0, bringing the fatty acid into complete solution by warming if necessary; adjust final volume to 100 ml. Store at 5°C.

Organic additions

Glucose, α-ketoglutaric acid, citric acid, L-cysteine hydrochloride, glutathione (reduced), L-asparagine, L-tryptophan, agar, purified starch (insoluble, i.e., not solubilized for chemical or enzymatic analyses)

Procedure for the preparation of 1 liter of agar medium

A. To a 500-ml graduated cylinder add and mix:

Solution 1 .... 250 ml
Bring volume to approximately 400 ml with distilled water.
Solution 2 .... 10 ml
Solution 3 .... 40 ml
Solution 4 .... 10 ml
Solution 5 .... 1 ml
Solution 6 .... 0.1 ml
Solution 7 .... 0.1 ml

B. Add and dissolve (in grams):

Citric acid 10.0
Glucose 10.0
α-Ketoglutaric acid 1.0
L-Cysteine hydrochloride 1.0
Glutathione (reduced) 0.5
L-Asparagine 0.1
L-Tryptophan 0.02

C. Neutralize carefully to pH 6.5 with 20% KOH; make to 500 ml and sterilize by filtration through a membrane filter (0.45-μm pore size) to give a 2× concentrated basal solution, i.e., part I.

D. To prepare 2× concentrated starch-oleic acid agar base, suspend 2 g of potato starch in 50 ml of distilled water and pour into 450 ml of boiling distilled water in a 2-liter flask; add 10 ml of solution 8; add 12.5 g of agar. Autoclave for sterilization at 121°C for 20 min. This is part II.

E. While still hot, add part I to part II, mix, and tube aseptically into screw-topped test tubes.

For the growth, sporulation, and maintenance of the mycelial phase (41)

The agar medium for the growth of the mycelial phase of *H. capsulatum* is prepared by using the same components and procedure as for the preparation of the yeast-phase medium except that the $ZnSO_4 \cdot 7H_2O$ and citric acid are deleted and 15.0 g of agar per liter of medium is added.

Notes. Screw-capped tubes are preferred to prevent drying of the agar slants during storage at 5°C. The media may be stored for 6 to 12 months at 5°C. If during this period the slants should turn yellow due to oxidation, fresh media should be prepared.

Once prepared, neither the "yeast-phase" nor "mycelium" agar can be remelted and repoured for use.

If liquid media are desired, delete agar, reduce the concentration of starch to 0.5 g/liter, and reduce the oleic acid to 1 mg/liter of final medium (i.e., use 1 ml of solution 8).

Although developed for *H. capsulatum*, the yeast-phase agar supports excellent growth of *H. duboisii*, *Blastomyces dermatitidis*, and *Sporotrichum schenckii*.

Indole-nitrite medium (9)
For anaerobes

Dehydrated indole-nitrate medium* .... 25 g
Distilled water .... 1,000 ml
Final pH 7.2 ± 0.1

Dissolve the ingredients by boiling for 1 min (with frequent mixing), adjust the pH, dispense 7 ml into 15- by 90-mm screw-capped tubes, autoclave at 121°C for 15 min, and cool.

See note on CHO fermentation broth for anaerobes. Place tubes in glove box, tighten caps, remove from glove box, and store at 4°C or ambient temperature.

Indole can be detected 24 h after good growth is evident in indole-nitrite liquid medium or in another noncarbohydrate medium containing tryptophan (e.g., plain Lombard-Dowell broth or chopped meat medium). Extract the indole by adding 1 dropperful (about 1 ml) of xylene. Add 1 dropperful of Ehrlich reagent. A red ring within 15 min is positive; no red is negative (9).

Nitrate reduction is determined 24 h after good growth in indole-nitrite medium. Add 1 dropperful (about 1 ml) of nitrite reagent A and 0.5 dropperful of nitrite reagent B to the broth culture. A red color within 5 min is positive (nitrite present). If no color is observed, add a pinch of zinc dust. A red color after zinc is negative; no color indicates reduction of nitrite to N_2 or another product (9).

Inhibitory mold agar
For primary recovery of fungi from clinical specimens; available from BBL

Tryptone .... 3 g
Beef extract .... 2 g
Yeast extract .... 5 g
Glucose .... 5 g

Soluble starch	2 g
Dextrin	1 g
Chloramphenicol	0.125 g
Salt A	10 ml
Salt C	20 ml
Agar	17 g
Distilled water	970 ml

Salt A

NaH_2PO_4	25 g
Na_2HPO_4	25 g
Distilled water	250 ml

Salt C

$MgSO_4 \cdot 7H_2O$	10 g
$FeSO_4 \cdot 7H_2O$	0.5 g
NaCl	0.5 g
$MnSO_4 \cdot 7H_2O$	2.0 ml
Distilled water	250 ml

Mix ingredients, and boil to dissolve agar. Adjust the pH to 6.7. Autoclave at 121°C for 15 min. Chloramphenicol is dissolved in 2 ml of alcohol (95%) and added to boiling medium. Dispense before or after autoclaving.

Iron-milk medium (9)
For anaerobes

Whole, nonhomogenized milk
Iron filings

Place a few iron filings in the bottom of 15- by 90-mm screw-capped tubes, add 7 ml of milk, autoclave at 121°C for 15 min, and cool.

See note on fermentation broth for anaerobes. Place tubes in glove box, tighten caps, remove, and store at 4°C or ambient temperature.

Iron-milk medium is used to detect proteolysis of milk proteins. Record C for clot or coagulation, G for gas production, D for digestion of milk proteins, and B for blackening. Hold for 7 days (or longer) before discarding as negative (9).

Jones-Kendrick pertussis transport medium

Soluble starch (BBL)	10.0 g
Yeast extract (Difco)	3.5 g
Heart infusion broth (Difco)	25.0 g
Agar (Difco)	20.0 g
Distilled water	1,000.0 ml
Activated charcoal powder (Norit)	4.0 g
Penicillin	300.0 U

1. Add starch, yeast extract, heart infusion, and agar to water, and boil to dissolve.
2. Add charcoal, mix well, and autoclave.
3. Cool to 50°C, add penicillin, and dispense into small bottles as slants. Cool, and seal tightly.
4. Store at 5°C. Stable for 2 to 3 months.
5. Use as a transport medium for pertussis by inoculating the surface of a slant, sealing, and sending to a reference laboratory.

Kanamycin-vancomycin (KV) blood agar (9)
For anaerobes

Prepare CDC anaerobe blood agar. After the sheep blood has been added, aseptically add 100 mg (base activity) of kanamycin (Bristol Laboratories) and 7.5 mg (base activity) of vancomycin (Lilly) per liter.

Kanamycin-vancomycin laked blood agar (KVLB)

Prepare brucella agar* and add 75 μg (base activity) of kanamycin per ml before autoclaving. Vancomycin (7.5 μg/ml), vitamin K_1 (10 μg/ml), and laked blood (5%) are aseptically added after autoclaving. Laked blood is prepared by freezing whole blood overnight and then thawing. Commercial injectable antibiotics cannot be used. Laboratory standard powders must be used.

2-Ketogluconate broth

Monopotassium phosphate	5.4 g
Potassium nitrate	2.0 g
Potassium gluconate	20.0 g
Distilled water	1.0 liter
Final pH 6.5	

Sterilize by filtration. Dispense aseptically, 1 ml per sterile 13- by 100-mm tube.

3-Ketolactonate medium

Yeast extract	1.0 g
Agar	2.0 g
Distilled water	100.0 ml

1. Mix and heat to obtain solution.
2. Dispense into 20- by 150-mm test tubes, using 20 ml per tube.
3. Autoclave at 121°C for 15 min.
4. While agar is still melted, aseptically admix 2.0 ml of 10% filtered aqueous lactose per tube.

Kligler iron agar*

Peptone or polypeptone	20.000 g
Meat extract (optional)	3.000 g
Yeast extract (optional)	3.000 g
Lactose	10.000 g
Glucose	1.000 g
Sodium chloride	5.000 g
Ferric ammonium citrate	0.500 g
Sodium thiosulfate	0.500 g
Agar	12 to 15 g
Phenol red	0.025 g
Water	1 liter
Final pH 7.4±	

1. Mix and heat with agitation until solution occurs.
2. Dispense for deep slants, and autoclave at 121°C for 15 min.

L-Cysteine-deficient growth medium (BCYEα-Lcys) (11)

ACES buffer (Research Organics)	10.00 g
KOH pellets (reagent grade, 85% KOH)	2.66 g

Activated charcoal (Norit A) .... 1.50 g
Yeast extract (Difco) .... 10.00 g
Agar (Difco) .... 17.00 g
α-Ketoglutarate monopotassium salt (Sigma) .... 1.00 g
Ferric pyrophosphate in 10 ml of distilled water (Sigma) .... 0.25 g
Distilled water .... 1.00 liter

1. Prepare this medium exactly the same way as BCYEα medium, but do not add L-cysteine. Note that less KOH is needed than in BCYEα medium.
2. Preincubate plates for 24 h at 35°C before using. Shelf life is 2 months at 5°C.

***Legionella pneumophila* media**

Charcoal-yeast extract diphasic blood culture medium (CYE-DBCM) (15)

Agar phase

**Activated charcoal (Norit SG) .... 2.0 g
Agar .... 17.0 g
Distilled water .... 500 ml

** Available through Sigma catalog no. C5510. Activated charcoal, washed with phosphoric and sulfuric acids.

Broth phase

Yeast extract .... 20.00 g
L-Cysteine hydrochloride · H_2O .... 0.40 g
$Fe(HO_3)_3 \cdot 9H_2O$ .... 0.10 g
Distilled water .... 500.0 ml

Prepare the agar first. Combine ingredients, boil, and dispense as 20-ml aliquots into each of a series of 125-ml Wheaton serum bottles. Stopper each bottle loosely with both a rubber stopper and a metal cap. Autoclave the bottles at 121°C for 20 min, cool to 50°C, remix the warm charcoal and agar thoroughly, and place the bottles at an angle so that an agar slant with a vertical height of 6.0 cm is formed. This procedure should leave a portion of the agar protruding above the 25 ml of liquid (broth plus specimen) that is added later.

Prepare the broth by first autoclaving the yeast extract and water at 121°C for 15 min. Allow to cool, and then add fresh sterile solutions of L-cysteine hydrochloride and ferric nitrate, in that order. Adjust the pH to 6.9 by adding 6.0 ml of 1 N KOH. Dispense the broth in 20-ml aliquots into Wheaton bottles with charcoal agar slants. Seal the bottles by crimping the metal caps over the rubber stoppers. Check for sterility by preincubation at 35°C for 2 days.

Editors' note. The authors recommend Difco ingredients.

Charcoal yeast extract (CYE) agar (14)

Yeast extract .... 10.00 g
**Activated charcoal (Norit SG) .... 2.00 g
L-Cysteine hydrochloride · H_2O .... 0.40 g
***Ferric pyrophosphate, soluble .... 0.25 g
Agar .... 17.00 g
Distilled water .... 1.0 liter

**Available through Sigma, catalog no. C5510. Activated charcoal, washed with phosphoric and sulfuric acids.
***Available upon request from Biological Products Division, Centers for Disease Control, Atlanta, GA 30333. This reagent must be kept dry and stored in the dark. Do not use if its color changes from green to yellow or brown.

Add all ingredients of CYE agar except L-cysteine hydrochloride and soluble ferric pyrophosphate to 980 ml of distilled water; dissolve by boiling, autoclave at 121°C for 15 min, and cool to 50°C in a water bath.

Prepare separate fresh solutions of L-cysteine hydrochloride (0.40 g in 10 ml of distilled water) and soluble ferric pyrophosphate (0.25 g in 10 ml of distilled water). Filter sterilize each solution separately. Add the L-cysteine hydrochloride to the basal medium first; then add the ferric pyrophosphate.

Adjust the pH with 4.0 to 4.5 ml of 1.0 N KOH so that the final pH of the medium is 6.90 ± 0.05. Use 1.0 N HCl when necessary.

Editors' note. The authors recommend Difco ingredients.

Feeley-Gorman (FG) agar (16)

Casein (acid hydrolysate) .... 17.50 g
Beef extractives .... 3.00 g
Starch .... 1.50 g
Agar .... 17.00 g
L-Cysteine hydrochloride · H_2O .... 0.40 g
**Ferric pyrophosphate, soluble .... 0.25 g
Distilled water .... 1.00 liter

** Available on request from Biological Products Division, Centers for Disease Control, Atlanta, GA 30333. This reagent must be kept dry and stored in the dark. Do not use if its color changes from green to yellow or brown.

Mueller-Hinton (MH) agar contains all the ingredients of FG agar except for L-cysteine hydrochloride and ferric pyrophosphate (soluble) and provides a source of casein, beef extractives, starch, and agar. (*Note.* Some lots of MH agar may vary too widely in formulation and stability to serve as a base for satisfactory FG agar.)

Add the casein (acid hydrolysate), beef extractives, starch, and agar (or their equivalent, 38 g of MH agar) to 980 ml of distilled water, autoclave at 121°C for 15 min, cool to 50°C in a water bath, and hold until L-cysteine hydrochloride and soluble ferric pyrophosphate are added.

Prepare separate fresh solutions of L-cysteine hydrochloride (0.4 g in 10 ml of distilled water) and soluble ferric pyrophosphate (0.25 g in 10 ml of distilled water). Filter sterilize each solution separately. Add the L-cysteine hydrochloride to the agar mixture first, then add the ferric pyrophosphate. Adjust the pH with either 1.0 N KOH or 1.0 N HCl so that the final pH of the medium is 6.90 ± 0.50.

Feeley-Gorman (FG) broth

FG broth has the same formulation as FG agar except that it lacks agar, and the concentration of soluble ferric pyrophosphate is 100 mg/liter (wt/vol). Dispense the broth in 2-ml aliquots into screw-capped tubes (13 by 100 mm).

MH-IH agar for *Legionella* species (16)

Component A

Mueller-Hinton agar*	38 g
Distilled water	490 ml

Component B

Hemoglobin powder*	10 g
Distilled water	490 ml

Component C

IsoVitaleX*	20 ml

Prepare components A and B separately, autoclave at 121°C for 15 min, cool to 50°C, combine A and B, and hold at 50°C in a water bath. Prepare component C by adding 10 ml of sterile distilled water to each of two vials containing lyophilized IsoVitaleX. Add the contents of both vials to the AB mixture.

Lipase test agar (Sierra medium) (48)

Peptone	10.0 g
Sodium chloride	5.0 g
Calcium chloride, monohydrous	0.1 g
Agar	15.0 g
Distilled water	1,000 ml
Final pH 7.4	

Mix, heat to boiling, and autoclave at 121°C for 15 min. Cool to 40 to 50°C.

Littman oxgall agar*

Oxgall	15.00 g
Peptone	10.00 g
Glucose	10.00 g
Agar, undried	20.00 g
(*or* Dried	16.00 g)
Crystal violet	0.01 g
Water	1.00 liter
Final pH 7.0 ± 0.2	

1. Mix and heat with agitation, and boil 1 min.
2. Autoclave at 121°C for 15 min.
3. Cool to 45 to 50°C, and add 3 ml of 1% aqueous solution of streptomycin.
4. Mix and dispense aseptically into sterile tubes or onto plates.

Littman oxgall agar* with birdseed extract

Prepare as described for Littman oxgall agar, but use only 800 ml of water and 200 ml of birdseed extract in the basic medium.

Birdseed extract is prepared as follows.

1. Grind 70 g of *Guizottia abyssinica* seeds in a blender.
2. Add 300 ml of water.
3. Autoclave at 115°C for 10 min.
4. Filter through gauze.

Lombard-Dowell (LD) agar (9)
For anaerobes

Pancreatic digest of casein	5.0 g
Yeast extract	5.0 g
Sodium chloride	2.5 g
L-Tryptophan	0.2 g
Sodium sulfite	0.1 g
**L-Cystine	0.4 g
**Hemin	10.0 mg
**Vitamin K_1, stock solution	1.0 ml
Agar	20.0 g
Distilled water	1,000.0 ml
Final pH 7.5	

** See CDC anaerobe blood agar for directions for preparing L-cystine, hemin, and vitamin K_1 solutions.

Dissolve the ingredients by heating, autoclave at 121°C for 15 min, cool, and dispense. If quadrant plastic plates (Presumpto) are used, aseptically dispense 5 ml into one quadrant of each plate and allow agar to solidify. Store Presumpto quadrant plates in refrigerator (4°C).

Lombard-Dowell bile (LD bile) agar (9)
For anaerobes

Prepare LD agar as described above, but add 1.0 g of D-glucose and 20 g of oxgall per liter of medium before autoclaving.

Lombard-Dowell (LD) broth medium (9)
For anaerobes

Trypticase (BBL)	5.0 g
Yeast extract (Difco)	5.0 g
Sodium chloride	2.5 g
Sodium sulfite	0.1 g
**L-Tryptophan	0.2 g
***Hemin (1% solution)	1.0 ml
***Vitamin K_1 (1% solution)	1.0 ml
Agar	0.7 g
Distilled water	1,000.0 ml

** Dissolve the L-tryptophan in 5 ml of 1 N NaOH before combining with the other ingredients.
*** For a stock solution of hemin, dissolve 1 g in 20 ml of 1 N NaOH, and add distilled water to make a final volume of 100 ml. For a 1% stock solution of vitamin K_1, suspend 1 g in 99 ml of absolute ethanol.

Heat to dissolve the ingredients, and adjust the pH. Dispense 7 ml into 15- by 90-mm screw-capped tubes, autoclave at 121°C for 15 min, and cool.

See note on CHO fermentation broth for anaerobes. Place tubes in glove box, tighten caps, remove from glove box, and store at 4°C or ambient temperature.

Lombard-Dowell egg yolk agar (LD EYA) (9)
For anaerobes

Prepare LD agar as described above, but add only 900 ml of distilled water, and add to it 2.0 g of D-glucose, 5.0 g of Na_2HPO_4, and 0.2 ml of a 5% aqueous solution of $MgSO_4$. Autoclave, and cool to 55°C. Add 100 ml of well-mixed egg yolk suspension. Mix well and dispense.

Lombard-Dowell esculin (LD esculin) agar (9)
For anaerobes

Prepare LD agar as described above, but add 1.0 g of esculin and 0.5 g of ferric citrate before autoclaving.

Lombard-Dowell neomycin egg yolk agar (NEYA) (9)
For anaerobes

Prepare LD EYA. After it is autoclaved and cooled to 50°C, aseptically add 100 mg (base activity) of neomycin sulfate (Lilly) per liter, mix, and dispense.

Determination of biochemical characteristics on quadrant plates

In 1977, Dowell and Lombard (8) developed a system for presumptive identification of gram-negative, nonsporeforming anaerobic bacilli which requires a minimal number of media and of biochemical tests. This system includes two blood agar plates (for determination of relation to oxygen, colony characteristics, microscopic features, and inhibition by penicillin, rifampin, and kanamycin), one tube of enriched thioglycolate medium (to determine growth characteristics and cellular morphology and to provide inoculum for other tests), and a quadrant plate, known as the Presumpto I plate, containing the following:

1. Lombard-Dowell agar—for indole disk test and growth control (to compare with growth on Lombard-Dowell bile agar)
2. Lombard-Dowell esculin agar—for determination of esculin hydrolysis, H_2S production, and catalase
3. Lombard-Dowell egg yolk agar—for determination of lipase, lecithinase, and proteolysis
4. Lombard-Dowell bile agar—for determination of growth in the presence of 20% bile (2% oxgall) and formation of an insoluble precipitate in the medium (under and surrounding growth)

Inoculate the above agar-based differential media with (i) an overnight thioglycolate or cooked meat-glucose broth culture from an isolated colony (24- to 48-h cultures are used if overnight growth is inadequate) or (ii) a turbid cell suspension (McFarland no. 1) in Lombard-Dowell growth broth without glucose.

Place 1 or 2 drops of broth culture or cell suspension on each quadrant of the Presumpto plate, using a capillary pipette, and streak three-fourths of each quadrant.

Place a sterile 0.25-in. (0.6-cm) blank disk near the outer periphery of the inoculum on the Lombard-Dowell agar quadrant for the indole test.

Moisten a sterile swab in actively growing thioglycolate broth culture, and evenly inoculate the surface of a brucella blood agar plate. Apply the following disks: penicillin, 2 U; kanamycin, 1,000 μg; and rifampin, 15 μg. Zones of inhibition of 12 mm or greater around the penicillin and kanamycin disks indicate susceptibility; zones of <12 mm indicate resistance. A 15-mm zone around the rifampin disk indicates susceptibility; <15 mm indicates resistance. If these disks are used, refer to the tables of Dowell and Lombard (8) or Koneman et al. (35). These antibiotic disks are available from BBL.

Incubate the above media in an anaerobic system at 35°C for 48 h. If an anaerobic glove box is used, inspect the media after 18 to 24 h of incubation. If good growth is obtained, the reactions can be read after overnight incubation.

Determination of results and test interpretation

Indole test. Add 2 drops of paradimethylaminocinnamaldehyde reagent to the disk on Lombard-Dowell agar. Blue is positive; any other color or no color is negative.

Lombard-Dowell bile agar. (i) Growth equal to or greater than growth on the Lombard-Dowell agar control is recorded as "E." Inhibition of growth on 20% bile medium (growth less than on the Lombard-Dowell agar control) is recorded as "I." (ii) Check for an opaque or whitish-yellow precipitate within the medium surrounding the colonies (known to be produced only by *Bacteroides fragilis* and some strains of *Bacteroides ovatus*; V. R. Dowell, Jr., personal communication).

Lombard-Dowell egg yolk agar. Lecithinase, lipase, and proteolytic activity are determined as for modified McClung-Toabe egg yolk agar (see above, reactions on egg yolk agar).

Esculin hydrolysis on Lombard-Dowell esculin agar. Browning or blackening of the medium indicates esculin hydrolysis. Loss of fluorescence compared with an uninoculated control observed under UV light (365 nm) indicates esculin hydrolysis.

H_2S production. Black colonies on Lombard-Dowell esculin agar indicate H_2S production.

Catalase. Expose the quadrant plate to air for 30 min. Flood the esculin agar quadrant with 3% H_2O_2. Vigorous sustained bubbling indicates catalase production. This sometimes takes 30 s to 1 min.

Lowenstein-Jensen medium*

Monopotassium phosphate, anhydrous	2.40 g
Magnesium sulfate · $7H_2O$	0.24 g
Magnesium citrate	0.60 g
Asparagine	3.60 g
Potato flour	30.00 g
Glycerol	12.00 g
Distilled water	600.00 ml
Homogenized whole eggs	1,000.00 ml
Malachite green, 2% aqueous	20.00 ml

1. Dissolve the salts and asparagine in the water.
2. Admix the glycerol and potato flour, autoclave at 121°C for 30 min, and cool to room temperature.
3. Scrub eggs, not more than 1 week old, in 5% soap solution, and then rinse thoroughly in cold running water.
4. Immerse in 70% ethyl alcohol for 15 min.
5. Break eggs into a sterile flask. Homogenize by hand shaking, and filter through four layers of gauze.
6. Add 1 liter of homogenized eggs to the potato-salt mixture.

7. Prepare the malachite green and admix thoroughly.
8. Dispense 6 to 8 ml into 20- by 150-mm screw-capped tubes.
9. Slant, and inspissate at 85°C for 50 min.
10. Incubate for 48 h at 37°C to check sterility, and store at 4 to 6°C with caps tightly closed.

MacConkey agar*

Bacto-Peptone or Gelysate (BBL)	17.000 g
Peptone: proteose or Polypeptone (BBL)	3.000 g
Lactose	10.000 g
Bile salts	1.500 g
Sodium chloride	5.000 g
Agar	13.500 g
Neutral red	0.030 g
Crystal violet	0.001 g
Distilled water	1.000 liter
Final pH 7.1	

Sucrose may be added to give a final concentration of 1%. Autoclave at 121°C for 15 min. Use in plates or as slants.

Malt extract agar*

Maltose	12.75 g
Dextrin	2.75 g
Glycerol	2.75 g
Peptone	0.78 g
Agar	15.00 g
Final pH 4.6	

Suspend 33.6 g of the dehydrated material in a liter of distilled water. Mix thoroughly. Heat with frequent agitation, and boil for 1 min. Sterilize by autoclaving at 121°C for 15 min. Aseptically dispense into plates before the agar solidifies.

Mannitol-salt agar USP*

Beef extract	1.000 g
Peptone or Polypeptone (BBL)	10.000 g
Sodium chloride	75.000 g
Mannitol	10.000 g
Agar	15.000 g
Phenol red	0.025 g
Distilled water	1.000 liter
Final pH 7.4	

Heat to dissolve, and autoclave at 121°C for 15 min.

Medium of Knisely for *Bacillus anthracis* (33)

Heart infusion agar (Difco; BHI agar)
Lysozyme (chicken egg white)
Thallous acetate [$Tl(OAc)_3$]
EDTA
Polymyxin

Add 55 g of the BHI agar to 1,000 ml of distilled water, and with the aid of pH paper, adjust the hydrogen ion concentration to pH 7.3 to 7.4. Autoclave, and cool to 48 to 54°C. Add (final concentration) polymyxin (30 U/ml), lysozyme (40 μg/ml), $Tl(OAc)_3$ (40 μg/ml), and EDTA (200 μg/ml). Carefully swirl the contents, and pour plates.

Medium (R-medium) for elaboration of toxins of *Bacillus anthracis* (43)

Amino acids (Trp, Gly, Cys, Tyr, Lys, Val, Leu, Ile, Thr, Met, Asp, Glu, Pro, His hydrochloride, Arg hydrochloride, Phe, Ser)
Glucose
Thiamine-hydrochloride, uracil, adenine sulfate
$CaCl_2 \cdot 2H_2O$, $MgSO_4 \cdot H_2O$, $MnSO_4 \cdot H_2O$, $NaHCO_3$, K_2HPO_4
Sodium hydroxide

To 500 ml of water add (milligrams in parentheses) Trp (35), Gly (65), Cys (25), Tyr (144), Lys (230), Val (173), Leu (230), Ile (170), Thr (120), Met (73), Asp (184), Glu (612), Pro (43), His hydrochloride (55), Arg hydrochloride (125), Phe (125), Ser (235), thiamine hydrochloride (1.0), glucose (2,500), $CaCl_2 \cdot 2H_2O$ (7.4), $MgSO_4 \cdot H_2O$ (9.9), $MnSO_4 \cdot H_2O$ (0.9), K_2HPO_4 (3,000), $NaHCO_3$ (8,000), uracil (1.4), and adenine sulfate (2.1). Adjust the pH to 8.0 with 5 N NaOH, and filter sterilize. Autoclave 500 ml of 3% (wt/vol) agar, and cool to 50°C. Mix filter-sterilized medium with agar, and pour plates. The medium is effective in promoting the elaboration of the exotoxins of *B. anthracis*.

Middlebrook and Cohn 7H10 agar*
7H10 agar

Stock solutions

Solution 1. Store at room temperature.

Monopotassium phosphate	15 g
Disodium phosphate	15 g
Distilled water	250 ml

Solution 2. Store at 4 to 10°C.

Ammonium sulfate	5.0 g
Monosodium glutamate	5.0 g
Sodium citrate · $2H_2O$ USP	4.0 g
Ferric ammonium citrate	0.4 g
Magnesium sulfate ($7H_2O$) ACS	0.5 g
Biotin, in 2 ml of 10% NH_4OH	5.0 mg
Distilled water, to	250.0 ml

Solution 3. Store at 4 to 10°C.

Calcium chloride · $2H_2O$ ACS	50 mg
Zinc sulfate · $7H_2O$ ACS	100 mg
Copper sulfate · $5H_2O$ ACS	100 mg
Pyridoxine hydrochloride	100 mg
Distilled water, to	100 ml

Solution 4. Glycerol reagent

Solution 5. Malachite green, 0.01% aqueous

1. To 975 ml of distilled water, add:

Solution 1	25 ml
Solution 2	25 ml
Solution 3	1 ml
Solution 4	5 ml

2. Adjust the pH to 6.6 by adding approximately 0.5 ml of 6 N HCl.

3. Add 2.5 ml of solution 5 and 15 g of agar.

4. Autoclave at 121°C for 15 min. Cool to 56°C, and add 100 ml of OADC Enrichment*:

Bovine albumin fraction V .... 50 g
Sterile saline, 0.85% .... 900 ml

Dissolve and add 30 ml of the following:

Oleic acid .... 0.6 ml
Distilled water, to .... 30.0 ml
Sodium hydroxide, 6 M .... 0.6 ml

Warm to 56°C; swirl until solution occurs.

5. Adjust the reaction to pH 7.0, measured electrometrically.

6. Add 40 ml of sterile 50% aqueous glucose. Sterilize by filtration.

7. Heat at 56°C for 1 h, incubate overnight, heat again at 56°C, and incubate again overnight at 37°C for sterility.

8. Add 100 ml of this solution to agar base.

9. Add also 2 ml of freshly prepared, membrane filter-sterilized catalase solution containing 1,000 μg/ml.

10. Dispense the complete medium completely into petri plates or tubes.

11. Store at 4 to 8°C; unsealed and unprotected containers should be used for no longer than 1 week.

12. During preparation and storage, protect from light.

7H11 agar*

Add 1 g of enzymatic casein hydrolysate per liter of 7H10 agar.

Selective 7H11 agar

Add the following antimicrobial agents per liter of 7H11 agar:

Carbenicillin .... 50 μg
Polymyxin B .... 200,000 U
Amphotericin B .... 10 μg
Trimethoprim lactate..................... 20 μg

Modified Diamonds medium for trichomonas culture

Trypticase (BBL) .... 20 g
Yeast extract .... 1.0 g
Maltose .... 0.5 g
L-Cystine hydrochloride .... 0.5 g
L-Ascorbic acid .... 0.02 g

Add distilled water to make 90 ml, and adjust the pH to 6.5 with 4 N NaOH or 1 N HCl. Autoclave, cool to 48°C, and add sodium penicillin G (100,000 U), streptomycin sulfate (0.15 g), and amphotericin B (200 μg) in 1 ml of water. (Final concentration of drugs is 1,000 U of penicillin, 1.5 mg of streptomycin, and 2 μg of amphotericin B per ml.) Add horse serum (10 ml, inactivated at 56°C for 30 min). Dispense in 5.0-ml quantities into sterile tubes. Store at 4°C for up to 14 days. Warm to 35°C before inoculation.

If usage rate is low, medium without horse serum may be frozen at −20°C. When needed, a tube of medium is thawed, and 0.5 ml of sterile horse serum is added.

Modified Thayer-Martin agar*

See Thayer-Martin agar.

Modified Wadowsky-Yee (MWY) medium (11, 61, 62)

Yeast extract .... 10.00 g
Agar (Difco) .... 17.00 g
Activated charcoal (Norit A) .... 1.50 g
ACES buffer (Research Organics) ..... 10.00 g
KOH pellets (reagent grade, 85% KOH) .... 2.90 g
α-Ketoglutarate, monopotassium (Sigma) .... 1.00 g
**L-Cysteine hydrochloride (ICN) .... 0.40 g
**Ferric pyrophosphate (Sigma) .... 0.25 g
Distilled water .... 1 liter
**Polymyxin B sulfate (Pfizer or Sigma) (as base) .... 50,000 U
**Vancomycin hydrochloride (Lilly) (as base) .... 1.00 mg
Glycine (Sigma) .... 3.00 g
**Bromothymol blue (Sigma) .... 10.00 mg
**Bromocresol purple (Sigma) .... 10.00 mg
**Anisomycin (General Diagnostics) .... 80.00 mg

** Added to medium after autoclaving at 50°C.

1. Prepare BCYEα as detailed above, with the following changes:
   a. Use 2.9 g of KOH pellets rather than 2.8 g.
   b. Add glycine to the mixture before autoclaving.

2. Add filter-sterilized antimicrobial agents separately at 50°C after autoclaving.

3. Add filter-sterilized dyes. Dissolve each dye separately in 0.1 N KOH in amounts that will yield 10 mg/ml (i.e., 1 g in 100 ml). Filter sterilize and add 1.0 ml of each to the medium.

4. Check pH. Should be between 6.85 and 6.95 at 40°C. Correct with 1 N HCl or KOH pellets, if needed.

5. Cool to 40°C and pour.

6. Preincubate the entire batch for 1 day at 35°C to detect contaminants.

7. Cool to room temperature, and then store noncontaminated plates at 5°C in plastic sleeves. Shelf life is 1 month at 5°C.

Motility medium (9)
For anaerobes

Motility medium (Difco)* .... 16.0 g
Nutrient broth* .... 4.0 g
Sodium chloride .... 1.0 g
Agar....................................... 4.0 g
Distilled water .... 1,000.0 ml
Final pH 7.2 ± 0.1

Heat to dissolve the above ingredients, adjust the pH, dispense 7 ml into 15- by 90-mm screw-capped tubes, autoclave at 121°C for 15 min, and cool.

See note on fermentation broth for anaerobes. Place tubes in glove box, tighten caps, remove from glove box, and store at 4°C or ambient temperature.

Motility test medium*

A. *For Enterobacteriaceae*

Beef extract .... 3.0 g
Peptone or Gelysate (BBL) .... 10.0 g
Sodium chloride .... 5.0 g
Agar .... 4.0 g
Distilled water .... 1 liter
Final pH 7.4

Dispense about 8 ml per tube. Autoclave at 121°C for 15 min.

B. *For nonfermenting gram-negative bacteria*

1. Method of Gilardi (19)

Pancreatic digest of casein USP .... 10.0 g
Yeast extract .... 3.0 g
Sodium chloride .... 5.0 g
Agar .... 3.0 g
Distilled water .... 1 liter
Final pH 7.2

Dispense 3.5 ml per 13- by 100-mm tube. Autoclave at 121°C for 15 min.

2. Method of Tatum et al. (54)

Tryptose or Biosate (BBL) .... 8.0 g
Sodium chloride .... 5.0 g
Agar .... 4.0 g
Nutrient broth .... 500 ml
Distilled water .... 500 ml

1. Mix, and heat to dissolve ingredients.
2. Dispense into test tubes, and autoclave at 121°C for 15 min.

C. *For Listeria monocytogenes* (31)

Beef extract .... 3.0 g
Peptone .... 10.0 g
Sodium chloride (NaCl) .... 5.0 g
Agar .... 4.0 g
Distilled water .... 1,000 ml

Before autoclaving, add 0.05 g of 2,3,5-triphenyltetrazolium chloride (TTC) (or 5.0 ml of a 1% TTC solution).

Mueller-Hinton agar*

Beef infusion, from .... 300.0 g
Acid hydrolysate of casein .... 17.5 g
Starch .... 1.5 g
Agar .... 17.0 g
Distilled water .... 1.0 liter

Dispense, and autoclave at 116 to 121°C for 15 min. Cool to 45 to 50°C and add blood, if desired.

Mycobiotic agar, mycosel agar
For primary recovery of dermatophytes

Soytone .... 10.0 g
Glucose .... 10.0 g
Agar .... 15.0 g
Cycloheximide, in 10 ml of acetone .... 0.05 g
Chloramphenicol, in 10 ml of 95% alcohol .... 0.05 g

Heat to dissolve the agar. Dispense, and autoclave at 121°C for 10 min.

Mycoplasma media
See H agar

Dialysate broth base

1. Dissolve 300 g of papaic digest of soy meal USP in 2 liters of water, and place in dialysis casing.
2. Autoclave 250 g of active dried yeast (e.g., Fleischman) in 2 liters of water for 5 min. Place in dialysis casing.
3. Dialyze both casings in 10 liters of water at 4°C for 2 days. Discard casings and contents.
4. Add 0.5% sodium chloride and 0.04% phenol red to dialysate. Adjust pH to 7.3 with 1 N sodium hydroxide.
5. Autoclave at 121°C for 15 min.

Diphasic H medium

Dispense H agar in 3-ml amounts into 16- by 125-mm screw-capped tubes, allow to solidify, and overlay with 3 ml of H broth. Store at room temperature. Tubes must have good closures to avoid excessive elevation of pH upon loss of CO_2 from some lots of horse serum.

Medium for production of complement-fixing antigen for *Mycoplasma pneumoniae*

Dialysate broth base .... 87 ml
Agamma horse serum .... 10 ml
Penicillin, 10,000 U/ml .... 2 ml
Thallium acetate, 3.3% .... 1 ml

1. Combine and dispense aseptically.
2. Inoculate with 1 to 5% actively growing cultures.
3. Agitate or stir with spin bars. Incubate for 5 to 8 days until haze is observed and pH begins to drop.
4. Harvest by centrifugation at 8,000 × *g*, and wash three times with saline.
5. Resuspend in a volume 1/100 to 1/1,000 of the original culture.

MES agar

Papaic digest of soy meal USP .... 20.00 g
Sodium chloride .... 5.00 g
2-(*N*-Morpholino)ethanesulfonic acid
(MES,* Calbiochem) .... 4.25 g
Water .... 1.00 liter
Agarose .... 10.00 g

1. Dissolve, adjust pH to 6.0 at 37°C, autoclave at 121°C for 15 min, and cool to 50°C.
2. To 65 ml of solution, add 10 ml of yeast dialysate, 25 ml of horse serum, and 2 ml of penicillin (10,000 U/ml).
3. Dispense into plates as for H agar.

Urea broth

Dialysate broth base .... 85 ml
Urea (1 M in water, filter sterilized) .... 0.5 ml
Horse serum .... 10 ml
MES buffer (1 M in water; adjust to pH 6.2 with 1 N NaOH at 37°C, and filter sterilize) .... 1 ml
Penicillin (10,000 U/ml)................... 2 ml
Sodium sulfite (100 mM in water, freshly prepared and autoclaved)................ 1 ml
Phenol red (1%) .... 0.1 ml

Dispense in 3-ml amounts into 16- by 125-mm test tubes.

This medium should not be stored for more than 3 days after the addition of sulfite. It is convenient to make up the complete medium without sulfite and to add an appropriate amount of sulfite before use.

Mycosel agar (Mycobiotic agar)*

Sabouraud dextrose agar................ 1.00 liter
Agar .... 5.00 g
Cycloheximide, in 10 ml of acetone ...... 0.50 g
Chloramphenicol, in 10 ml of 95% alcohol............................. 0.50 g

(Not to be used for isolation of zygomycetes.)

Nelson medium for *Naegleria fowleri*

Panmede (ox liver digest) .... 1.0 g
Glucose.................................. 1.0 g
Amoeba saline, to .... 1,000 ml

1. Dissolve the ingredients in amoeba saline (see nonnutrient agar plates).
2. Dispense into 16- by 125-mm screw-capped tubes, 10 ml per tube.
3. Autoclave for 15 min.
4. Add 0.2 ml of heat-inactivated fetal calf serum to each tube before inoculating the medium with amoebae.

Nitrate broth, enriched (54)

Heart infusion broth* or infusion (heart) broth .... 25 g
Potassium nitrate.......................... 2 g
Distilled water .... 1 liter

Dispense 4 ml per tube with inverted insert tubes. Autoclave at 121°C for 15 min.

Nitrate reduction broth (19)

Pancreatic digest of casein .... 10 g
Yeast extract............................. 3 g
Potassium nitrate.......................... 2 g
Distilled water .... 1 liter
Final pH 7.1

Dispense in 4-ml amounts into 13- by 100-mm test tubes containing Durham vials. Autoclave at 121°C for 15 min.

Nitrite reduction broth
For nonfermenting gram-negative bacteria

Heart infusion broth...................... 1 liter
Potassium nitrate, chemically pure .... 1 g
Adjust to pH 7.0

Dispense in 4-ml amounts into 15- by 125-mm tubes containing Durham vials. Autoclave at 121°C for 15 min.

Nonnutrient agar plates
For the isolation and culture of pathogenic free-living amoebae

Agar.................................... 1.5 g
Page amoeba saline .... 100 ml

Dissolve agar in the saline with heat, and sterilize at 15 lb/in^2 pressure in 15 min. Cool to 60°C and aseptically pour into plastic petri dishes. Use 20 ml for 100- by 15-mm dishes or 5 ml for 16- by 15-mm dishes. After the agar gels, store the plates in canisters at 4°C (refrigerator). Plates may be kept in the refrigerator for about 3 months.

Prepare Page amoeba saline as follows.

Sodium chloride (NaCl)................... 120 mg
Magnesium sulfate ($MgSO_4 \cdot 7H_2O$) .... 4 mg
Calcium chloride ($CaCl_2 \cdot 2H_2O$) .... 4 mg
Disodium hydrogen phosphate (Na_2HPO_4). 142 mg
Potassium dihydrogen phosphate (KH_2PO_4) .... 136 mg
Distilled water, to .... 1,000 ml

Dissolve the chemicals in water, and sterilize by autoclaving at 15 lb/in^2 for 15 min. The solution may be stored in the refrigerator for up to 6 months.

DL-Norleucine assimilation

See carbon assimilation medium.

Novy, MacNeal, and Nicolle (NNN) medium for *Leishmania* species and *Trypanosoma cruzi* (40)

Agar..................................... 7 g
Sodium chloride........................... 3 g
Distilled water .... 450 ml
Defibrinated rabbit blood (sterile)......... 150 ml

Mix agar, sodium chloride, and water, and bring to boiling to dissolve agar and salt. Autoclave at 121°C for 15 min. Cool to 52°C and add rabbit blood. Dispense 5-ml portions into sterile 13- by 100-mm screw-capped tubes, and slant the tubes in the refrigerator. Condensation of water at the bottom of the slant is desirable. If there is little moisture, a few drops of sterile water may be added with the inoculum. Medium should be tested for sterility by incubating a sample slant at 35°C for 48 h.

Nutrient agar*

Beef extract .... 3 g
Peptone or Gelysate pancreatic digest of gelatin .... 5 g

Agar .... 15 g
Distilled water .... 1 liter
Final pH 6.8

Autoclave at 121°C for 15 min.

Nutrient broth, standard II

Special peptone.......................... 8.6 g
Sodium chloride .... 6.4 g
Distilled water........................... 1.0 liter
Final pH 7.5 ± 0.1 at 37°C

Completely dissolve 15 g in 1 liter of freshly distilled or demineralized water. Sterilize in the autoclave (15 min at 121°C).

Nutrient broth, 6.0% NaCl

Prepare like nutrient agar but without agar plus 60 g of sodium chloride per liter.

Nutrient gelatin*

A. *For enterobacteriaceae and nonfermenting gram-negative bacteria*

Nutrient broth .... 1 liter
Gelatin .... 120 g

Autoclave at 121°C for 12 min.

Inoculate by stabbing with a wire, and incubate at 20°C for 30 days. Nutrient gelatin is recommended as the "standard" method in taxonomic work, since the rate of gelatin liquefaction is important in the characterization of certain groups and subgroups within the family *Enterobacteriaceae*. Some of the rapid methods are excellent for diagnostic work in which one is not interested in the rate of liquefaction. In those areas where the rate of liquefaction is of differential value (e.g., within the tribe *Klebsielleae*), positive tests obtained by the rapid methods should be repeated with the conventional methods. If the above-mentioned limitations are borne in mind, certain rapid methods can be recommended.

Gelatin liquefaction demonstration by rapid methods (34, 36)

Prepare nutrient gelatin, using 15 g/100 ml of distilled water. Add 3 to 5 g of powdered charcoal, mix thoroughly, cool, and pour into petri dishes or other flat containers to a depth of 3 mm. (Apply a thin film of petrolatum to containers first.) After medium has set, remove the sheet of gelatin and place it in 10% Formalin for 24 h. Cut the sheet into pieces 1 cm by 5 to 8 mm. Wrap in gauze, and wash in running tap water for 24 h. Place in wide-mouth, screw-capped jars or bottles, and cover with distilled water. Sterilize by exposure to flowing steam for 30 min on 3 successive days. After sterilization, the water may be decanted. Check sterility by placing pieces in tubes of nutrient broth.

Suspend the growth from an 18- to 24-h agar plate culture in 3 ml of 0.85% sodium chloride solution containing 0.01 M calcium chloride in a small test tube (e.g., 13 by 100 mm). The suspension should be very dense, since the rapidity of the reaction appears to be a function of density and temperature incubation. The agar plates should be thick; i.e., they should contain 35 to 40 ml of infusion agar medium. Three or four drops of a broth culture should be spread over the entire surface to obtain maximal confluent growth.

With aseptic precautions, add a piece of denatured charcoal to each dense suspension prepared as described above. Add 0.1 ml of toluene to each suspension, and shake the tube. Toluene appears to have an activating effect upon the reactions of strains that ordinarily are slow liquefiers of gelatin. Incubate at 37°C, and examine after 5 or 6 h and daily for 14 days. Positive reactions are indicated by the release of charcoal particles, which collect in the bottom of the tube.

Advantage may be taken of the bactericidal effect of toluene. Distribute 3-ml amounts of saline solution containing 0.01 M calcium chloride into small tubes, and autoclave at 121°C for 15 min. Place a piece of charcoal-gelatin and 0.1 ml of toluene in each tube, and stopper with corks that have been soaked in hot paraffin. Store until needed.

B. *For aerobic Actinomycetes*

Prepare as described above for medium A, but with demineralized water.

Inoculate just below the surface of the solidified medium. Incubate at room temperature or 37°C with an uninoculated control. When growth is sufficient, place tests and control in refrigerator until control solidifies; then observe tests for liquefaction.

C. *For Erysipelothrix*

Medium A is recommended.

D. *For miscellaneous gram-negative bacteria*

Heart infusion broth* or infusion (heart)
broth .... 25 g
Gelatin .... 120 g
Distilled water .... 1 liter

Heat to boiling to dissolve. Cool to 55°C, and adjust to pH 7.4. Dispense 5 ml per screw-capped 16- by 125-mm tube, and autoclave at 121°C for 15 min.

NYC medium

Basal medium

1. Agar* .... 20 g in 400 ml of distilled water

The solution is heated at 100°C until melted, e.g., by steaming in an Arnold unit.

2. Cornstarch ...... 1 g in 40 ml of distilled water

The solution is first mixed on a magnetic stirrer and then placed in the Arnold unit until homogeneous.

3. Proteose Peptone no. 3* .... 15 g
Dipotassium phosphate.................. 4 g

Monopotassium phosphate . . . . . 1 g
Sodium chloride . . . . . 5 g
Distilled water . . . . . 200 ml

Bring the components of step 3 to boiling on a heated magnetic stirrer.

4. Add the melted agar, cornstarch, and Proteose Peptone solution, and mix thoroughly on a heated magnetic stirrer.
5. Autoclave at 121°C for 15 min.
6. Allow to cool, and store at 4°C.
7. Final pH 7.1 ± 0.1.

Editors' note. The author recommends Difco ingredients.

Supplements

A. Yeast dialysate

Bakers' yeast . . . . . 908 g
Distilled water . . . . . 2,500 ml

Carefully mix the bakers' yeast to a smooth paste, autoclave for 10 min, and allow to cool. Place in dialysis tubing, and dialyze against 2 liters of distilled water in the cold for 48 h. Collect the dialysate, dispense in 25-ml aliquots, and autoclave at 121°C for 15 min. (A 25-ml sample of yeast dialysate is the amount necessary for 1 liter of final medium; aliquots can be prepared in any amount). The yeast dialysate is stored at 20°C. (It can be kept indefinitely.)

B. A 50% glucose solution distributed in 10-ml aliquots (for 1 liter of medium)

C. A 3% hemoglobin solution is prepared from packed erythrocytes (RBC). The RBC are kept after the plasma has been separated. They are not packed by centrifugation, but used as such. A 6-ml amount of sedimented RBC is added to 200 ml of sterile distilled water (for 1 liter of medium).

D. Horse plasma (citrated) preparation

Sodium citrate . . . . . 1,200 g
NaCl . . . . . 64 g
Water, to . . . . . 8 liters

Of this solution, 600 ml is placed in a receiving bottle, and blood is drawn to 6 liters to make a 10% final citrate concentration in the blood.

E. Antibiotic mixture

Vancomycin . . . . . 2 μg/ml
Colistin . . . . . 5.5 μg/ml
Amphotericin B . . . . . 1.2 μg/ml
Trimethoprim lactate . . . . . 3 μg/ml

1. Melt basal medium and allow to cool in a 55°C water bath.
2. The following additions are made for 1 liter of final medium:
   a. 120 ml of plasma
   b. 200 ml of a 3% RBC (hemolyzed) solution
   c. 10 ml of 50% glucose solution
   d. 25 ml of yeast dialysate
   e. 5 ml of antibiotic mixture
3. After the addition of the supplement to the basal medium, plates are poured and allowed to dry.
4. When dry, plates should be placed in plastic bags and stored in the refrigerator (4 to 8°C).

In these conditions, NYC medium will maintain its properties for about 2 to 2.5 months.

Oxidation-fermentation (OF) test medium
OF basal medium*

A. *For Corynebacterium, Enterobacteriaceae, Aeromonas, Pseudomonas, and nonfermenting gram-negative bacteria (19)*

Peptone, e.g., pancreatic digest of casein . 2.00 g
Sodium chloride . . . . . 5.00 g
Dipotassium phosphate . . . . . 0.30 g
Bromothymol blue . . . . . 0.03 g
Agar . . . . . 3.00 g
Distilled water . . . . . 1.00 liter
Final pH 7.1

1. Dispense 3 or 4 ml per 13- by 100-mm test tube. Autoclave at 121°C for 15 min.
2. Cool and add 10% glucose solution in distilled water for a final concentration of 1% glucose. Other carbohydrates may be substituted for glucose, if desired.
3. Dispense aseptically.

This medium aids in the differentiation of organisms that utilize carbohydrates oxidatively rather than fermentatively and, therefore, is helpful in the identification of pseudomonads and members of the tribe *Mimeae*. It also aids in the identification of organisms that do not utilize glucose in either way (e.g., *Alcaligenes* spp.). Inoculate (stab) lightly two tubes of medium from a young agar slant culture. Cover one of the tubes with a layer (about 5 mm) of sterile melted petrolatum or sterile paraffin oil. Incubate at 37°C, and observe daily for 3 or 4 days. Acid formation only in the open tube indicates oxidative utilization of glucose. Acid formation in both the open and the sealed tubes is indicative of a fermentative reaction. Lack of acid production in either tube indicates that the organism does not utilize glucose by either method.

B. *For nonfermenting gram-negative bacilli (54)*

Pancreatic digest of casein . . . . . 0.2 g
Phenol red, 1.5% aqueous . . . . . 0.2 g
Distilled water . . . . . 100.0 ml

1. Warm to dissolve the peptone, and adjust the pH to 7.3.
2. Dissolve 0.3 g of agar by heating.
3. Dispense 6 ml per 16- by 125-mm tube. Autoclave at 121°C for 15 min.
4. When basal medium is melted, add Seitz-filtered carbohydrate (glucose, D-xylose, D-mannitol, lactose, sucrose, or maltose) aseptically to a final 1% concentration.

P agar
For staphylococci

Peptone . . . . . 10 g
Yeast extract . . . . . 5 g

Sodium chloride	5 g
Glucose	1 g
Agar	15 g
Distilled water	1 liter

Final pH 7.5 (before autoclaving)

Autoclave at 121°C for 15 min.

Editors' note. The author recommends Difco ingredients.

Pablum cereal agar

Pablum precooked cereal	100 g
Agar	18 g
Chloramphenicol	50 mg
Distilled water	1 liter

Mix, and autoclave at 121°C for 15 min.

Paromomycin-vancomycin (PV) blood agar (9)

Prepare CDC anaerobe blood agar. After the sheep blood has been added, aseptically add 100 mg of paromomycin (Parke Davis) and 7.5 mg (base activity) of vancomycin (Lilly) per liter.

Pelargonate assimilation

See carbon assimilation medium.

Peptone-yeast extract-glucose (PYG) broth (9)
For anaerobes

Peptone*	20 g
Yeast extract	10 g
Cysteine hydrochloride	0.5 g
**Resazurin solution	4.0 g
***Salt solution (VPI)	40.0 ml
D-Glucose	10.0 g
Distilled water	1,000.0 ml

Final pH 7.2

** Resazurin solution: add 11 mg of resazurin to 44 ml of distilled water.
*** VPI salt solution (24):

$CaCl_2$	0.2 g
$MgSO_4$	0.2 g
K_2HPO_4	1.0 g
KH_2PO_4	1.0 g

Mix the above ingredients in water, adjust the pH, dispense 7 ml into 15- by 90-mm screw-capped tubes, autoclave at 121°C for 15 min, and cool. Place tubes in glove box, tighten caps, remove from glove box, and store at 4°C or room temperature (see note on CHO fermentation broth).

Mix $CaCl_2$ and $MgSO_4$ in 300 ml of distilled water. When dissolved, add 500 ml of distilled water. Swirl while slowly adding the remaining salts. Continue mixing until all the salts are dissolved. Add 200 ml of distilled water. Store at 4°C.

Peptone-yeast extract-glucose (PYG) medium
For Acanthamoeba

Proteose peptone*	20.00 g
Yeast extract*	2.00 g
$MgSO_4 \cdot 7H_2O$	0.980 g
$CaCl_2$	0.059 g
Sodium citrate · $2H_2O$	1.00 g
$Fe(NH_4)_2(SO_4)_2 \cdot 6H_2O$	0.02 g
KH_2PO_4	0.34 g
$Na_2HPO_4 \cdot 7H_2O$	0.355 g
Glucose	18.00 g
Distilled water, to	1,000 ml

Final pH 6.5 ± 0.2

1. Dissolve all ingredients, except $CaCl_2$, in about 900.0 ml of distilled water.
2. Add $CaCl_2$ while stirring.
3. Bring the volume to 1,000 ml.
4. Dispense into 16- by 125-mm screw-capped tubes, 5 ml per tube.
5. Autoclave for 15 min (15 lb/in^2).

Editors' note. The author recommends Difco ingredients.

Phenylethyl alcohol agar*
Phenylethanol agar*

Pancreatic digest of casein USP	15 g
Papaic digest of soya meal USP	5 g
Sodium chloride	5 g
Phenylethyl alcohol	2.5 g
Agar	15.0 g
Distilled water	1.0 liter

Final pH 7.3

Dissolve with heat, autoclave at 118°C for 15 min, cool, and pour the plates. Add 5% defibrinated sheep blood if desired.

For anaerobes

Prepare as described above, and add 5% defibrinated blood and vitamin K_1 (10 μg/ml) after autoclaving.

Phenylethyl alcohol (PEA) blood agar (9)
For anaerobes

Prepare CDC anaerobe blood agar, and add 2.5 g of phenylethyl alcohol (per liter) before autoclaving.

Pleisomonas differential agar (46)
Available from E. Merck AG, Darmstadt, Federal Republic of Germany

Peptone (tryptic)	7.5 g
Meat extract	7.5 g
NaCl	5.0 g
Bile salts	8.5 g
Brilliant green	0.00033 g
Neutral red	0.025 g
m-Inositol	10.0 g
Agar	13.5 g
Distilled water	1.00 liter

Final pH 7.2 ± 0.1

Autoclave.

Polysorbate (Tween) 80

Autoclave at 121°C for 20 min. Add 10 ml of polysorbate 80 to peptone agar, and pour the plates.

Potato-dextrose agar* (49)
Recommended to induce sporulation in all fungi

Potatoes (white, not new), diced	200 g
Glucose	10 g
Agar	18 g

1. Simmer the potatoes in water for 1 h. Filter through coarse paper.
2. Add the glucose and agar, dissolve by heat, and filter through cotton and gauze. Restore volume to 1 liter.
3. Tube, and autoclave at 121°C for 10 min.

Pril-xylose-ampicillin agar (44)

Nutrient agar plus:

Phenol red	25 mg
Ampicillin	30 mg
**Pril*	200 mg
Xylose	10 g
Distilled water	1.0 liter

**Böhme Fettchemie GmbH, Düsseldorf, Federal Republic of Germany.

Purple broth base*

Peptone: proteose or peptic digest of animal tissue USP	10.000 g
Beef extract	1.000 g
Sodium chloride	5.000 g
Bromocresol purple	0.015 g
Distilled water	1.000 liter
Final pH 6.8	

Autoclave at 121°C for not more than 15 min. Add sugars and alcohols to 1% (wt/vol).

Pyruvate utilization medium

Tryptone*	10 g
Yeast extract	5 g
Dipotassium phosphate	5 g
Sodium chloride	5 g
Sodium pyruvate	10 g
Bromothymol blue	104 mg
Distilled water	1,000 ml

Check the pH and adjust to 7.1 to 7.4, if necessary. Dispense into 13- by 100-mm screw-capped tubes. Autoclave at 121°C for 15 min.

A positive reaction is recorded when the indicator changes from green to definite yellow. Yellow-green indicates a weak reaction and should be regarded as negative utilization of pyruvate.

Rapid sugar test for *Neisseria* species (2, 3, 29)

A. Introduction

Neisseria gonorrhoeae is presumptively identified in the clinical laboratory by growth and colonial morphology on selective media, cell morphology and appearance on Gram stain, and oxidase reaction. It is confirmed either by the fluorescent-antibody test or by its reaction on various carbohydrate media. The rapid test is a time-saving modification of the standard procedure for determining carbohydrate fermentation of glucose, lactose, sucrose, maltose, and fructose. Although developed for the testing of neisseriae, it has been shown to be useful for other organisms such as *Haemophilus* spp., group DF-2, and *Kingella denitrificans*.

B. Reagents

1. Buffer-salt solution

KH_2PO_4	0.01 g
K_2HPO_4	0.04 g
KCl	0.80 g
Phenol red (1% aqueous solution)	0.40 g
Distilled water	100 ml

Adjust the pH to 7.0, filter sterilize, and store at 4°C in a sterile, screw-capped bottle.

2. Carbohydrate solutions

Prepare as 20% stock solutions in distilled water or broth (peptone, 10 g; meat extract, 3 g; sodium chloride, 5 g; distilled water, 1,000 ml). Adjust the pH to 7.0, and filter sterilize the solutions.

C. Procedure

1. Place 0.1 ml of the buffer-salt solution into each of five 10- by 75-mm sterile tubes.
2. Add 1 drop of the appropriate carbohydrate solution to each labeled tube.
3. Inoculate each tube with 1 drop of a very heavy 18- to 24-h culture suspension prepared in 0.35 ml of the buffer-salt solution.

Alternatively, a loopful of the suspected organism may be added directly into a tube of the buffer-salt-carbohydrate solution.

4. Incubate the tubes in a 35°C water bath for 4 h, and read at 30-min intervals.
5. A positive test is indicated by a yellow color (acid formation).

Regan-Lowe (RL) media for *Bordetella* species
For isolation of Bordetella pertussis and B. parapertussis

RL plates

Charcoal agar (Oxoid CM 119)	51 g
Distilled water	900 mg
Horse blood, defibrinated	100 ml
Cephalexin	0.040 g
Amphotericin B (optional)	0.050 g

1. Suspend dehydrated powder in water, and bring it to a boil to dissolve completely.
2. Sterilize by autoclaving at 121°C for 15 min.
3. Cool to 50°C and add defibrinated horse blood (to 10% wt/vol) and cephalexin (to 40 μg/ml final concentration), and optionally add amphotericin B (50 μg/ml final concentration).

4. Mix gently and dispense into sterile petri plates.
5. Plates may be stored sealed at 4°C for up to 1 week.

Regan-Lowe (RL) semisolid transport medium

1. Prepare Regan-Lowe (RL) agar as for plates, but use only 25.5 g of Oxoid CM 119 charcoal agar. After adding the blood and antibiotic supplements, aseptically dispense into small screw-capped vials to fill them half full.
2. Seal tightly and store at 4°C for up to 2 months.

Comments. The use of Oxoid CM 119 agar base is necessary for good results. The components of CM 119 in grams per liter are as follows: "Lab-Lemco" powder, 10; peptone, 10; starch, 10; bacteriological charcoal, 10; NaCl, 5; nicotinic acid, 0.001; agar, 12. Some American workers have successfully substituted defibrinated sheep blood for the horse blood on plates. However, use of sheep blood in transport medium is untested.

Rice grain medium

White rice (not enriched) .... 8 g
Distilled water.............................. 25 ml

Place in 125-ml Erlenmeyer flask, and autoclave at 121°C for 15 min.

This medium is recommended for the differentiation of *Microsporum* spp., *M. canis*, and *M. gypseum*, as well as other dermatophytes which grow and sporulate well on this medium. *M. audouini* produces only negligible growth.

Sabhi agar
For primary recovery of fungi from clinical specimens

Brain heart infusion broth .... 18.60 g
Calf brain, infusion from .... 50.00 g
Beef heart, infusion from.............. 62.50 g
Proteose or Gelysate (BBL) pancreatic
digest of gelatin .... 2.50 g
Sodium chloride........................ 1.25 g
Neopeptone .... 5.00 g
Disodium phosphate .... 0.625 g
Agar .... 7.50 g
Distilled water .... 1.00 liter

Dissolve, and autoclave at 121°C for 15 min.

Sabhi agar, modified

Brain heart infusion broth .... 18.60 g
Calf brains, infusion from .... 50.00 g
Beef heart, infusion from.............. 62.50 g
Proteose or Gelysate (BBL) pancreatic
digest of gelatin .... 2.50 g
Glucose .... 20.50 g
Sodium chloride........................ 1.25 g
Neopeptone .... 5.00 g
Disodium phosphate .... 0.625 g
Agar .... 7.50 g
Distilled water .... 1.00 liter

Dissolve, and autoclave at 121°C for 15 min.

Sabouraud dextrose agar*

Glucose .... 40 g
Neopeptone or Polypeptone (BBL) .... 10 g
(Pancreatic digest of casein USP.......... 5 g)
(Peptic digest of animal tissue USP .... 5 g)
Agar............................... 20 to 15 g
Demineralized water....................... 1 liter
Final pH 5.6

Heat to dissolve completely. Dispense into tubes (18 to 25 mm in diameter), and autoclave at 121°C for 10 min.

Sabouraud dextrose agar Emmons* (12)
Sabouraud agar Emmons

Glucose .... 20 g
Bacto-Peptone or Polypeptone (BBL)....... 10 g
(Pancreatic digest of casein USP.......... 5 g)
(Peptic digest of animal tissue USP .... 5 g)
Agar, not dried .... 20 g
(*or* Dried .... 17 g)
Distilled water .... 1 liter
Final pH 6.9

Dispense, and autoclave at 118 to 121°C for 10 min.

Sabouraud 2% dextrose agar
For subculture of fungi to induce sporulation

Glucose .... 20 g
Bacto-Peptone or Polypeptone (BBL)....... 10 g
(Pancreatic digest of casein USP.......... 5 g)
(Peptic digest of animal tissue USP .... 5 g)
Agar .... 17 g
Distilled water .... 1 liter
Final pH 6.9

Dispense, and autoclave at 188 to 121°C for 10 min.

Salmonella-Shigella (SS) agar*

Beef extract .... 5.000 g
Proteose or Polypeptone (BBL) .... 5.000 g
Lactose .... 10.000 g
Bile salts............................. 8.500 g
Sodium citrate .... 8.500 g
Sodium thiosulfate .... 8.500 g
Ferric citrate .... 1.000 g
Agar .... 13.500 g
Neutral red........................... 0.025 g
Brilliant green........................ 0.330 g
Distilled water .... 1.000 liter
Final pH 7.0

Heat to boiling. Do not autoclave. Cool to 42 to 45°C, and pour into plates. For slants, prepare the medium in a sterile flask, with sterile water. Dispense aseptically, 5 ml per sterile 16- by 125-mm tube.

Salt tolerance media
For nonfermenting gram-negative bacteria

1. Method of Tatum et al. (54): nutrient broth plus 65 g of sodium chloride per liter

2. Method of Gilardi (19): soybean-casein digest agar USP plus 65 g of sodium chloride per liter

Salt tolerance test for streptococci

Heart infusion broth .... 25 g
Sodium chloride .... 60 g
Indicator (1.6 g of bromocresol purple in
100 ml of 95% ethanol) .... 1 ml
Glucose .... 1 g
Distilled water .... 1,000 ml

Add all reagents together up to 1,000 ml; do not compensate for the volume loss caused by the sodium chloride. Final volume should be 1,000 ml.

Dispense into 15- by 125-mm screw-capped tubes, and autoclave at 121°C for 15 min.

A positive reaction is recorded when the indicator changes from purple to yellow or when growth is obvious even though the indicator does not change.

Selective medium for *Brucella* species

Heart infusion agar .... 40.0 g
Gelatin .... 1.0 g
Glucose .... 2.5 g
Distilled water .... 1,000.0 ml

Mix ingredients, autoclave for 15 min at 121°C, allow to cool to 50°C, and add 10 ml of sterile sheep blood and the following antibiotics:

Actidione (Upjohn), 0.2 ml of a 10-mg/ml stock solution; final concentration = 100 μg/ml of agar

Bacitracin (Upjohn), 1 ml of a 5,000-μg/ml stock solution; final concentration = 25 μg/ml of agar

Circulin (Upjohn), 0.3 ml of a 10-μg/ml stock solution; final concentration = 15 μg/ml of agar

Polymyxin B (Burroughs-Wellcome), 1.2 ml of a 1,000-μg/ml stock solution; final concentration = 6 μg/ml of agar

Selective 7H11 agar

See Middlebrook and Cohn 7H10 agar.

Semisynthetic medium for *Calymmatobacterium granulomatis*

See *Calymmatobacterium granulomatis* semidefined medium.

Serum dextrose agar

Oxoid blood agar base .... 40 g
Equine serum, inactive at 56°C for
30 min .... 50 ml
Glucose, 25% (wt/vol) autoclaved at 121°C
for 15 min .... 40 ml
Distilled water .... 1 liter
Final pH 7.4

Add Oxoid base to the water. Allow the mixture to stand for 10 min to keep the powder from caking. Apply gentle heat to aid solution before autoclaving at 121°C for 15 min. Cool to 50°C before adding serum and glucose solution. Pour plates or slants immediately.

7H11 agar

See Middlebrook and Cohn 7H10 agar.

Skirrow medium (50)

Blood agar base no. 2 (Oxoid CM 271) ...... 1 liter
Lysed defibrinated horse blood .... 50 ml
Vancomycin .... 10 mg
Polymyxin B .... 2,500 IU
Trimethoprim .... 5 mg

1. Melt blood agar base, and cool to 50°C.
2. Add blood and other ingredients to sterile molten agar.

Sodium hippurate broth

Heart infusion broth .... 25 g
Sodium hippurate .... 10 g
Distilled water .... 1,000 ml

1. Dispense into 15- by 125-mm screw-capped tubes, and sterilize in an autoclave for 15 min at 121°C.
2. Tighten caps to prevent evaporation.

Procedure

1. Inoculate with two or three colonies of beta-hemolytic streptococci, and incubate at 35°C for 20 h or longer.
2. Centrifuge the medium to pack the cells, and pipette 0.8 ml of the clear supernatant into a Kahn tube.
3. Add 0.2 ml of the ferric chloride reagent to the Kahn tube, and mix well. If a heavy precipitate remains longer than 10 min, the test is positive.

Soil extract agar
For demonstrating typical conidia of Histoplasma capsulatum and Blastomyces dermatitidis

Soil .... 500.0 g
Glucose .... 2.0 g
Yeast extract .... 1.0 g
Potassium phosphate .... 0.5 g
Agar .... 15.0 g
Tap water .... 1 liter

1. Mix 500.0 g of garden soil and 1 liter of tap water.
2. Autoclave for 3 h at 15 lb/in^2.
3. Filter through Whatman no. 2 filter paper. This is the soil infusion.
4. Add reagents and bring volume up to 1 liter.
5. Bring to a boil.
6. Dispense 7.0-ml aliquots into 16- by 125-mm test tubes.
7. Autoclave for 15 min at 15 lb/in^2.
8. Slant the test tubes.

Inoculate with hyphae of a test isolate, and incubate at 25°C. Typical conidia should be seen in tease mounts examined microscopically after 7 to 14 days of incubation. Make tease mounts in a bacteriological glove box or laminar-flow hood.

Soybean-casein digest agar* USP
Trypticase soy agar (BBL), tryptic soy agar (Difco), tryptone soya agar, casein soy peptone agar, etc.

Pancreatic digest of casein USP .... 15 g
Papaic digest of soy meal USP.............. 5 g
Sodium chloride........................... 5 g
Agar.................................... 15 g
Distilled water .... 1 liter
Final pH 7.3

Heat with agitation until the medium boils. Dispense, and autoclave at 118 to 121°C for 15 min.

Staib agar, modified (52)
Birdseed agar

Guizottia abyssinica seeds (niger or thistle seeds) .... 70.00 g
Creatinine.............................. 0.78 g
Glucose .... 10.00 g
Chloramphenicol (1 50-mg capsule) ...... 0.05 g
Agar .... 20.00 g
Distilled water.......................... 1.00 liter
Diphenyl............................ 100.00 mg

1. Grind seed powder in blender. Add 300 ml of water, and autoclave at 115°C for 10 min.
2. Filter through gauze, and bring the volume to 1 liter.
3. Add other ingredients except diphenyl, and autoclave at 121°C for 15 min.
4. Cool to 50°C.
5. Add diphenyl to 10 ml of 95% ethyl alcohol, and add aseptically to medium.
6. Stir and pour into plates.

Starch agar

Nutrient agar* .... 23 g
Potato starch............................ 10 g
Demineralized water....................... 1 liter

Dissolve nutrient agar medium in 500 ml of water. Dissolve starch in 250 ml of water by boiling. Combine and make up to 1-liter volume. Adjust the pH, and autoclave at 121°C for 30 min.

This medium is useful for differentiation of species of aerobic actinomycetes. Inoculate as for casein agar. When good growth is obtained, test for hydrolysis by flooding a small portion of the plate with Gram or Lugol iodine; this does not contaminate the plate, which may subsequently be retested if necessary. In a negative test, the agar immediately around the growth becomes (temporarily) dark blue; if it becomes red or remains unstained, partial or complete hydrolysis is shown.

Starch fermentation broth
For corynebacteria

A. Heart infusion broth with bromocresol purple

Heart infusion .... 5 ml
Demineralized water .... 200 ml
Bromocresol purple, 1% in ethanol .... 0.2 ml
Adjust to pH 7.8.

B. Starch 2%............................... 20 ml
Dissolve by bringing to boil with constant stirring.

1. Combine A and B and dispense into screw-capped tubes.
2. Autoclave at 121°C for 15 min.

Starch hydrolysis agar

Heart infusion agar .... 40 g
Soluble starch .... 20 g
Distilled water.......................... 1,000 ml

Warm to dissolve, and autoclave at 121°C for 15 min. Cool to 55°C, and pour into sterile petri dishes.

Hydrolysis of starch is determined by flooding the surface of the plate with Gram iodine 48 h after inoculation and incubation at 35°C. A zone of hydrolysis appears colorless, and a dark-blue to purple zone indicates that the starch has not been hydrolyzed.

Streptococcal growth medium
For growth at 10 and 45°C

Heart infusion broth .... 25 g
Glucose .... 1 g
Indicator (1.6 g of bromocresol purple in 100 ml of 95% ethanol).................... 1 ml
Distilled water.......................... 1,000 ml

Dispense in 5-ml amounts into 16- by 125-mm screw-capped tubes.

Autoclave at 121°C for 15 min.

A positive reaction is recorded when growth is indicated by a color change from purple to yellow or by frank growth in the tube.

Stuart *Leptospira* broth, modified*

Modified by the Department of Veterinary Medicine, Walter Reed Army Institute of Research, Washington, D.C.

Glycerol-asparagine-salt solution

Sodium chloride .... 1.930 g
Ammonium chloride .... 0.340 g
Magnesium chloride · $6H_2O$ .... 0.190 g
L-Asparagine............................ 0.130 g
Disodium phosphate .... 0.666 g
Monopotassium phosphate .... 0.087 g
Glycerol, optional .... 5.000 ml
Distilled water...................... 995.000 ml

Dissolve each ingredient separately in 100-ml portions of water. Mix and make up to 1 liter in water. Autoclave at 121°C for 15 min. Cool and add 100 ml of filter-sterilized rabbit serum, previously inactivated in a 56°C water bath for 30 min. Dispense.

Stuart transport medium*

Sodium thioglycolate .... 1.0 g
Sodium glycerophosphate............. 10.0 g
Calcium chloride .... 0.1 g
Methylene blue .... 0.002 g

Agar . . . . 3.0 g
Demineralized water. . . . . 1 liter
Final pH 7.3 ± 0.2 at 25°C

Sucrose agar
For glucan production

Heart infusion agar . . . . 40 g
Sucrose . . . . 50 g
Distilled water. . . . . 1,000 ml

Autoclave at 121°C for 15 min. Cool to 55°C and pour into 13- by 100-mm petri dishes (ca. 10 ml into each).

Glucan production typical of *Streptococcus sanguis* and *S. mutans* results in highly refractile-adherent or white-dry-adherent growth on the agar. Levan production typical of *S. salivarius* results in opaque, gummy, nonadherent growth. Typical *S. bovis* growth is similar to that of *S. salivarius* but is somewhat less gummy and rarely adheres to the medium. Large or small colonies that are mucoidal and nonadherent are considered negative or have no extracellular polysaccharide production.

Sucrose broth
For glucan production

Solution A

NIH thioglycolate broth . . . . 28.5 g
Dipotassium phosphate. . . . . 10.0 g
Sodium acetate . . . . 12.0 g
Distilled water. . . . . 500.0 ml

Solution B

Sucrose. . . . . 50.0 g
Distilled water. . . . . 500.0 ml

Sterilize solutions separately in an autoclave at 121°C for 15 min. Cool to 55°C, mix solutions, and dispense into 16- by 125-mm screw-capped tubes in 5-ml amounts.

Glucan production is indicated when the broth is partially or completely gelled—a typical *Streptococcus sanguis* reaction. Glucan production is also indicated when gelatinous, adherent deposits form on the bottom and walls of the growth tube—a typical *S. mutans* reaction. An increase in the viscosity of the broth indicates the production of slime (unknown polysaccharide) typical of *S. bovis*. Negative reactions are recorded when no gelling, deposit, or increase in viscosity occurs.

TCBS agar (thiosulfate-citrate-bile-sucrose agar)

Yeast extract . . . . 5 g
Proteose Peptone no. 3 . . . . 10 g
Sodium citrate . . . . 10 g
Sodium thiosulfate . . . . 10 g
Oxgall . . . . 8 g
Saccharose . . . . 20 g
Ferric citrate . . . . 1 g
Bromothymol blue. . . . . 0.04 g
Agar . . . . 15 g
Distilled water. . . . . 1 liter
Final pH 8.6 ± 0.2 at 25°C

Tetrazolium tolerance (TTC) agar

Soybean-casein digest agar USP plus 10 g of triphenyltetrazolium chloride per liter.

TGY agar*
Tryptone (tryptophan-peptone)-glucose-yeast agar

Pancreatic digest of casein USP . . . . 5 g
Glucose . . . . 1 g
Yeast extract . . . . 5 g
Dipotassium phosphate . . . . 1 g
Agar . . . . 15 g
Distilled water . . . . 1 liter
Final pH 6.8 to 7.0

Bring to a boil to dissolve agar. Dispense 5 ml per tube into 15- by 125-mm tubes. Autoclave at 121°C for 15 min, and slant.

Thayer-Martin agar*

Combine sterile solutions, with the agar base cooled to 50°C.

GC agar base* sterile, double strength, 100 ml (see chocolate agars). Add 2 g of agar and 0.5 g of glucose to the double-strength GC agar base before dissolving in 100 ml of water.

Hemoglobin,* 2% aqueous, 100 ml, or chocolated defibrinated blood, 5%

Antibiotic inhibitors* to give final concentrations per 100 ml of medium of: vancomycin, 300 μg; colistin, 750 μg; nystatin, 1,250 U; trimethoprim lactate, 500 μg

Chemical enrichment, e.g., 1% IsoVitaleX*

Vitamin B_{12} . . . . 0.010 g
L-Glutamine. . . . . 10.000 g
Adenine . . . . 1.000 g
Guanine hydrochloride. . . . . 0.030 g
p-Aminobenzoic acid. . . . . 0.013 g
L-Cystine . . . . 1.100 g
Glucose . . . . 100.000 g
Diphosphopyridine nucleotide
oxidized (coenzyme I) . . . . 0.250 g
Cocarboxylase . . . . 0.100 g
Ferric nitrate. . . . . 0.020 g
Thiamine hydrochloride. . . . . 0.003 g
Cysteine hydrochloride. . . . . 25.900 g
Distilled water . . . . 1.000 liter

Other supplements, e.g., supplement "B",* may be used instead of the defined chemical enrichment.

Also see Transgrow agar.

For modified Thayer-Martin agar,* add 2 g of agar and 0.5 g of glucose to the double-strength GC agar base before dissolving in 100 ml of water.

For Martin-Lewis agar, use 400 μg of vancomycin per 100 ml of medium, and substitute 200 μg of anisomycin for nystatin.

Thiogel medium (9)
For anaerobes

Thiogel medium (BBL)* . . . . 90 g
Distilled water. . . . . 1,000 ml

Preheat water to 50°C, add Thiogel medium, and let stand for 5 min. Boil for 1 min (with frequent mixing), and dispense 7 ml into 15- by 90-mm screw-capped tubes. Autoclave at 118°C for 15 min, and cool.

See note on fermentation broth for anaerobes. Place in tube in glove box, tighten caps, remove from glove box, and store at 4°C or ambient temperature.

Thiogel medium is used to test the ability of anaerobes to hydrolyze gelatin. Place culture and uninoculated control in a beaker of cold water in a refrigerator. Check for liquefaction when the control has solidified. The Centers for Disease Control manual states that reactions are usually complete by 7 days, but Thiogel may be incubated for up to 1 month before it is reported as negative (9).

Thioglycolate-bile broth

Thioglycolate medium without indicator* can be used to prepare 20% bile broth by adding 0.5 ml of a solution containing 40% oxgall and 2% sodium deoxycholate (sterilized by autoclaving) to 10 ml of thioglycolate medium.

Thioglycolate medium, Brewer modified*

Pancreatic digest of casein USP	17.500 g
Papaic digest of soybean meal USP	2.500 g
Glucose	10.000 g
Sodium chloride	5.000 g
Dipotassium phosphate	2.000 g
Sodium thioglycolate	1.000 g
Methylene blue	0.002 g
Final pH 7.2	

Dispense into test tubes, half full, and autoclave at 121°C for 15 min.

Thioglycolate medium without indicator,* plus hemin

Pancreatic digest of casein USP	17.00 g
Papaic digest of soy meal USP	3.00 g
Glucose	6.00 g
Sodium chloride	2.50 g
Sodium thioglycolate	0.50 g
Agar	0.70 g
L-Cystine	0.25 g
Sodium sulfite	0.10 g
Purified water	1.00 liter
Hemin	5.00 mg

1. Mix and heat with agitation to obtain solution. Omit hemin, if desired.
2. Dispense into tubes with calcium carbonate chips or powder, approximately 0.1 g per tube, to promote viability, spore formation, and maintenance of cultures.
3. Boil (or steam for 10 min), and cool to room temperature just before use.
4. Add filter-sterilized sodium bicarbonate to a concentration of 1 mg/ml and vitamin K_1 to a concentration of 0.1 g/ml.
5. Other supplements may be added, if desired, by introducing a pipette to the bottom of the tube and withdrawing the pipette as the supplement is added, e.g., 5% Fildes enrichment or 10% sterile animal serum. Do not shake or invert tubes.

Thioglycolate medium, supplemented

Use thioglycolate medium without indicator prepared according to the manufacturer's instructions. Add hemin (5 μg/ml) and vitamin K_1 (0.1 μg/ml). Dispense into tubes, each containing a marble chip (Fisher); fill the tubes two-thirds to three-fourths full. Autoclave as directed. Just before use, boil or steam for 5 min, cool, and supplement with normal rabbit or horse serum (10% vol/vol) or peptic digest of sheep blood (Fildes enrichment [5% vol/vol]*).

Tinsdale agar, Moore and Parsons, modified*

Proteose no. 3 or Thiotone (BBL) peptic digest of animal tissue USP	20.00 g
L-Cystine	0.24 g
Sodium chloride	5.00 g
Sodium thiosulfate	0.43 g
Agar, dried	14.00 g
(*or* Not dried)	20.00 g
Distilled water	1.00 liter
Final pH 7.4	

1. Heat with agitation and boil for 1 min. Dispense.
2. Autoclave at 121°C for 15 min. Cool to 56°C and, to each 100 ml of base, add:

Sterile serum, e.g., bovine	10 ml
Potassium tellurite, 1% aqueous	3 ml

3. Alternatively, the thiosulfate may be dissolved in 1.7 ml of water and added separately. It must be prepared fresh each time the medium is prepared. The cystine may be dissolved in 6 ml of 0.1 N HCl and added separately, in which case it may be necessary to add 6 ml of 0.1 N NaOH to ensure that the final pH is correct.

Toxigenicity test agar, Elek

Proteose or Thiotone (BBL) peptic digest of animal tissue USP	20.0 g
Maltose	3.0 g
Lactic acid	0.7 g
Sodium chloride	5.0 g
Agar	15.0 g
Sodium hydroxide, 40% aqueous	1.5 ml
Distilled water	1.0 liter

1. To 500 ml of water, add, with agitation, the peptone, maltose, lactic acid, and sodium hydroxide.
2. Heat to boiling, and filter through Whatman no. 2 filter paper.
3. Adjust the reaction of the filtrate to pH 7.8 by using 1 N HCl.
4. Add the agar and salt dissolved in 500 ml of water.
5. Mix, and dispense 10-ml quantities into screw-capped tubes.
6. Autoclave at 115°C for 10 min. Store at room temperature with caps closed.

Transgrow agar*

1. Prepare as for Thayer-Martin agar, except with added glucose (0.15%) and the agar increased to 2% in the GC agar base.

2. Gas the bottles with 20% CO_2 in air, and tighten caps securely.

3. Trimethoprim lactate (5 mg/liter) may be added to either Thayer-Martin or Transgrow agar, if desired, especially for the examination of rectal specimens.

Transport media for anaerobes (PRAS)

A. Without peptones

Ionagar 2 or equivalent .... 2.00 g
Resazurin solution .... 0.40 ml
See Cary and Blair transport medium, modified (PRAS)
L-Cysteine hydrochloride .... 0.05 g
Distilled water .... 100.0 ml

1. In a flask, boil all ingredients except cysteine.
2. When dissolved, gas with carbon dioxide.
3. Add cysteine. When dissolved, adjust the pH to 6.8 with 20% sodium hydroxide.
4. Cap the flask with a rubber stopper, pass into an anaerobic chamber, and dispense aseptically into tubes.

B. PYG medium

Peptone or Gelysate (BBL) pancreatic digest of gelatin .... 1.00 g
Yeast extract .... 1.00 g
Glucose .... 1.00 g
Resazurin solution .... 0.40 ml
Distilled water .... 100.00 ml
L-Cysteine hydrochloride .... 0.05 g
Salts solution .... 4.00 ml

1. Boil ingredients except cysteine until colorless.
2. Cool to 45°C while gassing with carbon dioxide.
3. Add cysteine. When dissolved, adjust the pH to 6.8 with 20% sodium hydroxide.
4. Cap the flask with a rubber stopper, pass into an anaerobic chamber, and dispense aseptically into tubes.

Note. Salts solution formula:

Calcium chloride .... 0.2 g
Magnesium sulfate .... 0.2 g
Dipotassium phosphate .... 1.0 g
Monopotassium phosphate .... 1.0 g
Sodium bicarbonate .... 10.0 g
Distilled water .... 1.0 liter

C. See Cary and Blair transport medium, modified (PRAS)

Triple sugar iron agar*

Pancreatic digest of casein USP .... 10.0 g
Peptic digest of animal tissue USP .... 10.0 g
or
Beef extract .... (3.00 g)
Yeast extract .... (3.00 g)
Peptone components .... (20.00 g)
Glucose .... 1.000 g
Lactose .... 10.000 g
Sucrose .... 10.000 g
Ferrous sulfate or ferrous ammonium sulfate .... 0.200 g
Sodium chloride .... 5.000 g
Sodium thiosulfate .... 0.300 g
Agar .... 12.000 g
Phenol red .... 0.24 g
Distilled water .... 1.000 liter

Dispense for 2.5-cm butts and 3.8-cm slants. Autoclave at 121°C for 15 min.

This medium is recommended for the detection of hydrogen sulfide production by *Enterobacteriaceae*. Inoculate by stabbing into the butt and streaking the slant. Incubate at 37°C, and observe daily for 7 days for blackening. Changes in pH in the butt and on the slant are recorded in 18 to 24 h only.

Trypticase soy agar

See soybean-casein digest agar USP.

Tyrosine or xanthine agar

Nutrient agar* .... 23 g
Tyrosine .... 5 g
(*or* Xanthine .... 4 g)
Demineralized water .... 1 liter

1. Dissolve the nutrient agar in the water.
2. Add tyrosine or xanthine and mix to distribute the crystals evenly.
3. Adjust the pH to 7.0, and autoclave at 121°C for 15 min.
4. Dispense onto plates, 20 ml per plate, with the crystals evenly distributed.

Recommended for the differentiation of species of aerobic actinomycetes. Use as for casein agar.

U agar

Papaic digest of soy meal USP .... 20 g
Sodium chloride .... 5 g
2-(*N*-Morpholino)ethanesulfonic acid (MES; Calbiochem) .... 4.25 g
Purified water .... 1 liter
Phenol red, 2% aqueous .... 2 ml
Agarose .... 10 g

1. Dissolve, adjust the pH to 5.5 at 37°C before adding agarose, autoclave at 121°C for 15 min, and cool to 50°C.
2. To 65 ml of solution, add 10 ml of yeast dialysate, 20 ml of horse serum, 0.3 ml of aqueous filter-sterilized 1 M urea, and 2 ml of penicillin (10,000 U/ml).
3. Dispense into plates as for H agar.

U broth

Papaic digest of soy meal USP .... 20 g
NaCl .... 5 g
Water .... 1 liter
Phenol red (2%) .... 2 ml

1. Dissolve, adjust the pH to 6.0, and autoclave at 121°C for 15 min. Cool.
2. To 75 ml of solution, add 10 ml of yeast dialysate, 10 ml of horse serum, 0.5 ml of 1 M aqueous filter-

sterilized urea, 2 ml of penicillin (10,000 U/ml), and 1 or 2 ml (the larger volume gives less pH change but longer viability) of 1 M MES (filter sterilized, adjusted to pH 6.0 at 37°C). This solution may be stored at 4°C. Just before use, add 1 ml of 100 mM sodium sulfite (aqueous solution, sterilized by autoclaving) per 100 ml of broth. The sterile stock sodium sulfite should be stored tightly capped in screw-capped tubes with phenol red indicator (the indicator turns from purple to orange upon oxidation to sulfate).

3. A diphasic U agar may be prepared by adding 3 ml of U agar to each tube and overlaying with 3 ml of U broth. This further enhances survival of *Ureaplasma urealyticum*.

Urea semisolid medium (9)
For anaerobes

Solution A

Thioglycolate without glucose or indicator*	9.6 g
Yeast extract*	0.8 g
Distilled water	400 ml

Dissolve the ingredients by heating, autoclave at 121°C, and cool at 60°C.

Solution B

Urea broth*	15.5 g
Distilled water	50.0 ml

Mix the urea broth and water, and filter sterilize.

Combine solutions A and B aseptically and mix. Dispense 7 ml into 15- by 90-mm screw-capped tubes.

See note on fermentation broth for anaerobes. Place tubes in anaerobic glove box, tighten caps, remove from glove box, and store at 4°C or ambient temperature.

Editors' note. The authors recommend Difco ingredients.

A deep red color is positive; no color is negative (2 to 5 days of incubation). If the phenol red indicator is reduced, add 2 to 3 drops of dilute phenol red to the culture on the last day of incubation.

Urease test agar (5)
Urea agar base, Christensen

A. *For Enterobacteriaceae*

Urea concentrate

Peptone or Gelysate (BBL) pancreatic digest of gelatin	1.000 g
Sodium chloride	5.000 g
Glucose	1.000 g
Monopotassium phosphate	2.000 g
Phenol red	0.12 g
Urea	20.000 g
Distilled water	100.000 ml

Adjust to pH 6.8 and sterilize by filtration. Dissolve 15 g of agar in 900 ml of water, and autoclave at 121°C for 15 min. Cool to 50 to 55°C in a water bath, and add 100 ml of sterile urea concentrate. Cool in a slanted position to form slants with deep butts.

Inoculate heavily over the entire surface of the slant, and incubate at 37°C. Examine at 2 and 4 h and after overnight incubation. Negative tubes should be observed daily for 4 days to detect delayed reactions given by members of certain groups other than *Proteus* spp. Urease-positive cultures produce an alkaline reaction evidenced by a red color.

B. *For aerobic actinomycetes*

Prepare agar base by dissolving 20 g of agar in 500 ml of demineralized water; dissolve the salts, peptone, and glucose in the remaining 500 ml of water. Combine, and adjust the pH to 6.9. To 1 liter of base cooled to 50°C, add 100 ml of 20% filter-sterilized urea solution. Dispense, slant, and inoculate as above.

Urea test broth* (52)

Urea solution

Yeast extract	0.100 g
Monopotassium phosphate	0.091 g
Disodium phosphate	0.095 g
Urea	20.000 g
Phenol red	0.010 g
Distilled water	1.000 liter

Sterilize by passing through a Seitz filter, and dispense 3-ml amounts into tubes. Alternatively, prepare basal medium in 900 ml of distilled water, and autoclave at 121°C for 15 min. After cooling, 100 ml of 20% filter-sterilized urea solution is added, and the medium is dispensed into sterile tubes.

Inoculate with three loopfuls (2-mm loop) from an agar slant culture, and shake to suspend the bacteria. Incubate in a water bath at 37°C, and read after 10 min, 60 min, and 2 h.

Vaginalis agar (V agar)

Columbia agar base*	10.00 g
Proteose peptone no. 3*	10.00 g
Distilled water	1.00 liter
Final pH 7.3	

1. Heat with frequent agitation, and boil for 1 min. Dispense, and autoclave at 121°C for 15 min.

2. Cool the sterile base to 45°C, and add 5% whole human blood.

Wickerham broths (63)

A. *For carbohydrate assimilation tests*

1. Use 100 ml of yeast nitrogen base (YNB),* 10×.
2. Add 10 g of carbohydrate (but add 20 g of raffinose). See auxanographic agar media.
3. Filter sterilize.
4. Add 0.5 ml of this concentrate to a tube containing 4.5 ml of sterile distilled water.

B. *For nitrate assimilation tests*

1. Use 100 ml of yeast carbon base (YCB),* 10×.
2. Add 0.78 g of potassium nitrate or peptone.

3. Sterilize by filtration, and follow the procedure for carbohydrate assimilation tests with Wickerham media.

Note. Inoculation for both tests should be with a suspension of starved yeast which gives at least 95% transmission (T) at 530 nm in a spectrophotometer. Incubate at 25 to 30°C with shaking. Examine tubes for growth as indicated by turbidity.

Wickerham media, modified

Bromocresol purple, 1.6%	1.0 ml
Deionized water	450.0 ml
Sodium hydroxide, 0.1 N	5.0 ml
Washed agar	10.0 g

Mix, and heat to dissolve. Cool to 45 to 50°C.

A. *For carbohydrate assimilation*

1. To 50 ml of YNB (10×), add 5.0 g of carbohydrate.
2. Admix to the melted basal medium.
3. Dispense in 5-ml amounts into sterile screw-capped tubes.
4. Autoclave at 115°C for 10 min. Cool in a slanted position.
5. Test each lot with standard control cultures of *Candida krusei, C. guilliermondii, C. pseudotropicalis,* and *Cryptococcus laurentii.*

B. *For nitrate assimilation*

1. To 50 ml of YCB (10×), add 0.5 g of peptone or potassium nitrate, and filter sterilize.
2. Admix to the melted basal medium.
3. Proceed as described above for steps 3, 4, and 5.

Wilkins-Chalgren agar*

Pancreatic digest of casein	10 g
Pancreatic digest of gelatin	10 g
Yeast extract	5 g
Glucose	1 g
Sodium chloride	5 g
L-Arginine (free base)	1 g
Sodium pyruvate	1 g
Hemin	5.0 mg
Vitamin K_1	0.5 mg
Agar	15 g
Distilled water	1,000 ml

Final pH 7.0 to 7.2

Mix and heat with agitation until dissolved. Dispense, and autoclave at 121°C for 15 min.

Wolin-Bevis agar (64)

Polysorbate (Tween) 80 (BBL)	3.00 ml
Glucose	0.25 g
L-Histidine hydrochloride	0.25 g
Ammonium sulfate	1.00 g
Monopotassium phosphate	1.00 g
Agar	20.00 g
Distilled water	1.00 liter

Dissolve, and autoclave at 121°C for 15 min.

Xanthine agar

See tyrosine agar.

Xylose-sodium desoxycholate-citrate agar (46)

Nutrient broth no. 2 (Oxoid)	12.5 g
Sodium citrate	5.0 g
Sodium thiosulfate	5.0 g
Ferric ammonium citrate	1.0 g
Sodium desoxycholate	2.5 g
Xylose	10.0 g
Neutral red (1% aqueous)	2.5 ml
Agar	12.0 g
Distilled water	1.0 liter

Heat to 100°C, simmer for 20 s, cool to 50°C, and plate.

Yeast ascospore agar (41)

Potassium acetate	10.0 g
Yeast extract	2.5 g
Glucose	1.0 g
Agar	30.0 g
Distilled water	1.0 liter

Dissolve, tube, and autoclave at 121°C for 15 min.

Ascospores are obtained in 2 to 6 days at room temperature (23 to 25°C).

Yeast carbon base* (YCB), 10×

Wickerham carbon base broth

Boric acid	0.500 mg
Copper sulfate	0.040 mg
Potassium iodide	0.100 mg
Ferric chloride	0.200 mg
Manganese sulfate	0.400 mg
Sodium molybdate	0.200 mg
Zinc sulfate	0.400 mg
Biotin	0.002 mg
Calcium pantothenate	0.400 mg
Folic acid	0.002 mg
Inositol	2.000 mg
Niacin	0.400 mg
p-Aminobenzoic acid	0.200 mg
Pyridoxine	0.400 mg
Riboflavin	0.200 mg
Thiamine hydrochloride	0.400 mg
L-Histidine hydrochloride	0.001 g
DL-Methionine	0.002 g
DL-Tryptophan	0.002 g
Potassium phosphate	1.000 g
Magnesium sulfate	0.500 g
Sodium chloride	0.100 g
Calcium chloride	0.100 g
Glucose	10.000 g
Water	1.000 liter

Final pH of the base 4.5±

Dissolve, and sterilize by filtration.
Dispense aseptically.

Yeast-dextrose agar

Dextrose .... 10.0 g
Yeast extract .... 10.0 g
Agar .... 15.0 g
Water .... 1,000.0 ml

Adjust the pH to 7.0.
Autoclave at 121°C for 15 min.

Yeast extract agar (12)
For identification of Histoplasma capsulatum, Blastomyces dermatitidis, and Coccidioides immitis

Yeast extract .... 1 g
**Buffer .... 2 ml
Agar .... 20 g
Distilled water .... 1,000 ml

** Dissolve 40 g of Na_2HPO_4 in 300 ml of distilled water, and then add 60 g of KH_2PO_4. The pH is 6.0. If necessary, adjust with 1 N HCl or NaOH. Adjust the volume to 400 ml with distilled water, and store at 4°C.

Boil into solution, and sterilize at 121°C for 15 min. Pour into sterile plastic petri dishes (35 ml per plate).

Yeast extract-phosphate agar
For primary recovery of dimorphic pathogenic fungi

Yeast extract .... 1 g
Phosphate buffer .... 2 ml
Agar .... 20 g
Distilled water .... 1 liter

Dissolve 40 g of Na_2HPO_4 in 300 ml of distilled water; add 60 g of KH_2PO_4. Adjust the pH to 6.0 with 1 N HCl or NaOH if necessary. Adjust the buffer volume to 400 ml with distilled water, and store at 4°C.
Dispense, and autoclave at 121°C for 15 min.
One drop of concentrated NH_4OH may be added to one side of the inoculated plate to decontaminate the sample.

Yeast morphology agar, buffered (buffered YMA)*

Ammonium sulfate .... 32.5 g
Asparagine .... 1.5 g
Glucose .... 10.0 g
L-Histidine monohydrochloride .... 10.0 mg
DL-Methionine .... 20.0 mg
DL-Tryptophan .... 20.0 mg
Biotin .... 2.0 µg
Calcium pantothenate .... 400.0 µg
Folic acid .... 2.0 µg
Inositol .... 2,000.0 µg
Niacin .... 400.0 µg
p-Aminobenzoic acid .... 200.0 µg
Pyridoxine hydrochloride .... 400.0 µg
Riboflavin .... 200.0 µg
Thiamine hydrochloride .... 400.0 µg
Boric acid .... 500.0 µg
Copper sulfate .... 40.0 µg
Potassium iodide .... 100.0 µg
Ferric chloride .... 200.0 µg
Manganese sulfate .... 400.0 µg
Sodium molybdate .... 200.0 µg
Zinc sulfate .... 400.0 µg
Potassium phosphate monobasic .... 1.0 g
Magnesium sulfate .... 0.5 g
Sodium chloride .... 0.5 g
Calcium chloride .... 0.1 g
Agar .... 18.0 g
0.01 M phosphate buffer, pH 7.0 .... 1.0 liter

1. Heat to boiling to dissolve.
2. Dispense, and autoclave at 121°C for 15 min. A comparable medium requiring the addition of buffer may be obtained commercially.

Yeast morphology agar, unbuffered (YMA)*
For agar dilution and diffusion disk testing with 5-fluorocytosine

Same as yeast morphology agar, buffered, except that 1 liter of distilled water is used in place of the 1 liter of 0.01 M phosphate buffer, pH 7.0. A comparable product is available commercially.
A similar prepared product (Buffered Shadomy Agar) containing phosphate buffer, pH 7.0 (0.33 g of Na_2HPO_4 and 0.92 g of KH_2PO_4 per liter, 1% glucose, 0.15% L-asparagine, and 100 µg of chloramphenicol per ml), also is available in 50-ml sterile volumes from A/S Rosco, 2630 Taastrup, Denmark.

Yeast nitrogen base,* 10×

Boric acid .... 500.0 µg
Copper sulfate .... 40.0 µg
Potassium iodide .... 100.0 µg
Ferric chloride .... 200.0 µg
Manganese sulfate .... 400.0 µg
Sodium molybdate .... 200.0 µg
Zinc sulfate .... 400.0 µg
Biotin .... 2.0 µg
Calcium pantothenate .... 400.0 µg
Folic acid .... 2.0 µg
Inositol .... 2,000.0 µg
Niacin .... 400.0 µg
p-Aminobenzoic acid .... 200.0 µg
Pyridoxine hydrochloride .... 400.0 µg
Riboflavin .... 200.0 µg
Thiamine hydrochloride .... 400.0 µg
L-Histidine monohydrochloride .... 10.0 mg
DL-Methionine .... 20.0 mg
DL-Tryptophan .... 20.0 mg
Magnesium sulfate .... 500.0 mg
Sodium chloride .... 100.0 mg
Calcium chloride .... 100.0 mg
Ammonium chloride .... 5.0 g
Monopotassium phosphate .... 1.0 g
Purified water .... 100.0 ml

Dissolve, sterilize by filtration, and dispense aseptically.

Yeast nitrogen base,* 10×, supplemented with asparagine and glucose
For susceptibility tests with yeasts and fungi

To yeast nitrogen base,* 10× (listed above), add:

L-Asparagine .... 1.5 g
Glucose .... 10.0 g

Dissolve, sterilize by filtration, and dispense aseptically. For use, dilute 1:10 in sterile water. For buffered 1× base, dilute 1:10 in sterile 0.01 M phosphate buffer, pH 7, instead.

Prepare the basic broth base in 10-fold concentration by adding 6.7 g to 100 ml of distilled water. Add the L-asparagine and glucose if required. Warm slightly to dissolve. Sterilize by filtration, and refrigerate for use as needed.

Company names and addresses

BBL Microbiology Systems, Cockeysville, Md.
Bristol Laboratories, Syracuse, N.Y.
Burroughs-Wellcome Corp., Research Triangle Park, N.C.
Calbiochem-Behring, La Jolla, Calif.
Difco Laboratories, Detroit, Mich.
EM Science, Cherry Hill, N.J.
Fisher Scientific Co., Pittsburgh, Pa.
Flow Laboratories, McLean, Va.
General Diagnostics, Anaheim, Calif.
ICN Pharmaceuticals, Inc., Cleveland, Ohio
Eli Lilly & Co., Indianapolis, Ind.
Merck & Co., Inc., Rahway, N.J.
Oxoid, U.S.A., Inc., Columbia, Md.
Parke, Davis & Co., Detroit, Mich.
Pfanstiehl Laboratories, Inc., Waukegan, Ill.
Pfizer Inc., New York, N.Y.
Research Organics, Cleveland, Ohio
Roche Laboratories, Nutley, N.J.
Sheffield Chemical, Union, N.J.
Sigma Chemical Co., St. Louis, Mo.
E. R. Squibb & Sons, Princeton, N.J.
The Upjohn Co., Kalamazoo, Mich.
Winthrop Laboratories, New York, N.Y.

LITERATURE CITED

1. **Blazer, M. J., I. D. Berkowitz, F. M. LaForce, J. Grovens, L. B. Reller, and W. L. Wong.** 1959. Campylobacter enteritis: clinical and epidemiological features. Ann. Intern. Med. **91**:179–185.
2. **Brown, W. J.** 1974. Modification of the rapid fermentation test for *Neisseria gonorrhoeae*. Appl. Microbiol. **27**:1027–1030.
3. **Brown, W. J.** 1976. Stability of working reagents for the modified rapid fermentation test (MRFT). Health Lab. Sci. **14**:172–176.
4. **Butzler, J. P., P. Dikeyser, M. Detrain, and F. Dehaen.** 1973. Related vibrio in stools. J. Pediatr. **82**:493–495.
5. **Christensen, W. B.** 1946. Urea decomposition as a means of differentiating *Proteus* and paracolon cultures from each other and from *Salmonella* and *Shigella* types. J. Bacteriol. **52**:461–466.
6. **Converse, J.** 1955. Growth of spherules of *Coccidioides immitis* in a chemically defined liquid medium. Proc. Soc. Exp. Biol. Med. **90**:709–711.
7. **Dowell, V. R., Jr., and T. M. Hawkins.** 1974. Laboratory methods in anaerobic bacteriology. CDC laboratory manual. Department of Health, Education, and Welfare publication (CDC) 78-8272. Center for Disease Control, Atlanta.
8. **Dowell, V. R., Jr., and G. L. Lombard.** 1977. Presumptive identification of anaerobic nonsporeforming gram-negative bacilli. Center for Disease Control, Atlanta.
9. **Dowell, V. R., Jr., G. L. Lombard, F. S. Thompson, and A. Y. Armfield.** 1977. Media for isolation, characterization and identification of obligately anaerobic bacteria. CDC laboratory manual. Department of Health, Education, and Welfare publication. Center for Disease Control, Atlanta.
10. **Dulaney, A. D., K. Guo, and H. Packer.** 1948. *Donovania granulomatis* cultivation, antigen preparation and immunological tests. J. Immunol. **59**:335–340.
11. **Edelstein, P. H.** 1984. Legionnaires' disease laboratory manual. National Technical Information Service, Springfield, Va.
12. **Emmons, C. W., C. H. Binford, J. P. Utz, and K. J. Kwon-Chung.** 1980. Medical mycology. Lea & Febiger, Philadelphia.
13. **Ewing, W. H., and B. R. Davis.** 1970. Media and tests for differentiation of *Enterobacteriaceae*. Center for Disease Control, Atlanta.

13a. **Facklam, R. R.** 1973. Comparison of several laboratory media for presumptive identification of enterococci and group D streptococci. Appl. Microbiol. **26**:138–145.

14. **Feeley, J. C., R. J. Gibson, G. W. Gorman, N. C. Langford, J. K. Rasheed, D. C. Mackel, and W. B. Baine.** 1979. Charcoal-yeast extract agar: primary isolation medium for *Legionella pneumophila*. J. Clin. Microbiol. **10**:437–441.
15. **Feeley, J. C., G. W. Gorman, and R. J. Gibson.** 1979. Primary isolation media and methods, p. 77–84. *In* G. L. Jones and G. A. Hebert (ed.), "Legionnaires": the disease, the bacterium and methodology. Center for Disease Control, Atlanta.
16. **Feeley, J. C., G. W. Gorman, R. E. Weaver, D. C. Mackel, and H. W. Smith.** 1978. Primary isolation media for Legionnaires disease bacterium. J. Clin. Microbiol. **8**:320–325.
17. **Fletcher, W.** 1928. Recent works on leptospirosis, Tsutsugamushi disease and tropical typhus in the Federated Malay States. Trans. R. Soc. Trop. Med. Hyg. **31**:265–268.
18. **Georg, L. K., L. Ajello, and C. Papageorge.** 1954. Use of cycloheximide in the selective isolation of fungi pathogenic to man. J. Lab. Clin. Med. **44**:422–428.
19. **Gilardi, G. L. (ed.).** 1978. Glucose nonfermenting gram-negative bacteria in clinical microbiology. CRC Press, Inc., West Palm Beach, Fla.
20. **Goldberg, J.** 1959. Studies on granuloma inguinale. IV. Growth requirements of *Donovania granulomatis* and its relationship to the natural habitat of the organism. Br. J. Vener. Dis. **35**:266–268.
21. **Gordon, R. E., and J. M. Mihm.** 1962. Identification of *Nocardia caviae* (Erickson) nov. comb. Ann. N.Y. Acad. Sci. **98**:628–636.
22. **Grayston, J. T., H. M. Foy, and F. E. Kenny.** 1969. The epidemiology of mycoplasma infections of the human respiratory tract, p. 651–652. *In* L. Hayflick (ed.), The Mycoplasmatales and L-phase of bacteria. Appleton Century Crofts, New York.
23. **Grove, D. C., and W. A. Randall.** 1955. The compilation of tests and methods of assay for antibiotic drugs. Food and Drug Administration: assay methods of antibiotics. Medical Encyclopedia, Inc., New York.
24. **Holdeman, L. V., E. P. Cato, and W. E. C. Moore (ed.).** 1977. Anaerobe laboratory manual, 4th ed. Virginia Polytechnic Institute and State University, Blacksburg.
25. **Hunter, G. W., III, W. W. Frye, and J. C. Schwartzwelder.** 1966. A manual of tropical medicine, 4th ed. W.B. Saunders Co., Philadelphia.
26. **Johnson, R. C., and V. G. Harris.** 1967. Differentiation of pathogenic and saprophytic leptospires. I. Growth at low temperatures. J. Bacteriol. **94**:27–31.
27. **Johnson, R. C., J. Walby, R. A. Henry, and N. E. Auran.** 1973. Cultivation of parasitic leptospires: effect of pyruvate. Appl. Microbiol. **26**:118–119.
28. **Kaper, J. B., H. Lockman, R. R. Colwell, and S. W. Joseph.** 1981. *Aeromonas hydrophila*: ecology and toxigenicity of isolates from an estuary. J. Appl. Bacteriol. **50**:359–377.
29. **Kellogg, D. S., Jr., and E. M. Turner.** 1973. Rapid fer-

mentation confirmation of *Neisseria gonorrhoeae*. Appl. Microbiol. **25:**550–552.

30. **Kenny, G. E.** 1973. Contamination of mammalian cells in culture with Mycoplasmata, p. 107–129. *In* J. Fogh (ed.), Contamination in tissue culture. Academic Press, Inc., New York.
31. **Kenny, G. E., and F. D. Cartwright.** 1978. Effect of urea concentration on growth of *Ureaplasma urealyticum* (T-strain mycoplasma). J. Bacteriol. **132:**144–150.
32. **Kimmig, J., and H. Rieth.** 1953. Antimykotia in Experiment und Klinik. Arzneim.-Forsch. **3:**267–276.
33. **Knisely, R. F.** 1966. Selective medium for *Bacillus anthracis*. J. Bacteriol. **92:**784–786.
34. **Kohn, J.** 1953. A preliminary report of a new gelatin liquefaction method. J. Clin. Pathol. **6:**249.
35. **Koneman, E. W., S. D. Allen, V. R. Dowell, Jr., and H. M. Somers.** 1983. Color atlas and textbook of microbiology, 2nd ed. J.B. Lippincott Co., Philadelphia.
36. **Lautrop, H.** 1956. A modified Kohn's test for the demonstration of bacterial gelatin liquefaction. Acta Pathol. Microbiol. Scand. **39:**357.
37. **Livingston, S. J., S. D. Kominos, and R. B. Yee.** 1978. New medium for selection and presumptive identification of the *Bacteroides fragilis* group. J. Clin. Microbiol. **7:**448–453.
38. **MacFaddin, J. F.** 1980. Biochemical tests for identification of medical bacteria, 2nd ed. Williams & Wilkins, Baltimore.
39. **McClary, D. O., W. L. Nulty, and G. R. Miller.** 1959. Effect of potassium versus sodium in the sporulation of Saccharomyces. J. Bacteriol. **78:**362–368.
40. **Novy, F. G., and W. J. MacNeal.** 1904. The cultivation of *Trypanosoma brusii*. J. Infect. Dis. **1:**1–30.
41. **Pine, L.** 1970. Growth of *Histoplasma capsulatum*. VI. Maintenance of the mycelial phase. Appl. Microbiol. **19:**413–420.
42. **Pine, L., and E. Drouhet.** 1963. Sur l'obtention et la conservation de la phase levure d'*Histoplasma capsulatum* et d'*H. duboisii*, en milieu chemiquement defini. Ann. Inst. Pasteur (Paris) **105:**798–804.
43. **Ristroph, J. D., and B. E. Ivins.** 1983. Elaboration of *Bacillus anthracis* antigens in a new, defined culture medium. Infect. Immun. **39:**483–486.
44. **Rogol, M., I. Sechter, L. Grinberg, and C. B. Gerichter.** 1979. Pril-xylose-ampicillin agar, a new selective medium for the isolation of *Aeromonas hydrophila*. J. Med. Microbiol. **12:**229–231.
45. **Schubert, R. H. W.** 1967. Das Vorkommen der Aeromonaden in oberirdischen Gewässern. Arch. Hyg. **150:**688–708.
46. **Schubert, R. H. W.** 1977. Ueber den Nachweis von *Plesiomonas shigelloides* Habs und Schubert, 1962, und ein Elektivmedium, den Inositol-Brilliantgrün-Gallesalz-Agar. E. Rodenwaldt-Archiv **4:**97–103.
47. **Shread, P., T. J. Donovan, and J. V. Lee.** 1981. A survey of the incidence of *Aeromonas* in human faeces. Soc. Gen. Microbiol. Q. **8:**184.
48. **Sierra, G.** 1957. A simple method for detection of lipolytic activity of micro-organisms and some observations on the influence of the contact between cells and fatty substrates. Antonie van Leeuwenhoek J. Microbiol. Serol. **23:**15–22.
49. **Skinner, G. E., C. W. Emmons, and H. M. Tsuchiya.** 1963. Henrici's molds, yeast and actinomycetes, 2nd ed. John Wiley & Sons, Inc., New York.
50. **Skirrow, M. B.** 1977. Campylobacter enteritis: a "new" disease. Br. Med. J. **2:**9–11.
51. **Smith, M. R., R. E. Gordon, and F. E. Clark.** 1952. Aerobic sporeforming bacteria, U.S. Department of Agriculture Monograph no. 16. Department of Agriculture, Washington, D.C.
52. **Staib, F.** 1962. Zur Kreatinin-Kreatin-Assimilation in der Hefepitz Diagnostik. Zentralbl. Bakteriol. Parasitenkd. Infektionskr. Hyg. Abt. 1 Orig. **191:**429–432.
53. **Sutter, V. L., V. L. Vargo, and S. M. Finegold.** 1975. Wadsworth anaerobic bacteriology manual, 2nd ed. University Extension, University of California, Los Angeles.
54. **Tatum, H. W., W. H. Ewing, and R. E. Weaver.** 1974. Miscellaneous gram-negative bacteria, p. 270–294. *In* E. H. Lennette, E. H. Spaulding, and J. P. Truant (ed.), Manual of clinical microbiology, 2nd ed. American Society for Microbiology, Washington, D.C.
55. **Totten, P. A., R. Amsel, J. Hale, P. Piot, and K. K. Holmes.** 1982. Selective differential human blood bilayer media for isolation of *Gardnerella (Haemophilus) vaginalis*. J. Clin. Microbiol. **15:**141–147.
56. **Trabulsi, L. R., and W. H. Ewing.** 1962. Sodium acetate medium for differentiation of *Shigella* and *Escherichia* cultures. Public Health Lab. **20:**137–140.
57. **Utz, C. J., and S. Shadomy.** 1977. Antifungal activity of 5-fluorocytosine as measured by disk diffusion susceptibility testing. J. Infect. Dis. **135:**970–974.
58. **Van den Bossche, H., H. G. Willemsens, and J. M. Van Cutsem.** 1975. The action of miconazole on the growth of *Candida albicans*. Sabouraudia **13:**63–73.
59. **Vera, H. D.** 1948. A simple medium for identification and maintenance of the gonococcus and other bacteria. J. Bacteriol. **55:**531–536.
60. **Vera, H. D., and D. A. Power.** 1980. Culture media, p. 965–999. *In* E. H. Lennette, A. Balows, W. J. Hausler, Jr., and J. P. Truant (ed.), Manual of clinical microbiology, 3rd ed. American Society for Microbiology, Washington, D.C.
61. **Vickers, R. M., A. Brown, and G. M. Garrity.** 1981. Dye-containing buffered charcoal-yeast extract medium for differentiation of members of the family *Legionellaceae*. J. Clin. Microbiol. **13:**380–382.
62. **Wadowsky, R. M., and R. B. Yee.** 1981. Glycine-containing selective medium for isolation of *Legionellaceae* from environmental species. Appl. Environ. Microbiol. **42:**768–772.
63. **Wickerham, L. J.** 1951. Taxonomy of yeasts. U.S. Department of Agriculture technical bulletin 1029. Department of Agriculture, Washington, D.C.
64. **Wolin, H. L., M. L. Bevis, and H. Laurora.** 1962. An improved synthetic medium for the rapid production of chlamydospores by *Candida albicans*. Sabouraudia **2:**96–99.

Chapter 112

Reagents and Stains

DONALD A. HENDRICKSON

Reagents and stains are explained in this chapter in the way the authors have presented the material in this Manual. Reagents and stains are presented in alphabetical order, with reagents first. When available, methodology is presented with the stains or reagents. Many procedures, however, are presented in the chapter in which their use is discussed. Please refer to specific chapters for the correct use of the reagents and stains. Additional information may be found in Chapter 98 of the 3rd edition of this Manual (17).

REAGENTS AND TEST PROCEDURES

Acetamide hydrolysis

See Chapter 29.

Nessler reagent

1. Solution A

Mercuric chloride .... 1 g
Distilled water .... 6 ml

Dissolve completely.

2. Solution B

Potassium iodide .... 2.5 g
Distilled water .... 6 ml

Dissolve completely and add solution A.

3. Solution C

Potassium hydroxide .... 6 g
Distilled water .... 6 ml

Dissolve completely and add to the mixture of solutions A and B. Add 13 ml of distilled water. Mix well, and filter before using.

Test procedure. Inoculate 1 ml of mineral base broth medium (carbon assimilation medium) supplemented with 0.1% acetamide. Incubate at 30°C for 24 h. Add 1 drop of Nessler reagent. The test is positive if a red-brown sediment appears. The sediment is due to the presence of ammonia from acylamidase action.

Andrade indicator (1)

Acid fuchsin .... 2.0 g
Distilled water .... 1 liter
Sodium hydroxide (1 N) .... 160.0 ml

1. Dissolve fuchsin in distilled water and add sodium hydroxide. If the fuchsin is not sufficiently decolorized after standing overnight, an additional 1 to 2 ml of alkali should be added. (The fuchsin should be slightly orange or darker than a faint straw color when brought up in a 10-ml pipette.)

2. Autoclave at 121°C for 20 min.

3. Andrade indicator should be aged for approximately 6 months before it is used, as it improves with aging.

Bile solubility test

Sodium deoxycholate .... 1 g
Distilled water, sterile .... 9 ml

Test procedure. To test for bile solubility, prepare two tubes, each containing a sample of fresh culture (a light suspension of the organism in buffered broth, pH 7.4). To one tube add a few drops of a 10% solution of sodium deoxycholate. A comparable volume of sterile physiological saline solution may be added to the second tube. If the cells are bile soluble, the tube containing the bile salt should lose its turbidity in 5 to 15 min and show an increase in viscosity concomitant with clearing.

Buffered saline

Sodium chloride (0.85%) is buffered to pH 7.2 with 0.067 M potassium phosphate mixture.

Catalase test

The organism to be tested should be grown on an agar slant heavily inoculated from a colony of the organism. The slant is usually incubated for 18 to 25 h at optimal temperature. To test for catalase, set the slant in an inclined position and pour 1 ml of a 3% solution of hydrogen peroxide over the growth. The appearance of gas bubbles indicates a positive test.

An alternative to conducting the test with a slant culture is to emulsify a colony in 1 drop of 30% hydrogen peroxide (superoxol) on a glass slide. Immediate bubbling indicates a positive catalase test. Extreme care must be exercised if a colony is taken from a blood agar plate. The enzyme catalase is present in erythrocytes, and the carry-over of blood cells with the colony can give a false-positive reaction.

Coagulase test

Tube test. To 0.5 ml of rabbit plasma, undiluted or diluted 1:4, add one loopful of growth from an 18- to 24-h-old agar culture, 0.5 ml of broth culture, or a single colony from a blood agar plate. Incubate in a water bath at 37°C, and examine the tubes at 6 and 24 h. Known coagulase-positive and coagulase-negative strains and, if possible, a weak coagulase producer must be set up as controls with each test.

A positive coagulase test is represented by any degree of clotting, from a loose clot suspended in plasma to a solid clot. It may be necessary to suspend a sterile loop in plasma to determine if clotting has occurred. The majority of coagulase-positive strains will produce a clot within the first 4 h, many within 1

h. False-positive tests may occur with mixed cultures or with pure cultures of some gram-negative rods, e.g., *Pseudomonas* spp., but the mechanism of clotting is different. Organisms which utilize the citrate used as the anticoagulant in the plasma will produce a clot. Therefore, the organism to be tested must first be determined to possess characteristics consistent with the genus *Staphylococcus*.

Slide test

1. Emulsify a colony in a drop of water on a glass slide to produce a dense, uniform suspension. If any evidence of autoagglutination is noted before the plasma is added, the culture is not suitable for the slide test.
2. Add one loopful or a drop of fresh plasma to the suspension, and mix by a continuous circular motion for 5 s.
3. A positive reaction is indicated by easily visible white clumps, which usually appear immediately or within 5 s.
4. Known coagulase-positive and coagulase-negative strains must always be set up in parallel.
5. All negative tests must be confirmed by the tube test.

Formalin buffer (10%) for parasitology

Formalin is an aqueous solution of formaldehyde, concentrated 37 to 40%. A 10% Formalin solution is about 3.7% formaldehyde.

Formaldehyde, 37% solution	100 ml
Water	900 ml
Na_2HPO_4	12 g
KH_2PO_4	3 g

The pH is approximately 7.4. Exact pH is not important.

Hippurate hydrolysis for *Legionella pneumophila* reagents (7, 11)

1. Sodium hippurate solution (1%)

Sodium hippurate	0.1 g
Sterile distilled water	10 ml

Mix in sterile test tubes and place 0.4-ml aliquots into small, sterile, screw-capped tubes. Freeze the solution at −20°C until it is used.

2. Ninhydrin solution (3.5%)

Ninhydrin	0.35 g
1-Butanol	5 ml
Acetone	5 ml

Mix butanol and acetone in a small, sterile, screw-capped tube. Add ninhydrin, mix, and store at room temperature in the dark.

Test procedure. Thaw the 1% sodium hippurate solution. To the solution add a loopful of organisms grown 1 to 4 days on BCYEα medium. Vortexing the mixture should result in a milky suspension. Incubate at 35°C in air overnight (18 to 20 h). Add 0.2 ml of 3.5% ninhydrin solution, mix gently, and reincubate at 35°C for 10 min. Remove the tube from the incubator and observe for color change in 20 min. Interpret the color as follows.

Purple	Positive
Very light purple	Weakly positive
Gray or light yellow	Negative

Indole test

A. *For Enterobacteriaceae (Chapter 24)*

Kovacs reagent (10)

Amyl or isoamyl alcohol	150 ml
p-Dimethylaminobenzaldehyde	10 g
Hydrochloric acid, concentrated	50 ml

Dissolve the aldehyde in alcohol, and then slowly add acid. The dry aldehyde should be light in color. Alcohols that result in indole reagents which become deep brown should not be used. The above-mentioned reagent is stable at room temperature and has a light color. Some authors recommend preparation of only small quantities, which are stored in a refrigerator when not in use.

Test procedure. Add about 0.5 ml of Kovacs reagent to a 40- to 48-h peptone-water culture incubated at 37°C, and shake the tube gently. A deep red develops in the presence of indole. Tests for indole may be made after 24 h of incubation, but if this is to be done, 1 or 2 ml of culture should be removed aseptically for testing. If the test is negative, the remaining portion of the culture should be reincubated for an additional 24 h.

B. *For nonfermenting gram-negative bacilli (Chapter 29)*

Ehrlich reagent

Ethyl alcohol, 95%	95 ml
p-Dimethylaminobenzaldehyde	1 g
Hydrochloric acid, concentrated	20 ml

Dissolve the aldehyde in alcohol, and then slowly add acid. The dry aldehyde should be light straw in color. Ehrlich reagent should be prepared in small quantities and stored in a refrigerator when not in use.

Test procedure. Add 1 ml of xylene to a 48-h 2% tryptone broth culture incubated at 35 to 37°C. Shake vigorously to extract the indole. Allow to stand 1 to 2 min for the xylene extract to layer on top. Add 0.5 ml of Ehrlich reagent down the side of the tube so that it forms a layer between the broth and the xylene. Do not shake the tube after the reagents are added. If indole is present, a red ring will develop just below the xylene layer.

Indophenol (cytochrome) oxidase test (8, 9)

See Chapter 29.

1. Solution A

α-Naphthol	1 g
Ethyl alcohol, 95 to 96%	100 ml

2. Solution B

p-Aminodimethylaniline hydrochloride .... 1 g
Distilled water .... 100 ml

(Solution B should be prepared frequently and stored in a refrigerator when not in use.)

Test procedure. The test is performed on nutrient agar slant cultures incubated at 37°C, or at a lower temperature if required. Add 2 or 3 drops of each reagent, and tilt the tube so that the reagents mix and flow over the growth on the slant. Positive reactions are indicated by the development of a blue color in the growth within 2 min.

Most positive cultures produce a strong reaction within 30 s. Any weak or doubtful reaction that occurs after 2 min should be ignored. Plate cultures may be tested by allowing an equal-parts mixture of the reagents to flow over isolated colonies.

KOH solution

Potassium hydroxide .... 10 or 20 g
Distilled water .... 100 ml

1. Mix the specimen (pus, exudate, tissue) with a drop of 10 or 20% solution on a clean slide, cover with a no. 2 cover slip (22 by 40 mm), and press gently to make a thin mount. Gentle warming may aid in clearing the mount. Viscid specimens may require overnight storage in a moist chamber. (Place the slide on applicator stick supports over moist filter paper in a petri dish, or place them in a screw-capped Coplin jar laid on its side.)

2. Scan under low power with reduced lighting. Switch to high power to check for the presence of suspected fungal elements.

Kovacs oxidase test (12)

The Kovacs oxidase test is recommended for nonfermenters and miscellaneous gram-negative bacteria.

1. Place filter paper in a petri dish, and saturate it with 0.5% tetramethyl-*p*-phenylenediamine hydrochloride.

2. With a platinum wire, pick a portion of the colony to be tested, and rub the colony on the filter paper.

A positive reaction is indicated by the appearance of a dark-purple color within 10 s. With this technique, there should be enough of the colony left for subculturing.

Alternate procedure. Place a few drops of the oxidase reagent directly onto colonies on a plate. Colonies producing oxidase become pink and then purple. The reagent does not interfere with the Gram stain, so purple colonies may be stained even though they are not viable. The oxidase reagent may be divided into small aliquots and stored at −20°C. Thaw one aliquot as needed, and do not use it for more than 1 day. Check the reagent daily with a known oxidase-positive organism. The Kovacs oxidase is more sensitive than the indophenol method.

Lactonate oxidation

Inoculate 2-ketolactonate agar plates and incubate them at 30°C for 24 to 48 h. Flood the plates with Benedict reagent. A positive test is indicated by the formation of a yellow ring of Cu_2O around the bacterial growth. This may take up to 1 h to develop.

Lactophenol cotton blue mounting solution

Phenol crystals .... 20 g
Lactic acid .... 20 ml
Glycerol .... 40 ml
Distilled water .... 20 ml

Dissolve the ingredients by heating the container in a hot-water bath. Add 0.05 g of cotton blue (Poirier blue).

Lectin solutions (5)

Dissolve lectins to a 1.0 mg/ml final concentration in 50 mM dipotassium phosphate adjusted to pH 7.3 with 0.1 N hydrochloric acid. Lectin solutions can be safely stored for several months at −20°C. *Glycine max* (soybean) or *Helix pomatia* (snail) lectin may be obtained from E-Y Laboratories, San Mateo, Calif.

McFarland nephelometer

1. In a rack, arrange 10 large test tubes, the same size as those used in the dilution of vaccines, and label them 1 to 10.

2. Add a 1% solution of anhydrous barium chloride and a 1% (by volume) cold solution of chemically pure sulfuric acid according to Table 1.

3. Seal the tubes and keep them in the refrigerator.

4. When the fine white precipitate of barium sulfate is shaken up well, each tube has a different density that corresponds approximately to the bacterial suspensions given in Table 1.

Methyl red test

Methyl red indicator

Methyl red .... 0.1 g
Ethyl alcohol, 95% .... 300.0 ml

Dissolve the dye in alcohol, and add sufficient distilled water to make 500 ml.

Test procedure. Inoculate buffered glucose-peptone broth lightly from a young agar slant culture. Incuba-

TABLE 1. Protocol for test tubes for McFarland nephelometer

Tube no.	Contents (ml)		Corresponding bacterial suspension per ml ($\times 10^8$)
	Barium chloride (1%)	Sulfuric acid (1%)	
1	0.1	9.9	3
2	0.2	9.8	6
3	0.3	9.7	9
4	0.4	9.6	12
5	0.5	9.5	15
6	0.6	9.4	18
7	0.7	9.3	21
8	0.8	9.2	24
9	0.9	9.1	27
10	1.0	9.0	30

tion at 37°C for 48 h is sufficient for the majority of cultures. Tests should not be made with cultures incubated less than 48 h. If the results are equivocal, repeat the test with cultures that have been incubated for 4 or 5 days. In such instances, duplicate tests should be incubated at 25°C.

Layer 5 or 6 drops of reagent for each 5 ml of culture. Reactions are read immediately. Positive tests are bright red, weakly positive tests are red-orange, and negative tests are yellow.

Methylene blue, acid buffered (pH approximately 3.8) for wet-mount staining of amoebic trophozoites

0.2 M solution of acetic acid (11.55 ml of glacial acetic acid per 1,000 ml of distilled water) .... 44.0 ml
0.2 M solution of sodium acetate (16.4 g of $C_2H_3O_2Na$ per 1,000 ml of distilled water) .... 6.0 ml
Distilled water .... 50.0 ml
Methylene blue (certified for use in histology) .... 0.06 g

Mix the acetic acid-sodium acetate solutions and the distilled water. Add methylene blue and dissolve it. The solution is stable at room temperature for a long time.

Methylene blue phosphate for filarial smears

Methylene blue chloride .... 1.0 g
Na_2HPO_4 (anhydrous) .... 3.0 g
KH_2PO_4 .... 1.0 g

Mix in a dry mortar. Dissolve in 1,250 ml of water (alternately, dissolve 1 g of the mixture in 250 ml of water).

Motility tests

For the *Enterobacteriaceae* (Chapter 24), a medium containing 0.4% agar is recommended. Inoculate by stabbing into the top of the column of medium to a depth of about 5 mm. Incubate at 35°C for 1 or 2 days. If the results are negative, follow with further incubation at 21 to 25°C for 5 days. For special purposes, such as enhancement of the motility and flagellar development in poorly motile cultures, it is often advisable to pass cultures first through a semisolid medium containing 0.2% agar tubed in Craigie tubes or in U tubes. Subsequent passages may be made in the 0.4% agar medium.

Aeromonas and *Pseudomonas* species are discussed in Chapters 25 and 30. Motility media containing agar concentrations higher than 0.3% produce gels through which many motile organisms cannot spread. Spreading in a semisolid medium is judged by macroscopic examination of the medium for a diffuse zone of growth emanating from the line of inoculation. Many aerobic pseudomonads fail to grow when they are deep in semisolid medium in a test tube. Organisms possessing "paralyzed" flagella are nonmotile and cannot spread in the medium. Some filamentous organisms spread in or on semisolid medium but are nonmotile and nonflagellated. Although cultures may grow at 37°C or at higher temperatures, the flagellar proteins of some organisms are not synthesized optimally at these temperatures; hence, motility medium should be incubated at temperatures near 18 to 20°C. These observations require a judicious interpretation of motility, and they limit, to some extent, the reliability of spreading in semisolid agar as the sole taxonomic criterion to delineate related species.

A deep layer of motility medium, 18 to 20 ml in a 100-mm-diameter petri dish, is useful for selecting strains from a predominantly nonmotile stock. The plate is inoculated in the center, and motile descendants are "fished" from the periphery of the giant colony after organisms have spread through the semisolid agar.

Nitrate reduction broth

Reagents

1. Solution A

Sulfanilic acid .... 8 g
Acetic acid, 5 N .... 1 liter

2. Solution B

N,N-Dimethyl-1-naphthylamine .... 6 ml
Acetic acid, 5 N .... 1 liter

(The 5 N acetic acid consists of 1 part glacial acetic acid to 2.5 parts distilled water.)

Although *N,N*-dimethyl-1-naphthylamine, unlike α-naphthylamine, has not been listed as a carcinogen by the Occupational Safety and Health Administration, Department of Labor, its structural similarity to α-naphythlamine would indicate that such safety precautions as the avoidance of aerosols, mouth pipetting, and contact with the skin should be followed.

For miscellaneous gram-negative bacteria (Chapter 28) and Pseudomonas spp. (Chapter 30)

Inoculate a fluid medium tubed with inverted Durham tubes, incubate it at 30°C (or 35°C), and examine it after 24 and 48 h for reduction of nitrate to nitrogen gas, which accumulates in the Durham tube. After 48 h, test for nitrite by the addition of 0.5 ml each of reagents A and B. A red color indicates a positive test, provided the uninoculated control medium is negative. Negative tests should be confirmed by the addition of zinc dust to convert unreduced nitrate to nitrite.

Nitrate disk test (23)

Disks

KNO_3 .... 30 g
$Na_2MoO_4 \cdot 2H_2O$ .... 0.1 g
Distilled water .... 100 ml

1. Dissolve the nitrate and molybdate completely, and then filter sterilize the solution by passing it through a membrane filter (0.45-μm pore size).
2. Dispense 20-μl quantities of the sterile solution to presterilized 0.25-in. (0.635-cm) filter paper disks.

Petri dishes (100 by 15 mm) are very convenient receptacles during the preparation of the disks.

3. After saturating the disks, cover them and allow them to dry at room temperature for 72 h.

Nitrate reagents

1. Solution A

Sulfanilic acid .... 0.5 g
Glacial acetic acid .... 30.0 ml
Distilled water .... 120.0 ml

2. Solution B

1,6-Cleve's acid (5-amino-2-naphthalenesulfonic acid) .... 0.2 g
Glacial acetic acid .... 30.0 ml
Distilled water .... 120.0 ml

Test procedure. The nitrate disk test is performed by removing the nitrate disk from the surface of a blood agar plate on which there is growth and placing it in a clean petri dish. Add 1 drop each of reagents A and B to the disks. Reduction of nitrate to nitrite is indicated by a pink to red color. If no color is seen within a few minutes, add a small amount of zinc dust and wait 5 min. Development of a red color indicates that nitrate was not reduced. If the disk remains colorless, nitrate was reduced beyond nitrite (interpreted as a positive test).

Nitrite reduction test for miscellaneous gram-negative bacteria (Chapter 28)

Inoculate inverted Durham tubes containing nitrite reduction broth. Incubate at 35°C, and examine after 24 and 48 h for reduction of nitrite to nitrogen gas, which accumulates in the Durham tubes. After 48 h, test for nitrite by the addition of 0.5 ml each of reagents A and B of nitrate reduction broth. The test is positive if there is no color change, provided a red color develops in the uninoculated control.

ONPG test (6)

Peptone water. Dissolve 1 g of peptone and 0.5 g of NaCl in 100 ml of distilled water. Autoclave at 121°C for 15 min.

ONPG solution. Dissolve 0.6 g of *o*-nitrophenyl-β-D-galactopyranoside (ONPG) in 100 ml of 0.01 M Na_2HPO_4. Sterilize by filtration; put in a sterile bottle. Store at 4 to 10°C, protected from light.

ONPG broth. Aseptically add 25 ml of ONPG solution to 75 ml of peptone water. Aseptically dispense in 0.5-ml amounts into sterile tubes (13 by 100 mm). This broth is stable for 1 month at 4 to 10°C, or longer if stored in the freezer. Do not use if yellow.

1. Inoculate 0.5 ml of ONPG broth with a heavy loopful of growth from triple sugar iron or Kligler iron agar.
2. Incubate in a heating block or water bath at 37°C for 1 h or more.
3. Results are interpreted as follows: Yellow color, positive; Colorless, negative.

This test cannot be performed with yellow-pigmented organisms.

Polyvinyl alcohol fixative for parasitology

Reagents

Polyvinyl alcohol (PVA) powder .... 50 g
Schaudinn stock solution (see Schaudinn fixative) .... 935.0 ml
Neutral glycerol .... 15.0 ml
Glacial acetic acid .... 50.0 ml

PVA (Evonal) powder may be purchased from Eastman Chemical Products, Inc., or Baker Co. Grades of high hydrolysis and low to medium viscosity are most satisfactory but must be pretested for adhesive property and lack of clumping.

Add the powder to the flask containing Schaudinn solution, acetic acid, and glycerol by slowly shaking in the powder while agitating the solution. Dissolve the powder by heating the solution in a 75°C water bath (do not boil). Shake the mixture frequently, or use a magnetic stirrer. When the powder is dissolved, remove the flask from the water bath and allow it to stand overnight. If a heavy precipitate forms during preparation or if the finished product is quite cloudy, discard and start again. PVA powder may be unsatisfactory.

Prepared PVA fixative may be purchased from several commercial sources including Marion Scientific, Kansas City, Mo.; Meridian Diagnostics, Cincinnati, Ohio; and Medi-Chem, Santa Monica, Calif. PVA fixative is also included in some commercial kits.

Shelf life varies with the specific lot and the quantity (larger amounts are satisfactory for a longer period). PVA fixative becomes cloudy or gels when it is unsatisfactory. Large volumes may be satisfactory for a year or more, whereas small volumes may be unsatisfactory after several months.

Rapid fermentation technique (RFT) solutions

1. Prepare a 10× buffered salts solution stock (10× BSS) as follows:

KCl .... 8.0 g
K_2HPO_4 .... 0.4 g
KH_2PO_4 .... 0.1 g
Distilled water .... 100 ml

Dissolve and filter sterilize.

2. Prepare a working dilution of BSS by diluting 10 ml of 10× BSS with 90 ml of distilled water. Add 0.5 ml of a 1% aqueous solution of phenol red, and filter sterilize. Aseptically pour into a sterile bottle, label, and store at 4°C.
3. Prepare 20% aqueous solutions of reagent grade glucose, maltose, fructose, sucrose, and lactose. Filter sterilize, and dispense 0.1-ml aliquots of each sugar into appropriately labeled Nunc tubes. Store in Nunc trays at 4°C.

Rapid pyrazinamidase test (20)

Substrate

Pyrazinamide .... 0.1 g
Sodium pyruvate .... 2.0 g

Agar .... 2.0 g
Distilled water .... 1,000 ml

Heat to melt the agar, dispense as 5-ml aliquots into screw-capped tubes (16 by 125 mm), and autoclave at 15 lb of pressure for 15 min.

Reagent

Ferrous ammonium sulfate (1%), freshly prepared

Test procedure. Use a swab to transfer a heavy inoculum from 24- to 72-h cultures from blood or chocolate agar plates to the top half of the agar tube. Incubate for 2 h at 37°C. Add 1 ml of freshly prepared reagent, and mix it in the top half of the tube. Within 1 to 5 min, a pink color develops for a positive test. A positive and negative control should be done.

Saponin solution (0.5%) for filaria concentration

Saponin powder .... 0.25 g

Add 50 ml of 0.85% saline. Mix carefully and thoroughly. Remove excess foam.

Note. Saponin (0.15 g) can be placed in 50-ml screw-capped centrifuge tubes and stored. Saline is added when saponin is needed.

Schaudinn fixative for fresh fecal films

Stock solution of saturated mercuric chloride solution with alcohol. Saturated mercuric chloride should be prepared several days before it is used. Add 47.5 g of mercuric chloride to 675 ml of distilled water, and heat to dissolve. Stopper tightly, let stand at room temperature overnight, and filter. Add 1 part 95% ethyl alcohol to 2 parts saturated mercuric chloride in stoppered container.

Schaudinn fixative. Add 1 part glacial acetic acid to 19 parts solution just before use (fixative should be prepared daily as needed). Usually 2.5 ml of glacial acetic acid is added to 48 ml of stock solution for each usage.

Slide culture technique

In the study of fungi, it is often necessary to observe the undisturbed relationship between reproductive structures and mycelium. This observation may be done by growing the fungi on glass slides in a moist chamber. The operator should be aware of infectious hazards of this procedure. This technique should never be used for the systemic pathogenic molds *Coccidioides immitis*, *Histoplasma capsulatum*, *Blastomyces dermatitidis*, and *Paracoccidioides brasiliensis*.

1. Place a piece of absorbent paper and a slide on a bent-glass rod in the bottom of a petri dish, add a cover slip, cover, and sterilize.
2. Prepare Sabouraud dextrose agar plates with about 15 ml of agar per plate. Allow to solidify and dry. Cut agar blocks about 1 cm square.
3. Using sterile technique, place a block of agar on the slide in a petri dish.
4. Inoculate the central portion of each of the four sides of the block with a small fragment of the fungus being studied.
5. Cover the inoculated block with a sterile cover slip.
6. Add 8.0 ml of sterile water to the bottom of the petri dish.
7. Incubate at 25°C until sporulation occurs. Maintain saturation of the paper throughout the incubation period. The slide preparation may be checked periodically under the low power of a microscope.
8. When sporulation is complete, carefully lift off the cover slip and lay it aside with the fungus growth up.
9. Lift the agar block from the slide and discard it.
10. Place a drop of lactophenol cotton blue in the center of growth on the slide, and cover it with a fresh cover slip. Place another drop in the center of growth on the cover slip, and drop the cover slip into place on another clean slide.
11. Blot away excess mounting fluid from the two preparations, and allow them to dry. Seal the edges with nail polish.

Sodium hippurate hydrolysis for streptococci

See Chapter 16.

Inoculate sodium hippurate broth with two or three colonies of beta-hemolytic streptococci, and incubate at 35°C for 20 h or longer. Centrifuge the medium to pack the cells, and pipette 0.8 ml of the clean supernatant into a Kahn tube. Add 0.2 ml of the ferric chloride reagent to the Kahn tube, and mix well. If a heavy precipitate remains longer than 10 min, the test is positive.

Ferric chloride reagent

$FeCl_3 \cdot 6H_2O$ .... 12 g
2% aqueous HCl .... 100 ml

The 2% aqueous HCl is made by adding 5.4 ml of concentrated HCl (37%) to 94.6 ml of H_2O.

Temperature requirements for nonfermentative bacteria (16)

Inoculate three slants of tryptone-glucose-yeast extract agar or Trypticase soy agar with one loopful of an overnight broth culture. Incubate one tube at 25°C, one tube at 37°C, and one tube at 42°C for 18 to 24 h. Examine the tubes for amount of growth.

Ureaplasma urease test reagent

Urea .... 0.6 g
$CuCl_2$ .... 1.1 g
Water .... 100 ml

Place several drops of reagent on an agar plate with visible or invisible colonies. A brown color surrounding *Ureaplasma urealyticum* colonies but not *Mycoplasma hominis* colonies results within 1 to 5 min on urea agar. The added urea also results in a red color from the phenol red indicator if colonies are present within 1 to 12 h. Colonies should be no more than 48 h old or should be held at 4°C until tested.

Voges-Proskauer test

Method 1

O'Meara reagent, modified

Potassium hydroxide .... 40.0 g
Creatine .... 0.3 g
Distilled water .... 100.0 ml

Dissolve the alkali in the water and add creatine. The reagent should be prepared frequently and should be refrigerated when not in use (13, 19). The reagent may be used for 2 to 3 weeks, but it deteriorates rapidly thereafter.

The Voges-Proskauer test is performed on the culture grown in buffered peptone-glucose broth incubated at 37°C for 48 h. Add 1 ml of reagent to 1 ml of culture, and place the mixture at 37°C or at room temperature, after shaking to aerate. A positive test is indicated by the development of a pink color. The test depends on the formation of acetylmethylcarbinol, which is oxidized in alkaline medium, in the presence of air, to form diacetyl. Diacetyl reacts with creatine to form the pink compound. Final readings are made after 4 h. If equivocal results are obtained, repeat the test with cultures incubated at 25°C.

Method 2, Barritt

1. Solution A

α-Naphthol .... 5 g
Ethyl alcohol, absolute .... 100 ml

2. Solution B

Potassium hydroxide .... 40 g
Distilled water .... 100 ml

Add 0.6 ml of solution A and 0.2 ml of solution B to 1 ml of culture. Shake well after the addition of each reagent. Positive reactions occur at once or within 5 min and are indicated by the production of a red color. The development of a copper color in some tests should be disregarded.

Method 3, Coblentz (4)

MR-VP medium .... 1.7 g
Distilled water .... 100.0 ml

Dispense in 2.0-ml volumes into 18- by 150-mm cotton-plugged tubes. Autoclave at 121°C for no longer than 10 min to prevent caramelization of the glucose.

Inoculate the medium with a massive inoculum, consisting of a loopful (2 to 3 mm) of an 18-h agar slant culture. Incubate at 30°C for 6 to 7 h, and add 0.6 ml of α-naphthol (5.0 g of α-naphthol in 100.0 ml of 95% ethyl alcohol) followed by 0.2 ml of 40% KOH containing 0.3% creatine. The tubes are then shaken vigorously for 30 s to 1 min.

Zinc sulfate solution for fecal parasite concentration

Zinc sulfate solution is prepared so that the final specific gravity is appropriate for the type of specimen to be examined (1.20 for Formalin-fixed specimens, 1.18 for fresh specimens).

1. Dissolve 386 g of reagent grade zinc sulfate ($ZnSO_4 \cdot 7H_2O$) in 400 ml of hot water, and cool to room temperature.
2. Place 400 ml of water in a 1,000-ml stoppered graduated cylinder.
3. Add zinc sulfate solution to the graduated cylinder, and mix.
4. Check the specific gravity with a hydrometer for heavy liquids (it should be over 1.20).
5. Add 10 ml of water for each 0.002 the specific gravity must be reduced, mix thoroughly, and check the specific gravity.
6. Repeat step 5 until the specific gravity is 1.20.
7. If the specific gravity is less than 1.195, dissolve as much zinc sulfate as possible in 50 ml of very hot water. Add some of this solution to raise the specific gravity.
8. After the specific gravity is correctly adjusted, allow the solution to stand overnight, and recheck.
9. Store at room temperature in a tightly stoppered container.
10. Recheck the specific gravity monthly.

STAINS

Capsule stain (Hiss method)

Mix a loopful of physiological saline suspension of growth with a drop of normal serum on a glass slide. Allow the smear to air dry, and heat fix. Flood the smear with crystal violet (1% aqueous solution). Steam the preparation gently for 1 min, and rinse with copper sulfate (20% aqueous solution). Capsules appear as faint blue halos around dark-blue to purple cells.

Dobell and O'Connor iodine solution for staining intestinal protozoa

Iodine (powdered crystals) .... 1 g
Potassium iodide .... 2 g
Distilled water .... 100 ml

Dissolve potassium iodide in water. Add iodine crystals, and shake thoroughly. The crystals may not dissolve completely; if they do not, filter or decant. Put a portion in a dropping bottle for daily use, and store the remainder in a brown bottle away from the light.

This is a weak iodine solution and should be prepared fresh every 2 to 3 weeks.

Flagellar stain (3)

See Chapter 29.

Reagents

1. Solution A

Basic fuchsin (certified for flagellar stain) ... 0.6 g
Ethyl alcohol, 95% .... 50.0 ml

Shake and let stand overnight to dissolve the fuchsin.

2. Solution B

Sodium chloride	0.37 g
Tannic acid	0.75 g
Distilled water	50.0 ml

Combine solutions A and B, and mix thoroughly. Adjust the pH to 5.0 with 1 N NaOH. If the solution is more acidic, flagella stain slowly and are difficult to detect. The stain may be used immediately, but it works better if held 2 to 3 days at 4°C. Store in a tightly stoppered bottle at 4°C. At 4°C the stain is stable for about 1 month. In the freezer, it will keep indefinitely. The stain deteriorates rapidly if held at room temperature. Allow any precipitate which forms to remain undisturbed. Use only the clear supernatant solution to stain flagella.

Test procedure. Make a slightly cloudy suspension in distilled water (not saline) from 18- to 24-h growth on agar plates or slants. Do not carry any agar over. Growth equivalent to a 1-mm colony is mixed with 3 ml of sterile distilled water. Emulsify the growth in a drop of water against the side of the tube, and then mix it with the remaining water.

Slides must be clean and grease free. Soak the slides 4 days in 3% concentrated HCl in 95% ethyl alcohol or acid dichromate solution. Rinse in 10 changes of tap water and 2 changes of distilled water, and air dry.

Pick up a clean slide with forceps and hold it for a few seconds in a blue Bunsen flame. When the slide has cooled, draw a thick wax pencil line on it to contain the stain in an area about 1 by 1.75 in. (ca. 2.5 by 4.4 cm).

Put a loopful of the suspension on one end of the slide. Tilt the slide, and allow the suspension to flow to the opposite end of the slide. Air dry on a level surface. Do not fix with heat.

Apply 1 ml of stain, warmed to room temperature, to the smear. Do not move the slide. A fine precipitate forms over the entire slide in 5 to 15 min. Once the precipitate has formed, wash the slide in a gentle stream of tap water. Do not tilt to rinse. Air dry, and examine under oil immersion.

For a control suspension, inoculate flagella broth with an actively motile bacterium, and incubate overnight at 25°C. Add 0.25 ml of Formalin to 5 ml of broth culture. Mix, and allow the mixture to stand for 15 min. Wash twice in distilled water. Resuspend the packed cells in enough distilled water to obtain a faintly turbid suspension.

Giemsa stain for chlamydiae (17)

See Chapter 85.

Stock solution

Giemsa powder	0.5 g
Methyl alcohol, absolute, acetone-free	33.0 ml

Mix thoroughly, allow to sediment, and store at room temperature.

Buffered water, pH 7.2

Solution 1. Prepare 0.067 M Na_2HPO_4 by adding 9.5 g of anhydrous salt to 1 liter of distilled water.

Solution 2. Prepare 0.067 M NaH_2PO_4 by dissolving 9.2 g of NaCl in 1 liter of distilled water.

Mix 72 ml of solution 1 with 28 ml of solution 2 and 900 ml of distilled water.

Working solution

Stock solution	1 part
Buffered water, pH 7.2	40 or 50 parts

Staining procedure. The smear is air dried, fixed with absolute methanol for at least 5 min, and again dried. It is then covered with the working Giemsa solution (freshly prepared the same day) for 1 h. The slide is then rinsed rapidly in 95% ethyl alcohol to remove excess dye, dried, and examined for the presence of the typical intracytoplasmic inclusion body. The elementary bodies stain purplish, whereas the initial bodies are slightly more basophilic and tend to stain bluer. The inclusions stand out against the grayish cytoplasm and in contrast with the pink nucleus of the cell.

This method gives excellent permanent preparations if a reliable brand of stain is used (for example, National Aniline and Chemical Company, Inc., New York, N.Y., or Gradwohl Laboratories, St. Louis, Mo.). Dilutions of the stock solution can be made with neutral distilled water (orange with neutral red or purple with hematoxylin), but buffered water solution is more reliable. Commercial cytological buffers may also be used. Any pH between 6.8 and 7.2 is acceptable (although the more basic side may be preferable) as long as it is kept constant to minimize tinctorial variation. There is some variability in currently available prepared stock Giemsa solutions, and these commercial products should be screened before being accepted for routine use.

Giemsa stain for demonstrating *Dermatophilus* spp. in paraffin sections (17)

Stock solution

Giemsa powder	4 g
Methyl alcohol	264 ml
Glycerol	264 ml

Working solution

Stock solution	1.25 ml
Methyl alcohol	1.50 ml
Distilled water	50.00 ml

Staining procedure. Pass the slide through xylol, absolute alcohol, and 95% alcohol. If the section was fixed in Zenker solution, remove the mercury precipitates by placing the section in iodine for 5 min and then in 5% sodium thiosulfate until clear. Wash in water, and rinse in distilled water. Flood with working Giemsa solution, and steam gently for 10 min. Wash in tap water. Differentiate in rosin alcohol (95% alcohol and a few drops of 10% rosin) to a macroscopic purplish-pink color; usually three swishes are sufficient. Pass through two changes each of absolute alcohol and xylol, and mount.

Giemsa stain for demonstrating *Dermatophilus* spp. in smears (17)

Giemsa blood stain (stock solution), Matheson, Coleman and Bell, 257 .... 1 ml
Distilled water .... 49 ml

Staining procedure. Fix the film with methyl alcohol for 30 s. Apply dilute stain for 45 min. Rinse with distilled water, and air dry.

Giemsa stain for malaria

Use a certified liquid or prepare as described below.

Giemsa stain, powdered (certified) .... 0.75 g
Methyl alcohol, pure .... 65.00 ml
Glycerol, pure .... 35.00 ml

Shake well in a bottle with glass beads. Keep tightly stoppered at all times. Filter if necessary.

Giemsa stain for parasites

Stock reagents

Giemsa stain powder (certified, Azure B) .. 600 mg
Methyl alcohol (acetone free, neutral) .... 50 ml
Glycerol (neutral, from freshly opened bottle) .... 50 ml

1. Place 600 mg of stain powder in a clean mortar.
2. Add part (about 10 ml) of the glycerol, and grind.
3. Pour off the top third into a clean 500- or 1,000-ml Erlenmeyer flask.
4. Add more glycerol, and repeat the grinding and decanting.
5. Repeat until most of the stain powder has been mixed with glycerol and the mixture has been poured into the flask.
6. Stopper the flask with a cotton plug, and then cap it with aluminum foil.
7. Place the flask so the bottom is flat in a water bath at 55 to 60°C for 2 h. Shake gently at half-hour intervals.
8. Add alcohol to the stain solution.
9. Store in a tightly stoppered dark bottle away from light.

Stock Giemsa stain, Azure B, may be purchased from scientific supply houses such as Harleco, Philadelphia.

Phosphate buffers

A. Stock buffers

1. Alkaline buffer, 0.067 M Na_2HPO_4 solution. Dissolve 9.5 g of Na_2HPO_4 in 1 liter of water.
2. Acid buffer, 0.067 M NaH_2PO_4 solution. Dissolve 9.2 g of $NaH_2PO_4 \cdot H_2O$ in 1 liter of water.

These buffers can be kept for a long period of time.

B. Buffered water (pH 7.0 to 7.2)

Acid buffer (NaH_2PO_4) .... 39 ml
Alkaline buffer (Na_2HPO_4) .... 61 ml
Distilled water .... 900 ml

Be sure glassware is clean!
Buffered water, if sealed, is stable for several weeks.

C. Triton X-100 stock solution (10%)

Triton X-100 .... 10 ml
Distilled water .... 90 ml

Triton X-100 is available from Rhon and Haas Co., Philadelphia, Pa., and Emulsion Engineering, Inc., Elk Grove Village, Ill.

Mix and store in a tightly stoppered bottle at room temperature; the solution will keep indefinitely.

D. Buffered water with Triton X-100 (10%)

The final concentration of Triton X is 0.01%. Add 1 ml of Triton X-100 stock solution to 1,000 ml of the buffered water described above.

Giemsa stain prepared in buffer with 0.01% Triton X-100 penetrates better and is recommended for staining blood and tissue slides. Buffered water without Triton X-100 is used for rinsing stained slides.

The pH (7.0 to 7.2) is critical, and if it is incorrect, it can be adjusted by adding small amounts of acid or alkaline buffer.

To make a working solution from stock alcohol solutions, dilute the stain at time of use with Triton X-100 buffered water. For blood films for malaria, dilute 1:40 with 60-min stain time. For impression smears or smears of sediment, dilute 1:20 with 30-min stain time. Rinse the stained slides briefly in buffered water without Triton X-100. (See Chapter 56 for procedure.)

Gimenez stain for chlamydiae—Gimenez modification of the Macchiavello technique (17)

Stock solutions

1. 10% (wt/vol) basic fuchsin in 95% ethanol .... 100 ml
4% (wt/vol) aqueous phenol .... 250 ml
Distilled water .... 650 ml
2. Sodium phosphate buffer solution (0.1 M) at pH 7.45. (Mix 3.5 ml of 0.2 M NaH_2PO_4, 15.5 ml of 0.2 M Na_2HPO_4, and 19 ml of distilled water.)
3. Aqueous malachite green oxalate, 0.8%.

To prepare a working solution of carbol-fuchsin, mix 4 ml of stock solution with 10 ml of buffer (pH 7.45); filter immediately, and filter again before each staining. The working solution remains satisfactory for about 40 h.

A very thin smear, air dried (heat fixation is not necessary for cytological reasons, but should be used for safety), is covered with the filtered carbol-basic fuchsin working solution and held for 1 to 2 min; after a thorough washing in tap water, it is covered with the malachite green solution for 6 to 9 s and again washed in tap water. The slides are finally dried with absorbent paper. Elementary bodies stain red; the background will be greenish.

Reagents

1. Carbol-basic fuchsin stock solution

10% basic fuchsin in 95% ethanol .... 100 ml

4% aqueous phenol . . . . 250 ml
Distilled water . . . . 650 ml

Mix well; incubate at 37°C for 48 h before use.

2. Stock buffers

a. 0.2 M NaH_2PO_4 . . . . 2.84 g in 100 ml of distilled H_2O
(*or* if $NaH_2PO_4 \cdot H_2O$ is used . . . 3.27 g in 100 ml of distilled H_2O)
b. 0.2 M Na_2HPO_4 . . . . 2.76 g in 100 ml of distilled H_2O

3. Buffer solution (0.1 M sodium phosphate, pH 7.45)

0.2 M NaH_2PO_4 . . . . 3.5 ml
0.2 M Na_2HPO_4 . . . . 15.5 ml
Distilled H_2O . . . . 19.0 ml

4. Carbol-basic fuchsin working solution

Carbol-basic fuchsin stock . . . . 4 ml
0.1 M sodium phosphate buffer solution, pH 7.45 . . . . 10 ml

Filter immediately and again before each use. Remains suitable for use for about 48 h.

5. Malachite green oxalate

0.8% solution in distilled water

Staining procedure

1. Air dry the smear and then heat fix it.
2. Filter working carbol-fuchsin onto the slide (through Whatman no. 2 filter paper), and let it stand for 1 to 2 min.
3. Wash the slide thoroughly with tap water.
4. Cover the smear with malachite green for 6 to 9 s.
5. Wash the slide thoroughly with tap water.
6. Cover the smear a second time with malachite green for 6 to 9 s.
7. Wash thoroughly with tap water.
8. Examine under an oil immersion objective with light microscopy.

(In properly prepared slides, *Legionella* sp. stains red, and the background material will appear green.)

Gram stain (2)

Modified Hucker crystal violet

1. Solution A

Crystal violet (certified) . . . . 2 g
Ethyl alcohol, 95% . . . . 20 ml

2. Solution B

Ammonium oxalate . . . . 0.8 g
Distilled water . . . . 80.0 ml

Mix solutions A and B. Store for 24 h before use. Filter through paper into staining bottle.

Gram iodine

Iodine . . . . 1 g
Potassium iodide . . . . 2 g
Distilled water . . . . 300 ml

Grind the dry iodine and potassium iodide in a mortar. Add water a few milliliters at a time, and grind thoroughly after each addition until solution is achieved. Rinse the solution into an amber glass bottle with the remainder of the distilled water.

Decolorizers

1. Slowest agent, ethyl alcohol (95%)
2. Fastest agent, acetone
3. Intermediate agent, acetone-alcohol (95% ethyl alcohol, 100 ml; acetone, 100 ml)

For an experienced worker, any one of the three decolorizing agents will yield good results.

Counterstain

1. Stock solution

Safranin O (certified) . . . . 2.5 g
Ethyl alcohol, 95% . . . . 100.0 ml

2. Working solution

Stock solution . . . . 10 ml
Distilled water . . . . 90 ml

Staining procedure. Flood the smear with crystal violet solution, and let it stand for 1 min. Wash the smear briefly with tap water, and drain off excess water. Flood the smear with iodine solution, and let it stand for 1 min. Wash with tap water, and decolorize until the dye doesn't run off the smear. Wash briefly with water. Counterstain with safranin for 10 s. Wash briefly with tap water, blot dry, and examine. Gram-positive organisms are blue; gram-negative organisms are red.

Gram stain for actinomycetes and other bacteria

Aniline-crystal violet solution

Aniline . . . . 40 ml
Water . . . . 1 liter
Crystal violet, saturated alcohol solution . 114 ml

Shake aniline and water in a closed container until they are mixed. Filter through four sheets of paper moistened with water. Add crystal violet solution to filtrate.

Gram iodine solution

Iodine . . . . 1 g
Potassium iodide . . . . 2 g
Water . . . . 300 ml

Grind iodine and potassium iodide in a mortar until they are well blended. Add water slowly to dissolve.

Safranin

Safranin, saturated alcohol solution (about 2.5 g/100 ml of 95% alcohol) .... 10 ml
Water .... 90 ml

Staining procedure. Stain with crystal violet for 2 min, wash, and dry. Apply Gram iodine for 1 min, wash, and dry. Decolorize with 95% alcohol for 30 s, and wash in running tap water. Counterstain with safranin for 30 s, and blot dry.

Gram-Weigert stain reagents

Eosin Y. Use 1 g of eosin Y in 100 ml of distilled water.

Crystal violet (gentian)

Solution 1. Combine 2 ml of aniline oil and 88 ml of distilled water; shake and filter.

Solution 2. Dissolve 5 g of crystal violet in 10 ml of 95% ethanol.

Combine solutions 1 and 2. Filter before use. The reagent will remain stable for 3 months.

Gram iodine solution. Dissolve 2 g of KI in a small amount of distilled water. Add 1 g of I_2 crystals, and bring the volume to 300 ml with distilled water.

Aniline oil and xylene. Use 1 part aniline oil and 1 part xylene. (*Note.* This must be pure aniline oil, not aqueous aniline oil.)

Grocott-Gomori methenamine silver nitrate stain for fungi in tissue sections

1. Solution A

Borax, 5% .... 8 ml
Distilled water .... 100 ml

2. Solution B

Silver nitrite, 10% .... 7 ml
Methenamine, 3%......................... 100 ml

Add equal parts of solutions A and B to make a working methenamine-silver nitrate solution. These solutions should be made up fresh. Counterstain with light-green stain.

Stock light-green solution

Light-green S.F. (yellow).................. 0.2 g
Distilled water.......................... 100.0 ml
Glacial acetic acid......................... 0.2 ml

Staining procedure. Deparaffinize tissue sections and rinse in distilled water. Oxidize in 5% chromic acid for 1 h. Wash in running tap water for a few seconds. Rinse in 1% sodium bisulfite for 1 min to remove residual chromic acid. Wash in tap water for 5 to 10 min. Wash in three or four changes of distilled water. Place in working methenamine-silver nitrate solution in oven (58 to 60°C) for 30 to 60 min. When section turns yellowish brown, use paraffin-coated forceps to remove the slide from the silver nitrate solution. Dip the slide in distilled water, and check with a microscope for adequate silver impregnation. Fungi should be dark brown at this stage. Rinse in six changes of distilled water. Tone in 0.1% gold chloride for 2 to 5 min. Rinse in distilled water. Remove unreduced silver with 2% sodium thiosulfate for 2 to 5 min. Wash in tap water. Counterstain for 1 min with a fresh 1:5 dilution of stock light-green solution in distilled water. Dehydrate, clear, and mount.

Hematoxylin and eosin stain for mycetoma granules and *Dermatophilus* in paraffin sections

Cut sections at 5 mm.

1. Two changes of xylol, for 2 min each
2. Two changes of absolute alcohol, for 1 min each
3. One change of 95% alcohol, for 1 min
4. One change of 90% alcohol, for 0.5 min
5. One change of 80% alcohol, for 0.5 min
6. One change of 60% alcohol, for 0.5 min
7. Two changes of distilled water or until slides have cleared
8. Harris hematoxylin with glacial acetic acid (5 ml of acetic acid–100 ml of hematoxylin), for 1 to 2 min
9. Rinse in distilled water.
10. Place in tap water, to which 20 to 40 drops of ammonium hydroxide has been added, for about 3 s (section will turn blue immediately).
11. Rinse in two changes of tap water to remove the ammonia.
12. Counterstain in picro-eosin solution for about 30 s.
13. Two changes of 95% alcohol, for 1 min each
14. Two changes of absolute alcohol, for 1 min in the first and 2 min in the last
15. Two changes of xylol, for 1 min each
16. Mount in neutral xylol-damar.

Note. At step 5, Zenker-fixed tissue is placed in Lugol iodine for 5 min, washed in tap water for 0.5 min, and placed in 5% sodium thiosulfate solution (hypo) for about 0.5 min or until color is removed. Then wash the sample in tap water for 1 min, rinse it in distilled water, triple the staining time in hematoxylin, and then stain the same as for other fixatives.

India ink mount for *Cryptococcus neoformans* capsules

1. Mix the specimen (pus, exudate, tissue, sputum, or sediment of centrifuged spinal fluid) with a small drop of India ink on a clean slide. Cover with a cover slip. The mount should be thin. Gentle pressure may have to be applied on the cover slip for pus, exudate, tissue, or sputum to obtain a thin mount. If India ink is too dark, dilute it to 50% with distilled water.
2. Scan under low power, with reduced lighting. Switch to high power to examine for the presence of

encapsulated cells. The mucoid capsules appear as a clear halo that surrounds the yeast cell or lies between the cell wall and the surrounding black mass of India ink particles. Capsules may be broad or narrow. The yeast cells may be round, oval, or elongate; buds may be absent, single, or rarely multiple. The buds may be detached from the mother cell but enclosed in a common capsule attached.

Kinyoun acid-fast stain

Kinyoun carbol-fuchsin

Basic fuchsin . . . . . 4 g
Alcohol, 95% . . . . . 20 g
Phenol crystals . . . . . 8 g
Distilled water . . . . . 100 ml

Acid-alcohol

Hydrochloric acid, concentrated . . . . . 3 ml
Ethyl alcohol, 95% . . . . . 97 ml

Methylene blue counterstain

Methylene blue . . . . . 0.3 g
Distilled water . . . . . 100.0 ml

Staining procedure. Flood the fixed smear with Kinyoun carbol-fuchsin, and let it stain for 2 min. Wash with tap water, and decolorize with acid-alcohol until the dye doesn't run off the slide. Wash with tap water, and counterstain for 20 to 30 s. Wash, blot dry, and examine. Acid-fast organisms stain red; the background and other organisms stain blue.

Kinyoun acid-fast stain, modified for actinomycetes

The solutions are the same as those described for Kinyoun acid-fast stain except that 2.5% methylene blue in 95% ethyl alcohol is used as the counterstain. Stain with Kinyoun carbol-fuchsin for 3 min without heat. Wash and decolorize for 5 to 10 s with acid-alcohol. Wash and counterstain for 30 s. Wash in water, and blot dry.

Lactophenol cotton blue mounting solution

Phenol crystals . . . . . 20 g
Lactic acid . . . . . 20 ml
Glycerol . . . . . 40 ml
Distilled water . . . . . 20 ml

Dissolve the ingredients by heating the container in a hot-water bath. Add 0.05 g of cotton blue (Poirier blue).

Loeffler alkaline methylene blue reagents

1. Solution A

Methylene blue . . . . . 0.3 g
95% ethanol . . . . . 30.0 ml

2. Solution B

Dilute potassium hydroxide (0.01%) . . . . . 100 ml

Dissolve the methylene blue in alcohol, and add the potassium hydroxide solution.

Staining procedure. Heat fix the smear with gentle heat. Flood the smear with the stain, and let it stand for 1 min. Wash briefly with tap water, blot dry, and examine.

Lugol iodine solution for staining intestinal protozoa

Stock solution (Lugol)

Iodine (powdered crystals) . . . . . 5 g
Potassium iodide . . . . . 10 g
Distilled water . . . . . 100 ml

The potassium iodide is dissolved in the distilled water, and the iodine crystals are added slowly and shaken until dissolved. Filter and place in a tightly stoppered bottle.

Working solution (D'Antoni)

Dilute the stock solution 1:5 with distilled water. Prepare a fresh working solution every 2 to 3 weeks.

Machiavello stain (modified) for chlamydiae (17)

Stock solutions

Basic fuchsin. Use 0.25 g in 100 ml of double-distilled water.
Citric acid. It is important to use a fresh solution daily. Use 0.5 g in 200 ml of double-distilled water.
Methylene blue. Use 1.0 g in 100 ml of double-distilled water.

Staining procedure. After drying in air, the smear or impression preparation is fixed by heat. The basic fuchsin solution, first passed through filter paper in a small funnel, is dropped onto the slide and left for 5 min before being quickly drained off. The slide is first washed in tap water and then dipped for a few seconds in the citric acid solution, best held in a Coplin jar. The slide is then washed thoroughly with tap water and stained with 1% methylene blue for 20 to 30 s; it is washed again in tap water and dried.

The citric acid solution must be fresh. Exposure to citric acid for more than a few seconds will decolorize the chlamydiae, and they will all stain blue. In a properly prepared slide, most elementary bodies will stain red against a blue background.

Methylene blue-phosphate stain for malaria

Methylene blue, medicinal . . . . . 1 g
Disodium phosphate, anhydrous . . . . . 3 g
Monopotassium phosphate, anhydrous . . . . . 1 g

All of the ingredients are thoroughly mixed in a dry mortar, and 1-g quantities are placed in well-stoppered vials. For use, 1 g is dissolved in 250 to 350 ml of distilled water. Filter if necessary.

M'Fadyean stain (15)

Reagents

Absolute methanol

Methylene blue solution (0.05 mg/ml in 20 mM potassium phosphate adjusted to pH 7.3)

Staining procedure. Smear the specimen on a slide, and air dry. Cover the smear with absolute alcohol for approximately 3 min, and air dry. The smear is then flooded with methylene blue solution for 30 to 45 s. Wash the slide gently with water, blot it dry, and examine under oil immersion. If *Bacillus anthracis* is suspected, all washings, blotting materials, and slides must be properly discarded and autoclaved.

Page amoeba saline

Sodium chloride (NaCl) .... 120 mg
Magnesium sulfate ($MgSO_4 \cdot 7H_2O$) .... 4 mg
Calcium chloride ($CaCl_2 \cdot 2H_2O$) .... 4 mg
Disodium hydrogen phosphate (Na_2HPO_4). 142 mg
Potassium dihydrogen phosphate
(KH_2PO_4) .... 136 mg
Distilled water .... 1,000 ml

Dissolve the chemicals in the water and sterilize by autoclaving at 15 lb/in^2 for 15 min. The solution may be stored in the refrigerator for up to 6 months.

Periodic acid-Schiff stain

This stain is recommended for observing fungi in tissues and smears (18).

Reagents

1. Basic fuchsin solution
Basic fuchsin.............................. 0.1 g
95% alcohol............................... 5.0 g
Water .... 95.0 ml

2. Zinc (sodium) hydrosulfite solution
Zinc (sodium) hydrosulfite .... 1.0 g
Tartaric acid.............................. 0.5 g
Water .... 100.0 ml

3. Saturated aqueous solution of picric acid (Do not let dry out.)

4. Light-green stain
Light green .... 1.0 g
Glacial acetic acid .... 0.25 ml
80% alcohol .... 100.0 ml

Skin scrapings. Spread thin scrapings on a slide coated with Mayer albumen, or coat the lesion with the fixative, scrape the surface scales off, and immediately press them on the surface of a clean slide. Heat gently, and check to see if the scrapings are fast on the slide before proceeding.

Staining procedure. Use 3 or 4 drops to cover the sample, or immerse the fixed scrapings as follows.

1. Immerse for 1 min in 95% alcohol.
2. Immerse for 5 min in 5% periodic acid.
3. Immerse for 2 min in basic fuchsin solution.
4. Rinse in tap water.
5. Immerse for 10 min in zinc (or sodium) hydrosulfite solution.
6. Rinse in tap water.
7. Counterstain with either saturated aqueous solution of picric acid for 2 min or with light-green stain for 5 s.
8. Rinse for a short time in tap water.
9. Dry well and observe under immersion oil; gently blot excess oil off slide before storage.
10. After step 8, dehydrate for about 10 s in 95% alcohol and for 1 min in 100% alcohol, rinse twice in xylol for about 1 min each time, and mount in Permount or other mounting medium.

The fungi stain a bright red or purplish red after periodic acid hydrolysis to release aldehydes that can combine with Schiff reagent. The carbohydrates in the cell walls take the red stain as a result of the reaction.

Tissue sections. Tissues should be fixed, dehydrated, embedded in paraffin, and sectioned by the routine method, as follows.

1. Place in xylol to deparaffinize.
2. Rinse in 100% alcohol.
3. Wash in distilled water.
4. Immerse for 10 min in 1% periodic acid.
5. Rinse in tap water for 5 to 10 min.
6. Immerse for 2 min in basic fuchsin solution.
7. Rinse in tap water for 30 s.
8. Immerse for 30 min (or possibly for up to 2 to 3 h for some material) in zinc (or sodium) hydrosulfite solution.
9. Rinse in tap water for 3 to 5 min.
10. Immerse for 2 min in light-green stain.
11. Rinse for a short time in tap water.
12. Dehydrate for about 10 s in 95% alcohol and for 1 min in 100% alcohol, rinse twice in xylol for about 1 min each time, and mount in Permount or other mounting medium.

Note. Hematoxylin and eosin (H and E)-stained slides may be restained by removal of the cover slip with xylol and rehydration. The slide is then placed in 1% periodic acid for 10 min, and the rest of the procedure is continued. This technique is useful when the H and E stain does not differentiate the fungus from the tissue sufficiently well.

Solutions in trichrome technique (20)

Iodine alcohol

Stock solution. Dissolve iodine crystals (approximately 1.0 g) in 20 ml of 70% alcohol by adding small amounts of alcohol while mixing. This procedure should produce a dark, concentrated solution. Store in capped brown bottles.

Working solution. Add stock solution to 70% alcohol until the color of strong tea is reached. The exact concentration of this solution is not important.

Acidified alcohol for trichrome staining. Add approximately 0.4 ml of glacial acetic acid to 100 ml of 90% ethyl alcohol.

Ethyl alcohol solutions. Solutions are aqueous dilutions of commercial 190-proof alcohol, which is 95%.

1. 70% solution

95% ethyl alcohol......................... 700 ml
Distilled water .... 250 ml

2. 90% solution

95% ethyl alcohol.......................... 900 ml
Distilled water............................. 50 ml

Carbol xylene. Carbol xylene is a 1:3 solution of carbolic acid and xylene.

Melted phenol crystals .... 250 ml
Xylene .... 750 ml

Liquify the crystals in a water bath, and measure in a warm graduated cylinder. Pour the liquid phenol into the xylene. Stir thoroughly. Store in tightly closed bottle.

Spore stain (Wirtz-Conklin)

Flood the entire slide with 5% aqueous malachite green. Steam for 3 to 6 min, and rinse under running tap water. Counterstain with 0.5% aqueous safranin for 30 s. Spores are seen as green spherules in red-stained rods or with red-stained debris.

Stain reagents for modified acid-fast stain for *Cryptosporidium* spp.

Kinyoun carbol-fuchsin reagent

Basic fuchsin............................. 4.0 g
95% ethanol .... 20.0 ml

Dissolve the dye in the alcohol by frequent agitation. Add to the dye solution a phenol-water mixture of:

Concentrated phenol .... 8.0 ml
Distilled water.......................... 100.0 ml

Distain reagent

Concentrated H_2SO_4 .... 1.0 ml
Distilled water........................... 99.0 ml

Trichrome stain reagents for parasites (Wheatley; 22)

Trichrome stain is quite stable and is used without being diluted. A good stain solution will be a very dark purple, almost black. Use only certified dyes.

Chromotrope 2R .... 6.0 g
Light-green SF .... 1.5 g
Fast-green FCF............................ 1.5 g
Phosphotungstic acid (CP) .... 7.0 g
Acetic acid (glacial) .... 10.0 ml
Distilled water .... 1,000.0 ml

Put the dry chemicals into a large dry flask (2 liters) and swirl gently to mix. Add the acetic acid slowly while mixing with a stirring rod to moisten all of the stain reagents. Completely cover. Allow the mixture to stand about 30 min to ripen. Add the distilled water slowly at first, with mixing, to dissolve all stain. Store in a tightly stoppered bottle at room temperature.

Prepared trichrome stain for parasitology can be purchased from commercial sources such as Meridian Diagnostics, Cincinnati, Ohio, and Remel (Regional Media Laboratories, Inc.), Lenexa, Kans.

Wayson stain

1. Solution A

Basic fuchsin (90% dye content)........... 0.20 g
Methylene blue (90% dye content)......... 0.75 g
Ethyl alcohol, 95% .... 20.00 ml

2. Solution B

Phenol, 5% .... 200.00 ml

Pour solution A slowly into solution B, and filter.

Staining procedure. Stain the smears for 10 to 20 s. Wash with water, and blot dry. The stain is especially useful for demonstrating polar staining.

Ziehl-Neelsen acid-fast stain for actinomycetes in tissues

Carbol-fuchsin solution

Phenol crystals, melted .... 2.5 ml
Alcohol, 95%.............................. 5.0 ml
Basic fuchsin.............................. 0.5 g
Distilled water.......................... 50.0 ml

Dissolve fuchsin in alcohol, add phenol and water, and let stand overnight. Filter through paper and then through a filter candle to remove any acid-fast bacilli. Store at room temperature, and filter before use.

Acid-alcohol, 3%

Hydrochloric acid, concentrated .... 3.0 ml
Alcohol, 70%............................. 99.0 ml

Working methylene blue solution

Methylene blue chloride .... 0.5 g
Glacial acetic acid........................ 0.5 ml
Distilled water.......................... 100.0 ml

Shake and filter twice.

Staining procedure. Any well-fixed tissue may be used. Cut sections at 4 to 6 μm. Deparaffinize sections through two changes of xylene, and run through absolute and 95% alcohols and distilled water as usual. Remove mercury precipitates through iodine and hypo solutions, if necessary. Stain with freshly filtered carbol-fuchsin for 10 min. Rinse well in tap water. Decolorize with 3% acid-alcohol until sections are pale pink. Wash thoroughly with running tap water for 8 min. Counterstain by dipping one slide at a time in working methylene blue solution for 15 to 30 s. Sections should be pale blue. Overstaining will mask bacilli. Wash with tap water and distilled water. Dehydrate with two changes of 95% alcohol and absolute alcohol; clear with two or three changes of xylene, and mount in Permount. Acid-fast actinomycetes are bright red, erythrocytes are yellowish orange, and other tissue elements are pale blue.

Ziehl-Neelsen stain, modified for staining *Brucella* spp.

Stock carbol-fuchsin

Basic fuchsin .... 1 g
Methyl alcohol, absolute .... 10 ml
Phenol, 5% .... 90 ml

Staining procedure. Stain for 10 min with a 1:10 dilution of stock carbol-fuchsin. Wash in tap water. Decolorize with 0.5% acetic acid for 20 to 30 s. Wash thoroughly. Counterstain with 1% methylene blue. Wash and blot dry. *Brucella* spp. stain red against a blue background.

Ziehl-Neelsen stain for mycobacteria

Carbol-fuchsin stain

Basic fuchsin.............................. 0.3 g
Ethyl alcohol, 95% .... 10.0 ml
Phenol, melted crystals .... 5.0 ml
Distilled water .... 95.0 ml

Dissolve the basic fuchsin in the alcohol; dissolve the phenol in the water. Mix the two solutions. Let stand for several days before use.

Acid-alcohol

Ethyl alcohol, 95%......................... 97 ml
Hydrochloric acid, concentrated .... 3 ml

Methylene blue counterstain

Methylene blue........................... 0.3 g
Distilled water......................... 100.0 ml

Staining procedure. Flood the entire slide with carbol-fuchsin, and heat slowly to steaming. Use low or intermittent heat to maintain steaming for 3 to 5 min, and then cool. Wash briefly with tap water, and decolorize until no more stain comes off. Wash with tap water, and counterstain for 20 to 30 s, wash, dry, and examine. Acid-fast organisms are red; the background and non-acid-fast organisms are blue.

LITERATURE CITED

1. **Andrade, E.** 1905–1906. Influence of glycine in differentiating certain bacteria. J. Med. Res. **14:**551–556.
2. **Bartholomew, J. W.** 1962. Variables influencing results, and the precise definition of steps in gram staining as a means of standardizing the results obtained. Stain Technol. **37:**139–155.
3. **Clark, W. A.** 1976. A simplified Leifson flagella stain. J. Clin. Microbiol. **3:**632–634.
4. **Coblentz, L. M.** 1943. Rapid detection of the production of acetyl-methyl-carbinol. Am. J. Public Health **33:**815–817.
5. **Cole, H. B., J. W. Ezzell, Jr., K. F. Keller, and R. J. Doyle.** 1984. Differentiation of *Bacillus anthracis* and other *Bacillus* species by lectins. J. Clin. Microbiol. **19:**48–53.
6. **Cowan, S. T., and K. J. Steel.** 1970. Manual for the identification of medical bacteria, p. 122. Cambridge University Press, Cambridge.
7. **Edelstein, P. H.** 1984. Legionnaires' disease laboratory manual. National Technical Information Service, Springfield, Va.
8. **Ewing, W. H., and J. G. Johnson.** 1960. The differentiation of *Aeromonas* and C27 cultures from *Enterobacteriaceae*. Int. Bull. Bacteriol. Nomencl. Taxon. **10:**223–230.
9. **Gaby, W. L., and C. Hadley.** 1957. Practical laboratory test for the identification of *Pseudomonas aeruginosa*. J. Bacteriol. **74:**356–358.
10. **Gadebusch, H. H., and S. Gabriel.** 1956. Modified stable Kovac's reagent for the detection of indole. Am. J. Clin. Pathol. **26:**1373–1375.
11. **Hébert, G. A.** 1981. Hippurate hydrolysis by *Legionella pneumophila*. J. Clin. Microbiol. **13:**240–242.
12. **Kovacs, N.** 1956. Identification of *Pseudomonas pyocyanea* by the oxidase reaction. Nature (London) **178:**703.
13. **Levine, M., S. S. Epstein, and R. H. Vaughn.** 1934. Differential reactions in the colon group of bacteria. Am. J. Public Health **24:**505.
14. **Mallory, F. B.** 1942. Pathological technique, p. 75. W. B. Saunders Co., Philadelphia.
15. **M'Fadyean, J.** 1903. A peculiar staining reaction of the blood of animals dead of anthrax. J. Comp. Pathol. **16:**35–41.
16. **Oberhofer, T. R.** 1979. Growth of nonfermentative bacteria at 42°C. J. Clin. Microbiol. **10:**800–804.
17. **Paik, G.** 1980. Reagents, stains, and miscellaneous test procedures, p. 1000–1024. *In* E. H. Lennette, A. Balows, W. J. Hausler, Jr., and J. P. Truant (ed.), Manual of clinical microbiology, 3rd ed. American Society for Microbiology, Washington, D.C.
18. **Peabody, J. W.** 1955. Demonstration of fungi by periodic acid-Schiff stain in pulmonary granuloma. J. Am. Med. Assoc. **157:**8858.
19. **Smith, M. R., R. E. Gordon, and F. E. Clark.** 1952. Aerobic sporeforming bacteria. U.S. Department of Agriculture Monograph no. 16. U.S. Department of Agriculture, Washington, D.C.
20. **Sulea, I. T., M. C. Pollice, and L. Barksdale.** 1980. Pyrazine carboxylamidase activity in *Corynebacterium*. Int. J. Syst. Bacteriol. **30:**466–472.
21. **Tatum, H. W., W. H. Ewing, and R. E. Weaver.** 1974. Miscellaneous gram-negative bacteria, p. 270–294. *In* E. H. Lennette, E. H. Spaulding, and J. P. Truant (ed.), Manual of clinical microbiology, 2nd ed. American Society for Microbiology, Washington, D.C.
22. **Wheatley, W. B.** 1951. A rapid staining procedure for intestinal amoebae and flagellates. Am. J. Clin. Pathol. **21:**990–991.
23. **Wideman, P. A., D. M. Citronbaum, and V. L. Sutter.** 1977. Simple disk technique for detection of nitrate reduction by anaerobic bacteria. J. Clin. Microbiol. **5:**315–319.

SUBJECT INDEX